Civil Engineer's Reference Book

Fourth Edition

Civil Engineer's Reference Book

Fourth Edition

Edited by
L S Blake

BSc(Eng), PhD, CEng, FICE, FIStructE
Consultant; formerly Director of the
Construction Industry Research and
Information Association

With specialist contributors

Butterworths

London · Boston · Singapore
Sydney · Toronto · Wellington

First published as *Civil Engineering Reference Book* in 1951
Second edition 1961
Third edition 1975
Fourth edition 1989

© **Butterworth & Co. (Publishers) Ltd, 1989**

British Library Cataloguing-in-Publication Data

Civil engineer's reference book.—4th ed.
  1. Civil engineering
  I. Blake, L. S. (Leslie Spencer)
  624
  ISBN 0-408-01208-0

Library of Congress Cataloging-in-Publication Data

Civil engineer's reference book.
  Includes bibliographies and index.
  1. Civil engineering—Handbooks, manuals, etc.
I. Blake, L. S. (Leslie Spencer), 1925–
TA151.C58 1989 624 88–30220
ISBN 0-408-01208-0

Typeset by Latimer Trend & Company Ltd, Plymouth

Printed in Great Britain by Anchor Press Ltd, Tiptree, Essex
and bound by Hartnoll Ltd, Bodmin, Cornwall

Preface to the fourth edition

The aim of this edition, as of those that preceded it, is to give civil engineers a concise presentation of theory and practice in the many branches of their profession. The book is primarily a first point of reference which, through its selective lists of references and bibliographies, will enable the user to study a subject in greater depth. However, it is also an important collection of state-of-the-art reports on design and construction practices in the UK and overseas.

First published in 1951, the book was last revised in 1975. Although civil engineering is not normally regarded as involving fast-moving technologies, so many advances have occurred in the theory and practice of most branches of civil engineering during the past decade or so that the preparation of a fourth edition became essential. Some of these advances have taken the form of improvements in earlier practices, for example in surveying, geotechnics, water management, project management, underwater working, and the control and use of materials. Other radical changes have resulted from the evolving needs of clients for almost all forms of construction, maintenance and repair. Another major change has been the introduction of new national and Euro-codes based on limit state design covering most aspects of structural engineering.

The fourth edition incorporates these advances and, at the same time, gives greater prominence to the special problems relating to work overseas, with differing client requirements and climatic conditions.

As before, careful attention has been given to the needs of the different categories of readers. Students and graduates at the start of their careers need guidance on the practice of design and construction in many of the fields of civil engineering covered in Chapters 11 to 44. The engineer in mid-career will also find these chapters valuable as presentations of the state of the art by acknowledged experts in each field, in addition to the references and bibliographies they contain for deeper study of specific problems. Chapters 1 to 10 provide engineers, at all levels of development, with up-to-date 'lecture notes' on the basic theories of civil engineering.

Although the book was primarily prepared for civil engineers in the UK and elsewhere in the world, members of other professions involved in construction—architects, lawyers, mechanical engineers, insurers and clients—will also benefit by referring to it.

I am most grateful to the authors who have contributed chapters. They are all engineers of considerable standing—consultants, contractors, research workers or academics—who have devoted a substantial amount of time to presenting their expert knowledge and experience for the benefit of the profession.

L.S. Blake
Bournemouth
November 1988

Contents

hardwood assemblies. Timber fastenings. Timber-framed construction. Repair and restoration. Termite-resistant construction. Storm-resistant construction. Earthquake-resistant construction. Design aids. References. Bibliography

17 Foundations Design

General principles. Shallow foundations. Deep foundations. Piled foundations. Retaining walls. Foundations for machinery. Foundations in special conditions. The durability of foundations. References. Bibliography

18 Dams

Definition. Brief history. Embankment dams. Concrete dams. Design concepts. Legislation. References. Bibliography

19 Loadings

Loading. Occupancy loads on buildings. Containers for granular solids. Road bridges. Railway bridges. Wind loading. Earthquake effects. References

20 Bridges

Plan of work. Economics and choice of structural system. Characteristics of bridge structures. Stress concentrations. Concrete slab decks. Skew and curved bridges. Dynamic response. Movable bridges. Items requiring special consideration. References. Bibliography

21 Buildings

Background. General management. Brief. The site. Landscape. Town planning. Public utility. Feasibility. Cost. Internal environment. Water supply, drainage and public health. Lifts, escalators and passenger conveyors. Energy. Building Regulations. Building security and control. Materials. Walls, roofs and finishes. Interior design and space planning. Structure. Tall buildings. References. Bibliography

22 Hydraulic Structures

Open channel structures. Enclosed flow. Spillways. Reservoir outlet works. Gates and valves. Cavitation. References

23 Highways

Introduction. Highway administration. Scheme preparation. Traffic appraisal. Environmental appraisal. Highway geometry. Earthworks. Drainage. Pavement design and surfacings. Lighting, signing, communications and safety. Specifications and materials testing. Roads and traffic in urban areas. Highway maintenance. Low-cost roads in developing countries. References

24 Airports

Introduction. Airport location. Standards. Airport concept and layout. Traffic forecasts. Aircraft pavements. Surface water drainage design. Ancillary services. Definitions. References. Bibliography

25 Railways

Earthworks and drainage. Ballast. Sleepers. Fastenings. Rails. Curved track. Welded track. Switches and crossings. Slab track. Track maintenance and renewal. Railway structures. Inspection and maintenance of structures. Research, development and international collaboration. Bibliography

26 Ports and Maritime Works

Siting of ports and harbours. Port planning. Navigation.

Design of maritime structures. Marginal berths. Piers and jetties. Dolphins. Roll-on roll-off berths. Loads. Fendering. Locks. Pavements. Durability and maintenance. References. Bibliography

27 Electrical Power Supply

Fuels. Plant layout, buildings layout and station siting. Power house steelwork. Roofs, walls, floors and ventilation. Turbo-generator support structures. Cooling-water systems. Natural-draught cooling towers. Chimneys. Nuclear reactors and reactor buildings. Hydro-electric power and pumped storage. Overhead transmission lines and supports. References. Bibliography

28 Water Supplies

Organization and management. Present consumption and estimated demand. Transmission and distribution of water. Measurement of flow in streams. Measurement of water in pipes. Service reservoirs. The underground scheme. Surface water schemes. Formation of reservoirs. Desalination. Treatment of water for potable supply. References. Bibliography

29 Sewerage and Sewage Disposal

Sewerage Introduction. Design of storm sewers. Sewage. Design of sewerage systems. Pumping sewage. Construction. Maintenance. **Sewage treatment** Introduction. Effluent disposal. Preliminary treatment. Primary treatment. Biological treatment. Tertiary treatment. Advanced treatment. Sludge treatment. Sludge disposal. Intermediate technology. References

30 Irrigation, Drainage and River Engineering

Part A: Irrigation and Drainage Irrigation—fundamental concepts. Irrigation methods. Drainage of agricultural land. **Part B: Land Drainage and River Engineering** Land drainage and flood alleviation. Hydrology. Channel regime. Sediment transport. Channel design. Channel improvements. Embankments. Detention basins, washlands and catchwater drains. Structures. Pumping. References

31 Coastal and Maritime Engineering

Tides. Waves. Exceptional water levels. Sea-bed and littoral sediments. Stratification and densimetric flow. Wave and current forces. Scaling laws and models. Surveys and data collection. Design parameters and data analysis. Materials. Sea-defence and coast protection works. Breakwaters. Sea-water intakes and outfalls. References

32 Tunnelling

The options for a tunnel route. Costs of tunnelling. Systematic site investigation. Tunnelling methods related to the ground. Tunnel construction. Aids to tunnelling. Ground movements. Tunnel design. References

33 Project and Contract Management

Introduction. Project and contract organization. Commercial considerations and cashflow. Construction planning. Cost estimating. Project appraisal. Engineering contracts. Contractual measurement and valuation. Project management. References. Bibliography

34 Setting Out on Site

Principles. Surveying instruments and their use in setting out. Working procedures. Site survey and preparations. Setting out

List of Contributors

Peter Ackers, MSc(Eng), CEng, FICE, MIWEM, MASCE
Hydraulics consultant

The late **J Allen,** DSc, LLD, FICE, FRSE
Emeritus Professor, University of Aberdeen

W H Arch, BSc(Eng), CEng, MICE

R W Barrett, MSc
Manager, Underwater Engineering Group, London

S C C Bate, CBE, BSc(Eng), PhD, CEng, FICE, FIStructE
Formerly at the Building Research Establishment, and later
consultant to Harry Stanger Ltd

B C Best, BSc
Consultant

Keith M Brook, BSc, CEng, FICE, FIHT
Wimpey Laboratories Ltd

Robert Cather, BSc
Arup Research and Development

Staff of **Central Electricity Generating Board**
Generation Development and Construction Division

G H Child, MSc, FGS
Keith Farquharson and Associates

C R I Clayton, MSc, PhD, CEng, MICE
Reader in Geotechnical Engineering, University of Surrey

D S Currie, FEng, FICE, MIMechE
Director of Civil Engineering, British Rail

R H R Douglas, BSc(Eng), CEng, FICE, FIHT, MConsE
Sir Frederick Snow and Partners

J B Dwight, MA, MSc, CEng, FIStructE, MIMechE
Emeritus Reader in Structural Engineering, University of
Cambridge

C J Evans, MA(Cantab), FEng, FICE, FIStructE
Wallace Evans and Partners

E V Finn, CEng, FICE, FIStructE, FRSH, MIWEM,
MConsE
Sir Frederick Snow and Partners

P G Fookes, DSc(Eng), PhD, BSc, CEng, FIMM, FGS
Consultant

Staff of **Goodfellow Associates Ltd**
Offshore and subsea technology

T R Graves Smith, MA, PhD, CEng, MICE
Department of Civil Engineering, University of Southampton

D J Irvine, BSc, CEng, FICE
Tarmac Construction Ltd

W M Jenkins, BSc, PhD, CEng, FICE, FIStructE
Emeritus Professor of Civil Engineering, The Hatfield
Polytechnic

T J M Kennie, BSc, MAppSci (Glasgow), ARICS, MInstCES
Lecturer in Engineering Surveying, University of Surrey

Philip King, BSc
Arup Research and Development

D J Lee, BSc Tech, DIC, FEng, FICE, FIStructE
Maunsell Group

T R Mills, CEng, MICE, FIDE
Chartered Engineer

Sir Alan Muir Wood, FRS, FEng, FICE
Consultant, Sir William Halcrow and Partners

F H Needham, BSc(Eng), CEng, ACGI, FICE, FIStructE
Formerly at the Constructional Steel Research and
Development Organisation

I K Nixon, ACGI, CEng, FICE, FGS
Consultant

D J Osborne, BSc(Eng), CEng, FICE, FIHT, MIWEM,
MBIM
Sir Frederick Snow and Partners

W Pemberton, BSc, CEng, FICE
Sir Murdoch MacDonald and Partners

A D M Penman, DSc, CEng, FICE
Geotechnical Engineering Consultant

F H Potter, BSc Tech, CEng, MICE, FIWSc, AMCT
Formerly of the Department of Civil Engineering, Imperial
College of Science and Technology

J L Pratt, BSc(Eng), CEng, MIEE, FWeldI
Formerly Research Manager, Braithwaite & Co. Engineers
Ltd

A Price Jones, BSc(Eng), MSc, CEng, MICE
Soil Mechanics Ltd

D W Quinion, BSc(Eng), CEng, FICE, FIStructE
Tarmac Construction Ltd

A L Randall, CEng, FIStructE
Formerly at the British Steel Corporation

B Richmond, BSc(Eng), PhD, FCGI, CEng, FICE
Maunsell Group

C E Rickard, BSc, CEng, MICE, MIWEM
Sir Murdoch MacDonald and Partners

J Rodin, BSc, CEng, FICE, FIStructE, MConsE
Building Design Partnership

B H Rofe, MA(Cantab), CEng, FICE, FIWEM, FGS
Rofe, Kennard and Lapworth

John H Sargent, CEng, FICE, FGS
Costain Group plc

R J M Sutherland, FEng, BA, FICE, FIStructE
Harris and Sutherland

F L Terrett, MEng, CEng, FICE, MConsE
Posford Duvivier

The late **A R Thomas,** OBE, BSc(Eng), CEng, FICE, FASCE
Formerly consultant, Binnie and Partners

P A Thompson, BSc(Eng), MSc, CEng, FICE, MIWEM
University of Manchester Institute of Science and Technology

M J Tomlinson, CEng, FICE, FIStructE, MConsE
Consulting engineer

J A Turnbull, CEng, FICE, FIHT, DipTE
Mott, Hay & Anderson

Cdr H Wardle
Subsea consultant, Sovereign Oil and Gas plc

Staff of **Watson Hawksley**
Consulting engineers

C J Wilshere, OBE, BA, BAI, CEng, FICE
John Laing Design Associates Ltd

T A Wyatt, PhD
Department of Civil Engineering, Imperial College of Science
and Technology

Hugh C Wylde
Independent management consultant

1

Mathematics and Statistics

B C Best BSc
Consultant

Contents

Mathematics

Statistics

Computers

MATHEMATICS

1.1 Algebra

1.1.1 Powers and roots

The following are true for all values of indices, whether positive, negative or fractional:

$$a^p \times a^q = a^{p+q}$$
$$(a^p)^q = a^{pq}$$
$$(a/b)^p = a^p/b^p$$
$$(ab)^p = a^p b^p$$
$$a^p/a^q = a^{p-q}$$
$$a^{-p} = (1/a)^p = 1/a^p$$
$$p\sqrt{a} = a^{1/p}$$
$$a^0 = 1$$
$$0^p = 0$$

1.1.2 Solutions of equations in one unknown

1.1.2.1 Linear equations

Generally $ax + b = 0$
of which there is one solution or root $x = -b/a$

1.1.2.2 Quadratic equations

Generally $ax^2 + bx + c = 0$
of which there are two solutions or roots

$$x = \frac{-b \pm \sqrt{(b^2 - 4ac)}}{2a} \tag{1.1}$$

where, if $b^2 > 4ac$, the roots are real and unequal, $b^2 = 4ac$, the roots are real and equal, and $b^2 < 4ac$, the roots are conjugate complex.

It is worth attempting to rearrange equations as, often, they can be put into a more familiar form simply by rearrangement, e.g.:

$$ax^{2m} + bx^m + c = 0$$

is a quadratic equation in x^m

while $a/x^2 + b/x + c = 0$

is the quadratic $cx^2 + bx + a = 0$

1.1.2.3 Cubic equations

Generally $x^3 + bx^2 + cx + d = 0$
If the substitution: $x = y - b/3$ is made the equation

becomes $y^3 + ey + f = 0$

where $e = (3c - b^2)/3$

and $f = (2b^3 - 9bc + 27d)/27$

now define

$$A = \left[-\frac{f}{2} + \left(\frac{f^2}{4} + \frac{e^3}{27} \right) \right]^{1/3}$$

$$B = \left[-\frac{f}{2} - \left(\frac{f^2}{4} + \frac{e^3}{27} \right) \right]^{1/3}$$

and the three roots, in terms of y are:

$$y_1 = [A + B]$$

$$y_{2,3} = [-(A + B)/2 \pm \sqrt{-3}(A - B)/2]$$

and in terms of x the three roots are:

$$x_{1,2,3} = y_{1,2,3} - \frac{b}{3}$$

1.1.2.4 Equations of higher degree

Equations of degree higher than the second (quadratic equations) are not solvable directly as the method of solving the cubic equation above shows. Generally recourse must be had to either graphical or numerical techniques.

If the equation be of the form:

$$F(x) = 0$$

e.g. $a_n x^n + a_{n-1} x^{n-1} \ldots + a0 = 0$

then plot the graph of $y = F(x)$ the values of x at which $y = 0$ are the roots or solutions to the equation. Frequently this graphical approach may be used fairly roughly (and therefore quickly) to obtain an estimate of a root. This estimate can then be improved by numerical means. For instance, values of $F(x)$ may be calculated for values of x close to that given as a root by the graphical method. The difficulty (which is not serious for hand calculations) is guessing by how much to adjust x to get $F(x)$ nearer to 0.

1.1.3 Newton's method

This is a method of step-by-step iteration in which an estimate of a root is refined.

Suppose that a_1 is an approximation to a root of an equation then, for small q:

$$F(a_1 + q) \simeq F(a_1) + qF(a_1)$$

So that if we assume $(a_1 + q)$ to be the better solution we are seeking, i.e.:

$$F(a_1 + q) = 0 \tag{1.2}$$

then:

$$q = \frac{-F(a_1)}{F'(a_1)} \tag{1.3}$$

and $a_2 = a_1 + q$ is a second and better approximation.

This is well illustrated by drawing a curve cutting the x-axis, assuming a value a_1 of x near to the intersection to have been found, drawing the ordinate to the curve $x = a_1$ and then constructing the tangent to the curve $y = F(x)$ at the point $x = a_1$.

The point $x = a_2$ where this tangent cuts the axis is plainly a better estimate of the intersection than is a_1.

This technique can be used successfully in automatic calculation on a computer. The problem then becomes that of determining when to stop the iteration process:

$$a_1, a_2, a_3 \ldots$$

which may be best done by stopping when the change between successive approximations, a_n and a_{n+1} becomes less than some small preset amount.

Graphical and numerical methods will generally be required to deal with transcendental equations although in some cases it may be more convenient to find the intersections of two graphs rather than try to compute where a more complicated graph cuts an axis

e.g. $x - \sin x = 0$

is best solved by plotting:

$y = x$

and $y = \sin x$

to find the intersection which will give an estimate which can be refined numerically.

1.1.4 Progressions

(1) Arithmetic progressions in which the difference between consecutive terms is a constant amount. Thus, the terms may be:

$a, a+d, a+2d, a+3d \ldots$

The nth term is $a+(n-1)d$ and the sum to n terms,

$$S_n = \frac{n}{2}\{2a + (n-1)\,d\} \tag{1.4}$$

(2) Geometrical progressions in which the ratio between consecutive terms is a constant. Generally terms are:

$a, ar, ar^2, ar^3 \ldots$

The nth term is ar^{n-1} and the sum of n terms is:

$$S_n = \frac{a(1-r^n)}{1-r} \tag{1.5}$$

If r is strictly smaller than 1 so $-1 < r < 1$, then r^n tends to zero as n becomes larger so that for such geometric progressions we can find the 'sum to infinity' of the series:

$$S_\infty = \frac{a}{1-r} \tag{1.6}$$

The geometric mean of a set of n numbers is the nth root of their product.

If we limit consideration to non-negative numbers then the arithmetic mean of a set of numbers will be greater than or equal to their geometric mean.

1.1.5 Logarithms

Logarithms, which, short of calculating machinery of some form, are probably the greatest aid to computation are based on the properties of indices.

Thus, if we consider logarithms to base a we have the following results:

$a^x = P$ is equivalent to $\log_a P = x$

$a^1 = a$ is equivalent to $\log_a a = 1$

$a^0 = 1$ is equivalent to $\log_a 1 = 0$

So that using rules for powers given on page 1/3:

If: $a^x = P$ and $a^y = Q$

then: $PQ = a^{x+y} H$

so: $\log_a PQ = x + y = \log_a P + \log_a Q$

Similarly: $\log_a (P/Q) = \log_a P - \log_a Q$

Also: $P^n = a^{nx}$

so: $\log aP^n = nx = n \log_a P$

In computation, it is generally convenient to use as base the number 10, i.e. in the expressions given above $a = 10$. However, in fundamental work or integration natural logarithms (also known as Napierian or hyperbolic logarithms) are generally used. These are logarithms to base e a transcendental number given approximately by:

$$e = 2.7182\,8 \tag{1.7}$$

and whose definition can be taken as: 'The value of the solution of the differential equation $dy/dx = y$ for $x = 1$.'

(Note the solution of $dy/dx = y$ is $y = e^x$.)

1.1.6 Permutations and combinations

If, in a sequence of N events, the first can occur in n_1 ways, the second in n_2, etc. then the number of ways in which the whole sequence can occur is:

$n_1 n_2 n_3 \ldots n_N$

1.1.6.1 Permutations

The number of permutations of n different things taken r at a time means the number of ways in which r of these n things can be arranged *in order*. This is denoted by:

$$^nP_r = n(n-1)(n-2) \ldots (n-r+1) = \frac{n!}{(n-r)!} \tag{1.8}$$

where $n! = n(n-1)(n-2), \ldots 3.2.1$ is called factorial n.
It is clear that:

$^nP_n = n!$

and that:

$^nP_1 = n$

If, of n things taken r at a time p things, are to occupy fixed positions then the number of permutations is given by:

$$^{n-p}Pr-p \qquad (1.9)$$

If in the set of n things, there are g groups each group containing $n_1, n_2 \ldots n_g$ things which are identical then the number of permutations of all n things is:

$$\frac{n!}{n_1!n_2!\ldots n_g!}$$

1.1.6.2 Combinations

The number of combinations of n different things, into groups of r things at a time is given by:

$$^nCr = \frac{n!}{r!(n-r)!} = \frac{^nPr}{r!} \qquad (1.10)$$

It is important to note that, whereas in permutations the order of the things does matter, in combinations the order does not matter. From the general expression above, it is clear that:

$$^nCn = 1$$

$$^nC_1 = n \qquad (1.11)$$

If, of n different things taken r at a time p are always to be taken then the number of combinations is:

$$^{n-p}Cr - p \qquad (1.12)$$

If, of n different things taken r at a time p are never to occur the number of combinations is:

$$^{n-p}Cr \qquad (1.13)$$

Note that combinations from an increasing number of available things are related by:

$$^{n+1}Cr = {}^nCr + {}^nCr - 1 \qquad (1.14)$$

also $\quad {}^nCr = {}^nCn - r \qquad (1.15)$

1.1.7 The binomial theorem

The general form of expansion of $(x+a)^n$ is given by:

$$(x+a)^n = {}^nC_0x^n + {}^nC_1 \times {}^{n-1}a^r + {}^nC_2x^{n-2}a^2 \ldots \qquad (1.16)$$

Alternatively this may be written as:

$$(x+a)^n = x^n + nx^{n-1}a + \frac{n(n-1)}{1.2}x^{n-2}a^2 + \frac{n(n-1)(n-2)}{1.2.3}x^{n-3}a^3 \qquad (1.17)$$

It should be noted that the coefficients of terms equidistant from the end are equal (since $^nCr = {}^nCn - r$).

1.2 Trigonometry

The *trigonometric functions* of the angle a (see Figure 1.1) are defined as follows:

$\sin a = y/r$	$\operatorname{cosec} a = r/y$
$\cos a = x/r$	$\sec a = r/x$
$\tan a = y/x$	$\cot a = x/y$

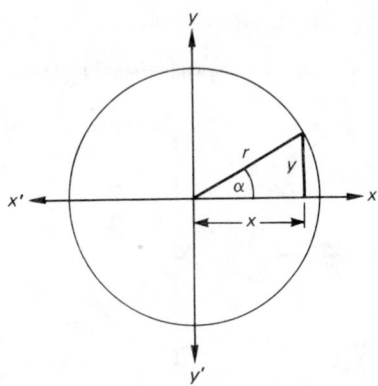

Figure 1.1 Trigonometric functions

These functions satisfy the following identities:

$$\sin^2 a + \cos^2 a = 1$$
$$1 + \tan^2 a = \sec^2 a$$
$$1 + \cot^2 a = \operatorname{cosec}^2 a$$

1.2.1 Positive and negative lines

In trigonometry, lines are considered positive or negative according to their location relative to the coordinate axes xOx', yOy', (see Figure 1.2).

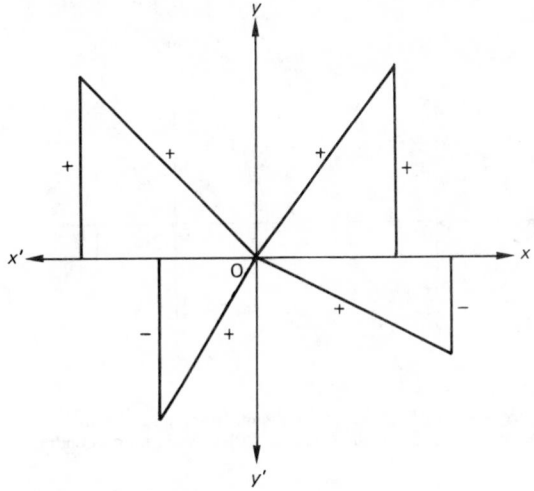

Figure 1.2 Positive and negative lines

1.2.1.2 Positive lines

Radial: any direction.
Horizontal: to right of yOy'.
Vertical: above xOx'.

1.2.1.3 Negative lines

Horizontal: to left of yOy'.
Vertical: below xOx'.

1.2.2 Positive and negative angles

Figure 1.3 shows the convention for signs in measuring angles. Angles are positive if the line OP revolves anti-clockwise from

Ox as in Figure 1.3a and are negative when OP revolves clockwise from Ox.

Signs of trigonometrical ratios are shown in Figure 1.4 and in Table 1.1.

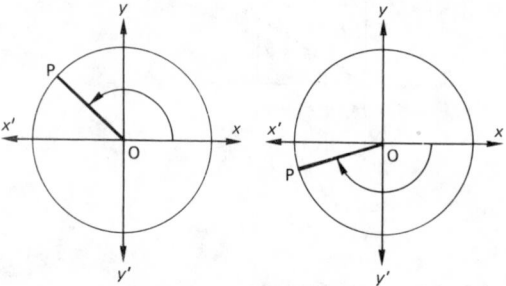

Figure 1.3 (a) Positive (b) negative angle

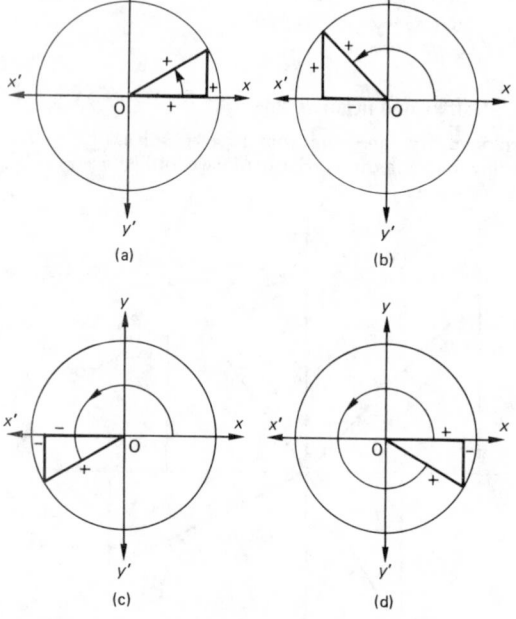

Figure 1.4 (a) Angle in first quadrant; (b) angle in second quadrant; (c) angle in third quadrant; (d) angle in fourth quadrant

1.2.3 Trigonometrical ratios of positive and negative angles

Table 1.1

Quadrant	Sign of ratio	
	positive	negative
First	sin	
	cos	
	tan	
	cosec	
	sec	
	cot	
Second	sin	cos
	cosec	sec
		tan
		cot
Third	tan	sin
	cot	cosec
		cos
		sec
Fourth	cos	sin
	sec	cosec
		tan
		cot

1.2.4 Measurement of angles

1.2.4.1 English or sexagesimal method

1 right angle = 90° (degrees)
1° (degree) = 60′ (minutes)
1′ (minute) = 60″ (seconds)

This convention is universal.

1.2.4.2 French or centesimal method

This splits angles, degrees and minutes into 100th divisions but is not used in practice.

1.2.4.3 The radian

This is a constant angular measurement equal to the angle subtended at the centre of any circle by an arc equal in length to the radius of the circle as shown in Figure 1.5.

$$\pi \text{ radians} = 180°$$

$$1 \text{ radian} = \frac{180}{\pi} = \frac{180}{3.141\,6} = 57° \, 17′ \, 44″ \text{ approximately}$$

Table 1.2

$\sin(-\alpha)$	$= -\sin\alpha$	$\tan(-\alpha)$	$= -\tan\alpha$	$\sec(-\alpha)$	$= \sec\alpha$
$\cos(-\alpha)$	$= \cos\alpha$	$\cot(-\alpha)$	$= -\cot\alpha$	$\mathrm{cosec}(-\alpha)$	$= -\mathrm{cosec}\,\alpha$
$\sin(90°-\alpha)$	$= \cos\alpha$	$\tan(90°-\alpha)$	$= \cot\alpha$	$\sec(90°-\alpha)$	$= \mathrm{cosec}\,\alpha$
$\cos(90°-\alpha)$	$= \sin\alpha$	$\cot(90°-\alpha)$	$= \tan\alpha$	$\mathrm{cosec}(90°-\alpha)$	$= \sec\alpha$
$\sin(90°+\alpha)$	$= \cos\alpha$	$\tan(90°+\alpha)$	$= -\cot\alpha$	$\sec(90°+\alpha)$	$= -\mathrm{cosec}\,\alpha$
$\cos(90°+\alpha)$	$= -\sin\alpha$	$\cot(90°+\alpha)$	$= -\tan\alpha$	$\mathrm{cosec}(90°+\alpha)$	$= \sec\alpha$
$\sin(180°-\alpha)$	$= \sin\alpha$	$\tan(180°-\alpha)$	$= -\tan\alpha$	$\sec(180°-\alpha)$	$= -\sec\alpha$
$\cos(180°-\alpha)$	$= -\cos\alpha$	$\cot(180°-\alpha)$	$= -\cot\alpha$	$\mathrm{cosec}(180°-\alpha)$	$= \mathrm{cosec}\,\alpha$
$\sin(180°+\alpha)$	$= -\sin\alpha$	$\tan(180°+\alpha)$	$= \tan\alpha$	$\sec(180°+\alpha)$	$= -\sec\alpha$
$\cos(180°+\alpha)$	$= -\cos\alpha$	$\cot(180°+\alpha)$	$= \cot\alpha$	$\mathrm{cosec}(180°+\alpha)$	$= -\mathrm{cosec}\,\alpha$

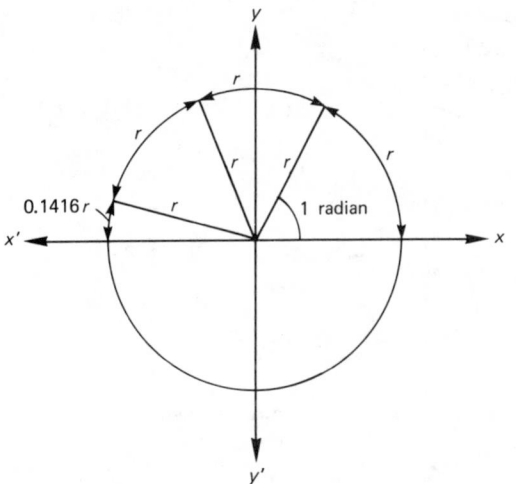

Figure 1.5 The radian

1.2.4.4 Trigonometrical ratios expressed as surds

Table 1.3

Angle in radians	0	$\frac{\pi}{6}$	$\frac{\pi}{4}$	$\frac{\pi}{3}$	$\frac{\pi}{2}$
Angle in degrees	0°	30°	45°	60°	90°
sin	0	$\frac{1}{2}$	$\frac{1}{\sqrt{2}}$	$\frac{\sqrt{3}}{2}$	1
cos	1	$\frac{\sqrt{3}}{2}$	$\frac{1}{\sqrt{2}}$	$\frac{1}{2}$	0
tan	0	$\frac{1}{\sqrt{3}}$	1	$\sqrt{3}$	∞

Table 1.3 gives these ratios for certain angles.

1.2.5 Complementary and supplementary angles

Two angles are complementary when their sum is a right angle; then either is the complement of the other, e.g. the sine of an angle equals the cosine of its complement. Two angles are supplementary when their sum is two right angles.

1.2.6 Graphical interpretation of the trigonometric functions

Figures 1.6 to 1.9 show the variation with α of sin α, cos α, tan α and cosec α respectively. All the trigonometric functions are periodic with period 2π radians (or 360°).

1.2.7 Functions of the sum and difference of two angles

$$\sin (A \pm B) = \sin A \cos B \pm \cos A \sin B$$
$$\cos (A \pm B) = \cos A \cos B \mp \sin A \sin B$$
$$\tan (A \pm B) = \frac{\tan A \pm \tan B}{1 \pm \tan A \tan B}$$

1.2.8 Sums and differences of functions

$$\sin A + \sin B = 2 \sin \tfrac{1}{2}(A + B) \cos \tfrac{1}{2}(A - B)$$
$$\sin A - \sin B = 2 \cos \tfrac{1}{2}(A + B) \sin \tfrac{1}{2}(A - B)$$
$$\cos A + \cos B = 2 \cos \tfrac{1}{2}(A + B) \cos \tfrac{1}{2}(A - B)$$
$$\cos A - \cos B = -2 \sin \tfrac{1}{2}(A + B) \sin \tfrac{1}{2}(A - B)$$
$$\sin^2 A - \sin^2 B = \sin (A + B) \sin (A - B)$$
$$\cos^2 A - \cos^2 B = -\sin (A + B) \sin (A - B)$$
$$\cos^2 A - \sin^2 B = \cos (A + B) \cos (A - B)$$

1.2.9 Functions of multiples of angles

$$\sin 2A = 2\sin A \cos A$$
$$\cos 2A = \cos^2 A - \sin^2 A = 2 \cos^2 A - 1 = 1 - 2 \sin^2 A$$
$$\tan 2A = 2 \tan A/(1 - \tan^2 A)$$
$$\sin 3A = 3 \sin A - 4 \sin^3 A$$
$$\cos 3A = 4 \cos^3 A - 3 \cos A$$
$$\tan 3A = (3 \tan A - \tan^3 A)/(1 - 3 \tan^2 A)$$
$$\sin pA = 2 \sin (p - 1) A \cos A - \sin (p - 2) A$$
$$\cos pA = 2 \cos (p - 1) A \cos A - \cos (p - 2) A$$

1.2.10 Functions of half angles

$$\sin A/2 = \sqrt{\left(\frac{1 - \cos A}{2}\right)} = \frac{\sqrt{(1 + \sin A)}}{2} - \frac{\sqrt{(1 - \sin A)}}{2}$$
$$\cos A/2 = \sqrt{\left(\frac{1 + \cos A}{2}\right)} = \frac{\sqrt{(1 + \sin A)}}{2} + \frac{\sqrt{(1 - \sin A)}}{2}$$
$$\tan A/2 = \frac{1 - \cos A}{\sin A} = \frac{\sin A}{1 + \cos A} = \sqrt{\left(\frac{1 - \cos A}{1 + \cos A}\right)}$$

1.2.11 Relations between sides and angles of a triangle (Figures 1.10 and 1.11)

$$\frac{a}{\sin A} = \frac{b}{\sin B} = \frac{c}{\sin C}$$

$$a = b \cos C + c \cos B$$

$$c^2 = a^2 + b^2 - 2ab \cos C \qquad (1.18)$$

$$\sin A = \frac{c}{bc} \sqrt{\{s(s - a)(s - b)(s - c)\}} \qquad (1.19)$$

where $2s = a + b + c$

Area of triangle $\triangle = \tfrac{1}{2}ab \sin C = \sqrt{\{s(s - a)(s - b)(s - c)\}}$

$$\tan \frac{A}{2} = \sqrt{\left\{\frac{(s - b)(s - c)}{s(s - a)}\right\}}$$

$$\cos \frac{A}{2} = \sqrt{\left\{\frac{s(s - a)}{bc}\right\}}$$

$$\sin \frac{A}{2} = \sqrt{\left\{\frac{(s - b)(s - c)}{bc}\right\}}$$

$$\tan \frac{B - C}{2} = \frac{(b - c)}{(b + c)} \cot \frac{A}{2}$$

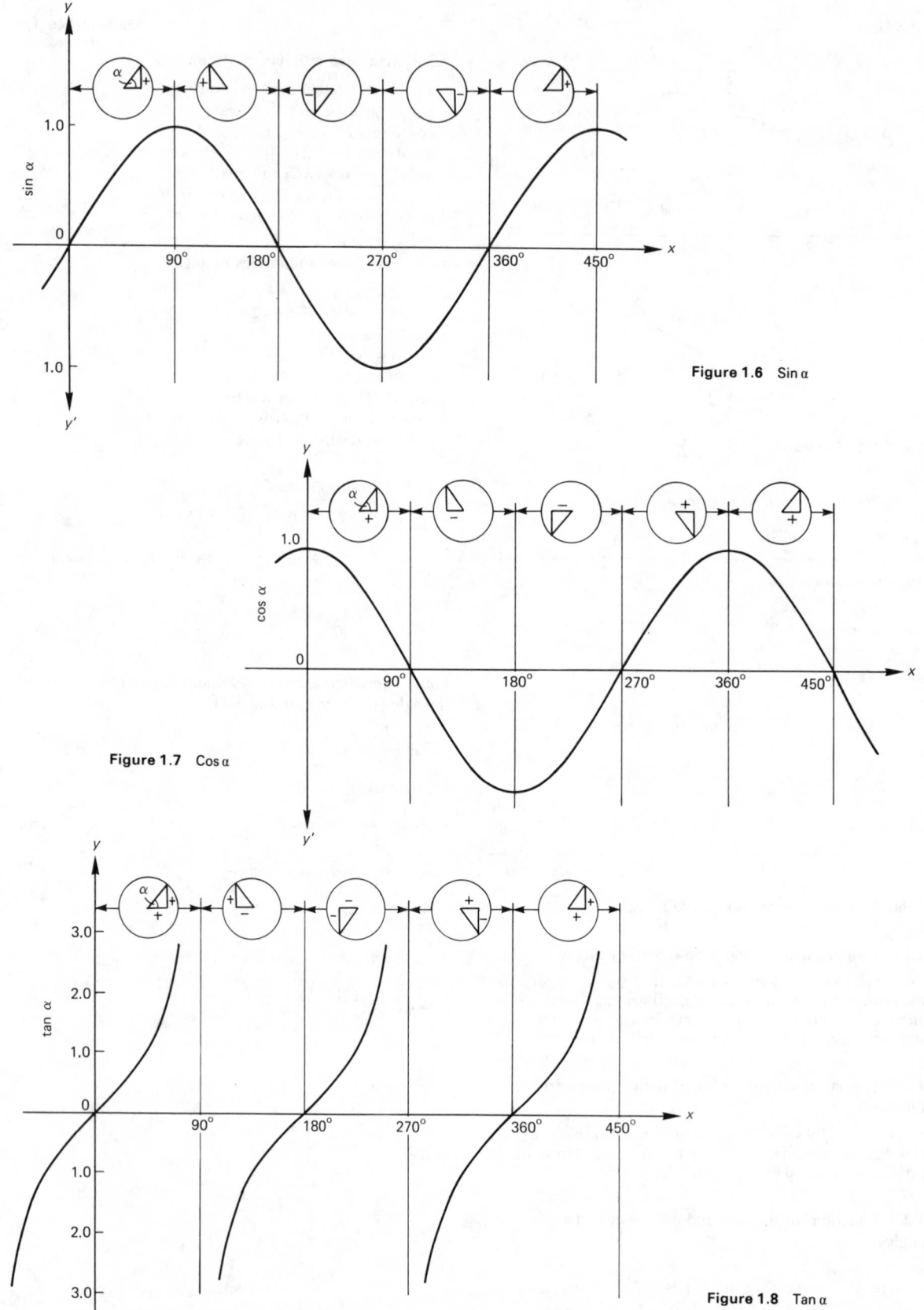

Figure 1.6 Sin α

Figure 1.7 Cos α

Figure 1.8 Tan α

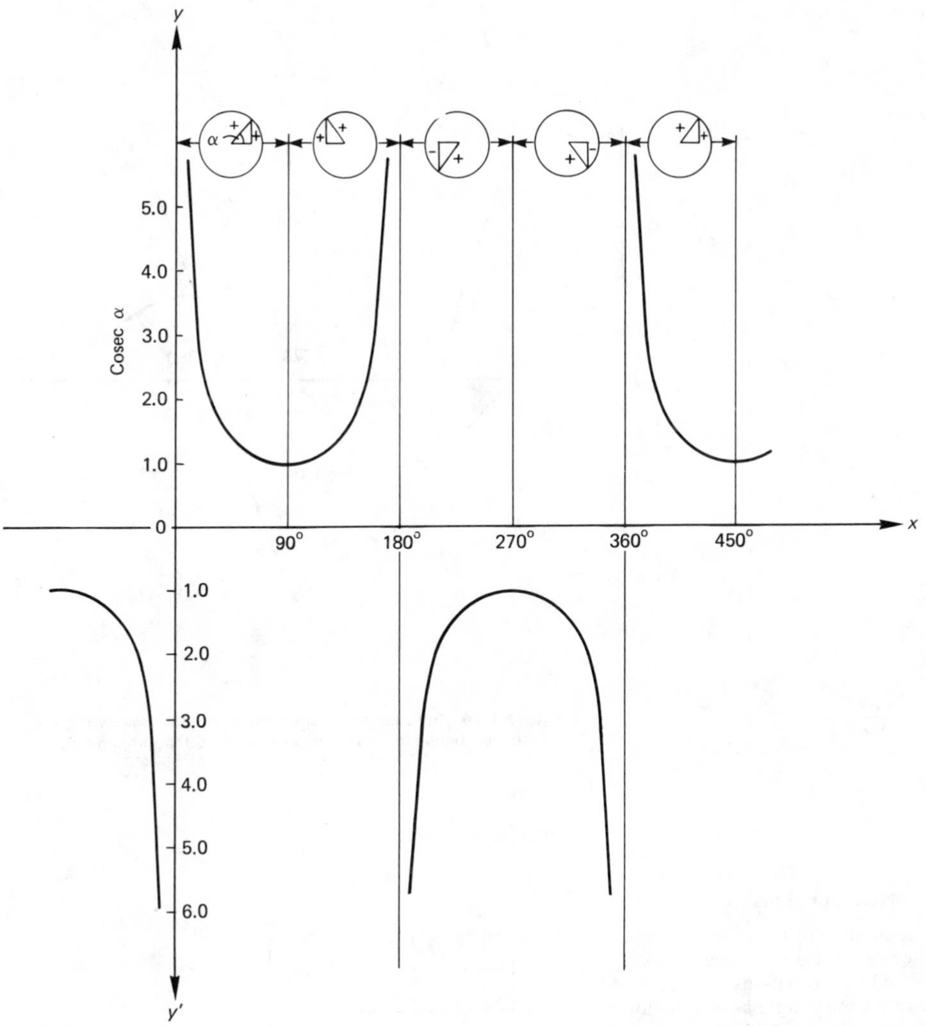

Figure 1.9 Cosec α

1.2.11.1 Any right angled triangle (Figure 1.12)

$a^2 + b^2 = c^2$; $A + B = 90°$; $\sin A = \cos B$; $\cot A = \tan B$ etc.

Area of $\triangle ABC = \frac{1}{2}ab = \frac{1}{2}bc \sin A = \frac{1}{2}ac \sin B$

1.2.11.2 Any equilateral triangle (Figure 1.13)

$a = b = c$; $A = B = C = 60° = \pi/3$

$$h = \frac{b\sqrt{3}}{2}; \quad \text{area} = \frac{b^2\sqrt{3}}{4}$$

Circumscribed circle radius $R = \frac{b\sqrt{3}}{3}$

Inscribed circle, radius $r = \frac{b\sqrt{3}}{6}$

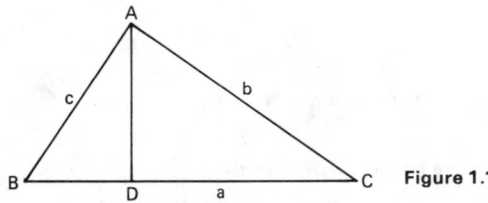

Figure 1.10

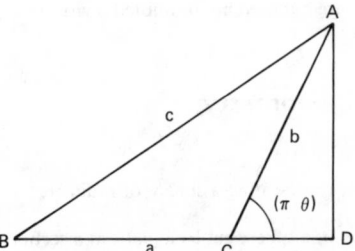

Figure 1.11

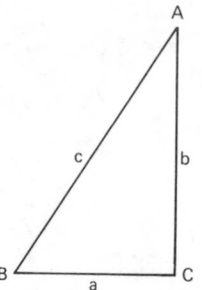

Figure 1.12

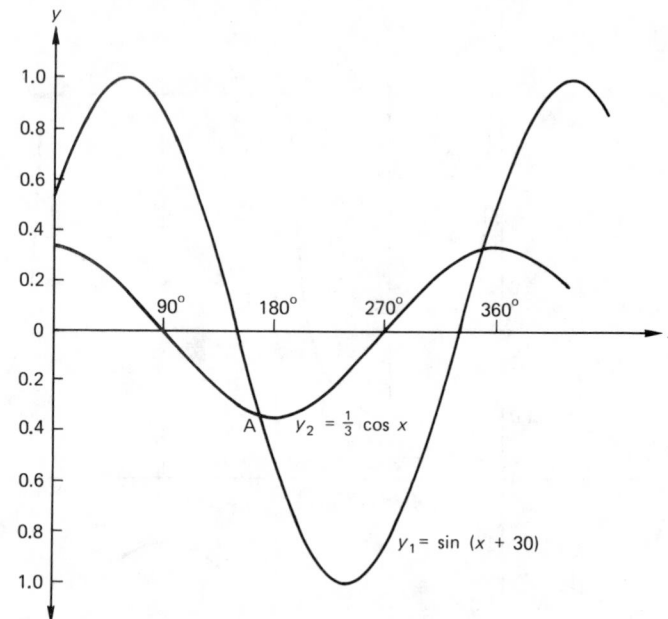

Figure 1.14   Solution of trigonometrical equations showing the intersection between $x=10$ and $x=\pi$ as $x=169°$ approximately

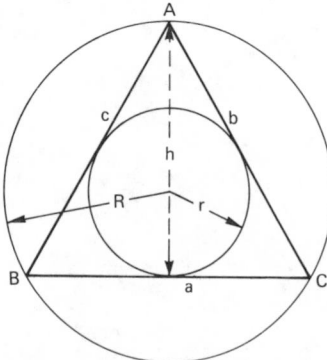

Figure 1.13

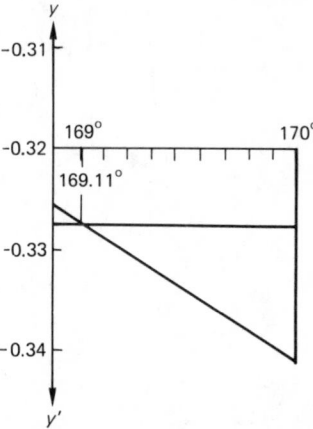

Figure 1.15   Enlargement at A of Figure 1.14

1.2.12 Solution of trigonometric equations

The method best suited to the solution of trigonometric equations is that described in the section on algebra which deals with the method of solving transcendental equations by means of graphs. The expression to be solved is arranged as two identities and two graphs drawn as shown in Figure 1.14. The points of intersection of the curves projected on to the coordinate axes give the values which will satisfy the trigonometric equation.

Example 1.1   Solve $\sin(x+30)=\frac{1}{3}\cos x$ for x between 0 and 2π.

Assigning values to x in Table 1.4 and calculating the corresponding values for $y=\sin(x+30)$ and $y=\frac{1}{3}\cos x$ gives the readings for plotting the curves in Figure 1.14.

Plotting the curves between $x=169°$ and $170°$ shows that the intersection is at $x=169.11°$ to the second approximation. Greater accuracy can be obtained by continuing the small range large scale plots of the type in Figure 1.15.

There is one further value of x between $x=300°$ and $360°$ which will satisfy the equation as can be seen on Figure 1.14.

1.2.13 General solutions of trigonometric equations

Due to the periodic nature of the trigonometric functions there is an infinite number of solutions to trigonometric equations. Having obtained the smallest positive solution, α, the general solution for θ is then given by:

$$\begin{array}{lll} if & a=\sin^{-1}x & then \quad \theta=n\pi+(-1)^n a \\ & a=\cos^{-1}x & \theta=2n\pi\pm a \\ & a=\tan^{-1}x & \theta=n\pi+a \end{array}$$

where θ and a are measured in radians and n is any integer.

1.2.14 Inverse trigonometric functions

Inverse functions of trigonometric variables may be simply defined by the example: $y=\sin^{-1}\frac{1}{2}$ which is merely a symbolic way of stating that y is an angle whose sine is $\frac{1}{2}$, i.e. y is actually 30° or $\pi/6$ in radian measure but need not be quoted if written as $\sin^{-1}\frac{1}{2}$.

1.3 Spherical trigonometry

1.3.1 Definitions

Referring to Figure 1.16, representing a sphere of radius r:

Small circle   The section of a sphere cut by a plane at a section not on the diameter of the sphere, e.g. EFGH.

Table 1.4

x	0	30	60	90	120	150	180
$y_1 = \sin(x+30)$	0.5	0.866	1.0	0.866	0.5	0	−0.5
$y_2 = \frac{1}{3}\cos x$	0.333	0.289	0.166 7	0	−0.166 7	−0.289	−0.333

x	210	240	270	300	330	360
$y_1 = \sin(x+30)$	−0.866	−1.0	−0.866	−0.5	0	+0.5
$y_2 = \frac{1}{3}\cos x$	−0.289	−0.166 7	0	0.166 7	0.289	+0.333

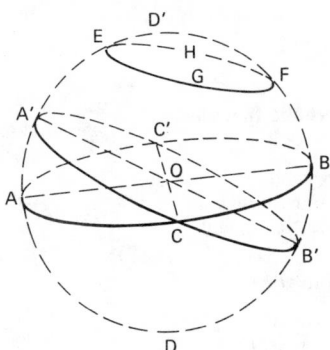

Figure 1.16 Sphere illustrating spherical trigonometry definitions

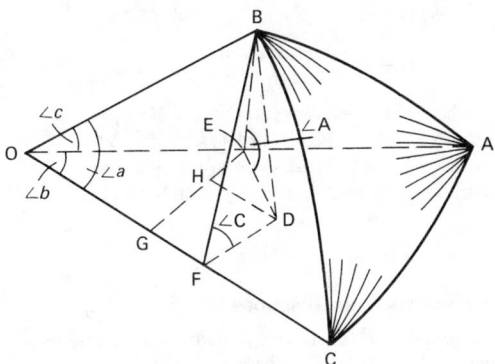

Figure 1.17 Spherical triangles

Great circle The section of a sphere cut by a plane through any diameter, e.g. ACBC'.

Poles Poles of any circular section of a sphere are the ends of a diameter at right angles to the section, e.g. D and D' are the poles of the great circle ACBC'.

Lunes The surface areas of that part of the sphere between two great circles; there are two pairs of congruent areas, e.g. ACA'C'A; CBC'B'C and ACB'C'A; A'CBC'A'.

Area of lune If the angle between the planes of two great circles forming the lune is θ (radians), its surface area is equal to $2\theta r^2$.

Spherical triangle A curved surface included by the arcs of three great circles, e.g. CB'B is a spherical triangle formed by one edge BB' on part of the great circle DB'BA the second edge

B'C on great circle B'CA'C' and edge CB on great circle ACBD'. The angles of a spherical triangle are equal to the angles between the planes of the great circles or, alternatively, the angles between the tangents to the great circles at their points of intersection. They are denoted by the letters C, B', B for the triangle CB'.

Area of spherical triangle $\text{CB'B} = (B' + B + C - \pi)r^2$.

Spherical excess Comparing a plane triangle with a spherical triangle the sum of the angles of the former is π and the spherical excess E of a spherical triangle is given by $E = B' + B + C - \pi$; hence, area of a spherical triangle can be expressed as $(E/4\pi) \times$ surface of sphere.

Spherical polygon A spherical polygon of n sides can be divided into $(n-2)$ spherical triangles by joining opposite angular points by the arcs of great circles.

$$\text{Area of spherical polygon} = [\text{sum of angles} - (n-2)\pi]r^2$$

$$= \frac{E}{4\pi} \times \text{surface of sphere.}$$

Note that $(n-2)\pi$ is the sum of the angles of a plane polygon of n sides.

1.3.2 Properties of spherical triangles

Let ABC, in Figure 1.17, be a spherical triangle; BD is a perpendicular from B on plane OAC and OÊD, OF̂D, OÊB, OF̂B, OĜE, DĤG are right angles; then BÊD = A and BF̂D = C are the angles between the planes OBA, OAC and OBC, OAC respectively. DÊH = CÔA = b also CÔB = a, AÔB = c, and since OB = OA = OC = radius r of sphere, OF = $r\cos a$, OE = $r\cos c$; then

$$\cos a = \cos b \cos c + \sin b \sin c \cos A$$
$$\cos b = \cos a \cos c + \sin a \sin c \cos B$$
$$\cos c = \cos a \cos b + \sin a \sin b \cos C$$

Also the sine formulae are:

$$\frac{\sin A}{\sin a} = \frac{\sin B}{\sin b} = \frac{\sin C}{\sin c}$$

and the cotangent formulae are:

$$\sin a \cot c = \cos a \cos B + \sin B \cot C$$
$$\sin b \cot c = \cos b \cos A + \sin A \cot C$$
$$\sin b \cot a = \cos b \cos C + \sin C \cot A$$
$$\sin c \cot a = \cos c \cos B + \sin B \cot A$$
$$\sin c \cot b = \cos c \cos A + \sin A \cot B$$
$$\sin a \cot b = \cos a \cos C + \sin C \cot B$$

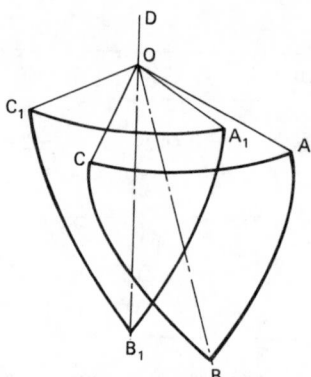

Figure 1.18 Polar triangles

1.4.1 Relation of hyperbolic to circular functions

$$\sin \theta = -i \sinh i\theta$$
$$\cos \theta = \cosh i\theta$$
$$\tan \theta = i \tanh i\theta$$
$$\operatorname{cosec} \theta = i \operatorname{cosech} i\theta$$
$$\sec \theta = \operatorname{sech} i\theta$$
$$\cot \theta = i \coth i\theta$$
$$\sinh \theta = -i \sin i\theta$$
$$\cosh \theta = \cos i\theta$$
$$\tanh \theta = -i \tan i\theta$$
$$\operatorname{cosech} \theta = i \operatorname{cosec} i\theta$$
$$\operatorname{sech} \theta = i \sec i\theta$$
$$\coth \theta = i \cot i\theta$$

In Figure 1.18, ABC, $A_1B_1C_1$ are two spherical triangles in which A_1, B_1, C_1 are the poles of the great circles BC, CA, AB respectively; then $A_1B_1C_1$ is termed the polar triangle of ABC and vice versa. Now OA_1, OD are perpendicular to the planes BOC and AOC respectively; hence $A_1\hat{O}D$ = angle between planes BOC and AOC = C. Let sides of triangle $A_1B_1C_1$ be denoted by $a_1b_1c_1$ then $c_1 = A_1\hat{O}B_1 = \pi - C$ also $a_1 = \pi - A$ and $b_1 = \pi - B$; $c = \pi - C_1$; $a = \pi - A_1$; $b = \pi - B_1$, and from these we get

$$\cos b = \frac{\cos B + \cos A \cos C}{\sin A \sin C} \qquad (1.20)$$

$$\cos a = \frac{\cos A + \cos B \cos C}{\sin B \sin C} \qquad (1.21)$$

$$\cos c = \frac{\cos C + \cos A \cos B}{\sin A \sin B} \qquad (1.22)$$

1.3.2.1 Right-angled triangles

If one angle A of a spherical triangle ABC is 90° then cos $a = \cos b \cos c = \cot B \cot C$

$$\cos B = \frac{\tan c}{\tan a}; \quad \cos C = \frac{\tan b}{\tan a}; \quad \sin B = \frac{\sin b}{\sin c};$$
$$\sin C = \frac{\sin c}{\sin a}; \quad \tan B = \frac{\tan b}{\sin c}; \quad \tan C = \frac{\tan c}{\sin b};$$
$$\cos B = \cos b \sin C; \quad \cos C = \cos c \sin B.$$

1.4 Hyperbolic trigonometry

The hyperbolic functions are related to a rectangular hyperbola in a manner similar to the relationship between the ordinary trigonometric functions and the circle. They are defined by the following exponential equivalents:

$$\sinh \theta = \frac{e^\theta - e^{-\theta}}{2} \qquad \operatorname{cosech} \theta = \frac{1}{\sinh \theta}$$

$$\cosh \theta = \frac{e^\theta + e^{-\theta}}{2} \qquad \operatorname{sech} \theta = \frac{1}{\cosh \theta}$$

$$\tanh \theta = \frac{\sinh \theta}{\cosh \theta} \qquad \coth \theta = \frac{1}{\tanh \theta}$$

1.4.2 Properties of hyperbolic functions

$$\cosh^2 \theta - \sinh^2 \theta = 1$$
$$\operatorname{sech}^2 \theta = 1 - \tanh^2 \theta$$
$$\sinh 2\theta = 2 \sinh \theta \cosh \theta$$
$$\cosh 2\theta = \cosh^2 \theta + \sinh^2 \theta$$

$$\operatorname{cosech}^2 \theta = \coth^2 \theta - 1$$

$$\tanh 2\theta = \frac{2 \tanh \theta}{1 + \tanh^2 \theta}$$

$$\sinh (x \pm y) = \sinh x \cosh y \pm \cosh x \sinh y$$
$$\cosh (x \pm y) = \cosh x \cosh y \pm \sinh x \sinh y$$

$$\tanh (x \pm y) = \frac{\tanh x \pm \tanh y}{1 \pm \tanh x \tanh y}$$

$$\sinh x + \sinh y = 2 \sinh \tfrac{1}{2}(x+y) \cosh \tfrac{1}{2}(x-y)$$
$$\sinh x - \sinh y = 2 \cosh \tfrac{1}{2}(x+y) \sinh \tfrac{1}{2}(x-y)$$
$$\cosh x + \cosh y = 2 \cosh \tfrac{1}{2}(x+y) \cosh \tfrac{1}{2}(x-y)$$
$$\cosh x - \cosh y = 2 \sinh \tfrac{1}{2}(x+y) \sinh \tfrac{1}{2}(x-y)$$

1.4.3 Inverse hyperbolic functions

As with trigonometric functions, we define the inverse hyperbolic functions by $y = \sinh^{-1} x$ where $x = \sinh y$:

Therefore: $x = (e^y - e^{-y})/2$

Rearranging and adding x^2 to each side:

$$e^{2y} - 2x \cdot e^y + x^2 = x^2 + 1$$

or: $e^y - x = \sqrt{(x^2 + 1)}$

and therefore: $y = \sinh^{-1} x = \log_e [x + \sqrt{(x^2 + 1)}]$ \qquad (1.23)

The other inverse functions may be treated similarly. We find:

$$\sinh^{-1} x = \log [x + \sqrt{(x^2 + 1)}];$$

$$\cosh^{-1} x = \log [x + \sqrt{(x^2 - 1)}];$$

$$\tanh^{-1} x = \tfrac{1}{2} \log \frac{1+x}{1-x};$$

$$\operatorname{cosech}^{-1} x = \log \frac{1 + \sqrt{(1+x^2)}}{x}$$

$$\operatorname{sech}^{-1} x = \log \frac{1 + \sqrt{(1-x^2)}}{x}$$

$$\coth^{-1} x = \tfrac{1}{2} \log \frac{x+1}{x-1}$$

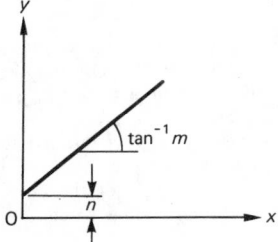

Figure 1.19 Straight-line equation $y=mx+n$

The relationships with the corresponding inverse trigonometric functions are as follows:

$$\sinh^{-1} x = -i \sin^{-1} ix$$
$$\cosh^{-1} x = i \cos^{-1} x$$
$$\tanh^{-1} x = -i \tan^{-1} ix$$
$$\sin^{-1} x = -i \sinh^{-1} ix$$
$$\cos^{-1} x = -i \cosh^{-1} x$$
$$\tan^{-1} x = i \tanh^{-1} ix$$

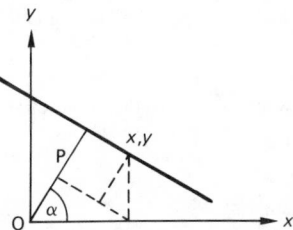

Figure 1.20 Straight-line equation $x \cos \alpha + y \sin \alpha = p$

1.5 Coordinate geometry

1.5.1 Straight-line equations

The equation of a straight line may be expressed as:

(1) $ax+by+c=0$ or $y = -\dfrac{a}{b}x - \dfrac{c}{b} = mx+n$ (1.24)

where a, b and c are constants and m is the slope of the line as shown in Figure 1.19.

(2) $\dfrac{x}{k} + \dfrac{y}{l} = 1$ (1.25)

where k is the intercept on the x axis and l is the intercept on the y axis.

(3) $x \cos a + y \sin a = p$ (1.26)

where p = length of the perpendicular from the origin to the line and a the inclination of this perpendicular to Ox in Figure 1.20.

The length d of a perpendicular (see Figure 1.21) from any point $(x'y')$ to a straight line is given by $(ax'+by'+c)/\sqrt{(a^2+b^2)}$ if the straight line equation is as given in (1), or $(x' \cos a + y' \sin a - p)$ if the straight line equation is as given in (3).

The equation of a straight line through one given point $(x'y')$ is $y - y' = m(x - x')$.

The equation of a straight line through two given points (Figure 1.22) $(x_1 y_1)(x_2 y_2)$ is:

$$\frac{y - y_1}{y_2 - y_1} = \frac{x - x_1}{x_2 - x_1}$$ (1.27)

The angle ψ between two straight lines (Figure 1.23) $y = m_1 x + n_1$ and $y = m_2 x + n_2$ is given by:

$$\tan \psi = \frac{m_1 - m_2}{1 + m_1 m_2}$$ (1.28)

For lines which are parallel $m_1 = m_2$.
For lines at right angles $1 + m_1 m_2 = 0$.

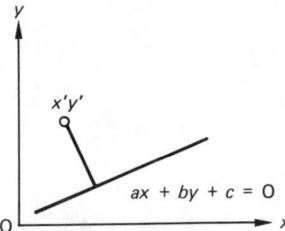

Figure 1.21 Perpendicular to straight line

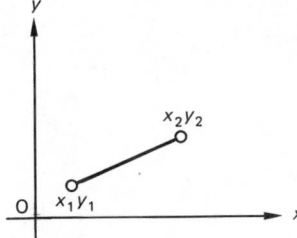

Figure 1.22 Straight line through two points

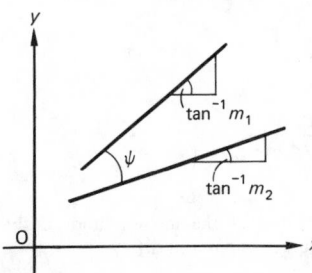

Figure 1.23 Angle ψ between two straight lines

1.5.2 Change of axes

Let the equation of the curve be $y=f(x)$ referred to coordinate axes Ox, Oy; then its equation relative to axes $O'x'$, $O'y'$ parallel to Ox, Oy with origin O' at point (r, s) is given by $y+s=f(x+r)$ in which x and y refer to the new axes.

If the equation of a curve is given by $y=f(x)$ referred to coordinate axes Ox, Oy, then if these axes are each rotated an angle ψ anti-clockwise about O, the equation of the curve referred to the rotated axes is given by $x\sin\psi+y\cos\psi = f(x\cos\psi-y\sin\psi)$.

1.5.2.1 Tangent and normal to any curve $y=f(x)$

The tangent PT and the normal PN at any point x_1y_1 on the curve $y=f(x)$ in Figure 1.24 are given by the following equations:

Tangent: $y-y_1=\dfrac{dy}{dx}(x-x_1)$ where $\dfrac{dy}{dx}=m=$ the slope of the curve at P

Normal: $(y-y_1)\dfrac{dy}{dx}+(x-x_1)=0$

If ϕ be the angle which the tangent at P makes with the axis of x, then:

$$\tan\phi=\frac{dy}{dx}; \cos\phi=\frac{dx}{ds}; \sin\phi=\frac{dy}{ds}$$

where s is the distance measured along the curve.

1.5.2.2 Tangent and normal to any curve $f(xy)=0$

The function is implicit in this case so that partial differential coefficients are employed in the equations for the tangent and for the normal at x_1 y_1.

Tangent: $(y-y_1)\dfrac{\partial f}{\partial y}+(x-x_1)\dfrac{\partial f}{\partial x}=0$

Normal: $\dfrac{(y-y_1)}{(\partial f/\partial y)}=\dfrac{(x-x_1)}{(\partial f/\partial x)}$

where $\dfrac{dy}{dx}=-\dfrac{\partial f}{\partial x}\Big/\dfrac{\partial f}{\partial y}$

1.5.2.3 Subtangent and subnormal to any curve $y=f(x)$

The subtangent is TQ and the subnormal is QN at any point $P(x_1y_1)$ on the curve $y=f(x)$ in Figure 1.24. Their lengths are given by:

Subtangent, $TQ=y_1\Big/\left(\dfrac{dy}{dx}\right)_1$

and subnormal, $QN=y_1\left(\dfrac{dy}{dx}\right)_1$

Example 1.2 Find the equation of the tangent and of the normal where $x=p$ on the curve $y=\cos\pi x/(2p)$

$$\frac{dy}{dx}=-\frac{\pi}{2p}\sin\frac{\pi x}{2p} \text{ and when } x=p, \sin\frac{\pi x}{2p}=1,$$

i.e. $\dfrac{dy}{dx}=-\dfrac{\pi}{2p}$ and $y=0$

Therefore:

the required equation of the tangent is $y=-\dfrac{\pi}{2p}(x-p)$

and the equation of the normal is $y=\dfrac{2p}{\pi}(x-p)$

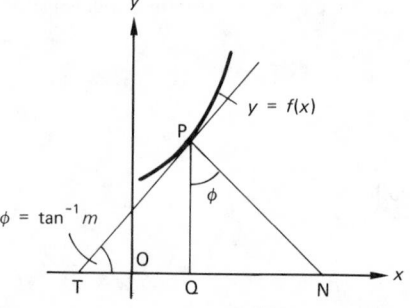

Figure 1.24 Tangent, normal, subtangent and subnormal to curve

1.5.3 Polar coordinates

The polar coordinates of any point P in a plane are given by r, θ where r is the length of the line joining P to the origin O and θ is the inclination of OP, the radius vector relative to the axis Ox (see Figure 1.25).

The relations between the rectangular coordinates x and y and the polar coordinates r and θ are:

$$x=r\cos\theta, \ y=r\sin\theta;$$

$$r=\sqrt{(x^2+y^2)}, \ \theta=\tan^{-1}y/x$$

If PT is a tangent to the curve at point P then:

$$\tan\phi=rd\theta/dr; \cot\phi=(1/r)(dr/d\theta);$$

$$\sin\phi=rd\theta/ds \text{ and } \cos\phi=dr/ds$$

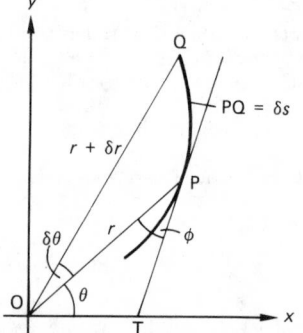

Figure 1.25 Polar coordinates

1.5.3.1 Polar subtangent and subnormal

In Figure 1.26 the polar subtangent is OR and the polar subnormal is OQ where QR is perpendicular to OP and their lengths are given by: polar subtangent $=r^2d\theta/dr$; polar subnormal $=dr/d\theta$.

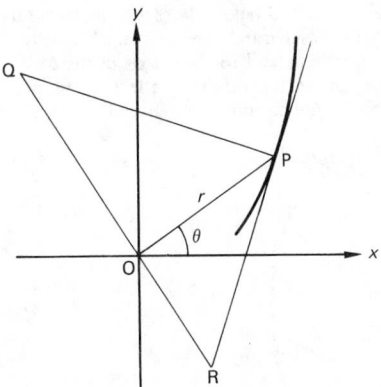

Figure 1.26 Polar subtangent and subnormal

1.5.3.2 Curvature

Let PQ in Figure 1.27 represent an elemental length δs of a given curve and PS, QT the tangents at the points P, Q then:

Curvature at $P = d\beta/ds$. For a circle centre at C, radius ρ, $ds = \rho\, d\beta$, i.e. curvature $= 1/\rho$.

Therefore: $\qquad \rho = \dfrac{ds}{d\beta} = \text{radius of curvature} \qquad (1.29)$

Putting $\beta = \tan^{-1}\left(\dfrac{dy}{dx}\right)$ and differentiating:

$$\text{Curvature} = \frac{d\beta}{ds} = \frac{1}{\rho} = \frac{\dfrac{d^2y}{dx^2}}{\left[1+\left(\dfrac{dy}{dx}\right)^2\right]^{3/2}}$$

$$\text{Radius of curvature } \rho = \frac{\left[1+\left(\dfrac{dy}{dx}\right)^2\right]^{3/2}}{\dfrac{d^2y}{dx^2}}$$

Where dy/dx is small (as in the bending of beams), the radius of curvature is given by:

$$\rho = \frac{1}{d^2y/dx^2} \qquad (1.30)$$

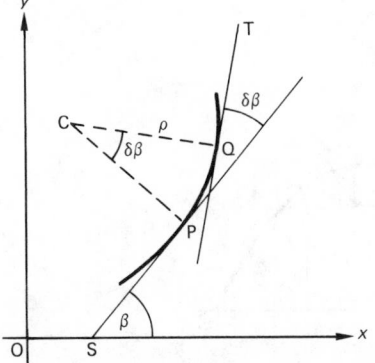

Figure 1.27 Curvature

Example 1.3 Find the radius of curvature at any point at the curve $y = a \cos x/a$:

$$\frac{dy}{dx} = \sinh \frac{x}{a}$$

Therefore:

$$\left[1+\left(\frac{dy}{dx}\right)^2\right]^{3/2} = \left[1+\sinh^2\frac{x}{a}\right]^{3/2}$$

$$= \left(\cosh^2\frac{x}{a}\right)^{3/2} = \cosh^3\frac{x}{a}$$

$$\frac{d^2y}{dx^2} = \frac{1}{a}\cosh\frac{x}{a}$$

Therefore:

$$\rho = \frac{a\cosh^3\dfrac{x}{a}}{\cosh\dfrac{x}{a}} = a\cosh^2\frac{x}{a} = \frac{y^2}{a}$$

1.5.4 Lengths of curves

1.5.4.1 General theory

From Figure 1.28:

$$ds^2 = dx^2 + dy^2$$

Hence:

$$ds = \sqrt{\left\{1+\left(\frac{dy}{dx}\right)^2\right\}}\,dx = \sqrt{\left\{1+\left(\frac{dx}{dy}\right)^2\right\}}\,dy$$

Therefore: $\qquad s = \displaystyle\int_a^b \sqrt{\left\{1+\left(\frac{dy}{dx}\right)^2\right\}}\,dx$

or: $\qquad s = \displaystyle\int_c^d \sqrt{\left\{1+\left(\frac{dx}{dy}\right)^2\right\}}\,dy$

For the evaluation of s for any given continuous function, use the first formula if x is single-valued, i.e. if one value of x corresponds to one point only in the function, e.g. Figure 1.29. If more than one point on the curve corresponds to one value of x, the second formula for a curve of the form shown in Figure 1.30, should be used.

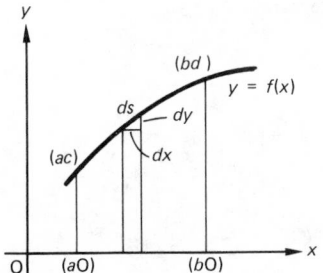

Figure 1.28

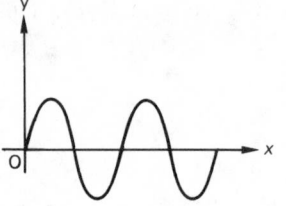

Figure 1.29

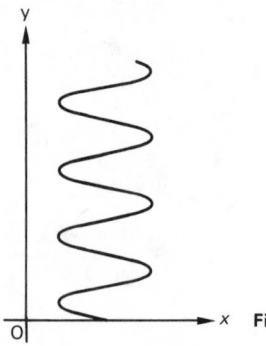

Figure 1.30

For polar coordinates, from Figure 1.31

$$ds = \sqrt{\{(\rho d\theta)^2 + (d\rho)^2\}} = \sqrt{\left\{\rho^2 + \left(\frac{d\rho}{d\theta}\right)^2\right\}} d\theta$$

$$s = \int_{\theta_1}^{\theta_2} \sqrt{\left\{\rho^2 + \left(\frac{d\rho}{d\theta}\right)^2\right\}} d\theta \qquad (1.32)$$

or:

$$s = \int_{\rho_1}^{\rho_2} \sqrt{\left\{1 + \left(\rho\frac{d\theta}{d\theta\rho}\right)^2\right\}} d\rho \qquad (1.33)$$

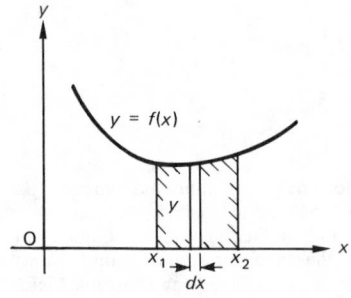

Figure 1.31

1.5.5 Plane areas by integration

See Figures 1.32 and 1.33.

1.5.5.1 General theory

From Figure 1.32, $A = \int_{x_1}^{x_2} y\,dx = \int_{x_1}^{x_2} f(x)\,dx$

1.5.5.2 Polar coordinates

From Figure 1.33, $dA = \frac{1}{2}\rho^2\,d\theta$

Therefore: $\qquad A = \frac{1}{2}\int\rho^2 d\theta = \frac{1}{2}\int\{f(\theta)\}^2 d\theta \qquad (1.34)$

(*Note.* For curve cutting x axis, equate $f(x)$ to zero, find values of x for $y = 0$ and integrate between these values for the area cut off by the x axis.)

When the area lies above and below the x axis integrate the positive and negative areas separately and add algebraically.

Where the area does not extend to the x axis in the case of cartesian coordinates, or to the origin in the case of polar coordinates, then double integration must be used.

Thus: $\quad A = \iint dx\,.\,dy\,.\iint\rho\,.\,d\rho\,.\,d\theta \qquad (1.35)$

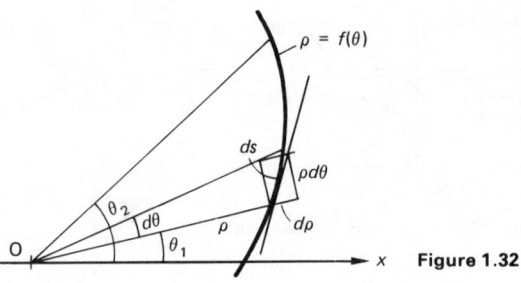

Figure 1.32

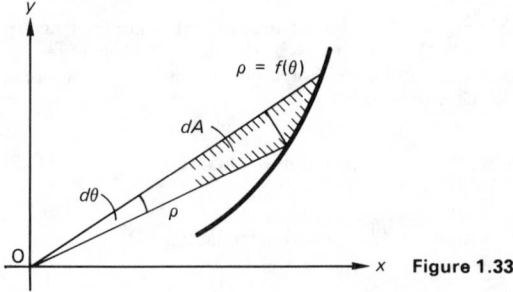

Figure 1.33

1.5.6 Plane area by approximate methods

See Figure 1.34.

1 *Trapezoidal rule*:

$$A = \frac{h}{2}\{y_0 + 2(y_1 + y_2 + \ldots + y_{n-1}) + y_n\} \qquad (1.36)$$

(2) *Durand's rule*:

$$A = h(0.4y_0 + 1.1y_1 + y_3 + \ldots + y_{n-2} + 1.1y_{n-1} + 0.4y_n) \qquad (1.37)$$

(3) *Simpson's rule* (n *made even*)

$$A = \frac{h}{3}(y_0 + 4y_1 + 2y_2 + 4y_3 + 2y_4 + \ldots + 2y_{n-2} + 4y_{n-1} + y_n) \qquad (1.38)$$

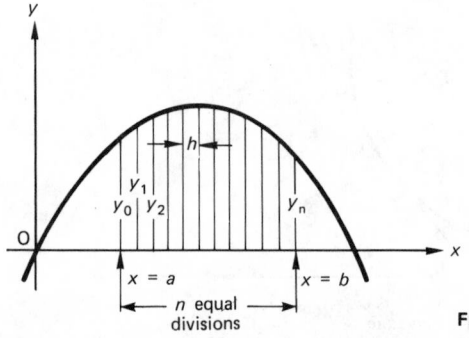

Figure 1.34

Of these, Simpson's is the most accurate. The accuracy is increased in all cases by increasing the number of divisions. Areas can often be determined more rapidly by plotting on squared paper and 'counting the squares' or by the use of a planimeter.

1.5.7 Conic sections

Conic sections refer to the various profiles of sections cut from a pair of cones vertex to vertex when intersected by a plane. Figure 1.35 shows a pair of cones generated by two intersecting straight lines AB, CD about the bisector EF of the angle between the lines.

Two straight lines. A section through the axis EF.

Circle. A section b–b parallel to the base of a cone.

Ellipse. A section c–c not parallel to the base of a cone and intersecting one cone only.

Parabola. A section d–d parallel to the side of a cone.

Hyperbola. A section e–e inclined to the side of a cone and intersecting both cones.

1.5.8 Properties of conic sections

A conic section is defined as the locus of a point P which moves so that its distance from a fixed point, the focus, bears a constant ratio, the eccentricity, to its perpendicular distance from a fixed straight line, the directrix.

Referring to Figure 1.36: the vertex of the curve is at V, the focus of the curve is at F, the directrix of the curve is the line DD parallel to yy'; the latus rectum is the line LR through the focus parallel to DD, $FL = FR = l$; the eccentricity of the curve is the ratio $FP/PQ = FV/VS = e$.

Then the curve is a parabola if $e = 1$, an ellipse if $e < 1$; and a hyperbola if $e > 1$. A circle is a particular case of an ellipse in which $e = 0$.

The polar equation of a conic is given by $l = \rho(1 - e \cos \theta)$ where ρ is the radius vector of any point P on the curve, θ the angle the vector makes with VX and l the semi latus rectum.

Parabola ($e = 1$)   (see Figure 1.36).

1.5.8.1 Equations

With origin at V and putting $a = VS = VF$ then for P at (x, y): $(x - a)^2 + y^2 = (x + a)^2$, i.e. $y^2 = 4ax$.

1.5.8.2 Tangents

Let PT be a tangent at any point P (x_1, y_1) then the equation of PT is given by:

$$y - y_1 = m(x - x_1) = (2a/y_1)(x - x_1)$$

or $yy_1 = 2a(x + x_1)$

since $d/dx(y^2) = 2y\, dy/dx = 4a$,

i.e. $m = dy/dx = 2a/y_1$ at P (x_1, y_1).

Alternatively, if any straight line $y = mx + c$ meets the parabola $y^2 = 4ax$ then $(mx + c)^2 = 4ax$ at the points of intersection and this expression will satisfy the condition for tangency if the roots of $m^2x^2 + 2(mc - 2a)x + c^2 = 0$ are equal, i.e. if $4(mc - 2a)^2 = 4m^2c^2$ or $c = a/m$ so that the equation for the tangent may be expressed as $y = mx + a/m$ for all values of m where $m = dy/dx$, and tangency occurs at the point $(a/m^2, 2a/m)$.

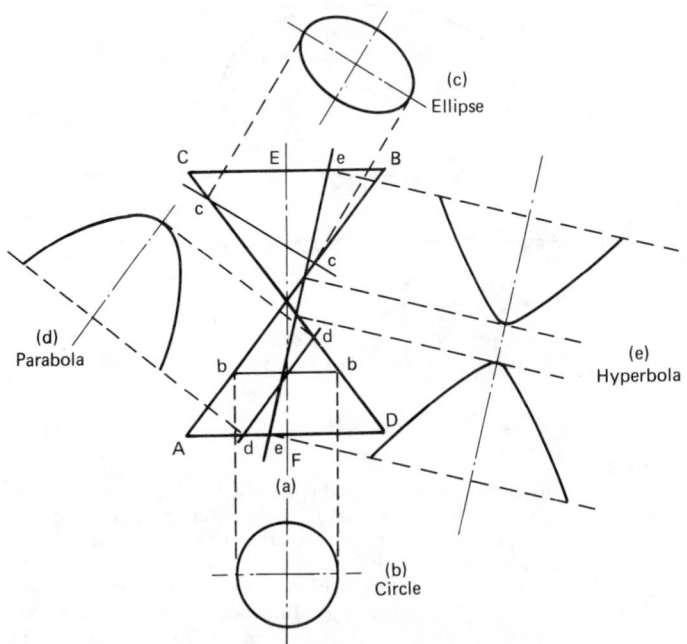

Figure 1.35   Circular cones generated by two intersecting straight lines

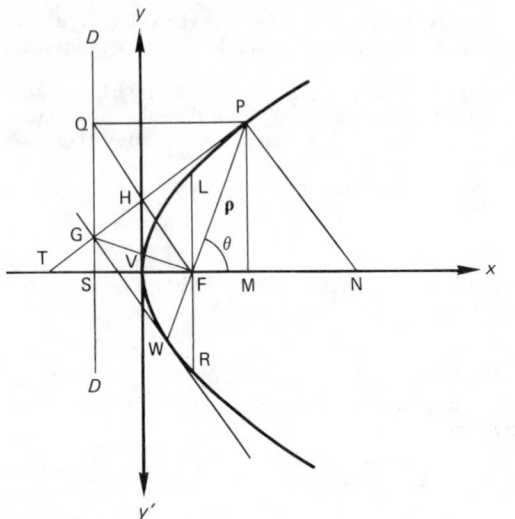

Figure 1.36 Properties of a conic section

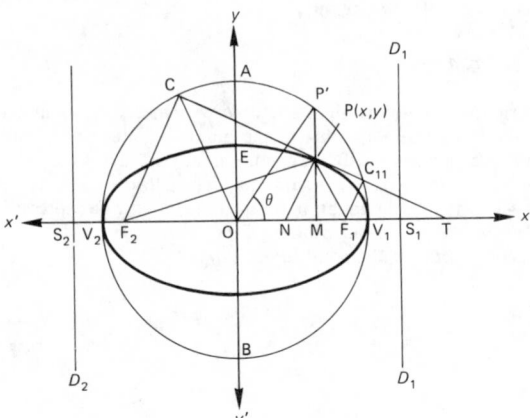

Figure 1.37 Ellipse in cartesian coordinates

1.5.8.3 Normal

Let PN be the normal at any point P $(x_1\, y_1)$; then the equation of PN is given by:

$$y - y_1 = -(y_1/2a)(x - x_1) \tag{1.39}$$

1.5.8.4 General properties

Tangents:

(1) The tangent PT bisects F$\hat{P}$Q.
(2) The tangents PG, GW where PW is a focal chord intersect at G on DD.
(3) The tangent PT intersects the axis of the parabola at a point T where TV = VM; TF = SM = PF.
(4) The angles GFP, PHQ and PGW are right angles.

Normals: any normal PN intersects VX at N where FT = FN.

Subnormals: the subnormal MN is a constant length, i.e. MN = FS = 2a.

1.5.8.5 Ellipse (e < 1)

Referring to Figure 1.37, F_1, F_2 and the foci; D_1D_1, D_2D_2 the directrices.

$$e = \frac{F_1V_1}{S_1V_1} = \frac{F_1V_2}{S_1V_2} = \frac{F_2V_1}{S_2V_1} = \frac{F_1P}{MS_1} = \frac{F_2P}{MS_2} = \frac{OF_1}{OV_1} = \frac{OF_2}{OV_2} = \frac{F_1F_2}{V_1V_2}$$

Let OV_1 the semi-major axis $= a$ and OE the semi-minor axis $= b$,

then $OF_1 = OF_2 = ae$ and $OS_1 = OS_2 = \dfrac{a}{e}$

also $F_1P = a - ex$; $F_2P = a + ex$ ∴ $F_1P + F_2P = 2a$

$$F_1E = eOS_1 = a; \ (OE)^2 = b^2 = (F_1E)^2 - (OF_1)^2 = a^2(1 - e^2),$$

or $e^2 = 1 - \dfrac{b^2}{a^2}$

Hence, as OM $= x$ and PM $= y$ we have the following.

1.5.8.5 Equation of ellipses

$$y^2 = a^2(1 - e^2) - x^2(1 - e^2)$$

or $\dfrac{x^2}{a^2} + \dfrac{y^2}{b^2} = 1$ in cartesian coordinates.

Substituting $\rho \cos a$ for x and $\rho \sin a$ for y (see Figure 1.38) in the above equation for an ellipse we have for the polar equation for an ellipse:

$$\frac{1}{\rho^2} = \frac{\cos^2 a}{a^2} + \frac{\sin^2 a}{b^2} \tag{1.40}$$

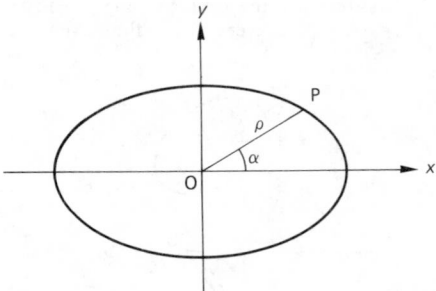

Figure 1.38 Ellipse in polar coordinates

1.5.8.6 Tangent

At any point P $(x_1\, y_1)$ on the ellipse
Let $f(xy) = 0$ represent the curve, then:

$$\frac{dy}{dx} = -\frac{\partial f}{\partial x} \Big/ \frac{\partial f}{\partial y} \tag{1.41}$$

Therefore $\partial f/\partial x = 2x/a^2$ and $\partial f/\partial y = 2y/b^2$ so that dy/dx at point $(x_1\, y_1)$ is given by $-b^2x_1/a^2y_1 = m$. Substituting this value of m in the equation of a straight line $(y - y_1) = m(x - x_1)$ we have the equation of tangent PT: $xx_1/a^2 + yy_1/b^2 = 1$.

Alternatively, the straight line $y = mx + c$ is a tangent to the ellipse $x^2/a^2 + y^2/b^2 = 1$ when the roots of $x^2/a^2 + (mx + c)^2/b^2 - 1 = 0$ are equal, i.e. when $c^2 = a^2m^2 + b^2$. Substituting we have for the equation of a tangent at any point P: $y = mx + \sqrt{(a^2m^2 + b^2)}$.

The equation to the tangent may also be written in the form:

$$\frac{x}{a}\cos\theta + \frac{y}{b}\sin\theta = 1. \tag{1.42}$$

The coordinates of the point of contact are $(a\cos\theta, b\sin\theta)$, θ being known as the eccentric angle (see Figure 1.37).

1.5.8.7 Normal

Substituting the value of m above in the general equation for the normal PN to a curve at point P $(x_1\ y_1)$ given by: $(y-y_1)m+(x-x_1)=0$ we have as the equation for the normal $(y-y_1)b^2/y_1=(x-x_1)a^2/x_1$.

1.5.8.8 General properties

(1) The circle AV_2BV_1 is termed the auxiliary circle (Figure 1.37).
(2) $OM \times OT = a^2$.
(3) $F_2N = eF_2P$.
(4) $F_1N = eF_1P$.
(5) PN bisects $\angle F_1PF_2$.
(6) The perpendiculars from F_1, F_2 to any tangent meet the tangent on the auxiliary circle.

1.5.8.9 Circle (c=0)

The circle may be regarded as a particular case of the ellipse (see above). The equation of a circle of radius a with centre at the origin is $x^2+y^2=a^2$ or, in polar coordinates, $\rho=a$.

The equation of the tangent at the point $(x_1\ y_1)$ is $xx_1+yy_1=a^2$, or, $y=mx+a\sqrt{(1+m^2)}$. The equation of the normal is $xy_1-yx_1=0$.

1.5.8.10 Hyperbola (e>1)

This is shown in Figure 1.39 where $F_1\ F_2$ are the foci, D_1D_1 and D_2D_2 the directrices and:

$$e=\frac{F_1V_1}{S_1V_1}=\frac{F_1P}{MS_1}=\frac{F_2P}{MS_2}=\frac{F_1V_2}{S_1V_2}=\frac{F_2V_1}{S_2V_1}=\frac{V_1V_2}{S_1S_2}=\frac{OV_1}{OS_1}=\frac{OV_2}{OS_2}$$

where O is the origin of the axes x and y.

Putting $OV_1=OV_2=a$ then $OF_1=OF_2=ea$ and $OS_1=OS_2=a/e$; also $F_1P=ex-a$ and $F_2F=ex+a$. Now $(F_1P)^2=(PM)^2+(F_1M)^2$, so $(ex-a)^2=y^2+(x-ae)^2$ which becomes $y^2=(e^2-1)x^2-(e^2-1)a^2$. Putting $(e^2-1)a^2=b^2$ then $y^2=(b^2/a^2)x^2-b^2$; therefore the equation of the hyperbola is given by $x^2/a^2-y^2/b^2=1$ in cartesian coordinates, or:

$$\frac{1}{p^2}=\frac{\cos^2\theta}{a^2}-\frac{\sin^2\theta}{b^2} \tag{1.43}$$

in polar coordinates.

Rearranging we have $y=b\sqrt{(x^2/a^2-1)}$, i.e. y is imaginary when $x^2<a^2$ and $y=0$ for $x=\pm a$. y is real when $x>a$ and there are two values for y of opposite sign.

1.5.8.11 Conjugate axis

The conjugate axis lies on yy' and is given by CC' where $OC=OC'=\pm b$.

1.5.8.12 Tangents

Let the straight line $y=mx+c$ meet the hyperbola $x^2/a^2-y^2/b^2=1$; then $x^2/a^2-(mx+c)^2/b^2-1=0$ will give the points of intersection. The condition for tangency is that the roots of this equation are equal, i.e. $c=\sqrt{(a^2m^2-b^2)}$ and the equation of the tangent is given by $y=mx+\sqrt{(a^2m^2-b^2)}$ at any point. Alternatively, the tangent to the hyperbola at $(x_1\ y_1)$ is given by $xx_1/a^2-yy_1/b^2=1$.

1.5.8.13 Normal

The equation for the normal at any point $(x_1\ y_1)$ on the curve is given by:

$$(y-y_1)b^2/y_1+(x-x_1)a^2/x_1=0. \tag{1.44}$$

1.5.8.14 Asymptotes

The tangent to the hyperbola becomes an asymptote when the roots of the equation $x^2/a^2-(mx+c)^2/b^2-1=0$ are both infinite, i.e. when $b^2-a^2m^2=0$ and $a^2mc=0$. Therefore: $m=\pm b/a$ and $c=0$. Substituting for m in $y=mx+c$ we have as the equation for an asymptote $y=\pm(b/a)x$. The combined equation for both asymptotes is given by:

$$\frac{x^2}{a^2}-\frac{y^2}{b^2}=0 \tag{1.45}$$

The equation of the hyperbola referred to its asymptotes as oblique axes is:

$$X.Y=\frac{a^2+b^2}{4} \tag{1.46}$$

1.5.8.15 General properties

(1) $F_2P-F_1P=2a$.
(2) The product of the perpendiculars from any point on a hyperbola to its asymptotes is constant and equal to $a^2b^2/(a^2+b^2)$.

1.5.8.16 Rectangular hyperbola

When the transverse axis V_1V (Figure 1:39) is equal to the conjugate axis CC' the hyperbola is a rectangular hyperbola, i.e. $a=b$ and the equation for the curve is given by $x^2-y^2=a^2$.

The equation for the asymptotes then becomes $y=\pm x$ which represents two straight lines at right angles to each other. The equation of the rectangular hyperbola referred to its asymptotes as axes of coordinates is given by $xy=$ constant.

1.5.8.17 General equation of a conic section:

The general equation of a conic section has the form:

$$ax^2+2hxy+by^2+2gx+2fy+c=0 \tag{1.47}$$

Let
$$D=\begin{vmatrix} a & h & g \\ h & b & f \\ g & f & c \end{vmatrix} \quad \text{and} \quad d=\begin{vmatrix} a & b \\ h & b \end{vmatrix}$$

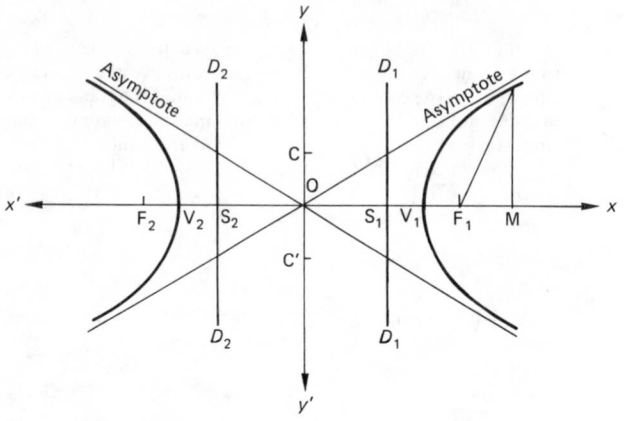

Figure 1.39 Hyperbola

Then, the general equation represents:

(1) An ellipse if $d > 0$;
(2) A parabola if $d = 0$;
(3) A hyperbola if $d < 0$;
(4) A circle if $a = b$ and $h = 0$;
(5) A rectangular hyperbola if $a + b = 0$;
(6) Two straight lines (real or imaginary) if $D = 0$;
(7) Two parallel straight lines if $d = 0$ and $D = 0$.

The centre of the conic $(x_0 y_0)$ is determined by the equations: $ax_0 - hy_0 + g = 0$, $hx_0 + by_0 + f = 0$.

1.6 Three-dimensional analytical geometry

1.6.1 Sign convention

1.6.1.1 Cartesian coordinates

This is shown in Figure 1.40, there being eight compartments formed by the right-angled intersection of three planes. The signs of x, y, z follow the convention that these are positive when measured in the directions Ox, Oy, Oz of the coordinate axes and negative when measured in the directions Ox', Oy', Oz' respectively.

1.6.1.2 Polar coordinates

The location of any point P in space (see Figure 1.41) is fully located by the radius vector $\boldsymbol{\rho}$ and the two angles θ and ϕ thus $(\rho\theta\phi)$. From Figure 1.41:

$$OP = \boldsymbol{\rho} = \sqrt{[(OD)^2 + (OB)^2 + (OC)^2]} = \sqrt{(x^2 + y^2 + z^2)}$$

and $\quad x = \boldsymbol{\rho} \sin\theta \cos\phi; \; y = \boldsymbol{\rho} \sin\theta \sin\phi; \; z = \boldsymbol{\rho} \cos\phi.$

1.6.1.3 Cylindrical coordinates

In this system the point P (Figure 1.41) is located by its perpendicular distance, z, from the x–y plane and the polar coordinates of the foot, A, of that perpendicular in the x–y plane. P is the point r, ϕ, z where $OA = r$.

1.6.1.4 Direction-cosines of a straight line

If the direction of the line OP in Figure 1.42 is determined by a,

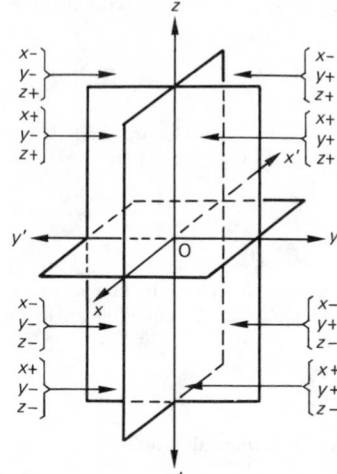

Figure 1.40 Sign conventions in analytical solid geometry

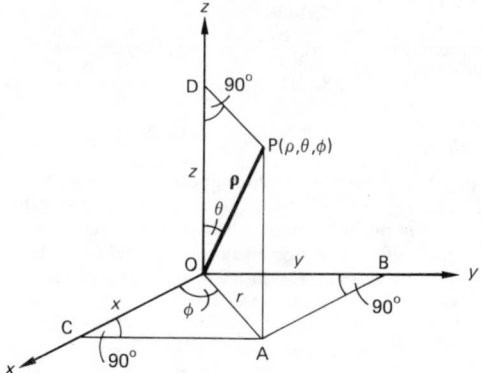

Figure 1.41 Polar coordinates in three dimensions

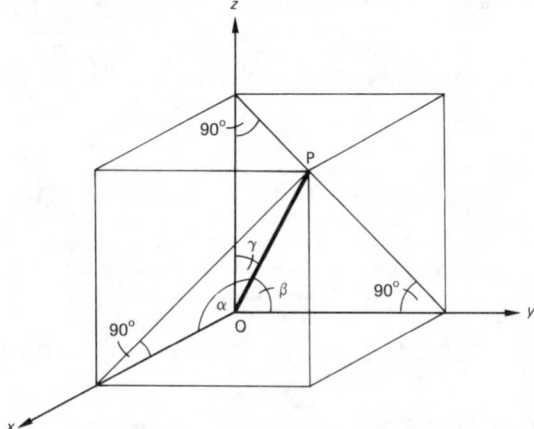

Figure 1.42 Direction-cosines

β, γ then the projections of a unit length of OP on to the axes Ox, Oy, Oz are given by $\cos\alpha$, $\cos\beta$, $\cos\gamma$ respectively, termed direction-cosines. Let $l = \cos\alpha$, $m = \cos\beta$, $n = \cos\gamma$ and $CP = \rho$; then $\rho^2(l^2 + m^2 + n^2) = x^2 + y^2 + z^2 = \rho^2$, i.e. $l^2 + m^2 + n^2 = 1$.

Also:
$$\sin^2 a + \sin^2 \beta + \sin^2 \gamma = (1-l^2)+(1-m^2)+(1-n^2)=2$$

Again if:
$$l:m:n=s:t:u \text{ then } \frac{l^2}{s^2}=\frac{m^2}{t^2}=\frac{n^2}{u^2}=\frac{l^2+m^2+n^2}{s^2+t^2+u^2}=\frac{1}{s^2+t^2+u^2}$$

i.e.:
$$l=\frac{s}{\sqrt{(s^2+t^2+u^2)}};$$

$$m=\frac{t}{\sqrt{(s^2+t^2+u^2)}};$$

$$n=\frac{u}{\sqrt{(s^2+t^2+u^2)}}$$

1.6.1.5 General equations

The expression $F(xyz)=0$ represents a surface of some kind and if we put $x=0$ the resulting equation is for a curve in the y–z plane; similarly, for $y=0$ the curve is in the x–z plane, etc. In general, any two simultaneous equations, $F(xyz)=0$, $F'(xyz)=0$ represent a line (either straight or curved) being the intersection of two surfaces. Any three such simultaneous equations represent a point (or several points).

1.6.2 Equation of a plane

The general equation of a plane is given by the expression $ax+by+cz+d=0$ (*abcd* being constants). By putting $y=0$, $z=0$ then $x=-d/a=a'$ which is the intercept of the plane on the x axis at a distance a' from the origin. Similarly, the intercepts on the y and z axes are b' and c'. Hence $a=-d/a'$; $b=-d/b'$; $c=-d/c'$ and substituting these values in the general equation for the plane we have the intercept equation for a plane as $x/a'+y/b'+z/c'=1$.

In Figure 1.43 let P be any point on the plane ABC and let OQ of length p be at 90° to the plane ABC; then if l, m, n are the direction cosines of OQ we have $p=lx+my+nz$, which is the perpendicular form of the equation to a plane. The various forms of the equation to a plane are interchangeable since:

$$p=-\frac{d}{\sqrt{(a^2+b^2+c^2)}}=la'=mb'=nc'$$

and
$$\frac{1}{a'^2}+\frac{1}{b'^2}+\frac{1}{c'^2}=\frac{1}{p^2} \qquad (1.48)$$

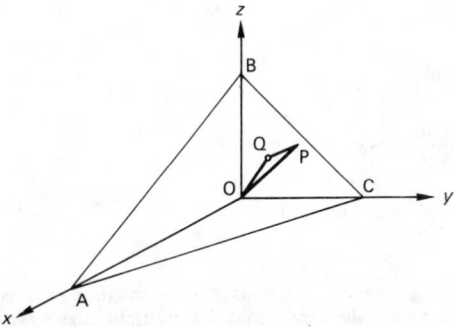

Figure 1.43 Equation of a plane

1.6.3 Distance between two points in space

Let the two points be $P(x_1y_1z_1)$; $Q(x_2y_2z_2)$. Assume origin shifted to P and axes kept parallel to original axes, then coordinates of Q relative to P are (x_2-x_1), (y_2-y_1), (z_2-z_1) and the length $PQ=r$, i.e. $r=\sqrt{[(x_2-x_1)^2+(y_2-y_1)^2+(z_2-z_1)^2]}$ and the locus of Q is a sphere if r is constant.

1.6.4 Equations of a straight line

Using direction cosines for PQ, $l=(x_2-x_1)/r$; $m=(y_2-y_1)/r$; $n=(z_2-z_1)/r$. If Q is taken as any point then the symmetrical equation of a straight line is given by $r=(x-x_1)/l=(y-y_1)/m=(z-z_1)/n$ and the coordinates of any point on the line are given by $x=x_1+rl$; $y=y_1+rm$; $z=z_1+rn$. For a line through the origin this becomes:

$$r=\frac{x}{l}=\frac{y}{m}=\frac{z}{n}$$

The equation of the straight line through the points $(x_1y_1z_1)$ and $(x_2y_2z_2)$ is:

$$\frac{x-x_1}{x_2-x_1}=\frac{y-y_1}{y_2-y_1}=\frac{z-z_1}{z_2-z_1}$$

1.6.4.1 Angle between two lines of known direction cosines

Let PA, QB be any two lines in space (Figure 1.44) and let P'O, Q'O be parallel to PA, QB respectively and having direction cosines $l_1m_1n_1$; $l_2m_2n_2$ respectively then $\cos a = l_1l_2+m_1m_2+n_1n_2$ where $a=\text{P'}\hat{\text{O}}\text{Q'}$.

1.6.4.2 The angle between two planes

Let the equations of the planes be:

$$a_1x+b_1y+c_1z+d_1=0$$

and:
$$a_2x+b_2y+c_2z+d_2=0$$

then the direction-cosines of the normals to these planes are:

$$\frac{a_1}{\sqrt{(a_1^2+b_1^2+c_1^2)}};\frac{b_1}{\sqrt{(a_1^2+b_1^2+c_1^2)}};\frac{c_1}{\sqrt{(a_1^2+b_1^2+c_1^2)}}$$

and:
$$\frac{a_2}{\sqrt{(a_2^2+b_2^2+c_2^2)}};\frac{b_2}{\sqrt{(a_2^2+b_2^2+c_2^2)}};\frac{c_2}{\sqrt{(a_2^2+b_2^2+c_2^2)}}$$

If θ is the angle between the planes, this is equal to the angle between the normals to these planes, i.e.:

$$\cos\theta=\frac{a_1a_2+b_1b_2+c_1c_2}{\sqrt{[(a_1^2+b_1^2+c_1^2)(a_2^2+b_2^2+c_2^2)]}} \qquad (1.49)$$

The planes are perpendicular to each other if $a_1a_2+b_1b_2+c_1c_2=0$. They are parallel if $a_1/a_2=b_1/b_2=c_1/c_2$.

1.6.4.3 The angle between a plane and a straight line

The angle θ between the plane $l_1x+m_1y+n_1z=p$ and the line $(x-x_1)/l_2=(y-y_1)/m_2=(z-z_1)/n_2$ is given by $\sin\theta=(l_1l_2+m_1m_2+n_1n_2)$.

1.6.4.4 Length of the perpendicular from a point $x_1 y_1 z_1$ to a plane

(1) Where the equation of the plane is the perpendicular form $lx + my + nz = p$ then the equation of a plane containing the point $(x_1 y_1 z_1)$ and parallel to the given plane is given by $lx + my + nz = p'$ where p and p' are the lengths of perpendiculars from the origin. Therefore required length of perpendicular is $p' - p = lx_1 + my_1 + nz_1 - p$, since the point $(x_1 y_1 z_1)$ lies on the second plane.

(2) Where the equation of the plane takes the general form $ax + by + cz + d = 0$ the length of perpendicular from point $x_1 y_1 z_1$ is given by:

$$\frac{ax_1 + by_1 + cz_1 + d}{\sqrt{(a^2 + b^2 + c^2)}}$$

In the above the equation of the perpendicular is given by:

$$\frac{x - x_1}{l} = \frac{y - y_1}{m} = \frac{z - z_1}{n}$$

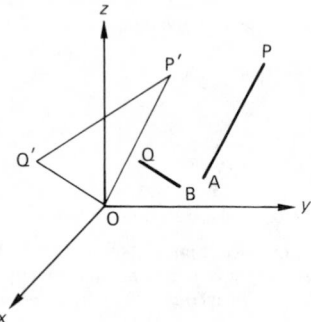

Figure 1.44 Angle between two lines of known direction-cosines

1.7 Calculus

The calculus deals with quantities which vary and with the rate at which this variation takes place.

Variables may be denoted by u, v, w, x, y, z and increments of these variables are denoted by $du, dv \ldots dz$. A simple example concerns the slope of a curve. Suppose that the curve is defined by some function:

$$y = f(x) \tag{1.50}$$

The slope at the point $x = x_1$ may be approximated to as follows. Let x_2 be close in value to x_1; then, provided the curve $f(x)$ is well behaved in the region of x_1, the line joining $f(x_1)$ to $f(x_2)$ is an approximation to the tangent to the curve at $x = x_1$. As x_2 is moved closer to x_1 the approximation becomes better and better, until, in the limit, when x_2 reaches x_1 the tangent (instead of the secant) is obtained and thereby the slope of the curve $y = f(x)$ is found at the point $x = x_1$. This process is known as 'differentiation'.

1.7.1 Differentiation

This process is used to find the value of dy/dx.
For combinations of functions u, and v of x:

$$\frac{d}{dx}(uv) = u\frac{dv}{dx} + v\frac{du}{dx} \tag{1.51}$$

$$\frac{d}{dx}\left(\frac{u}{v}\right) = \frac{v(du/dx) - u(dv/dx)}{v^2} \tag{1.52}$$

For polynomial functions, $y = ax^n$:

$$\frac{dy}{dx} = nax^{n-1} \tag{1.53}$$

The differentiation process may be carried out more than once. Thus:

$$\frac{d}{dx}\frac{(dy)}{dx} = \frac{d^2y}{dx^2} \text{ etc.} \tag{1.54}$$

As an example, if $y = f(x) = ax^4 + bx^3 + cx^2 + dx + c$

then: $\dfrac{dy}{dx} = f'(x) = 4ax^3 + 3bx^2 + 2cx + d$

$$\frac{d^2y}{dx^2} = f''(x) = 12ax^2 + 6bx + 2c$$

$$\frac{d^3y}{dx^3} = f'''(x) = 24ax + 6b$$

$$\frac{d^4y}{dx^4} = f^{iv}(x) = 24a$$

$$\frac{d^5y}{dx^5} = f^{v}(x) = 0$$

It is often convenient, when dealing with long, complicated expressions, to substitute a symbol for a part of a compound expression. Suppose we have:

$$y = f(x) \tag{1.55}$$

a complicated expression and we choose to make a substitution u then the differential, $f'(x)$ can be found from the rule:

$$\frac{dy}{dx} = \frac{dy}{du} \cdot \frac{du}{dx}$$

e.g.: $y = (x^4 + a^2)^6$

The substitution

$u = x^4 + a^2$ is appropriate

so: $y = u^6$

thus: $\dfrac{dy}{dx} = 6u^5$ and $\dfrac{du}{dx} = 4x^3$

so: $\dfrac{dy}{du} = \dfrac{dy}{du}\dfrac{du}{dx} = 6(x^4 + a^2)^5 4x^3$

$$= 24x^3(x^4 + a^2) \tag{1.56}$$

In the case of trigonometric functions the differentiation process can be obtained via the expressions for multiple angles (see section 1.2).

For, suppose:

$$y = \sin \theta$$

We let:

$$y + \delta y = \sin(\theta + \delta\theta)$$

$$= \sin\theta \cos\delta\theta + \cos\theta \sin\delta\theta$$

$$= \sin\theta + \cos\theta \cdot \delta\theta$$

So:

$$\delta y = \cos\theta \cdot \delta\theta$$

$$\frac{\delta y}{\delta\theta} = \cos\theta \text{ or } \frac{\delta y}{\delta\theta} = \cos\theta \qquad (1.57)$$

In cases where inverse trigonometric functions are involved, the principle of substitution is employed, for suppose:

$$y = \sin^{-1} u$$

then:

$$u = \sin y$$

so: $\dfrac{du}{dy} = \cos y = (1 - \sin^2 y) = (1 - u^2)$

so: $\dfrac{dy}{du} = 1 \quad \dfrac{du}{dy} = \dfrac{1}{(1-u^2)} \qquad (1.58)$

In cases where exponentiation is involved, the principle of substitution may again be employed:

For, suppose $\quad y = e^{3x^2/4}$

we write $\quad y = e^u$

where $\quad u = 3x^2/4$

and $\quad \dfrac{dy}{dx} = \dfrac{dy}{du} \cdot \dfrac{du}{dx} = e^u 6x/4$

$$= \frac{3x}{2} e^{3x^2/4} \qquad (1.59)$$

1.7.2 Partial differentiation

The dependent variable u may be a function of more than one independent variable, x and y, and we wish to find the rates of changes of u with respect to u and v separately. These rates of change, the partial differentials with respect to x and y are denoted by:

$$\frac{du}{dx} \text{ and } \frac{du}{dy}$$

In these processes the normal rules of differentiation are followed except that in finding du/dx, y is treated as a constant and in finding du/dy, x is treated as a constant.

The total differential of a function:

$$u = f(x, y)$$

where both x and y are functions of t is given by:

$$\frac{du}{dt} = \frac{du}{dx}\frac{dx}{dt} + \frac{du}{dy}\frac{dy}{dt} \qquad (1.60)$$

1.7.3 Maxima and minima

Maxima and minima of functions occur when the function has zero slope or first differential. Thus, in order to determine a maximum or minimum of a function $y = f(x)$:

we set: $\quad \dfrac{dy}{dx} = f'(x) = 0 \qquad (1.61)$

and solve this equation, say $x = x_1$.

To distinguish between maxima and minima it is necessary to evaluate:

$$\frac{d^2y}{dx^2} \text{ at the point } x_1$$

For a maximum: $\quad \dfrac{d^2y}{dx^2} < 0$

For a minimum: $\quad \dfrac{d^2y}{dx^2} > 0$

1.7.4 Integration

Integration is generally the reverse of the process of differentiation. It may also be regarded as equivalent to a process of summing a number of finite quantities but, in the limit the number of quantities becomes infinite and their size becomes infinitesimal.

By the reverse of the differentiation process the integral

$$\int ax^n \cdot dx = \frac{ax^{n+1}}{n+1} + c \qquad (1.62)$$

the c being an arbitrary constant which, for shortness, is frequently not written. This is called an indefinite integral as no range over which the integration is to be performed has been specified. If such a range is specified then we obtain the case of a definite integral, e.g.:

if $\quad F(x)\,dx = f(x)$

then $\quad \int_b^a F(x)\,dx = f(b) - f(a) \qquad (1.63)$

In geometrical terms, this integral represents the area bounded by the curve $y = F(x)$, the x axis, and the two lines $x = a$, $x = b$.

1.7.5 Successive integration

This is the reverse process from that of successive differentiation, each cycle of operations consisting of the integration of the function resulting from the immediately previous integration. In general terms instructions to carry out successive integration are expressed thus: $y = \iiiint f(x)\,dx,\,dx,\,dx,\,dx$, which means integration is to be successively carried out 4 times with respect to x.

Another form of successive integration is $v = \iiint f(x, y, z)\, dx, dy, dz$; referred to as a volume integral. A surface integral would take the form $s = \iint f(x, y)\, dx, dy$.

Example 1.4 Find a general expression for the deflection of a simple span girder of span l loaded uniformly by a load w per unit length of span given that $w = EI\, d^4y/dx^4$ and taking E and I as constant and x as measured from one end.

Load:

$$EI\frac{d^4y}{dx^4} = w$$

Shear:

$$EI\int\frac{d^4y}{dx^4}\,dx = EI\frac{d^3y}{dx^3} = wx + c_1 = wx - \frac{wl}{2}$$

$$\left(\text{Shear} = -\frac{wl}{2}\text{ for }x=0\right)$$

Bending moment:

$$EI\int\frac{d^3y}{dx^3}\,dx = EI\frac{d^2y}{dx^2} = \frac{wx^2}{2} - \frac{wlx}{2} + c_2 = \frac{wx^2}{2} - \frac{wlx}{2}$$

(B.M. $=0$ for $x=0$)

Slope:

$$EI\int\frac{d^2y}{dx^2}\,dx = EI\frac{dy}{dx} = \frac{wx^3}{6} - \frac{wlx^2}{4} + c_3 = \frac{wx^3}{6} - \frac{wlx^2}{4} + \frac{wl^3}{24}$$

$$\left(\text{Slope} = 0\text{ for }x = \frac{l}{2}\right)$$

Deflection:

$$EI\int\frac{dy}{dx}\,dx = EIy = \frac{wx^4}{24} - \frac{wlx^3}{12} + \frac{wl^3x}{24}$$

(Deflection $=0$ for $x=0$)

i.e. at any distance x from one end the deflection

$$y = \frac{1}{24}\frac{w}{EI}(x^4 - 2lx^3 + l^3x)$$

$$\iiint\frac{d^4y}{dx^4}\,dx\,dx\,dx\,dx = \frac{w}{24EI}(x^4 - 2lx^3 + l^3x)$$

which is in the general form.
For the mid-span deflection the range of integration is from $x=0$ to $x=\frac{1}{2}l$.

Hence:

$$\iiint_0^{l/2}\frac{d^4y}{dx^4}\,dx\cdot dx\cdot dx\cdot dx = \frac{wl^4}{24EI}\left(\tfrac{1}{16} - \tfrac{1}{4} + \tfrac{1}{2}\right) = 5\frac{wl^4}{384EI}$$

1.7.6 Integration by substitution

The integration of functions can often be simplified by substituting a new variable for a part or the whole of the original function, thereby reducing it to one of the standard forms.

Example 1.5 Find the value of $\int\sqrt{(3+x)}\,dx$

Let $\quad 3 + x = u$

Therefore $\quad dx = du$

so that

$$\int\sqrt{(3+x)}\,dx = \int u^{\frac{1}{2}}\,du = \tfrac{2}{3}u^{3/1} = \tfrac{2}{3}(3+x)^{3/2}\text{ or }\tfrac{2}{3}\sqrt{(3+x)^3}$$

Example 1.6 Find the value of

$$2\int\frac{dx}{e^{3x} + c^{-3x}}$$

Let $e^{3x} = v$; then $3e^{3x}\cdot dx = dv$, or $dx = dv/3v$.

Substituting

$$2\int\frac{dx}{e^{3x} + e^{-3x}} = \tfrac{2}{3}\int\frac{dv}{v(v + 1/v)} = \tfrac{2}{3}\int\frac{dv}{v^2 + 1} = \tfrac{2}{3}\tan^{-1}v = \tfrac{2}{3}\tan^{-1}e^{3x}.$$

Example 1.7 Find the value of $\int\sqrt{(1 - x^2)}\,dx$.
Put $x = \sin\theta$; then $\sqrt{(1 - x^2)} = \cos\theta$

Therefore

$$\int\sqrt{(1 - x^2)}\cdot dx = \int\cos\theta\cdot d\sin\theta = \int\cos^2\theta\cdot d\theta$$

$$= \int\frac{1 + \cos 2\theta}{2}\cdot d\theta$$

$$= \frac{\theta}{2} + \frac{\sin 2\theta}{4} = \frac{1}{2}\{\sin^{-1}x + x\sqrt{(1 - x^2)}\}$$

1.7.7 Integration by transformation

The integration of trigonometric functions can often be simplified by transformation into a standard form of integral.

TYPE

$$\int\sin^m\theta\cos^n\theta\,d\theta$$

Case 1: $m =$ positive odd integer, $n =$ any positive integer.

TRANSFORMATIONS

$$\int\sin^{m-1}\theta\sin\theta\cos^n\theta\,d\theta$$

$$= \int(1 - \cos^2\theta)^{(m-1)/2}\sin\theta\cos^n\theta\,d\theta$$

$$= -\int(1 - \cos^2\theta)^{(m-1)/2}\cos^n\theta\,d(\cos\theta)$$

Example 1.8 Solve $\int\sin^3\theta\cos^2\theta\,d\theta$

$$\int\sin^3\theta\cos^2\theta\,d\theta$$

$$= \int(1 - \cos^2\theta)\sin\theta\cos^2\theta\,d\theta$$

$$= \int\cos^2\theta\sin\theta\,d\theta - \int\cos^4\theta\sin\theta\,d\theta$$

$$= -\frac{\cos^3 \theta}{3} + \frac{\cos^5 \theta}{5}$$

Case 2: $m =$ any positive integer, $n =$ positive odd integer.

TRANSFORMATION

$$\int \sin^m\theta \cos^{n-1} \theta d\theta = \int (1 - \sin^2\theta)^{(n-1)/2} \sin^m \theta d (\sin \theta).$$

Example 1.9 Solve $\int \sin^2\theta \cos^3\theta d\theta$

$$\int \sin^2\theta \cos^3\theta d\theta = \int (1 - \sin^2\theta) \sin^2\theta \cos \theta d\theta$$

$$= \int \sin^2\theta \cos \theta d\theta - \int \sin^4\theta \cos \theta d\theta$$

$$= \frac{\sin^3\theta}{3} - \frac{\sin^5\theta}{5}$$

TYPE

$\int \tan \theta d\theta$ where n is an integer > 1.

TRANSFORMATION

$$\int \tan^{n-2} \theta \tan^2 \theta d\theta = \int \tan^{n-2} \theta (\sec^2 \theta - 1) \, d\theta$$

$$= \int \tan^{n-2} \theta . \tan \theta - \int \tan^{n-2} \theta . d\theta$$

TYPE

$\int \cot^n \theta d\theta$ where n is an integer > 1.

TRANSFORMATION

$$\int \cot^{n-2} \theta \cot^2 \theta d\theta = \int \cot^{n-2} \theta (\csc^2 \theta - 1) \, d\theta$$

$$= -\int \cot^{n-2} \theta . d\cot \theta - \int \cot^{n-2} \theta . d\theta.$$

TYPE

$\int \sec^n \theta d\theta$ where n is positive and even.

TRANSFORMATION

$$\int \sec^{n-2} \theta \sec^2 \theta d\theta = \int (\tan^2 \theta + 1)^{(n-2)/2} d \tan \theta.$$

TYPE

$\int \csc^n \theta d\theta$ where n is positive and even.

TRANSFORMATION

$$\int \csc^{n-2} \theta \csc^2 \theta d\theta = \int -(\cot^2 \theta + 1)^{(n-2)/2} \, d \cot \theta$$

TYPE

$\int \tan^m \theta \sec^n \theta d\theta$ where n is positive and even.

TRANSFORMATION

$$\int \tan^m \theta \sec^{n-2} \theta \sec^2 \theta d\theta = \int \tan^m \theta (\tan^2 \theta + 1)^{(n-2)/2} d \tan \theta.$$

TYPE

$\int \cot^m \theta \csc^n \theta d\theta$ where n is positive and even.

TRANSFORMATION

$$\int \cot^m \theta \csc^{n-2} \theta \csc^2 \theta d\theta = \int -\cot^m \theta (\cot^2 \theta + 1)^{(n-2)/2} d \cot \theta.$$

TYPE

$\int \tan^m \theta \sec^n \theta d\theta$ where m and n are odd.

TRANSFORMATION

$$\int \tan^{m-1} \theta \tan \theta \sec^{n-1} \theta \sec \theta d\theta$$
$$= \int (\sec^2 \theta - 1)^{(m-1)/2} . \sec^{n-1} \theta . d \sec \theta.$$

1.7.8 Integration by parts

The integration of functions can often be simplified by breaking up the function into two parts u and dv where u and v are the substituted variables in $\int u . dv = u . v - \int v . du$, the fundamental formula for integration by parts, $\int u . dv$ representing the function to be integrated. In applying this method of integration $\int v . du$ should not be more complex than $\int u . dv$. The integration of logarithmic, exponential, inverse trigonometric and products of algebraic expressions may be simplified by this procedure.

Example 1.10 $\int w \sin w \, dw$
Let $u = w$ and $dv = \sin w \, dw$ then $du = dw$ and $v = -\cos w$

Therefore:

$$\int w \sin w . dw = \int u . dv = -w \cos w + \int \cos w \, dw$$
$$= -w \cos w + \sin w$$

Example 1.11 $\int xe^x \, dx$
Let $u = x$ and $dv = e^x \, dx$ then $du = dx$ and $v = e^x$

Therefore:

$$\int xe^x \, dx = \int u . dv = xe^x - \int e^x \, dx = xe^x - e^x = e^x(x - 1)$$

Example 1.12 $\int \cos^2 \theta d\theta$
Let $u = \cos \theta$ and $dv = \cos \theta d\theta$ then $du = -\sin \theta d\theta$ and $v = \sin \theta$

Therefore:

$$\int \cos^2 \theta d\theta = \int u . dv = \cos \theta \sin \theta + \int \sin^2 \theta d\theta$$
$$= \frac{\sin 2\theta}{2} + \int (1 - \cos^2 \theta) \, d\theta$$

i.e.:

$$2\int \cos^2 \theta d\theta = \frac{\sin 2\theta}{2} + \theta \text{ hence } \int \cos^2 \theta d\theta = \frac{\sin 2\theta}{4} + \frac{\theta}{2}$$

Example 1.13 $\int \sec^3 \theta d\theta$
Let $u = \sec \theta$ and $dv = \sec^2 \theta d\theta$ then $du = \sec \theta \tan \theta d\theta$ and $v = \tan \theta$

Therefore:

$$\int \sec^3 \theta d\theta = \sec \theta \tan \theta - \int \tan^2 \theta \sec \theta d\theta$$

$$= \sec \theta \tan \theta - \int \sec^3 \theta d\theta + \int \sec \theta d\theta$$

i.e.:

$$2 \int \sec^3\theta d\theta = \sec\theta\tan\theta + \log_e\left\{\tan\left(\frac{\pi}{4} + \frac{\theta}{2}\right)\right\}$$

Therefore:

$$\int \sec^3\theta d\theta = \frac{1}{2}\left[\sec\theta\tan\theta + \log_e\left\{\tan\left(\frac{\pi}{4} + \frac{\theta}{2}\right)\right\}\right]$$

1.7.9 Integration of fractions

The integration of functions consisting of rational algebraic fractions is best carried out by first splitting the function into partial fractions. It is assumed that the numerator is of lower degree than the denominator; if not, this should first be achieved by dividing out. It may be shown that the prime real factors of any polynomial are either quadratic or linear in form. This leads to four distinct types of partial fraction solutions which are now described.

1.7.9.1 Fractions type 1

The denominator can be factored into real linear factors all different. The partial fractions are then of the form $a/(bx+c)$.

Example 1.14 $\int \dfrac{2x+3}{x^2-4}dx$

Now

$$\frac{2x+3}{x^2-4} = \frac{A}{x-2} + \frac{B}{x+2}$$

i.e.: $\quad 2x+3 = A(x+2) + B(x-2)$

i.e. $\quad A = \dfrac{7}{4}$ and $B = \dfrac{1}{4}$

Therefore:

$$\int \frac{2x+3}{x^2-4}dx = \frac{7}{4}\int\frac{dx}{x-2} + \frac{1}{4}\int\frac{dx}{x+2} = \frac{7}{4}\log_e(x-2) + \frac{1}{4}\log_e(x+2)$$

1.7.9.2 Fractions type 2

The prime factors of the denominator include quadratic functions and all factors are different. The partial fractions then include expressions of the form $(ax+b)/(cx^2+dx+e)$.

Example 1.15 $\int \dfrac{7x^2-3}{2x^3-3x^2+4x-6}\cdot dx$

Put $\quad \dfrac{7x^2-3}{2x^3-3x^2+4x-6} = \dfrac{Ax+B}{x^2+2} + \dfrac{C}{2x-3}$

i.e.: $\quad 7x^2-3 = (Ax+B)(2x-3) + C(x^2+2)$

$$= (2A+C)x^2 + (2B-3A)x - (3B-2C)$$

and therefore $\quad A=2,\ B=3,\ C=3$

Therefore:

$$\int\frac{7x^2-3}{2x^3-3x^2+4x-6}\cdot dx = \int\frac{2x+3}{x^2+2}\cdot dx + \int\frac{3}{2x-3}\cdot dx$$

$$= \int\frac{2x}{x^2+2}\cdot dx + \int\frac{3}{x^2+2}\cdot dx + \int\frac{3}{2x-3}\cdot dx$$

$$= \log_e(x^2+2) + \frac{3}{\sqrt{2}}\tan^{-1}\frac{x}{\sqrt{2}} + \frac{3}{2}\log_e(2x-3)$$

1.7.9.3 Fractions type 3

The denominator can be factored into real linear factors, some of which are repeated. The partial fractions then include expressions of the form $a/(bx+c)^n$.

Example 1.16 $\int \dfrac{3x^2+8x+16}{x^3+3x^2-4}dx = \int\dfrac{f(x)}{F(x)}dx$

$$F(x) = (x-1)(x+2)^2 \text{ and } \frac{f(x)}{F(x)} = \frac{A}{x-1} + \frac{B}{(x+2)} + \frac{C}{(x+2)^2}$$

Hence:

$$3x^2+8x+16 = A(x+2)^2 + B(x-1)(x+2) + C(x-1)$$

putting $x=1$ then $A=3$; $x=-2$ then $C=-4$; substitution gives $B=0$

Therefore:

$$\int\frac{f(x)}{F(x)}dx = 3\int\frac{dx}{x-1} - 4\int\frac{dx}{(x+2)^2} = 3\log_e(x-1) + \frac{4}{(x+2)}$$

1.7.9.4 Fractions type 4

The prime factors of the denominator include quadratic functions some of which are repeated. The partial fractions then include expressions of the form $(ax+b)/(cx^2+dx+e)^n$.

Example 1.17 $\int \dfrac{12x-1}{(x^2+1)^2(x+2)}dx = \int\dfrac{f(x)}{F(x)}dx$

$$\int\frac{f(x)}{F(x)}dx = \frac{Ax+B}{(x^2+1)^2} + \frac{Cx+D}{(x^2+1)} + \frac{E}{(x+2)}$$

i.e.

$$12x-1 = (Ax+B)(x+2) + (Cx+D)(x^2+1)(x+2) + E(x^2+1)^2.$$

Put $x=-2$; then $E=-1$.

Therefore:

$$x^4+2x^2+12x = Cx^4 + (D+2C)x^3 + (A+2D+C)x^2 + (2A+B+D+2C)x + 2(B+D)$$

Equating coefficients, we find $C=1$, $D=-2$, $B=2$ and $A=5$.

Therefore:

$$\int\frac{f(x)}{F(x)}dx = \int\frac{5x+2}{(x^2+1)^2}dx + \int\frac{x-2}{x^2+1}dx - \int\frac{dx}{x+2}$$

$$= \frac{5}{2}\int \frac{d(x^2+1)}{(x^2+1)^2} + 2\int \frac{dx}{(x^2+1)^2} + \frac{1}{2}\int \frac{d(x^2+1)}{x^2+1}$$
$$- 2\int \frac{dx}{x^2+1} - \int \frac{dx}{x+2}$$

$$= -\frac{5}{2}\int \frac{1}{x^2+1} + \frac{x}{x^2+1} + \int \frac{dx}{x^2+1} + \frac{1}{2}\log_e(x^2+1)$$
$$- 2\tan^{-1}x - \log_e(x+2)$$

$$= \frac{2x-5}{2(x^2+1)} + \log_e \frac{\sqrt{(x^2+1)}}{x+2} \tan^{-1}x$$

1.8 Matrix algebra

A matrix is an array of mn numbers in m rows and n columns

$$\begin{bmatrix} a_{11} & a_{12} & \cdots & a_{1n} \\ a_{21} & a_{22} & \cdots & a_{2n} \\ \vdots & & & \\ a_{m1} & a_{m2} & \cdots & a_{mn} \end{bmatrix}$$

The element in the ith row and jth column a_{ij} is called the (i, j)th element and the matrix is often denoted by $[a_{ij}]$ or A. When $m=n$ the matrix is square. An $m \times 1$ matrix is called a column vector or column matrix.

$$X = \begin{bmatrix} x_1 \\ x_2 \\ \vdots \\ x_m \end{bmatrix} \qquad (1.64)$$

A $1 \times n$ matrix is called a row vector

$$Y = [y_1, y_2 \ldots y_n] \qquad (1.65)$$

1.8.1 Addition of matrices

Two matrices may be added if and only if they are of the same order $m \times n$.

Then: $\quad A + B = [a_{ij}] + [b_{ij}] = [(a_{ij} + b_{ij})]$

i.e. the sum is formed by adding corresponding elements.

1.8.2 Multiplication of matrices

(1) By a scalar.
Any matrix may be multiplied by a scalar.

Then $\quad \lambda A = \lambda[a_{ij}] = [(\lambda a_{ij})] = A\lambda$

i.e. all the elements of A are multiplied by λ.

(2) Multiplication of two matrices.
Two matrices may be multiplied (A times B in that order) only if the number of columns of A is equal to the number of rows of B. If A is $[a_{ij}]$ of order $m \times n$ and B is $[b_{ij}]$ of order $n \times p$ then

$$AB = [a_{ij}][b_{ij}] = \left[\left(\sum_{k=1}^{n} a_{ik}b_{kj} \right) \right]$$

is of order $m \times p$.
It should be noted that, in general $AB \neq BA$.

1.8.3 The unit matrix

The unit matrix is a square matrix I for which:

$a_{ij} = 0$ for $i \neq j$
$a_{ij} = 1$ for $i = j$

1.8.4 The reciprocal of a matrix

The reciprocal matrix A^{-1} of A exists only if the determinant of A is nonzero and is given by:

$$AA^{-1} = I = A^{-1}A$$

1.8.5 Determinants

The determinant of a square matrix is defined as:

$$|A| = ||a_{ij}|| = \sum(\pm a_{1\alpha}a_{2\beta} \ldots a_{n\nu})$$

the summation of $n!$ terms being over all the arrangements $(\alpha, \beta, \ldots \nu)$ of the column suffixes and the sign $\pm$ being chosen according to whether the arrangement is even or odd.
In the simplest case,

$$\begin{vmatrix} a_{11} & a_{12} \\ a_{21} & a_{22} \end{vmatrix} = a_{11}a_{22} - a_{12}a_{21}$$

and from this can be developed the expressions for the expansion of determinants of higher order than the second. The minor of a_{ij} in A is the determinant of the matrix obtained by deleting the ith row and jth column of A. The cofactor A_{ij} of a_{ij} in A is $(-1)^{i+j} \times$ minor of a_{ij}.
Now the expression for a determinant is given by:

$$|A| = a_{i1}|A_{i1}| + a_{i2}|A_{i2}| \ldots + a_{in}|A_{in}|$$

The value of a determinant is unaltered by interchanging rows with columns. Interchanging either two rows or two columns changes the sign of a determinant. Thus, if either two rows or two columns are identical the determinant is zero.

1.8.6 Simultaneous linear equations

Simultaneous linear equations can be arranged in matrix form and their solution obtained via determinants

$$a_{11}x_1 + a_{12} \quad x_2 + \ldots + a_{in}x_n = b_1$$
$$\vdots \qquad \qquad \vdots \quad \vdots$$
$$a_{m1}x_1 + \ldots \qquad + a_{mn}x_n = b_m$$

may be written $\quad AX = B$

and now $\quad X = A^{-1}B$

alternatively $\quad x_j = \dfrac{|D_j|}{|D|}$

where D denotes the determinant $|a_{ij}|$ and D_j denotes the determinant D with the elements $a_{1j}a_{2j} \ldots a_{mj}$ replaced by $b_1 b_2 \ldots bm$.

STATISTICS

1.9 Introduction

Statistical techniques are used in engineering mainly in connection with the quality control of manufacturing of produced material and with the checking for compliance of such products, with whatever specifications or clauses are contained in the contracts covering their purchase and sale. In order to exercise quality control or to check for compliance it is necessary to make measurements of one sort or another. Now it is well established that the result of repeating a measurement (or of repeating an experiment) does not generally repeat the observation or original result. Further, repeat measurements will lead to further results and so appears the problem of variability.

Generally speaking, the variation in results arises both because the subjects of the measurement are themselves different and also because of errors introduced by the experiment or the measuring technique. Such variation is common experience in the measurement of, for example, the strengths of materials. It will often be desirable (if only from an economic viewpoint) to reduce the variation to as small an amount as can conveniently be arranged. However, it is not generally possible to reduce such variation to an unimportantly small value and so it becomes necessary to deal with the problem posed by the obtaining of different results from apparently identical experiments. It is to deal with the evaluation of such scattered experimental results that statistical techniques have been developed.

It is supposed that, were it possible to continue the experiments indefinitely, the results so obtained would cluster around some fixed value which would be the required value. (It is an implicit assumption that the indefinite series of experiments be conducted under identical conditions.) Since it is not possible to conduct indefinitely long experiments the problem becomes that of trying to determine, from a finite series of experiments, that fixed value (which is presumably the true value) about which the indefinite series of results would cluster. This attempt to determine is known as 'estimating', and while the use of that particular word does not imply that there has been any guesswork in obtaining it, there is an implication of uncertainty about the result. In statistical methods this uncertainty is calculated and specified in terms of confidence limits. A result obtained after statistical calculations should generally be given in terms of an estimate surrounded by confidence limits. Of course the more nearly certain we wish to be that the confidence limits contain the true value the wider those limits must be. In cases where the experiment or test is not aimed at the estimating of some particular quantity, the form of the estimate and confidence limit changes to one that such and such a result would not have arisen 'by chance' more than on so many per cent of occasions in an indefinitely long series of trials.

It is important that statistical results should be properly presented in the form of estimate and confidence limits: having decided upon such a form it is then sensible to use an appropriate precision for reporting the values. For example, when estimating the strength of concrete where an estimate might be of the form: $42 \pm 5 \, \text{N/mm}^2$ (at 95% confidence) there is clearly no point in reporting the result to several decimal places.

When an estimate of some quantity has been obtained, the interval between the confidence limits may be wider than it is desired they should be, in which case the interval may be narrowed by accepting a lower confidence. If this is not desirable it will be necessary to: (1) take more observations; or (2) improve the experimental techniques used to reduce the variability of the results.

It is important that the question of what is required by way of precision should be considered prior to an experiment so that the number of observations necessary to obtain the required precision may be assessed. In making that assessment it will be necessary to have information about the variability of parts of the experiment. This information may be available from previous experience, but if not it must be obtained by a pilot experiment.

It will be clear from the foregoing that any result which is obtained, being subject to error, may cause a wrong decision to be taken. Thus when dealing with, for instance, material to be checked for strength the contract for the supply of the material should indicate a test scheme to determine whether the strength of the material is correct or not.

Such a test scheme will involve experiments, and the possibilities for a wrong decision are:

(1) That the test will wrongly show as unsatisfactory, material with the correct strength. (This is known as the manufacturer's or supplier's risk.)
(2) That the test will wrongly show as satisfactory, material with an incorrect strength. (This is known as the consumer's risk.)

The performance of a test scheme is defined by its power and is represented by a graph showing, on one axis, the true value of the parameter in question (e.g. the strength of the material) plotted against the probability that material will pass the test and so be accepted. The calculation of such graphs is not a simple matter and requires full information about all aspects of the test scheme under consideration. The power curves of two test schemes represent, however, the only way in which the performance of the two schemes may be compared.

In the following sections are presented definitions of some of the terms used in statistical work, descriptions of statistical techniques and tests which may be used as a part of the experimenter's armoury of techniques and a description of central charts as a method of quality control. In the final section the references have, in the first cases, been selected for their readability as well as for their coverage of any particular point. Thus the works by Moroney[1] and Neville and Kennedy[2] are especially recommended as initial reading for anyone interested in statistical problems and techniques.

1.10 Definitions of elementary statistical concepts

1.10.1 Statistical unit or item

One of a number of similar articles or parts each of which may possess several different quality characteristics.

Example 1.18 A piece of glass tubing taken from a large number produced in quantity for which the diameter and other characteristics may be measured; a concrete cube for which the strength may be measured.

1.10.2 Observation – observed value

The value of a quality characteristic measured or observed on a unit.

Example 1.19 The diameter in millimetres of a piece of tubing; the strength of a concrete cube.

1.10.2.1 Sample

A portion of material or a group of units taken from a larger number which is used to obtain estimates of the properties of the larger quantity.

Example 1.20 Forty-eight pieces of tubing sampled from all the pieces produced during a day; the concrete cube made from a batch of concrete.

1.10.2.2 Random sample

A sample selected in such a manner that every item has an equal chance of inclusion.

1.10.2.3 Representative sample

A sample whose selection requires planned action to ensure that proportions of it are taken from different subportions of the whole.

Example 1.21 The forty-eight pieces of tubing selected two from every hour's production in one day; concrete cubes made, one from every batch, of a lot consisting of several batches.

1.10.2.4 Population

A large collection of individual units from one source. In particular circumstances this may be, for example, an output or batch: the bulk of material (concrete) or total collection of units (pieces of tube) produced by a set of machines or a factory in a specified time.

Example 1.22 Pieces of tubing made in a particular factory during a month; the concrete produced by a single plant during 1 day.

1.10.2.5 Statistic

A statistic is a quantity computed from the observations of a sample.

1.10.2.6 Parameter

A parameter is a quantity computed from the observations made on a sample. Thus, the value of a parameter for a population is estimated by the appropriate statistic for the sample.

1.11 Location

1.11.1 Measures

1.11.1.1 Arithmetic mean

The arithmetic mean, often called the 'mean' or the 'average', is the sum of all the observations divided by the number of observations:

$$\bar{x} = \frac{1}{n} \sum_{i=1}^{n} x_i \tag{1.66}$$

Example 1.23 $0.20\,(2.540 + 2.538 + 2.547 + 2.544 + 2.541)$
$= 2.542.$

1.11.1.2 Median

The value which is greater than one-half of the values and less than one-half of the values.

Example 1.24 The value 2.541 is the median of the above five numbers. (Had there been an even rather than an odd number of

numbers the median is the average of the two numbers either side of the median position.)

1.11.1.3 Midpoint or midrange

The value which lies half way between the extreme values.

Example 1.25 Using the numbers above the mid point is

$$0.5\,(2.538 + 2.547) = 2.542\,5$$

1.12 Dispersion

1.12.1 Measures

1.12.1.1 Range

The difference between the largest and the smallest values.

Example 1.26 $2.547 - 2.538 = 0.009.$

1.12.1.2 Deviation

The difference between a value and the mean of all the values.

1.12.1.3 Variance

The variance of a set of values is the mean squared deviation of the individual values and is normally represented by σ^2.

$$\sigma^2 = \frac{1}{n} \sum_{i=1}^{n} (x_i - \mu)^2 \tag{1.67}$$

where μ is the mean value.

A frequently occurring problem is that of estimating the main properties (the mean to describe the location and the variance to describe the dispersion) of a population by measurements (x_i) taken on a sample. From the measurements on the sample we can calculate the sample mean, $\bar{x}$ which is an estimate of the population mean μ. The sum of the squared deviations is smallest about the arithmetic mean; thus, for the population an estimate of variance using the sample mean and sample variance will be an underestimate. In cases where we wish to estimate population parameters from sample observations, a correction is made by using $(n-1)$ as divisor instead of n. Thus, the estimate of the population variance from observations x_i on a sample is:

$$\sigma^2 = \frac{1}{n-1} \sum_{i=1}^{n} (x_i - \bar{x})^2 \tag{1.68}$$

1.12.1.4 Standard deviation

The standard deviation is the square root of the variance.

$$s = \left[\frac{1}{n-1} \sum_{i=1}^{n} (x_i - \bar{x})^2 \right]^{1/2} \tag{1.69}$$

As in the case of variance, the divisor n is replaced by $(n-1)$ when working with sample observations to estimate a population standard deviation. The standard deviation has the same units as the original observations and their mean $\bar{x}$.

When carrying-out hand calculations, the identity:

$$\sum_{i=1}^{n} (x_i - \bar{x})^2 = \sum_{i=1}^{n} (x_i)^2 - n\bar{x}^2 \tag{1.70}$$

frequently saves effort. However, this method is not recommended for use on computers because of the danger of loss of accuracy when n is large and x_i has several significant figures.

1.12.1.5 Coefficient of variation

The coefficient of variation is the standard deviation expressed as a percentage of the mean. This is useful for dealing with properties whose standard deviation rises in proportion to the mean, for instance the strengths of concrete as measured by compressive tests on cubes.

1.12.1.6 Standard error

The standard error is the standard deviation of the mean (or of any other statistic). If in repeated samples of size n from a population the sample means are calculated, the standard deviation calculated from these means is expected to have a value:

$$Sm = \sigma/\sqrt{n} \qquad (1.71)$$

where σ is the standard deviation of the population.

An important result is that whatever the distribution of the parent population (normal or not) the distribution of the sample mean tends rapidly to normal form as the sample size increases.

1.13 Samples and population

1.13.1 Representations

1.13.1.1 Frequency

The number of observations having values between two specified limits. It is often convenient to group observations by dividing the range over which they extend into a number of small, equal, intervals. The number of observations falling in each interval is then the frequency for that interval. This allows a convenient representation of the information by means of a histogram.

1.13.1.2 Histogram or bar chart

A diagram in which the observations are represented by rectangles or bars with one side equal to the interval over which the observations occurred and the other equal to the frequency of occurrence of observations within that range (Figure 1.45).

1.13.1.3 Distribution curve

The result of refining a histogram by reducing the size of the intervals and correspondingly increasing the total number of observations. In the limit, when the intervals become infinitesimally small and the number of observations infinitely large, the tops of the rectangles of a histogram become a distribution curve (Figure 1.46).

1.13.1.4 Normal distribution (or Gaussian distribution)

A particular type of distribution curve given by:

$$y(x) = \frac{1}{\sigma(2\pi)^{\frac{1}{2}}} \exp\left\{ \frac{-\frac{1}{2}(x-\mu)^2}{\sigma^2} \right\}$$

$$(1.72)$$

where x is the observational scale value, μ the population mean and σ the population standard deviation.

These parameters of the distribution are estimated by the sample mean $\bar{x}$ and standard deviation s.

It has been found that a great many frequency distributions met with in practice fit quite closely to the normal distribution. However, one should beware of thinking that there is any law which says that this shall be so; it is simply a matter of experience. In circumstances where the observed frequency distribution does not appear to be normal it is often possible to transform the original data (e.g. by taking logarithms, square roots or squares) so that the transformed data is nearly normal. These two facts explain why so much of the effort in statistical theory has been devoted to treatment of normal-distribution problems.

For normal distributions the percentage of observations (in large samples) lying within certain limits of the observational scale are given in Table 1.5 and Figure 1.47.

1.14 The use of statistics in industrial experimentation

As has been stated, in experimental work units in a sample drawn from a parent population and the observations made on them are subject to error, and our task for which we use statistics is to make useful statements about the properties of the parent population. To achieve this, the most important statistics are the mean and the standard deviation. This section, therefore, considers the obtaining of sample means and standard deviations and confidence limits for them in situations where the parent population is normally distributed. Tests of significance for comparisons of means and variances are also described. Inevitably, only brief summaries are given and a study of standard works is advised before using the techniques on any important matters. As an alternative, the help of the statistical expert should be sought. If such assistance is to be obtained, it cannot be emphasized too strongly that it should be acquired right at the outset of the problem. It is rarely of much help to anyone (even though it happens only too frequently) for the statistician to be asked: 'Please tell me what these numbers show: they must mean something, I've collected so many, and they cost a great deal to obtain.'

1.14.1 Confidence limits for a mean value

If the form of the distribution were known together with the true mean μ and the standard deviation σ, then it is easy to make statements about the mean of a number of observations. If the population is normal then the mean $\bar{x}$ of a sample size n drawn randomly will, on average, satisfy:

$$\mu - \frac{3\sigma}{\sqrt{n}} < \bar{x} < \mu + \frac{3\sigma}{\sqrt{n}}$$

997 times out of 1000 (see Table 1.5).

Thus, if μ is actually unknown (and we are trying to estimate it) we may assert:

$$\bar{x} - \frac{3\sigma}{\sqrt{n}} < \mu < \bar{x} + \frac{3\sigma}{\sqrt{n}}$$

with 99.7% confidence. By this, we mean that if we go on making such assertions indefinitely we shall be wrong only 3 times in every 1000. We can make the containing interval narrower by reducing confidence so that we assert with 95% confidence that the limits for μ are $\bar{x} \pm 1.96\sigma/\sqrt{n}$ (Figure 1.48).

Very often we may be concerned with a limit on only one side, for instance we may require assurance that μ is greater than a certain value. Now, the probability of $\bar{x}$ falling above $\mu + 2\sigma/\sqrt{n}$

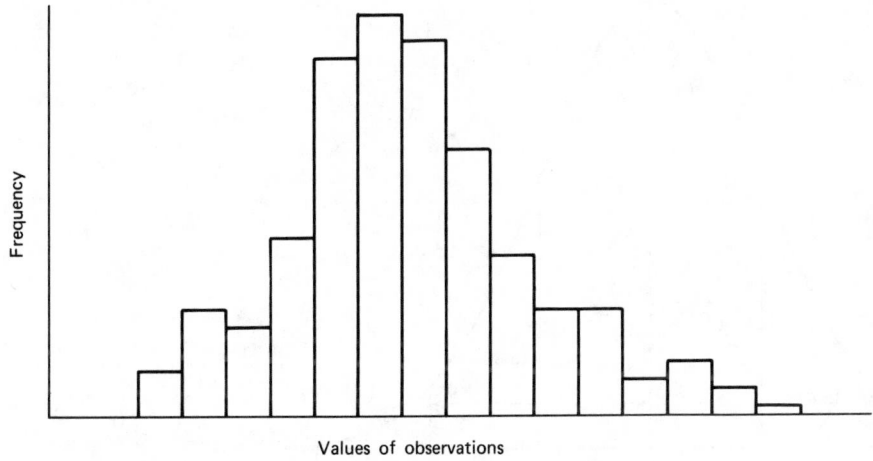

Figure 1.45 A histogram of observations from a sample

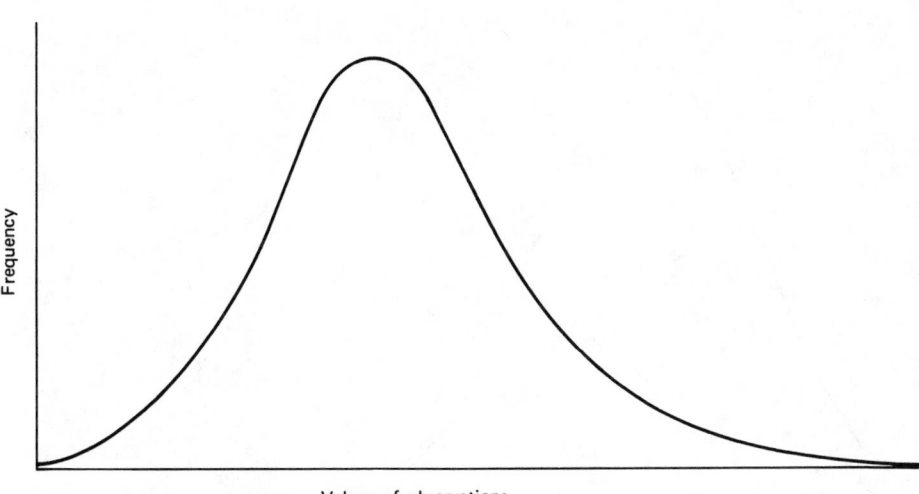

Figure 1.46 A continuous distribution curve

Table 1.5 The normal distribution

Range	Observations within range (%)
$\mu \pm \sigma$	68.27
$\mu \pm 2\sigma$	95.45
$\mu \pm 3\sigma$	99.73
$\mu \pm 1.96\sigma$	95
$\mu \pm 3.09\sigma$	99.8

is 2.27%. Thus, we may assert with 97.73% confidence that μ does not lie below $\bar{x} - 2\sigma/\sqrt{n}$.

Generally, the proportion of sample means $\bar{x}$ which exceed $\mu + u_a\sigma/\sqrt{n}$ is equal to a where u_a is the value given in a table of the normal distribution for a specified probability, say P. Because the distribution is symmetrical, a is also the proportion of sample means which are exceeded by $\mu - u_a\sigma/\sqrt{n}$. Thus, the whole range of values which μ may take is divided into three parts and three assertions can be made, to correspond one with each part:

(1) $\mu \geqslant \bar{x} - u_a\sigma/\sqrt{n}$ with confidence $100(1-a)\%$.
(2) $\mu \leqslant \bar{x} + u_a\sigma/\sqrt{n}$ with confidence $100(1-a)\%$.
(3) $\bar{x} - u_a\sigma/\sqrt{n} \leqslant \bar{x} + u_a\sigma/\sqrt{n}$ with confidence $100(1-2a)\%$.

This shows two sorts of statement, the single-sided (cases 1 and 2) and the double-sided (case 3). When using statistical tables it is important to check whether the tabulation is for single-tailed testing or two-tailed testing. (This description arises because cases 1 and 2 are, in the practical cases where a useful level of confidence is being used, representable as the two tails of a curve shaped like the normal distribution curve.)

In the discussion of confidence limits for the mean value μ of a population estimated by the mean of the sample $\bar{x}$ above it was assumed that the population standard deviation was known. Generally this will not be the case and μ will have to be estimated as s, a sample standard deviation and used in place of μ in the calculations above. The confidence limits for μ are now

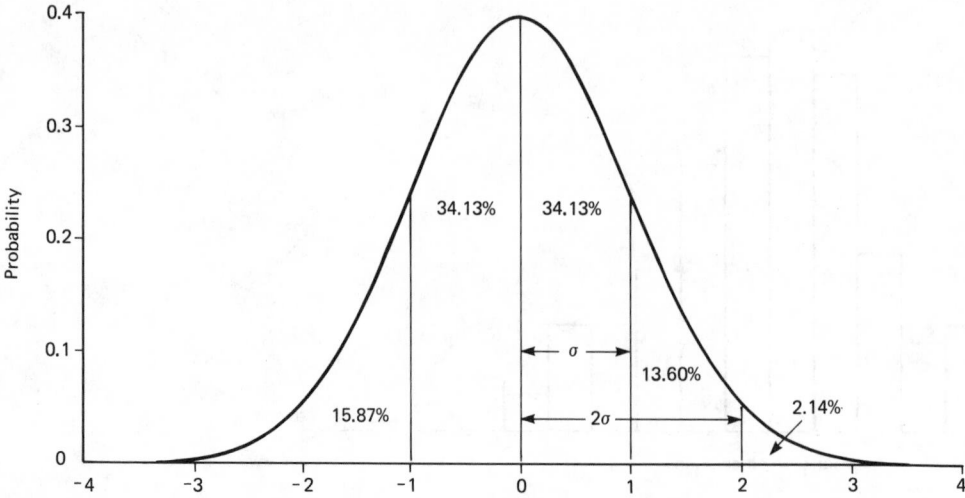

Figure 1.47 The normal distribution with limits

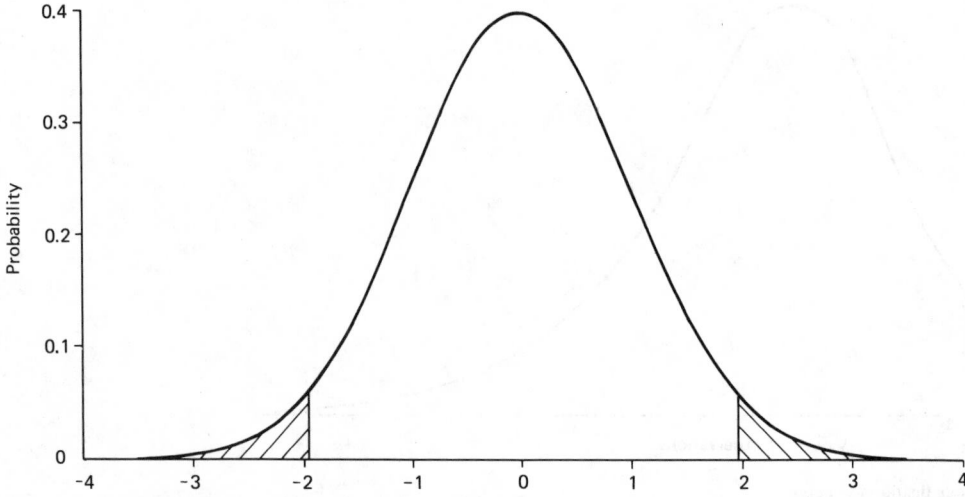

Figure 1.48 Diagrammatic representation of confidence limits
(with 95% limits shown)

wider because of the uncertainty about s and, instead of using u from the normal distribution curve it becomes necessary to use tables of student's t. The particular value of t to be used depends on how good the estimate s of σ is, which in turn depends upon the number of degrees of freedom in making the estimate. In the case of a standard deviation of n observations, the number of degrees of freedom is $(n-1)$. The number of degrees of freedom is generally denoted by ϕ. Some values of t are given in Table 1.6. In using this table the $100(1-2a)\%$ confidence limits are:

(1) Lower limit $\quad \bar{x} - t_a s/\sqrt{n}.$
(2) Upper limit $\quad \bar{x} + t_a s/\sqrt{n}.$

using the value of t_a for the appropriate number of degrees of freedom.

Table 1.6 Significance points of the t-distribution (single-sided)

ϕ	Probability: P				
	0.1	0.05	0.025	0.01	0.005
1	3.08	6.31	12.70	31.80	63.70
2	1.89	2.92	4.30	6.96	9.92
5	1.48	2.01	2.57	3.36	4.03
10	1.37	1.81	2.23	2.76	3.17
20	1.32	1.72	2.09	2.53	2.85
40	1.30	1.68	2.02	2.42	2.70
∞	1.28	1.64	1.96	2.33	2.58

1.14.2 The difference between two mean values

A problem which arises frequently is that of determining if the difference between two means has occurred by chance because of natural variation or whether there is a real difference. A real difference can only be asserted in the form of a statistical statement that the difference is significant at a certain level, i.e. there is a probability that there is a real difference. This is done by calculating a t statistic from information about the samples and comparing the result with the tabulated t values. The means, standard deviations and number of observations of the two tests are denoted by $\bar{x}_1$, $\bar{x}_2$, s_1, s_2, n_1 and s_2.

Calculate $\quad t = (\bar{x}_1 - \bar{x}_2)/s_p$

where $\quad s_p = \sqrt{\left[\left(\dfrac{1}{n_1} + \dfrac{1}{n_2} \right) \left(\dfrac{v_1 s_1^2 + v_2 s_2^2}{v_1 + v_2} \right) \right]}$

and $\quad v_1 = n_1 - 1, \; v_2 = n_2 - 1$

(*Note*: s_p is the pooled standard deviation for the samples 1 and 2.).

If this calculated value exceeds a tabulated value of t (for $\phi = v_1 + v_2$) then the difference is significant at the level determined by the probability heading the column of the t table.

As an example, consider the comparison of two testing machines for crushing concrete cubes. The machines are to be compared by making a single batch of twelve concrete cubes and testing six cubes on each machine. The results obtained are:

Machine 1   39.2   38.4   44.7   41.0   41.0   44.1
Machine 2   41.1   33.8   42.4   36.8   32.0   40.1

From these observations we can calculate:

Sample sizes $\qquad\qquad\quad n_1 = 6, \; n_2 = 6$
Sample means $\qquad\qquad\;\; \bar{x}_1 = 41.4, \; \bar{x}_2 = 37.7$
Sample standard deviations $\;\; s_1 = 2.54, \; s_2 = 4.19$

$$s_p = \sqrt{\left\{ \left(\frac{1}{6} + \frac{1}{6} \right) \left(\frac{5 \times 2.54^2 + 5 \times 4.19^2}{5 + 5} \right) \right\}} = 2.0$$

so: $\quad t = (41.4 - 37.7)/2.0 = 1.85$

The number of degrees of freedom $\phi = v_1 + v_2 = 10$.

From Table 1.6 it is seen that for $\phi = 10$ the single-sided 2.5% (or 0.025) point of the t distribution is 2.23 and so the calculated t value is not significant at the $2 \times 2.5\%$ level.
(*Note*: It is necessary to double the probability value from the table because the question: 'Are the testing machines different?' requires a two-sided test to be carried out. By contrast the question: 'Is mean 1 greater than mean 2?' would require a single-sided test.)

1.14.3 The ratio between two standard deviations

In a similar way in which it may be desired to compare two means, it may be desired to compare two standard deviations. Whereas means are compared by calculating their difference, standard deviations are compared by calculating the ratio of the variances (the square of the standard deviation) and comparing the ratio with tabulated values in an F test. In all such calculations the value obtained for the ratio must be greater than unity so that the larger standard deviation (say s_1) must be placed over the smaller (s_2) where s_1 and s_2 are sample standard

deviations and so estimates of the population standard deviations. Since we have of necessity $s_1 \geqslant s_2$ when we calculate

$$F = \frac{s_1^2}{s_2^2}$$

the F test is a one-sided test.

Values of F for comparison with the calculated value from the observed standard deviations are given in most statistical books.[2] Such tables are presented generally with one table for each specified probability and within such a single table the column headings are the values of v_1 (the number of degrees of freedom, $n_1 - 1$, of the smaller standard deviation estimate s_2).

By way of illustration, consider the example used above for the comparison of means. In that example the observations lead to:

$$n_1 = n_2 = 6$$

so: $\qquad v_1 = v_2 = 5$

$$s_1 = 2.54 \quad s_2 = 4.19$$

In this example $s_2 > s_1$ so the calculation of F is:

$$F = \left(\frac{4.19}{2.54} \right)^2 = 2.72$$

In the tables (e.g. in Neville and Kennedy[2]) the tabulated 1% confidence point of F is 10.97 and the 5% point is 5.05 both found for $v_1 = v_2 = 5$. Since the calculated F ratio is not greater than the tabulated values the conclusion to be drawn is that there is not strong evidence that population standard deviations are different.

1.14.4 Analysis of variance

If a manufacturing process or a testing scheme involves a number of independent factors, each of which contributes to the variability of the results, then the variance of the whole system is equal to the sum of the component variances. (Note that the variance must be added, not the standard deviation.) This additive property permits the technique of analysis of variance, which can take many forms depending on the structure of the process which is being analysed. One of the major difficulties of analysis of variance lies in deciding what form of structure is appropriate to the process being modelled by the analysis of variance. In the majority of cases which are not both simple and short it will be sensible for the arithmetic to be performed by computer. However, in simple and short situations the calculations may reasonably be undertaken by hand.

Probably the most commonly occurring simple situation is that of analysis to determine variance between and within batches. The methods are best described by an example. The example will be one in which concrete cubes are made batches (each of three cubes) and strength tested at (say) 28 days. The first step is to define the statistical model which is being used:

$$Y_{ij} = Y + A_i + E_{ij} \qquad (1.73)$$

where there are i batches each of j cubes, Y_{ij} is the observed strength of the jth cube in the ith batch, Y is the average strength (averaged over all tests), A_i is the difference between Y and the average strength of batch i, and E_{ij} is the difference between the jth cube of batch i and the average strength $Y + A_i$ of that batch.

If the data for four batches of cubes is:

19.8	21.1	19.8	(batch 1)
21.8	22.0	21.0	(batch 2)
21.2	21.5	21.2	(batch 3)
21.4	21.4	21.0	(batch 4)

it is found that $Y = 21.1$.

From this can now be found the sums of squares of the A_i and E_{ij}. Associated with each sum of squares is a number of degrees of freedom (as usual one less than the number of occurrences) so that dividing the sums of squares by the appropriate number of degrees of freedom gives the mean square. Thus is constructed an analysis of variance table as shown in Table 1.7.

The method of test is by F ratio so that the larger variance (the average of the sums of squares of errors) is divided by the smaller. Here, $1.07/0.23 = 4.61$ with 3 degrees of freedom for the column heading and 8 for the row heading when comparing with the tabulated F values. For $v_1 = 3$, $v_2 = 8$ the tabulated F value at 1% confidence level is 7.59. The observed value does not exceed this and so there is no assertion that can be made at the 1% level. However, the tabulated F value at the 5% confidence level is 4.07. The observed value exceeds this and so a result significant at the 5% level has been obtained. Thus, although there is not strong evidence there is some evidence of a real difference between batches.

In an example so small as this one the necessary arithmetic (especially if properly organized) may reasonably be tackled by hand. However, as can be deduced from examination of F tables, it is not always easy to get significant results with small experiments. Thus, the use of the technique will in many cases imply the use of a computer for handling the arithmetic. In such circumstances the engineer is likely to be using an existing computer program and need only concern himself with correctly presenting the data for the program to analyse and then with the interpretation of results and comparisons with tabulated F values. He has no need therefore to develop great skills in short-cut arithmetic methods.

Table 1.7

Model term	Sum of squares	Degrees of freedom	Mean square
A_i	3.2	3	1.07
E_{ij}	1.9	8	0.23

1.14.5 Straight-line fitting and regression

Experiments may be designed to examine whether two parameters are related. The circumstances may involve the effect on a property of a product of some parameter in the production process. In the experiment the parameter will be controlled or constrained to take a number n of prescribed values x_i over some range and the consequential observations y_i will be paired with them. The question now arises as to the 'best straight line' through the points x_i, y_i. It is assumed that the x_i values are error-free but that the observations y_i are subject to error. The method of obtaining the 'best' straight line in such circumstances is to choose the two parameters m and c of the straight line:

$$y = mx + c \tag{1.74}$$

in such a way that the sum of the squares of the errors in the y direction is a minimum. This is achieved by making:

$$m = \frac{n\Sigma xy - \Sigma x \Sigma y}{n\Sigma x^2 - (\Sigma x)^2} \tag{1.75}$$

and:

$$c = \frac{\Sigma x^2 \Sigma y - \Sigma x \Sigma xy}{n\Sigma x^2 - (\Sigma x)^2} \tag{1.76}$$

This line is called the line of regression of y on x and one of its properties is that it passes through the centroid $\bar{x}$, $\bar{y}$ of the observed points. The usual statistical question now arises concerning the confidence limits which should be applied to the calculated line which is an estimate of a relationship. To examine this problem the errors or deviations must be calculated. At every observation point x_i y_i which does not actually lie on the calculated line there is an e_i. The variance of y estimated by the regression line is then:

$$s_e^2 = \frac{\Sigma e_i^2}{v} \tag{1.77}$$

where v is the number of degrees of freedom.

Since calculation of m and c impose two restraints the value of v is given by:

$$v = n - 2 \tag{1.78}$$

The variance of the mean value $\bar{y}$ is given by:

$$s_{\bar{y}}^2 = \frac{s_y^2}{n} \tag{1.79}$$

so that the confidence limits for $\bar{y}$ are:

$$\bar{y} \pm ts_{\bar{y}} \tag{1.80}$$

where, just as for a sample mean, the value of t is found from tables using the appropriate number of degrees of freedom.

The variance of the slope m is given by:

$$s_m^2 = \frac{s_y^2}{\Sigma(x - \bar{x})^2} \tag{1.81}$$

and the confidence band for slope is given by:

$$m \pm ts_m \tag{1.82}$$

It may be necessary to compare one regression line with another, theoretical one, to see if there is any significant difference between the theoretical slope, m_0, and the observed slope m. This test is performed by calculating a t statistic:

$$t = \frac{m - m_0}{s_b} \tag{1.83}$$

and comparing with the tabulated values. Just as in the case of comparison by means of samples we can compare the slopes of two observed lines by replacing $m - m_0$ by $m_1 - m_2$ in Equation (1.83) and using a pooled standard deviation from the variances of the slopes of both lines in place of s_b. The number of degrees of freedom used in the t table will be $n_1 + n_2 - 4$.

1.15 Tolerance and quality control

Material is often manufactured for supply according to a specification which will include compliance clauses for the performance of the product. As an example, CP 110[3] lays down (in Section 6.8) certain strength requirements and also suggests a testing plan. The *Handbook* to that code[4] discusses the problems of compliance and shows how different forms of

testing plan after the operating characteristic of a test plan and so charge the risks run by the producer and by the customer. The customer has, in theory, the opportunity of reducing his risk by adopting a more vigorous testing plan. This, however, is likely to cost more and a customer may well deem this not worth while. The producer, on the other hand, must expect to have to meet the compliance clauses and needs to arrange his production methods so as to make a profit taking account of whatever limits or penalties may be imposed on him by the compliance clauses under which he has to operate. Thus, the manufacturer or producer is faced with a problem of how to control his product.

One example of a technique for exercising this control is shown by a system advocated for controlling the strength of ready-mixed concrete[5] by means of the cumulative sum chart which is an improved form of control chart especially developed and adapted to the problems of concrete manufacture.

In the process of manufacture and measurement of some property of the product natural variation will cause the results obtained to be distributed in some way. The problems facing the manufacturer are:

(1) To maintain adequate control over the process so that the variation in results does not become so large that an uneconomic number fall outside the specified tolerances.
(2) To detect any trend for the observations obtained to be moving out of the specified limits, sufficiently early to take useful corrective action.

As usual, samples are taken to estimate the properties of the parent population. To do this comparatively, many samples (25 or more) of comparatively small (but not less than about four and all the same) size are tested and the mean of the means $\bar{\bar{x}}$ used to estimate the population mean. The population standard deviation is estimated from the variance within samples, the average sample standard deviation from the average sample range.

Thus: $$\bar{\bar{x}} = \frac{\bar{x}_1 + \bar{x}_2 \ldots + \bar{x}_k}{k} \tag{1.84}$$

$$\bar{x} = s/\sqrt{n}$$

for k samples of size n.

Now a chart is drawn with time or sample number in the horizontal axis and observation values on the vertical axis. A line drawn at $\bar{\bar{x}}$ represents the target performance of the process and two surrounding lines at $\bar{\bar{x}} \pm 1.96/(\sqrt{n})s$ represent warning levels for the process while surrounding lines at $\bar{\bar{x}} \pm 3.09/(\sqrt{n})s$ can be regarded as action levels.

The choice of the figures 1.96 and 3.09 has been made on the assumption that the process is functioning in such a way that the specified tolerance limits are reasonable, i.e. they are not so stringent that the chance of the product meeting the requirements is not high while on the other hand the process is not so 'good' (in which case it may be unnecessarily expensive) that all the results obtained lie well within limits.

The design and use of control charts is a valuable use of statistical methods. Generally they are robust in the sense that their usefulness is little affected by factors such as non-normality of the basic data. However, for their efficient use in some area experience of the particular technology is desirable and for a better understanding of the possibilities of the techniques the reader is recommended to works by the British Ready Mixed Concrete Association[5] and Davies and Goldsmith.[6]

COMPUTERS

Computers and computing have made a substantial impact on most walks of life, civil engineering not excepted. The pace of development in computing is substantially greater than for any other area of activity in the engineering world. Although other subject areas are subject to bursts of activity from time to time, when research or some specific project provides the necessary spur, computers are developing rapidly all the time, whether the engineering world wishes it or not. In consequence a great many organizations find it difficult to keep abreast of what is available or of what might actually be of benefit to them in their work. This difficulty is not eased by the wide discrepancy between the useful life of most civil engineering work and the life of computers.

Although, in principle, computers are simple machines which can perform simple arithmetic and make simple decisions (according to a set of coded instructions—the program) that fact is ever more frequently masked by the use of sophisticated techniques which appear to make computers behave more and more like human beings, and able to undertake tasks previously the province of human effort.

1.16 Hardware and software

One of the most important distinctions which must be understood when considering computers is the difference between *hardware* and *software*. A simple criterion is to imagine that the hardware consists of the material pieces which one can see— boxes, wires, screens, discs, chips etc.—while the software comprises the instructions which the hardware obeys. It is in the nature of the general developments in society that the cost of making the hardware is, in real terms, falling all the time. This fall in cost comes about through better design of components, automated manufacturing techniques and so on. All this is similar to the developments which have been taking place in other fields of manufacturing.

The software, on the other hand, consumes human effort and imagination very intensively. It is not easy to improve the techniques of manufacture here! In consequence, the total cost of a computer installation—if it is regarded, for simplicity, as being composed of the two elements of hardware and of software—has changed considerably. In the early days, when computers were harnessed in working offices, the cost of the hardware was the major consideration and the software, if considered at all, tended to be something of an afterthought. Now we are recognizing that the software is, or ought to be, the major consideration. Once the major details of the software suitable for the envisaged tasks have been settled it is logical to search for the 'best' hardware solution which will accommodate the chosen software.

It is, of course, unlikely that any office, let alone organization, will wish to 'computerize' just one activity. It is normal for a great multitude of tasks to benefit from being done by machine rather than by man. In this event the choice of hardware will be constrained by a, perhaps wide, variety of software. This emphasizes the fact that computers should be thought of (in the hardware sense) as general-purpose machines.

1.17 Computers

The changes which have taken place in recent years encompass the change from remote 'batch' computing to 'personal' computing where every person who needs one in order to do his job appears to have access to one on his desk. In truth, this revolution has come about via an intermediate stage, i.e. the change from the large mainframe machines which, while they could perform several tasks apparently concurrently, had to be run by dedicated operators remote from the users, to the mini machines operated via remote terminals. With this scheme, the many users all feel (most of the time) that the computer is dedicated to them alone while, actually, the centre of the machine is servicing up to some tens of users and it is only the terminal which is dedicated to the individual user.

The development of the 'personal' computer has had a bigger impact on this situation than is at first sight obvious. Personal computers came about because the huge improvements in computer technology allowed the production of a machine which can sit, complete, on a desk, but which has power greater than the mainframe machines of a decade ago. (Those mainframes had required a large dedicated room, and air-conditioning, as well as operating staff.)

The presence of the computer on the desk, with an impressive array of available software, encouraged a situation in which the users were often repeating tasks being performed by colleagues (especially the inputting of data). One description of the development as it affected the functioning of an organization was that it was leading to near anarchy with little managerial control of what was happening to the benefit of the organization.

This independence of the personal computers has therefore been both a benefit and a source of difficulty. Trying to get the best of all possible worlds has led to much emphasis on communications. Here is meant the communication between different computers, generally communicating with other computers within the same organization, but sometimes further afield. Increasingly, for organizations of a certain size, the plan followed is one with a major computer at the heart of operations with a network of personal computers around it. These personal computers can be connected to the 'heart machine', or can be operated in stand-alone mode, at the will of the user. In these circumstances it becomes possible for the heart machine to be the repository of the valuable corporate data (which should then be held once only) to which the individual users can have access as and when their work demands it. The individual users can then use their own 'personal' data and run the programs of interest to them on their personal computer without affecting anyone else. Should an individual user have available something (be it data or be it a program) to which other users require access, this is arranged via the heart machine.

The organization and control of such an arrangement is not simple and produces interesting problems of a managerial and human nature. But it is now possible to make arrangements which seem to be getting near to providing a situation in which men, machines, and the organization can all work reasonably efficiently.

The providers of computing solutions (hardware and software) are, of course, in business. They will therefore advise potential customers of the benefits of the particular solutions they purvey. It is not easy for the (computing) lay person to judge the advice received from such quarters. It seems likely, therefore, that many organizations will be well advised to adopt the strategy of ensuring that they have in-house expertise to judge such matters. This notwithstanding the fact that, as time goes by, the purveyors of computing solutions are making greater emphasis of the idea that their solution needs no computer expertise. As in other walks of life, the lack of expertise in an activity in which one is engaged is likely to be costly.

Since the advent of the personal computer (able to double as a terminal) sitting on the desk, has come the mobile or portable computer. This is depicted as sitting on the knees of the user and working off batteries, thus freeing the user from the need for access to mains electricity. While an obvious early use was, for example, for salesmen to enter their transactions, other applications are being found. The collection of technical data on site is an obvious parallel to those activities in other fields, so such machines are proving useful to the engineer. It can reasonably be said that the use of the fruits of computer invention are limited only by human imagination. Though trite, this statement has considerable importance. It can be very difficult indeed to think of a really new way of achieving some objective: the straitjacket of 'we've always done it this way' can be extremely strong.

1.17.1 The use of computers by civil engineers

Although engineers have appeared, at times, to lag behind in the use of computers, they actually began using them at a very early stage. The first 'obvious' application lay in the solution of the many simultaneous linear equations to which many problems of structural analysis can be reduced. This mathematical problem had received much attention in an effort to speed, refine, and make more reliable, hand methods. The ability of the computer to perform repetitive tasks reliably shifted the search to making the preparation, and input, of the data describing the problem more robust. This search was hampered for some time by limitations of the hardware. However, the availability of substantially increased computing power eventually allowed the problem of the data to be encompassed as well as the problem of solving the equations and presenting the results.

This theme of the availability of increasing computing power allowing new tasks to be tackled has recurred frequently in the history of computing.

The use of computers for structural analysis represented the limit of activity in engineering for some time. However, developments of other machines for drawing or plotting prompted an attack on another phase of engineering design activity. Conceptually, the operation of a design project can be split into four stages: (1) the concept and choice of solution; (2) the analysis of the whole structure; (3) the design of individual members; and (4) the preparation of detailed drawings.

The contribution computers can provide to (1) above has only comparatively recently become apparent in terms of rearranging scheme drawings and the holding of base data. Stage (2) was covered by the early computing endeavours and (4), the detail drawings, became possible when the plotters, developed for aero work, for example, became cheap enough for use in civil engineering. The development of a package for the production of drawings of reinforced concrete details was a major breakthrough. In use the detailer has available the information arising from the design of members. Using a desktop computer, for example, he supplies data of the basic dimensions of the member and then, via a question-and-answer dialogue, supplies information to define the reinforcement detail. At all stages the information supplied by the user is checked for logical consistency, geometric compatibility and compatibility with appropriate code or standard documents. If an attempt is made to do something impossible or contrary to regulation, then the user is not permitted to proceed until the error has been rectified. By contrast, an attempt to do something which, according to standards incorporated with the program, is unusual will result in a message which the user can heed, or ignore, at will. Such interactive programs were impossible with the earlier batch machines.

Having been developed in modern environments, such a program is now expected to be very 'user friendly'. To take a cynical view, a user-friendly program is one for which the user has no need to consult the (written) user manual!

The effect of using such an aid is that a small team of detailers can become very much more productive, producing many more drawings per week than by manual means. Further, the fact that the data defining the drawings is stored means that, in the event of changes becoming necessary, the revised drawings can be produced very much more quickly (and reliably) than if done by hand.

An interesting sideline to the development of such a tool is the attitude taken to drawings. While some drawings are required to make an impression, and so are treated as works of art, the drawing of a reinforcement detail is just a technical necessity and it is used only by technical people and therefore may be less impressive. In consequence, detail drawings, produced on comparatively cheap dot matrix printers, have become quite acceptable. Only a few years ago even these technical drawings were also treated as works of art.

The third stage of the design office exercise, that of designing the individual members has also been solved. More than one approach has been adopted but this has the benefit of providing potential purchasers and users with competitive choice.

If we consider the three technical stages of the design office activity and the solutions listed above, we find the appearance of some more common occurrences such as the requirements for compatibility and the transfer of information between different stages of the work. In this example, the information from the global structural analysis is required by both the design and the detailing activities. Similarly, the information from the member design is required for the detailing phase. There are different views about the extent to which this information transfer can, or indeed should, be made automatic. At the time of writing the general feeling is to limit the amount of automatic transfer, it being held that the contribution of man is too difficult to codify and too valuable to lose. Such views have held sway before and have, eventually, been overturned. It seems probable that the developments in expert systems and other advanced computer technology may have the same overturning effect here in due course.

The production of the software for such systems represents a very substantial expenditure and it is important that the solutions developed should not be excessively dependent on particular hardware. In fact the drawing part of the solution has used plotters and more recently dot matrix printers. (A likely change is that laser printing will be a practical tool for such work.) The actual computers used have covered a wide range although the operating system used by the computer has been important. This is another area where the general developments in computing towards standardized operating and filing systems will make the transport of software solutions from machine-to-(often successive)-machine a comparatively painless task.

This example of the solution of an engineering set of problems—analysis, design, detailing—has been described at some length because it typifies the problems which will require consideration in some form almost whenever a computer solution to a problem is being sought. It is foolish to underestimate the benefits which the computer can bring, but it is important to be aware of, and consider properly, the problems and side issues which can arise. Proper treatment of such matters can sometimes bring unexpected benefits.

1.17.2 Nontechnical computing

Although, when first invented, computers were largely used by technical people to perform technical tasks, it has long been the case that the bulk of computer sales and use have been in commercial fields. For some time this affected the design of computers but the picture now is one of much more general application as computers are becoming the user's workhorse. It is not practicable to have many different computers to perform the many different tasks which an individual may tackle.

1.17.2.1 Spreadsheet

One example of the change of use to which software can be put is the spreadsheet. A spreadsheet is essentially a rectangular array of boxes identified by cartesian-type coordinates. A box may contain either a value or a formula. Such formulae may relate to the values held in other boxes. After entry of data and formulae the user will request that calculation of values be performed. For whatever reason an item of data, or a formula, may need to be altered. After the change a recalculation can be requested and will usually seem to be performed almost instantaneously. It does not matter, of course, to the computer, whether the change was a correction of an error or a change of mind on the part of the user. The traditional use of spreadsheets has been in financial areas where the slogan about answering 'What if?' questions was meant to appeal to those with responsibility for profit margins, etc. However, a spreadsheet is nothing more than a general-purpose organization of calculation: there is no reason why a spreadsheet should not be used to calculate a set of sine or of logarithm tables. Increasingly, engineers are finding that, if they think about a problem from a different angle, a different solution tool may come to their aid. The spreadsheet is one example. The use of an improved solution tool will not come about unless knowledge of the tool and its capabilities exists together with a knowledge of the tasks tackled by the organization. As has been mentioned above, the best results will come only when there is a proper awareness of requirements and capabilities.

1.17.2.2 Word processing

Probably the most widespread computer application now is that of word processing. Although, traditionally, an author has passed his original (e.g. manuscript or dictated tape) to another person who has sole responsibility for production of the typed form, this may well change. With the increasing use of computers many workers who at some stage take on the role of author, are becoming more or less keyboard-competent. Now, while it would be too much to claim that such authors can key as well as a professional typist, there is an increasing number of these 'sometime' authors who can type quickly enough to keep up with their own creative thought processes. Also, by using a few of the more basic capabilities of the word processing system the author can produce a good result that is well ordered and cogent (even if the spelling and layout may leave something to be desired) more quickly than with the older techniques. Even the problems of spelling and of layout can, in part, be tackled by the computer. It seems hardly likely that the secretary is under serious threat from such developments of author capability but the notion of the copy-typing task may well be one that will disappear. Provided reasonable control can be exercised over aspects of detail and over the proper use of an individual's time, there may well soon be a substantial increase in the number of engineers producing their own reports.

1.17.2.3 Networks

It is this chameleon-like behaviour of the user at his desk, wanting to be structural analyst, financial analyst, typist, etc. which makes the proper arrangement of personal computers that are able to double as terminals so important. While, in some circumstances, it may be suitable to have these personal computer networks connected to one another without the existence of any 'heart' machine, as described above, it seems likely that the more frequent situation will be one in which the central computer is needed to provide not only backup facilities but also the corporate data (details of cost rates for example) which many users may require.

1.17.3 Specific vs. general-purpose software

When seeking software there is sometimes a choice between a program which has been designed for a strictly limited and well-defined purpose on the one hand and a program designed to be general-purpose on the other hand. To illustrate this, consider the problem of producing drawings of reinforcement details as discussed above. The end product drawing is nothing more than a set of lines drawn on a piece of paper. Some of these lines are straight, some curved and some are in the form of characters. Thus, a general-purpose drawing package, capable of putting lines on the paper according to data instructions defining, for example, cartesian coordinates of end points of line segments, could produce the required drawing if the data are prepared. On the other hand, the special-purpose detailing program will require far less data in order to produce the same end result. It will, partly in consequence, be very much easier for humans to understand the data of the special-purpose version at a glance. They are thereby more able to spot mistakes and correct them.

On the other hand, the detailing program will be no use for the production of general-arrangement drawings or for a host of other tasks. It will generally be the case that the general-purpose program will feel, to the user, much more cumbersome, than a program built to specific purpose. When this occurs, most operators begin to feel that they are not properly in control and thereby become a little careless, allowing mistakes to creep in.

It is not practicable to produce special-purpose programs for all problems; there are too many problems. Indeed, even a special-purpose program will be, to some extent, general-purpose. (A detail drawing program will, for instance, be capable of drawing a wide variety of beam types, though it may not be capable of drawing a column.)

There is no universal answer to this choice problem. It is a question which can be resolved only by harnessing a proper awareness.

1.17.4 Computers and information

The last half decade has seen an explosion in the amount of data stored in computers. (Technically this has become possible as the cost of unit storage has reduced.) However, there is no point in storing data in a computer if it cannot be accessed with both speed and ease, and then manipulated to meet the need.

Computers traditionally have been regarded as 'unintelligent' so that information would, of necessity, be stored only in carefully prearranged patterns in order to permit subsequent location and retrieval. The argument has been that, if there is no pattern, retrieval will be impossible, so do not store. (There have inevitably been 'squirrels' who have adopted the policy of storing everything, in case it may be useful. This philosophy has not been regarded as generally cost-effective.)

Information, which can be expensive to collect and to keep, has typically been stored in large databases to which accredited users can gain access. These databases have generally been in very well-defined structural forms. In consequence access has, in general, been rapid. However, the design of the database envisages 'all' possible accesses which might be used in future. Now, however, increased machine speeds and the production of 'intelligent' software is cutting across these restrictions. The way ahead is not clear, nor is it likely to be quick because of the sheer volume of information which is available to man. However, there will be movement towards making information, generally, more easily available.

1.17.5 Computers and management

The size of projects in which mankind engages has increased enormously; so has the complexity. The management of projects (and the training of managers) has become a major problem.

Early tools to come to the aid of management have included bar-chart techniques, etc. The problem with most of these techniques is the volume of work necessary to cope with the inevitable alterations to the original plan. These alterations are liable to occur throughout the life of the project. Ideally, the manager would like to examine the effect of the change forced on him and then consider possible effects of changing his own plan for proceeding. Of course, the calculation power of the computer is the facility which makes such possibilities realistic.

However, this is only dealing with the techniques. There is also the problem of training managers, preferably without the trainees making mistakes (with very large cost consequences) on a real job. Developments in universities and research organizations have played a major part here.

1.17.5.1 Training games

These developments take the form of the simulation of a construction project, e.g. the construction of a manhole. This simulation is incorporated in a 'training game'. The game is set up by the tutor and included in it are details of the project and rates, e.g. for crane hire. Some of this information is made available to the player, who is invited to manage the construction. For each day's work his management will take the form of ordering types of labour and/or materials. The player has options, e.g. a cheap but not too reliable crane hire company, as opposed to a more reliable, more expensive one. The simulation makes available weather forecasts for the following day at the stage when the player is ordering. As in real life the weather is *generally* similar to the forecasts but differences do occur. With the labour and materials he has ordered, a certain amount of construction will get done in the day, and for this the player earns credit or payment. On the other side the labour and materials will be expensive. The actual progress of the work is subject to statistical interruptions whose level of occurrence is set by the tutor in advance. The objective for the players is to make a profit that is as large as possible.

The use of this training tool seems to be most effective when the player is actually a small team of about four students. The element of competition provided by three other teams working at the same time (but independently so they suffer different statistical 'accidents') increases the learning by sharing complementary experiences. There is clear evidence that this training is effective: it is certainly cheaper than making mistakes on a real job.

1.17.5.2 Project planning models

A further illustration of the way computers assist with tackling the unknown is to be found in a tool to be used in advance planning. For this application the project is modelled as a fairly conventional bar chart (possibly at more than one level). However, the model is not deterministic, i.e. it is recognized that when the chart is constructed, a bar is only a best advance guess and that it is subject, in the event, to variation. For many items (e.g. weather effects, rate of bricklaying, etc.) data is available about the variations which occur in practice. These variations are incorporated with the basic data. The best-guess bars represent just one way in which the project might be built. Changing one bar (within its allowed variation) produces another way the project might be built. What the computer does is to 'build' the project many hundreds of times allowing all the bars to vary stochastically. The result is an envelope of possible construction routes. Some will be quicker, others will be slower; some will be cheaper, others will be more expensive. Overall, however, the envelope will highlight potential holdups caused by delays, indicate cashflow requirements, etc. Clearly, the system can be run not only prior to construction but also during construction (when work already complete is, of course, no longer subject to variation). The tool ideally should be used collaboratively between client and contractor in a noncompetitive manner. At present, the world is far from ideal but there may be sufficient benefit for this route to appeal, especially to those involved in the very large projects for which it is best used. It is interesting that this potentially valuable exercise demands nothing more expensive than a fairly run-of-the-mill desktop personal computer in order to produce useful results.

References

1 Moroney, M. J. (1953) *Facts from figures.* 2nd edn. Penguin, London.
2 Neville, A. M. and Kennedy, J. B. (1964) *Basic statistical methods for engineers and scientists.* International Textbook Company, Scranton.
3 British Standards Institution (1985) *Structural use of concrete*, BS 8110 Part 1. BSI, Milton Keynes.
4 Rowe, R. E. *et al.* (1987) *Handbook to BS 8110 (Part 1), Structural use of concrete.* Chapman and Hall, London.
5 British Ready Mixed Concrete Association (1972) *Authorisation scheme for ready mixed concrete.* 2nd edn. BRMCA, Ashford.
6 Davies, O. L. and Goldsmith, P. L. (1972) *Statistical methods in research and production.* 4th edn. Oliver and Boyd, Edinburgh.

Bibliography

The following selection of works is not intended to be exhaustive. The literature of mathematics is vast and that relating to engineers is especially large. Thus, the reader should, in the first instance, turn to works with which he is already familiar as being the quickest way to find the answer to a problem. The list below includes books either of a wide range and general nature and intended for engineers or covering subject matter of recent development, and also some old books on specialist subjects.

Battersby, A. (1967) *Network analysis.* 2nd edn. Macmillan, London.
Douglas, A. H. and Turner, F. H. (1964) *Engineering mathematics.* Concrete Publications, London.
Hall, H. S. and Knight, S. R. (1892) *Higher algebra.* 4th edn. Macmillan, London.
Kreyszig, E. (1983) *Advanced engineering mathematics.* 5th edn. Wiley, New York.
Lamb, H. (1956) *An elementary course of infinitesimal calculus.* 3rd edn. Cambridge University Press, Cambridge.
Loney, S. L. (1922) *The elements of coordinate geometry.* Macmillan, London.
Morice, P. B. (1959) *Linear structural analysis.* Thames and Hudson, London.
Piaggio, H. (1950) *An elementary treatise on differential equations.* Bell, London.
Vine-Lott, K. M. (ed.) (1972) *Computers in civil engineering design.* National Computing Centre, Manchester.

2

Strength of Materials

T R Graves Smith MA, PhD, CEng, MICE
**Department of Civil Engineering,
Southampton University**

Contents

2.1 Introduction

The subject 'Strength of Materials' originates from the earliest attempts to account for the behaviour of structures under load. Thus the problems of particular interest to the first investigators, Galileo and Hooke in the seventeenth century, and Euler and Coulomb in the eighteenth,[1] were the very practical problems associated with the behaviour of beams and columns; at a somewhat later stage, general mathematical investigations of the behaviour of elastic bodies were made by Navier (1821) and Cauchy (1822). The theory of structures has subsequently developed so that it now includes many different and sophisticated fields of interest. Nevertheless, the topic 'Strength of Materials' traditionally covers those aspects of the theory that were the subject of the original research: the theory of bars and the general theory of elasticity. This chapter, therefore, is essentially a review of the main features of these two somewhat disparate theories, and contains some of the results that are of immediate importance to civil engineers.

2.2 Theory of elasticity

2.2.1 Internal stress

Internal stress is the name given to the intensity of the internal forces set up within a body subject to loading. Consider such a body shown in Figure 2.1(a) and an imaginary plane surface within the body passing through a point P. The internal forces exerted between atoms across this surface are represented in the expanded view of Figure 2.1(b). They are described by stress vectors (having the dimensions of force per unit area), and the particular vectors at P give a measure of the intensity of the internal forces at this point. They are denoted by σ and called *internal stress vectors*. If they are directed away from the material as in Figure 2.1(c) they are called *tensile*, and if towards the material *compressive*.

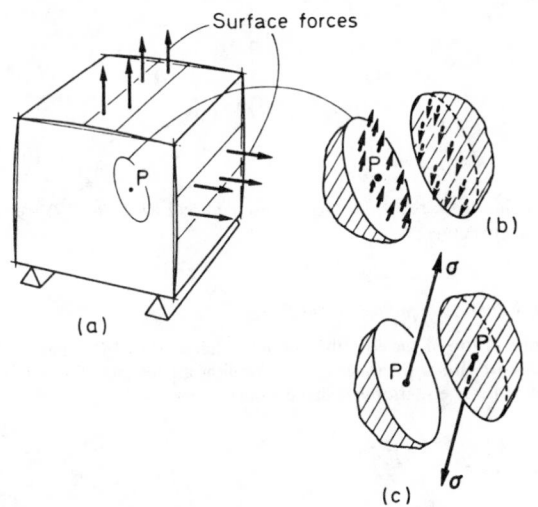

Figure 2.1

2.2.1.1 Components of stress

The complete state of stress at P is defined in terms of the internal stress vectors acting on three particular surfaces at P

called the *positive coordinate surfaces*. (The positive x coordinate surface is the surface parallel to the y–z plane of an x, y, z coordinate system, with the material situated so that a vector directed outwards from the material and normal to the surface is in the positive direction of the x coordinate line as in Figure 2.2.)

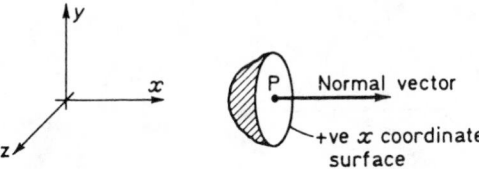

Figure 2.2

These internal stress vectors are distinguished by appropriate subscripts. Thus σ_x acts on the positive x coordinate surface, while σ_y and σ_z respectively act on the y and z surfaces. Their scalar components† are then denoted by two subscripts. Thus the components of σ_x are σ_{xx}, σ_{xy} and σ_{xz} and are shown in Figure 2.3(a). Similarly the components of σ_y are σ_{yx}, σ_{yy}, σ_{yz} and of σ_z are σ_{zx}, σ_{zy}, σ_{zz} as shown in Figure 2.3(b) and (c). σ_{xx}, σ_{yy} and σ_{zz} are called the *direct stress components* at P in the x, y and z directions respectively, while σ_{xy}, σ_{xz}, σ_{yx}, σ_{yz}, σ_{zx}, and σ_{zy} are called the *shear stress components*.

While the above notation is strictly logical and clarifies the basic concepts of stress, conventional engineering notation is somewhat different and emphasizes the physical differences between the components. Thus the direct stress components are written σ_x, σ_y and σ_z, while the shear stress components are written τ_{xy}, τ_{xz}, τ_{yx}, τ_{yz}, τ_{zx}, τ_{zy}. Except in section 2.2.1.3 (below), this latter notation is employed in the remainder of this chapter.

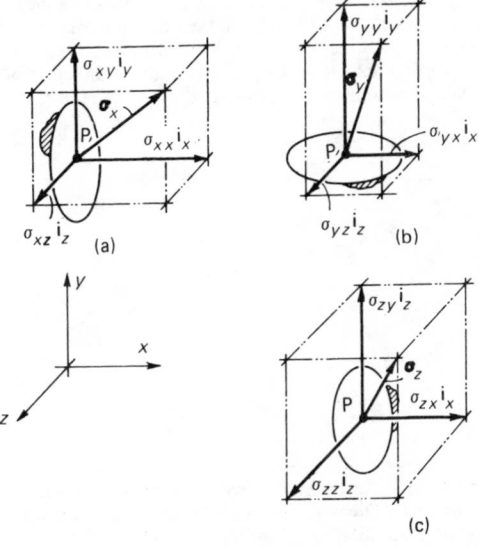

Figure 2.3

2.2.1.2 Stress on an arbitrary surface

Suppose a plane surface through P is defined in terms of the components n_x, n_y and n_z of the outward unit normal vector **n**, as in Figure 2.4. The stress vector σ_n acting on this surface is

† A vector **F** at P is equal to $F_x\mathbf{i}_x + F_y\mathbf{i}_y + F_z\mathbf{i}_z$, where F_x, F_y and F_z are the scalar components of **F**, and $\mathbf{i}_x$, $\mathbf{i}_y$ and $\mathbf{i}_z$ are unit base vectors parallel respectively to the x, y and z coordinate lines at P.

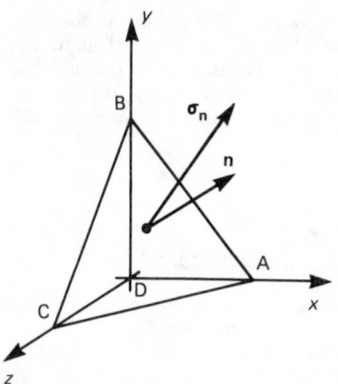

Figure 2.4

obtained in terms of the basic stress components defined in the previous section by considering the linear equilibrium of the differentially small trapezoidal element ABCD shown in the figure. Thus:

$$\sigma_{nx} = \sigma_x n_x + \tau_{yx} n_y + \tau_{zx} n_z \qquad (2.1)$$

$$\sigma_{ny} = \tau_{xy} n_x + \sigma_y n_y + \tau_{zy} n_z \qquad (2.2)$$

$$\sigma_{nz} = \tau_{xz} n_x + \tau_{yz} n_y + \sigma_z n_z \qquad (2.3)$$

where σ_{nx}, σ_{ny} and σ_{nz} are the components of σ_n.

2.2.1.3 Transformation of stress

Considering a new coordinate system x', y', z' rotated relative to the x, y and z system as in Figure 2.5, then the components of stress in the new system are defined as in section 2.2.1.1, so that $\tau_{x'y'}$ ($=\sigma_{x'y'}$), for example, is the component in the y' direction of the stress vector acting on the positive x' coordinate surface.

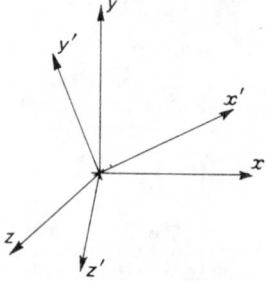

Figure 2.5

The components of stress in the two systems are related by equations of the following type (where for conciseness we employ the original notation of section 2.2.1.1):

$$\sigma_{x'y'} = \frac{\partial x}{\partial x'}\frac{\partial x}{\partial y'}\sigma_{xx} + \frac{\partial x}{\partial x'}\frac{\partial y}{\partial y'}\sigma_{xy} + \frac{\partial x}{\partial x'}\frac{\partial z}{\partial y'}\sigma_{xz}$$

$$+ \frac{\partial y}{\partial x'}\frac{\partial x}{\partial y'}\sigma_{yx} + \frac{\partial y}{\partial x'}\frac{\partial y}{\partial y'}\sigma_{yy} + \frac{\partial y}{\partial x'}\frac{\partial z}{\partial y'}\sigma_{yz}$$

$$+ \frac{\partial z}{\partial x'}\frac{\partial x}{\partial y'}\sigma_{zx} + \frac{\partial z}{\partial x'}\frac{\partial y}{\partial y'}\sigma_{zy} + \frac{\partial z}{\partial x'}\frac{\partial z}{\partial y'}\sigma_{zz} \qquad (2.4)$$

Equation (2.4) and eight similar equations formed by permuting x', y' and z' are called the *transformation equations of stress*. The partial derivatives in Equation (2.4) are called direction cosines, since $\partial y/\partial x'$, for example, is equal to the cosine of the angle between the y and x' coordinate lines.

2.2.1.4 Principal stresses

For a particular orientation of x', y' and z' it is found that all the shear stress components vanish, i.e. that the stress vectors $\sigma_{x'}$, $\sigma_{y'}$ and $\sigma_{z'}$ are directed at right angles to their respective coordinate surfaces. Calling this coordinate system X, Y and Z, the matrix of stress components takes the form:

$$\begin{matrix} \sigma_X & 0 & 0 \\ 0 & \sigma_Y & 0 \\ 0 & 0 & \sigma_Z \end{matrix}$$

The direct stresses σ_X, σ_Y and σ_Z are called the *principal stresses* at P, while the X, Y and Z coordinate lines are called the *principal directions of stress*.

The values of the principal stresses in terms of the stress components in the x, y and z system are equal to the three roots of the equation:

$$\sigma^3 - I_1\sigma^2 + I_2\sigma - I_3 = 0 \qquad (2.5)$$

where

$$I_1 = \sigma_x + \sigma_y + \sigma_z \qquad (2.6)$$

$$I_2 = \sigma_x\sigma_y + \sigma_y\sigma_z + \sigma_z\sigma_x - \tau_{xy}^2 - \tau_{yz}^2 - \tau_{zx}^2 \qquad (2.7)$$

$$I_3 = \sigma_x\sigma_y\sigma_z + 2\tau_{xy}\tau_{yz}\tau_{zx} - \sigma_x\tau_{yz}^2 - \sigma_y\tau_{zx}^2 - \sigma_z\tau_{xy}^2 \qquad (2.8)$$

The direction cosines of the Y coordinate line say, relative to the x, y and z coordinate lines (λ_{Yx}, λ_{Yy}, λ_{Yz}), are found by solving the equations

$$\begin{bmatrix} (\sigma_x - \sigma_Y) & \tau_{xy} & \tau_{xz} \\ \tau_{yx} & (\sigma_y - \sigma_Y) & \tau_{yz} \\ \tau_{zx} & \tau_{zy} & (\sigma_z - \sigma_Y) \end{bmatrix} \begin{bmatrix} \lambda_{Yx} \\ \lambda_{Yy} \\ \lambda_{Yz} \end{bmatrix} = 0 \qquad (2.9)$$

$$(\lambda_{Yx})^2 + (\lambda_{Yy})^2 + (\lambda_{Yz})^2 = 1 \qquad (2.10)$$

(Note that the three equations represented by Equation (2.9) are not independent.)

2.2.1.5 Internal equilibrium equations

Consideration of the equilibrium of a differentially small parallelepiped element of material surrounding an internal point P, leads to three equations of linear equilibrium:

$$\frac{\partial\sigma_x}{\partial x} + \frac{\partial\tau_{yx}}{\partial y} + \frac{\partial\tau_{zx}}{\partial z} + F_x = 0 \qquad (2.11)$$

$$\frac{\partial\sigma_y}{\partial y} + \frac{\partial\tau_{zy}}{\partial z} + \frac{\partial\tau_{xy}}{\partial x} + F_y = 0 \qquad (2.12)$$

$$\frac{\partial\sigma_z}{\partial z} + \frac{\partial\tau_{xz}}{\partial x} + \frac{\partial\tau_{yz}}{\partial y} + F_z = 0 \qquad (2.13)$$

and three equations of rotational equilibrium:

$$\tau_{xy} = \tau_{yx} \qquad (2.14)$$

$$\tau_{yz} = \tau_{zy} \qquad (2.15)$$

$$\tau_{zx} = \tau_{xz} \qquad (2.16)$$

In Equations (2.11 to 2.13), F_x, F_y and F_z are the components of any body force vector $\mathbf{F}$ (units: force per unit volume) acting at P. Note, for example, that a body force vector of magnitude (ρg)/unit volume is exerted by the Earth at all points within a body situated in its gravitational field, ρ being the local density of the body and g being the acceleration due to gravity.

The shear stress components τ_{xy} and τ_{yx} being equal, are called *complementary shear stresses*. It is apparent from Equations (2.14 to 2.16) that if a body is in equilibrium then only six of the nine stress components can take different values at any point.

2.2.1.6 Plane stress

For structures made of elements whose dimensions in the z direction are much smaller than the dimensions in the x and y directions, such as thin plate girders, slabs, shear walls, etc., the following assumptions can be made: (1) the stress components σ_z, τ_{yz}, τ_{xz} can be ignored; and (2) the stress components are uniform across the thickness of the element. That is, they are independent of z.

Such a state of stress is called *plane stress*.

For plane stress, the transformation Equations (2.4) take a simple and important form. Suppose the x', y', z' system is formed by a rotation of $\alpha°$ about the z axis anticlockwise from the reader's viewpoint, as in Figure 2.6. The transformation equations between σ_x, σ_y, τ_{xy} and $\sigma_{x'}$, $\sigma_{y'}$, $\tau_{x'y'}$, are then as follows:

$$\sigma_{x'} = \tfrac{1}{2}(\sigma_x + \sigma_y) + \tfrac{1}{2}(\sigma_x - \sigma_y)\cos(2\alpha) + \tau_{xy}\sin(2\alpha) \qquad (2.17)$$

$$\sigma_{y'} = \tfrac{1}{2}(\sigma_x + \sigma_y) - \tfrac{1}{2}(\sigma_x - \sigma_y)\cos(2\alpha) - \tau_{xy}\sin(2\alpha) \qquad (2.18)$$

$$\tau_{x'y'} = -\tfrac{1}{2}(\sigma_x - \sigma_y)\sin(2\alpha) + \tau_{xy}\cos(2\alpha) \qquad (2.19)$$

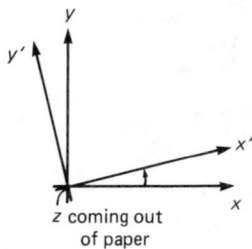

Figure 2.6

z coming out of paper

These equations can then be represented by the following graphical construction. Two axes are drawn, the vertical representing shear stress and the horizontal, direct stress, and a circle is constructed whose centre is at $(\sigma_x + \sigma_y)/2$ on the direct stress axis, and which passes through the point (σ_x, τ_{xy}) as in Figure 2.7. The line through the centre of the circle at an angle $2\alpha°$ *clockwise* to the line joining the centre and (σ_x, τ_{xy}) then intersects the circle at $(\sigma_{x'}, \tau_{x'y'})$. Produced backwards, it intersects the circle at a point whose abscissa is $\sigma_{y'}$. This construction was devised by Otto Mohr in 1882 and the circle is called *Mohr's circle of stress*.

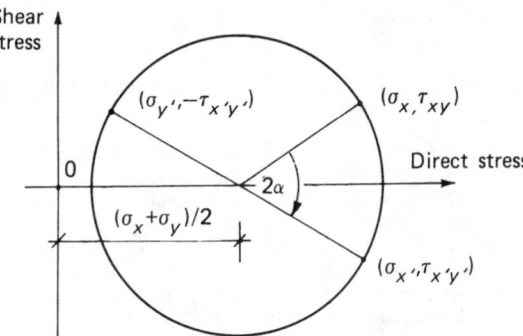

Figure 2.7 Mohr's circle of stress

2.2.2 Strain

Strain is the general name given to the deformation of a body subject to loading.

2.2.2.1 Displacements

A particular point P in a body before loading, occupies its initial position P_i say, and after loading its final position P_f. The line joining P_i to P_f is a vector which is denoted by $\mathbf{u}$ and called the *displacement vector* at P. In general, this vector varies continuously from point to point in the body, and its three components u_x, u_y and u_z are continuous functions of the coordinates of P.†

Consider two neighbouring points $P(x, y, z)$ and $P^*(x + dx, y + dy, z + dz)$ in the body. Then

$$du_x = \frac{\partial u_x}{\partial x}dx + \frac{\partial u_x}{\partial y}dy + \frac{\partial u_x}{\partial z}dz \qquad (2.20)$$

$$du_y = \frac{\partial u_y}{\partial x}dx + \frac{\partial u_y}{\partial y}dy + \frac{\partial u_y}{\partial z}dz \qquad (2.21)$$

$$du_z = \frac{\partial u_z}{\partial x}dx + \frac{\partial u_z}{\partial y}dy + \frac{\partial u_z}{\partial z}dz \qquad (2.22)$$

where the differentials du_x, du_y and du_z are the differences between the components of $\mathbf{u}$ at the two points. As such, these differentials can be regarded as the components of the vector giving the displacement of P^* *relative* to P.

2.2.2.2 Components of strain

In order to obtain a concise description of the deformation of the material at P it is convenient to define nine dimensionless components ε_{xx}, ε_{yy}, ε_{zz}, ε_{xy}, ε_{yz}, ε_{zx}, ω_{xy}, ω_{yz}, ω_{zx} by the following equations, called the *strain–displacement relations*:

$$\varepsilon_{xx} = \frac{\partial u_x}{\partial x}, \quad \varepsilon_{yy} = \frac{\partial u_y}{\partial y}, \quad \varepsilon_{zz} = \frac{\partial u_z}{\partial z} \qquad (2.23, 2.24, 2.25)$$

$$\varepsilon_{xy} = \frac{1}{2}\left(\frac{\partial u_x}{\partial y} + \frac{\partial u_y}{\partial x}\right), \quad \varepsilon_{yz} = \frac{1}{2}\left(\frac{\partial u_y}{\partial z} + \frac{\partial u_z}{\partial y}\right),$$

$$\varepsilon_{zx} = \frac{1}{2}\left(\frac{\partial u_z}{\partial x} + \frac{\partial u_x}{\partial z}\right) \qquad (2.26, 2.27, 2.28)$$

† In most cases $\mathbf{u}$ is so small that the coordinates of P do not change appreciably during the loading.

$$\omega_{xy}=\frac{1}{2}\left(\frac{\partial u_x}{\partial y}-\frac{\partial u_y}{\partial x}\right), \quad \omega_{yz}=\frac{1}{2}\left(\frac{\partial u_y}{\partial z}-\frac{\partial u_z}{\partial y}\right),$$

$$\omega_{zx}=\frac{1}{2}\left(\frac{\partial u_z}{\partial x}-\frac{\partial u_x}{\partial z}\right) \qquad\qquad (2.29, 2.30, 2.31)$$

The physical meaning of these components is clarified by considering the deformation of the rectangular element of material containing P shown in Figure 2.8(a) for each component in turn.

Thus if $\varepsilon_{xx}\neq0$, $\varepsilon_{yy}=\varepsilon_{zz}=\varepsilon_{xy}=\varepsilon_{yz}=\varepsilon_{zx}=\omega_{xy}=\omega_{yz}=\omega_{zx}=0$ then, by using Equations (2.20 to 2.22), it can be shown that the element deforms as in Figure 2.8(b). ε_{xx}, corresponding to this type of longitudinal deformation is called the *direct strain component* in the x direction at P. If it is positive it is called *tensile* and the element lengthens and if negative, it is called *compressive* and the element shortens. Similarly, the components ε_{yy} and ε_{zz} corresponding respectively to longitudinal deformation in the y and z directions are called the direct strain components in these directions.

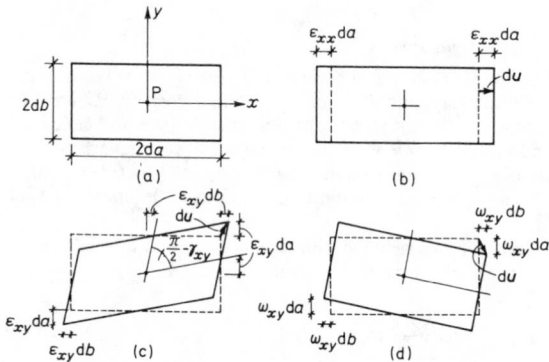

Note: deformation shown to an exaggerated scale

Figure 2.8

If $\varepsilon_{xy}\neq0$, $\varepsilon_{xx}=\varepsilon_{yy}=\varepsilon_{zz}=\varepsilon_{yz}=\varepsilon_{zx}=\omega_{xy}=\omega_{yz}=\omega_{zx}=0$ then $\partial u_x/\partial y=\partial u_y/\partial x=\varepsilon_{xy}$ and again by using Equations (2.20 to 2.22) it can be shown that the element deforms into a lozenge shape as in Figure 2.8(c). Deformation of this type is called shear strain, and ε_{xy} is called the *mathematical shear strain component* at P. The adjective 'mathematical' is used to distinguish between this and the engineering shear strain at P, which is denoted by γ_{xy} and is equal to the closure in radians of the angle between the x and y coordinate lines. From the geometry of Figure 2.8(c) we have

$$\gamma_{xy}=2\varepsilon_{xy} \qquad\qquad (2.32)$$

Similarly, the components ε_{yz} and ε_{zx} correspond to shear strain in the y–z and z–x planes respectively.

Finally, if $\omega_{xy}\neq0$, $\varepsilon_{xx}=\varepsilon_{yy}=\varepsilon_{zz}=\varepsilon_{xy}=\varepsilon_{yz}=\varepsilon_{zx}=\omega_{yz}=\omega_{zx}=0$ then $\partial u_x/\partial y=-\partial u_y/\partial x=\omega_{xy}$ and it can be shown that the element rotates without deformation about the z coordinate line as in Figure 2.8(d). ω_{xy} is called the *rotation* at P. Similarly ω_{yz} and ω_{zx} correspond respectively to local rotations about the x and y coordinate lines through P. These rotations are necessary in the theoretical discussion in order to define the displacement derivatives in Equations (2.20 to 2.22). However, since they do not define *deformation* directly, they are not considered further in elastic analysis.

As in the case of stresses, the conventional engineering notation for the strain components is somewhat different from

the above and the direct strain components are written ε_x ε_y and ε_z. Except in section 2.2.2.4 (below), this latter notation is employed in the remainder of the chapter, and the shear strains are described in terms of γ_{xy}, γ_{yz} and γ_{zx}.

In the majority of civil engineering structures, the strain components are very small, of the order of magnitude 10^{-3}. Thus, the deformation of the elements in Figure 2.8 is exaggerated. The strain–displacement relations in Equations (2.23 to 2.28) assume that the *displacements* are small. If this is not the case, nonlinear terms involving the products of the derivatives are included.[2] These nonlinear terms are significant in defining the buckling characteristics of thin elements in compression.[3,4]

2.2.2.3 Uniform strain

If the displacement components u_x, u_y and u_z are linear functions of the coordinates of P then the corresponding strains given by Equations (2.23 to 2.28) are uniform. The overall changes in the geometry of a body are then simply related to the strain components. Thus consider, for example, a line AB in or on the surface of the body which originally coincides with an x coordinate line. If the original length of AB is l and its increase in length is Δl, then:

$$\varepsilon_x=\Delta l/l \qquad\qquad (2.33)$$

2.2.2.4 Transformation of strain

Considering again a new coordinate system x', y', z' rotated relative to the x, y and z system as in Figure 2.5, then the components of strain in this new system are defined by strain–displacement relations similar to Equations (2.23 to 2.28). Thus $\gamma_{x'y'}(=2\varepsilon_{x'y'})$, for example, is given by:

$$\gamma_{x'y'}=\left(\frac{\partial u_{x'}}{\partial y'}+\frac{\partial u_{y'}}{\partial x'}\right) \qquad\qquad (2.34)$$

where $u_{x'}$, $u_{y'}$ and $u_{z'}$ are the components of the displacement vector **u** relative to x', y' and z'. The components of strain in the two systems are related by equations of the same type as Equation (2.4) (where again for conciseness we employ the original notation of section 2.2.2.2). Thus:

$$\varepsilon_{x'y'}=\frac{\partial x}{\partial x'}\frac{\partial x}{\partial y'}\varepsilon_{xx}+\frac{\partial x}{\partial x'}\frac{\partial y}{\partial y'}\varepsilon_{xy}+\frac{\partial x}{\partial x'}\frac{\partial z}{\partial y'}\varepsilon_{xz}$$

$$+\frac{\partial y}{\partial x'}\frac{\partial x}{\partial y'}\varepsilon_{yx}+\frac{\partial y}{\partial x'}\frac{\partial y}{\partial y'}\varepsilon_{yy}+\frac{\partial y}{\partial x'}\frac{\partial z}{\partial y'}\varepsilon_{yz}$$

$$+\frac{\partial z}{\partial x'}\frac{\partial x}{\partial y'}\varepsilon_{zx}+\frac{\partial z}{\partial x'}\frac{\partial y}{\partial y'}\varepsilon_{zy}+\frac{\partial z}{\partial x'}\frac{\partial z}{\partial y'}\varepsilon_{zz} \qquad (2.35)$$

The nine equations formed by permuting x', y' and z' in Equation (2.35) are called the *transformation equations of strain*.

2.2.2.5 Principal strains

For a particular orientation of x', y' and z', all the shear strain components vanish, and in most materials this orientation is the same as that of the principal directions of stress discussed in section 2.2.1.4. Calling the coordinate system X, Y and Z as before, the direct strains ε_X, ε_Y and ε_Z are called the *principal strains* at P.

The values of the principal strains are equal to the three roots of the equation:

$$\varepsilon^3 - E_1\varepsilon^2 + E_2\varepsilon - E_3 = 0 \tag{2.36}$$

where:

$$E_1 = \varepsilon_x + \varepsilon_y + \varepsilon_z \tag{2.37}$$

$$E_2 = \varepsilon_x\varepsilon_y + \varepsilon_y\varepsilon_z + \varepsilon_z\varepsilon_x - \tfrac{1}{4}(\gamma_{xy}^2 + \gamma_{yz}^2 + \gamma_{zx}^2) \tag{2.38}$$

$$E_3 = \varepsilon_x\varepsilon_y\varepsilon_z + \tfrac{1}{4}(\gamma_{xy}\gamma_{yz}\gamma_{zx} - \varepsilon_x\gamma_{yz}^2 - \varepsilon_y\gamma_{zx}^2 - \varepsilon_z\gamma_{xy}^2) \tag{2.39}$$

2.2.2.6 Compatibility equations

The three displacement components u_x, u_y and u_z can be eliminated from the six strain–displacement relations in Equations (2.23 to 2.28) to produce three equations called the *compatibility equations*, which must be satisfied by the strain components. This elimination can be done in different ways to produce different sets of equations. Two such are:

$$\frac{\partial^2 \varepsilon_x}{\partial y^2} + \frac{\partial^2 \varepsilon_y}{\partial x^2} - \frac{\partial^2 \gamma_{xy}}{\partial x \partial y} = 0 \tag{2.40}$$

$$\frac{\partial^2 \varepsilon_y}{\partial z^2} + \frac{\partial^2 \varepsilon_z}{\partial y^2} - \frac{\partial^2 \gamma_{yz}}{\partial y \partial z} = 0 \tag{2.41}$$

$$\frac{\partial^2 \varepsilon_z}{\partial x^2} + \frac{\partial^2 \varepsilon_x}{\partial z^2} - \frac{\partial^2 \gamma_{zx}}{\partial z \partial x} = 0 \tag{2.42}$$

$$\frac{2\partial^2 \varepsilon_x}{\partial y \partial z} - \frac{\partial}{\partial x}\left(\frac{\partial \gamma_{xy}}{\partial z} - \frac{\partial \gamma_{yz}}{\partial x} + \frac{\partial \gamma_{zx}}{\partial y}\right) = 0 \tag{2.43}$$

$$\frac{2\partial^2 \varepsilon_y}{\partial z \partial x} - \frac{\partial}{\partial y}\left(\frac{\partial \gamma_{yz}}{\partial x} - \frac{\partial \gamma_{zx}}{\partial y} + \frac{\partial \gamma_{xy}}{\partial z}\right) = 0 \tag{2.44}$$

$$\frac{2\partial^2 \varepsilon_z}{\partial x \partial y} - \frac{\partial}{\partial z}\left(\frac{\partial \gamma_{zx}}{\partial y} - \frac{\partial \gamma_{xy}}{\partial z} + \frac{\partial \gamma_{yz}}{\partial x}\right) = 0 \tag{2.45}$$

2.2.2.7 Plane strain

Plane strain is said to exist when the strain components ε_z, ε_{yz} and ε_{zx} are equal to zero. It occurs when $u_z = 0$ at every point within a

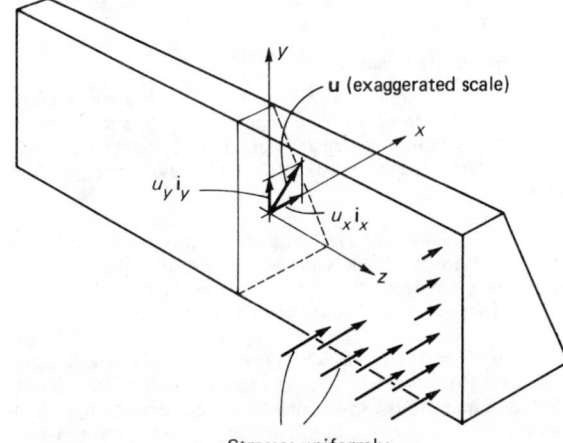

Stresses uniformly
distributed along length

Figure 2.9

region of a body. From symmetry this is the case in the central region of a body which: (1) is very long in the z direction; (2) is of uniform cross-section; and (3) is subjected to loading in the z plane that is uniformly distributed along its length (Figure 2.9). It can therefore occur in structures such as gravity dams, tunnel linings or retaining walls.

Considering again the new coordinate system x', y', z' formed by a rotation of $\alpha°$ anticlockwise about the z axis as in Figure 2.6, the transformation equations between ε_x, ε_y, ε_{xy} and $\varepsilon_{x'}$, $\varepsilon_{y'}$, and $\varepsilon_{x'y'}$ take the same form as Equations (2.17 to 2.19). These transformation equations are represented by a graphical construction called *Mohr's circle of strain*, whose function is the same as that of Mohr's circle of stress.

2.2.3 Elastic stress–strain relations

The relationship between the stress and strain components at a point in a body is a property of the particular material making up the body. For an isotropic elastic material the stress–strain relations are linear and are independent of the orientation of the x, y, z coordinate system. They take the following form:

$$\varepsilon_x = \frac{1}{E}[\sigma_x - v(\sigma_y + \sigma_z)] + \alpha\Delta T \tag{2.46}$$

$$\varepsilon_y = \frac{1}{E}[\sigma_y - v(\sigma_z + \sigma_x)] + \alpha\Delta T \tag{2.47}$$

$$\varepsilon_z = \frac{1}{E}[\sigma_z - v(\sigma_x + \sigma_y)] + \alpha\Delta T \tag{2.48}$$

$$\gamma_{xy} = \frac{1}{G}\tau_{xy}, \; \gamma_{yz} = \frac{1}{G}\tau_{yz}, \; \gamma_{zx} = \frac{1}{G}\tau_{zx} \tag{2.49, 2.50, 2.51}$$

where ΔT is the temperature change from some initial state. E and G are constants having the dimensions of force per unit area and are called *Young's modulus* and the *shear modulus* respectively, v is a dimensionless constant called *Poisson's ratio* and α is a constant having the dimensions °C^{-1} and is called the *temperature coefficient of expansion*. G in fact is related to E and v by the following equation:

$$G = E/2(1 + v) \tag{2.52}$$

Values of E, v and α for a variety of practical materials are given in Table 2.1.

The corresponding inverse stress–strain relations are found by solving Equations (2.46 to 2.51) for the stresses and are as follows:

$$\sigma_x = 2\mu\varepsilon_x + \lambda(\varepsilon_x + \varepsilon_y + \varepsilon_z) - (3\lambda + 2\mu)\alpha\Delta T \tag{2.53}$$

$$\sigma_y = 2\mu\varepsilon_y + \lambda(\varepsilon_x + \varepsilon_y + \varepsilon_z) - (3\lambda + 2\mu)\alpha\Delta T \tag{2.54}$$

$$\sigma_z = 2\mu\varepsilon_z + \lambda(\varepsilon_x + \varepsilon_y + \varepsilon_z) - (3\lambda + 2\mu)\alpha\Delta T \tag{2.55}$$

$$\tau_{xy} = \mu\gamma_{xy}, \; \tau_{yz} = \mu\gamma_{yz}, \; \tau_{zx} = \mu\gamma_{zx} \tag{2.56, 2.57, 2.58}$$

where for conciseness we employ the *Lamé constants* λ and μ defined in terms of E and v by the equations:

$$\lambda = vE/(1 + v)(1 - 2v) \tag{2.59}$$

$$\mu = E/2(1 + v) \tag{2.60}$$

Table 2.1 Properties of materials (representative)

Material	Density (kg/m³)	E (GN/m²)	μ	α (°C⁻¹)	Limit of proportionality (MN/m²)	Ultimate stress (MN/m²)	Uniform elongation
Mild steel	7840	200	0.31	1.25×10^{-5}	280	370	0.30
High-strength steel	7840	200	0.31	1.25×10^{-5}	770	1550	0.10
Medium-strength aluminium alloy	2800	70	0.30	2.3×10^{-5}	230	430	0.10
Titanium alloy	4500	120	0.30	0.9×10^{-5}	385	690	0.15
Magnesium alloy	1800	45	0.30	2.7×10^{-5}	155	280	0.08
Concrete	2410	25	0.20	1.2×10^{-5}	—	3 (tension) 30 (compression)	—
Timber (Douglas fir)	576	7 (with grain)		0.6×10^{-5}	43 (compression with grain)	52 (compression with grain)	—
Glass	2580	60	0.26	0.7×10^{-5}	—	1750	—
Nylon	1140	2	—	10×10^{-5}	77	90	1.00
Polystyrene (not expanded)	1050	4	—	10×10^{-5}	46	60	0.03
High-strength glass-fibre composite	2000	60	—	—	—	1600	—
Carbon fibre composite	1600	170	—	—	—	1400	—

The stress–strain relations hold for a wide range of stresses in most practical materials. They become invalid when the interatomic bonds in the materials break down, this process being called *yielding* or *fracture*. Yielding in steel can be demonstrated by the tensile test, where a known stress system $\sigma_x \neq 0$, $\sigma_y = \sigma_z = \tau_{xy} = \tau_{yz} = \tau_{zx} = 0$, called *uniaxial stress*, is induced in a specimen and the corresponding strain ε_x is measured. A typical plot of σ_x versus ε_x for a mild steel tensile specimen then takes the form shown in Figure 2.10(a). The initial straight section of the curve of slope equal to E corresponds to Equation (2.46), but at a certain stress of the order of 250 MN/m² the strain increases dramatically with little or no increase of load. This stress is called the *uniaxial yield stress* of mild steel. Subsequently, the stress–strain curve indicates that the specimen

supports larger stresses up to a maximum value of the order of 400 MN/m² which is called the *ultimate tensile stress*. The uniaxial stress–strain curve for an aluminium alloy specimen shown in Figure 2.10(b) does not display a marked yield stress and the material is linear elastic up to a stress called the *limit of proportionality* which again is of the order of 250 MN/m². Two other properties frequently quoted in engineering literature, the 0.2% *proof stress* and the *uniform elongation*, are shown in the figure. Values for the limit of proportionality, ultimate stress and uniform elongation are included in Table 2.1.

For accounts of yield criteria and plastic stress–strain relations corresponding to more general stress systems see, for example, Bisplinghoff et al,[5] and Prager and Hodge.[6]

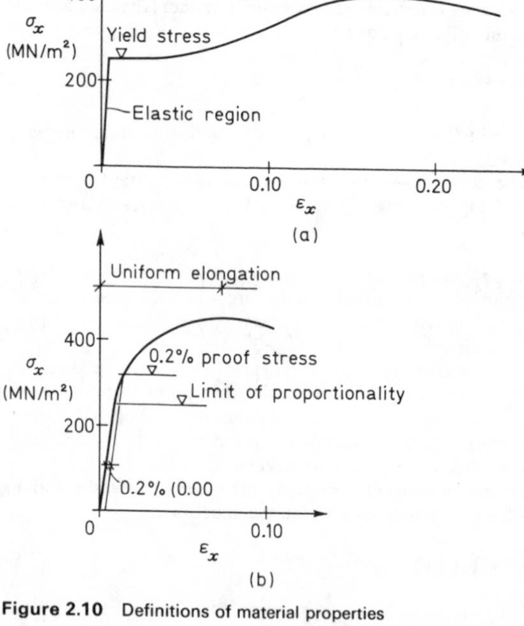

2.2.4 Analysis of elastic bodies

The internal equilibrium Equations (2.11 to 2.16), strain–displacement relations Equations (2.23 to 2.28) and the stress–strain relations Equations (2.46 to 2.51) are eighteen differential equations in the unknowns of the analysis problem, namely the nine stress components, the six strain components and the three displacement components. These equations must be satisfied subject to boundary conditions.

2.2.4.1 Boundary conditions

The boundary conditions at a point P on the surface of a body are expressed in terms of the components S_x, S_y and S_z of the surface stress vector **S** acting at P, and the components u_x, u_y and u_z of the displacement vector **u** of P. They are of three types, as follows.

Static boundary conditions. The three stress vector components at P are specified. Thus at an unloaded point on the boundary $S_x = S_y = S_z = 0$, while at a loaded point $S_x = k_1$, $S_y = k_2$, $S_z = k_3$, where k_1, k_2 and k_3 are known values at P.

Kinematic boundary conditions. The three displacement components at P are specified. Thus at a rigid support $u_x = u_y = u_z = 0$, while at a point whose displacements are constrained by, say, a screw jack $u_x = j_1$, $u_y = j_2$, $u_z = j_3$, where j_1, j_2 and j_3 are known values at P.

Mixed boundary conditions. Certain displacement and certain

Figure 2.10 Definitions of material properties

stress-vector components at P are specified simultaneously. For example, at the point P on the roller support shown in Figure 2.11, $S_x = 0$ and $u_y = u_z = 0$.

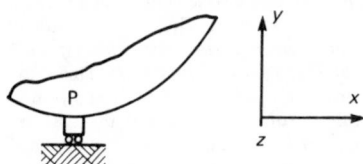

y

x

z

Figure 2.11

2.2.4.2 Solution in terms of displacements

A straightforward solution method involves treating the displacement components as the basic unknowns. The three linear equilibrium Equations (2.11 to 2.13) are expressed in terms of the displacements by using the stress–strain relations followed by the strain–displacement relations. The resulting differential equations in u_x, u_y and u_z are called the *Navier equations*. They are as follows:

$$\mu \nabla^2 u_x + (\lambda + \mu) \frac{\partial \Phi}{\partial x} + F_x = 0 \tag{2.61}$$

$$\mu \nabla^2 u_y + (\lambda + \mu) \frac{\partial \Phi}{\partial y} + F_y = 0 \tag{2.62}$$

$$\mu \nabla^2 u_z + (\lambda + \mu) \frac{\partial \Phi}{\partial z} + F_z = 0 \tag{2.63}$$

where

$$\nabla^2 u_x = \frac{\partial^2 u_x}{\partial x^2} + \frac{\partial^2 u_x}{\partial y^2} + \frac{\partial^2 u_x}{\partial z^2} \tag{2.64}$$

and:

$$\Phi = \frac{\partial u_x}{\partial x} + \frac{\partial u_y}{\partial y} + \frac{\partial u_z}{\partial z} \tag{2.65}$$

In order to solve these equations, the boundary conditions must all be expressed in terms of the displacements of the surface points. In the case of the static boundary conditions, equations for the internal stress components are obtained using Equations (2.1 to 2.3) with the components of σ_n replaced by the components of **S**. These are then converted to differential boundary conditions in displacements by again using the stress–strain and the strain–displacement relations. Thus at each internal point and each boundary point there are three simultaneous differential equations in the unknowns u_x, u_y and u_z. In most cases, a direct solution is obtainable only by a numerical procedure such as the finite-difference method.[7]

2.2.4.3 Solution in terms of stresses

A second solution method involves treating the nine internal stress components as the basic unknowns. Six equations in these unknowns are immediately available from the internal equilibrium Equations (2.11 to 2.16). A further three equations are obtained from the compatibility Equations (2.40 to 2.42) or

(2.43 to 2.45) by using the stress–strain relations to express them in terms of the stress components. The resulting equations are called the *Beltrami–Michell* equations and are as follows:

$$\nabla^2 \sigma_x + \frac{1}{(1+v)} \frac{\partial^2 \Theta}{\partial x^2} = \frac{-v}{(1-v)} \Psi - \frac{2 \partial F_x}{\partial x} \tag{2.66}$$

$$\nabla^2 \sigma_y + \frac{1}{(1+v)} \frac{\partial^2 \Theta}{\partial y^2} = \frac{-v}{(1-v)} \Psi - \frac{2 \partial F_y}{\partial y} \tag{2.67}$$

$$\nabla^2 \sigma_z + \frac{1}{(1+v)} \frac{\partial^2 \Theta}{\partial z^2} = \frac{-v}{(1-v)} \Psi - \frac{2 \partial F_z}{\partial z} \tag{2.68}$$

or:

$$\nabla^2 \tau_{xy} + \frac{1}{(1+v)} \frac{\partial^2 \Theta}{\partial x \partial y} = - \left(\frac{\partial F_x}{\partial y} + \frac{\partial F_y}{\partial x} \right) \tag{2.69}$$

$$\nabla^2 \tau_{yz} + \frac{1}{(1+v)} \frac{\partial^2 \Theta}{\partial y \partial z} = - \left(\frac{\partial F_y}{\partial z} + \frac{\partial F_z}{\partial y} \right) \tag{2.70}$$

$$\nabla^2 \tau_{zx} + \frac{1}{(1+v)} \frac{\partial^2 \Theta}{\partial z \partial x} = - \left(\frac{\partial F_z}{\partial x} + \frac{\partial F_x}{\partial z} \right) \tag{2.71}$$

where

$$\Theta = \sigma_x + \sigma_y + \sigma_z \tag{2.72}$$

$$\Psi = \frac{\partial F_x}{\partial x} + \frac{\partial F_y}{\partial y} + \frac{\partial F_z}{\partial z} \tag{2.73}$$

The only problems than can be solved directly in terms of stresses conveniently are those in which all the boundary conditions are static boundary conditions. In such problems, three equations in the internal stress components are obtained using Equations (2.1 to 2.3) and these, together with the three equations of rotational equilibrium and the three compatibility equations, provide the required nine equations at the boundary. In problems where displacements are specified at various boundary points, the corresponding boundary stresses cannot usually be obtained in advance of the solution except for those special cases where the body is externally statically determinate.

Direct solutions in terms of stresses can in principle be obtained using numerical procedures. However, many solutions, especially to two-dimensional problems,[2,8] have been obtained using stress functions which automatically satisfy the equilibrium equations.

2.2.5 Energy methods

2.2.5.1 Virtual work

Consider a body which is in equilibrium under surface stresses **S** over part of its surface and body forces **F**. Suppose the corresponding internal stress system is given by $\sigma_x, \sigma_y, \sigma_z, \tau_{xy}, \tau_{yz}, \tau_{zx}$. This is called an *equilibrium force system*.

Next consider an entirely independent system of displacements **u*** which vary continuously from point to point in the body and satisfy the kinematic boundary conditions. Suppose the corresponding strain system is given by $\varepsilon_x^*, \varepsilon_y^*, \varepsilon_z^*, \gamma_{xy}^*, \gamma_{yz}^*, \gamma_{zx}^*$. This is called a *compatible displacement system*.

The virtual work W_e^* done by the external forces **S** and **F**, supposing they were to move through **u***, is as follows:

$$W_e^* = \int_A (S_x u_x^* + S_y u_y^* + S_z u_z^*) \, dA$$
$$+ \int_V (F_x u_x^* + F_y u_y^* + F_z u_z^*) \, dV \tag{2.74}$$

where $\int_A(\ ) \, dA$ represents an integral taken over the loaded surface of the body, and $\int_V(\ ) \, dV$ represents an integral taken over its volume. By a purely mathematical operation[5] it can be shown that

$$W_e^* = W_i^* \tag{2.75}$$

where W_i^* is a quantity called the *internal virtual work* and is given by

$$W_i^* = \int_V (\sigma_x \varepsilon_x^* + \sigma_y \varepsilon_y^* + \sigma_z \varepsilon_z^* + \tau_{xy} \gamma_{xy}^* + \tau_{yz} \gamma_{yz}^* + \tau_{zx} \gamma_{zx}^*) \, dV \tag{2.76}$$

Equation (2.75) is called the *equation of virtual work*. Note that its derivation is independent of the nature of the stress–strain relations of the material making up the body.

2.2.5.2 Strain energy

Consider the body in equilibrium under **S** and **F** and suppose differential changes in the loading d**S** and d**F** occur causing corresponding differential changes in the *real* displacements d**u**. d**u** and the strains $d\varepsilon_x$, $d\varepsilon_y$, $d\varepsilon_z$, $d\gamma_{xy}$, $d\gamma_{yz}$, $d\gamma_{zx}$ can be regarded as a compatible system of displacements in Equation (2.75). The work terms on either side of Equation (2.75) are then differential quantities of real work caused by the loading change. In particular the internal work is given by

$$dW_i = \int_V (\sigma_x \, d\varepsilon_x + \sigma_y \, d\varepsilon_y + \sigma_z \, d\varepsilon_z + \tau_{xy} \, d\gamma_{xy} + \tau_{yz} \, d\gamma_{yz} + \tau_{zx} \, d\gamma_{zx}) \, dV \tag{2.77}$$

Using the elastic stress–strain relations it is possible to integrate Equation (2.77) to obtain the total internal work done on an elastic body from the initial state with zero stress to the final state with stresses corresponding to **S** and **F**. This internal work is found to be independent of the loading path to the final state and is called the *elastic strain energy U*. It can be expressed in three forms:

$$U = \int_V \frac{1}{2E} \left[(\sigma_x^2 + \sigma_y^2 + \sigma_z^2) - 2\nu(\sigma_x \sigma_y + \sigma_y \sigma_z + \sigma_z \sigma_x) \right.$$
$$\left. + 2(1+\nu)(\tau_{xy}^2 + \tau_{yz}^2 + \tau_{zx}^2) \right] dV$$

$$= \int_V \tfrac{1}{2}(\sigma_x \varepsilon_x + \sigma_y \varepsilon_y + \sigma_z \varepsilon_z + \tau_{xy} \gamma_{xy} + \tau_{yz} \gamma_{yz} + \tau_{zx} \gamma_{zx}) \, dV$$

$$= \int_V \left[\mu(\varepsilon_x^2 + \varepsilon_y^2 + \varepsilon_z^2) \right.$$

$$\left. + \frac{\lambda}{2}(\varepsilon_x + \varepsilon_y + \varepsilon_z)^2 + \frac{\mu}{2}(\gamma_{xy}^2 + \gamma_{yz}^2 + \gamma_{zx}^2) \right] dV \tag{2.78}$$

2.2.5.3 Principle of stationary total potential energy

The external work done by the loading in the previous subsection is given by:

$$dW_e = \int_A (S_x \, du_x + S_y \, du_y + S_z \, du_z) \, dA$$
$$+ \int_V (F_x \, du_x + F_y \, du_y + F_z \, du_z) \, dV \tag{2.79}$$

If the loading is *conservative*, so that all the loads on the body are independent of the displacements, it is possible to define a function V as follows:

$$V = U - \int_A (S_x u_x + S_y u_y + S_z u_z) \, dA - \int_V (F_x u_x + F_y u_y + F_z u_z) \, dV \tag{2.80}$$

so that the equation of virtual work for the differential change of the body in equilibrium takes the form:

$$d\Phi = 0 \tag{2.81}$$

Φ is called the *total potential energy* of the system of the body plus loads.

Equation (2.81) is the mathematical statement of the Principle of Stationary Total Potential Energy. Thus, *for a body in equilibrium, the total potential energy is stationary with respect to small changes in the actual displacements of the body*. This is the most important energy principle, and its method of application for the solution of structures involves expressing all the displacements of the structure in terms of a (usually limited) number of degrees of freedom. (This can be done exactly for frameworks, but only approximately for structures such as slabs.) The stationary position of the total potential energy is found by equating to zero the derivatives of Φ with respect to the degrees of freedom. The resulting equations are analogous to the stiffness equations in the stiffness method of structural analysis. They are solved for the degrees of freedom to yield the exact or approximate displacements of the structure corresponding to equilibrium.

If the structural displacements are assumed to be small so that the linear strain–displacement relations in Equations (2.23 to 2.28) are applicable, then it can be shown that the potential energy is a *minimum* for a structure in equilibrium.[9] The equilibrium is then said to be *stable*. If the displacements are not small, and the non-linear strain–displacement relations are used to obtain Φ, the equilibrium potential energy can either be a *minimum* or a *maximum*. In the latter case the equilibrium is said to be *unstable*. For certain values of load called the critical loads or *eigenvalues*, the equilibrium is *neutral*. This is indicated mathematically when the determinant of the coefficient matrix in the stiffness equations is zero. Extensive treatments of the eigenvalue problem have been given in many texts, e.g. by Croll and Walker[10] and by Thompson and Hunt.[11]

2.2.6 Measurement of stress and strain

2.2.6.1 Surface strain

The measurement of strain is usually limited to obtaining direct strains tangential to the surfaces of structures by means of mechanical or electrical strain gauges. If the complete state of tangential strain at a surface point is to be determined, then separate measurements of direct strain have to be obtained in three distinct directions at the point. In interpreting these measurements, we then use the fact that two of the principal directions of stress and strain are tangential to the surface whilst the third is normal to it. Thus using, for example, a 45° strain-gauge rosette, producing strain measurements ε_1, ε_2 and ε_3 as shown in Figure 2.12, it can be shown that the principal direction X is at $\theta°$ anticlockwise to the x coordinate line where:

$$\tan(2\theta) = \frac{(2\varepsilon_2 - \varepsilon_1 - \varepsilon_3)}{(\varepsilon_1 - \varepsilon_3)} \tag{2.82}$$

The two principal surface strains ε_X and ε_Y are then given by:

$$\varepsilon_X = \frac{(\varepsilon_1 + \varepsilon_3)}{2} + r \qquad \varepsilon_Y = \frac{(\varepsilon_1 + \varepsilon_3)}{2} - r \tag{2.83, 2.84}$$

where

$$r = \tfrac{1}{2}[(\varepsilon_1 - \varepsilon_3)^2 + (2\varepsilon_2 - \varepsilon_1 - \varepsilon_3)^2]^{1/2} \tag{2.85}$$

Example 2.1. The strains measured by the three gauges of the 45° rosette shown in Figure 2.12 are respectively:

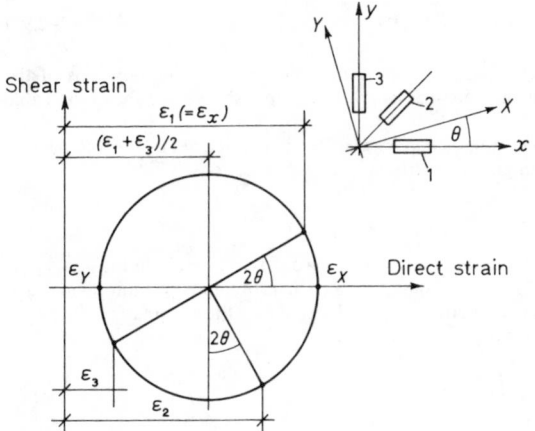

Figure 2.12

$$\varepsilon_1 = -5.0 \times 10^{-4} \quad \varepsilon_2 = +3.0 \times 10^{-4} \quad \varepsilon_3 = +1.0 \times 10^{-4}$$

What are the principal strains at the point and the orientation of the principal direction X, to the x coordinate line?
From Equation (2.85):

$$r = \tfrac{1}{2}[(-5.0 - 1.0)^2 + (2 \times 3.0 + 5.0 - 1.0)^2]^{1/2} \times 10^{-4}$$

$$= 5.8 \times 10^{-4}$$

Thus:

$$\varepsilon_X = 3.8 \times 10^{-4} \quad \varepsilon_Y = -7.8 \times 10^{-4}$$

From Equation (2.82):

$$\tan (2\theta) = -1.667$$

Thus:

$$2\theta = -59.0° \quad \text{or} \quad 121.0°$$

The ambiguity in the expression for θ is resolved by examining the position of the strains on the Mohr's circle of strain for the surface plane (Figure 2.13). Thus, it is clear that in this example, 2θ must be greater than 90°. The X coordinate line is therefore directed at 60.5° anticlockwise to the x coordinate line.

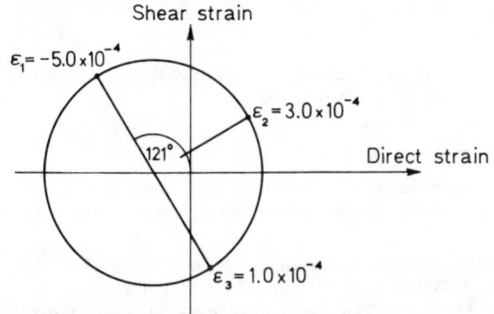

Figure 2.13

Another common layout for strain gauges is the 60° rosette shown in Figure 2.14. The principal direction X is then at $\theta°$ anticlockwise to the x coordinate line where:

$$\tan (2\theta) = \sqrt{3}(\varepsilon_2 - \varepsilon_3)/(2\varepsilon_1 - \varepsilon_2 - \varepsilon_3) \tag{2.86}$$

while the principal surface strains ε_X and ε_Y are given by

$$\varepsilon_X = \frac{\varepsilon_1 + \varepsilon_2 + \varepsilon_3}{3} + r \quad \varepsilon_Y = \frac{\varepsilon_1 + \varepsilon_2 + \varepsilon_3}{3} - r \tag{2.87, 2.88}$$

where

$$r = \tfrac{2}{3}(\varepsilon_1^2 + \varepsilon_2^2 + \varepsilon_3^2 - \varepsilon_1\varepsilon_2 - \varepsilon_2\varepsilon_3 - \varepsilon_3\varepsilon_1)^{1/2} \tag{2.89}$$

The complete state of surface stress corresponding to the strains measured above can be found from the stress–strain relations, noting that in the absence of surface loading the state of stress is one of plane stress.

2.2.6.2 The photoelastic method[12,13]

A good indication of the internal stresses in model structures can be obtained by making use of the property of certain materials such as glass and plastics, that they become double-refracting when subject to stress.

The apparatus for photoelastic stress analysis consists essentially of a light source L (Figure 2.15), a *polarizer* P, and an *analyser* A and the model M of photoelastic material, which is held in a reaction frame and subjected to loads. The lenses L_1 and L_2 are arranged so that a parallel beam of light passes through the model. An image containing bands of different colours then appears on the ground glass screen, these colours representing regions of equal principal stress difference $(\sigma_X - \sigma_Y)$ in the model. For further experimental and theoretical details see, for example, Hendry.[2]

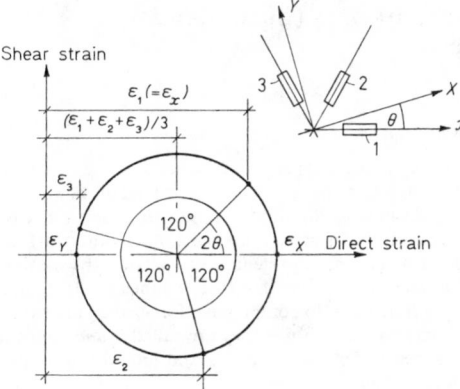

Figure 2.14

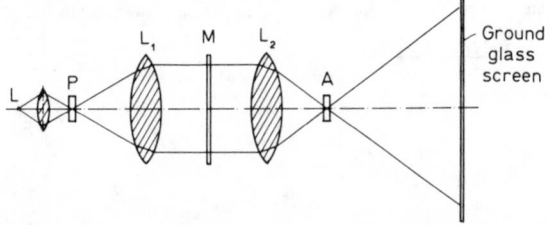

Figure 2.15

2.3 Theory of bars (beams and columns)

2.3.1 Introduction

A great many engineering structures contain components whose dimensions in two coordinate directions are small compared with their dimensions in the third. These components can be called *bars* as a means of general classification, although they are often given other names to denote the particular way they are loaded in structures. Thus if they are subjected to tensile forces they are called *ties*, to compressive forces they are called *struts* or *columns*, to lateral forces they are called *beams*, while if they are subjected to both compressive and lateral forces they are called *beam-columns*.

Structures completely composed of bars are called *frames*, and are either two-dimensional *plane frames*, or three-dimensional *space frames*.

This section reviews the engineering theory of straight bars of uniform cross-section.

2.3.2 Cross-section geometry

2.3.2.1 First moment of area

Consider a bar of some particular cross-sectional shape shown in Figure 2.16, and the two orthogonal axes y and z. (The choice of axes with y horizontal and z downwards, is quite arbitrary but has two advantages when applied to beams: (1) the displacements of a beam are usually vertically downwards, and therefore in the positive direction of z; and (2) as shown in section 2.3.5, a positive bending moment about the y axis causes tension on the bottom of the beam; and therefore positive stresses occur at points in the beam defined by positive values of z.) The first moment of area of the cross-section about the y axis G_y, is defined as the sum of the products obtained by multiplying each element of cross-sectional area dA by its distance z from the y axis. Thus:

$$G_y = \int_A z \, dA \tag{2.90}$$

Similarly:

$$G_z = \int_A y \, dA \tag{2.91}$$

The position of the *centroid* of the cross-section is such that the first moment of area about any axis passing through it is zero. Thus if C is the centroid in Figure 2.16, then

$$G_y = G_z = 0$$

From this it is clear that C must lie on any axis of symmetry of the section. The centroid can be located in general by selecting any two orthogonal axes y' and z'. The coordinates of the centroid relative to this system, y'_c and z'_c, are then given by:

$$y'_c = G_{z'}/A \qquad z'_c = G_{y'}/A \tag{2.92, 2.93}$$

where A is the area of the cross-section. The positions of the centroids of various cross-sectional shapes are shown in Table 2.2.

The *longitudinal axis* of the bar is defined as the line passing through the centroids of its cross-sections.

2.3.2.2 Moments of inertia

The *moment of inertia*† of the cross-section about the y axis I_y, is defined as the sum of the products obtained by multiplying each element of cross-sectional area dA by the square of its distance z from the y axis. Thus:

$$I_y = \int_A z^2 \, dA \tag{2.94}$$

Similarly:

$$I_z = \int_A y^2 \, dA \tag{2.95}$$

The *product of inertia*, I_{yz} is defined as:

$$I_{yz} = \int_A yz \, dA \tag{2.96}$$

where y and z are the respective distances of each element of area dA from the z and y axes.

The *polar moment of inertia* of the cross-section about the x axis, I_p, is defined as:

$$I_p = \int_A r^2 \, dA \tag{2.97}$$

where r is the distance of each element of cross-sectional area dA from the x axis. Note that since $r^2 = (y^2 + z^2)$

$$I_p = \int_A (y^2 + z^2) \, dA = I_z + I_y \tag{2.98}$$

If y' is an axis parallel to the centroidal axis y and distance c from it, then

$$I_{y'} = I_y + Ac^2 \tag{2.99}$$

The relationship in Equation (2.99) is known as the *parallel axis theorem*. This theorem facilitates the calculation of the moments of inertia of a complicated cross-section, for the section can be divided into separate simpler elements of area A_e say, whose moments of inertia I_{ye} about their *own* centroidal axes are known. If then c_e is the distance of an element centroid from the y axis, we have:

$$I_y = \sum_{\text{elements}} (I_{ye} + A_e c_e^2) \tag{2.100}$$

The moments of inertia about their centroidal axes, of various sectional shapes are given in Table 2.2.

2.3.2.3 Transformation of moments of inertia

Consider a new system of centroidal axes, y' and z', formed by a rotation of $\alpha°$ anticlockwise about the x axis as shown in Figure

---

† The term 'moment of inertia' is commonly used in engineering texts because the quantity I_y defined by Equation (2.94) is directly proportional to the mechanical moment of inertia about the y axis, of a thin lamina of the same shape as the cross-section. A more precise term for I_y is the 'second moment of area'.

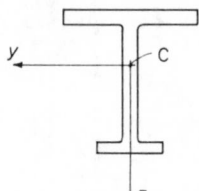

Figure 2.16

Table 2.2 Geometrical properties of plane sections

Section	Area A	Position of centroid C	Moments of inertia
(1) Rectangle	$A = bd$	$c = d/2$	$I_y = bd^3/12$ $I_z = db^3/12$
(2) Triangle	$A = bd/2$	$c = d/3$	$I_y = bd^3/36$ $I_z = db^3/48$
(3) Trapezium	$A = d(a+b)/2$	$c = d(2a+b)/3(a+b)$	$I_y = d^3(a^2 + 4ab + b^2)/36(a+b)$ $I_z = d(a^3 + a^2b + ab^2 + b^3)/48$
(4) Diamond	$A = bd/2$	$c = d/2$	$I_y = bd^3/48$ $I_z = db^3/48$
(5) Hexagon	$A = 0.866d^2$	$c = d/2$	$I_y = I_z = 0.0601d^4$
(6) Circle	$A = \pi r^2$ $= 3.1416r^2$	$c = r$	$I_y = I_z = \pi r^4/4$ $= 0.7854r^4$

Table 2.2 Geometrical properties of plane sections

Section	Area A	Position of centroid C	Moments of inertia
(7) Hollow circle	$A = \pi(r_1^2 - r_2^2)$ $= 3.1416(r_1^2 - r_2^2)$	$c = r_1$	$I_y = I_z = (\pi/4)(r_1^4 - r_2^4)$ $= 0.7854(r_1^4 - r_2^4)$
(8) Semicircle	$A = \pi r^2/2$ $= 1.5708r^2$	$c = 0.424r$	$I_y = [(\pi/8) - (8/9\pi)]r^4$ $= 0.1098r^4$ $I_z = \pi r^4/8$ $= 0.3927r^4$
(9) Ellipse	$A = \pi ab$	$c = a$	$I_y = (\pi/4)ba^3 = 0.7854ba^3$ $I_z = (\pi/4)ab^3 = 0.7854ab^3$
(10) Semi-ellipse	$A = \pi ab/2$	$c = 0.424a$	$I_y = 0.1098ba^3$ $I_z = 0.3927ab^3$
(11) Parabola	$A = 4ab/3$	$c = 2a/5$	$I_y = 0.0914ba^3$ $I_z = 0.2666ab^3$

2.17. Then the inertias $I_{y'}$, $I_{z'}$ and $I_{y'z'}$, being defined in the same way as I_y, I_z and I_{yz} in Equations (2.94 to 2.96), are related to I_y, I_z and I_{yz} by the equations:

$$I_{y'} = \tfrac{1}{2}(I_y + I_z) + \tfrac{1}{2}(I_y - I_z) \cos(2\alpha) - I_{yz} \sin(2\alpha) \qquad (2.101)$$

$$I_{z'} = \tfrac{1}{2}(I_y + I_z) - \tfrac{1}{2}(I_y - I_z) \cos(2\alpha) + I_{yz} \sin(2\alpha) \qquad (2.102)$$

$$I_{y'z'} = \tfrac{1}{2}(I_y - I_z) \sin(2\alpha) + I_{yz} \cos(2\alpha) \qquad (2.103)$$

Note that these transformation equations are similar in form to the transformation equations of plane stress in Equations (2.17 to 2.19), the difference being in the sign of α.

For a certain orientation of y' and z', the product of inertia $I_{y'z'}$ vanishes. Denoting these coordinates by Y and Z, then I_Y

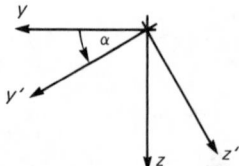

Figure 2.17

and I_Z are called the *principal moments of inertia* of the cross-section, and Y and Z are called the *principal axes*. Concerning their orientation, it can be shown in particular that one of the principal axes always coincides with an axis of symmetry in the section. Values of I_Y and I_Z for standard rolled sections are given in BS 4.[14]

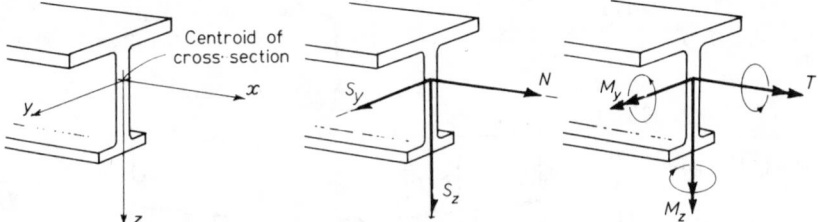

Figure 2.18

2.3.3 Stress resultants

The stresses acting across a particular cross-section of a bar under loads, are conveniently represented by their resultant forces and couples relative to the three coordinate axes x, y and z. Thus the resultants acting on the material of the bar on the negative[†] side of the cross-section are considered positive when acting in the directions shown in Figure 2.18 and are defined as follows:

Resultant	Defining equation	
Axial force N	$N = \int_A \sigma_x \, dA$	(2.104)
Bending moment about the y axis M_y	$M_y = \int_A \sigma_x z \, dA$	(2.105)
Bending moment about the z axis M_z	$M_z = -\int_A \sigma_x y \, dA$	(2.106)
Shear force in the y direction S_y	$S_y = \int_A \tau_{xy} \, dA$	(2.107)
Shear force in the z direction S_z	$S_z = \int_A \tau_{xz} \, dA$	(2.108)
Torque T	$T = \int_A (-\tau_{xy} z + \tau_{xz} y) \, dA$	(2.109)

These resultants are in equilibrium with the loads acting on that part of the bar which is on the negative side of the cross-section. Thus, if the bar is statically determinate, the resultants can be obtained directly by resolving and taking moments.

A *stress resultant diagram* represents the variation of the stress resultant with x for a specified bar loading. The diagram is drawn positive in the direction of the y and z coordinates. Thus given the beam subject to the vertical forces shown in Figure 2.19(a), the shear force (S_z) diagram and the bending moment (M_y) diagram take the form shown in Figures 2.19(b) and (c) respectively. Note that a positive bending moment M_y, causes tension on the bottom of the beam and therefore that the bending moment diagram is located on the tension side of the member. This orientation of the bending-moment diagram is very useful in reinforced concrete design leading to an immediate visual impression of where in the beam the tension reinforcement needs to be placed.

It is sometimes of interest in the case of beams to consider the value of a stress resultant (or any other parameter), at a particular point P in the beam, for various positions of a load moving across the beam. If, for example, we consider the bending moment about the y axis at P ($[M_y]_P$), caused by a unit vertical force at point x on the beam, then $[M_y]_P$ is a function of

[†] 'Negative' means the side in the negative direction of the x axis.

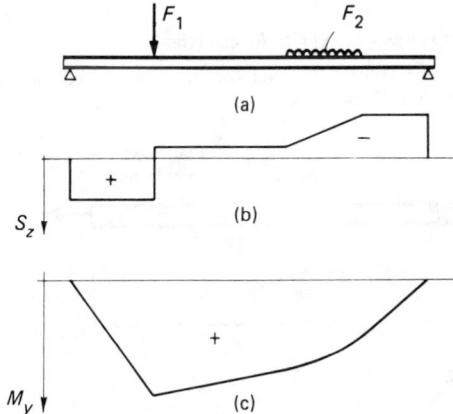

Figure 2.19

the coordinate x. The plot of $[M_y]_P$ versus x is called the *influence line* of M_y at P. Thus for the beam AB in Figure 2.20 the influence lines for $[S_z]_P$ and $[M_y]_P$ are as shown.

The stress resultants are not all independent of each other. Thus considering the rotational equilibrium about the y axis of a small element of a bar subject to a vertical distributed load q per unit length, as in Figure 2.21:

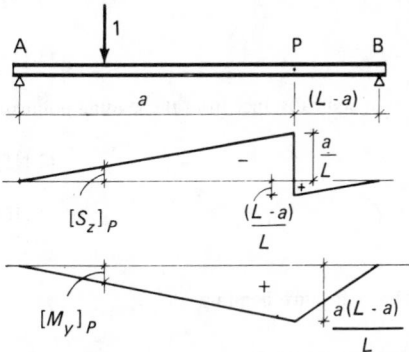

Figure 2.20

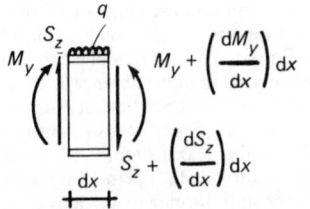

Figure 2.21

$$\mathrm{d}M_y/\mathrm{d}x = S_z \tag{2.110}$$

Further, considering vertical equilibrium:

$$\mathrm{d}S_z/\mathrm{d}x = -q \tag{2.111}$$

Whence, combining Equations (2.110) and (2.111) gives:

$$\mathrm{d}^2M_y/\mathrm{d}x^2 = -q \tag{2.112}$$

A similar set of equations relates M_z, S_y and the horizontal loading on the bar.

2.3.4 Bars subject to tensile forces (ties)

Consider a bar subject to axial forces N, produced by the loading shown in Figure 2.22.

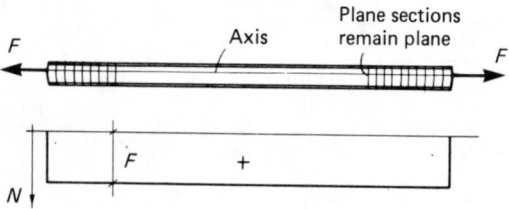

Figure 2.22

From the symmetry of the system at some distance from the loading points it can be *deduced* that plane sections originally normal to the longitudinal axis remain plane and normal to the axis after the deformation, while from the geometry of the bar, it can be assumed that the only nonzero component of stress is σ_x.[8]

The stress–strain relations corresponding to the uniaxial state of stress take the form:

$$\varepsilon_x = (\sigma_x/E) + \alpha\Delta T \tag{2.113}$$

$$\varepsilon_y = \varepsilon_z = -(\nu\sigma_x/E) + \alpha\Delta T \tag{2.114}$$

and it follows that at some distance from the loading points:

$$\sigma_x = N/A \tag{2.115}$$

$$\varepsilon_x = (N/EA) + \alpha\Delta T \tag{2.116}$$

2.3.5 Beams subject to pure bending

2.3.5.1 Beams symmetric about the vertical plane and subject to vertical loading

Consider a beam subject to a uniform bending moment M_y over part of its length, produced, for example, by the loading shown in Figure 2.23. (Note that Equation (2.110) implies that a uniform bending moment can only occur in the absence of shear forces.) From the symmetry of the system it can be *deduced* that: (1) the beam deforms in the vertical plane, and straight-line generators parallel to the longitudinal axis deform into segments of circles with a common centre; and (2) planes originally normal to the axis remain plane and normal to the axis after deformation.

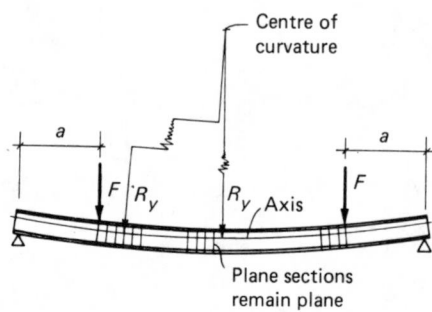

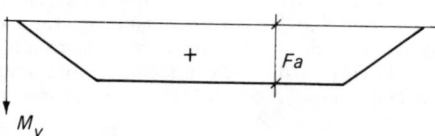

Figure 2.23

It can again be assumed that: (3) the only nonzero component of stress is σ_x.

The above three conditions are the fundamental assumptions made in the *engineering theory of the bending of beams*.

The surface containing those points in the beam at which $\varepsilon_x = 0$ is called the *neutral surface*. The intersection of the neutral surface with a cross-section produces a line called the *neutral axis*.

From the geometry of the deformation, the uniaxial stress–strain relations in Equations (2.113, 2.114), and the requirement of axial equilibrium ($N = 0$), it follows that:

(1) The neutral axis is given by the equation:

$$z = 0 \tag{2.117}$$

i.e. it is a horizontal straight line, coincident with the y coordinate line, and passing through the centroid of the section.

(2) $\sigma_x = \dfrac{M_y z}{I_y} \tag{2.118}$

and

$$\frac{1}{R_y} = \frac{M_y}{EI_y} \tag{2.119}$$

where R_y is the vertical radius of curvature of the beam axis.

2.3.5.2 Composite beams

Suppose the beam in the previous subsection is made of two materials of Young's modulus E_1 and E_2 respectively comprising areas A_1 and A_2 of the total cross-section, as in Figure 2.24. The three conditions of the engineering theory of the bending of beams discussed in the previous subsection still apply. It therefore follows that:

(1) The neutral axis is a horizontal straight line passing through a point C' called the *equivalent centroid of the cross-section*.

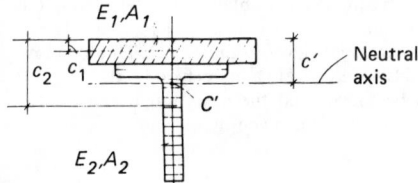

Figure 2.24

This is defined as being such that the first moment of *Young's modulus times area* about any axis passing through it is zero. Thus if c' is the distance of C' from the upper boundary of the beam and c_1 and c_2 are the distances of the respective centroids of the areas A_1 and A_2 from the upper boundary, then:

$$c' = \frac{E_1 A_1 c_1 + E_2 A_2 c_2}{E_1 A_1 + E_2 A_2} \tag{2.120}$$

$$(2)\ [\sigma_x]_{A_1} = \frac{M_y z}{I'_y} \qquad [\sigma_x]_{A_2} = \frac{E_2}{E_1} \frac{M_y z}{I'_y} \tag{2.121, 2.122}$$

and

$$\frac{1}{R_y} = \frac{M_y}{E_1 I'_y} \tag{2.123}$$

where $[\sigma_x]_{A_1}$ represents the axial stress in the area A_1, etc. I'_y is the *equivalent moment of inertia* of the cross-section defined as:

$$I'_y = \int_{A_1} (z^2)\, \mathrm{d}A_1 + \frac{E_2}{E_1} \int_{A_2} (z^2)\, \mathrm{d}A_2 \tag{2.124}$$

where $\int_{A_1} (\ )\, \mathrm{d}A_1$ represents an integral taken over the area A_1, etc. In the above equations, the coordinates are relative to axes y and z passing through the equivalent centroid of the section.

2.3.5.3 Reinforced concrete beams

A reinforced concrete beam behaves as a composite beam, except that where the concrete is in tension (i.e. below the neutral axis) its stress-bearing capacity is taken to be zero (Figure 2.25). Otherwise the conditions of the engineering theory of the bending of beams still apply.

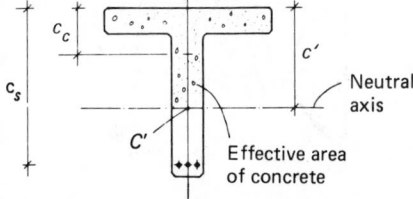

Figure 2.25

Let the subscripts c and s denote parameters associated respectively with the concrete and the steel. It then follows that:

(1) The neutral axis is a horizontal straight line passing through the equivalent centroid whose distance c' from the upper boundary of the beam is given by:

$$c' = \frac{E_c A_c c_c + E_s A_s c_s}{E_c A_c + E_s A_s} \tag{2.125}$$

(Note that since A_c and c_c are themselves functions of c', Equation (2.125) is an implicit equation.)

$$(2)\ [\sigma_x]_c = \frac{M_y z}{I'_y} \qquad [\sigma_x]_s = \frac{E_s}{E_c} \frac{M_y z}{I'_y} \tag{2.126, 2.127}$$

and

$$\frac{1}{R_y} = \frac{M_y}{E_c I'_y} \tag{2.128}$$

where

$$I'_y = \int_{A_c} (z^2)\, \mathrm{d}A_c + \frac{E_s}{E_c} \int_{A_s} (z^2)\, \mathrm{d}A_s \tag{2.129}$$

Example 2.2. A rectangular reinforced concrete beam with a single layer of reinforcement is shown in Figure 2.26. For this section:

$$c' = \frac{E_s A_s}{E_b b} \left[\left(1 + \frac{2E_c}{E_s} \frac{b(d-e)}{A_s} \right)^{1/2} - 1 \right] \tag{2.130}$$

$$I'_y = \frac{bc'^3}{3} + \frac{E_s}{E_c} A_s [d - (c'+e)]^2 \tag{2.131}$$

Note that the ratio $E_s : E_c$ is generally taken to be 15.

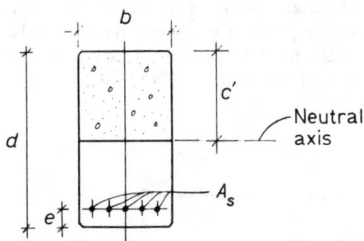

Figure 2.26

2.3.5.4 Beams of asymmetric section subject to both vertical and horizontal loading

Consider again a beam of homogeneous material. The general case of pure bending occurs when the beam is of asymmetric section and is subject to uniform bending moments M_y and M_z (Figure 2.27) over part of its length.

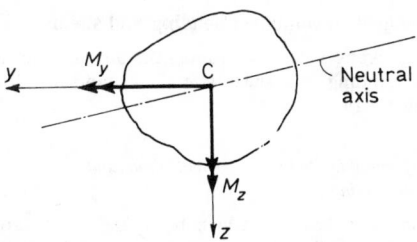

Figure 2.27

From the symmetry of the system it can be *deduced* that straight-line generators parallel to the axis of the beam deform into curves of constant horizontal and vertical curvature. The other conditions discussed in section 2.3.5.1 still apply.

From the geometry of the deformation, the uniaxial stress–strain relations and the requirement that $N=0$, it follows that:

(1) The neutral axis is given by the equation:

$$(M_y I_z + M_z I_{yz}) z - (M_z I_y + M_y I_{yz}) y = 0 \qquad (2.132)$$

i.e. it is a straight line passing through the centroid of the section, as shown in Figure 2.27.

(2) $\quad \sigma_x = \dfrac{(M_y I_z + M_z I_{yz}) z - (M_z I_y + M_y I_{yz}) y}{(I_y I_z - I_{yz}^2)}$

$$\qquad (2.133)$$

$$\frac{1}{R_y} = \frac{(M_y I_z + M_z I_{yz})}{E(I_y I_z - I_{yz}^2)} \quad \frac{1}{R_z} = \frac{-(M_z I_y + M_y I_{yz})}{E(I_y I_z - I_{yz}^2)} \qquad (2.134, 2.135)$$

where R_z is the horizontal radius of curvature of the beam axis.

Note:

(1) If the loading is vertical so that $M_z = 0$, Equation (2.135) indicates that the deformed beam is curved horizontally, i.e. $R_z \neq 0$.

(2) If y and z are principal axes, so that $I_{yz} = 0$, Equations (2.134, 2.135) indicate that the curvature about each axis is proportional only to the bending moment about that axis.

In some cases, where a standard commercial section is mounted obliquely, as in Figure 2.28(a) for example, I_y, I_z and I_{yz} will be known relative to the axes y', z', while the bending moments will be known about the axes y and z. In order to use the results in Equations (2.132 to 2.135) it is preferable to work in terms of the y' and z' axes and resolve the bending moments into equivalent moments about these axes, as in Figure 2.28(b).

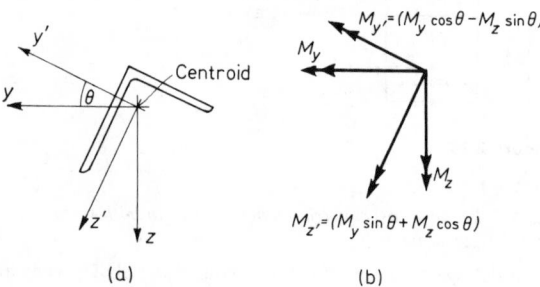

(a)

(b)

Figure 2.28

2.3.6 Beams subject to combined bending and shear

Practical loading arrangements on beams generally produce a combination of bending and shear stress resultants as, for example, in Figure 2.19.

2.3.6.1 Beams symmetric about the vertical plane and subject to vertical loading

Consider a point in a beam at which both M_y and S_z are nonzero. The presence S_z then implies the existence of the shear stresses τ_{xz} on the cross-section and corresponding shear strains γ_{xz}, and much of the symmetry of the deformation of a beam under a uniform bending moment is lost. In particular, plane sections no longer remain plane.

The following approximate analysis of the problem is due to St Venant.[15] It is assumed that the direct stresses σ_x and curvature $(1/R_y)$ are the same as they would be if M_y were acting alone. They are therefore given by Equations (2.118, 2.119). The

shear stresses in the beam are then obtained by considering the longitudinal equilibrium of the element of length dx shown shaded in the cross-sectional view of Figure 2.29. Thus employing Equations (2.110) and (2.118), namely $dM_y/dx = S_z$ and $\sigma_x = M_y z / I_y$, it can be shown that the *mean* longitudinal shear stress τ on the surface ABCD is given by:

$$\tau = \frac{S_z A_e c_e}{b_e I_y} \qquad (2.136)$$

where A_e is the cross-sectional area of the element, c_e is the distance of its centroid from the neutral axis, and b_e is the length of the curve joining AB (Figure 2.29).

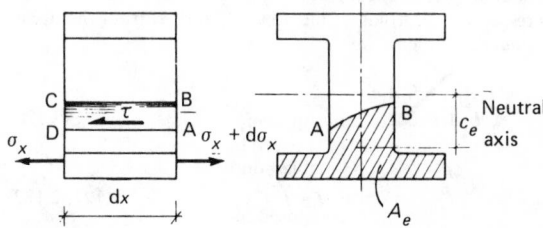

Figure 2.29

τ can then be related to the shear stresses τ_{xy} and τ_{xz} on the cross-section as follows. If the cut surface ABCD is a horizontal plane (i.e. it is a z-coordinate surface) then τ is the mean value of the shear stress component τ_{zx} on that surface. Whence, since $\tau_{zx} = \tau_{xz}$, it follows that τ is also the mean value of τ_{xz} on the line AB. For thin sections, we assume that τ_{xz} is uniformly distributed across the width so that:

$$\tau_{xz} = \tau \qquad (2.137)$$

Thus for the rectangular section shown in Figure 2.30, Equation (2.136) gives the following parabolic distribution of shear stress on the cross-section:

$$\tau_{xz} = \frac{3S_z}{2bd^3}(d^2 - 4z^2) \qquad (2.138)$$

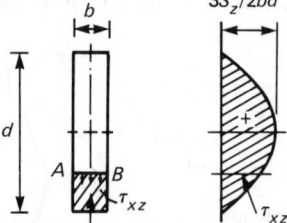

Figure 2.30

If the cut surface ABCD is a vertical plane (a y-coordinate surface) then τ is the mean value of the shear stress component τ_{yx} on that surface, or the mean value of τ_{xy} on the line AB. Thus for an I-section, the shear stresses in the flanges are as shown in Figure 2.31.

2.3.6.2 Composite beams

The existence of the longitudinal shear stress τ (Figure 2.29) is of

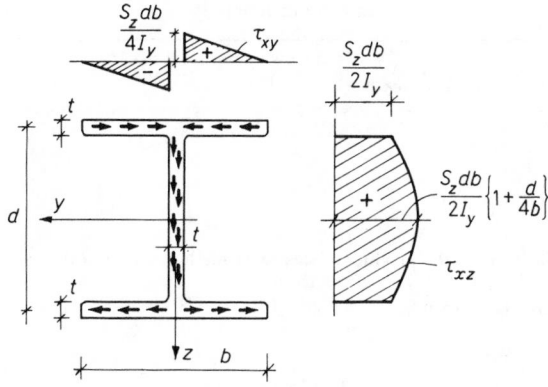

Figure 2.31

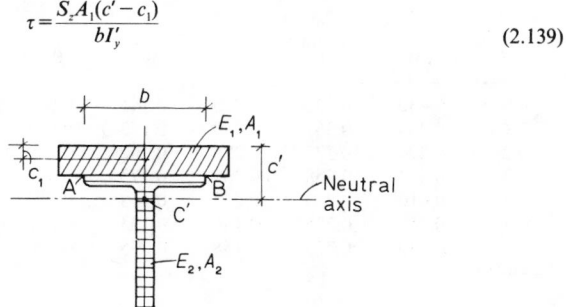

Figure 2.32

special significance in built-up composite beams, because this stress has to be transmitted between the separate components of the beams by means of suitable bonds such as welds, rivets or shear connectors.

Thus consider a beam composed of two materials of Young's modulus E_1 and E_2 respectively comprising areas A_1 and A_2 of the total cross-section (Figure 2.32). The position of the neutral axis and the equivalent moment of inertia of the cross-section are again given by Equations (2.120) and (2.124), whence, employing the assumptions of St Venant's theory, it can be deduced that the mean longitudinal shear stress at the interface AB is given by:

$$\tau = \frac{S_z A_1 (c' - c_1)}{b I'_y} \tag{2.139}$$

while the corresponding longitudinal shear force/unit length of beam F is given by:

$$F = b\tau \tag{2.140}$$

If the beam were composed, say, of a concrete slab connected to a steel T-section joist, then F would be transmitted by stud shear connectors of the type shown in Figure 2.33 which would be welded on to the top face of the T-section. Supposing that the factored shear strength of each connector were known experimentally to be F_s, then the connectors would need to be distributed at a concentration of F/F_s per unit length of beam.

2.3.6.3 The shear centre (beams asymmetric about the vertical plane)

In a beam of asymmetric cross-section the shear stresses given by St Venant's theory contribute to a torque T. Consider, for

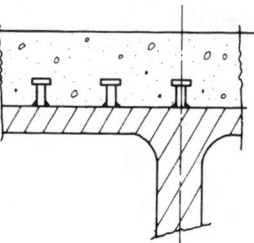

Figure 2.33

example, the shear stresses produced in the channel section shown in Figure 2.34. They are statically equivalent to the stress resultants S_f acting in the two flanges, and S_w in the web, where:

$$S_w = S_z \tag{2.141}$$

$$S_f = \frac{S_z b^2 dt}{4 I_y} \tag{2.142}$$

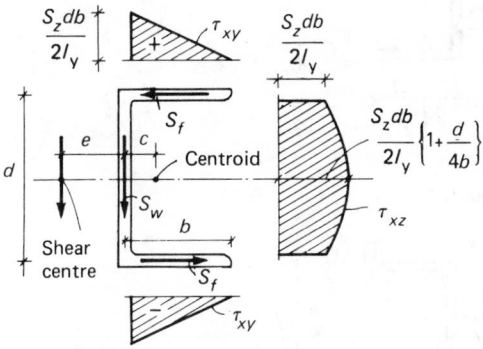

Figure 2.34

and because of the asymmetry of the section, they produce a torque T acting about the longitudinal axis of the channel given by

$$T = S_z c + \frac{S_z b^2 d^2 t}{4 I_y} \tag{2.143}$$

An important assumption of St Venant's theory is that the beam deflects vertically without twist. Thus, it can be deduced that if the loading on the beam is such that it produces the torque T, then twisting does not, in fact, occur. (If the loading did not produce T then some twisting of the beam would be necessary in order to modify the torque obtained in Equation (2.143).) T can be applied by positioning the vertical loading so that its resultant at any cross-section lies at a suitable distance from the centroid. Thus the torque in the channel can be produced by the loading shown in Figure 2.35. The point at

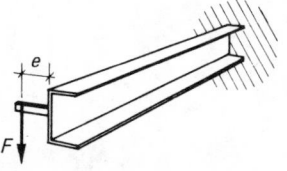

Figure 2.35

Table 2.3 Shear centres of the walled sections

Section	*Position of shear centre Q*

(1) Channel

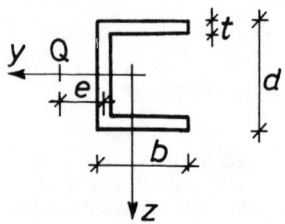

$$e = d\left(\frac{H_{yz}}{I_y}\right)$$

where H_{yz} is the product of inertia of the half section (above the y axis).
 If t is uniform:

$$e = \frac{b^2 d^2 t}{4 I_y}$$

(2) Lipped channel

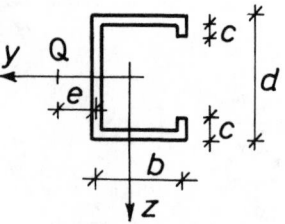

Values of (e/d)

	b/d				
c/d	1.0	0.8	0.6	0.4	0.2
0.0	0.430	0.330	0.236	0.141	0.055
0.1	0.477	0.380	0.280	0.183	0.087
0.2	0.530	0.425	0.325	0.222	0.115
0.3	0.575	0.470	0.365	0.258	0.138
0.4	0.610	0.503	0.394	0.280	0.155
0.5	0.621	0.517	0.405	0.290	0.161

(3) Hat-section

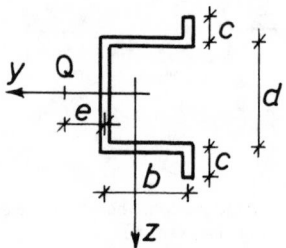

Values of (e/d)

	b/d				
c/d	1.0	0.8	0.6	0.4	0.2
0.0	0.430	0.330	0.236	0.141	0.055
0.1	0.464	0.367	0.270	0.173	0.080
0.2	0.474	0.377	0.280	0.182	0.090
0.3	0.453	0.358	0.265	0.172	0.085
0.4	0.410	0.320	0.235	0.150	0.072
0.5	0.355	0.275	0.196	0.123	0.056
0.6	0.300	0.225	0.155	0.095	0.040

(4) I-section

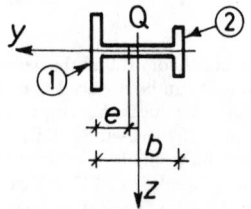

$$e = \frac{b I_2}{I_1 + I_2}$$

where I_1 and I_2 respectively denote the moments of inertia about the y axis of flange 1 and flange 2

(5) Split circle

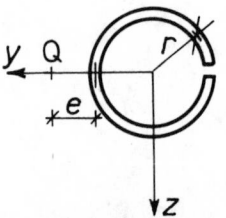

$$e = r$$

Section	Position of shear centre Q
(6) Z-section 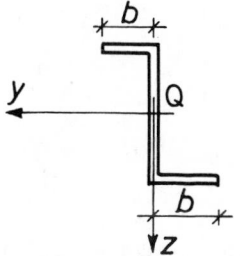	Shear centre coincides with centroid
(7) Sections with elements intersecting at a single point 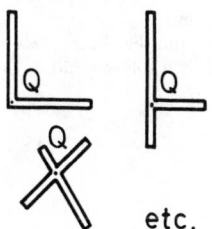 etc.	Shear centre lies at point of intersection

which the vertical resultant crosses the neutral axis is then called the *shear centre*, and for the channel section it is located at a distance e from the web (Figure 2.34) where

$$e = \frac{b^2 d^2 t}{4 I_y} \tag{2.144}$$

The positions of the shear centres of various cross-sectional shapes are shown in Table 2.3.

The *shear axis* of the beam is defined as the line passing through the shear centres of its cross-sections, and by definition, the resultants of all lateral forces acting on the beam must pass through this axis if the beam is to deflect without twist.

2.3.7 Deflection of beams

According to St Venant's theory, the curvature of a beam subject to combined bending and shear is given by Equation (2.119) thus: $1/R_y = M_y/EI_y$. Suppose u_z is the corresponding vertical deflection of the longitudinal axis of the beam, then from the geometry of the deformation (Figure 2.36), it can be shown that:

$$\frac{1}{R_y} = -\frac{d^2 u_z}{dx^2} \bigg/ \left(1 + \left(\frac{du_z}{dx}\right)^2\right)^{3/2} \tag{2.145}$$

In practice, the slopes of beams are extremely small and the denominator of the right-hand side of Equation (2.145) can be taken to be equal to unity, whence, combining Equations (2.119) and (2.145) gives the following differential equation:

$$\frac{d^2 u_z}{dx^2} + \frac{M_y}{EI_y} = 0 \tag{2.146}$$

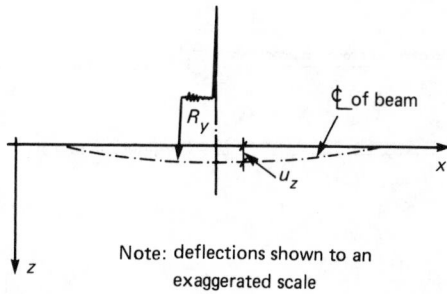

Note: deflections shown to an exaggerated scale

Figure 2.36

called the *differential equation of beams*. For statically determinate beams, where M_y can be found as a function of x, this second-order equation can be solved subject to boundary conditions by double integration. The solution $u_z(x)$ is then the deflected shape of a beam produced by the applied loading. Examples of the boundary conditions for particular cases are shown in Figure 2.37. Special techniques, such as the step function method[16a] and the moment-area method[15] have been devised to simplify the analysis.

The differential equation of beams can be expressed in two further forms using the results of Equations (2.110) and (2.112). Thus from Equation (2.110) we have:

$$\frac{d^3 u_z}{dx^3} + \frac{S_z}{EI_y} = 0 \tag{2.147}$$

while from Equation (2.112) we have:

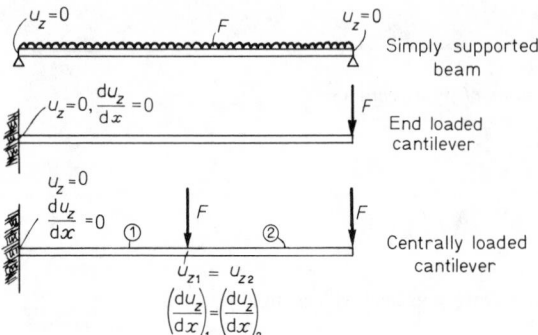

Figure 2.37

$$\frac{d^4u_z}{dx^4} - \frac{q}{EI_y} = 0 \tag{2.148}$$

Equation (2.148), expressing the deflections of beams in terms of the lateral loading, is directly equivalent to the three-dimensional Navier Equations (2.61 to 2.63), and can be solved if the boundary conditions are expressed in terms of the displacements. The solution of this equation as opposed to Equation (2.146), is necessary when a beam is statically indeterminate, i.e. when M_y cannot be found in advance. Examples of the required displacement boundary conditions for particular cases are shown in Figure 2.38.

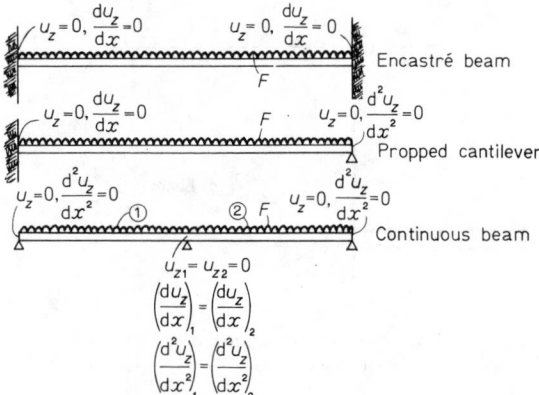

Figure 2.38

An interesting modification of Equation (2.148) occurs when a beam rests on an elastic foundation. Suppose the stiffness of the foundation is k per unit length of beam. Then in addition to the vertical applied loading q, the foundation resists the deflection of the beam with distributed forces equal to ku_z per unit length. Equation (2.148) then takes the form:

$$\frac{d^4u_z}{dx^4} + ku_z - \frac{q}{EI_y} = 0 \tag{2.149}$$

Examples of the solution of this equation are given by Hetényi.[16b]

2.3.8 Bars subject to a uniform torque

2.3.8.1 Bars of circular cross-section

Consider a bar subject to a uniform torque T produced, for example, by the loading shown in Figure 2.39.

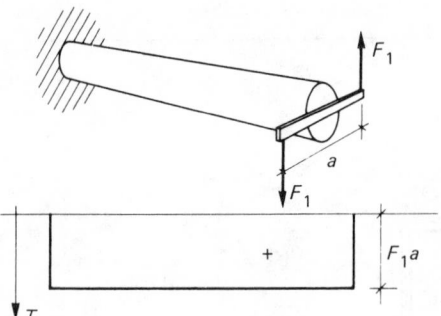

Figure 2.39

From the symmetry of the system it can be deduced that: (1) the bar twists about its longitudinal axis; (2) planes originally normal to the axis remain plane and normal to the axis and rotate like rigid laminae, and (3) the rotation θ of any plane is proportional to its distance along the beam.

From the geometry of the deformation and the shear stress–strain relations in Equations (2.49 to 2.51), it follows that:

$$\tau_{xt} = \frac{Tr}{J} \tag{2.150}$$

$$\frac{d\theta}{dx} = \frac{T}{GJ} \tag{2.151}$$

where τ_{xt} is the shear stress on the cross-section at a distance r from the axis, and tangential to the circle of radius r (Figure 2.40). J is a sectional constant, equal in this case to the polar moment of inertia I_p about the longitudinal axis.

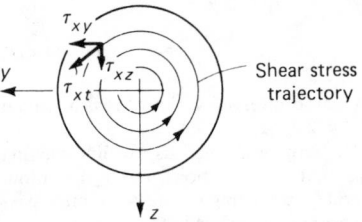

Figure 2.40

The quantity $d\theta/dx$ being the rate of change of rotation with x is called the *twist* of the bar, and is clearly uniform when the bar is subject to uniform torque.

2.3.8.2 Bars of arbitrary cross-section

The three assumptions of section 2.3.8.1 can be shown to lead to impossible values of τ_{xt} at the boundaries of an arbitrary section, since in order to satisfy longitudinal equilibrium conditions, τ_{xt} must be tangential to those boundaries (Figure 2.41).

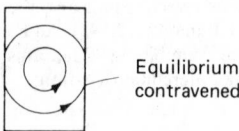

Figure 2.41

The theory for the analysis of bars of arbitrary section is again due to St Venant.[8] Thus the assumption in the previous subsection that plane sections remain plane is relaxed, and a point P is assumed to have an axial displacement u_x given by:

$$u_x = \frac{d\theta}{dx} \psi (y, z) \tag{2.152}$$

u_x is called the *warping* of the section, and is directly proportional to the twist, but is independent of x. The shear stresses τ_{xy} and τ_{xz} are then expressed in terms of a stress function $\phi(y, z)$ by the equations:

$$\tau_{xy} = \partial\phi/\partial z \quad \tau_{xz} = -\partial\phi/\partial y \tag{2.153, 2.154}$$

so that the internal equilibrium Equations (2.11) to (2.13) are identically satisfied. Satisfaction of the compatibility Equations (2.40) and (2.42) then leads to the following equation:

$$\frac{\partial^2\phi}{\partial x^2} + \frac{\partial^2\phi}{\partial y^2} = -2G \left(\frac{d\theta}{dx}\right) \tag{2.155}$$

Equilibrium conditions require that ϕ is constant along the boundaries of the section, and if the section is solid ϕ can be conveniently taken as zero along the boundaries, whence it can be shown that:

$$T = 2\int_A \phi \, dA \tag{2.156}$$

Equations (2.155) and (2.156) are solved simultaneously, either numerically, or experimentally,[8] and the shear stresses corresponding to T are obtained from Equations (2.153) and (2.154). The results can be expressed in the following form:

$$[\tau_{xb}]_{max} = T/k \tag{2.157}$$

$$d\theta/dx = T/GJ \tag{2.158}$$

where $[\tau_{xb}]_{max}$ is the maximum shear stress on the boundary of the section and is tangential to the boundary. k and J are constants, and their values for various cross-sectional shapes are shown in Table 2.4.

For the narrow rectangular section shown in Figure 2.42:

$$k = t^2 d/3 \quad J = t^3 d/3 \tag{2.159, 2.160}$$

and the maximum shear stress occurs along the boundaries of greatest length. These results can be used to determine the

(2) Equilateral triangle

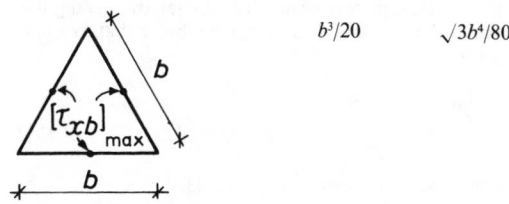

$b^3/20$	$\sqrt{3}b^4/80$

(3) Right isosceles triangle

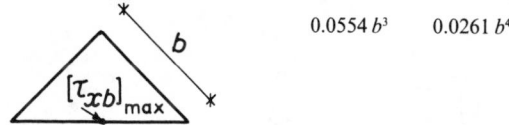

$0.0554\,b^3$	$0.0261\,b^4$

(4) Hexagon

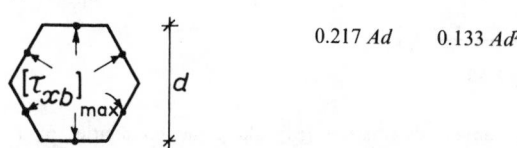

$0.217\,Ad$	$0.133\,Ad^2$

(5) Circle

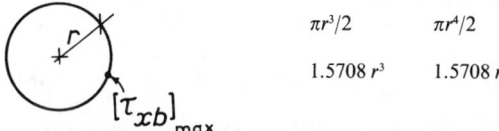

$\pi r^3/2$	$\pi r^4/2$
$1.5708\,r^3$	$1.5708\,r^4$

(6) Hollow circle

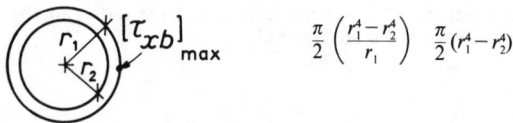

$\frac{\pi}{2}\left(\frac{r_1^4 - r_2^4}{r_1}\right)$	$\frac{\pi}{2}(r_1^4 - r_2^4)$

(7) Ellipse

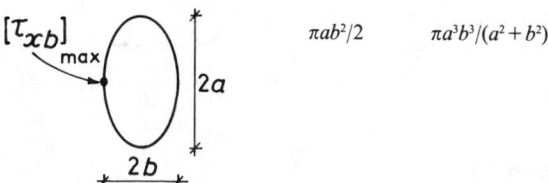

$\pi ab^2/2$	$\pi a^3 b^3/(a^2 + b^2)$

Table 2.4 Torsional constants

Section		k	J
(1) Rectangle	d/b		
	1.0	$0.208\,(b^2 d)$	$0.1406\,(b^3 d)$
	1.2	$0.219\,(b^2 d)$	$0.166\,(b^3 d)$
	1.5	$0.231\,(b^2 d)$	$0.196\,(b^3 d)$
	2.0	$0.246\,(b^2 d)$	$0.229\,(b^3 d)$
	2.5	$0.258\,(b^2 d)$	$0.249\,(b^3 d)$
	3.0	$0.267\,(b^2 d)$	$0.263\,(b^3 d)$
	4.0	$0.282\,(b^2 d)$	$0.281\,(b^3 d)$
	5.0	$0.291\,(b^2 d)$	$0.291\,(b^3 d)$
	10.0	$0.312\,(b^2 d)$	$0.312\,(b^3 d)$
	∞	$1/3\ (b^2 d)$	$1/3\ (b^3 d)$

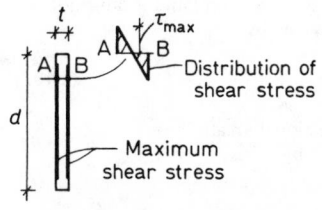

Figure 2.42

torsional properties of a thin-walled open-section bar, supposing that the cross-section can be divided into narrow rectangular elements of thickness t_e and d_e, for it can be shown that to a first approximation:

$$J = \sum_{\text{elements}} \frac{t_e^3 d_e}{3} \tag{2.161}$$

Thus for the I-section shown in Figure 2.43:

$$J = \frac{2d_f t_f^3 + d_w t_w^3}{3} \tag{2.162}$$

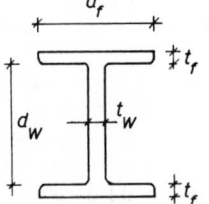

Figure 2.43

The maximum shear stress $[\tau_{xb}]_e$ along the boundaries of a particular element are given by:

$$[\tau_{xb}]_e = T/k_e \tag{2.163}$$

where

$$k_e = J/t_e \tag{2.164}$$

$[\tau_{xb}]_e$ however, is not the maximum shear stress on the cross-section, for this now occurs at the re-entrant corners. Thus, in a constant-thickness channel section (Figure 2.44(a)) $[\tau_{xb}]_{\text{max}}$ occurs at point P, and is related to $[\tau_{xb}]_e$ and the radius of the corner as shown in Figure 2.44(b).[8]

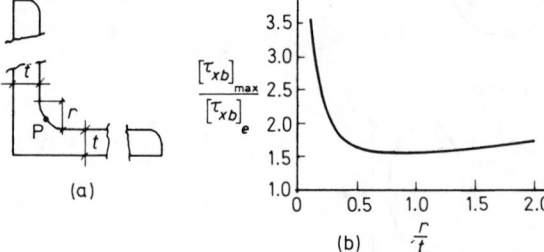

(a)

(b)

Figure 2.44

In a thin-walled closed-section bar, such as the tube of varying wall thickness t shown in Figure 2.45, the shear stress τ_{xb} is uniform across the thickness at any point and is tangential to the surface of the tube. It is given by:

$$\tau_{xb} = T/2At \tag{2.165}$$

where A is the gross cross-sectional area.
J in Equation (2.158) is given by:

$$J = 4A^2 \bigg/ \left(\oint \frac{\text{d}s}{t} \right) \tag{2.166}$$

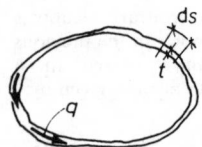

Figure 2.45

where $\text{d}s$ is an element of length round the tube (Figure 2.45). A further quantity q called the *shear flow* is defined at a point in the tube wall by the equation

$$q = \tau_{xb} t \tag{2.167}$$

It is then apparent from Equation (2.165) that the shear flow is independent of t.

In multicell thin-walled bars, as shown in Figure 2.46, the concept of circulatory shear flows q_1, q_2, q_3 is introduced, a concept which automatically satisfies the conditions of longitudinal equilibrium at junctions such as A. The shear flow at point B, for example, is then given by $(q_1 - q_2)$. The shear flows and the twist of the bar corresponding to a certain applied torque are calculated from the four simultaneous equations:

$$T = 2q_1 A_1 + 2q_2 A_2 + 2q_3 A_3 \tag{2.168}$$

$$\frac{\text{d}\theta}{\text{d}x} = \frac{1}{2A_1 G} \oint_1 \frac{q}{t}\,\text{d}s_1 = \frac{1}{2A_2 G} \oint_2 \frac{q}{t}\,\text{d}s_2 = \frac{1}{2A_3 G} \oint_3 \frac{q}{t}\,\text{d}s_3 \tag{2.169, 2.170, 2.171}$$

where $\oint_1$ represents the contour integral taken round cell 1, etc.

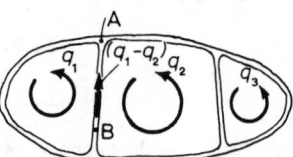

Figure 2.46

2.3.9 Nonuniform torsion

Nonuniform torsion in a bar is defined to occur when the twist $\text{d}\theta/\text{d}x$ varies along its length. This situation arises when the warping assumed in St Venant's theory is restrained at a rigid support, or when the torque exerted by the applied loading is nonuniform.

The nature of the modification necessary to St Venant's theory can be appreciated by considering the nonuniform torsion of the I-section cantilever shown in Figure 2.47. Since $\text{d}\theta/\text{d}x$ is not constant, the flanges are curved in the z plane. Considering the flanges as subsidiary beams, they contain shear forces $[S_z]_f$ which are related to this curvature. The torque T therefore includes an extra component $[S_z]_f d$. $[S_z]_f$ is given by Equation (2.147) as:

$$[S_z]_f = -E[I_y]_f \frac{\text{d}^3 u_z}{\text{d}x^3} \tag{2.172}$$

where $[I_y]_f$ is the moment of inertia of each flange about the y axis, and u_z is its displacement. Whence noting that:

$$u_z = \pm \frac{d}{2}\theta \tag{2.173}$$

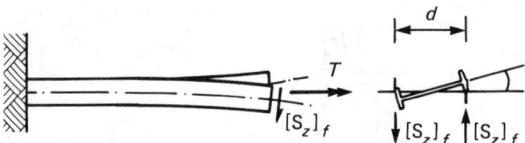

Figure 2.47

Table 2.5 Warping factors

Section	Γ
(1) I-section	$\frac{1}{24}b^3d^2t_1$
(2) Channel section	$\frac{1}{12}b^3d^2t_1\left(\dfrac{3bt_1+2dt_2}{6bt_1+dt_2}\right)$
(3) Z-section	$\frac{1}{12}b^3d^2t_1\left(\dfrac{bt_1+2dt_2}{2bt_1+dt_2}\right)$
(4) Thin-walled sections with elements intersecting at a single point	0

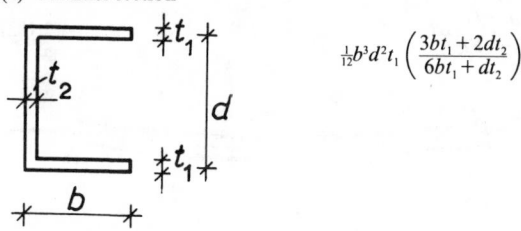

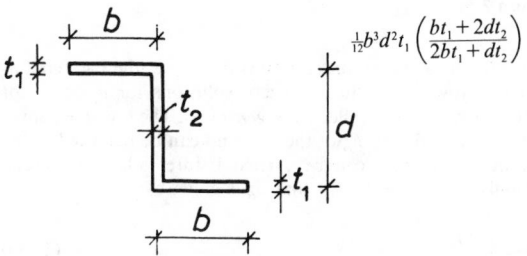

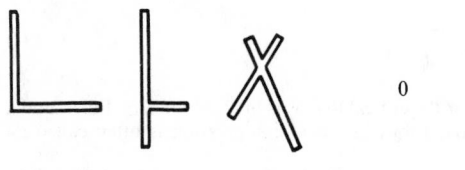

the additional torque component becomes:

$$-EI_y\frac{d^2}{4}\frac{d^3\theta}{dx^3}$$

where I_y is the *total* moment of inertia of the cross-section about the y axis. Combining this with the torque required for the uniform torsion of the bar, we obtain:

$$T=GJ\frac{d\theta}{dx}-EI_y\frac{d^2}{4}\frac{d^3\theta}{dx^3}\tag{2.174}$$

Equation (2.174) can be expressed in the more general form:

$$T=GJ\frac{d\theta}{dx}-EI\Gamma\frac{d^3\theta}{dx^3}\tag{2.175}$$

where Γ is a constant called the *warping factor*. Its values for various cross-sectional shapes are given in Table 2.5. The differential Equation (2.175) can be solved for various values of T applied to the bar, subject to boundary conditions in θ. Examples of these boundary conditions are shown in Figure 2.48.

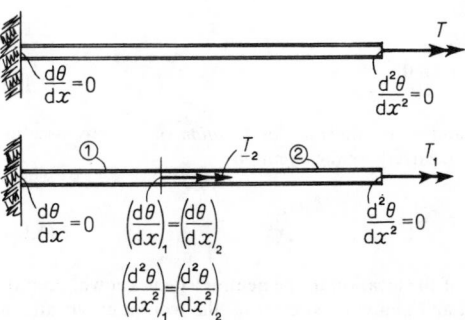

Figure 2.48

2.3.10 Bars subject to compressive forces (columns)

2.3.10.1 Short columns

If the geometry of a bar is such that its length is less than about 5 times its lateral dimensions, then it is usually stable under compressive forces. If therefore it is subjected to an axial compressive force F, then $N=-F$ and the corresponding stress σ_x is given by Equation (2.115) as: $\sigma_x=N/A$. If further, the bar is subjected to bending moments M_y and M_z acting about the principal axes y and z, then by superposition:

$$\sigma_x=\frac{N}{A}+\frac{M_yz}{I_y}-\frac{M_zy}{I_z}\tag{2.176}$$

and the neutral axis is given by the equation:

$$\frac{M_yz}{I_y}-\frac{M_zy}{I_z}+\frac{N}{A}=0\tag{2.177}$$

Combined compressive forces and bending moments occur in the bar if the compressive force F is eccentrically positioned as shown in Figure 2.49. Thus if the resultant due to F passes at a distance n and m from the y and z axes respectively, then:

$$N=-F \quad M_y=-Fn \quad M_z=+Fm\tag{2.178}$$

and:

$$\sigma_x=-F\left(\frac{1}{A}+\frac{nz}{I_y}+\frac{my}{I_z}\right)\tag{2.179}$$

The neutral axis is then given by the equation:

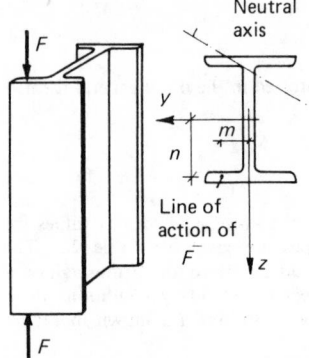

Figure 2.49

$$\frac{nz}{r_y^2}+\frac{my}{r_z^2}+1=0 \tag{2.180}$$

where r_y and r_z are the *radii of gyration* of the cross-section defined respectively by the equations:

$$r_y^2=\frac{I_y}{A}\quad r_z^2=\frac{I_z}{A} \tag{2.181}$$

Note that if the location of the neutral axis is known, then the maximum and minimum stresses on the section are located at those points which are at the greatest perpendicular distance from this axis. Their positions can easily be found graphically.

It is apparent from Equation (2.180) that the location of the neutral axis depends only on the coordinates n and m defining the eccentricity of F. If this eccentricity is such that the neutral axis falls outside the section, then the stress σ_x is negative (or compressive) at all points in the section. This situation arises if the stress resultant lies within an area called the *core* of the section. The dimensions of the cores of regular sections can be found analytically and some examples are given in Table 2.6.

Table 2.6 Cores of sections

Section

(1) Rectangle

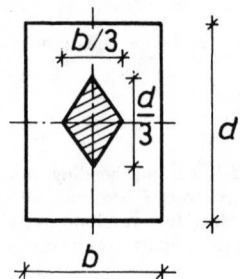

(2) Circle

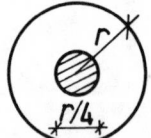

(3) I-section

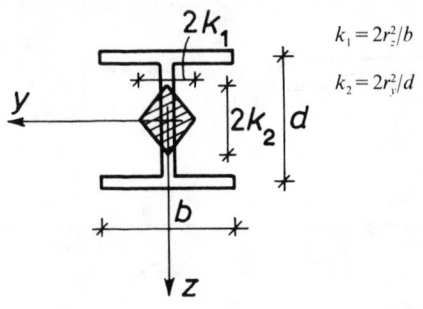

$k_1=2r_z^2/b$

$k_2=2r_y^2/d$

2.3.10.2 Long columns

If the length of a bar is greater than about 5 times its lateral dimensions, it can become unstable under compressive forces. Consider, for example, the pin-ended bar subject to an axial compressive force F shown in Figure 2.50. If u_z is the lateral displacement in the z direction of a particular cross-section, then the moment M_y exerted by F at the section is Fu_z. Thus from Equation (2.146) we have the differential equation

$$\frac{d^2u_z}{dx^2}+\frac{Fu_z}{EI_y}=0 \tag{2.182}$$

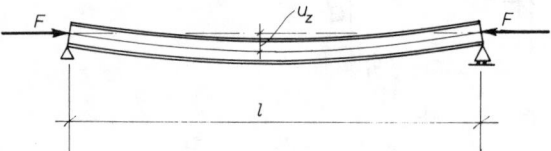

Figure 2.50

One solution of Equation (2.182) is $u_z=0$, i.e. the bar remains straight. However, further nonzero solutions for u_z occur for particular values of F called the *eigenvalues*. The lowest eigenvalue is the critical load F_{cr} of the bar, and can be regarded as the maximum load that can be carried before failure by lateral instability. It can be shown that F_{cr} is given by:

$$F_{cr}=\frac{\pi^2EI_y}{l^2} \tag{2.183}$$

while the corresponding deflected shape of the bar is sinusoidal and of arbitrary amplitude, taking the following form:

$$u_z=A\sin\left(\frac{\pi x}{l}\right) \tag{2.184}$$

The value for the critical load was first obtained by Euler, and a pin-ended bar subject to axial compression is often called an *Euler strut*.

Dividing Equation (2.183) by the area of the bar leads to the following expression for the *critical buckling stress* σ_{cr}:

$$\sigma_{cr}=\pi^2E/\lambda^2 \tag{2.185}$$

where $\lambda\ (=l/r_y)$ is called the *slenderness ratio*.

When, as in most cases, $I_y\neq I_z$, the strut buckles first about the minor principal axis, about that axis for which the moment of inertia of the section is a minimum.

The critical buckling loads of struts with other than pin-ended boundary conditions are given in Table 2.7. The corresponding *effective lengths* l_e are then defined so that the critical stresses can be given by an equation analogous to Equation (2.185) namely

$$\sigma_{cr} = \pi^2 E / \lambda_e^2 \qquad (2.186)$$

where $\lambda_e = l_e / r_y$. Values for l_e are included in the table.

Table 2.7 Critical buckling loads of struts

All struts are of length l; $I_z > I_y$

Lower end boundary condition	Upper end boundary condition	Mode	P_{cr}	l_e
(1) Hinge along y axis	Hinge along y axis	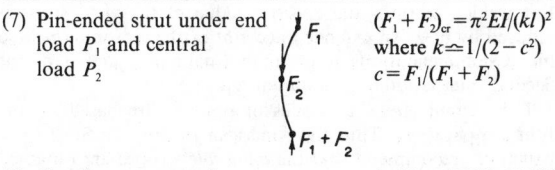 F	$\pi^2 E I_y / l^2$	l
(2) Clamped	Clamped	F F	$4\pi^2 E I_y / l^2$	$0.5\,l$
(3) Clamped	Hinge along y axis	F	$20.19\,E I_y / l^2$	$0.7\,l$
(4) Clamped	Free	F	$\pi^2 E I_y / 4l^2$	$2.0\,l$
(5) Hinge along z axis	Hinge along z axis		Smaller of $4\pi^2 E I_y / l^2$ or $\pi^2 E I_z / l^2$	$0.5\,l$
(6) Hinge along z axis	Hinge along y axis	F	$20.19\,E I_y / l^2$	$0.7\,l$

Special loading cases

(7) Pin-ended strut under end load P_1 and central load P_2		F_1 F_2 $F_1 + F_2$	$(F_1 + F_2)_{cr} = \pi^2 E I / (kl)^2$ where $k \simeq 1/(2 - c^2)$ $c = F_1 / (F_1 + F_2)$	
(8) Cantilever strut under uniformly distributed load q/unit length		q	$(q_{cr})l = \pi^2 E I / (1.122\,l)^2$	

The above type of buckling is called *flexural buckling*, and occurs when the cross-section of the strut has two axes of symmetry. For unsymmetrical sections, buckling may be accompanied by torsion as well as flexure, producing a correspondingly reduced critical load. Results for such cases are given by Bleich.[3]

2.3.10.3 Formulae for the strength of columns

The plot of σ_{cr} versus λ for various column lengths is the hyperbola shown in Figure 2.51. Clearly, when λ is very small, the critical stress becomes much greater than the yield stress σ_Y of the material, and the failure of the column is brought about by the yielding of the material rather than by flexural buckling. If the columns were perfectly straight and the axial load had no eccentricity then the ultimate stresses σ_u would be given by the upper curve in Figure 2.51, i.e. the elastic buckling hyperbola intersected by the horizontal 'squash' line. However, tests show that the strengths of real columns are considerably reduced by initial imperfections when $\sigma_{cr} \simeq \sigma_Y$ as indicated by the lower curve in the figure. The following semi-empirical formulae have been devised to account for this, giving the ultimate stresses of columns in terms of their geometrical and material properties.

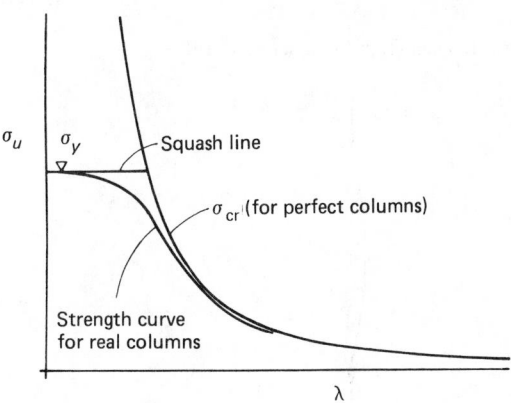

Figure 2.51

The Rankine formula.[17] A simple interaction formula relating σ_u, σ_Y and σ_{cr} is as follows:

$$\frac{1}{\sigma_u} = \frac{1}{\sigma_{cr}} + \frac{1}{\sigma_Y} \qquad (2.187)$$

gives:

$$\sigma_u = \frac{\sigma_Y}{1 + \dfrac{\sigma_Y \lambda^2}{\pi^2 E}} \qquad (2.188)$$

The interaction curve is tangential to the squash line at $\lambda = 0$, and to the buckling hyperbola at $\lambda = \infty$.

The Johnson parabola.[17] The formula:

$$\sigma_u = \sigma_Y \left(1 - \frac{\sigma_Y \lambda^2}{4\pi^2 E} \right) \qquad (2.189)$$

gives a parabolic interaction curve in the nonelastic range which is tangential to the squash line at $\lambda = 0$, and to the buckling hyperbola at the point $\sigma_{cr} = \frac{1}{2}\sigma_Y$.

The secant formula. The secant formula is derived assuming that the axial forces on the column have an initial *eccentricity e* (Figure 2.52(a)). In this case it can be shown that:

$$\sigma_u = \frac{\sigma_Y}{1 + \eta \sec\left[(\pi/2)\sqrt{(\sigma_u/\sigma_{cr})}\right]} \qquad (2.190)$$

where η is given by:

$$\eta = ec/r_y^2 \qquad (2.191)$$

c is the distance from the neutral axis to the extreme fibre of the section.

The Perry–Robertson formula. Assuming that the column has an initial *curvature* and that its maximum misalignment is e (Figure 2.52(b)), Ayrton and Perry derived the following formula:

$$\sigma_u = \frac{1}{2}[\sigma_Y + (1+\eta)\,\sigma_{cr}] - \left[\left(\frac{\sigma_Y + (1+\eta)\,\sigma_{cr}}{2}\right)^2 - \sigma_Y\sigma_{cr}\right]^{1/2} \qquad (2.192)$$

where η is again given by Equation (2.191).

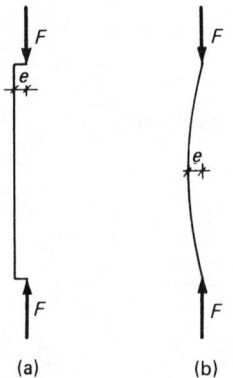

Figure 2.52

Robertson showed by experiment that a good but conservative prediction of the real strengths of columns can be obtained by making η proportional to λ, as follows:

$$\eta = 0.003\,\lambda \qquad (2.193)$$

Later experiments by Dutheil[17] led to the modified expression

$$\eta = 0.3\,(\lambda/100v)^2 \qquad (2.194)$$

2.3.10.4 Codes of practice for the design of columns

Section 2.3.10.3 summarizes the bases of simple empirical formulae for the strengths of columns. Current and projected codes of practice are somewhat more complicated, attempting to allow for the effects of variations in cross-sectional geometry and of residual stresses due to rolling and welding.

The *British* codes of practice are based on the Perry–Robert-

son formula. In the current standard for the design of steel bridges,[18] compression members are designed for η in Equation (2.191) which is linearly related to λ and a parameter α as follows:

$$\eta = 0 \qquad (\lambda < \lambda_0) \qquad (2.195)$$

$$\eta = 0.001\alpha\,(\lambda - \lambda_0) \quad (\lambda > \lambda_0) \qquad (2.196)$$

where λ_0 is the slenderness ratio below which the members are assumed to reach their full squash load. This is given as $0.2\,\lambda_1$ where $\lambda_1\ (= \pi\sqrt{E/\sigma_Y})$ is the slenderness ratio for which the critical stress is equal to the yield stress. Four curves for σ_u are presented, curves A, B, C and D corresponding to $\alpha = 2.5$, 4.5, 6.2 and 8.3 respectively. These are shown in a nondimensional plot in Figure 2.53. The curves appropriate for various cross-sections and fabrication methods are then selected according to

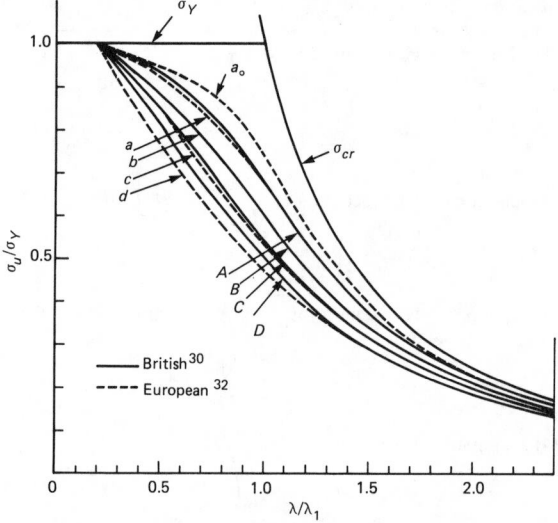

Figure 2.53 British and European column strength curves

Table 2.8.† The revised standard for steelwork in buildings,[19] adopts a similar approach, with slight differences in α for the different cases.

The *European Recommendations for Steel Construction*,[20] published by the European Convention for Structural Steelwork (ECSS) employ three basic column strength curves a, b and c again describing the strengths of groups of rolled and welded columns with various cross-sections. These curves are included as broken lines in Figure 2.53. The additional curves a_0 and d respectively deal with heat-treated sections in high-strength steel, and with sections with particularly thick plates (> 40 mm). For welded sections the effective value of the yield stress is reduced by 6%. An extended account of the reasoning behind the Recommendations is given in Chapters 2 and 3 of the Second International Colloquium report.[21]

The current *American* codes of practice are based on the Johnson parabola. Thus the American Institute of Steel Construction[22] recommend that the *allowable* stresses are obtained by dividing the interaction curve given by Equation (2.189) by a safety factor ϕ which depends on the slenderness ratio. Thus defining λ_2 to be the slenderness ratio for which $\sigma_{cr} = \frac{1}{2}\sigma_Y$, then

† Extracts from BS 5400: Part 3:1982 are reproduced by permission of the British Standards Institution, 2 Park Street, London, W1A 2BS from whom complete copies of the standard can be obtained.

Table 2.8 Selection of British column strength curves. British Standards Institution (1982) *Steel, concrete and composite bridges, BS 5400: Part 3.* BSI, Milton Keynes)

	Members fabricated by welding (excluding local welding of battens, lacing, etc.)	All other members (including stress relieved welded members)
$r_y/c \geqslant 0.7$	curve *B*	curve *A*
$r_y/c = 0.60$	curve *C*	curve *B*
$r_y/c = 0.50$	curve *C*	curve *B*
$r_y/c \leqslant 0.45$	curve *C*	curve *C*
All-rolled sections with flange thickness > 40 mm	curve *D*	
Hot-finished hollow sections	curve *A*	

Notes: (a) For intermediate values of r_y/c, linear interpolation may be used between the curves given.
   (b) *c* is defined as for Equation (2.191).

$$\phi = \tfrac{5}{3} + \tfrac{3}{8}\left(\frac{\lambda}{\lambda_2}\right)^4 - \tfrac{1}{8}\left(\frac{\lambda}{\lambda_2}\right)^{3/2} \quad (\lambda < \lambda_2) \tag{2.197}$$

$$\phi = \tfrac{23}{12} \quad (\lambda > \lambda_2) \tag{2.198}$$

For slender bracing and secondary members for which $\lambda > 120$, the allowable stresses may be divided by $(1.6 - \lambda/200)$, giving stresses similar to those of the Rankine formula. The Structural Stability Research Council (SSRC)[23] describe three column-strength curves (1), (2) and (3) each one representing the computed strength of a group of rolled or welded sections with realistic residual stresses and an initial bow of $l/1000$. These are shown in Figure 2.54.

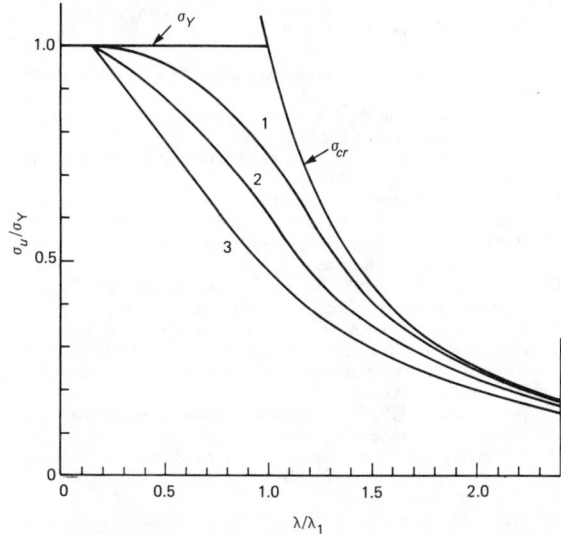

Figure 2.54 American column strength curves

2.3.11 Virtual work and strain energy of frameworks

The state of stress and strain at all points in a framework can be expressed in terms of the stress resultants at those points, using the appropriate equations of the previous sections and the stress–strain relations. The internal virtual work done in a framework corresponding to the general expression in Equation (2.76) is then given by:

$$W_i^* = \sum_{\text{bars}} \int_0^l \left(\frac{NN^*}{EA} + \frac{M_y M_y^*}{EI_y} + \frac{M_z M_z^*}{EI_z} + \frac{k_y S_y S_y^*}{GA} \right.$$

$$\left. + \frac{k_z S_z S_z^*}{GA} + \frac{TT^*}{G} \right) \, dx \tag{2.199}$$

where k_y and k_z are dimensionless form factors depending on the shape of the bar cross-section at each point in the framework. Values of the form factors for some common cross-sections are given in Table 2.9.

Table 2.9 Form factors

Section	k_y	k_z
1 Rectangle	1.20	
2 Circle	1.11	
3 Hollow circle	2.00	
4 I-section or hollow rectangle (approx.)	A/A_{flanges}	A/A_{webs}

Similarly the internal strain energy of a framework corresponding to the expression in Equation (2.78) is given by:

$$U = \sum_{\text{bars}} \int_0^l \left(\frac{N^2}{2EA} + \frac{M_y^2}{2EI_y} + \frac{M_z^2}{2EI_z} + \frac{k_y S_y^2}{2GA} \right.$$

$$\left. + \frac{k_z S_z^2}{2GA} + \frac{T^2}{2GJ} \right) \, dx \tag{2.200}$$

2.3.12 Note on the limitations of the engineering theory of the bending of beams (ETBB)

As noted in section 2.3.6, the basic assumptions of the ETBB, while quite correct when the beam is subject to pure bending, become invalid when the beam is also subject to shear. In particular, we can no longer assume that plane sections remain plane.

Some indication of the error involved in using the ETBB is obtained by analysing a thin-walled deep cantilever beam. Treating this as a plane stress problem, a complete solution is possible subject only to certain assumptions regarding the fixity at the encastre end.[8] Thus it can be shown that if the cantilever is loaded by a single vertical load *F* at its end so that the shear stress resultant is uniform along the length, the direct and shear stresses given by the ETBB are *exact*. However, the deflections u_x and u_z are given by:

$$u_x = \frac{F}{2EI_y}(-2lx + x^2)z + \frac{\nu F z^3}{6EI_y} - \frac{F z^3}{6GI_y} \tag{2.201}$$

$$u_z = \frac{F}{6EI_y}(3lx^2 - x^3) + \frac{\nu F}{2EI_y}(l - x)z^2 + \frac{F d^2 x}{8GI_y} \tag{2.202}$$

and the corresponding deflected shape of the beam is composed

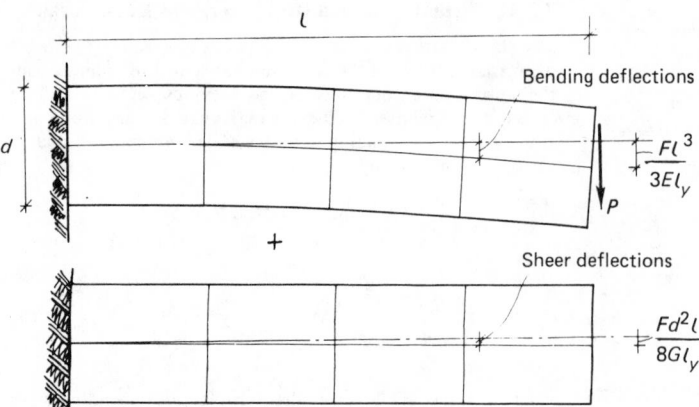

Figure 2.55

of two components as shown in Figure 2.55. One is the curved shape predicted by the ETBB, while the other is a linear vertical displacement due to the shear with the original plane cross-section taking up an S-shape in side view.

If the cantilever is loaded by a uniformly distributed load F/unit length so that the shear-stress resultant varies with x then the *stresses* given by the ETBB are also slightly inaccurate. However, it can be shown that the error is small, provided the span of the beam is large compared with its depth. Further the curvature of the beam is modified from Equation (2.119) to:

$$\frac{1}{R_y} = \left[\frac{M_y}{EI_y} + \frac{F}{EI_y} \frac{d^2}{4} \left(\frac{4}{5} + \frac{v}{2} \right) \right] \tag{2.203}$$

where the second term on the right-hand side represents the effect of the shear forces.

The preceding discussion concerns the behaviour of the webs of beams. However, in the flanges as well, it can be shown that plane sections no longer remain plane when beams are subject to shear. This phenomenon is called *shear lag*. It can be conveniently illustrated by the T-section cantilever shown in Figure 2.56(a). According to St Venant's theory (section 2.3.6), the forces in the flange are transmitted by longitudinal shear across the section A–B, so that the flange can be considered to behave like the cantilever plate shown in Figure 2.56(b) subjected to the uniformly distributed axial load along its centreline. It is then clear that the corresponding displacements u_x and the axial stress σ_x are nonuniform across the width of the flange. The

analysis of shear lag for practical cases is complex, and the topic is dealt with at some length by Williams.[24]

Further departures from the ETBB occur when beams become geometrically unstable. This instability can take the form of local compressive buckling of the flanges,[3] local shear buckling of the webs[3] and overall torsional buckling.[3]

References

1   Love, A. E. H. (1944) *The mathematical theory of elasticity.* (4th edn.), Dover, New York pp. 1–31.
2   Sokolnikoff, I. S. (1951) *Mathematical theory of elasticity.* (2nd ed.), McGraw-Hill, New York pp. 29–33; pp. 249–376.
3   Bleich, F. (1952) *Buckling strength of metal structures.* McGraw-Hill, New York; *ibid.*, pp. 104–147; *ibid.*, pp. 302–357; *ibid.*, 386–428; *ibid.*, 149–166.
4   Allen, H. G. and Bulson, P. S. (1980) *Background to buckling.* McGraw-Hill, Maidenhead.
5   Bisplinghoff, R. L., Mar, J. W. and Pian, T. H. H. (1965) *Statics of deformable solids*, Addison-Wesley, Reading, Mass., pp. 206–230.
6   Prager, W. and Hodge, P. G. (1951) *Theory of perfectly plastic solids*. Wiley, New York.
7   Dugdale, D. S. and Ruiz, C. (1971) *Elasticity for engineers.* McGraw-Hill, Maidenhead, pp. 155–194.
8   Timoshenko, S. P. and Goodier, J. N. (1951) *Theory of elasticity.* 2nd edn., McGraw-Hill, New York, pp. 29–130; *ibid.*, p.245; *ibid.*, pp. 258–315.
9   Washizu, K. (1968) *Variational methods in elasticity and plasticity.* Pergamon, Oxford.
10 Croll, J. G. A. and Walker, A. C. (1972) *Elements of structural stability*, Macmillan, London.
11 Thompson, J. M. T. and Hunt, G. W. (1973) *A general theory of elastic stability*, Wiley, London.
12 Frocht, M. M. (1941) *Photoelasticity*, 2 vols, Wiley, New York.
13 Hendry, A. W. (1966) *Photoelastic analysis*, Pergamon, Oxford.
14 British Standards Institution, (1980) *Structural steel sections*, BS 4: Part 1 BSI, Milton Keynes.
15 Timoshenko, S. P. (1955) *Strength of materials*, (3rd edn), Van Nostrand Reinhold, Princeton, p. 114; *ibid.*, pp. 149–170.
16a Case, J. and Chilver, A. H. (1971) *Strength of materials and structures*, Arnold, London, pp. 225–262.
16b Hetényi, M. (1946) *Beams on elastic foundations.* University of Michigan, Ann Arbor.
17 Godfrey, G. B. (1962) 'The allowable stresses in axially loaded steel struts', *Structural Engineer*, **40**, pp. 97–112.
18 British Standards Institution (1982) *Steel, concrete and composite bridges*, Part 3, *Code of practice for the design of steel bridges*, BS 5400. BSI, Milton Keynes.
19 British Standards Institution (1985) *Structural use of steelwork in building*, Part 1, *Code of practice for design in simple and continuous construction*, BS 5950. BSI, Milton Keynes.

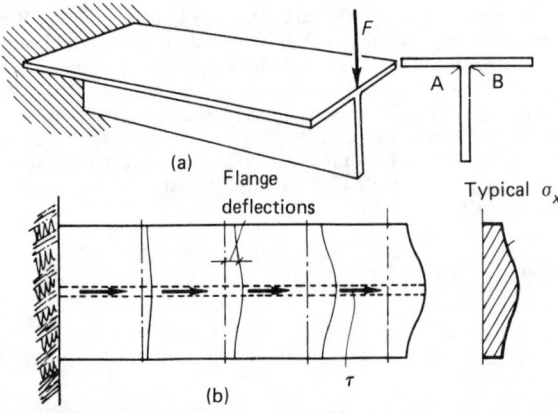

Figure 2.56

20 European Convention for Constructional Steelwork (1978) *European recommendations for steel construction*, ECCS, Milan, pp. 25–43.
21 *Manual on the stability of steel structures*, (1977) Introductory Report of the Second International Colloquium on Stability, ECCS/IABSE/SSRC/Col. Res. Committee of Japan, Tokyo (Sept. 1976).
22 American Institute of Steel Construction (1969). *Specification for the design, fabrication and erection of structural steel for buildings*, American Institute of Steel Construction, New York, pp. 5–16.
23 Johnston, B. G. (ed.), (1976) *Guide to stability design criteria for metal structures*, 3rd edn., Wiley/SSRC, New York, pp. 64–73.
24 Williams, D. (1960) *An introduction to the theory of aircraft structures*, Arnold, London, pp. 233–281.

Further reading

Den Hartog, J. P. (1952) *Advanced strength of materials*, McGraw-Hill, New York.

Dugdale, D. S. (1968) *Elements of elasticity*, Pergamon, Oxford.

Durelli, A. J., Phillips, E. A. and Tsao, C. H. (1958) *Introduction to the theoretical and experimental analysis of stress and strain*, McGraw-Hill, New York.

Graves Smith, T. R. (1974) *Stress and strain*, Chatto and Windus, London.

Green, A. E. and Zerna, W. (1968) *Theoretical elasticity*, 2nd edn., Oxford University Press.

Hetényi, M. (1950) *Handbook of experimental stress analysis*, Wiley, London.

Heywood, R. B. (1969) *Photoelasticity for designers*, Pergamon, Oxford.

Hoff, N. J. (1956) *The analysis of structures*, Wiley, New York.

Langhaar, H. L. (1962) *Energy methods in applied mechanics*, Wiley, New York.

Popov, E. P. (1952) *Mechanics of materials*, MacDonald, London.

Roark, R. J. (1954) *Formulas for stress and strain*, 3rd edn., McGraw-Hill, Maidenhead.

Shanley, F. R. (1957) *Strength of materials*, McGraw-Hill, New York.

Timoshenko, S. P. and Gere, J. M. (1961) *Theory of elasticity stability*, 2nd end., McGraw-Hill, New York.

3

Theory of Structures

W M Jenkins BSc, PhD, CEng, FICE, FIStructE
Emeritus Professor of Civil Engineering, School of Engineering, The Hatfield Polytechnic

Contents

3.1 Introduction

3.1.1 Basic concepts

The 'Theory of Structures' is concerned with establishing an understanding of the behaviour of structures such as beams, columns, frames, plates and shells, when subjected to applied loads or other actions which have the effect of changing the state of stress and deformation of the structure. The process of 'structural analysis' applies the principles established by the Theory of Structures, to analyse a given structure under specified loading and possibly other disturbances such as temperature variation or movement of supports. The drawing of a bending moment diagram for a beam is an act of structural analysis which requires a knowledge of structural theory in order to relate the applied loads, reactive forces and dimensions to actual values of bending moment in the beam. Hence 'theory' and 'analysis' are closely related and in general the term 'theory' is intended to include 'analysis'.

Two aspects of structural behaviour are of paramount importance. If the internal stress distribution in a structural member is examined it is possible, by integration, to describe the situation in terms of 'stress resultants'. In the general three-dimensional situation, these are six in number: two bending moments, two shear forces, a twisting moment and a thrust. Conversely, it is, of course, possible to work the other way and convert stress-resultant actions (forces) into stress distributions. The second aspect is that of deformation. It is not usually necessary to describe structural deformation in continuous terms throughout the structure and it is usually sufficient to consider values of displacement at selected discrete points, usually the joints, of the structure.

At certain points in a structure, the continuity of a member, or between members, may be interrupted by a 'release'. This is a device which imposes a zero value on one of the stress resultants. A hinge is a familiar example of a release. Releases may exist as mechanical devices in the real structure or may be introduced, in imagination, in a structure under analysis.

In carrying out a structural analysis it is generally convenient to describe the state of stress or deformation in terms of forces and displacements at selected points, termed 'nodes'. These are usually the ends of members, or the joints and this approach introduces the idea of a structural element such as a beam or column. A knowledge of the forces or displacements at the nodes of a structural element is sufficient to define the complete state of stress or deformation within the element providing the relationships between forces and displacements are established. The establishment of such relationships lies within the province of the theory of structures.

Corresponding to the basic concepts of force and displacement, there are two important physical principles which must be satisfied in a structural analysis. The structure as a whole, and every part of it, must be in equilibrium under the actions of the force system. If, for example, we imagine an element, perhaps a beam, to be removed from a structure by cutting through the ends, the internal stress resultants may now be thought of as external forces and the element must be in equilibrium under the combined action of these forces and any applied loads. In general, six independent conditions of equilibrium exist; zero sums of forces in three perpendicular directions, and zero sums of moments about three perpendicular axes. The second principle is termed 'compatibility'. This states that the component parts of a structure must deform in a compatible way, i.e. the parts must fit together without discontinuity at all stages of the loading. Since a release will allow a discontinuity to develop, its introduction will reduce the total number of compatibility conditions by one.

3.1.2 Force–displacement relationships

A simple beam element AB is shown in Figure 3.1. The application of end moments M_A and M_B produces a shear force Q throughout the beam, and end rotations θ_A and θ_B. By the stiffness method (see page 3/11), it may be shown that the end moments and rotations are related as follows:

$$\left. \begin{array}{l} M_A = \dfrac{4EI\theta_A}{l} + \dfrac{2EI\theta_B}{l} \\[2mm] M_B = \dfrac{4EI\theta_B}{l} + \dfrac{2EI\theta_A}{l} \end{array} \right\} \tag{3.1}$$

Or, in matrix notation,

$$\begin{bmatrix} M_A \\ M_B \end{bmatrix} = \frac{2EI}{l} \begin{bmatrix} 2 & 1 \\ 1 & 2 \end{bmatrix} \begin{bmatrix} \theta_A \\ \theta_B \end{bmatrix}$$

which may be abbreviated to,

$$\mathbf{S} = \mathbf{k}\theta \tag{3.2}$$

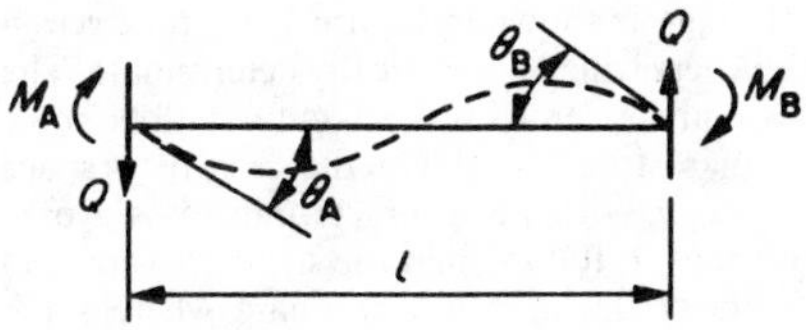

Figure 3.1

Equation (3.2) expresses the force–displacement relationships for the beam element of Figure 3.1. The matrices **S** and θ contain the end 'forces' and displacements respectively. The matrix **k** is the *stiffness matrix* of the element since it contains end forces corresponding to *unit* values of the end rotations.

The relationships of Equation (3.2) may be expressed in the inverse form:

$$\begin{bmatrix} \theta_A \\ \theta_B \end{bmatrix} = \frac{l}{6EI} \begin{bmatrix} 2 & -1 \\ -1 & 2 \end{bmatrix} \begin{bmatrix} M_A \\ M_B \end{bmatrix}$$

or

$$\theta = \mathbf{f}\mathbf{S} \tag{3.3}$$

Here the matrix **f** is the *flexibility matrix* of the element since it expresses the end displacements corresponding to *unit* values of the end forces.

It should be noted that an inverse relationship exists between **k** and **f**

i.e.

$$\mathbf{k}\mathbf{f} = I$$

or,

$$\mathbf{k} = \mathbf{f}^{-1} \tag{3.4}$$

or,

$$\mathbf{f} = \mathbf{k}^{-1}$$

The establishment of force–displacement relationships for structural elements in the form of Equations (3.2) or (3.3) is an important part of the process of structural analysis since the element properties may then be incorporated in the formulation of a mathematical model of the structure.

3.1.3 Static and kinematic determinacy

If the compatibility conditions for a structure are progressively reduced in number by the introduction of releases, there is reached a state at which the introduction of *one further* release would convert the structure into a mechanism. In this state the structure is *statically determinate* and the nodal forces may be calculated directly from the equilibrium conditions. If the releases are now removed, restoring the structure to its correct condition, nodal forces will be introduced which cannot be determined solely from equilibrium considerations. The structure is *statically indeterminate* and compatibility conditions are necessary to effect a solution.

The structure shown in Figure 3.2(a) is hinged to rigid foundations at A, C and D. The continuity through the foundations is indicated by the (imaginary) members, AD and CD. If the releases at A, C and D are removed, the structure is as shown in Figure 3.2(b) which is seen to consist of two closed rings. Cutting through the rings as shown in Figure 3.2(c) produces a series of simple cantilevers which are statically determinate. The number of stress resultants released by each cut would be three in the case of a planar structure, six in the case of a space structure. Thus, the degree of statical indeterminacy is 3 or 6 times the number of rings. It follows that the structure shown in Figure 3.2(b) is 6 times statically indeterminate whereas the structure of Figure 3.2(a), since releases are introduced at A, C and D, is 3 times statically indeterminate. A general relationship between the number of members m, number of nodes n, and degree of static indeterminacy n_s, may be obtained as follows:

$$n_s = \frac{6}{3}(m-n+1)-r \tag{3.5}$$

where r is the number of releases in the actual structure

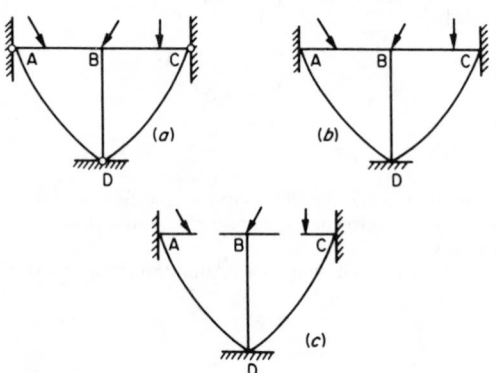

Figure 3.2

Turning now to the question of *kinematical determinacy*; a structure is defined as kinematically determinate if it is possible to obtain the nodal displacements from compatibility conditions without reference to equilibrium conditions. Thus a fixed-end beam is kinematically determinate since the end rotations are known from the compatibility conditions of the supports.

Again, consider the structure shown in Figure 3.2(b). The

structure is kinematically determinate except for the displacements of joint B. If the members are considered to have infinitely large extensional rigidities, then the rotation at B is the only unknown nodal displacement. The degree of kinematical indeterminacy is therefore 1. The displacements at B are *constrained* by the assumption of zero vertical and horizontal displacements. A *constraint* is defined as a device which constrains a displacement at a certain node to be the same as the corresponding displacement, usually zero, at another node. Reverting to the structure of Figure 3.2(a), it is seen that three constraints, have been removed by the introduction of hinges (releases) at A, C and D. Thus rotational displacements can develop at these nodes and the degree of kinematical indeterminacy is increased from 1 to 4.

A general relationship between the numbers of nodes n, constraints c, releases r, and the degree of kinematical indeterminacy n_k is as follows,

$$n_k = \frac{6}{3}(n-1)-c+r \tag{3.6}$$

The coefficient 6 is taken in three-dimensional cases and the coefficient 3 in two-dimensional cases. It should now be apparent that the modern approach to structural theory has developed in a highly organised way. This has been dictated by the development of computer-orientated methods which have required a re-assessment of basic principles and their application in the process of analysis. These ideas will be further developed in some of the following sections.

3.2 Statically determinate truss analysis

3.2.1 Introduction

A structural frame is a system of bars connected by joints. The joints may be, ideally, *pinned* or *rigid*, although in practice the performance of a real joint may lie somewhere between these two extremes. A *truss* is generally considered to be a frame with pinned joints, and if such a frame is loaded only at the joints, then the members carry axial tensions or compressions. *Plane* trusses will resist deformation due to loads acting in the plane of the truss only, whereas *space* trusses can resist loads acting in any direction.

Under load, the members of a truss will change length slightly and the geometry of the frame is thus altered. The effect of such alteration in geometry is generally negligible in the analysis.

The question of statical determinacy has been mentioned in the previous section where a relationship, Equation (3.5) was stated from which the degree of statical indeterminacy could be determined. Although this relationship is of general application, in the case of plane and space trusses, a simpler relationship may be established.

The simplest plane frame is a triangle of three members and three joints. The addition of a fourth joint, in the plane of the triangle, will require two additional members. Thus in a frame having j joints, the number of members is:

$$n = 2(j-3)+3 = 2j-3 \tag{3.7}$$

A truss with this number of members is statically determinate, providing the truss is supported in a statically determinate way. Statically determinate trusses have two important properties. They cannot be altered in shape without altering the length of one or more members, and, secondly, any member may be altered in length without inducing stresses in the truss, i.e. the truss cannot be *self stressed* due to imperfect lengths of members or differential temperature change.

The simplest space truss is in the shape of a tetrahedron with four joints and six members. Each additional joint will require three more members for connection with the tetrahedron, and thus:

$$n = 3(j-4) + 6 = 3j - 6 \qquad (3.8)$$

A space truss with this number of members is statically determinate, again providing the support system is itself statically determinate. It should be noted that in the assessment of the statical determinacy of a truss, member forces and reactive forces should all be considered when counting the number of unknowns. Since equilibrium conditions will provide two relationships at each joint in a plane truss (there is a space truss), the simplest approach is to find the total number of unknowns, member forces and reactive components, and compare this with 2 or 3 times the number of joints.

3.2.2 Methods of analysis

Only brief mention will be made here of the methods of statically determinate analysis of trusses. For a more detailed treatment the reader is referred to Jenkins[1] and Coates, Coutie and Kong.[2]

The *force diagram* method is a graphical solution in which a vector polygon of forces is drawn to scale proceeding from joint to joint. It is necessary to have not more than two unknown forces at any joint, but this requirement can be met with a judicious choice of order. The two conditions of overall equilibrium of the plane structure imply that the force vector polygon will form a closed figure. The method is particularly suitable for trusses with a difficult geometry where it is convenient to work to a scale drawing of the outline of the truss.

The method of *resolution at joints* is suitable for a complete analysis of a truss. The reactions are determined and then, proceeding from joint to joint, the vertical and horizontal equilibrium conditions are set down in terms of the member forces. Since two equations will result at each joint in a plane truss, it is possible to determine not more than two forces for each pair of equations. As an illustration of the method, consider the plane truss shown in Figure 3.3. The truss is symmetrically loaded and the reactions are clearly 15 kN each.

Consider the equilibrium of joint A,

vertically, $P_{AE} \cos 45° = R_A$; hence $P_{AE} = 15\sqrt{2}$ kN (compression)

horizontally, $P_{AC} = P_{AE} \cos 45°$; hence $P_{AC} = 15$ kN (tension)

It should be noted that the arrows drawn on the members in Figure 3.3 indicate the directions of forces acting *on the joints*. It is also seen that the directions of the arrows at joint A, for example, are consistent with equilibrium of the joint. Proceeding to joint C it is clear that $P_{CE} = 10$ kN (tension), and that $P_{CD} = P_{AC} = 15$ kN (tension). The remainder of the solution may be obtained by resolving forces at joint E, from which $P_{ED} = 5\sqrt{2}$ kN (tension) and $P_{EF} = 20$ kN (compression).

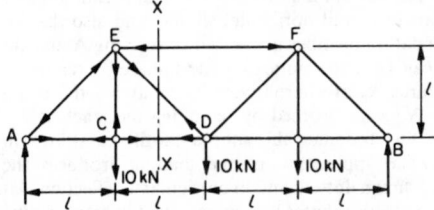

Figure 3.3

The *method of sections* is useful when it is required to determine forces in a limited number of the members of a truss. Consider, for example, the member ED of the truss in Figure 3.3. Imagine a cut to be made along the line XX and consider the vertical equilibrium of the part to the left of XX. The vertical forces acting are R_A, the 10 kN load at C and the vertical component of the force in ED. The equation of vertical equilibrium is:

$$15 - 10 = P_{ED} \cos 45° \quad \text{hence } P_{ED} = 5\sqrt{2} \text{ kN}$$

Since a downwards arrow on the left-hand part of ED is required for equilibrium, it follows that the member is in tension. The method of tension coefficients is particularly suitable for the analysis of space frames and will be outlined in the following section.

3.2.3 Method of tension coefficients

The method is based on the idea of systematic resolution of forces at joints. In Figure 3.4, let AB be any member in a plane truss, $T_{AB} =$ force in member (tension positive), and $L_{AB} =$ length of member.

We define:

$$T_{AB} = L_{AB} t_{AB} \qquad (3.9)$$

where $t_{AB} =$ tension coefficient.

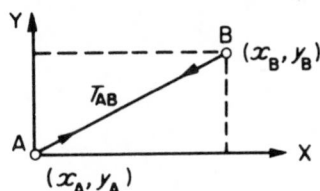

Figure 3.4

That is, the tension coefficient is the actual force in the member divided by the length of the member. Now, at A, the component of T_{AB} in the X-direction:

$$= T_{AB} \cos BAX$$

$$= T_{AB} \frac{(x_B - x_A)}{L_{AB}} = t_{AB}(x_B - x_A)$$

Similarly the component of T_{AB} in the Y-direction:

$$= t_{AB}(y_B - y_A)$$

At the other end of the member the components are:

$$t_{AB}(x_A - x_B), \; t_{AB}(y_A - y_B)$$

If at A the external forces have components X_A and Y_A, and if there are members AB, AC, AD etc. then the equilibrium conditions for directions X and Y are:

$$\left. \begin{array}{l} t_{AB}(x_B - x_A) + t_{AC}(x_C - x_A) + t_{AD}(x_D - x_A) + \dots + X_A = 0 \\ t_{AB}(y_B - y_A) + t_{AC}(y_C - y_A) + t_{AD}(y_D - y_A) + \dots + Y_A = 0 \end{array} \right\} \quad (3.10)$$

Similar equations can be formed at each joint in the truss. Having solved the equations, for the tension coefficients, usually a very simple process, the forces in the members are determined from Equation (3.9).

The extension of the theory to space trusses is straightforward. At each joint we now have three equations of equilibrium, similar to Equation (3.10) with the addition of an equation representing equilibrium in the Z direction:

$$t_{AB}(z_B - z_A) + t_{AC}(z_C - z_A) + \ldots + Z_A = 0$$

(3.11)

The method will now be illustrated with an example. The notation is simplified by writing AB in place of t_{AB} etc. A fabular presentation of the work is recommended.

Example 3.1. A pin-jointed space truss is shown in Figure 3.5. It is required to determine the forces in the members using the method of tension coefficients. We first check that the frame is statically determinate as follows:

Number of members = 6
Number of reactions = 9

Total number of unknowns = 15

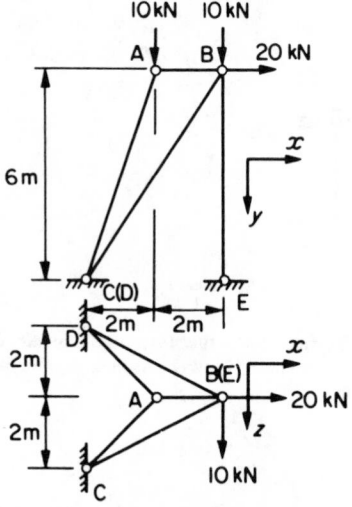

Figure 3.5

The number of equations available is 3 times the number of joints, i.e. $3 \times 5 = 15$. Hence, the truss is statically determinate. In counting the number of reactive components, it should be observed that all components should be included even if the particular geometry of the truss dictates (as in this case at E) that one or more components should be zero.

The solution is set out in Tables 3.1 and 3.2 where it should be noted that, in deriving the equations, the origin of coordinates is taken at the joint being considered. Thus, each tension coefficient is multiplied by the projection of the member on the particular axis.

The methods of truss analysis just outlined are suitable for 'hand' analysis, as distinct from computer analysis, and are useful in acquiring familiarity and understanding of structural behaviour. Much analysis of this kind is now carried out on computers (mainframe, mini- and microcomputers) where the stiffness method provides a highly organized and suitable basis. This topic will be further considered under the heading of the stiffness method.

Table 3.1

Joint	Direction	Equations	Solutions
A	x	$-2AC - 2AD + 2AB = 0$	$AC = AD = -\frac{10}{12}$
	y	$6AC + 6AD + 10 = 0$	$AB = -\frac{10}{6}$
	z	$2AC - 2AD = 0$	$-4BC - 4BD + \frac{10}{3}$ $+ 20 = 0$
			$2BC - 2BD + 10 = 0$
C	x	$-4BC - 4BD - 2AB$ $+ 20 = 0$	$BC = \frac{10}{24}$
	y	$6BC + 6BD + 6BE$ $+ 10 = 0$	$BD = \frac{130}{24}$
	z	$-2BD + 2BC + 10 = 0$	Hence $BE = -\frac{15}{2}$

Table 3.2

Member	Length (m)	Tension coefficient	Force (kN) (tension +)
AB	2	$-\frac{10}{6}$	-3.33
AC	6.62	$-\frac{10}{12}$	-5.52
AD	6.62	$-\frac{10}{12}$	-5.52
BC	7.48	$\frac{10}{24}$	$+3.12$
BD	7.48	$\frac{130}{24}$	$+40.5$
BE	6	$-\frac{15}{2}$	-45.0

3.3 The flexibility method

3.3.1 Introduction

The idea of statical determinacy was introduced previously (see page 3/4) and a relationship between the degree of statical indeterminacy and the numbers of members, nodes and releases was stated in Equation (3.5). A statically determinate structure is one for which it is possible to determine the values of forces at all points by the use of equilibrium conditions alone. A statically indeterminate structure, by virtue of the number of members or method of connecting the members together, or the method of support of the structure, has a larger number of forces than can be determined by the application of equilibrium principles alone. In such structures the force analysis requires the use of compatibility conditions. The flexibility method provides a means of analysing statically indeterminate structures.

Consider the propped cantilever shown in Figure 3.6(a). Applying Equation (3.5) the degree of statical indeterminacy is seen to be:

$$n_s = 3(2 - 2 + 1) - 2 = 1$$

(Note that two releases are required at B, one to permit angular rotation and one to permit horizontal sliding, and also that an additional foundation member is inserted connecting A and B.) The structure can be made statically determinate by removing the propping force R_B or alternatively by removing the fixing moment at A. We shall proceed by removing the reaction R_B. The structure thus becomes the simple cantilever shown in Figure 3.6(b). The application of the load w produces the deflected shape, shown dotted, and in particular a deflection u at the free end B. Note also that it is now possible to determine the bending moment at $A = wl^2/2$, by simple statical principles. The

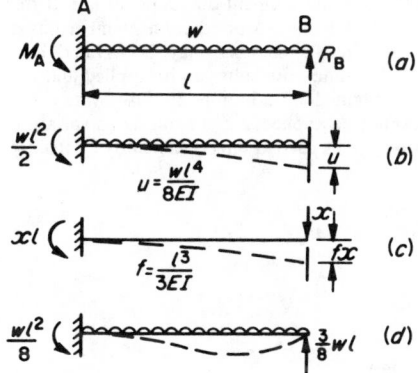

Figure 3.6 Basis of the flexibility method

$$\Delta_i = \int M \partial M / \partial F_i \; \frac{ds}{EI} \tag{3.13}$$

in which Δ_i is the displacement required, M is a function representing the bending moment distribution and F_i is a force, real or virtual, applied at the position and in the direction designated by i. It follows that $\partial M / \partial F_i$ can be regarded as the bending moment distribution due to unit value of F_i.

Consider the cantilever beam shown in Figure 3.7(a). Forces x_1 and x_2 act on the beam and it is required to determine influence coefficients corresponding to the positions and directions defined by x_1 and x_2. From now on we work with *unit* values of x_1 and x_2 and draw bending moment diagrams, as in Figure 3.7(b) and (c), due to unit values of x_1 and x_2 separately.

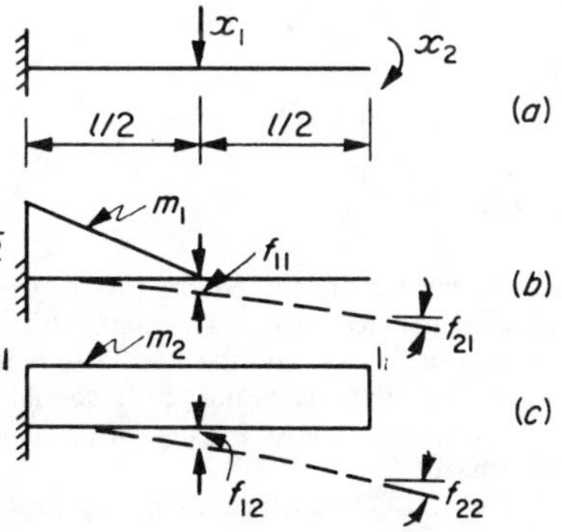

deflection u may be obtained from elementary beam theory as $wl^4/8EI$. We now remove the applied load w and apply the, unknown, redundant force x at B. It is unnecessary to know the sense of the force x; in this case we have assumed a downwards direction for positive x. The application of the force x produces a displacement at B which we shall call fx; i.e. a unit value of x would produce a displacement f. The compatibility condition associated with the redundant force x is that the final displacement at B should be zero, i.e.:

$$u + fx = 0 \tag{3.12}$$

and substituting values of u and f

$$x = -\tfrac{3}{8}wl$$

The process may be regarded as the superposition of the diagrams Figures 3.6(b) and (c) such that the final displacement at B is zero. The addition of the two systems of forces will also give values of bending moment throughout the beam, e.g. at A:

$$M_A = \frac{wl^2}{2} + xl$$

$$= \frac{wl^2}{2} - \tfrac{3}{8}wl^2 \qquad = \frac{wl^2}{8}$$

The actual values of reactions are as shown in Figure 3.6(d).

The displacement f is called a 'flexibility influence coefficient'. In general f_{rs} is the displacement in direction r in a structure due to unit force in direction s. The subscripts were omitted in the above analysis since the force and displacement considered were at the same position and in the same direction.

3.3.2 Evaluation of flexibility influence coefficients

As seen in the above example, flexibility coefficients are displacements calculated at specified positions, and directions, in a structure due to a prescribed loading condition. The loading condition is that of a single unit load replacing a redundant force in the structure. It should be remembered that at this stage the structure is, or has been made, statically determinate.

For simplicity we restrict our attention to structures in which flexural deformations predominate. The extension to other types of deformation is straightforward.[3] In the case of pure flexural deformation we may evaluate displacements by an application of Castigliano's theorem or use the principle of virtual work.[3] In either case a convenient form is:

Figure 3.7 Evaluation of flexibility coefficients

These are labelled m_1 and m_2. Consider the application of unit force at x_1 ($x_2 = 0$). Displacements will occur in the directions of x_1 and x_2. Applying Equation (3.13) the displacement in the direction of x_1 will be:

$$f_{11} = \int m_1 m_1 \frac{ds}{EI}$$

and in the direction of x_2:

$$\left. \begin{array}{c} \\ \\ \end{array} \right\} \tag{3.14}$$

$$f_{21} = \int m_2 m_1 \frac{ds}{EI}$$

Similarly, when we apply $x_2 = 1$, $x_1 = 0$, we obtain:

$$f_{22} = \int m_2 m_2 \frac{ds}{EI}$$

and:

$$\left. \begin{array}{c} \\ \\ \end{array} \right\} \tag{3.15}$$

$$f_{12} = \int m_1 m_2 \frac{ds}{EI}$$

The general form is:

$$f_{rs} = \int m_r m_s \frac{ds}{EI} \tag{3.16}$$

The evaluation of Equation (3.16) requires the integration of the product of two bending moment distributions over the complete structure. Such distributions can generally be represented by simple geometrical figures such as rectangles, triangles and parabolas and standard results can be established in advance. Table 3.3 gives values of product integrals for a range of combinations of diagrams. It should be noted that in applying Equation (3.16) in this way, the flexural rigidity EI is assumed constant over the length of the diagram.

We may now use Table 3.3 to obtain values of the flexibility coefficients for the cantilever beam under consideration. Using Equations (3.14) and (3.15) with Figures 3.7(b) and (c) we obtain:

$$f_{11} = \frac{1}{3} \cdot \frac{l}{2} \cdot \frac{l}{2} \cdot \frac{l}{2} \cdot \frac{1}{EI} = \frac{l^3}{24EI}$$

$$f_{21} = \frac{1}{2} \cdot \frac{l}{2} \cdot 1 \cdot \frac{l}{2} \cdot \frac{1}{EI} = \frac{l^2}{8EI}$$

$$f_{22} = l \cdot 1 \cdot 1 \cdot \frac{1}{EI} = \frac{l}{EI}$$

$$f_{12} = \frac{1}{2} \cdot \frac{l}{2} \cdot \frac{l}{2} \cdot 1 \cdot \frac{1}{EI} = \frac{l^2}{8EI}$$

It is seen that f_{21} and f_{12} are numerically equal, a result which could be established using the Reciprocal Theorem. This is a useful property since in general $f_{rs} = f_{sr}$ and the effect is to reduce the number of separate calculations required. It should be further noted that whilst $f_{21} = f_{12}$, f_{21} is an angular displacement and f_{12} a linear displacement.

The evaluation of the flexibility coefficients f_{rs} provides the displacements at selected points in the structure due to unit values of the associated, redundant, forces. Before the compatibility conditions can be written down, it remains to calculate displacements (u) at corresponding positions due to the actual applied load. The basic equation (Equation 3.13) is applied once more. Now the bending moment distribution M is that due to the applied loads and we will re-designate this m_0. As before, $\partial M / \partial F_i = m_i$, and thus:

$$u_i = \int m_0 m_i \frac{ds}{EI} \tag{3.17}$$

The table of product integrals, Table 3.3, can be used for evaluating the u_i in the same way as the f_{rs}.

Table 3.3

$$\frac{\text{Product integrals}}{(EI \text{ uniform})} \int_0^l m_r \, m_s \, ds$$

m_s \ m_r	a ▭	a ◺	a ▱ b
▭ c	lac	$\frac{1}{2}ac$	$\frac{l}{2}(a+b)c$
◣ c	$\frac{l}{2}ac$	$\frac{l}{3}ac$	$\frac{l}{6}(2a+b)c$
◥ c	$\frac{l}{2}ac$	$\frac{l}{6}ac$	$\frac{l}{6}(a+2b)c$
▱ c d	$\frac{l}{2}a(c+d)$	$\frac{l}{6}a(2c+d)$	$\frac{l}{6}\{a(2c+d)+b(2d+c)\}$
◠ c	$\frac{2}{3}lac$	$\frac{l}{3}ac$	$\frac{l}{3}(a+b)c$

In cases where the bending moment diagrams do not fit the standard values given in Table 3.3 or where a member has a stepped variation in EI, the member may be divided into segments such that the standard results can be applied and the total displacement obtained by addition. In cases where the standard results cannot be applied, e.g. a continuous variation in EI, the integration can be carried out conveniently by the use of Simpson's rule:

$$\int m_r m_s \frac{ds}{EI} \simeq \frac{a}{3}(h_1 + 4h_2 + 2h_3 + 4h_4 + \ldots + h_n)$$

where a = width of strip

$$h_i = \frac{m_r m_s}{EI} \text{ at section i.}$$

In using Simpson's rule it should be remembered that the number of strips must be even, i.e. n must be odd.

3.3.2.1 Sign convention

A flexibility coefficient will be positive if the displacement it represents is in the same sense as the applied, unit, force. The bending moment expressions must carry signs based on the type of curvature developing in the structure. Since the integrand in Equation (3.16) is always the product of two bending moment expressions, it is only the relative sign which is of importance. A useful convention is to draw the diagrams on the tension (convex) sides of the members and then the relative signs of m_r and m_s can readily be seen. In Figure 3.7(b) and (c), both the m_1 and m_2 diagrams are drawn on the top side of the member. Their product is therefore positive. Naturally, the product of one diagram and itself will always be positive. This follows from simple physical reasoning since the displacement at a point due to an applied force at the same point will always be in the same sense as the applied force.

3.3.3 Application to beam and rigid frame analysis

The application of the theory will now be illustrated with two examples.

Example 3.2. Consider the three-span continuous beam shown in Figure 3.8(a). The beam is statically indeterminate to the second degree and we shall choose as redundants the internal bending moments at the interior supports B and C. The beam is made statically determinate by the introduction of moment releases at B and C as in Figure 3.8(b). We note that the application of the load W now produces displacements in span BC only, and in particular rotations u_1 and u_2 at B and C. The bending moment diagram (m_0) is shown in Figure 3.8(c).

We now apply unit value of x_1 and x_2 in turn. The deflected shapes and the flexibility coefficients, in the form of angular rotations, are shown at (d) and (e). The bending moment diagrams m_1 and m_2 are shown at (f) and (g).

Using the table of product integrals (Table 3.3), we find:

$$EI f_{11} = \frac{2}{3}l$$

$$EI f_{22} = \frac{2}{3}l$$

$$EI f_{12} = EI f_{21} = \frac{l}{6}$$

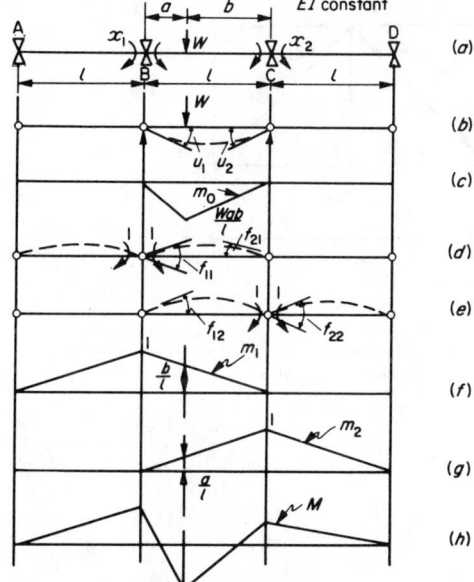

Figure 3.8 Flexibility analysis of continuous beam

$$EI u_1 = -\frac{a}{6}\left(1+\frac{2b}{l}\right)\frac{Wab}{l}-\frac{b}{3}\cdot\frac{b}{l}\cdot\frac{Wab}{l}$$

$$= -\frac{Wab}{6l}(a+2b)$$

and

$$EI u_2 = -\frac{Wab}{6l}(b+2a)$$

The required compatibility conditions are, for continuity of the beam:

at B, $f_{11}x_1+f_{12}x_2+u_1=0$
at C, $f_{21}x_1+f_{22}x_2+u_2=0$

or, in matrix form:

$$\mathbf{FX+U=0} \tag{3.18}$$

i.e.:

$$\frac{l}{6EI}\begin{bmatrix}4 & 1\\1 & 4\end{bmatrix}\begin{bmatrix}x_1\\x_2\end{bmatrix}=\frac{Wab}{16EIl}\begin{bmatrix}(a+2b)\\(b+2a)\end{bmatrix}$$

and the solutions are:

$$\begin{bmatrix}x_1\\x_2\end{bmatrix}=\frac{Wab}{15l^2}\begin{bmatrix}(2a+7b)\\(2b+7a)\end{bmatrix}$$

The actual bending moment distribution may now be determined by the addition of the three systems, i.e. the applied load and the two redundants. The general expression is:

$$M=m_0+m_1x_1+m_2x_2 \tag{3.19}$$

In particular:

$$M_B=x_1=\frac{Wab}{15l^2}(2a+7b)$$

$$M_C=x_2=\frac{Wab}{15l^2}(2b+7a)$$

and the bending moment under the load W is:

$$M_W=-\frac{Wab}{l}+\frac{b}{l}x_1+\frac{a}{l}x_2$$

$$=-\frac{2Wab}{15l^3}(4l^2+5ab)$$

The final bending moment diagram is shown in Figure 3.8(h).

Example 3.3. A portal frame ABCD is shown in Figure 3.9(a). The frame has rigid joints at B and C, a fixed support at A and a hinged support at D. The flexural rigidity of the beam is twice that of the columns.

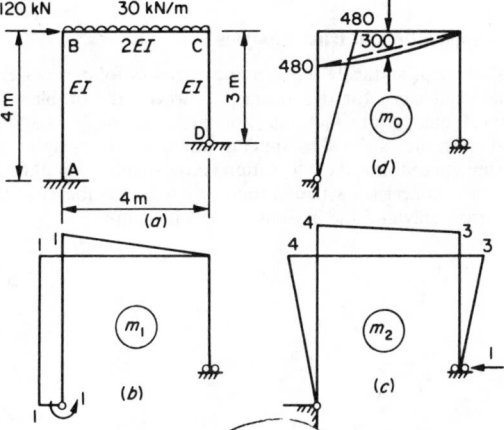

Figure 3.9

The frame has two redundancies and these are taken to be the fixing moment at A and the horizontal reaction at D. The bending moment diagrams corresponding to the unit redundancies, m_1 and m_2 and the applied load, m_0, are shown at (b), (c) and (d) in Figure 3.9.

Using the table of product integrals, Table 3.3, we obtain:

$$f_{11}=\int m_1^2\frac{ds}{EI}=\frac{14}{3EI}$$

$$f_{22}=\int m_2^2\frac{ds}{EI}=\frac{55}{EI}$$

$$f_{12}=f_{21}=\int m_1 m_2\frac{ds}{EI}=\frac{35}{3EI}$$

$$u_1=\int m_0 m_1\frac{ds}{EI}=-\frac{1320}{EI}$$

$$u_2=\int m_0 m_2\frac{ds}{EI}=-\frac{4600}{EI}$$

Thus the compatibility equations are:

$$\frac{1}{3}\begin{bmatrix}14 & 35\\35 & 165\end{bmatrix}\begin{bmatrix}x_1\\x_2\end{bmatrix}=+\begin{bmatrix}1320\\4600\end{bmatrix}$$

from which

$$x_1 = +157 \, \text{kNm}$$

and

$$x_2 = + \; 50 \, \text{kN}$$

The bending moment at any point in the frame may now be determined from the expression:

$$M = m_0 + m_1 x_1 + m_2 x_2$$

e.g.:

$$M_{BA} = 480 - 1(+157) - 4(+50) = 123 \, \text{kNm}$$

and

$$M_{CD} = 3x_2 = 150 \, \text{kNm}$$

3.3.4 Application to truss analysis

The analysis of statically indeterminate trusses follows closely on that established for rigid frames; however, the problem is simplified due to the fact that for each system of loading investigated, the axial forces are constant within the lengths of the members and thus the integration is considerably simplified. We are now concerned with deformations in the members due to axial forces only and the flexibility coefficients are:

$$f_{rs} = \sum p_r \, p_s \frac{l}{AE} \tag{3.20}$$

and

$$u_i = \sum p_0 \, p_i \frac{l}{AE} \tag{3.21}$$

in which the p_r system of forces is due to unit tension in the rth redundant member and similarly for p_s and p_i. The p_0 system of forces is that due to the applied load system acting on the statically determinate structure (i.e. with the redundant members omitted). Equations (3.20) and (3.21) should be compared with Equations (3.16) and (3.17) in the flexural case.

Example 3.4. The plane truss shown in Figure 3.10 has two redundancies which we will choose as the forces in members AE and EC. AE is constant for all the members and equal to $1 \times 10^6 \, \text{kN}$. The member EC is $l/10\,000$ short in manufacture and has to be forced into position. The member force systems p_0, p_1 and p_2 are found from a simple statical analysis and are listed in Table 3.4.

The flexibility coefficients may now be obtained as follows:

$$f_{11} = \sum p_1 p_1 \frac{l}{AE} = \frac{2l}{AE}(1 + \sqrt{2})$$

$$f_{22} = f_{11}$$

$$f_{12} = f_{21} = \sum p_1 p_2 \frac{l}{AE} = \frac{l}{2AE}$$

$$u_1 = \sum p_1 p_0 \frac{l}{AE} = \frac{Wl}{AE}(1 + 1/\sqrt{2})$$

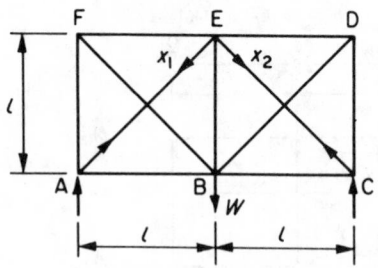

Figure 3.10

Table 3.4

Member	Length	p_0/w	p_1	p_2
AB	l	0	$-1/\sqrt{2}$	0
BC	l	0	0	$-1/\sqrt{2}$
CD	l	$-1/2$	0	$-1/\sqrt{2}$
DE	l	$-1/2$	0	$-1/\sqrt{2}$
EF	l	$-1/2$	$-1/\sqrt{2}$	0
AF	l	$-1/2$	$-1/\sqrt{2}$	0
FB	$\sqrt{(2)}l$	$1/\sqrt{2}$	1	0
BE	l	0	$-1/\sqrt{2}$	$-1/\sqrt{2}$
BD	$\sqrt{(2)}l$	$1/\sqrt{2}$	0	1
AE	$\sqrt{(2)}l$	0	1	0
EC	$\sqrt{(2)}l$	0	0	1

Ignoring, for the moment, the effect of the shortness in length of member EC, the compatibility equations are:

$$f_{11}x_1 + f_{12}x_2 + u_1 = 0$$

$$f_{21}x_1 + f_{22}x_2 + u_2 = 0$$

Clearly the symmetry will produce $x_1 = x_2$ and thus:

$$x_1 = x_2 = -W\frac{(2 + \sqrt{2})}{(5 + 4\sqrt{2})}$$

The effect of the prestrain caused by the forced fit of member EC may be obtained by putting:

$$U = -\begin{bmatrix} 0 \\ 10^{-4}l \end{bmatrix} \tag{3.22}$$

and then solving $\mathbf{FX} + \mathbf{U} = 0$ obtaining:

$$x_1 = \frac{-200}{(47 + 32\sqrt{2})} \, \text{kN}$$

$$x_2 = \frac{800(1 + \sqrt{2})}{(47 + 32\sqrt{2})} \, \text{kN}$$

The forces in the other members may now be obtained from $p = p_0 + p_1 x_1 + p_2 x_2$.

The sign of the lack of fit in Equation (3.22) should be studied carefully and it should be noted that the convention for the signs of forces is tension-positive throughout.

3.3.5 Comments on the flexibility method

For a more detailed treatment of the flexibility method the reader may consult any of the standard texts, e.g. Jenkins[1] and

Coates, Coutie and Kong.[2] The method has declined in popularity in recent years due to the widespread adoption of computerized methods based on *stiffness* concepts. In the context of automatic computation, the stiffness method, which will be considered in the next section, offers considerable advantages over the flexibility method. Methods based on flexibility offer some advantage for hand computation in structures with low (1 or 2) degrees of statical indeterminacy or with lack of fit, temperature change or flexible supports. The concept of flexibility influence coefficients is also useful in determining stiffness coefficients, e.g. in nonprismatic members.

3.4 The stiffness method

3.4.1 Introduction

This method has been very extensively developed in recent years and now forms the basis of most structural analysis carried out on digital computers. The method of 'slope-deflection' is an example of the application of the general stiffness method.

Consider the structure shown in Figure 3.11(a) which is fixed at A and C and has a rigid joint at B. The degree of kinematical indeterminacy, from Equation (3.6), is:

$$n_k = 3(n-1) - c + r$$

$$= 3(3-1) - 5 + 0$$

$$= 1$$

The five constraints are the zero displacements, three at C and two at B, related to the fixed point A. The single unknown

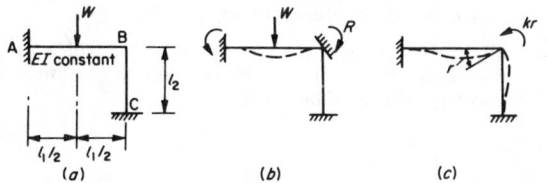

Figure 3.11 Basis of the stiffness method

displacement, nodal degree of freedom is, of course, the rotation of the joint B.

The procedure is to clamp the joint B so constraining the nodal degree of freedom r. On applying the load W, a constraining force, R, will be required at B to prevent the rotation of the joint. The constraining force R is now applied to the, otherwise unloaded, structure with its sign reversed and the nodal degree of freedom released. The result is a rotation of joint B through angle r. The external moment required to effect this rotation is kr where k is the stiffness of the structure for this particular displacement. Thus, for equilibrium:

$$kr = R \qquad (3.23)$$

From the table of fixed-end moments, Table 3.5:

$$R = \frac{Wl_1}{8}$$

and from the force–displacement relationships of Equation (3.1)

$$k = \frac{4EI}{l_1} + \frac{4EI}{l_2}$$

Thus:

$$4EI\left(\frac{1}{l_1} + \frac{1}{l_2}\right) r = \frac{Wl_1}{8}$$

Hence:

$$r = \frac{Wl_1^2 l_2}{32EI(l_1 + l_2)}$$

The member forces are now obtained by adding the two systems (b) and (c) in Figure 3.11, e.g.:

$$M_{BA} = \frac{Wl_1}{8} - \frac{4EI(r)}{l_1} = \frac{Wl_1}{8}\left(1 - \frac{l_2}{l_1 + l_2}\right)$$

$$= \frac{Wl_1^2}{8(l_1 + l_2)}$$

and

$$M_{BC} = -\frac{4EI(r)}{l_2} = -\frac{Wl_1^2}{8(l_1 + l_2)}$$

Note that in the above, clockwise moments are considered positive.

Table 3.5 Fix-end moments for uniform beams (clockwise moments positive)

M_{FL}	Loading	M_{FR}
$-\dfrac{Wab}{l}\left(\dfrac{b}{l}\right)$		$\dfrac{Wab}{l}\left(\dfrac{a}{l}\right)$
$-\dfrac{Wab}{2l^2}(a+2b)$		0
$-\dfrac{Wc}{12l^2}\left[12ab^2+c^2 (a-2b)\right]$		$\dfrac{Wc}{12l^2}\left[12a^2b+c^2 (b-2a)\right]$
$-\dfrac{wl^2}{12}$		$\dfrac{wl^2}{12}$
$-\dfrac{wl^2}{30}$		$\dfrac{wl^2}{20}$
$-\dfrac{5}{96}wl^2$		$\dfrac{5}{96}wl^2$
$\dfrac{Mb}{l^2}(2a-b)$		$\dfrac{Ma}{l^2}(2b-a)$
$\dfrac{M}{2l^2}(l^2-3b^2)$		0

3.4.2 Member stiffness matrix

In the stiffness method, a structure is considered to be an assemblage of discrete elements, beams, columns, plates, etc. and the method requires a knowledge of the stiffness characteristics of the elements. In the 'finite element' method (see page **3/14**) an artificial discretization of the structure is adopted. As an

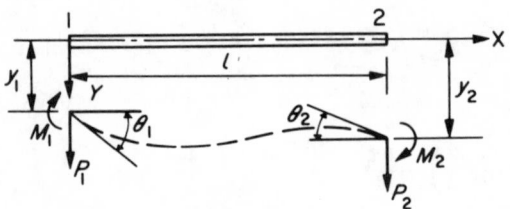

Figure 3.12 Structural beam element

example of the determination of stiffness influencing coefficients we shall consider the simple beam element shown in Figure 3.12. We neglect any axial deformation.

The expression for the bending moment in the beam with origin at end 1 and deflections y positive downwards is:

$$EI d^2 y/dx^2 = P_1 x - M_1$$

Integrating

$$EI dy/dx = \frac{P_1 x^2}{2} - M_1 x + C_1$$

$$= EI\theta_1 \text{ for } x = 0$$

Hence:

$$C_1 = EI\theta_1$$

$$= EI\theta_2 \text{ for } x = l$$

Hence:

$$EI(\theta_2 - \theta_1) = \frac{P_1 l^2}{2} - M_1 l \qquad (3.24)$$

Integrating again:

$$EIy = \frac{P_1 x^3}{6} - M_1 \frac{x^2}{2} + EI\theta_1 x + C_2$$

$$= EIy_1 \text{ for } x = 0$$

Therefore:

$$C_2 = EIy_1$$

$$= EIy_2 \text{ for } x = l$$

Hence:

$$EI(y_2 - y_1) - EI\theta_1 l = P_1 \frac{l^3}{6} - M_1 \frac{l^2}{2} \qquad (3.25)$$

Solving equations (3.24) and (3.25) for M_1 and P_1:

$$M_1 = \frac{4EI\theta_1}{l} + \frac{6EIy_1}{l^2} + \frac{2EI\theta_2}{l} - \frac{6EIy_2}{l^2} \qquad (3.26)$$

and

$$P_1 = \frac{6EI\theta_1}{l^2} + \frac{12EIy_1}{l^3} + \frac{6EI\theta_2}{l^2} - \frac{12EIy_2}{l^3} \qquad (3.27)$$

Two further relationships between the forces and displacements are obtained from statical equilibrium as follows:

For vertical equilibrium, $P_1 + P_2 = 0$

Hence:

$$P_2 = -P_1 \qquad (3.28)$$

Taking moments about end 1:

$$M_2 = -M_1 - P_2 l$$

$$= \frac{2EI\theta_1}{l} + \frac{6EIy_1}{l^2} + \frac{4EI\theta_2}{l} - \frac{6EIy_2}{l^2} \qquad (3.29)$$

Equations (3.26)–(3.29) may be combined in the matrix form:

$$\begin{bmatrix} M_1 \\ P_1 \\ M_2 \\ P_2 \end{bmatrix} = \frac{EI}{l^3} \begin{bmatrix} 4l^2 & 6l & 2l^2 & -6l \\ 6l & 12 & 6l & -12 \\ 2l^2 & 6l & 4l^2 & -6l \\ -6l & -12 & -6l & 12 \end{bmatrix} \begin{bmatrix} \theta_1 \\ y_1 \\ \theta_2 \\ y_2 \end{bmatrix}$$

or $\mathbf{S} = \mathbf{k}\Delta$ (3.30)

The matrix $\mathbf{k}$ is the stiffness matrix of the beam, and $\mathbf{S}$ and Δ are the matrices of member forces and nodal displacements respectively. Equation (3.30) expresses the force–displacement relationships for the beam in the *stiffness* form as distinct from the *flexibility* form. The symmetry of the matrix should be noted as consistent with the symmetry exhibited by flexibility coefficients (see page 3/9).

3.4.3 Assembly of structure stiffness matrix

The stiffness method involves the solution of a set of linear simultaneous equations, representing equilibrium conditions, which may be expressed in the form:

$$\mathbf{Kr} = \mathbf{R} \qquad (3.31)$$

Equation (3.31) is similar in form to Equation (3.23) with the important difference that now we are concerned with a multiple degree of freedom system as distinct from a single unknown displacement. $\mathbf{K}$ is the structure stiffness matrix, $\mathbf{r}$ is a matrix of nodal displacements and $\mathbf{R}$ a matrix of applied nodal forces.

The process of assembling the matrix $\mathbf{K}$ is one of transferring individual element stiffnesses into appropriate positions in the matrix $\mathbf{K}$. Naturally, this has been the subject of considerable organization for digital computer analysis and the subject is well documented.[3] Some aspects of a computerized approach will be considered later but the basic process will be illustrated here using a simple example. Consider the structure shown in Figure 3.13(a). The two beams are rigidly connected together at B where there is a spring support with stiffness k_s. End A is hinged and end C fixed. The structure has three degrees of freedom, rotations r_1 and r_3 at A and B and a vertical displacement r_2 at B. The stiffness matrix for each beam has the form of Equation (3.30) from which $\mathbf{k}$ may be written in the general form:

$$\mathbf{k} = \begin{bmatrix} k_{11} & k_{12} & k_{13} & k_{14} \\ k_{21} & k_{22} & k_{23} & k_{24} \\ k_{31} & k_{32} & k_{33} & k_{34} \\ k_{41} & k_{42} & k_{43} & k_{44} \end{bmatrix} \qquad (3.32)$$

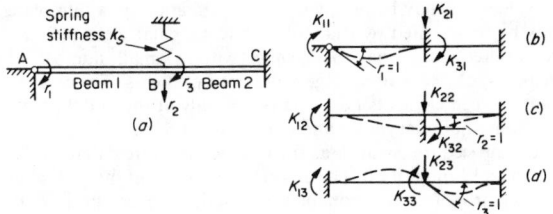

Figure 3.13

where $k_{11} = 4EI/l$; $k_{12} = 6EI/l^2$, etc.

We apply unit value of each degree of freedom in turn as shown in Figure 3.13(b), (c) and (d). It should be noted that when $r_1 = 1$ is applied, r_2 and r_3 are constrained at zero value and similarly with $r_2 = 1$ and $r_3 = 1$. The force systems necessary to achieve the unit values of the degrees of freedom are also shown at (b), (c) and (d). The equilibrium conditions are clearly:

$$K_{11}r_1 + K_{12}r_2 + K_{13}r_3 = R_1$$

$$K_{21}r_1 + K_{22}r_2 + K_{23}r_3 = R_2$$

$$K_{31}r_1 + K_{32}r_2 + K_{33}r_3 = R_3$$

i.e. $\mathbf{Kr} = \mathbf{R}$

where $\mathbf{R}$ is the matrix of applied loads. Clearly, the forces shown in Figure 3.13(b), (c) and (d) constitute the elements of the stiffness matrix $\mathbf{K}$ and this may now be assembled by inspection. Using the individual beam elements from Equation (3.30) with the notation of Equation (3.32):

$$\mathbf{K} = \begin{array}{|c|c|c|} \hline (k_{11})_1 & -(k_{12})_1 & (k_{13})_1 \\ \hline -(k_{12})_1 & (k_{44})_1 + (k_{22})_2 + k_s & (k_{23})_2 - (k_{14})_1 \\ \hline (k_{13})_1 & (k_{23})_2 - (k_{14})_1 & (k_{33})_1 + (k_{11})_2 \\ \hline \end{array} \tag{3.33}$$

and more specifically:

$$\mathbf{K} = \begin{array}{|c|c|c|} \hline 4\left(\dfrac{EI}{l}\right)_1 & -6\left(\dfrac{EI}{l^2}\right)_1 & 2\left(\dfrac{EI}{l}\right)_1 \\ \hline -6\left(\dfrac{EI}{l^2}\right)_1 & 12\left(\dfrac{EI}{l^3}\right)_1 + 12\left(\dfrac{EI}{l^3}\right)_2 + k_s & 6\left(\dfrac{EI}{l^2}\right)_2 - 6\left(\dfrac{EI}{l^2}\right)_1 \\ \hline 2\left(\dfrac{EI}{l}\right)_1 & 6\left(\dfrac{EI}{l^2}\right)_2 - 6\left(\dfrac{EI}{l^2}\right)_1 & 4\left(\dfrac{EI}{l}\right)_1 + 4\left(\dfrac{EI}{l}\right)_2 \\ \hline \end{array}$$

$$\tag{3.34}$$

3.4.4 Stiffness transformations

The member stiffness matrix $\mathbf{k}$ in Equation (3.30) is based on a coordinate system which is convenient for the member, i.e. origin at one end and X-axis directed along the axis of the beam. Such a coordinate system is termed 'local' as distinct from the 'global' coordinate system which is used for the complete structure. This subject is considered in detail in a number of

texts[2,3] and we shall give only a brief indication of the type of computation required.

Consider a three-dimensional coordinate system $\bar{X}\bar{Y}\bar{Z}$ (global) which is obtained by rotation of the (local) coordinate system XYZ. In the local system the force–displacement relationships for a beam element may be expressed in the partitioned matrix form:

$$\begin{bmatrix} \mathbf{S}_1 \\ \mathbf{S}_2 \end{bmatrix} = \begin{bmatrix} \mathbf{k}_{11} & \mathbf{k}_{12} \\ \mathbf{k}_{21} & \mathbf{k}_{22} \end{bmatrix} \begin{bmatrix} \mathbf{r}_1 \\ \mathbf{r}_2 \end{bmatrix} \tag{3.35}$$

in which the subscripts refer to ends 1 and 2.

The stiffness expressed in the coordinate system $\bar{X}\bar{Y}\bar{Z}$ may be obtained as follows:

$$\begin{bmatrix} \bar{\mathbf{S}}_1 \\ \bar{\mathbf{S}}_2 \end{bmatrix} = \begin{bmatrix} \lambda\mathbf{k}_{11}\lambda^T & \lambda\mathbf{k}_{12}\lambda^T \\ \lambda\mathbf{k}_{21}\lambda^T & \lambda\mathbf{k}_{22}\lambda^T \end{bmatrix} \begin{bmatrix} \bar{\mathbf{r}}_1 \\ \bar{\mathbf{r}}_2 \end{bmatrix} \tag{3.36}$$

in which λ is a matrix of direction cosines as follows:

$$\lambda = \begin{bmatrix} \lambda_{\bar{x}x} & \lambda_{\bar{x}y} & \lambda_{\bar{x}z} & 0 & 0 & 0 \\ \lambda_{\bar{y}x} & \lambda_{\bar{y}y} & \lambda_{\bar{y}z} & 0 & 0 & 0 \\ \lambda_{\bar{z}x} & \lambda_{\bar{z}y} & \lambda_{\bar{z}z} & 0 & 0 & 0 \\ 0 & 0 & 0 & \lambda_{\bar{x}x} & \lambda_{\bar{x}y} & \lambda_{\bar{x}z} \\ 0 & 0 & 0 & \lambda_{\bar{y}x} & \lambda_{\bar{y}y} & \lambda_{\bar{y}z} \\ 0 & 0 & 0 & \lambda_{\bar{z}x} & \lambda_{\bar{z}y} & \lambda_{\bar{z}z} \end{bmatrix} \tag{3.37}$$

where $\lambda_{\bar{x}x} = \cos \bar{X}OX$, etc.

3.4.5 Some aspects of computerization of the stiffness method

The remarkable increase in popularity of the stiffness method is due to the widespread availability of relatively cheap computing power. The method is of limited practical use *except* on computers. The stiffness method is eminently suitable for computers because the setting up of the data describing the structure and loading system to be analysed is a comparatively simple operation. Although there is then generally considerable numerical computation to do, this is done by the computer. Thus the human effort required is minimized and the likelihood of errors being made also reduced. With the phenomenal development of cheap and powerful microcomputers, which are quite suitable for analysing most 'run-of-the-mill' structures, it is quite likely that in the very near future almost all structural analysis will be carried out on computers.

It will be useful to look briefly at the more important aspects of adapting the stiffness method for use on computers. The method may be viewed as a succession of six stages:

(1) Define the nodal degrees of freedom of the structure (*n*) (Equation (3.6)), the nodal 'coordinates'. The total number determines the size of the structure stiffness matrix $\mathbf{K}$. The ordering is a matter of convenience but in some programs a judicial ordering of coordinates is necessary to reduce the 'band width' of $\mathbf{K}$. An array $\mathbf{K}$ ($n \times n$) is now generated in the computer and all elements are zeroed. This is necessary since component stiffnesses are going to be added-in to this array thus 'accumulating' the stiffnesses element by element.

(2) The individual structural elements are now defined and their force–displacement relationships expressed in stiffness matrices, $\mathbf{k}$ (Equation (3.30)); $\mathbf{S} = \mathbf{k}\Delta$. The dimensions of these matrices will depend on the type of element used but for most of the common elements (beam, column, pin-jointed truss member, etc.) the standard matrices are pub-

lished in the textbooks. The element stiffnesses are now transformed from local to global coordinates using matrix transformations as in Equation (3.36).

(3) The transformed stiffnesses are now transferred into appropriate locations of the structure stiffness matrix **K**. Suppose we are to transfer the stiffnesses of a particular element and suppose this element has two coordinates numbered 1 and 2. If the coordinates in the actual structure which correspond to 1 and 2 of the element are, say, i and j then the transfer of stiffness is carried out as follows:

$$\mathbf{k}_{11} \rightarrow \mathbf{k}_{ii}$$
$$\mathbf{k}_{12} \rightarrow \mathbf{k}_{ij}$$
$$\mathbf{k}_{21} \rightarrow \mathbf{k}_{ji}$$
$$\mathbf{k}_{22} \rightarrow \mathbf{k}_{jj}$$

There is considerable economy in organization and programming if the above procedure is applied to 'groups' of coordinates, e.g. *all* the displacements at one node. This can be achieved by *partitioning* the element stiffness matrices.

(4) Once **K** has been set up, the applied load matrix **R** is generated. This is simply a column matrix containing the applied (nodal) loads arranged in the same order as the nodal coordinates. If the structure is carrying loads other than at the defined nodes, then such loads must be converted to statically equivalent nodal loads. In rigid frames, for example, this is easily done using the standard values of 'fixed-end' effects. If a concentrated load does not coincide with the defined nodal coordinates then it is a simple matter, as an alternative, to introduce a node at the load point. This procedure, although it increases the size of the system to be solved, does have the advantage of yielding the displacements developing at the load point.

(5) The computer now solves the linear simultaneous equations (Equation (3.31)) **Kr** = **R** to produce the nodal displacements **r**.

(6) Lastly, the element forces are obtained from Equation (3.30) **S** = **kΔ**. In this last operation, some logical organization is clearly needed to extract the element nodal displacements **Δ** from the structure displacement **Sr**.

The foregoing is a description of the fundamental basis of the stiffness method applied on computers. Of course, it is possible to incorporate many refinements and devices to simplify the input and output, to check the results and to make changes in data without having to re-input all data.

In its most general form the stiffness method is used to analyse complex structures in which not only simple elements such as beams and columns are used but 'continua' such as plates and shells. This is the 'finite element' method which will now be examined briefly.

3.4.6 Finite element analysis

This extremely powerful method of analysis has been developed in recent years and is now an established method with wide applications in structural analysis and in other fields. Space permits only the most brief introduction here but the method is extensively documented elsewhere.[4-6] We have discussed the application of the stiffness method to framed structures in which the structural elements, beams and columns, have been connected at the nodes and the method observes the correct conditions of displacement compatibility and equilibrium at the nodes. The finite element method was developed, originally, in order to extend the stiffness method to the analysis of elastic continua such as plates and shells and indeed to three-dimensional continua. The first step in the process is to divide the structure into a finite number of discrete parts called 'elements'.

The elements may be of any convenient shape, e.g. a thin plate may be represented by triangular or rectangular elements, and the discretization may be *coarse*, with a small number of elements, or *fine*, with a large number of elements. The connection between elements now occurs not only at the nodal points but along boundary lines and over boundary faces.

The procedure ensures, as for framed structures, that equilibrium and compatibility conditions are satisfied at the nodes but the regions of connection between nodes are constrained to adopt a chosen form of displacement function. Thus, compatibility conditions along the interfaces between elements may not be completely satisfied and a degree of approximation is generally introduced. Once the geometry of the elements has been determined and the displacement function defined, the stiffness matrix of each element, relating nodal forces to nodal displacements, can be obtained. The remainder of the structural analysis follows the established procedures similar to those for framed structures. Naturally the best choice of element and discretization pattern, the precise conditions occurring at the interfaces and the accuracy of the solution, are matters which have received a great deal of attention in the literature.

A central stage in the process is the adoption of a suitable displacement function for the element chosen, and the subsequent evaluation of the element stiffnesses. This will be illustrated with one of the simplest possible elements, a triangular plate element for use in a plane stress situation.

3.4.6.1 Triangular element for plant stress

A triangular element ijk is shown in Figure 3.14. Under load, the displacement of any point within the element is defined by the displacement components u, v. In particular the nodal displacements are:

$$\Delta = \{u_i u_j u_k v_i v_j v_k\} \tag{3.38}$$

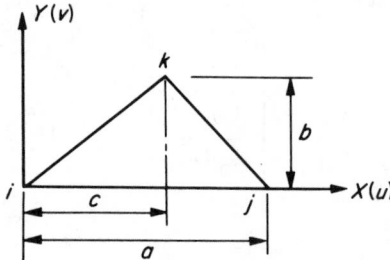

Figure 3.14

It is now assumed that the displacements u, v are linear functions of x, y as follows:

$$u = \alpha_1 + \alpha_2 x + \alpha_3 y$$
$$v = \alpha_4 + \alpha_5 x + \alpha_6 y \tag{3.39}$$

The nodal displacements Δ are now expressed in terms of the displacement parameters α, from Equations (3.39) and Figure 3.14:

$$
\begin{bmatrix} u_i \\ u_j \\ u_k \\ v_i \\ v_j \\ v_k \end{bmatrix}
=
\begin{bmatrix}
1 & 0 & 0 & 0 & 0 & 0 \\
1 & a & 0 & 0 & 0 & 0 \\
1 & c & b & 0 & 0 & 0 \\
0 & 0 & 0 & 1 & 0 & 0 \\
0 & 0 & 0 & 1 & a & 0 \\
0 & 0 & 0 & 1 & c & b
\end{bmatrix}
\begin{bmatrix} \alpha_1 \\ \alpha_2 \\ \alpha_3 \\ \alpha_4 \\ \alpha_5 \\ \alpha_6 \end{bmatrix}
\tag{3.40}
$$

or, $\Delta = \mathbf{A}\alpha$

The strains in the element are functions of the derivatives of u and v as follows:

$$\boldsymbol{\varepsilon} = \begin{bmatrix} \varepsilon_x \\ \varepsilon_y \\ \gamma_{xy} \end{bmatrix} = \begin{bmatrix} \partial u/\partial x \\ \partial v/\partial y \\ \partial u/\partial y + \partial v/\partial x \end{bmatrix} \quad (3.41)$$

$$= \begin{bmatrix} 0 & 1 & 0 & 0 & 0 & 0 \\ 0 & 0 & 0 & 0 & 0 & 1 \\ 0 & 0 & 1 & 0 & 1 & 0 \end{bmatrix} \begin{bmatrix} \alpha_1 \\ \alpha_2 \\ \alpha_3 \\ \alpha_4 \\ \alpha_5 \\ \alpha_6 \end{bmatrix} \quad (3.42)$$

i.e.:

$$\boldsymbol{\varepsilon} = \mathbf{B}\boldsymbol{\alpha} = \mathbf{B}\mathbf{A}^{-1}\boldsymbol{\Delta} \quad (3.43)$$

from Equation (3.40).

It should be noted that the matrix $\mathbf{B}$ in Equation (3.42) contains only constant terms and it follows that the strains are constant within the element.

The stress–strain relationships for plane stress in an isotropic material with Poisson's ratio v and Young's modulus E are:

$$\begin{bmatrix} \sigma_x \\ \sigma_y \\ \tau_{xy} \end{bmatrix} = \frac{E}{(1-v^2)} \begin{bmatrix} 1 & v & 0 \\ v & 1 & 0 \\ 0 & 0 & \frac{1}{2}(1-v) \end{bmatrix} \begin{bmatrix} \varepsilon_x \\ \varepsilon_y \\ \gamma_{xy} \end{bmatrix}$$

i.e.:

$$\boldsymbol{\sigma} = \mathbf{D}\boldsymbol{\varepsilon} = \mathbf{D}\mathbf{B}\mathbf{A}^{-1}\boldsymbol{\Delta} \quad (3.44)$$

Matrix $\mathbf{D}$ is the 'elasticity' matrix relating stress and strain. To obtain the element stiffness we employ the principle of virtual work and apply arbitrary nodal displacements $\overline{\boldsymbol{\Delta}}$ producing virtual strains in the element:

$$\overline{\boldsymbol{\varepsilon}} = \mathbf{B}\mathbf{A}^{-1}\overline{\boldsymbol{\Delta}} \quad (3.45)$$

The virtual strain energy in the element, from Equation (2.78) of Chapter 2, is:

$$\int_{vol} \overline{\boldsymbol{\varepsilon}}^T \boldsymbol{\sigma} \, \mathrm{d}V$$

where $V = $ volume of triangular element $= tab/2$, $t = $ thickness
Substituting for $\overline{\boldsymbol{\varepsilon}}^T$ and $\boldsymbol{\sigma}$ from Equations (3.45) and (3.44) respectively, the virtual strain energy is:

$$\int_{vol} [\mathbf{B}\mathbf{A}^{-1}\overline{\boldsymbol{\Delta}}]^T \mathbf{D}\mathbf{B}\mathbf{A}^{-1}\boldsymbol{\Delta} \, \mathrm{d}V$$

Now since all the matrices contain constant terms only and are thus independent of x and y, the expression for the virtual strain energy may be written:

$$\overline{\boldsymbol{\Delta}}^T \{[\mathbf{A}^{-1}]^T \mathbf{B}^T \mathbf{D}\mathbf{B}\mathbf{A}^{-1}V\}\boldsymbol{\Delta}$$

The external work is the product of the virtual displacements $\overline{\boldsymbol{\Delta}}$ and the nodal forces $\mathbf{S}$, hence equating external virtual work and internal virtual strain energy:

$$\overline{\boldsymbol{\Delta}}^T \mathbf{S} = \overline{\boldsymbol{\Delta}}^T \{[\mathbf{A}^{-1}]^T \mathbf{B}^T \mathbf{D}\mathbf{B}\mathbf{A}^{-1}V\}\boldsymbol{\Delta}$$

The virtual displacements are quite arbitrary and in particular may be taken to be represented by a unit matrix, thus:

$$\mathbf{S} = \{[\mathbf{A}^{-1}]^T \mathbf{B}^T \mathbf{D}\mathbf{B}\mathbf{A}^{-1}V\}\boldsymbol{\Delta}$$
$$= \mathbf{k}\boldsymbol{\Delta}, \text{ from Equation (3.30)}$$

Thus:

$$\mathbf{k} = [\mathbf{A}^{-1}]^T \mathbf{B}^T \mathbf{D}\mathbf{B}\mathbf{A}^{-1}V \quad (3.46)$$

Before the matrix multiplications required in Equation (3.46) can be performed we need to find $\mathbf{A}^{-1}$. This is easily determined as:

$$\mathbf{A}^{-1} = \frac{1}{ab} \begin{bmatrix} ab & 0 & 0 & 0 & 0 & 0 \\ -b & b & 0 & 0 & 0 & 0 \\ (c-a) & -c & a & 0 & 0 & 0 \\ 0 & 0 & 0 & ab & 0 & 0 \\ 0 & 0 & 0 & -b & b & 0 \\ 0 & 0 & 0 & (c-a) & -c & a \end{bmatrix}$$

Hence finally, with $|\lambda_1 = \frac{1}{2}(1-v)$ and $\lambda_2 = \frac{1}{2}(1+v)$ we obtain the stiffness matrix for the plane stress triangular element as shown in equation (3.47) below.

It is neither necessary nor economical to carry out these operations by hand; the computation of the element stiffness and, indeed, the entire computational process is easily programmed for the digital computer.

Computer 'packages' for finite element analysis of structures are highly developed, very powerful and readily available. Because of the comparatively heavy demands on computer storage, the use of the packages is generally confined to mainframe computers. A good example of a finite element system which is used very extensively is PAFEC.[6] The more important topics which should be studied in pursuing finite element analysis include: (1) shape (displacement) functions; (2) conforming and nonconforming elements; (3) isoparametric elements; and (4) automatic mesh generation.

$$\mathbf{k} = \frac{Et}{2(1-v^2)ab}$$

$b^2 + \lambda_1(c-a)^2$					
$-b^2 - \lambda_1 c(c-a)$	$b^2 + \lambda_1 c^2$	Symmetric			
$\lambda_1 a(c-a)$	$-\lambda_1 ac$	$\lambda_1 a^2$			
$-\lambda_2 b(c-a)$	$\lambda_1 cb + vb(c-a)$	$-\lambda_1 ab$	$\lambda_1 b^2 + (c-a)^2$		
$\lambda_1 b(c-a) + vcb$	$-\lambda_2 bc$	$\lambda_1 ab$	$-\lambda_1 b^2 - c(c-a)$	$\lambda_1 b^2 + c^2$	
$-vab$	vab	0	$a(c-a)$	$-ac$	a^2

$$(3.47)$$

3.5 Moment distribution

3.5.1 Introduction

Although the stiffness method, described in the previous section has the merit of simplicity, the solution of the equilibrium equations (3.31) is generally a matter for the digital computer since only for the simplest structures can a hand solution be contemplated. An alternative procedure which is eminently suitable for hand computation is the method of moment distribution which is essentially an iterative solution of the equations of equilibrium.

As in the general stiffness method, we first imagine all the degrees of freedom, joint rotations and joint translations, to be constrained. We ignore axial effects in members and consider flexure only. The constraints are imagined to be clamps applied to the joints to prevent rotation and translation. The forces required to effect the constraints are applied artificially and in the moment distribution processes these clamping forces are systematically released so as to allow the structure to achieve an equilibrium state. It is important to note that in the method as generally applied, the rotational joint restraints are relaxed by one process and the translational restraints by another. Finally the principle of superposition is used to combine the separate results.

It is necessary to assemble certain standard results before we can consider the actual process.

3.5.2 Distribution factors, carry-over factors and fixed-end moments

For the time being we confine our attention to prismatic members. The treatment of nonuniform section members will be touched on later.

Standard member stiffnesses are required and these are illustrated in Figure 3.15. The member end forces are those required to produce the deflected forms shown. Diagrams (a) and (b) relate to rotation without translation (sway), and diagrams (c) and (d) relate to sway without rotation. The results in diagrams (a) and (c) may be deduced from the stiffness matrix in Equation (3.30). The other results may be obtained easily from elementary beam theory, e.g. in Figure 3.15(b), taking the origin of x at the left-hand end and y positive downwards:

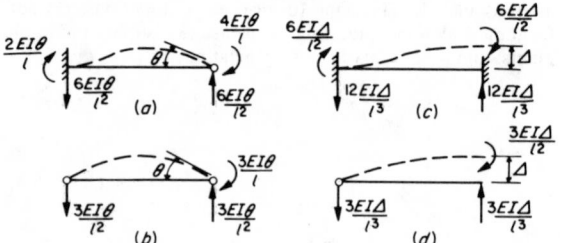

Figure 3.15

$$EI\,d^2y/dx^2 = \frac{Mx}{l},$$ where M is the moment, to be determined, at the right-hand end,

$$EI\,dy/dx = \frac{M}{l}\frac{x^2}{2} + C_1$$

$$= EI\theta \text{ for } x = l; \text{ hence } C_1 = EI\theta - M\frac{l}{2}$$

$$EIy = \frac{M}{l}\frac{x^3}{6} + \left(EI\theta - M\frac{l}{2}\right)x + C_2$$

$$= 0 \text{ for } x = 0; \text{ hence } C_2 = 0$$

$$= 0 \text{ for } x = l; \text{ hence, } M = \frac{3EI\theta}{l}$$

When loads are applied to members which are constrained at the joints, fixed-end moments are required to prevent the end rotations. This is another standard type of result which is required in the moment distribution method. Table 3.5 lists fixed-end moments for a selection of loading cases on uniform section beams. Again, these results may be obtained from elementary beam theory. It should be noted that the sign convention is that a moment is *positive* if tending to produce clockwise rotation of the end of the member at which it acts. This convention is different to, and should not be confused with, the sign convention for constructing bending moment diagrams which must be based on the curvature produced in the member.

As an illustration of the basic process, consider the structure ABC shown in Figure 3.11. This structure was analysed by the stiffness method previously. Joint B is considered to be clamped and thus a system of fixed-end moments is set up in member AB. The end moments in the members are shown in line 1 of Table 3.6. The constraining moment at joint B is seen to be $Wl_1/8$ clockwise and we imagine this moment to be removed by the application of a moment $-Wl_1/8$. The subsequent rotation of joint B, anticlockwise through angle θ, will develop moments in both members. Referring to Figure 3.15 the moments induced will be:

$$M_{BA} = -\frac{4EI\theta}{l_1}; \quad M_{AB} = -\frac{2EI\theta}{l_1}$$

$$M_{BC} = -\frac{4EI\theta}{l_2}; \quad M_{CB} = -\frac{2EI\theta}{l_2}$$

For equilibrium of joint B, the applied moment $-Wl_1/8$ must equal the sum of the moments absorbed by the two members meeting at the joint:

$$-\frac{Wl_1}{8} = -\frac{4EI\theta}{l_1} - \frac{4EI\theta}{l_2} = -4EI\theta\left(\frac{I}{l_1} + \frac{I}{l_2}\right)$$

and it is seen that the moment is 'distributed' to the members in proportion to their I/l values.

Thus:

$$M_{BA} = \frac{-Wl_1}{8}\frac{I/l_1}{(I/l_1 + I/l_2)} = \frac{-Wl_1}{8}\left(\frac{l_2}{l_1 + l_2}\right)$$

and:

$$M_{BC} = \frac{-Wl_1}{8}\frac{I/l_2}{(I/l_1 + I/l_2)} = \frac{-Wl_1}{8}\left(\frac{l_1}{l_1 + l_2}\right)$$

The moments induced at A and C are from Figure 3.15, one-half of those induced at B and the factor of one-half is termed the *carry over* factor. This set of moments is shown in line 2 of Table 3.6.

Joint B is now 'in balance' and since it was the only joint which was clamped we have reached an equilibrium state and no further distribution of moments is required. The final set of

Table 3.6

Stage	Operation	M_{AB}	M_{BA}	M_{BC}	M_{CB}
1	Fixed-end moments	$-Wl_1/8$	$+Wl_1/8$	0	0
2	Distribution at B	$-\dfrac{Wl_1}{16}\left(\dfrac{l_2}{l_1+l_2}\right)$	$-\dfrac{Wl_1}{8}\left(\dfrac{l_2}{l_1+l_2}\right)$	$-\dfrac{Wl_1}{8}\left(\dfrac{l_1}{l_1+l_2}\right)$	$-\dfrac{Wl_1}{16}\left(\dfrac{l_1}{l_1+l_2}\right)$
3	Total moments	$-\dfrac{Wl_1}{16}\left(\dfrac{2l_1+3l_2}{l_1+l_2}\right)$	$\dfrac{Wl_1}{8(l_1+l_2)}$	$-\dfrac{Wl_1}{8(l_1+l_2)}$	$-\dfrac{Wl_1}{16(l_1+l_2)}$

moments is obtained in line 3 of Table 3.6, by the addition of lines 1 and 2. This result is the same as that obtained from pure stiffness considerations. It should be noted that the zero sum of moments M_{BA} and M_{BC} indicates that joint B is in rotational equilibrium.

Two further points should be noted before we consider the moment distribution process in more detail. Referring to Figure 3.16, of the three members connected at joint A, member AD is hinged at the end remote from A whereas the other two members are fixed. Since D is hinged no moment can exist there and hence there is no carry-over to D. Furthermore, the moment–rotation relationship is different for a member pinned

at the remote end, as may be seen by comparing Figures 3.15(a) and (b). In calculating distribution factors this is taken account of by taking $\frac{3}{4}(I/l)$ for such members as compared with I/l for members fixed at the remote end.

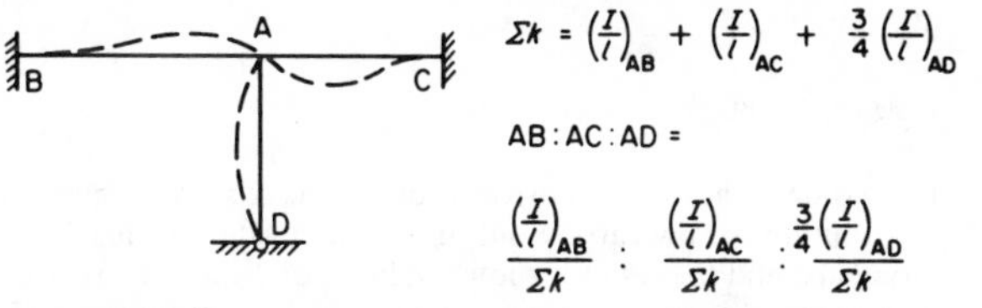

3.5.3 Moment distribution without sway

As an example of a structure with two degrees of freedom of joint rotation and no sway, consider the frame shown in Figure 3.17, EI (beams) $= 3 \times EI$ (columns).

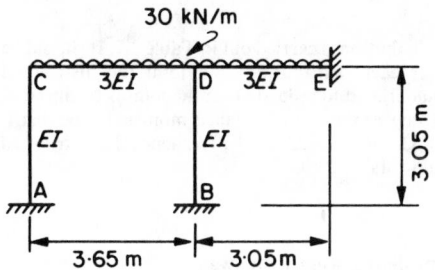

$$\Sigma k = \left(\tfrac{I}{l}\right)_{AB} + \left(\tfrac{I}{l}\right)_{AC} + \tfrac{3}{4}\left(\tfrac{I}{l}\right)_{AD}$$

AB : AC : AD =

$$\frac{\left(\tfrac{I}{l}\right)_{AB}}{\Sigma k} : \frac{\left(\tfrac{I}{l}\right)_{AC}}{\Sigma k} : \frac{\tfrac{3}{4}\left(\tfrac{I}{l}\right)_{AD}}{\Sigma k}$$

Figure 3.16 Distribution factors at typical joint

Figure 3.17

Table 3.7 Moment distribution for frame shown in Figure 3.17

	Joint	A	C		D			B	E
	Distribution factors		0.285	0.715	0.386	0.154	0.460		
	end moments	AC	CA	CD	DC	DB	DE	BD	ED
(1)	Fixed-end moments			−33.3	+33.3		−23.3		+23.3
(2)	Distribution at C		+9.5	+23.8					
(3)	Carry-over to A and D	+4.75			+11.9				
(4)	Distribution at D				−8.45	−3.38	−10.07		
(5)	Carry-over to C, B and E			−4.23				−1.69	−5.04
(6)	Distribution at C		+1.20	+3.03					
(7)	Carry-over to A and D	+0.60			+1.52				
(8)	Distribution at D				−0.59	−0.23	−0.70		
(9)	Carry-over to C, B and E			−0.30				−0.12	−0.35
(10)	Distribution at C		+0.09	+0.21					
(11)	Carry-over to A and D	+0.05			+0.11				
(12)	Distribution at D				−0.04	−0.02	−0.05		
(13)	Carry-over to C, B and E				May be neglected				
(14)	Total moments (kNm)	+5.40	+10.79	−10.79	+37.75	−3.63	−34.12	−1.81	+17.91

The fixed-end moments are, $(wl^2/12)$,

$$M_{FCD} = -30 \times \frac{3.65^2}{12}; \quad M_{FDC} = +30 \times \frac{3.65^2}{12} = 33.3 \text{ kNm}$$

$$F_{FDE} = -30 \times \frac{3.05^2}{12}; \quad M_{FED} = +30 \times \frac{3.05^2}{12} = 23.3 \text{ kNm}$$

and the distribution factors are:

at C, $CD:CA = \dfrac{3/3.65}{(1/3.05)+(3/3.65)} : \dfrac{1/3.05}{(1/3.05)+(3/3.65)}$

$= 0.715:0.285$

at D, DC:DB:DE =

$$\frac{3/3.65}{(3/3.65)+(1/3.05)+(3/3.05)} : \frac{1/3.05}{(3/3.65)+(1/3.05)+(3/3.05)} :$$
$$\frac{3/3.05}{(3/3.65)+(1/3.05)+(3/3.05)}$$

$= 0.386:0.154:0.460$

The moment distribution is carried out in Table 3.7. It should be noted that after each distribution at a joint the distributed moments are underlined to indicate that the joint is balanced at that stage. At step 4, the out-of-balance moment to be distributed at D is $+33.3 + 11.9 - 23.3 = +21.9$; hence the distributed moments should total -21.9.

3.5.4 Moment distribution with sway

This process will be illustrated with reference to Example 3.3 (page 3/9), for which the structure is shown in Figure 3.9. We first ignore any horizontal movement (sway) of the joints B and C and carry out a moment distribution.

The fixed-end moments are $wl^2/12 = \pm 40$ kNm; and the distribution factors are:

$BA:BC = \frac{1}{3}:\frac{2}{3}$

$CB:CD = \frac{2}{3}:\frac{1}{3}$ (noting $\frac{3}{4}I/l$ for CD)

The result of this (no sway) moment distribution is given in line 3 of Table 3.8. We now consider the horizontal equilibrium of the beam BC, Figure 3.18(a), and find that a force F_1 is required to maintain equilibrium. F_1 may be calculated by evaluating the horizontal shear forces at the tops of the columns as follows:

$$F_1 = 120 + \frac{(20+10)}{4} - \frac{20}{3} = 120.8 \text{ kN}$$

This force cannot exist in practice and what happens is that the beam BC deflects to the right and a new set of bending moments is set up with the effect that the out-of-balance horizontal force F_1 is removed. We consider the effect of this sway separately. Referring to Figure 3.18(b), a movement to the right of Δ, without joint rotation, requires column moments as shown. From Figure 3.15(c) and (d), these column moments are,

$$M_{FBA} = M_{FAB} = -6\left(\frac{EI}{l^2}\right)\Delta_{AB}$$

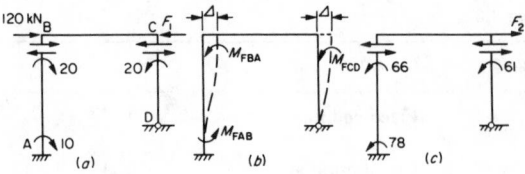

Figure 3.18

$$M_{FCD} = -3\left(\frac{EI}{l^2}\right)\Delta_{CD} \quad (\text{note } M_{FDC} = 0)$$

We cannot evaluate these moments unless Δ is known but we could proceed with an arbitrary value of Δ, and carry out a distribution to produce rotational equilibrium of the joints B and C. In fact, it is seen that any arbitrary values of moments can be used providing these are in the correct proportions between the two columns. The ratio in this example is:

$$AB:CD = \left(\frac{I}{l^2}\right)_{AB} : \frac{1}{2}\left(\frac{I}{l^2}\right)_{CD}$$

If we adopt

$$M_{FBA} = M_{FAB} = -90$$

and

$$M_{FCD} = -80$$

the moments are in the correct proportion. A second moment distribution is now carried out, using these values of fixed-end moments, and the result is shown in line 1 of Table 3.8. This set of moments is consistent with an applied horizontal force F_2, Figure 3.18(c), and:

$$F_2 = \frac{66+78}{4} + \frac{61}{3} = 56.3 \text{ kN}$$

Table 3.8

Joint		A	B		C	
End moments		AB	BA	BC	CB	CD
(1)	Arbitrary sway	-78	-66	$+66$	$+61$	-61
(2)	Corrected $[(1) \times \lambda]$	-167	-141	$+141$	$+131$	-131
(3)	No sway moments	$+10$	$+20$	-20	$+20$	-20
(4)	Final moments					
	$[(2)+(3)]$		-157	-121	$+121$	$+151$	-151

Now F_2 has to be scaled to equal F_1 and the scaling factor is $F_1/F_2 = \lambda = 120.8/56.3 = 2.14$.

The corrected moments are given in line 2 of Table 3.8 and the final moments are in line 4 obtained by adding lines 2 and 3.

3.5.5 Additional topics in moment distribution

Space has permitted only a brief introduction to the method of moment distribution. Additional topics which should be studied by reference to the standard texts,[3,4] are as follows:

(1) *Frames with multiple degrees of freedom for sway*. These are handled by carrying out an arbitrary sway distribution

for each sway in turn. Equilibrium conditions are then used to relate the out-of-balance forces and obtain the correction factors for each sway mode.

(2) *Treatment of symmetry.* In cases of symmetry the moment distribution process can be considerably shortened. Two cases arise and should be studied, systems in which it is known that the final set of moments is symmetrical and systems in which the final moments form an anti-symmetrical system.

(3) *Nonprismatic members.* If the flexural rigidity (*EI*) of a member varies within its length, then the effect is to change the values of end stiffnesses, carry-over factor and fixed end moments. A suitable general method for handling this situation is to evaluate end flexibilities by the use of Simpson's rule and then convert the flexibilities into stiffnesses.

3.6 Influence lines

3.6.1 Introduction and definitions

It is frequently necessary to consider loads which may occupy variable positions on a structure. For example, in bridge design it is important to determine the maximum effects due to the passage of a specified train or system of loads. In other cases the total load on a structure may be comprised of different loads which may be applied in various combinations and this again is a problem of variability of load or load position. The effect of varying a load position may be studied with the help of *influence lines*.

An influence line shows the variation of some resultant action or effect such as bending moment, shear force, deflection, etc. at a particular point as a unit load traverses the structure. It is important to observe that the effect considered is at a fixed position, e.g. *bending moment at* C, and that the independent variable in the influence line diagram is *the load position*. The following is a summary of influence line theory. For a more detailed treatment the reader should consult Jenkins.[1]

3.6.2 Influence lines for beams

Consider the simply-supported beam AB, Figure 3.19, carrying a single unit load occupying a variable position distant y from A. We require to obtain influence lines for bending moment and shear force at a fixed point X distant a from A and b from B.

If the unit load lies between X and B:

$$M_x = R_A \cdot a = 1 \frac{(l-y)}{l} a \tag{3.48}$$

If the unit load acts between A and X:

$$M_x = R_B \cdot b = 1 \cdot y/l \cdot b \tag{3.49}$$

Equations (3.48) and (3.49) are linear in y and when plotted in the regions to which they relate, form a triangle as shown in Figure 3.19(b). We note that, in both cases, substitution of $y = a$ gives $M_x = 1 \cdot ab/l$. Thus the influence line for M_x is a triangle with a peak value ab/l at the section X.

Turning now to the influence line for shearing force at X. For unit load between X and B:

$$S_x = R_A = \frac{l-y}{l} \tag{3.50}$$

(and now we have implied a sign convention for shear force

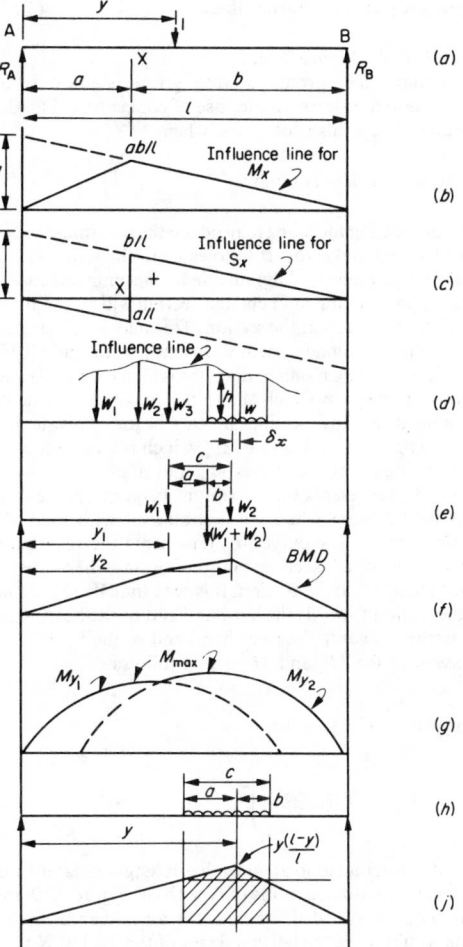

Figure 3.19 Influence lines and related diagrams for simply supported beams

namely that S_x is positive if the resultant force to the left of the section is upwards).

Where $y = a$, $S_x = b/l$

For unit load between A and X:

$$S_x = -R_B = -y/l \tag{3.51}$$

when $y = a$, $S_x = -a/l$

We note that Equations (3.50) and (3.51) give different values of S_x for $y = a$ and moreover the signs are opposite. This means that the shear force influence line contains a discontinuity at X as shown in Figure 3.19(c).

In using influence lines with a given system of loads and having determined the locations of the loads on the span, the total effect is evaluated as:

$$\sum(W \times \text{ordinate}), \text{ for concentrated loads} \tag{3.52}$$

and:

$$\int whdx = w \text{ (area under influence line)} \qquad (3.53)$$

for distributed loads (Figure 3.19(d)).

The maximum effect produced at a given position is of interest in the design process. In the case of concentrated loads, from Equation (3.52), this is obtained when:

$$\sum(W \times \text{ordinate}) \text{ is a maximum}$$

The process of locating the loads to produce the maximum value is best done by trial and error. It follows from the straight-line nature of a bending moment diagram due to concentrated loads, that the maximum bending moment at a section will be obtained when one of the loads acts at the section. This may be illustrated by reference to the two-load system shown at (e) in Figure 3.19. The shape of the bending moment diagram is as shown at (f) and at (g) is drawn a diagram which shows the maximum value of bending moment at any section in the beam. This is the *maximum bending moment envelope* M_{max} which is seen to consist of two intersecting parabolic curves M_{y1} and M_{y2}.

The curve M_{y1} represents the maximum bending moment at all sections in the beam when this is obtained with load W_1 placed at the section. The curve M_{y2} represents the maximum bending moment at all sections in the beam when this is obtained with load W_2 at the section. It is seen that W_1 should be placed at the section towards the left-hand end of the beam, and W_2 at the section towards the right-hand end of the beam.

The expressions for M_{y1} and M_{y2} are as follows:

$$\left.\begin{aligned} M_{y1} &= (W_1 + W_2)\frac{y_1}{l}(l - y_1 - a) \\ M_{y2} &= (W_1 + W_2)\frac{(l - y_2)}{l}(y_2 - b) \end{aligned}\right\} \qquad (3.54)$$

In the case of a distributed load which has a length greater than the span, then for an influence line of type (b) in Figure 3.19, the whole span would be loaded, whereas for an influence line of type (c) one would place the left-hand end of the load at X thus avoiding the introduction of a negative effect on the maximum positive value. For a short distributed load, as at (h), for maximum effect at y, the load must be placed so that the shaded area in (j) is a maximum.

The rule for this is:

$$y/l = a/c \qquad (3.55)$$

3.6.3 Influence lines for plane trusses

In the analysis of plane trusses, the influence line is useful in representing the variations in forces in members of the truss.

Figure 3.20(a) shows a Warren girder AB of span 20 m. For the unit load acting at any of the lower chord joints, the force in member 1 is:

$$P_1 = \frac{AR_A}{2\sqrt{3}}$$

The peak value occurs when the unit load is at C, and thus:

$$P_{1\,max} = \frac{2}{\sqrt{3}} \times \frac{4}{5} \times 1 = \frac{8}{5\sqrt{3}}$$

The influence line for P_1 is shown at (b).

For member 2, if the unit load lies between A and E, we take:

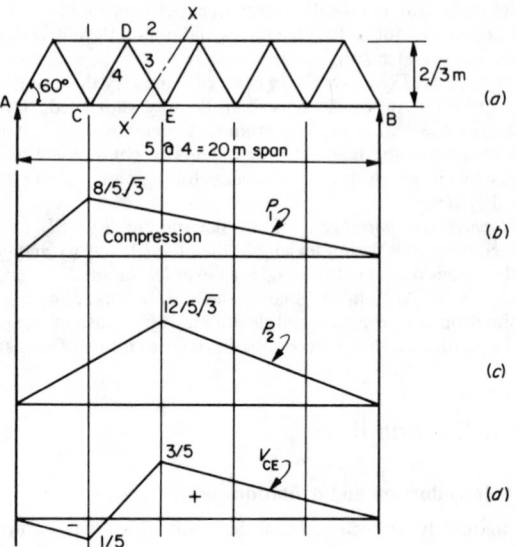

Figure 3.20 Influence lines for plane truss

$$P_2 = \frac{12R_B}{2\sqrt{3}}$$

or, if the unit load lies between E and B we take:

$$P_2 = \frac{8R_A}{2\sqrt{3}}$$

The result is a triangle with peak value $12/5\sqrt{3}$ at E, as shown in diagram (c).

It should be noted that both the P_1 and P_2 influence lines indicate compression for all positions of the unit load.

For members 3 and 4 it is useful to note that these members carry the vertical shear force in the panel CE, and we proceed by drawing the influence line for V_{CE} as at (d).

Considering now the force in member 3 and the section XX in diagram (a), it is clear that the relationship is:

$$P_3 = \frac{V_{CE}}{\sin 60°}$$

and that P_3 is tensile when V_{CE} is positive and compressive when V_{CE} is negative.

3.6.4 Influence lines for statically indeterminate structures

The use of influence lines in representing the effects of variable-position loads in statically determinate beams and trusses has been outlined. The concept is, of course, of general application. When dealing with statically indeterminate structures it is convenient to introduce some additional theorems to assist the analysis. It is possible to relate influence line shapes to deflected shapes of structures under particular forms of applied force. This involves an application of Mueller-Breslau's principle, which we shall look at in this section. The application of this principle can take the form of a model analysis, to which a simple form or model of the structure is made and particular distortions of the model produce scaled versions of influence lines.

With the enormous increase in computing power now available there is little need to use models in this way and it is generally more economical to produce influence lines by computer. It should be noted that it is always possible to construct influence lines by repeated analysis of the structure under a unit applied load, changing the load position for each analysis and thus producing a succession of ordinates to the influence line sought. This latter approach will be illustrated in section 3.6.8.

We now look at two important theorems concerned with influence lines.

3.6.5 Maxwell's reciprocal theorem

Consider the propped cantilever shown in Figure 3.21 to be subjected to a load W at A, producing displacements f_{11} and f_{21} as shown at (a), and then separately to be subjected to a moment M at B producing displacements f_{12} and f_{22} as at (b). Assuming a linear load–displacement relationship we may use the principle of superposition and obtain the combined effects of W and M by adding (a) and (b). Clearly it will be immaterial in which order the forces are applied. Applying W first and then M, the work done by the loads will be:

$$(\tfrac{1}{2}Wf_{11}) + (\tfrac{1}{2}Mf_{22} + Wf_{12}) \tag{3.56}$$

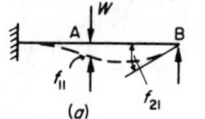 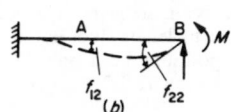

Figure 3.21

The first bracket in Equation (3.56) contains the work done during the application of W and the second bracket the work done (by both M and W) during the application of M.

In a similar way, if the order is reversed, the work done is:

$$(\tfrac{1}{2}Mf_{22}) + (\tfrac{1}{2}Wf_{11} + Mf_{21}) \tag{3.57}$$

From Equations (3.56) and (3.57) it is evident that:

$$Wf_{12} = Mf_{21} \tag{3.58}$$

If the applied actions are taken to have unit values, then Equation (3.58) simplifies to:

$$f_{12} = f_{21} \tag{3.59}$$

Equation (3.59) is a statement of Maxwell's reciprocal theorem. A more general theorem, of which Maxwell's is a special case, is due to Betti. This latter theorem states that if a system of forces P_i produces displacements p_i at corresponding positions and another set of forces Q_i, at similar positions to P_i, produces displacements q_i, then:

$$P_1 q_1 + P_2 q_2 + \ldots + P_n q_n = Q_1 p_1 + Q_2 p_2 + \ldots + Q_n p_n \tag{3.60}$$

3.6.6 Mueller-Breslau's principle

This principle is the basis of the indirect method of model analysis. It is developed from Maxwell's theorem as follows. Consider the two-span continuous beam shown in Figure 3.22(a). On removal of the support at C and the application of a unit load at C, a deflected shape, shown dotted in Figure

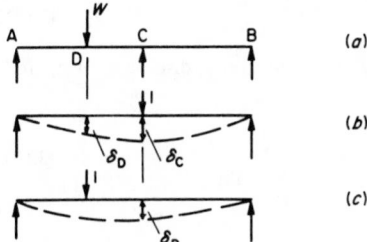

Figure 3.22

3.22(b), is obtained. If a unit load now occupies any arbitrary position D, as at (c), then from Maxwell's theorem the deflection at C will be δ_D. In other words, the deflected form (b) is the influence line for deflection of C.

Now the force at C to move C through $\delta_C = 1$

Hence, the force at C to move C through $\delta_D = 1 \times \delta_D/\delta_C$.

If a unit load acts at D, producing a deflection δ_D at C, then the upwards force needed to restore C to the level of AB is $1 \times \delta_D/\delta_C$. Hence, the reaction at C for unit load at D is $1 \times \delta_D/\delta_C$. Since D is an arbitrary point in the beam then it is seen that the deflected shape due to unit load at C, Figure 3.22(b), is to some scale, the influence line for R_C. The scale of the influence line is determined from the knowledge that the actual ordinate at C should equal unity. Hence, the ordinates should all be divided by δ_C.

This result leads to Mueller-Breslau's principle which may be stated as follows:

'The ordinates of the influence line for a redundant force are equal to those of the deflection curve when a unit load replaces the redundancy, the scale being chosen so that the deflection at the point of application of the redundancy represents unity.'

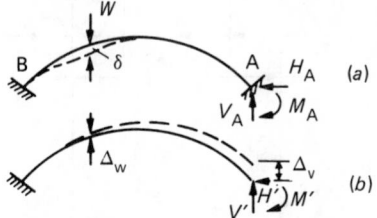

Figure 3.23

3.6.7 Application to model analysis

Consider the fixed arch shown in Figure 3.23(a). The arch has three redundancies which may be taken conveniently as H_A, V_A and M_A. We make a simple model of the arch to a chosen linear scale and pin this to a drawing board. End B is fixed in position and direction and the undistorted centreline is transferred to the drawing paper. We then impose a *purely* vertical displacement Δ_v at A and transfer the distorted centreline to the drawing paper. The distortion produced will require force actions at A, V', H' and M'. Let the displacement of a typical load point be Δ_w. Applying Equation (3.60) to the two systems of forces:

$$V_A(\Delta_v) + H_A(0) + M_A(0) + W(\Delta_w) = V'(0) + H'(0) + M'(0) + 0(\delta)$$

Hence:

$$V_A \Delta_v + W \Delta_w = 0$$

and if $W = 1$:

$$V_A = \frac{-\Delta_w}{\Delta_v} \qquad (3.61)$$

Similarly, we impose a *purely* horizontal displacement Δ_H and obtain:

$$H_A = \frac{-\Delta'_w}{\Delta_H} \qquad (3.62)$$

then a pure rotation θ and obtain:

$$M_A = -\frac{\Delta''_w}{\theta} \qquad (3.63)$$

In Equations (3.62) and (3.63) the displacements Δ'_w and Δ''_w represent the arch displacements due to the imposed horizontal and rotational displacements respectively. In each case the deflected shape, suitably scaled, gives the influence line for the corresponding redundancy.

3.6.7.1 Sign convention

The negative sign in Equations (3.61) to (3.63) leads to the following convention for signs. On the assumption that a reaction is positive if in the direction of the imposed displacement, then a load W will give a positive value of the reaction if the influence line ordinate at the point of application of the load is opposite to the direction of the load. This is evident in Figure 3.23(b) where the upward deflection Δ_w, being opposed to the direction of the load W, is consistent with a positive (upwards) direction for V_A.

3.6.7.2 Scale of the model

It should be noted that when using relationships (3.61) and (3.62) the ratios Δ_w/Δ_v and Δ'_w/Δ_H are dimensionless and thus the linear scale of the model does not affect the influence line ordinates. On the other hand, when using Equation (3.63) in obtaining an influence line for bending moment, Δ_w/θ has the dimensions of length and thus the model displacements must be multiplied by the linear scale factor.

In performing the model analysis, quite large displacements can be used providing the linear relation between load and displacement is maintained. Hence, the indirect method is sometimes called the 'large displacement' method.

3.6.8 Use of the computer in obtaining influence lines

With adequate computing facilities it is generally more economical to proceed directly to the computation of influence line ordinates by the analysis of the structure under a unit load, the unit load occupying a succession of positions. The actual method of analysis is immaterial but for bridge-type structures often the flexibility method offers some advantage especially if the structural members are 'nonprismatic'. An example of this type of computation is shown in Figure 3.24 where influence lines for bending moments at the interior supports of a five-span continuous beam are given. The beam is taken to be uniform in section over its length and, due to the symmetry of the spans, unit load positions need only be taken over one-half of the structure as shown.

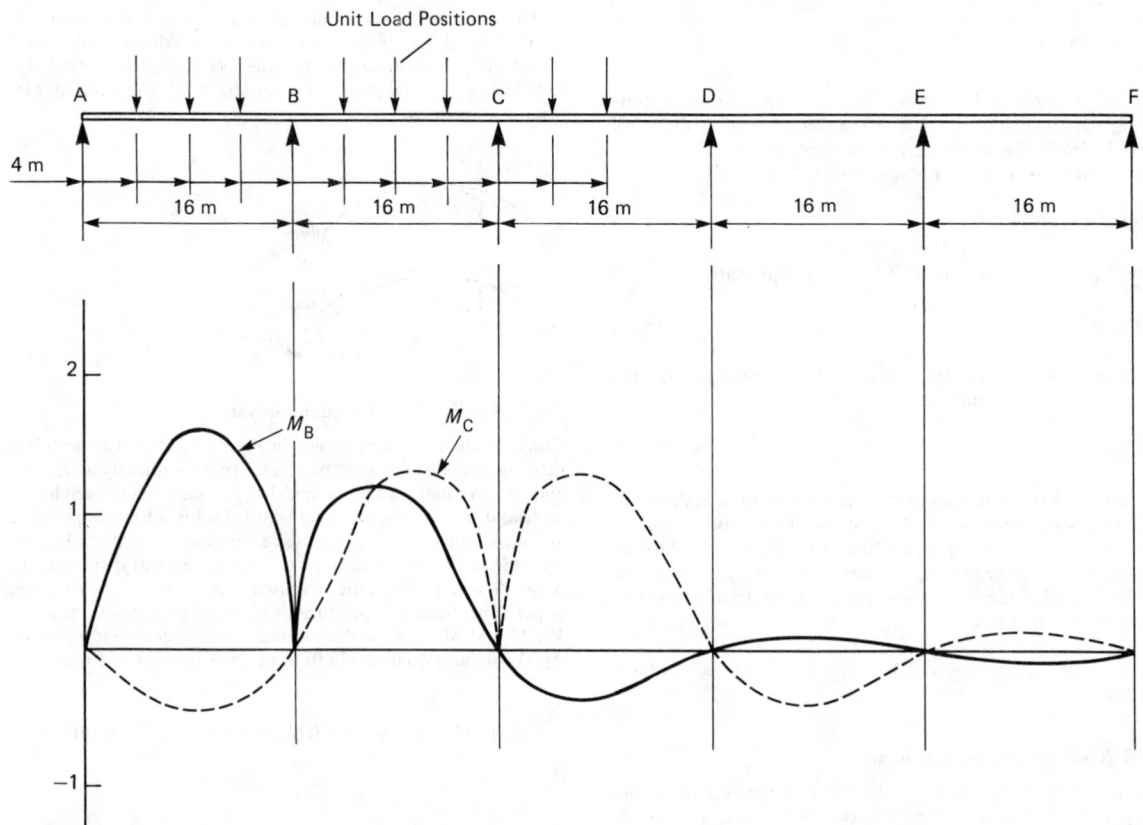

Figure 3.24 Influence lines for bending moments in a continuous beam obtained by computer analysis

3.7 Structural dynamics

3.7.1 Introduction and definitions

Structural vibrations result from the application of *dynamic* loads, i.e. loads which vary with time. Loads applied to structures are often time-dependent although in most cases the rate of change of load is slow enough to be neglected and the loads may be regarded as *static*. Certain types of structure may be susceptible to dynamic effects; these include structures designed to carry moving loads, e.g. bridges and crane girders, and structures required to support machinery. One of the most severe and destructive sources of dynamic disturbance of structures is, of course, the earthquake.

The dynamic behaviour of structures is generally described in terms of the displacement–time characteristics of the structure, such characteristics being the subject of vibration analysis. Before considering methods of analysis it is helpful to define certain terms used in dynamics.

(1) *Amplitude* is the maximum displacement from the mean position.
(2) *Period* is the time for one complete cycle of vibration.
(3) *Frequency* is the number of vibrations in unit time.
(4) *Forced vibration* is the vibration caused by a time-dependent disturbing force.
(5) *Free vibrations* are vibrations after the force causing the motion has been removed.
(6) *Damping.* In structural vibrations, damping is due to: (a) internal molecular friction; (b) loss of energy associated with friction due to slip in joints; and (c) resistance to motion provided by air or other fluid (drag). The type of damping usually assumed to predominate in structural vibrations is termed *viscous* damping in which the force resisting motion is proportional to the velocity. Viscous damping adequately represents the resistance to motion of the air surrounding a body moving at low speed and also the internal molecular friction.
(7) *Degrees of freedom.* This is the number of independent displacements or *coordinates* necessary to completely define the deformed state of the structure at any instant in time. When a single coordinate is sufficient to define the position of any section of the structure, the structure has a *single degree of freedom*. A continuous structure with a distributed mass, such as a beam, has an infinite number of degrees of freedom. In structural dynamics it is generally satisfactory to transform a structure with an infinite number of degrees of freedom into one with a finite number of freedoms. This is done by adopting a *lumped mass* representation of the structure, as in Figure 3.25. The total mass of the structure is considered to be *lumped* at specified points in the structure and the motion is described in terms of the displacements of the lumped masses. The accuracy of the analysis can be improved by increasing the number of lumped masses. In most cases sufficiently accurate results can be obtained with a comparatively small number of masses.

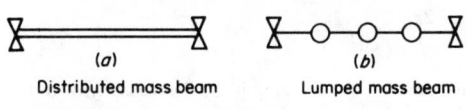

(a)
Distributed mass beam

(b)
Lumped mass beam

Figure 3.25

3.7.2 Single degree of freedom vibrations

The portal frame shown in Figure 3.26 is an example of a structure with a single degree of freedom providing certain assumptions are made. If it is assumed that the entire mass of

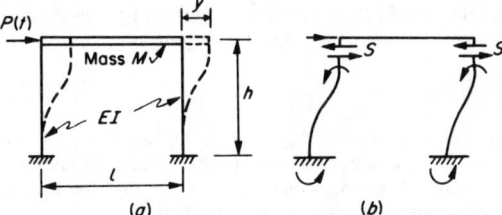

Figure 3.26

the structure (M) is located in the girder and that the girder has an infinitely large flexural rigidity and further, that the columns have infinitely large extensional rigidities, then the displacement of the mass M resulting from the application of an exciting force $P(t)$, is defined by the transverse displacement y. The girder moves in a purely horizontal direction restrained only by the flexure of the columns.

From Newton's second law of motion:

Force = mass × acceleration

i.e.:

$$\sum P = M\ddot{y} \tag{3.64}$$

Now from Figure 3.26(b), the force resisting motion is:

$$2S = 2\left(\frac{12EIy}{h^3}\right)$$

$$= 24\frac{EIy}{h^3} \tag{3.65}$$

Thus Equation (3.64) becomes:

$$P(t) - 24\frac{EIy}{h^3} = M\ddot{y}$$

or:

$$M\ddot{y} + 24\frac{EIy}{h^3} = P(t) \tag{3.66}$$

If the effect of damping is included then the equation of motion, Equation (3.66) is modified by the inclusion of a term $c\dot{y}$ where c is a constant. It should be noted that since the effect of damping is to resist the motion, then the term $c\dot{y}$ is added to the left-hand side of Equation (3.66). Thus:

$$M\ddot{y} + c\dot{y} + 24\frac{EIy}{h^3} = P(t) \tag{3.67}$$

Equation (3.67) may be generalized for any single degree of freedom structure by observing that the stiffness of the structure, i.e. force required for unit displacement horizontally, is given by:

$$k = 24\frac{EI}{h^3} \tag{3.68}$$

Combining Equations (3.67) and (3.68) we obtain the general single degree of freedom equation of motion:

$$M\ddot{y} + c\dot{y} + ky = P(t) \tag{3.69}$$

If in Equation (3.69) $P(t)=0$, we have a state of *free vibration* of the structure. The governing equation becomes:

$$M\ddot{y}+c\dot{y}+ky=0 \qquad (3.70)$$

The situation envisaged by Equation (3.70) would arise if the beam were given a horizontal displacement and then released. The resulting vibrations would depend on the amount of damping present, measured by the coefficient c.

The solution of Equation (3.70) is:

$$y=A_1 e^{\lambda_1 t}+A_2 e^{\lambda_2 t} \qquad (3.71)$$

where A_1 and A_2 are the constants of integration, to be evaluated from initial conditions, and λ_1 and λ_2 are the roots of the auxiliary equation:

$$M\lambda^2+c\lambda+k=0 \qquad (3.72)$$

or, substituting:

$$\left.\begin{array}{l} p^2=k/M \\ \text{and} \\ 2n=c/M \end{array}\right\} \qquad (3.73)$$

Equation (3.72) becomes:

$$\lambda^2+2n\lambda+p^2=0 \qquad (3.74)$$

Hence:

$$\lambda=-n\pm\sqrt{(n^2-p^2)} \qquad (3.75)$$

Four cases arise:

Case 3.1 $p^2<n^2$

Here (n^2-p^2) is always positive and $<n^2$ and thus λ_1 and λ_2 are real and negative.

Equation (3.71) takes the form:

$$y=e^{-nt}(A_1 e^{\sqrt{(n^2-p^2)}t}+A_2 e^{-\sqrt{(n^2-p^2)}t}) \qquad (3.76)$$

The relationship between y and t of Equation (3.76) is shown in Figure 3.27(a) and it is seen that the displacement y gradually returns to zero, no vibrations taking place.

Now, since $n^2>p^2$, then:

$$\frac{c^2}{4M^2}>\frac{k}{M}$$

or

$$c>2\sqrt{(Mk)} \qquad (3.77)$$

A structure exhibiting these characteristics is said to be *over-damped*.

Case 3.2 $p^2=n^2$

From Equation (3.75), $\lambda-n$ (twice) and hence,

$$y=e^{-nt}(A_1+A_2 t) \qquad (3.78)$$

Again, no vibrations result and Equation (3.78) has the form shown in Figure 3.27(a).

From Equation (3.73) the value of c for this condition is given by:

$$c_c=2\sqrt{(Mk)} \qquad (3.79)$$

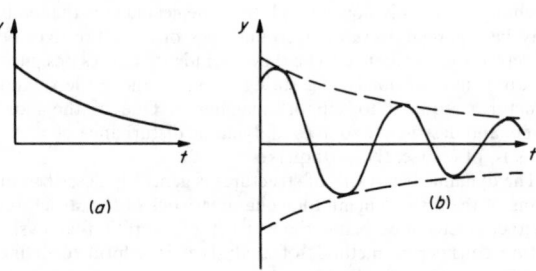

Figure 3.27

This is termed *critical damping* and the critical damping coefficient c_c is the value of the damping coefficient at the boundary between vibratory and nonvibratory motion. The critical damping coefficient is a useful measure of the damping capacity of a structure. The damping coefficient of a structure is usually expressed as a percentage of the critical damping coefficient.

Case 3.3 $p^2>n^2$

Here $c<c_c$ and the structure is *underdamped*.

From Equation (3.75), $\lambda=-n\pm i\sqrt{(p^2-n^2)}$

Hence:

$$y=e^{-nt}(A_1 e^{i\sqrt{(p^2-n^2)}t}+A_2 e^{-i\sqrt{(p^2-n^2)}t})$$

or, putting:

$$(p^2-n^2)=q^2$$

$$y=e^{-nt}(A_1 e^{iqt}+A_2 e^{-iqt})$$

or

$$y=e^{-nt}(A\cos qt+B\sin qt) \qquad (3.80)$$

A typical displacement–time relationship for this condition is shown in Figure 3.27(b).

An alternative form for Equation (3.80) is:

$$y=Ce^{-nt}\sin(qt+\beta) \qquad (3.81)$$

where C and β are new arbitrary constants

The period $T=\dfrac{2\pi}{q}=\dfrac{2\pi}{p\sqrt{\{1-(n/p)^2\}}}$

The period is constant but the amplitude decreases with time. The decay of amplitude is such that the ratio of amplitudes at intervals equal to the period is constant, i.e.:

$$\frac{y_{(t)}}{y_{(t+T)}}=e^{nT}$$

and $\log e^{nT}=nT=\delta$

δ is called the *logarithmic decrement*, and is a useful measure of damping capacity.

The percentage critical damping

$$= 100\frac{c}{c_c}$$

$$= 100\frac{\delta}{pT}$$

This is of the order of 4% for steel frames and 7% for concrete frames.

Case 3.4 $c=0$

In the absence of damping, Equation (3.70) becomes:

$$M\ddot{y} + ky = 0 \qquad (3.82)$$

The solution of which is:

$$y = A_1 e^{\lambda_1 t} + A_2 e^{\lambda_2 t}$$

where, from Equation (3.72):

$$\lambda_1 = ip$$

$$\lambda_2 = -ip$$

Thus:

$$y = A \sin pt + B \cos pt \qquad (3.83)$$

The period is, $T = \dfrac{2\pi}{p}$

where p is the *natural circular frequency*

The *natural frequency* is $f = \dfrac{1}{T} = \dfrac{p}{2\pi}$

3.7.3 Multi-degree of freedom vibrations

Vibration analysis of systems with many degrees of freedom is a complex subject and only a brief indication of one useful method will be given here. For a more comprehensive and detailed treatment, the reader should consult one of the standard texts.[7]

For a system represented by lumped masses, the governing equations emerge as a set of simultaneous ordinary differential equations equal in number to the number of degrees of freedom. Mathematically the problem is of the *eigenvalue* or *characteristic value* type and the solutions are the *eigenvalues* (frequencies) and the *eigenvectors* (modal shapes). We shall consider the evaluation of mode shapes and fundamental, undamped, frequencies by the process of matrix iteration using the flexibility approach (see page 3/6). The method to be described, leads automatically to the lowest frequency, the fundamental, this being the one of most interest from a practical point of view. The alternative method using a stiffness matrix approach leads to the highest frequency.

Consider the simply-supported, uniform cross-section beam shown in Figure 3.28(a). The mass/unit length is w and we will regard the total mass of the beam to be lumped at the quarter-span points as shown in Figure 3.28(b). We may ignore the end

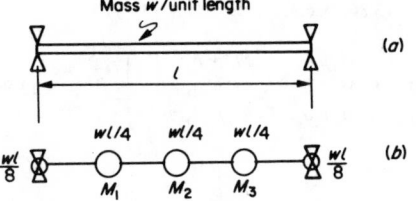

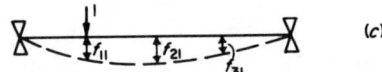

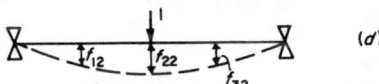

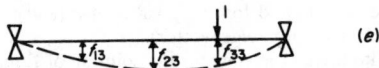

Figure 3.28

masses $wl/8$ since they are not involved in the motion, and consider the three masses

$$M_1 = M_2 = M_3 = wl/4.$$

The appropriate flexibilities, f_{ij}, are shown at (c), (d) and (e). Using the flexibility method previously described, we may obtain a flexibility matrix as follows:

$$\mathbf{F} = \begin{bmatrix} f_{11} & f_{12} & f_{13} \\ f_{21} & f_{22} & f_{23} \\ f_{31} & f_{32} & f_{33} \end{bmatrix} = \frac{l^3}{256EI} \begin{bmatrix} 3.00 & 3.67 & 2.33 \\ 3.67 & 5.33 & 3.67 \\ 2.33 & 3.67 & 3.00 \end{bmatrix} \qquad (3.84)$$

It should be noted that f_{ij} is the displacement of mass M_i due to unit force acting at mass M_j. Thus, if the forces acting at the positions of the lumped masses are $F_{1,2,3}$ and the corresponding displacements are $y_{1,2,3}$, then:

$$\left.\begin{array}{l} y_1 = f_{11}F_1 + f_{12}F_2 + f_{13}F_3 \\ y_2 = f_{21}F_1 + f_{22}F_2 + f_{23}F_3 \\ y_3 = f_{31}F_1 + f_{32}F_2 + f_{33}F_3 \end{array}\right\} \qquad (3.85)$$

For free, undamped vibrations, F_i is an inertia force $= -M_i \ddot{y}_i$.

Thus:

$$\left.\begin{array}{l} y_1 + f_{11}M_1\ddot{y}_1 + f_{12}M_2\ddot{y}_2 + f_{13}M_3\ddot{y}_3 = 0 \\ y_2 + f_{21}M_1\ddot{y}_1 + f_{22}M_2\ddot{y}_2 + f_{23}M_3\ddot{y}_3 = 0 \\ y_3 + f_{31}M_1\ddot{y}_1 + f_{32}M_2\ddot{y}_2 + f_{33}M_3\ddot{y}_3 = 0 \end{array}\right\} \qquad (3.86)$$

The solutions take the form:

$$y_1 = \delta_i \cos(pt + a) \qquad (3.87)$$

Hence:

$$\ddot{y}_i = -p^2 y_i \qquad (3.88)$$

Thus, Equations (3.86) become:

$$\left.\begin{array}{l}\delta_1 - f_{11}M_1p^2\,\delta_1 - f_{12}\,M_2\,p^2\delta_2 - f_{13}\,M_3\,p^2\,\delta_3 = 0 \\ \delta_2 - f_{21}M_1p^2\,\delta_1 - f_{22}\,M_2\,p^2\delta_2 - f_{23}\,M_3\,p^2\,\delta_3 = 0 \\ \delta_3 - f_{31}M_1p^2\,\delta_1 - f_{32}\,M_2\,p^2\delta_2 - f_{33}\,M_3\,p^2\,\delta_3 = 0)\end{array}\right\} \quad (3.89)$$

or:

$$\Delta = p^2\mathbf{FM}\Delta \tag{3.90}$$

where:

$$\Delta = \begin{bmatrix}\delta_1 \\ \delta_2 \\ \delta_3\end{bmatrix}; \qquad \mathbf{M} = \begin{bmatrix}M_1 & 0 & 0 \\ 0 & M_2 & 0 \\ 0 & 0 & M_3\end{bmatrix}$$

The unknowns in Equation (3.90) are the displacement amplitudes δ_i and the frequency p; p has as many values as there are equations in the system, and for every value of p (eigenvalue) there corresponds a set of y (eigenvector).

We adopt an iterative procedure for the solution of Equation (3.90) and first of all rewrite the equations in the form:

$$\mathbf{FM}\Delta = \frac{1}{p^2}\Delta \tag{3.91}$$

We start with an assumed vector Δ_0, thus:

$$\mathbf{FM}\Delta_0 = \frac{1}{p^2}\Delta_0$$

Putting $\mathbf{FM}\Delta_0 = \Delta_1$

$$\Delta_1 \simeq \frac{1}{p^2}\,\Delta_0 \text{ giving } p^2 \simeq \frac{\Delta_0}{\Delta_1}$$

We cannot form Δ_0/Δ_1 since each Δ is a column matrix, so we take the ratio of corresponding elements in Δ_0 and Δ_1 and form the ratio δ_0/δ_1. It is best to use the numerically greatest δ for this purpose.

Continuing the process:

$$\mathbf{FM}\Delta_1 \simeq \frac{1}{p^2}\Delta_1 \text{ giving } p^2 = \delta_1/\delta_2$$

$$= \Delta_2$$

and again:

$$\mathbf{FM}\Delta_2 \simeq \frac{1}{p^2}\Delta_2$$

$$= \Delta_3 \text{ giving } p^2 = \delta_2/\delta_3$$

It can be shown that this iterative process converges to the largest value of $1/p^2$ and hence yields the lowest (fundamental mode) frequency.

Applying the iterative scheme to the beam of Figure 3.28, and assuming:

$$\Delta_0 = \begin{bmatrix}1 \\ 2 \\ 1\end{bmatrix}$$

then, $\Delta_1 = \mathbf{FM}\Delta_0$

where $\mathbf{FM} = \dfrac{l^3}{256EI}\begin{bmatrix}3.00 & 3.67 & 2.33 \\ 3.67 & 5.33 & 3.67 \\ 2.33 & 3.67 & 3.00\end{bmatrix}\begin{bmatrix}wl/4 & 0 & 0 \\ 0 & wl/4 & 0 \\ 0 & 0 & wl/4\end{bmatrix}$

$$= \frac{wl^4}{1024EI}\begin{bmatrix}3.00 & 3.67 & 2.33 \\ 3.67 & 5.33 & 3.67 \\ 2.33 & 3.67 & 3.00\end{bmatrix}$$

Thus: $\Delta_1 = \dfrac{wl^4}{1024EI}\begin{bmatrix}12.67 \\ 18.00 \\ 12.67\end{bmatrix} = \dfrac{12.67wl^4}{1024EI}\begin{bmatrix}1.00 \\ 1.42 \\ 1.00\end{bmatrix}$

Hence: $p_1^2 = \dfrac{\delta_0}{\delta_1} = \dfrac{2 \times 1024EI}{12.67 \times 1.42wl^4}$

$$= 114\frac{EI}{wl^4}$$

A second iteration gives:

$$\Delta_2 = \mathbf{FM}\Delta_1 = \frac{wl^4}{1024EI}\begin{bmatrix}3.00 & 3.67 & 2.33 \\ 3.67 & 5.33 & 3.67 \\ 2.33 & 3.67 & 3.00\end{bmatrix}\frac{12.67wl^4}{1024EI}\begin{bmatrix}1.00 \\ 1.42 \\ 1.00\end{bmatrix}$$

$$= 12.67\left(\frac{wl^4}{1024EI}\right)^2\begin{bmatrix}10.54 \\ 14.91 \\ 10.54\end{bmatrix}$$

Hence:

$$p_2^2 = \frac{\delta_1}{\delta_2} = \frac{12.67 \times 1.42wl^4}{1024EI} \times \frac{1}{12.67(wl^4/1024EI)^2 \times 14.91}$$

$$= 97.5\frac{EI}{wl^4}$$

This result is very close to that produced by an exact method, i.e. $97.41EI/wl^4$.

3.8 Plastic analysis

3.8.1 Introduction

The plastic design of structures is based on the concept of a *load factor* (N), where

$$N = \frac{\text{Collapse load}}{\text{Working load}} = \frac{W_c}{W_w} \tag{3.92}$$

A structure is considered to be on the point of collapse when finite deformation of at least part of the structure can occur without change in the loads. The simple plastic theory is based on an idealized stress–strain relationship for structural steel as shown in Figure 3.29. A linear, elastic, relationship holds up to a stress σ_y, the *yield stress*, and at this value of stress the material is considered to be in a state of perfect plasticity, capable of infinite strain, represented by the horizontal line AB continued indefinitely to the right. For comparison the dotted line shows the true relationship.

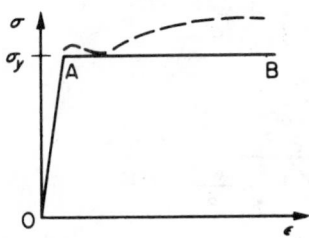

Figure 3.29

The term '*plastic analysis*' is generally related to steel structures for which the relationship indicated in Figure 3.29 is a good approximation. The equivalent approach when dealing with concrete structures is generally termed 'ultimate load analysis' and requires considerable modification to the method described here.

The stress–strain relationship of Figure 3.29 will now be applied to a simple, rectangular section, beam subjected to an applied bending moment M (Figure 3.30).

Under purely elastic conditions, line OA of Figure 3.29, the stress distribution over the cross-section of the beam will be as shown in Figure 3.30(b) and the limiting condition for elastic behaviour will be reached when the maximum stress reaches the value σ_y. As the applied bending moment is further increased, material within the depth of the section will be subjected to the yield stress σ_y and a condition represented by Figure 3.30(c) will exist in which part of the cross-section is plastic and part plastic. On further increase of the applied bending moment ultimately condition (d) will be reached in which the entire cross-section is plastic. It will not be possible to increase the applied bending moment further and any attempt to do so will result in increased curvature, the beam behaving as if hinged at the plastic section. Hence, the use of the term *plastic hinge* for a beam section which has become fully plastic.

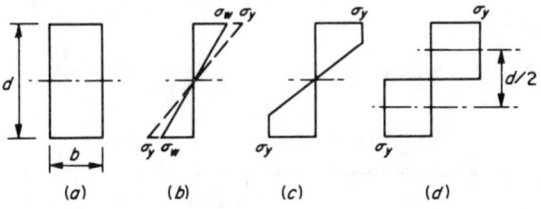

Figure 3.30

The moment of resistance of the fully plastic section is, from Figure 3.30(d):

$$M_p = b\frac{d}{2}\sigma_y\,\frac{d}{2} = \frac{bd^2\sigma_y}{4}$$

$$= Z_e\sigma_w \qquad\qquad (3.93)$$

where Z_p = plastic section modulus

In contrast, the moment of resistance at working stress σ_w is, from Figure 3.30(b):

$$M_w = b\frac{d}{2}\frac{\sigma_w}{2}\frac{2}{3}d = \frac{bd^2}{6}\sigma_w \qquad\qquad (3.94)$$

$$= Z_e\sigma_w$$

where Z_e = elastic section modulus

The ratio Z_p/Z_e is the shape factor of the cross-section. Thus the shape factor for a rectangular cross-section is 1.5.

The shape factor for an I-section, depth d and flange width b, is given approximately by:

$$\left(\frac{1 + x/2}{1 + x/3}\right)$$

where $x = \dfrac{t_w d}{2t_f b}$ and t_w and t_f are the web and flange thicknesses respectively

Values of plastic section moduli for rolled universal sections are given in steel section tables.

3.8.2 Theorems and principles

The definition of collapse, which follows from the assumed basic stress–strain relationship of Figure 3.29, has already been given. If the structural analysis is considered to be the problem of obtaining a correct bending moment distribution at collapse, then such a bending moment distribution must satisfy the following three conditions:

(1) *Equilibrium condition:* the reactions and applied loads must be in equilibrium.
(2) *Mechanism condition:* the structure, or part of it, must develop sufficient plastic hinges to transform it into a mechanism.
(3) *Yield condition:* at no point in the structure can the bending moment exceed the full plastic moment of resistance.

In elastic analysis of structures where several loads are acting, e.g. dead load, superimposed load and wind load, it is permissible to use the principle of superposition and obtain a solution based on the addition of separate analyses for the different loads. In plastic theory the principle of superposition is not applicable and it must be assumed that all the loads bear a constant ratio to one another. This type of loading is called 'proportional loading'. In cases where this assumption cannot be made, a separate plastic analysis must be carried out for each load system considered.

For cases of proportional loading, the uniqueness theorem states that the collapse load factor N_c is uniquely determined if a bending moment distribution can be found which satisfies the three collapse conditions stated.

The collapse load factor N_c may be approached indirectly by adopting a procedure which satisfies two of the conditions but not necessarily the third. There are two approaches of this type:

(a) We may obtain a bending moment distribution which satisfies the equilibrium and mechanism conditions, (1) and (2); in these circumstances it can be shown that the load factor obtained is either greater than or equal to the collapse load factor N_c. This is the 'minimum principle' and a load factor obtained by this approach constitutes an 'upper bound' on the true value.
(b) We may obtain a bending moment distribution which satisfies the equilibrium and yield conditions, (1) and (3), and in these circumstances it can be shown that the load factor obtained is either less than or equal to the collapse load factor N_c. This is the 'maximum principle' and its application produces a 'lower bound' on the true value.

It should be observed that whilst method (a) is simpler to use in practice, it produces an apparent load factor which is either correct or too high and thus an incorrect solution is on the unsafe side. A most useful approach is to employ both principles in turn and obtain upper and lower bounds which are sufficiently close to form an acceptable practical solution.

3.8.3 Examples of plastic analysis

This section contains some examples of plastic analysis based on the minimum principle. The method employed is termed the 'reactant bending moment diagram method'.

Example 3.5. The structure is a propped cantilever beam of uniform cross-section, carrying a central load W, as shown in Figure 3.31(a). The bending moment distribution under elastic conditions is shown in Figure 3.31(b) and it should be noted that the maximum bending moment occurs at the fixed end A.

As the load W is increased, plasticity will develop first at end A. As the load is further increased, end A will eventually become fully plastic with a stress distribution of the type shown in Figure 3.30(d) and the bending moment at A, M_A, will equal M_p the fully plastic moment of the beam. Further increase of load will have no effect on the value of M_A but will increase M_B until it also reaches the value M_p. The resulting bending moment distribution will now be as shown in Figure 3.31(c).

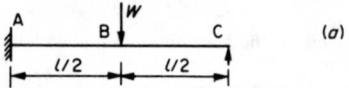

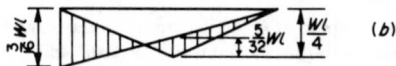

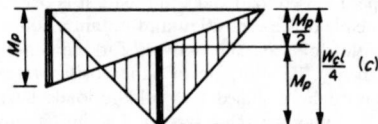

Figure 3.31

The geometry of the diagram produces a relationship between the load at collapse, W_c, and the plastic moment of resistance of the beam M_p, as follows:

$$\frac{W_c l}{4} = M_p + M_p/2$$

or:

$$W_c = 6\frac{M_p}{l} \tag{3.95}$$

If the working load is W_w then the load factor is given by:

$$N = \frac{W_c}{W_w} \tag{3.96}$$

Example 3.6. This is again a propped cantilever but here the load is uniformly distributed (Figure 3.32(a)). At collapse the bending moment diagram will be as shown in Figure 3.32(b) with plastic hinges at A and C. It should be noted that C is not at the centre of the beam. The location of the plastic hinge at C

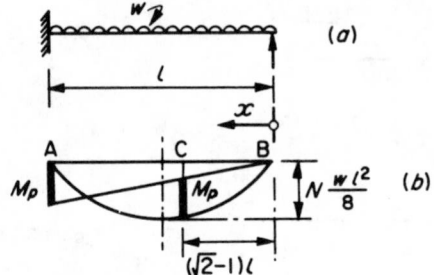

Figure 3.32

and the relationship between the load and the value of M_p may be obtained by differentiation as follows.

At C:

$$M_p = \left(N\frac{wlx}{2} - N\frac{wx^2}{2} \right) - M_p\frac{x}{l}$$

i.e.:

$$M_p = N\frac{wlx(l-x)}{2(l+x)} \tag{3.97}$$

$$\frac{dM_p}{dx} = N\frac{wl\{(l+x)(l-2x) - x(l-x)\}}{2(l+x)^2}$$

$$= 0 \text{ for } M_{p\,max}$$

Hence: $x^2 + 2xl - l^2 = 0$

i.e.:

$$x = l(\sqrt{2} - 1) = 0.414l$$

which locates the point C.

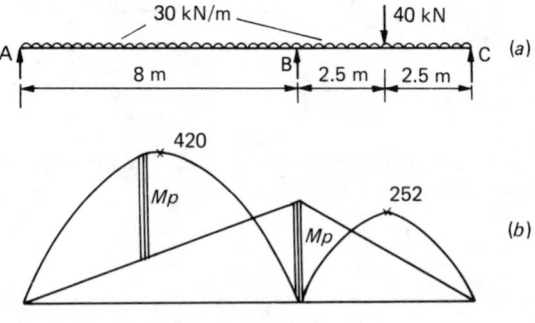

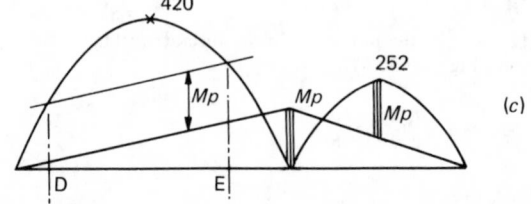

Figure 3.33

Also, substituting in Equation (3.97) for x:

$$M_p = \frac{Nwl^2(\sqrt{2}-1)}{2} \frac{}{\sqrt{2}}(2-\sqrt{2})$$

$$= \left(\frac{Nwl^2}{8}\right) 4(3-2\sqrt{2})$$

$$= 0.686 \left(\frac{Nwl^2}{8}\right)$$

Example 3.7. A two-span continuous beam is shown in Figure 3.33. The loads shown are maximum working loads and it is required to determine a suitable universal beam (UB) section such that $N = 1.75$ with a yield stress $\sigma_y = 250 \text{ N/mm}^2$. Effects of lateral instability are ignored for the purposes of this example.

With factored loads, the free bending moments are:

$$1.75 \times 30 \times \frac{8^2}{8} = 420 \text{ kNm}$$

$$1.75 \times 30 \times \frac{5^2}{8} + 1.75 \times 40 \times \frac{5}{4} = 252 \text{ kNm}$$

For collapse to occur in span AB, Figure 3.33(b)

$$420 \times 0.686 = M_p = 288 \text{ kNm}$$

For collapse in BC, assuming the span hinge in BC to occur at the centre (Figure 3.33(c)):

$$252 = \frac{3}{2}M_p; \quad M_p = 168 < 288$$

Hence the beam must be designed for $M_p = 288 \text{ kNm}$

$$= Z_p\sigma_y$$

Hence:

$$Z_p = \frac{288 \times 10^6}{250 \times 10^3} \text{ cm}^3 = 1152 \text{ cm}^3$$

From section tables, select 406×178 UB 60 ($Z_p = 1194 \text{ cm}^3$).

This design may be compared with elastic theory from which we obtain $M_{max} = 198 \text{ kNm}$, $Z_e = 1200 \text{ cm}^3$ (using $\sigma_w = 165 \text{ N/mm}^2$). A suitable section would be 457×152 UB 67 ($Z_e = 1250 \text{ cm}^3$) or, 406×178 UB 74 ($Z_e = 1324 \text{ cm}^3$).

The plastic design may be improved by choosing different sections for spans AB and BC:

For BC, $M_{PBC} = 168$ giving $Z_p = \frac{168}{250} \times \frac{10^6}{10^3} = 672 \text{ cm}^3$

Select 356×171 UB 45 ($Z_p = 773.7 \text{ cm}^3$)

For AB, $M_{PAB} \simeq 420 - \frac{1}{2}M_{PBC}$

$$= 420 - \frac{1}{2} \times \frac{773.7 \times 10^3 \times 250}{10^6}$$

$$= 420 - 96.7 = 323 \text{ kNm}$$

$$\therefore Z_p = \frac{323}{250} \times \frac{10^6}{10^3} = 1293 \text{ cm}^3$$

Select 406×178 UB 67.

The weights of steel used in the different designs may be compared.

First plastic design	780 kg
Elastic design	871 kg
Second plastic design	761 kg

As an alternative to the second plastic design the lower value of M_p could be used, based on collapse in BC (356×171 UB 45, $Z_p = 773.7$, $M_p = 193 \text{ kNm}$), and flange plates welded on to the beam in the region DE, Figure 3.33(c).

The additional M_p required at the plated section

$$= 420 - \frac{1}{2} \times 193$$
$$= 130 \text{ kNm}$$

Using plates 150 mm wide top and bottom, the plastic moment of resistance of the plates is approximately:

$$2 \left(150 \times t \times 250 \times \frac{356}{2}\right) \times 10^{-6}$$

$$= 13.4 \, t$$

where t = plate thickness in millimetres

Hence:

$$t = \frac{130}{13.4} \simeq 10 \text{ mm)}$$

Example 3.8. Here we consider the plastic analysis of a portal frame type structure as in Figure 3.34(a) and (b). At (a) the frame has pinned supports and at (b) fixed supports. A simple form of loading is used for illustration of the principles.

The frame is made statically determinate by the removal of H_A in both cases, and by the removal of M_A and M_E in case (b). The 'free' bending moment diagram is then as in diagram (c) and the reactant bending moment diagrams are as at (d) for H_A and at (e) for M_A and M_E combined. We now seek combinations of the diagrams which will satisfy the conditions of equilibrium, mechanism and yield (see page 3/27). We consider first the case of the two-hinged frame.

Diagram (f)

This is consistent with a pure sideway mode of collapse. From the geometry of the diagram:

$$M_p = \frac{Hh}{2} \tag{3.98}$$

The yield condition will be satisfied providing:

$$\frac{wl}{4} \leqslant \frac{Hh}{2} \tag{3.99}$$

Diagram (g)

This is a combined mechanism involving collapse of the beam and sideway. From the geometry of the diagram:

At D:

$$M_p = Hh \mp H_A h$$

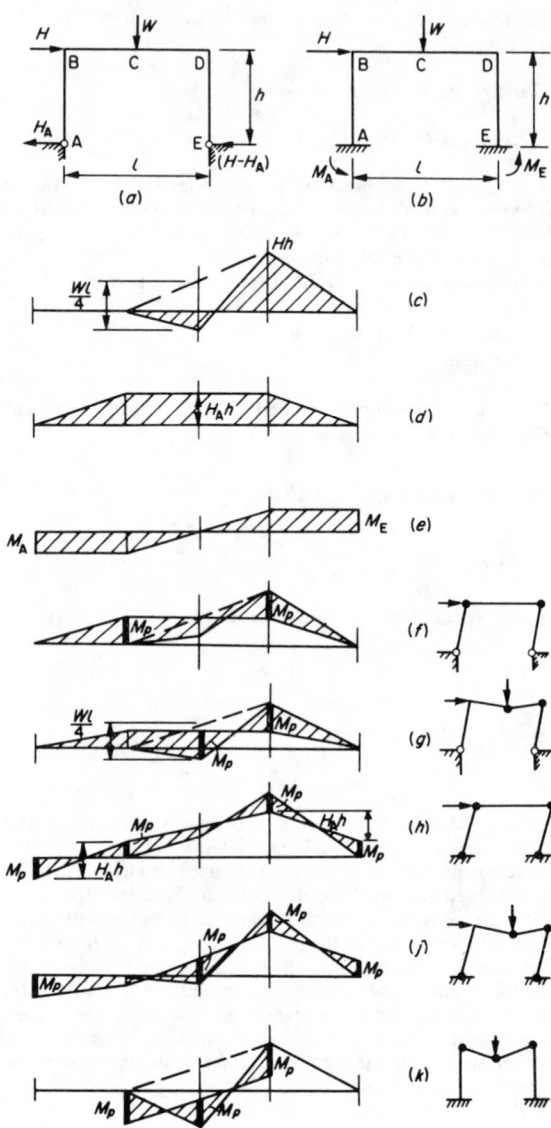

Figure 3.34

At C:

$$M_p = \frac{Wl}{4} - \frac{Hh}{2} \pm H_A h$$

Adding:

$$2M_p = \frac{Wl}{4} + \frac{Hh}{2}$$

or:

$$M_p = \frac{Wl}{8} + \frac{Hh}{4} \qquad (3.100)$$

In the case of the frame with fixed feet, there are three possible

modes of collapse. The corresponding bending moment diagrams are constructed at (h), (j) and (k) and the results are as follows:

Diagram (h):

$$M_p = \frac{H_A h}{2}$$

$$M_p = Hh - H_A h - M_p$$

Hence:

$$M_p = \frac{Hh}{4} \qquad (3.101)$$

Diagram (j):

$$M_p = \frac{Wl}{4} - \frac{Hh}{2} \pm H_A h$$

$$M_p = Hh \mp H_A h - M_p$$

Adding:

$$3M_p = \frac{Wl}{4} + \frac{Hh}{2}$$

or:

$$M_p = \frac{Wl}{12} + \frac{Hh}{6} \qquad (3.102)$$

Diagram (k)
This mode is the same as the collapse of a fixed end beam; the columns are not involved in the collapse apart from providing the resisting moment M_p at B and D. From the geometry of the diagram:

$$M_p = \frac{Wl}{8} \qquad (3.103)$$

Example 3.9. Here we consider a pitched roof frame, a structure which is eminently suitable for design by plastic methods. The frame is shown in Figure 3.35(a). The given loads are already factored and we are to find the required section modulus on the basis of a yield–stress $\sigma_y = 280$ N/mm², neglecting instability tendencies and the reduction in plastic moment of resistance due to axial forces.

The bending moment diagram at collapse is shown in Figure 3.35(b). The free bending moment diagram, EFGB, is drawn to scale after evaluating values of moment at intervals along the rafter members. The reactant line (H_A diagram) is then drawn by trial and error so that the maximum moment in the region BC is equal to the moment at D. This moment is the required M_p for the frame and is found to be:

$$M_p = 52 \text{ kNm} = \sigma_y Z_p$$

from which:

$$Z_p = \frac{52 \times 10^3 \times 10^3}{280 \times 10^3} = 186 \text{ cm}^3$$

Horne[8] and Baker and Heyman[9] should be consulted for a more

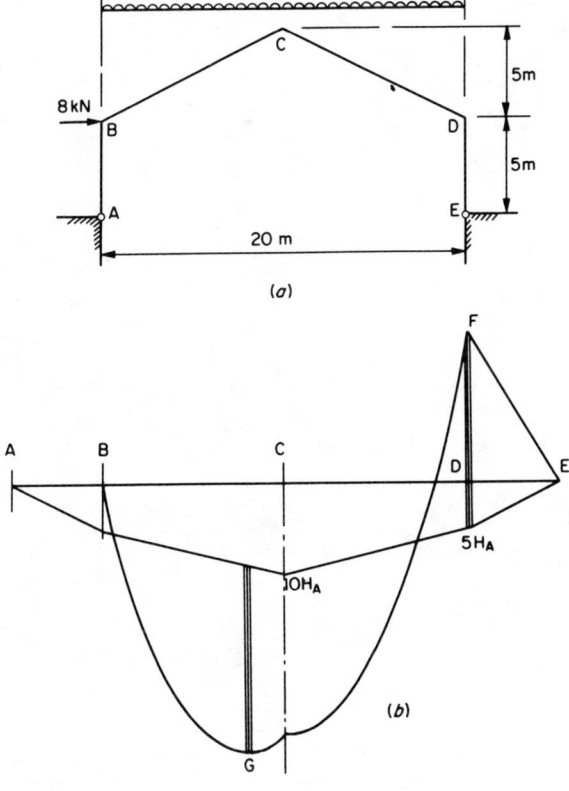

(5) Numbers of independent mechanisms.
(6) Shakedown.
(7) Effects of axial forces.
(8) Moment carrying capacity of columns.
(9) Behaviour of welded connections.

Figure 3.35

detailed study of plastic analysis. Among the topics deserving of further study are:

(1) Use of the principle of virtual work in obtaining relationships between applied loads and plastic moments of resistance.
(2) Effects of strain hardening.
(3) Evaluation of shape factors for various cross-sections.
(4) Application of the maximum principle in obtaining lower bounds.

References

1   Jenkins, W. M (1982) *Structural mechanics and analysis level IV/V*. Thomas Nelson/Van Nostrand Reinhold, London.
2   Coates, R. C., Coutie, M. G. and Kong, F. K. (1972) *Structural analysis*. Nelson, London.
3   Ghali, A. and Neville, A. M. (1978) *Structural analysis*. 2nd edn. Chapman and Hall, London.
4   Zeinkiewicz, O. C. (1977) *The finite element method*. McGraw-Hill, Maidenhead.
5   Desai, C. S. (1977) *Elementary finite element methods*. Prentice-Hall, New Jersey.
6   PAFEC (1978) PAFEC 75, Pafec Ltd, Pafec House, 40 Broadgate, Beeston, Nottingham NG9 2FW.
7   Warburton, G. B (1976) *The dynamical behaviour of structures*, 2nd edn. Pergamon Press, Oxford.
8   Horne, M. R. (1979) *Plastic theory of structures*, 2nd edn. Pergamon Press, Oxford.
9   Baker, J. and Heyman, J. (1969) *Plastic design of frames*. Cambridge University Press, Cambridge.

Bibliography

Bhatt, P. (1981) *Problems in structural analysis by matrix methods*. The Construction Press, London.
Cheung, Y. K. (1976) *Finite strip method in structural analysis*. Pergamon Press, Oxford.
Cheung, Y. K. and Yeo, M. F. (1979) *A practical introduction to finite element analysis*. Pitman, London.
Dawe, D. J. (1984) Matrix and finite element displacement analysis of structures. Clarendon Press, Oxford.
Graves Smith, T. R. (1983) *Linear analysis of frameworks*. Ellis Horwood, Chichester.
Harrison, H. B. (1980) *Structural analysis and design: some mini-computer applications*. Pergamon Press, Oxford.
Jenkins, W. M., Coulthard, J. M. and de Jesus, G. C. (1983) *BASIC computing for civil engineers*. Van Nostrand Reinhold, London.
McGuire, W. and Gallagher, R. H. (1979) *Matrix structural analysis*. Wiley, Chichester.

4

Materials

Philip King BSc
Robert Cather BSc
Arup Research and Development

Contents

4.1 Introduction

A good working knowledge of the materials used in civil engineering is very important to the engineer and in this book the characteristics and properties of many materials are described appropriately in other chapters as indicated below.

Material	Chapter
Soils	9
Rocks	8 and 10
Reinforcement	12
Steel	13
Aluminium	14
Bricks and masonry	15
Timber	16
Bituminous materials	23 (also 17 and 24)

This chapter is concerned with materials which are not covered elsewhere in the book and considers in detail only: concrete (pages **4**/3 to **4**/18), plastics and rubbers (pages **4**/18 to **4**/24) and paint (pages **4**/24 to **4**/26).

The authors gratefully acknowledge permission by Peter Pullar Strecker to include or update parts of his text from the 3rd Edition of the reference book (1974).

In the field of materials especially, the solution of problems often requires a full understanding of technologies outside the engineer's normal experience. Fortunately specialist help is usually readily available in the UK, although the enquirer does not always know where to look for it. Many sources are listed by the Construction Industry Research and Information Association (CIRIA)[1] *Guide to sources of construction information*. A selection of useful organizations and their addresses is as follows.

Aluminium Federation Ltd, Broadway House, Calthorpe Road, Five Ways, Birmingham B15 1TN.

Asbestos Information Centre, 40 Piccadilly, London W1V 9PA.

Association of Bronze and Brass Founders, 136 Hagley Road, Birmingham B16 9PN.

Brick Development Association, Woodside House, Winkfield, Windsor, Berks SL4 2DX.

British Aggregate Construction Materials Industries, 156 Buckingham Palace Road, London SW1W 9TR.

British Cast Iron Research Association, Alvechurch, Birmingham B48 7QB.

British Cement Association, Wexham Springs, Slough, Berks SL3 6PL.

British Ceramic Research Ltd, Queens Road, Penkhull, Stoke-on-Trent, Staffs ST4 7LQ.

British Constructional Steelwork Association Ltd, 35 Old Queen Street, London SW1H 9HZ.

British Glass Industry Research Association, Northumberland Road, Sheffield S10 2UA.

British Non-ferrous Metals Federation, 10 Greenfield Crescent, Edgbaston, Birmingham B15 3AU.

British Rubber Manufacturers' Association Ltd, 90–91 Tottenham Court Road, London, W1P 0BR.

British Standards Institution, 2 Park Street, London W1A 2BS.

British Steel Corporation, Corporate Research Laboratories, Swinden House, Moorgage, Rotherham S60 3AR.

British Wood Preserving Association, 150 Southampton Row, London WC1B 5AL.

Building Centres: London, Manchester, Bristol, Peterborough, Durham, Glasgow.

Building Research Establishment, Garston, Watford, Herts WD2 7JR.

Cement and Concrete Association, *see* British Cement Association.

Clay Pipe Development Association, Drayton House, 30 Gordon Street, London WC1H 0AN.

Concrete Pipe Association, 60 Charles Street, Leicester LE1 1FB.

Construction Industry Research and Information Association (CIRIA), 6 Storey's Gate, London SW1P 3AU.

Copper Development Association, Orchard House, Mutton Lane, Potters Bar, Herts EN6 3AP.

Flat Glass Council, 44–48 Borough High Street, London SE1 1XB.

Institution of Mining and Metallurgy, 44 Portland Place, London W1N 4BR.

Lead Development Association, 34 Berkeley Square, London W1X 6AJ.

National Physical Laboratory, Teddington, Middlesex TW11 0LW.

Paint Research Association, Waldegrave Road, Teddington, Middlesex TW11 8LD.

RAPRA Technology Ltd, Shawbury, Shrewsbury, Shropshire SY4 4NR.

Steel Construction Institute, Silwood Park, Ascot, Berks SL5 7QN.

Stone Federation, 82 New Cavendish Street, London W1M 8AD.

Timber Research and Development Association, Stocking Lane, Hughenden Valley, High Wycombe, Buckinghamshire HP14 4ND.

Zinc Development Association, 34 Berkeley Square, London W1X 6AJ.

4.1.1 Standards and codes of practice

British and some other standards and codes referred to in this chapter are listed separately in the bibliography.

4.2 Concrete

4.2.1 Cement

Hydraulic cement, i.e. a cement which hardens because of chemical reactions between the cement and water is the main, and often the only, binder used in concrete for civil engineering purposes. Portland cement or one of its variants is usually used, but high-alumina cement has advantages for some applications. The following list of cements is likely to be encountered in civil engineering. The relevant British Standards governing properties are given in the headings.

4.2.1.1 Ordinary Portland cement (OPC): BS 12

This is the most commonly used form of cement. It is made by heating together raw materials containing alumina and calcium. Clay and chalk or limestone are common sources. During the heating process the materials fuse to form clinker which is subsequently ground to a fine powder, gypsum usually being added at this stage to control the setting characteristics of the cement. Portland cements normally comprise four main phases or chemical compounds: tricalcium silicate, dicalcium silicate, tricalcium aluminate and calcium ferroaluminate. For convenience, these phases are usually given a shorthand notation of C_3S, C_2S, C_3A and C_4AF. This powder resulting from the grinding of clinker is the cement in its final form. The fineness of grinding, the raw materials and the conditions of the fusing process influence the nature and the reactivity of the cement, fine cement hardening more quickly than coarse cement of the same composition. The quality of British cement, although varying according to its source, usually exceeds the BS requirements by a considerable margin.

4.2.1.2 Rapid-hardening Portland cement (RHPC): BS 12

This is similar to OPC in composition but it is more finely ground. It gains strength more quickly than OPC, though the final strength is only slightly increased. Heat is generated more quickly during the hydration of the cement. This may have advantages in cold weather, or in precasting operations. The difference in strength development between OPC and RHPC has now become less marked.

4.2.1.3 Low-heat Portland cement: BS 1370

This cement is less reactive than OPC because it differs in composition, but it is nevertheless more finely ground than OPC. Heat is generated more slowly on hydration and lower concrete temperatures are reached. Early and eventual strengths are less than with OPC and the initial setting time is greater. This cement is made only to order in the UK.

4.2.1.4 Sulphate-resisting Portland cement: BS 4027

This cement is similar to OPC but the proportions of the cement phases are different and it is less prone to attack by sulphates principally by having a controlled low C_3A content. Heat may be generated more slowly than with OPC, but a little more quickly than with low-heat Portland cement.

4.2.1.5 Portland blast-furnace cement: BS 146

This cement is made by grinding together OPC clinker with granulated blast-furnace slag (see later). The granulated blast-furnace slag content must be less than 65% of the total weight. This cement is less reactive than OPC and gains strength a little more slowly. It has advantages in generating heat less quickly than OPC and in being more resistant than OPC to attack from sulphates. Portland blast-furnace cement is not widely available in the UK. (Low-heat Portland blast-furnace cement contains more slag but is manufactured only to order in the UK; BS 4246 governs its composition and properties.) Combination at the concrete mixer of Portland cement with ground granulated blast-furnace slag is more commonly used to achieve similar performance. By this method a wider range of OPC:slag ratios is readily achievable. These combinations are likely to be available in most parts of the UK.

4.2.1.6 Portland PFA cement: BS 6588

This cement is manufactured by intergrinding or combining at the cement plant pulverized fuel ash (PFA), complying with BS 3892, Part 1 (see later) with ordinary Portland cement. The PFA content should be between 15 and 35% by weight. The rate of strength development is slower than that of the respective Portland cement source. The cement may generate heat less quickly and be more chemically resistant in some circumstances.

Combination of PFA with ordinary Portland cement at the concrete mixer can produce concrete with a similar performance to that using this cement.

4.2.1.7 Pozzolanic cement with PFA as pozzolana: BS 6610

As for BS 6588 but the PFA content is between 35 and 50%. This cement is not referred to in BS 8110 or BS 5328 and is therefore unlikely to be used in reinforced concrete or other slender structural elements. The lower heat of hydration is useful property in massive structures.

4.2.1.8 White Portland cement

This cement is similar to OPC but with selected raw materials and processing to remove the normal OPC grey coloration; it would also comply with BS 12 for setting time and early and eventual strength.

4.2.1.9 Supersulphated cement: BS 4248

This cement is made from granulated blast-furnace slag, gypsum and not more than 5% of OPC clinker. It is more resistant to sulphate attack than sulphate-resisting cement, and it is not attacked by weak acids. This cement is much finer though less reactive than OPC, but eventual strengths are at least as high. It is not currently available in the UK. Good control of concrete mix is essential and its use has largely been superseded by other cement–slag combinations.

4.2.1.10 Water-repellent cement

This is made from OPC and stearates. It is used to reduce water permeability especially in screeds and rendering.

4.2.1.11 Masonry cement: BS 5224

This cement is made by mixing OPC with plasticizers and a fine powder (often whiting). It is used to give plasticity to bricklaying and rendering mortars, especially where the local sand is harsh.

4.2.1.12 High-alumina cement: BS 915

This cement is chemically different from OPC and its varieties. Concrete made with it has different properties from OPC concrete. High-alumina cement is very reactive and produces very high early strengths (the eventual strength may be reached in less than 1 day) but the initial setting is slower than with all varieties of Portland cement.

High-alumina cement is very resistant to attack from sulphates and is more resistant to acid attack than any variety of Portland cement but is attacked by alkalis. At temperatures above 700°C, high-alumina cement forms a ceramic bond with suitable aggregates and it can therefore be used for refractory concrete. Under moist conditions at temperatures of 40° to 100°C conversion takes place and high-alumina cement loses strength. Cement in this condition is less resistant to chemical attack.

It is widely believed that high-alumina cement should not be used in contact with hardened Portland cement. The scientific basis for this is, however, less well founded. Mixtures of unhardened Portland and high-alumina cements lead to very rapid 'flash' setting. This phenomenon has some practical applications where almost instantaneous setting is wanted, but the quality of the resulting concrete will be in most respects inferior to either Portland cement concrete or high-alumina cement concrete.

High-alumina cement concrete is not permitted for use in structural concrete in BS 8110. Applications such as floor toppings, hardstandings are still permissible.

4.2.1.13 Other cementing materials

Ground granulated blast-furnace slag. This is a by-product of the manufacture of iron from iron ore. The molten slag is removed from the furnace and quenched rapidly (granulation). Subsequent grinding can be either after combination with Portland cement clinker or more commonly of the granulated slag alone. The slag is composed mainly of calcium and magnesium silicates and alumino-silicates. Although some small strength gain or hardening would take place in water, the strengths developed are not likely to be sufficient for construction. Blending with a Portland cement produces a much faster and useful strength gain. Combinations of ground granulated blast-furnace slag and Portland cements have been used for many years both in the UK and overseas. An increase in the use and interest in these materials has taken place over recent years in the UK and BS 6699 gives composition and performance requirements. It is widely available in the UK.

Pozzolanas. Natural or artificial materials containing amorphous silica in a reactive form. The silica can react with lime to produce cementing compounds giving useful strength properties. This lime can be either hydrated lime or the calcium hydroxide produced during the hydration of Portland cements. The original pozzolana was volcanic ash from Pozzuoli, Italy. Using pozzolanas as a cementing component in Portland cement concretes can be useful to reduce heat of hydration or to improve resistance to some chemicals. Early age strength development may be affected unless the concrete is proportioned to allow for it.

Pulverized fuel ash (PFA). This is the most common pozzolana used in Portland cement concrete. It is electrostatically precipitated from the exhaust fumes of coal-fired power stations burning pulverized coal. It is widely available in the UK, and performance and compositional requirements are given in BS 3892, Part 1 (for use in structural concrete) and BS 3892, Part 2 (for miscellaneous uses in concrete).

Condensed silica fume. A high-purity silica pozzolana which has a very fine particle size much smaller than that of cement or PFA (mean particle size approximately 1 μm). Condensed silica fume is so fine it can be used to fill the interstices between cement particles and it reacts rapidly with the cement hydration products. Condensed silica flume is a by-product of the production of silicon and ferro-silicon being collected by cooling and filtering of furnace gases. Condensed silica flume can be used to produce very high strengths and good chemical resistance.

4.2.1.14 Non-UK standards

Many other national standards exist for Portland cements and combinations of Portland cements with blast-furnace slag or PFA. These standards cover similar ranges of materials to those in the British Standards given in the preceding pages although the overlap will not be complete for each country. Methods or terminology of classification vary for each country but common principles exist, e.g. sulphate-resisting cements are always low in C_3A content but the actual limiting value will be different.

Standards issued by the American Society for Testing Materials (ASTM) are widely used outside the US. Their standard C-150 has five main categories of Portland cement and a summary of these types is given in Table 4.1.

Other national standards for Portland cements which are likely to be encountered more widely are issued by Deutsches Institut für Normung (DIN) and in Japan as Japanese Industrial Standards (JIS). A wide range of cement specifications are incorporated within these standards and, hence, are not reproduced here.

4.2.2 Aggregates

Aggregates form more than three-quarters of the volume of concrete and the selection and proportioning of coarse and fine aggregates greatly influence the properties of both fresh and hardened concrete. The choice of grading, maximum aggregate size and aggregate:cement ratio are subjects for concrete mix design and are dealt with below. In this section the selection of aggregate type will be covered. Broadly, aggregates can be classified according to density as normal (particle density 2000 to 3000 kg/m³), lightweight (less than 2000 kg/m³) and heavy

aggregates (greater than 3000 kg/m³). Typical properties of concretes made with a range of aggregates are given in Table 4.2.

Table 4.1 Cement type classification in ASTM C-150

Type	Use	Special requirements
I	Where other special types not needed	
II	General use, moderate sulphate resistance or moderate heat of hydration	Max.C_3A (8%)
III	For high early strength	
IV	For low heat of hydration	Max. C_3S (35%) Min. C_2S (40%) Max. C_3A (7%)
V	For high sulphate resistance	Max C_3A (5%) Max. $C_4AF + 2 C_3A$ (20%)

4.2.2.1 Normal aggregates

These usually consist of natural materials, hard crushed rock or crushed or natural gravel and their corresponding sands, but artificial materials like crushed brick and blast-furnace slag can also be used. The specific gravity of these materials usually lies between 2.6 and 2.7. Because satisfactory concrete for most purposes can be made with a very wide range of aggregates, local sources of supply usually determine which aggregate will be used. Where very high strength, resistance to skidding, good appearance or other special properties are required, appropriate aggregates will have to be selected, preferably on the basis of previous experience.

For example, the low-speed skidding resistance of concrete roads is affected by the hardness of the sand but only slightly by the polished-stone value of the coarse aggregate. Thus, a hard sand should be chosen for concrete which is to form the surface of a concrete pavement.

Some aggregates have undesirable influences on important concrete properties or are themselves unsound. They should be used with caution, if at all. Examples are aggregates with high drying shrinkages, which may lead to poor durability in exposed concrete, aggregates which react with alkalis in the cement paste, aggregates which are readily oxidized, aggregates which can cause surface staining, and aggregates made from weathered, partially decomposed, rocks.

Other aggregates, although making reasonably satisfactory hardened concrete, for most purposes, may give the fresh concrete poor handling characteristics. Aggregates with flat, flakey, very angular or hollow particles tend to have this effect. In general, aggregates with well-rounded particles in the case of gravels, or near-cubical particles in the case of crushed rock, produce concrete with better workability and fewer voids than aggregates with angular particles.

Natural sands have advantages over crushed rock sands because their particles tend to be more rounded and they contain less very fine material (of 150 μm or less), but crushed rock sands may be preferable, e.g. where the grading of locally occurring natural sands is poor, where the colour of natural sands would be unsatisfactory in weathered concrete (many sands weather to a yellowish colour) or where resistance to slipping is important. General requirements for aggregates to be used in concrete are given in BS 882.

Table 4.2 Properties of concrete using different aggregates

Aggregate	Typical range of dry density		Compressive strength (N/mm²)	Drying shrinkage (%)	Thermal conductivity at 5% moisture content (W/m°C)
	Aggregate (kg/m³)	Concrete (kg/m³)			
Flint gravel or crushed rock	1350–1600	2200–2500	20–80	0.03–0.08	1.6–2.2
Crushed limestone	1350–1600	2200–2400	20–80	0.03–0.04	1.6–2.0
Crushed brick	1100–1350	1700–2150	15–30	—	0.85–1.50
Expanded clay, shale or slate and sintered pulverized fuel ash	300–1050	1350–1800	15–60	0.02–0.12	0.55–0.95
Foamed slag	500–950	1700–2100	15–60	0.04–0.10	0.85–1.40
Expanded clay, shale or slate and sintered pulverized fuel ash	300–1050	700–1300	2–7	0.03–0.07	0.24–0.50
Foamed slag	500–950	950–1500	2–7	0.03–0.07	0.30–0.65
Pumice	500–900	650–1450	2–15	0.04–0.08	0.21–0.63
Exfoliated vermiculite and expanded perlite	60–250	400–1100	0.5–7	0.20–0.35	0.15–0.39
Clinker	700–1050	1050–1500	2–7	0.04–0.08	0.35–0.65

4.2.2.2 Lightweight aggregates

These consist of various artificial and natural materials with specific gravities of between 0.1 and 1.2. They are used to make lightweight concrete for structural and insulating applications. In general, concrete made with lightweight aggregates has better fire resistance than dense concrete, but greater shrinkage and moisture movement.

Examples of lightweight aggregates are given below.

(1) *Sintered PFA* is made by heating pellets of PFA until they fuse to form hard spherical lumps.
(2) *Expanded clay, shale, slate and perlite* are made by heating suitable grades of these materials to their fusion temperature (about 1000°C) when they simultaneously fuse and are blown by gases generated within the material.
(3) *Pumice* is a natural lightweight aggregate consisting of a frothy volcanic glass.
(4) *Clinker* consists of fused lumps of fuel residues. To be suitable for use as a concreting aggregate it must be low in sulphates and residual fuel. Limits are given in BS 1156.
(5) *Foamed blast-furnace slag* is made by treating molten blast-furnace slag with water so that the steam which is generated blows the slag. Standards for this material are given in BS 877.
(6) *Exfoliated vermiculite* is made by heating vermiculite (a micalike mineral found in Africa and America) to a temperature of about 700°C when it expands to form a very light material.

Of these aggregates the sintered PFA, and the expanded clay, shale and slate and perlite are the most likely to be encountered.

4.2.2.3 Heavy aggregates

These consist either of natural or artificial materials and are used to make high-density concrete for radiation shielding or ballasting.

Examples of heavy aggregates are barytes, which is a naturally occurring rock consisting of 95% barium sulphate (specific gravity about 4.1; density of concrete up to 3700 kg/m³); iron ores such as magnetite, goethite, limonite and ilmenite (specific gravity about 3.4 to 5.3, density of concrete up to 4200 kg/m³); iron or steel shot (specific gravity 7.7; concrete density up to 5500 kg/m³); lead shot (specific gravity 11.4; concrete density up to 7000 kg/m³) and scrap-iron stampings and punchings. Provided the materials are sound and free from oil, satisfactory concrete of good structural strength can be made, especially if prepared by a method such as prepacking to avoid segregation. Consideration of the higher-density effect on mixing and batching facilities is important.

4.2.2.4 Contaminants, unsound aggregates and reactive aggregates

Aggregates may contain impurities which upset the hydration of the cement or coatings which interfere with bond, or the aggregates themselves may be unstable. To some extent, impurities and surface coating can be removed by suitable treatments, but aggregates which are unsound or reactive must be avoided.[2] Unsound or reactive particles may occur naturally with the aggregate source and may be detected by careful examination of the supply. It is also possible for a small percentage of contamination to occur during transportation or storage of aggregate.

Organic impurities. These may or may not delay or prevent the hydration of the cement and it is best to compare the strength of the concrete made with the contaminated aggregate with the strength of concrete made from similar but clean aggregate. Sugar, sugar-like substances and humic acid are among common contaminants which are known to retard or prevent cement hydration. Products of wood degradation such as 'cellibiose' have a similar effect.

Clay and fine material. These can contaminate aggregates either as a coating on the coarse aggregate or as a constituent of the fine aggregate. As coatings, these materials interfere with bond and therefore reduce concrete strength. As constituents of the mix they are less troublesome unless the quantity is great enough to require the addition of extra water to make the concrete workable. Clay, silt and fine material should not form more than 1% by weight of coarse aggregate, 3% by weight of gravel sand or 15% by weight of crushed rock sand (BS 882).

Salt is usually present in marine deposited or extracted aggregates and in small quantities it is harmless. Efficient washing of the aggregates before use in concrete is capable of reducing the salt to an acceptable level. The salt content should, however, be limited to the levels in Table 4.3 taken from BS 882:1983.

In addition to the limits given in Appendix C of BS 882, there is an overall limit given for the chloride ion from all sources calculated as a percentage by weight of cement given in Table 6.4 of BS 8110.

Table 4.3 Maximum chloride content of aggregates

Type or use of concrete	Maximum total chloride content expressed as percentage of chloride ion by mass of combined aggregate
Pre-stressed concrete ⎱ Steam-cured structural concrete ⎰	0.02
Concrete made with cement complying with BS 4027 or BS 4248	0.04
Concrete containing embedded metal and made with cement complying with BS 12	0.06 for 95% of test results, with no result greater than 0.08

Note:
Marine aggregate and some inland aggregate contain chlorides. Both should be selected carefully and may need efficient washing to achieve the limit required for use in pre-stressed concrete.

Nondurable particles. These are sometimes found in aggregates which are otherwise satisfactory. Examples of such particles are lumps of clay, shale, wood or coal. Being soft, they are easily eroded and will lead to pitting or spalling of the concrete surface. If more than about 5% of such particles are present in the aggregate they will also cause strength to be reduced. Although no limits are given for these in BS 882, generally such particles should not form more than 1% of the aggregate by weight. The actual significance of the particles in the structure will be affected by the nature of the structure, e.g. a concrete paving will be more affected by soft particles floating to the surface than will a wall.

Reactive particles. Reactive particles found in some aggregates may be soluble in, or react with, water or the hydrating cement paste. Mica and sulphates, e.g. gypsum, react with cement paste, and iron sulphides, e.g. pyrites and marcasite, react with air and water to form products which then react with the cement paste and cause staining or pop-outs.

Unsound material. This may form the whole of the aggregate or unsound particles may merely contaminate it. Unsoundness is the property of some aggregates to expand or contract excessively as a result of freezing and thawing, wetting and drying, or temperature changes. Such movements can be large enough to cause the aggregate itself to break down or they may disrupt concrete made with it. Examples of unsound aggregates are rocks with very high water absorption, porous cherts, limestones and other sedimentary rocks if they contain laminae of clay, and some shales. Foreknowledge of how such aggregates behave in concrete is the only reliable guide, but freezing and thawing tests may give some indication of an aggregate's unsoundness.

Reactive aggregates. Reactive aggregates are those which react chemically with the cement paste, the most common reaction being between reactive silica and alkalis (in the form of sodium and potassium ions). Reactive silicas occur in opaline and chalcedonic cherts, siliceous limestone, rhyolites, andesite and phyllites. The actual susceptibility of particular aggregate sources needs to be assessed by tests or previous experience. The silica forms a gel with the alkali and this gel expands continuously as it absorbs water, exerting enough force to disrupt the surrounding cement paste in some cases.[1] This phenomenon of alkali silica reactions is well known and recorded. It was first identified some 46 years ago by Stanton in the US. Since then, workers in other countries around the world notably Denmark, Iceland, Germany and South Africa have identified similar reactions. It was believed until recently that the combination of high alkali cements together with reactive aggregates did not occur in the UK. However, a number of cases of alkali silica reaction have now been reported in UK structures built over many years. It is not clear at this time what the extent of these occurrences are or what significance they will have in structural performance. Guidance is available on minimizing the risks of the reaction.[3]

4.2.3 Admixtures

Relatively small quantities of other materials called admixtures can be added to concrete to modify its properties in either fresh or hardened state. There are several classes of admixtures which are listed below.

The British Standard for admixtures BS 5075 is in separate parts for each class of admixture.

4.2.3.1 Water-reducing admixtures and workability aids (BS 5075, Part 1)

These materials are also commonly called plasticizers and have the effect of making concrete more workable for a given water content. They can also reduce the water:cement ratio for a constant workability and can therefore be used to improve strength development.

These materials can also entrain a little air in the concrete or, if used in too high a dosage, can cause retardation of the cement setting. If used as a result of trial mixes or in accordance with the manufacturer's recommendations these side-effects should not be significant under normal site conditions.

Plasticizers for mortars are used to give plasticity or cohesion. They function by entraining large amounts of air which, as a side-effect, reduces strength. This modification to mortars should be carried out using only admixtures specifically formulated for the particular use.

4.2.3.2 Superplasticizers and high-range water-reducing admixtures (BS 5075, Part 3)

These more specialized admixtures perform similar functions to normal plasticizers but with increased effectiveness. Very high workability or flowing concrete is a common application. Because of their very effective action on the fluid properties of the concrete, much closer control of the initial mix design and subsequent batching is needed to prevent excessive bleeding or segregation of the mix. Many of the general superplasticizing admixtures have a relatively limited activity and the concrete workability may fall back to normal levels after approximately 30 to 45 min.

4.2.3.3 Air-entraining agents (BS 5075, Part 2)

These are widely used admixtures, especially for paving concrete. Their importance is related to the capacity of concrete containing a small amount of air in the form of well-distributed small bubbles to have greater resistance to the destructive action of freezing and thawing when the concrete is saturated than similar concrete made without air-entraining agents. The freezing and thawing action is made more severe when de-icing salts are used, or can be brought on to the surfaces by vehicles. In such circumstances the use of air entrainment is strongly recommended in codes of practice. This increased durability is gained at the expense of some strength and it is therefore important to control the amount of entrained air between close limits.

The amount of air that will be entrained with a given addition of an air-entraining agent is influenced by the grading of the sand, the workability of the concrete, the type of mixer and the duration of mixing. Trial mixes are essential to establish how much of each agent is to be added. Frequent regular measurements must be made throughout the work to ensure that the correct air content is being maintained (see page 4/17). Some difficulty may be experienced when using fine sands, sands with an organic or carbon content or when PFA and ground granulated blast-furnace slag materials are incorporated in the mix constituents.

As well as being more resistant to damage from de-icing salts, air-entrained concrete is somewhat more cohesive than concrete made without an air-entraining agent and tends to have slightly higher workability, a factor which can be used partly to offset the strength reduction.

4.2.3.4 Accelerators and 'antifreezes' (BS 5075, Part 1)

These are used to hasten the hardening of concrete, particularly in cold weather. The term 'antifreeze' is misleading because these admixtures merely lessen the period when frost damage is likely; they do not prevent concrete from freezing. Since the prohibition of the use of chloride-based accelerators as a result of corrosion of embedded steel, other proprietary products, often based on calcium formate, have been developed. Such admixtures are much less efficient at accelerating the strength development and therefore are less attractive to use. There may also remain some uncertainty about the risks of inducing corrosion. Alternative procedures for protecting concrete or mortars from frost, such as heated materials and adequate protection for the formed work, may be preferable.

4.2.3.5 Retarders (BS 5075, Part 1)

These have the effect of delaying the onset of hardening and usually also of reducing the rate of the reaction when it starts. Ultimate strengths are unaffected by retardation for several hours but may be reduced if the addition of retarder is excessive. Accidental overdosage may cause retardation of a few days or it may prevent hardening altogether. The fear that this may happen is probably one of the reasons why retarders are seldom used in the UK. Nevertheless, retarders can be beneficial where large volumes of concrete have to be poured in one operation or where high ambient temperature conditions prevail which lead to rapid setting. Care must be taken in this situation that the rapid set is not the result of rapid moisture loss by evaporation. Trial mixes are essential to determine the dosage at which the retarder is to be used.

4.2.3.6 Mixed admixtures

Mixed admixtures containing a variety of materials are available. Examples are combinations of an air-entrainment admixture with water-reducing admixture, or water-reducing and retarding admixtures.

4.2.3.7 Other admixtures

These include waterproofers, viscosity modifiers, resin bonding agents, fungicides, etc. They may be useful for specific applications, but the claims made for them should be supported by impartial test results. This applies particularly to the permanence of the effects claimed.

Pigments may be incorporated in concrete mixes. If bright or pastel shades are wanted, white cement and light-coloured sand must be used for the basic concrete, but low-key colours and dark shades can be obtained with ordinary concrete. The pigments must be stable in cement, fast to light and resistant to being washed out by weathering. Requirements are given in BS 1014.

Although a number of organic pigments can be used in concrete, the most commonly used are iron oxides for red, brown, yellow and black, and chromium oxide for green. Synthetic iron oxides have better staining power than natural ones and are available in a greater colour range. Although more expensive than natural oxides, they may be cheaper in use. Carbon black gives a more intense black than iron oxide, but because it is often greasy it is difficult to disperse and has the reputation of being easily washed out. Pigment additions vary typically from about 2 to 10% or more by cement weight. Some strength reduction should be expected with the larger rates of addition.

4.2.4 Concrete mix design

4.2.4.1 General

The purpose of concrete mix design is to choose and proportion the ingredients used in a concrete mix to produce economical concrete which will have the desired properties both when fresh and when hardened. The variables which can be controlled are: (1) water:cement ratio; (2) maximum aggregate size; (3) aggregate grading; (4) aggregate:cement ratio; and (5) use of admixtures.

Interactions between the effects of the variables complicate mix design and successive adjustments following trial mixes are usually necessary. Experience built up by ready-mix concrete producers should enable them to produce suitable mix designs more quickly than this. Many different methods of mix design have been developed, one relatively simple method is given by Teychenné, Franklin and Erntroy.[5]

4.2.4.2 Water:cement ratio

Many of the most important properties of fully compacted hardened concrete and strength in particular are for normal concrete virtually decided by the water:cement ratio of the mix. The importance of this parameter is due to the fact that any excess of water over that needed to hydrate the cement (about 25% by weight) forms voids in the concrete, thus reducing its density. The reduced density leads to reduced compressive, tensile and bond strengths, lower durability, lower resistance to abrasion and greater permeability to water. Excess water cannot be eliminated altogether because it is needed to lubricate the mix and make it workable, but it should be kept to a minimum.

Figure 4.1 shows how strength is influenced by water:cement ratio and the first step in concrete mix design is to fix the water:cement ratio from a knowledge of the strength required. The shape of the curves will be similar for all types of Portland cements but the actual relationship between strength and water:cement ratio will be different for each cement source.

4.2.4.3 Workability

When the concrete is fresh it must be workable or fluid enough to be compacted easily under the conditions in which it will be

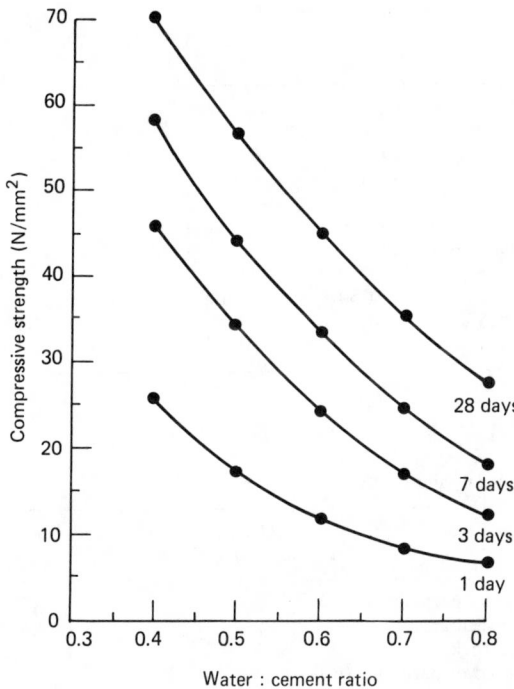

Figure 4.1 Influence of water:cement ratio on strength for typical UK OPC sources

Table 4.4 Suggested workabilities of concrete mixes

Workability	Suitable use	BS 1881 recommendation for method of measuring workability
Very low	Vibrated concrete in large sections	Vebe time
Low	Mass concrete foundations without vibration. Simple reinforced sections with vibration	Vebe time, compacting factor
Medium	Normal reinforced work without vibration, and heavily reinforced sections with vibration	Compacting factor, slump
High	Sections with congested reinforcements. Not normally suitable for vibration	Compacting factor, slump, flow
Very high	As for high workability plus large volume pours	Flow

placed. This is vitally important since loss of density has a very large effect in reducing strength. Table 4.4 gives suggested levels for workability suitable for different circumstances. Other factors may influence the selection of workability, e.g. some large foundations have been poured with very high workability to increase rates of placing. The cement paste is the lubricant which provides workability, but on grounds of economy as well as for technical reasons such as the limitation of shrinkage and thermal contraction, the amount of cement paste should be as small as possible. The next stages in the mix design process are intended to ensure that the mix will not be richer than is necessary.

4.2.4.4 Maximum aggregate size

Using a larger aggregate requires less fines to make up a volume of concrete; using a lower fines content will enable less cement and less water to be used for the same workability. The largest size of aggregate which can be used is governed by the dimensions of the section being cast and the spacing of the reinforcement. It is unusual to use aggregate with a maximum size of more than 20 mm for reinforced concrete, or with a maximum size greater than 25% of the section thickness for any work. Where large aggregates are to be used, consideration needs to be given to the use of appropriate-sized concrete testing equipment.

4.2.4.5 Overall grading

Sand and coarse aggregates frequently occur together, e.g. in gravels, but they seldom occur naturally in the best proportions for making concrete. Although all-in aggregates can be used, it is usually more satisfactory and more economical to separate sand from coarse aggregate and then recombine them in the required proportions. Further adjustments could be made by separating the aggregates into smaller-size groups and recombining them as required, but it is doubtful if this would repay its cost for most concreting requirements. Table 4.5 shows the

British Standard requirements for aggregate gradings. The criterion for determining what proportions of sand and coarse aggregate should be used is that the concrete shall be cohesive enough to resist segregation but not so oversanded as to require higher cement and water contents.

Fine sands provide more cohesion than coarse ones so less sand will be needed if it is fine. Very fine sand is not recommended for structural concrete unless tests have shown that concrete made with it is satisfactory. Very coarse sand may cause difficulties in surface finishing if floors or pavements are being constructed. To resist segregation, high workability mixes need more sand than low workability ones, and the proportion of sand must also be increased as the maximum aggregate size is reduced. Crushed rock coarse aggregates need more sand than rounded gravel aggregates.

Table 4.6 taken from BS 5328 gives an indication of sand:coarse-aggregate ratios considered suitable for a range of 'prescribed' mixes.

4.2.4.6 Cement content

When the water:cement ratio has been fixed, the related variable of cement content can be chosen to ensure that there will be enough cement paste to produce a workable mix. The cement content that will be needed to do this depends on the grading and shape of the aggregates, more cement being needed for mixes with a finer overall grading or a more angular coarse aggregate. Trial mixes will usually be needed before the choice of cement content is finally made. Cement contents considered suitable for a range of prescribed mixes are also included in Table 4.6 and may also form the starting point for trial mixes if the work by Teychenné et al.[5] is not available.

4.2.5 Properties of hardened concrete

4.2.5.1 Compressive strength

This depends on water:cement ratio, degree of compaction, the type of cement and aggregates used, the curing and the age of the concrete. To the extent that compressive strength reflects water:cement ratio and density, it is a good indicator of general

Table 4.5 British Standard requirements for aggregates[3]

(i) Coarse aggregate

| Sieve size (mm) | Percentage by mass passing BS sieves for nominal sizes | | | | | | | |
| | Graded aggregate (mm) | | | Single-sized aggregate (mm) | | | | |
	40–5	20–5	14–5	40	20	14	10	5*
50.00	100	—	—	100	—	—	—	—
37.50	90–100	100	—	85–100	100	—	—	—
20.00	35–70	90–100	100	0–25	85–100	100	—	—
14.00	—	—	90–100	—	—	85–100	100	—
10.00	10–40	30–60	50–85	0–5	0–25	0–50	85–100	100
5.00	0–5	0–10	0–10	—	0–5	0–10	0–25	45–100
2.36	—	—	—	—	—	—	0–5	0–30

*Used mainly in precast concrete products.

(ii) Fine aggregate

| Sieve size | Percentage by mass passing BS sieve | | | |
| | Overall limits | Additional limits for grading | | |
		C	M	F
10.00 mm	100	—	—	—
5.00 mm	89–100	—	—	—
2.36 mm	60–100	60–100	65–100	80–100
1.18 mm	30–100	30–90	45–100	70–100
600 μm	15–100	15–54	25–80	55–100
300 μm	5–70	5–40	5–48	5–70
150 μm	0–15*	—	—	—

*Increased to 20% for crushed rock fines, except when they are used for heavy duty floors.

Note:
Fine aggregate not complying with this table may also be used provided that the supplier can satisfy the purchaser that such materials can produce concrete of the required quality.

(iv) Clay, silt and dust

Aggregate type	Quantity of clay, silt and dust (max. % by mass)
Uncrushed, partially crushed or crushed gravel	1
Crushed rock	3
Uncrushed or partially crushed sand or crushed gravel fines	3
Crushed rock fines	15 (8 for use in heavy duty floor finishes)
Gravel all-in aggregate	2
Crushed rock all-in aggregate	10

Note:
The nature of the material passing the 75 μm BS 410 test sieve used in the decantation method differs between crushed rock and gravel or sand.

(iii) All-in aggregate

| Sieve size | Percentage by mass passing BS sieves for nominal sizes | | | |
	40 mm	20 mm	10 mm	5 mm*
50.0 mm	100	—	—	—
37.5 mm	95–100	100	—	—
20.0 mm	45–80	95–100	—	—
14.0 mm	—	—	100	—
10.0 mm	—	—	95–100	100
5.00 mm	25–50	35–55	30–65	70–100
2.36 mm	—	—	20–50	25–100
1.18 mm	—	—	15–40	15–45
600 μm	8–30	10–35	10–30	5–25
300 μm	—	—	5–15	3–20
150 μm	0–8†	0–8†	0–8†	0–15

*Used mainly in precast concrete products.
†Increased to 10% for crushed rock fines.

concrete quality and it is an easy property to measure with reasonable consistency. The cube crushing strength is consequently an important test of both the structural and general quality of the concrete (see page 4/16).

The development of compressive strength with age is greatly influenced by the temperature of the concrete, especially early in its life. Since the hydration of the cement itself generates heat, the temperature of the concrete is influenced not only by its initial temperature and the temperature of the surroundings, but also by the volume and shape of the section. Figure 4.2 indicates how the development of strength is related to the temperature of the concrete itself, and Table 12.4 in Chapter 12 relates the strength of various grades of concrete to the age at test. It should be remembered that because of these factors the properties and performance of the concrete in the structure will be different from the same concrete mix made into cubes which are stored and tested under controlled conditions.

The strength to which a concrete mix is designed depends on structural considerations and the fact that the concrete must be durable. Since there will be some variation in the quality of the concrete made on site and in the results of cube-crushing tests, the strength to which the mix is designed must exceed the strength actually needed by a safety factor which will depend on the degree of control which can be exercised over the concrete production process. These matters are discussed in Chapter 12; the question of the strength is also influenced by the overriding consideration that the concrete must be durable, and this factor often fixes the least cement content which can be used. This

Table 4.6 Cement contents and sand:coarse aggregate ratios considered suitable for a range of prescribed concrete mixes

Mass of dry aggregate to be used with 100 kg of cement

Grade of concrete (see Note 1)	Nominal maximum size of aggregate (mm)	40		20		14		10	
	Workability Range for standard sample (mm)	Medium 50–100	High 80–170	Medium 25–75	High 65–135	Medium 5–55	High 50–100	Medium 0–45	High 15–65
	Range for sample taken in accordance with 9.2 of BS 5328 (mm)	40–110	70–180	15–85	55–145	0–65	40–110	0–55	5–75
		(kg)	(kg)	(kg)	(kg)	(kg)	(kg)	(kg)	(kg)
C7.5P		1080	920	900	780	N/A	N/A	N/A	N/A
C10P		900	800	770	690	N/A	N/A	N/A	N/A
C15P	Total	790	690	680	580	N/A	N/A	N/A	N/A
C20P	aggregate	660	600	600	530	560	470	510	420
C25P		560	510	510	460	490	410	450	370
C30P		510	460	460	400	410	360	380	320

N/A not applicable
Source: BS 5328

Percentage by mass of fine aggregate to total aggregate

Grade of concrete	Nominal maximum size of aggregate (mm) Workability	40		20		14		10	
		Medium	High	Medium	High	Medium	High	Medium	High
C7.5P C10P C15P		30–45		35–50		N/A		N/A	
	Grading zone 1	35	40	40	45	45	50	50	55
C20P	2	30	35	35	40	40	45	45	50
C25P	3	30	30	30	35	35	40	40	45
C30P	4	25	25	25	30	30	35	35	40

N/A not applicable

Notes on the use of tables:

(1) The proportions given in the tables will normally provide concrete of the strength in newtons per square millimetre indicated by the grade except where poor control is allied with the use of poor materials.

(2) For grades C7.5P, C10P and C15P a range of fine-aggregate percentages is given; the lower percentage is applicable to finer materials such as zone F sand and the higher percentage to coarser materials such as zone C sand.

(3) For all grades, small adjustments in the percentage of fine aggregate may be required depending on the properties of the particular aggregates being used.

(4) For grades C20P, C25P and C30P, and where high workability is required, it is advisable to check that the percentage of fine aggregate stated will produce satisfactory concrete if the grading of the fine aggregate approaches the coarser limits of zone C or the finer limits of zone F.

conflict between specifying strength for structural considerations and minimum cement contents to satisfy durability considerations has in the past led to confusion. For externally exposed concrete ensure that the more onerous requirement (usually durability) is properly achieved. Table 4.7 gives the requirements of BS 8110.

4.2.5.2 Tensile and flexural strength

The tensile strength of concrete is much smaller than the compressive strength and is in any case usually effectively eliminated by cracking, whether this cracking is visible or not. Consequently the tensile strength of concrete is not usually taken into account for design purposes, though it can be important inasmuch as it influences the spacing and control of cracks in structures[6] and contributes to the flexural strength of concrete paving.

Tensile strength is measured either directly by testing bobbins or cylinders to failure in tension, or indirectly by the cylinder splitting test or flexural tests on concrete beams. Results from the latter test are referred to as 'modulus of rupture'.

Table 12.5 in Chapter 12 gives figures for tensile and flexural strengths for several grades of concrete, and testing is discussed on page 4/16.

While tensile and flexural strength both increase with increasing compressive strength, there is no fixed relationship between

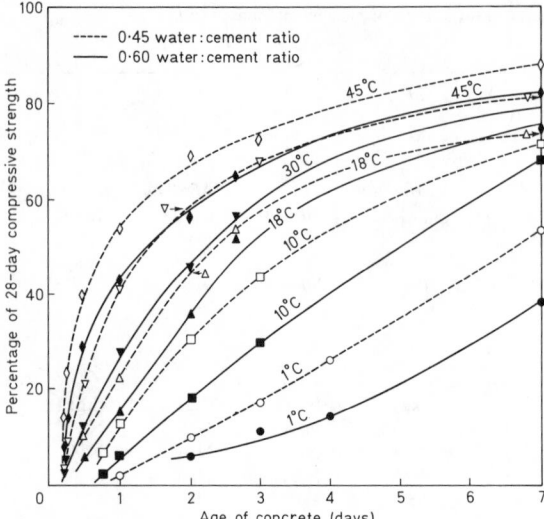

Figure 4.2 Influence of age and temperature on strength. (After Sadgrove (1970) *The early development of strength in concrete.* Construction Industry Research and Information Association)

them. Cylinder splitting tests have shown that the relationship is influenced by the nature of the aggregate, but that some surface characteristic of the aggregate, rather than whether the aggregate is crushed or rounded, is the cause of this influence.[8]

4.2.5.3 Elastic modulus

The elastic modulus of concrete is important in designing members to resist deflection, though concrete is not perfectly elastic and does exhibit significant creep behaviour. For design purposes, shrinkage, creep and elastic modulus are often allowed for together by designing on the basis of an 'effective modulus' which takes account of the three factors. This is discussed in Chapter 12.

The elastic modulus of concrete varies between about 7 and 50 kN/mm² depending on the strength of the concrete and the proportion and rigidity of the aggregate. The lowest figure would be applicable to low-strength concrete made with light-weight aggregate while normal structural concrete would have an elastic modulus of 25 to 30 kN/mm²; some values are given in Table 12.5 in Chapter 12.

The elastic modulus of concrete can conveniently be measured by vibrating a suitable specimen; the value for the modulus found in this way is termed the 'dynamic' modulus and is considerably higher than the static modulus because no creep occurs under the test condition.

4.2.5.4 Creep

'Creep' is the term given to the tendency for concrete to continue to strain over a period of time when the stress is constant. For design purposes, creep is allowed for by using an 'effective' modulus which takes account of both short- and long-term stress–strain relationships. This is covered in Chapter 12.

Factors which tend to increase creep are low strength, low ambient relative humidity, low-modulus aggregates, and high stressing. Methods for calculating creep deflections usually assume that creep is increased by early loading, but other investigations suggest that this effect may not be significant.[8]

Table 4.7 Minimum cement content and other requirements in Portland cement concrete to ensure durability under specified conditions of exposure (from BS 8110)

(i) Conditions of exposure

Environment	Exposure conditions
Mild	Concrete surfaces protected against weather or aggressive conditions
Moderate	Concrete surfaces sheltered from severe rain or freezing whilst wet
	Concrete subject to condensation
	Concrete surfaces continuously under water
	Concrete in contact with nonaggressive soil (see class 1 of Table 4.9)
Severe	Concrete surfaces exposed to severe rain, alternate wetting and drying or occasional freezing or severe condensation
Very severe	Concrete surfaces exposed to sea-water spray, de-icing salts (directly or indirectly), corrosive fumes or severe freezing conditions whilst wet
Extreme	Concrete surfaces exposed to abrasive action, e.g. sea-water carrying solids or flowing water with pH ⩽ 4.5 or machinery or vehicles

(ii) Minimum cement contents and other requirements for durability

Conditions of exposure	Nominal cover to reinforcement				
	(mm)	(mm)	(mm)	(mm)	(mm)
Mild	25	20	20*	20*	20*
Moderate	—	35	30	25	20
Severe	—	—	40	30	25
Very severe	—	—	50†	40†	30
Extreme	—	—	—	60†	50
Maximum free water:cement ratio	0.65	0.60	0.55	0.50	0.45
Minimum cement content (kg/m³)	275	300	325	350	400
Lowest grade of concrete	C30	C35	C40	C45	C50

*These covers may be reduced to 15 mm provided that the nominal maximum size of aggregate does not exceed 15 mm.
†Where concrete is subject to freezing whilst wet, air-entrainment should be used.
Note: This table relates to normal-weight aggregate of 20 mm nominal maximum size.

4.2.5.5 Shrinkage and moisture movement

Concrete shrinks when it dries. Part of this shrinkage, usually about 30% but sometimes as much as 60% is reversible, and is known also as 'moisture movement'. Shrinkage leads to cracking or distortion in members which are restrained or reinforced, though in this respect it is now considered to be less important than thermal movement (see section 4.2.5.6 below).

Shrinkage is increased with increasing cement content or the water content of the mix. High workability mixes shrink more than low workability mixes of the same strength. Aggregates with high elastic moduli are more effective in restraining shrink-age than low-modulus aggregates, this influence being virtually confined to the coarse aggregate. The phenomenon can be viewed simply as a two-component system – cement paste which tends to shrink and aggregate which tends to resist. Changing the balance of these two components will affect the overall magnitude of the shrinkage.

Figure 4.3 shows the influence of ambient relative humidity on the rate and amount of shrinkage. From the latter it can be seen that shrinkage is a more serious problem in dry countries or inside dry buildings than outside in the UK where the relative humidity usually exceeds 75%; indeed, where concrete remains permanently moist, it increases somewhat in volume.

Lightweight aggregates usually have less effect in restraining shrinkage than normal-weight aggregates, and where aggregate is absent, e.g. in aerated concrete, products have to be auto-claved to keep the shrinkage and moisture movement within reasonable limits.

4.2.5.6 Thermal movement

The linear coefficient of thermal expansion of concrete varies from about 5 to 15 microstrain per degree centigrade, depending on the richness of the mix and the coefficient of expansion of the aggregate. Rich mixes have higher coefficients than lean ones, and siliceous aggregates have higher coefficients than limestone and granite. In the same way as for shrinkage, thermal movements can be seen as the summation of properties of the two primary components cement paste and aggregates.

Since concrete tends to become heated when the cement hydrates, thermal contraction on cooling and hardening can set up enough stress on restrained members to cause cracking. Even if a reduced coefficient is used for immature concrete (to take creep into account) cooling strains in walls of normal thickness can reach 200 microstrain or more within a few days of the concrete being cast. Table 4.8 gives general figures for coefficient of thermal expansions for various aggregate types.

4.2.5.7 Durability

This important property of concrete has already been referred to on page **4**/11 where the minimum cement content needed for durability was mentioned in relation to conditions of exposure. Durability considerations will need to be of both the concrete itself and any embedded steel reinforcement. There is a great deal of information on the durability of concrete. Codes of practice are now focusing much more closely on this aspect of concrete performance; careful consideration of the relevant codes of practice are therefore necessary before producing a specification for the concrete. Special care must be taken when concrete is exposed to sulphates, acids or salts used for de-icing, or other aggressive chemicals.

In general, concrete which has low permeability will be much more durable than concrete which has high permeability and the effect may be so marked that it outweighs the influence of specially resistant cements. Well-compacted dense concrete con-

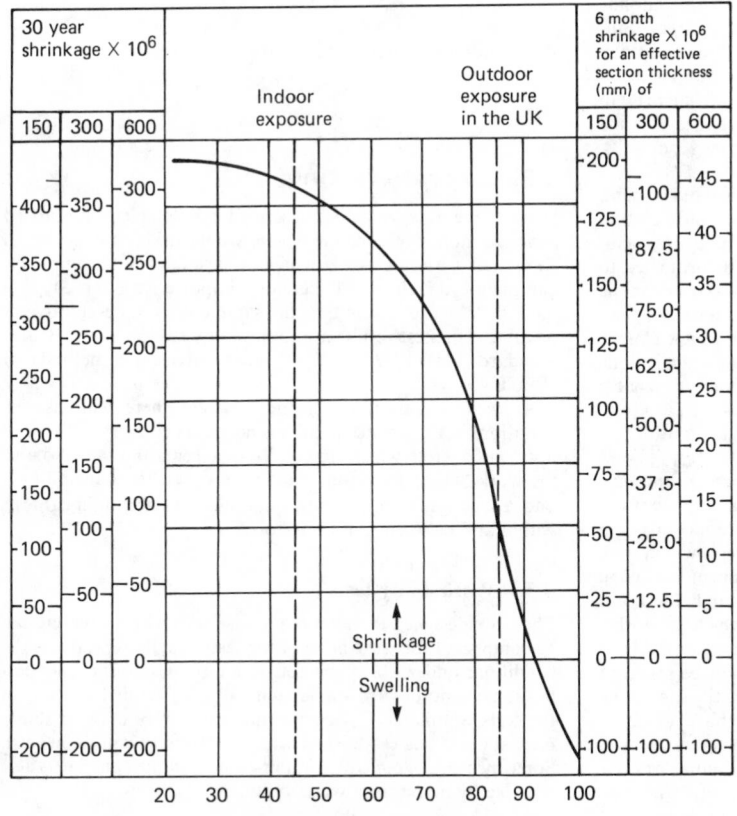

Figure 4.3 Drying shrinkage of normal-weight concrete (from BS 8110). The graph relates to concrete of normal workability with a water content of about 190 l/m³. Shrinkage may be regarded as proportional within the range of 150 to 230 l/m³

Table 4.8 Coefficients of thermal expansion

Coarse aggregate/rock group	Thermal expansion coefficient ($\times 10^{-6}/°C$) (microstrain/°C)	
	Rock	Saturated concrete
Chert or flint	7.4–13.0	11.4–12.2
Quartzite	7.0–13.2	11.7–14.6
Sandstone	4.3–12.1	9.2–13.3
Marble	2.2–16.0	4.4–7.4
Siliceous limestone	3.6–9.7	8.1–11.0
Granite	1.8–11.9	8.1–10.3
Dolerite	4.5–8.5	Average 9.2
Basalt	4.0–9.7	7.9–10.4
Limestone	1.8–11.7	4.3–10.3
Glacial gravel	—	9.0–13.7
Lytag (coarse and fine)	—	5.6
Leca (10 mm)	—	6.7

taining sufficient cement and no unnecessary water should always be used where durability is important. Additional measures may also be needed where exposure conditions are severe.

Sulphates in solution can attack cement paste if the concentration is sufficiently high. Sources of sulphate are calcium and magnesium sulphate present in some groundwaters, sulphates contained in sea-water and sulphates formed from sulphur dioxide present in the air in urban and industrial areas. Sulphates from the last two sources would be too dilute to attack good-quality concrete unless circumstances had allowed them to become concentrated by evaporation. This situation can arise in coastal splash and tidal zones, and on the undersides of units from which contaminated water drips, e.g. copings on walls. Table 4.9 gives the recommendations of BS 8110.

Acids of all kinds attack concrete made with Portland cement. Sources of acids are flue gases (if condensation occurs), carbon dioxide dissolved in water (moorland water is frequently acid) and acid formed from sewer gas. Concrete can be protected to some extent by applying acid-resisting coatings, and limestone concrete (curiously) has been found to be more resistant than other concrete possibly because the large area which can be attacked neutralizes the acid before much local damage is done to the cement paste alone. It is not clear how often serious acid attack of concrete actually occurs in service; however, a report by Eglington for CIRIA[10] gives a review of available information.

Freezing and thawing cycles attack poor concrete, but very good-quality concrete is resistant unless de-icing salts are used. Even good-quality concrete may have a more porous top surface as the result of waterbleed and evaporation which may be more vulnerable to frost action. Air entrainment (see page 4/8) has been found to provide protection, though there are different explanations of the mechanism by which it works. Where no de-icing salt is to be used but the concrete is liable to become frozen when wet, air entrainment may not be specified since concrete with a very low water:cement ratio should be satisfactory. However, the concrete will need to have a cement content in excess of 400 kg/m³ and a water:cement ratio less than 0.45. British Standard 8110 proposes a minimum concrete grade of C50 to ensure these requirements are met. It may be more practical and economic in these circumstances to use a lower-strength grade together with appropriate air-entrainment levels.

The corrosion of reinforcement in concrete is covered in Chapter 12.

4.2.6 Curing

If newly hardened concrete is to achieve its potential strength and durability, the hydration of the cement must be allowed to continue for as long as possible. The detrimental effects of inadequate curing on the durability of reinforced concrete may take many years to become apparent and therefore the relevance at the time of casting the concrete may be overlooked. For this purpose an excess of water must be present in the pores of the concrete and the act of ensuring that this is so is 'curing'. The excess water normally present in the concrete is enough to provide curing except in the case of very dry mixes, but near the surface of a member it will escape by evaporation unless this is prevented. Formwork is usually left in place long enough to provide initial curing, and where appearance and durability are not considered important, this may be enough in the UK climate. Where further curing is considered justified, spray-applied curing films or other means of preventing evaporation must be used. An alternative is to apply water to the surface for the curing period.[11] The curing of unformed surfaces should be commenced as soon as possible after placing the concrete to prevent rapid moisture loss and possible cracking in the plastic concrete.

4.2.7 Concreting in hot, arid climates

Reference should be made to Chapter 37 for the special requirements for mixes, production and curing in hot arid countries such as those in the Middle East.

4.2.8 Reinforcement and prestressing steel

These materials are covered in Chapter 12.

4.3 Concrete testing

Most of the tests which are described below have to be carried out on a sample of concrete which will inevitably be very small compared with the work which it is intended to represent. Sampling is therefore of the greatest importance and every care must be taken to ensure that the sample is as representative as possible if the test results are to have any real meaning. British Standard 1881:1983, Part 101 gives advice on methods of sampling.

As well as variations in the concrete there will also be variations in the test itself and it is necessary to carry out several tests on concrete which nominally represents the same part of the work. These variations have been called reproducibility R and repeatability r and can be quantified by careful repeat tests within and between test laboratories.

4.3.1 Workability tests

These are designed to measure the ease with which concrete can be compacted. Because none of the tests exactly reproduces the conditions under which concrete is compacted on site, each test has some limitations in applicability, though within limits any of the tests is suitable for monitoring uniformity of workability once site use has established what workability will be needed. Some measure of control of water content in the concrete is also possible by monitoring workability.

(1) Slump test (BS 1881:1983, Part 102):
    (a) *application*: quick approximate tests for medium and high workability concrete; suitable for site use; simple apparatus;

Table 4.9

Class	Concentration of sulphates expressed as SO₃			Type of cement	Dense, fully compacted concrete made with 20 mm nominal maximum size aggregates complying with BS 882 or BS 1047	
	In soil					
	Total SO₃	*SO₃ in 2:1 water:soil extract*	*In groundwater*		*Cement* content not less than*	*Free water cement* ratio not more than*
	(%)	(g/l)	(g/l)		(kg/m³)	
1	< 0.2	< 1.0	< 0.3	All cements listed in BS 5328 BS 12 cements combined with PFA† BS 12 cements combined with ground granulated blast-furnace slag†	— Moderate exposure; see Table 4.7	—
2	0.2–0.5	1.0–1.9	0.3–1.2	All cements listed in BS 5328 BS 12 cements combined with PFA† BS 12 cements combined with ground granulated blast-furnace slag†	330	0.50
				BS 12 cements combined with minimum 25% or maximum 40% PFA‡ BS 12 cements combined with minimum 70% or maximum 90% ground granulated blast-furnace slag	310	0.55
				BS 4027 cements (SRPC) BS 4248 cements (SSC)	280	0.55
3	0.5–1.0	1.9–3.1	1.2–2.5	BS 12 cements combined with minimum 25% or maximum 40% PFA† BS 12 cements combined with minimum 70% or maximum 90% ground granulated blast-furnace slag	380	0.45
				BS 4027 cements (SRPC) BS 4248 cements (SSC)	330	0.50
4	1.0–2.0	3.1–5.6	2.5–5.0	BS 4027 cements (SRPC) BS 4248 cements (SSC)	370	0.45
5	Over 2	Over 5.6	Over 5.0	BS 4027 cements (SRPC) and BS 4248 cements (SSC) with adequate protective coating	370	0.45

*Inclusive of PFA and ground granulated blast-furnace slag content.
†For reinforced concrete see 3.3.5; for plain concrete see 6.2.4.2 of BS 8110.
‡Values expressed as percentages by mass of total content of cement, PFA and ground granulated blast-furnace slag.
Notes:
Within the limits given in this table, the use of PFA or ground granulated blast-furnace slag in combination with sulphate-resisting Portland cement (SRPC) will not give lower sulphate resistance than combinations with cements to BS 12.
If much of the sulphate is present as low solubility calcium sulphate, analysis on the basis of a 2:1 water extract may permit a lower site classification than that obtained from the extraction of total SO₃. Reference should be made to BRE Current Paper 2/79 for methods of analysis, and to BRE Digests 250 and 222 for interpretation in relation to natural soils and fills, respectively.

(b) *special apparatus*: mould in shape of inverted cone frustum, flat baseplate;

(c) *method*: Concrete is compacted into the mould in three approximately equal layers with a 16 mm diameter tamping rod giving 25 tamps per layer. Top surface is struck off and finished with trowel. The mould is then lifted off vertically and the concrete is allowed to slump;

(d) *result*: difference in height between moulded and slumped condition measured to nearest 5 mm, *or* if total collapse or shear occur, these facts are recorded.

(2) Compacting factor (BS 1881:1983, Part 103):
(a) *application*: concrete of all workabilities; suitable for simple site laboratory;
(b) *special apparatus*: compacting factor apparatus consisting of two hoppers and a measuring cylinder fixed in vertical alignment; balance to weigh 25 kg to 10 g accuracy;
(c) *method*: Concrete is filled loosely into top hopper and allowed to fall to next hopper. Concrete from this hopper is then allowed to fall into the measuring cylinder. The surplus is struck off;
(d) *result*: ratio of the weight of concrete in the cylinder filled as above to the weight of concrete fully compacted into the cylinder.

(3) 'V-B' consistometer (BS 1881:1983, Part 104):
(a) *application*: concrete of all workabilities; suitable for large site laboratory;
(b) *special apparatus*: consistometer consisting of conical mould, cylindrical container, transparent disc kept horizontal by a guide and a vibrating table of specified size, frequency and amplitude, and stopwatch;
(c) *method*: the mould is placed in the cylinder and filled with concrete as for the slump test. The mould is then removed and the disc is allowed to rest on top of the slumped concrete. Vibration is then applied and allowed to continue until the underside of the disc is just covered with grout;
(d) *result*: vibration time in seconds to nearest 0.5 s.

(4) Ball penetration test (ASTM C360-82):
(a) *application*: similar to slump test;
(b) *special apparatus*: kelly ball consisting of 30 lb, 6 in diameter hemisphere with support frame and graduated scale;
(c) *method*: the frame is placed on the surface of the concrete, e.g. in a wheelbarrow, with the bottom of the hemisphere just touching the concrete. When the weight is released, the penetration into the concrete is measured;
(d) *result*: penetration in inches.

(5) Flow test (BS 1881:1984, Part 105):
(a) *application*: high-workability concrete;
(b) *special apparatus*: mould in shape of inverted cone frustum: hinged flat baseplate;
(c) *method*: concrete is compacted into cone on hinged baseplate. Cone mould lifted off and top half of hinged plate lifted and dropped through predetermined height (40 mm) fifteen times. Diameter of concrete 'cowpat' measured in two directions;
(d) *result*: diameter of flow, in millimetres.

Other workability tests. Other test methods have been developed but have not been widely accepted. One test method which may become more widely used has been developed by Tattersall[12] from laboratory research. This method is a so-called two-point method and attempts to measure more scientifically the rheological properties of the plastic concrete by measuring the torque required to turn an impeller at various speeds when immersed in the concrete.

4.3.2 Strength tests

Strength tests are designed to measure the potential strength of concrete when cured and tested in a standard manner; the actual strength in a structural member depends on compaction, curing and uniformity as well as the potential strength. It cannot therefore be measured except by tests on the member (for core tests and nondestructive tests see sections 4.3.5 and 4.3.6). The primary reason for these strength tests is to maintain control over the batching and mixing of the concrete supply, thereby checking compliance with specified requirements.

The main requirements of BS 1881 for strength tests on concrete specimens are summarized in Table 4.10.

4.3.3 Accelerated strength tests

The curing and testing regimes shown in Table 4.10 are for 'standard' control tests and most specifications have a requirement for strengths to be tested at 28 days. For the majority of construction work this is acceptable. There are circumstances, e.g., where construction is likely to be on a fast timetable, in which waiting for 28 days before confirming compliance of the concrete is not preferred. In this situation it is possible to use an accelerated curing regime to give strength testing at, say, 24 h. The elevated temperatures used for this curing do not produce a constant effect on concreting materials and therefore it is normally required to do correlation testing in advance of the main concreting work. British Standard 1881:1983, Part 112 gives some guidance on accelerated curing.

4.3.4 Tests on cores

Tests on cores cut with diamond-tipped core cutters are described in BS 1881:1983, Part 120. The diameters of the cores should conform with the diameters of cylindrical specimens (see table in BS 1881), but the length cannot usually be chosen. The ends of the cores must be flat and perpendicular to the axis (this can be achieved by capping or grinding). The cores must be soaked in water for at least 48 h before being tested and are then tested while still wet. The failure stress is calculated from the load × the correction factor for the length/diameter ratio, divided by the average actual area of cross-section.

4.3.5 Nondestructive strength tests

A wide range of tests have been developed to give an indication of strength in concrete structures. The most commonly used in the UK are the rebound (Schmidt) hammer and ultrasonic pulse velocity tests. Others include Windsor probe, pull out, internal fracture and break-off tests. Most of these tests measure the properties of a relatively small volume of concrete near the surface and all of them need calibration against concrete of known strength. A summary of nondestructive test methods is given in BS 1881:1986: Part 201.

Surface hardness tests measure the rebound of impact hammers (e.g. the Schmidt hammer test) or the depth of indentation. A large number of tests is needed for a satisfactory assessment because the closeness of the aggregate to the test point affects the results. British Standard 1881:1986 Part 202 gives details of these tests.

Ultrasonic pulse velocity tests depend on the relationship between transmission time and the density and elastic properties of the concrete, both of which are related to concrete strength. Guidance on the method and interpretation is given in BS 1881:1986 Part 203.

4.3.6 Tests on aggregates

British Standard 812 gives details of numerous tests on aggregates. A number of these are frequently used in concrete mix

Table 4.10 Strength tests for concrete specimens

	Crushing strength	Flexural strength	Indirect tensile strength
Specimen size (mm)	$150 \times 150 \times 150$	$150 \times 150 \times 750$	150 dia $\times$ 150
Specimen size (mm)[a]	$100 \times 100 \times 100$	$100 \times 100 \times 500$	
Rammer size (for hand compaction)[b]	$25 \times 25 \times 380$	$25 \times 25 \times 380$	$25 \times 25 \times 380$
Blows per layer (smaller specimens)	$\geqslant 35$ ($\geqslant 25$)	$\geqslant 150$ ($\geqslant 100$)	$\geqslant 30$
Rate of loading (smaller specimens)	0.2–0.4 N/mm²·s	0.06 ± 0.04 N/mm²·s	0.02–0.04 N/mm²·s
Test result	Failure stress (f_c)	Modulus of rupture (f_b)	Tensile strength (f_t)
Calculation	$f_c = \dfrac{\text{load}}{\text{nominal area}}$	$f_b = \dfrac{\text{load} \times \text{outer span}^c}{\text{breadth} \times \text{depth}^2}$	$f_t = \dfrac{2 \times \text{load}}{\pi \times \text{diameter} \times \text{length}}$
	(parts of broken beams may be used for crushing strength tests. The area of the platens is then the nominal area)	(for failure inside middle third)	

Notes:

[a]The smaller specimen size may be used where the maximum aggregate size is 25 mm or less.
[b]The weight of the rammer is 1.8 kg in each case.
[c]Outer span = 3 × inner span
Curing of specimens: until strong enough to be demoulded (usually after 24 h) the specimens are stored in their mould at a min RH of 90% and a temperature of 20°C ± 2°
(for specimens to be tested at 7 days or less) or 20°C ± 5° (for specimens to be tested at 7 days or more). After being demoulded they are stored in water at 20°C ± 2°.

design or for quality control. These are briefly summarized in Table 4.11.

4.3.7 Measurement of entrained air

It is important to control the air content of air-entrained concrete for the reasons given on page **4/8**. Entrained air is measured by compacting a sample of fresh concrete in three layers in a container of known volume (nominally 0.006 m³). Compaction must be sufficient to remove all entrapped air, but not so prolonged that entrained air is also removed. The container is then clamped to an airtight cover which incorporates a pressure gauge and a graduated sight tube. The space under the cover is filled with water and the vessel is pressurized with an air pump to compress the air contained in the concrete (the air pressure is usually about 1 atm). The change in volume is indicated by a drop in the water level in the sight tube which is calibrated directly in per cent of entrained air by volume. British Standard 1881:1983: Part 106 gives details of this test.

This test method is used as the basis for most entrained-air concrete specifications and is easily carried out on site. For development of admixtures and for demonstration of the effectiveness in resisting freezing and thawing the determination of the distribution of air bubbles in the hardened concrete may be preferred. This can be carried out using the methods described in ASTM C-457.

4.3.8 Analysis of fresh concrete

This is used to determine the proportions of the constituents and the grading of the aggregates before the concrete has hydrated sufficiently to bind the components together.

Several different approaches to this type of analysis have been developed. One method which was incorporated in BS 1881 comprised a set of sieves and used water wash to separate the concrete into material greater then 5 mm, between 5 mm and 150 μm, and less than 150 μm. These tests are not widely carried out on site and testing in a laboratory is difficult because of the need to retard the cement hydration.

A more convenient method of analysing fresh concrete quickly has been developed and is called the Rapid Analysis Machine (RAM). This machine separates-out the fine component of a concrete mix by elutriation in a water column, and by prior calibration an assessment of cement content can be made. As for accelerated strength testing this method can be very useful for construction work with a fast programme or where large volumes of concrete have to be placed.

Details of five methods of fresh concrete analysis are given in BSDD83:1983.

4.3.9 Analysis of hardened concrete

This is sometimes needed when a failure has occurred, or when, for some other reason, the constituents of the hardened concrete have to be determined. Full details of the methods used are given in BS 1881:Part 6:1971, although this method is shortly to be revised.

The chemistry of the analysis of hardened concrete is relatively straightforward; however, skill and experience are needed to ensure an accurate result is consistently achieved. This type of work should be carried out only in laboratories in which expertise is available and experience in interpreting results is possible.

Two main methods of analyses are: (1) determination of calcium oxide; or (2) of soluble silica. Both of these methods analyse a sample ground to a fine dust which is then dissolved in acid. Determination of either, or both, calcium oxide and soluble silica contents of the concrete can be related to the quantities in the original cement, assumed or from existing data, and hence give the cement content of the concrete. The analytical methods have inherent inaccuracies in them which can be defined and these should be investigated and accepted before an analysis is carried out.

Table 4.11 Tests on aggregates – summary of main tests in BS 812

Test	Property measured	Principle/apparatus/method
Sieve analysis	Aggregate grading, including clay and fine silt passing 75 µm	Dried sample of aggregate sieved over a number of test sieves conforming with BS 410; weight retained on each is measured
Sedimentation test	Proportion of clay, silt or dust	Fine material in suspension sampled with sedimentation pipette
Field settling test	Estimate of clay, silt or dust	Sample of aggregate shaken with salt solution; depth of material which has settled measured
Flakiness test	Percentage of flat particles	Sample of aggregate tested in specified slotted gauges
Specific gravity and water absorption	Specific gravity and percentage water absorption of coarse aggregates	Sample of saturated aggregate submerged – loss in weight indicates volume; sample dried to give dry weight and weight of absorbed water
Specific gravity and water absorption	Specific gravity and percentage water absorption of coarse aggregates	Sample of saturated aggregate submerged – displaced water indicates volume; remainder as above
Specific gravity and water absorption	As above, but for fine aggregates	Volume of sample measured by water displacement in a pycnometer: remainder of test as above
Bulk density	Bulk density and void volume of aggregate sample	Weight of aggregate required to fill container of known volume
Oven drying	Percentage moisture content	Weighed sample oven dried and reweighed
Siphon can	Percentage moisture content	Water volume determined by displacement in siphon can
Aggregate impact value	Resistance of aggregate to shock	Percentage of material passing 2.40 mm determined after specified impact test on aggregate sample
Aggregate crushing value	Resistance of aggregate to crushing	As above, but specified crushing test instead of impact
10% fines value	Resistance of aggregate to crushing	Determination of load required to produce 10% of material passing 2.40 mm in specified crushing test
Aggregate crushing strength	Compressive strength of rock	Crushing test on cylinder cut from rock sample
Aggregate abrasion value	Resistance of aggregate to surface wear	Determination of percentage loss in weight after specified lapping of aggregate sample

Aggregate type and grading are determined by breaking down a sample of concrete by heating it to 550°C for an hour or more. Cement is dissolved from the lumps of fine material with dilute hydrochloric acid, and a sieve analysis is carried out on the insoluble material which remains. Again, the method is difficult and rather approximate if the aggregates contain a substantial proportion of limestone.

The original water content is found by saturating a slice (sawn with a diamond saw) with carbon tetrachloride. This fills the pores left by the uncombined water and the volume of the pores is estimated from the weight gained. The combined water is found from the loss in weight on ignition of a sample of the concrete. The determination of original water content is also very approximate and, unless large variations in actual values are being sought, the test may not be beneficial.

4.4 Plastics and rubbers

4.4.1 Terminology

Standard definitions of terms relating to plastics (ASTM D883) includes the following.

Polymer A substance consisting of molecules characterized by the repetition (neglecting ends, branch junctions and other minor irregularities) of one or more types of monomeric units.

Plastic(s) A material that contains as an essential ingredient one or more organic polymeric substances of large molecular weight, is solid in its finished state and, at some stage in its manufacture or processing into finished articles, can be shaped by flow.

Rubber A material that is capable of recovering from large deformations quickly and forcibly, and can be, or already is, modified to a state in which it is essentially insoluble (but can swell) in boiling solvent, such as benzene, methylethylketone, and ethanol-toluene azeotrope.

A rubber in its modified state, free of diluents, retracts within 1 min to less than 1.5 times its original length after being stretched at room temperature (18 to 29°C) to twice its length and held for 1 min before release.

Elastomer A macromolecular material that at room temperature returns rapidly to approximately its initial dimensions and shape after substantial deformation by a weak stress and release of the stress.

Thermoplastic A plastic that repeatedly can be softened by heating and hardened by cooling through a temperature range characteristic of the plastic, and that in the softened state can be shaped by flow into articles by moulding or extrusion.

Thermoset A plastic that, after having been cured by heat or other means, is substantially infusible and insoluble.

4.4.2 Physical and chemical properties

4.4.2.1 Fusibility

Thermoplastics melt or soften when heated and return to their original state on cooling provided that they have not been degraded by overheating. Some thermoplastics, e.g. polystyrene, become very fluid when heated and can be used to make castings, others become soft and doughy but do not melt. These compounds, e.g. PVC, have to be shaped or formed under pressure.

4.4.2.2 Combustibility

All polymer materials should be considered combustible. However, the range of their behaviour in fire is wide. Some plastics based on, for example, chlorides, fluorides or formaldehyde, will be very difficult to ignite, and then will be self-extinguishing. Some plastics to which flame-retardant additives have been added may also behave in this way. On the other hand, some plastics which would normally be considered to be difficult to ignite or self-extinguishing may be rendered otherwise by the addition of combustible plasticizer. Plasticized PVC is an example of this.

4.4.2.3 Resistance to daylight and weathering

Ultraviolet light, present in daylight outside but effectively filtered-out by ordinary window glass, attacks many plastics and rubbers though some (acrylics for example), are largely unaffected. Those materials which are attacked can be made much more resistant by the incorporation of suitable pigments or ultraviolet absorbers in the formulation. The degree of attack naturally depends on the amount of ultraviolet light present, and performance data must relate to the appropriate conditions of exposure. Rain may leach out constituents of some formulations, and a few plastics and rubbers are not resistant to moisture.

4.4.2.4 Resistance to extremes of temperature

The flexibility of plastics compounds increases as the temperature rises and oxidation may degrade plastics and rubbers which are kept at high temperatures for long periods. Thermoplastics particularly are affected by temperature changes, and with some (bitumen is a familiar example), the ambient temperature range is enough to change them from the brittle to the fluid state. With others, e.g. polypropylene and thermosetting plastics, ambient temperature changes have little effect. These materials are stable at 100°C or more, and do not become brittle at normal low temperatures.

4.4.2.5 Thermal expansion

The coefficient of thermal expansion of polymers tends to be very high compared with conventional construction materials. Formulating compounds with high filler contents reduces this effect, but in design it must always be allowed for, e.g. by incorporating suitable movement joints and consideration when choosing fixing points. Rigid PVC formulations, such as those used for pipes and claddings, have coefficients of thermal expansion several times those of metals commonly used in construction.

4.4.2.6 Resistance to acids and alkalis

Polymers tend to be resistant to attack from acids and alkalis and are generally better than more common construction materials in this respect. The good chemical resistance of many polymers is made use of in formulating protective coatings and linings, but incorporating nonresistant fillers in compounds (chalk is a common filler for plastics) reduces or eliminates their resistance.

4.4.2.7 Resistance to oil and solvents

Polymers vary greatly in their resistance to oil and solvents. Many thermoplastics are attacked by a variety of solvents; thermosetting plastics and elastomers tend to be more resistant but may swell. Nylon and PTFE are notable among common thermoplastics for their solvent resistance. The solvent resistance of many polymers is highly specific, and polymers which are unaffected by one solvent may be readily attacked by another.

Applications of this behaviour of plastics that are worthy of note are: (1) joining by solvent welding; (2) the formulation of adhesives and paints; and (3) the possibility of incorporating 'plasticizers'. These are materials (usually liquids) which are compounded with certain plastics to make them more flexible – PVC is an example.

Problems associated with this behaviour are:

(1) Firstly, the migration of solvents and oils into or out of plasticized compounds. This occurs if the solvent or oil is miscible with the plasticizer in such compounds, even if the basic polymer would be immune from attack. Solvent welding and plasticizer migration are dealt with on pages 4/20 and 4/21.
(2) Environmental stress cracking and crazing of some polymers when stressed in an environment in which solvents are present. Polystyrene is an example of a material susceptible to this.

4.4.2.8 Resistance to oxidation and ozone

Some polymers are oxidized to a significant extent at high temperatures (the temperatures at which they fuse for example) and some (rubbers especially) are attacked at ambient temperatures by ozone. Formulation with suitable inhibitors can be used to make rubber and plastics compounds which are resistant to these effects.

4.4.2.9 Resistance to biological attack

Most polymers are immune from biological attack, though attack on ingredients used in the formulations of plastics compounds is not unknown. Casein (a protein) can be attacked; though not when it is crosslinked with formaldehyde. Borers, such as woodworm, have been known to make their way into plasticized PVC, but this is unusual and occurs only when the compound is in contact with some more palatable material. Rats sometimes bite through plastics water pipes (as they do through lead) but it is an uncommon hazard.

4.4.3 Mechanical properties

The mechanical properties of rubber and plastics compounds

are greatly influenced by both the basic polymer and by the other ingredients used in formulating the compound. The compounding and manufacturing process itself also influences the mechanical properties, especially where molecular orientation occurs.

Data on mechanical properties are thus very difficult to tabulate concisely, also because values vary so much with test conditions such as temperature, duration of loading and method of loading. For such reasons the data given in Table 4.12 are incomplete in some cases, and may appear to be very imprecise in many others.

4.4.3.1 Strength

Tensile and compressive strengths of plastics compounds vary over a wide range. High tensile strength is a property of polymers such as nylon and some forms (films and fibres) of polyester and polypropylene. The relatively low elastic modulus of many polymers makes the compressive strength more difficult to assess in practical terms. Thermosetting resins tend to have high compressive and tensile strengths, the latter being capable of being greatly increased by the incorporation of reinforcing fibres. Orientation in films and fibres is also a means of increasing strength.

4.4.3.2 Elastic modulus

Rubbers and thermosetting polymers behave elastically over a large part of their strain range, but thermoplastic polymers tend to strain irreversibly after a relatively small proportion of their ultimate strain. There are a number of exceptions to this general rule. Unmodified polystyrene is noted among thermoplastics for its exceptionally low strain at failure and it shatters easily. Thermosetting polymers tend to be less flexible than rigid thermoplastics and when broken they often show a brittle fracture.

4.4.3.3 Hardness and abrasion resistance

Rubber and plastics compounds are soft compared with most construction materials, though they are not necessarily easily abraded. Rubber and flexible thermoplastics are softer than rigid thermoplastics, thermosetting plastics being generally harder still. Abrasion resistance depends on several factors including hardness, elasticity, surface friction and the ability for abrasive particles to become embedded. Factors increasing abrasion resistance for some of these reasons tend to reduce it for others and it is a property which is difficult to predict without tests.

4.4.3.4 Creep

Strength and elastic modulus measured at high rates of loading are much higher than those which are obtained at very low rates of loading for most plastics compounds, though rubbers and thermosetting polymers are less prone to creep than thermoplastic polymers. When creep deflection is considered important, care must be taken to choose suitable compounds and to limit stresses to those which will not lead to unacceptable creep. Where loads are to be applied intermittently, creep is unlikely to be a problem as recovery can take place over a relatively long period. Creep in plastics increases greatly with higher temperatures.

4.4.4 Compounding, processing and fabrication

4.4.4.1 Compounding

Some of the ingredients which are used in formulating plastics compounds have been mentioned on page 4/19. Many polymers are used in commercial applications without addition, but the art or science of formulating PVC compounds suitable for particular applications is the converter's most important contribution in the manufacture of plastics articles and compounds based on this polymer. Guidance on formulation cannot be given here, but it is necessary that the engineer should understand that formulation is important.

4.4.4.2 Processing methods

There are many ways of making plastics compounds into useful articles or materials; some of the most usual methods are:

(1) Extrusion, where the compound is continuously forced through a die.
(2) Calendering, where the compound is forced between a series of rollers to form a sheet.
(3) Injection moulding, where the compound is forced into a die or mould.
(4) Spreading, where the compound (usually PVC) is spread on to a support (often temporary) to form a sheet.
(5) Casting, where the compound is allowed to flow into a mould under gravity or by centrifugal force.
(6) Dough moulding, where the compound is shaped under pressure by a die.
(7) Vacuum forming, where previously made sheet is shaped by being heated and forced on to an evacuated former under air pressure.

4.4.4.3 Influence of processing methods on properties

All processing methods except some used for thermosetting polymers need the polymer or compound to be heated and many thermoplastics compounds are degraded by prolonged heating. Thus, processing methods, like extrusion, which need the compound to be heated for only a short time have inherent technical advantages over methods like calendering where the compound may have to be kept hot over a long period.

Processing methods for compounds which do not become truly fluid on being heated shape the compound under pressure into a form which it will largely retain on cooling. However, some tendency to return to the unformed shape may remain and 'relaxation' of newly formed shapes (calendered sheet especially) in thermoplastics should be allowed for.

Where thermoplastics compounds are to be used at temperatures which even begin to approach the processing temperature, relaxation can be a severe problem. An example is vacuum-formed shapes which have been formed from sheet heated only enough to soften it slightly. Such shapes may relax enough to be considerably distorted even by the temperatures caused by sunshine on a summer's day.

4.4.4.4 Fabrication methods for materials and components

Materials made from thermoplastic polymers or compounds can be fabricated by heat or friction welding and, in the case of those which are soluble, by solvent welding. Mechanical methods of fabrication can also be used.

It is usually possible to find a solvent which can be used for solvent welding thermoplastics, though not all the solvents which attack a material are suitable for welding it. Important among thermoplastics which cannot be solvent welded are polyethylene, polypropylene, PTFE and nylon. The welding solvent may be modified by the addition of a separate polymer to make it tacky. This is useful in keeping joined parts in position while the solvent is doing its work. Properly made heat- or solvent-welded joints are often as strong as the parent material.

Materials made from thermosetting polymers and cross-linked rubbers cannot be welded, though many can be glued satisfactorily using thermosetting resin, or with some solvent-based adhesives made from other polymers. Best results are usually achieved if gluing is carried out as soon as possible after fabrication before cross-linking of the material is complete.

Glueing polymer materials together is likely generally to be less strong than welding. For this reason such applications as reservoir liners and waterproof membranes are far more reliable when welded. Often this is best carried out at works rather than on site. However, thermosetting adhesives can produce strong bonds. This is useful when other materials are involved, e.g. concrete, steel.

Fabrication can usually be limited to the joining of finished units because plastics materials are relatively simple to make in almost any shape, and even these can often have mechanical joints formed into them during manufacture.

4.4.4.5 Fabrication methods – direct fabrication from polymers and compounds: contact moulding

Manufacture of the material and fabrication into the required unit can often be combined into one operation. Glass-reinforced thermosetting polymers are often used in this way, and if the polymer can be cured under ambient site conditions fabrication on site is possible. When contemplating on-site fabrication of plastics components or the direct application of compounds such as, for instance, chemical-resistant epoxy surface coatings, it should be noted that full curing under ambient conditions (which might need to be modified by installing heating) will be needed. It sometimes happens that a compound which will harden under ambient conditions, and look as if it has cured fully, does not cross-link sufficiently to develop fully its desired properties of strength, durability and solvent-resistance.

On-site fabrication, or surface coating with thermosetting compounds, usually needs a curing agent to be added to the polymer shortly before fabrication. This is necessary because most compounds which will cure under ambient conditions would also cure during storage and could not be kept ready-mixed for more than a few hours. Exceptions include thermosetting compounds which cure through the absorption of atmospheric moisture and compounds whose storage life can be extended usefully by storing them under refrigeration.

Where heat can be applied to promote curing, ready-mixed thermosetting compounds which can be stored at ambient temperature are often convenient to use, since curing starts when heat is applied, and this time is under the fabricator's control. As well as thermosetting compounds which already contain the curing agent, pre-impregnated glass cloth can be fabricated in this way. This cloth is usually made with glass strand mats and compounds which have a high enough viscosity at ambient temperature to give a conveniently handled material. Before it is cured, the compound is in a thermoplastic condition, and the material can be shaped easily if it is slightly heated. Prolonged heating, or heating to a higher temperature, is then used to cure the compound after shaping and fabrication.

4.4.4.6 Application of plastics materials and components

The properties of plastics compounds described in the above sections should give the designer some guide on the virtues and limitations of the materials themselves, but in their application the interaction between plastics and other materials must also be taken into account. Two important limitations are the high coefficient of thermal expansion of plastics materials and the phenomenon of plasticizer migration.

In the case of flexible plastics, the high coefficient of thermal expansion causes few problems because the material's tendency to strain with temperature changes does not produce high stresses in the plastics materials or at the interface between plastics materials and the materials to which they are applied.

With rigid plastics materials, however, the stresses produced by restrained thermal expansion can be high enough to produce distortion, failure at the interfaces or even failure of the materials themselves. When designing fittings for rigid plastics components, provision must be made for thermal movement. Fixing through slotted holes or with clips is satisfactory provided that they are not fastened too tightly to allow free movement. Where weatherproofing has to be provided by plastics components, the design of joints which will remain weathertight while allowing movement is essential.

Plasticizer migration can be a serious problem when flexible thermoplastics containing plasticizers are bonded with adhesives which contain similar materials. In such cases the plasticizer and constituents of the adhesive diffuse into each other with the result that the plastics material may shrink and become brittle if there is a net loss of plasticizer, or soften excessively if there is a net gain and the adhesive may suffer similarly. The problem is best avoided by the choice of suitable adhesives, but where plasticized materials have to be applied over substrates into which plasticizer can migrate, e.g. over bituminous materials, a coating or an intermediate layer can be used to provide a barrier to the migration of the plasticizer.

4.4.5 Identification of polymers and plastics compounds

The suitability of any polymer or compound for any particular application will depend greatly on which compound is chosen, and it is therefore helpful to know how different compounds and polymers can be recognized. Although precise identification is often impossible without modern analytical equipment, a useful idea of the nature of the material can be obtained quite easily in many cases. The following is intended as a general guide.

(1) *Flexibility.* Rubbers can be bent without breaking or cracking and they snap back when released.
    Flexible thermoplastics can also be bent without breaking or cracking, though usually not as much as rubbers, and they do not snap back. Rigid thermoplastics can usually be bent a little, but continued attempts to bend them result in breaking or cracking. Polystyrene, however, is rigid and brittle unless modified. It cannot be bent. Thermosetting plastics are usually very rigid and break cleanly if an attempt is made to bend them.
(2) *Feel* is a difficult sensation to describe accurately, but polyethylene and PTFE have a waxy feel which other plastics do not have.
(3) *Bounce.* Most rubbers (but not butyl rubber) bounce.
(4) *Density.* A simple division can be made between polymers which float in water (a minority) and those which do not. (Table 4.12 lists specific gravities.)
(5) *Burning.* Many polymers support combustion and, of those that do, some burn with a smoky flame and others with a clear flame. Table 4.13 indicates behaviour on ignition.
(6) *Chemical tests.* Details of chemical tests are too long to be included here, but engineers who wish to carry out further tests for the identification of polymers and compounds will find that many of the simple tests can be carried out with rudimentary chemical knowledge and apparatus.

4.4.6 Foamed and expanded plastics

Thermal insulation is a very important application of plastics when they are in a foamed or expanded form. Very low bulk

Table 4.12 Properties and applications of some commonly used plastics and rubbers

Compound or polymer	Combustibility	Specific gravity	Fusing temperature (°C)	Maximum working temperature (°C)	Ultimate tensile strength (N/mm²)	Minimum working temperature (°C)	Tensile elastic modulus (N/mm²)	Tensile strain at failure (%)	Coefficient of thermal expansion (°C × 10⁻⁵)	Resistance to weathering	Resistance to acids and alkalis				Resistance to solvents					Typical applications	Relevant British Standards or codes of practice
											Concentrated inorganic acids	Diluted inorganic acids	Organic acids	Alkalis	Petrol	Paraffin, diesel and fuel oil	Aromatic and chlorinated solvents	Ethers, ketones and esters	Alcohols		
Acetal copolymers		1.41	160	80 to 120	35 to 80		1000 to 2000		8		−	−	−	+	+	+	① 0	+	+	Plumbing components, e.g. taps, door and window furniture	
Acrylic resins	② F C	1.18	100 to 120	80	40 to 70		3000	5	6	+	−	+	−	+	+	+	−	−	+	Moulded and shaped lights, e.g. rooflights and domelights, lighting fittings, sanitary ware	
Acrylonitrile-butadiene-styrene (ABS)	③ F S	1.10	−	80	40		400	1.5												Waste and drainpipes and fittings, pressure pipes and fittings	
Butyl rubber	F S	0.92	−	125	5 to 17	−50		500 to 800		+	+	+	+	+	+ to 0	+ to 0	−			Roof coverings, tank linings, BS 3227 damp-proof membranes, adhesives and mastics, sealants, bridge bearings	
Chlorosulphonated polyethylene (CSM)	B	1.10			14			300 to 500		+	+	+	+	+						Roof coverings, tank linings	
Epoxide resins	F S	1.25 to 1.30			55 to 70		5000	1.6 to 1.8		+	+	+	+	+	+	+	+		+	Adhesives, bedding and jointing mortars and grouts, concrete patching mortars; surface coatings	BS 4994 BS 6374
Melamine formaldehyde (laminates)	B	1.45		120	95		8000	0.7	3	0							①			Decorative laminates, cladding, surface coatings	BS 3794
Nylon	B S	1.10 to 1.40	220 to 265	80 to 120	50 to 100	−40	2000	75	8 to 10	+	−	−	−	+	+	+	0	+	+	Door and window furniture, cold water fittings, surface coatings, fairleads, ropes and straps	
Phenol formaldehyde (figures for	B	1.40 (1.40)	−	120 (120)(80)	50		7000	0.5 (3)	5 (3)	+	+ ⑦	+	+	−	+	+	+	0 to −	− to −	Laminates for roofing and walling panels, adhesives for timber, surface coatings. (See Table 4.13 for	BS 1203, BS 1204, BS 2572, BS 6374

Properties of plastics and rubbers (table; column headings appear on the facing page and are not printed on this page).

Material	Combustibility	Density (Mg/m³)	Other tabulated properties (in order of columns)	Uses	Relevant British Standards
Polyester (figures for laminates)	F, S (note ③)	1.1 (1.6)	—; 90 (90); to 20; 40 to 70 (≥300); 2000 (10 000); to 900; 2 (0.5); 5 to 10 (2); ④⑤ +	adhesives, surface coatings, bridge bearings; Surface coatings, as laminated material for: pipes, roofing and cladding, storage tanks, 'architectural' features	BS 3532, BS 4154, BS 4549, BS 4994, BS 6374
Polyethylene (low-density)	F (note ⑥)	0.91 to 0.94	110 to 125; 80; −60; 5 to 15; 100; 100 to 400; 14 to 24; ④ +	Damp-proof membranes, protective sheeting (temporary), cold-water supply pipes, cold-water storage tanks, drains	BS 1972, BS 1973, BS 3012, BS 4646, BS 6515
Polyethylene (high-density)	F (note ⑥)	0.94 to 0.97	105; 20 to 35; 1000; 50 to 200; ④ +		BS 4646
Natural rubber	F, S	0.93	70; −55; 20; up to 10; 500 to 800; —	Bridge bearings, adhesives, floor coverings	BS 1711, BS 6716
Polypropylene	F (note ⑥)	0.90 to 0.91	165 to 175; 120; 18 to 30; 1000 to 1500; 5; 9.0 to 13.5; ④ +	Drain and waste pipes and fittings, containers, pressure pipes, ropes, geotextiles	BS 3867, BS 3943, BS 4159
Polysulphide	F (note ⑧)	1.34	—; 95; 4 to 10; −50; 200 to 350; +	Flexible sealants	BS 4254, BS 5215
Polytetrafluoroethylene (PTFE)	R	2.15 to 2.24	325; 250; 12; 400; 150; 10; +	Bridge bearings, chemically resistant coatings, gaskets	BS 6564
Polyvinyl chloride (PVC) rigid	B, S (note ⑨)	1.39	75 to 85; 65; 35 to 55; −40; 5500; 10; 6; ④ +	Cold-water supply pipes, drains and wastes, rain-water goods, roofing and cladding, lining panels, ducting	BS 3505, BS 3506, BS 4203, BS 4346, BS 4514, BS 4576, BS 4660, BS 5481
Polyvinyl chloride (PVC) plasticized	B/F, S	1.30	60 to 85; 40 to 65; 10 to 25; 10; 200; ④ + to −	Roof coverings, waterstops, preformed seals, surface coatings, floor coverings	BS 2571, BS 3869

Notes:
(The data given in this table must be considered in relation to the descriptions and guidance given in the text.)
① Resistant to aromatic solvents.
② Burns with sweet 'gassy' smell.
③ Very sooty flame, penetrating 'styrene' smell.
④ If suitably pigmented or ultraviolet stabilized.
⑤ Flare-retardant grades are less resistant to weathering.
⑥ Burns, drips and smells like candle wax.
⑦ Oxidizing agents may attack.
⑧ Burns with sulphurous fumes.
⑨ Burns with sweetish acrid smell.

Key:
Combustibility:
F Flammable; burning continues after ignition
B Combustible but self-extinguishing
R Difficult to ignite
C Burns with clear flame.
S Burns with smoky flame.
Chemical resistance:
+ Resistant.
0 Some attack; use with caution if at all.
− Little or no resistance; unsuitable for use.

densities combined with sufficient strength for satisfactory handling and fixing can be obtained with these materials, and some of them have the additional advantage for low temperature insulation of low water vapour diffusance. Commonly used insulating materials made from plastics are listed in Table 4.13 together with their most important physical properties, typical applications and relevant British Standards and codes of practice.

4.4.6.1 Resin mortars

Resin mortars are usually based on epoxy or polyester. They can produce high strengths and are useful for setting-in and repair applications. When considering their use, a number of points should be borne in mind. In curing, these materials give out heat. The manufacturer's guidance should be sought as to the size of application at which this exotherm becomes too great. These materials are susceptible to creep under sustained load. In a fire, they will deteriorate and cease to perform their function. They do require care and proper conditions for their use on site, in particular: (1) correct proportioning of resin and hardener; (2) correct application and curing temperature; and (3) correct surface preparation.

4.4.6.2 Crack injection

There are a number of specialist contractors offering resin injection repair of cracked concrete. Such repairs are primarily used for restoring durability and should not be considered for structural use. Resins used are polyesters or epoxy of suitably low viscosity for penetration of hairline cracks. Different application techniques are offered by different contractors but most involve injection under pressure or application of a vacuum prior to injection. Again, care is needed to ensure correct proportioning and temperatures.

4.4.6.3 Geotextiles and liners

The key question is of course: If buried in the ground or underwater, how long will the material last? Plastics commonly used for these are: polypropylene, polyethylene polyester, PVC, polyamide, butyl, CPE and CSM. They will often be subject to oil, water, chemicals, biodeteriorating environments, and may be used in applications where there is a likelihood of mechanical damage or puncture. Rodents may be a problem. In general the materials will not usually be exposed to ultraviolet light, except perhaps where they emerge from the ground or water.

Guidance should be sought from manufacturers, but most of the information relating to the durability of the above materials can be found in Table 4.12.

4.5 Paint for steel

Painting of steelwork is covered by BS 5493:1977. Using this document, corrosion protection schemes can be chosen for most situations. This section discusses the key aspects of corrosion protection of steel and the coating types that will be met most frequently.

4.5.1 Zinc coatings

There are a number of ways in which a layer of zinc can be deposited on to the surface of steel: hot-dip galvanizing, metal spraying, sherardizing, and electro deposition. These will produce results of different thicknesses and composition through their thickness. It has been suggested that the life of the coat to first maintenance is proportional to the thickness. As such, the method of zinc deposition greatly influences the performance of the coating.

The mechanism by which zinc coating protects steel is that the difference in potential between the zinc coating and any of the steel surface which becomes exposed (scratches, etc.) and is subject to moisture which would otherwise cause it to corrode, causes an electric current to flow through the cell such that the more anodic metal, the zinc, will corrode preferentially. The zinc is lost sacrificially at the anode of the cell and the steel is cathodically protected.

Elsewhere, where the zinc coating remains intact over the steel and is not required to give cathodic protection yet, normal relatively slow corrosion rates of zinc will apply.

Some special cases should be mentioned. Firstly, zinc coating will not protect steel in hot water (in excess of about 60°C). Secondly, where zinc-coated steel is to be concreted it should be chromated beforehand. Consideration should also be given to the risks of hydrogen embrittlement and distortion when choosing the method of deposition.

4.5.2 Surface preparation

Surface preparation probably defines the quality of the overall scheme to a greater extent than any subsequent coats. It has a number of functions:

(1) Removal of millscale.
(2) Removal of existing paint.
(3) Removal of salts.
(4) Removal of contaminants, e.g. grease, dirt and weld fume.
(5) Roughening of the surface to improve adhesion.
(6) Preparation of galvanized or other zinc-coated surfaces to receive paint.

Removal of millscale. Millscale is a layer of dense oxide formed during the rolling processes at the steel mill. It is blue-grey in colour, reasonably shiny and initially is tightly adherent to the underlying steel. As such, it can easily be missed. It has the tendency to loosen and flake off with time even if coated. It is essential that it is removed before painting. Hand preparation alone (wire brushing, grinding, needleguns, etc.) is ineffective. If preceded by flame cleaning, however, it may be satisfactory. Blast cleaning is more efficient and reliable. There are a number of alternative forms. Possibilities exist for its use at works or on site.

Sand blasting is little used these days because of silicosis. Grit blasting is probably the most effective method but grit particles can get trapped in the surface of the steel. Shot blasting is common in automatic blasting plants there being problems recycling grit. It is not so efficient as grit for removing millscale because it tends to impact it into the surface.

Removal of existing paint. If paint is well adhered there is generally little point in removing it before overcoating unless an adhesion or interaction problem with the overlying coats is likely. Where existing paint is to be removed, possible methods are blast cleaning, water jetting, chemical paint strippers and hot-air strippers with mechanical cleaning. Removal of lead-based paints can be particularly problematic from a health and safety viewpoint.

Removal of salts. When rusting has occurred to the extent that even shallow pits have formed, hygroscopic iron salts will be present within the pits. These will cause the premature breakdown of paint films applied over them by underfilm corrosion. For their removal, hand preparation is useless. Normal blast cleaning is of limited effectiveness. Wet blasting or high-pressure water jetting after normal blasting are preferable.

Removal of contaminants. For oil or water-soluble contami-

Table 4.13 Properties of some foamed and expanded plastics

Expanded or foamed polymer	Bulk density (kg/m³)	Thermal conductivity (W/m°C)	Maximum working temperature (°C)	Water absorption at 7 days (% by vol.)	Vapour resistivity (MNs/g)	Combustibility	Typical applications	British Standards or codes of practice
Bead board polystyrene	16–24	0.033–0.035	80	2.5–3.0	100–600	*F	Lining walls and ceilings: insulating cold-water services; Integral wall, floor and roof insulation	BS 3837:1986 BS 6203:1982
Extruded polystyrene	32–40	0.033–0.035	80	1.0–1.5	1200–1800	*F	Similar to above	BS 3290:1960 BS 6203:1982
Expanded PVC	24–125	0.035–0.055	65	3.0–4.0	1000–1800	B	Lining walls and roofs: integral insulation in sandwich panels	BS 3869:1965
Foamed phenol formaldehyde	32–64	0.036	130	Depends on cell structure	50–250	R	Roof insulation under hot-applied finishes	BS 3927:1986
Foamed urea formaldehyde	8	0.038	100	High	20	R	In situ cavity wall insulation	BS 5617:1985 BS 5618:1985
Foamed polyurethane	32	0.020–0.025	100	2.5	400–600	*F	Lining walls and ceilings, integral insulation in sandwich panels, insulating pipes, sprayed on in situ insulation	BS 4841:1975

Key:
* Flame retardant grades are available
F Flammable and continues to burn after ignition
B Combustible but self-extinguishing
R Resistant to ignition

nants, washing with a detergent solution followed by thorough rinsing is preferable to solvent washing, which tends to spread the contaminant rather than remove it. Weld slag and fume may be removed by blast cleaning or mechanical abrasion.

Roughening of the surface. Surfaces can be either so smooth that many paints will not stick well to them or too rough so that peaks penetrate the first primer coat, possibly leading to spot rusting. In general, a peak-to-trough height between 50 and 100 μm is sensible.

4.5.3 Preparation of galvanized or other zinc-coated surfaces

Galvanized and other zinc-coated surfaces are inherently difficult to paint. If zinc has weathered sufficiently that significant surface roughening has occurred, then adhesion will be improved. In new construction this is rarely the case. Calcium plumbate primers were a traditional way of solving the problem; however, being slow drying, poisonous and not compatible with many modern paints, their use is much decreased.

The two most common solutions to the problem are: (1) the use at works of an etch primer, e.g. two-pack PVB etch primer (a polyvinyl butyral, zinc tetroxychromate, phosphoric acid etch primer); (2) the use on site of British Rail 'T-Wash' – a nonproprietary product available from almost any paint supplier. This etches the surface and, in so doing, discolours it.

A number of paint products have recently been introduced to the market, claiming to be directly applicable over such surfaces. Some of these appear promising.

4.5.4 Coating types

Primers. Commonly encountered types of primer are as follows.

(1) *Lead-based primers.* The use of red lead, metallic lead and calcium plumbate primers has reduced due to their toxicity.
(2) *Zinc chromate primers.* Zinc chromate primers are based on mixtures of zinc chromate and other pigments. A variety of media are possible. They perform comparatively poorly in industrial atmospheres and marine environments.
(3) *Zinc phosphate primers.* These are good general-purpose pigments for primers. They are available in a variety of media and give good performance in most environments.
(4) *Zinc-rich primers.* Zinc-rich primers are made from not less than 85% metallic zinc dust in styrene, chlorinated rubber or epoxide media. The dry film should contain at least 90% of zinc which is a high enough concentration to give electrical contact between the zinc and the steel. Cathodic protection therefore results, though it is believed that this mechanism of protection is soon replaced by the protective action of a dense layer of zinc and its corrosion products which form through the action of contamination by the atmosphere. Very good surface preparation is essential for these primers to ensure good electrical contact and continued adhesion. The solubility of zinc corrosion products formed on the surface of the primer makes it essential to wash primed surfaces very thoroughly before applying subsequent coats of the painting system.
(5) *Other primers.* A number of paint manufacturers recently have introduced to their range products especially formulated, it is claimed, to cope with poorly prepared steel surfaces, i.e. wire brushing only. Some of these appear promising.

Barrier and finishing paints. Barrier coatings, as their name suggests, should keep the harmful elements of the environment (moisture, ultraviolet, etc.) from the primer and from the steel. In some cases they also give the steelwork an attractive appearance. Where this is not the case, a separate finishing paint may often be used over the barrier coat, although this is not always possible; bitumen paints, for instance, cannot be overcoated successfully in such a way.

Lamellar pigments. Plate-like (lamellar) particles are much used for barrier coats. Examples are micaceous iron oxide, flake glass and flake aluminium. The overlapping of the pigment particles in the paint film provides a tortuous path for moisture to penetrate through its thickness. They can be used in a variety of media. Oleoresinous, two-pack epoxy or chlorinated rubber are common.

Paint media. Pigments, whether inhibitive, lamellar, decorative, reinforcing or with other purposes, are dispersed in media. Characteristics of types commonly met are given below.

(1) *Alkyd paints.* These are cheap and easy to use but, comparatively, not very long-lasting. They can be modified with silicone to produce more durable decorative properties.
(2) *Oleoresinous paints.* Linseed oil used on its own as a paint media is slow-drying but does have advantageous qualities. Alkyd and/or other resins are often incorporated with it in paint formulations to improve this and confer other properties. Phenolic resin is often used.
(3) *Chlorinated rubber paints.* These are solutions of chlorinated rubber and plasticizer in suitable solvents. They produce generally chemically resistant, highly impermeable coatings but they do have the disadvantage of being comparatively soft and remain susceptible to solvents. This can lead to some wrinkling during overcoating. Generally, however, they are easy to maintain.
(4) *Epoxy paints (two-pack).* These are very hard, chemically resistant and confer a very high degree of protection. However, they do require a high quality of surface preparation and can be difficult to overcoat at age, being smooth and hard. At low temperatures curing may be very slow.
(5) *Polyurethane paints (two-pack).* These are tough and chemically resistant and give a high degree of protection. Often their flexural and decorative properties are superior to those of an epoxy but the difficulties of overcoating them are greater. Also, their spray application has health hazards associated with it.
(6) *Acrylics, vinyls, acrylated rubber paints.* These are one-pack products that are similar to chlorinated rubber in performance although, in general, somewhat less soft.
(7) *Coal-tar pitch and bitumen paints.* These are based on solutions of these materials in solvents. Pigments and thickening agents have to be added to these solutions if the paints are to give adequate film thicknesses and be resistant to ultraviolet light. High-build paints can be formulated to give dry film thicknesses of up to 250 μm in one coat. These paints are normally black, but a limited range of dark colours can be produced. They are slow-drying.
(8) *Pitch/epoxide paints.* Pitch/epoxide paints are similar to coal-tar pitch paints, but have much better chemical and weathering resistance because of the addition of 30% or more of epoxide resin. These paints can be formulated to give dry film thicknesses of 250 μm or more in one coat and they give good protection in very severe exposure conditions. These paints are black or very dark in colour.

Standards and codes of practice referred to in Chapter 4

BS 12 1978: *Specification for ordinary and rapid-hardening Portland cement.*

BS 146:1973:Part 2: *Specification for Portland blast-furnace cement.*

BS 812:1984:Part 101: *Guide to sampling and testing aggregates.*

BS 877:1973 (1977):Part 2: *Specification for foamed or expanded blast-furnace slag lightweight aggregate for concrete.*

BS 882:1983: *Specification for aggregates from natural sources for concrete.*

BS 915:1972 (1983):Part 2: *Specification for high alumina cement.*

BS 1014:1975 (1986): *Specification for pigments for Portland cement and Portland cement products.*

BS 1165:1985: *Specification for clinker and furnace bottom ash aggregates for concrete.*

BS 1203:1979: *Specification for synthetic resin adhesives (phenolic and aminoplastic) for plywood.*

BS 1204:1979: *Synthetic resin adhesives (phenolic and aminoplastic) for wood.*

BS 1370:1979: *Specification for low-heat Portland cement.*

BS 1711:1975: *Specification for solid rubber flooring.*

BS 1881: *Methods of testing concrete:*

Part 6:1971: 'Analysis of hardened concrete'.

Part 101:1983: 'Method of sampling fresh concrete on site'.

Part 102:1983: 'Method for determination of slump'.

Part 103:1983: 'Method for determination of compacting factor'.

Part 104:1983: 'Method for determination of vebe time'.

Part 105:1984: 'Method for determination of flow'.

Part 106:1983: 'Methods for determination of air content of fresh concrete'.

Part 112:1983: 'Methods of accelerated curing of test cubes'.

Part 116:1983: 'Method for determination of compressive strength of concrete cubes'.

Part 117:1983: 'Method for determination of tensile splitting strength'.

Part 118:1983: 'Method for determination of flexural strength'.

Part 201:1986: 'Guide to the use of nondestructive methods of test for hardened concrete'.

Part 202:1986: 'Recommendations for surface hardness testing by rebound hammer'.

Part 203:1986: 'Recommendations for measurements of velocity of ultrasonic pulses in concrete'.

BS 1972:1967: *Specification for polythene pipe (type 32) for above ground use for cold-water services.*

BS 1973:1970 (1982): *Specifications for polythene pipe (type 32) for general purposes including chemical and food industry uses.*

BS 2571:1963: *Specification for flexible PVC compounds.*

BS 2572:1976: *Specification for phenolic laminated sheet and epoxide cotton fabric laminated sheet.*

BS 2752:1982 (1987): 'Specification for chloroprene rubber compounds'

BS 3012:1970 (1982): *Specification for low- and intermediate-density polythene sheet for general purposes.*

BS 3227:1980: *Specification for butyl rubber compounds (including halobutyl compounds).*

BS 3290:1960: *Specification for toughened polystyrene extruded sheet.*

BS 3505:1986: *Specification for unplasticized polyvinyl chloride (PVC-U) pressure pipes for cold potable water.*

BS 3506:1969: *Specification for unplasticized PVC pipe for industrial uses.*

BS 3532:1962: *Specification for unsaturated polyester resin systems for low-pressure fibre-reinforced plastics.*

BS 3794:1986: *Decorative laminated sheets based on thermosetting resins.*

BS 3837:1986: *Expanded polystyrene boards.*

BS 3869:1965: *Specification for rigid expanded polyvinyl chloride for thermal insulation purposes and building applications.*

BS 3892:1982:Part 1: *Specification for pulverized fuel ash for use as a cementitious component in structural concrete.*

BS 3892:1984:Part 2: *Specification for pulverized fuel ash for use in grouts and for miscellaneous uses in concrete.*

BS 3927:1986: *Specification for rigid phenolic foam (PF) for thermal insulation in the form of slabs and profiled sections.*

BS 4027:1980: Specification for sulphate-resisting Portland cement.

BS 4154:1985: *Corrugated plastics translucent sheets made from thermosetting polyester resin (glass fibre reinforced).*

BS 4203:1980 (1987): *Extruded rigid PVC corrugated sheeting.*

BS 4248:1974: *Specification of supersulphated cement.*

BS 4254:1983: *Specification for two-part polysulphide-based sealants.*

BS 4346:1969/70/82: *Joints and fittings for use with unplasticized PVC pressure pipes.*

BS 4514:1983: *Specification for unplasticized PVC soil and ventilating pipes, fittings and accessories.*

BS 4549:1970: *Guide to quality-control requirements for reinforced plastics mouldings.*

BS 4576:1970 (1982): *Specifications for unplasticized PVC rain-water goods.*

BS 4646:1970 (1982): *Specification for high-density polythene sheet for general purposes.*

BS 4660:1973: *Specification for unplasticized PVC underground drain pipe and fittings.*

BS 4841:1975: (1987 parts 1, 2, 3): *Rigid polyurethane (PUR) and polyisocyanurate (PIR) foam for building applications.*

BS 4994:1987: *Specification for design and construction of vessels and tanks in reinforced plastics.*

BS 5075: *Concrete admixtures:*

Part 1:1982: 'Specification for accelerating admixtures retarding admixtures and water-reducing admixtures'.

Part 2:1982: 'Specification for air-entraining admixtures'.

Part 3:1985: 'Specification for superplasticizing admixtures'.

BS 5139:1974: *Classification for polypropylene plastics materials for moulding and extrusion.*

BS 5215:1986: *Specification for one-part gun grade polysulphide-based sealants.*

BS 5328:1981: *Methods for specifying concrete, including ready-mixed concrete.*

BS 5481:1977: *Specification for unplasticized PVC pipe and fittings for gravity sewers.*

BS 5493:1977: *Code of practice for protective coating of iron and steel structures against corrosion.*

BS 5617:1985: *Specification for urea formaldehyde (UF) foam systems suitable for thermal insulation of cavity walls with masonry or concrete inner and outer leaves.*

BS 5618:1985: *Code of practice for thermal insulation of cavity walls (with masonry or concrete inner and outer leaves) by filling with urea formaldehyde (UF) foam systems.*

BS 6203:1982: *Guide to fire characteristics and fire performance of expanded polystyrene (EPS) used in building applications.*

BS 6374:1984:Part 3: *Specification for lining with stoved thermosetting resins. Lining equipment with polymeric materials for the process industries.*

BS 6515:1984: *Specification for polyethylene damp-proof courses for masonry.*

BS 6564:1985: *Polytetrafluoroethylene (PTFE) materials and products.*

BS 6588:1985: *Specification for Portland pulverized fuel ash cement.*
BS 6610:1985: *Specification for pozzalanic cement with pulverized fuel ash as pozzolana.*
BS 6699:1986: *Specification for ground granulated blast-furnace slag for use with Portland cement.*
BS 6716:1986: *Guide to properties and types of rubber.*
BS 8110:1985: *Structural use of concrete.*
BS DD 83:1983: *Assessment of the composition of fresh concrete.*

Other standards

ASTM C-150: *Standard specification for Portland cement.*
ASTM C-457: *Standard recommended practice for microscopical determination of air void content and parameters of the air void system in hardened concrete.*
ASTM D-883: *Standard definitions of terms relating to plastics.*

References

1 Construction Industry Research and Information Association (1984) *Guide to sources of construction information* (4th edn). CIRIA Special Publication No. 30. CIRIA, London.
2 Building Research Establishment (1980) *Materials for concrete,* BRE Digest No. 237 (2nd ser.). BRE, Garston.
3 British Standards Institution (1983) BS 882. *Specification for aggregates from natural sources for concrete.* BSI, Milton Keynes.
4 Cement and Concrete Association (1983) *Minimising the risk of alkali silica reaction: guidance notes.* C&CA, London.
5 Teychenné, D. C., *et al.* (1975) *Design of normal concrete mixes.* HMSO 1975.
6 Cement and Concrete Association (1958) *An introduction to concrete Eb1.* C&CA, London.
7 Sadgrove, B. M. (1970) *The early development of strength in concrete.* Construction Industry Research and Information Association Technical Note No. 12. CIRIA, London.
8 Chapman, G. P. (1968) 'The cylinder splitting test with particular reference to concrete made with natural aggregates', *Concrete*, **1**, 2.
9 Sadgrove, B. M. (1971) *The strength and deflection of reinforced concrete beams loaded at early age,* Construction Industry Research and Information Association Technical Note No. 31. CIRIA, London.
10 Eglington, M. S. (1975) *Review of concrete behaviour in acidic soils and groundwaters.* Construction Industry Research and Information Association Technical Note No. 69. CIRIA, London.
11 Birt, J. C. (1973) *Curing concrete: an appraisal of attitudes, practice and knowledge.* Construction Industry Research and Information Association Technical Note No. 43. CIRIA, London.
12 Tattersall, G. H. and Bloomer, S. J. (1979) 'Further development of the two-point test for workability and extension of its range', *Conc. Res.* **31**, 109, 202–210.

Bibliography

Brydson, J. A. (1989) *Plastics materials* (5th edn). Butterworth Scientific, Guildford.
Building Research Establishment (1977) *Durability and application of plastics.* BRE Digest No. 69 (2nd ser.) BRE, Garston.
Building Research Establishment (1979) *Cellular plastics for building.* BRE Digest No. 224 (2nd ser.) BRE, Garston.
Construction Industry Research and Information Association (1982) I. P. Haigh (ed.) *Painting steelwork.* CIRIA Report No. 93. CIRIA, London.
Construction Industry Research and Information Association (1984) *The CIRIA guide to concrete construction in the Gulf region.* CIRIA Special Publication No. 31. CIRIA, London.
Evans, U. R. (1960) *The corrosion and oxidation of metals.* Arnold, Glasgow.
Hall, C. (1981) *Polymer materials.* Macmillan, London.
Handbook on British Standard BS 8110:1985: Structural use of concrete: code of practice for design and construction. Palladin Publications.
Harrison, T. A. (1981) *Early-age thermal crack control in concrete.* Construction Industry Research and Information Association Report No. 91. CIRIA, London.
Lea, F. M. (1970) *The chemistry of cement and concrete* (3rd edn). Arnold, Glasgow.
Neville, A. M. (1981) *Properties of concrete* (3rd edn). Pitman, London.
Teychenné, D. C. (1978) *The use of crushed rock aggregates in concrete.* Building Research Establishment, Garston.
Wranglén, G. (1985) *An introduction to corrosion and protection of metals.* Chapman and Hall, London.

5

Hydraulics

The late J Allen DSc, LLD, FICE, FRSE
Emeritus Professor,
University of Aberdeen

Contents

5.1 Physical properties of water

5.1.1 Density

For most purposes in hydraulic engineering, the density of fresh water may be taken to be 1000 kg/m³. Correspondingly, the weight of 1 l is approximately 1 kg. In more precise work, usually of a laboratory or experimental nature, the variation of density with temperature may have to be taken into account in accordance with Table 5.1.

Table 5.1 Density of fresh water at atmospheric pressure

Temperature (°C)	Density (kg/m³)
0	999.9
4	1 000.0
10	999.7
20	998.2
30	995.7
40	992.2
50	988.1
60	983.3
70	977.8
80	971.9
90	965.3
100	958.4

The density of sea water depends on the locality but for general calculations the open sea may be assumed to weigh 1025 kg/m³. In a tidal river the density varies appreciably from place to place and time to time; it is influenced by the state of the tide and by the amount of fresh water flowing into the estuary from the higher reaches or from drains and other sources. At any one spot it may also vary through the depth of the water owing to imperfect mixing of the fresh and saline constituents.

5.1.2 Viscosity

Let us visualize a layer of a fluid as represented in Figure 5.1. The thickness of the layer is δy and particles in the plane AB are supposed to have a velocity v while those in CD have a different

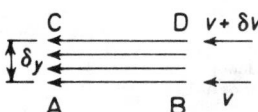

Figure 5.1 Layer of fluid illustrating laminar flow

velocity, say $v + \delta v$. The plan area of each plane, AB or CD, is a, say. Now the fluid bounded by AB and CD experiences a resistance to relative motion along AB analogous to shear resistance in solid mechanics. This force of resistance, divided by the area a, will give a resistance per unit area, or a stress f. Then:

$$f = \eta(dv/dy) \qquad (5.1)$$

as δy tends to zero, or as the layer assumes an infinitesimal thickness, so that dv/dy becomes the velocity gradient, i.e. the rate at which the velocity changes as we proceed outwards in a direction normal to the plane AB. η in Equation (5.1) is known

as the coefficient of viscosity. If force is defined by force = mass × acceleration then η will have the units of $f/(dv/dy)$, or:

$$([M] \times [L] \times [T^{-2}] \times [L^{-2}])/([L] \times [T^{-1}] \times [L^{-1}])$$

i.e. $[ML^{-1}T^{-1}]$, where $[M]$ represents mass, $[L]$ length, and $[T]$ time.

Thus if newtons (i.e. kilogram metres per squared second) are adopted for the force of resistance, metres for length, metres squared for area, and metres per second for velocity, then the coefficient of viscosity takes the units kilograms per metre second. For example, the numerical value of η in the case of water at 10° C is 0.001 31 kg/m s. This is the same as 0.0131 poise, i.e. 0.0131 g/cm s.

5.1.2.1 Kinematic viscosity

Kinematic viscosity v is defined as the ratio of the viscosity η to the density ρ of a fluid, or:

$$v = \eta/\rho \qquad (5.2)$$

It follows from this definition that if η is in kilograms per metre second and ρ in kilograms per cubic metre then v will be in square metres per second. Again, considering water at 10° C, v is 1.31×10^{-6} m²/s or 1.31×10^{-2} cm²/s.

Typical values of η and v, for both water and air, are given in Table 5.2, from which it will be seen that temperature has quite different effects on these two fluids.

Table 5.2 Viscosities of water and dry air at atmospheric pressure

Temperature °C	Water		Air	
	$10^3 \eta$ (kg/m s)	$10^6 v$ (m²/s)	$10^5 \eta$ (kg/m s)	$10^5 v$ (m²/s)
0	1.79*	1.79*	1.71*	1.32*
5	1.52	1.52	1.73	1.36
10	1.31	1.31	1.76	1.41
15	1.14	1.14	1.78	1.45
20	1.01	1.01	1.81	1.50
25	0.894	0.897	1.83	1.55
30	0.801	0.804	1.86	1.59
35	0.723	0.727	1.88	1.64
40	0.656	0.661	1.90	1.69
50	0.549	0.556	1.95	1.79
60	0.469	0.477	2.00	1.88
80	0.357	0.367	2.09	2.09
100	0.284	0.296	2.18	2.30

*To avoid any misinterpretation of the column headings note that, at 0° C, η for water is 1.79×10^{-3} kg/m s; v is 1.79×10^{-6} m²/s; η for air is 1.71×10^{-5} kg/m s; v is 1.32×10^{-5} m²/s.

5.1.3 'Non-Newtonian' fluids

Table 5.2 implies that for water and air, the coefficient of viscosity (sometimes called 'dynamic viscosity' or 'absolute viscosity') varies with temperature but otherwise is a constant coefficient in Equation (5.1), i.e. f is in simple proportion to dv/dy. This is true of very many other fluids, but there exist also the so-called non-Newtonian fluids. With some of these, η decreases as dv/dy increases; with others the reverse is true. Among the examples of substances which, in phases of their fluid state, behave in a 'non-Newtonian' way, are certain lubricants, e.g. grease used in bearings, plastics, and suspensions of particles.

5.1.4 Compressibility

In the vast majority of engineering calculations, water may be treated as an incompressible fluid. Exceptions arise when large and sudden changes of velocity occur, as in certain problems associated with the rapid opening or closing of a valve. If we imagine a mass of water to have its volume changed from V to $V - \delta V$ by an increase of pressure δp applied uniformly round its surface, then the bulk modulus of compressibility K is defined as the stress or pressure intensity δp divided by the volumetric strain produced by δp. This volumetric strain is $-\delta V/V$. Hence:

$$K = -\delta p \Big/ (\delta V/V) \qquad (5.3)$$

δV itself being treated as negative.

The value of K depends somewhat on the temperature and absolute pressure of the water, but in round numbers it is usually sufficiently accurate to take it as 2×10^9 N/m².

5.1.5 Surface tension

Surface tension is the property which enables water and other liquids to assume the form of drops, when it appears that the water is bounded by an elastic skin or membrane under tension. Another important manifestation is related to small waves or ripples where the form and motion are restricted or influenced by the tension in the surface. Surface tension depends on the liquid and gas in contact with one another. Suppose a portion of the liquid to have a bounding surface with radii of curvature in two mutually perpendicular directions R_1 and R_2 as in Figure 5.2. Then the excess of pressure intensity inside the boundary, over that outside it, is $\gamma(1/R_1 + 1/R_2)$, where γ is the surface tension, a force per unit length of the line to which it is normal.

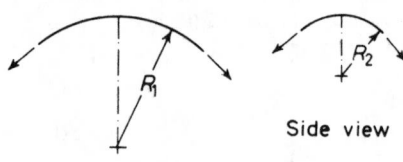

Side view

Front view

Figure 5.2 Bounding surface of a liquid

The surface tension of mercury in contact with air at 20° C is approximately 0.51 N/m.

Table 5.3 Surface tension of water in contact with air

(° C)	0	20	40	60	80	100
γ (N/m)	0.0756	0.0728	0.0700	0.0671	0.0643	0.0615

5.1.6 Capillarity

If a vertical tube is placed in a vessel containing water, the water will be drawn up the tube by capillary attraction. The angle α in Figure 5.3 is known as the angle of contact, and approaches the value zero for clean water in contact with a clean glass tube. In that case

$$h = (4 \times 10^6 \, \gamma/\rho dg) \text{ mm, approximately, or say } (4 \times 10^5 \, \gamma)/\rho d \qquad (5.4)$$

where γ is the surface tension in newtons per metre, ρ the density in kilograms per cubic metre, d the bore of the tube in millimetres, and g is in metres per second squared.

At 20° C, therefore, the elevation of the water in the glass tube amounts to about $30/d$ mm. It should be emphasized, however, that capillary attraction depends to a marked degree upon the state of cleanliness of the liquid and the tube.

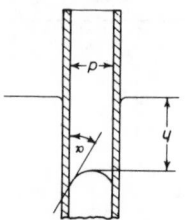

Figure 5.3 Capillary rise of water in a tube

If mercury is considered instead of water, there is a depression of the liquid in the tube as shown in Figure 5.4. The angle β is approximately 53° and h, at room temperature, is about $9/d$ mm.

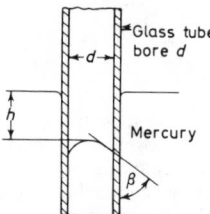

Figure 5.4 Capillary depression of mercury in a tube

Capillary attraction may be important in connection with the technique of measurement. For example, if the level of water in a tank is read, for convenience, on an external gauge as depicted in Figure 5.5, the bottom of the meniscus, or curved surface of the water in the gauge-tube, will stand higher than in the tank. Whether the effect is serious depends, of course, upon the standard of accuracy demanded, but it is generally advisable in such a case, or when using a differential gauge having two limbs, to use a tube not smaller than 9.5 mm bore, and as uniform as practicable throughout its length.

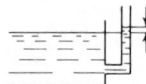

Figure 5.5 Effect of capillary attraction on measurement of height of water in a tank

5.1.7 Solubility of gases in water

At atmospheric pressure, water is capable of dissolving approximately 3, 2 and 1% of its own volume of air at temperatures of 0, 20 and 100° C respectively. Certain other gases, such as carbon dioxide, are dissolved in much greater volumes, but the presence of air alone, together with the phenomenon of vapour pressure of water, may lead to complications in certain pipelines or machines. To take an example, at the highest point of a siphon the pressure is below atmospheric and it is possible for air to come out of solution there which was originally dissolved in the water at atmospheric pressure. This accumulation of air may ultimately reduce the flow along the siphon very appreciably, or even break it entirely, unless precautions are taken to draw off the air and vapour as it collects. The suction-lift of pumps is also restricted in practice by the difficulty of release of air and generation of vapour so that, in practice, it is usual to limit the suction-lift to about 8.5 m instead of the full height of the water barometer, say 10.4 m.

5.1.8 Vapour pressure

If a liquid is contained within a closed vessel the space above it becomes saturated with its vapour and the space is subjected to an increase of pressure, which is the vapour pressure of the liquid at the temperature then obtaining.

Table 5.4 Vapour pressure p_v (N/m²) of water at various temperatures

(°C)	0	5	10	15	20	30	40
$10^{-4}p_v$	0.0610	0.0875	0.123	0.170	0.235	0.423	0.736

(°C)	50	60	70	80	90	100
$10^{-4}p_v$	1.23	2.00	3.09	4.76	7.00	10.1

To obtain p_v in metres of water at a given temperature, divide pv, (N/m²) by $g\rho$ where ρ (kg/m³) is given in Table 5.1. For example, at 100°C, p_v is $(10.1 \times 10^4)/(9.81 \times 9.58 \times 10^2)$, i.e. 10.8 m of water.

5.2 Hydrostatics

(1) A fluid at rest exerts a pressure which is everywhere normal to any surface immersed in it.
(2) The pressure intensity at a point P in a liquid is equal to that at the free surface of the liquid together with ρgh, where h is the depth of P below the free surface and ρ is the density of the liquid.

5.2.1 Force on any area

In many engineering problems all pressures are treated relative to atmospheric pressure as a datum. Adopting that system, consider the force exerted on an elementary, or infinitesimal, portion δA of an area A immersed in a liquid (see Figure 5.6).

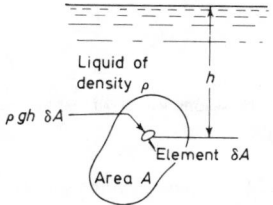

Figure 5.6 Elementary area δA immersed in liquid

The pressure intensity on δA is $p = \rho gh$. Hence, the force on δA is $\rho gh \cdot \delta A$, where h is the vertical depth of δA.

The total force on the whole area A of which δA is an element is the arithmetical sum of the forces on all its constituent elements, $\Sigma \rho gh \cdot A = \rho g\Sigma h \cdot \delta A$, assuming ρ to be constant throughout the liquid. Hence:

$$\text{total force} = \rho gH \cdot A \qquad (5.5)$$

where H is the vertical depth of the centroid of the whole area.

This total force is only equal to the resultant force if the area under consideration is a plane one. If the area is curved, then the forces acting on its elementary portions are not all parallel to one another so their simple arithmetic sum is not the same as their resultant.

5.2.2 Force on plane areas (Figure 5.7)

$$\text{Force on element} = (\rho gz \cos\theta)\delta A$$

Resultant force = total force in this case

$$= \Sigma (\rho gz \cos\theta)\delta A$$
$$= \rho g\bar{z}A \cos\theta \qquad (5.6)$$

where $\bar{z}$ is the inclined depth of centroid of A.

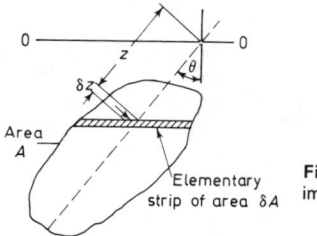

Figure 5.7 Plane area immersed in liquid

The resultant force will act through a point in the immersed area A known as its centre of pressure and such that its inclined depth Z is given by:

$$Z = \left(\int z^2\,dA\right)\Big/\bar{z}A = (\text{Second moment of area } A \text{ about } 00)/\bar{z}A$$
$$= I_{00}/Az \qquad (5.7)$$

or:

$$Z = k_{00}^2/\bar{z} \qquad (5.8)$$

where k_{00} is the radius of gyration about 00 and $k_{00}^2 = k^2 + \bar{z}^2$, where k is the radius of gyration about an axis through the centroid parallel to 00.

Examples 5.1 to 5.6. In the following examples C is the centroid, P the centre of pressure.

Example 5.1: Parallelogram (Figure 5.8(a)):

$$I_{00} = (bd^3/12) + bd(\bar{z})^2$$
$$Z = [bd^3/12 + bd(\bar{z}^2)]/bd\bar{z}$$
$$= (d^2/12 + \bar{z}^2)/\bar{z}$$
$$= \tfrac{2}{3}d \quad \text{if upper edge of parallelogram is in surface.}$$

Example 5.2: Circular area, diameter d (Figure 5.8(b)):

$$I_{00} = (\pi d^4/64) + (\pi d^2/4)\bar{z}^2$$

$$Z = \frac{(\pi d^4/64) + (\pi d^2/4)\,\bar{z}^2}{(\pi d^2/4)\,\bar{z}}$$
$$= (d^2/16 + \bar{z}^2)/\bar{z}$$
$$= 5d/8 \quad \text{if circle touches } 00.$$

Example 5.3: Triangular area, apex downwards (Figure 5.8(c)):

$$I_{00} = (bh^3/36) + (bh/2)\bar{z}^2$$
$$Z = \frac{(bh^3/36) + (bh/2)\,\bar{z}^2}{(bh/2)\,\bar{z}} = \frac{h^2/18 + \bar{z}^2}{\bar{z}}$$

where $\bar{z} = a + (h/3)$

$$Z = h/2 \text{ if } a = 0.$$

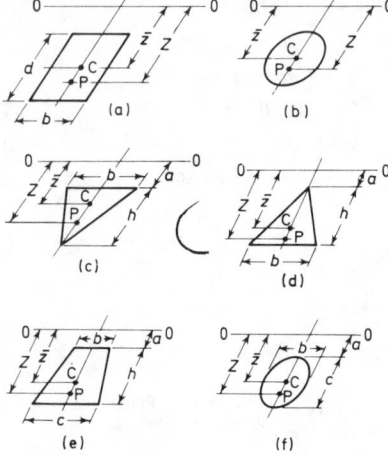

Figure 5.8 (a) Parallelogram; (b) circular area; (c) triangular area, apex downwards; (d) triangular area, apex upwards; (e) trapezium; (f) ellipse

Example 5.4: Triangular area, apex upwards (Figure 5.8(d)):

$$I_{00} = (bh^3/36) + (bh/2)\,\bar{z}^2$$

$$Z = \frac{h^2/18 + \bar{z}^2}{\bar{z}}$$

where $\bar{z} = a + (2/3)h$

$$Z = (3/4)h$$

if $a = 0$.

Example 5.5: Trapezium (Figure 5.8(e)):

$$Z = (k^2 + \bar{z}^2)/\bar{z}$$

where $k^2 = (h^2/18)[1 + 2bc/(b+c)^2]$

$$\bar{z} = \frac{h(2c+b)}{3(b+c)} + a$$

Example 5.6: Ellipse (Figure 5.8(f)):

$$Z = (k^2 + \bar{z}^2)/\bar{z}$$

where $k^2 = c^2/16$; $\bar{z} = a + (c/2)$

In the examples so far considered, the immersed areas have had a vertical plane of symmetry in which it is evident that the resultant force will act. All that has been necessary, therefore, was to determine the position of the resultant force in that plane of symmetry.

5.2.2.1 Force on an unsymmetrical plane area

Choose any convenient axes OX, OY. OX may be the line of intersection of the plane of the immersed area with the surface of the liquid. The elementary area δA has coordinates x and y relative to the chosen axes (Figure 5.9).

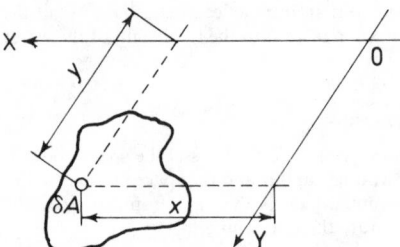

Figure 5.9 Unsymmetrical plane area

Let $\bar{x}$ and $\bar{y}$ be the coordinates of the centre of pressure of the whole area relative to these same axes. Then, by using the principle that the moment of the resultant force is equal to the sum of the moments of the individual elementary forces:

$$\bar{y} = (\Sigma y^2\,\delta A)/(\Sigma y\,\delta A) \qquad \bar{x} = (\Sigma xy\,\delta A)/(\Sigma y\,\delta A)$$

or:

$$\bar{y} = \frac{\iint y^2\,dx\,dy}{\iint y\,dx\,dy} \qquad \bar{x} = \frac{\iint xy\,dx\,dy}{\iint y\,dx\,dy} \tag{5.9}$$

But:

$$\Sigma y^2\,\delta A \quad \text{or} \quad \iint y^2\,dx\,dy = Ak^2$$

$$\Sigma y\,\delta A \quad \text{or} \quad \iint y\,dx\,dy = Ay_0 \tag{5.10}$$

where A is the total area, k its radius of gyration about OX and y_0 the ordinate of the centroid of the area.

5.2.3 Force on curved areas

The following examples will serve to illustrate some useful principles.

Hemispherical bowl, radius r, just full of water (Figure 5.10).

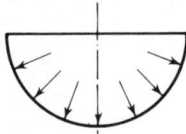

Figure 5.10 Hemispherical bowl just full of water

Total force = arithmetical sum of forces acting on the surface
= area × pressure intensity at centroid
= $2\pi r^2$ × density of water × depth of centroid × g
= $2\pi r^2 \times \rho \times (r/2)g$
= $\pi r^3 \rho g$

But the horizontal components of the corresponding forces on opposite sides of the vertical axis counterbalance one another, and:

Resultant force = weight of water contained
= volume of hemisphere × density of water × g
= $\frac{2}{3}\pi r^3\,\rho g$

Cylindrical vessel with hemispherical end, just full of water.

(1) (Figure 5.11(a)). Force on lid due to water = 0

Resultant force (vertical) on hemispherical base $= (\pi r^2 h + \frac{2}{3}\pi r^3)\rho g$

(2) (Figure 5.11(b)). Resultant force (vertical) on flat base $= \pi r^2 (h + r)\rho g$

Resultant force (vertically upwards) on dome $= \pi r^2 (h + r)\rho g - (\pi r^2 h + \frac{2}{3}\pi r^3)\rho g = \frac{1}{3}\pi r^3 \rho g$

(3) (Figure 5.11(c)). Horizontal force on either end $= \pi r^2 (r\rho)g = \pi r^3 \rho g$

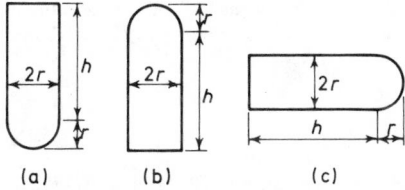

(a)   (b)   (c)

Figure 5.11 (a) Cylindrical vessel with hemispherical end just full of water; (b) the same, inverted; (c) the same, lying with axis horizontal

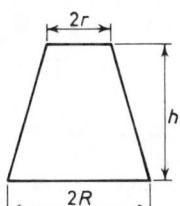

Figure 5.12 Truncated cone just full of water

Truncated cone, just full of water (Figure 5.12).

Resultant force (vertically downwards) on base $= \pi R^2 h\rho g$

Volume of water contained $= \frac{1}{3}\pi h(R^2 + Rr + r^2)$

Resultant force (vertically upwards) on curved side

$= [\pi R^2 h\rho - \frac{1}{3}\pi h(R^2 + Rr + r^2)\rho]g$

$= (\frac{2}{3}\pi R^2 h - \frac{1}{3}\pi Rrh - \frac{1}{3}\pi r^2 h)\rho g$

5.2.4 Buoyancy

A liquid of density ρ exerts a vertical upwards force $V\rho g$ on an immersed body of volume V (Figure 5.13). If the weight of the

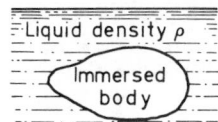

Figure 5.13 Body immersed in liquid

body is greater than $V\rho g$ it will sink. If the weight of the body is less than $V\rho g$ the body will float in such a way that the portion immersed has a volume V' which satisfies the following equation:

$$V'\rho g = \text{total weight of body} \tag{5.11}$$

Centre of buoyancy and metacentre. The centre of gravity of the displaced fluid is called the centre of buoyancy. When a body is floating freely, the weight of the fluid displaced equals the weight of the body itself, and its centre of buoyancy, for equilibrium, must be in the same vertical as the centre of gravity

of the body. The degree of stability for angular displacements involves the conception of the metacentre. In Figure 5.14, XX represents the vertical axis of symmetry of a floating body with a centre of gravity, owing to the distribution of its weight, we suppose to be at G. B is the centre of buoyancy.

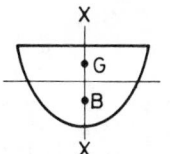

Figure 5.14 Centre of buoyancy

Let the body be displaced so that B_1 becomes the new centre of buoyancy, i.e. the centre of gravity of the liquid as displaced in the new position (Figure 5.15).

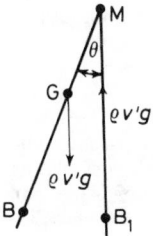

Figure 5.15 Metacentre

The new force of buoyancy acts vertically upwards through B_1, to intersect the deflected line BG in M, the metacentre. Strictly speaking, the metacentre is the position assumed by M as the angle of displacement θ tends to zero.

If M is above G, there will be a 'righting moment' $\rho V' \cdot GM \sin \theta \cdot g$.

The condition for initial stability, or stability during small displacements, is then that M shall be above G.

The metacentric height may be calculated from the equation:

$$GM = (I/V') \pm GB \tag{5.12}$$

where $I = Ak^2$: the plus sign is used if G is below B and the minus sign if G is above B.

A is the area of the water-line section and k is its radius of gyration about the axis Oy. V' is the volume of the immersed portion of body (shown in Figure 5.16).

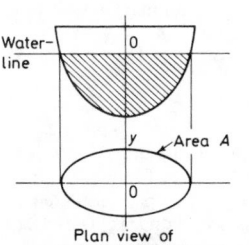

Plan view of waterline section

Figure 5.16 Metacentric height

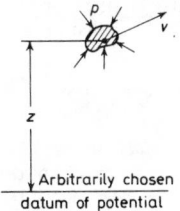

Figure 5.17 Mass of liquid subjected to p N/m^2 and moving with velocity v m/s

5.3 Hydrodynamics

5.3.1 Energy

A liquid possesses energy by virtue of the pressure under which it exists, its velocity and its height above some datum level of potential energy. These three forms of energy—pressure, kinetic and potential—may be expressed as quantities per unit weight of the liquid concerned. The result is the pressure, kinetic or potential head. Thus, referring to Figure 5.17, in which a mass of liquid is represented as subjected to a pressure p N/m², moving with a velocity v m/s and having its centre of mass at a height z above a datum of potential energy:

$$\text{total head} = (p/\rho g) + (v^2/2g) + z \qquad (5.13)$$

in metres, where ρ is the density in kilograms per cubic metre.

If a gas is considered, then account should be taken of its elasticity and the work done in compressing a given mass of it as it passes from a region of low to higher pressure.

5.3.2 Bernoulli's theorem

This states that in the streamline motion of an incompressible and inviscid fluid the total head remains constant from section to section along the stream tube, i.e.:

$$(p/\rho g) + (v^2/2g) + z = \text{constant} \qquad (5.14)$$

The idealized circumstances envisaged in this statement would seem, at first sight, to render the theorem useless for the solution of problems dealing with natural fluids, which are viscous, and especially when such fluids are not moving in streamlines, i.e. when the motion is turbulent, as it usually is in hydraulics. In fact, however, the theorem forms the basis of the majority of practical calculations if and when appropriate terms are added or coefficients introduced to allow for losses of head arising from various causes. The numerical values of these terms and coefficients are almost always the result of experiment or experience.

5.3.3 Streamline and turbulent motion

If we concentrate attention upon one point P in the cross-section of a pipe or channel along which a fluid is moving at a constant rate, the motion at P may be called 'streamline' if the velocity there is constant in magnitude and direction. On the other hand, the motion at P will be turbulent if the velocity there varies from time to time in magnitude and/or direction, despite the fact that the general rate of flow along the channel is constant. In this turbulent motion, the instantaneous velocity at P depends upon how the eddies are passing it at the moment under consideration. The eddies which characterize turbulent motion require energy for their creation and maintenance, and the law of resistance in streamline (sometimes called laminar) flow is quite different from that in turbulent flow.

5.3.3.1 Flow in pipes

Loss of head in smooth pipes The general equations for the loss of head in a pipe of uniform diameter d are as follows. Let h be the loss of head in metres, l the length of pipe considered in metres, v the mean velocity in metres per second $= Q/(\pi d^2/4)$, Q the rate of discharge (in cubic metres per second) and v the kinematic viscosity of fluid (in square metres per second). Then:

$$h = K(lv^{2-n}v^n)/gd^{3-n} \qquad (5.15)$$

where K is a coefficient.

Both K and n depend upon R_e the Reynolds number, vd/v. If R_e is less than 2100, $K = 32$ and $n = 1$.

These values give the equation for streamline flow, which may be deduced mathematically:

$$h = 32vlv/gd^2 \qquad (5.16)$$

Alternatively we may use the equation† commonly adopted by hydraulic engineers, i.e.:

$$h = flv^2/2gm \qquad (5.17)$$

in which m represents the hydraulic mean depth or the ratio of the area of section to the wetted perimeter. In a cylindrical pipe running full, $m = d/4$.

For values of R_e up to 2100:

$$f = 16(vd/v)^{-1} \qquad (5.18)$$

For values of R_e between 3000 and 150 000 (Davis and White[1]):

$$f = 0.08(vd/v)^{-0.25} \qquad (5.19)$$

Alternatively, if R_e exceeds 4000, the Prandtl equation may be used:

$$1/\sqrt{(4f)} = 2.0 \lg [R_e\sqrt{(4f)}] - 0.8 \qquad (5.20)$$

These relationships apply to smooth pipes of, say, glass, drawn brass, copper or large pipes with a smooth cement finish.

In calculating Reynolds numbers, the units in which v, d and v are measured should be consistent with one another; e.g. v in metres per second, d in metres and v in square metres per second. Values of f for smooth pipes in the equation $h = flv^2/2gm$ are plotted against $\lg vd/v$ in Figure 5.18.[2]

At a temperature of 15°C, v for water is 1.14×10^{-6} m²/s.

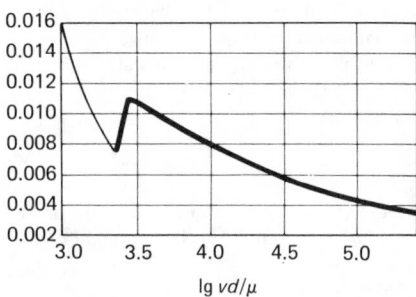

Figure 5.18 Values of f for smooth pipes. (After Stanton and Pannell (1914) Phil. Trans A, **214**; National Phys. Lab. Coll. Res., **11**)

5.3.4 Pipes of noncircular section

There is experimental evidence showing that if the flow is *turbulent*, the value of f for various shapes of section is approximately the same as for a cylindrical pipe at the same value of vm/v, where m again represents the hydraulic mean depth.

The critical value of vm/v, below which the motion is normally laminar, does depend to some extent, however, on the shape of the section. For a circular section it is 525 (i.e. $vd/v = 2100$). For rectangular sections the critical vm/v varies with the ratio of the lengths of the sides and has approximate values of 525 for a square section, 590 for a section having one

† Some writers prefer to use $h = \lambda lv^2/2gd$, rather than $4flv^2/2qd$, for cylindrical pipes. Their friction factor λ is then $4f$.

side 3 times the other, and 730 for a section in which the length of one side is large compared with the other.

During truly *laminar* or *viscous* motion, the loss of head h for various shapes of section is as follows (v being the mean velocity through the section):

Circular section (diameter d): $h = 32vlv/gd^2$

Rectangular section (one side $2a$, other side $2b$):

$$h = 3vlv \bigg/ gb^2 \left[1 - \frac{192}{\pi^5} \frac{b}{a} \left(\tanh \frac{\pi a}{2b} + \frac{1}{3^5} \tanh \frac{3\pi a}{2b} + \ldots \right) \right]$$

Square section (each side $2a$):

$$h = 7.12vlv/ga^2$$

Rectangular section having a large compared with b:

$$h \rightarrow 3vlv/gb^2$$

Circular annulus (mean velocity v through space of area $\pi(d_1^2 - d_0^2)/4$):

$$h = 32vlv \bigg/ gd_1^2 \left[1 + (d_0/d_1)^2 + \frac{1 - (d_0/d_1)^2}{\ln (d_0/d_1)} \right]$$

where d_1 is the outside diameter and d_0 the inside diameter.

5.3.5 Loss of head in rough pipes

Here the value of f also depends upon the ratio of the hydraulic mean depth to the height of the roughening projections from the wall, as well as upon the distribution and shape of these roughnesses. Figure 5.19 summarizes experimental results obtained by Nikuradse[3] with sand-roughened pipes, k being the mean size of the grain projecting from the wall.

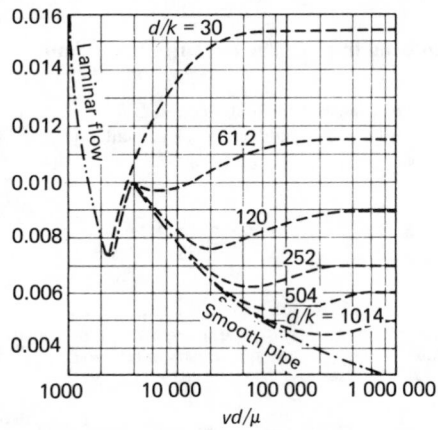

Figure 5.19 Values of f for rough pipes. (After Nikuradse (1933) *Forschungsh. Ver. dtsch. Ing.*, No. 361)

The general tendency is for f to become constant, for a given rough pipe, at sufficiently high Reynolds numbers. A commonly accepted explanation of this is that first of all the surface grains lie inside a very thin viscous layer at the wall of the pipe, even when the main motion is turbulent; at higher values of R_e,

however, they begin to project from this layer and to shed eddies for the maintenance of which additional energy is required.

Prandtl and von Kármán[4] have shown than Nikuradse's results may be made to lie within one band by plotting the quantity

$$1/\sqrt{(4f)} - 2 \lg (d/2k)$$

against a new Reynolds number V_*k/v, in which V_* represents $v\sqrt{(f/2)}$.

Again, if V_*k/v exceeds 60, f becomes constant for a given pipe, the flow then being 'fully turbulent' and the resistance proportional to v^2.

Under those conditions ($V_*k/v > 60$):

$$1/f = 16 \left(\lg \frac{3.7d}{k} \right)^2 \tag{5.21}$$

A pipe may be regarded as 'hydraulically smooth' if $V_*k/v < 3$.

To take a specific example, namely, $d/k = 252$: for values of the original Reynolds number vd/v less than about 11 500, f is the same as for a smooth pipe (see Figure 5.19), while if vd/v exceeds 250 000, f assumes a constant value of 0.007. Between the two there is a curve which represents a transition and which covers a wide range of Reynolds numbers (11 500 to 250 000 in the example $d/k = 252$).

A large proportion of the cases which occur in engineering will be found to lie within the zone of transition, for which Colebrook and White[5,6] have evolved the equation:

$$1/\sqrt{(4f)} = -2 \lg \left(\frac{k}{3.7d} + \frac{2.51}{R_e \sqrt{(4f)}} \right) \tag{5.22}$$

5.3.6 Formulae for calculating pipe friction (turbulent flow)

With the velocities commonly encountered in water pipes, the motion is turbulent. These velocities in fact are frequently within the range from 1.5 to 3.5 m/s, whereas in general the motion can only be expected to be streamline, considering water at ordinary temperatures, for velocities lower than $2.4/d$ m/s, where d is the pipe diameter (in millimetres).

Manning's formula. Among the many formulae which have been suggested from time to time, that of Manning[7] is much favoured:

$$v = (m^{2/3}i^{1/2}/n) \quad \text{(m/s)} \tag{5.23}$$

In this form it applies to pipes and open channels.

m is the hydraulic mean depth (in metres), i the virtual slope of the pipe (i.e. h/l) or the actual slope of the open channel under conditions of uniform flow. n depends upon the material of which the conduit is made.

Alternatively:

$$v = (m^{2/3}i^{1/2}/100n) \quad \text{(m/s)} \tag{5.24}$$

if m is expressed in millimetres.

For cylindrical pipes, the Manning formula in Equation (5.23) gives:

$$h = n^2 lv^2/m^{4/3} \tag{5.25}$$

But:

$$m = d/4$$

Hence:

$$h = n^2(4)^{4/3}(lv^2/d^{4/3}) = 6.35n^2(lv^2/d^{4/3}) \qquad (5.26)$$

If we write this in the form:

$$h = A(lv^2/d^{4/3})$$

the following are appropriate values of A for new pipes (see Table 5.5).

Table 5.5

Material	n	$A = 6.35 n^2$
Clean uncoated cast iron	0.013	0.001 1
Clean coated cast iron	0.012	0.000 92
Riveted steel	0.015	0.001 4
Galvanized iron	0.014	0.001 2
Brass, copper or glass	0.010	0.000 64
Wood-stave	0.012	0.000 92
Smooth concrete	0.012	0.000 92
Cement mortar finish	0.013	0.001 1
Vitrified sewer pipe	0.011	0.000 77

Comparing the formulae

$$h = 4flv^2/2gd \text{ and } h = Alv^2/d^{4/3}$$

it appears that

$$f = (4.91/d^{1/3}) A \qquad (5.27)$$

or

$$f = 31.2n^2/d^{1/3} \qquad (5.28)$$

where d is in metres. If d is in millimetres:

$$f = (49.1/d^{1/3}) A = 312n^2/d^{1/3} \qquad (5.29)$$

Hydraulic Research Papers, Nos 1 and 2 (Ackers[8]), first published by HMSO in 1958, contain not only a fascinating review of the resistance of fluids in channels and pipes but also tables and graphs for the use of designers. The results are derived from the formula of Colebrook and White in Equation (5.22) and values of the effective roughness dimension k are quoted for a great variety of commercial pipes, while in addition, a supplementary note provides information concerning the *actual* diameters of different classes of pipe in relation to their nominal bores.

5.3.7 Deterioration of pipes

Pipes deteriorate with age and to allow for this reduction in their carrying capacity Barnes[9] has suggested the following (see Table 5.6).

None of these values can be at all precise, since the reduction

Table 5.6

Type of pipe	Discharge for which to design, in terms of required discharge Q
Uncoated cast iron	1.55 Q
Asphalted cast iron	1.45 Q
Asphalted riveted wrought iron or steel	1.33 Q
Wood-stave	1.08 Q
Neat cement or concrete	1.06 Q

of carrying capacity must depend not only upon the material but also upon the nature and velocity of the water and upon the diameter of the pipe; an increased roughness due to tuberculation will be more troublesome, proportionately, in small- than in large-diameter pipes.[10]

5.3.8 Use of additives to reduce resistance

The literature dealing with this important subject has expanded greatly in the last 10 or 20 years and now includes a large number of papers giving information not only of laboratory studies but also of evidence adduced by full-scale applications.

Fortunately, the International Association for Hydraulic Research (IAHR) invited a panel or 'Task Force' consisting of R. H. T. Sellin, J. W. Hoyt, J. Pollert and O. Scrivener to compile a 'state-of-the-art review' and this has now been published in two parts in the Association's *Journal*.[11,12]

These reports deal largely with the addition of polymers and their observed effect in decreasing the resistance to the motion of the fluid. Attempts to explain the phenomenon are discussed and a wide range of applications are described: they include, *inter alia*, full-scale examples of sewers, oil pipelines, open channels, hydro-transport of solids, hydraulic machinery, ships and submerged bodies, in which significant reductions of resistance or increases of the carrying-capacity of pipes and channels have been recorded.

5.3.9 Losses of head in pipes due to causes other than friction

Sudden enlargement. With sufficient accuracy for practical purposes, loss due to sudden enlargement (Figure 5.20) $= (v_1 - v_2)^2/2g$

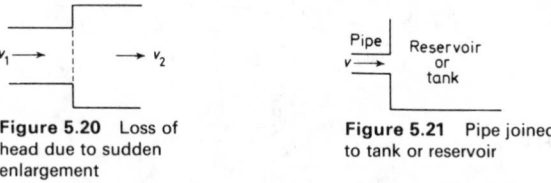

Figure 5.20 Loss of head due to sudden enlargement

Figure 5.21 Pipe joined to tank or reservoir

If the enlarged section is very large, as when a pipe is joined to a tank or reservoir (Figure 5.21):

$$\text{loss} = v^2/2g \qquad (5.31)$$

Gradual enlargement (Figure 5.22). Loss of head (including friction) may be taken as:

$$k(v_1 - v_2)^2/2g \qquad (5.32)$$

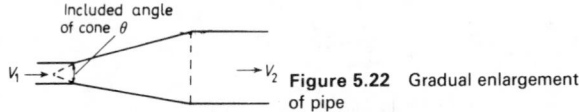

Included angle
of cone θ

Figure 5.22 Gradual enlargement of pipe

where k depends upon the angle of divergence θ in the following manner (Gibson[13]).

Table 5.7

θ (degrees)	2	5	10	15	20	40	60	90	120	180
k (circular pipe)		0.20	0.13	0.18	0.27	0.43	0.91	1.12	1.07	1.05	1.00

Sudden contraction. The loss due to sudden contraction is almost entirely due to the subsequent re-enlargement of the contracted stream (Figure 5.23). For practical purposes:

$$\text{loss} = (1/2)(v^2/2g) \tag{5.33}$$

and this may be taken as the immediate loss of head experienced as water flows from a reservoir into a pipe in which it attains a velocity v.

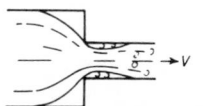

Figure 5.23 Loss of head due to sudden contraction

With a reasonably rounded entrance, the loss may be reduced to about:

$$(1/20)(v^2/2g)$$

If the pipe has a sharp entrance but projects into the reservoir and forms a re-entrant mouthpiece, the loss of head at the entrance is approximately $0.80\, v^2/2g$ (assuming that the pipe runs full).

5.3.10 Losses at pipe bends

Owing to the many variables involved (e.g. size of pipe, radius of bend, velocity of flow), it is impracticable at present to generalize with any degree of certainty, but the following data may be helpful in ordinary calculations (Figure 5.24).

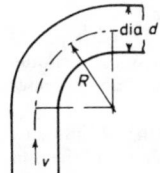

dia d

R

v

Figure 5.24 Pipe bend

Defining $Kv^2/2g$ as the loss in excess of that which would arise from friction in the same length of straight pipe, then approximate values of K are:

$R/d = 1$; $K = 0.50$ for either 90° or 180° bends;
$R/d = 2$ to 8; $K = 0.30$ for 90° and $K = 0.35$ for 180° bends.

For 90° elbows, $K = 0.75$ and for square, or sharp, elbows, $K = 1.25$.

The motion round the bend tends to take on the characteristics of a free vortex, having a greater velocity at the inside than at the outside. Correspondingly, the pressure at the inside is less than at the outside. Consider a section half-way round the bend and let $d = 2r$. If the discharge (volume per second) is Q, the velocities at the inside and outside of the bend are approximately:

$$v_i = \frac{Q}{2\pi(n-1)r^2[n-(n^2-1)^{1/2}]}$$

$$v_0 = \frac{Q}{2\pi(n+1)r^2[n-(n^2-1)^{1/2}]}$$

where $n = R/r$.

Correspondingly, the effect of the free vortex itself is to make the pressure head at the outside of the bend exceed that at the inside by an amount $(v_i^2 - v_0^2)/2g$.

5.3.11 Losses at valves

These depend, of course, upon the relative size and design, but the order of magnitude involved may be judged from Gibson's experiments[13] for:

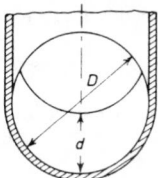

D

d

Figure 5.25 Circular sluice gate valve

(1) Circular sluice gate valves (Figure 5.25):

$$\text{Loss} = F(v^2/2g) \text{ where } v = \text{velocity in pipe} \tag{5.34}$$

Table 5.8

	F for d/D						
	0.2	0.3	0.4	0.5	0.6	0.8	1.0
$D = 50$ mm	30	11	4.2	2.1	0.9	0.22	0
$D = 600$ mm	36	11	3.0	1.6	1.0	—	—

(2) Butterfly valve (Figure 5.26):

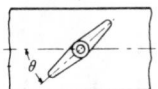

θ

Figure 5.26 Butterfly valve

Table 5.9

θ	5°	10°	20°	30°	40°	50°	60°	70°
F	0.24	0.52	1.54	3.9	10.8	32.6	118	751

5.3.12 Graphical representation of pipe-flow problems

This may be illustrated by the example of a pipeline joining two reservoirs (Figure 5.27), and for the sake of the example we will suppose that the pipe is enlarged somewhere along its length. Consider a particle of water to find its way, in effect, from the surface in the upper reservoir to that in the lower. The velocity of fall of the upper surface, or of rise of the lower one, may be treated as negligible in comparison with the velocity through the pipe.

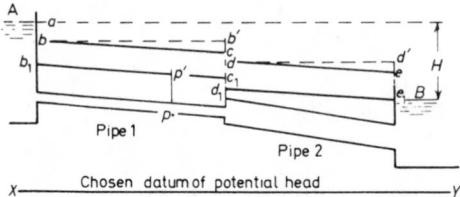

Figure 5.27 Graphical representation of pipe flow

Accordingly, the particle starting in surface A has originally no kinetic head. Taking atmospheric pressure as the datum of reference for pressure energy, it has also no pressure head. Its head is, in fact, entirely potential and may be represented diagrammatically by the height of A above any arbitrarily chosen datum of potential, XY. As the water enters the pipe there is a loss ab due to the sudden contraction: this is followed by a friction loss $b'c$ in pipe 1. Then comes the loss cd caused by the enlargement and this is followed by the friction loss $d'e$ in the second length of pipe.

The height of the line $abcde$ above XY therefore represents the total head.

Now drop bc by a distance bb_1, representing the kinetic head $v_1^2/2g$ in pipe 1, and de by a distance dd_1, representing the kinetic head $v_2^2/2g$ in pipe 2.

The height of $ab_1c_1d_1e_1$ above XY must then represent total head minus kinetic head, i.e. pressure plus potential head. $b_1c_1d_1e_1$ is then the line of virtual slope or hydraulic gradient of the pipe, and its height above the centreline of the pipe gives the pressure head in the pipe. thus, Pp' is the pressure head at P. If p' were below P, it would follow that the pressure in the pipe at P was below atmospheric and one of the advantages of this graphical method is to reveal clearly such points of suction.

Further, considering the bottom end of the pipeline, if the pipe 2 has a sharp, or flanged, connection to the reservoir, the whole of the kinetic head $v_2^2/2g$ will be lost in the creation of eddies at the enlargement. The point e_1 will then define the surface level in the lower reservoir, the total head under which flow is taking place being H. With a gradual enlargement, joining in a curved bell mouth to the lower reservoir, a proportion of $v_2^2/2g$, up to possibly $\frac{3}{4}(v_2^2/2g)$ might be regained and the level in B would be correspondingly higher than e_1 by this, say, $\frac{3}{4}(v_2^2/2g)$ amount of reconverted kinetic head.

5.3.13 Pipes in parallel

Consider two reservoirs connected by three pipes as shown in Figure 5.28. Pipe 1 is of diameter d_1, length l_1. Pipe 2 has diameter d_2, length l_2. Pipe 3 is of diameter d_3, length l_3. H is the loss of head in any one of the pipes. Neglecting end-effects:

$$H = \frac{4f_1 l_1 v_1^2}{2gd_1} = \frac{4f_2 l_2 v_2^2}{2gd_2} = \frac{4f_3 l_3 v_3^2}{2gd_3} \tag{5.35}$$

or, in the Manning form:

$$H = \frac{A_1 l_1 v_1^2}{d_1^{4/3}} = \frac{A_2 l_2 v_2^2}{d_2^{4/3}} = \frac{A_3 l_3 v_3^2}{d_3^{4/3}} \tag{5.36}$$

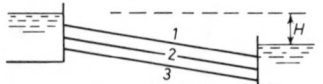

Figure 5.28 Pipes in parallel

The total rate of flow along the pipes:

$$= v_1(\pi d_1^2/4) + v_2(\pi d_2^2/4) + v_3(\pi d_3^2/4) + \ldots \tag{5.37}$$

5.3.14 Siphons

Consider two reservoirs connected by a siphon pipeline as shown in Figure 5.29a, Figure 5.29b being the diagrammatic equivalent.

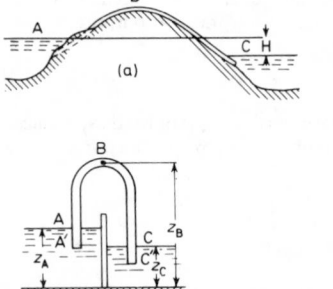

Diagrammatic equivalent of (a)
(b)　　　　　　　　**Figure 5.29** Siphon

If v is the velocity through the siphon, then $H=$ loss at entry + friction loss + loss at exit, i.e.:

$$H = 0.80(v^2/2g) + \text{friction loss} + (v^2/2g)$$

for submerged sharp entrance and exit ends of pipe. The end-effects are negligible in a reasonably long pipe.

Let F_1 be the friction loss of head in portion A'B of length l_1. Then pressure head at crown B is:

$$p_B/\rho g = z_A - z_B - \left(F_1 + 0.80\frac{v^2}{2g} + \frac{v^2}{2g}\right) \tag{5.38}$$

assuming a loss of $0.80(v^2/2g)$ at A'.

Hence p_B is negative, i.e. less than atmospheric.

In practice, to avoid undue difficulty arising from accumulation of air and vapour at B, the numerical value of $p_B/\rho g$ should not exceed 8.5 m of water, an absolute pressure of about 2 m of water being then left at the crown.

The analysis giving Equation (5.38) assumes that the pressure over the pipe-section at B is uniform. If the curvature of the pipe is pronounced, however (as it frequently is in the case of siphon-spillways), the free-vortex phenomenon mentioned earlier (section 5.3.10) becomes important. The difference of pressure between the inside and the outside of the bend has already been given for a *circular section*. In the case of a *rectangular section* having a breadth b, an inside radius r_i and an outer radius r_o, the pressure head at the outside of the bend exceeds that at the inside by an amount, due to the free vortex alone:

$$K^2/2g[(1/r_i^2) - (1/r_o^2)]$$

where $K = Q/b \ln r_o/r_i$ and Q is the volume flowing per second.

Difficulties may arise, therefore, in bends of severe curvature, through high suction at the inside of the bend.

5.3.15 Nozzle at the end of a pipeline (Figure 5.30)

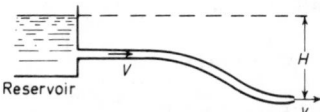

Figure 5.30 Nozzle at end of pipeline

Let H be the total head (in metres), V the velocity in the pipe (in metres per second), v the velocity in the jet (in metres per second), D the diameter of the pipe (in metres) and d the diameter of the nozzle end (in metres):

Discharge $= v(\pi d^2/4) \text{ m}^3/\text{s}.$ Then $v = V(D/d)^2$

Effective head behind nozzle $= H -$ loss at entrance to pipe $-$ friction in pipe. Hence $v^2/2g = H -$ friction head in pipe, neglecting entrance effect, or, more precisely:

$$v = C_N[2g(H - F)]^{1/2} \tag{5.39}$$

where F is the friction loss of head in pipe and C_N the coefficient of the nozzle.

Although C_N depends upon the Reynolds number vd/ν, for practical purposes it is usually sufficiently accurate to give it the value 0.98, either for elaborately streamlined nozzles or for a straight taper form ending in a cylindrical portion of length $d/2$ and diameter d.

Writing F in the form $F = 4flV^2/2gD$:

$$v^2 = \frac{2gC_N^2 H}{1 + [4flC_N^2/D(D/d)^4]} \tag{5.40}$$

The energy delivered in the jet $= \rho av^3/2 \text{ N m/s}$, where ρ is the density of water ($= 1000 \text{ kg/m}^3$) and $a = \pi d^2/4$.

This energy is a maximum, for a given H, if:

$$d^2 = \frac{D^2}{C_N}\sqrt{(D/8fl)} \tag{5.41}$$

The velocities are then such that the friction head F is very nearly equal to $H/3$.

5.3.16 Multiple pipes

Consider the example shown in Figure 5.31, in which it is desired to calculate the flow along the three pipes.

Height of A, B, C, J above some chosen datum of potential

$= z_A, z_B, z_C, z_J$ respectively.

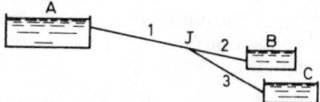

Figure 5.31 Multiple pipes

Neglecting losses other than those due to pipe friction F:

$$z_A = (p_J/\rho g) + (v_1^2/2g) + z_J + F_1 \tag{5.42}$$

$$z_B = (p_J/\rho g) + (v_2^2/2g) + z_J - F_2 \tag{5.43}$$

$$z_C = (p_J/\rho g) + (v_3^2/2g) + z_J - F_3 \tag{5.44}$$

Also:

$$v_1(\pi d_1^2/4) = v_2(\pi d_2^2/4) + v_3(\pi d_3^2/4) \tag{5.45}$$

Calling:

$$F_1 = (4f_1 l_1 v_1^2/2gd_1); \quad F_2 = (4f_2 l_2 v_2^2/2gd_2); \quad F_3 = (4f_3 l_3 v_3^2/2gd_3)$$

and assigning values to f_1, f_2, f_3, Equations (5.42) to (5.45) are sufficient for the determination of the four unknowns p_J, v_1, v_2, v_3, and if necessary the solutions may be modified by further calculation should the resulting v suggest values of f_1, f_2, f_3 different from those originally assumed.

The solution of (in this example) four simultaneous equations is, however, cumbersome and full of possibilities of arithmetical slips. A simpler and quicker method[14] is as follows.

If *assumed* total head difference between two ends of a pipe $= h$, then:

$$h = \frac{4flv^2}{2gd} = \frac{4fl}{2gd}\left(\frac{Q}{a}\right)^2 = KQ^2 \tag{5.46}$$

where Q is the rate of flow $= va$, a is area of section $= \pi d^2/4$, and $K = 2fl/gda^2$.

If the correct values of h and Q are $h + \delta h$ and $Q + \delta Q$, then $\delta h = 2h \cdot \delta Q/Q$ to a first approximation.

Now the error δQ for any one pipe is unknown, but the sum of the errors, $\Sigma(\delta Q)$, is known, being equal to the unbalanced flow, and:

$$\delta h = \Sigma(\delta Q)/\Sigma(Q/2h) \tag{5.47}$$

Example 5.7. $z_A = 30.50$; $z_J = 12.20$; $z_B = 6.10$; $z_C = 3.05$, all in metres.

(These might be levels with reference to ordnance datum, say.)

Pipe 1 3.0 km long, 600 mm diameter, $f = 0.007$
Pipe 2 1.5 km long, 300 mm diameter, $f = 0.007$
Pipe 3 1.5 km long, 300 mm diameter, $f = 0.007$

Let:

$$H = (p/\rho g) + (v^2/2g) + z$$

Steps in solution

(1) Assign some value to the total head at J. Evidently it must be less than that at A, which is 30.5 m. Hence we might first try $H_J = 20$ m, say.

(2) It then follows that:

$$F_1 = (4 \times 0.007 \times 3000v_1^2)/(2 \times 9.81 \times 0.600) = 30.5 - 20.0 = 10.5$$

Hence:

$$v_1 = \sqrt{(1.47)} = 1.21 \text{ m/s}$$
$$Q_1 = 0.342 \text{ m}^3/\text{s}$$

(3) Similarly, $F_2 = 20.0 - 6.1 = 13.9$
therefore:

$$\frac{4 \times 0.007 \times 1500v_2^2}{2 \times 9.81 \times 0.300} = 13.9$$

$v_2 = 1.40$ m/s

$Q_2 = 0.099\,0$ m³/s

$F_3 = 20.0 - 3.05 = 16.95$
therefore:

$$\frac{4 \times 0.007 \times 1500v_3^2}{2 \times 9.81 \times 0.300} = 16.95$$

$v_3 = 1.54$ m/s

$Q_3 = 0.109$ m³/s

Hence, our original assumption (that total head at $J = 20$ m) has led to an out-of-balance flow of $0.342 - (0.099\,0 + 0.109)$, or 0.134 m³/s. Hence:

$$\Sigma\delta Q = 0.134$$

(4) But $Q/2h$ for pipe $1 = 0.342/(2 \times 10.5) = 0.016\,3$

therefore $\Sigma(Q/2h) = 0.023\,1$

But $Q/2h$ for pipe $2 = 0.099\,0/27.8 = 0.003\,57$

But $Q/2h$ for pipe $3 = 0.109/33.9 = 0.003\,22$

so that $\Sigma\delta Q/[\Sigma(Q/2h)] = 0.134/0.023\,1 = 5.80$

(5) Now it is evident that we must aim at decreasing our original estimate of Q_1 while increasing those of Q_2 and Q_3.
We now try total head at J: $20.0 + 5.8 = 25.8$ m:

Our new estimate of
$$Q_1 = 0.342 \times \sqrt{\left(\frac{30.5 - 25.8}{30.5 - 20.0}\right)} = 0.229 \text{ m}^3/\text{s}$$

Our new estimate of $Q_2 = 0.099\,0 \times \sqrt{\left(\frac{19.7}{13.9}\right)} = 0.118$ m³/s

Our new estimate of $Q_3 = 0.109 \times \sqrt{\left(\frac{22.75}{16.95}\right)} = 0.126$ m³/s

and:

$$\Sigma\delta Q = 0.118 + 0.126 - 0.229 = 0.015$$

$Q/2h = 0.229/9.40 = 0.0244$ for pipe 1

$Q/2h = 0.118/39.4 = 0.0030$ for pipe 2

$Q/2h = 0.126/45.5 = 0.0028$ for pipe 3

Therefore:

$$\Sigma(Q/2h) = 0.030\,2$$

$$\frac{\Sigma\delta Q}{\Sigma(Q/2h)} = 0.015/0.030\,2 = 0.497$$

(6) Now make total head at J: $25.8 - 0.5 = 25.3$:

Q_1 then $= 0.342 \times \sqrt{(5.20/10.5)} = 0.241$

Q_2 then $= 0.099\,0 \times \sqrt{(19.2/13.9)} = 0.116$ ⎫
⎬ 0.241
Q_3 then $= 0.109 \times \sqrt{(22.25/16.95)} = 0.125$ ⎭

The flows are now in balance; the number of steps required in the process of successive approximation depends on the accuracy of the original guess at the total head of J.

Having obtained a sensibly accurate balance, we may if we choose carry out refined calculations based upon more acceptable values of f and including losses due to other causes such as the junction J, any bends, and so forth.

The example considered above is, however, comparatively simple. The more complicated cases frequently encountered in practice are nowadays analysed with the aid of analogue or digital computers.[15-18]

5.3.17 Flow measurement in pipes

5.3.17.1 The Venturi meter

The proportions shown in Figure 5.32 are not essential, but are fairly representative; the gently tapering divergent portion following the convergent tube (which is the real meter) is intended to minimize the overall loss of head due to eddies and friction between 1 and 3.

$$Q = \text{rate of flow} = va = C(\pi d^2/4)\left[\frac{2g(h_1 - h_2)}{(a_1/a_2)^2 - 1}\right]^{1/2} \tag{5.48}$$

where h_1 $(= p_1/\rho g)$ and h_2 $(= p_2/\rho g)$ are the pressure heads at sections 1 and 2 respectively.

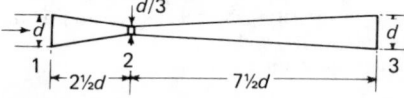

Figure 5.32 Venturi meter

If the throat diameter is too small, the pressure p_2 may be so low as to encourage release of air and vapour which will cause the flow to fluctuate and will introduce an uncertainty. To avoid this, it is advisable to use proportions which will not cause p_2 to be less than 21×10^3 N/m²; i.e. h_2 not less than 2.1 m of water (absolute).

Values of C. The value of C, the coefficient of the meter, is influenced by the Reynolds number and for the sort of designs of meter commonly used in practice, the following are approximately correct values[19] (Table 5.10).

5.3.17.2 Pipe orifice as meter

The pipe orifice as a measuring device is conveniently installed at a flange joint but has the disadvantage of creating an appreciable obstruction and consequent loss of head. In Figure 5.33, a and b are pressure tappings:

$$Q = CA\sqrt{2gh/[(D/d)^4 - 1]} \tag{5.49}$$

Table 5.10

Reynolds number vd/v as measured at throat	C
2 000	0.91
6 000	0.94
10 000	0.95
100 000	0.98
1 000 000	0.99

where $A = \pi D^2/4$, and h is the pressure head difference between points a and b. C is of the order of 0.61 for values of $(d/D)^2$ between 0.3 and 0.6, but it should be noted that the accuracy of the machining is of great importance, since the quantity $(D/d)^4$ occurs in the formula. Similarly, it is important to have a high degree of accuracy in the measurement of the diameter D of the pipe in which the plate is installed.

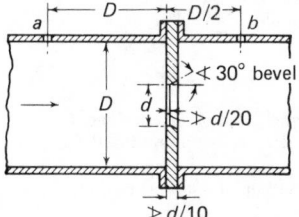

Figure 5.33   Pipe orifice as a meter

If a well-shaped convergent nozzle is used instead of a sharp-edged orifice plate, C has a value of 0.98 to 0.99 if $(d/D)^2$ does not exceed 0.2.

5.3.17.3 General notes on meters

The pressure holes used for meters or other purposes should be finished flush with the inside of the pipe. A reasonable length of straight pipe should precede the meter, and, though less important, should follow the meter.

For laboratory purposes it is always most satisfying to calibrate any meter *in situ* by comparison with the collection of a known weight of water in a measured time, but considerable accuracy may be expected from observance of the recommendations in the British Standard Code,[20] BS 1042:1943, which covers Venturi tubes, orifice plates and nozzles, and pitot tubes: it deals with gases as well as liquids. The US Standard[21] is *ASME Fluid Meters Report*.

5.3.18 Water hammer in pipes

If a valve is closed suddenly, successive masses of the water in the pipe are brought to rest; their kinetic energy is converted to strain energy and the effect is transmitted along the pipe with the velocity of sound waves in water. Some energy is, in fact, expended in stretching the pipe walls, thus reducing the water hammer pressure, but if this effect is neglected the rise of pressure p at the valve is given by:

$$p = v\sqrt{(K\rho)} \quad (\text{N/m}^2) \tag{5.50}$$

where v is the velocity of flow before the valve is closed (in metres per second), K the bulk modulus of compressibility of the water, equal to about 2×10^9 N/m^2 and ρ is the density of water, 1000 kg/m^3.

This formula leads to the result:

$$\text{Water hammer pressure} = 1.4 \times 10^6 \, v \quad (\text{N/m}^2) \tag{5.51}$$

where v is the original velocity of flow (in metres per second).

Pressures of this order of magnitude will result if a valve is closed in a time not exceeding $2l/V_p$ s, where $V_p = \sqrt{(K/\rho)}$ is the velocity of sound waves in the water and where l is the length of pipe (in metres). In round numbers, V_p may be taken as 1400 m/s.

If the time of closure exceeds $4l/V_p$ the stoppage becomes gradual. Supposing the valve to be then closed in such a manner as to cause a constant retardation α m/s^2 of the water column, the resulting rise of pressure will be $\rho l \alpha$ N/m^2, or $l\alpha/g$ m head of water.

5.3.19 Flow in open channels

5.3.19.1 Formulae for open channels

Consider the portion of an open channel shown in Figure 5.34. AB represents the surface of a stream: section A is distance l along the channel, section B a distance $l + \delta l$ along. h is the depth at A, $h + \delta h$ the depth at B. v is the mean velocity at A, r the depth of surface at A below some arbitrary datum and $(r + \delta r)$ the depth of surface at B below the same datum. m is the hydraulic mean depth, equal to the ratio of area to wetted perimeter, and f is the friction coefficient. Then:

$$dr/dl = (v/g)(dv/dl) + fv^2/2gm \tag{5.52}$$

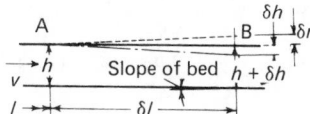

Figure 5.34   Portion of open channel

If the flow is uniform and the mean velocity constant from section to section along a channel of constant cross-section, this assumes the familiar form:

i = slope of bed = slope of water surface

= fall per unit length

$$= fv^2/2gm \tag{5.53}$$

alternatively, as in the Chézy equation, Equation (5.53) can be written:

$$v = c\sqrt{(mi)} \tag{5.54}$$

where c is known as the Chézy coefficient and is related to the friction coefficient f in the formula:

$$\text{friction head} = flv^2/2gm \quad \text{by} \quad c^2 = 2g/f \tag{5.55}$$

The numerical value of c depends on the units adopted; it has the units of $[L^{\frac{1}{2}}]/[T]$. Consequently, in the metre second system it is measured in metres$^{\frac{1}{2}}$ per second. On the other hand, f is dimensionless; it has the same numerical value in either system.

Chézy's c depends upon the nature of the channel and also upon the hydraulic mean depth of a channel of given material.

To some extent it also depends upon the mean velocity of the stream, although in most practical examples of open channel flow with which the engineer is concerned, this effect is of minor importance.

Although old fashioned in the sense that it dates back to the last century, a formula due to Bazin[22] is very reliable:

$$c = 158/(1.81 + N/\sqrt{m}) \quad (m^{\frac{1}{2}}/s) \tag{5.56}$$

the hydraulic mean depth m being measured in metres and N having the values given in Table 5.11.

Table 5.11

Class	N	Application
I	0.109	Smoothed cement or planed wood
II	0.290	Planks, bricks or cut stone
III	0.833	Rubble masonry
IV	1.54	Earth channels of very regular surface, or revetted with stone
V	2.35	Ordinary earth channels
VI	3.17	Exceptionally rough earth channels (bed covered with boulders) or weed-grown sides

As one of the many proposed alternatives to the Chézy–Bazin treatment ($v = c\sqrt{(mi)}$ where $c = [158/(1.81 + N/\sqrt{m})]$, the formula due to Manning[7] is much favoured and regarded by many as more convenient, though giving much the same result.

For the classes of channel already described in Table 5.11 in connection with Bazin's N, Manning's n may be taken as given in Table 5.12 and in Equation (5.23): $v = (m^{2/3}i^{1/2})/n$ (m/s) where m is in metres.

With rather more precision, the values of n given by Parker[23] are quoted in Table 5.13.

Table 5.12

Class	n
I	0.009 3
II	0.012 9
III	0.018 2
IV	0.022 5
V	0.025 8
VI	0.028 4

Table 5.13

Nature of channel	n
Timber, well planed and perfectly continuous	0.009
Planed timber, not perfectly true	0.010
Pure cement plaster	0.010
Timber, unplaned and continuous; new brickwork	0.012
Rubble masonry in cement, in good order	0.017
Earthen channels in faultless condition	0.017
Earthen channels in very good order or heavily silted in the past	0.018
Large earthen channels maintained with care	0.0225
Small earthen channels maintained with care	0.025
Channels in order, below the average	0.0275
Channels in bad order	0.030

Note that by comparing $v = c\sqrt{(mi)}$ with $v = (m^{2/3}i^{1/2})/n$ we may obtain the result:

$$c = m^{1/6}/n$$

or:

$$f = 2g/c^2 = 19.6n^2/m^{1/3} \tag{5.57}$$

Incidentally the formula $c = 20.7 + 17.7 \lg m/k$ m$^{\frac{1}{2}}$/s covers a remarkably wide range of both rough pipes and rough open channels.[24] Here k represents the size of roughening excrescences.

5.3.19.2 Form of channel for maximum v and Q

Q = rate of discharge (in cubic metres per second) = vA, where A is now the area of section (in square metres) and v is the mean velocity (in metres per second).

Adopting the Manning formula:

$$Q = vA = (Am^{2/3}i^{1/2})/n \tag{5.58}$$

for a given material of channel, and a given slope i, v is a maximum when m, or when A/P, is a maximum, P being the wetted perimeter.

For Q to be a maximum, however, $Am^{2/3}$ must be a maximum. Hence:

$$\left. \begin{array}{l} \text{for max } v, \; P\,dA - A\,dP = 0 \\ \text{for max } Q, \; 5P\,dA - 2A\,dP = 0 \end{array} \right\} \tag{5.59}$$

Examples 5.8 to 5.10.
Example 5.8: *Rectangular channel* (Figure 5.35).
$A = bd$; $P = b + 2d$. If A, n and i are fixed, maximum v and maximum Q will occur when $b = 2d$.

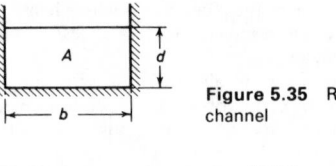

Figure 5.35 Rectangular channel

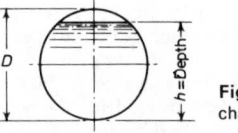

Figure 5.36 Trapezoidal channel

Figure 5.37 Circular channel

Example 5.9: *Trapezoidal channel* (Figure 5.36):
For maximum v and maximum Q with given A:

$$\sqrt{(1 + s^2)} = (b + 2sh)/2h$$

where $\tan \theta = 1/s$

This is satisfied if a semicircle can be drawn, centred in the water surface and touching both sides and bottom.

Example 5.10: Circular channel (Figure 5.37).
For maximum v, $h = 0.813D$; for maximum Q, $h = 0.938D$.

5.3.19.3 Resistance of natural river channels

This resistance is complicated by the losses of energy at bends and at relatively sudden changes of cross-sectional area. As these depend on the precise dimensions and shapes, it is quite impossible to generalize, but they are nevertheless important. For example, it has been shown that in a 13-km tortuous stretch of the River Mersey[25] the textural roughness of the bed and sides accounts for only 25 to 50% of the total loss of head depending on the rate of flow, the rest of the resistance being due principally to the bends.

Somewhat similar conclusions have been reached in a study of the River Irwell.[26]

5.3.19.4 Velocity distribution in open channels

Side and bottom friction cause the stream to be retarded. The highest velocity in any vertical at a particular section is usually found some distance below the surface; the mean velocity in a vertical line occurs at about 60% of the depth, whether the wind is blowing up- or downstream. This is the basis of one method of stream-gauging, in which the section is considered divided into strips of equal width and the velocity in these strips, or panels, is measured by current meter at 60% of the depth of each individual panel (Figure 5.38). The area of a strip, as found by sounding the bed, multiplied by the velocity so measured is assumed to give the flow through the strip and the addition for the total number of strips gives the flow through the whole section.

Figure 5.38 Measurement of velocity in open channel

5.3.19.5 Energy of a stream in an open channel, or 'specific energy'

If D is the depth and v the mean velocity, the energy head H_e, taking atmospheric pressure as the datum of pressure and the bottom of the channel as the datum of potential, is $D + v^2/2g$, or $D + Q^2/2gA^2$.

5.3.19.6 Critical depth

Critical depth is the depth at which maximum discharge occurs for a certain energy head, or, alternatively, the depth at which a given discharge takes place with the minimum energy head. It represents an unstable condition, often accompanied by water surface undulations. Under these conditions $D = v^2/g$ in a rectangular channel and Q then equals $(0.544b\sqrt{g})H_e^{3/2}$.

5.3.19.7 Nonuniform flow in open channels

If h is the depth at a distance l along the channel with slope i

$$\mathrm{d}h/\mathrm{d}l = \frac{i - (fv^2/2gm)}{1 - (v^2/gh)} \tag{5.60}$$

This condition of varying depth, even in a stream of constant width and constant rate of discharge, as here assumed, may be brought about by obstructions or irregularities.

h is now the depth actually found at a particular section, whereas H may be called the depth which would apply under conditions of uniform flow corresponding to the simple Equation (5.54). In other words, h becomes equal to H if $\mathrm{d}h/\mathrm{d}l = 0$.

Suppose now, in order to examine general trends, that the width of the stream is large compared with its depth. In such a case the hydraulic mean depth m is approximately the same as h, at any rate in channels having approximately uniform depth across their width. Then:

$$\mathrm{d}h/\mathrm{d}l = \frac{i - (fv^2/2gh)}{1 - (v^2/gh)}$$

and $Q = vbh$. Q is also equal to VbH, where V and H are the velocity and depth which would be obtained with uniform flow. It then follows that:

$$\mathrm{d}h/\mathrm{d}l = \frac{i[1 - (H/h)^3]}{1 - (2i/f)(H/h)^3} \tag{5.61}$$

5.3.19.8 Special cases of nonuniform flow

(1) *Sluice gate in channel with small slope and/or rough bed* $2i/f < 1$, $h^3 < (2i/f)H^3$ (Figure 5.39). $\mathrm{d}h/\mathrm{d}l$ becomes infinite when

$$h = (2i/f)^{1/3}H \tag{5.62}$$

A 'hydraulic jump' then results.

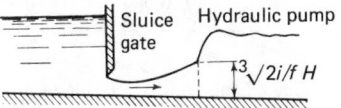

Figure 5.39 Sluice gate in channel, small slope and/or rough bed

(2) *Sluice gate in channel with steep slope and/or smooth bed* $2i/f > 1$, $h < H$. $\mathrm{d}h/\mathrm{d}l$ is again positive, but tends to zero, i.e. the depth increases gradually until it reaches that appropriate to uniform flow (Figure 5.40).

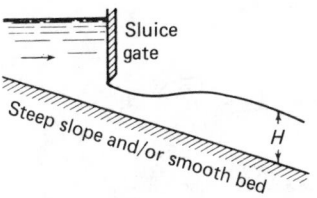

Figure 5.40 Sluice gate in channel, steep slope and/or smooth bed

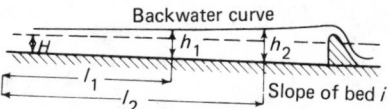

Figure 5.41 Weir or dam in channel, small slope and/or rough bed

(3) *Weir or dam in channel with small slope and/or rough bed* $2i/f < 1$, $h > H$ (Figure 5.41):

$$h_1 - h_2 = i(l_1 - l_2) + H[1 - (2i/f)][\Phi(y_1) - \Phi(y_2)] \tag{5.63}$$

where $\Phi(y) = $ backwater function $= -\int \mathrm{d}y/(y^3 - 1)$, and $y = h/H$.

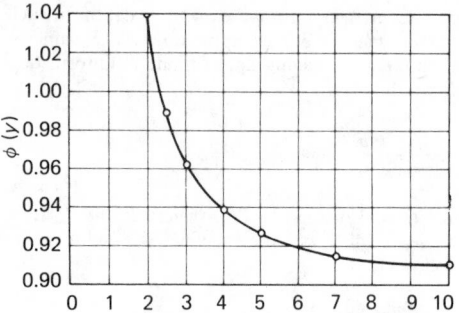

Figure 5.42 Backwater function

For values of the backwater function, see Figure 5.42.

Specific values are tabulated below:

y	1.000	1.005	1.010	1.015	1.020	1.050	1.100
$\Phi(y)$	∞	2.555	2.326	2.192	2.098	1.803	1.587

y	1.200	1.500	1.800	2.000	2.500	5.000	10.000
$\Phi(y)$	1.387	1.162	1.073	1.039	0.989	0.927	0.911

Example 5.11. The following example illustrates the application of the backwater function. A dam is built across a stream which was previously flowing with a depth of 1 m. The effect of the dam is to raise the level just behind it by 4 m. The slope of the bed is 1:2000 and $f=0.01$. What is the effect of the dam on the levels upstream? (Figure 5.43.)

$$i=1/2000; \ 2i/f=1/10; \ H=1 \text{ m}; \ H(1-2i/f)=0.9$$
$$h_2=\text{depth behind dam}=5 \text{ m}$$
$$y_2=h_2/H=5; \ \Phi(y_2)=0.927$$

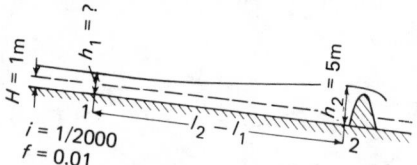

Figure 5.43 Application of backwater function

Therefore from Equation (5.63):

$$l=2000[4.166-h_1+0.9\,\Phi(y_1)] \tag{5.64}$$

where $l=l_2-l_1$; $h_1=$ depth at distance l from dam; $y_1=h_1/H=h_1/1$.

Now assign values to h_1 and calculate l from Equation (5.64). This process gives:

h_1 (m)	5	4	3	2	1.5	1.2
l (m)	0	2020	4070	6200	7420	8430

Inspection of the data given for this example will show that the water surface is virtually horizontal over the length extending 2000 m above the dam.

Although this theory of backwater is based upon the assumption of a rectangular channel of great width compared with its depth, and of f independent of depth, it nevertheless gives reasonably accurate results in practical cases provided that the value of H for the actual channel is known. Thus Gibson[27] has applied it to a circular conduit, in which the central depth was increased by a weir from its normal value H of 1.04 m to 1.67 m in the vicinity of the weir.

5.3.19.9 Flow over a horizontal bed

In this case, the slope i of the bed is zero and the simple Chézy Equation (5.54) ceases to be applicable. Instead of being uniform, the depth h decreases in the direction of flow and the fall of the water surface provides for the head required to overcome friction together with that needed to increase the kinetic head.

Considering such a rectangular channel of constant breadth b and writing $i=0$ in Equation (5.60):

$$dh/dl=-\left[\frac{(fv^2/2gm)}{1-(v^2/gh)}\right]$$

Let $Q=$ (constant) discharge (volume/s) supplied to the channel.

Then $v=Q/bh$

and:

$$dh/dl=f/[2gm(b^2h^2/Q^2-1/gh)]$$

If $Q_1=$ discharge per unit width:

$$dh/dl=-f/[2gm(h^2/Q^2-1/gh)]$$

Substituting $m=bh/(b+2h)$ and integrating between distances l_1 and l_2 along the channel in the direction of flow:

$$4Q_1^2fl/bg=4Q_1^2/g+b^3/2)\log_e\alpha-4/3(h_2^3-h_1^3)$$
$$+b(h_2^2-h_1^2)-b^2(h_2-h_1) \tag{5.65}$$

where $\alpha=(2h_2+b)/(2h_1+b)$ and $l=l_2-l_1$.

Even this complicated result assumes that f is constant all along the channel, whereas strictly f is influenced by the depth (which changes). More refined calculations may be made by considering small lengths with an adjustment for f from one to the next.

In the case of a comparatively broad channel, $m \to h$,

and:

$$dh/dl=-f/2g(h^3/Q_1^2-1/g)$$

from which:

$$l_2-l_1=\frac{1}{f}\left[\frac{g(h_1^4-h_2^4)}{2Q_1^2}-2(h_1-h_2)\right] \tag{5.66}$$

In practical cases, h_2 is often known from the conditions imposed at the end of the channel by, say, a weir. Another possibility is that the bed of the channel drops sharply at its end and the stream then issues as a waterfall. In such an example, the depth h_2 at or near the downstream end is $(Q_1/\sqrt{g})^{2/3}$, and by assuming a value for f, the depth h_1 at a distance l ($=l_2-l_1$) upstream can be estimated. Alternatively, the distances l

upstream of h_2 may be found for a succession of chosen values of h_1.

5.3.19.10 The hydraulic jump (sometimes called 'standing wave')

An hydraulic jump is illustrated in Figure 5.44.

$$h_1 - h_2[(h_1 + h_2)/2 - (h_1 v_1^2/h_2 g)] + (h_2 - d/2)d = 0$$

or, in a practically horizontal channel $(d=0)$:

$$h_2 = -h_1/2 + [(h_1^2/4) + (2h_1 v_1^2/g)^{1/2}]$$

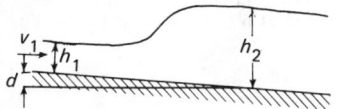

Figure 5.44 Hydraulic jump

and:

$$(h_2 - h_1) = [(h_1^2/4) + (2h_1 v_1^2/g)]^{\frac{1}{2}} - 3h_1/2 \qquad (5.67)$$

For information concerning the length of channel covered in forming the jump, see Allen and Hamid.[28]

5.3.19.11 The Venturi flume

The flume is a device for measuring rates of flow in an open channel and is not so liable to damage as a weir and does not offer the same obstruction to the flow. In order to preserve its surface and its hydraulic characteristics, it is sometimes lined with stainless steel.

It may be formed by inserting 'streamlined' humps on the sides (Figure 5.45a) and/or the bed (Figure 5.45b) of the channel.

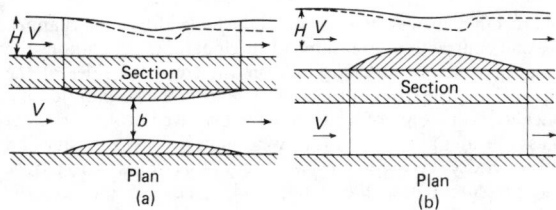

Figure 5.45 Venturi flume

If the discharge is 'free' as represented by the broken lines in Figure 5.45 and accompanied by the formation of a standing wave, the rate of discharge depends only upon the depth upstream of the constriction. With a throat of rectangular section and width b, Q is approximately $0.54g^{\frac{1}{2}}(H + V^2/2g)^{3/2}$, V itself of course depending upon H.

The general equation is:

$$Q = C_D\{b_2 d_2/[1 - (b_2 d_2/b_1 d_1)^2]^{\frac{1}{2}}\}[2g(d_1 - d_2)]^{1/2} \qquad (5.68)$$

where b_2 is the breadth at throat, d_2 the depth at throat, d_1 the depth upstream of the constriction, b_1 the breadth upstream of the constriction, and C_D is the coefficient of discharge.

For particular designs, C_D is best found by scale-model experiments.

Details as to proportions, shapes and types of flow may be found in papers by Engel.[29] See also Elsden.[30]

5.3.20 Orifices

5.3.20.1 Sharp-edged orifice (Figure 5.46)

Velocity at *vena contracta* $= C_v\sqrt{(2gh)}$, where C_v is the coefficient of velocity, about 0.985.

Neglecting air resistance:

$$x^2 = 4yhC_v^2 = 3.88yh \qquad (5.69)$$

Discharge $Q = C_v a_c\sqrt{(2gh)}$

$$= C_v C_c a\sqrt{(2gh)} \qquad (5.70)$$

where C_c is the coefficient of contraction, or:

$$Q = C'a\sqrt{(2gh)} \qquad (5.71)$$

where a is the area of the orifice and C' is the coefficient of discharge.

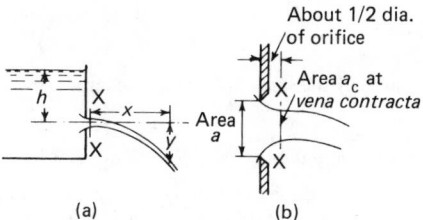

Figure 5.46 Sharp-edged orifice

Consideration of various published data[31] indicates that for orifices of 6.35 cm diameter or over, under heads of at least 0.43 m, $C' = 0.60$, provided $h \nless 3d$, where $d =$ diameter of orifice. Other typical results are given in Table 5.14.

It is doubtful whether a third significant figure is of any value, as a 1% error in measuring the mean diameter of the orifice at once makes 2% difference to the computed discharge. The absolute sharpness of the edge must also have some bearing upon the results.

Table 5.14 Values of C' for sharp-edged orifice

Head, h	d *of circular orifice, or side of square orifice*					
	0.64	1.27	2.54	6.35	15.2	30.5
(m)	(cm)	(cm)	(cm)	(cm)	(cm)	(cm)
0.12	—	0.64	0.63	0.61	—	—
0.18	0.66	0.64	0.62	0.61	0.60	—
0.31	0.65	0.63	0.62	0.61	0.60	0.60
0.61	0.63	0.62	0.62	0.61	0.60	0.60
1.22	0.62	0.61	0.61	0.61	0.60	0.60
3.05	0.61	0.61	0.61	0.60	0.60	0.60
15.24	0.60	0.60	0.60	0.60	0.60	0.60
30.48	0.60	0.60	0.60	0.60	0.60	0.60

The value of C' for a rectangular orifice appears to be somewhat higher than for a circular or square one of the same area. The difference amounts to about 2% for rectangles having a 4:1 ratio of sides and 4% for a ratio of 10:1.

5.3.20.2 Rounded or bell-mouthed orifice

For a design such as that shown in Figure 5.47, $C_c = 1$ and $C' = C_v$:

$$Q = 0.95(\pi d^2/4)\sqrt{(2gh)} \quad \text{to} \quad 0.99(\pi d^2/4)\sqrt{(2gh)} \tag{5.72}$$

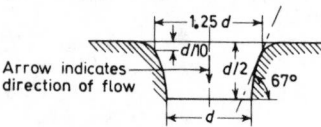

Figure 5.47 Rounded orifice

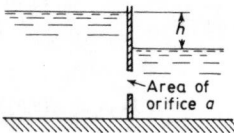

Figure 5.48 Submerged orifice

5.3.20.3 Submerged orifices

For a submerged orifice as shown in Figure 5.48:

$$Q = C'a\sqrt{(2gh)} \tag{5.73}$$

where C' is substantially the same as for free discharge into the atmosphere.

5.3.20.4 Time of discharge through an orifice

The equation for time of discharge through an orifice from a tank (Figure 5.49), without any simultaneous inflow is:

$$\mathrm{d}h/\mathrm{d}t = C'(a/A)\sqrt{(2gh)}$$

or

$$t_2 - t_1 = -\frac{1}{C'a\sqrt{(2g)}}\int_{H_1}^{H_2} Ah^{-1/2}\,\mathrm{d}h \tag{5.74}$$

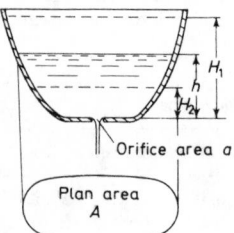

Figure 5.49 Discharge through an orifice

This can be solved if A can be expressed in terms of h, the instantaneous head at time t.

If A is constant, or independent of h:

$$[C'a\sqrt{(2g)}/A](t_2 - t_1) = 2(\sqrt{H_1} - \sqrt{H_2}) \tag{5.75}$$

treating C' as independent of h.

The time of discharge through a submerged orifice (Figure 5.50) is given by the equation:

$$(t_2 - t_1) = \frac{2}{C'(1/A_1 + 1/A_2)a\sqrt{(2g)}}(\sqrt{H_1} - \sqrt{H_2}) \tag{5.76}$$

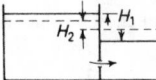

Figure 5.50 Discharge through submerged orifice

5.3.21 Weirs and notches

The term 'notch' is used for the smaller weirs common in laboratories, as distinct from outdoors.

5.3.21.1 Rectangular sharp-crested weir

For a rectangular weir as shown in Figure 5.51 in which $a \ll 4H$ and $c \ll 3H$, and H is measured at a distance 6 to 10 times H behind the weir, then:

$$Q = [0.410(2g)^{\frac{1}{2}}](b - H/10)H^{3/2} \tag{5.77}$$

This formula is probably accurate within 2% for all values of H from 0.08 to 0.61 m provided $b/H > 2$ and provided b is $\ll 0.305$ m.

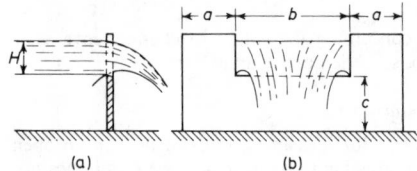

Figure 5.51 Rectangular sharp-crested weir

The effect of the velocity of the approaching stream is automatically allowed for in this formula, as in all others to be presented.

5.3.21.2 Suppressed rectangular weir

If a rectangular weir crest occupies the full width of the channel, the end contractions are suppressed. Under these conditions the 'nappe' or stream discharging over the crest, the sides of the channel and the front of the weir plate form the boundaries of a pocket of air, some of which may be dissolved in the turbulent mass of water at the downstream side of the weir on its downstream side and carried away (Figure 5.52). The effect of this would be to reduce the pressure below the nappe and, hence, to increase the

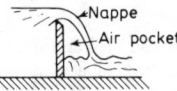

Figure 5.52 Suppressed rectangular weir

discharge for a given head. In itself this is no detriment but it introduces an element of uncertainty and of variation. To overcome this it is generally supposed that air vents should be provided through the sides of the channel in communication with the air-pocket with the object of maintaining atmospheric pressure and preserving a standardized condition.

Under such conditions (ventilated suppressed rectangular weir of height P m above the bed of the channel), the Rehbock formula[32] is perhaps accurate to within 1 or 2%. It reads:

$$Q = \tfrac{2}{3}\sqrt{(2g)}b\left(0.605 + \frac{1}{1050H - 3} + \frac{0.08H}{P}\right)H^{3/2} \quad (\text{m}^3/\text{s}) \tag{5.78}$$

Writing this as:

$$Q = Cb\sqrt{(2g)}H^{3/2} \tag{5.79}$$

values of C are as given in Table 5.15.*

Table 5.15 Values of C

P	H							
	0.06	0.15	0.31	0.46	0.61	0.91	1.22	1.52
(m)	(m)	(m)	(m)	(m)	(m)	(m)	(m)	(m)
0.15	0.436	0.461	0.512	0.565	0.617	0.724	0.831	0.936
0.31	0.425	0.434	0.459	0.486	0.511	0.564	0.617	0.672
0.61	0.421	0.422	0.433	0.446	0.458	0.484	0.510	0.537
0.91	0.418	0.417	0.423	0.431	0.440	0.458	0.475	0.492
1.52	0.416	0.413	0.416	0.421	0.426	0.436	0.446	0.457
3.05	0.415	0.411	0.411	0.412	0.416	0.421	0.425	0.431

5.3.21.3 The 90-degree vee-notch (sharp-edged)

Measuring the head with reference to the point v (Figure 5.53):

$$Q = 1.34H^{2.48} \quad (\text{m}^3/\text{s}) \tag{5.80}$$

over a wide range, H being in metres.

This notch is more convenient than the rectangular form for the measurement of small quantities but it should be remembered that an error of 1% in the measurement of H means 2.5% in the resulting estimate of Q, whereas with a rectangular notch the corresponding error is 1.5%.

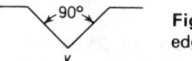

Figure 5.53 Sharp-edged 90° vee-notch

5.3.21.4 The Cippoletti weir (sharp-crested)

The discharge over this type of weir (Figure 5.54) is:

$$Q = 0.420\sqrt{(2g)}b[(H+h)^{\frac{3}{2}} - h^{3/2}] \quad \text{if } \tan\theta = \tfrac{1}{4} \tag{5.81}$$

where $h = v^2/2g$, v being the mean velocity in the approach channel.

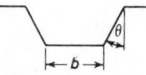

Figure 5.54 Sharp-crested Cippoletti weir

v cannot be allowed for until Q is known. Hence, as a first approximation, find Q from $Q = 0.420(2g)^{\frac{1}{2}}bH^{3/2}$. Then calculate $v = Q$/area of section of approach channel, and re-evaluate Q from Equation (5.81).

5.3.21.5 Weirs without sharp crests

Some typical examples are shown in Figure 5.55(a), (b), (c) and (d). The discharge $Q = C(2g)^{\frac{1}{2}}bH^{3/2}$.

5.3.21.6 Submerged weirs

If the downstream level rises above the crest of the weir, a 'drowned weir' results. The effect of this upon the discharge for a given upstream head is surprisingly small: in general, the reduction in discharge will not amount to more than 2 or 3% for 'downstream heads' or submergences up to 20% of the upstream head.

* $Q = 4.43CbH^{1/2}$ m³/s if b and H are in metres.

5.3.21.7 Time of discharge over a weir (Figure 5.56)

The time of discharge over a weir or spillway of length b may be calculated as follows.

(1) *With no inflow*:

$$A\,\delta H = -C(2g)^{\frac{1}{2}}bH^{\frac{3}{2}}\,\delta t$$

therefore:

$$\int_{t_1}^{t_2} dt = -\int_{H_1}^{H_2} \frac{A}{C(2g)^{\frac{1}{2}}b}H^{-\frac{3}{2}}\,dH$$

Or, time for head to fall from H_1 to H_2 is:

$$(t_2 - t_1) = t = -\frac{1}{b}\int_{H_1}^{H_2}\frac{A}{C(2g)^{\frac{1}{2}}}H^{-\frac{3}{2}}\,dH$$

This may be solved by splitting the change between H_1 and H_2 into stages over which mean values of A and C are applied.

If A and C are treated as constant:

$$t = \frac{2A}{bC(2g)^{\frac{1}{2}}}\left(\frac{1}{\sqrt{H_2}} - \frac{1}{\sqrt{H_1}}\right) \tag{5.82}$$

(2) *Reservoir with inflow as well as outflow* (Figure 5.56)

Let A be the surface area of reservoir, Q the inflow and h the instantaneous head over spillway of length b.

Then excess of inflow over outflow $= Q - C(2g)^{\frac{1}{2}}bh^{3/2}$. Hence:

$$dh/dt = [Q - C(2g)^{\frac{1}{2}}bh^{3/2}]/A$$

Let H be the head over spillway which would make the rate of outflow equal to the rate of inflow. Then:

$$Q = C(2g)^{\frac{1}{2}}bH^{3/2}$$

Let:

$$r = h/H \quad \text{and} \quad K_1 = C(2g)^{\frac{1}{2}}b.$$

The time taken for the head to change from h_1 to h_2 is given by:

$$t_2 - t_1 = (A/K_1\sqrt{H})[\Phi(r_2) - \Phi(r_1)] \tag{5.83}$$

In this equation, $r_1 = h_1/H$, $r_2 = h_2/H$ and Φ represents Gould's function[33] of h/H, i.e. of r.

Detailed values of $\Phi(r)$ for use when the time-interval is expressed in the form given by Equation (5.83) (where b as well as $C\sqrt{(2g)}$ is absorbed in K_1) have been calculated by Mathieson.[34]

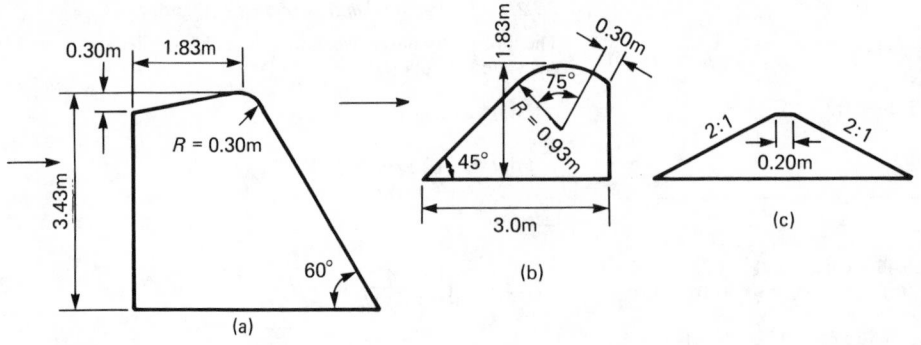

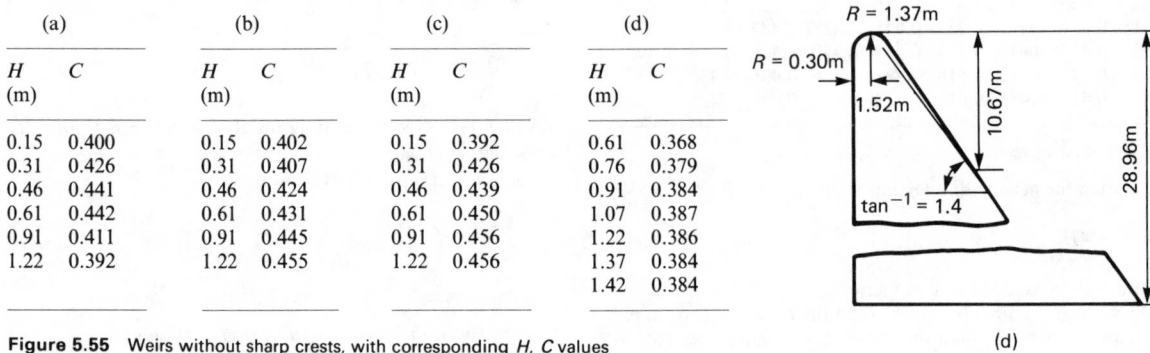

(a)		(b)		(c)		(d)	
H (m)	C	H (m)	C	H (m)	C	H (m)	C
0.15	0.400	0.15	0.402	0.15	0.392	0.61	0.368
0.31	0.426	0.31	0.407	0.31	0.426	0.76	0.379
0.46	0.441	0.46	0.424	0.46	0.439	0.91	0.384
0.61	0.442	0.61	0.431	0.61	0.450	1.07	0.387
0.91	0.411	0.91	0.445	0.91	0.456	1.22	0.386
1.22	0.392	1.22	0.455	1.22	0.456	1.37	0.384
						1.42	0.384

Figure 5.55 Weirs without sharp crests, with corresponding H, C values
See also *Hydraulics Research*, HMSO Annual Reports, e.g. 1964 p. 15 and 1965 p. 7 (the 'Crump' weir)

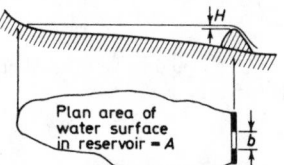

Figure 5.56 Discharge over a weir

When $r < 1$:

$$\Phi(r) = \frac{2}{3} \left\{ \ln\left[\frac{(r+\sqrt{r}+1)^{\frac{1}{2}}}{1-\sqrt{r}}\right] - \sqrt{3}\left[\tan^{-1}\left(\frac{2\sqrt{r}+1}{\sqrt{3}}\right) - \frac{\pi}{6}\right]\right\}$$

When $r > 1$:

$$\Phi(r) = \frac{2}{3} \left\{ \ln\left[\frac{(r+\sqrt{r}+1)^{\frac{1}{2}}}{\sqrt{r}-1}\right] - \sqrt{3}\left[\tan^{-1}\left(\frac{2\sqrt{r}+1}{\sqrt{3}}\right) - \frac{\pi}{2}\right]\right\}$$

Some of the values given by Mathieson are quoted below.

r	0	0.1000	0.2000	0.3000	0.4000	0.5000	0.6000
$\Phi(r)$	0	0.1013	0.2076	0.3220	0.4482	0.5920	0.7615

r	0.7000	0.8000	0.9000	0.9900	1.0100	1.0200	1.0400
$\Phi(r)$	0.9729	1.2619	1.7414	3.2925	4.4948	4.0426	3.5795

r	1.0600	1.1000	1.5000	2.0000	5.0000	10.0000	∞
$\Phi(r)$	3.3202	2.9838	1.9708	1.5730	0.9155	0.6376	0

Example 5.12. A reservoir has a surface area of 2.5 km². It is provided with an overflow weir of length 25 m, $C=0.400$. Initially there is a steady head of 0.25 m over the weir, but superimposed upon the discharge corresponding with this condition, additional flood water enters the reservoir as detailed below.

Time (h)	0	1	2	3	4	5	6	7	8
Additional inflow (m³/s)	0	10	35	50	40	20	10	0	−2.75

Investigate the variation of water level.

$$K_1 = Cb\sqrt{(2g)} = 0.4 \times 25.0 \times 4.43 = 44.3$$

$$A/K_1 = 2.5 \times 10^6/44.3 = 5.64 \times 10^4; \quad 3600K_1/A = 0.0637$$

Initial inflow = initial outflow = $44.3(0.25)^{3/2} = 5.53$ m³/s.

First hour

$$\text{Mean } Q = 5.53 + 5 = 10.53 \text{ m}^3/\text{s}$$
$$\text{therefore } H^{3/2} = 10.53/44.3 = 0.238$$
$$\text{and } H = 0.384 \text{ m}; \quad \sqrt{H} = 0.62$$
$$r_1 = h_1/H = 0.25/0.384 = 0.651; \quad \Phi(r_1) = 0.863$$
$$3600(s) = (5.64 \times 10^4/0.62)[\Phi(r_2) - 0.863]$$
$$\Phi(r_2) = 0.903; \quad r_2 = 0.67$$
$$h_2 = 0.67 \times 0.384 = 0.258 \text{ m}$$
$$= \text{head over spillway at end of 1 h}$$

Second hour

$$\text{Mean } Q = 5.53 + 22.5 = 28.03 \text{ m}^3/\text{s}$$
$$H^{3/2} = 28.03/44.3 = 0.634$$
$$H = 0.738 \text{ m}; \ \sqrt{H} = 0.859$$
$$r_1 = 0.258/0.738 = 0.35; \ \Phi(r_1) = 0.384$$
$$\text{therefore } 0.0637 = (1/0.859)[\Phi(r_2) - 0.384]$$
$$\Phi(r_2) = 0.438; \ r_2 = 0.392$$
$$\text{and } h_2 = 0.392 \times 0.738 = 0.289 \text{ m}$$
$$= \text{head at end of 2 h.}$$

Proceeding in this way, we obtain:

Hour	0	1	2	3	4	5	6	7	8
Head (m)	0.250	0.258	0.289	0.348	0.407	0.442	0.453	0.449	0.438

So the maximum head is 0.453 m at 6.10 h. This head would give a maximum *out*flow of $44.3(0.453)^{3/2}$, or $13.5 \text{ m}^3/\text{s}$, as compared with the maximum *in*flow of $55.5 \text{ m}^3/\text{s}$ at 3.10 h.

5.3.22 Impact of jets on smooth surfaces

5.3.22.1 *Single moving vane*

Let v be the absolute velocity of jet, u the absolute velocity of vane, assumed parallel to v, and a the area of section of jet (Figure 5.57).

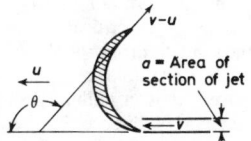

Figure 5.57 Impact of a jet on a single moving vane or series of moving vanes

Velocity of jet relative to vane $= v - u$.
Therefore mass striking vane $= \rho a(v - u)$.
This is unaltered in flow over the vane. (If roughness is taken into account, the final relative velocity $= k(v - u)$ where $k < 1$.)
Initial momentum per second of jet $= \rho a(v - u)v \leftarrow$
Final momentum per second of jet $= \rho a(v - u)[u + (v - u) \cos\theta] \leftarrow$
Therefore force exerted on vane $= \rho a(v - u)^2(1 - \cos\theta) \leftarrow = x$, say
Work done on vane $= \rho a(v - u)^2(1 - \cos\theta)u$

Initial kinetic energy of jet $= \rho a v^3/2$

Therefore efficiency $= [2(v - u)^2(1 - \cos\theta)u]/v^3$ (5.84)

Initial momentum per second of jet, in $\uparrow$ direction $= 0$
Final momentum per second of jet, in $\uparrow$ direction $= \rho a(v - u)^2 \sin\theta$
Therefore force exerted on vane in direction $\downarrow = \rho a(v - u)^2 \sin\theta = y$, say

Resultant force on vane $= \sqrt{(x^2 + y^2)}$ (5.85)

5.3.22.2 *Series of moving vanes*

Figure 5.57 also applies. Since successive vanes intercept the jet,

the mass of water striking them per second now is ρav. ρ is again the density of the water.

Force exerted in direction $\leftarrow = \rho av(v - u)(1 - \cos\theta) = x$, say
Work done on vanes $= \rho avu(v - u)(1 - \cos\theta)$
Efficiency $= 2u(v - u)(1 - \cos\theta)/v^2$ (5.86)

Force exerted in direction $\downarrow = \rho av(v - u)\sin\theta = y$, say (5.87)

Resultant force $= \sqrt{(x^2 + y^2)}$

5.3.22.3 *Cubical block resting on the bed of a stream (Figure 5.58)*

Let ρ' be the density of the material of the block (in kilograms per cubic metre), μ the coefficient of friction between the block and bed of stream and v the velocity of stream at height y above bed (in metres per second).

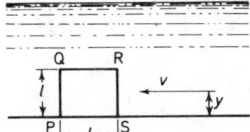

Figure 5.58 Impact of jet on cubical block resting on the bed of a stream

Resistance to overturning about $P = \{[(\rho' - \rho)l^4]/2\}g$

Resistance to sliding $= \mu(\rho' - \rho)l^3g$
where μ is the coefficient of sliding friction.

Force of impact of stream against face $RS = K'\rho l \int_0^l v^2 \, dy$

where K' is a coefficient, ~ 0.70, to allow for the fact that not all the forward momentum of the stream is 'destroyed'.
Therefore, block will overturn if:

$$K' \int_0^l v^2 y \, dy > \frac{g(\rho' - \rho)l^3}{2\rho} \quad (5.88)$$

or will slide if:

$$K' \int_0^l v^2 \, dy > \frac{\mu g(\rho' - \rho)l^2}{\rho} \quad (5.89)$$

These results neglect other effects such as reduction of pressure on top and lee faces. They serve to show, however, that the stability of the block is essentially dependent upon the way in which the velocity is distributed through the depth of the stream.
Experiments in laboratory flumes indicate that:

$\bar{v} = $ mean velocity of stream for which block overturns (in metres per second)
$\quad = $ rate of discharge divided by (area of section of channel $-$ area of section of block)

$$= \frac{L}{l}\left(5.52 - \frac{2930}{(32 + h/l)^2}\right)\left[\left(\frac{\rho' - \rho}{\rho}\right)l\right]^{1/2} \quad (5.90)$$

where L is the length of block measured parallel to direction of current (in metres), l the depth of block (in metres), h the depth of water above top of block (in metres), ρ' the density of material of block (in kilograms per cubic metre) and ρ the density of water (in kilograms per cubic metre).

Example 5.13. A concrete cube $1 \times 1 \times 1$ m of density 2400 kg/m³.

Total depth D (m)	Depth over cube h (m)	$\bar{v}$ (m/s)
2	1	3.34
3	2	3.53
4	3	3.70
6	5	3.99
11	10	4.56

For stability of more than one block, arranged in rows and tiers, or heaped at random, see Allen.[35]

5.3.22.4 Stability of flat beds and mounds of broken stone and sand

The paper just cited[35] suggests that, over a wide range including materials having 'equivalent cube lengths' of 0.098 to 26.2 mm and specific gravities of 2.016 to 3.89:

$$\bar{v} = k[(L+B)/2l]^{0.44}(\sigma'-\sigma)^{0.22}\,l^{0.22}\,M^{-0.22}\,h^{0.06}\,m^{0.22} \qquad (5.91)$$

where v is the mean velocity (in metres per second) of stream flowing over a flat bed, L is the average value of maximum length of individual grains, B the average value of maximum breadth of individual grains, l the length of side of a cube of the same weight as the average piece, σ' the specific gravity of material, σ the specific gravity of water (say unity), M the uniformity modulus of material, h the depth of stream, m the hydraulic mean depth and k is a constant.

k is equal to 0.067 for movement of the first few particles if L, B, l, h and m are all in millimetres.

The uniformity modulus M is defined as indicated in Figure 5.29.

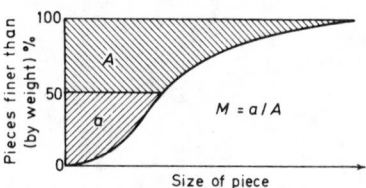

Figure 5.59 Definition of a uniformity modulus

Similarly, for mounds:

$$\bar{v} = 0.079(\sigma'-\sigma)^{0.27}\,l^{0.27}\,M^{-0.27}\,h^{0.21}\,m^{0.02} \qquad (5.92)$$

for initial disturbance.

$$\bar{v} = 0.083[(L+B)/2l]^{0.54}\,(\sigma'-\sigma)^{0.27}\,l^{0.27}\,M^{-0.27}\,h^{0.21}\,m^{0.02} \qquad (5.93)$$

for 'flattening the crest of the mound'.

Here $\bar{v}$ is the velocity (mean) of the stream flowing over the top of the mound and h the depth over the top (in millimetres).

Equations (5.92) and (5.93) apply to materials of irregular shape (sand or broken stone), but the case of mounds formed of *cubes* laid in random fashion has also been investigated by Allen.[35]

Example 5.14: 1 m cubes weighing 2400 kg so that $\sigma' = 2.4$.
Example 5.15: Broken stone, equivalent cube length $l = 1$ m, $\sigma' = 2.64$, $M = 0.9$

Depth over crest of mound laid in random fashion (m)	v for initial movement (m/s)	
	1 m cubes	Broken stone $l = 1$ m, $M = 0.9$
1	1.6	2.9
2	1.9	3.4
5	3.1	4.2
10	4.0	5.0

Note that σ', the true specific gravity, of silica or quartz sands is usually 2.64, or very nearly so.

5.3.23 Sediment transport

Considerable effort continues to be applied to the theoretical and experimental study of the transport of sand and suspended silt in both laboratory channels and natural waterways. One approach used for at least 100 years has been to try to relate the bed movement to the shear stress or tractive force per unit area exerted by the water on the bed of the channel, especially for the critical condition necessary to initiate motion of the grains. This stress is $\rho f v^2/2$, where f depends on the material which forms the bed and sides of the channel and on the hydraulic mean depth (see Equation (5.57)). The stress so defined has formed the basis of many theories and the means of interpreting observations of bed movement. The results cover a wide range of complexity, one of the most elementary being to write τ_c, the critical shear stress required to start movement of the sand, as proportional to the median diameter d_s of the grains. For example, $\tau_c = 1.2d_s$, where τ_c is in newtons per square metre if d_s is in millimetres. While such a simple relationship makes no allowance for the shape of the grains, or the variation in their individual sizes, or any cohesion such as is present with fine particles, it nevertheless gives the right order of magnitude of τ_c if d_s is not less than about 0.25 mm. Thus, for $d_s = 0.25$ mm, τ_c becomes 0.3 N/m². Equating this to $\rho f v^2/2$ and taking ρ for water as 1000 kg/m³, and supposing f to be of the order of 0.003 if the depth is, say, 1 m, it then follows that the mean velocity v in such a channel would be about 0.5 m/s. At this velocity, grains would begin to move.

Other researchers, such as G. M. White[36] and R. A. Bagnold,[37] have introduced a force of upward lift on a grain in addition to the longitudinal drag, while some methods of assessing the *quantity* of bed material transported by a given stream have been based on the amount by which the shear stress at the bed exceeds the critical shear stress τ_c. Yet another concept is to relate the sediment transport to the 'streampower' τ_v. More detailed accounts of these matters appear in many books and journals, e.g. a paper by P. A. Mantz.[38]

5.3.24 Vortices

5.3.24.1 Forced vortices
Forced vortices are of the type caused by stirring a liquid in a dish or by rotating the dish (Figure 5.60).

$$h = (\omega^2 r^2)/2g \qquad (5.94)$$

where ω is the angular velocity of rotation (rad/s).

5.3.24.2 Free vortices

Free vortices are of the type which results when a liquid flows

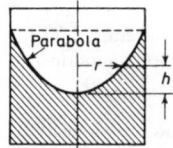

Figure 5.60 Forced vortex

through a hole in the bottom of a vessel. The water moves in stream lines spirally towards the centre, where an air-core tries to form. The coefficient of discharge through the orifice is greatly reduced as compared with its value in the absence of a vortex, i.e. with larger heads.

If the hole is now closed, the vortex motion persists for a time (until damped out by viscous resistance), the lower part having the characteristics of a forced vortex and the upper of a free cylindrical vortex (Figure 5.61):

$$c_1 - z = \text{constant}/2gr^2 \qquad (5.95)$$

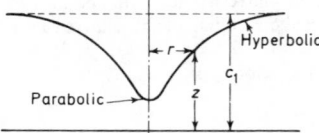

Figure 5.61 Free vortex

5.3.25 Waves

5.3.25.1 Waves of transmission or translation

Waves of transmission or translation are of the type formed when a sluice gate is suddenly opened to admit water to a channel, or when a tidal bore advances along a river.

Let h be the depth of the stream, v the velocity of the stream moving in the direction opposite to the wave, k the height of the wave crest above the surface of the stream and V the velocity of propagation of the wave. Then:

$$V = \left[\frac{2g(h+k)}{1 + h/(h+k)}\right]^{1/2} - v \qquad (5.96)$$

If k is small compared with h

$$V = [g(h+k)]^{1/2} - v$$

If k is very small compared with h

$$V = (gh)^{1/2} - v$$

5.3.25.2 Waves of oscillation

Waves of oscillation occur in comparatively deep water (Figure 5.62).

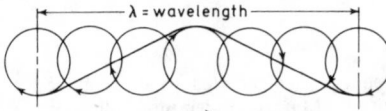

Figure 5.62 Waves of oscillation

Particles in the surface describe circular orbits, giving the appearance of a wave crest advancing with velocity $\sqrt{(g\lambda/2\pi)}$. The period of oscillation is $\sqrt{(2\pi\lambda/g)}$. In shallower water the

orbits become distorted into approximate ellipses – a condition intermediate between a wave of oscillation and one of translation – and finally the wave breaks on the shore.

5.3.25.3 Dynamic effects on hydraulic structures

These are not confined to the direct impact of streams or waves. They may also result from eddies giving rise to periodic forces and may become serious, e.g. with long, slender piles in deep tidal waters. An important occurrence of this phenomenon was investigated at Immingham Oil Terminal.[39-41]

5.3.26 Dimensional analysis

Dimensional analysis is a valuable tool in reducing the apparent chaos of experimental results involving many variables and also in the systematic design of experimental procedure or technique.

Consider the resistance R of a certain shape of submerged body. Let l be any representative dimension, such as the length, v the velocity relative to the stream, ρ the density of fluid, μ the viscosity of fluid and g the acceleration due to gravity.

Suppose R depends upon l, v, ρ, μ and g $\qquad$ (5.97)

Six quantities are involved here, and together they depend upon the three fundamental units or quantities of mass, length and time.

We therefore expect to find three dimensionless groups of the original quantities R, l, v, ρ, μ and g, related to one another.

Let N be a dimensionless group.

Choose any three quantities from Equation (5.97) such that together they include mass M, length L, and time T.

Suppose we choose R, l, ρ.

R (a force) has the dimensions of mass × acceleration, or $[MLT^{-2}]$

l has the dimensions of length $[L]$

ρ has the dimensions of $[ML^{-3}]$

Now write $N_1 = R^{a_1} l^{b_1} \rho^{c_1}$, the last quantity v being chosen at random from the remaining symbols of Equation (5.97).

Now N_1 is to have no dimensions in mass, length or time. Hence, dimensionally, $0 = [M^{a_1}][L^{a_1}][T^{-2a_1}][L^{b_1}][M^{c_1}][L^{-3c_1}][LT^{-1}]$.

Equating indices of $[M]$, $[L]$, $[T]$ in turn

for $[M]$, $0 = a_1 + c_1$
for $[L]$, $0 = a_1 + b_1 - 3c_1 + 1$
for $[T]$, $0 = -2a_1 - 1$

These simultaneous equations give $a_1 = -1/2$, $c_1 = 1/2$, $b_1 = 1$. Hence:

$$N_1 = \rho^{\frac{1}{2}} l v / R^{1/2}$$

Similarly, writing $N_2 = R^{a_2} l^{b_2} \rho^{c_2} \mu$ and remembering that the units of μ are $[M][L^{-1}][T^{-1}]$, we should find

$$N_2 = \mu / R^{\frac{1}{2}} \rho^{1/2}$$

Again, writing

$$N_3 = R^{a_3} l^{b_3} \rho^{c_3} g$$

we should get:

$$N_3 = R^{-1} l^3 \rho g$$

Now write:

N^1 = a function of N_2, $N_3 = \Phi(N_2, N_3)$

i.e.

$$\rho^{\frac{1}{2}} l v / R^{1/2} = \Phi(\mu/R^{\frac{1}{2}}\rho^{\frac{1}{2}}, l^3\rho g / R)$$

which result is not invalidated if we multiply any one of the dimensionless 'groups', N_1, N_2, N_3, by a function of itself or of one of the others, so long as we do not reduce the total number of such distinctive groups.

Therefore:

$$R/\rho l^2 v^2 = \Phi(v l \rho / \mu, v^2/lg) \tag{5.98}$$

In this, $v l \rho / \mu$ is the so-called Reynolds number, and v^2/lg is the so-called Froude number, although some writers define it as $v/\sqrt{(lg)}$.

For alternative methods of dimensional analysis, see Duncan[42] and Whittington.[43]

5.3.26.1 Application to scale models

The principle of dynamical similarity indicates that if a model is operated at a speed truly corresponding with the full size project, then:

$$R_1/R_2 = \rho_1 l_1^2 v_1^2 / \rho_2 l_2^2 v_2^2$$

where the suffix 1 refers to the full size and the suffix 2 to its model.

As far as the resistance of submerged bodies is concerned therefore, it follows from Equation (5.98) that the Reynolds number in the model should be the same as in the actual, and (at the same time) the Froude numbers should be the same in model and full-size project.

The one condition then requires:

$$v_2 = v_1(l_1/l_2)(\rho_1/\rho_2)(\mu_2/\mu_1) \tag{5.99}$$

and the other requires:

$$v_2 = v_1\sqrt{(l_2/l_1)} \tag{5.100}$$

These two conditions cannot, in general, be satisfied at one and the same time and one of the features of scale model technique is to choose one or other as the more important and then to discover, as a result of experiments on models of different scales, or by comparison of model results with prototype values, what is the inaccuracy or 'scale effect' caused by neglect of the other requirement. In more complicated examples, more than two requirements may ideally be necessary.

5.3.26.2 Ship models

In testing the resistance of a ship's hull, the model is towed at a speed given by $v_2 = v_1 \times (1/S)^{1/2}$, where $1/S$ represents the geometrical scale. Its total resistance R_T is then measured and from this is subtracted the skin-frictional resistance R_S as estimated from experiments on long 'boards' of similar surface roughness. The residue $(R_T - R_S)$ represents eddy and wave-making resistance R_E:

R_E multiplied by $\rho_1 l_1^2 v_1^2 / \rho_2 l_2^2 v_2^2$, or $(\rho_1/\rho_2)S^3$

then gives the eddy and wave-making resistance of the full-size ship at the corresponding speed $v_1 = v_2\sqrt{S}$, and to this is added the estimated skin friction of the ship at that speed.

5.3.26.3 Falling particles – terminal velocity

A particle descending through a fluid will accelerate until its effective weight is balanced by the resistance to its motion – it will then have attained its terminal velocity V_T.

Let ρ_1 = density of particle, ρ = density of fluid, η = viscosity of fluid, and R_T = force of resistance when velocity of particle reaches its terminal value V_T.

Then, once the terminal velocity is attained,

$$g(\rho_1 - \rho) \times \text{volume of particle} = R_T.$$

A general treatment of the resistance to relative motion of the solid through the fluid would involve considering inertia (arising from the property of mass possessed by the fluid) and viscosity (brought about by the molecular structure of the fluid).

In his classical study of a sphere moving through a fluid, Sir George Gabriel Stokes assumed that at sufficiently low velocities the inertia terms could be neglected and so established the result that under such conditions the resistance would be $3\pi\eta Vd$, where V is the velocity of the sphere of diameter d relative to the fluid.

Applying this to the terminal velocity V_T:

$$g(\rho_1 - \rho)[(1/6)\pi d^3] = 3\pi\eta V_T d,$$

from which $V_T = (gd^2/18)\left(\dfrac{\rho_1 - \rho}{\eta}\right)$.

Experiments confirm this provided that $V_T d\rho/\eta$ does not exceed approximately 0.5. Subject to that proviso, the formula for V_T may be used to determine: (1) the diameter of a small sphere from its observed terminal velocity in a fluid of known viscosity; and (2) the viscosity of a fluid by measuring the terminal velocity of a small sphere of known diameter.

In more general terms, the terminal velocity, even when $V_T d\rho/\eta$ exceeds 0.5, may be used as a measure of: (1) the size of particles in comparison with the well-established experimental results for spheres; and (2) the capacity of particles to be lifted into suspension by an upwards current.

(Incidentally, it will be seen that Stokes's $R = 3\pi\eta Vd$ is equivalent to $R/\rho l^2 V^2 = R/\rho d^2 V^2 = 3\pi/(\rho Vd/\eta)$, or, if A is the projected area of the sphere, $\pi d^2/4$, $R/\rho AV^2 = 12/(Vd/v)$.)

5.3.26.4 Models of sluice-gates, weirs, etc

In models of sluice-gates, weirs and spillways, the Froude number is to be adopted, provided the scale is well chosen. For example, a model of an overflow spillway for a dam may usually be relied upon if the lowest head on which deductions are to be based is not less than 6.4 mm in the model. Under heads smaller than this, surface tension and viscosity are generally responsible for appreciable scale effects. With this reservation, however, if we call h the head observed in the model and q the discharge, then the discharge of the full-size spillway under a head of Sh will be $qS^{5/2}$, the model scale being $1:S$.

5.3.26.5 River models

River models are frequently constructed to different scales horizontally and vertically. Many considerations influence the scales actually to be chosen. If the horizontal scale is $1:x$ and the vertical scale $1:y$, there will be a vertical exaggeration, or distortion, of scale, equal to x/y. This exaggeration is desirable in order to improve the prospects of: (1) the flow being turbulent in the model as it is in nature; (2) the water-surface slopes being reproduced; (3) bed material being shifted by the currents available in the model. The smaller the value of x, the smaller must be the exaggeration x/y. The nominal velocity scale for sensibly horizontal stream velocities in such a model is then $1:\sqrt{y}$ and the scale of time $1:x/\sqrt{y}$. If q is the rate at which water is fed to the model in order to simulate a flow Q in nature, then q should be equal to $Q/xy^{3/2}$. Some river models have, however, been operated with such flows and velocities, discovered by trial, as to give the proper gradients and bed movements irrespective of these ideal conditions which should, if practicable, be observed.

Sand or other material used on the bed of river models to give qualitative or approximately quantitative indications of scour is not necessarily reduced to scale. Frequently such materials are of the same order of size as in nature, the feasibility of this depending on the fact that the scouring property of a shallow stream is greater than that of a deep one of the same mean velocity.

5.3.26.6 Harbour models

Models of harbours specifically concerned with surface waves produced by storms have been successful with scales (undistorted) in the region of 1:50, 1:100 or 1:180, and with model waves in the exposed area outside the harbour works about 20 mm or more in height. The velocity scale of such a model having a geometrical scale of 1:100 would be $1:\sqrt{100}$; or 1:10, and its time scale would also be 1:10.

5.3.26.7 'Mathematical models'

Mathematical solutions to complex problems of fluid motion have developed on a wide front in recent years, in conjunction with the increasing availability of computers. They cover a range of problems embracing the analysis of data, the study of surges, the hydraulic behaviour of rivers and estuaries, sediment transport, etc. The representation or modelling by mathematical formulae and equations frequently contains terms and/or coefficients which have been derived from observations in the field, the laboratory, or on physical models. The choice of which kind of model to use – the mathematical or the physical – depends on the nature of the engineering problem concerned; in some cases, it is advantageous to combine the two methods, so treating one as complementary to the other. Recent literature provides a guide, and in the particular case of estuaries, the potential and the limitations of both kinds of model are discussed by McDowell and O'Connor.[44]

Notwithstanding the undoubted powers of the mathematical approach, engineers experienced in·these matters are likely to endorse the view of McDowell and O'Connor that 'the greatest single merit of physical models is their capacity to reproduce, on an easily observed scale, the intricate three-dimensional flow in a large estuary. They are, in consequence, an aid to thought and planning without equal. . . .' This applies not only to physical models of rivers and estuaries but equally to those of hydraulic structures, e.g. spillways, channels and other devices for the 'dissipation' of energy.

References

1   Davis, S. J. and White, C. M. (1929) 'A review of flow in pipes and channels', *Engng*, **78**, 71.
2   Stanton, T. E. and Pannell, J. R. (1914) *Phil. Trans. A*, **214**, 199; *National Phys. Lab. Coll. Res.*, **11**, 293.
3   Nikuradse, J. (1933) *Forschungsh. Ver. dtsch. Ing.*, No. 361.
4   Prandtl, L. and von Kàrmàn, T. (1933/1935) *Z. Ver. dtsch. Ing.*, **77**, 105 (1933); *Proceedings, 4th international congress on applied mechanics*, Cambridge (1935).
5   Colebrook, C. F. and White, C. M. (1937) 'Experiments with fluid friction to roughened pipes'. *Proc. R. Soc. A*, **161**, 367.
6   Colebrook, C. F. (1939) 'Turbulent flow in pipes, with particular reference to the transition region between the smooth and rough pipe laws', *J. Instn Civ. Engrs*, **11**, 133.
7   Manning, R. (1895) 'Flow of water in open channels and pipes', *Trans Instn Civ. Engrs, Ireland*, **20**, 161 (1891); **24**, 179.
8   Ackers, P. (1958) 'Resistance of fluids flowing in channels and pipes', *Hydraulics Research Papers*, Nos 1 and 2, HMSO, London.
9   Barnes, A. A. (1916) *Hydraulic flow reviewed*, Spon, London.
10 Colebrook, C. F. and White, C. M. (1937) 'The reduction of carrying capacity of pipes with age', *J. Instn Civ. Engrs*, **7**, 99.
11 Sellin, R. H. J., Hoyt, J. W. and Scrivener, O. (1982) 'The effect of drag-reducing additives on fluid flows and their industrial applications'. Part 1 'Basic aspects'. *J. Hydr. Res*, **20**, 1, 29.
12 Sellin, R. H. J., Hoyt, J. W., Scrivener, O. and Pollert, J. (1982) 'Present applications and future proposals', Part 2, *J. Hydr. Res*, **20**, 3, 235.
13 Gibson, A. H. (1910/1911) 'Loss of head in gradual enlargements', *Proc. R. Soc. A*, **83**, 366 (1910); 'Loss at enlargements and at valves', *Trans R. Soc. Edinburgh*, **48**, (1911).
14 Cornish, R. J. (1939) 'The analysis of flow in netwcrks of pipes', *J. Instn Civ. Engrs*, **13**, 147.
15 Skeat, W. O. and Dangerfield, B. J. (eds) (1969) *Manual of British water engineering practice* (4th edn) Vol. III, Institution Water Engineers, London, 7 pp. 168–174.
16 Stuckey, A. T. (1969) Methods used for the analysis of pipe networks, WWE.
17 Barlow, J. F. and Markland, E. (1969) 'Computer analysis of pipe networks', *Proc. Instn Civ. Engrs*, **43**, 249.
18 Al-Nassri, S. A. (1971) 'Flow in pipes and pipe networks', PhD thesis, University of Liverpool.
19 O'Brien, M. P. and Hickox, G. H. (1937) *Applied fluid mechanics*. McGraw-Hill, New York,
20 British Standards Institution (1943) *Flow measurement* (BS 1042). British Standards Institution, Milton Keynes.
21 American Society of Mechanical Engineers (1931) *Fluid meters report* (4th edn). ASME, New York.
22 Bazin, H. E. (1897) *Ann. des Ponts et Chaussées*, **4**, 20.
23 Parker, P. A. M. (1915) *The control of water*, Routledge, London.
24 Allen, J. (1943) 'Roughness factors in fluid motion through cylindrical pipes and through open channels', *J. Instn Civ. Engrs*, **20**, 91
25 Allen, J. (1939) 'The resistance to flow of water along a tortuous stretch of river and in a scale model of the same', *J. Instn Civ. Engrs*, **11**, 115.
26 Allen, J. and Shahwan, A. (1954) The resistance to flow of water along a tortuous stretch of the river Irwell (Lancashire) – an investigation with the aid of scale-model experiments', *Proc. Instn Civ. Engrs*, Part. III, **3**, 1, 144.
27 Gibson, A. H. (1924) *Hydro-electric engineering*, Blackie, London, Vol. I, p. 67.
28 Allen, J. and Hamid, H. I. (1968). The hydraulic jump and other phenomena associated with flow under rectangular sluice-gates', *Proc. Instn Civ. Engrs*, **40**, 345.
29 Engel, F. V. A. (1933/1934) 'Non-uniform flow of water', *The Engineer*, London, 21, 28 April, 5 May (1933); 'The venturi flume', *The Engineer*, London, 3 and 10 August (1934).
30 Elsden, O. (1964) 'Flow measurement' in: Guthrie Brown (ed.) *Hydro-electric engineering practice* (2nd edn) Chap. 2, Vol. I. Blackie, London and Glasgow.
31 Hamilton Smith, Jr (1886) *Hydraulics*.
    Bilton, H. J. I. (1908) *Victorian Inst. Engrs.*
    Judd, H. and King, R. S. (1906) *Am. Assoc. Adv. Sci.*
    Smith, E. S. and Walker, W. H. (1923) *Proc. Instn Mech. Engrs.*
    Bond, W. N. *An introduction to fluid motion*. Arnold, London.
32 Rehbock, T. (1912/1929) *Handbuch der Ingenieur wissenschaften*, Vol. I, Part 3/2 (1912); discussion in *Trans. Am. Soc. Civ. Engrs*, 93 (1929); also details in J. R. Freeman (ed.) *Hydraulic laboratory practice*. American Society Civil Engineers, (1929).
33 Gould (1901) *Engineering News*.
    Horton, D. F. (1918) *Engineering News Record*.
    Gibson, A. H. (1924) *Hydro-electric engineering*, Vol. I, Blackie, London and Glasgow, p. 75.
34 Mathieson R. (1953) 'The generalized Gould's function', *Proc. Instn Civ. Engrs*, Part III, **2**, 1, 142.
35 Allen, J. (1942) 'An investigation of the stability of bed materials in a stream of water', *J. Instn Civ. Engrs*, **18**, 1.

36 White, C. M. (1940) 'The equilibrium of grains on the bed of a stream'; *Proc. Roy. Soc. A*, **174**, 958, 322.
37 Bagnold, R. A. (1956) 'The flow of cohesionless grains in fluids'; *Phil Trans. Roy. Soc. A*, **249**, 964.
38 Mantz, P. A. (1983) 'Semi-empirical correlations for fine and coarse cohesionless sediment transport'; *Proc. Instn Civ. Engrs*, **75**, Part 2, 1.
39 Sainsbury, R. N. and King, D. (1971) 'The flow-induced oscillation of marine structures', *Proc. Instn Civ. Engrs*, **49**, 269.
40 Construction Industry Research and Information Association (1970/1971) Report Project 143, p.21.
41 British Transport Docks Board (1971) *Docks*, **8**, 11, 5.
42 Duncan, W. J. (1953) *Physical similarity and dimensional analysis*. Arnold, London.
43 Whittington, R. B. (1963) 'A simple dimensional method for hydraulic problems', *J. Hydr. Div. Proc. Am. Soc. Civ. Engrs*, **89**, No. HY5, 1.
44 McDowell, D. M. and O'Connor, B. A. (1977) *Hydraulic behaviour of estuaries*. Macmillan, London.

Bibliography

Addison, H. (1964/1940) *A treatise on applied hydraulics* (5th edn), Chapman and Hall, London; *Hydraulic measurements*, Chapman and Hall, London.
Allen, J. (1947) *Scale models in hydraulic engineering*, Longman, London.
Brown, J. Guthrie (1964) *Hydro-electric engineering practice* (2nd edn), vol. I, Blackie, London and Glasgow.
Duncan, W. J., Thom, A. S. and Young, A. D. (1960) *The mechanics of fluids*, Arnold, London.
Fox, J. A. (1977) *An introduction to engineering fluid mechanics* (2nd edn), Macmillan, London.
Francis, J. R. D. (1958) *A textbook of fluid mechanics for engineering students*, Arnold, London.
Francis, J. R. D. (1975) *Fluid mechanics for engineering students* (4th edn), Arnold, London.
Gibson, A. H. (1952) *Hydraulics and its applications* (5th edn), Constable, London.
Her Majesty's Stationery Office *Hydraulics research*. HMSO Annual Reports, London.
Institution Water Engineers (1969) *Manual of British water engineering practice* (3 vols) (4th edn). IWE, London.
Jaeger, C. (1956) in: Wolf (trans. and ed.), *Engineering fluid mechanics*, Blackie, London and Glasgow.
Jameson, A. H. (1937) *An introduction to fluid mechanics*, Longman, London.
King, H. W. (1954) in: Brater (ed.) *Handbook of hydraulics* (4th rev. ed.) McGraw-Hill, New York.
Lewitt, E. H. (1952) *Hydraulics and the mechanics of fluids* (9th edn), Pitman.
McDowell, D. M. and Jackson, J. D. (eds) (1970) *Osborne Reynolds and engineering science today*. Manchester University Press, Manchester, and Barnes and Noble, Inc. New York.
McDowell, D. M. and O'Connor, B. A. (1977) *Hydraulic behaviour of estuaries*, Macmillan, London.
Muir Wood, A. M. (1969) *Coastal hydraulics*, Macmillan, London.
Novak, P. and Cābéla, J. (1981) *Models in hydraulic engineering*, Pitman, London.
O'Brien, M. P. and Hickox, G. H. (1937) *Applied fluid mechanics*. McGraw-Hill, New York.
Pao, R. H. F. (1961) *Fluid mechanics*. Wiley, New York.
Parker, P. à. M. (1915) *The control of water*. Routledge, London.
Prandtl, L. (1952) *The essentials of fluid dynamics*. Blackie, London and Glasgow.
Raudkivi, A. J. (1967) *Loose boundary hydraulics*, Pergamon Press, Oxford.
Rouse, H. (1948) *Elementary mechanics of fluids*, Wiley, New York.
Vennard, J. K. (1962) *Elementary fluid mechanics* (4th edn). Wiley, New York.
Webber, N. B. (1979) *Fluid mechanics for civil engineers*, Chapman and Hall, London.

6

Engineering Surveying

T J M Kennie BSc, MAppSci (Glasgow), ARICS, MInstCES
Lecturer in Engineering Surveying,
University of Surrey

Contents

6.1 Introduction

The work of the land surveyor can be classified into three main areas of responsibility. Firstly, he is concerned with the recording of measurements which allow the size and shape of the Earth to be determined. Secondly, and primarily, he is involved in the collection, processing and presentation of the information necessary to produce maps and plans. Thirdly, he may be required to locate on the surface of the Earth the exact positions to be taken up by new roads, dams or other civil engineering works.

As a consequence of the diverse nature of the land surveyor's duties, several distinct branches of the subject have evolved.

6.1.1 Branches of surveying

Geodetic surveys are carried out on a national or international basis in order to locate points large distances apart. This type of survey acts as a framework for 'lower order' surveys. In order to ensure high accuracy, the effect of factors such as the curvature of the Earth on observations must be considered and the necessary corrections applied.

Topographic surveys are concerned with the small-scale representation of the physical features of the Earth's surface. Frequently, the data necessary for such an operation will be provided by the use of aerial photography. The science of taking measurements from photography in order to produce maps is known as photogrammetry. Topographic surveys are often the responsibility of a national organization such as, for example, the Ordnance Survey in the UK.

Hydrographic surveys, in contrast, involve the representation of the surface of the seabed. The end-product is normally a navigational chart. In recent years this branch has become increasingly important with the development of the offshore oil industry. In this case, in addition to the production of charts, the surveyor may be required to position large structures such as oil production platforms. This type of operation would normally necessitate the use of ground and satellite electronic position-fixing equipment.

Cadastral surveys relate to the location and fixing of land boundaries. In many countries in the world, e.g. Australia, the information supplied by the cadastral surveyor may be an integral part of a land registration system.

Finally, engineering surveys are required for the preparation of design drawings relating to civil engineering works such as roads, dams or airports. The surveys are normally at a large scale, with scales of 1:500 and 1:1000 being most common.

Many of these branches require highly specialized knowledge, beyond the scope of this chapter. In view of this, the aim in this chapter will be to discuss: (1) those aspects of the subject which are required in order to carry out simple surveys for engineering projects; (2) the processes involved in carrying out precise surveys for deformation monitoring projects; and (3) the use of computers in surveying for digital mapping and ground modelling.

6.1.2 Principles of surveying

In spite of the diverse nature of land surveying, it is possible to define certain basic principles which are common to all branches of the subject. These principles have proved over the years to be vital if accurate surveys are to be conducted.

The first and most important principle is the provision of an initial framework before observing and fixing the detail of a survey. This process is often known as providing control. It is essential to ensure that the positions of the control points are known to a higher order of accuracy than those of the subsidiary points. By satisfying this principle it is possible to ensure that errors, which inevitably occur, do not accumulate but are contained within the control framework.

A second and perhaps more obvious principle is that of planning. All too often it is tempting to rush into a survey without consideration for an overall plan. Of particular importance is the need to define a job specification. This is indispensable since the relationship between cost and accuracy is not linear and an increase in accuracy may have a disproportionate effect on cost. For example, if a distance of 500 m is to be determined to an accuracy of either 5 or 0.5 mm, the cost ratio of the respective accuracies may be of the order of 1:300. It is important, therefore, to choose techniques and instruments appropriate to the survey specification. Of equal importance is the need to plan the reconnaissance stage. Before starting a task it is essential to examine the area carefully, considering all the possible ways of doing the survey and then selecting the most suitable method. Remember, 'time spent on reconnaissance and planning is never wasted'.

A third principle is the need to ensure that sufficient independent checks are incorporated into the survey to eliminate or minimize errors. It is important that the checking system is included at all stages of the survey from fieldwork and computations to the final plotting. In addition, the checking system should be independent and not solely a repeat of the initial measurement. Examples of independent checks are:

Fieldwork:
- measure both diagonals of a quadrilateral
- measure distances in both directions
- measure angles using different parts of the theodolite circle

Computations:
- use the summation check on angle observations of geometric figures, e.g. sum of interior angles $(2n-4) \times 90°$
- levelling booking cross-checks

Plotting:
- plot positions of important points by using angles and distance and also using coordinates.

The final principle is that of safeguarding. Safeguarding is equally important at all stages of the survey, and refers to the process of ensuring that the survey results can be replicated if accidental, or other, damage occurs to the survey markers or field observations. Thus, it is important when constructing permanent survey markers to take 'witness or reference measurements' to points of prominent detail in the vicinity of the point. Linear measurements of this type enable the point to be relocated if it is damaged, or alternatively if it is difficult to find. The latter situation can often occur with road projects. In many instances there may be a gap of many years between initial survey and final setting-out. During this time the permanent survey markers may become overgrown with vegetation and hence difficult to locate. The use of witness marks and measurements can often be of crucial importance if the permanent marks are to be relocated.

Safeguarding of field observations is also of paramount importance. Thus, it is considered good practice to produce abstract sheets from the surveyor's fieldbook at the end of each day. These abstract sheets should summarize the major results from the fieldbook (e.g. rounds of angles, mean distance etc.) and should be carefully filed in the survey office. By such a process the possibility of several days' work being lost if the fieldbook is damaged or misplaced can be eliminated.

6.1.3 Errors in surveying

It is an unfortunate and often misunderstood fact that all measurements are affected by errors. So often, when confronted with the question: 'How accurate do you want the survey to be?', or 'How accurately do you want this point located?', the glib answer 'Exactly!', or 'Spot on!', is given as the reply by a prospective client. If, in addition, the question of errors is raised, it is quickly dispensed with the comment: 'Errors don't occur if you do it properly.' The answer is correct in one respect, i.e. in relation to mistakes, or, more correctly, gross errors. These should not occur if a survey is carried out according to the basic principles of surveying. However, other types of errors do occur which can be much more difficult to handle.

Systematic errors, as the name suggests, are errors which follow a pattern or system. Errors of this type are normally related to the variations in physical conditions which can occur when a measurement is made. For example, a steel tape is normally known to be a certain length at some standard temperature. If the temperature under which a measurement is made varies from this standard, a systematic error will occur. By knowing the coefficient of linear expansion a value for the expansion of the tape can be determined and a correction applied. Systematic errors whose effect can be modelled mathematically are, hence, eliminated.

Random errors, in contrast, do not follow a standard pattern and are entirely based on the laws of probability. These errors, or rather variations in measurement, will occur after gross and systematic errors have been eliminated. The measurement of a distance by taping can again be taken as an example. It is often not appreciated that the same distance measurement made under the same physical conditions with the same tape will produce different answers. Since it is assumed that the measurements will follow a normal distribution they can be examined using standard statistical techniques.

The following formula can therefore be applied to the analysis of random errors:

$$\text{arithmetic mean} = \bar{x} = \frac{\Sigma x_i}{n} \tag{6.1}$$

where $i = 1, 2, \ldots, n$ are the observed values and n denotes the number of observed values.

The arithmetic mean is significant because it is often taken to be the closest approximation to the 'true' value and as such is known as the most probable value (m.p.v.). The difference between the m.p.v. and the observed value is known as a residual (v).

A term often used in order to estimate the precision of a series of measurements is the standard error, where standard error of a single observation is:

$$\sigma_s = \pm \left(\frac{\Sigma v^2}{n-1} \right)^{1/2} \tag{6.2}$$

and standard error of the arithmetic mean is:

$$\sigma_M = \pm \left(\frac{\sigma_s}{n^{1/2}} \right) \tag{6.3}$$

For surveying purposes, the terms 'standard error', 'standard deviation' and 'root mean square (r.m.s.) error' are synonymous. All such terms are used to give an indication of the precision of the result, i.e. the degree of agreement between successive measurements. High precision may not be indicative of high accuracy, since accuracy is related to the proximity of the measurement to the true value. If, however, all the effects of the bias caused by systematic errors have been eliminated, these indices of precision may also be used as indices of accuracy.

For example, suppose an angle has been measured 9 times and the subsequent error analysis indicates that $\sigma_s = 3''$ and $\sigma_m = 0.81''$. What does this information tell us? Firstly, it indicates that the angle has been measured to a high precision. Secondly, it indicates that, statistically, there is a 68% chance or probability of the standard error of a single measurement being less than 3''. Furthermore, if one extends the confidence limit to a value 3 times the standard error, or 9'', then statistically the probability that the error will be less than 9'' is now 99.7% with only a 0.3% chance of the error being greater than 9''. This confidence limit is often applied as a rejection criterion to a group of observations. Any observation with a residual greater than 3 times the standard error may then be rejected, on the basis that it is highly unlikely that the variation is solely a consequence of random effects. Similar reasoning would apply to the standard error of the arithmetic mean. Further information on errors and their treatment can be found in Cooper[1] and Mikhail and Gracie.[2]

6.2 Surveying instrumentation

Surveying is essentially concerned with the direct measurement of three fundamental quantities: (1) the angle subtended at a point; (2) the distance between two points; and (3) the height of a point above some datum, normally mean sea-level. From the measurement of these three quantities, it is then possible to compute the three-dimensional positions of points.

With the exception of electronic methods of determining distance, the instruments used by the surveyor have not radically changed in principle for 40 to 50 years. The advances in technology may have reduced the size and increased the efficiency of the instruments, but the fundamental principles remain unchanged.

6.2.1 Angular measurement using the theodolite

The theodolite is used for the measurement of horizontal and vertical angles. In simple terms, a theodolite consists of a

Table 6.1 Characteristics of some modern theodolites

Type of theodolite:	1″ Precise	20″ Engineers	10′ Builders	Compass	Electronic
Typical example:	Kern DKM 2A-E	Sokkisha TM20ES	Zeiss (Ober.) TH51	Wild TO	Kern E-2
Country of manufacture:	Switzerland	Japan	W. Germany	Switzerland	Switzerland
Direct reading to	1″	20″	10′	1′	1″
By estimation to	0.1″	5″	1′	30″	—
Telescope magnification	32 ×	28 ×	20 ×	20 ×	32 ×
Telescope aperture (mm)	40	45	30	28	45
Sensitivity of plate					
Level per 2 mm run	20″	30″	45″	8′	—
Weight of instrument (kg)	6.2	4.2	2.2	2.9	8.7

telescope mounted on a platform which may be levelled to form a horizontal plane by means of a simple spirit bubble. Angles are measured by pointing the telescope at targets and establishing the difference between readings on a circular protractor mounted on the level platform.

There is a bewildering choice of theodolites available. Table 6.1 lists the characteristics of a selection of commonly available modern theodolites. The broad distinction can be made between those instruments which measure angles and those such as compass and gyro theodolites which measure bearings, relative to magnetic north and to true north respectively.

6.2.1.1 General construction of the theodolite

There are certain fundamental relationships and components which are common to all theodolites. Before examining the detailed construction of a modern glass arc theodolite, it is important to appreciate the geometrical arrangement of the axes of a theodolite, as illustrated in Figure 6.1.

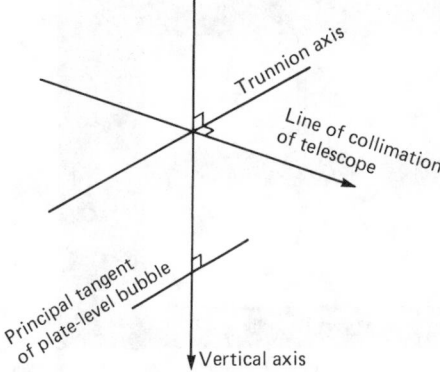

Figure 6.1 Theodolite axes

In this ideal arrangement the vertical axis is vertical, the trunnion axis is perpendicular to it and hence horizontal, and the line of collimation is perpendicular to the trunnion axis. Unfortunately, it is not possible during the manufacturing process to ensure that these orthogonal relationships occur exactly. Similarly, during use over a period of years, wear may occur which may also alter these conditions. The extent to which a theodolite fails to satisfy them can be measured by a series of instrument tests which may be carried out in the field. If, subsequently, the instrument is found to be out of adjustment, the instrument should be returned to the manufacturer or a specialist instrument technician for adjustment. Details of the field tests and methods of adjustment may be found in Cooper.[3] If, however, a modern theodolite is treated with care, and a suitable observational technique is employed, regular servicing should be all that is required in order to obtain good results.

The detailed construction of a modern 1 s precise theodolite is shown in Figures 6.2 and 6.3. Examination of these figures illustrates that the theodolite consists essentially of three distinct parts:

(1) *Base*. This consists of two main components: the tribrach and the horizontal circle. The tribrach can be screwed securely to the tripod and, by means of three footscrews, the instrument may be levelled. The circle is made of glass with photographically etched graduations. It is normally graduated in a clockwise manner. A circle-setting screw is also usually provided and the base will generally house an

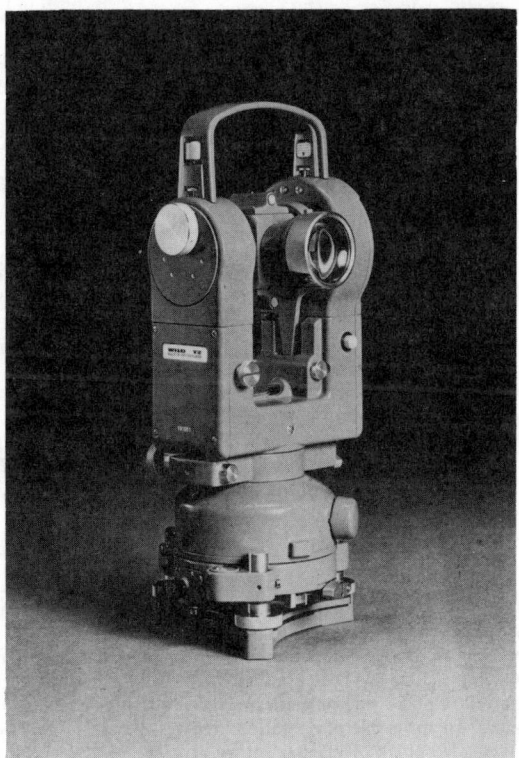

Figure 6.2 Wild T-2 one second theodolite (Wild Heerbrugg)

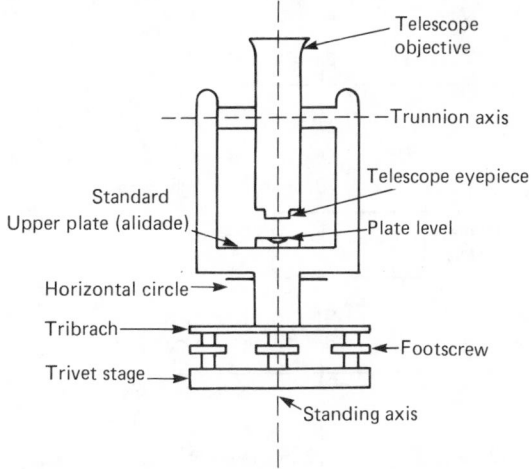

Figure 6.3 Construction of a theodolite

optical plummet. This consists of a small eyepiece with a line of sight which is deviated by 90° in order to point vertically down. By this process it is possible to centre the instrument precisely over a ground point. In some cases the optical plummet may be housed in the alidade.

(2) *Alidade*. This rotatable upper part of the theodolite may also be known as the upper plate. The alidade rotates about the vertical axis. Mounted on the alidade is the plate-level bubble which indicates whether the instrument is level. By

means of clamps and slow-motion screws it is possible to rotate and clamp the alidade relative to the base.

(3) *Telescope.* Attached to the trunnion axis of the theodolite is the telescope. The telescope magnifies the object and, by the use of cross-hairs, allows the exact bisection of the target. Focusing of the object and the cross-hairs is carried out using separate focusing screws. A further clamp and slow-motion screw allow precise pointing of the telescope in a vertical plane.

Angles of elevation or depression are measured using a vertical circle also attached to the trunnion axis. Prior to measuring a vertical angle it may be necessary to set the altitude bubble. However, most modern theodolites employ an automatic compensating mechanism. In these cases vertical angles may be recorded after the plate level has been set, without recourse to an additional bubble setting.

When the vertical circle is to the left of the telescope, the theodolite is in what is conventionally called the face left (FL) position. Conversely, when the vertical circle is to the right of the telescope as it views an object, the theodolite is in the face right (FR) position.

6.2.1.2 Circle reading

By projecting daylight through the standards of the theodolite, it is possible to illuminate the glass scale of both the horizontal and vertical circles.

In order to resolve a direction to a higher precision than that to which the circle has been graduated, an optical micrometer is employed. Optical micrometers are the modern equivalent of verniers. The principle of operation involves the use of a plane parallel-sided block of glass as shown in Figure 6.4. When the glass is in the normal position, as shown by position (a), light passing through will be uninterrupted. Rotation of the block of glass, however, produces a lateral shift of the incident beam as shown by position (b). This rotation is controlled by the

micrometer screw of the theodolite. Movement of this screw enables the observer to read, on an auxiliary scale, the lateral shift required in order to bring the image of the main-scale degree graduations into coincidence with the index marks which are built into the optical path. Using this technique it is possible to resolve directly to 20″ of arc if the micrometer is reading from one side of the circle. Resolution direct to 1″ is possible if a mean-reading optical micrometer is used. In this case, readings from two points diametrically opposite are meaned in order to eliminate the effects of any circle eccentricity.

Figures 6.5 and 6.6 illustrate two typical examples of the circle reading systems for both the single- and mean-reading optical micrometers.

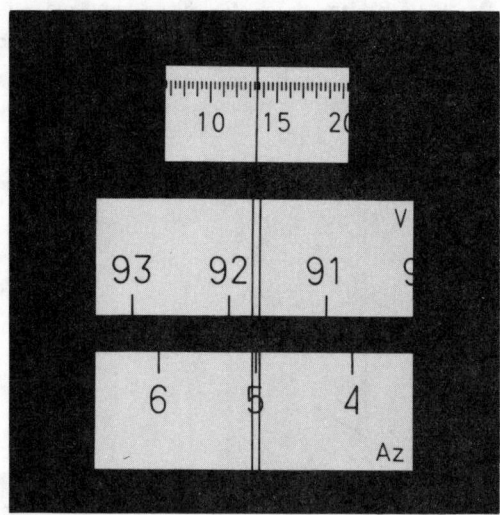

Figure 6.5 Single reading optical micrometer: circle reading Wild T-1A 05° 13′ 30″ (Wild Heerbrugg)

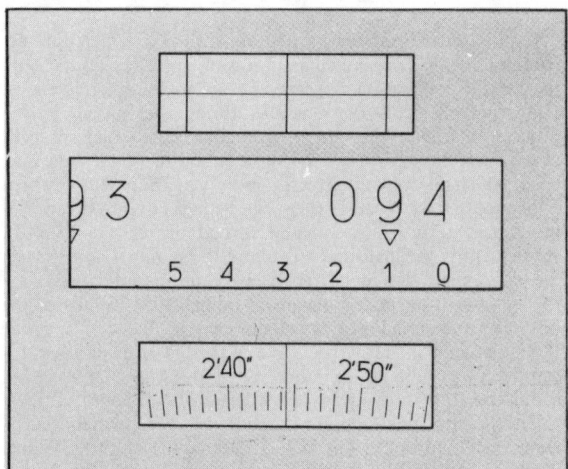

Figure 6.6 Mean reading optical micrometer: circle reading Wild T-2 94° 12′ 44.3″ (Wild Heerbrugg)

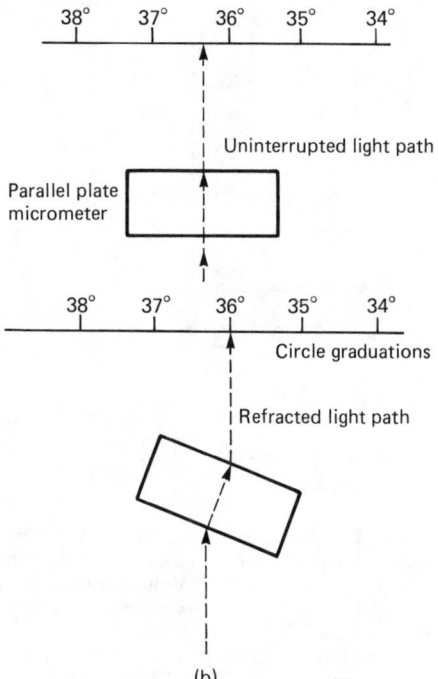

(b)

Figure 6.4 Parallel plate micrometer

6.2.1.3 Field procedure

Potentially the theodolite is a very precise instrument. It is, however, necessary to follow a strict procedure both in setting-up the instrument and in observing if this potential is to be realized. Incorrect use of a theodolite will undoubtedly result in poor results, regardless of how precise the instrument may be.

Setting-up. Setting-up a theodolite prior to observations being taken consists of three separate operations: centring, levelling and focusing.

Centring involves positioning the instrument exactly over a ground point. This may be achieved by means of either a plumb bob suspended from the instrument or a centring rod or an optical plummet. The process of centring and levelling should be considered as iterative in nature, becoming increasingly more precise after each operation.

Levelling the theodolite carefully is a necessary prerequisite for precise measurements. The following sequence of operations must be carried out in order to level a theodolite:

(1) Approximately level the instrument using the small circular bubble.
(2) Set the plate-level bubble parallel to any two footscrews, such as A and B in Figure 6.7(a). Rotate both footscrews together or apart until the bubble is in a central position.

(3) Rotate the alidade until the bubble is now approximately perpendicular to the initial position, as shown in Figure 6.7(b). Using footscrew C only, centralize the bubble.
(4) Return to the initial position and again centralize the bubble using footscrews A and B. Repeat (2) and (3) until the bubble is central in both positions.
(5) Rotate the alidade through 180° until the position shown by Figure 6.7(c) is achieved. If the bubble does not remain in a central position, move the bubble until it is in a position midway between a central position and its initial position.
(6) Rotate the alidade until the position illustrated by Figure 6.7(d) is achieved. Using footscrew C, move the bubble into the same position as in Figure 6.7(c). The bubble should then remain in the same off-centre position for any alignment of the alidade.

The final step before observations begin is to focus both the cross-hairs and the object to which observations will be made. It is important to ensure that both images appear clear and sharp. In addition, it is critical that parallax does not exist. Parallax refers to the apparent movement of the cross-hairs and objects relative to each other when the observer moves his head. It is caused by the image of the object not lying in the same vertical plane as the cross-hairs. If this occurs, the focusing operation must be repeated until it is eliminated.

Observational procedure. A strict observational procedure is essential if both human and instrumental errors are to be reduced to a minimum. Consider the problem of measuring the angle shown in Figure 6.8.

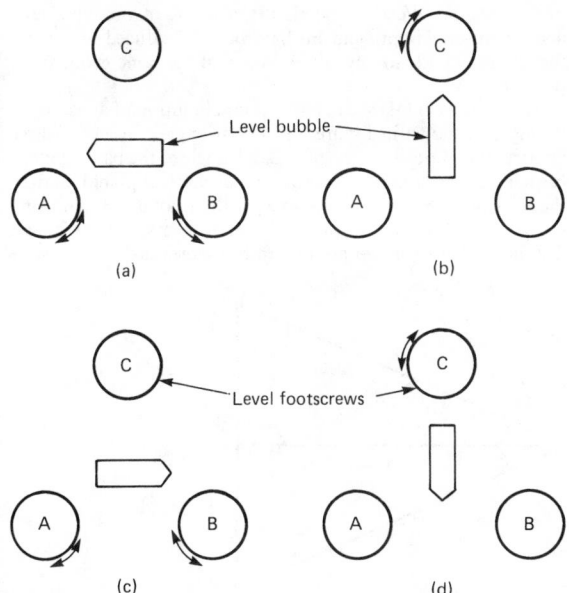

(a)   (b)

(c)   (d)

Figure 6.7 Levelling the theodolite

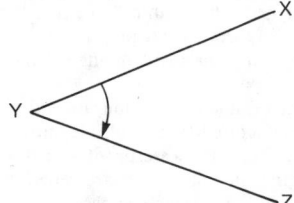

Figure 6.8 Angle measurement

The observational procedure which should be adopted is as follows. A booking procedure is illustrated by Table 6.2.

(1) Point with telescope in the FL position to the target at X and record the horizontal circle reading, e.g. 90° 20′ 30″.
(2) Point on FL to target Z and record, e.g. 130° 25′ 40″.

Table 6.2 Booking and reduction of theodolite readings

Station Y Observer ....Date ....

Height of Inst: ....Booker ....Weather................................

TO	FL			FR			MEAN			ANGLE			COMMENTS
	°	′	″	°	′	″	°	′	″	°	′	″	
Round 1													
X	90	20	30	270	20	40	90	20	35	40	05	10	
Z	130	25	40	310	25	50	130	25	45				
Round 2													
X	135	30	15	315	30	25	135	30	20	40	05	15	
Z	175	35	30	355	35	40	175	35	35				
Mean angle: 40° 05′ 13″													

(3) Change face to FR and point to target Z and record circle reading 310° 25′ 50″.
(4) Point on FR to target X and record direction 270° 20′ 40″.

In order to reduce the effect of instrument maladjustments to a minimum, the mean of the FL and FR minutes and seconds readings to the same point is averaged and the value entered in the mean column. The difference between the mean circle readings is then derived and entered in the angle column.

This constitutes one round of angles. The base setting screw should then be adjusted and the process repeated in order to increase the precision of the angle measurement. A minimum of two rounds is necessary for the least precise measurements; up to sixteen rounds may be required for very precise operations.

6.2.2 Distance measurement

The second fundamental quantity which it is necessary to measure is distance.

A wide variety of techniques can be used for the determination of distance. The general distinction can, however, be made between direct, optical and electronic methods. All of the techniques discussed are capable of varying levels of precision depending on the degree of sophistication of the instrumentation and the observational techniques adopted.

6.2.2.1 Direct distance measurement (DDM)

The simplest method of measuring distance is that of physically measuring the distance with a tape. In the past invar tapes were used for the precise measurement of baselines for triangulation networks. Nowadays, DDM is generally confined to either the precise measurement of short distances for setting-out or control purposes or the less precise measurement of the detailed dimensions of a building or land parcel.

There are basically two types of tape in common use. Fibreglass measuring tapes are manufactured from multiple strands of fibreglass coated with PVC. They are waterproof and normally either 30 or 50 m in length. Fibreglass tapes are generally used for detail measurements and have largely superseded the linen tapes which were available previously. For more precise measurements it is necessary to use steel bands. These are typically either 30, 50 or 100 m in length.

In order to obtain high precision with either type of tape, it is essential that it is periodically checked against a standard reference tape, the length of which is known to a higher order of accuracy than that of the tape being checked. If a significant variation exists, a standardization correction should be applied. In addition, it is vital that suitable attention is paid to the effect of variations in slope, temperature and tension which may necessitate appropriate corrections being applied to the measured distance. The corrections (C_1, C_2 and C_3 respectively) are:

Slope:

$$C_1 = -L(1 - \cos \theta) \qquad (6.4)$$

where θ = slope angle, and L = measured slope distance

or:

$$C_1 = -\Delta h^2 / 2L$$

where Δh = height difference between end-points

Temperature:

$$C_2 = \pm \alpha L(t_m - t_s) \qquad (6.5)$$

where t_m = measured temperature in the field, t_s = temperature at which the tape was standardized, usually 20°C, and α = coefficient of linear expansion (0.000 011 2 for steel bands)

Tension:

$$C_3 = \pm L(T_m - T_s)AE$$

where T_m = measured tension, T_s = standard tension, A = cross-sectional area of tape, and E = Young's modulus for the tape, typically 200 kN/mm² = 200 000 N/mm²

Miller[4] details the typical accuracy levels which can be achieved with steel tapes.

6.2.2.2 Optical distance measurement (ODM)

As an alternative to the direct method of measuring distance, it is also possible to measure distance indirectly by optical methods.

The development of ODM began over two centuries ago. James Watt is recorded as having used this approach in his survey of the West of Scotland in 1774. Although many instruments and improvements have been introduced since then, they all essentially involve the solution of the same geometrical problem.

All methods of ODM are based on the solution of an isosceles triangle, as shown in Figure 6.9. The triangle consists of three important components: the parallactic angle α, the base length B (which may be either in a horizontal or vertical plane), and D, the horizontal bisector of the base of the triangle. By knowing the relationship between the three components, the horizontal distance D between two points can be determined.

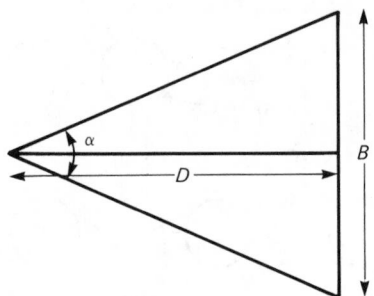

Figure 6.9 Optical distance measurement (ODM)

Two methods of solution are possible: either an instrument with a fixed parallactic angle is used and the variable base B is measured (Figure 6.10) or a base of fixed length is set up and the variable parallactic angle is measured (Fig. 6.11). In both cases, the variable quantity is proportional to the horizontal distance. By defining the mathematical relationship between the fixed and variable quantities it is therefore possible to determine the horizontal distance.

Tacheometry. The first approach described above (fixed angle, variable base) is commonly known as tacheometry or more correctly as vertical staff stadia tacheometry. It is normally used for the measurement of distance where a proportional error of between 1/500 and 1/1000 is acceptable, e.g. in picking-up survey detail points.

All modern theodolites have a diaphragm consisting of a main horizontal cross-hair and two horizontal stadia lines

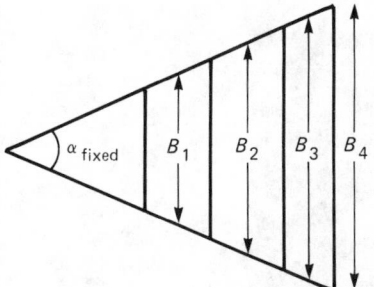

Figure 6.10 ODM: fixed angle, variable base

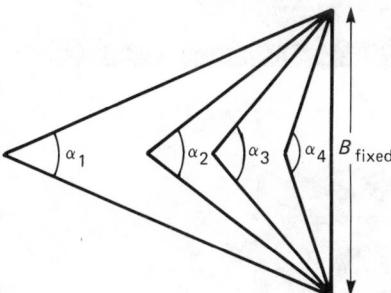

Figure 6.11 ODM: fixed base, variable angle

spaced either side of it. These stadia lines define the fixed parallactic angle. If the theodolite telescope is sighted on to a levelling staff and the readings of the outer lines noted, the difference in the readings, the staff intercept (s), will be directly proportional to the horizontal distance between the instrument and the staff. Generally, the distance between the stadia lines is designed in such a manner that the horizontal distance D_H between the instrument and staff is given by:

$$D_H = 100s \tag{6.7}$$

For inclined sights the geometry is as shown in Figure 6.12. Hence:

$$D_s = 100(s \cos \theta) \tag{6.8}$$

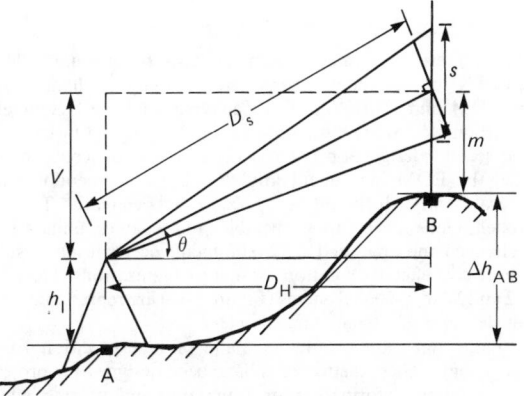

Figure 6.12 Stadia tacheometry: inclined sights

where D_s = slope distance, and θ = vertical angle measured by the theodolite. Therefore:

$$D_H = 100s \cos^2 \theta \tag{6.9}$$

$$\Delta h_{AB} = h_i + V - m \tag{6.10}$$

where Δh_{AB} = difference in height between A and B, h_i = height of instrument (trunnion axis to ground), m = middle hair reading, and V = difference in height between middle-hair reading and trunnion axis = $50s \sin 2\theta$ (6.11)

Several self-reducing tacheometers have also been designed. The main advantage of these instruments is their ability to compensate for the effect of the inclination of the theodolite telescope and, hence, allow the direct determination of horizontal distance without additional computation.

Two notable examples of this type of instrument are the Wild RDS vertical staff self-reducing tacheometer and the Kern DK-RT horizontal bar double-image self-reducing tacheometer. Details of the construction and use of these instruments may be found in Hodges and Greenwood,[5] and Smith.[6] In recent years, the manufacture of these precise optical devices has ceased, their place being taken by low-cost electronic measuring devices.

Subtense bar. The second approach (fixed base, variable angle) is commonly known as the subtense or horizontal subtense bar method. The method is normally confined to the measurement of distance for control purposes. Using this approach, distances may be determined with a proportional error of up to 1/10 000.

The instrumentation required consists of a subtense bar, normally 2 m long, and a one-second theodolite, such as the Wild T2. The bar has targets mounted at each end of an invar strip. The strip is protected by a surrounding aluminium strip in order to ensure that, for all practical purposes, the length of the bar remains constant at 2 m. The bar is set up and oriented at right angles to the line of sight of the theodolite, as shown in Figure 6.13.

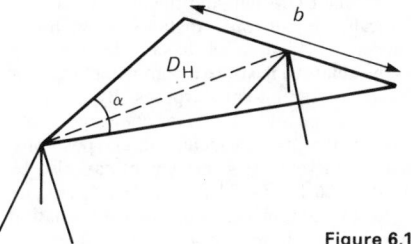

Figure 6.13 Subtense bar

The horizontal parallactic angle α is measured with the theodolite. Irrespective of the vertical angle to the bar, the horizontal distance is given by:

$$D_H = \tfrac{1}{2}b \cot(\alpha/2) \tag{6.12}$$

with $b = 2m$

$$D_H = \cot(\alpha/2) \tag{6.13}$$

For distances greater than 100 m it is advisable to subdivide the distance to be measured or, alternatively, to use the auxiliary base method (Hodges and Greenwood,[5] and Smith.[6]).

Subtense methods are also tending to be superseded by low-cost electronic methods. Nevertheless, many organizations still possess this type of equipment and for many projects it is a very suitable technique to adopt.

6.2.2.3 Electronic distance measurement (EDM)

Development. The first generation of EDM instruments was developed in the early 1950s. Typical of the early meters were the Swedish Geodimeter (GEOdetic DIstance METER) and the South African Tellurometer instrument. The former, an electro-optical instrument, used visible light measurement, whilst the latter used high-frequency microwaves. Both instruments were primarily developed for military geodetic survey purposes and had the ability to measure long distances, up to 80 km in the case of the Tellurometer, to a precision of a few centimetres. They were also, however, bulky, heavy and expensive in comparison to their modern-day equivalents.

During the late 1960s, developments in microelectronics and low-power light-emitting diodes led to the emergence of a second generation of EDM instruments. These electro-optical instruments utilized infra-red radiation as the measuring signal and were developed for the short range (< 5 km) market. In addition, they were considerably smaller, lighter and less expensive than their predecessors. Probably the best-known example is the Wild DI-10 Distomat.

The introduction of microprocessors into the survey world in the early 1970s led to the introduction of a third series of EDM instruments. With this group it became possible, not only to determine slope distance, but also to carry out simple computational tasks in the field. For example, the facility became available to compute automatically the corrected horizontal distance and difference in height between two points by manual input of the vertical angle read from the theodolite. Electronic distance measurement instruments of this type had also been reduced in size to the extent that the EDM unit could be theodolite-mounted. The Wild DI-3 is a typical example of this type of instrument.

The most recent short-range EDM instruments are similar to the previous group, but have several additional features worthy of mention. Firstly, the technology now exists to sense automatically the inclination of the EDM unit and therefore to be able to compute automatically the horizontal distance between two points. The Geodimeter 220 (Figure 6.14) has this facility. This instrument also has the ability to measure to a moving target, or track, a useful feature for setting-out purposes. By using an additional unit it is also possible to have one-way speech communication between the instrument and target positions, again valuable when setting-out. This instrument can also be connected to a Geodat 126 hand-held data collector (Figure 6.14), which is able to store automatically distance information from the EDM unit. Other relevant information (numeric or alphanumeric) can be input manually via the keyboard. The Geodimeter 220 has a range of 1.6 km with one prism and 2.4 km with three prisms determined to a standard error of ± 5 mm ± 5 parts per million (p.p.m.) of the distance.

The last development in the field of EDM instrumentation is the electronic tacheometer or 'total station'. The former term is more appropriate in view of the different interpretations, by the instrument manufacturers, of the term total station. In essence, an electronic tacheometer is an instrument which combines an EDM unit with an electronic theodolite. Hence, such instruments are capable of measuring, automatically, horizontal and vertical angles and also slope and/or horizontal distance. The majority also have the facility to derive other quantities such as heights or coordinates and store this data in a data collector. Two designs of instrument have evolved during the last 5 years.

(a)

(b)

Figure 6.14 (a) Geodimeter 220; (b) Geodat 126 data collector (Geotronics)

The first, the integrated design, consists of one unit which, generally, houses the electronic circle-reading mechanism and the EDM unit. The Wild TCI Tachymat and the Geodimeter 140 (Figure 6.15) are representative of this range of instrument. The second design approach is the modular concept. In this case, the EDM instrument and the electronic theodolite are separate units which can be operated independently. This approach tends to be more flexible and enables units to be exchanged and upgraded as developments occur; it may also be a more cost effective solution for many organizations. The Kern E-2 and Wild T-2000 Systems (Figure 6.16) are representative of this design of electronic tacheometer.

Finally, mention should be made of high-precision EDM instruments. These instruments have been designed for projects such as dam deformation or foundation monitoring where extremely high precision is necessary. Instruments which are

Figure 6.15 Electronic tacheometer, integrated design: Geodimeter 140

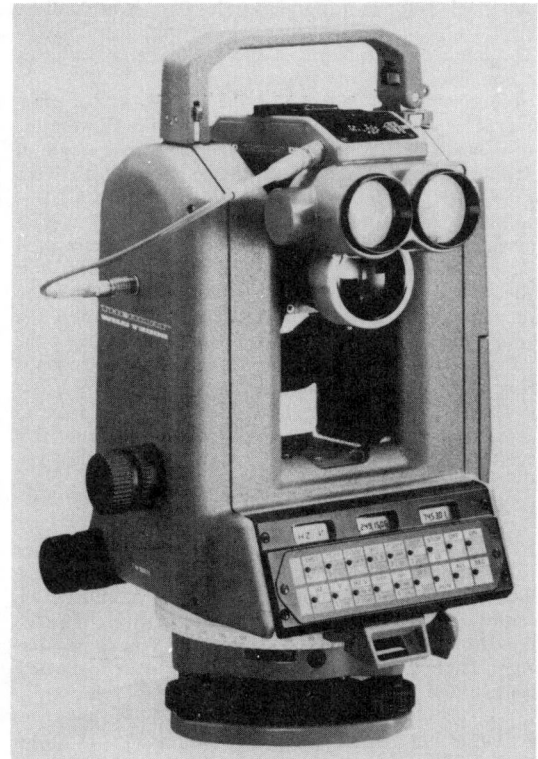

Figure 6.16 Electronic tacheometer, modular design: Wild T-2000 with DI4

Figure 6.17 High-precision EDM: Comrad Geomensor 204DME

10 km with a standard error of ± 0.1 mm ± 0.5 p.p.m. Further up-to-date technical information on many modern EDM instruments can be found in Burnside.[11]

Principle of measurement. Although there is a wide variety of EDM instruments on the market, they all measure distance using the same basic principle. This can be most clearly illustrated by means of the flow diagram (Figure 6.18), which relates specifically to electro-optical instruments.

An electromagnetic (EM) signal of wavelength equal to either 560 nm (visible light), 680 nm (HeNe laser) or 910 nm (infra-red) is generated. This signal is subsequently amplitude-modulated before being transmitted through the optical system of the instrument towards a retro-reflector mounted at the end of the line to be measured. The signal is then retro-reflected, or redirected through 180°, by a precisely ground glass corner cube. Cheaper acrylic corner cubes may also be used.[12,13] This reflected signal is consequently directed towards the receiving optical system. On entering the optical system of the instrument, the signal is converted by means of a photomultiplier into an electrical signal.

The next stage involves the measurement of the phase difference between the transmitted and received signals and the conversion of this information into distance. Figure 6.19 shows the path taken by an EM signal radiated by an EDM instrument together with the instantaneous phase of the signal. It is apparent that the distance X–Y–X travelled by the EM signal is equivalent to twice the distance to be measured. Also, this distance can be seen to be related to the modulation wavelength (λ) and the fraction of the wavelength ($\Delta\lambda$) by the following relationship:

representative of this design include the Tellurometer MA-100 Jaakola,[7] the Kern ME-3000 Mekometer (see Froome,[8] Meir-Hirmer,[9] and Murname[10]) and the Comrad Geomensor 204 DME (Figure 6.17). The latter instrument has a range of up to

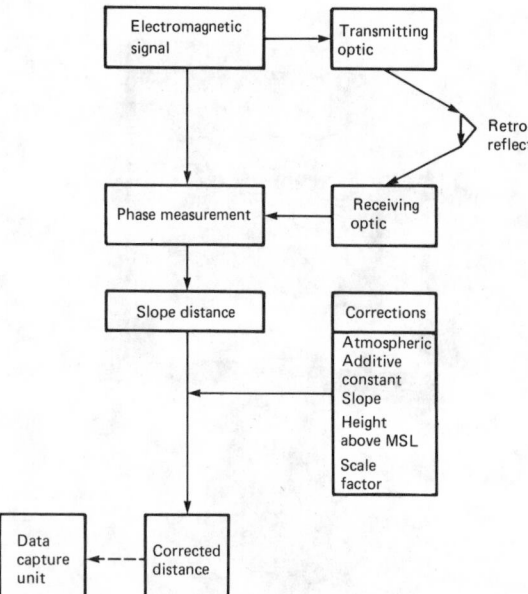

Figure 6.18 Principle of operation: electro-optical distance measurement

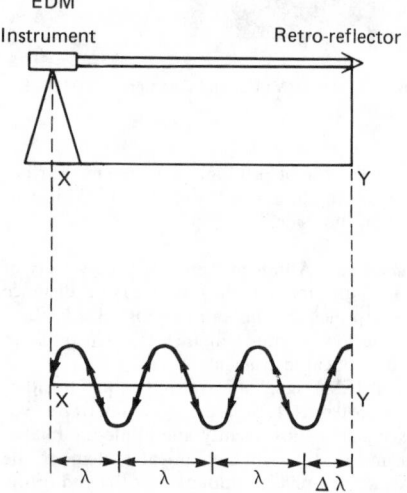

Figure 6.19 Double path measurement using EDM

$$2D = n\lambda + \Delta\lambda \qquad (6.14)$$

where n is an unknown integer number of wavelengths. The determination of the distance therefore involves resolving both $\Delta\lambda$ and n. Phase detectors are used to determine $\Delta\lambda$ which effectively measures the phase difference between the transmitted and received signal and, hence, allows the fractional part of the distance less than one full wavelength to be determined. The value of n can be determined by using two or more EM signals of slightly varying wavelengths. For example, assume $2D = 25.5$, $\lambda_1 = 2.5$ m and $\lambda_2 = 2.4$ m then:

$$2D = 2.5n_1 + 0.5 \qquad (6.15)$$

and

$$2D = 2.4n_2 + 1.5 \qquad (6.16)$$

Assuming $n_1 = n_2$ for short distances, solving for n leads to

$$n = 10 \qquad (6.17)$$

Substituting in Equations (6.15) or (6.16)

$$D = 12.75 \text{ m} \qquad (6.18)$$

This entire process is fully automatic in modern instruments, taking approximately 10 to 20 s to complete.

As with any other method of distance measurement, it is necessary to apply several corrections to the measured slope distance in order to determine the corrected horizontal distance. The first correction to be applied is the atmospheric or refractive index correction. Just as a steel tape varies in length with variations in temperature and pressure so, too, does the modulation wavelength of an EDM instrument. It is therefore necessary to measure the temperature, pressure and, in some cases, relative humidity during measurements. A correction is then applied to compensate for the variation in modulation wavelength caused by variations in atmospheric conditions.

Many instruments have the facility to compute automatically and apply this correction to observations directly in the field. Temperature, pressure and relative humidity readings are taken and the appropriate reading to be set on the refractive index correction dial is read from a nomogram.

A second important correction is the additive zero or prism constant. This correction represents the difference between the electro-optically determined distance and the correct length of line. It is a combination of the errors due to prism offset and the variation in the physical and electrical centres of the EDM instrument. Many manufacturers design their corner cube reflectors in order to eliminate this correction totally. However, if several different types of corner cube are being used, it is essential that a full field calibration be undertaken in order to determine the correction. (See Schwendener,[14] Ashkenazi and Dodson,[15] and Sprent and Zwart[16] for further details of the procedure for instrument calibration.) The slope correction is the same as for DDM. For distances measured above or below mean sea-level (MSL) a correction is necessary in order to reduce the distance to its equivalent at MSL. The correction (C_4) is given by:

$$C_4 = (-LH_m)/R \qquad (6.19)$$

where H_m is the mean height of the instrument and reflector above MSL and R is the radius of the Earth (6370 km). Finally, if the distance is to be used for computation of coordinates on the national grid, the horizontal distance at MSL must be multiplied by the local scale factor. For the Transverse Mercator projection, the local scale factor (F) may be approximately calculated from:

$$F = 0.999\,601\,27 + [1.228 \times 10^{-14} \times (E - 400\,000)^2] \qquad (6.20)$$

where E is the mean local national grid easting in metres of the line to be measured.

6.2.3 Height measurement using the level

The third and final quantity which is measured is height or, more correctly, height difference. This is achieved by means of a level.

The fundamental principle of the level is illustrated by Figures 6.20 and 6.21. Figure 6.20 represents the situation which normally exists when the level is initially set up. In this case, the standing axis of the level and the vertical do not coincide. Hence, the line of collimation of the level will not be horizontal. Figure 6.21 represents the geometrical arrangement of the axis when the instrument has been levelled using the procedure outlined in 'Setting up' in section 6.2.1.3. It can be seen that completion of this procedure ensures that, firstly, the standing and vertical axes are made coincident and, secondly, if the

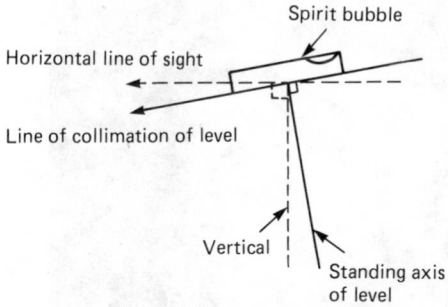

Figure 6.20 Geometry of the level axes: before levelling

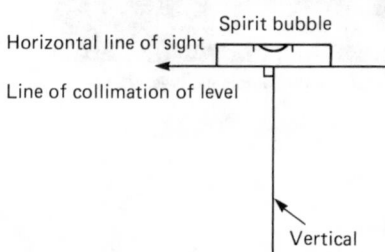

Figure 6.21 Geometry of the level axes: after levelling

instrument is in perfect adjustment, the line of collimation of the level is coincident with a horizontal line of sight.

Three distinct types of level are available for engineering survey purposes: (1) the dumpy level; (2) the tilting level; and (3) the automatic level.

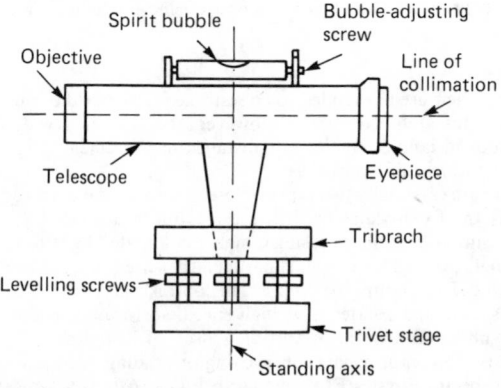

Figure 6.22 Level construction: dumpy level

6.2.3.1 Dumpy level

The dumpy level was so named because of the rather short telescopes which were used with early versions of this instrument.

The construction of a typical dumpy level is shown in Figure 6.22. The most distinctive feature of this type of level is that the axis of the telescope is fixed rigidly to the standing axis of the instrument. In order to satisfy the condition that both the vertical and standing axes are coincident, the standard levelling procedure outlined in section 6.2.1.3 is carried out. Rotation of the telescope will now define a horizontal plane.

In the past, this type of level was very popular for general engineering work. It has, however, been replaced in recent years by the automatic level.

6.2.3.2 Tilting level

The tilting level is a more precise instrument than the dumpy level. Figure 6.23 illustrates the main features. In contrast to the dumpy level, the telescope is not rigidly attached to the standing axis but is able to be tilted in a vertical plane about a pivot point X, by means of a tilting screw.

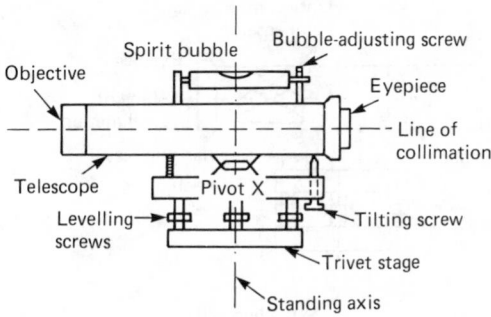

Figure 6.23 Level construction: tilting level

Prior to recording an observation, the instrument is approximately levelled. This is normally achieved by means of a 'ball-and-socket' arrangement and a small circular bubble. In order to set the standing axis exactly vertical, the tilting screw is turned and the main bubble altered until a coincident position (Figure 6.24), as viewed through a small auxiliary eyepiece, is achieved.

If the telescope is now rotated horizontally to sight a second or subsequent point, it is important to relevel the main bubble by means of the tilting screw.

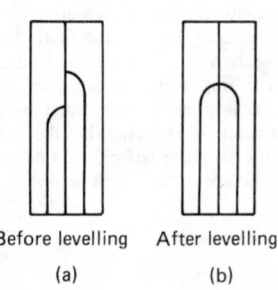

Before levelling   After levelling

(a)   (b)

Figure 6.24 Coincidence bubble-reading system

6.2.3.3 Automatic level

This type of level is not, as the name suggests, totally automatic. Human intervention is still necessary. However, one major source of human error, that of setting the bubble, is replaced by an automatic compensating system. In common with the tilting level, approximate levelling is still necessary. The tedious and error-prone bubble-setting process, however, is eliminated. As with the dumpy level, the instrument defines a horizontal plane when rotated. The automatic level therefore combines the speed of operation of the dumpy level with the precision of the tilting level.

Figure 6.25 illustrates the main components of this type of level. The essential feature of the instrument is the incorporation of an automatic optical–mechanical compensating mechanism. The use of such a system ensures that the line of collimation as defined by the centre cross-hair will trace out a horizontal plane irrespective of the fact that the optical axis of the instrument may not be exactly horizontal. It is, however, necessary to level the instrument approximately in order to ensure that the line of sight is within range of the compensating mechanism.

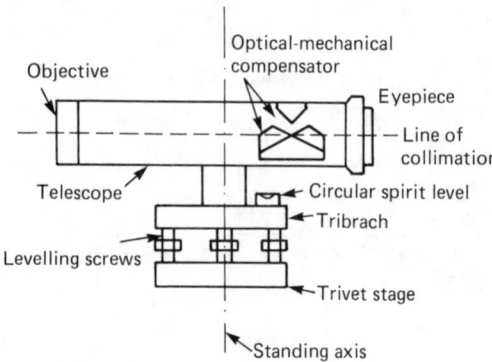

Figure 6.25 Level construction: automatic level

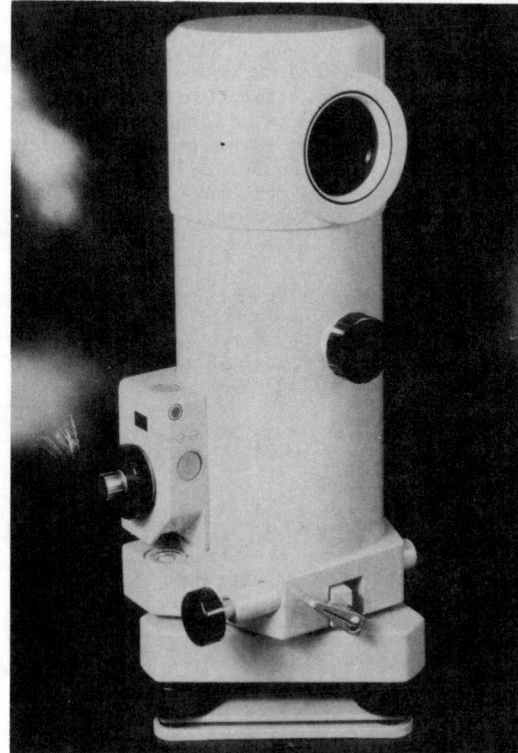

Figure 6.26 Zeiss (Jena) Ni 007 automatic precise level

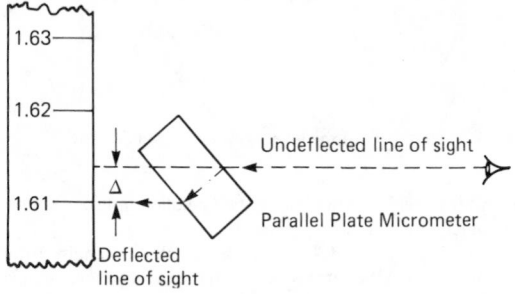

Reading = 1.61 + Δ

Figure 6.27 Operation of a parallel plate micrometer for precise levelling

For high-precision levelling, e.g. in order to detect the settlement of a building, a parallel plate micrometer (PPM), attached to the front of the objective of the telescope normally forms part of the construction of the level. Almost all precise levels in use nowadays are of the automatic design and, ideally, should be designed so that the PPM forms an integral part of the instrument, rather than being an 'add-on' attachment. One such instrument is the Zeiss (Jena) Ni 007 (Figure 6.26). This particular instrument also has an unusual compensating mechanism which results in the 'periscope'-type appearance of the instrument. The PPM operates by deflecting the line of sight to the nearest whole staff graduation, the amount of displacement which is required being measured by a micrometer. This value is then added to the staff reading to give the final staff reading. This is illustrated in Figure 6.27. Clearly in an operation such as precise levelling it is important to minimize the effects of systematic errors. This is partially overcome by a suitable field procedure,[17] and partially by ensuring that the staff is maintained at a constant length. In order to achieve this, an invar staff with stabilizing arms and a level bubble attachment is normally used.

6.2.3.4 Laser level

Lasers are monochromatic, coherent and highly collimated light sources, initially developed in the 1940s. Until relatively recently, their use has tended to be restricted to the field of pure scientific research. Nowadays, however, the laser is a widely used tool in land surveying for distance measurement, alignment,[18] and levelling purposes.

There are essentially two types of laser in use in civil engineering: (1) the fixed-beam; and (2) the rotating beam laser. The fixed-beam laser projects a single highly collimated light beam to a single point. This design is particularly suited to alignment problems. The rotating-beam laser, in contrast, takes the fixed-beam source and rotates it at high speed, so forming a plane (either in the horizontal or vertical sense), of laser light. This design is more appropriate for levelling or grading purposes.

The Spectra-Physics EL-1 shown in Figure 6.28 is a typical example of the laser levels currently in use. The laser beam in

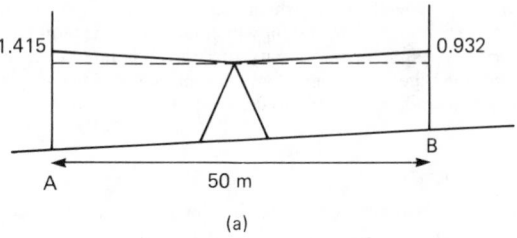

(a)

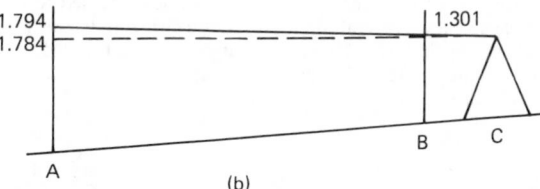

(b)

Figure 6.30 Two-peg test

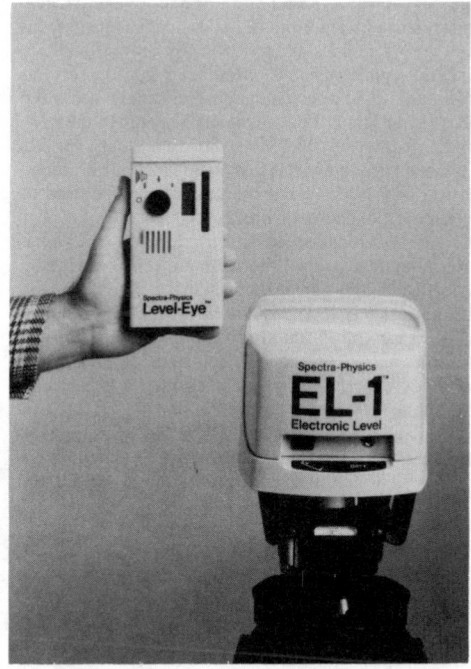

Figure 6.28 Spectra-Physics EL-1 (Spectra-Physics)

this case forms a 360° horizontal plane which is detected by a portable sensing device also shown in Figure 6.28. The laser unit automatically corrects for any error in level of the instrument, providing it has been roughly levelled to within 8° of the vertical. An accuracy of ± 5 to 6 mm per 100 m up to a maximum range of 300 m can be achieved with this type of instrument.

6.2.3.5 Collimation error

So far the assumption has been made that once the standing axis of the level has been set truly vertical, then the line of collimation will be horizontal. This may not always be the case.

If this condition does not occur, then a collimation error is said to exist. This is illustrated in Figure 6.29. If accurate levelling is to be achieved, it is essential that a regular testing procedure is established in order to check the magnitude of any collimation error that may exist.

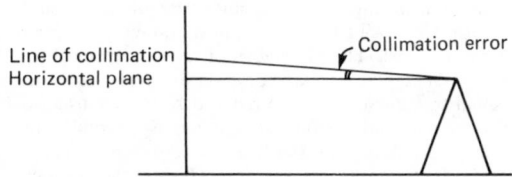

Figure 6.29 Collimation error

A common field procedure which can be used to test a level is known as the 'two-peg test'. The procedure is as follows:

(1) Set out two points A and B approximately 50 m apart, as shown in Figure 6.30(a). The level is set up at the mid-point of AB and levelled as in section 6.2.1.3.
(2) A reading is taken on to a staff held at points A and B and

the difference between the two readings calculated. This value represents the true difference in height between A and B. Any collimation error which exists will have an equal effect on both readings and, hence, will not affect the difference between the readings. In this case the difference in height is $1.415 - 0.932 = 0.483$ m.
(3) The instrument is now moved to a point C close to the staff at B (about 3 to 5 m away), as in Figure 6.30(b). The reading on staff B is recorded (1.301). If no collimation error exists, the reading on staff A should be equal to the reading on staff B $\pm$ the true difference in height as established in (2), i.e. $1.301 + 0.483 = 1.784$ m.
(4) The actual observed reading on staff A is now recorded (1.794). Any discrepancy between this value and that derived previously in (3) indicates the magnitude and direction of any collimation error. For example, in this case, the error would be $1.794 - 1.784 = 10$ mm per 50 m.

An error of up to 2 to 3 mm over this distance would be acceptable. If, however, the error is greater than this, the instrument should be adjusted. Unlike theodolite adjustments, this type of adjustment can normally be performed without any great difficulty by the engineer and the procedure is as follows.

For the dumpy and automatic level: alter the position of the cross-hairs until the centre cross-hair is reading the value which should have been observed from step (3) above. This is achieved by loosening the small screws around the eyepiece which control the position of the cross-hairs.

For the tilting level: again alter the position of the centre cross-hair until it is reading the value previously determined in (3), in this case by tilting the telescope using the tilting-screw. Unfortunately, this will displace the bubble. The bubble must, therefore, be centralized by means of the bubble-adjusting screw.

6.3 Surveying methods

6.3.1 Horizontal control surveys

Any engineering survey or setting-out project, regardless of its size, requires a control framework of known co-ordinated points. Several different control methods are available as described below. The choice of which method to use depends on many factors, e.g. the purpose for which the control is required, the accuracy required, the density of control points which is required, the type of equipment and computing facilities which are available and, lastly, the physical nature of the ground.

6.3.1.1 Triangulation

Triangulation is the oldest, and in the past was the most common, method of control for large civil engineering projects. The principles are well known and essentially involve the establishment of a measured baseline from which a network of triangles is formed, all of the angles of the triangle being measured. The development of EDM has, however, led to the establishment of several alternative control methods, such as trilateration and traversing. The introduction of EDM has therefore tended to make 'classical triangulation' obsolete as a method of control.

6.3.1.2 Trilateration

Trilateration is a method of establishing control which involves the direct measurement, normally using EDM, of all the sides of a network of triangles, in contrast to triangulation which involves the measurement of angles. Although the method has been used in this 'classical form', it does not offer any significant advantages over the method of triangulation. The method has therefore not become particularly common.

6.3.1.3 Traversing

A traverse is a method of establishing control by measuring the distance between successive points and also the horizontal angle between adjacent stations, as shown in Figure 6.31.

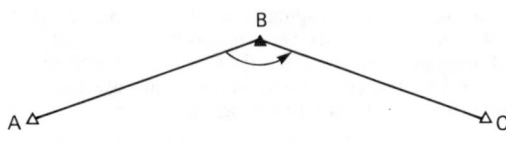

Figure 6.31 Traversing

The method is very popular for several reasons. Firstly, it is a more flexible method than triangulation. In the case of triangulation, the positions of the control stations must be chosen so that not only are they intervisible, but also that the triangles formed are well conditioned. For this reason the reconnaissance stage in triangulation projects is extremely important and often very time-consuming. In contrast, with traversing, much less attention has to be paid to the reconnaissance stage, since it is only required that adjacent stations be intervisible. This allows the surveyor much greater flexibility in the choice of control station positions. The stations can then be positioned in areas close to the detail to be picked up, or close to the project for which they are required. This can be of enormous benefit in areas which are either very flat or, alternatively, heavily forested. A second reason for the popularity of traversing is the computational simplicity, both of determining provisional coordinates and also of adjusting any misclosure which may exist. It should, however, be mentioned that in recent years more rigorous techniques of adjustment, based on the principle of least squares, have become more common for the adjustment of traverse.

Traversing does, however, suffer from one serious drawback: the lack of redundant observational data. As a consequence, the effect of small errors of measurement is not only difficult to detect but is also cumulative in nature. To counteract this problem, additional angle and distance observations are often taken in order to strengthen the control framework. In the past, this additional information tended to be used solely for the detection of gross errors. Nowadays, by using a suitable adjustment technique, these additional observations can be used to improve the precision of the coordinates.

The normal procedure adopted for traverse adjustment is firstly to determine and adjust the angular misclosure and, secondly, to determine and adjust the misclosure in easting and northing. The angular misclosure is determined, in the case of a closed polygon, by summing the internal angles. These should total $(2n-4)$ right angles, where n is the number of traverse sides. A misclosure of $>(20''\sqrt{n})$, for example, would not be acceptable for site traverses.

The question of which method to use for the adjustment of any misclosure in eastings and northings is a matter which has been examined by Schofield.[19] Traditionally, the Bowditch and Transit methods have been used.

$$\text{Bowditch:} \quad \Delta E_c = M_E \frac{d}{\Sigma d} \text{ and } \Delta N_c = M_N \frac{d}{\Sigma d} \tag{6.21}$$

$$\text{Transit:} \quad \Delta E_c = M_E \frac{\Delta E}{\Sigma|\Delta E|} \text{ and } \Delta N_c = M_N \frac{\Delta N}{\Sigma|\Delta N|} \tag{6.22}$$

where M_E, M_N, is the misclosure in easting, northing, d is the length of a traverse leg, $\Sigma|\Delta E|$, $\Sigma|\Delta N|$ is the absolute sum of the provisional ΔE, ΔN, and ΔE_c, ΔN_c is the correction to be applied to the provisional ΔE, ΔN.

For most small site traverses observed using a theodolite and steel tape both techniques will give acceptable results. Schofield,[19] however, also discussed the problems which arise when semi-rigorous methods of adjustment such as Bowditch and Transit are used to adjust modern EDM traverses. The main conclusion reached is that both methods are based on assumptions which are not applicable to EDM. For example, it is assumed that the expected error in an EDM measurement is proportional to the distance measured, which is clearly not true. It is suggested that all EDM-based traverses should be adjusted by a rigorous method such as variation of coordinates.

6.3.1.4 Survey networks

The combination of angle, distance and orientation measurements to form a control framework is now commonly referred to as a survey network. The advantages of such an approach are considerable. Firstly, the scale and orientation errors associated with classical triangulation can be reduced by the inclusion of additional distance and azimuth measurements. Secondly, the control framework does not suffer from the serious propagation of errors which can occur in traversing. Thirdly, the optimum number of observations (angular and distance) can be determined before the fieldwork commences, using computer simulation techniques. Finally, by using the method of variation of coordinates a least squares procedure can be used to determine the most probable values of the coordinates.

The principle of least squares states that the most probable value of a sample is that for which the sum of the squares of the residuals is a minimum. *Variation of coordinates* is a computational method, based on this principle, which is used for the determination of coordinates in a survey network.

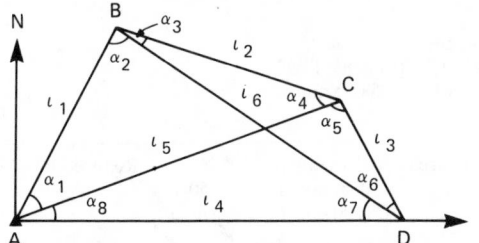

Figure 6.32 Adjustment of a braced quadrilateral by variation of coordinates

Consider Figure 6.32, which illustrates a small network in which all the eight angles $\alpha°1$ to 8 and all distances $l°1$ to 6 have been observed. Point A is fixed, the bearing to point D is fixed in order to orient the network and the a priori standard errors of the observations are estimated. There are, therefore, fifteen observations and six unknowns to be determined, i.e. the eastings and northings of points B, C and D. The method of solution is as follows: (1) assume provisional coordinates for all points. These may either be scaled graphically from a plan or computed using a selection of the measurements; (2) using the provisional coordinates, compute values for the angles $\alpha°1$ to 8 and the distance $l°1$ to 6; and (3) set up an observation equation for each individual observation.

For a distance l_{ij}:

$$-\cos \beta_{ij}dN_i - \sin \beta_{ij}dE_i + \cos \beta_{ij}dN_j + \sin \beta_{ij}dE_j$$
$$= (l^o_{ij} - l^c_{ij}) + v \tag{6.23}$$

or

$$P dN_i - Q dE_i + R dN_j + S dE_j = (l^o_{ij} - l^c_{ij}) + v \tag{6.24}$$

For bearing φ_{ij}

$$\frac{\sin \beta_{ij}}{l_{ij}}dN_i - \frac{\cos \beta_{ij}}{l_{ij}}dE_i - \frac{\sin \beta_{ij}}{l_{ij}}dN_j + \frac{\cos \beta_{ij}}{l_{ij}}dE_j$$
$$= (\varphi^o_{ij} - \varphi^c_{ij}) + v \tag{6.25}$$

For angle α_{ijk}:

$$\left(\frac{\sin \beta_{ik}}{l_{ik}} - \frac{\sin \beta_{ij}}{l_{ij}}\right)dN_i + \left(\frac{-\cos \beta_{ik}}{l_{ik}} + \frac{\cos \beta_{ij}}{l_{ij}}\right)dE_i$$

$$+ \frac{\sin \beta_{ij}}{l_{ij}}dN_j - \frac{\cos \beta_{ij}}{l_{ij}}dE_j - \frac{\sin \beta_{ik}}{l_{ik}}dN_k + \frac{\cos \beta_{ik}}{l_{ik}}dE_k$$

$$= (\alpha^o_{jik} - \alpha^c_{jik}) + v \tag{6.26}$$

For a position

$$dN_i = (N^o_i - N^c_i) + v \tag{6.27}$$

$$dE_i = (E^o_i - E^c_i) + v \tag{6.28}$$

where β_{ij} = direction of line ij, dN_i, dE_i = the unknowns, the corrections to the provisional coordinates, and v = residual.

These observation equations can be expressed in matrix form:

$$
\begin{array}{ccccccc}
A & & X & = & b & + & V \quad (6.29)
\end{array}
$$

$$
\underset{15}{\begin{bmatrix} \text{Matrix of} \\ \text{coefficients} \end{bmatrix}}^6 \underset{6}{\begin{bmatrix} \text{Vector of} \\ \text{unknowns} \end{bmatrix}}^1 \underset{15}{\begin{bmatrix} \text{Vector of} \\ \text{o–c terms} \end{bmatrix}}^1 \underset{15}{\begin{bmatrix} \text{Vector of} \\ \text{residuals} \end{bmatrix}}^1
$$

The solution is found by forming the normal equations.

$$A^T W A X = A^T W b \tag{6.30}$$

where W is the weight matrix, the diagonal elements of which are equal to $1/\sigma^2$ where σ refers to the a priori standard error of the observation.

The normal equations may then be solved using a Choleski's triangular decomposition method or by matrix inversion.

$$X = (A^T W A)^{-1} A^T W b \tag{6.31}$$

The column vector X therefore contains the corrections (dN_i, dE_i) to the provisional coordinates of the unknown points. Using these new values for the coordinates, the entire computational procedure is repeated until there is no further change in the coordinates.

Usually when this method is used a complete error analysis of the results is carried out. By this process information about the precision and reliability of the coordinates can be obtained. For further reading on this aspect of the method see Ashkenazi,[20,21] Ashkenazi et al.,[22,23] and section 3.4.2.

6.3.2 Detail surveys

After the main control survey has been observed and computed, a detail survey is carried out in order to locate the positions of features which are to be presented on the map.

6.3.2.1 Tacheometry

The most common method of carrying out a detail survey by ground methods at scales smaller than 1 : 500 is that of stadia tacheometry. The basic principles have been outlined in section 6.2.2.2.

In practice, the observational and booking procedure should be as follows. The theodolite should be set up and levelled over the point from which observations are to be taken. This point may coincide with one of the main control survey stations, or more commonly be one of a series of subsidiary control stations which have been established by traversing from the main control. By this process not only are any errors contained within the control network, but also the positions of the subsidiary control will be closer to the detail which is to be surveyed.

The circle of the theodolite should then be oriented until a horizontal circle reading of 0° coincides with the direction to an adjacent control point. This is known as the reference object (RO). This procedure simplifies the subsequent plotting of the field observations. The height of instrument should be recorded.

The staffman is then directed to the various points which are to be mapped. The staff is held vertical over the point and the stadia readings, vertical angle and horizontal circle reading to the point are observed from the theodolite. Several techniques for speeding-up the field recording of the stadia hair readings

Table 6.3 Booking and reduction of tacheometric readings

Observer:	TJMK		Station B
Booker:	DRG		Height of Instrument: 1.690 m
Date:	22/9/85		Reduced Level: 58.35 m

Point	Horizontal circle	Staff readings U M L	Staff intercept (S)	Vertical circle	Horizontal distance (D_H) ($100 \cos^2 \theta$)	V ($50\ s$ $\sin 2\theta$)	Reduced level
A(RO)	0° 00′	—	—	—	—	—	—
1	10° 30′	1.430 1.215 1.000	0.430	91° 00′	42.99	0.75	59.58

have been advocated.[24] Assuming that a pocket calculator, preferably programmable, will be used to reduce the observations, the most efficient field method is to set the bottom hair to the 1 m or 2 m point of the staff. Using this method, the mental calculation of the stadia intercept becomes much quicker.

The booking procedure for stadia observations is illustrated by Table 6.3. It is also essential to draw, whilst in the field, a good sketch-map of the area being surveyed. This is invaluable when the results are being plotted.

6.3.2.2 Chaining

This is the simplest form of detail surveying. The method involves measuring the lengths of the sides of a series of triangles or braced quadrilaterals. Points of detail are then picked-up by measuring offsets from these lines. The procedure is illustrated in Figure 6.33.

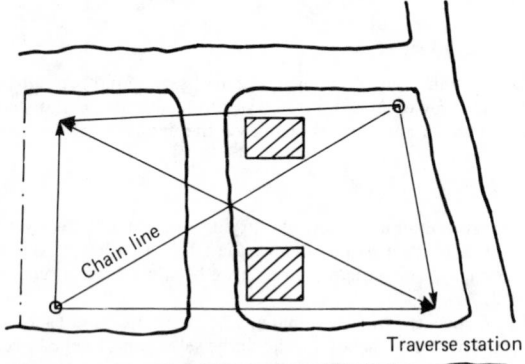

Figure 6.33 Chaining

In this case, two traverse stations from an existing control survey are within the boundary of the site which is to be surveyed.

The first stage in the survey involves breaking-down this existing control into either well-conditioned triangles or braced quadrilaterals. In this case, a braced quadrilateral is chosen. The inclusion of diagonal measurements provides an independent field check on the measurements. These 'chain lines' are normally measured with a steel band.

The second stage in the survey process involves the measurement of offset or ties from these chain lines in order to pick-up

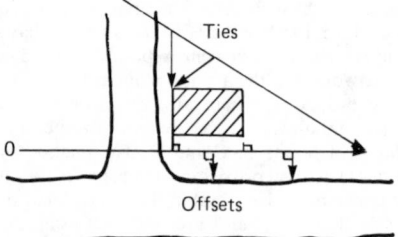

Figure 6.34 Offsets and ties

the survey detail. Figure 6.34 illustrates the distinction between offsets and ties.

Ties, which involve two measurements to one point, are normally employed when the offset distance is long and, hence, the accuracy with which the perpendicular to the chain line can be set out is low. Offset and tie measurements are normally made with a fibreglass tape.

A standard method of booking is conventionally adopted for chain surveying. Details of this and other points relating to chain surveying can be found in any of the standard surveying textbooks listed in the bibliography.

6.3.3 Vertical control surveys

In general, all civil engineering projects require not only planimetric control, but also vertical or height control points. The two most common methods of obtaining this height control are by levelling and by trigonometrical heighting.

6.3.3.1 Levelling

Levelling is the name given to the process of determining, by means of a surveyor's level, the height of a point above some datum, normally mean sea-level.

As is the case with horizontal control, it is important to work from 'the whole to the part'. It is therefore normal practice to design a levelling control framework in a hierarchical manner, in order to contain small errors. Typical maximum allowable misclosures for level loops in one such hierarchical design are shown in Table 6.4.

The basic principle of levelling involves taking horizontal backsight (BS), foresight (FS) and intermediate sight (IS) readings, as defined by the line of collimation of the level, on to vertical staves as shown in Figure 6.35. The difference between successive readings indicates the difference in height between points. By this process it is therefore possible to determine the reduced level (RL) of a series of points.

Table 6.4 Accuracy of levelling (*K* is the distance in kilometres)

Order	Maximum allowable misclosure (m)
1st	$0.004\sqrt{K}$
2nd	$0.008\sqrt{K}$
3rd	$0.012\sqrt{K}$
4th	$0.024\sqrt{K}$

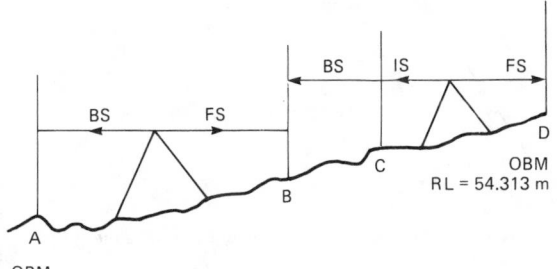

OBM
RL = 51 m

Figure 6.35 Levelling

The reduced level of a point is defined as its height above some datum. In the UK, the fundamental datum established by the Ordnance Survey is mean sea-level at Newlyn, Cornwall, known as Ordnance Datum. A further series of points of known height have been established throughout the UK and these are known as Ordnance bench-marks (OBM).

The results of a levelling operation are, by convention, booked in a standard manner. Two methods of booking are used. The rise and fall method (Table 6.5) is usually employed when running lines of levels between bench-marks in order to establish additional supplementary height points. The height of collimation method (Table 6.6), on the other hand, is more suitable for tasks such as recording cross-sectional information, in which many intermediate sights have been taken.

Contouring. A contour is an imaginary line joining points of equal elevation. The level can be used in a variety of ways for contouring. One of the simplest methods is by means of grid levelling.

With this technique, the area to be contoured is covered by an imaginary grid of lines forming squares of 10, 20 or 30 m. The level is set-up in a central position and levels are then taken to a site temporary bench-mark (TBM) and at the intersections of the grid lines, as shown in Figure 6.36.

Contours may then be interpolated either graphically or mathematically from the grid of levels.

Quantity determination. The method of grid levelling provides a convenient means of determining earthwork quantities of, for example, borrow pits.

The depth of cut (*h*) at each intersection point is established. This will be equal to the difference between the ground-level and the proposed formation level of the borrow pit. Figure 6.37 illustrates a simple example for a 20-m grid of levels.

The volume of excavation consists of a series of rectangular prisms each having a base area (*A*), in this case equal to 400 m². The total volume (*V*) for the general case is therefore:

$$V = \frac{A}{4}(\Sigma h_1 + 2\Sigma h_2 + 3\Sigma h_3 + 4\Sigma h_4) \tag{6.32}$$

Table 6.5 Rise and fall method of booking

	Backsight	Intermediate sight	Foresight	Rise	Fall	Reduced level	Remarks
	2.345					51.000	OBM
	1.935		0.632	1.713		52.713	B
		1.213		0.722		53.435	C
			0.335	0.878		54.313	OBM
Σ	4.280		0.967	3.313	0.000		

Checks: ΣBS $-\Sigma$FS $=3.313$ m $\quad \Sigma$RISE $-\Sigma$FALL $=3.313$ m.
First RL $-$ last RL $=3.313$ m

Table 6.6 Height of collimation method of booking

	Backsight	Intermediate sight	Foresight	*Height of Collimation (HC)		Reduced level (RL)	Remarks
	2.345			53.345		51.000	OBM
	1.935		0.632	54.648		52.713	B
		1.213				53.435	C
			0.335			54.313	OBM
Σ	4.280		0.967				

Checks: ΣBS $-\Sigma$FS $=3.313$ m $\quad$ First RL $-$ last RL $=3.313$ m
*Height of collimation $=$ reduced level $+$ backsight
RL $=$ HC $-$ FS or IS

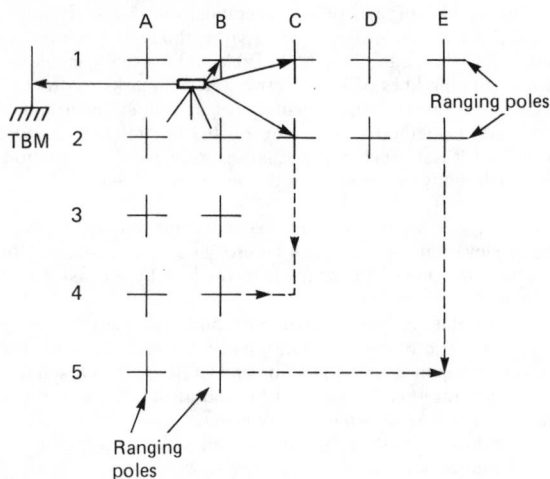

Figure 6.36 Grid levelling

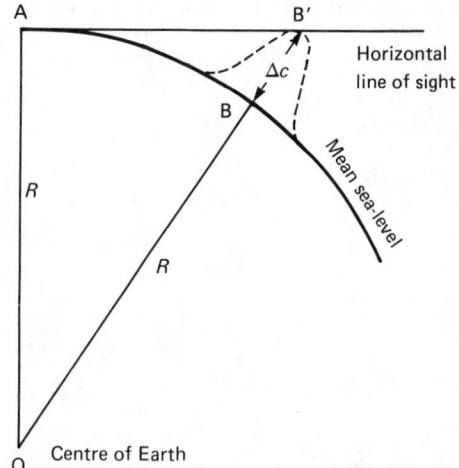

Figure 6.38 Trigonometrical heighting: Earth curvature

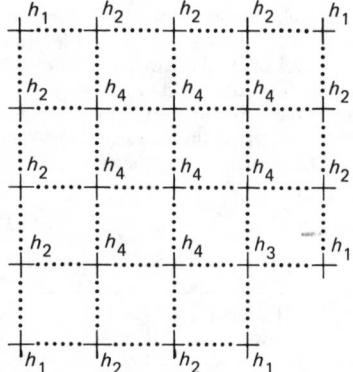

Figure 6.37 Quantity determination using a grid of levels

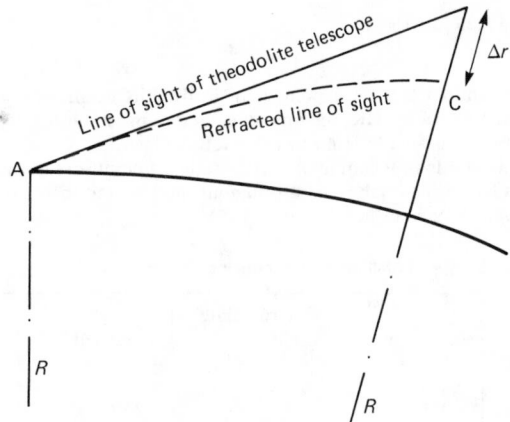

Figure 6.39 Trigonometrical heighting: atmospheric refraction

where h_1 = depth of cut used once, h_2 = depth of cut used twice, etc.

The volume determination assumes that the ground slope between grid intersections is constant. Clearly, therefore, by reducing the size of the grid, the accuracy of the quantity determination will be increased.

6.3.3.2 Trigonometrical levelling

For projects where the acceptable accuracy requirements of the height control are lower than would be obtained by levelling, the method of trigonometrical heighting is normally used.

The method is based on the measurement of the vertical angle between two points by means of a theodolite, together with either the slope or horizontal distance. By simple geometry, the difference in height can therefore be calculated. As the distance between the points increases, however, the effects of two phenomena, Earth curvature and refraction, become more significant. Figures 6.38 and 6.39 illustrate the highly exaggerated effect of these phenomena.

In Figure 6.38, a distant point B′ appears too low by an amount Δc. A positive Earth curvature is therefore necessary. It can also be seen in Figure 6.39 that the effect of refraction is to refract the line of sight of the telescope so that a distant point C

appears too high by an amount Δr. A negative refraction correction is therefore required.

It can be shown that the difference in height between A and C can be determined by the equation:

$$\Delta h_{AC} = d\tan\theta + (\Delta c - \Delta r) \tag{6.33}$$

where $\Delta c - \Delta r = \dfrac{d^2(1 - 2K)}{2R} \tag{6.34}$

and K = coefficient of refraction $\simeq 0.07$, R = radius of Earth = 6378 km, d = horizontal distance, and θ = vertical angle.

If $k = 0.07$ and d is in kilometres, then $\Delta c - \Delta r = 0.0675d^2$ m.

The results obtained using this simple approach can be increased significantly by observing vertical angles from both ends of a line. Reciprocal observations of this type, particularly if they are observed simultaneously, can eliminate completely the necessity for Earth curvature and refraction corrections. In this case, the difference in height between A and C can be computed by the general expression

$$\Delta h_{A_C} = \frac{d \tan \theta_A + \theta_c}{2} + \frac{(h_{S_A} - h_{T_C}) - (h_{S_C} - h_{T_A})}{2} \qquad (6.35)$$

where θ_A = vertical angle from A to C, θ_c = vertical angle from C to A, h_{S_A} = height of signal at A, h_{S_C} = height of signal at C, h_{T_A} = height of theodolite at A, h_{T_C} = height of theodolite at C

6.3.4 Deformation monitoring surveys

Deformation monitoring surveys using conventional land-surveying methods, photogrammetric survey methods or specialist geotechnical methods are used to quantify the amount by which an engineering structure has moved both vertically and horizontally over specific periods of time.

The information which surveys of this type provide can be of critical importance and may be used to indicate that either: (1) the structure is stable and consequently safe; (2) the structure is experiencing small random movements which are not imposing significant forces on the structure; (3) the structure is experiencing small localized systematic deformations, e.g. as caused by seasonal effects, which may or may not be significant; (4) lastly and most significantly, the structure is experiencing deformation which is increasing as a function of time. This may indicate that remedial measures have to be taken, in some cases immediately, in order to avoid catastrophic consequences.

For a variety of reasons, surveys of this type have been increasingly more important in recent years. Several interrelated factors have been responsible for this increase in interest. Firstly, for many types of structures it is now a mandatory element of the civil engineering process that a monitoring survey be commissioned. Notable in this respect in this country are, for example, reservoirs, which now have to be monitored following the implementation of the 1975 Reservoirs Act. A further example, from Switzerland, is the requirement by the Swiss Federal Government for all dams with a height greater than 15 m and a cubic capacity greater than 50 000 m³ to be monitored.[25] A second factor which has led to this increase in interest has been the speed of development both in the manufacture of precise survey instrumentation and also in high-speed, low-cost computing facilities. This, in conjunction with improvements in very elegant and highly sophisticated software for the design and analysis of surveying observations, now provides the surveyor and engineer with a highly accurate measurement system. Thirdly, the tolerances to which civil engineers are now designing and constructing many modern structures necessitates a much higher order of accuracy in the initial dimensional control and also in the subsequent deformation monitoring. Indeed, this factor may have acted as a 'springboard' for many of the developments in software and instrument design. It is also hoped that to some extent civil engineers are now more aware of the possibilities offered by modern survey techniques and are therefore requesting this type of survey more frequently than was the case in the past.

Most of the instruments which are used for deformation monitoring projects have been discussed in several of the previous sections of this chapter. It will therefore be the aim in this section to concentrate on two other important aspects of deformation monitoring surveys. The first is the design of suitable reference and monitoring points. Clearly, when very small displacements are being measured it is crucial that the points from which measurements are being recorded are not subject to movement. The second aspect which will be discussed is the computational processes associated with the horizontal control networks which are often used to quantify the extent of any structural deformation. Finally, it should be noted that deformation monitoring can often be carried out, in some cases more efficiently, using close-range photogrammetric techniques. Reference should be made to Chapter 7 for further details.

6.3.4.1 Design of reference and monitoring points

The full accuracy potential currently offered by modern surveying equipment for measuring small structural displacements can only be fully realized if care and attention is paid to the design of the reference points from which observations will be taken. Equally important is the need to design appropriate monitoring points to be placed on the structure under investigation.

Reference points. Two distinct types of survey reference point can be identified. The first, typically a survey pillar, forms the reference framework for the horizontal and vertical control measurements, whilst the second, a steel or concrete pile, forms the datum for levelling observations.

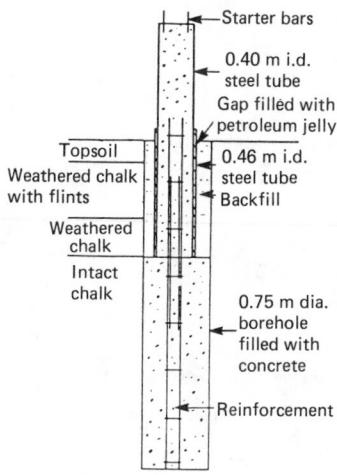

(a) Survey pillar (Penman and Charles[26])

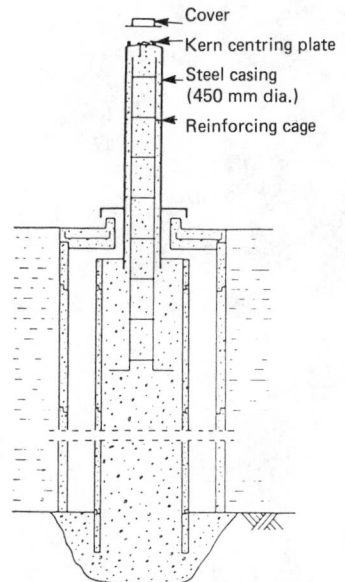

Scammonden reference monument

(b) Survey pillar (Penman and Charles[26])

Figure 6.40 Survey pillar designs

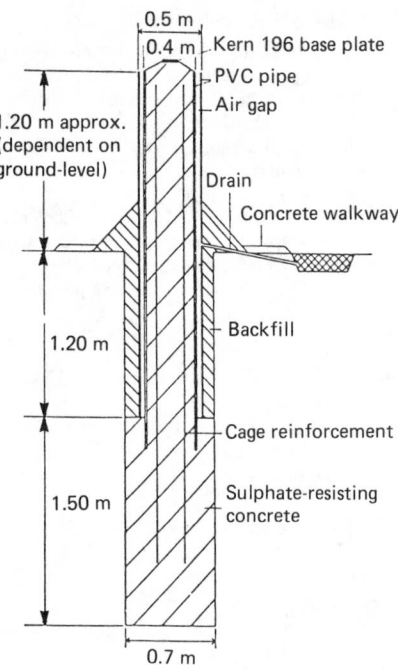

(c) Cross section of Base Line Pillar (Deeth *et al*[27])

Figure 6.40 (Continued) Survey pillar designs

Three distinct designs of survey pillar are illustrated in Figure 6.40. The main requirement in siting pillars is the need to ensure that they are founded on stable ground outside the zone of influence of the structure under investigation. It is, however, often difficult to assess whether this is the case before observations are recorded. It is therefore important to incorporate into the design an insulating gap which ensures that the central pillar is not in contact with the surrounding ground. This should ensure that the effects of diurnal or seasonal earth movements are minimized.

A further common design feature is the incorporation of some system of forced centring. The two most common forced centring systems for deformation monitoring are the Wild ball centring system, and the Kern system. Both designs are illustrated in Figure 6.41. Use of either system should ensure that errors from this source do not exceed ±0.1 mm.

This concept of an insulating sleeve around the reference point is also evident in the levelling datum designs illustrated in Figure 6.42. Figure 6.42(a) can be seen to consist of a steel foot driven into the ground at the bottom of a 5 m deep borehole. This steel foot is connected to a central rod which extends

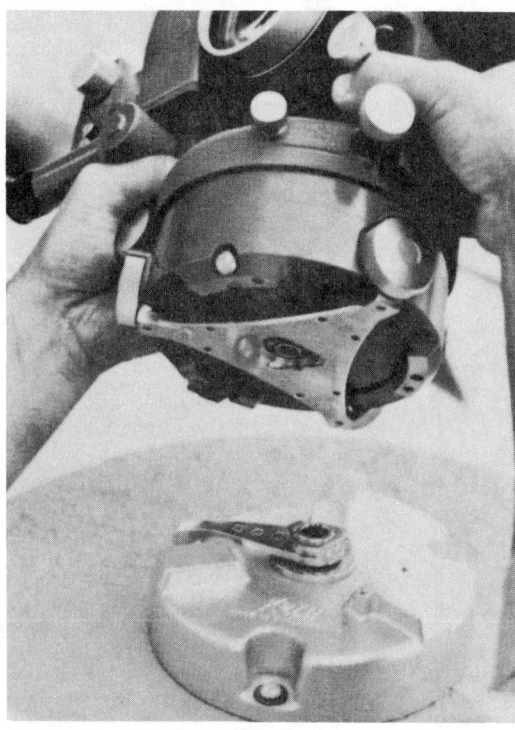

(a)

(b)

Figure 6.41 Pillar centering systems (a) Kern system; (b) Wild system

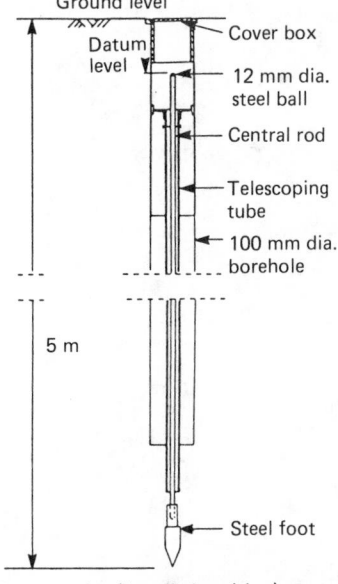

Ground level

Cover box

Datum level

12 mm dia. steel ball

Central rod

Telescoping tube

100 mm dia. borehole

5 m

Steel foot

Datum point (installed position)

(a) Datum (Penman and Charles[26])

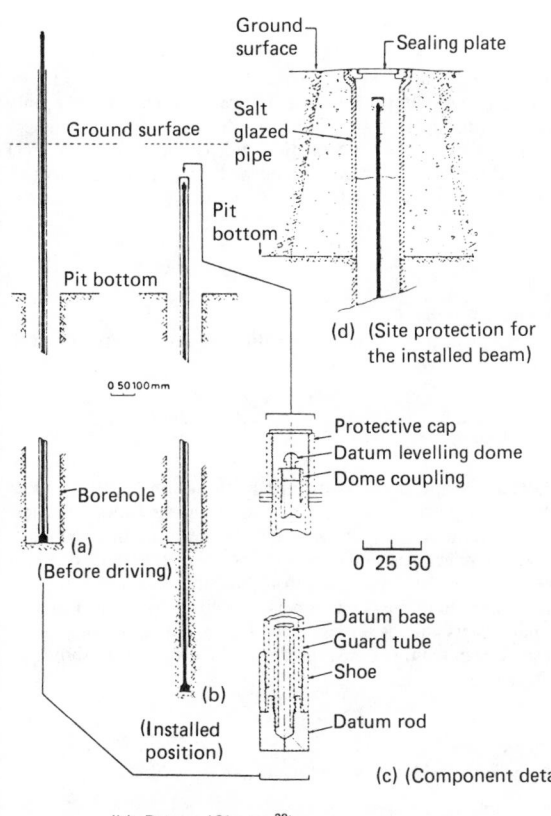

Ground surface

Sealing plate

Salt glazed pipe

Ground surface

Pit bottom

Pit bottom

(d) (Site protection for the installed beam)

0 50 100 mm

Protective cap
Datum levelling dome
Dome coupling

Borehole

(a)
(Before driving)

0 25 50

Datum base
Guard tube
Shoe

(b)
(Installed position)

Datum rod

(c) (Component detail)

(b) Datum (Cheney[28])

Figure 6.42 Levelling datum designs (from Penman and Charles,[26] and Cheney[28])

almost to ground-level. In order to minimize the potential effect of ground movement influencing the datum, the central rod is surrounded by a telescopic sleeve. An alternative design is illustrated in Figure 6.42(b). In this case, the inner datum rod consists of a 10 mm bore galvanized tube surrounded by a guard tube of 25 mm bore. The guard tube is driven into a 150 mm diameter hole which has been bored to a depth of 5.5 m. The datum point consists of a dome-shaped steel ball about 0.1 m below ground-level. The ball is covered by a protective cap and by a sealing plate (possibly a manhole cover), at ground-level. Further details relating to the installation procedure may be found in Cheney.[28] It is also important to install a sufficient number of datums to enable any settlement or uplift of the datums to be detected. Ideally, three should be installed.

Monitoring points.   The main requirement of a survey monitoring point is that it can be either permanently affixed to, or precisely relocated, on the structure being investigated. Again, two distinct types of monitoring point can be identified, those to which angle/distance measurements will be taken and those which will be used as precise levelling settlement points.

The measurement of angles and distances to monitoring points on the structure will normally require that the target points be designed so that they are capable of accepting both conventional survey targets and also corner cube reflectors. The simplest approach, and that commonly used for dam deformation work, is to build a series of pillars on the structure. The pillars are fitted with a suitable forced centring system which enables both survey targets and corner cube reflectors to be interchanged very accurately. An example of such a system is illustrated in Figure 6.43.[29] An alternative approach may be used in situations where the distances to be measured are short. In these cases it may be possible to permanently affix targets and reflectors to the structure. Details of one particular arrangement which involves the use of reflex acrylic reflectors is reported by Kennie.[13] Cheney[30] also discusses the use of permanently mounted reflectors, in this instance for use with the Kern Mekometer ME 3000.

For settlement measurements, one of the most common designs of monitoring point is that which has been designed by the Building Research Establishment (BRE). The components of the settlement system are illustrated in Figure 6.44. It can be seen that the system consists of four components: (1) a stainless steel socket 65 mm long by 22 mm diameter which is grouted into the structure; (2) a detachable settlement bolt; (3) a protective Perspex cap; and (4) an alloy wrench to allow removal of the cap. The main advantages of such a system are, firstly, a high degree of accuracy in relocating the settlement point; Cheney[28] states that the settlement bolts may be repositioned in the socket to within 0.03 mm. The second advantage is the level of protection offered to the settlement point from accidental or other damage.

6.3.4.2 *Computational processes*

Horizontal control surveys for deformation monitoring purposes are essentially no different from control surveys for other purposes (e.g. national control surveys). Whilst they may differ in size (generally much smaller), and accuracy requirements (generally much higher), the basic principles in terms of observational and computational techniques are common. It is not surprising to note, therefore, that in recent years the type of control survey generally adopted for deformation purposes consists of a mixed set of angle and distance observations or a 'survey network'. The advantages of such an approach in reducing scale and orientation errors have been discussed previously. However, the primary advantage of this approach occurs when the data are subjected to a rigorous least squares

Figure 6.43 Pillar target point (from Egger[29])

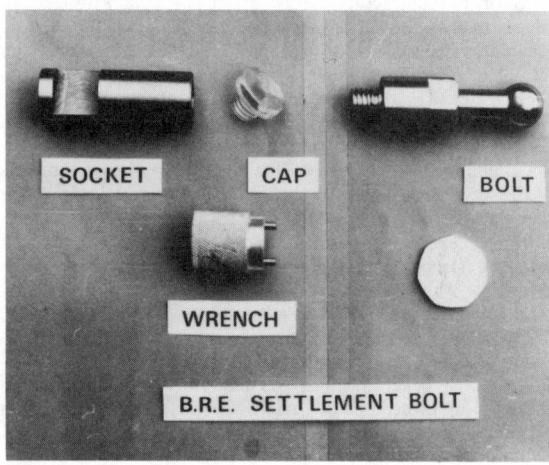

Figure 6.44 Building Research Establishment settlement bolt system

adjustment using the method of variation of coordinates (see section 6.3.1.4). This not only enables the most probable values of the positions to be arrived at, but also statistical data about the precision and reliability of the network to be determined. Furthermore, by interpreting these statistical indices and then the results of a network adjustment from two different epochs it is possible to evaluate the statistical significance of any changes which are observed. The various stages in the computational process are shown in Figure 6.45.

The various statistical indices which can be derived from a

variation of coordinates adjustment is well documented in the literature, e.g. Cooper.[1] It is therefore intended to make only brief mention of the primary features of each statistical index.

Three distinct types of statistical indices can be identified: (1) those which are concerned with network precision; (2) network reliability; and (3) for deformation analysis.

(i) Network precision.

Residuals
The residuals (**v**) of the observation equations are obtained from:

$$\mathbf{v} = \mathbf{AX} - \mathbf{b} \tag{6.36}$$

where A, **X** and **b** are as defined in Equation (6.29). In cases where a large number of observations exist, the ratio $\mathbf{v}/\sigma$, where σ refers to the standard error of the observation, may be used in order to reject suspect observations. For example, if $\mathbf{v}/\sigma$ is greater than 2.5, this indicates that there is only a 1.2% probability that this variation is caused by random observational effects and is more likely to be indicative of a poor observation. On this basis, therefore, the observation should be rejected.

Unit variance
The quantity σ_0^2 is computed from the expression:

$$\sigma_0^2 = \frac{\mathbf{v}^{\mathrm{T}} W v}{m - n} \tag{6.37}$$

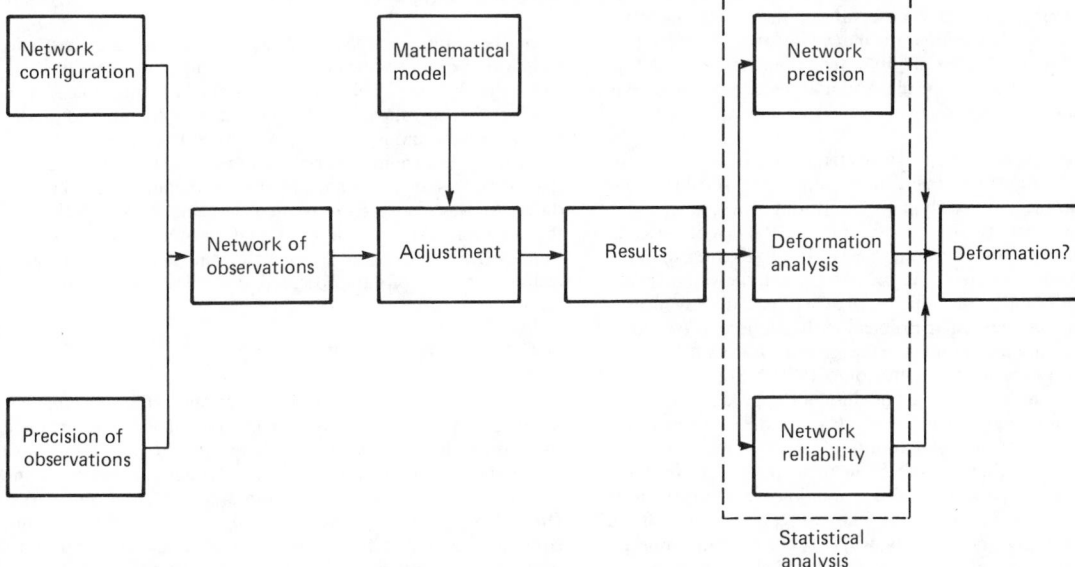

Figure 6.45 Stages of computation: deformation monitoring surveys

where m is the number of observations, n the number of unknowns and W is the weight matrix. It can be shown that, theoretically, σ_0 (the standard error of unit weight) should equal 1. A large departure from unity is usually indicative of some gross blunder, whilst a small departure may give an indication whether the initial a priori standard errors of the observations have been correctly estimated. In certain circumstances it is possible to correct the initial standard errors by a 'trial and error' process in order to bring the value of σ_0 closer to unity.

Variance – covariance matrix

The variance – covariance matrix (σ_{xx}), is an extremely useful by-product of the adjustment process. It is computed from the following expression:

$$\sigma_{xx} = \sigma_0^2 (A^T W A)^{-1} \qquad (6.38)$$

The diagonal elements of the matrix $(\sigma_{E_i}^2, \sigma_{N_i}^2, \ldots)$, refer to the variances or squares of the standard errors associated with the unknowns E_i, N_i, etc. It is therefore an extremely useful means of measuring the precision of the network. The off-diagonal elements, or covariances $(\sigma_{E_i,N_i}, \ldots)$ may be used in order to construct error ellipses.

Error ellipses

The 'absolute error ellipse' of a calculated point is an ellipse defined by σ_{min} and σ_{max} with the semimajor axis $a = \sigma_{max}$ and the semiminor axis $b = \sigma_{min}$. When drawn around a point the ellipse is generally interpreted as depicting the region within which there is 39% confidence that it contains the position of the point. The values of σ_{max} and σ_{min} are determined by the following expressions:

$$\sigma_{max} = \tfrac{1}{2}(\sigma_E^2 + \sigma_N^2) + [\tfrac{1}{4}(\sigma_E^2 - \sigma_N^2) + \sigma_{EN}]^{1/2} \qquad (6.39)$$

$$\sigma_{min} = \tfrac{1}{2}(\sigma_E^2 + \sigma_N^2) - [\tfrac{1}{4}(\sigma_E^2 - \sigma_N^2) + \sigma_{EN}]^{1/2} \qquad (6.40)$$

Two drawbacks exist with error ellipses of this type. Firstly the size of the error ellipse is not 'invariant' and depends on the position of the point which is chosen as the origin. Secondly, no information is given about any inter-station correlation which may exist. An alternative type of error ellipse which overcomes the first drawback is the 'relative error ellipse'. In this case the ellipse is computed on the basis of the variances and covariances of the differences in coordinates between points. It is therefore a measure of the relative positional accuracy between the points.

A posteriori standard errors

The a posteriori or post adjustment standard errors of the adjusted quantities are a further set of 'invariant' statistics which can be computed. For example, the a posteriori standard error associated with the adjusted distance between any two points in the network is given by:

$$\sigma_l^2 = [P\,Q\,R\,S] \begin{bmatrix} 4 \times 4 \text{ Sub-matrix} \\ \text{of } \sigma_{xx} \end{bmatrix} \begin{bmatrix} P \\ Q \\ R \\ S \end{bmatrix} \qquad (6.41)$$

where P, Q, R and S are the trigonometric functions defined in Equation (6.24).

(ii) Network reliability. Network reliability in this context is defined as the ability of the network to detect gross errors in the observed quantities. Ashkenazi[21] has suggested that the following statistic should be used in order to assess the reliability of the network:

$$\text{reliability} = \frac{\text{a posteriori standard error of the observation}}{\text{a priori standard error of the observation}}$$
$$(6.42)$$

Theoretically the greater the number of observations indirectly affecting a particular observation, the lower the ratio should be. Thus low ratios are indicative of high reliability, whilst a ratio of one would indicate complete unreliability. Ashkenazi suggests that the maximum acceptable value of this ratio should be 0.9

(iii) Deformation analysis. The statistical indices discussed so far relate to a single set of network observations. A much more common problem in deformation monitoring work, however, is the situation where two or more sets of observations of the same network, taken at different times, exist. In these circumstances, further analysis is required in order to determine whether any differences in coordinates which exist are statistically significant.

Currently this particular problem is the subject of considerable research interest. Following the second FIG symposium on Deformation Measurements in Bonn, 1978, a committee was established in order to investigate different approaches into the analysis of deformation measurements using the same measurement data. The provisional results of this committee's work were reported by Chrzanowski.[31] The primary problems which this group, and others, have been examining are, firstly, what is the most appropriate means of eliminating systematic effects from the data sets, and secondly, what are the most appropriate types of statistical test to apply to the recorded deformations in order to test their significance.

The first problem, that of eliminating systematic effects, such as a scale bias in EDM measurements, has been discussed by Ashkenazi *et al.*[23] and Ashkenazi.[21] In cases where a systematic bias is considered to be present in one set of data, the systematic bias parameters (scale, orientation and translation) may be determined by means of a least squares four-parameter Helmert transformation. By the use of this procedure it is therefore possible to eliminate from one data set a series of systematic effects which, if not eliminated, may be mistaken as systematic deformations.

The main dangers associated with the application of this technique to networks is that it may lead to the removal of a genuine systematic deformation under the mistaken believe that it is caused by observational error. It has therefore been suggested by Dodson,[32] that it is preferable for any systematic effects to be eliminated by careful instrument calibration, or by designing the network so that 'the influence of such effects does not influence the acceptance or rejection of any hypothesis of deformation'.

The second problem, that of devising the most appropriate statistical tests, has also been examined by Ashkenazi and Dodson.[33] The approach which they have adopted involves comparing the detected coordinate differences with the corresponding coordinate standard errors. The basis of this approach can be illustrated by considering two points P and Q in a fictitious network. If the position of P in the network is stronger (i.e. more accurately determined because of a good geometrical configuration and higher precision observations) than point Q, then quite clearly a smaller difference between two successive values of the coordinates of P, may be more significant than a larger difference corresponding to point Q. Thus periodic coordinate differences may only be considered to be significant if the standard error associated with the deformation at that point is also examined. Therefore, if:

$$\Delta^2 = (\Delta E^2 + \Delta N^2)^2 \tag{6.43}$$

where Δ is the computed deformation based on planimetric coordinate differences ΔE, and ΔN, and the standard error in Δ is given by:

$$\sigma_\Delta^2 = (\sigma_{\Delta E}^2 + \sigma_{\Delta N}^2)/2 \tag{6.44}$$

where $\sigma_{\Delta E} \approx \sqrt{2}\sigma_E$ and $\sigma_{\Delta N} \approx \sqrt{2}\sigma_N$ $\qquad$ (6.45)

then, the ratio Δ/σ may then be used in order to test the significance of any deformation. For example, if the ratio Δ/σ is 2 (95% probability level) then there is only a slight chance (19:1 against) that the movement has been caused by random observational errors and is much more likely to have been caused by movement of the point. In contrast, a ratio of 0.7 would indicate that there was an even chance (50% probability level) that the difference was caused by movement, it being equally likely that the variation was simply a manifestation of a random observational error. Ashkenazi[22,23] discusses the application of this technique to sets of data observed at Cattleshaw reservoir.

6.4 Computers in surveying

Computers have, throughout their development, been extremely important in the fields of surveying and mapping. Initially, their use was almost exclusively restricted to the 'number crunching' requirements of large organizations carrying out geodetic computations or the adjustment of major control frameworks. Operations of this type were carried out on large mainframe computers in batch mode. Whilst slow and cumbersome to operate by modern computing standards, these early computers offered enormous benefits to the surveyor: their computational speed, and their ability to deal with the application of rigorous computational techniques, such as the method of least squares, on a much greater scale than was possible previously. In recent years, the trend, in computational terms, has been towards the development of more 'user friendly' software which can be operated on smaller computers, particularly microcomputers (see Milne[34] and Walker and Whiting[35]).

Whilst the use of computers for solving mathematical problems in surveying continues to develop, the main thrust area of interest in recent years has been in the development of interactive graphics systems for processing and displaying surveying and mapping data. Thus, the emphasis is becoming concentrated more on the capability of the computer to store, search, retrieve and display digital map data, than on computational speed. Such systems offer considerable benefits both in terms of speed of access and flexibility of use, e.g. the ability to be able to display selectively map data at a user-defined scale. Indeed, there is little doubt that over the next decade the storage of map data in digital form will continue to expand at an ever-increasing speed as the costs of computer memory drop and the availability of data in digital form increases. The engineer will therefore come into increasing contact with these types of data, primarily at the design stage of a project, using large-scale digital maps and ground-modelling systems, but also at the project planning stage using data from land/geographic information systems.

6.4.1 Digital mapping and ground-modelling systems

In the UK, responsibility for national mapping lies with the Ordnance Survey. The Ordnance Survey are also involved in the production of digital maps (Thompson[36] and Logan[37]), and actively considering the possibility of producing a national digital topographic database (Thompson[38] and Rhind[39]). The task of compiling large-scale digital maps to cover the entire country is an enormous task and one which requires substantial resources if it is to be produced within a reasonable time-scale. To date, the success of the Ordnance Survey in providing large-scale digital mapping coverage has been very limited (10 to 15% coverage). It is hoped that as technology develops and, perhaps more significantly, as funding becomes more available, that this situation will improve. Nevertheless, in spite of the lack of

Ordnance Survey-derived data, many organizations are involved in the acquisition, processing and sale of digital survey data in the commercial sector. Several sophisticated suites of software have also evolved for the manipulation of this digital survey data, particularly in conjunction with civil engineering design information. Two examples which are representative of this range of software are MOSS and the Eclipse Interactive Ground Modelling System.

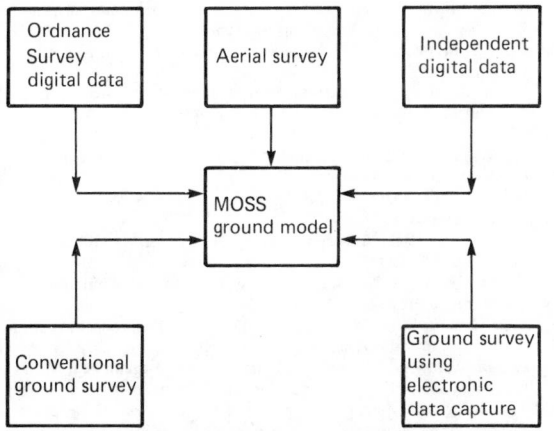

Figure 6.46 Alternative methods of creating a MOSS ground model

6.4.1.1 MOSS system

MOSS is a combined surveying and engineering design system which records information in the form of three-dimensional 'strings'. These 'strings' consist of a series of linked coordinated points representing features such as kerbs, roads, railway lines, or ground level detail, such as contours. Each string is given a label and by covering an area with a sufficient number of strings and/or point information it is possible to represent the terrain in digital form in the form of a MOSS 'model'. A wide variety of survey methods can be used in order to create a string ground-model, and Figure 6.46 illustrates a few of the options which are available. The design features within the system also enable a model of any proposed works to be generated. By comparison of both models it is therefore possible to generate other data such as earthwork quantities.

The MOSS system was initially launched in the mid 1970s by a consortium of several county councils within the UK. Although it was not initially developed with the production of large-scale plans as its primary aim, it has been upgraded and enhanced since its inception and is now widely used for large-scale survey purposes. Figure 6.47 is an example of the use of MOSS in this mode. MOSS is, however, much more widely used as an integrated survey and design system, rather than as a stand-alone survey system, and it has been used in this mode for a wide variety of projects including: highway design (Hougham[40] and Fawcett[41]), railway design (Bedingfield and Craine[42]) and the design of land reclamation schemes (Wilson[43]). A new interactive version of MOSS has recently been launched by MOSS Systems Ltd, a company which was formed by several members of the original consortium in 1983. This version will offer considerable benefits to its users, particularly in terms of speed and ease of use.

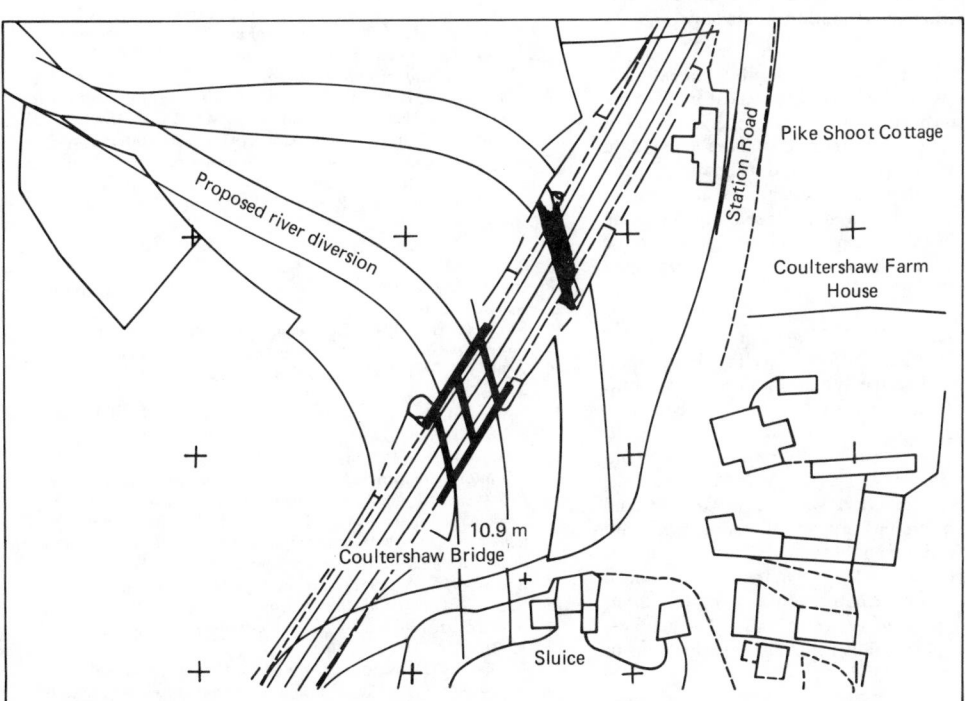

Figure 6.47 MOSS model taken from a graphics VDU showing an Ordnance Survey digital map, updated with a small road improvement, bridge detail, proposed river diversion, and culvert. (MOSS Systems Ltd, model courtesy of West Sussex County Council.)

6.4.1.2 Eclipse system

The Eclipse interactive ground modelling system is a further example of an integrated system for survey processing, digital mapping and engineering design. In common with the MOSS system the software is designed so that it can be installed on a variety of different computer systems. However, unlike MOSS which is designed for use with a variety of mainframe (IBM, ICL, etc.), minicomputers (DEC/VAX, Prime, etc.) and engineering workstations (Sun, Apollo, etc.), the Eclipse system has been designed primarily for Wang minicomputers and, recently, for the Wang PC and the IBM PC/AT.

The program modules within the system include software for processing and editing – interactively – survey control and detail observations. Data input can be by manual keyboard entry, or automatically by transfer from one of the electronic data collectors currently on the market. The design software includes facilities for road design, drainage design, building layout design and land reclamation design. Throughout, facilities exist for deriving additional information such as areas and earthwork quantities. For high-quality output a series of options within the draughting software enable the user to enhance cartographically the screen image in order to produce final contract drawings if necessary.

Other systems currently available which perform operations similar to those outlined include the Wild System-9, Intergraph, HASP, AXIS and ProSurveyor.

6.4.2 Land information systems

Land information systems (LIS), are currently one of the major growth areas in computing as applied to surveying and mapping. They are of particular relevance to engineers involved with feasibility and planning studies. The concept of a LIS can best be described by considering the definition offered by Andersson:[44]

> A land information system is a tool for legal, administrative and economic decision making, and as an aid for planning and development which consists, on the one hand, of a database containing spatially referenced land related data and, on the other hand, of procedures and techniques for the systematic collection, updating, processing and distribution of the data. The base of a LIS is a uniform spatial referencing system for the data in the system which also facilitates the linking of data within the system with the other land related data.

Two other terms, Geographic Information System (GIS) (Hallam[45]) and Urban Information System (UIS) (Parker and Bray[46]), are also used to describe systems which operate in a similar manner but within a regional or local urban area respectively.

All of these systems are very much in their infancy at present. However, many organizations concerned with, for example, public utilities (gas, electricity, waste water disposal) and land taxation/valuation/registration are already investing significant sums of money in the development of such systems. A review of the history and future possibilities for LIS in the UK is provided by Dale.[47] The LIS will have enormous impact on the amount of information available to the engineer and planner. The management and efficient use of this data is the challenge which has to be faced in the future.

6.5 Acknowledgements

The author would like to thank the following persons and organizations who helped in the preparation of this chapter:

Geotronics (UK) Limited, Moss Systems Limited, Spectra Physics (UK) Limited, and Wild Heerbrugg Limited for permission to reproduce photographs of their products. Also, to Mrs Veronica Brown for her valuable comments on the initial draft.

References

1   Cooper, M. A. R. (1987) *Control surveys in civil engineering.* Collins.
2   Mikhail, E. M. and Gracie, G. (1981) *Analysis and adjustment of survey measurements*, Van Nostrand Reinhold.
3   Cooper, M. A. R. (1982) *Modern theodolites and levels*, 2nd edn. Granada.
4   Miller, R. M. (1969) 'Accuracy of measurement with steel tapes', *Building Research paper CP51/69*, Building Research Station, Watford.
5   Hodges, D. J. and Greenwood, J. B. (1971) *Optical distance measurement.* Butterworths.
6   Smith, J. R. (1970) *Optical distance measurement*, Crosby Lockwood Staples.
7   Jaakola, M. (1971) 'Survey with the Tellurometer MA-100', *Survey Review*, **159**, 29–34.
8   Froome, K. D. (1971) 'Mekometer: EDM with submillimetre resolution, *Survey Review*, **161**, 98–112.
9   Meir-Hirmer, B. (1978) 'Mekometer ME3000. Theoretical aspects, frequency calibration, field tests', *Proc. Int. Symp. on EDM and the influence of atmospheric refraction.* IAGG, Wageningen, May.
10   Murmane, A. B. (1982) 'The use of the Mekometer ME3000 in the Melbourne and Metropolitan Board of Works', *Proc. of the 3rd Int. Symp. on deformation measurements by geodetic methods.* Budapest.
11   Burnside, C. D. (1982) *Electromagnetic distance measurement*, 2nd edn. Granada.
12   Kennie, T. J. M. (1983) 'Some tests of retroreflective materials for electro-optical distance measurement', *Survey Review*, **207**, 3–12.
13   Kennie, T. J. M. (1984) 'The use of acrylic retroreflectors for monitoring the deformation of a bridge abutment, *Civil Engineering Surveyor*, **9**, 6, 10–15.
14   Schwendener, H. R. (1972) 'Electronic distances for short range: accuracy and checking procedures', *Survey Review*, **164**, 273–281.
15   Ashkenazi, V. and Dodson, A. H. (1975) 'The Nottingham multi-pillar baseline', *Proc. of the 26th General Assembly of the Int. Ass. of Geodesy*, Grenoble.
16   Sprent, A. and Zwart, P. R. (1978) 'EDM calibration – a scenario', *Australian Surveyor*, **29**, 3, 157–169.
17   Ministry of Defence (1978) *Military engineering* Part 2: Field Survey, **29**, 1–8.
18   Murray, G. A. (1980) 'Lasers and Dinorwic', *Civil Engineering Surveyor*, **5**, 6, 6–11.
19   Schofield, W. (1979) 'The effect of various adjustment procedures on traverse networks', *Civil Engineering Surveyor*, **4**, 4, 13–19.
20   Ashkenazi, V. (1968) 'The solution and analysis of large geodetic networks', *Survey Review*, **146**, 166–173.
21   Ashkenazi, V. (1981) 'Least square adjustment: signal or just noise', *Chartered Land Surveyor*, **3**, 42–49.
22   Ashkenazi, V., Dodson, A. H. and Crane, S. A. (1980a) 'Monitoring deformations to millimetre accuracy', *Proc. of the FIG Commission 6 Symp. on engineering surveying.* London.
23   Ashkenazi, V., Dodson, A. H., Skyes, R. M. and Crane, S. A. (1980b) 'Remote measurement of ground movements by surveying techniques', *Civil Engineering Surveyor*, **5**, 4, 15–22.
24   Redmond, F. A. (1951) *Tacheometry.* The Technical Press.
25   Egger, K. (1983) 'Geodetic measurement and the unusual behavior of the Zeuzier arch dam', *Land and Minerals Surveying*, **1**, 10, 15–21.
26   Penman, A. D. M. and Charles, J. A. (1971) *Measuring movements of engineering structures*, Building Research Station Publication No. CP 32/71.
27   Deeth, C. P., Dodson, A. H. and Ashkenazi, V. (1979) 'EDM: accuracy and calibration', Symposium on EDM, Polytechnic of the South Bank. London.
28   Cheney, J. E. (1973) 'Techniques and equipment using the surveyor's level for the accurate measurement of building

movements, *Proc. of the British Geotechnical Society Symposium on Field Instrumentation*. London.

29 Egger, K. (1970) *Precision measurement with special reference to deformation of dams*, Kern Instrument Company.

30 Cheney, J. E. (1980) 'Some requirements and developments in surveying instrumentation for civil engineering monitoring', *Proc. of the FIG Commission 6 Symp. on Engineering Surveying*. London.

31 Chrzanowski, A. (1981) 'A comparison of different approaches into the analysis of deformation measurements', *Proc. of the FIG Conference*. Montreaux.

32 Dodson, A. H. (1984) 'Pre-analysis and design of a measurement scheme', *Land and Minerals Surveying*, **2**, 1, 13–19.

33 Ashkenazi, V. and Dodson, A. H. (1978) *Measuring deformations by surveying techniques*. Seminar, University of Nottingham.

34 Milne, P. H. (1984) *Basic programs for land surveying*. Spon.

35 Walker, A. S. and Whiting, B. M. (1983) 'Multitudinous micros and micros in mapping', *Land and Minerals Surveying*, **1**, 1, 34–42.

36 Thompson, C. N. (1978) 'Digital mapping in the Ordnance Survey 1968–78. *Proc. of the ISP Commission 4 Symp. on 'New Technology for Mapping'*. Ottowa.

37 Logan, I. T. (1981) 'Ordnance Survey digital mapping', *Proc. of the 1st UK National Conf. on Land Surveying and Mapping*. Paper G4.

38 Thompson, C. N. (1979) 'The need for a large-scale topographic database', *Proc. of the Conf. of Commonwealth Surveyors*. Cambridge.

39 Rhind, D. W. (1981) 'Digital data banks and digital mapping', *Proc. of the 1st UK National Land Surveying and Mapping Conference*. Paper G2.

40 Hougham, P. (1980) 'The application of MOSS to roadworks in a shire county', *Proc. 2nd Int. MOSS Conference*. Bournemouth, Paper 9.

41 Fawcett, D. S. (1980) 'The design of a complex minor interchange', *Proc. 2nd Int. MOSS Conference*. Bournemouth, Paper 8.

42 Bedingfield, P. G. and Craine, G. S. (1981) 'An automated survey and integrated design system', *Proc. of the 1st UK National Land Surveying and Mapping Conference*. Paper F3.

43 Wilson, P. (1980) 'Land reclamation using MOSS', *Proc. 2nd Int. MOSS Conference*. Bournemouth, Paper 4.

44 Anderson, S. (1981) 'LIS – what is that?', *Proc. of the FIG Congress*, Montreux, Paper 301.1.

45 Hallam, C. A. (1979) 'The USGS geographic information, retrieval and analysis system: overview', *Proc. of the 39th ACSM meeting*. Washington DC, 229–246.

46 Parker, D. and Bray, D. (1983) 'The surveyor and urban information systems', *Chartered Land Surveyor*, **4**, 3, 4–15.

47 Dale, P. (1984) 'Land information systems – which way in the UK?' *Land and Minerals Surveying*, **2**, 6, 313–316.

Bibliography

Bannister, A. and Raymond, S. (1986) *Surveying*, 6th edn. Pitman, 510pp.

Methley, B. O. F. (1986) *Computational models in surveying and photogrammetry*. Blackie.

Olliver, J. G. and Clendinning, J. (1978) *Principles of surveying*, Vols I and II, 4th edn. Van Nostrand Reinhold.

Schofield, W. (1984) *Engineering surveying 2*, 2nd edn. Butterworths, 276pp.

Shepherd, F. A. (1981) *Advanced engineering surveying*, Edward Arnold, 276pp.

Shepherd, F. A. (1984) *Engineering surveying*, 2nd edn, Edward Arnold, 370pp.

Uren, J. and Price, W. F. (1978) *Surveying for engineers*. Macmillan, 298pp.

Uren, J. and Price, W. F. (1984) *Calculations for engineering surveying*. Van Nostrand Reinhold, 309pp.

7

Photogrammetry and Remote Sensing

T J M Kennie BSc, MAppSci (Glasgow), ARICS, MInstCES
Lecturer in Engineering Surveying, University of Surrey

Contents

7.1 Introduction

Photogrammetry and remote sensing are two indirect methods of obtaining both quantitative and qualitative data about the Earth's surface or other features of interest. Since some debate exists regarding the demarcation between the two subjects, the following broad distinction will be used throughout this chapter. Photogrammetry will be considered to be concerned with the scientific methods of obtaining reliable measurements from ground or airborne imagery (primarily photographic) in order to produce a precise representation (graphical or digital) of the feature of interest. Remote sensing, in contrast, will be defined as encompassing those methods of detecting variations in radiant energy from the Earth's surface using airborne or satellite sensors. Interpretation (either visually or using computer techniques) of these recorded patterns is used to create thematic maps. The aim of this chapter will therefore be to: (1) provide an introduction to the fundamental principles of photogrammetry and remote sensing; (2) review the current state of development of instrumentation in both fields; and (3) examine the application of both techniques to problems in civil engineering.

7.2 Principles of photogrammetry

Since measurements may be taken from both air and ground images (normally photographs) two separate branches of the discipline are generally recognized: aerial and close range (or terrestrial) photogrammetry.

Aerial photogrammetry is a well-established technique in civil engineering for the production of topographic maps. Aerial photographs produced for such purposes can be obtained either with the optical axis of the camera pointing, nominally, vertically downwards so producing vertical aerial photographs or, with the axis intentionally tilted, to produce oblique aerial photographs. The former are almost exclusively used nowadays for photogrammetric purposes, although the latter have received a revival in recent years with the advent of analytical techniques. Vertical aerial photographs are normally acquired to achieve a 60% forward and a 20 to 25% lateral overlap in coverage. This enables a stereoscopic view of the terrain to be obtained and also a stereoscopic 'model' to be produced. The latter forms the basis of almost all of the photogrammetric instruments and techniques discussed in section 7.3.

Close-range, terrestrial or, to use a less attractive but commonly used alternative, nontopographic photogrammetry, is concerned with photographs taken on, or near the ground, rather than from an aerial platform. This branch of photogrammetry has developed considerably in the past 20 yr or so and has been applied to many problems in civil engineering. A selected number of these applications are presented in section 7.4.

A further distinction which is becoming used increasingly as a means of classification, both in aerial and close-range photogrammetry (CRP), is that between the analogue and analytical approaches to the subject. Apart from a few exceptions, until relatively recently photogrammetric instruments and techniques consisted of optical and mechanical analogue systems which were used for the production of a graphical end product (map, plan, elevation, section, etc.). However, the emphasis in photogrammetry is now moving towards analytical systems which use rigorous mathematical models to simulate the problem under investigation. By using these mathematical models in conjunction with modern computers, new photogrammetric instruments have evolved which offer significant advantages over their analogue counterparts in terms of speed of operation, accuracy and flexibility. Furthermore, the output products of these computer-based systems consist not only of graphical products, but also of other information sources such as digital terrain models.

7.2.1 Geometry of a single photograph

A photograph differs fundamentally in geometric terms from a map (unless the terrain is flat and the photograph vertical). Image displacement causes scale variations over the format due to the alteration in the position of points, compared with their corresponding map positions, as a result of the effects of ground relief and tilt of the photograph at the moment of exposure. The geometrical influence of both factors is illustrated by Figure 7.1 which shows that a tall tower is not imaged as a single point, as it would be on a map, but rather as a displaced line t–t and t′–t′ on the vertical and tilted photographs respectively. In this case, the displacement is greatest on the vertical photograph. A similar displacement will exist for all points above the datum level e.g. point X. It is also evident that the tilted and vertical photographs are coincident at point i, the isocentre, and that tilt displacement radiates from this point. This radial nature of the displacements can be used to devise a simple method of plotting from a pair of photographs.

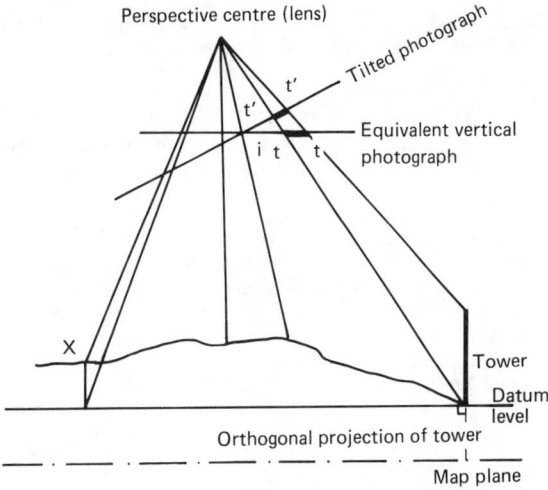

Figure 7.1 Relief and tilt image displacements on a vertical and near-vertical aerial photograph

The effect of image displacement on scale can be seen in Figures 7.2 and 7.3 which illustrate the separate results of tilt and variation in ground relief. It can be seen that, unlike a map which has a constant scale, the combined effects of relief and tilt produce a photograph which will be of constantly changing

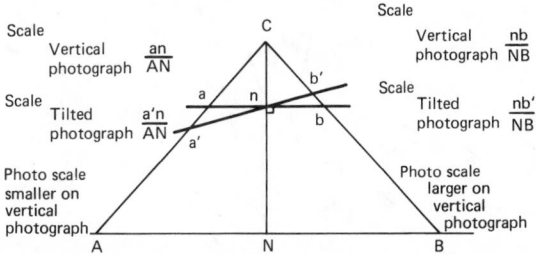

Figure 7.2 Scale changes on a vertical aerial photograph caused by variations in the elevation of the ground

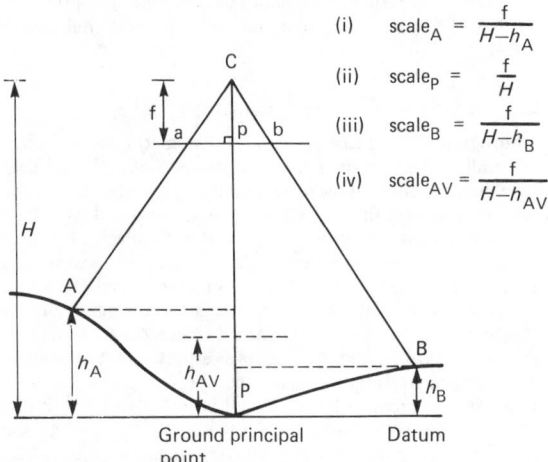

(i) $scale_A = \dfrac{f}{H-h_A}$

(ii) $scale_P = \dfrac{f}{H}$

(iii) $scale_B = \dfrac{f}{H-h_B}$

(iv) $scale_{AV} = \dfrac{f}{H-h_{AV}}$

Figure 7.3 Scale changes on a tilted and vertical aerial photograph

scale, although an average nominal value for the scale can be calculated (often referred to as the contact scale).

7.2.2 Stereoscopy and parallax

Stereoscopic viewing and the measurement of the perceived stereomodel is fundamental to photogrammetry and thus of great importance. If two photographs taken from different viewpoints of the same area are viewed simultaneously, the difference in position of a common image point on the two photographs results in a discrepancy in image coordinates and this leads to the concept of x parallax and y parallax.

All points appearing on successive overlapping photographs exhibit x parallax in the direction of flight and this is the principal reason why a stereomodel exists. It can be seen from Figure 7.4 that the combined movement of point A across the

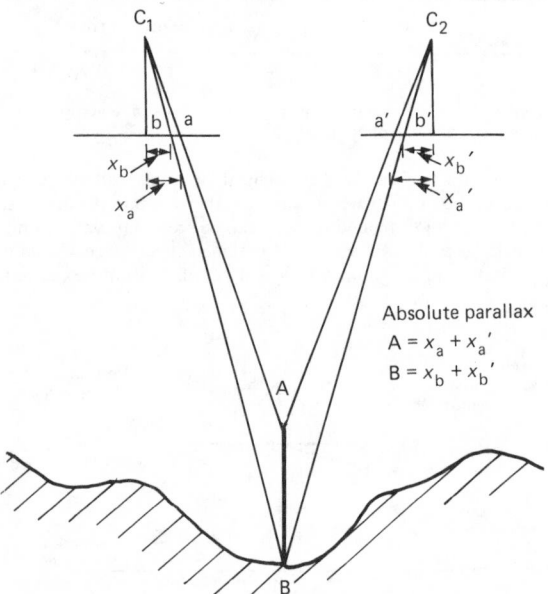

Figure 7.4 Variations in x-parallax on an overlapping pair of aerial photographs

focal plane of both photographs relative to the central axis is greater than for point B, thus point A is defined as exhibiting a greater absolute x parallax than B. It is also evident from the diagram that the absolute x parallax of a point is related to the height of that point and that it increases with increasing height. This simple concept forms the basis of the methods used to derive heights from aerial photographs. Section 7.3.3 outlines one such technique.

In contrast, y parallax is not related to height variations. It is produced if there is a difference in orientation of the two photographs at the time of exposure. The effect of y parallax on a stereomodel can be eliminated by reproducing the original camera orientations to ensure that corresponding rays from the two photographs will intersect. Such a process is termed the 'relative orientation' phase in setting up a stereoplotter.

7.2.3 Analytical photogrammetry

Many photogrammetric operations which were previously carried out by graphical or analogue methods are now being performed digitally using analytical instruments and computational techniques. Since most analytical methods involve some form of coordinate transformation, it is first necessary to examine the coordinate reference systems which are used. On the plane of the photograph, the reference system known as the 'image space' system is defined with respect to the principal point and the fiducial marks, as shown in Figure 7.5. It may also be important to take into account any variation in position which exists between the fiducial centre (the intersection of the fiducial marks) and the principal point (where the perpendicular line from the lens intersects the focal plane). This is defined by a camera calibration procedure. On the ground, positions are usually referred to a cartesian coordinate system, e.g. the National Grid. This is illustrated by Figure 7.6 in which X_c, Y_c and Z_c represent the 'object space' coordinates of the camera exposure station.

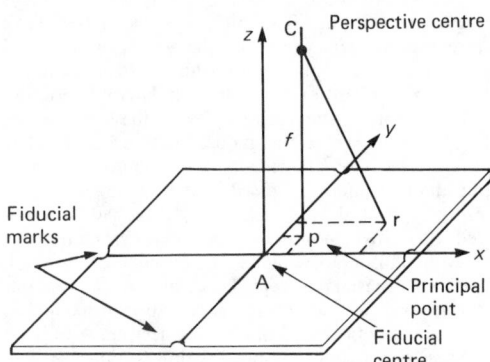

Figure 7.5 Image space coordinate reference system

An appreciation of the two basic assumptions which are used in order to define positions is also essential. Firstly, the line in space joining a point in the object space – the perspective centre of the camera lens and the position of the point on the photograph – is assumed to be a straight line, i.e. line p–C in Figure 7.6. In analogue stereoplotters, this assumption is imposed by optical projection or mechanical 'space rods'.

Analytically, it is expressed by means of the collinearity equations for a *single* photograph:

$$x_r = -f\frac{M_{11}(X_r-X_c)+M_{12}(Y_r-Y_c)+M_{13}(Z_r-Z_c)}{M_{31}(X_r-X_c)+M_{32}(Y_r-Y_c)+M_{33}(Z_r-Z_c)} \qquad (7.1)$$

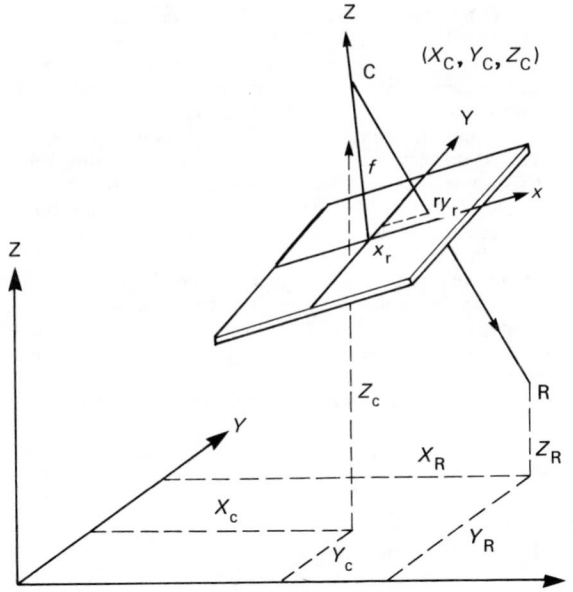

Figure 7.6 Object space coordinate reference system and collinearity condition

$$y_r = -f\frac{M_{21}(X_r - X_c) + M_{22}(Y_r - Y_c) + M_{23}(Z_r - Z_c)}{M_{31}(X_r - X_c) + M_{32}(Y_r - Y_c) + M_{33}(Z_r - Z_c)} \qquad (7.2)$$

where $M_{11} \ldots M_{33}$ represent terms of an orthogonal rotation matrix and the other terms are as defined in Figure 7.6. The collinearity equations are used extensively in analytical photogrammetry to relate image coordinates, ground coordinates and the coordinates of the perspective centre. The technique may then be used to derive the coordinates of a series of unknown positions.

A further analytical technique which is used by photogrammetrists is that defined by the coplanarity equations. Coplanarity equations can be established for a pair of photographs and they may be used to reconstruct the orientation of the cameras at the moment of exposure, i.e. an analytical method of relative orientation. In order to satisfy the condition, the two exposure stations C_1 and C_2, the object point R and its corresponding image points r_1 and r_2 are assumed to all lie in a common plane, known as the epipolar plane (Figure 7.7).

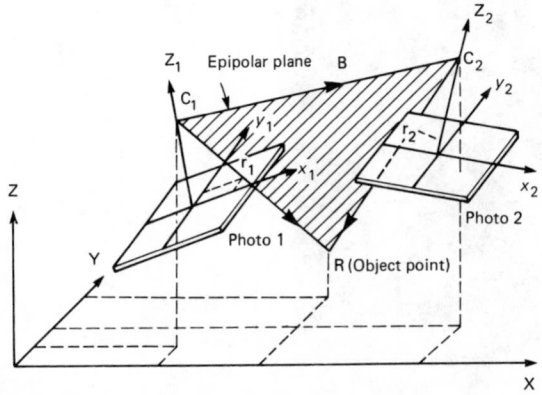

Figure 7.7 Coplanarity condition in analytical photogrammetry

7.3 Photogrammetric instrumentation

7.3.1 Cameras

Most photogrammetric measurements are recorded from photography produced by a high-quality camera specifically designed for photogrammetric purposes. The main feature which distinguishes this type of camera from others is that the internal geometrical characteristics are known precisely. Thus, data relating to the lens distortion, focal length and principal point location are established, and monitored periodically by a camera calibration procedure.

Both aerial and close-range cameras are used for data acquisition. Aerial mapping cameras are normally classified on the basis of the angular field of view of the lens, a parameter which relates directly to both the focal length (f) and the format size of the camera. The focal length of an aerial camera is defined as the distance from the optical centre of the lens to the image plane. For the general case of a 230×230 mm format, the classification shown in Table 7.1 can be produced. Normal angle (NA) photography is only occasionally used and it can be seen from Figure 7.8 that the principal advantage of the super wide

Table 7.1 Classification of aerial mapping cameras

Camera type	Approximate angular field of view (°)	Approximate focal length (mm)	Examples
Super wide angle (SWA)	120	90	Zeiss (Ober.) RMK A 8.5/23 Wild RC-10 (Super Aviogon lens) Zeiss (Jena) MRB 9/2323
Wide angle (WA)	90	150	Zeiss (Ober.) RMK A 15/23 Wild RC-8 Wild RC-10 (Universal Aviogon lens) Zeiss (Jena) MRB 15/2323
Normal angle (NA)	60	300	Zeiss (Ober.) RMK A 30/23 Wild RC-10 (Normal Aviogon lens)

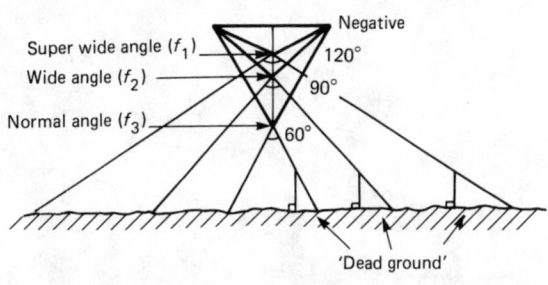

Figure 7.8 Variation in ground coverage of normal, wide and superwide angle cameras

angle (SWA) lens is its greater ground coverage for a given flying height. This is particularly important since it not only reduces the number of photographs required to cover an area, but also reduces the number of control points which are required for controlling the mapping. However, because SWA photography is more susceptible to the production of 'dead ground' (loss of detail because of the screening effect of features), a compromise solution is normally adopted by using a camera which enables wide angle (WA) photography to be produced.

Three distinct types of close range or terrestrial photogrammetric camera are currently available: (1) metric; (2) stereometric; and (3) nonmetric. The metric camera models include those in which a camera and a theodolite form one integrated unit, the classical phototheodolite, and those where the camera is a separate unit from the theodolite. The latter case tends to be more common and the UMK10/1318 is a typical example

(Figure 7.9). Details of the characteristics of several cameras of this type are listed in Table 7.2.

The second design, the stereometric system, consists of two matched camera units mounted on each end of a base bar, typically 0.4 to 3 m long. The camera axes are set at right angles to the bar and since the relative position and orientation of the cameras at the instant of exposure are known, the subsequent analysis phase is considerably simplified. The Zeiss (Oberkochen) range of cameras illustrated by Figure 7.10 are representative of this design. Further details of other cameras of this type are listed in Table 7.3.

A nonmetric camera is one which was not designed specifically for photogrammetry. Although such cameras offer advantages in terms of size and cost, their use is limited, unless analytical techniques are used to compensate for the unstable interior geometry of the camera. A varied selection of nonmetric cameras has been used for photogrammetric purposes including

Figure 7.9 Zeiss (Jena) UMK 10/1318 close-range camera

Table 7.2 Summary of main features of a selection of close-range photogrammetric cameras

Manufacturer	Model	Nominal focal length (mm)	Principal distance (F = fixed, V = variable)	Format (mm)	Minimum range (m)
Wild	P30	165	F	100 × 150	20
Wild	P31	45, 100, 200	F, V, V	100 × 150	7, 1.4, 8
Wild	P32	64	F	65 × 90	3.3
Officine Galileo	Verostat	100	V	90 × 120	2
Zeiss (Jena)	UMK 10/1318	100	V	130 × 180	1.4
Zeiss (Ober.)	TMK 6	60	V	90 × 120	5

(Adapted from Atkinson (1978) 'Photography in non-topographic photogrammetry'. In: Newman (ed.)
Photographic techniques in scientific research, vol. III. Academic Press, London)

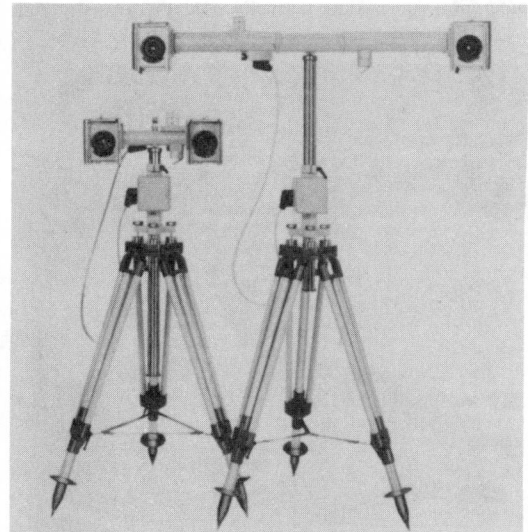

Figure 7.10 Zeiss (Oberkochen) SMK 120 and SMK 40 stereometric camera

medium-priced 'amateur' cameras, such as the Olympus OM-2, Canon TX, Minolta XG-7 and Rolleiflex SL66, as well as the more expensive 'professional' cameras such as the Linhof Technica and the Hasselblad 500EL.

7.3.2 Single-photograph-based instruments

In certain circumstances it may be appropriate to attempt to derive metric plan information from a single photograph. Reasonable results can be obtained provided that relatively flat ground has been imaged on an almost tilt-free photograph.

The camera lucida principle, whereby images of an existing map and the corresponding photograph are superimposed, has been adopted to produce a simple mapping instrument such as the Zeiss (Jena) Sketchmaster and the Bausch and Laumb Zoom Transfer Scope. Adjustments can be made either mechanically or optically in order to remove the effect of minor tilt distortions and to make a correction for small variations in scale between map and photograph. This type of instrument is of particular value for map revision. When detail shown on both photograph and map has been aligned, new details imaged on the photograph can be traced on to the map.

It is also possible to apply a digital approach to measurement on a single photograph using a monocomparator. This instrument essentially consists of a travelling microscope moving

Table 7.3 Summary of main features of stereometric close-range photogrammetric cameras

Manufacturer	Model	Nominal principal distance (mm)	Principal distance (F = fixed, V = variable)	Format (mm)	Focusing range
Wild	C40	64	F	65 × 90	1.5–7.0
Wild	C120	64	F	65 × 90	2.7–∞
Officine Galileo	Veroplast	100	V	90 × 120	2–∞
Zeiss (Jena)	SMK 5.5/0808	56	F	80 × 80	5–∞
Zeiss (Ober.)	SMK 120	60	F	90 × 120	5–∞

(Adapted from Atkinson (1978) 'Photography in non-topographic photogrammetry'. In: Newman (ed.)
Photographic techniques in scientific research, vol. III. Academic Press, London)

along orthogonal axes in order to determine image coordinates. Various refinements to the basic principle have been made; the Surveying and Scientific Instruments PI-1a, for example, employs a miniature CCTV observation system and provides digital readout with the MDR 1S/3 display, whilst the Zeiss PSK-1 instrument uses optical scales based on Moiré fringes. With the aid of a microcomputer and suitable software, a plot can be produced from such equipment.

7.3.3 Approximate solution stereoscopic instruments

When reconnaissance mapping or revision of existing maps from new photography is being carried out, it is often more cost effective to use a simple, approximate stereoscopic instrument rather than a sophisticated and expensive rigorous stereoplotter.

Several manufacturers produce stereoscopic versions of the monoscopic instruments which are based on the camera lucida principle, e.g. the Cartographic Engineering Stereosketch and the Bausch and Laumb Stereo Zoom Transfer Scope. The image of the stereoscopic view of an overlap is superimposed on the map image and photographic plan detail can be traced; no facilities are provided for height measurement. The Cartographic Engineering Radial Line Plotter (Figure 7.11) is a simple, lightweight instrument which can be used for small-scale planimetric mapping. The instrument design is based on the radial line assumption which states that angles measured at the

principal point are true if the photographic tilt and variation in ground relief are below some limiting value (normally 3° and 10% of the flying height). Other approximate instruments include the Zeiss (Oberkochen) G2 Stereocord and the Officine Galileo Stereomicrometer.

A parallax bar (Figure 7.12) is a portable device for obtaining measurements of differences in x parallax which can be used to calculate differences in height. Spot heights at changes of slope can hence be obtained and contours interpolated. The eye observes the measuring mark on the glass plates through a stereoscope. When adjusted to the correct x separation the two marks appear to fuse into a single point which may appear to 'float' above the level of the terrain. In order to make some correction for possible photographic tilt and variation in flying height for the stereopair, a minimum of five control points of known ground height are required for the overlap; ideally eight to ten points will be used. Each of the control points is observed in turn and the micrometer screw is turned to remove x parallax and, hence, bring the floating mark to ground level.

Any y parallax can be eliminated by adjusting the right-hand stage plate in the y direction. The micrometer readings are recorded for all the control points. In order to calculate spot heights, the mean flying height must be determined from scaled measurements on the photographs and the absolute parallax of one point must be determined using a travelling microscope. A computer program is then normally used to calculate crude

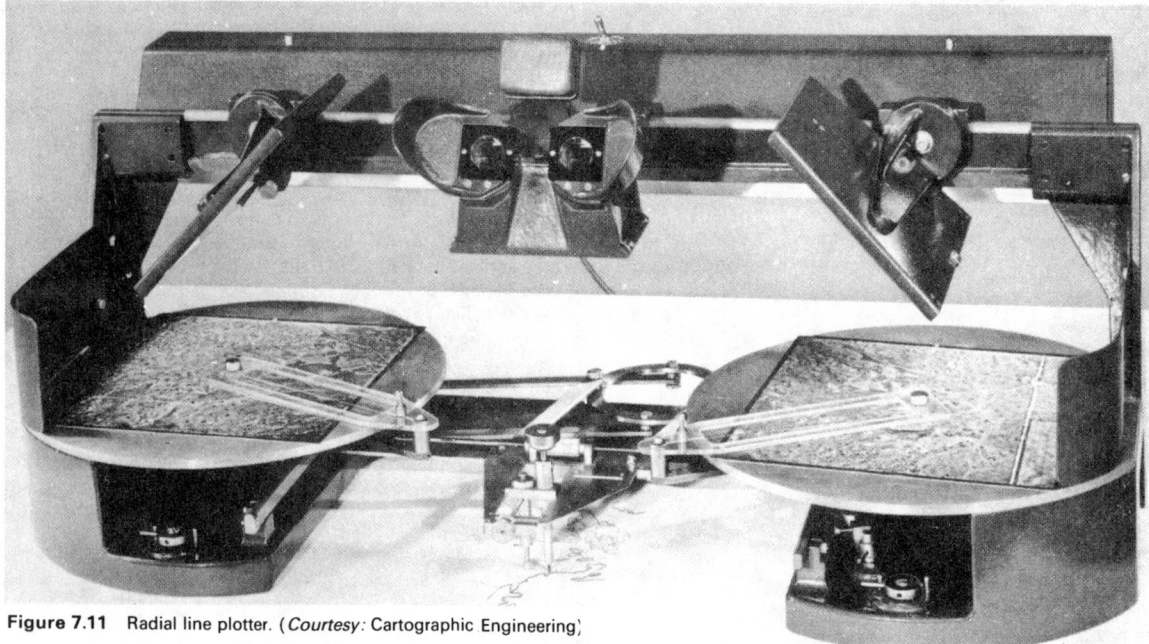

Figure 7.11 Radial line plotter. (*Courtesy:* Cartographic Engineering)

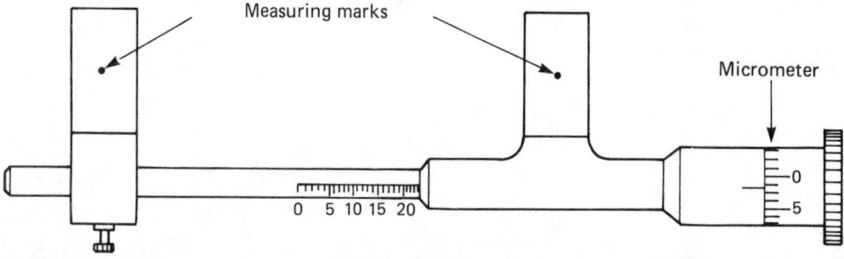

Figure 7.12 Parallax bar

heights from Equation (7.3) and to obtain corrected spot heights from Equation (7.4). If redundant control has been provided, residuals will be available and the heighting accuracy can be investigated.

$$h_B = h_A - \frac{(H - h_A)\Delta_{P_{AB}}}{p_A - \Delta_{P_{AB}}} \qquad (7.3)$$

where h_A = height of A (a reference point of known height), h_B = crude height of point B (point of required height), $\Delta_{P_{AB}}$ = difference in parallax bar readings between A and B (care must be taken with sign) and p_A = parallax of A.

$$h_{B^1} = h_B + a_0 + a_1 x + a_2 y + a_3 xy + a_4 x^2 \qquad (7.4)$$

where h_{B^1} = corrected height of point B and x, y are image coordinates of B, and $a_0 \ldots a_4$ are coefficients, constant for a given overlap. This method of height correction was first proposed by Thompson[1] and was subsequently modified by Methley.[2] Contours may be interpolated between spot heights, continuous reference being made to the stereomodel as they are drawn.

7.3.4 Rigorous solution stereoscopic instruments

7.3.4.1 Analogue stereoplotters

Analogue stereoplotters create and measure an exact three-dimensional model of the ground. Direct optical projection in the form of Multiplex equipment was the earliest method of realizing this objective. The Zeiss (Ober.) DP 2 stereoplotter (Figure 7.13) is a modern version of Multiplex equipment. Plotters such as these, which employ direct optical projection, have the disadvantage of relatively poor viewing conditions and an inflexible plotting scale, usually approximately double the photograph scale. However, their uncomplicated operation and

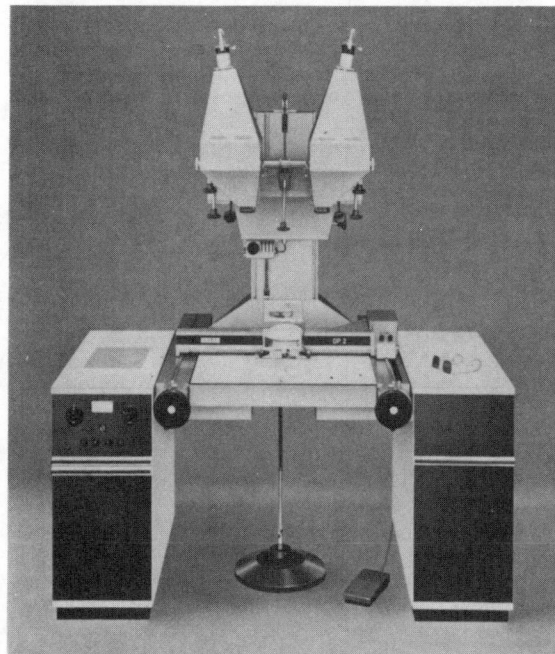

Figure 7.13 Zeiss (Oberkochen) DP-2 stereoplotter. (*Courtesy*: Zeiss (Oberkochen))

minimum of maintenance make them very suitable for medium-scale mapping and map revision, although in recent years they have tended to be superseded by mechanical and analytical instruments.

Mechanical rods replace light rays when a rigorous model is formed in instruments employing mechanical projection (Figure 7.14). Full-sized diapositives are placed in projector heads which can be tilted; a stereoview is obtained through a binocular eyepiece linked to viewing microscopes fitted to each projector. The movement of each microscope is communicated via a tie rod to a sleeve at the top of a space rod which pivots about a gimbal joint representing the projection centre. The two space rods take up the attitude of the space rays from the ground point to the camera stations; they normally intersect in a model point which is connected to a plotting pencil by gears. The principal distance of the taking camera is represented by the separation of the sleeve on the space rod which represents the image point and the gimbal joint; this distance can easily be varied so that mechanical projection plotters can accommodate photography taken with a wide variety of cameras. Considerable flexibility of plotting scales (up to 6 times the photographic scale) is provided by altering the gears linking the model point to the plotting point (Table 7.4).

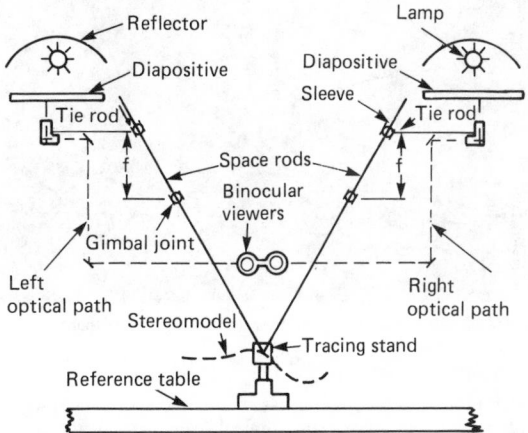

Figure 7.14 Mechanical projection stereoplotter

Table 7.4

Photographic scale	Flying height (m)	Enlargement factor	Mapping scale
1:3000	450	6	1:500
1:5000	750	5	1:1000
1:10 000	1500	4	1:2500
1:25 000	3750	2.5	1:10 000

Digital readout of model coordinates is now standard practice and recent developments include computer-assisted prompts, to guide the orientation procedures, and computer-aided drawing. The Wild A10 stereoplotter (Figure 7.15) is a typical example of a mechanical instrument.

7.3.4.2 Analytical stereoplotters

With the advent of digital computers, especially mini- and microcomputers together with electronic encoders for digitizing the output from stereoplotting equipment, a new generation of

Figure 7.15 Wild A-10 stereoplotter. (*Courtesy*: Wild Heerbrug)

photogrammetric stereoplotter has emerged: the analytical stereoplotter. With this type of instrument a mathematical model replaces the space rods and other mechanical linkages which are found in more conventional analogue stereoplotters. An instrument which is representative of this design is the Officine Galileo Digicart. This instrument consists of a stereoviewing system, linear encoders to digitally record the position of the carriages on which the photographs are mounted, a minicomputer to process the data, and a plotting table. Unlike a conventional analogue instrument where manual elimination of y-parallax is performed initially over the entire model, with this design any parallax existing within the model is continuously eliminated by a feedback loop which ensures that corrections are transmitted to the optics in real time (50 times a second) to provide a model free from parallax. Figure 7.16 illustrates the principle of operation.

The advent of analytical techniques has had a considerable influence on the practical application of photogrammetry to engineering problems. In particular, analytical techniques offer the following advantages over conventional analogue techniques:

(1) Higher accuracy of measurement can be achieved than was previously possible (up to $\pm 3\,\mu m$ at photoscale with suitable care).
(2) Photography taken with differing format sizes and focal lengths can be used. In certain cases, provided suitable

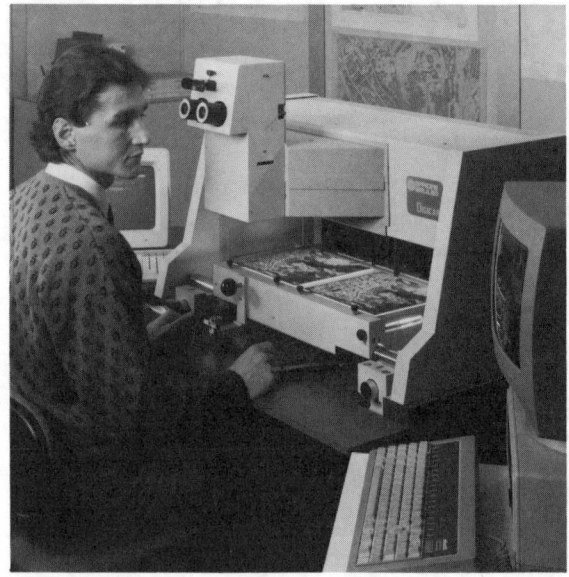

Figure 7.16 Officine Galileo Digicart analytical stereoplotter. (*Courtesy*: Officine Galileo)

precautions are taken to eliminate any systematic errors such as lens distortion, nonmetric cameras can be used.

(3) The setting up of any stereomodel presents little difficulty; most models can be set up in 10 to 15 min by an experienced operator, thus leaving more time for the actual measuring phase.

(4) Photography can be obtained in the field without too great a concern about the orientation of the camera. This can save a great deal of time (as much as 50% over the conventional approach). Where access is difficult, and vertical or normal case geometry is impeded, oblique photography may be taken. Such an approach would have been precluded using traditional methods.

(5) Data are captured in ground coordinates and stored on magnetic disks. This allows for plotting off-line on graphical plotters at any scale. Also, with data being collected digitally rather than graphically, much more flexibility exists in data handling, e.g. the data can be manipulated on a Computer Aided Drafting and Design (CADD) system.

7.3.4.3 Orientation of stereoplotters

Whatever type of equipment is used, before drawing can commence three stages of orientation must be carried out in order to set up the model correctly.

Inner orientation consists of the careful setting of the diapositives on glass stage plates. For analogue instruments, either the principal point or the fiducial marks on both diapositive and register glass must be aligned. Unless the principal distance is fixed (as is the case for direct optical projection instruments), this must be set to the value obtained from the camera calibration certificate. If an analytical plotter is used, the reference marker is taken to each fiducial mark in turn and the coordinates are recorded; the camera principal distance is entered into the computer as are any known lens distortion characteristics.

Relative orientation is next carried out in order to produce a true undistorted model. It does not require any reference to ground control and is achieved in analogue instruments by removing y parallax for clear points of detail selected at five standard positions. The entire model should then be clear of y parallax. For analytical models, y parallax is measured at five or more positions to enable the computer to deduce the movement of the servomotors subsequently required to view a parallax-free model.

Absolute orientation is required in order to scale the model and level it to ground datum. Scaling is basically accomplished by comparing a model distance with a known ground distance and adjusting the machine base in analogue instruments to achieve the required model scale. For heighting, the model and ground heights of a minimum of three points are compared and the model is rotated about the machine X and Y axes until differences in height agree. The machine datum will then be parallel to ground datum and its exact level can be determined. True ground heights can then be read directly from the instrument height scale. Analytically, photo coordinates and corresponding ground coordinates of three or more control points are compared in order to enable the required transformation coefficients to be computed.

The machine is now set up and plotting can commence. For plan detail, the operator moves the floating mark to follow image detail, keeping it in contact with the model surface by constantly adjusting its height. In correspondence, the plotting pencil traces out the detail on the map sheet. In order to plot a contour, the floating mark is kept at the required constant height and is moved in contact with the model surface so that the contour line is drawn on the plot. A similar procedure is followed for the analytical plotter as far as the movement of the floating mark by the operator is concerned, but computer-

assisted plotting offers additional options such as defining the end points of a straight line and instructing the pencil to join them.

The machine plot will only contain information that can be obtained directly from the stereopair and field completion is required to add ground detail which might be obscured on the photographs, e.g. within woodland or below the eaves of houses, and also to assist annotation of, for example, place names and road classifications.

7.3.4.4 Orthophoto systems

An orthoprojector is used to generate a photograph that is the geometrical equivalent of a map by a process of differential rectification. An exact model is set up in a stereoplotter, the effects of tilt being removed by the standard procedures of relative and absolute orientation. The operator scans the model systematically with the floating mark, adjusting the effective height of the mark continuously so that it remains on the surface of the ground. The orthoprojector operates in the darkroom conditions and holds a duplicate of one of the photographs of the stereopair. The tilt of the corresponding plotter projector, and the linear motion and height movement of the floating mark are transmitted either mechanically (Zeiss (Oberkochen) DP Ortho-3), optically (Wild PPO-8) or analytically (Zeiss (Oberkochen) Orthocomp Z-2) to the orthoprojector which allows exposure through a small slit on to light-sensitive paper. Thus a corrected print of the overlap area, known as an orthophotograph, is gradually built up as the slit moves in sympathy with the floating mark, the effects of height difference being removed by the continuous adjustment of the orthoprojector. Cartographic enhancement may be added to the print (e.g. contour lines, grid lines and place names), resulting in an orthophotomap.

7.4 The use of photogrammetry in civil engineering

7.4.1 Aerial photogrammetry

7.4.1.1 Topographic mapping

The use of aerial survey is generally considered to be the standard method of producing a topographic map or plan at scales smaller than 1:500. For scales greater than 1:500, ground survey would almost invariably be used.

The basic sequence of operations required for the production of a topographic map is shown by the flow diagram in Figure 7.17. The major air survey companies will have the equipment and manpower to carry out all the stages indicated but smaller concerns might, for example, subcontract the photography. It is very important that the exact requirements for the survey output should be known at the outset and that detailed discussion should take place between the civil engineer and the air survey company in order to identify any specific problems which might arise in connection with a particular project.

Flight planning. Before a survey flight can commence, a flight plan must be prepared detailing all the information required by the pilot, navigator and camera operator.

Aerial photographs which are to be used for mapping purposes must be taken in a regular sequence along parallel lines of flight. Navigation is usually visual by identifying landmarks shown on the map, mosaic or satellite image on which the

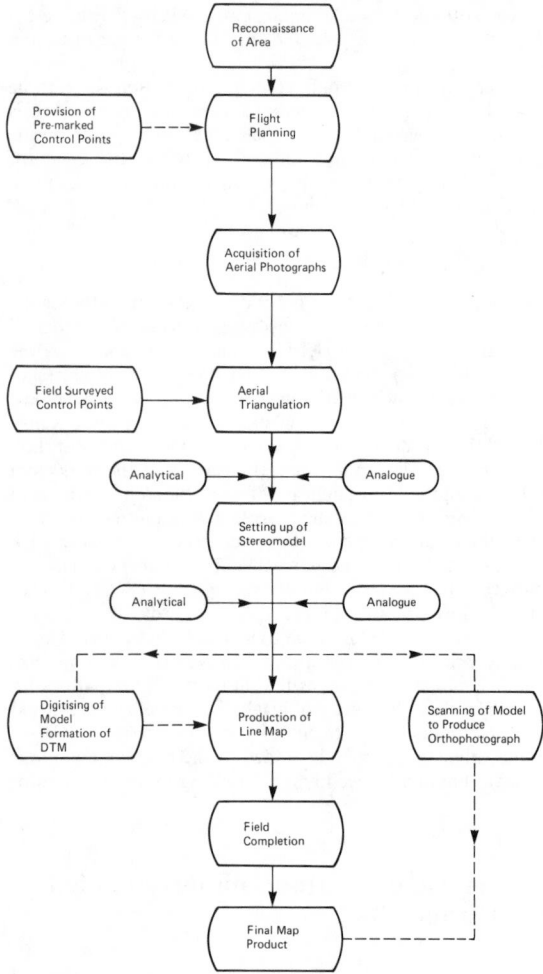

Figure 7.17 Flow diagram illustrating the sequence of photogrammetric operations involved in the production of a topographic map

required lines of flight have been marked. Increasingly, reliance may be placed on some form of radio navigation system. Flight lines will normally be drawn parallel to the longes edge of the area to be covered. However, if there is any marked grain of relief, this direction may be taken for the flight lines in order to attempt to maintain a constant scale along a given strip. Each line of flight produces a *strip* of photography, successive overlapping strips forming a *block* of photography.

Since a stereoscopic view of the entire area is needed, each ground point must be photographed from two camera stations giving a minimum overlap stipulation of 50% in the direction of the line of flight (the fore and aft direction). However, in order to give additional overlap to satisfy aerial triangulation requirements and to provide a safety margin, a value of 60% is usually specified.

A standard *Specification for vertical aerial photography* has been prepared by the Royal Institution of Chartered Surveyors[3] in collaboration with the British Air Survey Association. This specification is used by air survey companies throughout the world as a contract document for providing black-and-white aerial photography. The specification includes the air camera and its calibration, photographic coverage and operation, flying

conditions, the type of aerial film, processing and drying conditions, the resultant products to be supplied, and documentation and annotation.

Aerial triangulation. The object of aerial triangulation is to establish a network of control points with known ground coordinates which are required before a map can be plotted from aerial photographs. In order to perform absolute orientation when setting up a model in a stereoplotter, the minimum theoretical control required for each model is 2 points with known plan coordinates and 3 points with known heights (in practice, 3 plan points and 4 height points are usually considered to be the minimum). Thus, a large project would require the provision of an extensive amount of control; it would be very expensive and time-consuming if this had to be carried out exclusively by field survey methods. Aerial triangulation also has the advantages that it is easy to provide more than the bare minimum of control and it will usually be possible to obtain points in very convenient positions for their subsequent use in a stereoplotter.

Analogue methods of aerial triangulation have evolved steadily with the development of analogue stereoplotters. The method of aerial triangulation by independent models (AIM) is now used almost universally with modern analogue instruments. Since the technique requires model coordinates to be measured in an analogue plotter, a substantial amount of computation is required to obtain ground coordinates and it could therefore be described as a semi-analytical method.[4]

Whereas model coordinates measured in an analogue plotter form the basic data for the calculation of aerial triangulation by independent models, x, y coordinates measured either in a mono- or stereocomparator provide the raw data for the fully analytical methods. Much less observation time is required than with the equivalent analogue methods, but much greater emphasis is, however, placed on the data processing phase, both for the location of gross and systematic errors and for the computation of the coordinates. As outlined in section 7.2.3, it is possible to establish the mathematical equations of the bundle of rays from the optical centre of the camera lens to all image points observed on a given photograph by using the collinearity method. The combination of all the bundles of rays for the entire block can be used to carry out relative and absolute orientation simultaneously.[5] The calculated ground coordinates obtained by this method are adjusted to a best mean fit by least squares. It is clearly advantageous if both the necessary hardware and software are available to consider the block as an entity rather than to adjust a strip and then connect adjacent strips in a separate operation.

Having carried out aerial triangulation, each stereomodel is subsequently set up in turn using the orientation procedures mentioned in section 7.3.4.3. Plotting can then commence, so creating a topographic map or the operator may measure spot heights to form a digital terrain model.

7.4.1.2 Production of digital terrain models

The term digital terrain model (DTM) was originally coined in 1958 by Miller and La Flamme. Since then, the subject has developed considerably and is currently an area of widespread activity in surveying, geology and geophysics, civil and mining engineering and other disciplines in the earth sciences.

Although a variety of different techniques exist for the creation of DTMs in general, the distinction can be made between those which make use of height data which have been collected in a regular grid and those based on a network of randomly located height points. The former, simpler, approach is generally adopted when photogrammetric equipment is being used for the generation of the DTM. While the square grid is the most common form, rectangular-, hexagonal- and triangular-based grids are also used.

Although the regular grid is the simplest technique to adopt it has one serious limitation: the distribution of data points is not related to the characteristics of the terrain. If the data point sampling is conducted on the basis of a regular grid, then the density must be high enough to portray accurately the smallest terrain features present in the area being modelled. If this is done, then the density of data collected will be too high in most areas of the model, in which case there will be an embarrassing and unnecessary data redundancy in these areas.

One solution to this problem is to use progressive sampling and its development, composite sampling.[6] Instead of all points in a dense grid being measured, the density of the sampling is varied in different parts of the grid, being matched to the local roughness of the terrain surface. This approach has been widely implemented on a variety of grid-based terrrain modelling packages such as HIFI,[7] and SCOP.[8]

Terrain models derived by photogrammetric techniques are also used by modelling packages such as MOSS, ECLIPSE, and Intergraph DTMN. A comprehensive review of the use of such terrain modelling packages in surveying and civil engineering can be found in Petrie and Kennie.[9]

7.4.1.3 Monitoring

Aerial photogrammetry has been used extensively in the past as a means of monitoring changing conditions such as the depletion of coal stocks or the degree of coastal erosion occurring over a period of years. In more recent years, its use has also developed for monitoring the deformation of natural or man-made features.

Fraser[10] and Fraser and Gruendig[11] describe the planning and subsequent execution of a photogrammetric survey of the Frank landslide/rockslide region of Turtle Mountain in Alberta, Canada. A rockslide occurred in this area over 80 yr ago when more than 30 million m^3 of rock moved down the east face of the mountain covering an area 3 km square to a depth of 14 m. Since 1933, crack monitoring surveys of the area have been undertaken to determine the stability of the rock wedge forming the southern peak of the mountain. These surveys have been supplemented in recent years by an *in situ* monitoring program using crack motion detectors. Whilst these devices are capable of very high accuracy (± 0.1 mm/yr), they can only, realistically, be deployed over a very small area. Also, if hazardous movements are detected it may be considered unsafe to continue recording measurements. Aerial photogrammetry, by covering large areas in a noncontact manner, can overcome these limitations. Therefore, it was considered appropriate to evaluate the potential of photogrammetry for this project. Fraser[10] established that the optimum flight plan consisted of obtaining large-scale (1:2000) multiple photographs (10 to 15) of the area containing the thirty monitoring target points. Photography was obtained using a Wild RC8 camera (focal length 152 mm) on two occasions approximately 1 yr apart. A subsequent analysis of the results indicated that statistically significant deformations had occurred at eight of the thirty points, the maximum movement being 63 mm.

7.4.2 Close-range photogrammetry (CRP)

Whereas it is possible to identify a standard procedure for mapping from aerial photographs, each project which requires the use of close-range photographs will tend to be regarded as unique due to the wide variation in circumstances which can arise for both taking the photographs and the subsequent analysis. For example, the positioning of the cameras for photography needs to be adapted to the particular demands of

an individual task and may vary from a simulated aerial case with parallel, near vertical axes, through inclined and convergent axes to parallel horizontal axes. At the measurement stage, use of either analogue or analytical equipment may be specified according to the type of camera and whether a graphical or a numerical output is demanded.

Although some air survey companies also have a department which specializes in CRP, as do a few land survey organizations and universities, much of this type of work could be carried out by engineering firms themselves, perhaps within a research and development department as, for example, at Rolls Royce.[12] It is even more vital for CRP than for aerial photogrammetry that the engineer should be very closely associated with all stages of the work so that he can ensure that his exact requirements are met.

7.4.2.1 Monitoring

Deformation of structures. Close-range photogrammetry has been used widely in the field of structural engineering. The ability of the technique to record both the shape of a structure at one moment and the deformations which occur between two epochs have been exploited. Notable in the first case is the extensive work which has been carried out in architectural photogrammetry. In the latter case, comprehensive applications reviews have been written by Atkinson[13] and Cheffins and Chisholm.[14] A review of the various methods which can be adopted has been prepared by Cooper.[15] Cooper identifies four main methods including those which make use of a single camera (time parallax), controlled stereomodels, resection/intersection, and bundle adjustment. The latter is considered to be the most general case and most suitable for high-accuracy applications.

The range of large engineering structures which have been investigated by photogrammetric methods includes cooling towers, box girder bridges, elevation of St Paul's Cathedral, tower cranes, offshore platforms, and ships. In addition, photogrammetry has been used for the measurement and dimensional control of smaller structures such as microwave antennae, robots, and large compressors (see Welsh[16] for further details). Increasingly, the emphasis is being placed on the development of a 'real-time' photogrammetric system using video technology and solid-state cameras. Wong and Wei-Hsin[17] outline the development of such a system.

Earth and rockfill dam monitoring. The use of CRP for monitoring earth and rockfill dams has been discussed by Moore[18] and Brandenberger,[19] among others. In the case of Moore,[18] the author describes the use of a Wild P30 phototheodolite for monitoring the three-dimensional displacements of the Llyn Brianne rockfill dam, in mid Wales, during several stages of construction. The predicted displacements were in the range of ± 0.5 to 1.0 m, and whilst high absolute positional accuracy was deemed to be important, it was considered that the definition of the direction of movement was of equal importance. By using a Wild A7 stereoplotter, measurements were taken at over 80 targets, at differing levels of fill. The results indicated a range of displacements, from 0.1 to 0.6 m (with a standard error of ± 0.05 m).

Retaining wall monitoring. The use of a modified KA-2 aerial camera (focal length 610 mm, format 23×23 cm) to monitor the deflection of a gabion wall is described by Veress, Jackson and Hatsopoulos.[20] The wall, situated near Seattle, Washington State, was over 400 m long and varied in height from 2 to 18 m. The gabions were constructed of 1-m steel mesh rockfilled cubes. Photographs were obtained from fixed control points up to 1000 m from the wall. Over 100 target points were observed with an analytical plotter and ground coordinates computed. The wall was measured on eleven occasions. The authors suggest that CRP used in conjunction with an internal monitoring system such as an inclinometer would be an ideal system for monitoring new structures.

7.4.2.2 Slope/rock stability studies

One of the earliest reports of rockface mapping is given in Cheffins and Rushton.[21] This paper describes the procedures used to produce 1 : 50-scale contoured elevations of the north face of Edinburgh Castle Rock. The elevations were required by a team of engineering geologists who were carrying out rock bolting and grouting as stabilizing measures to reduce the chance of rock falls from the face. The survey was carried out by taking four pairs of overlapping photographs from a baseline approximately parallel to the face. The photographs were obtained using a Wild RC5A aerial camera mounted on a hydraulic platform. The elevations with 0.25-m contours were subsequently used, in conjunction with ultrasonic data about the joint structure, to plan the positions of the necessary boreholes for rock bolting.

Moore[22] discusses the use of a phototheodolite for mapping vertical faces, in this case in clay pits. The problem under investigation involved the determination of information about the continuity and spacing of major joints in the clay face. Photogrammetric techniques enabled a three-dimensional model of the spatial disposition of the joints to be constructed. Other references which mention the use of photogrammetry for slope and rockface stability monitoring include Torlegard and Dauphin[23] and Robertson et al.[24]

Landslides are a commonly occurring hazard in road design and construction. Heath, Parsley and Dowling[25] describe the use of a Wild P32 camera to obtain photographs of areas susceptible to landslipping in Columbia and Nepal. The photographs were observed using a Zeiss Topocart and Wild A40 Terrestrial Stereoplotter, and 1 : 200- to 1 : 10 000-scale contoured plots were produced. The plans were subsequently used to design remedial and preventative measures associated with the landslip areas. They also enabled a classification of landslide characteristics to be produced.

Kennie and McKay[26] discuss the use of CRP to monitor the erosion of chalk cliffs around a road tunnel in East Sussex. Control stations were monumented at both the top and base of the cliffs, ensuring that a minimum of six control points were visible within each stereomodel. The stations were surveyed by EDM and theodolite and targets positioned over the monuments for precise viewing and control pointing in the stereo model. Photography was taken with a Zeiss UMK 10/1318 camera using glass negatives for greater stability.

Due to the angle of slope of the cliff face from the camera a wide variation in photoscale resulted in this case, from 1 : 400 to 1 : 900. For each of the four faces under investigation, the control network was rotated about two points on top of the cliff thus forming a datum line parallel to the face itself. The shape of the cliff was then defined by contours at 0.25-m intervals of depth – i.e. as differences in the horizontal distance from this datum plane – rather than as more conventionally vertical elevations above a height datum. The accuracy of points shown on both the map and sections was within ± 5 cm of true ground position.

7.4.2.3 Tunnel profiling

Recent developments associated with the use of photogrammetry for producing tunnel profiles are described by Anderson and Stevens.[27] Anderson and Stevens outline the development and use of a 'mono-photogrammetric' tunnel profile measuring system. The two main elements of the system are a high-intensity light plane generator, which illuminates the section to be profiled, and a camera which records the line of light which is generated. By digitizing the line, applying corrections for lens distortion, and scaling the photographs, an accuracy of ± 10 mm is claimed. Furthermore, if the quoted progress rate of 100 sections per hour can be achieved in practice, the system would appear to be a highly cost-effective solution.

7.4.2.4 Laboratory-based applications

Close-range photogrammetry has also been applied to measurement problems in the laboratory; for example, Andrawes and Butterfield,[28] Wong and Vonderhoe,[29] and Davidson[30] have used the motion or 'false parallax' technique to measure the planar displacement fields associated with soil models. In the first two cases cited, the technique involved taking repeat photographs, from a single camera position, of a glass-sided tank which contained the soil under investigation. Photographs were taken using a nonmetric 35 mm camera before and after the sand within the tank was subjected to movement by a moving wall or wedge situated in the tank. By examination of enlargements of the photographs in an analogue stereoplotter, the relative positions of features could be measured and displacement contour diagrams produced. Analysis of these diagrams enabled the displacement component along the sand bed to be determined to within 5 μm. The latter two authors, in contrast, used analytical techniques to investigate the movement of soil particles around a model tunnel and soil penetration probe respectively.

7.5 Principles of remote sensing

The interpretation of aerial photography has for many years proved to be a valuable source of data for civil engineers. Until the early 1960s civilian 'remote sensing' was concerned primarily with the interpretation of such imagery. Since then, however, developments in orbiting satellites, sensor technology and computing have led to the creation of a discipline which now impinges on many areas in science and engineering. Although still in an embryonic state, particularly in the field of image processing, it has already proved to be a cost-effective method of investigating engineering phenomena of both large aerial extent, e.g. surface drainage, and those which are more localized, e.g. landslides, unstable land, etc.

7.5.1 The electromagnetic spectrum

Electromagnetic (EM) energy is all energy which travels in a periodic harmonic manner at the velocity of light. Electromagnetic energy is normally considered to consist of a continuum of wavelengths referred to as the EM spectrum (Figure 7.18).

It can be seen from Figure 7.18 that several regions of the EM spectrum are of particular importance in remote sensing. For example, the visible and reflected infra-red regions of the spectrum are important since they enable *reflected* solar radiation to be measured. In contrast, at longer wavelengths in the infra-red (8 to 14 μm waveband) the sensing of *emitted* thermal energy is of more significance. Measurements within the visible and infra-red (reflected and emitted) regions are considered to be *passive* in nature since the radiation being recorded occurs naturally. As the wavelength increases to the order of several millimetres it becomes more convenient *actively* to generate EM radiation of this wavelength and record the reflected radiation from the terrain. Thus, a typical *active* system would be side-

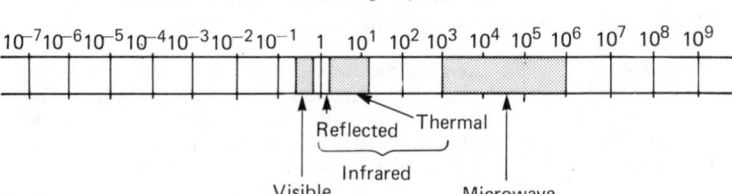

Figure 7.18 The electromagnetic spectrum

looking airborne radar (SLAR). It should be noted that instruments also exist for the measurement of *passive* microwave emission. However, since the emitted EM energy at this wavelength is very small, microwave radiometers are much less common than SLAR instruments.

Depending on the nature of the radiation being measured, it is possible to record the reflected or emitted energy by using either a lens and photographic emulsion or by using a linescanner and crystal detector. The geometrical distinction between the two approaches is illustrated in Figure 7.19. The primary advantage of using a linescanner approach is that it is possible to record radiation of wavelengths greater than about 0.9 μm. It is also possible to measure the variations in radiation within narrow spectral regions and to record directly these variations in digital form.

Mention should also be made here of the distinction between the terms 'photograph' and 'image'. A 'photograph' is an image which has been detected by photographic techniques and recorded on photographic film. In contrast, an 'image' is a more general term used to describe any pictorial representation of

detected radiation data. Therefore, although scanner data may be used to create a photographic product, this result is normally referred to as an 'image' since the original detection mechanism involved the use of crystal detectors creating electrical signals, rather than a lens focused on to photographic film.

7.5.2 Classification of remote sensing systems

Remote sensing systems can be classified using various criteria, such as sensor sensitivity range, mode of recording (photographic or scanning), mode of operation (active or passive) or type of sensor platform (aircraft or satellite). Table 7.5 provides a framework for the classification of data acquisition systems in remote sensing by using the latter two criteria; it also provides some common examples of each category.

Table 7.5 Classification of data acquisition systems

	Aircraft	*Satellite*
Passive systems	Wild RC-10 camera Daedalus 1268 MSS Daedalus 1230 Thermal Scanner Barr and Stroud IR18 TVFS	Spacelab metric camera NASA large-format camera Landsat MSS, RBV, TM SPOT MOMS
Active systems	Goodyear GEMS Westinghouse SLAR SAR 580	Seasat SIR A/B ERS-1 Radarsat

7.6 Data acquisition

Data can be acquired for remote sensing from a variety of aerial platforms, although fixed-wing aircraft and unmanned orbiting satellites are the most common. Both have specific and complementary advantages. Aircraft, for example, enable small localized phenomena to be investigated at high levels of resolution, whereas satellites enable wide synoptic views of the terrain to be obtained, often on a repeatable basis, but at much lower resolution.

7.6.1 Airborne systems

7.6.1.1 Photography

Vertical panchromatic aerial photography taken with a high-

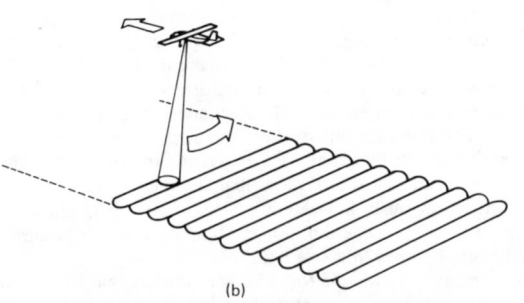

Figure 7.19 Geometry of data acquisition. (a) Camera; (b) linescanner

precision mapping camera is the most commonly available source of airborne remote sensing data (Figure 7.20(a) and Table 7.1). In addition to conventional aerial photography obtained using the system described above, it is also possible to utilize oblique aerial photography[31] (Figure 7.20(b)), low-cost or small-format photography,[32] or multiple-camera multispectral designs where each camera is filtered in order to record the reflected radiance in a particular waveband.[33] Alternatively, lower cost platforms such as microlight aircraft[34] or remotely piloted aircraft[35] may be used to obtain photography over a small site. A more recent development has been to produce photography from airborne video systems.[36]

Figure 7.20 Aerial photographs of Guildford Cathedral.
(a) Vertical; (b) oblique

7.6.1.2 Multispectral scanner

The principle behind the use of the multispectral scanner (MSS) concerns the detection of radiation from the Earth's surface. This radiance can be quantified by measuring the proportion of energy reflected or emitted by an element of the Earth's surface at various specific wavelengths. Each element is referred to as a pixel. An image can be created using an airborne linescanner by detecting the radiance in a linear pass across the ground perpendicular to the line of flight. Successive passes can be conducted at a rate commensurate with the aircraft flying speed, which can then be used to create an image. The image is therefore built up pixel by pixel in a sequential form, line by line.

Crystal detectors in the MSS transform the incident radiation into electrical signals which are recorded in digital form on magnetic tape.

The technique used in the MSS is to detect both radiation *emitted*, and solar radiation *reflected*, from the Earth's surface. These forms of radiation, within several different bands of wavelength, can be detected simultaneously using this method. The advantage of using airborne systems rather than satellite systems (such as Landsat MSS) is the much higher spatial resolution of the former (less than a metre) and the greater number of available spectral wavebands.

One MSS which has been used in the UK for a variety of engineering projects is the Daedalus AADS 1268 Airborne Thematic Mapper (ATM). This instrument has an instantaneous field of view (IFOV) of 2.5 mrad and the spectral specifications of the instrument are presented in Table 7.6. Data are recorded in flight on high-density digital tape and subsequently reproduced in a standard form on to computer compatible tape (CCT) for analysis on a digital image processing system. Figure 7.21 illustrates a typical example of imagery obtained with this type of scanner.

Table 7.6 Spectral bands for the Daedalus 1268 AADS ATM

Spectral band	Wavelength (µm)	Spectral region
1	0.42–0.45	Blue
2	0.45–0.52	
3	0.52–0.60	Green
4	0.605–0.625	Red
5	0.63–0.69	
6	0.695–0.75	Near IR
7	0.76–0.90	
8	0.91–1.05	
9	1.55–1.75	
10	2.08–2.35	Mid IR
11	8.50–13.00	Far (Thermal) IR

7.6.1.3 Thermal infra-red (IR) scanning

Thermal IR scanning techniques are similar to those utilized by MSSs but concentrate on a single wavelength within the far IR region of the spectrum. Two regions can be sensed by such instruments, corresponding to atmospheric 'windows' in the 3 to 5 µm and 8 to 14 µm regions. The former region is used primarily for examining very hot objects, e.g. forest fires, and is rarely used for engineering applications. The latter region, corresponding to the peak value of emitted radiation from the Earth, is of much greater importance.

The most common thermal IR sensor is the linescanner which is a widely used instrument based on a scan geometry identical to that for MSS devices. A cadmium mercury telluride (CMT) detector is used to record the temperature variations over the scene and a calibrated internal blackbody reference enables precise quantitative temperature variations to be measured. Several thermal IR linescanners are currently in operation in the UK, including a Daedalus 1230 dual-channel scanner and as mentioned previously a Daedalus 1268. Figure 7.22 shows a typical example of a thermal image obtained with this scanner which could be used to assess heat loss.

An alternative technique for sensing in the thermal IR region of the spectrum which has recently been evaluated for remote sensing applications in environmental engineering is the thermal video frame scanner (TVFS).[37] Several benefits accrue from the

Figure 7.21 Airborne multispectral scanner image of the Humber Bridge illustrating sediment transport patterns. (*Courtesy:* Huntings Geology and Geophysics)

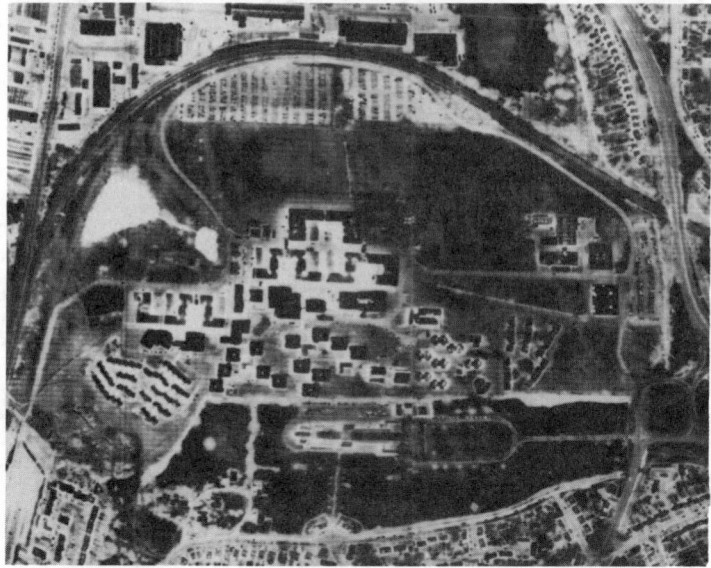

Figure 7.22 Thermal infra-red linescan image of the University of Surrey. (*Courtesy:* Clyde Surveys Ltd)

use of video-based instruments. These include system portability, almost real time verification of data capture, rapid turnaround (since there is no processing stage), low cost of operation (since the sensor can be operated from a light aircraft) and ease of duplication of data. Unlike linescanning systems, however, these instruments normally do not include a calibrated internal blackbody reference for quantitative temperature measurements. For some applications this could prove to be a severe limitation. However, for many engineering applications an assessment of the relative temperatures within a scene can be just as important as a detailed knowledge of the absolute ground temperatures. It should be noted that the quantitative determination of temperatures can, nevertheless, be carried out

by correlating grey scale levels and ground control points of known temperature.[38] One example of this type of instrument which is currently being used for remote sensing applications is the Barr and Stroud IR18 TVFS system.

7.6.1.4 Side-looking airborne radar (SLAR)

Side-looking airborne radar was first developed for military purposes in the early 1950s. It was not, however, until the early 1970s that it became available commercially. The main impetus for the development of SLAR arose out of its two significant advantages over optical photographic or scanning systems: (1) its ability to produce imagery both day and night; and (2) its ability to 'sense' through haze, smoke, cloud and even rain. These radar systems are consequently in great demand for the production of imagery in equatorial regions of the Earth's surface.

The main reason for these twin advantages stems from SLAR's use of microwaves as the imaging source. The image which is produced by SLAR is very different from that produced by conventional optical systems. The view which is obtained is a record of the Earth's reflective properties at microwave wavelengths. Consequently, the nature and intensity of the reflections by SLAR will be influenced by factors such as ground conductivity and surface roughness, which are much less significant with optical systems. In addition, the geometrical properties of the image differ significantly from a conventional aerial photograph.

The principle of operation of a typical SLAR system is illustrated in Figure 7.23.[39] It involves the measurement of the time interval between the transmission and reflection of a microwave pulse. This indirectly provides the range from the aircraft to the ground feature. Successive pulses can then be sensed sequentially so providing a radar picture of the terrain. Figure 7.24 is a large-scale image obtained using an airborne radar system.

Two distinct designs of SLAR exist. The earliest designs were termed real aperture radar (RAR) systems. Although able to sense through cloud their ground resolution was limited by the size of the antennae which could be mounted on the side of the aircraft. In order to overcome this limitation, an alternative design, synthetic aperture radar (SAR) has been developed. Using suitable computer processing such a system is able to provide much higher resolution imagery than that which can be provided by a RAR system. Table 7.7 outlines some of the characteristics of a selection of SLAR systems.

7.6.2 Satellite systems

Satellite remote sensing systems range from low-resolution systems such as the meteorological satellites to high-resolution photographic missions such as the recently flown metric camera on board the space shuttle. Although both have some application to civil engineering the low ground resolution of the former system and the limited coverage of the latter restrict their application considerably. Of much greater importance to civil engineers are the Earth resources satellites operating the visible, infra-red and microwave regions of the spectrum.

Figure 7.24 Side-looking airborne radar image

7.6.2.1 Earth resources satellites – passive

A selection of the characteristics of the most common Earth resources satellites is given in Table 7.8.

Landsat. The Landsat satellite system, previously known as the Earth Resources Technology Satellite (ERTS), was initiated in 1967 by NASA in conjunction with the US Department of the

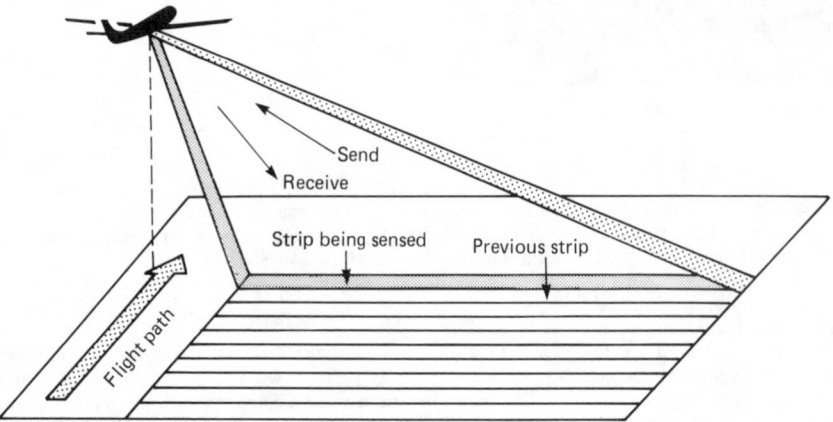

Figure 7.23 Side-looking airborne radar. (After Rudd (1974) *Remote sensing – a better view.* Duxbury Press, Massachusetts)

Table 7.7 Side-looking airborne radar systems

	Motorola AN/APS	Westinghouse	Goodyear GEMS	SAR-580
Type	RAR	RAR	SAR	SAR
Wavelength band	× (3 cm)	K(0.8 cm)	× (3 cm)	× (3 cm)
Resolution (m)	15	15	12	3

Table 7.8 Characteristics of passive Earth resources satellites

	Landsat's 1-5 multispectral scanner (MSS)	Landsat's 1-3 return beam vidicon (RBV)	Landsat's 4-5 thematic mapper (TM)	Modular optoelectronic multispectral scanner (MOMS)	Le Système Probatoire d'Observation de la Terre (SPOT)
Operated by (country)	EOSAT (USA)	EOSAT (USA)	EOSAT (USA)	DFVLR (W. Germany)	SPOT Image (France)
Date of launch	1972	1972–81	1982	1983	1986
Orbital altitude (km)	900	900	705	300	830
Ground resolution (m)	80	80, 30	30	20	10, 20
Spectral range (μm)	0.5–12.6	0.505–0.75	0.45–12.5	0.575–0.975	0.5–0.89
No. of wavebands	5	1	7	2	3
Further reading	NOAA[40]	NOAA[40]	NOAA[40]	—	Chevrel, Courtois and Wells[41]

Interior. The system was initially designed as an experiment in order to assess the feasibility of collecting Earth resource data from unmanned satellites. However, following the commercialization of remote sensing activities in 1985, the current operational activities of Landsat have been transferred to EOSAT, a joint venture formed by Hughes and RCA.

Three distinct sensors have been carried by Landsat: (1) a MSS; (2) a return beam vidicon (RBV); and (3) a thematic mapper (TM). The MSS is a linescanning device which uses an oscillating mirror to scan at right angles to the satellite flight direction. The IFOV of the sensors operating in the visible and near IR produce a resolution cell of approximately 56 × 79 m. The fifth channel has an IFOV of 0.258 mrad, or a ground resolution of about 235 m.

The MSS scans each line from west to east with the southward motion of the satellite providing the along-track progression of the scan lines (Figure 7.25).[40] Each Landsat MSS scene covers an area approximately 185 × 185 km. In view of the extremely high mirror oscillation rate which would be required using this approach if only one line was scanned, the system is designed to scan six lines simultaneously with each oscillation of the mirror. This results in an area 474 m × 185 km being recorded with each sweep.

A typical Landsat scene, 185 × 185 km, consists of 2340 scan lines with about 3240 pixels per line; each image therefore consists of over 7.5 million digital values. With four spectral observations per pixel, this means that over 30 million values have to be recorded for every Landsat scene.

The simplest Landsat MSS product is a photograph consisting of a black-and-white image for each spectral band (Figure 7.26). Although the resulting image is vaguely familiar, because

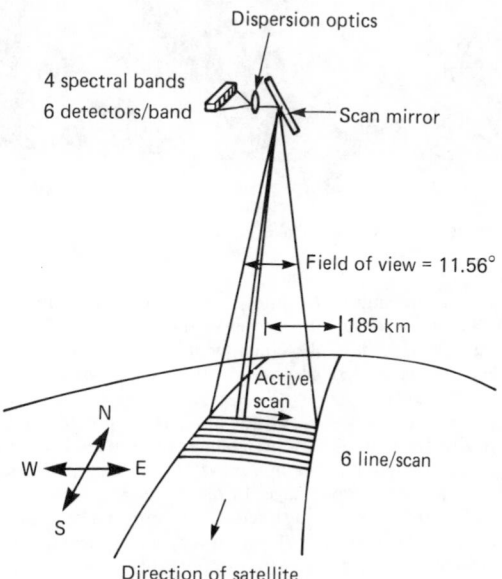

Figure 7.25 Landsat multispectral scanner[40]

it is similar to a conventional panchromatic aerial photograph, much of the information content of the image is lost. An alternative approach is to assign a different colour to each

Figure 7.26 A black-and-white image as presented by Landsat of Southeast England created by a MSS. (*Courtesy*: National Remote Sensing Centre)

spectral and superimpose these to produce a colour-composite image. Since the sensor also senses-in the reflected infra-red region of the spectrum, the consequent composite does not provide a true colour-coded image of the terrain but rather a 'false' colour image.

The second sensor, the RBV, was designed to provide high-resolution imagery suitable for mapping. In addition to having a high ground resolution (30 m on Landsat 3) it also had a réseau grid superimposed on the image. In recent years however, the advent of the TM has tended to reduce the importance of both MSS and RBV imagery, apart from forming a historical record.

The thematic mapper has been operational since 1984 and it provides data of higher spatial resolution (30 m) and finer spectral resolution (seven bands) than that available from the MSS or TM sensors. The comparative spatial resolution of the sensors is illustrated in Figure 7.27.

Modular optoelectronic multispectral scanner (MOMS). The modular optoelectronic multispectral scanner (MOMS) was designed by the West German MBB company for the German Aerospace Research establishment (DFVLR), primarily as a research system. The first satellite-borne imagery was obtained with the system during the seventh space shuttle flight in June 1983. The scanner has several unique design features of which the most significant are the dual lens system and four linear arrays of 1728 pixels which enable a continuous line of 6912 pixels to be swept out by the scanner (equivalent to 20 m on the ground).

A second-generation MOMS system is currently under development by MBB which will offer the possibility of obtaining stereoscopic imagery and which will also have an extended range of spectral bands.

Le Système Probatoire d'Observation de la Terre (SPOT). This is a commercial remote sensing system developed by the French Government and aerospace industry and operated worldwide by SPOT image. The sensors on board the satellite consist of two high-resolution visible (HRV) imaging instru-

TM (NO MAGNIFICATION)
CENTRAL LONDON FROM THE LANDSAT MULTI-SPECTRAL AND THEMATIC MAPPER SENSORS

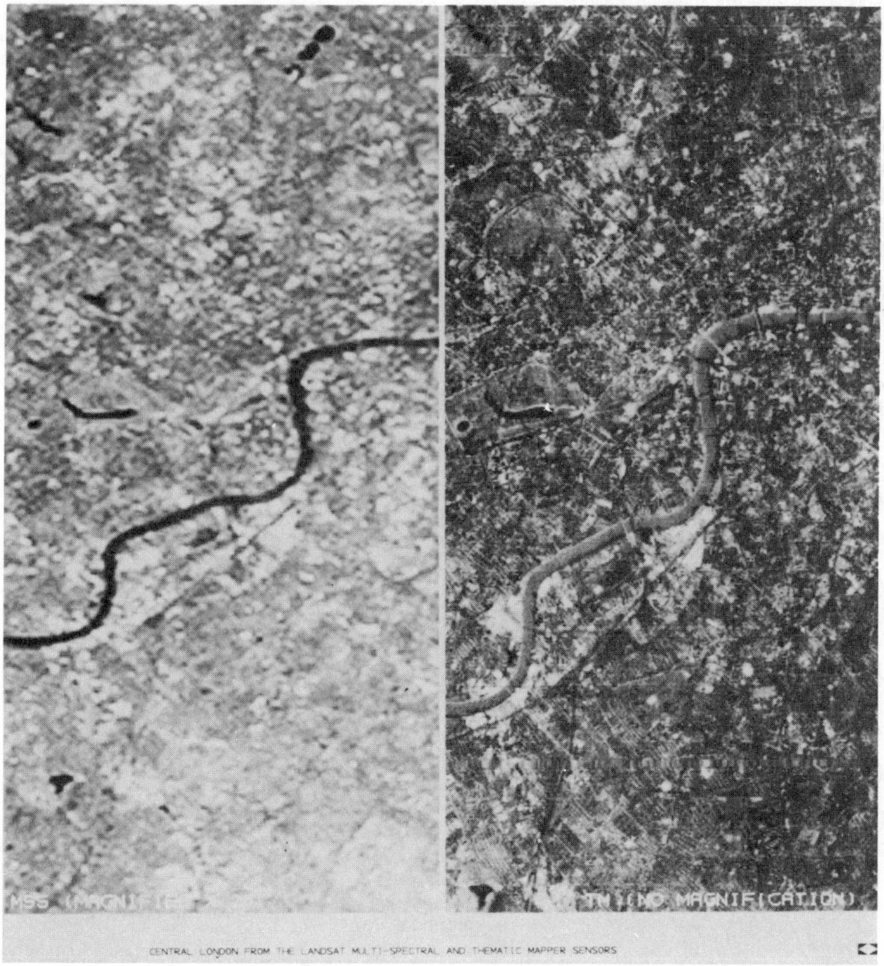

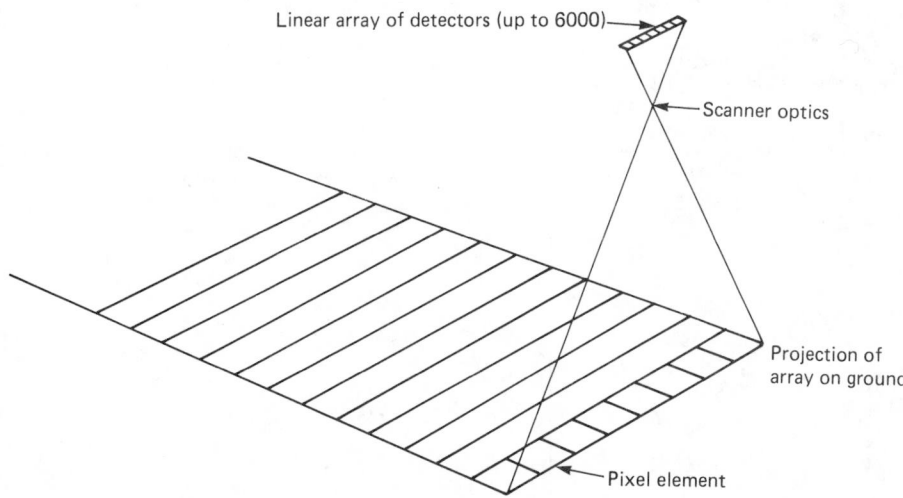

Figure 7.27 A comparative spatial resolution of the Landsat MSS (left) and thematic mapper sensors. (*Courtesy*: National Remote Sensing Centre)

Linear array of detectors (up to 6000)
Scanner optics
Projection of array on ground
Pixel element

Figure 7.28 SPOT: pushbroom scanner design

ments employing the multilinear array or 'pushbroom' design of MSS. This design of MSS differs from the optical–mechanical design discussed previously. In this case each line of the image is formed by measuring the radiances which are imaged directly on to a one-dimensional linear array of small detectors located in the instrument's focal plane. Each line is subsequently scanned electronically and the radiance values recorded on to magnetic tape. As before, successive lines of the image are produced by the forward motion of the satellite along its orbital path. This pushbroom concept is illustrated diagrammatically in Figure 7.28.

A further important feature of the SPOT system from an engineering point of view is its capacity to provide stereoscopic coverage of the Earth's surface from the lateral overlap between successive scenes. A rotatable mirror in the SPOT sensor package also permits scenes to be acquired over areas up to 400 km left or right of the normal vertical vantage point of the satellite. This feature will permit much easier acquisition of high-priority scenes (Figure 7.29). Figure 7.30 shows a typical example of a SPOT panchromatic image (10 m pixel).

7.6.2.2 *Earth resources satellites – active*

Following the success of the Seasat satellite SAR mission in 1978 (Table 7.9 and Figure 7.31), a great deal of interest has been shown in the development of further side-looking radar satellites. For example, the forthcoming ERS-1 satellite will be used to provide continuous monitoring of ocean parameters such as wave and wind height and pattern, which may help to im. rove the engineering design of oil platforms. The SAR data will also be used to aid the planning of shipping routes particularly in Arctic regions. Similarly, the Canadian Radarsat will provide imagery to enable the positions of icebergs to be monitored more precisely. This data will be used by oil tankers transporting oil and natural gas through the Northwest passages from oil fields in and around the Arctic islands of northern Canada.

7.7 Digital image processing (DIP)

Digital image processing is a crucial stage in the effective use of

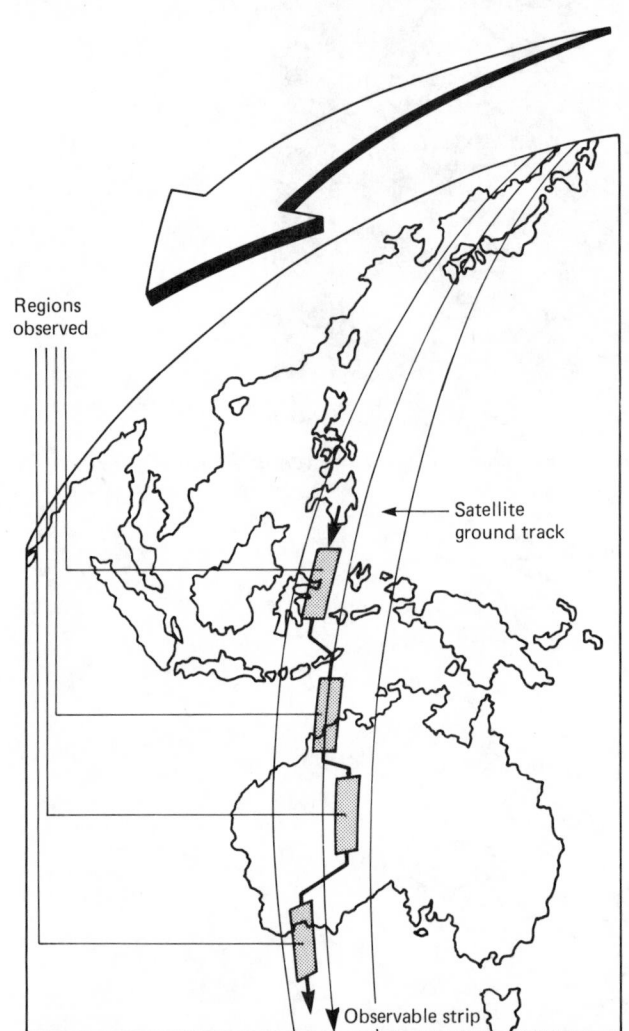

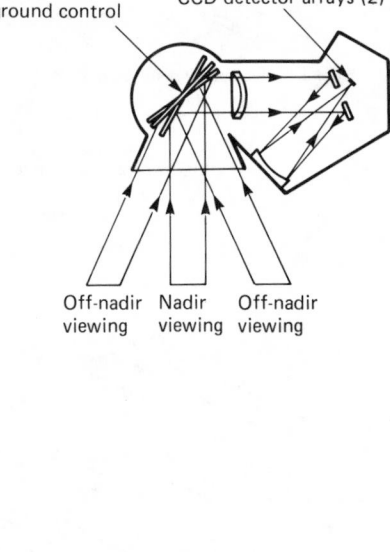

Figure 7.29 SPOT: nadir and off-nadir viewing. (*Courtesy*: SPOT Image)

Figure 7.30 SPOT: panchromatic image of part of Montreal (SPOT data copyright CNES, 1986, image provided by Nigel Press and Associates)

modern remote-sensing data. Although a great deal of information can be gleaned from the interpretation of a photographic image produced from, for example, Landsat, the full potential of the data can only be realized if the original data, edited and enhanced to remove systematic errors, is used with a suitably programmed image processing system.

7.7.1 Hardware

Until relatively recently, most DIP of remote sensing data was carried out on large, expensive, dedicated systems such as that illustrated in Figure 7.32. Such a system enables data to be read into the system from a magnetic tape reader, to be displayed on

a colour TV monitor and, after suitable processing, to be output to a colour filmwriter for the production of high-quality imagery. The trend, however, is towards smaller and cheaper image processing based on the new generation of 32-bit super mini-computers and 16-bit IBM-compatible personal computers.[42] Table 7.10 lists a selection of the remote-sensing systems currently available.

7.7.2 Software

The range of algorithms which are used to restore and classify remote-sensing data is extremely wide and varied. Furthermore, an appreciation of the choice of the most appropriate technique,

Table 7.9 Characteristics of a selection of active Earth resources satellites

	Seasat	*Shuttle imaging radar (SIR)-A*	*SIR-B*	*European radar satellite (ERS-1)*	*Radarsat*
Date of launch	1978	1981	1984	1989	1991
Operated by	NASA	NASA	NASA	European Space Agency	Canada
Altitude (km)	800	245	255, 274, 352	675	1000
Wavelength (mm)	230	230	540	540	540
Look angle (°)	20	50	57	—	30–45
Resolution (m)	25	40	30	25	30
Swath width (km)	100	50	20–25	80	150

Figure 7.31 SEASAT SAR image of the Tay estuary. (*Courtesy*: National Remote Sensing Centre)

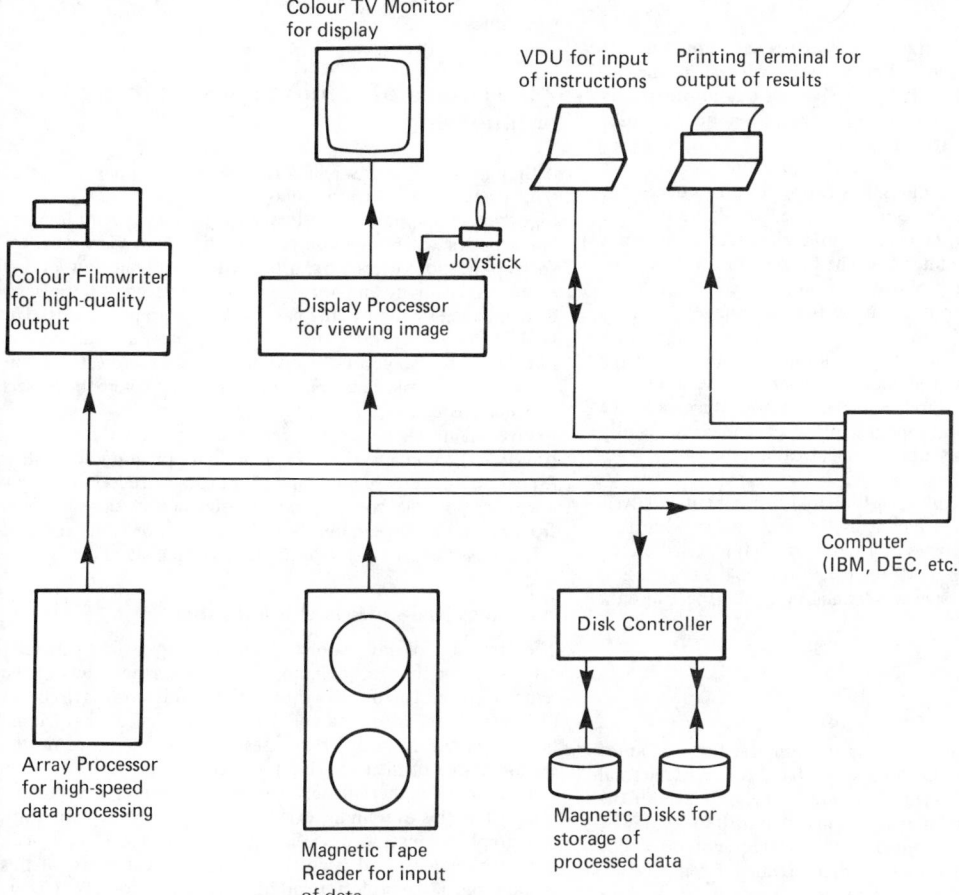

Figure 7.32 Digital image processing system

and the advantages and limitations of each technique, is a subject of some complexity beyond the scope of this chapter. Consequently, only a very brief description of the most commonly used techniques will be provided. For further details Bernstein,[43] Avery and Graydon[44] and Bagot[45] should be consulted.

7.7.2.1 Image restoration

Geometric correction. Several preliminary transformations are normally carried out on raw satellite data to reduce the effects of the Earth's rotation and curvature and variations in the attitude of the satellite. If a sufficient number of ground control points are available the imagery can be fitted to the control using a least squares technique. Such a procedure is also necessary if the remote sensing data is to be combined with existing map data.

Radiometric correction. Variations in the output from the detectors used in a MSS may cause a striping effect on the final image. Such an effect can be eliminated by a process termed 'destriping', which effectively interpolates the missing radiance values from the outputs of adjacent pixels. In some cases, complete scan lines may be missing from the data; preprocessing can also be carried out to overcome this type of problem.

Table 7.10 A selection of remote-sensing systems currently available

	Name of system	Company
Dedicated image-processing system	Gemstone 33	GEMS of Cambridge, UK
	I²S Model 75	Int. Imaging Systems, California, USA
	Vicom VDP Series	Vicom Systems, San Jose, USA
	Dipix Aries II	Dipix Systems, Ottawa, Canada
Personal computer-based systems	Diad Systems	Diad Systems, Edenbridge, Kent, UK
	LS10	CW Controls, Southport, Lancashire, UK
	Microimage	Terra Mar, California, USA
	ERDAS	Earth Resources Data Analysis System, Atlanta, USA
	ImaVision	RBA Associates, Ottawa

7.7.2.2 Image enhancement

Contrast stretching. Contrast stretching is a technique which is used to brighten an image and produce one which uses the full dynamic range of the data. It is often carried out automatically and is generally one of the first operations carried out by a user when viewing a new image.

Spatial filtering. Spatial filtering is carried out in order either to enhance boundaries between features (edge enhancement or high pass filtering) or to smooth and eliminate boundaries (smoothing or low pass filtering). The former approach is most suitable for enhancing geological features, whereas the latter may be used to reduce random noise from an image.

Image ratioing. Image ratioing is the process of dividing the radiance values of each pixel in one waveband by the equivalent values in another waveband. Ratioed images reduce radiance variations caused by local topographic effects and consequently may be used to enhance subtle spectral variations.

Principal components analysis. Principal components analysis is a statistical technique which is used to improve the discrimination between similar types of ground cover. It is based on the formation of a series of new axes which reduce the correlation existing between successive wavebands in a multispectral data set.

7.7.2.3 Image classification

Density slicing. Density slicing is the simplest form of automated classification. It enables a single band (often of thermal IR data) to be colour-coded according to the grey level of the individual pixels. Thus, for example, pixel values 0 to 19 may be coloured blue, 20 to 25 red and so on. This not only aids interpretation but if the density slices are related to temperature levels, the resulting image can be used quantitatively to assess the variations in temperature across the scene.

Supervised classification. In cases where ground data exists about the terrain under investigation, it is possible to use this data to 'train' the computer to perform a simplified type of automated interpretation. For example, a particular group of pixels may be known to represent an area of water. Examination of the maximum and minimum numerical values of the radiances in each of the spectral bands being used (seven for Landsat TM) would enable a 'spectral pattern' for water to be determined. If the computer then compared this multidimensional spectral pattern with the spectral pattern of every pixel in the scene it would be possible to display only those areas which satisfied the criteria set by the training sample. Consequently, all water areas over a large area (185×185 km for Landsat) could be determined.

The process outlined above refers to the simplest form of classifier, the 'box' classifier. Other more sophisticated techniques are also available.

Unsupervised classification. In unsupervised classification of multispectral data, no attempt is made to 'train' the computer to interpret common features. Instead, the computer analyses all the pixels in the scene and divides them into a series of spectrally distinct classes based on the natural clusters which occur when the radiances in n-bands are compared. The user may now analyse other evidence to determine the nature of each of the distinctive classes which have been identified by the computer. Unsupervised classification techniques are generally more ex-pensive in terms of processing requirements than supervised techniques.

7.8 The use of remote sensing in civil engineering

Although aerial and satellite remote-sensing imagery (other than black-and-white aerial photography) have been used for topographic mapping, the primary role of this type of imagery has been for the production of thematic maps. Generally, for such applications users are satisfied with a relatively low level of positional accuracy and are also willing to accept a lower level of completeness than is the case with topographic maps. The type of remote-sensing data which is appropriate will depend largely on the stage of the civil engineering project and the degree of economic development of the region where the project is being carried out.

Five main stages can be identified for a civil engineering project: (1) reconnaissance; (2) preliminary planning/feasibility; (3) design; (4) construction; and (5) post construction/maintenance. The potential role of remote sensing at each stage will vary from project to project, but the following sections indicate some of the possible uses of the techniques at each stage.

7.8.1 Reconnaissance level investigations

The first stage of any major civil engineering project generally involves some form of preliminary reconnaissance study of the region or project area. The primary aim of this study is to collect together all available data concerning the physical characteristics of the terrain in order to assess its likely influence on the overall design of the engineering works.

Traditionally, the engineer has carried out this reconnaissance stage by examining existing maps of the region. Both topographic maps and specialist thematic maps (e.g. geological, geomorphological and pedological) are of greatest use at this initial stage of the project. In addition, examination of engineering reports produced for previous projects in the region may also provide valuable reference material. The main problems associated with this approach are that: (1) in many regions existing maps may be at a very small scale ($< 1:500\,000$), be significantly out of date and, in some instances, may not exist at any scale; and (2) that whilst engineering reports may give valuable data about a specific site they may not be appropriate or representative of regional terrain conditions. Consequently, many engineering projects have involved an examination of satellite and aerial imagery at this early stage.

The use of Landsat MSS data for projects of this type has been reported by Beaumont.[46] In this case, use of Landsat data for water resources planning in northeast Somalia is reported. Visual interpretation of Landsat data enabled a series of transparent overlays to be produced illustrating drainage, surface water, groundwater potential and land capacity over the region.

The use of remote sensing for terrain evaluation purposes is also of considerable importance. Terrain evaluation is a form of thematic mapping in which the terrain is classified in a hierarchical manner into units having common landscape patterns and engineering characteristics. This process can be carried out very cost-effectively using satellite imagery in conjunction with aerial photography.[47]

Interpretation of Landsat data may also be useful for the provision of additional supplementary information on hydrological phenomena, e.g. flooding, water quality, groundwater potential and water depth. A more general review of hydrological applications can be found in Blyth.[48]

Landsat has also been used for estimating urban populations

over large areas,[49] and using multitemporal imagery for monitoring change, e.g. the expansion of urban areas, changes in soil erosion, varying river channels, the impact of desertification and so on.

7.8.2 Preliminary planning/feasibility

The aim of the preliminary planning/feasibility stage of a project is: (1) to select potential routes/sites; and (2) to select from these the best from the available options. Depending on the state of economic development of the region, it is likely that different factors will influence the choice of the optimum route/site. For example, in developing regions the location of low-cost construction materials such as calcrete may be important[50] or the identification of potential bridge crossing-points. Both activities can be carried out cost-effectively using satellite imagery in conjunction with aerial photography. A comprehensive review of the role of remote sensing for highway feasibility studies can be found in Lawrance and Beaven.[51]

7.8.3 Design

Remote sensing has a particularly valuable role to play at the design stage of a project. At this stage a detailed site investigation of the proposed site or route will be required, together with supplementary environmental information which may influence the design of the project being considered. A critical stage of a site investigation is the desk study. The desk study provides the engineer with an overview of the project area and is normally carried out using aerial photography. Not only can valuable information be obtained during this interpretation about possible engineering problems but it can also be very important for the planning of the subsequent ground investigations. A general review of the potential of aerial photography in this respect is provided by Dumbleton[52] and Matthews,[53] and a selected number of papers which consider the use of remote sensing for geotechnical investigations are presented in Table 7.11.

Of increasing importance in recent years has been the role of remote sensing in providing environmental engineering data which may provide useful supplementary data to the engineer to input into the design process. A comprehensive review of remote sensing in this context is provided by Mason and Amos.[54] Information which may be extracted from remote-sensing imagery (often thermal IR linescan) includes data on energy conservation (heat loss), power station cooling-water patterns, groundwater and spring detection, pipeline and field drainage patterns, hydraulic leakage from earth and concrete dams and the location of ice-prone regions on roads.

7.8.4 Construction

The role of remote sensing is much smaller during construction than at the preceding planning and design stages. Nevertheless, aerial photography can be used to provide a valuable historical record of construction activities and may also be used to plan construction activities. Singhroy[67] provides an interesting case-history concerning the use of large-scale colour IR aerial photography and aerial video during pipeline construction in Canada. Both types of data were used to plan construction activities, determine site conditions and assess environmental effects before, during and after construction activity.

7.8.5 Post-construction maintenance

As mentioned in the previous section, remote sensing can be used to assess the environmental impact of construction activities. Remote sensing may also be used to assess the use of a new road by conducting aerial traffic investigations.[68] Nevertheless, the use of remote sensing at this stage is again of much less importance than during the previous four stages which have been discussed.

In conclusion, Table 7.12 illustrates the recommended use of remote-sensing techniques at each stage of a civil engineering project.

7.9 Sources of remote-sensing data

7.9.1 Satellite data

The most general source of satellite data and information on the availability and cost of data are the national points of contact (NPOC) which exist in most countries throughout the world. In the UK, for example, the NPOC is the National Remote Sensing Centre, Royal Aircraft Establishment, Farnborough.

Several other organizations may have archives of satellite data or may act as distributors for Landsat or SPOT data. Table

Table 7.11 Remote sensing and site investigations – selected references

	Aerial photography	MSS	Thermal IR linescanning
Landslides/slope instability	Norman, Leibowitz and Fookes[55] Matthews and Clayton[31]	Liu[56]	Chandler[57]
Solution features	Norman and Watson[58]	Coker, Marshall and Thompson[59]	Kennie and Edmonds[60]
Derelict/contaminated land	Bullard[61]	Coulson and Bridges[62]	Ellyet and Fleming[63]
Surfacial/subsurface materials surveys	Caiger[64]	Lynn[65]	Singhroy and Barnett[66]

Table 7.12

Project phase	Photograph/ image scale	Satellite						Aircraft				
		Photog.	Landsat MSS	Landsat TM	SPOT	MOMS	Side-looking radar	Photog.	MSS	Thermal	SLAR	Video
Reconnaissance	1:1 000 000 to 1:50 000	***	****	***	***	***	***	**	—	—	—	—
Planning/feasibility	1:100 000 to 1:20 000	***	**	***	***	***	***	**	—	—	—	—
Design	1:20 000 to 1:2000	—	—	—	*	—	—	****	***	***	***	**
Construction	1:2000 to 1:500	—	—	—	—	—	—	****	*	*	*	***
Post construction/ maintenance	1:2000 to 1:500	—	—	—	—	—	—	****	*	*	*	**

**** Very useful   *** Very useful (but coverage limited)   ** Useful   * Of limited use   — inadequate

7.13 lists a selected sample of the organizations who provide remote-sensing services and may also hold copies of satellite data.

7.9.2 Aerial photography

As mentioned in Table 7.13, several of the commercial remote-sensing organizations offer facilities for obtaining airborne thermal and MSS data and should be consulted if such data are required. However, aerial photography remains the most popular source of data for remote-sensing applications in civil engineering. Table 7.14 lists some of the organizations which have archives of aerial photographs. In addition, several of the organizations listed in Table 7.13 may also have holdings of aerial photographs.

Table 7.13 Sources of airborne and satellite data

Name of Company/Organization	Based in	Comments
Clyde Surveys Ltd	Maidenhead, Berks, England	Operate Daedalus thermal scanner
DFVLR	Oberpfaffenhofen, W. Germany	Distributors of MOMS and metric camera data
EOSAT Corp.	Arlington, Virginia, USA	Responsible for operation of Landsat
GeoSurvey Ltd	East Molesey, Surrey, England	—
Huntings Geology and Geophysics Ltd	Borehamwood, Herts, England	Operate Daedalus airborne MSS
EROS Data Centre	Sioux Falls, S. Dakota, USA	Prime inter. distribution centre for Landsat
Nigel Press and Associates	Edenbridge, Kent, England	Distributors of SPOT data in UK
SPOT Image	Toulouse, Cedex, France	Main operator for SPOT
SPOT Image Corp.	Washington DC, Reston, Virginia, USA	Distributors of SPOT data in N. America

Table 7.14 Sources of aerial photography

Name of Organization	Based in:
Aerofilms Ltd	Borehamwood, Herts, England
BKS Surveys Ltd	Coleraine, Co. Londonderry, N. Ireland
Cartographical Services Ltd	Salisbury, Wilts, England
Central Register of Aerial Photography of N. Ireland (Ordnance Survey)	Belfast, N. Ireland
Central Register of Aerial Photography of Wales	Cardiff, Wales
ERSAC Ltd	Livingston, W. Lothian, Scotland
Scottish Development Department	Edinburgh, Scotland
J. A. Story and Partners	Mitcham, Surrey, England
Ordnance Survey and Overseas Survey Directorate	Southampton, Hants, England
Royal Air Force Film Library	Whitehall, London
University of Cambridge Committee for Aerial Photography	Cambridge, England

References

1 Thompson, E. H. (1954) 'Heights from parallax measurements.' *Photog. Rec.* **1**, 4, 38–49.

2 Methley, B. D. F. (1970) 'Heights from parallax bar and computer.' *Photog. Rec.* **6**, 35, 459–465.

3 Royal Institution of Chartered Surveyors (1984) *Specification for vertical aerial photography*, 2nd edn. The British Air Survey Association and the Land Surveyor's Division of the RICS, 13pp. Surveyors Publications, London.

4 Thompson, E. H. (1964) 'Aerial triangulation by independent models.' *Photogrammetria*, **19**, 7, 262–274.

5 Schut, G. H. (1980) 'Block adjustment by bundles.' *'Canadian Surv.*, **34**, 2, 139–151.

6 Makarovic, B. (1973) 'Progressive sampling for digital terrain models.' *ITC J.* 397–416.

7 Ebner, H. (1980) Hoffman-Wellenhof, B., Reiss, P. and Steidler, F. 'HIFI–a minicomputer program package for height interpolation by finite elements.' *14th Congress of the International Society of Photogrammetry*, Hamburg, Commission IV, 14pp.

8 Kosli, A. and Wild, E. (1984) 'A digital terrain model featuring varying grid size.' *Contributions to the XVth ISPRS Congress*, Rio de Janeiro, Institute of Photography, University of Stuttgart, **10**, 117–126.

9 Petrie, G. and Kennie, T. J. M. (1986) 'Terrain modelling in surveying and civil engineering.' *Proceedings, British Computer Society Displays Group meeting on state of the art in stereo and terrain modelling*, London, 32pp.; *Computer Aided Design*, **19**, 4, 171–188.

10 Fraser, C. S. (1983) 'Photogrammetric monitoring of Turtle mountain – a feasibility study.' *Photog. Engr. and Remote Sensing*, **49**, 11, 1551–1559.

11 Fraser, C. S. and Gruendig, L. (1985) 'The analysis of photogrammetric deformation measurements on Turtle mountain.' *Photog. Engr. and Remote Sensing*, **51**, 2, 207–216.

12 Stewart, P. A. E. (1979) 'X-ray photogrammetry of gas turbine engines at Rolls-Royce', *Photog. Rec.*, **9**, 54, 813–821.

13 Atkinson, K. B. (1976) 'A review of close-range engineering photogrammetry'. *Photog. Eng. and Remote Sensing*, **42**, 1, 57–69.

14 Cheffins, O. W. and Chisholm, N. W. T. (1980) 'Engineering and industrial photogrammetry'. In: K. B. Atkinson (ed.) *'Developments in close-range photogrammetry*, vol. I. Elsevier Applied Science, London, pp. 149–180.

15 Cooper, M. A. R. (1984) 'Deformation measurement by photogrammetry.' *Photog. Rec.*, **11**, 63, 291–301.

16 Welsh, N. (1986) 'Photogrammetry in engineering.' *Photog. Rec.*, **12**, 67, 25–44.

17 Wong, K. W. and Wei Hsin Ho, (1986) 'Close-range mapping with a solid state camera.' *Photog. Engr. and Remote Sensing*, **52**, 1, 67–74.

18 Moore, J. F. A. (1973) 'The photogrammetric measurement of the constructional displacements of a rockfill dam.' *Photog. Rec.*, **7**, 42, 628–648.

19 Brandenberger, A. J. (1974) 'Deformation measurements of power dams.' *Photog. Engr.*, **40**, 9, 1051–1058.

20 Veress, S. A., Jackson, N. C. and Hatsopoulos, J. N. (1980) 'Monitoring a gabion wall by inclinometer and photogrammetry.' *Photog. Engr. and Remote Sensing*, **46**, 6, 771–778.

21 Cheffins, O. W. and Rushton, J. E. M. (1970) 'Edinburgh Castle Rock, a survey of the north face by terrestrial photogrammetry.' *Photog. Rec.*, **8**, 46, 417–433.

22 Moore, J. F. A. (1974) 'Major mapping joints in the lower Oxford Clay using terrestrial photogrammetry.' *Q. J. Engrg. Geol.*, **7**, 57–67.

23 Torlegard, A. K. I. and Dauphin, E. L. (1975) 'Deformation measurement by photogrammetry in cut-and-fill mining.' *Proceedings of the ASP Symposium on close-range photographic systems*, University of Illinois, pp. 24–39.

24 Robertson, G. R., MacRae, A. M. R., Tribe, J., Sibley, D. W. and Smith, D. H. (1982) 'Use of photogrammetric methods for mine slope deformation surveys.' *4th Canadian Symposium on Mining and Deformation Monitoring*, Toronto.

25 Heath, W., Parlsey, L. L. and Dowling, J. W. F. (1978) 'Terrestrial photogrammetric surveys of unstable terrain in Columbia.' Transport and Road Research Laboratory, Publication No. LR876, TRRL, London.

26 Kennie, T. J. M. and McKay, W. M. (1986) 'Monitoring of geotechnical processes by close-range photogrammetry.' *Proceedings, 22nd Conference of the Engineering Group of the Geological Society*, Plymouth Polytechnic.

27 Anderson, H. and Stevens, D. (1982) 'Mono photogrammetric tunnel profiling.' *International Archives of Photographers' Commission 5 Symposium, York, Precision and speed in close-range photogrammetry*, **24**, 1, 23–30.

28 Andrawes, K. Z. and Butterfield, R. (1973) 'The measurement of planar displacements of sand grains', *Géotechnique*, **23**, 4, 571–576.

29 Wong, K. W. and Vonderohe, A. P. (1981) 'Planar displacement by motion parallax.' *Photog. Engr. and Remote Sensing*, **47**, 6, 769–777.

30 Davidson, J. L. (1985) 'Stereophotogrammetry in geotechnical engineering.' *Photog. Engr. and Remote Sensing*, **51**, 10, 1589–1596.

31 Matthews, M. C. and Clayton, C. R. I. (1984) 'The use of oblique aerial photography to assess the extent and sequence of landslipping at Stag Hill, Guildford.' *Proceedings 20th Regional Conference of the Engineering Group of the Geological Society of London*, Guildford, vol I, pp. 319–330.

32 Heath, W. (1980) *Inexpensive aerial photography for highway engineering and traffic studies.* Transport and Road Research Laboratory Supplementary Report No. 632, TRRL, London, 24pp.

33 Beaumont, T. E. (1977) *Techniques for interpretation of remote sensing imagery for highway engineering purposes.* Transport and Road Research Laboratory Report No. 753, 24pp.

34 Graham, R. W. and Read, R. (1984) 'Small format aerial photography from microlight platforms.' *J. Photog. Sc.*, **32**, 100–109.

35 Tomlins, G. F. (1983) 'Some considerations in the design of low cost remotely piloted aircraft for civil remote sensing applications.' *Can. Surveyor*, **37**, 157–167.

36 Meisner, D. E. and Lindstrom, O. M. (1985) 'Design and operation of a colour infra-red aerial video system.' *Photog. Engr. and Remote Sensing*, **51**, 5, 555–560.

37 Kennie, T. J. M., Dale, C. D. and Stove, G. C. (1986) 'A preliminary assessment of an airborne thermal video frame scanning systems for environmental engineering surveys.' *ISPRS, Commission VIII. Proccedings International Symposium on remote sensing for resources development and environmental management*, Enschede, Balkema Press, pp. 215–221.

38 Stove, G. C., Kennie, T. J. M. and Harrison, L. (1987) 'Airborne thermal mapping for winter highway maintenance using the Barr and Stroud IR18 thermal video frame scanner.' *Int. J. Remote Sensing*

39 Rudd, R. O. (1974) *Remote sensing–a better view*. Daxbury Press, Massachusetts.

40 NOAA *'Landsat data users' notes*. Issues 1–36.

41 Chevrel, M., Courtois, M. and Wells, G. (1981) 'The SPOT satellite remote sensing mission.' *Photog. Engrg and Remote Sensing*, **47**, 1163–1171.

42 Fearns, D. C. (1984) 'Microcomputer systems for satellite image processing.' *Earth Orient. Appl. Space Technol.*, **4**, 4, 247–254.

43 Bernstein, R. (1978) *Digital image processing for remote sensing*. IEEE Press, New York, 473pp.

44 Avery, T. E. and Graydon, L. B. (1985) 'Digital image processing.' In: *Interpretation of aerial photographs*, 4th edn, Chapter 15, Burgess Publishing Co., Minneapolis, pp. 451–536.

45 Bagot, K. H. (1985) 'Digital processing of remote sensing data.' In: T. J. M. Kennie and M. C. Matthews (eds) *Remote sensing in civil engineering*, Surrey University Press, London, pp. 87–105.

46 Beaumont, T. E. (1982) 'Land capability studies from Landsat satellite data for rural road planning in North East Somalia.' *Proceedings, OECD symposium on Terrain evaluation and remote sensing for highway engineering in developing countries*. Transport and Road Research Laboratory Report No. SR690, pp. 86–95.

47 Overseas Unit (1982) *Terrain evaluation and remote sensing for highway engineering in developing countries*. Transport and Road Research Laboratory Supplementary Report No. 690, 172pp.

48 Blyth, K. (1985) 'Remote sensing and water resources engineering.' In: T. J. M. Kennie and M. C. Matthews (eds), *Remote sensing in civil engineering*, Surrey University Press, London, pp. 289–334.

49 Forstner, G. (1983) 'Some urban measurements from Landsat data.' *Photog. Engr and remote sensing*, **49**, 12, 1693–1707.

50 Beaumont, T. E. (1979) 'Remote sensing for location and mapping of engineering construction materials in developing countries.' *Q. J. Engrg Geology*, **12**, 3, 147–158.

51 Lawrance, C. J. and Beaven, P. J. (1985) 'Remote sensing for highway engineering projects in developing countries.' In: T. J. M. Kennie and M. C. Matthews (eds), *Remote sensing in civil engineering*, Chapter 9, Surrey University Press, London, pp. 240–268.

52 Dumbleton, M. J. (1983) *Air photographs for investigating natural changes, past use and present condition of engineering sites*. Transport and Road Research Laboratory Report No. 1085, TRRL, London.

53 Matthews, M. C. (1985) 'Interpretation of aerial photography,' Chapter 8. In: T. J. M. Kennie and M. C. Matthews (eds) *Remote sensing in civil engineering*, Surrey University Press, London, pp. 204–239.

54 Mason, P. A. and Amos, E. L. (1985) 'Environmental engineering applications of thermal infrared imagery.' In: T. J. M. Kennie and M. C. Matthews (eds) *Remote sensing in civil engineering*, Chapter 10, Surrey University Press, London, pp. 269–288.

55 Norman, J. W., Leibowitz, T. H. and Fookes, P. G. (1975) 'Factors affecting the detection of slope instability with air photographs in an area near Sevenoaks, Kent.' *Q. J. Engrg Geol.*, **8**, 3, 159–176

56 Liu, J. K. (1985) 'Remote sensing for identifying landslides and for landslide prediction-cases in Taiwan.' *International Conference on advanced technology for monitoring and processing global environmental data*, Remote Sensing Society, London, 8pp.

57 Chandler, P. B. (1975) 'Remote detection of transient thermal anomalies associated with the Portuguese Bend landslide.' *Bull. Assoc. Engrg Geol.*, **12**, 3, 227–232.

58 Norman, J. W. and Watson, I. (1975) 'Detection of subsidence conditions by photogeology.' *Engrg Geol.*, **9**, 359–381.

59 Coker, A. E., Marshall, R. and Thompson, N. S. (1969) 'Application of computer-processed multispectral data to the discrimination of land collapse (sinkhole) prone areas in Florida.' *Proceedings 6th International Symposium on remote sensing of the environment*, Michigan, vol. I, pp. 65–69.

60 Kennie, T. J. M. and Edmonds, C. N. (1986) 'The location of potential ground subsidence and collapse features in soluble carbonate rocks by remote sensing techniques.' *Proceedings, American Society for Testing and Materials International Symposium on geotechnical applications of remote sensing and remote data transmission*. ASTM Special Technical Publication, pp. 206.

61 Bullard, R. K. (1983) *Abandoned land in Thurrock: an application of remote sensing*. Working Paper No. 8, North East London Polytechnic, London.

62 Coulson, M. G. and Bridges, E. M. (1984) 'The remote sensing of contaminated land.' *Int. J. Remote Sensing*, **5**, 4, 659–669.

63 Ellyet, C. D. and Fleming, A. W. (1974) 'Thermal infrared imagery of the burning mountain coal fire.' *Remote sensing of environment*, pp. 79–86.

64 Caiger, J. H. (1970) 'Aerial photographic interpretation of road construction materials in South Africa with special reference to its potential to influence route location in underdeveloped territories.' *Photogrammetria*, **25**, 151.

65 Lynn, O. W. (1984) 'Surface material mapping in the English fenlands using airborne multispectral scanner data.' *Int. J. Remote Sensing*, **5**, 4, 699–713.

66 Singhroy, V. and Barnett, P. (1984) 'Locating subsurface mineral aggregate deposits from airborne imagery.' A case study in southern Ontario, *International Symposium on remote sensing for exploration geology*, Colorado.

67 Singhroy, V. (1986) 'Case studies on the application of remote sensing data to geotechnical investigations in Ontario.' In: *Geotechnical applications of remote sensing and remote data transmission*. American Society for Testing and Materials Special Technical Publicaton No. 206.

68 Mountain, L. J. and Garner, J. B. (1981) 'Semi-automatic analysis of small-format photography for traffic control studies of complex intersections.' *Photogrammetric Rec.* **10**, 331–342.

Bibliography

Atkinson, K. B. (ed.) (1980) *Developments in close-range photogrammetry*, Elsevier Applied Science, London, 222pp.

Burnside, C. D. (1979) *Mapping from aerial photographs*. Granada, London, 304pp.

Colwell, R. W. (ed.) (1983) *Manual of remote sensing*, vols I and II, 2440pp.

European Space Agency (1984) *Remote sensing applications in civil engineering*. ESA, Publication No. SP–216, 198pp.

Karara, H. M. (ed.) (1979) *Handbook of nontopographic photogrammetry*. American Society of Photogrammetry, 206pp.

Kennie, T. J. M. and Matthews, M. C. (eds) (1985) *Remote sensing in civil engineering*, Surrey University Press, London, 356pp.

Kilford, W. K. (1979) *Elementary air survey*, 4th edn, Pitman, 345pp.

Lillesand, T. M. and Kiefer, R. W. (1979) *Remote sensing and image interpretation*. Wiley, New York, 611pp.

Moffit, F. H. and Mikhail, E. (1980) *Photogrammetry*, 3rd edn., Harper and Row, New York, 648pp.

Slama, C. C. (ed.), (1980) *Manual of photogrammetry*, 4th edn., American Society of Photogrammetry, 1056pp.

Wolf, P. R. (1983) *Elements of photogrammetry*, 2nd edn., McGraw-Hill, New York, 628pp.

8

Geology for Engineers

P G Fookes DSc(Eng), PhD, BSc, FIMM, FIGS
Consultant

Contents

This chapter introduces civil engineers to some basic geology and outlines the broad concepts of the subject.

Geology is concerned with the science of the Earth and the materials comprising the Earth. This includes *physical geology* or *geomorphology* (the surface form of the Earth), *palaeontology* (study of fossils), *stratigraphy* (the chronological sequence of rocks), *mineralogy* (study of minerals), *petrology* (study of the composition of rocks) and *structural geology* or *tectonics* (the broad structure of rocks). Together with newer and closely related branches such as geochemistry, geophysics or mathematical geology, and applied and biological aspects, the whole subject is rapidly developing and is now generally being called Earth Science.

Engineering geology is the branch of geology applied to civil engineering and, in Britain particularly, is applied to all aspects of foundation and excavation design, construction and performance. The extremes of the subject merge into the practices of soil mechanics, rock mechanics and some aspects of the extractive industries, as sand and gravel or opencast mining (Price[1]).

8.1 Basic geology

8.1.1 Introduction

Rock is strictly defined in geology as any natural solid portion of the Earth's crust which has recognizable appearance and composition. Some rocks are not necessarily hard, and in discussion a geologist may call peat or clay a rock as he would granite or limestone.

There are three major classes of rocks:

(1) *Sedimentary rocks* formed by the deposition of material at the Earth's crust, e.g. sandstone, clay.
(2) *Igneous rocks* formed from molten rock magma solidifying either at the Earth's surface or within the crust, e.g. basalt, granite (*s.l.*).
(3) *Metamorphic rocks* produced deep in the Earth by the transformation of existing rocks through the action of heat and pressure, e.g. marble, slate.

The interrelation and continual recycling of rock over long periods of geological time is illustrated in Figure 8.1.

8.1.2 Principles of stratigraphy

Sedimentary rocks cover some 75% of the Earth's land surface but form only a discontinuous and relatively thin cover to the underlying igneous and metamorphic rocks of the mantle.

The sedimentary layers (*strata*) normally lie one above another in order of decreasing age, but where there has been structural disturbance they are faulted and folded. Study of the strata in a particular area enables their sequence to be recorded, and this can then be compared with other local sequences. From such observations the general succession of sedimentary rocks over a wider area can be established: this has been done, for example, for nearly the whole of the British Isles. A list of strata for England and Wales was compiled by William Smith, 'the father of English geology'; in 1815 he produced the first simple coloured geological map of the country. As a result of his studies he stated two basic principles of stratigraphy, that 'the same strata are always found in the same order of superposition, and contain the same peculiar fossils'. These principles are still used to determine the relative ages of strata, i.e. in the order of superposition for an undisturbed series of sedimentary beds, the oldest (i.e. the first deposited) is at the bottom, and successively younger beds lie upon it. Sedimentary strata in different localities can usually be correlated by the diagnostic fossil remains they contain. Rapidly evolving fossils act as horizon markers so that a specimen of one of these enables the particular level of the rock outcrop in which it occurs to be identified in the geological column wherever in the world it is found.

The whole sequence of rocks comprising the geological column is broadly divided into the systems and groups shown in Table 8.1; this column applies particularly to British strata. The column shows the age of each group relative to the others, and was in use long before any of the recent radiometric methods of determining the absolute age in years was developed. Names of the geological systems, and of the larger groups are of worldwide application; they are also used to express the periods of time during which the rocks of the different systems were formed, e.g. the Jurassic system and the Jurassic period, or Mesozoic group and the Mesozoic era. The times of major mountain-building episodes (*orogenies*) and of phases of igneous activity in Britain are given in the third column of the table.

There are numerous further subdivisions down to 'zones' and even 'horizons', many of the smaller divisions being based on specific fossils.

In any given area the deposition of sediments was not continuous throughout the geological periods. There are breaks in the sequence of deposits, marked by *unconformities* which represent intervals of time during which there was no deposition

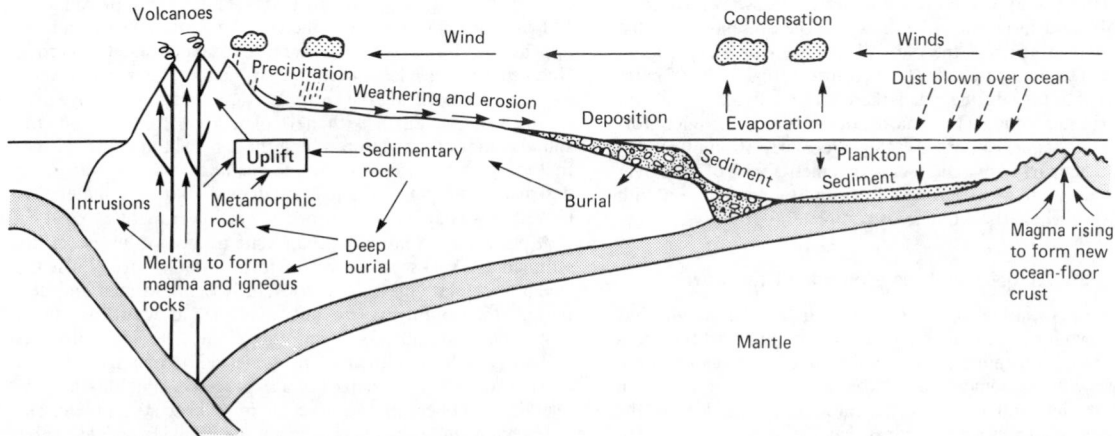

Figure 8.1 Diagrammatic representation of the long-term cycling of rocks. (After Bradshaw, Abbot and Gelsthorpe (1978) *The Earth's changing surface.* Hodder and Stoughton, London)

and erosion took place. The sea floors with their sediments were raised and became subject to erosion by wind and water. There were also periods of quiet sedimentation, when seas covered the land, and intervening episodes of disturbance when uplift and folding took place. This broad pattern of events – the transgression of the sea over the lands, then the regression of the sea, followed by orogenic upheaval – has been repeated many times throughout geological history (see Figure 8.2 which shows the typical simplified borehole sequence of such a chain of events).

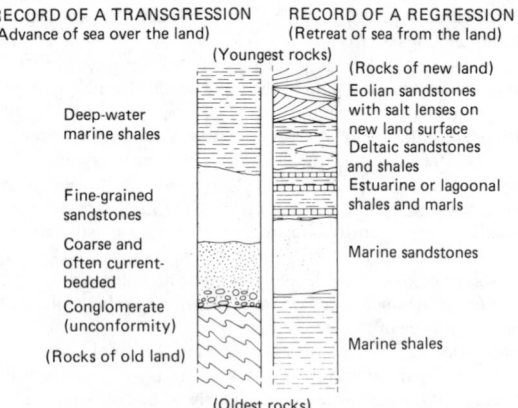

RECORD OF A TRANSGRESSION (Advance of sea over the land)

RECORD OF A REGRESSION (Retreat of sea from the land)

Figure 8.2 Marine transgression and regression as seen idealized in borehole core, tens of metres long. (After Read and Watson (1971) *Beginning geology*, 2nd edn. Macmillan/Allen and Unwin, London)

Unconformities are often marked by beds of pebble gravel, the beach deposits of a sea which gradually inundated the land during its submergence (see Figure 8.3). Examples of this are the pebbly quartzites at the base of the Cambrian, or the rounded flints at the base of the Eocene deposits of southeast England overlying the Chalk, both marking the oncoming of marine transgression. Boulder beds and hill or mountain screes formed on an old land surface during erosion, after uplift has taken place, may also be preserved as the lowest members of a newer series of rocks resting unconformably on older rocks; an example is the boulders and coarse sands at the base of the Torridonian in northwest Scotland which lie unconformably on an old land surface carved in the underlying Lewisian rocks.

An old land surface may be shown by the presence of a 'dirt bed' in which some of the old soil has been preserved, as at Purbeck, Dorset, or by other land-formed deposits. It indicates an interval of time during which there was locally no deposition of waterborne sediments. In marine deposits a minor unconformity (or nonsequence), representing a local cessation in deposition, can be marked by the absence of a metre or so of beds over a relatively small area. This can be found by comparison with other areas where the sequence is complete.

8.1.3 Plate tectonics and the evolution of the Earth

The close association of volcanic and earthquake activity has been known for some time but it is only during the last few years that it has been more or less understood. This association, together with the coincidence of young narrow fold mountain ranges on the continents, and trenches and ridges deep in the ocean basins also in narrow zones, has led to a new theory of Earth evolution known as plate tectonics. This idea was proposed in the late 1960s and has been received with widespread

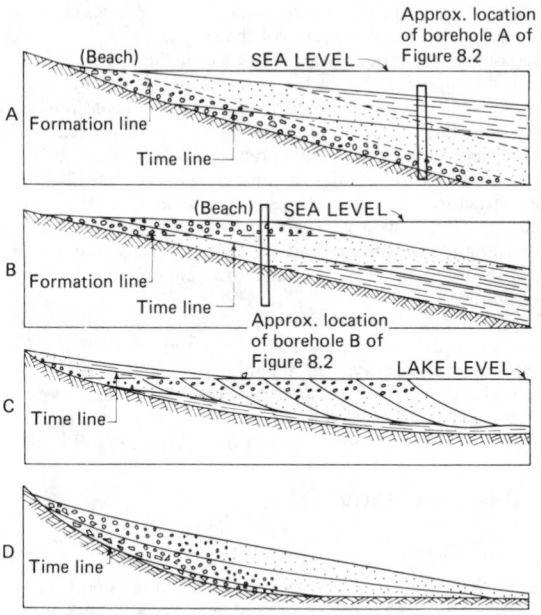

Figure 8.3 Examples of common marine and freshwater transgressions and regressions showing types and geometric distribution of sediment deposited. (After Lahee (1961) *Field geology*, 6th edn. McGraw-Hill, New York.) Lines parallel to lake and sea floors are time lines as they join sediment deposited contemporaneously. Lines essentially parallel to gravel, sand or clay deposits are formation lines. A, a marine transgression over the land; B, a marine regression from the land; C, a lake regression from the land; lake bottom muds are gradually covered by coarser sediments. Later transgression is shown left of a; D, an alluvial transgression by growth of a cone of river alluvium in mountainous area overlooking a desert plain

acceptance as more evidence has been found to fit the general model.

The concept suggests that the Earth's surface layers are divided into large segments or plates. Plates are approximately 100 km thick and therefore include the Earth's crust and the upper mantle, and measure several thousand kilometres across. One scheme considers there are six major plates and several smaller ones, covering the entire Earth. Plates slowly move over the face of the Earth with new plate rock formed from the solidification of slowly upwelling molten rock at the constructive margin as more new rock forms and travels towards the destructive margin where it is subducted, and rock material is moved downwards and returned to the lower mantle.

A plate may eventually accumulate a mass of lower density sedimentary rocks on its top to form a continent. Whilst the ocean-floor plate material is constantly being formed and destroyed, the continents are not consumed downwards at the destructive margin because their low density provides buoyancy. The continents are subjected to changes due to erosion and deposition by surface processes, but this has the overall effect of causing relatively light rocks to accumulate. The oldest known continental rocks are 3900 million yr old (Table 8.1) but nowhere are the ocean floor rocks known to be more than 200 million yr old.

Table 8.1 The geological column

Name of geological group or era		Name of geological system or period (ages in millions of years)	General nature of deposits, major orogenies, and igneous activity in Britain
CAINOZOIC	Quaternary	Recent ⎫ Pleistocene ⎬ (2)	Alluvium, blown sand, glacial drifts, etc. At least five ice ages separated by warmer periods. The Devensian (Weichselian or Newer Drift) is the last ice age
	Tertiary	Pliocene ⎫ Miocene ⎪ Oligocene ⎬ Eocene ⎭ (70)	Sands, clays, and shell beds *Alpine orogeny* *Igneous activity in Scotland and Ireland*
MESOZOIC (or Secondary)		Cretaceous Jurassic Triassic (225)	Sands, clays and chalk Clays, limestones, some sands Desert sands, sandstones and marls
PALAEOZOIC (or Primary)	Newer	Permian	Breccias, marls, dolomitic limestone *Hercynian orogeny* *Igneous activity*
		Carboniferous Devonian (and Old Red Sandstone) (c. 400)	Limestones, shales, coals and sandstones Marine sediments (Lacustrine sands and marls) *Igneous activity* Caledonian orogeny
		Silurian Ordovician Cambrian (c. 600)	Thick shallow-water sediments, shales and sandstones Older Volcanic activity in the Ordovician
PRE-CAMBRIAN		*Dalradian* ————— Schists ————— *Moinian* (740 +)	Schists and granulites
		Torridonian Uriconian Lewisian (3500 +)	Sandstones and arkoses Lavas and tuffs (Shropshire) *Pre-Cambrian orogenies* Orthogneisses, etc.

8.2 Geological description and classification of rock

Engineering classification of rock is discussed in Chapter 10, and engineering classification of soils in Chapter 9.

8.2.1 Sedimentary rocks

Sediments originate mainly from the weathering of all rocks, especially igneous rocks. Certain resistant minerals in igneous rocks such as quartz survive unchanged and are eventually incorporated in the new sediments; often they tend to be concentrated in certain types of sediment (e.g. sands). Other igneous minerals, such as the feldspars and ferromagnesian minerals, break down during weathering to give rise to new minerals and to colloidal and dissolved substances. The new minerals, chiefly clay-minerals, are concentrated in a second group of sediments (e.g. clays) and the colloidal matter, usually iron hydroxides, in a third. The substances taken into solution include calcium and magnesium salts which are precipitated by chemical and organic processes as carbonate rocks, and sodium

and potassium salts which may in certain circumstances crystallize out to give evaporites. Another group of sediments including coal and peat is produced by the piling up of decaying plant matter.

The products of weathering can be related, as is shown diagrammatically in Figure 8.4, into fairly distinct chemical and geological groups. This natural differentiation provides a simple classification of sediments into two broad groups:

(1) Detrital sediments made by the accumulation of fragmented particles of minerals or rocks, represented by (a) the pebbly rocks, and (b) the sands, made chiefly of inherited minerals or rocks, and (c) the clays made chiefly of new minerals.
(2) Chemical–organic sediments formed by the precipitation of material from solution or by organic processes, represented mainly by the limestones, the evaporites and the coals.

The sediments produced go on changing after deposition; e.g. they may be saturated by groundwater carrying salts in solution, or deformed by the weight of new sediments laid down on top of them. Changes produced by such means are called *diagenetic*

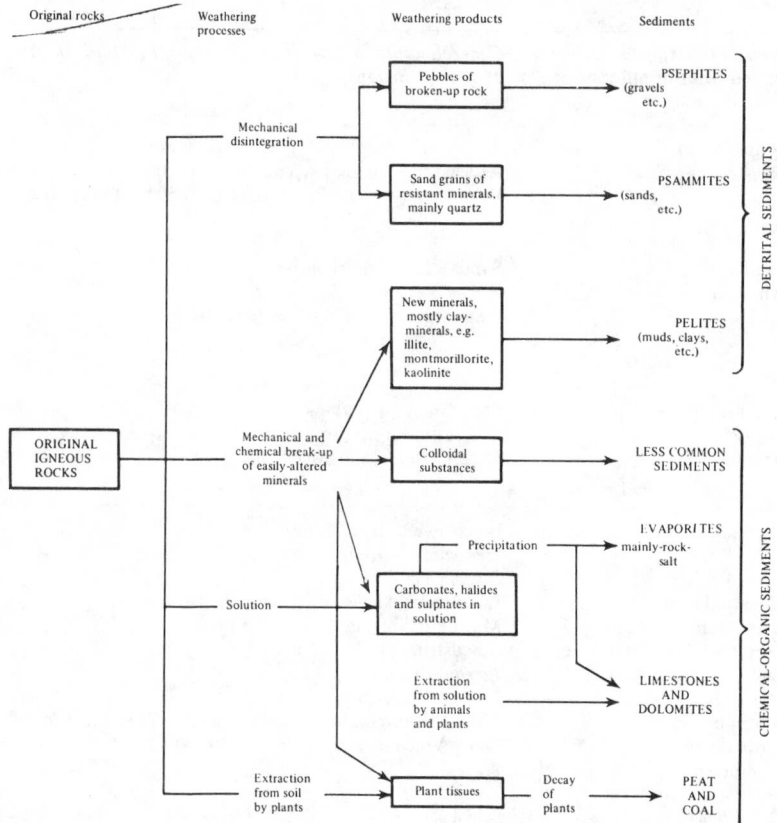

Original rocks — Weathering processes — Weathering products — Sediments

Figure 8.4 Sedimentary differentiation. (After Read and Watson (1971) *Beginning geology,* 2nd edn. Macmillan/Allen and Unwin, London)

changes and convert the *sediments* into consolidated or lithified (hardened) *sedimentary rocks*, e.g. a sand becomes a sandstone.

8.2.1.1 Deposition environments and textures of sedimentary rock

The characteristics and to a certain extent the engineering performance of recent sediments can be directly related to the environment occurring at their location of deposition, because the agents of deposition can still be seen in action. In the older sedimentary rocks, the environment of deposition can be reconstructed from the characters of the rocks themselves. The evidence for this reconstruction is provided by the *composition* and *texture* of the rock, the type of bedding, the fossil content and the relationship between any one bed and its neighbours. The sum of all these features decides its sedimentary *facies* and from this it is generally possible to deduce the conditions under which each rock was formed. This is summarized in Table 8.2.

The most obvious and characteristic feature of sedimentary rocks is *bedding,* i.e. the presence of recognizably different beds or strata in a sedimentary succession, and the presence within any one bed of depositional surfaces which are the bedding planes (see Figure 8.5).

Although many beds are homogeneous, some show considerable variation, especially *graded beds,* in which there is a passage from coarser to finer particles towards the top; lateral gradation may also be found. Thin laminae or layers, differing somewhat in colour or texture, may be present without causing a bed to lose its individuality. A bed is characterized by all of its

lithological features. These indicate that it was laid down in a particular environment, either uniform, or varying systematically. Although it may be arbitrary, some very thin strata may best be regarded as beds rather than as laminae within a bed. For example, in glacial varves each annual deposit of summer silt and winter clay is an individual bed even though its thickness is measured in millimetres, whereas sandy laminae in a graded greywacke are parts of the whole graded unit (see Figure 8.6).

In describing bedding it is necessary to distinguish firstly between bedding planes which are individual structures where each planar surface may be distinguished, and also depositional textures, which result from the parallel orientation of particles throughout a bed. Both are primary depositional features, and may be either parallel or inclined to the separation planes, bounding individual beds (Figure 8.6). In addition, various textures, the parallel orientation of mica-flakes, for example, may be induced by post-depositional effects such as consolidation. These are post-depositional *fabrics* but in many instances they are very difficult to separate from true depositional fabrics.

8.2.2 Igneous rocks

The important characteristics of igneous rocks are the chemical composition, the mineral composition and the texture.

8.2.2.1 Chemical composition

The chemical composition depends on the magma from which the igneous rock was derived. Some 99% of the various igneous

Table 8.2 Environments of deposition of sedimentary rocks

Environment of deposition			Common sedimentary rocks produced by the environment
SEA			
Shallow seas (continental shelf)	Littoral (beaches, sandbanks, tidal flats)		Conglomerate, sandstone, shale
	Neritic	{ Shelf seas in stable areas	Orthoquartzite, current-bedded sandstone, shale, organic and chemical limestones
		{ Restricted deep basins	Black shale
Deep seas		{ Geosynclinal seas in mobile belts	As for shelf seas with in addition greywackes
		{ Deep seas in stable areas	and other turbidite deposits
Abyssal seas			Calcareous ooze, siliceous ooze, Red Clay
LAND/SEA			
Deltas			Mainly sandstone, shale
Estuaries, lagoons			Shale
LAND			
Floodplain			Conglomerate, sandstone, shale
Lakes		{ with outlet to sea	Sandstone, shale, freshwater limestone
		{ in basins of interior drainage	Sandstone, shale, evaporates
Deserts			Sandstone, conglomerate, breccia
Piedmont (intermontane basins, alluvial fans)			Conglomerate, breccia, arkose, sandstone
Areas of glaciation			Tillite

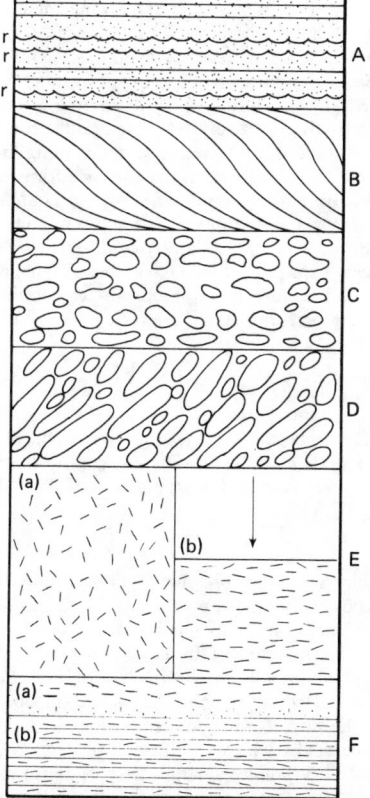

rocks are made up by combinations of only eight elements. Of these, oxygen is dominant, next is silicon and then aluminium, iron, calcium, sodium, potassium and magnesium. In terms of oxides, silica (SiO_2) is by far the most abundant, ranging from 40 to 75% of the total. The silica percentage therefore forms the basis of a fourfold chemical classification of the igneous rocks, the limits being given on Figure 8.7.

8.2.2.2 Mineral composition

Mineral composition depends largely upon the chemical composition. The chief minerals present will normally be silicates of the six common metal cations noted, together with quartz, when silica is present in excess. The minerals which actually form will be controlled by the silica percentage and the relative abundance of the cations. For example, silica-poor silicates such as olivine

Figure 8.5 Idealized types of sedimentary bedding. (After Sherbon Hills (1972) *Elements of structural geology*, 2nd edn. Chapman and Hall, London) A, sandstone with discrete bedding planes parallel to separation planes. Some beds ripple-marked (r); B, sandstone with discrete bedding planes inclined to separation planes (false or cross-bedding; an inclined deposition texture); C, conglomerate with long axes of pebbles approximately parallel to separation planes (a parallel depositional texture); D, edgewise conglomerate with long axes of pebbles inclined to separation planes (an inclined depositional texture); E(a), unconsolidated mud with random orientation of mica flakes and clay particles (a random depositional texture); E(b), consolidated clay or lithified mudstone with flaky particles approximately parallel, and parallel with separation planes (a parallel consolidation texture); F(a), mudstone with mica flakes deposited parallel to separation planes, but lacking discrete bedding planes (a parallel depositional texture, cf. C above); F(b), mudstone with mica flakes deposited parallel to separation planes, and showing discrete bedding planes. A thin bed of sandstone lies between the two mudstones

Approx. scale 1 m

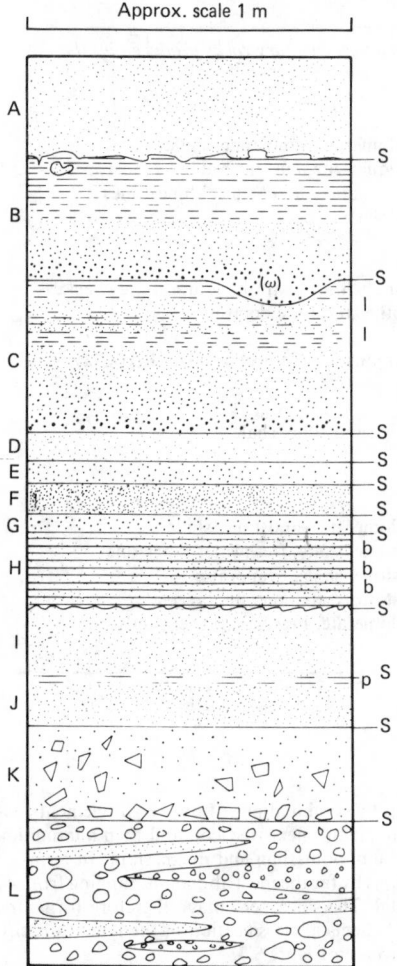

Figure 8.6 Idealized types of sedimentary beds. Beds are bounded by separation planes (S). A, uniform, massive sandstone with bottom structures at its base; B, simple graded bed, with uniform grading from coarse sandstone below to shale above and a washout (w); C, complex graded bed with thin sandstone laminae (l, l) in the shales; D, E, F, G, individual thin beds; H, single sandstone bed with discrete bedding planes (b, b, etc.); I, J, two sandstone beds separated by shale parting (p); K, heterogeneous bed of sandstone containing angular shale fragments; L, heterogeneous bed of conglomerate containing lenses of sand and gravel

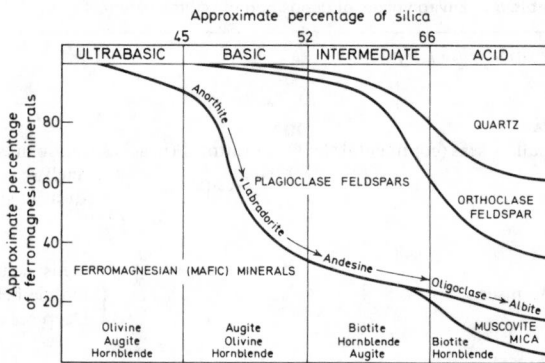

Figure 8.7 A classification of igneous rocks based on a silica percentage

will be most abundant in the ultrabasic and basic rocks and absent from the silica-rich acid rocks.

The chief minerals are quartz, orthoclase and plagioclase feldspars, micas, amphiboles, pyroxenes and olivines. Their distribution in the four chemical groups – ultrabasic, basic, intermediate and acid – established by silica percentage is shown diagrammatically in Figure 8.7. Many of the names given to igneous rocks are defined according to the presence of two or three particular minerals which are the essential minerals for that rock type. Other accessory minerals may also be present in small quantities, e.g. the essential minerals of granite are quartz, feldspar and mica; common accessories are zircon and iron oxide.

The predominant minerals of an igneous rock may determine its general appearance and it is usually possible to get some idea of its composition from its colour and density. Quartz is commonly colourless and transparent, feldspars opaque but pale coloured. Rocks made mostly of these minerals (i.e. acid and intermediate rocks) are therefore usually pale in colour and relatively light in weight. The coloured ferromagnesian silicates, olivines, pyroxenes and amphiboles, are abundant in basic and ultrabasic rocks which are usually dark and relatively heavy. Two important exceptions are very fine-grained or glassy rocks which tend to look dark whatever their composition, and weathering or other alteration which changes the colours of minerals. It is, therefore, usually necessary to look at fresh-broken surfaces to diagnose the parent rock type.

8.2.2.3 Texture

The texture of an igneous rock is shown by the arrangement of the constituent minerals and the relation of each mineral to its neighbours. The main textural character is the grain size and in a general way this depends on the rate of cooling of the magma. Coarse-grained rocks are the result of slow cooling which allowed time for the growth of large crystals; fine-grained rocks are produced by rapid cooling. By extremely rapid cooling, no time at all is given for crystallization and glasses are formed. *Holo-crystalline* rocks are entirely crystalline, *hypo-crystalline* are partly crystals, partly glass. A common distinctive texture is the *porphyritic* texture in which crystals of two different sizes occur: large *phenocrysts* are scattered through a finer-grained or glassy groundmass. The texture is an important controlling factor in the engineering performance of the rock.

8.2.2.4 Classification

Classification of the common igneous rocks is usually made on the basis of grain size and silica percentage as given in Table 8.3. The characteristic minerals of rocks of different compositions are shown in Figure 8.7 which should be studied with Table 8.3.

8.2.2.5 Form

A body of magma which is under pressure in the sial may be forced upwards intruding into the upper rocks of the crust. During the process of intrusion it may incorporate some of the rocks with which it comes into contact, by assimilation. In some cases it may also give off mobile fluids which penetrate and change the rocks in its immediate neighbourhood and mineralization may occur. If the intrusive magma cools at some depth below the surface, the rocks which result are called *plutonic* rocks and are coarsely crystalline; a large mass of this kind constitutes a major intrusion, e.g. a granite batholith which may

Table 8.3 A classification of igneous rocks on silica percentage and grain

Basic	Intermediate	Acid
Coarse-grained (plutonic) rocks. Grain size larger than about 5 mm. Liable to be brittle owing to presence of large crystals		
Gabbro	Syenite	Granite
Norite	Diorite	Granodiorite
(Not very common in the British Isles)	(Comparatively rare in the British Isles)	(Widely distributed in the British Isles)
Medium-grained (hypabyssal) rocks. Grain size between about 1 and 5 mm. Very frequently possess intergrown texture: include some of the best roadstones		
Dolerite	Porphyry	Microgranite
Diabase	Porphyrite	Granophyre
(Widely distributed in the British Isles)	(Fairly common in the British Isles)	(Fairly common in the British Isles)
Fine-grained (volcanic) rocks. Grain size below about 1 mm, i.e. below the limit of visible recognition. Similar to medium-grained rocks, but sometimes liable to be brittle and splintery		
Basalt	Trachyte	Rhyolite
Spilite	Andesite	Felsite
(Widely distributed in the British Isles)	(Not very common in the British Isles)	(Not very common in the British Isles)

←————————————— Continuous variation in properties ————————— ————————————————————————————→

Dark colour←—— ————————————————————————→Light colour

(Due to increase in ferromagnesian minerals)

High specific gravity←———————————————————————————————— ——————————————————→Low specific gravity
(2.9)   (Due to increase in ferromagnesian minerals)   (2.6)

have an aureole of thermally altered rock. When magma rises and fills fractures or other lines of weakness in the crust, it forms minor intrusions. These include *dykes*, which are steep or vertical wall-like masses, with more or less parallel sides, and *sills*, which are sheets of igneous rock intruded between bedding planes of sedimentary rocks and lying more or less horizontal. Dyke and sill rocks commonly have a fine-grained texture. Veins are smaller and irregular bodies of igneous material, filling cracks which may run in any direction.

Magma which rises to the Earth's surface and flows out as a lava, is called extrusive, and under these conditions it loses most of its gas content. These are the *volcanic* rocks, and since they have cooled comparatively quickly in the atmosphere they are frequently glassy (i.e. noncrystalline), or very fine-grained with some larger crystals.

These forms are summarized in Figure 8.8.

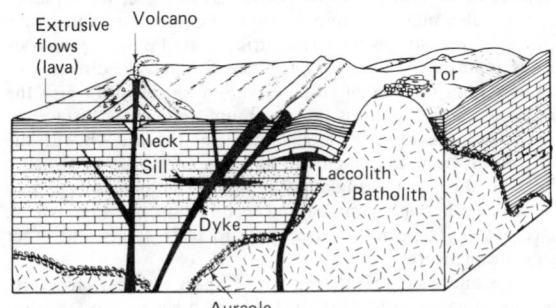

Figure 8.8 Idealized forms of intrusive plutonic rocks

8.2.2.6 Structure

The use of the term *structure* is reserved for more pronounced features of a rock than those described by the term 'texture'. In igneous rocks the structure may indicate a relative arrangement of different spatial features of the rock, both small (*microscopic*) and large (*macroscopic*). For example, gas bubble holes in an igneous rock may be characteristic of its structure. A *vesicular* structure is the presence of small holes, or vesicles, throughout the igneous rock, such as are found in pumices and some basalts. Holes larger than vesicles are *vugs* and are generally filled with minerals other than those forming the rock.

An important macroscopic structural feature is jointing of the rock. Joints are fractures and may be open or closed and run in various directions. They usually occur in more-or-less regular systems and may tend to break the rock into cubes or other regular blocks. This is an important engineering property and is discussed further later. Fractures or cracks are also macroscopic features and may run in any direction and may intersect each other at any angle. A fracture usually has an irregular surface in contrast to the planar or even surface of a joint.

8.2.2.7 Fabric

'Fabric' is a controversial term which sometimes is considered as a generalization of the term 'texture'. Here, igneous fabric denotes the spatial pattern of the rock particles which includes grain sizes and their ratios, grain shapes, grain orientation, microfracturing, packing and interlocking of particles, the character of the matrix, and so on, all of which help control the engineering performance of the rock.

8.2.3 Metamorphic rocks

Rocks formed by the complete or incomplete recrystallization, i.e. the change in crystal shape or in composition, of igneous or sedimentary rocks by high temperatures, high pressures, and/or high shearing stresses, are metamorphic rocks. A platy or foliated structure in such rocks indicates that high shearing stresses have been the principal agency in their formation.

Foliation is not always visible to the naked eye, but individual grains may exhibit strain lines when seen under the microscope.

Table 8.4 Metamorphic rock classification

Structure and texture	Composition	Rock name
FOLIATED OR PLATY	Various tabular and/or prismatic minerals (generally elongated)	Schist, some serpentines, slate, phyllite
MASSIVE:		
Banded, consisting of alternating lenses	Various tabular, prismatic, and granular minerals (frequently elongated)	Gneiss
Granular, consisting mostly of equidimensional grains	Calcite, dolomite, quartz, in small particles	Marble or quartzite

Metamorphic rocks formed without intense shear action have a massive structure. In Table 8.4 the most common metamorphic rocks are subdivided into two basic classes according to their structure. Foliated rocks usually have directional engineering properties.

8.2.4 A field identification of common rocks

Table 8.5 gives a simple field guide to the identification of the more common rocks. It is after the scheme by Krynine and Judd[2] for engineers with little training in geology and has been devised to present those features first seen when picking up a hand specimen. It is based primarily on texture and structure. They consider the scheme fits the most common occurrences of the rock but some variations will occur.

Textbooks such as the one by Lahee[3] give more specialized field identification techniques. Difficult or contentious identification should be carried out by a geologist who may require thin-section examination of a slice of the rock or even geochemical methods for complete identification.

8.2.5 Rock properties

Engineering characteristics of *rocks* are given more fully in Chapter 10 on rock mechanics and of *soils* in Chapter 9 on soil mechanics. Geological characteristics are given in the engineering geology and mineralogy and petrology textbooks listed in the bibliography.

Table 8.6 (from Shergold[4]) gives some general properties of common rocks and Table 8.7 (in part from Attewell and Farmer[5]) gives a range of mechanical properties of rocks identified by their British Standard (BS) 812:1951 trade group classification. This classification should be used with caution as the rocks listed in each group do not necessarily have close mechanical affinities. The results listed are probably on fairly fresh rock types, i.e. not weathered in the manner following.

8.3 Rock deformation in Nature – fractures and folds

When rocks of the Earth's upper mantle are subject to large stresses, they either break or bend with the production of fractures or folds. The kind of structure formed depends on the condition of the rocks and the rate at which deformation takes place. Most rocks are brittle at surface conditions and tend to fracture under stress though they may yield slowly by bending. At deeper levels where temperatures and pressures are high the majority of rocks become ductile and deform without breaking. Many special conditions at the Earth's surface cause minor fractures and folds, e.g. cooling of igneous lava, thermal stress by daily temperature changes, ground ice movement, and soil desiccation.

8.3.1 Joints

Joints are fractures without any displacement. They may appear to be somewhat random in direction, but a careful field examination will usually show that they have some relation to the host rock, e.g. with the bedding in sedimentary rock or with flow lines in igneous rock.*

In igneous rocks there are often three regular sets of joints (Figure 8.9). In an ideal situation one set lies approximately horizontal and parallel to the flow lines and is termed flat-lying. Another set, the cross-joints, is roughly perpendicular to the flow lines. The third set, the longitudinal joints, dips steeply and strikes parallel to the flow lines if the latter are projected to a plane surface such as a map.

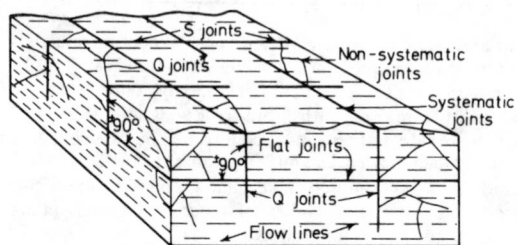

Figure 8.9 Block diagram of simple joint systems in an igneous rock. Systematic S joints are more commonly called longitudinal joints, and Q joints are more commonly called cross-joints perpendicular to the flow lines of the original molten rock

In sedimentary rocks, there are often two systems of mutually perpendicular joints, both perpendicular to the bedding plane.

Joints also may be grouped into strike joints and dip joints. Figure 8.10 illustrates the terms 'strike' and 'dip' where the rock bed is assumed to be an oblique plane. Strike is the direction of contour lines or lines of equal elevation on the surface of the rock mass, and the dip is the maximum slope of its surface. In Figure 8.10 the dip is the angle α made by the line AB with the horizontal. In measurements of dip, it is important to measure the 'true' dip, i.e. the angle located in a plane perpendicular to the strike; otherwise, a misleading *apparent* dip, β in Figure 8.10, is recorded. These terms also apply to beds, faults and other geometric features.

Joints and their orientation with respect to other structures have been widely studied in the field and it has been established that systematic joints usually show well-defined relationships to folds and faults which develop during the same tectonic episode.

The spacing of joints varies considerably and is of importance in engineering. Some rocks, such as sandstones and limestones in which the joints may be widely spaced, yield large blocks and

*Lines showing the flow of the originally liquid magma and indicated by the long axes of crystals

Table 8.5 Field identification of rocks (specimens should be unweathered and not altered in any way)

[LIGHT-COLOURED]

GRAINS OR CRYSTALS VISIBLE TO NAKED EYE

Angular particles				Rounded particles			Erratic large grains	
Large	Fine to medium	Very fine	Foliated or banded	Large	Fine to medium	Very fine	Rounded	Angular
Pegmatite Granite (+Q, +F)* Granodiorite (+Q, +F)* Monzonite (−Q, +F)* Syenite (−Q, +F)* Marble (reacts with HCl) Arkose (usually bedded)	Tuff (contains glasslike fragments)	Felsite* (rhyolite +Q and trachyte −Q)	Schist (shiny) Gneiss (may have sub-angular particles)	Conglomerate (+10% of grains over 2 mm diameter) Sandstone (bedded) (if it reacts to HCl = calcareous sandstone; if it gets slick when wet = argillaceous sandstone)	Quartzite (not friable and very hard)	Siltstone	Depositional breccia	Volcanic breccia and agglomerate fault breccia (may have clay)

NO GRAINS OR SPARSE CRYSTALS VISIBLE TO NAKED EYE

Glassy lustre	Dull lustre	Shiny lustre	Earthy appearance		Laminated					
			Spongy, light wt	Porous, moderate wt	Slick when wet	Not slick	Slick when wet	Not slick		
Quartzite	Felsites* (rhyolite, trachyte)	Schist (foliated)	Pumice Volcanic ash	Chalk (HCl reaction)	Shale	Shale Slate (dull) Phyllite (shiny)	Claystone Mudstone Serpentine (usually greasy and may be banded)	Reaction to HCl ⎯⎯⎯⎯ Limestone Chalk (earthy)	No reaction to cold HCl ⎯⎯⎯⎯ Dolomite	

[DARK COLOURED (DARK GREY OR GREEN TO BLACK)]

GRAINS OR CRYSTALS VISIBLE TO NAKED EYE

Angular particles		Rounded to subangular particles
Fine to medium	Very fine to glassy	Graywacke (fine- to medium-grained) Dark sandstones
Peridotite (−Q, −B)* Gabbro (−Q, −B)* Diorite (−Q, +B)* Dolerite (−Q, +B)*	Andesite* Basalt (usually vesicular)*	

NO GRAINS OR SPARSE CRYSTALS VISIBLE TO NAKED EYE

Glassy lustre	Dull lustre – laminated		Dull lustre – not laminated
	Slick when wet	Not slick	
Obsidian	Shale	Shale (flexible) Slate (brittle) (dull) Phyllite (shiny)	Basalt* Serpentine (usually greasy and may be banded)

*Rocks may contain occasional large crystals embedded in a very fine-grained matrix or occasional very large crystals in a medium-grained matrix – in either case the term 'porphyry' is appended to the rock name, e.g. syenite porphyry.

(+Q)=contains numerous white or colourless quartz crystals.
(−Q)=contains little or no quartz.
(+F)=contains numerous white to pink feldspar crystals.

(+B)=contains numerous flakes of black mica (biotite).
(−B)=contains little or no black mica.

Table 8.6 Summary of means and range of values for mechanical tests in each trade rock-group

Trade Group classification (BS 812:1951)		Aggregate* crushing value	Aggregate* impact value	Aggregate* abrasion value	Water* absorption (per cent)	Specific gravity	Polished-stone coefficient
Artificial	Mean	28	27	8.3	0.7	2.71	0.50
	Range	(15–39)	(17–33)	(3–15)	(0.2–1.8)	(2.6–3.4)	(0.35–0.60)
	Number of samples	55	18	18	19	19	9
Basalt	Mean	14	15	6.1	1.1	2.80	0.56
	Range	(7–25)	(7–25)	(2–12)	(0.0–2.3)	(2.6–3.0)	(0.45–0.70)
	Number of samples	123	79	65	68	68	25
Flint	Mean	18	23	1.1	1.0	2.54	0.35
	Range	(7–25)	(19–27)	(1–2)	(0.3–2.4)	(2.4–2.6)	(0.30–0.40)
	Number of samples	63	32	45	24	24	4
Granite	Mean	20	19	4.8	0.4	2.69	0.56
	Range	(9–35)	(9–35)	(3–9)	(0.2–0.9)	(2.6–3.0)	(0.45–0.70)
	Number of samples	41	32	28	16	16	13
Gritstone	Mean	17	19	7.0	0.6	2.69	0.69
	Range	(7–29)	(9–35)	(2–6)	(0.1–1.6)	(2.6–2.9)	(0.60–0.80)
	Number of samples	81	45	31	33	33	18
Hornfels	Mean	13	12	2.2	0.4	2.82	0.45
	Range	(5–15)	(9–17)	(1–4)	(0.2–0.8)	(2.7–3.0)	(0.40–0.50)
	Number of samples	28	24	13	15	15	4
Limestone	Mean	24	23	13.7	1.0	2.66	0.43
	Range	(11–37)	(17–33)	(7–26)	(0.2–2.9)	(2.5–2.8)	(0.30–0.75)
	Number of samples	164	61	34	42	42	33
Porphyry	Mean	14	14	3.7	0.6	2.73	0.51
	Range	(9–29)	(9–23)	(2–9)	(0.4–1.1)	(2.6–2.9)	(0.45–0.60)
	Number of samples	62	29	23	30	30	13
Quartzite	Mean	16	21	3.0	0.7	2.62	0.57
	Range	(9–25)	(11–33)	(2–6)	(0.3–1.3)	(2.6–2.7)	(0.45–0.65)
	Number of samples	57	37	29	21	21	8
All groups†	Mean	19	19	5.7	0.7	2.68	0.53
	Range	(5–39)	(7–35)	(1–26)	(0.0–3.7)	(2.3–3.4)	(0.30–0.80)
	Number of samples	724	370	311	313	313	134

*In these tests a numerically lower result indicates a better performance in the test.   †Including results from unclassified samples.

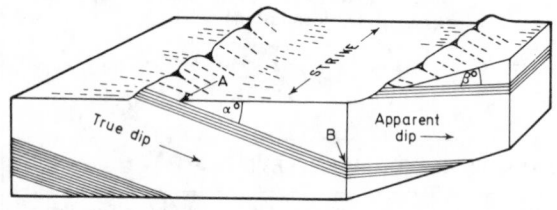

Figure 8.10 Idealized block diagram to show dip and strike relationships

may be suitable for masonry, for example, whereas other rocks may be so closely jointed as to break up into small pieces and may be suitable for aggregate or other purposes. Some joints in sedimentary rocks run only from one bedding plane to the next, but others may cross several bedding planes, and are called *master joints*.

The ease of quarrying, excavating or tunnelling in hard rocks largely depends on the regular or irregular nature of the joints

and their surface characteristics, e.g. attitude, size, frequency, openness and spacing. Joints and other fractures control groundwater and air flow in otherwise intact rock and help to promote rock weathering.

8.3.2 Faults

Faults are fractures in the crust along which there has been displacement of the rocks on one side relative to those on the other.

The surface on which movement takes place during faulting is the fault plane. It may be vertical, steeply inclined or gently inclined as with thrust faults. The intersection of a fault with the ground surface is known as the fault line or fault trace. The upper side of an inclined fault, and the rock which lies above it, is referred to as the hanging wall. Rock below it is the foot wall; dip faults strike parallel to the local direction of dip of the beds, strike faults are parallel to the strike and oblique faults cut across both strike and dip directions.

Movements on a fault may be in any direction. The displace-

Table 8.7 Typical rock strengths, porosity and bulk densities of rock materials

Rock	Strength N/mm^{-2}			Bulk density (Mg/m^{-3})	Porosity (n%)
	Compressive	Tensile	Shear		
Granite	100–250	7–25	14–50	2.6–2.9	0.5–1.5
Diorite	150–300	15–30	—	2.7–3.05	0.1–1.0
Dolerite	100–350	15–35	25–60	2.7–3.05	0.1–0.5
Gabbro	150–300	15–30	—	2.8–3.1	0.1–0.2
Basalt	150–300	10–30	20–60	2.8–2.9	0.1–1.0
Sandstone	20–170	4–25	8–40	2.0–2.6	5–25
Shale	5–100	2–10	3–30	2.0–2.4	10–30
Limestone	30–250	5–25	10–50	2.2–2.6	5–20
Dolomite	30–250	15–25	—	2.5–2.6	1–5
Coal	5–50	2–5	—	—	—
Quartzite	150–300	10–30	20–60	2.6–2.7	0.1–0.5
Gneiss	50–200	5–20	—	2.8–3.0	0.5–1.5
Marble	100–250	7–20	—	2.6–2.7	0.5–2.0
Slate	100–200	7–20	15–30	2.6–2.7	0.1–0.5
Rhyolite	—	—	—	2.4–2.6	4–6
Andesite	50–200	—	—	2.2–2.3	10–15

ment or slip is the sum of all the previous effects of movement and is shown by the relative positions on either side of the fault of two originally contiguous features as a bedding plane. The vertical component of the slip, taken by itself, is called the throw of the fault (see Figure 8.11).

Faults can be classified, according to the direction of movement that has taken place on them, into normal faults, reverse faults and transcurrent or strike–slip faults.

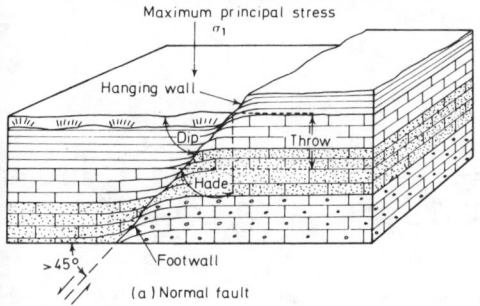

(a) Normal fault

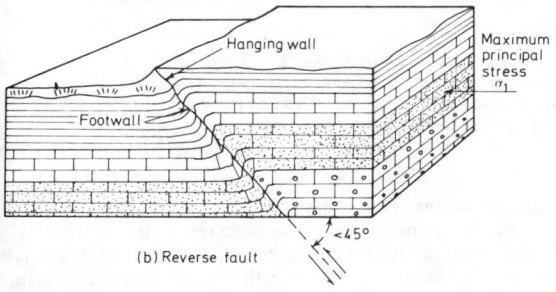

(b) Reverse fault

Figure 8.11

Normal faults. Normal faults (originally so-called because they are the normal type found in coalfields in the UK) are those in which the hanging-wall rocks have moved down the dip of the

fault plane. Small normal faults are extremely common in almost all geological situations and may even occur in Quaternary sediments. Large normal faults, occurring in groups, produce a considerable effect of lengthening and are especially common in the more stable areas of the Earth's crust. Groups of faults are arranged so that alternate dislocations dip in opposite directions and produce the effect of block faulting illustrated in Figure 8.12; the crust is separated into high blocks or *horsts* between outward-dipping faults and low blocks, troughs or *graben* between inward-dipping faults.

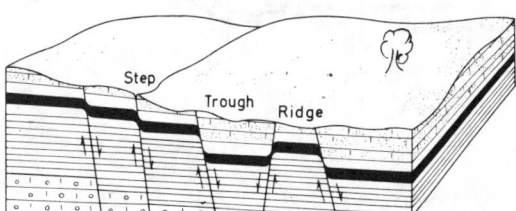

Figure 8.12 Idealized block diagram of some common fault groups. Note there is little effect on topography here as the surface bed is the same in all locations, but where difficult beds are exposed by the faulting, scarp topography may be found

Reverse faults. Reverse faults are those on which the rocks of the hanging wall move up the dip of the fault plane. They result in shortening across the fault and in duplication of strata; reverse faults with low dips are thrusts.

Transcurrent faults. These are wrench faults, tear faults or strike–slip faults on which horizontal movement takes place. The fault planes are almost vertical and the effect of faulting when seen on a map is to shift rocks laterally, even for many tens of kilometres. Examples of block diagrams to illustrate mapped outcrop patterns of faults are shown in Figure 8.13.

An example of the relationship between faulting and jointing in one complete episode is shown in Figure 8.14 from the textbook by Price.[6] Techniques and the use of stereographic projection in geology is given in Phillips.[7]

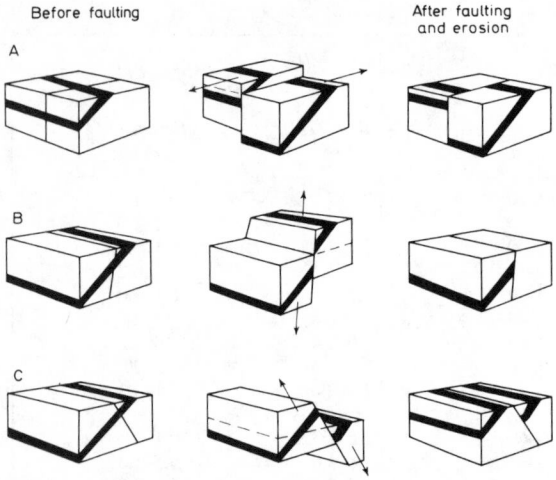

Figure 8.13 Idealized outcrop patterns of faulted beds. A, dip fault, i.e. movement in the dip direction; B, strike fault with downthrow in dip direction; C, strike fault with downthrow against the dip angle. (After Read and Watson (1971) *Beginning geology*, 2nd edn. Macmillan/Allen and Unwin, London)

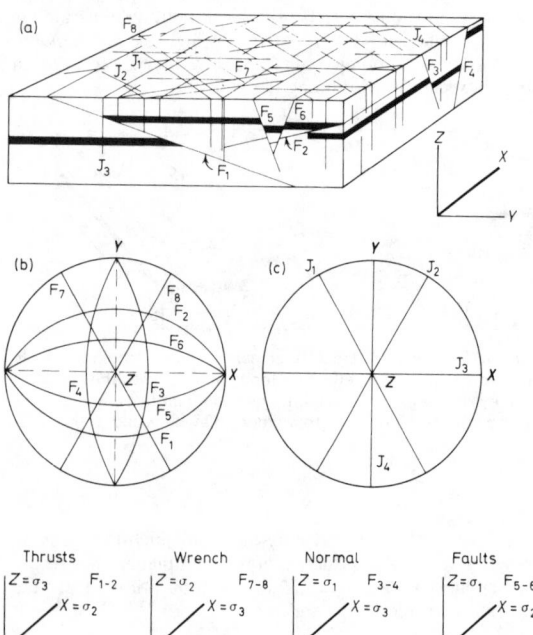

Figure 8.14 (a) Block diagram showing orientation of faults and joints in unfolded rocks which may result from various phases of compression and tension related to one complete tectonic episode; (b) stereogram of fault orientations shown in (a); (c) stereogram of joint orientation shown in (a); (d)–(g) orientation of stress fields when the various groups of faults were initiated. (Redrawn from Price (1966) *Fault and joint development in brittle and semi-brittle rock*. Pergamon Press, Oxford)

8.3.3 Folds

In geology weak rocks which deform under stress are termed *incompetent* whereas strong rocks that buckle and fracture are termed *competent*. These terms should not be confused, however, with similar terms describing the bearing capacity of foundation rocks.

A complete fold is composed of an arched portion, or *anticline*, and a depressed trough or *syncline* (Figure 8.15a). The highest point of an anticline is called the crest, and the inclined parts of the strata where anticline and syncline merge are the limbs of the fold. The youngest beds outcrop in the middle of a syncline and the oldest in the middle of an anticline.

The plane bisecting the vertical angle between equal slopes on either side of the crest line is the axial plane. Where this is vertical, as in Figure 8.15a, the fold is upright and symmetrical; where it is inclined the fold is asymmetrical (Figure 8.15b). Sometimes the middle limb has been brought into a vertical position by the compression which buckled the strata, and under still more severe conditions an *overturned fold*, or over-fold, is produced (Figure 8.15c). Here the middle limb is inclined in the same attitude as the axial plane, and the beds of which it is composed have a reversed dip, i.e. upper beds are now brought to dip steeply beneath lower beds, an inversion of the true sequence.

If the compression is so extreme as to pack a series of folds together so that their limbs are all virtually parallel and steeply dipping, the structure is referred to as *isoclinal folding*, i.e. all limbs have the same slope (Figure 8.15c).

Where the axial plane is inclined at a low or zero angle, the fold is said to be *recumbent* (Figure 8.15d), a type which is often found in intensely folded mountain regions such as the Alps.

The term *monocline* is for the kind of flexure which has two parallel gently dipping limbs with a steeper middle limb between them: it is in effect a local steepening of the dip in gently dipping (or horizontal) beds.

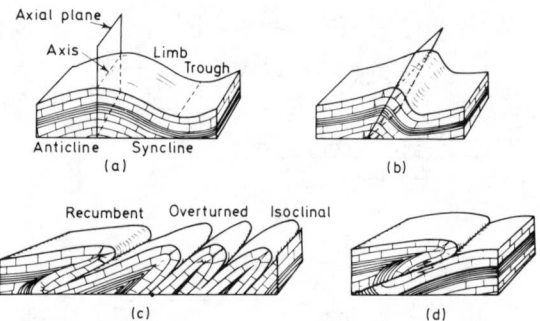

Figure 8.15 Idealized fold types. (a) Simple or gentle symmetrical; (b) simple or gentle asymmetrical; (c) tight assymetrical, recumbent, overturned and isoclinical; (d) recumbent passing into a thrust fault

The dimensions of anticlines and synclines vary between wide extremes, from small puckers millimetres across in sharply folded sediments, to broad archings of strata whose extent is measured in kilometres. The growth of such structures is, in general, a process which goes on slowly as stresses develop in any particular part of the Earth's crust; but superficial folds may develop in a comparatively short space of time, e.g. earthquake ripples forming quickly, in weak sediments or some types of hillcreep. Simple land topography largely controlled by folding is illustrated in Figure 8.16.

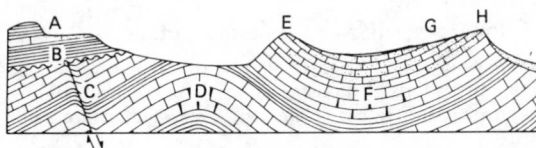

Figure 8.16 Simple fold forms and related topography. A, step topography; B, unconformity; C, normal fault; D, anticline; E, hog's back ridge; F, syncline; G, dip slope; H, scarp slope

8.3.4 Some engineering aspects of faults and folds

Any geological structure that influences one of the mass properties of the *in situ* rock, such as the strength, modulus of deformation or permeability, is highly significant. The most common structural features of significance are joints, bedding planes and foliation surfaces and 'shears' or faults. These are all planar or near-planar discontinuities, and have a strong anisotropic effect on the mass properties.

A search for discontinuities and other faults is not always

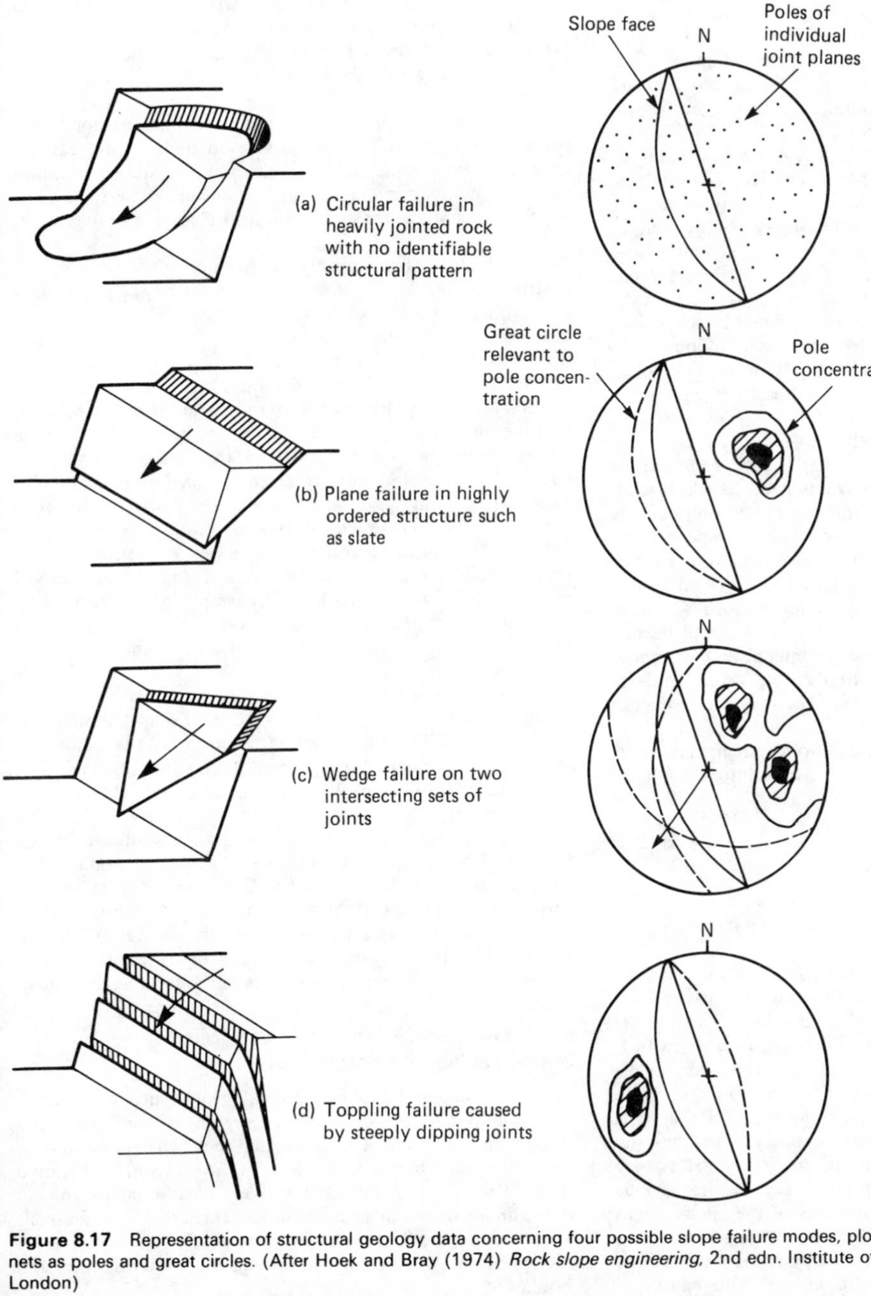

(a) Circular failure in heavily jointed rock with no identifiable structural pattern

(b) Plane failure in highly ordered structure such as slate

(c) Wedge failure on two intersecting sets of joints

(d) Toppling failure caused by steeply dipping joints

Figure 8.17 Representation of structural geology data concerning four possible slope failure modes, plotted on equatorial equal-area nets as poles and great circles. (After Hoek and Bray (1974) *Rock slope engineering*, 2nd edn. Institute of Mining and Metallurgy, London)

effective during site investigation, and significant faults, for example, are sometimes not discovered until construction or even afterwards. Stability of hillsides, cut slopes, quarry faces and so on may often be controlled by the geometric arrangement of joints and faults. (For examples see Figure 8.17.) Also the groundwater pattern may be controlled by the condition of the joints and faults whether they are open or closed or filled with debris or gouge and the persistence or continuity of such fractures may be important.

On large works the determination of whether a fault is *active*, *inactive* or *passive* may be important. Active faults are those in which movements have occurred during the recorded history and along which further movements can be expected any time (such as the San Andreas and some other faults in California). Inactive faults have no recorded history of movement and are assumed to be and probably will remain in a static condition. Unfortunately, it is not possible yet to state definitely if an apparently inactive fault will remain so. The fault may reopen, either because of a new stress accumulation in the locality or from the effect of earthquake vibrations.

From the alteration products of faulting, gouge is probably of the most concern in foundation problems. This is usually a relatively impervious clay-grade material and may hinder or stop the movement of groundwater from one side of the fault to the other and so create hydrostatic heads, e.g. if encountered in a tunnel. It may also reduce sliding friction along the fault plane. The presence of soft fault breccia or gouge may cause sudden squeezes in a tunnel that intersects the fault. Arch action of rocks in tunnels may be reduced by the presence of joints and faults. Rock falls on cuts and in tunnels, patterns of rock bolts, grout holes and so on are all controlled to a large extent by the joint and fault pattern.

In foundations, folds are generally not so critical as faults though they may give stability problems if their geometry is unfavourable. Occasionally, folds may influence the selection of a dam site; e.g. when the reservoir is located over a monocline containing pervious strata, there may be excessive seepage if the monocline dips downstream. If the monocline were to dip upstream, the reservoir might have little seepage providing the monocline contained some impervious layers such as shale which were not fractured in the folding. Serious water problems may arise in the construction and maintenance of tunnels intersecting synclines containing water-bearing strata. In deep cuts, analogous water problems arise that may create continuous maintenance problems.

Dipping beds, which must be part of a fold system, may cause stability problems if the dip is unfavourable into a cut face (Figure 8.17b).

8.4 Engineering geology environments

A geological environment is the sum total of the external conditions which may act upon the situation. For example, a 'shallow marine environment' is all the conditions acting offshore which control the formation of deposits on the sea bed: the water temperature, light, current action, biological agencies, source of sediment, sea bed chemistry and so on.

The concept of geological environment forms a suitable basis to study systematically the engineering geology of the deposits formed in or influenced by the various environments, as they condition the *in situ* engineering behaviour of the various deposits. A knowledge of the parameters of the environment enables predictions and explanations of the engineering behaviour to be attempted. Geomorphology is the study of the geology of the Earth's surface (see Fookes and Vaughan[8]).

8.4.1 Processes acting on the Earth's surface

A *landform* may be defined as an area of the Earth's surface differing by its form and other features from the neighbouring areas. Mountains, valleys, plains and even swamps are landforms.

The principal processes that are continually acting on the Earth's surface are gradation, diastrophism and vulcanism.

(1) *Gradation* is the building up or wearing down of existing landforms (including mountains), formation of soil and various deposits. Erosion is a particular case of gradation by the action of water, wind or ice.
(2) *Diastrophism* is the process where solid, and usually the relatively large, portions of the Earth move with respect to one another as in faulting or folding.
(3) *Vulcanism* is the action of magma, both on the Earth's surface and within the Earth.

With the exception of vulcanism and sometimes erosion, these processes may take hundreds and even millions of years to change the face of the Earth significantly. The sudden eruption of a volcano, for example, with the ensuing flow of lava or deposition of volcanic ash, can abruptly change land overnight.

Origin of soils. The majority of the soils are formed by the destruction of rocks. The destructive process may be physical, as the disintegration of rock by alternate freezing and thawing or day–night temperature changes. It may also be by chemical decomposition, resulting in changes in the mineral constituents of the parent rock and the formation of new ones.

Soils formed by disintegration and chemical decomposition may be subsequently transported by the water, wind or ice before deposition. In this case they are classified as alluvial, aeolian, or glacial soils and are generally called *transported* soils. However, in many parts of the world, the newly formed soils remain in place. These are called *residual* soils.

In addition to the two major categories of transported and residual soils, there exist a number of soils that are not derived from the destruction of rocks. For example, peat is formed by the decomposition of vegetation in swamps; some marly soils are the result of precipitation of dissolved calcium carbonate.

Soil-forming processes. There are very many and varied processes that take place in weathered rock and soils that affect the formation of soil profiles to varying degrees, but the major soil-forming processes are: (1) organic accumulation; (2) eluviation; (3) leaching; (4) illuviation; (5) precipitation; (6) cheluviation; and (7) organic sorting.

The soil-forming processes produce an assemblage of soil layers at horizons, called the *soil profile*. In its simplest it is categorized as three layers A, B and C but numerous varieties of this and many other soil classifications exist. Probably the most generally accepted one is that based on a geographical approach. This is the zonal scheme thought to reflect zones of climate, vegetation and other factors of the local environment.

8.4.2 Engineering significance of selected geomorphological environments

Much of what can be called 'classical' geotechnical engineering has developed in temperate climate regions of the Earth. As a result many of the concepts of soil and rock behaviour and their properties have been conditioned by the soil and rock found there. The climate and local geology play a major role in determining the local geotechnical characteristics of the soils and rocks. Figure 8.18 shows the generalized distribution of the four principal climatic engineering soil zones after Sanders and Fookes.[9]

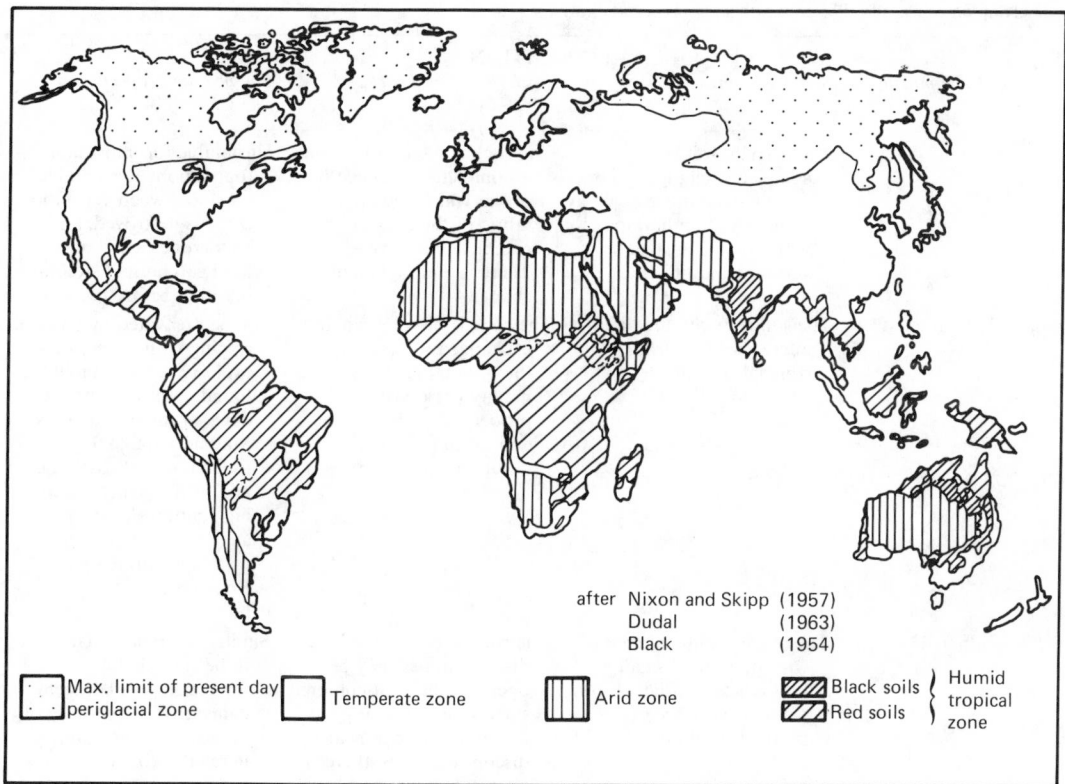

after Nixon and Skipp (1957)
Dudal (1963)
Black (1954)

⬚ Max. limit of present day periglacial zone	☐ Temperate zone	▥ Arid zone	▨ Black soils ▨ Red soils	Humid tropical zone

Figure 8.18 Generalized map showing present-day distributions of the four principal climatic engineering soil zones

The following section briefly describes the engineering geomorphology characteristics of the four principal climatic zones, together with other geological environments of particular interest.

8.4.2.1 Rock weathering

A rock can weather by physical breakdown without considerable change of its constituent minerals by disintegration. The residual or transported soil derived from this process consists of an accumulation of mineral and rock fragments virtually unchanged from the original rock. This type of weathering is found mainly in arid or cold climates. *Chemical decomposition* leads to the thorough alteration of a large number of minerals and only a few, among them quartz, may remain unaffected. The greater the percentage of weatherable minerals in the original rock, particularly the ferromagnesian minerals, the more conspicuous is the change from rock to soil. This type of weathering is generally found in warm and hot, wet climates and can lead to great thickness of weathered rock and soil. Biological weathering is generally of less importance than physical or chemical weathering and is a combination of biochemical and biophysical effects.

The weathering process produces a gradational and often quite irregular change in the rock from fresh some distance below ground surface to more or less completely weathered at the surface: Figure 8.19 shows somewhat schematically two examples of weathered rock profiles. Corestones need not always be present and in sedimentary rocks in particular are often missing.

There are several generalized weathering rock classifications for engineering purposes; the one reproduced here, Table 8.8, is from Fookes, Dearman and Franklin,[10] which discusses weathered rock mainly in the UK. For information on the engineering performance of weathered rock elsewhere, see also Little,[11] Deere and Patton[12] and Irfan and Dearman.[13]

8.4.2.2 Humid tropical residual soils

A residual soil is the end product (i.e. grades V and VI) of rock weathering. Different types of residual soils are produced in different environments (see Figures 8.20 and 8.21).

In tropical regions of high temperature and abundant surface water rock weathering it is at its most intensive. It is characterized by rapid breakdown of feldspars and ferromagnesian minerals, the removal of silica and bases and the concentration of iron and aluminium oxides. Kaolinite and related clay minerals form in well-drained areas. Because of the high iron concentration the resulting soils are usually red in colour. When dried, such materials may harden as a result of the cementing action by iron and aluminium oxides and they are commonly referred to as *laterites*. Large, fairly durable concretions may be formed this way to give lateritic gravels which are often important sources of aggregate for road construction and other uses, particularly in East and West Africa.

Tropical weathering of volcanic ash and rock leads to the formation of allophane and halloysite together with concentration of iron and aluminium oxides. Soils of this type, particularly common in the Far East, are known as *adisols*.

Nearer to the edges of the tropics to the north and south there

Table 8.8 Engineering grade classification of weathered rock

| | | FIELD RECOGNITION | | |
Grade	Degree of decomposition	Soils (i.e. soft rocks)	Rocks (i.e. hard rocks)	Engineering properties of rocks
VI	Soil	The original soil is completely changed to one of new structure and composition in harmony with existing ground surface conditions.	The rock is discoloured and is completely changed to a soil in which the original fabric of the rock is completely destroyed. There is a large volume change.	Unsuitable for important foundations. Unsuitable on slopes when vegetation cover is destroyed, and may erode easily unless a hard cap present. Requires selection before use as fill.
V	Completely weathered	The soil is discoloured and altered with no trace of original structures.	The rock is discoloured and is changed to a soil, but the original fabric is mainly preserved. The properties of the soil depend in part on the nature of the parent rock.	Can be excavated by hand or ripping without use of explosives. Unsuitable for foundations of concrete dams or large structures. May be suitable for foundations of earth dams and for fill. Unstable in high cuttings at steep angles. New joint patterns may have formed. Requires erosion protection.
IV	Highly weathered*	The soil is mainly altered with occasional small lithorelicts of original soil. Little or no trace of original structures.	The rock is discoloured; discontinuities may be open and have discoloured surfaces and the original fabric of the rock near the discontinuities is altered; alteration penetrates deeply inwards, but corestones are still present.	Similar to grade V. Unlikely to be suitable for foundations of concrete dams. Erratic presence of boulders makes it an unreliable foundation for large structures.
III	Moderately weathered*	The soil is composed of large discoloured lithorelicts of original soil separated by altered material. Alteration penetrates inwards from the surfaces of discontinuities.	The rock is discoloured; discontinuities may be open and surfaces will have greater discoloration with the alteration penetrating inwards; the intact rock is noticeably weaker, as determined in the field, than the fresh rock.	Excavated with difficulty without use of explosives. Mostly crushes under bulldozer tracks. Suitable for foundations of small concrete structures and rockfill dams. May be suitable for semipervious fill. Stability in cuttings may depend on structural features, especially joint attitudes.
II	Slightly weathered	The material is composed of angular blocks of fresh soil, which may or may not be discoloured. Some altered material starting to penetrate inwards from discontinuities separating blocks.	The rock may be slightly discoloured, particularly adjacent to discontinuities which may be open and have slightly discoloured surfaces; the intact rock is not noticeably weaker than the fresh rock.	Requires explosives for excavation. Suitable for concrete dam foundations. Highly permeable through open joints. Often more permeable than the zones above or below. Questionable as concrete aggregate.
I	Fresh rock	The parent soil shows no discoloration, loss of strength or other effects due to weathering.	The parent rock shows no discoloration, loss of strength or any other effects due to weathering.	Staining indicates water percolation along joints; individual pieces may be loosened by blasting or stress relief and support may be required in tunnels and shafts.

*The ratio of original soil or rock to altered material should be estimated where possible.

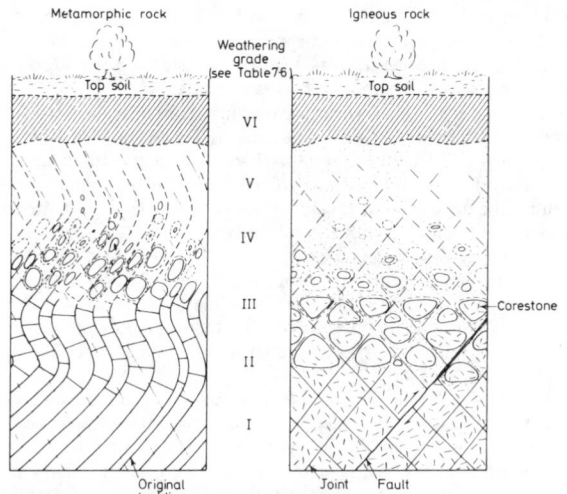

Figure 8.19 Diagrammatic weathering profile of an igneous and a metamorphic rock. (After Deere and Patton (1971) 'Stability of slopes in weathered rock.' *Proceedings, 4th Panamerican Conference*)

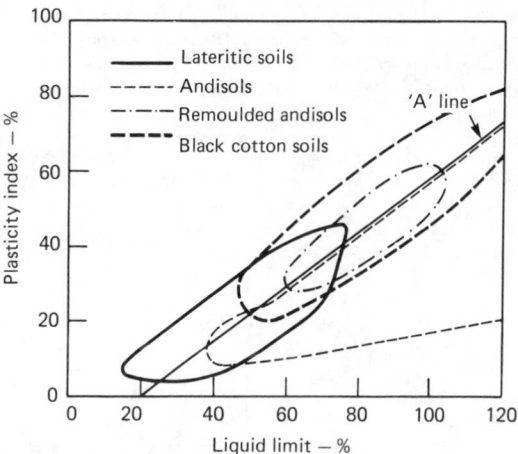

Figure 8.22 Approximate ranges of Atterberg limits for lateritic andisols and black cotton soils

is decreased rainfall, alternating wet and dry seasons and often poor drainage. Under these conditions smectite (montmorillonite) clays are found and the highly active *black cotton soils* develop. (See Figures 8.18, 8.21 and 8.22.)

In situ layers of tropical residual soils may be markedly nonhomogeneous. High geochemical activity in tropical environments can result in rapid changes in clay mineralogy and fabric both laterally and with depth. Cementation of particles into clusters and aggregates by sesquioxides and the hydrated state of some of the minerals are responsible for high void ratios (low densities), high strength, low compressibility and some-

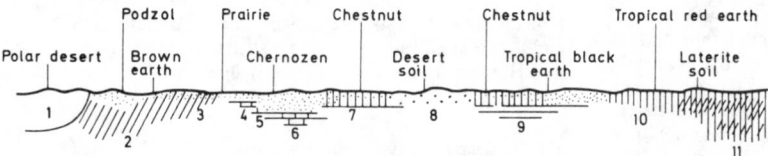

1. Frozen ground
2. Intense leaching and illuviation
3. Less leaching more mixing
4. Carbonate appears
5. Organic matter increases
6. Carbonate prominent
7. Organic matter decreases
8. Gypsum and salt accumulate
9. Organic matter increases
10. Ferrallitisation becomes dominant
11. Ferricrete formation

Figure 8.20 Diagrammatic cross-section of zonal soils from pole to equator. (After Ollier (1975) *Weathering*, 2nd edn. Oliver and Boyd, Cambridge)

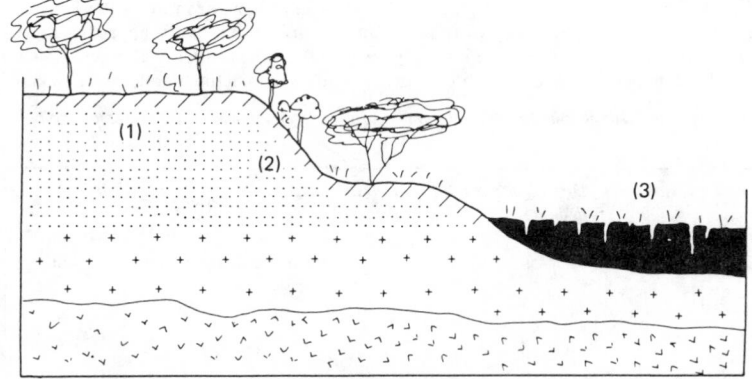

(1) Reddish-brown soil developed in deeply-weathered material beneath acacia grassland

(2) Reddish-brown soil with surface erosion developed in highly-weathered material

(3) Black soil of the depressions, fine-textured and salt-enriched in dry climates, deep cracking

Figure 8.21 Diagrammatic representation of topographic relationships of local soils in savanna lands of Africa. (After Thomas (1974) *Tropical geomorphology*. Macmillan, London)

Table 8.9 Physical properties of unremoulded, remoulded and sesquioxide-free lateritic soil (in Anon (1982) 'Engineering construction in tropical and residual soils'. *American Society of Civil Engineers, Geotechnical Engineering Division Special Conference,* Hawaii)

Property	Unre-moulded	Remoulded	Sesquioxide-free
Liquid limit (%)	57.8	69.0	51.3
Plastic limit (%)	39.5	40.1	32.1
Plasticity index (%)	18.3	28.0	19.2
Specific gravity	2.80	2.80	2.67
Proctor density (kN/m³)	13.3	13.0	13.8
Optimum moisture content (%)	35.0	34.5	29.5

times high permeability in relation to high plasticity and small particle size. Collapsing soils can occur. (See Table 8.9.)

Many of the red lateritic soils and adisols are susceptible to breakdown on manipulation which makes index property determination difficult as well as earthwork construction, changing a predominantly granular soil which excavates easily to a plastic mess that cannot be compacted easily. Irreversible changes in property of many tropical soils result from drying. (See, for example, Table 8.10.)

Most of the black cotton soils have high plasticity and marked swelling and shrinkage characteristics which in areas with marked dry and wet seasons cause special problems for foundations of structures and roads. Reference should be made to the appropriate literature for the particular soil type, though this itself may be difficult since in many of the published engineering articles the soil type may not be described sufficiently to characterize it. In addition, a great variety of different residual soils seem to be dubbed with the title of 'laterite' often quite erroneously. See Sanders and Fookes[9] for a general study of foundation conditions related to four principal climate zones: (1) periglacial; (2) temperate; (3) arid; and (4) humid tropical. See Deere and Patton[12] for an extensive treatise on slope stability aspects; Little[11] for 'laterites' and the *Proceedings* of the Institution of Civil Engineers[14] for a symposium on road and airfield construction in tropical soils; Chen[15] and Gidigasu[16] on swelling soils; and Anon[17] on engineering and construction.

8.4.2.3 Hot desert soils

Desert soils are formed in dry environments where the evaporation exceeds the precipitation and are generally associated with the world's hot deserts. Rainfall is low (say less than 150 mm per annum) and often seasonal. Physical weathering is dominant and the disintegration of the rock mainly results from insolation, but often other factors such as abrasion by windborne particles and salt weathering may contribute.

The products of this type of weathering are mainly of coarser-grained materials near hills or mountains. Parent materials of a high-silica content produce detrital sands and gravels, which, when sand is transported away from high land by wind, give sand-dune deposits and possibly loess or, when transported by water, give alluvial sands and gravels. Calcareous parent materials result in calcareous sands and gravels and evaporite salts are often present throughout the soil profile especially in internally draining areas as playa or salina flats.

Cooke *et al.*[18] relate potentially suitable sources of aggregates to some desert landforms as shown in Figure 8.23 and Table 8.11.

Fookes and Knill[19] (Figure 8.24) divided inter-montane desert basins into four sediment deposition zones which may be correlated with the degree of disintegration of the parent material.

Engineering problems provided by desert conditions are principally those related to the grading of the material. Coarse, angular, ill-sorted material generally occurs in zone II (Figure 8.24) and intermittent stream flow and occasional flash floods indicate that carefully designed drainage and runoff measures are required for engineering works. Better sorted and finer material occurs in zone III and the danger of sheet flood here may be greater. In zone IV, mobile sand dunes may require stabilization and loess soils can suffer metastable collapse on loading. Evaporite salts in the soil may cause expansive problems under thin pavements and attack concrete.

Desert engineering problems in general are discussed in Fookes and Knill[19] and Cooke *et al.*,[18] metastable loess soils in Holtz and Gibbs,[20] soluble salts in soils in Weinert and Clauss[21] and Fookes and French,[22] and construction materials by Fookes and Higginbottom[23] and Oweis and Bowman.[24]

8.4.2.4 Glacial soils ('drift' in the UK)

The five major glaciations of the Pleistocene period, begun about 2 million yr ago, constitute the last major episode in the shaping of much of the world's land surface. During each glaciation ice advanced over large areas of the northern and southern hemispheres. Post-glacial changes which have only occurred within the last 15 000 yr or so have been relatively limited. They are mostly confined to low ground, where alluvial deposits have tended to accumulate in response to the world-wide rise in sea-level caused by the latest recession of the ice sheets and glaciers. The deposits of one or more of the Pleistocene ice advances lie at the surface over, very approximately, 50% of the land area of the UK, whilst roughly another 10% is covered with post-glacial alluvium sometimes concealing glacial

Table 8.10 Effect of air-drying on index properties of a hydrated laterite clay from the Hawaiian Islands (In Gidigasu ((1975) *Laterite soil engineering*. Elsevier.)

Index properties	Wet (at natural moisture content)	Moist (partial air drying)	Dry (complete air drying)	Remarks
Sand content (%)	30	42	86	Dispersion prior to hydrometer test with sodium silicate
Silt content (%) (0.05–0.005 mm)	34	17	11	
Clay content (%) (<0.005 mm)	36	41	3	
Liquid limit (%)	245	217	NP	Soaking in water for 7 days did not cause
Plastic limit (%)	135	146	NP	regain of plasticity lost due to the air
Plasticity index (%)	110	71	NP	drying

Table 8.11 Major landforms as aggregate resources in hot deserts

---

Mountains
Including peaks, ridges, plateau surfaces, steep (excluding precipitous) slopes,* deep valleys and canyons, wadis, river terraces* and alluvial fans,* bounding scarp slopes.* Forms vary with rock type and the evolutionary history of the area

Pediments and alluvial fans
Rock pediment,† fan* and bajada,† with occasionally inselbergs* or salt domes° forming locally high ground

Plains
Occur downslope of pediments or alluvial fans without a distinct boundary and may include a whole variety of features including: alluvial† and colluvial plains,† wadi channels and flood plains, dune fields,° salt domes° inselbergs,* and extensive stone pavement surfaces°

Playa basins
Enclosed depressions receiving surface runoff from internal catchments or within escarpment zones.† They frequently contain lakes (either temporary or permanent), lake beaches, evaporite deposits° and may be strongly influenced by aeolian, fluvial and salt processes in their base zones

Coastal zones
These include beach ridges† (formed at periods of higher sea-level or during exceptional storms), sabkhas,° mud flats,° beach° and foreshore,° estuaries° and deltas°

---

*Normally a major source of aggregate, conditional on suitable mineralogy
†May be a reasonable source, depending on specific characteristics
°Normally should *not* be used for aggregates
(Symbols refer to Figure 8.23)

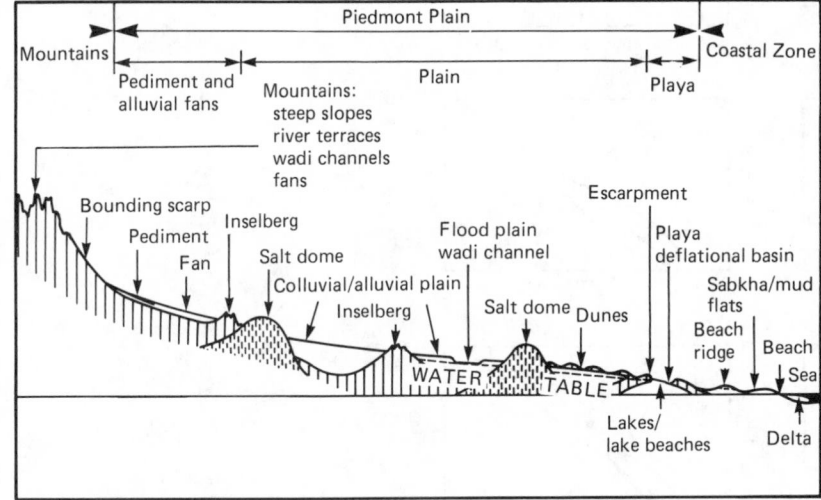

Figure 8.23 Some landforms of hot deserts and their potential suitability as sources of aggregates. (See Table 8.11 for explanation of terms)

---

materials at various depths. These deposits are generally known as 'drift' in the UK.

A large proportion of British site investigations therefore encounter glacial materials in one or more of their varying forms, and the property of rapid lateral and vertical change shown by some types of deposit has become notorious since construction has often revealed features undisclosed by the site investigation. It is commonly thought of as random and unpredictable but this is not always true. An understanding of the different facets of the glacial environment, each with its characteristic landforms, erosional processes and assemblages of deposits, can be of great value in predicting not only the range of variation but often also the actual location of anomalous geotechnical features.

Glacial till and outwash. Moving glaciers excavate soils and rocks in their paths which are carried along and released as the ice melts away. The material deposited directly by the glacier as it melts is called 'boulder clay' or much better *till*. If the ice front

remains more-or-less stationary for a long period of time, a considerable amount of till moraine may accumulate along the ice front. Sub glacial and other tills are also laid down in the form of extensive plains revealed as the glaciers retreat during periods of melting and are characterized by lack of stratification and large range in particle size (Figure 8.25). Many tills deposited by continental glaciers contain substantial amounts of clay-size particles and these are sometimes overconsolidated and form fairly stiff clays. Even though a deposit of till may be extensive in size and in texture, its strength may vary considerably from place to place.

Along the front of a glacier, water from the melting of the ice gathers to form large torrential streams which are capable of transporting great quantities of sediments. As the streams spread out over the plains most of the coarse sediment is deposited as *fluvioglacial* alluvium which has the characteristics of braided-stream deposits. This consists of granular soils with lenses of gravel, sand, or silt, which generally occur in front of a till moraine. In some localities extensive areas are covered by

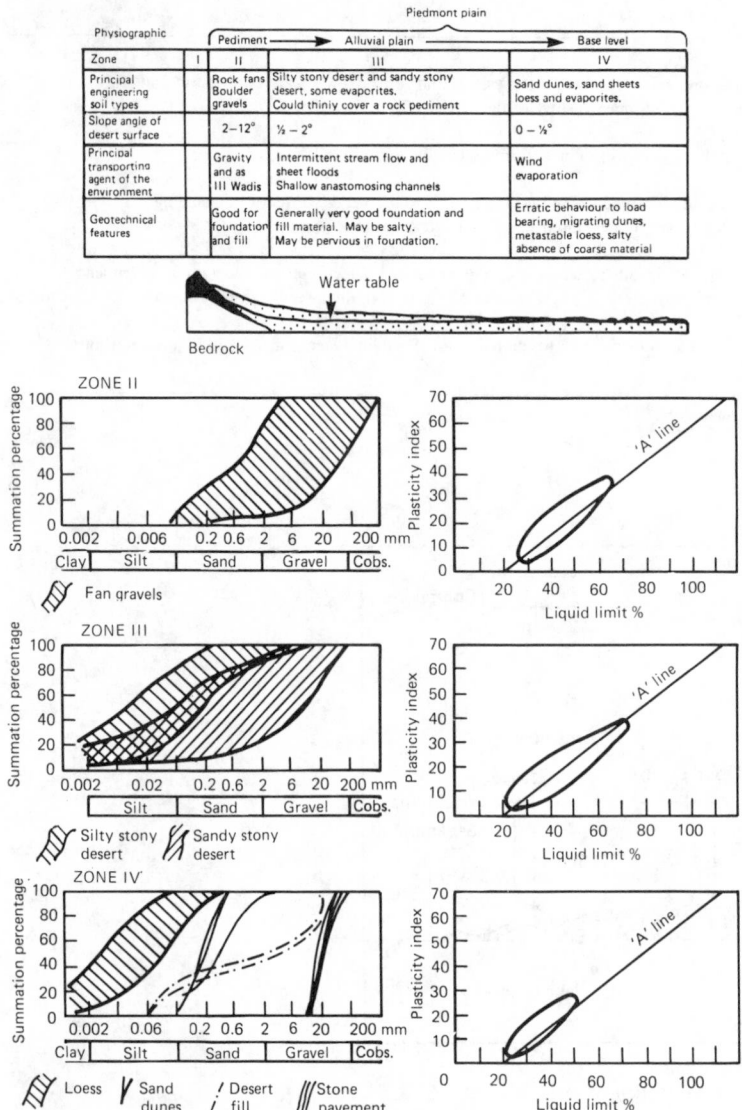

Physiographic		Pediment ⟶ Alluvial plain ⟶	Piedmont plain	Base level
Zone	I	II	III	IV
Principal engineering soil types		Rock fans Boulder gravels	Silty stony desert and sandy stony desert, some evaporites. Could thinly cover a rock pediment	Sand dunes, sand sheets loess and evaporites.
Slope angle of desert surface		2–12°	⅓ – 2°	0 – ⅓°
Principal transporting agent of the environment		Gravity and as III Wadis	Intermittent stream flow and sheet floods Shallow anastomosing channels	Wind evaporation
Geotechnical features		Good for foundation and fill	Generally very good foundation and fill material. May be salty. May be pervious in foundation.	Erratic behaviour to load bearing, migrating dunes, metastable loess, salty absence of coarse material

Figure 8.24 Idealized profile across mountain and plain desert terrain showing engineering zones I—IV and grading envelope, grading curves and Plasticity chart data from zones II, III and IV

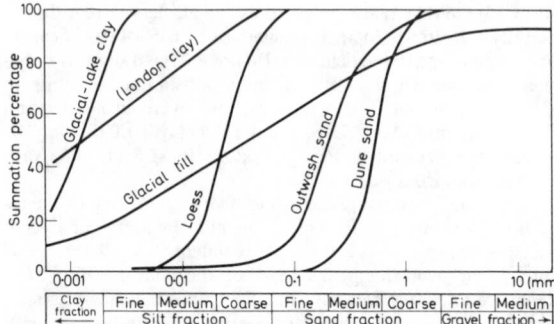

Figure 8.25 Examples of particle-size curves of some glacial soils and a London Clay for comparison

fluvioglacial deposits up to 30 m in thickness. In other areas they exist as thin lenses of limited lateral extent included between layers of till or peat.

Deposits of fluvioglacial soils may also occur in river valleys (valley trains) that once served as drainage outlets for glacial meltwater or locally in the form of ridges (eskers) and terraces (kames) as a result of various complications in the drainage system around glaciers. These fluvioglacial soils are composed primarily of silt, sand and gravel. Some of them are unstratified and others may exhibit irregular stratification. Glacial-lake deposits are often varved clays which exhibit characteristic thin stratification of silt and clay.

The principal engineering problem produced by the glacial environment is generally the difficulty of satisfactorily investigating the site. This is variously due to the rapidly changing soil type and engineering properties, large boulders in the till, lenses of clay, silt, sand or open gravel in other materials, concealed and weathered rockhead topography, structural disturbance of glacial deposits and complex groundwater conditions. Figure 8.26 shows some of the features associated with glaciers. Differential settlement may occur with heavy bearing structures and for water-retaining structures permeability may be a problem.

Linell and Shea,[25] symposium proceedings[26] and McGowan and Derbyshire[27] discuss geotechnical properties of glacial sediments.

8.4.2.5 Periglacial soils ('drift' and 'head' in the UK)

The term 'periglacial' is used to denote conditions under which frost action is the predominant weathering process. Mass transportation, wind action, or both, may occur, but only in association with very cold climatic conditions such as those near the margins of glacial ice. Perennially frozen ground (permafrost) is an important characteristic, but is not essential to the definition of the periglacial zone. The inner boundary of the zone is sharply defined by the current margin of the ice sheet, but the outer edge is gradational and the radial width of the periglacial zone is indefinite.

The distribution of permafrost may be strongly influenced by ground conditions and topographic features. The surface strata must be sufficiently porous or jointed to contain water, and their thickness must be greater than the potential thickness of the active layer. Permafrost may be thin or absent under surface

features such as large bodies of water but it is still extensive near the northern and southern polar ice caps, in parts of Canada, Siberia and elsewhere.

Most periglacial effects from the Pleistocene glaciations on the topography and surface deposits are well preserved in southern England, between the limits of the last (Devensian) glaciation and the loess belts of North and Central Europe. However, they are not restricted to this area but can be found over the whole of the UK since the periglacial zone moved northward in the wake of the receding ice sheet. In the unglaciated areas, the relationship of frozen-ground features to older or younger glacial drifts often establishes them as independent of specifically glacial processes.

The phenomena associated with the periglacial environment almost defy classification since they are essentially overlapping aspects of a continuously evolving situation. Factors such as surface relief, lithology and geological structure have an important influence on the purely climatic effect. Some of the features which may occur concurrently are shown in an idealized manner in Figure 8.27.

From the figure it can be seen that the engineering problems can be classed under three principal headings:

(1) *Superficial structural disturbance*, e.g. frost shattering, glacial shear, hill creep, ice wedges and involutions of chemical weathering.
(2) *Mass movements*, e.g. cambering and valley bulging, landsliding or mudflows.

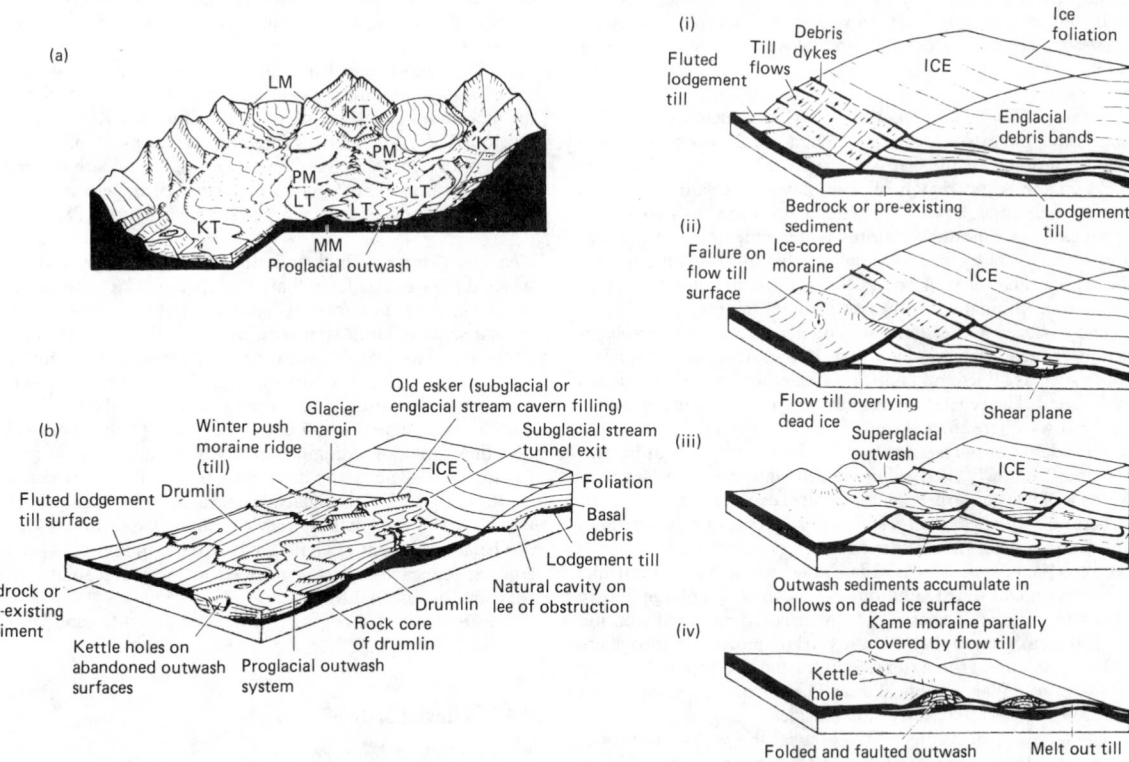

Figure 8.26 Schematic block diagrams showing three sediment associations and land systems. (a), glaciated valley system. LM – lateral moraine; MM – medial moraine (superglacial); LT – lodgement till; PM – push moraine; KT – kame terrace; (b), subglacial/proglacial system. Simple stratigraphy illustrated consists of outwash deposits on top of till resulting from single glacial advance followed by retreat; (c), superglacial system; progressive differential downwasting of ice margin produces ice-cored moraines between which meltwater streams flow and ultimately, an inversion of relief occurs as ice cores finally melt out. (After Derbyshire, Gregory and Hails (1979) *Geomorphological processes*. Butterworth, London, p. 312)

deflection indicates vertical strata; b, the outcrop which parallels

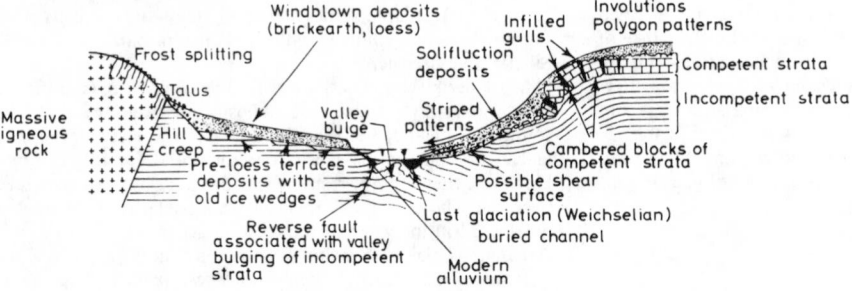

Figure 8.27 Idealized cross-section showing some periglacial features and deposits

(3) *Periglacial deposits*, e.g. loess, head or solifluxion soils.

A full discussion of these problems in the UK is given in Higginbottom and Fookes.[28] Weeks[29] discussed periglacial slope problems and Fookes and Best[30] discuss periglacial metastable soils.

8.4.2.6 Limestone landforms

Limestones usually develop more distinctive surface features of engineering relevance than other rock types, primarily as a result of its jointing, permeability and solubility in water containing carbon dioxide or humus acids. The lithology of the limestone is also significant, as strong, well-jointed limestones possess different features from those of weak limestones such as chalk.

Landforms on massive limestones. Rocks such as the Carboniferous Limestone of parts of England, the resistant Mesozoic Limestones of the Causses of Central France and similar rocks in the Karst region of Yugoslavia, are said to possess typical limestone landforms. The main process involved in producing the limestone features is the widening of fractures, joints and faults by solution, and it is helpful for this action if the water table is well below the surface to allow water to percolate continually downwards through the rock.

The surface of the limestone, due to the irregular solvent action of acid waters which pick out more readily attacked zones such as joints and faults, is often conspicuously furrowed and fretted. The vegetation, usually herbaceous plants, grows in the furrows where there is most likely to be a little soil, so that the real depth of the fretting may not be immediately apparent.

Joints are slowly enlarged by solution into holes, the shape of which will depend largely on the control exercised by the minor structural features of the rock. Two main types of solution holes are distinguished; funnel-shaped depressions with a hole at the centre (*doline, sink hole, swallow hole, swallet*) and shaft-like holes. With continued solution such holes may enlarge and in places several may coalesce to form larger, compound solution holes (*uvala*). Certain parts of the Carboniferous Limestone are dotted with grass-grown solution holes; the outcrop of the north of the South Wales coalfield near Penderyn, for example, has solution holes scattered over the hillside.

The largest depressions of Yugoslavia, the *poljes*, are probably not solution forms at all but tectonic depressions modified by solution of the limestone preserved in them.

If for some reason, such as the erosion of adjacent areas of impermeable rocks, the water table in a mass of limestone becomes lowered, the main underground channels will be displaced to successively lower levels. At the same time, general lowering of the limestone surface is thought to thin the rock above the underground caverns by solution loss so that eventually the roofs collapse and the drainage reappears at the surface in deep narrow gorges. Certain narrow valleys in Yugoslavia have been attributed to such cavern collapse and the same hypothesis has been applied to Cheddar Gorge in Britain. It must not be thought, however, that every narrow limestone valley is a collapsed cavern. The lowering of the water table in a limestone region is effected largely by the entrenchment of the valleys. During the process some rivers cut down their valleys more rapidly than others and, by the underground abstraction of drainage, the majority of the valleys become dry. The rivers which survive receive no surface tributaries but, instead, a supply of water from springs at river level. Streams having headwaters outside the limestone region will be assured of a supply of water, and consequently may survive as the main surface streams.

In the Yugoslavian Karst region, which is probably unique both on account of its area and because of the great thickness of its limestones, it is possible to formulate a Karst cycle of erosion. The cycle includes three important assumptions: (1) a thick and extensive mass of limestone; (2) an underlying impermeable stratum; and (3) a surface layer of impermeable rocks for the initiation of a stream pattern (see Figure 8.28).

Chalk landforms. Chalk and other weak limestones form relief which differs greatly from that developed on Carboniferous and similar massive limestones. Chalk does not often possess such a regular series of joints so there is usually little or no joint control of the relief comparable with that of Carboniferous Limestone districts, nor is the rock hard enough to be fretted at the surface by solution, nor usually is it strong enough to allow a development of large caves. A few caves and gaping fissures can occur, but they are not common, perhaps because the weight of fractured chalk above tends to close up any fissures widened by solution. In the UK, periglacial (freeze–thaw) disturbance of the upper part of the Chalk is very common (see section 8.4.2.5).

The general form of chalk landscapes, dominated by smooth convexo–concave curves is ascribed to the permeability of the rock and its residual soil.

See Dearman[31] for engineering data on limestone and Hobbs[32] for chalk.

8.4.3 Alluvial soils of rivers

8.4.3.1 Cycle of valley erosion

A river flowing in a valley erodes the material of its bed and local surface runoff contributes to the erosion of the walls of the valley. The eroded materials are transported in the form of sediment by the river and are eventually deposited.

Considered simply, rivers and the valleys along which they flow may be youthful, mature or old. At these three stages in the life of a river or valley its longitudinal profile, cross-section, and

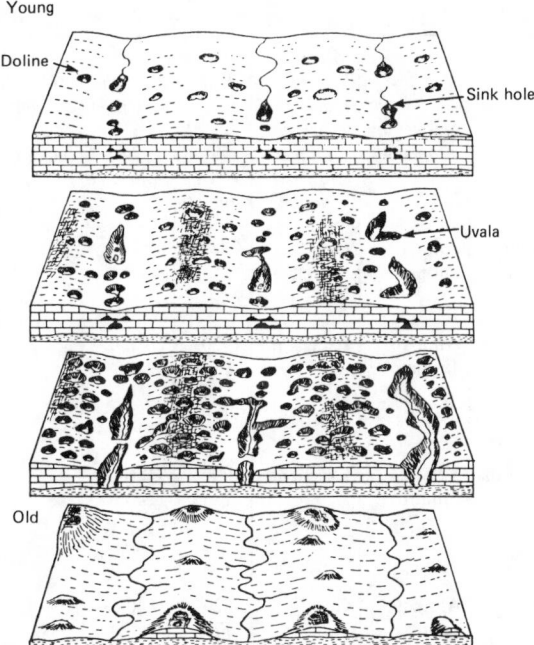

Figure 8.28 Simplified stages of the karst cycle of erosion (for exlanation, see text). The depth and size of the solution holes is greatly exaggerated in relation to the thickness of the limestone. (After Sparks (1972) *Geomorphology*, 2nd edn. Longman, London)

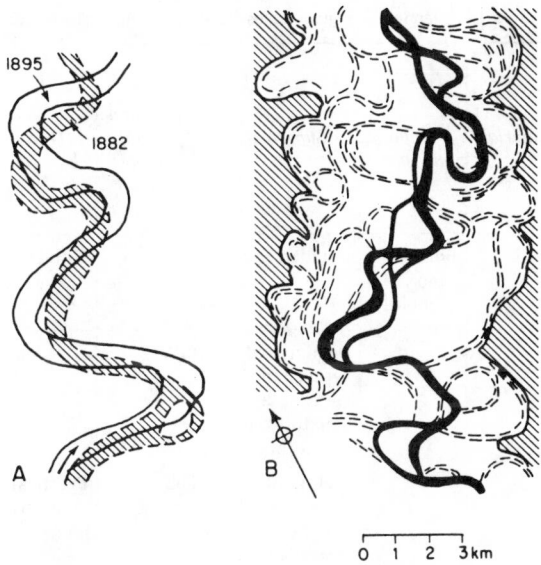

Figure 8.29 Meanders: A, downstream migration of meanders in the Mississippi; B, the river channel (black) and abandoned meanders of the Rhine

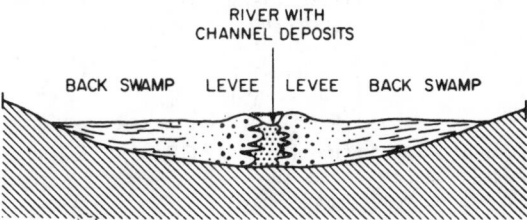

Figure 8.30 Deposits in the floodplain of a river

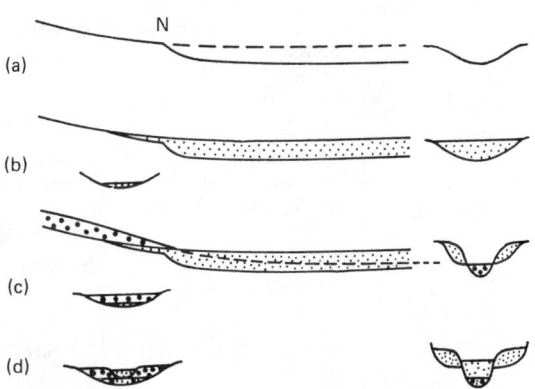

Figure 8.31 Long profiles of rivers with down-cutting and aggradation phases associated with changing sea-levels. Fine dots, alluvium of temperate stages; large dots, sands and gravels of cold stages: a, preglacial valley, river rejuvenated by low sea-level of glacial times. Nick-point marks the head of rejuvenation N; b, aggradation as a result of a rising interglacial sea-level; c, a further glacial low sea-level results in down-cutting in the lower parts of the valley and aggradation of outwash and weathering debris in the upper part; d, further aggradation during a second interglacial stage of higher sea-level

plan undergo gradual changes. At the *youthful* stage of a valley, its longitudinal profile is irregular and contains rapids, falls and even lakes because of local obstructions, such as hard rock strata, and its cross-section tends to be V-shaped. The plan of a youthful valley or river is somewhat angular or zigzag.

As erosion progresses, the river reaches *maturity*, irregularities gradually disappear and the plan acquires the shape of a smooth sinusoidal curve. The longitudinal profile also becomes reduced in gradient, decreasing gradually towards the mouth of the river. The valley at its mature stage is wide; its slopes are flatter than in its youth and often covered with talus (hillside rock debris).

Periodic floods contribute to the gradual widening of the valley until at its *old* age it becomes a wide floodplain. Between the floods, the old river meanders, changes its plan, but stays within a certain meander belt at the central part of the floodplain. In shifting from one location to another, a meandering river may leave behind oxbow lakes or abandoned oxbow-shaped depressions. Examples of a meandering river are the Thames, Rhine and Mississippi (see Figures 8.29 and 8.30).

During any period of geological time when climatic changes remain approximately constant, and in the absence of uplift (e.g. due to formation of folds) or change of base level (e.g. falling sea-level during a glacial period), downcutting of the river is slow enough for the lateral swinging of the river usually to make the valley wider than the channel itself. However, when the base level is lowered, the old floodplain is dissected and perhaps left in part at least, as a terrace, i.e. an abandoned floodplain. Rises in base level (e.g. the sea-level rising after a glacial period) causes the river to fill its channel (i.e. bury it) and to raise its bed level to keep pace with the rising base level. Figure 8.31 illustrates this by reference to a glacial cycle in the UK (e.g. the River Thames). More complex sequences can occur by the interacting of falling and rising base levels.

Table 8.12 Extended Casagrande classification of some alluvial soils

Material	Major divisions	Subgroups	Casagrande group symbol	Drainage characteristics	Potential frost action	Shrinkage or swelling properties	Value as a road foundation when not subject to frost action	Bulk unit weight before excavation (kg/m³)		Co-efficient of bulking (%)
								Dry or moist	Saturated	
Coarse-grained soils	Boulders and cobbles	Boulder gravels	—	Good	None to very slight	Almost none	Good to excellent	—	—	
	Gravels and gravelly soils	Well-graded gravel and gravel–sand mixtures, little or no fines	GW	Excellent	None to very slight	Almost none	Excellent	1920–2165	2080–2325	
		Well-graded gravel–sand mixtures with excellent clay binder	GC	Practically impervious	Medium	Very slight	Excellent	2000–2245	2160–2405	
		Uniform gravel with little or no fines	GU	Excellent	None	Almost none	Good	1520–1765	1840–2085	10–20
		Poorly-graded gravel and gravel–sand mixtures with little or no fines	GP	Excellent	None to very slight	Almost none	Good to excellent	1600–1845	1760–2005	
		Gravel with fines, silty gravel, clayey gravel, poorly-graded gravel–sand–clay mixtures	GF	Fair to practically impervious	Slight to medium	Almost none to slight	Good to excellent	1760–1925	1920–2085	
	Sands and sandy soils	Well-graded sands and gravelly sands, little or no fines	SW	Excellent	None to very slight	Almost none	Excellent to good	1840–2005	2000–2165	
		Well-graded sand with excellent clay binder	SC	Practically impervious	Medium	Very slight	Excellent to good	1920–2085	1920–2325	
		Uniform sands with little or no fines	SU	Excellent	None to very slight	Almost none	Fair	1520–1845	1840–2165	5–15
		Poorly-graded sands with little or no fines	SP	Excellent	None to very slight	Almost none	Fair to good	1440–1685	1520–1765	
		Sands with fines, silty sands, clayey sands, poorly-graded sand–clay mixtures	SF	Fair to practically impervious	Slight to high	Almost none to medium	Fair to good	1520–1765	1760–2005	

Table 8.12 *Continued*

Material	Major divisions	Subgroups	Casagrande group symbol	Drainage characteristics	Potential frost action	Shrinkage or swelling properties	Value as a road foundation when not subject to frost action	Bulk unit weight before excavation (kg/m³)		Coefficient of bulking (%)
								Dry or moist	Saturated	
Fine-grained soils	Soils having low compressibility	Silts (inorganic) and very fine sands, rock flour, silty or clayey fine sands with slight plasticity	ML	Fair to poor	Medium to very high	Slight to medium	Fair to poor	1520–1765	1600–1765	20–40
		Clayey silts (inorganic)	CL	Practically impervious	Medium to high	Medium	Fair to poor	1600–1765	1760–1925	
		Organic silts of low plasticity	OL	Poor	Medium to high	Medium to high	Poor	1440–1685	1520–1765	
	Soils having medium compressibility	Silty and sandy clays (inorganic) of medium plasticity	MI	Fair to poor	Medium	Medium to high	Fair to poor	1520–1765	1600–1765	—
		Clays (inorganic) of medium plasticity	CI	Fair to practically impervious	Slight	High	Fair to poor	1600–1765	1765–1925	
		Organic clays of medium plasticity	OI	Fair to practically impervious	Slight	High	Poor	1440–1685	1520–1765	
	Soils having high compressibility	Micaceous or diatomaceous fine sandy and silty soils, elastic silts	MH	Poor	Medium to high	High	Poor	<1680	<1925	—
		Clays (inorganic) of high plasticity, fat clays	CH	Practically impervious	Very slight	High	Poor to very poor	<1925	<1925	
		Organic clays of high plasticity	OH	Practically impervious	Very slight	High	Very poor	<1765	<1925	
Fibrous organic soils with high compressibility	Peat and other highly organic swamp soils		Pt	Fair to poor	Slight	Very high	Extremely poor	<1765	<1765	—

8.4.3.2 Alluvial soils

Eroded soil transported by water and deposited is alluvial (water-laid) soil, or alluvium. Immediately adjacent to the steep portion of the valley, boulders and coarser gravel might be expected and there will be a minimum of fine sizes. At a distance of several kilometres from the place of original erosion, fines may predominate.

Alluvial deposits are in many respects similar to glacial but are generally more stratified and their properties might be determined from fewer boreholes than under equal conditions in glacial soils. The alluvial deposits are somewhat heterogeneous. It is not unusual, for example, to find a bed of alluvial clay several metres long, although it may be fairly narrow and only a few tens of millimetres thick. Rather uniform sand and gravel beds of varying dimensions may be found and, although there may be lens-like inclusions of sand in gravel beds and vice versa, these deposits as a whole are fairly continuous.

Besides forming terraces and benches in the valley itself, deposition of alluvium also may occur on river plains and form relatively flat deposits. Large plains are not necessarily continuous but may be interrupted by isolated hills and occasional valleys. The sediment carried by a flow moving across a plain during a flood may be spread if the gradient of the stream decreases gradually and in this case a floodplain is formed. However, if the gradient decreases abruptly, a larger part of the sediment carried by the stream drops in one place and forms an alluvial fan, a broad cone with the apex at the point where the gradient breaks.

Particular cases of recent alluvium are organic silt and mud. These are fine outwash from hills and mountain ridges, deposited in estuaries and in the rivers flowing into them, especially in the lower reaches of these rivers. The greater part of the organic silt consists of angular fragments of quartz and feldspar, abundant sericite (fine mica), and clayey matter; numerous microorganisms are also present. In the natural state, organic silt is dark and smells unpleasant; after drying it can become light grey and lose its characteristic odour. Table 8.12 gives the Casagrande classification of soils deposited from river systems.

For further reading see the bibliography on general and physical geology, and for engineering behaviour see Chapter 9.

8.5 Geological maps

8.5.1 General geological maps

Maps of the British Geological Survey in the UK are published in two principal forms. 'Solid' maps show the rock outcrops as they would appear with the overburden removed. 'Drift' maps show overburden, usually with dotted lines to indicate the probable extent of the underlying outcrops. All show outcrop patterns, not just the actual exposure of the rock at the surface. As it is only exposures which can be seen, geological maps are necessarily in part conjectural.

Mapping is based on various techniques and the completed map is not simply a survey, but the sum of all the information gathered from various sources. The small-scale map, e.g. 1:625 000 or larger, is useful to obtain a general appreciation of the country over a relatively wide area, whilst the large-scale map, 1:50 000 or less, is more for detailed information. Regional guides and detailed memoirs are also published by the British Geological Survey as well as water and mineral memoirs and so on, to supplement their maps.

One of the first things to be considered on looking at a geological map is to determine whether the rocks shown are igneous, metamorphic or sedimentary, by checking the main outcrops shown against the key provided on the map. Assuming the rocks are sediments, the strike is usually the long axis of the

outcrop if it is a small-scale map. A more accurate picture is obtained by comparing the outcrop with the contours. By definition, in dipping beds the strike is the direction in which similar horizons of the strata are at the same elevation, so it follows that where the top (or bottom) boundary of a bed twice crosses the same contour line, the top of the bed will be at the same altitude at those two points and that a line drawn connecting them will show the strike direction.

The dip can be identified by noting the direction of dip arrows if these are shown, otherwise it can be deduced. The key will show which are the older of two successive beds; a boundary line on flat ground indicates that the older beds have emerged from below the newer beds, or in other words that the older beds are dipping towards the newer beds. If the boundaries follow contour lines, the beds are more-or-less horizontal, and if the boundaries cross contour lines at right angles, the beds are vertical. The relation of the outcrops to the contour lines in valleys is particularly helpful in diagnosing the general dip of the bed. (See examples in Figure 8.32.)

The degree of dip is obtainable by drawing parallel strike

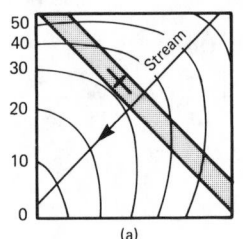 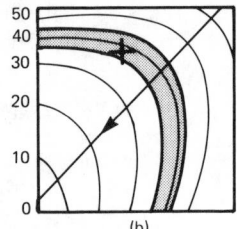

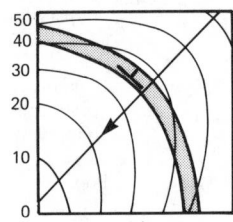 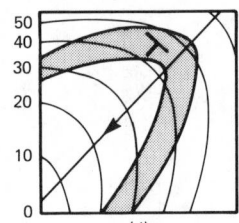

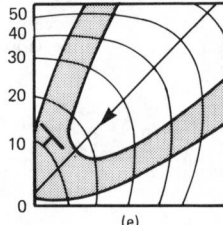

Figure 8.32 Outcrop patterns of a geological stratum related to topography: a, the outcrop crossing valley contours without defection indicates vertical strata; b, the outcrop which parallels topographic contours indicates horizontal strata; c, the outcrop which forms a blunt V pointing up a valley indicates a strata dipping upstream; d, the outcrop which forms a narrow V pointing up a valley indicates a strata dipping downstream but at an angle smaller than the valley gradient; e, the outcrop which forms a V pointing downstream across a valley indicates strata dipping downstream but at an angle greater than that of the valley gradient

lines, at different contour levels, using the same boundary between beds. Thus if a boundary between two beds crosses, say, the 100 m, 200 m and 300 m contours on each side of a valley, and the strike lines are drawn in at these levels, the distance between each strike line on the map is the distance over which the bed has dipped 100 m. The degree of dip may then be calculated or obtained graphically. Examples are given in Figures 8.33 and 8.34.

There are many other problems that can be solved by a study of geological maps, and textbooks dealing with this are given in the bibliography.

8.5.2 Special geological maps

8.5.2.1 Engineering geology maps

These are simplified maps in which the details and stratigraphy

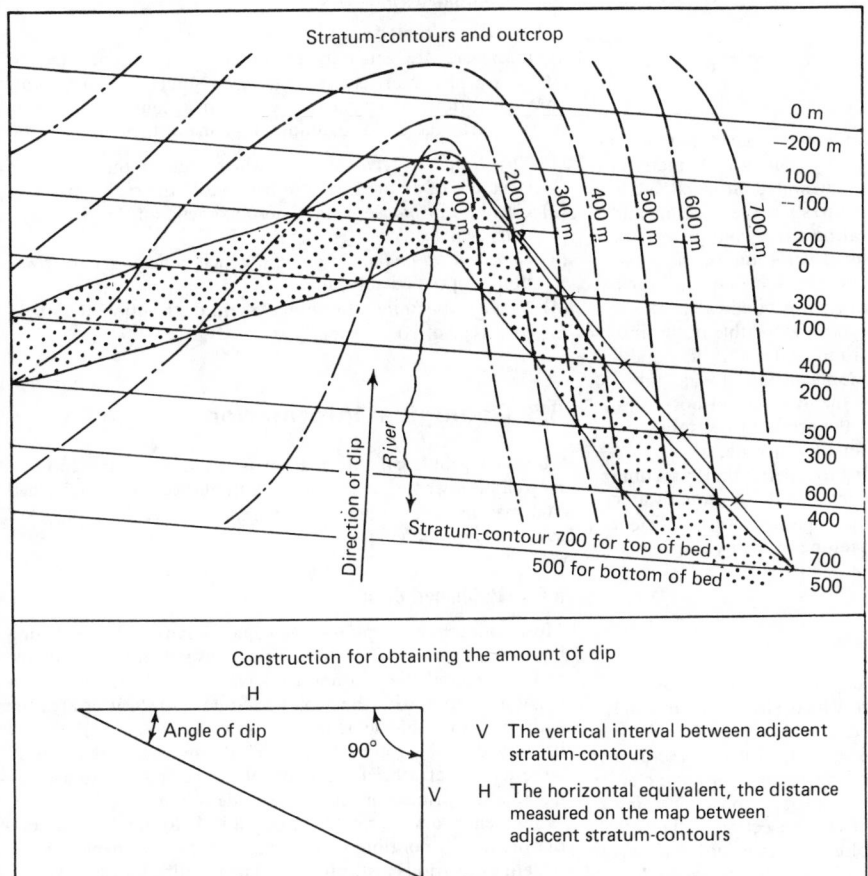

Figure 8.33 Stratum contours used to plot the outcrop of a bed and to calculate the dip. The construction is general, not specifically related to the example

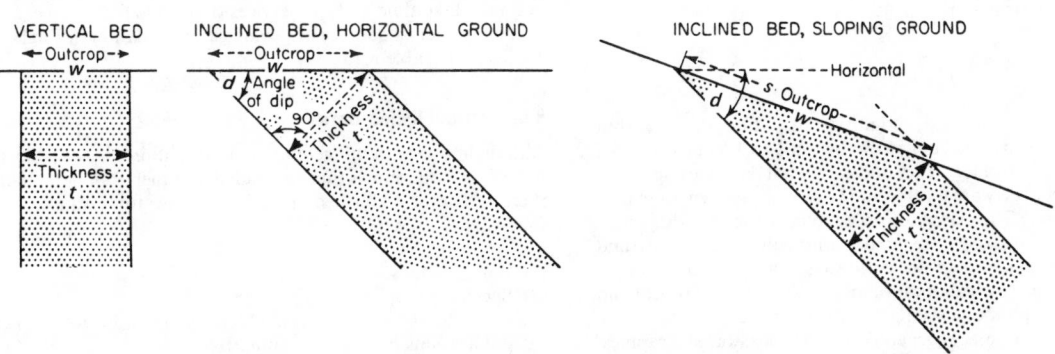

Figure 8.34 Width outcrop in relation to dip. Thickness *t*=outcrop width, w/cosec *d* −*s* where *s* is the angle of slope

are omitted as far as possible, and physical characteristics of the rocks and soils, and their uses, are given. The legends of such maps are generally prepared so as to give facts and inferences of importance to the engineer. Small-scale maps may, however, be over-simplified. The particular preparation of maps and plans in terms of engineering geology is described in a Geological Society report.[33] Such maps can also be produced by an engineering geologist from the standard geological maps, and Dumbleton and West[34] list references which describe the procedure. British Standard Code of Practice 5930[35] gives current terminology.

8.5.2.2 Hydrogeological maps

The purpose of hydrogeological maps is to enable various areas to be distinguished according to their hydrological character in relation to the geology. At present, only limited areas in the UK or elsewhere have been mapped, either to show the structure and distribution of the various water-bearing strata, and/or to show the depth, and sometimes the seasonal variation, of the water table. New maps being produced in the UK indicate, on a regional basis, the extent of the principal groundwater bodies, the scarcity of groundwater, the known or possible occurrence of artesian basins, areas of saline groundwater and the potability of groundwater. They also show, according to scale, information of a local character, such as the locations of boreholes, wells and other works, contours of the groundwater table, the direction of flow and variations in quality of water.

In general, any information leading to a better understanding of occurrence, movement, quantity and quality of groundwater in the preparation of such maps should be shown with sufficient geology to lead to a proper understanding of the hydrogeological conditions.

8.5.2.3 Landform maps

Land use and landform are interrelated. Some maps are available in the UK and elsewhere which show present land utilization. Other maps indicate the relative value of land for agricultural use. Where such maps exist, they may serve as a preliminary indication of land values and engineering characteristics (see Dumbleton and West[34] for information concerning England and Wales). Landform is itself a reflection of the type of climate and the nature of the geology. Patterns of landforms can be mapped as land systems, subdivided into facets and elements, which broadly relate to the geology of the ground and more particularly to its agricultural and engineering properties. The advantage of such maps is that they can be prepared easily from air photographs; the disadvantage is their small scale (usually 1:500 000 to 1:50 000).

8.5.2.4 Soil maps

Soil maps exist in many countries including the UK but generally are concerned only with the top 1 to 1.5 m of material at the Earth's surface. They indicate what kinds of soils are present, where they are located and to some extent what use they can best serve. The classification adopted varies with intended use. Most soil maps are prepared for agricultural purposes, and although there is no agreed worldwide system which will provide for the precise classification of all varieties of physical and chemical composition existing in the soils of the world, discrimination between units is generally on the basis of geographical association, parent material, texture, subsoil characteristics and drainage class. Some but not all such criteria have engineering significance.

8.5.2.5 Resource maps

These maps indicate the occurrence and distribution of economic rocks (e.g. gravel, limestone, minerals, coal, oil, etc.) They are often on a small scale and generalized, but may record useful, if rapidly outdated, production information. Maps showing economic deposits for construction (e.g. sand and gravel, limestone, etc.) are published for parts of the UK.

8.5.2.6 Subsidiary map types

All the above categories of maps are potentially useful to the engineer and are generally published by a national agency. There are also more specialized maps sometimes obtainable through various government or commercial sources which may meet a particular need. Examples of some of these are:

(1) *Structural and tectonic maps*, which indicate the geological structure of an area but not necessarily the rock types.
(2) *Geophysical maps*, such as aeromagnetic and gravity anomaly maps.
(3) *Geochemical and mineral maps*, which indicate the distribution of specified substances.
(4) *Single feature maps*, to emphasize the distribution of a single rock type or rock property.

8.6 Geological information

Although geological information is available in the form of maps and in written texts and both published and unpublished data may have to be acquired, *Military engineering*, vol. XV[36] suggests the following.

8.6.1 Published data

Most countries now publish geological maps with supporting literature. This basic literature, which is usually readily available and understandable to a non-geologist, may take the form of a memoir, dealing with the geology of one map sheet or area (or with one aspect of the geology of several map sheets); a book broadly describing the geology of a country or significantly large region; or a brief summary of the geology of an area or information printed on the reverse side of a map.

Supplementary literature, more difficult to obtain or assess, but often incorporating maps, generally exists as papers in the bulletins of various institutions and in scientific journals. Where the basic literature is inadequate, this supplementary literature could be used. However, problems are often caused by the magnitude of published data (some 50 000 geological papers are published annually) and sifting and extracting relevant information may take time and the services of a geologist.

Some principal sources of geological information in England are listed in Table 8.13.

8.6.2 Unpublished data

Although a wealth of geological information is available in published form, much that is detailed and therefore potentially useful is never published in full, such as, for example, the following.

(1) Borehole logs.
(2) Specialist reports.
(3) Field notebooks, maps and detailed records from which publications have been summarized.
(4) University theses and dissertations.

Because of its detail such information may be of great practical

Table 8.13 Sources of geological information in the UK

Location	Information available	Notes

General Information
British Geological Survey
Exhibition Road
South Kensington, London SW7 2DE
Tel. 01-589 3444
and/or Nickerhill, Keyworth,
   Nottingham NG12 5GG
(Regional offices in Edinburgh,
   Aberystwyth, Belfast, Exeter, Leeds
   and Newcastle)
	Maps and literature
	worldwide coverage
		Constituent body of Natural
		   Environment Research Council.
		   Specialist advisory service is also
		   available through staff members.
		Library of the Overseas Division
		   contains information on most
		   developing countries.
		Visits should be made by appointment

National Reference Library of Geology,
   Geological Museum,
   Exhibition Road,
   South Kensington, London SW7 2DE
	All scales of maps of Britain. Many
	   publications and maps of countries
	   worldwide. Comprehensive filing
	   system
		Available for copying. Also much
		   unpublished data such as large-scale
		   maps and borehole logs. Access free.
		   (Part of British Geological Survey)

Library of Geological Society of London,
   Burlington House,
   Piccadilly, London W1V 0JU
	Information on the UK and overseas
	   countries, also contains an engineering
	   geology section
		Consultation fee must be paid by
		   non-members

Libraries of British Museum,
   (Natural History)
   Cromwell Road,
   South Kensington, London SW7 5BD
	All British and many foreign publications
	   (including geological maps) relevant to
	   the work of the museum
		Access free

Geologists Association Library,
   University College,
   Department of Geology,
   Gower Street, London WC1E 6BT
	Information on the UK and overseas
	   countries
		Access free

Specialized Information
National Coal Board,
   Hobart House,
   Grosvenor Place, London SW1X 7AE
   (also area and regional offices)
	Detailed information about coalmining
	   areas	

Nature Conservancy (Geology and
   Physiography Section)
   Foxhold House, Thornford Road,
   Headley, Newbury, Berks
	Geology of conserved areas, sites of
	   special scientific interest, some
	   coastlines, new roads
		Constituent body of Natural
		   Environment Research Council

Institute of Hydrology
   Maclean Building,
   Crowmarsh Gifford, Wallingford,
   Berks (also Regional Water
   Authorities)
	Hydrogeological data	Constituent body of Natural
		   Environment Research Council

Institute of Mining and Metallurgy
   44 Portland Place, London W1N 4BR
	Information on the UK and overseas
	   countries	

Soil Survey of Great Britain,
   Rothamstead Experimental Station,
   Harpenden, Herts.
	Pedological soil surveys	Constituent body of Agricultural
		   Research Council

Macaulay Institute for Soil Research,
   Craigiebuckler, Aberdeen AB9 2QJ
	Information on the UK and overseas
	   countries	

Department of the Environment,
   Lambeth Bridge House
   London SE1 7SB
	Data on some resources and sites, UK
	   and overseas
		Minerals Division deals with geological
		   aspects of planning and environment,
		   also sand/gravel supplies

Transport and Road Research
   Laboratory,
   Crowthorne, Berks G11 6AU
	Geological information relevant to road
	   construction
		Department of the Environment

Table 8.13 *Continued*

Location	Information available	Notes
Water Resources Board, Reading Bridge House Reading RG1 8PS	Hydrogeological data	Department of the Environment
Building Research Station, Garston, Watford, Herts	Some geotechnical data relevant to building construction	Department of the Environment
Civil Engineering Laboratory, Cardington, Bedfordshire	Data relevant to foundations and to slope stability	Department of the Environment

use for engineering purposes, but problems may be encountered in locating it and securing permission for its release.

8.6.3 Books

The bibliographies of the Geological Society of America have been issued since 1933 and are available at most university or national reference libraries. They list all the geological information published about any country for a given year and in its present form have separate location and author indices. The value of the bibliographies lies in the detail of their references, but many of the publications listed may be difficult to obtain other than from the largest national lending libraries, e.g. the British Library Lending Division, Boston Spa, Wetherby, West Yorks LS23 7BQ, telephone (0937) 843434.

Many countries have an established Geological Society which publishes papers primarily related to the country. Copies of many of these papers are kept at the Geological Society of London. Major libraries throughout the world are listed in the *World of learning*, published annually by Europa Publications.

8.6.4 Institutions

Government institutes of geology and allied sciences provide the main reference source for geological information. They publish maps and literature, store unpublished data, and may provide a specialist advisory service through their staff members. Most countries now possess an equivalent to the British Geological Survey in the UK, although this may be called a Geological Survey Department, Bureau or Department of Mines and Mineral Resources. The work produced by such an institution will usually be in the language of that country and a technical translation of high quality may be essential.

Many universities have a department of geology or earth sciences and may possess information not available elsewhere, comprising research work in progress and unpublished theses, dissertations and reports. British universities have research interests overseas; details of staff and their research interest are published annually by the Department of Education and Science. The address of overseas universities and staff are given annually in the *World of learning*.

Local museums frequently serve as depositories for geological data, both published and unpublished. Certain national organizations in any country accumulate specialized geological information.

Similar organizations exist in some developing countries and for many of these areas particularly useful information may be contained in unpublished reports of oil companies, mining companies and civil engineering firms. The oil companies have produced geological maps for many otherwise unsurveyed areas and some of these have been published. A request for geological information, particularly about *near-surface* formations, may be received sympathetically depending on company policy and circumstances.

Besides the methods of obtaining geological information mentioned above, there are data retrieval organizations and geological abstracting services which will provide information for a fee.

Another source of information is satellite photographs. A catalogue and price list of satellite photographs can be obtained from Audio-visual Branch, National Aeronautics and Space Administration, Washington DC, US.

References

1. Price, D. G. (1971) 'Engineering geology in the urban environment', *Q. J. Engng Geol.* **4**, 191–208
2. Krynine, D. P. and Judd, W. R. (1957) *Principles of engineering geology and geotechnics.* McGraw-Hill, New York.
3. Lahee, F. H. (1961) *Field geology*, 6th edn. McGraw-Hill, New York.
4. Shergold, F. A. (1960) 'The classification, production and testing of roadmaking aggregates.' *Quarry Managers J.,* **44**, 2, 3–10.
5. Attewell, P. B. and Farmer, I. W. (1975) *Principles of engineering geology.* Chapman and Hall, London.
6. Price, N. J. (1966) *Fault and joint development in brittle and semi-brittle rock.* Pergamon Press, Oxford.
7. Phillips, F. C. (1971) *The use of stereographic projection in structural geology*, 3rd edn. Arnold, London.
8. Fookes, P. G. and Vaughan, P. R. (1986) *A handbook of engineering geomorphology.* Surrey University Press, London.
9. Sanders, M. E. and Fookes, P. G. (1970) 'A review of the relationship of rock weathering and climate and its significance to foundation engineering.' *Engng Geol.* **4**, 289–325.
10. Fookes, P. G., Dearman, W. R. and Franklin, J. A. (1971) 'Some engineering aspects of rock weathering with field examples from Dartmoor and elsewhere.' *Q. J. Engrg Geol.* **4**, 139–186.
11. Little, A. L. (1967) 'Laterites.' *Proceedings, 3rd Asian Regional Conference on Soil Mechanics Foundation Engineering.* Haifa, **2**, 61–71.
12. Deere, D. R. and Patton, F. D. (1971) 'Stability of slopes in weathered rock.' *Proceedings, 4th Panamanian Conference*, pp. 87–163.
13. Irfan, T. Y. and Dearman, W. R. (1978) 'Engineering classification and index properties of a weathered granite.' *Bull. Int. Assoc. Engng Geol.* **17**, 79–90.
14. Institution of Civil Engineers (1957) Symposium on airfield construction on overseas soils. *Proc. Instn. Civ. Engrs,* **8**, 211–292.
15. Chen, F. H. (1975) *'Foundations on expansive soils.* Elsevier, Amsterdam, p. 280.
16. Gidigasu, M. D. (1975) *Laterite soil engineering.* Elsevier, Amsterdam, p. 570.
17. Anon. (1982) 'Engineering construction in tropical and residual soils.' *Proceedings, American Society of Civil Engineers Geotechnical Engineering Division Special Conference.* Hawaii, p. 735.
18. Cooke, R. U., Brunsden, D., Doornkamp, J. C. and Jones, D. K. C. (1982) *Urban geomorphology in drylands.* Oxford, p. 324.

19. Fookes, P. G. and Knill, J. L. (1969) 'The application of engineering geology in the regional development of northern and central Iran.' *Engng Geol.* **3**, 81–120.

20. Holtz, W. G. and Gibbs, H. J. (1952) *Consolidation and related properties of laersial soils.* American Society of Civil Engineers Testing Material Special Technical Publication No. 126, pp. 9–33.

21. Weinert, H. H. and Clauss, M. A. (1967) 'Soluble salts in road foundations.' *Proceedings, 4th Regional Conference for Africa Soil Mechanics and Foundation Engineering.* pp. 213–218.

22. Fookes, P. G. and French, W. J. (1977) 'Soluble salt damage to surfaced roads in the Middle East.' *Highway Engineering,* **XXIV**, 12, 10–20.

23. Fookes, P. G. and Higginbottom, I. E. (1980) 'Some problems of construction aggregates in desert areas with particular reference to the Arabian peninsula.' *Proc. Instn Civ. Engrs,* Part 1, **68**, 39–90.

24. Oweis, I. and Bowman, J. (1981) 'Geotechnical considerations for construction in Saudi Arabia.' *Proc. Am. Soc. Civ. Engrs,* Paper 16092, 319–338.

25. Linell, K. A. and Shea, H. F. (1960) 'Strength and deformation characteristics of various glacial tills in New England.' *Proceedings, American Society of Civil Engineers Regional Conference on Shear Strength of Cohesive Soils.* pp. 275–314.

26. 'The engineering behaviour of glacial materials.' *Proceedings, Symposium on Midland Soil Mechanics and Foundation Engineering Society.* Birmingham 1975, (reprint by Geo Abstracts, Norwich, p. 275.)

27. McGowan, A. and Derbyshire, E. (1977) 'Genetic influences on the properties of tills.' *Q. J. Engng Geol.,* **10**, 389–410.

28. Higginbottom, I. E. and Fookes, P. G. (1970) 'Engineering aspects of periglacial features in Britain.' *Q. J. Engng Geol.,* **3**, 86–117.

29. Weeks, A. G. (1969) 'The stability of slopes in south-east England as affected by periglacial activity.' *Q. J. Engng Geol.,* **2**, 49–62.

30. Fookes, P. G. and Best, R. (1968) 'Consolidation characteristics of some late Pleistocene periglacial metastable soils of east Kent.' *Q. J. Engng Geol.,* **2**, 103–128.

31. Dearman, W. R. (1981) 'Engineering properties of carbonate rocks.' *Bull. Int. Assoc. Engng Geol.,* **24**, 3–17.

32. Hobbs, N. B. (1975) 'Factors affecting the prediction of settlement of structures on rock: with particular reference to the Chalk and Triass.' *Revised Paper Session IV: rocks,* 579–610. British Geotechnical Society, London.

33. Geological Society (1972) Geological Society Working Party Report on the preparation of maps and plans in terms of engineering geology.' *Q. J. Engng Geol.,* **5**, 4, 293–381.

34. Dumbleton, M. J. and West, G. (1970) 'Preliminary sources of information for site investigation in Britain.' *Ministry of Transport Road Research Laboratory Report* LR 403, Crowthorne, Berks, p. 100.

35. British Standards Institution (1981) *Site investigation.* Code of Practice 5930. BSI, Milton Keynes.

36. Hughes, N. F. (1977) 'Applied geology for engineers.' In: *Military Engineering,* vol. XV. HMSO, London.

Bibliography

Periodicals

Engineering Geology (Published quarterly by Elsevier).
Géotechnique (Published quarterly by Thomas Telford Ltd).
International Journal of Rock Mechanics and *Mining Sciences* (Published bimonthly by Pergamon).
Quarterly Journal of Engineering Geology (Published by Geological Society).
Rock Mechanics and Engineering Geology (Published quarterly by Springer-Verlag).

Dictionaries

Challinor, J. (1978) *Dictionary of geology,* 5th edn, University of Wales Press.
Whitten, D. G. A. and Brooks, J. R. V. (1972) *Dictionary of geology.* Penguin.

Dictionaries

Challinor, J. (1978) *Dictionary of geology,* 5th edn, University of Wales Press.
Whitten, D. G. A. and Brooks, J. R. V. (1972) *Dictionary of geology.* Penguin.

General and physical geology

Ager, D. W. (1975) *Introducting geology,* 2nd edn. Faber and Faber, London.
Blyth, F. G. H. and de Freitas, M. H. (1974) *A geology for engineers,* 6th edn. Arnold, London.
Bradshaw, J. J., Abbott, A. J. and Gelsthorpe, A. P. (1978) *The Earth's changing surface.* Hodder and Stoughton, London.
Bridges, E. M. (1970) *World soils* Cambridge University Press, Cambridge.
Dury, G. H. (1966) *The face of the Earth* (rev. edn). Penguin, London.
Fookes, P. G. and Vaughan, P. R. (1986) *A handbook of engineering geomorphology.* Surrey University Press, London.
Gass, I. G., Smith, P. J. and Wilson, R. C. L. (1972) *Understanding the Earth,* 2nd edn. Artemis Press.
Holmes, A. (1978) *Principles of physical geology,* 3rd edn. Nelson, London.
Ollier, C. D. (1975) *Weathering,* 2nd edn. Oliver and Boyd, Cambridge.
Read, H. H. and Watson, J. (1971) *Beginning geology,* 2nd edn. Macmillan/Allen and Unwin, London.
Shephard, F. P. (1973) *Submarine geology,* 3rd edn. Harper and Row, London.
Small, R. J. (1978) *The study of landforms,* 2nd edn. Cambridge University Press, Cambridge.
Sparks, B. W. (1972) *Geomorphology,* 2nd edn. Longman, London.
Thomas, M. F. (1974) *Tropical geomorphology.* Macmillan, London.
West, R. G. (1977) *Pleistocene geology and biology,* 2nd edn. Longman, London.

Engineering geology

Anon. (1976) *Engineering geological maps.* UNESCO Press, Paris.
Hoek, E. and Bray, J. (1974) *Rock slope engineering,* 2nd edn. Institute of Mining and Metallurgy, London.
Attewell, P. B. and Farmer, I. W. (1975) *Principles of engineering geology.* Chapman and Hall, London.
Brown, E. T. (ed.) (1981) *Rock classification, testing and monitoring.* ISRM Suggested Methods. Pergamon Press, Oxford.
British Standards Institution (1981) *Site investigation.* Code of Practice 5930. HMSO, London.
Derbyshire, E., Gregory, F. J. and Hails, J. R. (1979) *Geomorphological processes.* Butterworth, London, p. 312.
Hughes, N. F. (1977) 'Applied geology for engineers.' *Military Engineering,* Vol. XV. HMSO, London.
Stagg, K. G. and Zienkiewicz, O. C. (1968) *Rock mechanics in engineering practice.* Wiley, Chichester.
Legget, R. F. and Karrow, P. G. (1983) *Geology in civil engineering.* McGraw-Hill, New York.

Fieldwork and mapwork

Bennison, G. M. (1976) *An introduction to geological structures and maps,* 3rd edn. Arnold, London.
Blyth, F. G. H. (1976) *Geological maps and their interpretation,* 2nd edn. Arnold, London.
Lahee, F. H. (1961) *Field geology,* 6th edn. McGraw-Hill, New York.

Structural geology

Billings, M. P. (1972) *Structural geology,* 3rd edn. Prentice-Hall, Englewood Cliffs, New Jersey.
Hills, E. Sherbon (1972) *Elements of structural geology,* 2nd edn. Chapman and Hall, London.
Phillips, F. C. (1971) *The use of stereographic projection in structural geology,* 3rd edn. Arnold, London.
Price, N. J. (1966) *Fault and joint development in brittle and semi-brittle rock.* Pergamon Press, Oxford.
Ramsey, J. G. (1967) *Folding and fracturing of rocks.* McGraw-Hill, New York.

Mineralogy and petrology

Grim, R. E. (1968) *Applied clay mineralogy,* 2nd edn. McGraw-Hill, New York.

Hatch, F. H., Wells, A. K. and Wells, M. K. (1973) *The petrology of the igneous rocks*, 13th edn. Murby,

Hatch, F. H. and Rastall, R. H. (1978) *The petrology of the sedimentary rocks*, 6th edn. (rev. J. T. Greensmith). Murby,

Krumbein, W. C. and Sloss, I. L. (1963) *Stratigraphy and sedimentation*, 2nd edn. Freeman, New York.

Pettijohn, F. J. (1976) *Sedimentary rocks*, 3rd edn. Harper and Row, London.

Read, H. H. (1970) *Rutley's elements of mineralogy*, 26th edn. Murby

In addition to these general works, the following series are of interest to British engineers:

British Regional Geology: handbooks published by the Geological Museum, London, SW7.

British Palaeozoic Fossils ⎫ Handbooks published by the British
British Mesozoic Fossils ⎬ Museum
British Cainozoic Fossils ⎭ (Natural History) London SW7

Geologists' Association Guides: a series of excursion guides to selected British localities (available from The Scientific Anglican, 30/30A St Benedict's St, Norwich)

Publications of the Geological Society of London, Burlington House, Piccadilly, London W1V 0JU: various monographs and authoritative works

9

Soil Mechanics

C R I Clayton MSc, PhD, CEng, MICE
Reader in Geotechnical Engineering, University of Surrey

Contents

9.1 The basics of soil behaviour

In engineering terms, soil is the generally softer, weaker and more weathered material overlying rock. All soils consist of solid particles assembled in a relatively loose packing. The voids between the particles may be filled completely with water (fully saturated soils) or may be partly filled with water and partly with air (partly saturated soils).

Soil and rock materials can, very simply, be divided into the groups shown in Table 9.1.

The primary engineering problems which we attempt to solve in soil mechanics are those of predicting the strength, compressibility and time-dependent compression of soil materials. For this it is necessary to understand the principle of effective stress.

9.1.1 Effective stress

Soil can be considered as a two-phase system consisting of a solid phase—the skeleton of soil particles—and a fluid phase—water plus air in a partly saturated soil, and water alone in a saturated soil. It follows that the normal stress across a plane within a soil mass will have two components: (1) an intergranular pressure, known as the effective pressure or effective stress; and (2) a fluid pressure known as the pore pressure or neutral pressure u. The sum of these will constitute the total normal stress. The volume change characteristics and the strength of a soil are controlled by the effective stress, the pore pressure being significant only in so far as it determines the magnitude of the effective stress for a given total stress.

The simplest illustration of pore pressure and effective stress is given by consideration of the vertical stresses acting on a horizontal plane at a depth h under equilibrium conditions with a horizontal water table. The total vertical stress σ is given by the weight per unit area of soil and water above the plane:

$$\sigma = \gamma h \tag{9.1}$$

where γ is the bulk density of the soil, i.e. its total weight/unit volume (see section 9.7 for a definition of terms).

The pore pressure will be the water pressure, and if the plane is at a depth h_w below the water table then $u = h_w \cdot \gamma_w$.

The effective vertical stress is the difference between these:

$$\sigma' = \sigma - u \tag{9.2}$$

In partly saturated soils, there is a pore air pressure u_a as well as a pore water pressure u_w and the effective stress Equation (9.2) is then modified as follows:

$$\sigma' = (\sigma - u_a) + \chi(u_a - u_w) \tag{9.2a}$$

The parameter χ is related to the degree of saturation S_r. For full saturation, $\chi = 1$ and Equation (9.2a) reduces to (9.2). Equation (9.2a) is rarely used in practice.

A change in total stresses arising from a change in external loading conditions will give rise to a change Δu in pore pressure.

The excess pore pressure, positive or negative, will dissipate with time, the rate at which equilibrium pore pressure conditions are re-established being governed by the permeability of the soil. In coarse grained granular soils, such equilibrium conditions will be achieved immediately and changes in effective stress are equal to changes in total stress. At the other limit, with clays of low permeability, equilibrium conditions may take considerable time, up to tens of years, to be re-established. The relation between pore pressure change and the change in principal stresses can be expressed by the use of pore pressure parameters A and B. The basic relationship is in terms of the major and minor principal stresses, σ_1 and σ_3:

$$\Delta u = B[\Delta\sigma_3 + A(\Delta\sigma_1 - \Delta\sigma_3)] \tag{9.3}$$

It is also useful to relate the pore pressure change to the change in deviator stress $(\Delta\sigma_1 - \Delta\sigma_3)$ alone and also to the change in the major principal stress $(\Delta\sigma_1)$. For these purposes, two further parameters $\bar{A}$ and $\bar{B}$ are used as follows:

$$\Delta u = B \cdot \Delta\sigma_3 + \bar{A}(\Delta\sigma_1 - \Delta\sigma_3)$$
$$\text{or} \quad \Delta u = \bar{B} \cdot \Delta\sigma_1 \tag{9.4}$$

If the soil structure behaved in an elastic manner, the values of the pore pressures could be established theoretically, e.g. A would have a value of 1/3. However, soils behave nonelastically and A can have values ranging between $+1.3$ and -0.7 (values at failure in a triaxial compression test). Typical values of the pore pressure parameters are given by Bishop and Henkel.[2] For a full discussion of the parameters see Skempton.[1]

It is the effective stress, rather than the total stress, which controls the key properties of strength and compressibility and, to a certain extent, permeability.

9.1.2 Shear strength

Shear strength of a soil is commonly thought of as having two components: cohesion and frictional resistance. Clays are often described as cohesive soils in which the shear strength or cohesion is independent of applied stresses, and sands and gravels are described as noncohesive or frictional soils in which the shearing resistance along any plane is directly proportional to the normal stress across that plane:

$$s = p \cdot \tan\phi \tag{9.5}$$

The concepts of cohesion and friction were combined in Coulomb's equation for the shear strength of soil:

$$s = c + p \cdot \tan\phi \tag{9.6}$$

where c is the cohesion and ϕ is the 'angle of internal friction'.

Such simple concepts are, however, inadequate to deal with the complex problem of the shear strength of soils. The early history of the study of shear strength is somewhat confused.

Table 9.1 A simple grouping for soils and rocks

	Strength	Compressibility	Speed of drainage
Organic materials, e.g. peat	Very low	Very high	Generally rapid
Cohesive materials, e.g. clay	Low–medium	High	Slow
Granular soils, e.g. silts, sands and gravels	Medium–high	Low	Very rapid
Rocks	Very high	Very low	Often rapid

Attempts were made to represent the shear strength of a soil by the envelope to a Mohr circle diagram of stress,[3] the intercept on the vertical axis being taken as cohesion c, and the slope of the envelope being taken as the friction angle ϕ. It was found that, except in sands and gravels, the results for a given soil varied considerably depending on the test procedure used, particularly the rate of testing and the conditions of drainage of the specimens during test. However, following the realization that the strength of a soil is governed by the effective stress, it was possible to achieve a better understanding of the shear strength characteristics of soils. The shear strength can be expressed as:

$$\tau_f = c' + (\sigma - u) \tan \phi' \tag{9.7}$$

where c' and ϕ' are effective stress parameters, c' is the apparent cohesion, ϕ' the angle of shearing resistance and u the pore water pressure.

The Mohr circle diagram can be plotted in terms of effective stress, with c' as the cohesion intercept and ϕ' as the slope of the envelope (Figure 9.1).

In terms of effective principal stresses in the Mohr diagram the Coulomb failure criterion may be expressed as:

$$(\sigma_1' - \sigma_3') = \sin \phi'(\sigma_1' + \sigma_3') - 2c' \cos \phi' \tag{9.8}$$

and if c' is zero, then:

$$\sigma_3' = \sigma_1'(1 - \sin \phi')/(1 + \sin \phi') \tag{9.9}$$

which is the well-known Rankine failure condition.[4]

A special condition exists where the soil is loaded rapidly and no drainage is permitted. In this case, the shear strength is found to be independent of the total stress applied to the soil for a saturated unfissured clay. Changes in total stress lead to equal changes in pore water pressure u so that the effective stress in the soil remains unchanged. Because the soil is undrained its moisture content does not alter. Therefore, it is common practice to define the undrained shear strength of a clay at failure as:

$$\tau_f = c_u \quad (\phi_u = 0) \tag{9.10}$$

This parameter is used in so-called $\phi_u = 0$, 'short-term' or 'end of construction' analyses (see section 9.5).

9.1.3 Permeability

Permeability is that property of a soil which controls the rate of flow of water through the soil. In soil mechanics, permeability is defined by the equation derived from Darcy's law:

$$v = k \cdot i \tag{9.11}$$

where v is the superficial velocity of flow through the soil, i is the hydraulic gradient and k is the permeability.

k therefore has the dimensions of a velocity; it depends chiefly on particle size and grading, and to a lesser extent on the particle packing.

Typical values of permeability for soils range from 1×10^{-2} m/s for a coarse sand to 1×10^{-10} m/s for a clay. A very rough estimate of permeability for a relatively uniform sand can be obtained from Hazen's law:

$$k = D_{10}^2/100 \tag{9.12}$$

where D_{10} is the 10% size or effective size in millimetres and k is in metres per second.

The 10% size or effective size is the particle size at which the grading curve crosses the 10% line (Figure 9.2).

Research by Loudon[5] has shown that the permeability (in metres per second) of clean sand may be computed from simple soil tests, using Kozeny's formula, with an accuracy of ±20% (SI units):

$$kS^2 = 1.5 \times 10^{-4} \frac{n^3}{(1-n)^2} \tag{9.13}$$

He suggests an alternative formula which is easier to use and of equal accuracy:

$$\log_{10}(kS^2) = 1.365 + 5.15n \tag{9.14}$$

In both of these formulae, n is porosity and S denotes specific surface in square metres per cubic millimetre:

$$S = f(x_1 S_1 + x_2 S_2 + \ldots x_n S_n) \tag{9.15}$$

where f is an angularity factor, varying between 1.1 for a rounded sand and 1.4 for an angular sand, $x_1, x_2, \ldots$ is the

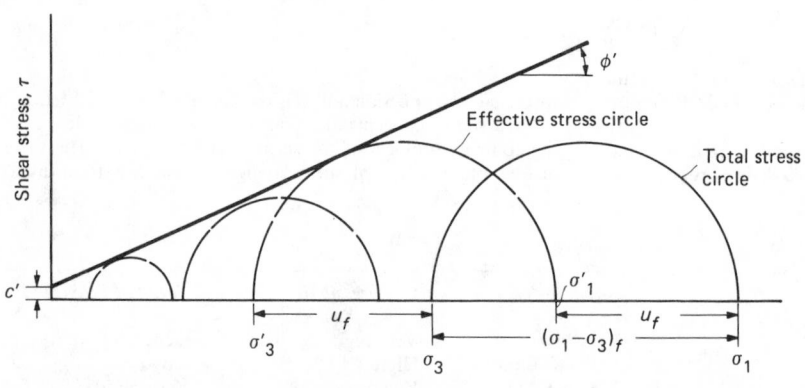

Figure 9.1 Mohr circle diagram

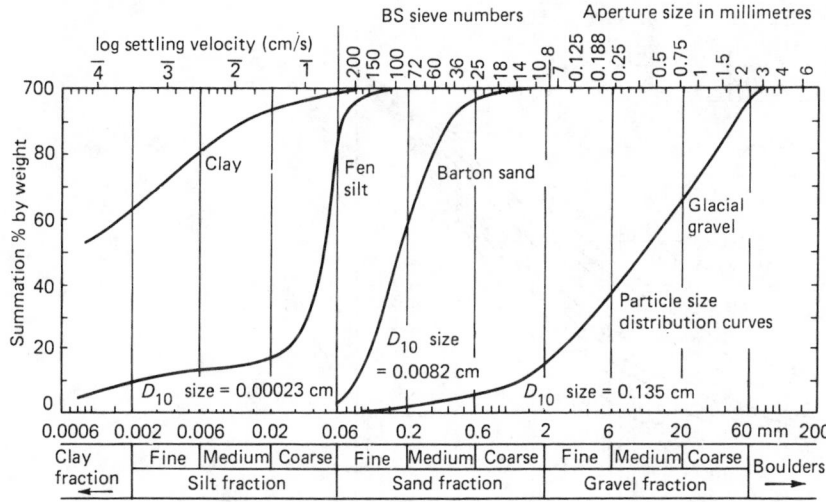

Figure 9.2 Particle size distribution curves (on standard sheet) showing D_{10} size. (The D_{10} size is now given in millimetres)

fraction in each sieve range and S_1, S_2, ... is the specific surface for each sieve range, from Loudon's tables. (Note that Loudon's tables are given in square centimetres per cubic centimetre units for S. It is therefore necessary to convert Loudon's values for S using $1 \text{ cm}^{-1} = 0.1 \text{ mm}^{-1}$.)

9.1.4 Consolidation

The ultimate change in volume of a soil occurring under a change in applied stress depends on the compressibility of the skeleton of soil particles. However, the water in the voids of a saturated soil is relatively incompressible and, if no drainage takes place, change in applied stress results in a corresponding change in pore pressure, and the volume change is negligible. As drainage takes place by flow of water from zones of high excess pore pressure to zones of less or zero excess pore pressure, and the excess pore pressures dissipate, the applied stress is transferred to the soil skeleton and volume change takes place. It is this volume change of cohesive soils resulting from dissipation of excess pore pressures which is known as consolidation.

A study of consolidation requires knowledge of the compressibility of the soil skeleton and of the rate at which excess pore pressures dissipate, which is related to the permeability. In Terzaghi's consolidation theory the relation between these factors can be expressed for the one-dimensional case as:

$$c_v \partial^2 u / \partial z^2 = \partial u / \partial t \tag{9.16}$$

where c_v is the coefficient of consolidation and is given by

$$c_v = k / m_v \cdot \gamma_w \tag{9.17}$$

u is excess pore water pressure, z the thickness of the stratum, t is time, and m_v is the modulus of volume compressibility, defined as:

$$m_v = - (\mathrm{d}e / \mathrm{d}p) / (1 + e_o) \tag{9.18}$$

where e is the void ratio and p the effective pressure.

The solution of the consolidation equation has been given by Taylor[3] and values have been tabulated for the degree of consolidation U against the time factor T, where:

$$T = c_v t / H^2 \tag{9.19}$$

and H is the length of the drainage path. Values of c_v and m_v are determined by laboratory tests known as oedometer, or consolidation, tests.

The relationship between the degree of consolidation, U, and the time factor T is dependent on the initial distribution of excess pore water pressure. Figure 9.3 gives a plot of U against t for various ratios of the initial excess pore pressure at the top and bottom of the compressible stratum u_1 / u_2. The cases given are all for single drainage.

For double drainage (i.e. where drainage can take place at the top and bottom of the layer) values corresponding to $u_1/u_2 = 1$ can be used for all ratios of initial pore pressure, but it should be noted that in the double drainage case, H is taken as only half of the layer thickness.

One-dimensional drainage is seldom fully realized in practice and for important calculations, particularly where the loaded area is small in comparison with the thickness of the compressible stratum, two- or three-dimensional consolidation should be considered.[6,7]

Furthermore, in many deposits, lateral permeability can be up to two orders greater than vertical owing to the presence of a laminar structure of thin layers and partings of silt and fine sand. This will have a very marked effect on rate of consolidation.[8] Problems involving a number of layers having different consolidation characteristics have been solved numerically using the finite-difference method.[9]

Methods of test to obtain the soil parameters described in the previous sections are detailed in Appendix 9.1.

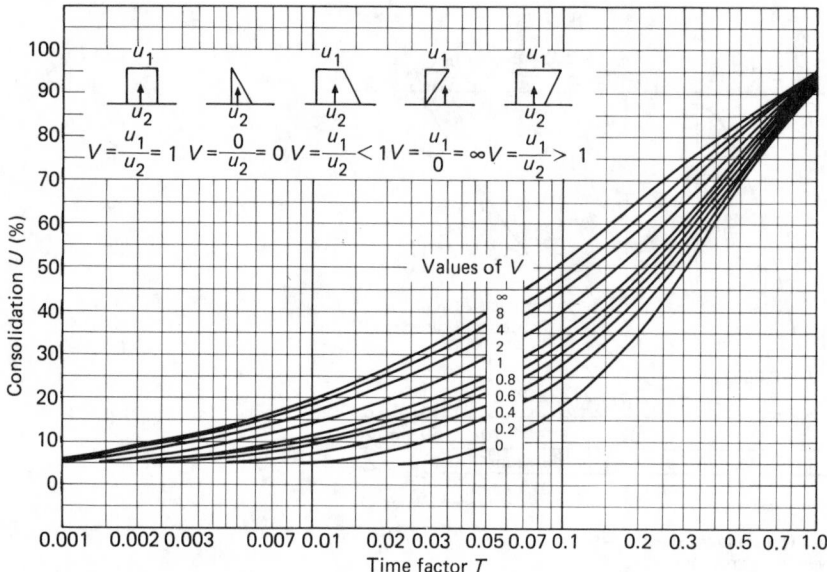

Figure 9.3 Values of time factor T

9.2 Design and limit states in soil mechanics and foundation engineering

Unlike virtually all other materials dealt with by civil engineers during the course of their work, soil is naturally occurring. It is inherently variable, not only from site to site but also at different levels and plan locations at any one site. The extent of its variability can be judged by examining the typical limits of some of its most important properties:

Undrained shear strength c_u	5–300 kN/m²
Coefficient of permeability k	10^{-2}–10^{-10} m/s
Coefficient of compressibility m_v	0.01–3.00 m²/MN

If the variability of soil is to be understood and allowed for, then a knowledge of geology and the processes leading to the formation and induration of the soil will be important. For any single soil type it is rarely reasonable to assume that the soil is uniformly variable, in the sense of conforming to a single Gaussian distribution for a given soil property, or that the full ranges of its properties are known.

The basic steps involved in soil mechanics design are:

(1) Arbitrary division of the soil into layers thought to have similar engineering behaviour. This process is carried out during site investigation (see Chapter 11, and Clayton, Simons and Matthews[10]) using sample description and classification testing. Approximate assessment of the principal soil mechanics parameters (e.g. undrained strength) for each soil group.
(2) Envisaging all the mechanisms by which the structure–soil combination may lead to a limit state for the structure (e.g. the structure may fail to perform as required because of foundation bearing capacity failure, excess differential settlement, chemical attack on foundations, etc.). Table 9.2 lists some of these factors as they affect low-rise construction.
(3) Testing of soil, both *in situ* and in the laboratory, to obtain detailed parameters suitable for sound geotechnical engineering calculations in order to assess the risk of 'failure' by all of the mechanisms envisaged in (2) above.

(4) Geotechnical calculation, associated with changes in structural design, to achieve maximum economy while avoiding unacceptable behaviour of the structure.

Table 9.2 Ground problems and low-rise building. (After Building Research Establishment (1987) *Site investigation for low-rise building: desk studies.* BRE Digest Number 318. HMSO, London)

Differential settlement or heave of foundations (or floorslabs)
Soft spots under spread footings on clays
Growth or removal of vegetation on shrinkable clays
Collapse settlements on pre-existing made ground
Mining subsidence
Self-settlement of poorly compacted fill
Floor slab heave on unsuitable fill material

Soil failure
Failure of foundations on very soft subsoil
Instability of temporary or permanent slopes

Chemical processes
Groundwater attack on foundation concrete
Reactions due to chemical waste or household refuse

Variations during construction
Removal of soft spots to increase depth of footings
Dewatering problems
Piling problems

Geotechnical engineers recognize an important division between soils which drain rapidly (e.g. silts, sands, gravels) and those which are slow-draining (e.g. clays and clayey soils). In the case of free-draining, noncohesive soils the important consequence of their rapid dissipation of excess pore water pressures is that they undergo increasing effective stress as load is applied; therefore, their strength increases as shear stresses brought about by loading also increase. The consequence is that when foundations are placed on granular soils, bearing capacity is rarely a problem.

Cohesive soils, on the other hand, do not drain rapidly; loading or unloading leads to excess pore water pressures which may take years to dissipate. For these materials the geotechnical engineer recognizes two types of loading which lead to critical conditions at two different times during the life of the structure.

Consider an element of soil beneath a foundation constructed on a clay foundation. As the loads are applied to the foundation during the relatively rapid process of constructing the superstructure of the building, the shear stress applied to the soil is increased. Because of the increase in load applied to the soil, excess pore pressures develop. Because clay is slow-draining, the excess pore pressures do not dissipate significantly, the effective strength does not change, and the shear strength of the soil remains constant. Thus, the factor of safety against bearing capacity failure decreases up to the end of the construction (Figure 9.4(a)), after which it increases. Normal practice, for almost all cases where a load increase is applied to the soil, is therefore to calculate a so-called 'short-term' factor of safety using the initial undrained shear strength of the soil combined with the structural loads and bearing pressures expected from the completed structure.

In those cases where unloading takes place (Figure 9.4(b)), the same logic leads to the conclusion that the critical time may be many years after the end of construction, in the 'long-term', once the pore pressures once again come to equilibrium. Examples of unloading situations include excavated slopes and retaining walls. In these cases, the shear strength used in calculations must take into account the changes in pore pressure and effective stress that have occurred. For this reason, long-term calculations are carried out using effective strength parameters.

9.3 Foundations

There are two ways in which a foundation can fail to perform satisfactorily: (1) by shear failure; and (2) by settlement. In the first case, a surface of rupture is formed in the soil, the foundation settles considerably and probably tilts to one side and heaving of the soil occurs on one or both sides of the foundation. In the second case, failure of the soil in shear does not occur, but the existing deformations are large enough to cause failure of the structure which the foundation is supporting.

Failure by settlement is therefore a function of the particular structure as well as the underlying soil. Skempton and Macdonald[11] have given a criterion for framed buildings based on angular distortion which is expressed by the ratio of differential settlement, δ, to the distance l, between two points, usually the column positions. From a detailed study of field data, a limiting value of $\delta/l = 1/300$ has been determined. More flexible structures, oil tanks for example, may undergo considerably greater settlements without sustaining damage. On the other hand, some sensitive machinery and stiff reinforced concrete slabs will tolerate very little settlement (see Chapter 17).

The ultimate bearing capacity of a foundation is the value of the net loading intensity at which the ground fails in shear. Before discussing bearing capacity, several definitions are necessary.

(1) The gross loading intensity, p, is the pressure due to the applied load and the total weight of foundation, including any backfill above the foundation.
(2) The net loading intensity, p_n, is the gross foundation pressure less the weight of material (soil and water) displaced by the foundation (and by the backfill above the foundation). Alternatively, the net pressure can be considered as equal to

the gross pressure less the total overburden pressure,
$$p_n = p - p_o.$$
(3) The safe bearing pressure is the ultimate bearing capacity divided by the factor of safety, $q_s = q_u/F$.
(4) The allowable bearing pressure, q_a, is less than, or equal to, the safe bearing capacity, depending on the settlements which are expected and which can be tolerated.

The term 'presumed bearing value' was introduced in CP 2004,[12] and was defined as the net loading intensity considered appropriate to the particular type of ground for preliminary design purposes. Table 9.3 gives the current presumed bearing values from BS 8004.[13]

9.3.1 Bearing capacity of shallow foundations

There are two groups of methods of determining ultimate bearing capacity: (1) analytical methods; and (2) graphical methods. The graphical methods are very flexible and will cover any conditions likely to be found in practice, but they are rather cumbersome in use. The analytical techniques, which are only strictly applicable in cases in which the soil is uniform, are quicker and easier to use, and therefore are the most often used.

The most general formula for the ultimate bearing capacity of a strip footing is that of Terzaghi,[14] which in terms of effective stress is:

$$q_u = p_o + c' \cdot N_c + p_o'(N_q - 1) + 0.5 \cdot \gamma \cdot B \cdot N_\gamma \qquad (9.20)$$

where p_o is the total vertical stress (overburden pressure) at foundation level, p_o' is the effective vertical stress at foundation level, c' is the effective cohesion intercept of the soil, γ is the bulk density of the soil, and N_c, N_q and N_γ are bearing capacity factors which depend upon the geometry of the foundation and the effective angle of friction (ϕ') of the soil.

For a circular footing of diameter D:

$$q_u = p_o + 1.3 \cdot c' \cdot N_c + p_o'(N_q - 1) + 0.3 \cdot \gamma \cdot B \cdot N_\gamma \qquad (9.21)$$

and for a square footing of width B:

$$q_u = p_o + 1.3 \cdot c' \cdot N_c + p_o'(N_q - 1) + 0.4 \cdot \gamma \cdot B \cdot N_\gamma \qquad (9.22)$$

Unfortunately, there is some disagreement in the literature on appropriate values of the bearing capacity factors. According to Vesic,[15] there is reasonable agreement on the values of N_c and N_q, but N_γ values vary considerably. Conservative values, which assume no friction between the soil and the underside of the foundation, are given in Table 9.4. These factors are only valid for foundations under uniformly distributed loads. If the foundation load is neither vertical nor central, then various factors must be applied. For the strip footing, for example, the equation for ultimate bearing capacity becomes:

$$q_u = p_o + c' \cdot N_c \cdot f_{c_i} + p_o'(N_q - 1) \cdot f_{q_i} + 0.5 \cdot \gamma \cdot B' \cdot N_\gamma \cdot f_{\gamma_i} \qquad (9.23)$$

B' is an effective (i.e. reduced) foundation width equal to:

$$B' = B - 2e$$

(Figure 9.5). R_h and R_v are the horizontal and vertical components of the resultant force on the foundation. Brinch Hanson[16] and Sokolovski[17] have proposed that:

$$f_{q_i} = \left[1 - \frac{R_h}{(R_v + B' \cdot c' \cdot \cot \phi')} \right]^2 \qquad (9.24)$$

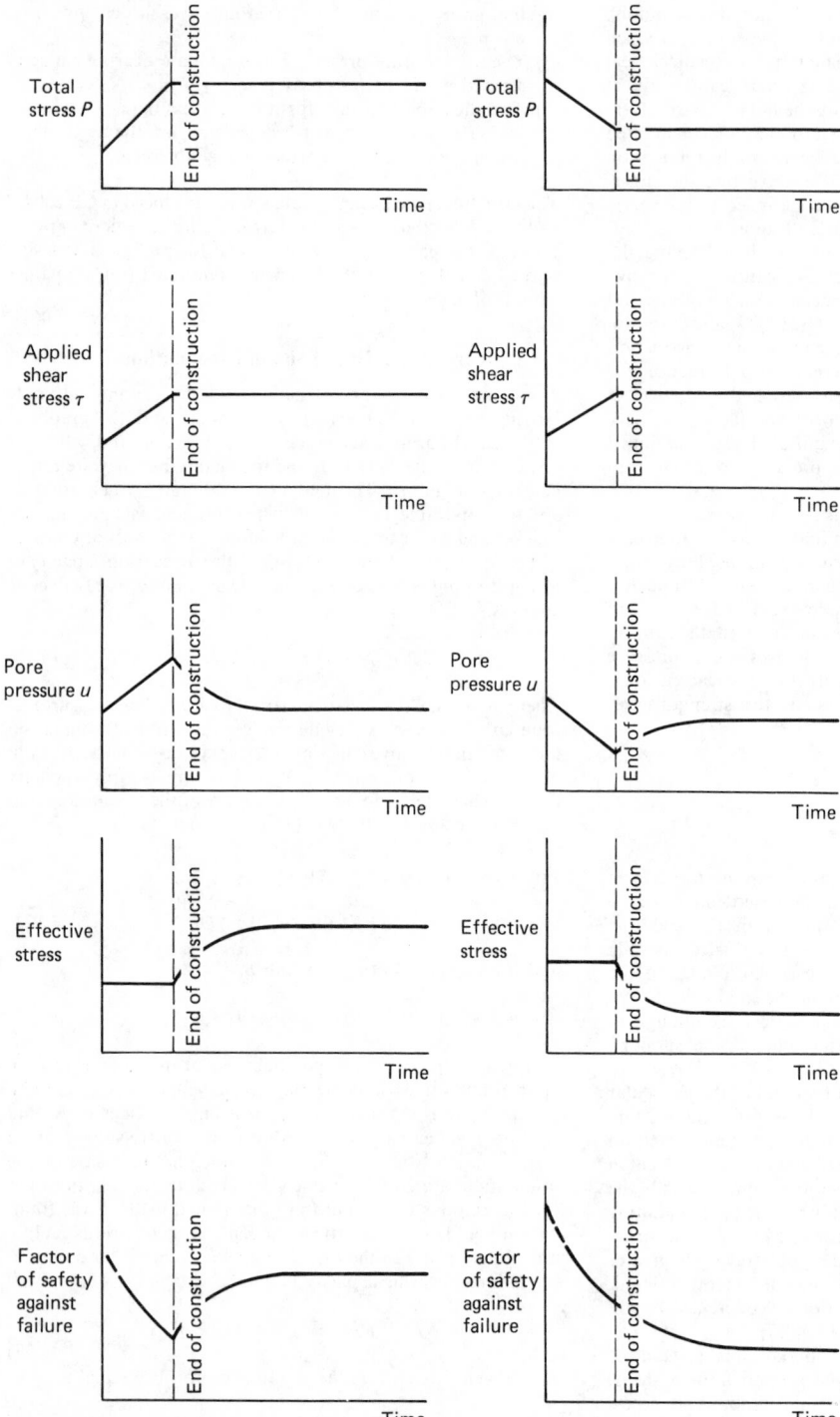

Figure 9.4 Load, pore pressure, shear strength and factor of safety. (a) For a clay beneath a foundation (loading case); (b) for a clay beneath a motorway excavation (unloading case). (After Clayton and Milititsky (1986) *Earth pressure and earth-retaining structures*. Surrey University Press, Glasgow)

Table 9.3 Presumed bearing values under vertical static loading. Refer to British Standards Institution (1986) *Foundations*, BS 8004 for further details)

Category	Types of rocks and soils	Presumed allowable bearing value		Remarks
		(kN/m²*)	(kgf/cm²* tonf/ft²)	
Rocks	Strong igneous and gneissic rocks in sound condition	10 000	100	These values are based on the assumption that the foundations are taken down to unweathered rock. (For weak, weathered and broken rock, see BS 8004)
	Strong limestones and strong sandstones	4 000	40	
	Schists and slates	3 000	30	
	Strong shales, strong mudstones and strong siltstones	2 000	20	
Noncohesive soils	Dense gravel, or dense sand and gravel	> 600	> 6	Width of foundation not less than 1 m. Groundwater level assumed to be a depth not less than below the base of the foundation. (For effect of relative density and groundwater level, see BS 8004)
	Medium dense gravel, or medium dense sand and gravel	< 200–600	< 2–6	
	Loose gravel, or loose sand and gravel	< 200	< 2	
	Compact sand	> 300	> 3	
	Medium dense sand	100–300	1–3	
	Loose sand	< 100	< 1	
		Value depending on degree of looseness		
Cohesive soils	Very stiff boulder clays and hard clays	300–600	3–6	Cohesive soils will undergo long-term consolidation settlement
	Stiff clays	150–300	1.5–3	
	Firm clays	75–150	0.75–1.5	
	Soft clays and silts	< 75	< 0.75	
	Very soft clays and silts	Not applicable		
Peat and organic soils		Not applicable		
Made ground or fill		Not applicable		

Note. These values are for preliminary design purposes only, and may need alteration upwards or downwards. No addition has been made for the depth of embedment of the foundation.
* 107.25 kN/m² = 1.094 kgf/cm² = 1 tonf/ft².

$$f_{c_i} = f_{q_i} - \frac{(1 - f_{q_i})}{N_c \cdot \tan \phi'}$$

$$f_{\gamma_i} = (f_{q_i})^{3/2}$$

For a granular soil, or when $c' = 0$, with a foundation approximately at ground level ($p_o = 0$), these equations reduce to:

$$q_{ult} = \frac{1}{2} \cdot \gamma \cdot B \cdot N_\gamma \left(1 - \frac{R_h}{R_v} \right)^3 \qquad (9.25)$$

and the vertical force per unit length of foundation at failure is then:

$$R_{v_{ult}} = q_{ult} \cdot B' \qquad (9.26)$$

For clays, the short-term or end of construction case is critical. For short-term, '$\phi_u = 0$' analysis, the ultimate uniform bearing pressure that a foundation can apply is:

$$q_{ult} = p_o + N_c \cdot c_u \qquad (9.27)$$

where p_o is the total vertical stress in the soil adjacent to the foundation, at footing level (as before), N_c is a bearing capacity factor, dependent on the geometry of the foundation, and c_u is the average undrained shear strength of the foundation equal to its width, from unconsolidated undrained triaxial compression tests on undisturbed samples.

Values of N_c for a strip footing are given in Table 9.5.

As before, this equation must be modified for nonvertical and eccentric levels. In this case:

$$q_u = c_u \cdot N_c \cdot f_{c_i} + p_o \qquad (9.28)$$

where

$$f_{c_i} = 1 - \frac{2 \cdot R_h}{B' \cdot c_u \cdot N_c} \qquad (9.29)$$

(See Figure 9.5.)

9.3.2 Bearing capacity of deep foundations

The Terzaghi bearing capacity equation has been generally accepted as a basis for the design of shallow foundations, but not for foundations where the depth greatly exceeds the width as in the case of piers and piles. Meyerhof[18] suggested that the shear surface beneath the base would return upwards and inwards to reach the shaft, and produced graphs of a general bearing capacity factor $N_{\gamma q}$ for various values of ϕ' and foundation depth ratio $D : B$. Large-scale experiments by Kerisel[19] have indicated that, in addition to ϕ', both depth and size influence the bearing capacity factor N_q. In a comprehensive study of deep foundations, Vesic[20] has observed that there is no evidence of failure surfaces reverting to the shaft.

Table 9.4 Bearing-capacity factors for shallow strip footage. (After Clayton and Milititsky (1986) *Earth pressure and earth retaining structures.* Surrey University Press, Glasgow)

Effective angle of friction of soil ζ' (deg.)	N_c^*	N_q^*	N_γ^*
0	5.14	1.00	0.00
15	10.98	3.94	2.65
20	14.83	6.40	5.39
25	20.72	10.66	10.88
30	30.14	18.40	22.40
35	46.12	33.30	48.03
40	75.31	64.20	109.41
45	133.88	134.88	271.76

* Bearing-capacity factors based on
$N_q = e^{\pi\tan\zeta'} \cdot \tan^2(45 + \zeta'/2)$
$N_c = (N_q - 1)\cot\zeta'$
$N_\gamma = 2(N_q + 1)\tan\zeta'$

Table 9.5 Bearing-capacity factors for clay in the short term. (After Skempton (1951) *The bearing capacity of clays.* Building Research Congress Division 1, Part 3, p. 180)

Ratio of depth to width D/B	Bearing-capacity factor N_c for	
	strip footings	circular and square footings
0	5.14	6.2
0.25	5.6	6.7
0.50	5.9	7.1
0.75	6.2	7.4
1.00	6.4	7.7
2.00	7.0	8.4
4.00	7.5	9.0

The design of deep foundations in granular materials is a highly complex matter. Terzaghi's bearing capacity factor N_q is too conservative for higher values of ϕ' and is independent of the depth:width ratio $D:B$. Berezantsev's[21] curves relating N_q to ϕ' (Figure 9.6) are now recognized as giving the best representation of the ultimate bearing capacity of deep foundations in terms of N_q and $D:B$, thus:

$$Q_u = A \cdot p_o' \cdot D \cdot N_q \tag{9.30}$$

where A is the base area of the pier or pile. See also Tomlinson[22] for a discussion on deep foundations.

For clays, in the short term, experience has shown that Skempton's bearing capacity factors are reasonable. An N_c value of 9 is therefore generally adopted (Table 9.5).

9.3.3 Settlement

The problem of predicting settlement is a three-part one: (1) predicting the maximum settlement; (2) predicting differential settlement; and (3) predicting the rate at which settlement will occur. Differential settlements cause the majority of damage to structures because of the bending movements that they induce in the structure, although excessive total settlements can lead to problems where services enter buildings. Unfortunately, differential settlements are usually very difficult to predict; for routine problems with flexible buildings, maximum differential settlements are often as much as 75% of the maximum total settlement, so that the designer will normally seek to control total settlement in order to avoid problems due to differential settlements. For heavily loaded, relatively rigid structures, some form of fairly complex computer analysis will be necessary if differential settlements are to be estimated.

The rate at which consolidation settlements occur is of particular importance to two types of problems:

(1) Multistorey structures, where the rigidity of the structure increases as its height (and hence its load) goes up. If consolidation is relatively rapid, then a higher proportion of differential settlement will occur whilst the structure is flexible, thus inducing less bending moment.
(2) Stage-constructed embankments, where it is necessary to allow the subsoil (i.e. beneath the embankment) to gain strength before the next lift of fill can be placed.

Different methods of settlement prediction are used depending principally upon the type of soil to be analysed, but also to a certain extent upon the type of test data used to characterize the compressibility of the soil. The principal soil types for which settlement methods exist are: (1) normally consolidated clays; (2) lightly overconsolidated clays; (3) heavily overconsolidated clays; and (4) sands and gravels.

The main types of data used to characterize the compressibility of soil are derived from: (1) oedometer tests; (2) the triaxial test; (3) stress path tests; (4) the standard penetration test; (5) the static (Dutch) cone test; and (6) plate loading tests.

9.3.3.1 Settlement of clays

An idealized representation of settlement of foundations in clays and of the heave on excavation is given in Figure 9.7. On excavation, heave occurs and the majority of this is recovered when the original total overburden pressure p_o is replaced. This recovery of heave is often neglected in settlement calculations. With further application of load, i.e. with increase in the net applied pressure, an immediate settlement occurs without volume change of the clay, followed by consolidation settlement as the excess pore pressures set up by the applied load are dissipated. In practice, the consolidation settlement starts immediately the net pressure is greater than zero, but at a very slow rate, so that it is convenient to ignore the consolidation settlement occurring during construction or, alternatively, to consider it as starting at halfway through the construction period. See Taylor[3] for a fuller treatment of this problem.

The net final settlement is the sum of the immediate settlement and the consolidation settlement:

$$\rho_{final} = \rho_i + \rho_c \tag{9.31}$$

The net settlement at any time t is given by:

$$\rho_t = \rho_i + \bar{U} \cdot \rho_c \tag{9.32}$$

where $\bar{U}$ is the degree of consolidation after time t.

Immediate settlement Immediate settlement occurs without change in the volume of the soil and as a result of the shear stresses imposed by the foundation or embankment loads. Because it involves a change in shape without a change in volume (and therefore no drainage of pore water is involved) immediate settlement takes place as the load is applied. The immediate settlement below the corner of a uniformly loaded rectangular area can be calculated from elastic theory using the Steinbrenner[23] equation (see also Terzaghi[24]):

$$\rho_i = \frac{3}{4} q \frac{B}{E} I_\rho \tag{9.33}$$

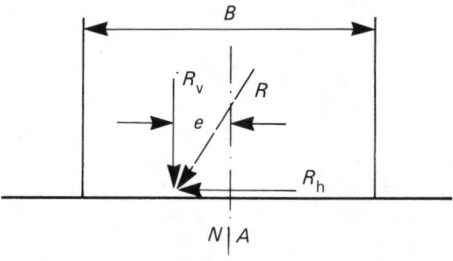

Figure 9.5 Bearing capacity for inclined eccentrically loaded strip foundations. Effective width $B' = B - 2e$. (After de Beer (1949) *Grondmekanica*. NV Standard Boekhandel, Antwerp; Meyerhof (1955) 'The bearing capacity of foundations under eccentric and inclined loads', *Proceedings, 3rd international conference on soil mechanics and foundation engineering*, Vol. III)

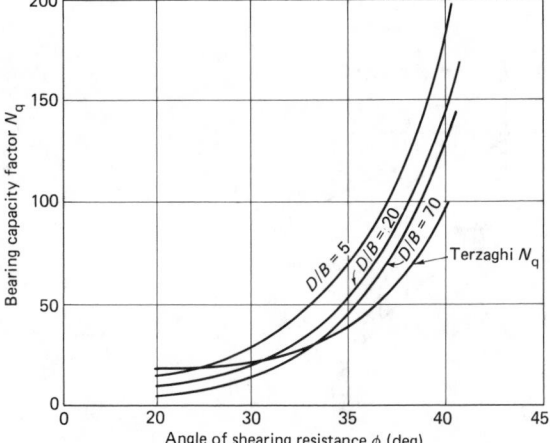

Figure 9.6 Berezantsev's bearing capacity factor N_q

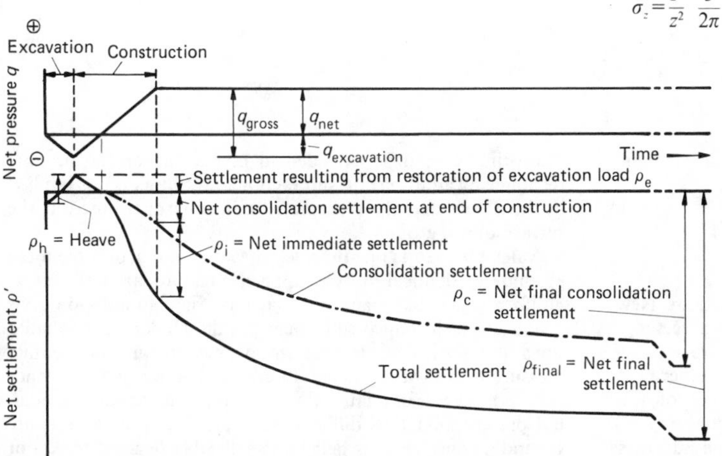

where E is Young's modulus for the clay, as measured by the appropriate tangent modulus of stress–strain curves from triaxial tests, and I_p is an influence factor which is a function of the length and width of the foundation and the thickness of the compressible layer below foundation level. Values of I_p for a Poisson's ratio value of 0.5 are given in Figure 9.8.

Settlements at other points below a rectangular area can be calculated by splitting the area into a number of rectangles and using the principle of superposition.

Consolidation settlement In practice, the contribution of consolidation to total settlement is calculated in different ways for clays with different stress histories.

Soft normally consolidated clays: oedometer data for normally consolidated clays should yield a straight line when plotted as a voids ratio (e) vs. logarithm of applied pressure ($\log_{10} p$) curve. Due to sample disturbance, they normally yield a curve at least in the early part of compression. To correct for this, the oedometer test should be continued until a void ratio of less than 0.42 times its initial value is reached. The 'virgin consolidation curve' can then be reconstructed by drawing a straight line through the point of the initial void ratio for the *in situ* effective vertical stress, p'_o, and the consolidation curve where $e = 0.42\varepsilon_o$. The slope of this straight line is termed the compression index, C_c (Figure 9.9(a)).

Consolidation settlement is calculated by dividing the compressible strata into layers (at least five layers are normally required to achieve adequate accuracy) and calculating the contribution of each layer. Thus:

$$\rho_c = \sum_{x=1}^{x=n} \frac{C_c}{(1+e_o)} \cdot z_x \cdot \log_{10}\left(\frac{P'_o + \Delta\sigma_v}{P'_o}\right) \tag{9.34}$$

where z_x is the thickness of the xth layer, $\Delta\sigma_v$ is the average vertical stress increase in the xth layer due to foundation or embankment loading, C_c is the compression index, e_o is the initial voids ratio of the xth layer, and P'_o is the initial effective vertical stress at the centre of the xth layer.

The increase in vertical stress at any level, due to a flexible foundation or embankment loading, can be obtained from elastic theory. Boussinesq[25] gave the following equation for the vertical stress increase at depth z due to a point load P on the surface of a semi-infinite solid:

$$\sigma_z = \frac{P}{z^2} \frac{3}{2\pi}\left(\frac{1}{[1+(r/z)^2]^{5/2}}\right) \tag{9.35}$$

Figure 9.7 Idealized representation of settlement of foundations in clay

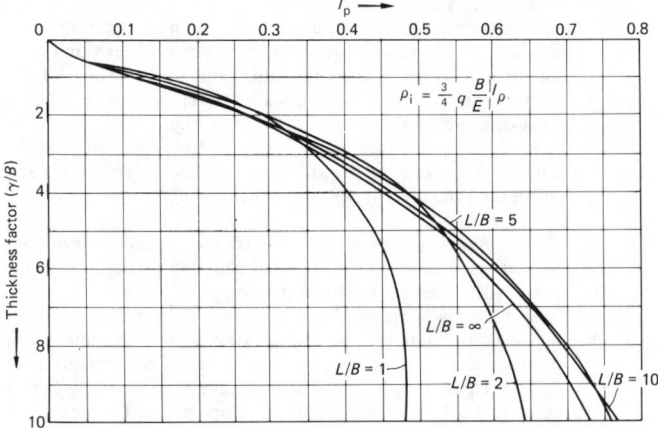

Figure 9.8 Steinbrenner's influence factors for loaded area $L \times B$ on compressible stratum of thickness Y

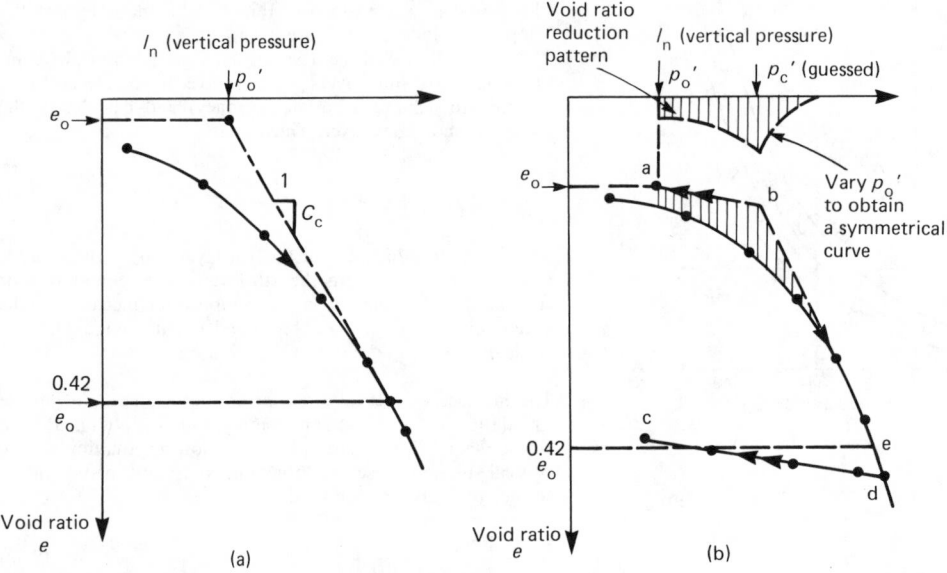

Figure 9.9 Schmertmann's methods of reconstructing the e vs $\log_{10} p'$ curve. (a) Normally consolidated soils; (b) lightly overconsolidated soils

where z and r are defined in Figure 9.10. This can be written:

$$\sigma_z = P \cdot K / z^2 \qquad (9.36)$$

Values of K are given in Table 9.6.

Equation (9.36) has been integrated and tabulated by Newmark[26] to give the pressure below the corner of a rectangle uniformly loaded at the surface, and the values are given in Table 9.7. In order to obtain the pressure below any other point it is necessary to regard that point as the corner of four adjoining rectangles (not necessarily the same shape or size), calculate the pressure below the corner of each and add these pressures. For example, the pressure below the centre of a rectangle is 4 times the pressure beneath the corner of a rectangle whose sides are half the sides of the original rectangle.

The principle can be extended to points outside the original rectangle by addition and subtraction of rectangles. It is implied in the above that the pressure is uniformly distributed at the surface of the ground.

A flexible load is one in which the contact reaction from the ground is identical to the applied pressure at each point. Uniformly flexible loads on relatively thick foundation soils produce a dish-shaped settlement profile. Under rigid foundations the settlement is uniform or planar but the contact pressure varies with soil type. Intermediate cases occur depending on the degree of relative rigidity of the structure and ground, but present analytical difficulties. In practice, foundations are generally considered as being either flexible (e.g. oil tanks) or rigid (e.g. high buildings on stiff rafts). The pressure distribution for the latter case has been worked out by Fox,[27] who gives a series of curves (Figure 9.11) for the mean vertical stress σ_z at a depth z beneath a rectangular area $a \times b$ uniformly loaded with a pressure q, the rectangle being on the surface.

For more important cases, it has been the recent practice to use the finite-element method to investigate soil–structure interaction, stress distribution within the structure and the ground, and the associated deformations. Alternatively, an excellent collection of elastic solutions has been made by Poulos and Davis.[28]

For lightly overconsolidated clays, where the maximum past vertical effective pressure, p_c', is somewhat in excess of its current value, p_o', sample disturbance and bedding effects are likely to mask the value of p_c'. The value of p_c' is required with some accuracy because the application of stress between p_o' and p_c' will lead to very little consolidation settlement, but once p_c' is exceeded then large consolidation settlements can be expected.

To overcome this problem, Schmertmann[29] has proposed the method of reconstruction of the e vs. $\log p'$ curve to find p_c', as shown in Figure 9.9(b).

The method is as follows:

(1) Plot (e_o, p_o'), point a.
(2) Estimate the maximum preconsolidation pressure p_c'.
(3) Draw a straight line through a, parallel to the rebound curve, c–d, to intersect the vertical line through p_c' at b.
(4) Through b draw a straight line to point e on the consolidation curve, where point e has a void ratio equal to $0.42e_o$.
(5) Construct the 'void ratio reduction pattern', which is the difference between the reconstructed curve abed and the curve obtained from the laboratory test.

Schmertmann suggests that the best estimate of p_c' is obtained when the most symmetrical voids ratio reduction pattern is obtained.

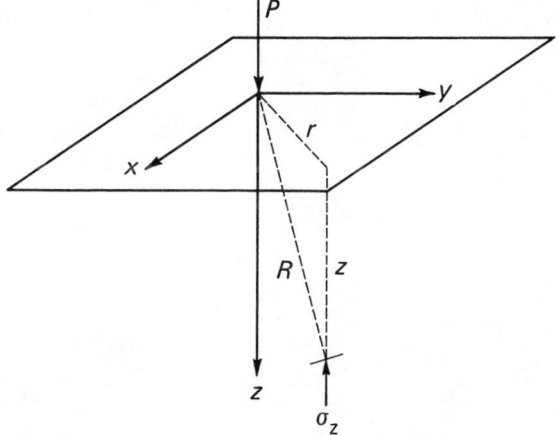

Figure 9.10 Diagram showing z and r in Boussinesq equation for concentrated load

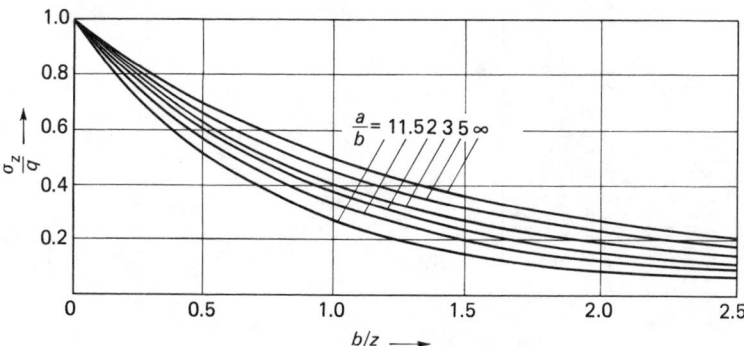

Figure 9.11 Mean pressure under a rigid foundation at ground level. (After Fox (1948) 'The mean elastic settlement of a uniformly loaded area at a depth below the ground surface', *Proceedings, 2nd international conference on soil mechanics and foundation engineering*, Vol. II)

Table 9.6 Values of coefficient K in Equation (9.36)

Ratio $\frac{r}{z}$	Coefficient K	Ratio $\frac{r}{z}$	Coefficient K	Ratio $\frac{r}{z}$	Coefficient K	Ratio $\frac{r}{z}$	Coefficient K	Ratio $\frac{r}{z}$	Coefficient K	Ratio $\frac{r}{z}$	Coefficient K	Ratio $\frac{r}{z}$	Coefficient K
0.00	0.477 5	0.60	0.221 4	1.20	0.051 3	1.80	0.012 9	2.40	0.004 0	2.84	0.001 9
								45	0.003 7	2.91	0.001 7
0.10	0.465 7	0.70	0.176 2	1.30	0.040 2	1.90	0.010 5	2.50	0.003 4	2.99	0.001 5
0.15	0.451 6	0.75	0.156 5	1.35	0.035 7	1.95	0.009 5	2.55	0.003 1	3.08	0.001 3
0.20	0.432 9	0.80	0.138 6	1.40	0.031 7	2.00	0.008 5	2.60	0.002 9	3.19	0.001 1
0.25	0.410 3	0.85	0.122 6	1.45	0.028 2	2.05	0.007 8	2.65	0.002 6	3.31	0.000 9
0.30	0.384 9	0.90	0.108 3	1.50	0.025 1	2.10	0.007 0	2.70	0.002 4	3.50	0.000 7
0.35	0.357 7	0.95	0.095 6	1.55	0.022 4	2.15	0.006 4	2.72	0.002 3	3.75	0.000 5
0.40	0.329 4	1.00	0.084 4	1.60	0.020 0	2.20	0.005 8	2.74	0.002 3	4.13	0.000 3
0.45	0.301 1	1.05	0.074 4	1.65	0.017 9	2.25	0.005 3	2.76	0.002 2	4.91	0.000 1
0.50	0.273 3	1.10	0.065 8	1.70	0.016 0	2.30	0.004 8	2.78	0.002 1	6.15	0.000 1
0.55	0.246 6	1.15	0.058 1	1.75	0.014 4	2.35	0.004 4	2.80	0.0002 1		

Table 9.7 Vertical pressure σ_z under corner of rectangle $a \times b$ loaded uniformly with intensity q. Table gives σ_z/q for values of $\alpha = a/z$ and $\beta = b/z$

α/β	0.1	0.2	0.3	0.4	0.5	0.6	0.7	0.8	0.9	1.0	1.2	1.4	1.6	1.8	2.0	2.5	3.0	4.0	5.0	6.0	8.0	10.0	∞
0.1	0.00470	0.00922	0.01320	0.01683	0.01980	0.02220	0.02420	0.02582	0.02700	0.02790	0.02930	0.03013	0.03061	0.03090	0.03110	0.03111	0.03145	0.03155	0.03156	0.03156	0.03156	0.03160	0.03160
0.2	0.00922	0.01799	0.02599	0.03288	0.03870	0.04350	0.04740	0.05040	0.05280	0.05470	0.05730	0.05890	0.05990	0.06060	0.06100	0.06106	0.06160	0.06180	0.06190	0.06200	0.06200	0.06200	0.06200
0.3	0.01320	0.02599	0.03740	0.04744	0.05599	0.06299	0.06866	0.07310	0.07660	0.07940	0.08320	0.08560	0.08710	0.08810	0.08870	0.08955	0.08985	0.09000	0.09010	0.09020	0.09020	0.09020	0.09020
0.4	0.01683	0.03288	0.04744	0.06024	0.07110	0.08010	0.08730	0.09310	0.09770	0.10130	0.10630	0.10940	0.11140	0.11260	0.11340	0.11453	0.11500	0.11532	0.11540	0.11542	0.11544	0.11545	0.11545
0.5	0.01980	0.03870	0.05599	0.07110	0.08400	0.09470	0.10340	0.11040	0.11580	0.12020	0.12630	0.13000	0.13240	0.13400	0.13500	0.13633	0.13683	0.13720	0.13740	0.13747	0.13750	0.13752	0.13755
0.6	0.02220	0.04350	0.06299	0.08010	0.09470	0.10690	0.11680	0.12470	0.13110	0.13610	0.14310	0.14750	0.15030	0.15210	0.15330	0.15480	0.15550	0.15600	0.15610	0.15620	0.15620	0.15620	0.15620
0.7	0.02420	0.04740	0.06866	0.08730	0.10340	0.11680	0.12770	0.13650	0.14360	0.14910	0.15700	0.16200	0.16520	0.16720	0.16860	0.17040	0.17110	0.17170	0.17190	0.17190	0.17200	0.17200	0.17200
0.8	0.02582	0.05040	0.07310	0.09310	0.11040	0.12470	0.13650	0.14610	0.15370	0.15990	0.16840	0.17390	0.17740	0.17970	0.18120	0.18320	0.18410	0.18470	0.18490	0.18500	0.18500	0.18500	0.18500
0.9	0.02700	0.05280	0.07660	0.09770	0.11580	0.13110	0.14360	0.15370	0.16190	0.16840	0.17770	0.18360	0.18740	0.18990	0.19150	0.19380	0.19470	0.19540	0.19560	0.19570	0.19580	0.19580	0.19580
1.0	0.02790	0.05470	0.07940	0.10130	0.12020	0.13610	0.14910	0.15990	0.16840	0.17520	0.18510	0.19140	0.19550	0.19810	0.19990	0.20240	0.20340	0.20420	0.20440	0.20450	0.20460	0.20460	0.20460
1.2	0.02930	0.05730	0.08320	0.10630	0.12630	0.14310	0.15700	0.16840	0.17770	0.18510	0.19580	0.20280	0.20730	0.21030	0.21240	0.21510	0.21630	0.21720	0.21750	0.21760	0.21770	0.21770	0.21770
1.4	0.03013	0.05890	0.08560	0.10940	0.13000	0.14750	0.16200	0.17390	0.18360	0.19140	0.20280	0.21020	0.21510	0.21840	0.22060	0.22360	0.22500	0.22600	0.22630	0.22640	0.22650	0.22650	0.22660
1.6	0.03061	0.05990	0.08710	0.11140	0.13240	0.15030	0.16520	0.17740	0.18740	0.19550	0.20730	0.21510	0.22050	0.22400	0.22680	0.23000	0.23110	0.23200	0.23240	0.23250	0.23260	0.23260	0.23260
1.8	0.03090	0.06060	0.08810	0.11260	0.13400	0.15210	0.16720	0.17970	0.18990	0.19810	0.21030	0.21840	0.22400	0.22740	0.22990	0.23330	0.23500	0.23600	0.23640	0.23670	0.23680	0.23680	0.23690
2	0.03110	0.06100	0.08870	0.11340	0.13500	0.15330	0.16860	0.18120	0.19150	0.19990	0.21240	0.22060	0.22680	0.22990	0.23255	0.23610	0.23780	0.23910	0.23955	0.23970	0.23980	0.23990	0.23990
2.5	0.03111	0.06106	0.08955	0.11453	0.13633	0.15480	0.17040	0.18320	0.19380	0.20240	0.21510	0.22360	0.23000	0.23330	0.23610	0.24000	0.24200	0.24300	0.24320	0.24330	0.24330	0.24330	0.24330
3	0.03145	0.06160	0.08985	0.11500	0.13683	0.15550	0.17110	0.18410	0.19470	0.20340	0.21630	0.22500	0.23110	0.23500	0.23780	0.24200	0.24500	0.24550	0.24610	0.24630	0.24650	0.24650	0.24650
4	0.03155	0.06180	0.09000	0.11532	0.13720	0.15600	0.17170	0.18470	0.19540	0.20420	0.21720	0.22600	0.23200	0.23600	0.23910	0.24300	0.24550	0.24790	0.24800	0.24820	0.24840	0.24840	0.24840
5	0.03156	0.06190	0.09010	0.11540	0.13740	0.15610	0.17190	0.18490	0.19560	0.20440	0.21750	0.22630	0.23240	0.23640	0.23955	0.24320	0.24610	0.24800	0.24900	0.24910	0.24920	0.24920	0.24920
6	0.03156	0.06200	0.09020	0.11542	0.13747	0.15620	0.17190	0.18500	0.19570	0.20450	0.21760	0.22640	0.23250	0.23670	0.23970	0.24330	0.24630	0.24820	0.24910	0.24940	0.24950	0.24950	0.24950
8	0.03156	0.06200	0.09020	0.11544	0.13750	0.15620	0.17200	0.18500	0.19580	0.20460	0.21770	0.22650	0.23260	0.23680	0.23980	0.24330	0.24650	0.24840	0.24920	0.24950	0.24970	0.24980	0.24980
10	0.03160	0.06200	0.09020	0.11545	0.13752	0.15620	0.17200	0.18500	0.19580	0.20460	0.21770	0.22650	0.23260	0.23680	0.23990	0.24330	0.24650	0.24840	0.24920	0.24950	0.24980	0.24990	0.24990
∞	0.03160	0.06200	0.09020	0.11545	0.13755	0.15620	0.17200	0.18500	0.19580	0.20460	0.21770	0.22660	0.23260	0.23690	0.23990	0.24330	0.24650	0.24840	0.24920	0.24950	0.24980	0.24990	0.25000

For heavily overconsolidated clays, the shear stresses which bring about immediate settlement also cause changes in pore pressure. This generally has the effect of reducing the consolidation settlement undergone by the structure relative to those that straightforward use of oedometer data would suggest. In practice, consolidation settlement is calculated by dividing the stressed zone of the overconsolidated clay into layers and calculating the average vertical stress change, as before, and then applying the equation:

$$\rho_c = \mu \sum_{x=1}^{x=n} m_v \cdot \Delta\sigma_v \cdot z_n \qquad (9.37)$$

where μ is the settlement coefficient proposed by Skempton and Bjerrum,[30] which is related to the pore pressure coefficient A by the equation:

$$\mu = A + a(1-A) \qquad (9.38)$$

where a is a coefficient depending on the geometry of the problem. The results for circular and strip footings, with various ratios of the thickness of the clay, z, to the width of the footing B, are given in Figure 9.12 in terms of the consolidation history.

9.3.4 Settlement of granular soils

Whilst, with clays, it is normally possible to obtain samples of sufficiently good quality to allow meaningful laboratory (oedometer) tests to be carried out, it is not feasible to obtain undisturbed samples of sand and gravels. In such cases, the engineer must rely upon *in situ* tests to obtain parameters from which to judge the compressibility of the ground. The most commonly used tests are penetration and plate loading tests.

At the outset, it is important to realize that neither the standard penetration test (SPT) nor the static cone test can provide an accurate measure of compressibility; both tests cause continuous soil failure as they penetrate the ground and are therefore more likely to give results related to the effective strength parameters and *in situ* stress levels of the soil. Nonetheless, the SPT is in widespread use. Recent evaluations of the accuracy of the numerous methods of estimating settlements from the SPT N value suggest that Schultze and Sherif's[31] method is one of the more accurate of those available.

Schultze and Sherif[31] used statistical methods to obtain a settlement equation from case-history data. The basic information for using the method is shown in Figure 9.13. Settlements are calculated from the equation:

$$p = \frac{s \cdot p}{N^{0.87}(1 + 0.4D/B)} \qquad (9.39)$$

where the granular soil extends for more than $2B$ below the foundation, p is the applied stress at foundation level, N is the average SPT N value over a depth of $2B$ below the foundation level, or d_s if the depth of granular soil is less than $2B$.

Methods based upon the cone test would be expected, in general, to give slightly more accurate results than those based upon the SPT, because of better standardization of the test and the lack of borehole disturbance. N values can be obtained from cone test data using the relationship:

$$N = q_c/107K \qquad (9.40)$$

where K can be obtained from Table 9.8.

While it is clear that the value of K increases with the particle size, comparison between tests on coarse materials becomes somewhat uncertain once the particle size becomes comparable with the diameter of the instruments, i.e. the penetrometer cone

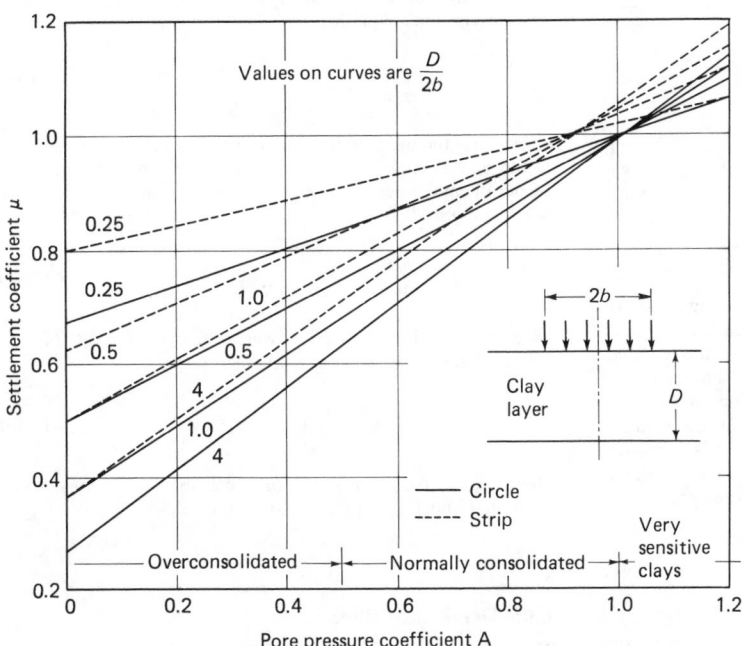

Figure 9.12 Settlement coefficient μ as a function of pore pressure coefficient A. (After Skempton and Bjerrum (1957) 'A contribution to the settlement analysis of foundations on clay', *Géotechnique*, **7**)

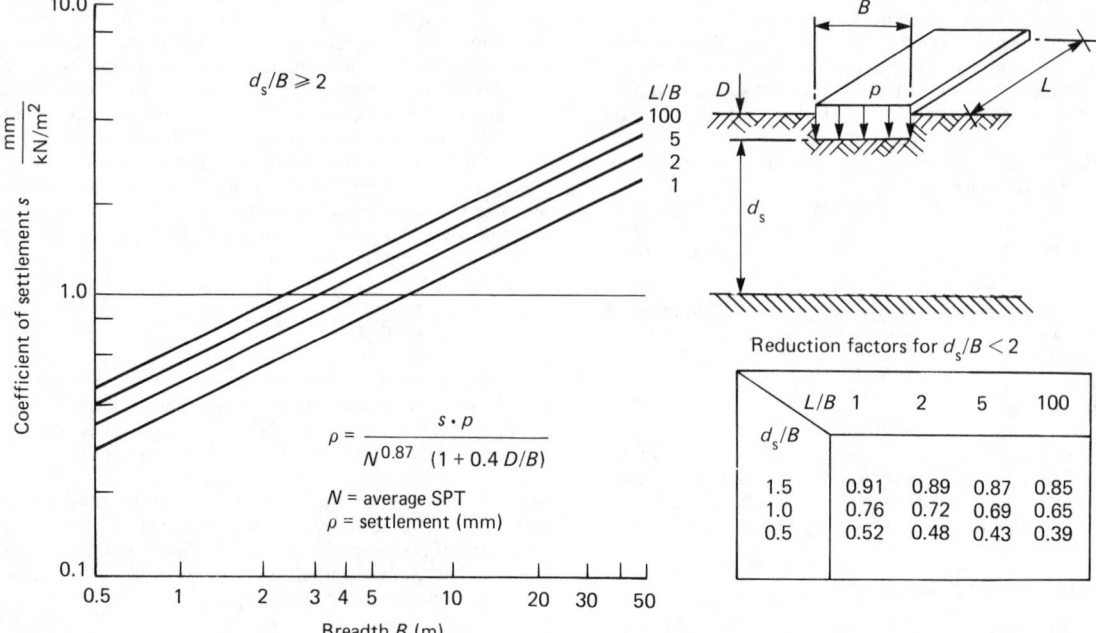

Figure 9.13 Schultze and Sherif's method. (After Schultze and Sherif (1973) 'Prediction of settlement from evaluated settlement observations for sand', *Proceedings, 8th international conference on soil mechanics and foundation engineering*, Vol. I)

Table 9.8 Values of coefficient K in Equation (9.40). (After Simons and Menzies (1977) *A short course in foundation engineering.* Butterworths, London)

Soil description	K
Sandy silt	2.5
Fine sand	4.0
Fine to medium sand	5.0
Sand with sandy gravel	8.0
Medium to coarse sand	8.0
Gravelly sand	8–18
Sandy gravel	12–16

and the SPT spoon. Difficulties also arise on making comparisons at the other end of the scale owing to the tendency of finer sands to 'pipe'; this results in an underestimate of the N value.

A number of methods exist to predict the settlements of foundations directly from cone data. According to Clayton, Simons and Matthews,[10] there is little difference in the accuracy of these methods. In one of the earliest, de Beer[32] has suggested that the compressibility C_s of a granular deposit could be related to the cone resistance q_c and the effective overburden pressure p'_o by the relation:

$$C_s = \frac{1.5 q_c}{p'_o} \qquad (9.41)$$

the compression of a layer of thickness H being given by the expression:

$$\rho = \frac{H}{C_s} \log_e \left(\frac{p'_o + \Delta\sigma_v}{p'_o} \right) \qquad (9.42)$$

where $\Delta\sigma_v$ is the increase in stress due to the net foundation pressure at the centre of the soil layer.

Schmertmann[33] proposed a different approach to the use of SPTs in the calculation of the settlement of footings on sands. He noted that the distribution of vertical strain under the centre of a footing on a uniform sand is not qualitatively similar to the distribution of the increase in vertical stress, the greatest strain occurring at a depth of about $B/2$.

Schmertmann's equation for calculating settlement is:

$$\rho = C_1 \cdot C_2 \cdot \Delta p \sum_0^{2B} \left(\frac{I_z}{E} \right) \Delta z \qquad (9.43)$$

where Δp is the increase in effective vertical pressure at foundation level, Δz is the thickness of layer under consideration, C_1 is the depth embedment factor, and I_z is the strain influence factor given in Figure 9.14.

$$C_1 = 1 - 0.5 \left[\frac{p'_o}{\Delta p} \right] \qquad (9.44)$$

where p'_o is the initial effective overburden pressure at foundation level and C_2 is the empirical creep factor,

i.e.: $\quad C_2 = 1 + 0.2 \log_{10} \left[\frac{t}{0.1} \right] \qquad (9.45)$

where t is the period in years for which the settlement is to be calculated and E is the deformation modulus:

$$E = 2 \cdot q_c \qquad (9.46)$$

9.3.5 Depth corrections

Where foundations are below ground level, and where elastic theory based upon a load applied at ground surface has been used to obtain foundation stress increases for settlement analysis, a correction factor is generally made to allow for the effect of the soil above foundation level. Traditionally the values of Fox[27] have been used; these lead to a maximum possible reduction in settlement (for deep foundations) of 50%. Because the figures are based upon elastic analysis they tend to overestimate the reductions to be applied. The more conservative values of Burland[34] are therefore preferred (Figure 9.15).

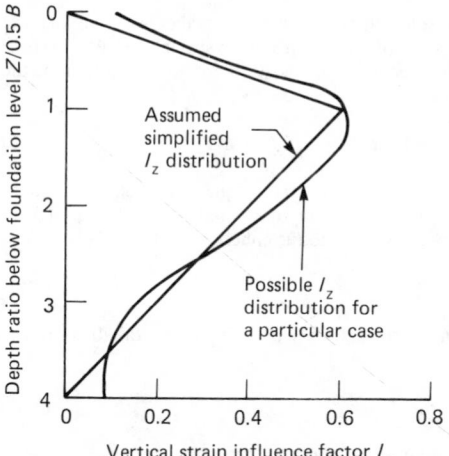

Figure 9.14 Variation of strain influence factor with depth. (After Schmertmann (1970) 'Static cone to compute elastic settlement over sand', *Proc. Am. Soc. Civ. Engrs*, **98**; Simons and Menzies (1977) *A short course in foundation engineering*. Butterworths, London)

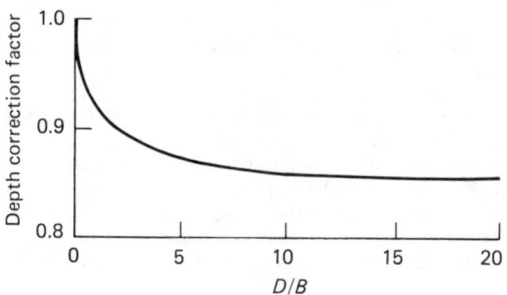

Figure 9.15 Depth correction factor. (After Burland (1970) *Proceedings, conference on* in situ *investigations in soils and rocks*. British Geotechnical Society, London)

9.4 Earth pressure

The pressure on a wall or structure in contact with soil is a complex matter, depending upon: (1) how the wall moves; (2) how the soil–wall system is placed; (3) the strength parameters of the soil; and (4) groundwater conditions.

Classical earth pressure analysis, which dates back to the eighteenth century, considers only the restricted cases of rigid walls which rotate about their base, either into or away from the soil. It is implicitly assumed that the soil is placed before the wall and that the wall can be placed without disruption to the soil. But in practice it is common to compact the soil behind a wall, and for the wall to slide or rotate in a way which is different to that implied by classical analyses. This leads to rather different pressure distributions. For a full discussion of earth pressure and the design of earth-retaining structures the reader is referred to Clayton and Milititsky.[35]

9.4.1 Active and passive conditions

The problem of earth pressure is the oldest soil mechanics problem. Active pressure is the lowest pressure which a retaining structure must be capable of resisting in order to prevent a soil mass from collapsing. The highest pressure which a structure can exert on a bank of earth without causing it to move in the direction of the pressure is the passive pressure or passive resistance. Examples of both are given in Figure 9.16.

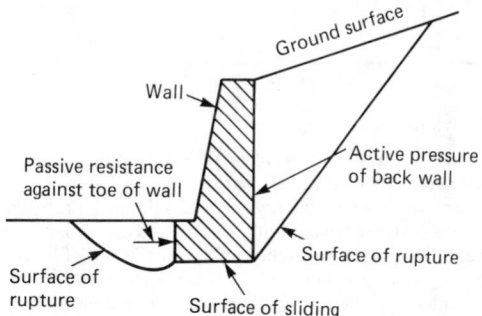

Figure 9.16 Active pressure and passive resistance

Below a level ground surface, the horizontal pressure is known as the 'pressure at rest'. This pressure lies between the active and passive pressures and is usually designated by the factor K_o which is the ratio of the horizontal and vertical effective pressures at any given depth. The factors K_a and K_p similarly relate the active and passive horizontal effective pressures to the vertical effective pressure.

The value of K_o depends on the depositional conditions and stress history of the ground. For loose sands the value of K_o is about 0.45, falling to about 0.35 in dense sands, following the relationship given by Jaky, $K_o = (1 - \sin \phi')$. A wider range is encountered in clays, typically 0.4 to 0.7, but in some heavily overconsolidated clays values considerably in excess of unity, and perhaps as large as 3, occur.

Tables 9.9 and 9.10 give typical values for K_a and K_p for cohesionless materials, vertical walls with horizontal ground where ϕ' is the angle of friction for the soil and δ' the angle of wall friction.

Experimental work by Rowe and Peaker[36] has emphasized the dominant role of the angle of wall friction, but has shown that the code values of K_p can be as much as 50% too high in loose and dense sands.

In order that the lateral pressure may change from the pressure at rest to either the active or passive value, movement must take place to mobilize shear forces. This generally occurs during the construction of the retaining structure.

Table 9.9 Values of K_a from CP2. (After Institution of Structural Engineers (1951) *Earth retaining structures*. ISE CP2, ISE, London)

	Values of ϕ'				
Values of δ'	25°	30°	35°	40°	45°
			Values of K_a		
0°	0.41	0.33	0.27	0.22	0.17
10°	0.37	0.31	0.25	0.20	0.16
20°	0.34	0.28	0.23	0.19	0.15
30°	—	0.26	0.21	0.17	0.14

Table 9.10 Values of K_p from CP2. (After Institution of Structural Engineers (1951) *Earth retaining structures*. ISE CP2, ISE, London)

Values of δ	Values of ϕ			
	25°	30°	35°	40°
			Values of K_p	
0°	2.5	3.0	3.7	4.6
10°	3.1	4.0	4.8	6.5
20°	3.7	4.9	6.0	8.8
30°	—	5.8	7.3	11.4

9.4.2 Active pressure

For an ideal material, the problem of determining the total active pressure is comparatively simple and is based on the wedge theory which was originally due to Coulomb[37] (1776) who solved it for a material having both friction and cohesion. Later workers omitted the cohesion, changed the coefficient of friction into the tangent of the angle of internal friction, which was taken as equal to the angle of repose, and extended the expression to include sloping walls, wall friction (anticipated in part by Coulomb), and inclined surcharges. Not all of these changes were improvements.

In the wedge theory it is assumed that the pressure on the wall is due to a wedge of earth which tends to slip down an inclined plane as shown in Figure 9.17. The forces acting on the wedge are also shown in the figure. The inclination of the plane BD is altered until the position which gives the greatest value for the force is found. This can either be done analytically or graphically.

The general formula for the total force over depth H exerted by a frictional material having no cohesion is:

$$P_a = \tfrac{1}{2}\gamma H^2 \left(\frac{K_a}{\sin a \cos \delta'} \right) \qquad (9.47)$$

where the coefficient of active earth pressure K_a is:

$$K_a = \frac{\sin^2(a+\phi')\cos\delta'}{\sin a \cdot \sin(a-\delta)\left\{ 1 + \left[\dfrac{\sin(\phi'+\delta')\sin(\phi'-\beta)}{\sin(a-\delta')\sin(a+\beta)} \right]^{1/2} \right\}^2} \qquad (9.48)$$

where ϕ' is the effective angle of friction, δ' the angle of wall friction and a, β and H are as shown in Figure 9.17.

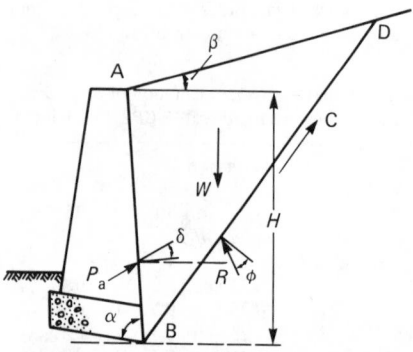

Figure 9.17 Coulomb or general wedge theory

For the case of a vertical wall and horizontal backfill, the values of K_a to be used in Equation (9.47) are as in Table 9.8. For the special case of no wall friction, vertical wall and horizontal backfill, this reduces to the Rankine formula:

$$P_a = \tfrac{1}{2}\gamma H^2(1 - \sin\phi')/(1 + \sin\phi') = \tfrac{1}{2}\gamma H^2 K_a \qquad (9.49)$$

No analytical solution exists for the general case of a soil having both friction and cohesion, but for the special case of a vertical wall, horizontal backfill and no wall friction or adhesion, the total pressure is given by:

$$P_a = \tfrac{1}{2}\gamma H^2 K_a - 2c' H \sqrt{K_a} \qquad (9.50)$$

For a purely cohesive material, Equation (9.50) reduces to $P_a = \tfrac{1}{2}\gamma H^2 - 2cH$ for the case of no wall adhesion, or if wall adhesion is taken as equal to the cohesion the formula becomes:

$$P_a = \tfrac{1}{2}\gamma H^2 - 2(\sqrt{2})cH \qquad (9.51)$$

The equation for the intensity of horizontal total stress at any level is:

$$p_a = K_a(\sigma_v - u) + u \qquad (9.52)$$

Since earth retaining structures generally are involved with unloading of the soil, it is normal to use effective strength parameters c' and ϕ' to determine K_a. The use of total stress analysis, even in the case of temporary works design, should be avoided. It is better to use an effective stress analysis with a relatively low factor of safety.

When using the earth pressure coefficient approach, the procedure is as follows:

(1) Establish a design profile from existing borehole records or available information.
(2) Determine ground levels, and levels of each soil layer.
(3) Determine the assumed position of the groundwater table (this may be different on either side of the wall); with good drainage, or a low groundwater table, there may be no influence from groundwater.
(4) Calculate the vertical total stress at ground level (which may not be zero, e.g. when there is a uniform surcharge), at groundwater level, and at the boundary of each soil type:

$$\sigma_v = \sum_0^n \gamma_n \cdot z_n + p_s \qquad (9.53)$$

where γ_n is the bulk weight of the soil in soil layer n, z_n is the thickness of soil layer n and p_s is the vertical uniform surcharge at ground level.
(5) Calculate the pore water pressure at each boundary. It is generally assumed that:

$$u = \gamma_w \cdot z \qquad (9.54)$$

where γ_w is the bulk unit weight of water (which can be taken as $10\,\text{kN/m}^3$) and z is the depth below groundwater level which implies no flow of water in the vicinity of the structure.
(6) Calculate the vertical effective stress at each level:

$$\sigma'_v = \sigma_v - u \qquad (9.55)$$

where σ'_v is the vertical effective stress, σ_v is the vertical total stress (see (4) above) and u is the pore water pressure (see (5) above).
(7) Calculate the horizontal effective stress *for each soil type*,

and each level. Where two soil types meet, the earth pressure coefficient will change. Using the value of vertical effective stress on the boundary, calculate two values of horizontal effective stress, and for each soil type:

$$\sigma'_{hn} = K_n \cdot \sigma'_v \qquad (9.56)$$

$$\sigma'_{h_{n+1}} = K_{n+1} \cdot \sigma'_v \qquad (9.57)$$

where σ'_{hn} is the horizontal stress in layer n, and $\sigma'_{h_{n+1}}$ is the horizontal stress level in layer $n+1$. K_n and K_{n+1} are the relevant earth pressure coefficients for layers n and $n+1$.

For passive pressures, the earth pressure coefficients are normally factored-down, typically by 2.

(8) Calculate the total horizontal stress at each level, for each soil type:

$$\sigma_h = \sigma'_h + u \qquad (9.58)$$

where σ_h is the horizontal total stress to be supported by the wall, σ'_h is the horizontal effective stress (see (7) above) and u is the pore water pressure (see (5) above).

9.4.2.1 Graphical method

A simple graphical solution can be used for walls with irregular outlines, for irregular backfills, external loads on the backfill, and variation in the properties of the backfill. The principle is that of the wedge theory, the most dangerous surface of slip being found by trial.

The polygons of forces are drawn for a number of wedges ABD_1, ABD_2, etc. as shown in Figure 9.18. The forces acting on each wedge are its weight W, the reaction R across the plane BD, the cohesion C acting up the plane BD, and the force P which is required to find. For limiting equilibrium, R is inclined at ϕ' to the normal to the plane BD, and $C = c \times BD$. If the direction of P is assumed, the polygon of forces can be completed, giving the value of P. It is convenient to plot all the polygons from a common origin of W as shown. The maximum value of P is then given by the point at which the envelope of the R lines cuts the line representing E. This P line can be drawn from the origin in any assumed direction, thus allowing for wall friction. Wall adhesion (cohesion) can be included by subtracting it from the weight of the wedge.

Alternatively, the force polygons can be drawn separately for each wedge and the value of P_a can be plotted above the position of the corresponding slip surface.

9.4.2.2 Surcharge loads

The effect of a distributed surcharge of magnitude p_s is to increase the earth pressure over the whole height of the wall by an amount $K_a p_s$ in the case of granular backfills and by an amount p_s in the case of cohesive backfills. The effect of a line load W_l can be estimated with sufficient accuracy for most designs by the construction shown in Figure 9.19.

The effect of a point load W_i on the backfill is more difficult to estimate. An approximate method given in CP2[38] is to assume that the load is spread through the backing at an angle of dispersion of 45° on each side of the load. The lateral pressure at any point due to the surcharge is then taken as K_a times the vertical pressure at the point. This method, however, tends to give results on the unsafe side. The following tentative approximate method is suggested. The line of action of the resultant force is obtained by a construction similar to that for a line load (Figure 9.19), the 40° line being constructed from the centre of the loaded area. It is assumed that, if the length of the loaded area be L and the distance between the back of the wall and the near edge of the area be x, the resultant lateral thrust will be distributed along a length of wall equal to $L + x$. Then if W_i be the load on the area, the resultant thrust per unit length of wall will be $K_a \cdot W_i(L + x)$.

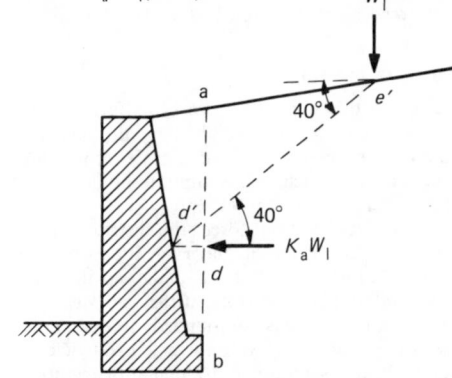

Figure 9.19 Method of estimating magnitude and line of action of pressure due to a line load

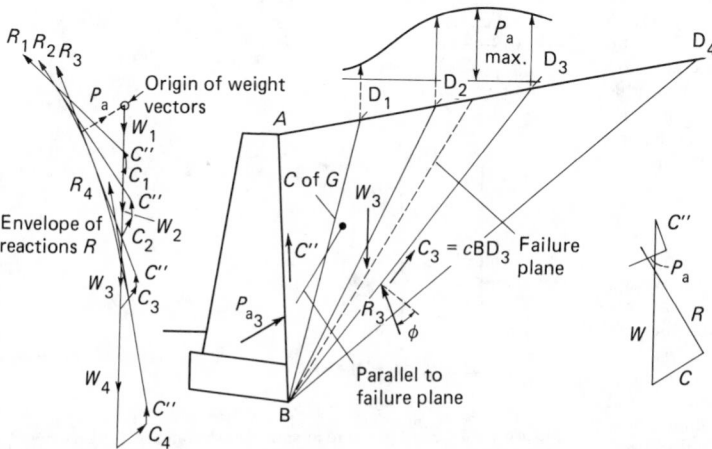

Figure 9.18 Engesser's method for determination of active pressure

9.4.3 Passive resistance

In the case of active pressure the assumption of a plane surface of failure gives results which are within 3 or 4% of the true value, and this is close enough for all practical purposes. With passive pressure, however, this may lead to results which differ significantly from the true values if wall friction is taken into account. Only for the case of no wall friction does the wedge theory give the true value. The passive pressure in this case is given by the formula:

$$P_p = \tfrac{1}{2}\gamma H^2 K_p + 2c'H\sqrt{K_p} \qquad (9.59)$$

Wall friction adds greatly to the passive resistance, but not so much as the wedge theory indicates; it is seldom, therefore, that it can be neglected. The wedge theory does not give the correct answer because the surface of failure is not a plane but is curved, as shown in Figure 9.17.

The effect of curved failure surfaces can be analysed using the log-spiral method, a graphical procedure which is given in detail by Terzaghi and Peck,[4] and which forms one basis for the computation of bearing capacity.

For the simple case of horizontal ground and vertical wall, in cohesionless soil, Table 9.3 gives values of the passive earth pressure coefficient P_p for use in the equation:

$$P_p = \tfrac{1}{2}\gamma H^2 K_p \sec \delta' \qquad (9.60)$$

For earth pressure coefficients for more complex geometrics and soil conditions, the reader is referred to Clayton and Milititsky.[35]

9.4.4 Distribution of pressure

It is generally assumed that pressure increases uniformly with depth. This is only true in certain special cases, although the assumption is not unreasonable in some other cases for which it is not strictly true. Cases in which the assumption should not be made are dealt with below.

The wedge theory gives the total force; the distribution of pressure cannot be obtained from this theory as it depends on the lines of action of the forces involved and on the way in which the wall yields and will only be triangular if the wall yields by turning about its base; if the wall is not founded on rock and is very rigid in itself this is usually the way in which it will yield. It is for this reason that the assumption of triangular distribution can often be made in practice. The centre of pressure on the wall using the wedge method can be estimated as shown in Figure 9.18.

In the case of cohesive soils, calculations indicate a zone of tension at the top of the wall. In this region there will be no pressure on the wall, and in practice deep tension cracks are often observed behind walls supporting cohesive soils. The value of this tension must not be subtracted from the pressure diagram. If it is possible for the tension cracks to become filled with water, the value of the water pressure must be included in the pressure calculations.

9.4.5 Strutted excavations

For excavations below groundwater level, a sheeted excavation is often used in preference to an open excavation with battered sides, particularly where space is limited and where the piles can be driven into a relatively impervious stratum to provide a cutoff against groundwater in overlying pervious strata. Such sheeting is usually braced by frames consisting of walings and struts. Calculations of the earth pressures on the sheeting follow the same lines as for retaining walls. However, during progressive excavation and placing of frames, deflections of the sheeting occur which lead to frame loads which are not in agreement with those calculated from the earth pressure diagram, assuming hinge points at all frame levels below the top frame. The load in the struts does not increase in general with depth below about one-quarter of the depth of the excavation.

For sands, Terzaghi has proposed an empirical design rule, based on a number of field observations, as shown in Figure 9.20(b); and the same rule, with a modification for use in clay, has been adopted in CP2.[38] Peck[39] and Tomlinson[40] have discussed this question and that of adjacent associated movements in some detail. Typical pressure distributions are given in Figure 9.20(b) and (c).

It should be emphasized that the redistributed pressure diagrams described above are, in effect, design devices for obtaining frame loads and do not necessarily imply any actual redistribution of earth pressure. In fact, Skempton and Ward[41] have described the results of strut and waling load measurements in a cofferdam at Shellhaven and have interpreted results to show that the frame loads can be accounted for in terms of deflections of the sheet prior to placing the struts in position and without the need for assuming any redistribution of earth pressure.

The two useful rules in determining levels at which frames should be placed have been given by Ward[42] as follows:

(1) In a deep deposit of normally consolidated clay the uppermost frame of struts should be placed across a cofferdam before the depth of excavation H_1 reaches a value given by $H_1 = 2c_u/\gamma$.
(2) The second frame of struts should be placed before the depth of excavation reaches a depth H_2 given by $H_2 = 4c_u/\gamma$.

Recent practice, particularly for larger excavations, is to avoid internal strutting by the use of ground anchors, placed at suitable levels as excavation proceeds, and which with advantage can be prestressed to stipulated loads. Reference should be made to articles by Littlejohn[43] for information on anchor design and construction.

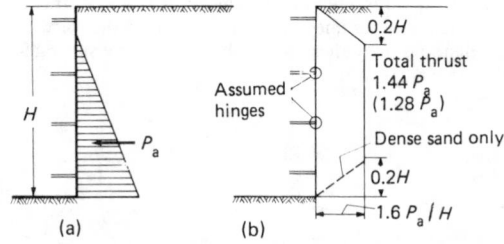

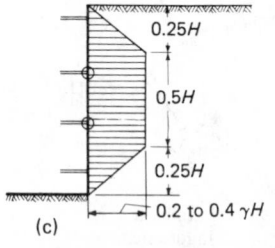

Figure 9.20 Earth pressure in strutted excavations. (a) Calculated earth pressure; (b) earth pressure distribution by CP2, based on Terzaghi's method; (c) stiff fissured clay. (After Terzaghi and Peck (1967) *Soil mechanics in engineering practice*. Wiley, New York)

9.4.6 Anchored bulkheads

An anchored bulkhead is usually in the form of a steel sheet-pile wall supported by ties at one level only and by passive pressure against the toe. However, anchored bulkheads may also be constructed with timber, precast reinforced concrete sheet piles, or continuous-bored piles. Calculations of active and passive earth pressures follow the same lines as for retaining walls but analysis of the stability of an anchored bulkhead requires the determination of bending moments in the piling and of the magnitude of the anchor pull.

The magnitude of maximum bending moment occurring in the bulkhead will be influenced by the relative rigidity of the soil and the section of the bulkhead.

The various dimensions and forces entering into a bulkhead calculation are indicated in Figure 9.21. A design procedure based on Rowe's method, and similar to that described by Terzaghi,[44] is given in sections 9.4.6.1 to 9.4.6.4.

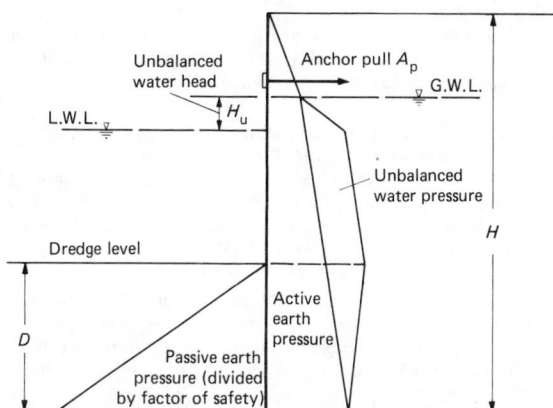

Figure 9.21 Dimensions and forces in anchored bulkhead calculation. Earth pressure diagrams illustrated are for homogeneous cohesionless soil. No pressures from surcharge loads have been shown

9.4.6.1 Forces acting on faces of bulkhead

The active and passive pressures are first calculated. The active pressure calculations must allow for the maximum possible unbalanced water pressure and for any surcharge in the form of distributed, line or point loads supported directly on the back-fill.

Strictly speaking, the water pressures should be calculated from a flow net (Figure 9.22(a)) taking into account the effects of stratification in the soils present. However, if the soils do not vary widely in their permeabilities it is sufficient to use the simplified pressure distribution shown in Figure 9.22(b). Allowance must also be made, where the passive pressure is provided by a permeable stratum, for a reduction in passive pressure due to seepage gradients (Figure 9.22(c)).

9.4.6.2 Computation of depth of penetration (free-end method)

A diagram of active and passive pressures is drawn as shown in Figure 9.21 for a trial penetration of the piling. The effects of any surcharge loads should also be included, as described on page **9**/19. Before passive pressures are plotted, a factor of safety F_p is applied. The choice of a value for this factor of safety for a given design depends on the accuracy of the data on which the earth pressure has been based, but in general it should not be less than 2.

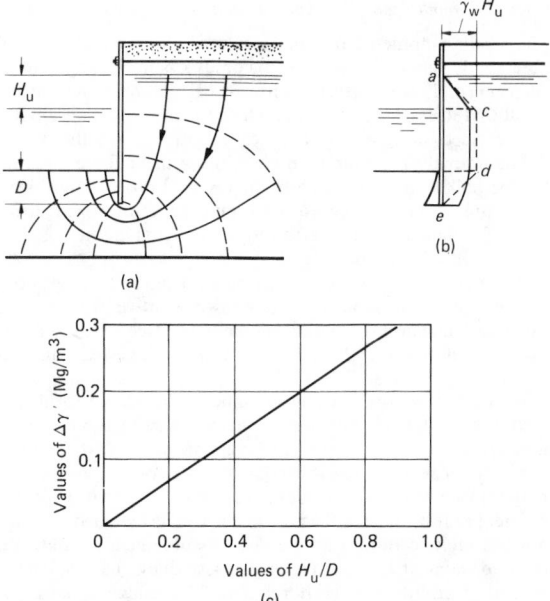

Figure 9.22 Unbalanced water pressure. (a) Flow net; (b) distribution of unbalanced water pressure; (c) average reduction of effective unit weight of passive wedge due to seepage pressure exerted by the upward flow of water. (After Terzaghi (1953) 'Anchored bulkheads', *Proc. Am. Soc. Civ. Engrs*, **79**)

The earth pressure diagram is then divided up into a number of convenient areas and the total load on each of these and its point of application is estimated. Moments of these loads are then taken about the line of the anchor pull. This is repeated for other trial depths of penetration until a depth giving zero total moment is obtained. Alternatively, the pressures below dredge level may be expressed in terms of the penetration D and the required depth found analytically. It is usual to increase the calculated depth of penetration by 20% to allow for the possibility of scour or overdredging.

9.4.6.3 Anchor pull

The anchor pull is determined by resolving horizontally all the forces acting on the bulkhead. There are, however, a number of factors which may lead to an anchor pull somewhat greater than that computed, and conservative design stresses should therefore be adopted.

Where the ties are taken back to blocks or beams the position of these should be such that no overlapping of active and passive zones occurs, as illustrated in Figure 9.23.

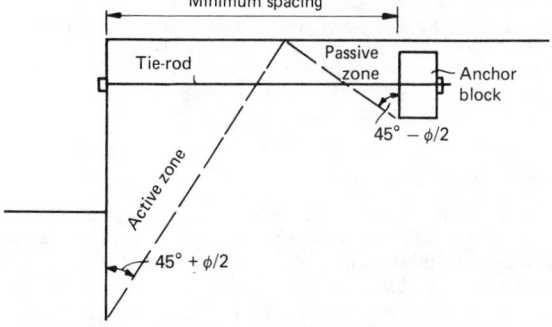

Figure 9.23 Minimum length of anchor ties

9.4.6.4 *Computation of maximum bending moments*

The bending moments are first calculated on the assumption of 'free earth support', and the maximum bending moment is determined. This is a straightforward calculation based on the resultant forces of the areas into which the pressure diagram has been divided, using either analytical or graphical methods.

The normal procedure is to construct a shear force diagram for the bulkhead, starting from the top and proceeding downwards only as far as is necessary to find the point of zero shear force. The maximum free earth support bending moment is then calculated at this point.

If the sheet piles are to be driven into fairly homogeneous stratum of clean sand with a known relative density, the calculated maximum bending moment for free earth support can be reduced on the basis of Rowe's investigations, as illustrated in Figure 9.24.

As a first step, the flexibility number is calculated for a trial section of bulkhead. For the calculated value of the flexibility number, a value of the moment reduction ratio M/M_{max} can be read off, depending on the relative density and, hence, the reduced moment can be obtained and compared with the moment of resistance of the piling. The trial is repeated until the most suitable section of the bulkhead is obtained, i.e. until the reduced moment is equal to, or just less than, the moment of resistance of the section. If required, the calculation can be extended to cover alternative construction materials.

If the sheet piles are to be driven into a homogeneous stratum of dense or medium-dense silty sand, the moment reduction curves for medium and loose sand should be used instead of those for dense and medium-dense sand. Sheet piles to be driven into loose silty sand should be calculated on the figure for free earth support, since compressibility of such sands may be high. Work by Rowe[45] has shown that in some cases moment reduction can be made where piles penetrate clay below dredge level, but this is rarely done in practice.

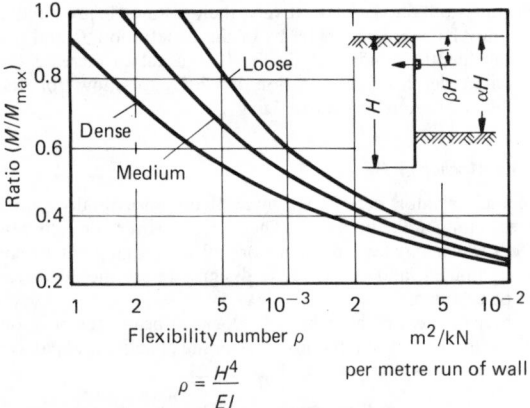

Figure 9.24 Relation between the flexibility number ρ of sheet piles and bending moment ratio M/M_{max} (logarithmic scale). (After Terzaghi (1953) 'Anchored bulkheads', *Proc. Am. Soc. Civ. Engrs*, **79**)

9.4.7 Overall stability

The design of a retaining wall, whether a mass wall or a sheet-pile wall, should always be considered from the point of view of overall stability, i.e. failure as a bank of earth. The forces are shown in Figure 9.25.

$$\text{Disturbing moment} = \Sigma\, TR + Pr \qquad (9.61)$$

$$\text{Resisting moment} = (\Sigma\, N \tan \phi' + c'L)R \qquad (9.62)$$

where L is the length of the arc over which c' acts and T and N are the tangential and normal components of W.

$$\text{Factor of safety} = \frac{(\Sigma\, N \tan \phi' + \Sigma\, c'L)R}{\Sigma\, TR + Pr} \qquad (9.63)$$

The methods described in section 9.5 are directly applicable.

9.5 The stability of slopes

The analysis of slopes is important because of the dangers to both structures and life that can be caused by two types of problem:

(1) Where construction or excavation causes stress changes in the soil which lead to failure in previously stable ground (the so-called 'first-time slide').
(2) Where construction or excavation reactivates movement on a pre-existing shear surface in the soil, usually part of an ancient and pre-existing landslide.

As with other areas of soil mechanics, an important distinction is made between short- and long-term conditions. Short-term conditions are the most critical when load increases are applied to the soil, but they may also be relevant for temporary works when load decrease takes place on clays, because in this situation the depressed pore pressures which are induced by excavation, for example, may not have time to dissipate. For short-term conditions the $\phi_u = 0$ analysis is applied, using the undrained shear strength C_u. In fissured clays, it is observed from back-analysed failures that the mobilized undrained shear strength at failure is considerably less than that measured from small-scale tests. Typically, the mobilized undrained shear strength is only 45 to 60% of that measured on samples of the order of 40 to 50 mm diameter, because these samples do not contain representative fissures (which are planes of weakness).

There is little available evidence of the rate at which negative pore pressures due to unloading decay, and therefore for how long the $\phi_u = 0$ analysis can reasonably be applied. As they do so, however, the soil swells and its strength decreases. In the long term, when pore pressures have equalized, peak effective strength parameters c' and ϕ' are used, provided the soil has not previously failed. If, on the other hand, it is known that a pre-existing shear surface exists (due to previous slope instability on the site) the residual strength parameters (c'_r and ϕ'_r must be used. These are normally derived either from stress reversal shear box testing or from ring shear testing.

If analysis is to be carried out in terms of effective stress, then the equilibrium pore pressures in the slope must be estimated. This is a complex matter, depending on climatic conditions, local groundwater conditions before construction, subsoil geometry and properties, and the effective stress dependency of soil permeability. In practice, in the UK, a value of r_u is normally selected on the basis of experience with monitored and back-analysed slopes. In the brown London Clay, a long-term value of $r_u = 0.3$ is sometimes used.

Stability analyses assume that shear strength is mobilized evenly over the slip surface, but in reality it is thought that progressive failure occurs. It is also known that while stability analysis is capable of giving a reasonable guide in terms of the factor of safety it yields to the stability of a slope, instability can occur at failure on a different surface to that predicted by analysis. A very useful review of European experience of slope stability problems and parameter selection is given by Chandler.[46]

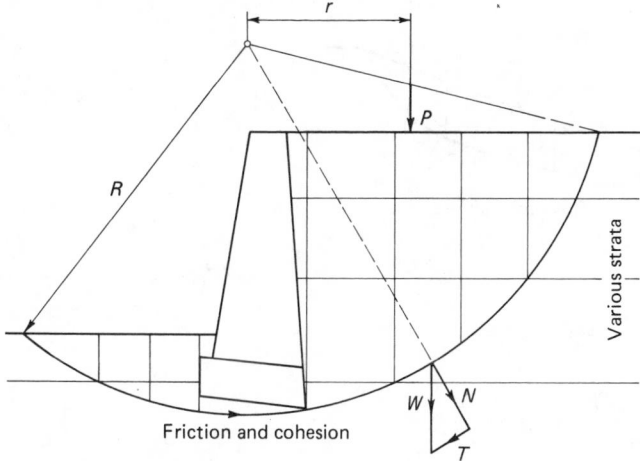

Figure 9.25 Overall stability of a retaining wall

9.5.1 Stability analysis

Slope stability problems in engineering works are usually analysed using limit equilibrium methods. Many such methods are available in practice and the most common ones call on the principle of slices. In this method, the failure mass is broken up into a series of vertical slices and the equilibrium of each of these slices is considered. This procedure allows both complex geometry and the variable soil and pore pressure conditions of a given problem to be considered.

The methods most commonly used are:

(1) The method of slices of Fellenius.[47]
(2) The modified, or simplified, method described by Bishop.[48]
(3) Janbu's generalized method of slices.[49]
(4) Morgenstern's and Price's method.[50]
(5) Spencer's method.[51]

The first two methods do not satisfy all the moment and force equilibrium equations and can only accommodate circular slip surfaces. The last three methods satisfy all equilibrium equations (although simplifying assumptions are required) and may be used to calculate the factor of safety along any shape of slip surfaces. Those methods which satisfy all the conditions of equilibrium, and Bishop's modified method, give accurate results which do not differ by more than 5% from the 'correct' answer, obtained by the log-spiral method.

The main conclusions of such studies can be found in La Rochelle and Marsal,[52] who prefer the simplified methods of Bishop and Janbu due to the simplicity of computer programming and the low cost of running such programs.

In practice, the design engineer must choose between circular and noncircular methods of analysis. Because of their relative simplicity, it is common to carry out routine slope-stability analyses using a circular failure surface. This assumption will be justified in relatively homogeneous soil conditions, since experience shows that the analysis can make good estimates of the factor of safety when failure is imminent. Noncircular analyses are used when: (1) a pre-existing surface has been found in the ground, and is known to be noncircular; and (2) circular failure is prevented, perhaps by the presence of a stronger layer of soil at a shallow depth. Under either of these conditions, the use of a circular shear surface will overestimate the factor of safety against failure.

9.5.1.1 Method of slices

In order to cater both for complex soil conditions and variable geometry it is common to divide the slipping mass of soil into slices (Figure 9.26). The factor of safety is normally defined in terms of the ratio between the average shear strength available on the shear surface and the average shear strength mobilized for stability, i.e.:

$$F = \frac{\bar{T}_{\text{available}}}{\bar{T}_{\text{mobilized}}} \tag{9.64}$$

where $F = 1$ at failure.

In order to obtain a solution to the problem, slope stability analyses examine the equilibrium of the mass of soil which is being considered. A number of different solutions, therefore, can be derived, depending on the approach taken in the analysis. Methods used to derive the basic equations are:

(1) Force equilibrium of a single slice.
(2) Moment equilibrium of a single slice.
(3) Force equilibrium of the total mass of soil above the slip surface.
(4) Moment equilibrium of the total mass of soil above the slip surface.

For example, Bishop's method combines (1) and (4), Janbu uses (1) and (3), and Morgenstern and Price use a combination of (1), (2) and (3). A solution cannot be obtained for the complete stability of the slipped mass without making simplifying assumptions, and therefore it is common to make assumptions concerning the interslice forces E and X (Figure 9.26), for example, in deriving a factor of safety for a circular failure surface. From the definitions of factor of safety:

$$\text{Available shear strength} = \frac{\text{peak shear strength}}{\text{factor of safety}} \tag{9.65}$$

For moment equilibrium about the centre of the circle:

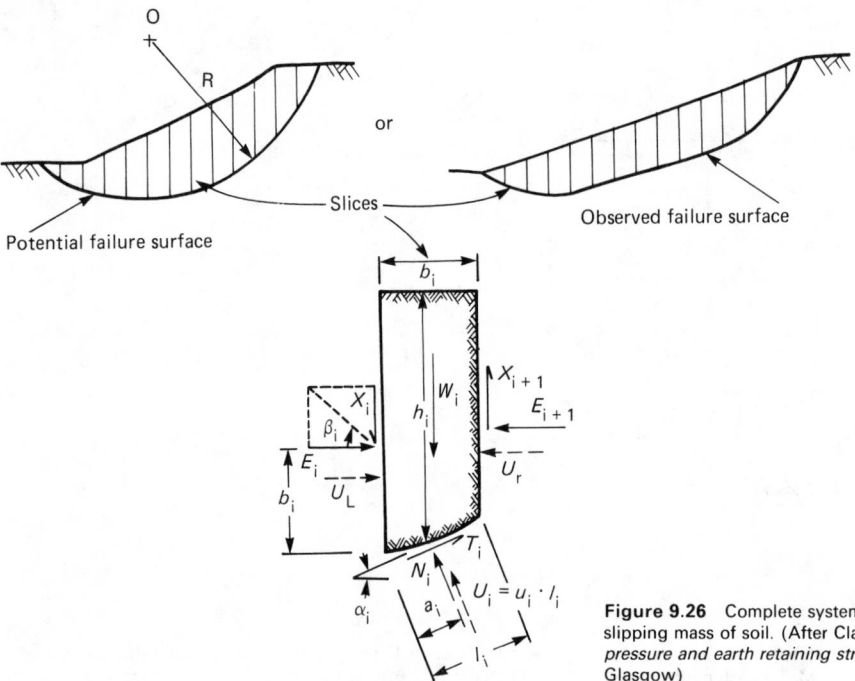

Figure 9.26 Complete system of forces on a single slice of a slipping mass of soil. (After Clayton and Mililitsky (1986) *Earth pressure and earth retaining structures*. Surrey University Press, Glasgow)

(moment of available shear strength on shear surface) =

(moment of weight of soil mass)

i.e. disturbing moments must equal restoring moments, at failure.

A summary of the most significant features of each method recommended is presented below. Slope stability analyses for earth retaining structures will normally be carried out in terms of effective stress using estimates of long-term, stabilized pore water pressure, since this will give the lowest factor of safety for an unloading case, e.g. excavation. Where load increase takes place on clays, e.g. for fill placed behind a sheet-pile wall, short- or intermediate-term stability analysis may give the lowest factor of safety.

9.5.1.2 Short-term stability analysis

Short-term, total, stress analysis is most conveniently carried out by the method of slices, since this method can take into account irregular ground surface profiles, changes in soil type, and variations in undrained shear strength around the slip surface. Partial submergence of slopes can also be analysed.

Interslice forces E_i, E_{i+1}, X_i, X_{i+1}, are ignored, and the strength at the base of each slice T_i is independent of the effective normal force N_i and is equal to the undrained shear strength of the soil, divided by the factor of safety.

For equilibrium:

$$R \Sigma W_i \cdot \sin a_i = \frac{R}{F} \Sigma C_{u_i} \frac{b_i}{\cos a_i} \qquad (9.66)$$

Therefore:

$$F = \frac{\Sigma C_{u_i} \cdot b_i / \cos a_i}{\Sigma W_i \cdot \sin a_i} \qquad (9.67)$$

The solution is obtained by dividing the slipping mass of soil into slices. For hand calculation, the use of 10 to 15 slices is recommended. Calculations are most easily handled in table form as suggested in Figure 9.27.

A search must be made for the slip surface which will yield the lowest factor of safety.

9.5.1.3 Long-term stability analysis: circular slips

Fellenius's method[47]   Like the short-term stability method described above, Fellenius's method for long-term analysis uses moment equilibrium of the slipped mass, and ignores interslice forces.

Considering the equilibrium of the ith slice, and resolving perpendicular to the slip surface,

$$W_i \cos a_i = N_i' + u_i \cdot l_i \qquad (9.68)$$

Therefore:

$$N' = W_i \cos a_i - u_i \cdot l_i \qquad (9.69)$$

By definition, $T_i = 1/F(c'l_i + N_i' \cdot \tan \phi')$ in terms of peak effective strength parameters and the weight of the slice.

$$W_i = \gamma \cdot h_i \cdot b_i \qquad (9.70)$$

Summing the moments for all slices:

$$\Sigma_{\text{disturbing moments}} = \Sigma W_i \cdot R \cdot \sin a_i = R \Sigma W_i \sin a_i \qquad (9.71)$$

$$\Sigma_{\text{restoring moments}} = \Sigma T_i R = \frac{R}{F} \Sigma c'l_i + N_i' \tan \phi')$$

$$= \frac{R}{F} \Sigma (c'l_i + (W_i \cos a_i - u_i l_i) \tan \phi') \qquad (9.72)$$

Slice no.	b (m)	α (deg.)	h (m)	C_u (kN/m^2)	γ_b (kN/m^3)	W	$W \sin \alpha$	$C_u \cdot b / \cos \alpha$
						Σ		Σ

Figure 9.27 Format for calculating slipping slices of soil

For equilibrium of the entire slipped mass:

$$R \Sigma W_i \sin \alpha_i = \frac{R}{F} \Sigma \left(c' l_i + (W_i \cos \alpha_i - u_i \cdot l_i) \tan \phi' \right) \quad (9.73)$$

Therefore:

$$F = \frac{\Sigma \left(c' l_i + (W_i \cdot \cos \alpha_i - u_i l_i) \tan \phi' \right)}{\Sigma W_i \cdot \sin \alpha_i} \quad (9.74)$$

As before, hand calculations are best carried out in tabular form.

A search must be made for the slip surface which gives the lowest factor of safety. Because of the simplifying assumption that the interslice forces are zero, or at least are in equilibrium for each slice, the Fellenius solution underestimates the factor of safety by between 5 and 20%. Errors are greatest when the variation of a is large, i.e. for deep circles. Results are conservative, but this may lead to uneconomical design. This method is not widely used in the UK, that by Bishop (see below) being preferred.

Bishop's method[48] As with Fellenius's method, Bishop's method uses force equilibrium of each slice and moment equilibrium of the entire slipped mass to derive an equation for the factor of safety. Unlike Fellenius's solution, however, the method does not entirely ignore interslice forces. The result is that the method is considerably more accurate than that of Fellenius.

As before, the factor of safety is defined in terms of the mobilized shear strength, i.e.:

$$T_i = \frac{1}{F}(c' l_i + N'_i \tan \phi'_i) \quad (9.75)$$

For force equilibrium within each slice, resolving vertically, and for moment equilibrium of the entire slip mass about zero:

$$F = \frac{1}{\Sigma W_i \sin \alpha_i} \Sigma \frac{1}{m_{\alpha_i}}$$
$$[c' l_i \cos \alpha_i + (W_i - u_i l_i \cos \alpha_i + (X_i - X_{i+1})) \tan \phi'_i] \quad (9.76)$$

where

$$m_\alpha = \frac{1}{\sec \alpha_i} \left[\frac{1 + \tan \phi' \tan \alpha_i}{F} \right] \quad (9.77)$$

The Bishop simplified solution assumes that $(X_i + X_{i+1}) = 0$ and thus:

$$F = \frac{1}{\Sigma W_i \sin \alpha_i} \Sigma \frac{1}{m_{\alpha_i}} [c' l_i \cos \alpha_i + (W_i - u_i l_i \cos \alpha_i) \tan \phi'_i] \quad (9.78)$$

Improvements on the accuracy achieved by Fellenius's solution derive from the fact that no assumptions are made with regard to the normal interslice forces E_i and E_{i+1}. Unfortunately, since m_α is a function of the factor of safety F (see Equation (9.77)), Equation (9.78) must be solved by assuming a trial factor of

safety, to input into the right-hand side of the equation, in order to evaluate the left-hand side of the equation. By iteration, the condition $F_{trial} = F_{calculated}$ is satisfied.

Convergence may be achieved rapidly by calculating the initial value of F_{trial} (to input into the right-hand side of the equation) using $m_a = 1$ for every slice. Adequate convergence occurs in hand calculations when F_{trial} and $F_{calculated}$ agree to two decimal places.

It is often convenient to express the pore water pressure at the base of each slice in a dimensionless form, where:

$$r_{u_i} = \frac{u_i}{\gamma \cdot h_i} \quad (9.79)$$

On this basis, Equation (9.8) becomes:

$$F = \frac{1}{\Sigma W_i \sin \alpha_i} \Sigma \frac{1}{m_{\alpha_i}} [c' l_i \cos \alpha_i + W_i (1 - r_{u_i}) \tan \phi'_i] \quad (9.80)$$

Hand calculations are carried out in tabular form as, for example, in Figure 9.28.

After iteration to determine the value of factor of safety for a given slip surface, the procedure must be repeated to locate the slip surface which will give the lowest factor of safety. This is normally achieved by: (1) moving the centre of the circle; or (2) changing the depth of the circle.

For a given depth of circle, contours of factor of safety are obtained as shown in Figure 9.29. The critical shear surface will give the lowest factor of safety.

9.5.1.4 Long-term stability analysis: noncircular slips

Janbu's method[49] Janbu's method is the only hand-calculation method in widespread use for estimating the stability of noncircular slips. The method uses the force equilibrium of the individual slice, coupled with the horizontal force equilibrium of the entire slip mass, to derive a solution. Interslice forces are assumed to be zero.

As with the other methods, the factor of safety is defined in terms of the mobilized shear strength of the soil, i.e.:

$$T_i = \frac{1}{F}(c'_i \cdot l_i + N_i \cdot \tan \phi'_i) = t_i \cdot l_i / F \quad (9.81)$$

for internal force equilibrium within each slice.

The factor of safety by this method is:

$$F = \frac{\Sigma \left(c'_i + (p_i - u_i) \tan \phi'_i \right) b \cdot (sec^2 \alpha_i / (1 + \tan \alpha_i \cdot \tan \phi' / F))}{\Sigma W \tan \alpha_i} \quad (9.82)$$

Once again, hand solution is carried out in tabular form.

Once the solution is achieved (by iteration, as with Bishop's method),[48] it is corrected by a factor introduced by Janbu to take account of the fact that interslice forces are ignored in the derivation (Figure 9.30):

$$F = f_0 \cdot F_{calculated} \quad (9.83)$$

Slice no.	Width b (m)	h (m)	γ (kN/m²)	r_u	c′ (kN/m²)	ϕ' (deg.)	α	W	W sin α
									Σ

c′b W(1 − r_u) tan ϕ'	M_α for $F = F_x$	(1) + (2) ÷ (3)
	Σ	

Figure 9.28 Format for recording hand calculations

where f_0 depends on the strength parameters used in the analysis, and also upon how curved or flat the shear surface is; f_0 varies between 1.0 and about 1.1.

Janbu's method is normally used where site observations and measurement indicate a noncircular failure surface. It is common to use this method when the geometry of the critical slip surface is known. If the position of a critical shear surface is unknown then trial surfaces must be used to determine the minimum factor of safety.

Morgenstern and Price's method.[50] Morgenstern and Price's method takes into account normal and shear forces on vertical planes within a slip mass bounded by a noncircular failure surface. The method appears at first sight to use slices as with the other methods previously discussed; in fact, the 'slices' are used to discretize the slip mass, allowing linear variations of a number of functions to be assumed within each slice.

The method is not suitable for hand calculation, but can be solved on relatively small desktop computers.

The equations of the method derive partly from considerations of equilibrium of an infinitely thin vertical slice. Moments are taken about the centre of the base of this infinitely thin slice, and further equations are developed by resolving normal and parallel to the slip surface. These equations are then used to numerically integrate across the slip mass.

A satisfactory solution is obtained when this integration yields zero lateral force, and zero moment about the slip surface, at the end of the last slice. The method therefore gives both moment and force equilibrium.

When using a computer program to make calculations by this method the engineer must not only define the geometry of the problem, soil parameters, and the desired division of the slip mass into 'slices', he must also determine the shape of the $f(x)$ distribution across the slip, where:

$$f(x) = \frac{1}{\lambda}\frac{X}{E} \qquad (9.84)$$

on any vertical plane within the slip mass. The program defines the available shear strength in terms of the factor of safety F and in common with other methods assumes that F is constant across the slip. The condition that zero thrust and moment should be obtained from integration across the slip mass is obtained using the Newton–Raphson method to converge on the values of F and λ.

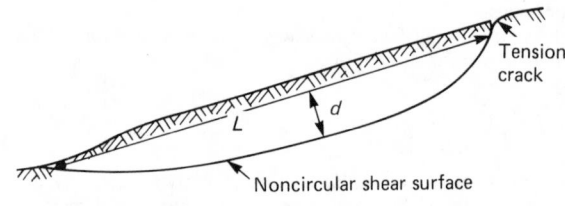

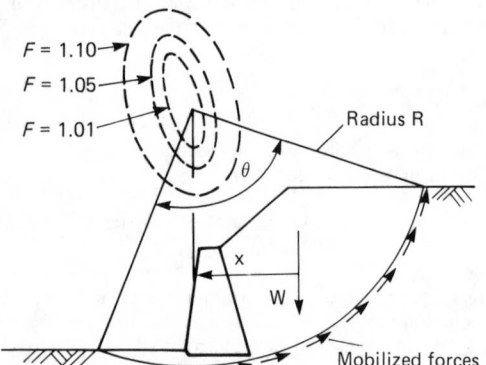

Figure 9.29 Contours of factor of safety. (After Clayton and Mililitsky (1986) *Earth pressure and earth retaining structures.* Surrey University Press, Glasgow)

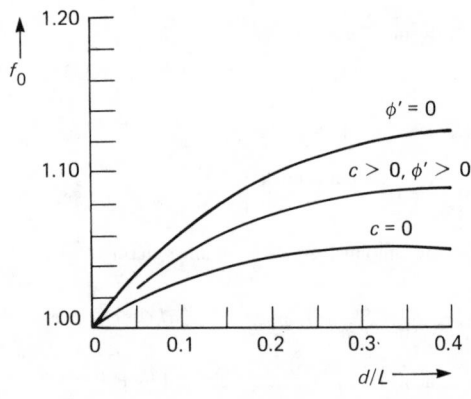

Figure 9.30 Correction factor f_0 for Janbu's slope stability method. (After Clayton and Mililitsky (1986) *Earth pressure and earth retaining structures.* Surrey University Press, Glasgow)

It is common practice to start the analysis by assuming a uniform distribution of $f(x)$ (say $f(x) = 1.0$) across the slip. Some authors have indicated that variations in the shape of the $f(x)$ distribution have little effect on the factor of safety determined by the analysis, but other experience suggests that this depends upon the geometry of the problem and the parameters used. In some cases the $f(x)$ distribution can affect the result significantly, and therefore different distributions should be used. These may take the form of either: (1) a half-sine curve; or (2) a distribution where $f(x)$ is proportional to the curvature of the slip surface.

The output from such a computer analysis should not be accepted without a check that:

(1) The shear stress on any vertical surface within the slip mass does not exceed the available shear strength.
(2) No tension is implied on any vertical surface within the slip mass. For this to be true the resultant force on any section should pass within the bounds of the ground surface and the slip surface.

It is therefore necessary to obtain values for the vertical shear force and position of the resultant thrust across the section in order to make these checks.

9.5.1.5 Flow slides

Not all slides are deep-seated shear slides. (For a classification of landslides see Skempton and Hutchinson.[53]) A fairly common type which should be mentioned is the flow slide. Flow slides generally take place in saturated masses of loose, fairly impervious soils such as fine sands, silts or silty clays. They can occur on quite flat slopes and can travel long distances, the soil and water flowing as a liquid mass. They are caused by a sudden reduction of the shear strength to zero, or close to it, by the transfer of all the pressure to the water in the voids, a process known as liquefaction. The chief factor here is the relative density of the soil, loose soils having a relative density of less than a critical value below which volume reduction occurs on shearing contrary to dense soils which tend to dilate, thus throwing a tension on to the pore water. Casagrande considers in Green and Ferguson[54] that sands having a relative density greater than 50% would be safe against liquefaction. Casagrande and others have drawn attention to a more general phenomenon called 'fluidization' where the fluid phase is mainly air. Such phenomena are not necessarily confined to fine grained soils. They can occur under a variety of circumstances, the chief of which is a sudden disturbance of a loose solid by a heavy shock or vibration. Drainage and compaction are the two main remedial and preventative measures.

In some clays, the existing undisturbed strength is much greater than the strength after remoulding. The ratio of these two strengths is called the sensitivity of the clay. In general, the undisturbed strength of soft clays is measured *in situ* by means of a vane test. Much work on this problem has been done in Norway and Sweden and is described by Skempton and Northey.[55] In England, sensitivities are usually below 10 and in these cases an analysis can be made using the undrained shear strength and the $\phi_u = 0$ method.

With clays of high sensitivity, i.e. over 20 and up to 100 such as occur in southern Norway and Sweden, experience shows that an analysis based on the $\phi_u = 0$ case and the undisturbed strength of the clay measured by a vane test, gives results which do not necessarily agree with practice (Bjerrum[56]). When failure takes place in a clay of high sensitivity, the result is in many respects similar to a flow slide described in the paragraph above, the clay becoming practically fluid and flowing through quite small apertures down flat valleys or hillsides for a long distance.

The formation of these very sensitive clays is believed to be due to the clay being laid down in salt water, then being raised above water level by isostatic readjustment, the salt later being leached out by the percolation of fresh water; thus, the liquid limit of the clay is reduced but the moisture content remains high. On disturbance, therefore, the moisture content greatly exceeds the liquid limit, and the clay acts as a heavy fluid. A description of the leaching process is given by Bjerrum.[57]

9.5.1.6 Protective and remedial measures

Certain protective and remedial measures can be taken to prevent a slip or to stabilize a slip which has occurred. Of these the most important is drainage. The majority of troubles are due to water; well-designed and adequately maintained drainage can prevent the ingress of water to a bank and so stop, or considerably delay, any softening which may take place and prevent the build-up of high pore pressures.

On sidelong ground, a drain should be installed at the top of the slope to catch surface water. Water from the slope itself should be caught in a toe drain and led away from the toe. On a long slope it may be necessary to catch surface water in herringbone drains or even to lead it to a longitudinal drain on a berm halfway up the slope. It is important to line the inverts of these drains.

A considerable degree of protection can be obtained by grassing the slope and the level area at the top where shrinkage cracks are likely to appear. Cracks once formed can become filled with water, the pressure of which exerts a considerable disturbing force on the bank. A good carpet of grass prevents not only drying and cracking but also surface erosion. Bushes and trees with strong root systems can also be of help, but trees which grow rapidly and absorb large amounts of water from the soil (e.g. poplars and elms) should be avoided.

Deeper drainage can be achieved by drilling horizontally into the slope—or in some cases adits can be used, either alone or in conjunction with vertical drainage holes. Counterforts are also used to improve drainage. These consist of trenches excavated back into the slope, backfilled with free-draining granular material.

In addition to drainage methods, the stability of deep-seated slides can be improved either by reducing the disturbing forces or by increasing the resisting forces by one or other of the following methods:

(1) Reducing the disturbing forces by loading the toe, or removing material from the top, or replacing material at the top by a lighter material, or altering the bank profile, e.g. introducing berms.
(2) Increasing the resisting forces by increasing the shear strength, by drying or adding frictional counterforts or keys through the slip surface.
(3) It may in some instances be possible to improve stability by introducing 'rigid' elements, such as sheet-piling driven through the toe, with or without ground anchors. However, great caution is required, since the rigid element will attract load and very high bending moments will develop.

9.6 Seepage and flow nets

A flow net is a graphical representation of the pattern of the seepage or flow of water through a permeable soil. It is possible, by means of a flow net, to calculate the hydrostatic uplift on a structure such as a dam or barrage, the amount of seepage

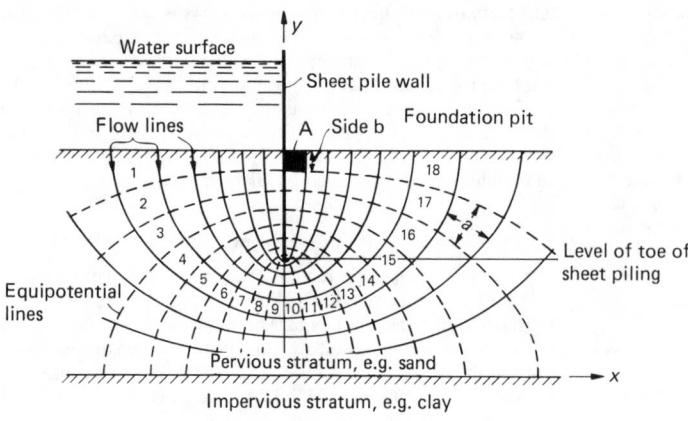

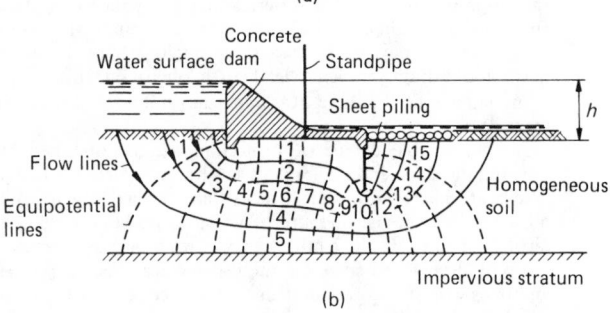

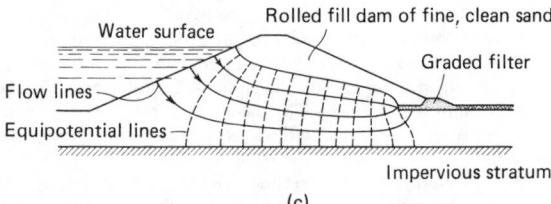

Figure 9.31 Examples of flow nets for simple cases. (a) Beneath sheet pile wall; (b) beneath concrete dam on sand with sheet pile cutoff wall; (c) through rolled fill dam with toe drain

through an earth dam or under a barrage, or estimate the probability of piping occurring in a cofferdam, for example.

A flow line is the path followed by a particle of water flowing through a soil mass. Flow lines are always smooth, even, curves as shown in Figure 9.31. An equipotential line is a line joining points at which the hydraulic head is equal; therefore, if standpipes are inserted into any two points on an equipotential line, the water will rise to the same level in each standpipe. Flow lines and equipotential lines are always at right angles to each other. For a discussion on the theory of flow nets, see Cedergren.[58]

9.6.1 Construction of flow nets

Four methods of constructing flow nets are in general use.

(1) *Mathematical.* For simple boundary conditions, the governing differential equation (Laplace) can be solved mathematically. Many computer-finite element packages now contain seepage programs which can be used to solve the more complex steady-state seepage problems, e.g. with layered soils and complex boundary conditions.

(2) *Electrical analogy.* The differential equation for flow nets is the same as that for flow of electricity, and flow nets can be drawn by using an electrical model and tracing lines of equal potential with a wandering probe. The soil is represented by a suitably shaped conducting paper, strips of copper represent water surfaces and the edges of the card represent an impervious surface. Once the equipotential lines are drawn, the flow lines can easily be drawn at right angles to them.

This method assumes that the permeability in both directions, horizontally and vertically, is the same. In the electrical resistance network method a scaled network of resistances of values proportional to the permeability is set up, the electrical potential at each node being a direct measure of the hydraulic potential at that point.

(3) *Hydraulic models.* An obvious approach is to construct a model of the problem in sand behind glass, to allow water to flow through it and to trace the flow lines by inserting a small drop of dye as it flows through the soil. This trace is then drawn on the glass with a wax pencil and the procedure repeated from a different point.

(4) *Graphical method.* After a little practice, it is quite possible to sketch a flow net for many problems, which is quite accurate enough for most practical purposes. The cross-section is drawn and the boundary conditions clearly marked. The flow net is then tentatively sketched in, bearing in mind that flow lines and equipotential lines are at right angles to each other, that flow lines always start at right angles to a free water surface and equipotential lines start or finish at right angles to an impervious surface. The number of flow and equipotential lines is chosen to divide the seepage area into shapes which are approximately square and which are bounded by two flow lines and two equipotential lines.

9.6.2 Examples of hydraulic problems by flow nets

9.6.2.1 Uplift pressure

In Figure 9.31(b), let the number of squares along a flow line be $n(=15)$ and the number along an equipotential line be $f(=5)$. Then if the total drop in head is h, the drop in head across each square is h/n, and at an imaginary standpipe through the concrete at the sixth equipotential line the loss in head will be:

$$6 \times h/n = 6h/15 = 0.4h \tag{9.85}$$

The uplift pressure at this point will be the remaining head times the density of water, i.e.:

$$(h - 0.4h)\gamma_w = 0.6h\gamma_w \tag{9.86}$$

Note that this result is independent of the number of squares in the flow net, since if $n = 30$ instead of 15, the borehole in the position shown would be on the twelfth equipotential and loss in head would be:

$$12h/30 = 0.4h$$

9.6.2.2 Hydraulic gradient and D'Arcy's law

The hydraulic gradient is defined as $\Delta h/\Delta l$ (i.e. the ratio of head loss to distance) and is related to the velocity of flow v and the permeability k by D'Arcy's law:

$$v = ki \tag{9.87}$$

9.6.2.3 Amount of seepage

The quantity of water Q flowing under the dam in Figure 9.31(b) is given by $Q = Atki$, where A is the area of flow, t is time, k is the coefficient of permeability and i is the hydraulic gradient. The hydraulic gradient i across any square of the flow net of side b is given by $i = h/nb$. The flow in unit time through the square is $Q = bkh/nb = kh/n$.

If the number of flow channels is f, the total flow is $Q = fkh/n$, and is independent of the size of the squares.

9.6.2.4 Factor of safety against piping

The factor of safety against piping is the ratio of the critical hydraulic gradient to the existing hydraulic gradient at exit.

In Figure 9.31(a) piping will occur at A when the upward force of the water issuing at A is greater than the effective weight of the particles.

The seepage force on the base of the last square is $\gamma_w ib^2 =$ effective weight of soil $= b^2 \gamma_w (G_s - 1)/(1 + e)$. Therefore, piping occurs when $i = (G_s - 1)/(1 + e)$.

For sand, G_s is about 2.7, and in the loose state e is about 0.7, giving:

$$i = (2.7 - 1)/(1 + 0.7) = 1 \tag{9.88}$$

i.e. the critical hydraulic gradient is about unity.

The exit hydraulic gradient at A in Figure 9.31(a) is $(h/n)/b$. Therefore, the factor of safety against piping is:

$$F = 1/(h/nb) = nb/h \tag{9.89}$$

Note that nb is independent of the number of squares.

It is generally considered that the value of the factor of safety against piping should be 4 or greater.

9.7 Definitions of terms used in soil mechanics

All soils consist of solid particles assembled in a relatively open packing. The voids may be filled completely with water (fully saturated soils) or partly with water and partly with air (partially saturated soils).

The relationships between void space and the volume occupied by the particles are fundamental and are characterized by the following definitions.

Porosity $n =$ volume of voids V_v/total volume of soil V_t.
Voids ratio $e = V_v$/volume of soil particles V_s.

Hence $\quad e = n/(1 - n) \quad$ and $\quad n = e/(1 + e) \tag{9.90}$

Degree of saturation $S_r =$ volume of water/V_v.
Water content $w =$ weight of water/weight of soil particles.

Hence, if G_s is the specific gravity of soil particles:

$$w = S_r e/G_s$$

For fully saturated soils, $S_r = 1$ and $w = e/G_s$.
Bulk density $\gamma =$ total weight of soil and water:unit volume W_t/V_t.

Hence $\quad \gamma = (G_s + S_r e)\gamma_w/(1 + e) \tag{9.91}$

where γ_w is the density of water (1 Mg/m³).

When $\quad S_r = 1, \ \gamma = (G_s + e)\gamma_w/(1 + e) \tag{9.92}$

dry density $\quad \gamma_d = W_s/V_t$.

Hence $\quad \gamma_d = G_s\gamma_w/(1 + e) \tag{9.93}$

and $\quad \gamma = \gamma_d(1 + w) \tag{9.94}$

Also $\quad n = 1 - \gamma_d/G_s\gamma_w \tag{9.95}$

and $\quad S_r = w/(\gamma_w/\gamma_d - 1/G_s) \tag{9.96}$

The submerged density, of soils below water table, is given by:

$$\gamma_s = (G_s - 1)\gamma_w/(1 + e)$$

$$= (G_s - 1)\gamma/G_s(1 + w) \tag{9.97}$$

Percentage air voids $\quad A =$ volume of air $\times 100/V_t$

$$A = 1 - \gamma_d(1 + wG_s)/\gamma_w G_s \tag{9.98}$$

All the foregoing definitions and relationships are in constant use in soil mechanics problems.

Appendix 9.1 Laboratory testing of soils

This appendix gives a brief outline of some of the main laboratory tests required to classify individual soils and to indicate their compaction and strength characteristics.

In order to obtain reliable results, it is essential to follow the recommendations in Chapter 11 with regard to sampling and then to follow closely the practices recommended in the appropriate standards for sample preparation, testing and reporting.

The outline in this section is based on British Standard BS 1377: 1975[59] but other national standards may be relevant for work in some countries, such as, for example, in:

(1) *Australia*:   AS 1289 – Methods of testing soils for engineering purposes.
(2) *France*:   Norme X31 – Qualité des sols.
(3) *West Germany*:   A series of DINs (Deutsche Normen), e.g. 18121–18127, 18130–18137 and 18196.
(4) *USA*:   A series of ASTM Standards.

Other standards may be applicable elsewhere in the world, and the reader should refer to the appropriate national standards office for further information. The reader should also ensure that the standard to which he works is the latest version. For example, it is known that BS 1377 is being revised and that a new version will be published after 1989.

A9.1.1 Soil classification: physical

A9.1.1.1 Liquid limit (LL): cohesive soils

The liquid limit is the moisture content at which a soil passes from the plastic to the liquid state.

The preferred method of determining the LL is now by use of a standard cone penetrometer, under a load of 80 g, on to prepared samples 55 mm diameter and 40 mm deep. A series of at least four tests on the sample at increasing moisture contents allows a linear plot between penetration and moisture content, from which the LL is interpolated as the moisture content corresponding to 20 mm penetration.

The alternative traditional, but less reliable, method is to use the Casagrande apparatus, in which the prepared soil sample is placed in a cup, grooved with a standard tool and then lifted mechanically and dropped on to a standard rubber block at a rate of 2 blows/s, until the two parts of the soil come together for a distance of 13 mm along the groove. The results of a series of at least four tests on the sample, at increasing moisture content, are then plotted with moisture content on a linear scale and number of blows on a logarithmic scale. The LL is interpolated as the moisture content corresponding to 25 blows.

A9.1.1.2 Plastic limit (PL): cohesive soils

The plastic limit is the moisture content at which a soil becomes too dry to be in a plastic condition, as determined by the PL test. The 20-g sample of soil, with moisture content sufficient for it to be moulded into a ball between the palms of hands, is rolled between fingers and palm until slight cracks appear on its surface. The sample is divided equally and each subsample divided into four for tests that each comprise forming a thread 6 mm diameter by rolling between first finger and thumbs, and rolling the thread on a glass plate, using fingertips with a uniform pressure, until the thread reduces to 3 mm or the number of rolling passes exceeds 10. Each subsample should be remoulded by finger and thumb, to reduce moisture, and the thread rolled on glass until a reduction to 3 mm is achieved within ten passes and, simultaneously, the thread shears transversely and longitudinally. The average of the moisture contents at which crumbling (shearing) occurs is the PL of the soil.

A9.1.1.3 Plasticity Index (PI): cohesive soils

The Plasticity Index of a soil is the numerical difference between the liquid and plastic limits of the soil, i.e. PI = LL − PL. Where PI is zero (i.e. LL = PL) or when the soil is insufficiently cohesive for a PL to be measured, the soil is termed nonplastic (NP).

A9.1.1.4 Classification: cohesive soils

The Casagrande classification of cohesive soils is illustrated in Figure A9.1.1, in which PI is plotted against LL. Above the A line are CL, CI and CH (low, intermediate and highly plastic clays) and below the line are ML, MI and MH (low, intermediate and highly plastic silts) or OL, OI and OH (low, intermediate and highly plastic organic clays). The Casagrande classification provides the engineer with a good indication of the characteristics of a soil both for design and construction purposes, but an extended classification system is given in Chapter 11, section 11.5.

A9.1.1.5 Particle size distribution: granular soils

In the majority of cases, wet sieving is required first to remove and record the loss of silt and clay-size particles. The BS sieves used in the test range from 75 mm to 63 µm and the results are recorded as the dry weight expressed as the percentage, by weight, of the total sample passing each sieve. Grading curves in the form shown in Figure A9.1.2, give cumulative percentages passing 12 or so of the BS test sieves between 75 mm and 63 µm. The procedure for washing and sieving and the use of sodium hexametaphosphate solution to help remove and to break down fine particles should be followed very closely.

From the particle size distribution curve the soil will be classified as gravel (G), sand (S), or silt (M) or combinations of these, and the shape of the curve indicates whether it is well graded (e.g. GW), poorly graded (e.g. GP) or silty (GM). Further information on soil classification is given in Chapter 11, section 11.5.

A9.1.1.6 Particle size distribution: fine grained soils

The particle size distribution of fine grained soils, e.g. silts and clays or fractions of these in coarser grained soils, is determined by sedimentation tests in which the material is brought into suspension in a solution of sodium hexametaphosphate and allowed to settle. The mass of solids in a given volume of solution is measured at a specific point, either by sampling (using a pipette or by the use of an hydrometer) at specific time intervals corresponding roughly to the equivalent particle diameters (e.g. 0.02 mm, 0.006 mm and 0.002 mm) for which information is required.

The basis of this test is Stokes's law, which gives the relationship for a spherical particle falling through a column of liquid:

$$v = \frac{d}{t} = \frac{2(\gamma_s - \gamma_1)\, gr^2}{9\eta}$$

where v is the velocity of the falling particle, d is the distance through which it falls in time t, γ_s is the density of the particle, γ_1 is the density of the liquid, r is the radius of the particle and η is the viscosity of the liquid. A monographic chart in BS 1377[59] facilitates the calculation of the equivalent particle diameter.

A9.1.1.7 The specific gravity of soil particles (G_s)

It is only rarely necessary to know the G_s of a soil for classification purposes, but values of G_s are required in the calculations and interpretation of some other test results and for design and construction purposes.

In essence, the determination of G_s is by measurement of the dry weight of a sample of soil, the weight of the same sample plus water required to fill a container and then the weight of water alone to fill the same container. For coarse grained soils, a gas jar of 1 litre is a suitable container but, for soil particles finer than about 2 mm, standard 50 ml density bottles (pycnometers)

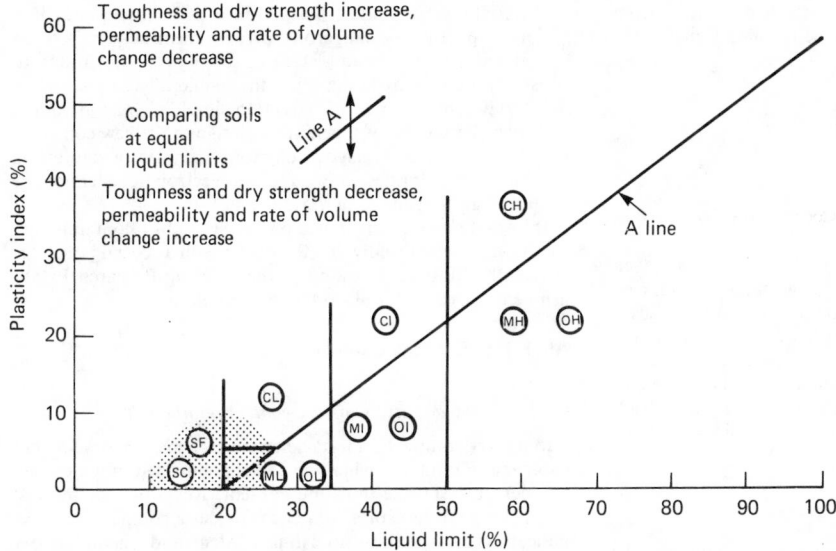

Figure A9.1.1 Plasticity chart used in the Casagrande soil classification. (After *Soil mechanics for road engineers* (1968). HMSO, London)

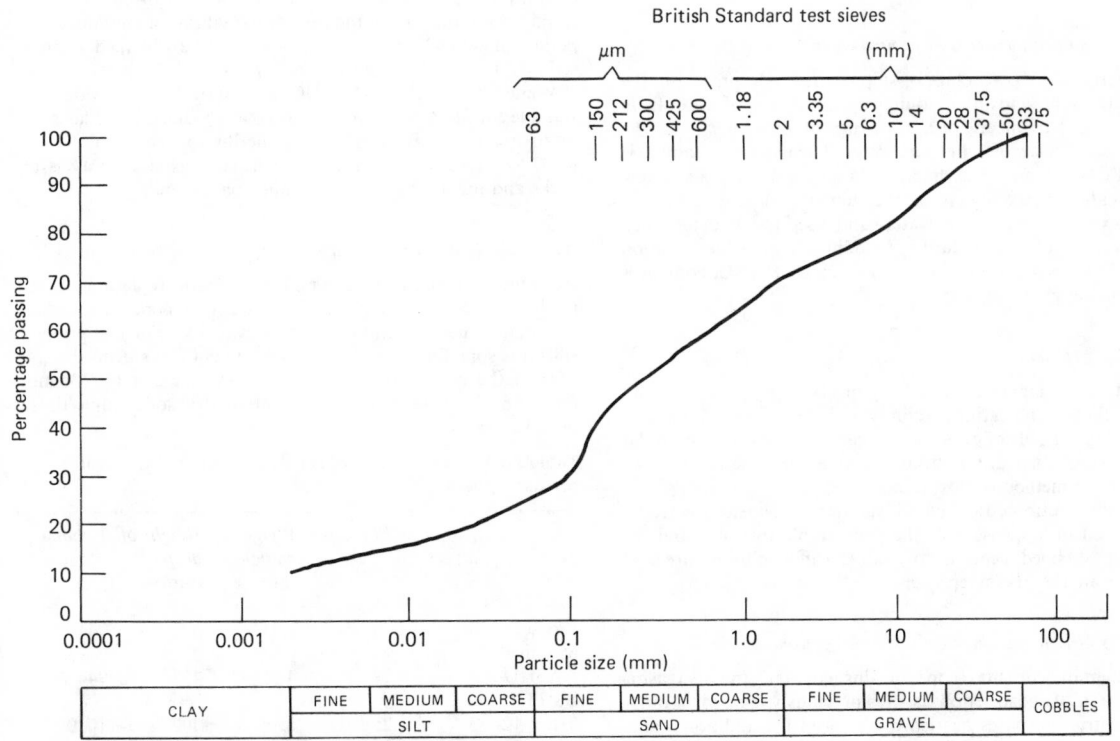

Figure A9.1.2 Particle size distribution chart. (After British Standards Institution (1975) *Methods of test for soils for civil engineering purposes.* BS 1377:1975. BSI, Milton Keynes)

are often used. In either case, care is required to exclude air from the samples, involving the use of air-free distilled water in the tests on fine grained soils.

A9.1.2 Soil classification: chemical

A9.1.2.1 Organic matter content

Organic content is expressed as the percentage by mass of organic matter present in the soil.

The organic matter in a sample of soil dried in an oven between 105 and 110° C is oxidized in a 500 ml conical flask by the addition of 10 ml potassium dichromate solution and 20 ml concentrated sulphuric acid. The sample of soil should weigh between 0.2 g for peaty soil and up to 5 g for a soil with low organic content. Processes of titration determine the volume of potassium dichromate used to oxidize the organic matter and, hence, the organic content. Soils containing sulphides or chlorides require some special treatment.

A9.1.2.2 Total sulphate content of soil

This is determined as the percentage mass of sulphate (as SO_3) present in the soil sample, represented by the material passing, or which can be crushed to pass, the 2 mm sieve. The sample for test is then pulverized until it passes the 425 µm sieve. Stones, other than gypsum, may be assumed to contain no sulphate.

The test procedure involves treatment of the sample with hydrochloric acid and ammonia to form an acid extract, which is boiled and barium chloride solution added, to form a precipitate of barium sulphate. The mass of the precipitate formed is determined by filtration and ignition, and gives a measure of the SO_3 present in the sample.

A9.1.2.3 Sulphate content of groundwater

To determine the percentage of sulphate (as SO_3) present in the soil water or groundwater requires the extraction of water from a soil sample by means of a centrifuge or other means or the collection of a sample of groundwater. The extract or sample is passed through an ion exchange column comprising a strongly acidic cationic exchange resin; the sulphate content of the soil water extract or the groundwater can be separately determined by titration of a standardized sodium hydroxide solution against the resultant solution. In both cases, the SO_3 content is expressed in grams per litre.

A9.1.2.4 pH value

The pH value indicates the acidity or alkalinity of a soil, with values above 7 indicating alkalinity and values below 7 increasing acidity. The pH of groundwater can be determined in similar ways, using either an electrometric or a calorimetric method. The former method is most usual.

For the electrometric method, standard pH meter electrodes are placed in suspension of the soil sample in water and the readings obtained record the pH value. Buffer solutions are used to calibrate the pH meter.

A9.1.2.5 Additional chemical tests for contaminants

Although outside the scope of this chapter, the increasing proportion of construction on sites previously used for a range of industrial purposes has led to the need for additional and, often, more extensive soil testing. The problems involved are referred to in Chapter 11 and attention is drawn to the bibliography in that chapter relating to contaminated sites.

On contaminated sites, the presence of methane and of a variety of potentially dangerous or toxic chemicals needs to be considered from the point of view of safety of workmen during construction. Corrosive materials that could affect the durability of materials in a new construction should also be identified and special precautions taken in the design of the new construction to limit such damage. Common examples of aggressive materials are chlorides, sulphates and electrolytic, chemical or bacteriological agencies of various types.

Information and experience on the subject of contaminated sites is increasing rapidly and the reader with a special interest is advised to consult, for example, the Building Research Establishment for the most up-to-date information.

A9.1.3 Soil compaction

A9.1.3.1 Dry density: moisture content relationships

Laboratory compaction tests determine the mass of dry soil per cubic metre obtained when the soil is compacted in a specified manner at a specific moisture content. Repetition of the test over a range of moisture contents provides a compaction curve indicating the optimum moisture content and maximum dry density obtainable for the compactive effort applied.

The original 'Proctor' test simulated the compactive effort of construction plant in the 1930s. Later, the modified American Society of State Highway Officials (AASHO) test was developed to model heavier construction plant and, in the UK, a vibratory test was introduced in 1967 to simulate the effect of heavy compaction by vibrating rollers and plates. A comparison of the British and American tests is shown in Table A9.1.1.

In each test, soil of predetermined moisture content is placed in a specified number of layers into a cylindrical mould, and each layer is compacted by a specified number of blows with a standard rammer or, in the case of the vibration method, for a period of 60 s. A typical compaction curve obtained from a series of tests is shown in Figure A9.1.3.

While the modified AASHO test or its BS equivalent is suitable for fine grained as well as coarse grained granular soils up to 20 mm, the BS vibratory method is applicable to soils up to 37.5 mm and is preferred for soils such as clean gravels or rocks and for uniformly graded and coarse sands.

A9.1.3.2 Measurement of dry density

Reference is made in Chapter 11 of the sand replacement test method for the measurement of dry density of compacted soil *in situ*, and to nuclear and other tests available. For fine grained soils it is sometimes more convenient to cut cores, trim the soil core to the cutter size (usually 100 mm diameter by 130 mm long) and to determine the weight of dry soil within those dimensions.

Table A9.1.1. Comparison of standard American and British compaction tests

Test	No. of layers	Blows per layer	Weight of hammer (kg)	Height of drop (mm)	Volume of mould (cc)
Proctor	3	25	2.5	305	944
Modified AASHO	5	27	4.55	457	944
BS 1377					
test 12	3	25	2.5	300	1000
test 13	5	27	4.5	450	1000
test 14 (vibratory)	3	60 s with 300–400 N down force			c.2300

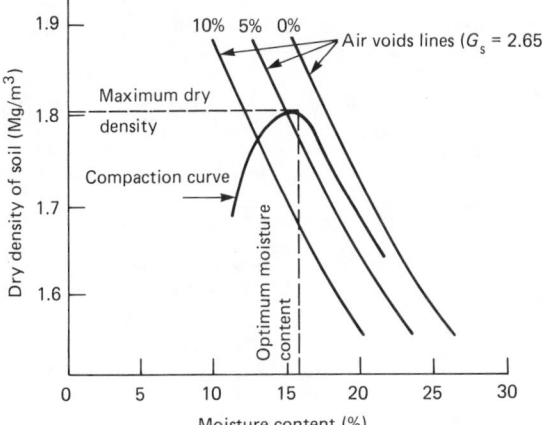

Figure A9.1.3 Typical compaction curve and the terms used. (After British Standards Institution (1975) *Methods of test for soils for civil engineering purposes*. BS 1377:1975. BSI, Milton Keynes)

An alternative laboratory method appropriate to lumps of soil, preferably approximately cubical or cylindrical, involves coating them with paraffin wax and subsequent determination of the soil volume by weighing the sample immersed in water or by a water displacement method.

A9.1.4 Strength tests

A9.1.4.1 Californian bearing ratio (CBR)

The Californian bearing ratio (CBR) test was developed in 1938 to evaluate Californian highway subgrade strengths and became the basis for the design of road and airfield pavements throughout the world. It is used both *in situ* and on prepared samples in the laboratory, but is limited to materials of particle sizes up to a maximum of 20 mm.

The test determines the relationship between force and penetration when a cylindrical plunger 1935 mm² in cross-section is pressed into soil at a given rate of 1 mm/min. For any given penetration, the ratio is expressed as a percentage of a standard force derived for crushed stone.

For CBR tests in the laboratory, the soil specimen is prepared at a predetermined moisture content and is compacted into a cylindrical mould 152 mm in diameter and 127 mm high either by continuous tamping, compression in three equal layers or dynamic compaction in layers, using either the rammers or the vibrating hammer used in the compaction tests. In all cases, the mass of soil poured into the mould is calculated as that required to provide the chosen dry density or air voids percentage on completion of compaction. These usually correspond to the optimum, determined from compaction tests, or are the values measured on the soil *in situ*.

Results of the test are plotted as shown in Figure A9.1.4. The CBR is calculated at penetrations of 2.5 and 5 mm, and the higher value taken. The standard forces corresponding to 100% CBR at these two penetrations are 13.24 and 19.96 kN respectively. The force–penetration curve is normally convex upwards, but curves beginning as concave upwards require correction by shifting the zero on the penetration axis.

A9.1.4.2 Other plate bearing tests

A variety of plate bearing tests has been used in site investigations and are mentioned in Chapter 11. However, for airfield

pavements and use in Westergaard's analysis of strains and deflections in concrete slabs, a test developed in the US[60] is used to measure the 'modulus of subgrade reaction'. The plate is usually 750 mm in diameter and a linear plot of settlement against pressure applied gives a curve which is convex upwards. The modulus of subgrade reaction k is calculated as:

$$k = p/1.27 \text{ g/mm}^2/\text{mm}$$

where p is the pressure required to cause settlement of 1.27 mm.

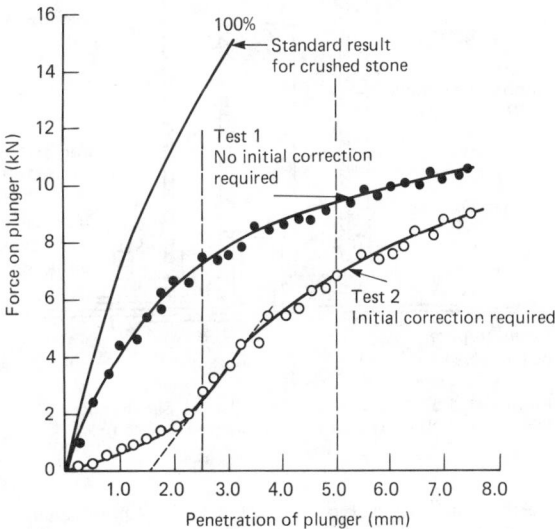

Figure A9.1.4 Typical California bearing ratio test results. (After British Standards Institution (1975) *Methods of test for soils for civil engineering purposes*. BS 1377:1975. BSI, Milton Keynes)

A9.1.4.3 Determination of shear strength

The vane test For *in situ* measurement of shear strength, a vane of cruciform section 50 or 75 mm in diameter and of length equal to twice the diameter is lowered into a borehole and pushed, without twisting, to a depth not less than 3 times the borehole diameter into the undisturbed soil. A torque head fitted to the top of the vane rods turns the vane at a rate between 10 and 20°/s until the soil is sheared.

For vanes with height twice the diameter, the vane shear strength S is given by:

$$S = \frac{M \times 10^6}{3.66D^3} \text{ kN/m}^2$$

where M is the torque to shear the soil in newton metres and D is the measured width of the vane in millimetres.

Direct shear test In this test, a soil sample is cut and trimmed carefully to fit closely into a metal box, either circular or square in plan, which is constructed to allow displacement along its horizontal midplane. The upper surface of the sample is confined by a normal load and a shear load is applied to the lower half of the box until the soil shears across the midplane.

The triaxial test The triaxial cell (Figure A9.1.5) provides the means of applying horizontal pressures to a cylindrical specimen by means of lateral hydraulic pressure in the cell chamber simultaneously with a vertical load applied by a ram. The

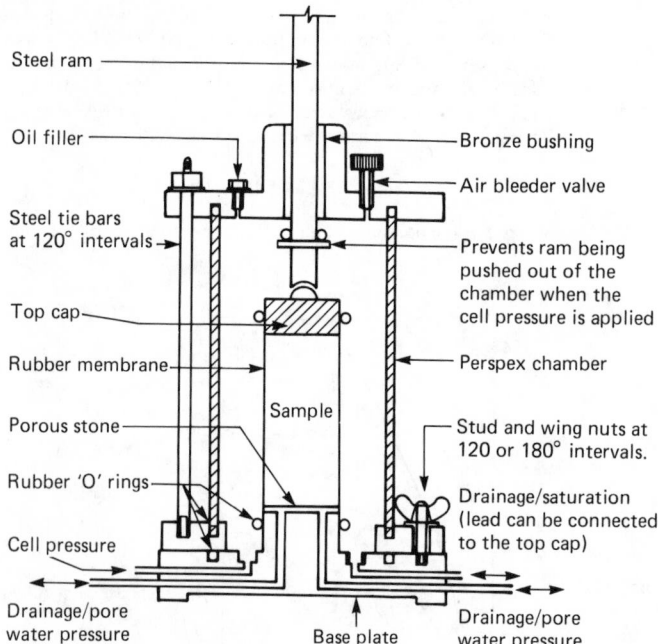

Figure A9.1.5 A triaxial cell. (After Clayton, Simons and Matthews (1982) *Site investigation*. Granada, London)

specimen in various types of test may be consolidated or unconsolidated, drained or undrained, and there is a facility to measure pore pressure.

The unconsolidated undrained compressive test is used to determine the undrained shear strength of undisturbed samples. At least three samples are tested over a range of cell pressures. At each pressure the ram force is increased and the vertical deformation recorded, until the maximum value of the stress had been passed or an axial strain of 20% reached. The principal stress difference (deviator stress $\sigma_1 - \sigma_3$) is calculated as the ram force divided by the cross-sectional area of the specimen. However, because the specimen diameter increases during the test, a corrected value of cross-sectional area is used in the calculation, such that:

$$A = A_0/(I - \varepsilon_a)$$

where A is the cross-sectional area of the specimen at a given time, ε_a is the measured axial strain at that time and A_0 is the initial cross-sectional area of the specimen.

The results of the test are plotted as curves of principal stress difference against strain. For conditions of maximum stress (i.e. failure) Mohr circles are plotted to give values for c_u, the apparent cohesion, and ϕ_u, the angle of shearing resistance. This is discussed further in section 9.1.2 and Mohr circles are shown in Figure 9.1.

A9.1.4.4 Penetration resistance N using the split barrel sampler

The penetration resistance N of a soil, relevant particularly to piling, is measured on site as the number of blows required to drive a standard sampler, fitted with a driving shoe of 35 mm internal, and 50 mm external, diameter, specific distances into a soil surface at the bottom of a borehole. The sampler is driven by a 65 kg hammer falling freely through a height of 760 mm.

The number of blows to penetrate 300 mm is termed the penetration resistance N. For gravelly soils, the driving shoe can be replaced by a 60° solid cone (BS 1377: 1975).

A9.1.5 Consolidation tests

A9.1.5.1 The oedometer test

One-dimensional consolidation properties can be determined as the magnitude and rate of consolidation of a disc of saturated soil confined laterally and subject to vertical pressure. The equipment consists of a ring, usually 76 mm in diameter and 19 mm high, to restrain the carefully cut soil disc, and a loading cell capable of being filled with water into which the ring is placed between porous plates. The loading device must be capable of maintaining constant load, giving pressures from the range 10, 20, 50, 100, 200, 400, 800, 1600 and 3200 kN/m². The compression movement is measured as the relative movement between the base of the cell and the loading cap.

Immediately after application of the initial load, water is poured into the cell and, if this causes swelling, the load is increased to the next stage until there is no swelling. The load is maintained for a period of up to 24 h and the results plotted as the compression movement against √time, and also as the movement against log–time. Methods are given in BS 1377[59] for calculating the consolidation coefficient from these graphs as c_v in square metres per year. Further information on the calculation of consolidation settlement is given in section 9.3.3 (page **9**/9).

A9.1.5.2 Triaxial dissipation test

An alternative to the use of an oedometer is a triaxial test, in which volume change is plotted against log–time and, in addition, pore pressure is measured at the base of the specimen. The compressibility measured as a result of triaxial dissipation is greater than that determined in the oedometer test.

Appendix 9.2 Pile capacities

Piles are used to transfer foundation loads to a deeper stratum when the surface soils are too weak or too compressible to carry the load without excessive settlement. Details of pile types and their design and use are given in Chapter 17. The reader's attention is drawn to the references in Chapter 17 for further information, particularly to BS 8004[13], Tomlinson[61,62] and to series of CIRIA/PSA piling guides.[63]

In this appendix, methods are given for estimating the carrying capacity of piles in various types of soil conditions.

A9.2.1 Groups of piles

A piled foundation generally consists of a group of several piles, the behaviour of which should be considered as an entity.

The piled group will consist of either point-bearing piles which transfer their load to a hard stratum of soil on which their points bear (e.g. piles to rock) or friction piles which transfer their load mainly by friction on the sides of the piles to a firm stratum into which they penetrate. Friction piles into a firm clay stratum will usually penetrate about 20 to 30 times the pile diameter into the clay. Piles driven through soft material into compact sand or gravel will usually penetrate about 5 times the diameter and will be partly point-bearing and partly frictional.

When friction piles are used in conditions in which the increase in strength of the soil with depth is only gradual, they must be of a length comparable to the size of the building to be of much advantage. This is shown in Figure A9.2.1, in which the stress distribution with and without piles is shown for two buildings of different widths but with piles of the same length. Unless the use of piles changes the stress pattern radically their use is probably not economic.

The design of a foundation on friction piles should always be checked from the point of view of overall stability, and assuming that the whole of the load is distributed uniformly over the area of the building and acts as a block foundation with its base at the foot of the piles. The friction round the circumference of the block of soil containing the piles should be subtracted from the foundation load.

If the foundation on friction piles is supported by a bed of clay, consolidation settlements will occur which can be estimated as described for deep foundations above.

A9.2.1.1 Single piles

Although the carrying capacity of a group of piles is not simply that of a single pile times the number of piles in the group, it is useful to know the load which a single pile will carry. Until the practice developed to treat pile groups as an equivalent deep raft, the capacities of groups were based on the sum of individual capacities with empirical efficiency ratios ranging from 1.0 for groups in sand down to 0.7 for widely spaced piles (i.e. 3 × diameter) in clay.

A9.2.1.2 Pile bearing on rock

Where the bedrock is massive and strong, bearing capacity is usually not a problem. However, if the rock surface is steeply sloping, it may be necessary to provide a driven pile with a special point to make sure that it is adequately toed-in. Bored piles into strong rock will only need a small penetration, say half to one pile diameter, in order to develop a working load equal to the maximum allowable working stress in the concrete – which is taken as 5000 kN/m². With small penetrations, particular care is needed to get a good contact between the pile toe and the rock.

With weaker and fractured rocks, it is still generally possible to develop the full allowable working stress in the concrete, but this will require penetration in order to carry some of the load in shear in the 'rock socket'. The required depth of penetration can be determined from a knowledge of the unconfined compression strength and fracture, or joint, spacing of the rock mass.[62]

A9.2.1.3 Piles bearing in deep deposits of sand and gravel

The ultimate point bearing capacity of a pile of end area A_B in sand may be estimated directly from the cone resistance C_r of the Dutch deep sounding test, thus:

$$Q_{uB} = A_B C_r \qquad (A9.2.1)$$

When using this method, however, it is necessary to take due account of the difference in scale between the cone (end area

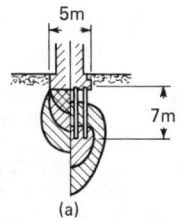

(a)

Vertical unit pressure in
% of load per unit of area
on foundation:

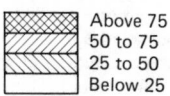

Above 75
50 to 75
25 to 50
Below 25

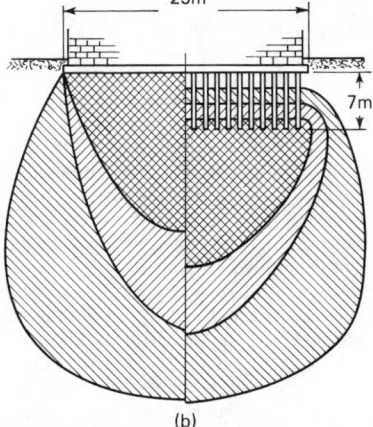

(b)

Figure A9.2.1 The increase of vertical pressure in soil beneath friction pile foundations having piles of equal lengths carrying equal loads. (a) Width of foundation small compared with pile length; (b) width of foundation large compared with pile length

10 cm²) and the pile by ensuring that the pile penetrates the bearing layer some 4 to 6 or more diameters for piles of less than 450 mm diameter. For larger piles, greater penetration ratios are necessary (see Tomlinson,[62] for further information on this problem). The bearing capacity can also be determined from Berezantsev's[64] curves. However, results of theoretical calculations may be unreliable, and past experience is a better guide.

In deeply embedded piles it is also necessary to take account of the side friction, a factor which depends upon the horizontal earth pressure coefficient K_s, the overburden pressure $\gamma'D$ and the angle of shearing resistance δ between the ground and the pile; thus the ultimate shaft resistance is given by:

$$Q_{us} = A_s \bar{K}_s (\gamma' \bar{D} \tan \delta) \tag{A9.2.2}$$

where A_s is the shaft area and $\bar{K}_s$ and $\gamma' \bar{D}$ refer to average values of K_s and $\gamma'D$.

The horizontal coefficient K_s depends not only upon the relative density and stress history of the deposit, but also upon the method of forming the pile. In bored piles it can be as low as $0.7 \times$ at-rest earth pressure coefficient, rising to twice this coefficient for large displacement. The value of δ depends upon the value of ϕ and the pile material; δ/ϕ varies from 0.50 to 0.80 for steel and precast concrete respectively (see Tomlinson[62] for a full discussion on the question of shaft resistance).

The frictional resistance of the ground for driven piles can also be determined by use of the friction sleeve adaptation to the Dutch deep-sounding apparatus. This gives the term $K_s\gamma'D$, tan δ throughout the depth of the deposit directly (see Chapter 11).

The ultimate bearing capacity of the pile is given by the sum of Equations (A9.2.1) and (A9.2.2):

$$Q_u = Q_{uB} + Q_{us} \tag{A9.2.3}$$

Frequently, pile bearing capacities have to be based on standard penetration test data. Meyerhof[65] has proposed the following empirical relationships for the components Q_u and Q_{us}:

$$Q_u = A_B(107)4N + A_s \frac{\bar{N}}{50}(107) \tag{A9.2.4}$$

where Q_u is in kilonewtons and A_B and A_s are in square metres and $\bar{N}$ is the average N value along the pile shaft.

An upper limit of 60 kN/m² is suggested for the shaft resistance.

It is suggested that the point bearing capacity term in equation (A9.2.4) might be written as follows, to take account of the actual soil type into which the pile is driven:

$$Q_{uB} = A_B(107)KN \tag{A9.2.5}$$

When it is possible to carry out a loading test on a full-scale pile this should be done, the test being carried to failure, i.e. until settlement continues under constant load.

The use of dynamic pile-driving formulae to calculate the ultimate load is not to be recommended for two reasons, namely: (1) the information is obtained too late (i.e. at the construction stage instead of the design stage unless previous expensive tests are carried out) in which case loading tests should be included; and (2) a very wide range of answers can be obtained depending on the formula used and the choice of constants in the formula.

Dynamic pile-driving formulae have their uses, however, and records of set and energy should always be kept as they are guides to the variation in ultimate loads over a site on which one or two loading tests have already been carried out. Engineers of wide experience can also make estimates of the load-carrying capacity from the results of a driving test, provided always that their experience was obtained with conditions similar to those relating to the test. The relation between the true ultimate load and that given by the dynamic formula is empirical and should be recognized as such in spite of the mathematical basis of Newtonian impact mechanics on which such formulae appear to be founded.

A9.2.1.4 Piles bearing in clay

A pile bearing in clay receives support from the adhesion along the shaft and from the resistance at its base, the relative contribution of these two components depending upon the strength–depth profile of the clay, the ratio of the length to diameter of the pile and the manner of its installation, i.e. whether bored, driven preformed, or cast-in-place. Skempton[66] has given the following expression for the bearing capacity of a bored pile in London Clay:

$$Q = 9c_u A_b + a\bar{c}_u A_s \tag{A9.2.6}$$

where a is an empirical factor which depends on the type of pile, the clay and its condition, and varies between 0.3 and 0.6 for a bored pile in London Clay, with a mean value of 0.45 for a well-constructed pile in typical unweathered clay, and $\bar{c}_u$ is the average shear strength along the pile shaft, with an upper limit of 100 kN/m² for the average ultimate skin friction resistance.

The bearing capacity of a bored pile can be increased greatly by constructing an enlarged base (belling or under-reaming). The value of a for under-reamed piles should not exceed 0.3.

The shear strength of the clay in commercial practice is determined on 38 mm specimens cut from 100-mm driven samples, and experience over the last decade has shown that, owing to the fissured nature of London Clay, this practice results in considerably higher strengths being used in the assessment of end bearing capacity than obtain in situ (see Whitaker and Cooke[67] and also Burland, Butler and Dunican[68]). It is necessary therefore to allow for this factor by introducing an empirical coefficient ω related to the base diameter B of the pile, thus:

$$Q = 9\omega c A_B + a\bar{c}_u A_s \tag{A9.2.7}$$

where $\omega = 0.8$ for $B < 1$ m and 0.75 for $B > 1$ m, and ac_u has an approximate limit of 100 kN/m².

In the event of the strength being determined on larger specimens or from in situ piston tests, upward modification in the value of ω is necessary. Load factors used in the design range from 2 for straight-shafted piles to 2.5 for base diameter less than 2 m as these values will generally restrict the short-term settlement of the single pile to less than 10 mm. With diameters larger than 2 m the working load should be checked by calculating the settlement using the curves relating Q_a/Q to settlement shown in Figure A9.2.2.

The range of a values discussed above is based on experience of London Clay. In dealing with other overconsolidated clays (e.g. the Lias, Kimmeridge and Oxford Clays and in weathered marls) some caution is needed in assigning a values, and where previous experience is not available, loading tests designed to establish a values should be undertaken.

Piles driven into soft sensitive clays result in loss of strength in the clay in contact with the pile, and thus test loading should be delayed for as long as possible after driving, preferably at least a month, to allow thixotropic strength to be regained.

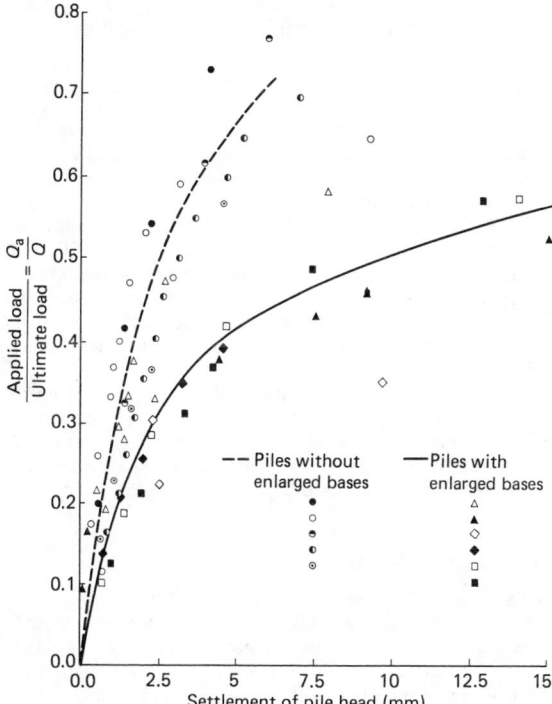

Figure A9.2.2 Q_a/Q against settlement in loading tests, showing the mean curves for piles with and without enlarged bases. (After Whitaker and Cooke (1966) 'An investigation of the shaft and base resistance of large bored piles in London Clay', *Proceedings, conference on large bored piles*. Institution of Civil Engineers, London)

Piles driven into stiff clay cause severe disturbance and fracturing to the clay and experience has shown that the effective cohesion can be extremely erratic, even on the same site (Tomlinson[62]). The lower limit of the adhesion factor, Tomlinson suggests, ranges from about 1.0 for soft clays to 0.2 for stiff and very stiff clays. When piles are driven to deep penetrations into stiff clays, the adhesion factor is influenced by the effective overburden pressure at any point in the pile shaft.[62]

Loading tests should be carried out whenever possible.

Piles driven into ground which is subject to consolidation from loading, or which is still settling under prior loading, are subject to downdrag forces (negative skin friction) which place an additional load on the toe. These forces can be severe and should be taken into account in the design and test loading. Negative skin friction can also occur, when driving piles through soft sensitive clays to bear on a harder stratum, owing to dissipation of the pore pressures induced by driving. Negative skin friction can be reduced by slicking the pile with the appropriate grade of bitumen. Johannessen and Bjerrum[69] recommend the evaluation of downdrag using effective stresses.

Appendix 9.3 Ground improvement

There are numerous cases in which the properties of naturally occurring soil or fill material can be improved or changed to help solve engineering problems arising either in temporary or permanent works. The methods of ground improvement cover a wide range of techniques – often referred to as geotechnical processes – and include compaction, moisture control, stabiliza-

tion, grouting and reinforcement. Reference should also be made to the use of geotextiles for reinforcement, separation and filtration in the ground and of mild steel reinforcement strips to produce reinforced earth structures with increased shear properties in embankments and fills. Various other processes have been developed and the whole subject of ground improvement deserves special study starting, for example, with Chapters 29 to 38 of the *Ground engineer's reference book*.[70]

This appendix lists and gives a brief description of the main ground improvement methods.

A9.3.1 Drainage and water lowering

Drainage systems to control groundwater in engineering works may include seepage reduction measures, such as impervious barriers of sheet piles, membranes or grouts, together with drains incorporating a filtering material, sometimes enveloped within a plastic fabric filter. In sands and gravels, water-lowering systems can be used to allow an excavation to be carried out in the dry or to reduce the water pressure on the sides and base of the excavation.

A9.3.1.1 Site investigation

A thorough site investigation to establish the hydrogeological characteristics of a site is an essential preliminary to the design of any groundwater lowering project. This is discussed further in Chapter 11.

9.3.1.2 Permeability and filters

The permeability (transmissibility divided by the depth of the aquifer) can be determined by pumping from a large-diameter well while monitoring the drawdown in a number of adjacent observation wells. The shape of the drawdown–time curve is matched to type curves derived theoretically, and the transmissibility and storage coefficients so deduced. A wide range of available type curves enables varying aquifers and hydrologic boundary conditions to be considered. Alternatively, for a fully penetrating well into an unconfined aquifer, permeability can be estimated from the equilibrium drawdown–distance curve, using Equation (A9.3.1). This latter method, which is the older of the two described, requires the establishment of equilibrium conditions, which may take many days; in contrast the time-variant method can be applied within a matter of hours after commencement of pumping.

Pumping tests give the most reliable value for k but an order of permeability can be obtained from grading curves (see Loudon[71]). Undistorted samples of fine sands are essential in order to see if the material is laminated – a fact which naturally can have a marked effect on the horizontal permeability.

Grading curves are also used to choose suitable sand as a filter medium using Terzaghi's empirical rule which states that the grading curve for the filter material should be the same shape as that for the material to be filtered, and:

$$\frac{D_{15} \text{ (filter)}}{D_{85} \text{ (soil)}} \leqslant 4 \text{ to } 5 \leqslant \frac{D_{15} \text{ (filter)}}{D_{15} \text{ (soil)}}$$

where D_{15} and D_{85} are the grain sizes corresponding to those at which 15% and 85% pass on the grading curves.

A9.3.1.3 Pumping capacity

The quantity of water to be pumped from a fully penetrating well into an unconfined aquifer can be calculated from the equation:

$$Q = \frac{\pi k (H^2 - h^2)}{\log_e R/A} \quad (\text{m}^3/\text{s}) \qquad\qquad (\text{A9.3.1})$$

where k is the permeability in metres per second, H is the depth from normal water level to the impermeable stratum in metres, h is the depth from lowered water level to the impermeable stratum in metres, R is the radius of cone of depression in metres and A is the radius of circle of area equal to area surrounded by wells in metres.

R can be obtained from the empirical relation

$$R = 30(H - h)\sqrt{k} \quad (\text{m}) \qquad\qquad (\text{A9.3.2})$$

The number of wells required can be obtained from the empirical relationship

$$Q = 3.63 \times 10^{-6} r_0 h_0 n k^{\frac{1}{2}} \quad (\text{m}^3/\text{s}) \qquad\qquad (\text{A9.3.3})$$

where r_0 is the radius of a well in metres. h_0 is the water level outside a well in metres, k the permeability in metres per second and n is the number of wells required.

A9.3.1.4 Well points, suction wells and deep wells

Three systems of water lowering are in common use and each has certain advantages and disadvantages.

Well points When the lowering of the water level required is 4.5 m or less, a well point system can be used. If the system operates very efficiently, greater lowering can be obtained, but this should not be assumed at the planning stage.

A well point is a metal tube about 50 mm diameter carrying a gauze filter about 1 m long at its lower end. New types of well points are now on the market with slotted plastic outer covering and metal centre tubes. The well points are jetted into the ground at intervals of 1 or 2 m and then connected to a header main through which the water is extracted by a well-point pump for exhausting the main and well points, and a centrifugal pump to remove the water.

Well points are cheap to install and, for a long progressive excavation such as a pipe or sewer trench, they are usually the most economical system. If greater lowering than 4.5 m is required, a two-stage system can be used.

A special trenching machine is now available for laying a horizontal porous pipe of 100 mm diameter at depths down to 5 m below ground level. A pipe up to a maximum length of about 230 m can be installed, and for continuous trench work the pipes are overlapped by about 4.5 m.

Shallow wells These are, in principle, the same as well points but the wells are bored into the ground. The wells are usually about 600 mm diameter with a 300 mm filter tube and a 75 to 100 mm diameter suction pipe. The space outside the 300 mm tube is filled with a gravel filter as the boring tube is withdrawn. Because of their greater diameter, the wells usually can be spaced about 10 to 15 m apart, and since there are many fewer connections than in a well-point system, the efficiency is greater. The wells are connected to a ring main and a well-point pump. Alternatively ordinary suction-lift pumps can be installed individually in the wells for small excavations.

The cost of pumping is much the same for a given lowering of the water level from either a well-point or a shallow well system, but the cost of installation of the shallow wells is higher. The shallow well installation is often to be preferred for an excavation of rectangular shape (as opposed to a long trench) where pumping must continue for many months, and in fine sands where a graded filter is necessary or in laminated soils where a definite vertical connection between the aquifers is required.

As in the case of well-points, two-stage systems can be used for a lowering of more than 4.5 m but, in general, deep wells will prove cheaper.

Deep wells With the deep-well system lowering of the water level can be achieved in one stage. This is because the pumps are placed at the bottom of the wells and deliver the water against pressure; there is no suction lift. The pumps used are electrically driven submersible pumps.

The well cannot be less than 450 mm diameter, which allows a 75 mm thickness of filter gravel, and if a two-stage filter is desired the well will be 600 mm diameter. For this reason the capacity of the wells is much greater than in the case of shallow wells and they can therefore be spaced further apart, in general up to 30 m, but this will vary considerably on different installations.

The cost of a deep-well system is high but it generally gives safe dry excavation and can often reduce the time of construction considerably. For safety, two independent sources of electric power must be provided for the pumps, since once they have started pumping it might be disastrous if they failed.

A9.3.1.5 Vacuum drainage

In soils of low permeability, such as coarse silts, drainage can sometimes be effected by sealing the wells or well points and exhausting the air from them. The pressure of the atmosphere then acts as a surcharge on the soil causing it to consolidate, and water is squeezed out of the soil into the filters of the wells. The amount of water removed is very small but the increase in strength of the silt is marked, and excavation is greatly facilitated.

A9.3.1.6 Electro-osmosis

Electro-osmosis is a further drainage process which can be used in silts. It is based on the principle that if a direct electric current is passed through the soil a flow of water takes place from anode to cathode. The cathode is made into a well and the water which reaches it is pumped out. The amount of water removed is small. The success of the method depends, as with the vacuum method, on the fact that the flow of water is away from the excavation, the free water surface is lowered and the water which remains in the soil above this surface is in tension, and that the capillary tensions add greatly to the strength of silt. To some extent also the water content of the soil is reduced, thus resulting in an increased strength. Electro-osmosis is an expensive process and should only be considered if more normal methods of construction are inapplicable. In many silts the vacuum method of drainage is probably nearly as effective and much cheaper.

A9.3.1.7 Settlements caused by water lowering

When the water level is lowered the effective weight of the soil between the original and the lowered water levels is increased because the buoyancy effect has been removed. Where the soil concerned is sand and gravel any settlements due to this increase in weight are normally small, but where silt, clay or peat occurs in the zone referred to, settlement will occur with time owing to the consolidation of this material under its own increased weight. Advantage is taken of this in the methods given in section A9.3.2 to accelerate settlement.

Before installing a groundwater lowering system it is essential, therefore, to consider what effect such settlements may have on structures within the zone of influence. Important structures will probably be founded below the compressible material, either directly or on piles, and will be unaffected. For structures founded on the compressible strata, it is necessary to calculate the probable settlement and to estimate what damage, if any, to the structure would result.

In order to limit the radius of influence of a groundwater lowering system, some of the pumped water can be 'recharged' or fed back into the aquifer by means of infiltration wells sited close to the structure below which it is desired to limit the potential settlement.[72]

A9.3.2 Vertical drains to accelerate settlement

Vertical drains are used to accelerate the settlement of layers of soft clay or silt under applied loads. In many cases settlement can be tolerated provided it occurs quickly, preferably during the construction period, e.g. road embankments on soft clay (Figure A9.3.1).

The rate at which a uniform thin clay layer consolidates is inversely proportional to the square of the drainage path, which is either the thickness or half the thickness of the layer depending on the drainage conditions. The principle of this method is to provide vertical drains in the clay. The drainage path is then reduced to half the spacing of the drains. When the load (e.g. a fill) is applied, the settlements take place quickly in, say, a few months instead of many months or even years. It is important to note that vertical drains do not reduce the amount of settlement.

Vertical drains are particularly effective in deposits where the horizontal permeability is high compared with its vertical permeability. However, care must be taken to avoid local reduction of horizontal permeability by the process of installation of the drain. In deposits of exceptionally high lateral permeability, vertical drains may not be necessary. It is important, therefore, to investigate the horizontal drainage characteristics with great care.[73]

The consolidation of the clay under load increases its strength, a fact which can sometimes be made use of by construction, in stages, of a fill which would cause foundation failure if placed in one operation.

A9.3.2.1 Vertical sand drains

Vertical sand drains usually have been formed by driving a hollow mandrel and filling the space formed by the forcibly displaced soil with sand as the mandrel is withdrawn. Jetting and augering methods have also been used. The drains are generally between 0.15 and 0.5 m in diameter at between 2 and 5 m centres and up to 30 m long, depending on conditions. A horizontal drainage blanket is required at ground level to link the vertical drains together before the fill is placed, unless the fill itself is permeable.

The disturbance of the foundation soil during the installation of vertical sand drains is an undesirable feature, particularly if the method of installation could cause remoulding of sensitive clays. Various attempts have been made to develop thinner drains to reduce disturbance, such as the band drains described in section A9.3.2.3.

A9.3.2.2 Prefabricated drains (sandwicks)

The original Kjellmann wick drain of treated cardboard has been replaced by the use of fabric stocking filled pneumatically with sand or grooved plastic cores with nonwoven textile or geotextile filter coverings. Several different designs of plastic drains are available.

A9.3.2.3 Prefabricated band drains

An extension and, in many respects, an improvement on the prefabricated sandwicks are the band drains[74] which consist of a flat core with internal drainage grooves surrounded by a filter

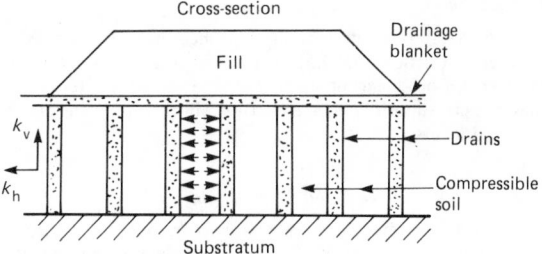

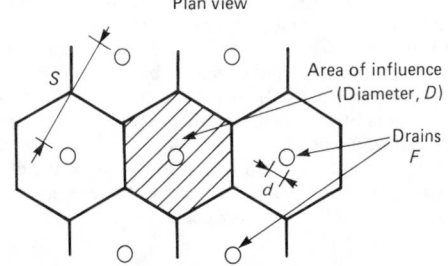

Figure A9.3.1 Typical vertical drain installation. (After Gambin (1987) 'Deep soil improvement', in: Bell (ed.) *Ground engineer's reference book* (Figure 36.6). Butterworth Scientific, Guildford)

cover, hence the name 'band'. The mandrel used to place these prefabricated bands may be circular, rectangular or of any other convenient section that reduces the disturbance of the surrounding soil. The equivalent diameters of band drains lie between 5 and 10 mm with a typical spacing of 1.2 to 1.5 m and a maximum length of 60 m (Figure A9.3.2).

The particular advantages of band drains are their simplicity, speed and cost of installation, together with minimum disturbance of the foundation soil.

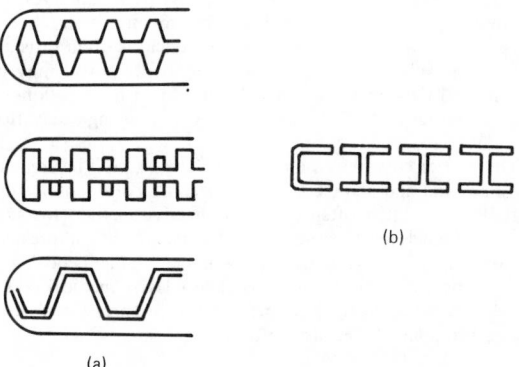

Figure A9.3.2 Examples of cross-sections of plastic band drains. (After Gambin (1987) 'Deep soil improvement', in: Bell (ed.) *Ground engineer's reference book* (Figure 36.10). Butterworth Scientific, Guildford)

A9.3.3 Exclusion of groundwater

Retaining systems excluding groundwater from a site, or from an area of a site, include sheet piled walls, *in situ* concrete diaphragm walls and contiguous bored pile diaphragm walls. The use of compressed air to exclude water from underground workings is another well-developed technique. Freezing and grouting are further possibilities which have the added potential advantage of strengthening the ground locally.

A9.3.3.1 Sheet piling

The use of standard interlocking steel sheet piles is described in Chapter 17 (section 17.3.2.5) in relation to the construction of basements. For all except shallow excavations, the interlocking piles require support either by struts, shores or by the use of ground anchors.

A9.3.3.2 In situ concrete diaphragm walls

Reinforced concrete diaphragm walls can be constructed to considerable depths by placing concrete by tremie tube in a narrow trench supported by bentonite mud. The trench is excavated in short panels, 2 to 7 m in length, using special cutters with circulating mud, or grabs and stationary mud. Junctions between panels are formed by means of steel tubes acting as end shutters. The mud displaced by the rising concrete is used in subsequent panels provided it is still in good condition.[75] It is possible to construct such walls in all types of ground, though difficulties have been known to occur when concreting in very soft alluvium owing to displacement of the clay under the high head of liquid concrete. In very permeable ground, significant mud losses may occur.

A9.3.3.3 Contiguous bored pile walls

Walls consisting of soldiers of bored piles can also be made with or without the use of mud depending upon the ability of the ground to support itself. The spacing can be varied and in the case of contiguous bored pile walls each is bored slightly into the completed adjacent pile while the concrete is still 'green'.

A9.3.3.4 Compressed air

The use of compressed air is well known as a construction expedient in underground work. It can be used in sands and gravels, silts and clays. The air pressure, acting on the surface of the soil in the excavation or, more correctly, on the water surfaces in the voids of the soil, prevents the flow of water through the soil and acts as a support. The air pressure theoretically must be equal to the water pressure. In practice, a pressure somewhat lower than the theoretical is often satisfactory. In gravels, the losses of air through the gravel may be serious and these can sometimes be cut down by injections of clay suspensions into the gravel before commencing excavation in order to reduce the permeability.

The cost of compressed-air working can be high in areas of silt, fine sand and some clays which require considerable support. However, it is often the most effective method for subaqueous tunnels in soft ground. Health hazards in compressed-air working, as in diving (Chapter 42), include the 'bends', and in the longer term, bone necrosis. The CIRIA medical code[76] should be applied, using the appropriate decompression procedure and equipment. See also section 17.3.4.4.

A9.3.3.5 Ground freezing

Another temporary method of preventing the access of groundwater to excavations and of strengthening the soil locally is by freezing the water. This is normally done by boring vertical holes into the ground, installing pipes in them and circulating brine or cryogenic liquids, cooled to below the freezing point of water, through the pipes. The freezing process is expensive and slow but once the water is frozen excavation can safely take place inside the frozen ring. The freezing must, of course, be continued until the permanent work is completed. One of the disadvantages of brine is that if a leak occurs in the pipes it will escape into the groundwater and it may then prove impossible

to freeze it. To overcome this objection the Dehottay process was introduced, in which liquid carbon dioxide is circulated instead of brine. More recently, liquid nitrogen has been used. The freezing process in general is applied to narrow, deep excavations such as mineshafts, but cases are on record of its use in foundation work (Figure A9.3.3).

A9.3.3.6 Grouting

Injection processes using various types of grout are dealt with separately in section A9.3.4 because of their wide applications in reducing permeability, increasing strength and reducing compressibility in soils and rocks.

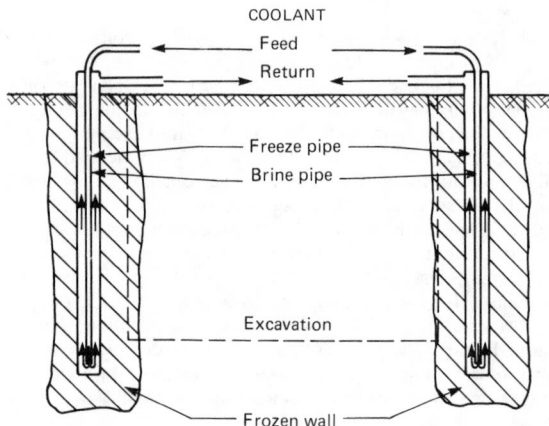

Figure A9.3.3 Scheme of ground freezing for support of an excavation. (After Jeffberger (1987) 'Artificial freezing of the ground for construction purposes', in: Bell (ed.) *Ground engineer's reference book*. Butterworth Scientific, Guildford)

A9.3.4 Injection processes: grouting

It is sometimes possible to change the properties of the ground encountered by injecting materials of various sorts into the voids of the soil. These changes include: (1) reduction in permeability; (2) increase in strength; and (3) decrease in compressibility, or a combination of these. A major use is for filling voids in mine workings and karstic limestone.

Cases in which the reduction in permeability is important include: (1) the formation of grouted cutoffs under dams; (2) grouting fissured rocks; (3) grouting sand and gravel to reduce air losses during construction work in compressed air; and (4) sealing gaps in sheet piling. The increase in strength is important in underpinning problems and in support of excavation in tunnelling. Injection processes can also be used to lift tanks and pavement slabs by hydraulic pressure, the grout later setting and supporting the structure in the raised position.

A9.3.4.1 Materials that can be grouted

Grouts can be injected into the fissures in a rock. This is probably one of the earliest applications of the process. Rockfill and rubble masonry can also be grouted. Cement grouts, containing sand or PF ash, are used in cases where the voids are fairly large.

Gravels and sands can be grouted successfully by a variety of different processes as described below, but clays and silts cannot because their voids are too small and their permeabilities too low. An exception to this is the use of the technique called claquage grouting, in which tongues of high-pressure grout penetrate planes and zones of weakness within the soil body.

A9.3.4.2 Grouting materials

Grouts generally consist of suspensions, solutions or aerated emulsions, the choice depending upon the nature of the work and the type of material to be grouted.

Suspensions The most commonly used grout is cement in water but, because of its high sedimentation rate, it can be relatively unstable, depending on the water/cement ratio. Pure cement grouts cannot be used for injecting sands or clays. Suspensions of cement with bentonite (e.g. in the proportions of 4 or 5:1) are more stable, easier to pump and, generally, produce a better result than pure cement grouts. If the voids to be filled are sufficiently large, sand is added to reduce shrinkage and cost. Another group of grouts comprises suspensions in solutions, e.g. in a solution of sodium silicate. An example would be a combined cement–bentonite–silicate grout.

Solutions There are several grouting processes in which solutions of chemicals based on sodium silicate are injected. The chemical processes can be used down to the fine sand range. They divide into the two-solution and the single-solution processes.

In the two-solution processes (Joosten and Guttman processes) the first chemical injected is sodium silicate, and this is followed immediately by calcium chloride or some such salt. The reaction is almost immediate and for this reason the solutions cannot penetrate far from the injection pipes which are therefore spaced at about 600 mm centres. The process gives considerable strength to the soil and also reduces the permeability to a very small fraction of its previous value.

In the single-solution processes, two chemicals are mixed before injection, possibly with a third chemical to delay the setting action for some time. The injection pipes can therefore be further apart. The processes reduce the permeability but do not give strengths comparable to the two-solution process.

In addition, a range of 'liquid' grouts is available, having acrylic, formaldehyde, lignin and epoxide bases. These grouts have low viscosities and therefore considerable penetration power, and achieve comparatively high strengths. They are, however, more expensive than the common grouts.

Aerated emulsions These are cement- or organic-based grouts into which a gas is emulsified. The properties of the resulting foam depend upon the distribution of the gas bubbles which, in turn, depends upon the materials and method of preparation. The foams are not particularly strong and this type of grout is used mainly as a filling.

Other types of grout Cement grouts with a low water/cement ratio (e.g. 0.4) are stable and pastelike in consistency. For grouting purposes, they can be made sufficiently fluid by the use of admixtures, such as plasticizers and swelling agents, or by vigorous stirring. 'Colgrout' is an example of the latter treatment and 'Prepakt' of the former. As the high potential strength of these low water/cement ratio grouts is often unnecessary, a large proportion of the cement can be replaced by pulverized fuel ash (PFA).

Bituminous emulsions can also be used as grouts, e.g. to reduce the permeability of soils down to the fine sand range.

A9.3.4.3 Methods of grout injection

In nearly all grouting work the injections are made by drilling or driving pipes into the ground and pumping the grout in under pressure through hoses attached to the pipes.

The spacing of the pipes varies widely with the process and the conditions, from 600 mm for the two-solution chemical process in sand, to about 3 m for clay injections in alluvium, and up to 6 m or more for cement grouts in fissured rocks.

When filling fissures in rock with cement grout it is usual to use piston pumps which will give a pressure up to 7500 kN/m². The same pumps can be used for clay injections with alluvial sands and gravels but the pressures must be controlled carefully in relation to the depth and nature of the overburden to avoid undue lifting of the ground surface. If the limitation of ground heaving is important, suitable instrumentation for the monitoring of heave may be necessary. In the Joosten and Guttman two-solution chemical processes the amount of grout required to fill the voids in the soil between injection pipes is pumped in with less regard to the pressure, subject to a maximum pressure of about 3000 kN/m². Piston pumps are used.

For very simple grouting jobs, a grout pan may be used. The cement grout is mixed in the pan by a paddle driven by hand or by a compressed air motor, an air pressure up to 750 kN/m² is then applied to the surface of the grout in the pan and this drives the grout through the hose into the injection pipe and so into the ground. This suffers the limitation that the injection pressures cannot easily be varied to suit the ground conditions.

For more complex jobs, sleeve grouting is frequently used using a 'tube-à-manchettes'. The system comprises a PVC tube of about 30 mm bore which is installed into a borehole of about 90 mm diameter and sealed into the ground with a relatively weak bentonite–cement sleeve grout. The tube is equipped at short intervals, normally 300 mm, with rubber sleeves covering perforations. An injection device is located against selected perforations in turn between upper and lower packers. The grout lifts the sleeve, fractures the sleeve grout and enters the ground. With this device it is possible to return to any position and regrout.

Grouting work in general is not simple and damage can be caused by the indiscriminate use of high pressures by inexperienced operators. The work should be planned and carried out by engineers and operators experienced in the use of grouting methods. The results need to be observed and monitored stage by stage.

A9.3.5 Reinforced soil

An example of reinforced soil as a constructional material consists of frictional soil backfill reinforced by linear flexible strips, usually placed horizontally. It was introduced by Vidal[77] in 1963 and has been since developed into a system comprising interlocking precast concrete or metallic wall-facing panels to which are fixed 5 mm thick galvanized ribbed mild steel strips, which provide the reinforcement. The facing plays no structural role, apart from helping to retain the backfill as it is compacted in layers and in locating the reinforcement strips in the backfill under compaction.

The performance of a reinforced soil structure depends on the friction developed between the soil and strip.

The choice of galvanized ribbed mild steel, instead of other metals or plastics, is based on durability, friction, creep and elastic property considerations.

Another form of reinforced soil is the use of woven plastic mesh placed on and wrapped around successive layers of compacted fill in embankment construction.

In Chapter 17, section 17.5.9, further information on both systems is given.

The range of uses for reinforced soil include retaining walls, sea walls, dams, bridge embankments and foundation slabs.

A9.3.6 Geotextiles

'Geotextile' is the generic name given to a wide variety of materials based on synthetic fibres, such as polyester, polyethylene, polypropylene, polyamine, etc. They may be woven, needle punched or formed into nets. Their potential applications in ground engineering are separation, filtration, drainage and reinforcement.

Most progress has been made in the use of these materials for filtration and drainage, but geotextiles are relatively new materials that are still developing. Their use needs to be considered in geotechnical applications.

A9.3.7 Ground anchors

Rock anchors and bolts are discussed in Chapter 10 but there are some similar requirements for anchorages and ties in ground engineering.

Diaphragm and pile walls are thin and generally require support which nowadays is provided by anchors, rather than strutting and bracing, as this facilitates excavation. However, for narrow excavations, strutting is often cheaper. With modern boring and injection techniques it is possible to install anchors at reasonably flat angles in all manner of soils.

The method comprises boring a hole using augers in clay and rotary percussion with water flushing and casing support in granular soils. Bar or strand is inserted into the hole and the predetermined anchor length grouted up with neat cement under pressure, the free end of the bar being sleeved off. The anchor can be stressed to loads in excess of the working load, if necessary, to test its capacity. When pulled to failure, special test anchors are installed.

The design procedure in clays is somewhat similar to that for bored piles. In sands, a semi-empirical approach is used based upon the density and grain size of the sand, the overburden pressure and the injection pressure.[78]

A9.3.8 Deep ground improvement

A variety of methods is available to improve the bearing capacity and decrease the compressibility of natural soils and manmade fills on site. They include preloading, vibro or dynamic compaction and the use of stone columns.

A9.3.8.1 Preloading

Improvement of soils by preloading is one of the techniques. The method is most applicable to loose sands, silts and waste materials. The types of work for which the method is most appropriate are those in which column loads will be relatively low, such as for embankments, low-rise buildings and light industrial developments (see Chapter 17, section 17.2.7).

The preload is applied by surcharging with imported fill, or water tanks, for the period of time necessary to achieve the required precompression. If it is required to accelerate the process, consideration should be given to the use of vertical drainage in conjunction with preloading. The surcharge load is restricted by the stability of the original ground but, if necessary, can be increased in stages as the ground improves with time.

A9.3.8.2 Vibrocompaction

Vibrocompaction (vibroflotation) is used for the deep compaction of cohesionless soils and fill materials to achieve improvements at depths to 20 to 30 m (see also Chapter 17, section 17.2.7). Increases achieved in density are greater for coarse grained than for fine grained material. The equipment consists

of a probe (vibroflot) of about 400 mm diameter, fitted with a vibrator giving horizontal amplitudes of 2 to 12 mm at between 30 and 50 Hz. The probe and its extension tubes are lowered by crane at penetration rates generally between 1 and 2 m/min until the required depth is reached. The tip of the probe has jetting holes for water supplied under pressure to assist penetration. Granular backfill is sometimes placed over the area of treatment and used as supplementary fill for the hole as the probe is withdrawn.

Other forms of vibrocompactors apply vertical vibrations using a vibrator, similar to that for piledriving, to drive a steel pipe of I-beam section into the material. However, this method is less effective in compacting the top 2 or 3 m (Figure A9.3.4).

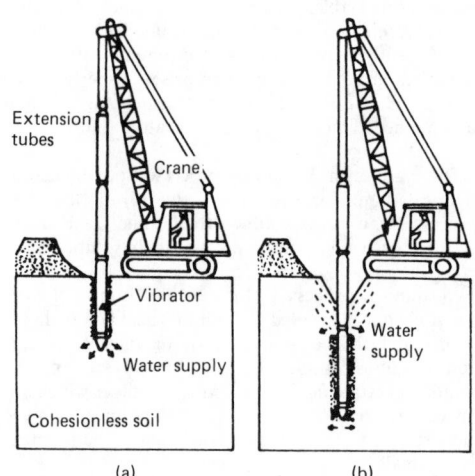

(a) (b)

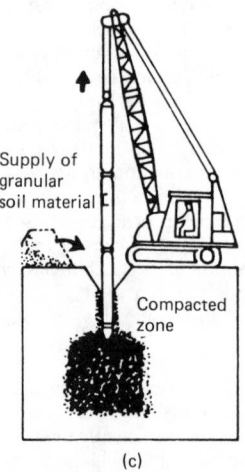

(c)

Figure A9.3.4 The vibrocompaction process. (After Gambin (1987) 'Deep soil improvement', in: Bell (ed.) *Ground engineer's reference book*. Butterworth Scientific, Guildford)

A9.3.8.3 Installation of stone columns

Cohesive materials and layered systems that are unsuitable for vibrocompaction can be reinforced by sand, gravel or stone columns formed by the vibrating probe or tubes driven by vibrators on top.

In partly saturated clays or fully saturated nonsensitive clays, the deep vibrating probe displaces the soil radially and the hole so formed is filled with well-graded 75 to 100 mm angular stone. No water is used in the probe, but compressed air is necessary for extraction of the probe. The fill is placed in layers, each layer being compacted by re-inserting the probe. The stone columns of 600 to 800 mm diameter allow partial mobilization of the passive resistance of the soil at small horizontal radial strains. In very soft cohesive soils (undrained shear strength 30 kN) the stone columns are formed by replacement of the soil under water pressure. The vibrating probe is operated in a similar way as for deep compaction, using the water jets, and the hole is washed out by repeated up and down movements of the probe. The stone column is formed and compacted as described earlier.

In very weak clays it is possible to couple three or four probes together to form a sufficiently stable hole to backfill with large self-supporting stone columns.

A9.3.8.4 Dynamic consolidation

A method of compaction introduced by Menard involves the dropping of a heavy weight, such as a concrete block or an assembly of plate steel of up to 20 t, from heights of about 20 m on to the ground. The compactive effect can reach depths in excess of 10 m. It is a method applicable particularly to the more freely draining soils and can strengthen cohesive soils although, with saturated cohesive soils, a combination of dynamic compaction and vertical sand drains may be necessary. It can also be used with manmade fills, including industrial wastes and some domestic waste tips (see Chapter 17, section 17.2.7).

Tamping weights can be selected to suit the depth and extent of improvement required. The site is first covered with a working blanket of free-draining material, about 1 m thick, and several impacts are applied at each centre in a predetermined grid. Several coverages of the area, including a perimeter strip, are required at intervals of up to several weeks, to allow the pore water pressure to dissipate.

A9.3.9 Shallow compaction

Shallow compaction refers to the compaction of material in layers, typically of 200 to 250 mm or less, in the construction of embankments, earth dams, pavement bases and sub-bases and in the process of fill, including the disposal of certain types of waste. Compaction increases the resistance to deformation, reduces permeability and increases the shear strength of a material. The principles of achieving a well-compacted material are illustrated by the laboratory soil compaction tests described in section A9.1.3. The maximum density achievable, measured either in terms of dry density or air voids content, depends upon the characteristics of the material, the moisture content at which it is compacted and the compactive effort applied, the latter depending on the number of passes as well as the weight or vibrating energy of the roller.

The most commonly used rollers were of the deadweight type (up to 20 t) including smooth-wheel, sheepsfoot or tamping rollers with variations such as grid rollers and pneumatic-type rollers which could apply higher local pressures. Vibrating rollers are now more common, ranging from self-propelled pedestrian-operated rollers to self-propelled and towed vibrating rollers of up to 20 t. Other important items of compaction equipment are vibrating plate compactors but, in addition, there are power rammers, dropping weight compactors and, more recently, an impact roller.

The choice of compaction equipment depends on the characteristics of the material to be compacted. Smooth wheel deadweight and vibrating rollers are suitable for most materials, but grid rollers and pneumatic tyred compactors are generally unsuitable on uniformly graded granular materials and silty clays.

Although the ideal is to bring the moisture content of material to be compacted to the optimum value determined by test or, preferably, by field compaction trial, adjustment of moisture content is difficult and sometimes impossible. In arid areas, compaction at moisture contents below optimum may have to be accepted. Where an adequate supply of water is available, the moisture content of the soil can be increased by mixing water into each layer using disc harrows or cultivators; however, quality of results may vary. In temperate and other countries with wet and dry seasons, earthmoving and compaction is usually confined to certain parts of the year.

A9.3.10 Soil stabilization

The term 'soil stabilization' is applied to a range of treatments which improve the properties of the existing ground materials, including changing their grading (mechanical stabilization) and chemical action (chemical stabilization). Ground freezing and grouting, described in section A9.3.3, are also forms of soil stabilization.

A9.3.10.1 Soil–cement

Soil stabilization includes treatments with cement, which can produce materials of considerable strength. For example, the flexural strength and elastic modulus of a fine grained soil–cement may be in the order of 0.5 to 1.5 MN/m^2 and 5 to 15 GN/m^2 respectively. For coarse grained soil–cement (e.g. cement-bound granular material) the strength and elastic properties approach those of lean concrete.

A9.3.10.2 Bitumen stabilization

Bitumen has also been used for stabilization in arid climates, but it has little application in moist soil conditions. Addition of bitumen to a granular soil or crushed rock improves its cohesion and resistance to water penetration. The most usual form of addition is bitumen emulsion but foam bitumen is a recent development.

A9.3.10.3 Lime stabilization

Lime stabilization is widely used, particularly in the warmer climates and developing countries. The reaction of lime with the clay content of a soil rapidly reduces the soil's plasticity; but the subsequent gain in strength of the soil–lime mix is slower and less than that obtained with cement. In some cases, lime is used to modify the properties of a soil rather than to produce a material with appreciable strength, i.e. soil modification rather than soil stabilization.

A9.3.10.4 Methods of construction

The most commonly used forms of stabilization – mechanical, cement, bitumen, lime – all require efficient mixing and compaction. Their main application is for road, airfield and other pavement areas, but soil–lime and soil–cement have been used to provide stable fill and embankments.

Mix-in-place methods, where the required thickness of the stabilized layer is less than about 200 to 250 mm, are generally the most appropriate, using single-pass or multipass machines with blades or tyres, similar to, but more sophisticated than, agricultural cultivators, or travelling mixers which pick up the soil from preformed windrows into cylindrical-drum mixers. Most of these specially developed machines have the capacity to disperse controlled amounts of water and to bring the mix to optimum moisture content; others each include a cement dispenser, otherwise the required amounts of cement and water are spread ahead of the mixer.

While it is possible to stabilize successive layers by mix-in-place methods, it requires considerable care and control. Mixing by central stationary mixing plant is often preferred, using either batch or continuous mixers fitted with paddles or blades to give a positive mixing action. Free-fall-type concrete mixers are not suitable. A range of spreading equipment is available to spread the mixed material *in situ* prior to compaction.

Compaction is usually by smooth wheel or vibrating rollers, but impact compactors are also available. Maximum density at optimum moisture content is the requirement for strength and durability (see section A9.3.9).

A9.3.10.5 Limitations to soil stabilization

Not all soils or other materials are suitable for stabilization and the following gives some indication of the main limitations.

Grading Uniformly or poorly granular soils which cannot be compacted adequately; stones should generally be less than 75 mm.

Plasticity Cohesive soils with LL > 45% cannot be mixed efficiently by most mix-in-place plant; soils containing more than about 10% plastic fines cement cannot be mixed effectively by most stationary mixers.

Organic content Peaty soils are unsuitable and soils with more than about 1% organic content give low strength with lime or cement.

Sulphate content Unsuitable for lime or cement stabilization without special precautions.

References

1 Skempton, A. W. (1954) 'The pore pressure coefficients *A* and *B*', *Géotechnique*, **4**, 143.

2 Bishop, A. W. and Henkel, D. J. (1957) *The measurement of soil properties in the triaxial test*. Edward Arnold, London.

3 Taylor, D. W. (1948) *Fundamentals of soil mechanics*. Wiley, New York.

4 Terzaghi, K. and Peck, R. B. (1967) *Soil mechanics in engineering practice*. Wiley, New York.

5 Loudon, A. G. (1952) 'The computation of permeability from simple soil tests', *Géotechnique*, **3**, 165.

6 Gibson, R. E. and Lumb, P. (1953) 'Numerical solution of some problems in the consolidation of clay', *J. Inst. Civ. Engrs*, **2**, Part 1, 182.

7 Davis, E. H. and Poulos, H. G. (1968) 'The use of elastic theory for settlement predictions under three-dimensional conditions', *Géotechnique*, **18**, 1.

8 Rowe, P. W. (1968) 'The influence of geological features of clay deposits on the design and performance of sand drains', *Proc. Instn Civ. Engrs*, supp. vol.

9 Schiffman, R. L. and Stein, J. R. (1969) *A computer program to calculate the progress of ground settlement*. University of Colorado Report Number 69–9a.

10 Clayton, C. R. I., Simons, N. E. and Matthews, M. C. (1982) *Site investigation*. Granada, London.

11 Skempton, A. W. and Macdonald, D. H. (1956) 'The allowable settlements of buildings', *J. Instn Civ. Engrs*, **5**, Part III, 3, 727.

12 British Standards Institution (1972) *Foundations*. CP 2004, BSI, Milton Keynes.

13 British Standards Institution (1986) *Foundations*. BS 8004, BSI, Milton Keynes.

14 Terzaghi, K. (1943) *Theoretical soil mechanics*. Wiley, New York.

15 Vesic, A. S. (1975) 'Bearing capacity of shallow foundations', in: Winterkorn and Fang (eds) *Foundation engineering handbook*. Van Nostrand Reinhold, New York.

16 Brinch Hansen, J. (1961) 'The ultimate resistance of rigid piles against transverse forces', *Dansk Geotechnisk Inst. Bull.*, **12**, 5.

17 Sokolovski, V. V. (1960) *Statics of soil media*. Butterworth, London.

18 Meyerhof, G. G. (1951) 'The bearing capacity of foundations', *Géotechnique*, **2**, 301.

19 Kerisel, J. (1961) 'Fondations profondes en milieux sableux', *Proceedings, 5th international conference on soil mechanics and foundation engineering*, Vol. II, p.73.

20 Vesic, A. S. (1967) *A study of the bearing capacity of deep foundations*. Georgia Institute of Technology, Georgia.

21 Berezantsev, V. G. (1961) 'Load bearing capacity and deformation of piled foundations', *Proceedings, 5th international conference on soil mechanics and foundation engineering*, Vol. II, p.11.

22 Tomlinson, M. J. (1986) *Foundation design and construction*. 5th edn. Longman Scientific and Technical, London.

23 Steinbrenner, W. (1934) 'Tafeln sur Setsungsberechnung', *Die Strasse*, **1**, 121.

24 Terzaghi, K. (1943) *Theoretical soil mechanics*. Wiley, New York.

25 Boussinesq, J. (1885) *Application des potentiels à l'étude de l'équilibre et du mouvement des solides élastiques*. Gauthier-Villars, Paris.

26 Newmark, N. M. (1935) *Simplified computation of vertical pressures in elastic foundations*. Engineering Experimental Station Bulletin Number 24. EES.

27 Fox, E. N. (1948) 'The mean elastic settlement of a uniformly loaded area at a depth below the ground surface', *Proceedings, 2nd international conference on soil mechanics and foundation engineering*, Vol. I, p.192.

28 Poulos, H. G. and Davis, E. H. (1974) *Elastic solutions for soil and rock mechanics*. Wiley, New York.

29 Schmertmann, J. H. (1953) 'Estimating true consolidation behaviour of clay from laboratory test results', *Proc. Am. Soc. Civ. Engrs*, **79**, 311.

30 Skempton, A. W. and Bjerrum, L. (1957) 'A contribution to the settlement analysis of foundations on clay', *Géotechnique*, **7**, 168.

31 Schultze, E. and Sherif, G. (1973) 'Prediction of settlement from evaluated settlement observations for sand', *Proceedings, 8th international conference on soil mechanics and foundation engineering*, Vol. I, p.225.

32 De Beer, E. (1948) 'Settlement records on bridges founded in sand', *Proceedings, 2nd international conference on soil mechanics and foundation engineering*, Vol. II, p.111.

33 Schmertmann, J. H. (1970) 'Static cone to compute elastic settlement over sand', *Proc. Am. Soc. Civ. Engrs*, **98**, SM3, 1011.

34 Burland, J. B. (1970) *Proceedings, conference on in situ investigations in soils and rocks*, discussion session A. British Geological Society, London, p.61.

35 Clayton, C. R. I. and Milititsky, J. (1986) *Earth pressure and earth retaining structures*. Surrey University Press, Glasgow.

36 Rowe, P. W. and Peaker, K. (1965) 'Passive earth pressure measurements', *Géotechnique*, **15**, 57.

37 Coulomb, C. A. (1776) *Essai sur une application des règles de maximis et minimis à quelques problèmes de statique, relatifs a l'architecture*. Mémorial de Mathématiques et de Physiques, Académie Royale des Sciences, Paris, p.343.

38 *Earth retaining structures*. Institution of Structural Engineers Code of Practice Number 2 (1951), ISE, London.

39 Peck, R. B. (1969) 'Deep excavations and tunnelling in soft ground', *Proceedings, 7th international conference on soil mechanics and foundation engineering*.

40 Tomlinson, M. J. (1970) 'Lateral support of deep excavations', *Proceedings, Institution Civil Engineers conference on ground engineering*, Thomas Telford, London, p.55.

41 Skempton, A. W. and Ward, W. H. (1952) 'Investigations concerning a deep cofferdam in the Thames estuary clay at Shellhaven', *Géotechnique*, **3**, 119.

42 Ward, W. H. (1955) 'Experiences with some sheet-pile cofferdams at Tilbury', *Géotechnique*, **5**, 327.

43 Littlejohn, G. S. (1970) 'Soil anchors', *Proceedings, Institution of Civil Engineers symposium on ground engineering*. Thomas Telford, London.

44 Terzaghi, K. (1953) 'Anchored bulkheads', *Proc. Am. Soc. Civ. Engrs*, **79**, 262.

45 Rowe, P. W. (1953) 'Sheet-pile walls in clay', *J. Instn Civ. Engrs,* **79**, 262.

46 Chandler, R. J. (1984) 'Recent European experience of landslides in overconsolidated clays and soft rocks', *Proceedings, 4th international symposium on landslides,* Toronto, Vol. I, p.61.

47 Fellenius, W. (1936) 'Calculations of the stability of earth dams', *Proceedings, 2nd congress on large dams,* Vol. IV, 445.

48 Bishop, A. W. (1955) 'The use of the slip circle in the stability analysis of slopes', *Géotechnique,* **5**, 7.

49 Janbu, N. (1954) *Stability analysis of slopes with dimensionless parameters.* Harvard Soil Mechanics, Series, Number 46.

50 Morgenstern, N. R. and Price, V. E. (1965) 'The analysis of the stability of general slip surfaces', *Géotechnique,* **15**, 79.

51 Spencer, E. (1967) 'A method of analysis for the stability of embankments assuming parallel interslice forces', *Géotechnique,* **17**, 11.

52 La Rochelle, P. and Marsal, R. J. (1981) 'Slope stability', *Proceedings, 10th international conference on soil mechanics and foundation engineering,* Vol. IV, p.485.

53 Skempton, A. W. and Hutchinson, J. N. (1969) 'The stability of natural slopes and embankment foundations', *Proceedings, 7th international conference on soil mechanics and foundation engineering.*

54 Green, P. A. and Ferguson, P. A. S. (1971) *On the liquefaction phenomenon.* Report of a lecture by A. Casagrande, *Géotechnique,* **21**, 197.

55 Skempton, A. W. and Northey, R. D. (1952) 'The sensitivity of clays', *Géotechnique,* **3**, 30.

56 Bjerrum, L. (1972) 'Engineering properties of normally consolidated clays'. Lecture at King's College, London.

57 Bjerrum, L. (1967) 'Engineering geology of Norwegian normally consolidated marine clays as related to settlement of buildings', *Géotechnique,* **17**, 81.

58 Cedergren, H. R. (1977) *Seepage, drainage and flow nets.* Wiley, New York.

59 British Standards Institution (1975) *Methods of test for soils for civil engineering purposes.* BS 1377. BSI, Milton Keynes. (A revised standard is due for publication after 1989.)

60 US Corps of Engineers (1943) 'Design of runways, aprons and taxiways at Army Air Force stations', *Engineering manual,* Chapter XX. US War Department, Office of the Chief Engineers, Washington, DC.

61 Tomlinson, M. J. (1986) *Foundation design and construction* (5th edn). Longman Scientific and Technical, London.

62 Tomlinson, M. J. (1986) *Pile design and construction practice* (3rd edn). Viewpoint Publications, London.

63 Construction Industry Research and Information Association/Public Services Agency (various dates). Piling guides Numbers 1 to 9. CIRIA, London.
Weltman, A. J. and Little, A. L. (1983) *A review of bearing pile types.* Piling Guide Number 1.
Thorburn. S. and Thorburn, J. Q. (1985) *Review of problems associated with construction of cast-in-place concrete piles.* Piling Guide Number 2.
Fleming, W. K. and Sliwinski, Z. J. (1986) *The use and influence of bentonite in bored pile construction.* Piling Guide Number 3.
Weltman, A. J. (1977) *Integrity testing of piles: a review.* Piling Guide Number 4.
Weltman, A. J. and Healy, P. R. (1978) *Piling in boulder clay and other glacial tills.* Piling Guide Number 5.
Hobbs, N. B. and Healy, P. R. (1979) *Piling in chalk.* Piling Guide Number 6.
Weltman, A. J. (1980) *Pile load testing procedures.* Piling Guide Number 7.
Healy, P. R. and Weltman, A. J. (1980) *Survey of problems associated with the installation of displacement piles.* Piling Guide Number 8.
Weltman, A. J. (1980) *Noise and vibration from piling operations.* Piling Guide Number 9.

64 Berezantsev, V. G. (1961) 'Load-bearing capacity and deformation of piled foundations', *Proceedings, 5th international conference on soil mechanics and foundation engineering,* Paris.

65 Meyerhof, G. G. (1956) 'Penetration tests and bearing capacity of cohesionless soils', *Proc. Am. Soc. Civ. Engrs,* **82**.

66 Skempton, A. W. (1959) 'Cast-in-situ bored piles in London Clay', *Géotechnique,* **9**.

67 Whitaker, T. and Cooke, R. W. (1966) 'An investigation of the shaft and base resistance of large bored piles in London Clay', *Proceedings, conference on large bored piles,* Institution Civil Engineers, London.

68 Burland, J. B., Butler, F. G. and Dunican, P. (1966) 'The behaviour and design of large diameter bored piles in stiff clay', *Proceedings, conference on large bored piles,* Thomas Telford, London.

69 Johannessen, I. J. and Bjerrum, L. (1965) 'Measurement of the compression of a steel pile to rock due to settlement of the surrounding clay', *Proceedings, 6th international conference on soil mechanics and foundation engineering,* Vol. II, Montreal 37.

70 Bell, F. G. (1987) *Ground engineer's reference book.* Butterworth Scientific, Guildford.

71 Loudon, A. G. (1952) 'The computation of permeability from simple soil tests', *Géotechnique,* **3**, 165.

72 Cashman, P. M. and Haws, E. T. (1970) 'Control of groundwater by water lowering', *Proceedings, conference on ground engineering.* Institution Civil Engineers, London.

73 Rowe, P. W. (1968) 'The influence of geological features of clay deposits on the design and performance of sand drains', *Proc. Instn Civ. Engrs* supp. vol.

74 Holtz, R. D., Jamiolkowski, M., Lancellotta, R. and Pedroni, S. *Performance of prefabricated band-shaped drains.* Construction Industry Research and Information Association. Butterworth Scientific, Guildford (to be published).

75 Littlejohn, G. S., Jack, B. and Sliwinski, Z. (1971) 'Anchored diaphragm walls in sand', *Gr. Engnrg,* **4**, 6, 18–21.

76 Construction Industry Research and Information Association (1982) *Medical code of practice for work in compressed air* (3rd edn). CIRIA Report Number R44, CIRIA, London.

77 Vidal, H. (1969) *The principle of reinforced earth.* Highway Research Records Number 282. Washington DC.

78 Hanna, T. H. (1980) *Design and construction of ground anchors* (2nd edn). Construction Industry Research and Information Association Report Number R65. CIRIA, London.

Bibliography

American Association of State Highway and Transportation Officials (1986) *Standard specifications for transportation materials and methods of sampling and testing.* AASHTO, Washington DC.

Baumann, V. and Bauer, G. E. A. (1974) 'The performance of foundations on various soils stabilized by the vibrocompaction method', *Can. Geol. J.,* **11**, 509–530.

Bishop, A. W. and Henkel D. J. (1957) *The measurement of soil properties in the triaxial test.* Edward Arnold, London.

Clayton, C. R. I., Simons, N. E. and Matthews, M. C. (1982) *Site investigation.* Granada, London.

Department of Transport (1986) *Specifications for highway works.* HMSO, London.

Head, K. H. (1986) *Manual of soil laboratory testing,* Vol. I. Pentech Press, London.

Institution of Civil Engineers (1983) *Proceedings, international conference on advances in piling and ground treatment for foundations.* Thomas Telford, London.

Jones, C. J. F. P. (1985) *Earth reinforcement and soil structures.* Butterworth Scientific, Guildford.

Acknowledgements

The author acknowledges the permission of Dr A. C. Meigh and Mr N. B. Hobbs to reuse some of the text and figures that they had prepared for the 3rd edition. Appendices A9.1, A9.2 and A9.3 include some text from the 3rd edition with additional and updated material prepared by the Editor.

10

Rock Mechanics and Rock Engineering

A Price Jones BSc(Eng), MSc, CEng, MICE
Head of Rock Mechanics, Soil Mechanics Limited

Contents

10.1 Introduction

10.1 The scope of rock mechanics and rock engineering

Rock mechanics is a term for science and engineering applied to rock masses. As such, the term has relevance in numerous fields such as the recovery of hydrocarbons in rock reservoirs, development of geothermal energy resources, studies of the Earth's crust, seismicity studies, as well as mining and civil engineering. The area of activity restricted to construction works which require or essentially comprise excavation into the surface of, or within, rock masses might appropriately be referred to as rock engineering. Theoretical methods of analysing the behaviour of rock, based on the understanding of certain material or mass properties, have advanced rapidly in recent years. This has been driven partly by the need for effective nuclear waste management strategies. In turn, engineering practice as applied to construction in rock is also advancing, albeit at a slower pace because design processes often depend upon the use of empirical rules based on established precedents and on the knowhow gained from practical experience.

Civil engineering works which impose significant foundation loads on to rock, such as dams or massive nuclear containment buildings, or which involve an excavation such as a pit or cavern, demand that rock engineering principles be applied in order to achieve a stable structure. In many cases, such as tunnels, deep road cuttings, hydro-powerhouse caverns, storage caverns and underground repositories, deep bunkers, existing cliffs or natural cavities, the 'structure' is largely formed of the rock mass and its 'design' will have to take into account the inherent flaws, variability and weaknesses of the natural constituents.

For many engineering projects the main steps followed are:

(1) Investigation to determine geotechnical properties.
(2) Classification and characterizing the site.
(3) Initial design.
(4) Excavation and support.
(5) Performance monitoring and design re-evaluation.

For the investigation, the compiling of relevant geological and geotechnical information is facilitated by the introduction of standard schemes for terminology and methods (see Chapter 8). A number of rock classification schemes assist in the assimilation of data and characterizing of sites (see section 10.3).

Design methods which make use of advanced numerical calculations are now widely available as computers have become commonplace, although they have not necessarily replaced other methods. Rock conditions can be modelled with some degree of realism in order to predict rock behaviour when numerical methods are appropriately applied. However, uncertainties will inevitably exist concerning the geology, the appropriateness of the design model, the parameters used (the scatter may reflect real variability or random errors), the loads applied and unforeseen factors. Various statistical methods and risk and reliability analyses may be useful as aids to assess confidence limits, but they do not replace the need to monitor actual performance. Further benefits from computer technology have been derived from using automated data acquisition and results analysis for the performance monitoring of rock structures which provides feedback for design assessments,[1] as shown on Figure 10.1.

Improvements in excavation methods are evident with increasing production rates as rippers, excavators, tunnel boring machines and roadheader machines have become powerful and resilient. Blasting practice has advanced through the continued development of hydraulic drilling machinery and a better understanding of explosives usage in obtaining the required fragmentation with the least damage to the surround-

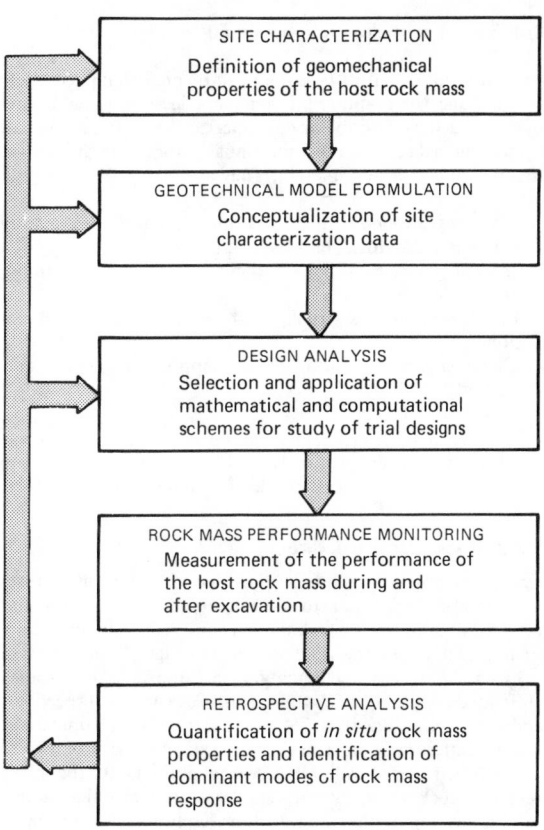

Figure 10.1 Components of a generalized rock mechanics programme. (After Brown (1985) 'From theory to practice in rock engineering'. *Proceedings, 4th international symposium on tunnelling 85.* Institute of Mining and Metals, London)

ing rock forming the structure. Particular advances have been made in construction procedures to form large excavations with the least disturbance to the surrounding rock by applying an improved understanding of rock mechanics stress analysis to the sequencing of excavation and support works. Support is provided either actively, i.e. with preload applied, or passively by rock reinforcement using dowels or rockbolts, rock anchors, shotcrete, concrete or steel supports or cast *in situ* concrete liners. These expedients may be used separately or in combination, or in conjunction with grouting or drainage measures.

10.1.2 Rock engineering principles

The principles of rock engineering have developed from the disciplines of civil and mining engineering, structural geology, engineering geology and, where possible, the practical application of analytical rock mechanics. All of these subjects are developing continually, with papers describing research, theory and practice being published frequently. Some relevant journals and text books are listed in the bibliography to this chapter together with abstract bulletins that relate to rock mechanics topics. The International Society for Rock Mechanics (ISRM) has established commissions to attempt the standardization of terminology and test methods and to investigate various topics including rock mechanics teaching, research and rock classification.

In addition to engineering knowhow, an overall appreciation of relevant aspects of geology (see Chapter 8), is an essential prerequisite to enable rock engineering principles to be applied effectively. The relevant *properties* of the rock essentially must be determined so that its *behaviour* can be predicted adequately and subsequently monitored. Parameters describing the following are needed to describe or characterize the rock mass for an engineering project and it is important to appreciate how these are linked together and interact. They are:

(1) *Rock mass structure*: intact rock and discontinuities (mainly strength and deformability).
(2) *Hydrology and void space*: groundwater flow, permeability and pressure.
(3) In situ *rock stresses*: principal stress magnitudes and directions.
(4) *Construction*: excavation and support sequence and methods.

A rock mechanics or rock engineering interaction matrix has been devised[2] and is shown in Table 10.1. These paremeters are further described below and are dealt with later in this chapter.

10.1.3 Rock mass structure

Geological appraisals allow the presence of important features of engineering relevance to be anticipated and, if peculiar unexpected conditions are later encountered, will serve as a warning that something is wrong. Knowledge of the origin of rock at a site, whether sedimentary, metamorphic or volcanic, will immediately suggest certain properties such as respectively bedding features, local zones of alteration or characteristic fracture patterns. Geological processes or upheavals known to have affected a site will all serve as pointers to the likely prevailing rock conditions. Examples would include folding, faulting, heating and cooling, tectonic compressive and tensile stresses, extreme loading by ice age cover, weathering, surface denudation relieving stresses, solutioning and chemical alteration causing volume and, hence, stress changes. Subsequent site investigations can thus be scheduled accordingly. Awareness during investigations or construction works of the 'expected but unpredictable', such as solution features in limestones, is also extremely helpful.

The geological appraisal might also assist at an early stage in the selection of an appropriate analysis method for designing the proposed works, in deciding on essential parameters to be measured and on whether or not material should be modelled as a continuum. For example, it will be known whether intact material is likely to be uniform, or zoned into bands or graded units by metamorphism or sedimentation such that possible anisotropic material properties can be predicted. Regular discontinuity patterns may exist as joint sets, bedding and cleavage planes, or irregular fracture orientations may be present due to faults and slump features. The likely planarity or undularity and persistence of discontinuities can sometimes be anticipated from their geological origin.[3,4] The mechanical behaviour of rock is controlled very largely by the presence and characteristics of the discontinuities contained within it.

The mechanical behaviour of intensely fractured rock can sometimes be approximated to that of a soil. At the other extreme, e.g. in very deep underground works where the rock is massive and the fractures confined, the rock can sometimes be considered as a continuous medium. More often, rock must be regarded as a discontinuum in which the mechanical properties of the discontinuities are of considerable relevance. Discontinuities also play a major role in determining water and stress environments in the mass.

Roughness, aperture, rock wall strength and *filling* in particular control the shear strength and deformability of fractures.[5] Even a thin weathered layer in a joint will considerably reduce the strength which is otherwise afforded by tightly interlocking roughness asperities.[6] Discontinuities that *persist* smoothly and without interruption over extensive areas, e.g. certain types of fault and bedding plane, offer considerably less resistance to shearing than discontinuities of irregular and interrupted pattern.[7] The *orientation* of fractures relative to the exposed rock surface is also critical in determining rock mass stability. Techniques are available for rapid assessment of measured fracture orientations using stereogrammetric methods for presentation of fracture orientations in statistical and graphical forms and for processing this type of data in rock mechanics calculations.[8–10] Fracture *spacing* (typically 20 to 2000 mm) and the *number of sets* of joints are important, since they determine the *block size* of the rock mass. Fracture spacing is relevant to problems of excavation stability and the design of bolted support systems, also in determining the ease or difficulty in rock excavation and in establishing its suitability for use as construction material.

10.1.4 Hydrogeology and void space

Intact rock material contains grains and intergranular pores

Table 10.1 Rock engineering interaction matrix. The four main parameters are drawn on the leading diagonal of the matrix and the cause–effect interaction of these is demonstrated by a clockwise motion to each off-diagonal element of the matrix. (After Hudson (1986) 'Rock engineering interaction matrix', *Symposium on rock joints*. Geological Society, London)

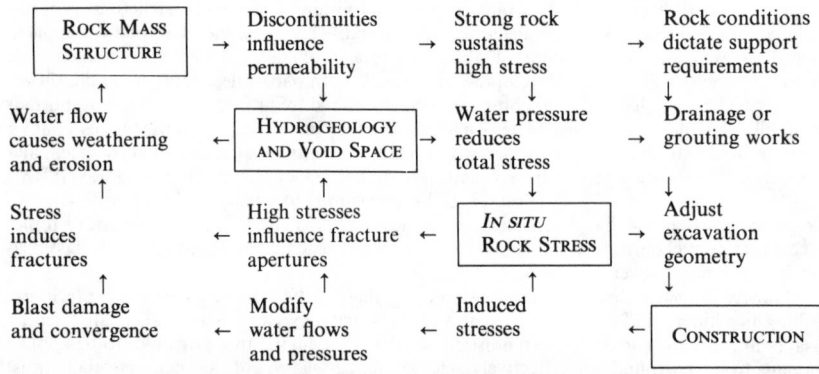

filled with air and water. The relative volumes and weights of these three constituents determine porosity, density, saturation and various related parameters as shown on Table 10.2. The presence of pores and pore water in the fabric of a rock material generally decreases its strength and increases its deformability. A small volume fraction of pores and the associated pore water pressure can produce an appreciable mechanical effect.[11-13]

The rock mass contains void space in the form of fissures as well as pores. The volume occupied by fissures is usually much less than that of pores, however, so that the porosity and bulk density of the rock mass and of intact material are usually similar. Fissures, however, have a much greater influence on mass permeability.[14] Fluid migration in rock masses is generally regulated by the intersecting network of conducting fissures which afford a less tortuous path for water flow than does a network of pores. It is possible usually to ignore the effect of intact material permeability in studies of water flow, except in some coarse-grained sedimentary rocks.

Construction works can cause the permeability of a rock mass to alter significantly depending on the deformations experienced by the joints. The hydraulic behaviour (i.e. conductivity) of rock joints is influenced by changes in effective normal stress because of the mechanical interaction between roughness, wall strength and aperture. The construction of dams, tunnels, caverns and slopes in jointed, water-bearing rock causes complex interactions between joint deformation and effective stress, so that deformation can take the form of normal closure, opening, shear and dilatation. The resulting changes in aperture can cause as much as three orders of magnitude changes in hydraulic conductivity of joints at moderate stress levels. This could influence both the design and scheduling (i.e. time of construction) of a grouted cutoff beneath a dam, for example.

The pressure of water in fissures imposes normal loads on the fissure walls and so forces are applied to the discrete rock blocks. These forces must be accounted for especially in stability analyses of rock slopes and in considering uplift forces in dam foundations. Groundwater pressure is measured with piezometers but, where fissure flow is expected, numerous measurements at specific locations are needed to establish the hydraulic gradients, i.e. spatial distribution of pressure differences. The use of specific drainage measures will reduce water pressures and thus generally improve the stability of rock structures.

10.1.5 Rock stress

The state of *in situ* stress refers to the stresses which exist naturally in a rock mass prior to any influence from engineering works. These stresses arise from gravitational and tectonic movements, crustal cooling and other major geomorphological influences. Complex stresses can develop in steep mountainous areas and near deeply incised river valleys. Given these possible origins, the state of *in situ* stress of a rock mass cannot readily be predicted and values can only be reliably obtained by direct measurement.[15] For major underground mining or civil engineering works, the *in situ* rock stress is a design parameter directly relevant to the layout, shape, size and orientation of excavations, to stability and support, and to sequence of excavation. Measured values of *in situ* rock stress are also useful for oilfield and geothermal development and studies related to evaluations of seismic hazard.

The state of stress existing at any point in a rock mass (or any material) is represented by a stress tensor which is a complete three-dimensional definition of stress magnitudes and directions. Local inhomogeneities or discontinuities within the rock mass can lead to variations in stresses such that measured values, besides showing some scatter, may be misleading in relation to the general regional stress field. In these circumstances it is necessary to consider the scale of the proposed excavations, the rock mass structure and the disposition of tests. It is also helpful to be aware of the possible constituents of the *in situ* stress field[16,17] which are shown in Figure 10.2 and Table 10.3.

Induced stresses (see Figure 10.3) are a function of the shape of the excavations and the previously existing *in situ* stresses. During early site investigations when direct access underground is not available, a practical method of measuring *in situ* stresses is hydrofracturing[18] which may be carried out in deep boreholes. When underground access via tunnels is available then stress

---

Table 10.2 Porosity and density definitions and interrelationship formulae

The components of a rock sample are defined as follows:

Grains: weight G_w, volume G_v
Pore water: weight W_w and volume W_v
Pore air: zero weight and volume A_v
Pores: volume $P_v = W_v + A_v$
Bulk sample: weight $B_w = G_w + W_w$
Bulk sample: volume $B_v = P_v + G_v$
Density of water: ρ_w = mass of water per unit volume

The physical properties – density, porosity, etc. – may be defined in terms of the above components as follows:

Water content	$w = (W_w/G_w) \times 100\%$	
Degree of saturation	$S_r = (W_v/P_v) \times 100\%$	
Porosity	$n = (P_v/B_v) \times 100\%$	
Void ratio	$e = P_v/G_v$	
Dry density of rock	$\rho_d = G_w/B_v$	(dry specific gravity $d_d = \rho_d/\rho_w$)
Density of rock	$\rho = (G_w + W_w)/B_v$	(specific gravity $d = \rho/\rho_w$)
Saturated density of rock	$\rho_s = (G_w + P_v\rho_w)B_v$	(saturated specific gravity $d_s = \rho_s/\rho_w$)
Grain density	$\rho_g = G_w/G_v$	(grain specific gravity $d_g = \rho_g/\rho_w$)

Having defined the three properties, water content, porosity and dry density, the remaining properties may be calculated from the following interrelationships:

$S_r = 100(w\rho_d)/n\rho_w$
$e = n/(100 - n)$
$\rho = (1 + w/100)\rho_d$
$\rho_g = \rho_d(1 - n/100)$

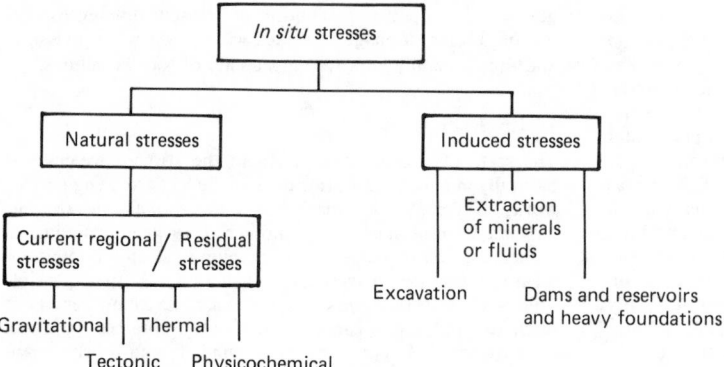

Figure 10.2 *The constituents of* in situ *stress fields. (After Hyett, Dyke and Hudson (1986) 'A critical examination of basic concepts associated with the existence and measurements of* in situ *stress'.* Proceedings, international symposium on rock stress, *Sweden)*

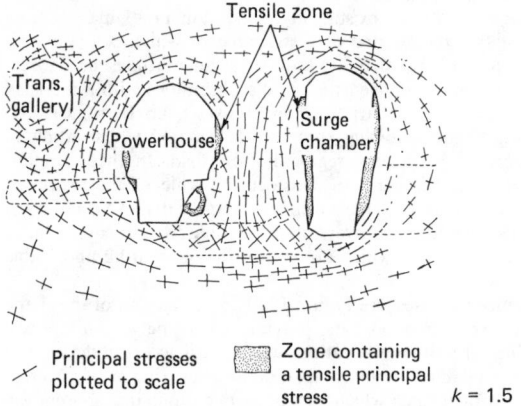

Figure 10.3 *Induced stresses caused by excavation. (After Benson, Murphy and McCreath (1970)* Modulus testing of rock at the Churchill Falls underground powerhouse, Labrador. *American Society for Testing and Materials. Special Technical Publication Number 477, pp.89–116.*

relief methods, such as overcoring[19,20] may be used to determine *in situ* stress. Provided sufficient care is exercised in making such measurements, the results are usually adequate for design purposes.

Other less direct methods may be used to determine natural stresses on a wider regional scale. These include the use of various geological indicators (e.g. foliations, cleavage, lineations, stylolites, slickensides, etc.) earthquake focal mechanisms[21] and borehole breakout studies.[22] Favourable comparisons have been obtained for direct measurements by overcoring and stress orientations from earthquake focal mechanism solutions in a seismically active area.[23]

10.2 Rock tests

Table 10.4 illustrates the various categories of testing, currently presented as *Suggested methods* by the ISRM.[24] Tests for classification and characterization of rock (index tests) are used for rock quality description and mapping and are generally quick and relatively cheap. Also, proposed construction methods can be evaluated from index test results, for example abrasiveness and hardness are relevant to tunnelling machine

Table 10.3 Glossary of *in situ* stress terms. (After Hyett, Duke and Hudson (1986) 'A critical examination of basic concepts associated with the existence and measurements of *in situ* stress'. *Proceedings, international symposium on rock stress.* CENTEK, Sweden)

Natural stress: The stress state which exists in the rock prior to any artificial disturbance. The stress state is the result of various events in the geological history of the rock mass. Therefore, the natural stresses present could be the product of many earlier states of stress. Synonyms include 'virgin', 'primitive', 'field' and 'active'.

Induced stress: The natural stress state as perturbed by engineering.

Residual stress: The stress state remaining in the rock mass, even after the originating mechanism(s) has/have ceased to operate. The stresses can be considered as within an isolated body that is free from external tractions; neither are they caused by the action of body forces or thermal gradients, etc. Sometimes the word 'remanent' is used as a synonym for 'residual'.

Tectonic stress: The stress state due to the relative displacement of lithospheric plates.

Gravitational stress: The stress state due to the weight of the superincumbent rock mass.

Thermal stress: The stress state set up as a result of temperature variation.

Physicochemical stress: The stress state set up as a result of chemical alteration and/or physical changes in the rock, e.g. recrystallization, absorption of water and fluctuation of groundwater levels.

Paleostress: A previously active *in situ* stress state no longer in existence. It can be considered as old and no longer present; whereas a residual stress is old and remains. Paleostress can be inferred from geological structures and from crystallography but cannot be measured.

Near field stress: The stress state perturbed by a heterogeneity (usually caused by engineering activities, e.g. a tunnel as a low-modulus inclusion).

Far field stress: A stress state which is not perturbed by a heterogeneity.

Regional stress: The stress state in a relatively large geological domain.

Local stress: The stress state in a small geological domain – usually of the dimensions of an engineering structure.

Active stress: A stress state with an associated strain energy state.

Table 10.4 Test categories published, or scheduled to be published, as *Suggested methods* by the International Society of Rock Mechanics. (After Brown (1981) *Rock characterization, testing and monitoring*. Pergamon, Oxford)

Field index tests for characterization	*Field design tests*
Discontinuity orientation	Deformability using a plate test
Discontinuity spacing	Deformability plate test down a borehole
Discontinuity persistence	Deformability radial jacking test
Discontinuity roughness	Deformability flexible borehole jack
Discontinuity wall strength	Deformability rigid borehole jack
Discontinuity aperture	Deformability flatjack
Discontinuity filling	Deformability *in situ* uniaxial triaxial test
Discontinuity seepage	Shear strength – direct shear
Discontinuity number of sets	Shear strength torsional shear
Discontinuity block size	Piezometric head (3 methods)
Discontinuity drill core recovery/RQD	Permeability transmissivity (5 methods)
Geophysical logging of boreholes	Flow velocity logs
Seismic refraction (2 methods)	Flow velocity – tracer dilution
Acoustic logging	Flow paths using tracers (4 methods)
Seismic measurements between boreholes	Stress determination – flatjack
Sonic log	Stress determination – surface coring
Caliper log	Stress determination – 'doorstopper'
Temperature log	Stress determination – strain gauge cell
SP log	Stress determination – USBM-type gauge
Resistivity logs (2 methods)	Stress determination – hydraulic fracturing
Focused current logs	
Induction log	*Field monitoring*
Gamma ray log	Movements: probe inclinometer
Neutron log	Movements: fixed-in-place inclinometer
Gamma-gamma log	Movements: tiltmeter
	Movements: borehole extensometers
Field 'quality control tests'	Movements: convergence meter
Rockbolt anchor strength	Movements: joints and faults
Rockbolt tension (torque wrench)	Movements: survey triangulation
Rockbolt tension (load cells)	Movements: survey levelling
Cable anchor tests	Movements: survey offset
Shotcrete: visual assessment	Movements: survey EDM
Shotcrete: pull tests	Vibration and blast monitoring
Shotcrete: box mould tests	Pressure – hydraulic cells
Shotcrete: core tests	Rock stress variations
Gas level measurements	Pendulum and inverted pendulum
	Strains in linings and steel ribs
Laboratory index tests for characterization	*Laboratory 'design tests'*
Water content	Triaxial strength
Porosity/density (4 methods)	Direct tensile strength
Void index (quick absorption)	Indirect (Brazil) tensile strength
Swelling pressure	Direct shear test (+ field method)
Swelling strain (2 methods)	Permeability
Slake-durability	Time-dependent and plastic properties
Uniaxial compressive strength	
Uniaxial deformability (E, v)	
Point load strength index	
Resistance to abrasion (Los Angeles test)	
Hardness (Schmidt rebound)	
Hardness (Shore scleroscope)	
Sound velocity	
Petrographic description	

selection. Engineering design tests provide data specifically required for design calculations and often may be complex and expensive. The quality control and monitoring categories of test relate to specific materials, rockbolts and shotcrete, and to methods of observation. They are of relevance to practical rock engineering, e.g. tunnelling or landslide studies, and assist in the development of practices based on empirical knowledge and provide feedback for comparisons with design calculations. Research tests devised for academic investigations into rock properties and behaviour fall beyond the scope of this section. Tests should be selected to suit a specific and well-defined requirement. For example, shear strength parameters or rock mass deformation moduli may be required for design using either limiting equilibrium methods or the finite element method respectively. Some of the more commonly used tests are summarized in Table 10.5[25] and some are shown in Figures 10.4 to 10.7. Test procedures mentioned below may generally be found in the ISRM *Suggested methods.*[24]

10.2.1 Laboratory index tests

10.2.1.1 Petrographic description

This is performed by microscopic examination of a thin section of the rock material to obtain information relevant to its mechanical behaviour such as mineral composition, grain size, texture, fabric (e.g. anisotropy), degree of alteration or weathering, and microfractures. The method does not provide a sufficiently accurate estimate of volumetric pore content (i.e. porosity), but can provide supplementary information on shape, size and interconnection of pores. The examination can be carried out only by a trained petrographer.

The information obtained can be useful in assessing abrasive properties. In particular, quartz content and intensity of cementation have been studied[26] and a cementation coefficient has been devised in an attempt to quantify the information (see Table 10.6).

10.2.1.2 Water content, porosity, density, absorption and related properties

These laboratory tests require measurement of the volumes and weights of rock constituents and of the bulk sample (Table 10.2). Bulk volume can be measured directly from caliper measurements on specimens of regular geometry, alternatively using displacement of a fluid such as mercury or water, or from buoyancy measurements using Archimedes' principle. Grain volume can be found by crushing the specimen to a powder. Pore volume is obtained using a water saturation technique or in Boyle's law gas-pressure cells. A simple quick absorption method is available that gives a void index for rocks that do not appreciably disintegrate when immersed in water. The test calls for a minimum of equipment and may be suitable for rock classification purposes as an index of porosity, or degree of weathering or alteration.

10.2.1.3 Swelling, slake durability index and durability tests

Clay-bearing rocks (shales, mudstone and some weathered igneous rocks) can swell or disintegrate when relieved of *in situ* confining stresses and when exposed to atmospheric wetting and drying. Swelling tests[24] similar to soil consolidation tests can be used to measure swelling pressure or strain during wetting of the specimen; alternatively, the swelling of an unconfined rock cube or cylinder can be measured as an index property. A quick slake durability index test can be used to measure the disintegration of rock subjected to wetting and atmospheric weathering. These tests are index tests, best used in classifying and comparing one rock with another. The swelling strain index, for example, should not be taken as the actual swelling strain that would develop *in situ*, even under similar conditions of loading and water content.

Sodium sulphate soundness and other salt crystallization tests[27] subject rock specimens to the disruptive action of salt crystals growing from solution in the pore space, and are appropriate in assessing the durability of aggregates and building stones exposed to saline conditions. Freezing and thawing tests[28] are available to measure susceptibility to frost damage.

10.2.1.4 Hardness, abrasion and attrition

The hardness and abrasiveness of rock are dependent on the type and quantity of the various mineral constituents of the rock and the bond strength that exists between the mineral grains. Quartz content in particular can be a useful guide. Tests for each property have been developed to simulate or to correlate with field experience. Many of the tests now used for rock have been adapted from highway materials, concrete and metals testing.

Abrasion tests measure the resistance of rocks to wear. These tests include wear when subject to an abrasive material, wear in contact with metal and wear produced by contact between the rocks. Abrasiveness tests can also measure the wear on metal components (e.g. tunnelling machine cutters) as a result of contact with the rock. These tests can be grouped in three categories: (1) abrasive wear impact tests; (2) abrasive wear with pressure tests; and (3) attrition tests.

The first category includes the Los Angeles test (in which the rock sample and an abrasive charge of steel spheres are rotated

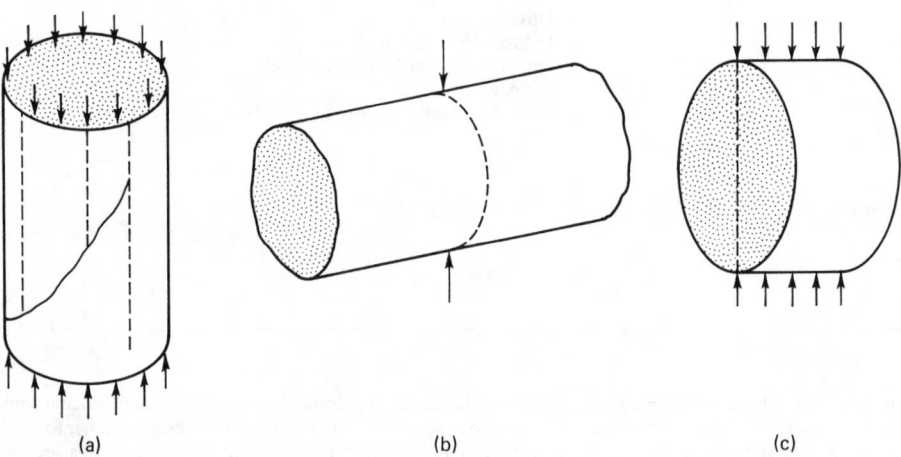

Figure 10.4 Strength index tests showing the loading configuration, specimen shape and failure pattern in: (a) The uniaxial compressive strength test; (b) the diametral point-load test; (c) the Brazilian test

Table 10.5 Summary of rock mechanics field and laboratory tests. (After Hoek (1977) 'Rock mechanics laboratory testing in the context of a consulting engineering organization', *Int. J. Rock Mech. Min. Sci.,* **14**)

Text	Purpose	Specimen preparation	Comments
(1) Point load test	Originally used to determine the tensile strength of rock (Brazilian test); now used as an index test or to estimate the uniaxial compressive strength of rock material	None	Simple, quick and inexpensive field test which gives reliable results provided that it is used on brittle, isotropic rock cores
(2) Slake durability test	Determination of the rate of breakdown of rock material under varying moisture-content conditions	None	Oven drying and weighing of sample adds complexity to simple test equipment
(3) Field shear test	Determination of shear strength of rock discontinuities in small field specimens	Specimens have to be mounted in plaster or cement with discontinuity surface horizontal	Simple test giving reasonable values for angle of friction within limitations of sample size
(4) *In situ* shear test	Shear strength determination on undisturbed samples	Area surrounding specimen has to be cleared and load frame mounted parallel to discontinuity surface	Specimen preparation relatively expensive and load limitation restricts use of apparatus to weak materials such as coals
(5) Large specimen shear tests	Determination of shear strength of discontinuities in samples of reasonable size under laboratory conditions	Trimming of specimens is usually required and specimens have to be mounted in plaster or cement with discontinuity horizontal	Tests are relatively expensive and are normally only justified on large jobs where improved accuracy of results can be used to design steeper slopes
(6) Triaxial testing of granular materials	Determination of shear strength of broken rock or weathered rock	None	Maximum lump size is limited by size of triaxial cell
(7) Uniaxial compression tests	Determination of elastic moduli and unconfined compressive strength of rock cores	Careful preparation of end of core specimens essential	Most commonly quoted rock strength value and one of the most difficult to obtain economically
(8) Triaxial compression tests	Determination of the triaxial unconfined compressive strength of rock cores	Careful preparation of ends of core specimens essential and mounting of specimen in cell critical	Only usually justified for special projects concerned with underground excavation design
(9) Anisotropic triaxial compression tests	Determination of triaxial compressive strength variations with direction in schistose rock materials	Core must be drilled at specified angles to schistosity directions. End preparation essential	Only usually justified for underground excavation design in highly anisotropic materials such as slate
(10) Triaxial shear test	Determination of shear strength of discontinuities in rock core with controlled pore-pressure conditions	As for (9) Discontinuity must be inclined in core specimen	Most effective means for carrying out shear tests with controlled pore pressure Platens must be free to move laterally to accommodate component of shear displacement and resulting mechanical complexity limits popularity of test
(11) Creep tests	Determination of deformation and failure characteristics of rocks which exhibit marked time-dependent behaviour	Preparation of core specimen ends critical and accurate control of temperature usually required	Normally only justified when designing underground excavations in evaporites such as salt or potash
(12) Stiff machine tests	Determination of the post-failure deformation behaviour of rock materials	Very accurate specimen end preparation essential	Information required for the design of yielding pillars in underground mines
(13) Scaled material model test	Simulation of behaviour of complete rock structure/ applied load interaction	Choice and preparation of model materials and loading system critical if similitude requirements are to be satisfied	Normally only justified by research institutes for national scale problems

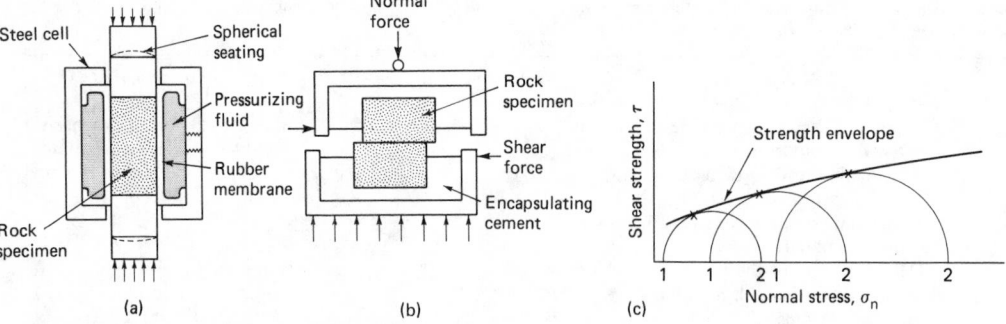

Figure 10.5 Triaxial and direct-shear tests in the laboratory. (a)
The triaxial test is most often used for intact rock; (b) the shear test
is used for determining the strength of joints or weakness planes;
(c) the Mohr–Coulomb diagram is used for plotting the results from
direct-shear tests or triaxial tests in order to obtain a strength
envelope for the rock

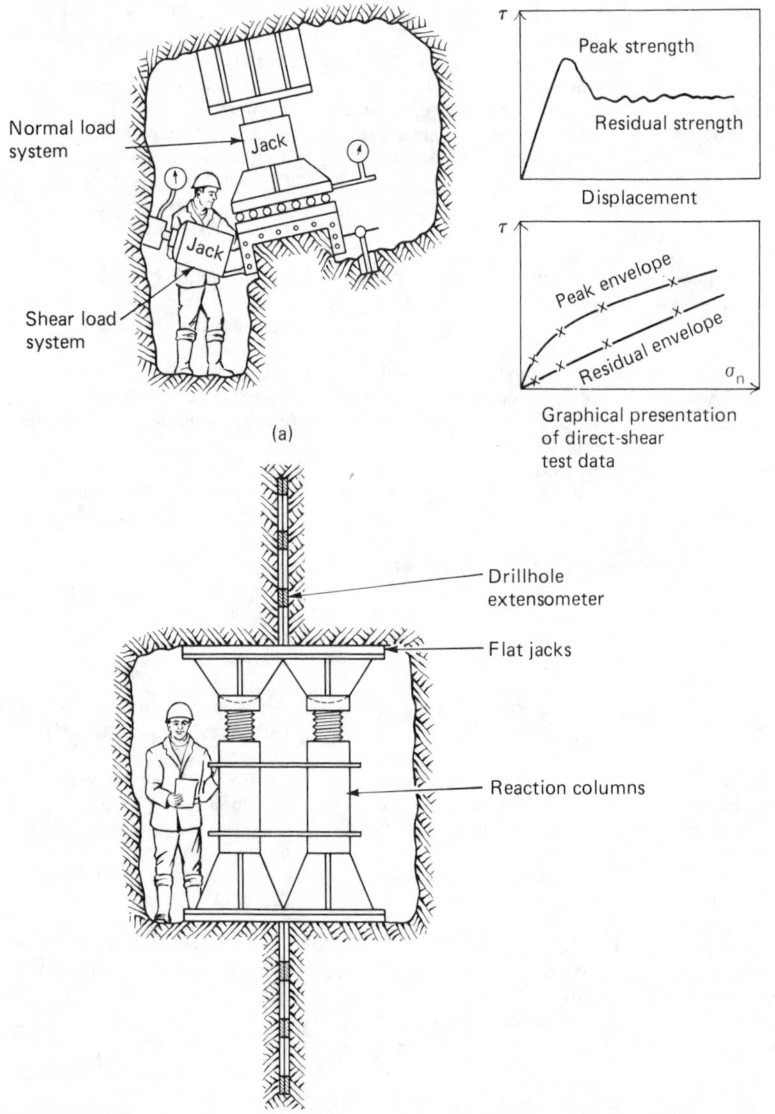

(a)

(b)   **Figure 10.6** (a) Direct-shear test; (b) deformability test

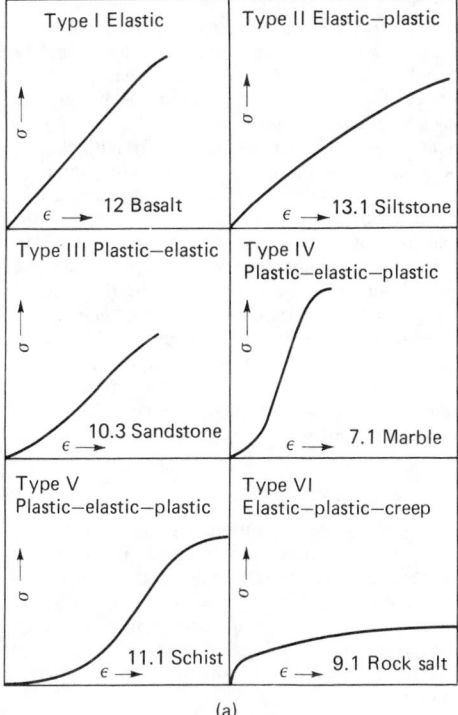

(a)

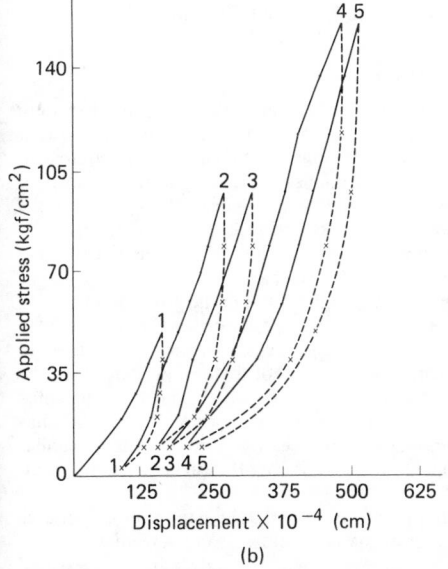

(b)

Figure 10.7 Stress–strain curves for laboratory and field deformability tests. (a) Laboratory stress–strain curves showing the influence of rock type on the shape of the curve. (After Hendron (1968) 'Mechanical properties in intact rock', in: Stagg and Zienkiewicz (eds) *Rock mechanics*. Wiley, Chichester, pp.21–53); (b) result of a field plate loading test on granulite rock at Monar dam, Scotland, showing the effect of repeated cycling of the applied load. A deformability modulus must be selected to suit the problem, and depends on the stress level, number of cycles, and whether load is increasing or decreasing

Table 10.6 Cementation coefficient. (After McFeat-Smith (1977) 'Rock property testing for the assessment of tunnelling machine performance'. *Tunnels & Tunnelling*, **9**, 3, 29–33)

(1) Noncemented rocks or those having greater than 20% voids
(2) Ferruginous cement
(3) Ferruginous and Clay cement
(4) Clay cement
(5) Clay and calcite cement
(6) Calcite (or Halite) cement
(7) Silt; Clay or Calcite with quartz overgrowths
(8) Silt with quartz overgrowths
(9) Quartz cement, quartz mozaic cements
(10) Quartz cement with less than 2% voids

in a test drum), a sand blast test[29] (where the surface of the test sample is abraded by an air blast containing abrasive powder) and the Burbank test (in which the sample abrades a metal testpiece). In the second category is the Dorry test (in which the test specimen is pressed against a rotating steel disc in the presence of an abrasive powder) and the Taber abraser test (in which an NX core sample is rotated against an abrasive wheel). Weight loss of both provides indices of abrasiveness. Various tests have been devised to determine the wear on rock drilling bits. Finally, attrition tests examine the wear produced merely by the rubbing of surfaces together. The Deval test is similar to the Los Angeles test but without the abrasive charge of steel spheres.

Hardness is not regarded as a fundamental material property because a quantitative measure of hardness depends on the type of test employed. Three types of test have been used: (1) indentation test; (2) dynamic or rebound test; and (3) scratch test.

A number of indentation tests normally used on metals have been variously applied to rocks and include Vickers hardness test and the Knoop test. A test used more extensively in the UK coal measures rocks is the National Coal Board's cone indenter test.[30] A scale of hardness and some typical values are illustrated in Table 10.7.[26]

The Shore scleroscope measures mineral hardness by recording the rebound height of a small diamond-tipped plunger after its impact with the rock surface. The Schmidt rebound hammer works on a similar principle but uses a larger plunger so that a measure of 'average rock hardness' is obtained. Care must be taken in standardizing the methods for holding the specimen and for preparing the surface at the point of impact. However, the instrument is portable and has frequently been used in the field. The use of the Schmidt hammer has been adopted particularly for the assessment of the strength of rock on joint surfaces (see section 10.3.3 below). Moh's scale of hardness commonly employed in petrographic studies is based on a scratch test

Table 10.7 Scale of hardness for the standard cone indenter test. (After McFeat-Smith (1977) 'Rock property testing for the assessment of tunnelling machine performance'. *Tunnels & Tunnelling*, **9**, 3, 29–33)

Standard hardness	Description
0–1.00	Soft
1.01–1.80	Moderately soft
1.81–2.50	Moderately hard
2.51–4.00	Hard
4.01–6.00	Very hard
6.00–	Extremely hard

applied to minerals. Micro-indentation tests are also available to measure mineral hardness under the microscope for purposes of mineral identification.

A number of derived indices have been devised based on the results of the individual tests mentioned above, and attempts have been made to correlate these with construction or excavation performance.

For the Shore scleroscope[26] the coefficient of plasticity:

$$K = \frac{H_2 - H_1}{H_1} \times 100\%$$

where H_1 is average Shore hardness and H_2 is Shore hardness after 20 tests on the same point.

In some respects this is a measure of 'work-hardening'.

Similarly, for the Schmidt hammer,[31] the deformation coefficient:

$$D = \frac{R_2 - R_1}{R_2} \times 100\%$$

where R_1 is the initial rebound number at a point and R_2 is the rebound number after 20 tests on the same point

Also for the Schmidt hammer when used on rock exposures of jointed weathered rock:[32]

$$\text{Reduction Index} = \frac{R_1 - R_F}{R_1} \times 100\%$$

where R_1 is the average rebound number from intact areas of the rock exposure free of discontinuities, and R_F is the mean value of the rebound number from randomly located tests on an exposure of fractured rock (such as a tunnel face).

To assess the tenacity of rock against the effort of rock cutters or picks such as on tunnelling machines, a number of parameters are obtained from *rock cutability tests* (or core grooving tests).[33] The test may be carried out on either core samples or on block samples. Four cuts are normally made in the rock sample at a constant depth of 5 mm with a tungsten carbide chisel-shaped tool of a standard geometry and composition, mounted on an instrumented shaping machine. The shaping machine is instrumented with a strain-gauged triaxial force dynamometer. For the standard instrumented cutting test, forces are analysed in the cutting and normal directions, since sideways forces are balanced due to the symmetrical design of the cutting tool. The strain gauge output from the dynamometer is recorded together with other information, such as weight of debris and length of cut. The analysis provides the following cutting parameters:

(1) Cutting and normal mean, mean peak and peak force components acting on the cutting tool.
(2) *Specific energy*. This is defined as the work done by the resistance forces within the rock against the cutter tip per unit volume of rock excavated.
(3) *Cutting wear*. This is the weight of tungsten carbide lost by the cutting tool during the four experimental cuts. Alternatively, cutting wear may be expressed in millimetres of wear-flat generated per metre of cut.

10.2.1.5 Strength tests on intact rock

The *uniaxial (unconfined) compression test* is mainly used for strength classification.[24,34] Rock cylinders are crushed by axial loading between steel platens to give compressive strength, defined as the ratio of failure force to specimen cross-sectional area (Figure 10.4). The length:diameter ratio for specimens should be between 2.0 and 2.5 to avoid the confining effect due to platen friction, and the end faces of the cylinder should be machined flat and parallel. Alternative crushing tests on cubes have been used for classification of rock aggregates but are not recommended in other applications.

The *point-load strength test* is an alternative strength classification.[35] Rock in the form of either core or irregular lumps is tested by compressing the specimen between conical platens, and a strength index is obtained as the ratio of failure load to the square of the distance separating the platen contact points. The test does not require machined specimens, employs portable testing equipment, and may be carried out in the field to produce a strength log for rock core. The *Brazilian test*[24,36] employs line loading of machined discs rather than point loading. The test provides an index of the tensile strength, which may also be determined by a direct tensile test.[34] *Impact tests* employing a pendulum or falling weight[37] are sometimes used for strength classification of rock aggregates.

10.2.1.6 Index tests on chalk

Chalk requires special consideration as an engineering material[38] and there are specific tests available.[39] The saturated moisture content is a measure of the porosity, and the higher the value the more water that can be released to generate an unworkable slurry when chalk lumps are broken down by compaction equipment. The chalk crushing value is a measure of the resistance to mechanical breakdown of pieces of chalk.

10.2.2 Field index tests

The quantitative description of discontinuities in rocks masses is in this category (Table 10.4), and this aspect is discussed below in section 10.3.3 and in Chapter 8. Further rock mass index properties can be assessed from the following methods.

10.2.2.1 Geophysical methods

Seismic, resistivity, magnetic and gravimetric techniques[40] were developed for mineral and oil prospecting, but have been successfully adapted to the more detailed surveys required in civil engineering. The techniques are used mainly to map rock structure but can also be used to give index values related to the mechanical character of rocks and soils.

Refraction seismic surveys measure the velocity of sound emitted usually by an explosive charge and received by one or more geophones after travelling through the ground. Velocity is usually highest in rocks of low porosity or when pores are filled with water. Therefore velocity measurements can be used to indicate the spacing and openness of fissures, particularly if the velocity of sound *in situ* is compared with that through unfissured laboratory specimens of the same rock. The method has proved useful in mapping the quality of rock in dam foundations and abutments and in assessing the effectiveness of grouting treatment.[41]

Hammer seismograph equipment, where the explosive source is replaced by a sledgehammer blow, is convenient for small-scale engineering surveys. Sound velocity can be measured between adits or, using a piezoelectric or sparker source, can be measured between drillholes or in a single drillhole, to give a more complete picture of rock quality variation.

Sound velocity values can be used in deriving dynamic elastic parameters (Young's modulus and Poisson's ratio) for the rock,[42] but the calculated values are usually quite different from the 'static' values measured, for example, by plate testing. The reason is that the rock strains produced by a travelling stress wave are of much smaller amplitude and higher stress rate and gradient than in static loading. Sound velocity can, however, be used as an index to rock deformability if a suitable correlation is established.

Field resistivity, magnetometry and *gravimetric surveys* are generally only used for structural mapping, and then in relatively few applications in comparison with seismic methods. However, these and other techniques are becoming increasingly used in *downhole geophysical logging*.[43,24] Instrumental probes are lowered to the base of a drillhole on a wireline and then drawn up the hole producing a record or 'log' of various rock properties. Common instruments measure sound velocity, spontaneous electric potentials in drilling fluid and natural or induced nuclear radiation. The drillhole logs can be used to evaluate rock porosity, density, saturation, clay content and other mechanically relevant information. These methods of geophysical logging of boreholes are summarized in Table 10.8. Caliper logging, which shows variations in hole diameter, has particular significance to the study of *in situ* stress-related 'breakouts', mentioned above in section 10.1.5.

10.2.3 Design tests

10.2.3.1 Permeability tests

The majority of flow through the rock mass usually occurs along fissures rather than pores, so that tests on *laboratory specimens* usually have limited significance. Such tests are, however, useful if the rock is very porous. Often they employ air or an inert gas rather than water as the test fluid, in order to speed up the testing procedure.

Permeability can be measured in the field using a *packer test* where a section of borehole is isolated by pneumatically inflated packers.[44,45] Tests in rock require packers that are long in relation to the test section and, preferably, also to the fracture spacing in the rock. Borehole permeability tests can be carried out by 'pumping in' under conditions of either constant or falling head, or by 'pumping out'. A graph of flow rate against test pressure is usually linear at lower pressures where laminar flow occurs, but may become nonlinear at higher pressures owing to turbulence or to the effect of water pressures in increasing the width of existing fissures. A summary of the main permeability test methods is given in Table 10.9. Analysis of test results and water flow problems in general should account for the influence of both total stress and pore pressure on rock permeability. Anisotropic permeability may require that tests be carried out with holes drilled at different inclinations.

10.2.3.2 Strength testing for design

Triaxial strength tests[46] are most often required for stress/ strength design particularly of deep underground excavations. Cylindrical specimens are tested in the laboratory by applying a confining pressure to the curved surface of the specimen, using a pressurizing fluid confined in a triaxial cell and separated from the specimen by a flexible membrane to avoid generation of pore pressures. The confining pressure is radially symmetric and provides the (equal) intermediate and minor principal stresses acting on the specimen. The major principal stress is usually applied with a hydraulic ram acting through steel platens along the axis of the cylinder. This stress is increased, usually at constant confining pressure, until the specimen fails. The test is repeated for a range of confining pressure in order to define a 'strength criterion' for the material (Figure 10.5 and section 10.3.2).

Direct-shear tests are usually employed to provide data for limit equilibrium analyses, particularly for analysis of the stability of rock slopes, dam foundations and abutments.[47] They are best suited to testing a well-defined discontinuity such as a joint, bedding plane or the interface between concrete and rock rather than to testing of intact material. Rock samples containing joint planes can be tested in the field, or in a laboratory, using a portable shear box.[9,48] Typically an *in situ* test block with dimensions $700 \times 700 \times 350$ mm is first isolated by sawing or line drilling. Stress is applied normal to the discontinuity to be tested, and a force is then applied to shear the discontinuity (Figure 10.6). To avoid generation of tension at the heel of the test block, the shearing force may be inclined so as to pass through its centre of area. A graph showing shear displacement as a function of shear force is used to find values of 'peak' and 'residual' strength. Further tests at different normal stress values allow strengths to be plotted as a function of normal stress. The shearing resistance of the discontinuity, thus evaluated, may be applied in limit equilibrium design calculations. The strength of discontinuities in rock is discussed in more detail in section 10.3.3.

10.2.3.3 Deformability testing for design

Deformability test data are required for designs involving analysis of stresses and displacements in rock, e.g. analysis of foundation settlements on softer rocks, analysis of stresses around underground openings or design of tunnel linings. The most commonly used design methods assume linear elastic behaviour for the rock and require values for the elastic parameters (Young's modulus and Poisson's ratio).

Laboratory deformability tests can be used in situations where the rock material is likely to be much more deformable than the discontinuities, e.g. in soft rocks and at depths where the discontinuities are tight. A cylindrical specimen is loaded along its axis, and strains occurring in the central third of the specimen are compared with the corresponding levels of stress. Strains are typically measured with electric resistance strain gauges or transducers bonded or clamped to the specimen surface. The stress–strain curves (Figure 10.7) are typically nonlinear, reflecting inelasticity due to the closing of pores at low stress levels and to the generation of failure cracks at higher levels of stress.[49] Elastic parameters can be obtained by approximating the slopes of these curves to straight lines over the restricted ranges of stress that are of relevance to the problem. Hysteresis effects (where the curve for unloading does not correspond to the loading curve) are observed in both laboratory and field tests and are probably due to friction acting on the surfaces of cracks and pores. The deformability of rock samples during and after 'failure' (i.e. beyond peak strength) can be examined using 'stiff' testing machines, and the relevance of such 'post-peak' behaviour is reviewed in section 10.3.2.

Rock fissures and joints are typically more open near the surface owing to weathering and stress relief, and in such materials laboratory tests can indicate that rocks are less deformable than is really the case. Field tests are designed to affect as large a volume of rock as possible in order to fully reflect the contribution of fissures to deformability.

The most economic field deformability tests are carried out in drillholes using a jack or *dilatometer*.[50,51] This instrument applies a radial pressure through either a rigid split cylinder or a flexible membrane, and the resulting radial displacements may then be analysed in terms of elastic parameters. In softer ground these tests can also give an approximation to the strength of the rock.

Plate-loading tests[17] (Figure 10.6) can be used to obtain deformability values that reflect the behaviour of a larger volume of rock. Tests using smaller-diameter plates in drillholes have the advantage of simplicity when a large number of strata

Table 10.8 Geophysical (borehole logging) methods. (After International Atomic Energy Agency (1982) *Site investigations for repositories for solid radioactive wastes in deep continental geological formations.* IAEA, Geneva)

Mode	Name of geophysical log	Basis of method	Application
Electrical (requires uncased boreholes and presence of borehole fluid, except where indicated)	(1) Spontaneous potential (SP)	Measurement of variations in natural potentials of lithologies in borehole	Lithological variations – presence of conductive or oxidizing/reducing minerals, dykes, etc.; delineation of shale–sandstone sequences. Usually run in association with normal resistivity
	(2) Normal resistivity	Measurement of formation resistivity using electrode spacings from a few to about 6 m	Variations in lithology, porosity and groundwater salinity; formation penetration depends on the spacing of the electrodes; fracture analysis
	(3) Focused resistivity laterologs and (guard logs)	Measurement of formation resistivity by focusing the current into thin sheets	Porosity and groundwater salinity; indication of occurrence of fractures, and conductive or oxidizing/reducing minerals. Higher resolution than normal resistivity logs
	(4) Induction	Measurement of formation conductivity by the measurement of secondary magnetic fields within the formation generated by an alternating magnetic field created by a borehole sonde	Measurement of porosity in dry sections of borehole indication of occurrence of fractures (can be made in dry boreholes)
	(5) Microresistivity (micrologs and microlaterologs)	Measurement of small volumes of the formation immediately adjacent to the borehole wall. Done with miniature focused resistivity logs	Porosity and lithological variations; fracture analysis relative permeability (with mud-drilling fluids only)
	(6) Dipmeter	Specialized use of microresistivity measurements; data from 4 probes are fed to a computer for the computation of dips of bedding planes and fractures	Determination of the dip (inclination) and strike of planar structures (bedding planes, fractures, etc.) intersected in the borehole
	(7) Radar	Measurement of reflected pulses of induced electromagnetic energy	Gives 3-dimensional picture of inhomogeneities inside a rock mass within a radius of approximately 80 m of a borehole. In development stage
	(8) Fluid resistivity	Measurement of variations in resistivity (conductivity) of borehole fluids	Groundwater salinity; elevations of groundwater in-flow. Important for corrections to other logs
Radiometric (can be run in cased or uncased boreholes with or without a borehole fluid except where indicated)	(1) Natural gamma	Measurement of natural gamma radiation by a scintillation counter located in a sonde	Radioactive elements are normally concentrated in clays and shales so the logs give data on lithological variations
	(2) Gamma-gamma	Measurement of reflected radiation from induced radiation by gamma scource	Bulk density. Results can be disturbed by presence of high natural gamma background
	(3) Neutron-neutron	Measurement of neutrons emitted from hydrogen atoms in the rock or pore fluid. The hydrogen atoms capture neutrons from a neutron source and emit neutrons with a different energy	Porosity. Results can be interpreted in terms of water content, both free water and in minerals
	(4) Neutron-gamma	Measurement of gamma emitted from hydrogen atoms in the pore fluid. The hydrogen atoms capture neutrons from a neutron source and emit gamma radiation	Porosity. Results can be interpreted in terms of water content. Requires uncased borehole
	(5) Spectral gamma	Measurement of natural gamma radiation over a spectrum of energies	Determination of lithological and mineralogical variations, including detection of potassium, uranium and thorium. May provide interrelationship of different fissure systems
Acoustic (requires uncased borehole, preferably with a borehole fluid)	(1) Acoustic or sonic	Measurement of time required for sound to travel a certain distance through formations surrounding a borehole	Used to measure local sonic velocity. The response for a given formation depends on its lithology, porosity and fracturing. Provides a check on cement emplacement behind casing

Mode	Name of geophysical log	Basis of method	Application
	(2) 3-dimensional sonic	Measurement of time required for reflection of induced sonic energy	Determination of number and extent of fractures
	(3) Sonic waveform	Depiction of individual waveforms at discrete depths	Determination of number and extent of fractures
	(4) Downhole seismic	Measurement of surface signal recorded by geophone clamped to side of borehole	In combination with seismic data collected from surface geophones yields information on subsurface geology without the need to rely on closely spaced boreholes
	(5) Televiewer	Measurement of time for reflection of high-frequency acoustic pulses from borehole surfaces. Records a continuous image of a borehole wall	Determination of fractures along the borehole
Miscellaneous (uncased boreholes)	Television camera	Gives visual image of borehole wall, capable of recording image on videotape for playback	Examination and measurement of fractures and other discontinuities. Requires clear borehole fluid
	Caliper	Measurement of borehole diameter. Mechanical arms in contact with the borehole wall give a log of borehole diameter	Identifies fractures at the borehole; identifies cavitation or closure of certain formations intersected by the borehole; provides corrections to other logs for hole diameter variations; checks condition of casing
	Fluid temperature	Direct measurement of temperature variations of borehole fluid	Heat flow within a particular region. Under pumping and nonpumping conditions can be used to derive information on groundwater flow
	Flowmeter	Direct measurement of the flow of water, in a vertical direction, within the borehole	Determination of vertical hydraulic gradients, groundwater conditions, levels and amounts of inflow of groundwater; identifies casing leaks and condition of screens
	Gravimetric	Measurement of *in situ* rock density	Improves interpretation of regional data on rock density
	Magnetic	Measurement of magnetic intensity and/or orientation	Detection of magnetite or pyrrhotite-rich layers or bodies; improved interpretation of regional magnetic data

are to be tested. The trend, however, is to employ loaded areas of increasingly large diameter. Tests at diameters in excess of 1 m require *flat jack techniques*[52] in order to ensure uniformity of loading and to facilitate application of the high forces that are required. Flat jacks comprise two thin steel plates circumferentially welded and inflated by hydraulic pressure. Pressure is applied in increments and the corresponding rock displacements are recorded by extensometers either at the surface or at depths beneath the loaded plate or flat jack. Reaction can be provided by the opposite wall of an adit or test chamber, or methods are available where a flat jack is grouted into a slot machined into the rock face. Even larger volumes of rock can be tested by *radial jacking in a test chamber* using flat jacks distributed around the circumference of a ring beam, or by applying hydraulic pressure to a section of adit that has been lined and plugged with concrete.

In situations where rock behaviour cannot reasonably be assumed to be elastic, because of viscous effects, the creep or time-dependent behaviour of the rock must be measured.[53] These properties are of greatest relevance to geological processes involving crustal deformations over long periods of time and in engineering their relevance is restricted to certain rock types such as evaporates, salt deposits and clay-bearing rocks under conditions where flow and squeezing of rock is likely.

10.2.4 Rock as an engineering material

Rock materials used in engineering construction include building stone, riprap and rockfill, also concrete and road aggregates.

Rocks suitable for building stone are typically homogeneous and have well-defined, planar and persistent discontinuities. Materials with a fracture spacing of about 1000 mm are suitable for monumental stone, and smaller sizes may be useful for general-purpose building. Facing stone, flags and slates may be naturally flaggy, with fracture spacing in two orthogonal directions much greater than in the third direction, or may be sawn from blocks of more cubic shape. Appearance of the stone and ease of dressing are also important. Weathering deterioration of building stones is more often associated with incipient weakness planes (e.g. cleavage or poorly cemented bedding) than with the intact material, although porous materials in particular are subject to the action of frost and salt crystallization.[54]

Rockfill and riprap selection[55] is based on similar considerations although there is more flexibility in shapes, sizes and heterogeneity of materials, and visual appearance is seldom of great importance. Large-sized material is essential for marine works to inhibit removal by tide and wave forces. An optimum size grading is essential for adequate placement and compaction and to achieve size density and placed permeability. This is most

Table 10.9 Summary of permeability test methods. (After International Atomic Energy Agency (1982) *Site investigations for repositories for solid radioactive wastes in deep continental geological formations.* IAEA, Geneva)

Type of test	Description	Duration	Results obtained
1 FIELD TESTING			
1.1 Withdrawal	Removal of water from wells		
1.1.1 Single well	Pumping from single well; measurement of rate of withdrawal and drawdown	Minutes–hours	Transmissivity
1.1.2 Multiple wells	Pumping from single well; measurement of discharge and drawdown from pumped well; measurement of drawdown in one or more radially spaced observation wells	Hours–weeks–months	Transmissivity; storage coefficient (related to porosity (n)); identification of aquifer boundaries including aquitards and aquicludes
1.2 INJECTION	Injection of a measured volume of water into a well		
1.2.1 Slug test	Injection of measured volume of water into well and measurement of rate of recovery to pretest water level	Minutes–hours	Transmissivity
1.2.2 Single packer	Injection of measured volume of water at predetermined pressure into well below the level of an emplaced packer; measurement of water flow and pressure changes	Minutes–hours	Transmissivity; size of fracture apertures
1.2.3 Double packer	Injection of measured volume of water of predetermined pressure into interval of well between packers; measurement of water flow and pressure changes	Minutes–hours	Permeability; size of fracture apertures
1.2.4 Pulse test	Injection of small volume of water into well interval isolated by double packers; measurement of hydraulic pressure decay	Minutes–hours	Permeability; size of fracture apertures
1.2.5 Tracer tests	Injection of identifiable materials into single or multiple wells; may be employed in zones of a well isolated by packers; periodic sampling of adjacent wells to determine tracer concentrations	Hours–months	Groundwater velocity fracture interconnectivity; aquifer anisotropy; hydraulic gradient direction; sorption characteristics of aquifer

economically achieved if the natural fracture spacing is such as to give approximately the correct block shapes and size grading without appreciable secondary blasting and crushing. Tests to evaluate the suitability of building stone or rockfill might, for example, include point-load strength evaluation both parallel and perpendicular to weakness planes, porosity, density measurements, and evaluation of cementing materials and rock texture by an examination of hand specimens and thin sections under the microscope. Tests on small pieces of rock cannot usually be used to predict deterioration caused by extensive planes of weakness, and an examination of the weathering of rock that has been exposed for a number of years in a quarry or rockface can often provide the answer.

Concrete and road aggregate[56] can take the form of either natural gravels, artificial materials or crushed rock. For the latter, the crushability of the potential source of aggregate may be of even greater importance than its properties in use. The material should break readily into approximately cubic frag-

ments without an excess of fines. Brittle, dense and crystalline materials are better from this point of view than porous or friable rocks. Surface roughness is advantageous for a satisfactory bond between the aggregate and cement or bitumen. A summary of relevant test methods is given in Table 10.10.

The mineralogy of the rock may also affect this bond, but probably to a lesser extent than roughness. Porosity is a major factor; some of the porous, yet strong, limestones give excellent bonding characteristics. Road surfacing materials also require to be resistant to polishing, the best polishing resistance being afforded by rocks with minerals of contrasting hardness or rocks with grains that are plucked rather than worn smooth by traffic. Rock constituents that are undesirable in that they react chemically with cement or bituminous substances can often be detected by a mineralogical examination, but tests on concrete made from the aggregates are usually also required to evaluate this hazard.[57]

Table 10.10 Summary of tests for aggregates. (After Collis and Fox (1985) *Aggregates: sand, gravel and crushed rock aggregates for construction purposes.* Geological Society, London)

Physical tests:

(1)	Aggregate grading	BS 882 and 1201:1973
		BS 812:1975
		ASTM Designations C33 and 136
(2)	Aggregate shape, angularity, sphericity, roundness, surface texture	
		BS 812:1975:Part 1
(3)	Relative density, bulk density, unit weight	BS 812:1975
		ASTM Designations C29 and 127
(4)	Water absorption	BS 812:1975
		ASTM Designations C127 and 128
(5)	Aggregate shrinkage	BRS Digest 35
	Petrographic examination	ASTM Designation C295

Mechanical tests:
Strength

(1)	Aggregate impact value	BS 812:1975
(2)	Aggregate crushing value	BS 812:1975
(3)	10% fines value	BS 812:1975
(4)	Franklin point load test	ISRM 1985
(5)	Schmidt rebound number	Duncan 1969

Durability

(1)	Aggregate abrasion value	BS 812:1975
(2)	Aggregate attrition value	BS 812:1943
(3)	Los Angeles abrasion value	ASTM Designation C131
(4)	Polished stone value	BS 812:1975
(5)	Slake durability value	Franklin 1970
(6)	Sulphate soundness	ASTM Designation C88

Chemical tests:

(1)	Chloride content	BS 812:Part 4:1976
(2)	Sulphate content	BS 1377
(3)	Organic content	BS 1377
(4)	Adhesion tests	HMSO 1962

10.3 Characterizing rock mass properties

10.3.1 Rock mass classification

Rocks may be classified using geological names only, but this approach can mislead because the names are sometimes general and depend on properties that are of little engineering significance. For example, 'granite' can be a crumbly sand or a broken rubble rather than the monolithic material implied by the name. Shales, mudstone and limestone can also exhibit an extremely broad range of engineering properties. On the other hand, there are over 2000 igneous rock names in existence, reflecting minor mineralogical changes that are usually mechanically insignificant.

Various index tests can be introduced to supplement the classification, but the samples tested in the laboratory usually represent only a very small fraction of the total volume of the rock mass. Since often only those specimens which survive the collection and preparation process are tested, the results of these tests could represent a highly biased sample.

The classification can be made more realistic by including properties characteristic of *in situ* conditions as well as hand-specimen conditions, e.g. characteristics of the fractures and discontinuities. Fracture spacing has been used together with intact strength in a number of rock classification schemes[58,59] and others have been formulated on the basis of Young's modulus and uniaxial compressive strength tests[60] (Figure 10.8) and using porosity, density and crystallinity.[61] One objective of a rock classification scheme is to allow compilation of maps and cross-sections to show the 'geotechnical' or 'geomechanical' (rather than geological) variations with depth and extent to facilitate design and to simplify analytical models. For example, a finite element mesh constructed to represent variations in geomechanical properties is likely to be more appropriate for design analysis than one based solely on geological boundaries.

A number of rock mass classification systems have been developed in an attempt to provide guidance on the properties of rock masses upon which the selection of tunnel support systems can be based. Details of the more useful systems and of how to apply them have been compiled by Hoek and Brown.[62]

A simple yet useful rock tunnel classification system based on combinations of different parameters considered to be the main causes of ground instability was devised by Terzaghi[63] for cases where steel arch supports were to be used, but the system is purely descriptive. There is no objective assessment of rock quality, although later Deere[64] and Cording, Hendron and Deere[65] further developed Terzaghi's classification to include rock quality designation (RQD).

Other systems have been devised, such as the concept of rock structure rating (RSR)[66] which takes account of the local geological structure, the pattern and orientation of joints relative to the tunnel direction and groundwater and joint condition. The RSR is then related empirically to a support requirement in terms of rib ratio (RR) for steel arch supports.

The most useful of these classifications take account of the interrelation of several key rock properties and relate them empirically to precedent case-histories. The two foremost classifications are those published by Bieniawski[67] and by Barton, Lien and Lunde.[68] These classifications include: (1) information on the strength of the intact rock materials; (2) the spacing, number and surface properties of the structural discontinuities as well as allowances for the influence of groundwater; and (3) *in situ* stresses and the orientation and inclination of dominant discontinuities.

Bieniawski developed the Council for Scientific and Industrial Research (CSIR) scheme in South Africa for classifying rock mass stability using input parameters including RQD, state of weathering, rock strength, joint spacing, separation and continuity, and groundwater flow. This scheme also considers the orientation of discontinuities with respect to the excavation, and it rates rock quality using a 'merit points' system. Graphical relationships between this classification, the span of the excavation and the time which may elapse before an unsupported length of tunnel would start to fail (the 'stand-up time') allow empirical correlations to be made between these different factors for any span up to 20 m.

The Norwegian Geotechnical Institute (NGI) method of Barton, Lien and Lunde[68] is based on an evaluation of about 200 Scandinavian tunnel or cavern case-histories and proposes an index, Q, describing 'tunnelling quality'.

The numerical value of index Q is defined by an expression in terms of a derived 'joint structure number' J_n, 'joint roughness number' J_r, 'joint alteration number' J_a, 'joint water reduction factor' J_w, and 'stress reduction factor' (SRF) as follows:

$$Q = \frac{\text{RQD}}{J_n} \times \frac{J_r}{J_a} \times \frac{J_w}{\text{SRF}}$$

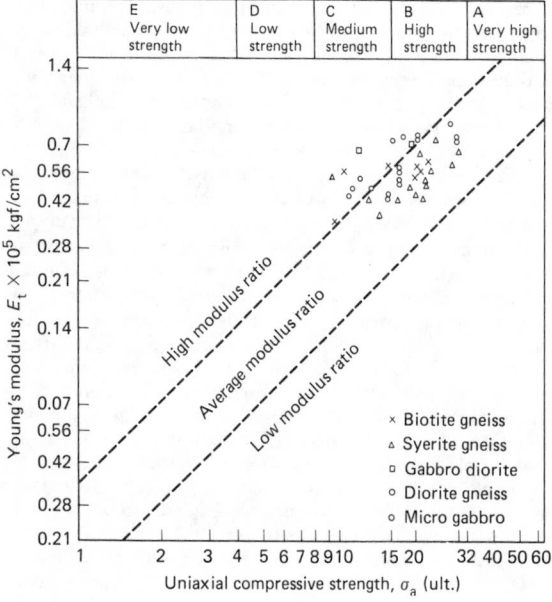

(a)

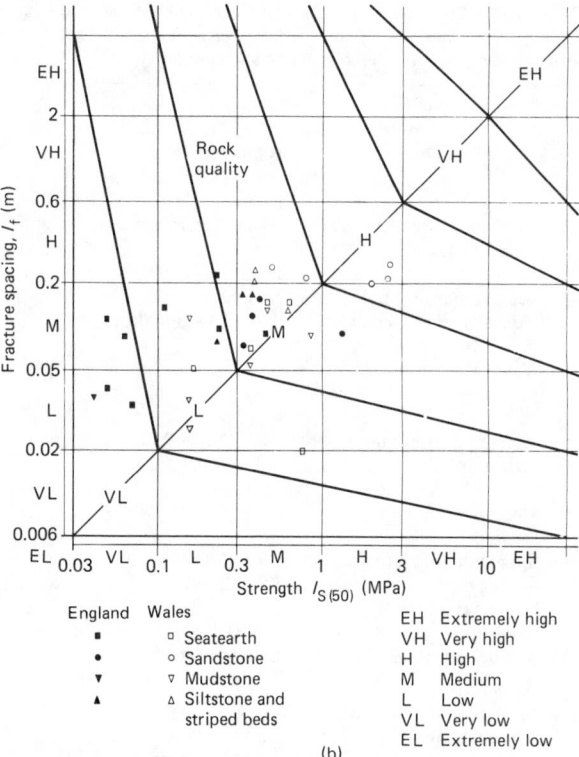

England Wales | | EH Extremely high
■ | □ Seatearth | VH Very high
● | ○ Sandstone | H High
▼ | ▽ Mudstone | M Medium
▲ | △ Siltstone and | L Low
| striped beds | VL Very low
| | EL Extremely low

(b)

Figure 10.8 Rock classification methods. (a) Rock classification using laboratory measurements of Young's modulus and uniaxial compressive strength (σ_a). (After Deere (1968) 'Geological considerations', in: Stagg and Ziekiewicz (eds) *Rock mechanics.* Wiley, Chichester, pp.1–20). The plotted results relate to specimens from the Churchill Falls underground powerhouse (see: Benson, Murphy and McCreath (1970) *Modulus testing of rock at the Churchill Falls underground powerhouse.* American Society for Testing and Materials. Special Technical Publication Number 477); (b) rock classification using observations of fracture spacing and point-load strength. The plotted results relate to specimens of Coal Measures rock core. (After Franklin, Broch and Walton (1971) 'Logging the mechanical character of rock', *Trans Inst. Min. Metall.,* **80,** A1–A10; discussion, **81,** A43–A51 (1972))

The value of Q can be related to width of span and support requirements, using tables and graphs based on precedent practice for several categories for rock mass quality. The scheme includes a factor, excavation support ratio (ESR), which varies for temporary or permanent openings and is similar to the concept of factor of safety in engineering works. The index Q is essentially a function of three parameters which represent:

(1) Block size RQD/J_n.
(2) Inter-block shear strength J_r/J_a.
(3) Active stress J_w/SRF.

These rock mass classification systems have proved to be very useful practical engineering tools not only because they provide a starting point for the design of tunnel support but also because the properties of the rock mass must be examined in a very systematic manner. The familiarity and understanding gained from this systematic study are themselves of great value to enable sound engineering judgements to be made.[69]

Caution should be exercised, however, when considering the use of these classification schemes in rock types significantly different from those from whence the case-history information is taken.

10.3.2 Strength, deformation and failure criteria

In recent years, rock mechanics has advanced such that the demands on testing work have changed. From merely providing information to classify and to put into groups rocks on a quantitative as well as descriptive basis, the results of tests must now contribute input parameters for some very powerful analytical methods. These analyses require realistic constitutive relations for stress, strain and failure criteria in order to be useful. They draw upon quantitative data from laboratory tests, field tests, descriptive observations or the classification schemes mentioned in section 10.3.1.

Deformability tests can be used to determine the stress–strain relationship (i.e. modulus and Poisson's ratio) for rocks prior to the yield point or the peak strength being attained. In addition, the development of servo-controlled stiff testing machines[70] has permitted the determination of the complete *post-peak* stress–strain curve for rocks. This information is important in the design of underground excavations since the properties of the 'failed' rock surrounding the excavations have a significant influence upon the stability of the excavations. The term 'failure' is sometimes taken to mean the attainment of the peak strength. However, when the rock is confined, the *post-peak behaviour* is also relevant, and failure may alternatively mean the rock can no longer sustain the forces applied to it. This may involve considerations other then peak strength, e.g. excessive deformation may be a more appropriate criterion of 'failure'.[71]

Post-peak stress–strain curves are used, for example, in rock-support interaction analysis, described in section 10.4.7. Also various types of elastic–plastic, or elastic–strain softening and brittle behaviour can be applied to the analysis of 'yield zones' around tunnels.[72]

The stress–strain behaviour of jointed rock masses has been studied theoretically to determine equivalent overall elastic constants for use in design analyses taking into account the presence of numerous joint sets,[73] although the modulus of rock masses in relation to empirical classification schemes has also been studied. This information is summarized in Figure 10.9, which is a useful guide to selecting initially a modulus of deformation of a rock mass for preliminary calculations.

Rock mass strength concepts have been examined using several theoretical formulations based on expressions of three-dimensional stresses or equations of strain energy. Alternatively, a number of empirical approaches have been used which in general are more useful in solving practical engineering problems.[74] The simpler failure criteria use the maximum (major) and minimum (minor) principal stresses in their formulation, generally ignoring the effects of the intermediate principal stress. Some of the criteria provide an insight into theoretical behaviour, especially in the study of *fracture mechanics*, but they rarely agree with experimental results for a broad range of rocks or rock mass characteristics.[75]

Determination of the strength of a rock mass by laboratory testing is generally not practical (Table 10.11). Hence, this strength must be estimated from geological observations and from test results on individual intact rock pieces removed from the rock mass. This question has been discussed extensively by Hoek and Brown[62] who used the results of theoretical and model studies and the limited amount of available strength data, to develop an empirical failure criterion for jointed rock masses.

The empirical equation of Hoek and Brown's strength criterion which describes a Mohr failure envelope is:

$$\sigma_1' = \sigma_3' + (m\sigma_c\sigma_3' + s\sigma_c^2)^{1/2}$$

where σ_1' is the major principal effective stress at failure, σ_3' is the minor principal effective stress or the confining pressure in the case of a triaxial test, σ_c is the uniaxial compressive strength of the intact rock material, and m and s are empirical constants.

The equation may be normalized with respect to σ_c by division to give:

$$\sigma_{1n}' = \sigma_{3n}' + (m\sigma_{3n}' + s)^{1/2}$$

which is a very useful form when comparing the shape of these Mohr failure envelopes for different rocks (Figure 10.10).

The criterion thus contains three constants m, s and σ_c. Both m and s depend on the properties of the rock and the extent to which it is broken before the application of failure stresses. The value of s is zero for a broken rock with no tensile strength and no cohesion, ranging up to 1 for intact rock. It is found that m has values characteristic of certain rock types. The higher values (15 to 25) are associated with brittle igneous rocks with high friction angles such as granites, and lower values (3 to 7) with more ductile rocks such as limestones. The uniaxial compressive strength σ_c was adopted for the criterion as it is such a widely quoted index of rock property and can be reasonably estimated at the initial stages when perhaps no firm data is available. The three constants can be determined statistically from experimental data, although values for m and s have been

Table 10.11 Summary of range of rock mass characteristics. (After Hoek (1983), 'Strength of jointed rock masses', *Géotechnique*, **33**, 3)

	Description	Strength characteristics	Strength testing	Theoretical considerations
	Hard intact rock	Brittle, elastic and generally isotropic	Triaxial testing of core specimens in laboratory relatively simple and inexpensive and results usually reliable	Theoretical behaviour of isotropic elastic brittle rock adequately understood for most practical applications
	Intact rock with single inclined discontinuity	Highly anisotropic, depending on shear strength and inclination of discontinuity	Triaxial testing of core with inclined joints difficult and expensive but results reliable. Direct shear testing of joints simple and inexpensive but results require careful interpretation	Theoretical behaviour of individual joints and of schistose rock adequately understood for most practical applications
	Massive rock with a few sets of discontinuities	Anisotropic, depending on number, shear strength and continuity of discontinuities	Laboratory testing very difficult because of sample disturbance and equipment size limitations	Behaviour of jointed rock poorly understood because of complex interaction of interlocking blocks
	Heavily jointed rock	Reasonably isotropic. Highly dilatant at low normal stress levels with particle breakage at high normal stress	Triaxial testing of undisturbed core samples extremely difficult due to sample disturbance and preparation problems	Behaviour of heavily jointed rock very poorly understood because of interaction of interlocking angular pieces
	Compacted rockfill	Reasonably isotropic. Less dilatant and lower shear strength than *in situ* jointed rock but overall behaviour generally similar	Triaxial testing simple but expensive because of large equipment size required to accommodate representative samples	Behaviour of compacted rockfill reasonably well understood from soil mechanics studies on granular materials
	Loose waste rock	Poor compaction and grading allow particle rotation and movement resulting in mobility of waste rock dumps	Triaxial or direct shear testing relatively simple but expensive because of large equipment size required	Behaviour of waste rock adequately understood for most applications

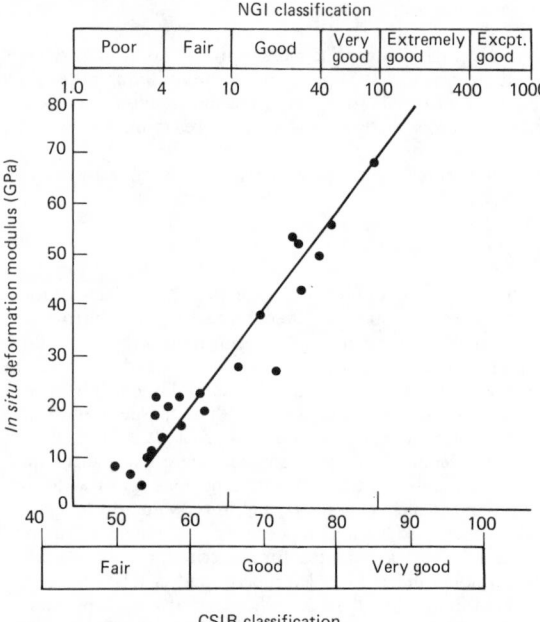

Figure 10.9 The relationship between *in situ* deformation modulus of rock masses and rock mass classification. (After Hoek and Brown (1980) *Underground excavations in rock*. Institution of Mining and Metallurgy, London)

proposed based on the rock mass classification systems and these are shown in Table 10.12. The effect of comparative size of the proposed excavation and the spacing of discontinuities in the rock mass (see Figure 10.11) is important to the design. Table 10.12 should be used only for rock masses containing several joint sets or for intact rock. There are other specified procedures necessary to apply this method to anisotropic or schistose conditions influenced by dominant discontinuities.

This criterion gives reasonably good estimates for disturbed hard rock masses and tends to be conservative in tightly interlocking hard rock masses. A similar empirical criterion by Johnston[75] appears to extend the principle to intact materials ranging down to soft rocks and firm clays.

10.3.3 Properties of joints and discontinuities

Standard methods of observing, recording and describing the characteristics of discontinuities mentioned above in sections 10.1.3 and 10.1.4 have been put forward by the ISRM[24] in which many practical aspects are described. A joint survey of a rock mass can be carried out either in a *subjective* (biased) manner where only those discontinuities which appear to be important are described, or in an *objective* (random) manner where all joints within a delineated area are described. A subjective survey is best applied where structural patterns can be identified clearly and will save time and effort, whereas an objective survey is time-consuming even though automatic data processing may be used to advantage to analyse all the data. The principal survey methods are:

(1) Outcrop or excavation (e.g. tunnel wall) exposure description.
(2) Drillcore or drillhole wall description.
(3) Terrestrial photogrammetry.

For method (1) 'scanlines' are often used to define a sample of

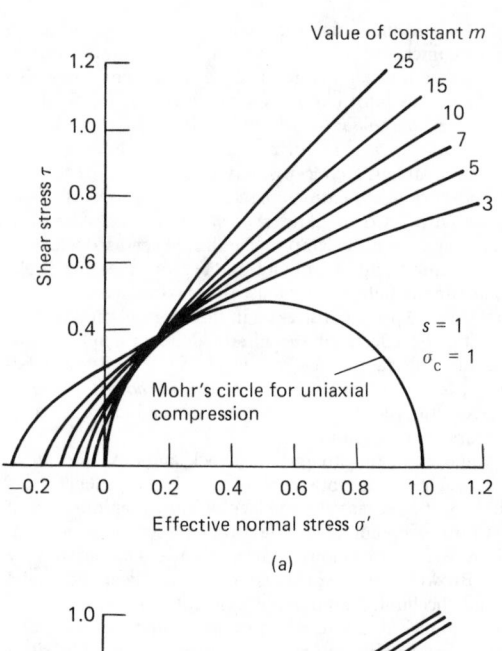

(a)

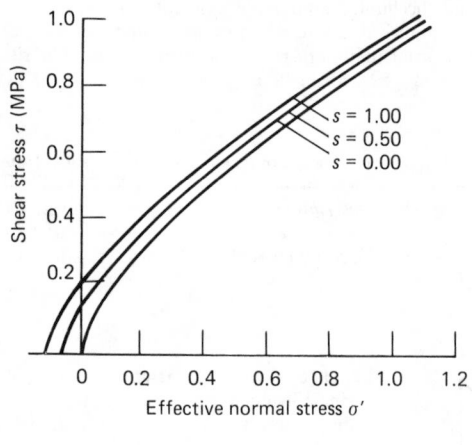

(b)

Figure 10.10 The influence of the values of *m* and *s* in Hoek and Brown's failure criterion. (a) Influence of the value of the constant *m* on the shape of the Mohr failure envelope at different effective normal stress levels; (b) influence of the value of the constant *s* on the shape of the Mohr envelope at different normal stress levels. (After Hoek (1983) 'Strength of jointed rock masses', *Géotechnique*, **33**, 3, 187–223)

the exposure and as such are similar to the linear sampling of the rock mass afforded by drillholes in method (2). The statistical significance of these samples in relation to the three-dimensional rock structure, the nature of probable errors and statistical methods of analysing the various data have been examined.[8,10]

There are three fundamental properties of discontinuities which significantly affect the rock mass and they are: (1) strength (or weakness); (2) deformability (or stiffness); and (3) conductivity (or permeability).[76] They are all interrelated, or coupled and they are stress-dependent. The parameters' strength and stiffness are strongly stress-dependent and may vanish under tensile stress. When under compression, however, they vary between fairly well-understood limits. The parameter which varies most of all under varying compression and shear is the joint aperture and, hence, the hydraulic conductivity.

Table 10.12 Approximate relationship between rock mass quality and material constants m and s. (After Hoek (1983) 'Strength of jointed rock masses', *Géotechnique*, **33**, 3)

Empirical failure criterion $\sigma_1' = \sigma_3' + (m\sigma_c\sigma_3' + s\sigma_c^2)^{1/2}$ σ_1' = major principal stress σ_3' = minor principal stress σ_c = uniaxial compressive strength of intact rock m, s = empirical constants		Carbonate rocks with well-developed crystal cleavage, e.g. dolomite, limestone and marble	Lithified argillaceous rocks, e.g. mudstone, siltstone, shale and slate (tested normal to cleavage)	Arenaceous rocks with strong crystals and poorly developed crystal cleavage, e.g. sandstone and quartzite	Fine-grained polymineralic igneous crystalline rocks, e.g. andesite, dolerite, diabase and rhyolite	Coarse-grained polymineralic igneous and metamorphic crystalline rocks, e.g. amphibolite, gabbro, gneiss, granite, norite and quartzdiorite
Intact rock samples Laboratory size samples free from pre-existing fractures		$m=7$ $s=1$	$m=10$ $s=1$	$m=15$ $s=1$	$m=17$ $s=1$	$m=25$ $s=1$
Bieniawski (1974b) (CSIRO)* rating	100					
Barton et al. (1974) (NGI)† rating	500					
Very-good-quality rock mass Tightly interlocking undisturbed rock with rough unweathered joints spaced at 1–3 m		$m=3.5$ $s=0.1$	$m=5$ $s=0.1$	$m=7.5$ $s=0.1$	$m=8.5$ $s=0.1$	$m=12.5$ $s=0.1$
Bieniawski (1974b) (CSIRO)* rating	85					
Barton et al. (1974) (NGI)† rating	100					
Good-quality rock mass Fresh to slightly weathered rock, slightly disturbed with joints spaced at 1–3 m		$m=0.7$ $s=0.004$	$m=1$ $s=0.004$	$m=1.5$ $s=0.004$	$m=1.7$ $s=0.004$	$m=2.5$ $s=0.004$
Bieniawski (1974b) (CSIRO)* rating	65					
Barton et al. (1974) (NGI)† rating	10					
Fair-quality rock mass Several sets of moderately weathered joints spread at 0.3–1 m, disturbed		$m=0.14$ $s=0.0001$	$m=0.20$ $s=0.0001$	$m=0.30$ $s=0.0001$	$m=0.34$ $s=0.0001$	$m=0.50$ $s=0.0001$
Bieniawski (1974b) (CSIRO)* rating	44					
Barton et al. (1974) (NGI)† rating	1					
Poor-quality rock mass Numerous weathered joints at 30–500 mm with some gouge. Clean, compacted rockfill		$m=0.04$ $s=0.00001$	$m=0.05$ $s=0.00001$	$m=0.08$ $s=0.00001$	$m=0.09$ $s=0.00001$	$m=0.13$ $s=0.00001$
Bieniawski (1974b) (CSIRO)* rating	23					
Barton et al. (1974) (NGI)† rating	0.1					
Very-poor-quality rock mass Numerous heavily weathered joints spaced at 50 mm with gouge. Waste rock		$m=0.007$ $s=0$	$m=0.010$ $s=0$	$m=0.015$ $s=0$	$m=0.017$ $s=0$	$m=0.025$ $s=0$
Bieniawski (1974b) (CSIRO)* rating	3					
Barton et al. (1974) (NGI)† rating	0.01					

*CSIRO Commonwealth Scientific and Industrial Research Organization.
†NGI Norwegian Geotechnical Institute.
(For references used in this table, consult source, above.)

Stresses applied normal to the plane of the discontinuity (normal stresses) will cause closure (i.e. reduction of the aperture) and the 'normal stress' to 'normal displacement' ratio is the 'normal stiffness'. Shear stresses cause shear displacements and this ratio is the shear stiffness; shear displacements, however, usually result in dilatation (i.e. increase in aperture) of the discontinuity. The above are all governed by the rock-to-rock contact of asperities across discontinuities which is described by the 'roughness' and 'waviness', 'wall strength' of the discontinuity surfaces and the friction angle.

An important factor affecting shear strength is the magnitude of the effective normal stress σ_n' acting across the joint. In many rock engineering problems the maximum effective normal stress will lie in the range 0.1 to 2.0 MPa for those joints considered critical for stability.

It has been customary to fit Coulomb's linear relation $\tau = c + \sigma_n \tan \phi$ to the results of shear strength investigations on rock joints (τ is peak shear strength, c is the cohesion intercept and ϕ is the friction angle). If this equation is applied to the results of shear tests on *rough joints*, under both high normal

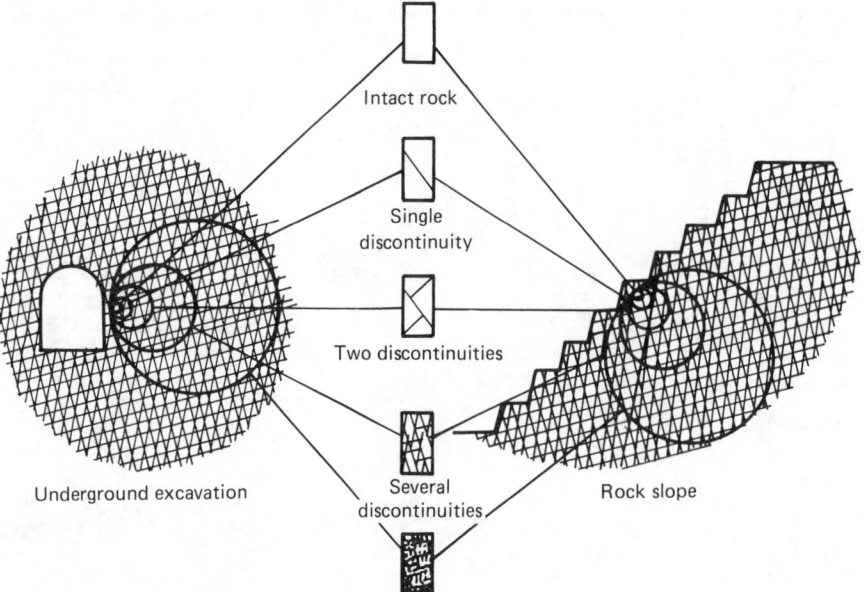

Figure 10.11 A simplified representation of scale on the type of rock mass behaviour model which should be used in designing underground excavations or rock slopes. (After Hoek (1983) 'Strength of jointed rock masses', *Géotechnique,* **33**, 3, 187–223)

stress and low normal stress, one finds the former recording a cohesion intercept of tens of megapascals and a friction angle of perhaps only 20°, while the latter has a friction angle of perhaps 70° and zero cohesion. The peak shear strength envelopes for *nonplanar* rock joints are strongly curved.

This behaviour can be explained[77] by assuming two mechanisms, illustrated for practical situations in Figure 10.12(a):

(1) Dilatation, in which the moving block rides up and over the surface irregularities – assumed to be applicable when the normal stress on the surface is small.
(2) Shear or fracture through the irregularities resulting in a planar surface and nondilatational slip – assumed to be applicable when the normal stress on the surface is large.

These ideas suggest a bilinear strength envelope such as illustrated in Figure 10.12(b). The angle ϕ represents the friction angle between clean planar rock surfaces and the angle i is the angle of dilatation.

For the first mechanism the apparent friction angle is $(\phi + i)$ and the joint behaviour is described by the relation:

$$\tau = \sigma_n \tan (\phi + i)$$

It should be noted that extrapolation back to low values of stress of the upper part of the bilinear strength envelope results in gross overestimation of strength represented by the cohesion intercept, c.

Empirical methods of quantifying roughness and wall strength and utilizing them in shear strength relations were developed by Barton and Choubey.[78] They proposed that the peak shear strengths of joints τ in rock could be represented by the empirical relation:

$$\tau = \sigma_n' \tan \left[\text{JRC} \log_{10} \left(\frac{\text{JCS}}{\sigma_n'} \right) + \phi_r' \right]$$

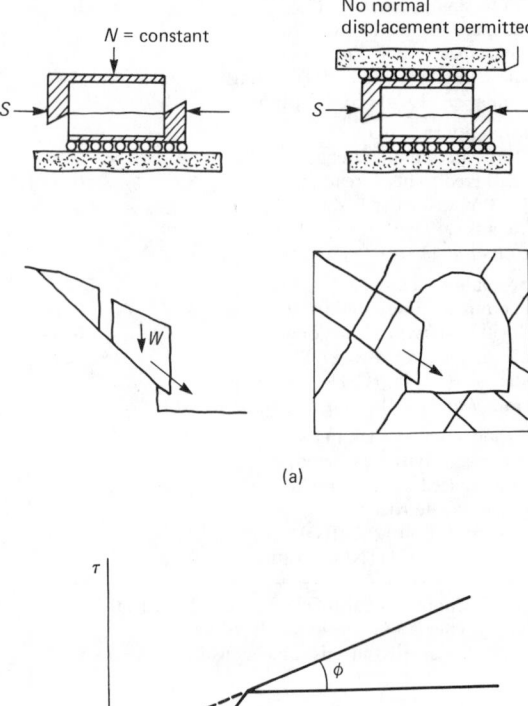

(a)

(b)

Figure 10.12 (a) Examples of two types of shear behaviour on joints; (b) bilinear shear strength envelope

where σ_n' is effective normal stress, JRC is joint roughness coefficient on a scale of 1 for the smoothest to 20 for the roughest surfaces, JCS is joint wall compressive strength and ϕ_r' is drained, residual friction angle.

This equation suggests that there are three components of shear strength – a frictional component given by ϕ_r', a geometrical component controlled by surface roughness (JRC) and an asperity failure component controlled by the ratio (JCS/σ_n'). As Figure 10.13 shows, the asperity failure and geometrical components combine to give the net roughness component i. The total frictional resistance is then given by $(\phi+i)$. The equation and Figure 10.13 show that the shear strength of a rough joint is both scale-dependent and stress-dependent.

The JRC and ϕ_r' can be obtained indirectly from extremely simple tilt tests using pieces of intact and jointed core, as shown in Figure 10.14. The ideal sample would be jointed axially, but routine testing can also include obliquely jointed samples, as typically recovered from a drilling program. The JRC, ϕ_r' are obtained from such tests as follows:

$$JRC = \frac{a - \phi r'}{\log(JCS/\sigma'_{n_0})}$$

where α is the tilt angle when sliding occurs and σ'_{n_0} is the corresponding value of effective normal stress when sliding occurs (weigh upper sample, correct for cos α, measure joint area).

The value of JRC typically varies from 0 to about 15, 0 corresponding to residual nondilatant joint surfaces for which:

$$\alpha = \phi_r'$$

The residual friction angle ϕ_r' may be lower than the *basic friction angle* ϕ_b obtained from the tilt tests on core cylinders (lowest sketch Figure 10.14) due to weathering or alteration effects. A simple empirical relation was developed to estimate ϕ_r' from ϕ_b using the Schmidt hammer. It is based on rebound tests on both the unweathered, dry rock (rebound R) and on the weathered and saturated joint wall (rebound r):

$$\phi_r' = (\phi_b - 20) + 20\frac{r}{R}$$

The *Schmidt hammer* test is also used to estimate the JCS of the fresh or altered joints in the saturated state, if this is appropriate to *in situ* conditions. The relationship derived by Miller[79] (Figure 10.15) is used to convert the rock density and the rebound r to estimate the compression strength JCS. Full details of the above characterization tests were given by Barton and Choubey.[78]

The three parameters (JRC, JCS, ϕ_r) are all that are needed to develop shear strength, displacement, dilatation and normal stress-closure curves for any given joint. Coupling conductivity with these processes requires additional information concerning the initial joint aperture but similar methods are available for such analyses.[7]

Discontinuities may contain infilling materials such as clay 'gouge' in faults, silt in bedding planes, weak low-friction minerals such as chlorite or stronger materials such as quartz or calcite in veins. Clay gouge infill will decrease both joint stiffness and shear strength, low friction mineral coatings will decrease friction angles (hence shear strength), while vein materials like quartz will increase the strength and stiffness. Occasionally the last category mentioned is described as a 'healed joint' but this term is not encouraged here. Clearly, major joints infilled with weak materials such as clay or silt are of particular concern as they may present the dominant failure mechanism. These weak materials are usually analysed as soils using an effective stress Coulomb relationship. The behaviour of filled discontinuities often has two characteristics, the first reflecting the deformability of the infill material without rock-to-rock contact, and the second reflecting the deformability and shear strength of rock asperities in contact. Swelling clay infill is dangerous as it can develop high swelling pressures and will lose strength on swelling.

10.4 Design methods

The purpose of designing a rock structure (e.g. a tunnel, cavern, open cut or foundation) is to decide the size and shape of the excavation, to determine whether measures to improve the rock conditions such as grouting, rock reinforcement, anchoring or drainage are needed, whether rock support such as shotcrete, mesh, rockbolts, steel arches, or concrete liners are required and to select and design the appropriate systems as necessary.

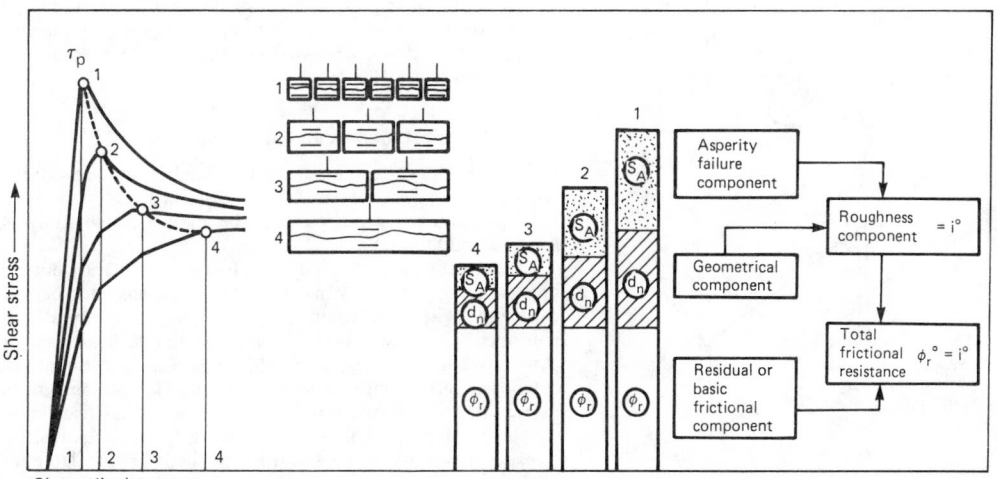

Figure 10.13 An illustration of the size-dependence of shear stress-deformation behaviour for nonplanar joints. (After Bandis, Lumsden and Barton (1981) 'Experimental studies of scale effects on the shear behaviour of rock joints'. *Int. J. Rock Mech. Min. Sci.,* **18**)

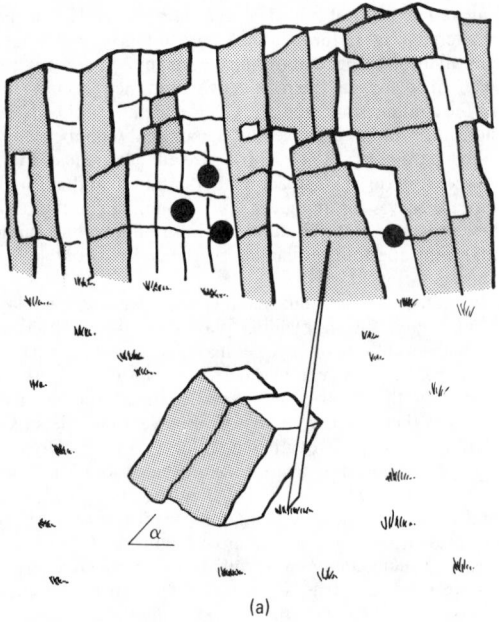

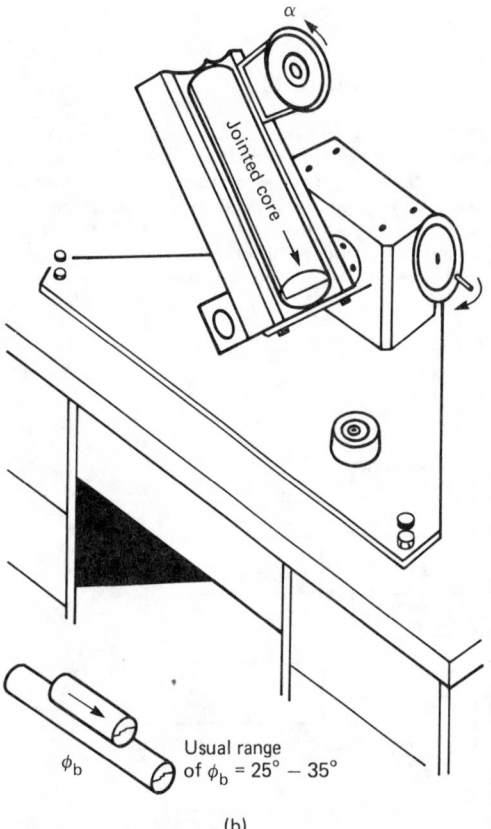

Figure 10.14 Tilt tests for obtaining joint roughness and basic friction parameters. (After Barton (1986) 'Deformation phenomena in jointed rock'. *Géotechnique*, **36**, 2)

The analysis of a rock structure should not start without first preparing a complete statement of the factors involved. These usually include the geometry and intended purpose of the structure together with the main elements of the rock engineering matrix described previously. That is, the structure of the rock, mechanical properties of intact rock materials and the discontinuities, the nature of water and stress conditions in the mass and the proposed construction/excavation method. It is not usually possible to take each factor into full account, so that analysis is often based on simplified mathematical or physical 'models'. The choice of simplifying assumptions and the errors that these assumptions are likely to introduce are matters for engineering judgement, and it is often advisable to check a design by carrying out more than one type of analysis wherever possible. 'Failure' of the structure needs to be defined (e.g. excessive displacements) and potential failure mechanisms need to be identified so that design calculations are addressed to these specifically. The adequacy of the initial design can be checked by instrumentation and monitoring and modified if necessary.

Design methods for the stability of the rock structure consider either the resulting *stresses* in the surrounding rock or the *displacements* which are induced or which become kinematically feasible, e.g. wedge failure on joint planes. In applying methods of stress analysis, judgement is required in deciding whether or not the rock should be regarded as a continuum, and then whether as an infinite space or half space. Various conditions of anisotropy and of inelastic behaviour can be simulated with some models. In discontinuous media, methods are available to analyse the displacements, provided parameters describing rock joint behaviour can be determined.

10.4.1 Empirical methods

A useful first approach is design by 'rule of thumb' using design principles that experience has shown to give satisfactory results. Where no precedent exists simple rules can be established by undertaking a programme of field observations to determine relationships between 'cause' and 'effect'. This is more likely to succeed if it is preceded by an attempt to establish, using theory and simple methods, those parameters that might prove fundamental to the design, and the trends to be expected.[9,80] Empirical design rules are usually only safe to apply in the context for which they were originally formulated, and extrapolation can be unreliable, particularly if the method has no theoretical basis.

An example of empirical methods based on precedent practice is the classification of rock in relation to support requirements for underground openings, namely the Q system, RSR and others (see section 10.3.1).

10.4.2 Physical models and analogue methods

Physical, as opposed to mathematical, models can be used in a laboratory analysis of stresses or displacements in a rock structure, and can sometimes also be applied to the study of fracture and failure.[81]

Elastic models may be used to analyse stresses mainly but are applicable where the rock may reasonably be assumed to behave elastically. The most common type, a *photoelastic* model, is machined from a stress-birefringent material such as glass or plastic. When loaded and viewed in polarized light the model exhibits coloured fringes (isochromatics) that follow contours of maximum shear stress in the model. Black fringes (isoclinics) visible in plane polarized light indicate the principal stress directions. Stress-freezing and other methods are available for analysis of problems in three dimensions. *Other types of elastic model* have also been used, e.g. metal sheet with resistance strain gauges or moire fringe grids mounted on the surface. Elastic modelling has been largely superseded in most applications by

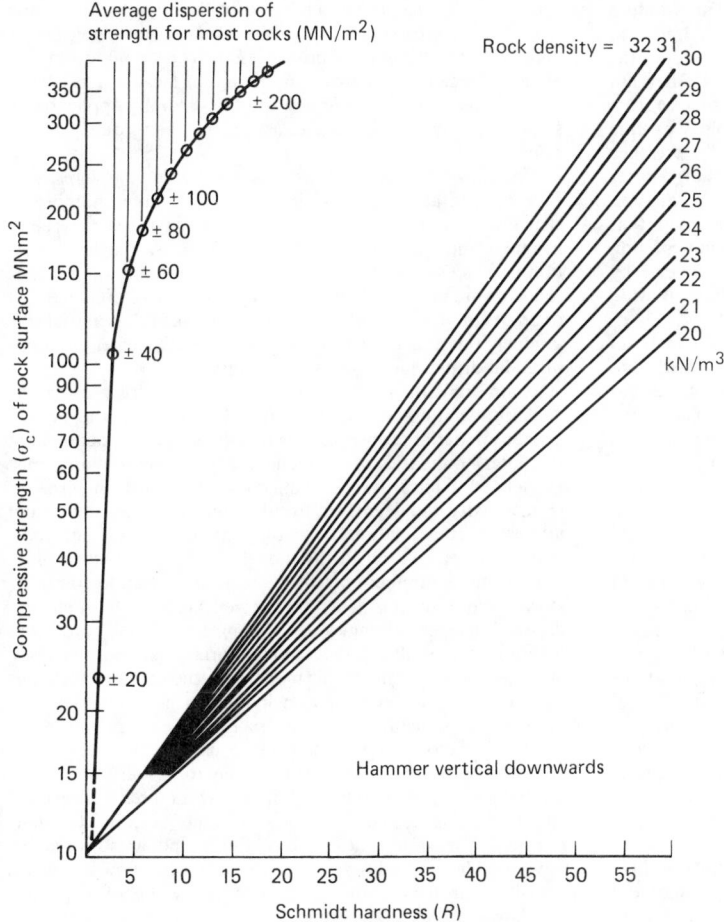

Figure 10.15 Correlation chart for the Schmidt hammer, relating rock density, compressive strength and rebound number. (After Miller (1965) 'Engineering classification and index properties for intact rock'. PhD thesis, University of Illinois)

computer analyses that afford greater flexibility, require fewer simplifying assumptions and less preparation time, and can also solve inelastic problems. Photoelastic models, however, can still be useful in presenting a visual and easily understood representation of stress distributions, and can cope easily with complex geometrical configurations.

Inelastic physical models and *block models* to study displacements are built from materials chosen, according to principles of dimensional similarity, to scale-down prototype properties such as density, strength and deformability.[82] Physical models have been used, for example, to investigate the behaviour of rock slopes, underground excavations at various stages of construction, and subsidence above mine workings.[83] Gravity can be simulated by building the model lying on a flat surface with a movable backing sheet which can be drawn in the downward direction: the friction forces on the blocks simulate gravity forces. Such models are known as 'base-friction models'.[84] Simple and approximate physical models can be valuable at the early stages of analysis in helping to visualize possible kinematic mechanisms and in formulating the problem, but care is needed to select appropriate scaling factors.

The equations governing electric potential differences and currents are analogous to the equations governing stress distributions or water flow in the rock mass. Thus, stress or water

problems can be simulated and solved by electric analogue methods.[85,86] *The conducting paper method*, of limited accuracy and flexibility but simple, uses an impregnated paper with probes to monitor surface potential differences. *The resistance network method* uses a grid of interchangeable or variable resistors, can solve anisotropic and heterogeneous problems and is more accurate, but has the disadvantage of restricting measurements to a limited number of nodal points. Analogues other than electric ones are possible, e.g. the Hele–Shaw method uses the flow of viscous fluids between closely spaced parallel plates.[87]

10.4.3 Numerical methods of analysis

Analysis of stresses, strains and displacements based on principles of classical stress analysis and continuum mechanics[88] assume that the material is continuous throughout and that conditions of equilibrium and compatibility of displacements are satisfied for given boundary conditions. The constitutive equation, i.e. the relationship between stress, strain and time for the rock mass, must be known or assumed in order to formulate the problem. This relationship can in theory be established by testing, although in practice, tests serve only to measure the parameters in an idealized constitutive equation such as one of linear elasticity. A satisfactory constitutive equation should account for rock behaviour both before, during and after failure in intact material. In most rock structures, zones of fractured rock can develop owing to the induced stresses exceeding the rock strength (zones of 'overstress') which, because they are confined by unfractured material, do not lead to collapse. It is important to recognize when it is or is not reasonable to assume that a problem may be analysed as a continuum. The presence of fractures or discontinuities does not invalidate the premises of continuum mechanics provided that a constitutive equation can be formulated for an 'element' or test specimen that incorporates a large number of such discontinuities. Soil materials, for example, contain discrete grains bounded by discontinuities but can be tested in this way, thus allowing continuum mechanics to be applied to soil problems.

Having formulated an appropriate equation for the material the problem is solved taking into account the geometry and boundary conditions for the rock structure, and ensuring that conditions of equilibrium and compatibility of displacements are satisfied. Particular solutions which are most useful in considering tunnels or underground openings are those for circular holes in stressed elastic media. Equations for analysing thick-walled cylinders or circular holes in an infinite elastic solid (Kirsch equations) will determine the radial and tangential stresses around the surface and within the rock and the displacements surrounding the opening.[89] A wide range of *exact* or *'closed form' solutions* are available for solving two-dimensional problems of various geometries, particularly for linear elastic behaviour, but few solutions have been formulated for three-dimensional problems. Boussinesq and Cerruti give solutions for normal and shear point-loading on a three-dimensional elastic half-space that can be used to build up, by a process of simple superposition, the distribution of stresses and displacements for any system of applied loads.[90–91] Savin[92] gives closed-form solutions for stress concentrations around holes in an elastic plate; these and other solutions are reviewed by Obert and Duvall.[93] In practice, it is unnecessary to derive solutions to particular problems because published collections exist of most analytically tractable problems. Such a collection by Poulos and Davis[94] is most thorough. It is important, however, to ensure that an appropriate solution is being applied to each design problem.

10.4.4 Computational methods of analysis

For many design problems in rock mechanics it is necessary to seek a more detailed understanding of stress distribution than can be obtained by superposition of standard analytical solutions. Conditions of complex geometry, rock mass anisotropy, nonlinear constitutive behaviour and nonhomogeneity require more versatile methods of solution. To solve the many stress analysis problems for which no solution in closed form is available one must resort to *numerical approximation methods* for solving the continuum mechanics equations. These methods are now widely used since computers have become generally available. There are two categories of computational methods of analysis, namely differential methods and integral methods (see later). For the first of these the problem is divided up into a set of discrete elements and the solution based on numerical approximations of the governing equations, i.e. the differential equations of equilibrium and the stress–strain–displacement relations.

The *finite difference (relaxation) method*[90] has had a long-established use in civil engineering. Partial differential equations that define material behaviour and boundary conditions are replaced by finite-difference approximations at a number of discrete points throughout the rock mass. The resulting set of simultaneous equations is then solved. A finite difference computer program is available specifically designed for modelling

soil and rock behaviour and is based on the Lagrangian method (FLAC).[95] It is capable of modelling elastic, anisotropic-elastic and elasto-plastic material properties and can simulate joint slip planes. Its main advantage is in solving for large displacements and collapse due to plastic flow, and is generally applicable to slope and foundations analyses as well as underground excavations.

The *finite element method*[96,97] is similar in many respects, except that the rock mass is subdivided into a number of structural components or interacting elements that may be of irregular and variable shape (Figure 10.16). A judicious selection of element is critical to the efficiency of computation. The elements are assumed to be interconnected at a discrete number of points on their boundaries, and a function is chosen to define uniquely the state of displacement within each element in terms of nodal displacements at element boundaries.

Strain may then be defined and, hence, stress using the constitutive equation for the material. Nonlinear and heterogeneous material properties may readily be accommodated, but the outer boundaries of the model (or 'problem domain') must be defined arbitrarily. The boundary conditions, in terms of relative fixity and degrees of freedom, may influence the area of interest to the analysis depending upon how distant these boundaries can be set in the model. There are practical limitations on the number of elements used in relation to computer storage, and the elements themselves need to be well-conditioned shapes (triangles or rectangles). Nodal forces are determined in such a way as to equilibriate boundary stresses, and the stiffness of the whole model may then be formulated as the sum of contributions from individual elements. The response of the structure to loading may then be computed by the solution of a set of simultaneous equations. Finite element computer programs have been written for a variety of rock mechanics problems, to tackle both two- and three-dimensional situations, elastic, plastic, and viscous materials, and to incorporate 'no-tension zones', joints, faults and anisotropic behaviour. The method is also used to solve water-flow problems, heat-flow problems, and an even wider scope of situations unrelated to rock engineering.

Analytical techniques, which are based on continuum idealization, are not always suitable for jointed rock problems. Numerical methods such as the finite element method or finite difference method in which rock masses are simply considered as elastic or elasto-plastic continua, are not suitable to model the geometric irregularity in natural jointing.

Several methods have been proposed to simulate such discontinuous media. They are divided into two groups: (1) 'equivalent' continuum analyses in which the jointed rock mass is represented by a homogeneous, anisotropic and continuous medium;[98,99] and (2) the methods which can deal with discontinuities directly and can express positively the behaviour of discontinuous rock masses.

Finite element techniques using joint elements[100] have been used in the analysis of certain problems, particularly in configurations involving a relatively small number of major faults or joints. However, for models with dense jointing, a large number of degrees of freedom are required.

The *distinct element method* or *dynamic relaxation method*[101] allows a problem to be formulated assuming rock blocks to be rigid, with deformation and movement occurring only at the joints and fissures so that for this type of analysis in its simplest form no information is needed on the deformability and strength of intact rock. The method is a discontinuum modelling approach which is suitable in cases where the behaviour of the rock mass is dominated by the properties of joints or other discontinuities. In such cases the discontinuity stiffness (i.e. force/displacement characteristics) is much lower than that of intact rock. Calculations are based on laws relating forces and

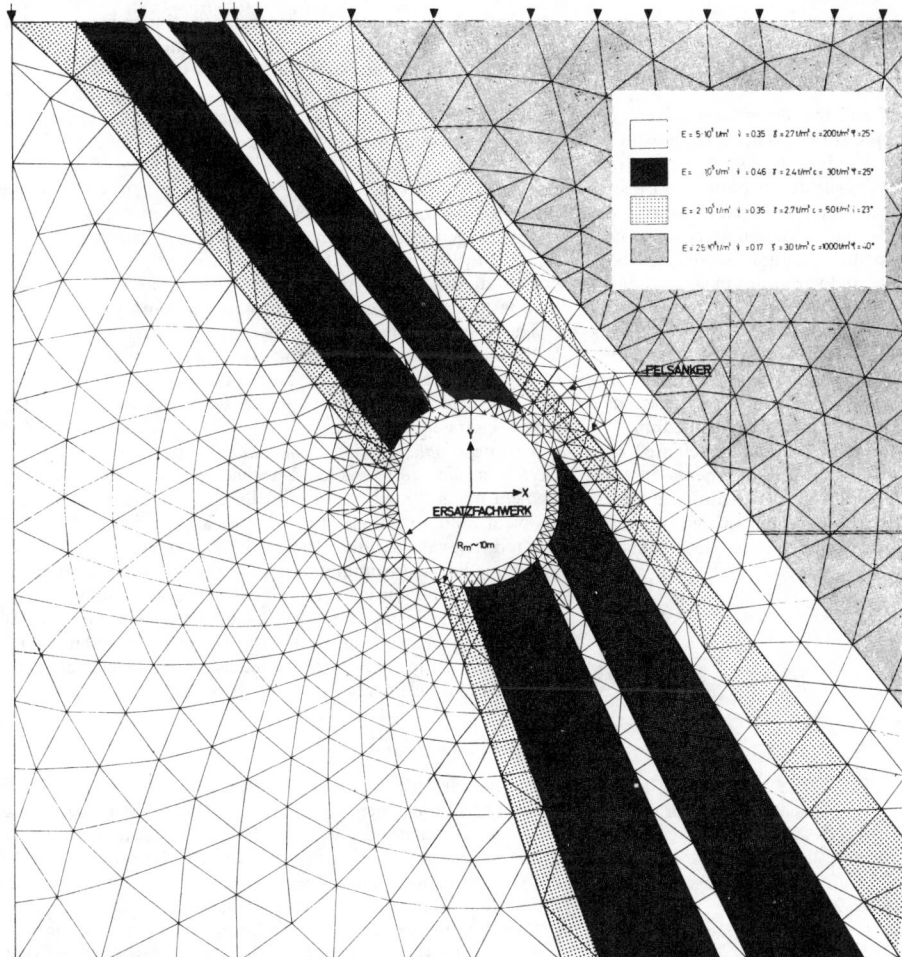

Figure 10.16 Finite-element method. An example showing finite-element mesh for the analysis of stresses acting on a pressure tunnel lining. Elements of varying stiffness have been used to simulate rock zones of varying competence. (Grob, *et al.* (1970) *Proceedings, 2nd international conference on rock mechanics,* Belgrade. Paper 4–69)

displacements between blocks (e.g. laws of elasticity or friction) and on the laws of motion (e.g. creep, viscosity or Newton's laws). Behaviour of the model is constricted to be compatible with boundary force or displacement conditions. Large movements can be modelled – not normally possible with any accuracy using a finite-element method. The method of computation is ideally suited for considering the development of rock movements incrementally with time (Figure 10.17). The distinct element method first described by Cundall[101] treats the rock as an assemblage of blocks interacting across deformable joints of definable stiffness. It is a development of the relaxation method and the dynamic relaxation method described by Otter, Cassell and Hobbs.[102] A force–displacement relationship governs interaction between the blocks and laws of motion determine block displacements caused by out-of-balance forces. Several forms of distinct element codes have been developed to cover a variety of *in situ* conditions. The Universal Distinct Element Code (UDEC), has been developed recently which provides, in one code, all the capabilities that existed separately in previous

programs. Features exist for modelling variable rock deformability, nonlinear inelastic behaviour of joints, plastic behaviour and fracture of intact rock, and fluid flow and fluid pressure generation in joints and voids. An automatic joint generator produces joint patterns based upon statistically derived joint parameters. The program can simulate the influence of the far-field rock mass for both static and dynamic conditions as it is coupled to a boundary element program which represents the effects of a static, elastic far-afield response, and nonreflecting boundary conditions are available for dynamic simulations.[103] The technique has three distinguishing features which make it well suited for discontinuum modelling:

(1) The rock mass is simulated as an assemblage of blocks which interact through corner and edge contacts.
(2) Discontinuities are regarded as boundary interactions between blocks; joint behaviour is prescribed for these interactions.
(3) The method utilizes an explicit time-stepping algorithm

which allows large displacements and rotations and general nonlinear constitutive behaviour for both the rock matrix and the joints.

The *boundary element method*[104] and the *displacement discontinuity method*[105] are integral methods of stress analysis, in which the problem is specified and solved in terms of forces (or tractions) and displacements on the surface or boundary of the model. The boundary (such as a tunnel perimeter) is divided into discrete elements whilst the far-field boundary may be infinite (or semi-infinite). Thus 'discretization' errors are restricted to the problem boundary, and variations in stresses and displacements are fully continuous. The field equations at a point within the continuum are satisfied exactly, and errors are associated with the approximations occurring at the boundary only. The boundary forces or tractions determine the stresses in the surrounding medium which are evaluated using expressions for stress components at any point in an infinite medium.[106] Elastic displacements around the excavation are calculated by making use of standard solutions for displacements in an infinite medium due to point or line loads. The methods are not particularly well suited to heterogeneous, anisotropic or nonlinear material behaviour. A reasonably clear and concise

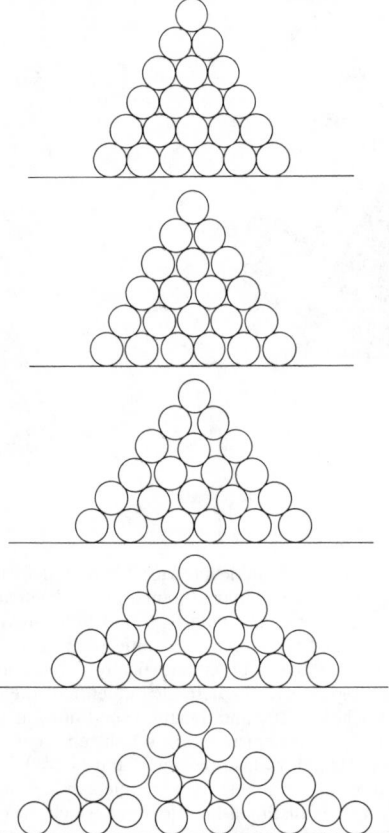

Figure 10.17 The dynamic relaxation method. The figure shows progressive collapse of a stack of cylinders, with displacements computed as a function of time using the dynamic relaxation method. Similar calculations can be used to show the collapse of rectangular blocks such as comprise a rock mass. (After Cundall (1971) 'A computer model for simulating progressive large-scale movements in blocky rock systems'. *Proceedings, international symposium on rock mechanics rock fracture*, Nancy. Paper 2–8)

description of the method is presented in the appendices of a practical manual on underground excavations.[62] This further provides a collection of stress distributions calculated using the boundary element method around single openings of various shapes within different stress fields. Reference to these is useful as a first assessment of stress concentration around proposed excavations.

More recently[107] a boundary element formulation has been presented for modelling structural discontinuities, joints, faults and heterogeneous rock. The model is divided into regions, each one homogeneous, separated by interfaces which can represent discontinuities. The solution, however, is an iterative approximation to account for nonlinear joint equations.

The displacement discontinuity method is particularly suited to the analysis of tabular openings (such as coalmines). It is able to work in three dimensions, data input is relatively easy, and will model multiple seam-mining layouts or folded or faulted single-seam deposits.

Various *coupled computational analyses* (or 'hybrid' methods) have been devised to make advantageous use of the boundary element integral method for modelling the far-field region of a problem, and to couple this with an appropriate differential method (relaxation, finite element or distinct element) to model the immediate surroundings of the excavation. A domain of complex behaviour is thus embedded in an infinite elastic continuum. Lorig and Brady[108,109] have described the coupling of the discrete element method with boundary elements, and Ushijima and Einstein[110] a three-dimensional finite element code with boundary elements.

10.4.5 Slope design

In *the limiting equilibrium method*[111] a rock mass is considered under conditions where the mass is on the point of becoming unstable. The method gives no information on magnitudes of displacement or on rock behaviour prior to failure, so that the design calculations cannot readily be checked by instrumentation and monitoring of rock movements.

Equilibrium is examined by relating the shear and normal forces on the sliding surface to the sliding resistance of that surface. Shear tests are necessary to evaluate sliding resistance, but otherwise a constitutive equation for the rock mass is not required. The geometry and position of the sliding surface must be predicted in advance, and for this reason the method has been most commonly applied to slope stability problems where the sliding surface is more readily predicted than in underground situations. The development of computers has made it possible to use conveniently some very powerful limit equilibrium methods.[69]

Slope design[9] usually employs limiting equilibrium analysis. A first step is to assess whether any kinematic mechanisms of potential slope failure are likely to be more closely approximated by a plane failure, a sliding wedge, rotational slip or a toppling failure model, and to identify the beds, joints or faults that could conceivably control such a failure, as illustrated in Figure 10.18. Throughgoing discontinuities such as faults, beds or older pre-existing failure surfaces are likely to be of considerably greater significance than impersistent or rough features. The presentation and analysis of geological structure data using the method of stereographic projections (stereonets)[10] is invaluable for such an assessment. Clearly, the collection in the field of the geological data relevant to the problem is of paramount importance, as described in Chapter 8.

Quick and approximate calculations at this stage help to assess whether there is indeed a problem, and whether a more detailed analysis is justified. These can employ hand calculations or design charts[9] using data for rock strength and water pressures estimated after examining the rock *in situ*. Worst and

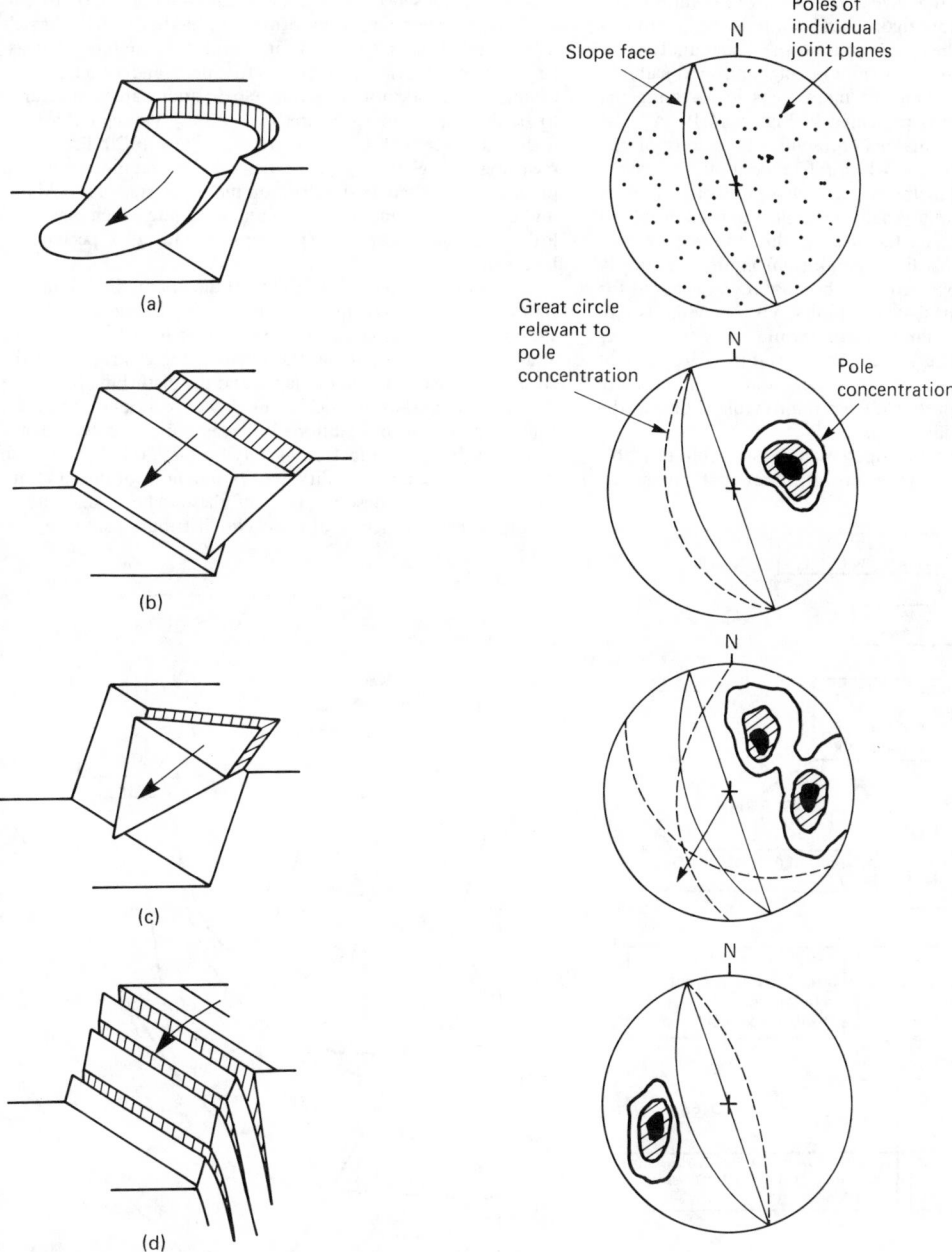

Figure 10.18 Representation of structural data concerning four possible slope failure modes, plotted on equatorial equal-area nets as poles and great circles. (a) Circular failure in heavily jointed rock; (b) plane failure in highly ordered structure such as slate; (c) wedge failure on two intersecting sets of joints; (d) toppling failure caused by steeply dipping joints. (After Brown (1981) *Rock characterization, testing and monitoring*. Committee Testing Methods International Society Rock Mechanics. Pergamon, Oxford)

best estimates can be used to give upper and lower bounds in the stability calculations. More rigorous calculations then require *in situ* measurement of shear strength or the back analysis of existing slides, water pressure monitoring and permeability testing. A flowchart showing the main steps in assessing the stability of a rock slope is presented in Figure 10.19. A most comprehensive practical manual[9] presents full details of the methods mentioned above, to which reference should be made.

The two-dimensional methods of analysis most often used in soil mechanics should not normally be applied to rock problems although Hoek[74] describes the use of the nonvertical slice method of Sarma.[112] This limit equilibrium method is ideally suited to many rock slope problems because it can account for specific structural features such as faults. Vector methods[9] are particularly suited to the limiting equilibrium analysis of three-dimensional wedges. The kinematics of stability are also of greater relevance in rock than in soil, and techniques are available for selecting probable from improbable slides on the basis of kinematic considerations.[113]

Natural or excavated rock slopes might be shown to be stable overall whilst the possibility remains of minor rockfalls occur-ring due to loosened blocks. These may be controlled by full stabilization methods, or by protection methods such as catch fences and ditches to arrest or retard the tumbling blocks. Experimental work, notably by Ritchie[114] and others,[115] has examined the trajectory of falling blocks and enabled guidelines to be developed for slope–toe rock traps. A design chart for ditch and fence rock traps is shown in Figure 10.20. Protection measures are generally not expensive but require continual maintenance, whereas stabilization measures such as rockbolt-ing, buttressing, trimming, mesh and shotcrete which may need little maintenance can be very expensive to install, especially for high slopes.

Owing to the inherent variability of the orientation of discontinuities, even though they occur in 'sets', the design of rock slopes may in some circumstances lend itself to a probabilistic analysis.[116-118] For example, the resisting forces are due to the shear resistance of the joint planes and the disturbing forces are due to the weight of the rock block or wedge, both of which are functions of joint orientation. Whereas normal 'deterministic' analysis is based on a factor of safety for the ratio of resisting to disturbing forces, probability density functions or distributions can be derived to describe each of these. The probability of failure is then a function of these two distributions.

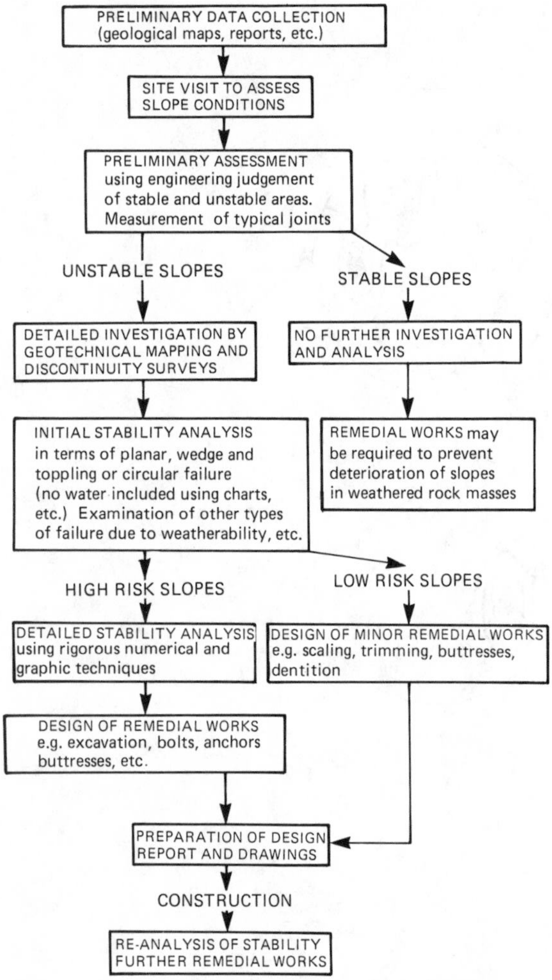

Figure 10.19 Flowchart for rock slope stability assessment and remedial works. (After Powell and Irfan (1986) *Slope remedial works in weathered rocks for differing risks. Rock engineering and excavation in an urban environment.* Institute Mining and Metallurgy, Hong Kong)

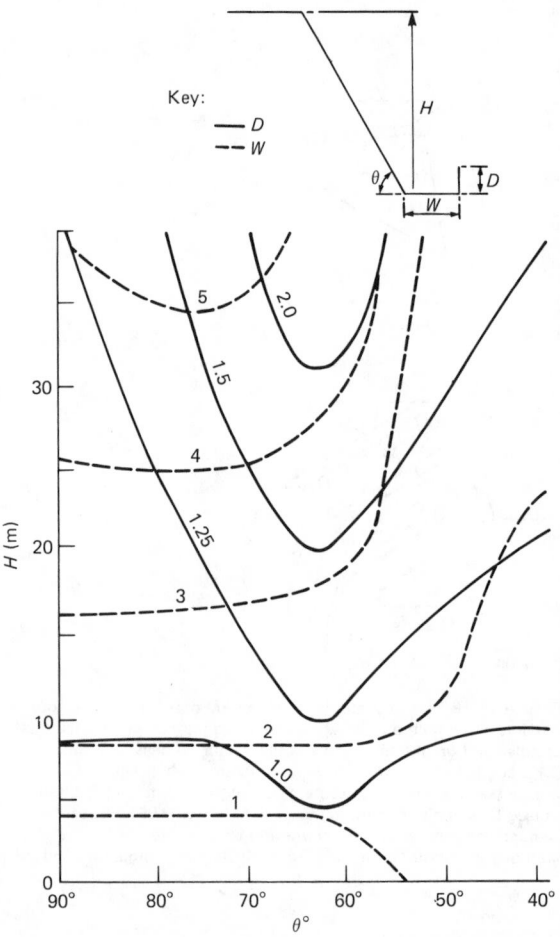

Figure 10.20 Rock trap guidelines. (After Whiteside (1986) Discussion, *Proceedings, symposium on rock engineering in an urban environment.* Institute of Mining and Metallurgy, Hong Kong)

10.4.6 Foundation design

Rocks generally have a high allowable bearing pressure which may be reduced by the presence of weak layers, discontinuities and weathering. The allowable bearing pressure depends on the compressibility and strength of the rock mass and the permissible settlement of the structure.

Detailed design calculations for foundation rock are generally required only if the rock is weak and/or broken or the loading is unusually high, and in these cases the problems are usually associated with settlement prediction rather than foundation bearing failure. The compressibility of the rock mass is related to the strength and modulus of intact rock, the lithology, and the frequency, nature and orientation of the discontinuities. Guidance on allowable bearing pressures and a method of calculation may be obtained from the British Standard Code of Practice.[119] These values are not necessarily suitable for very large heavy foundations nor for structures sensitive to settlement which should be considered having regard for the size of foundation, the variation in strength and nature of fracturing both with depth and laterally, and the variation in modulus with intensity of stress.[120]

Typical problems that require a more detailed study include the design of end-bearing piles or caissons carried to rock, particularly when the depth of overlying less competent materials is such as to require a minimum diameter of excavation with correspondingly high bearing pressures. The design of rock-socketed piers has been reviewed by Rowe and Armitage[121] and they have produced a number of design charts for their proposed method. Piling in chalk has been reviewed comprehensively by Hobbs and Healy.[122] Certain types of structure are particularly vulnerable to differential settlements, and others are particularly massive so as to impose high foundation loads (e.g. arch dams and the heavier types of nuclear reactor)[120,123] and in these instances rock foundations require careful design.

Site exploration is primarily aimed at locating suitable foundation levels, and the relative, rather than the absolute, competence of strata. Rock-quality maps can be useful in making this choice. The depth of rock weathering (Chapter 8) is often of particular significance. Approximate allowable bearing pressures can be estimated empirically[119] and often at this stage the foundation design can be modified or improved by grouting.

More detailed analyses of foundation behaviour may employ closed-form elastic or plastic solutions or the various computational methods of analysis described in section 10.4.4. These require information of rock deformability that is usually obtained by *in situ* plate-loading tests or from borehole dilatometer tests. Seismic refraction and other geophysical methods may be useful to assess the character of the rock mass on a large scale. Foundations on argillaceous rocks can be subject to plastic deformation under high contact stresses and may require a study of long-term (time-dependent) behaviour. Stability analyses – taking into account particularly any uplift forces due to water pressure – in addition to settlement calculations are necessary when designing dam foundations and abutments, and when a foundation is situated above a rock slope. Limiting equilibrium methods are appropriate for these analyses, although rock foundation–structure interaction analyses using computational methods of analysis are appropriate for any major structures.

10.4.7 Design of underground openings

Underground openings for civil engineering purposes require perhaps the most rigorous use of rock mechanics practice in their design and construction. They are so demanding because of the severe consequences of being unsuitable or unsafe for their intended purpose. Tunnels and caverns for, say, railway systems or hydro-electric power complexes are not only occupied by the public in some cases, but even minor instability or surface ravelling of small blocks is wholly intolerable to the function of the works. The responsibility on the designer given the natural variability of the materials is very great and it is also necessary to recognize the essential requirements of the various uses to which caverns may be put. In the above examples stability in every sense is essential but the effects on the groundwater regime may be of lesser importance. There are some caverns used for storage of water, oil or gas, and mining openings in which limited minor instability might be permissible, but to maintain, for example, an unlined gastight cavity, it is essential that drainage of the surrounding rock pore water and fissure water does not occur.

The most demanding of purposes for caverns for the designer is the storage of radioactive waste, and this requirement is one explanation for the recent most rapid advance in rock mechanics field experimentation and development of understanding. Stability is essential and must be considered both for the very long term and in relation to extreme thermal effects and radionuclide absorption on rock mass properties, as well as to the dynamic disturbing forces of possible future seismic events. The prediction of groundwater flow around and away from nuclear repositories must consider the stress and thermal effects on hydraulic conductivity of joints and fissures, and the hydrothermal migration or convection effects on groundwater. Such details can be examined theoretically using the computational methods described above, but the determination of realistic input parameters remains a major problem.

The shapes and sizes of underground excavations are often dictated by economic and functional considerations, but their precise location and orientation should be adjusted to suit ground conditions wherever possible. Optimum orientation requires a knowledge of the geological structure, rock mass structure and also of the directions and relative magnitudes of principal stresses in the ground prior to excavation, and so detailed layouts of proposed works should be held in abeyance until after the investigation stage if possible.

A guide to the most important steps in the stability design of underground openings is given in Chapter 1 of a comprehensive manual by Hoek and Brown[62] and full details of the various methods are given. In the design process it is necessary to characterize and zone the rock mass 'geo-mechanically', to determine constitutive relations and strength criteria for each zone, and then bearing these and the proposed geometry of the opening in mind to select the appropriate method of analysis which will render where potential zones of instability lie. The criteria for support design must then be decided upon, e.g. it is common to provide rockbolts of sufficient length and number to support the deadweight of tension zones or 'overstressed' rock determined by stress analysis by anchoring back into 'sound' rock as illustrated in Figure 10.21. The definition of relevant 'failure' criteria which may be influenced by water pressure or seepage considerations, dynamic seismic forces or limiting displacements is a significant design input. Each failure state must then be tested for various stages of the excavation progress because areas of stress concentration will vary. When the excavation sequence and support element dimensioning are decided, verification of performance monitoring is necessary.

A useful design method which is valuable in developing an understanding of the mechanics of rock support is known as *rock-support interaction analysis.*[62] Although the method makes numerous simplifying assumptions (e.g. a circular excavation in a uniform *in situ* stress field is assumed) the principles may potentially be extended to more general cases. The method analyses stress and displacement in the surrounding rock and in the support elements, taking into account rock mass properties, the *in situ* stress, the development of a 'zone of plastic failure'

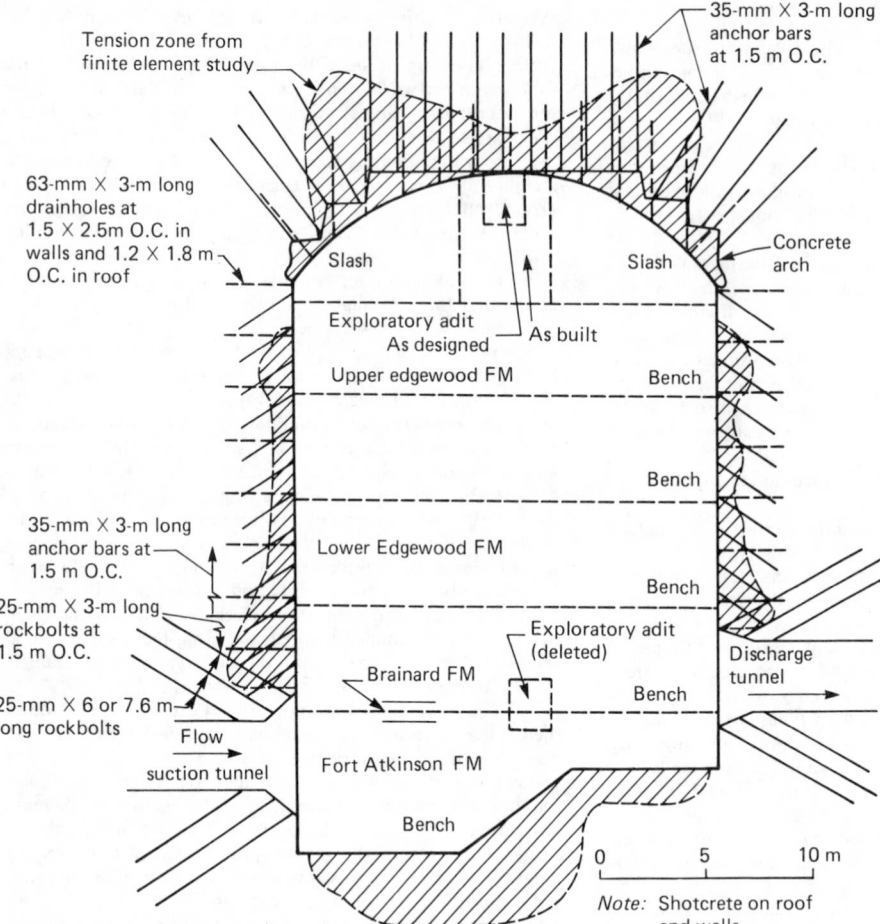

Figure 10.21 Example of rockbolt support for a major cavern. (After Cikanek and Goyal (1986) 'Experiences from large cavern excavation for TARP'. *Proceedings, symposium for large rock caverns*, Helsinki. Pergamon, Oxford)

around the opening, the stiffness of the support and the timing of its installation after excavation. Whenever an excavation is made there will be inward radial movements (convergence) of the surrounding rock which, in practice, are not instantaneous. In the meantime, supports such as rockbolts or arches are installed and, as convergence continues, so the supports will provide reaction forces. The characteristics of the rock are represented by a 'ground response curve' and the support pressure by a 'support reaction line'. Methods of response curve calculation which make use of nonlinear peak strength and residual strength criteria, and the method for pressure tunnel design have been described.[124,125]

Certain types of civil engineering excavation may give rise to subsidence problems, e.g. unsupported chambers for storing water and gas. Furthermore, the civil engineer is often affected in his surface construction operations by mining excavations beneath the site. Subsidence can be predicted to some extent although, since the phenomenon is essentially time-dependent, the analysis is complex and often based on empirical observations.[126]

10.5 Construction methods and monitoring

10.5.1 Excavation

Processes of rock fragmentation are known collectively as comminution processes. In spite of a considerable amount of research aimed at improving these techniques the gap between theory and practice is still great and an empirical approach is more often used. Much research has been directed towards understanding the mechanisms of fragmentation during drilling, in order to improve the design of conventional mechanical bits (diamond bits, percussion or rotary drag bits) and to develop new ways of drilling such as by water-jet cutting, flame cutting and pellet impact. Mechanical drilling techniques suffer from energy losses due to inefficient transfer of energy from the bit to the rock. The drilling process ideally should produce fragments small enough to be flushed from the drillhole but not so intensely crushed as to absorb a considerable proportion of the input energy. Other factors such as drilling rate and for core drilling, the quality of core recovery are of primary importance.

The theory and practice of blasting are reviewed in detail by Langefors and Kihlstrom[127] and details of techniques and explosives are given in Chapter 32.

Blast pressure in a drillhole can exceed 100 000 atmospheres. This pressure tends to shatter the area adjacent to the drillhole and a stress wave is generated that travels outward from the hole at a velocity of 3000 to 5000 m/s. The leading front of the stress wave is compressive, but is closely followed by tensile stresses that are responsible for the major part of rock fragmentation. When the stress wave is reflected from a nearby joint surface or exposed rock surface it again gives rise to tensile stresses which may cause scabbing of the superficial rock. The stress wave is the initial cause of fracturing, gas pressure serving to widen and extend the cracks previously generated.

Variables in designing a blasting pattern include the degree of fragmentation required (size of fragments), the explosive used, the diameter, inclination and method of loading the drillhole, the burden (distance from the drillhole to the free face) and the spacing between holes. Also the sequence of firing can be varied (e.g. with delayed charges) to minimize vibration levels and unwanted rock damage and to give a more efficient pattern of rock removal. To predict and control excessive blast vibrations and assess their likely effects involves the use of an attenuation law for vibrations through the ground. Various theoretical relationships exist between instantaneous charge weights, distance and vibration intensity but frequently site-specific experiments are necessary.[128] A well-designed blast gives maximum yield, controls the size and shape of fragments (critical to subsequent crushing processes for aggregate production and to utilization of material in rockfill constructions), controls the throw and scatter of fragments, and minimizes the amount of drilling and explosive required.

In road cuts, blasting is complicated by the continually varying height of bench; heights of more than 10 m are usually blasted in more than one lift. In foundation blasting it is particularly important to control throw and also vibration levels, usually by means of short-delay multiple-row blasting with small-diameter drillholes and reduced charges.

In tunnel blasting (Chapter 32) the first holes to be detonated should create an opening towards which the rest of the rock is successively blasted. The holes of this 'cut' are usually arranged in a wedge, fan or cone pattern. The remainder of the round is designed to leave the intended excavation contour undamaged. In long excavations of diameter larger than 8 m the upper section is often removed first, followed by removal of the remaining bench in one operation after installation of roof support. This can give a more economic result and also facilitates both mucking out and the installation of support. Several successive benches are excavated for caverns, as in Figure 10.21.

Smooth-wall blasting is a comparatively recent innovation that greatly improves rock stability and at the same time reduces the amount of concrete required for lining the excavation. The techniques are well proven[129] but not as widely used as they might be. The greater level of control needed may represent an additional cost, but experience of projects in which carefully controlled blasting has been used generally shows that the amount of support can be reduced significantly. The overall cost of excavation and support thus is lower than in the case of poorly blasted excavations. In *presplit blasting* cracks for the final contour are created prior to firing holes for the rest of the pattern. Spacing for the contour holes is typically 10 to 20 times the hole diameter, and holes are loaded with a reduced charge density. The contour holes should be ignited simultaneously.

Machine excavation causes very little disturbance to the surrounding rock. The development of larger and more efficient ripping and tunnelling machines together with the continuing increase in manpower costs has resulted in the increasing use of mechanical excavation as an alternative to blasting. Ripping can be highly competitive particularly in the larger-scale surface mining operations.

The mechanics of rock excavation with a ripper or with the cutting head of a tunnelling machine are in some respects similar to those of drilling, although on a larger scale the natural fractures and planes of weakness in the rock play an increasingly important role. Also a tunnelling machine, unlike a drill bit, cannot be changed at will to suit rock conditions. Machines where the spacing, size and type of cutting disc or pick as well as the thrust and speed of cut can be varied, may be desirable but the construction of machines of such wide versatility is not generally feasible. Hence, the considerable capital investment associated with the purchase of ripping and, particularly, tunnelling plant requires a careful study of rock conditions prior to selecting a machine. Rock quality classifications can help in making this choice when used in association with site investigation and geological studies.[130] Fracture spacing is often the most relevant property to note, since it determines the sizes of rock block to be excavated. Intact material strength, abrasiveness and specific energy are also of relevance. Sonic velocity mapping has sometimes been used to assist in assessing the state of fracturing of the ground and, hence, its rippability.

10.5.2 Rock support methods

Unstable rock conditions can very often be improved by rock-bolting and support, grouting or drainage. The cost is frequently offset by the benefits, e.g. a rock slope can in some cases be steepened by 10° or more if an efficient drainage system is used; in deep rock cuts this appreciably reduces excavation costs.

10.5.2.1 Rockbolting and anchoring

A rockbolt assembly usually comprises a bar with an anchor at one end and a faceplate assembly (faceplate, nut and wedge or spherical washer) at the other. The anchor may be mechanical or grouted, and the bar may be substituted by a cable to achieve greater lengths and loads. For rockbolts to be used for permanent civil works a careful study of corrosion resistance is essential.

The *dowel*, or *fully bonded rockbolt*, comprises a bar installed in a drillhole and bonded to the rock over its full length. A cement-grouted bar can be installed by packing the drillhole with lean quick-set mortar into which the bar is driven. The 'Perfo' system uses a split perforated sleeve to contain the mortar, which is extruded through the perforations as the bar is driven home. Fully bonded resin anchors usually employ a polyester or epoxy resin and catalyst in the form of cartridges that are ruptured and mixed by a bar driven into the drillhole with a small rotary drill.

Point-anchored rockbolts comprise a bar anchored over a comparatively limited length. The slot and wedge system uses a bar with a longitudinally sawn slot. A wedge inserted into the slot expands the slotted section against the rock when the bar is driven home. Sliding wedge anchors employ a pair of wedges drawn over each other when the bolt is rotated. Expansion shell anchors employ two or more wedges or 'feathers' that are expanded by a threaded cone. Explosive anchors have also been used in softer rocks and comprise a split tube, sections of which are driven into the rock by an explosive detonated within the tube. The resin point anchor is identical to the fully bonded resin anchor described above, except that a limited quantity of resin is used to provide an anchor only at the base of the drillhole.

Point-anchored rockbolts must essentially be *tensioned* to work efficiently, since the action of opposing anchor and bearing plate forces effectively tightens the superficial zone of loose rock. This zone can then make a significant contribution to the support of rock at greater depth. Dowels cannot gain from tensioning during installation but tension is induced when the rock begins to move and dilate. Rockbolts should be installed as soon as feasible after excavation, before the rock begins to move with consequent loss of interlocking resistance.

Grouting of point-anchored rockbolts reinforces the anchor, protects against bolt corrosion, and is particularly necessary in softer rocks where point anchorages are seldom reliable in the long term. Grout should be injected at the lowest point in the drillhole, using a bleeder tube to remove air. Resin-bonded bolts can be arranged so that a fast-setting resin is introduced first in the drillhole, followed by slower setting resin cartridges. The bolt is tensioned when the point anchor provided by the fast-setting cartridge has gained sufficient strength, and the slow-setting cartridges then polymerize to grout the remainder of the bolt length. The use of fully resin-bonded wooden, plastic or glass-fibre rockbolts has the advantage that the bolts will not damage excavating machinery and so can be used if necessary for temporary support at, say, a tunnel portal or advancing face.

Friction anchored rockbolts develop anchorage loads along their full length within the drillhole. The *split-tube bolt* is hammered into a drillhole of slightly smaller diameter than the bolt itself. The bolt, which is manufactured from sprung steel and has a C-shaped cross-section, recoils against the drillhole wall. The *inflatable bolt* is manufactured from malleable steel as a reniform tube which is inserted into the drillhole. It is then inflated to take up the form of the drillhole by the use of water at high pressure.

Rockbolts should be field-tested in the rocks in which they are to be installed in order to establish their design performance. A bolting pattern may then be designed on the basis of test results, taking into account the rock structure and the size and shape of the slope or underground excavation.[62,131]

10.5.2.2 Sprayed concrete and other lining methods

A first essential in rock support is to prevent even small quantities of material from ravelling from the rock face, since this can lead to general loosening and progressive failure. The size of rockbolt faceplates is often adjusted to minimize ravelling, and *wire mesh* or *ribs* can be installed beneath the plates to give added protection. *Sprayed concrete* (gunite or shotcrete) can be used to supplement bolting and mesh, or may under some circumstances provide adequate support on its own,[132] it is a particularly appropriate technique for preventing the slaking deterioration of mudstones and shales. A thin sprayed concrete lining, unlike more rigid methods of support, will crack to reveal zones of instability before they develop fully, allowing the placing of additional local support. Sprayed linings as thin as 5 to 10 mm have been used effectively in some instances. Several proprietary systems for a more rigid tunnel lining using, for example, interlinked expanded metal sheeting and pumped or sprayed concrete, are in use and the methods are described in greater detail in Chapter 32.

10.5.2.3 Grouting

The injection of a grout into the rock mass so that air or water in fissures is replaced by a solid material or gel will inhibit percolation of water and may also provide added rigidity and possibly strength. Injection into rock normally requires a grout consisting of a mixture of Portland cement and water. Sand, clay, or other inert materials may be added to reduce cost provided that these filler grouts can flow, without undue segregation, through the sizes of fissure present in the rock. High grouting pressures can be necessary for adequate grout emplacement, depending on rock mass permeability and on the fluidity of the grout. Grouting at high pressure can result in hydraulic fracturing and lifting of beds. Although this assists emplacement it can be detrimental to the resulting strength of grouted fissures, can result in damage to nearby structures, and in the worst case can itself initiate rock collapse. The efficient grouting of narrow fissures can require grouts of greater than usual fluidity, and in these cases chemical grouting materials can be used. Permeability tests are usually needed to select appropriate grouting pressures and materials. Grouting is commonly used to reduce leakage beneath dams and into tunnels or excavations beneath the water table. It is also used in association with drainage measures to control uplift and pore pressures in dam foundations and abutments, and to consolidate loose rock in foundations or in the vicinity of an excavation. *Consolidation grouting* (distinct from the term 'consolidation' as applied in soil mechanics) serves to improve rock strength but, more importantly, it considerably reduces rock deformability. Cementitious materials are usually required. *Grout curtains* on the other hand are used to reduce permeability (e.g. beneath a dam) and may employ lower-strength gel-type materials. Temporary control of water flow, and temporary rock mass consolidation, can sometimes be provided by *freezing techniques*, used, for example, in the driving of shafts through highly fractured rock. The efficiency of rock grouting is usually assessed by monitoring of the grouting operation itself, by visual inspection, or using sonic velocity techniques.[41]

10.5.2.4 Drainage

A drainage system installed around a rock structure under construction can have the immediate advantage of improving working conditions (e.g. in a tunnel or cut excavation) but its principal objective is usually to reduce water pressures within the rock mass and hence improve its stability.[133,134] Groundwater will present a problem either by the seepage of copious quantities into excavations leading to inundation of the works, or by the destabilizing effects of water pressure acting in pores, joints and fissures. The former is more likely in permeable formations and the latter in formations of lower permeability. Hence, the control of seepage by grout injection usually results in a build-up of water pressures which must be relieved by suitable drainage methods to ensure stability. Sprayed concrete linings, for example, must usually be provided with drain holes if they are to remain intact under conditions of high water pressure.

The design of drainage systems requires a comparison of water pressures and flow paths in the drained and undrained structure (Figure 10.22). Darcy's law – that the flow velocity is proportional to the change in pressure per unit distance (hydraulic gradient) along the flow path – can be assumed to be valid in most cases. Flow through the rock may be solved relatively simply using graphical methods (flow-net sketching), analogue methods or numerical techniques as discussed in section 10.4. The graphical methods are particularly suitable for an initial examination of the problem, and can be relatively accurate given some practice in flow-net construction and reliable data on rock conditions.[135]

10.5.3 Monitoring

Instrumentation and monitoring gives a check on the design and its inherent assumptions and simplifications. Alternatively, it can be used in cases where a detailed analytical design is not justified by the nature of the problem, but when rock stability remains in question. The object of performance monitoring is to give sufficient warning of adverse or unpredicted behaviour to

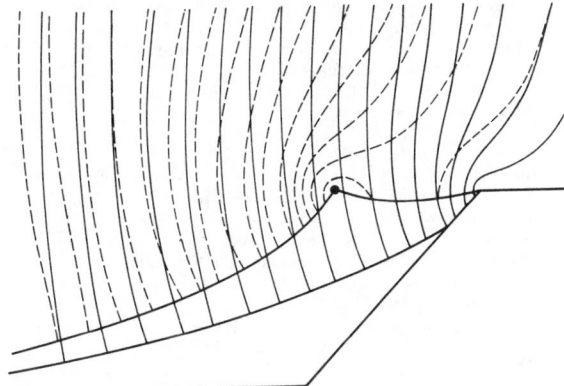

Figure 10.22 Comparison between drawdown curves and equipotential lines for a 45° slope in isotropic material with and without drainage *(after Sharp et al. Ref 133)*

allow timely remedial action, and to assess the effect of remedial works.

10.5.3.1 Movement monitoring

Conventional survey techniques (e.g. precise levelling and tri-angulation) provide the simplest and perhaps the most reliable control but their accuracy is not always sufficient for detection of movements smaller than a few millimetres.

Electronic distance measuring methods are capable of good accuracy, with the added advantages of speed and of a range greater than 1 km.

Photogrammetric techniques (ground photogrammetry in particular) are particularly useful for monitoring large rock slopes. Although their accuracy is usually not better than 100 or 200 mm, depending on the object distance, movements can be detected at every visible point even though they do not coincide with prelocated survey markers. *Lasers* can be used to detect changes in alignment, and *surface extensometers* to monitor the development of tension cracks, tunnel convergence, and the superficial movements of rock slopes.

Movement monitoring devices can also be installed in boreholes to supplement, or sometimes to replace, surface monitoring. *Borehole extensometers* measure changes in length of the borehole. Multiple position extensometers are available to measure differential movement at as many as ten or fifteen anchors at varying depth. Simple settlement devices often work on a hydraulic or 'U-tube' principle. Multiple-position instruments may employ rods or wires and either a mechanical or an electric transducer system for recording movements.

Borehole inclinometers record changes in borehole inclination. Moving-probe inclinometers employ a capsule or probe that travels along the borehole which for this purpose is cased with flexible plastic or aluminium grooved tubing. The probe may comprise, for example, a cantilever pendulum, a short length of pivoted rod, an inertial system or a system where the position of a bubble in a 'spirit level' is monitored electrically. Inclinometers of the fixed-position type usually employ a system of pivoted rods anchored at various positions along the borehole. Another device, the *shear strip*, comprises a set of electric resistors connected in parallel at regular intervals along the length of a printed circuit conductor and is used to detect the depth at which the strip, grouted into the borehole, is sheared by ground movements.

Blasting, earthquake and vibration studies require monitoring of *dynamic movements* with an array of geophones located in drillholes or at the rock surface. Seismic arrays can be used either to locate the sources of 'rock noise' (e.g. in monitoring landslide movements or rockburst phenomena) the development of hydraulic fractures, or to record the waveforms associated with blast or earthquake tremors, traffic or machinery vibrations.

10.5.3.2 Water pressure and flow

Piezometers (water-pressure measuring devices) are installed in drillholes to provide information for design analysis, or to record changes in water pressures so as to monitor one of the major causes of instability. They may take the form of *simple standpipes* where pressure is measured as a change in water level using a probe lowered into the standpipe tube. Artesian pressures and local pressure anomalies are common in rock, and the drillhole must usually be sealed with grout over its complete length except for the test sections at which pressures are to be measured. Several piezometers recording pressures at test sections of different elevation may be installed in a single drillhole.

Standpipe piezometers are suitable only for pressure monitoring in near-vertical drillholes and also have a comparatively slow response to water pressure fluctuations. This response can be improved by use of a *pressure transducer* method. Pressure transducers can be used in drillholes at any inclination and usually employ a flexible diaphragm exposed to water pressure on one side. They incorporate a device for measuring diaphragm deflection that typically uses bonded resistance strain gauges or a vibrating-wire system. Pressure sensors which determine back-pressures or balancing pressures across diaphragms and which use pneumatic or hydraulic readout equipment are also commonly used.

Water-flow monitoring is most often required in connection with reservoir leakage problems, pollution and hydro-geological studies, but a knowledge of flow velocities, directions and the identity of particular fissures carrying the majority of flow is often needed for other types of problem. Flow directions and velocities can be evaluated using radioactive or dye tracer techniques where a concentrated tracer is injected into one drillhole and the time taken for the tracer to appear in nearby observation holes recorded. Flow can also be measured in a single drillhole by observing the rate of dilution of a tracer or saline solution. Flow velocities and directions can also be logged in the drillhole by a probe incorporating a small turbine or electrically heated wires, the object being to find horizons of greatest permeability prior to installation of piezometers or to permeability testing.

10.5.3.3 Stress measurement

Rock stress measurements have the object of evaluating the natural stress field for purposes of design (see section 10.1) or of monitoring stress changes (i.e. induced stresses) for purposes of control and warning. Overcoring methods make the assumption that rock core is relieved of its *in situ* stress by core drilling and will return to its initial unstressed configuration. The expansion of the rock core is measured by first bonding a stressmeter device into a small-diameter drillhole, recording initial readings, and then drilling a concentric hole of larger diameter to remove an annulus of core with the stressmeter inside.[136] The stress prior to overcoring is then computed from initial and final strain measurements assuming elastic properties for the rock.

Borehole stressmeters have the advantage of measuring stresses sufficiently deep in the rock, away from disturbing influences due to the ground surface or the presence of excavated cavities. However, they need considerable expertise in their installation. They can be installed and left in place without overcoring if the object is to measure stress changes.

A further method for stress measurement is that of *hydrofracturing*, in which fluid is pumped at high pressure into a section of the drillhole isolated between two inflatable packers, and an accurate record kept of the relationship between pressure and flow volume. A sudden increase in volume at constant pressure indicates that the rock has been fractured (often this involves joint or bed separation rather than fracture of intact material) and gives an estimate of the pre-existing rock pressure normal to the fracture plane. The direction of the fracture plane must be known for meaningful interpretation of results.

Flat-jack techniques can also be used to measure rock stresses in close proximity to an excavation. A pattern of displacement measuring points is bonded to the rock surface, and after taking initial readings a slot is cut between the points. A flat jack is then grouted into the slot and inflated until the initial displacement values have been reinstated. The pressure at which this is achieved gives a measure of the rock stress that existed perpendicular to the plane of the slot prior to slot cutting. Slots at various positions and orientations around the excavations are required, and the evaluation of stresses must depend on assumptions as to the effect of the excavation on the natural stress field.

10.5.3.4 Pressure and load

Pressures on retaining walls and tunnel linings are usually monitored with hydraulic flat jacks or load cells. The flat jack or load cell may be connected to a pressure gauge as a sealed system so that changes in pressure are recorded directly, or there may be provision for inflation of the jack prior to taking readings, so that a null displacement condition is achieved.

Load cells (force transducers) can be used to record tension in rockbolts and anchors, compression in ribs, steel sets and prop supports, or can be incorporated into walls and tunnel linings as a means of measuring pressure. Essentially they use a 'proving ring' principle with a semi-rigid member to carry the load to be measured, and a means for monitoring the strain in this member. Electric-resistance load cells use one or more metal columns on to which the strain gauges are bonded. Hydraulic load cells employ a sealed hydraulic capsule with measurement of internal fluid pressure. Rockbolt and anchor load cells are available that work on an electric-resistance or vibrating wire principle, use a photo-elastic glass plug or employ a sandwich of rubber between metal plates whose separation is measured with a micrometer. Load cells often incorporate a spherical seating to ensure that the measured load is coaxial with the cell.

Acknowledgements

This chapter has been revised to incorporate the many advances in rock mechanics and rock engineering that have taken place since the 3rd edition of this book was published in 1975, but the author acknowledges that some parts of the chapter were based on the 3rd edition text prepared by Dr J. A. Franklin, then of Rock Mechanics Ltd and now Professor of Earth Sciences at the University of Waterloo, Ontario, and President of the International Society for Rock Mechanics. The author is grateful for the encouragement of the Geotechnical Consultancy Group of Soil Mechanics Ltd.

References

1 Brown, E. T. (1985) 'From theory to practice in rock engineering'. 19th Sir Julius Wernher Lecture. *Proceedings, 4th international symposium on tunnelling 85*. Institution Mining and Metallurgy, London.

2 Hudson, J. A. (1986) 'Rock engineering interaction matrix'. *Symposium on rock joints*. Geological Society, London.

3 Price, N. J. (1966) *Fault and joint development in brittle and semi-brittle rock*. Pergamon, Oxford.

4 Ramsay, J. G. (1967) *Folding and fracturing of rocks*. McGraw-Hill, New York.

5 Barton, N. R. (1973) Review of a new shear strength criterion for rock joints. *Engng Geol.*, **8**, 287–332.

6 Brekke, T. L. and Selmer-Olsen, R. (1966) 'A survey of the main factors influencing the stability of underground constructions in Norway'. *Proceedings, 1st international congress on rock mechanics*, Lisbon, Vol. II, pp.257–260.

7 Barton, N., Bandis, S. and Bakhtar, K. (1985) 'Strength, deformation and conductivity coupling of rock joints'. *Int. J. Rock Mech. Min. Sci.*, **22**, 3, 121–140.

8 Priest, S. D. and Hudson, J. A. (1976) 'Discontinuity spacings in rock'. *Int. J. Rock Mech. Min. Sci.*, **13**, 135–148.

9 Hoek, E. and Bray, J. W. (1977) *Rock slope engineering* (2nd edn). Institute of Mining and Metallurgy, London.

10 Priest, S. D. (1985) *Hemispherical projection methods in rock mechanics*. Allen & Unwin, Hemel Hempstead.

11 Serafim, J. L. (1968) 'Influence of interstitial water on the behaviour of rock masses', in: Stagg and Zienkiewicz (eds) *Rock mechanics*. Wiley, pp.55–97.

12 Murrell, S. A. F. (1963) 'A criterion for brittle fracture of rocks and concrete under triaxial stress and the effect of pore pressure on the criterion', in: C. Fairhurst (ed.) *Rock mechanics*. Pergamon, Oxford, pp.563–577.

13 Morgenstern, N. R. and Phukan, A. L. T. (1966) 'Nonlinear deformation of sandstone'. *Proceedings, 1st international congress on rock mechanics*, Lisbon, pp.543–548.

14 Snow, D. T. (1970) 'The frequency and apertures of fractures in rock'. *Int. J. Rock Mech. Min. Sci.*, **7**.

15 Fairhurst, C. (1986) 'In-situ stress determination – an appraisal of its significance in rock mechanics'. *Proceedings, international symposium on rock stress*. CENTEK, Sweden.

16 Hyett, A. J., Dyke, C. G. and Hudson, J. A. (1986) 'A critical examination of basic concepts associated with the existence and measurements of *in situ* stress'. *Proceedings, international symposium on rock stress*. CENTEK, Sweden, p.387.

17 Benson, R. P., Murphy, D. K. and McCreath, D. R. (1970) 'Modulus testing of rock at the Churchill Falls underground powerhouse, Labrador. Determination of the insitu modulus of deformation of rock', American Society for Testing and Materials Special Technical Publication Number 477, pp.89–116.

18 Haimson, B. C. (1978) 'The hydrofracturing stress measuring method and recent field results'. *Int. J. Rock Mech. Min. Sci.*, **15**, 167–178.

19 Leeman, E. R. and Hayes, D. J. (1966) 'A technique for determining the complete state of stress in rock using a single borehole'. *Proceedings, 1st international congress rock mechanics*, Lisbon, Vol. II, pp.17–24.

20 Worotnicki, G. and Walton, R. J. (1976) 'Triaxial "hollow inclusion" gauges for determination of rock stresses *in situ*'. *Proceedings, International Society Rock Mechanics* symposium, supplement, pp.1–8.

21 McKensie, D. P. (1969) 'The relation between fault plane solutions for earthquakes and the direction of principal stresses'. BSSA, **59**, 2.

22 Klein, R. J. and Barr, M. V. (1986) 'Regional state of stress in western Europe'. *Proceedings, international symposium on rock stress*. CENTEK, Sweden, p.33.

23 Price Jones, A. and Sims, G. P. (1984) '*In situ* rock stresses for a hydroelectric scheme in Peru'. *Proceedings, international symposium International Society Rock Mechanics*, Cambridge.

24 Brown, E. T. (1981) *Rock characterization, testing and monitoring*, Committee for Testing Methods, International Society Rock Mechanics. Pergamon, Oxford.

25 Hoek, E. (1977) 'Rock mechanics laboratory testing in the context of a consulting engineering organization'. *Int. J. Rock. Mech. Min. Sci.*, **14**, 93–101.

26 McFeat-Smith, I. (1977) 'Rock property testing for the assessment of tunnelling machine performance'. *Tunnels & Tunnelling*, **9**, 3, 29–33.

27 De Puy, G. W. (1965) 'Petrographic investigations of rock durability and comparisons of various test procedures'. *J. Am. Ass. Engng. Geol.*, **2**, 31–46.

28 American Society for Testing and Materials (1969) *Concrete and mineral aggregates*. ASTM testing and materials standards, Part 10.

29 Verhoef, P. N. W. (1987) 'Sandblast testing of rock'. *Int. J. Rock Mech. Min. Sci.* Technical note, **24**, 185–192.

30 National Coal Board (1977) *The cone indenter test*. MRDE Handbook No. 5. NCB Mining Department.

31 Fowell, R. J. and McFeat-Smith, I. (1976) 'Factors influencing the cutting performance of a selective tunnelling machine'. *Proceedings, symposium on tunnelling 76*, London.

32 Young, R. P. (1978) 'Assessing rock discontinuities', *Tunnels & Tunnelling*, **10**, 5, 45–48.

33 McFeat-Smith, I. and Fowell, R. J. (1977) 'Correlation of rock properties and the cutting performance of tunnelling machines'. *Proceedings conference on rock engineering*. Newcastle, pp.581–602.

34 Hawkes, I. and Mellor, M. (1970) 'Uniaxial testing in rock mechanics laboratories'. *Engng Geol.*, **4**, 3, 177–285.

35 International Society for Rock Mechanics (1985) 'Suggested method for determining point load strength', *Int. J. Rock Mech. Min. Sci.*, **22**, 51–60.

36 Mellor, M. and Hawkes, I. (1971) 'Measurement of tensile strength by diametral compression of discs and annuli'. *Engng Geol.*, **5**, 173–225.

37 Ramsey, D. M. (1965) 'Factors influencing aggregate impact value in rock aggregate'. *Quarry Mngrs J.*, **49**, 129–134.

38 Jenner, H. N. and Burfitt, R. H. (1974) 'Chalk as an engineering material'. *Proceedings, Institution Civil Engineers Brighton Meeting*, Thomas Telford, London.

39 Ingoldby, H. C. and Parsons, A. W. (1977) *The classification of chalk for use as a fill material*. Transport and Road Research Laboratory Publication Number LR 806.

40 Griffith, D. H. and King, R. F. (1965) *Applied geophysics for engineers*. Pergamon, Oxford.

41 Knill, J. L. (1970) 'The application of seismic methods in the prediction of grout take in rock'. *Proceedings, 1969 conference on in situ investigations in soils and rocks*, British Geotechnical Society, London.

42 Masuda, H. (1964) 'Utilization of elastic longitudinal wave velocities for determining the elastic properties of dam foundations rock'. *Proceedings, 8th international congress on large dams*, Vol. I, pp.253–272.

43 Kennett, P. (1971) 'Geophysical borehole logs as an aid to ground engineering'. *Ground Engng*, **4**, 5, 30–32.

44 Muir Wood, A. M. and Caste, C. (1970) '*In situ* testing for the Channel tunnel'. *Proceedings, 8th conference on in situ investigations in soils and rocks*, pp.109–116. British Geotechnical Society, London.

45 British Standards Institution (1981) *Code of practice for site investigations*. BS 5930. BSI, Milton Keynes.

46 Franklin, J. A. and Hoek, E. (1971) 'Developments in triaxial testing technique'. *Rock Mech.*, **2**, 223–228.

47 Kutter, H. K. (1971) 'Stress distribution in direct shear test samples'. *Proceedings, international symposium on rock mechanics and rock fracture*, Nancy.

48 Ross-Brown, D. M. and Walton, G. (1975) 'A portable shear box for testing rock joints'. *Rock Mech.*, **7**, 3, 129–153.

49 Hendron, A. J. (1968) 'Mechanical properties of intact rock', in: Stagg and Zienkiewicz (eds). *Rock mechanics*. Wiley, pp.21–53.

50 Rocha, M. and Silveira, A. (1970) 'Characterization of the deformability of rock masses by dilatometer tests'. *Proceedings, 2nd congress on rock mechanics*, Belgrade, Vol. I, pp.2–32.

51 Finn, P. S. (1984) New developments in pressuremeter testing. *Gr. Engng*, **17**, 5.

52 Rocha, M. (1970) *New techniques in deformability of the in situ rock masses. Determination of the in situ modulus of deformation of rock*. American Society for Testing and Materials Technical Publication Number 477, pp.39–57.

53 Meigh, A. C., Skipp, B. O. and Hobbs, N. B. (1973) 'Field and laboratory creep tests on weak rocks'. *Proceedings, 8th international conference on soil mechanics and foundation engineering*.

54 Honeyborne, D. B. and Harris, P. B. (1958) 'The structure of porous building stone and its relation to weathering behaviour'. *Proceedings, 10th symposium Colston Reservoir Society*, Bristol, Butterworth Scientific, Guildford, pp.343–354.

55 Marachi, N. D., Chan, C. K. and Seed, H. B. (1972) 'Evaluation and properties of rockfill materials'. *J. Soil Mech. Found. Div., Am. Soc. Civ. Engrs*, **98**, **SM1**, 95–114.

56 Collis, L. and Fox, R. A. (1985) *Aggregates: sand, gravel and crushed rock aggregates for construction purposes*. The Geological Society, London.

57 Bloem, D. L. (1966) *Concrete aggregates – soundness and deleterious substances*. American Society for Testing and Materials Special Technical Publication Number 169-A, pp.497–512.

58 Ward, H. H., Burland, J. B. and Gallois, R. W. (1968) 'Geotechnical assessment of a site at Mundford, Norfolk, for a large proton accelerator'. *Géotechnique*, **18**, 399–431.

59 Franklin, J. A., Broch, E. and Walton, G. (1972) 'Logging the mechanical character of rock'. *Trans Inst. Min. Metall.*, **80**, A1–A10; disc **81**, A43–A51.

60 Deere, D.U. (1968) 'Geological considerations', in: Stagg and Zienkiewicz (eds) *Rock mechanics*. Wiley, pp.1–20.

61 Duncan, N. (1966) 'Rock mechanics and earthworks engineering'. *Muck Shifter*, parts 1–8.

62 Hoek, E. and Brown, E. T. (1980) *Underground excavations in rock*. Institution of Mining and Metallurgy, London.

63 Terzaghi, K. (1946) 'Rock defects and loads on tunnel supports', in: Proctor and White (eds) *Rock tunnelling with steel supports*. Commercial Shearing and Stamping Co., pp.15–99.

64 Deere, D. U. (1964) 'Technical description of rock cores for engineering purposes'. *Rock Mech. & Engng Geol.*, **1**, 1, 17–22.

65 Cording, E. J., Hendron, A. J. and Deere, D. U. (1971) 'Rock engineering for underground caverns. *Proceedings, symposium on underground rock chambers*, Phoenix. American Society Civil Engineers, New York, pp.567–600.

66 Wickham, G. E., Tiedemann, H. R. and Skinner, E. H. (1972) 'Support determination based on geological predictions'. *Proceedings, 1st North American rapid excavation and tunnelling conference*, American Institute Mechanical Engineers, New York, pp.43–64.

67 Bieniawski, Z. T. (1976) Rock mass classification in rock engineering. *Proceedings, symposium on exploration for rock engineering*, Johannesburg. Vol. I, pp.97–106.

68 Barton, N. R., Lien, R. and Lunde, J. (1974) 'Engineering classification of rock masses for the design of tunnel support'. *Rock Mech.*, **6**, 189–239.

69 Hoek, E. (1986) 'Practical rock mechanics – developments over the past 25 years'. *Proceedings, symposium rock engineering in an urban environment*. Institute Mining and Metallurgy, Hong Kong.

70 Hudson, J. A., Crouch, S. L. and Fairhurst, C. (1972) 'Soft, stiff and servocontrolled testing machines: a review with reference to rock failure'. *Engng Geol.*, **6**, 155–189.

71 Stacey, T. R. (1981) 'A simple extension strain criterion for fracture of brittle rock. *Int. J. Rock Mech. Min. Sci.*, **16**, 469–474.

72 Brown, E. T. and Bray, J. W. (1983) 'Ground response curves for rock tunnels. *J. Geotech. Engng*, **109**, 15–39.

73 Gerrard, C. M. (1982) 'Elastic models of rock masses having one, two and three sets of joints'. *Int. J. Rock Mech. Min. Sci.*, **19**, 15–232.

74 Hoek, E. (1983) 'Strength of jointed rock masses'. *Géotechnique*, **33**, 3, 187–223.

75 Johnston, I. W. (1985) 'Strength of intact geomechanical materials'. *J. Geotech. Engrg*, **111**, 6.

76 Barton, N. R. (1986), 'Deformation phenomena in jointed rock'. *Géotechnique*, **36**, 2, 147–167.

77 Patton, F. D. (1966) 'Multiple modes of shear failure in rock'. *Proceedings, 1st congress international society rock mechanics*, Lisbon, Vol. I, pp.171–187.

78 Barton, N. and Choubey V. (1977) 'The shear strength of rock joints in theory and practice'. *Rock Mechanics*, Vol. X, Springer-Verlag, pp. 1–54.

79 Miller, R. P. (1965) 'Engineering classification and index properties for intact rock'. PhD thesis, University of Illinois.

80 Cording, E. J., Hendron, A. J. and Deere, D. U. (1971) 'Rock engineering for underground caverns'. *Proc. Am. Soc. Civ. Engrs* (National Water Resources Meeting, Phoenix).

81 Fumagalli, E. (1968) 'Model simulation of rock mechanics problems', in: Stagg and Zienkiewicz (eds) *Rock mechanics*. Wiley, pp.353–384.

82 Stimpson, B. (1970) 'Modelling materials for engineering rock mechanics'. *Int. J. Rock. Mech. Min. Sci.*, **7**, 1, 77–121.

83 Sutherland, H. J., Hecker, A. A. and Taylor, L. M. (1984) 'Physical and numerical simulations of subsidence above high-extraction coal mines'. *Proceedings, International Society Rock Mechanics symposium on design performance of underground excavation*, Cambridge. ISRM, London.

84 Bray, J. W. and Goodman, R. E. (1981) 'The theory of base friction models'. *Int. J. Rock Mech. Min. Sci.*, **18**, 453–468.

85 Herbert, R. and Rushton, K. R. (1966) 'Groundwater flow studies by resistance networks'. *Géotechnique*, **16**, 53–75.

86 Wilson, J. W. and More-O'Ferrall, R. C. (1970) 'Application of the electric resistance analogue to mining operations'. *Trans Inst. Min. Metall.*, **79**, Sect. A.

87 Santing, G. (1963) *The development of groundwater resources with special reference to deltaic areas. UNESCO Water Resources Series 24*, pp.85–87.

88 Fung, Y. C. (1965) *Foundations of solid mechanics*. Prentice-Hall, New Jersey.

89 Coates, D. F. (1970) *Rock mechanics principles*. Mines Branch Monograph Number 874, Department of Energy and Mines Resources, Canada.

90 Timoshenko, S. P. and Goodier, J. N. (1951) *Theory of elasticity* (3rd edn). McGraw-Hill.

91 Lysmer, J. and Duncan, J. M. (1969) *Stresses and deflections in foundations and pavements* (4th edn) Department of Civil Engineering. University of California, Berkeley.

92 Savin, G. N. (1961) *Stress concentration around holes*. Pergamon, Oxford.

93 Obert, L. and Duvall, W. I. (1967) *Rock mechanics and the design structure in rock*. Wiley, New York.

94 Poulos, H. G. and Davies, E. H. (1974) *Elastic solutions for soil and rock mechanics*. Wiley, New York.

95 Marti, J. and Cundall, P. A. (1982) 'Mixed discretization procedure for accurate solution of plasticity problems'. *Int. J. Num. Methods in Engrg*, **6**, 129–139.

96 Zienkiewicz, O. C.and Cheung, Y. K. (1967) *The finite element method in structural and continuum mechanics*. McGraw-Hill, Maidenhead.

97 Brebbia, C. A. and Connor, J. J. (1973) *Fundamentals of finite element techniques*. Butterworth Scientific, Guildford.

98 Duncan, J. M. and Goodman, R. E. (1968) *Finite element analysis of slopes in jointed rocks*. US Army Engineer Waterways Experimental Station Report Number 568–3. Vicksburg.

99 Morland, L. W. (1976) 'Elastic anisotropy of regularly jointed media'. *Rock Mech.*, **8**.

100 Goodman, R. E., Taylor, R. L. and Brekke, T. L. (1968) 'A model for the mechanics of jointed rock'. *Proc. Am. Soc. Civ. Engrs*, **SM3**: 637–657.

101 Cundall, P. A. (1971) 'A computer model for simulating progressive large-scale movements in blocky rock systems'. *Proceedings, international symposium on rock mechanics and rock fracture*. Nancy, Paper 2–8.

102 Otter, J. R. H., Cassell, A. C. and Hobbs, R. E. (1966) 'Dynamic relaxation'. *Proc. Instn Civ. Engrs*, **35**, 633–665.

103 Lemos, J. V., Hart, R. D. and Cundall, P. A. (1985) 'A generalized distinct element, program for modelling jointed rock mass: a keynote lecture'. *Proceedings, international symposium on fundamentals of rock joints*.

104 Lachat, J. D. and Watson, J. O. (1977) 'Progress in the use of boundary integral equations, illustrated by examples'. *Computer Meth. Appl. Mech. & Engng*, **10**, 273–289.

105 Sinha, K. P. (1979) 'Displacement discontinuity technique for analysing stresses and displacements due to mining in seam deposits'. PhD thesis, University of Arizona.

106 Brady, B. H. G. and Brown, E. T. (1985) *Rock mechanics for underground mining*. Allen & Unwin, Hemel Hempstead.

107 Crotty, J. M. and Wardle, L. J. (1985) 'Boundary integral analysis of piecewise homogeneous media with structural discontinuities'. *Int. J. Rock Mech. Min. Sci.*, **22**, 6.

108 Lorig, L. J. and Brady, B. H. G. (1982) 'A hybrid discrete element – boundary element method of stress analysis', in: *Issues in rock mechanics: Proceedings, 22nd symposium on rock mechanics*. Society Mining Engineers/American Institute Mining, Metallurgical and Petroleum Engineers, pp.628–636.

109 Lorig, L. J. and Brady, B. H. G. (1984) *A hybrid computational scheme for excavation and support design in jointed rock media. Design and performance of underground excavation*. International Society for Rock Mechanics, Cambridge.

110 Ushijima, R. S. and Einstein, H. H. (1985) Application of 3-D coupled finite element – boundary element method. *Proceedings, symposium on rock masses*, American Society Civil Engineers, Denver.

111 Morgenstern, N. R. (1968) 'Ultimate behaviour of rock structures' in: Stagg and Zienkiewicz (eds) *Rock mechanics*. Wiley, Chapter 8.

112 Sarma, S. K. (1979) 'Stability analysis of embankments and slopes'. *J. Geotech. Engrg Div. Am. Soc. Civ. Engrs*, **105, GT12**, 1511–1524.

113 Hueze, F. E. and Goodman, R. E. (1971) *A design procedure for high cuts in jointed hard rock – three-dimensional solutions*. US Bureau of Reclamation. Final Report Contract 14–06 D–6990, University of California, Berkley.

114 Ritchie, A. M. (1963) *Evaluation of rockfall and its control*. *Highway Research Record*, **17**, pp.13–28.

115 Whiteside, P. G. D. (1986) Contribution to discussion, *Proceedings, symposium on rock engineering and excavation in an urban environment*. Institute Mining and Metallurgy, Hong Kong.

116 McCracken, A. and Jones, G. A. (1986) 'Use of probabilistic stability analysis and cautious blast design for an urban excavation'. *Proceedings, symposium on rock engineering and excavation in an urban environment*. Institute Mining and Metallurgy, Hong Kong.

117 Harr, M. E. (1977) *Mechanics of particulate media: a probabilistic approach*. McGraw-Hill, New York.

118 Priest, S. D. and Brown, E. T. (1982) 'Probabilistic stability analysis of variable rock slopes'. *Trans Instn Min. Metall.*, **92**, A1–A12.

119 British Standards Institution (1986) *British Standard code of practice for foundations*. BS 8004. BSI, Milton Keynes.

120 Hobbs, N. B. (1974) *Settlement of foundations on rock. Proceedings, conference on the settlement of structures*, general report.

121 Rowe, R. K. and Armitage, H. H. (1987) 'A design method for drilled piers in soft rock'. *Can. Geotech. J.*, **24**, 1, 126.

122 Hobbs, N. B. and Healy, P. R. (1979) *Piling in chalk*. Construction Industry Research and Information Association. Report Number PG6. CIRIA, London.

123 Rocha, M. (1956) 'Arch dam design and observations of arch dams in Portugal'. *Proc. Am. Soc. Civ. Engrs*, **82**, 997.

124 Brown, E. T. and Bray, J. W. (1982) 'Rock-support interaction analysis for pressure shafts and tunnels'. International Society Rock Mechanics symposium, Aachen.

125 Brown, E. T. and Bray, J. W. (1983) 'Characteristic line calculations for rock tunnels'. *J. Geotech. Engng Am. Soc. Civ. Engrs*, **109**, 15–39.

126 National Coal Board (1975) *Subsidence: engineer's handbook*. NCB Mining Department, London.

127 Langefors, U. and Kihlstrom, B. (1963) *The modern technique of rock blasting*. Wiley.

128 Dowding, C. H. (1985) *Blast vibration and control*. Prentice-Hall, New Jersey, pp.297.

129 Holmberg, R. and Persson, P. A. (1980) 'Design of a tunnel perimeter blasthole pattern to prevent rock damage'. *Trans. Inst. Min. Metall.*, **89**, A37–A40.

130 Tarkoy, P. J. and Henderson, A. J. J. (1983) *Rock hardness index properties and geotechnical parameters for predicting tunnel boring performance*. Final Report NSF Research Grant GI-36468, NTIS, Springfield.

131 Stillborg, B. (1986) *Professional user's handbook for rock bolting*. Rock and Soil Mechanics Series, Vol. XV, Trans-tech Publications.

132 American Concrete Institute (1976) 'Shotcrete for ground support'. *Proceedings, conference of the American Society Civil Engineers*, ACI Publication Number SP–54, ACI, New York.

133 Sharp, J. C., Hoek, E. and Brawner, C. O. (1972) 'Influence of groundwater on the stability of rock masses: 2-drainage systems for increasing the stability of slopes'. *Trans. Inst. Min. Metall.* **81**, A113–120.

134 Morgenstern, N. R. (1970) 'The influence of groundwater on stability'. *Proceedings, 1st international conference on stability*, pp.65–82.

135 Cedergren, H. R. (1967) *Seepage, drainage and flow nets*. Wiley, New York.

136 Price Jones, A., Whittle, R. A. and Hobbs, N. B. (1984) 'Measurements of *in situ* stresses by overcoring. *Tunnels and Tunnelling*, **16**, 1.

Bibliography

American Society of Civil Engineers *Journal*, Soil Mechanics Foundation of the Engineering Division. ASCE, New York.

Geological Society *Quarterly Journal of Engineering Geology*. GS, London.

Institution of Civil Engineers *Géotechnique*. Thomas Telford, London.

International Journal of Rock Mechanics and Mining Sciences. Pergamon, Oxford.

International Society for Rock Mechanics (1986) *Report on ISRM fields of activities*. ISRM Secretariat, Lisbon.

Rock Mechanics Information Service *Geomechanics Abstracts*. Pergamon, Oxford.

Textbooks

Brown, E. T. (1981) *Rock characterization testing and monitoring*. International Society for Rock Mechanics *Suggested methods*. ISRM Commission on Testing Methods. Pergamon, Oxford.

Brown, E. T. and Brady, B. H. G. (1985) *Rock mechanics for underground mining*. Allen & Unwin, Hemel Hempstead.

Coates, D. F. (1965) *Rock mechanics principles*. Mines Branch Monograph Number 874. Department of Mines and Technical Surveys, Ottawa.

Goodman, R. E. (1980) *Introduction to rock mechanics*. Wiley, Chichester.

Hoek, E. and Bray, J. W. (1977) *Rock slope engineering* (2nd edn). Institute of Mining and Metallurgy, London.

Hoek, E. and Brown, E. T. (1980) *Underground excavations in rock*. Institute of Mining and Metallurgy, London.

Jaeger, J. C. and Cook, N. G. W. (1979) *Fundamentals of rock mechanics* (3rd edn). Chapman & Hall, London.

Obert, L. and Duvall, W. I. (1967) *Rock mechanics and the design of structures in rock*. Wiley, Chichester.

11

Site Investigation

I K Nixon ACGI, FICE, FGS
Consultant

G H Child MSc, FGS
Associate, Keith Farquharson and Associates

Contents

11.1 Preliminary assessment

11.1.1 General

Site investigation in the overall sense is the process by which the various factors influencing the selection and use of the most appropriate location for a project are evaluated. Identification of the primary factors aids the initial selection of the site. Thus whereas topography and the geology determine the site of a dam, minimal environmental pollution requirements often define the location of an airport and preferential government aid that of a new industrial development. Each of the primary factors should be considered in sufficient depth to disclose any adverse item that may be critical before proceeding in detail to examine the technical feasibility of using the site. Guidelines for this preliminary assessment are given below.

11.1.2 Environmental considerations

This refers to the local conditions and resources both natural and those already existing including the infrastructure. A preliminary site reconnaissance should be carried out as early as possible utilizing available data, in order to consider the surroundings in relation to the project. Aerial photographs can be a valuable aid. The principal topics in this group are given below:

(1) *Topography* – Suitability of surface features at the site on land or over water.
(2) *Public and private services* – Availability of a suitable workforce, transportation facilities for access, water supply, power and telecommunications, sewerage and drainage, disposal of wastes.
(3) *Living amenities* – Facilities available or required for accommodation during construction and afterwards. The extent and standard of the community services either existing or planned.
(4) *Geology* – The local ground conditions at the site and in the surrounding area would normally be indicated sufficiently at this stage from the geological survey maps where available, otherwise by a site visit and/or an enquiry addressed to the local public authority. Check for adverse natural conditions such as unstable ground, underground caverns and subsidence potential.
(5) *Construction materials* – If *in situ* deposits are of interest refer to the geology or earlier uses of the site (e.g. old tips). For new roads in virgin country use aerial survey and remote sensing.
(6) *Hydrology* – Surface and groundwater conditions, river and tide levels, currents and stream flow, flood levels and drainage conditions. Periodic occurrence of springs.
(7) *Other uses of site area* – Past, current and proposed other uses at and around the site, such as mine workings (underground or open-cast), tunnels and underground bulk storage. Former industrial areas, refilled gravel pits, refuse tips, reclamation, waste and spoil dumps, buried pipelines, services, drains, pollution, radioactivity and other hazards. Ecological and conservational impacts. Consider both the site and its surroundings.
(8) *Meteorology* – Regional temperature, rainfall, humidity, prevailing winds and fog. Seasonal effects. Local microclimatical conditions before and after construction of the project.
(9) *Earthquakes and ground tremors* – See Chapter 8.

11.1.3 Administrative considerations

National or local plans for development or redevelopment, where they exist, should be inspected. Any restrictions such as those pertaining to access, noise, atmospheric pollution and site rehabilitation should be examined. The existence of mine workings, mineral rights, ancient monuments, burial grounds and rights of light, support and way, including any easements, should be established.

Proposal for development with outline plans should be submitted to the local planning authority and an application made for approval of use of the site before the preparation of a detailed scheme. It may be necessary to present evidence at a public enquiry.

11.1.4 Financial considerations

Although outside the scope of this chapter, another preliminary consideration of a proposed project should be a cost/benefit study covering both initial capital cost for construction along with subsequent running costs, wherein financial and economic factors, together with any social or amenity benefit are considered with respect to project feasibility and its alternatives.

In cases where the cost of the project may be significantly influenced by the ground conditions, such as for dams and major highways, including where comparisons are needed between alternatives, it would normally be advisable to extend the initial feasibility stage to include the preliminary appreciation of the site at least and its alternatives as described in section 11.2.

11.1.5 Environmental surveys

Engineering projects cannot properly be evaluated in isolation from their environment.

With minimal infrastructure there is the opportunity to select the most suitable site from a relatively simple environmental study of the natural features and facilities in the locality in order to consider the best compromise of aspects related, say, to topography, geology, biology and meteorology. As the intensity of local development and/or conflicting interests increases, site location becomes more restricted, and before it is examined in detail it may be necessary for impact studies to be made of the effects of the proposed development on the local human, natural and man-made environment or vice versa. Such studies take on a special significance where a public inquiry may ensue before final approval for a project can be given.

Some examples of major considerations that may be involved include: (1) preservation of natural vegetation, wildlife and land quality; (2) preservation of areas of archaeological, historical, or other special interest; (3) prevention of pollution of atmosphere in quality or by noise; (4) prevention of contamination of surface water, the ground or groundwater; (5) prevention of erosion on land, siltation and scour by water action; (6) preservation of social, public and private amenities; and (7) acceptable disposal or re-use of waste materials.

The principle to be adopted for undertaking an environmental survey should be to consider, comprehensively, all the pertinent factors whether or not at first sight they appear relevant having regard to the possible influence of the project. Full advantage should be made of available data and to ensure that sufficient territory has been taken into account, aerial photography generally provides the most convenient means of studying local topographical conditions that may be affected. Moreover, aerial photographs and multi-spectral techniques (satellite imagery) often reveal features not otherwise easily discernible, such as man-made buried workings and morphological changes. Trained interpreters, properly briefed, are able to extract much information.

A bibliography on this subject is given at the end of this chapter.

11.2 Site examination

11.2.1 General

Once the basic feasibility of a site for a new project has been established, the next step is to undertake a more detailed examination of the site itself to further the assessment of the relevant aspects required for the design and construction of the project.

11.2.2 Topographical surveys

The first stage in a detailed examination is to prepare an accurate survey from which plots on any required scale may be made. All levels should be referred to a reliable datum and the site preferably related to the national mapping and levelling system. Where derelict land, underground cavities or old workings have been identified, they should be included in this survey.

On large sites a local graticule of coordinates should be established, employing permanent beacons for more detailed surveys, extensions and setting out the works. Aerial photography can be advantageous, particularly on extended sites such as cross-country highways. Vertical photographs with stereoscopic overlap permit measurements to be made provided there is ground control. Photographs can be rectified to produce a true map and are then known as orthophotos. Such photographs can have contours plotted directly on to them. All aerial photographs can be produced as black and white with or without enhanced tones or infra-red-sensitive. Natural colour or false colour may also be used to identify special features. Existing air-photo cover sufficient for preliminary purposes may sometimes be available from the photo libraries of air survey companies and a few other organizations.

11.2.3 Hydrographic surveys

Hand-sounding may be sufficient for small areas of work. For larger works in tidal areas, deep water and high flows, more reliable methods will be required. Echo-sounding may be employed to provide a bed profile and under favourable conditions would be more convenient and accurate than hand-soundings. Surface profiling may be combined with sub-surface work by employing continuous seismic profiling or side-scan sonar systems, although the accuracy would be less. A comprehensive description of hydrographic surveying is given in the two standard works in the bibliography at the end of this chapter.

Photogrammetry is very useful for surveying coastal and inter-tidal zones. Offshore rocks, coral pinnacles, islands, sandbanks, shelves and buoys are easily located. Techniques are also available for reliably measuring nearshore depths. The disadvantages include unseen underwater hazards and drying-out areas which cannot usually be delineated.

Multi-spectral techniques may sometimes reveal detailed hydrological information not otherwise visible.

Estimates of peak flood levels should involve specialist advice and may require an extended survey far beyond the boundaries of the site.

11.2.4 Ground investigation

This refers to the collection and interpretation of data on the ground conditions at and surrounding the site for the design, construction, operation and maintenance of the project. Further details are given in the remainder of this chapter.

11.3 Principles of ground investigation

11.3.1 Primary objectives

Particular primary objectives of a ground investigation are to:

(1) Ascertain geological conditions at the site and groundwater hydrology to assess general suitability and for geotechnical study.
(2) Collect geotechnical data on relevant formations for quantitative design study of permanent and temporary works.
(3) Consider changes in ground stability and groundwater regime after construction due to the structure and/or future changes in ground such as mining or seismic activity.
(4) Evaluate effects of alternative excavation and construction methods, also temporary works.

In the case of existing structures, other factors that may be involved are:

(5) Need to ascertain reasons for structural defects, instability or failure.
(6) Consideration of remedial measures.

11.3.2 Contaminated site hazards

Additional investigation work, sometimes allied to that referred to above, is required when there has been some earlier use of a site that may have given rise to some significant disturbance or change in the conditions. A particularly difficult case is when the ground has become chemically contaminated with toxicants. This is a growing problem in developed countries.

The presence of chemical contaminants or ground liable to subsidence creates risks to personnel, for which reason special precautions are necessary during the investigation.

Objectives in this case, which may concern risks to construc-

Figure 11.1 Ground investigation using percussion boring equipment on waste tip containing hazardous materials. (*Courtesy*: Wimpey Laboratories Ltd)

tion workers, eventual users, animals, plants and building structures, are:

(1) To identify the types, extent and importance of the hazards, so that an assessment of their potential dangers to personnel, plants and/or the proposed end-use of the site can be made.
(2) To advise on suitable remedial measures to overcome identified hazards such as:

    (a) Settlement problems, e.g. subsidence due to decomposition, weathering and natural compaction, leaching and sudden collapse;
    (b) Obstructions, e.g., old foundations, piles, buried sea-walls.
    (c) Other problems which include:

        (i) fire, smoke, noxious fumes, gases and explosions from combustible material, microbial reaction of organic matter, volcanic areas;
        (ii) deleterious attacks on personnel from toxic powders, asbestos, fibres, liquids, explosive and asphyxiant gases, radioactivity and biological contamination;
        (iii) deleterious effects on the growth of plants or to the safe consumption of edible plant material;
        (iv) deleterious attacks on construction materials from residual chemicals (see also corresponding problem with natural ground as described in section 11.3.7.5(5) aggressive ground and groundwater);
        (v) pollution of streams and aquifers and the control of leachates, which involve the determination of the type of contaminant, its source and drainage plume. Wind action on contaminated dusts.

Table 11.1 indicates the range of potentially hazardous areas that may or may not include toxic materials. Future changes that possibly might occur in the ground at the site need also to be considered. Examples of this include underground mining, tunnelling and underground storage.

Table 11.1

Major landfills	Derelict works	Old workings and cavities
Domestic waste	Gasworks	Underground workings
Colliery waste	Sewage works and sludge disposal	for coal, stone, lime and flints, etc.
Ash (PFA) and clinker	Ferrous and non-ferrous works	Opencast workings
Slurry lagoons		Metalliferous mines
Chemical wastes	(smelting, refining	Abandoned shafts and adits
Metallurgical slag	and processing)	Cellars and basements
Hospital wastes	Pickling tanks	Sewers and tunnels
Scrapyards	Plating works	Salt mines
Industrial fill	Chemical works	Underground storage
Radioactive waste	Tanneries	Wells and tanks
Backfilled quarries and pits	Oil refineries	

11.3.3 Governing factors and limitations

Sufficient knowledge and experience exist of the difficulties in predicting ground conditions locally without a proper study, and the inherent weaknesses in many soils and rocks, to provide a justification for ground investigations in order to ensure safe, practical and economic designs.

Neither surface inspection nor information from outside the site is usually sufficient to provide reliable data on the ground conditions below the site so that exploration penetrating into the ground is used, at points on the site and related to the project.

The intensity of the investigation depends upon the character and variability of the ground as well as the magnitude of the project. The investigation depends upon the collection of representative data at sufficient points of exploration to enable the relevant geotechnical properties to be inferred for any part of the project.

The wide variety in ground conditions coupled with the range of design and construction problems to be solved make the subject complex, so that precise rules on the manner and extent of any study are not possible. Both experience and judgement are necessary. Too little investigation may not reveal a potential hazard, or involve extra costs for safety, while too much would be uneconomical.

An investigation should be planned and executed sufficiently far in advance of the commencement of design and construction to allow for a full study and the most effective use of the conclusions.

Codes of practice for ground investigation are now available in a number of countries and where appropriate should always be studied as they are likely to embody important local experience. Some are referred to in the bibliography under 'Main Investigation' given at the end of this chapter.

11.3.4 Cost

The cost of the investigation cannot be measured solely according to the size of the site or the magnitude of the project. It also depends upon having a knowledge of project details together with as much information as is available on the ground conditions. Even so, adjustments may still arise as the investigation proceeds depending upon whether simpler or more complex geotechnical solutions to those originally contemplated are appropriate.

Particular conditions may exist at a site that will involve higher costs than normal even for the same kind of development. One reason for this would be the presence of naturally occurring 'problem' soils or rocks, e.g. exceptionally weak soils such as peats and unstable material such as loess. The site may have become contaminated. Another reason would be the location of the site in a high seismic risk area or a cavernous region.

Because of the wide variety of soil and rock conditions many different investigation techniques have been developed varying in range of application and accuracy depending on general and particular requirements.

As an approximate guide, ground investigation may cost about 0.1 to 0.5% of the capital cost of new works and about 0.1 to 2% of earthworks and foundation costs although exceptionally the cost may be several times these ranges.

Sometimes the cost may be related to an overall cost saving; more often, though, the value of the investigation lies in the assurance against costly over/under design, unforeseen ground conditions with consequential delays in construction and poor in-service performance.

11.3.5 Ground investigation stages

The investigation should be a systematic expansion in knowledge of the ground conditions, directed towards solving the geotechnical problems. It is convenient to distinguish three stages in the complete process: (1) preliminary appreciation; (2) main investigation; and (3) construction review. Each of these stages is described in the following sections and is embodied in Figure 11.2 which presents in outline the sequence of

Contaminated derelict sites	Project conception	Geotechnical operations
Consult recorded data, site visit	Preliminary assessment	Geological outline
Consult specialist	Environmental/impact surveys	Geomorphological study
	Detailed topographical/ hydrographical survey of site/alternatives	
Specialist decides safety precautions and makes site visit	Preliminary appreciation Desk study and site reconnaissance leading to PRELIMINARY REPORT on: ground conditions, possible engineering problems and main investigation programme with estimated costs	Surface inspection, possibly with engineering geologist. Very occasionally borings/ pits, or overwater geophysics
Plan sampling patterns and methods	Main investigation SELECT RESOURCES for field exploration (1) In-house contribution (2) Contract fieldwork only, or (3) Contract fieldwork, testing and analysis. CHECK: staff competence, equipment adequacy and fieldwork supervision responsibility. Ensure flexibility of programme and methods.	Plan timing access, availability of resources, etc.
Maintain safety precautions	FIELDWORK EXECUTION Regular liaison between person in charge of investigation, field	Borings/pits etc., geophysics, *in situ* testing, instrumentation
Fieldwork	supervisor, geologist and engineering design.	
Laboratory tests	Minimal delays for submission of preliminary results. Samples to laboratory for testing leading to	Laboratory testing by stages with regular reviews
Specialists analysis and report	Record of results: FACTUAL REPORT ENGINEERING DESIGN REPORT	
Check sampling and testing	Construction review Compare predicted conditions with ground revealed in excavations, samples from bored piling, pile tests borrow pit conditions, trial embankments RECORD DATA AND ENTER ON DRAWINGS	Inspection, check results, full-scale trials instrumentation

Figure 11.2 Sequence of operations for ground investigation

events from commencement of the ground investigation to the completed development.

The particular investigation where a site is possibly contaminated is conveniently and economically carried out at the same time as the geotechnical study, as some of the processes are of mutual benefit, and knowledge of the results has an equal bearing on the predictions to be made.

11.3.6 Preliminary appreciation

Selected works on this subject are given in the bibliography, at the end of this chapter, particularly in the section headed 'Main Investigation'.

The preliminary appreciation of the available data on a site is an invaluable first stage in a ground investigation. By this means, full benefit is taken of experience to prevent wasteful exploration work and normally it provides a sound approach for the planning and commencement of the detailed study which should be considered as a separate exercise.

The time required to search and the amount of available data to be studied often cannot be predicted so that it is generally not a suitable subject for competitive bidding, especially on a lump-sum basis.

In order to proceed it is necessary to have some knowledge of the project besides knowing its location. Minimal information initially would be the approximate overall size, layout and purpose of the principal structural units. It needs to be sufficient to establish the main geotechnical problems in relation to the ground and site conditions revealed by the appreciation. However, at an early stage and before deciding upon the detailed plan for the main investigation, it will be advisable to have more particulars such as loadings, floor levels and settlement tolerances of buildings, to enable the scope of the testing to be determined. Clearly, the more detailed information that is available at this stage the more effective the planning of the investigation.

11.3.6.1 Objectives

The preliminary appreciation consisting of both a desk study and site reconnaissance, which should be properly recorded for future reference with a description of the project, should have the following objectives:

For natural ground:

(1) As clear a conception as is possible of the ground structure and groundwater regime underneath and neighbouring the site; the formations present and likely to be affected; their degree of complexity, the presence of problem soils and natural hazards, e.g. seismic activity, and subsidence; the possible alternatives in interpretation of the data and the probable degree of accuracy of each.
(2) The principal ground engineering problems anticipated and possible solutions in the design and for the construction, e.g. need for piling, excavation below water table, stability of natural slopes, suitable fill for embankments, slopes for cuttings (see list of primary objectives in section 11.3.1).
(3) Detailed proposals for the main investigation having regard to the factors outlined in the next section, including the methods to be used and the amount of work necessary in each. The budget cost and possible extent of contingencies that may be required having regard to the probable degree of accuracy of the preliminary information.

Additionally, for contaminated and derelict land:

(4) A carefully executed survey of available evidence on the earlier uses of the site and its surroundings to assess what is possibly present that is hazardous, where it is and how much (see list of potential hazardous areas earlier in section 11.3.2).
(5) Precautions to safeguard personnel and equipment during site visits and the investigation work.
(6) Main investigation programme, kinds of specialist services and personnel required.

Where the available data on a site are found to be inadequate to indicate the course of the main investigation or where important inconsistencies arise at this stage that raise doubts on the best exploratory method to be employed, it may be necessary to make a preliminary survey by carrying out a limited amount of field and laboratory work for the preliminary appreciation. Such an approach could apply to any type of investigation. Whilst the amount of such work should be kept to a minimum for economical reasons, it is important that this initial work is of a sufficiently high quality to enable the preliminary appreciation to be soundly based and that it is carried out sufficiently far in advance of the main investigation in order to provide adequate time to consider the results and to make the appropriate arrangements.

One particular form of a preliminary survey, well suited to investigation along a line, such as for a highway, railway or pipeline, is to utilize a system of land form mapping, sometimes called 'land surface evaluation' or 'terrain evaluation'. Predictions on conditions below ground-level, however, should always be checked, e.g. by geophysics.

The preliminary survey for an offshore structure is usually made by geophysics in the absence of drilling data and at the same time an inspection of the seabed topography is made to select a suitable site.

11.3.6.2 Desk study

The collection of available evidence on the site that may be relevant to the investigation and project is conveniently referred to as the 'desk study'. The time spent on this exercise will depend upon the amount of recorded data, the complexity of the site conditions and the magnitude of the project. Where the amount of information is significant, study of this prior to the initial site reconnaissance can be of advantage.

The sources of the information will vary according to the country in which the site is located. The following list represents the better-known potential sources, to act as an *aide-mémoire* in the search. For major projects, important evidence may be available in another country.

(1) Topographical maps. Past surveys can help identify filled land, subsidence and erosion.
(2) Geological maps and memoires. Mining records. The latest information is often only in manuscript form.
(3) Local administrative authority and museum records. Personal visits are likely to be the best approach. Enquire for national code of practice on ground investigation. Building by-laws. Regional code of practice for seismic areas. Records of unstable ground or flooding. Always ask about earlier uses. Archival search and specialist consultants can be helpful where mining was done.
(4) Aerial photographs and satellite imagery (the latter requires specialist interpretation). Commercial and government sources. Photography from model aircraft and balloons can be useful as accurate scaling is generally not needed.

Whilst aerial photography is excellent to gain an overall view and for interpreting surface features, e.g. old slips or swamps, caution should be exercised when predicting conditions below ground-level as these can only be inferred and changes may occur at very shallow depths. Experience aids interpretation.

(5) Previous investigations or construction work at or near the site. Valuable source in developed countries.
(6) Agricultural data. Often confined to surface data but can show local variations.
(7) Public services, water, electricity, sewerage, etc.

Advantage should also be taken of the growing body of recorded soil and rock mechanics' information on the more important ground formations throughout the world, e.g. laterites, decomposed granite and London Clay. These offer useful quantitative data for making tentative predictions on the engineering characteristics of the ground as well as suggesting the particular problems pertaining to the formations in question that have been noted from experience and which might not necessarily be revealed by an individual investigation at a site but nevertheless may deserve safeguards in the design. For example, where cavities (natural or man-made) may possibly be present as in chalk or limestone formations, whether or not any are revealed, foundations should be reinforced against local collapse and in certain cases permanent precautions taken during the subsequent use of the site against initiating subsidence by restricting the use of soakaways and garden hoses and by employing flexible joints on services.

Where the site is potentially contaminated, the desk study can be done in parallel with that for the natural ground and similar sources of information are used. However, specialists should interpret the hazards likely to be present and provide general guidance on the procedures to be used during the main investigation.

11.3.6.3 Site reconnaissance

Prior knowledge and interpretation of the available data on a site greatly aids the value to be gained from a preliminary inspection. It should best be done on foot to make a thorough visual examination on the topographical and geotechnical features; to note the layout of the project in relation with the ground and services, also to ascertain the overall suitability. The reconnaissance is very helpful for planning the main investigation, particularly with respect to the methods and means of access for rigs and the larger items of plant.

For all major projects an inspection should also be made by an experienced engineering geologist to interpret the geological features and to assess their engineering implications.

In the case of contaminated sites, particularly where chemical contamination is suspected, a specialist, preferably knowledgeable in the suspected contaminants, should be included in the team for the reconnaissance. Moreover, whatever the hazard it is vital that there is prior consultation between all members of the team to ensure that every reasonable precaution is taken not to endanger their health or safety during the site visit.

Guidelines for undertaking a site reconnaissance include:

Preparation – Take, if possible, project layout to scale, local and district map or chart, geological data, aerial photos and notebook and, if needed, simple surveying equipment, compass and geological hammer. For soil sampling, take hand auger, plastic bags and labels.

Ensure permission has been obtained to make the visit.

General – Position the project and assess effect on existing boundaries, topography and geology. Check access. Consider effect of earlier uses of site. Check services that may be available.
Ground – Study surface features in relation to available data. Note differences. Sample main soil types for simple identification. Study neighbouring geological features and existing structures. Inspect vegetation and make deductions on the soils present. Note presence of any problem soils such as peat, weak clays or loose sands. Look for chemical waste, significant odours, discoloured soil and blighted vegetation.
Main investigation – Inspect access to and around the site. Look for possible obstructions, such as power cables, buried pipelines and services, fences, etc. Seek water sources and power supply if required. Select location of offices, sample store and laboratory. Consider accommodation and communications.

11.3.7 Main investigation

11.3.7.1 General

The object is to develop, in sufficient detail, the initial concept of the ground and groundwater conditions formed from the preliminary appreciation to enable a final choice of site and layout to be made; a safe and economic design to be prepared of the works, with alternatives where appropriate; potential construction problems anticipated and hazards identified. This will almost invariably entail the use of specialized equipment in the field to establish the geological structure, soil and rock types and groundwater conditions, with *in situ* and laboratory tests, in conjunction with experience to assess the values of the engineering parameters.

The investigation should be a reiterative process whereby information gathered in the early stages is used in the checking of the preliminary appreciation and in the directing of the later stages. It is not unusual for a preliminary appreciation to be found inaccurate and it is important therefore that the investigation programme should be flexible enough and, at all stages, to permit, if needed, changes to be made in the amount, location and type of investigation methods and tests employed.

The scope of the main investigation involves taking into account a number of considerations as set out in sections 11.3.7.2 to 11.3.7.6, together with the selection of the appropriate methods of ground investigation.

The ground conditions usually determine the methods of field exploration and sampling but the numbers and types of tests needed are usually governed as much by the requirements of the project as by the ground conditions. All methods of investigation have their limitations and these must be continually borne in mind (see section 11.4).

The main investigation will usually consist overall of field and laboratory work, the results of which will be under continual scrutiny. Upon completion, all the results, their interpretation together with the conclusions should form the basis of a formal report.

11.3.7.2 Types of main investigation

Initial considerations affecting the scope of the main investigations are the influence of the basic engineering requirements as indicated below.

Whilst the majority of investigations concern new works, there are a number of other types each with particular requirements as referred to below.

New works. Attention always needs to be given to every aspect particularly where 'greenfield' sites are concerned, including the effects on adjacent properties. It may be necessary to consider alternative locations. The least favourable conditions should be taken into account for design purposes and to safeguard subsequent performance.

The type of new work can strongly influence the quantity and quality required for the main investigation, e.g. nuclear power stations, large industrial developments, petrochemical complexes and other potentially hazardous sites would require detailed investigation of the locations of the key plant sites to ensure that any hazard to the environment is limited to an acceptably low level. Major water-retaining structures could also come into the same category because of the potential hazard to the environment. It is usually necessary for the more sensitive projects to carry out a seismic risk analysis even if the site is located in a relatively low-risk area.

At the other end of the scale, single or small groups of houses, minor extensions to existing structures, small sewers and pipelines may only require relatively rudimentary investigation with shallow auger holes, trial pits and visual observations.

Extensions to existing works. Data from previous investigations should be sought and used together with design and construction experience of the existing works and their subsequent performance. It will be necessary to consider the effect of the new works on the old as this may influence radically the type of foundation to be employed. See also remarks in *Safety of existing works* above.

Damaged works. It is very important initially to establish the causes of the problems, so that with the collection of other relevant information required for the design of the remedial works, the outcome is that a more detailed investigation is often required than for a new project of similar size. Litigation is another reason for an extended investigation to provide ample justification for the redesign.

Measurements should be made to check for continuing movement.

Safety of existing works. This situation can occur where there has been a change of use entailing heavier loading conditions or ground conditions encountered which were not anticipated in the original design. Where safety is concerned, the key factor is to establish all the possible problems that could adversely affect it and where conditions indicate a marginal situation a careful and often detailed investigation is needed, including monitoring of the performance of the works.

Materials for constructional purposes. In soils, pits or, better still, trenches are generally more suitable than boreholes since these enable more detailed examination of local variations, can give some indication of excavation problems, and allow for larger samples for testing. Classification testing generally needs to be more detailed than for foundation investigations.

Rock classification should be on the basis of the size of material needed for its proposed use to take account of its actual jointing and planes of fracture. Drillings into bedrock should be planned to determine joint structure to predict excavation costs and mode of extraction. A blasting trial should be considered.

11.3.7.3 Lateral extent of exploration

Fieldwork, in conjunction with available geological information, is used to explore the ground conditions at points distributed over the plan area of the project and extending at least up to its boundaries. Where the project could affect or be affected by adjacent areas or structures outside the site boundary, the points of exploration should cover these areas also. For example, proposed basements may extend below adjacent existing foundations or there may be sloping ground just beyond the site boundary. This principle should be followed even if the effect is only temporary, e.g. during construction.

By carrying out exploration initially at widely spaced points the overall conditions become known at an early stage. Further exploration can then proceed within this framework by comparison of the results at two or more points. There are dangers in assuming that an investigation at a point is representative of some undefined area all around it and there is no indication of the dip of bedding. Investigation should normally be made at the extremities of a structure, with additional intermediate exploration if variable conditions exist, and at points of concentrated loading.

Spacing of the points of exploration depends on the interaction of such factors as the type of ground conditions, the significance of the project, the requirements for the investigation, the relative merits of methods of investigation and their availability. At the lower end of the scale, several machine-dug trial pits or hand-auger holes could be more appropriate for a small investigation on reasonable ground conditions than the cost equivalent in rig boreholes. However, in general terms boreholes (or their equivalent) are often as close as 10 to 30 m, with not less than three per 200 to 900 m^2 of project plan area. With increasing number of points of exploration the intensity of investigation tends to be reduced where the ground conditions are relatively uniform.

Large complex projects require the points of exploration to be concentrated in the areas of the more significant units, e.g. deep basements, high retaining walls, large, tall or heavily loaded structures, liquid retaining structures, structures sensitive to

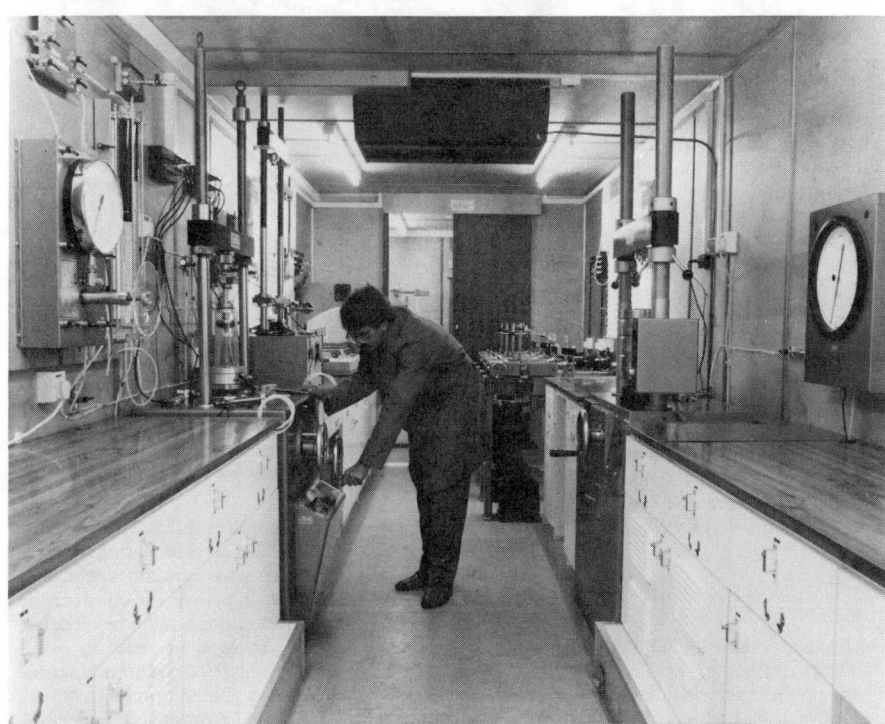

Figure 11.3 A site laboratory usefully minimizes sample disturbance by reducing delays in testing and transport. This illustration shows the interior of one of four mobile air-conditioned laboratories, belonging to the China National Coal Development Corporation, that incorporates computer-controlled triaxial, oedometer and shear box equipment. (*Courtesy*: ELE International Ltd)

Figure 11.4 Rotary core drilling showing sealing of rock core in plastic tubing upon removal from double tube core barrel

movement, units with a potential hazard risk. For minor units located within a complex, fewer points of exploration are used provided they reveal ground conditions consistent with interpolations between the major areas of investigation.

Very large projects such as reservoirs, stockyards, spoil tips and reclamation schemes should be subjected initially to a broad study using a gridded pattern of points of exploration at, say, 300 m centres.

Linear structures such as pipelines, channels, roads, railways, airport runways and tunnels have points of exploration located along the centreline with some straddling it to detect lateral variations. Structures en route are investigated as separate units; otherwise, points of exploration are spaced about 50 to 200 m apart.

Construction on sidelong ground, that might entail retaining walls, or where there is an actual or potential slope stability problem, would usually include three to five or more points of exploration on line in the critical direction across, as well as beyond, the area.

11.3.7.4 Depth of exploration

In principle, the investigation should extend to such a depth that it identifies all the strata and groundwater regimes that will be affected by or will affect the project and provides sufficient data for design and construction. It is advisable at an early stage of the main investigation to establish or confirm the overall ground profile to the maximum depth required at least at one point under each major structure. Provided ground conditions are consistent and satisfactory it may not be necessary for the full

profile to be established at all exploration points. Any excavation work (surface or underground) that is planned should always be adequately supported by investigation beyond its full depth.

Where bearing capacity and settlement are the controlling factors, as for single or multiple foundations, the investigation should extend to a depth at which the increase in vertical stress caused by the foundations will have negligible effect on the ground. This is usually taken as that depth at which the increase in vertical stress is less than 10% of the applied bearing pressure and less than 5% of the effective vertical stress in the ground. Where loadings are not known, the initial depth of exploration should be at least one-and-a-half times the width of the building, not less than 10 m unless very strong ground is encountered at shallow depth precluding any problem. Such a stratum should be investigated to a depth of at least 3 m. Where this stratum is rock, any very weathered zone should be fully penetrated to

Figure 11.5 Location of gravel-filled channel under St James's Park, London, using ground radar (electro-magnetic profiling) equipment. (*Courtesy*: Wimpey Laboratories Ltd)

ensure an improving profile with depth and that a boulder has not been mistaken for bedrock. Where loadbearing piles or other deep foundations, cantilever walls, ground anchors or other similar forms of temporary or permanent construction may be employed, the depth of exploration should be reckoned below the lowest possible founding level in order to assess overall stability and settlement. The depth of exploration should also extend below the depth of any proposed ground treatment such as freezing, chemical injection or dynamic compaction. Exploration depth below excavations and basements should be assessed by the change in vertical stress criteria as for foundations given above. If artesian or sub-artesian conditions are suspected the depth should extend to below the aquifer or to

Figure 11.6 Self-boring pressuremeter in operation complete with on-site data acquisition and monitoring system. (*Courtesy*: Soil Mechanics Ltd)

such a depth below which such conditions if they existed would not be significant.

Some relaxation of the above depth guides are permissible for high fills provided the ground conditions are shown to be satisfactory and settlement is not a problem. Side slopes should be investigated for stability by exploring to depths of half to one-and-a-half times the side slope width with the greater factor for the steeper slopes. Slope stability problems should be investigated to depths below any potential failure surface or to a hard stratum below the slope toe.

Where ground permeability is an important factor, as for dams and water-retaining structures, the depth of exploration should be sufficient to enable flow nets to be drawn, i.e. about 1 to 2 times the height of retention or half the base width, whichever is the greater. Where a vertical cut-off could be considered greater depths may be necessary.

Highway and airfield pavements require investigation to about 3 m depth below subgrade in cuts or ground-level under low fills unless the ground is very weak, such as in peaty areas when exploration needs to extend through the weak material.

Lightly loaded areas may involve an extended depth of exploration to penetrate all weak and compressible ground and particularly the zone of ground influenced by seasonal climatic changes and vegetation. In temperate conditions such as in Britain the zone affected by seasonal wetting, drying and frost may only extend to 1 or 2 m depth in open sites. However, the effect of tree roots can extend to about 5 m. In hotter, drier, climates the zone can extend to 20 to 30 m under trees. In arctic climates the temperature effect can extend to depths of 15 to 30 m.

11.3.7.5 Natural problem conditions

Particular types of natural ground conditions, representing problem conditions, are known from experience to require more careful exploration and testing because of the difficulties they can cause. Some examples are:

(1) Organic soils, peat, soft alluvial clays and silts, black cotton soil, quick clays leached by freshwater percolation, sensitive clays that weaken significantly on disturbance, swelling (expansive) and shrinking clays which are markedly affected by changes in moisture content.

(2) Weak granular soils such as dune (rounded grain) sand; or very loose sand which can settle significantly when subject to minor vibrations from, say, nearby pile driving. Earthquakes in some cases have caused spontaneous liquefaction in such soils.

(3) Metastable soils, such as loess, having been deposited in an exceptionally loose state and which can collapse on saturation leading to catastrophic settlement. In the dry state, such soils are very stable. This can be misleading during investigation and more so because collapse can take place in the boring process.

(4) Duricrusts. Hard crust sometimes occurs, normally near the top of a particular soil profile, beneath which much weaker soils exist. One example is the caprock at the top of a lateritic soil profile, others are lava flows and basaltic layers. The thickness and strength of the crust is often extremely variable.

(5) Aggressive ground and groundwater, that may contain constituents such as gypsum which attack Portland Cement concrete; or electrolytic, chemical or bacteriological agencies that attack metals, particularly cast iron. Saline ground, groundwater, as well as sea water and soft water may also require special consideration.

(6) Permafrost and frost-susceptible ground. Dealing with permanently frozen ground is a subject requiring special expertise, while frost-susceptible ground includes silts,

chalk and some shales which expand or disintegrate on freezing due to the development of internal ice lenses.

(7) Noxious and explosive gases are occasionally present in soils, rock and groundwater. These may cause a hazard during construction and in unlined excavations such as tunnels. Methane is soluble in groundwater, increasingly so above ambient pressure, and it has been known to cause explosions in boreholes and underground workings. A certain type of bacteria exists naturally in some ground, that can deplete the oxygen supply in poorly ventilated underground areas.

(8) Rocks subject to rapid weathering or swelling (see Chapters 9 and 10).

(9) Unstable profiles, geological or topographical anomalies, faults, ancient slip planes and extended discontinuities. Buried channels which may affect the project. Inclined drill holes may be of value to assist in the location of faults.

(10) Ground liable to subsidence such as in cavernous limestone and chalk areas, or collapse associated with valley cambering.

11.3.7.6 Contaminated site surveys

It is essential that a most thorough and careful preliminary appreciation has first been made to ascertain as much information as is available on the previous long-term history of the site and its surroundings to give the best indication of what potential hazards to expect from all previous uses. The next most important stage is to assess the immediate risks to personnel and plant in order to decide the safety precautions to be taken during the main investigation. Established procedures often exist for these precautions and specialist advice should be sought according to the risks. There may be national legal obligations to be complied with such as the Health and Safety at Work Act.

In the case of physical hazards, such as the location of derelict underground structural work or the size and extent of old mine workings, the problems are similar to natural ones and the main investigation represents simply an extension of the use of the established methods of ground exploration with the intensity of the points of investigation being programmed according to the available information. A grid pattern of boreholes is often used with the spacing being steadily reduced until sufficient confirmation of the hazard and its extent is obtained. Geophysics may also be employed with advantage.

Where contaminants are suspected, in liquid, gas, solid, bacteriological or radiological form, experience shows that investigation can become very complex to locate and identify what is the amount of risk, so that suitable specialists should be engaged to determine land quality and the precautions necessary for its safe development. The principal objective is to locate those parts of the site or its surroundings where concentrations of contaminants remain that are sufficiently high to impose a risk to the development envisaged.

To give some idea of what may be involved a systematic sampling strategy is usually employed in conjunction with a thorough visual inspection of ground and vegetation, including noting unusual smells. Machine-excavated pits (entry may be dangerous and backfilling should be prompt) are preferred to boreholes for ascertaining what variations there are in the concentration of the contaminants from one part of the site to another, laterally and in depth. Places with the highest levels are the important ones and probably would be sampled in a second stage in more detail. Initial sampling of soil and groundwater, might be on a 25 to 100 m grid and, say, at 1-m intervals in depth depending upon the kind of development. Sampling should be ample, it may include taking vegetation for test, and could involve special precautions to obviate contamination

from the container. Surface water would need to be sampled. Where groundwater contamination is involved geological and hydrological surveys become more important with a careful assessment of ground permeability.

The wide range of contaminants makes testing a specialist subject, the more so because the quantities may be small. Inorganic, organic, bacteriological and radiological testing may be involved so that a comprehensive study becomes multidisciplinary although chemical testing usually predominates. While some guidelines have been proposed on the level of concentration that involves precautions for the more common contaminants it will vary with the kind of development proposed and considerable judgement is involved. In fact, the presence of contamination does not necessarily mean a problem exists and many contaminated sites can be safely re-used.

11.3.7.7 Interpretation and the geotechnical report

Interpretation of the data should be a continuous process from the commencement of the investigation, leading firstly to reliable ground and groundwater profiles, then realistic values for the ground characteristics and ultimately solutions where possible to the ground engineering and site problems. As the fieldwork explores the stratigraphy, in situ tests are carried out and samples taken for examination and laboratory work. The types of sampling and testing are chosen compatible with the methods of exploration (see section 11.4) and the engineering problems. The amounts are based on previous experience of what is appropriate to give sufficient information. In view of the fundamental importance of the fieldwork, an experienced geotechnical engineer or engineering geologist should be employed full-time on site to supervise. Changes and adjustments in procedures can also be effected more competently and economically. Such a person, or a trained assistant, should personally inspect all samples and plot the logs. These should represent what the actual ground conditions are considered to be, at that point of exploration, weighing all the evidence from the boring-records, tests and sample descriptions, taking account of the inherent disturbance that sometimes occurs due to the boring and sampling operations.

The report will be the only lasting record of the investigation and therefore should contain a statement of the purpose of the investigation, a plan showing the site and its location, surface conditions, earlier uses, existing structures and topographical features, time of the fieldwork and for whom it was carried out. Along with the description of the proposed works would be given a summary of the local geology and a full record of the types and results of the field and laboratory work. Information from boreholes and trial pits should be recorded graphically and in cross-sections. Classification tests should be used to check the sample descriptions making due allowance for sample disturbance. Full correlations should be made with the geological information. Test results including water-level records should be tabulated and where appropriate plotted graphically. Up to this point a description of the interpretation of the ground conditions with a record of the results of the investigation is generally referred to somewhat irrationally as a factual or, more appropriately, descriptive report and may complete the work of a ground-investigation contractor. (Such reports should always be provided complete to all tenderers for the construction work.)

Access difficulties sometimes mean that it is not possible to investigate at the preferred locations. Such situations are undesirable and should be recorded in the report on the main investigation.

Where an engineering interpretation is required, the terms of reference should be recorded, with information on proposals supplied by the client. The derivation of values of ground

Table 11.2 Exploration methods on land

Method	Geology	Technique	Applications and limitations
Boreholes	Clays, silty clays and peats.	Hand or power auger boring (single blade or continuous spiral). Usually without addition of water.	Shallow reconnaissance. Power operation fast. Limited to non-caving ground except for power-operated hollow continuous augers.
	As above, also silts and sands	Wash boring, with water or drilling mud.	Preliminary exploration, with disturbed sampling, frequently includes SPT. Unsatisfactory for precision work but inexpensive.
	As above with gravel, occasional cobbles, and boulders also decomposed rocks.	Light cable percussion boring with casing. (Shell, auger and clay cutter cable-operated boring tools)	Standard for soil exploration. Water added below water table to stabilize base of boring. Before core sampling cohesive soils, the borehole should be properly cleaned out.
	As above and up to moderately weak rocks	Non-coring drilling, with pneumatic chisel or rotary tricone bit.	Limited to location of hard ground (check for presence of boulders), cavities, or testing, at pre-arranged levels, in suitable soils.
	All rocks and occasionally soils	Rotary core drilling, usually with water flush. Drilling mud stabilizes wall and counters stress relief at base. Alternatives: air flush, foam and other liquid additives.	Standard for rock exploration. Reliability depends on correct selection of core barrels, bits and flush fluid. Water table observations difficult. For proving rock at base of cable percussion boring with casing use pendant drilling attachment. For clays, sands and very weathered rocks, use triple tube retractor barrels.
Pits and shafts	Clays and peats	Excavation by hand, power grab, or auger with support as required. Tractor-mounted hydraulic back hoe excavators particularly suitable.	Detailed study of local soil variations. Direct access gives best opportunity for inspection of ground in situ, presence of stratification and thin clay layers. Depth usually limited by problems of groundwater lowering.
	Silts, sands and gravels	Close timbering or piling, groundwater lowering essential below water table.	
	Weak to moderately weak rocks	Hand excavation, power grab, or auger.	Detailed study of bedrock conditions, weathering, fissures and joints. Depth usually limited as above.
Trenches	All soils above water	Excavation usually by machine such as hydraulic powered excavators. Support as required.	Exploration of borrow areas. Direct access with extended inspection of lateral variations.
Adits	All soils and rocks	Appropriate forms of hand excavation and timbering as for tunnelling.	Established method for detailed exploration of dam abutments and underground structures. Sub-surface exploration of steeply inclined rock strata.

Prime safety precautions: (Boreholes)
- Overall collapse above cavities, buried mine shafts, etc.
- Gas and fires from peat beds and organic fills.
- Hand and head injuries.

Prime safety precautions: (Pits and shafts)
- Beware collapse of walls
- Asphyxiation without ventilation
- Quick condition in bottom
- Refer to BS 5573*

Safety: See above (Trenches)

Safety: As for tunnels (Adits)

Notes: (1) Locate pits and trenches outside proposed foundation areas to obviate soft spots under structures.

(2) Seal boreholes with impermeable backfill where it is necessary to prevent access for groundwater upwards or downwards into excavations.

(3*) BS 5573:1978 Code of Practice for Safety Precautions in the Construction of Large-diameter Boreholes for Piling and Other Purposes (formerly CP 2011).

parameters from the investigation data for design purposes should be explained with an assessment of reliability. The interpretation may take the form of recommendations or comments on a client's proposals, and may be qualified and subject to confirmation by further work. Various topics may be referred to depending on the project but could include comments upon: pad and raft foundations, working loads for piles, earth pressures for retaining walls, flotation of basements, the need for special construction techniques (e.g. anchors, groundwater control, chemical treatment), chemical attack on foundations, pavement design, temporary and permanent stability of slopes, subsidence, methods of excavation and filling, sources of construction materials. For further information see bibliography under 'Main Investigation'.

11.3.8 Construction review

The degree of confidence placed in the conclusions and recommendations in any investigation must recognize that they are based on a first-hand knowledge of only a minute proportion of the ground influencing or influenced by the project. Accordingly, during construction the results of the investigation should be verified. In simple cases, this will consist of comparing the conditions revealed in any excavations with the predicted soil profile. Significant differences that arise may require design amendments, possibly after further investigation. Such differences should be recorded properly for use later when modifications may be introduced or extensions added.

In the case of specialist geotechnical processes, e.g. piling, grouting, ground anchors and diaphragm walls, check tests may sometimes be required to compare local conditions with design criteria established from the main investigation. This third stage would also normally include full-scale trials made at the commencement of the contract.

The full extent of the bedrock structure can only be seen properly in an excavation and full allowance should be provided in the design for all reasonable eventualities. This applies to earthworks and especially dam construction where modifications in the design to suit the conditions revealed is frequently normal practice.

Groundwater observations may have to extend over a wet season or even several years and into the construction period. Records are also needed when groundwater lowering is being used to note its effect on the excavation work and outside the site where it may affect water supply and cause ground settlement.

Instrumentation measurements are often usefully continued throughout and after construction to observe the performance of the project, particularly where, because of complex ground conditions, predictions from tests are less reliable than usual. This is particularly desirable in the case of, say, dams where instrumentation can provide the only possibility of an early warning of the onset of unacceptable conditions.

In some cases, where the interaction of the project and ground conditions is a complex one, the construction review becomes more fundamental in the solution of the problem. This is considered further in section 11.4.2.5.

Table 11.3 Over-water ground exploration

Method	Technique	Applications and limitations
Cable percussion boring with casing. Conventional core rotary drill with rods to prove bedrock	As on land, from fixed platform or from floating pontoon, barge or ship fixed in position by anchors. (Wash boring with *in situ* sampling and testing may suffice for preliminary surveys.)	*River, lake and coastal structures in water depths up to about 50 m.* Sampling and borehole tests as on land. Proving bedrock facilitated by using pendant attachment mounted on boring casing to combat wave and tidal effects.
Rotary wireline drilling through guide tube from surface vessel or platform	Extension of above technique with cable extractable rotary core barrels. Vessels typically 50 m long for water depths up to 200 m. Greater depths require dynamic positioning gear.	*Widely used for offshore platforms* using percussion wash boring through sediments and rotary coring through rock. Heave compensation needed for rotary drilling from floating craft, or use pendant attachment.
Submerged rotary drills operated by divers	Hydraulic power from support vessel anchored above, using wireline coring technique and *in situ* testing.	*Projects in harbours and open water depths up to 40 m.* Total penetration typically 20–60 m. Maximum current 2–3 knots. At depth around 20 m may require 4–6 divers. At depths around 30/40 m may require minimum of 8 divers.
Submerged remote-controlled rotary corers and seabed samplers	Power supply and control via umbilical cable from support vessel to seabed unit with rotating head. Some incorporate magazine of drill pipes to increase penetration.	*Offshore structures.* Low penetration units limited to 5 m. Larger corers designed for penetrations around 50 m or more in water depths 200 m and more. Core sizes typically 50–100 mm dia.
Flexodrilling	Power supply, flushing media and remote control via special flexible non-rotating cable on the lower end of which is a motor-driven rotary drill that may incorporate coring and non-coring facilities.	*Coastal and offshore structures.*

Table 11.4 Sampling methods on land

Source	Geology	Disturbed		Undisturbed	
Boreholes	Clays, silty clays and peats	Hand auger	Normally representative of composition for classification, but unreliable for examination of structure.	Open tube samplers Area ratio (AR)	General purpose 100 mm dia × 450 mm long, heavy duty < 30% A Suitable for local stratigraphical identification and soil mechan testing on cohesive soils and weak rocks excluding pore-water pressure measurement on softer materials. Thin walled samples < 10% AR. 75 to 250 mm dia. better for soft and firm without stones.
		Clay cutter	As above, but liable to more mixing.		
	As above, also silts and sands	Shell	Standard for non-cohesive strata to examine composition (particle size and distribution). Best when whole contents of shell is emptied into tank and allowed to settle before taking representative sample from sediment.	Piston samplers	Less disturbance and better recovery than for open tube samplers. Fixed piston superior to free piston. Non-cohesive strata retain only within mud filled borehole. Improved quality helpful when testing soft recent clays and for effective stress analysis. Reliabi aids studies of specific horizons. Sample diameters range from 250 mm and lengths up to 1 m.
		Powered auger	Liable to considerable disturbance and mixing except when conditions in depth are very uniform.	Continuous samplers (usually commenced from ground surface)	(a) Delft 29 mm dia. (Nylon stocking) rapid method with individ samples up to 18 m long in recent alluvium (Dutch cone resista below 10 MN/m²) for stratigraphical identification.
		Water flush	Liable to serious disturbance and mixing (strata identification only).		(b) Delft 66 mm dia. (Nylon stocking) as for 29 mm sampler, also all standard soil mechanics testing.
		Standard penetration test sampler (SPT)	Provides small specimens of both cohesive and non-cohesive soils for classification purposes but is not normally suitable for retaining structural features.		(c) Swedish 68 mm dia. (Steel foils) individual samples up to abou 29 m in soft recent alluvial clays and laboratory strength tests correspond to *in situ* vané results. Can also be used in silts and sands of medium and low density.
		Flow-through sampler	Self-contained incremental sampling technique commenced from ground surface for strata identification. Size of sample similar to that of an SPT.	Compressed air sampler (60 mm dia.)	For recovery of silt and sand from above or below water table wi use of mud, to study laminar structure and composition, densit and permeability.
				Rotary core barrels (Total volume sampling)	Triple tube types (see below under rocks) including those with spring-loaded inner barrel (retractor type, Mazier) and face discharge tungsten carbide bits with removable inner liner usua plastic, e.g. Mylar. Synthetic polymer flushing fluid can be advantageous using low flow rates. Larger core sizes preferable typical 100 mm nom.
	Gravel, cobbles and boulders	Shell	Standard for gravel, but grading may be unreliable.		No common method in use, although injection of chemical grout been tried, also freezing where saturated.
		Power auger	Specimens up to gravel size may be recovered without reliance on source.	Rotary core barrels	Double and triple tube types to core boulders.
	Weak rocks (including *hard clays*)	Auger	Sample identification generally misleading due to remoulding which produces a weaker material.	Driven samplers	Shatter during driving causes serious structural disturbance which affect results of soil mechanics tests.
		Air flush (vacuum recovery)	Possible study of mineral composition above water table.	Rotary core barrels	Double and triple tube types (see below and above under fine grai soils) and pitcher sampler.
		Flow through sampler	See note above.		
	All rocks	Water flush	Rock sludge samples provide opportunity for identification by microscope when conditions are uniform if no core is recovered.	Rotary core barrels	*Single tube.* Simplest type suitable only for massive uniformly stro rock.
					Double tube types support and protect core during drilling.
					Inner tube rigid: least likely to jam but liable to cause serious sa disturbance in variable and broken rock.
					Inner tube swivel: *internal discharge:* adversely affects core recove variable and broken rock which is minimised w discharge is below core lifter.
					face discharge: although expensive is considere best method to minimise losses in variable and broken rock.
					Triple tube types provide extra split inner tube which assists in rer of core from barrel with least disturbance. Other special barrels in *spring loaded inner barrel* which extends to protect core in weak la *Wire line barrels* provide facility to withdraw and return inner ba and core from bottom of hole independently of outer barrel and t *Water flush* is generally used to cool bit and remove cuttings. *Air* requires special equipment to maintain air speeds, can have advan when coring above the water table. *Mud flush* can be helpful to re erosion of core. *Foam flush* helps reduce bit wear in hard rocks an increases speed. Rock cutting is usually with diamond bits but tur carbide inserts are applicable for uniform soft rocks. *Chilled steel* is used only for large diameter cores (over 150 mm dia.) when som loss is acceptable. Fissures must be grouted to prevent loss of sho
					Suitable ancillary equipment as well as skilful operation are essent for good core recovery and the greater the complexity in ground conditions the higher the degree of skill required. The more broke ground the shorter each drill run should be to ensure good recove The core should be preserved in 'lay-flat' plastic tubing for labora testing or in polyurethane foam as described below.
Pits, trenches & adits	Clays and peats, silts, sands and gravels. Up to moderately strong rock.	Hand excavation	For identification purposes particularly useful to study local variations and anomalies. Ensure fresh *in situ* surface is exposed before sampling.	Open tube and piston samplers. Block samples	See notes for boreholes. Offers opportunity for horizontal and inc as well as vertical tube samplers, silts and sands as well as clays. Ensure fresh surface is exposed before sampling. Hand-cut *block samples* of self-supporting soil or weak rock, carefully cut and trin *in situ* to provide undisturbed sample with minimal disturbance. Samples, often 150 mm cube are coated in wax reinforced with m as each face is exposed or wrapped in foil and encapsulated in polyurethane foam.
Groundwater		Bail out borehole or pool, sample after the water has returned to its former level. Rinse the container thoroughly beforehand, preferably using water from test source. Ensure surface or rain water has not diluted water to be tested.			

11.4 Methods of ground investigation

11.4.1 General

Primarily, the very large number of methods in use can be divided into two groups; those that rely upon samples and those that provide *in situ* measurements. In any major investigation, both are normally required, the relative amounts depending upon the available information and the magnitude of the project. To aid basic selection, it is important to note that first the stratigraphy and the groundwater conditions must be interpreted adequately, before it is possible to decide upon the required engineering properties. In certain circumstances the general ground profile is reasonably well known in advance, in which case it is often possible to depend solely upon *in situ* measurements to interpret the soil conditions and determine the properties (see also section 11.4.3.1).

The merit of sampling is the opportunity it gives for direct inspection, although sample disturbance must always be taken into account. In the laboratory, more control is possible in the boundary conditions during a test, and adjustments can be made prior to the main test, e.g. in stress conditions or moisture content where these are of consequence in the design. Long-term tests are also more conveniently carried out.

11.4.2 Stratigraphical methods

11.4.2.1 Geological mapping

The structure in depth is inferred from mapping of the surface features. This gives a general indication of the ground conditions and, where there are numerous surface features to aid identification, may provide a very good indication of the structure. However, it may fail to reveal comparatively minor geological features, which have a decisive influence on the project. This method is more fully discussed in Chapter 9.

11.4.2.2 Exploration by boreholes, pits, trenches and adits

Geological mapping should always be supplemented by exploration using boreholes, pits, trenches and adits in which the ground in depth is exposed for direct examination and representative samples retained for identification and laboratory testing (section 11.4.3.2). The various techniques of exploration and their application to the ground conditions are given in Table 11.2, for use on land, from platforms or staging and, in Table 11.3, in offshore situations where floating craft are necessary (section 11.4.4).

Table 11.5 Deep-water soil sampling

Method	Technique	Applications and limitations
Dredges	Sheet metal or chain bag with open end, towed under way.	Disturbed samples of loose material from surface. *Reconnaissance only.*
Grabs	Variety of methods force jaws to close before withdrawal on winch rope.	*Bulk samples of bed material.* Unconsolidated clays, silts and sands. Limited use in compact soils and gravels. Difficult to use in rough conditions.
Drop or gravity corers	Similar to open-tube soil sampler with weight at the top end, and trip device for release at pre-selected distance above bed. Normally with plastic liner. Some types are driven in with an explosive charge. *Free-fall corers* for deep water obviate the need for winch and use release of buoyant chamber after sampling to return core to water surface.	*Open-drive tube* samples of soft cohesive soils up to few metres long. Fine non-cohesive soils recovered using tulip-type core catchers. Tube dia. typically 40–120 mm. Can be used at depths exceeding 400 m.
Piston corers	Drop corers in the form of a free-piston sampler. Overall weight 300–1500 kg	Offers opportunity to recover longer and better cores typically up to 10 m, than with drop corer. Operation simple but requires large boom and calm conditions to handle heavy weight on cable. Can be used at depths exceeding 400 m.
Vibrocorers	Core barrel, with cutter and catcher, is vibrated from the bed by hydraulic, pneumatic or electric motors, all mounted in a frame. May also rotate. Sometimes with piston and plastic liners.	*Widely used technique.* Lengths 2–10 m, dia. 100–300 mm. Cores subject to disturbance. Sampling possible in most unconsolidated sediments up to gravel size, as well as stiff clays and soft chalk. Time per core typically 2/3 h. Maximum current about 2 knots.
Air lift	A vertical pipe from the bed to above the water surface has injected into it, close to its base, compressed air which induces a strong upward flow that lifts the sediment to the surface.	*Large bulk samples* of unconsolidated sediments. Samples very disturbed and suffer same limitations as those from wash borings.

Besides the ground conditions, information must also be obtained, where possible, on the groundwater, e.g. the level at which it is struck, and presence of artesian conditions. However, such observations can be affected by the exploration method and it is advisable, therefore, to use observation wells or piezometers, which also take account of tidal and seasonal variations in level, in order that measurements may be made from time to time to establish the worst conditions. When drilling in rock, levels at which the circulating water fails to return should be noted as these denote the existence of open fissures.

Sufficient samples of the right size and type should be taken in order to fully represent the ground being investigated. In soils, this means that each stratum should be sampled at regular intervals, typically every metre, over its whole depth. Samples are either 'disturbed', i.e. taken from the spoil from the borehole, pit, etc. and not, therefore, representative of the soil structure, or 'undisturbed', i.e. showing the undisturbed soil structure. The latter, however, are still subject to some disturbance depending upon the method of sampling used. Further information on this and the methods used for both disturbed and undisturbed sampling is given in Table 11.4 for land and Table 11.5 for over water situations.

The size of the standard undisturbed sample is normally sufficient for the usual laboratory tests, although larger-diameter or block samples are sometimes required.

The size of the disturbed sample should be governed by the nature of the soil and the type and number of tests which are to be made upon it. Typical sizes are as shown in Table 11.6.

Table 11.6

Purpose of sample	Type of soil	Minimum amount of sample required (kg)
Soil identification natural moisture content and chemical tests	Cohesive soils and sands	1
	Gravelly soils	3
Compaction tests	Cohesive soils and sands	12
	Gravelly soils	25
Comprehensive examinations of construction materials including soil stabilization	Cohesive soils and sands	25–45
	Gravelly soils	45–90

11.4.2.3 Exploration by penetration tests

Advantage of these less expensive and quicker methods may be taken on occasion in preference to boring or pitting, to determine sufficient information of the ground formations as well as their engineering properties, albeit in an empirical form. The methods available for testing soils and their relative merits are set out in Table 11.7.

11.4.2.4 Geophysical methods

The techniques used are *in situ* methods of measuring contrasts in particular physical properties of strata and, hence, determining the stratigraphy and occasionally the water table. Where appropriate, the techniques represent a valuable and economic means of extending ground profile information outwards from a point of exploration. The methods are summarized in Tables 11.8 and 11.9. Further information is given in the selected bibliographies at the end of this chapter. Although there are no

great difficulties in carrying out the site measurements, excluding adverse marine conditions, experience and a knowledge of the geology are essential for interpreting the data correctly. The results should always be checked with some form of direct exploration, such as rotary core-drilling.

11.4.2.5 The observational method

Sometimes, because of the complexity of the ground conditions or a need for some flexibility in the project plan, as often required with a dam, it is not economically feasible to assess completely the problems after the main investigation. However, by adjusting the construction programme, it is possible to monitor the construction so that the design can be checked and modified where necessary. A good example of this observational method is the construction of road embankments over soft ground, where the preliminary assessment indicates a very low factor of safety. Construction is monitored, the design checked and, if necessary, side slopes, rate of earthmoving, etc. adjusted.

The method is equally applicable to investigations other than those for new works and especially for investigations into failures. It should be noted that successful application of the method to any project depends upon obtaining reliable relevant field data, which, in turn, requires the correct field instrumentation. Basically, instrumentation is to enable measurements to be made of displacement, earth pressure and pore-water pressure. Two important points need always to be borne in mind: firstly, to select the simplest form of apparatus consistent with the required accuracy and, secondly, always to make provision for some breakdowns due to the difficulties arising during the installation and subsequently as a result of the severity of the operating environment.

11.4.3 Measurement of engineering properties

11.4.3.1 In situ testing and instrumentation

Normally a main advantage of *in situ* testing over laboratory work is that the ground under test is less disturbed, and occasionally this includes the retention of the natural *in situ* ground stress pattern. The amount of ground tested by each measurement may also be larger than would otherwise be economically possible to test in the laboratory. Against this, the boundary conditions are no longer precise compared with those in a laboratory. The method may also relate to a direction of testing different from that which will be subsequently imposed.

The various methods of *in situ* testing in soils and instrumentation with their main applications and limitations are set out in Tables 11.7 and 11.10. Exploration and *in situ* testing in rocks is described in Chapter 10.

11.4.3.2 Laboratory testing of representative samples

The samples obtained from the exploration are generally tested in the laboratory to assist in the identification of strata and to determine their relevant engineering properties. The various laboratory tests are summarized in Table 11.11 with their application to routine engineering problems. Compared to *in situ* testing, laboratory testing is under controlled boundary conditions and to defined testing procedures. Moreover, there is no doubt as to the soil type, state and structure under examination.

11.4.3.3 Geophysical methods

Although most geophysical work falls into the category of 'stratigraphical methods', referred to above, some techniques can be used for ascertaining certain engineering properties as given in Tables 11.7 and 11.10.

Table 11.7 *In situ* testing in soils for foundations

Location	Method	Technique	Applications and limitations
Normally on land or from platform	**Borehole tests** Standard Penetration Test (SPT)	Standardized intermittent dynamic test. Provides small disturbed sample excepting gravel when solid cone is used. Maintain positive head of water or drilling mud.	*Most widely used preliminary field test.* Relative densities. *Bearing values of non-cohesive soils.* Unreliable in gravel. Correction applied in fine-grained soils. Indicates settlement of spread footings in granular soils. Aids estimation of liquefaction potential.
	Pressuremeter test	Lateral pressure and deformation tests from an expanding cell. Types include: – *Menard.* Used in pre-formed hole. Slotted steel casing for gravel. – *Stuttgart.* Split metal cylinder expanded hydraulically. – *Stressprobe.* Pressed below borehole and takes core. – *Camkometer.* Self-boring with pore pressure cell. – *Marchetti* dilatometer (DMT). Steel plate containing stress cell on one side, pushed edgewise into soil.	*Strength and deformation properties in most fine-grained soils.* Direct bearing values, but may not correspond to vertical loading. Less expensive than vertical loading tests and larger volume stressed than in laboratory test. Camkometer most sensitive, where boring is possible, able to measure k_0 and effective stress.
	Downhole bearing test	Plate loading test on base of borehole. Alternatively, simple screw plate augered through disturbed zone.	In-situ *bearing values for clays.* Not commonly used. Baseplate restricted to above the water table.
	Hydraulic fracturing	In hydraulic piezometers	Measurement of minor stress.
	Penetration tests (continuous record) Simple probe	Driving a rod by drop hammer or pneumatically.	Location of hard ground beneath weak strata. Beware of boulders and influence of friction of the rods.
	Dynamic Probing (DP)	Standardized dynamic cone penetration testing procedures.	Bearing values where local specialized experience exists. Mainly used in non-cohesive soils. Caution needed at depths when rod friction may be high.
	Weight Sounding Test (WST)	Standardized procedure using dead weights on screw point, via rods followed by rotation to provide profile of half-turns/0.2 m.	*Well-established Scandinavian technique.* Inexpensive. Used in most soils except dense sediments and compact layers. Applied to footings and pile design, also compaction control.
	Cone Penetration Test (CPT)	Standardized procedure using shielded cone rod with slow constant rate of penetration. Local friction just behind cone also measured. Piezometer can be fitted in cone. Adaptable for small piston sampling at selected levels. Electric cones preferable to mechanical. Special equipment measures tilt, temperature and density. Piezocones measure pore-water pressure.	*Bearing value and length of piles in silts and sands.* More rapid and less costly than boreholes, suited to generally known conditions. Indicates soil types. Empirical formula for foundation design in sands and clays. Penetration affected by coarse-grained soils and cemented layers. Strength relationships also vary considerably in cohesive soils. Results in weak clays and silts can be suspect, particularly with the mechanical equipment. Accessory measurements include: (a) inclinometer to observe verticality, (b) temperature, e.g. beneath permafrost or cold stores, (c) acoustic for qualitative enhancement to differentiate soil types.
	Static-dynamic penetration tests	CPT procedures with dynamic sounding in dense layers or for extra penetration.	Investigating coarse soils and compact layers. Should be complemented by boreholes.

Location	Method	Technique	Applications and limitations
On land	**Independent tests** Vane test	Direct penetration from surface and in boreholes or pit.	Undrained shear strength, for sensitive clays with cohesion up to $100\,kN/m^2$. Cross-check results, beware of silt or sand pockets and fibrous peat.
	Plate bearing tests	Incremental load/deformation test with plate encastré. Ensure plate unaffected by test load support.	*Bearing value of 'stoney' clays, weak and weathered rocks for foundation design.* Test by boring for softer deposits at depth.
	Load settlement test	Waste skip or metal tank incrementally loaded with settlement observations typically over 1–6 months period.	Immediate and short-term settlement data on fill or recent alluvium. Correlate with borehole data when extrapolating results.
On land or over water	Pile tests	Loading, pulling and lateral as required. (a) Maintained load method (ML) (b) Constant rate of penetration method (CRP) (c) Equilibrium load method (EL). Requires fairly even temperatures and leakproof ram. In all types of tests, end-load can be measured separately by load cell.	Pile design. Ratio of settlement in sands between individual test and group suggested by Skempton. ML method represents conventional technique. CRP method is very quick for load-carrying behaviour. EL method is compromise for determining load-carrying behaviour quickly. Load increment is applied and load system sealed so that as settlement occurs load decreases until equilibrium is reached.
Offshore marine	**Cable-operated equipment from vessel** (for wireline operation)		
	Penetrometers	Short-drive dynamic and static devices. Capacity 3–10 m approximately.	Intermittent empirical resistance diagrams. CPT (see above) very widely used. Results can be affected by the drilling operations.
	Pressuremeters	Operation may be below borehole casing, remote from seabed unit, or direct via seabed probe. Menard commonly used, but stressprobe has specially suited features.	*Used in most soil conditions,* excluding only the coarse types. For design of piles and spread foundations.
	Vane	Remote-controlled torque measuring device lowered to base of borehole, acting on short-drive vane rod.	Intermittent strength determinations in very weak cohesive soils, particularly where undisturbed sampling proves difficult.
	Logging	Seismic, nuclear, dip, calipers.	
	Seabed equipment Remote-controlled static cone penetration test	Capacity typically about 30 m. Usually Dutch electric cone.	*Preferred method for determining mechanical properties of sands and consolidated clays.* Unit may sink into very weak seabeds. Very dense or coarse soils penetrated only a few metres.

Notes: (1) Test equipment should be regularly recalibrated and these results should be available on site.
(2) Seabed reaction frame may facilitate testing offshore with cable-operated equipment.

11.4.3.4 Model and prototype tests

It may be necessary to carry out full-scale trials in the field or model tests in the laboratory to check the parameters used in the preliminary analyses. For example, trial embankments may be constructed in the field on soft ground to check for stability or settlement, pile-loading tests carried out to measure shaft adhesion and/or end bearing, or compaction trials to test the suitability of fill.

Table 11.8 Geophysical methods on land

Method	Technique	Applications and limitations
Electrical resistivity	The form of flow of an induced electric current is affected by variations in ground resistivity, due mainly to the pore or crack water. Current is passed through an outer pair of electrodes whilst the potential drop is measured between the inner pair. Extension of direct measurements of porosity, saturation and permeability. Analysis is most often done by theoretical curve-fitting techniques.	Simplest and least expensive form of geophysical survey. – Location of simple geological boundaries: *depth to bedrock beneath clay, and water bearing granular stratum over clay* (sub-surface saline bodies). – *'Expanding' electrode* method for changes in sequence with depth. Limited to 3 or 4 layers of similar thickness. – *Constant separation* method for lateral delineation of boundaries, e.g. *location of faults, dykes, shafts and caverns.* Reliability affected by metal pipes, electrical conductors, complex and sloping strata, railway lines and power cables.
Seismic	The speed of propagation of an induced seismic impulse or wave is affected by the dynamic elastic properties and density of the ground. Impulse generated by falling weight and on open sites by explosive charges. *'Refraction'* method with single shots concerns travel times of refracted waves which travel through sub-strata and are rebounded to the surface. Valid only when seismic velocities increase with depth. Short separate traverses used to check this. *'Reflection'* method with single shots concerns the directly reflected impulses from horizons of abrupt increase in seismic velocity.	Most highly developed form of geophysical survey. Can be quite accurate under suitable conditions, particularly for horizontally layered structures. – Also for ground vibration problems. *Determination of depth of bedrock beneath sands and gravel with low water table.* Variations laterally in rock, also buried channels and domes. – Direct evidence of seismic velocities in refracting strata. – For checking effectiveness of cement grouting of rock. – Interpretation generally possible only for depths greater than is normally required for civil engineering on land.
Gravimetric	The Earth's natural gravitational field is affected by local variations in ground density. Measurements are made of differences between stations in the vertical component of the strength of gravity, which is then accurately corrected for latitude, height and topography to reflect only changes due to sub-surface geology. Precise topographical survey of exact station positions is necessary to obtain reliable results as differences are small.	– The interpretation of regional geology, without depth control, mainly where some geological information is already available. – For distinguishing local anomalies such as *buried rock faces in infilled quarries, large faults.* Also for positioning buried channels, large cavities and old shafts. – Fitting techniques based on simplified structures can be applied for studying anomalies.
Magnetic	Many rocks are weakly magnetic and the strength varies with the rock type depending upon the amount of ferromagnetic minerals present. This modifies the Earth's field. Surveys are similar to those for gravity measurements. Although the fieldwork is simpler, the interpretation is more difficult.	– For locating the hidden boundaries between different types of crystalline rock and positions of faults, ridges, dykes and *large ferrous ore bodies.* – *Field detectors for locating buried pipes and cables.*
Ground radar	An impulse radar system, representing the electromagnetic equivalent to echo-sounding. Variants including mapping and profiling techniques.	– For locating very shallow buried anomalies, such as disused hidden shafts and cavities, and simple shallow geological boundaries. – Unsuitable if clay topsoil or otherwise relatively high conducting surface layer present.
Borehole logging	The application of geophysical methods in boreholes.	Electrical and sonic methods to distinguish between strata especially where core recovery is difficult. See also Tables 11.7 and 11.10.

Notes: (1) All methods rely upon strong contrast in density, void space or resistivity across the boundaries to be identified. Transitional zones lead to uncertainty.
(2) Results should not be considered in isolation but correlated always with direct exposures, e.g. boreholes.
(3) Value of the results is very dependent upon use of the appropriate method or a combination of complementary methods, the specialist experience used in the analysis and the amount of geological knowledge that is available.
(4) Assessment of engineering parameters with geophysics is given in Table 11.10.

In the laboratory, dams, embankments and cuttings can be modelled and tested in a centrifuge and problems of permeability and seepage can be investigated in a flow tank. Model piles and footings can also be tested.

11.4.4 Ground investigations over water

Although in simple cases a land-type investigation is used with the additional facilities needed for access and the depth of water, as the difficulties with these two added complications increase so

Table 11.9 Over-water geophysics

Method	Technique	Applications and limitations
Echo-sounding	The times taken for a short pulse of high-frequency sound to travel from a source normally on the vessel's hull, vertically down and back to a detector via reflecting surfaces at and beneath the seabed.	– A *continuous water depth profile*. – Qualitative interpretation to limited depths of boundaries of higher density material beneath soft seabed deposits.
Side-scan sonar	Directional echo-sounding analogous to oblique aerial photography. Transducer source normally trailed at an elevation of 10–20% of maximum range.	– *Quantitative guide to position and shape of surface anomalies such as rock outcrops, wrecks and pipelines.* Qualitative guide to material on seabed.
Continuous seismic reflection profiling	The reflected wave trace of the seabed and underlying strata from a high rate of acoustic (sonar) impulses of short period for high resolution. *Pingers* are of frequencies typically 3–7 kHz penetrating a few tens of metres in soft silts, less in sands, few metres in stiff clay and none in compact gravel. *Boomers* 400 Hz–3 kHz penetrate over 100 m in soft sediments and only a few tens of metres in gravels and stiff clays. *Sparkers* 200–800 Hz of high resolution can penetrate more than 1 km in overburden or rock. *Air guns* are for much deeper penetrations. Optimum towing arrangements for sensors and detectors are adjusted to suit noise characteristics of survey vessel, its speed and the tides.	– *Valuable complementary aid to exploratory borings* for intermediate interpretation of stratigraphical horizons, location of *aggregate deposits, buried pipelines*, etc. within range of the impulses. Also applicable beneath rivers and lakes. Velocities of transmission obtained by seismic refraction for quantitative analysis. – *Type of acoustic source must be suited to local ground conditions.* Unable to distinguish between different formations with similar geophysical responses, e.g. coarse boulder clay and very weathered bedrock. – *Ineffective in water depths less than about 2 m.* Also noise of rough seas can cause signal losses.
Magnetic	See Table 11.8. Sensor trailed close to seabed and behind towing vessel, if iron about 2 ship lengths or more.	– Location of local buried structural changes with strong magnetic contrasting *dykes,* also *buried iron vessels* and pipelines.
Radioactive	Counter for detection is towed at or close to the seabed for location of geological anomalies. Alternatively *artificial isotopes can be introduced* for subsequent detection.	– Strong natural contrast between granite and basalt. Tracers indicate effluent dispersion and sediment mobility. Probes give estimate of seabed density.
Gravimetric	See Table 11.8. Seabed or ship-borne.	– Unlikely to be justified.

Note: Tidal corrections and good survey control essential, including, when appropriate, electronic navigation.

it becomes necessary to employ specially adapted techniques as given in Tables 11.3, 11.5, 11.7 and 11.9.

Selection of a suitable means for supporting the boring plant at the exploration positions is of fundamental importance in order that the work can be carried out safely and with minimal delays. It is generally not economical where craft are needed to provide a method that will permit working to continue during adverse weather and tidal conditions. Alongside jetties or river banks it is often possible to use a staging, but great care is essential to safeguard against overturning a cantilever platform, particularly during withdrawal of the casing. In protected waters, scaffold platforms or small pontoons are adequate. Larger pontoons or dumb barges usually suffice in more open conditions in the better weather or in estuaries near a safe haven, although the alternative of a jack-up pontoon platform may offer attractive advantages including overcoming difficulties due to tidal conditions. Offshore in the more exposed locations self-propelled craft, sometimes of special design and of sufficient size to remain stable are advisable. Adequate stability against the normal wave and tidal conditions with floating craft can be significantly improved by using ballast. At least four anchors suited to the seabed are essential to maintain position, preferably with an additional two in the direction of the tides or main current flow. In deep water over about 80 m special craft with computer-controlled thrust devices are used.

Separate facilities are also required in the form of an auxiliary vessel or helicopter for transporting personnel, materials and for visits to the boring-platform. Generally, an auxiliary vessel is also used for positioning dumb craft and to help with the time-consuming operation of handling and laying the anchors. Another use would be for sounding and geophysical surveys.

Position-fixing of the points or lines of exploration must be reliable and properly related to permanent stations. Sextants and theodolites are employed for simple cases very near land. Offshore electronic methods provide an accuracy of about 3 m with a range of up to 50 km. Seabed markers may also be employed.

Due consideration needs to be given to shipping, harbour and other regulations with respect to carrying out the investigation at the site and for permission to use the overwater facilities proposed.

Offshore investigations, say for oil production platforms,

Table 11.10 *In situ* testing and field instrumentation for earthworks, groundwater and other purposes

Nature of works	Geology	Technique		Applications and limitations
Earthworks, soil and rock slopes	Soils	*In situ* shear strength.	Normally undrained direct shear test.	Undrained *in situ* shear strength.
		Plate loading.	See static loading test below.	Refined cycled test for modulus or simple load test for bearing capacity.
		Sand replacement.	(Also water balloon device). Calibrate sand at natural humidity.	Bulk density during construction. Standard techniques unsuitable in coarse non-cohesive material: then use water replacement.
		Water replacement.	Water filled pit, lined with plastic.	
		Nuclear devices at surface.	Radioactive sources and counting unit.	Bulk density (preferably by attenuation method) and *in situ* moisture content.
		Nuclear density probe.	Usually back-scatter method with radioactive isotopes.	Bulk density measurements above and below water table, with casing if required.
		Proctor needle.	In earthwork construction.	Field control and consistency of fine-grained soils.
		In-situ CBR.	In earthwork construction and for roads.	Only appropriate in clay soils and subject to climatic changes.
		Piezometers.	High air value for partially saturated soils.	
		Total pressure cells.	Require very careful positioning.	Total earth pressure against sub-structures and within a soil mass.
		Settlement and heave instruments.	Types: Water, mercury, magnetic ring, buried plates, rods and notched tubes.	Total and relative settlement.
		Conventional survey methods.	Laser, photogrammetry.	Total and relative surface movement.
	Soils and rocks	Inclinometers and deflectometers. Extensometers.	Portable and installed.	Creep and slip detection. Expansion due to relief of stress and across tensile zones arising from differential settlement.
Groundwater permeability, etc.	Soils and rocks	Observation wells and piezometers.	Use effective filter, test regularly and seal from extraneous infiltration.	Level of water table, artesian and sub-artesian conditions.
		Rapid-response recorders.	Electric or pneumatic transducer type.	Tidal measurements, effect of surges, rainstorms and earthquakes.
	Clays and silts	Constant head seepage tests.	*In situ* measurement of permeability.	Coefficient of consolidation. (Certain advantages over laboratory tests.) Test is time-consuming. Groundwater must be at equilibrium at start of test.
	Sands and gravels	Pumping tests.	Pump to equilibrium conditions measuring transients during draw-down and recovery. Use at least two lines of observation wells.	Best form of test for natural permeability measurement. Transient measurements provide storage coefficient.
		Two-well pumping test.	Established technique.	Estimation of difference between horizontal and vertical permeability.
		Radioactive tracers.	Various	
		In situ permeability.	Careful shelling beforehand. Make both rising and falling head tests.	Local measurement of *in situ* permeability either through base of borehole or after placing coarse filter and withdrawing casing. Treat results with caution. A considerable number of tests are required to compensate for scatter.
		Infiltration above the water table.	Soakaway design.	
		Electrical resistivity.	Four electrodes. Wenner or Schlumberger configuration.	Extension of direct measurement of porosity, degree of saturation, and permeability.
	Rocks	Formation tests.	Expanding packers isolate zone under test.	Joint seepage and condition of joints by measuring flow under varying pressures, rising and falling.
Foundations for dynamic loads	Soils and rocks	Static loading test	Extra sensitive plate test cycled over expected stress range to give a modulus of reaction.	'Spring constant' for foundation design.
		Dynamic loading test.	Small vibrators mounted on soil to give resonance response.	Values of dynamic modulus. Poisson's ratio and damping.
		Seismic velocity measurements.	In various modes.	Dynamic and possibly static moduli for small strains.
Miscellaneous	Soils	Thermocouples and thermistors.		Ground temperature of coal tips on fire, beneath boilers and refrigeration plant.
		Electrical resistivity.	Four electrode system. Wenner configuration or two electrode probe.	'Apparent' resistivity for corrosion survey.
		Corrosion probe. (redox potential)	Short circuit current between reference cell and platinum electrode to earth.	Measure of oxygen in soil to assess microbial corrosivity.
		Stray current measurement.		For corrosive effect.
	Soils and rocks	Periscope calipers and borehole cameras, with video-tape recording.		Defining cavities, fractures, etc.
	Rocks	Noise detectors.	Considerable amplification required.	Incipient ground movement at faults, slopes, tunnels.

Note: Test equipment should be regularly recalibrated and these results should be available on site.

Table 11.11 Laboratory tests

Category	Test		Remarks	Category	Test		Remarks
		*				*	
Identification and classification	Moisture content	FR	standard for all fine soils	Strength	Quick undrained triaxial compression	F	bearing capacity and short term stability. Variations include specimen sizes, single/multi-stage, quick/slow tests.
	Atterberg limits	F					
	Particle size distribution	FC	standard for all coarse soils				
	Particle density (specific gravity)	FR	used in conjunction with other tests		Uniaxial compression	R	bearing capacity
	Linear shrinkage and shrinkage limit	F	shrinkage/swell behaviour		Shear vane	F	soft soils (not peats) bearing capacity
	Saturation moisture content	R			Shear box	C	bearing capacity of recompacted soils.
Compaction	In-situ density	FR	used with other tests particularly strength and deformation			F	peak and residual effective strengths.
					Slow triaxial compression with/without pore-pressure measurements.	C	effective strength and pore pressure parameters for long term stability.
	Compacted density-moisture content	FCR	standard, heavy and vibrating plate compaction tests for all fill materials				
					Ring shear	F	residual strengths.
	Maximum and minimum density	C	used with in-situ density to indicate relative density	Deformation	Oedometer consolidation	F	drained modulus
					Triaxial consolidation	FR	
	Moisture condition value	FCR	suitability of fill for compaction. Adapted for chalk as the crushability test		Rowe cell consolidation	F	
					Cyclic undrained triaxial	FR	undrained modulus
Pavement design	California bearing	FCR	pavement thickness	Chemical corrosivity	Bacteriological content		Risk of attack on buried metals.
	Frost heave test	FCR	susceptibility to frost heave		Redox potential		
					Resistivity		
Permeability	Constant head	C			Organic contents		Risk of attack on buried concrete
	Variable head	FC			pH value		
					Sulphate content		
	Triaxial consolidation	F			Carbonate content		
					Chloride content		
	Rowe cell consolidation	F			Methane content		Health risk
					Full chemical analysis		
Erodability	Pinhole test	F					

*Legend F = fine soils (clays and silts); C = coarse soils; R = soft rocks, mainly cohesive

Notes:
1. Every test specimen should have a complete soil rock description in order to assist interpretation of the test result.
2. Peat generally treated as fine soil

Table 11.12 Identification and description of soils

	Basic soil types	Particle size (mm)	Visual identification	Particle nature and plasticity[1]	Composite soil types (mixtures of basic soil types)			
Very coarse soils	BOULDERS	— 200	Only seen complete in pits or exposures	Particle shape:	Scale of secondary constituents with coarse and very coarse soils. Term either before or after principal constituent			
	COBBLES	— 60	Often difficult to recover from boreholes	Angular Subangular Subrounded Rounded Flat Elongated	Term before	Principal	Term after	Approximate
Coarse soils	GRAVELS	Coarse — 20	Easily visible to naked eye; particle shape can be described; grading can be described			SAND, GRAVELS, COBBLES OR BOULDERS	With a little (sand*) or occasional (cobbles†)	<
		Medium — 6			Slightly (sandy*)			
		Fine — 2		Texture:	–(sandy*)		With some (sand*) or some (cobbles†)	5 2
	SANDS	Coarse — 0.6	Visible to naked eye; very little or no cohesion when dry; grading can be described	Rough Smooth Polished	Very (sandy*)		With much (sand*) or many (cobbles)	2 4
		Medium — 0.2			—		and (sand*) or and (cobbles†)	
		Fine — 0.06			* Fine or coarse soil type as appropri† Very coarse soil type as appropriate+ Or described as fine soil depending mass behaviour			
Fine soils[2]	SILTS	Coarse — 0.02 Medium — 0.006 Fine — 0.002	Only coarse silt barely visible to naked eye; exhibits little plasticity and marked dilatancy; slightly granular or silky to the touch. Disintegrates in water; lumps dry quickly; possesses cohesion but can be powdered easily between fingers.	Non-plastic or low plasticity	Scale of secondary constituents with fine soils. Term either before or after principal constituent.			
					Term before	Principal	Term after	Approximate
	CLAYS		Dry lumps can be broken but not powdered between the fingers; they also disintegrate under water but more slowly than silt; smooth to the touch, exhibits plasticity but no dilatancy; sticks to the fingers and dries slowly; shrinks appreciably on drying, usually showing cracks. Intermediate and high plasticity clays show these properties to a moderate and high degree, respectively.	Intermediate plasticity (Lean clay)	Slightly (sandy*)	CLAY OR SILT	With a little (sand*)	<
					(Sandy*)		With some (sand*)	
					Very (sandy*)		With much (sand*)	>
				High plasticity (Fat clay)	* Coarse soil type as appropriate Or described as coarse soil dependi on mass behaviour			
Organic	ORGANIC CLAY, SILT OR SAND	Varies	Contains substantial amounts of organic vegetable matter.		Examples of composite types (indicating preferred order for description) Loose, brown, subangular, very sandy, coarse GRAVE with small pockets of soft grey clay.			
	PEATS	Varies	Predominantly plant remains usually dark brown, or black in colour, often with distinctive smell; low bulk density.		Firm, brown, thinly laminated SILT and CLAY. Dense, light brown, clayey, fine and medium SAND.			

Notes: (1) Plasticity, compaction and strength can be defined in more detail, see text.

(2) Fine soils. Rarely exist as either silt or clay, but as mixtures, typified by intermediate behaviour. Such soils may be described as above, i.e. an appropriate plasticity or in such terms as silty CLAY, very clayey, SILT, etc.

Compactness/strength		Structure			Colour
Term	Field test	Term	Field identification	Interval scales	
Loose	By inspection of voids and particle packing	Homo-geneous	Deposit consists essentially of one type	Scale of bedding spacing	Red Pink Yellow Brown
Dense		Inter-stratified	Alternating layers of varying types or with bands or lenses of other materials. Interval scale for bedding; bedding spacing may be used	Term — Mean spacing (mm)	Green Blue White Grey Black, etc.
				Very thickly bedded — Over 2000	
				Thickly bedded — 2000–600	
Loose	Can be excavated with a spade; 50 mm wooden peg can be easily driven.	Hetero-geneous	A mixture of types	Medium bedded — 600–200	Supplemented as necessary with
Dense	Requires pick for excavation, 50 mm wooden peg hard to drive			Thinly bedded — 200–60	Light Dark Mottled, etc.
		Weathered	Particles may be weakened and may show concentric layering	Very thinly bedded — 60–20	
Slightly cemented	Visual examination; pick removes soil in lumps which can be abraded			Thickly laminated — 20–6	and
				Thinly laminated — Under 6	Pinkish Reddish Yellowish Brownish, etc.
Soft or loose*	Easily moulded or crushed in the fingers	Fissured	Break into polyhedral fragments along fissures. (Interval scale for spacing of discontuities may be used)	Scale of spacing of other discontinuities	
Firm or dense*	Can be moulded or crushed by strong pressure in the fingers			Term — Mean spacing (mm)	
Very soft*	Exudes between fingers when squeezed in hand	Intact	No fissures	Very widely spaced — over 2000	
Soft*	Moulded by light finger pressure	Homo-geneous	Deposit consists essentially of one type	Widely spaced — 2000–600	
				Medium spaced — 600–200	
Firm*	Can be moulded by strong finger pressure	Inter-stratified	Alternating layers of varying types. Interval scale for thickness of layers may be used	Closely spaced — 200–60	
Stiff*	Cannot be moulded by fingers Can be indented by thumb.			Very closely spaced — 60–20	
Very stiff*	Can be indented by thumb nail	Weathered	Usually has crumb or columnar structure	Extremely closely spaced — under 20	
* Mainly silt † Mainly clay					
Firm	Fibres already compressed together	Fibrous	Plant remains recognizable and retains some strength.		
Spongy	Very compressible and open structure				
Plastic	Can be moulded in hand, and smears fingers	Amor-phous	Recognizable plant remains absent		

have become a specialist activity for a limited number of organizations in view of the particular personnel and plant requirements. The relevant selected bibliography deals with the subject in some detail.

11.4.5 Personnel

Competent execution of an investigation not only requires selection of the appropriate methods and equipment but the use of properly trained and experienced persons to carry out the field and laboratory work. In charge of a major investigation there should be a suitably qualified senior engineer whose control should cover all aspects of the work. Throughout the investigation this person should meet the senior design engineer regularly.

Site work in the field should be fully supervised by a resident qualified geotechnical engineer or engineering geologist, who should describe all samples and prepare the logs. Operatives carrying out boring and testing should have been trained and should execute their work according to standardized procedures, to include full documentation of the results.

11.4.6 Contracts

Where specialist contractors are to be employed for the main investigation they should be selected on the basis of having adequate resources in the following respects: (1) specialist equipment for the field and laboratory work, with trained personnel for its operation; (2) experience in executing investigations of equivalent magnitude and to the required quality; and (3) professional staff for supervision and interpretation with knowledge and experience covering the range of techniques likely to be employed.

Although supervision may sometimes be supplied separately the intricacies of much ground investigation continues to demand an adequate level in the knowledge and experience of the contractor in order to execute the work competently and to produce reliable results.

There should be discussions before the contract is signed in order that the contractor may become acquainted with the results of the preliminary appreciation and have an opportunity to contribute from their experience on the formulation of an efficient and adequate programme for the field and laboratory work. They will visit the site except in the simplest of cases in order to plan their work.

Model conditions of contract and specifications exist (see bibliographies) that are specifically suited for ground investigations. Legal advice should always be sought if changes are contemplated in the wording of the clauses.

Flexibility should be an essential feature of the contract to deal with unexpected requirements and changes in the project. Moreover, a close working relationship should be formed between the client's engineering representative and the specialist geotechnical contractor for the provision of detailed and continuous communication to provide the best opportunity for a successful investigation. Ground investigation work awarded purely on the basis of the most competitive price for individual operations or the overall content has strict limitations in scope, may suffer in quality and could actually be misleading.

11.5 Description of soils and rocks

11.5.1 General

Soils and rocks need to be described in terms which readily convey their engineering properties. As these are largely physical properties they are first assessed from a visual inspection of samples or exposures using a few simple hand-tests. Samples or areas of exposures described as similar are grouped together as forming a geotechnical unit often with relatively uniform properties throughout, e.g. a stratum on a log. This grouping forms the basis for the selection of samples for laboratory testing and/or zones for *in situ* testing to obtain representative values of the engineering properties of each unit. A final assessment is then made taking into account the test results. For descriptions to fulfil their purposes they must be to a common system accepted by all concerned.

11.5.2 Systematic soil description

A universally accepted system does not exist at present so the following is presented as probably the most widely used. Non-organic soils are primarily described in terms of particle size, plasticity, compactness/strength, structure, and organic soils in terms of strength and structure (see Table 11.12). The particle size limits of the basic soil types are defined by various national standards which are all similar and broadly correspond to significant changes in the engineering properties. Soils are usually a wide-ranging mixture of particle sizes, with engineering properties largely controlled by the finer particles, particularly where they are clay-size and even where that constituent is a relatively minor percentage. The influence of the fines content varies widely with percentage, mineralogy, fabric and the particular engineering property under consideration.

For descriptions to be complete all the information indicated by the column headings in Table 11.12 should be recorded, either assessed visually, or as determined by the appropriate field and/or laboratory tests. Some typical descriptions are shown in the table embodying the results of particle-size distribution tests (examples of which are given in Chapter 9), standard penetration test (SPT) *N*-values, and shear strength tests (see Tables 11.13 and 11.14 respectively). These three tables are based on British practice.

The degree of plasticity of the amount passing 425 μm sieve fraction can be further described by liquid and plastic limits and

Table 11.13 State of compaction of coarse soils

State of compaction	SPT N value* (blows per 300 mm penetration)	Approximate relative density (%)
Very loose	0–4	0–15
Loose	4–10	15–35
Medium dense	10–30	35–65
Dense	30–50	65–85
Very dense	Over 50	85–100

* Standard penetration test *N* values may be corrected for overburden pressure, etc. in refined analyses.
SPT method using triggered hammer and free fall.

Table 11.14 Consistency of fine soils

Consistency	Undrained shear strength range (kN/m²)	Equivalent SPT N value (very approximate)
Very soft	Under 20	Under 2
Soft	20–40	2–4
Firm	40–75	4–8
Stiff	75–150	8–15
Very stiff	Over 150	Over 15

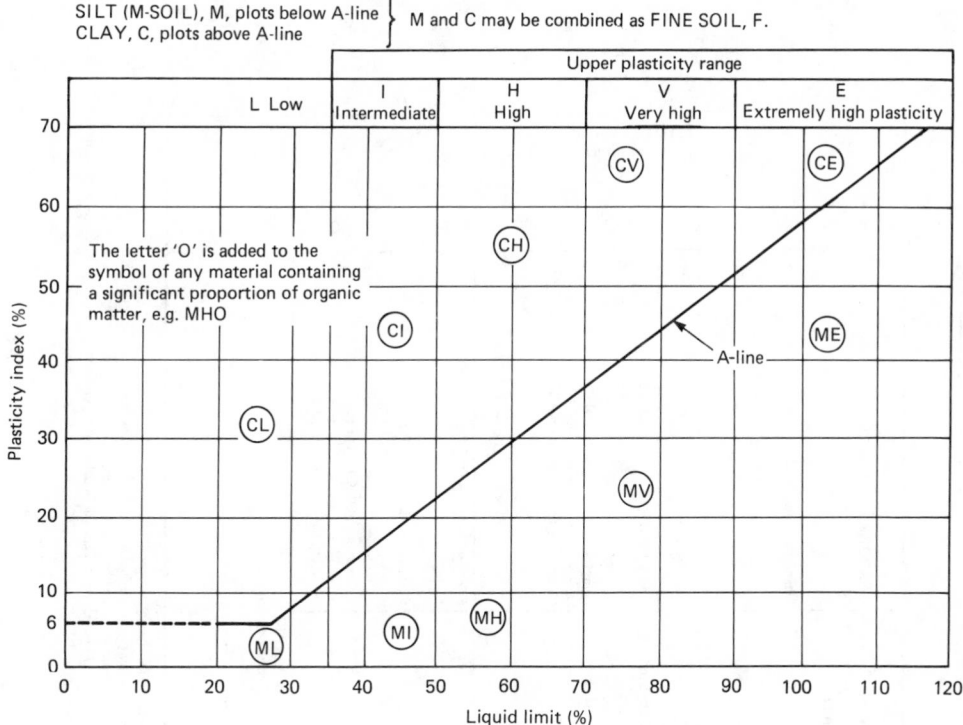

SILT (M-SOIL), M, plots below A-line ⎱ M and C may be combined as FINE SOIL, F.
CLAY, C, plots above A-line ⎰

Figure 11.7 Plasticity chart for the classification of fine soils (measurements made on material passing a 425 μm BS sieve). (After BS 5930:1981.)

the plasticity chart (see Figure 11.7). Soils which plot below the A-line are predominantly silt and those above predominantly clay. High organic content moves the plots significantly to the right.

In the British Isles the principal non-organic soil types usually occur as siliceous coarse soils and silts and as alumino-siliceous clays, but worldwide many chemically and mineralogically different soils occur giving rise to particular mechanical and chemical characteristics (see Chapter 9). For such soils the standard description would need to be modified, e.g. for carbonate soils (and rocks) (see Table 11.15).

11.5.3 Classification of soils

Soils may be classified on any basis relevant to a particular project or problem, e.g. strength, aggressiveness to concrete, permeability, compaction. Probably the most widely used is the American Unified Soil Classification System (USCS) (see Figure 11.8). The British Soil Classification System for Engineering Purposes (BS 5930), of which Figure 11.7 is a part, is an expansion of the USCS system but as yet less widely used. Both classification systems assess soils objectively on the basis of specific tests, and therefore sometimes a classification may be at variance with a description which can be more subjective.

11.5.4 Systematic rock description

Rock description is generally more complex than soil description: not only is a description of the intact rock material required but a complete description of the rock mass *in situ* (i.e. structure and discontinuity pattern) is usually essential, together with the weathering state of the material and mass. The preferred scheme of description is set out in Table 11.16, together with definitions of terms used. Because weathering can be

caused by mechanical and/or chemical agencies acting in varying degrees and ways on the intact rock and along discontinuities it has to date proved impossible to propose a single universally applicable scale of weathering, but this table includes a general scale which can be used. Rock colours should be in accordance with the scheme for soils.

Grain size can be assessed visually as for soils, but as grain size can imply a particular rock type, some guidance is given in Table 11.17. The texture refers to individual grains and their arrangement, the rock fabric. Structure is the inter-relationship of textural features. Sometimes texture and structure are implied in the rock name and therefore need not be described separately or structure may be better described as a minor lithological characteristic. Spacing of structural and lithological features should be described and oriented by direction and dip. Comments on degree of openness of discontinuities, irregularity of their surfaces and type of infilling should be included, as well as the size and shape of rock blocks.

Apart from the descriptive assessment, mechanical or fracture logging should be undertaken to complete a rock core or exposure log (see Table 11.16). For further comments on this topic see Chapter 9.

11.5.5 Boreholes and trial pit logs

Typical borehole and drillhole logs are shown in Figures 11.9 and 11.10 respectively, with standard legends in Figure 11.11 and a trial pit log in Figure 11.12. Apart from the soil and rock descriptions, each aims to present all factual data which may assist at design and construction stages. The log itself should take into account all disturbance caused by the ground exploration technique and sampling and thus be an interpretation of what that core or block of ground was and what the groundwater conditions were before exploration at that location.

Table 11.15 Classification of carbonate soils and rocks (based on Middle East experience)

Additional descriptive terms based on origin of constituent particles — Increasing grain size of particulate deposits →

Origin of constituent particles: Not discernible | BIOLASTIC (organic) | OOLITE (inorganic) | SHELL (organic) | CORAL (organic) | ALGAL (organic) | PISOLITES (inorganic)

TOTAL CARBONATE CONTENT % (constituent particles plus matrix)

Grain size boundaries: 0.002 mm | 0.060 mm | 2 mm | 60 mm

Not discernible (and biolastic) descriptive nomenclature by grain size and carbonate content

Carbonate content	Clay (< 0.002 mm)	Silt (0.002–0.060 mm)	Sand (0.060–2 mm)	Gravel (2–60 mm)
Non-indurated				
90%	CARBONATE MUD	CARBONATE SILT	CARBONATE SAND	CARBONATE GRAVEL
50%	Clayey CARBONATE MUD	Siliceous CARBONATE SILT[1]	Siliceous CARBONATE SAND[1]	Mixed carbonate and non-carbonate GRAVEL[2]
10%	Calcareous CLAY	Calcareous SILT[1]	Calcareous silica SAND[1]	
—	CLAY	SILT	silica SAND	GRAVEL
Slightly indurated				
90%	CALCILUTITE (carb. clayst.)	CALCISILTITE (carb. siltst.)	CALCARENITE (carb. sandst.)	CALCIRUDITE (carb. conglom. or breccia)
50%	Clayey CALCILUTITE / Siliceous CALCILUTITE	Siliceous CALCISILTITE	Siliceous CALCARENITE	Conglomeratic CALCIRUDITE[2]
10%	Calcareous CLAYSTONE	Calcareous SILTSTONE	Calcareous SANDSTONE	Calcareous CONGLOMERATE
—	CLAYSTONE	SILTSTONE	SANDSTONE	CONGLOMERATE OR BRECCIA
Moderately indurated				
90%	Fine-grained LIMESTONE	Fine-grained LIMESTONE	Detrital LIMESTONE	CONGLOMERATE LIMESTONE
50%	Fine-grained Argillaceous LIMESTONE	Fine-grained Siliceous LIMESTONE	Siliceous detrital LIMESTONE	Conglomeratic LIMESTONE[2]
10%	Calcareous CLAYSTONE	Calcareous SILTSTONE	Calcareous SANDSTONE	Calcareous CONGLOMERATE
—	CLAYSTONE	SILTSTONE	SANDSTONE	CONGLOMERATE OR BRECCIA
Highly indurated (50%)	CRYSTALLINE LIMESTONE OR MARBLE (tends towards uniformity of grain size and loss of original texture). Conventional metamorphic nomenclature applies in this section.			

Strength and induration

Degree of induration	Non-indurated	Slightly indurated	Moderately indurated	Highly indurated
Approximate unconfined compressive strength	Very soft to hard (< 36 to > 300 kN/m²)	Hard to moderately weak (0.3–12.5 MN/m²)	Moderately strong to strong (12.5–100 MN/m²)	Strong to extremely strong (70 to > 200 MN/m²)

Notes:
(1) Non-carbonate constituents are likely to be siliceous apart from local concentrations of minerals such as feldspar and mixed heavy minerals.
(2) In description the rough proportions of carbonate and non-carbonate constituents should be quoted and details of both the particle minerals and matrix minerals should be included.
(3) The preferred lithological nomenclature has been shown in block capitals; alternatives have been given in brackets and these may be substituted in description if the need arises.
(4) Calcareous is suggested as a general term to indicate the presence of unidentified carbonate. Where applicable, when mineral identification is possible calcareous referring to calcite or alternative adjectives such as dolomitic, aragonitic, sideritic etc. should be used.

UNIFIED SOIL CLASSIFICATION INCLUDING IDENTIFICATION AND DESCRIPTION

Major divisions	Group symbols	Typical names	Laboratory classification criteria
COARSE GRAINED SOILS — More than half of material is *larger* than No. 200 sieve size U — **GRAVELS** — More than half of coarse fraction is larger than No. 4 sieve size — **CLEAN GRAVELS** (Little or no fines)	GW	Well graded gravels, gravel-sand mixtures, little or no fines	$C_u = \dfrac{D_{60}}{D_{10}}$ Greater than 4; $C_c = \dfrac{(D_{30})^2}{D_{10} \times D_{60}}$ Between 1 and 3
	GP	Poorly graded gravels, gravel-sand mixtures, little or no fines	Not meeting all gradation requirements for GW
GRAVELS WITH FINES (Appreciable amount of fines)	GM	Silty gravels, poorly graded gravel-sand-silt mixtures	Atterberg limits below 'A' line or PI less than 4 — Above 'A' line with PI between 4 and 7 are *borderline* cases requiring use of dual symbols
	GC	Clayey gravels, poorly graded gravel-sand-clay mixtures	Atterberg limits above 'A' line with PI greater than 7
SANDS — More than half of coarse fraction is *smaller* than No. 4 sieve size — **CLEAN SANDS** (Little or no fines)	SW	Well graded sands, gravelly sands, little or no fines	$C_u = \dfrac{D_{60}}{D_{10}}$ Greater than 6; $C_c = \dfrac{(D_{30})^2}{D_{10} \times D_{60}}$ Between 1 and 3
	SP	Poorly graded sands, gravelly sands, little or no fines	Not meeting all gradation requirements for SW
SANDS WITH FINES (Appreciable amount of fines)	SM	Silty sands, poorly graded sand-silt mixtures	Atterberg limits below 'A' line or PI less than 4 — Above 'A' line with PI between 4 and 7 are *borderline* cases requiring use of dual symbols
	SC	Clayey sands, poorly graded sand-clay mixtures	Atterberg limits above 'A' line with PI greater than 7

Determine percentages of gravel and sand from grain size curve. Depending on percentage of fines (fraction smaller than No. 200 sieve size) coarse grained soils are classified as follows:—
Less than 5% — GW, GP, SW, SP.
More than 12% — GM, GC, SM, SC.
5% to 12% — *Borderline* cases requiring use of dual symbols

Use grain size curves in identifying the fractions as given under field identification

Field identification procedures for fraction smaller than No. 40 sieve size

Major divisions	Group symbols	Typical names	Dry strength (crushing characteristics)	Dilatancy (reaction to shaking)	Toughness (consistency near plastic limit)
FINE GRAINED SOILS — More than half of material is *smaller* than No. 200 sieve size — **SILTS AND CLAYS** — Liquid limit less than 50	ML	Inorganic silts and very fine sands, rock flour, silty or clayey fine sands with slight plasticity	None to slight	Quick to slow	None
	CL	Inorganic clays of low to medium plasticity, gravelly clays, sandy clays, silty clays, lean clays	Medium to high	None to very slow	Medium
	OL	Organic silts and organic silt-clays of low plasticity	Slight to medium	Slow	Slight
SILTS AND CLAYS — Liquid limit greater than 50	MH	Inorganic silts, micaceous or diatomaceous fine sandy or silty soils, elastic silts	Slight to medium	Slow to none	Slight to medium
	CH	Inorganic clays of high plasticity, fat clays	High to very high	None	High
	OH	Organic clays of medium to high plasticity	Medium to high	None to very slow	Slight to medium
HIGHLY ORGANIC SOILS	Pt	Peat and other highly organic soils	Readily identified by colour, odour, spongy feel and frequently by fibrous texture		

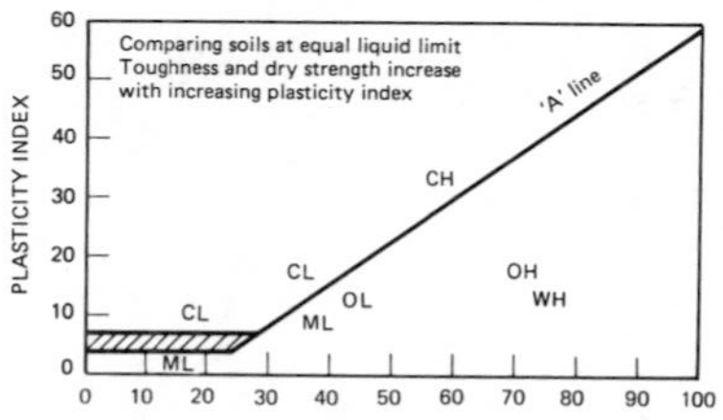

(The No. 200 sieve size is about the smallest particle visible to the naked eye)

(For visual classifications the ¼ in. size may be used as equivalent to the No. 4 sieve size)

Information required for describing soils

Give typical name; indicate approximate percentages of sand and gravel; max size, angularity, surface condition, and hardness of the coarse grains; local or geological names and other pertinent descriptive information; and symbol in parentheses.

For undisturbed soils add information on stratification, degree of compactness, cementation, moisture conditions and drainage characteristics.

EXAMPLE:—
Silty sand, gravelly; about 20% hard, angular gravel particles ½ in. maximum size; rounded and subangular sand grains coarse to fine; about 15% non-plastic fines with low dry strength; well compacted and moist in place; alluvial sand; (SM)

Give typical name; indicate degree and character of plasticity, amount and maximum size of coarse grains; colour in wet condition, odour if any, local or geological names, and other pertinent descriptive information, and symbol in parentheses.

For undisturbed soils add information on structure, stratification, consistency in undisturbed and remoulded states, moisture and drainage conditions.

EXAMPLE:—
Clayey silt, brown; slightly plastic; small percentage of fine sand; numerous vertical root holes; firm and dry in place; loess; (ML)

Plasticity chart — Liquid limit vs Plasticity index. 'A' line. Regions: ML, CL, OL, MH, OH, CH, CL-ML. "Comparing soils at equal liquid limit Toughness and dry strength increase with increasing plasticity index". For laboratory classification of fine grained soils.

Boundary classifications—Soils possessing characteristics of two groups are designated by combinations of group symbols. For example GW–GC, well graded gravel-sand mixture with clay binder.
All sieve sizes on this chart are US standard

Figure 11.8 Unified soil classification chart. (*After Earth Manual* (1974), US Department of Interior Bureau of Reclamation)

Table 11.16 Definitions of terms used in description of rock

Description of cores and exposure carried out in general accordance with the recommendations set out in British Standards Institution Code of Practice for Site Investigations (BS 5930:1981). The description of weathered state is based on the Working Party Report on the Preparation of Maps and Plans in Terms of Engineering Geology in *Quarterly Journal of Engineering Geology*, **5**, 316–317, 1972.

Scheme of description
Colour/grain size*/texture and structure/weathered state/minor lithological characteristics/ROCK NAME/strength/other characteristics and properties.

* Grain size often implied by the rock name.

<table>
<tr><td rowspan="40" style="writing-mode: vertical-lr">A. Definition of terms used to describe cores and exposures</td></tr>
<tr><td colspan="3">Spacing of structural and lithological features</td></tr>
<tr><td>(Mean) spacing</td><td>Bedding plane spacing term*</td><td>Discontinuity spacing in one dimension</td></tr>
<tr><td>> 2000 mm</td><td>Very thickly</td><td>Very widely</td></tr>
<tr><td>600–2000 mm</td><td>Thickly</td><td>Widely</td></tr>
<tr><td>200–600 mm</td><td>Medium</td><td>Moderately widely</td></tr>
<tr><td>60–200 mm</td><td>Thinly</td><td>Closely</td></tr>
<tr><td>20–60 mm</td><td>Very thinly</td><td>Very closely</td></tr>
<tr><td>6–20 mm</td><td>Thickly laminated (sedimentary and metamorphic rocks)
Narrow (metamorphic rocks and igneous rocks)</td><td>Extremely closely</td></tr>
<tr><td>< 6 mm</td><td>Thinly laminated (sedimentary and metamorphic rocks)
Very narrow (metamorphic and igneous rocks)</td><td></td></tr>
<tr><td></td><td colspan="2">* Bedding plane spacing describes thickness of lithological variations within rock. Frequency of 'bedding plane joints' described separately using discontinuity spacing terms.</td></tr>
</table>

Weathered state	
Term	Description
Residual soil	Rock is discoloured and completely changed to a soil. Original rock texture and structure is completely destroyed.
Completely weathered	Rock is discoloured and changed to a soil but original textures and structure is mainly preserved. There may be occasional corestones.
Highly weathered	Rock is discoloured; discontinuities may be open and stained. Rock texture and structure near discontinuities may be altered. Alteration may penetrate deeply, corestones are still present.
Moderately weathered	Rock is discoloured; discontinuities may be open and stained with alteration starting to penetrate inwards. Intact rock is noticeably weaker than fresh rock.
Slightly weathered	Rock may be slightly discoloured, particularly adjacent to discontinuities, which may be open and slightly stained. Intact rock is not noticeably weaker than fresh rock.
Fresh	Rock shows no discolouration, loss of strength or other weathering effects.
	Additional term faintly weathered often used when staining is present but limited to surface of major discontinuities. *In moderately and highly weathered rocks, the ratio of original rock to weathered rock should be estimated where possible.*

Strength classification		
Term	Compressive strength (MN/m²)	Field identification
Very weak	< 1.25	Crumbles easily in hand
Weak	1.25–5	Thin slabs or edges broken by light hand pressure
Moderately weak	5–12.5	Thin slabs or edges broken by heavy hand pressure
Moderately strong	12.5–50	Broken by light hammer blows
Strong	50–100	Broken by heavy hammer blows
Very strong	100–200	Rock chipped only by heavy hammer blows
Extremely strong	> 200	Rock rings on hammer blows

Table 11.16 *(continued)*

	Fracture state (assessed natural fractures only unless otherwise specified)
B. Additional terms used to describe rock cores	**Term** — **Definition** Total core recovery (TCR%) — Percentage ratio of core recovered (both solid and non-intact) to the total length of core run. Solid core recovery (SCR%) — Percentage ratio of solid recovered to the total length of core run. Solid core is here defined as pieces with at least one full diameter, but not necessarily with a full circumference measured along axis of the core. Rock quality designation (RQD%) — Percentage ratio of total length of solid core pieces, each greater than or equal to 100 mm in length, to the total length of core run. Fracture spacing (l_s mm) — Average spacing of natural fractures over core lengths of reasonably uniform characteristics. Minimum/average/maximum of dimensions often quoted. The term non-intact (NI) is used when the core is recovered in a broken or fragmented state.

	Size and shape of rock blocks		
C. Additional terms used to describe rock exposures	**First term**	**Maximum dimension**	**Second term**
	Very large	> 2000 mm	Blocky
	Large	600–2000 mm	or
	Medium	200–600 mm	Tabular
	Small	60–200 mm	or
	Very small	< 60 mm	Columnar
	In addition the following indices can be measured:	R_f (mm) the assessed dimension of individual blocks in geological units of reasonably uniform characteristics. I_f mm a linear measure of spacing of fractures usually along scan lines. (Similar to I_s as measured in cores.) Both indices are quoted as minimun/average/maximum dimensions.	

Orientation of linear and planar features
Orientation described by direction of dip as three digit compass bearing relative to stated north point/two digit angle of dip from horizontal (e.g. 035°/15°).

Table 11.17 Aid to identification of rocks for engineering purposes

Grain size (mm)	Bedded rocks (mostly sedimentary)									
	Grain size description				At least 50% of grains are of carbonate		At least 50% of grains are of fine-grained volcanic rock			
20 —	RUDACEOUS		CONGLOMERATE Rounded boulders, cobbles and gravel cemented in a finer matrix Breccia Irregular rock fragments in a finer matrix			Calcirudite		Fragments of volcanic ejects in a finer matrix Rounded grains AGGLOMERATE Angular grains VOLCANIC BRECCIA	SALINE ROCKS Halite Anhydrite
6 —										
2 —										
0.6 —	ARENACEOUS	Coarse	SANDSTONE Angular or rounded grains, commonly cemented by clay, calcitic or iron minerals Quartzite Quartz grains and siliceous cement Arkose Many feldspar grains Greywacke Many rock chips		LIMESTONE and DOLOMITE (undifferentiated)	Calcarenite		Cemented volcanic ash TUFF	Gypsum
0.2 —		Medium							
		Fine							
0.06 —	ARGILLACEOUS		MUDSTONE	SILTSTONE Mostly silt	Calcareous mudstone		Calcilu-tite	CHALK	Fine-grained TUFF
0.002 —			SHALE Fissile	CLAYSTONE Mostly clay			Calcilu-tite		Very fine-grained TUFF
Amorphous or crypto-crystalline			Flint: occurs as bands of nodules in the Chalk Chert: occurs as nodules and beds in limestone and calcareous sandstone						COAL LIGNITE
			Granular cemented – except amorphous rocks						
			SILICEOUS			CALCAREOUS		SILICEOUS	CARBON-ACEOUS

SEDIMENTARY ROCKS

Granular cemented rocks vary greatly in strength, some sandstones are stronger than many igneous rocks. Bedding may not show in hand specimens and is best seen in outcrop. Only sedimentary rocks, and some metamorphic rocks derived from them, contain fossils.

Calcareous rocks contain calcite (calcium carbonate) which effervesces with dilute hydrochloric acid.

Grain size boundaries approximate

Igneous rocks: generally massive structure and crystalline texture				
Grain size description				
	Pegmatite			Pyroxenite
COARSE	GRANITE[1]	Diorite[1,2]	GABBRO[1,2]	
MEDIUM				Peridotite
FINE	These rocks are sometimes porphyritic and are then described, for example, as porphyritic granite			
COARSE				
MEDIUM	Microgranite[1]	Microdiorite[1,2]	Dolerite[3,4]	
FINE	These rocks are sometimes porphyritic and are then described as porphyrites			
	RHYOLITE[4,5]	ANDESITE[4,5]	BASALT[4,5]	
	These rocks are sometimes porphyritic and are then described as porphyries			
	Obsidian[5]	Volcanic glass[6]		
	Pale ◄——— Colour ———► Dark			
	ACID Much quartz	INTER- MEDIATE Some quartz	BASIC Little or no quartz	ULTRA BASIC

(vertical label: Increasing grain size)

Metamorphic rocks	
Foliated	Massive
GNEISS Well-developed but often widely spaced foliation sometimes with schistose bands. Migmatite Irregularly foliated; mixed schists and gneisses	MARBLE QUARTZITE Granulite HORNFELS Amphibolite
SCHIST Well-developed undulose foliation; generally much mica	Serpentinite
PHYLLITE Slightly undulose foliation; sometimes 'spotted'	
SLATE Well developed plane cleavage (foliation) Mylonite Found in fault zones, mainly in igneous and metamorphic areas	
CRYSTALLINE	
SILICEOUS	Mainly SILICEOUS

IGNEOUS ROCKS

Composed of closely interlocking mineral grains.
Strong when fresh; not porous.

Mode of occurrence: (1) Batholiths; (2) Laccoliths;
(3) Sills; (4) Dykes; (5) Lava flows; (6) Veins

METAMORPHIC ROCKS

Generally classified according to fabric and
mineralogy rather than grain size.

Most metamorphic rocks are distinguished by
foliation which may impart fissility. Foliation in
gneisses is best observed in outcrop. Non-foliated
metamorphics are difficult to recognize except by
association.

Most fresh metamorphic rocks are strong although
perhaps fissile.

Name of company: A N Other Ltd		Borehole No. 1 Sheet 1 of 1				
Equipment and methods: Light cable tool percussion rig. 200 mm dia.hole to 7 m. Casing 200 mm dia. to 6 m		Location No:6155 QUAGMIRE MOOR FARTOWN				
Carried out for: Smith, Jones & Brown		Ground level 9.90 m (Ordnance datum)	Coordinates: E 350 N 901	Date: 17 − 18 June 1974		

Description	Reduced level	Legend	Depth & thickness	Samples/tests				Field records
				Depth	Sample Type	No.	Test	
Made Ground (sand, gravel, ash, brick and pottery)	9.40		(0.50) 0.50	0.20	D	1		
Made Ground (red and brown clay with gravel)	9.10		(0.30)					
Firm mottled brown silty CLAY (Brickearth)			0.80	0.70−1.15	U D	2 3		24 blows*
	7.90		(1.20) 2.0	1.20				
Stiff brown sandy gravelly CLAY (Flood Plain Gravel)			(1.65)	2.10−2.55 2.55	U D	4 5		50 blows
	6.25		3.65	3.60−4.05 3.65	D U	6		No recovery
Medium dense brown sandy fine to coarse GRAVEL (Flood Plain Gravel)			(1.65)	4.00−4.30			S N27	
				4.00−5.00	B	7		
	4.60		5.30	5.00−5.30			S N15	
Firm becoming stiff to very stiff fissured grey silty CLAY with partings of silt (London Clay)				5.30	D	8		Standpipe inserted 5.30 m below ground level
			(2.15)	6.00−6.45	U	9		35 blows
Water level observations during boring	2.45		7.45	7.00−7.45	U	10		44 blows
				End of borehole				

Date	Time	Depth of hole (m)	Depth of casing (m)	Depth to water (m)	Remarks
17 Jun 18 Jun	1615 0800 1130	1.50 1.50 3.70	1.00 1.00 1.00	Dry Trace 3.30	Water encountered after 15 min
28 June	1430 1000 1015	7.45 − −	5.50 − −	Dry 2.36 2.46	} Stand pipe

SPT: Where full 0.3 m penetration has not been achieved, the number of blows for quoted penetration is given (not N−value).
Depths: All depths and reduced levels in metres, Thicknesses given in brackets in depth column.
Water: Water level observations during boring are given on last sheet of log.

Sample/test key
D Disturbed sample
B Bulk sample
W Water sample
▮ Piston (P), tube (U) or core sample; length to scale
S Standard penetration test
C Standard cone penetration test
V Vane test
C Core recovery (%)
r Rock quality Designation (RQD%)

Remarks
Water added to facilitate boring from 4.00 m to 5.30 m Borehole back-filled with natural spoil from 7.00 m to 5.30 m, gravel to 0.80 m, clay to 0.50 m, a concreted cock box to ground level.

*Blows to drive U100

Logged by: ABC

Scale:

As drawn

Figure 11.9 Typical log of data from a light cable percussion borehole. (After BS 5930:1981)

Name of company: A N Other Ltd		Borehole No. 14 Sheet 1 of 4		
Equipment and methods: Rotary coring, water flush and with diamond bits. PWF bit to 8.8 m and HWF beyond.		Location No: 65117 LUKE STREET UPHILL		
Carried out for: Smith, Jones & Brown		Ground level 125.3 mm O.D.	Coordinates: E 295 N 635	Date: 8 – 18 March 1975

Main description	Detail	Reduced level	Legend	Depth & thickness	Samples/tests Depth	Sample Type	No.	Test	Field records
							C	r	Drilling & casing progress / Water recovery
Yellow brown clayey gravelly SAND (Glacial drift)				(2.1)	0–1.7		0	0	Flush return normal to 9 m, where it ceased. Normal flush restored when borehole cased to 8.8 m
		123.2		2.1					
Firm to stiff reddish brown sandy silty CLAY gravelly and with cobbles (Glacial drift)				(1.1)	1.7–4.6		93	0	
		122.1		3.4					
Black friable coaly SHALE (Glacial drift?)	Grey clay and mudstone at base			(0.6)					
		121.5		4.1					
Grey thinly bedded sandy MUDSTONE, moderately weak (Middle Coal Measures)	Thin bands of fine grained grey argillaceous sandstone and occasional ironstone nodules			(1.8)					8 Mar
		119.7		5.6					
Light grey thinly bedded to medium bedded fresh fine grained SAND-STONE, moderately strong (Middle Coal Measures)	Cross bedded frequently fissured and with bands of sandy dark grey mudstone			(4.1)	4.6–7.2		93	17	
					7.2–8.8		100	32	9 Mar
		115.6		9.7	8.8–10.3		93	25	
Grey friable CLAY - (old mine workings - Crank coal)	Some broken coal in mudstone								

SPT: Where full 0.3 m penetration has not been achieved, the number of blows for the quoted penetration is given (not N–value). Depths: All depths and reduced levels in metres. Thickness given in brackets in depth column. Water: Water level observations during boring are given on last sheet of log.	Sample/test key D Disturbed sample B Bulk sample W Water sample ▮ Piston (P), tube (U) or core sample; length to scale S Standard penetration test C Cone penetration test V Vane test C Core recovery (%) r Rock quality Designation (RQD%)	Remarks Borehole cased to 19 m	Logged by: ABC Scale: As drawn

Note: TCR, SCR, RQD and I$_f$ recorded as required.

Figure 11.10 Typical log of data from a rotary drillhole. (After BS 5930:1981)

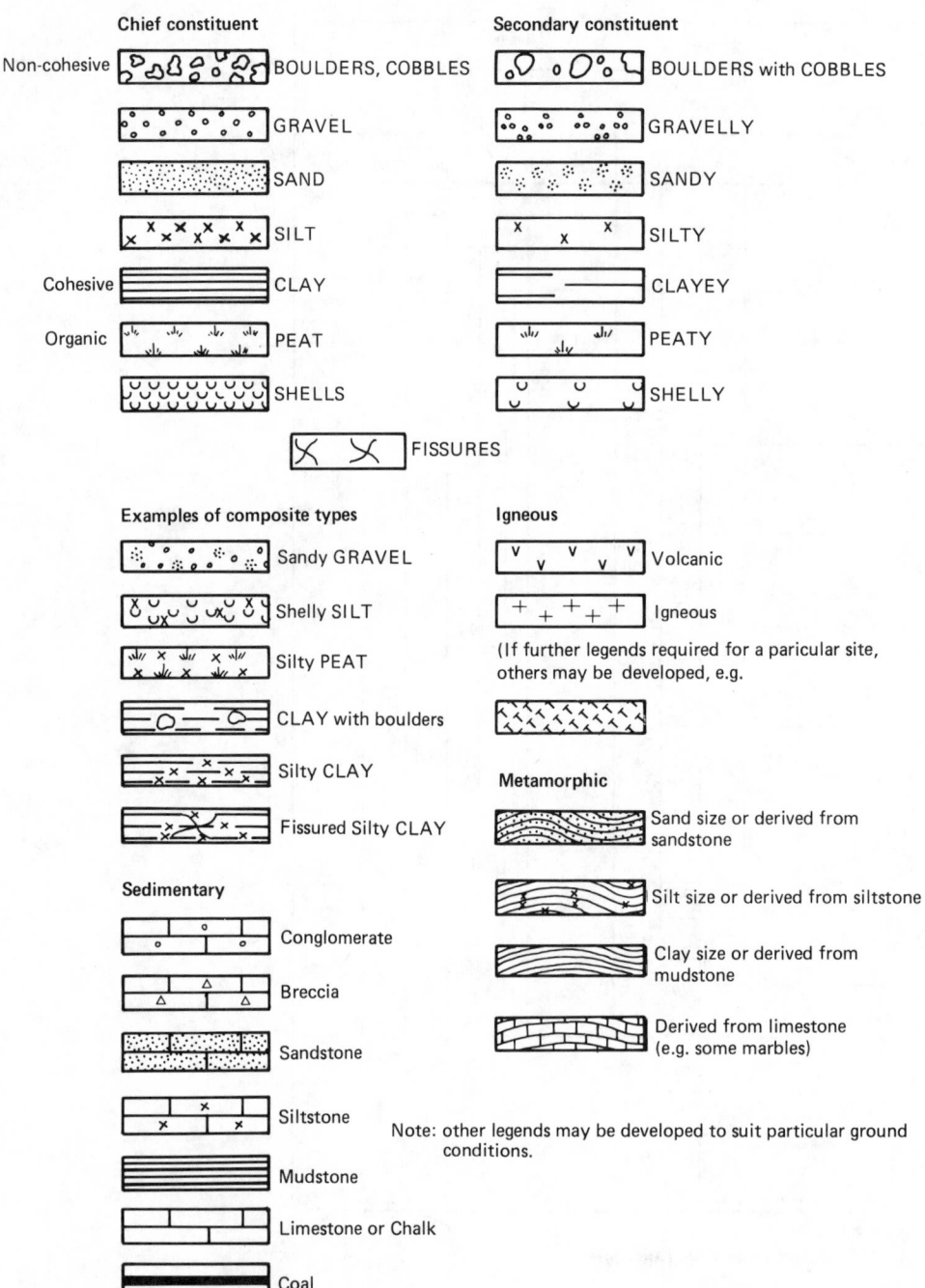

Chief constituent

Non-cohesive — BOULDERS, COBBLES

GRAVEL

SAND

SILT

Cohesive — CLAY

Organic — PEAT

SHELLS

FISSURES

Secondary constituent

BOULDERS with COBBLES

GRAVELLY

SANDY

SILTY

CLAYEY

PEATY

SHELLY

Examples of composite types

Sandy GRAVEL

Shelly SILT

Silty PEAT

CLAY with boulders

Silty CLAY

Fissured Silty CLAY

Sedimentary

Conglomerate

Breccia

Sandstone

Siltstone

Mudstone

Limestone or Chalk

Coal

Igneous

Volcanic

Igneous

(If further legends required for a paricular site, others may be developed, e.g.

Metamorphic

Sand size or derived from sandstone

Silt size or derived from siltstone

Clay size or derived from mudstone

Derived from limestone (e.g. some marbles)

Note: other legends may be developed to suit particular ground conditions.

Figure 11.11 Standard legends

Little Springs, Tiverton
Trial Pit No. 1 : WSW Face

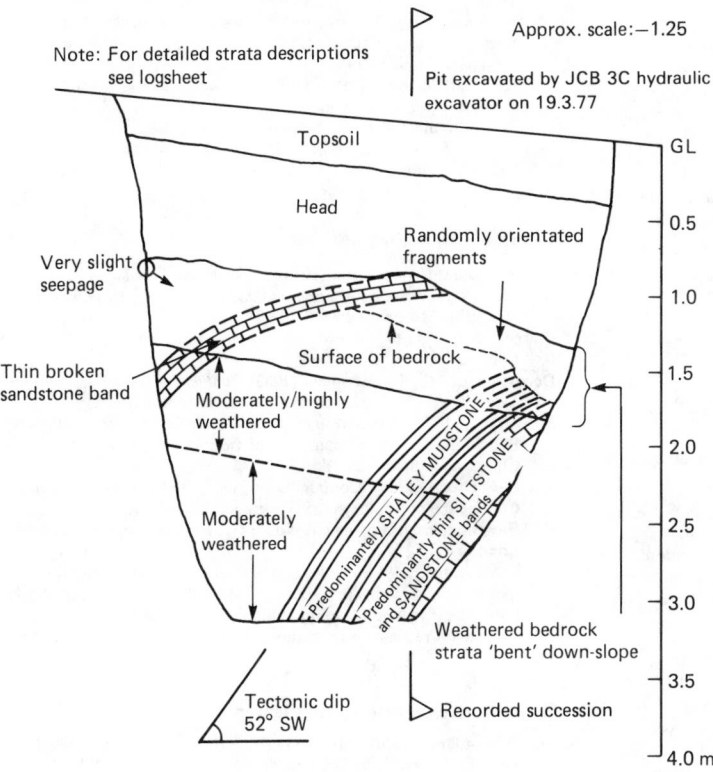

Approx. scale:—1.25

Note: For detailed strata descriptions
see logsheet

Pit excavated by JCB 3C hydraulic
excavator on 19.3.77

Trial pit no. 1 (GL:76.51 m AOD)

Depth (m)	Description
GL–0.30/0.40 0.30/0.40 to 1.00/1.30	Ashy TOPSOIL with a little brick gravel. Firm to stiff becoming firm, red-brown becoming orange-brown silty CLAY with a little fine to coarse gravel sized fragments of various rock types including shaley mudstone, siltstone and subround quartz cobbles. (HEAD)
1.00/1.30–1.60/1.80	Stiff orange-brown and grey mottled shaley CLAY with some (25% increasing to 50%) gravel and angular cobbles and cobble sized blades of fine grained sandstone and shaley mudstone. Fragments random but striking 140/320 and dipping 37° SW at base. (COMPLETELY/HIGHLY WEATHERED BEDROCK)
1.60/1.80–2.40	Very weak dark grey, stongly stained orange brown, moderately to highly weathered thickly laminated shaley MUDSTONE and occasional very thin beds of fine grained sandstone. Degenerating to very stiff shaley clay in parts strike 115/295 and Dipping 56° SSW. (UPPER CARBONIFEROUS BEDROCK)
2.40–3.10	—Transition— Moderately weak to moderately strong, brownish-grey moderately weathered thinly interbedded shaley MUDSTONE and SILTSTONE. Bedding strikes 130/310° and dips 55° SW. (UPPER CARBONIFEROUS BEDROCK)

Notes:
(1) Pit orientated SSE—NNW and 3.5 m by 1 m in plan
(2) Very slight ground-water seepage at 1.3 m depth and southern
end of pit and 'sweating' between approximately 1.5 m and
2.5 m depth. After 6 hours pit dry.
(3) Sides of pit shored and pit descended to examine strata
(4) Pit terminated in hard digging
(5) Recorded succession as shown in sketch.
(6) Hand Shear Vane Testing

At: 0.75 m   85kN/m^2 1.50 m   120kN/m^2
1.25 m   65kN/m^2 2.00 m   115kN/m^2

Could not penetrate pit sides below 2 m.

(7) Bulk disturbed samples recovered at 1.0 m, 1.7 m, and 2.3 m depth.
(8) Weather: fine and dry

Note on logging:
Where justified, each face may be shown separately and samples located on each face.

Figure 11.12 Typical log of data from a trial pit

Selected bibliographies

Environmental surveys

Canter, L. W. and Hill, L. G. (1981) *Handbook of variables for environmental impact assessment*. Ann Arbor Science, Inc., Michigan.
Gunnerson, C. G. and Kalbermatten, J. M. (eds) (1979) 'Environmental impacts of international civil engineering projects and practices'. American Society of Civil Engineers' National Convention, California, Oct. 1977. American Society of Civil Engineers, New York.
Lacy, R. E. (1976) *Climate and building in Britain*, HMSO, London.
United States Department of the Interior (1978) *Land use and land cover information and air quality planning*. Geological Survey Prof. Paper 1099B. United States Government Printing Office.

Hydrographic surveys

Ingham, A. E. (1975) *Sea Surveying*. 2 Vols. Wiley, London.
Hydrographer of the Navy (1982) *Admiralty manual of hydrographic surveying*. Admiralty, London.
British Standards Institution (1984). BS 6349 *Code of practice for maritime structures*. Part 1: 'General criteria'. BSI, Milton Keynes.

Preliminary appreciation

Amos, E. M., Blakeway, D. and Warren, C. D. (1984) *Remote sensing techniques in civil engineering surveys*, Twentieth Regional Meeting, Engineering Group, Geological Society of London.
Beaumont, T. E. and Beavan, P. J. (1977) *The use of satellite imagery for highway engineering in overseas countries*, **SR279**. Transport and Road Research Laboratory, Crowthorne.
Dumbleton, M. J. and West, G. (1976) *Preliminary sources of information for site investigations in Britain*, **LR403**. Transport and Road Research Laboratory, Crowthorne.
Dumbleton, M. J. and West, G. (1974) *Guidance on planning, directing and reporting site investigations*, **LR625**. Transport and Road Research Laboratory, Crowthorne.
Dumbleton, M. J. (1983) *Airphotographs for investigating natural changes, past use and present conditions of engineering sites*, **LR1085**. Transport and Road Research Laboratory, Crowthorne.
Geological Society of London Working Party (1982). 'Land surface evaluation for engineering practice', *Q. J. Eng. Geol.* **15**, 265–316.
Mollard, J. D. (1962) 'Photo analysis and interpretation in engineering geology investigations: a review'. From: *Reviews in engineering geology*. The Geology Society of America, New York.

Main investigation and methods of ground investigation

Bell, F. G. (ed.) (1987) *Ground engineers reference book*. Butterworth Scientific, Guildford.
British Standards Institution (1981) 'Code of Practice for Site Investigations' (formerly CP 2001), BS 5930. Milton Keynes.
Clayton, C. R. I., Simons, N. E. and Mathews, M. C. (1983) *Site investigation – a handbook for engineers*. Granada, London.
Cottington, J. and Akenhead, R. (1984) *Site investigation and the law*. Thomas Telford, London.
Dumbleton, M. J. and West, G. (1974) *Guidance on planning, directing and reporting site investigations*, **LR625**. Transport and Road Research Laboratory, Crowthorne.
Hanna, T. H. (1985). 'Field instrumentation in geotechnical engineering'. Trans. Tech. Pub. POB **266. D3392.** Clausthal-Zellerfeld Federal Republic of Germany.
Institution of Civil Engineers (1983) *ICE conditions of contract for ground investigation*, Thomas Telford, London.

National Research Council of Canada (1975) *Canadian manual on foundation engineering*. Ass. Com. on National Building Code. Ottawa.
Peck, R. B. (1969) 'Advantages and limitations of the observational method in applied soil mechanics', *Géotechnique* **19**, 2, 171–187.
Sanglerat, G. (1979). *The penetrometer and soil exploration*. Elsevier Scientific. Amsterdam. (Includes proposed European Standards (CPT, DPT and SPT).)
Uff, J. F. and Clayton, C. R. I. (1986) *Recommendations for the procurement of ground investigation*. Construction Industry Research and Information Association, London.
Weltman, A. J. and Head, J. M. (1983) *Site investigation manual*. SP25. CIRIA, London.

Contaminated sites

British Standards Institution (1984) *Draft British Standard Code of Practice for the identification and investigation of contaminated land*. BSI, Milton Keynes.
Cairney, T. (ed.) (1987) *Reclaiming contaminated land*. Blackie, London.
Department of the Environment (1983) *Guidance on the assessment and redevelopment of contaminated land*. HMSO, London.
Kelly, R. T. (1979) 'Site investigation and materials problems'. Paper B2, Conference on Reclamation of Contaminated Land. Society of Chemical Industry, London.
Kelly, R. T (ed.) (1984) 'Contaminated land. The London experience'. Conference proceedings, 25 November 1983. London Environmental Supplement No. 7. Greater London Council, London.
Smith, M. A. (ed.) (1985) 'Contaminated land: reclamation and treatment'. Plenum Press. London. (Results of NATO/CCMS pilot study by seven leading industrialized countries. Includes chapter on rapid on-site methods of chemical analysis.)

Ground investigations over water

British Standards Institution (1984) *Draft British Standard Code of Practice for fixed offshore structures* (rev. of BS 6235:1982). (Includes discussion on site investigation and environmental data.) BSI, Milton Keynes.
Carter, P. G., Pirie, R. M. and Sneddon, M. (1984) 'Marine site investigations and BS 5930'. Proceedings, Twentieth Regional Meeting Engineering Group of Geological Society of London. Vol. 1, pp.86–92.
St John, H. D. (1980) *A review of current practice in the design and installation of piles for offshore structures*. Department of Energy, offshore technical paper. CIRIA, London.
Tirant le, Pierre (1976) *Seabed reconnaissance and offshore soil mechanics for the installation of petroleum structures*. Trans. J. C. Ward. Graham and Trotman, London.

Laboratory and *in situ* tests

American Society for Testing and Materials (Annual). 'Soil and rock; building stones; peats', Section 4, Vol. 04.08. *Annual book of ASTM standards*. Philadelphia.
British Standards Institution (1975) *Methods of test for soils for civil engineering purposes*. BS 1377. (under revision 1987). BSI, Milton Keynes.
British Standards Institution (1975). *Methods of test for stabilized soils*. BS 1924. (under revision 1987). BSI, Milton Keynes.
Head, K. H. *Manual of soil laboratory testing*. (1980) Vol. 1. *Soil classification and compaction tests*; (1982) Vol. 2. *Permeability, shear strength and compressibility tests*; (1986) Vol. 3. *Effective stress tests*. Pentech Press, Plymouth.

12

Reinforced and Prestressed Concrete Design

S C C Bate CBE, BSc(Eng), PhD,
C Eng, FIStructE, FICE
Formerly at the Building Research
Establishment and later, currently Consultant
to Harry Stanger Ltd

Contents

12.1 Introduction

The design of reinforced and prestressed concrete has been increasingly codified during the past 40 years. Before the Second World War, recommendations for design had been published in the UK in a Code of Practice prepared by the Department of Scientific and Industrial Research, which was issued in 1934,[1] and in the Building By-laws of the London County Council of 1938.[2] After the war, the DSIR Code was revised and became the British Standard Code of Practice, CP 114, in 1948.[3]

The Institution of Structural Engineers published its First Report on Prestressed Concrete in 1951,[4] which gave design procedures for prestressed construction. This report was subsequently revised and issued as BS Code of Practice, CP 115, in 1959.[5] The BS Code of Practice for the Design of Precast Concrete, CP 116,[6] appeared in 1965 and supplemented the two earlier BS codes. By this time, a number of codes dealing with specialized forms of concrete construction were being prepared. Codes of Practice 114, 115 and 116 have been updated from time to time and are currently adopted as deemed-to-satisfy documents in the Building Regulations.

An important innovation took place in 1972 when a unified BS code, CP 110, for the structural use of concrete[7] was published. This code, which it was intended should supersede Codes 114, 115 and 116, introduced a new feature in design, namely limit state design in which account was taken directly of the possibility of failure or unserviceability occurring during the life of the structure being designed. The particular factors considered included the risks resulting from variability of the materials, inaccuracy in design assumptions and construction, variability of loading and the incidence of accidental damage. Whilst the approach to design was modified, many existing methods of analysis and calculation were retained. Provision was made for the incorporation of new data on loading and materials, and on structural performance and methods of construction as they became available. The basis for this approach had been developed by the European Committee for Concrete assisted by the International Federation for Prestressing who published a jointly prepared code[8] in 1978 having previously issued separate codes. This code was used in the production of a code for the European Economic Community.[9]

Code of Practice 110 did not replace the earlier Codes 114, 115 and 116, which still remain in force, but it has now been revised as BS 8110.[10] The approach adopted in CP 110 has been retained and the content has been brought up to date. In addition, a manual[11] has been prepared by the Institution of Structural Engineers conforming with its recommendations but presented in simpler form and dealing with a more limited range of construction. The guidance given in this chapter is related directly to the contents of BS 8110.

Whilst these developments were taking place in the UK, somewhat similar changes were occurring elsewhere. Some idea of the differences between the recommendations adopted in the UK and elsewhere are given in Table 12.1, which makes some comparisons between BS 8110, the American Building Code, ACI 318-83[12] and the EEC Code.

12.1.1 Definitions

This chapter is concerned with the basic approach to design of reinforced and prestressed concrete. It deals with both cast-in-place and precast concrete whether reinforced or prestressed. It includes information on the use of plain or deformed steel reinforcing bars and with tendons which may be either pretensioned or post-tensioned. In this context some definitions and an indication of limitations may be useful.

(1) Reinforcement which is used to provide the tensile compo-

nent of internal forces in reinforced concrete, generally consists of one of three types of material: plain round mild-steel bar produced by hot-rolling; plain square or plain chamfered square twisted mild-steel bar which has had its yield stress raised by cold-working; ribbed bars, which may be hot-rolled from steel with high yield stress or cold-worked by twisting from hot-rolled mild-steel.

Since steel reinforcement can only develop an effective tensile force by extension of the concrete by cracking, there is a limit on the maximum strength of steel that can be used. In general the yield stress should not exceed 500 N/mm² although higher strength steels may be used if particular care is taken to avoid excessive cracking or deflection.

(2) Tendons are used to impart a prestress to concrete before service loads are applied which offsets the tensile stresses which will later result from the application of these loads. Tendons are usually comprised of plain, indented or deformed cold-drawn carbon steel wire, of seven-wire or nineteen-wire strand spun from one or two layers respectively of cold-drawn carbon steel wire around a core wire, or of high-tensile alloy steel bar. The strength of steel used must be high enough for it to be extended sufficiently to avoid excessive loss of tension due to elastic contraction, creep and shrinkage of the concrete. In general it is not of lower tensile strength than about 1000 N/mm².

(3) In prestressed concrete, prestressing may be effected by pretensioning or post-tensioning the tendons. Pretensioned tendons are stressed before the concrete is cast. They are stretched either between temporary anchorages placed sufficiently far apart for a number of moulds to be assembled in line around the tendons, i.e. the 'long-line' method, or between the ends of specially strong moulds, i.e. the 'individual' mould method; in each case, concrete is then cast and allowed to harden before the tendons are released from their temporary anchorages. The methods are best-suited to mass production in the factory and usually use wire or the smaller sizes of strand as tendons.

With post-tensioning, however, the tendons are stressed after the concrete has hardened and are usually accommodated in ducts within the concrete being held at their ends by anchorages, of which there are various proprietary types. Subsequently the ducts are grouted with cement grout to protect the tendons from corrosion. This method is mostly applied to site construction and tends to use tendons of relatively large size.

12.2 Behaviour of structural concrete

The characteristics of concrete that have conditioned its development as a structural material are its high compressive strength and relatively low tensile strength. In consequence its use for flexural members did not become practicable until it was discovered that steel reinforcement could be cast in the concrete to carry the bending tensile stresses whilst relying on the concrete to carry the bending compressive stresses. Experiment showed that mild steel, when present in the tension zone in relatively small amounts, provided a material with characteristics for deformation and strength which complemented those for concrete and provided a practical form of construction. Early research workers concluded that the presence of the steel increased the extensibility of the concrete. Later experiments showed, however, that this was not so. It then became clear that as the tensile stress in the steel of a beam increased beyond a small amount, which is appreciably less than that developed under service loading, cracks developed in the concrete. These cracks were controlled in width and numbers by the position of the reinforcement relative to the concrete surface and by the size

Table 12.1 Notes on different Codes of Practice (British, American and EEC)

(A) BS 8110 – *Structural use of concrete*[10]	(B) ACI 318 – *Building code requirements for reinforced concrete*[12]	(C) Eurocode No. 2 – 'Common unified rules for concrete structures'[9]

Status

(A) This national code was prepared by the British Standards Institution, an organization with some direct support from Government, and accepted as providing conformity with British Building Regulations, but not in itself mandatory, other authenticated design procedures may be acceptable.

(B) This national code was prepared by the American Concrete Institute; it is used extensively in State regulations for building control. It is widely recognized internationally and is adopted in part or wholly in the codes of a number of other countries.

(C) The code has been prepared by the Commission of the European Community for use in member countries, and is one of a number now being produced to deal with all common materials and forms of construction. It is likely to be adopted for building control in those countries and will be recognized as satisfying the requirements of national regulations. The code has drawn on the work of international organizations, which are supported worldwide, and hence it is likely to have an important influence on the formulation and revision of codes in other countries outside as well as inside the Community.

Design procedure

(A) Limit state procedures (described in this chapter) are adopted following closely the 1964 Recommendations of the European Committee for Concrete, which were used subsequently in developing the EEC Code; the basic approach in the two codes is therefore very similar.

The ultimate limit states include strength and stability under dead and imposed loads, wind loads, and earth and water pressure for which partial safety factors are defined depending on load groupings, and the effects of accidental loading and damage. Durability and fire resistance are not treated as limit states but are included in the design process, the former being given more emphasis than in previous codes. As far as possible the analysis of structures is based on ultimate behaviour but, where methods have not been developed, elastic analysis is accepted. The strength of sections is based on the strength of the materials, as reduced by partial safety factors, and compatibility between stress and strain using idealized stress–strain relationships. Simplifying assumptions relating to these stress–strain relationships are allowable for many types of construction.

The serviceability limit states include deflection and cracking under dead, imposed and wind loads, appropriate partial safety factors for combinations of loading being given with limitations on deflection and crack width. Where necessary, allowance is required for the effects of shrinkage and creep and of temperature change. For common types of construction, limits on deflections are imposed by placing limits on span : depth

(B) The objectives of the ACI Code are similar to those of (A) but the means of achieving them are different.

Design requires consideration of ultimate strength but a single factor is defined for relating strength to the loads to be supported instead of adopting the combined effects of partial safety factors for both loading and strengths of materials, as in (A). The principles for calculating the strength of sections are otherwise similar in requiring compatibility between stress and strain. For flexure, the strength of concrete is defined in terms of 85% of the cylinder strength reducing as strength increases instead of 67% of the cube strength irrespective of strength, as in (A) and there are slight differences in the shape and extent of the stress–strain curve assumed; precautions are introduced to avoid brittle compression failures.

Serviceability with respect to deflection is ensured either by limiting span : depth ratios or by checking that the long-term deflections do not exceed defined limiting values. Cracking is controlled by limitation of calculated crack width and provision of reinforcement for both reinforced and prestressed concrete. In (A), on the other hand, cracking of prestressed concrete may be controlled by limitation of tensile stress, the nominal tensile stress being related to amount and distribution of secondary reinforcement.

The ACI Code has an appendix which gives an alternative method of design for reinforced concrete which is based on permissible stresses in the materials.

(C) Since the developments of the BS code and of the EEC code have drawn on a common source, the two codes, as already noted, have a common basis. However, the former was drafted by a committee with a British background in design and in the development of codes, whereas the latter has incorporated multi-national European experience and there are therefore a number of differences. Also in Britain, codes are generally regarded as advisory whereas in Europe they are mandatory. The main differences between the two codes are, however, in detail, the EEC Code tending to be more precise.

Thus, the definition of limit states, the loads to be considered, and the strengths to be adopted with their relevant partial safety factors are closely similar.

The EEC code is, however, based on cylinder strengths of concrete which gives rise to some differences when compared with the BS code. The simplified assumptions for calculation of flexural strength for each code, for example, show an apparent ratio of cylinder strength to cube strength of 0.89, which is appreciably higher than for most experimental data.

Table 12.1 Notes on different Codes of Practice (British, American and EEC)—*continued*

(A) BS 8110 – *Structural use of concrete*[10]	(B) ACI 318 – *Building code requirements for reinforced concrete*[12]	(C) Eurocode No. 2 – 'Common unified rules for concrete structures'[9]
ratios, and cracking may be controlled for reinforced concrete by defining the form that the reinforcement should take. Since many practical engineers have pressed for the retention of permissible stress methods of design, the previous code applicable for reinforced concrete[3] has been retained in use. It seems likely, however, that it will be withdrawn in the longer term.		

Concluding comment: Of necessity, these comparisons are very limited and superficial in character but should serve to show that developments in codes proceeding currently in different countries have much in common. This trend is likely to increase through the medium of the extensive international collaboration that now takes place.

of bars used. Thus with closely spaced bars near the surface, a large number of small cracks would develop, but with large widely spaced bars, the cracks would be fewer in number and much larger for the same stress in the steel. If the stress in the steel were increased the size of the cracks increased and their size was little influenced by the surface roughness of the steel, although at one time it was thought that roughening of the surface resulted in appreciably smaller cracks of larger numbers. It was eventually established that the main benefit of using bars with a roughened surface was in developing good end-anchorage.

Because steel needs to extend to develop stress and hence causes cracking and deformation of the concrete, there is a limit to the strength of steel that can be used efficiently for reinforcement, since unsightly cracking, which could lead to severe corrosion in adverse conditions and unacceptable deflections, must be avoided. The use of steel in prestressed concrete, where the stress in the steel is imposed before the concrete member is subjected to external load, avoids this problem, since the initial tensile force is developed without extending the concrete, and so no upper limit is imposed on the strength of steel that can be employed. This was not, however, appreciated in the early development of prestressed concrete. Then, steel of relatively low strength was used with a small initial tension. The experimenters found that, although this was effective at the start, the initial prestress disappeared with time. Eventually, however, it was established that this nonelastic behaviour was limited in extent and that if a sufficiently large elastic extension was imparted to the steel, the nonelastic effects of creep and shrinkage of the concrete did no more than reduce the prestress by an acceptable amount. Although for a time there was a tendency to underestimate the losses of prestress due to contraction of the concrete and to ignore creep in the steel tendons, research has now, however, clearly set the limits on what needs to be considered in design.

The performance of reinforced concrete and prestressed concrete beams under increasing load is characteristically different since cracking develops in different ways in each form of construction. This is illustrated by the results of tests on beams in each form of construction as illustrated in Figures 12.1 and 12.2.

Examined in more detail the deformation of the reinforced concrete beam under load is linear until cracking occurs; thereafter it approximates to a linear relationship until the steel yields as cracking becomes more extensive for beams of normal design. Subsequent deformation leads to the development of a hinge with continued yielding of the steel accompanied by damage to the concrete. This deformation continues at approximately constant moment until a stage is reached where the resistance reduces. The occurrence of this stage is influenced by the amount of transverse shear reinforcement in the section.

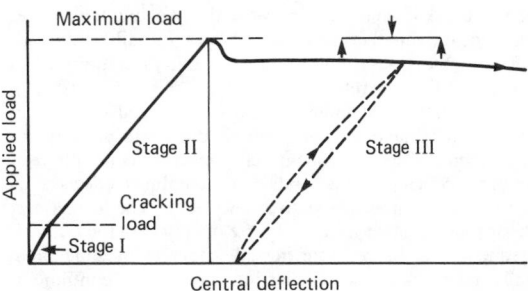

Figure 12.1 Relationship between applied load and deflection for a reinforced concrete beam showing recovery and reloading

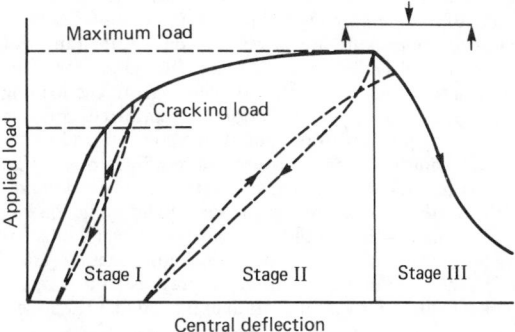

Figure 12.2 Relationship between applied load and deflection for a prestressed concrete beam showing recovery and reloading

The prestressed concrete beam, however, remains uncracked usually until the service load is exceeded, and in this range its deformation is elastic. Once cracking has occurred deformation increases disproportionately rapidly with increasing load as cracks widen until the maximum load is reached. Subsequently there is a rapid reduction in resistance. Since the prestressed concrete beam is usually uncracked under service conditions its stiffness is greater than that of reinforced concrete beams of the same overall depth.

In continuous construction subjected to applied loads of short duration, deformation of both reinforced concrete and prestressed concrete members is elastic or effectively elastic until service loads are exceeded. With further loading, as the applied moment at any section approaches the resistance moment at that section, there is a tendency for the moment to be relaxed and redistributed to sections that are less seriously stressed. Thus a loaded beam, built in at each end, may reach its maximum resistance moment at mid-span before the maximum resistance moments at the supports are attained; a hinge then forms at midspan with the applied moment there remaining sensibly constant whilst the applied moments at the supports increase until hinges form at the supports. The beam has then reached its maximum carrying capacity. The capability of reinforced and prestressed concrete beams for rotation at hinges is limited, however, and restrictions therefore need to be placed on allowances in design for redistribution of moment. These allowances are smaller for prestressed concrete sections than for reinforced concrete sections since their rotational capacities are smaller.

Under long-term loading, the deflection of reinforced concrete beams increases usually to about 2 or 3 times the initial deflection. Although the initial deflection is primarily influenced by the amount of steel in the section and its stress, the subsequent deflection is largely the result of creep of the concrete, breakdown of bond between the steel and the concrete in the tension zone between cracks which initially stiffens the beam, and the effect of the reinforcement in restraining the shrinkage of the concrete.

Since prestressed concrete is usually uncracked under long-term load the initial deflection is mainly due to the deformation of the concrete. The subsequent deflection results mainly from creep of the concrete and depends on the combined effects of the prestress and the stresses due to applied load. The former tend to deform the member in the opposite direction to the latter. In consequence, a loaded prestressed concrete member may initially have an upward deflection which can continue to develop upwards or downwards depending on how heavily it is loaded.

Under cyclic loading, reinforced concrete members usually fail in fatigue by fracture or yield of the reinforcement. The properties of most reinforcing steels, provided that they are free from welded connections, are, however, such that the ranges of stress experienced under service loading determined for static conditions are usually within the fatigue range. Cyclic loading leads to some increase in deflection of reinforced concrete members partly due to deformation of the concrete and partly due to breakdown of bond between cracks. Since prestressed concrete is uncracked under normal static service load conditions, the fluctuations of stress in the steel under cyclic loading are small. Fatigue failure of the steel only occurs when substantial cracks have developed and deflections are generally unacceptable. The effect of cyclic loading on prestressed concrete is to increase deflection by a small amount, i.e. 20 to 30% largely as a result of creep of the concrete. Large numbers of repetitions within the normal range of service loading do not reduce the ultimate strength of prestressed or reinforced concrete. Because of its freedom from cracking, prestressed concrete behaves better than reinforced concrete under severe cyclic loading and has therefore been used extensively for railway sleepers.

Resistance of beams to impact is indicated by the energy absorbed in deforming which is given by the area of the load deflection curves. Referring again to Figures 12.1 and 12.2, the deformation of prestressed and reinforced concrete beams has been defined in three stages. In stage I, deformation is elastic and largely recoverable; in stage II, deformation is in part elastic but accompanied by cracking and is partly recoverable; whilst in stage III, deformation is mainly due to permanent damage to the materials. Since stages I and II represent the largest amounts of absorbed energy for prestressed concrete, this material has a considerable capacity for recovery after impact. For reinforced concrete, the energy absorbed in stage III is substantially greater than in the other two stages. Thus, reinforced concrete does not show much recovery after impact but has a high ultimate impact resistance which is appreciably higher than that for prestressed beams designed for the same static loads. Prestressed concrete

beams are, however, better in resisting repetitions of relatively light impacts with little residual damage.

So far, performance has been considered mainly in terms of bending conditions, but conditions of direct stress in compression exist in columns and walls. In such construction, unless high bending moments are also likely to occur, prestressed concrete would be unsuitable and reinforced concrete should be used with the steel acting in compression. For columns, transverse steel in the form of links is essential to contain the longitudinal steel and ensure ultimate resistance to strains in excess of those causing failure of plain concrete. Evidence from long-term tests also shows that the effect of creep of the concrete in a column under load is to raise the stress in the longitudinal steel to its yield stress and hence there is a need to retain it in its correct alignment. Walls when lightly reinforced are slightly weaker than walls without reinforcement and they can therefore only be treated as reinforced when the longitudinal reinforcement exceeds a specific minimum.

Other aspects of behaviour which are of importance are shear and torsion. In each case if these cause failure, the mode of failure tends to be brittle and less ductile than bending failures. Hence in design, the procedure is to avoid such failure by the inclusion of sufficient transverse reinforcement to ensure bending or compression failure in the event of severe overloading.

Members subjected solely to tension are relatively rare. If they are of reinforced concrete, then the role of the concrete is to protect the reinforcement which is designed to take the whole tensile force. In prestressed members, however, the precompressed concrete can sustain the tension until the load exceeds the cracking loading when the behaviour reverts to that of reinforced concrete with the steel carrying the whole of the tension, stiffened to some extent between cracks by the concrete.

For most building structures, the Building Regulations define fire resistance requirements, which are expressed in terms of a required endurance under service load when components are subjected to a standard heating regime. Both reinforced concrete and prestressed concrete are primarily influenced in their behaviour in fire by the behaviour of the steel at high temperature; as its temperature is raised its strength and yield characteristics are reduced. For reinforcing steels the rate of reduction in strength is lower than for steels used in tendons and hence greater amounts of protection are needed for prestressed concrete. This may take the form of concrete cover and the optional addition of insulating material. It is often easier, however, to provide the greater thicknesses of cover needed for tendons without loss of efficiency than that needed for reinforcement, since the positioning of tendons is governed by different requirements.

The need to provide adequate durability also affects the amount of cover required to the reinforcement or tendons. As concrete ages, carbon dioxide in the air causes carbonation of the concrete which, as it progresses, reduces its capacity for inhibiting rusting of the steel. For dense concrete the rate of progress is very low but, since defects exist, experience has shown that a greater thickness of concrete is required to prevent spalling of the concrete caused by expansion of the corrosion products on rusting. Cover requirements also affect the width of cracks that are likely to occur and hence need attention in dealing with serviceability.

These characteristics of the behaviour of both reinforced and prestressed concrete are considered in more detail in presenting design procedures.

12.3 Philosophy of design

The early developments of the design of reinforced concrete were crystallized in this country by the issue in 1934 of Recommendations for a Code of Practice[1] prepared by a committee set up by the Department of Scientific and Industrial Research. It was based on the premise that the stresses in the steel and concrete should not exceed certain permissible values, related to the strengths of the materials by safety factors, when the structure was subjected to the maximum loads that it would need to carry in service. The materials were assumed to behave elastically and compatability of strains between steel and concrete was ensured by assigning a value for the ratio of their moduli of elasticity. Some account was taken of the inelastic effects of creep of concrete by adopting a low value for the modulus of elasticity of concrete in determining the modular ratio for use in the design calculations. No account was taken of the effects of shrinkage and no estimate was made of the ultimate strength of the structure. When the British Standards Institution issued its first Code for Reinforced Concrete, CP 114,[3] in 1948, it followed the same general approach. In the revision in 1957, however, there was an alternative method for design in flexure which limited the stresses to the same permissible values as for elastic design but assumed that they were distributed as at failure and avoided the use of the modular ratio; this was therefore a form of ultimate strength design.

Limitations on the permissible stresses in the steel and on span:depth ratios were imposed to guard against excessive deflection or cracking. Thus it could be argued that CP 114 provided for safety against failure and for the avoidance of unserviceability.

The earliest formal presentation of a design procedure for prestressed concrete was contained in the First Report on Prestressed Concrete[4] published by the Institution of Structural Engineers in 1951. Many of the recommendations in that report found their way into the British Standard Code of Practice for Prestressed Concrete, CP 115,[5] issued in 1959. It conformed with CP 114 in the sense that it was based primarily on the limitation of stresses to permissible values related to the strengths of the materials with the object of preventing cracking and avoiding excessive deflection. It also provided for the calculation of ultimate strength and introduced separate requirements for minimum load factors for the dead and imposed loads.

Thus, when the drafting of CP 110[7] commenced in 1964 it had already been demonstrated that there were a number of limiting conditions or limit states which had to be considered by the designer in the overall conception of structural safety and adequacy. These were primarily limits of collapse, deformation and cracking, but other matters such as the effects of vibration, of fatigue, of deterioration with time or as a result of fire, needed attention in the design process.

A further major change in the content of structural codes first introduced in CP 110 in 1972 was the move towards considering the coordinated design of the structure as a whole for safety and serviceability rather than the separate design of its component parts with only limited appreciation of their interaction. This development has become necessary partly as a result of the evolution of design philosophy and partly because the utilization of the materials has become more onerous following the general increase in the levels of stress in both concrete and steel under service conditions.

12.3.1 Criteria for limit state design

The aim in limit state design is to codify the procedures normally adopted by engineers in the design of structures to provide safe, serviceable and economic construction with a reasonable degree of certainty, and to do this with a better appreciation of the margins of safety and of ignorance involved. As far as possible, it takes into account the variations likely to occur in the loads on the structure and in the strength of the

materials of which it is comprised; it can allow for inadequacies of construction and methods of analysis, and should lead to design being more closely related to the risk of occurrence of specific conditions of failure and unserviceability.

For the purposes of design, both loads and strengths are expressed in terms of characteristic values. For loads, these are defined loads with a small but acceptable risk that they will be exceeded in service; they are given in the British Standard loadings for buildings,[13] in BS 5400 for highway bridges and in other standards for other construction. To meet the needs of limit state design, there has been a move in recent years away from specifying loads as maximum values and towards expressing them in terms of their likelihood of occurrence where possible determined from observations of their imposition on structures (see Chapter 19).

The characteristic values of loads allow for normally expected variations in loading but not for: (1) unforeseen loading effects; (2) lack of precision in design calculations; (3) inadequacies in the methods of analysis; and (4) dimensional errors in construction which alter the assumed positions or directions of loads and their effects, e.g. incorrect positioning of reinforcement and inaccurate alignment of columns in successive storeys. The values for loads used in design are therefore increased by partial safety factors to cater for these effects and to provide the margin of safety appropriate to the need for ensuring that a particular limit state is not reached. Thus, for conditions of failure, higher values are used than for those of serviceability. Where a combination of loads is assumed to be acting, the partial safety factors for each source of loading are smaller since the simultaneous occurrence of high values for each load is less likely. The loads for use in the design are therefore the sums of the products of the appropriate characteristic loads and their partial safety factors for the limit states and combination of loads being considered. For simplicity, the structural code for concrete, BS 8110,[10] reduces the number of situations needing consideration to a minimum, as will be seen later.

Characteristic values for the strengths of materials are usually given in the relevant standard or code. Research on materials shows that their strengths conform reasonably closely to a normal distribution, and their characteristic strengths can therefore be stated as follows:

Characteristic strength = mean strength $- k_1 \times$ standard deviation or:

$$f_k = f_m - k_1 \times \sigma_f \tag{12.1}$$

k_1 is usually given a value of 1.64, which ensures for a normal distribution that not more than 5% of strengths are less than the characteristic strength. This definition of strength has been adopted in British Standards for both steel and concrete.

The magnitude of the loads used in design is therefore increased by factors, partial safety factors for loads, to cater for these effects and to provide a margin of safety appropriate to the need for ensuring that any particular limit state is not reached. Thus, when envisaging conditions of failure, higher values for the factors are adopted than when considering serviceability.

The strengths of the materials used in the design calculations are those defined in the specification for the structure, which are checked by physical tests. The strengths of the materials as they exist in the structure, however, are likely to differ from those determined from test specimens and some allowance is also required for changes or deterioration with time. Partial safety factors for the materials are therefore introduced and the strengths taken for design are the characteristic strengths divided by a partial safety factor, γ_m, which has a value depending on the limit state being considered and the nature of the material, being less for steel than for concrete.

An idealized and simplified situation for a homogeneous material is illustrated in Figure 12.3. The provisions for safety outlined so far then require:

$$F_k \cdot \gamma_f \leqslant f_k / \gamma_m \qquad (12.2)$$

where F_k = the characteristic load

This conforms reasonably closely with what has now become accepted practice in the recent revisions of British Standards codes, and was first adopted in CP 110 in 1972. Current thought, however, accepts the view that a further partial safety factor should be introduced to take account of the nature of the construction and its behaviour under overload conditions, e.g. whether it is capable of sustaining large deformations and so giving warning of the imminence of collapse, and of the seriousness of failure in terms of the risks to health, life and property. This factor, γ_c, might have a value of less than 1 for temporary construction not normally occupied by human beings but of more than 1 for buildings with large spans used for public assemblies. Thus design would then require:

$$F_k \cdot \gamma_f \cdot \gamma_c \leqslant f_k / \gamma_m \qquad (12.3)$$

For an idealized situation, the global factor of safety relating characteristic loads to characteristic strength is then $\gamma_f \cdot \gamma_c \cdot \gamma_m$.

If the concept of relating the factors of safety to the nature of the construction is not followed, then the global factor is $\gamma_f \cdot \gamma_m$. Since reinforced concrete and prestressed concrete are composite materials, the value of the global factor for each limit state cannot be expressed as simply as this; it is dependent on the interaction between steel and concrete, each of which has a different value for γ_m. Also, γ_f cannot be given a single value for each limit state since the partial safety factors for dead, imposed, wind and other loads may differ and change with different combinations of loads. Hence, only upper and lower values for the global factor can be defined which makes comparison of the new Code with earlier or other codes imprecise. Nevertheless, in preparing the new Code, the aim has been to avoid substantial changes in the dimensions of the resulting structures whilst at the same time obtaining more consistent levels of safety and leaving room for development on more rational lines in the future.

It is convenient to divide the limit states to be considered in design into two kinds, namely those concerned with collapse and those concerned with serviceability. Limit states of collapse deal with overturning of the complete structure, failure of the whole or a large part of the structure as a result of overstressing of a number of sections or buckling of a number of compression members or as a result of a serious accident; the effects of fire and fatigue may also be included. Deflection, cracking, deterioration, corrosion and vibration are all aspects of serviceability and require limits of acceptability to be set for consideration. In the Code for Structural Concrete the limit states specifically dealt with are ultimate conditions in general, and deflection and cracking under the heading of serviceability. The criteria defining the serviceability limits are set out in Table 12.2.

The partial safety factors γ_f to be used with the characteristic loads for dead, imposed and wind loads obtained from CP 3, Chapter V, or other appropriate specification, are set out in Table 12.3 with notes on interpretation for ultimate and serviceability limit states. The combinations of loading to be taken are those which create the most severe conditions within the limits specified.

The partial safety factors for materials, γ_m, for the limit states considered are given in Table 12.4 also with notes on their interpretation.

The Code for Structural Concrete has special provisions to satisfy the requirement that, when a building suffers accidental damage, the amount of damage caused shall not be inconsistent with the original cause. It would seem reasonable to apply this same approach to other structures where safety and avoidance of excessive damage are necessary considerations in the event of accidents. To achieve this in buildings, attention should be given to the choice of an appropriate plan form since this may have a large influence on the mode of collapse as a result of an accident. When it is necessary to consider the effects of excessive loads outside those normally likely to be experienced or the residual strength of a structure after accidental damage the value of γ_f

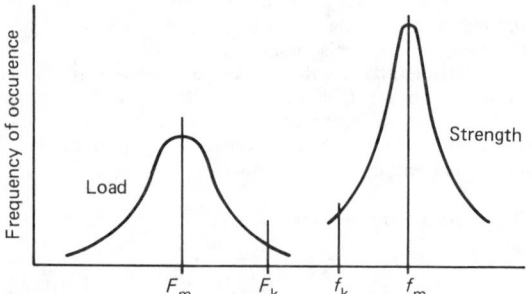

Figure 12.3 Idealized relationship between load and strength for a structure

Table 12.2 Limits for serviceability conditions

Limit state	Reinforced concrete	Prestressed concrete
Cracking	Controlled by detailing rules for sizing and spacing reinforcement (BS 8110, Part 1) For unusual structures or conditions, more specific recommendations are given in BS 8110, Part 2	Class 1: No flexural tensile stress Class 2: Flexural tensile stresses permitted but no cracking Class 3: Nominal flexural tensile stresses adopted to limit cracking to not more than 0.1 mm for severe exposures, e.g. sea water and otherwise not more than 0.2 mm (BS 8001, Part 1)
Deflection	Normally controlled by rules for span:depth ratio (BS 8110, Part 1). Exceptionally (BS 8110, Part 2), under vertical loads, not more than span/250: or for brittle finishes and partitions, not more than span/500 or 20 mm, or for nonbrittle materials, not more than span/350 or 20 mm	Normally controlled by limitations of stresses under service loadings for cracking considerations (BS 8001, Part 1) Exceptionally (BS 8001, Part 2), as for reinforced concrete but also applied to upward deflections

Table 12.3 Partial safety factors γ_f for loads and load effects

	Limit state design loads[a]		
	Ultimate		Serviceability
Load combination			
Dead and imposed load	$\begin{Bmatrix} 1.4\,G_k + 1.6\,Q_k \\ 1.0\,G_k \end{Bmatrix}$	See note (1)	$\begin{Bmatrix} 1.0\,G_k + 1.0\,Q_k \\ 1.0\,G_k \end{Bmatrix}$
Dead and wind load	$\begin{Bmatrix} 1.0\,G_k + 1.4\,W_k \\ 1.4\,G_k + 1.4\,W_k \end{Bmatrix}$	See note (2)	$1.0\,G_k + 1.0\,W_k$
Dead imposed and wind load	$1.2\,G_k + 1.2\,Q_k + 1.2\,W_k$		

[a]The figures given in the table are the values for the partial safety factors γ_f.

Notes:

(1) The minimum load for this combination should not be less than $1.0\,G_k$. When alternate spans are considered loaded in the design of continuous beams, for example, the loaded spans should be assumed to carry $1.4\,G_k + 1.6\,Q_k$ and the 'unloaded' spans to carry $1.0\,G_k$.

(2) The most serious load condition will usually occur when the design dead load is taken as $1.0\,G_k$, but for certain cantilevered structures, for example, a more serious situation may exist when the design dead load for part of the structure is $1.4\,G_k$.

Table 12.4 Partial safety factors for materials, γ_m

Material	Limit state values for γ_m		
	Ultimate	Serviceability	
		Deflection	Cracking
Concrete[e] in bending and compression	1.5[a]	1.0[c]	1.3[d]
Steel	1.15[b]	1.0	1.0

[a] This value is related to the standards of workmanship and supervision advocated in the code for the production of concrete. If these standards are not applied, a higher value should be used. It relates primarily to compressive strength of concrete.

[b] This value is for reinforcement in tension or tendons. For reinforcement in compression it is increased to $1.15 + f_y/2000$.

[c] Calculations of deflection are based on the characteristic strength of the material and therefore the modulus of elasticity of concrete derived for this strength is less than the mean value for the component or structure which strictly speaking would be more relevant. This slightly conservative approach is justified in the interests of simplicity.

[d] This higher value for γ_m is selected for all calculations of stress for class 2 prestressed concrete.

[e] For shear without reinforcement, γ_m should be 1.25 and for bond at the ultimate limit state should be 1.4.

can be taken as 1.05 for those loads likely to be experienced. In these circumstances also, the values for γ_m for steel and concrete may be taken as 1 and 1.3 respectively. The wind loading should be taken as one-third the characteristic wind load. These low values for the factors are acceptable because the loading considered will not be experienced by most buildings and it would therefore be uneconomic to design for it to be sustained without damage.

12.3.2 Characteristics of materials

The grades of concrete used for reinforced and prestressed concrete construction in the Structural Concrete Code are expressed as the characteristic strengths determined from 28-day tests on cubes; they are given in Table 12.5 with their application and properties relevant to design, including the increase in cube strength with age. No data are given for lightweight aggregate concrete since its properties are dependent

on density in addition to strength as well as on the type of aggregate. The figures for flexural and indirect tensile strength refer to concretes made with smooth gravel aggregates; for crushed rock aggregates of tough texture, tensile strengths for the same grades of concrete would be somewhat higher. Generally, the minimum grade of concrete for reinforced concrete will be grade 25; there are, however, areas in Britain where the natural aggregates are not of high enough quality for concrete to meet this grade even though its cement content is sufficient to conform with requirements for durability. Unless there are special needs, grades stronger than grade 40 are unlikely to be used for reinforced concrete. When lightweight aggregate is used, a lower grade, grade 15, is acceptable for reinforced concrete but it is preferable to use a higher grade for the lightweight aggregates of higher strength. No upper limit needs to be set on the strength for prestressed concrete and higher grades than grade 60 may therefore be used, but only special circumstances would justify the much greater cost and need for control and supervision.

Calculations for conformity with ultimate and serviceability limit states require the strength and deformation characteristics for concrete to be defined in numerical terms. In particular, data are required on the relationships between stress and strain in compression under short-term loading and on creep and shrinkage when serviceability in the longer term is being considered. These aspects of behaviour are dealt with in section 12.4 and are simplified for design later in this section, but it must be recognized that there are substantial variations in the behaviour of concrete, depending on its constituent materials and environment, and that the values given for calculation should only be adopted if more reliable data are not available.

The strength properties of steel reinforcement and steel tendons are defined in British Standards which are summarized in Tables 12.6 and 12.7. For reinforcing bars of hot-rolled steel the characteristic strength is derived from the yield stress, but for cold-worked bars or wire reinforcement, it is derived from the 0.2% proof stress. The characteristic strength of tendons for prestressed concrete, however, is derived from their ultimate tensile strengths. In each case these are the relevant strengths for calculating ultimate strength for structural concrete members. Also in each case, the conformity with the specified characteristic strength is determined by ensuring that not more than two in forty consecutive results of tests made during the production of the steel falls below the specified value.

The design calculations for serviceability of structural concrete require information on the modulus of elasticity of steel.

Table 12.5 Grades and properties of structural concrete

Grade[a] (characteristic strength 28 days) (N/mm²)	Cube strength[a] (N/mm²) at the age of:						Flexural strength at 28 days (N/mm²)	Indirect tensile strength at 28 days (N/mm²)	Modulus[a] of elasticity at 28 days (kN/mm²)	Use[a]
	7 days	28 days	2 mths	3 mths	6 mths	1 year				
15	—	15	—	—	—	—		—	—	Reinforced concrete with lightweight aggregate
20	13.5	20	22	23	24	25	2.3	1.5	24	
25	16.5	25	27.5	29	30	31	2.7	1.8	25	Reinforced concrete with natural dense aggregates
30	20	30	33	35	36	37	3.1	2.1	26	Prestressed concrete for post-tensioning, ≮ 15 N/mm² at transfer
40	28	40	44	45.5	47.5	50	3.7	2.5	30	Prestressed concrete for pretensioning, ≮ 15 N/mm² at transfer
50	36	50	54	55.5	57.5	60	4.2	2.8	32	

[a]Recommendations in the Code of Practice for Structural Concrete.

The values adopted in the new code are: for reinforcement for all types of loading 200 kN/mm², and for short-term loading for wire and strand of small diameter 200 kN/mm² and for alloy bars and strand of large diameter 175 kN/mm².

In prestressed concrete, considerations of serviceability require allowance not only for the effects of creep and shrinkage of the concrete but also relaxation of the tendons which may modify the prestress conditions substantially. Appropriate requirements are incorporated in the standards which therefore provide guidance on values for relaxation to be used in design.

The stress–strain characteristics for concrete and steel may be needed for calculations of the deformation of structural members under short-term loading or for assessing ultimate strength.

These are given in Figure 12.4 for concrete, in Figure 12.5 for reinforcement and in Figure 12.6 for tendons.

In interpreting these curves, the value of γ_m appropriate to the limit state being considered should be obtained from Table 12.4. The values for the modulus of elasticity given in these figures should not be used for estimating the required extension of tendons. These data should be obtained from stress–strain curves for actual material being stressed, which are supplied by the manufacturers.

The creep and shrinkage characteristics of concrete are considered in section 12.4. Where it is necessary to calculate long-term deformation, the effects of creep can be conveniently allowed for by adopting an effective modulus:

Table 12.6 British Standards for reinforcing bars for concrete

Type of steel[a]	Specified characteristic strength[b] (N/mm²)	Elongation at fracture (%)	Diam. for 180° bend test (no. of bar diam.)	Upper limit for: carbon content (%)	sulphur content (%)	phosphorus content (%)
BS 4449:1978 Hot-rolled steel bars for the reinforcement of concrete						
250 Grade	250	22	2	0.25	0.06	0.06
460 Grade	460	12	3	0.40	0.05	0.05
BS 4461:1978 Cold-worked steel bars for the reinforcement of concrete						
460 Grade	460	12	3	0.25	0.06	0.06
BS 4482:1969 Hard-drawn mild steel wire	485	—	Rebend test	0.25	0.06	0.06

Notes: [a]Preferred sizes:

BS 4449 }
BS 4461 } 8, 10,12,16, 20, 25, 32 and 40 mm diameter, if a smaller size is required then 6 mm, if larger 50 mm diameter

BS 4482 – 5, 6, 7, 8, 10 and 12 mm diameter.

[b]The characteristic strength is the yield stress below which not more than 5% of results should fall

Table 12.7 British Standards for prestressing tendons for concrete

Type of steel	Range of sizes available (dia. in mm)	Range of specified characteristic breaking load (kN)	Other information					
BS 5896:1980 High-tensile steel wire and strand for the prestressing of concrete								
Cold-drawn steel wire in mill coils	3–5	12.2–30.8	Relaxation: 1000 h					
			60%	breaking load*			8%	
			70%				10%	
Stress relieved and may be crimped or indented and treated to reduce relaxation	4–7	21.0–64.3	Relaxation:1000 h					
					Class	1	2	
			60%	b.l.		4.5%	1.0%	
			70%			8.0%	2.5%	
			80%			12.0%	4.5%	
Strand seven wire stress-relieved			Relaxation: 1000 h					
					Class	1	2	
standard	9.3–15.2	92–232	60%	b.l.		4.5%	1.0%	
super	8.0–15.7	70–265	70%			8.0%	2.5%	
drawn	12.7–18.0	209–380	80%			12.0%	4.5%	
BS 4757:1971 Nineteen wire strand for the prestressing of concrete								
as spun strand normal relaxation	25.4–31.8	659–979	Relaxation: 1000 h 60%	b.l.	9%,	70%	b.l.	14%
strand	18.0	370	60%		7.0%,	70%		12%
low-relaxation strand	18.0	370	60%		2.5%,	70%		3.5%
BS 4486:1980 Hot-rolled and hot-rolled and processed high-tensile alloy steel bars for the prestressing of concrete								
hot-rolled	20–40	325–1300	Relaxation: 1000 h all bars 60%	b.l.				1.5%
hot-rolled and processed	20–32	385–990	70%					3.5%
			80%					6.0%

*The breaking load for relaxation testing is the actual breaking load

Note: These values for relaxation at 1000 h apply in temperate climates and are those obtained at 20°C. When the prestressing steel is used at higher temperatures to prestress concrete, these percentage values should be increased. The increase may be as much as 2% for each 10°C increase in temperature.

$$E_{c,\text{eff}} = E_{ci}/(1 + \phi_c E_{ci})$$ (12.4)

where E_{ci} is the short-term modulus of elasticity of concrete and ϕ_c is the creep of concrete under a unit stress of 1 N/mm^2.

The effects of shrinkage may be treated by assuming that the concrete contracts without a change in stress except for that caused by the effect of the change in strain on the stress in the steel. Some readjustment of strains then becomes necessary to balance the forces in the cross-section by assuming that the concrete is stressed under this strain in proportion to the effective modulus.

12.4 Analytical and design procedures

12.4.1 Objectives

A recent trend in the approach to the initial design is to place much greater emphasis on the requirements for the durability of construction, since experience has shown that deterioration is a more serious cause of failure and of high maintenance costs than shortcomings in the structural calculations. It has therefore become more necessary to treat compliance with requirements for the quality of concrete, as placed in the construction and for

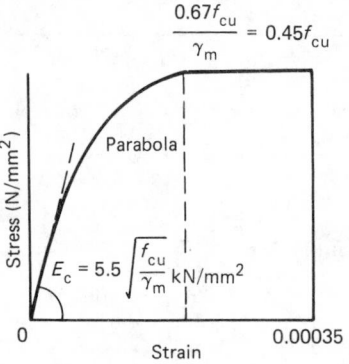

Figure 12.4 Short-term stress–strain curve for concrete

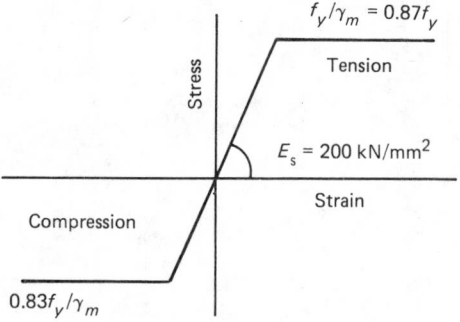

Figure 12.5 Short-term stress–strain curve for reinforcement

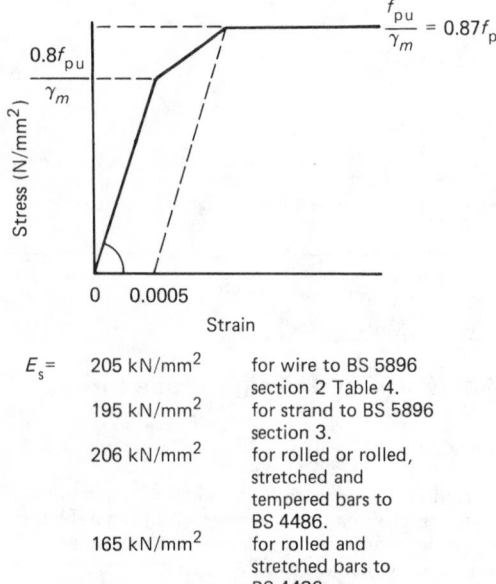

$E_s =$	205 kN/mm^2	for wire to BS 5896 section 2 Table 4.
	195 kN/mm^2	for strand to BS 5896 section 3.
	206 kN/mm^2	for rolled or rolled, stretched and tempered bars to BS 4486.
	165 kN/mm^2	for rolled and stretched bars to BS 4486

Figure 12.6 Short-term stress–strain curve for normal and low relaxation tendons

the protection of embedded steel by adequate concrete cover or other means, as being at least as important as compliance with requirements derived from design calculations. Whilst it is not practicable to define requirements for durability in terms of a limit state, it is nevertheless an aspect of the overall design process requiring primary attention.

For somewhat similar reasons, more care is now given to the requirements for fire resistance and information is presented in the Code (Part 2) which can be used for an analytical approach as an alternative to satisfying somewhat arbitrary requirements for concrete cover in order to obtain the necessary fire grading.

Inevitably, structural calculations continue to be a major part of design. Whilst the principles of limit state design require all possible limit states to be examined in the design of a particular structure, part of the purpose of the Code is to provide guidance on containing the effort required in design within reasonable limits without overlooking significant features, i.e. limit states. In doing this, the Code relies on the experience of the designer to ensure that the interpretation is sensible in each instance.

12.4.2 General assumptions

For most forms of concrete construction, with the possible exception of slabs, it is most convenient at the present time to base all design on elastic analysis of the structural system. The analysis would then apply directly to the serviceability limit states of deflection and cracking and, with some limited redistribution of moments and shear forces to the ultimate state. For slabs, other than one-way spanning slabs, it will usually be more satisfactory to use yield line methods or the strip method for ultimate design. For most construction it will usually be preferable to determine conformity with the serviceability limit states by using the arbitrary rules given in the Code for span:depth ratios and reinforcement detailing instead of calculating deflections and widths of cracks.

The Code recommends procedures in Part 1 for the detailed design of beams, solid slabs supported by beams or walls, flat slabs, columns, walls, staircases and bases, which are given at some length, and generally apply to both reinforced concrete and prestressed concrete. More information dealing with ultimate strength, serviceability and deformation due to creep, shrinkage and temperature effects is contained in Part 2. In this relatively brief summary, it is only possible to cover the more basic recommendations, and detailed design therefore requires reference to the main documents.

Other methods of analysis and design, where experimental procedures are used to develop the theoretical approach or to determine performance, are acceptable but will normally only be employed for specially complex structures or where repetition justifies more refinement than is obtained by established methods of calculation. The assessment of stresses in the region of load concentrations or of holes in continuous construction may be determined by photoelastic procedures. Model testing using special materials or scaled concrete has found applications in developing design methods, e.g. in the design of concrete box-girder bridges and pressure vessels for nuclear power stations. In precast concrete construction particularly, the behaviour of joints can only be established by tests on full-scale assemblies. It may also be economic to derive the dimensions of precast components for mass production by testing successively refined prototypes to obtain the final form; this approach applies particularly in dealing with the requirements for fire resistance. The interpretation of test data for design requires the special care of experienced engineers since tests cannot embrace all the loads and load effects that may need to be sustained and the circumstances that exist in actual structures cannot necessarily be fully reproduced experimentally. When test results are applied, therefore, there is a need to show convincingly the justification for departures from established practice, especially so, if these lead to less conservative design. If test data are applied in contexts for which they were not originally sought even more caution is necessary.

For the purpose of analysis, the Code offers three alternative methods for estimating beam and column stiffness: (1) the concrete section; (2) the gross section; and (3) the transformed section. The concrete section is the whole concrete section excluding the reinforcement, the gross section is the whole concrete section including the reinforcement allowing for the modular ratio, usually taken as 15, and the transformed section is the section of concrete in compression together with the reinforcement again allowing for the modular ratio. Generally, the concrete section is most convenient for use in design. For checking existing structures or for design in special circumstances, it would be more appropriate to use the transformed section for reinforced concrete; in construction where flexural cracking has occurred, however, the actual stiffnesses obtained by this assumption will be greater since the concrete exerts some tensile stiffening in the regions between cracks through bond with the reinforcement. The appropriate section for checking the design of existing or special prestressed concrete structures is the gross section since cracking does not usually occur with elastic deformation under service loads even for class-3 prestressed concrete.

12.4.3 Robustness

The Building Regulations require that all buildings of more than four storeys in height should be designed to resist accidental damage. The Regulations require that these buildings should be capable of sustaining removal of a structural member without excessive collapse resulting or should be able to withstand an internal pressure of $34 \, \text{kN/m}^2$ without collapse.

The layout of the structure and its general form should not be sensitive to accidental damage whatever the cause. It is more realistic to interpret this as meaning that, in the event of an accident, the resulting damage should not be disproportionate to the magnitude of the cause. Where impact from vehicles is a possibility, buildings should be protected by barriers, such as bollards or earth banks. Greater margins should be allowed in design when the occupancy of a building may result in a greater than normal risk of accident, e.g. in flour mills and bonded stores.

Provisions envisaged in the Code go further in some respects in dealing with the effects of accidents than the Regulations require. The recommendations for robustness deal with both expected and accidental forms of loading, and include the following:

(1) All buildings should be so designed that all dead, imposed and wind loads are safely transmitted to the foundations.
(2) All buildings should be capable of withstanding a horizontal design ultimate load applied at roof and each floor level simultaneously corresponding to 1.5% of the dead-weight of the structure between the mid height of the storey below and mid height of the storey above for floors, and the surface for the roof. This, in effect, sets a lower limit for wind loading for the first two combinations of loading in Table 12.3.
(3) All buildings should be tied with effectively anchored and continuous reinforcement which is capable of withstanding the notional forces outlined in the following paragraphs. This reinforcement may consist of bars provided to resist stresses due to normal loads, which may be ignored for this purpose, and it may be assumed to be stressed up to its characteristic strength.

Buildings of four storeys or less require tying horizontally

in two directions approximately at right angles with internal ties and peripheral ties.

Internal ties, which should be anchored to the peripheral ties and should be accommodated in the beams or slabs, should be capable of resisting a notional force of: $[(g_q + q_k)/7.5](\rho/5)F_t \, \text{kN/m}$ or $1F_t \, \text{kN/m}$ width, whichever is the greater, where $(g_q + q_k)$ is the sum of the average characteristic dead and imposed load in kilonewtons per square metre, and ρ is the greater of the distances in metres between the centres of supporting columns, frames or walls of any two adjacent floor spans parallel to the tie. F_t is the lesser of $(20 + 4n_0)$ or 60, n_0 being the number of storeys. The spacing of the ties should not be more than 1.5ρ.

Peripheral ties should be provided at each floor and roof level and be capable of withstanding a notional force of not less than $1F_t \, \text{kN}$. They should be located within 1.2 m of the edge of the building.

Horizontal ties to external columns and walls should be provided for each external column, in two directions for corner columns, and for each metre length of external wall at each floor and at roof level. The notional force considered should be the greater of the following: $2F_t$ (or $(F_t/2.5) \times$ ceiling height m) kN, or 3% of the total design ultimate load carried by the column or wall at that level.

Buildings of five storeys or more require additional provision for robustness, which usually will be met by the inclusion of vertical ties in all walls and columns. These should be designed for a notional force corresponding to the maximum design ultimate dead and imposed load received by the column or wall from any one storey or roof.

(4) Where there are key elements in a building design, the failure of which might cause extensive collapse, their design should take their importance into account if their use cannot be avoided. Where vertical ties cannot be provided (see (3) above), provision should be made for bridging by the structure above in the event of their removal.

The purpose of these recommendations is to ensure that all structures are insensitive to damage from localized disturbances. It is therefore important in providing ties, for bridging or any other action, that the arrangements are sound engineering.

12.4.4 Beams and slabs

The effective span (l) of beams or slabs, which are simply supported, is taken as either the distance between the centres of bearings or the clear distance between supports plus the effective depth, whichever is the smaller. For continuous members, however, the effective span is the distance between the centres of the supports. Whilst for a cantilever which forms part of a continuous beam or slab, it is to the centre of the support, but for an isolated cantilever the effective span is to the centre of the support plus half the effective depth.

The effective width of a flange to a T-beam may be taken as the smaller of the width of the web plus one-fifth of the distance between points of zero moment or the actual width. Similarly, the effective width of flange for an L-beam is taken as the smaller of the width of the web plus one-tenth of the distance between points of zero moment; for continuous beams the distance between points of zero moment may be assumed to be $0.7 L$.

The lateral stability of beams may need attention, usually by providing for adequate restraints and stiffness. The limits between lateral restraints for simply supported beams or continuous beams should not exceed $60b_c$ or $250b_c^2/d$, where d is the effective depth and b_c the breadth of the compression face midway between supports. For cantilevers restrained only at the support, its length should not exceed $25b_c$ or $100b_c^2/d$.

The following loading conditions should usually be considered in the design of continuous beams and slabs: (1) the design ultimate load of $1.4G_k + 1.6Q_k$ on all spans; and (2) the design ultimate load as (1) on alternate spans with $1G_k$ on intermediate spans. When moments at sections are determined by elastic analysis, the maximum moment may be reduced by redistribution provided that the calculated depth of the neutral axis is not greater than $(\beta_b - 0.4)d$ where d is the effective depth and β_b is:

$$\frac{\text{moment at the section after redistribution}}{\text{moment at the section before redistribution}} \not> 1$$

and that the resistance moment at any section is not less than 70% of the moment at that section from elastic analysis.

12.4.5 Continuous and two-way solid slabs

Slabs which are continuous in extent in one or two directions may be designed as simply supported, provided that continuous ties that may be required for overall stability of the structure are incorporated in the construction. In such cases, cracking will develop in the top surface of the floors at their supports and some provision will be needed for dealing with this in applying floor finishes.

Where slabs are required to span in one direction over a number of supports, they should be designed for moments and shears, calculated in similar manner to those for continuous beams.

If solid slabs are required to span in two directions, yield line analysis or the strip method of design may be used. British Standard 8110, however, gives simple methods for the design of rectangular slabs for simply supported two-way panels and two-way continuous or restrained slabs.

12.4.6 Flat slab construction

Flat slab construction usually consists of a slab which spans between columns in two directions without supporting beams. Drops may be provided over the columns by increasing the depths of the slab and sometimes the column heads may be flared to reduce shear stresses. The slabs may be solid or ribbed in two directions.

British Standard 8110 offers a method of design but does not exclude the use of other methods such as finite element analysis or other procedures. In the BS method, it is assumed that the slab is supported by a rectangular grid of columns in which the ratio of the longer spans to the shorter spans is not greater than 2. The slabs are divided longitudinally and transversely into column strips and middle strips; the columns and column trips are designed as frames spanning in each direction. Each frame is then analysed elastically; a simplified method is given for the situation where the structure is braced against lateral loading and the column grid has a regular layout. Procedures are given for determining the widths of column strips and for the treatment of drops.

12.4.7 Frames

The loads to be adopted in the design of frames with their factors have already been given in Table 12.3. When considering the ultimate limit state, the forces, shears and moments calculated for design should be the worst combinations of loading regarded as feasible. British Standard 8110 gives some simplified procedures, which may be used for a number of common forms of construction. These analyse frameworks by breaking them down into subframes and make some provision for

redistribution of moments. Two types of frame are dealt with – the no-sway frame, in which bracing, such as shear walls and lift or stair wells, are used to restrain sidesway, and sway frames, in which the frame itself provides the lateral restraint. For the latter, the amount of moment redistribution allowed is restricted with further restrictions on frames of four or more storeys in height to avoid excessive deflection and the possibility of frame instability.

12.4.8 Columns and walls

The determination of the loads and moments on columns is given in BS 8110 to which reference should be made for details. A column is described as slender when the ratio of the effective length to the corresponding breadth with respect to either axis is greater than 12 (10 for lightweight aggregate concrete); if the ratio is less than 12, the column is said to be short. The effective length is dependent on the length of the column and on the degree of restraint at the top and bottom connections with the structure. Generally, the slenderness ratio for a column should not be greater than 60. A distinction is made between braced and unbraced columns, a column being described as braced when the lateral stability of the whole structure is ensured by providing walls or bracing to resist all horizontal forces.

The procedures for dealing with walls in BS 8110 have much in common with those for columns. A concrete component is defined as a wall when the greater of the lateral dimensions is at least 4 times the smaller. For plain walls, however, the ratio may be less (since columns without reinforcement are not recognized) and reduction factors are then applied. To be described as a reinforced wall, the area of vertical reinforcement should not be less than 0.4% of the cross-sectional area of concrete; if the amount of reinforcement is less, the wall should be designed as a plain wall. Some reinforcement may be required in plain walls to control cracking. A stocky wall is one in which the ratio of effective length to thickness does not exceed 12 (10 for lightweight concrete), otherwise the wall should be treated as being slender. As for columns, the effective length is dependent on the height and conditions of end-restraint. Methods for calculating the loads and moments on walls (as for columns) are also given in some detail in BS 8110 to which reference should be made.

Provided the recommendations in the British Standards are followed, the deflections of columns and walls should not be excessive.

12.5 Reinforced concrete

12.5.1 General

In the design of reinforced concrete to meet the requirements of the Code, BS 8110, it will usually be most appropriate to consider the ultimate limit state first and then check the design against the requirements for cracking and deflection. This might be inappropriate in exceptional circumstances, e.g. where steels of characteristic strengths in excess of 500 N/mm² are being used or where spans were exceptionally long: in these cases cracking or deflection might govern design. In the sections that follow, design will be treated on the assumptions that normal conditions obtain. For these the Code gives simplified treatments for dealing with both cracking and deflection. It also gives methods more suited to the exceptional cases for which reference to the Code should be made.

12.5.2 Beams

12.5.2.1 Bending

Ultimate resistance in bending is calculated by assuming that:

(1) Sections which are plane before bending remain plane after bending.
(2) Stresses in the concrete may be determined using the stress–strain curve in Figure 12.4 (as assessed in the preparation of the design charts in Part 3 of the BS 8110), or may be taken as uniformly distributed across the most stressed 90% of the compression zone as indicated in Figure 12.7(a) with a value of $0.67f_{cu}/\gamma_m$, i.e. $0.45f_{cu}$ for deriving simplified formulae. Ultimate compressive strain in the concrete for analysis of sections is 0.0035.
(3) The strength of the concrete in tension is ignored.
(4) The stress in the steel is derived from the stress–strain relationships in Figure 12.5 with a value not greater than f_y/γ_m, i.e. $0.87f_y$ in tension and not greater than $0.83f_y/\gamma_m$ in compression, i.e. $0.72f_y$.

The simplified assumptions may be used to derive design formulae, which are shown in Figure 12.7(a–d). For beams reinforced in tension only:

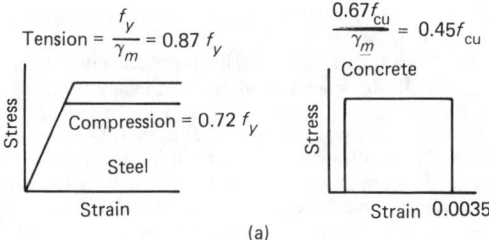

(a)

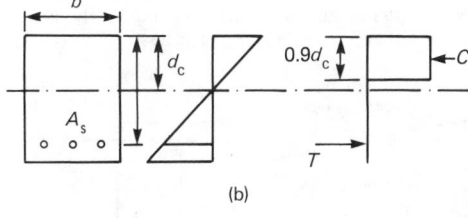

(b)

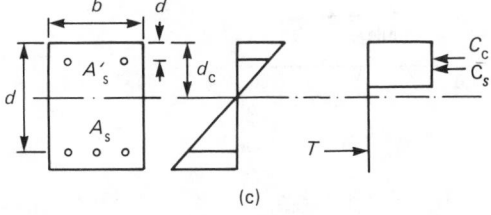

(c)

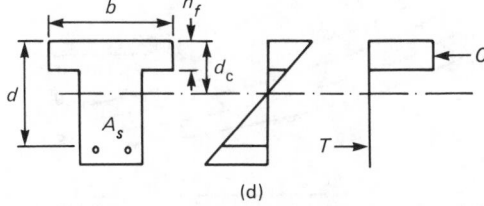

(d)

Figure 12.7 Flexural strength of beams – approximate methods.
(a) stress–strain curves assumed;
(b) beams reinforced in tension only;
(c) beams reinforced in tension and compression;
(d) flanged beams

$C = 0.4f_{cu}bd_c$ but not greater than $0.2f_{cu}bd$

T not greater than $0.87f_yA_s$ \hfill (12.5)

If d_c is not to exceed $0.5d$ as a practical limit, then:

$$M_u = 0.87f_yA_sd(1 - 0.97f_yA_s/f_{cu}b_d)$$

and not greater than $0.156f_{cu}bd^2$ \hfill (12.6)

For beams reinforced in tension and compression:

$C_c = 0.4f_{cu}bd_c$ but not greater than $0.2f_{cu}bd$ \hfill (12.7)

$C_s = 0.0035[(d_c - d_1)/d_c]A'_sE_s$ but not greater than $0.72A'_sf_y$ \hfill (12.8)

$T = 0.0035[(d - d_c)/d]A_sE_s$ but not greater than $0.87f_yA_s$ (12.9)

If d_c is not greater than $0.5d$ and d' is not greater than $0.5d_c$ where

$$d_c = [(T - C_s)/C_c]d \hfill (12.10)$$

then:

$$M_u = C_c(d - 0.45d_c) + C_s(d - d_1) \hfill (12.11)$$

For flanged beams:

If $h_f < 0.9d_c < d/2$ then

$$C = 0.45f_{cu}bh_f \hfill (12.12)$$

$$T = 0.87f_yA_s \hfill (12.13)$$

$M_u = 0.87f_yA_s(d - h_f/2)$ but not greater than $0.45f_{cu}bh_f(d - h_f/2)$ \hfill (12.14)

provided that moment redistribution is restricted to not more than 10%. For full moment redistribution, considered earlier, either the more complex stress–strain relationships should be used in the calculations or, more readily, the Code design charts should be employed. This also applies when the form of section cannot be readily dealt with by the simple formulae.

12.5.2.2 Shear

The resistance of beams in shear is calculated for the ultimate limit state. The procedure generally takes account of the contribution of the concrete as being additional to that of the shear reinforcement. The amount of shear reinforcement required is governed by the nature of the structural member and the level of shear stress in the concrete v in relation to the design shear strength of the concrete v_c. The shear stress, v, is given by

$$v = V/(b_v \cdot d)$$

where V = shear force due to ultimate loads, b_v = breadth of the section or the mean, breadth of the web for flanged beam, and d = effective depth \hfill (12.15)

The design shear strength of the concrete v_c is dependent on the strength of the concrete, the proportion of longitudinal reinforce-

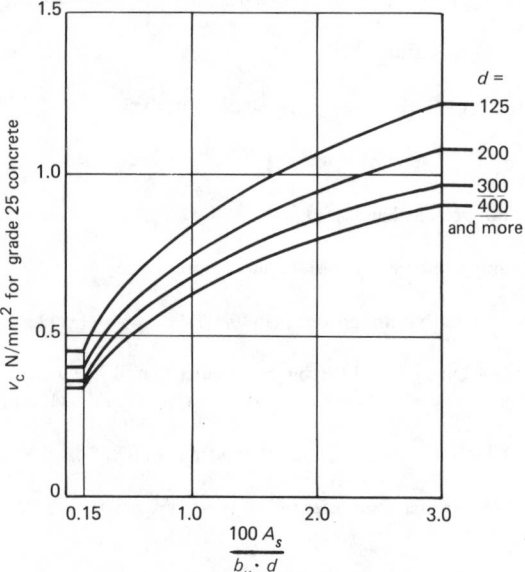

For other grades of concrete multiply v_c for grade 25 concrete

by:
    1.06 for grade 30 concrete
    1.12 for grade 35 concrete
    1.17 for grade 40 concrete and stronger grades.

Figure 12.8 Design shear stress for concrete beams v_c

ment and the effective depth. Values for v_c for grade 25 concrete are given in Figure 12.8 with factors for determining v_c for other grades.

The situations considered in the Code are as follows:

(1) Where v is less than $\frac{1}{2}v_c$ and members are of no structural importance, no shear reinforcement is required. If the members are of structural importance, minimum shear reinforcement, as in (2) should be provided.
(2) Where v is greater than $\frac{1}{2}v_c$ but less than $v_c + 0.4$, the area of reinforcement required A_{sv} should not be less than $(0.4b_v \cdot s_v)/0.87f_{yv}$

where s_v = spacing between links, and f_{yv} = characteristic strength of links $\not> 460$ N/mm^2.

The links should be positioned throughout the length of the beam, spaced not further apart than in (3).
(3) Where v is greater than $v_c + 0.4$ but less than $0.8\sqrt{f_{cu}}$ or 0.5 N/mm^2, whichever is the less (this limit is the limit for all beams), the amount of shear reinforcement required in the form of links is not less than:

$$A_{sv} = \frac{b_v v s_v (v - v_c)}{0.87 f_{yv}} \qquad (12.16)$$

The spacing of the links longitudinally should not exceed $0.75d$ and transversely not more than d with no tensile reinforcement more than 150 mm from the vertical leg of a link. Alternatively, up to 50% of these links may be replaced by bent-up bars, which should be bent up at an angle of not less than 45° with a longitudinal spacing of not more than $1.5(d - d_l)$ reduced correspondingly if the angle is increased.

12.5.2.3 Deflection

The accuracy of any calculation of deflection is dependent on the extent to which the conditions of loading are known both with respect to position and duration, and to which the assumptions made in design conform with the behaviour of the structure in reality. Apart from the dead load on the structure which may be known with reasonable accuracy, the imposed load that is actually applied may be unpredictable. The structure itself may have non-loadbearing components such as floor screeds and partitions which make a substantial contribution to its stiffness. The characteristics of the concrete may also not be known precisely since these are dependent on the different constituents actually used and provide additional uncertainty.

In most cases, therefore, it is not practical to calculate long-term deflections and this is recognized in the Code by giving a method of complying with the limit state requirements for deflection which take a number of features into account and define limits for span:depth ratios.

In defining these limits it is assumed that deflection of beams is primarily influenced by the conditions of support, the shape of the section, the proportions of tension and compression reinforcement in the section and their levels of stress under service loading. These features are dealt with by introducing modifying factors given with the basic span:depth ratios in Figure 12.9(a–d). To determine the limiting span:depth ratio for spans up to 10 m, a value for the ratio is obtained from (a), which is multiplied by the modification factor for tension reinforcement from (c) and, if appropriate, by the modification factor for compression reinforcement from (d). Figure 12.9(b) is used to derive the service stress in the steel required in the use of (c). The values given were developed for the Code to meet the requirement that the total deflection will not exceed span/250 and that the deflection after completion of finishes and partitions will not exceed span/350 or 20 mm, whichever is less.

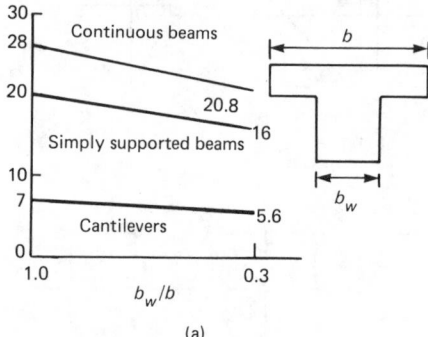

(a)

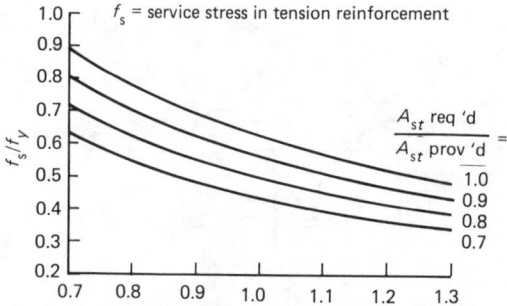

$$\beta = \frac{\text{moment at the section after redistribution}}{\text{moment at the section before redistribution}}$$

(b)

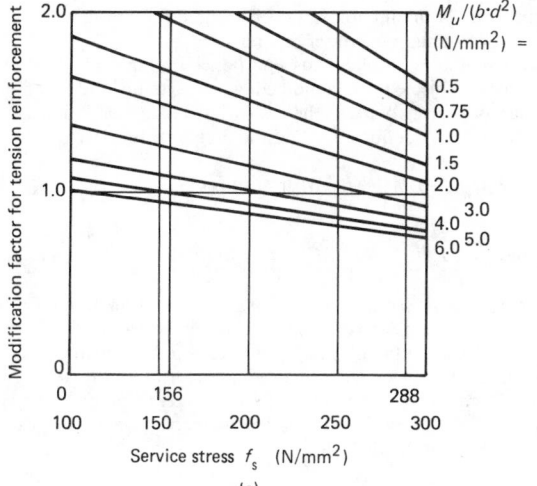

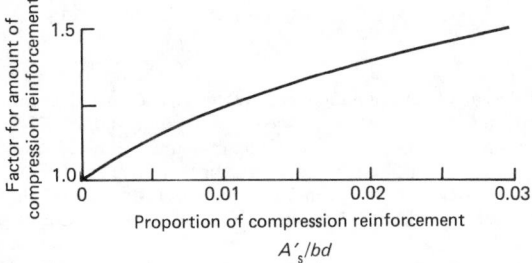

Figure 12.9 Factors for determining limiting span : depth ratios
(a) basic span: depth ratios;
(b) service stress for use in (c);
(c) modification factor for tension reinforcement;
(d) modification factor for compression
reinforcement

For spans greater than 10 m where limitation of deflection is not necessary to avoid damage to finishes and partitions, the limiting span:depth ratios obtained above may still be used, but if such damage is not acceptable the limiting span:depth ratio should be reduced by multiplying by a factor 10/span. For a cantilever with a span greater than 10 m, the deflection should be calculated as indicated in Part 2 of the Code.

12.5.2.4 Control of cracking

The width of cracks at a particular location in a flexural member is dependent on a large number of parameters of which the following have been found by experimental investigations to be the most important:

(1) The distance from the nearest reinforcing bar spanning the crack.
(2) The distance from the neutral axis of the section.
(3) The mean strain at the level of the section considered.

These investigations, which showed that the surface characteristics of the bars have only a relatively small effect, have led to the derivation of the formulae recommended in Part 2 of BS 8110 for use in special circumstances when the calculation of crack width is necessary. For most construction, satisfaction of the requirements for the cracking limit state is provided by meeting

the detailed needs for distribution of reinforcement in the concrete section with respect to location and spacing, which are dealt with later in sections 12.5.8 and 12.5.9.

In all construction, thorough moist curing of the concrete plays an important part in minimizing the extent of cracking due to drying shrinkage.

12.5.3 Slabs

The flexural strength of slab sections, including ribbed and flat slab sections, is treated in the same manner as for beam sections dealt with previously, design being based on derived moments. The moments, and also the shear forces, resulting from concentrated and distributed loads should be determined by elastic analysis or by yield-line or strip methods provided that these latter methods give a ratio of span-to-support moments similar to that obtained by elastic analysis. Rules are also given in the Code for the distribution of concentrated loads and for loading on slabs continuous over a number of bays when it is usually sufficient to assume that the most severe loading occurs with all spans fully loaded; this may not apply when cantilever spans are included. The design of two-way spanning slabs is covered in substantial detail with methods for calculating moments and shear forces and with requirements for the distribution of reinforcement between middle and edge strips and of reinforcement for resisting torsion at corners.

The shear stress in a solid slab should also be calculated as for a beam. The value of v for width b and an effective depth d is given by:

$$v = V/(b \cdot d) \qquad (12.17)$$

should not exceed the lesser of 5 N/mm² or $\sqrt{f_{cu}}$ whatever shear reinforcement is provided. The recommendations for design shear stress for beams v_c shown in Figure 12.8 also apply for solid slabs and the following situations are considered:

(1) Where v is less than v_c, no shear reinforcement is required.
(2) Where v is greater than v_c, but less than $v_c + 0.4$, minimum links are required with a cross-sectional area of A_{sv} of $(0.46b \cdot s_v)/f_{yv}$ where s_v and f_{yv} are the spacing of the links and yield stress of steel as for beams.
(3) Where v is greater than $v_c + 0.4$, the amount of shear reinforcement required in the form of links is not less than

$$A_{sv} = [b \cdot s_v \cdot (v - v_c)]/0.87 f_{yv} \qquad (12.18)$$

Alternatively, these links may be partly or completely replaced by bent-up bars. The spacing of links or bent-up bars need not be less than d.

Since it is difficult to bend and fix reinforcement for slabs with a depth of less than 200 mm, such slabs should be designed to avoid the need for shear reinforcement.

For most design, the deflection of solid slabs should be controlled by restrictions on span:depth ratio as for beams using the data in Figure 12.9(a–c). For two-way slabs, the span used in the calculations should be the shorter span.

Cracking is normally controlled by conforming with requirements for spacing reinforcement given on pages 12/19 to 12/21.

A convenient form of floor is the cast *in-situ* ribbed, hollow block or voided floor. The ribs may be connected by a structural topping of concrete of the same grade as that of the ribs or by a nonstructural topping not necessarily of the same grade. The ribs of floors with structural topping may be formed by solid or hollow blocks or formers, which can contribute to the structural strength provided that they are made of concrete or burnt clay (complying when appropriate with BS 3921) with a characteristic strength in the direction of compressive stress in the floor of 14 N/mm² or more. The spacing of *in-situ* ribs should not be greater than 1.5 m and their depth without topping should not be more than 4 times their width. The minimum thicknesses of structural topping required are related to the form of construction as follows:

(1) When the clear distance between ribs is not more than 0.5 m and permanent blocks are jointed with cement:sand mortar not leaner than 1:3 or weaker than 11 N/mm² – 25 mm.
(2) When the clear distance between ribs is not more than 0.5 m but the permanent blocks are not jointed with cement:sand mortar – 30 mm.
(3) All other slabs with permanent blocks – 40 mm or one-tenth of the clear distance between ribs whichever is more.
(4) All slabs without permanent blocks – 50 mm or one-tenth of the clear distance between ribs whichever is more.

If it is impracticable to provide sufficient reinforcement to develop the full support moment for continuous ribbed slabs, they may be designed as simply supported with not less than 25% of the mid-span reinforcement for the adjacent spans over the supports to restrict cracking; this reinforcement should extend for 15% of the span into the adjacent spans. When calculating the ultimate resistance moment of the section, the compressive stress in the blocks may be assumed to have a value of 0.3 times the specified characteristic strength. Design for shear follows that for solid slabs, the width of the section being taken as the width of the rib plus the thickness of the walls of hollow blocks or plus half the depth of the rib for solid blocks. The depth:span ratio of ribbed slabs should meet the requirements for the control of the deflection of beams; the thickness of the walls of hollow blocks may be added to the thickness of the ribs in making this check.

12.5.4 Columns

The Code draws particular attention to the need when commencing the design of columns to consider the dimensional requirements for cover for durability and for cover and minimum dimensions for fire resistance. Minimum amounts of reinforcement are given in Table 12.9.

Moments forces and shears in columns are derived by the analytical and design procedures considered earlier. For most construction with braced columns, i.e. where the structure is fully braced against lateral loading, the ratio of effective height to minimum breadth will not exceed 12 and the columns may be treated as short columns.

For short-braced axially loaded columns the ultimate load is derived from the assumptions made for beams and is given by:

$$N = 0.45f_{cu}A_c + 0.72A_{sc}f_y \quad (12.19)$$

but to allow for inaccuracy in construction it is reduced by about 10% to give the relationship in the Code:

$$N = 0.4f_{cu}A_c + 0.67A_{sc}f_y \quad (12.20)$$

where f_{cu} is the characteristic strength of concrete, f_y is the characteristic strength of steel in compression, A_c the area of concrete and A_{sc} the area of steel in compression.

If the braced short column has an approximately symmetrical arrangement (i.e. within 15% of span) of uniformly loaded beams, then loading may be treated as axial using a reduced value for N to deal with the small moments induced:

$$N = 0.35f_{cu}A_c + 0.60f_yA_{sc} \quad (12.21)$$

This formula should not, however, be used for unsymmetrically loaded columns, e.g. corner columns.

When these simplified assumptions are applied to short columns subjected to combined axial loading and bending about one or two axes, the following recommendations are made in the Code for adjusting the moments for design:

For M_x/h, M_y/b, $\quad M'_x = M_x + \beta(h/b)M_y \quad (12.22)$

and for M_y/b, M_x/h, $\quad M'_y = M_y + \beta(b/h)M_x \quad (12.23)$

where M_x and M_y are the estimated design ultimate moments about the x and y axes respectively, and M'_x and M'_y are the corresponding ultimate design moments for use in the design calculations; h and b are the overall dimensions of the rectangular columns at right angles to the x and y axes respectively, and β is a factor with values given below in relation to the axial ultimate design load, N.

N/bhf_{cu}	0.2 or less	0.3	0.4	0.5	0.6	0.7 or more
β	0.90	0.65	0.53	0.40	0.28	0.15

The design of long columns receives extensive coverage and the details are not readily amenable to abbreviation. Two situations are, however, considered, namely braced and unbraced columns, and for each category effective lengths are defined in terms of the conditions of end-restraint and the corresponding additional moments for use in design are developed.

The deflection of short columns (and braced long columns) do not need to be checked since they will normally be within acceptable limits. Cracks are unlikely to occur in columns when the design ultimate axial load is greater than $0.2f_{cu} \times$ the net cross-sectional area of the column: if bending predominates, cracking should be considered as for beams.

12.5.5 Walls

12.5.5.1 Reinforced concrete walls

A wall is usually defined as a vertical loadbearing member with a length exceeding 4 times its thickness. The method of design of of reinforced concrete walls is generally similar to that for columns, the treatment of stocky (effective length of 12 or less) and slender walls corresponding to that for short and long columns respectively.

Where a braced stocky wall cannot be subjected to significant moments, its ultimate load is given by:

$$N = 0.4f_{cu}A_c + 0.67A'_sf_y \quad (12.24)$$

This is the same as the formula for columns and includes a reduction to allow for the effects of constructional tolerances.

If the spans on either side of a wall do not differ by more than 15% and are uniformly loaded, then it may be assumed that loading is axial and:

$$N = 0.35f_{cu}A_c + 0.60A'_sf_y \quad (12.25)$$

12.5.5.2 Plain concrete walls

For stocky braced plain walls, the ultimate load per unit length, n_w is:

$$n_w = (h - 2e_x)_w f_{cu} \tag{12.26}$$

where e_x is the resultant eccentricity of load at right angles to the plane of the wall, * is a coefficient with a value of 0.3 reduced by a factor varying linearly between 1.0 and 0.8 as the length reduces from 4 to 1 times its thickness.

Reinforcement may be needed in plain walls to control cracking due to flexure or drying shrinkage; it should not be less in each direction than 0.25% for 460-grade, nor 0.30% for 250-grade, steel.

12.5.6 Bond and anchorage

Bond and anchorage as distinct from cracking in reinforced concrete are affected substantially by the surface characteristics of the reinforcement. The BS Code recognizes three types of bar surface, i.e. plain round bars, type 1 deformed bars (which are usually twisted bars of square or chamfered square cross-section) and type 2 deformed bars which are usually of round cross-section with transverse ribs.

Where there are rapid changes in stress in the longitudinal direction over a short length or changes in the depth of the section, excessive local bond stresses at ultimate should be avoided by making sufficient provision for the anchorage of bars on each side of critical sections.

At the end of any bar, a sufficient length should be provided to anchor the tensile or compressive force in the bar. The length is found by dividing the force in the bar by the product of the ultimate average bond stress (f_{bu}) and the perimeter of the bar or group of bars; the perimeter of a group of bars is taken as that of a single bar of equal cross-sectional area. f_{bu} is assumed to be uniform along the bond length and it is obtained as follows:

$$f_{bu} = \beta \sqrt{f_{cu}} \tag{12.27}$$

where β = the bond coefficient with the values given in Table 12.8.

Table 12.8 Values of the bond coefficient β

Bar type	Bond coefficient β	
	Tension	Compression
Plain bars	0.28	0.35
Type 1 deformed bars	0.40	0.50
Type 2 deformed bars	0.50	0.63
Fabric	0.65	0.81

For beams where the minimum amount of link reinforcement for shear is not required, the values of β for plain bars should be used for deformed bars, too. This restriction does not apply to slabs.

For hooks conforming with BS 4466, the anchorage provided should be the smaller length of 24 times the bar size or 8 times the internal radius of the hook but not less than the length of the bar in the bend and the straight part of the hook. For 90° bends, the anchorage provided should be the smaller length of 12 times the bar size or 4 times the internal radius of the bend but not less than the length in the standard 90° bend. The radius of the bend is limited to twice the bend test radius in the appropriate British Standard or by the ultimate bearing stress in the concrete.

Bearing stress $= F_{bt}/r\phi$

and not exceed $1.5 f_{cu}/(1 + 2\phi/a_b)$ \hfill (12.28)

where F_{bt} is the tensile force in the reinforcement, r is the internal radius of the bend, ϕ is the size of the bar or the equivalent size for a group of bars and a_b is the distance between bars perpendicular to the bend or the cover + ϕ for bars adjacent to the face of the member.

Links should be anchored by being passed through at least 90° round a longitudinal bar not less than its own size and continued for a length of at least 8 times its own size. Again, the internal radius of the bend should not be less than twice the bend test radius.

12.5.7 Cover

Concrete cover provides protection to the steel against corrosion and against too-rapid heating in the case of fire. Thus, the conditions of exposure and the requirements for fire resistance have a major influence in determining the amount of cover provided in design. Other factors are the dimensions of the reinforcing bars, the nature of the aggregate and the quality of the concrete.

For natural aggregate concretes, the nominal concrete cover to all reinforcement should not be less than that given for the appropriate condition of exposure and grade of concrete in Figure 12.10.

12.5.8 Spacing of reinforcing bars

The spacing and location of reinforcement must conform with the design requirements and must also allow proper compaction of the concrete to safeguard the protection of the steel against corrosion. The spacing must also be such that it tends to inhibit the spread of cracking and conforms with the needs for satisfying the criteria for the limit state of cracking.

In general, the maximum size of the aggregate governs the minimum spacing of bars, but when the size of the largest bars is 5 mm greater than that of the aggregate the spacing should not usually be less than the bar size. Bars or groups of bars should be located in horizontal layers with the gaps between the bars or groups in each layer in-line vertically.

Limitations on spacing are shown in Figure 12.11. The maximum distance between bars in tension is defined in the Code as a simple method of controlling crack width. For beams, the horizontal distance between bars is given in Figure 12.12(a). These requirements also apply to slabs except:

(1) When the slab is less than 200 mm thick, or 250 mm thick if f_y is less than 460 N/mm^2, or where the reinforcement is less than 0.3%.
(2) When the amount of reinforcement is less than 1% the spacing given in Figure 12.12(a) may be increased by dividing this spacing by the percentage of steel.

These recommendations relate to bars which, in size, are at least 0.45 times the size of the largest bar in the section and do not apply for particularly aggressive environments when f_y has a higher value than 300 N/mm^2.

The amount and spacing of side reinforcement required for beams of greater depth than 750 mm is illustrated in Figure 12.12(b).

12.5.9 Laps and joints

Bars can be lapped, welded or joined with mechanical devices to obtain continuity but joins should be located away from points of maximum stress. Load may be transferred in compression by cutting the ends of the bars square and holding them in direct alignment by a steel sleeve.

In general, the lap length should not be less than the greater of

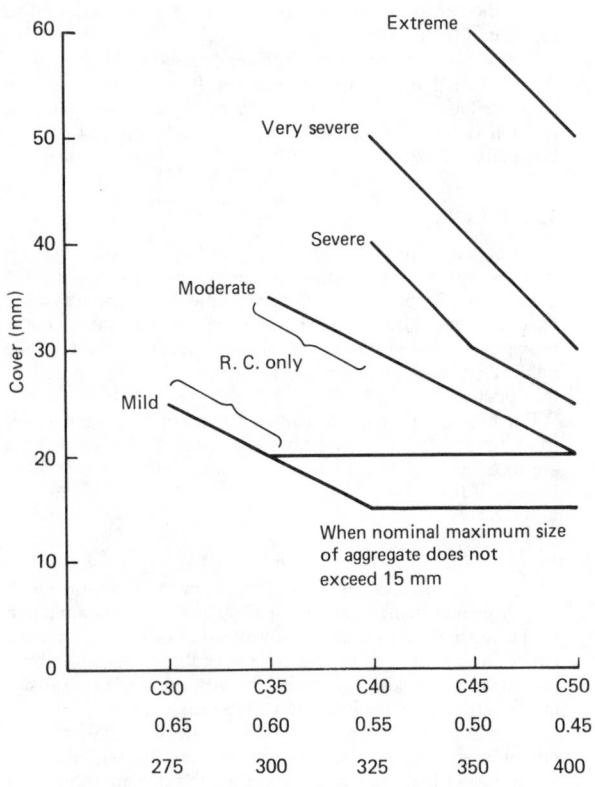

Figure 12.10 Cover to reinforcement and tendons

Conditions of exposure

Mild — Concrete surfaces protected against weather or aggressive conditions

Moderate — Concrete surfaces sheltered from severe rain or freezing whilst wet; subject to condensation, continuously under water or in contact with soil (SO_3 content less than 0.2%)

Severe — Concrete surfaces exposed to driving rain, alternate wetting and drying and occasional freezing, severe condensation of flowing water

Very severe — Concrete surfaces exposed to sea water spray, directly or indirectly to de—icing salts, corrosive fumes or freezing conditions whilst wet

Extreme — Concrete surfaces exposed to the abrasive action of sea water or flowing water with PH equal to, or less than, 4.5

For concrete of C40 grade subjected to very severe exposures and for concrete of C45 grade subjected to extreme and very severe exposures and in each case to freezing whilst wet, air entrainment should be used

Lowest grade of concrete

Maximum free water : cement ratio

Minimum cement content — kg/m^3

Note: Where low workability concrete is used in a precast factory, the minimum cement content may be reduced by up to 10% provided that the corresponding water : cement ratio is reduced by the same percentage.

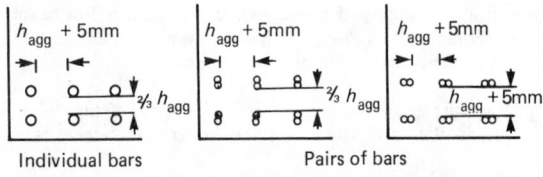

Individual bars   Pairs of bars   Bundled bars

Figure 12.11 Bar spacing – h_{agg} is the maximum size of coarse aggregate

either 15 times the bar size or 300 mm. For tension reinforcement, the lap length should not be less than that required for anchorage in tension, but when the cover is less than 2ϕ and, either the bar is at the top of the member as cast or at the corner of a section, or the clear distance between adjacent laps is less than the greater of 75 mm or 6ϕ, the length should be multiplied by 1.4. If both conditions apply, the length should be multiplied by 2. For compression laps, the length should not be less than 1.25 times the required anchorage in compression. Lap lengths may be based on the size of the smaller bar when two sizes of bar are joined. If the size of a bar at a lap is greater than 20 and the cover is less than 1.5 times the size of the smaller bar, transverse links should be used; they should not be smaller than one-quarter the size of the smaller bar with a spacing of not less than 200 mm.

British Standard 8110 includes recommendations for joining bars by welding but stipulates that these should not be at bends and that, where possible, they should be staggered between parallel main bars. Where tests have shown that the strength of the welded bar is not less than that of the parent bar, joints in compression may be designed for the full strength of the joined bars and joints in tension for 80% of the strength of the joined bars. Welding should not normally be used when the stress in the bar is predominantly cyclic.

12.5.10 Curtailment and anchorage of bars

In principle, except at the ends of members, all reinforcement should extend beyond the point where it is no longer needed, i.e. where the resistance moment for the continuing reinforcement is equal to the design moment. The amount of this extension should not be less than either the effective depth of the member

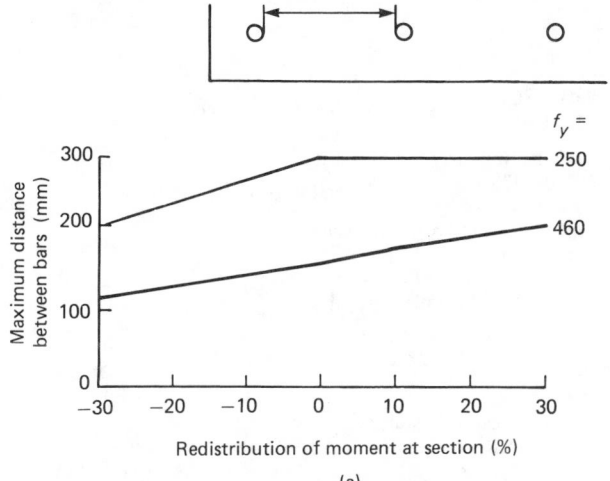

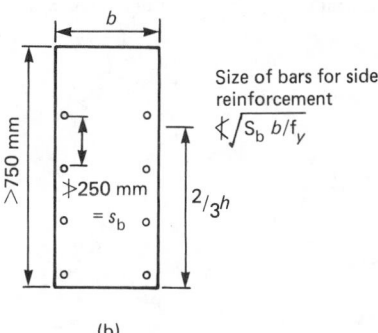

Size of bars for side
reinforcement
$\not< \sqrt{S_b\, b/f_y}$

Figure 12.12 Reinforcement for crack control.
(a) maximum distances between bars in beams and
slabs; (b) minimum amounts and maximum
spacing of side reinforcement for beams of 750
mm depth or more

or 12 times the bar diameter. Since the tension zone in the concrete may be cracked under service loading, special provisions for anchoring the bars in this region are necessary. These are met by one of the following:

(1) The extension should be an anchorage length.
(2) The shear capacity of the section where the bar stops should be twice that required.
(3) The continuing bars at this section provide twice the flexural strength required.

Each tension bar should be anchored at the end of a simply supported member by one or other of the following effective anchorage lengths:

(1) 12 times the bar diameter beyond the centre of the support, no bend or hook starting before $d/2$ from the face of the support.
(2) 12 times the bar diameter $+ d/2$ from the face of the support.
(3) For slabs, the greater in length of one-third the width of the support or 30 mm beyond the centre of the support provided that the design ultimate shear stress at the face of the support is not greater than half that allowed.

Items (1) and (2) above may be applied to hooked or bent bars whilst item (3) refers to straight bars. Simplified rules for application to the common cases of uniformly loaded beams and slabs are given in BS 8110 to which reference should be made.

In heavily reinforced members, curtailment of bars should be staggered to avoid undue cracking which might otherwise occur if a number of bars were stopped at almost the same position.

Despite these recommendations for anchorage and curtailment, the provision of ties required for the overall robustness of the structure should ensure their continuity and effective connection at changes of direction. At corners, the ties should extend 12 bar diameters beyond all the bars of the transverse ties or an effective anchorage length beyond their centreline.

12.5.11 Limits on the amount of reinforcement

For a concrete structure to be regarded as properly reinforced it

is necessary to have reinforcement crossing all sections which could otherwise develop fracture planes and lead to failure. The Code only exempts certain columns and plain walls from this requirement. Generally it sets out lower limits for structural members which are listed in Table 12.9. These are supplementary to the amounts of steel required to provide stability in the event of partial damage (see page 12/13) and are contributory to these requirements.

Small amounts of vertical steel in walls do not contribute to the strength of the wall and hence the minimum percentage is 0.4% except when fire resistance is required when the minimum is 1%. For axially loaded reinforced walls the steel may be placed in one layer and in that case transverse links are not necessary, but if two layers are used the Code requires transverse links.

To avoid difficulty in compaction of concrete, upper limits are set on the amounts of steel in the section, and these are also shown in Table 12.9.

12.6 Prestressed concrete

12.6.1 General

The primary objective in prestressing is to avoid excessive cracking and deflection whilst at the same time enabling high-strength materials, particularly high-tensile steel, to be used efficiently in construction. The main criteria governing the design of prestressed concrete are therefore characteristics of serviceability rather than ultimate strength.

In setting out the criteria for serviceability of prestressed concrete in Table 12.2, three classes of structure have been identified but no indication was given, nor is it given in the Code, for what purposes these different classes of structure should be used. They nevertheless represent a logical progression from reinforced concrete construction which is likely to be cracked under service loading through class 3 and class 2 to class 1 prestressed concrete construction which is not only completely free from cracking but free from flexural tensile stresses under service conditions.

Where there are particularly adverse conditions of exposure or where cyclic or dynamic loading is severe, it may be appropri-

Table 12.9 Maximum and minimum requirements for reinforcement

Member	Maximum	Minimum*
Tie	—	250G–0.80% 460G–0.45% of total area of concrete
Beams – tension reinforcement	4% of gross cross-sectional area of concrete	Rectangular sections – 250G–0.24% ⎫ 460G–0.13% ⎬ of total area of concrete (applies to each direction in slabs) Flanged beams – webs $\dfrac{\text{breadth of web}}{\text{breadth of beam}}$

<table>
<tr><td></td><td></td><td>< 0.4</td><td>≥ 0.4</td></tr>
<tr><td>250G –</td><td></td><td>0.32%</td><td>0.24%</td></tr>
<tr><td>460G –</td><td></td><td>0.18%</td><td>0.13%</td></tr>
</table>

of area – breadth of web × effective depth
Flanged beams – flanges at supports of continuous beams

<table>
<tr><td></td><td>T-beams</td><td>L-beams</td></tr>
<tr><td>250G –</td><td>0.48%</td><td>0.36%</td></tr>
<tr><td>460G –</td><td>0.26%</td><td>0.20%</td></tr>
</table>

of area – breadth of web × effective depth

Member	Maximum	Minimum*
Beams – compression reinforcement	4% of gross cross-sectional area of concrete	Rectangular sections – 0.2% of total area of concrete Flanged beams – flange in compression 0.4% of breadth × depth of flange Flanged beams – web in compression 0.2% of breadth of web × effective depth
Beams – shear	Limited by limit on shear in beams	Limitations on the amount of shear reinforcement are given in the sections shear in beams and slabs. Transverse reinforcement in flanges of flanged beams – 0.15% of the longitudinal cross-sectional area of the flange positioned near the top surface.
Columns	Vertically cast – 6% Horizontally cast – 8% At all laps – 10% of the gross cross-sectional area of concrete	Rectangular sections – 0.4% of total area of concrete. When all or part of the reinforcement is in compression in a column (or beam), ties or links are required at least one-quarter the size of the largest compression bar but not smaller than 6 mm at a spacing not more than 12 times the size of the smallest compression bar. Each corner bar and alternate bars should be tied by links and no bar should be more than 150 mm from a tied bar.
Walls	4% of total area of concrete	0.4% of total area of concrete. For up to 2% of compression bars, horizontal bars are needed not less than 6 mm size – 250G 0.30% or 460G 0.25% of concrete area – evenly spaced.

Table 12.9 Maximum and minimum requirements for reinforcement—*continued*

Member	Maximum	Minimum*
		For more than 2% of compression bars, links should go through the wall not less than one-quarter size of largest compression bars or 6 mm size, at a spacing of not less than twice wall thickness horizontally and vertically and also vertically not more than 16 × the bar size. No compression bar not enclosed by a link should be more than 200 mm from an enclosed bar

*250G and 460G refer to the grade of steel, no grade is given for compression steel and either may be used.

ate to use class 1 structures. Where, on the other hand, these effects do not exist and cracking is acceptable, class 3 structures would be more appropriate. For general purposes, however, including water-retaining structures, class 2 structures offer most advantages being free from cracking but more economical than class 1 structures.

Since the serviceability requirements tend to dominate the design process rather than ultimate strength as in reinforced concrete (and in some prestressed concrete class 3 construction), the procedures for calculating stresses due to the prestress and likely service loadings in relation to serviceability will be given first. The main advantages of prestressing and its most important applications are seen in flexural members and main attention will therefore be given to beams and slabs with secondary attention to ties and columns subjected to bending; little advantage is gained by prestressing members subjected mainly to compression and such columns and walls are not therefore considered.

In dealing with serviceability, different sets of conditions need to be examined:

(1) During the imposition of the prestress, the stresses in the materials should not exceed certain values determined by the need to avoid failure during the transfer operations or excessive loss of prestress due to creep effects in the materials.
(2) After the losses of prestress have occurred due to creep and shrinkage of the concrete and relaxation of the steel, the remaining prestress should be sufficient to ensure that the limit state for cracking does not occur under the appropriate design loads.
(3) During none of these stages should the deflections exceed the limits set.

Additionally, attention has to be paid to the secondary effects that arise in anchoring prestressing tendons and finally to the need to meet the ultimate loading conditions.

12.6.2 Prestress and serviceability

12.6.2.1 General

In assessing the conditions of prestress, it is sufficient to assume that the concrete deforms elastically under short-term loading, that creep of concrete can be treated by adopting an effective modulus for the concrete (see page 12/10), and that shrinkage is uniform across the section.

12.6.2.2 Prismatic members

The stress conditions in a uniform member, subjected to external moment M and external force F with a prestressing force P with eccentricity to one axis only of e are shown in Figure 12.13.

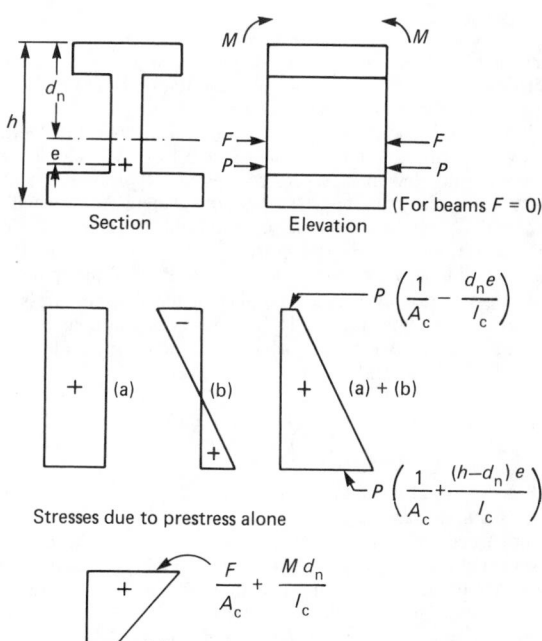

Stresses due to prestress alone

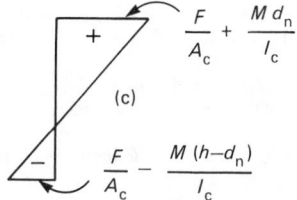

Stresses due to external moments and forces

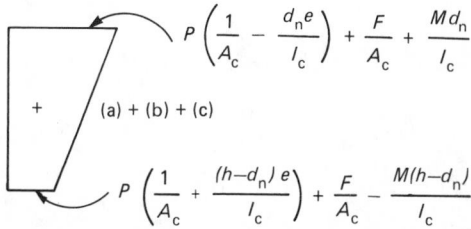

Stresses due to external moments and forces and prestress

Figure 12.13 Elastic analysis of prestressed concrete sections. M, external moment; F, external force; P, prestressing force; A_c, area of concrete; I_c, second moment of concrete area; A_{ps}, area of tendons

Steel is normally used to impose the prestressing force although it can be applied by jacks or by the use of other materials. Such applications are, however, so rare that they are not considered further and this section is therefore concerned only with design for pretensioning and post-tensioning with steel tendons.

For pretensioning, the stress in the steel immediately after transfer, f_{p2}, is less than the initial stress, f_{p1}, by an amount corresponding to the relaxation of the steel at that stage and to the elastic contraction of the concrete adjacent to the steel; the stress is then:

$$f_{p2} = f_{p1} - \Delta f_{p1} - \frac{E_s}{E_{c1}} \left[f_{p2} A_{ps} \left(\frac{1}{A_c} + \frac{e^2}{I_c} \right) - \frac{M_g e}{I_c} \right] \qquad (12.29)$$

where Δf_{p1} is the relaxation of steel before transfer, M_g is the moment due to the proportion of dead load effective at transfer, E_s and E_{c1} are the moduli of elasticity of concrete and steel respectively, A_c and I_c are the area and second moment of area respectively of the concrete section and e is the eccentricity of the tendons.

The final term in the equation is the stress in the concrete adjacent to the steel multiplied by the modular ratio, to give the equivalent change in stress in the steel.

For beams of uniform section with steel at constant depth along the beam, the moment due to dead load, M_g, should be ignored, since the most severe conditions of prestress occur away from mid-span near the supports where M_g is small. Where the beams are of nonuniform section, the eccentricity of the steel is normally reduced towards the ends and checks on the stress conditions are required at several sections in the span.

In post-tensioning where a number of tendons are stressed successively, only the first tendon to be stressed contracts by the full amount of the elastic shortening of the concrete and the stress in the steel after transfer is then given by:

$$f_{p2} = f_{p1} - \frac{E_s}{2 E_{c1}} \left[f_{p2} A_{ps} \left(\frac{1}{A_c} + \frac{e^2}{I_c} \right) - \frac{M_g e}{I_c} \right] \qquad (12.30)$$

Δf_{p1} is not included since there is little relaxation of steel between stressing and anchoring.

In structures with post-tensioned tendons, the dead-load moment effective at transfer may be large and may include dead load from additional superstructure which is temporarily propped but becomes effective on stressing the tendons. This dead-load moment has an important influence on design since it can, if properly manipulated, lead to improvements in efficiency.

The effect of time on the deformation of concrete and steel is taken into account in the following expressions which give the stresses in the tendons. For pretensioning:

$$f_{p3} = f_{p1} - \Delta f_p - S E_s - \left(\frac{E_s}{E_{c1}} + \phi E_s \right)$$

$$\left[f_{p3} A_{ps} \left(\frac{1}{A_c} + \frac{e^2}{I_c} \right) - \frac{M_g e}{I_c} \right] \qquad (12.31)$$

For post-tensioning:

$$f_{p3} = f_{p1} - \Delta f_p - S E_s - \left(\frac{E_s}{2 E_{c1}} + \phi E_s \right)$$

$$\left[f_{p3} A_{ps} \left(\frac{1}{A_c} + \frac{e^2}{I_c} \right) - \frac{M_g e}{I_c} \right] \qquad (12.32)$$

where Δf_p is the total relaxation loss in the steel, S is the shrinkage strain and ϕ is the creep strain for unit stress.

Using these formulae, the value of P is obtained from:

$$P = f_p A_{ps} \qquad (12.33)$$

where f_p is the stress in the tendon at the stage considered. P is then substituted in the appropriate expressions in Figure 12.13 to obtain the required stress conditions immediately after transfer and subsequently under the loads for serviceability limit states.

In members subjected to bending in one direction only it will be normal to locate the centre of the prestressing tendons at an eccentricity which will provide maximum compression at what will become the tension face with a small amount of tension or compression at what will become the compression face under load. For members subjected to loads from any transverse direction, the tendons will be placed concentrically to give a uniform prestress. The formulae given apply to either case.

12.6.3 Losses of prestress

In making these calculations for the stress conditions during manufacture and in service, quantitative allowances must be made for the elastic contraction, shrinkage and creep of concrete, and relaxation of the steel. These characteristics are variable and are much influenced by the nature of the materials used, methods of production and the service conditions. Part 1 of BS 8110 gives values for the calculation of these losses for general use, while Part 2 amplifies this information for special circumstances; it is recommended that specialist literature should be consulted for very unusual conditions of temperature or exposure.

The shrinkage of concrete, so far as it affects the loss of prestress, depends on the quality of the concrete, the size of the component, the nature of the aggregate, age at transfer and the conditions of exposure. It is usually reasonable to assume that the shrinkage of concrete may be taken as 100×10^{-6} for external exposure in the UK and 300×10^{-6} for indoor conditions.

Experimental evidence shows that the creep of concrete is proportional to the applied stress for the stresses generally applied during transfer and, as for shrinkage, is considerably affected by circumstances. For the calculation of the loss of prestress, it is convenient to define the amount of creep as a multiple of the elastic contraction at transfer, and to adopt a factor of 1.8 for transfer within 3 days, reducing to 1.4 for transfer at 28 days for outdoor exposure in the UK. These values may also be used for class 1 and 2 structures for internal conditions. Further advice for other conditions and for class 3 structures is given in Part 2 of BS 8110; for other problems specialist publications should be consulted.

If it is necessary to estimate the amount of creep at some intermediate stage in the life of a structure, it is often sufficient to assume that about half occurs during the first month after transfer and that about three-quarters of the total occurs during the first 6 months following transfer.

The relaxation of prestressing tendons due to creep of steel is dependent on the type of steel and method of manufacture. The relevant BS Standards (Table 12.7) require a 1000-h test for relaxation of tendons at different levels of initial stress as part of the acceptance test procedure. The relaxation loss used in design is obtained from the value in the manufacturers' test certificate corresponding to the initial prestress multiplied by a relaxation factor. The values in BS 8110 are quoted in Table 12.10. If, at the design stage, the steel supplier is not known, it will usually be sufficient to base calculations on specified values.

For many forms of repetitive construction, once the losses of prestress have been established, it may be possible to express the total loss as a percentage of the initial stress in the steel at transfer. Values for total loss, due to elastic contraction, shrinkage and creep of concrete, and relaxation of steel, of 20% for

Table 12.10 Relaxation factors for different types of tendons

Type of tendon	Relaxation factor	
	Pretensioning	Post-tensioning
Wire and strand		
relaxation class 1	1.5	2.0
relaxation class 2	1.2	1.5
Bar	—	2.0

pretensioning and 15% for post-tensioning for an initial stress in tendons of 70% of their characteristic strength have been found to be appropriate. If such an assumption is made, however, detailed refinement in design should not be attempted.

Other sources of loss also need to be taken into account with post-tensioning. These arise through the movement of the tendons in the anchorage during the process of transfer, which needs to be determined by measurement and should be given by the manufacturer of the system, and through the development of friction between the tendon and its surroundings.

The profiles of the cables or bars may be curved to provide for counteracting variation of moment due to dead and imposed loads or due to continuity. As a result, friction develops during stressing between the cables or bars and the inner surfaces of ducts or tendon deflectors. The amount of friction depends on the construction of the cable, the materials in sliding contact and the angular displacement. For long cables, the actual profiles are likely to deviate from their correct position to such an extent that they have an effective additional curvature, which causes considerable frictional effects. Then the force in a tendon, P_x, at a distance x from the jacking point is given by:

$$P_x = P_o \exp - [(\mu x/r_{ps}) + Kx] \qquad (12.36)$$

where P_o is the force in the tendon at the jacking end, μ is the coefficient of friction from Table 12.11, r_{ps} is the radius of curvature, x/r_{ps} is the angle of deviation over length x and K is the constant and the form of tendon and duct which has a usual value of 33×10^{-4}/m but may be reduced to 17×10^{-4}/m for rigid sheaths or rigidly fixed duct formers or to 25×10^{-4}/m for greased strands in plastic sheaths.

Table 12.11 Values for coefficient of friction μ

Condition	μ
Lightly rusted strand on unlined concrete duct	0.55
Lightly rusted strand on lightly rusted steel duct	0.30
Lightly rusted strand on galvanized duct	0.25
Bright strand on galvanized duct	0.20
Greased strand on plastic sleeve	0.12

12.6.4 Stress limitations at transfer and for serviceability conditions

Limits need to be set on the stresses in the steel and concrete at transfer to ensure that the deformation of the materials is not excessive since this would lead to high losses of prestress, severe cracking and undue distortion of components. Limits also need to be imposed on the stresses likely to occur in service to keep deflections within acceptable bounds and to control cracking to required limits.

All calculations of stresses for these two sets of conditions are based on the assumptions that the section is uncracked and that the strains due to applied stresses are proportional to those stresses.

The stress in tendons during the initial stressing operations should not normally be more than 75% of the characteristic strength but may be as much as 80% provided special care is taken. The stress at transfer should not normally be more than 70% and never more than 75% of the characteristic strength.

The allowable maximum limit for compressive stress in concrete at transfer is $0.5f_{ci}$ at the extreme compression face or $0.4f_{ci}$ for a nearly uniform prestress, where f_{ci} is the strength at the time of transfer.

For the serviceability limit state, the compressive stress in the concrete should not be more than $0.33f_{cu}$ at the extreme compression face but for continuous construction this limit may be raised to $0.4f_{cu}$ within the negative moment zone. The stress in direct compression should not be greater than $0.25f_{cu}$.

Flexural tensile stresses in concrete are defined according to the class of structure decided at the outset of design. For class 1 structures, the maximum tensile stress at transfer is limited to 1 N/mm^2 and no tensile stresses are allowed for serviceability limit states.

For class 2 structures, in which some flexural tensile stresses are allowed up to the tensile strength of the concrete for pretensioning and up to 0.8 times the tensile strength of the concrete for post-tensioning, the tensile strength is assumed to be $0.45\sqrt{f_{ci}}$ for transfer and $0.45\sqrt{f_{cu}}$ for the serviceability limit states. Where a design service load is only likely to be rarely imposed and the concrete is normally stressed in compression so that any cracks that might occur are closed, the allowable tensile stress may be increased by 1.7 N/mm^2 provided that pretensioned tendons are well distributed throughout the concrete stressed in tension and post-tensioned tendons are supplemented by secondary reinforcement.

Although cracking is permitted in class 3 structures, the section is assumed to be uncracked and limits are set for notional tensile stresses for use in calculations for service loading to impose some restriction on the widths of cracks. At transfer, the limits set for tensile stresses are, however, the same as those for class 2 structures. The values for allowable notional tensile stresses are obtained from Table 12.12, which are multiplied by the factors in Table 12.13 to allow for the effect of the depth of section on cracking.

Table 12.12 Class 3 members – limits for notional tensile stresses

Group	Limiting crack width (mm)	Design stress (N/mm^2) for concrete of grade		
		30	40	50 and over
Pretensioned tendons	0.1	—	4.1	4.8
	0.2	—	5.0	5.8
Grouted post-tensioned tendons	0.1	3.2	4.1	4.8
	0.2	3.8	5.0	5.8
Pretensioned tendons distributed in tensile zone and close to the concrete tension face	0.1	—	5.3	6.3
	0.2	—	6.3	7.3

Table 12.13 Class 3 members – depth factors

Depth of member including composite members (mm)	Factor
200 and under	1.1
400	1.0
600	0.9
800	0.8
1000 and over	0.7

These stresses may be exceeded in certain circumstances for class 3 structures as indicated in BS 8110.

12.6.5 Beams

12.6.5.1 Flexural strength

The methods of calculation of the ultimate flexural strength of prestressed concrete beams are similar to those for reinforced concrete beams with the additional need that allowance must be made for the effect of the condition of prestress. The assumptions made are:

(1) Sections which are plane before remain plane after bending.
(2) The stresses in the concrete may be determined from the stress–strain curve in Figure 12.4 or, more normally, may be taken as uniformly distributed across the compression zone to a depth of 0.9 times that of the neutral axis with a value of $0.45f_{cu}$ for deriving simple formulae as in Figure 12.14(a). As for reinforced concrete, the ultimate compressive strain for the concrete is taken as 0.0035.
(3) The tensile strength of the concrete is ignored.
(4) The strains at ultimate in pretensioned tendons and in post-tensioned and bonded tendons are assumed to conform generally with the strains in the concrete so that the stresses may be determined from the stress–strain relationships for steel given in Figure 12.6 making allowance for the initial stress condition in the steel after all losses. As a simple alternative, the stresses at ultimate may be obtained from the curves in Figure 12.14(b) where allowance is made for the strength of the steel and the concrete, their respective proportions and the initial stress in the steel.
(5) The strains at ultimate in unbonded post-tensioned tendons do not conform directly with the compressive strains in the adjacent concrete but are directly influenced by the increase in separation between the end anchorages. A method of calculation of the stress in the steel is given in BS 8110 or the stress may be determined by test or analysis.
(6) Any additional steel reinforcement close to the tension face should be assumed to be stressed to its characteristic yield stress at ultimate. Such reinforcement in the compression zone should normally be ignored.

If the compression zone at failure is not rectangular in shape, the ultimate strength should be calculated from first principles using the stress–strain relationships shown in Figures 12.4 and 12.6.

12.6.5.2 Deflection

Control of deflection in the design of reinforced concrete beams is governed in BS 8110 by limitations on span:depth ratio, but the method is not appropriate for prestressed concrete beams. For normal construction no specific requirement is given, the limitations on stresses for service conditions usually being sufficient to avoid deflections becoming excessive.

For special construction, where checks are required however,

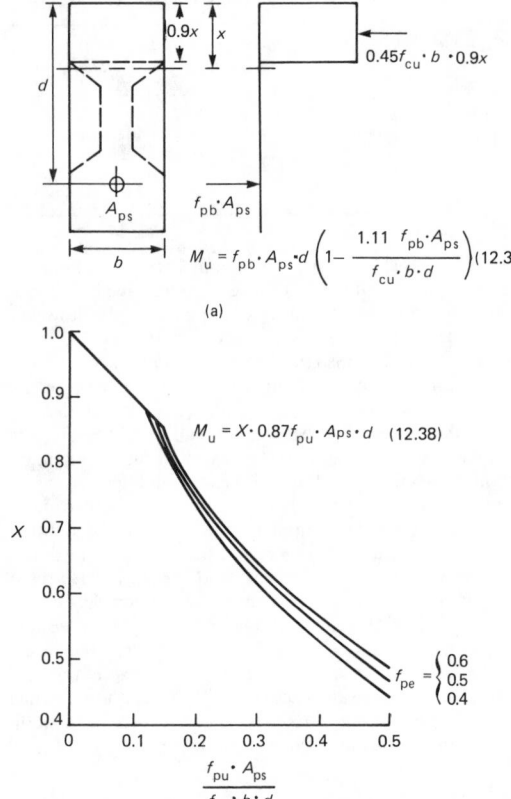

$$M_u = f_{pb} \cdot A_{ps} \cdot d \left(1 - \frac{1.11 \ f_{pb} \cdot A_{ps}}{f_{cu} \cdot b \cdot d}\right) \quad (12.37)$$

(a)

$$M_u = X \cdot 0.87 f_{pu} \cdot A_{ps} \cdot d \quad (12.38)$$

$$f_{pe} = \begin{cases} 0.6 \\ 0.5 \\ 0.4 \end{cases}$$

(b)

Figure 12.14 Flexural strength of beams – approximate method (pretensioning and post-tensioning with bond)

some guidance is provided in Part 2 of BS 8110. It should then be assumed in the calculation of the short- and long-term deflections of class 1 and class 2 prestressed beams that behaviour is elastic and that the properties of the section are those for the concrete with the deformation characteristics appropriate to the nature of the loading. For long-term loading, an effective modulus should be used in the calculations, which may be derived from the data given on creep. The same approach to the calculation of deflection may be adopted for class 3 prestressed beams provided that the section is not cracked. If, however, it is cracked under the load being considered, deflection is more likely to require limitation and it should then be calculated from the moment–curvature relationship determined from the properties of the materials and the characteristics of the section.

It should be noted that for members, such as precast mass-produced units with pretensioned steel with a uniform eccentric prestress along their length, the upward deflection at transfer and later in service, if the permanent loading is light, may need to be checked by calculation to ensure that it is not excessive. If such members are heavily loaded, it should be remembered that the regions near their ends are subjected to the effects of reversed bending due to the prestress which will tend to reduce the central deflection.

12.6.5.3 Shear

Shear in prestressed concrete beams needs only to be considered for ultimate conditions. Then, sections subjected to shear remote from regions of maximum bending are likely to be uncracked but those subjected to both bending and shear will usually be cracked in flexure. These two situations give rise to substantially different distributions of stress and therefore require different methods of analysis; each is dealt with in BS 8110. In each case, the contribution of the concrete to shear strength is calculated and may be taken into account when provision is made for shear reinforcement.

Firstly, irrespective of the situation and amount of shear reinforcement, a limitation is set on the maximum shear that a section may sustain. This maximum shear strength is defined by limiting the maximum shear stress $(V/b_v \cdot d)$ for cracked or uncracked sections to the lesser of $0.8\sqrt{f_{cu}}$ or 5 N/mm^2.

For uncracked sections, the ultimate resistance of the concrete (V_{co}) is given by:

$$V_{co} = v_{co} \cdot b_v \cdot h \qquad (12.39)$$

where v_{co} is the ultimate shear stress that can be sustained by the uncracked concrete and is given in Figure 12.15; it is expressed in terms of the grade of concrete and compressive stress at the centroid due to the prestress $(f_{cp}) \cdot b_v$ is the breadth of the section or of the web for T- and I-sections and h is the overall depth of the section

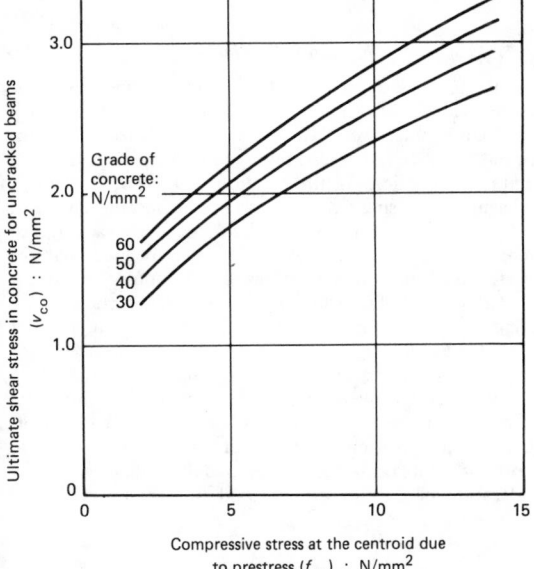

Figure 12.15 Ultimate shear stress in concrete for uncracked beams

The shear at the ends of units with pretensioned steel should be determined at a distance equal to the height of the centroid of the section above the soffit from the edge of the bearing. If this position is within the transmission length, the value of f_{cp} should be reduced by multiplying by the factor $r(2-r)$ where r is the distance of the section from the end of the unit as a fraction of the transmission length.

The ultimate shear strength of the cracked concrete section is given by:

$$V_{cr} = \left[1 - 0.55 \left(\frac{f_{pe}}{f_{pu}}\right)\right] v_c \cdot b_v \cdot d + M_0 \cdot (V/M) \qquad (12.40)$$

but:

$$V_{cr} \not< 0.1 b_v \sqrt{f_{cu}} \qquad (12.41)$$

where v_c is given in Figure 12.8 and is the same as for reinforced concrete (the cross-sectional area of reinforcement should then be the sum of the areas of tendons and reinforcement in the tensile zone). f_{pe} is the effective prestress in the tendons but not more than $0.6f_{pu}$, M_0 is the moment to produce zero stress at depth d in the beam, and V and M are the shear and the moment at ultimate respectively

The calculated shear resistance of the concrete alone (V_c) for uncracked sections should be taken as that given by V_{co} above. For cracked sections, however, both V_{co} and V_{cr} above should be calculated and V_c assigned the lower value.

Since, if it were to occur, failure in shear of prestressed concrete beams without secondary reinforcement might take place suddenly with little warning, it is usually recommended that the provision of shear reinforcement should be conservative. Hence, BS 8110 recommends that reinforcement should only be omitted from prestressed concrete beams when the shear strength of the concrete is more than the required resistance. In such circumstances, reinforcement may be omitted from components of minor structural significance or from those which have been proved by tests. Shear reinforcement is not required in other beams when the shear strength provided by the concrete is calculated to be twice that required.

The Code gives the following guidance on the calculation of amount of shear reinforcement required. When the shear resistance required is less than $V_c + 0.4 b_v \cdot d$, the shear reinforcement in the form of links should not be less than:

$$A_{sv}/s_v = (0.4b_v)/(0.87f_{yv}) \qquad (12.42)$$

If the shear resistance required (V) is greater than $V_c + 0.4 b_v \cdot d$, the shear reinforcement required in the form of links should not be less than:

$$A_{sv}/s_v = (V - V_c)/(0.87f_{yv} \cdot d_t) \qquad (12.43)$$

where A_{sv} is the cross-sectional area of two legs of a link, s_v is the link spacing, f_{yv} is the characteristic strength of the reinforcement but not more than 460 N/mm^2 and depth d_t is of the section from the compression face to the centroid of the tendons or to the longitudinal reinforcement if greater.

Links should pass round longitudinal bars or tendons of larger diameter than the link at the corners of the tensile zone and as close as the cover requirements permit to the tensile and compression faces. They should be anchored firmly and enclose all longitudinal tendons and reinforcement. The spacing of the links should not be more than $0.75d_t$ or 4 times the web thickness. The maximum spacing should be reduced to $0.5d_t$ when V exceeds 1.8 times V_c. Lateral spacing should not exceed d_t.

12.6.6 Other forms of member

12.6.6.1 Slabs

The design of slabs in prestressed concrete adopts the methods for dimensioning prestressed concrete beams. No shear reinforcement is required provided that V is less than V_c.

For two-way slabs, the stresses under service conditions should be determined by elastic analysis while the methods of design used for reinforced concrete slabs should be used for ultimate conditions.

12.6.6.2 Columns

It will usually only be appropriate to use prestressed concrete columns when eccentricities of load are high and so require high bending stresses to be resisted. British Standard 8110 recommends that columns with a mean precompression of less than 2 N/mm² should be treated as reinforced concrete columns.

The analysis of columns in prestressed concrete frameworks should, in general, be similar to that adopted for reinforced concrete but modified to take account of their different moment/curvature characteristics.

12.6.6.3 Tension members

Although prestressed concrete is seldom used for tension members, it is well suited for the purpose.

For serviceability limit state of cracking, the tensile stresses should be limited to possibly half those for class 2 construction. Ultimate strength should be calculated by assuming that both tendons and any secondary reinforcement are stressed to 0.87 times their respective characteristic strengths.

12.6.7 Requirements for tendons and reinforcement

12.6.7.1 Transmission lengths for pretensioning

With pretensioning, it is usual to anchor the tendons by bond with the concrete. In this case, the stress in the steel and the prestress in the concrete builds up along the length of the member near its ends. Consequently, in dealing with shear in this region some allowance must be made for this build-up, as already explained in the section on shear.

The transmission length is dependent on a number of factors including the strength of the concrete, the degree of compaction, the pretension in the tendons, the nature of the tendon, its size, and surface characteristics. Since the concrete near the top of a member is often less well compacted than that near the bottom, the transmission length for tendons near the top is usually greater. Values for the transmission length, obtained from measurements in the field and in the laboratory, show that very substantial variations do occur. Where possible, therefore, transmission lengths should be determined for the particular conditions of production that obtain in the precast works. Where such information is not available, the method of calculation recommended in BS 8110 should be used. The transmission length is calculated as follows:

$$l_t = (K_t \cdot \phi)/N f_{ci} \tag{12.44}$$

where K_t is 600 for plain or indented wire or crimped wire with a small offset; 400 for crimped wire with an offset greater than $0.15\,\phi$; 240 for standard or super seven-wire strand; 360 for drawn seven-wire strand; ϕ is the nominal diameter of the tendon; and f_{ci} is the strength of the concrete at transfer

Within the anchorage region, the pretensioned tendons should be distributed as uniformly as possible across the section to avoid the development of unnecessary bursting stresses between groups of tendons. Reinforcement of the end-regions may be necessary; it should enclose all the tendons and some guidance on the amount required may be obtained from that needed in end-blocks.

12.6.7.2 End-blocks for post-tensioning

The anchorages of post-tensioned tendons at the ends of members give rise to high bursting stresses in the concrete. Recommendations for the design of end-blocks may be made by the manufacturers and, if so, should be followed.

If they are not made the recommendations in the Code should be followed. These give a method of estimating the bursting forces for square blocks which may be applied also to rectangular blocks or combinations of rectangular blocks to cover irregular shapes of anchorage. The basic formula is:

$$F_{bst}/P_k = 0.32 - 0.30(y_{po}/y) \tag{12.45}$$

for:

$$0.3 < y_{po}/y < 0.7$$

where F_{bst} is the bursting force, P_k is the maximum prestressing force during jacking when tendons are grouted, y_{po} is the half-side of the loaded square end-plate when y is the half-side of the square block

The reinforcement required in each direction should be located transversely in the distance $0.2\,y_{po}$ to $2\,y_{po}$. For rectangular end-blocks, the treatment should be similar giving different amounts of reinforcement in each direction. Circular bearing plates should be treated as square plates of the same area. The reinforcement should consist of spirals or closed links stressed to not more than 200 N/mm². Where the whole anchorage region is comprised of a number of individual anchorages additional reinforcement will be required to enclose the whole.

For unbonded tendons, the calculations should be based on the characteristic strength of the tendons and reinforcement assumed to be stressed to 0.87 times its characteristic strength.

12.6.7.3 Proportions of prestressing steel

The characteristics of prestressed concrete of being crack-free under normal conditions and of exhibiting extensive cracking and deflection under overloads have been emphasized as advantages, but these are not necessarily obtained with all prestressed concrete construction. Brittle fracture will occur if there is insufficient prestressing steel in the section to sustain the tensile stresses transferred to it on the development of cracking. To prevent this from taking place, it is recommended that the calculated ultimate moment of resistance should exceed the calculated moment of resistance of the uncracked section corresponding to a maximum flexural tensile stress of $0.6\sqrt{f_{cu}}$ after allowing for losses.

Brittle fracture can also occur by crushing of the concrete in 'over-reinforced' members. It may be experienced with precast members of inverted T-section which are designed for incorporation in composite construction but which are susceptible to this form of failure during erection. Difficulty can be avoided by consideration of erection stresses and proper supervision of construction.

12.6.7.4 Cover and spacing for tendons

In general the requirements for concrete cover for tendons and for their spacing are governed by the same needs as for reinforced concrete.

The cover of concrete necessary to protect steel against different conditions of exposure is given in Figure 12.10 (page 12/20) and applies to both tendons and reinforcement. The concrete should not be less than grade 30 and, for concretes of higher grades than 50, there should be no reduction in the thickness of the cover. The cover to ducts for post-tensioned tendons should not be less than 50 mm. Since prestressing steels are more sensitive to the effects of heat than reinforcing steels, the requirements for concrete cover for the protection of tendons in fire are more onerous and are dealt with later on pages 12/32 and 12/33.

Experience has shown that ends of individual pretensioned tendons do not require concrete cover for protection and may be left cut flush with the end of the member. Where post-tensioned tendons are positioned outside the member, they are normally protected with added concrete to provide the cover; some attention should be given, however, to the way in which this is done since the development of possible paths for penetration of moisture to the steel must be avoided.

There should be sufficient space between tendons to allow proper compaction of the concrete and the rules applied to reinforced concrete should therefore apply. Large ducts, however, provide more difficulties than are experienced with large reinforcing bars and careful attention is needed in detailing. If these ducts are required in thin diaphragms or webs then care must also be given to the avoidance of bursting the concrete and possibly to the provision of additional reinforcement. Additional reinforcement may also be required in members with pretensioned steel if the individual tendons are in separate groups to prevent longitudinal splitting at the ends where transmission of the prestress by bond is developed.

12.6.7.5 Secondary reinforcement

Reinforcement is required in prestressed concrete to meet requirements for resisting shear, to permit higher tensile stresses in class 3 structures, to prevent bursting in the region of endanchorages, to reinforce thin webs and to retain concrete cover in place for longer periods of fire resistance. Any longitudinal reinforcement provided for these purposes can be taken as contributing to ultimate strength as can the ties needed to ensure stability in the event of accidental damage.

Reinforcement may also be desirable for members prestressed by post-tensioning to restrict cracking after casting caused by restraint of the formwork due to shrinkage or cooling of the concrete.

12.7 Precast and composite construction

12.7.1 General

The previous sections have dealt with concrete construction without specific reference to the method of making the concrete. Those sections are generally applicable when the concrete is cast *in situ*. Not uncommonly, however, it is economic or convenient to precast concrete in the factory and use it in construction with site-cast concrete to form composite members or structures. This section deals with the special needs of using precast concrete in construction which are generally additional to those already given for cast *in situ* construction.

Precast members must be handled, and possibly transported, stored and erected, at an early age. The strength of the concrete must therefore be sufficient, not only to satisfy the normal design requirements for the finished structure, but also to sustain these constructional operations without damage. It is therefore not unusual for precast members to be made with concrete of a much higher strength than would be used if they were to be cast *in situ*. Characteristic strengths as high as 50 or 60 N/mm^2 at 28 days may be obtained since the manufacturing requirement may be for 10 N/mm^2 or more at 24 h.

Since the strength and robustness of concrete construction depends mainly on the combined action of concrete in compression and steel in tension at all sections, the presence of joints between members requires special attention from the designer to provide continuity of reinforcement and between members in the development of overall stability. To help in achieving this aim, the responsibility for design should also be vested in one engineer to ensure that, not only are the individual units

adequate for their purposes, but that the overall performance of the structure is also adequate.

12.7.2 Structural connections between units

The recommendations for the stability of precast construction conform generally with those for *in situ* construction. The first requirement in the design of structural joints is therefore to provide for transmitting the tie forces required for stability. In providing this reinforcement, it should be ensured that the arrangements for anchorage are realistic and can be obtained with normal site workmanship. Particular account needs to be taken of the requirements in BS 8110 with respect to location and anchorage. Bars must almost inevitably be lapped and anchored in cast *in situ* concrete or mortar and the Code gives detailed recommendations on conditions, relevant dimensions and form of secondary link reinforcement that should be provided. These details are not reproduced here and reference should be made to BS 8110.

When the joint is not required to transmit horizontal forces or moments, the detailing should be such that these effects are not, in fact, transmitted or, if that is not possible, that any unintended transmission should not lead to any untoward cracking or local damage. It must be recognized that shrinkage and creep effects can lead to the development of substantial restraints in structures and so cause cracking, and that these effects can be more serious in precast structures, particularly when the units are prestressed.

In dealing with the transmission of forces and moments at connections, the normal procedures for calculation for reinforced concrete, prestressed concrete or structural steel, should be used. Where special difficulties arise, tests should be made to assess both strength and mode of failure.

Attention needs to be paid to the protection of joints against the weather and corrosion and against fire to make sure that the performance of the joint under these conditions is not inferior to that of the rest of the structure. Emphasis has already been placed on the importance of detailing joints in such a way that they can be made properly on the site and effectively inspected. To achieve this, any projecting bars, ribs or fins should be sufficiently robust to avoid damage in transit and erection and sufficiently accurately located and dimensioned for ease of casting and assembly. The space between members being jointed should be sufficient to allow filling of the joint without difficulty and for subsequent checks on workmanship. Further, to make sure that joints are made properly, written instructions should be given to the site supervisor giving full details and sequence for jointing with information on what should be done in the event of misfits. The instructions should also contain information on the making of the joints in relation to progress of construction since stability during erection may well depend on the extent of the completion of joints in other parts of the structure.

Continuity of reinforcement in precast construction may require special consideration. If sockets or slots are left for lapping or anchoring reinforcement they should be sufficiently large and of a form likely to achieve their object with a suitable form of surface for developing bond between the infill concrete, mortar or grout and the precast units. Sleeves or threaded anchors for connecting reinforcement may be used as well as welded connections with structural steel sections. Test data for many of these types of connections are available.

12.7.3 Beams, slabs and frames

Where the continuity of beams, slabs and frames is obtained by the appropriate positioning and anchoring of steel reinforcement or tendons, the amount of redistribution of moments adopted for *in situ* construction may also be assumed for precast construction.

12.7.4 Floor slabs

Precast slabs are often used in floor construction. Units are usually placed side by side with an *in situ* topping and have provision for the transmission of transverse shear between units. Where these floors are needed to support partitions or other concentrated loads it may be assumed, subject to certain provisos in BS 8110, that the width of floor supporting the load may be up to 3 times the width of the units together with their joints for unreinforced topping and up to 4 times for reinforced toppings.

12.7.5 Bearings

The design of bearings in precast concrete construction has received detailed scrutiny in recent years, since a number of roof failures have been attributed to their inadequacy. The function of the bearing is to provide support for beams and slabs without cracking or spalling of the concrete under the relatively high stresses due to rotational effects needing to be sustained, whilst ensuring at the same time that unintended relative linear movement at the bearing does not take place. If rotational effects are likely to change the point of application of loading on a member or longitudinal restraint and likely to induce substantial stresses, these should be allowed for in design. Constructional inaccuracies are usually adjusted at the bearings and so it is also essential that proper allowance is made for accommodating building tolerances. A further requirement is that reinforcement in the supporting component should have an effective overlap of the member being supported. These aspects are dealt with in greater detail in BS 8110, but the necessary design calculations are based on the following limitations:

(1) The net bearing width, i.e. in the direction of the span of the supported member after allowing for inaccuracies, should not be less than 40 mm; for isolated members, it should not be less than 60 mm.
(2) The effective bearing length, i.e. the transverse direction to span of the supported member, should not be taken as greater than the effective length, half the effective length + 100 mm or 600 mm, whichever is the less.
(3) The ultimate bearing stress should not exceed $0.4f_{cu}$ for concrete directly to concrete, $0.6f_{cu}$ for a bedded bearing on concrete or $0.8f_{cu}$ for concrete bearing on steel where the dimension of the steel does not exceed 40% that of the concrete.

The corbel is a common form of support for beams on columns, which BS 8110 defines as a short cantilever beam, and provided the limiting dimensions given in Figure 12.16 are adopted it may be designed as an inclined strut. The tie reinforcement should be capable of sustaining an additional horizontal force equal to at least half the vertical force and compatibility between the strains in the concrete and the steel tie should be checked at the root of the corbel. The tie should be anchored at the front of the corbel by fabricating the reinforcement as a closed loop. The depth of the corbel at the face of the support should be determined from considerations of shear. Reinforcement should also be provided in the form of horizontal links with a total cross-sectional area of half that of the main tie over the upper two-thirds of the corbel at its root to control cracking under service conditions.

Where precast floor units or other units are supported by beams, a concrete nib may be required along the sides of the beams, which may need to be continuous. This may be designed

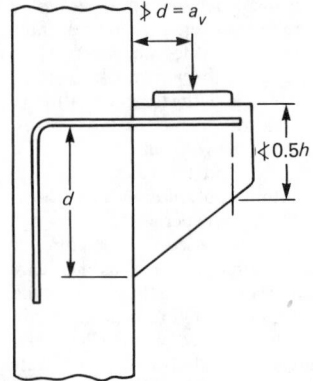

Figure 12.16 Dimensions of corbels

as a cantilever assuming that the load acts at the outer edge of the loaded area, i.e. the edge for a nib without a chamfer, the edge of the chamfer or the edge of a bearing pad. The eccentricity of the load is then taken as the distance to nearest vertical link reinforcement in the beam, which should be anchored in the compression zone of the beam. The reinforcement in the nib should extend as near to the front face of the nib as requirements for cover permit and it should be anchored there by being in the form of a closed loop. Shear should be checked and the shear stress should not exceed normal values for reinforced concrete multiplied by a factor equal to twice the effective depth divided by the eccentricity.

12.7.6 Composite concrete construction

Precast members of reinforced or prestressed concrete are commonly used in composite construction, particularly beams in combination with a structural cast *in situ* concrete topping to form floor slabs. Design of composite sections generally follows the methods developed for sections in reinforced or prestressed concrete with the exception that, when considering serviceability, some allowance should be made for differential shrinkage between the precast and cast *in situ* concrete.

With regard to serviceability, the recommendations for calculation of deflection and cracking normally apply with some additional provisions. For precast prestressed units in composite members, the compressive stress in the concrete may be up to 50% greater than for noncomposite conditions, provided that the ultimate mode of failure of the member is tensile in form and, when the member is effectively continuous, the compressive stress at the end of the prestressed unit within the transmission length for the tendons may be ignored. Also, for prestressed precast units with compositely cast *in situ* concrete, the flexural tensile stress under composite loading in this concrete should not exceed 3.2 N/mm^2 for 25 grade concrete, 3.6 N/mm^2 for 30 grade concrete, 4.4 N/mm^2 for 40 grade concrete and 5 N/mm^2 for 50 grade concrete; these values may be increased by up to 50% so long as there is a corresponding numerical reduction in the tensile stress permitted in the prestressed concrete unit. For normal conditions of use, it will usually be sufficient to assume that the differential shrinkage between cast *in situ* and precast concrete is 100×10^{-6}, otherwise the advice in Part 2 of the Code should be taken.

The flexural strength should be checked for ultimate loading conditions by the methods given for noncomposite sections. Particular attention, however, needs to be paid to the horizontal shear strength at the interface between the precast and the cast *in situ* concrete under ultimate loads. The Code limits the horizontal shear stress for design to values which depend on the nature of the interface and the amount of vertical reinforcement. The horizontal shear force is calculated from the ultimate bending moment and is taken as the total compressive force when the interface is in the tension zone or as the compressive force in that part of the section above the interface when this is in the compression zone. The ultimate shear stress, v_h, for use in design is then obtained by distributing the shear force as given above over the contact area between the point of maximum positive or negative moment and the point of zero moment and then adjusting the value for stress, so calculated, in proportion to the vertical shear stress in the member. Values for v_h are given in Table 12.14. When nominal links are provided across the interface, they should have a cross-sectional area of not less than 0.15% of the contact area. When the horizontal shear stress exceeds that in Table 12.14, the reinforcement required, which should be anchored on both sides of the interface, A_h, in equal millimetres is given by:

$$A_h = (1000b \cdot v_h)/0.87f_y \qquad (12.46)$$

The thickness of the structural topping should, in general, be greater than 40 mm and nowhere less than 25 mm, and special care should be exercised in placing to obtain effective adhesion.

12.8 Structural testing

In design and construction in reinforced and prestressed concrete the testing of components and structures can play an important role. It may take the form of:

(1) Model or full-scale testing to aid in the evolution of improved analytical and design methods.
(2) Development testing of prototypes of components or ancillary equipment for performance or feasibility of construction procedure.
(3) Check testing of factory production as a control on quality of output.
(4) Investigations of structural adequacy of construction which for some reason may have become suspect.

Model or full-scale testing to obtain a better understanding of structural behaviour has been used for many types of construction, as already mentioned. The tests may relate to solving general structural problems or may deal with the design of specific structures. In either case, testing requires sophisticated backing of experimental facilities and staff and can be undertaken only by universities and technical colleges, government laboratories and established industrial research organizations. The planning of the experiments and the interpretation of the

data are largely matters for those with the expert knowledge and experience.

The development testing of components possibly for mass production in the factory or of ancillary equipment such as bridge bearings, prestressing jacks and structural connections is the direct concern of designers and specialist contractors. The objects may be to produce economic design, to simplify procedures or to ensure that details in design give the performance required. Where concern is with repetitive production or procedures, statistical methods of analysis should be used; where it is with structural performance some care may be necessary to be sure that, in isolating the problem for experimental evaluation, it has not been so simplified that its solution is irrelevant.

Testing as a control procedure has been used very widely by industry for many years and has found application in the production of precast concrete components. It may take the form of nondestructive testing of a proportion of production possibly with a smaller proportion being tested to failure. The types of test and the procedures are covered by British Standards for a large number of products, such as kerbs, paving slabs, pipes and lamp standards.

The Code of Practice, BS 8110, deals with some of the aspects of testing concerned with checking concrete quality and refers to the cutting of cores and the use of gamma radiography, of ultrasonic tests, of covermeters and of rebound hammers. These methods provide some check on construction when the quality of the work is in doubt and are used to give guidance on the need for structural tests on components and structures. Such checks may become necessary because of faults in construction, because the structure has been damaged possibly by fire or because of a change of occupancy.

The Code contains recommendations for load tests on structures or more usually on parts of structures to provide a check on serviceability or strength. It should be noted that when a part of a structure, e.g. part of a floor, is to be tested, attention should be paid to the possibility that some proportion of the test load may be supported by other parts of the structure not being subjected to test. To avoid the effects of lateral transfer of load in tests on part of a floor, the width tested should be at least as great as the span. The deflections and strains on the centreline transverse to the direction of the span may then be assumed to be unaffected.

The test loading recommended in BS 8110 is given as not normally less than the characteristic dead and imposed loads combined, and not greater than either the characteristic dead load with 1.25 times the characteristic imposed load or 1.125 times the sum of the characteristic dead and imposed loads, whichever is the greater. This level of loading is sufficient for checks on serviceability limit states but insufficient for a direct check on ultimate strength. In general, it is seldom feasible to make a direct assessment of compliance with ultimate limit state requirements, since to do so would almost certainly render the

Table 12.14 Permissible design ultimate shear stress (v_h) (N/mm²)

Grade of in situ *concrete*		25	30	40 and over
Vertical reinforcement	Type of surface			
No links	1	0.4	0.55	0.65
	2	0.6	0.65	0.75
	3	0.7	0.75	0.8
With nominal links projecting into cast *in*	1	1.2	1.8	2.0
situ concrete	2	1.8	2.0	2.2
	3	2.1	2.2	2.5

Notes: Surface type (1) As cast with a rough finish or open-textured as extruded.
           (2) Brushed, screeded or rough tamped, i.e. surface deliberately roughened but not exposed aggregate.
           (3) Laitance removed by washing or surface treated with retarder and cleaned.

structure unserviceable, but some information on ultimate strength may be inferred from ancillary measurements such as deflection or strain at loads somewhat greater than the characteristic dead and live loads. Some caution is needed, however, in interpreting such information since it may not be possible always to rule out the sudden development of an unexpected mode of failure, such as shear failure when bending failure is expected.

At least two tests should be made applying the load in increments, and an interval of at least 5 min should be allowed after each increment before measurements of deformation are taken with at least a gap of 1 h between tests. Sometimes, it may be preferred to leave the maximum load in position for 24 h. Whenever deformation measurements are to be made in tests, provision should be made for recording the environmental temperature and, if there are likely to be variations, the relative humidity, since changes in these conditions may affect the results.

Results should be compared with the design calculations, and it may therefore be necessary to establish the strengths of the materials, determined in the way described in the appropriate British Standard, and to allow for any features of the structure that differ from what was assumed in design. Cracking and deflection in the tests should be consistent with the design calculations and, after the second loading, the recovery should be at least 75% for reinforced concrete and class 3 prestressed concrete structure and at least 85% for other prestressed concrete construction. After the tests, the structure should be examined to determine whether there are any unexpected signs of distress.

For prestressed concrete structures any departure from linearity of the load–deflection curve when plotted will give some indication of the level of prestress existing in the structure. This information is particularly useful in the case of structures damaged by fire since the retention of a high proportion of the prestress indicates that the tendons have not lost strength.

Other forms of test may be necessary to establish the structural performance of possibly an unusual form of construction or of precast concrete components as part of a quality assurance scheme, when the test schedule should be agreed by all concerned. Should it be necessary as part of such a system to test components to failure, the Code recommends that the ultimate strength of the units should be not less than 5% greater than the design ultimate load and the deflection under the design ultimate load should not exceed 2.5% of the span.

12.9 Fire resistance

The Building Regulations define requirements for health and safety in general and set amongst other matters the provisions that must be made to ensure safety in buildings in the event of fire. The Regulations deal with the prevention of the spread of fire, means of escape and facilitating the fighting of fires. They set out the maximum size of buildings or compartments into which buildings should be divided according to the class of occupancy and height, and then define the required fire endurance for the size of building and occupancy. The endurance is determined by tests conforming with the requirements of BS 476.

Owing to lack of comprehensive information on overall behaviour of structures it is assumed that, if each component part of the structure, walls, floors, columns and beams, is designed for the required fire resistance, the whole structure will have the necessary fire resistance. In general, this assumption would appear to be conservative but for certain types of structure, e.g. unbraced tall framed buildings, it might be conceivable for a fire to become widespread in one storey and so lead to instability of the columns at that level as their stiffness reduces with increasing temperature. Isolated fires would probably not produce this effect since the requirements for ties to prevent excessive damage due to accidents would enable the structure to bridge over any individual failures. The possibilities should, however, be given some consideration in design.

Both the Building Regulations and the Code give requirements for the minimum dimensions of members for various periods of fire resistance for concrete floors, walls, beams and columns. These are based on the results of fire tests in which the members were tested in a furnace heated at a rate laid down as a standard time–temperature curve. Results of these tests show that a number of factors may have an important influence although not all can yet be taken into account in design: these include the effects of size and shape of member, the properties of the different types of steel and concrete that may be used, the protection afforded to the steel, the level of stress that must be sustained by the steel (which is dependent on the proportion of the service load carried at the time of the fire) and the degree of restraint provided by the rest of the structure.

Parts 1 and 2 of the Code offer three methods of designing for fire resistance, the simplest method being presented in Part 1 where brief requirements for cover and minimum dimensions of members are tabulated. Some of this information has been extracted in Tables 12.15 and 12.16. Both parts of the Code should, however, be consulted for further details.

The factors, which affect the limitations defined in the tables apart from cover and size, are the degree of exposure of the member, whether it is continuous or simply supported, the type of steel used, and the maximum size and nature of the aggregate. Clearly, if a column is incorporated in construction in such a way that the full effects of a possible fire would not be experienced, a smaller amount of concrete cover is needed to control the rise in temperature of the steel. Continuity of construction allows some redistribution of moments and forces from weakened regions to those less severely heated and, hence, the temperature rise in the steel is less critical in its effect on fire resistance.

The steels used for prestressing are more sensitive to loss of strength at higher temperatures than those used as reinforcement. For example, cold-drawn wire or strand starts to lose strength at about 150° C and is reduced to half strength at about

Table 12.15 Fire resistance – minimum dimensions of reinforced and prestressed concrete members

Fire resistance (h)	Min. width of beams (mm)	Min. width of ribs (mm)	Min. thickness of floors (mm)	Min. width of columns 100% exposed (mm)	50%
0.5	200	125	75	150	125
1.0	200	125	95	200	160
1.5	200	125	110	250	200
2.0	200	125	125	300	200

Table 12.16 Fire resistance – nominal cover requirements for all steel including links

Fire resistance	Nominal cover (mm)						
			(SS = Simply supported: C = continuous)				
	Beams		Floors		Ribs		Columns
(h)	(SS)	(C)	(SS)	(C)	(SS)	(C)	
Reinforced concrete							
0.5	20*	20*	20*	20*	20*	20*	20*
1.0	20*	20*	20	20	20	20*	20*
1.5	20	20*	25	20	35	20	20
2.0	40	30	35	25	45+	35	25
Prestressed concrete							
0.5	20*	20*	20	20	20	20	—
1.0	20	20*	25	20	35	20	—
1.5	35	20	30	25	45+	35	—
2.0	60+	35	40	35	55+	45+	—

Notes:
*These thicknesses of cover may be reduced to 15 mm provided that the maximum size of aggregate does not exceed 15 mm.
+Additional measures are required to avoid spalling referred to below.

400° C, whereas the corresponding temperatures for reduction in the yield stress for reinforcement are 300 and 550° C. Hence, the requirements for cover in prestressed concrete are more onerous than for reinforced concrete.

Spalling of concrete in fire may be experienced when the initial water content of the concrete is greater than 2–3% for dense aggregates and becomes increasingly likely when the thickness of concrete cover exceeds about 40–50 mm. It is most common with siliceous aggregates such as flint gravel, much less so with limestone aggregate and rare with lightweight aggregates, for which the requirements for concrete cover may be reduced. Spalling is also aggravated if thermal expansion of the member is restrained leading to high stresses in the concrete and if the shape of the cross-section leads to concentrations of stress under severe temperature gradients. Its effect is to expose the main steel to the full effects of the fire and so result in premature failure but its effect may be determined by full-scale fire-testing or may be offset by appropriate detailing in design.

The measures that can be taken to avoid spalling include the application of a plaster or vermiculite concrete finish, the provision of some form of cladding such as a suspended ceiling, and the incorporation of a wire fabric mesh within the cover but conforming with the cover requirements for durability or the use of lightweight aggregates.

References

1 Department of Scientific and Industrial Research (1934) *Recommendations for a code of practice for the use of reinforced concrete in buildings*. Department of Scientific and Industrial Research, HMSO, London.
2 London County Council (1938) *Construction of buildings in London*. London County Council.
3 British Standards Institution (1948) 'The structural use of reinforced concrete in buildings'. CP 114, British Standards Institution, Milton Keynes (rev. 1957).
4 Institution of Structural Engineers (1951) *First report on prestressed concrete*. Institution of Structural Engineers, London.
5 British Standards Institution (1959) 'The structural use of prestressed concrete in buildings.' CP 115, British Standards Institution, Milton Keynes.

6 British Standards Institution (1967) 'The structural use of precast concrete in buildings.' CP 116, British Standards Institution, Milton Keynes.
7 British Standards Institution (1972) 'The structural use of concrete.' CP 110, Part 1, British Standards Institution, Milton Keynes.
8 International Federation for Prestressing (1978) *CEB/FIP Model code concrete structures*. European Committee for Concrete, Brussels.
9 Commission of the European Communities (1984) *Industrial processes – building and civil engineering:* Eurocode No. 1 – 'Common unified rules for different types of construction and material'; Eurocode No. 2 – 'Common unified rules for concrete structures.' Commission of the European Communities, Brussels.
10 British Standards Institution (1985) *The structural use of concrete*. BS 8110, Parts 1, 2 and 3. British Standards Institution, Milton Keynes.
11 Institution of Structural Engineers (1985) *Manual for the design of reinforced concrete building structures*. Institution of Structural Engineers, London.
12 American Concrete Institute (1983) *Building code for reinforced concrete*. ACI 318–83. American Concrete Institute, Washington DC.
13 British Standards Institution (1984) *Design loading for buildings*. BS 6399, Part 1: 'Code of practice for dead and imposed loads'. British Standards Institution, Milton Keynes.

Further reading

British Standards Institution (1972) *Code of basic data for the design of buildings*. BS CP 3, Chapter V, Part 2: 'Wind loading'. (Revision pending). British Standards Institution, Milton Keynes.
Concrete Society and Institution of Structural Engineers (1985) *Standard method of detailing structural concrete*. Draft for discussion. Concrete Society and Institution of Structural Engineers, London.
Institution of Structural Engineers (1980) *Appraisal of existing structures*. Institution of Structural Engineers, London.
Read, R. E. H., Adams, F. C. and Cooke, G. M. E (1980) *Guidelines for the construction of fire-resisting structural elements*. Department of the Environment, HMSO, London.

13

Practical Steelwork Design

F H Needham BSc(Eng), ACGI,
FIStructE, FICE
and

A L Randall CEng, FIStructE

Contents

13.1 Standards for the design of structural steelwork

13.1.1 British Standards and codes of practice

The British Standards and codes of practice for the design of structural steelwork have been reviewed over the past two decades and new codes, based on limit-state philosophy, have been developed. Some of the new codes have been published since 1980 and others are due to be issued after the publication of this book.

The earlier standards, which have served the industry well for a considerable number of years, are based on elastic analysis. Before being withdrawn they will continue to be available for a transition period of some years from the date of issue of the new codes.

It should be noted that, whereas many current British Standard (BS) codes of practice have the prefix CP in front of the number, more recent issues have BS numbers which the British Standards Institution (BSI) intends to adopt for all future codes. The use of the separate CP numerical series will therefore gradually disappear.

The philosophies behind the new steelwork documents differ considerably from those which have gone before. In particular they will require limit-state design to be used, entailing predictions of collapse loads and limits of serviceability and the adoption of partial load factors for different classes of loading taking into account, perhaps somewhat crudely, the statistical probabilities of different classes of loading occurring simultaneously.

The reader may wonder what the difference is between a standard specification and a code of practice. According to the BSI (see the 1981 BS guide *A standard for standards*), a specification is a detailed set of requirements to be supplied by a product, material or process indicating, wherever appropriate, procedures for checking compliance with these requirements. The function of a specification is to provide a basis for understanding between the purchaser and supplier, and the text is usually written with this interface in mind.

The main function of a code of practice is to recommend good accepted practice as followed by competent practitioners. Codes bring together the results of practical experience and research in a form that enables immediate use to be made of proven developments and practices. Codes tend to be complex documents, in many cases almost resembling textbooks – not only do they recommend good practice but some also indicate practices to be avoided.

Specifications may be looked upon as mandatory documents and codes as only advisory ones. In the UK, however, the code requirements ultimately become mandatory too, since many structural codes are called up, or referred to, in the national Building Regulations.

The following lists some of the more important standard specifications and codes of practice currently in use (1988) for the design of steelwork for structural applications.

(1) BS 153:1972 *Steel girder bridges*:
   Part 1, 'Materials and workmanship'.
   Part 2, 'Weighing, shipping and erection'.
   Part 3A, 'Loads'.
   Part 3B, 'Stresses'.
   Part 4, 'Design and construction'.

It should be noted that this standard is no longer cleared for use in the UK by the British Department of Transport, and BSI have withdrawn it; however, it must be recognized that some overseas countries may still refer to it.

(2) British Standard 449 *The use of structural steel in building*:

Part 1, 1970: 'Imperial units'
PD 3343, Supplement No. 1 to BS 449:1970, Part 1, 'Recommendations for design'.
PD 4064, Addendum No. 1 (1961) to BS 449:1970, Part 1, 'The use of cold-formed steel sections in building'.
Part 2, 1969: 'Metric units': Addendum No. 1 (1975) to BS 449:1969, Part 2, 'The use of cold-formed steel sections in building'.

Part 1 of this standard is, according to BSI, obsolete although it is still referred to both in the UK and, particularly, in overseas countries still using imperial units.

The Supplement No. 1, being an extract from the final report of the Steel Structures Research Committee in 1936, is now a rather dated document that is rarely referred to in present-day designs.

The Addendum No. 1 (1961) to BS 449:1970, Part 1, covers cold-formed sections and is another obsolescent imperial document which the BSI have not yet withdrawn.

Addendum No. 1 (1975) to BS 449:1969, Part 2 is simply the metricated equivalent of the imperial version. Ultimately, BS 5950, Part 5, when published, will supersede it.

British Standard BS 449:1969, Part 2 is a metricated version of Part 1 of 1970. This will ultimately be superseded by BS 5950, Parts 1 and 2; meanwhile the BSI have indicated that there should be a transition period prior to its withdrawal.

(3) Code of Practice 117 'Composite construction in structural steel and concrete'.

Part 1, 1965: 'Simply supported beams in building'.
Part 2, 1967: 'Beams for bridges'.

Both parts of this code of practice are in imperial units. Part 1 will ultimately be superseded by BS 5950, Part 3. Part 2 has been superseded by BS 5400:1979, Part 5, and has therefore been withdrawn by the BSI.

(4) British Standard 5400 *Steel, concrete and composite bridges*.

Part 1, 1978: 'General statement'.
Part 2, 1978: 'Specification for loads'.
Part 3, 1982: 'Code of practice for design of steel bridges'.
Part 5, 1979: 'Code of practice for design of composite bridges'.
Part 6, 1980: 'Specification for materials and workmanship: steel'.
Part 9, 1983: 'Bridge bearings': Section 9.1, 1983 'Code of practice for design of bridge bearings', and Section 9.2, 1983 'Specification for materials, manufacture and installation of bridge bearings'.
Part 10, 1980: 'Code of practice for fatigue'.
Part 10C, 1980: 'Charts for classification of details for fatigue (these are large wall charts of the details presented in Part 10)'.

This standard, which has been written in metric units, generally has the approval of the British Department of Transport. Part 5, as written, is not entirely acceptable and modifications to the requirements are to be published shortly. It supersedes BS 153 (all parts) and CP 117, Part 2.

In order to assess the effectiveness and soundness of the newly drafted standard, the Department of Transport financed an extensive calibration exercise from which much valuable experience was gained. These standards were therefore 'tried and tested' before publication.

(5) British Standard 5950 *The structural use of steelwork in building*.
   Part 1, 1985: 'Code of practice for design in simple and continuous construction: hot-rolled sections'.
   Part 2, 1985: 'Specification for materials, fabrication and erection: hot-rolled sections'.

Part 3, 'Code of practice for design in composite construction'.

Part 4, 1982: 'Code of practice for design of floors with profiled steel sheeting'.

Part 5, 'Code of practice for design in cold-formed sections'.

Part 6, 'Code of practice for design in light gauge sheeting, decking and cladding'.

Part 7, 'Specification for materials and workmanship: cold-formed sections'.

Part 8, 'Code of practice for design of fire protection for structural steelwork'.

Part 9, 'Code of practice for stressed skin design'.

Most parts of this standard are in the drafting stage. Only Parts 1, 2, 4 and 5 had been published by 1988.

British Standard 5950 is being written in metric units and will be based on limit-state philosophy where appropriate. Parts 1 and 2 will eventually supersede BS 449, Part 2. With a view to 'debugging' the newly drafted Parts 1 and 2, the UK Department of the Environment financed a large-scale calibration exercise to verify its suitability prior to finalization and publication.

Part 3 will supersede CP 117, Part 1, and Part 5 will replace Addendum No. 1 to BS 449:1969.

Parts 4, 6, 7, 8 and 9 cover subjects which have not previously been included in British Standards.

In the field of structures for agricultural purposes, the relevant standard is BS 5502:1978–86, Parts 1, 2 and 3 'Code of practice for the design of buildings and structures for agriculture'. Each part is published in a number of separate sections covering, for example, materials, design, fire protection, insulation, etc.

On the subject of overhead runway beams, the current standard, namely BS 2853:1957 *The design and testing of overhead runway beams*, is a very outdated document, written in imperial terms and containing information on steel material and some steel beam sizes long-since withdrawn from production. It does not include details of the parallel flanged universal I-shaped sections specified in BS 4, which present certain design problems that do not arise with the older tapered flange joist sections. Apart from the general design philosophy contained therein, only the section dealing with testing requirements is currently pertinent. For more detailed information on this subject, it is recommended that the reader refers to runway manufacturers' publications.

13.1.2 European and international standards and codes of practice

Eurocode 1: *Basic principles*.
Eurocode 3: *Steel structures*.
Eurocode 4: *Composite steel and concrete structures*.
International Organization for Standardization (ISO): *Steel and aluminium structures*.

Part 1: 'Steel – material and design'.
Part 2: 'Steel – fabrication and erection'.

These standards and codes have been published only in draft form. In many respects they will be similar to BS 5400 and 5950. As they are currently at the drafting stage, considerable committee work is still anticipated before the documents are finally issued. It is well known how long it takes and how difficult it is to reach agreement on the contents of standards and codes within any one country. The reader will readily appreciate, therefore, how much more difficult drafting will be, and how it will be extended over longer periods, before agreement is reached both within Europe and internationally.

The reader may be interested to know why the UK is so concerned with the drafting of Eurocodes and international standards for structural steelwork. Basically, they become harmonization documents for each country's national standards and regulations governing the construction industry both in Europe and worldwide. In the case of Europe it is anticipated that, once published, Eurocodes will be called up in EEC framework directives and thus override national standards.

13.2 Steel as a structural material

13.2.1 Fundamentals of the steelmaking process

It is axiomatic that the designer of any engineering undertaking ought to have a fair understanding of the nature of the material he proposes to use. This ideal does not always obtain in steelwork or other building materials. To provide some grasp of the varying characteristics of steel and the origin and nature of possible defects, it is necessary to consider the manufacturing processes by which steel plates and sectional shapes are made.

The production of structural plates and sections is a three-stage process, namely: ironmaking, steelmaking and rolling. Ironmaking is performed in a blast furnace and consists of chemically reducing iron ore, using coke and crushed limestone. It is essentially a continuous process. The resulting material, cast iron, is high in carbon, sulphur and phosphorus. Steelmaking, on the other hand, is a batch process and consists of refining the iron to reduce and control carbon, sulphur and phosphorus and also to add controlled proportions of manganese, chromium, nickel, vanadium, niobium, etc. where necessary, depending on the grade of material to be produced. During this century, the technique of steelmaking has undergone vast changes in scale and new processes have been developed continually to meet the demands of speed, quantity and quality. Today, however, there are only two major steelmaking processes: (1) electric arc; and (2) basic oxygen (BOS). The latter method is really an enlarged and refined development of the old basic Bessemer process, now generally obsolete, and is some 15 times faster than the open hearth process, which is also obsolescent.

Comparatively little iron is allowed to solidify, the metal mostly being tapped and transferred directly, in the liquid state, to the steelmaking furnace. The 'melt', so called, in a steel furnace may well be several hundred tonnes, economy deriving from bulk production. This in large measure explains why small quantities of specially alloyed steels are expensive.

The steelmaking process may last an hour or more, during which chemical change is taking place. Samples are taken at intervals and analysed for composition in a laboratory. During the minutes which this takes the chemical process continues, and it remains a matter of nice judgement when to stop it by cutting the oxygen and tapping the melt into a teeming ladle. Once tapped into the ladle, a further sample analysis is made, the 'ladle analysis', which is taken as a record of the whole melt; however, it must be recognized that there may be some variability in the dispersion throughout the melt, which explains why samples of a part of the product may show slight variations from the ladle analysis. At this stage most of the slag, being lighter than the steel, rises to the surface of the ladle or is left behind in the furnace.

Next, the steel is poured, or 'teemed', from the ladle into moulds to form ingots (Figure 13.1) into specially shaped castings, or directly into slabs, blooms or billets by the continuous casting process (Figure 13.2)

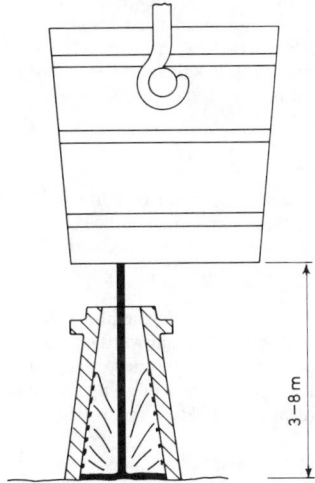

Figure 13.1 Teeming

3–8 m

13.2.1.1 Ingots

Not so long ago, a 10 t ingot was considered big; today, one of 40 t is common. Defects that concern the structural engineer may occur at this stage. In the first place, the steel may be poured from as high as 6 or 9 m into the bottom of the mould, splashing up the sides. Some drops, which 'freeze' instantly on contact with the relatively cold mould, may not remelt when the surface level rises to encompass them, and may even not fully forge into the body of the steel on rolling. This results in surface imperfections, which are mostly of little importance, apart from appearance. More seriously, oxidation inevitably occurs at the free surface of liquid metal, and some slag may still remain in the melt.

Most of this nonmetallic material floats to the surface of the ingot before solidification, but some may remain in the body of the steel, leading to internal laminations after rolling (Figure 13.3).

Ingots free of slag inclusions can be produced by special processes, such as uphill teeming (Figure 13.4) but inevitably are more expensive. For the bulk of heavy-engineering purposes, therefore, one must expect a small amount of laminations. As

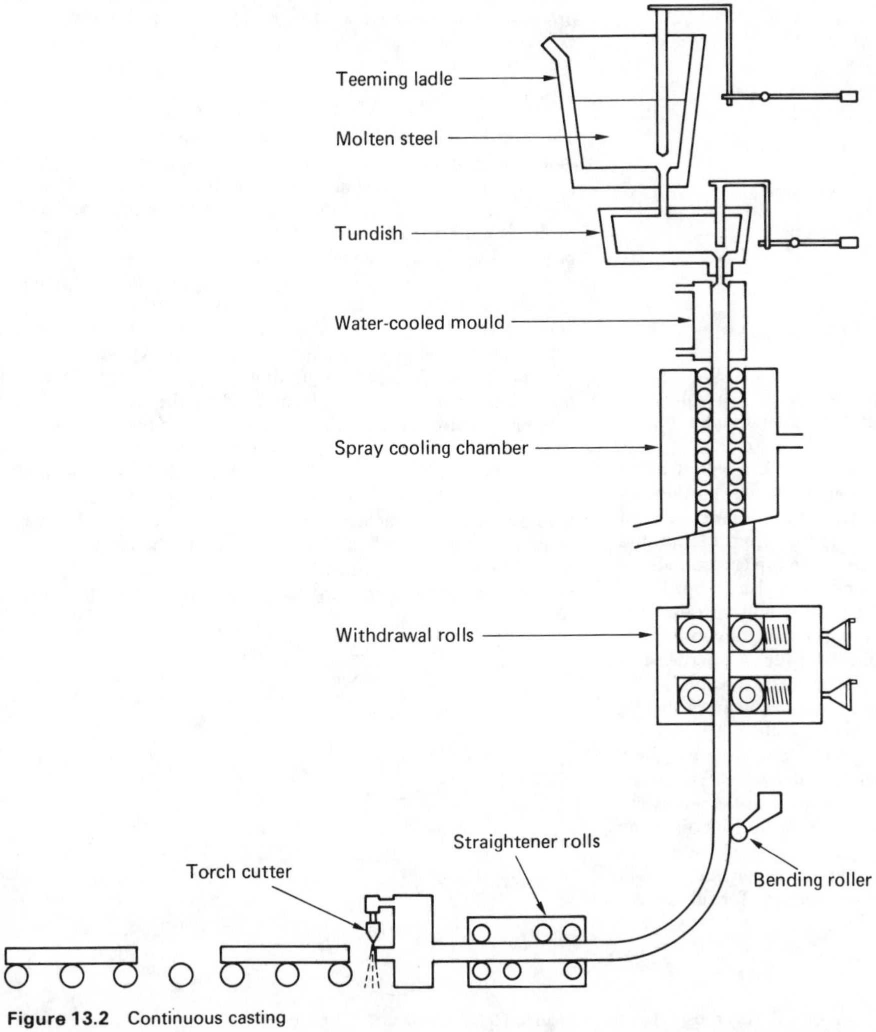

Teeming ladle

Molten steel

Tundish

Water-cooled mould

Spray cooling chamber

Withdrawal rolls

Straightener rolls

Torch cutter

Bending roller

Figure 13.2 Continuous casting

Figure 13.3 Inclusions

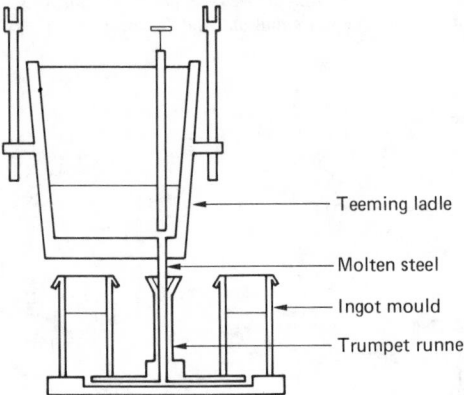

Figure 13.4 Uphill teeming

with knots in wood, what matters is where they are, how big they are and whether they render the material unfit for its intended purpose.

After the ingot has solidified and the mould has been stripped off, the ingot is transferred to a soaking pit, in which other hot ingots are stacked, to ensure even distribution of heat. After a period of time the red-hot ingot is removed from the pit and passed to the primary rolling mill. Here, the top end containing the slag puddle is cropped. The amount to be cropped is a matter of nice judgement on the part of the mill operator, the ingot passing towards him top end first. If he crops too much, the yield (i.e. the proportion of rollable product) is reduced; if too little, end piping or lamination will be present. Naturally, he errs on the safe side, but mistakes can occur sometimes leading to laminations in the end bar or two of a rolling. The ingot is now rolled down in a primary cogging or blooming mill by a series of passes to and fro between the rolls, which are closed slightly between each pass, reducing the girth of the metal and elongating it (Figure 13.5). The hot semi-finished blooms may be allowed to cool at this stage or be further rolled down to billets or narrow slabs. Some of these products may then be sold under the description of 'semi-finished products' for subsequent reheating and finish-rolling.

13.2.1.2 Special castings

Steel castings of various shapes for structural purposes vary in size and are usually required for special applications, such as bridge bearings, crane hooks, rope saddles, rope sockets and special jointing components for structural frames, etc. The molten steel is teemed into sand moulds which are broken up, when the steel has solidified, to remove the castings.

13.2.1.3 Continuous casting

Hitherto, before molten steel could be rolled or formed into plates, sheets, sectional shapes, bars, etc., it had to be cast into ingots which were then reheated in furnaces (or soaking pits) to bring them to a uniform temperature suitable for rolling semi-finished products in a primary mill.

In the continuous-casting process, the liquid metal is teemed direct into a casting machine (Figure 13.2 shows the process diagrammatically) which produces billets, blooms, slabs and dog-bone blooms for sectional shapes, instead of going through the ingot casting stage before being reheated and rolled in a primary mill.

The development of continuous casting has therefore eliminated much of the primary process and, whilst it has not yet superseded all steel-product making, it is used for a large proportion of the production of structural-steel products. Some of the larger-sized shapes, however, are still produced by the ingot route. A further economy of the continuous-casting process is the increased yield of usable product from a given weight of steel.

13.2.1.4 Finishing process

The semi-finished product, produced by either the conventional ingot route or the modern continuous-casting process, is then passed to a finishing mill, a universal beam mill, plate mill, etc. where it is further reduced in size by rolling through a series of reversals until it reaches its final shape which can, for example, be plate, sheet, joist, channel, universal beam, universal column, angle, square or round. Plate is rolled in a two-stand reversing mill which uses two rollers (Figure 13.5) and in which the gap between the rolls can be reduced progressively, the final gap determining the finished thickness. For very heavy work or for wide plates, two further back-up rollers may be incorporated to prevent the working rollers from bending (Figure 13.6). The maximum width available clearly depends upon the width of the widest mill which, in the UK, is currently 3.9 m.

Joists, channels and angles are also rolled in two-stand reversing mills similar to those for plates except, of course, that the desired shape is achieved by machined grooves in the rolls (Figure 13.7 shows the roll shape for forming channel sections). The sections are obtained by being passed through a succession of grooves, whose shapes change progressively from the billet

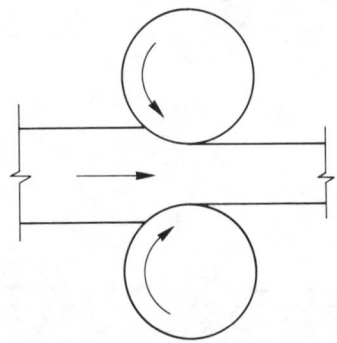

Figure 13.5 Two-stand reversing mill

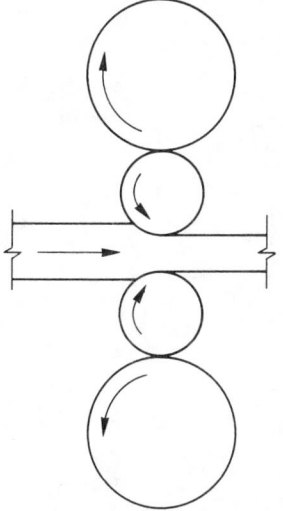

Figure 13.6 Use of back-up rollers

Figure 13.8 Universal mill

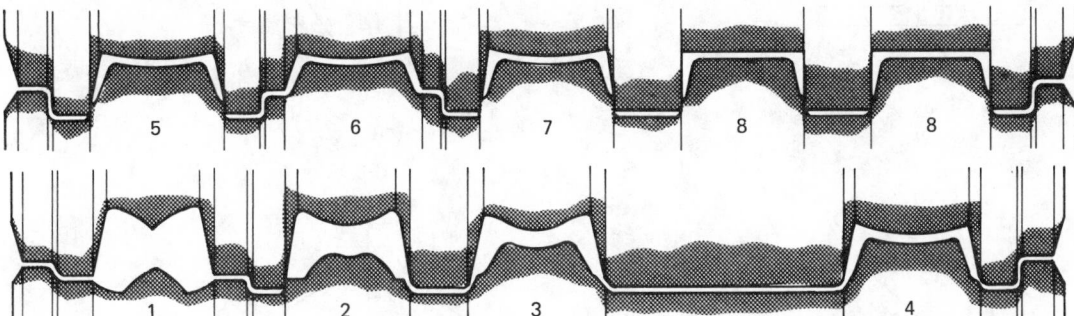

Figure 13.7 Roll shapes and forming sequence for channel sections

shape to that of the final section. Clearly, different sets of rolls are needed for each different sectional shape and size. For maximum economy, roll-changing must, however, be kept to a minimum and the tonnage output of any one rolling kept as high as possible.

Universal beams and columns are rolled in a different kind of mill, known as a universal mill, so called because for any one serial size of a sectional shape, it can roll a variety of final weights. Roll-changing is only necessary when the serial size needs to be changed. Figure 13.8 shows the principle of this type of mill from which it will be seen that different sizes within a serial range can be produced by varying the spacings between both the vertical and horizontal rolls. It is particularly import-ant to note that for universal sections the dimension between the inside faces of the flanges is constant for any one serial size and not the overall dimensions as with rolled joists and channels. The web thicknesses, flange widths and thicknesses can all be varied.

Irrespective of the type of rolling mill employed, the total number of passes depends upon the final required thickness of product, and the cooling rate increases as the thickness is reduced. Thin sections, therefore, go through their final pass much cooler than thick sections, and in so doing take up a degree of work-hardening not evident in thick sections. Since, for any grade of steel, it is desirable to have as little variation in yield strength with thickness as possible, and as alloying ele-ments are expensive, the steelmaker aims to keep additions to a

minimum if thin sections are being rolled. Likewise, thick sections are likely to contain alloying elements near to the maximum for the particular specification. Consequently, the material of which thin sections are made is inherently easier to weld than thick sections, apart from thermal problems which are likely to arise with thick material.

After rolling, the hot bars or plates are sawn to length and transferred to a cooling bank. (It is to be noted that at this stage the steel is most unlikely to be straight or flat.) The rate of cooling of different parts of the plate or section will vary, depending upon its exposure. For instance, the toes of flanges and the centre portion of a deep thin web of an I-section will cool faster than the thicker material at the junction of the web and flanges, and it is this which creates the pattern of residual stresses (Figure 13.9a). The residual stress patterns for channels, angles and plates are shown in Figure 13.9(b)–(d). When the cold material is cold-straightened, sometimes by rolling and sometimes by pressing, dependent on the shape, this affects the level and distribution of residual stresses. The straightening process has a stress-relieving effect and so a bar which needs extensive treatment will finish with a low level of residual stress, while the rare bar which remains straight when cooled will have the highest level. The presence of residual stresses is most clearly seen when a member is cut longitudinally, as when an I-member is slit into two Ts, which curve noticeably on division owing to redistribution of longitudinal stresses. Plate material also shows similar behaviour to a smaller extent.

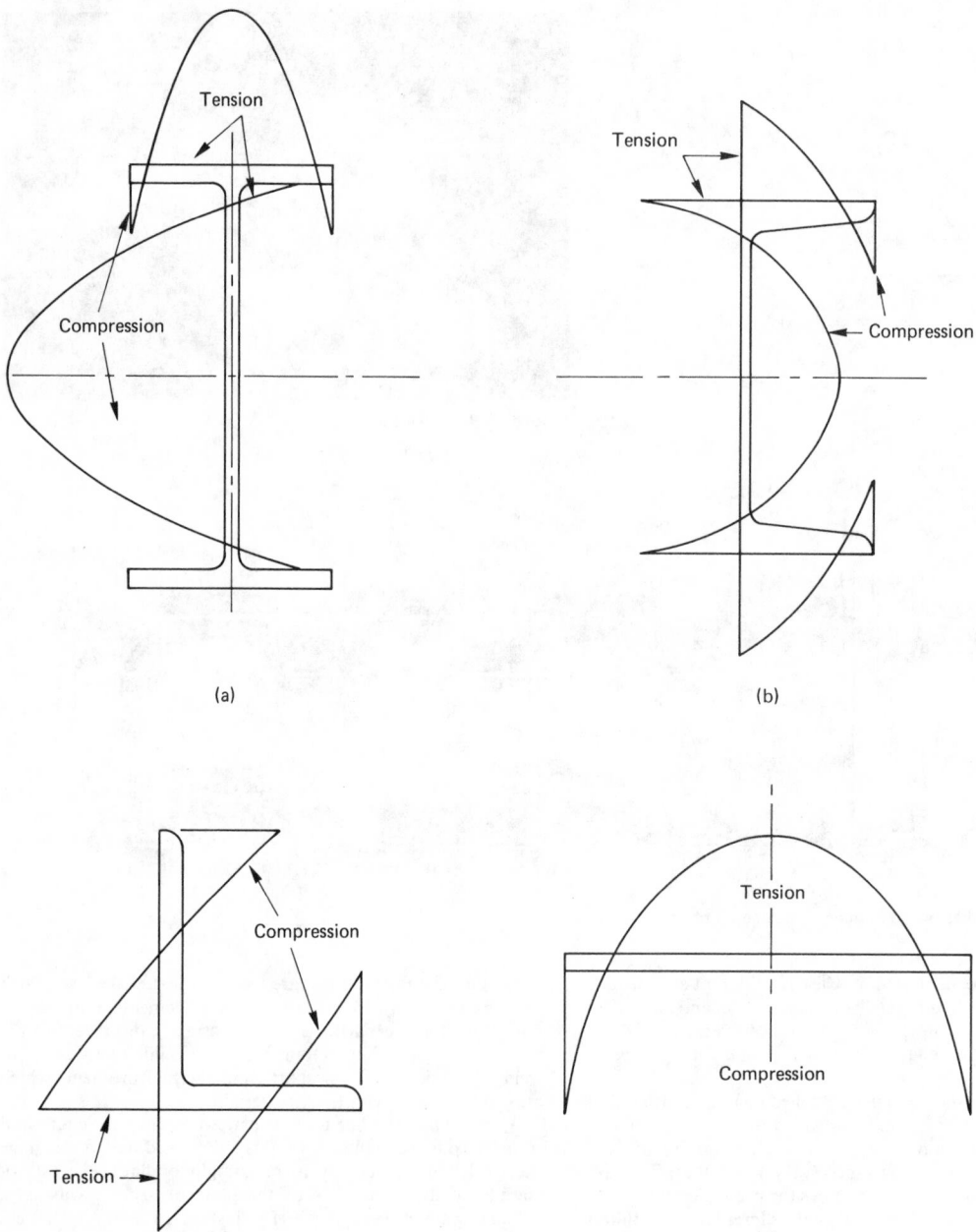

Figure 13.9 Residual stress patterns. (a) Universal sections and joists; (b) channels; (c) angles; (d) flats and plates

13.2.2 Fundamental properties of structural steel

Figure 13.10 indicates the familiar tensile stress–strain curve for steel. The properties of usual concern to the the design engineer are the value of the yield strength (R_e) and the gradient of the elastic portion, i.e. the modulus of elasticity, the value of which may be taken as between 200 and 210 kN/mm². Fortunately or otherwise (some think otherwise) the second property varies little from one grade of steel to another and cannot be controlled. It is to be noted that the finite value of the tensile

strength (R_m) sometimes incorrectly referred to as the ultimate tensile strength (UTS) is of little significance in structural design, since a structure can have manifestly failed at a stress much lower than yield stress. The gradient of the line immediately after yield (i.e. the strain-hardening modulus) can be of significance in plastic design as it can affect moment/rotation behaviour. But the property which is at least as significant from a structural point of view as the yield strength is the ductility, expressed by the length of the horizontal portion of the curve. It is this plateau of ductility which enables the steel to relieve and redistribute residual stresses arising from cooling and welding.

Another factor to be recognized is that the yield and tensile strength values quoted in standards usually refers to a specimen cut in the rolling direction. A similar specimen cut transversely to the direction of rolling would show lower characteristics and, for thick plate, through-thickness strength can be markedly less.

Having referred to residual welding stresses some remarks concerning their mode of origin and magnitude seem appropriate. During welding, the heat put into the heat affected zone (HAZ) causes the zone to try to expand, but this expansion is prevented by the surrounding cold material. The HAZ therefore yields in compression and becomes effectively shorter and, when cooled, attempts to contract further causing stress reversal. The weld and the HAZ is then in a state of residual tensile stress, with a corresponding residual compressive stress in the main bulk of the material maintaining equilibrium. It can be assumed that, as welded, all weld metal and HAZ material is in a state of tensile stress equal to the yield stress. These residual stresses can be relieved, by heating in a stress-relieving furnace or by local heating with electric mats or by proof loading to a degree greater than that anticipated in service. It is here that the property of ductility comes into play. For the majority of structural purposes, stress-relieving is too expensive and occurs fortuitously on the first application of severe load, during which the 'peaks' of the residuals are cut off, so to speak. This explains the partly inelastic initial deflection of a heavily welded member. Recoveries of initial deflection of 80 to 90% have been observed, after which behaviour is fully elastic.

The foregoing underlines the necessity for achieving adequate ductility in the weld metal. The attainment of strength usually presents no problem, owing to the quenching effect of fairly rapid cooling. This quenching effect can be too severe tending to produce brittleness as, for example, in multipass welds in thick material. Therefore, preheating and, in extreme cases, postheating, are needed in order to slow the cooling rate in the interests of achieving ductility.

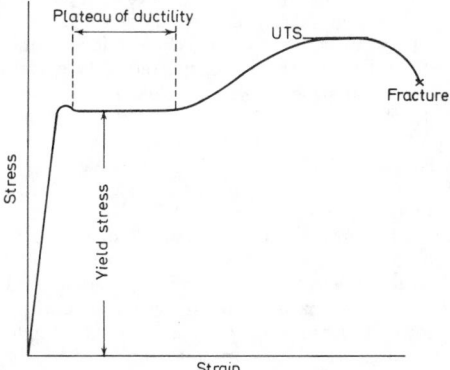

Figure 13.10 Tensile stress–strain diagram (mild steel)

13.2.3 Notch ductility

Let us now consider the mechanism of ductility. We recognize that uniaxial extension is accompanied by transverse contraction. If biaxial tensions are present, only one dimension remains to provide elongation, i.e. a reduction in thickness. If triaxial tensions are present, no distortions are possible, and apparent strength is much greater than the yield strength shown by a uniaxial tensile test piece. (This is the converse effect of triaxial compressions in soils or concrete.) However, a marked loss of

ductility accompanies this increased tensile strength. We then have the situation where a material demonstrably ductile can behave in a brittle fashion under certain stress conditions. The material itself does not distinguish between the origins of different stresses, so that the situation where residual stresses in two mutually perpendicular directions exist together with applied stress in the third is to be avoided. This situation occurs in positions such as joints, where three-dimensional continuity is accompanied by restraint against contraction.

Stresses at the tip of a crack subjected to tension perpendicular to the plane of the crack are essentially triaxial which, as we have seen, leads to brittleness. The property of notch ductility or toughness is the ability to resist the propagation of such cracks under tensile stress.

One must recognize that, even with the closest of control over an industrial process, some defects will occur. It therefore follows that no weld can ever be perfect and may contain slag inclusions, porosity and cracking, and so for design purposes one must assume that welds contain cracks which are undetectable. We have therefore to face the fact that adequate notch ductility is vital in any welded structure. The problem, however, is one of measurement.

The generally accepted measure of notch ductility is given by the Charpy impact test in which a 10 mm square notched test piece 100 mm long is struck and broken by a pendulum and the loss of kinetic energy measured. The specimen is supported at both ends and struck in the centre opposite the notch, putting the notch tip into tension. The radius at the notch tip is 0.25 mm which is very much greater than that at a crack tip. The specimen is also quite small relative to actual structures and, of course, not subject to residual stresses.

If, however, standard material is tested over a range of temperatures, and energy value is plotted against temperature, a curve of the form shown in Figure 13.11 results. Steel shows the characteristic of high notch ductility at high temperatures and a marked fall in notch ductility at low temperatures. The temperature, or rather range of temperature, over which this transition takes place is known as the 'transition temperature'. The transition temperature can be altered by heat treatment of the steel, quenching tending to raise the level, and annealing, or normalizing, tending to lower it. For a given steel, strength can be increased at the expense of notch ductility and transition temperature, and vice versa. For any particular requirement a balance has to be struck.

It is important to recognize, though, that the Charpy test has its limitations. Firstly, the specimen is small and does not reveal the inherent increase of brittleness with thickness due to triaxial

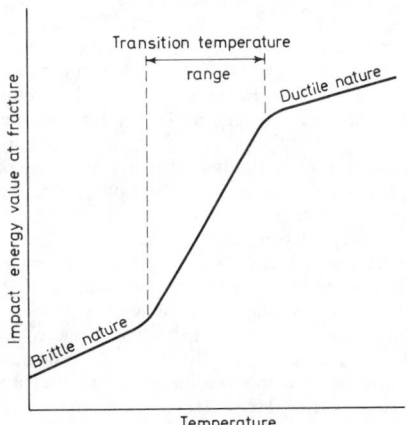

Figure 13.11 Impact–transition temperature curve

effects. Secondly, the notch tip is relatively blunt and, finally, the strain rate is high. High strain rate leads to loss of ductility, so the combination of these effects means that a finite value of a Charpy test does not give any accurate guide as to the actual in-service behaviour of a particular structure or detail. What it does provide is a quality control standard at the steelworks by which a batch of steel can be compared with another of known performance.

Other tests giving much closer correlation to in-service performance have been evolved, notably the Wells wide plate test, the crack opening displacement test and the Pellini drop weight tear test.

Factors leading to possible brittleness, as stated in the first paragraph of this section, include three-dimensional continuity or restraint, such as is usually present in welded components of thick material, coupled with loading giving high strain rates. But in addition to these, three features are essential for a brittle fracture to occur, namely: (1) a notch or severe stress concentration; (2) a tensile stress; and (3) a service temperature below the transition temperature. If any one of these three features is absent, brittle fracture cannot occur. The task of the designer, therefore, is to make a careful assessment of the particular situation in order to determine the degree of risk. In the majority of structural configurations, the risk is slight but where it is not, such factors as are under the designer's control, namely the detailed design and the material quality, in that order, must be modified. It is true to say that, of the failures that have occurred, more could have been prevented by good detailed design than by varying the steel quality.

For a fuller treatment of the subject of brittle fracture and testing procedures the reader should refer to material prepared by Richards[1] at the Welding Institute.

13.2.4 Fatigue

Failure from brittle fracture is sudden and temperature-dependent and can occur at any time in the life of a structure (though usually early owing to the effect of in-service stress relieving). Failure due to fatigue, however, is quite different. It arises principally due to the cumulative effect of pulsating or alternating stresses, and propagates slowly, often over a period of years. As one would expect, initiation occurs at a stress raiser – such as a sharp notch or crack – and its rate of growth is dependent both on the magnitude of the stress variation and the total number of cycles of stress. Whilst fatigue crack propagation occurs slowly it must be appreciated that it is nonductile in nature, i.e. no significant deformation takes place, and therefore detection of cracking can be difficult.

In considering the design of a structural element, the designer's first task must be to assess whether the live loading is essentially static or dynamic. In the majority of cases of building steelwork (with the notable exception of gantry girders and other lifting gear) the loading is essentially static and therefore fatigue presents no problems. Design may, therefore, proceed on the usual working stress basis. However, in bridgework, crane structures and their supports, and other structures subject to dynamic loading such as wind-induced vibrations, prevention of fatigue failure may determine the design.

As indicated earlier, stresses can be fluctuating above a certain minimum level or alternating or somewhere between the two. What matters is the stress range, or amplitude and the mean stress. It has been established that if one does a series of tests on identical specimens, varying the stress range S and recording the number N of cycles to failure and the results are plotted of S against log N, a straight line results as shown in Figure 13.12. This shows that, as the stress range is reduced, so the number of cycles to failure is increased. A family of curves can be derived for different levels of mean stress. If the proposed

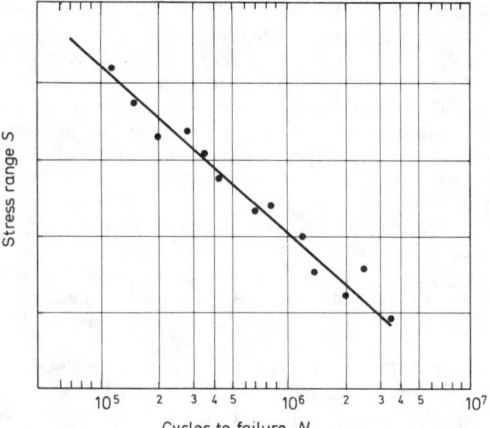

Figure 13.12 Fatigue curve

life is known, then a permissible stress range can be found and compared with that permitted under static conditions.

If we now study the effect of varying the form of specimen to include on the one hand plain rolled plate and on the other a plate with discontinuities, such as holes, notches or welded attachments, another phenomenon presents itself, i.e. that the presence of discontinuities significantly reduces the fatigue life – the more severe the discontinuity, the greater the effect. A plain-rolled plate at a stress of 250 N/mm² will give a fatigue life of 2×10^6 cycles in pulsating tension, i.e. at 2×10^6 cycles, plain grade 43 steel plate has a strength equal to the yield stress. (Note that there is no particular magic about 2×10^6 cycles. It may be a more than adequate life or it may be grossly insufficient in any particular case.)

A detail with a marked discontinuity can show a strength at 2×10^6 cycles, as low as 30 or 40 N/mm². As a result of extensive experimental work at the Welding Institute it has been possible to classify various structural details in terms of their severity as regards fatigue. These are clearly presented in the fatigue clauses of BS 5400:1980, Part 10, to which reference should be made.

For crane structures designed to BS 2573 *Rules for the design of cranes* (Part 1, 1983 and Part 2, 1980), it should be noted that the fatigue rules are still based on BS 153 rules, not those in the new BS 5400, Part 10.

It may seem curious that fatigue can be a consideration when the applied loading is wholly compressive. This is because, as explained earlier, the presence of high residual stress due to welding can cause local reversals. Thus, a fatigue crack in a zone of high residual tensile stress, not overcome by the applied compressive stress, will propagate until it reaches a zone where the total stress is wholly compressive. Care must also be taken to see that zones of high local stress are not overlooked as, for example, the severe stress occurring in the top flange-to-web joint in a gantry girder immediately below a local wheel load.

In considering fatigue one recognizes that, in practice, stress variations are mostly random in nature and rarely of a uniform amplitude. If a picture of the stress spectrum can be arrived at, either by prediction or by measurement, cumulative damage can be assessed by applying Miner's rule, which states that the fatigue damage at any particular stress level is directly proportional to the number of cycles of that stress applied and accumulates linearly until failure occurs. This leads to the relation:

$$\sum \frac{n}{N} = \frac{n_1}{N_1} + \frac{n_2}{N_2} + \frac{n_3}{N_3} + \ldots + \frac{n_r}{N_r} = 1 \qquad (13.1)$$

Failure is predicted in terms of n_r the number of cycles of a given magnitude and type actually occurring during a given period, and N_r the number of cycles of that type required to produce failure in a constant amplitude test. Failure is assumed to occur for k different types of loading when the summation:

$$\frac{k}{r} = 1 \frac{(n_r)}{(N_r)} = 1 \tag{13.2}$$

Full details are given in a publication by the Welding Institute[2] and by Gray and Spence.[3]

It should also be emphasized that fatigue behaviour cannot be assessed with any great accuracy, there being a wide scatter evident in test results. This arises partly because the degree of severity at a discontinuity, such as a butt weld or fillet weld, depends upon the weld shape and the presence or otherwise of weld undercutting. Thus the skill of the individual welder enters into the picture, which is, of course, a variable.

The Welding Institute also publish a valuable booklet edited by Richards[4] which is worthy of study if fuller information is needed.

13.2.5 Engineering judgement

In the matter of fatigue and brittle fracture, it should be appreciated that in practice these factors are not dominant in the majority of structural configurations. When they are, however, they can be absolutely critical. The safety and life of the structure thus depends on the engineering judgement of the designer in his assessment of the various factors involved. This cannot be divorced from considerations of the consequences of failure, which may vary from simple inconvenience to catastrophic failure. Accordingly, the designer must therefore err on the optimistic or pessimistic side.

Published Document PD 6493:1980 *Guidance on some methods for the derivation of acceptance levels for defects in fusion welded joints* – provides guidance concerning the present state of knowledge on specifying acceptance levels for welds by making use of fracture mechanics methods. It covers brittle fracture, fatigue, yielding, corrosion fatigue, stress corrosion, creep, etc. and includes details on the simplified treatment of the use of fracture mechanics methods to establish acceptance levels based on fitness for purpose. The BSI issued the document as a published document rather than a British Standard because further research is required before the available information can be codified. Feedback on the published document is therefore encouraged to further such information.

13.3 Available structural steel material

13.3.1 British material standards for structural steels

In 1968 a new British Standard, BS 4360, *Weldable structural steels* was published, which superseded all the previous standards for structural steels. So important had welding become as a fabricating technique that it became necessary for all steels marketed for structural purposes to be of weldable quality. At the same time a new classification system was introduced whereby steels were graded according to their tensile strength and new higher-strength steels were introduced. Since 1968, BS 4360 had been revised on three occasions in 1972, 1979 and 1986; these revisions have not only incorporated changes resulting from developments in steelmaking practice but take cognizance of users' latest design requirements.

The current version of BS 4360 is the 1986 edition. It presents compositions, tensile strengths, yield strengths and notch ductility (or Charpy impact) properties for five principal grades,

namely 40, 43, 50, 55 and WR 50 (these numbers roughly corresponding to the tensile strengths expressed in kilograms force per square centimetre) for plates, rolled sections and hollow sections. It also shows the nearest equivalent Euronorm 25 and ISO 630 grades where applicable (see section 13.3.2).

The WR 50 group covers seven grades of weather-resistant weldable structural steels, generally based on the familiar 'Corten' steels, whose properties fall close to the grade 50 group as present specified.

The dimensional tolerances, i.e. rolling margins, for plate and flat material are given, and for all products details of all mills tests and their frequencies. These include tensile tests, bend tests, flattening tests for hollow sections and, where the specified properties require, Charpy impact tests. Other matters dealt with include identification, marking, provision of test certificates and inspection.

The three principal documents are:

(1) BS 4360:1986 *Specification for weldable structural steels.*
(2) BS 1449:1983 *Steel plate, sheet and strip.*
    Part 1: 'Specification for carbon and carbon manganese plate, sheet and strip'.
(3) BS 2989:1982 *Specification for continuously hot-dip zinc coated and iron–zinc coated steel: wide strip, sheet/plate and slit wide strip.*

The latter specification is used primarily for light-gauge cold-formed sectional shapes such as profiled metal decking and cold-formed channel purlins and side rails. See sections 13.5.1 and 13.5.10 for further details.

13.3.2 European and international material standards for structural steels

13.3.2.1 European standards

Euronorm	EU 25-72	*Structural steels for general purposes.*
	EU 113-72	*Special-quality weldable structural steels grades and quality. General provisions*

Both EU 25 and EU 113 are now being reviewed by the relevant EEC working group to produce a single European standard (EN).

	EU 130-77	*Cold-rolled noncoated mild unalloyed steel flat products for cold forming – Quality standard.*
	EU 131-77	*Cold-rolled noncoated mild unalloyed steel flat products for cold-forming. Tolerances on dimensions and shape.*
	EU 137-83	*Plates and wide flats made of weldable fine-grained structural steels in the quenched and tempered condition. Part 1, 'Technical delivery conditions – General requirements'.*
	EU 139-81	*Cold-rolled uncoated nonalloy mild steel narrow strip for cold-forming. Quality standard.*
	EU 140-81	*Cold-rolled uncoated steel narrow strip. Dimensions, tolerances on dimensions, shape and mass.*

EU 142-79   *Continuous hot-dip zinc-coated unalloyed mild steel sheet and coil for cold-forming. Quality standard.*

EU 143-79   *Continuous hot-dip zinc-coated unalloyed mild-steel sheet and coil for cold-forming. Tolerances on dimensions and shape.*

EU 147-79   *Continuous hot-dip zinc-coated unalloyed steel sheet and coil with specified minimum yield strengths for structural purposes. Quality standard.*

EU 148-80   *Continuous hot-dip zinc-coated unalloyed steel sheet and coil with specified minimum yield strengths for structural purposes. Tolerances on dimensions and shape.*

EU 155-80   *Weathering steels for structural purposes. Quality standards.*

13.3.2.2 International standards

ISO 630:1980   *Structural steels.*

ISO 1035/1:1980   *Hot-rolled steel bars – dimensions.* Part 1, 'Dimensions of round bars'.

ISO 1035/2:1980   *Hot-rolled steel bars.* Part 2, 'Dimensions of square bars'.

ISO 1035/3:1980   *Hot-rolled steel bars.* Part 3, 'Dimensions of flat bars'.

ISO 1035/4:1982   *Hot-rolled steel bars.* Part 4, 'Tolerances'.

ISO 1052:1982   *Steels for general engineering purposes.*

ISO 4950/1:1981   *High-yield strength flat steel products.* Part 1, 'General requirements'.

ISO 4950/2:1981   *High-yield strength flat steel products.* Part 2, 'Products supplied in the normalized or controlled-rolled condition'.

ISO 4950/3:1981   *High-yield strength flat steel products.* Part 3, 'Products supplied in the heat-treated (quenched and tempered) condition.*

ISO 4951:1979   *High-yield strength steel bars and sections.*

ISO 4952:1981   *Structural steels with improved atmospheric corrosion resistance.*

ISO 4995:1978   *Hot-rolled steel sheet of structural quality.*

ISO 4996:1978   *Hot-rolled steel sheet of high-yield stress structural quality.*

ISO 4997:1978   *Cold-reduced steel sheet of structural quality.*

ISO 4998:1977   *Continuous hot-dip zinc-coated carbon steel sheet of structural quality.*

ISO 5951:1980   *Hot-rolled steel sheet of higher yield strength with improved formability.*

ISO 5952:1983   *Continuously hot-rolled steel sheet of structural quality with improved atmospheric corrosion resistance.*

ISO 6316:1982   *Hot-rolled steel strip of structural quality.*

ISO 6930:1983   *High-yield strength flat steel products for cold-forming.*

ISO 7778:1983   *Steel plate with specified through-thickness characteristics.*

13.3.3 Economic considerations

One of the principal concerns of the designer is the matter of economy, and this must be considered when specifying the grade of steel to be used. Clear guidance on this issue is somewhat sparse. Not only is the pricing structure of the various grades complex and subject to periodic change but even if material costs are known, no clear-cut answer can be given in most instances. This is largely because more than half the cost of a finished steel structure derives from fabrication and erection costs. A paper by Needham[5] sets out in qualitative terms guidance for engineers in achieving the economic use of structural steelwork. The higher grades of steel require more carefully controlled, and therefore more costly, welding procedures, applied to a reduced total tonnage. Therefore, estimated rates per tonne can be misleading, total costs being the criteria. In particular, fabricators tend to load tender rates when steels with which they are not familiar are specified. Suffice it to say that in many structural applications the high-tensile steels possess potential which has not as yet been fully exploited.

Over the years, numerous attempts have been made to draw realistic cost comparisons between components and structures built from the various forms of structural material available. It has long been accepted that many steel-framed structures show considerable advantages over other forms of building both in economy and speed of construction. One of the most recent comparisons[6] is worthy of study.

If an engineer needs rather special properties outside those in the published standards, and the bulk tonnage required is likely to warrant the making of a special cast (say 250 t and upwards), he should explore the possibilities with the steelmakers.

Also worthy of mention is the possibility of supplying, for structural purposes, quenched and tempered (QT) steels of tensile properties much higher than grade 55. Such QT steels are only on the fringe of the structural market and are expensive but, for particular requirements, such as for offshore oil and gas platforms, these steels are available against a special order.

13.4 Available structural steel shapes

13.4.1 British Standards for structural shapes

As already mentioned in section 13.3, only the dimensional rolling tolerances for plates, wide flats, universal wide flats and flat, round and square bars are specified in BS 4360. For sectional shapes the sizes, sectional properties and dimensional rolling tolerances are specified in the standards published for the different profiles. These are BS 4 *Structural steel sections*, and BS 4848, *Specification for hot-rolled structural steel sections*, each of which is published in several parts.

BS 4:1980, Part 1 'Specification for hot-rolled sections' (as amended subsequently) covers universal sections (beams and columns), joists, structural Ts cut from universal beams and column sections, rolled Ts and channels.

BS 4:1969, Part 2, 'Hot-rolled hollow sections' was withdrawn in 1979 and replaced by BS 4848:1975. The sizes given in BS 4, Part 1, although in metric terms, are based on the old imperial sizes which have been in existence for many years. This range of sections is known as a 'soft' metric series. On the other hand a 'hard' metric series produced in true metric sizes (not simply by metricating imperial sizes) has been under consideration by the ISO for some 25 years, but to date agreement has not been reached between the world's numerous steelmaking companies. It seems, therefore, that the current ranges will be with us for some considerable time to come.

In the case of hollow sections, angles and bulb flats, these have been metricated, their sizes and properties have been published under BS 4848 *Hot-rolled structural steel sections*. The only parts issued so far are:

Part 2, 1975 'Hollow sections'.
Part 4, 1972 (1986) 'Equal and unequal angles'.
Part 5, 1980 'Bulb flats'.

These three parts are based on, and generally in accordance with, the equivalent international standards, namely ISO/R 657/14 for hollow sections, and ISO/R 657/1 and 657/2 for equal and unequal angles and Euronorms, namely EU 56-77 for hot-rolled equal angles, EU 57-78 for hot-rolled unequal angles and EU 67-78 for bulb flats. Although the ISO range of angles was revised in 1983, the UK steelmakers do not intend amending the range of sizes currently given in BS 4848: 1986, Part 4.

It should be noted that considerable variation in tolerance levels exists between the various national specifications with regard to cross-section length and area/mass tolerances of steel sections. For example, BS 4 and BS 4848 call for a ±2.5% tolerance on the specified mass per unit length, whereas on the continent of Europe and worldwide (see the equivalent Euronorms and international standards) the mass/area tolerance is usually not only greater (or wider) but is based on a total bar length, not a unit 'metre' length as in the UK. Generally, British steel plates and sections have tighter tolerances than most foreign products!

In the field of structural cold-formed steel products the two main uses are channel shapes for purlins and side rails, and profiled sheets for cladding, roofing and flooring. This type of product lends itself to many different shapes and sizes and, whilst standard ranges are produced by numerous cold-forming companies, individual 'one off' shapes can be obtained (at a cost!) if especially required for a project.

British Standard 2994:1976 *Specification for cold-rolled steel sections*, gives dimensions, properties and dimensional tolerances of a basic range of shapes, namely angles, channels, Ts, Zs, etc.

Euronorm 162-81 *Cold-rolled steel sections produced on machines for cold-forming sections* gives details of some six shapes available in Europe.

No standards are yet available on metal decking profile shapes.

Various hot-rolled structural shapes are available in accordance with the national standards of many countries throughout the world and the details of some of these are also covered in relevant Euronorms (for the ECSC) and international standards (worldwide). As an aid to identifying appropriate national structural steelwork specifications and to assist in the determination of equivalent sections where conditions dictate that sections other than those specified may be used, the British Constructional Steelwork Association published a summary[7] of various national steelwork design codes, steel material specifications and dimensions and properties of sections. This publication is currently under review for revision.

13.4.2 Euronorms for structural shapes

Euronorms are not yet widely used but the following lists the relevant documents.

EU 19:57	*IPE joists with parallel flanges. Dimensions.*
EU 24:62	*Standard beams and channels. Rolling tolerances.*
EU 29:81	*Hot-rolled plates 3 mm thick and above. Tolerances on dimensions, shape and mass.*
EU 34:62	*Broad beams with parallel sides. Rolling tolerances.*
EU 44:63	*Hot-rolled IPE joists. Rolling tolerances.*
EU 53:62	*Broad flanged beams with parallel sides. Dimensions.*
EU 54:80	*Small hot-rolled steel channels.*
EU 55:80	*Hot-rolled equal flange Ts with radiused root and toes in steel.*
EU 56:77	*Hot-rolled equal angles (with radiused root and toes).*
EU 57:78	*Hot-rolled unequal angles (with radiused root and toes).*
EU 58:78	*Hot-rolled flats for general purposes.*
EU 59:78	*Hot-rolled square bars for general purposes.*
EU 60:77	*Hot-rolled round bars for general purposes.*
EU 67:78	*Hot-rolled bulb flats.*
EU 162:81	*Cold-rolled steel sections (produced on machines for cold-forming). Technical conditions of delivery.*

13.4.3 International standards for structural shapes

A limited range of sizes from these ISO standards have been adopted in the UK and form the basis for the structural shapes given in BS 4 and BS 4848.

ISO/R 657/1:1968	*Dimensions of hot-rolled steel sections.* Part 1, 'Equal-leg angles – metric series – dimensions and sectional properties'.
ISO/R 657/2:1968	*Dimensions of hot-rolled steel sections.* Part 2, 'Unequal-leg angles – metric series – dimensions and sectional properties'.
ISO 657/5:1976	*Hot-rolled steel sections*, Part 5, 'Equal-leg angles and unequal-leg angles tolerances for metric and inch series'.
ISO 657/11:1980	*Hot-rolled steel sections*. Part 11, 'Sloping flange channel sections (metric series) – dimensions and sectional properties'.
ISO 657/13:1981	*Hot-rolled steel sections*. Part 13, 'Tolerances on sloping flange beam, column and channel sections'.
ISO 657/14:1982	*Hot-rolled steel sections*. Part 14, 'Hot-finished structural hollow sections – dimensions and sectional properties'.
ISO 657/15:1980	*Hot-rolled steel sections*. Part 15, 'Sloping flange beam sections (metric series) – dimensions and sectional properties'.
ISO 657/16:1980	*Hot-rolled steel sections*. Part 16, 'Sloping flange column sections (metric series) – dimensions and sectional properties'.
ISO 657/19:1980	*Hot-rolled steel sections*. Part 19, 'Bulb flats (metric series) – dimensions, sectional properties and tolerances'.
ISO 657/21:1983	*Hot-rolled steel sections*. Part 21, 'T-sections with equal depth and flange width – dimensions'.

13.5 Types of steel structure

It is helpful in this section to review the classes of structure in which structural steelwork finds its principal applications. Accurate statistics of use are elusive, the best available being recorded by the British Steel Corporation (BSC). However, the rolling mills produce a much greater tonnage of plates and sections than are revealed in construction statistics. Much of this difference is accounted for by sales to stockholders, the ultimate destiny being obscure, the motor industry for truck and trailer bodies, crane manufacturers, pressure vessels, scaffolding, etc. There remains, though, an inexplicable discrepancy in the figures.

Below is given a schedule of types of structure, followed by guidance on some of them which it is hoped will prove of assistance in the early stages of design.

(1) Light industrial buildings, e.g. factories, warehouses, etc. which are mostly single-storey structures.
(2) Heavy industrial buildings, e.g. power stations, steelworks, mill buildings, etc.
(3) Multistorey buildings, e.g. offices, hotels, flats and industrial structures.
(4) Industrial plant, e.g. bunkers, hoppers, conveyor structures, tanks, petrochemical plant, etc.
(5) Single-storey institutional and commercial buildings, e.g. hypermarkets, sports stadia, etc.
(6) Pressure vessels.
(7) Short- and medium-span bridges.
(8) Major bridges.
(9) Other miscellaneous structures, e.g. electricity transmission towers, television masts, railway electrification structures, temporary bridges, space frame roofs, offshore and maritime structures, including oil and gas platforms, etc.
(10) Use of profiled metal decking in some of the types of structure noted above.

It is to be noted that tubular structures have not been shown as a separate category since circular and rectangular hollow sections (CHS and RHS) have been widely used alongside more traditional sectional shapes for many years. Indeed, these shapes are now accepted throughout the industry as one of the arrows in the quiver, and their full versatility is being exploited more and more. The same principle applies to sectional shapes cold-rolled from strip.

13.5.1 Light industrial buildings

It is in this application that there exists the greatest competition between fabricators, which is reflected in design. There is plenty of scope for new ideas and changes in techniques as indicated by the swing from roof trusses to portal frames. Plastic design has been widely adopted for design of portals, but it is worth pointing out that if absolute economy is desired (and the roof space is not needed for stacking goods, etc.) tied portals show benefit, i.e. using a horizontal slung tie between the knees at eaves level. Controversy continues as to whether haunched or

Figure 13.13 Typical portal frame building

uniform section portals are most economic. Whilst material costs tend to increase, so too does the cost of labour, but at a faster rate. Therefore, in today's conditions and for the foreseeable future the amount of fabrication should be minimized even if this means an increase in material content. Tapered members are often used but, in the majority of cases, at the expense of overall economy.

A paper by Horridge and Morris[8] gives details of an extensive investigation into the relative costs of the more popular structural forms and should be helpful to design engineers in determining the most appropriate arrangements.

Whilst not condemning the pursuit of economy in the design of frames it must be stressed that it is unwise to economize on wind and rafter bracing. There have been too many instances of frames moving or collapsing before the cladding has been fixed. In this context, interesting work has been done by Professor E. R. Bryan at Salford University on the stiffening and bracing effect of cladding.[9] Even when his methods of stressed skin design are contemplated, adequate reusable erection bracing must be included.

It is not often appreciated that the cost of purlins and side cladding rails form significant portions of the total. Two points should be considered here: (1) if the structural designer has freedom in fixing the spacing of trusses or frames it is worth designing the purlins and rails first and stressing them fully by adjusting their spans; (2) cold-formed sections almost invariably show cost advantages over hot-rolled angle or channel purlins or side rails, largely because most cold-formed sections are symmetrical and consequently show greater strength/weight than angle shapes.

Whilst cold-formed sections are almost universally used today, it is sometimes worth considering the use of hot-rolled symmetrical shapes, such as small-sized universal beams, joists or channel sections, particularly if the first procedure mentioned above is adopted, i.e. maximum spacing of supports to utilize maximum strength. In addition, purlin cleats can be dispensed with as these purlin sections can be attached directly to the frames.

In light factory construction, increasing emphasis is being put on the benefit to productivity of a pleasant environment, and this tends to influence design. Tubular space frames have much to offer here and should be considered. On the other hand, for high-bay warehouses and warehouses designed round the characteristics of the forklift truck, internal aesthetics are insignificant and structural capability is all-important. Benefit can be derived by integrating the roof and wall structure with the internal racking.

13.5.2 Heavy industrial buildings

In considering this group one must emphasize that great accuracy of calculation and sophisticated techniques of analysis are largely misplaced. Highly competitive designs which give rise to a host of minor troubles to the user are not preferred to relatively crudely designed structures of a high standard of robustness. The principal loading arises from the plant to be accommodated and, whilst in an ideal world the plant design should be finalized before the structural engineer starts, in practice this situation rarely obtains. Instead, one can expect loading figures given in the first instance to be subject to almost continuous and sometimes radical change during the development of the project, even up to the erection stage. Accordingly, the wisdom of providing an additional margin of stress in design to accommodate some of these probable loading changes will be evident. There is also the strong possibility of structural requirements changing during the lifetime of the building, and the designer will not be blessed if a succession of strengthening operations is needed subsequently.

In this class of structure there is not a great deal of scope for originality on the part of the designer. For the most part, plant requirements dominate and the structural layout is not decided – it occurs. It is important, nonetheless, to ensure adequate economy by studying actual needs. For example, a steel melting shop and a power station turbine house may demand cranes of similar capacity but, whereas the ladle crane may make a dozen passes a day at maximum load, the turbine house crane only needs its full capacity for the replacement of a turbo-alternator every few years. Clearly, different levels of fatigue are present in these instances, which results in quite different standards for structural adequacy.

Figure 13.14 Anchor steelworks, Scunthorpe

13.5.3 Multistorey buildings

Steelwork for this application faces the severest competition from concrete, so much so that a decade or so ago it seemed as if it had almost been ousted from the scene, except for a few very special applications. This had largely come about as a result of fire regulations. There are signs, however, which indicate that with some of the newer concepts, particularly for very high-rise prestige office buildings, steelwork offers economic and practical solutions. Two factors have brought about this situation. Firstly, the fire regulations require the lift and stair well enclosures to be separated and protected from the spread of fire from the occupancy floors. Secondly, the current wind code CP3:1972, Part II, Chapter V requires larger transverse forces to be taken by the building than pertained a decade ago. What better solution could there be than to have a reinforced concrete core containing the lifts and services which is itself capable of resisting transverse forces and surrounding it with a simple steel frame supporting only vertical forces? In certain cases, the height and plan form of the core may require vertical post-tensioning to enable lateral forces to be resisted. This post-tensioning can be provided by the surrounding office floors if these are suspended from the top of the core. Several examples of such structures have been constructed; these have the added advantage, in city centre developments, of keeping all the foundation work within the plan area of the building.

Two developments have taken place in multistorey buildings in recent years. Firstly, the gas explosion at Ronan Point in 1968 has led to the general recognition that structures need to have adequate three-dimensional coherence if they are to be sufficiently resistant to accidental damage. This philosophy is now embodied in section A3 of the Building Regulations,[10] which states that in the event of misuse or accident, damage should not be disproportionate to the cause. In truth, with very few

exceptions, any normal well-designed steel-framed multistorey building will be found, on analysis, to comply automatically with the statutory requirements for peripheral and transverse tying forces, although in design it is now necessary to check this. If difficulty is found in compliance, it should be taken as an indication that the basic structural anatomy is capable of improvement.

Figure 13.15 National Westminster Bank

Secondly, partly arising from the wide acceptance of sheet steel composite floors (treated in detail in section 13.5.10) and the savings in construction times thus brought about has come a wider recognition of the economic savings accruing from rapid construction generally. Given that immediate occupation is anticipated, high interest rates dictate that the fastest method is, in the long run, the cheapest. This recognition has led to some increase in the use of steelwork for multistorey buildings.

Having dealt shortly with the latest developments one must state that now, and in the future, the bulk of multistorey steel frames will be discontinuous column and beam frames designed on simple methods. That is to say that beams are assumed to be simply supported and columns are assumed to carry only such bending moments as arise from the end reaction of the beams acting at a given eccentricity from the column centreline. This method is easy and quick to apply and has stood the test of time in that failures in such structures are extremely rare indeed, and if they do occur it is usually due to instability during erection. However, it is important to realize that the stress values arrived at in calculations bear little relationship to those actually experienced by the frame. This is due primarily to the omission of any consideration of joint stiffness between beam and column. In practice a true simple support is very rare and, even with the simplest connection, some moment transference will take place between them. In the event of both beams and columns being encased in concrete, the degree of joint stiffness will be considerable. Thus, the simple design method tends to overestimate beam moments and underestimate column moments.

This does not matter, certainly so far as internal columns supporting beams of approximately equal span and loading are concerned, but outer columns (particularly corner columns) carrying asymmetrical moments deserve some thought, certainly if uncased. Therefore, in a highly competitive design it is quite safe to trim the design of beams and internal columns to the bone, even to accepting some degree of overstress, but leaving the face and corner columns conservatively designed.

The results of tests on full-scale structures (apart from tests on isolated elements) have indicated a very considerable margin of strength in the columns, well in excess of any predictions made by elastic design. Clearly, further research is needed before this can be codified into a simple design method.

13.5.4 Industrial plant

In many respects this group is a speciality and indeed has its own trade association. Design problems are likely to arise in many respects, notably in connection with working stresses and whether those specified in BS 449 and BS 5950 are applicable and appropriate. Judgement is required, particularly in respect of the possibility of overloading, accidental or deliberate. Clearly, tanks cannot be overloaded, and a slender load factor may be appropriate, whereas in some other applications normal load factors will result in structures insufficiently robust for their purpose.

In design, fatigue may be a consideration in dynamically loaded structures such as conveyor frames, gantry structures, etc. Reference to the BS 153, Part B or BS 5400, Part 10 rules will indicate whether this is a governing factor.

Figure 13.16 Plate girder for British Steel Corporation project, Scunthorpe, being welded

In such work, maintenance costs to prevent corrosion can be heavy. It is not sufficient to design and detail a steel structure and then ask oneself what means of protection should be applied. It may by then be too late since moisture traps may have been built in and certain areas may be inaccessible for painting after erection, e.g. back-to-back angles. Care must be taken in detailing to ensure that water cannot collect in puddles. If this situation is unavoidable, drain holes should be provided. The BSC publish two useful booklets[11,12] which give simple guidance. Chandler and Bayliss[13] provide a practical guide to current knowledge on coatings and processes involved in achieving sound protection of steelwork. It should be borne in mind that hot-dip zinc coating gives excellent protection at modest cost and that the trend is for longer baths enabling greater lengths and larger areas of steel to be treated.

13.5.5 Single-storey institutional and commercial buildings

In recent years a considerable increase in steel use has occurred in such buildings as leisure centres, swimming baths, cash and carry warehouses, sports stadia, hypermarkets, exhibition halls, etc. to the extent of becoming a new classification. A study reveals a very wide variety of different structural forms, including columns and beams, column and truss, arches, domes, space frames, etc. and it is clear that potential exists for the use of imagination and inventiveness. However, they mostly share two characteristics: (1) that they are single-storey; and (2) require the roofing of large areas, mostly uninterrupted by internal columns. As such, the latter presents common problems, including the disposal of large quantities of rainwater, the provision of services of considerable total weight often within the roof structure and, not least, the accommodation of temperature movements. This last is known to have caused problems in the plumbing of columns, when cladding sensitive to such movement was adopted. Provision must be made for movement joints in the cladding of both roofs and walls, and also possibly in the structure itself. In the absence of the latter in a large building it is futile to call for accuracy in plumb which is less than the anticipated total expansion.

Agreement as to total tolerance must be obtained at the design stage or disputes may subsequently arise. Further, lining and levelling of the final structure, before fixing the cladding, is best done at night, when the structure has settled to an even temperature with no differential between the shaded side and that exposed to strong sunlight.

13.5.6 Pressure vessels

The design of pressure vessels forms the subject of a separate British Standard BS 5500:1985 *Specification for unfired fusion welded pressure vessels.* The requirements laid down therein are considerably more stringent than for normal steelwork, both as regards material quality and fabrication, and inspection procedures are correspondingly more severe. That this is right is undeniable, having regard to the catastrophic consequences of possible failure, but it does lead to fabrication costs greatly in excess of the usual. Specialist plant is needed, such as that necessary to bend thick plates and spin dished ends.

In addition to the basic BS 5500 standard, the BSI also publish a series of *Enquiry cases* under the same number, which are supplied free of charge to subscribers of the BS 5500 updating service. These documents, which are numerous, provide information on a wide variety of pressure vessel problems.

In 1982, the Pressure Vessels Quality Assurance Board (PVQAB) came into being; its prime purpose is to provide the pressure vessel industry with a means of recognition, both nationally and internationally, of the quality of their products. The Board was originally set up under the aegis of the Institution of Mechanical Engineers with the object of providing a central UK authority capable of approving and certifying the quality assurance systems of pressure-vessel manufacturers, their products, and the material suppliers. It has now been taken over and is being run by Lloyd's. It is a similar scheme to that operated for many years by The American Society of Mechanical Engineers (ASME).

13.5.7 Short- and medium-span bridges

Competition with concrete is very severe. It will frequently be found that the most economic solution is arrived at by using steel beams or girders acting compositely with a reinforced concrete deck. In design it is no longer adequate to consider statical transverse distribution. An analysis in accordance with the theories of Hendry and Jaeger[14] or Morice and Little,[15] or a finite-element grillage analysis will demonstrate great economy in the sizes of beams at the expense of a minimum addition to the transverse reinforcement in the slab. If full transverse distribution is assumed it is usual to provide continuous transverse top steel in the slab.

A number of model tests have been carried out on composite bridges designed to these principles (see also section 13.7.4.5). Departure from strict linearity occurs at around twice working load, which indicates reasonable economy with adequate safety. Destructive tests, however, give enormous margins of strength at collapse, ultimate load factors of 10 being common.

Figure 13.17 Clunie Bridge, Pitlochry

In one such test which the authors witnessed, an ultimate load factor of 7 was achieved with a system in which the transverse top slab steel was discontinuous, i.e. over the beams only, calling into question the necessity for continuous top steel.

Girders may be rolled universal beams, or purpose-built plate girders, either with equal flanges or with a top flange smaller and narrower than the bottom. In the latter case, the intermediate design stage must be examined when the laterally unrestrained beam is supporting the weight of the wet concrete. It is rare for such systems to be designed as propped, and thus composite action is assumed for live load only.

The theoretical advantage of unequal flanges giving minimum weight of steel will prove largely illusory if a rolled beam could have been used as an alternative. Minimum fabrication leads to minimum total cost. A steel bridge design guide by Nash[16] illustrates the application of the new steel bridge code, BS 5400:1982, Part 3.

It is in the kind of bridge discussed, where the girders are largely protected from driving rain, that the 'Cor-ten' or weathering steels are becoming popular. At small extra cost, bridges can be built which will be effectively maintenance-free during their life. Depending on the corrosion environment, the Department of Transport require a sacrificial 'skin' of either 1 or 2 mm on the exposed surfaces. To facilitate the use of rolled beams, modified tables of properties, omitting this 'skin' have been published by Constrado.[17]

13.5.8 Major bridges

It is here that the designer has greatest scope for imagination and inventiveness, as is illustrated by the wide variety of elegant structures which have been built in recent years throughout the world. Steel dominates the scene here, largely due to its strength:weight ratio, since the greater part of the load sustained by a large-span bridge is its own self-weight. It is, unfortunately, given to few of us to have a hand in the design of

such works but, with increasing complexity, design becomes a matter of teamwork, and no one name can any longer be attached to one structure. On the other hand, the hazards are great, as events over the years have shown.

The considerable effort put into box-girder research following the collapse of four box-girder bridges in the early 1970s has been distilled into the new steel bridge code, BS5400:1982, Part 3.

13.5.9 Other miscellaneous structures

Within this group one should include electricity transmission towers (themselves something of a speciality), lighting masts, television masts, railway electrification structures, temporary bridges, footbridges, offshore structures, etc.

Structures frequently are evolved which could not be visualized by the drafting committees responsible for British Standards, e.g. railway electrification structures supported only on one side of the track, gallows style. In this instance, tension in the conductor wires can produce considerable torsion in the column and severe moments arise when the track is on a curve. Accordingly, strict application of codes of practice for design needs to be tempered by engineering judgement, and it is arguable whether the existing British Standards are really applicable in such cases.

Likewise, difficulties have been known to arise in transmission towers. Even greater line voltages have necessitated taller and taller towers. Some years ago, 30 m was a tall tower; today, 75 m is not uncommon and for long river crossings even greater heights are needed. It is not sufficient for the designer merely to extrapolate from existing designs. There will always be a tendency for the designer to assume that what is flat on his drawing, will also be planar when erected. This may be a reasonable assumption up to a certain size, but ceases to be accurate to a larger scale. Self-weight deflections of long members of lattice structures can become significant, particularly if the member is intended to act as a strut.

There was an obvious need for a rational approach to the design of lattice towers and masts which led the BSI to set up a steering committee in 1970 with the task of preparing a code of practice. The result of the drafting committee's deliberation is the recently published BS 8100:1986 *Lattice towers and masts*, which interestingly does not supersede any current standard. It is published in two parts, namely:

Part 1, 'Code of practice for loading'.
Part 2, 'Guide to background and use of Part 1'.

This standard defines procedures for the determination of loading for the design or appraisal of free-standing towers of lattice construction up to 300 m in height.

Part 2 includes two worked examples cross-referenced to the relevant clauses of the code. Also available is a draft for development DD 133:1986, *Code of practice for strength assessment of members of lattice towers and masts* which has been issued in this form to enable industry to provide practical feedback in the light of experience so that ultimately a British Standard code under BS 5950 can be published.

Developments in the North Sea opened up a vast potential market for steel structures. The sizes of many of those built so far beggar description, and the limit has by no means been reached. Again, it was arguable whether BS 449 really applied and the design engineer was in many ways out on a limb and had to work from first principles. To start with, the assessment of loading was far from straightforward and much had to be learned about the effect of wave forces on structures of various shapes. Temperature also had to be considered and the principles of notch ductility and fatigue mentioned in sections 13.2.3

and 13.2.4 had to be observed. From this it follows that a structure satisfactorily designed for use in a tropical sea may not be suitable, indeed may be downright dangerous, in Arctic conditions. The problems, therefore, required a conservative approach, tempered in the light of experience.

Figure 13.18 Electricity transmission tower. (*Courtesy*: Central Electricity Generating Board)

Actual failure of some offshore structures has provided much-needed information and, having spelt out some fundamental lessons, one hopes that the experience gained will result in safer structures in the future, although no doubt there are other traps awaiting the unwary. As designers tend to overlook stability questions on landbased structures in the part-erected condition, so also is there a danger of failing to recognize the situation when an offshore structure is being towed, usually on its side, to its final location. Buoyancy considerations arise which warrant more than cursory attention.

13.5.10 Use of profiled metal decking in some of the types of structure noted in sections 13.5.1 to 13.5.9

In 1983, the first part of the new BS 5950 *The use of structural steelwork in building* was published; it was, in fact, numbered Part 4, 'Code of practice for the design of floors with profiled steel sheeting'. This code is intended to assist both structural designers and the construction industry in general in the selection of steel deck floors for the most appropriate applications, and it is based on a recommendation made in 1970 by Committee 11 of the European Convention for Constructional Steelwork (ECCS). In North America, floors of the type described in BS 5950, Part 4 have been incorporated in the construction of multistorey buildings since the early 1930s. In the early 1960s this kind of floor was introduced into Western Europe where they are now widely used. Various interim recommendations

have been produced in North America and Europe covering their analysis and design, but no standard specifications or codes of practice have yet been published, although in the US a standard has been in the drafting stage for some years.

Profiled decking can be used noncompositely in the form of permanent shuttering or, more usually, compositely with concrete for roof and floor slabs. This type of construction is being specified and used more each year because of its economic and performance advantages over other more traditional forms of floor construction. The deck provides the positive reinforcement for the slab and at the same time replaces the temporary shuttering usually associated with reinforced concrete construction. Once placed, the decking provides both an immediate working platform for subsequent trades on the level on which it is laid and cover for the operatives working on the floors below. Thus, construction can proceed on several levels simultaneously whilst conforming more readily to the requirements of the Health and Safety at Work Act 1974.

Figure 13.19 Metal decking

The use of this type of floor is not limited to steel structures only, since it is equally suitable for use in structures of reinforced and precast concrete, timber or masonry.

The designer should appreciate that it is misleading simply to compare costs of floors using profiled steel sheeting with other more conventional forms of floor construction, since changes in overall weight, stiffness and stability, composite action, speed of construction, etc. will influence the cost relating to other parts of the supporting structure and foundations.

Recent building projects have shown quite clearly that the overall building costs of structures using steel deck floors are considerably lower than those of structures utilizing other forms of construction. The reasons for these cost reductions have been investigated in depth and confirmed in the findings of a report by Walker and Gray.[6]

The advantages of using profiled steel decking may therefore be summarized for both composite and noncomposite floors as follows:

(1) Steel decks provide permanent formwork and do not normally require the use of additional temporary formwork or propping.
(2) The low weight of the steel deck unit means they can be manhandled with comparative ease, thus reducing or eliminating mechanical handling costs.
(3) Faster construction times are obtained with improved safety.
(4) Maximum efficiency and cost-effectiveness is obtained where regular grids are adopted, although the designer has freedom to arrange the structural support framing to accommodate irregular shapes.
(5) For floors acting compositely there is an efficient use of the two basic construction materials of steel and concrete, since the steel deck acts as both formwork and tensile reinforcement.

13.6 Overall structural behaviour

Having referred frequently to British Standards for design, i.e. BS 449, BS 153, BS 5400, BS 5500 and BS 5950, a word of caution is warranted. These documents, as all British Standards, are drafted by committees of experienced people drawn from all quarters, charged with the task of formulating design rules in the light of the state of knowledge at the time, incorporating the best modern practice and exploiting the latest research. This is no easy task. They recognize that many of those who would use the documents would not understand, indeed would not want to understand, the background theory behind many of the requirements. Inevitably, many areas of doubt and uncertainty are encountered and differences of opinion expressed. At the end of the day, they rightly tend on the conservative side in such situations though, in one or two instances, subsequent research has indicated that their rules were not sufficiently cautious in the earlier standards. However, these documents have stood the test of time and, by and large, when properly interpreted result in structures in which safety and economy are reasonably in balance. But they tended to lay undue emphasis on the determination of exact values of permissible stresses and insufficient attention (certainly in the case of BS 449) to questions of overall structural behaviour and stability, particularly during erection.

The young designer can be forgiven for supposing that, if all members and connections in a structure are assessed and designed strictly in accordance with the British Standard, then the complete structure will be adequate for its task. In the great majority of cases this will be so, but in rare instances it has been found to be a fatal error. In a number of cases the designer has, unconsciously perhaps, relied on the stiffening effect of the cladding to provide overall stability and a collapse has occurred, sometimes under the weight of an erection crane before the cladding has been fixed. What one does not know, and cannot assess, is how many structures are satisfactorily in service which were within a hair's breadth of collapse at some stage during erection. As an illustration, it may seem an economic solution in the design of a frame for a multistorey building to have a number of load-carrying plane frames, interconnected with light ties. In the absence of temporary bracing, such a structure relies for stability on the stiffness of the joints at the ends of the tie beams. Such joints, being on nominally unloaded or light members, are often regarded as being unimportant and are sometimes detailed as simple connections, instead of joints

which have some moment capacity. Thus, in the unclad state the frame possesses insufficient longitudinal stability. This illustration is but one of many which could have been quoted but it is hoped that the point has been driven home. Overall supervision of the design of both members and connections by a competent engineer, and the intelligent assessment of the problems by the erection supervisor, should be sufficient to prevent such mishaps.

The stiffening effect of cladding, which has been referred to, is very great indeed but, except in the case of very simple structures, it is likely to remain impossible to assess in the foreseeable future. Because of this and because of the accidental and unquantifiable composite action which occurs between steel members and other materials, the actual stresses experienced in service are likely to be very different and usually much lower than those calculated. This fact, coupled with the fact that loadings which are specified, particularly wind loadings, are at best only general estimates, means that great nicety of calculation is largely misplaced. Thus, the pursuit of notional absolute economy is a cardinal error if it means that under pressure of time the consideration of overall behaviour is omitted.

A related problem frequently encountered and on which little guidance is available, concerns the amount of sway or horizontal deflection permitted at the top of a tall multistorey building under wind loading. It has been common to limit such sway to 1/500 of the height, calculated on the stiffness of the bare steel frame. This largely begs the question, since most of the wind force would be absent without cladding. With it in place, the deflection must be considerably reduced in consequence of its stiffening effect. What is being considered here is not structural safety – that is not in question – but serviceability. In the absence of any knowledge at all of cases where discomfort has been caused to occupants or damage to finishes through excessive sway, the authors conclude that the above notional limit is conservative. Clearly, long-term observations are called for which may take years to yield worthwhile data. Meanwhile, it is suggested that if the above limitation is deemed acceptable for dwellings with curtain walling, then relaxation of the limitation would seem appropriate in respect of two factors: (1) other uses; and (2) stiffer cladding. Again, in city centres, the sheltering effect of neighbouring buildings may be taken into account in assessing wind forces. Is one to take into account the possibility of some, or all, of them being demolished some time in the future? Such considerations need careful assessment, and authoritative advice.

13.7 Design of structural components

There would seem little point in including worked examples in this section, in view of the fact that the old design standards are unlikely to be used for most new projects, whilst designs to the new standards are doubtless covered in numerous recent publications such as that by the Steel Construction Institute.

13.7.1 Importance of correct loading assessment

Great care should be taken in the assessment of dead and live loading, as mistakes at this stage can make all subsequent calculations abortive and, in extreme cases, result in completed members being scrapped. This has been known to happen and unfortunately not all that infrequently. Some common causes of mistakes in this matter are: (1) failing to make adequate allowances for finishes as, for instance, in providing screeding to falls in a roof slab to provide adequate drainage; (2) failure to make proper allowance for surges in gantry structures; (3) incorrect interpretation of wind loading requirements; and (4) effect on pressure of flow in granular materials, etc. Safety requires that a full loading schedule for the intended structure should be drawn up and checked independently before the detailed design of structural frames and members is commenced.

13.7.2 Determination of structural layout

It is all too rarely, in building structures at least, that the structural engineer is summoned sufficiently early in the planning of a project to be able to contribute to the overall efficiency of the building by influencing the choice of layout. The steelwork designer more often is set an almost impossible problem by a form of building already finalized by others with little thought being given to structural needs. Between these two extremes fall the majority of problems. Thus, frequently it comes about that the structural layout merely 'occurs' and the steelwork design engineer has little room to manoeuvre in planning such matters as column layout, overall beam depth, pitch of roof, etc. Where such freedom does exist, however, and time permits, the structural designer should explore the effect upon overall economy of various arrangements before any one particular layout is finally selected.

13.7.3 Calculations

All calculations for a project should be neatly prepared in logical sequence, properly indexed and cross-referenced. It should be remembered that in most cases schemes including calculations have to be submitted to a local authority for approval. Incomprehensible papers invite criticism and do not smooth the road to acceptance. Additionally, mistakes are more easily spotted if the work is clearly presented, and the effect of structural alterations, either during the contract period or perhaps at a much later date, can be more readily followed.

13.7.4 Consideration of individual members

13.7.4.1 Angle members

Angle members, e.g. purlins, cladding or sheeting rails, bracing members, truss components, usually act as simple beams, over one or two spans, or pin-ended struts and ties. Within the stress rules laid down, having regard to deduction for holes and eccentricity of connections, etc. design is a matter of trial and error and simple arithmetic. Certain practical points, though, are worth bearing in mind. Typical mistakes made by designers in the past include: (1) the transport of double-angle rafters in too-long lengths, resulting in extensive damage in transit; (2) the design of diagonal bracing to resist longitudinal surge in a gantry, which although structurally adequate, was much too flimsy (in this case, when examined, the bracing looked far too light, and since the gantry served a scrapyard it was obvious that the bracing members would be quickly damaged); (3) long span purlins and side rails which, prior to placing cladding, required sag-rods; and (4) angles have sometimes been used for unloaded (tie) members in frameworks when heavier I or ⌐ sections, having greater inertia, would have been more sensible.

13.7.4.2 Simply supported beams

Such members fall into two categories: (1) laterally restrained; and (2) laterally unrestrained. In (1), direct design is possible and simple arithmetic will usually be enough to give the required section modulus, either elastic or plastic depending on the design method. The necessary lateral support is often provided either by the load itself or connecting load-bearing beams, a floor of steel chequer plate, profiled steel sheeting, *in situ* or precast concrete or timber suitably secured in some way to the beam. In the case of long spans, it is sometimes economic to provide restraint in the form of light tie beams rather than

design the main beam as an unrestrained member. The laterally unrestrained beam is designed to a lower working stress depending on a number of factors: (1) section shape; (2) slenderness ratio; (3) effective length; and (4) torsional restraint at the supports. The designer must first assess the effective length of the compression flange in lateral buckling and both BS 449 and BS 5950 provide guidance which in turn depends partly on torsional restraint at the supports. A trial section is then selected, the slenderness ratio $l:r$ (length:radius of gyration) calculated, the shape parameter allowed for and a permissible stress derived. Thus, design is a trial-and-error method.

It must be understood that the assignment of an effective length is a somewhat arbitrary process and therefore it follows that permissible stresses based on this value are themselves somewhat notional. Again, great accuracy in calculating actual stresses is therefore unnecessary.

One word of warning: the safe load tables for beams refer to beams fully laterally restrained, and therefore beam sizes cannot be selected direct from these tables if there is any doubt whatever on this issue. Examples of unrestrained beams include wall-bearing beams not at a floor level (as, for instance, at the rear of a lift shaft), beams supporting runways or hoists and, of course, the worst possible case – gantry beams – where the load, far from restraining the compression flange, can impart disturbing horizontal forces due to surge and impact.

13.7.4.3 Plate girders

This is a subject on which whole books have been written. At one time most steel bridges were built of plate girders in one form or another but in recent years box girders have entered the scene in a big way. Plate girders have also themselves been the subject of rapid development and change, owing to the almost complete supersession of riveting in favour of welding. Nowadays most plate girders consist principally of three plates. Variations include asymmetrical girders with larger tension flanges than compression for use in composite action with a concrete deck, and girders which are the opposite, where the top compression flange is wider than the tension flange particularly for use in unrestrained conditions or to resist surge in the case of gantry girders. Plate girders can also have continuous or curtailed flanges. The wisdom of curtailing flanges depends on the size of the girder, and to an extent the form of the loading or rather the shape of the bending moment diagram or envelope. It can be said in general terms that, if the size of the girder is such that flanges can be supplied in one piece, it is never economic to curtail the flanges by introducing butt welds in them, with all that this entails in terms of machined preparations, possible preheating, welding and final nondestructive testing. It is most important for designers to recognize that minimum weight does not necessarily mean minimum cost, and it is in the latter that most clients are really interested. Lower weight constructions usually require a more expensive fabrication.

It is rare in the case of rolled beams that the web requires stiffening to resist buckling tendencies, and then only at the supports and points of concentrated loads. In plate girders, however, the web thickness is decided by the designer and is not simply presented as one of the dimensions of a section of adequate bending resistance. Thus, the web can be much thinner relative to the girder depth than in the case of rolled beams, bringing in its train the necessity to provide web stiffeners. Stiffeners do not have to be of great size, except in the case of bearing stiffeners, and those under a concentrated load. Quite small restraining forces are needed to keep an initially flat plate in that condition. Thus, if appearance requires it, stiffeners can be provided on one side of the web only.

Whilst many rolled beams are used as simply supported members, worthwhile economy can be derived particularly in plate girders by making them continuous or semicontinuous, i.e. alternate cantilevers and suspended spans. This creates a situation over the supports where maximum bending moment and maximum shear force act together, a circumstance which requires special consideration of combined stresses.

Plate girders with thin webs often display waves in the webs during fabrication owing to the shortening of the weld and HAZ in the web-to-flange joint. This can make the subsequent fitting of the stiffeners difficult. Where possible, therefore, the stiffeners should be welded to the web beforehand, but this precludes the use of automatic welding, which is a disadvantage. Thus, it can come about that it is not necessarily economic to design with the bare minimum web thickness, and additional web area does also add to the bending resistance.

If it is necessary to make a full-strength splice in a plate girder the web and flange welds should coincide. With the flanges butted for welding there should be a clearance in the web to allow for shrinkage when welding the flanges, otherwise the waviness mentioned above will again be evident. The reason the welds should coincide is that there will inevitably be a slight difference between the two web depths, making fit-up difficult. Only by great good luck will the fit be exact.

13.7.4.4 Columns

Columns can be continuous as, for example, in a multistorey building, or discontinuous, as in a column-and-truss shed. They may be laterally restrained at intervals by other members, or entirely unrestrained throughout their lengths (or heights). These factors are taken into account in the design of an axially loaded column in accordance with BS 449 by assigning an 'effective length' somewhat arbitrarily as a first step in design. Thereafter, a section is selected for trial, its slenderness ratio, i.e. effective length divided by the relevant radius of gyration, calculated, and from this a permissible stress found from a column-stress formula. Such a formula is usually presented in the form of tables or curves. In BS 5950, a similar procedure is adopted, with the exception that four different column curves are presented. The choice of curve will depend on the section shape, it having been established experimentally that upon this depends the level of residual stress, which influences column behaviour.

There are many stress formulae similar to those of BS 449 used in various codes throughout the world. Most of them incorporate some factor to cover notional initial imperfections, and none of them is exact, since they are all attempts to express mathematically what is in essence a naturally occurring situation. Further, all of them are based on the behaviour, under laboratory tests, of pin-ended struts, and it should be understood that a truly pin-ended strut is a very rare occurrence in structural engineering, since a special bearing would be necessary at each end to meet this requirement.

Almost equally rare is a truly axially loaded column, and some bending moment is almost invariably imparted by the connecting members, about one or both axes, at one or both ends. The values of these bending moments may be insignificant or they may be such that the stresses induced thereby dominate the situation and render the axial stress insignificant. In order to take account of this, some form of stress interaction formula is used. The permissible stress in bending is established in the same manner as for a laterally unrestrained beam. It can be argued that this manoeuvre is not accurate, as indeed it is not, but it has the virtue of simplicity and can be readily applied to a variety of cases. Suffice it to say that it has stood the test of time and results generally in safe columns.

Effective lengths, or lengths between points of contraflexure, have also been seen to vary with the magnitude of the axial load; they also vary with the loading pattern. Thus, in a multistorey

frame, variations of loading can induce conditions ranging from full double curvature to full single curvature, i.e. effective lengths of $0.5l$ and $1.0l$ (Figure 13.13(a) and (b)). One discounts the possibility of 'chequerboard' loading and adopts an intermediate value, say $0.7l$ or $0.85l$.

The foregoing refers to elastic design of columns and, as indicated, the accepted methods err on the side of conservatism. Plastic design methods cannot as yet be universally adopted, and form the subject of widespread research. The first practical design method, for columns in portal frames, was evolved by Professor M. R. Horne. An extended version of Professor Horne's method is given by Morris and Randall.[19] It is in column design that the greatest potential exists for achieving economy through research in pursuit of realistic design methods.

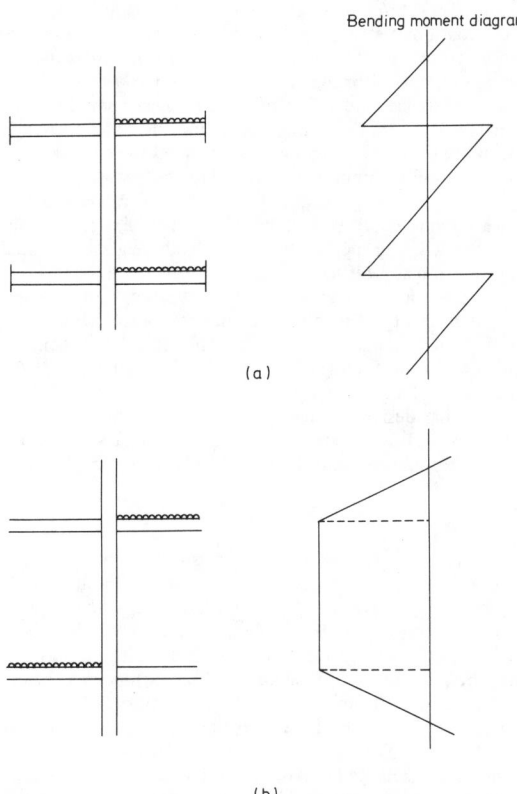

Figure 13.20 Column bending moments

13.7.4.5 Box girders

In the early 1970s nationwide research was undertaken at the behest of the Merrison Committee into a number of governing parameters in box-girder design in order to establish new design rules following the failures of certain bridges, as noted in section 13.5.7. Much of the information published in the public press was misleading and some was downright wrong. It may therefore be helpful to spell out some of the first principles in order to put matters into perspective.

Firstly, box girder construction is not new; some structures built in Victorian times incorporated steel box girders, albeit of riveted construction. The particular virtue of box construction is its much greater torsional resistance compared with normal single web girders. For this reason, welded box girders have been used for a number of years in crane construction, particularly heavy electric overhead travelling (EOT) cranes subjected to racking and surge forces. Of course, with the complete changeover from riveting to welding since the Second World War has come the greater ease with which one can tailor a member to suit a need. It is worth remembering that many hundreds of welded box-girder bridges have been successfully put into service, on the Continent and elsewhere, and none has failed in service.

Clearly, therefore, such structures could be designed successfully. Had it not been for the pressure to build the cheapest possible structure (which in the eyes of some wrongly equates to minimum weight[9] (see section 13.7.4.3)) there would have been no great problems. It was in pursuit of maximum economy, particularly in plate thicknesses, that the trouble arose.

With the exception of the long suspended span, it is true to say that box-girder construction is almost always more expensive than using discrete single web girders, with or without cross-beams. For a straight bridge this is always so – it is only when the structure is subject to high torsional forces as, for instance, in a curved bridge, that the particular virtue of high torsional resistance renders box construction economic. For the most part it has been the clean appearance and good corrosion performance of boxes which have determined the choice.

The Merrison Committee report led to the promulgation of the Interim Design and Workmanship Rules (IDWR)[20] which in turn were replaced by BS 5400:1982, Part 3. Experience has shown that when properly applied this code leads to cheaper box girders, albeit a bit heavier.

Turning to detailed design, one must accept that it has become a speciality of its own. The determination of approximate web and flange sizes to suit the calculated bending moments and shear forces is not of itself difficult, if a bit tedious in the absence of a computer program. But it is in the refinement of the outline design in regard to such matters as support diaphragms, web and flange stiffening, torsional stresses, the effect of shear lag, etc. that the process becomes complicated.

13.7.4.6 Simple bridges

In recent years, there has been some resurgence of the use of steelwork in bridges and viaducts of small and medium span. In some cases alternative designs submitted at the tender stage have demonstrated significant savings and have thus been adopted. In these situations, the cheapest solution has frequently been found to consist of continuous universal beams spaced at 2.5 to 3.0 m apart, acting compositely with the reinforced concrete deck. In all cases the transverse distribution capacity of the deck has been fully utilized by way of a grillage analysis. It is to be noted that design to the new bridge code, BS 5400:1982, Part 3 has increased somewhat the maximum spans for which rolled beams can be acceptable from 22 m to the order of 25 m. For greater spans, welded plate girders of constant depth, at somewhat greater spacing, have been adopted, again continuous and composite. The common characteristic in these successful alternatives has been great simplicity of detail, so as to reduce the workmanship content to the minimum.

Bridges with gracefully curved lower flanges are undoubtedly elegant, but do entail greater workmanship content. Splices have to be carefully detailed and made, and diaphragm bracings contain few common parts. However, only the client authority can indicate where the balance is to be drawn between elegance and economy. Current evidence suggests, rightly or wrongly, that greater emphasis is placed on lowest first cost. This is not to say, though, that parallel girders cannot of themselves be visually attractive, given cleanliness of detail.

13.7.4.7 Square hollow section (SHS) members

Circular hollow sections (CHS) have been available for many years, supplemented in the 1960s by square and rectangular hollow sections (RHS) in a large range of sizes. The use of these members is accelerating annually. Initially inhibited by a high price, ex-mill, compared with traditional sections, this factor is of decreasing importance having regard to the greater rise in fabrication labour rates and, architecturally, the cleaner lines presented.

Although it has long been appreciated that a CHS is technically the most efficient form of strut, the greater ease with which connections can be effected with the square and rectangular shapes has led to their wide acceptance. For example, to fabricate a lattice girder from circular sections requires the use of a profiling machine to generate the interpenetration curves necessary to give a fit-up adequate for welding, whereas only straight cuts are needed for square hollow sections (SHS) or RHS.

Care must be taken, however, in detailing the joints in such structures. It is unwise to use hollow sections for the secondary internal members which are significantly smaller than the main chords, unless some form of stiffening can be used, since the effect of applying a load to a small area of a flat face can lead to premature distortions in the main member. Such mistakes can lead to yielding occurring at or below working load.

Square hollow sections are widely used for space structures, which have aesthetic appeal, but the problems of jointing are considerable. Consequently, the cost of jointing dominates. In order to overcome this difficulty, the Tubes Division of the BSC developed and patented a special joint for space frames which is called the Nodus joint.

It is unfortunate that the way tubular fabrication has developed into a specialization has tended to lead to structures which are either all-tubular or all-traditional. Clearly, the greatest potential for both types of section will be realized only when they are fully blended.

The versatility of SHS is demonstrated by the very wide range of applications, from steel furniture to aircraft hangar roofs. On this topic, a notable success was achieved with the construction of the 'jumbo jet' hangar at Heathrow Airport in 1969 which broke new ground on two counts: (1) it was the largest space frame built at that time, the fascia girder carrying the doors spanning 130 m (the roof behind supports heavy loads from servicing equipment); (2) it was the first large structure in the UK to make use of BS 4360 Grade 55 steel of 448 to 463 N/mm² yield strength which was successfully shop- and site-welded in large thicknesses, notably in the main chords which were formed from curved plate into hollow sections 2.74 m in diameter. This structure clearly pointed the way for others and since its construction many equally imposing large-span structures have been built.

13.7.4.8 Cold-formed members

As indicated earlier, cold-rolled purlins have already established a wide market. Other shapes are used in the industrialized building field and in proprietary components such as lightweight lattice beams. These forms of structural sections have certain advantages: (1) when the loading is light, as in the case of purlins and cladding rails, the section can be structurally more efficient than a hot-rolled section owing to the limited number of smaller sizes of the latter; (2) they can be made from tight-coated galvanized strip which, with a subsequent paint coat, gives excellent corrosion protection.

It is not always appreciated that a cold-forming mill is a relatively cheap piece of equipment, only a fraction of the cost of a hot-forming mill. Thus, if a reasonable market is anticipated the designer has the freedom to design his own 'tailor-made' section to suit his particular requirement. This is not to say that it would be economic to design sections for a particular building but it would be for a range of standard buildings. There are several manufacturers willing to undertake such work and to give guidance.

Figure 13.21 Cold-forming mill

In the design of sections, prevention of local buckling is the prime consideration, and the derivation of exact solutions can be most complex. To prevent local buckling the edges of members are frequently lipped, inwards or outwards, and comparatively slender webs are often formed with a ridge or groove, longitudinally. In deriving a section it is often helpful to ensure that sections will 'nest'. Apart from saving space in transport it also ensures that the minimum damage will occur in transit, to which cold-formed sections are otherwise somewhat sensitive.

It is worth pointing out that, in the manufacturing process, since it is done cold, a significant amount of work-hardening takes place, particularly in those zones bent to a small radius. Thus, a section tends to be stronger than calculated on the basis of the yield strength of the flat strip. This effect is not at present taken into account in design and, hence, one has more margin of stress than might be supposed. When published, BS 5950, Part 5 will give much-needed guidance on this subject.

13.7.4.9 Gantry girders

These members are being considered separately since they present unique problems, not found in other members. One must admit that there is a degree of irrationality in British Standards, as it would seem logical for gantry girders to be designed to the same stresses and safety factors as the cranes they support. However, this is not the case and cranes are designed with higher margins than the supporting structure. Of course, a line of demarcation has to be drawn somewhere and one can put up a good case for making this between the moving and the static structure, which is the situation which obtains. Nonetheless, this brings in its train certain difficulties:

(1) For insurance purposes EOT cranes are subjected to a test load higher than the normal crane capacity. If this should be done with the crane midway between columns, with the crab at the end of its cross-travel, flange stresses in the gantry girder may well be approaching yield stress.
(2) Because the operatives are aware of the test overload requirement, a blind eye is often turned on the specified safe working load (SWL) when a particularly heavy load has to be lifted, and again it is often not possible to estimate the weight of a complicated piece of machinery or other load accurately. These factors lead to occasional overloading of the gantries.
(3) In many applications, notably in steelworks, it is not uncommon for a crane to be uprated by retesting, with little thought being given to the supporting structure.

In view of all this, it is most unwise to design gantry girders, or

indeed their supporting columns, to fine limits. Prudence suggests that a fair margin of stress be left, consistent with reasonable economy. These comments apply with particular force to the heavier types of crane supported by plate girders, and less so to lighter cranes carried on rolled beams.

Turning to the principles of design, BS 449 and BS 5950 lay down certain factors for longitudinal and transverse surge and vertical impact. These represent a reasonable average, although one suspects that the factors tend to be too conservative for heavy cranes and the opposite for light cranes. The treatment of load combinations is given, vertical loading being taken together with surge either along the rail or transverse to it.

In calculating vertical bending moments, one must know not only the load to be lifted but also the crane characteristics in terms of self-weight, crab weight, end-carriage wheel spacing, etc. Envelope diagrams of shear and bending moment should be derived. In this context it is worth bearing in mind that two cranes often run on the same track. One therefore has to take into account how closely they can be spaced. Sometimes long buffers are fixed to the cranes in order to prevent the two cranes from running on to one girder. But beware, it has been known for such buffers to be removed by the operatives because they are inconvenient!

The worst condition, which usually governs the design, is that with maximum vertical bending moment and transverse surge. In the calculation of stresses it is usual to consider that only the top flange resists the transverse surge, this being somewhat conservative. If a rolled beam can be used it will offer by far the cheapest solution. The next-best alternative is a rolled beam as a core section with either a flat plate or a toe-down channel connected to the top flange to accommodate surge stresses. If calculations suggest that the bottom flange also should be reinforced, one should then turn to a tailormade plate girder, since there will be no difference in the number of main weld runs and the section will be lighter. Finally, very heavy gantries are often built in two parts connected at intervals, i.e. main girder to resist vertical loads and horizontal surge girder for transverse loads, this latter often serving also as a maintenance walkway.

Final selection of section size is a trial-and-error process which can be very tedious. As a first trial one could assume a span:depth ratio of about 15 and a top flange with 30 to 50% greater cross-sectional area than the bottom. At this stage it is worth examining the shear situation at the supports in order to determine an appropriate web thickness, which will affect the overall moment of resistance. Ideally, one should arrive at a section where maximum permissible bending stresses are approached simultaneously in top and bottom flanges. This can only rarely be achieved and in any case it is an illusion to calculate stresses to a greater degree of accuracy than one's real knowledge of the loads.

Finally, certain detail points should be considered. Experience with the earliest welded gantry girders was unfortunate, since little regard had been paid to the local effect of rolling wheel loads. In the absence of accurate web-to-flange fit-up, the fillet welds were required to transmit the whole of the local compressive stresses under the wheel. This was sometimes compounded by a slightly eccentric rail, creating a rocking effect to the top flange. The result was early fatigue failure of the web-to-flange fillet welds. Several solutions have subsequently demonstrated their effectiveness: (1) the rail can be mounted on a resilient pad to give overall rather than point contact on the high spots; (2) where possible, the weld position can be moved to a zone of lower local stress by using a T-section top flange, formed by using a universal T-section, i.e. by splitting a universal beam, the web plate being butt-welded to the stalk; and (3) if this is not possible, the web-to-flange weld should be made by a full-strength double-V butt.

The effect of misalignment should also be considered. With

the best of efforts, no foundation can be guaranteed free from some degree of settlement in the lifetime of a building and, if magnified by height, some realignment of the track may be needed. This is greatly facilitated if there is some means provided at the girder supports for transverse adjustment. If this is not provided, the only solution is to realign the rail on the top flange, leading to eccentricity of loading. Some degree of eccentricity is, however, inevitable, owing to lack of fit of the rail, even with a pad, and the wheels not running on the crown of the rail. For this reason, web stiffeners should not be spaced too widely apart and it may be desirable to introduce short intermediate stiffeners supporting the top flange.

Various means have been used for securing the rails on gantries. For light work, bolts through both rail and girder flanges suffice, but these should not be at too wide a spacing, otherwise each bolt will in turn suffer the benefit of full transverse surge. Rail clips are more popular, can be purpose-made, and offer the possibility of adjustment without the necessity to slot holes. Direct welding of rail to flange has been tried, often unsuccessfully. This is mostly due to the fact that rail steels have a high manganese content to give resistance to wear and require a welding technique foreign to many fabricating shops. Also, rail replacement is a most difficult operation!

13.7.4.10 Curved steel sections

Recent developments in section bending have opened up new scope for imaginative structural design using curved steel sections. There are several specialist section benders whose technical brochures give useful information. Briefly, some modern bending rolls are now capable of maintaining the geometric shape of sections during bending without web and flange buckling occurring. The entire range of UK structural sections, and most continental sizes, are now readily available curved about either the major or minor axes. There are obviously limits to the minimum radius for each section, and also cold-bending results in work-hardening, leading to greater strength but reduced ductility, whilst reducing substantially residual rolling

Figure 13.22 Lee Valley Ice Centre. (*Courtesy*: Angle Ring Co. Ltd)

stresses. These factors must be taken into account in structural design, e.g. elastic design must be adopted but, even so, such members allow great architectural and engineering freedom. Typical applications include whalings to support steel sheet piling; domed and vaulted roofs for atria, arcades, etc. arched lintels and strengthening or support for masonry arches. A striking example is the roof of the Lee Valley Ice Centre, in which $533 \times 210 \times 122$ kg/m universal beams, curve to 24-m radius and form nine 40-m span three-pinned arches.

13.8 Methods of design

The current British Standards for the design of structural steelwork, BS 449 and BS 5400, are based on elastic design principles, except that BS 449 has a 'let-out' clause which permits other proven methods of design, or design on an experimental basis. This allows plastic design to be adopted where this is possible, e.g. in continuous beams and single-storey portal frames, but does not lay down any design criteria or even a recommended overall load factor. The new BS 5950 is based on limit-state philosophy, and sets out values for partial safety factors for use in plastic design.

Briefly, elastic design is based on the philosophy of permissible working stresses for different situations, these being some proportion of yield stress in the case of tie members or laterally restrained beams, and some proportion of the critical buckling load in the cases of laterally unrestrained beams and columns. A similar principle applies to shear stresses. Unfortunately, these proportions or so-called 'safety factors' are not stated but are implicit in the permissible stresses specified. If one delves deeply enough one can establish that the safety factors differ from one member to another, which is irrational to say the least. It is this anomaly which the drafting committees for BS 5950 were anxious to eradicate.

If the basic aims of structural design are considered, it becomes clear that structures should have an adequate margin against collapse and not become unserviceable (due perhaps to excessive deflection) under normally anticipated service loading. The only real purpose in calculating levels of stress assumes that from this the margin against failure can be predicted. And here we must come to define failure. The principles of elastic design implicitly define failure as either the attainment of first yield in an extreme fibre in a tension situation, or the attainment of the critical buckling stress in a compression or shear situation. So far, this appears to be rational, and it is on such lines that the design of the majority of building steelwork and all bridge steelwork is based. It is argued that if a known factor against first yield is ensured, a safe structure results, as indeed it does, and that what happens under a greater load than that necessary to achieve yield is of no consequence.

Such philosophy has been upset with increasing knowledge of actual behaviour gained from practical research. Firstly, it is now widely recognized that although members may be elastically designed many parts of a structure may reach yield stress, and indeed physically yield, on the first application of working load. A typical example of such a situation is the traditional end seating bracket and top cleat beam-to-column connection. Simple design assumes this to be a pinned connection and that the top cleat is simply a stabilizing fixing, not intended to transmit a fixed-end-moment. For the design assumption to be realized, some end rotation must take place, causing permanent but local yielding in the top cleat. Yield stress is often attained in many places, particularly if account is taken of residual cooling and welding stresses. But this is not to say the structure is unsafe.

Secondly, it has been established experimentally that all structures possess a margin of strength considerably greater than that calculated on the elastic basis. That is to say, on yielding, stresses will be redistributed extensively before a structure shows permanent deformation such as to render it unserviceable. What now appears to make the elastic philosophy irrational is that this extra margin of strength differs from one situation to another as, for instance, a laterally restrained beam compared to a slender strut. Redefining failure as excessive and unacceptable permanent deformation, ultimate load philosophy (plastic design in the case of steelwork) attempts to quantify this additional margin of strength in different situations, and to utilize it as part of the overall margin against failure, with the aim of achieving economy.

The principles of plastic design were originated by Sir John Baker at Cambridge, developed there by him, with others, and later taken up extensively at Manchester and Lehigh Universities, and elsewhere. As a result, plastic design can now be readily applied to continuous beams and portal frames of various shapes. The majority of steel portal frames built in this country are now designed plastically. The justification, if any is needed, for the extensive research which is proceeding lies in the fact that plastic design is by far the most accurate method of predicting that load at which real structural failure will occur.

As stated, plastic design consists of recognizing, and quantifying, the additional margin of strength in bending, beyond the attainment of first yield in an I-section. It has been established experimentally that before a beam or structure can be made to collapse it must be transformed from a structure to a mechanism. This occurs due to the formation of 'plastic hinges' at points of severe moment, which occur only when the whole depth of a member has reached yield stress, compressive on one side of the neutral axis and tensile on the other, i.e. no portion of the member depth remains elastic. The principles of design of continuous beams and portal frames are clearly set forth in some Constrado publications and other works. The Constrado brochure by Morris and Randall,[19] last published in 1979, together with its supplement (which incorporates design charts extracted and metricated from earlier publications[21]) has been published in various editions during the past two decades. It has been an invaluable source of information on the subject of plastic design worldwide, and contains details of numerous structural forms and design processes including details of two design methods applicable to multistorey frames.

A Constrado monograph by Horne and Morris,[22] and a paper by Morris[23] provides the most up-to-date information available on the subject of plastic analysis.

Having established that a structure has a known margin against collapse one has satisfied the strength criterion, and the value of stresses in various parts under working load are of little significance provided overall and local stability requirements are satisfied. Serviceability conditions, i.e. elastic deflections under working load, may however, not have been satisfied, and in effect one has to do an elastic analysis in order to establish the position, which lengthens the design process somewhat. Computer programs now exist for portal frames which, for given loading conditions, will select a section found plastically and provide values for elastic deflections.

Fully rigid multistorey frames, however, present considerable problems. Under sway conditions a large frame may require very many hinges to form before a mechanism is created and at any intermediate stage the frame is part elastic and part plastic. Proper understanding and control of stability considerations is essential and as yet rules of thumb suitable for codes of practice cannot be produced.

As an intermediate step it was proposed, many years ago, that frames could be designed as 'semi-rigid', i.e. with certain connection requirements satisfied, part transference of moment from beam to column could be assumed. This never found favour, owing to a complicated design process, but a gross

simplification of it is allowed in BS 449 and BS 5950, Part 1 whereby beam moments may be reduced by 10% provided the columns are designed to resist such extra moment and are structurally cased. This is a swings-and-roundabouts situation, not leading to any great economy.

13.9 Partial load factors

Alongside the development of ultimate load philosophy, i.e. plastic design, has come consideration of appropriate load factors, i.e. the determination of the right margin necessary against collapse. As indicated earlier, there are inconsistencies in present practice, using the principles given in BS 449.

Clearly, a structure must never closely approach collapse conditions in service, but have some margin. This margin is to allow for a number of uncertainties, including design inaccuracies, variations in material strengths, fabrication errors, lack-of-fit, foundation settlement, residual stresses and errors in assumed loading. Most of these uncertainties will never be quantified, but must be allowed for by a global factor, with two exceptions: (1) material strength which can be treated statistically; and (2) errors in assumed loading. Strength data sufficient for a rational statistical treatment has been accumulated over the past decade or so. This, together with loading data which has also been reviewed, has resulted in a new BS 6399 *Loading for buildings*. The BS 6399:1984, Part 1 'Code of practice for dead and imposed loads', which replaces CP 3:1967, Part 1, Chapter V, enables a more satisfactory design treatment to be adopted than hitherto according to BS 449, i.e. the concept of partial safety factors has been introduced. Different factors are to be applied to dead, imposed and wind loading, etc. by which the specified loads are to be multiplied. These factored loads are summated and the structure designed to be on the point of collapse under such factored loading. There is thus no fixed value for an overall safety factor, since it will depend on the proportions of the loadings from the several sources.

In the design of a building, for instance, one can calculate and control self-weight or dead load quite closely. The magnitude of applied load is very much less certain, and with possible change of use in the lifetime of a building, control is difficult, even by legislation. Further, naturally occurring loading such as wind loading and snow loading entail the consideration of the likely frequency of attainment and the probability of this occurring simultaneously with maximum service loading. One must also consider in continuous or semi-continuous structures that dead load may have a counterbalancing effect to certain live-load situations. Since it is possible for self-weight to be overestimated it is therefore necessary to apply minimum as well as maximum dead-load factors.

A rational approach, therefore, is to adopt maximum and minimum dead-load factors acting either alone or in combination with different factors for imposed and wind loads, the values of the latter depending on whether they act together or separately. This procedure has been adopted in BS 5950 and is fully documented in Part 1 of that standard.

The effect of this, however, will be to complicate the design process somewhat, but the benefit will be a much greater consistency in safety margins. When applied it will appear to make little difference to the run-of-the-mill structure previously designed to BS 449, but for the type of structure in which one kind of loading dominates, e.g. dead load or wind load, significant differences will be apparent.

13.10 Limit-state design

The fundamental objectives of structural design are to provide a structure which is safe and serviceable in use, economical to build and maintain, and which satisfactorily performs its intended function. All design rules, whatever the philosophy, aim to assist the designer to fulfil these basic requirements. However, as mentioned in section 13.9, it must be appreciated that design procedures are formulated to produce a satisfactory structure without necessarily representing its exact behaviour.

The design rules given in the various codes and standards represent the consensus of opinion of many experienced engineers. The rules, however, are not able to cover in detail every situation which designers may encounter, and so judgement must be exercised in their interpretation and application.

By using limit-state philosophy the design engineer has to consider two possible conditions of failure:

(1) *The serviceability limit state*, i.e. when a structure, although standing and experiencing safe stresses, will not be of use due to, say, excessive deformations.
(2) *The ultimate limit state*, i.e. when the structure has reached the point when it is unsafe for its intended purpose, and catastrophic failure is about to take place or already has taken place.

The appropriate factors of safety must therefore be determined, and whilst the primary object is to ensure that it is consistent throughout the structure, it must be recognized that parts of the framework are more sensitive than others to failure.

The change from working stress analysis, according to BS 449, to limit-state philosophy in accordance with BS 5950, Part 1, will produce certain changes in the structural design of a comparable framework. Hence, consideration must be given to economic as well as safety aspects by those responsible for establishing safety levels.

Safety can seldom, if ever, be absolute, but an increase in the level of safety will almost invariably be accompanied by an increase in the construction cost. The public, however, are extremely sensitive to the risk of failure and unserviceability but, whilst extreme care must be exercised, it should not be at the expense of pricing the structure such that it becomes uneconomic to build. An excellent résumé on this topic is given by Tordoff.[24]

13.11 Corrosion protection

The British Standard document giving guidance on this topic is BS 5493 *Code of practice for protective coating of iron and steel structures against corrosion.* It gives details of how to specify a chosen protective system, how to ensure its correct application, and how to maintain it. Many textbooks have also been written on the subject, to which reference may be made.[13,25] For specialist advice the BSC runs a Corrosion Advice Bureau for dealing with ad hoc problems. The BSC also publishes various leaflets on the subject of corrosion protection of structures, which may be obtained from BSC sections and commercial steels.

Generally, one must admit that, at one time, corrosion protection was given far-too-scant attention, as regards the effect of both structural detail and protective treatment. In recent years great advances have been made in both directions. Welding has relieved a lot of detail difficulties, and sophisticated surface and paint treatments have been developed which promise long repaint lives.

Whilst major exposed structures such as big bridges, where repainting is expensive, rightly receive 'Rolls-Royce' treatment, it should be remembered that such may not be appropriate in each and every case. It seems to the authors that in some respects the pendulum has swung too far the other way, some engineers specifying treatments and inspection standards not

warranted in many cases. Once again, judgement enters the picture and in any design it must be first assessed whether a problem exists at all. In the case of steelwork which can be guaranteed to be kept dry (say internal to a heated building) little problem exists. If exposed within the building, painting is usually carried out for cosmetic reasons only.

Steelwork exposed to the elements presents an entirely different problem which should receive attention before design commences, let alone detailing. Again, although account should be taken of the probable life of the structure, its use, atmospheric environment and location before coming to a judgement, structures in the public eye rightly demand the full treatment. Industrial structures of a temporary or semi-temporary nature do not warrant expensive treatment, particularly if they are likely to be subject to accidental damage.

The treatment appropriate to a particular case can thus vary from nothing to the blast-cleaned, metal spray and four-coat paint system. Further information on methods of treatment is given in Chapter 4.

As mentioned in section 13.5.7 it may sometimes be advantageous to consider the use of weathering steel to solve the corrosion problem as an alternative solution to protecting steel from corrosion. Whilst a sacrificial surface can sometimes be allowed, as required by the Department of Transport for bridge works, weathering grade steels painted subsequently usually provide a more durable protective system than ordinary grade steels.

When using universal sections requiring a sacrificial weathering allowance, the design properties are obviously different from those given in BS 4 and other informative documents. In this respect Constrado produced a useful brochure[26] which provides properties adjusted to give a 1 or 2 mm weathering allowance. The use of this brochure is not, however, necessarily confined to bridgework only.

13.12 Detailed design

Ideally, design should be carried out only by those familiar with shopfloor problems and procedures, but this is a counsel of perfection. Where doubt exists an approach to a fabricator is worth while.

What are referred to here are the difficulties which arise because a designer may sometimes interpret technical information literally without recognizing its limitations. The tables published in BS 4:1980 *Structural steel sections*, Part 1; BS 4848 *Specification for hot-rolled structural steel sections*, Parts 2, 4 and 5 and in various handbooks[18, 27–29] giving dimensions and web and flange thicknesses, are average values only. All steel members are subject to weight rolling margins of ± 2.5%. Sections can also be out of square and the limits of tolerance on shape are given in the relevant publications. Whilst it would be unfair to assume that all member sizes and shapes are at the tolerance limits, it is also unfair to suppose that all dimensions are strictly accurate to three significant figures. One must have some regard to the possibility of members being slightly out of true. No bar is ever completely straight, no plate flat, and no flange at right angles to its web. Fortunately, in most cases the work can be forced into alignment (albeit introducing locked-in stresses) but occasions can sometimes arise where this is not possible. When this occurs it becomes necessary to use packing pieces, occasionally needing costly machining which cannot be charged for. The designer's aim, therefore, must be to eliminate the need for accurate fit-up and to use machining only as a last resort.

Another point frequently overlooked is the matter of accessibility for welding. The easiest and therefore the best fillet weld results when the electrode, either manual or automatic, can be offered at 45°. The limits for satisfactory work are roughly 120 and 60° and outside these limits poor welds result since the electrode must be bent, breaking the flux coating and introducing many stops and starts. The designer cannot be expected to foresee all possible difficulties, but having made some attempt he should retain an open mind and be prepared to consider sequences and procedures suggested from the shop floor.

Handling and floor space also deserve some thought. For straightforward fabrication one should aim to keep complicated weldments small so that they can be handled readily to enable all welding to be done downhand. A complicated end detail to a long member makes this difficult if not impossible. If design can be effected such that all that long members need is to be cut to length and drilled, probably on an automatic machine, competitive tender sums will be offered. This is at least partly due to the fact that shop-floor space used is kept to a minimum and hence throughput can be high.

Where possible, repetition should be the aim. Money will not be saved if in the pursuit of imaginary economy a great variety of member sizes is used for broadly similar loading conditions. An example, for instance, occurred in a bridge consisting of 54 girders, all of which were different. Admittedly, this is an extreme case but some saving through repetition must have been possible. Since building structures generally offer the greatest scope for repetition, this should not be overlooked at the design concept stage.

13.13 Connections

The greater part of the cost of a steel structure is in the connections, whether they be bolted or welded. Simplicity therefore must be the keynote, with the greatest standardization possible if economy is to result. Typical examples of a great variety of connections are illustrated in the *Steel designers' manual*.[30] Suffice it to say here that the general trend is to use shop welding and site bolting. Site welding tends to be very expensive and should be considered only if extensive work is in hand as, for instance, in a big bridge or long pipeline, whilst shop bolting is usually more expensive than welding.

13.13.1 Welding

The British Standards for welding of greatest concern to the structural engineer are:

BS 5135:1984 *Specification for the process of arc welding of carbon and carbon manganese steels.* This standard supersedes BS 1856:1964 and BS 2642:1965 which have been withdrawn by the BSI.

BS 639:1976 *Covered electrodes for the manual metal-arc welding of carbon and carbon manganese steels.*

There are many other current British Standards covering various welding processes, inspection procedures, and welding of special alloy steels and other materials. Indeed, so extensive is the coverage that to the uninitiated great difficulty may be experienced in selecting the appropriate standard or most effective procedure. However, excellent guidance is available from the Welding Institute which publishes a series of booklets, some of which have already been mentioned, but Richards[4] is particularly recommended. It is couched in easily understood language and defines the fundamentals and points out pitfalls for the unwary. Armed with such guidance, an attempt can be made to propose details and procedures for a particular case, but an open mind should be retained for ideas and proposals from the welding engineers responsible for carrying out the work.

The PD 6493:1980 referred to in section 13.2.5 is also invaluable, as is BS 4870:1981 *Specification for approval testing of welding procedures*, Part 1, 'Fusion welding of steels'. This standard gives details of various processes, types of test weld and test pieces and recommended test procedures.

13.13.2 Bolting

The various types of bolts in structural use and the respective British Standard to which they are made, or governing their use, are as follows (the user must ensure that the version incorporating the latest amendments is used):

(1) *Black bolts*: BS 4190:1967 *Specification for ISO metric black hexagon bolts, screws and nuts.* Two obsolete standards covering imperial sizes, namely BS 325:1947 and BS 916:1953 have not yet been withdrawn by BSI.

(2) *High-tensile bolts*: BS 3692:1967 *Specification for ISO metric precision hexagon bolts, screws and nuts. Metric units.*

(3) *High strength friction grip (HSFG) bolts*: BS 4395:1969: *High strength friction grip bolts and associated nuts and washers for structural engineering*, Part 1, 'General grade'; Part 2, 'Higher-grade bolts and nuts and general-grade washers'; and Part 3, 'Higher-grade bolts (waisted shank) nuts and general grade washers'. BS 4604, *Specification for the use of high strength friction grip bolts in structural steelwork. Metric series*, Part 1, 'General grade'; Part 2, 'Higher grade (parallel shank)'; and Part 3, 'Higher grade (waisted shank)'.

Whilst at one time the most popular structural bolt was the black bolt to BS 4190 (Grade 4.6), in recent years there has been a move towards the much wider use of Grade 8.8 bolts covered by BS 3692. But since these bolts are mostly used in clearance holes, the shank diameter precision is not exploited. Consequently, manufacturers now supply bolts of Grade 8.8 strength grade to BS 4190 tolerances, specifically for structural purposes. These Grade 8.8 bolts are of material properties comparable to HSFG bolts to BS 4395, Part 1, but whereas the latter are supplied with Grade 10 nuts, the former come with Grade 8 nuts. Further, the HSFG nut is thicker in order to limit thread stresses during tightening. The result is that bearing and shearing values for Grade 8.8 bolts are greater than slip values for HSFG bolts, and the designer must decide whether some slight initial movement is admissible – usually it is in building structures, except where reversals of load may occur. Thus, in these cases Grade 8.8 bolts offer economy compared to HSFG bolts, particularly since they do not require controlled tightening or special tools. One would suggest, however, that the greater strength in tension offered by HSFG bolts, due to their thicker and harder nuts, justifies their adoption in wholly tension situations. An amendment in 1982 to BS 4604, Part 1 permits such bolts to be used without controlled tightening in these circumstances, which is quite acceptable for most building structures.

In the past, turned and fitted bolts were used only where accurate fit-up was essential. These are not now used having been superseded by HSFG bolts.

Turned barrel bolts are bolts in which the machined shank is of a larger diameter than the protruding threaded end, being shouldered at the spigot. They are for situations in which it is necessary to secure the bolt effectively without gripping the work and exerting pressure between the plies. Such a situation occurs in an expansion joint where one of the holes is slotted to allow for movement.

High strength friction grip bolts are now widely used, having already become popular in the mid 1960s. Both general- and higher-grade bolts, as the name implies, resist shear through the interface friction arising from bolt tension. The coefficient of friction to be assumed is called the slip factor, and the strength of a bolt is calculated as bolt tension × (slip factor/load factor) × number of interfaces. It is assumed that bolts are tightened-up to their proof load in tension and, to ensure this, alternative methods of tightening are specified in BS 4604. These are the part-turn method and the torque-control method. In the former the bolts are brought up hand-spanner tight, to bring the surfaces into contact, the nuts and threads marked, and tightening is then continued by a predetermined amount depending on dimensions. The latter method depends upon the use of either a manual or power-driven tool preset to slip at a particular torque value, which can be adjusted to suit the case. Of the two methods, the former is the more accurate and reliable, but tedious, whilst the latter is much quicker but less accurate owing to the fact that torque and bolt tension do not necessarily relate exactly, being dependent on thread fit.

Additionally, and not referred to in any British Standard, since there is only one manufacturer, is the use of 'Coronet' load indicating washers. These are washers, with raised nibs, to be inserted under the bolt heads. These raised nibs are flattened when the specified shank tension has been attained during nut-tightening. Since they do not require special calibrated tools, nor any marking procedure, they are very much simpler to use than either of the other two methods. Accordingly, their use now accounts for some 80% of cases. It is important, however, that pattern tightening of bolt groups is retained, or relaxation of the bolts tightened first will occur. Also, proper inspection during and after bolting is still essential.

The slip factor normally adopted is 0.45 for untreated surfaces, but BS 4604 *Specification for the use of high strength friction grip bolts in structural steelwork, metric series* gives details of a slip factor test to determine the value in other cases. It also recommends a load factor in design of 1.4. British Standard 449 refers to BS 4604 whereas BS 153 called for higher load factors depending on load combinations. In the new bridge and buildings codes (BS 5400 and 5950) the load factors are more closely related.

The reason for the apparently low load factor of 1.4 lies in the fact that a bolt possesses a margin of strength after slip has occurred, when the bolt starts to act in bearing as well as friction.

It might be supposed from the apparently full coverage in British Standards outlined above that all outstanding problems regarding HSFG bolts had been solved. Unfortunately, this is far from being the case. In the first place, post-slip strength is uncertain and clearly depends to some extent on the thickness of plies. Secondly, the test to determine slip factors consists of four bolts in line, two either side of the joint, with double cover plates putting the bolts into double shear. The effect of eccentricity arising in single-shear conditions is not determined. More particularly, though, it has been established that very large or long joints do not behave in the simple fashion assumed. For instance, a very long cover plate acting in tension transverse to its length also develops longitudinal tension. The Poisson's ratio effect in this biaxial tension situation brings about a reduction in thickness and, thus, bolt relaxation. This can reduce bolt strengths by as much as 20%. Similar biaxial tension effects can occur in large joints in lattice girders, particularly when more than three plies are involved. Indeed, the effect of a large number of plies is completely unknown.

Another difficulty arises in large joints, i.e. that if faces are not machined truly flat an indeterminate amount of bolt tension is used to bring the surfaces in contact, even with the best fabrication. An extreme case would occur in a splice in a box girder. If the overall widths and depths of the two lengths do not coincide exactly towards the corners, proper contact cannot be made without the use of machined packings. Proper contact would be made only towards the middle of the faces where biaxial tensions exist. Large joints therefore need careful thought and cannot be treated by applying rule-of-thumb methods. It should be noted that where HSFG bolts are used, machining of faying surfaces is detrimental since the slip factor will be considerably reduced due to the smoother surface; hence more or larger-sized bolts will be required in the joint.

Two papers by Needham[31,32] on the subject of connections are available. They cover in some detail the basic principles and design philosophy underlying the design of both welded and bolted joints. They emphasize the need for joint design to be consistent with assumed structural behaviour.

13.14 Inspection of structural steelwork during construction

Any inadequacies of materials and structural frameworks of building and bridge structures always causes concern, especially among owners who are frequently faced with the high costs of remedial work.

Many of the defects producing constructions not complying with the design specification may be found to originate from one or more of the following:

(1) *Material*: supply of incorrect grade or out-of-standard material – wrong sizes, etc.
(2) *Detail design*: poor and inaccurate details.
(3) *Fabrication*: incorrect material and size selected. Inadequate or incorrect assembly of joints – use of wrong-grade weld material or bolts.
(4) *Site erection*: misplacing of similar sized but different-grade sections. Wring-grade weld material or bolts. Lack of fit – overstraining of components. Poor foundation connections; omission of, or inadequate, bracing.

Thus, from the outset, the design concept and process should include considerations of:

(1) Availability and reliability of proposed materials.
(2) Special requirements for control of quality and speed of fabrication and erection.
(3) The degree of supervision and inspection likely to be present during fabrication and erection.
(4) The type of contract and its effect on the design/construction process.

It is important to recognize the contractual relationship as defined in the contract documents and that legal responsibility for satisfactory erection rests with the contractor. It is necessary to have good site management by way of planning at all stages, including delivery sequence, laying out of stock and positioning of cranes. Correct setting out is a prerequisite for satisfactory completion of a project, followed by inspection of steelwork on delivery and during erection, temporary and permanent bracing, lining and levelling and examination of all connections.

Two publications on this subject are by the Institution of Structural Engineers[33] and by Needham.[34]

References

1 Richards, K. G. (ed.) (1971) *Brittle fracture of welded structures.* The Welding Institute, London.
2 Richards, K. G. (ed.) (1969) *The fatigue strength of welded structures.* The Welding Institute, London.
3 Gray, T. F. and Spence, J. (1982) *Rational welding design*, 2nd edn. Butterworth Scientific, Guildford.

4 Richards, K. G. (ed.) (1967) *The weldability of steel*. The Welding Institute, London.
5 Needham, F. H. (1977) 'The economics of steelwork design'. *Struct. Engr*, **55**, 9.
6 Walker, H. B. and Gray, B. A. (1985) *Steel-framed multistorey buildings – the economics of construction in the UK*, 2nd edn. Constrado, London.
7 British Constructional Steelwork Association (1983) *International structural steelwork handbook* (No. 6). BCSA, London.
8 Horridge, J. F. and Morris, L. J. (1986) 'Comparative costs of single-storey steel-framed structures'. *Struct. Engr*, **64A**, 7.
9 Bryan, E. R. (1972) *The stressed skin design of steel buildings* (Constrado monograph). Crosby Lockwood Staples, London.
10 Her Majesty's Stationery Office (1985) *The building regulations*. HMSO, London.
11 British Steel Corporation (1983) 'Protection of steel from corrosion', in: *Steel protection guide*. BSC, London.
12 British Steel Corporation (1982) 'Interior environments', in: *Steelwork corrosion protection guide*. BSC, London.
13 Chandler, K. A. and Bayliss, D. A. (1985) *Corrosion protection of steel structures*. Elsevier Applied Science, London.
14 Hendry, A. W. and Jaegar, L. G. (1958) *The analysis of grid frameworks and related structures*. Chatto and Windus, London.
15 Morice, P. H. and Little, G. *The analysis of right bridge decks subjected to abnormal loading*. Cement and Concrete Association Publication No. D6/11. CCA, London.
16 Nash, G. F. J. (1984) *Composite universal beam – simply supported span*. Constrado, London.
17 Constrado (1983) *Weather-resistant steel for bridgework*. Constrado, London.
18 British Constructional Steel Association (1985) *Guide to BS 5950*, vol. 1 *Section properties – member capacities*. Constrado, British Steel Corporation and BCSA, London; British Constructional Steelwork Association (1986) *Guide to BS 5950*, vol. 2 *Worked examples*. Steel Construction Institute, London.
19 Morris, L. J. and Randall, A. L. (1975) *Plastic design*. Constrado, London.

20 Merrison, A. W. (1973) *Inquiry into the basis of design and method of erection of steel box-girder bridges. Report of the committee into interior design and workmanship rules*, Parts 1, 2, 3 and 4. Department of the Environment/Scottish Development Office/Welsh Office. HMSO, London.
21 Constrado (1979) *Plastic design supplement*. Constrado, London.
22 Horne, M. R. and Morris, L. J. (1981) *Plastic design of low-rise frames* (Constrado monograph). Granada, London.
23 Morris, L. J. (1983) 'A commentary on portal frame design'. *Struct. Engr*, **59A**, 12; discussion, **61A**, 6 and 7.
24 Tordorf, D. (1983) *Introduction to the limit-state design of structural steelwork*. Constrado, London.
25 Fancutt, F., Hudson, J. C. and Stanners, J. F. *Protective paintings of structural steel*. Chapman and Hall, London.
26 Constrado (1983) *Weather-resistant steel for bridgework*. Constrado, London.
27 British Steel Corporation and British Constructional Steel Association (1982) *The sections book*. BSC and BCSA, London.
28 British Constructional Steel Association (1978 and 1973) *Structural steelwork handbook: properties and safe load tables for sections to BS 4:1978; Structural steelwork handbook: properties and safe load tables for metric angles to BS 4848:1973*. Constrado and BCSA, London.
29 Steel Construction Institute (1986) *A checklist for designers*. SCI, London.
30 *The steel designer's manual*, 4th edn. Crosby Lockwood, London.
31 Needham, F. H. (1980) 'Connections in structural steelwork for buildings'. *Struct. Engr*, **58A**, 9.
32 Needham, F. H. (1983) 'Site connections to BS 5400, Part 3'. *Struct. Engr*, **61A**, 3.
33 Institution of Structural Engineers (1983) *Inspection of building structures during construction*. ISE, London.
34 Needham, F. H. (1981) 'Site inspection of structural steelwork'. *Proc. Instn Civ. Engrs*, **70**, Part 1.

14

Aluminium

J B Dwight MA, MSc, CEng, FIStructE, MIMechE
Fellow of Magdalene College and
Emeritus Reader in Structural Engineering,
University of Cambridge

Contents

14.1 Introduction

14.1.1 History

Although the most abundant metal in the Earth's crust, aluminium was ranked as a precious metal until 1890. In that year the modern electrolytic method of smelting was invented, which transformed the status of aluminium as an industrial metal. Today, it is second-cheapest, after steel, among the metals suitable for structural use. Its volume usage roughly equals that of all the other nonferrous metals put together.

The first strong alloy ('Duralumin') was developed in 1905, which made possible the structural use of aluminium in the German Zeppelins of the First World War. Between the wars its use was developed in aircraft, leading to a vast increase in aluminium production during the Second World War. This was accompanied by a dramatic decrease in cost relative to other metals. After 1945 there was great pressure to develop fresh outlets for aluminium and many new markets were found. By now it is well established in a wide range of industries. Aerospace accounts for a fairly small, but important, part of the total tonnage.

The use of aluminium for civil engineering structures was pioneered in the US during the early 1930s, the first epic example being several 45 m dragline jibs used on the Mississipi's levees. This was followed by the replacement of the steel deck of the Smith Street Bridge in Pittsburgh by an aluminium one in 1933, thus uprating its load capacity. This deck lasted for over 40 yr until replaced by a second aluminium one. Today, aluminium is acknowledged as a general structural material. It is chosen for main structures in situations where its special properties – low density and nonrustability – justify the extra metal cost compared with steel. Figure 14.1 shows a large aluminium roof structure built in Malaysia. This structure is mechanically jointed. Welding is also now widely used as, for example, in the aluminium military bridge shown in Figure 14.2.

A much greater tonnage of aluminium is consumed in secondary structural applications, such as maintenance gantries, glazing bars, window frames, curtain walling, shopfitting, prefabricated buildings, greenhouses, balustrades, crash-barriers, road signs and lamp posts. It is also used widely in the form of profiled sheeting for the cladding of buildings.

Figure 14.1 Aluminium roof of 50 m diameter structure for the Selangore State Mosque, Malaysia. Tubular construction, employing special extruded node units for the joints. Tubes in 6061-T6 alloy, and all other extrusions in 6082-T6. (Triodetic system. *Courtesy*: British Alcan Aluminium plc)

Figure 14.2 Rapidly erectable aluminium bridge for the British Army, built of specially designed extruded sections. The prefabricated units are of welded construction in 7019-T6 alloy. (Crown copyright)

14.1.2 Comparison with steel

The following is a crude statement of how aluminium differs from steel as a structural material (G = good, B = bad):

Light:	one-third the density	G
Nonrusting:	seldom needs painting	G
Extrusion process:	design your own sections	G
Fabrication:	generally easier	G
Brittle-fracture:	not susceptible	G
Expensive:	about 3 times the cost by volume	B
Deflections:	E one-third that of steel	B
Fatigue:	more susceptible	B
Buckling:	more critical	B
Ductility:	tends to be lower	B
Welded strength:	suffers from heat-affected zone (HAZ) softening	B

Other differences from steel are:

Thermal expansion:	twice that of steel
High conductivity (electrical and thermal)	

All aluminium alloys have a rounded stress–strain curve (Figure 14.3), in which they resemble cold-rolled rather than hot-finished steel. Yield is defined in terms of the 0.2% proof stress.

14.2 Production of structural material

14.2.1 Primary production

Aluminium is obtained from the ore bauxite, the first stage being to extract pure alumina. The smelter comprises lines of relatively small furnaces ('pots'), in which the alumina powder is dissolved in liquid cryolite and smelted electrolytically using big carbon electrodes. The output of the smelter is pure aluminium ingot, a major item in the cost of which is the electricity. The ingot is shipped to secondary plants where it is remelted and alloyed to produce wrought ('semi-fabricated') products, in the form of plate, strip, sheet, sections and tube.

14.2.2 Wrought products

14.2.2.1 Flat material

Mill practice for this is much as for steel. Continuously cast slabs are hot-rolled to produce plate, which may then be cold-reduced to make strip, from which sheets are cut. The coiled strip can be roll-formed to produce cold-rolled sections (less common than in steel) or profiled sheeting.

14.2.2.2 Sections

Hot-finished sections are made by a radically different technique from steel, namely extrusion. Cast billets are heated and inserted into a horizontal extrusion press, where they are forced through an aperture in a die, the emerging sections having a cross-section determined by the shape of the aperture. The process is highly flexible, and the die charge is minute compared with the cost of rolls for a steel section. It is common practice for a customer to order new sections to meet his special requirements. The complexity of possible profiles is almost unlimited. Hollow shapes are readily extruded.

14.2.2.3 Tubes

Normally tubes are produced as hollow extrusions. Thin tubing, e.g. irrigation pipe, can be formed from strip by welding. Alternatively, thin tubing for precision use may be produced as drawn tube, at greater cost.

14.2.3 Castings and forgings

Aluminium castings (see section 14.4.4.5) may be employed in conjunction with wrought material, typically for small fittings and attachments. They may be sand-cast, or else chill-cast (for larger quantities). Aluminium forgings can fill a similar role when better strength and ductility are needed.

14.3 Control of strength

14.3.1 Heat-treatable and nonheat-treatable material

Nearly all the aluminium alloys, unlike steel, are unacceptably weak in the hot-finished state. There are two main kinds of alloy: (1) 'heat-treatable'; and (2) 'nonheat-treatable'. The producer strengthens the former by heat treatment, and the latter by cold-working. The heat-treatable alloys are generally the stronger, but less tough, and are more often the choice for main structural use. The nonheat-treatable alloys typically appear in the form of sheet, for which the necessary cold-working is provided during manufacture by the reduction in the rolling operation. Either type can be softened again by annealing.

It is essential for extrusions to be in heat-treatable alloy, since there is no way of cold-reducing them during manufacture. Drawn tube can be of either type.

14.3.2 Heat treatment

The strengthening process for heat-treatable alloys consists of quenching ('solution treatment') followed by ageing. The quench has little immediate effect, but with time the metal will gradually harden at room temperature, reaching its final strength after several days. Such material is said to be 'quenched and naturally aged'.

The ageing process is speeded up usually by heating the metal in a furnace for some hours at about 150 to 200°C ('precipitation treatment'). Such material is stronger than if naturally aged. It can be described as 'quenched and artificially aged', or more commonly as 'fully heat-treated'.

The quench ideally takes place from a carefully controlled temperature in the region of 500°C. The resultant distortion has to be corrected, usually by stretching (before artificial ageing). With wide, thin, extrusions distortion is a major problem, and

may well dictate the thickness and, hence, economy of a design. A common practice with the 6000-group alloys is to spray-quench thin extrusions as they emerge from the die, which is more economic and causes less distortion than if they were heated and quenched in a tank as a subsequent operation. For very slender profiles in the 6063 alloy it is even possible to turn off the water entirely and rely on an air-quench at the die, thereby reducing distortion even more. Ideal quenching is obviously not achieved with air-quenching and the resulting material has reduced properties.

14.3.3 Cold-working

Nonheat-treatable aluminium material is strengthened by means of cold-working applied during manufacture. This is possible for products that are cold-reduced in bringing them to their final thickness, i.e. sheet and drawn tube. It is also possible, to a lesser degree, for plate at the lower end of the thickness range ('cold-rolled plate').

The required properties are achieved by careful control of the reduction passes and of interpass annealing. It is common to refer to cold-worked material as being one-quarter, one-half, three-quarters or fully hard, as an indication of its 'temper', i.e. of the extent to which it has been strengthened. The strongest temper is fully hard, but a lower temper may be called for when formability is a factor. For the stronger of the nonheat-treatable alloys the fully-hard temper is not offered.

14.4 Alloys

14.4.1 Alloy numbering system

The specification of aluminium materials has been much simplified by the recent worldwide adoption of the US numbering system. Engineers should abide by this and use no other.

A given alloy, i.e. composition, is referred to by a four-digit number, the first digit of which indicates the alloy group to which it belongs. The alloys are grouped according to main alloying elements as follows, the groups of interest to civil engineers being given asterisks:

Heat-treatable alloys:	2000 group	Copper
	*6000 group	Magnesium plus silicon
	*7000 group	Zinc
Nonheat-treatable alloys:	1000 group	(Pure)
	*3000 group	Manganese
	4000 group	Silicon
	*5000 group	Magnesium

Apart from the alloy it is necessary to specify the condition (heat treatment or temper). This is done by means of appropriate symbols written after the alloy number. For heat-treatable alloys:

T6	Fully heat-treated, i.e. quenched and artificially aged
T5	Air-quenched and artificially aged (extrusions)
T4	Quenched and naturally aged

For nonheat-treatable alloys:

H12 or H22	Quarter-hard temper
H14 or H24	Half-hard temper
H16 or H26	Three-quarter-hard temper
H18 or H28	Fully hard temper

For all alloys:

O Annealed
F As-extruded or as-rolled

Thus, typical material specifications would read as follows:

6082-T6   An alloy with a particular aluminium–magnesium–silicon composition, in the fully heat-treated condition

3103-H14 An alloy with a particular aluminium–manganese composition, in the half-hard temper

In the temper designation for nonheat-treatable alloys the first digit after the H (1 or 2) is of academic interest to the average user; it merely shows whether the material has been cold-reduced to the final temper, or has been partly annealed after the last pass. What matters is the ensuing digit (2, 4, 6 or 8) which indicates the actual hardness.

The F-condition is ill-defined. It essentially refers to hot-finished material (extrusion, plate) that has received no further treatment, the properties of which cannot be specified closely.

14.4.2 Selected alloys – properties

14.4.2.1 Strength values

Table 14.1 gives a short list of structural aluminium materials that are of interest in civil engineering. The quoted mechanical properties are based on BS 1470[1] (flat products) and BS 1474[2] (extruded sections). The reader is urged to refer to these or other national standards for fuller information.

14.4.2.2 Physical properties

Approximate values roughly applicable to all aluminium alloys are as shown in Table 14.2. Weight of aluminium material may be estimated using the following formulae:

Weight of section (N/m) = 0.027 × area in square millimetres
Weight of plate (N/m²) = 27 × thickness in millimetres

Table 14.2

Density	2.7 g/cm³
Modulus of elasticity E	70 kN/mm²
Shear modulus	26 kN/mm²
Poisson's ratio μ	0.33
Linear expansion coefficient	24×10^{-6} per °C
Melting point	660°C

14.4.3 Heat-treatable alloys

14.4.3.1 2000-group

This group of alloys, sometimes referred to as 'Duralumin', is typified by a high copper content (around 4%). It includes most of the strong alloys used for aircraft, an example being 2014-T6 with a tensile strength approaching 450 N/mm². These alloys are seldom used outside the aerospace industry, because of their low ductility in the T6 condition, higher cost, inferior corrosion resistance and nonweldability. They have to be fabricated with great care. To reduce corrosion it is possible to use them in the form of 'clad sheet', a product with rolled-on pure aluminium facings.

14.4.3.2 6000-group

This very important group, covering aluminium alloyed with magnesium and silicon, essentially comprises two basic grades of alloy: (1) a stronger; and (2) a weaker grade. The stronger comes in slightly different versions in different parts of the world, the European version, 6082, being broadly similar to the 6061 more commonly used in North America. Material of this type in the T6 condition may be regarded as the 'mild steel' of aluminium, and is the commonest choice for general structural use. It has a 0.2% proof stress about equal to the yield of mild steel, although with a lower tensile strength and less ductility. It is readily welded, but with nearly 50% loss of strength in the heat-affected-zone (HAZ) (Figure 14.3). A particular feature is the ease with which these alloys can be extruded into thin intricate sections.

The second type of alloy in the group is 6063, which is considerably weaker. This is the extrusion alloy *par excellence*

Table 14.1 Selected structural alloys

Approximate % composition	Alloy and condition	Form	t_m (mm)	Minimum properties			Design stresses	
				f_0 (N/mm²)	f_u (N/mm²)	elong. (%)	p_o (N/mm²)	p_a (N/mm²)
Heat-treatable								
Zn4.0, Mg2.0	7019-T6	E, P	—	330	380	8	330	355
Mg0.9, Si1.0, Mn0.7	6082-T6	E, S	20	255	295	8	255	275
		P	25	240	295	8	240	265
Mg0.7, Si0.4	6063-T6	E	—	160	185	8	160	170
Mg0.7, Si0.4	6063-T5	E	25	110	150	8	110	125
Nonheat-treatable								
Mg4.5, Mn0.7	5083-0	P	—	125	275	14	105	145
Mg4.5, Mn0.7	5083-H22	P	6	235	310	8	235	270
Mg2.0, Mn0.3	5251-H24	S	6	175	225	5	175	200
Mn0.6, Mg0.5	3105-H18	S	3	190	215	1–2	190	200

Notes:
f_0 = 0.2% proof stress, f_u = tensile strength.
E = extrusion, P = plate, S = sheet
t_m = maximum thickness for which stresses are valid.
The following are similar to the 6082 alloy, but are slightly weaker: 6061, 6081, 6181, 6261, 6351.
The alloy 3103 is similar to 3105, but again slightly weaker.

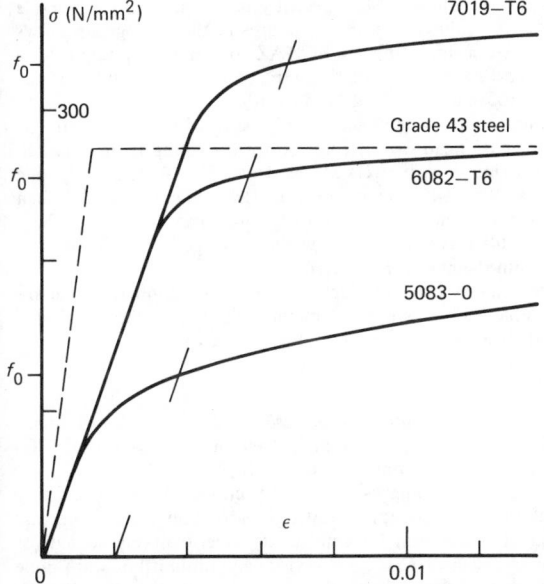

Figure 14.3 Typical stress–strain curves. f_0=0.2% proof stress

(even better than 6082) and is the automatic choice for slender complex architectural shapes, such as window sections and curtain-wall mullions, where stiffness rather than strength is important. Another feature is its smooth surface finish (with a well-made die). Extrusions in 6063 can be produced in the normal T6 or else in the weaker T5 condition (air-quenched), the latter being suitable for very slender profiles that would not otherwise be feasible; 6063 is not supplied in the form of plate or sheet.

14.4.3.3 7000-group

This group, comprising aluminium alloyed with zinc and magnesium, was originally developed in the form of ultra-strong materials for use in military aircraft, having tensile strengths exceeding 500 N/mm². Of value to the civil engineer are the less-strong alloys 7020 and 7019, which are of special interest for welded construction. With these, the heat-affected material adjacent to a weld gradually regains strength over a period at room temperature, and after a month gets back to 75% or more of the full T6 properties. Both alloys extrude nearly as well as 6082. The weaker version 7020 has comparable strength to 6082 in the T6 condition, while 7019 is up to 30% stronger.

Material of 7000-type is susceptible to stress corrosion, when the amount of the alloying elements exceeds a critical level. The alloy 7020 was developed and standardized in Europe with this danger taken into account. As a result it is safe, but hardly any stronger than the cheaper and more readily available 6082. The stronger 7019 version would seem more attractive to a designer. However, 7019, being more highly alloyed, is closer to the critical level for stress corrosion. It was developed in the UK for military bridges, in which form some 15 000 t have been used satisfactorily. But this success was only achieved by very careful control of fabrication procedures. It is essential for an intending user to realize that 7019 is not as simple to fabricate as 6082 or 6061, and to seek advice before doing so.

14.4.4 Nonheat-treatable alloys

14.4.4.1 1000-group

This comprises nominally pure aluminium with different levels of guaranteed purity, material too weak for serious structural use. The cheapest version is 1200 with a minimum aluminium content of 99.0%. Higher purities are available, and in the annealed condition these can provide a valid alternative to lead as a soft flashing material.

14.4.4.2 3000-group

This covers material with manganese as the main alloying element. The two common versions are 3103 and 3105, of which 3105 is slightly the stronger. Used in the fully hard H18 temper, they represent the standard type of material used for profiled aluminium sheeting as employed for cladding of buildings.

14.4.4.3 4000-group

The only interest here in this minor group (aluminium plus silicon) is that it includes one type of weld filler wire.

14.4.4.4 5000-group

This comprises a range of alloys having varying amounts of magnesium as the main alloying element. They are characterized by their ductility and toughness. They are generally unsuitable for use as extruded sections.

The most important in structural terms is the strongest (5083), a plate material. Until recently this was supplied either in the annealed 0 condition, or else in the indeterminate F condition (as-hot-rolled). Its use was confined to low-stress applications, where toughness rather than high yield was needed as, for example, for the entire superstructure of the liner *Queen Elizabeth II*. Material 5083-0 has too low a proof stress (only 125 N/mm²) for use in highly stressed structures; 5083-F will often have a proof stress 40% higher, but this cannot be guaranteed and the designer must still work to the low 0-condition properties. Recently, 5083 plate in thicknesses up to 6 mm has become available in the H22 temper; in this condition it becomes much more attractive, its properties matching those of 6082-T6 for which it is a valid replacement.

Other 5000-group alloys in decreasing order of strength are 5154A, 5454 and 5251. They are typically used in the medium tempers, where their combination of formability and toughness makes them suitable for boatbuilding and sheet metal fabrications generally.

14.4.4.5 Casting alloys

A useful casting alloy contains aluminium with a nominal 12% Si (known as LM6 in the UK). This has a tensile strength roughly comparable to that of 6063-T6, but with a proof stress 50% lower. It has excellent foundry characteristics and good ductility. An alternative is the Al–5% Mg alloy (known as LM5) which is better in terms of surface finish, but which can only be cast into simple shapes; it is less ductile and slightly weaker than LM6.

14.4.5 Alloy selection – summary

For highly stressed welded construction the ideal choice is 7019-T6, because of its good strength and less severe degree of HAZ softening. But it is vital to realize that this is not a material for amateurs in anything but the simplest fabrications because of the latent risk of stress corrosion if correct procedures are not followed.

A more common choice is 6082-T6 or 6061-T6, which is somewhat weaker but more straightforward to fabricate. HAZ softening at welds is more severe than with 7019, and this calls for ingenuity in the location and design of joints.

For welded stiffened plating where toughness is needed, the normal choice would be 5083-F plate welded to 6082 (or 6061)-T6 extruded stiffeners. At thicknesses up to 6 mm the plate is available as 5083-H22 with higher and more precise properties.

For triangulated structures (trusses, space frames) the normal choice is 6082 (or 6061)-T6. Mechanical joints are often used, instead of welding, to avoid the problem of HAZ softening. For extruded members whose design is governed by stiffness rather than strength, as for intricate architectural profiles, the natural choice is 6063-T6. If the section is on the limit of feasibility due to its slenderness, the weaker condition 6063-T5 has to be used instead.

Profiled sheet used for the cladding of buildings is normally supplied in 3105-H18 or the slightly weaker 3103-H18.

Sheet metal fabrications of secondary importance can be readily made in 5251-H14 which is ductile and formable. Pure aluminium in the form of 1200-H14 can be useful for unstressed sheet-metal work. Pure aluminium is also employed, in higher purities, for chemical plant (as an alternative to stainless steel) and for electrical conductors (busbars and transmission lines).

A good general-purpose casting alloy is the Al–12% Si (known as LM6).

14.5 Fabrication

14.5.1 Cutting and forming

14.5.1.1 Cutting and machining

Thin material can be sheared like steel, but more readily. For thicker material cold-sawing is used, with either a circular or band saw. Aluminium (except in the softest tempers) can be sawn faster than steel, especially if suitable coarse-toothed saws are used.

Aluminium is also more readily machined than steel, and it is not unusual in design to employ extrusions incorporating attachment flanges which are machined away over the greater length of the member. (Stiffened panels in aircraft are often machined out of the solid.)

Ordinary flame-cutting is unsuitable for aluminium, because of the ragged edge produced. Instead, one can employ plasma-arc cutting, an adaptation of the tungsten inert gas (TIG) welding process.

14.5.1.2 Bending

The heat-treatable alloys in the full strength T6 condition are less easily manipulated than steel. They will only accept a small deformation when bent cold, due to their lower ductility. Heating, on the other hand, disturbs the heat treatment and causes severe softening. One solution is to form the material in the more ductile T4 condition and then bring it up to the full T6 strength by subsequent artificial ageing in a low-temperature furnace.

With the nonheat-treatable alloys the practice for forming is more as for steel. Cold-bending is employed when possible, the temper of the material being selected to suit the severity of the bend. Springback is more than for steel. For severe manipulations it is possible to apply local heating with a gas flame, the necessary temperature being 450 to 500°C. Great care is necessary to avoid overheating the aluminium, since there is no colour change at this temperature. Temperature-sensitive crayons may be used; alternatively one can rub a pine stick on the heated area and see if it leaves a mark.

14.5.2 Mechanical joints

14.5.2.1 Riveting

For many years, riveting was the normal means of making shop joints in aluminium. More recently there has been a wholesale move to welding, even for structures in the 6000-group alloys which are severely affected by HAZ softening. Riveting is little used and rivets have become hard to get. One wonders if the swing to welding has not been overdone.

Small solid rivets would usually be in 5154A alloy with the small 'pan' head driven cold. Squeeze riveting is preferred to hammering. Larger rivets can be driven hot. Alternatively, one can use 6082-T4 rivets (or equivalent) which have been held in a refrigerator since quenching, to suppress natural ageing. These are readily driven cold, after which they age-harden in position to attain their proper T4 strength.

Proprietary fasteners such as 'Pop' and 'Chobert' rivets are available for joints in sheet-metalwork. These are suitable for blind riveting, i.e. from one side, and are quick to use.

14.5.2.2 Bolting

Aluminium structures can be assembled using either ordinary bolting (dowel action) or high-strength friction-grip (HSFG) bolting (friction action).

Ordinary bolding is used with clearance or close-fitting reamered holes as appropriate, possible bolt materials being: 6082-T6 aluminium (or equivalent), steel (suitably coated) or stainless steel (316S16 or 304S15). Aluminium bolts are none too good in tension, especially in fatigue. On the other hand it may be difficult to get steel bolts with a coating of sufficient durability to match that of the aluminium, unless they are painted. The ideal answer is stainless steel, which is usually worth paying for.

In recent years it has become acceptable to employ HSFG bolting for aluminium, taking care with the protection of the steel bolts. Bolt material (high yield steel) and torqueing procedures follow HSFG practice in steel. Proper attention must, of course, be paid to the condition of the contact surfaces, which should be grit-blasted. The slip resistance can be improved by applying epoxy resin (HSFG bolting is not recommended for use on plates having a 0.2% proof stress under 230 N/mm^2).

14.5.2.3 Screwing

Tapped holes in aluminium tend to be unsatisfactory. Patent stainless-steel thread-inserts are available, which give good service on parts that have to be screwed and unscrewed repeatedly.

14.5.3 Welding

14.5.3.1 Welding processes

Alloys in all groups except 2000 are readily welded. Unfortunately, welding is accompanied by local HAZ softening. This occurs to a greater or lesser degree depending on the parent alloy (see section 14.7.4), except with annealed material.

The standard arc-welding process is manual inert gas (MIG), using d.c. current. This is similar to CO_2 welding of steel except that the shielding gas is argon (or helium in North America). It is easy to operate and ideal for positional welds. It can be used on thicknesses down to about 2 mm. With the MIG process, aluminium can be welded as easily as steel, after an initial training period. Current settings are higher and deposit areas tend to be greater.

For thin work the TIG process is used instead of MIG. In this the arc is struck from a nonexpendable tungsten electrode, the filler wire being held in the left hand. This is an a.c. process which needs more skill than MIG. It is slower and causes more distortion.

Aluminium can be spot-welded, but with higher energy inputs than for steel.

14.5.3.2 Filler wire

Simplified recommendations for selection of arc welding filler wire material are shown in Table 14.3. For further information refer to BS 3019 or 3571.[3]

Table 14.3

Parent alloy group	Filler composition
6000*	5% Si (4043A), or 5% Mg (5056A, 5356)
5000 or 7000	5% Mg (5056A or 5356)
3000 or 1000	Parent composition

Note:
*When welding 5000 to 6000 use the 5% Mg wire.

14.5.4 Adhesive bonding

Aluminium is eminently suitable for glued joints using epoxy resin, a technique successfully used for lamp posts and other components. The epoxies are attractive because of their ability to tolerate poor fit-up. Shear strengths up to 15 N/mm^2 can be developed, but it is essential to guard against premature failure due to peeling from the end of a connection. An extruded tongue-and-groove feature is often a good way of preventing this.

The resin can be used cold or, alternatively, can be hot-cured to give improved strength. In the latter case the curing temperature is the same as that needed for artificial ageing. Thus, with heat-treatable alloys it is economic to order the material in the T4 condition, and rely on the hot-curing operation to harden the aluminium (up to T6).

14.5.5 Use of extruded sections

14.5.5.1 Availability

The relatively low cost of extrusion dies often makes it economic to design one's own section or 'suite' of sections to suit the job in hand. The use of such sections can reduce fabrication costs and produce an improved final product provided, of course, the quantities are sufficient.

Extrusion is mainly confined to the 6000 and 7000 alloy groups, the order of merit for extrudability being: (1) 6063; (2) 6082 or 6061; and (3) 7019 or 7020. Complex sections, including hollows, are produced in all of these. Extrusions are also possible in 2014 (high-strength) and 5083 (high-ductility), but with severe limitations on profile and at much higher cost.

Hollow sections are normally produced using a 'bridge die' in which a mandrel, defining the internal shape, is supported on feet locating on the body of the die (which defines the outer shape). Since the hot plastic metal has to flow around these feet and reunite, the final section contains longitudinal welds. These cannot be seen and, in the vast majority of applications, are quite acceptable. But there are some situations where they would be regarded as a potential danger. Hollow sections extrude more slowly than nonhollows, and thus cost more per kilogram; the die charge is also higher.

Apart from custom-made profiles, the designer has a wide range of conventional sections from existing dies to choose from, such as channels, angles, T- and I-sections and boxes. Stockists hold these, usually in 6082-T6 or equivalent.

Sections are extruded in long lengths and can be supplied up to 20 m long to meet special needs. The normal limit on length is much less than this and is dictated by handling and transport.

14.5.5.2 Limiting dimensions

Sections generally are available up to about 300 mm wide from small and medium extrusion presses. With large presses, using special die assemblies, it is possible to extrude sections up to 600 mm wide, depending on the shape. But relatively few mills contain such equipment.

The designer often wants a section to be as thin as possible, for economy. In 6063 alloy the lower limit on thickness can very roughly be taken as the lesser of 1.0 mm and width/120. In 6082 (or equivalent) the corresponding values are 1.5 mm and width/80, while in 7019 they are somewhat more. Sections of 6063 at the limit of slenderness can be supplied in the T5 condition (air-quenched) to reduce the amount of post-extrusion straightening needed to correct distortion.

14.5.5.3 Section design

Figure 14.4 shows a few of the devices that can be incorporated in the design of extruded shapes. Figure 14.4(a) shows a lipped channel space-frame chord, which is a more efficient shape than a plain (unlipped) channel, having greatly increased local buckling resistance. The planking section (Figure 14.4(b)) incorporates various features, including integral stiffeners, interlock, and anti-slip surface. Planking sections, first developed as flooring for trucks, have also been employed in bridge decks and (after piercing) for open-work flooring. Figure 14.4(c) shows a double-sided planking section, again interlocking.

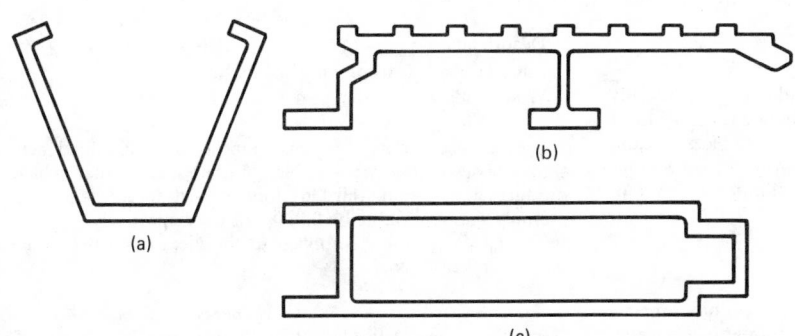

Figure 14.4 Examples of extruded sections

14.6 Durability and protection

14.6.1 Unpainted use of aluminium

14.6.1.1 The corrosion process

Atmospheric corrosion of unprotected aluminium proceeds by localized pitting, a radically different process from the rusting of steel. The oxide corrosion products formed at the pits are voluminous, giving an exaggerated impression of the actual damage. The rate of attack, defined by the depth of pitting, becomes stifled by the corrosion products and slows down after the first 2 or 3 yr. In outdoor sites the corrosion is less when the surface is regularly washed or rained on.

Corrosion failures, on the rare occasions that they happen with aluminium, usually stem from contact with other materials (see section 14.6.3).

14.6.1.2 Durability rating

Aluminium will usually last for ever unprotected, even out of doors. The decision whether or not to paint in an exposed environment depends on the durability rating of the alloy as shown in Table 14.4.

Table 14.4

Rating	Alloy groups	Whether to paint
A	1000, 3000, 5000	Usually no need
B	6000	Only necessary when exposed to severe industrial or marine environment
C	2000, 7000	Generally necessary, except in dry unpolluted situation

14.6.2 Protective systems

14.6.2.1 Conventional painting

When an aluminium structure has to be painted, it is important that the priming and subsequent coats contain no copper, mercury or graphite, and preferably no lead. A zinc chromate priming coat is recommended.

14.6.2.2 Powder coating

In recent years, powder coating has become an economic process for the coating of aluminium components, on a mass-production basis, and has to some extent replaced anodizing. The powder is sprayed on and stoved, the resulting coat having a more even thickness than with solvent-based paint. Components are often powder coated for purely decorative purposes.

14.6.2.3 Anodizing

This is a process whereby the inherent oxide film is artificially increased electrolytically, the minimum oxide thickness for 'architectural anodizing' being 25 μm. This gives a pleasant satin appearance, which will last for years *if regularly washed*. Colour anodizing is also available, but only in a limited number of shades.

14.6.3 Contact with other materials

When aluminium is in direct contact with certain other metals under moist conditions, the adjacent aluminium gets eaten away. This is known as 'electrolytic' or 'galvanic' corrosion.

Failure to take suitable precautions is likely to cause serious trouble.

Such corrosion occurs when aluminium is in contact with steel (other than stainless) or cast iron and, more severely, with copper, brass and bronze. The attack can be stopped by preventing direct contact, either by means of bituminous paint, or preferably with an interposed tape or gasket. Electrolytic corrosion need not be a problem if suitable precautions are taken.

With copper the electrolytic effect is so strong that water dripping off a copper roof on to aluminium sheeting will quickly perforate the aluminium, because of dissolved copper ions. The action between aluminium and lead is only slight. When aluminium and zinc are in contact it is the zinc that suffers. Galvanized bolts in an exposed aluminium structure tend to lose their protection more quickly. Aluminium that is to be embedded in concrete should be protected with bituminous paint; otherwise it will suffer attack while the concrete is 'green'.

14.7 Structural calculations

14.7.1 Principles of design

14.7.1.1 Codes of practice

At the time of writing (1987) existing codes for aluminium design are in the process of being redrafted into limit state format. In the UK, BS 8118 *Structural use of aluminium* which is near to publication and due to replace CP 118:1969, will be in two parts. Part 1: 'Code of practice for design' and Part 2: 'Specification for materials, fabrication and protection'. The simplified design rules given below have been broadly based on the draft to Part 1, which is still subject to possible alteration.

14.7.1.2 Basic requirements

All structures should satisfy: (1) the ultimate, and (2) the serviceability limit state. Fatigue may also be a factor (see Section 14.7.7).

14.7.1.3 Ultimate limit state

Every component, i.e. member, joint, must satisfy the following:

Action under factored loading ≤ factored resistance

where *action* means moment or force, as appropriate, *resistance* means ability to withstand that action, *factored loading* is nominal loading × γ_f and *factored resistance* is calculated resistance/γ_m.

The partial factor γ_f, applied to the nominal working loads, has basic values as follows:

Dead loads	1.20
Imposed loads (except wind)	1.33
Wind loads	1.20

However, when more than one imposed or wind load acts simultaneously it is permissible, in the case of that which produces the second, third or fourth most severe action, to multiply the basic value by 0.8, 0.6 or 0.4 respectively.

The partial factor γ_m, applied to the ideal calculated resistance, is taken thus:

	Members	Joints
Nonwelded construction	1.20	1.25
Welded construction	1.25	1.30

14.7.1.4 Limiting design stresses

Resistance calculations (see sections 14.7.2 to 14.7.5) involve the use of limiting stresses p_o and p_a, listed in Table 14.1 for selected alloys. p_o is usually taken to be equal to the guaranteed 0.2% proof stress. However, a reduced value is taken for materials having a high ratio of ultimate to proof stress, such as 5083-0, for which the stress–strain curve tends to be more rounded. This is to prevent plastic deformation at working load.

14.7.1.5 Combined actions

When a member carries simultaneous axial load and moment, the ultimate limit state is satisfied if:

$$P/P_R + M/M_R \leqslant 1.0$$

where P and M are actions arising under factored load and P_R and M_R are the separate factored resistances.

14.7.1.6 Serviceability limit state

The requirement is that recoverable elastic deflection under nominal (unfactored) loading should not exceed the specified limiting value. In view of the lower modulus it is common to accept larger deflections in aluminium than those normal with steel.

14.7.2 Section classification

14.7.2.1 Compact and slender sections

The first step in checking a member for the ultimate limit state, except in simple tension, is to establish whether it has a *compact* cross-section. If, instead, it is of *slender* section, the resistance will be reduced by premature failure due to local buckling.

14.7.2.2 Classification for axial load or moment

The plate elements comprising a section are of two basic sorts: *outstand* and *internal* (Figure 14.5). The procedure for classifying the section is as follows:

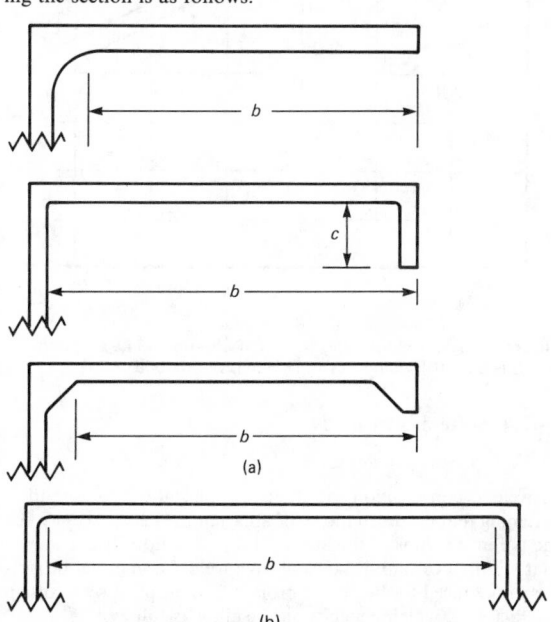

Figure 14.5   Plate elements as considered for local buckling.
(a) Outstand (plain and reinforced): (b) internal element

(1) Determine the parameter β/ε for each of the elements comprising the section (except the tension flange of a beam). β depends on the width:thickness ratio $b:t$ as follows, with b measured to the toe of the root fillet (if any):

Plain outstand element, uniform
   compression $\qquad\qquad\qquad\qquad\quad \beta = b:t$
Internal element, uniform compression $\quad \beta = b:t$
Web of beam, neutral axis at centre $\quad \beta = 0.35b:t$
and $\varepsilon = \sqrt{250/p_o}$, with p_o in newtons per square millimetre (Table 14.1).

(2) Classify the individual elements, according to the value of β/ε, in Table 14.5.

Table 14.5

		Outstand	Internal
Element in strut	Compact	$\leqslant 7(6)$	$\leqslant 22(18)$
	Slender	$> 7(6)$	$> 22(18)$
Element in beam	Fully compact	$\leqslant 6(5)$	$\leqslant 18(15)$
	Semi-compact	$\leqslant 7(6)$	$\leqslant 22(18)$
	Slender	$> 7(6)$	$> 22(18)$

The values in brackets represent the tighter limits applicable to *welded* elements.

(3) The classification of the section is then taken as that of the least favourable element.

14.7.2.3   Reinforced outstand elements

The ability of outstands to resist local buckling can be increased by stiffening the free edge with a lip or bulb (Figure 14.5). For such an element, if reinforced by a standard lip of thickness t equal to that of the plate, a more favourable value of β may be taken as follows:

$$\beta = (b:t)\{1 + 0.03(c:t)^3\}^{-1/3} \qquad (14.1)$$

where c is the internal lip height (Figure 14.5).

If c is large, there is the chance of the lip itself buckling prematurely as a plain outstand, and this should be checked. With any other shape of reinforcement, β should be found by replacing it with an equivalent standard lip (thickness t), the inertia of which about the mid-plane of the plate is the same as that for the actual reinforcement.

(*Note*: In channel-section struts it is immaterial if lips face in or out. But in a beam any lip on the compression flange must be inward facing to be effective.)

14.7.2.4 Classification for shear force

This depends on the depth:thickness ratio $d:t$ of the web or webs and on ε (defined in section 14.7.2.2), as follows:

Compact $\qquad\quad d:t \leqslant 49\varepsilon$
Slender $\qquad\qquad d:t > 49\varepsilon$

14.7.3 Resistance of the cross-section

14.7.3.1 Axial load resistance

The factored resistance P_R of an unwelded section is found as follows:

Tension: $P_R = $ lesser of $p_a A_n/\gamma_m$ and $p_o A/\gamma_m$ (14.2)

Compression (with overall buckling prevented):

Compact, unwelded section $P_R = p_o A/\gamma_m$ (14.3a)

Other sections $P_R = p_o A_e/\gamma_m$ (14.3b)

where p_a, $p_o =$ limiting design stresses (Table 14.1), A, $A_n =$ gross and net section areas, $A_e =$ area of *effective* section (see sections 14.7.3.3, 14.7.4.4) and $\gamma_m =$ partial safety factor (see section 14.7.1.3).

14.7.3.2 Bending moment resistance

The moment resistance M_R of an unwelded section in the absence of lateral–torsional buckling, is found thus:

Fully compact, unwelded section $M_R = p_o S/\gamma_m$ (14.4a)

Fully compact, welded section $M_R = p_o S_e/\gamma_m$ (14.4b)

Semi-compact, unwelded section $M_R = p_o Z/\gamma_m$ (14.4c)

Other sections $M_R = p_o Z_e/\gamma_m$ (14.4d)

where S and Z are plastic and elastic section moduli and S_e, Z_e are the same for the *effective* section (see sections 14.7.3.3 and 14.7.4.4).

14.7.3.3 Effective section

For sections classified as slender (see section 14.7.2.2) the effect of local buckling is catered for by basing the section properties (A_e and Z_e) on an *effective* section, instead of the true one. In unwelded construction the effective section is found by taking a thickness of k_L times the true thickness for any slender element within the section. k_L is read from Figure 14.6, the quantities β and ε needed to enter which are as defined in sections 14.7.2.2 and 14.7.2.3.

The effective section to be used for welded members, to allow for HAZ softening at welds, is defined in section 14.7.4.4.

14.7.3.4 Shear force resistance

The factored shear resistance V_R is found thus for sections having unwelded compact webs, normally orientated:

$$V_R = 0.6 p_o A_v/\gamma_m \qquad (14.5)$$

where A_v is the web area.

For unwelded webs classified as slender (see section 14.7.2.4) the following formula may be used:

$$V_R = \frac{600 \times 10^3 A_v}{(d:t)^2 \gamma_m} \qquad (14.6)$$

For welded webs, refer to section 14.7.4.4. Note that Equation (14.6) becomes oversafe if applied to very slender stiffened webs.

14.7.3.5 Moment and shear combined

The moment resistance is unaffected by the presence of a shear force V not exceeding half the value of V_R. For higher values of V, M_R becomes reduced as follows:

$$M_R = M_{R_0}\{1 - 8(V/V_R - 0.5)^3\} \qquad (14.7)$$

where M_{R_0} is the factored resistance in the absence of shear.

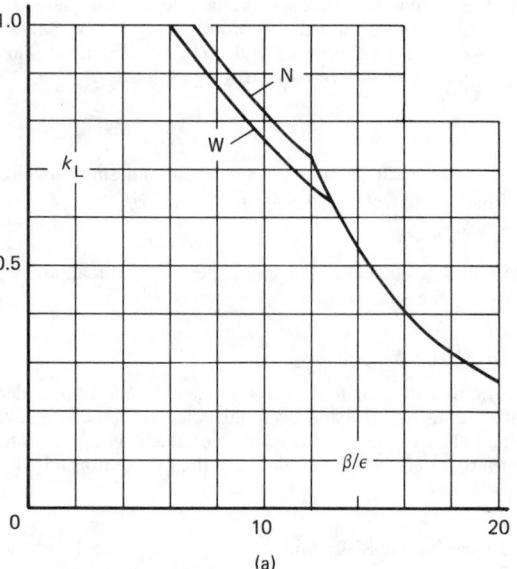

(a)

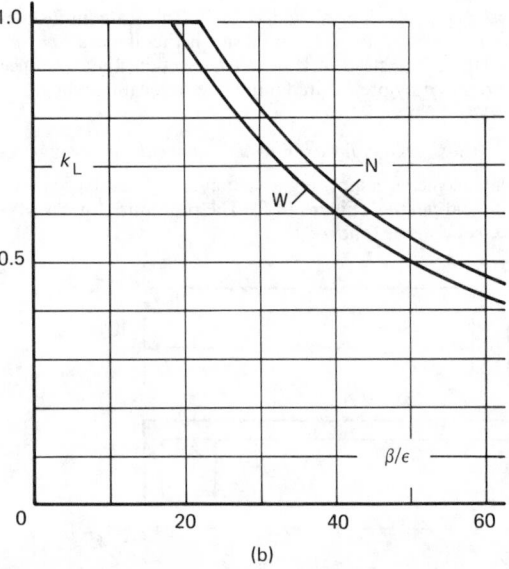

(b)

Figure 14.6 Local buckling factor k_L for slender plate elements. (a) Outstand: (b) internal. N=unwelded, W=welded

14.7.4 Softening at welds

14.7.4.1 Severity of softening

In welded construction the designer must allow for the local softening that occurs in the HAZ adjacent to welds, except when the parent metal is in the annealed (0) condition. It is assumed that within a certain distance of each weld the material properties are reduced to the parent properties multiplied by a softening factor k_z, which depends on the alloy as follows:

7000-group, T6 condition $k_z = 0.75$

6000-group, T6 condition $\quad k_z = 0.55$

5000-group $\quad k_z = \dfrac{\text{tensile strength in 0 condition}}{\text{tensile strength in temper used}}$

14.7.4.2 Extent of the HAZ

The area over which the material properties are thus reduced (the HAZ) is assumed to extend a distance z from a weld, measured: (1) transversely from the centre of a butt weld or the root of a fillet; and (2) longitudinally from the end of any weld. Provided welding is by the MIG process, with rigorous thermal control (see below), z may be generally taken as the lesser of two values found as in Table 14.6, where t_1 is the average thickness of the plates joined (but not exceeding $1.5t_2$), and t_2 is the thinnest plate thickness. But note that these values become unreliable if: (1) t_2 exceeds 25 mm; or (2) longitudinal welds have a total deposit area exceeding 3% of the gross area of the section for 7000- or 5000-group alloys, or 4% thereof for 6000-group.

Table 14.6

	Butt	Fillet
7000-, 5000-groups	$4.5t_1$ or 35 mm	$(4.5t_2^2/t_1)$ or 35 mm
6000-group	$3t_1$ or 25 mm	$(3t_2^2/t_1)$ or 25 mm

It is important to exercise rigorous thermal control during welding to limit the extent of the HAZ. The values of z given above are only valid if the metal temperature adjacent to a weld at the start of deposition, of any pass, does not exceed 40°C (for 7000- and 5000-group parent alloys) or 50°C (for 6000). If these temperatures are exceeded, the predictions in section 14.7.4.2 will underestimate the affected area. Also, with 7000-group material the softening factor k_z may drop below 0.75.

14.7.4.4 Effective section of welded members

To allow for the effects of HAZ softening, the true section is replaced by an effective one, which is assumed to have full parent properties throughout, but with reduced thickness in the HAZs. The resistance is then found generally as in section 14.7.3, using section properties based on the effective section:

(1) *Compact sections.* The effective section is obtained by taking an assumed thickness in the HAZ equal to $k_z t$ instead of the true thickness t.
(2) *Slender sections.* For any plate element that is both slender and affected by welding, the assumed thickness is taken as the lesser of $k_z t$ and $k_L t$ in the HAZ and as $k_L t$ elsewhere in that element. The rest of the section is treated according to (1) above or section 14.7.3.3 as appropriate.

The shear resistance of a welded web of slender proportions may be taken as the lower of two values: (1) based on Equation (14.5) with HAZ effects allowed for in the calculation of A_v using *compact sections* in (1) above; and (2) based on Equation (14.6) with HAZ effects ignored.

14.7.5 Buckling

14.7.5.1 Buckling of struts

There are two possible modes of overall buckling to be con-

sidered in axial compression: (1) *flexural*; and (2) *torsional*. Torsional buckling tends to become critical for thin open sections such as angles and channels.

The factored resistance for either mode is taken as the basic resistance P_R of the section (Equation (14.3a) or (14.3b)) times the factor k_c which is read from Figure 14.7. In order to enter the figure the quantity ε is found as follows, with p_o in newtons per square millimetre:

Compact, unwelded section $\qquad \varepsilon = \sqrt{(250/p_o)}$ $\qquad$ (14.8a)

Other sections $\qquad \varepsilon = \sqrt{(250A/P_R)}$ $\qquad$ (14.8b)

When considering ordinary flexural buckling, the slenderness parameter λ is simply the effective slenderness ratio $k_L{:}r$ as used in conventional steel design.

For torsional buckling, λ may be obtained from the general expression:

$$\lambda = \pi\sqrt{(EA/P_{cr})} \qquad (14.9)$$

where P_{cr} is the elastic critical load for torsional buckling of a strut, as given in textbooks, allowing for interaction with flexure when necessary.

For struts of plain angle section (unreinforced) the torsional buckling check may be waived when the section is compact (section 14.7.2.2). If the angle section is slender, it can be assumed that torsion is adequately covered by simply taking P_R based on the effective section, with $k_c = 1$.

(*Note*: The use of Figure 14.7 may tend slightly to overestimate buckling strength of struts that are: (1) welded; or (2) of very asymmetric section (buckling axis much nearer to one edge than the other).)

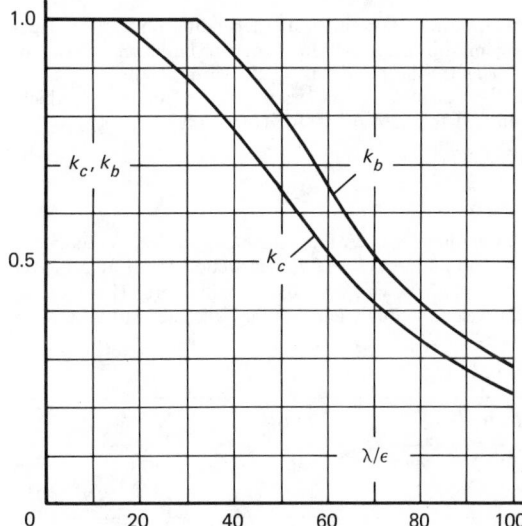

Figure 14.7 Overall buckling factors k_c (struts) and k_b (beams)

14.7.5.2 Lateral–torsional buckling of beams

The factored moment resistance for a member prone to lateral–torsional buckling is taken as the basic resistance M_R of the section (based on Equation (14.4a), (14.4b), (14.4c) or (14.4d)) times the factor k_b which is again read from Figure 14.7. In entering the figure ε is now found as follows, again with p_o in newtons per square millimetre:

Fully compact, unwelded section $\varepsilon = \sqrt{(250/p_o)}$ (14.10a)

Other sections $\varepsilon = \sqrt{(250S/M_R)}$ (14.10b)

The slenderness parameter λ may be obtained from the following general expression:

$$\lambda = \pi\sqrt{(ES/M_{cr})}$$

where M_{cr} is the elastic critical moment for lateral–torsional buckling, as given in textbooks. Alternatively, for beams of conventional I-, channel, T-shapes, λ may be found using the appropriate steel data.

14.7.6 Connections

14.7.6.1 Ordinary riveting and bolting

Limiting stresses for aluminium fasteners, to be used in conjunction with an appropriate value of γ_m (see section 14.7.1.3), are given in Table 14.7 in newtons per square millimetre. The corresponding bearing stress on the ply is taken as $2p_a$ (see Table 14.1). Limiting stresses for steel fasteners should normally be taken as $0.7p_y$, $2p_y$ and p_y respectively for shear, bearing and tension, where p_y is the yield stress.

Table 14.7

Fastener	Shear	Bearing	Tension
5154A rivet	125	400	—
6082-T4 rivet	95	310	—
6082-T6 bolt	170	550	220

14.7.6.2 Friction grip bolting

The factored resistance in shear (depending on friction capacity), again with an appropriate γ_m, may be based on a slip factor of 0.3. This is valid provided: (1) the surfaces are grit-blasted; (2) the bolt diameter is not less than the combined ply thickness; and (3) the 0.2% proof stress of the ply material is not less than 230 N/mm².

14.7.6.3 Welded joints

Suitable limiting stresses for weld metal, used in conjunction with an appropriate value of γ_m (see section 14.7.1.3), are given in Table 14.8 in newtons per square millimetre. These assume that the welds are sound, and that the right filler wire is used (see section 14.5.3.2).

Table 14.8

Parent alloy	Tension	Shear
7019, 7020, 5083	240	170
6082, 6061, 5251	150	100

14.7.7 Fatigue

Fatigue calculations are generally based on stress arising under nominal working loads (unfactored). In the new aluminium codes (e.g. BS 8118) fatigue data will be presented in terms of stress range, following steel practice. Details are classified broadly as in steel. For a given detail the stress range to be used in design, corresponding to a given number of cycles, is about one-third of the corresponding stress range for steel. It is largely independent of the alloy used.

References

1 British Standards Institution (1972) *Specification for wrought aluminium and aluminium alloys for general engineering purposes – plate, sheet and strip.* BS 1470. BSI, Milton Keynes.
2 British Standards Institution (1972) *Specification for wrought aluminium and aluminium alloys for general engineering purposes – bars, extended round tube and sections.* BS 1474. BSI, Milton Keynes.
3 British Standards Institution (1985) *Specification for manual inert gas welding of aluminium and aluminium alloys.* BS 3571, Part 1. BSI, Milton Keynes.

Bibliography

British Standards Institution (1972) *Specification for profiled aluminium sheet for building.* BS 4868. BSI, Milton Keynes.
British Standards Institution (1972) *Specification for wrought aluminium and aluminium alloys for general engineering purposes – drawn tube.* BS 1471. BSI, Milton Keynes.
British Standards Institution (1972) *Specification for wrought aluminium and aluminium alloys for general engineering purposes – rivet, bolt and screw stock.* BS 1473. BSI, Milton Keynes.
British Standards Institution (1974) *Specification for anodic oxide coatings on wrought aluminium for external architectural applications.* BS 3987. BSI, Milton Keynes.
British Standards Institution (1976) *Specification for performance and loading criteria for profiled sheeting in building.* BS 5427. BSI, Milton Keynes.
British Standards Institution (1983) *Specification for filler rods and wires for gas-shielded arc welding.* BS 2901, Part 4. BSI, Milton Keynes.
British Standards Institution (1984) *Specification for powder organic coatings for application and stoving aluminium extrusions, sheet and preformed sections for external architectural purposes and for the finish on aluminium alloy extrusions, sheet and preformed sections coated with powder organic coatings.* BS 6496. BSI, Milton Keynes.
British Standards Institution (1984) *Specification for liquid organic coatings for application to aluminium extrusions, sheet and preformed sections for external architectural purposes, and for the finish on aluminium alloy extrusions, sheet and preformed sections coated with liquid organic coatings.* BS 4842. BSI, Milton Keynes.
Mazzolani, F. M. (1985) *Aluminium alloy structures.* Pitman, London.
Narayanan, R. (ed.) (1987) *Aluminium structures.* Elsevier Applied Science, London.
Robertson, I. and Dwight, J. B. 'HAZ softening in welded aluminium, *3rd international conference on aluminium weldments*, Munich.

15

Load-bearing Masonry

R J M Sutherland
FEng, BA, FICE, FIStructE

Contents

15.1 Introduction

In this chapter the word 'masonry' has been used to describe either brickwork or blockwork as well as natural stone. Today, natural stone is seldom used except as a decorative facing and the advice which engineers need on masonry relates primarily to brickwork and concrete blockwork.

The use of brickwork in the UK has changed quite markedly in the 20 to 25 years following its rebirth as a structural material in the early 1960s. Full exploitation of its strength for slender load-bearing walls in high-rise flats has now been halted by the sharp social reaction against this form of building. In the domestic field what once looked like the greatest justification for 'engineered' brickwork has given way to the limited demands of traditional housing. However, there are now new challenges at least as great as those of high-rise housing.

The structural use of concrete blockwork largely dates from the 1960s and its fortunes have followed the same path as those of brickwork.

One of the new structural challenges for masonry lies in the construction of buildings for sport, education, manufacturing and storage. Here the economy of masonry is being used increasingly in unframed buildings, often single-storey, with larger spans than in domestic construction, taller walls and few partitions or returns to brace the whole assembly. As a result of these changes, today's engineering problems with masonry in buildings are largely wind resistance, overall stability and composite action with floors and roofs. Crushing strength takes second place.

Another field where masonry is finding increasing favour is in the cladding of framed construction especially in large industrial units built in steelwork. In this case not only are there problems of the lateral strength of large thin panels but there are complex questions of movement and of the compatibility of the different materials.

All these are very much engineering problems and not matters of architectural opinion.

In civil engineering, the once dominant place of masonry was taken about a century ago first by mass concrete and then by reinforced concrete. Concrete may be more in keeping with a mechanized age than a labour-dominated material like masonry but its appearance is increasingly being criticized and now doubts are arising as to its durability, especially when reinforced. What is more, with growing appreciation of the structural performance of masonry, especially when reinforced or prestressed, concrete has a rival both in slenderness and load-bearing capacity. This makes masonry particularly attractive for structural use in retaining walls, bridge abutments and other civil engineering works, particularly in areas which are visually sensitive. There is also a good case for the revival of the masonry arch.

Engineers need to keep in touch with developments in the use of masonry. Today it is not just a craft material for houses or decorative facings, as was thought 30 years ago, but a major structural element and one benefiting increasingly from engineering understanding.

15.2 Material properties

Before embarking on any structural design in masonry it is important to distinguish between the physical properties of the different materials of which the units are made and to appreciate the limitations of each.

Table 15.1[1] shows the types of masonry units normally available with their materials, sizes, unit strengths and approximate share of the UK market. It also gives the numbers of the current British Standards which define the acceptable quality of each.

Table 15.2 gives some indication of the dimensional stability of each type of masonry unit, i.e. its response to changes in temperature, load and moisture content. Equivalent figures are also given for other materials commonly used in construction. The most essential factors to note are the initial moisture movements:

(1) Clay units are fired at a high temperature and expand, for the most part irreversibly, as they take up moisture from the atmosphere. The expansion is greatest immediately after firing but continues at a diminishing rate for effectively about 10 to 20 years.
(2) Concrete units (bricks or blocks) are cast wet and shrink as they dry out, again largely irreversibly. The shrinkage is greatest immediately after casting but continues at a diminishing rate for effectively about 10 to 20 years.
(3) Calcium silicate bricks, which are of sand and lime, hydrated, pressed and then autoclaved, behave similarly to concrete units.

Not only are the initial moisture movements generally greater than any subsequent cyclic ones due to change of atmospheric conditions, but *those of clay and concrete are of comparable magnitude and in opposite directions.*

This simple distinction between the behaviour of clay and concrete has frequently been ignored in the past with results which have sometimes caused major disruption. Today, now that the different properties of the materials are better understood, there is a tendency to over-react to the problems of movement and sometimes to take precautions which are unnecessary and could even be harmful. The question of precautions against movement is discussed in section 15.7.

15.3 Codes of practice

In the UK, the accepted guidance on the way in which masonry should be designed is given in the British Standard Code of Practice BS 5628. The first part of this Code dealing with unreinforced masonry was issued in 1978.[2] This part is the successor to the greater part of the earlier code CP 111 and deals essentially with walls and piers.

The second part of BS 5628[3] which covers the structural use of reinforced and prestressed masonry was not published until 1985. It makes good the wholly inadequate treatment of reinforced masonry in CP 111 and also puts prestressed masonry on an 'official' basis for the first time. This part of the Code covers the design of all types of spanning structures in masonry as well as walls and piers.

The third part of BS 5628,[4] also published in 1985, gives advice on various aspects of detailing with masonry and on workmanship, durability and similar topics. It could be said to be more architecturally slanted than the first two parts of this Code and is the successor to the earlier Code CP 121.

Since the issue of all three parts of BS 5628, masonry in the UK has been on a parallel basis to concrete in up-to-date and officially recognized design methods. This does not mean that all an engineer needs to know about masonry is in the three parts of this Code – far from it. However, anyone designing masonry structures, in the UK at least, should be aware of the contents of this Code and, whether experienced in masonry or a newcomer to it, would do well to consult the handbooks to Parts 1 and 2. References to these handbooks and to a selection of other publications on the structural design of masonry are given at the end of this chapter.[5,6]

While BS 5628, together with the relevant material standards, will be used as anchor points for the advice in this chapter, this Code should be used for checking design rather than as a

Table 15.1 Types of masonry units normally available.

	Material and manufacture	Normal (actual) dimensions of unit (mm)	Type of unit	Characteristic strength of unit (N/mm²)		Approximate share of UK market 1985 (10⁶m² of wall)
				Range in British Standard	Range commonly used	
Clay brick (BS 3921)	Clay fired generally at > 1000°C to achieve ceramic bond	Standard 215 × 102.5 × 65 high	Solid, frogged or perforated	7–100	14–100	58*
Calcium silicate brick (BS 187)	Sand and lime; hydrated, pressed and autoclaved	Metric modular (small demand) 190 × 90 × 65 (BS 6649)	Solid, or frogged	14–48.5	20.5–34.5	1.75
Concrete (BS 6073)	Aggregate and cement hydrated and moulded with pressure and vibration		Solid or frogged	7–40	7–40	4.5
Aggregate Concrete block (BS 6073)		Varies widely: length 390–590 height 140–290 thickness 60–250	Solid or hollow	2.8–35	3.5–21	Dense 30.4 lightweight 22
Autoclaved (aerated) concrete block (BS 6073)	Cement and ground sand or PFA with aerating agent hydrated and moulded in large blocks and then cut		Solid only	> 2.8	2.8–7.0	23

* Equivalent based on 102.5 mm wall

starting-point. The wide variety of forms of structural brick-work and blockwork make their use even more of a design matter, needing individual judgement, than almost any other material.

It is the aim in this chapter to point to these design aspects and to emphasize both the great opportunities for the use of masonry and some of the pitfalls, rather than to provide another handbook to the BS Code.

Reference is made throughout this chapter to BS 5628.

Readers working in countries other than the UK will need to be aware of the local codes, which may differ quite markedly from BS 5628. The following information may be helpful in this respect.

US
The most widely used code in the US is the Uniform Building Code. Chapter 24 of the 1985 edition deals with masonry on a linear elastic (working stress) basis.

Canada
The current code CAN-S304-M84 issued by the Canadian Standards Association covers both design by rules and design by full engineering analysis. This is still on a working stress basis. A limit state code is planned for 1990.

Australia
A unified code incorporating AS1640-1974 (SAA Brickwork Code) and AS1475 (SAA Blockwork Code, Part 1: unrein-forced blockwork and Part 2: reinforced blockwork) is about to be issued. This is written in ultimate strength format and will be converted to a full limit state form in the next edition.

New Zealand
References to existing codes may be misleading but two new codes are in draft DZ 4229 for masonry not requiring specific design and DZ 4210 for designed masonry.

The information given above is considered as a starting-point only. Readers are advised to check directly with the appropriate authority in each country.

15.4 Limit state principles

The design guidance in BS 5628: Parts 1 and 2 for unreinforced, reinforced and prestressed masonry follows the same limit state principles, with partial safety factors, as are used with reinforced

Table 15.2 Dimensional stability of masonry compared with reinforced concrete and steel

	Coefficient of thermal expansion (per °C × 10⁻⁶)	Movement as result of 20°C change (%)	Unrestrained drying shrinkage (partly reversible) (%)	Unrestrained moisture expansion (%)	Elastic modulus (kN/mm²)	Creep with time: creep factor = final strain/elastic strain (for stress ≤ 0.5 × ult.)
Clay brickwork	5–8	0.010–0.016	Shrinkage of mortar allowed for in expansion figures (right)	Depends on type of clay and firing temperature. Probably 0.02–0.12%. Precise figures uncertain. Too few long-term tests	4–26	1.2–4.0
Calcium silicate brickwork	8–14	0.016–0.028	0.01–0.04 (BS limit 0.04)	—	14–18	Approximately 2.5
Aggregate concrete blockwork*	6–12	0.012–0.024	0.02–0.06 (BS limit 0.09 maximum)	—	4–25	2.0–7.0
Aerated concrete blockwork	Approximately 8	0.016	0.02–0.09 (BS limit 0.09 maximum)	—	1.5–4.0	No test results available
Reinforced concrete	7–14	0.017–0.028	0.02–0.10	—	15–36	1.0–4.0
Steelwork	Approximately 12	0.024	—	—	175–210	—

* Concrete bricks similar

and prestressed concrete. Most engineers in the UK are now familiar with these principles but, regrettably, there is still a lack of uniformity in the terminology used in the different BS material codes.

In BS 5628, the phrases 'design load' and 'design strength' are used to denote the factored loads and strengths which need to be compared to check adequacy. Thus, for the ultimate limit state, if γ_f is the partial factor of safety for loading and γ_m is the partial factor of safety for material strength, adequacy is achieved if:

$$\frac{\text{Ultimate (characteristic) strength}}{\gamma_m} \geqslant \text{Characteristic load} \times \gamma_f$$

In all cases (direct load, bending, shear, etc.) the partial factors of safety are expressed separately in BS 5628 and never lost within the characteristic values quoted.

The same principle and the same terminology are used in BS 5628 for the serviceability limit states and for precautions against disproportionate collapse following a major explosion or other accident, but in these cases the partial factors of safety are different.

Table 15.3 shows the principal factors for each limit state and how these compare with the factors of safety used in BS 8110[7] for concrete. With unreinforced masonry the serviceability limit states of deflection and cracking are seldom if ever relevant but,

when considering the behaviour of reinforced or prestressed masonry sections in bending, they can be vital.

15.5 Unreinforced masonry

15.5.1 The mechanism of failure in compression

Rather than just accept the characteristic strengths and 'Code' factors of safety for masonry, designers are advised to consider what influences its strength and to try to visualize the actual mechanism of failure.

Table 15.4 lists some of the major factors affecting the strength of a masonry wall.

The mechanism of failure seems to be generally agreed. Because the mortar is almost always weaker than the masonry units it tends to be squeezed out of the joints. This movement of the mortar is restrained by the bricks or blocks, which are thus subjected to lateral tensile stresses which lead first to splitting and finally to collapse. This mechanism is shown diagrammatically in Figure 15.1.

Even under absolutely uniform downward loads, masonry walls – brick or block – fail first due to vertical splitting. This is virtually universal. With brickwork, the wall strength averages about 0.3 times (0.15 to 0.45 times) the brick strength and with

Table 15.3 Partial factors of safety (material) for masonry compared with concrete (BS 5628 and BS 8110)

BS 5628: 1978, Part 1	(Unreinforced masonry)	Ultimate limit state	Accidental damage	Serviceability limit state	Notes
γ_{mm} (compression)	*Control level*				
	Manufacturing and site special	2.5	1.25		
	Manufacturing normal and site special	2.8	1.4		
	Manufacturing special and site normal	3.1	1.55	—	
	Manufacturing and site normal	3.5	1.75		
γ_{mv} (shear)	—		2.5	1.25	—
γ_m (wall ties)	—		3.0	1.5	—
BS 5628:1985, Part 2 (Reinforced and prestressed masonry)					
γ_{mm} (compression)	*Control level*				
	Manufacturing and site special	2.0	1.0	1.5	
	Manufacturing normal and site special	2.3	1.15	1.5	
γ_{mv} (shear)	—		2.0	1.0	—	No shear reinforcement assumed
γ_{mb} (bond to steel)	—		1.5	1.0	—
γ_{ms} (steel reinforcement)	—		1.15	1.0	1.0
BS 8110:1985 Parts 1 and 2 (Concrete)					
γ_m (compression or bending)	—		1.5	1.3	1.05
γ_m (shear)	—		1.25	—	1.05
γ_m (bond to steel)	—		1.4	—	1.05
γ_m (steel reinforcement)	—		1.15	1.0	1.05

Note: Partial factors of safety for load (γ_f) with masonry similar to those for concrete (basically 1.4 for dead load and 1.6 for superimposed load with variations for combinations and different limit states)

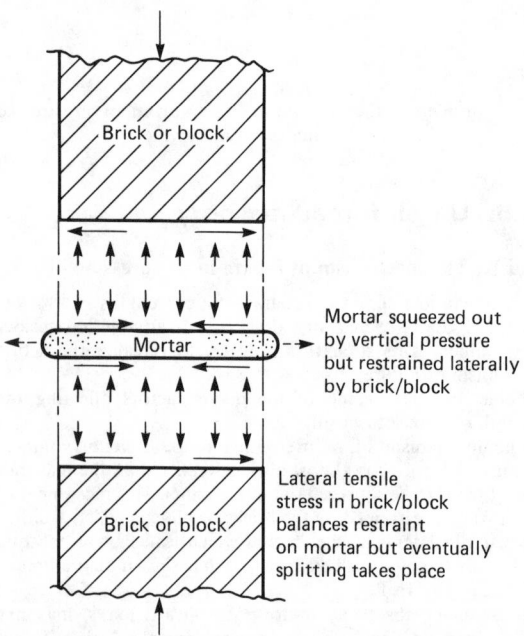

Figure 15.1 Simplified failure mechanism for vertical loads on masonry

Mortar squeezed out by vertical pressure but restrained laterally by brick/block

Lateral tensile stress in brick/block balances restraint on mortar but eventually splitting takes place

concrete blockwork it averages about 0.8 times the block strength.[1]

The difference between the apparent performance of bricks and blocks in walls compared with their individual strengths is primarily due to the shape of the units. The 'cube' strength of the concrete in the blocks is generally well below the equivalent strength of the fired clay in bricks but, when tested as units, the taller blocks fail at a stress nearer to that in a wall than the squat bricks, which are more fully restrained laterally by the platens of the testing machine. This is shown in Figure 15.2. The logical climax is that a storey-high unit should fail at the same load as the wall into which it is built.

Figure 15.3 shows the relationship, as given in BS 5628, Part 1, of the characteristic compressive strength of different types of masonry to that of the individual masonry units. This follows the principles outlined above.

Other important factors affecting the capacity of a wall or pier to support vertical loads, apart from those shown in Table 15.4, are slenderness, eccentricity and concentration of loading.

Table 15.4 Major factors influencing the strength of masonry walls

Variable factor	Effect on wall strength	
	Brickwork	Concrete blockwork
Strength of masonry unit	Most dominant factor: wall strength proportional to square root of brick strength	Most dominant factor
Strength of mortar	Not very significant: wall strength proportional to cube root of mortar strength for middle range of brick strengths	Little effect on wall strength
Thickness of mortar bed	Fairly critical: 17 mm bed instead of 10 mm gives 30% reduction; with ground faces and no mortar, wall strength approaches brick strength	No experimental data; effect probably less significant than with bricks
Geometry of masonry units	Ratio of wall strength to brick strength little affected whether wirecut, deeply frogged or perforated	Ratio of wall to block strength about 0.8 Ratio of wall to block strength reduced to about 0.5 with normal bond because cross-webs do not line up; higher with stack bond
Bond	English (50% cross-bonded material)⎫ Flemish (33% cross-bonded material)⎬ No noticeable difference in strength Stretcher (100% cross-bonded material) — Up to 40% stronger than English or Flemish Collar jointed (steel ties only between skins of stretcher bond but no cavity) — 10–15% weaker than English or Flemish	Seldom used other than in stretcher bond (or in stack bond with reinforcement in horizontal joints)
Poor filling of bed joints	Tests show 30% reduction in strength common, and more possible	No equivalent tests
Poor filling of perpendicular joints	No reduction found in tests even with wholly unfilled perpendiculars	Effect of not filling perpendicular joints at all is small

Source: Sutherland (1981) 'Bride and block masonry in engineering'. *Proc. Instr. Civ. Engrs*, **70**, Table 2.

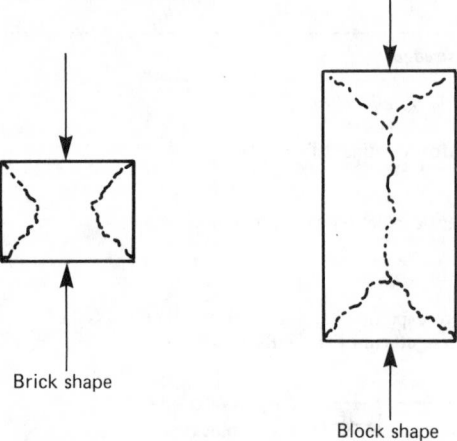

Figure 15.2 Effect of platen restraint on failure of bricks and concrete blocks under test

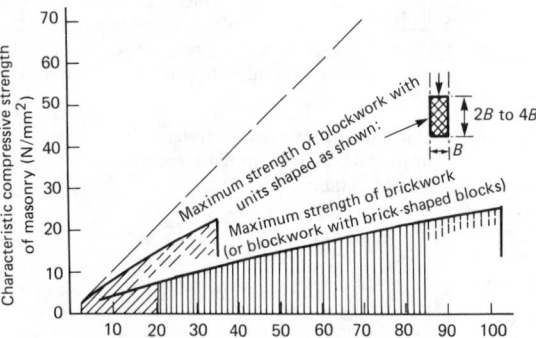

Figure 15.3 Relationship of characteristic compressive strength of masonry to the compressive strength of the units (BS 5628)

15.5.2 Slenderness

Tests on walls in recent years have shown that slenderness is less of a problem than it was thought to be 30 years ago and, in successive British Codes, reductions in load-carrying capacity for slenderness have tended to become smaller.

Reduction factors for slenderness tabulated in BS 5628 permit walls with a slenderness (effective height/thickness) of up to 27. Thus, for instance, allowing for end fixity a half-brick (102.5-mm thick) wall could be 3.7 m high and still carry 40% of the axial load appropriate for a low-height wall of the same thickness. Such a slim wall, as shown in cross-section in Figure 15.4, must lack robustness and most designers would prefer not to extend slenderness to this limit even with purely axial loading.

15.5.3 Eccentricity of loading

Simple eccentricity can be dealt with conservatively by using the appropriate reduction factor for slenderness and eccentricity in BS 5628 and assuming that the eccentricity at the top of any wall reduces to zero at the next level of lateral support below as shown in Figure 15.5(a). In practice, the eccentricity at the bottom of the wall is likely to be as in Figure 15.5(b) which is normally more favourable as it tends to counteract any further eccentricity applied at that level.

As an alternative to this simple procedure, the whole structure can be analysed rigorously as a frame.

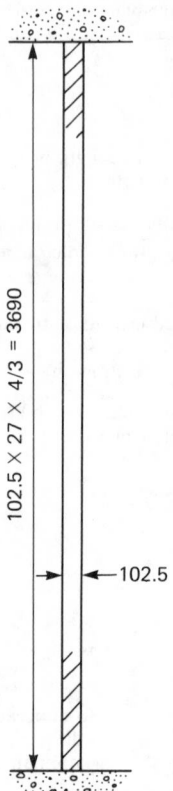

Figure 15.4 Maximum slenderness of wall (BS 5628)

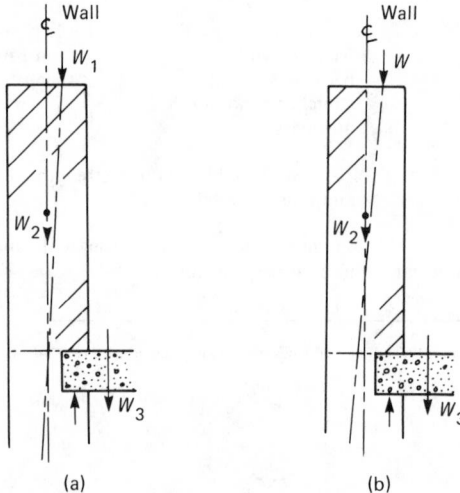

Figure 15.5 Line of thrust due to eccentric loading: (a) simplified assumption on line of thrust (BS 5628); (b) more likely line of thrust

Unless loaded heavily enough to provide fixity as in Figure 15.6(a), the rotation of the ends of floor slabs can cause unsightly cracking as in Figure 15.6(b). This is particularly likely with long-span concrete roofs or upper-floor slabs supported on masonary walls. In such locations the cracking can usually be eliminated, or at least controlled, without over-stressing the masonry, by allowing rotation as shown in Figure 15.6(c).

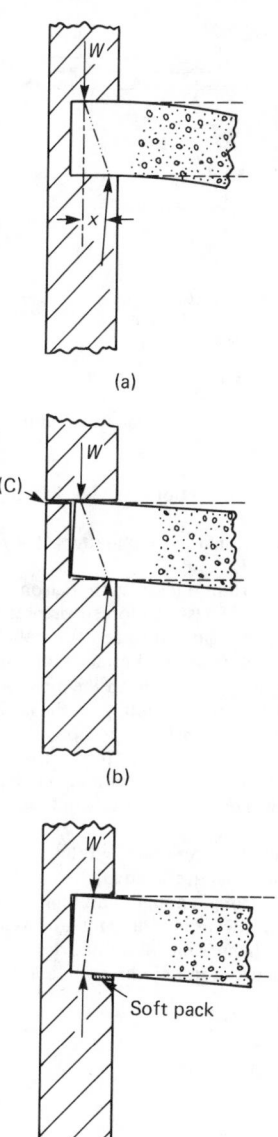

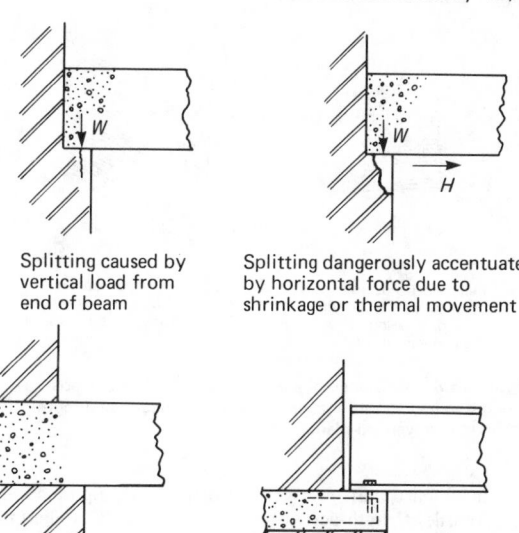

(a) Splitting caused by vertical load from end of beam

Splitting dangerously accentuated by horizontal force due to shrinkage or thermal movement

(b) Beam ends carried well back along supporting wall

Figure 15.7 Concentrated loads on ends of walls or on stiffening piers

Figure 15.6 Avoidance of cracking of masonry walls due to rotation of floor slabs at supports. (a) Slab rotation resisted by couple Wx. No cracking of masonry. (b) Load W not great enough to resist slab rotation. Upper wall lifted by rotation of slab and crack induced at point (C). (c) Load W not great enough to resist slab rotation, but soft pack as shown permits rotation without cracking of masonry walls

15.5.4 Concentrated loads

Traditionally, the need to spread concentrated loads from columns or the ends of beams has been met by using padstones of greater strength than the basic masonry of the walls or by laying local courses of stronger brick or stone. Today, this practice has been formalized in rules such as those in BS 5628. The problem is one of splitting and is most serious at the ends of walls or at piers where the splitting is most likely to lead to failure.

Such splitting could also be caused – or accentuated – by horizontal forces due to thermal or other movements. This is shown in Figure 15.7(a). It is something which is not explicitly covered in most Codes of Practice but which designers should consider. With long-span beams especially, the bearing should be carried back as far as practicable or a reinforced padstone should be used as indicated in Figure 15.7(b).

15.5.5 Lateral loads on masonry panels

The design of masonry walls so that they can be shown to resist wind forces is one of the major areas of doubt in the treatment of the material.

The uncertainty is greatest with panel walls with no vertical load except their own weight, but it is present with most thin or lightly loaded walls.

Following an extensive series of tests by the British Ceramics Association[8] some partly empirical and partly theoretical guidance has been incorporated in BS 5628, Part 1. This can be used to demonstrate adequacy in many of the most common situations but with vertical spanning in particular it is restrictive compared with what has been common practice for many years.

Where it can be shown to be appropriate, design for arching is a very good method of proving lateral stability; it, too, is recognized in BS 5628.

Real walls do sometimes blow out, or over, and those which do always seem to have a slenderness or lack of restraint far outside recommended limits. Some unlikely 'successes' may well be due to arching – even unexpected arching – and some to much higher tensile strengths in mortar joints than those generally assumed. In many cases the full 'Code' wind forces may never have occurred and may never occur in the future.

Much research is still being carried out on the resistance of masonry panels to lateral loads and it is hoped that further guidance may be given in future amendments to BS 5628. In the meantime, it is worth keeping in touch with the results of this research.

Two points of detail which designers would do well to remember are these:

(1) Whatever the published bending strengths, bending in a horizontal plane as in Figure 15.8(a) is much more reliable

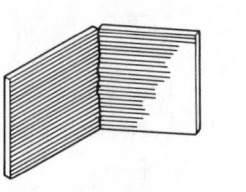 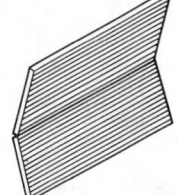

(a) Failure depends on breaking of the masonry units or shearing of the bed joint mortar

(b) Failure depends only on the tensile strength of the bed joint mortar

Figure 15.8 Resistance of masonry panels to lateral loads: (a) Strength in bending in horizontal plane; (b) greater and more certain than in vertical plane

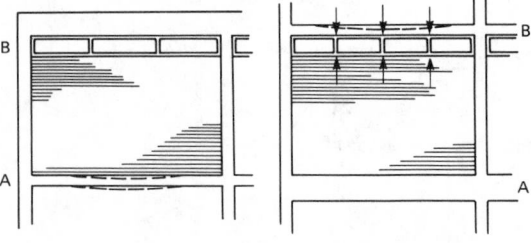

Worse than planned: creep deflection of slab at level A reduces panel to one way span

Better than planned: creep deflection of slab at B loads panel (even say through window) and improves lateral stability

Figure 15.9 Possible effects of deformation on lateral support for masonry cladding

in practice than in a vertical plane as in Figure 15.8(b). Tensile strength across bed joints can easily be disrupted especially during construction after the mortar has set but before it has gained its full strength.

(2) The edge support to panels will often be altered – sometimes for better and sometimes for worse – by the deformation of the structure, as can be seen from Figure 15.9. It is important to consider such possible movements and to make sure that adequate fixings are used at all edges assumed to provide restraint, even if there are relative movements due to deflection, shrinkage or settlement.

15.5.6 Stability and robustness

The great bulk of the advice in the world's masonry codes is devoted to stresses and compressive strength with only passing exhortations to consider stability. This seems curious when one considers that almost all failures of masonry structures – few in practice – have been due, not to overstressing, but to some form of instability. The reason may be that attempts to codify stability, while helpful in some circumstances, have tended to

lead to anomalies, unreasonable restrictions, or even new dangers, in others.

Stability and robustness are best seen as design matters. They need thought rather than rules.

Stability is a particularly important factor with masonry because of its low tensile strength. Unless reinforced or prestressed, masonry should generally be planned to rely for stability on gravity forces and on the friction induced by such forces.

With today's thin walls, this stability depends on the interaction of walls, floors and roofs. A lateral force acting on the face of a wall, such as that due to wind, is transferred to floors and roofs, which act as rigid horizontal plates and in turn transfer the force to the ground through shear walls running in the direction of the force. This is shown diagrammatically in Figure 15.10.

Designers should always follow the forces through this route to make sure that all the connections are adequate, that the floors and roofs are stiff and strong enough and that the 'racking' shear strength of the shear walls is great enough. Even if the structure is fully sheltered from wind it is important to consider a nominal lateral force acting in any direction and to

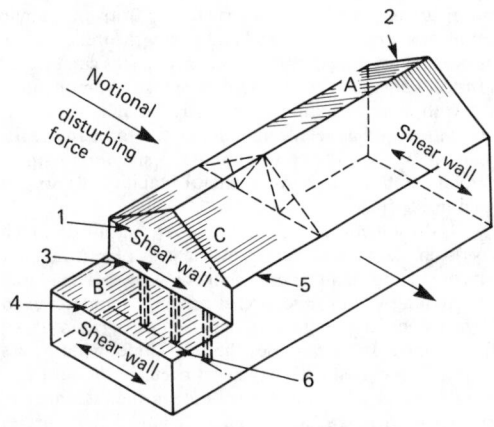

Roof braced in either horizontal planes (B) and (C) or on slope (A) and horizontal plane (B) to form effectively horizontal plates

Critical connections, or joints, to transfer force to the ground

Laterally loaded wall to horizontal plates: Joints 5 and 6
Horizontal plate to shear walls: Joints 1,2 and 4
Shear wall to lower horizontal plate: Joint 3

Figure 15.10 Diagrammatic representation of transfer of disturbing force on masonry structure to provide stability

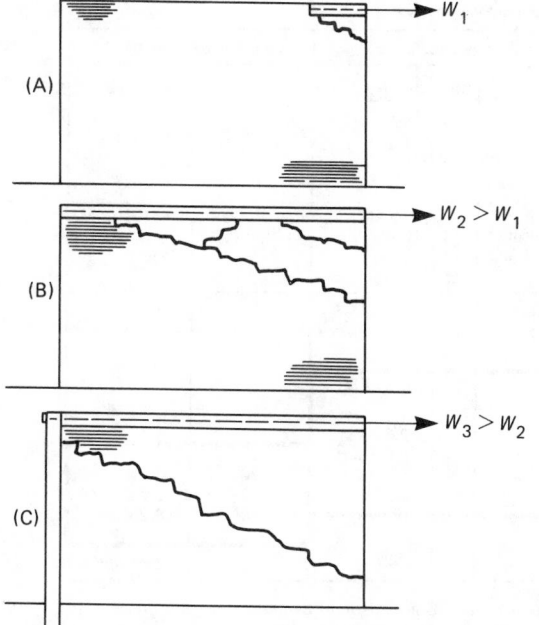

Figure 15.11 Relative strengths of three possible methods of connecting on to a shear wall

follow this to the ground. British Standard 5628 recommends a nominal lateral force of 1.5% of the dead load above the level being considered. A larger force may sometimes be preferable.

One of the most difficult questions with masonry is the assessment of the strength of connections. Figure 15.11 shows three means of connecting on to a shear wall. Designers would do well to opt for the strongest connection which is compatible with cost and other requirements rather than accept the lowest level of apparent adequacy.

There are few absolutes in design for stability. The skill lies in

balancing requirements which are often in conflict. This is illustrated in Figure 15.12. There is no virtue in just achieving what appears to be adequate stability if an unnoticeable change can make this much more certain. Accidental forces and material defects are largely unpredictable.

15.5.7 Accidental forces

Design to allow for accidental forces is just a particular case of design for stability. Rules introduced following the partial collapse of the large panel concrete structure at Ronan Point in 1968 have largely blinded engineers to the broad issue of accidental forces. These rules, which apply only to buildings of five storeys and above, were made to guard against gas explosions and do not in themselves ensure safety against all hazards.

The thinking was rightly to limit damage but the tying forces introduced in the structural codes to satisfy the rules may in some cases actually spread the damage. This is illustrated in Figure 15.13 where continuous vertical ties could actually cause progressive collapse as was shown in tests on a quarter-scale model at the Building Research Establishment.[9]

Accidental damage can be limited by planned sacrifice or by greater structural strength. The choice is a design matter although the solution must satisfy the broad functional requirements of the Building Regulations in most practical cases.

15.6 Reinforced and prestressed masonry

15.6.1 General

Much of the previous section on unreinforced masonry still applies but, when reinforced or prestressed, the scope for the use of masonry is greatly extended. Reinforced or prestressed beams can be made, as well as walls, following exactly the same general principles as those used with concrete. This has been demonstrated in tests and in practice.[10] The question to be decided is when such forms are advantageous, and this can be done only with some knowledge of the possible structural performance of

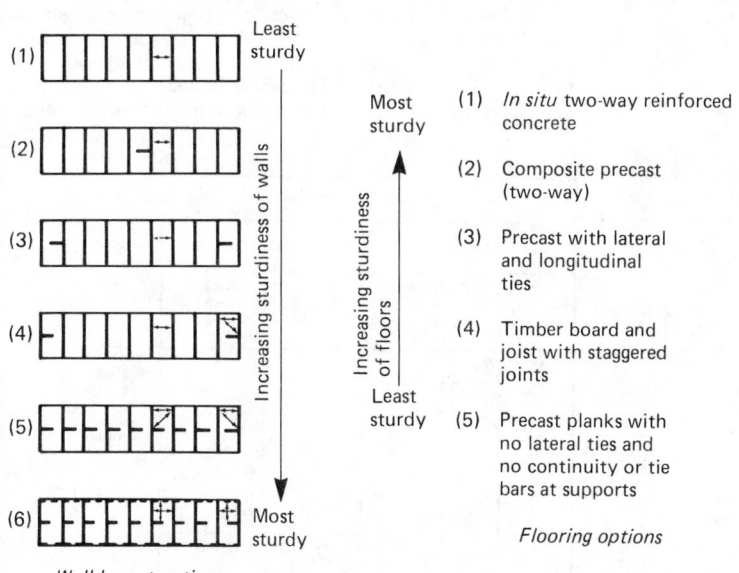

Wall layout options

(1) *In situ* two-way reinforced concrete

(2) Composite precast (two-way)

(3) Precast with lateral and longitudinal ties

(4) Timber board and joist with staggered joints

(5) Precast planks with no lateral ties and no continuity or tie bars at supports

Flooring options

Figure 15.12 Comparison of wall and floor options for simple masonry cross-wall construction

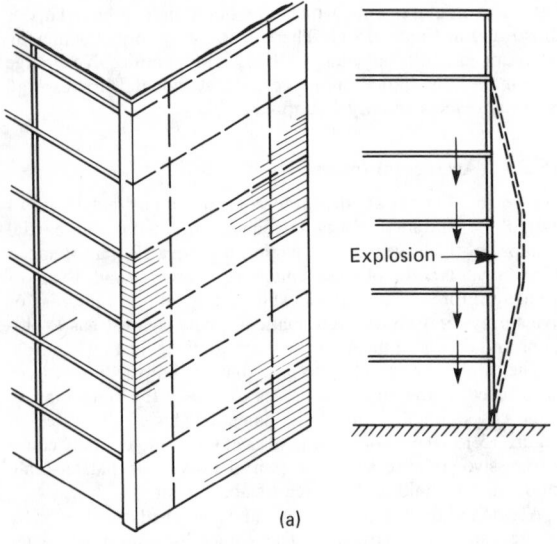

(a)

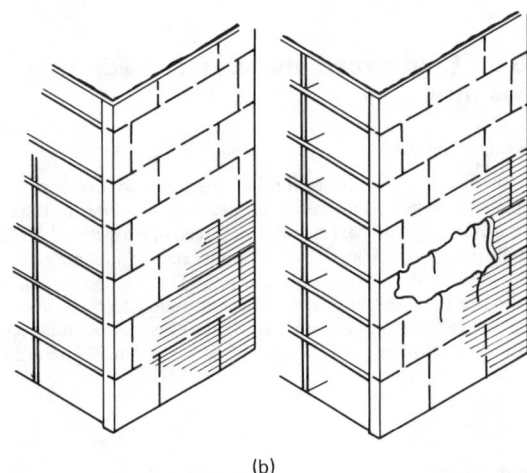

(b)

Figure 15.13 Diagrammatic representation of how simple code recommendations can spread accidental damage. (a) Horizontal ties broken by explosive force. Continuous vertical ties (as codes) keep well intact which bows out dropping ends of several floors above and below. (b) Independent staggered ties would allow local sacrifice and confine damage

reinforced or prestressed masonry. It is convenient, in this context, to compare the properties of masonry with those of concrete.

15.6.2 Structural performance of reinforced masonry

Figure 15.14 shows the relative bending strengths of reinforced brickwork (BS 5628, Part 2) and reinforced concrete in accordance with BS 8110. It can be seen that with moderate brick strengths the bending capacity of brickwork matches that required for most reinforced concrete and that, even with the lowest brick strength likely to be used (20 N/mm² unit strength), there is a very useful level of bending strength. A similar

Design resistance moments

Brick (BS 5628:Part 2)			Concrete (BS 8110:1985)	
Mortar (i) : special mfg. control		$y \boxed{} \dfrac{x}{x} \dfrac{y}{} = 0.6$	$m = 0.156\, bd^2\, f_{cu}$	
$M_d = 0.2\, bd^2\, f_k$				
Unit strength (N/mm²)	f_k N/mm²	DESIGN R.M. M_d (N.mm)	CHAR. CUBE STRENGTH f_{cu} (N/mm²)	DESIGN R.M. m (N. mm)
7	3.4	$0.7 \times bd^2$		
10	4.4	$0.9 \times bd^2$		
20	7.4	$1.5 \times bd^2$	10	$1.6 \times bd^2$
35	11.4	$2.3 \times bd^2$	15	$2.3 \times bd^2$
50	1.5	$3.0 \times bd^2$	20	$3.1 \times bd^2$
100	2.4	$4.8 \times bd^2$	30	$4.7 \times bd^2$
			45	$6.2 \times bd^2$
			50	$7.8 \times bd^2$

Figure 15.14 Comparison of bending strength of reinforced brickwork and reinforced blockwork

comparison can be made with concrete blockwork although the highest practical bending strength with blockwork is not so great.

With shear, the comparison is not so favourable to masonry as it is with bending. Figure 15.15 shows – again for brickwork – that without shear reinforcement the shear strength of all strengths of brickwork is well below that of reinforced concrete; the same is true of concrete blockwork. It is worth noting that the shear strength is virtually independent of the strength of the masonry units, being dependent on the tensile strength of the mortar and its bond to the units.

One advantage of prestressing over reinforcing with any material is that the prestress reduces the principal tensile stress and thus increases the shear capacity. With masonry this is a very real advantage but, curiously, no reference has been made to it in BS 5628, Part 2. The omission does not mean that designers cannot take advantage of this property of prestressing.

Design shear strength				
Brick (BS 5628:Part 2)		Concrete (BS 8110)		
Unit strength (N/mm²)	Design shear strength (N/mm²)	Char. cube strength (N/mm²) f_{cu}	Design strength (N/mm²)	Notes:
All	0.35 → 0.7 ($\gamma_m = 2.0$) ($\gamma_m = 2$)			(1) No shear steel
	0.175 → 0.35	25	0.34 – 1.22	(2) Shear strength varies with proportion of main steel
	In some cases with short shear spans this can be increased to a max. of 0.87	30	0.36 – 1.27	
		40	0.40 – 1.43	
		Note: Factor of safety of 1.25 included for concrete		

Figure 15.15 Comparison of shear strength of reinforced brickwork and reinforced concrete

It has been an accepted feature of prestressed concrete for at least 40 years.

Compared with reinforced masonry, there has been little practical experience with prestressed masonry and there is a school of thought which considers that it would have been best to omit it altogether from BS 5628 at this time. Nevertheless, it has been used successfully, as is shown later in this chapter, and may be used more in the future.

15.6.3 Uses for reinforced masonry

Not only is the shear strength of masonry low, but it is very difficult to incorporate shear reinforcement in it to increase this. For both these reasons, masonry compares unfavourably with concrete for use in beams except possibly in special cases such as that of a deep beam within the plane of a wall. However, reinforced masonry comes into its own in laterally loaded walls where the low shear strength is seldom a major problem.

Masonry is essentially a wall material, or one for arches and vaults. Traditionally, it used to depend for stability on its mass but today, with reinforcement, laterally loaded walls can be made of comparable slenderness to those in reinforced concrete. Further, while concrete walls, if visible, are increasingly being faced with masonry, reinforced masonry has its elegance built into the structure. Figure 15.16 shows a number of ways of reinforcing masonry walls using standard bricks and blocks.

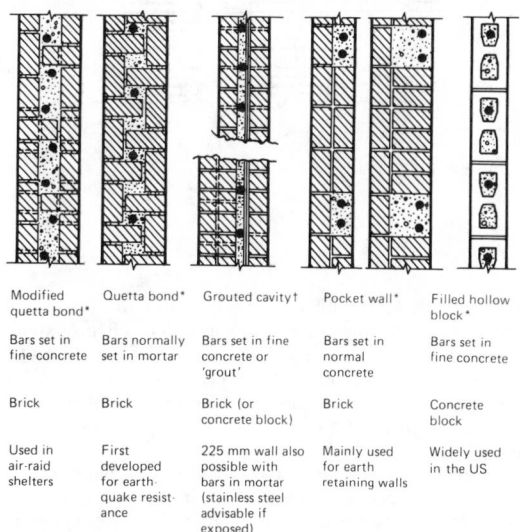

Vertical-spanning reinforced masonry

Modified quetta bond*	Quetta bond*	Grouted cavity†	Pocket wall*	Filled hollow block*
Bars set in fine concrete	Bars normally set in mortar	Bars set in fine concrete or 'grout'	Bars set in normal concrete	Bars set in fine concrete
Brick	Brick	Brick (or concrete block)	Brick	Concrete block
Used in air-raid shelters	First developed for earth-quake resistance	225 mm wall also possible with bars in mortar (stainless steel advisable if exposed)	Mainly used for earth retaining walls	Widely used in the US

* Essentially vertical span but secondary horizontal reinforcement often used in bed joints
† Full two-way span possible (or horizontal only). (Grout is the term used in the US for fine high slump concrete.)

HORIZONTAL-SPANNING REINFORCED MASONRY

Normally used for light loads only (except grouted cavity type).
Reinforcement normally small bars or mesh set in bedding mortar.
Equally suitable for brickwork and blockwork.

Figure 15.16 Typical methods of reinforcing masonry walls

15.6.4 Durability of reinforced masonry

As with reinforced concrete, durability is a factor which is receiving increasing attention today. British Standard 5628, Part 2 gives clear and full recommendations on cover to normal carbon steel and to galvanized reinforcement in masonry for different levels of exposure.

With stainless steel reinforcement there is no need for any cover to the steel specifically for durability. Although the supply cost of the steel is several times that of normal carbon steel, the percentage extra on the whole project tends to be very small once fixing and all the other costs of the construction are included.

Stainless steel is being used increasingly for ties and fixings in masonry as well as for reinforcement and is no longer the exorbitantly expensive material it used to be.

15.7 Dimensional stability of masonry

The order of unrestrained movements of clay and concrete products is shown in Table 15.2. Because of the restraints which exist to some extent in all real buildings these movements tend to be less than one would calculate from the tabulated figures. The vital question is how much less, and the answer must be that at present we do not know.

British Standard 5628, Part 3 recommends an allowance of 1 mm of movement per metre for clay brickwork, with movement joints a maximum of 15 m apart. For calcium silicate bricks joints are recommended at 7.5 to 10 m and at 6 m for concrete bricks and blocks.

It is worth remembering what these joints are for. In the case of clay brickwork they are needed primarily for expansion and thus, to be effective, must be wide enough and filled with something soft enough to allow the expansion to take place. In the case of calcium silicate bricks and all concrete units the joints are needed mainly for shrinkage and effective sealing becomes most important.

As in the case of movement joints in concrete structures, there is some evidence that the recommended cure for movement problems has not always been effective. The cure may even introduce new problems of maintenance. A recent study for the Construction Industry Research and Information Association (CIRIA)[11] showed that, both for clay and concrete units, adherence to the spacing then recommended in CP 121 failed to eliminate noticeable defects in a significant number of cases, while in quite a large proportion of other cases no noticeable defects were found even in walls well beyond recommended limits of unbroken length.

The problem of movement is too complex for simple rules. What is more, in many cases it may be most economical, and in the long term most satisfactory, to reduce the number of joints, risk some cracking and repoint after a few years.

Designers would do well to study case histories, observe real buildings and then try to recognize the situations where movement may be serious and those where damage, if it occurs, is only of a cosmetic nature. Figures 15.17 and 15.18 show some key factors but these should only be considered as examples.

15.8 Application of masonry and scope for future use

15.8.1 High-rise (small-cell) residential buildings

Unreinforced masonry has formed the sole vertical support to residential buildings of up to at least 18 storeys while with vertical reinforcement it has been used in blocks of over 20 storeys even in seismic zones. In most cases the design has been dominated by the need for resistance to lateral forces and the assessment of interaction between floors and walls to achieve this. Perhaps surprisingly, tension has often proved more of a problem than compression.

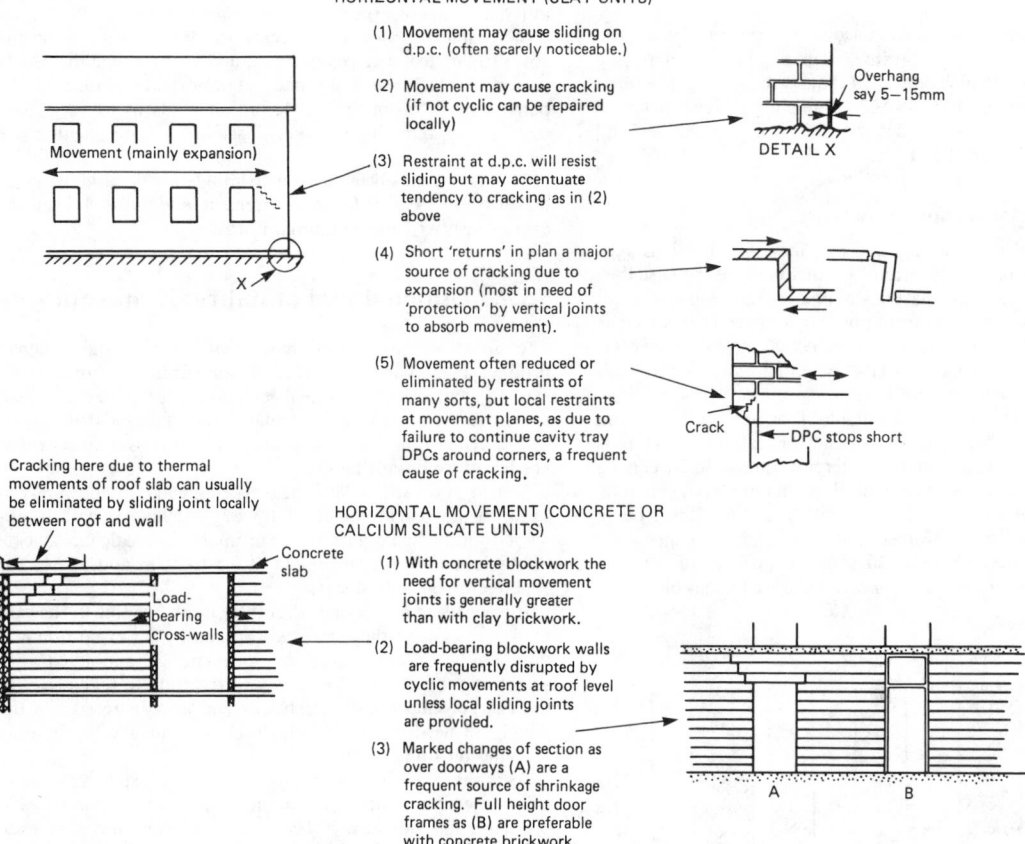

Figure 15.17 Typical serviceability problems due to horizontal movements in masonry

With the increased development of computer programs in the last 15 years the design of such buildings should be much easier than it was in their heyday in the mid 1960s. There are also some signs of a revival in the popularity of high-rise flats and hostels.

15.8.2 Low-rise (large-cell) buildings

The challenge of extending the economic use of masonry, as proved in domestic construction, to larger-cell buildings such as sports halls or warehouses has been mentioned at the beginning of this chapter.

Here the downward loads tend to be small but the walls may need to span vertically 2 or 3 times as far as in housing. This has been achieved in frameless construction with deep ribs at regular intervals, emphasized architecturally, or by making the walls cellular. In such cellular walls – popularly known as 'diaphragm' walls – the masonry is placed in the most efficient way to resist lateral forces (Figure 15.19a). The walls are usually designed to span from the ground to a roof which is braced to transfer the load to shear walls as already discussed. Sometimes such walls are prestressed with vertical cables either bonded or in voids as shown in Figure 15.19(b), or they can be reinforced. Care is needed during construction to make sure that the walls do not fail in wind before the bracing of the roof is provided.

15.8.3 Boundary walls

Masonry has proved itself for boundary walls over a very long period. Such walls, either of constant section or with wholly inadequate stiffening piers, frequently defy all probability of stability but have given good service for decades or centuries; some such walls survive even in spite of considerable bulges or tilts. Nevertheless, from time to time they do blow over and there is no justification for building unstable boundary walls today.

Stability can be achieved by an irregular planform or by reinforcing vertically or prestressing or by any combination of these. Reliance on the tensile strength of the masonry across the bed joints is unwise except on a very small scale. The forms of reinforcement shown in Figure 15.16 can be used or stable walls of very slender sections can be built with special or cut bricks as shown in Figure 15.20. The use of stainless steel is advisable in most boundary walls because of their extreme exposure.

Shear strength is virtually never a problem with boundary walls.

15.8.4 Retaining walls

Reinforced masonry has proved to be particularly suitable for retaining walls, either on a small scale associated with housing or on major civil engineering works. It is hard to understand why it has not been used more. Reinforced concrete walls are often faced in masonry for appearance. Why not use the masonry structurally? The answer to this must be that it is largely habit which prevents engineers from thinking of reinforced masonry, or fear which leads to unduly high pricing of

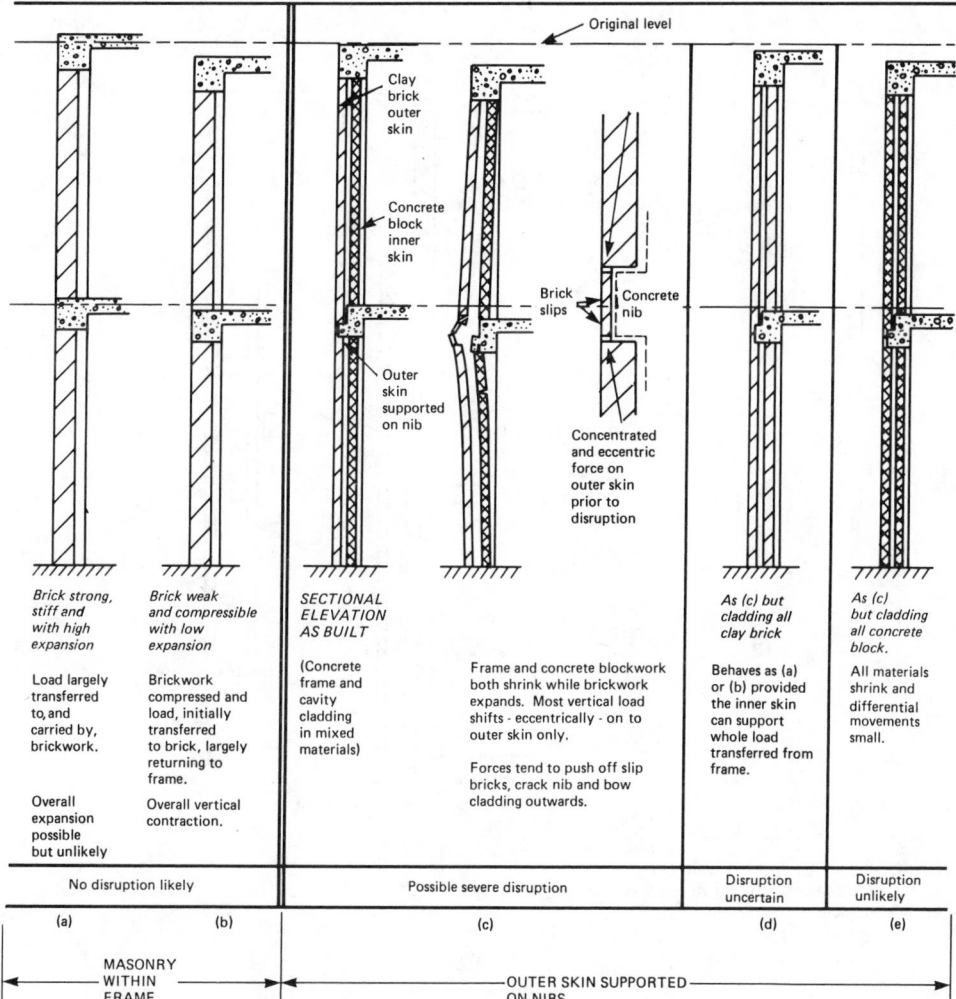

Figure 15.18 Relative movements of masonry cladding and concrete frames: diagrammatic representation based on two storeys. Real problems tend to be confined to multi-storey building

the slightly unfamiliar. However, today there is enough evidence of successful construction of retaining walls in masonry to prove that they are easy to build, and perform well.

Brick retaining walls have been built successfully in Quetta bond, in grouted cavity construction and using the 'pocket wall' technique. With concrete blockwork the fixing of reinforcing bars in the filled hollows in the blocks has become quite common. The planforms of such walls are shown in Figure 15.16.

Unlike boundary walls, retaining walls normally need only resist lateral forces in one direction. Thus, for cantilevers the reinforcement should be as near to the loaded face as possible. The pocket-type of wall is ideal in this respect. Figure 15.21 shows how such walls are formed, with the thickness increasing to match the increasing bending moment and the reinforcement as close to the rear face as practicable. In some cases the steps in thickness have been repeated two or three times.

In pocket-type walls the reinforcement is surrounded in dense concrete whose compaction can be checked once the small back shutter to the pocket is removed. Thus, the durability is equiva-

lent to that of reinforced concrete but the compressive force is resisted not by the concrete but by the brickwork.

Pocket-type retaining walls have been built in the US with heights of up to 7.3 m.[12] There are now quite a few major walls of this type in the UK used for bridge wing walls and similar purposes. One British pocket-type retaining wall approximately 4 m high was monitored for 517 days after which the deflection was only 16 mm. As expected, the movement was apparently continuing but tailing off.[13]

The design of pocket-type walls is covered in BS 5628, Part 2 which deals with concrete cover, pocket spacing and workmanship as well, of course, as with structural design. In some circumstances, the characteristic shear strengths in BS 5628, Part 2 may prove a restriction on the performance of such walls, although tests on actual walls have almost all shown failure in bending. There is a good case for revising the shear clauses in BS 5628 especially in relation to retaining walls.

One objection which is sometimes raised to the use of masonry retaining walls on civil engineering projects is speed of building. This objection may or may not be real but with

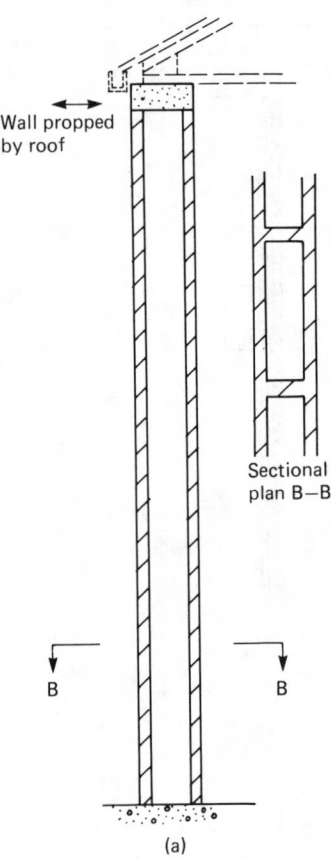

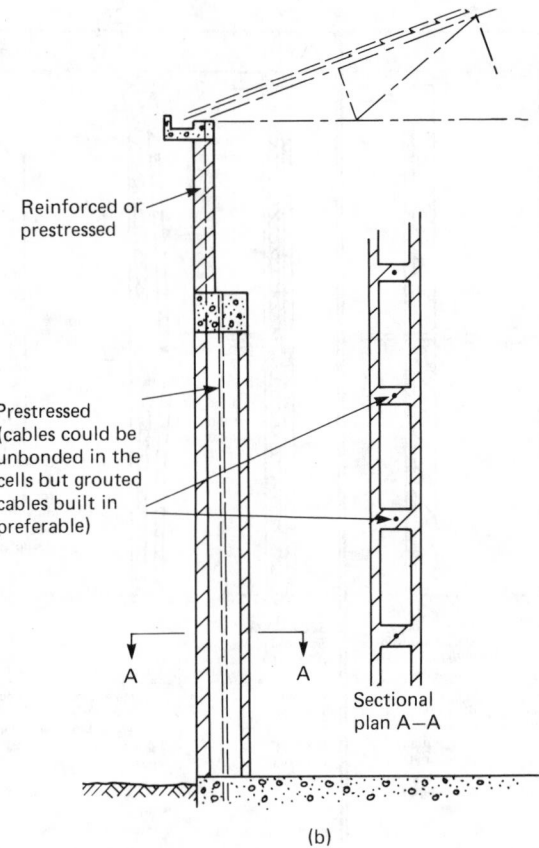

Figure 15.19 Typical forms of cellular (diaphragm) wall. (a) Unreinforced (wall spans from ground to roof). (b) Prestressed (or reinforced). Wall acts as vertical cantilever

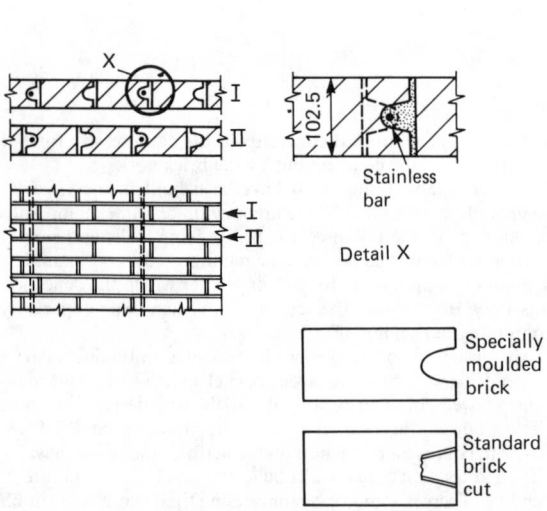

Figure 15.20 Thin but stable boundary walls formed with special or cut bricks (concrete blocks also as in Figure 15.16)

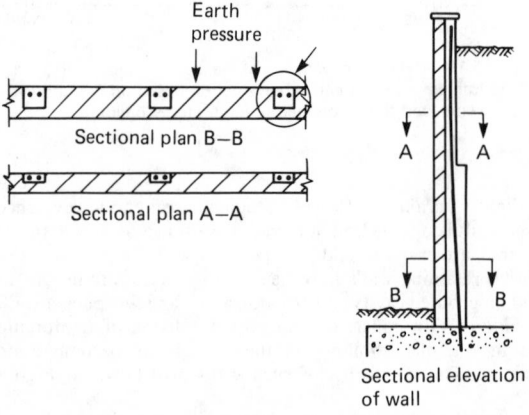

Figure 15.21 Typical pocket-type retaining wall

pocket-type walls, prefabrication is a very real possibility. This was demonstrated by a trial some years ago.[1]

15.8.5 Bridges

Masonry bridges have generally proved more durable than

those of iron, steel or concrete. There are thousands of them under roads and railways and over canals which have had minimal maintenance during 100 years or more. They have adapted themselves to settlement without distress and still look attractive. Tests have been made in recent years, both in the laboratory[14] and by the Transport and Road Research Laboratory on actual arch bridges. Our understanding of the behaviour of masonry bridges is better today than it ever was but in spite of this we continue to use slab or beam bridges under highways which are subject to corrosion due to rain and de-icing salt.

In the future, there seems a clear case for using more masonry arches, possibly combining mass concrete with brickwork. Such arches would be particularly appropriate for medium spans where piped culverts are too small but major long spans are not needed. Useful guidance on the appraisal of existing masonry bridges is given in the Departmental Standard BD 21/84 and the associated Advice Note BA16/84 both issued by the Department of Transport.[15] These documents are also relevant, at least in part, to new construction.

15.9 Conclusions

With the publication of all three parts of BS 5628, the UK is probably leading the world in recognized guidance on masonry design. Further, in research, the UK has taken a leading role for 20 years. It is now up to design engineers to make full use of this rediscovered material.

15.10 Acknowledgements

Tables 15.1 to 15.3 were first published in the Author's paper[1] but have been brought up to date. Figures 15.1, 15.3, 15.16 and 15.21 and parts of Figures 15.19 and 15.21 have been adapted with the permission of Thomas Telford Ltd from those in this paper. Figures 15.10, 15.11, 15.12 and 15.13 have been adapted with the permission of the Institution of Structural Engineers from those already published in the Author's paper.[9] The Author wishes to thank those concerned for permission to reproduce this material.

References

1 Sutherland, R. J. M. (1981) 'Brick and block masonry in engineering.' *Proc. Instn Civ. Engrs*, Part 1, **70**, 31–63. [Tables 15.1, 15.2 and 15.3 are updated versions of tables first published in the above paper.]

2 British Standards Institution (1978) *Code of practice for use of masonry*. BS 5628, Part 1: 'Unreinforced masonry.' BSI, Milton Keynes.

3 British Standards Institution (1985) *Code of practice for use of masonry*. BS 5628, Part 2: 'Structural use of reinforced and prestressed masonry.' BSI, Milton Keynes.

4 British Standards Institution (1985) *Code of practice for use of masonry*. BS 5628, Part 3: 'Materials and components, design and workmanship.' BSI, Milton Keynes.

5(a) Haseltine, B. A. and Moore, J. F. A. (1981) 'Unreinforced masonry.' In: R. G. D. Brown (ed.) *Handbook to BS 5628: structural use of masonry*, Part 1: Unreinforced Masonry Brick Development Association, Windsor. (b) Roberts, J. J., Edgell, G. J. and Rathbone, A. J. (1986) 'Palladian.' *Handbook to BS 5628*, 'Structural use of reinforced and prestressed masonry.' Viewpoint Publication No. 13.028, London.

6(a) Hendry, A. W. (1981) *Structural brickwork*. Macmillan, London. (b) Curtin, W. G. *et al*. (1982) *Structural masonry designers' manual*. Granada, St Albans; (c) Gage, M. and Kirkbride, T. (1980) *Design in blockwork*, 3rd edn. Architectural Press, London. (d) Orton, A. (1986) *Structural design of masonry*. Longman, Harlow.

7(a) British Standards Institution (1985) *Structural use of concrete*. BS 8110, Part 1: 'Code of practice for design and construction.' BSI, Milton Keynes; (b) British Standards Institution (1985) *Structural use of concrete*. BS 8110, Part 1: 'Code of practice for special circumstances.' BSI, Milton Keynes.

8 West, H. W. H., Hodgkinson, H. R. and Haseltine, B. A. (1977) 'The resistance of brickwork to lateral loading; Part 1: Experimental methods and results of tests on small specimens and full-sized walls.' *Struct. Engr* **55**, 10, 411–421.

9 Sutherland, R. J. M. (1978) 'Principles for ensuring stability.' *Symposium on stability of low-rise buildings of hybrid construction*. Institution of Structural Engineers, 5 July 1978, London. pp. 28–33.

10 Bradshaw, R. E., Drinkwater, J. P. and Bell, S. E. (1983) 'Reinforced brickwork in the George Armitage office block, Robin Hood, Wakefield.' *Struct. Engr*. **61A**, 8, 247–254.

11 Construction Industry Research and Information Association (1987) Movement and cracking in long masonry walls. CIRIA Practice note. (To be published.)

12 Abel, C. R. and Cochran, M. R. (1971) 'A reinforced brick masonry retaining wall with reinforcement in pockets.' In: H. W. H. West and K. H. Speed, (eds) *SIBMAC Proceedings, International Brick Masonry Conference*. British Ceramic Research Association, Stoke-on-Trent. pp.295–298.

13 Maurenbrecher, A. H. P. (1977) *A pocket-type reinforced brickwork retaining wall*. Structural Clay Products, Potters Bar, SCP Publication No. 13.

14 Sawko, F. and Towler, K. (1982) 'Load-bearing brickwork: structural behaviour of brickwork arches.' *Proc. Br. Ceramic Soc.*, **30**, 7, 160–168.

15(a) Department of Transport (1984) *The assessment of highway bridges and structures*. Roads and Local Transport Directorate. Advice Note No. BA16/84. HMSO, London. (b) Department of Transport (1984) *The assessment of highway bridges and structures*. Roads and Local Transport Directorate. Departmental Standard BD21/84. HMSO, London.

16

Timber Design

F H Potter BSc Tech, CEng, MICE, FIWSc, AMCT
Senior Tutor, Imperial College of Science and Technology

Contents

16.1 Introduction

Timber is one of the finest structural materials: it has a high specific strength, can be easily worked and jointed and does not inhibit design. Like most other structural materials it suffers attack causing deterioration (corrosion, weathering and biodeterioration) but once the material is known and the causes understood, effective preventative measures can be taken easily and economically.

Design is thus a confluence of specification, structural analysis, detailing and protection, each of which is of equal importance if an effective design is to be achieved.

Nowadays, structural design covers a wider range of components than ever before, for the intense wind loadings in high-rise building coupled with large glazed areas has meant that much window joinery is now subject to structural design. In addition, the effects of wind loadings together with the requirements of the Building Regulations and the newer building shapes has meant that even in low-rise buildings, components which once were built must now be designed.

Timber is thus used for a wide variety of structural purposes, either on its own or in combination with one of the 'heavier' materials. It can take extremely simple forms such as solid beams, joists and purlins or can be used in the more recent forms of glued–laminated construction or plywood panel construction. These latter forms allow the design of exciting and economic structural shapes, the variety of which may be judged from Tables 16.1 and 16.2.

Some of the characteristics of timber may be found in Table 16.3, whilst general properties are given in other publications.[1-3]

16.2 Design by specification

Essentially, this is a prescription of fitness for use under service conditions and requires the choice not only of an appropriate material but also of its condition, use and protection.

The success of the specification will depend upon its interpretation; standard glossaries are available for timber and woodwork,[4] nomenclature of timber[5] and preservative treatment.[6]

Functional and user needs will dictate the choice of material based on the following factors.

16.2.1 Species and use

Very many timbers are structurally useful, whereas usefulness for joinery purposes is often more restrictive. Where timber is used for structural joinery, the combination of requirements may be even more restrictive.

Table 16.3 lists most of the timbers and their characteristics for which working stresses are given in BS 5268: Part 2,[7] whilst BS 1186[8] indicates the joinery use of specific species. A comprehensive guide to West African species and their uses, both structural and joinery, is given in pamphlets issued by the United Africa Company.[9]

Flooring, particularly industrial flooring, has particular requirements and recommendations for suitable timbers for these

Table 16.1 Roof selection

Division	Subdivision	Construction	Minimum support conditions	Maximum economic spans (m)	Fastenings
Beams		Solid timber	Vertical support at ends	6	None
		Laminated, either vertically or horizontally, depending on size		24	Glue
		I or box sections: flanges solid or laminated. Webs plywood or diagonally boarded		30	Glue and/or nails
Arches		Laminated horizontally	Vertical and horizontal support at ends	46	Glue and/or nails for laminating
		I or box sections: flanges laminated horizontally. Webs diagonally boarded		46	Connectors for site joints
Portals		Laminated horizontally		24	Glue and/or nails for laminating
		I or box sections: flanges solid or laminated. Webs plywood or diagonally boarded	Vertical and horizontal support at ends	46	Connectors for site joints
Trusses	Belgian	Solid timber	Vertical support at ends	12 / 24	Nails and/or glue / Connectors
	Warren	Solid timber		12 / 30	Nails and/or glue / Connectors
	Bowstring	Laminated chords. Solid webs		46	Glue and/or nails for laminating / Connectors at joints

(After: L. G. Booth, *Engineering*, 25 March 1960)

Table 16.2 Roof selection

Division	Subdivision		Construction	Minimum support conditions	Maximum economic sizes (m)	Fastenings
Plates	Flat		Membrane formed from plywood or layers of diagonal boarding A single-skin structure may have stiffening ribs	Vertical support at corners	12 × 12	Nails and/or glue
	Folded		A double-skin structure will have spacing ribs Edge beams and end diaphragms required	Vertical support at corners	18 × 9	Membrane with nails and/or glue Diaphragms with nails or connectors
Singly curved shells	Circular cylindrical		Membrane formed with layers of diagonal boarding. May have stiffening ribs Edge beams required End diaphragms required	Vertical support at corners	30 × 12	Membrane with nails and/or glue Beams (see Table 16.1) Diaphragms with nails or connectors
Doubly curved shells	Spherical dome		Boarded membrane with or without laminated ribs Laminated ring beam	Ring beam to be supported at intervals	30 dia.	Membrane with nails and/or glue Ribs and ring beam glued
	Hyperbolic paraboloid		Boarded membrane with laminated edge beams	Vertical support only at low points, if columns tied together. Otherwise buttresses at low points	21 × 21	Membrane with nails and/or glue Edge beams glued
	Elliptical paraboloid		Boarded membrane with laminated tied arches along edges	Vertical support at corners	24 × 24	Membrane with nails and/or glue Tied arches with glue and connectors
	Conoid		Boarded membrane with laminated tied arches on ends Edge beams required	Vertical support at corners	30 × 9	membrane with nails and/or glue Beams (see Table 16.1) Tied arches with glue and connectors

(After: L. G. Booth, *Engineering*, 25 March 1960)

requirements are given in Princes Risborough Laboratory (PRL) Technical Note No. 49.[10]

16.2.2 Availability and sectional properties

Availability is equally important, and Table 16.3 indicates the availability of the structural timbers from the viewpoints of supply and length. The geometric properties of sawn and precision timber to be used in design are also given in BS 5268: Part 2. Guidance on the usefulness of worldwide species may be found,[11] whilst the available sizes for hardwoods are given in BS 5450.[12]

16.2.3 The movement of timber

Even with dried timber, changes in atmospheric conditions will result in a varying moisture content which will induce fluctuating dimensional changes in the timber, known as 'movement'. The variation can be designed-for quite simply but some knowledge of the degree of possible movement is helpful. Some indication can be obtained from Table 16.3, whilst further information can be obtained from the publications of the PRL.[2,3,13,14]

16.2.4 Moisture content and end use

Every species of timber will achieve a fairly steady moisture content for a particular environment – the equilibrium moisture content. The PRL has established moisture contents for various environments.[12] Greater reliability can be achieved by drying timbers to these moisture contents before construction.

Table 16.3 Characteristics and availability of some structural timbers

Standard name	Approx. density at M/C 18% (kg/m³)	Natural durability	Resistance to preservative treatment	Moisture movement	Working quality	Availability		Relative price
						Supply	Normal length (m)	
SOFTWOODS (IMPORTED)								
Douglas fir–larch	590	Moderately	Resistant	Small	Good	Good	4.20–4.80	Medium
Hem–fir	530	Not	Resistant	Medium	Good	Good	4.20–4.50	Low
Parana pine	560	Not	Moderately	Medium	Good	Good	3.60–3.90	Low
Pitch pine	720	Durable	Resistant	Medium	Good	Good	4.50–9.00	Medium
E. redwood	540	Not	Moderately	Medium	Good	Good	1.50–7.00	Low
E. whitewood	510	Not	Resistant	Small	Good	Good	1.50–7.00	Low
Canadian spruce–pine–fir	450	Not	Very	Medium	Good	Good	2.40–5.10	Low
W. red cedar	380	Durable	Resistant	Small	Good	Good	2.40–7.30	Low
SOFTWOODS (HOME GROWN)								
Douglas fir	560	Moderately	Resistant	Small	Good	Fair	1.80–4.50	Low
Larch (E—Japan)	560	Moderately	Resistant	Medium	Good	Good	1.80–3.60	Medium
Scots pine	540	Not	Moderately	Medium	Good	Good	1.80–3.60	Low
European spruce	380	Not	Resistant	Small	Good	Good	1.80–3.60	Low
Sitka spruce	400	Not	Resistant	Small	Good	Good	1.80–3.60	Medium
Corsican pine	510	Not	Moderately	Small	Good	Fair	1.80–3.60	Medium
HARDWOODS (IMPORTED)								
Abura	590	Perishable	Moderately	Small	Good	Good	1.80–6.00	Low
African mahogany	590	Moderately	Extremely	Small	Medium	Good	1.80–7.30	Medium
Afrormosia	720	Very	Extremely	Small	Medium	Good	2.40–7.30	Med high
Greenheart	1 060	Very	Extremely	Medium	Difficult	Good	4.80–9.00	High
Gurjun/Keruing	720	Moderately	Resistant	Large	Medium	Good	1.80–7.30	Low
Iroko	690	Very	Extremely	Small	Medium	Good	1.80–6.00	Medium
Jarrah	910	Very	Extremely	Medium	Difficult	Good	1.80–8.40	Med high
Karri	930	Durable	Impermeable	Large	Difficult	Good	1.80 up	Med high
Opepe	780	Very	Moderately	Small	Medium	Good	1.80–6.00	Medium
Red meranti	540	Moderately	Resistant	Small	Good	Good	1.80–7.30	Low
Sapele	690	Moderately	Resistant	Medium	Good	Good	1.80 up	Medium
Teak	720	Very	Extremely	Small	Medium	Good	1.80 up	High
HARDWOODS (HOME GROWN)								
European ash	720	Perishable	Moderately	Medium	Good	Fair	1.80 up	Low
European beech	720	Perishable	Permeable	Large	Good	Good	1.80 up	Medium
European oak	720	Durable	Extremely	Medium	Medium	Good	1.80 up	Medium

16.2.5 Working properties

Ease of fabrication is indicated in Table 16.3, although more detailed information may be found in PRL publications.[2,3]

16.2.6 Natural resistance to attack

Timber has a widely varying resistance to attack by fungi, insects, marine borers and termites. Fungi will not normally attack timber having a moisture content lower than 20% but a timber's ability to resist fungal attack is classified according to Table 16.4.

The natural durability of some structural timbers is given in Table 16.3. Information on further timbers will be found in PRL

Table 16.4 Durability classification of the heartwood of untreated timbers

Grade of durability	Approximate life in ground contact (yr)
Very durable	More than 25
Durable	15–25
Moderately durable	10–15
Nondurable	5–10
Perishable	Less than 10

Technical Note No. 40[15] and the *Handbooks* on softwood and hardwoods[2,3] whilst further advice on the control of decay will be found in PRL Technical Notes 29, 44 and 57.[16–18]

Termite attack and its prevention are dealt with by the PRL[19] in which the following timbers are mentioned as being generally resistant: iroko, opepe, Californian redwood and teak. Other timbers are given in the *Handbooks* on softwood and hardwoods.[2,3] Marine borers are a hazard in the sea or brackish waters and PRL Leaflet No. 46[20] gives advice on the protection of timbers against this attack. Highly resistant timbers suitable for marine works are: greenheart, pyinkado, turpentine, totara, jarrah, basralocus and manbarklak.

16.2.7 Preservative treatment

The sapwood of all timbers is liable to attack by fungi and insects but it is often possible to obtain a more attack-resistant structure by pressure-impregnating nondurable or perishable timbers than by using durable species. Indeed, it is sometimes more economic also.

The amenability of timbers to preservative treatment is given in Table 16.3 and is related to the following classification:

Permeable:	Easily treated by either pressure or open tank.
Moderately resistant:	Fairly easy to treat by pressure, penetration 6 to 20 mm in 2 to 3 h.
Resistant:	Difficult to impregnate, incising often used. Penetration often little more than 6 mm.
Extremely resistant:	Very little penetration can be achieved even after prolonged treatment.

Further information and guidance on satisfactory types and methods of treatment may be found in publications of the British Standards Institute (BSI)[21–23] and the Timber Research and Development Association (TRADA). The economics of timber preservation is discussed in *Timberlab 17.*[24]

16.2.8 Fire resistance

Although timber ignites spontaneously at about 250°C, ignition is a function of the external surface area to the total volume of timber and the rate of charring does not significantly increase with a rise of temperature. The rate of charring is generally taken as about 0.5 mm/min (western red cedar 0.85 mm/min, dense hardwood 0.42 mm/min), but perhaps the most important factor is that the structural properties of uncharred material remain virtually unchanged.

Thus, if adequate protection against combustion is provided, timber is one of the safest structural materials in a severe fire. These measures are usually: (1) the provision of sacrificial material; (2) chemical impregnation; and (3) protective covering.[25–27]

16.3 Stress grading and permissible stresses

Timber is a natural organic material and therefore is subject to wide variability because of environmental, species and genetic effects. This variability affects both visible quality and strength.

If, for any particular property and species only one design stress were specified, this would have to be set so low (to allow for variability) that the material would have a very limited

structural application. In consequence, a number of stress grades have been adopted, leading not only to a more economic use of the material but also to a higher yield of structurally useful material.

There are two main methods of stress grading for solid timber: (1) visual grading; and (2) mechanical grading. Each requires a different procedure.

16.3.1 Visual stress grading

In visual grading, data obtained from clear material (straight grained and free from knots and fissures) are analysed statistically for each species and basic stresses for each property are devised. These basic stresses are then reduced by factors which take account of the strength-reducing effects of the permissible growth characteristics for each stress grade.

At the present time, there are two sets of quality requirements for visual stress grading in this country, one for softwoods and one for hardwoods.

The first set is given in BS 4978.[28] Besides setting the requirements for two grades for solid softwood timber construction (SS and GS), this standard restates the requirements for laminating timber grades.

The second set is given in BS 5756[29] for tropical hardwoods (HS grade). In the current edition of BS 5268:Part 2 home-grown hardwoods have been deleted, but it is probable that these will be reintroduced.

16.3.2 Mechanical stress grading

Mechanical stress grading[30] is a method of non-destructive testing each piece to be graded. The piece is bent under a constant central load over a constant short span. The strength of the material can then be calculated accurately from the resultant deflection. Four grades are presented (M75, M50, MSS and MGS) and the grade stresses for the dry condition only are tabulated in BS 5268:Part 2. At present, machine-grade stresses are limited to six softwood species for which control information is available. However, it will be possible to machine-stress-grade other timbers in accordance with BS 4978.[28]

16.3.3 Glued–laminated timber grades

In glued–laminated members, the presence of strength-reducing characteristics will have a smaller effect than in solid timber, since the probability of identical structural defects appearing in identical positions in adjacent laminations is very small. British Standard 5268:Part 2, therefore allows higher grade stresses for glued–laminated timbers, these being obtained by applying tabulated modification factors to the grade stresses for each species.

16.3.4 Strength classes of timber

For the first time, BS 5268:Part 2 introduces the concept of strength classes for timber. Softwood species–grade combinations for strength classes, graded to BS 4978 are tabulated in Tables 3, 4 and 5 of that standard, whilst species groupings of hardwoods graded to BS 5756 are tabulated in Table 7 of that standard for the higher strength classes.

The concept, similar to the older species groupings, is that a strength class rather than a species may be specified. However, sometimes there are advantages in specifying a particular species and grade where the grade stress is higher than the strength class stress.

16.3.5 Permissible stresses

Permissible design stresses for both solid and laminated timber components are governed by the type of component, the conditions of service and the type of loading. They are obtained from grade stresses by applying the appropriate modification factors.

16.4 Design – general

Design in timber is similar to that in any other structural material as long as timber's peculiar qualities are acknowledged; indeed, these qualities can be exploited by resourceful designers. Timber is idealized as an orthotropic material, but in practice, only two directions need be considered: that parallel to the grain (along the trunk) and that perpendicular to the grain. Most strength properties, of both timber and joint fasteners, vary according to these directions and the variation has been found to follow the Hankinson relationship:

$$N = \frac{PQ}{P \sin^2 \theta + Q \cos^2 \theta}$$

where θ is the angle between directions of load and grain, N the stress at θ to the grain, P the stress parallel to the grain, and Q the stress perpendicular to the grain

from which intermediate stress or strength values can be calculated. This is not normally required for solid beams, joists and columns where only the major directions are used, but is often met where members intersect at joints. The stresses given in BS 5268:Part 2 are for permanent loading and increased values are allowed for short- and medium-term loads. This Code of Practice governs general design, but additional information is available.[31-41]

In the past, design has been inhibited by the relatively short lengths of timber available (Table 16.3 indicates availability). However, the production of durable resin adhesives has led to new construction techniques and structural forms being developed. Glued–laminated timber in which thin laminae are glued together to form structural components of almost any shape or length is a common reality, whilst structural plywood can be combined with either solid or glued–laminated timber to produce composite components which are lightweight, reliable and pleasing. The design in these forms is more complex than in solid timber but information on a wide variety of structural forms can be found,[42-70] whilst advice on the selection of a particular form is given in Tables 16.1 and 16.2. General design advice is provided by TRADA.[71]

16.5 Design in solid timber

Since permissible stresses are maxima there may be some advantage in using structural hardwoods or the higher-grade-stress softwoods whenever stress governs design. However, if deflection governs, there will be no advantage in using these more expensive materials unless the moduli of elasticity are sufficiently high. A possible exception is keruing (*dipterocarpus* spp.) whose current cost is roughly similar to that of softwoods. Some indication of price is given in Table 16.3.

As design in solid timber is limited by the maximum size of timber available, this has led to the development of many types of girder framework: however, where there is sufficient headroom, trussed beams can give an economic solution for heavy loads and long spans.[31,35,41]

In BS 5268:Part 2, minimum sizes are specified and the geometric properties tabulated in Appendix D of BS 5268:Part 2 are based on those minimum sizes.

Further reductions in section should be made for notches, mortices and bolt, screw and connector holes. Modification factors may also be required for the length and position of bearing, the shape of a beam and its depth if greater than 300 mm, whilst for compression members, combined factors are given for both slenderness and loading.

Lateral stability is important both for deep beams and for compression members, and in built-up members web stiffeners should be provided wherever concentrated loads occur.

General design data are available[71,73] applicable to particular structural forms[45,50-61,65-67] whilst design aids have been published for solid beams, portal frames and trussed rafters.

16.6 Glued–laminated timber assemblies

Glued–laminated timber is essentially a built-up section of two or more pieces of timber whose grains are approximately parallel and which are fastened together with glue throughout their length. This enables the properties of timber to be regulated to some degree and provides structural sizes and shapes which would not be possible in solid timber. Variation in section is possible, whilst high-grade material can be placed in zones of high stress and low-grade material in zones of low stress. All softwoods glue well and are generally preferred in the UK, although occasionally there can be some advantage in using wholly hardwood laminae.

Construction may use either vertical or horizontal laminations.

With vertical laminations, the zones of equal stress are shared between the laminations so that the strength of a beam can be said to be the sum of the individual laminations. This load-sharing concept has led to the grade–stress modification factors tabled in BS 5268:Part 2 which give higher permissible stresses for the laminated beam.

Horizontally laminated beams have been permitted since 1967 but the philosophy for behaviour is entirely different from that for vertical laminations. A beam will consist of material containing knots whose presence will affect the strength ratio of the beam. Since the knot effect will vary according to the sizes of knots and the number of laminations, BS 5268:Part 2 tables *basic stress* modification factors according to these variables.

Since curved laminated beams are fabricated by bending the individual laminations, fabrication stresses are induced which depend upon the degree of curvature, the thickness of the lamination and the species of timber. Therefore, modification factors to be applied to the grade bending stresses for different values of t/R are specified in BS 5268:Part 2.

The production of long laminations depends upon the use of efficient methods of end jointing. Where the efficiency of an end joint is known, the laminations containing them can be included when calculating the section properties, but where efficiencies are not known, the laminations containing the end joints must be omitted when calculating section properties. Efficiency ratings for plain scarf joints and for finger joints are given in BS 5268:Part 2: these are used to modify the basic stresses to give the maximum stresses to which any particular lamination may be subjected. British Standard 5291[74] governs finger joints in structural softwood. Butt joints do not transmit load and should only be used in zones of zero or very low stress.

Apart from the consideration of end joints and curvature, design is similar to that for solid timber,[46,64,75,76] whilst design aids are noted for glulam beams.

16.7 Plywood and tempered hardboard assemblies

Plywood is a type of glued–laminated construction in which the laminae are formed from thin flat veneers of timber. These veneers are produced by the rotary cutting of logs and are laid alternately at right angles in an odd number of layers. Since both the shrinkage and strength of timber differ according to the grain direction, the type of construction gives greater dimensional stability and tends to equalize the strength properties in both major directions of the plywood sheet.

There are two distinct design philosophies: (1) the North American approach which only considers the 'parallel plies', i.e. those plies whose grain lies in the direction of the load (this approach is based on the basic stresses and moduli for solid timber); and (2) the Finnish and British approach, known as the 'full cross-section' approach, in which grade stresses and moduli for the sheet materials have been determined from tests. In BS 5268:Part 2 all the grade stresses and moduli are for full cross-section, but it is well to remember that many North American design manuals will be based on the 'parallel-plies' approach. Plywood is a strong, durable and lightweight structural material which can be used to produce exciting structural shapes.[44,47–49,61–63] Design data are available for a variety of constructions[77–79] whilst design aids are available for stressed skin panels and portal frames.

Perhaps plywood's most useful property is that of providing excellent shear resistance for a small cross-section, although it is well to remember that lateral stability constraints may be required.

Tempered hardboard is a durable compressed fibreboard for which BS 5268:Part 2 now gives grade stresses for use in structural components instead of plywood.

larly true in timber for which highly efficient methods of transferring tensile loads have been developed only during the past 50 yr. Split-ring and tooth-plate connectors are now available which have load capacities much greater than those for nailed and bolted joints. A comparative indicator of fastener efficiency and the required member sizes is given in Table 16.5.

The strength of mechanical fasteners depends upon member size and thickness and the spacing of the fasteners. British Standard 5268:Part 2 tabulates permissible values of a wide range of variables, whereas some manuals prefer a presentation as a series of design curves.[31,34,35,40]

However, the major advance in fastening techniques has been in glued joints. Early glues were unreliable, deteriorating quickly, but the present phenolic and resorcinol resins are so durable that the risk of delamination has been almost entirely eliminated, even under extreme exposure. Unfortunately, gluing still requires controlled conditions and its application to site work is still limited.

Since the shear strength of adhesives is usually higher than that of timber, a fastener efficiency of 100% can be achieved. Nevertheless, it is important to remember that glues seldom have a good tensile strength, so that they should be stressed in shear as much as possible.

Information is available on gluing,[80] the requirements for adhesives,[81] and the compatibility between glues and preservatives.[82] The permissible stresses for glued joints are the shear stresses for the timber;[7] however, regard must be paid not only to the variation of that shear strength but also to the possibility of differential shrinkage and stress concentrations in the joint.

The type of fastener chosen will depend upon the skills and equipment available, possible fabrication conditions, relative costs and whether or not it is necessary to take down and reassemble the structural components.

16.8 Timber fastenings

Available methods of jointing are perhaps the most important criteria for the design of structural components. This is particu-

Table 16.5 The relative strengths of timber joints

Comparison – axial compression in GS/M50 (SC3) European redwood

| Type & dia. (mm) | FASTENER | | | TIMBER | |
	No.	Capacity (kN)	Size (mm)	Effective Area (mm²)	Capacity (kN)
NAILS					
3.75	63	40.3	47 × 145	5581	40.7
8.00	21	42.7	72 × 145	8712	63.6
SCREWS					
8.43	16	42.2	60 × 145	6677	48.7
BOLTS					
M8	30	41.3	44 × 169	5676	41.4
M12	13	40.3	60 × 145	6540	47.7
TOOTH PLATE					
2/51 mm dia. + M12 bolt	5	40.5	60 × 145	6540	47.7
2/64 mm dia. + M12 bolt	5	43.5	60 × 169	7980	58.3
SPLIT RING					
2/64 mm dia. + M12 bolt	3	49.7	60 × 145 or 72 × 120	6800 6740	49.6 49.2

Assumptions: (1) three member joints loaded to 40 kN in axial compression parallel to grain
(2) timber to timber joints
(3) GS/M50 European redwood
Grade stress Table 8 SC3 6.8 N/mm². Grade stress Table 9 M50 7.3 N/mm². Using Table 9 value, required timber area = 5479 mm²

EXAMPLES OF THE DESIGN OF A SIMPLE TENSION SPLICE JOINT

LOAD CAPACITY: 25 kN   DURATION: MEDIUM TERM

TIMBER: EUROPEAN REDWOOD   GRADE: M50

EXPOSURE CONDITION: DRY

(1) *REQUIRED TIMBER SECTION*

Dry exposure condition grade stresses, Tension //g

SC3 (**TABLE 8**): 3.2 N/mm^2 × 1.25 = 4.0 N/mm^2

European redwood (**TABLE 9**): 4.0 N/mm^{2*} × 1.25 = 5.0 N/mm^2

Maximum permissible timber stress = 5.0 N/mm^2

Section = $\dfrac{25000}{5}$ = 5000 mm^2

allow 10% reduction in effective section at joint

SAY 41 × 145 mm planed = 5950 mm^2 (**TABLE 99**)

(2) *NAILED JOINT*

CHOICE OF NAIL DIAMETER

Possible joint thickness = 3 × 41 mm = 123 mm

Maximum available stock lengths:

4 and 4.5 mm ϕ: 100 mm   5 mm ϕ: 125 mm

Standard thicknesses for members in double shear:

4 mm	ϕ: 0.7 × 44 = 31 mm	(splice 35 × 145)
4.5 mm	ϕ: 0.7 × 51 = 36 mm	(splice 35 × 145)
5 mm	ϕ: 0.7 × 57 = 40 mm	(splice 41 × 145)

CLAUSE 41.4.2   **TABLE 57**

Required nail lengths:

4 mm ϕ : 2 × 35 + 41 = 111 mm

4.5 mm ϕ : 2 × 35 + 41 = 111 mm

lengths not available

5 mm ϕ : 2 × 41 + 41 + = 123 mm*

DESIGN OF JOINT (5 mm nails)

Basic single shear lateral load capacity, dry exposure:

5 mm, SC3: 635 N (**TABLE 57**)

Multiple shear factor (**CLAUSE 41.42**)

0.9 × number of shear planes provided each member is thicker than 0.7 × standard point size penetration

Permissible joint load (**CLAUSE 41.8**)

= basic × K_{48} × K_{49} × K_{50}

duration of load   moisture content   number of nails
(medium term: 1.12)   (dry: 1.0)   (<10 'line': 1.0)

= 635 × (2 × 0.9) × 1.12 × 1.0 × 1.0 − (assumed)

= 1280.16 N.

Number of 5 mm nails required = $\dfrac{25}{1.28}$ = 20

TABLE 56 SPACING, modified by 0.8 (**CLAUSE 41.3**)

TRY 4 × 5 pattern (20 nails)

```
│25 × 40 × 40 × 40 × 25│   undrilled width ⩾ 170 mm
│    ×    ×    ×    ×   │
│    ×    ×    ×    ×   │   joint length 4 × 40 + 2 × 40 = 240 mm
│    ×    ×    ×    ×   │
│25 × 12 × 12 × 12 × 25│   drilled width ⩾ 86 mm*
```

Effective area = (145 − 4 × 5) 41 mm^2 (**CLAUSE 41.2**)

= 5125 mm^2

Effective timber load capacity = 25.625 kN

Therefore the nailing pattern is acceptable, but requires predrilling

Alternatively, try 3 × 7 pattern (21 nails, but easier to control), to avoid cost of pre-drilling

```
│         ×    ×    ×   │
│25 × 40 × 40 × 25│        undrilled width ⩾ 130 mm*
│         ×    ×    ×   │
│         ×    ×    ×   │   joint length 5 × 40 + 2 × 40 = 320 mm
│         ×    ×    ×   │
│         ×    ×    ×   │
│         ×    ×    ×   │
```

Effective area = (145 − 3 × 5) 41 = 5330 mm^2 (**CLAUSE 41.2**)

Effective timber load capacity = 26.65 kN

Fastener capacity = 21 × (2 × 0.9) × 1.12 × 1.0 × 635 kN
= 26.9 kN

(3) *BOLTED JOINT*

CHOICE OF BOLT DIAMETER

Number of lines of bolts (*n*) possible in a 145 mm wide member

= $\dfrac{\text{required timber capacity}}{\text{effective area} \times \text{permissible stress}}$

M10 bolt, *n* = 2.3 lines

For any larger diameter bolts, only one line of bolts would be possible.

DESIGN OF JOINT

Basic single shear lateral load parallel to grain, (**TABLE 67**), member 41 mm thick.

M10, 1.28 kN: M12, 1.84 kN: M16, 3.15 kN (interpolated)

Double shear factor (**CLAUSE 43.4.2**) 2.0

Permissible joint load (double shear)

= 2 basic × K_{55} × K_{56} × K_{57}

medium term = 1.25   number 'in line'

m/c dry = 1.0

Number of bolts required

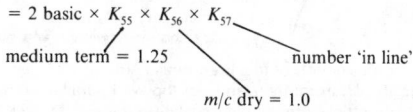

M10, 7.8: M12, 5.44: M16, 3.17:

bolts 'in line' and $(K_{57}) = [1 - \frac{3(n-1)}{100}]$ for $n < 10$

M10, 4 (.91): M12, 6 (.85): M16, 4 (.91)

revised number of bolts $= \dfrac{\text{original number}}{K_{57}}$

M10, 8.6: M12, 6.4: M16, 3.5:

Effective timber capacity = effective area × permissible stress

M10, 25.63 kN: M12, 27.26 kN: M16, 26.45 kN

BOLTS REQUIRED

M10, 9 (in two lines): M12, 7 (one line): M16, 4 (one line)

Compare with Timber Research and Development Association (1986) *Design aid DA1.* p. 25.

16.9 Timber-frame construction

It is estimated that the major use of structural timber will be in the housing field.

In high-rise construction, timber will play a supplementary role to the heavy material, being used for partitions, infill panels and floor and roofing systems. In this role, timber's ready adaptability to prefabrication is a great benefit.

In low-rise construction, on the other hand, timber is increasingly being used to provide the structural skeleton for the building; indeed, at the present time, timber-frame construction constitutes some 20% of all house construction. The method of construction is a simplification and refinement of that which has been used successfully for many centuries, but which is equally well applicable to many other uses besides housing.

The structural form is that of a free-standing skeleton for which standard details have been produced.[50-52,56,57,59] The basis of the skeleton is the stud-framed panel for which designs are described.[53,55,60]

Of especial importance is the structure's ability to withstand

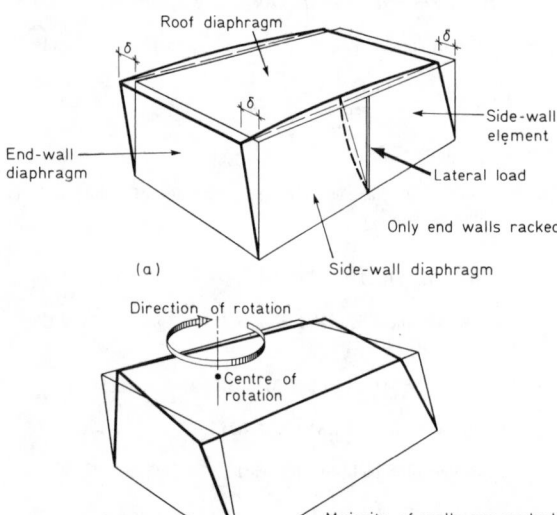

(a)

(b)

Figure 16.1 Deformation of idealized house structure under lateral loading. (a) Uniform translation of top wall parallel to the lateral load; (b) rotation after lateral translation, caused by lack of symmetry

lateral loading and, particularly, that resistance to planar deformation of a wall panel known as its racking resistance. Figure 16.1 shows the deformation of an idealized frame structure under lateral loading. Figure 16.1(a) shows the uniform translation of the walls which occurs when there is complete symmetry of both structure and loading. This hardly ever occurs and this lack of symmetry causes a rotation in addition to the translation (Figure 16.2(b)). The calculation of racking resistance is described by TRADA[58] and will also be dealt with by BS 5268:Part 6 (currently being written).

16.10 Repair and restoration

An increasing amount of work is being carried out on the repair and restoration of old timber-framed buildings. The work is specialized and requires not only sound structural assessment of the existing building but also a good understanding of the methods used in its construction and of acceptable methods of repair and restoration.

The methods of repair may be either by replacement of the damaged joints or members in the traditional manner or by the use of resin or stainless steel rods and resin fillers.

Brunskill[72] describes the traditional methods of timber-frame construction whilst Charles and Charles[83] give restoration case studies. Conservation and restoration are reported on by Fielden,[84] Gifford,[85] Ministry of Public Buildings and Works,[86] and repair by Avent et al.,[87-88] Oates and Richards[89] and Powys.[90]

16.11 Termite-resistant construction

There are two basic kinds of termite, which are mainly found in tropical and sub-tropical areas. The subterranean termite (white ant or wet-wood termite) has the larger distribution, builds huge nests, farms fungi and needs to maintain contact with the damp earth. Attack is from the ground.

There are many thousands of species of subterranean termites and a species of timber providing good termite resistance in one area may not be resistant in another. Resistive construction depends upon correct detailing, ground poisoning and preservative treatment.

The dry-wood termite, on the other hand, is less widely distributed, has fewer species, and flies, mates and deposits its eggs in timber to continue the life cycle.

Fly screens and preservative treatment give the best protection for new structures, whilst fumigation may be required when infestation is found in existing buildings. The recognition, control and detailing required are given by Harris[91] and Sperling.[92]

16.12 Storm-resistant construction

Wind loading is one of the commonest and most variable types of loading that can occur and ranges from normal wind loading to tornadoes.

Tropical storms have wind velocities in the range 55 to 117 km/h and damage is usually caused by flooding, including that produced by waterborne detritus.

Hurricanes are more severe and damage is caused by wind action, flooding and flying debris. Air circulation is counterclockwise with a damage area having a diameter of 50 to 160 km. Wind speeds are commonly 120 to 190 km/h with gusting up to 320 km/h. Rainfall, normally 125 to 250 mm, occasionally reaches 750 mm.

Tornadoes are unquestionably the most devastating of wind storms. The vortex is much smaller than that of a hurricane, causing intense damane from wind action and large pieces of flying debris.

Generally, storm-resistant structures require strength linked to rigidity with interconnected members and components securely fastened together. The Southern Pine Association gives valuable advice.[93]

16.13 Earthquake-resistant construction

Horizontal and vertical movement of the Earth's surface, caused by earthquakes, results in forces being generated by the inertia of the mass of the structure. The magnitude of these inertial forces varies directly as the mass of the structure and in consequence heavy structures are more severely loaded than are lightweight ones. Indeed, for timber structures, the forces may be little higher than those produced by normal wind loading. However, unlike storm-resistant structures, those for earthquake-resistant construction require strength linked to flexibility. Several publications give outlines of sound practice.[94-96]

16.14 Design aids

There are three areas in which design aids can make a valuable contribution to the design process. These are: (1) rapid preliminary design either for comparison or estimation of cost; (2) routine elementary design; and (3) complex design processes for which the design time can be reduced drastically.

Nomograms and load–span tables have been used for many years, but the application of computer programming has extended considerably the use of design aids.

The bibliography to this chapter indicates some of the design aids which are now available for structural design in timber.

References

General properties of timber

1 Dinwoodie, J. M. (1981) *Timber, its nature and behaviour.* Von Nostrand Reinhold, New York.
2 Princes Risborough Laboratory (1972) *Handbook of hardwoods.* Building Research Establishment, Garston.
3 Princes Risborough Laboratory (1977) *Handbook of softwoods.* Building Research Establishment, Garston.

Glossaries

4 British Standards Institution (1972) *Glossary of terms relating to timber and woodwork,* BS 565. BSI, Milton Keynes.
5 British Standards Institution (1974) *Nomenclature of commercial timbers,* BS 881 and 589. BSI, Milton Keynes.
6 British Standards Institution (1968) *Glossary of terms relating to timber preservatives,* BS 4261. BSI, Milton Keynes.
7 British Standards Institution (1984) *Code of practice for permissible stress design, materials and workmanship,* BS 5268: Part 2. BSI, Milton Keynes.

Species, use and availability

8 British Standards Institution (1971) *Quality of timber,* BS 1186: Part 1. BSI, Milton Keynes.
9 United Africa Company (1971) *West African hardwoods,* Parts 1 and 2. UAC, London.
10 Princes Risborough Laboratory (1971) *Hardwoods for industrial flooring,* Technical Note No. 49, PRL, Building Research Establishment, Garston.

11 Timber Research and Development Association (1979/1980) *Timbers of the world,* Vols 1 and 2. Longman, London.
12 British Standards Institution (1984) *Sizes of hardwoods and methods of measurement,* BS 5450. BSI, Milton Keynes.

Moisture content, moisture movement

13 Princes Risborough Laboratory (1969) *The movement of timbers,* Technical Note No. 38. Building Research Establishment, Garston.
14 Princes Risborough Laboratory (1971) *Flooring and joinery in new buildings.* How to minimize dimensional changes. Technical Note No. 12. Building Research Establishment, Garston.

Natural durability and the protection of timber

15 Princes Risborough Laboratory (1975) *The natural durability classification of timber.* Technical Note No. 40. Building Research Establishment, Garston.
16 Princes Risborough Laboratory (1968) *Ensuring good service life for window joinery.* Technical Note No. 29. Building Research Establishment, Garston.
17 Princes Risborough Laboratory (1971) *Decay in buildings – recognition, preservation and cure.* Technical Note No. 44. Building Research Establishment, Garston.
18 Princes Risborough Laboratory (1972) *Timber decay and its control.* Technical Note No. 57. Building Research Establishment, Garston.
19 Princes Risborough Laboratory (1965) *Termites and the protection of timber.* Leaflet No. 38. Building Research Establishment, Garston.
20 Princes Risborough Laboratory (1950) *Marine borers and methods of preserving timber against their attack.* Leaflet No. 46. Building Research Establishment, Garston.

Preservative treatment

21 British Standards Institution (1975) *Guide to the choice, use and application of wood preservatives.* BS 1282. BSI, Milton Keynes.
22 British Standards Institution (1977) *Preservative treatments for construction timbers.* BS 5268; Part 5. BSI, Milton Keynes.
23 British Standards Institution (1978) *Code of practice for preservation of timbers. Section 7 timber for use in prefabricated building in termite-infested areas.* BS 5589. BSI, Milton Keynes.
24 Tack, C. H. (1969) 'The economics of timber preservation.' *Timberlab 17,* Princes Risborough Laboratory, Building Research Establishment, Garston.

Fire resistance

25 British Standards Institution (1978) *Fire resistance of timber structures.* BS 5268: Part 4. BSI, Milton Keynes.
26 Fire Research Station (1970) *Fire and the structural use of timber in buildings.* BSI, Milton Keynes.
27 Wardle, T. M. (1966) *Notes on the fire resistance of heavy construction.* New Zealand Forestry Service Information Series 53.

Stress grading

28 British Standards Institution (1978) *Timber grades for structural use.* BS 4978. BSI, Milton Keynes.
29 British Standards Institution (1980) *Specification for tropical hardwoods graded for structural use.* BS 5756. BSI, Milton Keynes.
30 Curry, W. T. (1969) 'Mechanical stress grading of timber'. *Timberlab 18,* Building Research Establishment, Garston.

Design

Textbooks, etc.

31 American Institute of Timber Construction (1985) *AITC timber construction manual.* Wiley, New York.

32 Baird, J. A. and Ozelton, E. C. (1986) *Timber designer's manual.* Granada, London.

33 Breyer, D. E. and Ank, J. A. (1980) *Design of wood structures.* McGraw-Hill, New York.

34 Hansen, H. J. (1962) *Modern timber design.* Wiley, New York.

35 Karlsen, G. G. (1967) *Wooden structures.* Mir, Moscow.

36 Laminated Timber Institute of Canada (1980) *Timber design manual* (metric edn). LTIC, Ottawa.

37 Leicester *et al.* (1974) *Fundamentals of timber engineering,* Parts 1 and 2: 24 lectures given by officers of the Division of Building Research, CSIRO, Victoria, Australia.

38 Mettem, C. J. (1986) *Structural timber design and technology.* Longmans.

39 Oberg, F. R. (1963) *Heavy timber construction.* The Technical Press, London.

40 Pearson, R. G., Kloot, N. H. and Boyd, J. D. (1967) *Timber engineering design handbook.* Jacaranda, Melbourne.

41 US Department of Agriculture (1974) *Wood handbook.* Handbook No. 72, US Printer's Office, Washington, DC.

Arches and portal frames

42 Burgess, H. J. (1970) *Exploiting geometrical symmetry in timber structures.* Timber Research and Development Association, High Wycombe.

43 Council of Forest Industries of British Columbia (1972) *Portal frame manual.* COFI, London.

44 Kharna, J. and Hooley, R. F. (1965) *Design of fir plywood panel arches.* Council of Forest Industries of British Columbia Report No. TDD–43. COFI, London.

45 Timber Research and Development Association (1969) *Ridged portals in solid timber.* TRADA E/IB/18, London.

46 Wilson, T. R. C. (1939) *The glued laminated wood arch.* United States Department of Agriculture Technical Bulletin No. 691.

Barrel vaults

47 Kharna, J. (1964) *Design of fir plywood barrel vaults.* Council of Forest Industries of British Columbia Report No. TDD–40. COFI, London.

Folded plates

48 Council of Forest Industries of British Columbia (1969) *Fir plywood folded plate design.* COFI, London.

Formwork

49 Council of Forest Industries of British Columbia (1967) *Fir plywood concrete form manual.* COFI, London.

Housing

50 Anderson, L. O. (1970) *Wood frame house construction.* US Department of Agriculture Handbook No. 73. US Government Printing Office, Washington, DC.

51 Council of Forest Industries of British Columbia (1978) *Timber frame construction – a guide to platform frame construction.* COFI, London.

52 Council of Forest Industries of British Columbia (1977) *Load-bearing timber-framed walls.* Construction data. COFI, London.

53 Council of Forest Industries of British Columbia (n.d.) *Wind loading calculation for a typical timber-frame house.* COFI, London.

54 Canadian Wood Council (1977) *Canadian wood construction data files.* CWC, Ottawa.

55 Swedish/Finnish Timber Council (1976) *Timber stud walls of Swedish redwood and whitewood.* SFTC, Retford.

56 Swedish/Finnish Timber Council (1981) *Principles of timber-framed construction.* SFTC, Retford.

57 Timber Research and Development Association (1980) *Timber-frame housing manual.* Construction Press, TRADA, London.

58 Timber Research and Development Association (1980) *Calculating the racking resistance of timber-framed walls.* Wood Information Sheet No. 1–18. TRADA, High Wycombe.

59 Timber Research and Development Association (1981) *Introduction to timber-frame housing.* Wood Information Sheet No. 0–3, TRADA, High Wycombe.

60 Timber Research and Development Association (1981) *Timber-frame housing, structural recommendations.* Construction Press, TRADA, High Wycombe.

Plyweb beams

61 Burgess, H. J. (1970) *Introduction to the design of ply-web beams.* Timber Research and Development Association Note No. E/IB/24. TRADA, High Wycombe.

62 Council of Forest Industries of British Columbia (1963) *Nailed fir plywood web beams.* COFI, London.

63 Council of Forest Industries of British Columbia (1970) *Fir plywood web beam design.* COFI, London.

Shells

64 Keresztcsy, L. O. (1966) 'Interconnected, prefabricated laminated timber diamond type shell', *Proceedings, International Conference for Space Structures.* Surrey University, Guildford.

65 Tottenham, J. (1958) *The analysis of hyperbolic paraboloid shells.* Timber Research and Development Association Note No. E/RR/5. TRADA, London.

66 Tottenham, H. (1959) 'Analysis of orthotropic cylindrical shells.' *Civ. Engrg.*

67 Council of Forest Industries of British Columbia (1971) *Fir plywood stressed skin panels.* COFI, London.

68 Finnish Plywood Development Association (1970) *Design data for stressed skin panels in Finnish birch plywood.* Technical Bulletin No. 11 (M). FPDA, Welwyn Garden City.

69 Wardle, T. M. and Peek, J. D. (1970) *Plywood stressed skin panels: Geometric properties and selected designs.* Timber Research and Development Association Report No. E/IB/22. TRADA, High Wycombe.

Trussed rafters

70 British Standards Institute (1985) *Code of Practice for trussed rafter roofs.* BS 5268: Part 3. BSI, Milton Keynes.

General design data

71 Timber Research and Development Association (1967) *Design of timber members.* TRADA, High Wycombe.

72 Brunskill, R. W. (1985) *Timber building in Britain.* Gollancz, London.

73 Timber Research and Development Association (1986) *Design examples to BS 5268: Part 2, 1984.* DA1, TRADA, High Wycombe.

74 British Standards Institution (1984) *Specification for finger joints in structural softwood.* BS 5291. BSI, Milton Keynes.

75 Curry, W. T. (1955) *Laminated beams from two species of timber. Theory of design.* Princes Risborough Laboratory Special Report No. 10. HMSO, London.

76 Freas, A. D. and Selbo, M. L. (1954) *Fabrication and design of glued laminated wood structural members.* US Department of Agriculture Technical Bulletin No. 1069. US Government Printing Office, Washington, DC.

77 Council of Forest Industries of British Columbia (1972) *Canadian fir plywood data for designers.* COFI, London.

78 Council of Forest Industries of British Columbia (1971) *Plywood construction manual.* COFI, London.

79 Finnish Plywood Development Association (1964) *Finnish birch plywood handbook*, FPDA, Welwyn Garden City.

Glues for structural components

80 Princes Risborough Laboratory (1967) *The gluing of wooden components*. Technical Note No. 4. Building Research Establishment, Garston.
81 Knight, R. A. G. and Newall, K. J. (1971) *Requirements and properties of adhesives for wood*. Bulletin No. 20, Building Research Establishment, Garston. HMSO, London.
82 Princes Risborough Laboratory (1968) *Gluing preservative-treated wood*. Technical Note No. 31. Building Research Establishment, Garston.

The repair and restoration of timber structures

83 Charles, F. W. B. and Charles, Mary (1984) *Conservation of timber buildings*. Hutchinson, London.
84 Feilden, B. M. (1982) *Conservation of historic buildings*. Butterworth, London.
85 Gifford, E. W. H. and Taylor, P. (1964) 'Restoring old structures'. *Struct. Engr*, **42**, 10, 332–334.
86 Ministry of Public Buildings and Works (1965) *Notes on the repair and restoration of historic buildings*. HMSO, London.
87 Avent, R. R., Emkin, L. Z., Howard, R. H. and Chapman, C. L. (1970) 'Epoxy-repaired bolted timber connections'. *J. Struc. Div. Am. Soc. Civ. Engrs*, **102**, 821–838.
88 Avent, R. R., Sanders, P. H. and Emkin, L. Z. (1979) 'Structural repair of heavy timber with epoxy'. *For. Prod. J*. **29**, 3, 15–18.
89 Oates, D. W. and Richards, M. (1984) 'Timber engineering *in situ*.' *Record of British Wood Preserving Association Annual Convention 1984*, Paper 8, pp. 76–88.
90 Powys, A. R. (1981) *Repair of ancient buildings*. Dent and Sons, London (1929). Reprinted by Robert Maclehose, Glasgow.

Termite-resistant construction

91 Harris, W. V. (1971) *Termites – their recognition and control*. Longman, London.
92 Sperling, R. (1976) *Termites and tropical building*. Overseas Building Note No. 170. Building Research Establishment, Garston.

Storm-resistant construction

93 Southern Pine Association (1970) *How to build storm-resistant structures*. The Association, New Orleans.

Earthquake-resistant construction

94 Building Research Establishment (n.d.) Building in earthquake areas. Overseas Building Note No. 143, BRE, Garston.
95 Architectural Institute of Japan (1970) *Design essentials in earthquake-resistant structures*, Chapter 4, 'Wooden structures'. Elsevier, Amsterdam.
96 Takayame, K., Hisadat and Ohsaki, Y. (1960) 'Behaviour and design of wooden buildings subject to earthquakes'. *Proceedings, 2nd World Conference on Earthquake Engineering*, Tokyo.

Bibliography

General

Burgess, H. J. (1971) 'Design aids including computer programmes'. Paper No. WCH/71/5/8 World Work, Consultation Housing, Vancouver. (General appraisal of the development of design aids by Timber Research and Development Association.)

Finnish Plywood Development Association (1972) *Design for roof structures in Finply*. Technical Publication No. 17. FPDA, Welwyn Garden City. (Includes standard designs for box and I-beams, stressed-skin panels, portal frames and gussetted trusses.)
Timber Research and Development Association (1984) *The structural use of hardwoods*. Wood Information Sheet 01–17. TRADA, High Wycombe. (Span tables for 65-grade Keruing for floor and roof joists, purlins, ply-box beams and two-hinged portals.)
United Africa Company (1972) *Guide to the use of West African hardwoods*. (Load-span charts and tables for beams, joists, purlins, studs and ridged portal frame members.) (Universal span charts for any timber, grade and load duration together with simplified tables.)

Solid timber beams and joists

Burgess, H. J. and Masters, M. A. (1976) 'Span charts for solid timber beams'. Timber Research and Development Association Publication No. TBL 34. TRADA, High Wycombe.
Burgess, H. J. (1971) *Further applications of TRADA span charts*. Timber Research and Development Association Publication No. TBL 42. TRADA, High Wycombe.
Burgess, H. J. (1984) *Span tables for floor joists to BS 5268*. Timber Research and Development Association Publication No. DA 3.84. TRADA, High Wycombe.
Burgess, H. J. (1985) *Joist span tables for domestic floors and roofs to BS 5268*. Timber Research and Development Association Publication No. DA 6. TRADA, High Wycombe.
Burgess, H. J., Collins, J. E. and Masters, M. A. (1972) *Use of the TRADA universal span chart for a range of load cases*, Timber Research and Development Association Publication No. TBL 47. TRADA, High Wycombe.
Timber Research and Development Association (1984) *Span tables for floor joists to BS 5268: Part 2, Processed timber sizes to DA 3*. TRADA, London.
Timber Research and Development Association (1985) *Joist span tables for domestic floors and roofs* – processed timber sizes to DA 6. TRADA, London.
Hearmon, R. F. S. and Rixon, B. E. (1970) *Limiting spans for machine stress-graded European redwood and whitewood*. Princes Risborough Laboratory, Timberlab 30. HMSO, London. (Span tables.)
Council of Forest Industries of British Columbia (1973) *Hem-fir*. (Load-span tables for floor, roof and ceiling joists for a wide variety of distributed and concentrated loads.) COFI, London.

Glued–laminated timber beams

British Woodworking Federation (1967) *Span–load tables for glued–laminated softwood beams*. BWF, London.

Plywood box and web beams

Council of Forest Industries of British Columbia (1968) *Computer analysis of plywood web beams*. (A Fortran IV program for the analysis of both symmetrical and unsymmetrical beams.) COFI, London.
Council of Forest Industries of British Columbia (1968) *Fir plywood web-beam selection manual*. (Tabulates the properties of 4000 standard glued beams.) COFI, London.
Council of Forest Industries of British Columbia (1971) *Nailed fir plywood web-beams*. (Load span–deflection tables for twenty-four standard beams.) COFI, London.
Timber Research and Development Association (1984) *Load tables for nailed ply box beams to BS 5268: Part 2*. TRADA Publication No. DA 4, TRADA, High Wycombe.
Timber Research and Development Association (1984) *Load tables for glued ply box beams to BS 5268: Part 2*. TRADA Publication No. DA 5. TRADA, High Wycombe.

Portal frames

Burgess, H. J. *et al.* (1970) *Span tables for ridged portals in solid timber.* Timber Research and Development Association Publication No. E/IB/17. (Selection tables for portal member sizes for five different timbers.)

Council of Forest Industries of British Columbia (1972) *Portal frame manual.* (Design and selection manual.) COFI, London.

Stressed-skin panels

Finnish Plywood Development Association (1970) *Design data for stressed skin panels.* Technical Publication No. 11 (M). FPDA, Welwyn Garden City.

Wardle, T. M. and Peek, J. D. (1970) *Plywood stressed skin panels.* Timber Research and Development Association Publication No. E/IB/22. (Geometric properties and selected designs.) TRADA, High Wycombe.

Abbreviations and useful addresses

AITC   American Institute of Timber Construction, 1100, 17th Street NW, Washington DC 20036, USA.

BRE   Building Research Establishment, Garston, Watford, Hertfordshire.

BSI   British Standards Institution, 2 Park Street, London, W1A 2BS.

BWF   British Woodworking Federation, 82 New Cavendish Street, London, W1M 8AD.

BWPA   British Wood Preserving Association, 62 Oxford Street, London, W1N 9WD.

CITC   Canadian Institute of Timber Construction, 100 Bronfon Avenue, Ontario, Canada.

CPA   Chipboard Promotion Association, 50 Station Road, Marlow, Bucks, SL7 1NN

COFI   Council of Forest Industries of British Columbia, Tileman House, 131–133 Upper Richmond Road, Putney, London, SW15 2TR.

CWC   Canadian Wood Council, 701–710 Laurier Avenue West, Ottawa, Canada K1P SV5

FIDOR   Fibreboard Development Association, 1 Hanworth Road, Feltham, Middlesex, TW13 5AF.

FINPLY   Finnish Plywood Development Association, P.O. Box 99, Welwyn Garden City, Herts, A16 0HJ.

PRL   Princes Risborough Laboratory (Timberlab), Building Research Establishment, Aylesbury, Bucks, and at Building Research Station, Garston, Watford, Herts.

S/FTC   Swedish/Finnish Timber Council, 21 Carolgate, Retford, Notts, DN22 6BZ.

TRADA   Timber Research and Development Association, Stocking Lane, Hughenden Valley, High Wycombe, Bucks.

UAC   United Africa Company (Timber) Ltd., United Africa House, Blackfriars Road, London, SE1.

17 Foundations Design

M J Tomlinson
FICE, FIStructE, MConsE

Contents

17.1 General principles

17.1.1 The function of foundations

Foundations have the function of spreading the load from the superstructure so that the pressure transmitted to the ground is not of a magnitude such as to cause the ground to fail in shear, or to induce settlement of the ground that will cause distortion and structural failure or unacceptable architectural damage. In fulfilling these functions the foundation, substructure and superstructure should be considered as one unit. The tolerable total and differential settlement must be related to the type and use of the structure and its relationship to the surroundings. Foundations should be designed to be capable of being constructed economically and without risk of protracted delays. The construction stage of foundation work is not infrequently subjected to delays arising from unforeseen ground conditions. The latter cannot always be eliminated even after making detailed site investigations. Thus, elaborate and sophisticated designs and construction techniques which depend on an exact foreknowledge of the soil strata should be avoided. Designs should be capable of easy adjustment in depth or lateral extent to allow for variations in ground conditions and should take account of the need for dealing with groundwater.

Foundation designs must take into account the effects of construction on adjacent property, and the effects on the environment of such factors as piledriving vibrations, pumping and discharge of groundwater, the disposal of waste materials and the operation of heavy mechanical plant.

Foundations must be durable to resist attack by aggressive substances in the sea and rivers, in soils and rocks and in groundwaters. They must also be designed to resist or to accommodate movement from external causes such as seasonal moisture changes in the soil, frost heave, erosion and seepage, landslides, earthquakes and mining subsidence.

17.1.2 General procedure in foundation design

The various steps which should be followed in the design of foundations are as follows.

(1) A *site investigation* should be undertaken to determine the physical and chemical characteristics of the soils and rocks beneath the site, to observe groundwater levels and to obtain information relevant to the design of the foundations and their behaviour in service. The general principles and procedures described in Chapter 11 should be followed.

(2) The *magnitude and distribution of loading* from the superstructure should be established and placed in the various categories, namely:
   (a) dead loading (permanent structure and self-weight of foundations);
   (b) 'permanent' live loading, e.g. materials stored in silos, bunkers or warehouses;
   (c) intermittent live loading, e.g. human occupancy of buildings, vehicular traffic, wind pressures;
   (d) dynamic loading, e.g. traffic and machinery vibrations, wind gusts, earthquakes.

(3) The *total and differential settlements* which can be tolerated by the structure should be established. The tolerable limits depend on the allowable stresses in the superstructure, the need to avoid 'architectural' damage to claddings and finishes, and the effects on surrounding works such as damage to piped connections or reversal of fall in drainage outlets. Acceptable differential settlements depend on the type of structure; a framed industrial shedding with pin-jointed steel or precast concrete elements and sheet metal cladding, for example, can withstand a much greater degree of differential settlement than a 'prestige' office building with plastered finishes and tiled floors.

(4) The most suitable *type* of foundation and its *depth* below ground level should be established having regard to the information obtained from the site investigation and taking into consideration the functional requirements of the substructure, e.g. a basement may be needed for storage purposes or for parking cars.

(5) Preliminary values of the *allowable bearing pressures (or pile loadings)* appropriate to the type of foundation should be determined from a knowledge of the ground conditions and the tolerable settlements.

(6) The *pressure distribution* beneath the foundations should be calculated based on an assessment of foundation widths corresponding to the preliminary bearing pressures or pile loadings, and taking into account eccentric or inclined loading.

(7) A *settlement analysis* should be made, and from the results the preliminary bearing pressures or foundation depths may need to be adjusted to ensure that total and differential settlements are within acceptable limits. The settlement analysis may be based on simple empirical rules (see Chapter 9) or a mathematical analysis taking into account the measured compressibility of the soil.

(8) Approximate *cost estimates* should be made of alternative designs, from which the final design should be selected.

(9) *Materials* for foundations should be selected and concrete mixes designed taking into account any aggressive substances which may be present in the soil or groundwater, or in the overlying water in submerged foundations.

(10) The *structural design* should be prepared.

(11) The *working drawings* should be made. These should take into account the constructional problems involved and, where necessary, should be accompanied by drawings showing the various stages of construction and the design of temporary works such as cofferdams, shoring or underpinning.

17.1.3 Foundation loading

A foundation is required to support the dead load of the superstructure and substructure, the live load resulting from the materials stored in the structure or its occupancy, the weight of any materials used in backfilling above the foundations, and wind loading.

When considering the factor of safety against shear failure of the soil (see Chapter 9) the dead loading together with the maximum live load may be either a statutory or code of practice requirement, e.g. the requirements of the BS *Code of practice for loading*, BS 6399, or it may be directly calculated if the loads to be applied are known with some precision.

With regard to wind loading the BS *Code of practice for foundations*, BS 8004 states:

Where the foundation loading beneath a structure due to wind is a relatively small proportion of the total loading, it may be permissible to ignore the wind loading in the assessment of allowable bearing pressure, provided the overall factor of safety against shear failure is adequate. For example, where individual foundation loads due to wind are less than 25% of the loadings due to dead and live loads, the wind loads may be neglected in this assessment. Where this ratio exceeds 25%, foundations may be so proportioned that the pressure due to combined dead, live and wind loads does not exceed the allowable bearing pressure by more than 25%.

When considering the long-term settlement of foundations, the live load should be taken as the likely realistic applied load over the early years of occupancy of the structure. Consolidation settlements should not necessarily be calculated on the basis of the maximum live load.

Loadings on foundations from machinery are a special case which will be discussed in section 17.6.

17.1.4 The design of foundations to eliminate or reduce total and differential settlements

The amount of differential settlement which is experienced by a structure depends on the variation in compressibility of the ground and the variation in thickness of the compressible material below foundation level. It also depends on the stiffness of the combined foundation and superstructure. Excessive differential settlement results in cracking of claddings and finishes and, in severe cases, to structural damage. Where the total settlements are expected to be small, cracking and structural damage can be avoided by limiting the total settlement. For example, if the total settlement of buildings on isolated pad foundations is limited to about 25 mm the differential settlement is unlikely to cause any significant damage. Buildings on rafts can usually tolerate somewhat greater total settlements. Where total settlements are expected to be appreciably greater than 25 mm the effects of differential settlement should be considered in relation to the type and function of the structure. These effects are discussed comprehensively by Padfield and Sharrock[1] who tabulate acceptable deflection limits as shown in Table 17.1.[2]

Differential settlement may be eliminated or reduced to a tolerable degree by one or a combination of the following measures:

(1) Provision of a rigid raft either as a thick slab, or with deep beams in two directions, or in cellular construction.
(2) Provision of deep basements or buoyancy rafts to reduce the net bearing pressure on the soil (see sections 17.3.2.1 and 17.3.3).
(3) Transference of foundation loading to deeper and less compressible soil by basements, caissons, shafts or piles (as described in sections 17.3 and 17.4).

(4) Provision of jacking pockets within the substructure, or brackets on columns from which to re-level the superstructure by jacking.
(5) Provision of additional loading on lightly loaded areas by ballasting with kentledge or soil.
(6) Ground treatment processes to reduce the compressibility of the soil.

17.2 Shallow foundations

17.2.1 Definitions

British Standard 8004 defines shallow foundations as those where the depth below finished ground level is less than 3 m and which include many strip, pad and raft foundations. The code states that the choice of 3 m is arbitrary, and shallow foundations where the depth:breadth ratio is high may need to be designed as deep foundations.

(1) *A pad foundation* is an isolated foundation to spread a concentrated load (Figure 17.1).
(2) *A strip foundation* is a foundation providing a continuous longitudinal bearing (Figure 17.2).
(3) *A raft foundation* is a foundation continuous in two directions, usually covering an area equal to or greater than the base area of the structure (Figure 17.3).

17.2.2 Foundation depths

The first consideration is, of course, that the foundation should be taken down to a depth where the bearing capacity of the soil is adequate to support the foundation loading without failure of the soil in shear or excessive consolidation of the soil. The minimum requirement is thus to take the foundations below loose or disturbed topsoil, or soil liable to erosion by wind or flood. Provided these considerations are met the object should then be to avoid too great a depth to foundation level. A depth greater than 1.2 m will probably require support of the excavation to ensure safe working conditions for operatives fixing

Table 17.1 Limiting values of distortion and deflection of structures. (After Tomlinson (1986) *Foundation design and construction* (5th edn.). Longman Scientific and Technical)

Type of structure	Type of damage	Limiting values			
		Values of relative rotation (angular distortion), β			
		Skempton and MacDonald[3]	Meyerhof[4]	Polshin and Tokar[5]	Bjerrum[6]
Framed buildings and reinforced load-bearing walls	Structural damage	1/150	1/250	1/200	1/150
	Cracking in walls and partitions	1/300 (but 1/500 recommended)	1/500	1/500 (0.7/1000 to 1/1000 for end bays)	1/500
		Values for deflection ratio Δ/L			
		Meyerhoff[4]	Polshin and Tokar[5]	Burland and Wroth[7]	
Unreinforced load-bearing walls	Cracking by sagging	0.4×10^{-3}	$L/H = 3:0.3$ to 0.4×10^{-3}	At $L/H = 1$: 0.4×10^{-3} At $L/H = 5$: 0.8×10^{-3}	
	Cracking by hogging	—	—	At $L/H = 1$: 0.2×10^{-3} At $L/H = 5$: 0.4×10^{-3}	

Note: The limiting values for framed buildings are for structural members of average dimensions. Values may be much less for exceptionally large and stiff beams, or columns for which the limiting values of angular distortion should be obtained by structural analysis.

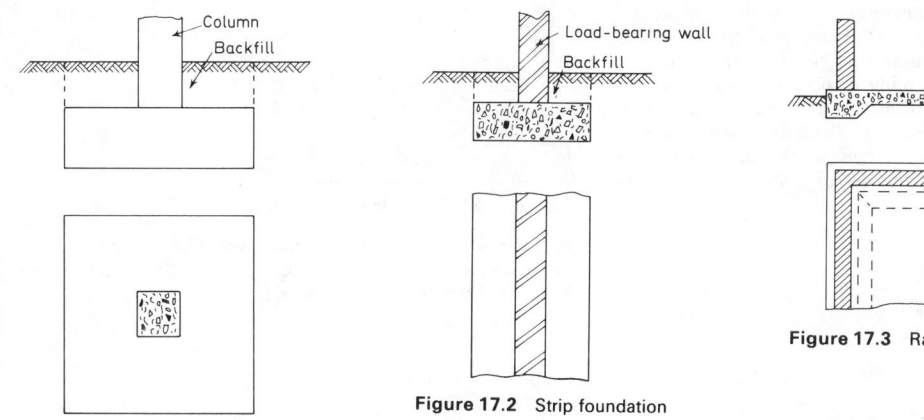

Figure 17.1   Pad foundation

Figure 17.2   Strip foundation

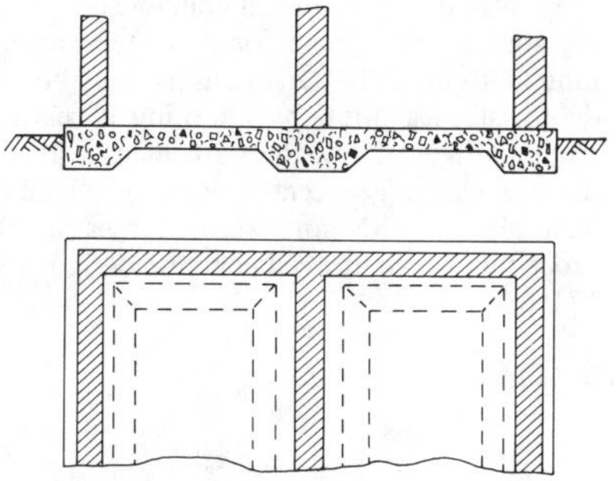

Figure 17.3   Raft foundation

reinforcing steel or formwork, which adds to the cost of the work. If at all possible the foundations should be kept above groundwater level in order to avoid the costs of pumping, and possible instability of the soil due to seepage of water into the bottom of an excavation. It is usually more economical to adopt wide foundations at a comparatively low bearing pressure, or even to adopt the alternative of piled foundations, than to excavate below groundwater level in a water-bearing gravel, sand or silt.

Apart from considerations of allowable bearing pressures, shallow foundations in clay soils are subject to the influences of ground movements caused by swelling and shrinkage (due to seasonal moisture changes or tree root action), in cohesive soils and weak rocks to frost action, and in most ground conditions to the effects of adjacent construction operations such as excavations or pile-driving.

It is usual to provide a minimum depth of 500 mm for strip or pad foundations as a safeguard against minor soil erosion, the burrowing of insects or animals, frost heave (in British climatic conditions other than those sites subject to severe frost exposure), and minor local excavations and soil cultivation. This minimum depth is inadequate for foundations on shrinkable clays where swelling and shrinkage of the soil due to seasonal moisture changes may cause appreciable movements of foundations placed at a depth of 1.2 m or less below the ground surface. A depth of 0.9 to 1 m is regarded as a minimum at which some seasonal movement will occur but is unlikely to be of a magnitude sufficient to cause damage to the superstructure or ordinary building finishes.[8]

Movements of clay soils can take place to much greater depths where the soil is affected by the drying action of trees and hedges, and in countries where there is a wide difference between the rainfall in the dry season and wet season.[9] Permafrost (permanently frozen ground) has a considerable influence on foundation depths.

Consideration should be given to the stability of shallow foundations on stepped or sloping ground. Analyses as described in Chapter 9 should be made to ensure that there is an adequate safety factor against a shear slide due to loading transmitted to the slope from the foundations.

The depth of foundations in relation to mining subsidence problems is discussed in section 17.7.2.

17.2.3 Allowable bearing pressures

Allowable bearing pressures (see definition in Chapter 9) for shallow foundations may be based on experience, or for prelimi-

nary design purposes on simple tables of presumed bearing values for a standard range of soil and rock conditions.

Where appropriate, more precise allowable bearing pressures for shallow foundations on cohesionless soils may be obtained from empirical relationships based on the results of *in situ* tests made on the soils (Chapter 11). In the case of shallow foundations on cohesive soils, the allowable bearing pressures may be obtained by applying an arbitrary safety factor to the ultimate bearing capacity calculated from shear strength determinations on the soil (Chapter 9). Where settlements are a critical factor in the design of foundations, detailed settlement analyses will be required based on the measured compressibility of the soil (Chapter 9).

17.2.4 Description of types of shallow foundations

17.2.4.1 Pad foundations

Pad foundations (Figure 17.1) are suitable to support the columns of framed structures. Pad foundations supporting lightly loaded columns can be constructed using unreinforced concrete, in which case the depth is proportioned so that the angle of spread from the base of the column to the outer edge of the ground bearing does not exceed 1 vertical:1 horizontal (Figure 17.4). The thickness of the foundation should not be less than the projection from the base of the column to its outer edge, and it should not be less than 150 mm.

Pad foundations to be excavated by a powered rotary auger should be circular in plan, so providing a self-supporting excavation in firm to stiff cohesive soils and weak rocks. Square or rectangular foundations can be excavated by mechanical grabs or backacters. The designs should not require the bottom to be trimmed by hand to a regular profile (Figure 17.4). This necessitates operatives working at the bottom of excavations in confined conditions, and for safety reasons the sides of excavations deeper than 1.2 m may have to be supported.

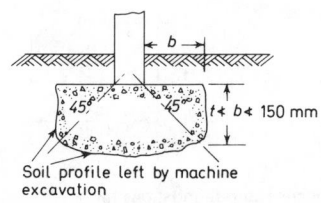

Figure 17.4   Proportioning of unreinforced concrete foundations

Savings in the volume of concrete can be obtained by providing steel reinforcement for pad foundations where heavy column loads are to be carried, and it may be advantageous to save depth of excavation by adopting a relatively thin base slab section (Figure 17.5). Reinforcement is also necessary for foundations carrying eccentric loading which may induce heavy bending moments and shear forces in the base slab. The procedure for reinforced concrete design is described in section 17.2.6.

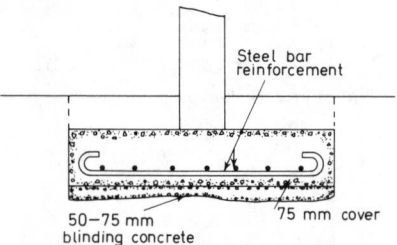

Figure 17.5 Reinforced concrete strip foundation

17.2.4.2 Strip foundations

Strip foundations are suitable for supporting load-bearing walls in brickwork or blockwork. The traditional form of strip foundation is shown in Figure 17.6(a). The concrete-filled trench foundation (Figure 17.6(b)) is suitable for stable soils in level ground conditions but should not be used where substantial swelling of clay soils may occur owing, say, to removal of trees or hedges. The swelling is accompanied by horizontal thrust on the foundation followed by movement of the foundation and superstructure. Strip foundations are also an economical method of supporting a row of closely spaced columns (Figure 17.7).

As a general rule, the thickness of unreinforced strip foundations should not be less than the projection from the base of the wall and not less than 150 mm. Where foundations are laid at more than one level, at each change of level the higher founda-

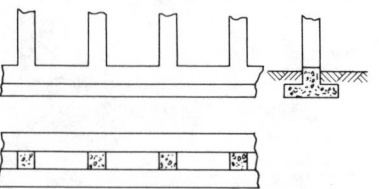

Figure 17.7 Strip foundation for closely spaced columns

tion should extend over and unite with the lower one for a distance of not less than the thickness of the foundation and not less than 300 mm (Figure 17.8).

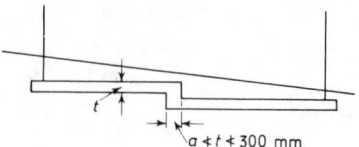

Figure 17.8 Stepping of strip foundations

The excavations for strip foundations are normally undertaken by a backacter machine, and it is usually possible to trim by the machine bucket to a rectangular bottom profile.

Reinforcement can be provided to strip foundations to enable savings to be made in the volume of concrete and also in foundation depths owing to the lesser required thickness of the base slab. Reinforcement is also necessary to enable the foundations to bridge over weak pockets of soil to minimize differential settlement due to variable loading conditions, e.g. when a strip foundation is provided to support a row of columns carrying different loads.

The procedure for the design of reinforced concrete foundations is described in section 17.2.6. In nonaggressive soil conditions a concrete mix consisting of 1 part of ordinary Portland cement to 9 parts of combined aggregate is suitable for unreinforced concrete strip foundations. The design of concrete mixes suitable for aggressive soil conditions is described in section 17.8.4.

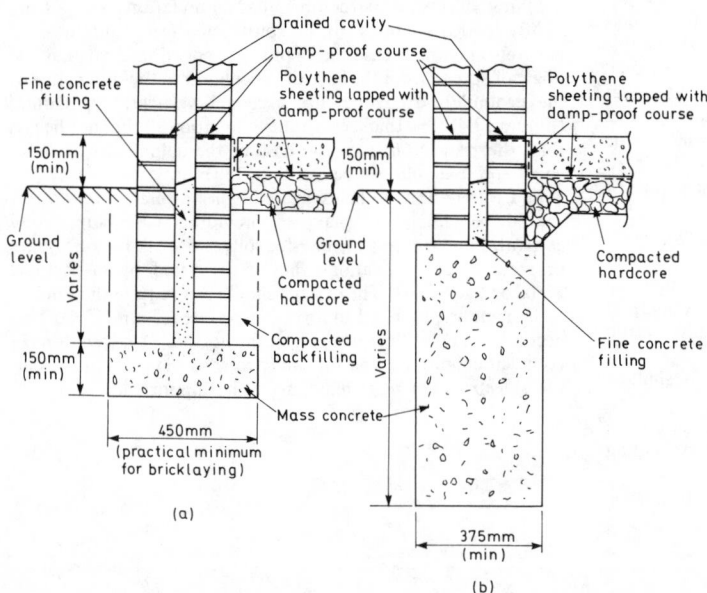

Figure 17.6 Unreinforced concrete strip foundations for load-bearing walls. (a) Traditional; (b) concrete-filled trench

17.2.4.3 Raft foundations

Raft foundations are a means of spreading foundation loads over a wide area thus minimizing bearing pressures and limiting settlement. By stiffening the rafts with beams and providing reinforcement in two directions the differential settlements can be reduced to a minimum.

Edge beams and internal beams can be designed as 'upstand' or 'downstand' projections (Figure 17.9). Downstand beams save formwork and allow the rafts to be concreted in one pour. However, the required trench excavations may not be self-supporting in loose soils and there are difficulties in maintaining the required profile in water-bearing ground. Upstand beams are required where rafts are designed to allow horizontal ground movements to take place beneath them, as in mining subsidence areas (section 17.7.2.3).

Raft foundations, in order to function as load-spreading substructures, must be reinforced and concrete mixes must be in accordance with code of practice requirements for reinforced concrete (BS 8110). Special mixes may be required in aggressive soil conditions.

17.2.5 Shallow foundations carrying eccentric loading

The soil adjacent to the sides of shallow foundations cannot be relied on to provide resistance to overturning moments caused by eccentric loading on the foundations. This is because in clays the soil is likely to shrink away from the foundation in dry weather and, in the case of cohesionless soils, excavation and subsequent backfilling will cause loose conditions around the sides. It is therefore necessary to check that the soil beneath the foundation will not be overstressed or suffer excessive compression under the unequal bearing pressures induced by the eccentric loading.

The pressure distribution beneath an eccentrically loaded foundation is assumed to be linear. For the pad foundation shown in Figure 17.10(a) where the resultant of the overturning moment M and the vertical load W falls within the middle third of the base:

Maximum pressure

$$q_{max} = \frac{W}{BL} + \frac{My}{I} \qquad (17.1)$$

For a centrally loaded pad foundation this becomes:

$$q_{max} = \frac{W}{BL} + \frac{6M}{B^2L} \qquad (17.2)$$

The minimum bearing pressure is given by:

$$q_{min} = \frac{W}{BL} - \frac{6M}{B^2L} \qquad (17.3)$$

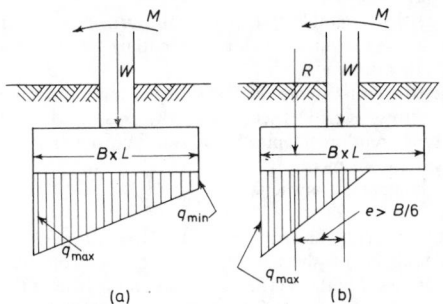

Figure 17.10 Eccentrically loaded foundations. (a) Resultant within middle third; (b) resultant outside middle third

When the resultant W and M falls outside the middle third of the base, Equation (17.3) indicates that tension theoretically occurs beneath the base. However, tension cannot develop and redistribution of bearing pressure will occur as shown in Figure 17.10(b). The maximum bearing pressure is then given by:

$$q_{max} = \frac{4W}{3L(B-2e)} \qquad (17.4)$$

In Equations (17.1) to (17.4) W is the total axial load on the column, M is the bending moment on the column, y is the distance from the centroid of the pad to the edge, I is the moment of inertia of the plan dimensions of the pad, e is the distance from the centroid of the pad to the line of action of the resultant loading.

The maximum bearing pressure q_{max} should not exceed the allowable bearing pressure appropriate to the depth and width of the foundation, but the effective width for consideration of settlement in cohesionless soils (see Chapter 9) can be taken as one-third of the overall width for the pressure distribution shown in Figure 17.10(b) for a triangular distribution of pressure.

17.2.6 The structural design of shallow foundations

17.2.6.1 Pad and strip foundations

The following steps should be taken in the structural design of a pad foundation.

(1) Calculate the base area of the foundation by dividing the total net load by the allowable bearing pressure on the soil, taking into account any eccentric loading.
(2) Calculate the required overall depth of the base slab at the point of maximum bending moment.

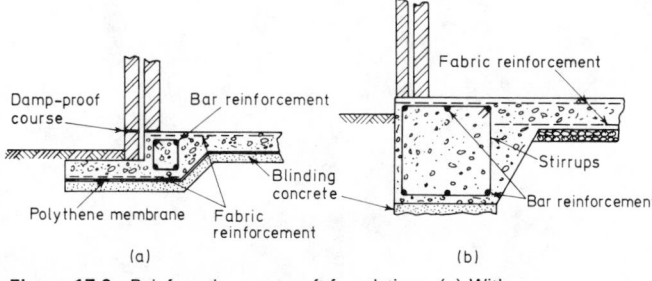

Figure 17.9 Reinforced concrete raft foundations. (a) With upstand beam; (b) with downstand beam

(3) Decide on either a simple slab base with horizontal upper surface or a sloping upper surface, depending on the economics of construction.
(4) Check the calculated depth of the slab by computing the beam shear stress at critical sections on the assumption that diagonal shear reinforcements should not be provided.
(5) Design the reinforcement.
(6) Check the bond stress in the steel.

The main reinforcement, consisting of bars at the bottom of the base slab, is designed on the assumption that the projection behaves as a cantilever with its critical section on the face of the column (Line X–X in Figure 17.11), and with a loading on the underside of the cantilever equal to net bearing pressure under the worst conditions of loading, i.e. maximum eccentricity if the loading is not wholly axial. In Figure 17.11, the bending moment at the face of the column is given by:

$$M_b = \frac{q \times b^2 \times L}{2} \qquad (17.5)$$

For pads of uniform thickness, the critical section of shear is along a vertical section Y–Y extending across the full width of the pad at a distance from the face of the column as defined in clause 3.4.5.8 of BS 8110. It is also necessary to check the punching shear along a critical peripheral section at a distance 1.5 times the thickness of the pad from the faces of the column. If the shear stress or punching shear stress exceed permissible limits they should be reduced by increasing the effective depth of the pad. Shear reinforcement in the form of stirrups or inclined bars should be avoided if at all possible.

Strip foundations are designed in the same manner, the critical sections for bending moment and shear being as shown in Figure 17.11.

17.2.6.2 Raft foundations

Rafts are provided on compressible soils, and particularly on soils of variable compressibility. Thus, wherever rafts are needed from the aspect of soil compressibility, some settlement is inevitable, either in the form of dishing (on soils of uniform compressibility) or hogging (where the compressibility of the soil or the thickness of the compressible layer varies across the raft) or twisting where the compressibility conditions are irregular.

Distortion of a raft will also occur as a result of variation in the superimposed loading. The magnitude of dishing, hogging or twisting, i.e. the angular distortion of the raft, will depend on the stiffness of the raft and of the superstructure. Only in the case of a uniformly loaded raft on a soil of uniform compressibility can the raft be designed as an inverted floor, either in slab and beam construction or as a stiff slab (Figure 17.3). In all other cases the design is a complex process of redistributing column load bending moments and shears by the amount calculated from a consideration of the stiffness of the substructure and superstructure and the settlement of the soil. The starting point is always the theoretical total and differential settlements calculated by the soil mechanics engineer on the assumption of a fully flexible foundation. Flexibility of the raft is desirable to keep bending moments and shears to a minimum, but if the raft is too flexible there will be excessive distortion of the superstructure.

Analysis of the complex interaction between the raft structure and a subgrade soil undergoing elastic or plastic deformation lends itself to computer methods for solution. A report by the Institution of Structural Engineers[10] discusses the problems involved in computer analysis. Reference may also be made to the work of Hooper[11] and Poulos and Davis.[12] Where settlements are expected to be fairly small, the complexities of raft design can be avoided by designing the substructure as a series of touching but not interconnected pad or strip foundations.

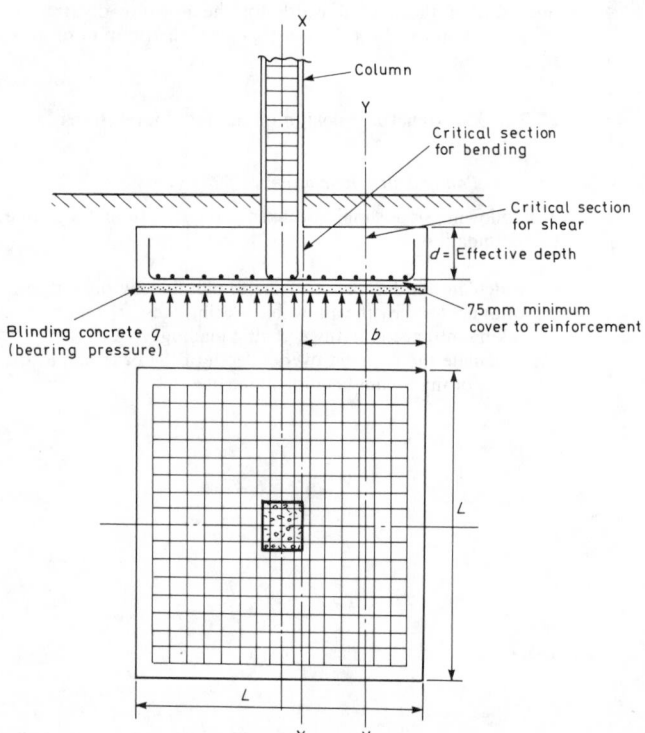

Figure 17.11 Reinforced concrete pad foundations

This will greatly reduce the amount of reinforcement required to resist the high bending moments and shears which occur in the short stiff members of a raft with close-spaced columns.

17.2.7 Ground treatment beneath shallow foundations

If the ground beneath a proposed structure is highly compressible it may be economical to adopt shallow foundations in conjunction with a geotechnical process to reduce the compressibility of the ground as an alternative to deep foundations taken down to a stratum of lower compressibility. Geotechnical processes which may be considered are:

(1) Preloading.
(2) Injection of cement or chemicals.
(3) Deep vibration.
(4) Dynamic compaction.

(See also Chapter 9.)

Preloading Preloading consists of applying a load to the ground equal to, or greater than, the proposed foundation loading so that settlement of the ground will be complete before the structure is erected. The method is applicable to loose granular soils or granular fills, where the settlement will be rapid. It is generally unsuitable for soft clays where shear failure may occur under rapid application of preload and, because of the long-term character of consolidation settlement, the preloading would have to be sustained over a long period to be effective. Preloading is most economical over a large area where the granular material such as gravel or colliery waste can be provided in bulk and moved progressively across a site using earthmoving machinery.

The injection of cement or chemicals Injection of cement or chemicals is suitable for treatment of loose granular soils or fills where the particle size distribution of the materials is suitable for the acceptance of grouts. The effect of injecting cement or chemicals is to replace the void spaces by relatively incompressible material, thus greatly reducing the overall compressibility of the ground mass.

Cement or chemicals used for injection are costly and the process is not normally recommended for dealing with large foundation areas or deep compressible strata. The process is usually restricted to small-scale application beneath important structures such as complex machinery installations. It is also employed as a remedial treatment to arrest the excessive settlement of foundations.

Unslaked lime can be mixed with soft clays by rotary drilling equipment to form load-bearing columns of stabilized soil.[13] These are suitable for the foundations of light buildings provided that minor settlements are acceptable.

Deep vibration Deep vibration methods comprise the insertion of a large vibrating unit into the soil for the full depth required followed by its slow withdrawal. Granular material is fed into the depression surrounding the vibration unit as it is withdrawn, and the unit is re-inserted several times to form a cylinder of densely compacted soil mixed with the imported material. By adopting close-spaced insertions on a grid pattern beneath loaded areas or in single or double rows beneath strip foundations, the whole mass of compressible soil can be compacted to a reasonably uniform state, thus reducing the total and differential settlements beneath the applied loading.

In the 'vibroflotation' process the vibratory unit is assisted in its insertion by water jetting. During withdrawal the direction of the jets is reversed to consolidate the added materials. In the 'vibro-replacement' process no water jetting is used, the vibratory unit resembling a large poker vibrator. Compressed air is used to assist penetration of the vibratory unit in the vibro-displacement process.

The depth of treatment is limited to the maximum depth to which the vibratory unit can be inserted which, with the most powerful units assisted by water jetting, is about 20 to 30 m. The process has been used to advantage in compacting very loosely placed brick rubble and building debris filling on urban redevelopment sites. Houses can then be built on conventional strip foundations on the fill which has been compacted to a reasonably uniform state of density. The process may not be suitable if the debris contains a high proportion of timber or other organic or soluble materials which may decay or dissolve over a period of years, resulting in further settlement of the fill.

Dynamic compaction This consists of dropping a heavy weight on to the surface of the soil to compact and consolidate the weaker upper layers. Commonly weights of 15 to 20 t are dropped from heights of about 20 m to achieve useful compaction of the soil over a depth of about 10 m. Tamping is usually undertaken on a rectangular grid at points spaced 5 to 10 m apart. About five to ten blows of the tamper are applied to each grid point and the resulting craters are backfilled with granular material. Successive passes are then applied to the same or intermediate grid points until the desired standard of compaction has been achieved. The process is suitable for free-draining coarse granular soils, rockfill, refuse tips and industrial waste tips. Fill material in waste tips should not contain appreciable quantities of biodegradable or soluble substances.

The deep vibration and dynamic compaction processes have been reviewed comprehensively by Greenwood and Kirsch.[14]

17.3 Deep foundations

17.3 Definitions

Deep foundations are required to carry loads from a structure through weak compressible soils or fills on to stronger and less compressible soils or rocks at depth, or for functional reasons. The types of deep foundations in general use are as follows.

(1) Basements.
(2) Buoyancy rafts (hollow box foundations).
(3) Caissons.
(4) Cylinders.
(5) Shaft foundations.
(6) Piles.

Basements These are hollow substructures designed to provide working or storage space below ground level. The structural design is governed by their functional requirements rather than from considerations of the most efficient method of resisting external earth and hydrostatic pressures. They are constructed in place in open excavations.

Buoyancy rafts (hollow box foundations) Buoyancy rafts are hollow substructures designed to provide a buoyant or semi-buoyant substructure beneath which the net loading on the soil is reduced to the desired low intensity. Buoyancy rafts can be designed to be sunk as caissons (see below): they can also be constructed in place in open excavations.

Caissons Caissons are hollow substructures designed to be constructed on or near the surface and then sunk as a single unit to their required level.

Cylinders Cylinders are small single-cell caissons.

Shaft foundations These are constructed within deep excavations supported by lining constructed in place and subsequently filled with concrete or other prefabricated load-bearing units.

Piles Piles are relatively long and slender members constructed by driving preformed units to the desired founding level, or by driving or drilling-in tubes to the required depth – the tubes being filled with concrete before or during withdrawal – or by drilling unlined or wholly or partly lined boreholes which are then filled with concrete. Piles form a large group within the general classification of deep foundations and will be described separately in section 17.4.

17.3.2 The design of basements

17.3.2.1 General

Basements are constructed in place in open excavations. The latter can be excavated with sloping sides, or with ground support in the form of sheeting or sheet piling. The choice of either excavation method depends on the clear space available around the substructure and the need to safeguard existing structures adjacent to the excavation. It may be economical to use the permanent retaining walls as the means of ground support as described in section 17.3.2.3. A circular shape to a basement can save construction costs where ground support is required, as cross-bracing to support the sheeted sides may not be needed. A circular plan should always be considered for structures such as underground pumping stations.

The walls of basements are designed as retaining walls subjected to external earth pressure and water pressure. The methods of calculating earth pressure on retaining walls are described in Chapter 9. If no groundwater is encountered in site investigation boreholes it must not be assumed that there will not be any water pressure. For example, where backfill is placed between the walls of a basement and the sides of an excavation in clay soil a reservoir will be formed in which surface water running across the site will collect and a head of water will progressively rise around the walls. Such accumulations of water will not occur in permeable soil or rock formations in which the rate of downward seepage exceeds the inflow from surface water.

The floors of basements are designed to resist the upward earth pressure and any water pressure. The basement slabs span between the external walls or cross-walls or between ground beams placed along the lines of the interior columns. Alternatively, they can be designed as flat slabs propped at column and wall positions. They act as raft foundations subjected to bending moments and shears induced by differential settlements. The results of the site investigation will normally provide estimates of total and differential settlement on the alternative assumptions of a rigid raft (heavy beam and slab construction) or a fully flexible raft (thin flat slab construction). It is then a matter for the structural designer's judgement to assess the degree of flexibility of the raft and its interaction with the superstructure for the particular design under consideration. The complexities of this assessment have already been discussed in section 17.2.6.2. Particular points to be taken into consideration with basement floor designs are noted below.

Basements constructed in water-bearing strata may become buoyant if the groundwater level in the excavation around the completed (or partly completed) structure is allowed to rise to its normal rest level. At this stage there may not be sufficient loading from the superstructure to prevent uplift occurring. Therefore care should be taken to keep the excavation pumped down until the structural loads have reached the stage when uplift cannot occur.

17.3.2.2 Design of basement floors

Basement floors founded on rock or other relatively incompressible soils will not undergo appreciable downward movement due to elastic or consolidation settlement of the subgrade material. Then differential settlements will be negligible and it will be necessary only to design the floor to resist upward water pressure. If no water table exists or cannot develop in the future then columns and walls can be designed with independent foundations, the floor slab being only of nominal thickness (Figure 17.12).

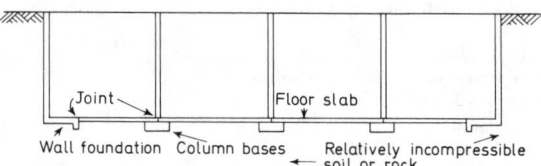

Figure 17.12 Basement floor founded on relatively incompressible stratum

Where appreciable total and differential settlements of the substructure can occur the basement floor should be designed as a stiff raft, either in slab and beam construction (Figure 17.13(a)) or as a flat slab (Figure 17.13(b)). Design practices are similar to those described in section 17.2.6.2 for surface rafts.

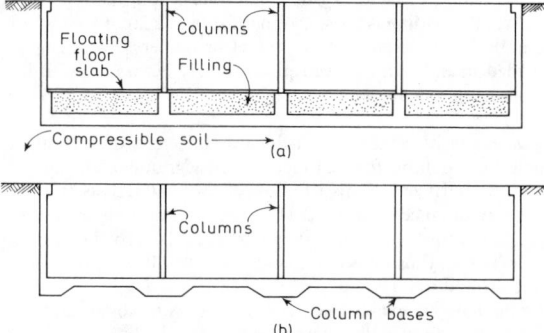

Figure 17.13 Basement floor founded on compressible stratum

When basements are supported on piles and settlements are expected in the pile group, i.e. where the piles terminate on compressible soils, some loading will be transferred to the underside of the floor slab. The magnitude of the pressure which develops will depend on the amount of settlement of the piles, the amount of heave of the base of the excavation due to relief of overburden pressure, the amount of heave and reconsolidation of the soil due to the installation of the piles and the time interval between completion of the excavation (including final trimming and removal of heaved soil) and the time when yielding of the piles commences due to superstructure loading. In all cases where there is potential transfer to the underside of the floor slab, or where hydrostatic pressure has to be resisted, the piled raft (Figure 17.14(a)) is the appropriate form of construction. The problems of load sharing between the piles and basement slab of a piled raft have been reviewed by Padfield and Sharrock[1] and by Hooper.[15]

Where the piles are terminated on rock or other relatively incompressible material and there is no hydrostatic pressure, there will be no load transfer to the floor slab, the latter being only of nominal thickness (Figure 17.14(b)). This assumes that ground heave causing uplift on the underside of the slab has ceased and that the heaved soil has been stripped off before placing the floor concrete.

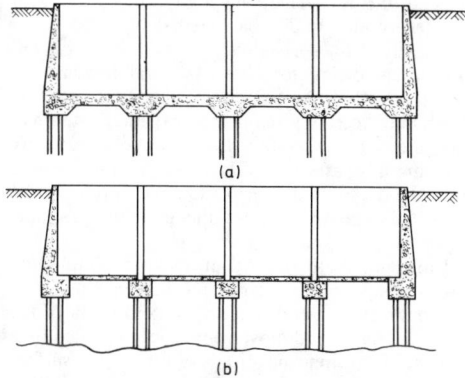

(a)

(b)

Figure 17.14 Piled basement floors. (a) With load transfer to floor slab; (b) with no load transfer to floor slab

17.3.2.3 Design of basement walls

Although the exterior walls of basements are supported by the ground-floor slab of the main structure and any intermediate subfloors in deep basements, they should be designed as free-standing cantilever retaining walls (Figure 17.15). This is because the supporting floors are not usually constructed until the final stage of the work (a special method of supporting the external walls of deep basements is shown in Figure 17.21, page **17/13**). Similarly, the foundation slab of the retaining wall should not be dependent on its connection to the basement floor slab for stability.

The structural form of the retaining wall is governed to some extent by the ground conditions and by the need or otherwise for waterproofing treatment (see below). Thus, the sloping back and projecting heel shown in Figure 17.15(a) require additional width of excavation, the cost of which may outweigh the increase in concrete volume required by a wall of uniform thickness (Figure 17.15(b)). In stable ground it may be possible to undercut the excavated face to form the heel enlargement. The wider excavation required for the sloping back wall (Figure 17.15(a)) may be needed in any case to allow room for applying a waterproof asphalt layer, whereas the vertical back requires either an enlarged excavation or the construction of a separate vertical backing wall on which to apply asphalt.

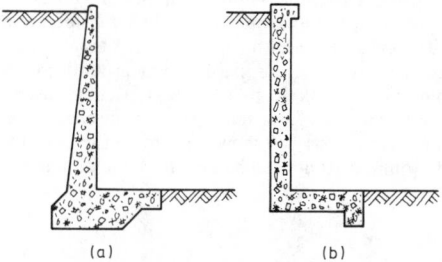

(a)

(b)

Figure 17.15 Basement floors. (a) With sloping back and heel; (b) with vertical back and no heel

The basement walls can be constructed as diaphragm walls by excavating a narrow trench by a mechanical grab using bentonite to support the excavation (Figure 17.16). The excavation is taken out in alternate panels 3 to 6 m long between guide walls. The level of the guide walls should be such that there is at least a 1-m head of bentonite slurry above the highest groundwater level. A preassembled reinforcing cage is lowered into the bentonite-filled trench and then concrete is placed by tremie pipe. The intermediate panels are then constructed in a similar

manner. Diaphragm walls are designed as retaining walls using conventional methods for calculating earth pressure (Chapter 9). However, they cannot usually be designed to act as cantilever walls at the final stage of excavation, and they require to be propped by shores (or held at the top or intermediate levels by ground anchors) as described in section 17.3.2.5.

Contiguous bored pile walls faced with reinforced concrete can also be used for basements (see Figure 17.43(f), page **17/24**).

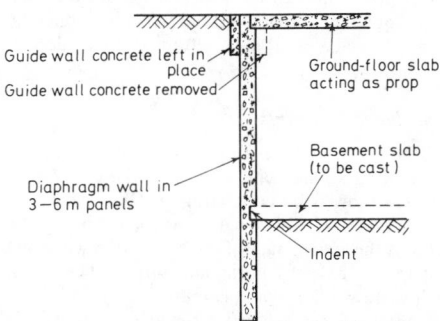

Figure 17.16 Diaphragm wall construction

17.3.2.4 Waterproofing basements

Watertightness of a basement can be obtained either by relying on impervious concrete and leaktight joints, or by providing an impermeable membrane in the form of trowelled-on asphalt tanking or preformed sheathing material. Neither method is entirely satisfactory.

If complete watertightness is required for functional reasons in a basement it is probable that the asphalt tanking method has a slight advantage compared with relying on the concrete alone, as tanking is a distinct operation carried out by skilled operatives, and the work can be restricted to favourable weather conditions and subjected to intensive supervision; whereas if the concrete alone is to be relied upon for watertightness, the concreting operations proceed in stages over a long construction period, in all weathers, with comparatively unskilled labour, and in congested situations, thus making close supervision difficult at all times.

Asphalt tanking or self-adhesive plastics sheathing is laid on blinding concrete beneath the basement floor and may be applied either to the exterior of the retaining walls if space is available around the excavations or, in restricted space conditions, it can be applied to a vertical backing wall before constructing the main wall (Figure 17.17). It is useless to apply tanking to the interior of the structural wall as the water pressure will merely force it off. Tanking applied to the exterior of the retaining wall should be protected by a 100-mm thick backing wall (in a manner similar to that shown in Figure 17.17) to prevent damage by sharp objects in the backfill materials.

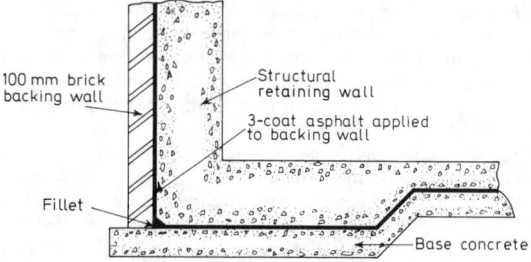

Figure 17.17 Asphalt tanking to basement

Asphalt tanking is covered by BS 988 and BS 1162 for limestone aggregate and natural rock asphalt aggregate respectively. The tanking should be applied in three coats to a total thickness of not less than 27 mm for horizontal work and 20 mm for vertical work. Other points of workmanship are covered in CP 102. An alternative to asphalt tanking is the use of Volclay panels. These consist of fluted cardboard slabs. The flutes are filled with bentonite which swells when wetted to form a permanent flexible gel.

Pumps keeping down the groundwater level around the excavation should not be shut down until the structural concrete walls have been concreted and have attained their design strength.

17.3.2.5 Construction of basements

If space around the substructure permits, the most economical method of constructing a basement is to form the excavation with sloping sides, followed by concreting the floor slab and then the retaining walls. If the space is restricted it will be necessary to support the vertical face of the excavation with steel sheet piling (Figure 17.18) or by horizontal timber sheeting in conjunction with vertical soldier piles (Figure 17.19). The sheet piling method is suitable for soft or water-bearing ground where continuous support is necessary and where it is desired to maintain the surrounding groundwater table at its normal level to safeguard existing structures. Horizontal sheeting can be used in 'dry' ground conditions, or where drainage towards the excavation can be permitted. In the latter case, hydrostatic pressures do not develop with correspondingly reduced loads to be carried by the bracing system.

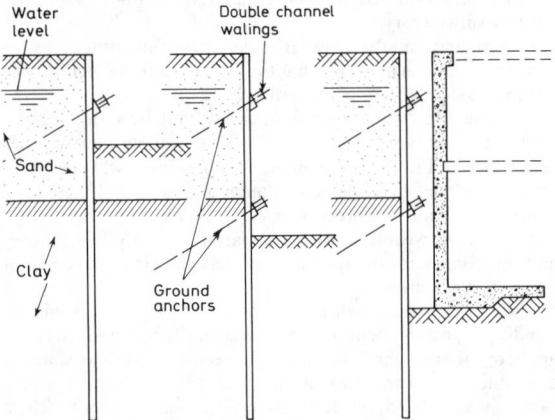

Figure 17.18 Excavation supported by tied-back sheet piling

The bracing system required to support sheeting to excavations of moderate width (say up to 30 m) can be in the form of horizontal struts and walings restrained against buckling by king piles and vertical cross-bracing (Figure 17.20). The struts can be preloaded by jacking to minimize inward movement of the sides. Where wide excavations have to be supported it is preferable to use a system of ground anchors (shown in various stages of construction in conjunction with sheet piling in Figure 17.18) or raking shores (shown in conjunction with horizontal sheeting in Figure 17.19).

Ground anchors have the advantage of providing a clear working space within the excavation and they can conveniently provide a preloading force to minimize inward movement, but there may be problems with existing sewers or other obstructions preventing their installation; also, it may be impossible to obtain wayleaves from surrounding property owners. Raking shores obstruct the working space and require substantial

bearing blocks at the toe. These may give difficulties with maintaining waterproofing in thin basement slabs.

Inward movement of the sheeted sides of an excavation will take place inevitably owing to relief of lateral pressure on removal of the excavation, the compression of the supporting struts (or stretch and creep of ground anchors) and the thermal movements of the support system if the work is properly designed and carefully executed. The inward movement is proportional to the depth of the excavation and appears to be independent of the type of soil and the particular support system.

The inward movements of strutted or anchored diaphragm walls in a wide range of soil types have been shown by observation to be in the general range of 0.05 to 0.6% of the excavation depth.[2] The inward movement is accompanied by a vertical settlement of the same magnitude of the ground surface close to the perimeter of the excavation. The settlement is about half this maximum value at half the excavation depth from the face and falls to a negligible amount at a distance of 3 or 4 times the excavation depth from the face.

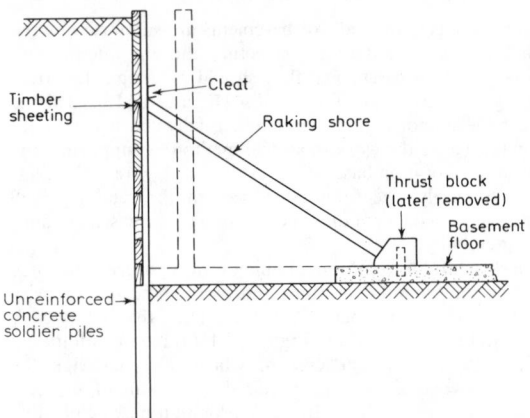

Figure 17.19 Excavation supported by soldier piles and sheeting

Where sheet piling is supported by berms of soft clay sloping not steeper than 2 horizontal:1 vertical, observations have shown a maximum inward deflection of about 2% of the excavation depth.[2]

If there are existing structures within a distance of 3 times the excavation depth from the excavation line then consideration will have to be given to the need for underpinning them before excavation commences. For reasonably good ground conditions, underpinning is unlikely to be needed if the existing structures are not nearer than a distance equal to the excavation depth. For example, Figure 17.20 shows the order of settlements of the ground around a 10-m deep basement. A building in the

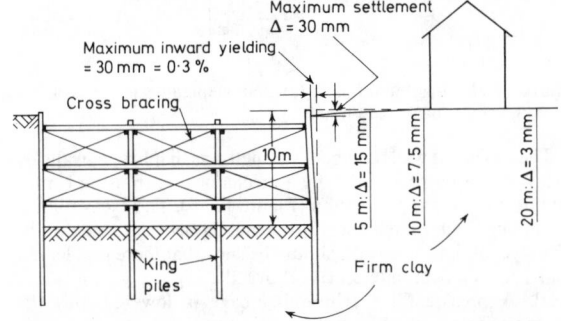

Figure 17.20 Bracing to wide excavation (also showing inward movement)

position indicated would not need to be underpinned. Consideration should be given to the comparative cost of repairs to make good cracking caused by small settlements and that of underpinning, bearing in mind that underpinning operations are themselves usually accompanied by some small settlement.

The various stages of excavation of a four-level deep basement using ground anchors to support the upper two levels and the basement floors to support the lower levels of a diaphragm wall are shown in Figure 17.21. Excavation is undertaken beneath the completed floors and openings are left for removal of spoil. The permanent columns supporting the basement floors are set in drilled holes before commencing the excavation. The inherent stiffness of a diaphragm wall combined with preloading of ground anchors, say to 50% higher than the calculated working load, reduces to a minimum (but does not eliminate) inward yielding of the wall.

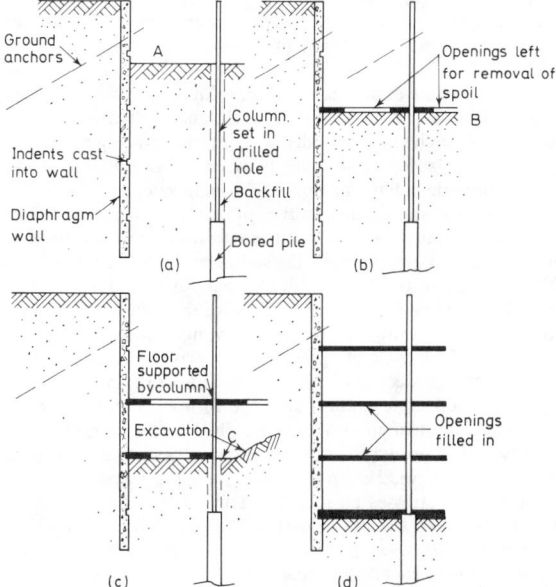

Figure 17.21 Construction of deep basement. (a) Excavation to level A and ground anchors installed; (b) excavation to level B and floor slab cast; (c) excavation to level C and further floor slab cast; (d) completed excavation with all basement floor slabs cast

17.3.3 Buoyancy rafts (hollow box foundations)

The substructure should be as light as possible consistent with the requirement of stiffness. A cellular ('egg box') construction is suitable. This structural form does not normally allow the substructure to be used for any purpose other than its function as a foundation element.

A cellular buoyancy raft may be designed as a caisson (Figure 17.22) which is an economical method of sinking for soft ground conditions, but ground disturbance during sinking can result in some settlement. A buoyancy raft should preferably be constructed within an open excavation. If necessary, the cells may be constructed in individual small areas or strips which are subsequently bonded together. By limiting the area of the excavation in this way, the heave and subsequent reconsolidation of a soft clay can be minimized to a marked degree.

Although considerable gain in uplift can be obtained if buoyancy rafts are designed as watertight structures, there are practical difficulties in achieving this. The space within the cells of a buoyancy raft is normally unoccupied and, if leaks occur, either through the substructure or from fracture of water pipes

within the structure, the flooding of the cells may remain undetected. While the cells can be interconnected and provided with a drainage sump and automatic pumping arrangements there can be no certainty that these arrangements will be maintained in a sound working condition throughout the life of the supported structure. Therefore, unless drainage by gravity to an existing piped system is possible, the net bearing pressures beneath the buoyancy raft should be calculated on the assumption that the cells will become flooded to the level at which gravity drainage can be assured. As noted in section 17.3.2.4, the tanking of a buoyancy raft with asphalt does not give any guarantee of lasting watertightness.

Pipes carrying potentially explosive gases should not be routed through the cells of a buoyancy raft. Leakage of gas into the unventilated cells could remain undetected with a consequent risk of an explosion from accidental ignition.

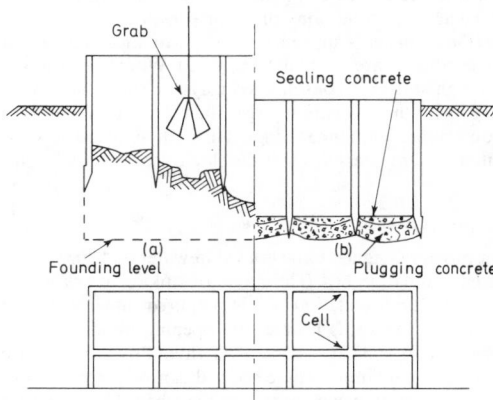

Figure 17.22 Caisson-type cellular buoyancy raft

17.3.4 Caisson foundations

17.3.4.1 General

The types of caisson foundation are:

(1) *A box caisson*, which is closed at the bottom but open to atmosphere at the top.
(2) An *open caisson*, which is open both at the top and bottom.
(3) A *compressed air* or *pneumatic caisson*, which has a working chamber in which air is maintained above atmospheric pressure to prevent the entry of water and soil into the excavation.
(4) A *monolith*, which is an open caisson of heavy mass concrete or masonry construction containing one or more wells for excavation.

The allowable bearing pressures beneath caissons are calculated by the methods described in Chapter 9. However, allowance must be made for the disturbance which may occur during the installation of the foundation. These factors are noted in the following subsections which describe the design and construction methods for the various types.

Caissons are often required to carry horizontal or inclined loads in addition to the vertical loading. As examples, caisson piers to river bridges have to carry lateral loading from wind forces on the superstructure, traction of vehicles on the bridge, river currents, wave forces and sometimes floating ice or debris. Caissons in berthing structures have to be designed to withstand impact forces from ships, mooring-rope pull, and wave forces. Methods of calculating the bearing pressures beneath eccentrically loaded foundations are described in section 17.2.5. A caisson will be safe against overturning provided that the bearing pressure beneath its edge does not exceed the safe bearing capacity of the foundation material, but it is also necessary to ensure that tilting due to elastic compression and consolidation of the foundation soil or rock does not exceed tolerable limits.

The walls of caissons are frequently subjected to severe stresses during construction. These stresses may arise from launching operations (when caissons are constructed on a slipway and allowed to slide into the water), from: (1) wave forces when floating under tow or during sinking; (2) racking due to uneven support whilst excavating individual cells; (3) superimposed kentledge; and (4) the drag effects of skin friction.

Lateral pressures on the external walls of caissons initially may be relatively low, corresponding to active pressure of soil loosened by the sinking process. However, with time the loosened soil will reconsolidate and, because the walls may be rigid and unyielding the conditions of earth pressure 'at rest' may develop (the coefficients appropriate to 'active' or 'at rest' earth pressure conditions are stated in Chapter 9). Where caissons are sunk through stiff over-consolidated clays or shales it may be necessary to cut the excavation larger than the plan dimensions of the foundation. With time the soil will swell to fill the gap and substantial swelling pressures may develop on the external walls.

17.3.4.2 Box caissons

Box caissons are designed to be floated in water and sunk on to a prepared foundation bed. The stages of sinking are shown in Figure 17.23. The foundation bed is prepared under water by divers, and the caisson is lowered by opening flood valves to allow the unit to sink at a controlled rate. Box caissons are suitable for site conditions where the bed can be prepared with little or no excavation below the sea- or river-bed. Thus, they are unsuitable for conditions where scour can undermine a shallow foundation. They are also unsuitable for conditions where scour can occur during the final stages of sinking by the action of eddies and currents in the gap between the base of the caisson and the bed material as the gap diminishes. For founding on soft clay or in scouring conditions, box caissons can be sunk on to a piled raft constructed underwater, but this method is normally more expensive than adopting an open-well caisson.

Box caissons can be of relatively light reinforced-concrete construction, since they are not subjected to severe stresses during sinking. Light construction is desirable to give the required freeboard whilst floating. After sinking they can be filled with mass concrete or sand if dead weight is required for the purpose of increasing the resistance to overturning or lateral forces.

17.3.4.3 Open caissons

Open caissons are designed to be sunk by excavating while removing soil beneath them through the open cells. They are designed in such a manner that the dead weight of the caisson together with any kentledge which may be placed upon it exceeds the skin friction of the soil around the walls and the resistance of the soil beneath the bottom (cutting) edges of the walls. To aid sinking, the soil may be excavated from beneath the cutting edges, or kentledge may be placed on the top of the walls to increase the dead weight. The skin friction around the external walls can be reduced considerably by injecting a bentonite slurry above the cutting edge between the walls and the soil. On reaching founding level, mass concrete is placed to plug each cell after which any water in the cells can be pumped out and further concrete placed to form the final seal. The portions of the cells above the sealing plugs can be left empty, or they can be filled with mass concrete, sand, or fresh water depending on the function of the unit and the allowable net bearing pressure. The stages of sinking are shown in Figure 17.24.

The lower part of an open caisson is known as the *shoe*. This is usually of thin mild steel plating stiffened at the edges with steel tees or angles and provided with internal bracing members. Concrete is placed in the space between the skin plates of the shoe to provide ballast for sinking through water and thereafter more concrete and further strakes of skin plating are added to obtain the required downward forces to overcome skin friction and the bearing resistance of the soil beneath the cutting edges. While the top of the shoe is still above water level, formwork is assembled and the walls extended above the shoe in reinforced concrete. The formwork is usually arranged in lifts of about 1.5 m and a 24-h cycle of operations comprises grabbing to sink 1.5 m, erecting steel skin plating or formwork in the walls, placing the concrete and striking the formwork. Sinking proceeds steadily throughout this cycle. Thick walls are needed for rigidity and to provide dead weight. As well as being reinforced to withstand external earth and hydrostatic pressures, they must resist racking stresses and vertical tension stresses. The latter may occur when the upper part of the caisson is held by skin friction and the lower part tends to fall into the undercut and loosened zone beneath the shoe.

The form of construction, incorporating a shoe fabricated in steel plating, is the traditional method of design, which provides optimum conditions for control of sinking at all stages. However, the introduction of bentonite injection techniques to aid sinking has improved the control conditions making it possible to design caissons entirely in reinforced concrete and enabling them to be sunk to great depths. Circular caissons were sunk to depths of as much as 105 m below the bed of the Jamuna River.[16]

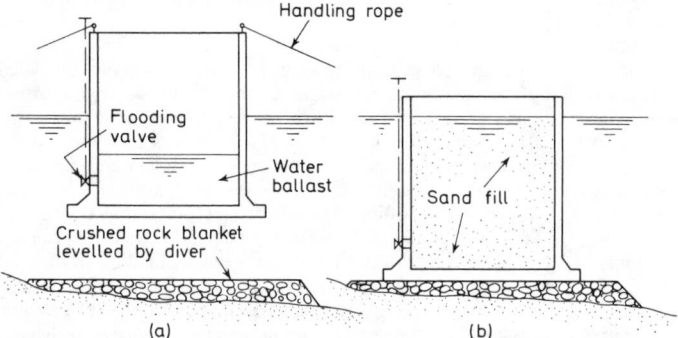

Figure 17.23 Stages in sinking a buoyancy raft. (a) Flooding valve opened to admit water ballast; (b) caisson sunk in final position

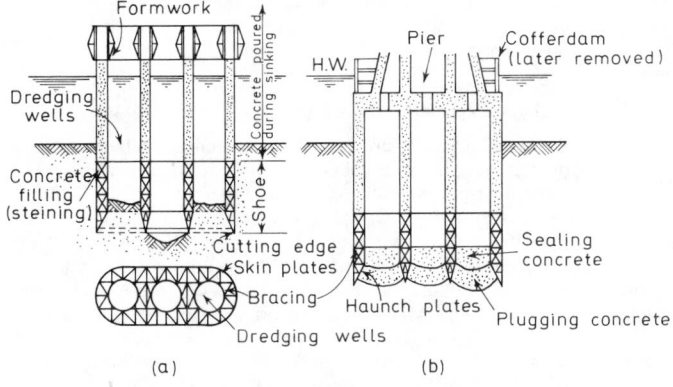

Figure 17.24 Stages in sinking an open caisson. (a) Grabbing from cells and concreting in walls; (b) plugging and sealing concrete in place with caisson at final level

Some typical values used to give a rough guide to skin friction are shown in Table 17.2.[17]

Table 17.2 (After Terzaghi and Peck (1967) *Soil mechanics in engineering practice*. Wiley)

Type of soil	Skin friction (kN/m²)
Silt and soft clay	7–30
Very stiff clay	50–200
Loose sand	10–35
Dense sand	30–70
Dense gravel	50–100

The soil is excavated from within the cells and, where necessary, from below cutting edge level by mechanical grab. In uncemented granular soils, the spoil can be removed by an airlift pump. On reaching founding level any kentledge placed on the walls is removed to arrest sinking and mass concrete is quickly placed at and below cutting edge level in the corner cells to provide a bearing on which the caisson comes to rest. The remaining outer cells are then plugged with concrete followed by completion of excavating and plugging of the inner cells. The concrete plugs are placed under water and after the concrete has hardened the cells are pumped out and further sealing concrete is placed.

Accuracy in the positioning of caissons and control of verticality while sinking are necessary. Various methods of achieving these are:

(1) Sinking between moored pontoons (Figure 17.25).
(2) Sinking within a piled enclosure (Figure 17.26).
(3) Sinking through a sand island (Figure 17.27).

The choice of method depends on the site conditions, i.e. the depth of water, degree of exposure, and velocity of sea or river currents. It also depends on the number of caissons to be sunk on any particular project. The cost of an elaborate floating sinking set as shown in Figure 17.25 is justified if spread over a number of sinking sites. Lowering during sinking can be achieved by using suspension links and jacks (Figure 17.26) by lowering from block and tackle (Figure 17.25) by free sinking with the use of guides (Figure 17.27) or by the controlled expulsion of air from the cells in conjunction with air domes (Figure 17.28).

Open-well caissons are best suited to sinking in soft or loose soils to reach a founding level on stiff or compact material, i.e.

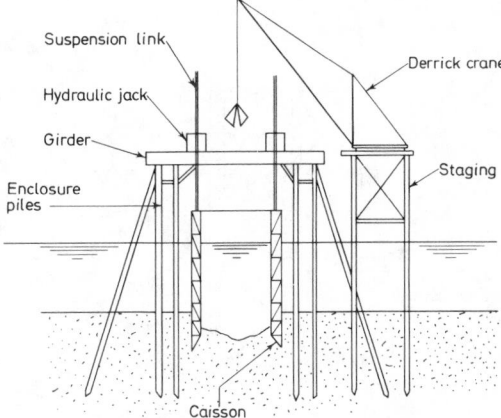

Figure 17.26 Lowering caisson from piled staging

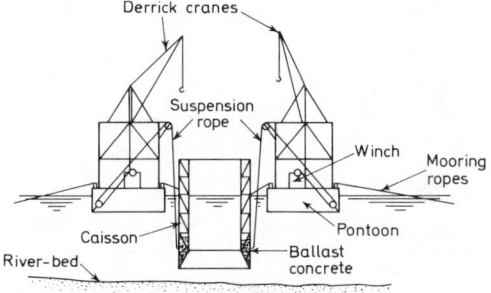

Figure 17.25 Lowering caisson from pontoons

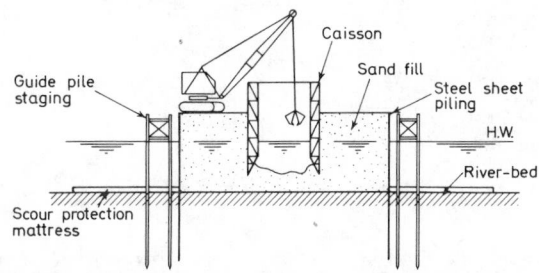

Figure 17.27 Sinking caisson through a sand island

through materials which can be dredged readily and are free of obstruction such as boulders, tree trunks or sunken vessels. They are unsuitable for ground containing obstructions which cannot be broken out from beneath the cutting edge, and are also unsuitable for sinking on to an irregular rock surface. Problems also arise when founding on weak rocks. Grabbing through water causes softening and breakdown of the rock, making it difficult to judge when a satisfactory bearing stratum has been reached and to clean the rock surface to receive the concrete plug.

Removal of soil from within or below the cells of an open caisson causes quite appreciable loss of ground, i.e. the total volume of soil excavated exceeds the volume displaced by the caisson. Open caissons are therefore unsuitable for sinking close to existing structures.

Some of the difficulties mentioned above can be overcome by providing an open caisson with air domes. These are provided with airlocks and are designed to be placed over individual cells as required. Having placed a dome on top of a cell, compressed air is introduced to expel water, after which workmen can enter through an airlock to remove obstructions or to prepare the bottom to receive the sealing concrete. There are limits to the air pressure under which operatives can work in this manner (see section 17.3.4.4). Air domes provided on all cells can be used as the means of floating an open caisson to the sinking site and for controlling its vertical aspect during sinking by varying the rate of expulsion of air from individual cells. Caissons designed in this way are known as flotation caissons. A design used for the Tagus River bridge[18] is shown in Figure 17.28. The cutting edge of this caisson was 'tailored' to suit the profile of the rock surface on which the caisson was landed. The domes of flotation caissons are not normally provided with an airlock. After they have been removed, grabbing proceeds in the normal way for open well caissons.

17.3.4.4 Pneumatic caissons

Pneumatic caissons are designed to be sunk with the assistance of compressed air to obtain a 'dry' working chamber. The general arrangement is shown in Figure 17.29. The caisson consists of a single working chamber surrounded by the shoe with its cutting edge, and a heavy roof. Walls are extended above the shoe in the form of double steel skin plating with mass concrete infilling. The height of the walls depends on the weight required to provide sinking effort and the need to provide freeboard when sinking through water. The airshaft extends from the working chamber to the full height of the caisson and it is surmounted by a combined manlock and mucklock. As the names imply, the former is used for access and egress by operatives and the latter for removal of spoil in crane buckets. The manlocks must at all times be above the highest tide or river flood levels, with due allowance being made for rapid sinking in soft or loose soils.[19]

Work in pneumatic caissons is regulated by the statutory regulations governing working conditions in compressed air. The regulations require 0.3 m^3 of fresh air per minute per person in the working chamber *at the pressure in the chamber*. The air is supplied from stationary compressors powered by diesel or electric motors. Standby power must be available if the site conditions are such as to endanger life or property if the main supply fails. To improve working conditions and to reduce the incidence of caisson-sickness the air supply should be treated to

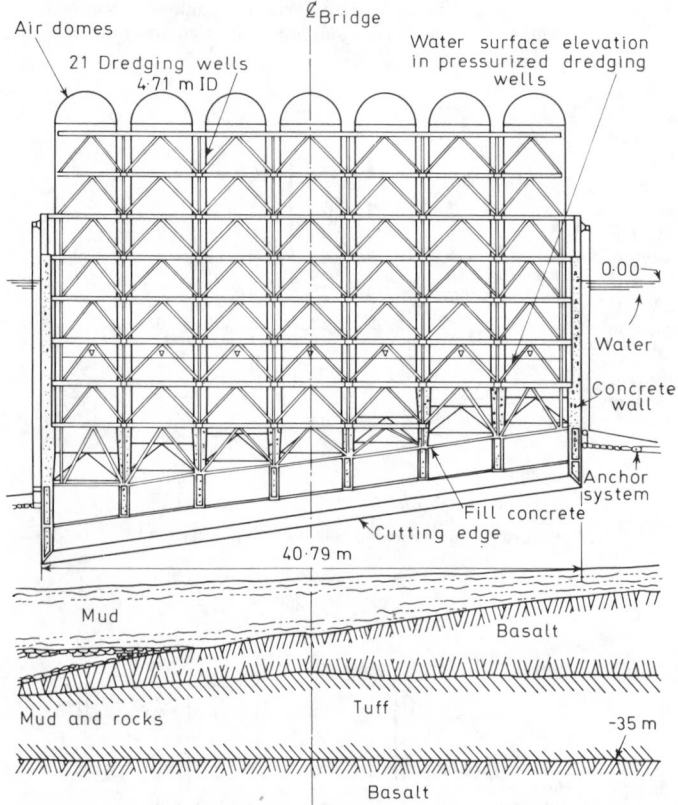

Figure 17.28 Flotation caisson for the Tagus River bridge. (After Riggs (1965) 'Tagus River Bridge – tower piers', *Civ. Engng* (USA) (Feb.) 41–45)

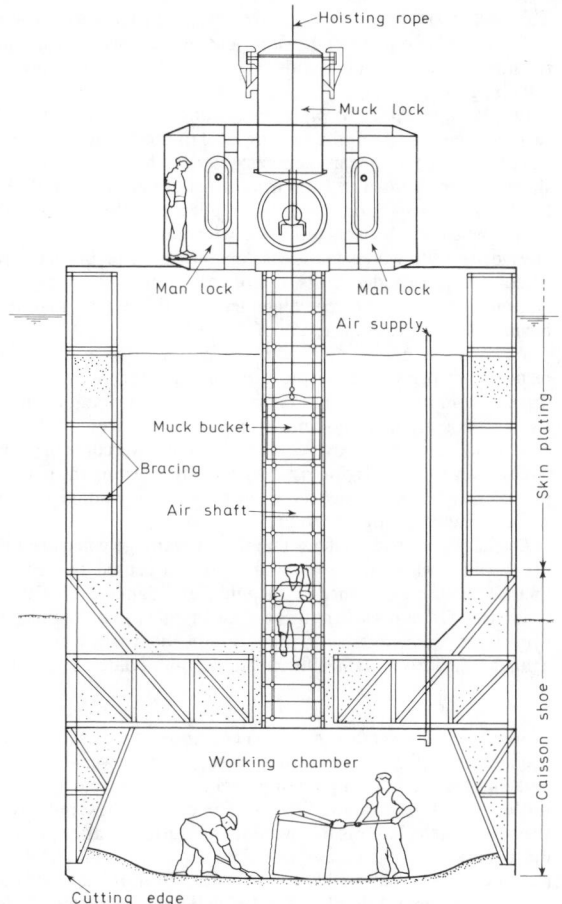

Figure 17.29 Compressed-air caisson. (After Wilson and Sully (1949) *Compressed air caisson foundations.* Works Construction Paper Number 13, Institution Civil Engineers)

warm it for working in cold weather and to cool it for hot-weather working. In tropical climates the air should be dehumi-dified to keep the wet bulb temperature at less than 25° C. In very permeable ground the escape of air into the soil beneath the working chamber may cause too great a demand on the air supply. This can be reduced by pregrouting the ground with clay, cement or chemicals.

If the dead weight of the caisson, together with any added kentledge, is insufficient to overcome the skin friction, the effective sinking weight can be increased temporarily by 'blow-ing down' the caisson. This involves removing the operatives from the working chamber, then reducing the air pressure by about one-quarter of the gauge pressure.

On nearing founding level, concrete blocks are placed on the floor of the working chamber and the roof is allowed to come to rest on them. The working chamber is then filled with concrete and the airshaft and airlocks removed.

The pneumatic caisson is suitable for sinking close to existing structures since the excavation is not accompanied by loss of ground. It is also suitable for sinking in ground containing obstructions, and for founding on an irregular rock bed. Pneu-matic caissons have the severe limitation that the depth of sinking cannot exceed a level at which the required air pressure to exclude water from the working chamber exceeds the limit at which operatives can work without danger to their health. A pressure of 345 kN/m² is considered generally to be a safe maximum but stringent medical precautions and supervision are

required at all stages of the work.[20] The high cost of compressed-air sinking generally precludes pneumatic caissons for all but special foundations where no alternatives are feasible or eco-nomically possible.

17.3.4.5 Monoliths and cylinders

Monoliths are open caissons of reinforced concrete or mass concrete construction (Figure 17.30) and are mainly used for quay walls where their heavy weight and massive construction are favourable for resisting the thrust of the filling behind the wall and for withstanding the impact forces from berthing ships. Because of their weight they are unsuitable for sinking through deep soft deposits. Their design and method of construction generally follow the same principles as those for open caissons in section 17.3.4.3.

Open caissons of cylindrical form and having a single cell are sometimes referred to as cylinder foundations.

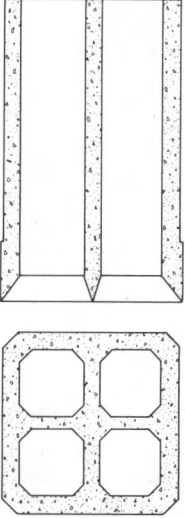

Figure 17.30 Concrete monolith

17.3.4.6 Shaft foundations

Where deep foundations are required for the heavily loaded columns of a structure it may be desirable to sink the foundation in the form of a lined shaft excavated by hand or by mechanical grab. This type of foundation is similar to the large bored pile as described in section 17.4.3.1 but its distinguishing characteristic is the construction of the lining in place, taken down stage-by-stage as the shaft is deepened. The shaft foundation would be selected in cases where the required diameter was larger than the capacity of the large-bored-pile drilling machine, in ground containing boulders or other obstructions which could prevent machine drilling or caisson sinking, and in localities where specialist pile-drilling plant is not available but where labour for hand excavation can be provided from local resources.

Shaft foundations can be of any desired shape but the cylindrical form is the most convenient since internal bracing is not required. The lining can consist of mass concrete placed *in situ* behind formwork (Figure 17.31(a)) or bolted precast con-crete, steel or cast-iron segments (Figure 17.31(b)). The *in situ* concrete lining is suitable for relatively dry ground which can stand without support for a height of about 1.5 m. Segmental lining can be used in water-bearing ground which can stand unsupported for the height of a segment. Cement grout must be injected at intervals into the space between the back of the segments and the soil. This is necessary to prevent excessive flow of water down the back of the lining, and also to support the segments from dropping under their own weight augmented by downdrag forces from the loosened soil. The collar at the top of the shaft is also required to support the lining.

Shaft foundations may be constructed as a second stage after first sinking through soft or loose ground as a caisson (Figure 17.31(a)) or at the base of a sheet piled cofferdam (Figure 17.31(b)).

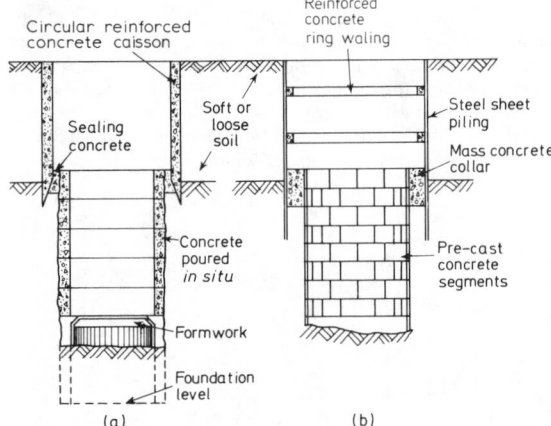

Figure 17.31 Shaft foundations. (a) With mass concrete lining constructed below caisson; (b) with precast concrete segmental lining constructed below a sheet-piled cofferdam

17.4 Piled foundations

17.4.1 General descriptions of pile types

There is a large variety of types of pile used for foundation work.[21] The choice depends on the environmental and ground conditions, the presence or absence of groundwater, the function of the pile, i.e. whether compression, uplift or lateral loads are to be carried, the desired speed of construction and consideration of relative cost. The ability of the pile to resist aggressive substances or organisms in the ground or in surrounding water must also be considered.

In BS 8004, piles are grouped into three categories:

(1) *Large displacement piles*: these include all solid piles, including timber and precast concrete and steel or concrete tubes closed at the lower end by a shoe or plug, which may be either left in place or extruded to form an enlarged foot.
(2) *Small displacement piles*: these include rolled-steel sections, open-ended tubes and hollow sections if the ground enters freely during driving.
(3) *Replacement piles*: these are formed by boring or other methods of excavation; the borehole may be lined with a casing or tube that is either left in place or extracted as the hole is filled.

Large or small displacement piles In preformed sections these are suitable for open sites where large numbers of piles are required. They can be precast or fabricated by mass-production methods and driven at a fast rate by mobile rigs. They are suitable for soft and aggressive soil conditions when the whole material of the pile can be checked for soundness before being driven. Preformed piles are not damaged by the driving of adjacent piles, nor is their installation affected by groundwater.

They are normally selected for river and marine works where they can be driven through water and in sections suitable for resisting lateral and uplift loads. They can also be driven in very long lengths.

Displacement piles in preformed sections cannot be varied readily in length to suit the varying level of the bearing stratum, but certain types of precast concrete piles can be assembled from short sections jointed to form assemblies of variable length. In hard driving conditions preformed piles may break causing delays when the broken units are withdrawn or replacement piles driven. A worse feature is unseen damage particularly when driving slender units in long lengths which may be deflected from the correct alignment to the extent that the bending stresses cause fracture of the pile.

When solid pile sections are driven in large groups the resulting displacement of the ground may lift piles already driven from their seating on the bearing stratum, or may damage existing underground structures or services. Problems of ground heave can be overcome or partially overcome in some circumstances by redriving risen piles, or by inserting the piles in prebored holes. Small-displacement piles are advantageous for soil conditions giving rise to ground heave.

Displacement piles suffer a major disadvantage when used in urban areas where the noise and vibration caused by driving them can cause a nuisance to the public and damage to existing structures. Other disadvantages are the inability to drive them in very large diameters, and they cannot be used where the available headroom is insufficient to accommodate the driving rig.

Driven and cast-in-place piles These are widely used in the displacement pile group. A tube closed at its lower end by a detachable shoe or by a plug of gravel or dry concrete is driven to the desired penetration. Steel reinforcement is lowered down the tube and the latter is then withdrawn during or after placing the concrete. These types have the advantages that: (1) the length can be varied readily to suit variation in the level of the bearing stratum; (2) the closed end excludes groundwater; (3) an enlarged base can be formed by hammering out the concrete placed at the toe; (4) the reinforcement is required only for the function of the pile as a foundation element, i.e. not from considerations of lifting and driving as for the precast concrete pile; and (5) the noise and vibration are not severe when the piles are driven by a drop hammer operating within the drive tube.

Driven and cast-in-place piles may not be suitable for very soft soil conditions where the newly placed concrete can be squeezed inwards as the drive tube is withdrawn causing 'necking' of the pile shaft, nor is the uncased shaft suitable for ground where water is encountered under artesian head which washes out the cement from the unset concrete. These problems can be overcome by providing a permanent casing. Ground heave can damage adjacent piles before the concrete has hardened, and heaved piles cannot easily be redriven. However, this problem can be overcome either by preboring or by driving a number of tubes in a group in advance of placing the concrete. The latter is delayed until pile driving has proceeded to a distance of at least 6.5 pile diameters from the one being concreted if small (up to 3 mm) uplift is permitted, or 8 diameters away if negligible (less than 3 mm) uplift must be achieved.[22] The lengths of driven and cast-in-place piles are limited by the ability of the driving rigs to extract the drive tube and they cannot be installed in very large diameters. They are unsuitable for river or marine works unless specially adapted for extending them through water and cannot be driven in situations of low headroom.

Replacement piles or bored piles These are formed by drilling a borehole to the desired depth, followed by placing a cage of steel reinforcement and then placing concrete. It may be necessary to support the borehole by steel tubing (or casing) which is driven down or allowed to sink under its own weight as the borehole is drilled. Normally the casing is filled completely with easily workable concrete before it is extracted, when the concrete slumps outwards to fill the void so formed.

In stiff cohesive soils or weak rocks it is possible to use a rotary tool to form an enlarged base to the piles which greatly increases the end-bearing resistance. Alternatively, men can descend the shafts of large-diameter piles to form an enlarged base by hand excavation. Reasonably dry conditions are essential to enable the enlarged bases to be formed without risk of collapse.

Care is needed in placing concrete in bored piles. In very soft ground there is a tendency to squeeze of the unset concrete, and if water is met under artesian head it may wash out the cement from the unset concrete. If water cannot be excluded from the pile borehole by the casing, no attempt should be made to pump it out before placing concrete. In these circumstances the concrete should be placed under water by tremie pipe. A bottom-opening skip should not be used. Breaks in the concrete shafts of bored piles may occur if the concrete is lifted when withdrawing the casing, or if soil falls into the space above the concrete due to premature withdrawal of the casing.

Bored piles have the advantages that their length can be readily altered to suit varying ground conditions, the soil or rock removed during boring can be inspected and if necessary subjected to tests, and very large shaft diameters are possible, with enlarged base diameters up to 6 m. Bored piles can be drilled to any desired depth and in any soil or rock conditions. They can be installed without appreciable noise or vibration in conditions of low headroom and without risk of ground heave.

Bored piles are unsuitable for obtaining economical skin friction and end bearing values in granular soils because of loosening of these soils by drilling. However, stable conditions can be achieved if the pile borehole is supported during the drilling operation by a bentonite slurry. Boring in soft or loose soils results in loss of ground which may cause excessive settlement of adjacent structures. They are also unsuitable for marine works.

17.4.2 Details of some types of displacement piles

17.4.2.1 Timber piles

In countries where timber is readily available, timber piles are suitable for light to moderate loadings (up to 300 kN). Softwoods require preservation by creosote in accordance with BS 913. If this is done they will have a long life below groundwater level but are subject to decay above this level. Where possible, pile caps in concrete should be taken down to water level (Figure 17.32(a)). If this is too deep, a composite pile may be installed, the upper part above water level being in precast concrete or concrete cast-in-place jointed to a timber section (Figure 17.32).

To prevent damage to timber piles during driving, the head should be protected by a steel or iron ring, and the toe by a cast-iron shoe (Figure 17.33(b)).

British Standard 8004 requires that the working stresses in compression on a timber pile do not exceed those tabulated in BS 5268 for compression parallel to the grain for the species and grade of timber used, due allowances being made for eccentricity of loading, nonverticality of driving, bending stresses due to lateral loads, and reductions in section due to drilling lifting holes or notching the piles. The working stresses of BS 5268 may be exceeded while the pile is being driven.

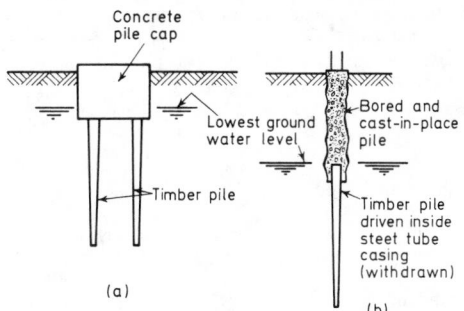

Figure 17.32 Methods of avoiding decay in timber piles

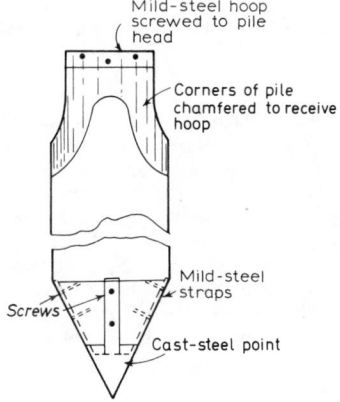

Figure 17.33 Protecting the head and toe of a timber pile

17.4.2.2 Precast and prestressed concrete piles

Precast reinforced concrete piles may not be economical for use in land structure because a considerable amount of steel reinforcement is needed to withstand bending stresses during lifting and subsequent compressive and tensile stresses during driving. Precast concrete piles are also liable to damage on handling and during driving in hard ground. However, the reinforcement may be needed for resisting lateral forces on the pile, e.g. for resisting impact forces on wharves or jetty piling. Much of this reinforcement is not required once the pile is in the ground.

The effect of prestressing of solid or hollow concrete piles in conjunction with high-quality concrete is to produce a unit which should not suffer hair cracks while being lifted or transported and therefore should produce a more durable foundation element than the ordinary precast concrete pile. This is advantageous in aggressive ground conditions. However, prestressed concrete piles are liable to crack during driving and require careful detailing of reinforcement and precautionary measures during driving to ensure concentric blows of the hammer and accurate alignment in the leaders of the pile frame.

The maximum pile lengths for main reinforcement of various diameters are listed in Table 17.3. These lengths allow for the pile to be lifted at the head and toe.

The pile lengths were based on a characteristic stress in the steel of 250 N/mm² and concrete having a characteristic strength of 40 N/mm². British Standard 8004 requires lateral reinforcement in the form of hoops or links to resist driving stresses, the diameter of which shall not be less than 6 mm. For a distance of 3 times the width of the pile from each end the volume of the lateral reinforcement should not be less than 0.6% of the gross volume. In the body of the pile the lateral reinforcement should not be less than 0.2% of the gross volume spaced at a distance of

Table 17.3 Maximum pile lengths for given reinforcement

Bar diameter for 4 bars (mm)	300 mm pile (m)	350 mm pile (m)	400 mm pile (m)	450 mm pile (m)
20	9.0	8.5	—	—
25	11.0	10.5	10.0	9.5
32	—	13.0	12.5	12.0
40	—	—	15.5	15.0

not more than half the pile width. The transition between close spacing at the ends and the maximum spacing should be made gradually over a length of about 3 times the width. A typical precast concrete pile of solid section designed for fairly easy driving conditions and the minimum transverse reinforcement required by BS 8004 is shown in Figure 17.34. Other recommendations are:

(1) *Reinforcement*: to comply with BS 4449 and 4461.
(2) *Concrete mixes*: for hard to very hard driving conditions and all marine work use cement content of 400 kg/m^3. For normal or easy driving use cement content of 300 kg/m^3.
(3) *Concrete design*: stresses due to working load, handling and driving not to exceed those in BS 8110 or CP 116.
(4) *Cover to reinforcement*: to comply with BS 8110:Part 1, Table 3.4.

Where piles are driven through hard ground which must be split to achieve penetration or ground containing obstructions liable to damage the toe of a pile, a cast-steel or cast-iron shoe should be provided as shown in Figure 17.35(a). For driving on to a sloping hard rock surface a rock point should be provided as shown in Figure 17.35(b) to prevent the toe skidding down the slope. A shoe need not be provided for easy to fairly hard driving in clays and sands when the pile may have a flat end or be terminated as shown in Figure 17.35(c).

The recommendations of BS 8004 for prestressed concrete piles are as follows:

(1) *Materials*: to be in accordance with BS 8110 or CP 115.
(2) *Design*: maximum axial stress 0.25 × (28-day works cube stress less prestress after losses). The stress should be reduced if the ratio of effective length:least lateral dimension is greater than 15.

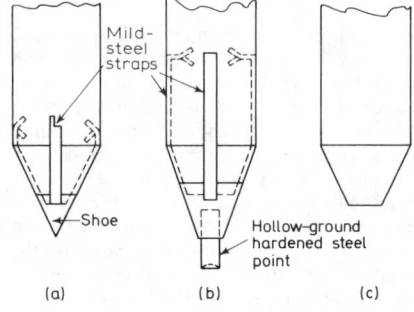

Figure 17.35 Design of toe for precast or prestressed concrete pile

Static stresses produced by lifting and pitching not to exceed values given in Tables 1 and 2 of CP 115 using in Table 2 of that code the values relating to loads of short duration.

(3) *Prestress*: minimum prestress is related to ratio of weight of hammer:weight of pile thus:

Ratio	0.9	0.8	0.7	0.6
Minimum prestress for normal driving (N/mm^2)	2.0	3.5	5.0	6.0
Minimum prestress for easy driving (N/mm^2)	3.5	4.0	5.0	6.0

The minimum prestress for diesel hammers should be 5.0 N/mm^2

(4) *Lateral reinforcement*: mild steel stirrups not less than 6 mm diameter spaced at a pitch of not more than side dimensions less 50 mm. At top and bottom for length of 3 times side dimension stirrup volume not less than 0.6% of pile volume.

(5) *Cover*: as for precast concrete piles (see section 17.4.2.2).

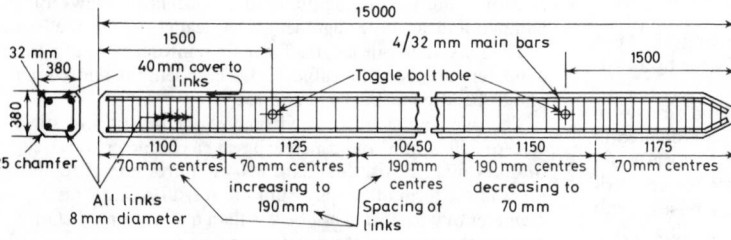

Figure 17.34 Design of precast concrete pile suitable for fairly easy driving conditions and for lifting at third point from one end or at positions shown

To minimize damage to pile heads during driving, precast concrete or prestressed concrete piles should be driven with timber or plastic packing between the helmet and the hammer. The hammer weight should be roughly equal to the weight of the pile and never less than half its weight. The drop should be 1 to 1.25 m. Particular care is necessary when driving with a diesel hammer when an uncontrollable sharp impact can break the pile if the toe meets a hard layer. Drop hammers or single-acting hammers are preferable for these ground conditions.

A typical prestressed concrete pile designed to the above recommendations is shown in Figure 17.36.

lateral forces and to buckling. They are advantageous for marine work. They can be lengthened by welding on additional lengths as required and cut-off sections have scrap value. If a small displacement is needed to minimize ground heave the H-section can be used or tubular piles can be driven with open ends and the soil removed by a drilling rig.

Various types of steel pile are shown in Figure 17.38. Reference should be made to the British Steel Corporation's handbook for dimensions and properties of the various sections. British Standard 8004 requires steel piles to conform to BS 4360, grades 43A, 50B or other grades to the approval of the engineer.

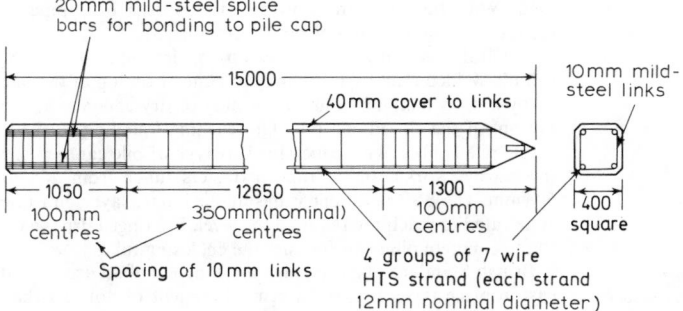

Figure 17.36 Design of prestressed reinforced concrete pile

17.4.2.3 Jointed precast concrete piles

One of the drawbacks of ordinary precast or prestressed concrete piles is that they cannot be readily adjusted in length to suit the varying level of a hard-bearing stratum. Where the bearing stratum is shallow a length of pile must be cut off and is wasted. Where it is deep the pile must be lengthened with an inevitable delay in the process of splicing on a new length. This drawback can be overcome by the use of precast concrete piles assembled from short units. Two principal types are available. The West's pile (Figure 17.37(a)) consists of short cylindrical hollow shells made in 380, 405, 445, 510, 535 and 610 mm outside diameters. The shells are threaded on to a steel mandrel which carries a shoe at the lower end. The driving head is designed to allow the full weight of the drop hammer to fall on the mandrel while the shells take a cushioned blow. Shells can be added or taken away from the mandrel to suit the varying penetration depths of the piles. On completion of driving, the mandrel is withdrawn, a reinforcing cage is lowered down the shells and the interior space filled with concrete. Care is needed with this type of pile in driving through ground containing obstructions. If the mandrel goes out of line there is difficulty in withdrawing it and the shells may be displaced. The shells are also liable to be lifted due to ground heave in firm to stiff clays. Piles driven in groups should be prebored for part of their length or the order of driving arranged to minimize ground heave.

The other type comprises solid square or hexagonal section precast units with locking joints which are stronger than the concrete section. The joints are capable of withstanding uplift caused by ground heave. The lengths are manufactured to suit the requirements of the particular job and additional short lengths are locked on if deeper penetrations are required. Piles of this type include the West's Hardrive, the Herkules and Balken sections.

17.4.2.4 Steel piles

Steel piles of tubular, box, and H-section have the advantages of being robust and easy to handle and can withstand hard driving. They can be driven in long lengths and have a good resistance to

The stress under the working load should be limited to 30% of the yield stress except where piles are driven through relatively soft soils to an end bearing on dense soils or sound rock, when the allowable axial working stress may be increased to 50% of the yield stress.

Slender-section steel piles driven in long lengths are liable to go off-line during driving. It is desirable to check them for

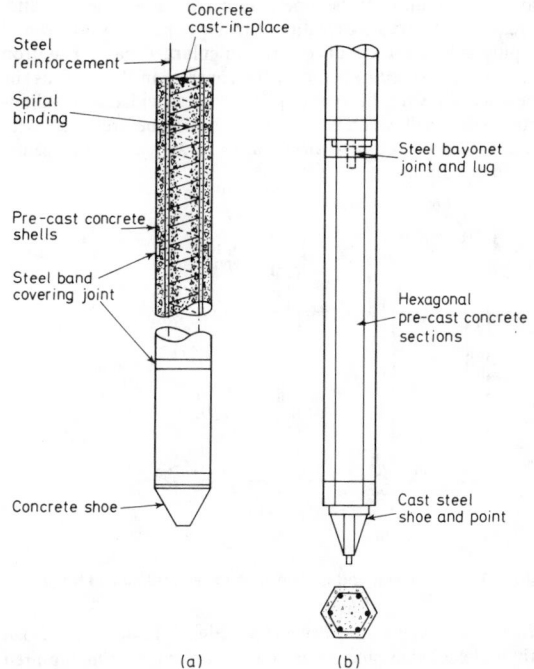

Figure 17.37 Jointed precast concrete piles. (a) West's shell pile; (b) Herkules pile

curvature after driving by inclinometer (a small-diameter tube can be welded to the web of an H-section pile for this purpose). If H-piles or unfilled tubular piles have a curvature of less than 360 m they should be rejected. Tubular piles need not be rejected if they are designed to be filled with concrete capable of carrying the full working load.

Steel piles are liable to corrosion where oxygen is available, e.g. above the soil line or above water level, but allowance can be made for corrosion losses within the useful life of the structure or special protection can be provided (see section 17.8.3).

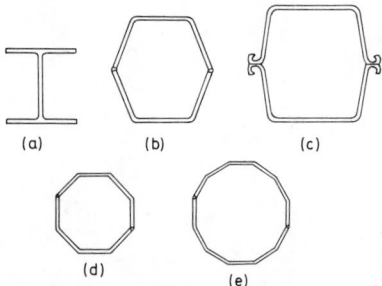

Figure 17.38 Steel-bearing piles of various types. (a) Universal bearing pile (UBP); (b) Rendhex foundation column (obsolete); (c) Larssen box pile; (d) Frodingham octagonal pile; (e) Frodingham duodecagonal pile

17.4.2.5 Driven and cast-in-place piles

There is a wide range of types of proprietary driven and cast-in-place piles in which a steel tube is driven to the required penetration depth and filled with concrete. In some types the tube is withdrawn during or after placing the concrete. In other types the tube of a light steel shell is left permanently in place.

In one type (Figure 17.39) a drop hammer acts on a plug of gravel at the bottom of the tube. This carries down the tube and, on reaching the bearing stratum, further concrete is added and the plug is hammered out to form an enlarged base. The drop hammer is also used to compact the concrete in the shaft as the tube is withdrawn. This type of pile can be provided with a light-section steel shell which is placed in the tube before filling with concrete to provide a permanent casing to withstand 'squeezing' ground conditions.

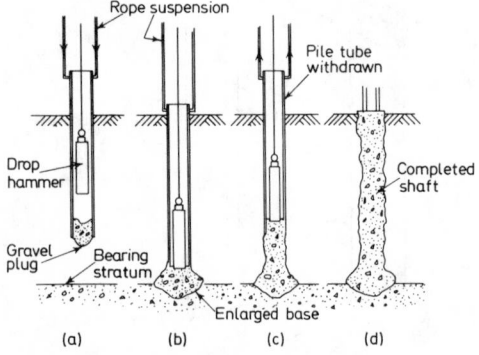

Figure 17.39 Driven and cast-in-place pile (end closed by gravel plug)

In another type a steel drive tube (Figure 17.40) is provided with a detachable steel shoe and is driven to the required penetration by a drop hammer or diesel hammer acting on top of the tube. A reinforcing cage is then placed in the tube and concrete is poured before or during withdrawal of the tube.

Driven and cast-in-place piles of the types described above are cast to nominal outside diameters ranging from 250 to 750 mm. Their lengths are limited by the capacity of the rig to pull out the drive tube to a maximum of about 40 m.

In the Raymond Step Taper Pile light gauge steel shells of progressively reducing diameter are driven to the required depth on a mandrel. The latter is then withdrawn and the shells are filled with concrete. Placing concrete in the shells should be delayed until ground heave has ceased when driving these piles in groups. Ground heave can be reduced by preboring. When the required pile length exceeds the limits of the available equipment to drive an all-shell pile, a pipe step-taper pile may be used. With this type the bottom unit consists of a pipe of constant 273 mm section of the required length.

The BSP cased pile system consists of driving a fairly light spirally welded steel tube either by a hammer on top of the pile or by a drop hammer acting on a plug of dry concrete at the bottom of the closed-end pile. On reaching founding level the whole pile is filled with concrete. This type of pile can be used for marine works. Inside tube diameters range from 245 to 508 mm. The BSP cased pile is unsuitable if hard layers must be penetrated to reach the required toe level. Prolonged driving on to the concrete plug can fracture the enclosing tube.

British Standard 8004 requires the concrete of all driven and cast-in-place types to have a cement content of not less than 300 kg/m³. The average compressive strength under working loads shall not exceed 25% of the specified 28-day works cube strength. Care should be taken to ensure that the volume of concrete placed fills the volume of the soil displaced by the drive tube or the volume of shells left in place. This is a safeguard against caving of the ground while withdrawing the tube or collapse of shells.

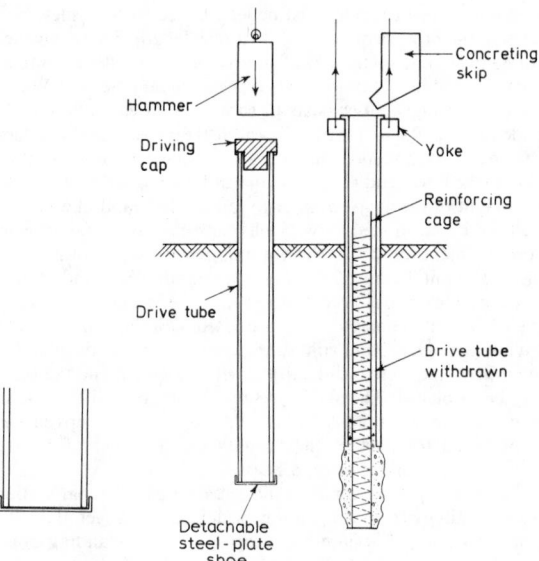

Figure 17.40 Driven and cast-in-place pile with detachable shoe

17.4.3 Types of replacement piles

17.4.3.1 Rotary bored piles

If the soil is capable of remaining unsupported for a short time the pile borehole can be drilled by a rotary spiral plate or bucket auger. Support to soft, loose or water-bearing superficial soil deposits in the upper part of the pile borehole can be provided by a length of temporary casing which is driven down to seal into a stiff cohesive soil in advance of the drilling operation. The borehole is continued in the stiff cohesive soil or weak rock without support by temporary casing unless it is desired to enter the hole for visual inspection of the base or to enlarge the base by manual excavation. In these cases it is necessary to give temporary support by full-length lining tubes which are suspended from the ground surface. After completion of drilling and cleaning the bottom of the borehole the reinforcing cage is inserted and concrete is placed by discharging it from a hopper at the mouth of the hole. An easily workable self-compacting mix with a slump of 125 to 150 mm is used.

Base enlargements can be formed in stiff cohesive soils and weak rocks by a rotary under-reaming tool provided that the borehole is reasonably dry.

Where groundwater seepages enter the borehole below the level of the temporary casing in quantities which cause accumulations at the bottom of the hole of more than a few centimetres in 5 min, no attempt should be made to bale out the water which should be allowed to rise to its standing level. The concrete should then be placed under water through a tremie pipe. The mix should have a slump of 175 mm or more and a minimum cement content of 400 kg/m^3.

In 'squeezing' soils or in ground contaminated by substances aggressive to concrete, light steel or plastic tubing can be used as a permanent sheathing to the concrete in the pile shaft.

In water-bearing soils and rocks and in cohesionless soils, support to the pile boreholes can be provided by a bentonite slurry. The concrete in the pile shaft is placed through the slurry by tremie pipe.

Rotary augers can drill to depths of up to 60 m with shaft and base diameters up to 5 and 6 m respectively.

Safety precautions in bored piling work are covered by BS 5573.

17.4.3.2 Percussion-bored piles

In ground which collapses during drilling, requiring continuous support by casing, the pile boring is undertaken by baling or grabbing. For small-diameter (up to 600 mm) piles the tripod rig is used to handle the drilling tools and to extract the casing. For large-diameter piles a powered rig which combines a casing oscillator and a winch for handling grabbing and chiselling tools is used to drill to diameters of up to 1.5 m and depths of 50 m or more. *Barrettes* are rectangular- or cruciform-shaped piles formed by excavating under a bentonite slurry by a trenching grab, followed by placing the concrete through the slurry by tremie pipe. Barrettes are suitable for deep foundations carrying high lateral forces, e.g. in retaining walls.

Problems of placing concrete in difficult conditions, e.g. in 'squeezing' ground, can be overcome in special cases by placing concrete under compressed air with the assistance of an airlock on top of the casing, i.e. the Pressure pile, or by placing precast concrete sections in the casing and injecting cement grout to fill the joints between and around the sections while withdrawing the casing (the Prestcore pile).

17.4.3.3 Auger-injected piles

A continuous-flight auger is used to drill the pile borehole to the required depth. A sand–cement grout or concrete is then pumped down the hollow stem of the auger as it is being withdrawn. The reinforcement cage is lowered down the shaft after the auger has been fully withdrawn. Presently available rigs can drill to diameters in the range of 300 to 750 mm and to depths of up to 25 m. The auger-injected pile is suitable for most soils. The process is virtually vibration-free which makes it suitable for use close to existing structures.

17.4.3.4 Concrete for replacement piles

The cement content should not be leaner than 300 kg/m^3. The average compressive stress under the working load should not exceed 25% of the specified works cube strength at 28 days. British Standard 8004 permits a higher allowable stress if the pile has a permanent casing of suitable shape.

The concrete should be easily workable and capable of slumping to fill all voids as the casing is being withdrawn without being lifted by the casing. If a tremie pipe is necessary for placing concrete under water the mix should not be leaner than 400 kg of cement per cubic metre of concrete and a slump of 175 mm is suitable.

17.4.4 Raking piles to resist lateral loads

Where lateral forces are large it may be necessary to provide raking piles to carry lateral loading in compression or tension axially along the piles. Arrangements of raking pile foundations for a retaining wall and a berthing structure are shown in Figures 17.41(a) and (b) respectively.

Raking piles should not have a rake flatter than 1 in 3 if difficulties in driving are to be avoided, but flatter rakes are possible with short piles. It is not easy to install driven and cast-in-place or bored piles on a rake.

Methods of calculating the ultimate capacity and deflection of piles under horizontal loading are given by Tomlinson[23] and Elson,[24] but load testing is necessary if deflections are critical.

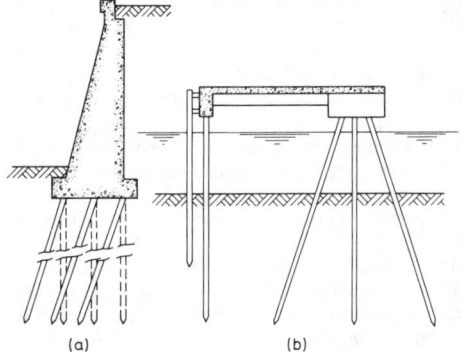

Figure 17.41 Raking piles to resist lateral loads. (a) Beneath retaining wall; (b) in a marine berthing structure

17.4.5 Anchoring piles to resist uplift loads

Piles can be anchored to rock by drilling in a steel tube with an expendable bit at its lower end. Grout is injected through the tube to fill the annulus to form an unstressed or 'dead' anchor. Alternatively, a high-tensile steel rod or cable can be fed into a predrilled hole. It is stressed by jacking from the top of the pile. In the second method the upper part of the anchor should be prevented from bonding to the grout by surrounding the greased metal with a plastic sheath. This is to ensure mobilization of the uplift resistance of the complete mass of rock down to the bottom of the anchorage. Methods of calculating this resistance are described in Chapter 10.

17.4.6 Pile caps and ground beams

A pile cap is necessary to distribute loading from a structural member, e.g. a building column, on to the heads of a group of bearing piles. The cap should be generous in dimensions to accommodate deviation in the true position of the pile heads. It is usual to permit piles to be driven out of position by up to 75 mm and the positioning of reinforcement which ties in to the projecting bars from the pile heads should allow for this deviation. Caps are designed as trusses or beams spanning the pile heads and carrying concentrated loads from the superimposed structural member.[25] The heads of concrete piles should be broken down to expose the reinforcing steel which should be bonded into the pile cap reinforcement. The loading on to steel piles can be spread into the cap by welding capping plates to the pile heads or by welding on projecting bars or lugs as shear keys. A three-pile cap is the smallest which can be permitted to act as an isolated unit. Single- or two-pile caps should be connected to their neighbours by ground beams in two directions or by a ground slab. A system for the standardization of pile-cap dimensions has been described by Whittle and Beattie.[26]

Piles placed in rows beneath load-bearing walls are connected by a continuous cap in the form of a ground beam (Figure 17.42). In the illustration the ground beam is shown as constructed over a compressible layer such as cellular cardboard designed to prevent uplift on the beam due to swelling of the soil, and the pile is sleeved over its upper part to prevent uplift within the zone of swelling. The ground beam should be designed to resist horizontal thrust from the swelling clay. As an alternative to sleeving the upper part of the pile it may be preferable to provide for uplift by increasing the length of the shaft.

17.4.7 Testing of piles

Tests to determine the integrity of the shafts of concrete piles can be made by nondestructive methods described by Weltman.[27] In soils where time effects are not significant in determination of bearing capacity a reasonably accurate prediction of ultimate bearing capacity and settlement can be made by measurements of strain and acceleration under hammer impact at the time of driving.[28]

Loading tests on piles may be needed at two stages: (1) to verify the carrying capacity of the piles in compression, uplift or lateral loading; and (2) to act as a proof load to verify the soundness of workmanship or adequacy of penetration of working piles.

For first-stage testing either the constant rate of penetration (CRP) test or the maintained load (ML) method may be used. The latter is to be preferred if information on the deflection of the pile under the working load, or at some multiple of this load, is needed.

For proof loading of working piles the ML test should be made. It is not usual to apply a load of more than 1.5 times the working load in order to avoid overstressing the pile.

The procedures for the CRP and ML tests are described in BS 8004.

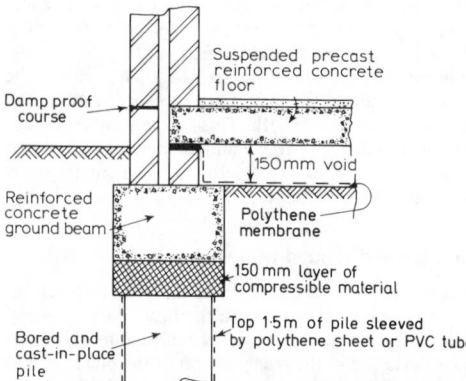

Figure 17.42 Ground beam for piles carrying a load-bearing wall

17.5 Retaining walls

17.5.1 General

This section covers the design and construction of free-standing or tied-back retaining walls. The design of retaining walls for basements, bridge abutments and wharves is described in section 17.3.2.3, in Barry[29] and in Chapters 23 and 26 respectively.

Free-standing or tied-back retaining walls can be grouped for design purposes as follows:

(1) Gravity walls which rely on the mass of the structure to resist overturning (Figure 17.43(a)).
(2) Cantilever walls which rely on the bending strength of the cantilevered slab above the base (Figure 17.43(b)).
(3) Counterfort walls which are restrained from overturning by the force exerted by the mass of earth behind the wall (Figure 17.43(c)).
(4) Buttressed walls which transmit their thrust to the soil through buttresses projecting from the front of the wall (Figure 17.43(d)).
(5) Tied-back diaphragm walls which are restrained from overturning by anchors at one or more levels (Figure 17.43(e)).
(6) Contiguous bored pile walls (Figure 17.43(f)).

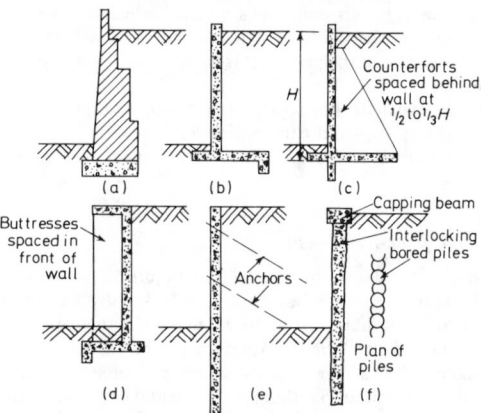

Figure 17.43 Types of retaining wall. (a) Gravity wall; (b) cantilever wall; (c) counterfort wall; (d) buttressed wall; (e) tied-back diaphragm wall; (f) cantilevered wall contiguous bored pile wall

It is assumed that sufficient forward movement of free-standing walls takes place to allow the earth pressure behind the walls to be calculated as the 'active pressure' case (see Chapter 9). Where the foundation of the wall is at a shallow depth below the lower ground level, the passive resistance to overturning or sliding is neglected since it may be destroyed by trenching in front of the wall at some future time.

The forces acting on a gravity and simple cantilever wall are shown in Figures 17.44(a) and (b). The force R is the resultant of the active earth pressure P_A and the weight of the wall W and backfill above the wall foundation. The surcharge on the fill behind the wall is allowed for when calculating P_A but is not included in the weight W. To prevent overturning of the wall the resultant R should cut the base of the wall foundation within its middle third, i.e. the eccentricity must not exceed $B/6$.

Having determined the position and magnitude of R, the bearing pressures at the toe and heel of the base are determined as described in section 17.2.5. These should not exceed the allowable bearing pressure of the ground, and the settlement at the toe should be within tolerable limits. Then the resistance to sliding of the base should be determined. If this is inadequate the base should be widened or taken down to a depth where the passive resistance in front of the wall may be safely mobilized (Figure 17.45).

Hydrostatic pressure behind the retaining walls should be avoided by the provision of a drainage layer behind the wall combined with weepholes and a collector drain (as shown in Figure 17.47).

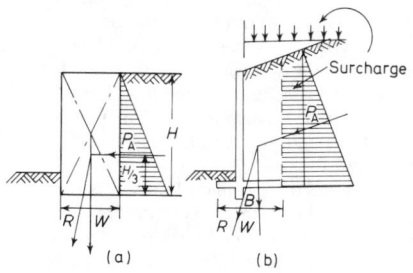

Figure 17.44 Forces acting on a free-standing retaining wall. (a) Simple gravity wall; (b) cantilever wall

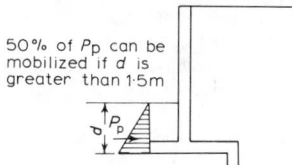

50% of P_p can be mobilized if d is greater than 1·5m

Figure 17.45 Passive resistance at toe of retaining wall

17.5.2 Gravity walls

Typical designs for gravity walls in brickwork, mass concrete and cribwork, are shown in Figure 17.46(a) and (b). Walls of these types are economical for retained heights of up to 2 to 3 m, or up to 5 m for cribwork walls. The width of the base should be about 0.40 to 0.65 times the overall height. For walls designed to present a 'vertical' appearance the front face should be battered back slightly say to 1 in 24 to allow for the inevitable slight forward rotation. The sloping wall and base (Figure 17.46(b)) provides the best alignment to resist earth pressure. Vertical joints in brick walls should be at 5 to 18 m and in concrete walls at 20 m centres or at some convenient length for a day's 'pour' of concrete. A preformed joint filler strip in bituminized fibre or PVC may be used.

Gravity walls of a type similar to that shown in Figure 17.46(b) can be built up from gabions (rectangular wire baskets filled with graded stone).

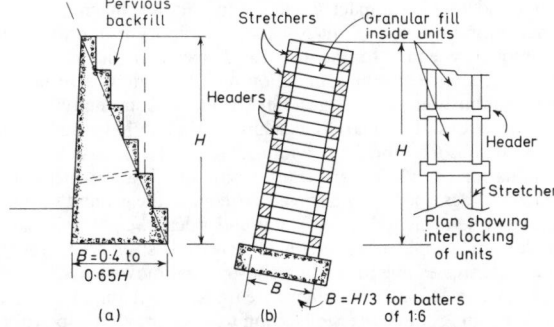

Figure 17.46 Designs for gravity retaining walls. (a) Mass concrete; (b) cribwork

17.5.3 Cantilevered reinforced concrete walls

A typical design for a cantilevered wall is shown in Figure 17.47. The projection of the base slab in front of the wall may be omitted if the wall face forms the boundary of the property but this arrangement should be avoided if at all possible because of the high pressure on the soil at the toe and the consequent risk of excessive forward rotation.

The design shown in Figure 17.47 is economical for heights of 4.5 to 6 m. The counterfort or buttressed types (see sections 17.5.4 and 17.5.5 respectively) should be used for higher walls.

The width of the base should be from 0.40 to 0.65 times the overall height of the wall. The minimum wall thickness should be 150 mm for single-layer reinforcement and 230 mm for front and back reinforcement. Although economy of concrete can result from progressive reduction in thickness of the wall section from the base to the top, a uniform thickness will give the lowest overall cost for walls up to 6 m high. A sloping or stepped-back face may show savings for higher walls.

The base slab thickness should equal the wall thickness at the stem of the latter. The projection in front of the wall should be about one-third the base width.

Expansion joints should be provided at spacings determined by the estimated thermal movement. A spacing of from 20 to 30 m is suitable for British conditions. The reinforcement should not be carried through these joints. Vertical contraction joints are required at 5 to 10 m spacing. The reinforcement may be carried through the contraction joints or stopped on either side. Where possible, construction (daywork) joints should coincide with expansion or contraction joints. The minimum cover to the reinforcing steel, appropriate in each case to the exposure conditions, is shown in Figure 17.47.

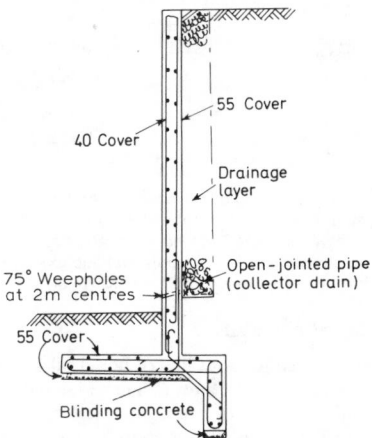

Figure 17.47 Design for reinforced concrete cantilever wall

17.5.4 Counterfort walls

The wall slab of counterfort retaining walls spans horizontally between the counterforts except for the bottom 1 m which cantilevers from the base slab. The counterforts are designed as T-beams of tapering section, and they are usually spaced at distances of one-third to one-half the height of the wall. The base of the counterfort must be well tied into the base slab. The latter acts as a horizontal beam carrying the surcharge load of the backfill and spanning between counterforts or from back beam to front beam. The counterforts transmit high bearing pressures to the ground at their front ends and may require piled foundations or a stiff front beam to distribute the pressure along the front of the wall.

17.5.5 Buttressed walls

Buttressed walls are economical for walls higher than 6 m designed to be cast against an excavated face, whereas the counterfort wall is more suitable where the ground behind the wall is to be raised by filling. The wall slab spans horizontally between the buttresses except for the bottom 1 m which cantilevers from the base slab. The buttresses act as compression members transmitting loading to the base slab or to piles on weak ground.

17.5.6 Tied-back diaphragm walls

The stages in constructing a tied-back wall in the form of a diaphragm wall are shown in Figure 17.48. In a stage I excavation the wall must be designed to cantilever from the stage I excavation level. For stage II excavations the wall spans between the anchorage level and the soil at the excavation line, similarly at stage III. At the latter stage the passive resistance of the soil in front of the buried portion must be adequate to prevent the wall moving forward at the toe, and the pressure beneath the base of the wall due to the vertical component of the anchor stress must not exceed the allowable bearing pressure of the soil.

The use of the tied-back wall as a basement retaining wall is described in section 17.3.2.5 and the design of ground anchors is discussed in Chapter 9. Guidance on the design of retaining walls of this type is given by Padfield.[30]

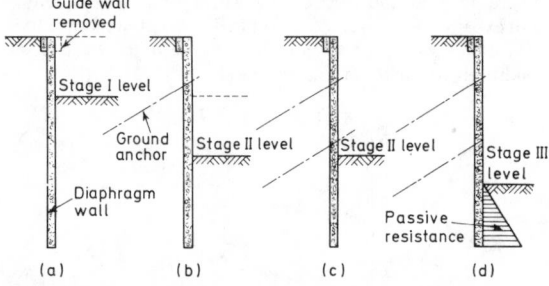

Figure 17.48 Stages in constructing a tied-back diaphragm wall. (a) Excavating to first stage in preparation for installing top-level ground anchors; (b) top-level anchors installed, excavation to second stage in preparation for installing bottom-level anchors; (c) bottom-level anchors installed; (d) excavation for third (final) stage

17.5.7 Contiguous bored pile walls

Retaining walls formed by a continuous line of bored piles can be designed as simple cantilever structures (Figure 17.43(f)) or as tied-back walls. Walls of this type are economical to construct by rotary auger drilling methods (see section 17.4.3.1) in self-supporting ground above the water table. In these conditions the piles can be installed merely as abutting units.

In water-bearing cohesionless soils the piles must interlock. If this is not done water and soil will bleed through the gaps causing loss of ground behind the wall. Interlocking is done by drilling and concreting alternate piles; then, by using a chisel to drill in the space between these piles, forming a deep groove in each of the latter. The drilled-out space is then filled with concrete to form the continuous wall. Construction in this manner is likely to cost more than the diaphragm wall.

17.5.8 Materials and working stresses

Concrete mixes and the quality of bricks or blocks should be selected as suitable for the conditions of exposure, attention being paid to frost resistance. Information on the durability of these materials in aggressive conditions is given in section 17.8.

The materials and working stresses for reinforced concrete should be in general accordance with BS 8110.

17.5.9 Reinforced soil retaining walls

Retaining walls can be constructed from soil which is reinforced to resist the internal tensile stresses which are induced by the horizontal movement towards the retained face of the wall.

There are two principal types of reinforced soil wall. In Figure 17.49(a), granular fill is brought up in compacted layers, each layer being reinforced by horizontal metal or plastic ties spaced at predetermined horizontal and vertical intervals. The vertical or steeply inclined face of the soil wall is retained by cladding panels which are secured to the ends of the ties. These panels may be constructed in precast concrete, metal or plastics and they can be preformed to a patterned profile to give a decorative effect to the finished wall.

In Figure 17.49(b), granular fill is placed on sheets of woven plastic mesh and compacted to form a thick bottom layer. The leading edge of the mesh is then folded back over the fill layer and a second sheet is placed on it followed by a second and successive layers of fill, each layer being partly wrapped by the sheets of mesh. The latter act as horizontal reinforcement restraining the fill from spreading outwards and as a means of retaining the steep outer face of the wall. Protection to the face can be given by precast concrete blocks, hand-placed stone pitching or turf. The mesh is designed to have tensile strength principally in the direction of horizontal forces induced by earth movements.

Reinforced soil walls for temporary works have been constructed by layers of scrap motor tyres lashed together by wire rope with granular fill placed in layers in the interstices between the tyres.

Reinforced soil retaining walls have the advantage of a high degree of flexibility which makes them suitable for retaining the face of deep cuttings where considerable heave and lateral movement may take place as a result of stress relief after excavating for the cutting. Walls of this type are also suitable for use in mining subsidence areas and in retaining the toe of embankments built on sloping ground.

The principles of reinforced soil have been stated by Jones.[31]

17.6 Foundations for machinery

17.6.1 General

In addition to their function of transmitting the dead loading of the installation to the ground, machinery foundations are subjected to dynamic loading in the form of thrusts transmitted by the torque of rotating machinery or reactions from reciprocating engines. Foundations of presses or forging hammers are subjected to high impact loading and rotating machinery induces vibrations due to out-of-balance components vibrating at a frequency equal to the rotational speed of the machine. Thermal stresses in the foundation may be high as a result of fuel combustion, exhaust gases or steam, or from manufacturing processes. Foundation machinery should have sufficient mass to absorb vibrations within the foundation block, thus eliminating or reducing the transmission of vibration energy to surroundings; they should spread the load to the ground so that excessive settlement does not occur under dead weight or impact forces and should have adequate structural strength to resist internal stresses due to loading and thermal movements.

Machinery foundation blocks are frequently required to have large openings or changes of section to accommodate pipework or other components below bedplate level. These openings can induce high stresses in the foundation block due to shrinkage

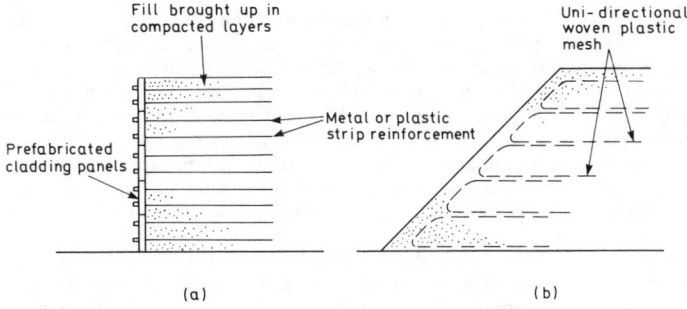

Fill brought up in compacted layers

Uni-directional woven plastic mesh

Metal or plastic strip reinforcement

Prefabricated cladding panels

(a) (b)

Figure 17.49 Reinforced soil construction. (a) Gravity-type retaining wall reinforced with strips of metal or plastic; (b) embankment reinforced wit woven plastics mesh

combined with other effects. Abrupt changes of section should be avoided, and openings should be adequately reinforced.

17.6.2 Foundations for vibrating machinery

When the frequency of a foundation block carrying vibrating machinery approaches the natural frequency of the soil, resonance will occur and the amplitude may be such as to cause excessive settlement of the soil beneath the foundation, or beneath other foundations affected by the transmitted wave energy. This is particularly liable to occur with foundations on loose granular soils. Knowing the weight of the machine and its foundation and the vibration characteristics of the soil, it is possible to calculate the resonant frequency of the machine–foundation–soil system. The frequency of the applied forces ideally should not exceed half of this resonant frequency for most reciprocating machines and should be at least 1.5 times the resonant frequency for machinery having frequencies greater than the natural frequency. If the applied frequencies are within this range there is a danger of resonance and excessive amplitude. These criteria are recommended by Converse[32] who describes various mathematical theories for calculating natural frequency and amplitude, and tabulates recommended ratios of foundation weight: engine weight for various types of machinery. The aim in design generally is to provide sufficient mass to absorb as much of the energy as possible within the foundation block and to proportion the block in such a manner that energy waves are reflected within the mass of the block or transmitted downwards rather than transversely in order not to affect adjacent property. In some cases it may be advantageous to mount the foundation block on special mountings such as rubber carpets or rubber–steel sandwich blocks.

17.6.3 Foundations for turbo-generators

The foundation blocks for large turbo-generators are complex structures subjected to periodic reversing movements due to differential heating and cooling of the concrete structures, moisture movements related to ambient humidity, steam and water leakage and to dynamic strains within the elastic range. They are also subjected to progressive movements resulting from long-term settlements of the foundation soil and from shrinkage and creep of concrete. These movements may be of sufficient magnitude to cause misalignment of the shafts of the machinery.[33]

17.7 Foundations in special conditions

17.7.1 Foundations on fill

If granular fill can be placed in layers with careful control of

compaction, the resulting settlement due to the foundation loading and the settlement of the fill under its own weight will be small. Provided the fill has been placed on a relatively incompressible stratum the settlement of the structure will be little if anything greater than would occur with a foundation on a reasonably stiff or compact natural soil.

However, in most cases of construction on filling, the material has probably not been placed under conditions of controlled compaction but has been loosely end-tipped, and the age of the fill may not be known with certainty. However, it is usually possible to obtain a good indication of the constituents of the fill and its state of compaction from observations in boreholes and trial pits (preferably the latter). From these observations an estimate can be made of the likely remaining settlement due to consolidation of the fill under its own weight and that of the superimposed loading. Reference should be made to *Building Research Establishment Digest* Number 274[34] for information on the amount and rate of settlement of various types of fill material.

For shallow granular fills, strip or pad foundations are suitable for most types of structure. For deeper granular fills which have not had special compaction, it will be necessary to use raft foundations for structures which are not very sensitive to differential settlement, or piled foundations for structures for which small settlements must be avoided.

Ordinary shallow foundations can be used on hydraulically placed sand fill where this can consolidate by drainage but not when the fill has been allowed to settle through water. Piled foundations are necessary for structures on hydraulically placed clay fill or on domestic refuse.

Raft or piled foundations can be avoided on loose granular fills if one of the ground treatment processes described in section 17.2.7 is adopted.

Where piled foundations are used in fill areas, consolidation of the fill and of any underlying natural compressible soil will cause dragdown forces on the pile shafts which must be added to the working load from the superstructure.

Where bored piles are used through the fill the dragdown or negative skin friction forces may be very high, and for economy it may be desirable to minimize the dragdown by adoption of slender preformed sections, e.g. high-strength precast concrete, or to surround the pile shaft with a sleeve or a layer of soft bitumen.

Fills consisting of industrial wastes may contain substances which are highly aggressive to buried concrete or steelwork.

17.7.2 Foundations in areas of mining subsidence

17.7.2.1 General

Cavities are formed where minerals are extracted from the

ground by deep mining or pumping. In time, the ground over the cavities will collapse wholly or partly filling the void. This leads to subsidence of the ground surface. Movements of the surface may be large both in a vertical direction and in the form of horizontal ground strains and the foundations of structures require special consideration to accommodate these movements without resulting damage to the superstructure. The majority of foundation problems in the UK are due to coalmining, but subsidence can occur due to extraction of other minerals such as brine.

In the nineteenth century and earlier, coal was extracted by methods known variously as 'pillar-and-stall', 'room-and-pillar' and 'bord-and-pillar'. The galleries were mined in various directions from the shaft followed by cross-galleries leaving rectangular or triangular pillars of coal to support the roof above the workings (Figure 17.50).

The current method of coalmining is by 'longwall' methods in which the coal seam is extracted completely on an advancing face (Figure 17.51). The amount of subsidence at ground level is less than the thickness of coal extracted owing to bulking of the collapsed strata.

The problems of foundations of buildings on old mine workings are discussed by Healy and Head.[35]

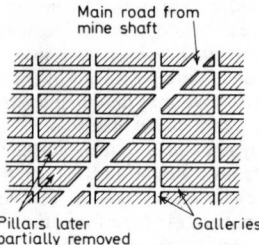

Figure 17.50 'Pillar-and-stall' mineworkings

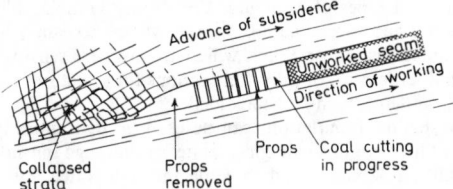

Figure 17.51 Extraction of coal by the longwall method

17.7.2.2 Foundation design in areas of pillar-and-stall workings

The risk of collapse depends on the conditions of the 'roof' over the workings. Where this consists of weak or broken rock, stage collapse will occur at some time and the void formed will gradually work its way up to the ground surface to form a 'crown hole' (Figure 17.52(a)). If, however, the roof is a massive sandstone it will bridge over the cavity for an unlimited period of years (Figure 17.52(b)). However, the pillars of coal may suffer slow deterioration at an unpredictable rate.

In considering the design of foundations over workings of this type an appraisal is made of the general geological conditions. Where the collapse of overburden strata or coal pillars could result in severe local surface subsidence, precautions against these effects must be taken. Methods which may be considered are:

(1) Filling the workings by injection techniques; or
(2) Constructing piled or deep shaft foundations to a founding level below the workings.

Method (1) is used where the workings are at such a depth that method (2) is uneconomical. No attempt is made to locate individual galleries but the area of the structure is ringed by a double row of injection holes at close spacing. Gravel or a stiff sand–cement grout is fed down these holes to form a barrier in the voids of the worked seam. Holes are then drilled on a nominal grid in the space within the barrier and low-cost materials are fed down these holes to fill all accessible voids. These materials may consist of sand–pulverized fuel ash–water slurry, or a lean sand–pulverized fuel ash–cement grout.

Where deep shaft or piled foundations (method (2)) are used the shaft is sleeved where it passes through the overburden to prevent transference of load to the foundation in the event of subsidence. The outer lining forming the sleeve must be strong enough to resist lateral movement caused by subsidence. Where structures are to be built on soft compressible soils overlying mine workings, piled foundations bearing on a thin cover of rock strata above the workings must not be used since the toe loading from the piles may initiate subsidence. Buoyancy raft foundations should be used (see section 17.3.3).

17.7.2.3 Foundation design in areas of longwall workings

In the case of current or future workings, subsidence is inevitable and the degree to which precautions are taken in foundation design depends on the type and importance of the structure under consideration.

It will be seen from Figure 17.53 that as the subsidence wave crosses a site the ground surface is first in tension and then in compression. As subsidence ceases, the residual compression strains die out near the surface. The simplest form of construction is a shallow reinforced concrete raft. This is usually adopted for houses for which the cost of repairs due to distortion of the raft can be kept to a reasonable figure.

Points to note in the design of raft foundations are:

(1) The underside of the raft should be flat, i.e. it should not be keyed into the ground.
(2) A slip membrane is provided beneath the raft to allow ground strains to take place without severe compression or tension forces developing in the substructure.
(3) Reinforcement is provided in the centre of the slab to resist bending stresses caused either by hogging or sagging.

A raft may not be suitable for heavy structures such as bridges or factories. In these cases, the principle to be adopted is to use bearing pressures as *high* as possible, so minimizing the foundation area and, hence, the horizontal tension and compression forces transmitted to the superstructure. If the layout permits, the structure should be supported on only three bases to allow it to tilt without distortion.

Trenching around a structure can be used to reduce compressive strain but this method is ineffective in countering tension strains.

17.7.2.4 Foundations adjacent to existing shafts

Foundation problems may arise owing to the collapse of deteriorated shaft linings followed by surface subsidence. The type of material for filling the shaft should be ascertained. If it is granular, the loose material and any cavities can be consolidated by injection of a low-cost grout. If the infill consists of clay, grouting may be ineffective. However, in this case the shaft may be capped with a reinforced-concrete slab. The latter method should be adopted only if the shaft lining is sound and durable over its full depth. If not, or if for reasons of safety the condition of the lining cannot be ascertained, the shaft should be surrounded by a ring of bored piles or by a diaphragm wall taken down to a stable stratum.

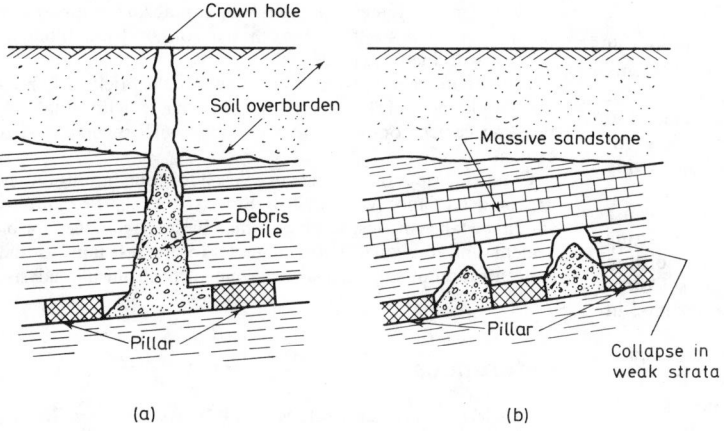

Figure 17.52 Subsidence due to collapse of cavities in mineworkings. (a) Weak strata over coal seam; (b) strong 'roof' over coal seam

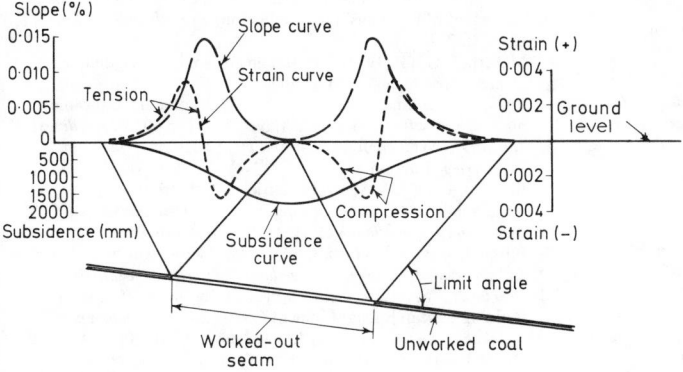

Figure 17.53 A form of subsidence above longwall workings

17.8 The durability of foundations

17.8.1 General

Foundation materials are subjected to attack by aggressive compounds in the soil or groundwater, living organisms and mechanical abrasion or erosion. The severity of the attack depends on the concentration of aggressive compounds, the level of and fluctuations in the groundwater table or the variation in tidal and river levels and on climatic conditions. Immunity against deterioration of foundations can be provided to a varying degree by protective measures. The protection adopted is usually a compromise between complete protection over the life of the structure, and the cheaper partial protection while accepting the possible need for periodic repairs or renewals. Problems of durability of a wide range of materials and the appropriate protective measures have been reviewed by Barry.[29] Methods of protection of some foundation structures are described in the following sections.

17.8.2 Timber

Timber piles are liable to fungal decay if they are kept in moist conditions, i.e. above the groundwater level. Piles wholly below the water level, if given suitable preservative treatment, can perform satisfactorily for a very long period of years. Properly air-seasoned timber, if kept wholly dry, i.e. moisture content less than 22%, will also remain free of decay for an indefinitely long period. The best form of protection against fungal attack and termites is pressure treatment with coaltar creosote or the copper (chrome) arsenic-type waterborne preservative. Creosote protection should be applied in accordance with BS 913 and the waterborne type to BS 4072.

Timber piles in marine structures are liable to destruction by molluscan and crustacean borers which inhabit saline or brackish waters. Although preservative treatment gives some protection against these organisms, the longest life is given by a timber known to be resistant to their depredations. Greenheart, jarrah and blue gum are suitable for cold European waters. In other countries, billian in the China seas, turpentine in New South Wales, black cypress and ti-tree in Queensland, spotted gum in Tasmania and teak in India have been found to have some immunity.

Protection can be given by jacketing piles in concrete before driving, or by gunite mortar after installation. Concrete can also be used in land foundations either in composite concrete–timber piles, or in deep pile caps down to groundwater level (see Figure je17.32, page **17/19**).

17.8.3 Metals

Protection can be given to steel piles by impervious coatings of bitumen, coaltar, pitch or synthetic resins but these treatments are not effective for piles driven into the ground since the coatings are partly stripped off. It is the normal practice to provide sufficient cross-sectional area of steel to allow for wastage over the useful life of the structure while still leaving enough steel to keep the working stresses within safe limits.

In undisturbed cohesive soils corrosion is negligible because of the absence of oxygen. There may be some local pitting corrosion near the ground surface where the capillary moisture zone is mobile and replenished with oxygenated waters. Corrosion generally is low or negligible below groundwater level in natural soils, again because of the absence of oxygen.

Morley[36] quotes a corrosion rate of 0.08 mm/yr for unprotected low-alloy steel in static sea-water, and 0.1 to 0.25 mm/yr in the splash zone. He quotes average corrosion rates of 0.1 to 0.2 mm for corrosion in an industrial atmosphere in the UK.

In severe conditions, i.e. in polluted ground, it may be necessary to adopt a system of cathodic protection. Steel piles in river and marine structures can be protected above the soil line by heavy coatings of coaltar, bituminous enamel, epoxy pitch and vinyl pitch. However, these coatings are liable to damage by floating objects or barnacle growth and cathodic protection is necessary in marine structures if a long life is desired.

Cast iron has a similar corrosion resistance to mild steel and protective coatings provide the best method of treatment of substructures such as cylinder foundations constructed from cast-iron segments.

17.8.4 Concrete

The principal cause of deterioration of concrete in structures below ground level is attack by sulphates in the soil or groundwater. Sulphates occur naturally in some soils and in peats. They occur in sea-water at a concentration of about 230 parts per 100 000 which is greatly in excess of the figure regarded as marginal between nonaggressive and aggressive. However, because of the inhibiting effect of the chlorides in sea-water the sulphates do not cause an expansive reaction to normal Portland cement concrete if it is of good quality and well compacted. However, it is a good idea as a precaution to use sulphate-resisting cement or Portland blast-furnace cement in *reinforced* concrete structures immersed in sea-water.

Concentrations of sulphates may be high in industrial wastes, particularly in colliery wastes and some blast-furnace slags. Where fill material contains industrial wastes a full chemical analysis should be made to identify potentially aggressive compounds.

The precautions to be taken to protect concrete substructures are listed in Building Research Establishment Digest Number 250[37] and also in BS 8004 but these recommendations do not give much consideration to the workability required for the recommended concrete mixes for the particular placing conditions. Guidance on this aspect is given by Tomlinson.[2]

In normal climatic conditions in the UK, concrete at a depth greater than 300 mm is unlikely to suffer disintegration due to frost expansion. In severe conditions of exposure a dense concrete mix should be used with a water:cement ratio of less than 0.5. If the ratio is between 0.5 and 0.6 there is a risk of frost attack and above 0.6 the risk becomes progressively greater.

The required cover of steel reinforcement to prevent corrosion of the steel for various exposure conditions is listed in BS 8110.

17.8.5 Brickwork

Bricks with a high absorption should be avoided since they are liable to frost disintegration, and they can absorb sulphates or other aggressive substances from the soil or from filling-in contact with the brickwork.

In sulphate-bearing soils or groundwater the brickwork mortar should be a 1:3 cement:sand mix made with sulphate-resisting cement or in severe conditions with supersulphated cement.

Concrete bricks or blocks may be used for foundations if they are in accordance with British Standard 1180. Precautions should be taken against sulphate attack by specifying the type of cement and the quality of concrete to be resistant to the concentration of sulphates as determined by chemical analysis.

References

1 Padfield, C. J. and Sharrock, M. J. (1983) *Settlement of structures on clay soils.* Construction Industry Research and Information Association Special Publication Number 27/PSA (Civil Engineering Technical Guide Number 38) pp. 67–70.

2 Tomlinson, M. J. (1986) *Foundation design and construction*, (5th edn), Longman, Scientific and Technical, Harlow.

3 Skempton, A. W. and MacDonald, D. H. (1956) 'The allowable settlement of buildings' (and discussion) *Proc. Instn Civ. Engrs*, **5** (Part 3), 727–784.

4 Meyerhof, G. G. (1947) 'The settlement analysis of building frames', *Struct. Engnr*, **25**, 9, 309.

5 Polshin, D. E. and Tokar, R. A. (1957) 'Maximum allowable nonuniform settlement of structures', Vol. I, p.402 *Proceedings, 4th International conference on soil mechanics and foundation engineering*, London.

6 Bjerrum, L. (1963) 'Allowable settlement of structures', *Proceedings, 3rd European conference on soil mechanics and foundation engineering*, Vol. II, pp.16–17. Wiesbaden.

7 Burland, J. B. and Wroth, C. P. (1975), 'Settlement of buildings and associated damage', *Proceedings, conference on settlement of structures*, Cambridge, Pentech Press, London, p.611–54.

8 Building Research Establishment (1980) *Low-rise buildings on shrinkable clay soils* (Part 2) Digest Number 241, BRE, Watford.

9 Driscoll, R. (1983) 'The influence of vegetation on the swelling and shrinkage of clay soils in Britain', *Géotechnique*, **33**, 93–105.

10 Institution Structural Engineers (1978) *Structure–soil interaction*, ISE, pp.43–57.

11 Hooper, J. A. (1984) 'Raft analysis and design – some practical examples', *Struct. Engnr*, **62A**, 8.

12 Poulos, H. G. and Davis, E. H. (1974) *Elastic solutions for soil and rock mechanics.* Wiley, New York.

13 Bredenberg, H. and Broms, B. B. (1983) 'Lime columns as foundations for buildings', *Proceedings, Conference on advances in piling and ground treatment for foundations.* Institution Civil Engineers, pp.95–100.

14 Greenwood, D. A. and Kirsch, K. (1983) 'Specialist ground treatment by vibratory and dynamic methods', *Proceedings, Conference on advances in piling and ground treatment for foundations*, Institution Civil Engineers, pp.17–45.

15 Hooper, J. A. (1979) *Review of behaviour of piled raft foundations*, Construction Industry Research and Information Assocation Report 83. CIRIA, London.

16 Chandler, J. A., Peraine, J. and Rowe, P. W. (1984) 'Jamuna River, 230 kV Crossing, Bangladesh: Construction of Foundations', *Proc. Instn. Civ Engrs*, **76**, 1, 965–984.

17 Terzaghi, K. and Peck, R. B. (1967) *Soil mechanics in engineering practice* (2nd edn), John Wiley, Chichester, p.563.

18 Riggs, L. W. (1965) Tagus river bridge – tower piers', *Civ. Engng* (USA) (Feb.) 41–45.

19 Wilson, W. S. and Sully, F. W. (1949) *Compressed air caisson foundations*, Institution Civil Engineers. ICE, London. Works Construction Paper Number 13.

20 Walker, D. N. (1982) *Medical code of practice for work in compressed air*, Construction Industry Research and Information Association Report Number 44 (3rd edn) CIRIA, London.

21 Weltman, A. J. and Little, J. A. (1977) *A review of bearing pile types.* Construction Industry Research and Information Association Report Number PG1. CIRIA, London.

22 Cole, K. W. (1972) 'Uplift of piles due to driving displacement', *Civ. Engng and Pub. Works Rev.*, **67**, 788, 263–269.

23 Tomlinson, M. J. (1986) *'Pile design and construction practice'* (3rd edn) Viewpoint Publications, London.

24 Elson, W. K. (1984) *Design of laterally loaded piles*, Construction Industry Research and Information Association Report Number 103. CIRIA, London.

25 Clarke, J. L. (1973) 'Behaviour and design of pile caps with four piles', Cement and Concrete Association Report Number 42.489. C & CA, London.

26 Whittle, R. T. and Beattie, D. (1972) 'Standard pile caps', *Concrete*, **6**, 1, 34–36 (January) and **6**, 2, 29–31 (February).

27 Weltman, A. J. (1977) *Integrity testing of piles – a review*. Construction Industry Research and Information Association Report Number PG4.

28 Goble, G. G. and Rausche, F. (1979) 'Pile driveability predictions by CAPWAP', *Proceedings, Conference on numerical methods in offshore piling*, Institution of Civil Engineers, London, pp.29–36.

29 Barry, D. L. (1983) *Material durability in aggressive ground*, Construction Industry Research and Information Association, CIRIA, London. Report Number 98.

30 Padfield, C. J. and Mair, R. J. (1984) *Design of retaining walls embedded in stiff clays*, Construction Industry Research and Information Association Report Number 104. CIRIA, London.

31 Jones, C. J. F. P. (1985) *Earth reinforcement and soil structures*. Butterworth. CIRIA, London.

32 Converse, F. J. (1962) 'Foundations subjected to dynamic forces', In: *Foundation engineering*, McGraw-Hill, Maidenhead, pp.769–825.

33 Fitzherbert, W. A. and Barnett, J. H. (1967) 'Causes of movement in reinforced turbo-blocks and developments in turbo-block design and construction', *Proc. Instn Civ. Engrs*, **36**, 351–393.

34 Building Research Establishment (1983) *Fill*, Part I: 'Classification and load-carrying characteristics', BRE Digest Number 274, BRE, Watford.

35 Healy, P. R. and Head, J. M. (1984) *Construction over abandoned mine workings*, Construction Industry Research and Information Association Special Publication Number 32.

36 Morley, J. (1979) *The corrosion and protection of steel piling*. British Steel Corporation, Report NumberIV T/CS/1115/1/79/C.

37 Building Research Establishment (1981) *Concrete in sulphate-bearing soils and groundwater*. BRE Digest Number 250, BRE, Watford.

Bibliography

CODES OF PRACTICE AND STANDARDS

INSTITUTION OF STRUCTURAL ENGINEERS
CP 2:1951 *Earth retaining structures*
BRITISH STANDARDS INSTITUTION
BS 6399 Loading for buildings
CP 101 *Foundations and substructures for non-industrial buildings of not more than four stories*
CP 102 'Protection of buildings against water from the ground'
BS 8110, *The structural use of concrete*
CP 112, Part 1 1967: Part 2 1971
BS 4978, *Timber grades for structural use*
BS 8004, *Foundations*
BS 5573, *Safety precautions in the construction of large-diameter boreholes for piling and other purposes*

BS 449, *The use of structural steel in building*
BS 913, *Pressure creosoting of timber*
BS 988, 1076; 1097; 1451, *Mastic asphalt for building (limestone aggregates)*
BS 1162, 1410; 1418, *Mastic asphalt for building (natural rock asphalt aggregates)*
BS 1180, *Concrete bricks and fixing bricks*
BS 4072, *Wood preservation by means of waterborne copper/chrome/arsenic compositions*
BS 4360, *Weldable structural steels*
BS 4449, *Hot-rolled steel bars for the reinforcement of concrete*
BS 4461, *Cold-worked steel bars for the reinforcement of concrete*
BS 5930, *Code of practice for site investigations*
BS 6031, *Code of practice for earthworks*

18

Dams

A D M Penman DSc, CEng, FICE
Geotechnical Engineering Consultant

Contents

18.1 Definition

In the UK, the name 'dam' is given to a civil engineering structure built across a valley to form an artificial lake as a reservoir of water. There are numerous variants. Some reservoirs are formed on relatively flat land by building long dams to encircle the required areas. Others are built to store materials other than water. In South Africa and some other countries the word 'dam' is used for the reservoir which is retained by a 'wall' or 'dam wall'.

18.1.1 Types of dam

Dams are separated into two main types by the choice of material used for their construction: (1) embankment; and (2) concrete dams.

(1) *Embankment dams* are made from nonorganic particulate material excavated from the Earth's surface local to the dam site and used more or less as excavated. They are subdivided into earthfill and rockfill dams, although many embankment dams contain both types of fill. Further subdivisions can be made, according to the material used, to make the waterproof element, e.g. central clay core, sloping clay core or upstream membrane of asphalt or reinforced concrete.

(2) *Concrete dams* are made from a carefully selected and processed harder fraction of this material, bound together and strengthened by an hydraulic cement. They are subdivided according to their mechanism for remaining stable.

  (a) *gravity dams*: these are the simplest because they rely on gravitational force to oppose the overturning moment caused by the pressure of the reservoir water on their upstream faces.

  (b) *hollow gravity dams*: these require less concrete and therefore cost less to construct. Foundation requirements are more critical.

  (c) *buttress dams*: these also require less concrete than gravity dams. The buttresses support the upstream face of a buttress dam. The upstream edges of the buttresses are commonly widened so that they join, forming the contiguous buttress dam. As an alternative, the upstream face may consist of small arches between buttresses, *forming a multi-arch dam.*

  (d) *arch dams*: these may be constructed as a whole in one large arch, spanning the valley sides and relying on them to carry the very large thrusts caused by reservoir water pressure. This type is the most sophisticated of the concrete dams and may be subdivided into single-curvature and double-curvature, according to whether the vertical section is straight, or is curved to further reduce bending moments in the concrete.

18.2 Brief history

Dams have made a major contribution to the development of our civilization. Their earliest role was to provide storage for irrigation water; now, they also provide hydro-electric power and water for industry and large cities.

Early dams were all of the embankment type, of necessity built from the earth and stones found at the site. Helms (quoted by Kerisel)[1] states that the oldest dam in the world is at Jawa in Jordan, dating from about 4000 B.C. and was built of earth with a masonry facing. Perhaps the second-oldest is the Sadd el-Kafara on the Wadi el-Garawi near Helwan in Egypt, built about 2900 B.C. It was 11 m high with upstream and downstream rubble masonry walls, each 24 m wide at their bases, separated by a central earthfill section 36 m wide.

Rao[2] reports that there was a tradition of dam building in India where it was once considered as one of the seven meritorious acts which a man ought to perform during his lifetime. During the period of British tenure, many embankment dams were constructed by traditional methods and were accepted as a means of famine relief, giving employment to thousands. According to Buckley,[3] the completed schemes were not only profitable, but brought happiness and contentment to the people by ensuring reliable crop production.

In the UK the industrial revolution required water for transport, industrial processes and a growing population. In the eighteenth century, dams were built to store water for canals; during the nineteenth, the majority were for water supply, and early in the twentieth century, dams were built specifically to provide power for aluminium smelting in Scotland and Wales.

Before 1800, almost all the world's dams were of the embankment type. During the nineteenth century, concrete technology and methods of structural analysis were developed. The many sites then available with sound rock foundations at shallow depth created increasing interest in concrete dams.

The increasing size of the human population of the world, together with ever-rising demands for irrigation water and power, caused rapid increase in the number of dams built. Figure 18.1 shows, from 1800 to the present, the increase in world population, together with the number and height of embankment dams. Although the numbers of dams increased in response to the rise in population, the number of embankment dams fell during the second half of the nineteenth and early twentieth centuries due to the number of concrete dams being built in that period. The proportion of the total that were embankment dams being built during any 5-yr period fell to a minimum of about 30% by the end of the first quarter of this century.

Since that period, the proportion has increased until currently more than 80% of dams being built are embankment dams. The highest dam in the world (Nurek, 300 m) is of the embankment type and is soon to be exceeded by another (Rogan) which will be 325 m high when complete. Both are in the Soviet Union.

This reversal of trend can be attributed to:

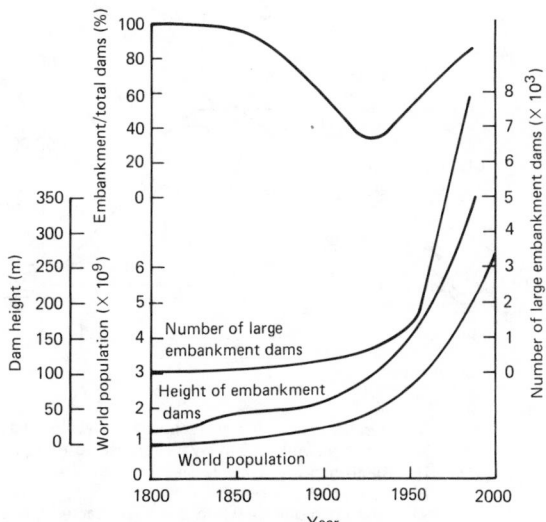

Figure 18.1 Numbers of people and embankment dams, 1800–1985. Curves also show heights of embankment dams and their proportion of total world dams

(1) Improved understanding of the behaviour of embankment dams due to advances in the art and science of soil mechanics since publication of Terzaghi's *Erdbaumechanik* in 1925.

(2) Increased capacity of earthmoving machinery after the introduction of the internal combustion engine and caterpillar tracks.

(3) Reduction in the number of sites with bedrock at shallow depth suitable for concrete dams.

(4) Increasing cost of labour.

At present, even at sites with strong rock near the surface suitable for concrete dams, e.g. the Lesotho Highlands schemes, embankment dams are often found to be cheaper. They can tolerate relatively poor foundation conditions and, when combined with their low price, this makes them a most attractive option.

18.3 Embankment dams

18.3.1 Introduction

Embankment dams can be subdivided according to type and position of the waterproof element. In this section, a description of the salient features of each type will precede an actual example.

18.3.2 Rockfill dam with upstream reinforced concrete membrane

This type is currently rising in prominence and so will be described first. (This does not mean that it is more important than other types of embankment dam.)

The Foz do Areia (160 m) on the River Iguaçu in Brazil is the world's highest dam of this type and is a typical example (Figures 18.2 and 18.3). It was required for hydro-electric power and river control. The region of the dam site is made up of medium to thick basaltic flows of 25 to 55 m depth. As is common with volcanic rocks, the interface between successive flows forms a zone more pervious than the massive basalt. The site investigation showed that about 70% of total volume was predominantly dense basalt with about 30% consisting mostly of basaltic breccia. Table 18.1 gives some geomechanical properties of the rocks.

More than 14 million m³ of rock excavation was required for tunnels, power station and spillway chute. This excavated rock was used as fill for the dam. It was placed in layers up to 1.6 m thick, sluiced with water at a rate of 25% volume of placed rock and compacted by four passes of a 10-t vibrating roller. Details of layer thickness and zones of rockfill are given in Figure 18.2, showing a cross-section of the dam.

Fragmentation of the hard rock by blasting produced a fill that was rather too uniform in size. To reduce compressibility under the upstream membrane, a transition zone of rockfill crushed to a specified grading with a maximum particle size of 150 mm was placed in 400 mm layers, compacted by the main 10-t vibrating rollers. The upstream face of this zone was smoothed and coated with a bitumen emulsion covered with sprayed sand to prevent erosion and facilitate compaction. This was achieved by six passes with the smooth 10-t roller pulled by cable up and down the slope, vibration only being used while moving upwards.

Great care was exercised in constructing the plinth to ensure a satisfactory water-tight connection between upstream membrane and valley sides. The plinth section is shown in Figure 18.4.

The reinforced concrete membrane was made 800 mm thick at the base, tapering to 300 mm at the top of the dam. The main part of it was cast as slabs 16 m wide, placed in slipforms. Before slipforming could start, the bottom piece of each strip, at its junction with the plinth, had to be formed in fixed shutters to produce a square end for the slipforms.

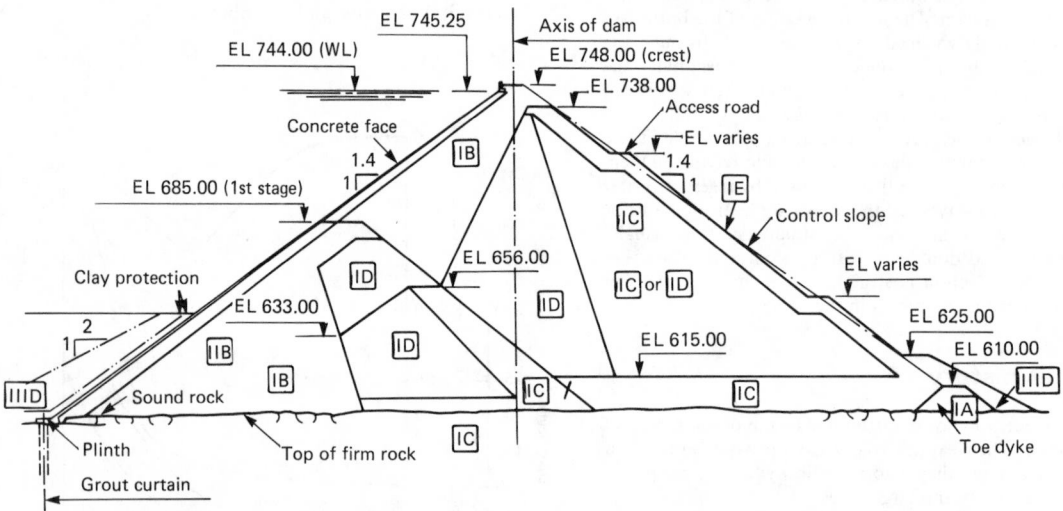

1C   Basalt rockfill. 1.6 m layers compacted four passes 10 t vibrating roller water 0.25 fill volume
1B   Basalt rockfill. 0.8 m layers compacted four passes 10 t vibrating roller water 0.25 fill volume
1E   Basalt rockfill. selected pieces placed
1A   Basalt rockfill. dumped
1D   Basalt and breccia 0.8 m layers compacted four passes 10 t vibrating roller water 0.25 fill volume
11B Well-graded crushed basalt maximum size 150 mm 400 mm layers
111D Impervious earthfill 300 mm layers

Figure 18.2 Foz do Areia: cross-section

Figure 18.3 Foz do Areia

Table 18.1 Foz do Areia: geomechanical properties of bedrock

		Dense basalt	Basaltic breccia
Specific gravity		2.80	2.30
Porosity (%)		1.30	11.80
Compressive strength (MN/m²)	Dry	233	37
	Saturated	190	25
Modulus of elasticity (MN/m²)	Dry	66 690	25 500
	Saturated	64 725	23 550
Soundness test			
Sodium sulphate (% of loss)	Coarse aggregate	2	50
	Fine aggregate	5	35
Los Angeles abrasion test			
(Type E grading, 1000 rev.) (%)		11	20

Construction of the dam began early in 1977 and the dam had reached full height in 1979. Impounding began in March 1980. Details of the design and performance of this dam have been given by Pinto, Materon and Marques.[4]

A list of six other concrete-faced, compacted-rockfill dams and four proposed dams is given in Table 18.2.

18.3.3 Rockfill dam with upstream asphaltic membrane

The superior flexibility of asphalt makes it appear attractive as material for an upstream membrane. Its use avoids the construction of contraction joints and the reinforcement of a concrete membrane, but considerable care has still to be exercised at the plinth, where excessive relative movements can tear an asphaltic membrane and permit leakage. Asphaltic membranes have been used on dams up to 83 m high. Slopes vary from about 1:1.5 to 1:2 (Figures 18.5 and 18.6).

Winscar dam (54 m) was constructed on the River Don in South Yorkshire between 1972 and 1975. The site is in the Millstone Grit series which includes alternating bands of sandstone and shale. The shale was moderately strong *in situ*, but deteriorated rapidly on exposure and had to be immediately covered with a fine granular fill where it was encountered at formation level. The sandstone, particularly those seams known as the Huddersfield White rock, was most stable and was used for the rockfill.

The pronounced fissure structure of this rock tended to control the shape and size of the pieces produced in the quarry by blasting. Fragmentation produced plenty of the finer sizes and the resulting rockfill was relatively well graded. It was compacted in 1.7 m layers by four passes of a 13.5 t vibrating smooth roller. Water was added to the rockfill when necessary to bring the water content up to 6%.

A selected finer fraction of the fill was placed under the 1:1.7 upstream slope and rolled with the 13.5-t roller, hauled by cable from the crest, but without use of vibration. The surface was sprayed with a tackcoat of bitumen before being covered with a levelling layer of porous asphalt placed by a paving machine. It was followed by two layers of dense asphalt, placed with staggered joints and compacted by small, smooth vibrating rollers. It was covered with a sealcoat of bitumen and treated with a white finish above low-water level as a protection from sunshine. Figure 18.6 shows conditions while the membrane was being placed.

18.3.4 Rockfill dam with central asphaltic core

In the absence of any suitably impervious fill that could be used for a core, asphalt has been used to construct a central water-proof element.

The practice may be said to have started with the construction of the 45 m high Vale de Gaio dam in Portugal in 1948. A tightly packed rubble stone wall was formed to a slope of 1 (verti-

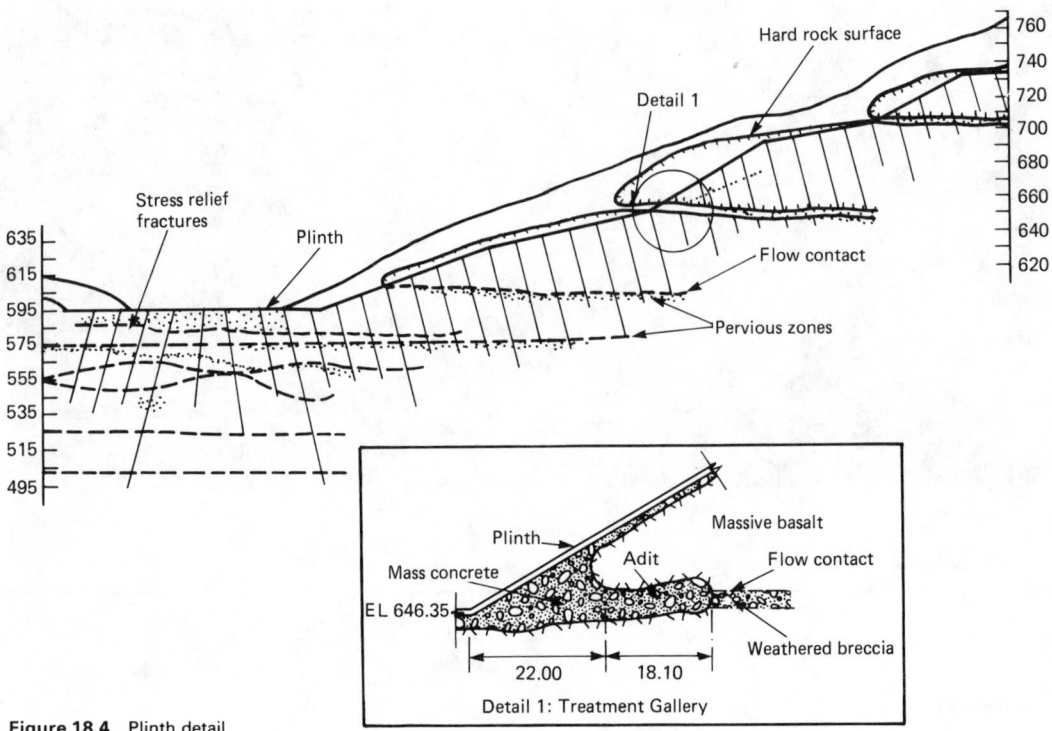

Figure 18.4 Plinth detail

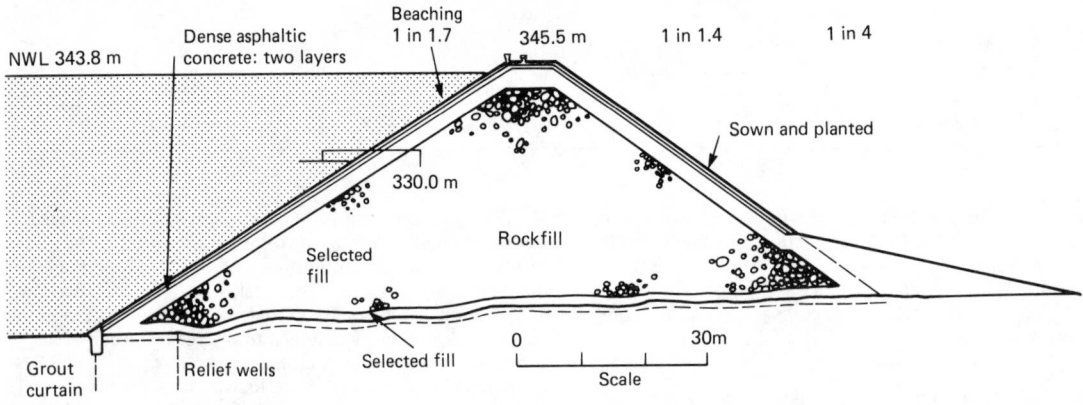

Figure 18.5 Winscar dam: cross-section

Table 18.2 Concrete-faced compacted-rockfill dams

Name	Country	Height (m)	Year completed
Foz do Areia	Brazil	160	1980
New Exchequer	USA	150	1966
Salvajina	Colombia	148	1984
Alto Anchicaya	Colombia	140	1974
Khao Laem	Thailand	130	1984
Shiroro	Nigeria	125	1984
Miel 1	Colombia	185	Proposed
Segredo	Brazil	145	Proposed
Xingo	Brazil	140	Proposed
Ita	Brazil	125	Proposed

Figure 18.6 Construction of asphaltic membrane: Winscar dam

cally):0.8 (horizontally) against the downstream fill and smoothed with a lean concrete that acted as a drainage layer. Asphalt with a maximum particle size of 9 mm was then placed under timber shuttering to a thickness of 200 mm, tapering to 100 mm near the top of the dam. As the upstream shoulder was raised, so the shuttering was moved up to form the sloping core.

According to Steffen,[5] this method of construction was not popular. In 1954, the 58 m high Henne dam in Germany was built with a 1 m wide sloping core formed between shutters. A soft bituminous concrete overfilled with bitumen and filler was stiffened by adding clean, dry stones that were vibrated into the mix. The shutters were moved up at each lift. This technique was used to form cores for several dams in Germany, Austria and France until the end of the 1960s.

A similar, but perhaps simpler, approach was used in Norway in 1969 for building a vertical asphaltic core without the use of special plant. The idea was to construct the core of aggregate and then fill the voids with bitumen. To economize on bitumen the voids should be small, but to obtain deep penetration they should be large and, preferably the aggregate should be warm. Laboratory tests showed that a rounded aggregate graded from 19 to 76 mm, not colder than 3° C, was penetrated to a depth of 20 to 30 cm by bitumen poured at 160° C.

A core was built by this method in the 12 m high approach dam at the south end of the spillway structure of the 67 m high Grasjø dam near Trondheim. Timber shutters were used to produce a core width of 0.5 m and the aggregate was compacted with adjoining filter material. A 180- to 200-penetration bitumen at 170° C was poured into the voids in the core aggregate. The core, as shown in Figure 18.7a and b, was extended to the

very top of the dam to ensure that the head of bitumen was higher than the head of water in the reservoir to prevent it from being displaced by the water pressure (hydraulic fracture) and so that the bitumen could be topped-up if required. Electrical bitumen pressure gauges were installed in the core during construction. Measured pressures were less than the hydrostatic bitumen pressure, as shown in Figure 18.7c. As the reservoir rose to top water-level, the bitumen pressure increased to values shown in Figure 18.7d and remained above water pressure. In a description of this work, Kjaernsli and Sande[6] say that the performance of the core during the first 2 yr of operation was satisfactory with negligible leakage through the dam.

Machines were developed in Germany in the early 1960s to place narrow cores of dense bitumen–concrete. Since 1962, at least 23 dams have been built with machine-placed bitumen–concrete central cores. Those higher than 50 m are listed in height order in Table 18.3.

The first, completed in 1962, was the Dhünn valley rockfill dam (35 m high), described by Breth.[27] It was founded on river gravel over a bedrock of greywacke sandstone and argillaceous slate: the shoulder fill was crushed slate placed in 0.6 m layers and compacted by a 7 t sliding vibrator. Concern was expressed that the asphaltic core should be compatible with the slate rockfill. A system of measurement was therefore devised (Figure 18.8) consisting of a vertical inspection shaft lined with concrete rings each separated by a few centimetres to allow for settlement, built with the dam and situated 6 m downstream of the core at about the major section. Six horizontal cross-pipes, spaced 5.5 m vertically, facilitated observation of the downstream face of the core.

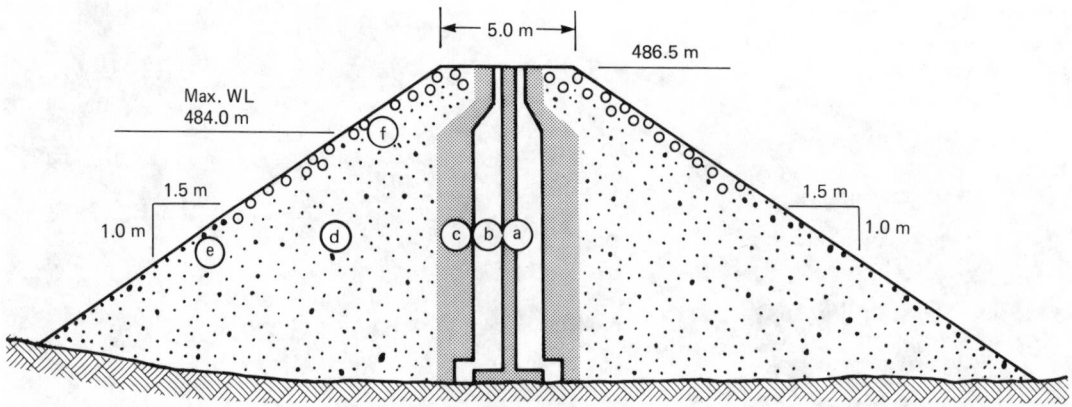

a, Asphaltic core; b, filters; c, transition zone; d, rockfill placed in
1.5 m layers and sluiced; e, riprap; f, larger-size riprap for wave
protection

(a)

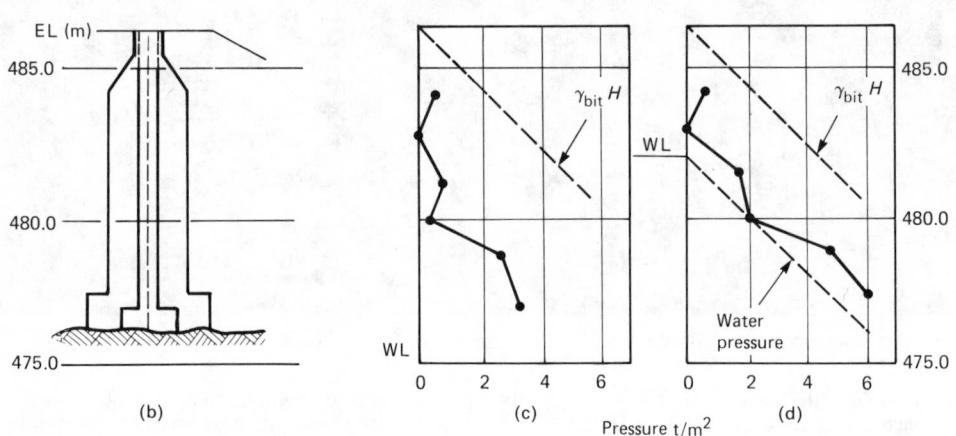

Figure 18.7 (a) Dam section; (b) Asphaltic core and filter;
(c) and (d) bitumen pressures in the core wall

Table 18.3 Dams (> 50 m) with machine-placed bitumen–concrete central cores

Dam	Height (m)	Core thickness (cm) (max.) (min.)	Completion date	Country
High Island – east	105.0	120–80	1977	Hong Kong
High Island – west	96.0	120–80	1977	Hong Kong
Finstertal	92.0	70–50	1981	Austria
Kleine Kinzig	67.5	65–50	1981	West Germany
Dhünn main dam	62.5	60	1981	West Germany
Megget	58.0	90–60	1983	UK
Wiehl main dam	54.0	60–50	1971	West Germany

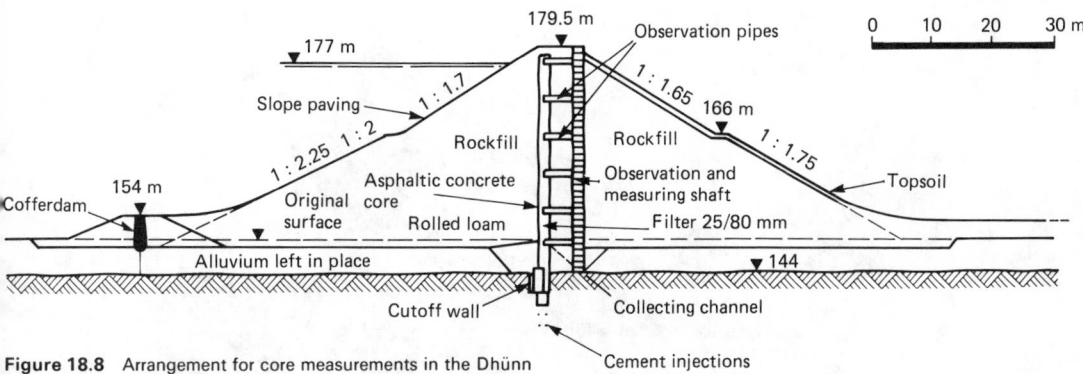

Figure 18.8 Arrangement for core measurements in the Dhünn dam

The very bottom of the core was made 1 m wide where it was in contact with a concrete groutcap, but the main core was 700 mm wide over the lower part, 600 mm in the middle section and 500 mm at the top. The results of measurements published by Lohr and Feiner[7] are given in Table 18.4. As the reservoir was filling for the first time, the upper part of the core moved upstream presumably because of the effect of wetting the upstream rockfill, but subsequently the horizontal movements have been in a downstream direction. Some part of this downstream movement could be caused by spread of the core under vertical loading. Core settlement at each measuring point tended to exceed settlements of the fill slightly, indicating that the core settlements were caused by self-weight rather than downdrag by the fill.

Example 18.1

In the UK, two dams have been built with this type of core; (1) the Sulby rockfill dam (60 m) on the Isle of Man, completed in 1982; and (2) the gravel-fill Megget dam (56 m) in Scotland, completed in 1983.

The Sulby dam. (Figure 18.9) was originally intended to be built in two stages. The first was to be a 35 m high structure with a central core wall of machine-placed bituminous concrete. The second stage, which was expected to be built at some future date to raise the height to 60 m, was to consist of rockfill placed downstream of the first dam and fitted with an upstream membrane of bituminous concrete (b) connected to the top of the central core wall (a).

While the 35 m high dam was under construction, it was realized that it made economic sense to build both stages at the same time, and so the contractor continued with the 60 m high dam (68 m above lowest foundation) using an upstream membrane for the upper part.

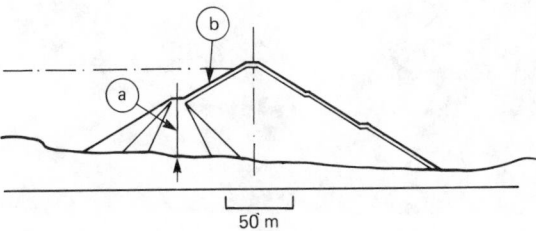

Figure 18.9 Sulby dam: cross-section.
a, Central core of bituminous concrete; b, membrane of bituminous concrete

The vertical core was placed by a Teerbau machine, which used steel sideplate shutters to form the core to the required width of 750 mm with vertical sides. Hot asphaltic mix was dropped into the hopper and fed down to the shutters as the machine advanced along the dam, the machine's tracks bearing on the transition material, adding a lift of 200 to 250 mm to the core. A second part of the machine, linked to the first, placed transition material on either side over a width of about 1.5 m. After it had passed, the machine left a level compacted surface of transition material with the asphaltic core in the middle. Dam fill was brought up on either side to support the transition material.

Example 18.2

The Megget dam. The vertical core of this dam was formed with a Strabag machine which also travelled with its tracks bearing on a 1.5 m width of transition material on either side of the core. The machine (Figure 18.10) had a long steel nose which projected over the core and contained preheating equipment. Hot asphaltic mix passed from a hopper through an adjustable

Table 18.4 Measured movemements of the asphaltic core of Dhünn main dam. (After Lohr and Feinner (1973) 'Asphaltic concrete cores: experiences and developments'. *Transactions, 11th international congress on large dams,* Madrid)

Position	Base	1	2	3	4	5	6	Crest
Height of measuring position above base (m)		3.4	8.9	14.4	19.9	25.4	30.9	35
Settlement during construction to 1962 (mm)		35.0	60.0	130.00	170.0	100.0	20.0	
Movements during operation to 1970 – vertical (mm)		50.0	78.0	200.0	290.0	253.0	202.0	
Horizontal (mm) maximum upstream on impounding		0.0	0.0	0.0	8.0	14.0	20.0	
Total downstream to 1970		33	39	81	71	63	52	

Figure 18.10 Strabag machine placing core at Megget dam

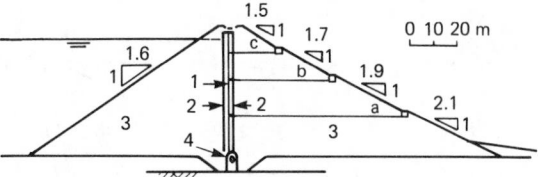

Figure 18.11 Major section of Megget dam, showing the positions of horizontal plate gauges a, b and c where (1) is the asphaltic concrete core, (2) the transition zones, (3) gravel fill, and (4) the control gallery

aperture to give the desired width of core and was immediately supported by transition material that had been dumped over the nose. As the machine advanced, guide plates spread the fine granular transition material which was compacted, together with the core leaving, as with the Teerbau machine, a firm level surface to the 200 to 250 mm lift of core plus transition.

This method of construction gave an inverted 'fir tree'-shaped edge to the core. Although the aperture gave the design width to the bottom of the lift, compaction caused the upper part to spread slightly and press into the transition fill. Thus, the design width of 900 mm at the base, reducing in steps to 600 mm in the upper part of the dam, was never decreased: it was slightly exceeded at the top of each lift.

The surface of each lift was kept clean by a strip of tarpaulin laid over it. This was very effective and even allowed core placement to continue immediately after snowfall. The machine was guided by a string pegged out at a specified distance upstream of the core. The Megget core was curved in plan and the pegs were positioned with the aid of a theodolite-mounted electronic distance-measuring device stationed over a reference point above crest level on the left abutment.

Shoulder fill was a well-graded gravel that was placed in 400 mm layers and compacted with four passes of a 5.5 t vibrating roller. This produced a very high density and a very stiff material. Numerous instruments installed in the dam during construction included horizontal plate gauges taken through the downstream fill to touch the core and thereby measure any movements that occurred. Figure 18.11 is a section of the dam showing the positions of these gauges, and the movements observed during first filling of the reservoir are given in Figure 18.12. A comparison between observed and predicted deformations of this dam have been given by Penman and Charles,[8] who conclude that the asphaltic concrete core acted as a thin diaphragm. It had little effect on construction deformations and during reservoir impounding simply transmitted the increased lateral thrust caused by the impounded water on to the fill of the downstream shoulder. During dam construction, there were virtually no downstream movements of the downstream face of the core, indicating that, unlike a clay core, it did not exert a large lateral thrust on the gravel fill because of its own weight.

18.3.5 Rockfill dam with central clay core

At sites where there is a source of suitable clay as well as rock, this type of dam may prove more economic than a rockfill dam with an upstream membrane, particularly if delivery of cement or asphalt to the site would be expensive. This design avoids the detailed handwork and special machinery required for slipform work or placing and compacting asphalt. It also obviates the dangers of damaging, concentrated movements near the plinth of upstream membranes, and the very high hydraulic gradient under the plinth that could cause internal erosion in some bedrock formations.

The width of the rolled clay core is usually between $0.5H + C$ to $0.33H + C$ where H represents height and C is the minimum width at dam crest. This provides a theoretical hydraulic gradient of 2 or 3 along the contact between the core and the foundation or abutment on which it rests.

In order to ensure good contact and an absence of cracks or fissures which would allow a passage for water under the clay, the cleaned formation is often coated with a layer of concrete over the area of contact with the clay core. The foundation is often sealed by grout injected through boreholes to form a grout curtain under the contact area to act as a below-ground cutoff.

Example 18.3

The Llyn Brianne dam (91 m) is an example of this type of dam

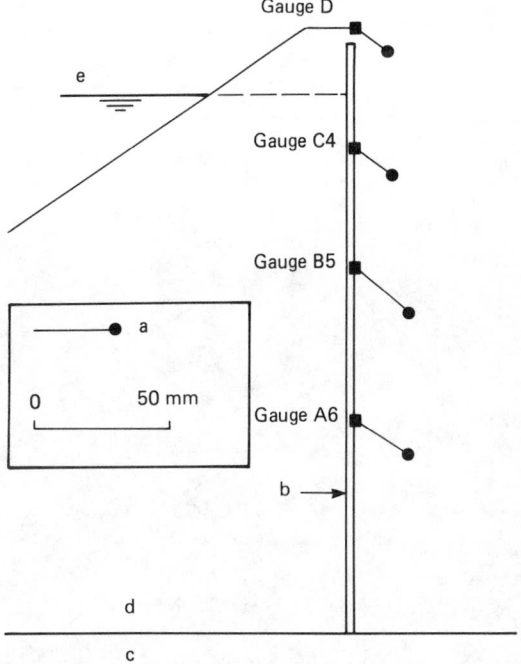

Figure 18.12 Downstream movements of the core on impounding. a, Observed movements; b, asphaltic concrete core; c, foundation; d, gravel fill; e, reservoir water level

(Figures 18.13 and 18.14). The dam site, on the River Towy in central Wales, is in a slatey, argillaceous rock which fragmented into plate-shaped pieces when won by blasting in the quarry. A trial embankment showed that placing and compaction prevented any preferred orientation, broke up some of the pieces and produced a dense fill.

The stripped bedrock was blanket-grouted to 10 to 15 m depth over the core contact area and coated with a pneumatically applied mortar skin 50 mm thick. In addition, a grout curtain was formed to a depth of 45 to 75 m under the centreline.

The rockfill to form the shoulders was spread in 0.5 m layers and compacted on every second layer, i.e. on a 1 m layer, by four passes of either an 8.6 or 13.5 t smooth vibrating roller.

Riprap on the upstream slope consisted of oversize pieces of the rockfill 1 to 1.5 m size, bulldozed out from the general fill.

18.3.6 Earthfill dam – homogeneous section

Dams built almost entirely from one type of fill, without

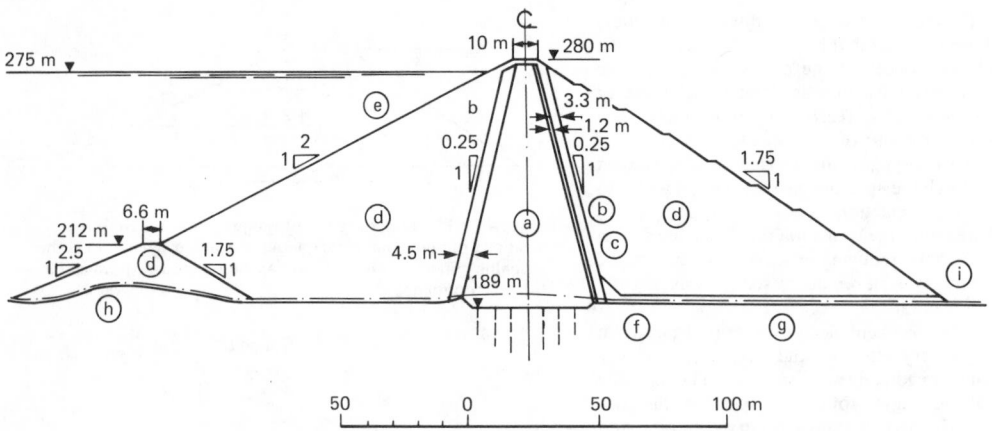

Figure 18.13 Llyn Brianne: cross-section.
(a) Clay core; (b) transition; (c) filter; (d) rockfill; (e) riprap;
(f) rockfill drain in river channel; (g) excavated level bedrock;
(h) cofferdam; (i) original ground level

Figure 18.14 Llyn Brianne dam

provision for neither a less pervious core nor more stable shoulders, became popular in the Americas as the size and power of earthmoving machinery developed in the 1920s and 1930s.

A danger quickly recognized was that, if under full reservoir, the phreatic line could cut the downstream slope and local slips develop, leading to backsapping and eventual destruction of the dam.

The US Bureau of Reclamation installed standpipe piezometers in the mid 1930s to check on the positions of phreatic surfaces and in this way revealed the presence of construction pore pressures. Placement at water contents below Proctor optimum substantially reduced construction pore pressure but produced a relatively stiff fill. Differential settlements could cause cracking: examples of cracked and failed dams have been given by Sherard.[9]

A solution was provided by Terzaghi in his design of the Vigario dam in 1947 by using a central vertical core of filter material to drain leakage and prevent the phreatic surface reaching the downstream slope. Examples of this design are shown in Figure 18.15.

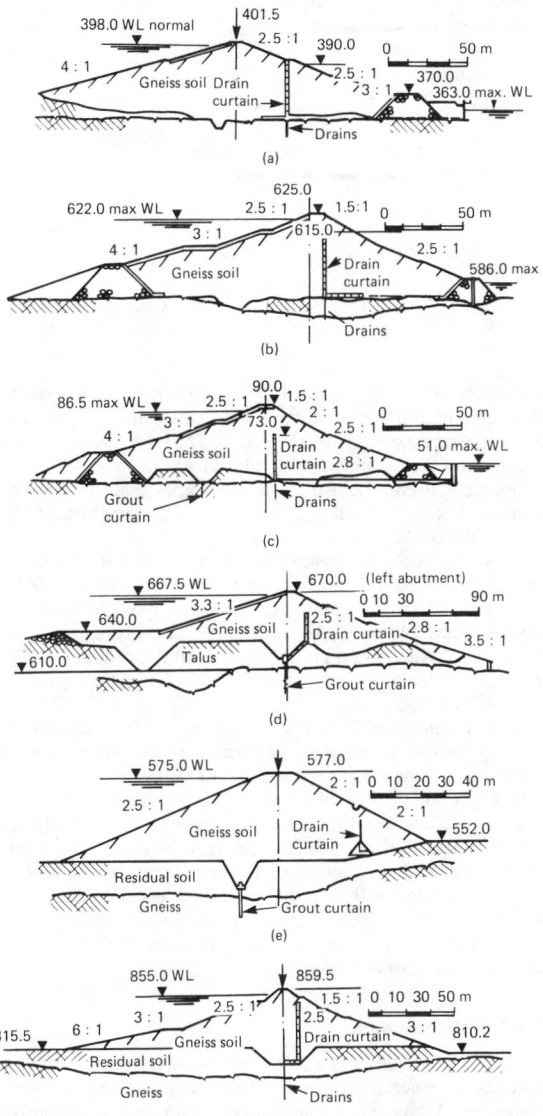

Figure 18.15 Homogeneous earth dams with Brazilian section. (a) Vigario Dike; (b) Santa Branca; (c) Ponte Coberta; (d) Euclides da Cunha; (e) Limoeira; (f) Graminha

18.3.7 Earthfill dam with central clay core

Central cores of puddled clay were used in the traditional British dam in the nineteenth century. It had an upstream slope of 1 in 3 and a downstream slope of 1 in 2.5. The puddled clay core was usually taken down in trench to form a below-ground water-stop.

The fissures, bedding planes, silt layers, etc. found in deposits of clay can give it a relatively high *in-situ* permeability. The action of puddling destroys this fabric and, by addition of water

where necessary, the strength was usually reduced sufficiently ($c_u = 10$ to 15 kN/m²) to enable the puddled clay to be compacted down to an air void of about 5% by the heeling of the puddling gang.

Increasing cost of labour and unsuitability of compacting machinery of the time for working with puddled clay caused the change to rolled clay cores. The era of British puddled clay cores came to an end during the 1950s, although there were still dams occasionally built with these cores until 1970.

The use of earth rather than rockfill for the shoulders poses the problem of construction pore pressures. In areas of high rainfall and/or when the borrow material is wet, compression of the fill under self-weight as height is increased can produce undesirable pore pressure which, in the extreme, may endanger stability by preventing the required increase of effective stresses within the fill.

This problem is overcome by use of drainage layers placed in the fill during construction. These reduce the length of the drainage path which the pore water must traverse to escape. The time required for dissipation of pore pressure is proportional to the square of the length of the drainage path, so drainage layers are particularly effective in reducing pore pressures in shoulder fill during construction.

Granular material for the drainage layers is often graded so that it acts as a filter to prevent loss of fines from the fill, but this aspect is not of prime importance in the downstream shoulder, where volume of escaping pore water is unlikely to cause significant particle migration. In the upstream shoulder, however, there is a danger that fluctuations of reservoir level could produce damaging flows if the layers do not act as effective filters.

Example 18.4

The Backwater dam (43 m) was constructed during the period 1964–69 (Figure 18.16). At the site, boulder clay overlies a schistose grit and micaceous schist. The embankment was built from compacted boulder clay placed in layers with a slight outward fall to help shed rainwater. Granular drainage layers were placed at about 7 m vertical intervals.

The core was also constructed of boulder clay, carefully selected in the borrow pits to contain least stones and most clay. It was placed over a grout curtain cutoff on to a prepared surface. It was separated from the downstream shoulder by a drainage filter which connected to the drainage layers and the main underdrainage layer separating shoulder fill from foundation.

18.4 Concrete dams

18.4.1 Introduction

Use of certain volcanic ashes by the Romans to cement together sand and gravel into a reconstituted rock is often regarded as the first concrete. The aggregates used for modern concrete, such as well-graded sands and gravels, and hard rock that has been crushed, sieved and graded to a designed grading, would form excellent fill, even without the addition of cement. The additional strength imparted to it by the cement enables less volume to be used, so redressing the high cost of production.

18.4.2 Gravity dams

The simplest type of concrete dam has a gravity section, i.e. it is heavy enough not to be overturned by horizontal thrust from the reservoir water. The foundations have to be relatively strong to support the large weight and they should not be subject to

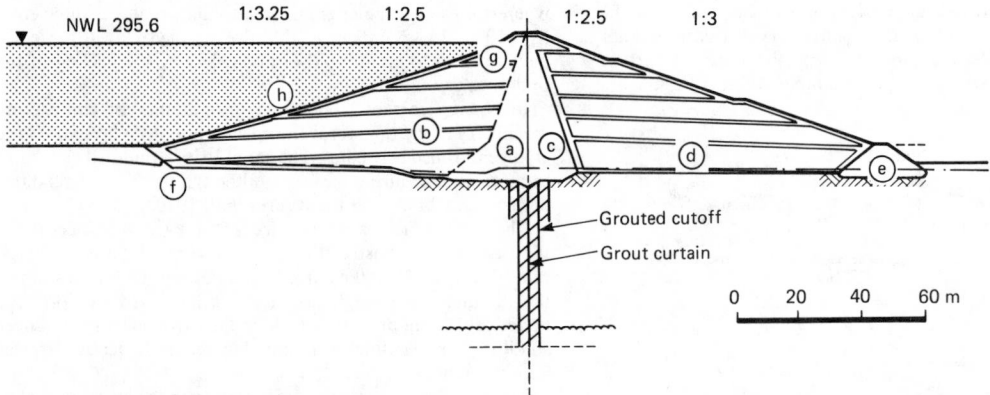

Figure 18.16 Backwater dam: section showing drainage layers.
(a) Glacial till core; (b) glacial till shoulder; (c) chimney drain;
(d) drainage layers; (e) rubble toe; (f) spoil; (g) concrete block;
(h) riprap

long-term settlements that could be caused by consolidation of an underlying clay.

The heat liberated by hydration of the cement in concrete can produce damaging temperature rises in large masses of concrete. As with the problem of dissipating construction pore pressures in earthfill shoulders, so consideration has to be given to limiting the temperature rise which will occur in a gravity dam as it is being built. The counterparts of drainage layers are layers of cooling pipes placed in the mass concrete as it is being cast. A refrigeration plant is used to lower the temperature of brine circulated through the cooling pipes.

It is usual to construct a gravity dam as several separate sections or, sometimes, as large monoliths. When temperatures have fallen sufficiently so that little more cooling shrinkage is likely to occur, the joints between the sections are grouted or the monoliths are connected with infilling sections to form the complete dam.

Temperature rise can be reduced by slow rates of construction, use of slow-setting, coarse-ground cement and use of relatively inert cement replacements such as power station fly ash.

Drainage is essential in a gravity dam to limit uplift pressures. In effect, the upstream face forms the waterproof element, supported by the remainder of the massive dam. Reservoir water percolating along the interface between dam and foundation or along any of the horizontal lift joints could develop destabilizing uplift pressures if not safely drained away. Drainage galleries and shafts are formed in the body of the dam. A gallery is usually provided close to the foundation on the upstream side from which either additional drainage or grouting holes can be drilled into the foundation if found to be necessary during reservoir operation.

Compressive stress in the concrete forming the upstream face should always exceed reservoir water pressure to avoid tensile cracking.

Example 18.5

The Grand Dixence dam (285 m) was constructed during the period 1953–62 (Figures 18.17 and 18.18). This remained the world's highest dam for 18 yr until it was exceeded by the Russian Nurek (300 m) embankment dam.

The Grand Dixence dam was built in the Swiss Alps across the River Dix on sound bedrock. Its 695 m crest length was divided into 16 m blocks connected by two copper sealing strips to form a continuous, watertight upstream face. The blocks,

particularly in the lower part of the dam where it was very wide (201 m maximum) were further divided by joints to allow for some shrinkage movements on cooling. The joints were grouted up only after the concrete temperature had fallen below 6° C.

Cooling was effected with layers of pipes placed on every 3.2 m lift. About 1000 km of 20 mm bore pipes were installed to circulate the cooling water.

The waterproof, upstream face was made with concrete containing 250 kg of cement per cubic metre so as to be watertight and frost-resistant. Concrete in other parts of the dam contained 140 to 300 kg/m³ of cement according to calculated stresses. Air-entraining agent was used to produce 3 to 4% air void in order to improve workability. Maximum aggregate size was 120 mm.

The concrete was placed with the aid of four cableways spanning the valley over the dam. On the right bank they were attached to mobile carriages anchored to rails so that the cables could be brought over almost every part of the dam. The cableway buckets could carry 6 m³ of concrete. After discharge from the buckets, the concrete was spread by small bulldozers. It was compacted with vibrating pokers: frames of five pokers were attached to the blades of other bulldozers so that they could be moved about and lowered into the concrete. At the time, this was considered to be the first use of earthmoving machinery on a concrete dam.

18.4.3 Rollcrete dams

The term 'rollcrete' was coined by Lowe[10] to name a new approach to concrete placement. Increasing labour costs were making the labour-intensive concrete dams less competitive than the embankment dam, even on sites suitable for a concrete dam. To help redress the balance, Lowe proposed using a relatively dry, lean concrete placed by earthmoving machinery and compacted by smooth vibrating rollers. His first use of the material was to construct the core of the cofferdam for Shihmen dam, Taiwan in 1961–62.

At the same time, a similar approach was being used in Italy. Gentile[11] designed the 175 m high Alpe Gera dam for placement by earthmoving machines. He used a blastfurnace cement to reduce heat of hydration, but also provided an adequate number of open contraction joints to allow for shrinkage. The gravity section, shown in Figure 18.19, was designed only to support the forces imposed by the reservoir; it was not intended to be watertight. A 3 mm thick steel plate was used to form the waterproof element on the upstream face of the dam. Maximum

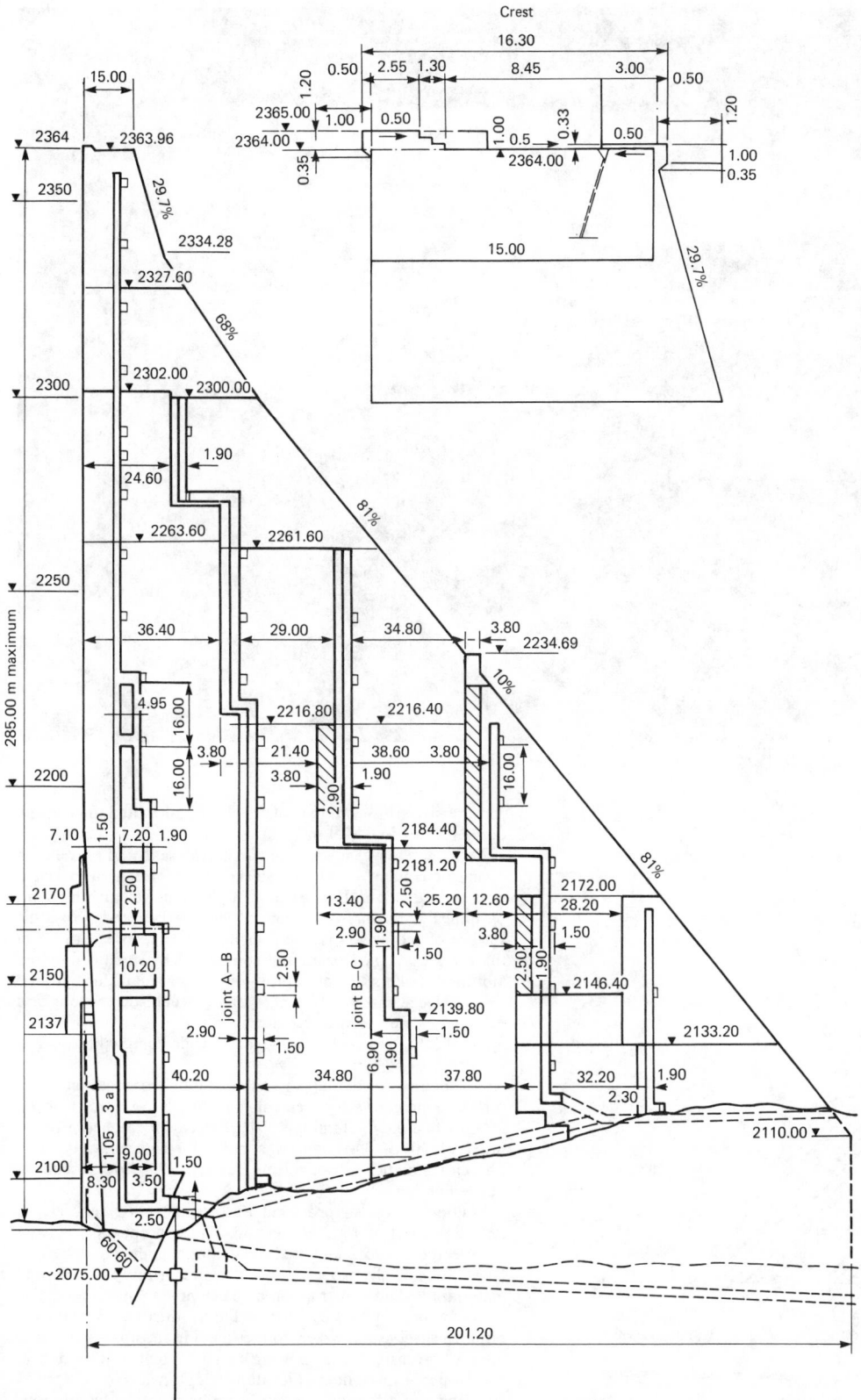

Figure 18.17 Grand Dixence dam: cross-section

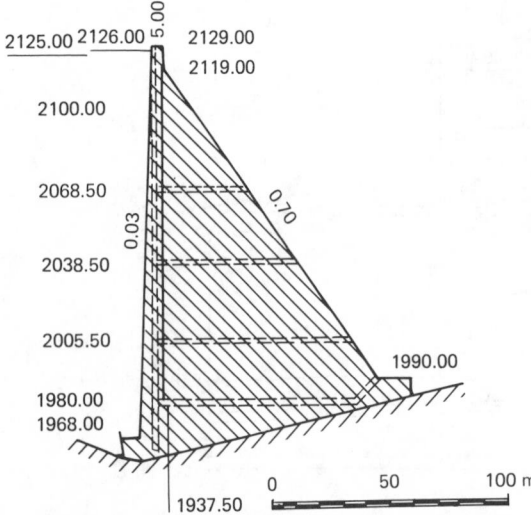

Figure 18.18 Grand Dixence dam

compressive stresses in the concrete were calculated to be in the range 3.25 to 5.60 MN/m^2.

The concrete was made from an alluvial sand and gravel with a cement content of 115 to 300 kg/m^3 according to expected stresses in the dam. It was transported from the batching plant by funicular railway to placement level and carried across the surface of the new concrete in 6 m^3 dumpers, spread to 0.8 m thickness by angle dozers and compacted by tractor-mounted vibrating pokers. Contraction joints were cut through the freshly placed concrete by a manganese steel blade 3 m long and 1 m high carried on a wheeled chassis and pressed down with a force of 2.5 t by two hydraulic jacks while being vibrated at a frequency of 50 Hz.

No cooling was employed and maximum measured temperature in concrete with a cement content of 150 kg/m^3 reached 35° C. Concrete containing 115 kg/m^3 cement reached a maximum of 30° C. The dam was built during three-and-a-half 6-month summer periods from August 1961 to the end of September 1964.

Wallingford[12] also proposed the use of rolled concrete to reduce the cost of a gravity section. This concept was followed by research studies, described by Moffat,[13] at the University of Newcastle into the properties of dry lean concrete as a potential material for dam construction by earthmoving machinery. The dam section proposed by Moffat (Figure 18.20) used a poured-bitumen sheet as the waterproof element in the upstream facing made of precast units. Following a study of temperature rises in the Upper Tamar dam, Dunstan[14,15] outlined a dam section containing low cement content concrete placed in continuous 0.6 m layers between upstream and downstream facings. These

Figure 18.19 Alpe Gera dam; rollcrete with sheet steel waterproof element

facings were to be laid as horizontally slipformed kerbs, with a sheet of reinforced butyl rubber inside the upstream facing as the waterproof element. Dunstan's further work[16] led him to use almost an excess of fly ash to ensure that voids between larger particles were completely filled, and in order to improve workability, reduce water content to very small amounts and produce a concrete of low permeability.

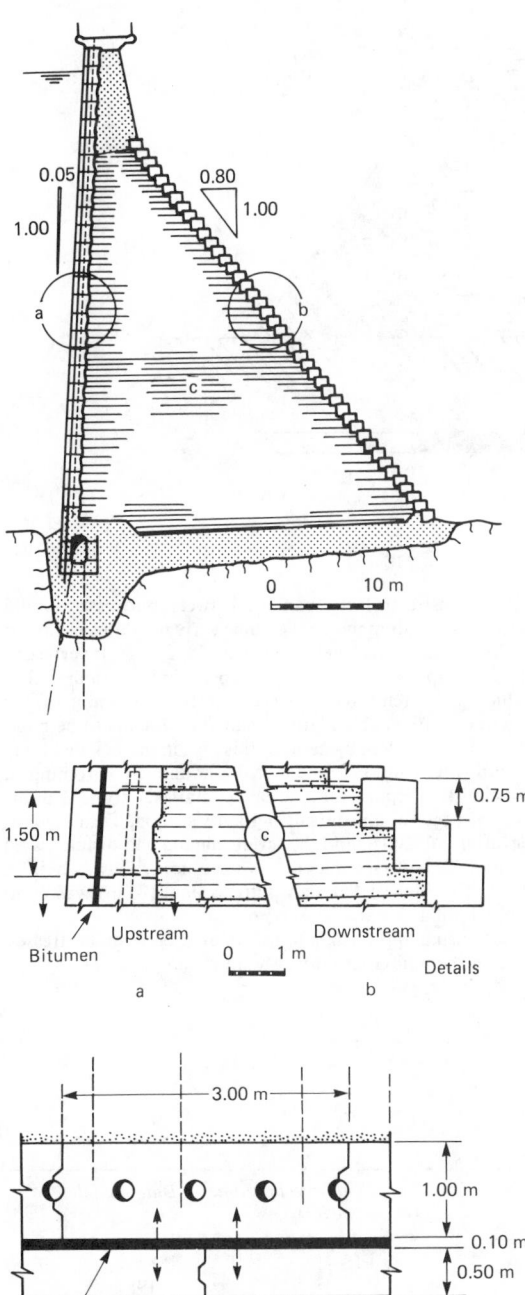

Figure 18.20 Proposed dry lean concrete dam with bituminous waterproof element. (After: Moffat (1973) 'A study of dry lean concrete applied to the construction of gravity dams'. *Transactions, 11th international congress on large dams*, Madrid)

Dunstan supervised construction of a trial bank at the site of Wimbleball dam in 1979. The bank was contained between slipformed kerbs (Figure 18.21). The cementitious content of the concrete was 0.75 fly ash and 0.25 cement. Each cubic metre of concrete contained 85 kg cement. It was compacted in 0.3 m layers by a 7 t duplex vibrating roller. The very cohesive mix did not segregate, but flowed under the vibrating action of the roller. This produced a hard surface on which the laser-guided, offset slipformer could immediately run to lay a further lift of kerb. Several different time intervals were used between lifts, and subsequent tests on cores which were axially drilled for water-pressure tests showed no leakage at the lift joints. The work was in preparation for the Milton Brook dam but financial restrictions caused postponement of construction.

In the US, roller compacted concrete has been used to build several dams (Table 18.5). One of the first was the 37 m high Willow Creek dam, completed in 1982. The construction joints proved to be fairly porous, causing the downstream face to be fairly wet, giving a clear indication of reservoir level. The joints have since been grouted to reduce leakage.

Rapid construction is one of the attractive features of roller-compacted concrete dams: some times of construction are given in Table 18.5. As an example of the concrete mix, that used for the Upper Stillwater dam contained only 77 kg cement per cubic metre, with 170 kg fly ash and 107 kg water.

In Japan, roller compaction was used to build the 89 m high Shimajigawa dam completed in 1980. Previously a base pad over the fractured rock foundation for the concrete gravity Okawa dam (75 m) used the roller-compacted dam method in 1978. The mat had an average thickness of 25 m and its volume of 300 000 m³ was placed in 9 months.

The use of roller compaction for these dams has been followed by its use for the construction of the 100 m high Tamagawa dam which has the fairly large (for a gravity dam) volume of 1.14×10^6 m³. The maximum size of aggregate, previously limited to 80 mm, has been increased to 150 mm. A section of the dam is shown in Figure 18.22 and the mixes used in the various parts of the dam are given in Table 18.6. The rolled concrete was placed in 0.75 m thick layers and compacted by twelve passes of a Bomag BW-200 vibrating smooth roller weighing 7 t.

Other Japanese roller compacted dams are listed in Table 18.7.

In South Africa, rollcrete was used for the first time to construct the lower half of the concrete section of the 52 m high Braam Raubenheimer dam in eastern Transvaal. The concrete gravity section is 33 m high and contains 20 000 m³ of rollcrete placed in 1984.

It has also been used to construct the Zaaihoek dam and the 30 m high De Mist Kraal weir on the Little Fish River in the Eastern Cape. Figure 18.23 shows a section of this weir. Rapid construction was achieved, with 34 000 m³ of rollcrete being placed in 26 days during the winter of 1986. The contents of each cubic metre are given in Table 18.8.

The 36 m high gravity section of the Arabie dam on the Olifants River, 27 km north of Marble Hall was constructed of rollcrete, completed in 1987.

18.4.4 Buttress dams

The weight and, therefore, volume (and cost) of a gravity dam can be greatly reduced by retaining only the upstream face and supporting it by buttresses instead of the whole mass of the gravity section. To utilize some of the reservoir water pressure to resist overturning moment, the upstream face is usually sloped. The water pressure acting on it has a vertical component that replaces some of the weight of a gravity section as well as the horizontal component producing the overturning moment.

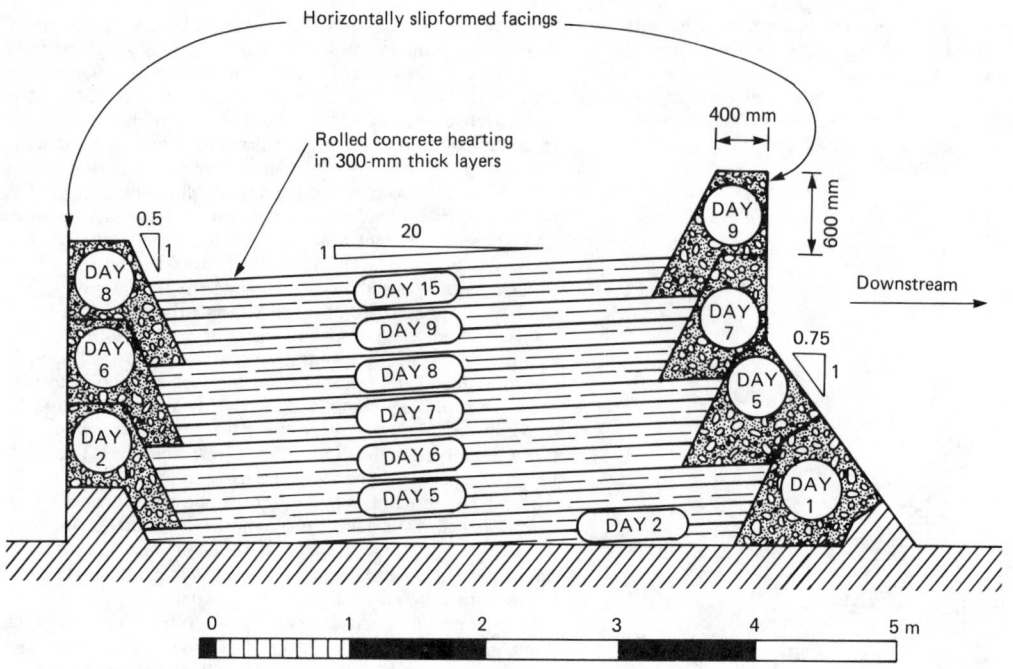

Figure 18.21 Milton Brook trial bank: roller-compacted concrete between slipformed kerbs

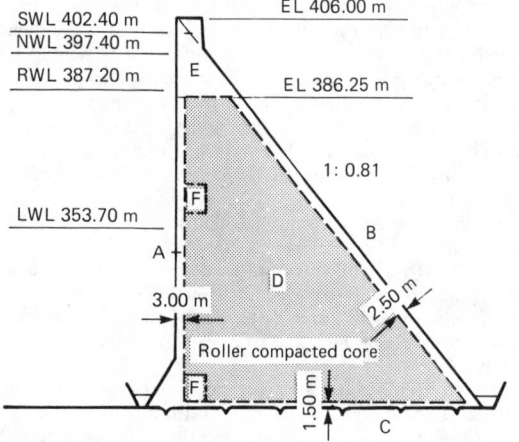

Figure 18.22 Tamagawa roller-compacted dam

A plane upstream face, spanning between buttresses, would have to resist bending moments. Some early milldams in timber had a plane deck at a relatively flat slope, e.g. 1 vertical:3 horizontal supported on wedge-shaped timber frames. The introduction of reinforced concrete at the beginning of the twentieth century enabled larger, plane-faced dams to be built. Ambursen designed many dams of this type in the US, including the record-breaking 41 m high La Prele dam in Wyoming in 1909. Schnitter[17] reports that the dam contained only 43% of the concrete needed for an equivalent gravity section. The dam is at an elevation of 1600 m and the severe climate had disintegrated more than 20% of face thickness when a new slab was built in 1977–79. The highest flat slab buttress dam (83 m) was completed in 1948 at Escaba in Argentina.

A more usual approach is to use an arch between buttresses, producing the multiarch buttress dam (Example 18.6).

Table 18.5 Roller-compacted concrete dams in the USA

Name of dam	Height (m)	State	Construction time (weeks)	Date completed
Upper Stillwater	87	Utah	18	1987
Pamo Valley	80	California	—	c. 1987
Elk Creek	76	Oregon	—	Designed 1983
Willow Creek	52	Oregon	21	1982
Galesville	51	Orgeon	6	1985
Monksville	46	North East	—	
Middle Fork	38	Western Colorado	7	· 1984

Table 18.6 Tamagawa dam – concrete mix designs

Position	Grade	Aggregate size (max.) (mm)	Cement (kg/m³)	Fly ash (kg/m³)	Water (kg/m³)
Upstream face	A	150	168	72	115
Downstream face	B	150	154	66	112
Foundation contact	C	150	126	54	108
Main body – rolled	D	150	91	39	95
Crest section	E	150	112	48	106
Reinforced openings	F	80	189	81	138

Table 18.7 Roller-compacted dams in Japan

Name of dam	Height (m)	Volume (m³ × 10³)	River	Construction
Sakaigawa	115.0	626	Sakai	1988–92
Tamagawa	100.0	1140	Tama	1983–87
Shimajigawa	90.0	324	Shimaji	1977–80
Asahi Ogawa	84.0	350	Ogawa	1986–89
*Shin-Nakano	74.9	201	Kameda	1979–82
Mano	69.0	212	Mano	1985–88
Shiromizugawa	54.5	312	Shiroizu	Under construction
†Pirica	40.0	360	Shiribeshi	1982–88

*Dam raised with roller-compacted concrete
†Composite with rockfill

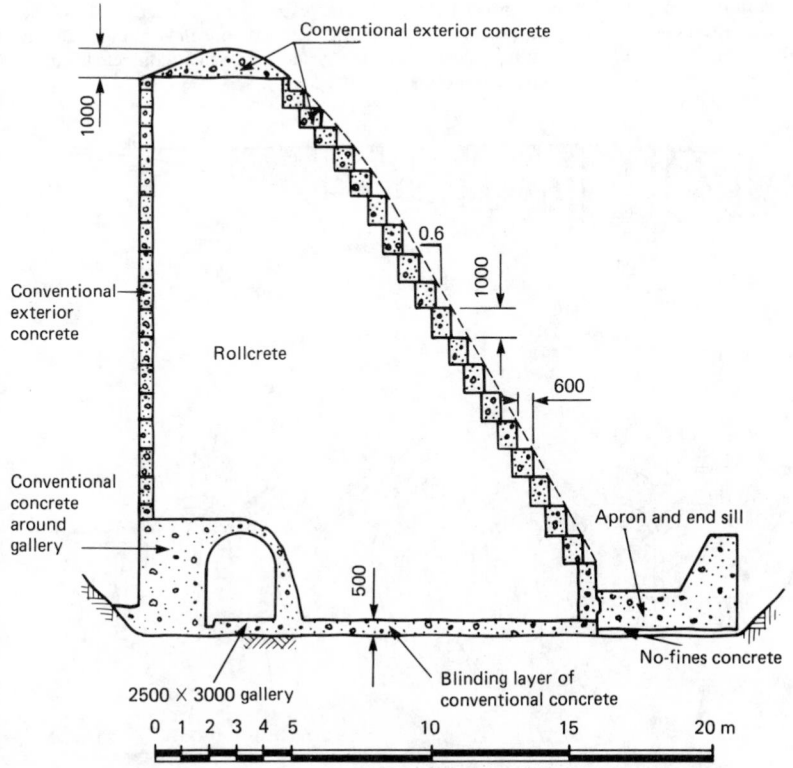

Figure 18.23 De Mist Kraal weir. Typical section through spillway

Table 18.8 Rollcrete mix for de Mist Kraal

Material	Quantity (kg)
Water	105
Portland cement	58
Fly ash	58
Sand	736
Aggregate (75–37.5 mm)	805
Aggregate (37.5–19 mm)	537
Aggregate (19–9.5 mm)	268
Aggregate (9.5–4.75 mm)	121
Conplast air-entraining agent	99 cm³

Compressive strengths	(MN/m³)
7 days	8.1
28 days	13.3
1 yr	25.0

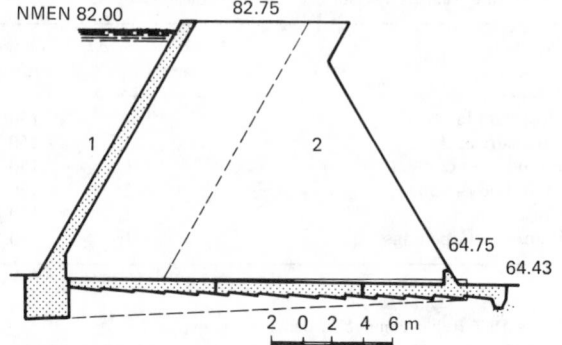

Figure 18.25 Meicende dam: cross-section, where (1) is the arch thickness, and (2) is the abutment between arches

Example 18.6

The Meicende dam northwest Spain (20 m) completed 1961 (Figures 18.24 and 18.25). Bedrock is granite and valley floor contained outcrops of sound granite but some decomposed rock and a lode of feldspar aplite. Use of the multiarch design enabled buttresses to be positioned off areas of poorest quality.

The final design provided a maximum height of 20 m with: 11 circular arches, each of 11 m internal radius and 1 m thick; 12 buttresses, 2.5 m thick at 22 m centres, with arch springing face at a slope of 1 vertical : 0.577 horizontal.

In contiguous buttress dams, the upstream face may be formed by widening the upstream edges of each buttress so that they touch and are sealed by suitable waterstops, instead of using a flat slab or multiple arches.

Example 18.7

The Itaipu dam, Parana river (196 m) constructed 1975–82 (Figure 18.26, showing the double buttress of hollow gravity block in river channel, and Figure 18.27, showing the diamond-headed buttress of the main dam).

This is an outstanding example of a contiguous buttress concrete dam. It crosses the frontier between Brazil and Paraguay, storing water and providing head for the world's largest hydro-electric installation, designed for an output of 12 600 MW.

The river section hollow-gravity dam, composed of double-buttress monoliths and the diamond-headed buttresses, is founded on basaltic flows 15 to 50 m thick. There are sub-horizontal discontinuities at different levels and special treatment was required to increase shear strength and reduce compressibility. Treatment included consolidation and contact grouting, special drainage systems, as well as concrete keying at some weak points.

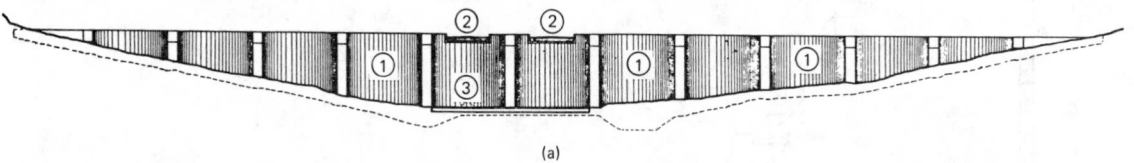

(a)

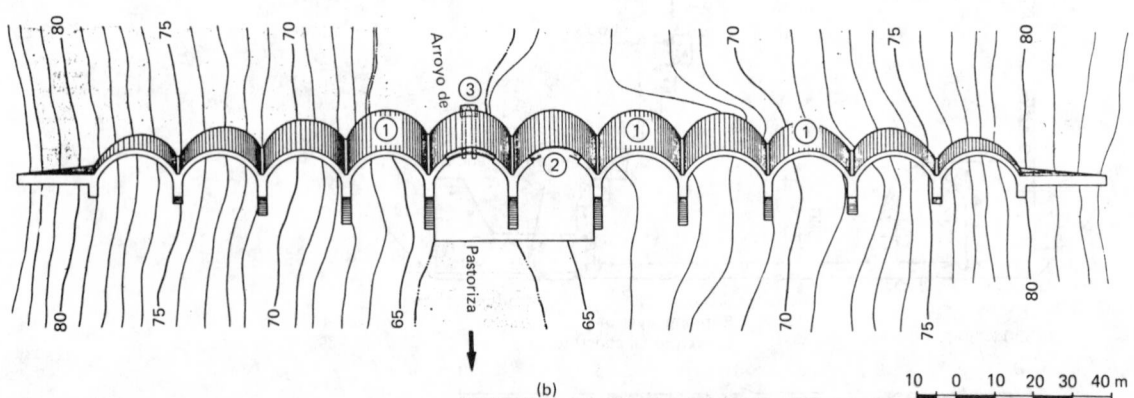

(b)

Figure 18.24 Meicende dam. (a) Elevation; (b) plan, where (1) is an arch, (2) are spillways, and (3) is the intake

A preloading test was made by flooding the space between the upstream coffer dam and the main dam. This enabled water to be raised almost 100 m against the major section of the dam. Good agreement was found between observed and predicted deformations. It lent reassurance that the structure would behave, as it did, in a satisfactory manner during the very rapid, irreversible impounding which took only 14 days. The main concrete section of the dam is flanked by embankment dams, forming a total length of 7772 m. Individual lengths are given in Table 18.9.

Table 18.9

	(m)
Hollow gravity of double buttress monoliths	612
Diamond-headed buttress	1450
Mass gravity section	532
Rockfill embankment	1984
Earthfill embankment	3194

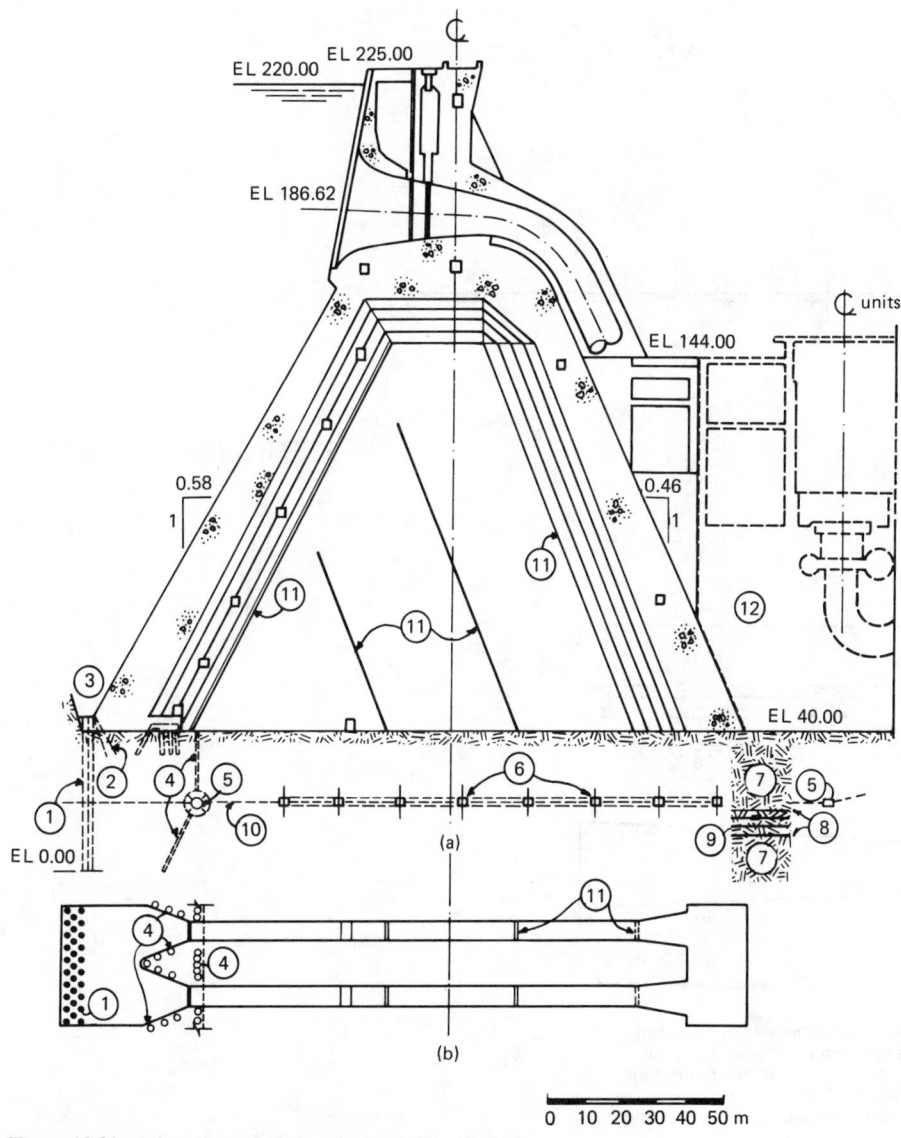

Figure 18.26 Itaipu dam: double buttress monolith of hollow gravity section in river channel. (a) Section; (b) plan, where (1) is the grout curtain, (2) contact grouting, (3) consolidation and contact grouting, (4) drainage holes, (5) drainage tunnel, (6) shear keys, (7) dense basalt, (8) breccia, (9) vescicular amygdaloidal basalt, (10) discontinuities, (11) contraction joints, and (12) is the powerhouse

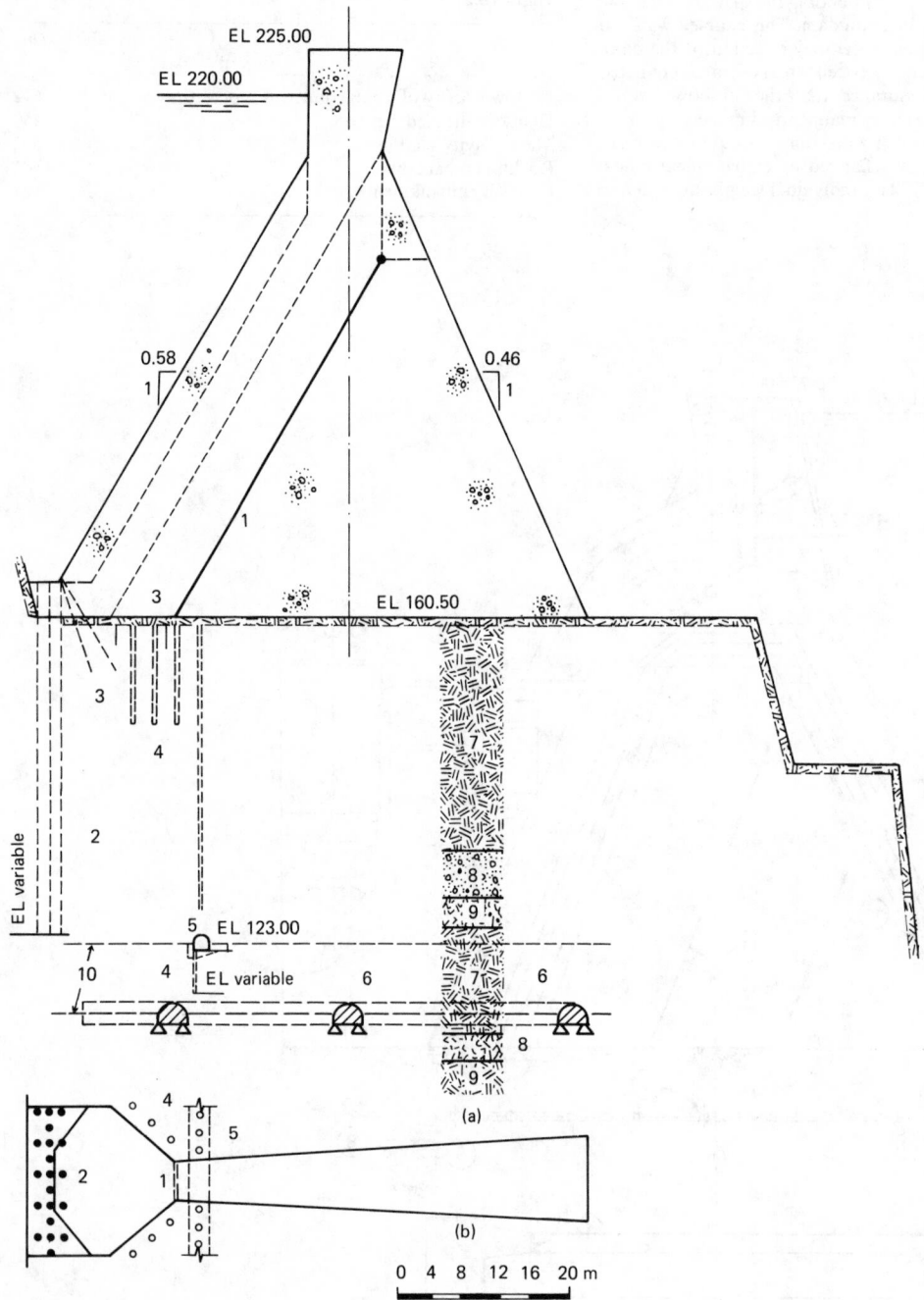

Figure 18.27 Itaipu dam: diamond-headed buttress of main dam. (a) Section; (b) plan, where (1) is the contraction joint, (2) grout curtain, (3) consolidation grouting, (4) drainage holes, (5) drainage tunnel, (6) shear keys, (7) dense basalt, (8) breccia, (9) vescicular amygdaloidal basalt, and (10) represents discontinuities EL 125 and EL 112

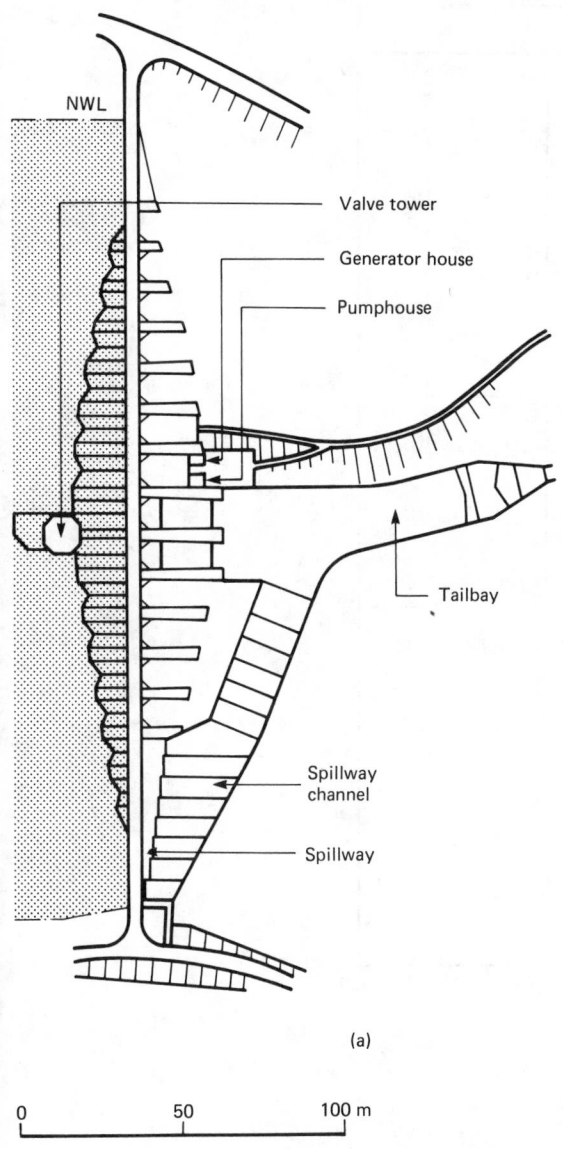

(a)

0   50   100 m

Example 18.8

The diamond-headed buttress – Wimbleball dam, southwest England (63 m) built 1975–79 (Figure 18.28).

The foundation consists of highly folded and contorted sandstones and siltstones with intercalated beds of slatey shales and mudstones. To limit underseepage, grouting was extended to 40 m below formation level and relief wells were drilled between the buttresses. Fly ash was used in the concrete to reduce the heat of hydration. Contraction joints between diamond heads were sealed by a 300 mm rubber waterbar upstream of a 200 mm Paracore plug.

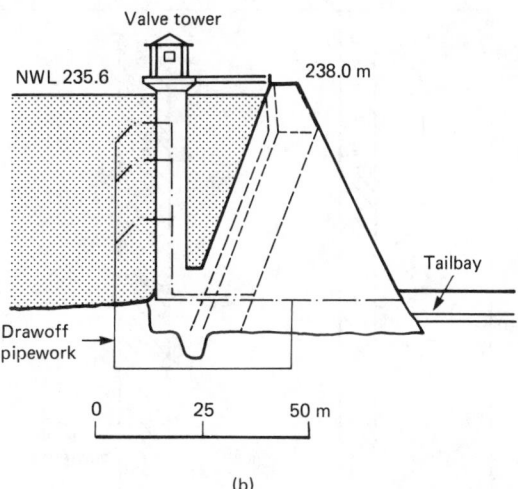

(b)

Figure 18.28 Wimbleball dam: a diamond-headed buttress. (a) General arrangement; (b) section through dam and valve tower

The diamond heads, forming the upstream face of the dam, sloped at 1 vertical : 0.383 horizontal.

The heights achieved by the various types of buttress dam are shown in Figure 18.29.

18.4.5 Arch dams

An arch dam can be likened to an arch bridge lying on its side with the abutments acting as springings for the arch. The arch transfers pressure from the reservoir water to thrust on the abutments, which must be strong enough to carry the thrust without permitting unacceptable deformations. The joints and bedding planes in many bedrocks provide potential weaknesses. Valley formation releases horizontal stresses and can be accompanied by strains and movements such as cambering and valley bulging which can loosen the rock mass. Great care is needed to ensure that the rock forming the abutments can accept both the magnitude and direction of the thrust that will come from the arch dam. As the reservoir filled for the first time, the inability of the left abutment rock structure to support the thrust from the 61 m high Malpasset dam brought about its collapse on 2 December 1959 causing serious damage to the downstream town of Fréjus, and killing 421 people.

18.4.5.1 Arch gravity dam

A mass gravity dam requires much less weight to keep it stable if it is built curved in plan (just as a piece of cardboard will stand on its edge if it is curved). Stresses thrown on to the abutments are less than with a thin arch dam but, unlike a true gravity dam, no separate section would be able to support the reservoir water pressure without the benefit of the arch action.

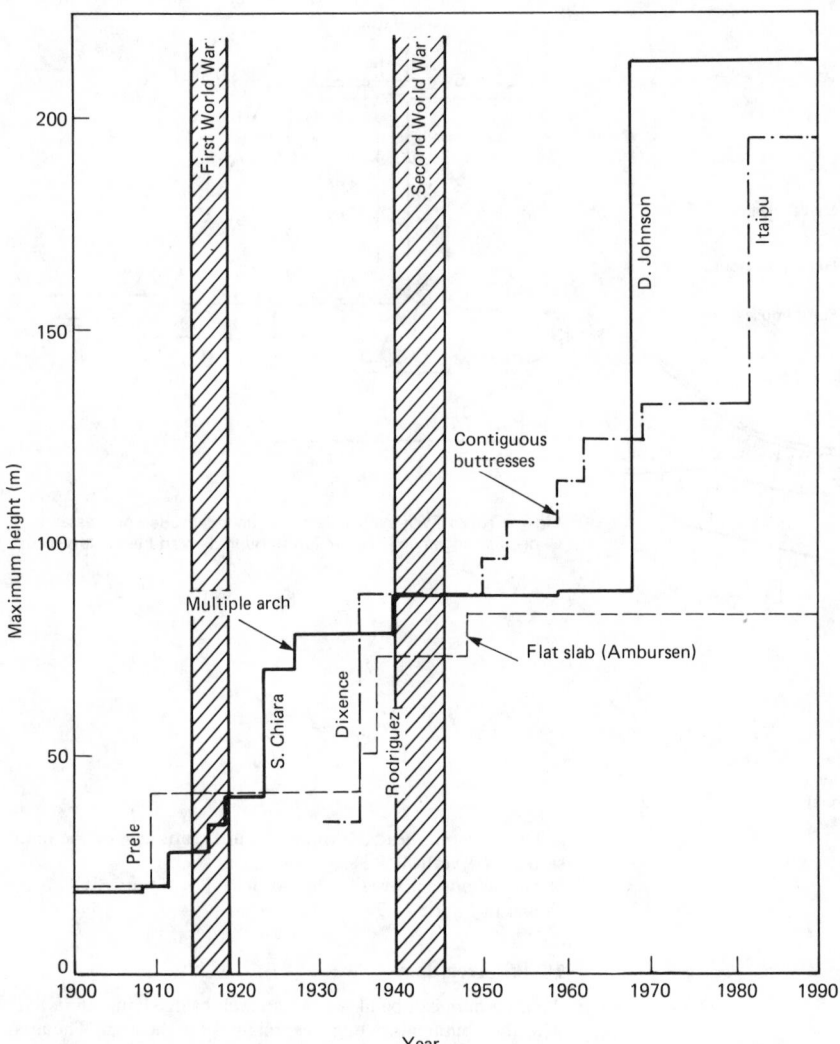

Figure 18.29 Maximum heights of buttress dams since 1900

Example 18.9

The Pieve di Cadore dam, Italy (112 m) constructed 1946–49 (Figures 18.30 and 18.31). This dam is founded in dolomitic limestones of the upper Trias. The modulus of elasticity of the rock lay in the range 2000 to 3000 MN/m²: this was improved by grouting to 5000 to 6000 MN/m². The lowest part of the gorge was filled in with a mass-concrete plug to a height of 57 m. The arch gravity dam above this level was constructed in 33 monoliths, each about 12 m wide.

The upstream face, varying in thickness from 1.5 m at the top to 4 m at the base, was made with concrete containing 250 kg of cement per cubic metre. The remaining major part contained 200 kg/m³.

At the interface between the two concretes, there was a drainage system of 300 mm diameter pipes over the whole height of the dam.

To allow cooling and shrinkage, alternate monoliths were constructed ahead of the others. On completion, and after sufficient cooling, the 34 vertical construction joints were grouted and coated with bituminous material to protect any tensile cracking that might develop.

18.4.5.2 Double curvature arch

To avoid tensile stresses in an arch resisting reservoir water pressure, some curvature is desirable in the vertical section. With a perfect shape, putting the concrete into pure compression, the section could be relatively thin without exceeding the safe working compressive strength of the concrete.

A model using an elastic sheet in pure tension, bulging under the water pressure, gives the shape in mirror image for pure compression. In practice, some abutment deformation, temperature stresses and uplift pressures from water in the foundation interface and joints in the concrete affect the pure compressive stress ideal and require greater thickness to be used.

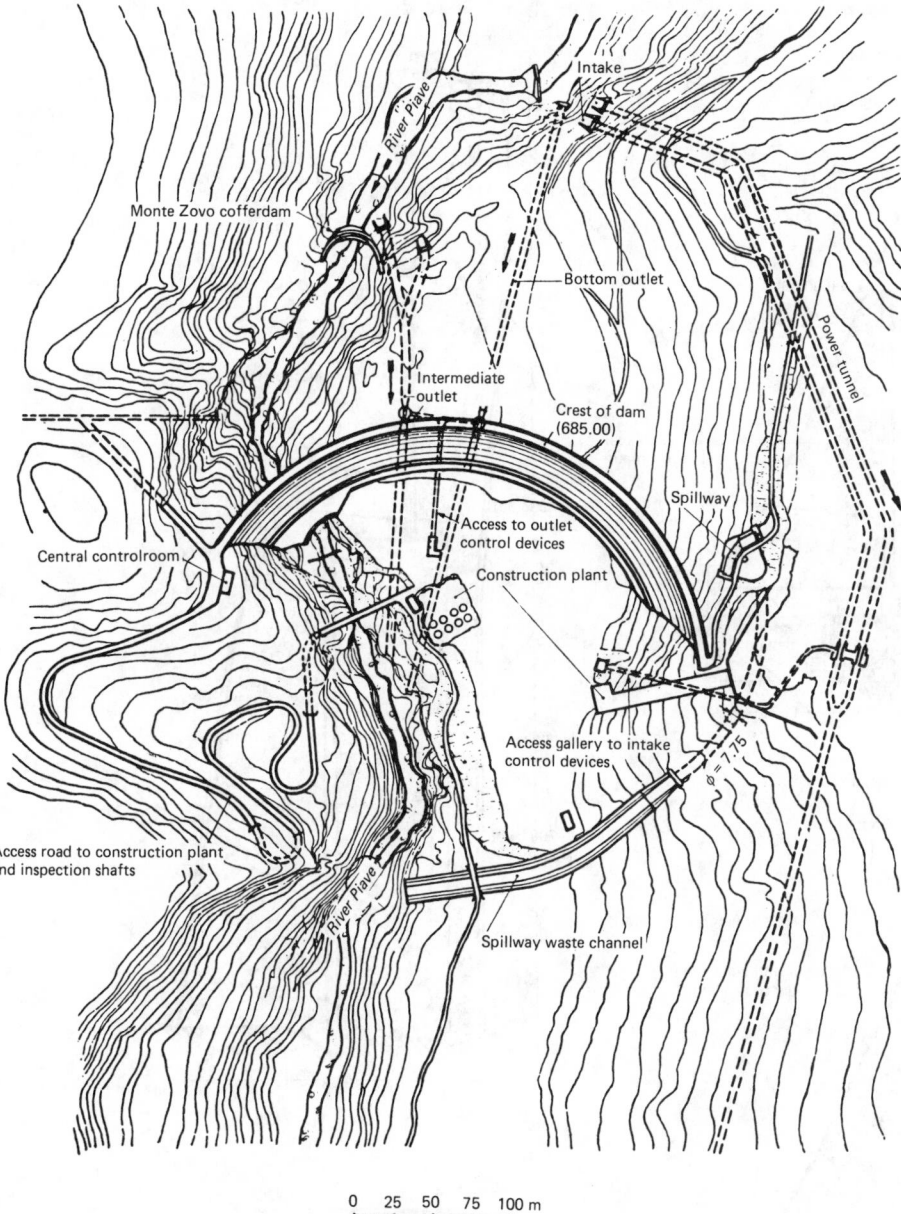

Intake

River Piave

Monte Zovo cofferdam

Bottom outlet

Intermediate outlet

Crest of dam (685.00)

Power tunnel

Spillway

Central controlroom

Access to outlet control devices

Construction plant

Access gallery to intake control devices

ø = 7.75

Access road to construction plant and inspection shafts

River Piave

Spillway waste channel

0   25   50   75   100 m

Figure 18.30   Pieve di Cadore dam: plan

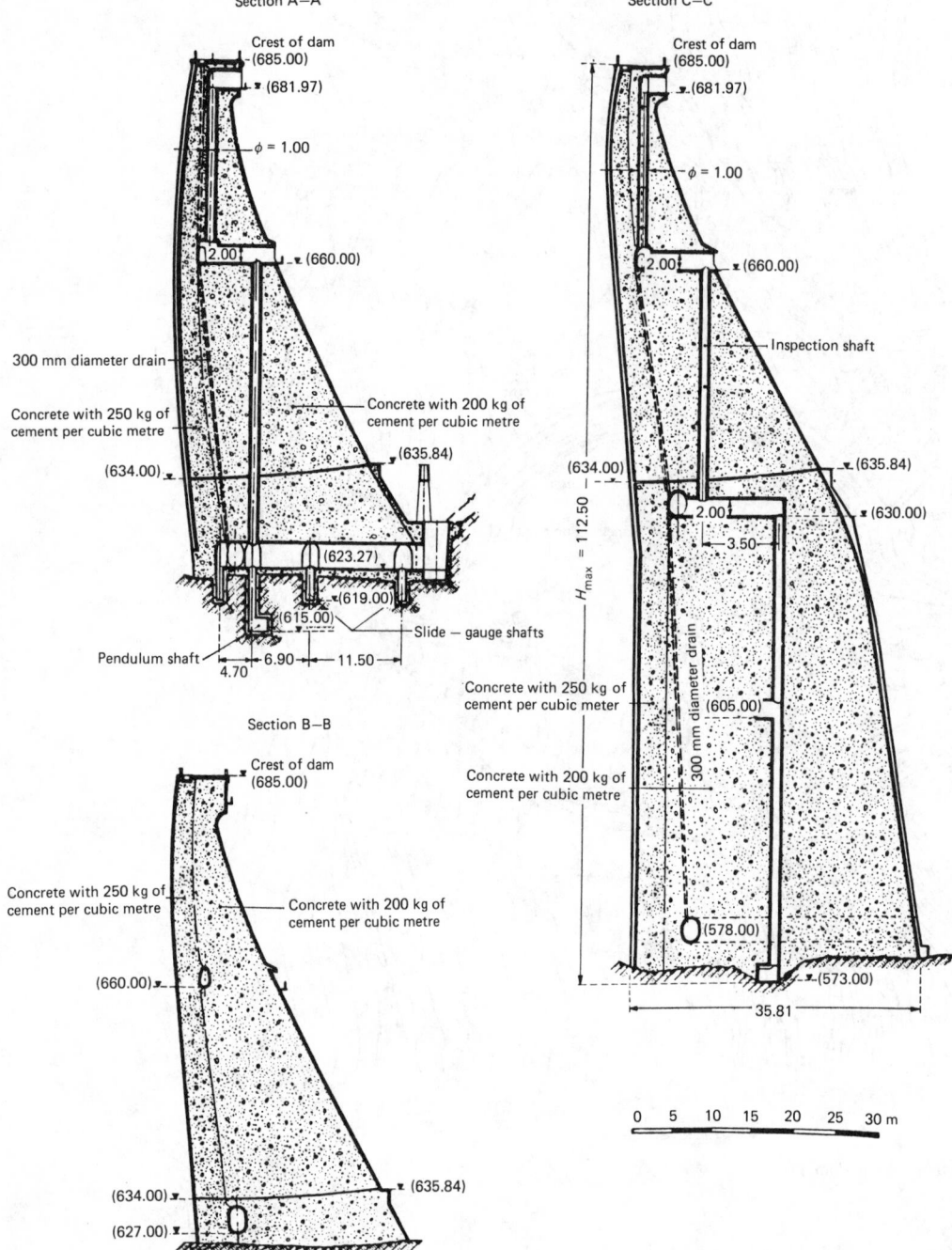

Figure 18.31 Pieve di Cadore dam: cross-sections

Example 18.10

Neves dam, Italy (94.7 m) constructed 1960–63 (Figures 18.32 and 18.33). Neves dam is founded in a gneissic granite formation. Design, based on a mathematical model using finite element techniques, was based on a modulus of elasticity for the rock of 14 000 MN/m² and a value of 28 000 MN/m² for the concrete in the lower third of the dam, 30 000 MN/m² for the upper part. The radius of the arch increased with height as shown in Figure 18.34.

The dam was built in 12 sections, each about 15 m long with concrete containing 240 kg/m³ of pozzolanic cement.

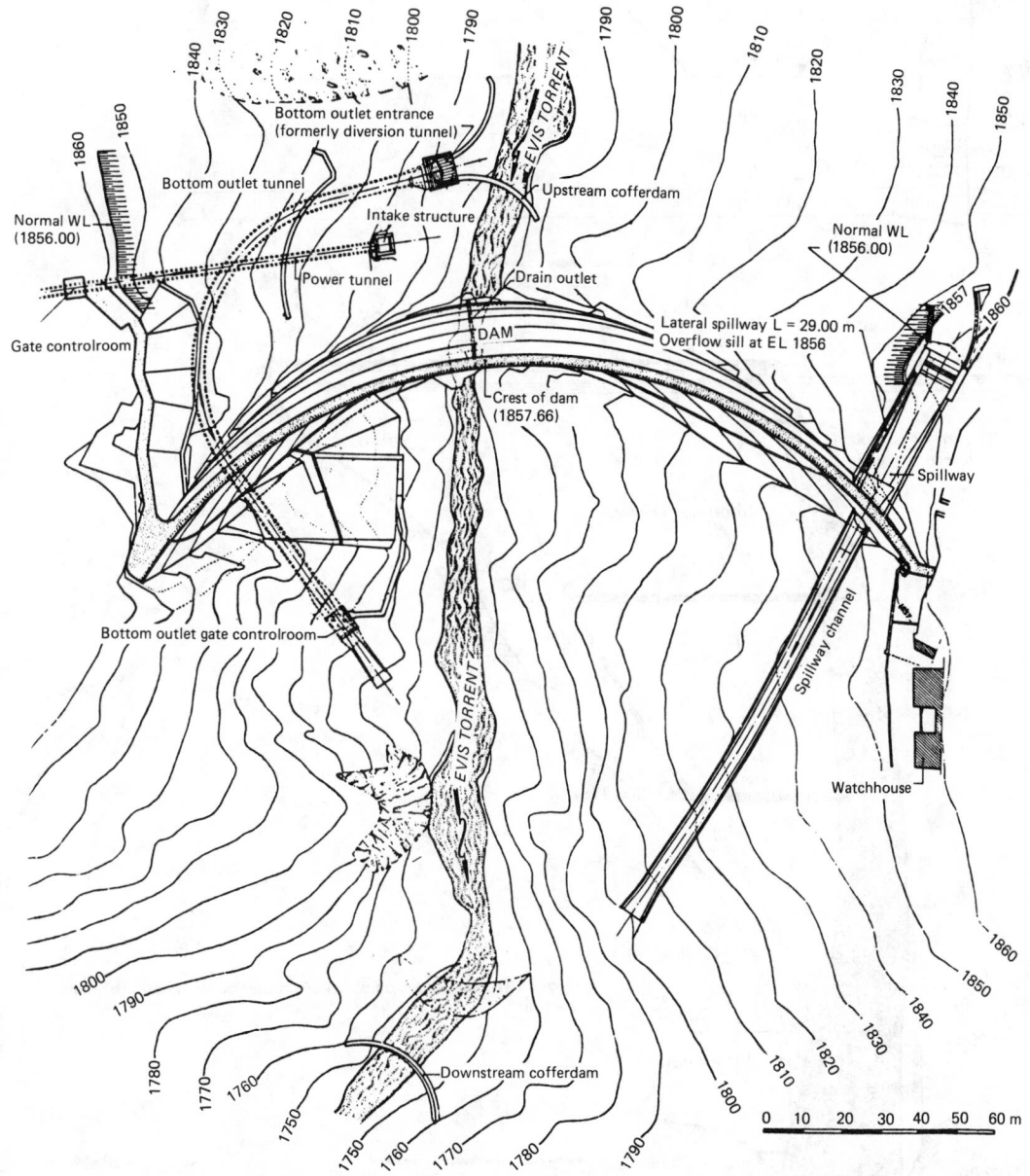

Figure 18.32 Neves dam: layout

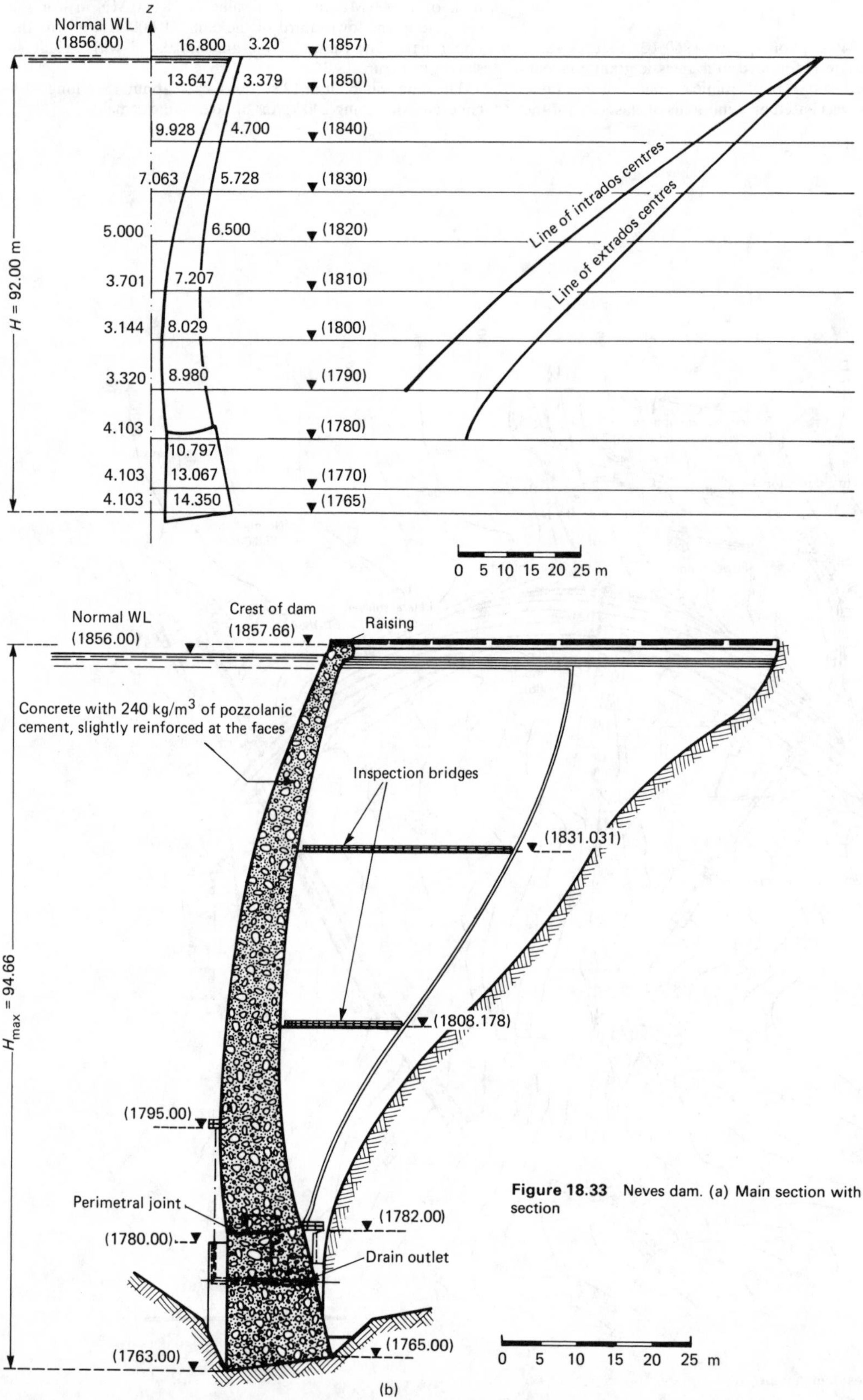

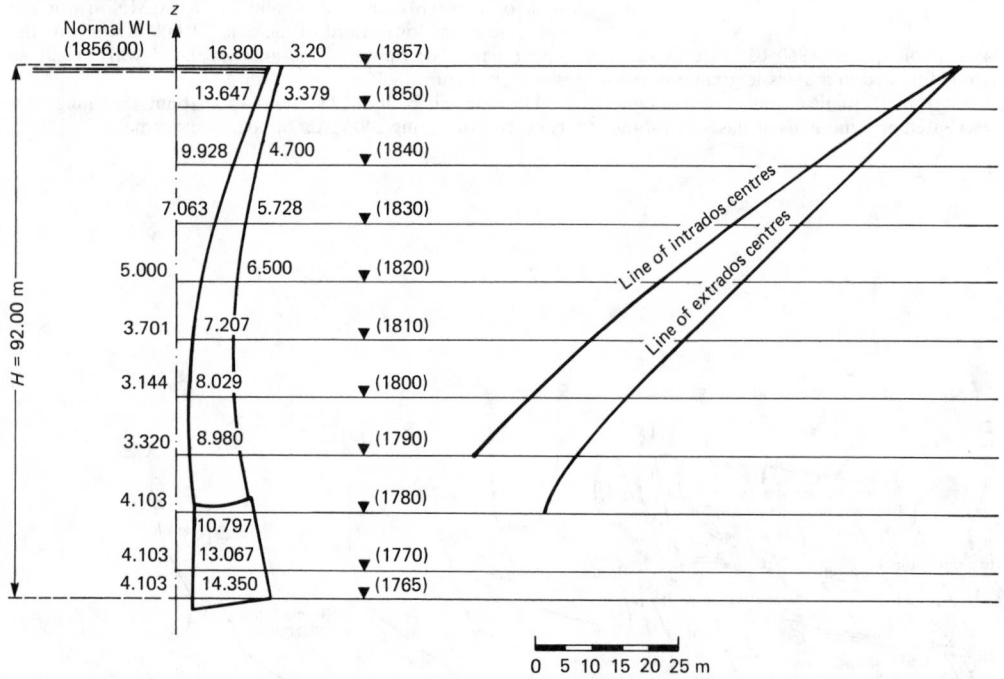

Normal WL (1856.00)

16.800	3.20	(1857)
13.647	3.379	(1850)
9.928	4.700	(1840)
7.063	5.728	(1830)
5.000	6.500	(1820)
3.701	7.207	(1810)
3.144	8.029	(1800)
3.320	8.980	(1790)
4.103		(1780)
	10.797	
4.103	13.067	(1770)
4.103	14.350	(1765)

$H = 92.00$ m

Line of intrados centres

Line of extrados centres

0 5 10 15 20 25 m

Normal WL (1856.00)

Crest of dam (1857.66)

Raising

Concrete with 240 kg/m³ of pozzolanic cement, slightly reinforced at the faces

Inspection bridges

(1831.031)

(1808.178)

$H_{max} = 94.66$

(1795.00)

Perimetral joint

(1780.00)

(1782.00)

Drain outlet

Figure 18.33 Neves dam. (a) Main section with data; (b) main section

(1763.00)

(1765.00)

0 5 10 15 20 25 m

(b)

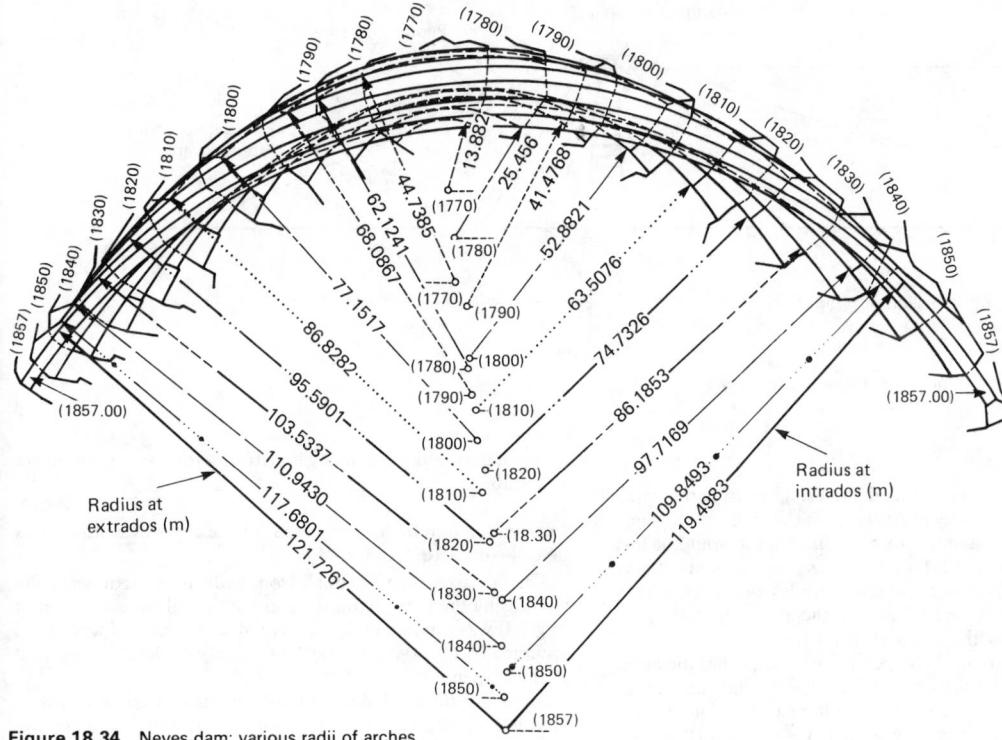

Figure 18.34   Neves dam: various radii of arches

18.5 Design concepts

18.5.1 Embankment dams – homogeneous section

Two important points must be borne in mind:

(1) It is essential to prevent the phreatic surface from reaching the downstream slope. This is most effectively achieved by placing a filter drain to separate the upstream and downstream shoulders (see Figure 18.15, page **18**/13).
(2) Foundation deformations and settlements over abutment irregularities such as those indicated in Figure 18.35 may create tensile strains that can cause visible cracks and may permit leakage to the central drain.

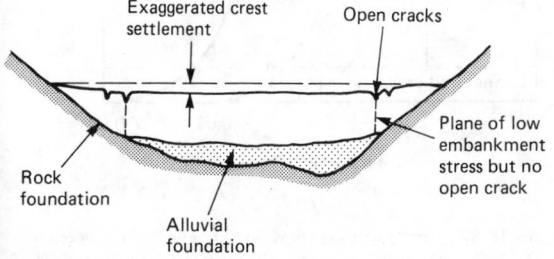

Figure 18.35   Zones of tensile strain

18.5.2 Embankment dams with central clay cores

The traditional British section had upstream slopes of 1:3 and downstream slopes of 1:2.5 (Figure 18.36).

18.5.2.1 Shoulders

The function of shoulders is to support the clay core. They must resist sliding along the foundation and any plane through the fill against horizontal thrust from the core.

Slope stability should be checked along a slip surface that might pass through core and foundation, or within the fill (see Chapter 9).

High construction pore pressures in the shoulders should be avoided by:

(1) Use of permeable fill.
(2) Provision of drainage layers.
(3) Placement of fill at a water content which allows a maximum dry density to be developed by the compacting machines to be used.

A transition is required between permeable shoulder fill and semi-impervious core material. This is often achieved by selecting borrow material so that the coarsest is placed in the outer portions of the shoulders, with the finest fraction adjacent to the core. It is good practice to place a graded filter between core and shoulder fill.

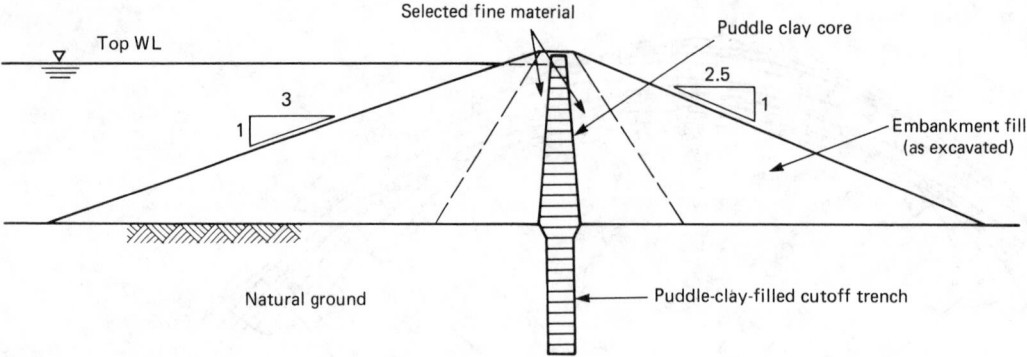

Figure 18.36 Traditional puddled clay core dam

18.5.2.2 Filters

The aim is to use a well-graded material, with pores small enough to prevent entry by particles from the core. At the same time, the filter material must not be so fine that it would be lost in voids in the shoulder fill. To satisfy these requirements, it may be necessary to use several different grades of material in a composite filter with finest next to the core graduating to coarsest in contact with the shoulder fill.

Traditional filter design is based on the concept that the pores in a granular material are only about one-fifth the diameter of the particles. Thus, the grain size of a filter can be about 5 times that of core material. In practice, filter material is well graded and the controlling size is often taken as the maximum of the smallest 15% of the mixture. When a sieving analysis has been made, producing a grading curve of the type shown in Figure 18.37, the controlling size is D_{15}.

A filter rule is:

(1) D_{15} maximum of the filter must be equal to or less than 5 times D_{85} minimum of the core (or smaller-size filter zone).
(2) D_{15} minimum of the filter must be equal to or greater than 5 times D_{15} maximum of the material to be protected to provide adequate permeability.

Specifications usually contain other restraints such as maximum and minimum sizes, absence of gap grading, etc.

The smallest particle in the clay core may be of clay size, i.e. < 0.002 mm and it is unlikely to be practical to provide a theoretically correct filter to trap this size of particle. Fortunately, most clayey fills that are used for cores are well-graded materials containing coarse sand or even pebbles. The clay particles tend to floc (unless they are dispersed by the chemical composition of the reservoir water) so that it is only necessary to filter the floc size.

Even though, initially, some clay flocs may pass into the pores of the filter, the slightly larger sizes of the clay core cannot pass into the pores and, after an initial slight loss of material, a finer filter of core material builds up against the provided filter. It is necessary to ensure that the grading of the core material will allow this to happen and that the various zones of the filter will prevent loss of filter material into the shoulder fill.

Vaughan and Soares[18] have developed a method for designing filters by using their permeability to indicate pore size. The permeability of a filter is proportional to the square of its pore size and if retained particle size is used to represent this, then:

$$k = Ad^x \qquad (18.1)$$

where k represents filter permeability, d, particle size, A is a

constant depending on other geometric factors, and x is a power of about 2.

Tests with a number of clays used for the cores of British dams gave $A = 6.1 \times 10^{-6}$ and $x = 1.42$.

The floc size d can be found from hydrometer sedimentation tests, using the local, natural water with a soil concentration of 25 g/l. It was found that floc size could be determined with equal accuracy by observing the settling velocity of the clearing front of the suspension.

Filters for the small Ardingly dam in Sussex were designed by this method. Sedimentation tests with river water gave a floc size of 6 to 15 µm, with an average of 10 µm. The above values of A and x gave a required permeability for the filter $k = 1.6 \times 10^{-4}$ m/s. A natural, medium-sized sand was found with $D_{50} = 0.4$ mm and $D_{15} = 0.23$ mm. Its permeability was 0.9×10^{-4} m/s and it has been used successfully.

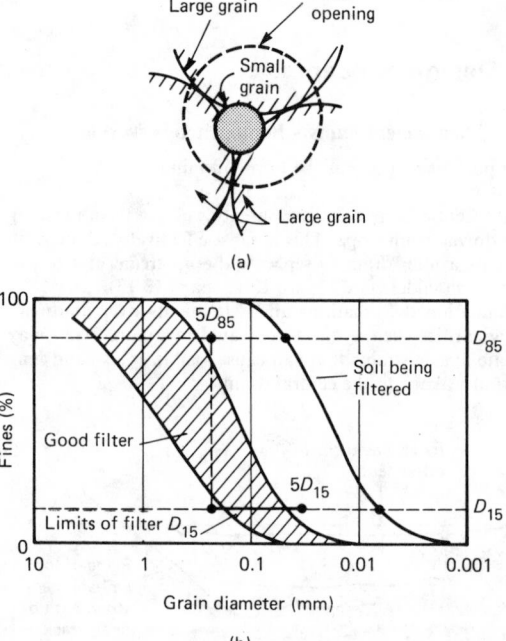

Figure 18.37 Grading curves showing filter rule. (a) Large grains screen small grains at filter opening; (b) grain size criteria for soils used as filters

Further work by Vaughan and Soares indicated values as high as $A = 6.7 \times 10^{-6}$ and $x = 1.52$. The small difference between these sets of values may be regarded as the current latitude in this new design method.

Drainage layers under and in the shoulder fill can be designed as filters to prevent fill finding its way into them. This aspect is not important on the downstream side where water flows are likely to be small. In the upstream side, however, it may be desirable to design the drains as filters, particularly when the reservoir will be subjected to repeated large fluctuations, as in the case of a pumped storage hydro-electric scheme.

18.5.2.3 Clay core

The permeability of particulate materials is dependent on stress history, as well as, for example, grain shape, size and grading.

A heavily overconsolidated clay may show a low permeability, e.g. 1×10^{-9} cm/s when an intact sample is tested in the laboratory. Field tests from piezometers usually show much higher values due to the joints, fractures, bedding planes, silt inclusions, etc. which form the fabric of the clay.

When excavated and compacted into a rolled clay core, intact fragments may retain the low permeability of the intact sample, but the mass permeability of the core may be controlled more by the joint surfaces between fragments. Compaction to optimum density will ensure that these joints are not only tightly closed, but carry across them a prestress induced by the compaction machinery – a stress which will be increased as the height of fill rises.

Hydraulic fracture. Water from the reservoir tends to percolate along a myriad joints. If its pressure exceeds the total stress acting across the joints, they may be forced apart, producing a fracture that will penetrate into the core until it meets a zone of higher stresses.

The amount by which the joints are forced open depends on confining conditions and the compressibility of the compacted clay. The fissure may not become very wide and as water migrates into the pores of the clay forming its walls, reducing the effective stresses to zero, the clay will swell and may close the fissure, if the reservoir water pressure ceases to rise.

A fissure which extends until it finds an outlet may permit sufficient flow to cause erosion which will increase fissure width, thereby further increasing flow and leading to piping. Rate of erosion may depend on how well the clay will disperse into the reservoir water. The degree of dispersion of a clay may be measured by the Volk[19] double hydrometer test or the Sherard *et al.*[20] pinhole test. A fairly clear indication may be obtained, however, from the ball test which used to be carried out on clay samples to check their suitability for use in a puddled clay core.

Ball test. Samples of the clay at the proposed placement water content were rolled by hand into balls about 50 mm diameter. These were carefully placed in a bucket of river water (to be the reservoir water) and left for 24 h. The degree of dispersion could be judged from the amount and fineness of the material that had fallen away from each ball. In the extreme, a ball could disintegrate into a loose heap and produce a muddy suspension covering the bottom of the bucket. A ball of nondispersive clay would remain intact in clear water.

Wet seams. Materials of low plasticity, such as silts, erode readily. If a fissure develops slowly, without appreciable flow, i.e. before it has reached an outlet, then, under the release of effective stress, the silt can expand into the water to loosely fill the fissure. As it progresses through a wide core, under the rising reservoir pressure, it may fill up with loose silt until, when it reaches an outlet, the permeability of the loose silt may be sufficiently low to limit flow to nonerodable amounts. The resulting seam of loose, saturated core material may be found if dry excavations are made into the core. Boreholes drilled with flush water will not disclose such seams, although when they are reached, loss of washwater can be expected.

Sherard[21] has described wet seams found in the cores of Manacouagan 3, Yard's Creek and Teton dams. He believed they originated in hydraulic fractures and proposed the above mechanism to account for their development. A more detailed discussion about the wet seams found in the Teton core during the post-failure investigation were given by Sherard.[22] The wet seams in the main body of the Teton core were not the cause of failure.

Total pressures in a core. The risk of hydraulic fracture can be reduced by increasing total stresses in the core. The maximum total stress on a horizontal plane could not be expected to exceed the overburden pressure. More usually it is much less.

It requires relatively little differential settlement between core and shoulders to develop the full shear strength of the clay. A narrow core can resemble material in a silo and its attachment to the silo walls supports some of its weight. This silo, or arching action, reduces the vertical total stress in the core, below its potential value of $\sigma_v = \gamma h = \sigma_o$ the overburden pressure, where σ_v represents the vertical total stress, γ the bulk density of core material, and h the height from considered point to crest.

If the core is assumed to have vertical sides, as indicated in Figure 18.38, then a guide to the vertical total stress at any level in the core can be obtained from:

$$\sigma_v = h\left(\gamma - \frac{c_u}{a}\right)$$

where a represents the half width of the core and c_u the undrained shear strength of the core.

In an axial direction, where the core is restrained by the valley sides, the total pressure σ_a, which may control hydraulic fracture, is given by:

$$\sigma_a = K_0 (\sigma_v - u) + u$$

where K_0 represents the coefficient of earth pressure at rest, and u the pore pressure in the core.

In order to maximize σ_v, a dense material should be used for the core and adjustments made by reducing c_u and increasing a to obtain desired values.

In general, reducing c_u increases K_0 so that maximum values of σ_a can be obtained with suitably low values of c_u.

An increased water content will reduce c_u and increase $\bar{B}$ (see Chapter 9), thereby increasing construction pore pressure u. This increase in the value of u also helps to increase σ_a.

It is desirable to have the clayey core material in such a condition that, at the end of construction, piezometric level in the core is above crest level. This may fall by dissipation as reservoir level rises and an ideal solution would be for the piezometric level to have fallen to top-water level in time to meet the reservoir water at that level.

18.5.2.4 Below-ground cutoff and core contact area

Traditional British treatment at a site on alluvium was to excavate a below-ground trench along the core centreline, taken down to a suitably impervious strata.

The trench was filled with concrete to form an impervious cutoff and this was taken up a short distance into the clay core. The top of the wall was sometimes shaped like a spearhead so that it would push up into the clay as the core settled. To avoid flow of water along the interface between concrete and clay it

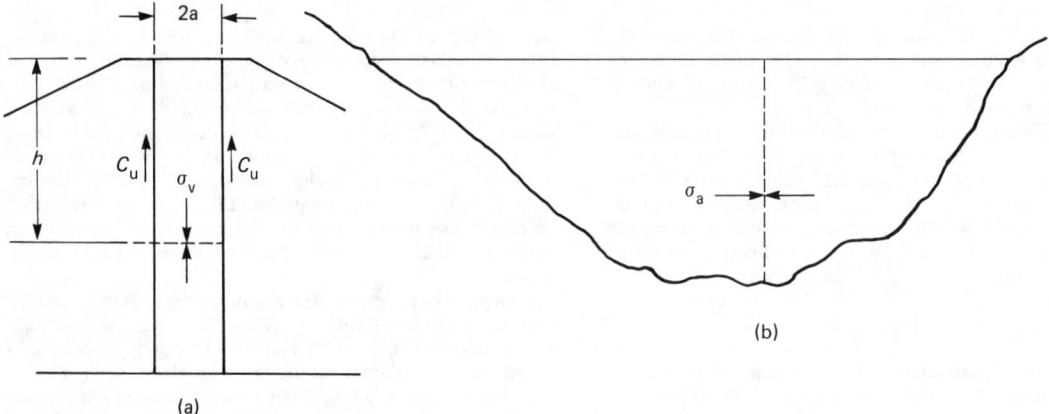

Figure 18.38 Silo or arching action in a clay core. (a) Total vertical stress at any level in a vertical core; (b) core restrained by valley sides

was necessary to have a total pressure across the interface in excess of reservoir pressure.

The deep trench required a strong support system which made the whole operation expensive. A more modern alternative is to support the trench with a slurry during excavation and place the concrete by tremie-pipe, displacing the slurry from the bottom upwards.

Since the purpose is to stop water flow under the dam, there is no need of a strong concrete wall. To allow for consolidation of the ground under dam weight, a compressible concrete is sometimes used to fill the trench.

Another method is to form the wall with contiguous bored piles constructed of concrete. This can be made by boring holes for alternate piles and then, after they have been cast, bore between them, cutting a small crescent from each.

The most common below-ground treatment is the injection of grout from small-diameter boreholes with the aim of forming a continuous wall of ground, made watertight by the grout. The type and extent of treatment depends on ground conditions. Sometimes a grout curtain is formed from one line of holes along the core centreline, but it is more usual to have three lines. The grouted width is usually increased near ground surface to at least core base width by blanket grouting from shallow holes, as indicated in Figure 18.39.

Grout can be injected as the hole is drilled, but it is more common to drill to full depth and grout in stages from the bottom. It is now usual practice to use perforated tubes covered with rubber tubes (*tube à manchette*) which are grouted into the bored hole with a weak grout. A double packer is then passed into the steel tube so that grout can be injected horizontally at any level. The rubber tube acts as a flap valve and allows the perforated steel tube to be washed out, so that further grouting can be carried out if necessary.

A test for determining whether grouting is necessary and to check the effectiveness of grouting was devised by Lugeon.[23] This requires injection of water in a borehole at a pressure of 10 bar. Flow is measured and expressed as litres per metre length of hole. A flow of 1 litre per metre per minute (referred to as a Lugeon) is regarded as the limit before grouting is needed.

This test, intended for bedrock, is dangerous because the top-of-hole pressure of 10 bar will exceed overburden pressure to a depth of about 50 m. Many sites show much higher Lugeon values over the upper 20 to 30 m, and this is often taken to indicate the degree of weathering, but it can also be due to excessively high water-pressure. Borehole diameter was not specified by Lugeon so it is difficult to convert Lugeon values to *in situ* permeability. In bedrock, water loss is along fissures, so that a determination of *in situ* permeability which would be comparable with a sand or gravel is not possible.

Centre line

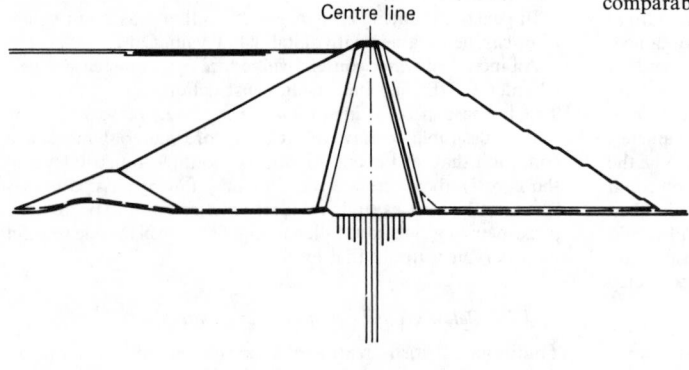

Figure 18.39 Grout holes under a clay core

Today, the danger of causing hydraulic fracture in the ground through use of excessive pressures is generally recognized, so although tests carried out are still referred to as Lugeon tests, water pressures are related to overburden pressure. This also applies to grout pressures. It is very easy to discharge large volumes of grout into the ground, but it may be opening fissures rather than sealing them.

A great deal of money can be wasted on extensive preconstruction grouting programmes on a dam site. Tests to measure *in situ* permeability should be designed on soil mechanics principles (see Chapters 9 and 11) and be carried out during the site investigation. If the results indicate that dam underflow will not cause unacceptable water loss or dangerous build-up of pressure under the downstream shoulder, consideration should be given to omitting a grout curtain.

In the interests of security, an inspection/drainage gallery should be built in a trench at ground surface to form part of the base contact area for the core.

On completion of the dam, the underflow condition can be assessed from instrument readings as the reservoir fills, when any necessary extra drainage or grouting can be carried out from the gallery. Much higher grouting pressures can then be used because the foundation is held down by the weight of the dam.

It is unfortunate that financial arrangements can seldom be made that will allow construction of a below-ground grout curtain after dam completion. If the cost of the grouting programme is not included in the estimated cost of the dam, it is extremely difficult to obtain money at the end of construction. It is also impossible to avoid the notion that the design has been a failure if grouting is found to be necessary as the reservoir fills for the first time.

18.5.2.5 Rockfill

Earthmoving machinery can handle rockfill containing pieces of 1.5 to 2 m. Heavy vibrating rollers can compact layers of 2 m thickness and these sizes have commonly been used in rockfill dams.

The rockfill should be well graded with a tendency towards an excess of the smaller sizes to ensure that the large pieces are fully bedded and do not touch. In general, most of the material from a quarry can be accepted. A limit to the amount and size of fines is when the *in situ* permeability of the rockfill is reduced to 1×10^{-5} m/s. A soft rock such as a sandstone may produce an ample quantity of the finer material, whereas a very hard rock such as basalt may be short of fines. Because of this, the situation often arises in which a rockfill of soft rock suffers less deformation than one composed of hard rock.

Placement. Segregation may be reduced by tipping the rockfill 4 or 5 m back from the advancing edge of the layer, then bulldozing it over to the level of the new layer. The larger pieces fall to the bottom and are covered by fines which produce a smooth surface that is kind to the placing machinery and makes good contact with a smooth vibrating roller, transmitting the compaction energy into the fill. Provided there are sufficient of the small fines, the voids between large pieces become completely filled and pressed into the lower surface so that surfaces between layers cannot be detected except when markers (such as coloured sand) are used. Water content should be sufficient for workability, typically 5 to 10%. An excess of water is not harmful because the rockfill should be sufficiently permeable for it to drain freely.

Compaction control. It is not practical to control rockfill compaction by *in situ* density measurements. Specification is usually of method (size of roller, number of passes, etc.) but roller performance can be measured with a compactionmeter which will show when a desired compaction has been achieved. An apparatus designed by Geodynamik in conjunction with

Dynapac Research has been described by Forssblad[24] and Thurner and Sandstrom.[25]

Stability of rockfill. The failure envelope for rockfill is curved. A typical example is shown in Figure 18.40.[27] The value of shear strength can be expressed by:

$$\tau = A(\sigma')^b$$

where τ represents the shear strength of the rockfill in kilonewtons per square metre, σ' represents the normal effective stress in kilonewtons per square metre and A and b are constants.

Values of A and b found from tests on several rockfills are given in Table 18.10.

Testing equipment used to obtain the parameters A and b requires to be fairly large, but clearly it is impractical to test the full-size rockfill. It has been found that samples obtained by sieving off the larger sizes give results more representative of the whole rockfill than samples with parallel, scaled-down grading curves. Test samples of 300 mm diameter and 700 mm height should be regarded as a minimum. Density and water content should be as close as possible to those expected in the field.

Design of slopes can be simplified by use of stability numbers given by Charles and Soares.[26] They have shown that the factor of safety:

$$F = \frac{\Gamma A}{(\gamma H)^{(1-b)}}$$

where Γ represents the stability number, H the height of the slope, and γ the bulk density of the rockfill.

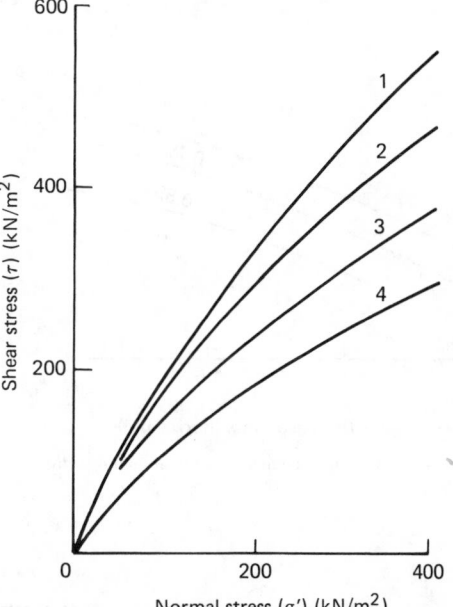

Figure 18.40 Curved failure envelope: rockfill, where (1) is sandstone, (2) slate, (3) is also slate, and (4) is basalt. (After: Charles and Watts (1980) 'The influence of confining pressure on the shear strength of compacted rockfill'. *Géotechnique*, **30**, 4)

The stability number for a given slope (1 vertical : x horizontal) is given in Figure 18.41.

As a guide to the confining pressures to be used in the tests for the determination of A and b, Charles and Soares have provided design curves to give values for σ'_m, the maximum normal stress on the critical failure surface. The curves are reproduced in Figure 18.42.

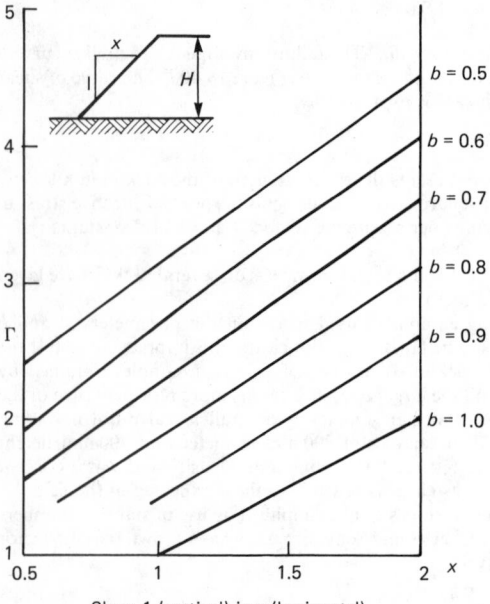

Figure 18.41 Stability numbers: rockfill slopes

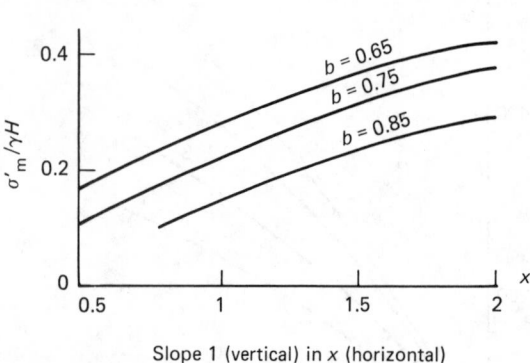

Figure 18.42 Maximum normal stress on critical failure surface

18.5.3 Roller-compacted concrete

A well-graded sandy gravel composed of rounded particles forms an excellent shoulder fill for embankment dams. It readily compacts to a maximum density and is usually sufficiently permeable to be classed as rockfill. Addition of some cementitious materials can both increase its strength so that it can be placed at steeper slopes, and reduce its permeability so that it becomes suitable for the construction of gravity section dams. By restricting heat liberation during hydration, temperature rises can be kept low enough so that long, continuous lengths will not develop shrinkage cracks. This enables it to be placed like earthfill. For a given height of dam, less volume of fill is required than with earth or rockfill, so more rapid construction can be achieved.

Mixes. Cement-stabilized earth and lean mix have been used to construct road sub-bases for half a century. Rollcrete used by Lowe[10] for the core of the Shihmen cofferdam in Taiwan was made from a concrete-type aggregate to a grading as shown in Figure 18.43. It had a maximum size of 76 mm and contained only 53 kg Portland cement and 53 kg fly ash per cubic metre (traditional concrete would contain about 300 kg Portland cement per cubic metre). Lowe used the modified American Association of State Highway Officials (AASHO) test (see Chapter 9) to determine optimum placement water content. He found this also gave maximum compressive strength, as shown in Figure 18.44.

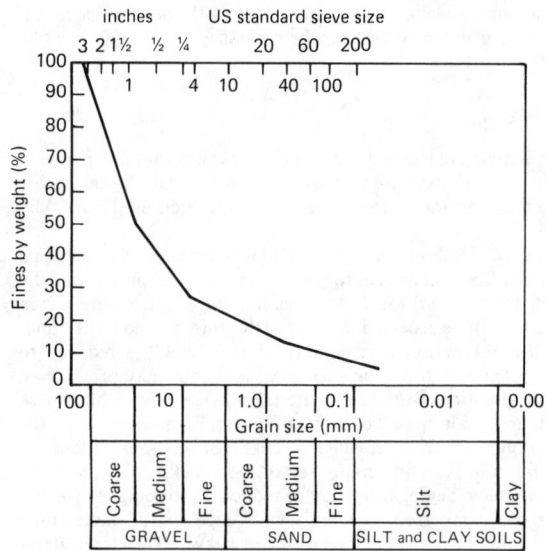

Figure 18.43 Rollcrete for Shihmen core: grading curve

Table 18.10 Values for the parameters A and b for several compacted rockfills

Rock type	A	b	Reference
Diorite	2.0	0.870	Marsal[28, 29]
Silicified conglomerate	2.6	0.846	Marsal[28,29]
Pizandarau sand and gravel	2.2	0.876	Marsal[28]
Nelzahualcoyotl conglomerate	2.1	0.881	Gamboa and Benassini[30]
Malpaso conglomerate	3.8	0.808	Marsal[29]
Carboniferous sandstone	6.8	0.670	Charles and Watts[27]
Palaeozoic slate	5.3	0.750	Charles and Watts[27]
Palaeozoic slate (weathered)	3.0	0.770	Charles and Watts[27]
Basalt	4.4	0.810	Charles and Watts[27]

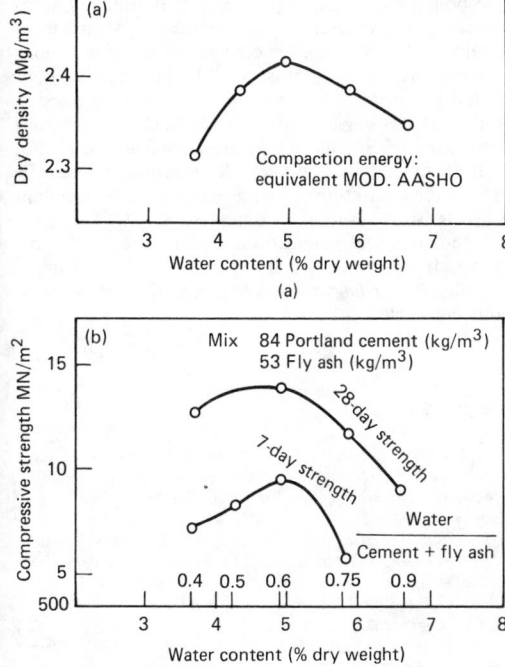

Figure 18.44 Rollcrete and water content. (a) Density; (b) compression strength

The mix used for Willow Creek dam (51.5 m), the first American dam built with roller compacted concrete and completed in 1982, contained 36.3 kg Portland cement and 14.5 kg fly ash per cubic metre. Segregation at the base of each layer produced porous joints which had to be grouted at a later stage.

The British approach outlined by Dunstan[16] has been to fill completely the voids in the aggregate to both minimize permeability and improve workability at low water contents. The air:voids ratio in a compacted fine aggregate passing a 5 mm sieve is typically 0.35:0.38.

A paste of this volume is therefore required to fill the voids completely. A study of the behaviour of joints between successive layers showed that even higher proportions of paste were advantageous.

The Milton Brook trial bank used a crushed limestone aggregate and an excess of paste with very low water content.

In Japan, the roller compacted dam method of construction has tended to follow the American roller compacted concrete approach, namely of a dry, lean-mix concrete. Fly ash is used to reduce heat of hydration but in general much less has been used than at the Milton Brook trial. Mixes that have been used for the hearting of several roller-compacted dams are given in Table 18.11. To allow for some shrinkage, transverse contraction joints, made by vibrating a cutting plate into freshly placed layers, have been spaced typically at 15 m.

Consistency tests. Concrete for roller compaction is too dry for the traditional slump test. Instead, vibrating equipment is used. The Japanese consistency meter is a steel cylinder of 480 mm diameter and 400 mm height which is filled with the mix and mounted on a vibrating table. It has a transparent plastic plate/piston, loaded to 200 kN, on the surface of the mix. The table vibrates with an amplitude of 1 mm at a frequency of 4000 c/min. The vibrating compaction value is the time, in seconds, for the cementitious paste to have risen so as to make complete contact with the underside of the transparent plastic plate.

There is a close relation between number of passes required to compact the mix with a vibrating roller and the vibration compaction value.

In designing a mix, the test can be used to adjust sand content, water content, etc. to give optimum values. The effect of varying sand content is shown in Figure 18.45.

Dam section. In a dam section, two approaches may be used:

(1) Placing a waterproof element i.e. steel sheet, bituminous membrane, or dense concrete with contraction joints containing waterstops, on the upstream face. With this arrangement, the main body of rollcrete does not have to have low permeability and can be provided with open contraction joints.

(2) In the homogeneous section, upstream and downstream faces may be formed by horizontally slipformed kerbs, precast units being held on by reinforcement laid into the rollcrete, etc. The rollcrete mix must be given low permeability by use of excess paste content – this will also enable tight joints to develop between successive layers. As with homogeneous embankment dams, drainage should be provided to prevent water reaching the downstream face.

There may be an advantage in sloping both faces: this may simplify provision of finishes to the faces and will assist stability by giving a vertical component to the thrust from the reservoir water.

Galleries. Drainage/inspection galleries can be formed:

(1) With traditional concrete, using formwork.

Table 18.11 Mixes for the hearting of several roller-compacted dams

Name	Aggregate size (max.) (mm)	Coarse aggregate (> 5mm)	Fine aggregate (< 5mm)	Cement	Fly ash	Water	Date
		(kg/m³)					
Shimajigawa	80	1476	749	91	39	105	1977–80
Ohkawa	80	1500	686	96	24	102	1978–79
Shin-Nakano	150	1468	685	84	36	90	1979–82
Pirica	80	1588	668	84	36	90	1982–88
Tamagawa	150	1544	657	91	39	95	1983–87
Mano	80	1552	726	96	24	102	1985–88
Asahiogawa	80	1500	706	96	24	102	1986–89
Sakaigawa	80	1182	752	84	36	105	1988–92

(2) As precast units, placed by crane.
(3) As sandbags and loose fill.

The first two systems interfere with placement of rollcrete and cause time delays. In the third system, the shapes of the galleries are defined by laying sandbags in the rollcrete and filling the space between with loose fill (sand and gravel or other suitable fill). In this way, the placing machinery can work over the galleries without interruption. When the rollcrete work is complete, the uncemented fill is dug out to form the galleries.

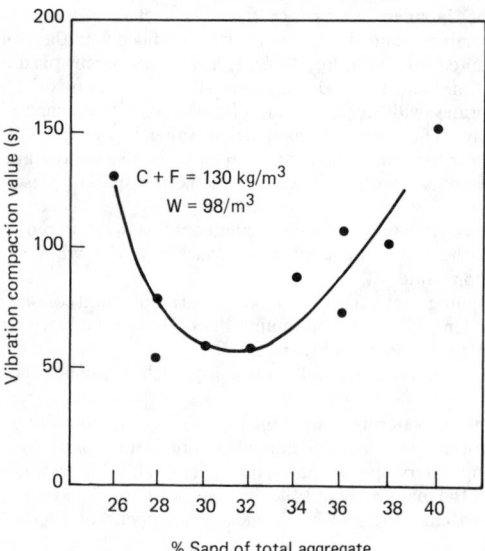

Figure 18.45 Sand : aggregate ratio and vibration compaction values

18.6 Legislation

Dams are subject to legislation in most countries. In the interests of public safety, there are usually conditions imposed on the qualifications of engineers permitted to design and supervise the construction of dams. There is also often a system of inspection of dams during their operation.

In the UK, dams retaining a reservoir of more than $24\,000$ m^3 are controlled by the Reservoirs Act 1975. This supersedes the Reservoirs (Safety Provisions) Act 1930, which required engineers to be approved and registered by the Secretary of State before they were permitted to design and supervise dam construction. Dams had to be inspected at certain times and at least every 10 years, when a report had to be made available to interested parties.

The new Act continues this general principle, but provides powers to implement recommendations for repairs or modifications contained in the Inspector's report. Appointments to a panel are for 5 years only and all reservoirs must be registered and continuously supervised by a qualified civil engineer, in addition to being inspected periodically by an independent engineer.

18.7 Further reading

The International Commission on Large Dams (ICOLD), holds congresses every 3 yr in various parts of the world. The transactions of these congresses contain a wealth of information and form milestones along the path of developing technology. The breadth of subject discussed at each congress is kept within reasonable bounds by being addressed to only four questions chosen internationally prior to each congress. Abstracts of ICOLD publications covering the contents of the transactions are available in two volumes. A list of all ICOLD publications, which includes special volumes such as *Lessons from dam incidents* and *World register of dams* as well as numerous bulletins covering specific subjects, is available from the central office of ICOLD in Paris. In the UK, information may be obtained from the Institution of Civil Engineers, where meetings are held by the British National Committee of ICOLD.

A few of the numerous publications on dams are listed in the bibliography. Innumerable papers relating to dams can be found in the *Proceedings* of civil engineering institutions throughout the world.

References

1 Kerisel, J. (1985) 'The history of geotechnical engineering up until 1700'. *Proceedings, 11th international conference on soil mechanics and foundation engineering.* San Francisco. Golden Jubilee Volume, pp. 3–93.
2 Rao, K. L. (1951) 'Earth dams ancient and modern in Madras state'. *Transactions, 4th International Congress on large dams*, New Delhi, vol. 1, pp. 285–301.
3 Buckley, R. B. (1898) 'Discussion on reservoirs in India'. *Proc. Instn. Civ. Engrs*, **132**, 213–217.
4 Pinto, N. L. de S., Materon, B. and Marques, P. L. (1982) 'Design and performance of Foz do Areia concrete membrane as related to basalt properties'. *Transactions, 14th international congress on large dams*, Rio de Janeiro, vol. 4, pp. 873–906.
5 Steffen, H. (1982) *Bituminous cores for earth and rockfill dams.* International Commission on Large Dams, Bulletin No. 42, ICOLD, Paris.
6 Kjaernsli, B. and Sande, A. (1973) *A new waterproofing technique for Norwegian dams.* Norwegian Technical Institute, Publication No. 98, pp. 1–4.
7 Lohr, A. and Feiner, A. (1973) 'Asphaltic concrete cores: experiences and developments'. *Transactions, 11th international congress on large dams*, Madrid, vol. 3 pp. 827–42.
8 Penman, A. D. M. and Charles, J. A. (1985) 'A comparison between observed and predicted deformations of an embankment dam with central asphaltic core'. *Transactions, 15th international congress on large dams*, Lausanne, vol. 1, pp. 1373–89.
9 Sherard, J. L. (1973) 'Embankment dam cracking', in: *Embankment dam engineering.* Casagrande volume, Wiley, Chichester, pp. 271–353.
10 Lowe, J. (1962) Discussion on 'The use of rollcrete in earth dams' (unpublished discussion). *1st American Society of Civil Engineers Water Resources Engineering Conference*, Omaha (reproduced in part in *Proceedings, international conference on rolled concrete for dams* (1981), pp. W1–W5, London). Construction Industry Research and Information Association, London.
11 Gentile, G. (1964) 'Study, preparation and placement of low-cement concrete, with special regard to its use in solid gravity dams'. *Transactions, 8th international congress on large dams*, Edinburgh, vol. 3, pp. 259–77.
12 Wallingford, V. M. (1970) 'Proposed new techniques for construction of concrete gravity dams'. *Transactions, 10th international congress on large dams*, Montreal, vol. 4, pp. 439–52.
13 Moffat, A. I. B. (1973) 'A study of dry lean concrete applied to the construction of gravity dams'. *Transactions, 11th international congress on large dams*, Madrid, vol. 3, pp. 1279–99.
14 Dunstan, M. R. H. and Mitchell, P. B. (1976) 'Results of a thermocouple study in mass concrete in the Upper Tamar dam'. *Proc. Instn Civ. Engrs*, Part 1 **60**, 27–52.
15 Dunstan, M. R. H. (1976) 'The Upper Tamar dam' (discussion). *Proc. Instn Civ. Engrs*, Part 1 **60**, 670–71.
16 Dunstan, M. R. H. (1981) *Rolled concrete for dams: a resumé of laboratory and site studies of high fly-ash content concrete.* Construction Industry Research and Information Association Report No. 90, CIRIA, London.

17 Schnitter, N. J. (1984) 'The evolution of buttress dams' in: *Intnl Wat. Pow. and Dam Constn.*, **36**, 6, 38–42 and **36**, 7, 20–22.

18 Vaughan, P. R. and Soares, H. F. (1982) 'Design of filters for clay cores of dams'. *Proceedings, Am. Soc. Civ. Engrs Geotech. Engrg Div.*, **108**, 17–31.

19 Volk, G. M. (1937) 'Method of determination of the degree of dispersion of the clay fraction of soils'. *Proc. Soil Sc. Soc. America.*

20 Sherard, J. L., Dunnigan, L. P., Decker, R. S. and Steele, E. F. (1976) 'Pinhole test for identifying dispersive soils'. *J. Geotech. Engng Div. Am. Soc. Civ. Engrs*, **102**, GT1, 69–85.

21 Sherard, J. L. (1985) 'Hydraulic fracturing in embankment dams', in: R. L. Volpe and W. E. Kelly (eds), *Seepage and leakage from dams and impoundments*, American Society of Civil Engineers, pp.115–41.

22 Sherard, J. L. (1987) 'Lessons to be learned from the Teton dam failure', in: *Engineering geology – special issue on dam failures*. Workshop on dam failures, Purdue. Elsevier, London.

23 Lugeon, M. (1933) 'Barrages et geologie'. Bulletin Technique Suisse Romande, Lausanne. Publication No. 58, 225–40.

24 Forssblad, L. (1980) 'Compaction meter on vibrating rollers for improved compaction control'. *Proceedings, international conference on compaction*, Paris.

25 Thurner, H. and Sandstrom, A. (1980) A new device for instant compaction control. *Proceedings, international conference on compaction*, Paris, pp. 611–614.

26 Charles, J. A. and Soares, M. M. (1984) 'Stability of compacted rockfill slopes'. *Géotechnique*, **34**, 1, 61–70.

27 Breth, H. (1964) 'Measurements on a rockfill dam with bituminous concrete diaphragm'. *Proceedings, 8th international conference on large dams*, Edinburgh, vol. 2, pp. 305–315.

28 Marsal, R. J. (1967) 'Grain forces in noncohesive soils'. *3rd Panamerican conference on soil mechanics and foundation engineering*, Caracas, vol. 1, p. 227.

29 Marsal, R. J. (1973) 'Mechanical properties of rockfill', in: *Embankment dam engineering*. Casagrande volume. Wiley, Chichester.

30 Gamboa, J. and Benassini, A. (1967) 'Behaviour of Netzahualcoyotl dam during construction'. *Proc. Am. Soc. Civ. Engrs*, **93**, SM4, 211.

31 Charles, J. A. and Watts, K. S. (1980) 'The influence of confining pressure on the shear strength of compacted rockfill'. *Géotechnique*, **30**, 4, 353–67.

Bibliography

American Society of Civil Engineers (1967). *Design criteria for large dams*. ASCE, New York.

American Society of Civil Engineers (1974) 'Inspection, maintenance and rehabilitation of old dams' *Proceedings Engineering and Foundation Conference*, ASCE, New York.

Balasubramaniam, A. S., Yudhbir, Tomiolo A. and Younger, J. S. (eds) (1982) *Geotechnical problems and practice of dam engineering*. Balkema, Rotterdam.

International Commission on Large Dams (1974). *Lessons from dam incidents*. ICOLD, Paris.

Oliver, H. (1975) *Damit*. Macmillan, South Africa.

Reservoirs Act 1975. An Act to make further provision against escapes of water from large reservoirs or from lakes or lochs artificially created or enlarged, Chapter 23. HMSO, London.

Sherard, J. L., Woodward, R. J., Gizienski, S. F. and Clevenger, W. A. (1963) *Earth and earth-rock dams*. Wiley, Chichester and New York.

Sowers, G. F. and Sally, H. L. (1962). *Earth and rockfill dam engineering*. Asia Publishing House, Bombay.

Thomas, H. H. (1976). *The engineering of large dams*. Wiley, Chichester.

19

Loadings

T A Wyatt PhD
Department of Civil Engineering, Imperial College

Contents

19.1 Loading

The process of structural design usually leads to some criterion of acceptability based on comparing the maximum predicted action of loads with an assured value of structural resistance. The assessment of the loading is thus as important as the structural analysis proper, although it has tended in the past to receive much less critical attention. This lack of attention has been fostered by a tendency for design loadings to be specified by clients or by governmental authority in broad terms to a degree of rigidity that leaves little freedom of choice to the designer.

Virtually all structural loadings are subject to some degree to statistical uncertainty; in other words, the maximum load that will act on any given structure during its life cannot be precisely known in advance, even if the probability of exceeding any particular value is known. In conjunction with the statistical uncertainty in the actual strength of any structure, the problem of safety is essentially probabilistic: a satisfactory design is one that limits the chance of occurrence of a load exceeding the actual strength to an acceptably low value (indeed, probably very small indeed, so that both its calculation and assessment of its significance are rather difficult), but does not make this event strictly impossible.

The format advocated by the International Standards Organization[1] suggests identification of a 'characteristic load' based on a statistical description of the maximum value of the load to occur in the design lifetime. The characteristic load may be taken as the expected maximum value (i.e. the mean of the values that might be observed in the life of an 'ensemble' of structures of the given type), or preferably as a value having a lower chance of occurrence, such as the expected maximum plus (say) one standard deviation of the variation across the ensemble. The characteristic value is then augmented by a partial safety factor to produce the 'design load' (F_d, say). It is often not easy to assess the relationship of these values to a 'nominal load', e.g. the value to be quoted on a notice of permitted loading. Indeed, the term 'nominal load' is best avoided, as international usage permits it to be applied to a very arbitrarily-assessed base value.

The check relevant to any limit state is then applied, taking account of the design load, a similarly-defined design material strength (f_d, say; characteristic value divided by a partial factor for the given material), and other factors. Typically, for the ultimate limit state the design condition can be separated into an action effect function S and a resistance function R and expressed by:

$$\gamma_n S(F_d, a_d, \gamma_{Sd}) \leqslant R(f_d, a_d, \gamma_{Rd}, C) \qquad (19.1)$$

where γ_n is a factor reflecting the importance of the structure and the consequences of failure; γ_{Sd}, γ_{Rd} are coefficients reflecting uncertainty in the relationship of the loads effect to the loads, and the structural resistance to the material strength ('model uncertainties'); a_d are geometric parameters of the structure; and C are any additional constraints operative. (Subscript d signifies 'design' values.)

Unfortunately, these definitions are often difficult to apply, and there are considerable variations of interpretation and application among current codes of practice. The British Standard for dead and imposed loads on buildings,[2] which covers an extremely wide range of occupancy loadings for buildings, including stadia and car parks, does not explicitly consider the probability level, or frequency of occurrence, of the loadings specified. The corresponding standard for wind loading on buildings[3] gives the parameters of a statistical extreme-value distribution that can be used to estimate the strength of the storm having any required low probability of being exceeded during the design life of the structure. It should be noted, however, that this is not the only source of variability or uncertainty in wind loading; further uncertainty is introduced in making allowance for the effect of terrain (ground roughness) and topography (ground contours) modifying the basic storm wind speed, for the various effects of gusts and in estimation of force coefficients, etc. It is common to base design on the storm having a return period equal to the design life of the structure. It is a property of the postulated extreme-value distribution that the probability of this value being exceeded at least once in the design life is 0.63; the load defined in this way is thus a statistically based characteristic value, but one having a rather high probability level. This approach is followed in the British Standard specification for loading on lattice towers[4] which has the additional feature of indicating variation of the associated load factor according to the function and consequences of failure of the tower.

Recent UK practice for traffic loads on bridges has been to specify[5] values having a relatively low probability of exceedance, although other countries (notably the North American specifications, which also have wide international influence) commonly refer to 'nominal' vehicles or arrays of vehicles. Thus, despite efforts at harmonization,[1] it remains necessary to caution the reader that load specifications are not freely interchangeable between the structural design specifications with which they interact, and great caution must be used to interpret the probability level (sometimes explicit, but still more commonly not explicitly stated) associated with each loading specification.

The objective in this chapter is to give some guidance on the fundamental characteristics of various types of loadings. No attempt is made to summarize specifications in detail, nor to give densities of building or other materials such as would normally be found in a data handbook, but rather to present background material and to highlight features where a lack of appreciation of fundamental characteristics could lead to misuse of specified values. Shortage of space has prevented discussion of certain difficult specific problems, such as the probability of simultaneous occurrence of high wind loading and ice accretion on slender structures. On the problem of snow loading, a comprehensive British Standard[6] has only recently appeared, which has a format compatible with the wind loading.[3] The British Standard for agricultural buildings[7] also makes useful recommendations in this field, and the French specification (Règles Neige–Vent) includes a helpful commentary.

19.2 Occupancy loads on buildings

The floor loadings specified for office or residential buildings have remained little changed for many years and are undoubtedly based as much on experience that the accepted values lead to a satisfactory level of safety as on detailed knowledge of actual loads in service. Two major surveys, covering office and retail premises respectively, are, however, now available.[8] The raw observations of actual weight loadings in the office premises were first used to determine the actual average load over notional 'bays' of various sizes (irrespective of the real structural systems of the floors concerned); the relative frequency of finding a 'bay' to be subject to any given load is shown by Figure 19.1, for three selected sizes of bay. The most remarkable characteristic is the wide range of loads observed, even when the average is taken over quite a large region of floor. The average observed value (including personnel or other 'mobile' loads) was about 0.62 kN/m^2, leaving lowest basement floors out of account, but loads considerably in excess of 2.5 kN/m^2 were observed, even among values averaged over bays of 100 m^2.

Values of load having 99% and 99.9% probability of not

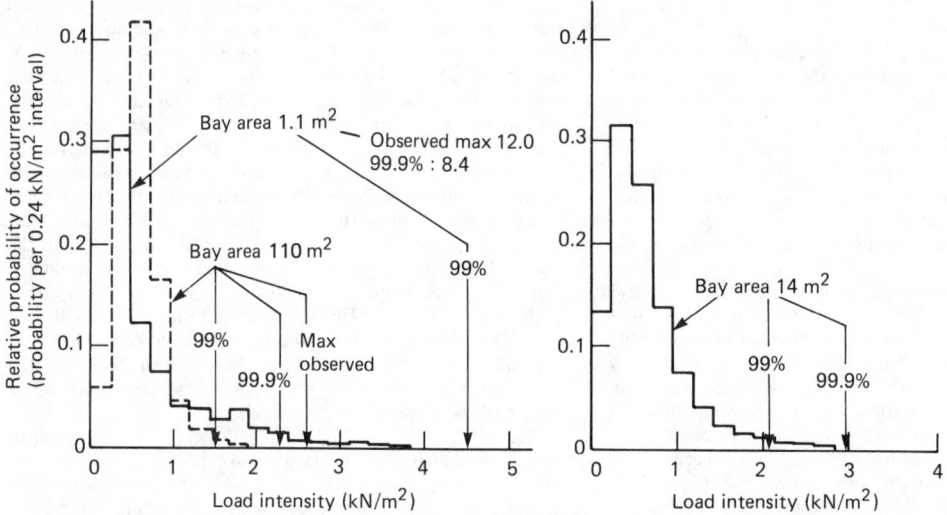

Figure 19.1 Observed local intensities in office buildings (ground floors and basements excepted)

being exceeded in the design life (which has been taken as equivalent to twelve complete changes of load such as would occur on a change of occupancy of the premises) for office premises taken from the International Standards Organization, document DIS 2394, are given in Table 19.1.

Table 19.1 Average load intensities in office premises corresponding to 99% and 99.9% probabilities of not being exceeded with 12 changes of occupancy (excluding ground floors and lowest basements)

Area of bay (m²)	1.1	5.2	14	31	111	192
Load for 99% probability (kN/m²)	9.4	4.3	3.2	2.6	2.15	1.7
Load for 99.9% probability (kN/m²)	17.4	5.3	4.3	3.5	3.2	2.3

The UK Code BS 6399 Part 1 (1984)[2] specified 2.5 kN/m² for general office premises, with a moderate reduction permissible when designing beams or further members supporting areas greater than 40 m². It is difficult to relate the above results to the standard format, as the sensitivity of the values to the probability considered is such that the usual partial load factors applied to *strengths* would no longer adequately fulfil their role of assuring a consistent level of safety between different structural types or materials. Furthermore, the dependence of this sensitivity on the size of bay suggests that the partial factor applied to the load would have to vary with the area. This problem may prove to be better treated by a more advanced probabilistic specification format.

Broadly, however, simple replacement of the existing code values by the 99% probability values with an appropriate reduction of partial load factor (say 1.25 in place of 1.6) would give a more rational balance of safety against size. It should be noted that a moderate improvement of safety with increasing size is desirable, in view of the likelihood of more serious consequences following the failure of a large bay. The variation of design load with bay size would thus be much more than hitherto accepted. The extent to which the average load intensity on any bay should be modified according to the shape of the influence function for the structural effect under consideration,

to allow for local concentrations of high intensity within the bay area, is also discussed in the reference quoted.

The office occupancy load survey also permits some general observations on the nature of the load. The occurrence of high values of loading was commonly associated with shelving, often in conjunction with filing cabinets. The relatively frequent change of building occupation is an important factor, and it is suggested that it is unwise to assume that these heavy items will be restricted to particular floor zones throughout the life of a building. The loads resulting from computer equipment have been shown not to require special consideration.

The survey of retail premises by Mitchell and Woodgate mentioned above shows a clear distinction between sales zones and non-sales zones; the latter were particularly important in food retailing, amounting to roughly half the area of such premises and subject to much heavier loading than the actual sales zones. Books and ironmongery also showed heavily laden storage areas, but with these exceptions the distinctions between trades were not very important. Taking all trades together, the result obtained for the load intensity having 99.9% probability of not being exceeded in fourteen changes of the 'fixed' loads, including an allowance for the weight of persons in the bays concerned, is shown in Table 19.2.

Table 19.2 Average load intensities in retail premises corresponding to 99.9% probability of not being exceeded with fourteen changes of occupancy irrespective of trade

Area of bay (m²)	1.1	5.6	15	28
Load on sales areas (kN/m²)	9.1	5.4	4.0	3.3
Load on non-sales areas (kN/m²)	18.2	10.8	7.7	6.3

The statistical variability was rather less wide than in the case of office premises, and the reduction of load intensity as the area increases was also less marked; for sales zones there was virtually no further reduction beyond 28 m² area, but insufficient evidence was available for larger areas in non sales zones.

For buildings for special purposes the lessons of the possibility of wide statistical variation should be borne in mind and in particular the possibility of change in use, unless the structural layout imposes positive constraints on the use that would ensure qualified professional consideration being given prior to any major change. One important practical example of such constraint is the multi-storey (or any other roofed) car park, where restricting headroom to approximately 2 m effectively ensures that vehicles heavier than private cars are excluded.

19.3 Containers for granular solids

The general problem of forces in a body of granular material is more appropriately classified in the field of soil mechanics rather than loading, but some important special factors can occur, particularly in the form of large transient forces or dynamic effects during the discharge of material from bunkers or silos. The terminology is somewhat imprecise, with no strict distinction between these two terms; both can be described as bins (a common usage in the US). A hopper may be a container with inclined walls only (i.e. an inverted cone or pyramid), or the section with inclined walls forming the base of a parallel-sided bin.

It is also necessary to distinguish between 'mass flow' and 'funnel flow' when discharging. In mass flow the movement of material towards the outlet is uniform across the cross-section with the exception of a fairly localized 'boundary layer' adjacent to the walls, whereas in funnel flow movement is localized in a relatively narrow pipe or core which is replenished from the top. The former behaviour is often called for when storing perishable material to ensure that material is discharged in approximately the same order as it was loaded, but has the disadvantage of being associated with considerable increases during discharge in the loads acting on the walls of deep bins.[9] These increases are imperfectly understood; they appear to be rather inconsistent, and although often referred to as dynamic loads they are not generally true inertial effects.

The most common basis for design of deep bins is the theory of Janssen, the horizontal load p_h at a depth h below the free surface being related to a material parameter k by the equation:

$$p_h = \frac{\rho R}{\mu'}(1 - e^{-h/c}) \qquad (19.2)$$

where ρ is the density of material, R the ratio of area to perimeter of bin cross-section (one-quarter of diameter for circular bin), μ' the coefficient of friction of material on wall, and $c = R/k\mu'$.

In Janssen's derivation, k is the ratio of the horizontal pressure to the vertical pressure in the active state, given by $k = (1 - \sin\phi)/(1 + \sin\phi)$, ϕ being the angle of internal friction of the material. This angle may in practice be somewhat less than the angle of repose; some values are given in Table 19.3. The coefficient of friction of these materials on a concrete wall is

Table 19.3 Granular materials

	Angle of friction, ϕ	Density, ρ (10^3 kg/m^3)
Gravel	35°–45°	1.6–2.2
Coal	20° (fines) to 40° (washed coal)	0.9
Grain	30°	0.5 (oats) to 0.8 (wheat)
Cement	10°–18°	1.4

about 0.5, rather less on a steel wall. Experimental results generally imply a rather lower value of k than given by the above (i.e. smaller loads near the top but asymptotically the same at greater depths). The larger value $k = 0.5$ suggested by Reynolds would seem to include some allowance for the dynamic effects, although the increase of load thus predicted does not entirely conform to the description that follows.

The dynamic effects during discharge may increase the effective horizontal loading by a factor as great as 3, and many failures have been reported as thus caused. According to Jenike[10] (who also gives an excellent bibliography) the explanation is that the condition at rest is approximated by the 'active' state with the major pressure nearly vertical (as in Janssen's theory), but that on withdrawal of material from below there is vertical expansion producing a 'switch' to a 'flow' condition with the major pressure nearly horizontal. This approximates to the passive pressure state, corresponding to arching across the bin. Once this state is established each 'arch' has only to support its own weight and the horizontal loading is again similar to the Janssen theory, but at the instant of 'switch' the top of the arching region has to give some support to the 'active state' material above, producing a very high horizontal loading locally. The vertical expansion required to cause the switch is very small, so that the switch generally propagates rapidly upwards and the strength provided must at all points cater for the corresponding concentrated load. At the time of writing these theories had not been fully verified and demonstrated. The Russian specification (see Jenike[10]) suggests that the basic value given by the Janssen formula should be doubled over the lower 65% of the height of deep bins to allow for this dynamic effect. The lower 15% of height of circular bins with flat floors are also exempted from the dynamic loading, because of the formation of a dead zone of inert material. The specification of the American Concrete Institute for grain silos requires allowance only where the outlet is markedly eccentric, a condition which can give rise to severe 'ovalization' loading; qualitative warning is given about dynamic effects in other materials.

Dynamic effects are relatively small in 'funnel flow', but the design features necessary to ensure funnel flow are not fully established. Funnel flow is assured if the depth does not exceed 1.3 to 1.5 diameters, the lower limit applying to grain silos. A perforated tube or a lattice tower placed over the discharge orifice also prevents the type of mass flow that can cause dynamic loading. Projecting circumferential fins on the inside face of the wall are of rather less certain action. The problem of very fine materials such as cement has been discussed by Leonhardt et al.[11]

19.4 Road bridges

Probabilistic considerations are also important in traffic loading on road bridges. Where the loaded length is sufficient to admit more than one loaded vehicle, the governing condition will occur when the traffic is brought to a standstill, minimizing the separation between vehicles. The maximum load intensity of heavy goods vehicles is very much larger than the load intensity of light goods vehicles and private cars, so the effective loading is strongly influenced by the degree of 'dilution' of the heavy vehicles. There is generally a clear gap of vehicle weights between about 25 kN and 50 kN; for census purposes, the UK licensing limit of 3 t unladen is convenient (above this limit, 'heavy goods vehicles', 'HGV'), and 12 000 lb laden (53 kN) is common in North America.

Recognition of these factors led to the British HA loading specification. When formulated in 1954,[12] it was presumed on the basis of common experience that stationary traffic implied traffic congestion and thus in turn a time of high traffic demand

and consequently a high proportion of light vehicles. Three maximum load-intensity vehicles (at that time, 24 t four-axle vehicles about 7.2 m long for bulk loads such as heavy liquids or powders) were considered to occur consecutively followed by a greatly diluted vehicle sequence, giving a lane load falling from about 30 kN/m to about 5 kN/m. Specifications in other countries followed similar lines, albeit with a less dramatic decrease of load intensity with increasing span.

When the Motor Vehicle (Construction and Use) Regulations were amended in 1964 to permit total vehicle weights exceeding 24 t, *minimum* permissible values of wheelbase were specified for such vehicles, such that the heaviest (32 t) vehicles would give a load intensity of about 25 kN/m. This was designed to protect bridges against overloading. However, it rapidly became apparent that very large numbers of maximum-weight vehicles were being put into service, and the totally revised British Standard for bridges issued in 1978[5] included a modification to the HA loading such that the lane load for very long spans did not fall below 9 kN/m.

The Construction and Use Regulations were further eased in 1978 and, more significantly in terms of loading, in 1982 (effective 1983) when the maximum permitted vehicle weight became 370 kN (38 t) for a vehicle of length between 12 and 15.5 m. It was no longer possible to require (or, indeed, to permit) a corresponding increase of vehicle length. A substantial proportion of the total traffic can now be expected to contribute 25 kN/m, and several vehicle configurations are permitted to reach 30 kN/m, to which allowance may be added for overloading.

Recent re-examination of the bridge specifications, backed up by quantified statistical reliability analysis, has drawn attention to the possible importance of obstructions to traffic flow that may occur even at times of low flow, e.g. due to inadequately secured loads. The increased journey speeds permitted by motorways have greatly reduced the motivation to night-time travel, given the relatively short travel distances within the UK, and the residual night-time traffic commonly has a very high proportion of large heavy vehicles. Examples have been recorded on British motorways of 1-h traffic counts in the very early morning with well over 80% heavy goods vehicles, and thus over 90% heavy goods vehicles by length in any stationary queue that might have formed from this traffic. Furthermore, such 'off-peak' traffic has a significantly higher proportion of loaded vehicles than the long-term average. The current *average* weight of a four-axle articulated vehicle in these circumstances is over 20 kN/m. Taking both traffic constitution and the probability of blockage into account, attention has now focused on the period around 6 a.m., when larger flows occur, but still with typically 60 to 70% heavy goods vehicles.

As a result of this reappraisal, the basic unit lane load

$$W = 36/L^{0.1} \text{ (kN/m)} \qquad (19.3)$$

has been put forward for loaded lengths exceeding about 50 m, where L is the loaded length in metres. At (for example) $L = 400$ m, this gives 19.8 kN/m, which is more than twice the value given by the existing BS 5400 or the preceding BS 153. A coexistent point load ('knife-edge load') of 120 kN is proposed (unchanged), which now has rather little influence on the total load effect. A value of this order would be specified for new designs for the British Department of Transport. However, the probabilistic description of the maximum load events now envisaged differs considerably from that considered when the various load factors were selected for the existing specification, and the final outcome remains to be seen.

For short loaded lengths, the governing case in the UK has hitherto generally been the HB loading, representing special-purpose vehicles which travel under supervision. It may, however, prove necessary to include a new approach to the HA loadings for short lengths in future specifications.

Recent thinking in North America has recognized similar trends, although the result has not yet been as severe as outlined above. Vehicle weights have shown the same inexorable upwards trend, and great concern has also been expressed about vehicle overloading. The latter is perhaps exacerbated by discrepancies between individual state/province regulations, and widespread use of 'citizens' band' radio has impeded enforcement. It has been suggested by Buckland *et al.*[13] that a maximum truck weight of 530 kN should currently be considered. In terms of current influence and importance, the 1979 Ontario Bridge Code should be cited, together with the AASHTO Standard Specification for Highway Bridges as revised in 1977. Buckland *et al.* give much background to recent proposals by the American Society of Civil Engineers.[14] It appears, however, that the proportion of vehicles approaching maximum weight is relatively low in North America, and the postulated overall average weight of the heavy vehicle component of traffic is only about 10 kN/m. The overall average of heavy goods vehicles in the UK is similar, but the weights in critical conditions range much higher, as noted above.

Buckland *et al.* also define clearly their assumptions concerning the frequency of traffic blockage and on the behaviour of traffic when only a fraction of the carriageway width is obstructed. These factors have a strong influence on the appropriate value of loading for design (including the relevant load factors), especially when the diurnal pattern of variation of traffic is taken into account, but there is very little published information. In the authors' opinion, this is the aspect of bridge loading most worthy of further studies.

Typical vehicle configurations in the UK together with the idealized vehicle used in the AASHTO specification, are shown in Figure 19.2. An international comparison of specifications has been published by the Transport and Road Research Laboratory.[15]

The fatigue-check count of load cycles, being to a first approximation a long-term average, is relatively insensitive to the time-of-day factors which have emerged in the reassessment of maximum loading discussed above. Excellent guidance on the cycle count for British traffic conditions is given in appendices to the British specification.[16]

Concern over the dynamic effects of traffic on highway bridges has perhaps receded in recent years. As already noted, the occurrence of the maximum total load on any span greater than about 20 m requires that the traffic shall be stationary. For shorter spans, allowance should be made for dynamic augmentation of the load effect. Some consideration of public reaction to perceived motion of bridges caused by the passage of traffic may also be desirable.

The dynamic action of the load is predominantly a question of the movement of the vehicle on its springs (or of the 'unsprung weights' of wheels and axles on the tyres); the dynamic effect of the addition of the weight of the vehicle *per se* arising from the time it takes to travel from the end of the span to somewhere near midspan is negligible, presuming the vehicle to be running smoothly on a smooth road. The excitation is therefore at a vehicle natural frequency; the fundamental is typically about 1.4 Hz for commercial vehicles, which corresponds to a suspension having a static deflection of about 150 mm. Only rarely is the fundamental frequency significantly lower than this, although this may not remain true if the trend continues towards self-levelling suspensions that can have a much larger equivalent static deflection; it might be considered possible for the vehicle to be resonant with a structural frequency within the range from 1 to 3 Hz, the latter value being limited to older or only part-laden vehicles. The natural frequency of the unsprung masses on the tyres is in the range from

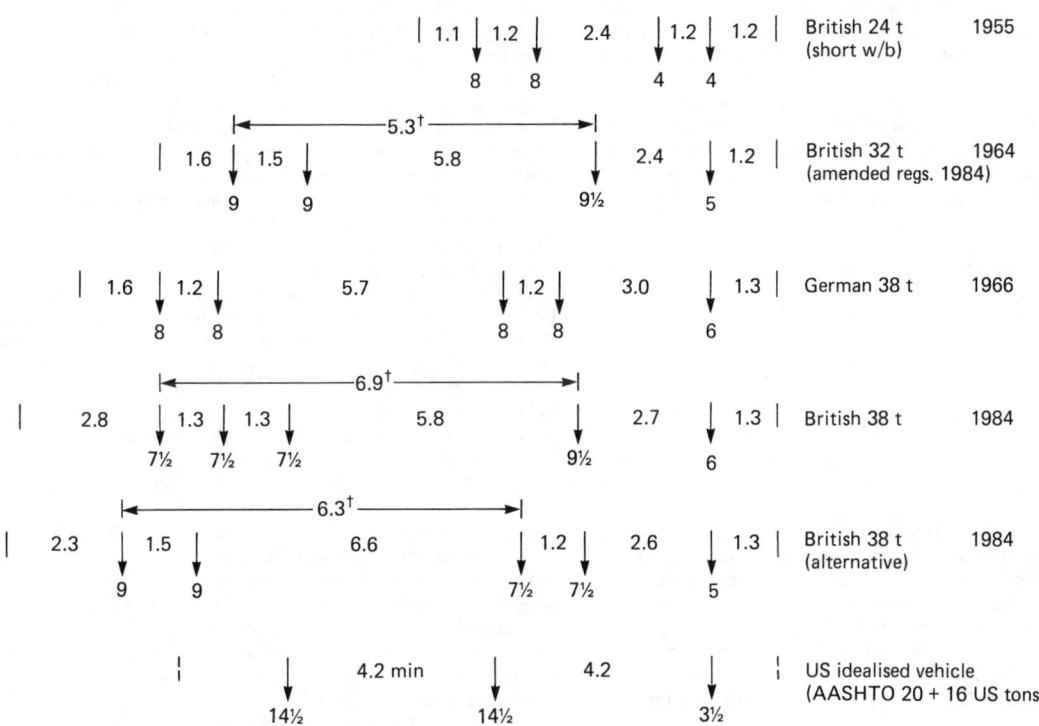

Figure 19.2 Typical road vehicles. Masses in tonnes. Dimensions in metres. Dimensions marked thus † are critical minimum values in regulations effective May 1984

8 to 14 Hz, above the range of basic frequencies of the whole bridge, but possibly significant for deck units.

It is useful to distinguish two classes of possible oscillation of the vehicle which then leads to excitation of the structure: (1) passage over a single severe road surface irregularity leading to a large vehicle motion which is then damped out by the vehicle dampers; or (2) a more random motion caused by the succession of small imperfections in the surface. The former is believed to be the governing factor on most bridges, associated usually with the joint between abutment and bridge. The first pulse (half cycle) of excitation to the bridge can be taken as applied at a distance from the bump equal to the distance travelled by the vehicle in one-quarter of the natural period, and subsequent pulses clearly progress across the span but rapidly diminish in amplitude; a vehicle damping of 15% of critical damping can be assumed, so that each half-cycle has an amplitude 0.6 of the preceding one. The amplitude of the first pulse can conservatively be taken from Table 19.4.

Table 19.4 Dynamic loading caused by single major surface irregularity

Speed of vehicle (m/s)	10	20	30
Amplitude of first load pulse / Weight of vehicle	0.25	0.4	0.6

This description of the dynamic excitation may be useful when it is desirable to assess a design with unusual dynamic parameters, or to assess the resonance effects on spans exceeding 20 m from the viewpoint of user perception. For the governing maximum total load effect on short spans, specifications are commonly based on a simple impact factor approach based on a generalization of practical experience ignoring the specific dynamic characteristics of the structure. The British specifications have hitherto included an allowance of 25% (impact factor 1.25) on the maximum effect of one axle load. A recent TRRL report[17] on measurements of short-span motorway bridges has indicated higher values, which are likely to be imposed on updating the short-span HA loading. In view of the sensitivity to vehicle speed (cf. Table 19.4), the HB load is not considered to be affected.

It remains to consider the possible reaction of users to any noticeable oscillation, especially if pedestrians have access to the bridge. A suggested approach is to consider the response of the system to a unit sequence of load pulses corresponding to a vehicle with suspension resonant with the bridge and thus to deduce the magnitude of excitation (measured by the amplitude of the first pulse) necessary to induce a response that would be considered unwelcome by a typical pedestrian, say, an oscillation building rapidly to an acceleration amplitude of 0.1 g, subsequently decaying at the relatively slow natural damping of the bridge. By comparison of the critical excitation with the nature of the expected commercial traffic density and its speed, Table 19.4 will enable an estimate to be made of the proportion of pedestrians using the structure who would regard the motion as unpleasant.

The kinematics of a person walking are such that the centre of mass of the body moves vertically over a range of some 30 mm during each pace. This clearly results in a cyclic variation of the vertical load imposed on the bridge deck, with an important Fourier component at the walking-pace frequency. This is not large enough to have a serious effect on a massive highway bridge, but requires serious consideration for light, long-span footbridges. As there is a positive tendency for walkers to synchronize their pace with perceived motion of the deck such as to augment the motion, if unfavourable response is possible it is likely to be developed fairly often and to constitute a significant problem of acceptability to the user.

A good design check procedure is given as Appendix C to the current British Standard.[5] This is based on an exciting force of amplitude 180 N irrespective of frequency: the author believes that it would be worth taking account of the typical variation of walking kinematics with pace as suggested in Table 19.5.

Table 19.5 Footbridge dynamic excitation

Walking description	Frequency Pace		Force amplitude
	(Hz)	(mm)	(N)
Leisurely	1.6	870	140
Brisk	1.9	970	240
Very hard	2.2	1040	370

Because this problem is associated with substantial resonant build-up of response, adding damping may be an effective and economical counter-measure. Friction dampers have been found to be useful, probably because they result in a change of resonant frequency with amplitude which may not be followed by the walker, losing synchronism. The possibility of wilful stronger excitation should also be considered, particularly to ensure that the structure is adequately located on its supports.

19.5 Railway bridges

Train weights and loadings are generally closely under the control of the owner, and a single train can extend to cover almost any loaded length of practical interest. The statistical variability of the estimate of the maximum loading that will occur on any specific structure is thus relatively small, but because trains crossing the bridge at speed may frequently approach close to the maximum weight, dynamic effects are important.

Design loadings may be in the form of an idealized train, specifying axle loadings and spacings (the body of the train

apart from the locomotives is commonly taken as a uniformly distributed load (UDL)); or as an equivalent UDL tabulated as a function of span. In either case, most existing specifications are based on steam-locomotive practice and somewhat outdated freight vehicle types, with the weight per unit length of the locomotives about twice that of the trailing vehicles. With modern traction, the disparity between locomotive and trailing load intensities is much less, although when two diesel locomotives run coupled together a very sharp load concentration arises from the two bogies coming adjacent to the coupling. Typical modern rolling stock for European standard-gauge railways is illustrated in Figure 19.3.

The UK specification[5] now includes, as the RU railway loading, the recommendations of the International Union of Railways (UIC) for the European Region. This is clearly described, with diagrams of the governing vehicles, in Appendix D to the specification. Loadings for urban systems, including light rail transit (LRT), are usually prepared on the basis of the anticipated rolling stock. Allowance should be made for maintenance traffic and for future changes. The British Standard gives an RL loading based on London practice as an illustration.

Dynamic effects on railway bridges were fully studied in Britain during the 1920s, with particular reference to the 'hammer blow' caused by steam locomotives. The report[18] is an excellent exposition of the factors involved, although large unbalanced reciprocating masses are a thing of the past and big changes have also affected the relevant bridge parameters; natural damping is commonly much lower, and resonant (natural) frequencies have also fallen.

A more recent investigation focused on the ratio of the peak measured bending stress to the corresponding value calculated statically from the nominal train weights, for a large number of short spans in Britain,[19] gave results summarized in Figure 19.4. The most important dynamic excitation here was probably bouncing or other oscillation of the rolling stock on its suspensions; some departures of the actual train weights from the nominal values are presumably included in the results shown.

Another dynamic excitation that has received much attention in recent years as train speeds have increased but bridge natural frequencies have fallen is the rate of application of the load to the span; it may be that somewhat excessive attention has been paid to this factor, taking note that the trend to increase train speeds appears now to have fallen away. On the other hand,

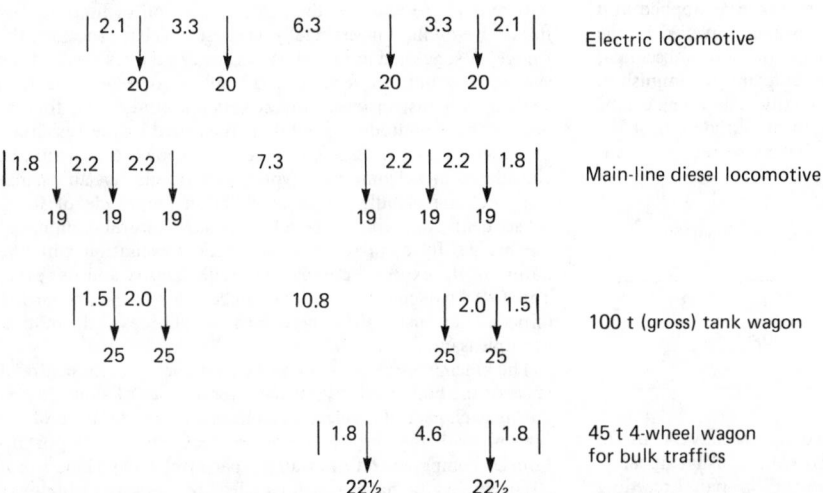

Figure 19.3 Railway vehicles. Masses in tonnes. Length and spacing in metres

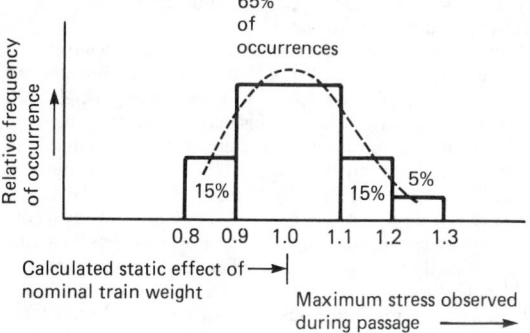

Figure 19.4 Histogram of measurements of maximum stress caused by passage of train (or single vehicle) at speed, compared with the nominal effect of the load

assumptions of improved track and vehicle suspension may have been overoptimistic, when the whole stock over a long period of time must be covered. The normalized speed parameter

$$K = V/2nL$$

where V is the train speed, n is the bridge natural frequency and L is the span, compares the time over which the load builds up (presuming the train to be longer than the span) with K being one-half of the natural period of oscillation. The UIC has recommended[20] the formula:

$$\phi' = K/(1 - K + K^4)$$

as an empirical bound to this effect, where ϕ' is an impact factor such that the maximum total bending moment is $(1 + \phi')$ times the maximum static value. Two-thirds of this value is taken for design shear force. The gross loading recommended by the UIC is an envelope comprising this impact factor, with the addition of a term making allowance for vehicle dynamic response to track irregularities (in two grades), for a variety of trains having different maximum speeds according to type, and bridge natural frequencies in the practical range of current construction. The British Standard RU loading is based thereon.

19.6 Wind loading

Wind is by its very nature a dynamic loading. It is obvious that the gustiness always noticeable in strong winds will cause a fluctuation of the loading; but in addition to this action, aerodynamic instability of the flow pattern round the structure, or interaction between motion of the structure and the flow, may cause periodic fluctuation of the loading that can result in serious oscillation of the structure. The gust action is most important as regards excitation in the downwind direction and increases rapidly with wind speed, so that this is normally the action governing the 'static' strength required. The instabilities usually cause maximum motion perpendicular to the flow, and may have their most serious effect at relatively moderate wind speeds, so that these may be most important in respect of fatigue damage, comfort of occupants, or in some cases deflection serviceability criteria. The instability excitation is usually only significant on slender structures, but in both cases it is generally true that reducing either the natural frequency or the natural damping markedly increases the risk of serious dynamic res-

ponse. Current trends in design are thus forcing designers to pay much more attention to these problems.

The basis for calculation of the required strength is the equation:

$$p = \tfrac{1}{2} C_p \rho V^2 \tag{19.4}$$

in which p is the pressure on the structure (e.g. N/m²), ρ is the density of air, 1.23 kg/m³, and V is the wind speed (e.g. m/s).

C_p, the 'pressure coefficient', is dependent on the shape of the body. A complete analysis thus requires: (1) analysis of local meteorological records and extrapolation to determine the strength of wind having a given low probability of being exceeded; (2) a 'model' of the gusts, involving definition of their fluctuation in space (area of influence of any one gust) and time (dynamic effects); (3) knowledge of the pressure coefficient for the given shape; and (4) dynamic analysis of the structure to determine the maximum value of stress in any selected structural element.

19.6.1 Meteorological data

The strength of the wind is usually conveniently expressed by its hourly mean value ($\bar{V}$, say), because at the peak of a major storm (very localized tornado phenomena excepted) the gusts can be treated as a 'stochastic' (random) process that is 'stationary' (having constant statistical properties although the instantaneous values at any point are changing) over such a period. Furthermore, in step (4) the fluctuations in the response prove to be sufficiently rapid that an hour provides a large sample, and the maximum reached in the sample is then relatively insensitive to the actions of chance: thus, to estimate the overall maximum response having a given probability of occurrence, take the hourly mean wind having that probability and multiply the mean response by the 'expected' value (average that would be found from a number of statistically similar samples) of the ratio of the peak gust response to the mean. The probability of a worse condition arising from a particularly adverse low-probability gust action in an hour of lower mean speed can be neglected.

The method usually adopted for extrapolation of meteorological records to predict the wind speed having the selected low probability of occurrence is to take the maximum values from each year of the available record, and fit a Fisher-Tippett Type-I extreme value distribution. Shellard[21] initiated the application of this method in the UK. The main difficulty is that a long run of reliable and consistent records is required, and serious distortion can occur if there is a systematic change within the duration of the record, e.g. due to change or resiting of the anemometer or even to change of its exposure.

Important progress has recently been made in improving the application of the extreme value distribution by careful correction of the raw data from meteorological stations for variation of the terrain and/or topography as a function of wind direction, by optimization of the number of storms that can be regarded as contributing usefully to the extreme value distribution (i.e. including more than simply the single biggest value each year) and by application of Lieblein's method of parameter estimation.[21]

The high-level winds which are basically dependent on the synoptic meteorology are greatly modified in their influence on any practical civil engineering structure by the roughness of the ground, averaged over many kilometres of the approach of the wind. Indeed, in strong winds the gustiness is regarded as wholly caused by the mechanical disturbance of the flow by ground obstructions.[22] Ground roughness is assessed empirically by reference to qualitative descriptions of the terrain; the usual parameter by which this is expressed is now generally the *roughness length* z_0, although this can readily be related approximately to the power law index α which is more familiar to most practising engineers (see below).

Meteorological records are usually corrected to the value applicable at a height of 10 m above ground, and are most often from sites in open terrain. British specifications are now being based on a typical inland open terrain with hedgerows and scattered trees, $z_0 = 0.03$ m. A logarithmic formulation is preferred for scientific purposes to describe the variation of wind speed with height, but the simple empirical power law remains useful and convenient:

$$\bar{V}_z = \bar{V}_{10}(z/10)^\alpha \tag{19.5}$$

in which $\bar{V}_z$ is the hourly mean speed at z (m) above ground and α is the power law index.

The wind speed at height 10 m as a function of terrain can be expressed by a factor R, such that:

$$\bar{V}_{10} = R\bar{V}_B \tag{19.6}$$

where $\bar{V}_B$ is the value for the basic open terrain. Gustiness is expressed by the root mean square value $\sigma(V)$, or by the intensity of turbulence $I = \sigma(V)/\bar{V}$; suffices are used to indicate height above ground.

A concise summary of these terrain-dependent parameters is given in Table 19.6. Further guidance on terrain classification is given in British Standard Specification 8100,[4] Engineering Sciences Data Unit items 82026 and 83045,[23] and items 74030/1 and 75001.[24] For terrain Classes IV and V the height should be measured from a substitute datum, 2 and 10 m respectively above the actual general level of the ground surface. It is the author's opinion that designers should not assume values based on any greater roughness (e.g. for city centres), because the terrain in such locations is inevitably 'heterogeneous', i.e. the roughness cannot be regarded as uniform over sufficient distances to achieve the equilibrium statistical pattern of turbulence and the flow incident to any particular structure will be influenced by specific neighbouring features.

Table 19.6 Terrain parameters

	Category and description	z_0 (m)	α	R	I_{10}
(I)	Sea coasts	0.003	0.13	1.2	0.15
(II)	Open country, exceptionally few obstacles	0.01	0.14	1.1	0.17
(III)	Basic British terrain; arable farmland with some hedges and isolated trees	0.03	0.165	1.0	0.19
(IV)	Farms with small fields, many hedgerows, trees	0.1	1.19	0.86	0.21
(V)	Extensive suburbs or mixed forest	0.3	0.23	0.72	0.25

Early descriptions of gust structure[25,26] suggested that $\sigma(V)$ was invariant with height. The most systematic investigation to date of gust structure, sponsored in the UK Construction Industry Research and Information Association (CIRIA) with results published in 1981,[27] showed that over a substantial part of the height range of practical importance, $\sigma(V)$ decreases with increasing height above ground. For very tall structures this is worth taking into account by use of the detailed results cited. The values shown above apply when the wind is sufficiently strong that mechanical mixing breaks down any thermal effects,

say $\bar{V}_{10} > 10$ m/s in British latitudes although possibly higher where solar radiation is stronger.

Topographic features have a substantial effect on wind speeds near the ground. Near the crest of hills (ridges, escarpments) the rate of decrease of wind speed with decreasing height (cf. parameter α) is greatly reduced. This may considerably increase wind loads in such locations, and reference should be made to the 1985 amendment to the general UK Code[3] or to the UK specification for towers.[4] Broadly, these presume that the hourly mean speeds are increased but that the superimposed gusts are unchanged by the topography. Earlier design proposals giving only a modified effective ground-level are non-conservative and should not be used.

The maximum instantaneous value of wind load on a large structure is dependent on the correlation of gust speeds over the area of the structure, and possibly on some time-averaging to allow for an approximation to the steady flow pattern to be established. The effective correlation of gust speeds is a function of the size of the structure relative to the cross-wind integral scale(s) of the turbulence, which express the effective dimensions of gusts in the horizontal and vertical directions. The corresponding along-wind scale parameter is also important; this can be expressed as a length (L_1 is a common notation) or as a timescale (T, say) on the basis

$$L_1 = \bar{V}T \tag{19.7}$$

The appropriate practical values of these scale parameters are still subject to argument. There has been a progressive increase in the values put forward as representative of very strong winds, and it seems likely that consensus may be reached on values not greatly different from those discussed in Papers 4 and 7 of a report by CIRIA.[27] The timescale (T) is thus about 8 s at 10 m above ground, increasing upwards approximately in proportion to $\bar{V}$ for the terrain roughness in question.

The computation of the effective correlation of gust action should also take account of the shape of the structural influence line for the load effect in question, which expresses the relative sensitivity to gusts affecting various parts of the structure. The result is an 'aerodynamic admittance' (J, say) which in its simplest form expresses the root mean square (r.m.s.) fluctuation $\sigma(F)$ of some load effect F by reference to $\sigma(V)$ and the respective mean values $\bar{F}$, $\bar{V}$:

$$\frac{\sigma(F)}{\bar{F}} = 2J\frac{\sigma(V)}{\bar{V}} \tag{19.8}$$

The factor 2 reflects the quadratic relationship between wind force and wind speed. This relationship becomes more complicated for vertical structures because of the changes in the respective parameters with height, but the additions have little effect on the practical results. Results are given in Paper 7 of the report by CIRIA.[27]

Such complete solutions may be required for structures with irregular or otherwise special influence lines, e.g. where the wind on some parts of the structure normally acts to reduce the net wind load effect, as may occur in the torsion of tall buildings, and in some bracings of towers of 'Eiffelized' profile. The UK specification for towers[4] provides design charts based on these complete solutions. For many cases, simpler generalized design formulations are sufficient,[28,29] commonly based on the assumption of a 'triangular' influence line, such as applies for the wind moment on a cantilever structure. The static response considered thus far constitutes the 'background excitation' in the terminology of these references.

A further simplification can be made by visualizing gusts as eddies carried along by the mean flow. The size of a gust will thus be in proportion to its duration measured at a point in free flow. The UK Code[3] has used this basis for many years, computing the equivalent loading for structures of dimension exceeding 50 m by using the 15 s gust wind speed. The 15 s gust is approximately equal to the mean speed plus $2.2\sigma(V)$. The increase of the consensus estimate of the gust scale parameters referred to above means that this may give a somewhat low estimate for slender lattice or 'line-like' structures, although remaining acceptable for 'solid' structures where the correlation effects are multi-dimensional.

To include the effect of fluctuation with time, recourse is made to power spectrum analysis, and the wind speed is subjected to a form of Fourier analysis, using an integral transform in place of the familiar Fourier series. In this way, the speed is represented not as a series of discrete harmonic components each having an identifiable amplitude, but as an integral of infinitesimal components over a continuous range of increments of frequency. This does not imply that identifiable periodicity exists in the wind, but does give a measure of the extent to which a structure of any given natural frequency would pick-up excitation. The wind speed spectrum $S^v(n)$ is found to have a universal shape (in the appropriate nondimensional form) for any height or terrain, as shown in Figure 19.5. A power spectrum portrays the distribution of the *square* of the quantity considered with frequency, so the units of $S^v(n)$ are $(m/s)^2/Hz$: to cover the wide frequency range of the natural wind a logarithmic scale of frequency (n, say) is preferred and $nS^v(n)$ is then plotted so that areas on the plot retain their significance as the distribution of the square of the gust speed fluctuations, $nS^v \, d(\log n) = S^v \, dn$. The use of the spectrum has been explained by Davenport[25,28] and space here permits only to point out that the spectrum can be operated upon by frequency-dependent functions expressing the correlation of the gusts over the structure and the dynamic magnification in terms of response in each natural mode of the structure. The correlation of any infinitesimal frequency component (measured by the 'normalized co-spectrum') follows similar rules to the correlation of the gross gust speed except that the characteristic longitudinal dimension is now the wavelength corresponding to the given frequency, $\bar{V}/n$. The effective crosswind dimension of frequency component n (twice 'lateral scale') can be taken as about $2\bar{V}/9n$.

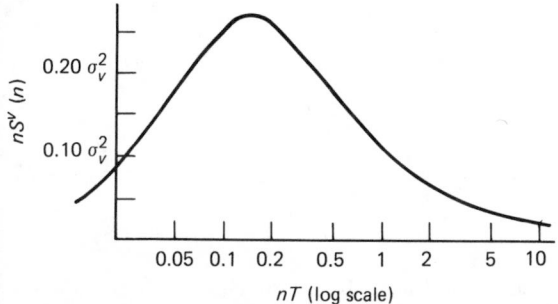

$$nS^v (n)$$

0.20 σ_v^2

0.10 σ_v^2

0.05 0.1 0.2   0.5   1   2   5   10

nT (log scale)

Figure 19.5 Wind gust speed spectrum. Universal nondimensional form for strong winds. S^v, power spectrum of wind; σ_v^2, variance of wind speed due to gustiness; n, frequency (Hz); T, time-scale of turbulence

Inspection of Figure 19.6 and the substitution of dimensions, such as for a typical example $\bar{V} = 25$ m/s, lowest natural frequency $n_1 = 0.5$ Hz, size of structure 50 m, shows that the resonant frequencies lie well on the diminishing upward tail of the gust spectrum and that the 'resonant gust' is small (here $2\bar{V}/9n = 11$ m) compared with the size of the structure. For most cases the dynamic effect is not large and is deemed to be covered by the load factor, but it is clear that the effect is sensitive to the

value of frequency in relation to size, as well as to structural damping, and investigation is advisable if either of these factors is suspected to be lower than is usual. If such study is made it is the opinion of the writer that a 10 to 20% trade-off from the load factor is justifiable. A study of this kind (or better) is obligatory in Canada for buildings exceeding 120 m in height. For structures of simple shape and simple dynamic first mode shape, tabulated solutions are available;[28,29] these are based on a marginally different algebraic form for the spectrum, but also imply a smaller scale of turbulence than the values discussed above (it may be noted that Davenport's normalized independent variable $\tilde{n} = n\mathcal{L}/\bar{V}_{10}$ is equivalent to $12nT$ in Figure 19.6). This has the effect of increasing the predicted dynamic response.

The small size of the 'resonant gust' also permits drastic simplification of the 'exact' procedure, giving a straightforward explicit formulation for the variance of the resonant contribution to structural response.[27] The variances (mean square) of the static and resonant components are additive.

The power-spectrum method of analysis is directly applicable to some problems involving deflection or vibration amplitude criteria, although consideration should also be given to the possibility of oscillation perpendicular to the wind direction as discussed below, particularly for slender solid prismatic bodies, such as very tall buildings having a uniform cross-section in plan.

19.6.2 Force and pressure coefficients

Little has been said in the above about the pressure coefficient C_p or the corresponding force coefficients C_D, C_L that express the total force resolved into components parallel and perpendicular to the wind direction (or the 'body axes' $C_x C_y C_z$ maintaining constant direction as the wind direction varies). Several collections of these factors can be used for reference, including Hoerner's wide-ranging book,[30] the current UK and other national codes, Cowdrey[31] for modern bridge sections, Cohen and Perrin[32] for lattice masts, and the present author[33] for antenna structures. Very large numbers of ad hoc tests have been reported, notably by the National Physical Laboratory (Teddington, Middlesex, UK).

For prismatic shapes having sharp edges the coefficients are truly independent of wind speed in the practical range, whereas for rounded sections there is generally a sharp change in the flow pattern when the speed reaches a 'critical Reynolds number', which is usually about 10^5: in normal environmental conditions Reynolds number $= 6 \times 10^4 Vd$, where V is the speed in m/s and d the diameter in metres. The essential feature of the change is that the point at which the flow separates sharply from the surface moves towards the back of the body, accompanied by a reduction of drag in the so-called supercritical range. This change is provoked at a lower Reynolds number by roughness of the body surface, and by turbulence in the incident flow. At still higher 'postcritical' or 'transcritical' Reynolds numbers, notably corresponding to the design conditions for large cylindrical structures including chimneys, the drag coefficient increases once more, and this effect also is sensitive to roughness. The most refined procedure is to relate the force coefficients to an 'effective Reynolds number', computed to take account of these effects.[34]

Most wind-tunnel tests in the past have been conducted in a smooth flow, with the 'gustiness' reduced as far as possible. Turbulence has deliberately been introduced in relatively few cases, and its effect is by no means fully understood. The rear-face suction ('base pressure') is principally affected and in the case of prisms of substantial 'depth' in the downwind direction (square, or downwind greater than crosswind dimension) the result is to reduce the loading (care is necessary when reading test reports as an *increase* in the absolute value of base pressure

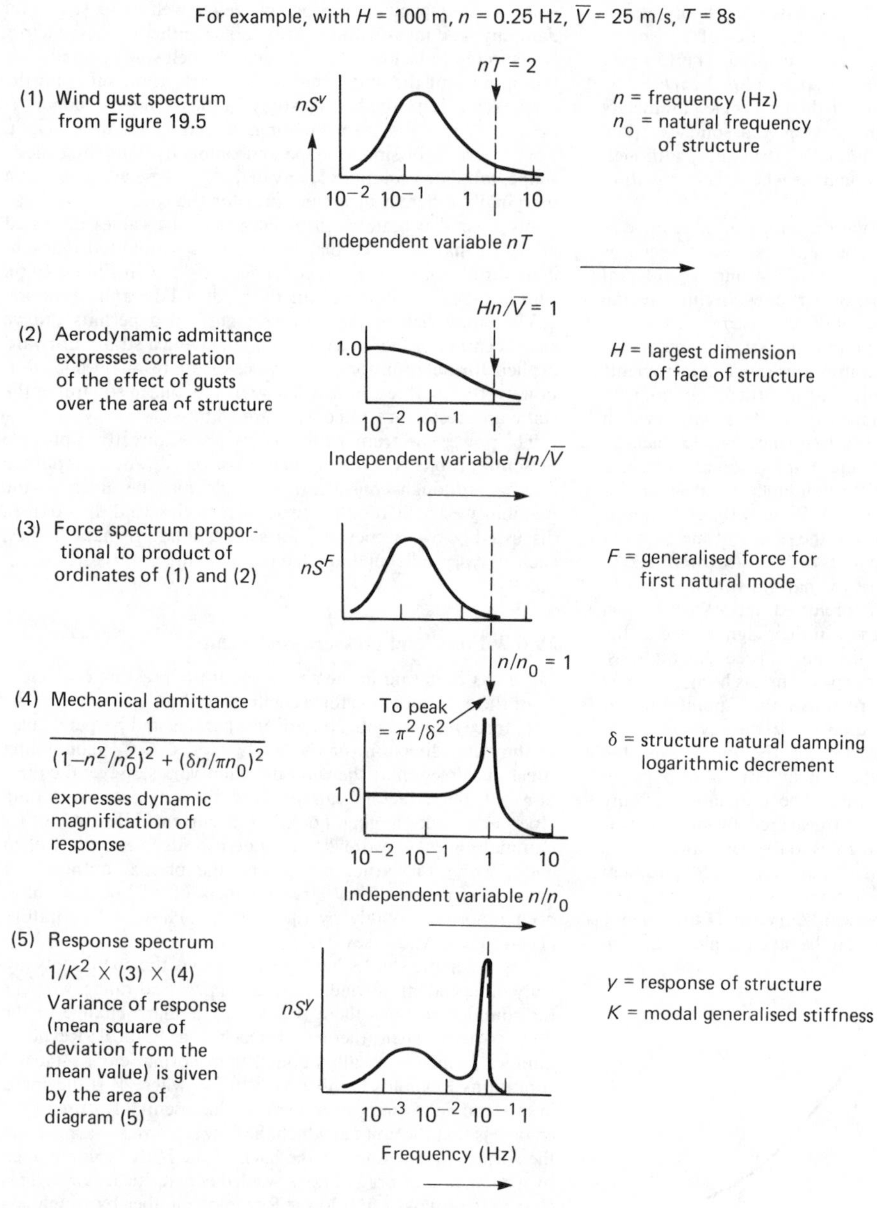

For example, with $H = 100$ m, $n = 0.25$ Hz, $\overline{V} = 25$ m/s, $T = 8$s

(1) Wind gust spectrum
from Figure 19.5

nS^V

$nT = 2$

Independent variable nT

n = frequency (Hz)
n_0 = natural frequency
of structure

(2) Aerodynamic admittance
expresses correlation
of the effect of gusts
over the area of structure

1.0

$Hn/\overline{V} = 1$

Independent variable $Hn/\overline{V}$

H = largest dimension
of face of structure

(3) Force spectrum proportional to product of
ordinates of (1) and (2)

nS^F

F = generalised force for
first natural mode

(4) Mechanical admittance

$$\frac{1}{(1-n^2/n_0^2)^2 + (\delta n/\pi n_0)^2}$$

expresses dynamic
magnification of
response

$n/n_0 = 1$

To peak
$= \pi^2/\delta^2$

1.0

Independent variable n/n_0

δ = structure natural damping
logarithmic decrement

(5) Response spectrum

$1/K^2 \times$ (3) $\times$ (4)

Variance of response
(mean square of
deviation from the
mean value) is given
by the area of
diagram (5)

nS^y

Frequency (Hz)

y = response of structure
K = modal generalised stiffness

Figure 19.6 Illustration of steps in power spectrum analysis of gust action

corresponds to decrease of load): the substantial reduction in the force coefficients specified for *buildings* in the current UK Code[3] by comparison with earlier editions is in recognition of this. Conversely, for prisms of small downwind dimension (approaching a plate) the load may be slightly increased. These effects are unfortunately scale-dependent, as they are caused by the small rapid fluctuations corresponding to 'gust wavelengths' smaller than the cross-section of the structure. For small bodies such as members in lattice structures the corresponding components of turbulence in the natural wind may not then be sufficient to cause the reduction of coefficient referred to above, but it is not possible to reproduce the natural turbulence to a sufficiently large scale in a wind tunnel to determine the limiting conditions. A further reason for distrusting many early wind-tunnel results is that the effect of 'blockage' is much greater than

was appreciated until about 1950; this is the increase in the apparent drag coefficient resulting from the constriction of the flow in the tunnel where it must pass the test body.

An important factor which is often misunderstood is the 'shielding effect', the fact that the drag force on (for example) a square lattice tower is less than the sum of the forces that would be predicted for the faces considered each in isolation. This is related to the flow 'seeing' the shape of the structure as a whole, and thus as the total 'density' of the structure increases, the incident velocities local to individual elements are reduced. It is thus not necessary that elements should be exactly 'shadowed'. Formulae for shielding factors commonly give the total load on two parallel frames as $(1 + \eta)$ times the load for one frame independently, although the total load is in fact more nearly equally divided; for example:[35]

$$\eta = 1.0 - 1.7(\phi - 0.1\sqrt{(B/D)}) \text{ (with } B/12D < \eta < 1.0) \qquad (19.9)$$

where ϕ is the solidity ('shadow fraction') of one frame and B, D are the width and spacing of frames respectively (e.g. $B/D = 1$ for a square tower).

For lattice towers it is generally satisfactory to neglect the bracings in planes parallel to the wind direction when calculating the load for the case of wind direction perpendicular to one face. For square towers the maximum load component resolved perpendicular to a face is about 10% higher, and the diagonal-incidence load about 20% higher, than the normal-incidence value. The principal difficulty lies in assessing the 'ancillaries', the ladders, cable runs, etc. which commonly substantially increase the wind resistance. To a first approximation, if the tower face solidity exceeds 0.15, judgement can be used to estimate the fraction of the area of such elements that should be added to the basic windward face area. Detailed guidance is given in the UK specification.[4]

The suction on the rear face referred to above is much less in cases with a three-dimensional flow pattern (such as short prismatic structures where flow can pass over the end as well as the sides) than strictly two-dimensional flow past a very long prism. This can be expressed by an 'aspect-ratio factor', the aspect ratio being defined as the ratio of the longest dimension of the face presented to the wind to the lesser dimension (twice this value if flow can pass one end only, as in the case of a chimney) and in the absence of specific information this can be taken as reducing the overall drag coefficient by the ratio k_1, shown in Table 19.7.

Table 19.7

Aspect ratio	∞	40	20	10	5	2	1
Reduction factor k_1	1.00	0.90	0.80	0.72	0.67	0.63	0.60

It will be noted that the effect is important even for quite large aspect ratios. A small obstruction to end flow, such as the supporting post of a signboard, is not significant, but a larger obstruction such as the deck of a bridge covering the end of a supporting pier destroys the effect.

19.6.3 Wind-excited oscillation

The flow pattern passing a slender prismatic body commonly has unsteady features, often leading to the regular build-up of vortices in the wake which, on reaching a limiting size, are swept away by the stream. Unless this occurs simultaneously symmetrically on both sides (which is unusual) the result is a periodic fluctuation of the aerodynamic force in the crosswind direction. The frequency at which this process takes place depends on the cross-section shape, and is expressed nondimensionally by the Strouhal number, S:

$$S = nD/V \qquad (19.10)$$

where n is the frequency (Hz), D the cross-section dimension (usually that perpendicular to flow) and V the flow velocity.

Thus, as the velocity increases, so does the vortex-shedding frequency, and a critical condition is likely to arise when the frequency is resonant with the structure natural frequency N (say). For this reason the velocity may be expressed nondimensionally as the 'reduced velocity', V_R:

$$V_R = V/ND \qquad (19.11)$$

The critical condition is clearly $V_R = 1/S$. For most cross-section shapes giving trouble in practice V_R lies between 5 and 8; 5 for circular or roughly circular cross-sections (decagon, etc.), 6 to 8 for rectangular sections, up to higher values for rectangular sections where the downwind dimension is large (e.g. wind perpendicular to the *narrow* face of a slab block) although the excitation then becomes relatively weak. Sufficient data on Strouhal numbers, or critical values of reduced velocity, are available to permit prediction of the critical speed for most cases, but the strength of excitation is much less regular and often requires ad hoc wind-tunnel testing if the critical speed falls within the practical range of speed for the site. The strength of excitation is often sensitive to detail in the cross-section shape, e.g. details at the leading edge of prismatic bridge cross-sections. Test results are conveniently expressed by relating the steady-state amplitude to the damping nondimensional parameter $2m\delta_s/\rho D^2$, known as the Scruton number, in which m is the mass per unit length, δ_s the structure damping as a logarithmic decrement, and ρ the density of air.

The problem is often important in the case of circular sections, such as chimneys. Excitation is reduced by presence of the free end and is thus not normally significant for chimneys of height less than eight diameters. The excitation is also Reynolds number-dependent; below the critical Reynolds number the response is typically very steady as long as the wind speed is critical. In the region of the critical Reynolds number (2×10^5 to 4×10^5) the excitation is very weak. At higher values, excitation is again present but is more random, vortices of relatively short length along the body forming and shedding with poor correlation along the length, producing a response in which the natural frequency predominates but the amplitude is continuously modulated. In this region the test results are usually quoted as an r.m.s. value, either as r.m.s. displacement, or the r.m.s. amplitude which is $\sqrt{2}$ larger. The random modulation means that allowance must be made for values significantly larger than the r.m.s.; it may be assumed that the amplitude has a Rayleigh probability distribution. In smooth-flow tests, if the r.m.s. response amplitude exceeds about 0.015D, the motion acts dramatically to improve the correlation of shedding along the length of the structure, and the response increases sharply;[36] the values thus predicted would be unacceptable to most massive structures. In turbulent flow this effect still applies, but the transition is much less sharp.

In cases such that resonance would occur at a low wind speed such that the standard mechanically induced turbulence cannot be presumed with complete confidence (limiting value perhaps $\bar{V}_{10} = 10$ m/s, see above), smooth flow should be considered. In such cases, and also where the critical condition would give a sub-critical Reynolds number (e.g. tubular structural members), or amplitudes exceeding (say) 5% of the diameter (e.g. slender steel chimneys), prediction may be based on a postulated regular harmonic excitation, characterized by a coefficient of fluctuating lift $\tilde{C}_L$. $\tilde{C}_L$ is generally taken in the range 0.25 (post-critical) to 0.4 (sub-critical).

For large chimneys the critical resonance is commonly at a speed such that turbulent flow is assured, in the post-critical Reynolds number range. Practical experience is gradually accumulating to support the use of a power-spectrum representation of the basic exciting force (characterized by the r.m.s. value $\sigma(\tilde{C}_L)$) and a negative aerodynamic damping substituted for the lock-on effect of motion of the structure.[37] Simple design formulae can be derived from this model.[38,39]

The analogy between the lock-on effect and negative aerodynamic damping has the effect that response may be very sensitive to the structural damping. It has long been recognized that unsatisfactory behaviour is unlikely in the post-critical range if the Scruton number is greater than 20.

Methods have been developed to reduce the excitation by

aerodynamic means. The addition of helical 'strakes' to chimneys is now quite widely practised,[40] and can effectively eliminate response but at the expense of considerably increased drag. An alternative[40] is the 'perforated shroud', rather more expensive to make, but minimizing the drag penalty. Responses in modes other than the fundamental have rarely been reported for massive structures, but should be considered. They have been observed on guyed cylinders and are common on tensioned cables; the singing of telegraph wires is of this type, and has led to the whole class of vortex-induced oscillation being referred to as 'Aeolian'.

The shedding of vortices from a structure can also have very serious effects on other structures nearby downwind. This effect can extend over quite substantial separations,[41] up to at least $10D$. For simple cases, such as pairs of similar chimneys, the ad hoc test technique is straightforward, simulating the upwind element by a fixed model in the wind tunnel. For complex cases, such as slender tower-type buildings in a city-centre environment, very complex procedures are required, including simulation of the general incident turbulence, which can be undertaken only at a very few specialist research centres.

There are several other forms of aerodynamic excitation of oscillation in a crosswind direction. The best known arises from a 'negative lift slope' condition, and is often referred to as 'galloping' from the very large amplitude oscillations suffered by electricity transmission lines when their aerodynamic coefficients are modified by the shape of ice accretions. This behaviour can briefly be explained as follows: downwards motion (say) of a horizontal prismatic structure causes the incidence of a horizontal wind to appear as if inclined upwards when viewed from the structure, and if this causes a *decrease* in the lift force (positive upwards), the motion is reinforced. In most cases this is not the case, but it does occur over a limited range of incidence angle for a square prism (and a range of other rectangular prisms)[42] and over quite a wide range of angle for D-section or pear-shape cross-sections as on power lines with ice accretion. Clearly, a true circular section is immune. In the simplest case, where a principal elastic plane of the structure coincides with motion perpendicular to the wind (i.e. excitation in that direction causes motion exactly in that direction) and the variation of lift coefficient (C_L) with angle of incidence (α) is closely linear in the region of the mean incidence, the critical wind speed for the onset of oscillation is given by:

$$V = \frac{4mN\delta_s}{\rho D_0 (dC_L/d\alpha)} \qquad (19.12)$$

where D_0 is the dimension used in defining C_L. The other variables are as previously defined.

Galloping is distinguished by the direct proportionality of the critical wind speed to the structural damping. If this speed is exceeded, the amplitude grows to a large value, limited only by the curvature of the C_L versus α relationship. Methods for taking this into account are given in the references quoted. In contrast, vortex-shedding excitation is distinguished by a peak of response at the speed corresponding to $V_R = 1/S$, although the amplitude may increase again to higher values at substantially higher wind speeds in a turbulent wind. In response to vortex shedding, the amplitude (rather than critical speed) is damping-dependent, commonly roughly inversely proportional; although there may be a controlling damping value above which oscillation is virtually entirely suppressed, or marking a sharp change in amplitude as related to the effect of motion on correlation of shedding discussed above.

More complicated behaviour, possibly involving the phenomenon of flutter, with or without interaction with the two mechanisms already described, can lead to strong oscillation of sections akin to airfoils (having relatively large dimensions in

the plane of the wind), such as slender bridges. Interim design rules sponsored by the British Department of Transport, together with extensive background discussion and commentary, are given by the Institution of Civil Engineers.[43]

19.7 Earthquake effects

The effect of earthquakes on civil engineering structures is primarily a question of the dynamic response of the structure excited by motion of the ground; in general, it is the horizontal components of ground acceleration that govern, although increasing attention is being paid to the effect of the vertical component of ground motion on such cases as large-span sheds having only small live load effects from other causes. The ground motion is normally assumed to be the same at all points on the foundation of the structure; it is not within the practical power of the engineer to deal with the possibility of major relative movement on some fault-line passing within the foundation, apart from site investigation to minimize the risk of building over an existing or incipient fault where movement can be expected.

The equations of motion of the masses of a structure excited by ground motion can be manipulated to give exactly the same differential equations for the displacements *relative to the ground* (and thus the strains in the structure) as for the case of the structure on a fixed base, subjected to horizontal loads applied to every mass equal to the product of the mass and the ground acceleration. A simple basis for design is thus to express an acceleration as a fraction of the acceleration of gravity (g) and to design for this fraction of the weight of the system, treated as a horizontal loading. Due to the dynamic nature of the problem, however, this equivalent acceleration is not simply equal to the maximum ground acceleration but will depend on the natural frequency (or natural period) of the structure and on the history of the ground motion extending over some time prior to the instant when maximum relative displacement is found to occur. For a given ground motion it is a straightforward matter to solve the equations of motion numerically and record the maximum response; repeating this process for single-degree-of-freedom structures of varying natural frequency (or period) leads to the *spectrum* of the earthquake. The so-called velocity spectrum, S_v, is the most commonly given form; the equivalent acceleration for design is ωS_v, where ω is the 'circular' natural frequency, rad/s. The spectrum is also dependent on the natural damping of the structure. The maximum of ωS_v typically occurs at a frequency of the order of 3 Hz; the structures of fundamental natural frequency below 3 Hz are progressively relatively less sensitive to earthquakes, although the effect of higher modes may become significant.

Unfortunately, the prediction of a ground motion to form a reasonable design basis for any specific structure is subject to many uncertainties. An earthquake occurs when strain energy gradually built up in the Earth's crust is suddenly released by movement on some fault plane. The energy released is measured by the *magnitude* of the earthquake, whereas its effect at some point on the ground is the *intensity* at that point. A rather crude single-parameter measure of intensity is given by scales such as the Modified Mercalli or Rossi-Forel ratings, which are based mainly on an only roughly quantified description of the human sensation or structural damage experienced (or expected).[44] The intensity experienced at a given distance from an earthquake of given magnitude depends greatly on the subsoil or shallow-rock conditions and considerably worse ground motion can be experienced where a thick layer of low-density low-stiffness material overlies heavier, stiffer, material. The duration and frequency content (and thus the shape of the spectrum) can also vary greatly even for cases where the overall intensity rating would be similar; a motion of given intensity recorded close to a low-magnitude shock would be shorter and have higher predominant frequencies by comparison with motion of the same intensity recorded distant from a high-magnitude shock. In the case of energy release from long faults the movements may be progressive along the fault, again leading to considerable differences in duration and frequency content from point to point.

The final factor to be introduced before describing the most useful approaches to design is that experience has shown that for most structures and in most regions where earthquake is a major design consideration, it would be highly uneconomic to base design on an 'elastic' or 'no significant damage' criterion. For most structures the aim must be to prevent major failures causing collapse and loss of life, while making use to the full of the possibility of inelastic structural behaviour resulting in dissipation of energy that is to a substantial degree analogous to increased structural damping. The obvious exceptions to the application of this principle are cases where even moderate damage must be prevented, such as nuclear reactor containment vessels, or buildings housing vital post-disaster services.

The most widely used format for a design code incorporating the factors described above is exemplified by the Unified Building Code of the US. The total horizontal load (base shear) V is given by:

$$V = ZISKAW \qquad (19.13)$$

in which W is the weight of the structure. Z, the zoning factor, reflects the basic seismicity of the region, modified where necessary by the soil effect factor, S. Factor I permits allowance to be made for the significance of possible failure of the structure, whether as a result of the importance of the structure to post-earthquake services, or the severity of consequential risks in the event of failure. Factor K expresses the capacity of the structure for inelastic energy dissipation, varying from 0.67 to 1.33 (1.5 for exceptional cases), with low values for 'brittle' structural forms. Factor A represents the spectrum $(2\pi n S_v)$, a simple, perhaps crude, approximation is generally specified, allowing a reduction as a function of predicted fundamental frequency where this is less than about 0.5 Hz. The UBC currently suggests $A = 0.07 n^{1/2}$ (but not more than 0.12) for the US.

The total force (V) is then distributed over the structure in proportion to the product of the mass and the mode shape function for the first mode (the latter is often approximated by direct proportionality to the height above the ground). It has been noted above that slender tall structures may also show significant higher-mode response, and this is most liable to increase stresses near the top (a so-called 'whiplash' effect); an added proportion of the total load, perhaps 15%, may thus be required to be applied at the highest point.

When it is desired to give more detailed consideration to the behaviour of the structure in the inelastic range, the 'reserve energy' technique is simple to apply and can quickly give very useful guidance and economy in design. To proceed to greater detail requires ad hoc computer step-by-step solution of the response to a given ground motion; this is increasingly commonly done in both US and Japan, and is general practice in the latter country for buildings exceeding fifteen storeys. Two important points must be noted. Firstly, that most of the available ground motion records to input to this procedure were obtained at a substantial distance from a large shock, so that special consideration is necessary for sites in a region where more localized energy release is typical (producing a higher characteristic frequency in ground motion) as well as sites on soft subsoil (possibility of lower frequencies as well as overall magnification). Secondly, any one record is but one chance example of the superposition of ground-wave motions of considerable complexity. Although the broad statistical properties of the ground motion are thus generally representative, the actual net peak response of one specific structure will vary greatly owing to the random factors in this superposition. One technique is to generate artificial ground motion sequences, all having the same broad statistical properties, so that the calculated maximum responses can be averaged (or the value for any given probability of occurrence selected). A somewhat more crude method to make use of a single record is to repeat analysis with a scale factor applied to the mass of the structure to modify the natural frequency. Averaging the responses obtained over a range of (say) ±30% of frequency greatly reduces the probable error due to the random factors.

Finally, it is worth repeating that design to ensure ductility can give much more benefit for a given cost than directly increasing strength. Good design keeps to simple shapes and simple structural forms to reduce the risk of large-scale 'stress concentrations' which would arise, for example, between two wings of a building having different natural frequencies. The conference proceedings that include Blume's analysis of response[45] is strongly recommended for further reading; this has been followed by a consensus guide.[46] A wider reference handbook is also available.[47]

References

1 International Standards Organization (1984) Draft International Standard ISO/DIS 2394, *General principles on reliability for structures*. International Standards Organization (UK agent, British Standards Institution, Milton Keynes).
2 British Standards Institution (1984) British Standard Specification BS 6399, *Design loading for buildings*: Part 1 'Code of practice for dead and imposed loads', BSI, Milton Keynes.
3 British Standards Institution (1972) British Standard Code of Practice CP3: Chapter V, Part 2, 'Wind loads', as revised 1985, BSI, Milton Keynes.
4 British Standards Institution (1985) British Standard Specification BS 8100, *Lattice towers and masts*: Part 1, 'Loading'. Also Part 2, 'Commentary'. BSI, Milton Keynes.
5 British Standards Institution (1978) British Standard Specification BS 5400, *Steel, concrete and composite bridges*: Part 2, 'Specification for loads'. BSI, Milton Keynes.
6 British Standards Institution (1986) British Standard Specification BS 6399 *Design loading for buildings*: Part 3, 'Snow loading'. BSI, Milton Keynes.
7 British Standards Institution (1980) British Standard Specification BS 5502, *Agricultural buildings*: section 1.2, 'Design contruction and loading'. BSI, Milton Keynes.
8 Mitchell, G. R. and Woodgate, R. W. *Floor loading in office buildings – the results of a survey*, Building Research Station Current Paper Number 3/71; *Floor loading in retail premises – the results of a survey*, Building Research Station Current Paper Number 25/71. BRS, Garston.
9 Turitzin, A. M. (1963) 'Dynamic pressure of granular material in deep bins', *Proc. Am. Soc. Civ. Engrs*, **89**, ST2 (Apr.).
10 Jenike, A. W. and Johansen, J. R. (1968) 'Bin loads', *Proc. Am. Soc. Civ. Engrs*, **94**, ST4 (Apr.).
11 Leonhardt, F. *et al.* 'The safe design of cement silos', Cement and Concrete Association translation Number 94. CACA, London.
12 Henderson, W. (1954) 'British highway bridge loading', *Proc. Instn Civ. Engrs*, **3** Part 2 (June).
13 Buckland, P. G. *et al.* (1980) 'Proposed vehicle loading of long-span bridges', *Proc. Am. Soc. Civ. Engrs.*, **106**, ST4 (Apr.).
14 American Society of Civil Engineers (1981) 'Recommended design loads for bridges', *Proc. Am. Soc. Civ. Engrs*, **107**, ST7 (July).
15 Thomas, P. K. (1975) 'A comparative study of highway bridge loading in different countries', Transport and Road Research Laboratory Supplementary Report Number 135 UC. TRRL, Crowthorne.
16 British Standards Institution (1980) British Standard Specification BS 5400, *Steel, concrete and composite bridges*: Part 10, 'Fatigue'. BSI, Milton Keynes.

17 Page, J. (1976) *Dynamic wheel load measurements on motorway bridges*, Transport and Road Research Laboratory Report Number LR722. TRRL, Crowthorne.

18 *Report of the Bridge Stress Committee* (1928) HMSO, London.

19 'Discussion on the basis of the revised fatigue clause for BS 153', *Proc. Instn Civ. Engrs*, **27** (Feb. 1964) .

20 'Loads to be considered in design of railway bridges'. Recommendation ref. 776–1, International Union of Railways, Paris.

21 Shellard, H. C. (1958) 'Extreme wind speeds over Great Britain and Northern Ireland', *Met. Mag.*, **87**.

22 Cook, N. J. (1982) 'Towards better estimation of extreme winds' *J. Wind Engrg and Ind. Aerodyn.* **9** (Sept.) or *The designer's guide to wind loading of building structures*, Part I (1985), Butterworths, London.

23 Engineering Sciences Data Unit (1982/83) ESDU data items 82026 and 83045, 'Strong winds in the atmospheric boundary layer', ESDU, London.

24 Engineering Sciences Data Unit (1974/75) ESDU data items 74030/1 and 75001. *Characteristics of atmospheric turbulence near the ground*, ESDU, London.

25 Davenport, A. G. (1961) 'The application of statistical concepts to wind loading of structures', *Proc. Instn Civ. Engrs*, **19**.

26 Construction Industry Research and Information Association (1971) *The modern design of wind sensitive structures*. CIRIA, London.

27 Construction Industry Research and Information Association (1981) *Wind engineering for the eighties*. CIRIA, London.

28 Davenport, A. G. (1967) 'Gust loading factors', *Proc. Am. Soc. Civ. Engrs*. **93**, ST3 (June).

29 National Building Code of Canada.

30 Hoerner, S. F. (1965) *Fluid-dynamic drag*. Published by S. F. Hoerner.

31 Cowdrey, C. F. (1971) 'Time average aerodynamic forces on bridges', NPL *Aero Rep*. 1327; continuation NPL *Mar. Sci. Rep*. 1–72, 1972.

32 Cohen, E. and Perrin, H. (1957) 'Design of multi-level guyed towers – wind loading'. *Proc. Am. Soc. Civ. Engrs*. 83 ST5 (Sept.).

33 Wyatt, T. A. (1964) 'The aerodynamics of shallow paraboloid antennas', *Ann. N.Y. Acad. Sci.* **116,** 1.

34 Engineering Sciences Data Unit (1980/81) 'Mean forces, pressures and flow field velocities for circular cylindrical structures', Data items 80025 and 81017. ESDU, London.

35 Scruton, C. and Newberry, C. W. (1963) 'On the estimation of wind loads for building and structural design', *Proc. Instn Civ. Engrs*, **25** (June).

36 Wootton, L. R. (1969) 'The oscillations of large circular stacks in wind', *Proc. Instn Civ. Engrs*, **43**, (Aug.).

37 Vickery, B. J. and Clark, A. W. (1972) 'Lift or across-wind response of tapered stacks', *Proc. Am. Soc. Civ. Engrs*, **98**, ST1 (Jan.).

38 Vickery, B. J. and Busu, T. I. (1984) 'The response of reinforced concrete chimneys to vortex shedding', *Engineering Structures*, **6,** (Oct.).

39 Wyatt, T. A. *et al.* (1985) 'The treatment of cross-wind excitation of chimneys proposed for the British Standard Draft for Development for reinforced concrete chimneys', *Proc. 5th International Chimney Congress*, CICIND, Essen.

40 Walshe, D. E. and Wootton, L. R. (1970) 'Preventing wind-induced oscillations of structures of circular section', *Proc. Instn Civ. Engrs, 47,* (Sept.).

41 Whitbread, R. E. and Wootton, L. R. (1967) *An aerodynamic investigation for tower blocks for Pink Shek Estate, Hong Kong*, NPL *Aero Special Rep*. Number 002.

42 Novak, M. (1972) 'Galloping oscillations of prismatic structures', *Proc. Am Soc. Civ. Engrs*. EMI (Feb.).

43 Institution of Civil Engineers (1981) *Bridge aerodynamics*. Thomas Telford, London.

44 Neuman, F. (1962) 'Seismic forces on engineering structures', *Proc. Am. Soc. Civ. Engrs*. **88**, ST2 (Apr.).

45 Blume, J. A. (1972) 'Analysis of dynamic earthquake response', ASCE/IABSE Conference, *Planning and design of tall buildings* (Lehigh). Paper Number 1b/6.

46 American Society of Civil Engineers (1983) *Tall buildings – criteria and loading*, Vol. CL of Tall Building Monograph, ASCE, New York.

47 Wiegel, R. L. (ed.) (1970) *Earthquake engineering*. Prentice-Hall, New Jersey.

20

Bridges

D J Lee BScTech, DIC, FEng, FICE, FIStructE

B Richmond BSc(Eng), PhD, FCGI, FICE
Maunsell Group

Contents

20.1 Plan of work

The design of bridges requires the collection of extensive data and from this the selection of possible options. From such a review the choice is narrowed down to a shortlist of potential bridge designs. A sensible work plan should be devised for the marshalling and deployment of information throughout the project from conception to completion. Such a checklist will vary from project to project but a typical example might be drawn up on the following lines.

(1) *Feasibility phase*:

(a) data collection;
(b) topographical and hydrographical surveys;
(c) hydrological information;
(d) geological and geotechnical information;
(e) site investigation requirements for soil and rock evaluation;
(f) Meteorological and aerodynamic data;
(g) assembly of basic criteria;
(h) likely budget.

(2) *Assembly of design criteria*:

(a) data and properties on the material to be used including steel, concrete, aluminium, timber, masonry, etc.;
(b) foundation considerations;
(c) hydraulic considerations, flood, scour;
(d) loading and design criteria;
(e) clearances height and width (such as for navigation, traffic);
(f) criteria for gradients, alignment, etc.;
(g) hazards such as impact, accident;
(h) proximity to other engineering works, etc.;
(i) functional requirements;
(j) transportation and traffic planning;
(k) highway and/or railway engineering aspects;
(l) drainage requirements;
(m) provision for services (water, sewage, power, electricity, telephone, gas, communications links, etc.);
(n) design life and durability considerations.

(3) *Design phase*:

(a) choice of bridge;
(b) detailed design of bridge including foundations, substructure and superstructure;
(c) production of drawings and documentation, etc.;
(d) preparation of quality assurance plan;
(e) estimation of cost and programme.

(4) *Construction phase*:

(a) contractual matters;
(b) construction methods;
(c) budget and financial control;
(d) quality control;
(e) supervision of construction;
(f) commissioning;
(g) operating, inspection and maintenance schedules for each part of the work.

(5) *Performance phase*:

(a) obligations of owner;
(b) management of facility;
(c) inspection, maintenance and repair;
(d) rehabilitation and refurbishment requirements (change of loading, widening, change of use and durability aspects);
(e) decommissioning and demolition.

Such a project list serves to highlight the various and sometimes conflicting requirements of a bridge project, and those aspects where the bridge designer should seek the approval of the client throughout all the stages of a project for a truly successful collaboration.

This chapter covers the selection and analysis of bridge superstructures and attempts to relate the most frequently used bridging materials – steel and concrete.

As extensive treatment as possible is given to box girder analysis, an important aspect of modern bridge construction. Information about individual bridges will be found in the bibliography. Reference to these specific examples will assist an understanding of the historical background and the existing state of the art. A good general review of the structural form of bridges is given by Beckett,[1] whilst a sensitive aesthetic assessment is provided by Mock.[2]

Masonry arches and steel trusses have not been dealt with but interesting examples of these types of bridges are contained in the reference list.

The principles developed in this chapter for open or closed sections are applicable to trussed structures if suitable modifications are made to allow for shear behaviour of the truss system.

Thus, the authors hope that there is adequate information in this chapter to make preliminary assessments for most modern bridge designs by methods which enable the essential natures of structural behaviours to be perceived and which can be developed to detailed analyses without the necessity of revising basic principles.

20.2 Economics and choice of structural system

Cost comparisons which would make it possible to arrive at the most economical choice of material, structural form, span, etc. have been sought for many years by bridge engineers, but since the costs of any one bridge depend on the circumstances prevailing at that time, the information is always imprecise. Cost data must be up to date and sufficiently detailed to allow adjustments to be made for changed circumstances. It is the changes in these factors which lead to new methods of construction and new structural systems; a major change of this kind has been that involving box girders, plate girders and trusses.

A very early steel box girder bridge, the Britannia Bridge,[3] built by Stephenson over the Menai Straits (main spans 140 m, completed in 1850) was very successful and was in regular use for railway trains until it was damaged by fire. Each span was lifted into place in its entirety by hydraulic jacks. The advantages of truss construction were, however, sufficient to convince engineers for the next 100 years that box structures were not economical, though plated structures were used in the form of I-beams for smaller spans and lighter loads. The steel box girder re-emerged as a structural system for bridges after the Second World War, although short-span multicellular bridges in reinforced concrete had been used for short spans in the 1930s. In 1965 a large proportion of structures other than short spans were built as box structures of one form or another. A greater degree of selectivity then began to emerge and open cross-sections, even for substantial spans, were again being used provided no problems of aerodynamic stability arose. The use of plate girders has been further encouraged by the reaction caused by failures of steel box girder bridges but it seems likely that a balanced view of the merits of various forms of construction will prevail.

Figure 20.1[4] shows the possible cross-sections for bridge structures which can include truss systems if the plane of each triangulated panel is represented by either a web or flange member. The significance of box structures in a more general sense now becomes clear. It is the open cross-section that is a particular, although important, form of construction, whereas

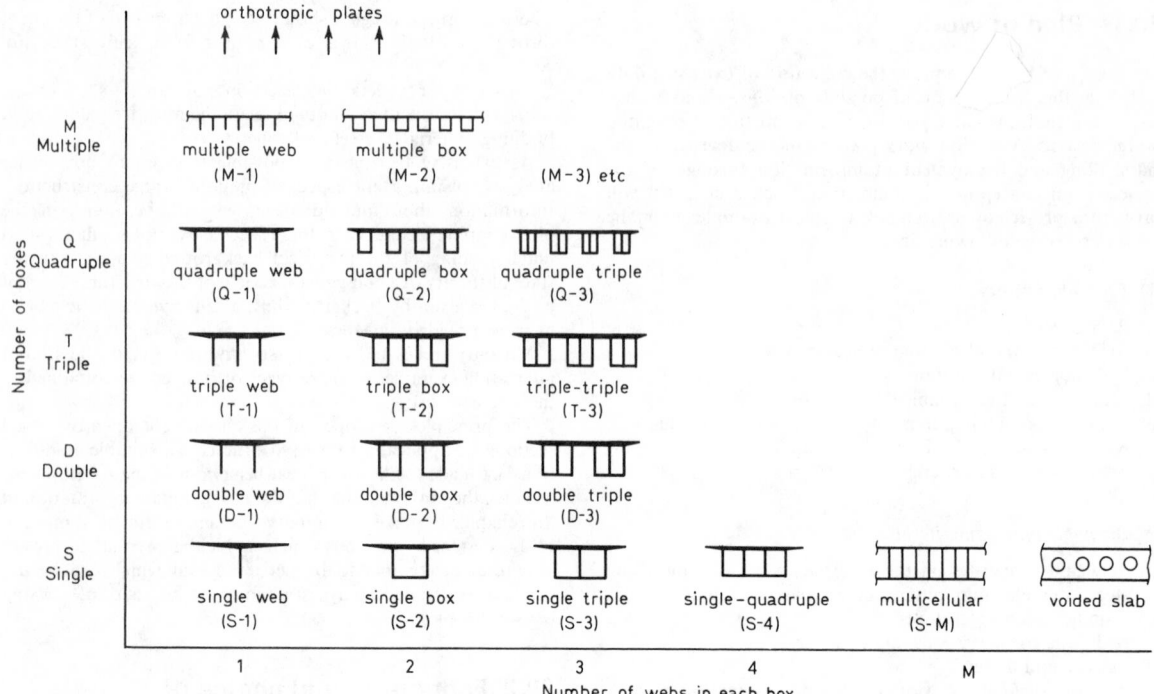

Figure 20.1 Classification of bridge-deck classifications. (After Lee (1971) 'The selection of box-beam arrangements in bridge design', *Developments in bridge design and construction.* (Crosby Lockwood).

the box system is perhaps a misleadingly simple description of the general range of structures.

The most basic structural dimension for a given span affecting both the least-cost and the least-weight methods of measuring efficiency is the effective lever arm of the structure for resisting bending moments resulting from the vertically acting forces from self-weight and imposed loads and vertical components of the support reactions. In bridges which depend on horizontal reactions from the ground, this distance is the rise of an arch above its foundations, or the dip of a suspension cable between towers. If the supports are at different levels, the dip or rise is measured vertically from the chord joining the supports.

The high strength:weight ratio of steel wire and favourable price:strength ratio results in dip:span ratios of 0.1 being suitable for even the longest suspension bridges (Table 20.1). The shallow cable has a higher tension which improves its capacity for carrying uneven loads without large deflection and increases its natural frequency of vibration. The cost of the cable alone is not, however, sufficient to reach conclusions on economics, since the cost of foundations to anchor the cables is substantial and varies with the ground conditions.

The lower strength:weight ratios of steel in compression and concrete combined with the destabilizing effect of the compressive force of the thrust lead to the rise:span ratios being considerably higher on average (Tables 20.3 and 20.4). Good foundations and the requirements of local topography may lead to reduced ratios, and arches – such as at Gladesville,[5] which are in flat country and yet have the roadway running above the arch rib – and the requirement for a low rise to minimize the cost of approach embankments.

The depth between compression and tension flanges is the lever arm of a simply supported beam structure, such as a truss, plate girder or box girder. If the structure is continuous at both ends, the sum of the depths at the centre span and one of the supports is the lever arm (Tables 20.6 and 20.7).

Table 20.1 The world's leading suspension bridges

Name of bridge	Year	Main span (m)	Cable sag (m)	Span/ sag	Location
Humber	1981	1410	125	11.3	Humber River
Verrazano					
Narrows	1964	1298	117	11.0	New York Harbor
Golden Gate	1937	1280	145	8.8	San Francisco
Mackinac Straits	1957	1158	108	10.76	Michigan
Minami Bisan-					
Seto (Road/Rail)	u.c.				
	(1988)	1100			Inland Sea of Japan
2nd Bosphorus	1988	1090			
Bosphorus	1973	1074	93.4	11.5	Ortakoy, Turkey
George					
Washington	1932	1067	96	11.1	Hudson River, New York state
Tagus	1966	1013	106	9.5	Lisbon
Forth	1964	1006	91	11.0	Queensferry
Kita Bisan-Seto					
(Road/Rail)	u.c.				
	(1988)	990			Inland Sea of Japan
Severn	1966	988	82	12.0	Beachley, UK
Ohnaruto		876			Naruto, Japan
Tacoma Narrows					
II	1950	853	87	9.8	Puget Sound, Washington
Lions Gate	1938	846			Vancouver

u.c. = under construction

Table 20.2 Leading cable-stayed bridges

Name	Location	Year	Main span length (m)	Span arrangement	Planes	Cables Arrangement	Material	Function	Special notes
Annacis	Vancouver, Canada	1986	465	Sym	2	MF	St/C	Road	
Hooghly	Calcutta, India	u.c.	457	Sym	2	F	St/C		
Barrios de Luna	Sierra Cantabrica, Spain	w	440	Sym	2	F	C	Road	
Hitsuishijima	Iwakurojima (two bridges)	(u.c. 1987)	420	Sym	2	MF	St	Road and Rail	Part of Kojma-Sokaido
Saint Nazaire	Loire estuary, Brittany France	1974	404	Sym	2	F	St	Road	
St Johns River	Jacksonville, Florida, US		400			F	C		
Rande	Vigo Estuary, Spain	1978	400	Sym	2	F	St	Road	
Luling	Mississippi River, Louisiana, US	1982	372	Sym	2	F	St	Road	
Dusseldorf Flehe	W. Germany	1979	368	Ass	1	H side span, MF main span	St/C	Road	Multiple side span anchor piers
Tjörn	Askeröfjord, Sweden	1981	366	Sym	2	MF	St/C	Road	Replaced steel arch demolished by ship collision
Sunshine Skyway	Florida, US	1987	366		1		C	Road	
Yamatogawa	Osaka, Japan		355	Sym	1	H	St	Road	
Duisberg–Neuenkamp	Rhine River, Duisberg–Moers, W. Germany	1970	350	Sym	1	MF	St	Road	
Jindo	S. Korea	1985	345	Sym	2	F	St	Road	
Westgate	Yarra River, Melbourne, Australia	1978	336	Sym	1	DF	St/C	Road	
Brazo Largo	Guagu, Argentina	1977	330	Sym	2	F	St	Road and Rail	Connected by long embankment
Zarate	Palmas, Argentina	1977	330	Sym	2	F	St	Road and Rail	
Posadas–Encarnacion	Paraguay, Argentina		330				C	Road and Rail	
Kohlbrand	Hamburg, W. Germany	1974	325	Sym	2	MF	St		
Knie	Rhine River, Dusseldorf, W. Germany	1969	320	Ass	2	H	St	Road	Multiple side span anchor piers
Brotonne	Seine River, Rouen, France	1977	320	Sym	1	MF	C	Road	
Bratislava	Danube River, Czechoslovakia	1971	316	Ass	1	S side span, F main span	St	Road	Two unequal spans, backward leaning tower
Erskine	Clyde River, Scotland	1971	305	Sym	1	S	St	Road	
Severins	Cologne, W. Germany	1959	302	Ass	2	F	St	Road	
Dnieper	Kiev, Soviet Union	1976	300	Ass	2	MF	St/C	Road	Two unequal spans
Pasco Kennewick	Washington State, US	1978	299	Sym	2	F	C	Road	
Neïwied	Rhine River, W. Germany		292	Ass	1	MF	St	Road	Two unequal spans with longitudinal A frame tower
Deggenau	Danube River, W. Germany	1975	290	Ass	1	F	St	Road	Two unequal spans
Coatzacoalcos II	Mexico	1984	288	Sym	1	MF	C	Road	
Kurt Schuhmacher	Rhine River, Mannheim Nord, W. Germany	1971	287	Ass	2	F	St	Road and tramway	Two side span anchor piers

Table 20.2 (cont)

Name	Location	Year	Main span length (m)	Span arrangement	Cables		Material	Function	Special notes
					Planes	Arrangement			
Wadi-el-Kuf	Beida, Libya, N. Africa	1971	282	Sym	2	S	C	Road	Articulated
Leverkusen	Rhine River, W. Germany	1965	281	Sym	1	H	St	Road	
Friedrich-Ebert	Rhine River, Bonn Nord, W. Germany	1967	280	Sym	1	MF	St	Road	
Dolsan	S. Korea	u.c.	280	Sym	2	F	St		
Speyer	Rhine River, W. Germany	1975	275	Ass	1	S side span, F main span	St	Road	
East Huntingdon	Ohio River, US	u.c.	274	Ass	2	MF	C	Road	Two unequal spans
Tiel	Waal River, Holland	1972	267	Sym	2	F	C	Road	
Theodor Heuss	Dusseldorf, W. Germany	1958	260	Sym	2	H	St	Road	
Oberkassel	Dusseldorf, W. Germany	1976	258	Ass	1	H	St	Road and streetcar	Multiple side-span anchor piers
Rees	W. Germany	1967	255	Sym	2	H	St	Road	
Save	Belgrade, Yugoslavia	1978	254	Sym	2	MF	St	Railway	
Papineau	Montreal, Canada	1969	251	Sym	1	F	St	Road	
Suchiro	Tokushima, Japan	1976	250	Sym	1	MF	St		
Manuel Belgrano	Parana River, Corrientes, Argentina	1972	245	Sym	2	F	C	Road	Articulated
Kessock	Inverness, Scotland	1982	240	Sym	2	H	St	Road	
General Rafael Urdaneta	Lake Maracaibo, Venezuela	1962	235	Sym	2	S	C	Road	Multiple spans. Articulated
Wye	Beachley, Wales	1966	235	Sym	1	S	St	Road	
Penang Crossing	Malaysia	1980	225	Sym	2	H	C	Road	
Luangwa	Zambia		223						
Rokko Island Double-decked	Kobe, Japan	1977	220	Sym	2	MF	St	Road	
Hawkshaw	New Brunswick, Canada	1969	217						
Toyosato	Yodo River, Osaka, Japan	1970	216	Sym	1	MF	St	Road	
Onomichi	Hiroshima Pref., Japan	1968	215	Sym	2	F	St	Road	
Polcevera Creek	Genoa, Italy	1967	210	Sym	2	S	C	Road	Multiple spans. Articulated
Albert Canal	Godsheide, Belgium	1977	210	Sym	2	MF	St	Road	
Batman	Tamar River, Tasmania	1968	206	Ass	2	S side span, F main span	St	Road	Two unequal spans. Forward leaning tower
Arno	Florence, Italy	1977	206	Sym	2	S side span, F main span		Road	Towers lean backwards
Stromsund	Sweden	1955	183	Sym	2	DF	St/C	Road	
Adhamiyah	Baghdad, Iraq	1984	182	Ass	1	H	St/C	Road	Two side-span anchor piers
New Galecopper	Rhine Canal, Amsterdam, Holland	1971	180	Sym	1	S	St	Road	Twin bridges skew spans
Maxau	Rhine River, W. Germany	1967	175	Ass	1	MF	St	Road	Two unequal spans
Ganter	Simplon Pass, Valais, Switzerland	1980	174	Sym	2	S	C	Road	Cables enclosed in web extensions. Curved side spans

Table 20.2 (cont)

Name	Location	Year	Main span length (m)	Span arrange-ment	Planes	Cables Arrangement	Material	Function	Special notes
North Elbe	Hamburg, W. Germany	1962	172	Sym	1	ST	St	Road	
Daikoku	Yokohama, Japan	1974	165	Ass	2	MF	St		
Massena	Paris, France	1971	162	Sym	1	MF	St	Road	
Steyregger Donau	Linz, Austria	1979	161	Ass	2	S	St/C	Road	Two unequal spans
Kamatsugawa	Japan	1971	160	Sym	1	H	St		
Ishikara–Kako	Hokkaido, Japan	1975	160	Sym	2	F	St	Road	
Arakawa	Tokyo, Japan	1970	160	Sym	1	H	St	Road	
George Street	River Usk, Newport, Wales	1964	152	Sym	2	H	St/C	Road	
Sancho el Major	Rio Ebro, Castejon, Spain		146			MF	C		
Metten	Danube River, W. Germany		145		1	S	C		
Magliana	Tiber River, Rome, Italy	1967	145	Ass	2	S	C	Road	Curved in plan. Two unequal spans. Backward leaning towers
Dnieper	Kiev, Soviet Union	1964	144	Sym	2	F	C	Road	
Maya	Kobe, Japan	1966	139	Ass	1	MF	St		Two unequal spans
Ludwigshafen	W. Germany	1968	138	Eq	2	F	St	Road	Four-leg A-frame tower
Sitka Harbour	Alaska, US	1972	137	Sym	2	S	St/C	Road	
Danube Canal	Vienna, Austria	1975	119	Sym	2	S	C		
Second Main Bridge	Frankfurt, W. Germany	1972	148	Ass	2	H	C	Road and rail	Articulated main span connects to fin back. Three anchored side spans
Tarano	Alba, Italy	1983	114	Ass	1	S side span, F main span	St	Road	Two unequal spans. Backward sloping towers
Harmsen	Rotterdam, Holland	1968	108						
Bridge of the Isles	Montreal, Canada	1967	105	Eq	2	S	St/C	Road and rail	
St Florent	River Loire, France	1969	104	Eq	2	F	St/C		
Julicherstrasse	Dusseldorf, W. Germany	1963	99	Sym	1	S	St	Road	

Ass – asymmetric; C – concrete; DF – double fan; Eq – two equal; F – multiple fan; H – harp; MF – modified fan; S – single; St – steel; ST – star; St/C – composite steel and concrete; Sym – symmetric; u.c. – under construction

Table 20.3 The world's leading steel arch bridges

Name of bridge	Span (m)	Rise (m)	Rise/span	Year	Location
River Gorge	518			1977	West Virginia, US
Bayonne	504	81	0.161	1931	New York, New York, US
Sydney Harbour	503	107	0.212	1932	Sydney, Australia
Fremont*	383			u.c.	Portland, Oregon, US
Port Mann*	366	76	0.208	1964	Vancouver, Canada
Thatcher†	344			1962	Balboa, Panama
Laviolette†	335			1967	Trois Rivières, Canada
Zd'ákov	330	42.5	0.129	1967	Lake Orlik, Czechoslovakia
Runcorn–Widnes	330	66.4	0.202	1961	Mersey River, England
Birchenough	329	65.8	0.200	1935	Sabi River, Rhodesia
Glen Canyon	313			1959	Arizona, US
Lewiston–Queenston	305	48.4	0.159	1962	Niagara River, N. America
Hell Gate	298			1917	New York, New York, US

Other steel arch bridges of interest

Name of bridge	Span (m)	Rise (m)	Rise/span	Year	Location
Rainbow	289	45.7	0.158	1941	Niagara Falls, N. America
Fehmarnsund*	249	43.6	0.175	1963	Fehmarnsund, W. Germany
Adomi (Volta)	245	57.4	0.234	1957	Adomi, Ghana
Kaiserlei*	220			1964	Frankfurt-am-Main, W. Germany

u.c. = under construction
*Tied arch   †Cantilever arch

Table 20.4 The world's leading concrete arch bridges

Name of bridge	Span (m)	Rise (m)	Rise/span	Year	Location
Krk II	390			1980	Adria, Yugoslavia
Gladesville	305	40.8	0.134	1964	Sydney, Australia
Rio Paraná	290	53.0	0.183	1965	Paraná River, Brazil–Paraguay
Bloukrans	272			1983	Cape Province, South Africa
Arrabida	270	51.9	0.192	1963	Portugal
Sandö	264	40.0	0.151	1943	Angerman River, Sweden
Shibenik	246			1967	Krka River, Yugoslavia
Fiumarella	231	66.1	0.286	1961	Catanzaro, Italy
Novi Sad	211			1961	Danube River, Yugoslavia
Linenau	210			1967	Bregenz, Austria
Van Stadens	200			1971	Van Stadens Gorge, S. Africa
Esla	192			1942	Esla River, Spain
Groot River	189			1983	Cape Province, South Africa
Rio das Antas	180	28.0	0.156	1953	Brazil
Traneberg	178	26.2	0.147	1934	Stockholm, Sweden
Plougastel (Albert Louppe)	173	33	0.190	1930	Elorn River, France
Selah Creek	168			1971	Yakima, Washington, US
Bobbejaans	165			1983	Cape Province, South Africa
La Roche–Guyon	161	23.0	0.143	1934	France
Cowlitz River Bridge	158				Mossyrock, Washington, US
Caracas–La Guaira	152	39.0	0.257	1952	Caracas, Venezuela
Puddefjord	145			1956	Norway
Podolska	145			1942	Czechoslovakia

Other concrete arch bridges of interest

Name of bridge	Span (m)	Rise (m)	Rise/span	Year	Location
Revin–Orzy	120	10.0	0.083		Meuse River, France
Glemstal	114	27.1	0.238		Stuttgart, W. Germany
Slängsboda	111	12.0	0.108	1961	Stockholm, Sweden

u.c. = under construction

Table 20.5 The world's leading truss bridges

Name of bridge	Span (m)	Year	Location
Quebec Railway	549	1918	Quebec, Canada
Forth Railway	2 × 521	1890	Queensferry, Scotland
Minato	510	1974	Japan
Delaware River	501		Chester, Penn–Bridgeport, New Jersey, US
Greater New Orleans	480	1958	New Orleans, Louisiana, US
Howrah	457	1943	Calcutta, India
Transbay	427	1936	San Francisco, California, US
Baton Rouge	376	1968	Baton Rouge, Louisiana, US
Tappan Zee	369	1955	Tarrytown, New York, US
Longview	366	1930	Columbia River, Washington, US
Queensboro	360	1909	New York, US
I Carquinez Strait	2 × 335	1927	San Francisco, California, US
II Carquinez Strait	2 × 335	1958	San Francisco, California, US
Second Narrows	335	1960	Vancouver, Canada
Jacques Cartier	334	1930	Montreal, Canada
Isaiah D. Hart	332	1967	Jacksonville, Florida, US
Richmond–San Rafael	2 × 326	1956	San Pablo Bay, California, US
Grace Memorial	320	1929	Cooper River, South Carolina, US
Newburgh–Beacon	305	1963	Hudson River, New York, US
Auckland Harbour	244	1959	Auckland, New Zealand

Table 20.6 Some of the world's leading steel girder bridges

Name of bridge	Span (m)	Depth (d) at midspan (m)	Depth (d₂) at pier (m)	$\frac{d_1 + d_2}{Span}$	Year	Type	Location
Niteroi	300	7.4	12.9	0.068	1974	B	Rio de Janeiro, Brazil
Sava I	261	4.6	9.8	0.055	1956	P	Belgrade, Yugoslavia
Zoo	259	4.5	10.0	0.056	1966	B	Cologne, W. Germany
Sava II	250				1969	B	Belgrade, Yugoslavia
Koblenz	235					B	Rhine River, W. Germany
Foyle	234				1984	B	Londonderry, N. Ireland
San Mateo–Hayward	228	4.6	9.2	0.060	1967	B	California, US
Hochbrücke 'Radar Insel'	221	5	9.5	0.066		P	Nord–Ost see Canal, W. Germany
Moselle	219					B	Moselle Valley, W. Germany
Milford Haven	213	5.9	5.9	0.055		B	Pembroke Dock, Wales
Fourth Danube	210				1970	B	Vienna, Austria
Martigues	210				1976	Portal B	France
Düsseldorf–Neuss	206	3.3	7.8	0.054	1951	B	Düsseldorf, W. Germany
Wiesbaden–Schierstein	205	4.4	7.4	0.057		P	Rhine River, W. Germany
Europa	198	7.7	7.7	0.078	1964	B	Sill Valley, Austria
Köln–Deutz	185				1948	B	Rhine River, W. Germany
Poplar Street	183	6.2	7.6	0.070	1967	B	St Louis, Mississippi, US
Italia	175	8.5	8.5		1969	B	Lao River, Italy
Avonmouth	174	2.6	7.6	0.059	1974	B	Gloucestershire, England
Friarton	174	2.7	7.5	0.059	1978	B	Perth, Scotland
Gemersheim	165	9.1	5.4	0.058	1971	B	Rhine River, W. Germany
Speyer	163	3.4	6.40	0.060	1956	B	Rhine River, W. Germany
Concordia	160	4.9	4.9	0.060	1967	B	Montreal, Canada
New Temerloh	151	3.7	5.9	0.064	1974	B	Temerloh, Malaysia
Other steel girder bridges of interest							
Calcasieu River	137	2.1	7.0	0.078	1963	P	Louisiana, US
St Alban	135	2.8	9.3	0.062	1955	P	Basel, Switzerland
Amara	82	3.7	12.1	0.087	1958	B	Tigris River, Iraq

u,c. = under construction
Bridge type: B box girder, P plate girder

Table 20.7 Some of the world's leading concrete girder bridges

Name of bridge	Span (m)	Depth (d) at midspan (m)	Depth (d₂) at pier (m)	$\frac{d_1 + d_2}{Span}$	Year	Type	Location
Gateway	260	4.0	14.0	0.069	1986	C	Brisbane, Australia
Hikoshima	236					C	
Urato	230	4.0	12.5	0.072	1972	C	Shikoku, Japan
Three Sisters	229				u.c.	C	Potomac River, Washington, DC, US
Bendorf	208	4.4	10.4	0.071	1965	C	Bendorf, W. Germany
Orwell	190					C	Ipswich, England
Manazuru	185	3.1	10.0		u.c.	C	Japan
Brisbane Water	183					C + SS	New South Wales, Australia
Gardens Point	183					C	Brisbane Australia
Redheugh	160						Newcastle upon Tyne, England
Amakusa Nakana	160	3.0	10.0				Japan
Medway	152	2.2	10.8	0.086	1963	C + SS	Rochester, England
Neckarsulm	151	4.2	7.4	0.078	1968	C	Neckarsulm, W. Germany
Moscow River	148				1957	CG	Soviet Union
Amakusa	146				1966	C	Japan
Kingston	143	2.4	10.0	0.087	1970	C	Glasgow, Scotland
Victoria	142				1970	C + SS	Brisbane, Australia
Tocantins	142				1961	C	Tocantins River, Brazil
Bettingen	140	3.0	7.0	0.089		C	Main River, W. Germany
Don	139				1964	C	Rostow, Soviet Union

Table 20.7 (cont.)

Name of bridge	Span (m)	Depth (d) at midspan (m)	Depth (d_2) at pier (m)	$\dfrac{d_1 + d_2}{Span}$	Year	Type	Location
Pine Valley	137				u.c.	CG	California, US
Alnö	134				1964	C	Alnösund, Sweden
Öland	130				1972	C	Kalmar Sound, Sweden
			Other concrete girder bridges of interest				
Worms	114	2.5	6.5	0.079	1952	C	Rhine River, W. Germany
Koblenz	114	2.7	7.2	0.087	1954	C	Moselle River, W. Germany
Nötesund	110	2.2	5.7	0.072	1966	C	Orust, Sweden
Siegtal	105	5.8	5.8	0.110	1969	CG	Eiserfeld, W. Germany
Chillon Viaduct	104	2.2	5.6	0.072	1973	CG	Chillon, Switzerland
Narrows	97	2.2	4.2	0.068	1959	C + SS	Perth, Australia
Benjamin Sheares	84				1981	C + SS	Singapore
Oleron	79	2.5	4.5	0.089	1966	CG	Rochefort, France

u.c. = under construction

C – concrete; C + SS – concrete with suspended span; CG – continuous girder

The cable-supported bridge can be seen as either a suspension bridge or a continuous beam with the effective depth at the supports equal to the height of the tower. Figure 20.2 shows the various arrangements of cables that are used, and various finished bridges are shown in Figures 20.3 to 20.9.

The choice of span depends on the foundations, depth of water and height of the deck but, in many cases, other requirements – such as navigation clearances – dictate the minimum span. It is usually only shorter spans where, proportionately at any rate, there is considerable variation possible. It has been claimed in the past that at the most economic span of a multispan structure, the cost of foundations equals the cost of the superstructure less the basic deck structure costs. The assumptions necessary for this to be valid are that the cost of superstructure per unit length should increase linearly with span and that that of the substructure should vary inversely with span. The slopes of the respective cost–span curves are then equal and opposite at the point of intersection of the curves provided any constant costs in both foundations and superstructure are first subtracted. If the cost of the superstructure is assumed to increase proportionately to the square root of the span, however, the same approach requires that half the superstructure cost should equal the foundation cost. In modern structures it is difficult to separate the costs of the basic deck system from the total of the multispan structure.

The well-known rule – that for maximum economy the total area of the flanges of a beam should equal the area of the web – is a more useful guide. Table 20.8 shows that for a given web thickness and a total area of cross-section of $1.0t$ the maximum section modulus is at a depth of 0.75 where the total flange area is one-third the web area, but at a depth of 0.5 where the flange and web areas are equal the section modulus is only 11% lower. A shallower beam is usually more economical because a simpler web is then possible provided the shear force can be carried. Fabrication, transportation and erection are also less costly.

Table 20.9 shows the types of standardized precast concrete beams that are appropriate to various parts of the short-span range. Apart from the cost advantages of standardization and factory production, which may be offset by higher overheads and transport costs, there are the following advantages.

(1) Estimates of cost more reliable.
(2) Speed of construction.
(3) No temporary staging required.
(4) Sample beams can be tested to demonstrate level of prestress and ultimate strength.

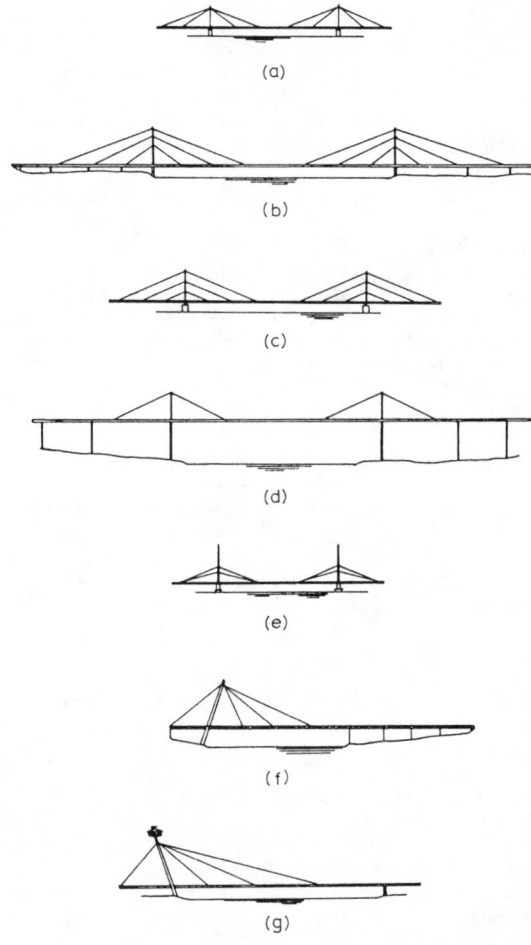

Figure 20.2 Examples of different cable systems (scale: approximately 1/10 000). (a) Fan (Stromsünd); (b) modified fan (Duisberg–Neuenkamp); (c) harp (Theodor Heuss); (d) single cable (Erskine); (e) star (Norderelbe); (f) asymmetric systems (Batman); (g) Bratislava. (*Courtesy*: Polensky and Zöllner)

Figure 20.3   Concrete girder bridge, Bettingen, Frankfurt-am-Main

Figure 20.4   Steel girder bridge, Rio-Niteroi, Brazil. (*Courtesy*: Redpath Dorman Long and the Cleveland Bridge and Engineering Co. Ltd)

Figure 20.5   Concrete cable-stayed bridge, Tempul Aqueduct, Spain. (*Courtesy*: Torroja Institute, Madrid)

Figure 20.6 Steel trussed cable-stayed bridge. Batman Bridge, Tasmania. (*Courtesy*: Maunsell and Partners)

Table 20.8

Depth	0.5	0.6	0.7	0.75	0.8
A_f	$0.25t$	$0.2t$	$0.15t$	$0.125t$	$0.1t$
Section modulus Z	$0.167t$	$0.180t$	$0.187t$	$0.1875t$	$0.187t$

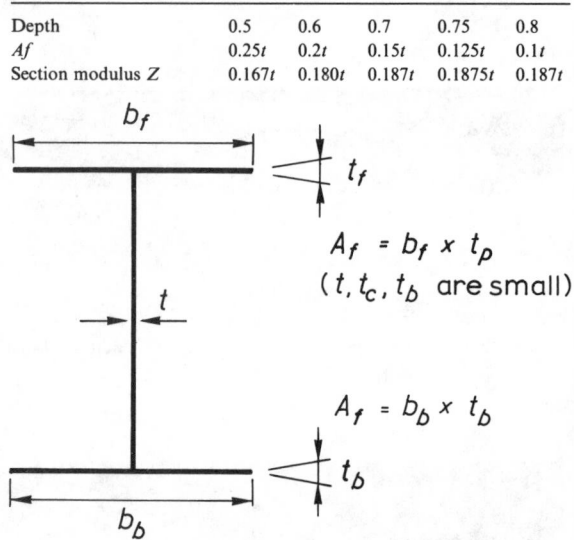

$$A_f = b_f \times t_p$$
$$(t, t_c, t_b \text{ are small})$$

$$A_f = b_b \times t_b$$

Note: Total cross-section throughout $= 1.0t$

Figure 20.7 Concrete-arched bridge, Gladesville, Sydney. (*Courtesy*: G. Maunsell and Partners)

Figure 20.8 Humber suspension bridge. (*Courtesy*: Freeman Fox and Partners)

Figure 20.9 Annacis cable-stayed bridge, Vancouver. (*Courtesy*: Buckland and Taylor Ltd)

In simple right spans, the system chosen, apart from span, depends on construction depth limitations, difficulties of access and, of course, prevailing prices. For example, the top hat beam system[6] is suitable for restricted access and small construction depths. The U-beam system[7] is suitable for similar conditions but requires an increased depth. At the greater depth it is more economical. An advantage of torsionally stiff structures of this type, particularly when they are designed to be spaced apart in the transverse direction, is that they can readily be fanned out to support the structures with complex plan forms that are now common.

The standard concrete beams are essentially a series of elements that can be placed across the complete span, requiring only simple shuttering to support the transversely spanning top slab. Diaphragm beams at the supports are required and occasionally intermediate diaphragms may be provided.

Steel beams can be used as an alternative form of construction in the same span range. Either a series of I-sections or small box girders can be used.

Table 20.9 Precast concrete bridge beams

Type of beam	Name of beam	Classification (as Figure 20.1)	Span (m)	Beam section
I	C & CA I-section beam	M-1	12–36	
Inverted T	C & CA inverted T-beam for spans from 7–16 m	Orthotropic slab	7–16 m	
Inverted T (M range)	MoT/C & CA prestressed inverted T-beam for spans from 15 to 29 m	(a) T-beam M-1 (b) Pseudo box S-M	15–29	
Box	C & CA box section beam	S-M	12–36	
U	U-beam	M-2	15–36	

Section through part of typical deck	*Remarks*
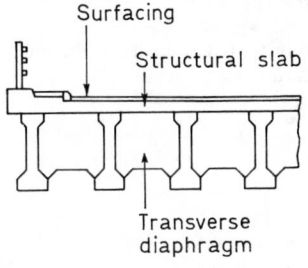	20 standard sections (I1–I20) Holes for transverse reinforcement provided at 30–50 mm centres
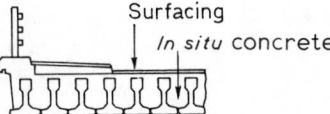	7 standard sections (T1–T7)
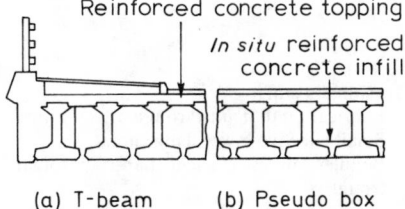	10 standard sections (M1–M10)
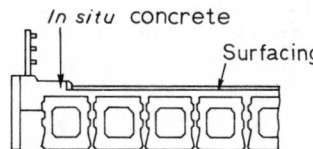	17 standard sections (B1–B17) Transverse prestress used to give optimum load distribution
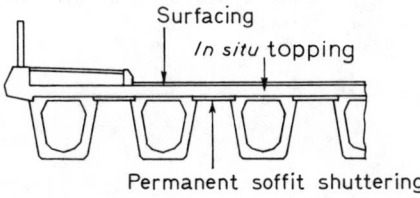	12 standard sections (U1–U12)

Table 20.10 Longitudinal stiffeners for orthotropic decks

Type of stiffener	Classification (as Figure 18.1)		Remarks
Flat	M-1, etc.		Torsionally weak. Easily spliced. Poor transverse load distribution. Earliest form.
Bulb flat	M-1		Torsionally weak. Easily spliced. Poor transverse load distribution. Bulb flats difficult to obtain. Out of date.
Trapezoidal trough	M-2, etc.		Torsionally stiff. Fabrication difficult through cross-frames. Relatively popular.
V-trough	M-2		Torsionally stiff. Fabrication difficult through cross-frames. Small effective lower flange area. Popular but less efficient than trapezoidal trough
Wineglass	M-2		Torsionally stiff. Very complicated fabrication. Expensive.
(cut from universal beam)	M-1		Easily spliced. Requires large cutout in cross-frame. Torsionally weak. Inefficient.

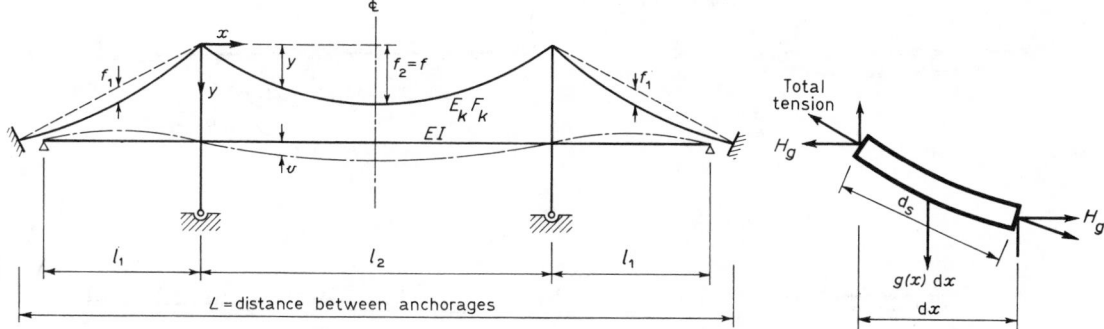

Figure 20.10 Suspension-bridge notation

Precast or prefabricated elements can be made as transverse rather than longitudinal elements and then joined together on site by prestressing in concrete structures or welding or bolting in steel structures. This approach, sometimes known as segmental construction, was used for the structures of Figure 20.11(d), (e), (f), (h), (i), (j) and (k). It was also used for the steel structures of Figure 20.11 (l) and (m). The remaining steel structures shown in Figure 20.11(n) to (r) were constructed by a similar process but with the subdivision taken a stage further. Each transverse slice was built up on the end of the cantilevering structure from several stiffened panels.

In situ concrete, reinforced or prestressed, can be used to form complete spans in one operation or else the cantilevering approach can be used. In the latter case, the speed of construction is limited by the time required for the concrete to reach a cube strength adequate for the degree of prestress necessary to support the next section of the cantilever and the erection equipment. Segmental methods of construction[8] avoid such delays. In shorter spans, provided that the restrictions on construction depth are not too severe, *in situ* concrete structures can be built economically using the cross-section of Figure 20.11(c). The simple cross-section[9] was developed to suit the use of formwork which, after supporting a complete span, could be moved rapidly to the next span. The resulting machine is only economical for multispan structures.

The stiffened steel plates (Table 20.10) are used for deck systems of long-span, and movable, bridges in order to reduce the self-weight of the structure.

20.3 Characteristics of bridge structures

The following theories have been chosen and developed for their value in demonstrating the principal characteristics of various types of bridge structure. Other methods of calculation, based on finite elements, for example, may be more accurate and more economical in certain circumstances. The theories are, however, linked to the main structural properties of the bridge types considered and are meant to assist the process of synthesis necessary before detailed calculations begin. The concepts described are also useful for idealizing structures when using computer programs and for interpreting and checking the computer output.

20.3.1 Theory of suspension bridges and arch bridges

The basic theory of arch and suspension bridges is the same and the equation derived below for suspension bridges is applicable to arches if a change in sign of H and y is made.

20.3.1.1 Suspension bridges with external anchorages

The dead load of the cable and stiffening girder is supported by the force per unit length of span produced by the horizontal component of the cable force and the rate of change of slope of the cable:

$$H_g y''(x) + g = 0 \tag{20.1}$$

where y, etc. are shown in Figure 20.10.

For a parabolic shape of cable corresponding to constant intensity of load across the span l, $y''(x) = -8f/l^2$ and:

$$H_g = gl^2/8f \tag{20.2}$$

The cable tension increases under live load $p(x)$ to:

$$H = H_g + H_p \tag{20.3}$$

The increase in support from the cable is $-[Hv''(x) + H_p y''(x)]$ where $v(x)$ is the vertical deflection of the cable and stiffening girder. The stiffening girder contributes a supporting reaction per unit length of $[EIv''(x)]''$ and adding the cable and stiffening girder contributions and equating them to the intensity of the applied load gives:

$$[EIv''(x)]'' - Hv''(x) = p(x) + H_p y'' \tag{20.4}$$

The term $H_p y''$ is added to the live load in order to show that the equation can be represented physically by the substitute structure of Figure 20.12. y'' is $-8f/l^2$ and therefore represents a force in the opposite direction to the live load.

H_p depends on the change in length of the cable and if Δdx is the horizontal projection of the change in length of an element ds then for fixed anchorages:

$$\int_0^L \Delta dx = 0 \tag{20.5}$$

Integrating along the cable and allowing for a change in temperature of ΔT gives:

$$\int_0^L \Delta dx = H_p \frac{L_k}{E_k F_k} \pm a_T \Delta T L_T + y'' \int_0^L v(x)dx = 0 \tag{20.6}$$

Approximate values of L_k and L_T are (see Figure 20.13):

$$L_k \simeq \left(1 + 8\frac{f^2}{l^2} + \frac{3}{2}\tan^2 v_0\right) + \frac{S_1}{\cos^2 v_1} + \frac{S_2}{\cos^2 v_2} \tag{20.7}$$

$$L_T \simeq \left(1 + \frac{16}{3}\frac{f^2}{l^2} + \tan^2 v_0\right) + \frac{S_1}{\cos v_1} + \frac{S_2}{\cos v_2}$$

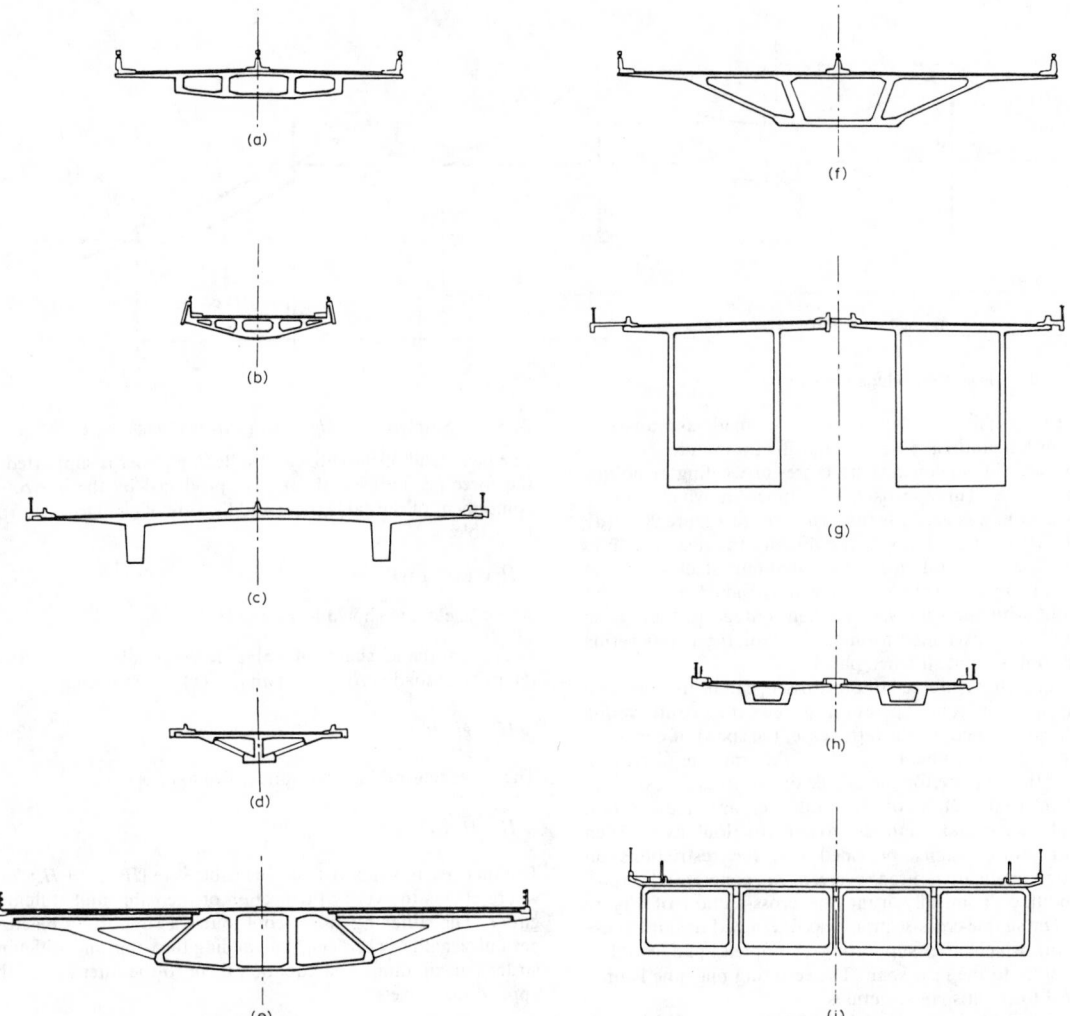

Figure 20.11 Elevated roadways. (a) Westway, Section One; (b) Tunnel relief flyover, Liverpool; (c) Vorlandbrucke Obereisesheim; (d) Illtal; (e) West Gate approach viaducts; (f) Westway, section five; (g) Bendorf, section at pier; (h) Mancunian Way; (i) Gladesville; (j) London; (k) Narrows; (l) Annacis; (m) Severn; (n) Europa; (o) Duisberg-Neuenkamp; (p) Concordia; (q) Kniebrücke; (r) Sava I; (s) Zoo

Equations (20.4) and (20.6) must be satisfied simultaneously and, although this makes the problem nonlinear, the correct value can be satisfactorily determined by interpolation by solving for two assumed values of H. Each assumed H gives an incorrect solution to Equation (20.6) and, assuming the error varies linearly, the correct value of H can be found. For each assumed value of H, the structure behaves as a simple beam and influence lines can be constructed for bending moments, etc., and for $\int_0^L v(x)\mathrm{d}x$. Hawranek and Steinhardt[10] suggest that for a particular loading case the bending moment and shear forces be found from both sets of influence lines as well as the $\int_0^L v(x)\mathrm{d}x$ values. H is found by interpolation and then the final bending moments and shears are found by interpolating between the two sets of values already found from the influence lines.

Typical results for a continuous stiffening girder are shown in Figure 20.14.

The above treatment follows that given by Hawranek and Steinhardt[10] who also give a comprehensive set of standard solutions for the substitute girder. The result quoted below illustrates the form the solutions take. Using:

$$\mu^2 = H/EI$$

For the load case of Figure 20.15, deflections as a function of x are given by:

$$v(x, \xi) = PG(x, \xi)$$

$$= P\frac{l}{H}\left[\frac{x}{l}\left(1 - \frac{\xi}{l}\right) - \frac{\sinh \mu x \sinh \mu(l-\xi)}{\mu l \sinh \mu l}\right] \quad \text{for } \xi \geqslant x$$

$$(20.8)$$

$$v(x, \xi) = PG(x, \xi)$$

$$= P\frac{l}{H}\left[\frac{\xi}{l}\left(1 - \frac{x}{l}\right) - \frac{\sinh \mu \xi \sinh \mu(l-x)}{\mu l \sinh \mu l}\right] \quad \text{for } \xi \leqslant x$$

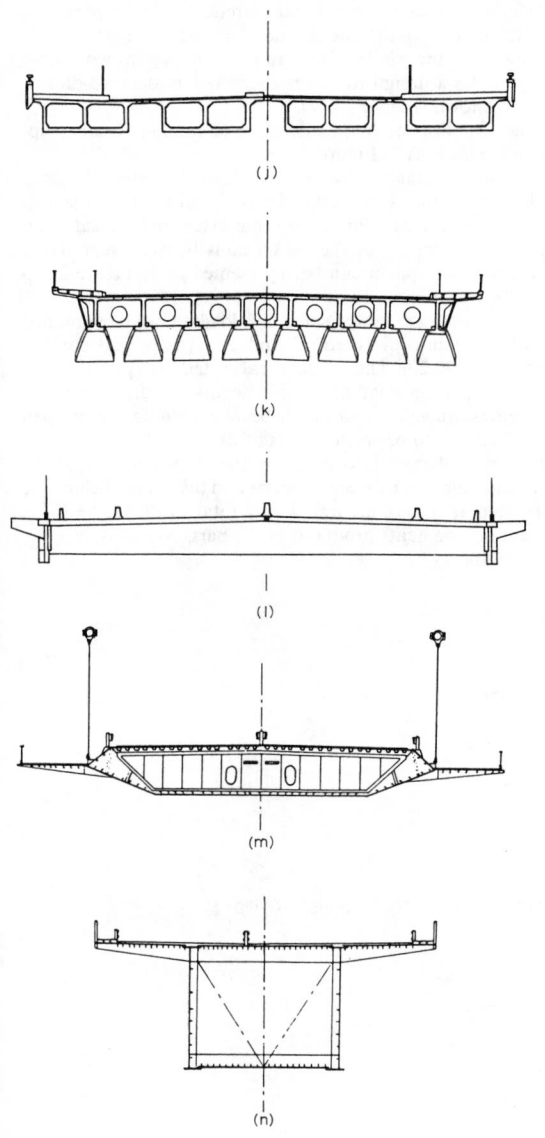

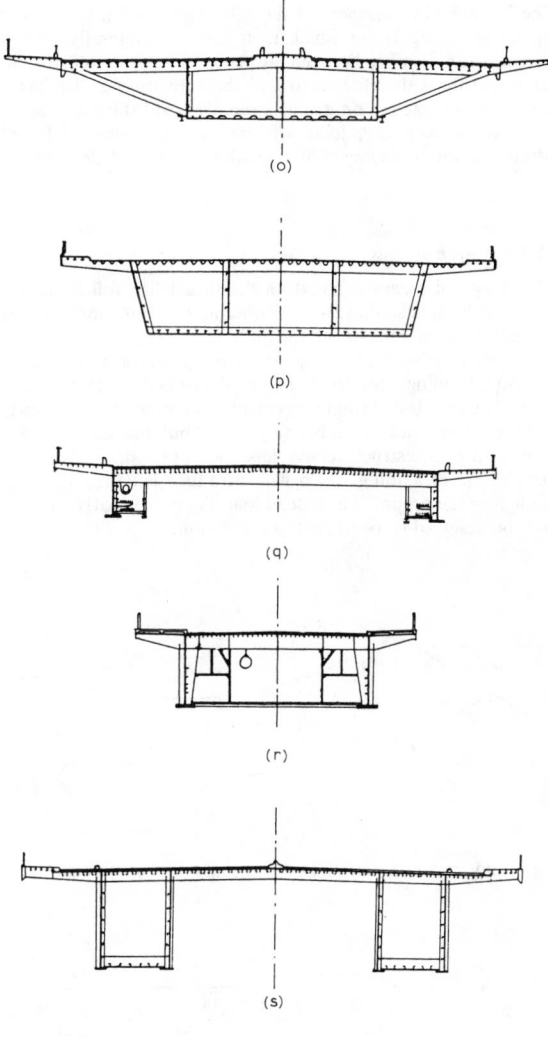

G(x, ξ) is known as a Green's function.

And:

$$F(\xi) = \int_0^L v(x)\mathrm{d}x$$

$$= P\frac{l^2}{H}\left[\frac{\xi(l-\xi)}{2l^2} - \frac{1}{(\mu l)^2}\left(1 - \frac{\cosh\mu(l/2-\xi)}{\cosh\mu(l/2)}\right)\right]$$

$$(20.9)$$

Computers can be used to analyse suspension bridges either by following the above approach or by means of standard framework programs provided the interaction of axial loads and deflections is allowed for. In other words, the change in geometry of the cable is considered. In some programs the axial loads must be stated as part of the data in the same way that H is

assumed in obtaining a solution to Equation (20.4). In others, an interactive process produces the correct axial forces. The structure solved can include the actual system of suspenders, tower properties, etc. or can be a very simple solution of the substitute structure of Figure 20.12.

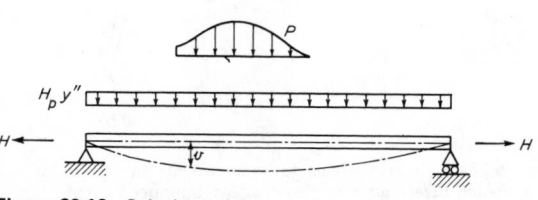

Figure 20.12 Substitute girder

20.3.1.2 Self-anchored suspension bridges

The horizontal component of the cable tension can be resisted by the stiffening girder which then acts as a laterally loaded compression member between suspenders. The net tension on the structure is therefore zero and the substitute girder has a zero axial load acting on it. The structure is substantially linear in its response to live load whereas the externally anchored bridge has an increasing stiffness with increasing deflection.

20.3.1.3 Arch bridges

The design of arches is based on the thrust line following the shape of the arch so that there is either no bending moment or a reduced bending moment in the arch member.

The shape of arch can only satisfy one condition of loading without bending moments being developed. Temperature changes, creep, foundation movements and imperfections must, however, introduce some bending in all but the three-hinged arch. In a bridge structure, live loading will produce a varying distribution of loading which will introduce bending. Clearly, the higher the proportion of dead load the more nearly can the arch be designed to be in pure compression.

The most common shapes are the circular arch, the parabolic arch and more recently the inclined leg frame (Figure 20.16). Loadings over the whole of (c) can be examined in two stages, which enables a design to be produced before detailed dimensions are known (Figure 20.17).

Arches for bridges frequently have continuous beams supporting the deck as in Figure 20.18.

At the design stage, since the dead load carried by the deck will depend on the construction method and the thrust jacked into the arch, the structure is to some extent determinate but, clearly, some bending of the deck beams between supports is introduced. The system can be represented as in Figure 20.19 where $H \simeq wl^2/8f$.

At the preliminary design stage, live loading can be examined by splitting it into symmetrical and antisymmetrical components (Figure 20.20). The symmetrical system will produce to a first approximation small bending moments and the asymmetrical system is equivalent to a simple beam with half the arch span (zero thrust due to opposite effects of load).

The local effects of loading on the deck can always be examined as a beam between columns and the overall behaviour can be seen as that of an arch with a total EI of $EI_{deck} + EI_{arch}$. The bending moments produced in the parts will be in proportion to stiffness.

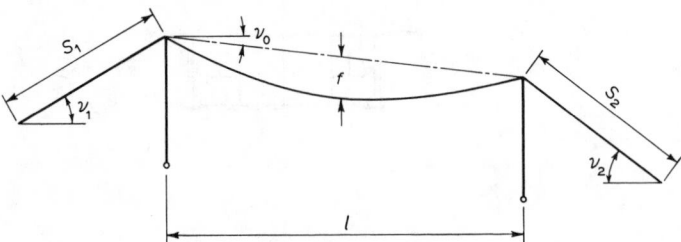

Figure 20.13 (After Hawranek and Steinhardt (1958) *Theorie und Berechnung der Stahlbrücken.* Springer-Verlag)

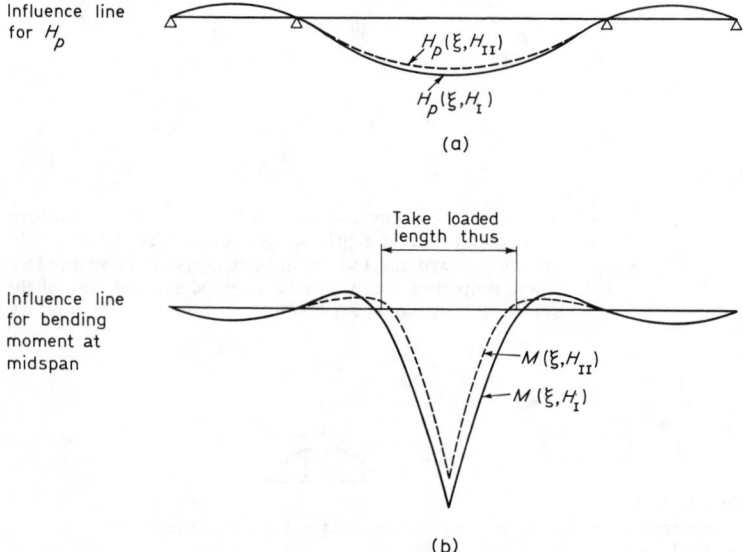

(a)

(b)

Figure 20.14 Influence lines. (a) Influence line for H_p; (b) for bending moment at midspan. (After Hawranek and Steinhardt (1958) *Theorie und Berechnung der Stahlbrücken.* Springer-Verlag)

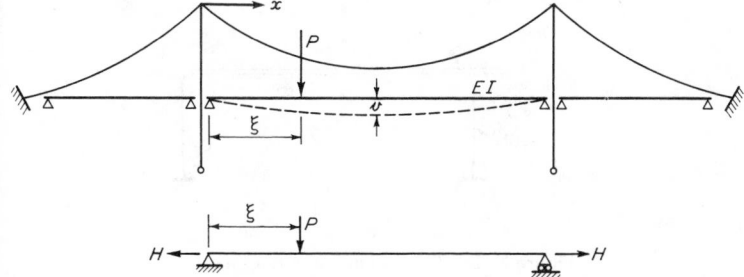

Figure 20.15 (After Hawranek and Steinhardt (1958) *Theorie und Berechnung der Stahlbrücken.* Springer-Verlag)

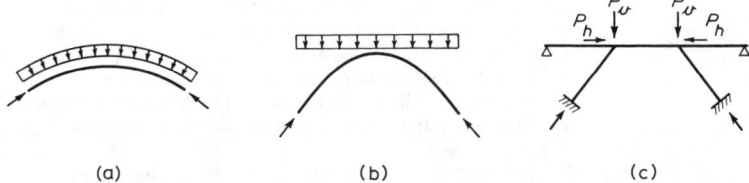

Figure 20.16 Arches with the loading that produces no arch member-bending. (a) Circular arch; (b) parabolic arch; (c) inclined leg frame

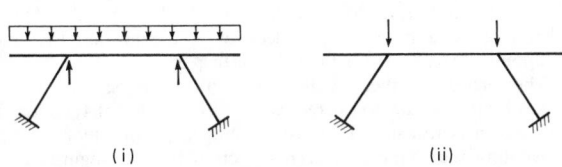

Figure 20.17 Two-stage analysis of inclined leg frame

Table 20.11

f_t		0	0.1	0.2	0.3	0.4	0.5	
For a two hinged arch		39.4	35.6	28.4	19.4	13.7	9.6	$H_{cr} = \dfrac{kEI}{l^2}$
For a fixed arch		80.8	75.8	63.1	47.9	34.8	24.4	

Arches may require correction for deflections since the thrust will magnify these in the same way that a strut is affected by end load. A two-hinged arch with a uniform loading buckles as in Figure 20.21 and to a first approximation the effective length of the strut is $l/2$. The magnification of moments can be found as for a strut using $1/(1-H/H_{cr})$ as the factor. Critical loads are available for a variety of cases as in Table 20.11.

Equation (20.4) for suspension bridges is applicable to arches with changes of sign in H and y. The nonlinear effects of deflections can be determined using that equation or the substitute beam in compression instead of tension. Computer programs can be used to analyse arches in the way already described for suspension bridges but if H/H_{cr} is small it is unnecessary to use programs which allow for changes of geometry.

Out-of-plane deflections can cause buckling or significant stresses in single arches or arches not braced laterally. This effect can be investigated readily using a grillage programme which allows for the interaction of axial loads and transverse deflections. The plane of the grillage must be considered as the plane of the arch for that purpose.

20.3.1.4 Tied arches

In the tied arch, the thrust is balanced by tensile forces in the stiffening girder which simplifies the in-plane behaviour of the structure since there is no net thrust on the structure. In-plane buckling is thus prevented but out-of-plane buckling is still possible.

20.3.2 Bridge girders of open section

The cross-girder and bracing members are shown as broken lines in Figure 20.22 since they do not affect the girder under twisting loads unless it has torsionally stiff members.

To find q (Figure 20.23) it is necessary to consider the equilibrium equation of the top flange (cf. simple beam theory). It is assumed that the shear force on the top flange is zero as shown and that the second moment of area of the top flange about a vertical axis is I_T.

Taking moments in a horizontal plane:

$$bq\,dx = \frac{(\sigma + d\sigma)}{b/2} I_T - \frac{\sigma}{b/2} I_T$$

$$= d\sigma \frac{2I_T}{b}$$

Therefore:

$$q\,dx = d\sigma \frac{2I_T}{b^2}$$

This is the same equation as that used for simple beams if their top flange area A is made equal to $2I_T/b^2$. It follows that q can be found by considering an equivalent simple beam with A_T replacing the deck and acted on by shear forces Q (Figure 20.24). Note that Q is not altered by horizontal shears and is therefore the same as in a simple beam loaded by W.

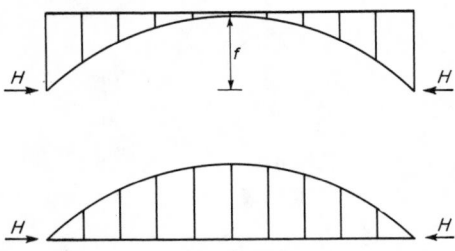

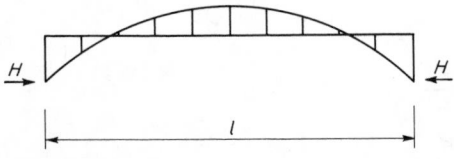

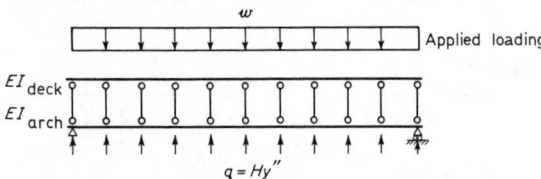

Figure 20.18

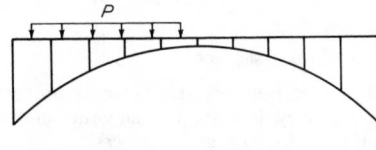

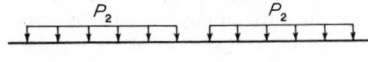

$q = Hy''$

Figure 20.19 Application of substitute girder

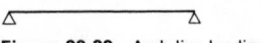

Figure 20.20 Arch live loading

Figure 20.21 Buckling of a two-hinged arch

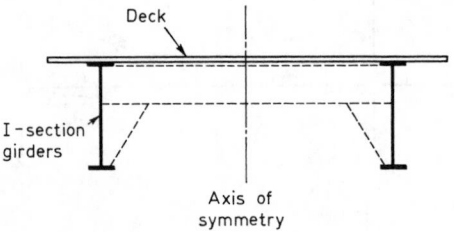

Figure 20.22 Girder of thin-walled open section

Any twisting load can be referred to the web positions giving the loads to be applied to the effective girder of Figure 20.24. The bending moments produced in the effective girder, acting as a single beam with the span of the actual structure, are applied to the effective girder cross-section. The second moment of area I_{eff}^v is used to find the longitudinal stresses between the top and bottom flanges. The remainder of the top flange stresses can be found by assuming a linear variation of stress between the web–flange junctions.

More complex loads (Figure 20.25) require consideration of horizontal forces and it is convenient to note that the centre of rotation (or shear centre) of the cross-section under purely twisting loads is a distance y_1 (Figure 20.26) above the top flange. This can be seen if it is recognized that longitudinal strain distribution and, hence, curvature results in the ratio of deflections in Figure 20.26 to be $w/v = 2y_1/b$ or $w = 2(y_1/b)v$. Therefore, at a height above deck of y_1 the normal to the midpoint of the deck must cut the vertical axis.

The vertical members of the cross-section may be individual boxes or thick-walled concrete webs with significant torsional stiffness in both cases. If transverse bracing is provided, thus preventing the shape of the cross-section from changing, the rotation of the boxes will equal that of the deck. This effect can be included to give the following governing equations with the deflections shown in Figure 20.25 which include a vertical displacement of the whole cross-section z. I_z and I_y are the usual second moments of area about a horizontal and vertical axis respectively. GK is the sum of the torsional stiffness of the whole cross-section:

$$\left.\begin{array}{l} EI_z z^{iv}(x) = P_1(x) \\[2ex] EI_{eff}^v v^{iv} - GK\dfrac{2}{b^2}v'' = P_1\dfrac{e_2}{b} + P_2\dfrac{(e_1 + y_1)}{b} \\[2ex] EI_y\left(w^{iv} - \dfrac{2y_1}{b}v^{iv}\right) = P_2 \end{array}\right\} \quad (20.10)$$

Equations (20.10) are not put forward for solution as a set of differential equations but as a description of the various mechanisms involved. The second is similar to the suspension bridge Equation (20.4), since $GK(2/b^2)v''$ can be compared with the Hv'' term. There is no term corresponding to $H_p y''$. The equivalent structure is, therefore, a beam of stiffness EI_{eff}^v under axial tension $GK(2/b^2)$.

The deflections and bending moments will be duplicated correctly by such a structure but the stresses do not, of course, require a direct contribution from the imaginary tensile force.

20.3.3 More general behaviour of suspension bridges and arches

The above treatment of a girder can be extended to include the usual twin cables of a suspension bridge. The positions of the

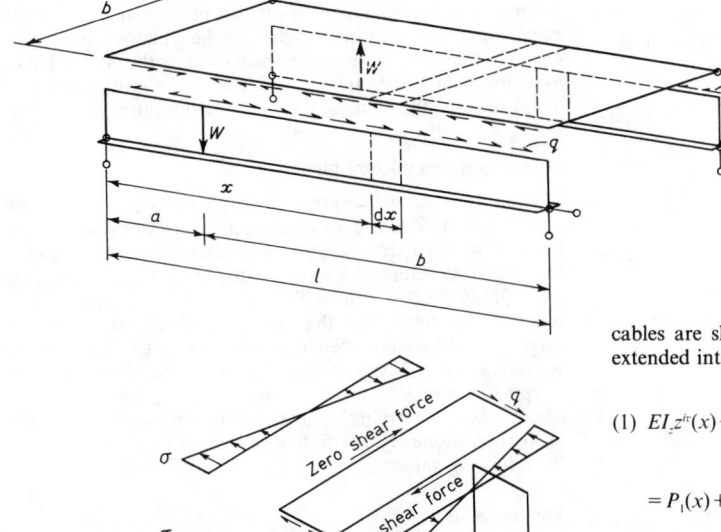

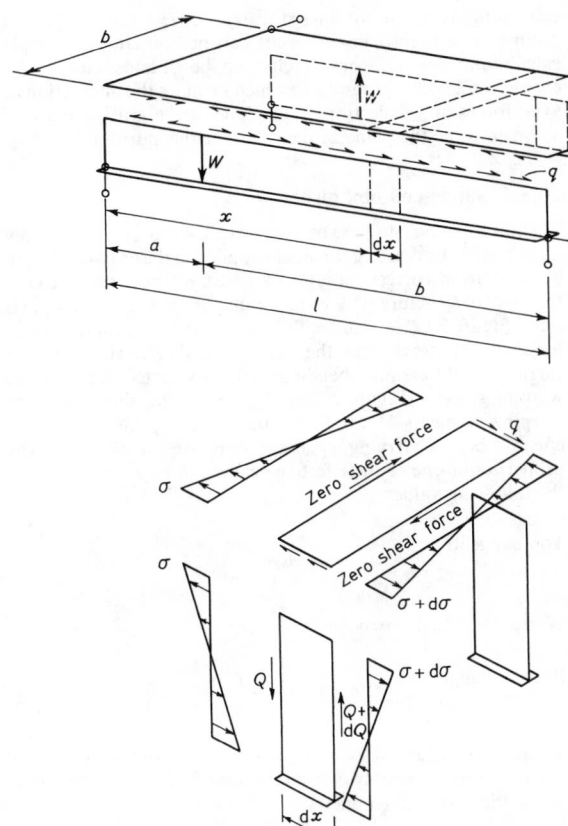

Figure 20.23 Loading and stresses on thin-walled open section

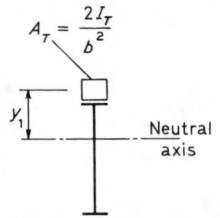

$$A_T = \frac{2I_T}{b^2}$$

Neutral axis

Flexural rigidity $= EI^v_{eff}$

Figure 20.24

cables are shown in Figure 20.27. Equations (20.10) become extended into:

(1) $EI_z z^{ir}(x) - (H_1 + H_2)z''(x) - (H_1 - H_2)\dfrac{2e}{b}v''(x)$

$$= P_1(x) + (H_{p1} + H_{p2})y''$$

(2) $EI^v_{eff} v^{ir}(x) - GK\dfrac{2}{b^2}v''(x) - (H_1 + H_2)\dfrac{2e^2}{b^2}v''(x) - (H_1 - H_2)\dfrac{e}{b}z''(x)$

$$(20.11)$$

$$= P_1\frac{e_2}{b} + P_2\frac{e_1}{b} + P_2\frac{y_1}{b} + (H_{p1} + H_{p2})y''\frac{e}{b}$$

(3) $EI_y\left(w^{ir}(x) - \dfrac{2y_1}{b}v^{ir}(x)\right) = P_w$

The terms underlined are fairly small and can be neglected, which leads to (1) and (2) being independent equations in z and v.

Further, if purely torsional loading is assumed, $H_{p1} \simeq H_{p2} = H_p$ and writing $H_1 + H_2 = 2H$ the second equation becomes:

$$EI^{eff}_v v^{ir}(x) - \left(GK\frac{2}{b^2} + 2H\frac{2e_2}{b^2}\right)v''(x)$$

$$= P_1\frac{e_2}{b} + P_2\frac{(e_1 + y_1)}{b} + 2H_p\frac{e}{b}y'' \qquad (20.12)$$

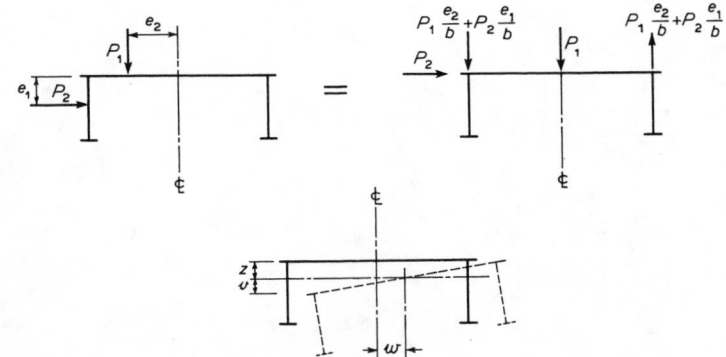

Figure 20.25 Overall loading and displacement

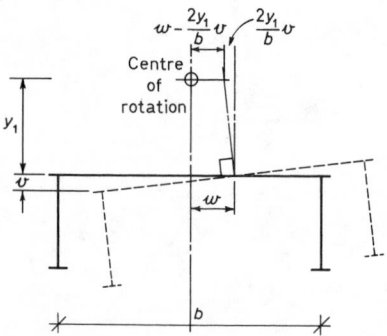

Figure 20.26 Twisting about centre of rotation

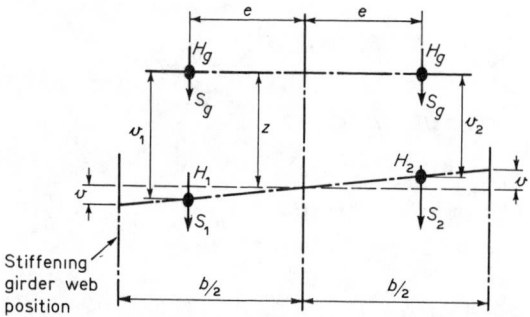

Figure 20.27 Change in cable position

For each cable the condition $\int_0^L \Delta \, dx = 0$ must be satisfied leading as before to the correct value of H_p. Equation (20.12) is clearly of the same form as the ordinary suspension bridge equation for vertical deflections except that EI^v_{eff} replaces EI and $GK(2/b^2) + 2H(e^2/b^2)$ replaces H. Therefore, the same results can be used to solve this equation.

The above equations may be used to investigate aerodynamic and other vibrational effects as well as live loading. The above treatment follows that of Hawranek,[10] but the effective beam concept has been used in order to give the equations more physical significance. Hawranek works in terms of a warping function and relates loads to the shear centre. Hawranek and Steinhardt give, however, a number of useful results for the natural frequency of various systems.

Horizontal deflections due to wind may produce a significant lateral component of the cable force but the complete properties of the structure must be known before the effect can be determined, e.g. the tower stiffness influences this type of

behaviour. It is not considered in the above equations. For a given system, approximate results can be estimated by simple calculations but rigorous results can be obtained using, for example, a grillage programme which includes the interaction of axial forces and deflections. The plane of the grillage must be assumed to be the plane of the cable for this purpose.

20.3.4 Single-cell box girder

The full torsional stiffness of a single cell, $4A^2G/\oint(1/t)\,ds$ is only mobilized if the twisting forces are applied in the cross-section in a distribution corresponding to a constant shear flow around the box. A structure with effective diaphragms at the supports only (Figure 20.28(a)) and with hinges at the long edges, has no torsional stiffness under the twisting loads shown. They are carried in differential bending. The associated stresses are warping stresses (Figure 20.29). Figure 20.28(a) shows how the warping moments M_0 are found and Figure 20.28(c) shows the effective beams carrying equal and opposite values of M_0. The properties of one of the effective beams can be calculated from the following values:

Top flange area

$$A_T = \frac{2I_T}{b_T^2}$$

Web as in actual box beam

Bottom flange area

$$A_B = \frac{2I_B}{b_B^2}$$

where I_T and I_B are, respectively, the second moments of area of the top flange assembly and the bottom flange assembly about the vertical axis of symmetry.

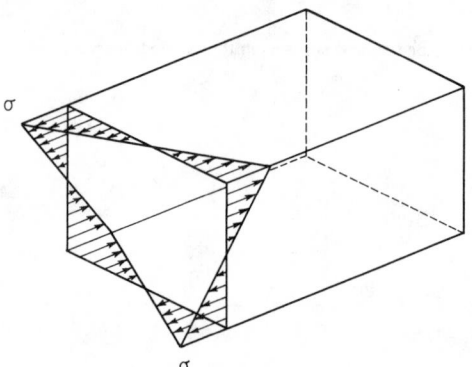

Figure 20.29 Warping stress distribution

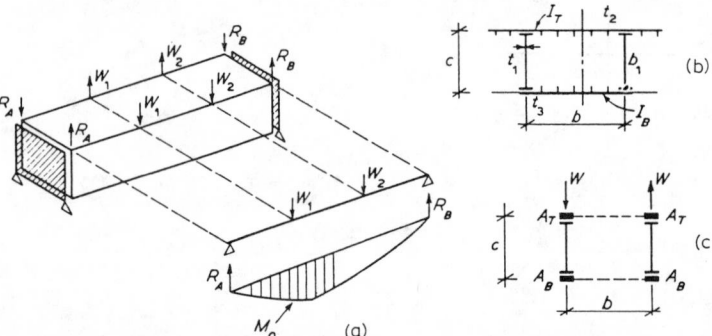

Figure 20.28 Equivalent I-beam

The warping stresses in the effective beam are determined as in normal beam theory and the remaining warping stresses in the flanges can be found by assuming a linear variation of stress across the top and bottom flanges.

Complex box structures can be solved most economically by suitable finite-element programs. Elaborate calculations by alternative means are not justified, but in order to understand the behaviour of box structures or for structures where an approximate result shows that further investigation is unnecessary, the following methods are of value.

20.3.5 Boxes with discrete diaphragms

Steel boxes usually have a number of diaphragms formed from a lattice system or solid plate and sometimes unbraced frames. The following influence-coefficient approach is one of several methods of dealing with such structures acted upon by twisting loads.[11,12]

The structure is rendered statically determinate for twisting loads by releasing the warping moments at each diaphragm thus permitting relative warping rotations to occur (Figure 20.30). Each bay of the box between diaphragms acts in a similar way to the box of Figure 20.28. At each diaphragm, however, instead of the twisting moments being taken by the supports, they are fed through the diaphragms into the box as pure torsion. The diaphragms then act as a series of elastic supports to the equivalent beam as shown in Figure 20.31 except that at the initial stage the beam acts as if hinged at each spring and diaphragm and is therefore determinate.

The stiffness of the diaphragms k is found from the relationship between the distortional force system, Figure 20.32(a), which is the distortional component of Figure 20.32(b), and the distortional deflection system shown in Figure 20.33(a).

$$\frac{P}{v} = \frac{\text{applied distortional forces}}{\text{distortional deflection}}$$

The deflections of the diaphragms or springs induce relative rotations at each hinge. The rotations at the releases are increased by the warping produced by the torque fed into the box at each diaphragm and the effect of the load between diaphragms. Denoting θ_1 as the relative rotation due to spring deflections and local loads, the total relative rotation is:

$$\theta_r = \theta_1 + \frac{Q_{r+1}}{\bar{f}_{r+1}} - \frac{Q_r}{\bar{f}_r} \tag{20.13}$$

where Q is the torque T divided by b, and $\bar{f}$ is the shear stiffness linking torsion and warping rotation:

$$\bar{f} = \frac{G}{E} \frac{8c^2}{b/t_3 + b/t_2 - 2c/t_1} \tag{20.14}$$

where r and $r+1$ refer to bays between diaphragms as in Figure 20.30.

The influence coefficients for solving the series of compatibility equations are:

$$E\theta_{r,r-2} = \frac{1}{a_{r-1}a_rk_{r-1}}$$

$$E\theta_{r,r-1} = \frac{a}{6I_r} - \frac{1}{a_rf^*_r} - \frac{1}{a_r}\left[\left(\frac{1}{a_r}+\frac{1}{a_{r-1}}\right)\frac{1}{k_{r-1}}\right.$$
$$\left. + \left(\frac{1}{a_r}+\frac{1}{a_{r+1}}\right)\frac{1}{k_r}\right]$$

$$E\theta_{r,r} = \frac{a}{3I_r} + \frac{a_{r+1}}{3I_{r+1}} + \frac{1}{a_rf^*_r} + \frac{1}{a_{r+1}f^*_{r+1}} + \frac{1}{a^2_rk_{r-1}}$$
$$+ \left(\frac{1}{a_r}+\frac{1}{a_{r+1}}\right)^2\frac{1}{k_r} + \frac{1}{a^2_{r+1}k_{r+1}}$$

$$E\theta_{r,r+1} = \frac{a_{r+1}}{6I_{r+1}} - \frac{1}{a_{r+1}f^*_{r+1}} - \frac{1}{a_{r+1}}\left[\left(\frac{1}{a_{r+2}}+\frac{1}{a_{r+1}}\right)\frac{1}{k_{r+1}}\right.$$
$$\left. + \left(\frac{1}{a_{r+1}}+\frac{1}{a_r}\right)\frac{1}{k_r}\right] \tag{20.15}$$

$$E\theta_{r,r+2} = \frac{1}{a_{r+2}a_{r+1}k_{r+1}}$$

where the effective distortional bending stiffness of the box in bay r is EI, and an additional f^* is a shear stiffness for the warping rotation produced by distortional shears in the box:

$$f^* = \frac{G}{E} \frac{8c^2}{b/t_3 + b/t_2 + 2c/t_1} \tag{20.16}$$

The various results given above can be found from concepts of virtual work using the mechanism shown in Figure 20.34 consisting of a series of shear webs and booms of axial stiffness. It can be used to obtain more general results such as those for boxes of trapezoidal cross-section.[11,13]

The warping produced by torsion results in stresses only if there is a change in torsion and therefore incompatible warping. Longitudinal stresses act to remove the lack of continuity. An upper bound estimate of these stresses can be made by assuming that the cross-section cannot deform. The warping moments produced by a change in torque $T = Pb$ is:

$$X = \frac{P}{2} \frac{\sqrt{f^*I_\theta}}{f} \exp\left[-\sqrt{(f^*/I_\theta)}x\right] \tag{20.17}$$

where x is the distance from the cross-section at which the change occurs.

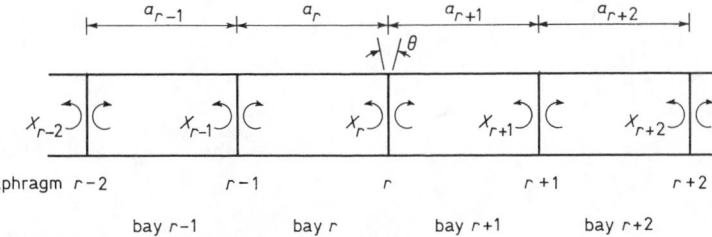

Diaphragm $r-2$ $r-1$ r $r+1$ $r+2$

bay $r-1$ bay r bay $r+1$ bay $r+2$

Figure 20.30 Box with discrete diaphragms

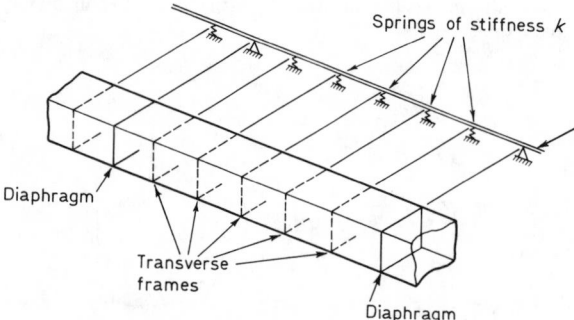

Figure 20.31 Equivalent beam on elastic supports

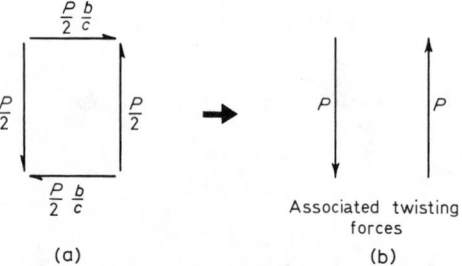

Figure 20.32 Distortional force system

The distribution of shear forces associated with warping stresses can be found from the warping stress distribution (Figure 20.35). Starting from the edge of the cantilever and assuming σ is the change in longitudinal stress over a unit length:

$$q = \int_0^x \sigma t \, ds$$

where q is the shear flow τt.

The complementary shear to q is on the face of the cross-section. The shear at the centre of the top flange can be assumed to be zero at the first stage of the calculation, which enables a simple shear system to be found. A pure torsional shear flow must be added to remove any component of pure torsion acting on the cross-section.

20.3.6 Box beams with continuous diaphragms

The frame action of the webs and flanges in concrete boxes provides a continuous resistance to distortion; consequently special diaphragms are not usually necessary except at disturbances such as bearings and other support points.

Smaller steel box beams which rely on the stiffness of the sides plus the frame action of web and flange stiffness or larger box beams with special frames which leave the interior of the box unobstructed, have similar characteristics to the concrete boxes. Frames are generally flexible compared with braced or plate diaphragms but stiff frames can be made which will have properties that can only be explored fully by a treatment which is suitable for discrete diaphragms. The validity of the following approach for steel boxes can be determined from the half wavelength which should be greater than twice the spacing of the frames.

The effect of twisting loads P applied at the corners of the web on the vertical deflections caused by distortional bending of the box and allowing for the diaphragm action of the cross-section is:[14]

$$v = \frac{P}{2a\beta} e^{-ax} \left\{ \beta \left[\frac{\lambda^2}{k} + \frac{1}{2E} \left(\frac{1}{f^*} - \frac{1}{f} \right) \right] \cos \beta x \right.$$

$$\left. + a \left[\frac{\lambda^2}{k} - \frac{1}{2E} \left(\frac{1}{f^*} - \frac{1}{f} \right) \right] \sin \beta x \right\} \quad (20.18)$$

assuming that the distance to a support is infinitely long. EI_θ is the effective distortional bending stiffness of the box, and k is the diaphragm stiffness per unit length. λ is defined by:

$$\lambda^4 = \frac{k}{4EI_\theta}$$

and:

$$a = (\lambda^2 + k/4Ef^*)^{1/2}, \quad \beta = (\lambda^2 - k/4Ef^*)^{1/2}$$

The generalized warping stress resultant is:

$$x = \frac{P}{2a\beta} e^{-ax} \left[\beta \left(\frac{1}{2} + \frac{\lambda^2 I_\theta}{f} \right) \cos \beta x \right.$$

$$\left. + a \left(-\frac{1}{2} + \frac{\lambda^2 I_\theta}{f} \right) \sin \beta x \right] \quad (20.19)$$

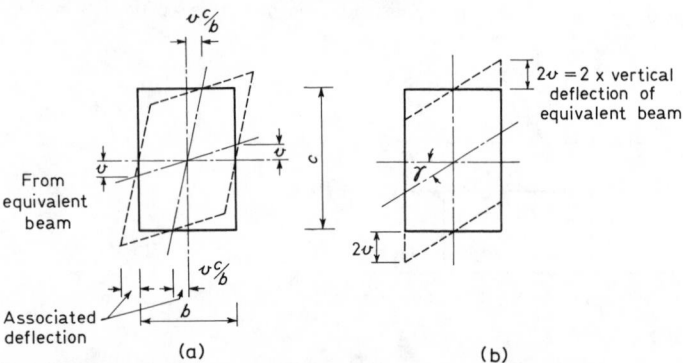

Figure 20.33 Cross-section distortion

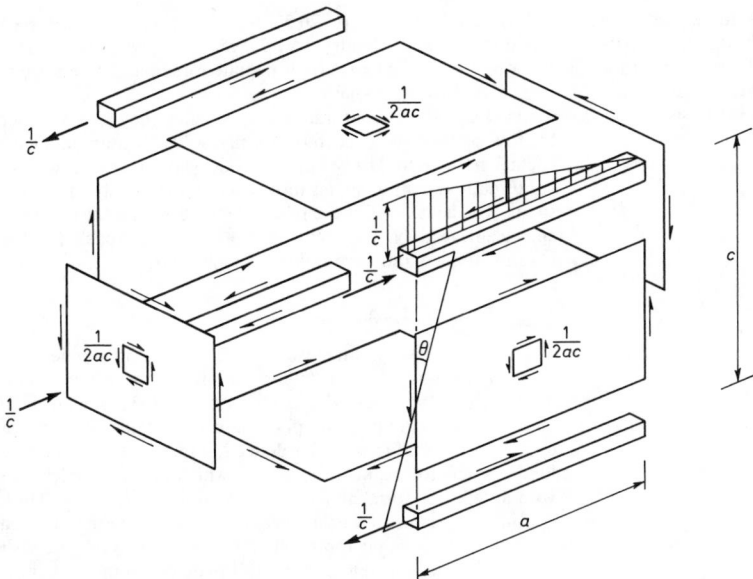

Figure 20.34 Exploded box girder

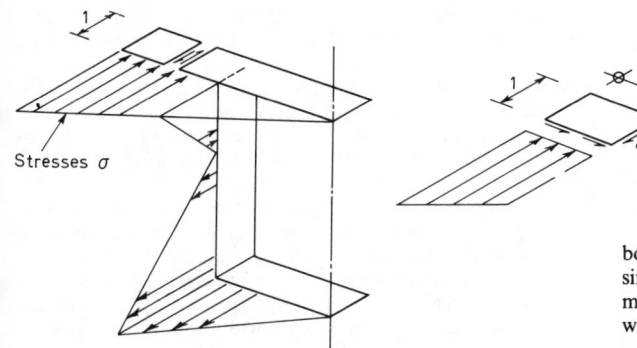

Figure 20.35 Half box

If the properties of the box are within the range $\lambda^2 - (k/4Ef^*) < 0$ then $\bar{\beta} = (k/4Ef^* - \lambda^2)^{1/2}$ replaces β and hyperbolic functions are used.

The half wavelength is $\pi/2\beta$.

In many cases, the effect of shear is not considerable and $\alpha = \beta = \lambda$. The equations then simplify into the standard beam on elastic foundation results. The above equations include, however, the effect of the change in torque due to the twisting loads P and in order to correct the results of beam on elastic foundation theory the warping moment of Equation (20.17) should be added. The correction is likely to be of most significance in boxes which are much wider than their depth.

The above equations are valid for boxes of trapezoidal cross-sections if the concepts are generalized as was illustrated by Dalton and Richmond.[13]

20.3.7 Box girders with cantilevers

The advantages of box girders over structures of open cross-section are sometimes only marginally linked to structural efficiency. However, where a compact structure is to carry a much wider deck, producing a large cantilevered section, the torsional stiffness and strength of the box is of primary importance. The importance of the interaction of the cantilever and the

box and the magnitude of the stresses is correspondingly great since any loss of torsional strength and stiffness could result in a major increase in the shear stresses and longitudinal stresses with a corresponding reduction in the load factor.

The three types of cantilever in Figure 20.36 show:[15]

(1) Transmission of torque Pe into twisting couple which must be resisted by diaphragm action.
(2) Cantilever bracket which, in the position shown, produces horizontal loads which are in the correct ratio to the vertical shear P to give a torsional shear flow without diaphragm action. In a concrete box the transverse moments in walls of the structure would be small.
(3) The relative stiffness of the web results in most of the cantilever moment being taken by the web which produces horizontal forces corresponding to a brace at any position of the load. The transverse moments in the walls are therefore reduced rather than increased as the eccentricity increases. If $e = b_T/2 + b_B/2$ they are practically zero.

In concrete boxes where the only diaphragm stiffness is due to transverse bending of the walls, except at bearings, the beneficial effects described in (2) and (3) are of considerable importance. The reduction in transverse moments has been described but another important benefit is the reduction in the shear forces in the outer web.

20.3.8 Multiple web girders of open cross-section

The choice of an individual main girder or beam as the member to carry say a concentrated load applied to it with the distribution of that load as the next, correcting, operation is more appropriate to smaller spans where typically there are large numbers of main beams and the width of the bridge is comparable with the span. In most shorter span structures the width is in fact more than the span which lends further weight to the argument. The following section on harmonic analysis adopts this approach.

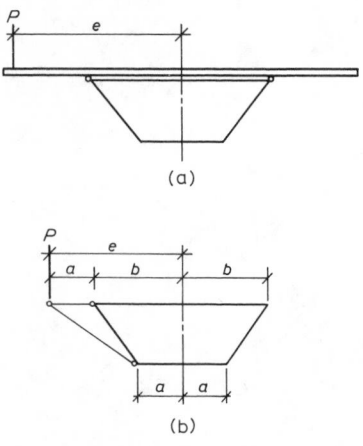

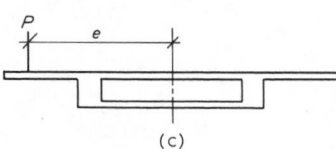

Figure 20.36

20.3.8.1 Harmonic analysis

The interaction between a series of separate simply supported beams of constant cross-section can be investigated for arbitrary loading using harmonic analysis. The sine series is the most suitable approach because any load which varies sinusoidally over the span in a complete number of halfwaves produces a deflected profile for each beam of the same form but of varying magnitudes. This result is justified provided that it is recognized that the interacting forces between the beams will be proportional to the transverse deflected form and will be also vary sinusoidally.

The interaction between the beams is dependent on the ratio of the transverse to longitudinal stiffness:

$$a = \frac{12}{\pi^4} \left(\frac{L}{h} \right)^4 \frac{D_y}{D_x} \qquad (20.20)$$

where D_y and D_x are stiffnesses of the equivalent orthotropic plate in the transverse and longitudinal directions.

For a single half-wave loading on one beam the distribution coefficients giving the fraction of the load taken by each beam have been calculated for a number of different systems by Hendry and Jaeger.[16] Figure 20.37 shows the coefficients for a five-beam bridge with beam 2 loaded (Figure 20.38).

The coefficients given are for the first harmonic only. Coefficients for subsequent harmonics can be found by varying a as appropriate for the shorter wavelength. Alternatively, if a sufficiently close approximation is given by the first harmonic alone for distribution to the unloaded beams, the behaviour of the loaded beam is given by its 'free deflection' curve less that which has been distributed to the other beams.

The results in Figure 20.37 are for a system with zero torsional rigidity; Hendry and Jaeger also give results for a torsionally rigid system.[16] Intermediate torsional stiffnesses can be analysed by interpolation.

Fixed-ended and continuous beams which they also consider by this method may be more easily solved by an influence coefficient method. Hinge releases at the supports can convert a continuous system of beams into two or more simply supported spans. The behaviour of the released structure and the influence coefficients can be found using the above approach for the loading applied and each influence coefficient.

20.3.8.2 Eigenvalue methods

The above method of analysing a grillage of beams under out-of-plane loading is a particular example of a more general method, i.e. the eigenvalue approach. Whereas a sinusoidal variation of loading of simply supported beams of constant cross-section produces similar deflection forms, there are eigen-load systems for beams of all forms which produce deflections with a deflected form similar to the load-intensity curve. Thus, continuous beams of varying cross-section can be investigated by such an eigenvalue approach. Where the transverse member is a continuous concrete slab or steel plate or comprises a large number of transverse beams it may be assumed that a continuous transverse medium is the most appropriate physical model. In some circumstances – where there are a small number of transverse members or other significant variations from a constant transverse medium – the eigenload system becomes a series of discrete loads on the main girder corresponding to the positions of the transverse members. The discrete approach may also be the most appropriate representation of a continuous transverse system in order to facilitate the analysis of more complex systems by numerical techniques.

Longer span bridge superstructures are nearer in behaviour to a single beam formed from the aggregate of the individual longitudinal main beams, slabs and plates of the complete bridge cross-section. Consequently, there are both conceptual and numerical advantages in analysing such structures as single aggregate beams with subsequent correcting operations to allow for the deformations of the transverse beams, slabs and diaphragms which contribute to the transverse stiffness of the superstructure.

The eigenvalue approach can be used to determine the necessary corrections by considering the characteristics of the transverse structural system of the bridge thereby turning the method described above in section 20.3.8.1 through 90°. The five-beam system for which the distribution coefficients were given in Figure 20.36 for beams of zero torsional stiffness can be used as an example of this approach.

Figure 20.39(a) shows a point load A acting at any spanwise position on the central beam of a set of girders which form any system of spans and have any support condition provided that both spans and supports are the same for all girders. Figure 20.39(b) is then the load distribution required to produce the effect of all beams acting as an aggregate beam to carry the point load.

The correcting systems are shown in Figure 20.39(c) and (d) and are eigenloads or vectors $\mathbf{p}_1$ and $\mathbf{p}_2$, of the transverse beams or other transverse system. The particular system shown is valid for a transverse beam or slab system that is constant in its flexural properties across the width of the five beams. The eigenload systems are both self-equilibrating and $\mathbf{p}_1$ and $\mathbf{p}_2$ can be calculated readily from statics such that (b), (c) and (d) are equivalent to (a). The eigenloads themselves are found by choosing a pattern of point loads that produces the same ratio of deflection of the transverse system at each beam position relative to the respective component of the eigenload system.

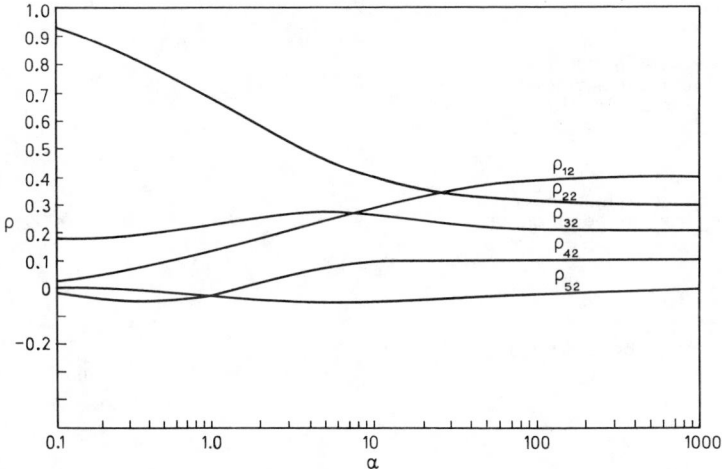

Figure 20.37 Distribution coefficients for a five-beam bridge, beam two loaded. (After Hendry and Jaeger (1958) *The analysis of grid frameworks and related structures*. Chatto and Windus)

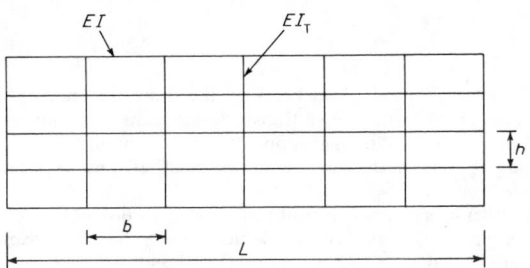

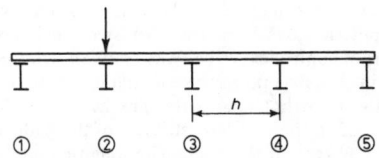

Figure 20.38 Five-beam bridge. (After Hendry and Jaeger (1958) *The analysis of grid frameworks and related structures*. Chatto and Windus)

Each beam is thereby effectively supported by a spring of equal stiffness when the complete system is loaded by one of the eigensystems. The spring stiffnesses for a transverse medium of flexural stiffness D per unit length for each beam is:

$$\frac{12\omega_1 D}{h^3} \quad \text{for eigenload system } \mathbf{p}_1 \tag{20.21}$$

and:

$$\frac{12\omega_2 D}{h^3} \quad \text{for eigenload system } \mathbf{p}_2 \tag{20.22}$$

where $\omega_1 = 0.0759016$ and $\omega_2 = 2.35267$

Thus, the beam systems under eigenload systems $\mathbf{p}_1$ and $\mathbf{p}_2$ can be analysed by considering for each load system the behaviour of the appropriate elastically supported beam. The essential

feature of the method is, of course, that only one beam has to be investigated for each load system. Where the transverse medium of stiffness D is appropriate, that single beam can be analysed using beam-on-elastic foundation theory. The supports and continuity can be allowed for as appropriate. Consequently, the two correcting solutions are readily produced either as exact solutions or as a means of assessing the characteristics of the structural system.

Grillage programs enable solutions to be obtained by computer with the advantage that complex geometries can be simulated without difficulty. In both cases, it is necessary to estimate the effective top flange unless the more complex form of beam and slab program using finite elements is used.

20.3.9 Multiple single-cell box beams

A series of box beams connected by a top deck and, in some cases, cross-beams and stiffening diaphragms, can be analysed by various approaches. The grillage approach using a computer is not necessarily suited to all problems of this type but it is discussed first because, in determining the properties of the members, the essential mode of behaviour of this form of structure emerges.

Figure 20.40 shows part of a typical cross-section which could represent a series of concrete main longitudinal beams spanning 20 and 60 m or the trough stiffeners on a steel deck system spanning 4 m.

In such systems, the interaction between the beams is through the deck slab or deck plate only if no special cross-beams, etc. are provided. The magnitude of the interacting forces will be mainly dependent on the overall deflections of the beams and so the distortional stiffness of the individual boxes will almost equal the frame stiffness of the sides. The distortional bending or warping stiffnesses of the boxes may be assumed to be nil except for local wheel loads, which will be mentioned later.

Through isolating one of the boxes and its share of the deck slab, its behaviour can be considered further (Figure 20.41). A unit value of the antisymmetrical component of the vertical shearing forces will act as shown. The twisting effect will be resisted by pure torsion if there is no significant distortional resistance of the box beam except for diaphragm action. The cross-section, acting as a frame, is loaded by the distortional component of the twisting load. The net effect can be obtained by the device illustrated. The pin-jointed bars are placed so as to apply a pure torsional shear flow to the cross-section. They also prevent only a pure rotation of the cross-section since distor-

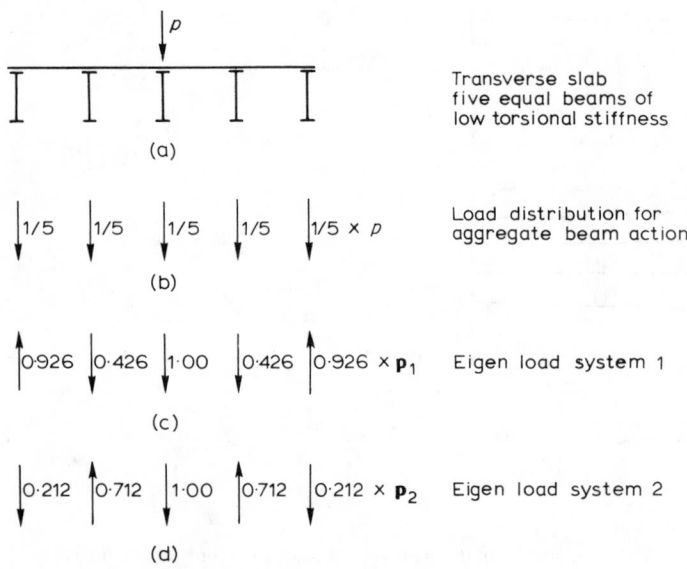

The three load systems (b)+(c)+(d) are equivalent to the single load p in (a) if $\mathbf{p}_1 = 0.3247$ and $\mathbf{p}_2 = 0.4752$

Figure 20.39 (a) Point load A acting on central beam; (b) load distribution acting as aggregate for point load p; (c) eigenloads for system 1; (d) eigenloads for system 2

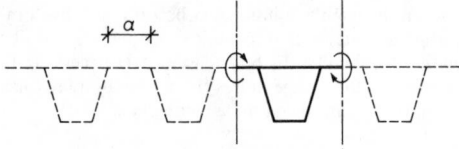

Figure 20.40 Interconnected multiple box girders

tional deflections do not have components in the direction of the restraints. A simple plane-frame analysis of the system gives the deflections and, hence, effective stiffness of a transverse member cantilevering out a distance $(a+b)/2$ from a grillage element with the torsional and flexural stiffnesses of the box. The antisymmetrical moments and symmetrical shears and moments can be applied to find the appropriate stiffnesses and, if the simplest form of grillage is used, a single compromise value must be chosen. The one based on antisymmetrical shears alone has been found to give results which compared well with a three-dimensional finite element simulation of a series of concrete boxes at 2-m centres.

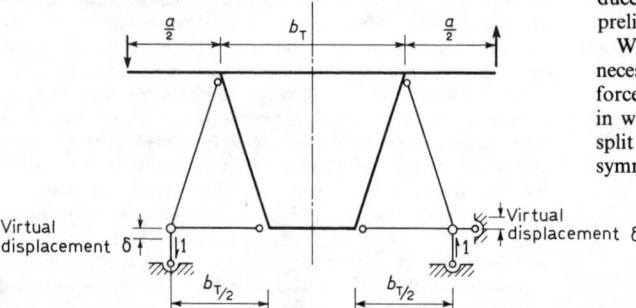

Figure 20.41 Torsional support system for slice of beam to give frame stiffness

The effect of local wheel loads on the transverse moments must be added to the above transverse moments by assuming that the boxes provide rigid supports. The maximum moments in the slab were, in the case mentioned, unaffected by the local slab loading.

Another effect already mentioned is the distortion of the box due to wheel loads applied on one side only. In fact, for boxes that are spaced at up to 2-m centres, the wheel loads are spread over a width sufficient to make the highest loads fairly symmetrically disposed about an individual box. An allowance can be made, however, by using the single cell theory to calculate stresses which are superimposed on those described above.

A similar approach has been used for steel deck systems, assuming points of contraflexure halfway between stringers but, instead of treating the structure as a series of discrete beams, it is transformed into an orthotropic plate. The transverse flexural stiffness is included in the torsional stiffness of the plate and is, therefore, taken as zero in the plate. The longitudinal flexural stiffness is determined in the usual way. Transversely, the deflections of the plate are represented as a sine series in order to solve the plate equation for wheel loading. A large number of terms in the series are required because of poor convergence which, together with the difficulties in obtaining detailed stress values other than longitudinal ones, from the solution, make the method of limited value. Graphs have, however, been produced[17] for the wheel loading used in the US which are useful for preliminary estimates.

Where a small number of large box girders are used, it may be necessary to allow for the various components of the interacting forces more exactly. An example of this is shown in Figure 20.42 in which the nonuniform component of load on two boxes is split into three load systems with the properties of either symmetry or antisymmetry. Releasing the forces at the centre of the connecting slab or cross-girder produces a lack of compatibility in each case. The influence coefficients are found from the unit forces of Figure 20.43(a) which relate to the compatibility equations including u_a and u_c and Figure 20.43(b) for u_b. δ represents the overall bending deflection of the box beam, $a\theta$ is the deflection produced by torsional rotation and δ_c is the deflection of the cross-girder or deflection of the slab. δ_c will be found from Figure 20.43(c) if it is a concrete box of the type already discussed, and θ_c similarly from Figure 20.43(d). Sinusoidally varying forces can be used for boxes without discrete diaphragms except at the supports, otherwise the influence coefficients can be related to individual cross-members.

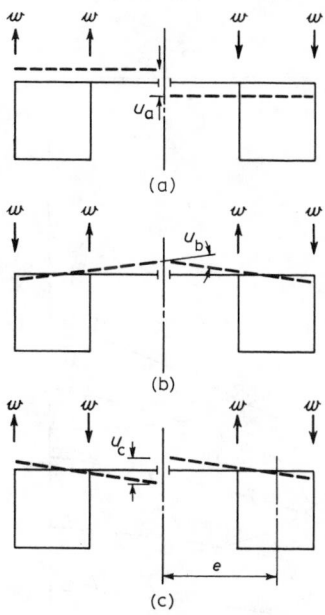

(a)

(b)

(c)

Figure 20.42 Interconnected boxes

20.3.10 Multicellular bridge decks

Bridge structures similar to the top-hat beam deck[6] which has 115 mm thick webs and no diaphragms between supports, have cross-sections which are relatively flexible in transverse shear. The usual grillage or orthotropic plate approach in which shear deformations are neglected is consequently invalid. The following treatment[6] is also relevant to cellular steel decks which may be even more flexible owing to higher web depth:thickness ratios.

Transverse shears are carried by the Vierendeel frame action (Figure 20.44) and the flexibility of the frame can be simulated by an equivalent web area of the transverse beams:

$$A_w = (E/G)\frac{12h/d}{(dh/2I_1)+(h^2/I_2)} \tag{20.21}$$

The grillage program used must include the effects of deflections due to shear strains. It is necessary to differentiate between rotations of initially horizontal and initially vertical lines when shear deformations are considered. In the grillage program used for Table 18.9 structure the rotation of vertical lines was the variable used. The flexural parameters are derived in the usual way but the torsional stiffnesses of the grillage members require further consideration.

20.3.11 Symmetrical loading

Loads disposed symmetrically in the transverse sense produce only relatively small transverse movements. True torsion is absent but Figure 20.45 shows that transverse members of grillage will be subjected to twisting which in the actual structure is simply a set of shear strains leading to shear transfer between the beams in a horizontal plane. This is sometimes referred to as shear lag. If the shear lag is small the shear transfer is high and the whole flange will be stressed uniformly. If the transverse members are assigned a stiffness per unit longitudinal distance of $h^2t/2$, this effect will be simulated. The torsional stiffness of the longitudinals is largely immaterial since they do not rotate significantly.

20.3.12 Antisymmetrical loading

Loads which cause twisting produce rotations of the cross-section which are the mean of a rotation of a vertical and horizontal line. The vertical line component is given directly by the grillage. The rate of change of rotation of a horizontal line is

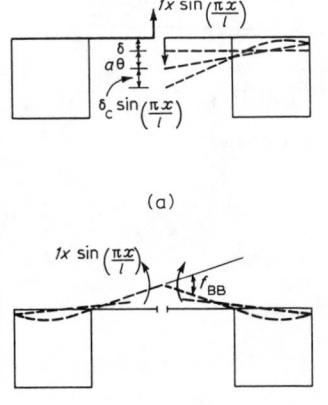

(a)

(b)

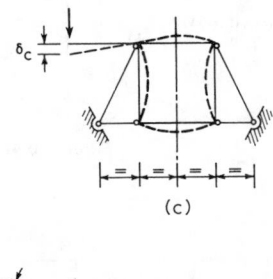

(c)

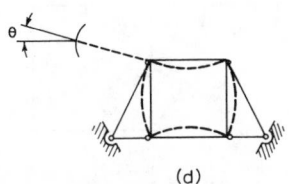

(d)

Figure 20.43 Unit loads and couples applied at releases

almost equal to the rate of change of rotation of a vertical line considered in the longitudinal and transverse direction respectively, since the true webs only undergo small shear strains compared with the frame of the cross-section.

Therefore, by allocating half the torsional stiffness GK of the cross-section to longitudinals and the remainder to transverse members, the mean rotation will be used as required.

GK for the whole structure is: $4GA^2/\oint(\mathrm{d}s/t)$. Therefore, the torsional stiffness of members divided by the spacing should be $(4GA^2/\oint(1/t)\mathrm{d}s)/2b$ for both sets of members.

Comparisons have shown that solving the structure in two stages using the different torsional properties described above does give good agreement, but it is clearly preferable to have one set of properties for any load case. It has been found that adopting $h^2t/2$ as the torsional stiffness for transverse members and dividing the remainder, $4GA^2/\oint(1/t)\mathrm{d}s-(h^2t/2)b$ amongst the longitudinals is a satisfactory compromise for all loading cases.

20.3.13 Design curves

Design curves have been obtained using the above approach for cellular decks constructed using precast 'top-hat' beams. They are of value for other cellular decks of similar proportions for preliminary design studies.

Figure 20.46 gives curves for HB coefficients (see Chapter 19). The range of decks covered is for spans from 60 ft to 120 ft (18.3 to 36.6 m) and deck widths from 30 ft to 90 ft (9.1 to 27.4 m), but the curves can be extrapolated to include values outside these figures and approximate solutions can also be derived for skew decks.

It is important to note that the curves have been derived by means of a grillage representation. Consequently, the application of the curves and the results obtained from them relate to the grillage solution and are subject to its conditions and limitations. The curves provide a coefficient per foot width of beam of the total moment M_L. The coefficients are plotted against the breadth: span ratio $B:L$ and are dependent upon the span L in conjunction with the edge stiffness ratio I and also upon the distance D from the centre of the outer wheels to the edge of the deck.

The values of k_L and k_I are derived from the intersection of $B:L$ and L, and k_p from the intersection of $B:L$ and D. The design live load moment for a composite top-hat beam is then given by:

$$M = k_L k_p[1 - k_I(I-1)]bM_L \tag{20.22}$$

where b is the beam width of either the edge beam, which may be asymmetrical, or the adjacent inner beam.

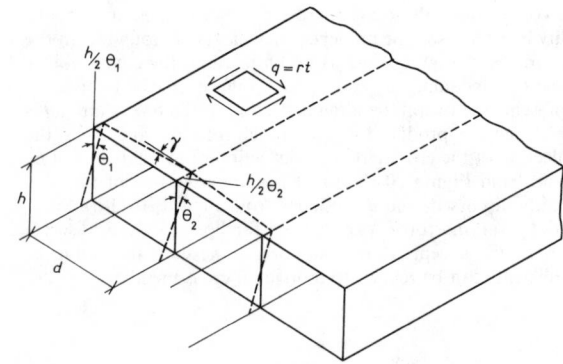

Figure 20.45 Shear strain due to differential longitudinal deflections

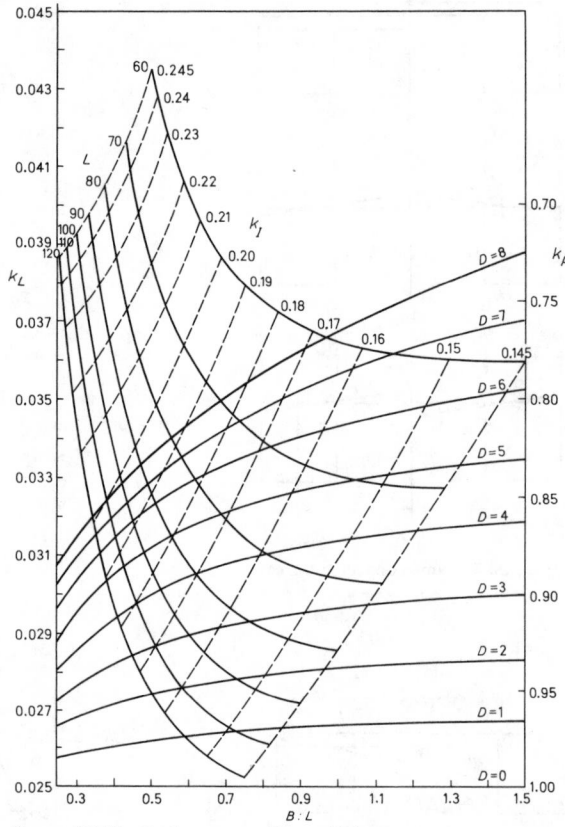

Figure 20.46 Design curves, HB coefficients

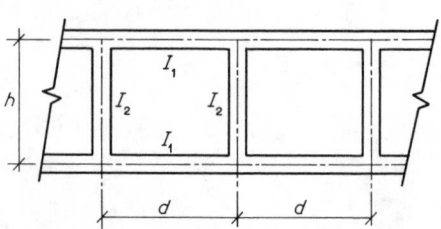

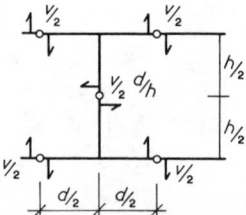

Figure 20.44 Transverse frame

20.4 Stress concentrations

Sudden changes in loads or in the shape of a structure produce stresses that cannot be calculated by normal beam theory. Concentrated loads such as the reactions at bearings and holes cut out of flanges are obvious examples of such changes but alterations in the direction of flanges, variations in thickness or width may produce significant effects. The introduction of strengthening members such as diaphragms or stiffening around holes are examples of situations in which the stress concentrations may weaken the structure instead of adding strength, if the full implications of the addition are not considered.

20.4.1 Shear lag due to concentrated loads

Beams with wide flanges are subject to shear lag effects, since the longitudinal stresses at points remote from the webs must be generated by the shear stress field across the flange. At sudden changes in shear, due to concentrated loads, the necessary changes in the longitudinal stresses in the flange require differential longitudinal movements in the transverse direction which produces the shear stress field in the flange. In other words, a longitudinal strain variation and therefore a stress variation across the flange is produced which is called the shear lag effect. It is important to note that transverse stresses are associated with shear lag as can be demonstrated by a consideration of the statics of the shear stress field alone.

Shear lag is most pronounced in girders of rectangular cross-section. The effect of cambering the flanges in the convex sense is to reduce shear lag. In the extreme case of a circular cross-section, shear lag does not occur, provided that all cross-sections remain circular. This is because the web of a rectangular box can deform in shear without shear deformations of the flange being necessary, whereas the circular beam can only deform in shear if all parts of the beam are subjected to shear strains. The distribution of shear strain in the latter case, which depends on purely geometrical considerations, agrees with the shear stress distribution produced by the longitudinal bending stresses of normal beam theory. Differential longitudinal movements are not required and no shear lag occurs. Figure 20.47 shows a substitute structure which enables shear lag effects to be found from standard formulae or relatively simple calculation. It is also useful for evaluating more precise results. The area A_F is the area of the edge member plus one-third of the web area between the flange and neutral axis. The area A_L is equal to half the area of the plate plus longitudinal stiffeners and b_s is given by Kuhn[18] as:

$$b_s = [0.55 + (0.45/n^2)] \qquad (20.23)$$

where n is the number of stringers in one-half of the flange. The actual distance of the centroid of the half-flange to the web is b_c whereas b_s is the distance which, together with the shear stiffness of the actual plate per unit width, simulates the shear lag property of the flange relative to the webs.

The above calculation gives the increase in stress at the web flange junction, $\Delta\sigma_F$, above normal beam theory. Figure 20.48 shows how the stress across the flange may be found assuming a cubic law variation.

The effects of shear lag are usually more important in steel box girders as the webs are more widely spaced than in concrete structures. Also, the thick diaphragms at bearings in concrete structures reduce the longitudinal stresses.

20.4.2 Changes in thickness and cut-outs

The change in thickness of a whole flange causes changes in the shear stresses between web and flange as well as local effects at the interface between the different flanges. In steel it is usual to taper the thicker plate to reduce the effects of fatigue and the possibility of brittle fracture. In some cases only part of the flange is thickened in areas of concentrated load such as forces from supporting cables or prestressing cables. The effect of such thickening is to tend to concentrate all flange forces in that part of the flange which must be allowed for by either gradually tapering out the increased area or by carrying the greater thickness through to a more lowly stressed region. A premature end to the reinforcement may overload the connecting unreinforced section.

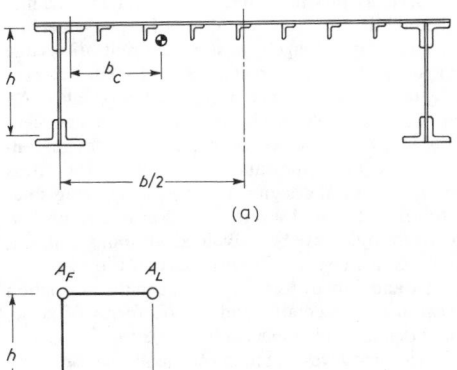

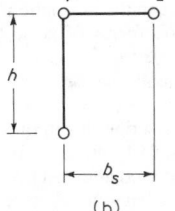

Figure 20.47 Transformation of actual into substitute beam cross-section. (After Kuhn (1956) *Stresses in aircraft and shell structures.* McGraw-Hill)

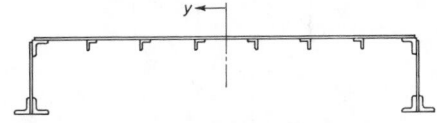

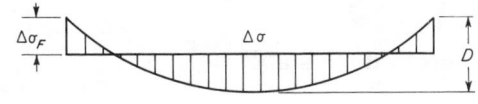

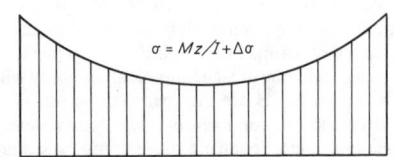

Figure 20.48 (After Kuhn (1956) *Stresses in aircraft and shell structures.* McGraw-Hill)

The reinforcement required for cut-outs must be continued or tapered for similar reasons but the need to do so is more obvious. The transverse and shear stresses associated with cut-outs are, however, also of considerable importance. In steel structures they are likely to cause buckling, fatigue and brittle fracture problems, whereas in concrete structures they cause cracking. It is, therefore, necessary to connect diaphragms, etc. to the structure by much more than nominal amounts of reinforcing steel. It is instructive to note with reference to steel structures that a number of large tankers have experienced local failures at cut-outs owing to the effects described above. It seems likely that the extrapolation of design knowledge from smaller structures was not backed-up with sufficient research into the complex stress systems produced and the associated buckling phenomenon. Similarly, the causes of failure of several steel box girder bridges have mainly been due to stress concentrations due to cut-outs in stiffeners, support reactions, and a cut-out produced by partly unbolting a main compression flange splice. An earlier failure of a plate girder bridge due to the stress concentrations produced by a flange cover plate completes an unanswerable case for the importance of allowing for stress concentrations in structural design. The basic engineering solution to this problem is to avoid severe stress concentrations and all those mentioned could have been avoided without significant cost or difficulty. Some degree of stress concentration is, however, unavoidable and only by using test data can the distinction be drawn between the acceptable and unsafe forms of structures, structural details, and associated stress levels.

The stress levels themselves due to known loads can be found with considerable accuracy using, for example, two- and three-dimensional finite element methods. Kuhn[18] describes ingenious methods for the approximate analysis of shell structures with cut-outs as well as the shear lag approach described above. They give a valuable insight into the structural behaviour of such systems but are more expensive to use than computer-based techniques using finite elements.

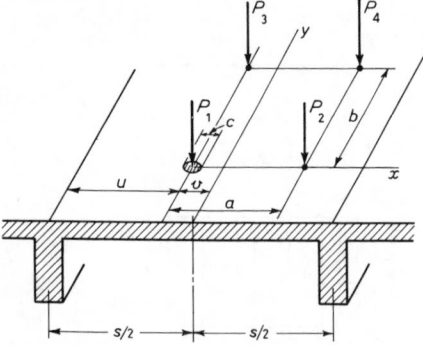

Figure 20.49 Slab supporting wheel loads

20.5 Concrete deck slabs

The usual form of deck system is a concrete slab spanning transversely between longitudinal beams which are often the main members of the bridge. Cross-girders can be used to produce a longitudinally spanning slab or the slab can be supported on a series of stringers spanning cross-girders. The simplest system is generally the most economical and only where there are special requirements are stringers and cross-girders used. In large steel trusses, for example, the loads must be carried to intersection points requiring more complex systems.

The slab must be designed for three different modes of behaviour:

(1) Local flexure due to the transfer of wheel loads to the adjacent beam members.
(2) Flexure due to relative movements of various parts of structure.
(3) In-plane stresses due to beam action of main and secondary members of structure.

Local stresses can be found by assuming that all supporting members are rigid when evaluating the slab moments and shears due to wheel loading. The remaining effects can then be found by applying loads equal to the reactions of the supporting members to those members. It is important that the latter loads should be statically equivalent to the vehicle loading, but the exact spanwise distribution is not usually required.

There are several publications giving influence surfaces for local slab bending moments for several types of support, i.e. simply supported on four sides, cantilever slabs, fixed on four sides.[19,20] A well-known treatment is by Westergaard[21] which uses Nádai's equations to obtain the bending moment under a wheel load:

$$\left.\begin{array}{c} M_x \\ M_y \end{array}\right\} = \frac{(1+\mu)P}{4\pi}\left[\ln\left(\frac{4s}{\pi c_1}\cos\frac{\pi v}{s}\right) + \frac{1}{2}\right] \pm \frac{(1+\mu)P}{8\pi} \quad (20.24)$$

The meaning of the symbols is given in Figure 20.49 except for c_1 which is the equivalent diameter of the loaded area. According to Westergaard, the equivalent diameter c_1 is expressed with satisfactory approximation by the following formula, applicable when $c < 3.45sh$:

$$c_1 = 2[(0.4c^2 + h^2)^{1/2} - 0.675h] \quad (20.25)$$

He derived Figure 20.50 for the case of a wheel at the centre of a simply supported span, for a Poisson's ratio of 0.15, using Equation (20.25). The fixed edge values of M_x and M_y, respectively M'_x and M'_y, are given in the same figure. A conservative value of the equivalent width of slab carrying the load is:

$$b_e = 0.58s + 2c$$

which was used to derive the result:

$$M_{0x} = \frac{Ps}{2.32s + 8c} \quad (20.26)$$

The above result was used by Henderson[22] to show that the abnormal vehicle loading of 90 t on two axles 1.83 m apart, does not exceed the design uniformly distributed loading curve moments by more than 20%, which is within the overstress allowance permitted at that time.

The bending moments due to wheels at other points on the slab can be found from influence surfaces of the kind shown in Figure 20.51 which are also taken from Westergaard's paper.[21]

Influence surfaces for cantilever slabs of varying depth have been produced by Homberg and Ropers.[23] These show that longitudinal and transverse sagging moments under the wheel may be as large as the root moment. It should be noted that the root moment is often at the thickest part of a cantilever slab.

The assumption of a fixed edge at the root and a simple free edge at the tip of the cantilever are generally made in influence surfaces. In fact, edge beams are usually provided and the root may be supported by a deck slab and a web with significant flexibility. The root moment is nonuniform and its distribution is affected by that flexibility. Treating the first harmonic of the fixed edge moment as a fixing moment, a moment distribution process can be used to allow for the flexibility, but it is usually sufficiently accurate to consider one joint. The stiffness of the

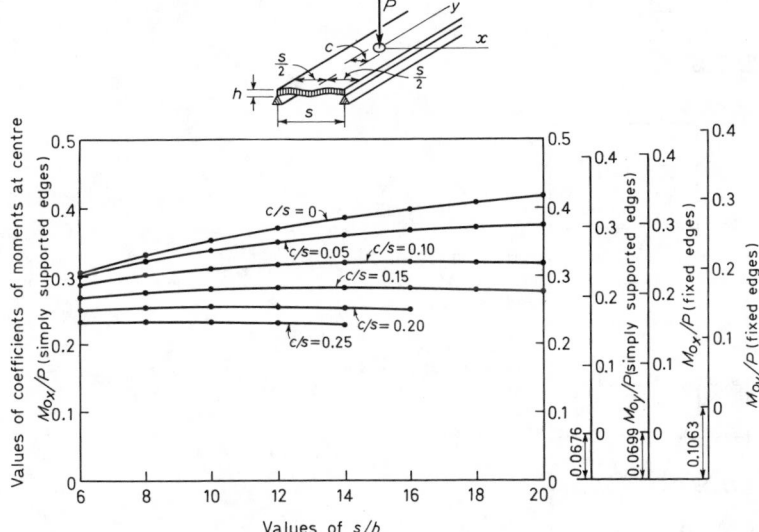

Figure 20.50 Coefficients of bending moments M_{0_x} and M_{0_y} in directions x and y respectively, produced at centre of slab by a central load P distributed uniformly over the area of a small circle with diameter c

flange, web and cantilever under sinusoidal edge moments must be used. The edge beam may also produce a significant change in the distribution of moments at the root. Both effects produce changes in longitudinal and transverse sagging moments in the cantilever.

20.6 Skew and curved bridges

Modern highway alignments result in many bridges with skewed supports, curvature in plan, and variation in width.

20.6.1 Skew

The primary shears and moments in bridge structures with angles of skew up to 15° approximate to those in a similar system with zero skew. A wider range of structures, provided that they approximate to slabs, are covered by the influence surfaces given by Rüsch and Hergenröder.[24] Most structures outside the range of the above approaches can be analysed by the grillage methods such as those mentioned in the previous sections.

Even where the skew is small, the behaviour of the structure near the bearings, particularly at the obtuse corner, requires special consideration, e.g. the interaction of the bearing diaphragm with the structure, uplift at bearings, transverse moments in the deck and shear distribution. These effects depend on the detailed form of the structure and the articulation of stiffness of the bearings and piers. In torsionally stiff structures, with twin bearings at both ends of a span, skew may produce significant end fixity with correspondingly high loads on the inner bearings (obtuse corners) and low loads on the others.

20.6.2 Curved in plan

Curved, torsionally stiff, structures supported against torsion at each vertical support can be analysed for the effects of bending continuity as if they were straight beams provided that the following requirements are satisfied:[25]

$\gamma < 1$ and $\alpha < 30°$
$\gamma < 5$ and $\alpha < 20°$
$\gamma < 10$ and $\alpha < 15°$

where $\gamma = EI/GK$ and α is the angle subtended by one span. The resulting bending moments are within 6% of curved beam solutions.

The solution found by that approach allows only for the vertical loads. The twisting moments produced by eccentricity of the loads relative to the shear centre must be considered separately. The vertical loads alone, however, produce torsional moments in the structure and graphs by Garret and Cochrane[25] enable these to be found from the continuity moments. The effects of applied torques for the ranges of structures given above can be estimated from considerations of statics and relative stiffnesses. More generally, the standard theory of curved beams can be applied to curved bridge structures or grillage programs can be used.

In all curved structures the longitudinal stresses produce lateral loads due to the curved path they follow. The effect of these loads, including the local loads caused by the curvature of prestressing cables, must always be given careful consideration. The resulting forces are resisted by the frame action of the cross-section producing deformations of the cross-section. In exceptionally thin-walled concrete box structures or steel box structures without special frames or diaphragms, these deformations can result in significant changes in the primary bending moments.[26] In the usual type of box structure, however, it is possible to consider this effect separately and, provided that the forces acting on the cross-section allow for curvature, reasonable results can be obtained by assuming the box is straight and using theories such as those described above.

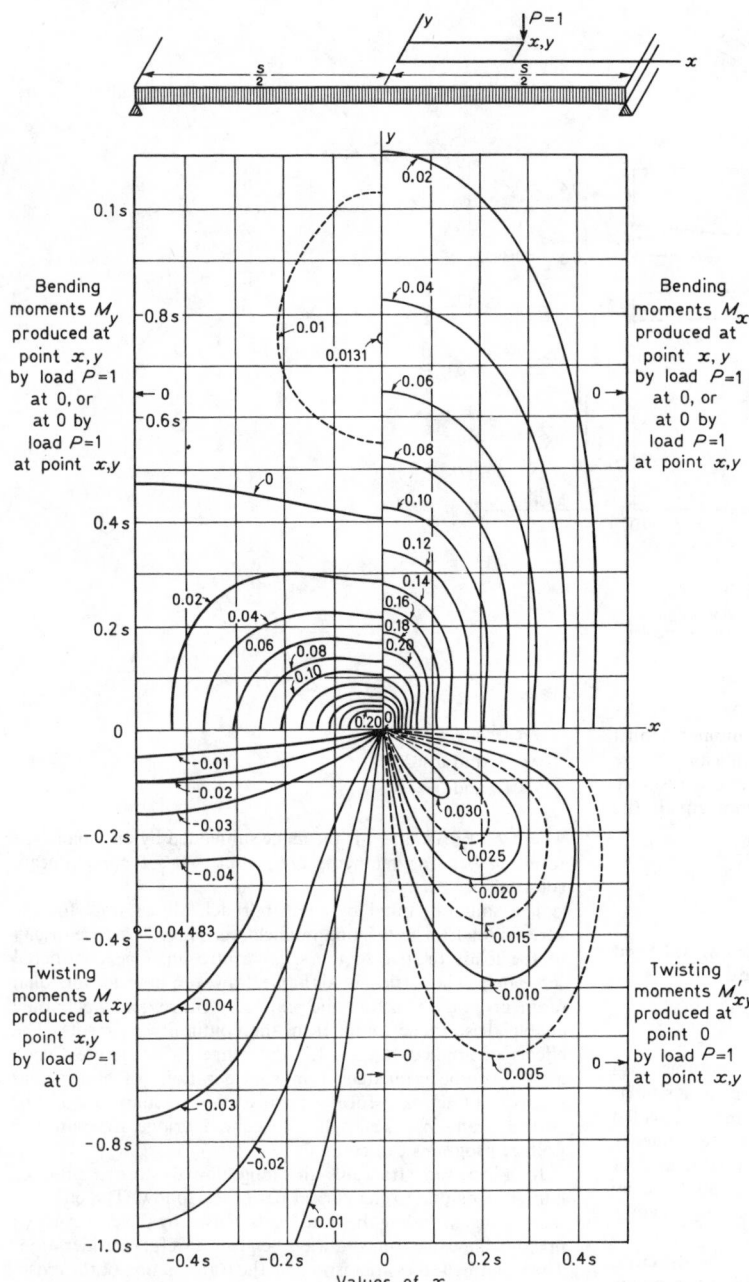

Figure 20.51 Contour lines of surfaces representing moments. Poisson's ratio ν=0.15

20.7 Dynamic response

The stresses produced by the dynamic response of a bridge to a vehicle traversing it are covered by loading specifications. Unusual structures may require a special investigation but the main problem will be that of defining the vehicle size, speed and frequency of occurrence. Damping, though difficult to quantify, is not likely to be important in this type of problem since the main effects are usually short-term ones where damping is of little significance.[27] The prolonged oscillations produced by a series of vehicles acting in phase with each other is highly improbable at high live-load stress levels.

The effect of vibration on the user of a bridge is, potentially, more important since small accelerations, 0.02g, can produce discomfort in pedestrians and occupants of stationary vehicles. Occupants of moving vehicles cannot distinguish bridge movements as they are masked by the normal oscillations of the vehicle. Figure 20.52 from Leonard[28] shows limits proposed by various workers in the industrial field compared with those found by the Road Research Laboratory from the reactions of

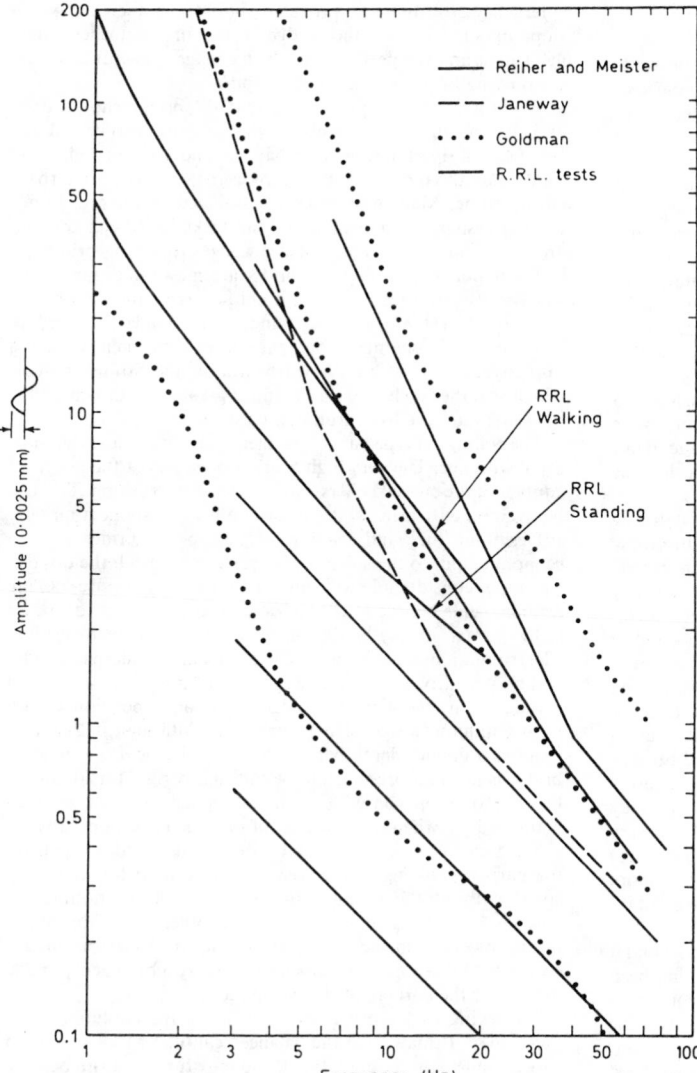

Figure 20.52 Comparison of tolerance levels to vibration

people using a footbridge which has been excited to a steady-state resonant motion. Short periods of oscillation much greater than the limits in the figure are likely to be acceptable to bridge users if they understand that it is part of the normal behaviour of the structure. The stationary car occupant is a different case, since resonance can produce oscillations of increased amplitude if the natural frequency of the vehicle, 1–3 Hz, and the bridge are nearly the same. Wyatt has commented, however, that for spans over 30 m:

> ... an obvious design aim would be to avoid the frequency range of 1–3 Hz. It has, however, also been shown that as a result of 40 years' design progress the economic structure is likely to have a frequency in this range and it is tantamount to throwing away this progress if higher frequencies are specified. Any specification which imposes a limit on frequency is thus strongly to be deplored.

Wind-excited oscillations are known to be of primary importance in long-span structures of the suspension bridge and cable-braced type. What is not clear is the significance of this phenomenon in more modest spans with or without cable supports. Theoretical investigations of the problem must allow for effects of vortex shedding on the particular cross-section in question; similarly, the changes in lift characteristics with rotation and deflection. Structural and aerodynamic damping are both important and, of course, the elastic characteristics of the structure must be known.

The behaviour of long-span structures is usually determined with the aid of wind-tunnel tests on models of segments of the deck and stiffening girder. Spring supports can reproduce to scale the natural frequency calculated for the prototype. The use of wind tunnels will continue to be important for major bridges where the economic advantages of determining the most suitable deck system are considerable.

In the intermediate range of spans the data for various shapes of cross-section from wind-tunnel tests are likely to be sufficient in the near future to permit an analytical estimate of the effect of wind without being unduly conservative.

20.8 Movable bridges

The principles associated with the various types of movable bridge are well established but increasing knowledge, particularly in the fields of control machinery and materials technology, has made new applications possible. The selection of the type of movable bridge is largely governed by the nature of the site. The relative priority of the two thoroughfares, usually a road or railway over a waterway, will help to establish the acceptable closed position headroom and also the speed and conditions under which the bridge will be required to operate.

Where an existing bridge is to be replaced, the condition of the existing foundations will influence the choice of type for the new bridge. For example, it would normally be uneconomic to excavate a tail chamber for a trunnion bascule bridge out of existing heavy concrete foundations.

There is a need for speed during those phases of construction that will obstruct navigation channels. This obstruction can be minimized by finally assembling the bridge as a few large items; the assembly of the leaves of one modern bascule bridge has been done in the bridge open position.

Minimum deadweight has the obvious advantage of reducing the required capacity of the bridge machinery; an orthotropic steel deck is used on most modern bridges although care is needed to ensure adequate adhesion between the wearing surface and the steel deck. A flexible PVC surface has been used.

The machinery to operate a movable bridge is as important as the structure itself. Particular requirements are that it should provide high torque at low speeds for starting, very slow inching speeds – to give final alignment for swing bridges and for landing on bearings for other types – and precise control at all times. The final drive to the movable structure has traditionally been rack-and-pinion, but many modern continental examples use a hydraulic linear actuating cylinder. A factor which may continue to favour the rack and pinion is the need for a very precise and rigid control of travel; also the braking system must be completely separated from the hydraulics and it is more difficult to arrange brakes for a sliding system than for rotary pinions.

A bascule bridge must be landed without violent impact and it is desirable to ensure a positive reaction in doing so. It has been customary to provide special inching motors which come into operation at the end of travel, with limit switches to stop the motors just before the bearing pads are reached. However, many modern examples are now using infinitely variable speed systems incorporating hydraulic motors or hydraulic transmission to remove the need for separate motors. Several modern Dutch bascule bridges incorporate the so-called 'snail' feature which is a mechanical device fitted on the rack-and-pinion drive used for slowing-down the movement of the bridge and locking it.

The normal requirement that a movable bridge should continue to function in all circumstances necessitates complete reliability of its machinery. Extensive duplication of machinery is incorporated, together with the provision of auxiliary motors so that the bridge can still operate, at reduced speed if necessary, with some of its machinery out of action for servicing or due to failure. In the event of failure of the main power supply, a standby power source, such as a diesel generator, may be provided; in some instances provision is made for hand operation.

A movable bridge out of control can do enormous damage; a braking system must be provided to ensure that this never occurs. This system, too, will require extensive duplication as well as being of the 'fail-safe' type. It must be adequate for the strongest anticipated wind loading for, even if the wind is sufficient to prevent the bridge from being moved, the brakes must be able to hold it with complete safety. Some form of emergency stop must be provided should a sudden danger to shipping occur while the bridge is moving. Another safety measure is an overspeed switch which engages the brakes should the moving bridge speed-up unacceptably.

While it is obviously desirable that the bridge should open and close as quickly as possible, attention must also be given to the speed of operation of crash barriers and warning lights and hooters as these operations occupy a major part of the road-closure time. Machinery must be provided to operate bridge locking systems such as nose-and-tail locks for bascule bridges and nose support jacks and centre wedges for swing-bridges.

The major types of movable bridge are shown in Figure 20.53. The Strauss, overhead counterweight Scherzer, and drawbridge have all their parts above ground. This avoids the need to provide a tail chamber, but care must be taken to avoid obtrusiveness. This is particularly true of the Strauss bascule which is rather inelegant and requires greater depth behind the quayside than the fixed trunnion bascule.

The rolling lift type has the problem of rolling tracks deteriorating with age. The very high bearing pressures at the points of contact of the curved rollers may lead to local crushing. This can be overcome by using wider tracks on heavier support girders.

Designing for wind loading on the opened bridge sets an economic limit to the single-leaf bascule span. With the double-leaf type, considerable care must be taken in the nose-locking arrangements between the two leaves; if the bridge is to carry a railway track it may be difficult to obtain a satisfactory joint. The trunnion bascule is the type used in many modern examples and may be driven in several ways. The rolling lift bascule gives a wider clearance with the bridge in the open position than a fixed trunnion bridge of the same span, although it therefore requires a greater depth behind the quay. The leaf of a bascule bridge is normally designed to be sufficiently rigid torsionally to be able to be opened with the drive applied to only one of its main girders, when the other set of machinery is inoperative.

The vertical lift-bridge sets a headroom limit and is expensive for narrow crossings. However, it can be used for very long spans without the nose-locking problems of the double-leaf bascule. The lifting machinery may be either at the top of the lifting towers or in the piers; a mechanical or electrical linkage connects the separate motors to ensure synchronized parallel motion of the corners of the lifting section.

The swing-bridge often provides the cheapest solution for a given span. It may be of the balanced cantilever type or have a shorter tail span ('bobtail' type), and may turn on a rim bearing or a central pivot bearing. Its main disadvantage is the need to protect the bridge in the open position. Space must be provided at the quayside to lay the span in the open position; this may not be possible, especially where there are adjacent locks requiring a clear quay for handling mooring ropes. The retractable bridge is not often used as it requires a suitable approach to accommodate the span in the open position and heavy rolling or sliding ways.

20.9 Items requiring special consideration

This chapter has described in general terms the types of bridges used throughout the world, primarily of steel or composite construction or a combination of both types, together with general principles of analysis which enable the preparation of the main members of the design to be established correctly. Space precludes treatment of many interesting aspects of bridge design.

The authors would like to draw attention to several points which, from experience, are likely to give rise to difficulties

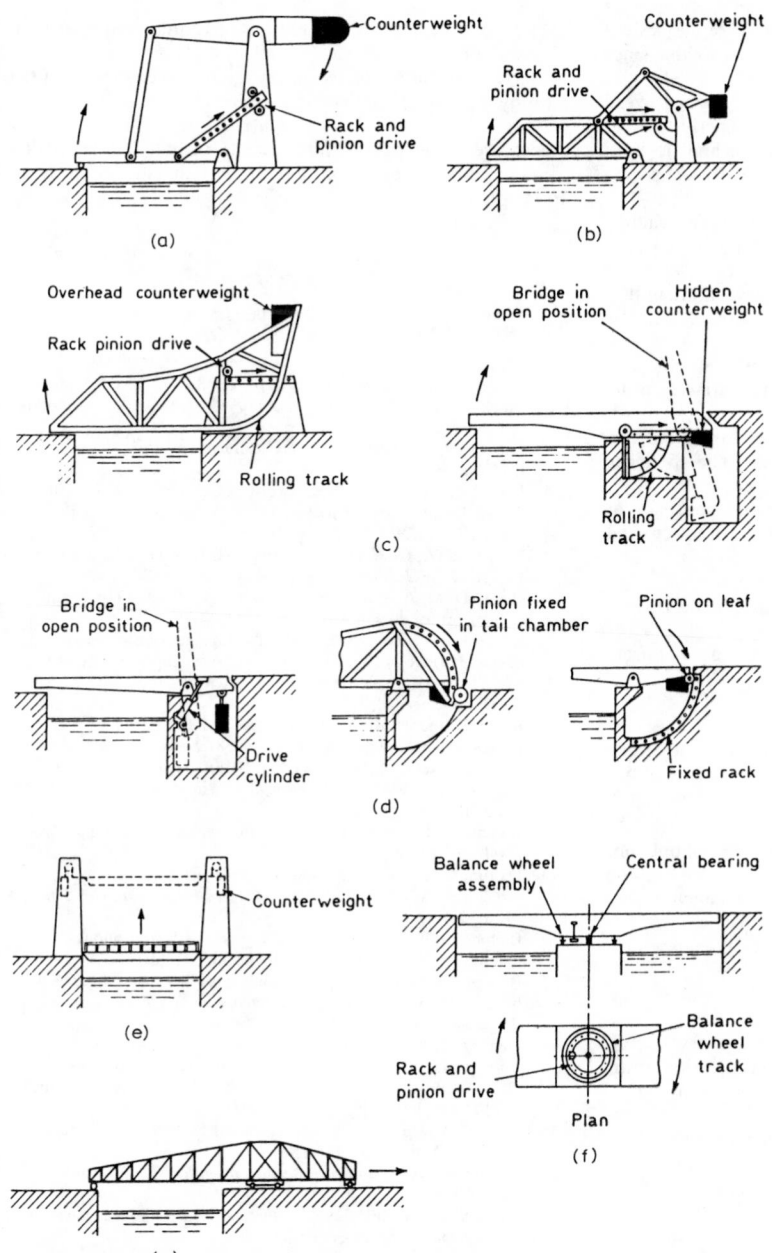

Figure 20.53 Types of movable bridge. (a) Drawbridge; (b) Strauss; (c) rolling lift (Scherzer) bascules; (d) trunnion bascules; (e) vertical lift; (f) swing; (g) retractable

unless adequate consideration is given to them. Many are obvious but it is a regrettable fact that it is obvious points which give rise to recurring difficulty.

It is essential that adequate time and effort is devoted to the checking of any bridge design by a senior engineer fully experienced in the type of work involved. Much time can be wasted in elaborate analysis either by slide-rule or computer when the simplest elementary assessment of some details will demonstrate inadequacies or inaccurate conception of the mode of behaviour. It is important for bridge engineers to have some grasp of three-dimensional actions of structures rather than a preoccupation with cross-sections, plans and elevations.

To this end, the engineer is encouraged to make simple models of paper or cardboard so that the implications of what is apparently an obvious decision can be grasped and interpreted. The relationship of frame members in space, the junctions of beams and columns, the implications of running bridge girders into diaphragms and end blocks are typical examples.

The following checklist may perhaps reinforce the plan of work and help bridge engineers to reduce the number of errors or difficulties on site.

(1) Check all the detailed items of the bridge structure and not just the main structural members.

(2) Check the basic principles of reinforced concrete design used. The same applies to steelwork details for welding and bolting.

(3) Can the reinforcement be fixed by a man of normal size?

(4) Can the concrete be placed and vibrated properly?

(5) Can welds be made in a suitable sequence and do bolt spanners, etc. foul other completed parts of the construction?

(6) The geometric calculations should be checked. The leading dimensions on the general arrangement drawings cause considerable confusion if they are not correct.

(7) Secondary forces and stresses should not be forgotten. It is better to consider them by elementary methods than to omit taking them into account.

(8) Do not forget constructional tolerances which will cause lack of fit and possibly additional loads and stresses due to eccentricity.

(9) Particular attention should be paid to bearing design and details for drainage and expansion joints. The specifications for these items should not be skimped.

(10) Remember that lateral loads are generated from lack of straightness in tension. These are often forgotten and include things as widely diverse as:

    (a) substantial lateral loading on elements of cellular construction;

    (b) change of direction of thrust members as at portal frame joints;

    (c) haunches on the soffit of a beam;

    (d) overall curvature of an arch;

    (e) the behaviour of prestressing cables in plan and elevation imposes a local loading as well as the effect of the line of pressure.

(11) Post-tensioned prestressed concrete anchorages should not only be considered in terms of bursting, splitting and the length of transmission of force for the single anchorages, but in terms of sets or the whole group.

(12) Openings for services and access for maintenance and inspection.

(13) Anchors require careful detailing. Holes, ties, etc. become stress raisers and they must be detailed to suit.

(14) Temporary openings and drain-holes necessary during construction should receive adequate consideration at an early stage, for the removal of items of plant, jacks, falsework, shuttering, lifting gear, etc.

(15) Engineers sometimes apply normal beam theory without considering whether this is applicable. Many elements act as deep beams, i.e. brackets, corbels, halved joints, anchorages, tiebacks. They require application of a specific approach which adequately caters for behaviour which would actually occur in service either in bending or in shear or torsion.

(16) Consideration should be given to the weathering of a structure in terms of orientation appearance and durability.

(17) The designer should consider whether the concept of erection is clear in his mind and what the effect would be of lack of compliance by the contractors.

(18) Are there any unforeseen events which would provoke an unfortunate chain of circumstances such as traffic collision on any parts of the structure, crane booms and dump trucks inadvertently striking part of the bridge during construction, fires, etc?

(19) When inspecting bridges under construction, keep a look-out for things which are not in accordance with the intention of the design from a safety point of view.

(20) Is the articulation of the structure as the designer intended?

(21) Is reinforcement in the bottom of cantilevers rather than the top?

(22) Has distribution or secondary reinforcement been omitted by mistake at the detailing stage?

These are some of the many items included in a quality assurance scheme and which should be extended to all the items in the plan of work.

References

1 Beckett, D. (1969) *Bridges*. Hamlyn, London.

2 Mock, B. (1949) *The architecture of bridges*. Museum of Modern Art, New York.

3 Fairburn, W. (1849) *Conway and Britannia tubular bridges*. London.

4 Lee, D. J. (1971) 'The selection of box beam arrangements in bridge design', *Developments in bridge design and construction*, Crosby Lockwood, London.

5 Baxter, J. W., Gee, A. F. and James, H. B. (1966) 'Gladesville Bridge', *Proc. Instn. Civ. Engrs*, **34**, June.

6 Hook, D. M. A. and Richmond, B. (1970) 'Precast box beams in cellular bridge decks', *Struct. Engr*, **48**, March.

7 Chaplin, E. C. *et al*. (1973) 'The development of a precast concrete bridge beam of U-section', *Struct. Engr*, **51**, 383–388.

8 Lee, D. L. (1970) 'Western Avenue extension, the design of section 5', *Struct. Engr*, **48**, 109–120.

9 Beyer, E. and Thul, H. (1967) 'Hochstrassen', Baton-Verlag.

10 Hawranek, A. and Steinhardt, O. (1958) *Theorie und berechnung der Stahlbrücken*, Springer-Verlag, Berlin.

11 Resinger, F. (1959) *Der dünnwandige Kastenfrager, Köln*, Stahlban-Verlag.

12 Richmond, B. (1966) 'Twisting of thin-walled box girders', *Proc. Instn. Civ. Engrs*, **33**, 659–675.

13 Dalton, D. C. and Richmond B. (1968) 'Twisting of thin-walled box girders of trapezoidal cross section', *Proc. Instn. Civ. Engrs*, **39**, 61–73.

14 Richmond, B. (1969) 'Trapezoidal boxes with continuous diaphragms', *Proc. Instn. Civ. Engrs*, **43**, 641–650.

15 Richmond, B. (1971) 'The relationship of box beam theories to bridge design', *Proceedings, Conference on developments in bridge design and construction*, Crosby Lockwood.

16 Hendry, A. W. and Jaeger, L. G. (1958) *The analysis of grid frameworks and related structures*, Chatto & Windus, London.

17 *Design manual for orthotropic steel plate deck bridges*, Am. Inst. Steel Constructors, New York (1963)

18 Kuhn, P. (1956) *Stresses in aircraft and shell structures*, McGraw-Hill, New York.

19 Pucher, A. (1964) *Influence surfaces of elastic slabs*, Springer-Verlag, Vienna.

20 Krug, S. and Stein P. (1969) *Influence surfaces of orthogonal on isotropic plates*, Springer, Berlin.

21 Westergaard, H. M. (1930) 'Computation of stresses in bridge slabs due to wheel loads', *Public Roads*, **11**, 1.

22 Henderson, W. (1954) 'British Highway Bridge Loading', *Proc. Instn. Civ. Engrs*, **3**, Part II.

23 Homberg, H. and Ropers, W. (1963) 'Kragplatten mit verändlicher Dicke', *Beton and Stahlbetonbau*, March.

24 Rüsch, H. and Hergenröder, A. (1961) *Influence surfaces for moments in skew slabs*, (Cement and Concrete Association translation) Munich.

25 Garrett, R. J. and Cochrane, R. A. (1970) 'The analysis of prestressed beams curved in plan with torsional restraint at the supports', *Struct. Engr*, **48**, 128–132.

26 Dabrowski, R. (1968) *Curved thin-walled girders, theory and analysis*. (Cement and Concrete Association translation) Springer-Verlag, Berlin.

27 Biggs, J. M. *et al*. (1959) 'Vibration of simple-span highway bridges', *Trans. Am. Soc. Civil Engrs*, **124**.

28 Leonard, D. R. (1966) 'Human tolerance levels for bridge vibrations', *Road Research Laboratory Report No. 34*.

Bibliography

Amman, O. H. *et al.* (1933) 'George Washington bridge', *Trans Am. Soc. Civ. Engrs,* **97,** 1–442.

Amman, O. H. *et al.* (1966) 'Verrazano Narrows bridge', *Proc. Am. Soc. Civ. Engrs J. Constr. Div,* **92,** (CO2), 1–192.

Anderson, J. K. (1964) 'Runcorn–Widnes bridge', *Proc. Instn Civ. Engrs,* **29,** 535–570.

Anderson, J. K. (1965) 'Tamar bridge', *Proc. Instn Civ. Engrs,* **31,** 337–360.

Anderson, J. K. and Brown, C. D. (1964) 'Design and construction of the Kingsferry lifting bridge, Isle of Sheppey', *Proc. Instn Civ. Engrs,* **28,** 449–470.

Anderson, J. K. *et al.* (1965) 'Forth Road bridge', *Proc. Instn. Civ. Engrs,* **32,** 321–512.

Andrew, C. E. (1947) 'Unusual design problems – second Tacoma Narrows bridge', *Proc. Am. Soc. Civ. Engrs,* **73,** 10, 1483–1497.

Anon. (1918) 'The Quebec bridge', *The Engineer,* 138–140.

Anon. (1928) 'Design of the 1675 ft Killvankull steel arch bridge', *Engng News Record,* 873–877.

Anon. (1952) 'Spectacular Venezuelan concrete arch bridge', *Engng News Record,* 28–32 (11 Sept.).

Anon. (1964) 'San Mateo–Haywood bridge', *California Highways and Pub. Wks* (Sept., Oct.).

Anon. (1969) 'The Batman bridge, Tasmania', *Building with steel.* British Constructional Steelwork Association, p. 69.

Anon. (1970) 'Rolling falsework arch cuts construction time', *Engng News Record,* 32–33 (26 March).

Anon. (1971) 'Record prestressed box girder span under way', *Engng News Record,* **14** (24 June).

Anon. (1972) 'Inverted suspension span is simple and cheap', *Engng News Record* (11 May) 27–31.

Baldwin, R. A. and Woolley, C. W. (1966) 'Cumberland basin bridges, scheme 2: construction'. *Proc. Instn. Civ. Engrs,* **33,** 289–312.

Baur, W. (1969) 'Die Durchstichbrücke Neckarsulm', *Beton u. Stahlbetonbau,* **64,** 3, 57–61.

Baxter, J. W., Birkett, E. M. and Gifford, E. W. H. (1961) 'Narrows Bridge, Perth, Western Australia', *Proc. Instn. Civ. Engrs,* **20,** 39–84.

Baxter, J. W., Lee, D. J. and Humphries, E. F. (1972) 'Design of Western Avenue extension (Westway)', *Proc. Instn. Civ. Engrs,* **50,** 177–218.

Beyer, E. (1970) 'Die Kniebrücke in Düsseldorf, *Stahlbau,* **39,** 6, 185–189.

Beyer, E. and Ernst, H. J. (1964) 'Brücke Jülicher Strasse in Düsseldorf', *Bauingenieur,* **39,** 12, 469–477.

Beyer, E., Grassl, H. and Wintergerst, L. (1958) 'Nordbrücke Düsseldorf', *Stahlbau,* **27:** 1–6 (Jan.); 57–62 (Mar.); 103–107 (Apr.); 147–154 (June); 184–188 (July).

Beyer, E. and Thul, H. (1967) *Hochstrassen,* Beton-Verlag GMBH, Berlin.

Biggart, A. (1887) 'The erection of the Forth bridge', *The Engineer,* 438–439.

Bignall, V. (1977) *Catastrophic failures (Westgate Bridge Collapse* pp. 127–165). Open University Press, Milton Keynes.

Bill, M. and Maillart, R. (1969) *Bridges and constructions* (3rd edn). Architectural Publishers Artemis, Zürich.

Bingham, T. G. and Lee, D. J. (1969) 'The Mancunian Way elevated road structure', *Proc. Instn Civ. Engrs,* **42,** 459–492.

Borelly, W. (1970) 'Die Nordbrücke Mannheim – Ludwigshafen im Bau', *Stahlbau,* **39,** 5, 156–157.

Boynton, R. M. and Riggs, L. W. (1966) Tagus river bridge, *Civ. Engng* (New York), **36,** 2, 34–45.

Bradley, J. N. (1978) *Hydraulics of bridge waterways.* US Department of Transportation (Federal Highway Administration Hydraulic Design Series No. 1 (rev. March 1978).

Brebbia, C. A. and Connor, J. J. (1973) *Fundamentals of finite element techniques,* Butterworths, London.

British Railways (1976) *Assessment of the live load-carrying capacity of underbridges.* BR, London.

Brown, C. D. (1965) 'Design and construction of George Street bridge over River Usk, at Newport, Monmouthshire', *Proc. Instn Civ. Engrs,* **32,** 31–52.

Brummer, M. and Hanson, C. W. (1956) 'Development and design of Walt Whitman bridge', *Proc. Soc. Am. Civ. Engrs. J. Struct. Div,* **82,** (ST4).

Building Research Establishment (1979) *Bridge foundations and substructures* BRE Report. HMSO, London.

CEB–FIP (1978) *Model code for concrete structure* (3rd edn).

Chettoe, C. S. and Henderson, W. (1957) 'Masonry arch bridges: a study', *Proc. Instn Civ. Engrs,* **7,** 723–762.

Clark, L. A. (1983) *Concrete bridge design to BS 5400* and supplement.

Constrado (1983) *Weather-resistant steel for bridgework. Section properties of universal beams with 1 and 2 mm weathering allowances.* Constrado, London.

Coombs, A. S. and Hinch, L. W. (1969). 'The Heads of the Valleys road', *Proc. Instn Civ. Engrs,* **44,** 89–118.

Covre, G. and Stabilini, P. (1970) 'Steel spans over three bays of the Italia viaduct on the Salerno–Reggio Calabria motorway', *Acier-Stahl-Steel,* **35,** 7–8, 307–313.

Daniel, H. (1965) 'Die Bundesautobahnbrücke über den Rhein bei Leverkusen', *Stahlbau,* **34,** 33–36 (Feb.); 83–86 (March); 115–119 (Apr.); 153–158 (May); 362–368 (Dec.).

Daniel, H. (1971) 'Die Rheinbrücke Duisberg-Neuenkamp, Planung, Bau wettbewerb und seine Ergebnisse', *Stahlbau,* **40,** 7, 193–200.

Department of Transport (1979 *Standard bridges publicity brochure.* HMSO, London.

Department of Transport (1982) *Damages to low bridges, bridge height gauges.* Report of working party of November 1982. HMSO, London.

Department of Transport (1984) *Bridge inspection guide.* HMSO, London.

Diamant, R. M. E. (1963) 'Maracaibo bridge', *Civ. Engng* (London), **58,** 681, 482–484.

Erde, J. M. (1972) 'Lowestoft double-leaf trunnion bascule bridge'. *Civ. Engng & P.W. Rev.,* **67,** 790, 465–469.

Faber, L. (1964) 'Die Europabrücke, Uberbau', *Stahlbau,* **33,** 7, 193–199; 'Der Unterbau der Europabrücke', *Bautechnik,* **41,** 7, 217–227.

Fairhurst, W. A. and Beveridge, A. (1965) 'Superstructure of Tay road bridge', *Struct. Engr,* **43,** 3, 75–82.

Fairhurst, W. A., Beveridge, A. and Farquhar, G. F. (1971) 'The design and construction of Kingston bridge and elevated approach roads, Glasgow'. *Struct. Engr,* **49,** 1, 11–33.

Faltus, F. and Zeman, J. (1968) 'Die Bogenbrücke über die Moldau bie Zdakov', *Stahlbau,* **37,** 11, 332–339.

Farraday, R. V. and Charlton F. G. (1983) *Hydraulic factors in bridge design.* American Society of Civil Engineers, New York.

FIP (1984) *Practical design of reinforced and prestressed concrete structures based on the CEB–FIP Model Code (MC78).* Thomas Telford, London.

Farago, B. and Chan, W. W. (1960) 'The analysis of steel decks with special reference to highway bridges at Amara and Kut in Iraq', *Proc. Instn. Civ. Engrs,* **16,** 1–32.

Feidler, L. L. (1962) 'Erection of Lewiston–Queenston bridge', *Civ. Engng* (New York), **32,** 11, 50–53.

Feige, A. (1966) 'The evolution of German cable-stayed bridges and overall survey', *Acier-Stahl-Steel,* **31,** 12, 523–532.

Feige, A. and Idelberger, K. (1971) 'Long-span steel highway bridges today and tomorrow', *Acier-Stahl-Steel,* **36,** 5, 210–222.

Finch, R. M. and Goldstein, A. (1959) 'Clifton bridge, Nottingham: initial design studies and model test: design and construction', *Proc. Instn. Civ. Engrs,* **12,** 289–316, 317–352.

Finsterwalder, U. and Schambeck, H. (1965) 'Die Spannbetonbrüke über den Rhein bei Bendorf', *Beton u. Stahlbetonbau,* **60,** 3, 55–62.

Freeman, R. Sr. (1934) 'Sydney harbour bridge, design of structure and foundations', *Proc. Instn. Civ. Engrs,* **238,** 153–193.

Freudenberg, G. (1968) 'Die Doppel klappbrücke (Herrenbrücke) über die Trave in Lubeck', *Stahlbau,* **37,** 10, 289–298.

Freudenberg, G. (1970) 'Die Stahlhochstrasse über den neuen Hauptbahnhof in Ludwigshafen/Rhein', *Stahlbau,* **39,** 9, 257–267.

Freudenberg, G. (1971) 'The world's largest double-leaf bascule bridge over the Bay of Cadiz', *Acier-Stahl-Steel,* **36,** 11, 463–472.

Freudenberg, G. and Rotha, O. (1966) 'Die Zoobrücke über den Rhein in Köln', *Stahlbau,* **35,** 225–235 (Aug.); 269–277 (Sept.); 337–346 (Nov.).

Freyssinet, E., Muller, J. and Shama, R. (1953) 'Largest concrete spans of the Americas', *Civ. Engng* (New York), **23,** 41–55.

Gee, A. F. (1971) 'Cable-stayed concrete bridges', in: K. C. Rockey, J. L. Bannister and H. R. Evans (eds) *Developments in bridge design and construction.* Crosby Lockwood, London.

Gifford, E. W. H. (1962) 'The development of long-span prestressed concrete bridges', *Struct. Engr,* **40,** 10, 325–335.

Gill, R. J. and Dozzi, S. (1966) 'Concordia orthotropic bridge: fabrication and erection', *Engng J.*, **49**, 5, 10–18.

Gimsing, N. J. (1983) *Cable-supported bridges – concept and design.*

Gowring, G. I. B. and Hardie, A. (1968) 'Severn Bridge: foundations and substructure', *Proc. Instn Civ. Engrs*, **41**, 49–68.

Gutfleisch, W. and Krüger, U. (1972) 'Der Stahlüberbau der Rheinbrücke Gemersheim', *Stahlbau*, **41**, 2, 33–40.

Guyon, Y. (1957) 'Long-span prestressed concrete bridges constructed by the Freyssinet system', *Proc. Instn. Civ. Engrs*, **7**, 110–179.

Hardesty, S. (1941) 'The Rainbow bridge at Niagara Falls', *Civ. Engng* (New York), **11**, 9, 532–534.

Hartwig, H. J. and Hafke, B. (1961) 'Die Bogenbrücke über den Askerofjord', *Stahlbau*, **30**, 289–303 (Oct.); 365–377 (Dec.).

Havemann, H. K., Aschenberg, H. and Freudenberg, C. (1963) 'Die Brücke über die Norderelbe in Zuge der Bundesautobahn Südliche Umgehung Hamburg, *Stahlbau*, **32**, 193–198 (July); 240–248 (Aug.); 281–287 (Sept.); 310–317 (Oct.).

Heckel, R. (1971) 'The fourth Danube bridge in Vienna – damage and repair', in: K. C. Rockey, J. L. Bannister and H. R. Evans (eds) *Developments in bridge design and construction.* Crosby Lockwood, London.

Hedifine, A. and Mandel, H. M. (1971) 'Design and construction of Newport Bridge', *Proc. Am. Soc. Civ. Engrs. J. Struct. Div.*, **97**, ST 11, 2635–2651.

Hess, H. (1960) 'Die Severinsbrücke Köle', *Stahlbau*, **29**, 8, 225–261.

Higgins, G. E. (1983) *The lightweight aggregate concrete bridge at Redesdale.* Transport and Road Research Laboratory Report No. LB 788. HMSO, London.

Hilton, N. and Hardenberg, G. (1964) 'Port Mann bridge, Vancouver, Canada', *Proc. Instn Civ. Engrs*, **29**, 677–712.

Hoppe, C. (1963) 'Die erste Strassenbrücke über den Panama-Kanal', *Der Bauingenieur* (Berlin), 5, 205–206.

Huet, M. (1960) 'Le Pont de Tancarville', *Soc. Ingénieurs Civ. France Mem*, **113**, 5, 23–42.

Hyatt, K. E. (1968) 'Severn bridge: fabrication and erection', *Proc. Instn Civ. Engrs*, **41**, 69–104.

Institution of Structural Engineers (1984) *The art and practice of structural design.* 75th anniversary international conference.

IABSE (1982) *Maintenance repair and rehabilitation of bridges*, Final Report.

IABSE (1983) *Ship collision with bridges and offshore structures.* IABSE colloquium, Copenhagen: Introductory report No. V41; Preliminary Report No. V42.

Karol, J. (1963) 'Calcasieu river bridge'. *Welding J*, **42**, 11, 867–870, 877–880.

Kerensky, O. A., Henderson, W. and Brown, W. C. (1972) 'The Erskine bridge', *Struct. Engnr*, **50**, 4, 147–170.

Kerensky, O. A. and Little, G. (1964) 'Medway bridge: design', *Proc. Instn Civ. Engrs*, **29**, 19–52.

Kier, M., Hansen, F. and Dunster, J. A. (1964) 'Medway bridge: construction', *Proc. Instn Civ. Engrs*, **29**, 53–100.

Klingenberg, W. (1962) 'Neubau einer Hängebrücke über den Rhein bei Emmerich', *Bauingenieur*, **37**, 7, 237–239.

Kondo, K., Komatsu, S., Inoue, H. and Matsukawa, A. (1972) 'Design and construction of Toyosato–Ohhashi bridge', *Stahlbau*, **41**, 6, 181–189.

Lacey, G. C., Breen, J. E. and Burns, N. H. (1971) 'State of the art for long span prestressed concrete bridges of segmental construction', *J. Prestressed Concrete Inst.*, **16**, 5, 53–77.

Lee, D. J. (1967) *The design of bridges of precast segmental construction.* Concrete Society Technical Paper, No. PCS 10.

Lee, D. J. (1967) *Prestressed concrete elevated roads in Britain.* Concrete Society Technical Paper No. PCS 12.

Lee, D. J. (1970) 'Prestressed concrete in Britain: bridges 1966–1970', *Concrete*, **4**, 6, 227–248.

Lee, D. J. (1971) *The theory and practice of bearings and expansion joints for bridges.* Cement and Concrete Association, London.

Lemieux, P., Kalnavarns, E. and Mordnz, N. (1968) 'Trois-Rivières bridge, design and construction', *Engng J.* (Montreal).

Leonhardt, F. (1983) *Bridges.* Architectural Press, London.

Leonhardt. F., Baur, W. and Trah, W. (1966) 'Brücke über den Rio Caroni, Venezuela', *Beton u. Stahlbetonbau*, **61**, 2, 25–38.

Lippert, E. (1965) 'Die Bauausführung der Rheinbrücke Bendorf, Los 1', *Beton u. Stahlbetonbau*, **60**, 4, 81–91.

Modjeski and Masters (1960) *Greater New Orleans bridge over Mississippi river.* Final Report to Mississippi River Bridge Authority.

Murphy, F. (1959) 'Building the world's highest arch span', *Civ. Engng* (New York), **29**, 2, 86–89.

Nash, G. F. J. (1985) Bridges to BS 5400. *Tables and graphs for simply supported beams and slab design.* Constrado, London.

New, D. H., Lowe, J. R. and Read, J. (1967) 'The superstructure of the Tasman bridge, Hobart', *Struct. Engnr*, **45**, 2, 81–90.

O'Connor, C. (1971) *Design of bridge superstructures.* Wiley, New York.

Organisation for Economic Cooperation and Development (1983) *Bridge rehabilitation and strengthening.* OECD, Brussels.

Plowden, D. (1974) *Bridges, the spans of North America.*

Podolny, W. J. R. (1986) *Construction and design of cable-stayed bridges* (2nd edn). Wiley, New York.

Prestressed Concrete Association (1984) *Prestressed concrete bridge beams* (2nd edn.). PCA, London.

Purcell, C. (1934) 'San Francisco–Oakland Bay bridge', *Civ. Engng* (New York), 183–187.

Radojković (1966) 'The evolution of welded bridge construction in Jugoslavia', *Acier-Stahl-Steel*, **31**, 12, 533–541.

Rawlinson, Sir Joseph and Stott, P. F. (1962) 'The Hammersmith flyover', *Proc. Instn Civ. Engrs*, **23**, 565–600.

Roberts, Sir Gilbert (1968) 'Severn bridge: design and contract arrangements', *Proc. Instn. Civ. Engrs*, **41**, 1–48.

Roberts, G. and Kerensky, O. A. (1961) 'Auckland Harbour bridge – Design', *Proc. Instn. Civ. Engrs*, **18**, 459–478.

Schafer, G. (1957) 'The new highway bridge over the Sava between Belgrade and Zemun (Yugoslavia), *Acier-Stahl-Steel*, **22**, 5, 213–218.

Schöttgen, J. and Wintergerst, L. (1968) 'Die Strassenbrücke, über den Rhien bei Maxau', *Stahlbau*, **37**, 1–9 (Jan.), 50–57 (Feb.).

Schröter, H. J. (1968) 'Hängebrucke über den Kleinen Belt in Dänemark', *Stahlbau*, **37**, 4, 122–124.

Schröter, H. L. (1970) 'Zwie neue stählerne Hochbrücken in Norddeutschland', *Stahlbau*, **39**, 10, 314–316. (Oct.).

Scott, P. A. and Roberts, G. (1958) 'The Volta bridge', *Proc. Instn. Civ. Engrs*, **9**, 395–432.

Schields, E. J. (1966) 'Poplar Street bridge – design and fabrication', *Civ. Engng* (New York), **36**, 2, 52–55.

Shirley-Smith, H. (1964) *The world's great bridges* (rev. edn.). Phoenix House, London.

Shirley-Smith, H. and Freeman, R., Jr. (1945) 'The design and erection of the Birchenough and Otto Beit bridges, Rhodesia', *Proc. Instn Civ. Engrs*, **24**, 171–208.

Smith, H. S. and Pain, J. F. (1961) 'Auckland Harbour bridge – construction', *Proc. Instn Civ. Engrs*, **18**, 459–478 (Apr.).

Sorgenfrei, O. F. (1958) 'Greater New Orleans bridge completed', *Civ. Engng* (New York), **28**, 6, 60 (June); also **28**, 2, 96 (Feb.).

Stein, P. and Wild, H. (1965) 'Das Bogentragwerk der Fehmarnsundbrücke', *Stahlbau*, **34**, 6, 171–186.

Steinman, D. B., Gronquist, C. H., Joyce, W. E. and London, J. (1959) 'Mackinac bridge', *Civ. Engng* (New York), **29**, 1, 48–60.

Stellman, W. L. O. (1966) 'Brücke über den Rio Paraná in Foz do Igaçú Brasilien', *Beton u. Stahlbetonbau*, **61**, 6, 145–149.

Strauss, J. B. (1938) 'The Golden Gate bridge, Golden Gate bridge and highway district'.

Swan, R. A. (1972) *A feature survey of concrete box spine-beam bridges.* Cement and Concrete Association Publication No. TR 469. CCA, London.

Talati, J. B., Holloway, B. G. R. and Chapman, R. G. (1971) 'A twin leaf bascule bridge for Calcutta', *Proc. Instn. Civ. Engrs*, **48**, 285–302.

Thoma, W. and Perron, M. (1963) 'Pont de Revin-Orzy', *Construction*, **18**, 9, 333–337.

Thul, H. (1966) 'Brückenbau (Beiträge der Deutschem Gruppe der FIP, "Bedeutende Spannbetonbauten" zum V. Internationalen Spannbeton-Kongress)', *Beton u. Stahlbetonbau*, **61**, 5, 97–115.

Thul, H. (1972) 'Entwicklungen im Deutschen Schragseilbrückenbau', *Stahlbau*, **41**, 6, 161–171.

Timoshenko, S. and Woinowsky-Krieger, S. (1959) *Theory of plates and shells* (2nd edn), McGraw-Hill, New York.

Tordoff, D. (1985) *Steel bridges: the practical aspects of fabrication which influence efficient design.* British Concrete and Steel Association, London.

Troitsky, M. S. (1977) *Cable-stayed bridges: theory and design.* Crosby Lockwood Staples, London.

Van Neste, A. J. (1970) 'Ten years of steel bridges at Rotterdam', *Acier-Stahl-Steel*, **35**, Part 1, 343–348 (July–Aug.), (Part 2, 388–396 (Sept.).

Vavasour, P. and Wilson, J. S. (1966) 'Cumberland basin bridges scheme planning and design', *Proc. Instn Civ. Engrs,* **33,** 261–288.

Virola, J. (1967) 'The proposed Ahashi Straits bridge, Japan, compared with other great suspension bridges', *Acier-Stahl-Steel,* **32,** 3, 113–116.

Virola, J. (1968) 'World's greatest suspension bridges before 1970'. *Acier-Stahl-Steel,* **33,** 3, 121–128.

Virola, J. (1969) 'The world's greatest cantilever bridges', *Acier-Stahl-Steel,* **34,** 4, 164–170.

Virola, J. (1971) 'The world's greatest steel arch bridges', *International Civ. Eng,* **2,** 5, 209–224.

Walther, R. (1969) 'Spannbandbrücken' *Schweizerische Bauzeitung,* **87,** 8, 133–137; (1971) English trans., *International Civ. Eng,* **2,** 1, 1–7.

Ward, A. and Bateson, E. (1947) 'The new Howrah bridge, Calcutta, design of structure, foundation and approaches', *Proc. Instn. Civ Engrs,* **28,** 167–236.

Weitz, F. R. (1966) 'Entwicklungstendenzen des Strahlbrückenbaus am Beispel der Rheinbrücke Wiesbaden–Schierstein', *Stahlbau,* **35,** 289–301 (Oct.), 357–365 (Dec.).

West, R. E. (1971) 'New Manchester road bridge in the Port of London', *Proc. Instn Civ. Engrs,* **48,** 161–194.

Wittfoht, H. (1970) 'Die Siegtalbrücke Eiserfeld in Zuge der Autobahn Dortmund–Giessen', *Beton u. Stahlbetonbau,* **65,** 1, 1–10.

Wittfoht, H. (1971) 'Spannbeton-Kongress 1970 (Bericht), Arbeitssitzung V. Bemerk enswerte Bauwerke-Brücken', *Beton u. Stahlbetonbau,* **66,** 2, 25–31.

Wittfoht, H. (1984) *Building bridges.* Beton-Verlag, Berlin.

Wittfoht, H., Bilger, W. and Schmerber, L. (1961) 'Neubau der Mainbrücke Bettingen', *Beton u. Stahlbetonbau,* **56,** 85–96, 114–122.

Woodward, R. J. (1981) *Conditions within ducts in post-tensioned prestressed concrete bridges.* Transport and Road Research Laboratory Publication No. LR 980. HMSO, London.

Zeman, J. (1967) 'A 1083-ft span steel arch bridge in Czechoslovakia', *Proc. Instn Civ. Engrs,* **37,** 609–631.

21

Buildings

J Rodin BSc, CEng, FICE, FIStructE, MConsE
Building Design Partnership

Contents

The design of the total building including its internal and external environment has traditionally been the responsibility of the architect but this is now so complex a task that, except for the simplest of buildings, a multidisciplinary involvement is necessary whereby engineering, surveying and other specialist skills are integrated with those of the architect to achieve consistent quality throughout the project.

Internal form and environment will be determined by the functional requirements of the occupying organization, the space needed to meet these functional requirements and the required comfort levels in regard to such items as noise, temperature, humidity and lighting. The external form and environment will be determined by the characteristics of the site and adjacent buildings. Influencing all aspects will be the constraints arising from time and cost, town planning and building regulations.

21.1 Background

Architects look to the civil and structural engineer for a positive contribution to the building design; from concept to completion, with understanding of the basic objectives of the project and with sensitivity and inspiration in their realization. The technical and economic solution to a predetermined problem, on its own, is no longer sufficient. The primary responsibility of the civil and structural engineer will be to ensure the safety and rigidity of the building but, with the architect, he can make a creative contribution to the building form, the spaces within it and its impact, visually and psychologically.

The potential freedom of building layout and expression which stemmed from the development of the structural frame took a surprisingly long time to be understood and put into practice. Framing techniques were available from the mid nineteenth century but, in the main, they were used simply to support buildings of predetermined form and style into which the required functions had to fit. It was not until a few leaders of architectural thought and practice adopted a more rational approach to design that the potential of the structural frame was grasped and put to good effect. For those who understood and wanted it, there was now much greater freedom of internal planning and external expression; and for the rationalists, form could more easily follow function.

The design of the Bauhaus building in Dessau by its founder Walter Gropius was a turning point in bringing logic into the design of buildings. It was the first major building to derive its form not from the irrational imposition of style, symmetry and proportion, but from the requirements of function and structure. Its character came from intrinsic materials and design detail, not from applied decoration; its subtlety of form and space from an ordered solution to the planning problems, not from some preconceived design formula or style. It was a major demonstration of a rational design approach founded upon the working out of solutions from first principles. The architectural features of the Bauhaus building became popular among progressive architects: assymetry, rectangular forms, lightness of the external wall, space and precision, all, in a way, reflections of the contradictory combination of freedom and discipline afforded by the sensible use of structure.

For form to follow function became the natural starting point for design; indeed, it seemed strange that it could ever have been thought otherwise. Later experience showed that a too-rigid adherence to this principle leads to a too-'tailored' building unable to respond to changing need and that a loose-fit approach is advantageous. The introduction of the structural frame allows building expression to be whatever is wanted and acceptable. Structure and building services may be expressed or hidden. Height is no longer a problem if it is acceptable to the planners and is economically viable. Almost any clear span is achievable. In short, the constraints are no longer technical; given the resources, design options are now almost limitless.

The more significant question has become: How is this technical freedom to be applied? Buildings are for people, to provide them with shelter, comfort, spiritual uplift and psychological support, and to accommodate the sophisticated processes that are part of modern life. Changing expectations of people and social relationships have become major determinants of the volume and nature of building. Communication systems of all types have changed remarkably. Science-based industries of unimagined complexity now exist requiring extreme levels of environmental control. These changes have led to the need for completely new types of building.

The design and construction itself is complex, requiring great skills of co-ordination and management. Functional requirements in many building projects are now so diverse that specialist input and understanding are required to establish the brief before building design can commence. Building materials, methods and forms of contract are diverse and changing, as are the constraints of cost and time, town planning and building regulation. The finished building itself is complex and highly serviced; and it often requires sophisticated building control and security systems to ensure satisfactory and safe performance. Cost in use, maintenance and energy consumption, have become as important considerations as first cost.

The diversity and depth of these aspects of building design and construction cannot be covered in a single chapter of a book devoted primarily to civil engineering practice. What follows is an introduction to the subject, to help the civil and structural engineer see his contribution better in the context of building design and construction as a whole.

References are in the main to UK practice but most aspects are, in principle, applicable generally.

21.2 General management

The procedures for handling large-scale building projects as opposed to civil engineering projects are complicated by the larger number of individual professional parties involved and by the large amount of legislation on permissions and approvals. The handling of such projects in the UK has been studied by the Royal Institute of British Architects (RIBA).[1] A similar publication relating to US practice has been produced by the American Institute of Architects.[2]

The overall procedures for the organization of building projects are covered in another publication produced by the RIBA.[3] Table 21.1, taken from that publication, shows the twelve discrete stages into which the project can be divided and briefly indicates the contents of each stage and the parties directly involved. Full details of the work required from each of the several professions and contractors at each stage are shown in separate diagrams. For example the detailed breakdown of Stage C, Outline Proposals, is shown in Table 21.2 in which column 5 details the input required from the civil and structural engineer.

21.3 Brief

Buildings are either purpose-built for a particular user or are speculative. In either case, the first step is to compile an agreed brief setting out the basic requirements of the project covering:

Table 21.1 Outline plan of work. (After Royal Institute of British Architects (1973) *Plan of work*. RIBA, London).

Stage	Purpose of work and decisions to be reached	Tasks to be done	People directly involved	Usual terminology
(1) INCEPTION	To prepare general outline of requirement and plan future action.	Set up client organization for briefing. Consider requirements, appoint architect.	All client interests, architect.	BRIEFING
(2) FEASIBILITY	To provide the client with an appraisal and recommendation in order that he may determine the form in which the project is to proceed, ensuring that it is feasible, functionally, technically and financially.	Carry out studies of user requirements, site conditions, planning, design, and cost, etc. as necessary to reach decisions.	Clients' representatives, architects, engineers, and quantity surveyor according to nature of project.	
(3) OUTLINE PROPOSALS	To determine general approach to layout, design and construction in order to obtain approval of client on outline proposals and accompanying report.	Develop the brief further. Carry out studies on user requirements, technical problems, planning, design and costs, as necessary to reach decisions.	All client interests, architects, engineers, quantity surveyor and specialists as required	SKETCH PLANS
(4) SCHEME DESIGN	To complete the brief and decide on particular proposals, including planning arrangement appearance, constructional method, outline specification, and cost, and to obtain all approvals.	Final development of the brief, full design of the project by architect, preliminary design by engineers, preparation of cost plan and full explanatory report. Submission of proposals for all approvals.	All client interests, architects, engineers, quantity surveyor and specialists and all statutory and other approving authorities.	

Brief should not be modified after this point.

Stage	Purpose of work and decisions to be reached	Tasks to be done	People directly involved	Usual terminology
(5) DETAIL DESIGN	To obtain final decision on every matter related to design, specification, construction and cost.	Full design of every part and component by collaboration of all concerned. Complete cost checking of designs.	Architects, quantity surveyor, engineers and specialists, contractor (if appointed).	WORKING DRAWINGS

Table 21.1 (continued)

Any further change in location, size, shape, or cost after this time will result in abortive work.

Stage	Purpose of work and decisions to be reached	Tasks to be done	People directly involved	Usual terminology
(6) PRODUCTION INFORMATION	To prepare production information and make final detailed decisions to carry out work.	Preparation of final production information, i.e. drawings, schedules and specifications.	Architects, engineers and specialists, contractor (if appointed).	
(7) BILLS OF QUANTITIES	To prepare and complete all information and arrangements for obtaining tender.	Preparation of bills of quantities and tender documents.	Architects, quantity surveyor, contractor (if appointed).	
(8) TENDER ACTION	Action as recommended in paras 7–14 inclusive of *Selective tendering**	Action as recommended in paras 7–14 inclusive of *Selective tendering**	Architects, quantity surveyor, engineers, contractor, client.	
(9) PROJECT PLANNING	Action in accordance with paras 5–10 inclusive of *Project management**	Action in accordance with paras 5–10 inclusive of *Project management**	Contractor, subcontractors.	SITE OPERATIONS
(10) OPERATIONS ON SITE	Action in accordance with paras 11–14 inclusive of *Project management**	Action in accordance with paras 11–14 inclusive of *Project management**	Architects, engineers, contractors, subcontractors, quantity surveyor, client.	
(11) COMPLETION	Action in accordance with paras 15–18 inclusive of *Project management**	Action in accordance with paras 15–18 inclusive of *Project management**	Architects, engineers, contractor, quantity surveyor, client	
(12) FEEDBACK	To analyse the management, construction and performance of the project.	Analysis of job records. Inspections of completed building. Studies of building in use.	Architects, engineers, quantity surveyor, contractor, client.	

*Publication of National Joint Consultative Council of Architects, Quantity Surveyors and Builders.

Table 21.2 Stage C: Outline proposals – plan of work for design team operation

(To determine general approach to layout, design and construction, in order to obtain authoritative approval of the client on the outline proposals and accompanying report.)

Col. 1 Client function	Col. 2 Architect management function	Col. 3 Architect design function	Col. 4 Quantity surveyor function	Col. 5 Engineer, civil and structural, functions	Col. 6 Engineer, mechanical and electrical, functions	Col. 7 Contractor (if appointed) function	Col. 8 Remarks
(1) Contribute to meeting: note items on agenda in col. 8.	(1) Organize design team. Call meeting to discuss directive prepared in stage B, action 9 (col. 2): establish responsibilities, prepare plan of work and timetable for stage C. (See col. 8 for items for agenda for meeting.)	(1) Contribute to meeting: note items on agenda in col. 8.	(1) Contribute to meeting: note items on agenda in col. 8.	(1) Contribute to meeting: note items on agenda in col. 8.	(1) Contribute to meeting: note items on agenda in col. 8.	(1) Contribute to meeting: note items on agenda in col. 8.	ITEMS FOR AGENDA FOR MEETING: (1) *State objectives and provide information: (a) brief as far as developed; (b) site plans and other site data; (c) restate cost limits or cost range, based on client's brief; (d) timetable; and (e) agree dimensional method.* (2) *Determine priorities.* (3) *Define roles and responsibilities of team members and methods communication and reporting* (4) *Define method of work, tender procedure and contract arrangements.* (5) *Agree drawing techniques.* (6) *Agree systems of cost and engineering checks on design.* (7) *Agree type of bill of quantities.* (8) *Agree check list of actions to be taken.* (9) *Agree programming and progressing techniques.*
(2) Provide all further information required by architect. Assist as required in all studies carried out by members of design team. Initiate and conclude according to timetable, any studies that are required within own organizations. Make decisions on all matters submitted for decision relevant to stage C.	(2) Elicit all information relevant to stage C by questionnaire, discussion, visits, observations, user studies, etc. Initiate studies by consultants and client as required. Maintain and coordinate progress throughout this stage.	(2) Carry out studies relevant to stage C, e.g.: (a) study published analyses of similar projects, visit if possible; (b) study circulation and space association problems; and (c) try out detail planning solutions and study effect of planning and other controls.	(2) Carry out studies relevant to stage C, e.g.: (a) Obtain all significant details of client's requirements relevant to cost and contract information on site problems, etc.; and (b) re-examine, supplement and confirm cost information assembled in stage B.	(2) Carry out studies relevant to stage C, e.g.: (a) site surveys, soil investigation; and (b) complete questionnaires on structural and civil requirements.	(2) Carry out initial studies relevant to stage C, e.g.: (a) environmental conditions, user and services requirements, appraise M and E loadings on an area or cube basis; and (b) consider possible types of installation and analyse capital and running costs, possible sizes and effects of major services installations, main services supply requirements.	(2) Carry out studies relevant to stage C, e.g visit site and investigate: (a) ground conditions, access and availability of services for construction; (b) local labour situation; and (c) local sub-contractors and suppliers to assess quality reliability, production potential and price level, etc.	
		(3) In consultation with team assimilate information obtained in action 2, and produce diagrammatic analyses, discuss problems.	(3) Outline design implications of cost range or cost limit.	(3) Advise architect on, for example: (a) types of structure; (b) methods of building; (c) types of foundation; and (d) roads, drainage, water supply, etc.	(3) Advise architect on design implications of studies made, e.g.: (a) factors which would influence efficiency, and cost of engineering elements, i.e. site	(3) Advise architect on findings and also on: (a) approximate times for construction of alternative methods; and (b) effect of construction times	

Table 2.1.2 (continued)

Col. 1 Client function	Col. 2 Architect management function	Col. 3 Architect design function	Col. 4 Quantity surveyor function	Col. 5 Engineer, civil and structural, functions	Col. 6 Engineer, mechanical and electrical, functions	Col. 7 Contractor (if appointed) function	Col. 8 Remarks
					building aspect and grouping, optimum construction parameters, etc.; (b) possible services solutions and ramifications of them; and (c) regulations and views of statutory authorities.		
		(4) Try out various general solutions; discuss with team; modify as necessary, and decide on one general approach. Prepare outline scheme, indicating, for example, critical dimensions, main space locations and uses and pass to team.	(4) Collaborate in preparation of outline scheme. Prepare quick cost studies of alternative structural and services solutions, and advise on economic aspects of solutions.	(4) Collaborate in preparation of outline scheme, prepare notes and sketches, consider alternatives, agree decision on general approach, and record details of alternative plans and assumptions.	(4) Collaborate in preparation of outline scheme, check that services decisions remain valid; record details of alternative plans and assumptions.	(4) Collaborate in preparation of outline scheme: continue to advise on time and cost implications of alternative designs or methods. Record details of proposals and assumptions.	
		(5) Assist quantity surveyor in preparation of outline cost plan; discuss and decide on cost ranges for main elements, and method of presentation of estimate to client.	(5) Confirm cost limit or give firm estimate based upon user requirements and outline designs and proposals. Prepare outline cost plan in consultation with team, either from	(5) Provide quantity surveyor with information for outline cost plan, with sketches on which to base estimate, and agree quantity surveyor proposals	(5) Provide quantity surveyor with cost range information for outline cost plan, and agree quantity surveyor proposals: interpret agreed standards by illustration.	(5) Provide quantity surveyor with information affecting price levels, for outline cost plan and agree quantity surveyor proposals.	

Table 21.1 (continued)

Col. 1 Client function	Col. 2 Architect management function	Col. 3 Architect design function	Col. 4 Quantity surveyor function	Col. 5 Engineer, civil and structural, functions	Col. 6 Engineer, mechanical and electrical, functions	Col. 7 Contractor (if appointed) function	Col. 8 Remarks
			comparison of requirements with analytical costs of previous projects or from approximate quantities based on assumed specification.				
	(6) Compile dossiers provided by team members on final (or alternative) sketch designs, recording all assumptions, and issue to all members of the team.	(6) Contribute to design dossiers, assemble all sketches and note all relevant assumptions.	(6) Record basis of estimate to contribute to design dossiers.	(6) Compile dossier of essential data collected in actions (2) to (5) above.	(6) Compile dossier of essential data collected in actions (2) to (5) above.	(6) Compile dossier of basic cost information agreed with quantity surveyor and architect.	The report includes: (a) the brief as far as it has been developed; (b) an explanation of the major design decisions; and (c) firm estimate with outline cost plan.
	(7) Prepare report as coordinated version of all members' reports, including fully developed brief.	(7) Contribute to preparation of report.	(7) Contribute to preparation of report.	(7) Contribute to preparation of report.	(7) Contribute to preparation of report.	(7) Contribute to preparation of report.	
(8) Receive architect's report; consider, discuss and decide outstanding issues. Give instructions for further action.	(8) Present report to client; discuss and obtain decisions and further instructions.						

(1) Purpose, function and scope including limitations of cost and time; proposed activities and organization including numbers and types of people concerned, internal and external service requirements, particular systems such as document retrieval, special functional requirements such as security.

(2) Design factors and required standards covering internal and external environment; spatial requirements, organizational relationships and required groupings affecting layout.

(3) Internal and external traffic and required access for pedestrians, vehicles and materials.

(4) Factors affecting type of construction, expansion, alteration, change of use, life.

(5) Phasing required.

(6) Special sensitivities or critical functions.

Of primary importance is the building use and the associated schedule of basic accommodation including the number and nature of the intended occupants. By adding allowances for circulation, services, plant, toilet and ancillary accommodation, a close assessment of the gross floor area can be made and thereby the size of building determined. By considering the relationships between the different activities, the optimum grouping of the spaces provided for them can be analysed in preparation for their translation into a physical plan to suit the particular site.

A user client may have special requirements: most buildings are expected to have a useful life of 60 to 100 years, but in some cases, a more limited life span may be envisaged dictating a light form of construction which can be demolished and replaced easily and cheaply. Alternatively, a client may require a robust building shell of long life in which internal adaptation can be carried out to suit a later, and perhaps unknown, alternative use. Substantial mechanical and electrical service requirements, as occur in hospitals and some specialist laboratories and factories, may dominate the design leading, perhaps, to the incorporation of near-storey-height service floors alternating with the functional floors.

21.4 The site

Early site appraisal is vital. Suitability for the purpose intended requires consultation with various planning authorities to confirm zoning and land use definition. Access for vehicles, people and goods must be checked and the availability of public transport and future road or transport links determined. Increasingly, good access to major international air and rail termini or proximity to the national road network is a prerequisite of a site.

Subsoil deficiencies and underground service easements may present difficulties in development. Investigation of old mineral workings (e.g. brick clay, salt, sand and gravel extraction), coalmines, shafts and wells should be undertaken, particularly if such work is known to have occurred in the vicinity. Evidence of filling should be investigated and dated. A subsoil survey should be recommended to the client (together with a cost estimate) early in the life of the project to identify the underlying conditions which may ultimately influence the building location, arrangement and cost.

The local climate requires early checking: high wind speeds will involve special stiffening; atmospheric pollution or salt-laden coastal winds will require the selection of suitable materials and careful detailing of exposed building elements. Excessive external noise from major roads, railways or airports may necessitate soundproofing in the building or sound screening between the building and the noise source.

Confined city sites introduce problems such as: (1) delivery and storage of building materials and components; (2) the threat of restrictions or stoppages arising from local objection to construction noise; and (3) protection of adjoining property which may need underpinning and should be surveyed for dilapidations before work commences on site.

21.5 Landscape

The landscape is the setting to which new developments must relate, therefore its consideration is vital at the outset of each project. Landscape and civil engineering bear a close affinity, due to a mutual and direct concern with land form and natural resources. All but the most cosmetic landscape treatment involves civil engineering considerations. Landscape considerations include feasibility studies, environmental assessments, public inquiries, erosion control, reclamation, restoration, conservation, transportation, industry, commerce, natural heritage and the landscape related to all types of buildings both exterior and interior.

The quality of the landscape is now an essential constituent of the planning consent process. Early site appraisal should include an analysis of the landscape or urban space. Among the factors to be considered are geology, topography, soil, microclimate, drainage, land use, artefacts, vegetation and visual analysis. The effects of the interaction of these factors should be considered in relation to the development. A skilful appraisal will lead to the establishment of sound principles, which will enhance the less favourable aspects of the site whilst conserving the best.

On a yet broader scale, a full environmental assessment leading to designs which cause the least damage socially, aesthetically and to our natural resources, would extend to a large team including other specialists.

Reclamation of abandoned industrial and domestic wasteland is an area in the re-creation of the environment where engineering and landscape are inseparably combined. Such operations can restore the form of the landscape, provide new sites for housing, industry and recreation, and create new habitats, from wetlands to woodlands.

There are numerous factors concerning planning and design which will be important to the landscape architect, the civil engineer and the architect. These factors include planning for vehicles, finished levels and materials, economic cut-and-fill and the integration between the hard paved areas of the scheme and its immediate environment. Close collaboration between the professions is therefore needed to achieve an economic and sympathetic design.

As part of the site investigations, soil tests should be taken to assess biological qualities and should include horizon depth, soil type, texture, moisture content and pH. It is advisable to obtain a chemical analysis from an approved laboratory to assess deficiencies.

Planning the site operations to achieve the best results involves many decisions related to the landscape. Vegetation and topsoil are delicate natural resources which are easily damaged by thoughtless construction techniques. Their value must be assessed at the outset by a specialist, and if considered of value they should be protected carefully and retained. Topsoil must not be mishandled, as compaction and poor storage can render it useless as a growing medium. Drainage and grading should also be considered regarding any vegetation to be retained.

Working areas should be kept to a minimum in order to leave the maximum undisturbed area and avoid the replacement or restoration of topsoil and subsoil. Excavation, compaction, changes in water table and in finished level within the root spread of trees, should be avoided. The canopy will suffer in proportion to the amount of root damage sustained and similarly the stability, appearance and life expectancy of the tree will also be affected. On no account should the level of soil adjacent to the trunk be changed. Where trenching is unavoidable within the root spread, hand digging and retention of roots will be advantageous.

Where trees are an important feature close to buildings, roads or drainage foundations should be designed to withstand the effects of root growth or moisture movement.

The structural requirements and economic viability of planted areas within the building which are conceived as roof gardens, terraces or interior gardens, must be considered early in the design process together with the client's understanding of the long-term maintenance commitment. In addition, drainage, access and the sequence of construction are significant factors. The growing medium and planting should either be installed as the very last of the building operations, or adequately isolated and protected from further construction activities. All planting, but especially interior planting, can be affected adversely and even destroyed by subsequent operations such as the repair of faulty tanking, the installation of lighting and irrigation or the grinding of materials such as marble and terrazzo. It is clear, therefore, that where planting is part of a design concept it needs careful integration into the building process.

Management of the landscape in the long term is essential and should be discussed at the earliest opportunity, preferably when the brief is being formulated to ensure the wellbeing of the newly created environment and that a succession of planting is provided for the future.

Figure 21.1 and the accompanying text describe the careful integration of an important headquarter building within a beautiful parkland setting. The aim was to provide a headquarters which would give high quality conditions for work, training and recreation. The new building was planned to have minimal impact upon the local environment, and to ensure that its landscaped surroundings would enhance working conditions. At the same time it had to cater for the latest demands of information technology, ensuring that the layout and fabric of the building were flexible enough to accept inevitable future change.

While the briefing process was underway, surveys were carried out on site conditions, tree planting and acoustic aspects of the location. It was seen as vital to respond to the exceptional natural quality of the site, and to the architectural qualities of the two main existing buildings there – the listed Fulshaw Hall, and Harefield House.

The form of the new headquarters evolved from these considerations as a low-lying building, tucked into the landscape on a slope of land across the lake from Fulshaw Hall. Car parking is provided discreetly to the south. The three-storey construction, pitched roofs, and brick and slate materials of the building help further to establish it as a worthy neighbour to the hall.

The plan provides outer and inner bands of office space, linked at intervals to create enclosed courtyards. Its external appearance is of a series of linked pavilions, sweeping round in a gentle curve that focuses upon the hall itself. The western end of the building surmounts a landscaped terrace facing the entrance from the A34 road.

Inside, circulation is provided by a pedestrian mall on the middle level of the inner band, facing the park. Vertical access is via stair towers at back and front. All the offices are fitted out with a raised floor to accommodate all cabling and air handling needs. The offices are 12 m wide along the bands, and 15 m wide along the links, so providing a good level of natural light. The planning module is a highly flexible 1.5 m allowing practically any type of interior fit-out. Uplighters bounce light off an

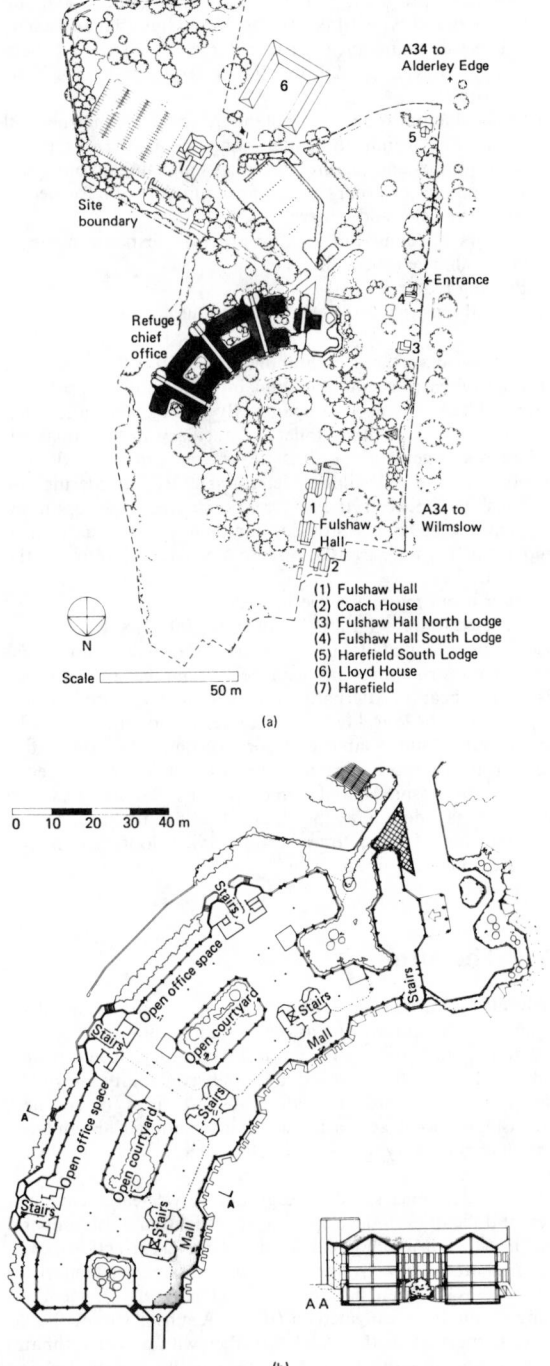

(1) Fulshaw Hall
(2) Coach House
(3) Fulshaw Hall North Lodge
(4) Fulshaw Hall South Lodge
(5) Harefield South Lodge
(6) Lloyd House
(7) Harefield

(a)

(b)

Figure 21.1 Integration of building and landscape. (a) Plan of the Fulshaw Hall site; (b) disposition of spaces

acoustically treated structural ceiling, providing glare-free conditions and easy replanning of working spaces.

A 1000 m² computer suite is provided, together with dining room, coffee lounge and kitchens. Although the main spaces are left open, cellular offices may be provided as required throughout the plan.

21.6 Town planning

Most development and construction work is governed by the Town and Country Planning Act 1971. Section 22 of the Act defines development as: 'the carrying out of building, engineering, mining or other operations or the making of any material change of use of buildings or other land'. With some exceptions (mostly under the General Development Orders 1977 and 1981) permission to undertake any development is required from the local planning authority.

Other planning powers are concerned with individual buildings listed as of special architectural or historic interest (where consent is required for any works of demolition or alteration), conservation areas, advertisements, caravan sites, tree preservation, national parks and the countryside.

County and district planning authorities prepare structure (broad policy) and local plans against which applications for planning permission are judged. If planning permission is refused or conditions are imposed upon the permission, the applicant has the right of appeal to the Secretary of State for the Environment, and such appeals may be heard at a local public inquiry.

' Early consultation with the local planning authority (the district or borough council) is recommended when advice will be given on the need to obtain planning permission, the scale of fees charged and the adopted planning policies which should be taken into account.

In some cases the local authority may provide access to grants available for special types of development. These include derelict land grant for approved ground restoration works and urban development grants for joint public/private sector funding of approved inner areas projects. In addition, most authorities offer grant, loan, site and premises assistance to encourage economic development in their area.

Major civil engineering projects such as oil refineries, power stations, radioactive toxic and dangerous waste treatment and disposal, iron and steelworks, asbestos extraction, chemical plants, motorways, ports and airports are all listed in an EEC Directive as likely to require an environmental impact assessment.

Further details of the planning legislation will be found in a work by Telling.[4]

21.7 Public utility

Once an outline brief exists and a site is under consideration the various public utility organizations (PO, gas, electricity, water authorities) should be consulted to determine the availability of their various services.

21.8 Feasibility

The compatibility of brief and site with the external constraints in their varying forms logically leads to the preparation of a feasibility study. This is normally the first design exercise and provides the design team with an opportunity to explore the problem, propose solutions, cost the alternatives and identify options for the client. Presentation of a preferred option with objective data supporting the preference completes the first stage and forms the basis for the final design.

21.9 Cost

Cost is an important factor at all stages of the design process. Alternative design solutions or materials must be considered carefully to ensure that cost is within budget, that money is allocated in a balanced way to best suit the client's needs and that, throughout the project, good value is obtained for the money spent. The most significant decisions affecting cost occur in the concept and outline planning stage.

Of first importance is the economic use of space in the proposed building. Although the basic range of accommodation is fixed, considerable additional space is required for circulation and access, stores, plant rooms and toilet facilities. This additional space, sometimes called 'balance area', can vary considerably according to the layout adopted and should be kept to the minimum by efficient planning of staircases and service ducts, grouping of toilet facilities and a restriction on the area of circulation routes. The economic planform will also aim at reducing the ratio of external wall area to total floor area thus saving expensive wall materials and reducing heat losses (or gains) and, hence, minimising the installation and running costs of the heating, ventilation or air-conditioning systems. The reduction of storey heights to a minimum will have similar cost benefits but could affect significantly the building's future adaptability.

It is usual to prepare a cost plan for the project in elemental form. Initially it is a cost estimate based on the preferred scheme and structural system together with a specification covering the main building elements. In the long term it forms a cost structure for monitoring the cost effect of changes and the detailed development of the design. The cost plan should state whether it provides for price inflation to tender stage or building completion, or is based upon rates current at date of estimate.

Major elements should be kept in reasonable balance, e.g. the use of an expensive cladding material could leave too little money for the remainder of the work resulting in a visually pleasing but operationally unsuccessful building. The cost plan is an excellent means of checking the balance between the different elements of structure, finishes and services though the relative percentages of the overall cost will vary from case to case according to the type of building and its user requirements.

While the capital construction cost of a building is of primary importance, other costs will also be significant and could affect design. The annual running cost is one such item and services installations, particularly, should be considered in terms of operational as well as initial cost. Similarly, the use of an expensive but hardwearing material may be justified in terms of subsequently reduced outlay on cleaning or maintenance. Discounting techniques and, possibly, tax considerations are necessary to make true cost assessments of such comparisons.

The total cost of a building project will also include expenditure on land, borrowed capital and the fitting out of the completed building, compensation to adjoining owners and other associated costs as well as legal and design consultant's fees and expenses. In some cases, the earlier a development can be occupied the better the cost advantage to the client. The construction method and programme are then significant and may affect the design form. It is often possible to assess the financial advantage of early completion and by comparative financial analysis to justify additional construction cost to shorten the construction period. Similarly, value engineering can be applied to ensure that optimum arrangements are adopted to meet the client's objectives.

21.10 Internal environment

21.10.1 Thermal environment

The required comfort conditions and tolerances are determined by the intended function of the space concerned.

Thermal comfort depends on a complex of inter-related factors: air temperature, ventilation rates, relative humidity and mean radiant temperature of the enclosing space. Mean radiant temperature is generally a function of enclosure construction, although the form of heating can have an influence. All other factors are determined by the air-conditioning system. Many attempts have been made to devise indices which will represent in one figure the composite effect of the different variables, such as equivalent temperature (T_{eq}) and corrected effective temperature (CET). The former incorporates three of the basic variables: (1) air temperature; (2) mean radiant temperature; and (3) rate of air movement; the CET adds relative humidity. For the purpose of design calculations, however, the generally accepted index is resultant temperature, which is the mean of the air temperature and the mean radiant temperature.

Internal design temperatures for air-conditioned buildings in this country are usually of the order of 20° C in winter and 22° C in the summer; relative humidity values are usually kept within limits depending upon the spaces served, the types of system, condensation considerations and the enclosure construction. Glass area and type, especially large single glazed windows, has an appreciable effect on mean radiant temperature and also restricts the permitted humidity level in cold weather.

21.10.1.1 Site and climate

Internal thermal control will also be influenced by external seasonal temperatures, relative humidity, wind velocities and direction, air quality (industrial smoke pollution, etc.), solar orientation and latitude and relation of the site to surrounding locality and adjacent buildings.

In other than air-conditioned buildings, external temperature related to occupancy levels and internal heat gains determine the amount of external ventilation air to be introduced. Where windows can be opened, however, occupant behaviour tends to be the dominant influence. In air-conditioned buildings ventilation air quantity can be related to external temperature and relative humidity, but this is dependent on the type of air-conditioning system. In warm summer conditions, the amount of ventilation air has a direct effect on refrigeration loads, but at other times of the year, cool outside air can be introduced beneficially to offset internal heat gains.

Excessive infiltration through openings such as doors, window gaps, etc. can reduce performance seriously and increase operating costs; satisfactory sealing is necessary as are effective measures to reduce the stack effect (flow of air up stair and lift areas) which grows in significance with increasing building height.

Solar penetration into the building is determined by latitude and season and the resulting heat gain can be serious. Methods of control include internal or external louvres and blinds, special heat-absorbing and reflecting glasses, small glass areas and various forms of external shading structure.

21.10.1.2 Building function and form

Thermal design is affected by the energy-producing elements within the building: human, mechanical and electrical. Building configuration, size and proportion and construction of the building shell influence the adaptability and capacity of the system to cope with external environmental changes. The proportion between interior space which is independent of external effects and perimeter space which is not, is important. External conditions penetrate a building to approximately 6 m: this perimeter zone will require a system which can quickly adapt to rapid variations in the heating or cooling loads. In contrast, load changes in interior spaces are usually less rapid and represent a predominantly cooling requirement.

21.10.2 Air-conditioning

Natural ventilation has certain potential drawbacks: (1) noise infiltration through open windows; (2) overheating during summer due to solar and internal heat gains; (3) excessive infiltration of outside air resulting in uncontrollable internal air movement; and (4) ineffective ventilation beyond about 5 m from the perimeter with attendant overheating.

Mechanical ventilation solves only a few of these problems. Noise and outside air infiltration are reduced as windows are opened less frequently. Increased air movement during warmer weather can alleviate discomfort to some degree.

Overheating and high humidity can, however, occur due to the inability of the system to supply air at the correct thermal condition. This inability is overcome by the inclusion of refrigeration, thereby changing the system from mechanical ventilation to air-conditioning.

Air-conditioning provides a controlled internal thermal environment which is largely independent of the external conditions or of any changes in the internal load conditions. Planning and configuration of the building will be influenced by the provision of air-conditioning. Deep space can be created with the knowledge that a satisfactory internal thermal environment will be achieved. Similarly, nonopening windows avoid infiltration problems which are accentuated with increased building height.

Moisture control and filtration of the incoming air are integral parts of full air-conditioning giving a cleaner, healthier and more comfortable atmosphere compared with ventilation by natural methods. Redecorating costs and absenteeism may be reduced and working efficiency increased.

21.10.2.1 Air-conditioning systems

Many types of air-conditioning systems are available and can be classified into three basic groups: (1) 'centralized'; (2) 'decentralized'; and (3) 'self-contained' systems; some solutions are combinations of these three.

Centralized systems. Centralized systems are:

(1) Systems where air is processed at a central plant and distributed for use without further treatment:
  (a) single-duct all-air systems using high-, medium- or low-velocity distribution;
  (b) double-duct all-air systems using high-, medium- or low-velocity distribution with local terminal mixing units (referred to as dual-duct systems).
(2) Systems where air is processed at a central plant, but with final heat addition or subtraction at the point of use:
  (a) single-duct all-air reheat/recool systems, using high-, medium- or low-velocity air distribution with associated heating and/or cooling water distribution;
  (b) perimeter induction air/water systems using high-, medium- or low-velocity primary air distribution with secondary heating and/or cooling water distribution on a two-, three- or four-pipe principle.

Decentralized systems. Decentralized systems are:

(1) Systems where a liquid medium is distributed from a central

point to units which condition air locally: some such systems also have a supplementary primary air supply from a central plant to the unit or space:

(a) room fan coil unit air/water system with two-, three- or four-pipe water distribution and local outside air connections;
(b) as for (a) but with supplementary primary air from central plant;
(c) localized zone air-handling unit all-air systems with associated heating/cooling water distributions and with low-velocity air distribution to conditioned spaces from the units;
(d) radiant ceiling systems supplied with heating/cooling water distribution and supplemented with separate single-duct all-air system.

Self-contained systems. Self-contained systems are systems where self-contained air-conditioners process and supply air at the point of use.

Each system has merits and limitations. The simpler low-velocity all-air single-duct systems require a large amount of duct space and are not a practical solution where a large number of zones of varying use are to be served. In these cases a system which can respond to these variations is required. One of the following systems would be appropriate. Double-duct all-air systems mix air from separate hot and cold distribution ducts using ceiling- or sill-mounted terminal mixing boxes. This system is very adaptable, but the combination of two supply ducts plus a return air duct requires considerable service space, even when using high- and medium-velocity distribution.

The induction unit discharges primary air supplied from the central plant through high-pressure nozzles and this induces air from the space into the unit which then mixes with the primary air before discharging back to the space; temperature control is achieved by a heating/cooling coil. Space is saved because the air is distributed at high velocity. The basic difference between two-, three- and four-pipe associated water distribution systems is that the latter two can provide, at the point of use, the simultaneous facility for either heating or cooling, while the two-pipe system is restricted at any one time to one or the other.

Fan coil systems incorporate a heating/cooling coil and a circulating fan. Primary air can be ducted direct to the units from a central system or discharged to the space independently or alternatively, each unit can draw in air direct from outside.

Radiant heating/cooling ceilings, when used with a supplementary air system, can provide an effective environment although their adaptability to meet rapid fluctuations in heating and cooling loads is limited.

Self-contained packaged air-conditioning units are usually restricted to smaller specialized projects.

21.10.2.2 Air-conditioning – distribution and integration

Considerable duct distribution space is required and air outlets and extracts are often incorporated in the detailing of light fittings and suspended ceilings. From the earliest stages, therefore, the air-conditioning system should be integrated into the total planning and detail design process of both the building elements and the structure.

Perimeter units can be served from a network of air ducts or water pipes concentrated in zones near the outer wall, within the under-sill or ceiling void for horizontal piping or ducts and within structural column enclosures for vertical distribution. Alternatively, the perimeter area may be served from the central core with ducts and pipes accommodated above a false ceiling, within a structural hollow floor or beneath a raised floor.

In areas where little flexibility for changing use is required, a totally integrated solution using the structure to accommodate

air and water distribution may produce some economies including reduced storey height. Where a high degree of flexibility is required as, for example, in open-plan buildings, ceiling distribution on a modular basis for interior zones and sill or ceiling distribution for the perimeter becomes essential and a false ceiling is required, the ceiling space being used to accommodate the ducts and pipes.

The above systems can be described as fully ducted. There are two other basic air-supply and exhaust methods using the ceiling space as a large duct or plenum:

(1) *Negative plenum*: air is extracted into the plenum through outlets in the false ceiling which are usually part of the light fittings. Air supply is ducted to diffusers or slots incorporated in the ceiling design.
(2) *Positive plenum*: the plenum is used as the supply duct, air being forced through ports in the false ceiling. Extracted air is ducted from terminals usually incorporated in the light fittings.

When the air is exhausted through the light fittings it cools and, hence, increases the efficiency of the light source: it also removes excess heat (arising from high light levels) which can be transferred for use elsewhere, e.g. the perimeter area, but is more commonly vented to the exterior. The outlets require careful design coupled with adequate ceiling height, 4 to 5 m if possible, to prevent downdraughts.

The completely ducted system has fewer thermal problems, but occupies more space and is more expensive. The plenum systems substantially reduce duct requirements, but are less efficient; they also require careful control of temperature to prevent condensation and, sometimes, the incorporation of insulation on the underside of the structural floor to confine the plenum effects to the storey intended.

21.10.3 Accommodation of building services

Services can occupy 15% or more of the volume of a building and their distribution through the building is critical to its performance and flexibility. The organizing of space for services is thus of vital importance both in the strategic planning and detail design stages of the building. The servicing systems may be given direct expression or be entirely hidden within the overall form and finishes of the building.

The strategic planning of the services installations involves the optimization of the location and size of plant room spaces and the distribution systems linking them with the building areas being serviced, coupled with their integration with the structural and architectural elements. Frequently, there is pressure on the design team to minimize the space occupied by the services as the result of planning height restrictions or on grounds of economics. This can prove a false economy as such an approach can affect significantly future flexibility in the use of the building.

Plant rooms should be positioned as close as possible to the centre of gravity of the areas they serve to keep maximum duct sizes to a minimum and should be readily accessible to connecting ductwork without impediment from adjacent structure. The impact of weight, noise or vibration on adjacent elements or building functions should be considered. In general, service runs should not be more than 25 m from the point of origin and, vertically, plant rooms should not serve floors more than ten storeys away. Plant rooms should be sensibly proportioned avoiding L-shapes and long thin spaces. Clear height generally has to be to the underside of structural beams and if possible the plant room space should be column-free.

Frequently several plant rooms are required covering the following items.

(1) *Boilers and refrigerators*: commonly referred to as the energy centre.
(2) *Air handling*: fans, heating and cooling coils and filters.
(3) *Water*: storage tanks.
(4) *Sprinklers*: storage tanks.
(5) *Cooling towers*: serving the cooling plant.
(6) *Lifts*: motors, winding gear or pumps.
(7) *Electrical*: switchroom or substation or standby generator.
(8) *Telecommunications*: telephone and data transmission equipment.

Some of these items must not be incorporated in the same plant space. Examples are water and electricity, refrigeration machines (chillers) and boilers to avoid toxic fumes from refrigerator gas coming into contact with boiler flames.

Access for installations, repair and maintenance must be incorporated and construction problems including speed should be taken into account in the siting of these elements especially when the installation of plant is on the critical path to completion.

In air-conditioned buildings, the air intake must be separated from the air discharge. The top of the building is often the preferred location for the air-handling plant particularly with all air systems involving recirculated air. In very large buildings, a number of air-handling plant rooms distributed through the building provide greater flexibility and less inroad into usable building volume.

The combined plant room area typically ranges from 4 to 15% of total floor area depending upon building type. Some plant area requirements are as follows:

	(%)
Hospitals/laboratories	9–15
Swimming pools/ice rinks	5–12
Shopping centres	5–8
Theatres/concert halls	9–11
Air-conditioned speculative offices	6–9
Residential/hotels	4–5
Factories/warehouses	3–4

Special cases can lead to even greater plant areas when environmental control is required to extremely fine limits, e.g. in pharmaceutical or semiconductor production facilities.

The incorporation of the horizontal services within the ceiling and floor construction is a vital element in the efficient design of the building particularly when overall floor depth is critical. Many different arrangements have been developed around both steel and concrete structural elements with the objectives of keeping floor depth to a minimum yet providing easy access and flexibility for future change. Some typical arrangements and corresponding floor depths are shown in Figure 21.2. In some cases complete storeys may be given over to services distribution.

The location of the plant rooms and the location of pipe and duct runs can have a critical impact on structural arrangement and detail. The co-ordination of structural penetrations is an important task for the structural engineer and timely receipt of relevant information from the services engineers is vital. In certain cases, duct and plant room walls may be subjected to positive or negative pressure which the structural engineer may need to take into account. Enclosure materials and construction would need to be appropriately airtight.

Modern office design has to cater for widespread use of the computer. Space for cabling and easy access for modification or extension are essential ingredients for good design catering for both immediate and long-term requirements. At the same time, increased space is needed for air ducting to deal with the higher heat loads generated. The growing impact of this heat-generat-ing equipment on the total heat load that has to be dealt with by the air-conditioning system is illustrated in Figure 21.3. Figures 21.4 (a)–(c) show diagrammatically three methods used for the incorporation of air-conditioning and cabling in the present-day electronic office.

Figure 21.4(a) shows a conventional-sandwich ceiling and raised floor. In this arrangement, which is favoured by most speculative developers, the air supply and removal and the general lighting are incorporated in the space between the structure and a suspended ceiling. All cable services are in the elevated floor usually between 75 and 150 mm deep. In some cases, cables are run in hollow cells in the structural floor deck. This arrangement separates service systems cleanly but costs more.

Figure 21.4(b) shows a total ceiling servicing using 'stalactites'. In this arrangement, a little more depth is added to the ceiling space and facilities provided for easy and frequent access so that heavy cabling can be accommodated in the ceiling space. The cabling is brought to the workstation down partitions, columns or free standing 'power poles'. This is the lowest cost option but is not much used in new design outside the high-technology industries. However, the increasing shift back to cellular offices coupled with the arrival of slimmer, more flexible, data cabling could make this solution more acceptable.

Figure 21.4(c) shows the total floor servicing using 'stalagmites'. In this arrangement, all the services and cabling are incorporated in the floor void. Uplighting is bounced off the ceiling helping to provide glare-free background lighting for visual display unit (VDU) working, and is augmented where required by task lighting. Air can be supplied, under occupants' control, through desks and removed through heat-producing equipment and light fittings. Partitions sit between the heavy floor and the solid ceiling giving better sound insulation. The exposed structural ceiling acts favourably as a heat sink helping to even-out internal temperatures. Removal of a small proportion of overhead stale air can be effected through uplighter units or in voids at walls or around structural columns.

21.10.4 Heating/cooling generation

Arrangements for the heating and cooling generating plant will depend on a number of general and localized factors: (1) availability, suitability, and economic costs associated with the utilization of fuel and power; (2) resources peculiar to the site; and (3) utilization of recoverable energy associated with the heating and cooling systems installed within the building.

Fuel and power considerations are complex and include a detailed appraisal of operating and capital costs for various fuel alternatives (coal, gas and oil) and power. Boiler plants incorporating combined dual-firing burners suitable for gas (town or natural) and oil can offer attractive capital and operating cost characteristics combined with greater flexibility.

Heat-recovery systems have been gaining popularity. A common arrangement is to utilize low-grade heat being rejected from refrigeration machines. Another is to transfer heat extracted from the interior of deeply planned areas, which have to be cooled, to spaces requiring a heating load, such as perimeter zones, during winter and certain mid-season periods.

On larger specialized projects, total energy is finding an application. This is based on the concept that the total energy requirement, in all its forms, can be provided from a single fuel source. These systems incorporate electrical generation with heat being produced as a byproduct. Refrigeration, which can be met by either electricity or heat, is usually a complementary part of such an integrated energy system.

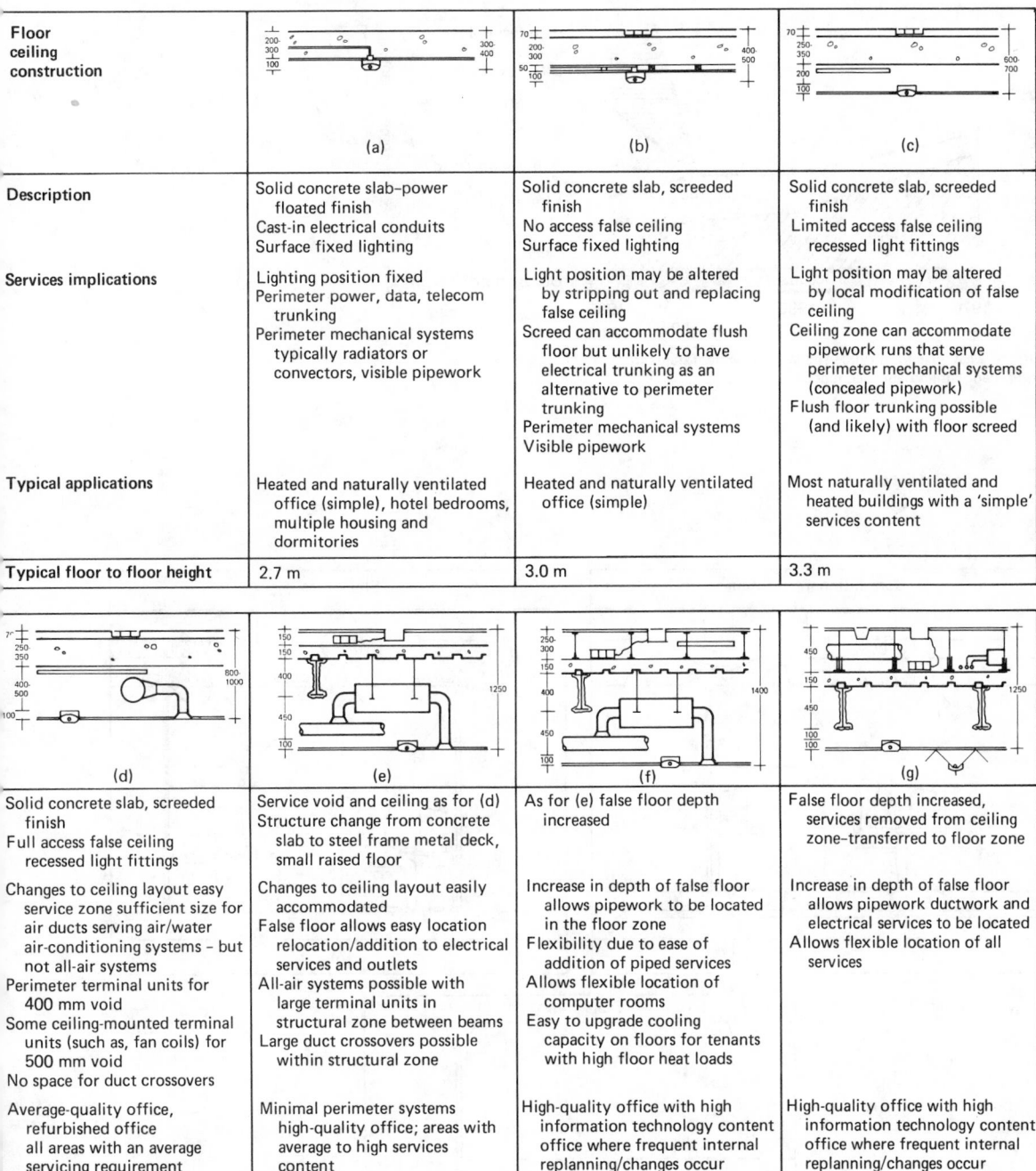

Floor ceiling construction	(a)	(b)	(c)
Description	Solid concrete slab–power floated finish Cast-in electrical conduits Surface fixed lighting	Solid concrete slab, screeded finish No access false ceiling Surface fixed lighting	Solid concrete slab, screeded finish Limited access false ceiling recessed light fittings
Services implications	Lighting position fixed Perimeter power, data, telecom trunking Perimeter mechanical systems typically radiators or convectors, visible pipework	Light position may be altered by stripping out and replacing false ceiling Screed can accommodate flush floor but unlikely to have electrical trunking as an alternative to perimeter trunking Perimeter mechanical systems Visible pipework	Light position may be altered by local modification of false ceiling Ceiling zone can accommodate pipework runs that serve perimeter mechanical systems (concealed pipework) Flush floor trunking possible (and likely) with floor screed
Typical applications	Heated and naturally ventilated office (simple), hotel bedrooms, multiple housing and dormitories	Heated and naturally ventilated office (simple)	Most naturally ventilated and heated buildings with a 'simple' services content
Typical floor to floor height	2.7 m	3.0 m	3.3 m

(d)	(e)	(f)	(g)
Solid concrete slab, screeded finish Full access false ceiling recessed light fittings	Service void and ceiling as for (d) Structure change from concrete slab to steel frame metal deck, small raised floor	As for (e) false floor depth increased	False floor depth increased, services removed from ceiling zone–transferred to floor zone
Changes to ceiling layout easy service zone sufficient size for air ducts serving air/water air-conditioning systems – but not all-air systems Perimeter terminal units for 400 mm void Some ceiling-mounted terminal units (such as, fan coils) for 500 mm void No space for duct crossovers	Changes to ceiling layout easily accommodated False floor allows easy location relocation/addition to electrical services and outlets All-air systems possible with large terminal units in structural zone between beams Large duct crossovers possible within structural zone	Increase in depth of false floor allows pipework to be located in the floor zone Flexibility due to ease of addition of piped services Allows flexible location of computer rooms Easy to upgrade cooling capacity on floors for tenants with high floor heat loads	Increase in depth of false floor allows pipework ductwork and electrical services to be located Allows flexible location of all services
Average-quality office, refurbished office all areas with an average servicing requirement	Minimal perimeter systems high-quality office; areas with average to high services content	High-quality office with high information technology content office where frequent internal replanning/changes occur	High-quality office with high information technology content office where frequent internal replanning/changes occur
3.6 m	3.9 m	4.2 m	3.9 m

Figure 21.2 Options for horizontal service distribution showing increasing size and complexity of service zone planning as sophistication increases provision. (After *Architect's Journal* **183**, 9, p.62 (1986))

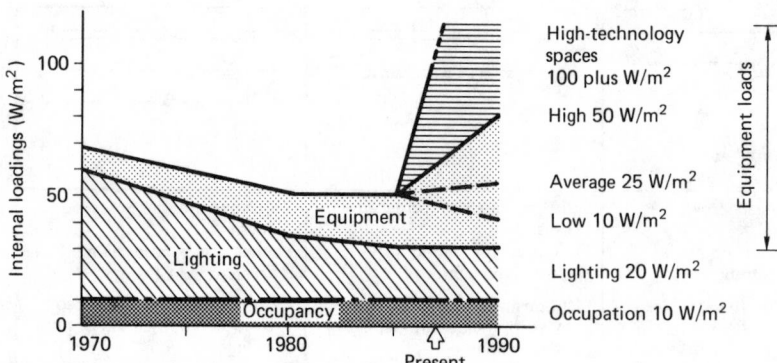

Figure 21.3 Trend in office space internal heat gains from equipment lighting and occupancy

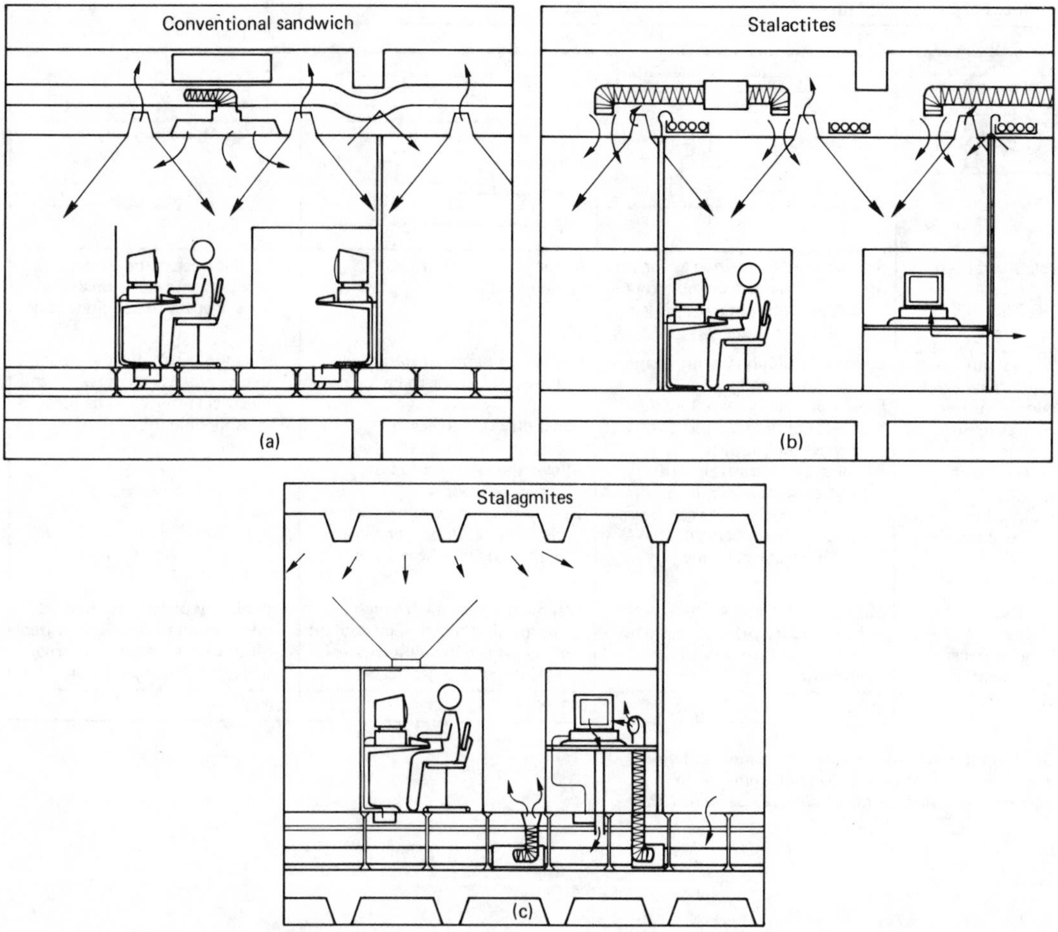

Figure 21.4 Incorporation of air-conditioning and cabling in the electronic office. (a) Conventional sandwich; (b) stalactites; (c) stalacmites

21.10.5 Thermal insulation

The object of thermal insulation, together with heating, is to obtain, irrespective of the prevailing weather conditions, a near-constant internal temperature determined by requirements of human comfort and satisfactory conditions for manufacturing processes or storage of goods. Adequate insulation is needed to avoid excessive expenditure on heating plant and fuel. The insulation and heating of buildings for human occupation are normally designed to maintain a temperature of 16 to 21° C according to use when the outside temperature is $-1°$ C. Good thermal insulating materials generally are those which entrap air, such as lightweight concrete, wood-wool slabs, glass or mineral fibre wool.

Calculation of the thermal transmittance, or U value, of a wall, floor or roof is carried out by adding together the thermal resistances of the materials and surface coefficients and taking the reciprocal of the answer to provide the thermal transmittance of the composite construction:

$$U = 1/R \tag{21.1}$$

$$R = R_{s_1} + R_{s_2} + \frac{(L_1)}{(K_1)} + \frac{(L_2)}{(K_2)} \text{ etc.} + R_a + R_h \tag{21.2}$$

where R is the total thermal resistance of the structure, R_{s_1} is the surface resistance of the inner face, R_{s_2} is the surface resistance of the outer face, R_a is the resistance of the air cavity if present, R_h is the resistance of unit material such as hollow block where resistance per unit thickness does not apply and L/K is the resistance of one layer of material or thermal conductivity K and thickness L.

To evaluate the total heat loss from a room or building the thermal transmittance of the walls, windows, floor and ceiling must be calculated and allowance made for the losses involved in heating-up the ventilating air and the structure when heating is intermittent.

Structural members penetrating the full thickness of a wall produce 'cold bridges', locally reducing the thermal resistance and internal surface temperatures, with consequent added risk of condensation. In such cases the cold bridge should be reduced in width or eliminated by appropriate insulation.

Ureaformaldehyde foam is sometimes used to fill the cavity in cavity-wall construction resulting in an almost 80% reduction in heat loss through the walls. Double glazing, as well as reducing heat loss, has advantages in increasing the temperature on the inside window surface and may improve internal comfort conditions.

Condensation problems have increased due to new methods of building, standards of heating and control of ventilation, and changing family habits which have led to intermittent heating coupled with the generation of more moisture inside the dwelling. Old buildings, particularly domestic ones, usually had open fires and flues and windows were generally less well-fitting resulting in natural, if draughty, ventilation which got rid of moisture-laden air and avoided condensation on cold walls and windows. Condensation in modern buildings can be avoided by adequate combination of insulation, heating and ventilation.

The amount of moisture which air can hold, increases with the temperature and when it can hold no more water it is said to be saturated and the relative humidity is 100%. The temperature at which air with any particular moisture content is saturated is called the dewpoint and if that air falls on a surface which is colder than the dewpoint, condensation will occur. Another object of thermal insulation, in conjunction with heating and ventilation, is to ensure that the inside surfaces of walls, floors, ceilings, roof and, if possible, windows, are kept above the dewpoint.

Moisture-laden air can pass through a porous wall or roof construction and condense inside where it meets a temperature below the dewpoint. Figure 21.5,[5] shows the relationship between the local material temperature and dewpoint through the cross-section of varying arrangements of a composite external wall, for given internal and external air temperatures and moisture contents. By appropriate positioning of a vapour barrier and combination of materials forming the wall, the local dewpoint can be kept above the local temperature and condensation avoided. Temperature drops across the section are determined by the proportional thermal resistances of the materials, surfaces and airgap; dewpoints are obtained by first determining the local vapour pressures from the proportional vapour resistances and then converting these to their respective dewpoint temperatures.

21.10.5.1 Estimation of condensation risk

At any point where the computed temperature is lower than the computed dewpoint temperature, condensation can occur in the conditions assumed. In the worked example, liquid may form in a position where, clearly, it can reduce the effectiveness of insulation and it is likely also to put the nearby timber at risk of rot. As in illustration of the effect of structural detailing, Figure 21.5(b) shows the construction reversed and free from risk in the same surrounding conditions. Slight modifications shown in Figure 21.5(c) and (d) are sufficient, however, to limit the potential risk by using materials that modify the vapour pressure gradient.

21.10.6 Lighting

Three types of lighting are used: (1) daylight; (2) daylight integrated with electric lighting; and (3) electric lighting. Good daylighting is more than the provision simply of large windows. Optimum size, shape and position of windows is a function not only of the required lighting levels, but also of the resulting eye adaptation conditions, sky glare and external view. In addition, heat loss or solar gain, ventilation, noise transmission, privacy and the shading effects of adjacent buildings, present or future, must be taken into account. Side-lit rooms often appear badly illuminated because of the contrast between the areas adjacent to and those remote from the windows, even though working illumination levels may be adequate throughout.

At one time, daylighting appeared cheap and its real cost went unquestioned. The present position is different: modern light sources cost less and are more efficient while the true cost of daylight is recognized in terms of added cost in construction, maintenance, heat loss or gain and, in urban areas, the inefficient use of the available site area. Simultaneously, the expected standards have increased in both quantity and quality and, in modern buildings, daylighting would not be relied upon as the sole source of light even during periods of good outdoor light.

By introducing electric lighting of a colour to blend with daylight it is possible to provide adequate illumination over the whole working area without a sense of deprivation of daylight. Moreover, such arrangements – known as permanent supplementary artificial lighting of interiors (PSALI) – can be applied without visual discomfort over areas much greater than can be lit by daylight alone, irrespective of the prevailing outdoor light; its added cost must be weighed against the direct and indirect costs of higher ceilings and bigger windows, reduced floor space for lightwells, and/or restricted useful depth of rooms.

The current quest for saving energy has stimulated research into methods of securing greater penetration of daylight into buildings. One such method involves the use of carefully machined acrylic prisms sandwiched between sheets of glass

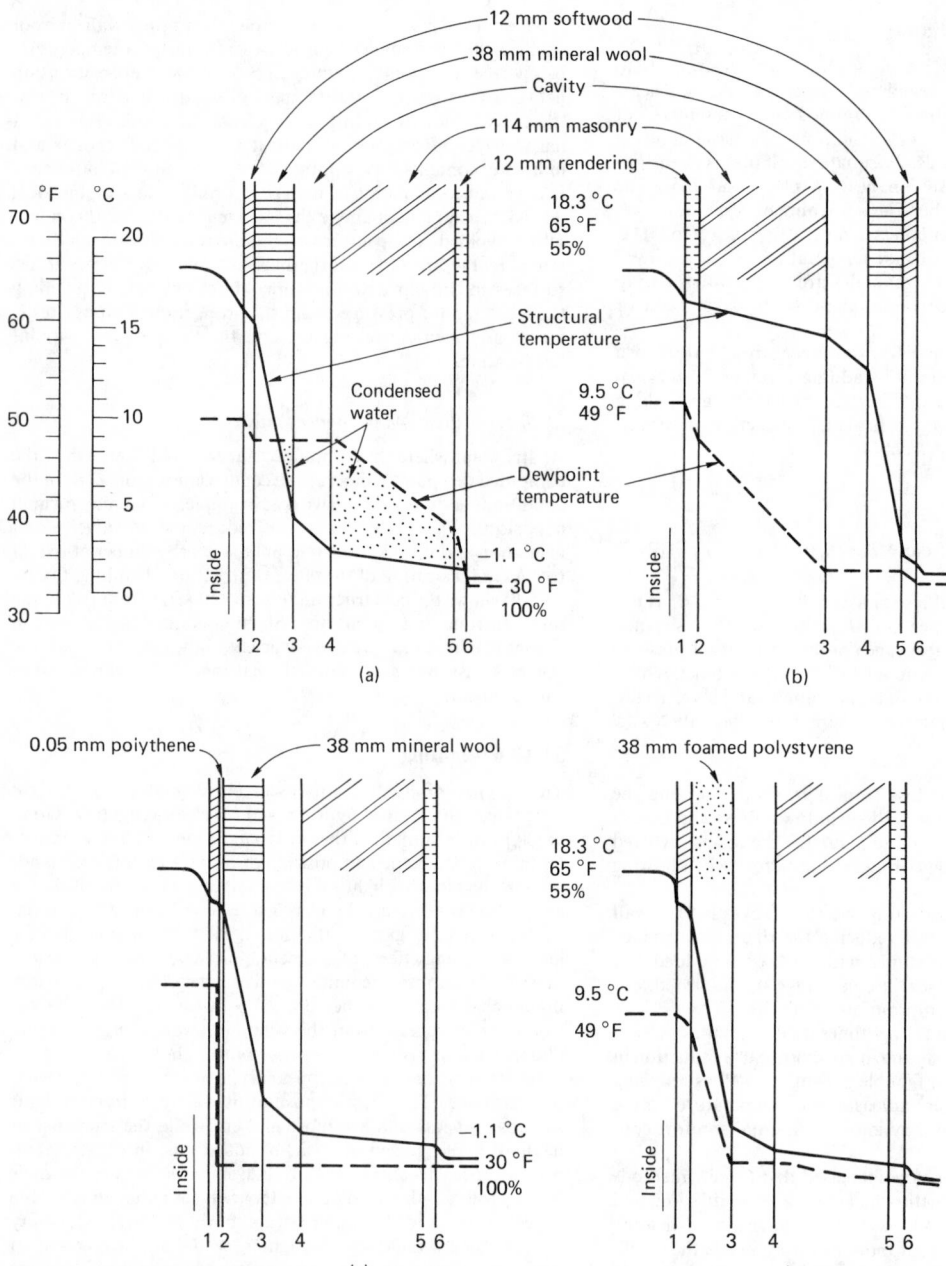

Figure 21.5 Prevention of condensation in wall cavities. (After Building Research Establishment (1979) *Thermal, visual and acoustic requirements in buildings*. Digest No. 91 (2nd edn). BRE, Watford)

attached to the exterior of the building as a form of shading. The prisms redirect the Sun's rays parallel to the ceilings within the building whilst blocking sky glare. The ceilings are specially shaped to divert the parallel beams and provide the uniform illumination on the working plane. Another approach uses heliostats to direct the Sun's rays into a hollow square acrylic pipe which, using the principle of total internal reflection in a manner similar to fibre optics, can feed light down the risers of a

building to illuminate the inside. This method has the advantage that as daylight fades, artificial light sources can irradiate the same piping.

The use of external automatic sunblinds has had limited acceptance, mainly due to high costs both initial and subsequent. Inevitably, the continued operation of electro-mechanical devices such as these, subject to external forces, is difficult. More promising is the development of special glasses similar to the familiar photochromic but whose light transmission may be varied reliably by the application of an electrical potential.

Quality of the electric light is as important as quantity and design should take into account: (1) brightness and colour patterns; (2) directional lighting where appropriate; (3) control of direct or reflected glare from light sources; (4) colour rendering; and (5) prevention of excessive contrast between adjacent areas.

The most common light sources are tungsten lamps and fluorescent tubes with a growing acceptance of high-pressure discharge lamps. Tungsten lamps are common in domestic and decorative installations, but are inefficient in their light output and are generally uneconomic for the lighting levels required in most modern buildings. However, at a time when the new compact-source fluorescent lamps seemed likely to oust tungsten lamps even in the home, a specialized form – the low-voltage reflector tungsten halogen lamp – is growing in popularity, especially for display purposes. Their small size, longer life, improved efficacy and excellent colour rendition compared with standard tungsten lamps have tended to outweigh high capital cost and the inconvenience of the stepdown transformers and heavy cabling.

Fluorescent tubes are the most commonly used, but can take up considerable amounts of ceiling space. High pressure discharge lamps provide similar benefits of efficiency and long life, but more closely approach a point source, permitting greater freedom in ceiling design. The ability to accommodate an economic light fitting will depend upon the planning and structural grids. When these are not appropriate to the light fitting, the lighting system will be expensive in itself and may also cause extra cost in removing the unwanted heat.

The light fittings have to be spaced carefully to provide adequate lighting levels over the whole working plane. Due to the physical discomfort which can be caused by the brightness of the light source, careful attention must be given to the prevention of direct or reflected glare. Glare standards exist for most types of working environments and the glare characteristics of lamp fittings and control diffusers are readily available.

The rationale behind such lighting layouts has always been the ensurance of a high degree of uniformity so that any location of the working plane will be served adequately. The basic inhumanity of such schemes, together with the absurdity of lighting circulation spaces to the same level as the task, has resulted in the growing popularity of uplighting where the lighting plane is illuminated indirectly by light bounced off a reflecting surface, usually the ceiling. As in most spheres of life, high quality is difficult to reconcile with efficiency and the cost of a superior working environment is increased consumption, typically 16 W/m² for 400 lux on the desk. Possibly the greatest single factor behind the popularity of uplighting is the expansion of the use of VDUs and word processors where, unlike most other forms of lighting, a correctly designed indirect scheme can limit tiring and distracting reflections from the screen.

The varying colour qualities and corresponding luminance efficiencies of the available light sources have an important bearing, not only on the visual environment, but also on the degree of heating or air-conditioning that may be required. The colour appearance of a light source is always cause for much subjective judgement and prejudice. Daylight cannot be used as a reference value since its spectral composition shifts throughout the day. Indeed, what is wrong with light sources, the purists insist, is that their colour appearance does not noticeably change. Fluorescent tubes can now be had in a bewildering range of phosphors equally able to imitate tungsten lamps or cold north light. The triphosphor tubes now make it possible to have both excellent colour rendition and high efficacy. The most promising light source for commercial interiors is the high-

pressure sodium lamp, which is able to better the fluorescent tube on most counts. However, its colour appearance even in the de luxe form remains controversial.

In some buildings, the energy for lighting can be a substantial part of the total required for all purposes. Since most of that provided for light appears as heat the possibility exists of using this as a major, and perhaps the only, source of internal heating; alternatively, the extra heat load may prove an embarrassment to the air-conditioning system. In either case the lighting must be treated as an integral part of the total environmental design.

Having selected the most efficient light source and used it in the most effective luminaire, the remaining part of the energy equation is control of the running hours. In many situations, people switch lighting on but never off so the advent of remote controls providing automatic switching is beneficial. Generally, such controllers operate either on a time basis or in response to some local stimulus. Their switching programmes may be held in their memories for as much as a year ahead with all holidays and weekends catered for. The instructions in the form of codes are transmitted along dedicated hard wiring or even over the supply cables themselves to the luminaires which are equipped with decoders enabling them to respond to one or several instructions. Local overriders often in the form of hand-held infra-red transmitters enable the central instructions to be modified. Less extensive forms of automatic lighting control take the form of presence detectors which switch off after a preset period, as the result of high daylight levels, or in the absence of people. The detecting principle may be either acoustic or infra-red.

21.10.6.1 Lighting for various categories of building

Speculative offices. Such buildings are generally leased without lighting fittings to avoid inhibiting either the letting pattern or the tenant's partition layout. Where lighting fittings are supplied, the preference is often for surface-mounted hot-cathode fluorescent tube units with prismatic light controllers. Lighting levels are currently in the region of 400 lux.

Offices: purpose design. In keeping with the design standards recommended in the Chartered Institute of Building Services Engineers (CIBSE) code for interior lighting,[6] average levels of 500 to 750 lux are usual, depending on the task. Such levels using combinations of light controllers with 'batwing' and asymmetric distributions may be had for as low as 10 W/m² but at the cost of inflexible and regimented workstation layouts. It is now possible to simulate lighting effects by means of models and artificial skies but this is best used where the budget will permit the purchase of purpose-designed luminaires. Much interest is being focused on the introduction of high-frequency control gear for fluorescent tubes which, for example, would reduce the loss on a 1500 mm tube from 13 to 5 W with gains in freedom from flicker and with silent operation.

Offices: burolandschaft. The gentle modulation of light and shadow produced by uplighting is particularly apt for this form of office. Using either metal halide or high-pressure sodium discharge lamps, uplighting brings good colour rendition, high efficacy, low maintenance and lack of glare, either direct or reflected. It saves the cost of a discrete lighting circuit since uplighting is usually fed from the small power points installed in the floor. The design process involved in an uplighting scheme is still unfamiliar to many, being task-related rather than building-related, and this unfamiliarity has tended to limit its more general acceptance.

Hospitals. The difficulties of reconciling the lighting needs in wards of patients who may either be lying supine or sitting up in their beds has led to separate systems being installed. In the latter case, wall-mounted units are preferred and these are often incorporated into continuous horizontal trunking runs which may contain other services such as oxygen, sound broadcasting, nurse call systems, etc. The former requirement is met by fluorescent fittings generally of the suspended pattern. There are many specialized considerations, such as operating theatres and anaesthetics rooms where totally enclosed, noise-proof fluorescent fittings sealed into the ceiling structure provide general illumination whilst shadowless operating-table lighting fittings incorporating tungsten light sources produce intensities up to 10 000 lux in the operating area.

Housing. Whilst tungsten fittings are still the norm for the home, the advent of compact fluorescent lamps with their significant economic advantages and tungsten-like colour appearance may change this. More sophisticated forms of lighting control, such as touch dimmers and infra-red switching, are now available and are beginning to be installed.

Schools. Cost considerations usually dictate surface-mounted fluorescent fittings with prismatic light controllers with levels in the region of 600 lux. In rooms where the seating has a fixed orientation, directional fittings may be used.

Industrial buildings. When ceiling heights are below about 4 m, fluorescent fittings are still the most-used light source. Above this, high-pressure mercury or sodium discharge lamps in reflector fittings are used with a wide range of distribution curves, both symmetrical and asymmetrical. The colour rendition of mercury fluorescent, mercury halide or high-pressure sodium light sources are satisfactory, but care has to be exercised in machine shops because of stroboscopic effects.

Car parks. The majority of multideck car parks use bare fluorescent tubes in fittings with moisture-proof lampholders and glassfibre or PVC-coated bodies. In the larger open car parks, increasing use is being made of high mast lighting.

Museums and art galleries. The lighting of museums and art galleries should be designed principally to meet the requirements of conservation, display and specialized study. Apart from atmospheric pollution, the main destructive agents will be the ultraviolet and infra-red content of light. Natural light is the worst offender with discharge sources such as fluorescent tubes, with high-pressure sources coming second. All three require careful filtering before they can be used to illuminate any exhibits containing organic materials or pigments.

Even tungsten halogen sources are suspect because of ultraviolet energy. The usual formula is a blend of tungsten display fittings giving a restrained average illumination plus fluorescent tubes with ultraviolet filtering. Deterioration of organic materials is a product of the intensity of the harmful wavebands and the length of exposure. The use of presence detectors – which ensure that exhibits are only illuminated for the period when there are people to see them – would be of value.

21.10.7 Noise

The control of noise requires consideration of its nature, source and mode of transmission. Typically, the main problems are: (1) reduction of noise to an acceptable level for efficient working; and (2) effective noise barriers for privacy. Problems of sound insulation and sound absorption are involved.

The main source of external noise is air or road traffic; penetration is reduced by double glazing (cavity preferably not less than 200 mm), minimum window area and heavy wall construction. In extreme cases, windows must be kept permanently closed and the building air-conditioned.

Internally, structural walls and floors are generally of sufficient mass to provide effective barriers against airborne sound but impact sound is not reduced by mass alone and a resilient material must be added to provide adequate total sound insulation. The lighter building elements, such as suspended ceilings or demountable partitions, do not provide good sound insulation. Continuity of sound insulation, where it is required, is important; a sound-insulating wall would need to extend through the void above a suspended ceiling, for example, unless the ceiling is itself a good sound insulator.

The use of sound-absorbing surface materials and shapes is effective in reducing the ambient noise level and may be so successful in *burolandschaft* offices that a degree of manufactured ambient sound may be needed to mask and, hence, reduce the disturbance from local intermittent noise.

Appropriate planning and detailing of the building is vital to the elimination of noise problems and the establishment of privacy. Wherever possible, areas requiring low noise levels should be divorced from noisy areas such as plant rooms, loading bays and lift motor rooms. Many items of mechanical and electrical equipment produce airborne noise which can pass along air-conditioning or ventilation ducts which then require silencer units. Equipment located in occupied rooms must be selected with appropriate low noise characteristics; in certain cases, especially on high-pressure systems, secondary silencer units are required. Rotating or reciprocating plant should be isolated from the structure to prevent structure-borne noise or vibrations. The increase in plant noise within buildings is increasingly a factor in modern design, requiring specialist advice.

Rooms with a high level of sound within them do not require such a good standard of insulation from adjoining rooms of similar level, but low-tolerance rooms will require a high standard. Figure 21.6 gives an indication of sound reduction levels for different room tolerances.

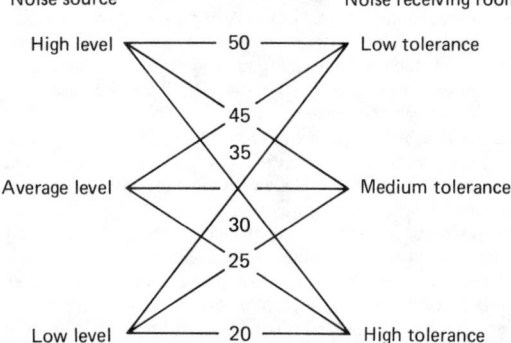

Figure 21.6 Sound reduction levels for various room tolerances. (After Parkins, Humphreys and Cowell (1979) *Acoustics, noise and building* (4th edn). Faber and Faber)

The sound reduction of dense walls varies with the sound frequency and with the weight of wall. At 550 Hz, the sound reduction is as follows:

Weight (kg/m²)	3	6	12	25	50	100	200	400	800	1000
Sound reduction (dB)	20	24	28	32	36	40	44	49	54	55

For a cavity wall, a reduction value corresponding to the combined weight of the two leaves is used and to this is added the additional assistance provided by the cavity which varies with its width as follows:

Air space (mm)	30	40	50	60	80–100	150	200
Added sound reduction (dB)	6	8	9	10	12	10	6

If the wall contains a door, the equivalent resistance is an intermediate value between those for the wall and door, dependent upon the relative decibel values and areas. For a brick wall of say 46 dB and door of 20 dB and the wall 10 times the area of the door, the equivalent sound reduction values (obtained from charts, e.g. Neufert[7]) is 30 dB. The full insulation value is obtained only if all holes, e.g. for services, are sealed; even very small openings such as keyholes and open joints represent serious sound leaks and must be taken into account in the design if good insulation is to be achieved.

21.11 Water supply, drainage and public health

21.11.1 Water supply

Potable water supplies are generally supplied from the local water undertaking's mains, the local water companies being required under the Water Act, 1945 and EEC directives to supply consumers with a potable supply. The conditions are based on the Model Water Byelaws of 1982, the purpose of which is to prevent waste, undue consumption, misuse and contamination. Water charges may be based on rateable value, assessed annual consumption or on metered consumption.

In the UK, storage provision for cold water for purposes other than drinking is normal and is provided for convenience in the event of mains failure. British Standard Code of Practice 310 schedules the amount of water storage required based upon occupancy (or number of fittings) and building type. Water storage is ideally located at roof level and below the available mains head to minimize operating and maintenance costs and to avoid pumping. A major revision to the various existing water services codes of practice is BS 6700.

It is increasingly found that water mains have insufficient head to deliver water to the upper levels of buildings without the aid of supplementary boosting. The method of boosting should take into account the location of storage, the possible need for any intermediate storage, pressure limitations or requirements in the distribution system, routing, quantities and usage of water. The two most common methods are direct centrifugal pumps serving high-level storage tanks or pneumatic pressure cylinders to boost the available mains pressure; the latter avoids the need for, but does not preclude, the use of high-level storage tanks. Break-cisterns are often required at ground level to cushion demand and very high buildings require break-pressure cisterns restricting gravity drops to about 30 m.

The distribution pipework generally separates cold- and hot-water service feeds and is preferably arranged to provide hot and cold water to the fitments at equal pressures. The routing should take into consideration maintenance, the requirements for draining down, protection against back siphonage and insulation against freezing and condensation.

Most large buildings have extended hot-water distribution systems served by a central heating plant which generally also provides the space heating. A central plant offers economies of scale and uses less fuel than a system of dispersed boilers. The boiler water is kept separate, the hot-water supply being heated by means of heat-exchange coils in calorifiers located in proximity to the outlets being served. Deadlegs need to be avoided wherever possible. Intermediate calorifiers can be located to act as break-pressure cisterns.

21.11.2 Fire installations

Water for fire-fighting purposes in buildings is separated from general water usage and is required for the hose reels, wet risers and sprinklers.

Consultations with the local fire authorities are required to ensure that storage and system duties are met. A number of packaged pumping units are available on the market for hydraulic hose reel installations. Wet risers are a fire authority requirement in tall and large-volume buildings. Sprinklers may be a requirement of the local fire authority or the building owner's insurance company. In the UK, most installations are required to comply with the 29th edition of the Fire Officers' Committee *Rules*[8] which have very specific water flow/pressure requirements and can involve large bulk water-storage requirements, dependent upon the fire risk hazard category. Specialist advice should be sought on these installations.

21.11.3 Water treatment

The growth of the electronics and pharmaceutical industries has expanded the need for water-quality levels far in excess of those supplied by the statutory authorities and special advice should be sought. In hospitals, additional chemical treatment may be required to reduce the rise of disease transmission through the water system.

21.11.4 Drainage

The aim of a well-designed building drainage, sanitation and rainwater installation is to convey foul waste and rainwater efficiently to the sewer or outfall without nuisance or risk to health and self-cleansing. The layout should be as simple and direct as possible and in accordance with the requirements of BS Code of Practice 8301:1985 'Building drainage', BS 572:1978 *Sanitary pipework*, and BS Code of Practice 6367:1985 'Drainage of roofs and paved areas'.

21.11.4.1 Design considerations

The practice of combining soil and rainwater pipes within a building is extremely unwise and the connection of the two systems, even with a combined sewer system, should be located externally, preferably at the last manhole before discharging to the sewer. Soil and waste stacks should be as vertical as possible with the minimum number of offsets. Particular care should be taken with discharges from kitchens, laboratories and disposal units. Separate systems should be provided for activities involving chemical and radioactive effluents. Ventilation pipes are required to maintain a balanced air pressure throughout the soils and waste system. All access locations for rodding should be reviewed in design and located to enable easy maintenance. Ground-floor fittings should be discharged direct to drains and separate from upper-floor fittings. Consideration should be given to draining basement levels via pumps to reduce the risk of flooding in the event of sewer back-up. In selecting pipework materials, consideration should be given to such items as noise, fixings, condensation and material damage in addition to the general material performance criteria.

All sanitary appliances need to be trapped to prevent sewer and drain smells entering the building. Precautions are required to prevent the seals being broken by siphonic action or plug pressure generated within an adjoining stack. Traps can be protected against these dangers by design or by the incorporation of secondary venting immediately behind the trap. Generally, the provision of sanitary appliances should accord with BS 6465:1984, Part 1.

21.11.5 Public health

The importance of providing a wholesome drinking water

supply and an efficient system of sanitation within buildings cannot be stressed strongly enough and improvements in the standards of installation and design must continually be sought to avoid the risk of infection and the creation of health hazards. The inter-relationship of these aspects with the other building services, particularly air-conditioning, is of growing importance as more is understood of the nature and transmission of diseases such as legionnaires. Specialist advice is available from the Department of Health and others on the precautions required.

21.12 Lifts, escalators and passenger conveyors

Many modern buildings are dependent upon lifts and thus demand high standards of performance and reliability from the drive and control systems. The advent of microprocessor controls has meant greater flexibility and quicker response to changing traffic conditions since a wide variety of inputs, such as car positions and car loading, even system failures or the number of people waiting at each landing, can be scanned many times a second. This continuous updating is used to secure the optimum lift performance. Equally, lift-drive systems are benefiting from the use of electronic speed-control techniques which will enable the robust and simple a.c. motor to replace the d.c. motor with its higher maintenance costs. Bulky worm gearing has been used traditionally to reduce the rotational speed of the traction motor but helical gearing with its superior mechanical efficiency and compact dimensions is now being considered.

With the issue of the various parts of BS 5655 for lifts and BS 5656 for escalators, the UK lift industry is now closely aligned with European standards. Only small national differences remain. Although not mandatory, these British Standards define standards of safe and practicable transport for buildings.

High-rise buildings may call for special solutions to the transport needs. One approach is the provision of 'shuttle' lifts where there are common liftshafts shared by two cars. One car covers the zone from ground level to an interchange floor or 'sky lobby', rising nonstop. The second car covers the zone from the sky lobby to the top floor served. A variation is where the shuttle cars are built as double-deckers, serving two levels at a time. Here, of course, two sky lobbies are required.

Undoubtedly, the type of lift attracting the most interest at present is the 'wallclimber' and its close relative, the panorama lift. The wallclimber lifts move on guides attached to the exterior elevation of a building and generally are found only in congenial climates. Panorama lifts resemble more closely a conventional lift but with the car projected through the shaft wall opposite the lift entrance. In both cases there is an emphasis on concealing mechanism and providing the largest practicable area of glass in the car construction. The current popularity of the atrium has added further impetus to the use of such lifts.

A lift pit is required at the bottom of every lift well of depth determined by lift speed; no occupied space is permitted beneath unless special provisions are incorporated to strengthen the pit bottom and lift safety gear. Lift motor rooms should be restricted to lift machinery and associated equipment. The lift well enclosure, pit and motor room form part of the building construction and may require particular construction as a 'protected shaft' passing between fire compartments.

Escalators have a much greater carrying capacity, but can only be used between two floors. Their use is mainly in high-flow areas with a limited number of floors. Capacity is varied by width and speed and can exceed 10 000 persons per hour.

Passenger conveyors are used basically for horizontal movement but increasing use is being made of them on shallow inclines to replace pedestrian ramps. They are used in transport terminals and interchanges.

21.13 Energy

The ready availability of cheap energy and the technology to control the internal environment meant that, for a long time, energy aspects were not a primary consideration in the design of buildings. Generally more effort was put into saving initial cost than into saving energy. All this changed radically when energy costs escalated in the mid 1970s. Today, energy aspects are a fundamental consideration in the design of buildings.

In the UK, analysis showed that buildings use half the nation's energy and that potentially 30% of this could be saved. In new building the potential saving is even greater. For example, a study of the energy used in hospitals[9] showed that savings of 50% and more could be achieved without any major change in hospital standards or building techniques.

Given a reasonable payback period, investment to reduce energy is sound economics. The problem is deciding what is an appropriate payback period. Various energy accounting methods exist. One approach is to compare the primary energy saved with the added primary energy needed to effect the saving. Alternative methods use a traditional financial approach, but these were sensitive to future fuel prices, inflation and interest rates. In the end, economic analysis is seen as a tool to sharpen judgement, particularly when comparing options within the resources available for investment. Other things being equal, priority should be given to those options which would be more difficult to introduce when the building is in occupation. Figure 21.7 is an interesting way of illustrating the combined impact of an energy-saving measure on cost of construction and cost in use. The most attractive measures reduce both initial and running costs of the building as a whole.

Heat loss considerations are important components of building regulations in most countries. In some cases these rely on specific thermal properties for the building fabric. More advanced regulations call for examination of the thermal performance of the building as a whole, thus encouraging the innovative skills of architects and engineers.

Existing buildings are being adapted to the new energy situation by energy conservation through insulation, the introduction of more sophisticated control systems or by the introduction of new plant.

In new construction, energy-conscious design is now the norm through building form and fabric and through the installations and controls provided. It is not enough simply to provide more thermal insulation; a whole set of measures is required to obtain the optimum solution.

The principal steps in producing an energy-efficient design are as follows:

(1) Computer analysis of local daily and seasonal weather conditions to optimize peak and long-term energy requirements.
(2) Analysis of building orientation, shape, height and construction to control heat gain or loss and the use of internal thermal capacity to reduce peak and total energy demands.
(3) Incorporation of heat conservation and recovery by transfer from points of surplus to points of need and reclamation of waste heat.
(4) Incorporating sophisticated control systems sensitive to variable building use and external weather conditions, to ensure that energy is injected only when needed.
(5) Where applicable using combined heat and power plants so that waste heat may be put to good use.

To save energy buildings are designed so that air-conditioning is unnecessary unless some special factor predominates such as wind, noise or fumes, or client requirement. Walls and roofs are used as the primary climatic modifiers with the environmental

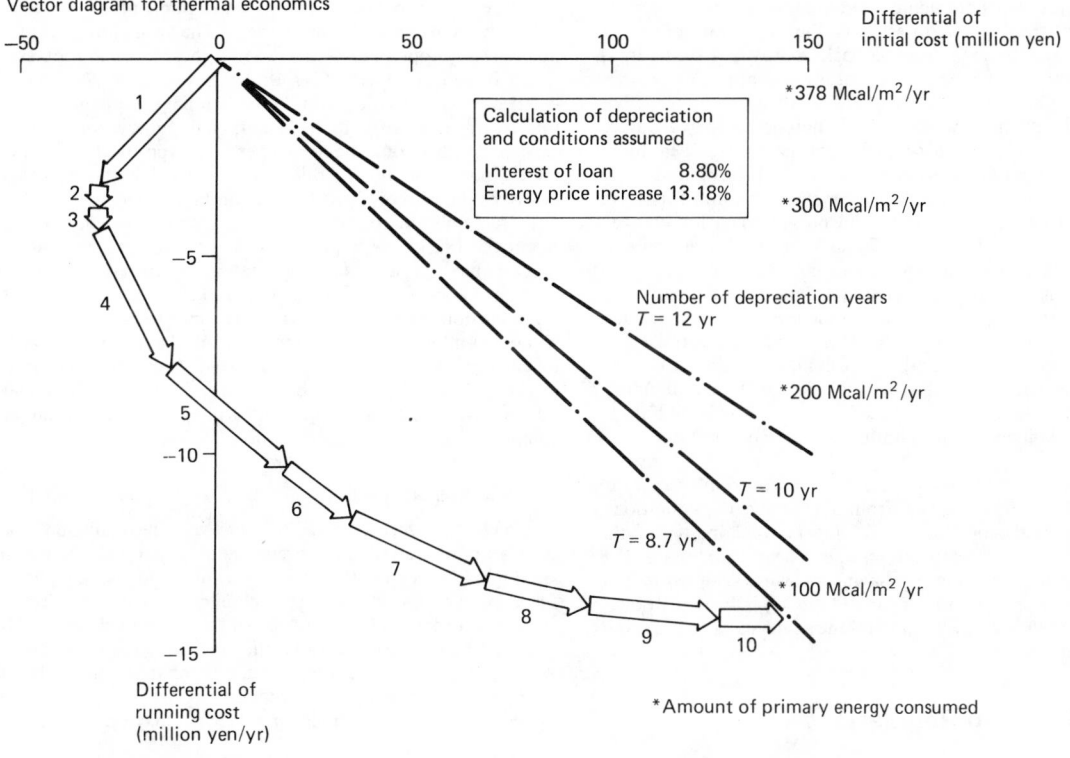

Vector diagram for thermal economics

Differential of initial cost (million yen)

−50   0   50   100   150

*378 Mcal/m²/yr

Calculation of depreciation and conditions assumed

Interest of loan   8.80%
Energy price increase 13.18%

*300 Mcal/m²/yr

Number of depreciation years
T = 12 yr

*200 Mcal/m²/yr

T = 10 yr

T = 8.7 yr

*100 Mcal/m²/yr

Differential of
running cost
(million yen/yr)

*Amount of primary energy consumed

(1) Architectural planning:
    (a) optimum building direction;
    (b) partial underground;
    (c) twin core;
    (d) reduction in floor height;
    (e) cubical structure;
    (f) 6 other methods.

(2) Reduction in power for plumbing facilities:
    (a) water-saving toilet;
    (b) local domestic hot water supply;
    (c) 7 other methods.

(3) Reduction in ventilation:
    (a) local ventilation;
    (b) double-staged use of conditioned air;
    (c) 2 other methods.

(4) Reduction in thermal load:
    (a) use of thermal wheel exchanger;
    (b) control on natural free cooling;
    (c) volume control on minimum outside air intake;
    (d) adoption of outside-air intake through underground pipes;
    (e) use of non-leaky damper;
    (f) 7 other methods.

(5) Insulation, shading device, and ventilation of building:
    (a) reduction in glazing areas;
    (b) use of insulated window shutters;
    (c) use of external louvre blinds;
    (d) adoption of double skin;
    (e) tilting outer glass of double skin;
    (f) 10 other methods.

(6) Reduction in power for fans and pumps:
    (a) adoption of VAV system;
    (b) adoption of large temperature differential system;
    (c) control on number of operating pumps;
    (d) 7 other methods.

(7) Reduction in lighting power:
    (a) task/ambient lighting;
    (b) dimming control on lighting in perimeter zone;
    (c) turning off light during lunch hour;
    (d) 8 other methods.

(8) Upgrade of efficiency:
    (a) adoption of heat reclaim system;
    (b) adoption of thermally-stratified heat storage tank;
    (c) optimal control on starting operation;
    (d) upgraded insulation around mechanical system;
    (e) 9 other methods.

(9) Active solar:
    (a) direct utilization of solar energy for air-conditioning
        and heating;
    (b) earth heat storage of solar energy;
    (c) 3 other methods.

(10) Reduction in electric power:
    (a) improvement of power factor;
    (b) control on number of operating transformers;
    (c) solar photovolatic power generation system without
        battery unit;
    (d) 5 other methods.

Figure 21.7 Economics of energy-saving devices in Ohbayashi's energy conservation building. (*Courtesy*: IABSE PERIODICA, IABSE, CH-8093, Zurich)

engineering systems providing fine-tuning to match local need. Elevational treatment is directed to the beneficial control of solar gain, the windows being set back to provide shading from the high Sun in summer, but permitting warmth to enter from the low Sun in winter. Glass remains an attractive material but is now less extensively used and is double- or triple-glazed. Passive solar heating can be an integral part of the heating in major buildings. There is now less general lighting and more local or task lighting with automatic systems of control sensitive to external daylight conditions and programmed for planned internal use. There is greater emphasis on variable air systems which use less energy and are more tailored to provide local ventilation needs. Heat reclamation techniques are now widely applied, particularly those based on the use of heat pumps and exhaust-air-inlet heat exchange. Heat from lights and office equipment is reclaimed, and 'run-around' systems are used to transfer heat from the hot, to the cool, side of a building. Energy-consciousness has extended from individual buildings to groups of buildings incorporating complementary energy requirements.

The impact on the structure arises mainly from changes in building form which result from the above considerations, coupled with the possible use of the thermal inertia of the structure to stabilize temperatures or as a heat store. Figure 21.8 illustrates the range of energy-saving measures adopted for the Tokyo Electrical Power Company (TEPCO), Ohtsuka Branch, Tokyo, while Figure 21.9 compares the achieved savings against those estimated.

12.14 Building Regulations

Building control in England and Wales is governed by the Building Act 1984 and the Building Regulations 1985. Scotland and Northern Ireland have separate systems of control.

Building work is defined as: (1) the erection or extension of a building; (2) the material alteration of a building; (3) the provision, extension or material alteration of a controlled service or fitting; and (4) work required on a material change in use. Certain small buildings and extensions and buildings for certain purposes are exempt from the regulations.

The regulations are silent on when repair work becomes subject to control. However, there comes a point in some cases when so much has to be done to repair or replace that the local authority could reasonably require the regulations to be applied.

21.14.1 Procedures

The new regulations contain an important innovation in that the proposed building work may be supervised either by the local authority or by an 'approved inspector'. Under local authority supervision, a further choice is available of depositing either full plans or a 'building notice' which contains much less information.

Full plans may be accompanied by a certificate of compliance with regulation requirements relating to structural stability and/or energy conservation. Such certificates can be given only by an 'approved person' and must be accompanied by a declaration that an approved insurance scheme applies. The details of who will qualify as an 'approved person' have yet to be resolved but it is expected that the relevant professional institutions will become the approving bodies for individuals wishing to undertake this work. When full plans are deposited, the local authority must pass or reject them within 5 weeks, or 2 months if the developer agrees. The 'building notice' option is simpler but is not applicable to shops or offices or any building work subject to the requirement for means of escape in the event of fire. The 'building notice' contains a short description of the work, a block plan and proposals for drainage. The local authority does not issue any approval but may wish to check work in progress.

The Building (Approved Inspectors) Regulations 1985 set out detailed procedures for private certification by approved inspectors as an alternative to local authority supervision. Under private certification, the developer and approved inspector jointly serve an 'initial notice' on the local authority. This describes the proposed works and can be rejected by the local authority only on certain prescribed grounds specified in the Approved Inspectors Regulations. The 'initial notice' must be accompanied by a declaration that an approved scheme of insurance applies to the work. If the local authority does not reject the notice within 10 days it is presumed to have accepted it without conditions. On acceptance, the local authority's powers to enforce the regulations are suspended and the approved inspector becomes responsible for inspecting the plans and building work and, on completion, issuing a final certificate of compliance.

21.14.2 Appeals procedure

When building work is alleged to contravene the regulations, the local authority can require its removal or alteration by serving a Section 36 notice on the building owner who has a right of appeal to a magistrates' court. The building owner may elect to obtain an expert's report from a 'suitably qualified person'. If accepted by the local authority, the Section 36 notice would be withdrawn and the building owner reimbursed his costs. If rejected, the report may be produced to the court and if the appeal is successful the building owner will recover his costs.

21.14.3 Approved Documents and mandatory requirements

The new regulations are much shorter and simpler since the technical details are now contained in a set of nonstatutory Approved Documents which give practical guidance on ways of meeting the regulation requirements. The obligation is to meet the requirement. The Approved Document may be used in whole or in part, or some other arrangement may be adopted provided the basic requirement is met. Use of the Approved Document would tend to be regarded as evidence of compliance with the regulations. If some other method is used, the reverse would apply and demonstration of compliance will be required.

Means of escape from fire, however, are covered by mandatory requirements although the local authority may agree to relax them in particular circumstances. The requirements are covered by the *Rules*.[10] A fire certificate is required for certain uses of building. Such buildings, in addition to means of escape, must include provision for fire alarms and fire-fighting equipment. Currently included are certain factories, offices, shops, railway premises, hotels and boarding houses. The Health and Safety Executive have similar responsibilities for special risk premises such as nuclear installations, buildings at the surfaces of mines and large chemical or petrochemical plants where the processes carried out affect general fire precautions.

Heat losses from certain classes of building are subject to minimum mandatory requirements. Four procedures for meeting the requirements are allowed:

(1) Specified insulation thickness.
(2) Specified U values to take account of full construction features.
(3) Calculated trade-off within glazed and solid areas and in the case of dwellings between solid areas and windows.
(4) Calculated energy use in buildings other than dwellings, allowing heat gains to be set off against heat losses.

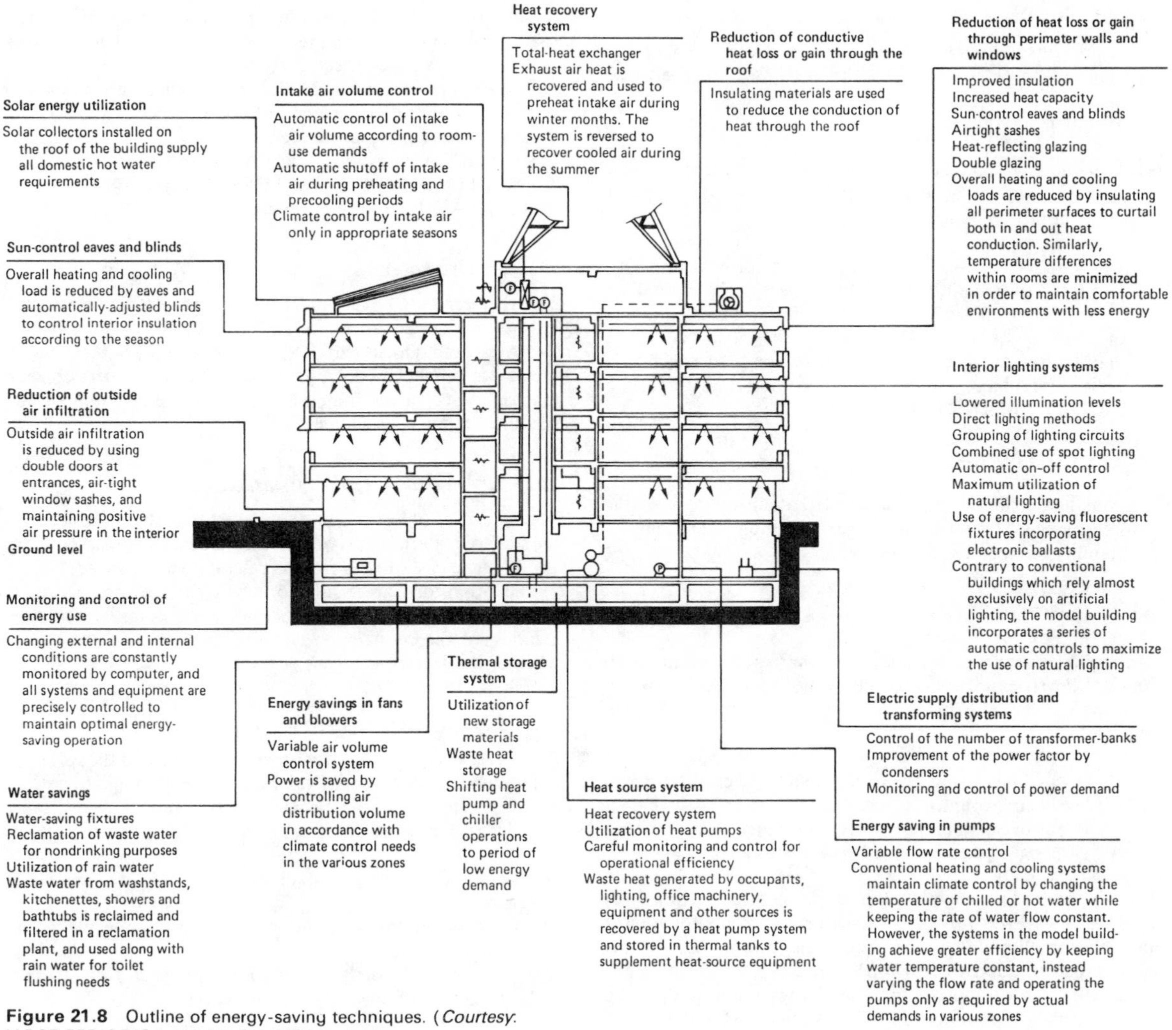

Heat recovery system

Total-heat exchanger
Exhaust air heat is recovered and used to preheat intake air during winter months. The system is reversed to recover cooled air during the summer

Reduction of conductive heat loss or gain through the roof

Insulating materials are used to reduce the conduction of heat through the roof

Reduction of heat loss or gain through perimeter walls and windows

Improved insulation
Increased heat capacity
Sun-control eaves and blinds
Airtight sashes
Heat-reflecting glazing
Double glazing
Overall heating and cooling loads are reduced by insulating all perimeter surfaces to curtail both in and out heat conduction. Similarly, temperature differences within rooms are minimized in order to maintain comfortable environments with less energy

Intake air volume control

Automatic control of intake air volume according to room-use demands
Automatic shutoff of intake air during preheating and precooling periods
Climate control by intake air only in appropriate seasons

Solar energy utilization

Solar collectors installed on the roof of the building supply all domestic hot water requirements

Sun-control eaves and blinds

Overall heating and cooling load is reduced by eaves and automatically-adjusted blinds to control interior insulation according to the season

Reduction of outside air infiltration

Outside air infiltration is reduced by using double doors at entrances, air-tight window sashes, and maintaining positive air pressure in the interior
Ground level

Interior lighting systems

Lowered illumination levels
Direct lighting methods
Grouping of lighting circuits
Combined use of spot lighting
Automatic on-off control
Maximum utilization of natural lighting
Use of energy-saving fluorescent fixtures incorporating electronic ballasts
Contrary to conventional buildings which rely almost exclusively on artificial lighting, the model building incorporates a series of automatic controls to maximize the use of natural lighting

Monitoring and control of energy use

Changing external and internal conditions are constantly monitored by computer, and all systems and equipment are precisely controlled to maintain optimal energy-saving operation

Energy savings in fans and blowers

Variable air volume control system
Power is saved by controlling air distribution volume in accordance with climate control needs in the various zones

Thermal storage system

Utilization of new storage materials
Waste heat storage
Shifting heat pump and chiller operations to period of low energy demand

Electric supply distribution and transforming systems

Control of the number of transformer-banks
Improvement of the power factor by condensers
Monitoring and control of power demand

Water savings

Water-saving fixtures
Reclamation of waste water for nondrinking purposes
Utilization of rain water
Waste water from washstands, kitchenettes, showers and bathtubs is reclaimed and filtered in a reclamation plant, and used along with rain water for toilet flushing needs

Heat source system

Heat recovery system
Utilization of heat pumps
Careful monitoring and control for operational efficiency
Waste heat generated by occupants, lighting, office machinery, equipment and other sources is recovered by a heat pump system and stored in thermal tanks to supplement heat-source equipment

Energy saving in pumps

Variable flow rate control
Conventional heating and cooling systems maintain climate control by changing the temperature of chilled or hot water while keeping the rate of water flow constant. However, the systems in the model building achieve greater efficiency by keeping water temperature constant, instead varying the flow rate and operating the pumps only as required by actual demands in various zones

Figure 21.8 Outline of energy-saving techniques. (*Courtesy*: IABSE PERIODICA, IABSE, CH-8093 Zurich)

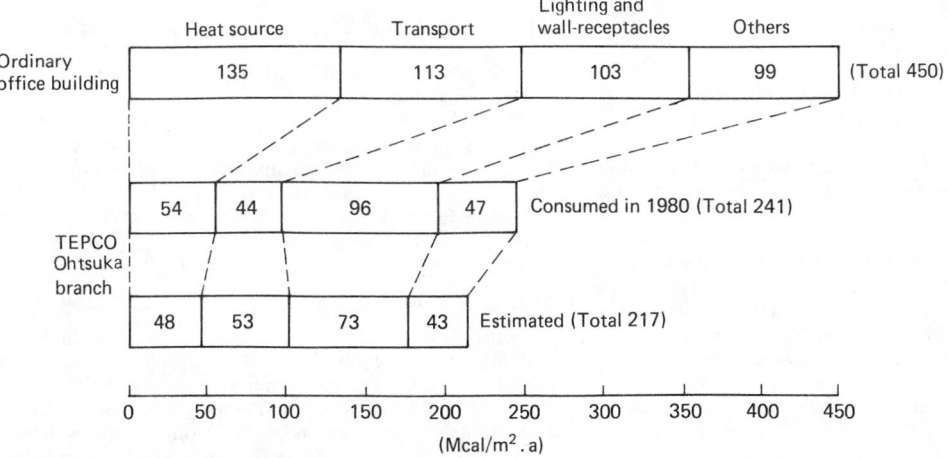

	Heat source	Transport	Lighting and wall-receptacles	Others	
Ordinary office building	135	113	103	99	(Total 450)
TEPCO Ohtsuka branch	54	44	96	47	Consumed in 1980 (Total 241)
	48	53	73	43	Estimated (Total 217)

0   50   100   150   200   250   300   350   400   450

(Mcal/m². a)

Figure 21.9 Energy consumption in the Tokyo Electrical Power Co., Ohtsuka Branch, Tokyo. (*Courtesy* IABSE PERIODICA, IABSE, CH-8093 Zurich)

21.14.4 Structure

There are three requirements: A1 concerned with loading; A2 with ground movement; and A3 with disproportionate collapse.

A1: Loading.

(1) The building shall be so constructed that the combined dead, imposed and wind loads are sustained and transmitted to the ground:
    (a) safely; and
    (b) without causing such deflection or deformation of any part of the building, or such movement of the ground, as will impair the stability of any part of another building.
(2) In assessing whether a building complies with (1) regard shall be had to the imposed and wind loads to which it is likely to be subjected in the ordinary course of its use for the purpose for which it is intended.

A2: Ground movement

The building shall be so constructed that movements of the subsoil caused by swelling, shrinkage or freezing will not impair the stability of any part of the building.

A3: Disproportionate collapse.

The building shall be so constructed that in the event of an accident the structure will not be damaged to an extent disproportionate to the cause of the damage. This requirement applies only to:

    (1) A building having five or more storeys (each basement level being counted as one storey).
    (2) A public building the structure of which incorporates a clear span exceeding 9 m between supports.

The Approved Document's guidance is in three sections: (1) Section 1 gives sizes for certain structural elements for houses and other small buildings, covering walls, floors, roofs, chimneys and strip foundations: (2) Section 2 lists codes and standards which are appropriate for all buildings and which may be used to satisfy the requirements of regulations A1 and A2; and (3) the final part deals with disproportionate collapse and quotes codes and standards which may be used in designing to meet the requirement of A3 for a building having five or more storeys, provided the recommendations on ties, and the effect of misuse or accident, are followed. Structural work of concrete is covered by BS 8110:1985, Parts 1 and 2. For steel it is BS 5950:1985, Part 1, and the accident loading referred to in clause 2.4.5.5 should be chosen having particular regard to the importance of the key element and the consequences of its failure; the key element should always be capable of withstanding a load of at least 34 kN/m^2 applied from any direction. British Standard 5628:1978 is quoted as the relevant standard for structural work of masonry.

By way of 'additional information', the Approved Document states that the structural failure of any member not designed as a protected key element or member, in any one storey, should not result in failure of the structure beyond the immediately adjacent storeys or beyond an area within those storeys of: (1) 70 m^2; or (2) 15% of the area of the storey, whichever is less.

Protected key elements or members are single structural elements on which large parts of the structure rely, i.e. supporting a floor or roof area of more than 70 m^2 or 15% of the area of the storey, whichever is less, and their design should take their importance into account. The least loadings they have to withstand are described in the codes and standards listed.

The Approved Document offers no specific guidance on widespan public buildings of less than five storeys. Clearly, such buildings do require special care in design and construction to reduce risks to human safety and the designer should consider carefully structural behaviour and consequence in the event of an unforeseen hazard or accident.

Structural matters are referred to additionally in other Approved Documents, e.g. in AD7 ('Materials and workmanship') and AD B2/3/4 ('Fire spread').

21.14.5 Fire spread

The requirements cover internal fire spread (surfaces) in B2, internal fire spread (structure) in B3, and external fire spread in B4. Approved Document B/2/3/4 describes how the requirements can be met by controlling aspects of the construction and materials used in the building, e.g. fire resistance and surface spread of flame characteristics.

21.14.5.1 Internal fire spread (surfaces)

The spread of fire over a surface is restricted by provisions for the surface material to have low rates of surface spread of flame, and in some cases to restrict the rate of heat produced.

The provisions are made for walls and ceilings and vary according to the use of the building or compartment, the location of the room or space concerned, and in some cases whether the surface is a wall or ceiling.

21.14.5.2 Internal fire spread (structure)

Premature failure of the structure is prevented and the spread of fire within a building restricted by specifying minimum periods of fire resistance for the elements of structure. Fire resistance includes requirements for one or more of the following:

(1) Resistance to collapse (stability – applicable to load-bearing elements).
(2) Resistance to fire penetration (integrity – applicable to fire-separating elements).
(3) Resistance to the transfer of excess heat (insulation – applicable to fire-separating elements).

Structural stability under fire conditions necessitates consideration of the following aspects of design:

(1) The coincident effects of dead, imposed and wind loads.
(2) The effect of heat on the structural elements.
(3) The provision of structural integrity to resist the effect of fire-induced movements.

The minimum period of fire resistance is set out relevant to the building use and depends upon the height of the building and on the size of the building or compartment. In basements, the provisions are generally more onerous in view of the greater difficulty of dealing with a fire.

21.14.5.3 Compartmentation

The spread of fire can also be restricted by subdividing the building into compartments of restricted floor area and cubic capacity, by means of compartment walls and compartment floors. The degree of subdivision depends upon the use of the building and in some cases its height. In single-storey buildings, the life risk from fire involving the whole building is generally less than that for a multistorey building. Thus, compartmentation in single-storey buildings applies only to those with a significant sleeping risk. There are, however, provisions for compartmentation of most multistorey buildings. Other forms of compartmentation apply between adjoining buildings and between a small garage and a house to which it is attached or forms part. As the minimum period of fire resistance increases with size of building or compartment it may be advantageous to provide compartments of smaller size than specified. Similarly, where a building is used for more than one purpose, it may be desirable to separate by compartmentation the use which has the more onerous fire-resistance requirement.

In order for compartments to be effective, junctions between the different elements enclosing the compartment must be protected. Similarly, any openings connecting one compartment with another should not present a weakness. Any spaces connecting the compartments need to be protected to restrict fire spread. These are termed 'protected shafts'.

21.14.5.4 Concealed spaces and fire stopping

Hidden voids provide a ready route for smoke and flame spread, e.g. above a suspended ceiling or in a roof space. As any spread is concealed, it presents a greater danger than would a more obvious fire weakness in the construction. Provision is therefore made to restrict the hidden spread of fire in concealed spaces by closing the edges of cavities, interrupting cavities which could form a pathway around a fire barrier and subdividing extensive cavities. Similarly, pipe or cable penetrations through a fire protection building element should be sealed appropriately.

21.14.5.5 External fire spread

The construction of external walls and the separation between buildings to prevent external fire spread are closely related, and many of the provisions specified are related to the distance of the wall from the boundary.

Whether a fire will spread across an open space between buildings, and the consequences if it does depend on: (1) the size of the fire; (2) the risk it presents to people in the other buildings; (3) the distance between the buildings; and (4) the fire protection given by their facing sides.

There are provisions limiting the extent of openings in external walls to reduce the risk of fire spread by radiation. Less onerous provisions for separation apply where compartmentation exists. Provisions for roof construction and coverings vary with the size and use of the building and the proximity of the roof to the site boundary.

21.14.5.6 Structural integrity of the building as a whole

The required level of performance will be met if the guidance in AD A1 and AD A2 and the *Guidelines*[11] are followed.

21.14.5.7 Varying the provisions

Where the fire provisions are thought to be unduly restrictive, the fire safety of the building as a whole may be considered and guidance is given about varying the provisions. Factors to be taken into account include: (1) whether the building is new or existing; (2) the construction and fire properties of the materials; (3) fire hazard and fire load; (4) space separation from boundaries and other buildings; (5) means of escape; (6) ease of access for fire-fighting; and (7) the provision of any compensatory features such as sprinklers or other automatic fire-detection systems. For example, the maximum compartment size in shops may be doubled when a sprinkler installation is fitted.

Shopping centres pose special problems and alternative measures and additional compensatory features to those set out in the Approved Document may be appropriate. These include:

(1) Unified ownership and control of fire prevention measures.

(2) Adequate means of escape and smoke control.
(3) Sprinkler protection of all areas of fire load.
(4) Automatic fire alarm system.
(5) Good access for fire fighting.
(6) Materials of limited combustibility and fire spread.
(7) Generally 2 h fire resistance of structure (4 h in basement).
(8) Floors generally of compartment construction.
(9) Walls between shop units of compartment construction.
(10) Compartmentation between large shop units (over 3700 m²) and a mall, and between opposing shop units (over 2000 m²) and a mall. Fire shutters may be used for this purpose.

The above items are not exhaustive but draw attention to the need to consider proposals as a comprehensive fire-safety package.

21.14.6 Other Approved Documents

Other Approved Documents are as follows.

C1/2/3	'Site preparation and contaminants'.
C4/	'Resistance to weather and ground moisture'.
D1–	'Cavity insulation'.
E1/2/3	'Airborne and impact sound'.
F1	'Means of ventilation'.
F2	'Condensation'.
H1/2/3/4	'Drainage and waste disposal'.
J1/2/3	'Heat-producing appliances'.
K1/2/3	'Stairways, ramps and guards'.
L2/3/4/5	'Conservation of fuel and power'.
Part 7	'Materials and workmanship'.

21.15 Building security and control

With the increased size and complexity of buildings, systems designed to monitor and control the mechanical and electrical installation, fire protection and escape, burglary, assault and emergency communication have become very important.

In tall buildings, and other major complexes, the most important security requirement is the fire-safety system. In addition to the structural precautions of fire protection and compartmentation, special systems are required to monitor and control: (1) fire detection and suppression; (2) movement and protection of people; (3) smoke control including pressurization and barriers; (4) safe places of refuge; and (5) emergency arrangements and communication.

In major buildings, these arrangements are integrated with those required to monitor and control the heating, ventilation and air-conditioning systems and other aspects of security within a single electronic system. The computer monitors all significant local conditions and appropriate action is taken. Such measures for security and control could, for example, bring in the use of: (1) heating, ventilation and air-conditioning plant and equipment to suit internal and external conditions or programmed requirements; (2) data collection for maintenance and resource management, particularly energy use and analysis, programmed responses to suit anticipated emergencies, e.g. defining smoke-free zones and escape routes in the event of fire; and (3) security interlocks, surveillance and access control.

The terms energy management system (EMS), building automation system (BAS) and building management system (BMS) are used to describe these systems. The EMS controls the environmental functions, the BAS the technical automatic controls, and the BMS includes such matters as status reports on environmental conditions, lifts and the location of people for security purposes. All these aspects influence and are influenced by the overall building designs.

The problems of security are by no means limited to major building complexes. Studies of housing estates with very different crime rates but with comparable density, size and tenant income, demonstrate clear relationships with specific building design characteristics, the nature of the surrounding areas and whether these are under the ready surveillance of the inhabitants. This has given rise to the design concept of 'defensible space' whereby tenants can act as their own 'policemen' simply because they identify with a particular space and easily monitor what is going on.

21.16 Materials

The principal materials used in buildings are concrete, steel, brick and masonry, and timber: each has its own developing technology. Other materials include aluminium, various alloys, glass, plastics and rubber. When used structurally, the essential properties concern strength, rigidity, durability and fire resistance. Relevant general properties concern hardness, thermal characteristics of insulation and expansion, weight, uniformity, appearance and workability. All these may be affected by changing temperature, humidity and weather. The choice of material for a particular building element will be determined by suitability for the intended purpose, cost and availability and compatibility with other materials.

21.16.1 Concrete

As a building and structural material, concrete is durable and relatively impermeable. It is readily available, and with the use of different cements, aggregates, forms and surface treatments it can provide a wide range of strengths, densities and finishes. Its mass automatically provides good airborne sound insulation and high thermal capacity, and with appropriate constituents it is resistant to chemical attack.

Lightweight concrete, in both structural and nonstructural elements, is used when weight is at a premium or when its better fire and thermal resistance is required. Precast applications include building blocks, fire casings, and floor, roof and cladding units. It is used *in situ* for floor slabs, screeds and generally for reinforced or prestressed structures. Its reduced stiffness should be allowed for in design.

Specially dense concrete is used as kentledge or for radiation shielding.

Normal-weight concrete is used unreinforced in mass foundations, gravity retaining walls, screeds and blinding, and precast as paving flags, kerbs and building blocks. For structural work it is normally reinforced with bars of high-tensile or mild steel, fibre-reinforced (with steel, glass or plastic), or prestressed. Fibre reinforcement gives improved impact resistance and can allow thinner sections; current applications are precast cladding, pipes and pile shells. Prestressing is used in precast floor units and in long-span beams as a means of controlling deflection or reducing member depth, and has special application in hanging and transfer structures.

Concrete cladding to buildings is generally precast, and may be nonstructural or structural. Panels may incorporate windows or doors, and may be of sandwich construction with a layer of thermal insulation. Many surface finishes are available including exposed aggregate, mosaic or tiles, and profiles may be plain or sculptured. Great care is required in detailing the profile and surface finish to control staining and to ensure satisfactory weathering.

The choice between precast and *in situ* construction will depend on cost, speed, access and availability of labour. Precasting minimizes the on-site work, but requires good accuracy and suitable lifting equipment. *In situ* work requires more on-site labour, but can be speeded by using prefabricated formwork systems and concrete pumps.

Where reinforced concrete is exposed to the weather, to water or to the ground, durability must be considered. Protecting the reinforcement from corrosion requires an adequate cover of densely compacted concrete with sufficient cement in the mix to maintain the steel in an alkaline environment. Where this cannot be achieved, other measures such as coating the concrete surface, or using galvanized or stainless steel reinforcement, may be needed. Chlorides in concrete must be restricted as they promote corrosion of the reinforcement; they may arise from the use of marine aggregates or from the unwise use of certain admixtures.

At the time of writing (1986), several cases of alkali–silica reaction (ASR) are being reported. This is a reaction which occurs between the alkali in the cement and certain types of aggregate, leading to expansion and loss of strength. It appears to be restricted to high-cement mixes, but there are as yet no reliable tests for aggregates. Although this problem is rare, advice on the latest knowledge should be sought before committing the mix design on large or important concrete elements, particularly if the concrete will be exposed to weather or water.

Glass-reinforced cement (GRC) is a relatively new material which utilizes alkali-resistant glasses developed in the 1970s. A typical mix will contain cement, sand, glass fibre and water, with cement:sand-ratio of 2:1, a water:cement ratio of 0.3:1, and 5% by weight of glass fibre. Careful detailing is needed to accommodate initial drying shrinkage, which is higher than for normal concrete. When young, GRC is a tough material which can deform without breaking. As it ages and weathers, the toughness reduces to a stable level after 2 to 5 yr. Several methods of casting GRC are used, but the normal method is to spray the constituents on a mould to form a layer 6 to 10 mm thick. This is demoulded and carefully cured. Simple shapes can be formed by folding flat sheets before they harden; complex shapes require purpose-made moulds. Glass-reinforced cement shares many applications with glass-reinforced plastic (GRP) but the GRC is heavier, stiffer and has better fire resistance. The main applications are drainage pipes for use underground, permanent formwork for concrete, roof tiles, street furniture and cladding panels. Cladding panels can be formed as a single skin of GRC, or as a sandwich of two skins with insulation between.

21.16.1.1 Fire protection

Fire protection of concrete structures is based upon the provision of adequate thickness of construction and adequate cover to the reinforcement or prestressing tendons. Lightweight concrete has an improved fire resistance because of its better thermal resistance and with some artificial lightweight aggregates is virtually free of spalling during a fire.

21.16.2 Steel

In one form or another steel is extensively used in practically all buildings. As a structural element it is available in a wide range of section and composition to suit the particular requirements of stress, deflection, corrosion or jointing technique: (1) it is of high strength; (2) in itself, occupies little space and is prefabricated for easy and rapid erection on site; and (3) it readily lends itself to extension or alteration. Sections can be cold-curved to form arches or rings. It has two disadvantages: (1) fire; and (2) corrosion. Several methods exist to overcome its fire sensitivity – various coatings are applied to resist corrosion and some steels (Corten) can be left exposed without treatment. Castellated beams are useful to reduce deflection and to provide holes for the passage of services. Hollow sections, of various wall thicknesses, are used in tubular structures, columns, trusses, space frames. Combined sections are commonly used and composite action, via shear connectors, with concrete floor construction can be advantageous. Sheet steel applications include roof and wall cladding, ducting and in floor slab construction, acting compositely with *in situ* concrete topping.

21.16.2.1 Fire protection

Fire protection of structural steel has moved towards the use of boarded lightweight encasements, sprayed surface materials and intumescent coatings. *In situ* concrete casing tends to slow down the construction process, although this can be overcome by precasting the concrete surround leaving only the junctions to be dealt with on site.

Boarded systems. A variety of boards are available in thicknesses from 6 to 80 mm giving fire periods up to 4 h. They are generally manufactured from vermiculite or mica using cement and/or silicate binders and are particularly suitable for column protection and for 'all dry' construction.

Spray systems. A wide variety of lightweight materials are available generally based on vermiculite plus a binder (often cement) or mineral fibres. It is generally applied direct to the steel surface, although in some situations it is applied to an expanded steel lathing to form a hollow box protection. Fire periods of up to 4 h can be achieved; mesh reinforcement may be required for the longer periods.

Intumescent coatings. These thin film coatings or mastics swell under the influence of heat and flames producing an insulating layer 50 times thicker than the original. Fire periods up to 2 h can be achieved. A variety of products are available; not all are suitable for damp environments and only a limited number are approved for the fire protection of columns.

Fire-protection thickness. The thickness required for most steel members given in manufacturers' literature has been given by the Association of Fire Protection Contractors and Constrado.[12] This generally relates the thickness of protection to both the fire resistance period and the H_p/A value of the steel section, where H_p is the perimeter exposed to the fire and A the cross-sectional area. The lower this value, the lower the heating rate.

Structural fire engineering. Whilst it is convenient in most situations to adopt the regulatory approach to fire resistance, in special cases a more fundamental approach may be appropriate and acceptable to the authorities concerned. Fire engineering is directed towards a more accurate assessment of the fire protection required by considering the significance and severity of a real fire in the building and the response of the structure as a totality to it. The process involves the consideration of:

(1) The heating rate and maximum temperature in the compartment – related to the fire load, ventilation and insulation of lining materials.
(2) The temperature rise in the structural member – related to location, weight per metre and perimeter of the structural member exposed along with any fire protection applied and, in the case of beams, their height above the fire.
(3) The stability of the structure – related to the load applied, the grade of steel and the effects of any composite action, restraint, continuity and movement.

Such considerations are particularly applicable to buildings where the function and fire load are unlikely to change. Examples are: schools, offices, hospitals, sports stadia, public assembly buildings and transport terminals. The concept is most cost effective when the analysis leads to the acceptance of the bare structure without fire protection.

The fire load is a measure of the amount of material available to burn and is calculated as weight × calorific value of the contents and building materials used in each compartment. The resulting figure is often converted to the equivalent quantity of wood having the same total heat content. In calculating the heating rate inside the compartment, the available ventilation and insulating properties of the compartment envelope are taken into account. Well-ventilated fires are shorter and hotter and an insulated envelope retains the heat. On the basis of this information, it is possible to predict the likely heating cycle within the compartment which can then be related to the heating strength/deformation curves for the steel.

21.16.3 Brick and masonry

Brick and masonry have the advantages of a long heritage of experience and simple construction based on traditional skills; building plant costs are low, but the labour content is high and not always in sufficient supply. The common materials are bricks and concrete blocks of various types, finishes and strength, and natural or reconstituted stone. The main applications occur in load-bearing walls and piers, particularly in low- and medium-rise buildings, and as cladding. Internally they are used as the inner skin of cavity construction or as partitions and, within limits, may be used to brace framed construction. Reinforcement can be added in the horizontal joints to produce beam action or vertically, in piers, for tying purposes.

In general, the massive nature of these materials provides good sound and thermal insulation together with good compressive strength and durability. Movement due to shrinkage in the case of concrete bricks/blocks and expansion in the case of fired clay bricks, temperature and moisture change, and chemical action must be allowed for and provision made in design against progressive collapse. The design of brickwork has become more sophisticated both structurally and architecturally in keeping with the swing back to more traditional forms of construction.

21.16.4 Timber

The main advantages are: (1) it is readily available in a wide variety of types and section; (2) it is light in weight; and (3) it is easily worked, employing traditional skills. Typical applications are in floors, roofs, framing to light buildings, cladding, wall and ceiling construction and in surface finishings. It is often used in temporary buildings and for temporary works and formwork.

Size, form and consistency limitations have been reduced by the introduction of glued laminates and special fastening systems have made larger-scale structures possible. Combustibility, rot and insect infestation can be retarded by chemical impregnation while treatment with steam or ammonia gas introduces flexibility. Temperature and moisture movements remain problems. Fire resistance of exposed timber members can be assessed from the rate of charring. This varies with different kinds of timber but is generally about 0.6 mm/min on each exposed face.

21.17 Walls, roofs and finishes

21.17.1 External finishes, materials and weathering

The design of the external fabric requires a knowledge of the behaviour of the materials and elements of construction and includes consideration of weathering and water-shedding characteristics. External materials must be durable under the influence of climatic extremes and the local environmental conditions including wind velocity and prevailing direction, and whether coastal, urban, industrial or rural. The cost of maintenance, and accessibility for maintenance are also important considerations. The major functional requirements for the external wall include heat and sound insulation and damp-proofing.

Design elements range from screws to complete assemblies. Deterioration due to weathering may be aesthetic or functional and may be visible or concealed, and design details should be such as to avoid structural deterioration, especially in concealed situations, and should be designed bearing the possible colour change or staining effects of weathering in mind.

The weathering characteristics of the main materials used externally are described in detail elsewhere.[13]

Concrete finishes depend upon the moulds used, the material properties and surface treatment. Blemishes such as surface airholes cannot always be avoided and untreated smooth surfaces generally weather badly, although when, with hard concretes of plain or white mixes, the surface laitance is ground off and sealed, good results can be obtained. Patterning and texture can provide interesting finishes, as with rough board markings, or ribbed surfaces which may be hammered or tooled in various ways. Deep patterning can also be very effective. Exposed aggregate finishes are available from a wide variety of processes and aggregates, and these generally weather well.

Brickwork is traditionally used as external walling and a variety of colours and textures are available in facing bricks. Attention must be given to weathering performance, porosity and freezing effects, efflorescence, sulphate attack, etc. When associated problems are recognized, solutions are available. Details must be designed to accommodate relative movement of the brickwork and other building elements.

Timber and timber products may be used as cladding, but colour changes usually occur on exposure to light and water, which can also lead to damage such as splitting, warping and dirt penetration. Various preservation treatments are available, including the use of varnishes, synthetic resinous clear finishes, opaque paints and applied film overlays.

With the use of metals externally, their particular properties as to electrochemical corrosion in relation to weathering and the problem of bimetallic electrolytic action need to be understood. Aluminium, bronze and copper weather well and are used in roofing, cladding, window framing and flashing applications. Lead may be used in sheet form for special roof covering and more frequently for flashings. Zinc provides useful coatings and flashings. Outstanding durability can be obtained with the use of stainless steel, and low-alloy steels of good weathering properties are available.

Curtain walling can provide an economic form of cladding and glazing, with advantages of lightness, thinness (as affecting usable floor area), flexibility of fenestration and speed of erection without external scaffolding. It is provided in two main types: (1) unit assemblies in which self-supporting panels are prefabricated including glazing and solid infilling with interlock or lap joints; and (2) part assemblies in which frame members are erected and the glass and solid sheets added. Weather resistance and connection to the structure must be adequate to meet high local wind pressures and conditions of driving rain. Sealing methods include the use of mastics, gaskets, cover tapes and spring strips. Thermal movements may be very large since the panels have low heat capacity and respond rapidly to changes in temperature, giving rise to differential movements between one part of the curtain walling and another, and between the curtain walling and the structure.

Different types of glass are available including: (1) clear, coloured, or opaque; (2) heat absorbing, filtering or reflecting; (3) toughened, single or bonded into insulated sandwich construction.

A wide range of plastics for external use is available, with different resistances to ultra-violet light, temperature, water, oxygen, micro-organisms, atmospheric pollution and loading. These include PVC, used for rainwater goods, glassfibre-reinforced polyester resins forming sheet or shell products, polymethyl methacrylate providing transparent sheets of high strength and durability and the phenolic and amino resins for laminated sheets. Polymer films may be applied to other materials such as boards or metals to improve durability.

21.17.2 Floor, ceiling and wall finishes (internal)

Such finishes may be integral with the structure or applied. Type of usage, cost, chemical resistance, aesthetic requirements, fire resistance or 'spread of flame' requirements, maintenance, are some of the factors influencing selection.

Floor finishes integral with concrete rely on good workmanship: power-float finish, use of hardeners, dust inhibitors, waterproofers, or the application of granolithic concrete finishes to 'green' concrete. Timber and metal decking can also be in the 'integral' category. Applied floor finishes can vary from simple sheet or tile materials stuck (or laid loose in some cases) to the structural slab, with or without levelling screeds. Damp-proof membranes or vapour barriers may be required for slabs on the ground, depending on the type of finish to be applied and/or the groundwater conditions. Screeds may need to be of adequate thickness to allow for the running of service conduits, or thickened, reinforced and isolated by insulation in the case of floors to be heated or used for impact sound insulation. Raised floors to accommodate electrical or other services have become increasingly popular in modern office buildings.

Integral wall finishes result from the use of controlled shuttering on concrete work, fair-faced brick or blockwork or self-finished plane or profiled sheet materials. Applied wall finishes can be basically divided into wet and dry applications, plastering – by hand or spray – being typical of the former, and dry-lining such as plasterboards and proprietary insulation boards, acoustic finishes, being among the final finishes that may be required.

Integral ceiling finishes result from the use of untreated structural soffits such as concrete, metal or timber. Applied finishes may be divided into direct and suspended, the former, as indicated, being the application of wet or dry 'lining' or finishing direct to the structural soffit, such as paint, plaster, sprayed finishes, plasterboards, acoustic insulation boards or tiles. Suspended ceilings can be used to conceal structural members, to provide space for services, to reduce room heights for functional or aesthetic reasons, to provide a grid for flexible layout of partitioning. Such ceilings may be partly or wholly demountable for access to services or may be 'monolithic', e.g. plaster on expanded metal.

Building regulations must be referred to when considering internal finishes, requirements for resistance to fire being particularly stringent in areas such as staircases and circulation spaces, but also applicable to other areas and varying according to building use, area and volume.

21.17.3 Roofs

Roofs must keep out the weather, be durable and structurally stable, provide heat insulation in most cases and in certain others provide light and ventilation. The choice of roof structure will generally be determined by the general form of the building and the activities for which it is designed. Unlike floors, there is not usually any restriction on the depth of a roof and this gives a wide flexibility for economical and appropriate solutions. Sometimes the roof structure will be outside the main building.

A roof must carry its own weight together with imposed loads of roof finish and usually insulation, snow, the effects of wind, normal maintenance and often plant. It must resist excessive deflection or distortion which, though not leading to collapse, may damage decorations and services and if visible lead to lack of confidence and anxiety for the occupants. In accommodation for sedentary work or living, heat insulation, lighting, ventilation and sound insulation are important.

In general, the spacing of supports should be as close as possible consistent with present or possible future use.

Short-span roofs below 7.6 m are generally used for houses, blocks of flats, many multistorey buildings and some warehouses. On houses, roofs are often traditional in design. Sheet materials allow a lower pitch but the uplift effect of wind is important. Flats and multistorey buildings are normally roofed with a concrete slab similar to the floor construction.

Medium-span roofs 7.6 to 24 m are generally used for industrial buildings, warehouses, transit buildings, etc. Here, intermediate supports are often a nuisance. Appropriate systems are precast or prestressed concrete beams or steel trusses, lattice girders and portal frames.

Long-span roofs over 24 m are for exhibition halls, industrial buildings, leisure buildings, sports stadia and transport buildings. Many of these buildings require roofs which only keep the elements off the occupants. Systems would include steel lattice girders, space frames, roofs supported by suspended cables, prestressed concrete, arched construction, concrete folded plates and hyperbolic paraboloids.

Roof coverings include slates and tiles for houses, sheet materials flat or profiled, asphalt, felt, new materials based on synthetic rubber, plastics, sprayed-on materials and glass. It must always be remembered that provision must be made for roof drainage.

Fire spread is important in relation to roof coverings and is covered by BS 476:1975, Part 3.

Thermal insulation is often required to conserve heat in the buildings and is covered by the Building Regulations, but it is also important to reduce solar gain and avoid excessive expansion in the structure of the roof, which sometimes distorts the structural frame and outside walls. A reflective external finish to the roof also assists.

Condensation is a serious problem in roofs. Where thermal insulation is provided below the roof deck at ceiling level, cross-ventilation should be provided above the insulation. In some situations, a vapour check is necessary on the warm side of the insulation (the face nearest the inside of the building). Condensation is a subject on which a great deal of research has been carried out in recent years and deserves careful study.

21.17.4 Partitions

Partitions divide large areas into individual spaces for specific purposes such as stores, offices, etc. and separate circulation from working or living areas. The type of partition is determined by requirements of acoustic or thermal insulation, security, privacy, fire resistance and flexibility of planning. When the partitions are structural, brick or blockwork or concrete are commonly used; however, partitioning is generally kept separate from the structures.

Commonly used partitions, in increasing weight, are: (1) light framing with infilling of glass or building board; (2) plasterboard dry partition panels; (3) woodwool and compressed straw building slabs; (4) sandwich composite panels; (5) precast autoclaved concrete panels; (6) light to dense blockwork; and (7) brickwork and concrete.

21.18 Interior design and space planning

Space planning and interior design are the link between the design of the building itself and its eventual use by the occupants. In new building, space planning features at the initial briefing stages, and later in the completion and fitting out of the building for use. During the life of the building many changes are likely to occur in the utilization of the space provided and the building design must allow for this. A structural engineering input is important not only in new building design but also in advising on space utilization in existing construction.

In achieving an efficient, flexible and visually attractive environment the space planner has to balance the client's needs against the restrictions imposed by the nature of the building and the requirements of the statutory bodies. Structural aspects can have a profound effect on the success of the design and the following factors should be considered:

(1) Structural grid related to floor size and shape and intended use.
(2) Actual location of columns and beams.
(3) Form of the structure and its integration into the building fabric including services.
(4) Floor to ceiling heights.
(5) Ability to make satisfactory fixings to the structure.
(6) Floor loadings.

Modular co-ordination in building has not been completely successful. Planning grids have never entirely resolved the conflict between, for example, the sizes of workstations and cellular enclosures on the one hand and the sizes of building boards, ceiling tiles and partitioning systems on the other. Tartan grids are more economic for the internal fitting-out process as there is less need to cut partitions for ceiling tiles. Open-plan arrangements within most buildings do not produce much difficulty but modular co-ordination can become a problem in cellular office accommodation. Offices, for instance, are nearly always 10, 15 or 20 m² in area according to the status of the occupier. Building components are 1200, 900 or 300 mm. The most widely used structural grids are 6.0 and 7.2 m; 6.0 m can be subdivided to provide 4×1.5 m or 5×1.2 m window bays while 7.2-m grids give rise only to 1.2 m window modules. Although 1.2 m is very suitable for building boards and partitions it does not lend itself to the provision of individual offices – two window bays gives an office only 2.44 m wide or less. The 1.5 m module is superior in this respect and can accommodate smaller components such as ceiling tiles of 300 mm width; larger components of 1.2 m or 900 mm cannot be installed economically. Perimeter details can be critical: a continuous flush surface allows easy partition connection, avoids cutting around spandrel profiles and service runs, which is both costly and unsightly, and maintains continuity of acoustic insulation. It is more expensive to produce a flush window wall and it is not attempted in speculative buildings. The form of the building often dictates the space-planning principles. Few buildings today are of the traditional narrow form providing two rows of rooms divided by a corridor; most are wider and the most effective use of the space uses a mixture of cellular and open-plan accommodation.

For reasons of cost, buildings often have restricted floor-to-ceiling heights on, or slightly above, the statutory minimum. Alterations to the air-conditioning system are frequently necessary to service individual rooms or to deal with the 'wild heat' produced by the new technologies. Such alterations cause difficulties if the ceiling height is too low or if deep beams have to be traversed; the formation of bulkheads causes problems of another kind. The major problem for the interior designer today lies with the distribution of power, telecommunications and data cables. Speculative offices and older building stock rarely have sufficient capacity to deal with the plethora of wires and the new local area networks. The best solution appears to be the installation of suspended floors fed by extra vertical risers but again the floor-to-ceiling height must be sufficient to allow for this. If it proves impossible then the floor construction should allow for extensive trunking runs in the screed.

Fixings to an existing structure can sometimes prove difficult and in some cases it has not been possible to fix the supports for a plasterboard ceiling to a slab. The strength of the slab gives rise to quite a different set of problems. Even with the growth of information technology there still seems to be more and more paper produced and the filing and storage of bulk paper is dependent upon the floor loading. The mandatory minimum of $2.5 \, kN/m^2$ is rarely enough – especially when the weight of a suspended floor forms part of the live load. Even normal filing units require $5 \, kN/m^2$ and compactors, mechanical and power files, safes, etc. can create serious difficulties. It is often possible to position the storage on beam on lines or directly adjacent to the core walls where maximum strength occurs, but these limitations are detrimental to the functional efficiency of the layout. A value of $4 \, kN/m^2$ live load plus $1 \, kN/m^2$ for partitions, etc. should be achieved if possible.

Similar cautions should be voiced concerning the strength of roofs. It is common to site additional service plants in such locations and the strength of the roof construction is of prime importance.

21.19 Structure

The design of building structures is an iterative process by which the type, shape, dimensions, materials and location of the various structural elements are initially chosen as a first approximation; loads are then determined and the design developed by a process of adjustment and verification that structural performance will be satisfactory. The structure must also satisfy the functional needs of the building, site factors and the many technical requirements concerned with the safety, health, comfort and convenience of the occupants.

21.19.1 Structural behaviour

Assessment of structural behaviour must cover: (1) 'serviceability limit states'; and (2)'ultimate limit states'.

21.19.1.1 'Serviceability limit states'

These are concerned with acceptable vibration, horizontal and vertical deflections and structural cracking and the compatibility of these with the secondary elements supported by the structure, such as partitions, cladding, finishes.

21.19.1.2 'Ultimate limit states'

These are concerned with the provision of adequate reserves of strength to cater for variations in materials, structural behaviour, loading and consequences of failure. Partial factors are used for this purpose as follows:

γ_m allows for variations in strength and is the product of:
   γ_{m_1} to take into account the reduction in strength of materials in the structure as a whole, as compared with the control test specimen; and
   γ_{m_2} to take account of local variations in strength due to other causes, e.g. the construction process.

γ_f allows for variability of loads and load effects and is the product of:
   γ_{f_1} to take account of variability of loads above the characteristic values used in design;
   γ_{f_2} to allow for the reduced probability of combinations of loads; and
   γ_{f_3} to allow for the adverse effects of inaccuracies in design assumptions, constructional tolerances such as dimensions of cross-section, position of steel and eccentricities of loading.

γ_c takes into account the particular behaviour of the structure and its importance in terms of consequential damage, should failure occur. It is the product of γ_{c_1} and γ_{c_2} where:
   γ_{c_1} takes account of the nature of the structure and its behaviour at or near collapse (whether brittle and sudden or ductile and preceded by warning) and the extent of collapse resulting from the failure of a particular member (whether partial or complete); and
   γ_{c_2} takes account of the seriousness of a collapse in terms of its economic consequence and dangers to life and the community.

Relevant structural codes do not give values for the subcomponents (γ_{m_1}, γ_{m_2}, etc.) quoting only global values for γ_m and γ_f, which vary with the circumstances and load combinations being considered. However, the subcomponent definitions are useful reminders of the variables that need to be taken into account.

21.19.1.3 Hazards

Building structures may be subjected to such hazards as: (1) impact from aircraft or vehicular traffic; (2) internal or external explosion caused by, for example, gas or petrol vapour or by sabotage; (3) fire; (4) settlement; (5) coarse errors in design, detailing or construction; and (6) special sensitivities, e.g. as to acceptance of movement or differential movement or as to conditions of elastic instability, not appreciated or allowed for in design. Hazards involving risk of collapse or damage may also be introduced during design, construction or service. They derive from mistakes, ignorance or omission, inadequate communication or organizational weakness.

These hazards cannot be quantified except in special circumstances. However, for buildings of five or more storeys, the Building Regulations requirements concerning progressive collapse provide a general level of protection whereby the stability of a building is not put excessively at risk as a result of local structural damage arising from whatever cause. In cases of known risk the special requirements should be included in the design brief.

Many methods are available for confining the effects of accidental damage to the immediate locality of the incident. These include designing to accept the forces involved, the provision of alternative paths for the loadings, 'fail-safe' and 'back-up' structures. Research has been carried out on partial-stability conditions, whereby the remaining components of the building framework are capable of bridging or stringing over an area of total local damage by beam, catenary or membrane action.

Statutory requirements as to fire resistance and means of escape are devised to ensure continued stability for sufficient time to permit evacuation of the building and fire-fighting to protect adjoining property.

The introduction of new methods and materials requires careful consideration of the structural response to all the events that may occur during manufacture, construction and life of the structure, not just those idealized in design procedures, codes or standards.

It is very important to recognize that hazards exist outside the range of conditions normally considered in design; they must be eliminated or the structure designed so that their consequence is acceptable. The alternative is to accept that a particular hazard is so remote a risk that it can be ignored. This conclusion involves not just the statistical assessment of the hazard risk itself but careful examination of the consequence should it occur, since, though the hazard risk itself may be constant, the consequence in one type of building or structure compared with another may be catastrophically different. This applies not only to the protection of human life but also to particular functions, the continuation of which may be of paramount importance to the building user.

Four basic philosophies exist, aimed at reducing hazards or their consequence:

(1) The probabilistic approach.
(2) Discernment of proneness to hazard.
(3) Alternative paths and partial stability considerations.
(4) The hazard–consequence relationship.

These philosophies can be applied individually or in combination as a means of risk control or as a tool for risk comparison between alternative courses of action.

The probabilistic approach (1) recognizes the statistical hazard of adverse combinations of high load and low strength and provides a method of comparative measurement of safety. It is the basis of current limit-state codes. Proneness to hazard (2) recognizes the 'climate' in which errors, misjudgements or accidents occur and is an aid to discerning potential hazard, e.g. a new form of construction being hurriedly adopted under political and/or commercial pressure. Considerations of alternative path/partial stability aspects (3) of a proposed structure, encourages recognition and, hence, elimination of excessive sensitivity to local damage from unforeseen hazards. The hazard–consequence approach (4) seeks to identify the potentially serious consequence and directs resources and attention to the most vulnerable aspects determining future structural performance.

21.19.1.4 Structural tests

The behaviour of the structure is normally assessed by analytical methods but may also be estimated by tests on prototypes or models or by a combination of analysis and experiment. Prototype testing is sometimes used, for example, in precast concrete construction where the accuracy of the design assumptions may be in question or, in cases of repetitive application, where a better or more reliable understanding of structural behaviour may lead to economy. Model testing may be used to determine internal forces or stresses, and in special cases photoelastic analysis may be used to check complex local stress conditions, e.g. around service openings in major structural members.

21.19.2 Robustness

Recent experience has demonstrated the vital importance of the quality of robustness in determining the long-term performance and adaptability of buildings and their structures. These qualities can often be incorporated in the building structure with little, if any, extra cost if appropriate consideration is given in the early stages of design. These desirable qualities include:

(1) An ability to cope with hazards in an acceptable way, i.e. the building and its structure have been consciously designed so that damage would not be disproportionate to the cause.

(2) There is provision to eliminate, or reduce to acceptable levels, the risk or consequence of hazards not allowed for in (1) above.
(3) The structure is not sensitive to:
   (a) marginal departures from the design assumptions;
   (b) local defects or movement;
   (c) environmental change.
(4) The structure does not deflect or vibrate to an extent that alarms the occupants or disturbs intended function.
(5) There is an inbuilt ability to cope with remodelling or increased loading to suit changing use.
(6) The structure is readily buildable and not unduly dependent upon perfect compliance with the specification for workmanship and materials or future maintenance.
(7) The structure is such that early warning would occur before serious defect or collapse.

A very important aspect of robustness is the ability of the structure to cope with change. Very high demand and limited resources in the 1960s coupled with the Modern movement approach to functional design led to 'tailor-made' buildings with very little thought or scope for future change. It is now realized that functional requirements are changing rapidly in many forms of building. Ample spaces for services and the facility to change them are, for the building owner and user, welcome additions to the general robustness of the structure, as is the ability to accept higher loading and some cutting or rearrangement to accommodate additional lifts or service runs. Loading is particularly significant. In many buildings the cost of the structure is small in relation to the value of the finished building. The relative cost of providing for additional loading may be slight and a sound investment for an uncertain future. In the same way, ample plant room space and extra storey height can prove invaluable for the incorporation of additional building services.

A robust structure, however well designed, also needs a robust management and communication system for its production. Analysis of past failures confirms that they result primarily from lack of perception and poor communication rather than from insufficient knowledge of behaviour or circumstance. There are many links in the tortuous chain required to produce a robust building. One of the most significant is that between the designer and builder including those responsible for site inspection and supervision. Two particular aspects emerge from these considerations: (1) buildability; and (2) quality assurance. A readily buildable structure is an essential start to ensuring good quality. Designers should bear in mind the problems of construction and seek advice from a contractor during the early stages of design whenever possible. Conversely, supervisory staff should understand the structural behaviour assumed in design, not only of the structure as a totality on completion but also of parts of the structure during construction of the whole. They should direct particular effort to the more important aspects of quality control and seek to create the right climate on site – to do a good job even when not supervised and to get it right first time. All this is helped by the production of an 'inspection brief' prepared by the design team to ensure the effective deployment of the supervisory staff. Such a document should consolidate all the work leading up to the site start, define responsibilities, confirm and clarify relevant documentation and lines of communication, alert people to critical aspects and co-ordinate specialist inputs, all as a formal handover from those responsible for design to those responsible for construction.

21.19.3 Wind effects on buildings

The airflow around a building is affected by the adjacent land

and building complex and by the shape and size of the building itself, the roof type, the position and size of overhangs, the area and location of openings and the direction of the wind. Account needs to be taken of the loading effect of wind turbulence on the building as a whole (and perhaps during construction) and on components, the local wind environment in the vicinity of buildings, the general weather-tightness, natural ventilation and the air pollution around buildings.

Code of Practice 3:1967. Part 2, Chapter V gives the method to be used for assessing wind loads on individual buildings and takes account of the location of the building, the topography, the ground roughness, the building size, shape and height and a statistical factor related to the probability of given wind speeds occurring during the specified life of the building. For groups of buildings, particularly those including tall buildings, the environmental effects are frequently studied by means of model tests in wind tunnels.

21.19.4 Movement

The problem of movement in buildings is not so much the determination of its absolute value, but more that of achieving a compatibility of movement between parts. Without this, cracking and other disturbances are inevitable and are likely to recur even after repair; in such cases, the accurate assessment of movement serves little purpose. On the other hand, the simple recognition that relative movement will occur, coupled with a broad assessment of its significance, are essential first steps in establishing a compatible design in which the actual amount of movement is relatively unimportant.

When significant parts of a building tend to move appreciably relative to each other, they can be separated into independent blocks. Within each block differential movement will occur between elements, e.g. between the structure and partitions, and also between different parts or different materials forming a particular element.

Factors affecting the division of a building into blocks include: (1) differential foundation settlement due to load or soil variations, changes in foundation type or major intervals in construction; (2) longitudinal movements due to changes in temperature, shrinkage or prestress (immediate and long-term); (3) abrupt changes in building plan or floor or roof level; and (4) abrupt differences in structural stress. Within each block, the most common movement problems are: (1) partitions damaged by floor deflection or relative longitudinal movement; (2) crushing or buckling of cladding and partitions due to relative vertical movement between them and the structure; and (3) separation of surfacings due to movement relative to the backing material.

Structural joints include: (1) hinge details, to permit rotation; (2) expansion and contraction joints, commonly used with flexible seals; and (3) complete separation, e.g. where double columns are introduced. Joints between nonstructural elements allow for expansion or contraction or lozenging by providing appropriate horizontal and vertical gaps between the elements which are sealed with a flexible material or by cover strips secured to one side of the joint. Joints exposed to the weather require special attention in detailing and manufacture and in the choice of sealing materials. The open drain joint coupled with an effective air and water back-seal has proved a successful and reliable joint.

Wherever possible, restraints to longitudinal movement should be avoided by appropriate design and location of wind stiffening cores or bracing walls. For example, in rectangular buildings it is frequently appropriate to locate a service core at one end, providing rigidity in two directions, and a cross-wall at the other which provides the necessary torsional stiffness but permits free movement in the longitudinal direction. Temperature movements are minimized by keeping the structure within the insulation envelope.

21.19.5 Structural arrangement

A great variety of structural arrangement is used in practice depending upon the planning, functional, aesthetic and economic requirements of the building and site. Even for similar buildings, the relative priorities attached to the individual factors affecting structural decisions will vary depending upon the particular circumstances and the views of the client. General rules governing structural arrangement cannot be given, but in most cases the following principles are valid: (1) vertical loads should be transmitted along the shortest and most direct path to the supporting ground; (2) when minimum structural sections are dictated by nonstructural requirements, e.g. sound insulation, they should be deployed to gather extra load; (3) vertical load-bearing elements should be stacked directly over each other; (4) transfer structures, including vertical ties, should be used only when justified; and (5) structural layout should be regular to increase repetition of identical building components and improve construction rhythm.

The structural arrangement, construction method and material may be chosen on the basis of some overriding consideration, such as the provision for future alteration or extension, passage for services, speed of construction, availability of labour and materials or difficulty of access. Where alternatives are equally appropriate, the choice is generally made by cost comparison, but this must take into account all aspects affecting the total cost of the building including, where appropriate, running and maintenance costs.

The primary structural systems available for spanning vertical loads across space are; (1) the catenary, acting in tension; (2) the arch, in compression; and (3) the beam, in bending, item (3) being of most importance in buildings. Columns and walls are the commonly used members for vertical load support, and on occasion tension members are used to suspend lower work from high-level beam or cantilever construction. Frame structures are of two basic types, those in which horizontal forces are taken by shear walls or bracing and those in which the frames, comprising columns rigidly jointed to beams or slabs, are designed to accept horizontal as well as vertical loading.

Structures relying solely on frame action for stability become increasingly inefficient with height and reach a normal practical limit of about 15 to 18 storeys. For tall framed buildings, sway limitations (which must take account of increased lateral displacement due to the action of vertical loads on frames deflected by wind loading) are such that much larger and stiffer members than necessary for vertical loads alone would need to be used.

21.19.6 Resistance to vertical load

21.19.6.1 Columns and walls

Column and load-bearing wall positions are determined mainly by the building use. Where large clear spans are not necessary and regular and permanent space division is required as, for example, in multistorey flats, load-bearing walls are commonly adopted. Column spacing, in conjunction with the floor construction, will be affected by the available structural depth and the necessary provisions for the passage of mechanical and electrical services. Concrete columns may be of any reasonable shape; in tall buildings the section may be kept constant for construction convenience or reduced at upper levels to save usable floor area. When floor space is particularly valuable, spiral reinforcement or solid steel sections may be used or tension supports may be provided; the latter may be prestressed, in stages, to minimize the required sectional area and eliminate cracking.

21.19.6.2 Floors

In addition to their structural function, floors may need to provide impact and airborne sound insulation, thermal insulation and appropriate fire resistance depending upon their location in the building and the building type. For longer items, deflection and vibration will also be important considerations.

In situ concrete floors. Reinforced and prestressed concrete flat-slab construction has the advantages of: (1) minimum structural depth; (2) adaptability to irregular arrangements of columns or walls; and (3) not requiring a suspended ceiling (but if one is provided gives complete freedom for the passage of services). Solid reinforced concrete construction, 150 to 250 mm thick, may be used for spans up to about 7 m, or greater if post-tensioned. Punching shear at the columns, midspan deflection and the size and location of openings require special attention. When larger spans or weight reductions are required, the slab may be coffered to provide one- or two-way spanning: standard or purpose-made plastic, steel or fibreglass moulds are available and can produce a visually attractive self-finished ceiling. Alternatively, permanent cavity formers may be left in position. Beam and slab construction, at the expense of greater floor depth and slower construction, has the advantages of: (1) longer spans; (2) ready provision for large openings, e.g. for stairs and lifts; (3) adaptability to varying size and building shape; and (4) relatively light weight. It is most economic when large repetitive areas, or a heavy loading is required. *In situ* beam and slab construction may be the only valid method of construction for complex shapes and areas.

Precast floors. These have the advantages of speed of erection and quality and accuracy of manufacture; they are economic for medium to large spans particularly where layouts are straightforward and repetitive. They may be used in conjunction with steel or concrete frames in addition to load-bearing wall construction. The largest use of precast flooring is in the form of hollow or solid slabs, reinforced or prestressed, with widths varying normally from 300 mm to 2.7 m and up to 7 or 8 m in the case of large-panel construction where such slabs may incorporate openings, ducting and floating screeds. Precast slabs may be designed to act compositely with the supporting concrete or steel beams. For longer spans, single or double T-beams, which combine floor slab and beam are available; they are connected by welding, bolting or *in situ* jointing to provide secondary load distribution and equalize deflection.

Composite floors. These rely on the composite action between an *in situ* concrete topping and precast concrete soffit elements, which may take the form of precast concrete ribs, planks or slabs incorporating the tension element of the composite slab and having projecting reinforcement or other appropriate interface to ensure composite action. Alternatively, permanent steel shuttering may form the soffit. These floors are easily erected, do not need shuttering and provide the shallow depth of *in situ* slabs. A further example of composite action is that between concrete floor slabs, *in situ* or precast, and steel beam or frame construction.

21.19.6.3 Transfer structures

It is frequently found, in multistorey construction, that the special functions of the lower storeys require an arrangement of columns or bearing walls very different from that required for the efficient support of the superstructure above. A transfer structure is then required to transmit the typical floor column loads to the fewer but larger supports beneath. Often a very substantial structure involving storey height beams is required – this should be taken into account in the early stages of design

since it may provide a suitable location for plant. An alternative is to place the transfer structure at roof or intermediate upper-floor level and to suspend the lower structure by means of steel or prestressed concrete hangers.

21.19.7 Resistance to horizontal load

In low- to medium-rise buildings, the structural system is designed primarily to resist the vertical loads and is then checked for lateral forces which may be taken by moment-resisting frame action, braced frames or shear walls conveniently located around lift shafts or stair wells. Simple shear walls provide the necessary horizontal restraint for 'pin-jointed' frameworks which are often convenient and economic. A minimum of three bracing walls are required so disposed as to provide: (1) resistance in each of two directions at right angles; (2) an overall torsional resistance; and (3) minimum restraint to thermal or similar movement.

21.19.8 Multistorey construction

Steel and concrete are the primary structural materials and both can be used in a variety of forms. The choice of material and form will be dependent upon many factors particularly relevant to the building project, such as:

(1) Required speed of construction.
(2) Integration of services.
(3) Adaptability for future change of use.
(4) Lead times for delivery or construction.
(5) Fire protection.
(6) Impact on foundation feasibility and cost.
(7) Buildability and dependence upon workmanship and materials.
(8) Stability during construction.
(9) Road access for delivery and erection.
(10) Off site/on site labour availability.
(11) Dependence upon supplier.
(12) Impact on other trades.
(13) Securing early watertightness.
(14) Cost.

A key aspect is the provision for services either by complete separation from, or integration within, the structural elements. Integration involves deep, perforated structural elements and requires very careful planning and co-ordination. Separation requires zoning outside the structural elements, the configuration of primary and secondary beams to provide these obstruction-free zones is then a very important part of the structural concept.

The common forms of concrete and steel multistorey framed construction are given in Figures 21.10 and 21.11.

Structural continuity is easily effected and commonly adopted in *in situ* reinforced concrete construction, increasing both stiffness and economy. In precast or steel construction, continuity is more difficult to achieve and the theoretical savings in structural material are offset by the complications in construction and reduced future adaptability. Repetition, simplicity and continuity of erection are more important in achieving economy than marginal savings in material obtained by a too-tailored approach in design. However, with modern methods of fabrication, 'specials' can be introduced economically provided the needs of quantity, repetition and simplicity of erection can still be met. The optimum structure will also take into account its impact on other building elements and on the construction process as a whole.

A number of buildings require a special structural response outside the commonly used systems referred to above. Architec-

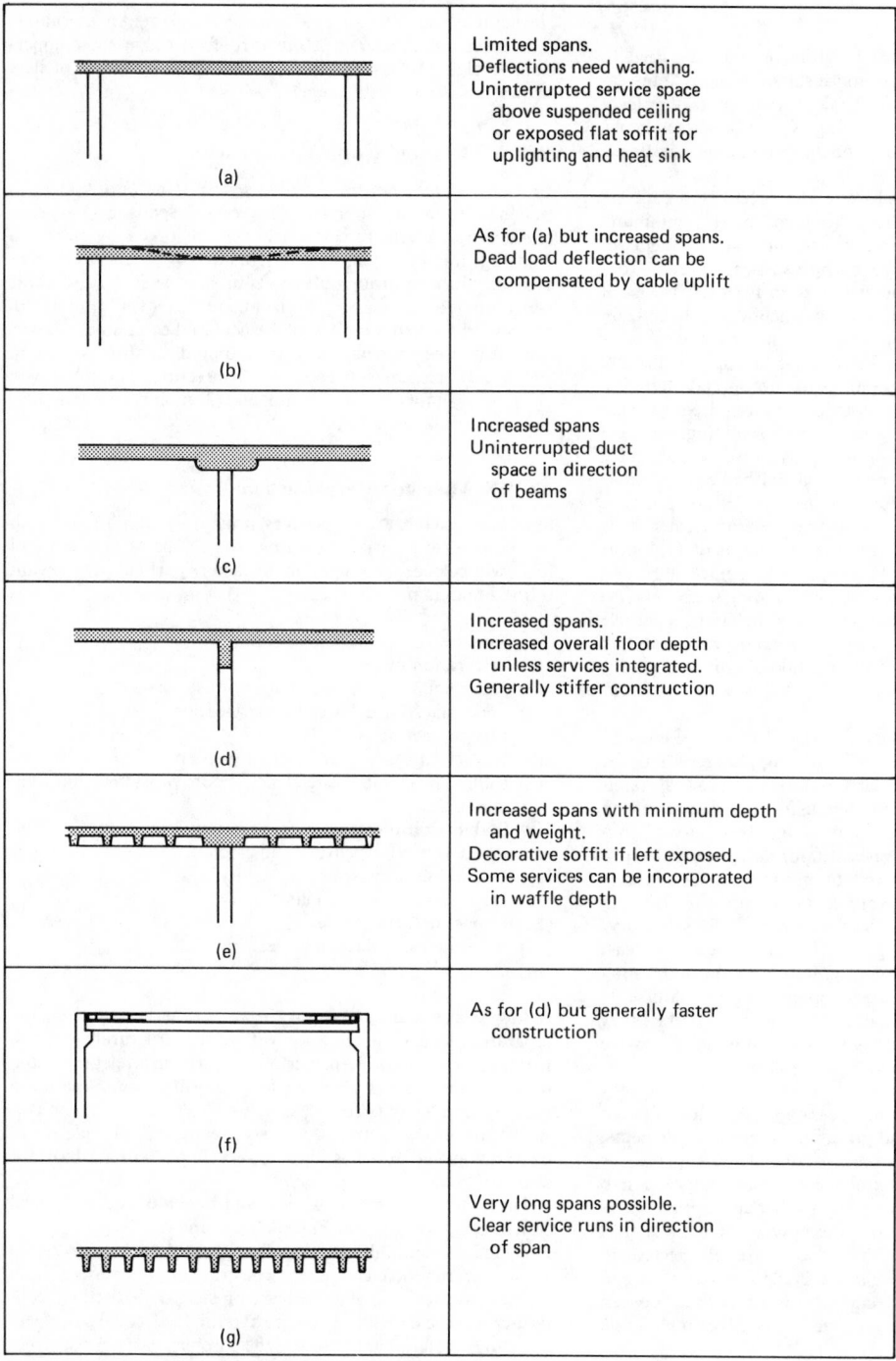

(a)	Limited spans. Deflections need watching. Uninterrupted service space above suspended ceiling or exposed flat soffit for uplighting and heat sink
(b)	As for (a) but increased spans. Dead load deflection can be compensated by cable uplift
(c)	Increased spans Uninterrupted duct space in direction of beams
(d)	Increased spans. Increased overall floor depth unless services integrated. Generally stiffer construction
(e)	Increased spans with minimum depth and weight. Decorative soffit if left exposed. Some services can be incorporated in waffle depth
(f)	As for (d) but generally faster construction
(g)	Very long spans possible. Clear service runs in direction of span

Figure 21.10 Concrete construction. (a) *In situ* reinforced concrete slab; (b) *in situ* prestressed concrete slab; (c) flat slab with shallow beams, reinforced or prestressed concrete; (d) *in situ* reinforced or prestressed concrete beam or slab; (e) waffle reinforced or prestressed concrete slab; (f) precast beam and slab; (g) precast T on double T

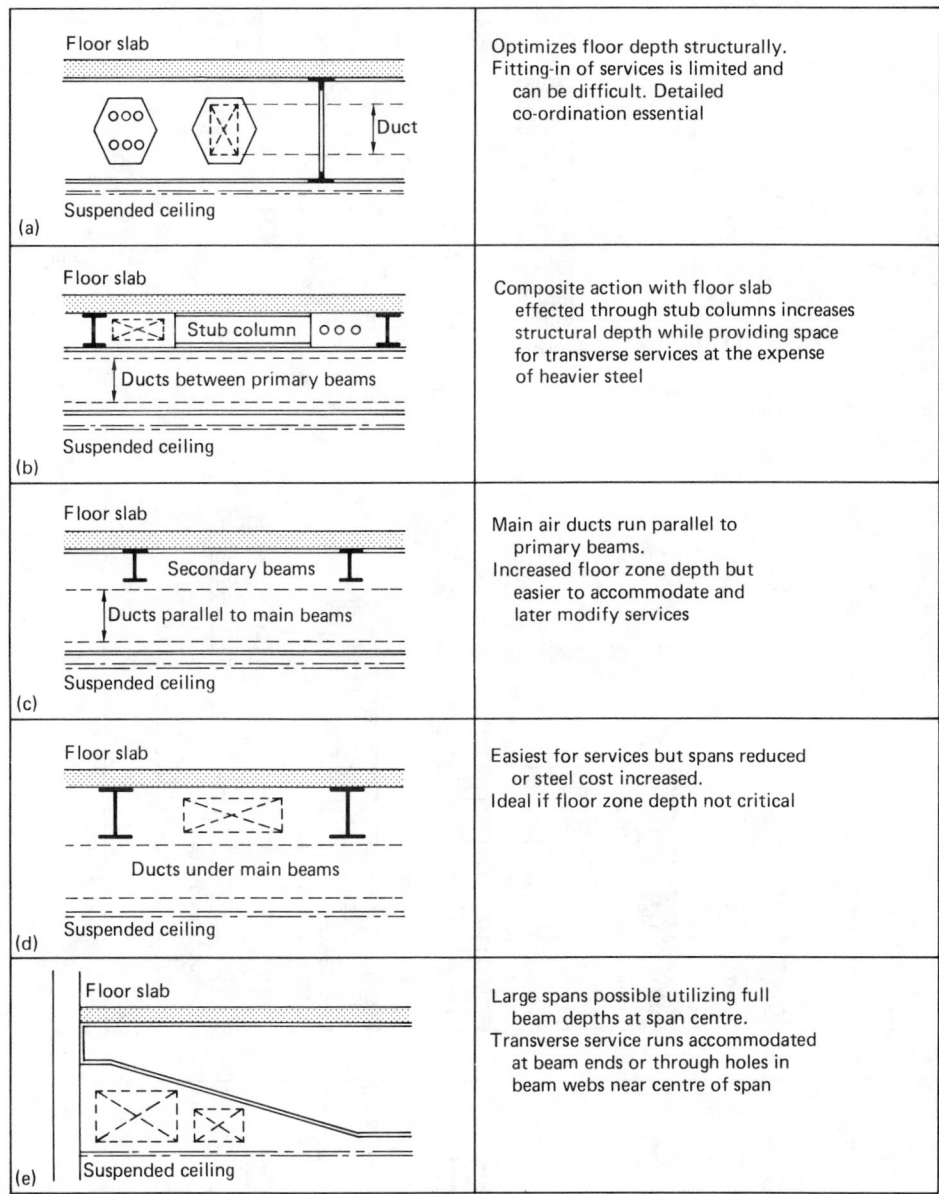

Figure 21.11 Floor slabs may be *in situ* or precast concrete and are frequently designed to act compositely with the beams. Profiled steel decking with through-welded stud shear connectors supporting an *in situ* lightweight or normal concrete topping can provide a fast and practical composite floor construction. (a) Castellated deep beams; (b) shallow beams with stub columns; (c) deep primary beams with shallow secondaries supporting slab; (d) shallow primary beams at close centres; (e) tapered beams

tural form, large spans or great height, or special physical criteria, e.g. seismic, may dictate a totally different approach.

21.20 Tall buildings

With increasing height, resistance to horizontal forces begins to dominate the design and may add substantially to the total cost.

In addition to structural safety, sway limitations must be satisfied in terms of horizontal accelerations as well as actual movement. In principle, the lateral resistance may be provided by frames in bending, braced frames or by shear walls, as in lower structures, but the greater magnitude of the forces and movements necessitates a more sophisticated approach. The various systems of resistance to horizontal forces in tall buildings are illustrated in Figure 21.12.

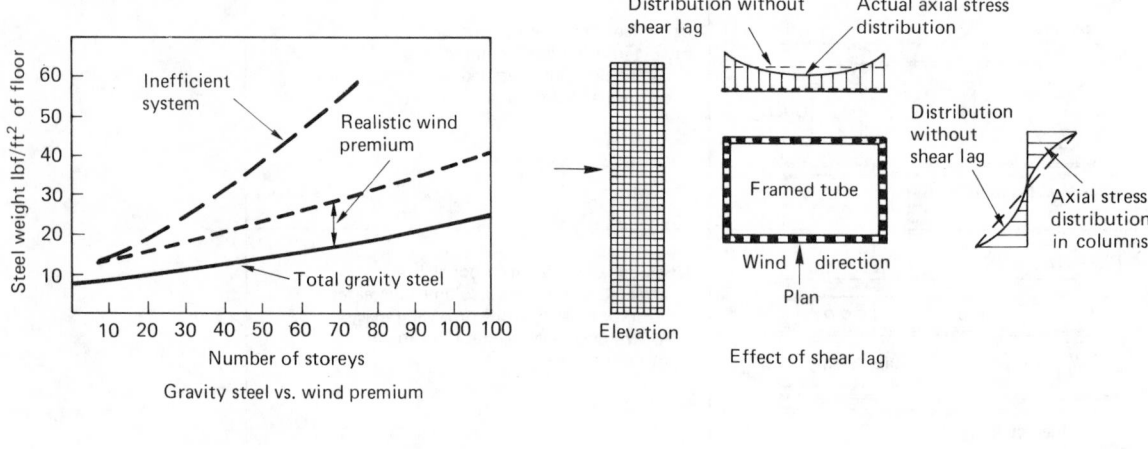

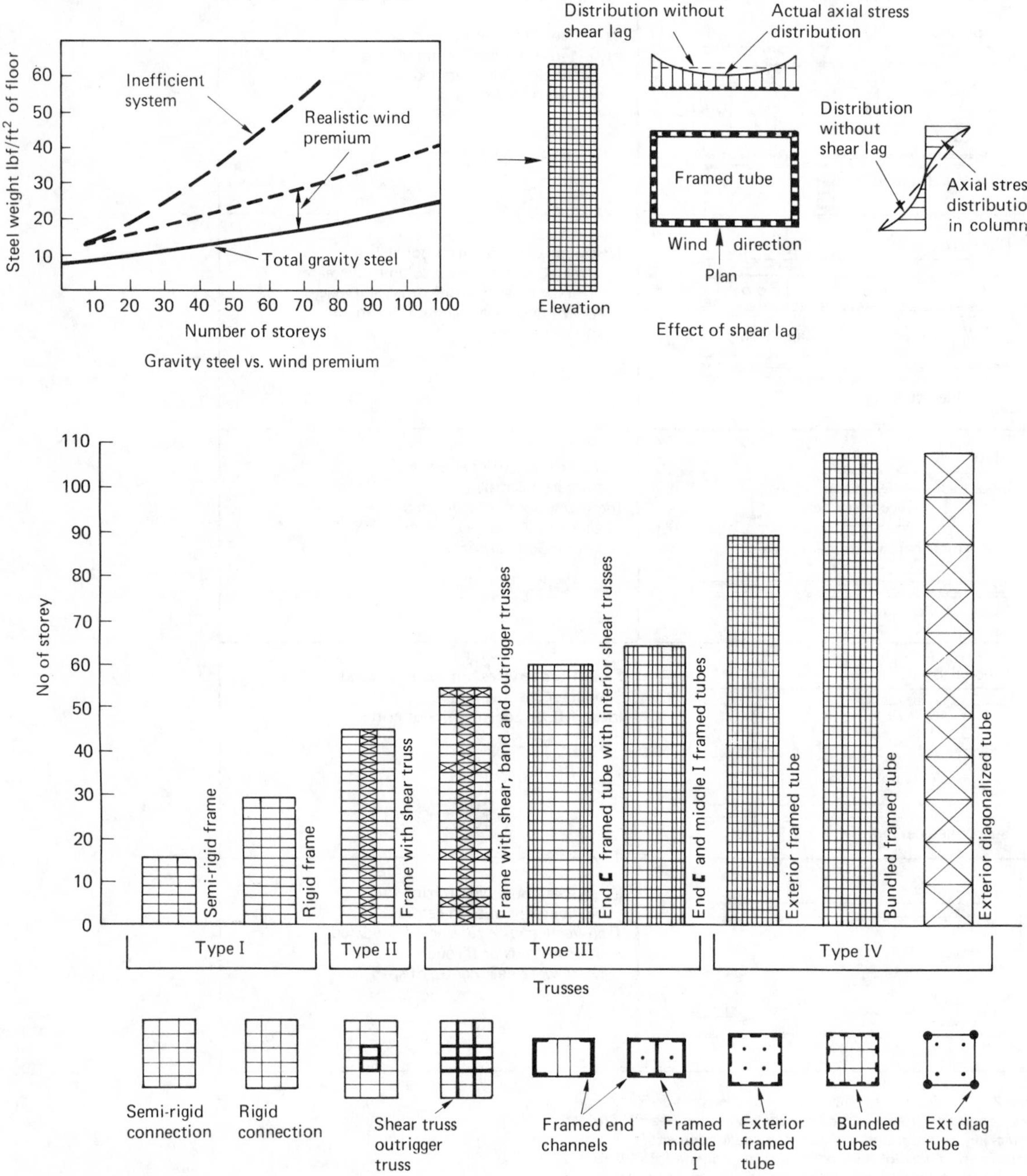

Elevation

Effect of shear lag

Gravity steel vs. wind premium

Trusses

Figure 21.12 Systems of resistance to horizontal forces in tall buildings. (After Iyengar (1972) 'Preliminary design and optimization of steel building systems', *Conference on planning and design of tall buildings.* ASCE/IABLE)

21.20.1 Frames in bending

Internal frames are comparatively inefficient and flexible as a result of the planning necessity for a wide spacing of internal columns and limited floor beam depth. On the other hand, exterior frames, formed in the plane of the external wall, may have closely spaced columns connected by deep spandrels. In this way, the entire perimeter of the building may be developed as a major lateral load-resisting system referred to as the 'boxed frame' or 'framed tube'. Subject to 'shear lag' considerations, the building walls act respectively as the webs or flanges of a box section cantilevering from the foundation. To allow for 'shear lag', two channel sections may be considered operative in place of the complete box. Deep spandrels, although advantageous, are not essential and apartment buildings of up to 46 storeys have been constructed in the US using part of the adjacent flat slab floor as the beam continuous with the closely spaced external columns.

21.20.2 Braced frames

Braced frames usually incorporate single- or double-diagonal braces or K-bracing within the beam and column framework and may be used internally, around service cores or in the external wall. When used externally, intermediate transfer structures may be incorporated to transmit a major proportion of the total vertical load to the corner columns. Such transfer structures may also support and, hence, separate, different configurations of internal supports required when the function varies between different vertical zones of the building. Such arrangements have a major impact on the internal planning and external appearance, and have important relevance to the planning and construction of very tall buildings using steelwork. Similar external truss action may be achieved in concrete by blocking out windows to form solid and continuous diagonal members or, for limited height, by using precast cladding forming a multiple-diagonal system. When they can be accommodated, internal trusses can bring into action lengths of external wall otherwise rendered ineffective by 'shear lag'. By alternating the plan position of such internal trusses, from storey to storey, the structural span of the floors may be reduced to half of the architectural planning bay by providing additional hanging supports from the trusses in the storey above.

21.20.3 Shear walls and cores

Shear walls may be internal or external or may surround internal service areas to form cores; their location and dimensioning are major design elements since they seriously impinge on internal planning and may affect external appearance. In the early formative stages of design, quick structural appraisal of alternative shear walls will be required followed by careful design and analysis of the final arrangement.

In office buildings, the service core – which includes lifts, stairs, ducts and toilets – can occupy 20% or more of the total floor area while fire and sound insulation require this area to be bounded by heavy wall construction. These conditions naturally lead to the use of the service core as a major vertical wind brace. However, away from the core area, open office space is generally required and even if partitions are used they would be demountable to allow for future alteration; internal bracing walls are therefore a planning impediment in offices and are generally avoided. External bracing walls, however, are often used, generally in conjunction with internal cores. In housing or hotels, partitions are normally fixed, need to be heavy for sound insulation and are regularly spaced; they therefore provide many convenient locations for internal bracing walls.

When shear walls alone are used, the general structural

requirements are: (1) at least three must be provided of which at least two must be parallel and widely spaced, to provide torsional resistance, with the third at right angles; (2) the centroid of the shear walls should be close to the centre of gravity of the loading; and (3) walls likely to need very large openings should be avoided if alternatives are available, since their stiffness and, hence, load-resisting contribution will be diminished substantially.

Walls with openings produce a stiffness intermediate between that of the total combined length acting as a monolithic wall and the sum of the stiffness of the parts acting separately, depending upon the relative size and location of the openings. Normal analysis assumes that all the shear walls or cores act from a completely rigid foundation such that relative rotation or vertical movement does not occur. Since even small relative movements could seriously invalidate the design, it is important that this assumption is realized in the foundation design or, if this is not practical or economic, the shear wall system should be designed to suit. In general, the total horizontal load is distributed between the shear walls in proportion to their relative stiffnesses taking into account any eccentricity of the applied load. The floor system then acts as a horizontal diaphragm equalizing horizontal displacement and rotation at each floor level.

Shear walls and cores are almost invariably constructed in concrete and are often slipformed. Precast large panels have been successfully used, particularly in high-rise blocks of flats, with the combined functions of load-bearing walls and vertical wind braces. The joints between the panels and the lintels over openings require particular consideration in the design of such structures.

The use of shear walls or cores is an economic and efficient method for resisting large horizontal forces but, in most cases, deflection limits their use to below 30 to 40 storeys. However, if the shear wall, and building, are shaped in plan along their length, a vertical shell or folded plate action could be developed permitting greater heights; otherwise, a 'boxed frame' or one of the combined systems described below is required to control deflection. Another limitation of shear walls, particularly if lightly reinforced, arises from the possibility of brittle failure which could make them unsuitable for seismic structures. However, by suitable framing around the shear wall, the necessary ductile behaviour can be obtained to absorb the considerable strain energy arising from an earthquake.

21.20.4 Combined systems

Internal cores may be used in conjunction with external moment-resisting frames so that the substantial overturning resistance of the façade frame is combined with the excellent shear resistance of the core to form a highly efficient total system known as 'tube in tube'. This arrangement still relies upon closely spaced external columns; when widely spaced external columns are required, a beneficial interaction between the core and the external columns can still be obtained by connecting the two with deep stiff beams rigidly connected to the core and located at convenient levels (roof and service floors). In this arrangement, the core continues to take the shear but the overturning resistance of the full building depth is called into play and deflection is reduced. Another advantage of this system is that it can help control the effects of differential expansion or contraction of the external columns.

Figure 21.13 illustrates the application of the bundled tube principle to the 110-storey Sears building in Chicago. The faces of each tube are stiffened by deep beams and columns in vierendeel action. The tubes are bundled to their maximum effect at ground level but are dropped off with increasing height to suit both structural and internal planning requirements. The internal framing at the junctions between tubes reduces shear lag in the long faces of the multiple tubes as shown in Figure 21.13(b).[14] Maximum effect would be achieved with diagonal bracing incorporated in the tube faces.

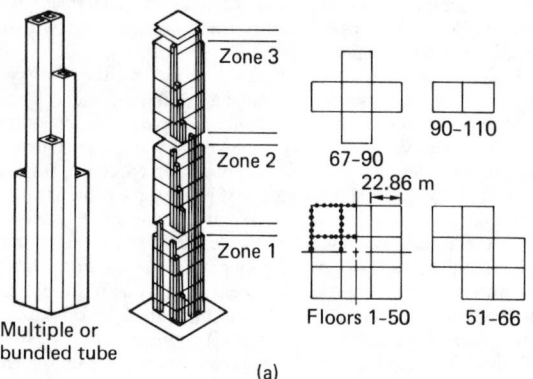

Multiple or bundled tube

(a)

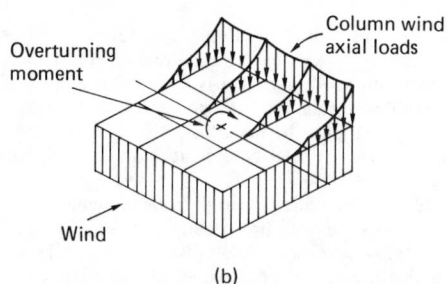

(b)

Figure 21.13 The Sears building in Chicago. (After Fischer (ed.) (1980) *Engineering for architecture.* McGraw-Hill, New York)

21.20.5 Vertical movement

Another aspect distinguishing tall buildings is the need to consider the possibility of differential vertical movements due to temperature or stress and, in the case of concrete structures, those due to creep and shrinkage. The movement is most marked between the internal structure and the external columns, particularly if the latter are totally or partly outside the external wall and glazing. It affects most the external cladding and partitions located at right angles to the external wall, as well as any linking structural element.

The effects are best controlled by attempting to achieve uniformity of stress and exposure (including insulation where necessary) and uniform surface:volume ratios for the concrete elements to reduce differential shrinkage or creep. When the problem is particularly severe, the building can be divided into two or more sections by incorporating intermediate transfer structures, in effect producing horizontal expansion or contraction joints. Alternatively, the movement can be restrained by stiff beams connecting the external columns to the core. Another approach is to freely permit the movement and incorporate appropriate movement details in the structure, partitions and finishings.

21.20.6 Lateral movement and dynamic effects

Wind loading on tall buildings is not simply a matter of statics. As with long-span bridges, aerodynamic effects and the dynamic response of the structure under variable wind loading are important considerations. The primary considerations are: (1)

vortex shedding, arising from the plan shape of the building; (2) the extent of horizontal movement and related accelerations that can be tolerated; and (3) the means of dissipating the wind energy imparted to the building, i.e. damping.

The damping characteristics of the early tall buildings of traditional construction were good. The exterior walls and internal partitions helped to dissipate the wind energy imparted to the building by friction between their parts as they moved relative to each other under wind action. The façades were generally textured, creating turbulence, and sculptured, thus reducing the risk of transverse oscillation from vortex shedding. With the development of simpler shapes and taller, more flexible buildings, these damping qualities have been diminished. Damping through the structure itself is small since, in this context, building structures are very elastic. Damping through the nonstructural elements is difficult to control and can be expensive in maintenance and repair. This leaves aerodynamic damping, through form and cladding texture, or specially designed mechanical damping as the means of absorbing energy and controlling movement to within acceptable limits. The alternative is to increase mass but this can be very expensive.

Acceptable lateral movement is not simply a matter of deflection but more of people's perception of, and psychological response to, movement. Deflection is a product of stiffness and can be controlled, at a cost, through structural quantity and configuration. However, people's reaction to sway is related more to acceleration than actual amount of movement and, more particularly, to the rate of change of acceleration. Increasing the structural stiffness reduces the deflection but does not affect acceleration since the frequency is increased. Indeed, the more critical rate of change of acceleration is also increased. Reducing stiffness increases sway and the relative movement between building elements as well as visual disturbance. Thus, in some cases the only two practical methods of dealing with this problem are to increase the mass or improve the damping characteristics of the building. Increasing the mass is expensive. Aerodynamic damping is possible if suitable building forms are acceptable. Where this is not possible, or in cases of extreme height, additional damping can be provided by incorporating inertia elements strategically located within the building, or energy-absorbing devices at points of movement within the structure. Such damping has the effect of reducing both the amount of movement and the acceleration.

An example incorporating inertia elements is the Citycorp Center in New York which has a tuned mass damper to slow down and reduce movement due to wind. The system is housed in the roof structure and consists of a 400-t concrete inertia block mounted on a 'frictionless' bearing which is free to remain still when the building starts to move under wind action. The mass is connected to a system of pneumatic springs and dashpots in which the energy is absorbed by oil. The dynamics of the damping system are adjustable to suit the building's actual dynamic characteristics. Fail-safe devices are incorporated to ensure that the movement of the mass relative to the building is kept within predetermined limits. Analysis and wind-tunnel testing indicated that this device would reduce the acceleration under wind loading by 38%.

The other methods of damping were adopted at the World Trade Center, also in New York. The simple geometric shape of the twin 110-storey towers was such that the vibration due to vortex shedding could not be discounted. To combat this danger, the building corners were chamfered, so modifying vortex generation, and a viscoelastic damping system introduced. Some 10 000 such dampers were built into each tower, consisting of steel connections incorporating viscoelastic material between the floor trusses and columns.

Figure 21.15 illustrates diagrammatically the dampers incorporated in the 110-storey World Trade Center in New York.

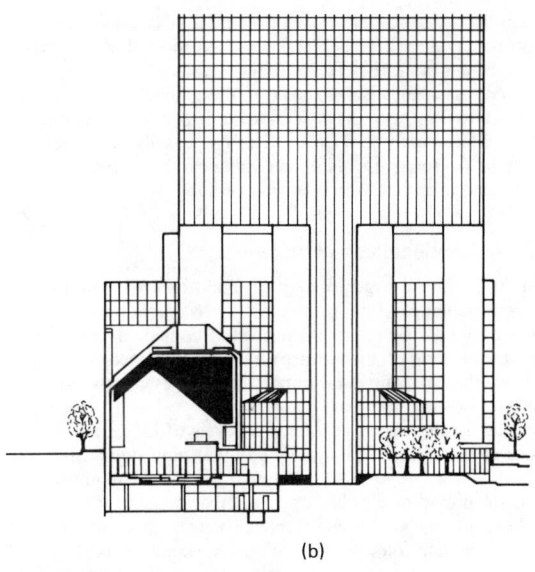

(b)

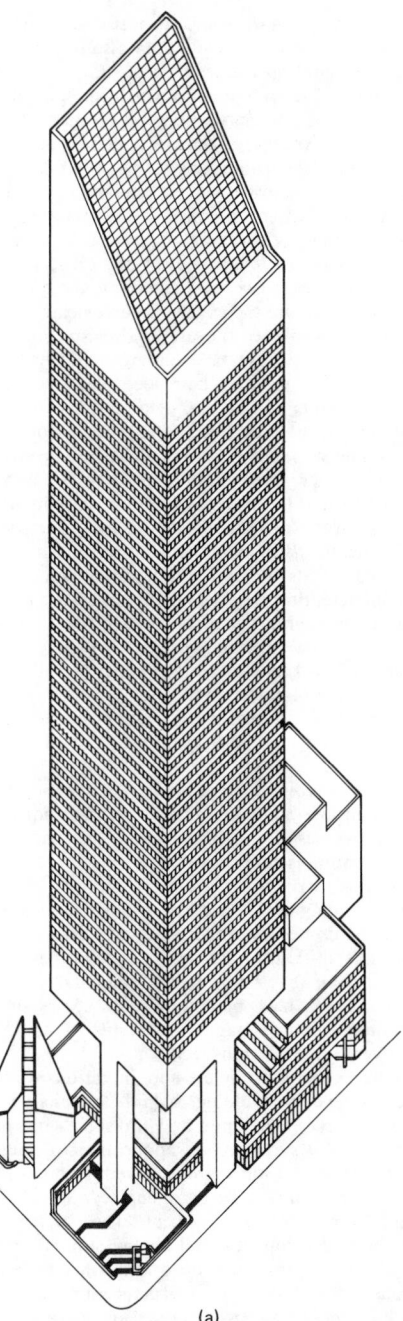

(a)

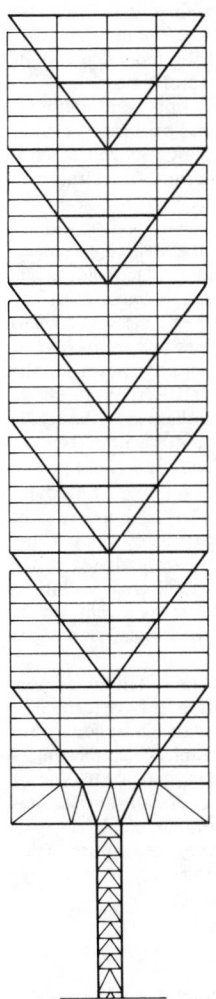

(c)

Figure 21.14 The Citicorp Center, New York. (a) The tower; (b) open space at the bottom of the tower; (c) braced external wall structures. (After Fischer (ed.) (1980) *Engineering for architecture*, McGraw-Hill, New York)

The damping unit comprises two Ts bonded to a central plate by viscoelastic material. The plate is attached to the bottom chord of the beams and the Ts to the columns. Wind energy is absorbed by shear displacement of the viscoelastic material. Figure 21.16 illustrates the general arrangement of a tuned mass damper while Figure 21.17 illustrates graphically the beneficial impact of damping on building oscillation. Similar principles can be used to control the vibration of light structures.

21.20.7 Additional considerations

In addition to the wind loading on buildings, the wind effect between buildings and, particularly, that produced by the presence of a tall building, has proved of considerable concern. Wind-tunnel tests on models are an essential element of design, to check the environmental impact created by a new building within an existing complex.

Special arrangements in the lower levels of tall buildings are encouraged in some locations by city planning policy. In New York, for example, a 20% additional floor area is given as a bonus for providing a public open space or plaza in front of the building; this has now been extended to encourage covered and secure pedestrian freeways through the buildings. Incentives are also given to encourage a richer mixed use of tall buildings, e.g. by incorporating a theatre when the building is in a theatre district, and shops within a shopping district. The intention is to produce a lively combination of activities and more demand for transport and services over a longer period of time.

These aspects have given rise to another important consideration in the design of both high- and low-rise buildings, i.e. their treatment at ground level. For example, the Citicorp Center in New York incorporates at its base a minicentre for culture and commerce including a church, a theatre, a room for jazz performances and a complex of international food boutiques. To accommodate these, the bottom of the structure consists only of the centre core and massive columns placed not at the corners but at the mid-point of each side. In most cases the bridging requirements to achieve these open spaces at ground level are met by a straightforward transfer structure in steel or concrete. This is sometimes placed at roof or upper level and the building beneath hung from it. In other cases, more subtle bridging means have been employed to make use of the whole superstructure, acting as a total bridging entity within itself.

Figures 21.14(a) and (b) illustrate how open space was created at the bottom of the Citicorp Center tower and utilized for religious and cultural purposes. Figure 21.14(c) illustrates the braced external wall structures. Each eight-storey tier is structurally independent with loads from each tier gathered to the four exterior mast columns via the diagonal truss members. Wind shear in each eight-storey tier is taken by a central core structure but the overturning moment is then transferred to the external masts.

In the urban context, the placing and form of tall buildings is of great importance. The architect and engineer have learned to solve the technical and aesthetic problems of tall buildings but generally as isolated elements. As one example which recognizes the very great positive or negative impact that a building of exceptional height may have, the City of San Francisco has adopted an urban plan which sets down fundamental principles and critical urban design relationships to govern future major development. This not only protects community interests but also gives reliable guidance to design.

21.21 Special structures

Special structures include means of covering or enclosing large areas without internal supports (by means of beam, membrane, tension, skeletal or pneumatic structural action), space frames and tall buildings, each of which has its own particular field of application and specialized technology. They also include cases where the normal assumptions regarding structural action may not apply. For example, deep beams, in which the span:depth ratio is small (less than 5:1) behave differently from shallower beams and the normal theory of flexure does not apply. In portal frames, in which the beam spans are large in relation to the column lengths, the distribution of bending moments is highly sensitive to the relative stiffnesses of the members, and the normal assumptions of stiffness of reinforced concrete sections (whether cracked or uncracked) or of steel sections at yield stresses are insufficiently accurate. Similarly, the geometry of the system may exaggerate the effects of movement.

Membrane structures involve the use of thin surfaces geometrically arranged to support vertical loading mainly by forces parallel to the surface. They may be folded plates or singly curved, as in barrel vaults and arches, or doubly curved as in domes, hyperbolic paraboloids and other special forms of curved surface. While secondary bending effects are present at right angles to the surface, the main internal forces are parallel to it and it is essential that the surface and boundary conditions are geometrically correct for resisting the loading. The simple example is the dome, in which compressive forces occur within the dome, and ring tensions or external restraints are required at the perimeter. Concrete, timber and sometimes steel are used in forming long-span membrane structures; they use materials efficiently, but their economic viability depends mainly on the workmanship and labour required. Folded plates provide functional and aesthetically interesting roofs, in which normal flexural action occurs; additional considerations of edge support, end shear, buckling and distribution of out-of-balance loading also apply.

Tension structures involving cable-supported sheeting have been used in exhibition buildings and in sports stadia; the structural system involves steel cables acting in catenary from which decking or tenting is supported.

Skeletal space-frame structures are used in the form of plane grids of rectangular, diagonal ('diagrid'), triangular or hexagonal pattern, arches, domes and other structures analogous to membrane structures, in that the geometric shape of the surface is such that the principal resisting forces act parallel to the surface. These structures have been applied to sports stadia, exhibition buildings, aircraft hangars, terminals and other places requiring very large column-free areas.

Pneumatic construction uses air pressure in various ways to stabilize the membrane of the building. Such structures are light, economical, easy to erect, dismantle and transport, and have proved to be practically successful in application to housing temporary exhibitions, warehousing and covering sports halls and other specialist buildings. The basic engineering principle employed is that the membrane can accept tensile stresses and will fold when not in tension. Internal air pressure is used to maintain membrane tensions when dead and other loads are imposed. Various types of pneumatic structure have been developed, including the basic air-supported membranes and inflated dual-walled and ribbed structures, and various hybrid types. The larger spans are achieved by the use of arched or domed forms, and cables, cable nets and internal membrane walls are used to control shapes and improve stiffness.

21.22 Foundations

Foundations must safely transfer loading from the superstructure to the ground without excessive absolute or differential settlement. The foundation type will depend upon the nature of both the underlying soil and the superstructure since the two must be compatible as far as the settlement characteristics are concerned. In some cases, to economize in foundations, the

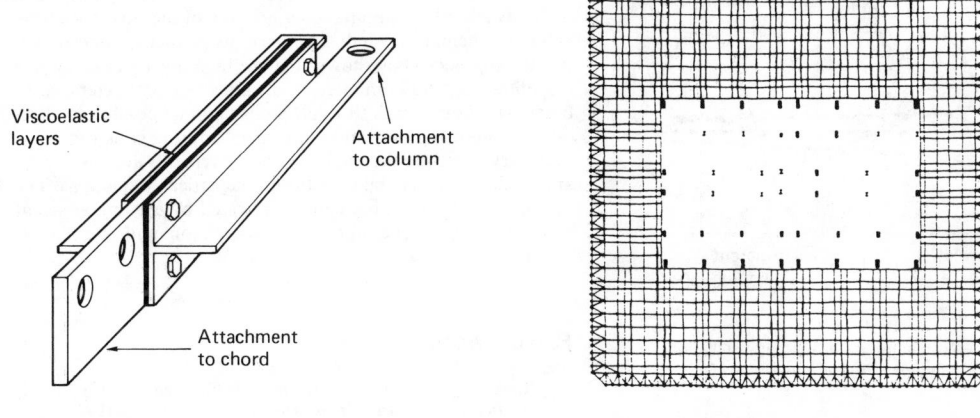

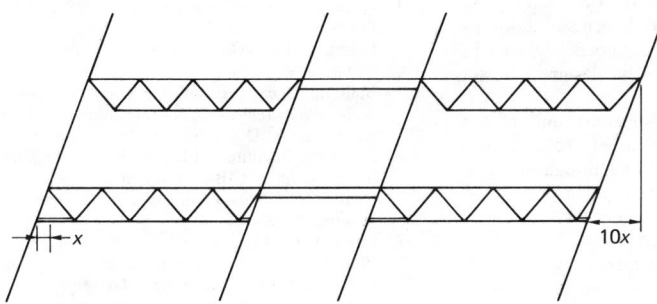

Figure 21.15 Dampers used in the World Trade Center, New York. (After Fischer (ed.) (1980) *Engineerng for architecture*. McGraw-Hill, New York)

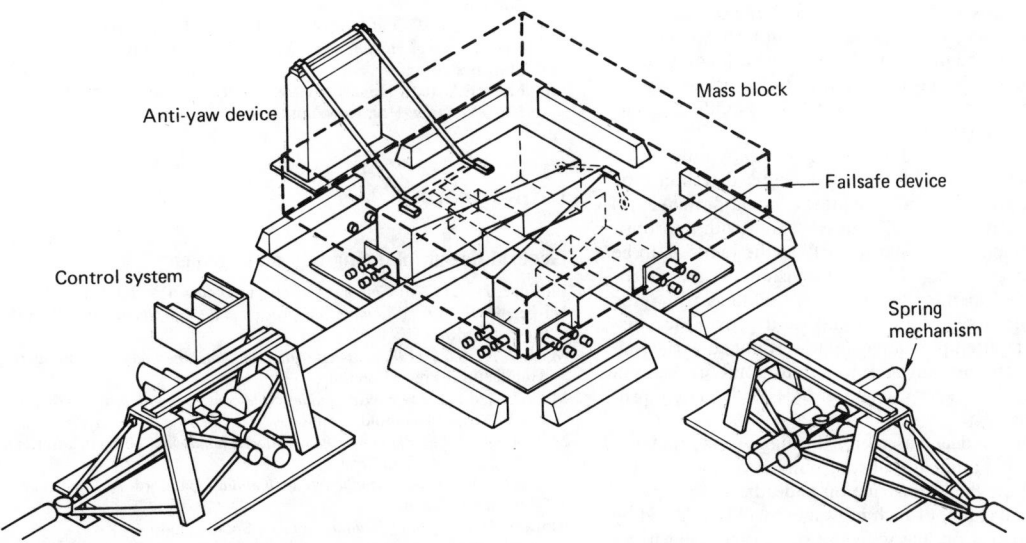

Figure 21.16 General arrangement of a tuned mass damper. (After Fischer (ed.) (1980) *Engineering for architecture*, McGraw-Hill, New York)

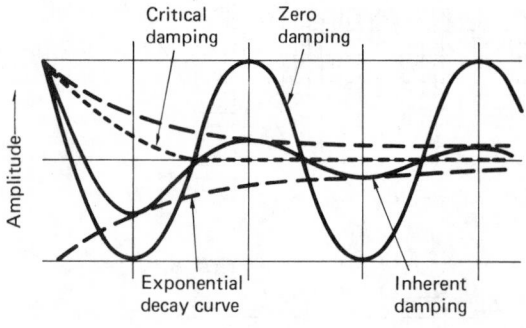

Figure 21.17 The impact of damping on building oscillation. (After Fischer (ed.) (1980) *Engineering for architecture*, McGraw-Hill, New York)

superstructure will be designed to allow for differential settlement by the incorporation of, for example, jointed construction in the structure, cladding and finishings. In other cases, a type of foundation will be adopted which limits settlements to amounts acceptable to the previously determined superstructure. In special instances, the superstructure may be designed to act compositely with the foundation.

In all cases the designs of the superstructure and of the foundations are interdependent; knowledge of the soil conditions is thus essential at all stages of the design process, beginning with at least broad local knowledge prior to land acquisition. Such information may also influence the location of particular buildings on the site.

The risk of mining subsidence must be assessed at an early stage and the appropriate course of action decided, e.g. whether to seal and grout the workings, pile through the workings, design the superstructure to accommodate subsidence, or to accept the risk without special action.

The main types of building foundation are individual pad footings, strip footings, rafts, piers and deep footings, deep-spread foundations and piles. Basement construction is useful in reducing settlements and in controlling differential settlements between parts of buildings of different height.

Ground floor slabs may be suspended, but normally bear directly on the soil; they are required to span local weaknesses or voids due to settlement and to transfer local loading to an appropriate area of the ground. Their design is largely an empirical process; specification, workmanship, joints and surface treatments are important.

Basement construction and ground floor slabs must resist penetration of water or water vapour to a degree determined by the building use; provisions may include: (1) land drainage; (2) good-quality concrete of suitable thickness and with appropriate workmanship and details particularly at the joints as to be adequately waterproof; and (3) the provision of internal or external waterproof membranes. In some buildings it may be more economic to accept some leakage and provide for discharging it than to attempt completely watertight construction. The presence of water may give rise to serious uplift problems in basements during or after construction unless adequately provided for in design.

The construction of deep basements through ground material other than rock is bound to cause ground movement which could affect both the new construction and adjacent property. The magnitude and extent of such movement will be affected by the method of excavation, the sequence of basement construction and any dewatering that may be required.

Top-down methods of basement construction incorporating diaphragm walls or secant piling can be effective in controlling movement. Where speed of construction is important, such methods also allow simultaneous erection of the superstructure. While mathematical modelling exists to predict ground movements, experience has shown that the most reliable indications are obtained by a combination of modelling and reference to basement construction through similar ground conditions. Frequent monitoring of ground movements during construction is necessary to ensure that no excessive movements are developing and to check the reliability of the mathematical predictions. The extent of movement that can be tolerated by existing adjacent property will depend upon its construction and foundation system.

References

1 Royal Institute of British Architects (1980) *Handbook of architectural practice and management*, 4th rev. edn. RIBA, London.
2 American Institute of Architects (1976) *Current techniques in architectural practice.* AIA, Washington.
3 Royal Institute of British Architects (1973) *Plan of work.* RIBA, London.
4 Telling, A. E. (1986) *Planning law and procedure*, 7th edn. Butterworth, London.
5 Building Research Establishment (1979) *Thermal, visual and acoustic requirements in buildings.* Building Research Establishment Digest No. 91. BRE, Watford.
6 Chartered Institute of Building Services Engineers (1984) *Code for interior lighting.* CIBSE, London.
7 Neufort, E. (1980) *Architects' data*, 2nd edn. Crosby Lockwood, London.
8 Fire Officers' Committee *Rules*, 29th edn. FOC, London.
9 Building Design Partnership, Aarens, Burton and Karalec, and Gifford and Partners (1981/82) *Low-energy hospital study.* Report compiled on behalf of the Department of Health and Social Security. HMSO, London.
10 Her Majesty's Stationery Office (1985) *The Building Regulations: mandatory rules for means of escape in case of fire.* HMSO, London.
11 Her Majesty's Stationery Office (1982) *Guidelines for the construction of fire-resisting structural elements.* HMSO, London.
12 Association of Fire Protection Contractors/Constrado *Fire protection of structural steel in buildings.* AFPC/Constrado, London.
13 Simpson, J. W. and Horrobin, P. J. (1970) *Weathering and performance of building materials.* Manchester University Press, Manchester.
14 Fischer, E. (ed.) *Engineering for architecture – architectural record 1980.* McGraw-Hill, New York.

Bibliography

General design planning and management

Eldridge, H. J. (1976) *Common defects in building.* HMSO, London.
Harper, D. R. (1979) *Building: the process and the product.* Construction Press, London.
Martin, D. (ed.) (1985) *Specification building methods and products.* Architectural Press, London.
Mills, D. (ed.) (1985) *House's guide to the construction industry*, 9th edn. Van Nostrand Reinhold, London.
Mills, E. D. (1985) *Planning: the architect's handbook*, 10th edn. Butterworth Scientific, Guildford.
Osbourn, D. (1986) *Introduction to building.* Batsford Academic and Educational, London.
Ransom, W. H. (1981) *Building failures.* Spon, London.
Reid, E. (1984) *Understanding buildings.* Construction Press, London.
Rich, P. (1982) *Principles of element design*, 2nd edn. Longman, Harlow.
Saxon, R. G. (1986) *Atrium buildings development and design*, 2nd edn. Architectural Press, London.

Cost planning and control

Bathurst, P. E. and Butler, D. A. (1980) *Building cost control techniques and economics*. Heinemann, London.

Cartlidge, D. P. and Mehrtens, I. N. (1982) *Practical cost planning*. Hutchinson, London.

Seeley, I. H. (1983) *Building economics*. Macmillan, Basingstoke.

Stone, P. A. (1983) *Building economy*. Pergamon Press, Oxford.

Internal environment

British Standards Institution (1975) *Code of basic data for the design of buildings: the control of condensation in buildings*. BS 5250. BSI, Milton Keynes.

British Standards Institution (19??) *Sound insulation and noise reduction in buildings*. BS CP3, Part 2. BSI, Milton Keynes.

Chartered Institute of Building Services Engineers (1984) *Guide to current practice*. CIBSE, London.

Faber, O. and Kell, J. R. (1979) *Heating and air-conditioning of buildings*, 6th edn. Architectural Press, London.

Lord, P. and Templeton, D. (1986) *The architecture of sound*. Architectural Press, London.

Parkins, P. H., Humphreys, H. R. and Cowell, J. R. (1979) *Acoustics, noise and building*, 4th edn. Faber and Faber, London.

Energy

Kasabov, G. (ed.) (1979) *Buildings: the key to energy conservation*. RIBA Energy Group, London.

Regulations

Her Majesty's Stationery Office (1985) *Manual to the Building Regulations*. HMSO, London.

Her Majesty's Stationery Office (1985) *The Building Regulations: Approved Documents*. HMSO, London.

Fire building security and control

Hopf, P. S. (1979) *Handbook of building security planning and design*. McGraw-Hill, Maidenhead.

Vincent, G. S. and Peacock, J. (1985) *The automatic building*. Architectural Press.

Structure and tall buildings

American Society of Civil Engineers (1985) *The engineering aesthetics of tall buildings*. ASCE, New York.

Institution of Engineers (1984) *Proceedings, international conference on tall buildings*. IE, Singapore.

22

Hydraulic Structures

The late A R Thomas OBE, BSc (Eng),
CEng, FICE, FASCE
Formerly consultant,
Binnie and Partners

Peter Ackers MSc (Eng), CEng, FICE,
MIWEM, MASCE
Hydraulics consultant

Contents

22.1 Open channel structures

22.1.1 Basic concepts

22.1.1.1 The Bernoulli theorem and critical flow

Two important concepts in the hydraulics of flow through structures are the Bernoulli and pressure–momentum theorems. The former (see page **5**/8) expresses conservation of energy, and when applied to straight-line flow in an open channel, taking bed level as reference level, may be expressed as:

$$H = d + \alpha V^2/2g \qquad (22.1)$$

where H is the specific energy head, d the depth of flow above the bed, α coefficient, V the mean velocity and g the gravitational constant

Where the flow is curvilinear, depth will vary across the channel and d is a mean value. Under normal conditions of flow in wide uniform channels, $\alpha = 1.02$ for smooth boundaries but higher for rough boundaries. For example, if $n/d^{1/6} = 0.0225$ (where $n =$ Manning's roughness factor) $\alpha = 1.12$. In order to simplify calculations where velocity head is relatively small, α is often assumed to be unity. Head loss must be allowed for in the value of H. For channels of rectangular cross-section, Equation (22.1) can also be expressed as:

$$H = d + \alpha q^2/2gd^2 \qquad (22.2)$$

where q is the discharge per unit width of channel Q/B where Q is the total discharge and B the width

To derive d from known H and q, with $\alpha = 1$ Figure 22.1 may be used.

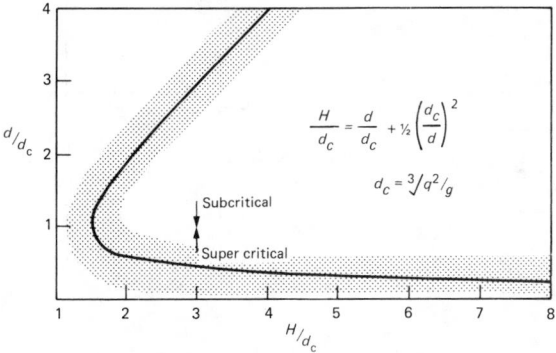

Figure 22.1 Specific energy of flow in open channels. Depth of flow d may be determined from specific energy head H and discharge per unit width q

In a channel of rectangular cross-section with horizontal bed and $\alpha = 1$ (as, for example, immediately downstream of a contraction)

critical velocity $V_c = (gd)^{1/2}$ $\qquad (22.3a)$

and critical depth

$$d_c = V_c^2/g = (q^2/g)^{1/3} \qquad (22.3b)$$

In the more general case, applicable to non-rectangular channels of slope angle θ:

critical velocity $V_c = \left(\dfrac{gd_m \cos\theta}{\alpha} \right)^{1/2}$ $\qquad (22.3c)$

For critical depth in circular and horseshoe-shaped channels see Figure 22.19 (page **22**/17).

22.1.1.2 Froude number

$F = V/(gd)^{1/2}$ is a useful indicator of the stability of free surface flow. When $F < 1$, the flow is subcritical; when $F = 1$ it is critical and when $F > 1$, supercritical. As F approaches unity from either direction, the flow becomes unstable and surface waves may develop. Surface undulations may occur in subcritical flow when F exceeds 0.5.

22.1.1.3 The pressure–momentum theorem

Unlike the Bernoulli theorem, this applies whether there is head loss or not. It follows from Newton's second law and can be expressed as:

$$P = M_2 - M_1 = \frac{w}{g} Q(V_2 - V_1) \qquad (22.4)$$

where P is the resultant force on a mass of fluid over a specified length, M_1 and M_2 represent momentum at entry and exit, w is the specific weight of fluid, Q the constant discharge and V_1 and V_2 are the flow velocities at entry and exit

P usually is the resultant of fluid pressures and boundary pressures in the direction of flow.

22.1.1.4 Hydraulic jump

This is a relatively abrupt change in flow depth when the flow changes from supercritical to subcritical as described on pages **5**/17 to **5**/19 and illustrated in Figure 22.2. Except at the limiting condition when both depths are critical, it involves a head loss, dissipated in extra turbulence. In Figure 22.1 it can be represented by a transfer from a point on the supercritical curve to a lower point on the subcritical curve. It may be stationary or moving. Its character and movement can be determined by application of the pressure–momentum equation (Equation (22.4)). In a rectangular channel of width B and horizontal bed, $P_1 = \frac{1}{2}Bd_1^2$ at entry and $P_2 = \frac{1}{2}Bd_2^2$ at exit, where d_1 and d_2 are depths; no other pressures have components in the direction of flow. If pressure plus momentum of the supercritical flow $(P_1 + M_1)$ exceeds the pressure plus momentum of the subcritical flow $(P_2 + M_2)$, the jump will move downstream, if they are equal the jump will be stationary and if $(P_2 + M_2)$ exceeds $(P_1 + M_1)$ the jump will move upstream.

For a stationary jump in a horizontal rectangular channel, the relationship between upstream and downstream depths is:

$$\frac{d_2}{d_1} = \sqrt{(0.25 + 2F_1^2)} - 0.5 \qquad (22.5)$$

where d_1 and d_2 are the conjugate depths, i.e. the depths of flow upstream and downstream of the jump, respectively, and F_1 is the Froude number upstream of the jump

A number of laboratory tests have shown close conformity to this relationship.

The jump height, $d_j = (d_2 - d_1)$, on a horizontal floor may be determined from Figure 22.2, which may be extended by use of Equations (22.1) and (22.5). The length of a jump cannot be precisely defined but is approximately 5 to $8 \times d_j$, the greater factor applying to lower Froude numbers.[1]

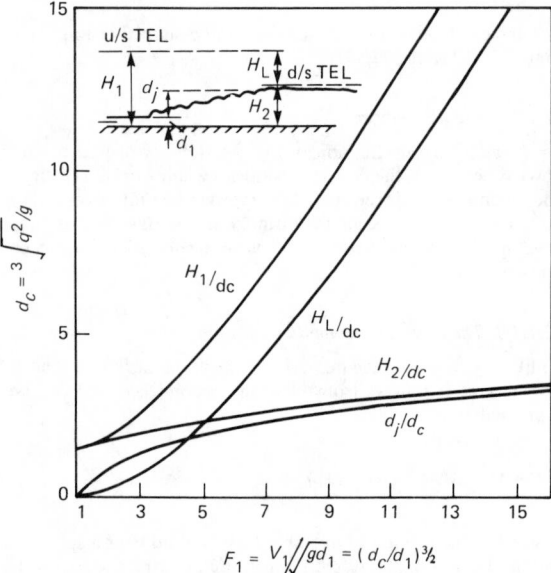

Figure 22.2 Hydraulic jump relationships for horizontal or gently sloping beds. (After Thomas (1958) Discussion on Bradley and Peterka (1957) op. cit. *Proc. Am. Soc. Civ. Engrs*, **84**, HY2, Paper 1616)

Equation (22.5) and Figure 22.2 give results with little error in channels with beds sloping at 10% or less, but with steeper slopes the components of vertical pressures have significant effect.

In channels which are not of rectangular section the jump may be distorted in plan, but the pressure–momentum equation (22.4) can be applied to the whole cross-section. Several methods for calculating the conjugate depths in channels of various shapes are available.[2-5]

22.1.2 Transitions

22.1.2.1 Subcritical flow

In channels of variable cross-section, Equation (22.1) or Figure 22.1 may be used to determine depth of flow, provided changes are sufficiently gradual to avoid significant head loss. In converging flow, q and hence d_c increase with the reduction in width. Therefore with subcritical flow and constant specific energy H, it is evident from Figure 22.1 that d reduces. As examples, a channel may be contracted at a bridge and allowed to expand downstream, or a gated regulator may have a raised sill. In both cases the surface is depressed in the contraction. Provided the flow remains subcritical the process is reversible in a downstream expansion. If, however, a contraction reduces the depth to the critical value, any further contraction has the effect of raising the upstream head, because critical depth is the mini-

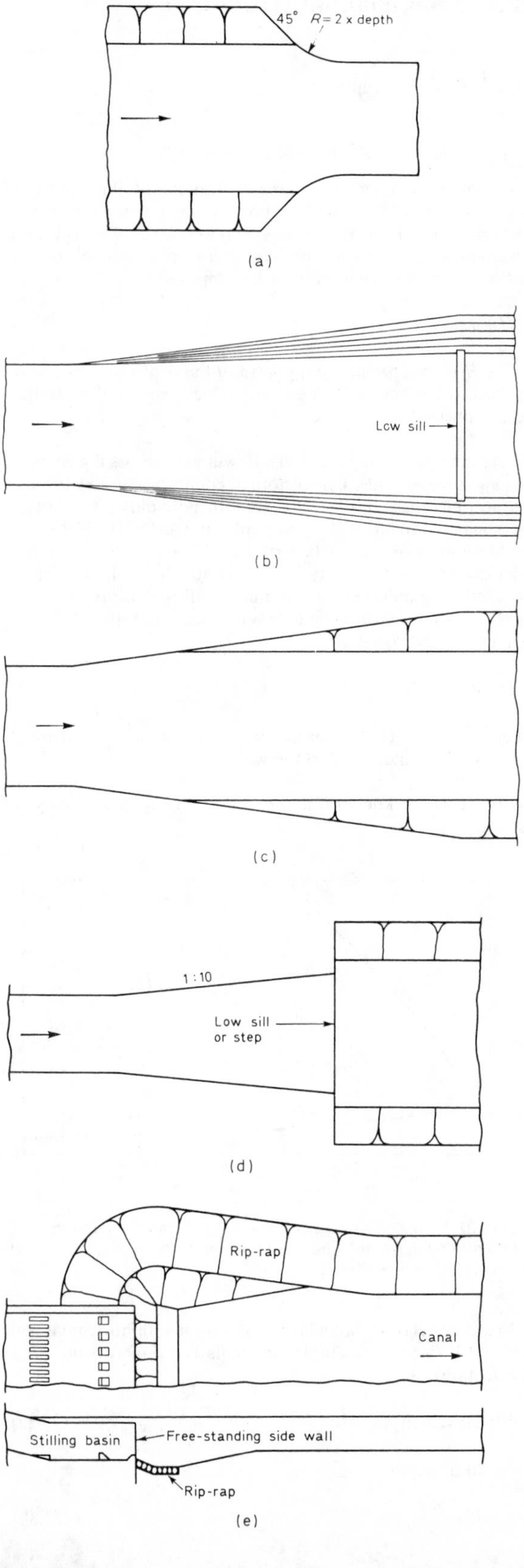

Figure 22.3 Typical transitions for subcritical flow. (a) Contraction from sloping to vertical sides; (b) warped expansion; (c) expansion with vertical sides; (d) short expansion; (e) example of transition from stilling basin to canal in erodible material

mum depth possible for any given specific head (see Figure 22.1). The result is a rise in upstream water level, the excess head generates supercritical flow downstream of the throat, or section of maximum contraction, and is lost in a hydraulic jump where the flow changes back to subcritical. The throat is then acting as a 'control'. If head loss is to be avoided, the Froude number should not be allowed to approach close to unity.

Convergences for subcritical flow may be rapid but external angles in the side walls should be avoided by the use of large-radius curves, as shown in Figure 22.3a. Diverging channels in subcritical flow are liable to result in separation of flow from one or both side walls unless expansion is gradual. Side expansions of 1:10 are usually satisfactory. Sharper divergences may be followed in some conditions;[6] expansion is assisted by a rising floor, baffle blocks or a raised sill downstream and by a hydraulic jump. The expansion ratio is also a factor – see page **22/17** – where expansions in enclosed flow are discussed. Some examples of diverging transitions are shown in Figure 22.3b to 22.3e. Figure 22.3e illustrates a transition from a drop structure to the canal beyond.

Changes of direction cause head loss because of the secondary flow which distorts the flow pattern; the flow near the bed is deflected more sharply than the surface flow. If the bend is very sharp there may be complete separation at the inner boundary. These effects may be minimized by adopting a large radius for the bend. In rectangular channels with depth: width ratio of 0.6 to 1.2, Shukry[7] found that head loss became minimal with radius 3 × width. In channels with erodible boundaries, unless bank protection is provided, the minimum radius depends on the velocity and erodibility of bank material. On irrigation canals in India the radius is generally 20 to 30 × surface width.

22.1.2.2 Transitions – supercritical flow

The problems here are different from those discussed so far. Whereas in subcritical flow, pressure changes can be transmitted laterally from the side walls to the whole flow, inducing change of depth or direction, in supercritical flow the velocity of transmission of a small disturbance or wave is less than the flow velocity. The result is that a change in direction of a side wall creates an oblique shock wave which is reflected from side to side downstream.

Convergences and divergences should be very gradual. Figure 22.4 shows the shock waves created by a convergence. A sharp convergence may cause high-velocity flow to ride up and overtop the wall. It is therefore preferable, if possible, to locate convergences and other changes in wall direction where the velocity is low, e.g. at the upstream end of a chute, and maintain a straight chute where velocity is high. It may, however, be possible to use lateral inclination of the bed, e.g. superelevation, to assist in convergence or divergence. Where shock waves are unavoidable, they will occur in a zigzag pattern for some distance downstream owing to reflection from side walls. The side walls should therefore be high enough to contain them at

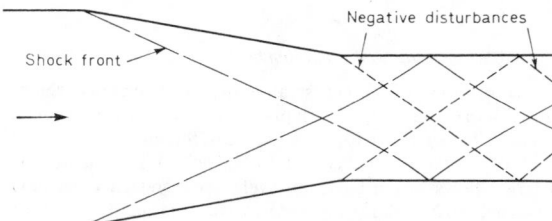

Figure 22.4 Example of shock waves at convergence in supercritical flow. (After Ippen *et al.*, (1951) 'High-velocity flow in open channels.' *Trans. Am. Soc. Civ. Engrs*, **116**, Paper 2434)

points of reflection. Sloping side walls, as in trapezoidal channels, are particularly vulnerable. Methods are available for the calculation of pattern and height of shock waves in simple cases and for minimizing their effects.[9] Scale models may also be used.

Long-radius bends are preferable to short radius, especially where overtopping is a danger. Knapp[9] recommends compound curves for the side walls of bends, with radius $2r$ in the approach and exit over a length of $B/\tan \beta$ in each case, where r is the radius of the centreline of the main curve, B is the channel width and $\sin \beta = F$, the Froude number. This arrangement creates counter waves which tend to neutralize the shock waves generated by the main curve, so reducing disturbances downstream.

22.1.3 Weirs and flumes

22.1.3.1 General

Weirs are used to control flow or water levels, or to measure flow. They range from low walls across streams to the spillway crests of high dams.

The basic equation for free flow over weirs is:

$$q = c(2g)^{\frac{1}{2}}H^{1/2} \tag{22.6}$$

where q is the discharge per unit width, c is a discharge coefficient, g the gravitational acceleration and H the total head level upstream above weir crest, normally taken as $h_1 + V_1^2/2g$, where h_1 is the upstream depth of flow above weir crest level and V_1 is the mean velocity of approach

Equation (22.6) can be derived from Equations (22.2) and (22.3) assuming critical flow and applying a coefficient c to take account of departure from flow on a horizontal bed. The coefficient depends on the shape of the weir and, in general, it varies with head over the weir; only in a few special cases is it constant. There are many weir profiles, each with different characteristics in relation to discharge coefficient and modularity. Weir flow is said to be 'modular' or 'free flow' when it is unaffected by tailwater level. The point at which a rising tailwater begins to affect the upstream head or flow is termed the 'modular limit', expressed as the ratio of downstream to upstream depth above crest level. Values of the coefficients of weirs of many different profiles have been published, e.g. by King and Brater[9] (see also section 22.5). In this section, some types in general use are considered as follows.

Sharp-crested weirs. These are formed of metal plates and are used for precise measurements of flow. Flow over weirs with narrow crests having rectangular upstream corners is effectively sharp crested, with a coefficient c approximately 0.406, provided the nappe springs clear and is fully vented.

Triangular profile weirs. These have sensibly constant coefficients throughout their modular range; no venting is required and the coefficient is greater than that of a sharp-crested weir. For example, the Crump weir (Figure 22.7), with 1:2 upstream and 1:5 downstream slope, has a free-flow coefficient c of 0.442 and a modular limit (within 1% of discharge) of 0.74. Weirs of this type are widely used for measurement of stream flows.

Trapezoidal profile weirs. Trapezoidal profile weirs have flat upstream and downstream slopes and narrow horizontal crests, formed by the gate sill, are often used in gated controls and barrages (see, for example, Figure 22.10). They have a free-flow coefficient which is variable but generally exceeds 0.383 and under drowned conditions the afflux is small.

Broad-crested weirs. These have horizontal crests wide enough for parallel flow effectively to develop. Control is then at the point of critical depth so that $c = 1.70$. To ensure that this condition applies and c is constant, the upstream edge should be rounded to avoid the formation of a roller above crest level. In practice, the value of c is 1 to 3% lower due to friction loss. If the downstream floor falls at a gentle slope, say 1:10, the modular limit is between 0.7 and 0.8. Broad-crested weirs have been extensively used for flow measurement and for proportional distribution of flow at dividing points in irrigation systems.

Free-nappe profile weirs. Free-nappe profile weirs with profile according to the shape of an undernappe of flow over a sharp-crested weir (Figure 22.5) have been widely used for overflow spillway crests. The standard profile is one with vertical upstream face and weir height P large compared with head over crest, H. The profile varies with smaller values of P/H and sloping upstream faces. This profile has the advantages that c is comparatively high for the profile discharge (i.e. the discharge corresponding to the nappe profile used), subatmospheric pressures do not develop within the range up to profile discharge, no venting is required and the flow characteristics are well documented and predictable.

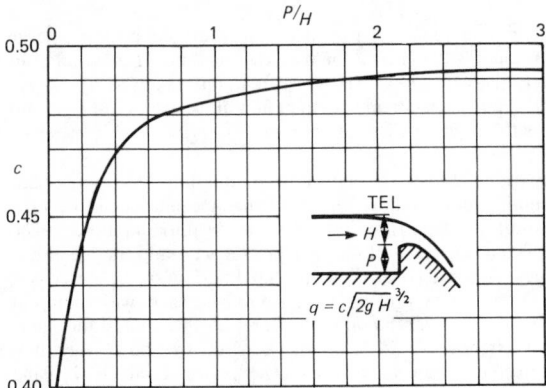

Figure 22.5 Discharge coefficient of free-nappe weirs at design discharge. (Based on USBR data; US Department of the Interior (1960) *Design of small dams*. Denver, Colorado)

The coefficient c of the standard weir at profile discharge is shown in Figure 22.5. By adopting a profile discharge lower than the maximum discharge, a higher coefficient is obtained at flows exceeding profile discharge.[10] Discharge in excess of the profile discharge causes pressures on the face of the weir to fall below atmospheric in the vicinity of the crest where the curvature is sharp.[11] This is usually acceptable provided that the structure is safe against uplift, and a reasonable margin of pressure is allowed above cavitation level to allow for fluctuations.

Profile coordinates have been published[12] from which weirs of standard profile and some variations can be designed.

Sharp side contractions at the abutments of weirs reduce the discharge capacity locally. They should be curved as in Figure 25.3a. Piers have a similar effect, to avoid which spillway piers are often extended upstream, so that the contraction at the pier noses occurs in a region of lower velocity.

22.1.3.2 Submerged weirs

The effect of a tailwater level above the modular limit is to raise the upstream water level for a given discharge. The degree to which the upstream head or discharge is affected depends on the weir profile: moreover in certain ranges of submergence the flow

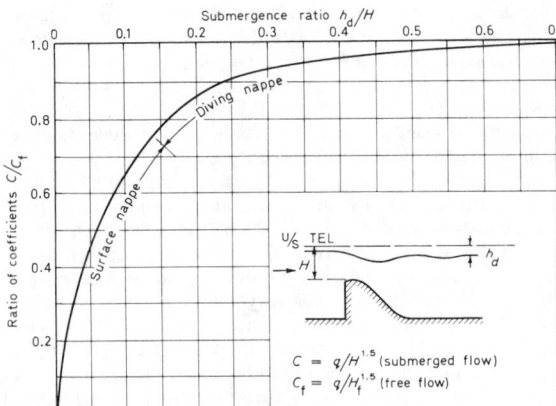

Figure 22.6 Free-nappe profile weirs. Effect of tailwater level on discharge coefficient. (Based on USBR data; US Department of the Interior (1960) *Design of small dams*. Denver, Colorado)

pattern is uncertain and may change from diving nappe, which follows the downstream weir face, to surface nappe, which separates near the weir crest, a roller developing beneath. Observations of discharge related to upstream and downstream heads or water levels therefore cannot be regarded as of general application. Nevertheless, good indications can be obtained. Figure 22.6 shows the effect of submergence on standard free-nappe profile weirs[12] and Figure 22.7 the effect on Crump triangular profile weirs.[13]

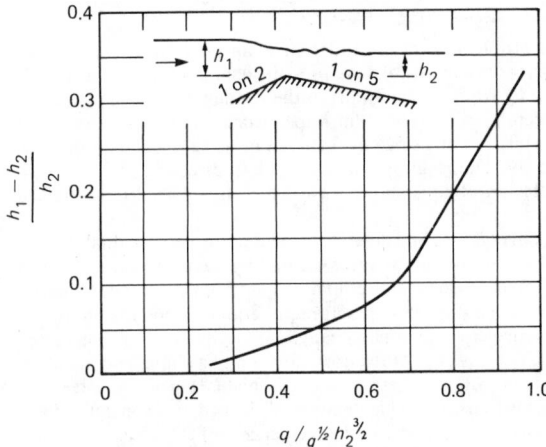

Figure 22.7 Afflux at submerged Crump weirs. (Based on data of White (1971) 'The performance of two-dimensional and flat-vee triangular profile weirs.' *Proc. Instn Civ. Engrs*, Paper 7350S.)

22.1.3.3 Measuring weirs and flumes

For laboratory and other small-scale measurements, sharp-crested weirs consisting of thin plates in the form of rectangular or Vee-notch weirs are found convenient. Standard formulae or tables of discharge for these are available.[9,14] For measurement of larger flows in the field, however, sharp-crested weirs have drawbacks, particularly the need to vent the nappe, the head difference required to ensure modular conditions and the effect of accretion of upstream bed level following the erection of a gauging weir.

Weirs of several other types have been thoroughly investigated and are subjects of national and international standards. A comprehensive account of the performance and use of the main types of weir and flumes is given by Ackers *et al.*[14] Bos[15] has reviewed a wide range of devices capable of use for flow measurement. Those most useful in a civil engineering context depend on the creation of critical flow. This can be induced by providing a sill or weir, or by contracting the width, or by a combination.

The critical flow formula in a rectangular cross-section channel, Equation (22.6), is the most basic formula for flow measurement by weirs and flumes, and applies to free flow, i.e. when the critical flow at the crest of a weir or in the throat of a flume is not drowned by the tailwater level exceeding the modular limit (see para 22.1.3.1). The cross-section of flow may also be made non-rectangular – V-shape, U-shape or trapezoidal – for particular applications, but then adjustment has to be made to the flow formula of Equation (22.6). For precise measurement with weirs and flumes, allowance for boundary layer development and other secondary effects has to be made.[14]

Unless there is already a local drop in level, the introduction of a measuring device will result in a rise in upstream level, though this may be quite small if a device with high modular limit is chosen, or a Crump weir with crest tapping which can be used when drowned by high tailwater level.[13] Where the range of discharge is large and it is desired to obtain an accurate measurement of low flows, a stepped weir may be used, consisting of a short weir at low level for the low flows flanked by longer weirs at a higher level. Alternatively, a flat Vee-weir may be used with crest tapping for submerged conditions.[13]

In the UK, broad-crested weirs with a round nose and Crump weirs have been accepted as standard.[16] In the US, Parshall measuring flumes have been widely used.[17] These were designed with plane surfaces so that they might be easily constructed of wood or concrete. *c* is not constant but calibration formulae and tables are available. Where the stream to be gauged carries appreciable bed load, a critical-depth flume with a flat or nearly flat bed at the channel bed level is desirable. The bed load can then pass through without excessive accretion upstream, though there may be some at the sides. A measuring flume of this type is shown in Figure 22.8. The degree of contraction sufficient to ensure modular flow can be checked by comparing calculated upstream water levels (using *c* = 1.66) with existing tailwater levels. The broad-crested weir coefficient is applicable, adjusted for head loss upstream of the location of critical depth.[14]

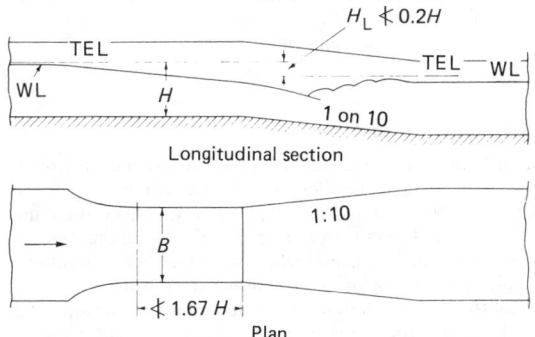

Figure 22.8 Measuring flume with flat floor for debris-laden flow

Structures of many other types are used for flow measurement, mostly depending on the critical depth principle or on orifice control as, for example, devices on irrigation canal outlets.[18]

22.1.4 Control weirs and barrages

22.1.4.1 Gated weirs

Weirs are used to control the water levels of a river or canal, for such purposes as diversion of flow into canals, extraction of water by pumping, creating head for hydro-electric power or maintaining a required depth of water for navigation. A fixed weir also raises flood levels, which may not be acceptable. A gated weir, or barrage, however, does not have this drawback if the gate sill is level with the river bed, or on a low weir crest. The gates are kept closed during low flows, maintaining the required upstream water level, but opened as may be necessary to pass floods. The range of water level is thus much less than with a simple weir, and the gates can be operated to maintain constant water level over a wide range of flow. Types of gates are described on pages **22**/11 to **22**/28.

The choice of crest profile depends on the circumstances. For example, a weir with a free-nappe profile is suitable where the crest is to be above the upstream channel bed and there is considerable head difference from upstream to downstream. On the other hand, a low crest with flat triangular profile is better suited where, at high rates of flow, the afflux or rise of upstream water level due to the weir must be kept to a minimum.

22.1.4.2 Control structures in alluvial rivers

Whereas structures in rivers with rocky beds and banks can often be of simple design, with an upstream cutoff wall into the rock and a basin or bucket energy dissipator downstream, the design of control structures in alluvial rivers requires consideration of many other factors.

Firstly, the site and orientation of the structure in relation to the river channel pattern is most important and generally should take priority over other considerations. Alluvial rivers without constraint by structures, training works or outcrops of rock or clay, may change course over a period of years, forming new patterns of river channels. The history of a river course is a good guide to such tendencies. The site for a control structure should be a stable one in the long term, i.e. it should remain operative despite changes in the channel pattern over a number of years, maintained if necessary with the aid of training works. Where a weir or barrage is used for diversion or abstraction of water it is usually desirable to ensure that the quantity of sediment in the water abstracted is a minimum. The best location for the offtake with this in view is generally on the outside of a bend, and the training works should be located to maintain the approach channel accordingly. This consideration applies even where special arrangements are made for sediment exclusion.

A typical barrage forming the headworks of an irrigation canal system on a large river in Pakistan is shown in Figure 22.9. A weir or barrage may occupy only a small part of the width of river channel and floodplain. For example, in India and Pakistan it is general practice to make the width of waterways between abutments equal to or rather greater than the width of Lacey regime channel $4.8Q^{1/2}$ where Q is the maximum design discharge in cubic metres per second.[19] Flanking bunds or embankments are then required extending from the abutments to high ground on either side. Where flood levels are being raised by the control, marginal bunds or flood embankments are often provided extending upstream on each bank. To prevent oblique approach, protect the bunds and avoid outflanking; guide banks are required extending upstream from the abutments (see Figure 22.9). In stable rivers these may be quite short, but where there may be wide swings in the river course they are generally approximately equal in length to the width of waterway between them. In addition, in rivers of this type, spurs or groynes may be provided upstream, but these may cause further

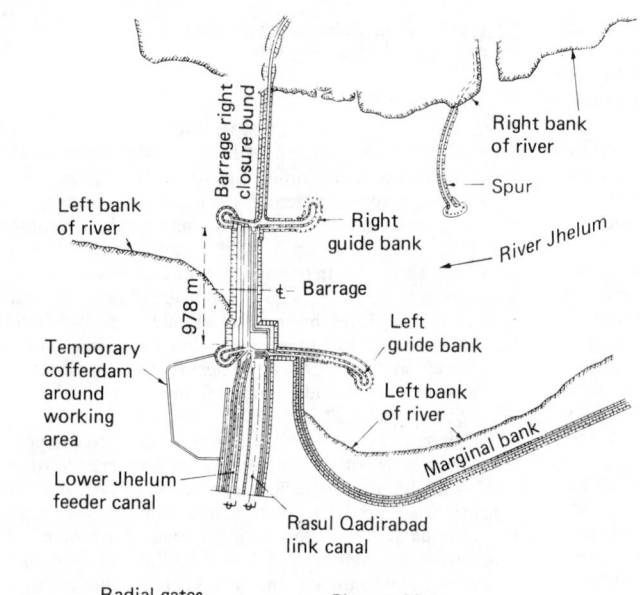

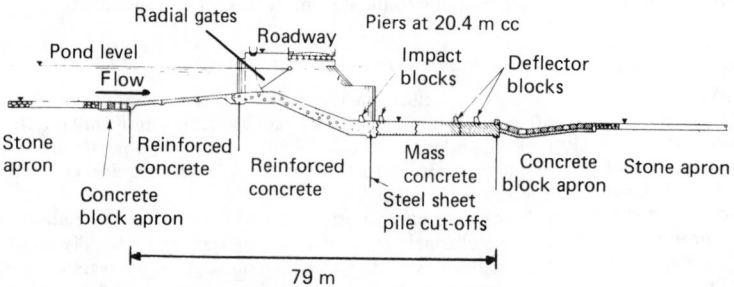

Figure 22.9 Rasul barrage on the river Jhelum, Pakistan. General layout (top), longitudinal section (above). Note the flow from right in layout (top) and from left in section. (Consulting Engineers: Coode and Partners)

trouble unless correctly located. Model tests are desirable before construction. Similar measures are used to train alluvial rivers at bridges. The guide banks and spur heads are protected against scour, by rip-rap or concrete slabs (see pages **22**/15).

Low-level sluices provided in the weir or barrage, generally adjoining the canal regulator, have three functions: (1) they discharge river flow during construction at a low level; (2) during operation of the works they enable the approach to the regulator to be sluiced at intervals to remove deposit of sediment deposit; and (3) if kept open during a flood they draw the main stream towards the canal regulator, thus maintaining a deep channel for water to gain access to the intake during the dry season. To fulfil these functions the sill should be well below the canal regulator sill level and the sluices should have sufficient capacity to influence flood flow distribution. A divide wall is often provided normal to the weir between undersluices and weir to enable the canal to draw supplies from a pocket of low-velocity water, the undersluices being kept closed. A divide wall also facilitates the sluicing operation. If the canal must operate continuously, control of coarse sediment can be provided by tunnels beneath the level of the canal regulator sill, which draw off the bed load and discharge it downstream.[20]

Downstream of the weir and undersluices, a floor is provided to protect the foundations against scour (Figure 22.9). The drop in water level across the weir or undersluices is accompanied by the formation of a hydraulic jump, except possibly at high flood flows when it may be drowned. A flexible apron of loose stone

or concrete blocks is beneficial as an extension to the floor.

For design of floor and apron see page **22**/15. To allow for nonuniform discharge distribution, the design discharge per unit width of floor should exceed the mean by an allowance depending on the approach conditions. In India and Pakistan a factor of 20% has generally been added for alluvial rivers but in extreme conditions it should be higher, e.g. where curvature of approach could cause a high concentration.

22.1.4.3 Irrigation canal structures

Canal head regulators are usually located immediately upstream of a weir or barrage (see Figure 22.9). On alluvial rivers the intake should be well above the sill of the undersluices. A stilling basin of sufficient depth, to ensure that the hydraulic jump is retained within it, is essential where the canal bed is erodible, and is also generally provided where the canal is lined.

Where the general ground slope exceeds the design slope of a canal, falls or drop structures are required at intervals to dissipate the excess head and lower the canal to conform to the ground level. Falls are designed in a similar way to weirs, with ungated crest and stilling basin. To reduce cost, the width of waterway is often made less than the width of canal. The upstream contraction presents little difficulty, but the downstream expansion must be gentle to avoid asymmetrical flow downstream (see page **22**/4).

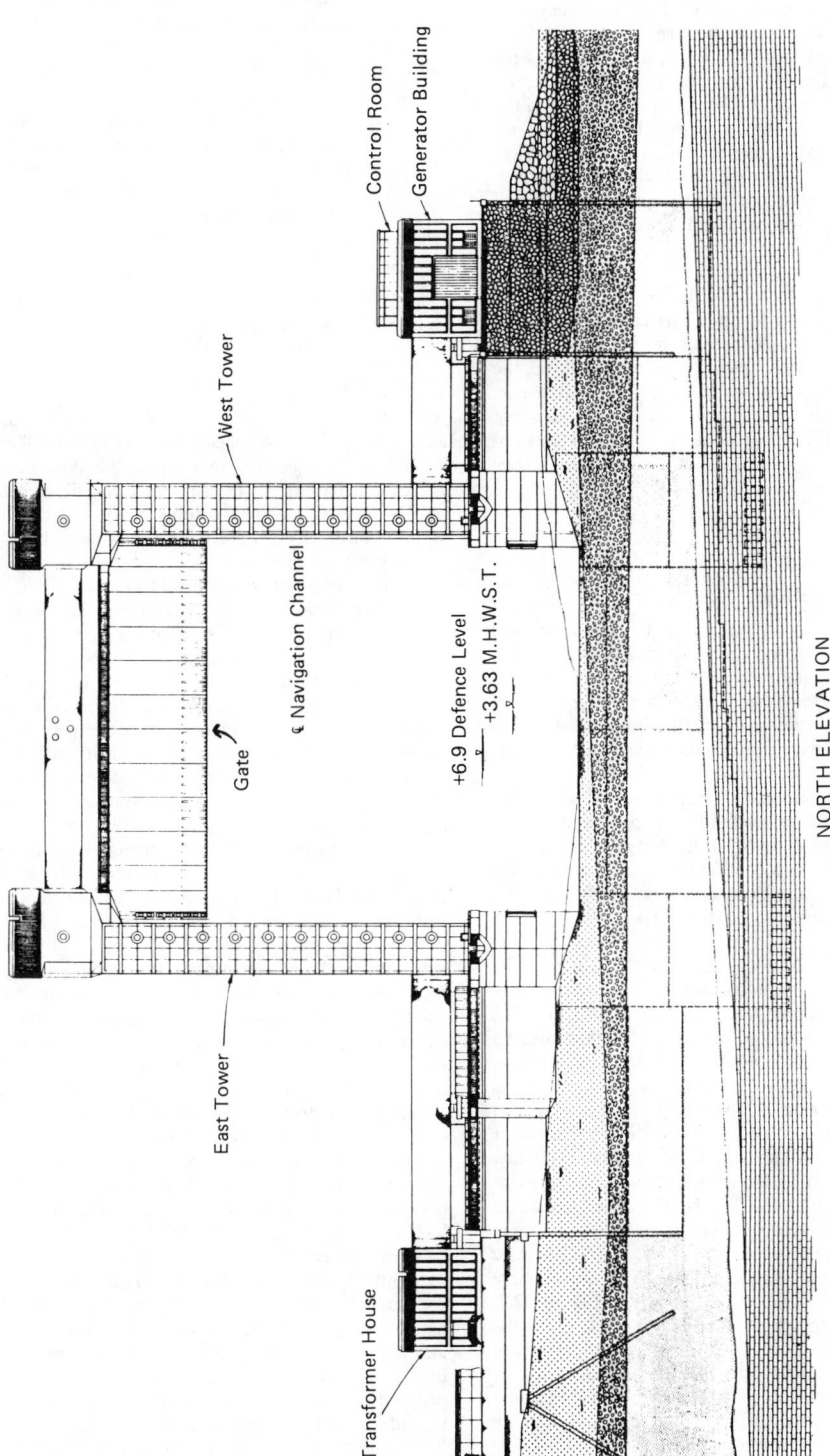

East Tower

Transformer House

West Tower

Control Room

Generator Building

Gate

₡ Navigation Channel

+6.9 Defence Level
+3.63 M.H.W.S.T.

NORTH ELEVATION
(LOOKING DOWNCREEK)

Figure 22.10 Tidal barrier, Barking Creek. (Consulting Engineers: Binnie and Partners)

22.1.4.4 Tidal barriers

The risk of serious flooding from tidal surges penetrating inland via estuaries and tidal inlets has led to several major schemes for tide-excluding barriers. These are in the form of gates, perhaps single gates for schemes of modest size but multiple gates for major estuaries. Navigation is often the controlling feature determining the necessary span, the elevation of the sill and the clearance under any structure spanning over the waterway.

Very large vertical lift gates have been used as tidal barriers and one example, at Barking Creek in the Thames Estuary,[21] is illustrated in Figure 22.10. The gate normally rests at the top of its support towers, thus providing clearance for navigation by medium-sized ships.

Rising sector gates do not require a high supporting structure because they normally rest below the bed of the navigation channel. The main gates of the Thames Barrier are of this type and their operating mechanism is such that they can be rotated through 180° from their normal position in their sills on the estuary bed to raise them above water level for maintenance. They turn through 90° to close the barrier against the tide (Figure 22.11). The rising sector gates in the four main spans of the Thames Barrier have a span of 61 m and effectively they form box-girders between their end wheels. There are six subsidiary gates with spans of 31.5 m.[22]

22.1.5 Permeable foundations

Special consideration is required if a hydraulic gradient will exist across a structure founded on permeable materials. Examples include weirs, regulators across canals, barrages and tidal barriers. Two important requirements are that every part of the structure must be safe against uplift pressures beneath and that underflow or seepage through the permeable materials should be controlled so that there is no failure by 'piping'. Where a continuous impermeable stratum is within reach, underflow can be prevented by a line of sheet piles or a curtain wall intersecting it, or possibly by grouting, but the sealing must be perfect. If, however, the permeable materials are too deep for this treatment, the floor must be safe against uplift pressures exceeding the tailwater level acting on the underside of the structure throughout.

Uplift depends on the hydraulic gradient of flow through the material beneath the work, reducing from the upstream water level to the downstream water level. Its distribution may be affected considerably by the nonuniformity of the materials so a prior investigation of the character of the material, its uniformity and the existence of strata of different permeability is necessary. The floor upstream of a weir or gates is safe against uplift because of the water load above but the downstream floor is particularly vulnerable at times of high upstream and low downstream water levels. Measures to reduce uplift pressures on the downstream floor include the lengthening of the upstream floor and provision of transverse lines of sheet piling upstream or beneath the weir, both serving to lengthen the effective seepage path, and provision of relief drains. Typical protective measures beneath a gated structure are shown in Figure 22.9. 'Piping' consists of the removal of foundation material by the flow of seepage water. It can occur at the tail end of a structure where the underflow emerges and is a potential cause of undermining and ultimate failure of the structure. It is caused by excessive exit gradient. Information on flow nets to determine uplift pressures and exit gradient is given in Chapter 9.

It is usual to protect against piping, where the foundation material is granular, by providing coarser filter material to intercept the seepage over its exit area. This is generally covered by loose stone or other protection against scour, but in case this

should fail, other measures are needed to reduce the exit gradient. Such measures include the lengthening of the structure and the provision of transverse lines of sheet piling to reduce the overall hydraulic gradient, provision of relief drains and the provision of a curtain wall or line of sheet piling at the tail end of the floor. The last is most important to avoid a locally steep gradient and protect the floor from undermining by scour, but it should not be too deep because it increases uplift beneath the floor. The upstream or central sheet piling should extend laterally into the flanking embankments, and lines of piling are carried around as may be necessary to intersect seepage paths and box in the foundations. For general design procedures, reference may be made to Haigh,[20] Leliavsky[23] and Foy and Green.[24]

22.1.6 Energy dissipation

22.1.6.1 Stilling basins

At weirs, barrages, sluices, spillways, tunnel outfalls, canal falls and in general where a sharp fall occurs in total energy level, a stilling basin is needed to contain the flow in the region of energy dissipation. This is especially important where the channel bed is erodible. The surplus energy may be dissipated by water spilling into a pool, which may be in bed rock, or lined with rip-rap or concrete.

In most cases the energy head to be dissipated is sufficient to create supercritical flow, defined on page **22**/3. A hydraulic jump is then generally the most effective and economical way of dissipating the surplus energy. The object is to provide a stilling basin lined with nonerodible material, usually concrete, deep enough to retain the jump over the whole range of flow conditions and long enough for the eddies generated in the jump to be reduced to an acceptable intensity before reaching the channel downstream. The minimum depth is thus related to the characteristics of the jump while the minimum length is related also to the degree of stilling required. Where the channel bed is erodible, a greater length of basin is generally required than where it is in rock or is concrete-lined. In the basin, chute blocks, baffle blocks or piers are often provided to help to stabilize the jump and reduce the length of basin required.

As shown earlier, the stability of a hydraulic jump is expressed by the pressure–momentum equation (Equation (22.5)) representing the condition at which the jump is at its limit of stability, i.e. any increase in discharge or upstream head would cause 'sweep-out' or movement of the jump downstream and possibly out of the basin. In the design of stilling basins, however, the quantities which are known are usually the discharge, head drop and tailwater level and it is required to determine the basin floor level. Equation (22.5) therefore cannot be applied directly, but the maximum acceptable floor level can be easily found with the aid of Figure 22.2. The procedure is to compute upstream and downstream total energy levels (water level + velocity head), compute $H_L = H_1 - H_2$ (see Figure 22.2), compute critical depth d_c by Equation (22.3a), compute H_L/d_c, read off H_2/d_c directly beneath H_L/d_c, i.e. for same F_1, and compute H_2. This gives the minimum depth of basin floor beneath tailwater total energy level. It applies to a plain floor and may be reduced by 10 to 20% if chute blocks and/or baffle blocks and end sill are provided. However, it is often the practice to provide the full depth and consider the blocks to provide a safety margin in addition. It is usually necessary to determine minimum basin depth for several discharges throughout the range, because the most severe case is not always with the maximum discharge. When determining q in cases of nonuniform distribution across the basin it may be necessary to use a value rather higher than mean $q = Q/B$, where Q is the total

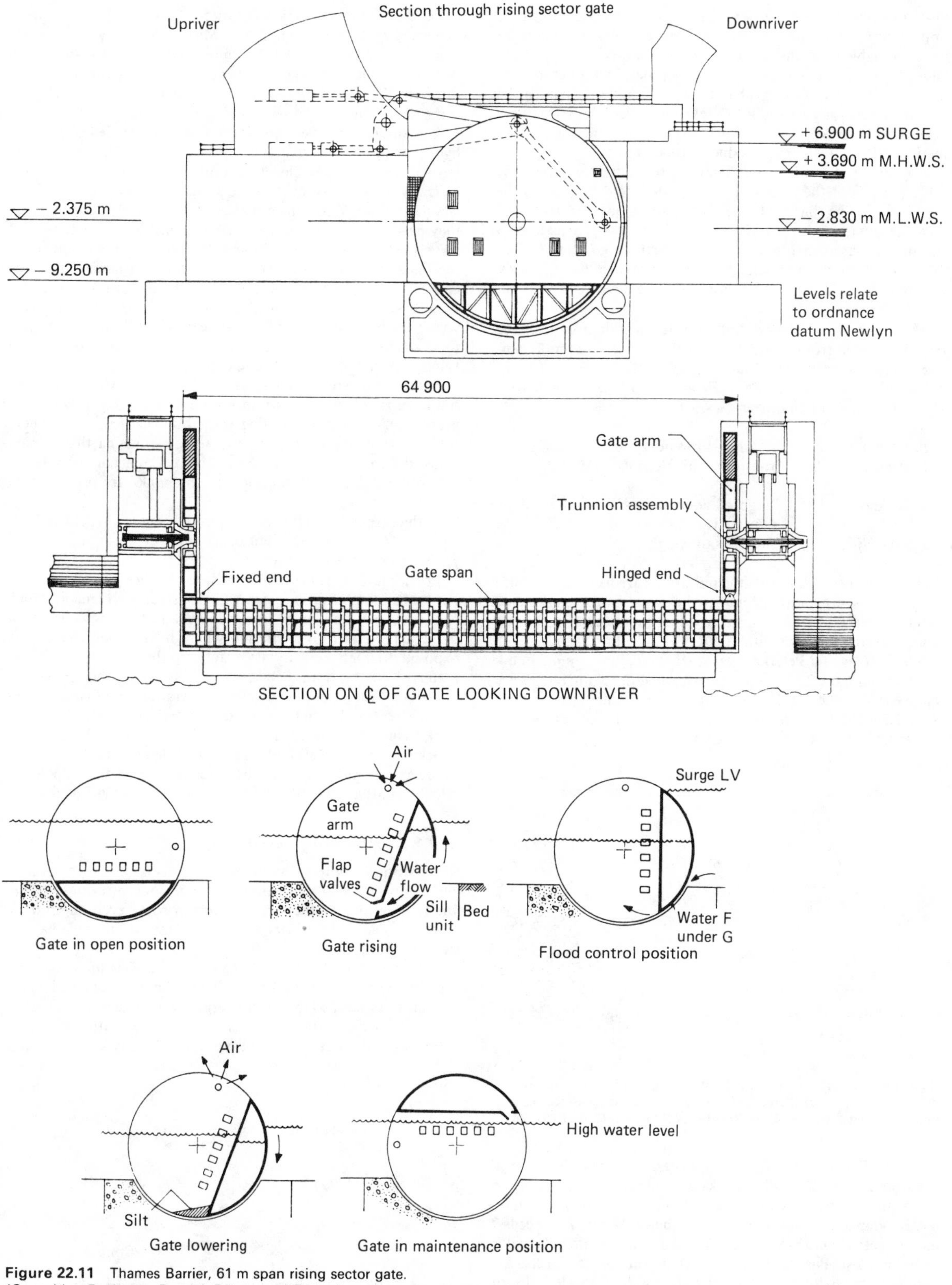

Section through rising sector gate

Upriver Downriver

▽ + 6.900 m SURGE

▽ + 3.690 m M.H.W.S.

▽ − 2.375 m

▽ − 2.830 m M.L.W.S.

▽ − 9.250 m

Levels relate
to ordnance
datum Newlyn

64 900

Gate arm

Trunnion assembly

Fixed end Gate span Hinged end

SECTION ON ℄ OF GATE LOOKING DOWNRIVER

Air

Gate
arm

Surge LV

Flap
valves Water
flow

Sill Bed
unit

Water F
under G

Gate in open position Gate rising Flood control position

Air

High water level

Silt

Gate lowering Gate in maintenance position

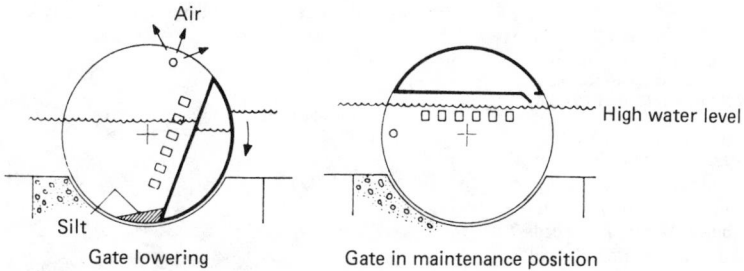

Figure 22.11 Thames Barrier, 61 m span rising sector gate.
(Consulting Engineers: Rendel, Palmer and Tritton)

discharge and B the width. Tailwater level is clearly of critical importance for the stability of the jump and it is necessary to have a reliable stage discharge curve, with allowance for future changes as, for example, due to channel bed degradation downstream. The lowest probable levels should be used. In the case of basins for gated spillway releases, where discharge may be increased rapidly over a short period, allowance should be made for low tailwater levels due to time lag.

The length of basin required cannot be defined so precisely. On a plain floor the length of a jump may be 4 or 5 times the depth d_2 in the basin. If residual eddies can be tolerated downstream because the bed is not erodible or is protected by a flexible apron, as in Figure 22.12, a length of $4d_2$ may suffice. Where chute blocks and baffle blocks are provided in such cases, a length of $2.5d_2$ is sometimes considered adequate (but see below).

Many standard designs of hydraulic jump stilling basins have been developed from model tests, one of the most comprehensive being that of Bradley and Peterka.[25,26] Four types of jump were defined according to the Froude number F_1, each with somewhat different characteristics, namely:

F_1 from 1.7 to 2.5	Pre-jump, low energy loss
F_1 from 2.5 to 4.5	Transition, rough pulsating water surface
F_1 from 4.5 to 9.0	Range of good jumps least affected by tailwater variations
F_1 exceeding 9.0	Effective but rough

If F_1 is in the range 2.5 to 4.5 the pulsations are likely to produce surface waves which are propagated downstream. The Froude number is generally determined by other factors, but if there is any choice it is clearly desirable for it to be within the range 4.5 to 9.0. Bradley and Peterka's basin III for F_1 between 4.5 and 9 is shown in Figure 22.12. The dimensions of the chute blocks are made equal to the depth d_1 and those of the baffle blocks range from $1.3d_1$ for $F_1 = 4$ to $3d_1$ for $F_1 = 14$. The height of end sill ranges from $1.2d_1$ for $F_1 = 4$ to $2d_1$ for $F_1 = 14$.

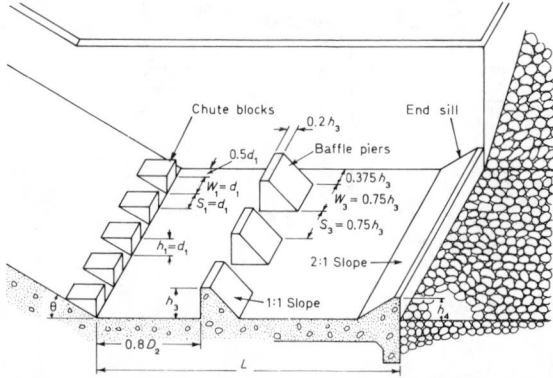

Figure 22.12 US Bureau of Reclamation stilling basin, type III. (After Beichley (1978) 'Hydraulic design of stilling basin for pipe or channel outlets.' USBR Water Resources Research Report No. 24)

Where F_1 is between 2.5 and 4.5 (basin IV) the chute block height is $2d_1$ and the baffle blocks are omitted or, according to Bhowmik,[27] a special arrangement of blocks and deflector may be provided to give improved jump stability. Where F_1 exceeds 9 (basin II) the baffle blocks are omitted and a dentated end sill is recommended. Basins II and IV, having no baffle blocks, are required to be longer than basin III, with floor lengths of approximately $4d_2$. In the case of high head structures, if the velocity much exceeds 15 m/s, chute blocks and baffle blocks are

liable to be damaged by cavitation. They can be omitted or protected by steel cladding, as at Mangla Spillway.[28]

Erosion of bed and banks immediately downstream of the stilling basin can be a serious problem, whether the head drop through the structure is great or small – see remarks on transitions, page 22/4.

A normal cause of erosion is the residual turbulence from the hydraulic jump. This may scour the bed beneath the level of the basin floor, so a flexible protection such as rip-rap is needed which will adjust its level to the scoured bed downstream of it (see page 22/8). When the banks are formed of erodible material they need slope protection to guard against local velocities and wave wash. In the case of weirs and barrages on alluvial rivers the banks are carried downstream a short distance – perhaps equal to a quarter of the width of river channel (see Figure 22.9). A loose stone apron is provided at the toe. In the case of canals where the banks are erodible, the slope protection is continued for a distance in which the surface waves will be reduced and velocity distribution will become normal.

A layout of stilling basin and canal banks which has been found satisfactory is shown in Figure 22.3e. The gently diverging side walls are free-standing at their downstream ends, where they consequently do not have to serve as high earth-retaining walls; the channel downstream is widened to accommodate the side rollers which will develop and the banks are protected by rip-rap.

In the case of small flows, shorter and simpler structures have been used, e.g. the straight-drop spillway basin of the US Department of Agriculture.[29]

For large flows and high heads, experience has shown that hydraulic jump basins are generally satisfactory. Damage which has occurred has been due mainly to the basin being of inadequate depth, to cavitation where baffle blocks have been exposed to high velocity flow and to abrasion due to loose materials in the basin.[30,31] In some cases these materials may have remained from river diversion operations but in other cases bed material and even rip-rap has been carried into the basins by backwash. There have also been instances of vibration and shock due to flow instability. In large-scale basins it is especially necessary to guard against flow separation at the side walls, which can be a cause of both these last effects and of backwash.

22.1.6.2 Bucket energy dissipators

The hydraulic jump stilling basins described above are effective but costly, especially for high-discharge concentrations. Where the foundations of the structure are in rock, even an erodible rock, a much higher degree of residual turbulence may be acceptable.

A submerged roller bucket (see Figure 22.13) is suitable over a wide range of Froude numbers. The bucket is placed well below the tailwater level so that a submerged roller forms in the bucket and exit velocities are not excessive. Compared with a hydraulic jump basin, it is deeper but shorter and generally less costly; but

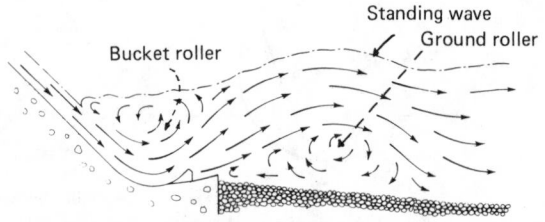

Figure 22.13 Submerged roller bucket – Angostura-type slotted bucket. (After Beichley and Peterka (1959) 'The hydraulic design of slotted spillway buckets.' *Proc. Am. Soc. Civ. Engrs*, **85**, HY10)

the range of tailwater level for satisfactory operation is limited, which precludes its use in some cases. Rules for design have been given by McPherson and Karr[32] and by Bleichley and Peterka[33] who found that slotted buckets were superior to plain buckets.

22.1.6.3 Terminal structures for pipes and valves

High-velocity jets from pipes and terminal valves have considerable erosive power, even on hard rock. Means of protection include the use of valves which disperse the jet in the air, e.g. the cone valve, or valves which project the jet some distance, where a plunge pool can be provided, or structures devised to contain the jet and allow most of the energy to be dissipated before discharge into an erodible channel.

Figure 22.14 shows an impact stilling basin developed by the US Bureau of Reclamation (USBR)[5] for pipe and open-channel outlets with discharges up to 10 m³/s and velocities up to 9 m/s. It may also be considered for terminal valves within the limits

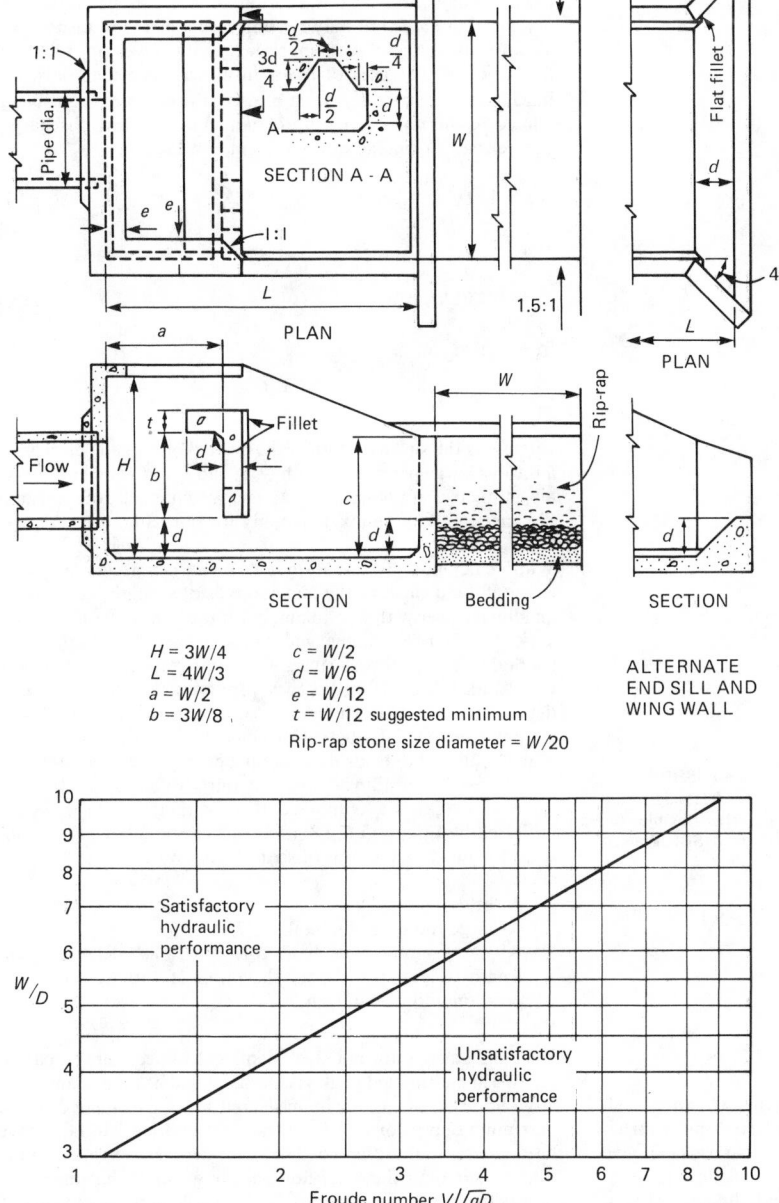

$H = 3W/4$ $\quad c = W/2$
$L = 4W/3$ $\quad d = W/6$
$a = W/2$ $\quad\quad e = W/12$
$b = 3W/8$ $\quad t = W/12$ suggested minimum

Rip-rap stone size diameter = $W/20$

ALTERNATE
END SILL AND
WING WALL

W = inside width of basin
D = square root of area of flow entering basin
V = velocity of flow entering basin

Tailwater depth uncontrolled

Figure 22.14 US Bureau of Reclamation impact-type energy dissipator – basin VI. (After Beichley (1978) 'Hydraulic design of stilling basin for pipe or channel outlets.' USBR Water Resources Research Report No. 24)

stated. The required dimensions may be obtained from Figure 22.14.

A special basin has been developed by the USBR for hollow jet valves.[5,34] Basins have also been used for cone valves. A very effective energy dissipator for pipe outlets is a vertical well in which the pipe outlet is deeply submerged at a short distance above the bottom. Figure 22.15 shows a USBR type of well.[35]

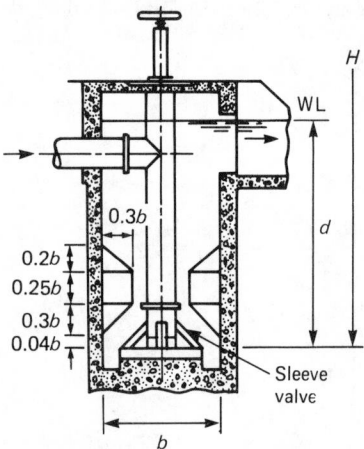

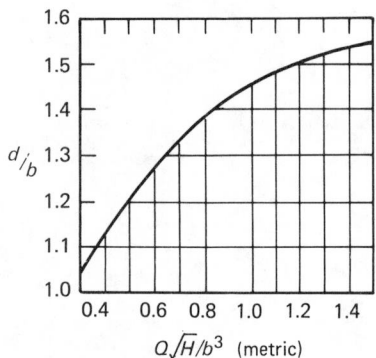

Figure 22.15 Vertical stilling well with sleeve valve (USBR design). Well is of square section in plan with corner fillets as shown. Q is discharge, H is head above pedestal. (After Beichley (1978) 'Hydraulic design for pipe or channel outlets.' USBR Water Resources Research Report No. 24)

Regulation is provided by the sleeve valve at the pipe outlet, operated from above (see page **22**/31).

22.1.7 Scour and erosion

22.1.7.1 Depth of scour at structures

Apart from scour downstream of stilling basins, structures such as bridges, jetties, groynes and constrictions forming obstructions to flow in rivers and channels with erodible beds can give rise to scour due to disturbance of the normal flow pattern. Scour can also be caused by oblique flow at the upstream of control structures such as barrages and regulators. It is generally required to estimate the depth of scour so that adequate protection can be provided or so that the foundations can be located at sufficient depth to avoid the possibility of undermining.

Local scour results from the deflection and, hence, concentration of flow caused by an obstruction. The depth of scour depends on the shape of the obstruction, its orientation to the flow, the channel cross-section and discharge, the character of the erodible bed material, the sediment in transport and the time history of the flow. With so many variables it is not surprising that there is no single formula available for calculation of scour. In the case of important works it is usual to carry out model tests.

It is also possible to determine the order of magnitude and probably the upper limit of scour depth by comparison with depths observed in actual cases, providing a useful check on model results or a fair indication in other cases. Local experience is a guide but may not embrace the highest discharges. To apply historical data from elsewhere it is necessary to adjust for scale. In the cases of rivers in alluvial bed materials, the Lacey regime formulae[19,36] can be used, the depth of scour being related to the normal depth of channel of the same discharge. The relevant formulae in the present context are:

$$B = 4.8Q^{1/2} \tag{22.7}$$

and

$$d = 0.47(Q/f_L)^{1/3} \tag{22.8}$$

from which can be derived

$$d = 1.34q^{3}/f_L^{1/3} \tag{22.9}$$

where B is the surface width, d the mean depth, Q the discharge, q the discharge per metre width Q/B, and f_L is a sediment factor, all in metric units relating to stable channels of constant discharge. f_L may be taken as unity for fine sand

Width calculated by Equation (22.7) with Q = design discharge is a useful indicator of the maximum bridge length required for an alluvial river with floodplain, but if the banks are of cohesive materials, the river channel width may be less; Nixon[37] found the average widths of British rivers to be approximately $3Q^{1/2}$, where Q is bank-full discharge. Lacey proposed that the maximum depths of scour at sharp bends in alluvial rivers could be taken as approximately $2d$, where d is calculated from Equation (22.8). Inglis[38] collated data of deep scour observed at structures and training works in alluvial rivers at thirty different locations in India and Pakistan, compared them with the normal depths indicated by Equation (22.8) and reached the following conclusions for maximum depth of scour below water level:

(1) At bridge piers, $2d$.
(2) At large radius guide banks, $2.75d$.
(3) At spurs along river banks, $1.7d$ to $3.8d$, depending on length of spur projection, sharpness of curvature of flow, position and orientation.

Here d is Lacey's normal mean depth calculated from Equation (22.8) using estimated peak discharge. It will be appreciated that large flood flows cannot be measured but are estimated, while maximum depths of scour are transient and may, in fact, have been greater than observed. The scour depths are related to the total rather than the local flow on the grounds that the scour results from the concentration of the whole flow. In the case of braided rivers, allowance could be made for the division of total flow into several channels. Scour depth in rivers in gravel and boulders would be less than indicated but the difference may be small. Scour depths in cohesive materials could be less because of the time required to reach maximum scour.

For the purpose of design of aprons upstream and downstream of barrages in northern India, maximum depth of scour below water level was taken as $1.5d$ at the upstream end of the hard floor and $2d$ at the downstream end of the basin. Here d was calculated from Equation (22.9) using mean q.

Scour at bridge piers has been studied in some detail in models, scour depth being related to discharge per unit width and sometimes expressed as depth below upstream bed level.[39-41] To apply such relationships, the discharge concentration, which can in the worst case greatly exceed the average, has to be estimated, and the upstream depth has then to be calculated for the corresponding flood condition. The latter can be done using Equation (22.9) which is likely to give a conservative value because of time lag and sediment load. The calculation should be checked by use of the appropriate Inglis factor above.

22.1.7.2 Protection against scour

This generally consists of one of the following materials.

Boulders. Boulders from the river bed which are generally rounded and therefore less stable than quarried stone of similar weight.

Rip-Rap. Rip-rap, or pitching of quarried stone, is widely used. In some cases it is hand-packed, especially on side slopes which are expected to remain as placed without settlement, but with increasing use of mechanical equipment it is more often placed in a random manner. This is also preferable in locations where it is expected to settle or move down due to scour. On side slopes the thickness of rip-rap should be sufficient to accommodate the biggest stones without large gaps – at least 1.5 × median stone diameter – and an underlayer or filter of smaller stone is generally provided to prevent the base material from being washed out by wave action. In the case of bed protection, surfaces not subject to scour may be treated in the same way, but at transitions from stilling basins and in general where the channel bed may scour beneath the apron level, the volume of rip-rap should be sufficient to protect a slope at the angle of repose of the rip-rap on the bed material (for a sand bed generally 1:2) extending from the apron level to the level of anticipated deepest scour. For this purpose the rip-rap may be laid on a prepared slope or it may be laid in a horizontal apron which it is assumed will settle to a slope when scour occurs. A margin should be allowed for uneven settlement.

The size of rip-rap which will remain stable may be estimated from Figure 22.16. Sixty per cent by weight of the material should be equal to or larger than the size shown. In the case of rigid structures it may be dangerous to rely on loose stone protection; it is generally best to provide foundations at low levels beneath possible scour. If stone or concrete blocks are used to protect existing structures they should be placed as low as possible beneath normal bed level.

Derrick stone. Derrick stone is stone in blocks too heavy to be placed by bulk handling and which therefore requires individual placing. It is usually placed on an underlayer of graded rip-rap.

Concrete blocks, slabs or units of various shapes. As the density of concrete is less than that of stone, larger and heavier blocks are required than the corresponding stone sizes. Concrete blocks are used in locations where stone of suitable quality and weight is not available or is too costly. Concrete blocks or slabs on edge, e.g. 2 m wide × 0.5 m long × 1 m deep, have been used successfully for flexible aprons downstream of barrages in rivers with sand beds. Concrete slabs are used in slope protection but require good compaction of fill beneath to avoid uneven settlement. Concrete units of special shapes have been developed

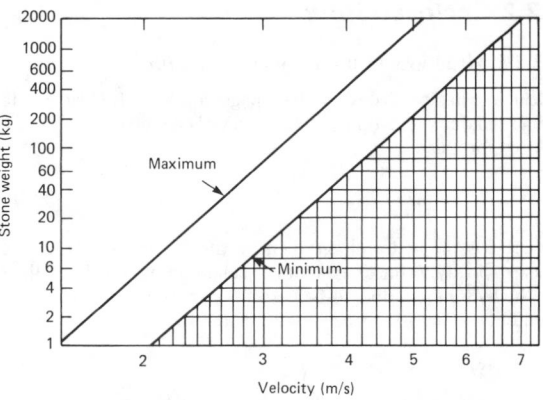

Figure 22.16 Stability of loose rock in flowing water. Graph relates to rock of specific gravity 2.65. For other specific gravity rock weight = weight shown × $1.7s/(s-1)^3$ where s = specific gravity. Use minimum weight graph in normal flow and maximum weight graph for very turbulent flow. (After US Army Corps of Engineers (1952–70) Hydraulic design criteria. US Army Engineer Waterways Experiment Station, Vicksburg, Mississippi)

which require less concrete than do concrete cubes for the same duty; some of these are extensively used in coastal protection and can also be used in river channel works.

Gabions. Gabions, consisting of wire crates containing boulders or broken stone, generally wired together to form an apron, form an economical temporary protection against erosion and have been used in permanent works, though the wire crates may be subject to corrosion. The standard metric size is 2 × 1 × 1 m but thinner mattresses are available. They offer great resistance to removal by flow and a gabion apron has considerable flexibility in adjusting to scour, though less than an apron of rip-rap.

Asphalt. Asphalt provides a smooth impervious cover but does not have much flexibility.

Sheets of polypropylene. These, and other synthetic materials, woven to a fine mesh provide an effective filter layer over sand and have been used to form thin mattresses, with pockets filled with cement grout, for side slope protection.

Brushwood fascine mattresses. These form a traditional protection consisting generally of willow twigs bound in bundles and formed into longitudinal and lateral layers bound together before it is launched by weighting with stone and sinking into place, which is still used and has a considerable life under water.

Vegetation. Vegetation, in particular certain grasses and shrubs, when established above normal water level, can protect a bank against occasional high level wave wash or even shallow overtopping.

For protection of formed banks in cut or fill, any of the above materials would be suitable (see also Chapter 18) subject to adequate protection against scour of the toe of the bank or the channel bed near it. This may be provided by a line of sheet piling at the toe or by a flexible apron laid horizontally which will subside and protect the underwater slope when scour occurs. Quarried stone rip-rap is usually used for the apron where available.

22.2 Enclosed flow

22.2.1 Head loss in large conduits and tunnels

Head loss in pipes is dealt with on pages **5**/8 to **5**/11. Head loss in large conduits and tunnels may similarly be estimated by the Darcy formula:

$$i = \lambda V^2/2gD = \lambda V^2/8gm \qquad (22.10)$$

where i is the hydraulic gradient, λ the friction factor, V the mean velocity, D the diameter of circular conduit flowing full, or m the hydraulic radius to be used for part full and noncircular conduits. λ and Manning's n are related by:

$$n = \lambda^{1/2} D^{1/6}/10.8 = \lambda^{1/2} m^{1/6}/13.6 \qquad (22.11)$$

In nearly all actual cases of large conduits the boundary cannot be classed as smooth or rough but falls within the transition region. λ therefore depends on the effective roughness and on the Reynolds number VD/v or $4Vm/v$, where v is the kinematic viscosity (for values of v see pages **5**/8 to **5**/10). Although many types of roughness are composite, e.g. smooth concrete with projections due to formwork joints, and therefore the equivalent sand roughness concept is not completely representative, it does provide a method of predicting the friction factor, based on recorded experience. In the case of new works this depends on the type of forms used, quality of workmanship and degree to which projections are ground down. Deterioration occurs with age and use. A steel lining may corrode and be roughened by tuberculation, as for pipes. Concrete inverts may be roughened by abrasion during river diversion. There may be deposits due to leaching through joints and cracks in a concrete lining, even vegetation and animal growths, while the deposit of slime by untreated water is commonplace.

Typical values of equivalent sand roughness k for new surfaces, based mainly on Ackers[42] and USBR experience,[43] are given in Table 22.1.

It is more difficult to predict the friction factor in a tunnel after many years of use; the best guide is often obtained from actual measurements in similar tunnels under similar conditions. Observations in many tunnels have been published.[43–45] The effect of slime has been studied by Colebrook.[45]

Table 22.1

Surface	k range (mm)
Asbestos	0.012 to 0.015
Spun bitumen lined	0.0 to 0.030
Spun concrete lined pipes	0.0 to 0.030
Uncoated steel	0.015 to 0.060
Coated steel	0.03 to 0.15
Rivetted steel	0.3 to 6.000
Wood stave, planed planks	0.2 to 1.5
Concrete:	
against oiled steel forms with no surface irregularities	0.04 to 0.25
against steel forms, wet mix or spun precast pipes	0.3 to 1.5
against rough forms, rough precast pipes or cement gun	0.6 to 2.0
smooth trowelled surface	0.3 to 1.5
Glazed brickwork	0.6 to 3.0
Brick in cement mortar	1.5 to 6.0
Ashlar and well laid brickwork	1.5
Rough brickwork	3.0

When a suitable k value has been determined, the relative roughness k/D or $k/4m$ can be calculated and a value of λ determined from Figure 22.17 which is based on the Karman–Nikuradse–Prandtl formulae with Colebrook–White transitions described on page **5**/9. Some examples of large-conduit observations are shown in Figure 22.17.[43,44] Figures 22.18 and 22.19 show characteristics of circular and horseshoe conduits flowing part full.

The surface of conduits for high-velocity flows should be to a very high standard of finish to avoid damage by cavitation – see page **22**/7.

22.2.2 Unlined and lined-invert tunnels in rock

Excavated rock surfaces are very rough and the hydraulics are complicated by 'overbreak', i.e. excavation beyond the minimum required by the specification. Rahm found a relation between the variation in cross-sectional area and the friction

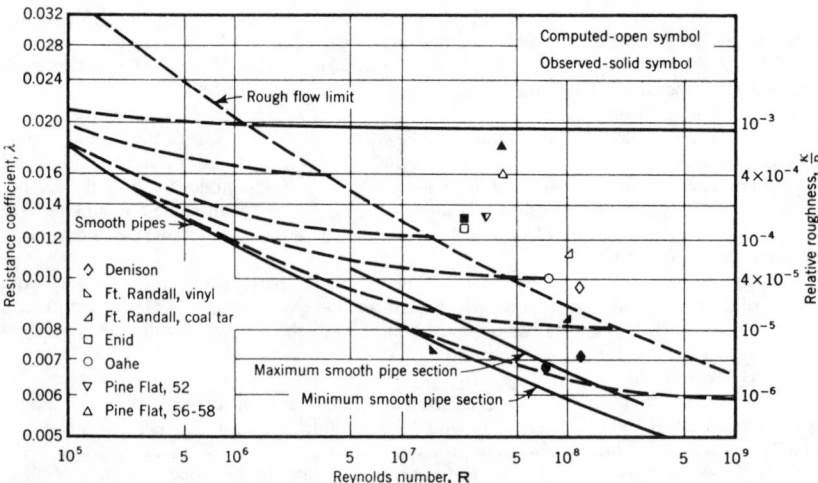

Figure 22.17 Head loss in uniform conduits. Open symbols, computed; solid symbols, observed. (After US Army Corps of Engineers (1952–70) *Hydraulic design criteria*. US Army Engineer Waterways Experiment Station, Vicksburg, Mississippi)

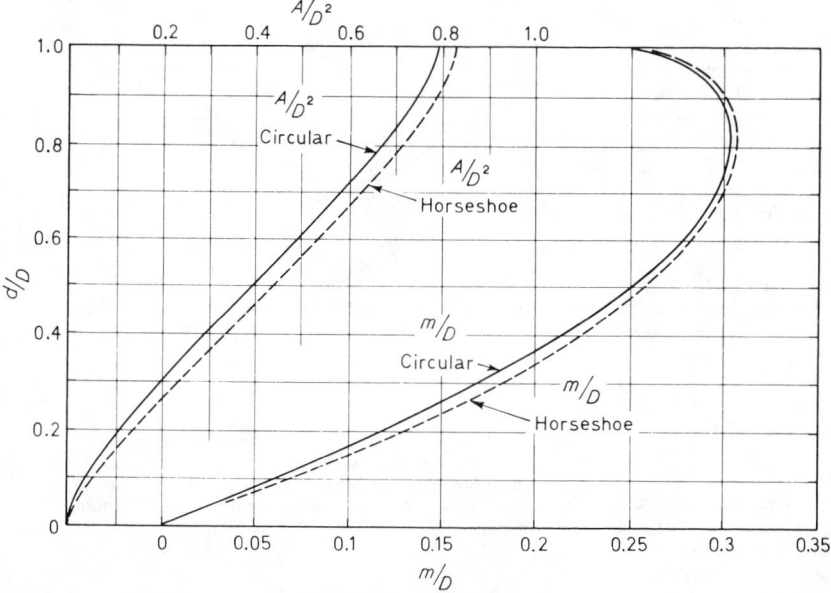

Figure 22.18 Area and hydraulic radius of conduits, part full. For key, see Figure 22.19

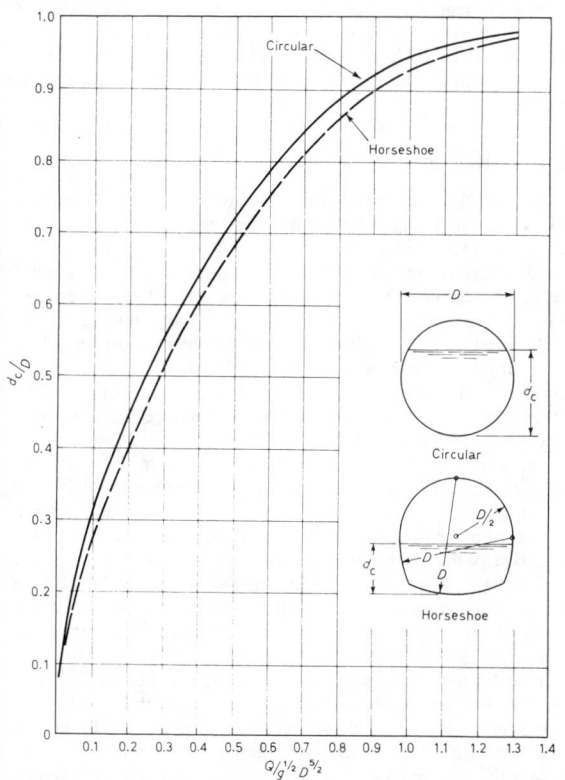

Figure 22.19 Critical depth in circular and horseshoe conduits

22.2.3 Transitions and bends

Transitions may be from circular to noncircular sections or vice versa, or from one circular section to another of different diameter. In conduits for high-velocity flows, transitions are generally gradual to avoid flow separation and possibly cavitation. It is also necessary to adopt moderate rates of expansion if head is to be conserved and instability of flow downstream avoided. Circular sections can be merged into rectangular or horseshoe sections without double curvature and avoiding sharp local divergences. Figure 22.20 shows head loss in diffusers of circular section; curves of similar pattern but slightly differing values apply to diffusers of rectangular section.[47] It will be seen that for expansion ratios of 2 or more the head loss may be considerable unless the angle of divergence is small. Where rapid expansion is required, divide walls may be used so that the

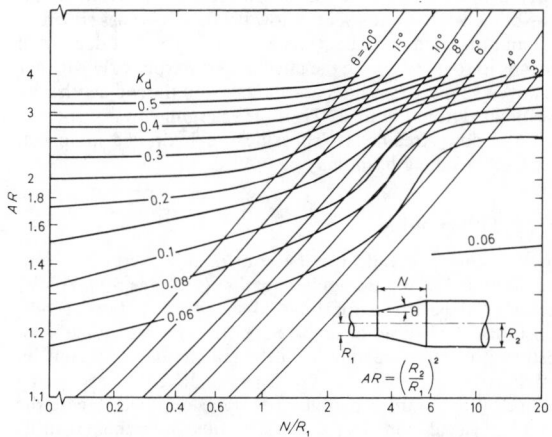

Figure 22.20 Head loss coefficient K_d of conical diffusers with tailpipe. (After Miller (1971) *Internal flow. A guide to losses in pipe and duct systems*. British Hydro-mechanics Research Association, Cranfield)

factor. The subject was further developed by Colebrook[45] and again by Wright,[46] who showed that the resistance of unlined tunnels can be reduced considerably by providing a concrete invert.

flow is carried in a number of ducts, each of which is a reasonably efficient diffuser while the overall angle of expansion may be as much as 90°. For head loss in a sudden enlargement, see Chapter 5, pages **5**/10 and **5**/11; expansions of this type are sometimes used for energy dissipation in closed conduits. Contractions may be more rapid than diffusers, but to avoid head loss due to the formation of a *vena contracta* in the downstream conduit, it is desirable to provide a rounded external angle between the transition and conduit at a radius of at least one-sixth diameter. In high-velocity flow this is an area of potential cavitation and the radius should be greater.

Head losses at bends in large conduits are similar to those in pipe bends (see pages **5**/11). A compromise then has often to be reached between the greater head loss of a short-radius bend and the greater cost of long-radius bend; bend radii of between 1.5 and 3 diameters are often adopted. The flow instability induced by a bend persists for a distance of many diameters downstream and may affect the performance of turbines or pumps.

Bifurcations and manifolds, dividing the flow, for example, for two or more machines, are generally designed with great care to achieve a smooth change of velocity, absence of swirl and minimum head loss. Model tests with air are useful to indicate flow pattern and pressure drop in closed conduit transitions; relatively low pressures are used to avoid compressibility effects.

Transitions leading from subcritical flow in open channels to closed conduit flow may, where the approach velocity is low, be designed on the same principles as apply to intakes from reservoirs (see page **22**/25). Sharp corners lead to separation and the formation of a *vena contracta* with head loss; this may be avoided by providing a rounded or bellmouth entry. With higher approach velocity the transition should be more gradual, with curves of larger radius. To avoid the formation of a hydraulic jump with resulting air entrainment the design should be such that the contact between free surface and roof occurs where the flow is subcritical, preferably with Froude number well below unity.

22.2.4 Exits

If the exit of a conduit is fully or partially submerged, head loss can be reduced by providing a gradual expansion, which can often be extended in the tail channel. If the conduit exit is not submerged, a free surface may develop some distance upstream of the exit portal, even when the conduit is flowing under pressure. The depth of the exit depends on downstream conditions but where tailwater level is low the flow becomes supercritical and the conduit exit acts as a control. The end depth still depends to some extent on the tail channel, particularly whether the flow is supported at the sides and bed, but the end depth may be estimated from Figure 22.21. If the emerging flow is supercritical and downstream flow subcritical, a hydraulic jump will occur and a stilling basin may be needed.

22.2.5 Flow routing

Flow through conduits can be routed and energy gradient plotted by use of the Bernoulli equation (see page **5**/8). Allowance should be made for head loss due to friction, bends, transitions and hydraulic jumps. Subcritical flow is routed in an upstream direction starting from the tail channel or a control, and supercritical flow in a downstream direction, using step methods if necessary. Computer programs exist which ease the burden of calculation. To locate a hydraulic jump the pressure–momentum theorem can be used, taking account of the slope of the conduit and the pressure against the conduit roof if submerged downstream. The method is described by Kalinske and Robertson.[48] Critical depths in circular and horseshoe conduits

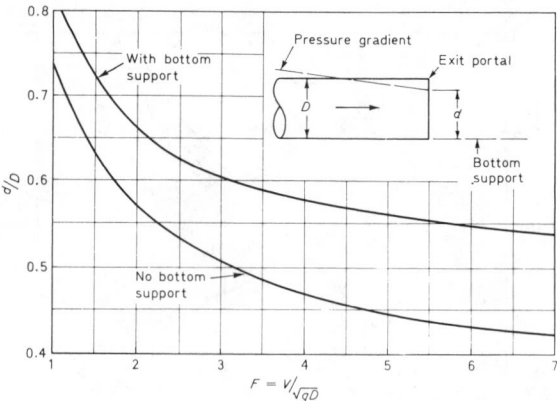

Figure 22.21 Exit depth in circular conduits. V is the mean velocity in the conduit flowing full (based on USWES data). (After US Army Corps of Engineers (1952–70) *Hydraulic design criteria*. US Army Engineer Waterways Experiment Station, Vicksburg, Mississippi)

may be determined from Figure 22.19, and diagrams facilitating computation of jumps in conduits of circular and other cross-sections have been published.[2-4] In cases where hydraulic jumps might occur in closed conduits with undesirable results, due to additional head loss or air (see below), it is recommended that the routing be repeated for several discharges using both high and low values of head loss coefficients and upper and lower limits of tailwater rating curve, to obtain a complete account of the flow.

22.2.6 Drop shafts

Sometimes flow has to be dropped from a high-level system to a low-level system, e.g. from shallow sewers to deep interceptors, from river intakes in mountains to a water-transfer tunnel, from the drainage of an open-pit mine to an adit from an adjacent valley. An economic solution is to use a shaft, but there are problems associated with air entrainment and release in any shaft system, especially if the base of the shaft is submerged by the hydraulic conditions in the low-level system. These problems can be minimized by generating a vortex at the top of the shaft, either by a scroll-shaped inlet chamber (Figure 22.22a) or by a tangential vertical slot (Figure 22.22b).

The vortex action ensures that the flow down the shaft will cling to the walls. This has the advantage of minimizing air entrainment and encouraging the return of air back up the centre to the head of the shaft, and at the same time maximizes head dissipation in the shaft by wall friction. The vortex motion is persistent: it will continue for the full length of fairly deep shafts provided the entry is well designed. The theory of the scroll inlet is given by Ackers and Crump,[49] and the slot inlet has been investigated by Eppema, Jain and Kennedy.[47]

If the bottom of the shaft is submerged, as in Figure 22.22c, it will be necessary to provide an air-release chamber. If the full-bore shaft velocity exceeds about 0.5 m/s, bubbles will be carried down with the flow. Problems – perhaps serious ones – could arise if this entrained air was allowed to travel along the tunnel system (due to potentially explosive blowout further downstream) and hence a stilling chamber should be provided to allow the entrained air to separate and rise to the crown of the chamber where the bubbles will coalesce to return via the vent pipe.

For unsubmerged conditions, Figure 22.22a illustrates a type of collecting chamber at the base of the shaft found suitable for

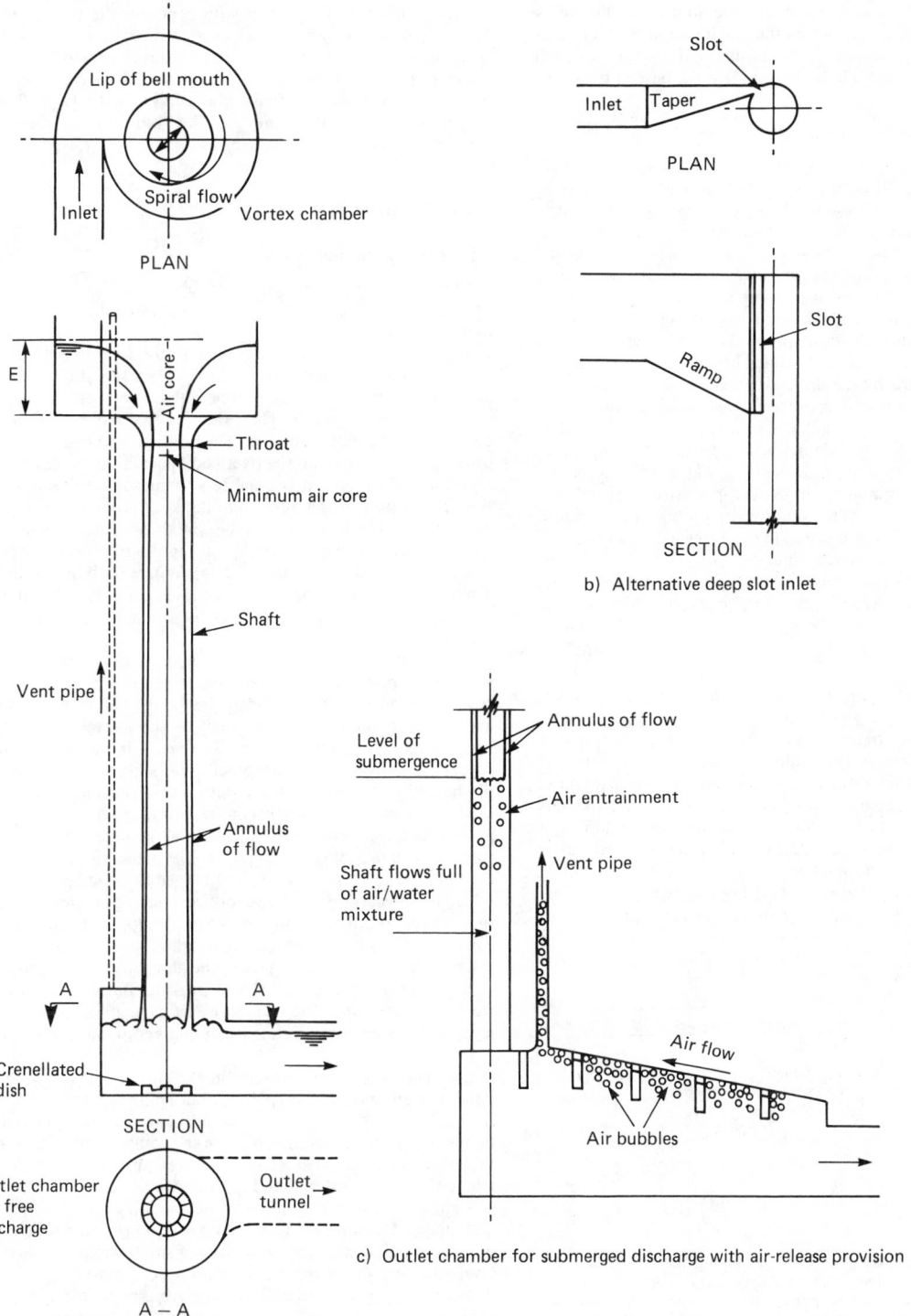

Figure 22.22 Vortex drop. Alternative forms: (a) normal sewage structure; (b) alternative deep slot inlet; (c) outlet chamber for submerged discharge with air-release provision

sewerage systems. With deep shafts, the annulus of flow may reach terminal conditions where the gravitational component is equalled by the friction at the shaft walls, and so there is a limit to the amount of energy to be dissipated at the base of the shaft.

22.2.7 Air problems in conduits

Air entrained at high velocity releases through gates and valves into conduits, e.g. at outlets from reservoirs, air entering from drop shafts or junctions and air entrained at hydraulic jumps, can lead to dangerous air and cavitation problems unless the conduits are adequately vented. Air can also collect and restrict the flow of water. Air release valves, often combined with vacuum relief to admit air if pressure falls below atmospheric, are therefore provided at high points. Vents are often provided in horizontal tunnels downstream of junctions where entrained air may enter. Air which has collected beneath the soffit tends to be carried forward by the flow, even against a small gradient, but with a variable flow may move upstream and downstream at different times. At vertical shafts in pressure conduits and at deeply submerged exits the intermittent escape of air produces shock waves due to slap on the soffit as water replaces the air. This effect can be minimized by vents for controlled air release.

Hydraulic jumps entrain air and when a jump in a conduit is in contact with the soffit much of the air is released downstream. Following model tests in a conduit with various slopes by Kalinske and Robertson[48] and others, and several observations at full scale, the US Army Corps of Engineers[11] use the formula:

$$\beta = 0.03(F_1 - 1)^{1.6} \qquad (22.12)$$

which gives higher values than found in the model tests to allow for scale effect. Here β is the air:water ratio Q_B/Q_W, $F_1 = V_1/\sqrt{(gd_e)}$, V_1 is the upstream velocity and d_e the effective upstream depth (= water area:surface width). A particular application of these formulae is the estimation of air demand downstream of gates or valves located in closed conduits, where high-velocity flow at part openings is transformed to full-conduit flow through a jump (see Figure 22.23). Full-scale observations in three different cases showed that with rectangular gate openings peak demand occurred at 60 to 85% opening, with a secondary peak at about 5%. Further analysis has been provided by Sharma.[50]

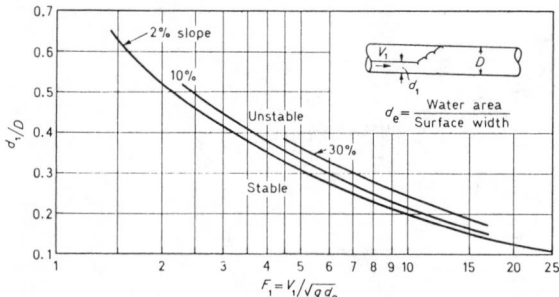

Figure 22.23 Stability of entrained air downstream of hydraulic jumps in circular conduits. (After Kalinske and Robertson (1943) 'Closed conduit flow. Symposium on entrainments of air in flowing water.' *Trans. Am. Soc. Civ. Engrs,* **108**, Paper 2205, 1435)

The air pumped by the jump may be carried downstream by the full-bore flow but, if the velocity is insufficient for this, air will collect immediately downstream of the jump and when a quantity of air has accumulated it will 'blow back' through the jump. Figure 22.23 shows the limiting conditions for the air just carried by the flow, as found by Kalinske and Robertson.[48]

Sailer[51] compared these curves with conditions in a number of full-scale inverted siphons and found verification in that five cases where blowback had occurred were represented by higher values of $(F_1 - 1)$ than shown by the curves, while others giving no trouble were on or below the curves. With large flows, blowback through the jump is, like 'blowout' at the exit, explosive and potentially dangerous.

22.3 Spillways

22.3.1 Purpose and types

A spillway is provided to remove surplus water from a reservoir and thus protect the dam and flanking embankments against damage by overtopping.

The best type and location of a spillway depends very much on the topography and geology of the dam site and adjoining area, and on the type of dam. Where the dam is of concrete or masonry founded on hard rock, the spillway may be within the dam, consisting either of a high-level overflow or of submerged orifices, discharging into the river bed beneath. In the case of an earth or rockfill dam, it is usual to site the spillway away from the deepest part of the dam; high flanking ground or a saddle away from the dam site can be suitable locations where a spillway channel may be excavated and control structure provided (see, for example, Figure 22.24). Where the dam is built in a narrow gorge and there is no suitable separate site for the spillway, a side-channel spillway is often adopted (Figure 22.25).

If control is by a fixed ungated weir, the maximum retention level of the reservoir is the weir crest level; at times of spill the reservoir level rises and sufficient freeboard has to be allowed above maximum water level, which is the level at which the design maximum flood discharge is released. In the case of gate-regulated spillways, on the other hand, flood flows can be discharged with reservoir at retention level, which need never be exceeded. For a given dam height, retention storage can thus be greater but, because there is less flood storage, the spillway capacity also may have to be greater. The gates, however, enable the reservoir to be drawn down in advance of a flood peak, given adequate forecasting. Low-level orifices, having greater capacity than required for purposes of normal supply, have greater capability than has a gated crest overflow in drawing down a reservoir in the event of damage to the dam, an important aspect in areas where earthquake risk is present. But crest overflow weirs have a greater rate of increase of capacity as a reservoir level rises above normal, thus providing additional safety margin.

Cost is a major consideration in the choice between a regulated and an unregulated spillway, but spillways without gates have advantages in respect of reliability, absence of mechanical maintenance problems and no power requirements. They are therefore often adopted at remote sites and for small dams where the cost of gates would not be justified.

Siphon spillways carry some of the advantages of both gated and ungated spillways. They can be designed to prime and operate to maximum discharge within a small range of reservoir level and they are automatic, with no moving parts.

Another type of spillway, particularly suited for use with earth or rockfill dams is the bellmouth or 'morning glory' spillway, which can be built quite independently from the dam, and which is described further in section 22.3.5. If the reservoir is for water supply, the bellmouth and shaft are often combined in the same structure as a drawoff tower and the low-level tunnel can be used for river diversion during construction, as discussed and illustrated in section 22.4.1.

In many cases it is advantageous to provide more than one spillway. Instead of relying on a single spillway to control all

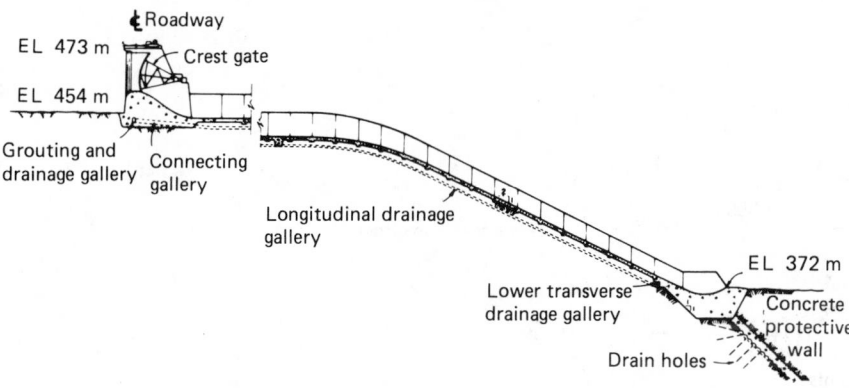

Figure 22.24 Chute spillway, Tarbela Dam, Pakistan. (Consulting engineers: Tippetts–Abbott–McCarthy–Stratton)

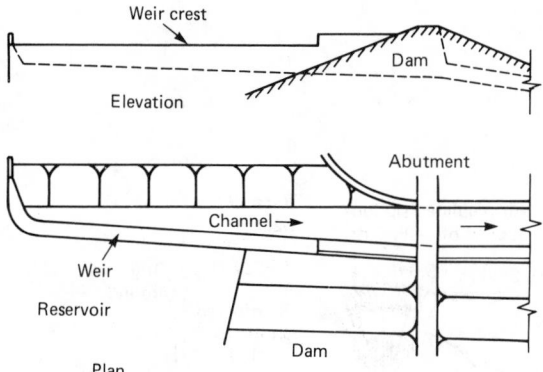

Figure 22.25 Side-channel spillway

floods up to catastrophic, it may be safer and more economical to provide a main or service spillway, fully regulated and capable of controlling all floods up to perhaps a 20- or 50-yr return period, and a secondary or emergency spillway which will bring the combined capacity to the catastrophic level. The capacity of the emergency spillway should be adequate to control normal inflows alone for a reasonable period should the main spillway be damaged.

22.3.2 Channel spillways

The simplest form of spillway, consisting of a channel excavated in rock, is often used for small reservoirs and for emergency spillways. The control is usually provided by a hard sill or weir at the entrance. The channel downstream should be given sufficient slope to ensure that the weir will not be drowned by backwater effect. If the rock is erodible, a curtain wall with bed protection or stilling basin is needed to avoid erosion undermining the downstream end of the weir.

In the case of emergency spillways, a 'fuse plug' is often provided, consisting of an erodible bank across the channel. Its crest is below the dam crest level but above normal operating level. When overtopped it quickly erodes down to the level of the hard sill, bringing the full discharge capacity of the channel into operation. To avoid excessive draw-down of the reservoir, emergency spillways of this type are preferably wide and shallow. Bed and bank protection are usually minimal but it is essential that there is no risk of erosion progressing upstream and breaching the reservoir rim beneath the level of the hard sill.

22.3.3 Weirs

These may be any of the types described earlier on page **22**/5 but are generally of the free-nappe profile type. It is usual for gates to close on to the weir face slightly downstream of the highest point of the crest, so that the jet at small openings will be projected downwards. Even so, sub-atmospheric pressures can develop, and for this and other reasons it is often desirable to avoid prolonged releases under high heads with small gate openings. Provision of separate sluices or valves for small releases is preferable.

Piers affect the rating curve, as described earlier (page **22**/6). A typical rating curve is shown in Figure 22.26 (AB).

A side-channel spillway consists of a weir discharging into a parallel channel, as in Figure 22.25 where the weir is aligned on a ground contour normal to the axis of the dam. The channel must be of sufficient width and depth to allow for energy dissipation and the generation of exit flow without drowning the weir. Calculation is based on the momentum principle.[12,52] Side-channel weirs have been investigated by el-Khashab and Smith.[53]

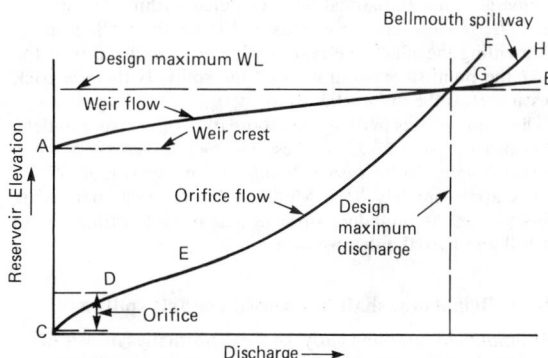

Figure 22.26 Typical rating curves for high-level weir, deep orifice and bellmouth spillways

22.3.4 Low-level outlets

These may discharge through a dam or original ground into conduits, chutes or stilling basins. They are generally of rectangular section and regulated by radial gates (see, for example, Figure 22.27). When the reservoir is drawn-down so that the orifices are not fully submerged, flow is of the free-surface weir

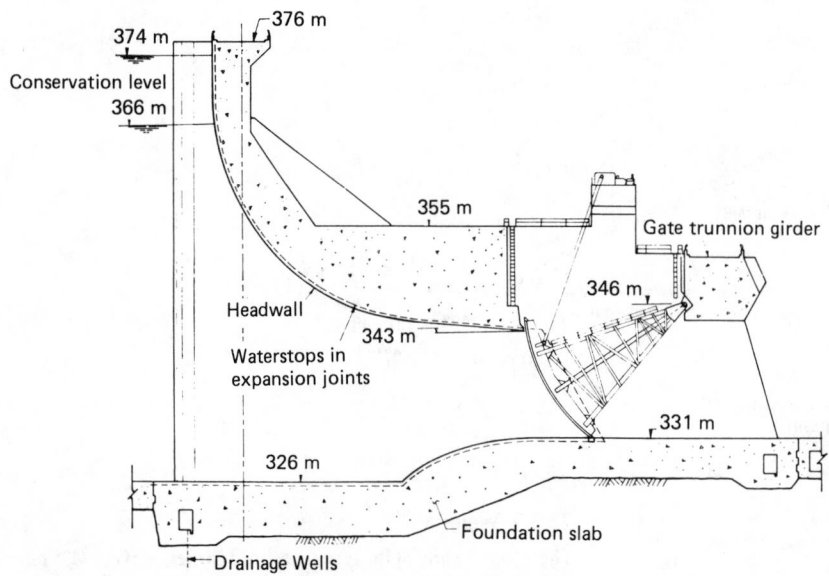

Figure 22.27 Main spillway, Mangla Dam, Pakistan. Section through gate structure. (Consulting Engineers: Binnie and Partners in association with Harza Engineering Co., and Preece, Cardew and Rider)

type. The rating curve of an orifice spillway with gates fully open therefore consists of two main parts, as may be seen from the example in Figure 22.26. The lower part CD relates to weir flow and the upper part EF to submerged orifice flow. Between the two is a transition DE. The discharge for weir flow is $Q = c_1(2g)^{\frac{1}{2}}BH_1^{1.5}$ where Q is the discharge, c_1 is the weir coefficient, B is the width and H_1 is the upstream head above the sill. c_1 is generally a variable.

In the range of orifice flow, $Q = c_2 B \sqrt{(2gH_2)}$ where c_2 is the coefficient of discharge and H_2 is the effective head below reservoir level. If the jet springs clear, so that atmospheric pressure obtains around the whole periphery, H_2 is best measured from the centre of the jet at its exit from the orifice. If the jet emerges on a horizontal floor confined within vertical side walls, H_2 is more correctly measured from the soffit level as representing the effective elevation plus pressure head over the jet at the point of separation from the soffit. If there is back pressure H_2 is the differential head.

The requirements in design are those of high-pressure outlets, described on page **22**/27. If these are met the coefficient can approach unity. In the case of Mangla spillway[54] (Figure 22.27) it was approximately 0.95. Model tests are used to indicate pressures on the boundary surfaces and provide rating curves for full and partial gate openings.

22.3.5 Bellmouth, shaft and closed-conduit spillways

A bellmouth or 'morning glory' spillway normally consists of an overflow weir, circular in plan, but in some cases multisided, and a vertical shaft discharging into a tunnel or culvert carried through high ground with outfall into the downstream river. The weir can be provided with gates or siphons – Figure 22.28 shows an example of the latter.

The hydraulics of bellmouth and closed-conduit spillways are complicated by the number of potential controls and the entrainment and release of air. At low flows the bellmouth crest provides weir control; at a higher stage the throat at the foot of the bellmouth can exert orifice control; the bend at the foot of the shaft leading into the tunnel can also exert orifice control

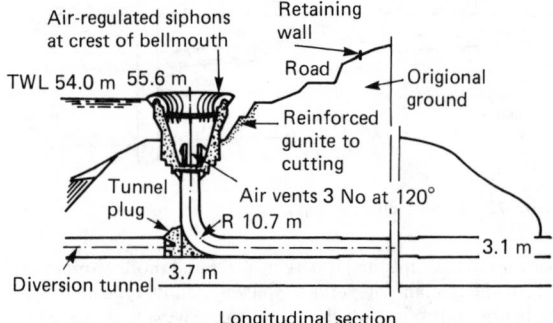

Figure 22.28 Siphon bellmouth spillway, Shek Pik Reservoir, Hong Kong. (Consulting Engineers: Binnie and Partners)

and if the tunnel flows full this may well control the discharge. To avoid instability due to controls operating intermittently, the range of each control should be clearly defined with stable transitions from one to another. It is best to reduce the number of potential controls to one or at most two. The weir (or siphons) provide the primary control and it is normal practice for the design maximum discharge to be reached in this range, with an adequate margin, before the weir is drowned by 'gorging' in the shaft and bellmouth due to controls in the system downstream. Similar considerations apply to spillways which are not of the bellmouth type but which have weir, gate or siphon as a primary control, discharging into shaft and tunnel. If, however, use is to be made of flood storage in the reservoir at higher levels, a bellmouth spillway may be allowed to become completely submerged.

As in the case of straight weirs, sharper curvature raises the coefficient of discharge. Profiles based on the shape of under-nappe of a jet have been designed for weirs circular in plan.[12,55] In several cases of bellmouth spillways measures were necessary to reduce swirl and prevent vortex formation particularly at the highest flows, when it could greatly reduce discharge capacity. Vortex flow is induced by asymmetrical approach in the reservoir; anti-vortex measures usually consist of crest piers or vanes, or a single divide wall on a diameter extending from below crest level to above maximum reservoir level.[28,56] When the jets from opposite sides of the shaft intersect, the entrainment of air can result in negative pressures on the walls. Vents may be necessary to avoid instability and vibrations.

A typical rating curve is shown in Figure 22.26. Weir control is represented by the curve AG. At low flows there is a free surface flow in the bend and tunnel but with rising discharge and downstream conduit not flowing full the bend begins to act as an orifice with water level rapidly rising in the shaft. When it reaches crest level it begins to drown the weir flow. The bellmouth is then said to be 'gorged'. This is represented by the intersection of the two curves at G, above which the bend assumes control of the rate of flow. The rating curve of the spillway is therefore AGH with a short transition at G representing drowned weir control and bend orifice control at H.

If the bend is too sharp, flow from the bend to the conduit is very disturbed; if it is too easy the downstream culvert may flow full; a bend radius of 1.5 to 2 times the diameter is generally satisfactory.

For proper control of flow at the bend and smoother flow in the conduit a deflector may be placed on the inside wall at the upstream of the bend.[57] Where the conduit is used for river diversion during construction, a properly shaped bend can later be formed when the diversion intake is plugged. Unless the conduit is very short, flow with a free surface is desirable with sufficient air space above for entrained air to be released without trouble. Sufficient slope should be provided to ensure that the depth does not exceed the desired limit. As the result of a model study, Mussalli and Carstens[57] recommend upper limits for the proportion of water flow in such conduits, ranging from 97% of the area, when the Froude number is 2, to 50% when it is 8.5. Where the velocity is high enough to entrain air it is desirable to provide an air vent at or near the bend. With high velocities it is also best to avoid bends and other conditions downstream which could cause a hydraulic jump to form in the conduit.

It is evident that the concrete surface at the base of the shafts of bellmouth spillways can be subjected to high impact loads by water, possibly with ice and logs, spilling from a great height. In some cases steel or cast-iron lining has been provided in this area, but from a survey of sixteen bellmouth spillways, of which eight with unlined concrete inverts had undergone a fair test, Bradley[56] found no erosion of a serious nature. Dense concrete with smooth surface finish is called for here and in the conduit.

22.3.6 Siphon spillways

Compared with a free-surface weir, flow through a siphon reaches a high rate of discharge per unit width with only a small rise of reservoir level needed to prime the siphon. This permits a higher retention level for a given maximum water level, or alternatively a higher concentration of flow in a restricted width.

Reservoir retention level is equal to siphon crest level. As the reservoir level rises, the action of a siphon passes through the following successive phases: (1) weir flow, when water spills at low depth over the crest; (2) priming phase, when air is being extracted from the crown of the siphon; and (3) fully primed siphonic flow. When the reservoir falls, (3) gives way to (4), a depriming stage, when air is admitted in sufficient volume to break the siphonic action and the action vellerts to (1), weir flow. In recent years, many air-regulated (or partialized) siphons have been built. In these, phase (3) consists of two parts: in (a) when priming has occurred the entry of air continues so that the flow consists of an air–water mixture. The air intake is so designed that the volume of air admitted is insufficient to break the siphon (except at low flows) but is controlled by very small variations in reservoir level. As the reservoir rises further the

volume of air is reduced until stage (b) is reached when the siphon flows 'blackwater', i.e. with no entrained air. This performance is illustrated in the typical stage : discharge function in Figure 22.29a. The advantage of air regulation is that, whereas without it the siphon on priming runs directly to a high blackwater discharge which if in excess of inflow will draw the reservoir down and lead to intermittent priming and depriming, an air-regulated siphon will remain in the fully primed phase over a wide range of flow, with continuous discharge matching the rate of inflow. Examples of air-regulated siphons are Eye Brook,[58] Shek Pik[59] (Figure 22.28) and Plover Cove[60] (Figure 22.29b).

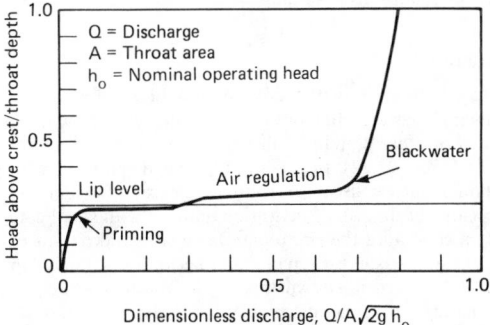

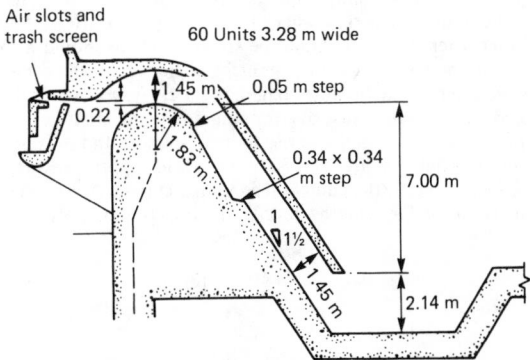

Figure 22.29 Air-regulated siphon spillway. (a) Typical stage: discharge curve; (b) Section through Plover Cove siphon spillway, Hong Kong
(Consulting Engineers: Binnie and Partners)

Spillway siphons are generally designed to prime automatically when the reservoir has risen to a level such that: (1) an upstream air inlet is submerged; (2) the siphon outlet is sealed by a deflected jet or a downstream weir; and (3) the flow is sufficient to entrain and remove air from the crown of the siphon. Various priming devices have been used[61] and the priming depth above crest level is in some cases as little as one-sixth of the throat diameter.

The blackwater discharge capacity of a siphon can be expressed as:

$$Q = cA\sqrt{(2gH)} \qquad (22.13)$$

where c is a coefficient allowing for head losses, A is the cross-sectional area of the flow at exit and H is the head from upstream reservoir to effective exit level – usually the downstream lip of the hood

Ackers and Thomas[62] reviewed the design and operation of siphon spillways. The value of c obtained in the Plover Cove siphon model (Figure 22.29b)[60] was 0.68. That of the model of Shek Pik bellmouth siphon (Figure 22.28), where the shape was radial and no sealing weir was provided,[59] was 0.66.

Surface waves in the reservoir are an important factor in siphon design. Model tests showed that despite provision of baffles, waves caused surging in the siphon but the air intakes could be designed to counterbalance the effects of surging and wave wash. The head in siphons is usually limited to about 7 m to avoid cavitation at the crest. Tests on the Plover Cove siphons showed that wave action resulted in transient pressures below average pressures, but a total head of 7.3 m was still feasible. Each case should, however, be examined in the light of the particular conditions obtaining.

22.3.7 Chutes

Chutes may be built into the downstream faces of concrete dams; longer chutes are often provided to convey the flow from side channel or other flanking spillways to the river bed downstream (see Figure 22.24). In general, high head spillways with chutes should not be used for routine releases of water for supply, because of the risk of cavitation damage with small gate openings, also because the chute may have to be taken out of service in the dry season for repairs, and because low flows can create problems of erosion downstream. The gradient is likely to be steep enough to generate high-velocity flow. As lateral changes of direction could result in overtopping of the side walls, any essential changes in the alignment should be made near the control structure where the velocity is relatively low and thereafter the chute should be straight. There should also preferably be no changes of alignment in the side walls where flow is supercritical because diagonal shock waves would be created which might cause overtopping downstream.

High-velocity flow can give rise to high pressure and the most careful precautions are necessary to avoid uplift pressures developing beneath the chute slabs. A high standard of surface finish is called for and the profile should contain only very

gradual curvature. Joints between slabs should be keyed and bridged by flexible water stops sealed at intersections. Projections at the joints facing upstream should not be allowed but offsets facing downstream up to 12 mm are often accepted or even specified. Drains are generally provided beneath and parallel to all joints, so that in the event of leakage, uplift pressure cannot build up. For additional protection, chute slabs are often anchored to the foundation rock. Special care is needed where the chute rests on jointed or fissured rock because pressure can be transmitted from leakage at a higher level despite underslab drainage. Chutes at a steep slope are especially vulnerable and may call for deep anchors. Stability should be checked for all possible modes of failure.

The problems of cavitation caused by high-velocity flow are reviewed on page **22**/17. Severe damage has occurred on some chute spillways which have been operated at high velocity and hence a careful assessment of the cavitation risk is needed for any chute spillway, especially if the total fall exceeds 50 m. Cavitation damage can be avoided if natural aeration of the flow from its free surface is high enough to provide several per cent of volumetric air concentration at the bed of the chute in the region of high velocity.

The development of flow down the spillway is illustrated in Figure 22.30. A layer of slower-moving fluid influenced by friction at the solid boundary grows in thickness beyond the crest until it occupies the full depth of flow. This defines the 'point of inception' of air entrainment by turbulence at the surface. The entrained air diffuses down within the partially aerated region of flow to occupy the full depth of flow, usually some considerable distance down the spillway. It is only in this fully aerated region that the natural aeration at the solid boundary can be sufficient to prevent cavitation damage if the cavitation index (see page **22**/32) drops below the safe limit. The likelihood of cavitation damage therefore depends on whether velocities rise too high as the flow accelerates down the chute before there is sufficient aeration at the bed. This is most likely at high discharge intensities. A method of calculation of the

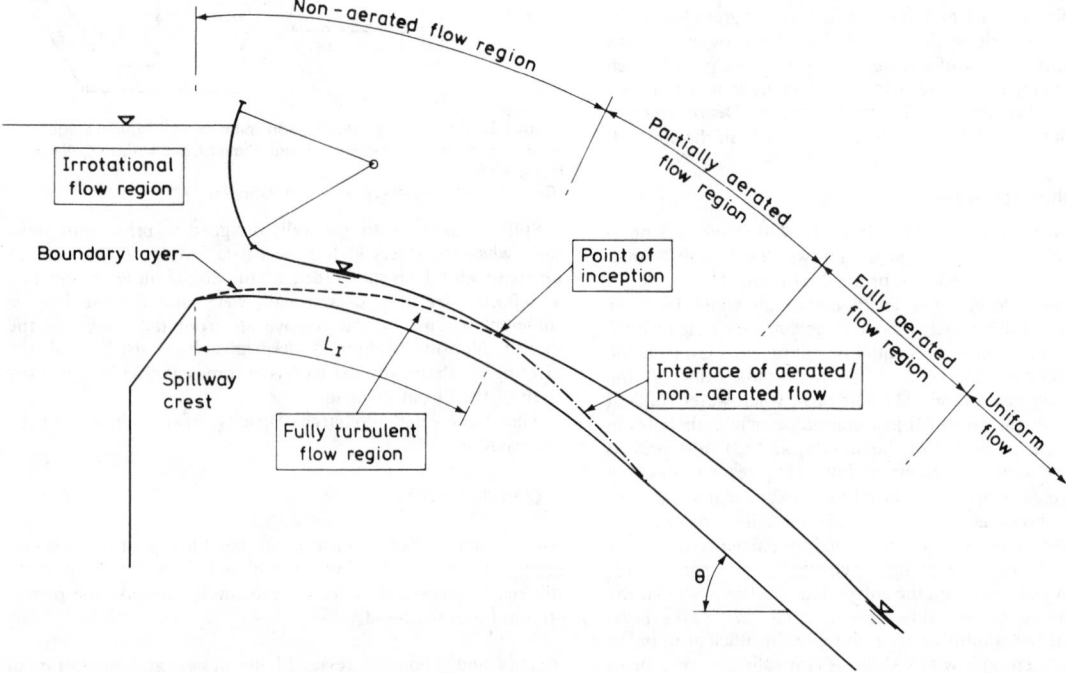

Figure 22.30 Development of flow down a typical spillway

growth of boundary layer is given by Wood, Ackers and Loveless.[63] If natural aeration is not sufficient, purpose-built air-entraining slots should be provided, fed from ducts at either side.[64,65]

Calculation of depth of flow in chutes may be done by a step method beginning at the top, using curves showing energy against depth for the required discharge per unit width, similar to that of Figure 22.1. The calculation should be performed with two roughness coefficients, one representing a maximum, for use in determining side-wall height, and the other representing a minimum, for use in the design of energy-dissipating works. The height of side walls should include an allowance for bulking due to air entrainment. The US Corps of Engineers[11] provide a design curve based on observed data, with the equation:

$$c = 0.436 \log_{10}(S/q^{\frac{1}{5}}) + 0.971 \tag{22.14}$$

where c is the ratio of air volume to air-plus-water volume, q the discharge per foot width in square metres per second, and S the sine of the angle of chute inclination

22.3.8 Energy dissipation

The energy to be dissipated at the outfall from a spillway is very considerable. The means of protection of the dam and other structures from its erosive action depend largely on the rate of discharge and its head, the erodibility of the materials of the river bed and surrounding ground and on the proximity of the dam.

In the case of rivers in alluvium or other easily erodible ground, a stilling basin designed to contain a hydraulic jump is often provided. This may be of the rectangular type or a submerged roller bucket (see sections 22.1.6.1 and 22.1.6.2). The former is generally more efficient but more costly, especially in the case of large structures where deep retaining walls would be required. Where the river bed material is rock a 'ski jump', trajectory or 'flip' bucket is generally provided. This is elevated above maximum tailwater level, so that the jet trajectory carries the water into a plunge pool some distance from the bucket. Dissipators of this type are generally less costly and are suitable where the bedrock is resistant enough so that erosion does not progress back and endanger the foundations of the dam or other structures. Ski jumps have also been used where the river bed is of alluvium, and the structures are suitably protected against erosion.

The radius of flip buckets is not critical provided it is large relative to the depth of flow. The exit angle is important as it determines the throw distance; the angle is generally between 20 and 40° and the theoretical throw distance is given by x in the formula:

$$\frac{x}{h_v} = \sin 2\theta + 2 \cos \theta \left(\sin^2 \theta + \frac{y}{h_v} \right)^{1/2} \tag{22.15}$$

where h_v is the velocity head at the bucket lip, θ the bucket exit angle, measured from the horizontal, y the vertical height of bucket lip above tailwater[11]

However, because of air resistance and internal shear the jet tends to break up and diffuse so that the actual trajectory distance may be 10 to 30% less than indicated by the formula. Elevatorski[5] quotes examples of models and full-scale spillways.

Flip buckets are not necessarily of circular profile, nor axisymmetrical. There are several instances of buckets composed of flat deflecting surfaces, some designed to deflect to one side to suit downstream requirements.[66]

In the design of the side walls, allowance must be made for the additional lateral pressure due to centrifugal force. This may be calculated by methods of Gumensky,[67] Balloffet[68] or the US Army Corps of Engineers.[11] To avoid cavitation damage it is usual to avoid the use of teeth, and in some cases the lip edge is protected by stainless steel.

The size and depth of the plunge pool depend primarily on the discharge concentration and the characteristics of the materials which are eroded. In the formation of a deep pool the eroded material is lifted out by the flow. Incoherent alluvial materials are readily removed and form flat side slopes. Rock disintegrates into fragments by transient pressures in the joints and the fragments are reduced by abrasion until small enough to be removed.[73,74] A plunge pool in rock has relatively steep side slopes and is less extensive in plan than one in alluvium. The erosion is not always confined to the plunge pool because the action of the jet creates large horizontal eddies which can extend back to the chute. Small flows and flows at low heads have shorter trajectories or may not be sufficient to sweep out of the bucket but spill over the lip, causing erosion beneath. In such cases special protection is needed (see, for example, Figure 22.23). Mason[75] has reviewed experience of energy dissipation works at dam outlets.

22.4 Reservoir outlet works

22.4.1 Intakes

The type of intake for drawing water from a reservoir depends on the type of dam and on the purpose of the supply. The velocity may be low and against a back pressure as, for example, in intakes for domestic water supply, and into penstocks for power generation, or it may be high, e.g. in spillways and into diversion tunnels during construction. With high velocity, special problems concerned with head loss and cavitation arise. If the dam is of concrete, the intakes may be located in the dam. If the dam is of earth or rockfill, a separate intake structure may be built (see Figure 22.31), leading into a tunnel, or a free-standing drawoff tower may be provided, sometimes combined with a shaft spillway as in Figure 22.32. A free-standing tower is particularly suitable where drawoff is required at several levels, as where the water is for domestic supply. In such cases a bottom drawoff or 'scour' sluice is generally provided; this is opened at intervals to prevent sediment from building up a deposit in the immediate vicinity of the lowest drawoff to supply.

Deep intakes have advantages in that they will remain submerged at low reservoir levels, are less affected by vortices and are less susceptible to obstruction by ice and floating trash. Against these, the gate structure is more costly and access to the screens for cleaning more difficult.

A square edge or small radius edge to an orifice would result in flow separation and a *vena contracta*, so orifices and sluice entrances, whether circular or rectangular in section, are usually shaped to a bellmouth. The head loss associated with the formation of a *vena contracta* at a circular orifice can be greatly reduced by providing a simple bellmouth, as shown in Figure 22.33a, but for high velocities the curvature should be less to avoid low pressures which might result in cavitation damage. Compound curves of two or more radii and elliptical curves are often suitable profiles. A typical example of an elliptical profile is shown in Figure 22.33b. With this profile the minimum pressure at the boundary is approximately $0.1V^2/2g$ below the corresponding pressure in parallel flow in the orifice downstream where V is mean velocity. As this is a mean pressure, lower pressures may occur owing to fluctuations. For very high velocities this may not be acceptable and the profile may be

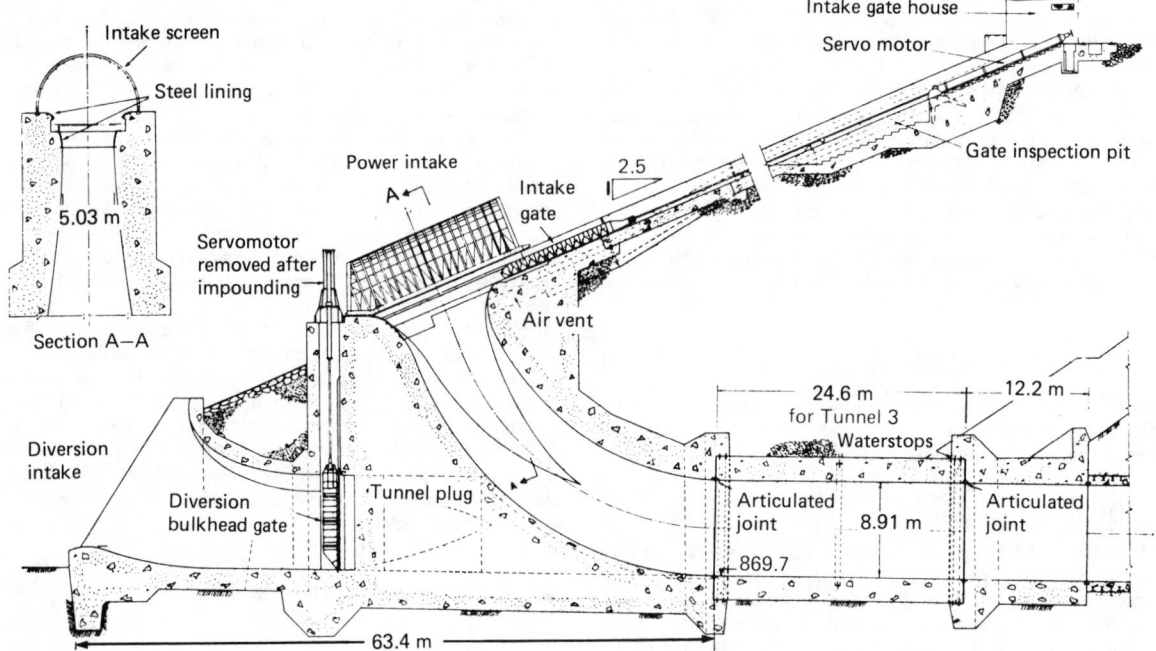

Figure 22.31 Intakes at Mangla Dam, Pakistan. Longitudinal section. (Consulting Engineers: Binnie and Partners in association with Harza Engineering Co., and Preece, Cardew and Rider)

based on the profile of a jet springing from a sharp-edged orifice[72,73] or may be compound elliptical.[11] In the case of important works, especially with high velocities, intake entry curves are usually tested in hydraulic models. Control may be by gate or valve, located at the intake or in a pressure conduit. Radial gates in rectangular orifices, as seen in Figure 22.27, are particularly suited for flood releases. If the gates are to be used for regulation at part openings under heads exceeding 10 m, the inverts immediately downstream are generally lined with steel as protection against cavitation. A second gate is often provided upstream of each service gate for emergency closure and to allow maintenance work on the service gate to proceed when the reservoir is at a higher level. This is generally a vertical-lift gate closing on to a steel sill but requiring side slots. The latter contribute to head loss and, where cavitation is a danger, require special design as, for example, is illustrated in Figure 22.34.

Where outlets fill a vital role in the safety of works, the possibility of failure or malfunction of a gate or valve should be considered and alternative measures provided for an emergency.

22.4.2 Vortices

Though a slight surface swirl may be of no consequence, a vortex with an air core extending to an intake can be harmful in reducing the discharge capacity of the intake, causing gate vibration or resulting in admission of air to pumps or turbines. Any tendency for a vortex to form in a model test should be carefully investigated because vortices form more readily and develop further at full scale.

A free vortex tends to form in accelerating flow towards a region of low pressure, as at a submerged intake. It is facilitated by boundary geometry consistent with vortex shape, and by an initial circulation in the reservoir, and is more marked the greater the pressure drop to the outlet relative to the depth below surface. Vortex action is reduced by deeper submergence of the intake, by reduced velocity at the intake and by obstructions to the rotation, such as horizontal grids and projecting walls.[74,75]

Vortices are also a problem in pump sumps where even a slight swirl may affect pump efficiency and more refined measures are needed. Velocity of approach is generally limited to 1 m/s, but eddies can still form at points of separation. Sharp wall angles and regions of dead water should be avoided and expansions should be gradual. General guidelines are available[76–78] but model tests are often needed to determine optimum pump sump geometry.

There are situations where vortices may be used to advantage, as in the vortex drop structures described earlier.

22.4.3 Screens

Screens or trash racks are provided at intakes to hydro-electric plants, pumps and water-treatment works. Log booms are often placed upstream, but it is generally required to intercept small debris and possibly fish. The spacing of the bars may be 2 to 20 cm depending on the duty. The main requirements in design are that the bars are stiff enough not to vibrate and are arranged for easy cleaning. As a general guide for screen area the mean velocity is usually limited to 0.6 m/s or less. Vibration is avoided if the dimensions of the bars are such that the natural frequency of the bars is higher than the forcing frequency. Screens may be fixed and cleaned by raking or lifted above water for cleaning. For ease of cleaning from above, the vertical bars generally project upstream of the lateral bars.[79,80]

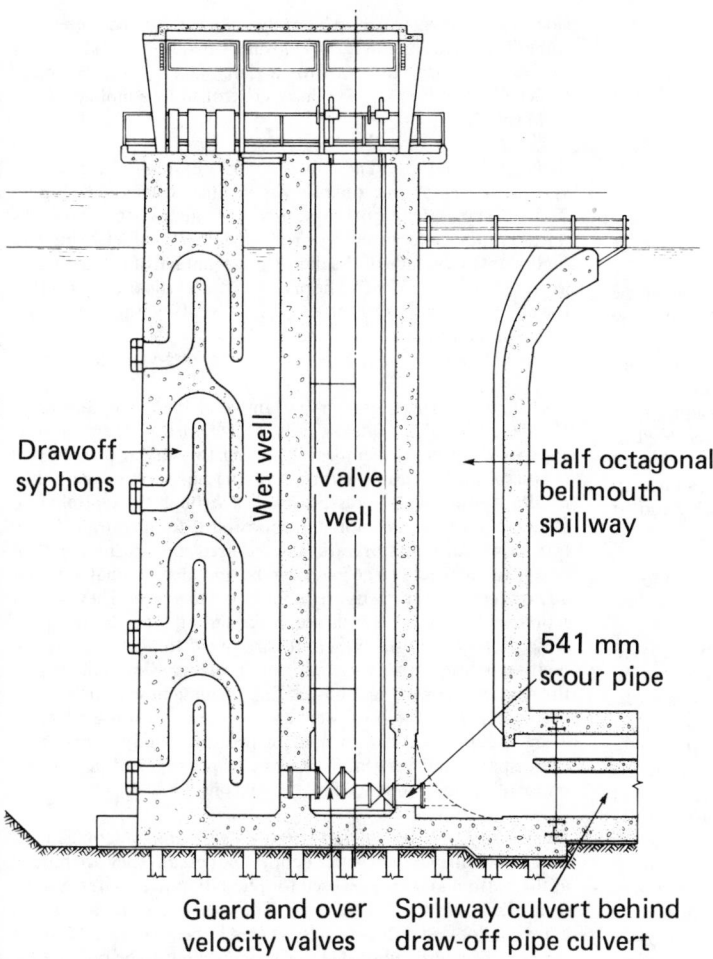

Figure 22.32 Combined drawoff tower and spillway, Seletar
Reservoir, Singapore. Air-controlled siphons are used instead of
valves for drawoff purposes. (Consulting Engineers: Binnie and
Partners, Malaysia)

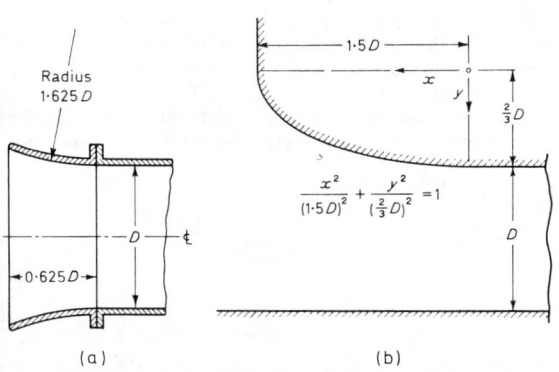

Figure 22.33 (a) Simple bellmouth; (b) elliptical roof profile for
conduit intake with parallel sides and horizontal floor. (After US
Army Corps of Engineers (1952–70) *Hydraulic design criteria*. US
Army Engineer Waterways Experiment Station, Vicksburg,
Mississippi)

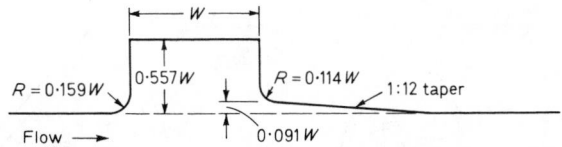

Figure 22.34 Typical gate slot with downstream offset to
minimize cavitation. (After US Army Corps of Engineers (1952–70)
Hydraulic design criteria. US Army Engineer Waterways Experiment
Station, Vicksburg, Mississippi)

22.5 Gates and valves

22.5.1 Gates

22.5.1.1 Uses and types

Gates are used to control flow in open channels or closed conduits by restricting or closing the waterway. They may be required:

(1) For regulating the flow, when they must be capable of operating at any required degree of opening.
(2) For emergency or guard purposes, when they must be capable of closing under any condition of runaway flow which could occur.
(3) As bulkhead gates for closing a conduit for inspection, maintenance or construction works. When they are permanent installations, such gates are generally designed to open and close only under balanced pressures, but when used to close diversion tunnels during construction works, closure may be against a considerable flow. Stop logs are similar in function to bulkhead gates, but are in smaller units handled individually and placed above one another.

The types generally used in outlets from reservoirs and in spillways, barrages and canals are as follows.

Vertical lift gates. Vertical lift gates are supported by guides in slots at the side walls of the conduit. They may open by raising or by lowering; in some cases they are in two or even three parts, each operating independently. They may have sliding contact with the guides or may have wheels (fixed-wheel gates) or a moving train of rollers (Stoney gates). They have seals at the sides and (in orifices or closed conduits) also at the top and generally close on to a steel sill with an inset compressible seal if required. Advantages of vertical lift gates are their simplicity and ease of maintenance; disadvantages are the requirement of slots in the side walls and limitations of loading on axles at very high heads. Sliding gates can be used for high heads but require correspondingly powerful actuators. However, sliding gates have been installed for heads as high as 200 m, carrying a water load exceeding 1000 t.[81] A disadvantage in the use of vertical lifting crest gates for spillways, barrages and canals is the requirement of a high superstructure. In the case of conduits the hydraulic disadvantage of side slots in high-velocity flow has been overcome by the introduction of 'jet flow' gates; in these, narrow side-slots are provided but an upstream contraction causes the flow to spring clear of the slots, re-attaching to the

side walls downstream. Caterpillar-track-mounted gates are sometimes used for emergency closure, operating on flat bearing faces on the upstream face of a dam. One gate is generally sufficient for a number of orifices, controlled by a mobile gantry from the top.

Inclined lift gates. These are similar in many respects to vertical lift gates but operate on inclined tracks (see Figure 22.31). They are sometimes used for guard or emergency purposes at intakes in earth or rockfill dams, the track being laid on the upstream face of the dam. An advantage of this gate is its low cost compared with alternatives of a vertical lift gate in a tower in the reservoir or in a gate shaft within the dam. Disadvantages may include the remoteness and inaccessibility of the gate in case of emergency and the long-vent shaft.

Radial (or tainter) gates. An example of these may be seen in Figure 22.27. The gate skin is of cylindrical shape and is supported on cross-members spanning two radial arms which rotate on short axles extending from the side walls or piers. The resultant of the water pressures passes through the centreline of the axles creating no moment opposing gate operation; therefore powerful actuators are not required and in the event of power failure quite large gates can be operated manually. Other advantages are simplicity, reliability and low cost. They do not require side slots; side seals are of the sliding type and the gates close on to a steel sill. Where it is required to allow passage of floating debris, or sensitive control of reservoir level, the top of the gate may consist of a hinged flap opening downwards. They are widely used for both weir and orifice control in spillways. They are also sometimes used in pressure conduits but need more space than vertical lift gates and problems of access and removal for maintenance have to be considered.

Hinged leaf, bascule or flap gates. These are sometimes used for crest control where water depth is not great. They are hinged at the bottom and may be used for regulation with water spilling over them; they need venting. They have the advantage of allowing floating debris to pass at small openings but, although they can be of curved profile, the weir crest has to be rather wide to provide the recess. On many European rivers, bascule gates are used for regulating upstream water level, operated by hydraulic actuators located below the weir crest. Hinged gates can be made to open automatically by a simple mechanical device when the upstream water level rises to a given height. Gates hinged at the top are also used where the whole assembly is retractable to allow the passage of ships.

Drum and sector gates. Examples of these can be seen in Figure 22.35. These are crest gates which open downwards, retracting into a recess in the crest. They may be hinged on the upstream (drum) or downstream (sector). The upper surface can be shaped to suit the weir profile when fully open. In the examples shown, the gate consists of a watertight vessel con-

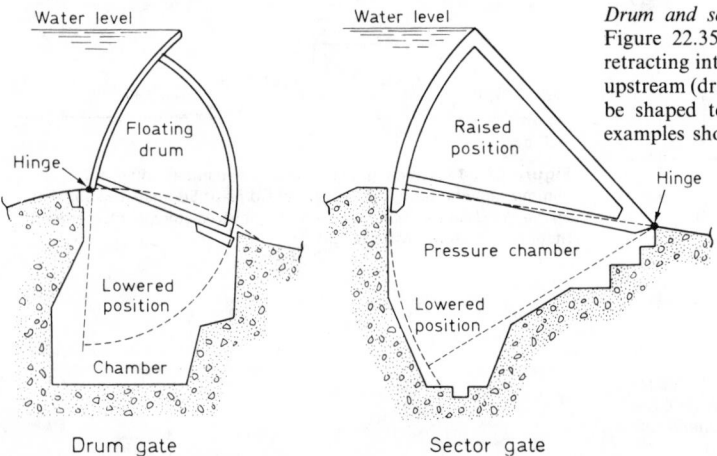

Figure 22.35 Drum and sector gates

trolled by application of headwater pressure beneath; it is sealed at the hinge and gate seat. These gates can be arranged to operate automatically by the upstream water level. (The sector gates of the Thames Barrier are illustrated in Figure 22.11, page **22**/11.)

Bear-trap gates.   When raised, bear-trap gates are in the form of a flat 'A', with upstream and downstream leaves forming the two legs, hinged at the bottom, with seals at the hinges and the apex. The gate is raised by admitting water under pressure from the headwater. When lowered the upstream leaf overlaps the downstream leaf and both fold flat. Bear-trap gates have for long been used in Europe and the US for river regulation.

Rolling gates.   Rolling gates consist of a roller with toothed-wheel meshing with an inclined toothed rack at each end. The gate is rotated by a chain and accordingly moves up and down the racks. Roller gates have been used for river regulation.

Cylinder or ring gates.   Cylinder or ring gates moving on vertical axes have been used as crest gates on bellmouth spillways and for bottom outlets.[81] Some of the former open by being lowered vertically into a recess in the weir crest, controlled by water pressure, others and the bottom outlet gates by being lifted from above. Lateral control of the gate motion is provided by guides.

22.5.1.2   Partial gate openings

These are orifices of which three sides are the fixed boundaries and the fourth is the gate lip. When unsubmerged, the jet springs clear from the gate lip forming a *vena contracta*, the dimensions of which depend on the shape of the gate skin and lip and on the upstream profile of the sill. It is therefore usual to calibrate gates by model tests. Details of calibration of various gates are available.[81–84] Where gates are partially submerged downstream, calibration is complicated by the additional variable, and is also less reliable because of possible variation of flow pattern. Submergence may also lead to vibration problems.

22.5.1.3   Vibration

In general, vibration is a result of resonance where the frequency of a pulsating force is equal or nearly equal to the natural frequency of a flexible part of the structure. Gates are liable to vibrate when: (1) overtopped and not adequately vented; (2) when significant flow occurs both over and under a gate; (3) when the gate is partially or fully submerged; (4) when the location of flow separation is unstable; or (5) there is flow re-attachment. Vibration by the two latter causes may occur at the bottom edge of a lifting or radial gate which should therefore be designed so that the flow separates at a sharp edge and cannot become re-attached by contact elsewhere. This is not always possible at small openings, so that vibration may occur in a limited range of opening. Flexible seals are a potential cause of instability and are often omitted from the bottom edges of gates for this reason. Many cases of gate vibration and remedial methods have been described.[85,86]

22.5.1.4   Downpull and upthrust forces

These forces can act on the upper and lower edges of a gate as well as on the face. They affect the operating forces required and gates are often designed so that the resultant force is of assistance. In particular, if the top edge of a lifting gate is subjected to static pressure whereas the bottom edge is at atmospheric pressure because the lip is on the upstream edge, the resultant downpull assists in gate closure, a safety measure

in case of power failure. Pressures measured on bottom edges of various shapes are available.[87]

22.5.1.5   Gate seals

Where a small amount of leakage can be tolerated, as in most works in the open, the seal at the bottom of a lifting or radial gate is usually metal to metal, between the gate edge and a steel sill set flush in the floor. A bottom slot would fill with debris and a projecting rubber seal may vibrate, but if leakage is to be minimal a rubber seal may be inset flush in the floor. Side and top seals can be metal to metal but with close tolerances these may be costly. Flexible rubber seals are therefore frequently used, in the form of strip tightly clamped with small projection, or moulded into bulbous shapes (e.g. music-note type) and arranged to be held in contact by the water pressure. As frictional resistance between metal and rubber seal increases with pressure, brass cladding is often used for sliding seals where the head exceeds 60 m. For very high heads, metal-to-metal contact may be required.

22.5.2   Valves

22.5.2.1   Uses and types

Valves are used to regulate flow or pressure in pipes and conduits or to close them against flow, often at high pressure. A service valve, whether for regulation or closure, is generally protected by an upstream gate or guard valve which can be closed against flow to isolate the regulating valve for maintenance or repair, and to prevent leakage if the regulator is not adequately sealed. Valves may be 'in-line', i.e. with pipeline upstream and downstream, or 'terminal' at the discharge end of a pipeline. Variations in valve design are numerous; only a few types which are normally used in reservoir outlet and hydro-electric power systems are described below. Of these, gate, spherical, butterfly, needle and tube valves are generally used in-line, while needle, tube, hollow jet and sleeve valves are used as terminal regulators.

The discharge capacity of a valve may be expressed as:

$$Q = CA\sqrt{(2gH_L)} \tag{22.16}$$

where Q is the discharge, C the coefficient, A the cross-sectional area of the valve at entry and H_L the head loss across the valve

Gate or sluice valves.   In their simplest form (Figure 22.36) these consist of a sliding leaf in a valve body with side slots, thus resembling a vertical lift gate with operating rod sealed to contain the pressure. The sealing contact is metal to metal and the leaf is usually wedge shaped to provide tight sealing when fully closed. Parallel guides prevent vibration at part openings, but gate valves are not suited for regulation except at low or moderate pressures. A bypass is usually provided to balance pressures but a valve for guard or emergency duty may have to close under unbalanced pressure. Advantages of gate valves include the simplicity of design and low head loss when fully open. Disadvantages are the considerable power required to operate under unbalanced heads, cavitation damage to the slots at high velocity and damage by abrasion of the sealing faces if sediment is carried in the flow. The main disadvantages of slots are overcome in the 'ring-follower' valve in which the gate when raised is followed by a cylindrical ring which effectively covers the slots. It requires a cavity for the ring beneath the conduit and is suitable only for guard purposes.

For regulation under high heads a special type of gate valve, termed the 'jet-flow' gate valve, has been devised by the USBR to operate free of cavitation. The flow is expanded, then sharply

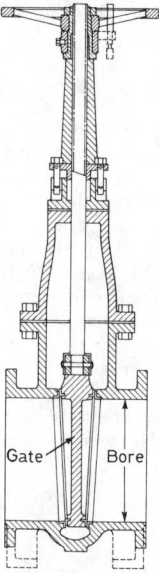

Figure 22.36 Typical gate valve, wedge type. (*Courtesy*: J. Blakeborough and Sons Ltd)

contracted, at the boundaries upstream of the gate slot and springs clear into a vented surround downstream. Valves of this type have been successfully used in sizes up to 2.44 m diameter and under heads to 120 m.[58]

Spherical or rotary plug valves. These are used for guard and on–off duties. They generally consist of a short length of tube of the same diameter as the conduit and length about the same dimension, which is in line with the conduit in the open position and is rotated through 90° to effect a closure. This is enclosed in a body of roughly spherical shape. The great advantage of this valve is that it offers no obstruction to the flow when in the fully open position. Hydraulic characteristics are given by Guins.[89]

Butterfly valves. Butterfly valves (Figure 22.37a) are widely used as guard or isolating valves but under low- or medium-pressure differentials they can be used for regulation. The blade or disc is mounted on a shaft, or on two stub axles; rotation by hydraulic piston and crank is simple and direct. The obstruction caused by the blade in the fully open position inevitably results in head loss and downstream turbulence. The former is reduced by adopting a slim blade of greater diameter than that of the pipe. However, a slim blade may not have the strength to resist high-pressure loading, so to meet this need, blades in some valves consist of two thin parallel discs rigidly connected by structural members parallel to the flow; these have great strength while offering less resistance to the flow than corresponding solid blades. Where the valve is in a terminal position, or is guarding a terminal valve, the pressure of the surrounding fluid may be near atmospheric; this also calls for slim blades to avoid cavitation at high velocities. The resultant torque due to fluid pressure is always acting to close the valve, but for safety when closed it is best for the lower half of the disc (if the axis is horizontal) to close in the direction of flow. Guard valves are often designed to be opened under balanced pressure, for which a bypass valve is provided, but to close against full flow in case of emergency.

Butterfly valves present special sealing problems because of the axles but this problem has been overcome and very low leakage rates have been achieved[88] with rubber seals mounted on

disc or body. In some valves the axles are offset to facilitate replacement of the sealing ring. For very high heads, rubber is not suitable and metal-to-metal seals are required.

Hydraulic and torque characteristics of valves with blades of various shapes are available.[89–91]

Needle valves. Needle valves have long been used for precise flow regulation in terminal locations. They consist of a needle or tapered plunger which moves axially within an orifice forming part of the valve body (see Figure 22.37b). The plunger is located by guides and operated by screw or hydraulic pressure. In the Larner–Johnson valve, actuation is by the pressure difference across the valve controlled by a pilot valve in the nose of the plunger. The plunger and body are precisely shaped so that the flow accelerates through the valve, to assist actuation, while avoiding cavitation under operating conditions. The discharge coefficient for a fully open valve is from 0.4 to 0.72 depending on the throat area ratio.

Tube valves. Tube valves which were used on some USBR reservoir outlets resemble needle valves but the part of the needle downstream of the sealing ring is omitted. This reduced cavitation problems but vibration was experienced with some valves. This valve has been satisfactorily operated fully submerged. The discharge coefficient, when fully open, based on valve outlet diameter is about 0.6. An in-line regulator of similar general form is shown in Figure 22.37c.[91] This has tubular ports at its discharge end which direct jets towards the centre of the downstream pipe where excess head is dissipated, thus avoiding cavitation damage. The port openings are regulated by an axial movement of the plunger.

Hollow jet valve. This was developed by the USBR as successor to needle and tube valves and has been generally satisfactory as a high-pressure terminal regulator. It resembles a needle valve with the downstream half of the needle omitted, while the remaining half advances upstream to seal against a ring on the body (Figure 22.37d). The flow is deflected by the surrounding tubular body and is trajected in the form of a jet with nearly parallel sides and hollow centre. Pressure is admitted to the plunger to assist operation, which is mechanical. When fully open the valve normally has a coefficient of discharge of 0.7 based on the outlet diameter.[92,93] Hollow jet valves by the USBR are stated to operate satisfactorily when partially submerged up to centre level, but should not be operated fully submerged.[88] The trajectory can be calculated approximately by the mechanics of a projectile. Though some aeration and dispersion of the jet occurs the jet fallout is concentrated in a relatively small area and in some cases erosion is a problem. A stilling basin has been developed for this valve.[5,34]

Fixed cone-sleeve valves. Also known as Howell–Bunger valves (Figure 22.38), these are widely used in free-discharge terminal applications, including pressure relief for turbines. They have a tubular body on the outside of which is a cylindrical sleeve. This is operated by screws or hydraulic servomotors and retracts to form an opening through which the discharge occurs, deflected outwards by the fixed cone. Discharge coefficient at normal maximum opening is approximately 0.85.[94]

Asymmetry of approach flow may lead to vibration; the valve should not be located too close to a bend in the conduit. A tapered contraction from conduit to valve assists to stabilize the flow. The sharply increasing diameter as the water leaves the valve forces the jet to break up, which is excellent for energy dissipation but sometimes creates problems due to fallout from drifting spray. The limits of the trajectory can be calculated by assuming projectiles at the upper and lower points and two sides of the jet, leaving the valve at velocity equal to $(2gH)^{1/2}$ where H

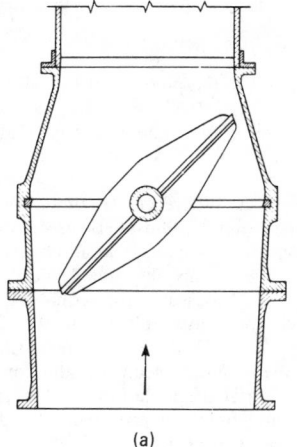

(a)

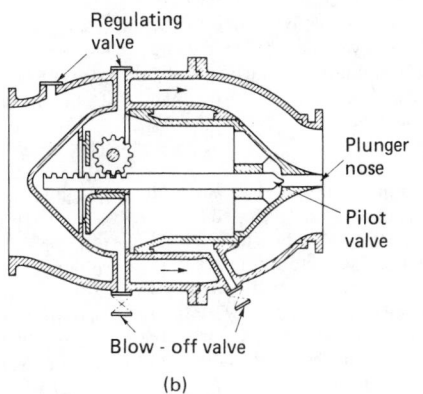

Regulating valve

Plunger nose

Pilot valve

Blow - off valve

(b)

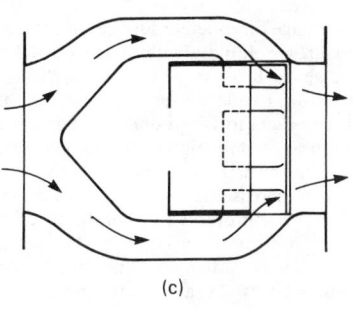

(c)

Figure 22.37 (a) Example of butterfly valve. The valve diameter exceeds the conduit size to allow for the obstruction of waterway by the blade; (b) Larner–Johnson needle valve, with internal pilot valve control; (c) in-line regulating valve (*Courtesy*: Glenfield and Kennedy Ltd); (d) hollow-jet valve (USBR design)

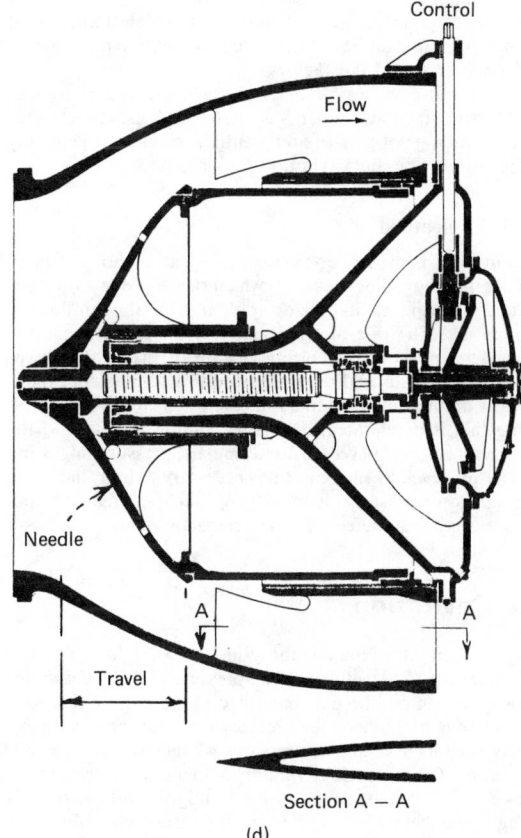

Control

Flow

Needle

A   A

A   A

Travel

Section A – A

(d)

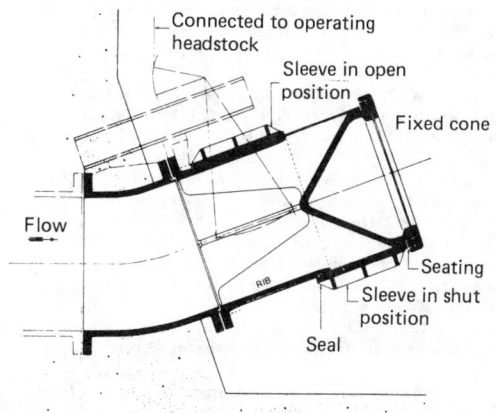

Connected to operating headstock

Sleeve in open position

Fixed cone

Flow

Seating

Sleeve in shut position

Seal

Figure 22.38 Fixed-clone sleeve valve. (*Courtesy*: Glenfield and Kennedy Ltd)

is the pressure plus velocity head in the valve body, and initial trajectory according to the angle of the cone, normally 45° to the axis. Fallout distance is, however, appreciably reduced by air resistance and affected by wind. In some installations a cylindrical hood is provided to restrict the dispersion of the jet, which then becomes tubular in form. To avoid vibration and failure by fatigue this should be rigid and is often of steel with concrete surround. Because of the large air demand, free access of air is essential.

The advantages of sleeve valves are simplicity and relatively low cost, low actuating power and the energy-dissipating characteristics of the jet. They have been made in sizes up to 3.850 m diameter and (smaller valves) for heads up to 280 m. Several cases of damage due to vibration have been recorded, particularly fatigue failure of the ribs attaching the fixed cone to the body, but this weakness has been overcome by increasing rigidity and avoiding causes of instability, in particular by fairing the leading edges of the ribs. Valves of this type have been operated fully submerged but trouble has occurred with partial submergence.

Submerged sleeve valves. Located in a stilling well, a valve of this type is an excellent terminal regulator. As developed by the National Engineering Laboratory, the valve has an internal sleeve sliding in a perforated cylinder. The perforations result in numerous small jets, which can be stilled in a small chamber, and the perforations can be graded to obtain any desired discharge/stroke characteristic. In some valves large ports are provided with the latter object but causing less obstruction and the full opening of the sleeve can be utilized, without perforations or ports. In this case the discharge coefficient is greater but a larger stilling well is needed. Dimensions of stilling wells providing adequate energy dissipation may be determined from Figure 22.15, but shallower, wider wells of about equal volume also have been found satisfactory.

Multijet sleeve valves have also been developed as energy dissipators in pipelines.[95] Energy is dissipated by small jets and, over a wide range of pressure differential, the valve is generally free from vibration and cavitation problems.

22.5.3 Air demand

Air vents are provided downstream of gates and valves in conduits, to relieve low pressures which develop due to regulation and to avoid cavitation or column separation following valve closure. The rate of air demand of a hydraulic jump is discussed on page 22/20. During closure of the valve, downstream pressure falls and the water standing in the vent pipe is gradually drained. At the same time the water in the conduit, if flowing full, is decelerated, but if the conduit is long and the valve closes before the vent is admitting air, pressure might fall to cavitation level. In such cases it is necessary to limit the speed of valve closure, especially when approaching final closure. Pressure can be estimated by a step calculation.

22.6 Cavitation

In high-velocity flow and regions of low pressure, local pressure may approach the level of vapour pressure causing cavitation bubbles to form and become entrained in the flow. When these reach regions of higher pressure they collapse, producing extremely high local transient pressures which can damage solid boundaries. Examples of cavitation damage have occurred in high-velocity flow past rough surfaces and joints in concrete, in stilling basins below high spillways, downstream of submerged orifices and in valves and rotary pumps. Metal becomes damaged by pitting; concrete begins to disintegrate.

Cavitation in valves, pipe fittings and pumps is characterized by a high-pitched tapping sound and the discharge coefficient or pump efficiency may be affected. Cavitation damage usually occurs immediately downstream of the point of lowest pressure but if the cavities are formed away from solid boundaries as, for example, by fluid shear downstream of a submerged orifice, they can be carried some distance in the flow and in some cases collapse harmlessly in the fluid, but if the collapse occurs near or on a solid boundary this can be damaged. The collapse of a diversion tunnel at Tarbela dam was attributed to this cause.[96]

An index of cavitation potential is the cavitation number:

$$\sigma = \frac{H_2 - h_v}{H_T - H_2} \quad \text{or} \quad \frac{p_2 - p_v}{\rho V^2/2} \tag{22.17}$$

where H_2 is the pressure head and p_2 the pressure at the point concerned, h_v the vapour pressure head, H_T the total (static + velocity) head, ρ the density of water, p_v the vapour pressure and V_2 the mean or relevant velocity

The heads and pressures appear as differences so should be related to the same datum. If the datum is atmospheric pressure, h_v will be negative, usually taken as -10 m at sea-level. For the second expression the pressures are usually absolute. The cavitation number represents the ratio of pressure drop required to initiate cavitation to the velocity head available and therefore indicates the potential for cavitation. The number at which cavitation occurs depends on the boundary geometry and flow pattern. If the number for the onset of cavitation in a given situation is known through research or experience it can be used to test whether cavitation will occur over a range of velocities and pressures and (subject to small variations due to scale effect) in situations of geometrical similarity but different absolute dimensions, as in scale models.

The risk of cavitation damage on concrete surfaces in contact with high-velocity flow increases rapidly with increasing velocity, velocities in the range 20 to 40 m/s being of particular concern. It is necessary to provide a very smooth finish; specifications often call for offsets to be ground to a flat slope such as 1:30.[95] In stilling basins below high spillways, concrete baffle blocks may be protected by steel cladding. Use of special additives in the concrete can increase its resistance to damage. Damage can also be alleviated or prevented by aeration as, for example, provided by suction of air at offsets through special ducts.[64,65] In the case of vertical and radial gates and gate valves, mild cavitation may occur with $\sigma = 2.0$ and more severe cavitation with $\sigma = 1.0$. In valves of other types the critical value of σ differs in different designs and with the amount of opening. Butterfly valves were found to have incipient cavitation characteristics with $\sigma = 1.5$ for 30° opening but 3.9 for 80° opening.[97]

References

1 Bakhmeteff, B. A. and Matzke, A. E. (1936) 'The hydraulic jump in terms of dynamic similarity.' *Trans. Am. Soc. Civ. Engrs,* **101** Paper 1935, 530–647.
2 Stevens, J. C. (1933) 'The hydraulic jump in standard conduits.' *Civ. Engrg* (New York), **3**, 000–000.
3 Massey, B. S. (1961) 'Hydraulic jump in trapezoidal channels – an improved method.' *Water Power,* **13**, 232.
4 Silvester, R. (1964) 'Hydraulic jump in all shapes of horizontal channels.' *Proc. Am. Soc. Civ. Engrs,* **90**, HY1, 23–55.
5 Elevatorski, E. A. (1959) *Hydraulic energy dissipators.* McGraw-Hill, New York.
6 Simmons, W. P. (1964) 'Transitions for canals and culverts.' *Proc. Am. Soc. Civ. Engrs,* **90**, HY3, **115**, 115–153.
7 Shukry, A. (1950) 'Flow around bends in an open flume.' *Trans. Am. Soc. Civ. Engrs,* **115**, Paper 2422, 751–788.
8 Ippen, A. Y., Knapp, R. T., Rouse, H. and Hsu, E. Y. (1951) 'High velocity flow in open channels – a symposium.' *Trans. Am. Soc. Civ. Engrs,* **116**, Paper 2434, 268–295.
9 King, H. W. and Brater, E. F. (1963) *Handbook of hydraulics,* 5th edn. McGraw-Hill, New York.
10 Rouse, H. (1938) *Fluid mechanics for hydraulic engineers.* McGraw-Hill, New York.
11 US Army Corps of Engineers (1952–70) *Hydraulic design criteria.* US Army Engineer Waterways Experiment Station, Vicksburg, Miss.
12 US Department of the Interior Bureau of Reclamation (1960) *Design of small dams.* Denver, Colorado.
13 White, W. R. (1971) 'The performance of two-dimensional and flat-vee triangular profile weirs.' *Proc. Am. Soc. Civ. Engrs,* Paper 7350S.

14 Ackers, P., White, W. R., Perkins, J. A. and Harrison, A. J. M. (1978) *Weirs and flumes for flow measurement*. Wiley, Chichester and New York.

15 Bos, M. H. (ed.) (1978) *Discharge measurement structures*. International Institute for Land Reclamation and Improvement, Wageningen, The Netherlands.

16 British Standards Institution (1969) *Methods of measurement of liquid flow in open channels, long-base weirs*, BS 3680, Part 4B. BSI, Milton Keynes.

17 Parshall, R. L. (1936) *The Parshall measuring flume*. The Colorado Agricultural Experimental Station, Fort Collins, Colorado, Bulletin 423.

18 Thomas, C. W. (1960) 'World practices in water measurements at turnouts.' *Proc. Am. Soc. Civ. Engrs*, **86**, IR2, 29.

19 Lacey, G. (1958) 'Flow in alluvial channels with sandy mobile beds.' *Proc. Instn Civ. Engrs*, **9**, 145–164.

20 Haigh, F. F. (1941) 'The Emerson barrage.' *J. Inst. Civ. Engrs*, **2**, 107–152.

21 Gerrard, R. T., Long, J. J. and Shah, H. H. (1982) 'Barking Creek tidal barrier.' *Proc. Instn Civ. Engrs*, **72**, 533–562.

22 Clark, P. J. and Tappin, R. G. (1978) 'Final design of Thames Barrier gate structures.' In: *Thames Barrier design*. Paper 7. Institution Civil Engineers, London.

23 Leliavsky, S. (1935) *Design of dams for percolation and erosion*. Chapman and Hall, London.

24 Foy, Sir T. and Green, H. S. (1969) 'Barrages and dams on permeable foundations.' In: C. V. Davis and K. E. Sorensen (eds) *Handbook of applied hydraulics*, 3rd edn, Chapter 17. McGraw-Hill, New York.

25 Bradley, J. N. and Peterka, A. J. (1957) 'The hydraulic design of stilling basins.' *Proc. Am. Soc. Civ. Engrs*, **83**, HY5, Papers 1401–1406; discussion 1958.

26 Peterka, A. J. (1978) *Hydraulic design of stilling basins*. US Department of the Interior, Bureau of Reclamation. Water Resources Research Report No. 24.

27 Bhowmik, N. G. (1975) 'Stilling basin design for low Froude number.' *Proc. Am. Soc. Civ. Engrs*, **101**, HY7, 901–915.

28 Binnie, G. M. (1938) 'Model experiments on bellmouth and siphon bellmouth overflow spillways.' *J. Instn. Civ. Engrs*, **10**, 65.

29 Donnelly, C. A. and Blaisdell, F. W. (1965) 'Straight drop spillway stilling basin.' *Proc. Am. Soc. Civ. Engrs*, **91**, HY3, 101.

30 Berryhill, R. H. (1957) 'Stilling basin experiences of the Corps of Engineers'. *Proc. Am. Soc. Civ. Engrs*, HY3, **83**, paper 1264.

31 Berryhill, R. H. (1963) 'Experience with prototype energy dissipators.' *Proc. Am. Soc. Civ. Engrs*, **89** HY3, 181.,

32 McPherson, M. B. and Karr, M. H. (1957) 'A study of bucket-type energy-dissipator characteristics.' *Proc. Am. Soc. Civ. Engrs*, **83**, HY3, Paper 1266.

33 Beichley, G. L. and Peterka, A. J. (1959) 'The hydraulic design of slotted spillway buckets.' *Proc. Am. Soc. Civ. Engrs*, **85**, HY10, 1–36.

34 Beichley, G. L. and Peterka, A. J. (1961) 'Hydraulic design of hollow-jet valve stilling basin.' *Proc. Am. Soc. Civ. Engrs*, **87**, HY5, 1–36.

35 Burgi, P. H. 'Hydraulic design of stilling wells.' *Proc. Am. Soc. Civ. Engrs*, **101**, HY7, 801–816.

36 Lacey, G. (1929–30) 'Stable channels in alluvium.' *Proc. Instn Civ. Engrs*, **229**, 259–292.

37 Nixon, M. (1959) 'A study of the bank-full discharges of rivers in England and Wales.' *Proc. Instn Civ. Engrs*, **9**, 145–164.

38 Inglis, Sir Claude (1949) *The behaviour and control of rivers and canals*. Central Waterpower Irrigation and Navigation Research Station, Poona, India, Part II, p. 327.

39 Laursen, E. M. (1962) 'Scour at bridge crossings.' *Trans. Am. Soc. Civ. Engrs*, **127**, Part I, 166–180.

40 Neill, C. R. (1965 and 1967) 'Measurements of bridge scour and bed changes in a floating sand-bed river.' *Proc. Instn Civ. Engrs*, **30**, 415–421; discussion on above, *Proc. Instn Civ. Engrs*, **36**, 397–436.

41 Jain, S. C. (1981) 'Maximum clear-water scour around circular piers.' *Proc. Am. Soc. Civ. Engrs*, **107**, HY5, 611–626.

42 Ackers, P. (1958) *Resistance of fluids flowing in channels and pipes*. Hydraulics Research Paper No. 1. HMSO, London.

43 Bradley, J. N. and Thompson, L. R. (1962) *Friction factors for large conduits flowing full*. US Department of the Interior Bureau of Reclamation, Water Resources Engineering Monograph 7 (revised).

44 American Society of Civil Engineers (1965) 'Factors influencing flow in large conduits.' Task Force on Flow in Large Conduits, *Proc. Am. Soc. Civ. Engrs*, **91**, HY6, Paper 4543, 123–152.

45 Colebrook, C. F. (1958) 'The flow of water in unlined, lined and partly lined rock tunnels.' *Proc. Instn Civ. Engrs*, **11**, 103–132.

46 Wright, D. E. (1971) *The hydraulic design of unlined and lined-invert rock tunnels*. Construction Industry Research an Information Association, Report No. 29. CIRIA, London.

47 Eppema, R., Jain, S. C. and Kennedy, J. F. (1982) *Hydraulic design of drop structures: a state of the art review*. Iowa Institute for Hydraulic Research, Iowa. Limited Distribution Report No. 98.

48 Kalinske, A. A. and Robertson, J. M. (1943) 'Closed conduit flow. Symposium on entrainments of air in flowing water.' *Trans Am. Soc. Civ. Engrs*, **108**, Paper 2205, 1435.

49 Ackers, P. and Crump, E. S. (1959–60) 'The vortex drop.' *Proc. Instn Civ. Engrs*, **65**, 16, 433–442.

50 Sharma, H. R. (1976) 'Air entrainment in high head gated conduits'. *Proc. Am. Soc. Civ. Engrs*, **102**, HY11, 1629–1646.

51 Sailer, R. E. (1955) 'San Diego aqueduct.' *Civ. Engrg* (New York), 268.

52 American Society of Civil Engineers (1963) 'Task force on hydraulic design of spillways: progress report.' *Proc. Am. Soc. Civ. Engrs*, **89**, HY4, Paper 3573, 117.

53 El-Khashab, A. and Smith, K. V. H. (1976) 'Experimental investigation of flow over side weirs.' *Proc. Am. Soc. Civ. Engrs*, **102**, HY9, 1255–1268.

54 Binnie, G. M., Gerrard, R. T., Eldridge, J. G., Kirmani, S. S., Davis, C. V., Dickinson, J. C., Gwyther, J. R., Thomas, A. R., Little, A. L., Clark, J. F. F. and Seddon, B. T. (1967) 'Engineering of Mangla.' *Proc. Instn Civ. Engrs*, **38**, 343–544.

55 Wagner, W. E. (1954) 'Morning-glory shaft spillways: determination of pressure-controlled profiles.' *Proc. Am. Soc. Civ. Engrs*, **80**, 432, 1–38.

56 Bradley, J. N. (1954) 'Morning-glory shaft spillways: prototype behaviour.' *Proc. Am. Soc. Civ. Engrs*, **80**, Paper 431, 1–33.

57 Mussallii, Y. G. and Carstens, M. R. (1969) *A study of flow conditions in shaft spillways*. Georgia Institute of Technology, Atlanta, Georgia WRC-0669.

58 Oliver, G. C. S. 'Eye Brook reservoir spillway.' *J. Instn W. Engrs*, **13**, 205.

59 Fellerman, L. (1965) *Models tests of Shek Pik siphon spillway, Tung Chung Water Scheme*. British Hydromechanics Research Association, RR841.

60 Hydraulics Research Station (1971) *Plover Cove Reservoir: air-regulated siphon spillway*. Report No. EX539.

61 Charlton, J. A. (1962) *Self-priming siphons – an appraisal*. British Hydromechanics Research Association Publication No. SP725. BHRA, Cranfield.

62 Ackers, P. and Thomas, A. R. (1975) 'Design and operation of siphons and siphon spillways: air-regulated siphons for reservoir and head-water control.' *Symposium of the design and operation of siphons and siphon spillways*. British Hydromechanics Research Association, Cranfield.

63 Wood, I. R., Ackers, P. and Loveless, J. (1983) 'General method for the critical point on spillways.' *Proc. Am. Soc. Civ. Engrs*, *HY2 308–312*.

64 Oskolnikov, A. G. and Semenkov, V. M. (1979) 'Experience in design and maintenance of spillway structures in large rivers in the USSR.' *Proceedings of the 13th Congress on Large Dams*, vol. III, p. 789. International Commission on Large Dams, New Delhi.

65 Beichley, G. H. and King, D. H. (1967) 'Cavitation control by aeration of high-velocity jets'. *Proc. Am. Soc. Civ. Engrs*, **101**, HY7, 829–846.

66 Rhone, T. J. and Peterka, A. J. (1959) 'Improved tunnel spillway flip buckets.' *Proc. Am. Soc. Civ. Engrs*, **85**, HY12, 53–76.

67 Gumensky, D. B. (1953) 'Design of side walls in chutes and spillways.' *Proc. Am. Soc. Civ. Engrs*, **79**, 1–7, 175.

68 Balloffet, A. (1961) 'Pressures on spillway flip buckets.' *Proc. Am. Soc. Civ. Engrs*, **87**, HY5, 87–98.

69 Gunko, F. G. (1965) 'Research on the hydraulic regime and local scour of river bed below spillways of high head dams.' *Proceedings, International Association for Hydrographical Research*, Paper 1, p.50.

70 Akhmedov, T. K. L. (1968) 'Local erosion of fissured rock at the downstream end of spillways.' (Trans. from Russian.) *Hydrotech. Const., Am. Soc. Civ. Engrs,* No. 9, 821.

71 Mason, P. J. (1982) 'The choice of hydraulic energy dissipator for dam outlet works based on a survey of prototype usage.' *Proc. Instn Soc. Civ. Engrs,* Part I, **72,** 209–219.

72 Rouse, H. (ed.) (1950) *Engineering hydraulics.* Wiley, New York, p.32.

73 Joglekar, D. V. and Damle, P. M. (1957) 'Cavitation-free sluice outlet design.' *Proc. Int. Assoc. Hyd. Res.,* Paper B2.

74 Denny, D. F. and Young, G. A. J. (1957) 'The prevention of vortices and swirl at intakes.' *Proc. Int. Assoc. Hyd. Res.,* Paper C1.

75 Anwar, H. O. (1968) 'Prevention of vortices at intakes.' *Water Power,* **20,** 393.

76 Hydraulic Institute (1975) *Hydraulic Institute Standards,* 13th edn. Hydraulic Institute, Cleveland, Ohio, pp.108–115.

77 Prosser, M. J. (1977) *The hydraulic design of pump sumps and intakes.* British Hydromechanics Research Association, Cranfield.

78 Sweeney, C. E., Elder, R. A. and Hay, D. (1982) 'Pump sump design experience: summary.' *Proc. Am. Soc. Civ. Engrs,* **108,** HY3, 361–377.

79 Sell, L. E. (1971) 'Hydro-electric power plant trashtrack design.' *Proc. Am. Soc. Civ. Engrs,* **97,** PO1, 115–121.

80 American Society of Civil engineers (1959) 'Design and maintenance of intakes, racks and booms.' Committee on Operation and Maintenance of Hydro-electric Generating Stations of the Power Division; *Proc. Am. Soc. Civ. Engrs,* **85,** PO5, 71–87.

81 Bleuler, W. (1963) 'Sluice gates.' *Water Power,* **15,** 460.

82 Bradley, J. N. (1954) 'Rating curves for flow over drum gates.' *Trans. Am. Soc. Civ. Engrs,* **119,** Paper 2677.

83 Toch, A. (1953) 'Discharge characteristics of Tainter gates.' *Proc. Am. Soc. Civ. Engrs,* **79,** Paper 295.

84 Anwar, H. O. (1964) 'Discharge coefficients for control gates.' *Water Power,* **16,** 152.

85 Simmons, W. P. (1965) 'Experiences with flow-induced vibrations'. *Proc. Am. Soc. Civ. Engrs,* **91,** HY4, 185–204.

86 Schmidgall, T. (1972) 'Spillway gate vibrations on Arkansas river dams'. *Proc. Am. Soc. Civ. Engrs,* **98,** HY1, 219–256.

87 Colgate, D. (1959) 'Hydraulic downpull forces on high head gates.' *Proc. Am. Soc. Civ. Engrs,* **85,** HY11, 39–52.

88 Kohler, W. H. and Ball, J. W. (1969) 'High-pressure outlets, gates and valves.' In: C. V. Davis and K. E. Sorensen (eds) *Handbook of applied hydraulics,* 3rd edn, section 21. McGraw-Hill.

89 Guins, V. G. (1968) 'Flow characteristics of butterfly and spherical valves.' *Proc. Am. Soc. Civ. Engrs,* **94,** HY3, 675–690.

90 McPherson, M. B., Strausser, H. S. and Williams, J. G. (1957) 'Butterfly valve flow characteristics.' *Proc. Am. Soc. Civ. Engrs,* Paper 1167, 1–28.

91 Miller, E. (1968) 'Flow and cavitation characteristics of control valves.' *J. Inst. W. Engrs,* **22,** 7.

92 Thomas, C. (1955) 'Discharge coefficients for gates and valves.' *Proc. Am. Soc. Civ. Engrs,* **81,** Paper 746, 1–26.

93 Lancaster, D. M. and Dexter, R. B. 'Hydraulic characteristics of hollow jet valves.' *Proc. Am. Soc. Civ. Engrs,* **85,** HY11, 53–63.

94 Elder, R. and Dougherty, G. B. (1952) 'Characteristics of fixed dispersion cone valves.' *Proc. Am. Soc. Civ. Engrs,* **78,** Paper 153, 1–21.

95 Ball, J. W. (1976) 'Cavitation from surface irregularities in high velocity.' *Proc. Am. Soc. Civ. Engrs,* **102,** HY9, 1283–1297.

96 Kenn, M. J. and Garrod, A. D. (1981) 'Cavitation damage and the Tarbela Tunnel collapse of 1974.' *Proc. Instn Civ. Engrs,* Part I, **70,** 65–89.

97 Tullis, J. P. and Marschner, B. W. (1968) 'Review of cavitation research on valves.' *Proc. Am. Soc. Civ. Engrs,* **94,** HY1, 1–16.

23

Highways

J A Turnbull CEng, FICE, FIHT, DipTE
Mott, Hay & Anderson

Contents

23.1 Introduction

The system of administering highways varies considerably from country to country, depending upon economic and political considerations as well as on the customs and laws of each country. There are differences also in highway design, construction and maintenance practices, related to climate, traffic intensity and available resources.

Motorways and other national routes of strategic importance are in most countries the responsibility of a government (federal) department and are funded by central government. Routes of major importance, including those feeding into the national system, are the responsibility of local authorities (counties in the UK, states in the US). Local roads are generally the responsibility of local authorities and may range from residential roads to roads linking local towns and communities to urban motorways.

Highway design and construction practices are developing all the time and lessons learned from experience and research quickly pass from one country to another although, of course, the success or failure of any new feature may take several years to establish in practice. The relevant national government department usually takes a leading role in developing national standards. There are, in addition, a number of formal and informal means of collaboration and exchange of information between highway engineers and materials and other specialists, that are gradually bringing greater uniformity of practice.

It is beyond the scope of this chapter to provide a comprehensive review of all international practice; it is therefore based on UK practice with occasional reference to other countries.

The problems of constructing low-cost roads in developing countries (described in section 23.14) and the methods of traffic and environmental appraisal (sections 23.4 and 23.5) and the engineering design aspects all have international applications.

23.2 Highway administration

23.2.1 Highway authorities

Highway authorities in the UK exercise powers under highway and traffic legislation and have various public responsibilities and duties. The highway authorities in the UK are:

(1) For trunk roads including inter-urban motorways in:

    (a) England – Secretary of State for Transport via the Department of Transport;
    (b) Scotland – Secretary of State for Scotland via the Scottish Development Department;
    (c) Wales   – Secretary of State for Wales via the Welsh Office.

(2) For principal and other roads including urban motorways – county councils and district councils in England and Wales, regional councils in Scotland, and London boroughs.

23.2.2 Agent authorities

Certain local highway authorities (normally county councils, large district councils and London boroughs), act as agents for the respective Secretary of State, on maintenance and other works on trunk roads and motorways, including studies, design and supervision of construction.

23.2.3 Consulting engineers

Consulting engineers are often appointed by the Secretaries of State to undertake various studies, design and supervision of construction for new works and improvement. In England,

consulting engineers have also been appointed to manage maintenance.

23.2.4 Financial arrangements

The financial responsibility for the construction, improvement and maintenance of trunk roads and motorways lies with the Secretaries of State. On non-trunk roads, the local highway authority has this responsibility with certain direct and/or indirect contributions from central government.

23.2.5 Highway legislation

Highway law for England and Wales was first codified in 1835 and since then there has been much legislation, most of which was consolidated in the Highways Act of 1980. The *Encyclopedia of highway law and practice*[1] is a useful reference and is updated three times per annum. The Highways Acts give powers to protect future trunk and motorway routes and lay down statutory procedures which safeguard the rights of individuals affected by proposed roads. Highway authorities have power to acquire land compulsorily for highway purposes but, if there is objection to a local authority scheme, the Secretary of State must, in general, confirm the Compulsory Purchase Order. The Land Compensation Act 1973 provides the machinery to compensate those whose properties are affected adversely by highway schemes. The main planning legislation as it affects highways was re-enacted in the Consolidated Town and Country Planning Act 1962 with further amendments in the Acts of 1968, 1971 and 1983. The Local Government Act 1985 allowed the transfer of certain functions to London boroughs and district councils following the abolition of the Greater London Council and metropolitan county councils respectively.

23.2.6 Statutory undertakers' apparatus within the highway (electricity, gas, water, telephone, etc.)

The legal position in the UK regarding apparatus owned by statutory undertakers laid in public highways is defined in the Highways Act 1980, the Pipelines Act 1967 and the Public Utilities Street Works Act 1950.

(1) *All-purpose roads.* For all-purpose roads, the statutory bodies have rights under the Highways Acts to install apparatus within the highway boundaries subject to the approval of the Highway Authority for the actual location of the service.

(2) *Motorways.* With the exception of British Telecom, statutory undertakers have no rights to lay apparatus on, under or over land along the route of a motorway, although permission may be granted under special conditions.

(3) *Public Utilities Street Works Act 1950.* This defines the procedure to be adopted wherever works are carried out in a highway. A brief outline of the four parts of the Act and of the general procedure to be adopted by highway authorities and undertakers is as follows:

    Part 1 (sections 3 to 14 and first, second and third schedules) protects the highway authorities responsible for streets, etc. when statutory undertakers exercise their own statutory powers to open streets, etc.;

    Part 2 protects statutory undertakers where their apparatus is affected by road, bridge or other highway works;

    Part 3 is a miscellaneous section and deals with the effect of one statutory undertaker on another, road closures, etc.;

Part 4 covers financial provisions, special application of the Act in London and Scotland, etc.

The Act was reviewed in 1984 by a committee headed by Professor Horne of Nottingham University and a report submitted to the Government.

Costs arising from the diversion of existing apparatus owned by statutory undertakers which must be altered owing to road-works fall on the highway authority, except that the Public Utilities Street Works Act 1950 provides for offset payments in respect of: (1) betterment, i.e. where different or improved apparatus is requested by the Undertaker; and (2) deferment of renewal of the apparatus.

23.2.7 Department of Transport publications

The Department of Transport currently issues a variety of documents which form the background to UK highway design and construction. These fall into the following categories:

(1) Circular roads – generally relating to legal and administrative procedures.
(2) Trunk Road Management and Maintenance Notices (TRMMs).
(3) Technical Memoranda Bridges (BEs) (issued until mid 1978).
(4) Departmental Standards – Bridges and Structures (BDs).
(5) Departmental Advice Notes – Bridges and Structures (BAs).
(6) Technical Memoranda – Highways (Hs) (issued until mid 1978).
(7) Departmental Standards – Highways (HDs).
(8) Departmental Advice Notes – Highways (HAs).
(9) Departmental Standards – Traffic Engineering and Control (TDs).
(10) Departmental Advice Notes – Traffic Engineering and Control (TAs).

23.2.8 Road research

Research on highway engineering and associated subjects in the UK is focused on the Transport and Road Research Laboratory (TRRL) based at Crowthorne, Berkshire, and at Livingston, West Lothian.

A number of external advisory committees exist to advise on the various aspects of research. Apart from work carried out at the TRRL, research is sponsored at universities and other research establishments and contracts are conducted in co-operation with industry. Results of research are issued as: (1) road notes (to 1981); (2) laboratory reports and supplementary reports (to 1985); (3) research reports, contractor reports and application guides (from 1985).

23.2.9 Maps

23.2.9.1 Ordnance Survey

The Ordnance Survey produces maps for official and public use and also offers many other services which are by-products of its normal work of triangulation, levelling, aerial and field surveys, drawing, printing and publication. Their Central Register of Aerial Photography maintains an index of all aerial photographs in the UK.

The Ordnance Survey publishes large-scale maps at three scales:

1:1250 These are all based on the national grid.

1:2500 These may be either national grid or county series mostly based on the national grid.
1:10 000 Based on the national grid.

The 1:1250 and 1:2500 maps represent the features of the ground to scale but the 1:10 000 maps are generalized.

23.2.9.2 British Geological Survey

Geological maps and memoirs issued by the British Geological Survey give information on solid rock formations and drift deposits throughout the UK.

23.2.10 Research and technical guidance in other countries

Many countries have organizations providing research results and technical guidance. Some examples are:

Transportation Research Board (TRB), Washington DC.
American Association of State Highways and Transportation Officials (AASHO), Washington DC.
Federal Highways Administration (FHWA), Washington DC.
Laboratoire Central des Ponts et Chaussées, Paris.
Centre de Recherches Routières, Brussels.
Australian Road Research Board, Vermont S., Victoria.
Bundesanstalt für Strassenwesen, Bergisch Gladbach.
National Institute for Transport and Road Research, Pretoria.

A useful source of reference is the publications of the Permanent International Association of Road Congresses (PIARC).

23.3 Scheme preparation

23.3.1 Introduction

Policy for roads in England[2] states that the objectives of building roads are: to assist economic growth by reducing transport costs; to improve the environment by removing through traffic (especially lorries) from unsuitable roads in towns and villages; and to enhance road safety. All scheme assessments should be made with these objectives in mind, together with the requirements that any investment should provide value for money as stated in the Department of Transport *Traffic appraisal manual*.[3]

Although this chapter deals primarily with the stages adopted in the UK in the evolution of major new routes for rural trunk roads or motorways the principles are equally applicable to other countries. Many described could be applied to urban and smaller scale inter-urban schemes including 'on-line' and junction improvements.

23.3.2 The scheme identification study

The Department of Transport commissions a scheme identification study in order to ascertain whether a scheme should be included in the national trunk roads programme. The planning brief requests the investigation of environmentally feasible options for the route and defines the scope of the study. The report is normally based on a collection of available data together with limited surveys and is a greatly reduced version of the technical appraisal report which is described in the following section 23.3.3. If a viable route or routes is/are identified in the scheme identification study the scheme is added to the national trunk roads programme, which is published periodically.[2]

23.3.3 The technical appraisal report

The next stage is the technical appraisal report which forms Part 1 of the preliminary report. The technical appraisal reviews, modifies and examines in greater detail the scheme identification study option(s) together with any route alterations not previously considered. The normal technical appraisal format is as follows:

(1) Introduction.
(2) Planning brief.
(3) Existing conditions:

    (a) description of built-up areas;
    (b) existing highway network;
    (c) traffic;
    (d) accidents;
    (e) topography, land use, property and industry;
    (f) climate;
    (g) drainage;
    (h) geology;
    (i) mining;
    (j) public utilities;
    (k) environmental status.

(4) Planning factors.
(5) Description of alternative schemes:

    (a) general;
    (b) alternative scheme A;
    (c) alternative scheme B;
    (d) other alternatives.

(6) Traffic analysis:

    (a) traffic data;
    (b) traffic analysis;
    (c) road layout and standards;
    (d) conclusions.

(7) Economic assessment:

    (a) application of cost-benefit analysis program (COBA);
    (b) networks and printouts;
    (c) discussion of COBA results.

(8) Appraisal framework:

    (a) framework including environmental aspects;
    (b) summary of consultation with public bodies;
    (c) comparison of alternative schemes.

(9) Programme.
(10) Conclusions and recommendations;

    (a) alternatives for public consultation;
    (b) preferred solution.

Detailed cost estimates are also provided.

It can be seen that considerable amounts of data have to be collected and analysed in order to make an assessment of alternative schemes and reach conclusions. The following notes expand on existing conditions (item (3) above).

Data collection should include the routes of public footpaths and bridleways through the alternative route corridors. County councils maintain definitive maps.

(1) *The existing highway network* (item (3)(b) above) comprises an extensive existing route inventory including:

    (a) vertical and horizontal alignment standards;
    (b) cross-section information, including edge details;
    (c) signing and road-marking, including speed limits;
    (d) lighting and safety fences;

    (e) junctions and frontage access.

(2) *Land use* (item (3)(e) above) should include land classified under the Ministry of Agriculture, Fisheries and Food system.[4] The reports which accompany the classification maps are also useful sources of climatological data.

(3) *Geology* (item (3)(h) above) is covered by solid/drift geology maps and memoirs published by the British Geological Survey. Further preliminary sources of information are published by the TRRL.[5]

(4) *Environmental status* (item (3)(k) above) covers a variety of features namely:

    (a) areas of outstanding natural beauty, the national landscape classification and various regional classifications;
    (b) sites of special scientific interest, nature reserves and other areas of ecological value defined by the Nature Conservancy Council;
    (c) ancient monuments and archaeological sites;
    (d) conservation areas and listed buildings;
    (e) country parks, national parks, public open spaces and common land.

(5) Planning factors relate primarily to policies to be found in Government white papers, regional plans, structure plans, local/district plans and transport policies and programmes. Reports of survey, the data which local authorities use in formulating or revising policies, can be useful sources of information.

(6) Traffic analysis is described in section 23.4.

(7) Economic assessment is described in section 23.4.6.

(8) Appraisal framework is a convenient tabular means of comparing alternative schemes. The use of frameworks was recommended by the independent Advisory Committee on Trunk Road Assessment in 1977.[6] The Committee became the Standing Advisory Committee on Trunk Road Assessment (SACTRA) and recommended a detailed approach to the preparation of frameworks in 1979.[7] Department of Transport publications formalized their use as a decision-making tool and as a means of factual presentation to the public.[8]

In their 1986 report,[9] SACTRA suggested that frameworks could be more effectively applied. They recommended developing the framework method of presenting data with the ultimate aim of producing a framework (Table 23.1) called an assessment summary report which would detail the objectives and summarize to what extent those objectives would be achieved by alternative schemes. This approach was regarded as being appropriate to both urban and inter-urban schemes. The Government has accepted this recommendation in principle.[10]

23.3.4 The Landscape Advisory Committee

Before the completion of the technical appraisal report, the Department of Transport seeks comments on the alternatives from a wide range of public bodies including the Countryside Commission, English Heritage and the Nature Conservancy Council. The Department of Transport also refers the scheme to the independent Landscape Advisory Committee. The Landscape Advisory Committee normally conducts site visits along the alternative routes and makes recommendations on route modifications to minimize the impact of schemes on the landscape.

23.3.5 Public consultation

Following the completion of the technical appraisal report, viable schemes are explained at a public exhibition, normally held at a number of venues in the affected area. Rejected

Table 23.1 The types of measures likely to be presented under each heading in the assessment summary

Assessment heading	Sub-headings (examples only)	Information (examples only)
Objectives and problems statement	National and local objectives with statements of problems directly relevant to each. Scheme objectives	Description of how each option satisfies
Options and consultations statement	Type of issue raised, e.g. mobility, access, circulation, parking, safety	Points raised for and against each option in consultation
Traffic appraisal	Changes in flows, journey times and accidents	Types of traveller, journey, purposes, times of day; immediate and longer-term effects
Economic evaluation	Cost benefits to travellers, transport operators and government	Net present value and first year rate of return
Environmental and social impact	Noise, air pollution, community severance and townscape	Immediate and long term effects on people, buildings and sites

schemes may also be included to indicate the breadth of the study.

Brochures containing plans and questionnaires to invite the views of the public, are distributed. The report is written-up as Part 2 of the preliminary report to the following format:

(1) Introduction.
(2) General:
    2.1 consultation arrangements;
    2.2 attendance at exhibition;
    2.3 effectiveness of consultation.
(3) Local preference:
    3.1 questionnaire;
    3.2 written comments.
(4) Main factors.
(5) Non-local views:
    5.1 questionnaire;
    5.2 other comments.
(6) Other information.
(7) Special considerations.
(8) Other routes suggested.
(9) Summary of results.
(10) Conclusion.

23.3.6 The scheme assessment report and recommendation

Parts 1 and 2 of the preliminary report are distilled to create this report which forms Part 3. It recommends a route for publication and is formated as follows:

(1) Introduction.
(2) Summary of existing conditions.
(3) Summary of planning factors.
(4) Summary of do-nothing consequences.
(5) Summary of alternative schemes.
(6) Summary tables of traffic, economics, costs, environmental and design data.
(7) Summary of public consultation.
(8) Conclusion.
(9) The recommended route. } in confidence
(10) Confidential consultations. }

23.3.7 The preferred route

If the Secretaries of State for Transport and the Environment accept the recommendation, the preferred route is published and the planning authorities protect the line.

23.3.8 The preliminary design

A site investigation and topographical survey (normally an aerial survey) are carried out along the preferred route corridor, following the preferred route announcement. Preliminary design is carried out, usually to 1/2500 scale.

23.3.9 The publication of orders

Before publication of orders can proceed, the Department of Transport's internal procedures require a final check on the viability of the scheme. This is presented as the order publication report which includes detailed cost breakdowns, scheme justification and traffic and economic appraisals.

The publication of a scheme under the Highways Act 1980 requires draft line, slip road, side road and, if appropriate, detrunking orders, together with order plans.

Within a short period of order publication, the Department of Transport normally holds a public exhibition where the published proposals are explained in detail. Amelioration measures such as noise barriers and landscaping are illustrated and use is made of photomontages, aerial photo-mosaics and topographical models to aid lay understanding of the proposals.

23.3.10 The public inquiry

During the specified periods for comment, the Department of Transport is likely to receive many representations and objections. Statutory objectors, including owners losing land to the scheme, can force an inquiry. The Secretaries of State for Transport and Environment are likely in any case to exercise their discretion to hold an inquiry if there are many objections from nonstatutory objectors, be they individuals or organizations.

Between publication and inquiry, the Department of Transport normally consults the Royal Fine Art Commission on the appearance of proposed structures, particularly if they are to be located in environmentally sensitive areas.

An independent inspector appointed by the Lord Chancellor's Office hears evidence at the public inquiry and reports to the Secretaries of State, who issue their decision letter in due course. If the scheme is accepted, it goes ahead, sometimes with modifications which require further orders to be published and, possibly, another public inquiry may result.

'Made' orders are ultimately published, incorporating any modifications discussed at inquiry and approved by the Secretaries of State.

23.3.11 The detailed design

Following the issue of the decision letter, contracts are prepared and let for detailed topographical and soil surveys. Flying for the aerial survey is at a lower altitude to produce 1/500- or 1/1000-scale contoured mapping. The site investigation takes account of the finalized route location and types of structures proposed. Detailed design is carried out and land reference plans and schedules are prepared showing scheme requirements and land ownerships.

23.3.12 The compulsory purchase order

A draft compulsory purchase order is published under the Highways Act 1980 and Acquisition of Land Act 1981. Compulsory purchase order plans and schedules are prepared. A further public inquiry may follow where the land requirement for the scheme may be questioned, but not the need for the scheme itself.

If the decision is to confirm the compulsory purchase order, a made order will be published as before. Notices to treat and notices to enter are issued before construction can begin. The former requires plans for each individual/organization with an interest in the land being taken. Land interest plans are also required for negotiation and conveyancing.

The works commitment estimate has to be approved, prior to inviting tenders, by the Department of Transport for budgeting purposes.

The stages in the development of a trunk road scheme are shown in Figure 23.1.

23.3.13 Local authority procedures

23.3.13.1 Planning

Although many of the elements of local authority scheme preparation are similar to those of the Department of Transport, there are differences. Structure plans generally give the first indication of new highway schemes although, historically, many routes were the subject of development plans and have been protected for a considerable time.

Local or district plans define the land to be reserved for highway proposals. Such land can be reserved by a planning and transportation committee resolution. Public inquiries or 'examinations in public' are normally held at draft structure plan and local plan stages.

Transport policy programmes are for local authorities, analogous to the Government's *Policy for roads* document[2] but are in the wider context of transportation as a whole. Transport policy programmes may also be designed to assist transport supplementary grant applications for specific schemes.

23.3.13.2 Orders

Local authorities normally publish a scheme via an application for planning permission under town and country planning legislation followed by publication of a side roads order and compulsory purchase order under the Highways and Land Acquisition Acts.

Objections to the planning application may result in the Secretaries of State 'calling in' the application and deciding to hold a public inquiry.

23.4 Traffic appraisal

The aim in traffic appraisal is to assess current flows on existing roads and predict future flows on both existing and new roads in a chosen network. This information is crucial in determining whether a scheme is justified on operational and economic grounds.

23.4.1 The *Traffic appraisal manual*

The methods of traffic appraisal recommended by the Department of Transport are set out in its *Traffic appraisal manual*.[3] This was produced following the SACTRA forecasting recommendations[11] that the Department should produce such a manual for inter-urban road appraisal. The principles encompassed by the manual, however, are equally applicable to urban schemes and local authority projects. Further information concerning urban roads can be obtained from the recommendations on urban road appraisal by SACTRA[9] and *Roads and traffic in urban areas*.[12]

Details on the various steps to be considered in carrying out a traffic appraisal are clearly set out in the *Traffic appraisal manual* and include: (1) study area definition; (2) sources of traffic data and survey methods; (3) model selection, construction and validation; (4) assessment of errors and uncertainty; and (5) forecasting.

Much original work relating to traffic appraisal techniques originated in the US and is being documented in numerous publications by the Transportation Research Board and Federal Highways Administration.

23.4.2 Study area definition

A traffic study should be tailored to the level of detail and accuracy appropriate to be taken. The study area should not be made unnecessarily large or over-complicated.

23.4.3 Traffic data

The accuracy of a traffic appraisal is dependent upon the quality and reliability of the input traffic data. Existing traffic information is obtained from sources such as traffic counts, origin–destination[13] and travel time surveys. These provide information, by time of day, on how many and what types of vehicles are using the road network. Methods of data collection are discussed in detail in Chapter 65 of the *Traffic appraisal manual*. Additional information concerning the use of video techniques is contained in the *Video system of traffic analysis* manual.[14]

Traffic in urban areas may need to be described in the context of all travel within the area, including journeys by public transport and on foot. This information can be obtained from sources such as household interview and other land-use surveys.

Traffic varies widely over the hours of the day and throughout the year, and information therefore is required on seasonal variations. Details of daily and seasonal variability, by road type, are provided in the *Traffic appraisal manual* and *COBA user manual*.[15] Local information on seasonal variability can be obtained from automatic traffic count data. The Department of Transport and local authorities operate long-term monitoring programmes for the major road network which provide additional sources of local reference data.

23.4.4 Model choice

There are several different types and variation of traffic model available for use. These range from simple growth-factor-based techniques to comprehensive network models requiring large amounts of survey data. Figure 23.2 provides an outline of the processes involved in a full modelling exercise. Such models are complex, require large amounts of data and are time consuming to analyse; however, they can be simplified to suit less demanding situations.

For less complex schemes, including those for traffic manage-

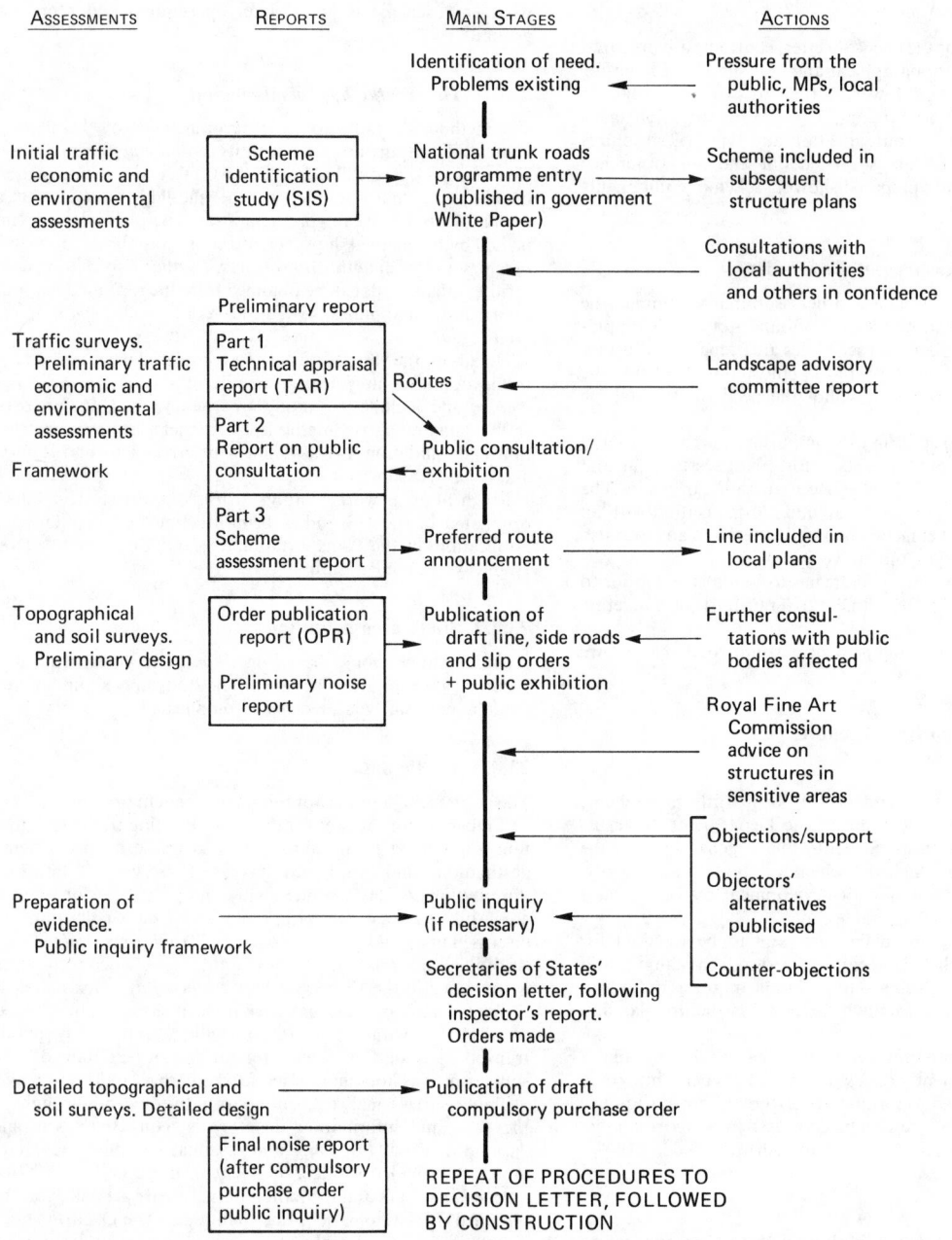

Figure 23.1 Stages in the development of a trunk-road scheme

ment and control, a localized area traffic model may be sufficient to provide the information required. There are several models of this kind available, each with their own special capabilities and limitations. Such models include TRANSYT,[16] SATURN,[17] CONTRAM[18] and TRAFFICQ.[19]

23.4.5 Traffic forecasts

Traffic forecasts are required in order that estimates can be obtained for:

(1) Economic evaluation – normally assessed over a 30 year period (the assumed useful life of a scheme).
(2) Environmental appraisal.
(3) Selection of design standards – it is normal to design schemes to cope with future traffic levels 15 years after scheme opening (the design year).
(4) Provision of information to satisfy certain statutory obligations, particularly those concerned with noise.

Traffic forecasts are based on existing traffic levels which are

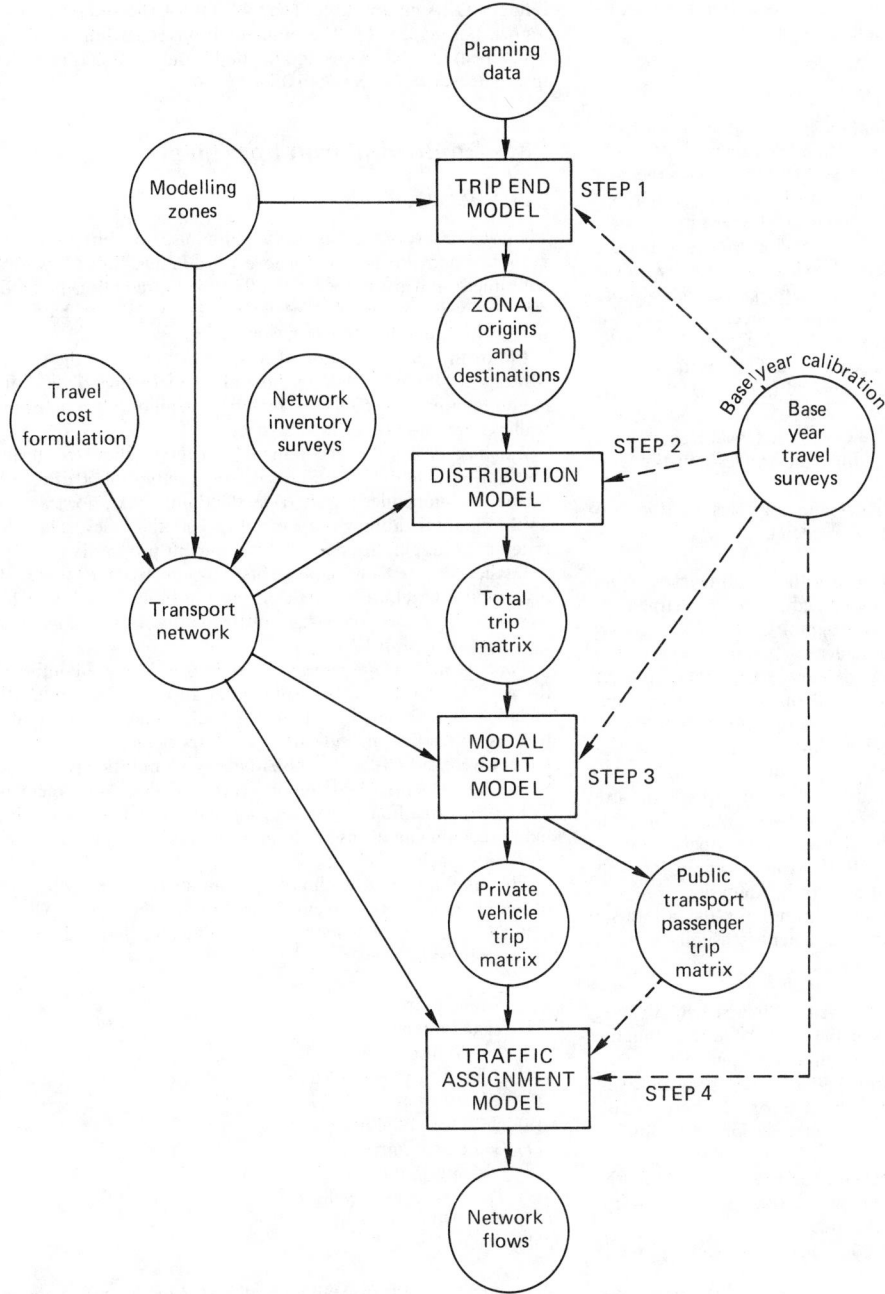

Figure 23.2 Summary of a serial traffic model

subsequently projected in line with national and local planning projections. In its simplest form a single growth rate may be applied to the entire area under consideration. Such an approach may be used for the design of an isolated junction or minor road improvement. For more complicated schemes, where it would be inappropriate to assume an equal growth of trips from all zones in the study area, a method of differential growthing should be applied.

National projections of traffic growth are provided in *National road traffic forecasts*.[20] These are provided as an average national growth rate (per year) for each of a series of vehicle types. The estimates are based on such variables as personal income, GDP per capita and petrol prices. In view of the inevitable uncertainty in forecasting, the Department of Transport employs alternative sets of forecasts, one high and the other low. The range is intended to provide a realistic but cautious view. In addition, the Department of Transport maintains two national submodels: (1) the national car ownership model; and (2) the national trip-end model. These reflect changes in car ownership and land-use over time at a local county and district level.

Detailed planning information within each county and district is available from their structure plans.

23.4.6 Economic evaluation

The purpose of economic evaluation is to estimate the travel-related benefits which might result from the provision of a new or improved transport facility and compare those with the cost of provision. The benefits relate only to those items which are relatively easy to value, namely savings in travel time, vehicle operating costs and accidents. The cost estimates generally taken into account are those elements which fall directly on the financing authority. These costs relate to scheme preparation, land acquisition, construction and maintenance.

The inputs to the analysis are crucially dependent on the output of the traffic forecasting and modelling exercise. Two basic modelling procedures are in general use:

(1) Assuming a fixed demand matrix, i.e. travel demands in terms of origins and destinations and modes and times of travel remain unchanged.
(2) Allowing variations in the demand matrices for the two situations, with and without the scheme.

The fixed trip matrix assumption has the advantage of being relatively simple to apply. It is embodied in the Department of Transport's cost-benefit analysis program COBA.[15] User costs are assessed over 30 years (the assumed useful life of the scheme) and are discounted back to a present value year to obtain the net present value (NPV) for the scheme. Traffic user costs resulting from delays during construction and future traffic-related maintenance works cannot be assessed directly. These costs are estimated separately using the Department of Transport program QUADRO 2 (queues and delays at roadworks)[21] and are incorporated into the cost-benefit analysis. Where incident delays are also likely to be significant (e.g. in the operation of a road tunnel) then the program CIDEL (calculation of incident delays)[22] may be used as an addition to the analysis.

The COBA program is applied throughout the appraisal process. The output in terms of cost-benefit is usually used to contribute to the following decisions:

(1) To assess the need for an improvement in a specified route. The improvement could take the form of upgrading the existing road or providing an entirely new one.
(2) The determination of the priority that should be given to an individual scheme, by comparing its economic return with those from other schemes both in the region and country-wide.
(3) The optimal timing of a scheme, taking into account the merits of staged construction and the timing of other road-improvement proposals in the area.
(4) The selection of potential solutions to present at public consultation.
(5) The selection of the preferred option by ministers after public consultation.
(6) The determination of the optimal link design standards, and the initial assessment of optimal junction designs, by the comparison of economic returns from the feasible alternatives.

Naturally the extent to which a full COBA can be undertaken, or is worth while, will depend on the stage reached in the assessment procedure, the data available and the particular decisions to be taken. As scheme preparation proceeds, more refined economic analyses become possible.

Some urban schemes may cause changes to travel behaviour which extend beyond simple reassignment of trips and thus render COBA inapplicable. The SACTRA *Urban road appraisal* report[9] has made several recommendations regarding economic assessment in urban areas and further useful reading is provided in *Roads and traffic in urban areas*.[12]

23.5 Environmental appraisal

23.5.1 History

The 1963 report of the Committee on Noise[23] proposed the first noise standard in the UK for levels inside dwellings. Research continued, and a number of traffic noise prediction methods were evolved. The method in *Design bulletin 26*[24] was based on that of the Building Research Station.

Following the publication of *New roads in towns*,[25] the Urban Motorways Project Team considered a wider range of environmental factors. Their report[26] included techniques for assessing traffic noise and visual impacts.

Planning and noise[27] recommended a 'good standard' inside dwellings of 40 dB(A) L_{10} 18 h with a maximum of 50 dB(A). It further recommended 70 dB(A) as the limiting exterior (façade) level. Planning authorities currently use these levels as the criteria for judging residential development proposals.

Lassière covered a wide range of environmental effects including visual intrusion evaluated on the solid angle method. This was later redefined as visual obstruction in the *Manual of environment appraisal*.[8]

Route location with regard to environmental issues[28] continued the work of developing environmental appraisal techniques. It included an assessment method for traffic-generated air pollution using lead as an indicator. The report and its discussion papers were submitted to the Advisory Committee on Trunk Road Assessment, whose report[6] reviewed the Department of Transport's method of appraising trunk-road schemes and made recommendations, including the adoption of the framework approach described earlier.

The development of techniques culminated in the publication of the *Manual of environmental appraisal*[8] by the Department of Transport in 1983. The manual gives guidance on the appraisal of the following impacts:

(1) Traffic noise.
(2) Visual impact.
(3) Air pollution.
(4) Community severance.
(5) Effects on agriculture.
(6) Heritage and conservation areas.
(7) Ecological impact.
(8) Disruption due to construction.
(9) Pedestrians and cyclists.
(10) View from the road.
(11) Driver stress.

A Council of the European Community directive was made on 27 June 1985 on the assessment of the effects of certain public and private projects on the environment.[29] A government guideline is being prepared for its use and is expected to be issued in mid 1988.

23.5.2 Vibration and low-frequency sound

Traffic-induced vibration is not an effect covered by the *Manual of environmental appraisal*.[8] However, it is invariably part of general environmental objections to road proposals.

Vibrations can be experienced in two ways:

(1) Airborne low-frequency sound (infrasound) can cause the

vibration of parts of a building, such as suspended floors, windows and doors. Although infrasound itself is below the audible range of sound, its effects on the building may cause its occupants to fear structural damage. However, no cases of damage due to infrasound have been established.

(2) Noticeable structural vibration can be generated by heavy vehicles running over irregularities in the road surface. If the pavement is adequately maintained such vibration is unlikely to occur.[30, 31]

23.5.3 Urban road appraisal

The report on *Urban road appraisal*[9] by SACTRA found that the *Manual of environmental appraisal*[8] needed some changes of emphasis to make it more suitable for application to the full range of urban schemes (Table 23.2).

Table 23.2 Environmental and social impacts: summary of changes proposed for the assessment of urban main-road schemes

Type of impact list in the Manual of environmental appraisal	*Changes proposed for assessment of urban main roads*
Traffic noise	Important; add night-time noise where relevant
Visual impact	Include with consideration of townscape
Air pollution	Can be perceived to be important
Community severance	Important; further research needed
Effects on agriculture	Rarely important
Conservation areas	Include with consideration of townscape and site-specific impacts
Ecological impact	Rarely important
Disruption due to construction	Important
Accidents	Important
View from the road	Omit
Driver stress	Omit
Social consequences of blight	Add description where relevant
Development potential	Add description where relevant

23.5.4 Noise

Noise is the most noticeable type of pollution and is a basis for numerous objections to road proposals. Traffic-noise measurement and forecasting is as a result the most long-standing of assessment tools representing, until recently, a surrogate for other forms of environmental effect.

The *Noise insulation regulations*[32] made in 1973, revised in 1975, introduced the entitlement to double glazing related to noise levels of 68 dB(A) or above, subject to certain criteria.

23.5.4.1 Traffic noise calculation

The *Noise insulation regulations*[32] specify that traffic noise is to be calculated in accordance with *Calculation of road traffic noise*.[33] The *Regulations* require the usage of the L_{10} level which is the mean sound level exceeded for 10% of the 18-h day

between 0600 and 2400. L_{10} is calculated in A-weighted decibels taking account of the response of the human ear.

The basic 18-h source level is computed from:

$$L_{10} = 28.1 + 10 \log Q \text{ dB(A)} \tag{23.1}$$

where Q is the number of vehicles per 18-h day in the design year based on high-growth economic assumptions.

Calculation of road traffic noise[33] applies corrections to the basic level for speed, gradient and percentage of heavy goods vehicles and provides methods for assessing attenuation between source and receiver.

Changes in noise level can be calculated from the ratio of traffic flows, namely: change $= 10 \log (Q_1/Q_2)$ (this expression being derived from Equation (23.1)). This, for example, gives a 3 dB(A) increase or decrease when doubling or halving the flow respectively.

Noise attenuation being independent of the source level, it is apparent that changes in the noise environment adjacent to existing roads can be established simply and quickly from a knowledge of the change in traffic flows.

23.5.4.2 Construction noise

The Control of Pollution Act 1974 empowers local authorities to set construction noise limits. The preferred procedure under section 61 of the Act is for the contractor to apply for a consent-to-work agreement. This consent is deemed on Department of Transport contracts to have been granted by virtue of compliance with the specification. The limits are normally agreed with the appropriate district council's environmental health officer before tenders are invited.

Construction noise is evaluated or measured in terms of $L_{A_{eq}}$, the equivalent continuous noise level. Noise-analysing equipment is available to measure this and L_{10} levels, in addition to other noise classifications.

In order to estimate likely construction noise levels, assumptions need to be made about the contractor's method of work. Noise control on construction and open sites is covered in BS 5228.[34]

23.5.4.3 Tasks in relation to scheme stages

The assessment of noise is a progressive process as a road scheme develops. A few ambient levels are measured in the earliest stages or prevailing levels calculated from existing traffic flows, and dwellings are counted in distance bands from alternative routes. Relief or increases along existing roads may be assessed from traffic flow ratios as previously described.

Some levels are forecast and compared at public consultation and orders publication exhibitions with existing levels at sample locations. The need for noise barriers and/or earthwork banks should be established as early as possible, together with the land-take implications of such ameliorative measures. Noise effects will need to be re-evaluated as the scheme is refined and developed.

Detailed evidence on existing and future noise levels will need to be prepared and presented at the public inquiry. Preliminary and final noise reports are required by the Department of Transport at order-publication-report and post-compulsory-purchase-order-inquiry stages respectively. Interim reports may also be required.

Statutory maps and lists are prepared in accordance with the *Noise insulation regulations*[32] after completion of the final report. A period of 6 months follows, during which householders can appeal against omission from the insulation list. The insulation of windows for eligible applicants is carried out by the local authority where possible in advance of scheme construction.

There is a second appeal period of 12 months after the opening of the new road. After this period, claims for compensation under Part 1 of the Land Compensation Act 1973 are considered.

In evaluating such claims, the district valuer would require, for example, information on noise increases attributable to the new road.

23.6 Highway geometry

23.6.1 Introduction

The purpose of design is to achieve an acceptable level of performance in average conditions in terms of traffic safety, operation, and economic and environmental effects. Attempts have been made in recent years to introduce much greater flexibility into highway design by regarding the standard as the starting point for the assessment of schemes. No minimum level of service to be provided is stipulated and each decision in an assessment has to reflect value for money in economic and environmental terms. This can result in different standards of road for the same flow of traffic in different parts of the country. For example, in an area of difficult, hilly terrain where construction costs are above average, a single carriageway may be the only economic possibility, but in flat open country it may be possible to justify a dual carriageway for the same amount of traffic. These aspects are covered in TD 9/81, *Road layout and geometry – highway link design.*[35]

23.6.2 Design speed

The speed of traffic varies according to the impression of constraint that layout, geometry, junction frequency and flow per lane impart to the driver. Design speed is, by definition, the estimate of the speed that traffic will be likely to adopt with the alignment configuration proposed.

For new rural roads, design speeds are derived from Figure 1, Part B of TD 9/81.[35]

For urban roads, design speed is selected with reference to the speed limit envisaged for the road but with some allowance for speeds in excess of the limit and is derived from Table 2, Part B of TD 9/81.[35]

23.6.3 Road width and capacity

23.6.3.1 Rural roads

The recommended procedure for the selection of carriageway width is as follows:

(1) Forecast the high and low growth 24 h annual average daily traffic flows for year 15 after opening of the road (the design year).
(2) Compare these flows with the flow ranges given for various carriageway widths in Table 2 of TD 20/85, *Traffic flows and carriageway width assessment.*[36] The flow ranges reproduced in Table 23.3 are in condensed form as in paragraph 4.6 of TA 46/85, *Traffic flows and carriageway width assessment for rural roads.*[37] No upper limit is shown for D3AP and D4M because these are the widest standards for all-purpose roads and motorways.
(3) Select for assessment those carriageway widths within those flow ranges where either or both forecast flows lie.
(4) Consider whether any local factors, such as unusually high or low construction costs, severe environmental effects and major network changes during the evaluation period suggest

Table 23.3 Carriageway widths related to annual average daily traffic flows in the design year

Carriageway standard		Annual average daily traffic flow in year 15
Single 2 lane	S2	Up to 13 000
Wide single 2 lane	WS2	10 000–18 000
Dual 2 lane all-purpose	D2AP	11 000–46 000
Dual 3 lane all-purpose	D3AP	40 000 and above
Dual 2 lane motorway	D2M	28 000–54 000
Dual 3 lane motorway	D3M	50 000–79 000
Dual 4 lane motorway	D4M	77 000 and above

that different widths outside those flow ranges should be assessed.
(5) Carry out economic assessments for the selected width(s) taking account of the implications of width on future maintenance costs. Economic assessment normally is carried out using the computer program COBA,[15] and the effects of future maintenance are assessed with the computer program QUADRO 2.[21]
(6) Enter all relevant economic and environmental factors into an assessment framework prepared as described in the *Manual of environmental appraisal.*[8]
(7) Select the optimal width. Guidance on making the decision can be found in TA 30/82: *Choice between options for trunk road schemes.*[38]

A particular carriageway width will not necessarily be the optimum along the whole length of a scheme. If two-way flows change by more than about 15% between adjacent lengths of a route, a change of road standard should be assessed.

23.6.3.2 Urban roads

Design flows for urban roads are the peak hourly flows in year 15 after opening. Peak hourly flow is the highest flow for any hour of the week averaged over 13 consecutive weeks during the busiest period of the year. This is normally Friday evening between 5 and 6 p.m. during June, July and August.

Tables A and B of TD 20/85[36] (shown here as Tables 23.4 and 23.5 respectively) relate peak hourly flows to carriageway widths for one- and two-way urban roads. Table C (shown here as Table 23.6) gives the adjustment to be made to peak hourly flows where the proportion of heavy goods vehicles is greater than 15%.

For all-purpose urban roads which are not grade-separated, it is often the junctions on the route which determine the capacities of the roads. Care should be taken to select carriageway widths with maximum design flows which do not exceed the capacity of the junctions under peak traffic conditions.

23.6.4 Alignment

A hierarchy of design criteria related to design speed is defined in TD 9/81.[35] The hierarchy is:

(1) Desirable minimum or above.
(2) Relaxations.
(3) Departures.

Design within at least desirable minimum standards will produce a high standard of road safety and should be the initial objective. However, where strict application of these standards would lead to disproportionately high construction costs or

Table 23.4 Design flows on two-way urban roads

Road type		Two-lane carriageway Peak hourly flow (vehs/h, both directions of flow)					Undivided carriageway Peak hourly flow (vehs/h, both directions of flow) 4 lane			6 lane	Dual carriageway Peak hourly flow (vehs/h, both directions of flow) Dual 2 lane		Dual 3 lane
		6.1 (m)	6.75 (m)	7.3 (m)	9 (m)	10 (m)	12.3 (m)	13.5 (m)	14.6 (m)	18 (m)	dual 6.75 (m)	dual 7.3 (m)	dual 11 (m)
A	Urban motorway											3600	5700
B	All-purpose road, no frontage access, no standing vehicles, negligible cross traffic			2000		3000	2550	2800	3050		2590*	3200*	4800*
C	All-purpose road, frontage development, side roads, pedestrian crossings, bus stops, waiting restrictions throughout day, loading restrictions at peak hours	1100	1400	1700	2200	2500	1700	1900	2100	2700			

60/40 directional split can be assumed.
*Include division by line of refuges as well as central reservation; effective carriageway width excluding refuge width is used.

Table 23.5 Design flows on one-way urban roads

Road type		Carriageway width (veh/h one direction of flow)					
		6.1 (m)	6.75 (m)	7.3 (m)	9 (m)	10 (m)	11 (m)
(1)	All-purpose road, no frontage access, no standing vehicles, negligible cross-traffic		2950	3200			4800
(2)	All-purpose road, frontage development, side roads, pedestrian crossings, bus stops, waiting restrictions at peak hours	1800	2000	2200	2850	3250	3550

Table 23.6 Adjustment of flows for heavy vehicle content

The recommended flows allow for a proportion of heavy vehicles equal to 15%. No allowance will need to be made for lower proportion of heavy vehicles; the peak hourly flows at the year under consideration should be reduced when the expected proportion exceeds 15% by:

Heavy vehicle content	Total reduction in flow level (vehs/h)		
	Motorway and dual carriageway all-purpose road (per lane)	10 m wide and above single-carriageway road (per carriageway)	Below 10 m wide single-carriageway road (per carriageway)
15–20%	100	150	100
20–25%	150	225	150

Table 23.7 Design parameters, related to design speeds

Design speed (km/h)	120	100	85	70	60	50
(1) *Stopping sight distance* (m)						
A1 desirable minimum	295	215	160	120	90	70
A2 absolute minimum	215	160	120	90	70	50
(2) *Horizontal curvature* (m)						
B1 minimum R* without elimination of adverse camber and transitions	2880	2040	1440	1020	720	510
B2 minimum R* with superelevation of 2.5%	2040	1440	1020	720	510	360
B3 minimum R with superelevation of 3.5%	1440	1020	720	510	360	255
B4 desirable minimum R with superelevation of 5%	1020	720	510	360	255	180
B5 absolute minimum R with superelevation of 7%	720	510	360	255	180	127
B6 limiting radius with superelevation of 7% at sites of special difficulty (category B design speeds only)	510	360	255	180	127	90

Design speed (km/h)	120	100	85	70	60	50
(3) *Vertical curvature*						
C1 Full overtaking sight distance overtaking crest K value	*	400	285	200	142	100
C2 desirable minimum* crest K value	182	100	55	30	17	10
C3 Absolute minimum crest K value	100	55	30	17	10	6.5
C4 Absolute minimum sag K value	37	26	20	20	13	9
(4) *Overtaking sight distance*						
D1 Full overtaking sight distance (m)	*	580	490	410	345	290

*Not recommended for use in the design of single carriageways

severe environmental impact, relaxations and departures should be considered.

Design parameters related to design speeds are reproduced from Table 3, Part B of TD 9/81[35] as shown in Table 23.7. Desirable minimum values represent the comfortable values dictated by design speed while absolute minimum values are identical to the desirable values for one step below the design speed. For horizontal radius of curvature an additional lower-level limiting radius is equivalent to a further step below the design speed.

Co-ordination of vertical and horizontal alignments is desirable for aesthetic reasons, but for single carriageways the need to design for adequate overtaking is more important.

23.6.5 Horizontal alignment

Transition curves are used to ease the change between a straight and a circular curve or two circular curves where the difference in radii is large. They are normally of spiral or clothoid form described by Equation (23.2):

$$RL = A^2 \tag{23.2}$$

where R is the radius of horizontal curve in metres; L is the length of clothoid in metres; and A is a constant which controls the scale of the clothoid.

In practice, the length of transition curves is often determined by the distance required to accommodate the introduction of superelevation.

Comprehensive transition tables and a treatise on their theory and use can be found in *Highway transition curve tables* (metric) compiled by the County Surveyors' Society.[39]

23.6.5.1 Camber and superelevation

On straight sections of road and those with radii greater than those shown in line B1 of Table 23.7 the crossfall or camber applied from the centre of the road to the outer channel should be 2.5%. For curves of lower radii, superelevation is required to counteract the effects of centrifugal force on vehicles. Superelevation is normally limited to 7%. Care should be taken when applying superelevation that large flat areas of carriageway are not created as they do not drain.

23.6.5.2 Stopping sight distance

The measurement of the stopping sight distance is described in paragraph 2.1.2, Part B of TD 9/81.[35] Absolute minimum stopping sight distance for the design speed should at least be provided on all dual- and single-carriageway roads. Paragraph 1.5.2, Part B of TD 9/81[35] describes cases where relaxations below desirable stopping sight distance are not permitted. Notice should be taken of obstruction to sightlines when measuring stopping sight distances. These include cutting slopes, safety fences, bridge parapets and supports. Widening of verges and central reserves may be necessary to achieve stopping sight distance, and Figure 23.3 shows how it is measured.

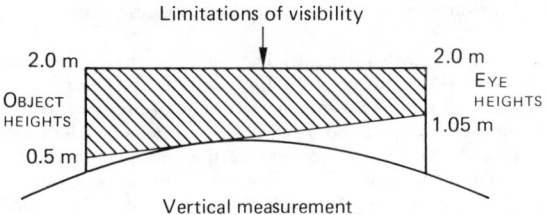

Figure 23.3   Measurement of the stopping sight distance

24.6.5.3 Full overtaking sight distance

Full overtaking sight distance is the distance required for overtaking vehicles by using the opposing traffic lane on single-carriageway roads. It should be checked in both the horizontal and vertical planes. (See paragraph 2.2.2, Part B of TD 9/81[35] for the measurement of full overtaking sight distance.) As with the stopping sight distance, widening of verges and central reserves may be required. Figure 23.4 shows the measurement of full overtaking sight distance.

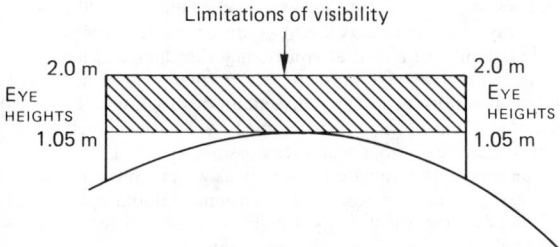

Figure 23.4   Measurement of the full overtaking sight distance

23.6.6 Vertical alignment

23.6.6.1 Gradients

In hilly terrain, gradients steeper than the desirable maximum can make significant savings in construction and environmental costs at the expense of higher user costs, but there is also a progressive decrease in safety with increasing gradients, and a gradient steeper than 8% should be considered as a departure from standard. Desirable maximum gradients can be seen in Table 23.8.

Table 23.8 Desirable maximum gradients

Road	Desirable maximum gradient (%)
Motorway	3
All-purpose dual carriageway	4
All-purpose single carriageway	6

For effective drainage a minimum gradient of 0.5% should be maintained wherever possible.

23.6.6.2 Curves

Vertical curves should be provided at all changes in gradient.

Curve radii should be large enough to provide for comfort and for stopping sight distance. Since vertical curve radii tend to be large, curves are described by their K value (Table 23.7) where $100K = R$; also, $L = K(g_1 - g_2)$.

The curvature of crest curves, where visibility between the driver and a stationary object is obstructed by the intervening road, should be large enough to provide at least absolute minimum stopping sight distance.

At sag curves, visibility is not obstructed and design is based on comfort criteria, except that roads of design speed 70 km/h or below in unlit areas should have sag curves designed to ensure that headlights illuminate the road surface for at least absolute minimum stopping sight distance. Relevant K values for hog and sag curves are given in Table 23.7.

Vertical curves are calculated as follows:

$$h_2 = h_1 + \frac{g_1 x}{100} - x^2 \frac{(g_1 - g_2)}{200L} \tag{23.3}$$

which is an approximation of a circular arc to a simple parabola, where h_2 is the level of a point on the curve; h_1 is the level at the start of the curve; g_1 is the approach gradient; g_2 is the following gradient (to be added algebraically); x is the distance from the start of the curve; and L is the length of the curve.

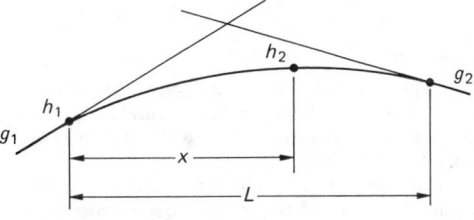

Figure 23.5   A diagram of a vertical curve

23.6.6.3 Climbing lanes

Additional uphill climbing lanes may be provided on gradients greater than 2% and longer than 500 m where they can be justified on economic and environmental grounds. Criteria for their provision and layout are described in section 5, Part B of TD 9/81.[35]

23.6.7 Carriageway layout

23.6.7.1 Single carriageways

Single two-lane carriageways should be designed with the objectives of safety and uncongested flow in mind. Clearly identified overtaking sections for either direction of travel are required to be provided frequently so that vehicles can maintain the design speed in off-peak conditions. Provision of full overtaking sight distance along the whole length of a route often generates high cost/environmentally undesirable layouts. The alternative of clearly identifiable overtaking sections interspersed with clearly nonovertaking sections is frequently more cost-effective. Part C of TD 9/81[35] defines overtaking and nonovertaking sections and sets out minimum requirements for the percentage of overtaking opportunity for different categories of single carriageway.

Vertical alignment should be co-ordinated with the horizontal alignment to ensure the most efficient overtaking provision.

Road markings, in accordance with the *Traffic signs manual*,[40] should be used to indicate nonovertaking sections to drivers.

23.6.7.2 Dual carriageways and motorways

All-purpose dual carriageways and motorways should be designed to permit light vehicles to maintain the design speed.

At the lowest level, dual carriageways should have a vertical alignment following the topography with horizontal alignment phased to match and at grade junctions.

On motorways, horizontal and vertical curves should be as generous as possible.

Complete elimination of access other than at interchanges and service areas and prohibition of usage by pedestrians and certain types of vehicle result in high speeds throughout. Design speed should be 120 km/h for rural motorways.

23.6.7.2 Changes in width

Changes from dual to single carriageway are a potential hazard and changes in width should be made clear to drivers by signing and marking. Lengths of dual carriageway within a generally single-carriageway road, or vice versa, should be at least 2 km, and preferably 3 km, long.

The cross-sectional layout of a carriageway should be such as to indicate to the driver the standard of road on which he is travelling. Department of Transport Standard TD 27/86, *Cross-sections and headroom*[41] and the *Highway construction details*[42] give typical cross-sections for all classes of trunk road. Figures 23.6 and 23.7 show typical cross-sections for rural motorways and all-purpose roads.

23.6.7.3 Urban roads

It is not possible to tabulate overall layout characteristics for roads in urban areas in the same way as for rural areas, as the constraints of the existing urban fabric will result in designs tailored to meet the specific site requirements.

Urban standards embracing mandatory speed limits, design speeds generally 85 km/h and below and reduced cross-section design, are more conducive to safe conditions where the surrounding development is very much of an urban nature. Figures 23.8 and 23.9 show typical cross-sections for major urban roads.

There is usually less scope for co-ordinating geometric features in urban, than in rural, areas and junction choice may be limited by land availability. Urban designs based solely on maximum peak-hour demand may result in schemes of high cost and environmental impact. A balance must be achieved between increased delay to vehicles at peak hours and the cost of providing for the highest level of demand.

The increased frequency of junctions in urban areas means that the usual method for the calculation of speeds on links for economic analysis may not always be appropriate. In particular, the COBA program[15] was not designed specifically for urban evaluation. Alternative methods of appraisal for urban schemes are discussed in *Roads and traffic in urban areas*.[12]

23.6.7.4 Residential roads

Differences between residential and other types of roads are not only of scale. Residential roads are an integral part of housing layouts where an environment free from traffic nuisance is of prime importance, and in the patterns of movement around buildings the needs of pedestrian safety and convenience should be given priority in design over the use of vehicles.

Layout considerations for residential roads are discussed in *Design bulletin 32*.[43] Some county councils have their own specific requirements for residential roads. An example is the Hampshire County Council publication *Roads in residential areas*.[44] When designing roads for residential areas, care should be taken that sufficient accesses and turning bays are provided for service and delivery vehicles including removal vans, refuse lorries and fire engines. *Swept paths*[45] and *Designing for deliveries*[46] provide useful data on turning requirements.

23.6.8 Computer programs

A variety of computer programs is available to assist the engineer in the design and assessment of highways. The two program suites most often used for highway design are:

(1) BIPS 3 (British Integrated Program System) for highway design, (Department of the Environment and County Surveyor's Society), a comprehensive suite of programs covering vertical and horizontal alignment, superelevation, ground models, cross-sections, earthworks, preliminary design and automatic plotting and generation of perspectives.
(2) MOSS (MOdelling SystemS) (Moss Systems Ltd), a complete highway design system based on the use of string models for surfaces, features and utility networks, e.g. electricity.

For economic assessment, COBA[15] and QUADRO 2,[21] referred to in sections 23.4 and 23.6, are used.

23.6.9 Interchange and junction design

23.6.9.1 Junction strategy

The aim of junction strategy should be to provide drivers with layouts which have consistent standards and are therefore unlikely to cause confusion.

The simplest strategy is to have motorways where all interchanges should be grade-separated. On all-purpose dual carriageways junctions may be at-grade or grade-separated but design should be aimed at consistency throughout a length of route. Single carriageways are not normally grade-separated but, as junctions generally represent an obstruction to overtaking, maximum efficiency is obtained by locating junctions in nonovertaking sections wherever possible.

The emphasis in junction selection, as with highway design, is on flexibility and the economic, environmental and operational effects of a junction strategy should be assessed before making the final decision with the aid of a framework.[8]

23.6.9.2 Selection and assessment

In order to generate alternative junction options for assessment frameworks, trial traffic flows are needed for detailed design. These are called 'design reference flows'. The range of design reference flows used for the generation of options at a site should be wide enough to embrace junction designs which will be optimal economically and operationally but not so wide as to produce junction designs which would clearly be under or over provision. They should also encompass the range of turning movements that can be expected at a site. Junction options should not be assessed on a single set of turning movements.

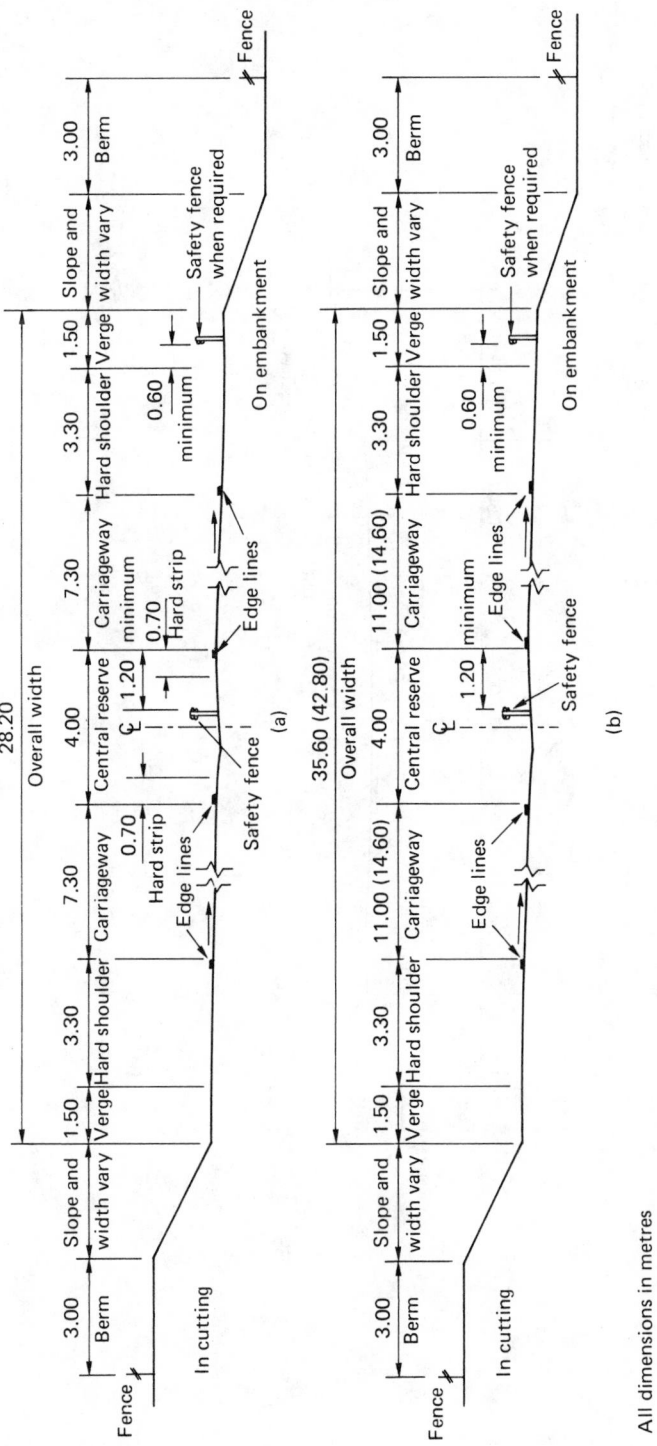

Figure 23.6 Typical cross-sections of rural motorways. (a) Dual two-lane motorway; (b) dual three-lane motorway. (A dual four-lane motorway is the same as for the three-lane, but the uprated dimensions are shown in brackets for carriageway and overall width.) (After Department of Transport (1987) *Highway construction details*. HMSO, London)

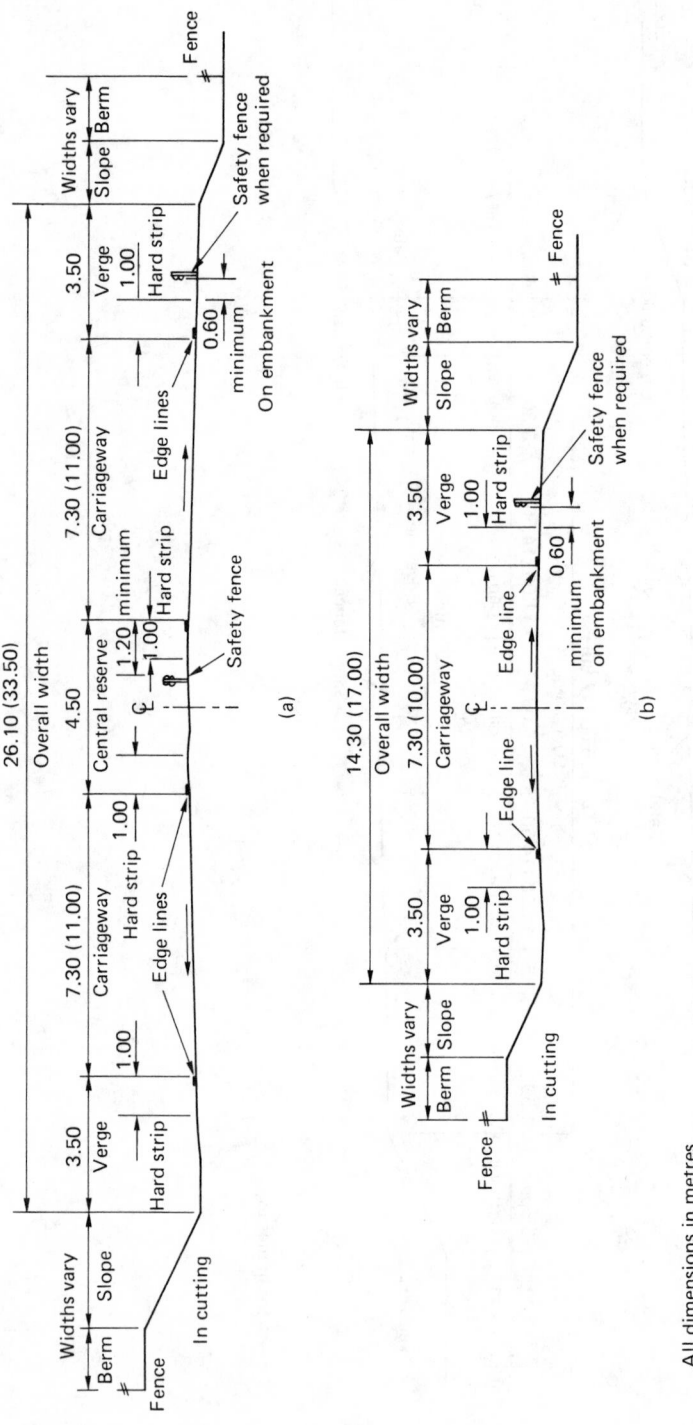

Figure 23.7 Typical cross-sections for rural all-purpose roads. (a) Dual two-lane all-purpose road (the dual three-lane all-purpose road carriageway and overall width are shown in brackets); (b) single 7.3 m road (the wide single 10 m carriageway road is the same except for the uprated dimensions for carriageway and over all widths). (After Department of Transport (1987) *Highway construction details.* HMSO, London)

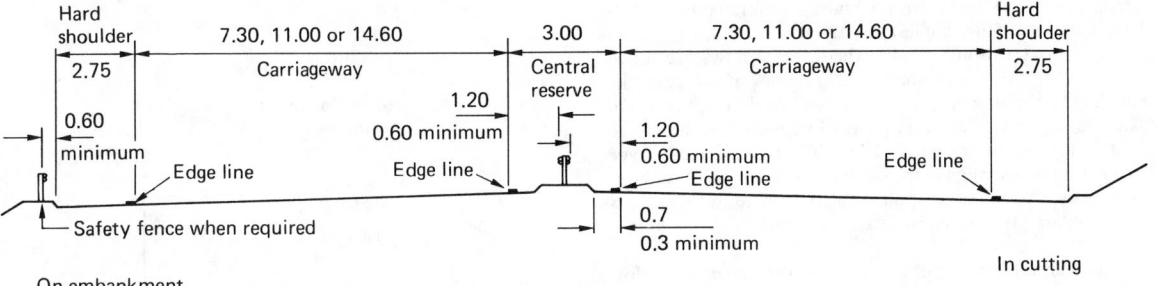

All dimensions in metres

Figure 23.8 Typical cross-sections of urban motorways (up to 85 km/h design speed). (After Department of Transport (1987) *Highway construction details*. HMSO, London)

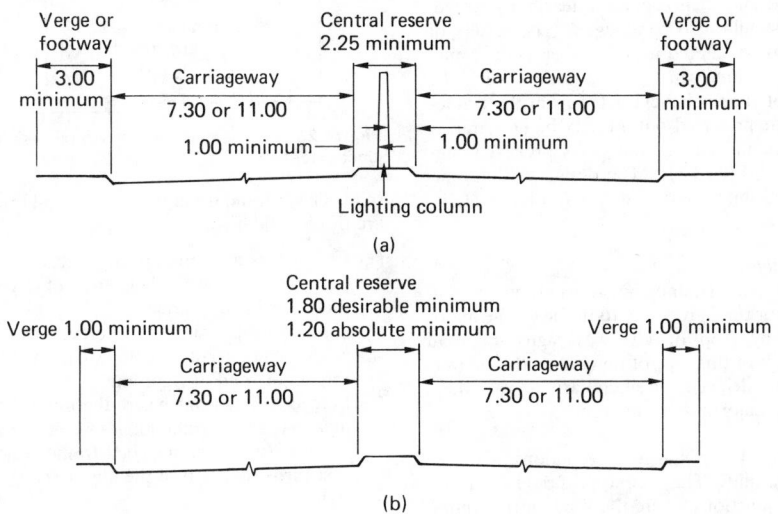

(a)

(b)

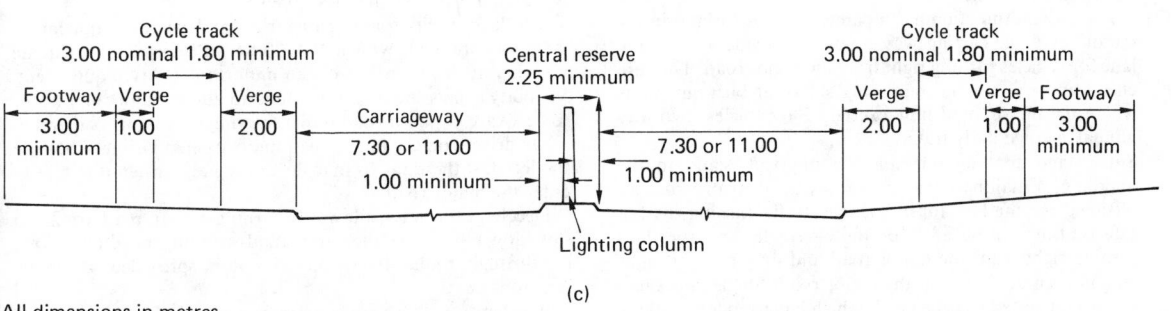

(c)

All dimensions in metres

Figure 23.9 Typical cross-sections of urban all-purpose roads (up to 85 km/h design speed). (a) With footways; (b) without footways and with lanterns bracketed from buildings or suspended; (c) with footways and cycle tracks. (After Department of Transport (1987) *Highway construction details*. HMSO, London)

Design reference flows are an hourly flow rate. For inter-urban roads the 50th highest hourly flows are likely to be suitable. On recreational roads, where traffic flows are much greater in the peak season than at other times of the year, the 200th highest hour will generate viable junction options. On urban roads, flows do not vary greatly by season and designs to the 30th highest hour are likely to be justified. Road type classification is described in the *COBA user manual*.[15] Peak hour flow factors which can be applied to the annual average daily traffic flow are given in Appendix D14 of the *Traffic appraisal manual*.[3]

Capacity at a roundabout is defined as the maximum inflow from an entry when traffic flow at that entry is sufficient to cause continuous queueing in the approach road. Major/minor junction flows are considered to be at capacity when there is continuous queueing feeding a particular turning movement. Not all movements at a junction need to be at capacity for the junction to be considered at capacity.

The reference flow:capacity (RFC) ratio is an indicator of the likely performance of a junction and should be calculated or computed for each trial design. An entry RFC ratio of 85% means theoretically that queueing will be avoided in the chosen design year peak hour in five out of six cases. Designs with an RFC ratio of this order are likely to be economically justifiable. Once RFC ratios have been calculated, certain trial junctions may be seen to be unsuitable and rejected. The remaining junctions can be refined and their economic feasibility tested. Environmental effects should also be assessed. The results of these evaluations should be entered into an assessment framework.[8, 9, 38] The various options can be compared in pairs, the significant advantages of each being extracted from the framework and examined. The preferred option may be determined from a process of sequential elimination. This process is described in Appendix 3 of TA 23/81.[47] The design–evaluation–decision process may be summarized as shown in Figure 23.10.

23.6.9.3 At-grade junctions

Major/minor junctions. Major/minor junctions are most common. Traffic on the minor road gives way to traffic on the major road and is controlled by 'stop' or 'give way' signs and road markings. The advantage of this type of junction is that major-road through-traffic is not delayed. On single carriageways there are three basic types of major/minor junction:

(1) Simple junctions are T- or staggered junctions with no 'ghost' or physical islands. They are appropriate for most accesses and minor junctions where the flow on the minor road does not exceed about 300 vehicles, two-way average daily traffic.
(2) A ghost island junction has a painted hatched island in the middle of single-carriageway roads to provide a diverging lane for vehicles turning right from the major road. They are cheap and effective in improving safety at busy junctions where the minor road flow exceeds 500 vehicles, two-way annual average daily traffic.
(3) Single lane dualling introduces a physical island in the middle of a major single-carriageway road. It provides an offside diverging lane for major road traffic turning right, a safe central waiting area for those vehicles and the ones turning right from the minor road and only one through-lane in each direction on the major road. Single-lane dualling is best-suited to rural roads which have good overtaking capacities between junctions.

On continuous dual carriageways, major/minor junctions are formed by widening the central reserve to form a diverging lane and waiting space for vehicles turning right from the major

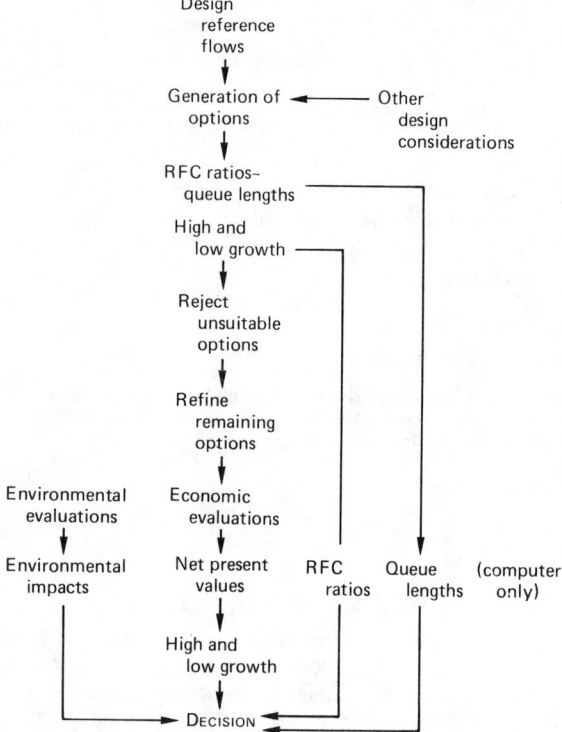

Figure 23.10 The decision-making process for interchange and junction design

road. They should never be used on dual three-lane roads. There are three basic layouts:

(1) Crossroads are only appropriate where there are very low minor road flows. They are not recommended for new junctions in rural areas.
(2) T-junctions are suitable wherever a minor road terminates at the major road. The minor road should preferably join the major road at right angles.
(3) Staggered junctions where the minor roads should join the major road at right angles wherever possible. Right/left staggers (where minor road traffic crossing the major road first turns right out of the minor road) are preferred to left/right staggers.

Advice on size and layout of major/minor junctions can be found in TA 23/81[47] and TA 20/84.[48]

The layout, siting and geometric standards of major/minor junctions are dealt with in TA 20/84.[48] Provision of adequate visibility at these junctions is an important safety requirement. Visibility splays should be provided to allow major road traffic to be aware of minor road traffic entering the major road and to allow drivers on the minor road unobstructed visibility both to the left and the right so that they may judge when it is safe to enter the major road.

Deceleration lanes allow left turning major road traffic to slow down and leave the major road without impeding following-through traffic. They should not be provided at simple junctions.

Acceleration lanes allow minor road traffic to accelerate before joining the major road traffic. They should only be used at dual-carriageway junctions with a design speed of 85 km/h or above and a heavy, minor road left-turning demand. Equations for the calculation of turning stream capacities at major/minor junctions are given in TA 23/81.[47] Queueing and capacity calculations are tedious and may be more easily computed with the PICADY[49] and MIDAS[50] programs.

Roundabouts. Roundabouts are equally suitable for both urban and rural areas. They are normally the safest form of at-grade junction over a wide range of entry flows and approach speeds and are particularly effective where a heavy right turn occurs. Their main disadvantages are that they cause delay to all traffic and often require more land than major/minor junctions.

There are three basic types of roundabout:

(1) Normal – a roundabout with a one-way circulatory carriageway around a kerbed island 4 m or more in diameter. Approaches are usually flared to allow multiple vehicle entry.
(2) Mini – a roundabout with a one-way circulatory carriageway around a flush or slightly raised circular marking less than 4 m in diameter and with or without flared approaches.
(3) Double – an individual junction with two normal or mini roundabouts either contiguous to, or connected by, a central link road or kerbed island.

Normal roundabouts perform best with only three or four entries; for more entries, double roundabouts should be considered. Mini roundabouts are particularly useful for improving existing urban junctions where space is limited.

Provision of adequate visibility at roundabouts is important. Drivers should have clear visibility from the 'give way' entry line both to the right and ahead as far as the previous or next entry or for a distance of 50 m. Drivers should also have clear visibility to the next entry or 50 m when circulating. Geometric design of roundabouts is covered by TD 16/84[51] and TA 42/84.[52]

As with major/minor junction manual calculation of queues, capacity and delays at roundabouts is tedious and again computer programs ARCADY[49] and MIDAS[50] are used for operational assessment. Capacity equations are given in TA 23/81.[47]

Traffic signals. Traffic signals reduce conflict between traffic streams and increase safety at junctions by separating traffic movements in time and regulating their position on the road. They are usually more economical in their use of road space than roundabouts, providing similar capacity but allowing more flexibility in layout and land-take; although vehicle-actuated, they produce overall delays to traffic in off-peak periods.

Traffic signals in the UK are provided under powers contained in the Road Traffic Regulation Act and must comply with current directions issued by the Department of Transport.[53] The use of signals is not recommended at junctions where the 85 percentile approach speed on any arm exceeds 65 km/h. It is difficult to design efficient signalling for junctions with five or more arms; for these reasons, traffic signals are normally used at urban and suburban junctions.

As with other at-grade junctions the manual design and operational assessment of traffic signals is complex, and there are several computer programs available to assist calculation, including: OSCADY,[54] which aids optimization of signal timings and models the operation of isolated junctions; SIGCAP,[55] SIGSET[56] and LINSIG,[57] which deal with more complicated junctions; and TRANSYT,[16, 58] which is primarily designed to optimize signal networks.

Guidance on junction layouts and general principles of control by traffic signals and signals on high-speed roads is given in TA 12/81,[59] TA 16/81[60] and TA 18/81.[61]

Pedestrians and cyclists. When considering junction design, provision should be made for pedestrians and cyclists wherever there are sufficient numbers to merit it. Advice is given in

TA.20/84[48] for major/minor junctions, TA 42/84[52] for roundabouts, and TA 15/81[62] for signalized junctions.

Long vehicles. Long vehicles can have problems in negotiating junctions, especially at sites where the road area is restricted by buildings. In a computer program called TRACK,[45] a library of vehicle-swept paths is available. These can be used to check that a junction design can accommodate turning movements of long vehicles. The Freight Transport Association also publishes a guidance document for long vehicles.[46]

Grade separation. The complexity and conflict between traffic movements at the intersection of two or more major roads can be reduced by providing for traffic on several different levels. This 'grade-separation' of traffic allows the heaviest flows to pass unhindered through the junction whilst lighter flows, including all turning movements, are dealt with on separate levels above, below or both.

Even where the topography is favourable, grade-separation usually involves high capital cost and therefore economic, operational and environmental assessments should always be made to ensure that it is justified.

The lowest physical level of a multi-level interchange ideally should be used by the heaviest traffic flow as this:

(1) Reduces visual and noise impact.
(2) Allows high-speed traffic leaving the major road to be slowed naturally by the up-grade of exit sliproads.
(3) Aids acceleration of traffic entering the major road on the down-grade of the entry sliproads.
(4) Often reduces the cost of structures.

Advice on the selection of standards and design procedures for all types of interchange is contained in TA 48/86[63] and TD 22/86.[64] Examples of grade-separated junctions and interchanges are shown in Figure 23.11.

Grade-separated interchanges fall into two different categories: (1) those which permit complete free-flow conditions; and (2) those which impose the need for certain streams of traffic to give way to others.

Free-flow interchanges are the most complex and expensive form of junction. They provide uninterrupted movement for vehicles moving between principal routes by the use of link roads with a succession of merging/diverging lanes and weaving sections. When designing merging and diverging lanes the worst combinations of design year hourly flows for the principal routes and merge/diverge should be used.

Non-free-flow interchanges include those between motorways where a roundabout is incorporated and those which involve two level intersections where the main route is connected by sliproads to the minor route at grade. These latter intersections include diamond and dumb-bell layouts and the use of roundabouts.

23.7 Earthworks

23.7.1 Planning

General advice is given in BS 6031.[65] Further advice will be available in an earthworks advice note to be published by the Department of Transport by the summer of 1988. Reference should also be made to Chapters 9, 11 and 18 of this book.

23.7.2 Desk study

The desk study should be undertaken as a part of route selection to investigate likely ground conditions and reveal potential problem areas such as landslip zones, mining areas, extensive

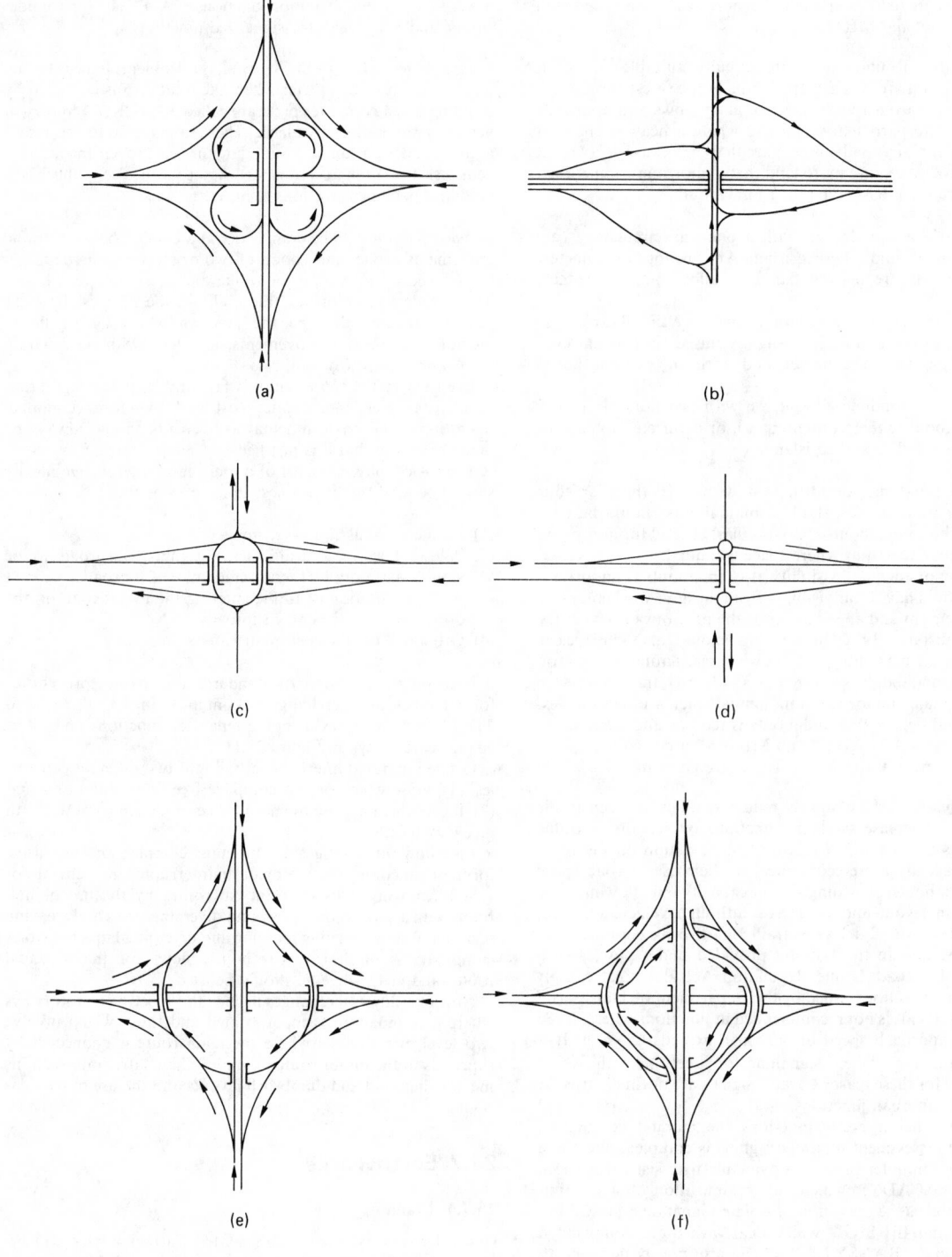

(a)

(b)

(c)

(d)

(e)

(f)

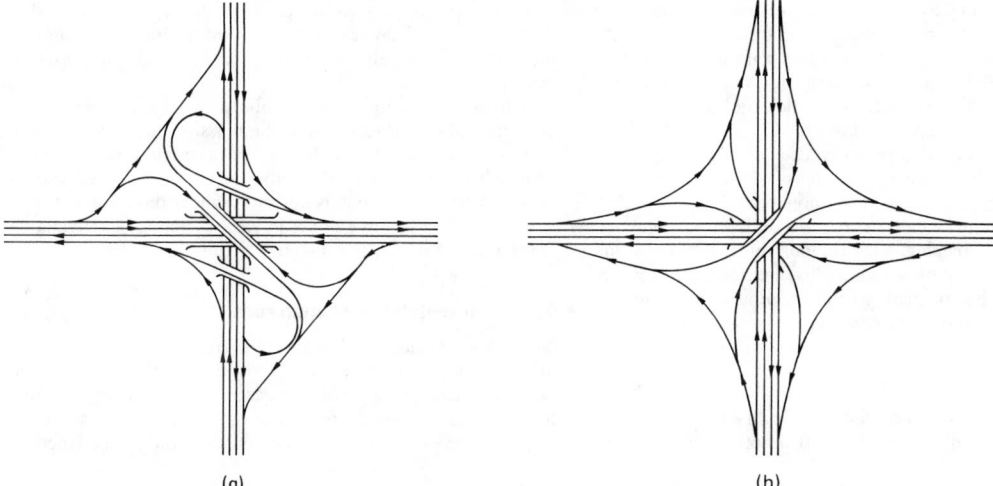

(g)   (h)

Figure 23.11 Examples of grade-separated junctions and interchanges. (a) Cloverleaf; (b) diamond; (c) two-level roundabout; (d) dumbell; (e) three-level roundabout; (f) two-level interchange; (g) three-level interchange; (h) four-level interchange. (After Department of Transport (1987) *Highway construction details*. HMSO, London)

marshes, etc. and to provide initial guidance on design parameters. Sources include geological maps and memoirs, agricultural surveys, aerial photographs and records of public and water authorities. An extensive list of sources is published by the TRRL.[66]

23.7.3 Ground investigations

A ground investigation should be carried out following route selection which may be in several stages. Further information on site investigation is given in Chapter 11. The investigation should be designed to confirm anticipated ground conditions and clarify areas of doubt. It should enable parameters to be established for: (1) determining the acceptability of excavated materials; (2) the stability of slopes of cuttings; (3) the stability and settlement characteristics of fill materials; (4) for determining the necessity for drainage measures; and (5) for the structural and pavement foundation conditions.

23.7.4 Highway alignment

The route should be chosen where practicable to avoid difficult soil conditions.

The road profile should aim to provide a balance between cut and fill, taking into account anticipated volumes of unacceptable material and earthworks side-slopes. Since these may be inter-related, an early assessment of likely acceptability limits should be made. Cut-and-fill diagrams can be used to highlight material distribution and to indicate haul lengths. Likely haul routes should be investigated, particularly in environmentally sensitive areas.

Contract documents should state the quantities and classes of materials in individual cut-and-fill areas to enable the contractor to select the appropriate plant. The contract commencement date should be chosen to allow the maximum time for earth-moving, taking into account local climatic conditions and the contract period.

23.7.5 Slopes of cuttings

Slopes in dry granular soils are stable to any height provided the slope is flatter than the angle of repose of the material. This varies with the angularity of the grains and the density. (See also Chapters 9 and 11.)

Groundwater control measures are necessary below the water table, particularly with fine sands and coarse silts in which running sand conditions may develop. Control measures comprise counterfort drains, cutoff drains or a drainage layer on the face of the cutting, the drainage media being graded to suit the subsoil or protected by a filter membrane to avoid migration of fines. Similar considerations apply in the regard to drainage and the road foundations.

For cohesive soils, slope stability varies with time. Excavation of the cutting results in pore water suction which draws-in additional moisture resulting in long-term swelling and softening. The rate of loss of strength depends on permeability, and for some stiff fissured clays[67] it can take up to 90 years to achieve equilibrium.

Stability analysis should be in terms of effective stress parameters derived from consolidated triaxial testing. However, the results should be treated with caution, since results obtained from small samples frequently do not accurately represent the properties of the soil mass *in situ*.

Information on equilibrium groundwater levels and resultant pore pressures needs to be assessed for this kind of analysis.

Rock slopes require a detailed knowledge of the geology, the joint bedding and dip and the piezometric level.

23.7.6 Slopes of embankments

As for cuttings, embankment slopes in dry granular materials are stable if the slope is flatter than the natural angle of repose of the material. However, the stability of the existing ground under the applied load due to the embankment needs to be considered.

When cohesive materials compact during construction, increased pore pressures result; these dissipate with time resulting in consolidation and an increase in strength. The critical period is thus usually during, and soon after, construction and analysis can be based on the results of quick undrained triaxial tests without regard to pore pressure.

The question of how representative the results of tests on small undisturbed samples are of the mass of remoulded compacted fill should receive careful attention.

Some stiff over-consolidated clays are difficult to compact and may swell after placing, with the possibility of long-term failure. Where the material strength is greater than 150 to 200 kN/m², the ability of plant to fully compact the material, and its acceptability should be considered.

23.7.6.1 Slope stability analysis

The most widely used method of slope stability analysis in the UK is that developed by Professor Bishop[68, 69] of Imperial College, London, based on circular slip surfaces. For slope failures in relatively homogeneous soils, Bishop's simplified method can give reliable answers but in other circumstances errors can be serious.

A method of solution for a multiple-wedge problem, incorporating shear surfaces within the mass of potentially slipping material, has been found by Dr Sarma at Imperial College.[70] This is being implemented for the Department of Transport in program NCSLIP which is currently under development.

Guidance on suitable slopes in various materials is available[71] from which Table 23.9 is taken.

23.7.7 Acceptability of materials

Most road schemes within the UK are constructed in accordance with the Department of Transport specification[72, 73] which defines material in terms of its acceptability. Upper and lower limits of the various properties of materials have to be defined for each class of material in the contract, the limits depending

Table 23.9 Geology and maximum slopes of cuttings and embankments. (After Parsons and Berry (1985) 'Slope stability problems in ageing highway earthworks', *Symposium on failures in earthworks*. Thomas Telford, London)

Geology	Maximum slope (v : h)		
	Height 0–2.5 m	2.5–5 m	More than 5 m
Cuttings:			
Gault Clay	1 : 3.5	1 : 4	1 : 4
Oxford Clay	1 : 2.5	1 : 3	1 : 3.5
Reading Beds (cohesive)	1 : 4	1 : 4	1 : 4
Reading Beds (noncohesive)	1 : 2.5	1 : 2.5	1 : 3
Lower Coal Measures	1 : 1.75	1 : 2	1 : 2.5
Plateau Gravel	1 : 2.5	1 : 2.5	1 : 3.5
Boulder Clay	1 : 4	1 : 4	1 : 4
Middle Coal Measures	1 : 2	1 : 2.5	1 : 3
Valley Gravel	1 : 2.5	1 : 2.5	1 : 2.5
London Clay	1 : 3.5	1 : 3.5	1 : 3.5
Clay with flints	1 : 2	1 : 2	1 : 2
Glacial Gravel	1 : 2	1 : 2	1 : 2
Chalk	1 : 1.5	1 : 2	1 : 2
Lower Greensand	1 : 1.75	1 : 1.75	1 : 1.75
Kimmeridge Clay	1 : 2.5	1 : 2.5	—
Kellaways Clay	1 : 2	1 : 3	1 : 3.5
Cornbrash	1 : 1.5	1 : 1.5	1 : 1.5
Upper Coal Measures	1 : 1.75	1 : 1.75	1 : 2
Embankments:			
Gault Clay	1 : 2.5	1 : 3	1 : 3.5
Reading Beds (cohesive)	1 : 3	1 : 4	1 : 4
Reading Beds (noncohesive)	1 : 1.75	1 : 1.75	1 : 1.75
Kimmeridge Clay	1 : 2.5	1 : 3.5	1 : 3.5
Oxford Clay	1 : 3	1 : 3.5	1 : 3.5
London Clay	1 : 2	1 : 3	1 : 3
Middle Coal Measures	1 : 2	1 : 3	1 : 3
Clay with flints	1 : 2	1 : 3	1 : 3.5
Boulder Clay	1 : 3	1 : 3	1 : 3
Chalk	1 : 2	1 : 2	1 : 2
Lower Greensand	1 : 2	1 : 2	—
Glacial Gravel	1 : 2	1 : 2	1 : 2
Coral Rag	1 : 2	—	—
Great Oolite Clay	1 : 1.75	1 : 1.75	1 : 1.75
Lower Coal Measures	1 : 2	1 : 2	1 : 2

Maximum slopes required to restrict failure rates to below 1% within 10 to 22 years of construction.
The use of flatter slopes offers the possibility of returning land to agriculture and reducing land-take. However, the land may be of lower agricultural grade and may not be acceptable to farmers.

on the use it is intended to make of the material. The moisture content limits are influenced by shear strength/slope stability requirements, and the need to limit settlements and provide an adequate subgrade.

Limits of acceptability for granular materials are normally related to the optimum moisture content as measured in BS 1377,[74] Test 12, although Test 13 represents better the performance of modern compaction plant. Typical limits would be 2 to 4% below optimum moisture content (Test 12) to 0 to 2% above optimum moisture content.

The main criterion for cohesive materials is the shear strength, the limits normally being from 35 to 50 kN/m². This is not usually specified directly. Instead, correlations are used between strength and properties such as the California bearing ratio (CBR), the moisture condition value (MCV) or moisture content : plastic limit ratio.

The MCV test[75, 76] is more reliable[77] with cohesive than with granular materials.

Suitability of rockfill is dealt with in the TRRL Research Report 60,[78] and the suitability of chalk in LR 806[79] and LR 112.[80]

Maximum use of site materials can be achieved by layering[81] wet and dry materials, incorporating drainage layers to speed settlement and dissipation of pore pressures, lime stabilization[82] of cohesive materials and by using poorer materials in landscape areas. Consideration should be given to specifying a better-quality material within a metre of the subgrade as dealt with in section 23.9.

23.7.8 Compaction

Increasing compaction of material results in an increase in strength leading to improved side slope and subgrade stability, together with reduced permeability and risk of settlement within the fill.

The state of compaction achieved is influenced by the moisture content, the compaction energy applied and the thickness of layer.

For a given compactive effort, material compacted in a dry state achieves a low dry density with a high proportion of air voids. When compacted in a wet state, dry density is low, with a low air-void content but also a high moisture content. Between these extremes is an optimum moisture content for the compaction energy used at which the maximum attainable dry density will be achieved.

The lower the acceptable dry density, the greater the range of moisture contents over which the material is acceptable, but the performance requirements should be considered to ensure that settlement criteria are not exceeded, side slopes are stable and subgrades are of adequate strength. Low moisture contents may result in under-compaction with voids at the bottom of the layer. This can only be checked by digging trial pits if under-compaction is suspected, but compactive effort can be increased or the layer thickness reduced to compensate.

High moisture contents may result in over-compaction which can be relieved by reducing the compactive effort, but the high moisture content will only be reduced by consolidation leading to long-term settlements.

The limiting factor on acceptability of material in shallow-fill areas, may be the question of whether it will be easy or difficult to operate earthmoving plant. Studies of the effect of soil condition on plant operation have been made by the TRRL.[83, 84]

Subgrades. The strength of the subgrade is a major factor in determining the required pavement thickness. The subgrade strength is assessed in terms of CBR and two design conditions need to be considered: (1) the construction condition; (2) the in-service condition. The former is usually the most critical. The

CBR can be obtained by testing samples recompacted to *in situ* density, and is dependent on the moisture content; a method for estimating soil moisture under various design conditions based on soil suction is published by the TRRL.[84]

However, the assessment given can be very negative for construction conditions and the TRRL[85] gives guidance on what are considered to be realistic values for various soil types and conditions as shown in Table 23.10.

A capping layer should be used to avoid thick sub-bases where subgrade CBR is less than 5%, as described in section 23.9. A better-quality fill may additionally be specified within a metre of subgrade. In some areas the sub-base may comprise cement-stabilized material.[86, 87]

23.7.9 Testing

23.7.9.1 Acceptability testing

The testing for acceptability is concerned both with granular and cohesive soils.

(1) *Granular soils*:   Density/moisture content relationship. Optimum moisture content curve is time-consuming to carry out. Moisture condition value is a rapid test but less accurate.

(2) *Cohesive soils*:   Moisture content : plastic limit ratio. Plastic limit test is time-consuming and liable to error. Moisture content value is rapid and more reliable with cohesive materials.

Moisture contents can be determined rapidly by microwave drying but results should be verified in cases of doubt by normal oven drying.

A number of special fills require grading tests.

23.7.9.2 Compaction testing

Regular density tests are required to check for compliance with an end-result specification. These may be cores, sand replacement tests or density meter tests. For a method specification, occasional tests are made to check that consistent results are being achieved. A trial area is advisable for comparison purposes. A careful assessment should be made of the potential output of compaction plant compared to that of the earthmoving fleet and checks made on speed of compactors, depths of layers and frequency of vibration if relevant.

23.7.9.3 Geotextiles[88]

There are four basic types of geotextiles.

(1) *Non-woven fabrics*:   used for general ground reinforcement and layer separation (e.g. between sub-base and soft subgrade). The fabric density can vary. Permeability varies inversely with density and strength.

(2) *Woven fabrics*:   used where higher strengths are required for ground reinforcement. The warp and weft are generally of different strengths but two-way fabrics are available.

    For special applications very wide woven textiles are now available.

(3) *Geogrids*:   comprise extruded sections welded together, or punched and stretched plastic sheet. Used as grids for soil reinforcement.

(4) *Directionally-structured-filament fabrics*:   manufactured by inter-meshing individual filaments on a multineedle machine producing a lacelike structure of regular-sized

Table 23.10 Equilibrium suction-index CBR values. (After Transport and Road Research Laboratory (1984) *The structural design of bituminous roads*. TRRL Report No. LR 1132.)

Type of soil	Plasticity index	High water table						Low water table					
		Construction conditions						Construction conditions					
		poor		average		good		poor		average		good	
		Thin	Thick	Thin	Thick	Thin	Thick	Thin	Thick	Thin	Thick	Thin	Thick
Heavy clay	70	1.5	2	2	2	2	2	1.5	2	2	2	2	2.5
	60	1.5	2	2	2	2	2.5	1.5	2	2	2	2	2.5
	50	1.5	2	2	2.5	2	2.5	2	2	2	2.5	2	2.5
	40	2	2.5	2.5	3	2.5	3	2.5	2.5	3	3	3	3.5
Silty clay	30	2.5	3.5	3	4	3.5	5	3	3.5	4	4	4	6
Sandy clay	20	2.5	4	4	5	4.5	7	3	4	5	6	6	8
	10	1.5	3.5	3	6	3.5	7	2.5	4	4.5	7	6	>8
Silt*	—	1	1	1	1	2	2	1	1	2	2	2	2
Sand (poor graded)	—	←					20						→
Sand (well graded)	—	←					40						→
Sandy gravel (well graded)	—	←					60						→

*Estimated assuming some probability of material saturating

holes regularly spaced within the mesh. Both permeability and strength can be varied and the fabric constructed in a range from fine meshes to large-diameter reinforcement grids. They are nonfraying, uniform in structure, and are more expensive than some alternatives but capable of being matched accurately to the required task.

23.8 Drainage

23.8.1 Sources of rainfall data

Preliminary estimation of runoff rainfall intensities are given in Table 7 of *A guide for engineers to the design of storm sewer systems*.[89] More accurate intensities related to a particular location can be obtained from two other sources.[90, 91]

More comprehensive information including storm profiles related to location is available from the Meteorological Office in Bracknell. (See also Chapters 28 and 30 of this book.)

23.8.2 Storm return periods

The basic return period used for most highway schemes is 1 year, but drainage systems should be checked to see that water will not surcharge on to the carriageway surface in the event of at least a 5 year storm.

Outfalls from low points in cuttings should be designed for a storm frequency of 2 to 5 years, depending on the size of the cutting and the road classification. Some designers apply a return period of 2 years to all carriageway cross-drains, but in flat areas care should be taken to ensure that the resulting increase in pipe size does not fall below the self-cleansing velocity.

The effect of a scheme on the overall hydrology of an area will need to be assessed in conjunction with the local water authority from whom a formal drainage consent will normally be required. Sensitive areas should be identified at an early stage and design storm return periods agreed. These will usually vary between 10 and 100 years, depending on the locality.

Culverts carrying existing watercourses under main roads are typically designed for a storm frequency of 10 years with no entry surcharge and 100 years utilizing all the available surcharge at entry. These figures may depend on the type and size of the upstream catchment, and on the local water authority's requirements.

23.8.3 Estimation of runoff from carriageways and associated earthworks

23.8.3.1 The Rational (Lloyd–Davies) Method

This is probably the method most widely used for estimating peak flows for urban catchments of up to 10 ha. The description of the method to be utilized is set out in *Road note 35*[89] and is based on the equation:

$$Q = 2.78CiA$$

where Q is the runoff in litres per second; C is the impermeability coefficient; i is the rainfall intensity in millimetres per hour; and A is the drained area in hectares.

Typical impermeability coefficients for highways are shown in Table 23.11.

Table 23.11 Impermeability coefficients

Type of surface	Coefficient C
Paved areas	1.0
Grass verges and central reserves	0.2–0.5
Slopes up to 1:2	0.4–0.7
Steep slopes	0.7–1.0

23.8.3.2 The Wallingford Modified Rational Method[91]

This is an updated version of the Lloyd–Davies procedure, the principal difference being the inclusion of a routing coefficient which allows for variations in rainfall during the time of concentration and is applicable to catchments up to 150 ha. Both Rational and Wallingford methods can be used manually, but computer programs are also available.

23.8.3.3 The Transport and Road Research Laboratory Hydrograph Method

This method is applicable to catchments of up to 50 km^2 and is more sophisticated than the Rational Method in that it takes account of variation in rainfall intensity and distribution within the catchment area, and predicts runoff volume and hydrograph shape as well as peak discharge. The number of calculations required is such that a computer program is essential, but the principals of the method are explained in *Road Note 35*.[89]

The input data required are similar to that for the Rational Method, except that the rainfall data must be in the form of a series of storm profiles for a range of return periods.

23.8.3.4 The Wallingford Hydrograph Method

This is based on a similar concept to the TRRL method, but has been extended to incorporate separate modelling of the surface-water and pipe-flow phases.

It is important to remember that all the above techniques have been developed for use in urban areas, and this should be borne in mind when designing rural highway drainage. Comparison of the design simulations with observed events for all of the methods has shown a wide scatter in their accuracy in predicting peak discharge or runoff volume, with standard deviations of up to 30%. These procedures, including the Wallingford Modified Rational Method[91] and the TRRL Hydrographic Method are available in a software package called WASSP, which represents the most up-to-date techniques currently available for urban drainage design.

23.8.3.5 Estimation of runoff from overland flow

The Rational Method may be used if the catchments draining towards the road are relatively small and well-defined. The *Handbook of steel drainage and highway construction products*[92] contains a monograph for estimating the time of concentration for overland flow. Care should be taken in selection of a value for the impermeability coefficient, since it is the only manipulative factor in the formula. Typical values can be found in Table 4.2 of the *Handbook*[92] and Table 4.1 of *Surface water sewerage*.[93]

Where several subcatchments are contributing to the flow at a particular point, their runoff hydrographs should be plotted on a common time-base and the ordinates added to produce the combined hydrograph. The times of concentration of the subcatchments are likely to be different, so the combined hydrograph should be plotted for values of T_c ranging from the shortest to the longest, bearing in mind that, for the shorter times, some of the subcatchments will only be contributing a proportion of their area.

The runoff hydrograph for a subarea can be approximated by drawing a triangle with base length $2 \times T_c$ and height equivalent to peak flow estimated from the Rational Method equation.[89]

A TRRL report[94] sets out a method for estimating the runoff from natural catchments up to 10 km^2 in area. It is only applicable to catchments with underlying deposits of clay and times of concentration up to 60 h.

The *Flood studies report* and its guide *Methods of flood estimation*[96] summarizes an extensive research programme on flood-prediction techniques.

Two basic methods for estimating design floods are put forward, namely the statistical analysis of peak flows, or the unit hydrograph synthesis of the flood corresponding to a design storm. Both methods are clearly explained step by step, but the most appropriate method for highway drainage will in most cases be the unit hydrograph approach.

23.8.4 Hydraulic design

23.8.4.1 Culverts

An effective culvert design must satisfy a number of criteria. The *Handbook of steel drainage and highway construction products*,[92] Chapter 4, section C, contains a useful checklist.

There are two main types of culvert flow: (1) inlet; and (2) outlet control. A full explanation of these flow conditions is given in the *Hydraulic engineering circular No. 5*.[97]

Where the upstream catchment contributing to a culvert is difficult to define, design flows can be estimated by examining existing hydraulic structures such as culverts and weirs and calculating their maximum discharge using relevant flow equations. These are listed below for the more common structures.

Examination of culverts. The capacity should be checked using the methods in *Hydraulic engineering circular No. 5*.[97]

Examination of weirs.

Board-crested	$Q = 1.71BH^{1.5}$
Sharp-crested	$Q = 1.85BH^{1.5}$
Side–straight channel	$Q = 1.83B^{0.83}H^{1.67}$
Side–constricted channel	$Q = 1.83B^{0.9}H^{1.6}$

where Q is the discharge in cubic metres per second; B is the length of weir in metres; and H is the head above weir in metres.

Examination of orifices. General equation is $Q = mA\sqrt{2gH}$

where Q is the discharge in cubic metres per second; A is the area of orifice in square metres; g is 9.806 in metres per second squared; H is the head above centre of orifice in metres; and m is 0.62 for sharp edged orifice or penstock, 0.86 for opening in chamber well, 0.81 for short length of pipe.

A more comprehensive list can be found in Figure 38 of *Guide to design of storage ponds*.[98]

23.8.4.2 Pipes

The flow in an unsurcharged pipe is dependent on its gradient, diameter and friction coefficient k_s. This information is presented in chart form[99] and as tables.[100] Recommended values of k_s are listed in those documents.

The minimum velocity in a pipe flowing full should not normally be less than 1 m/s to prevent deposition of solids. Where flat gradients make this difficult to achieve, the minimum velocity may be lowered to 0.6 m/s providing measures are taken to exclude solids from the water by using trapped gullies, catchpits and stilling basins as necessary. Geotextile fin drains when used as subsoil drains are also an effective means of excluding grit.

There is normally no upper limit to the velocity of flow in a pipe system, but it may often be necessary to reduce it to an acceptable limit at outfalls in order to prevent erosion in downstream channels. This can be done by providing energy dissipators or a stilling basin at the outfall. Where manholes are located on pipe runs with high discharge and velocity, special attention should be paid to the safety of inspection/maintenance operatives.

Recent trends indicate an increasing use of plastics for the manufacture of pipes. These have very low friction coefficients which help to reduce silting and allow flatter gradients. However, care must be taken in the design of systems using permitted alternative materials ranging from plastics to porous concrete because the difference in friction coefficients will have a significant effect on the times of flow in the pipes and, hence, on the time of concentration.

The effect of surcharging on discharge through a pipe is described in a paper by Colyer.[101] The ratio of surcharged flow to normal flow is given by the equation:

$$Q_s/Q_0 = 1 + (\Delta_y/\Delta_H)$$

where Q_s is the surcharge flow in cubic metres per second; Q_0 = unsurcharged pipe − full flow in cubic metres per second; Δ_y = surcharge head in metres; and Δ_H = difference in level between upstream and downstream pipe inverts.

23.8.4.3 Open channels

The fundamental formula for open channel flow is Chezy's equation:

$$Q = AC\sqrt{mI}$$

where Q is the discharge in metres per second; A is the cross-sectional area of flow in square metres; C is Chezy's constant; m is the hydraulic mean depth = A/(wetted perimeter) in metres; and I is the slope in millimetres per metre.

This equation was improved by Manning to give the most commonly used formula for open channel flow:

$$Q = (A/n)\, m^{2/3} I^{1/2}$$

Typical values of Manning's n are given in Tables 4.4 and 4.5 of the *Handbook of steel drainage and highway construction*.[92] Maximum permissible velocities above which erosion is likely to occur in unlined open channels are set out in Table 6.6 of a work by Bartlett.[93]

The maximum discharge along a rectangular channel will occur when the depth of flow is equal to half the width of the channel, and along a trapezoidal channel when the hydraulic mean depth is equal to half the depth of flow, i.e. when the sides and base are tangential to a semicircle drawn with its centre on the waterline.

23.8.5 Flood storage

The three basic criteria to be considered in the design of a flood storage pond are: (1) the maximum outflow into the existing watercourse; (2) the design storm; and (3) the water area and depth available for storage. The types of storage pond are listed in Table 23.12.

The first two criteria should be determined in consultation with the local water authority, but the third will be governed by site conditions.

Table 23.12 Types of flood storage pond

Type of pond	Description
Onstream	Dry weather flow passes through storage area
Offstream	Dry weather flow bypasses storage area
Dry	Storage area is dry under dry weather flow conditions
Wet	Storage area contains water at all times
Wet/dry	Storage area is part wet/part dry under dry weather flow conditions

Storage ponds of the dry type in rural areas should be used whenever possible so that the storage area can remain in use as land for recreation or grazing for most of the year. Restrictions on space may make this impractical in urban areas, in which case wet ponds will be needed. Careful design should ensure that these will enhance the environment and provide recreational facilities.

The basic design principle is to calculate the flow into and out of the storage area for small increments in time over the duration of the design storm. Comprehensive guidance on the design of flood storage ponds is available.[98] Figure 23.12 illustrates the principle of flood storage.

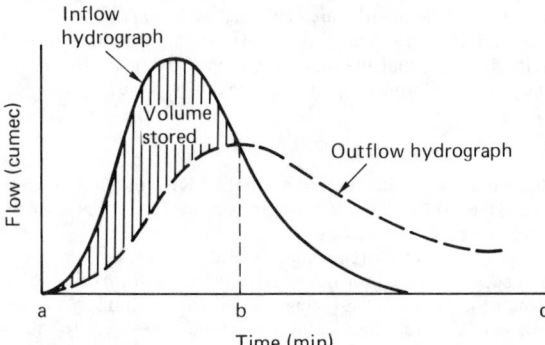

Figure 23.12 Stormwater storage by hydrography (weir outlet). (After Construction Industry Research and Information Association (1980) Technical Note No. 100. CIRIA, London)

23.8.6 Drainage details

A comprehensive set of drainage details is contained in *Highway construction details*.[42] Typical carriageway edge and drainage details are shown in Figures 23.13 to 23.16.

The current details include the use of geotextile fin drains for groundwater control and sub-base drainage, with surface water runoff being collected by a roadside channel. This system has several advantages over traditional methods, and should help to prolong the useful life of roads while being easier to maintain. A guide to the design of subsoil drains can be found in TRRL Report LR 110.[102]

The minimum longitudinal gradient at the road edge should be limited to 0.5% provided there is a crossfall of at least 2.5%. Where superelevation is being applied or reversed, particularly against the grade, care should be taken to ensure that the combination of longitudinal gradient and crossfall will provide a minimum resultant fall of about 1%. If this cannot be achieved, a rolling crown may need to be constructed, but this should be avoided if possible.

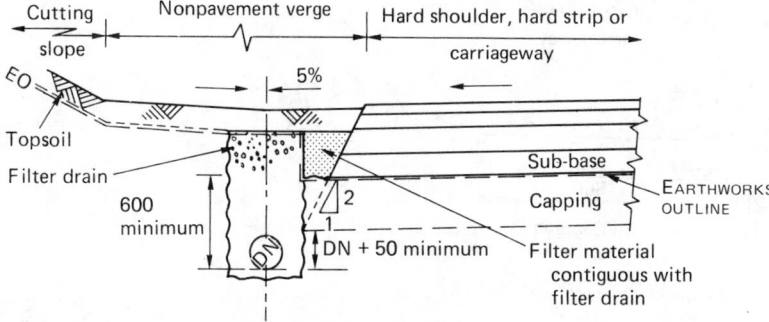

Figure 23.13 Edge-of-pavement and drainage detail in a cutting (flexible carriageway). (After Department of Transport (1987) *Highway construction details*. HMSO, London)

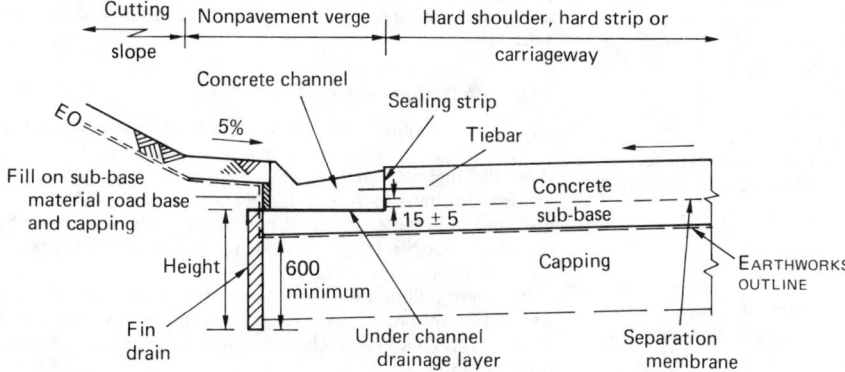

Figure 23.14 Edge-of-pavement and drainage detail in a cutting (rigid carriageway). (After Department of Transport (1987) *Highway construction details*. HMSO, London)

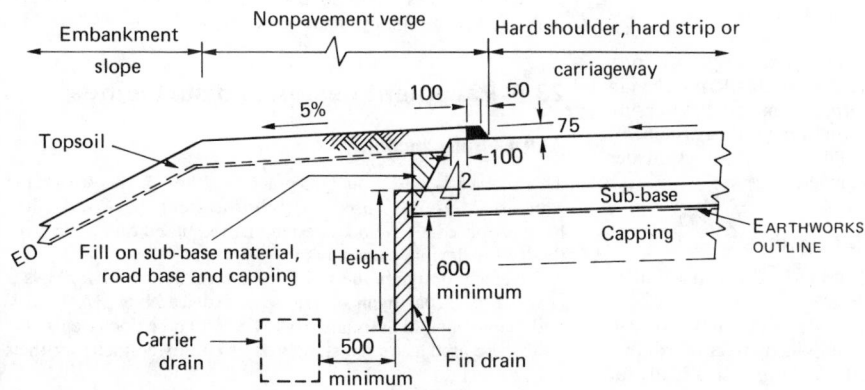

Figure 23.15 Edge-of-pavement and drainage detail on an embankment (flexible carriageway). (After Department of Transport (1987) *Highway construction details*. HMSO, London)

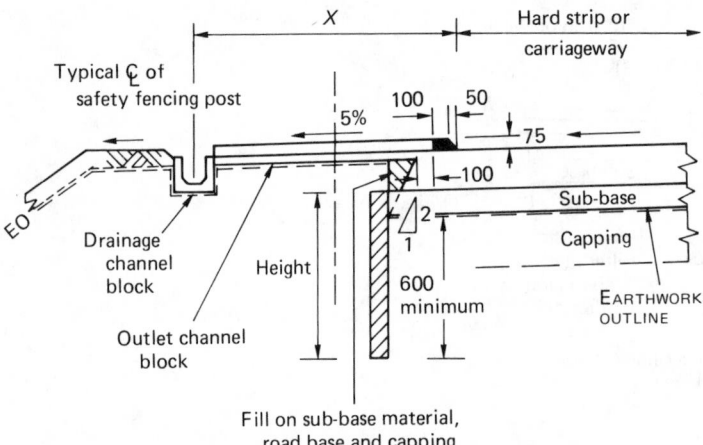

Figure 23.16 Edge-of-pavement and drainage detail on an embankment (rigid carriageway). (After Department of Transport (1987) *Highway construction details*. HMSO, London)

On straight-kerbed roads with normal 2.5% crossfall but gradients less than 0.5%, slotted channel blocks can be installed to provide continuous edge drainage.

Steep gradients should ideally have at least 2.5% crossfall in order to prevent surface water from running nearly parallel to the road.

Computer-aided design (CAD) packages include the facility to plot surface contours, and this should be used for checking surface water paths and gradients in problem areas. The following expression for the depth of water on a road surface is given in TRRL Report LR 236[103]:

$$d = 0.0474 \, LI \, X^{0.2}$$

where d is the depth of water in millimetres; L is the drainage path length in metres; I is the rainfall intensity in millimetres per hour; and X is the drainage path slope (1 in X).

Gully spacing on kerbed roads may be designed using TRRL Report LR 277[104] for roads on gradients above 0.33% and TRRL Report LR 602[105] for level or nearly level roads.

Roadside channels can be designed using either Manning's equation or the nomograph in Figure 4.53 of the *Handbook of steel drainage*.[92] A second nomograph in Figure 4.55 of the *Handbook* covers the design of slotted channel blocks.

Soakaways can be used where there is no outfall into which surface water can be discharged providing the underlying soil is sufficiently permeable. A guide to the design of soakaways is given in BRE Digest 151.[106]

Headwalls and outfall structures should be designed to prevent scour in the existing watercourse. Care should be taken in the detailing and selection of construction materials to ensure that the structure does not detract from the local environment.

Grilles should be fitted to all outfalls with openings greater than 450 mm diameter in order to prevent access.

23.8.7 Structural design of buried pipelines

More detailed information on the design and construction of buried pipelines is given in Chapter 40.

Pipes up to 3 m diameter with cover depths from 0.6 to 10 m can be designed by reference to the simplified tables of external loads on buried pipelines.[107] These tables give design loads for pipes under main roads and under fields. More complex load cases may be analysed using the *Guide to design loadings for buried rigid pipes*.[108]

Designs for pipes greater than 900 mm diameter associated with motorways and trunk roads must be submitted to the Department of Transport as part of their technical approval procedure.[109] Other roads come under the jurisdiction of local authorities who usually will require the same procedure to be followed.

23.8.8 Pollution control

Discussions should be opened with the local water authority at an early stage in the design, at which sensitive watercourses should be identified and preventative measures agreed.

Surface water runoff from highways contains both solid and dissolved pollutants. The solid pollutants can be settled-out by introducing catchpits into the drainage system and, if necessary, a stilling pond at the outfall.

Dissolved pollutants cannot be removed, and so the only sure way to prevent pollution of pure streams in sensitive areas such as fish farms or watercress beds is to avoid discharging highway runoff in those areas.

Accidental spillages of liquids on to the highway can be contained in interceptor chambers which are usually located at the outfall. These are available as prefabricated units or can be constructed *in situ*, and normally consist of either single or multicell chambers with a manually operated pen stock valve on the outlet. The storage volume of the chamber will depend on the size of the highway catchment and the requirement of the local water authority. It may be necessary in some cases to provide a bypass route so that the outfall can be reopened immediately the spillage has been contained.

23.9 Pavement design and surfacings

23.9.1 Introduction

This section is based on UK practice although the AASHTO standards[110] are also used widely throughout the world. Shell have proposed a pavement design method based on elastic layer theory for flexible pavements.[111, 112]

Pavement design in the UK for trunk roads and motorways is covered by Department of Transport Advice Note HA 35/87[113] and Departmental Standard HD 14/87.[114] These documents also tend to be used as design standards by county councils for their major schemes.

Charts for the design of rigid and flexible pavements in relation to initial traffic flows of commercial vehicles per day are provided in HD 14/87.[114] A commercial vehicle is defined as having over 15 kN unladen vehicle weight and the traffic flow should be the best estimate of the one-way 24 h annual average daily traffic flow of commercial vehicles per day in the opening year of the new road.

The design charts allow for a 2% per annum traffic growth rate over the design period. Advice on required pavement thicknesses in relation to higher growth rates can be obtained from the Department of Transport. The serviceable life of all types of pavement, with appropriate maintenance, should be 40 years. This section on pavement design and surfacings includes charts from HD 14/87[114] and some examples of typical details abstracted from the numerous drawings contained in the *Highway construction details*.[42] The clause numbers referred to in the charts and details are clause numbers in the Department of Transport *Specification for highway works*.[72]

23.9.2 Capping layer and sub-base

Capping layer and sub-base thicknesses for flexible and composite pavements (Figure 23.17), rigid and rigid composite pavements (Figure 23.18), in relation to subgrade CBR values are provided in HD 14/87[114] and are designed to achieve a satisfactory working platform for constructing subsequent pavement layers; they are thus independent of the traffic loading after completion. (See section 23.7 for guidance on CBR values for construction conditions.)

If the subgrade is frost-susceptible the total thickness of non frost-susceptible capping, sub-base and other pavement layers should be not less than 450 mm. An extra thickness of sub-base or capping would be necessary to achieve this.

Cement-bound sub-bases are used for rigid and rigid composite construction to minimize the risk of water penetration at slab joints and permit better compaction of the concrete slab. A depth of 150 mm of sub-base is employed above the capping layer where the subgrade CBR requires a capping layer, or directly above sub-grade where the CBR is above 15%. The specifications for the various cement-bound materials shown in the figures as CBM1, C10, etc. are given in the series 1000 clauses of the *Specification for highway works*.[72]

23.9.3 Flexible and flexible composite pavements

23.9.3.1 General

Flexible construction is defined by HD 14/87[114] as consisting of surfacing and road base materials that are bound with bituminous binder whilst flexible composite construction has a road base or lower road base of cement-bound material.

In order to achieve a 40-year service life for the above types of pavement, two-stage construction is required. Initially the pavement is designed for the predicted traffic loading over the first 20 years.

Major maintenance in the form of overlaying and/or partial reconstruction would then be required to carry the loading over the remaining 20-year period. Pavement strengthening is recommended when 15% of its length has reached the critical condition.[113, 115]

23.9.3.2 Design

Charts 3, 12 and 13 of HD 14/87,[114] (Figures 23.19, 23.20 and 23.21) give design thicknesses for flexible pavements with traffic loadings between 100 and 10 000 commercial vehicles per day. Flexible composite designs are either determinate or indeterminate and take account of the gradual deterioration of the cement-bound road base. Charts 4 and 5 of HD 14/87[114] (Figure 23.22 and Table 23.13) give design thicknesses and materials.

The longer life indeterminate designs are generally intended to be used for traffic loadings greater than 1000 commercial vehicles per day where traffic delays during maintenance are likely to be severe. Determinate designs are intended for flows below 1000 commercial vehicles per day. However, the choice is essentially an economic decision and an indeterminate life design may be appropriate for a lower flow in certain circumstances.[114]

Designs for traffic flows below 100 commercial vehicles per day are covered in TRRL Report LR 1132[85] which gives road-base and surfacing thicknesses for the predicted cumulative number of standard axles during the design life of a pavement. A standard axle load is defined as 80 kN and the cumulative number is normally expressed in millions of standard axles (or m.s.a.). It is prudent to test the sensitivity of the pavement design to variations in initial traffic flows, and traffic growth rates, particularly if flows cannot be forecast with a reasonable degree of confidence.

23.9.3.3 Materials

The UK is unusual in using hot-rolled asphalt with precoated chippings rolled in as a wearing course for the higher traffic loadings. Many countries use asphaltic concrete. Hot-rolled asphalt is essentially a gap-graded material depending on the hardness of the matrix for its mechanical stability, whereas asphaltic concrete depends primarily on a well-graded aggregate for its strength and uses the bitumen as a binder.

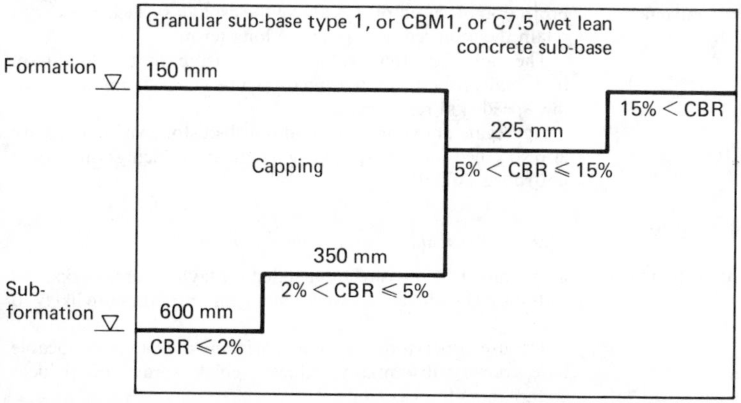

Granular sub-base Type 2 may be used as a sub-base for design traffic loadings less than 400 commercial vehicles per day at opening. Granular sub-base Type 2 shall have a CBR of 30% or more when tested in accordance with Clause 804.

Notes:
(1) Dimensions shown refer to the thickness of the layer they are assigned to.
(2) If the sub-grade is frost-susceptible the total thickness of non frost-susceptible capping, sub-base, and pavement layers as defined in Clauses 602 and 705 shall be not less than 450 mm.

Figure 23.17 Capping layer and sub-base thicknesses for flexible and composite pavements. (After Department of Transport (1987) *Structural design of new road pavements*. HMSO, London)

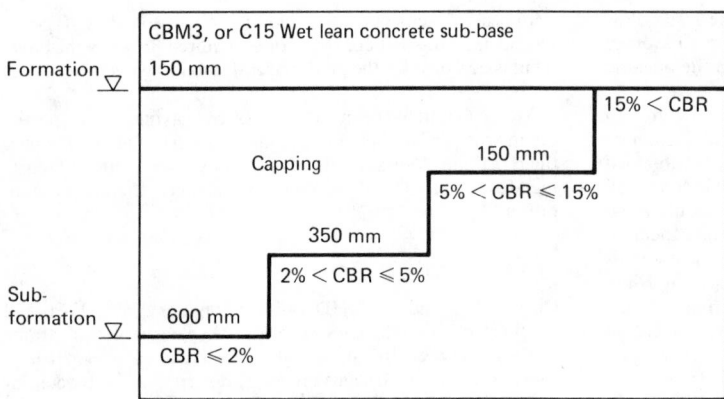

Formation ▽

CBM2 or C10 wet lean concrete may be used as sub-base for design traffic loadings less than 700 commercial vehicles per day at opening.

Notes:
(1) Dimensions shown refer to the thickness of the layer they are assigned to.
(2) If the sub-grade is frost-susceptible the total thickness of non frost-susceptible capping, sub-base, and pavement layers as defined in Clauses 602 and 705 shall be not less than 450 mm.

Figure 23.18 Capping and sub-base thicknesses for rigid and rigid composite pavements. (After Department of Transport (1987) *Structural design of new road pavements.* HMSO, London)

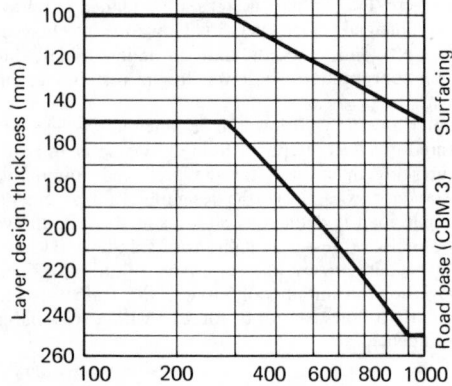

Design traffic loading
(commercial vehicles per day at opening in one direction)

Figure 23.19 Design thickness of materials for flexible composite pavements with determinable life. (After Department of Transport (1987) *Structural design of new road pavements.* HMSO, London)

Hot-rolled asphalt is normally batched to a design mix to clause 911. However, a less economic recipe mix to clause 910 is available for use in certain circumstances.

Alternative surfacings for more lightly trafficked pavements are as follows.

(1) Dense bitumen macadam to clause 912.
(2) Dense tar surfacing to clause 913.
(3) Cold asphalt to clause 914.
(4) Open textured bitumen macadam to clause 916.
(5) Open textured tarmacadam to clause 917.

Open-textured macadams are normally surface-dressed to seal the pavement.

Local authorities may have differing requirements as to which materials are acceptable.

23.9.3.4 Surface characteristics

Riding quality of a new pavement is governed by clause 702 of the *Specification for highway works*[72] which specifies surface level tolerances and limits the number of irregularities exceeding certain values.

Specification requirements for aggregate properties and texture depths[116] states that the skidding resistance of a road surface is determined by two basic characteristics: (1) its microtexture, providing friction with the vehicle tyre; and (2) its macrotexture, necessary at high vehicle speeds to facilitate rapid drainage of water from the surface of the road in contact with the tyre and to utilize the hysteresis effects in the tyre tread rubber to absorb some of the kinetic energy of the vehicle.

The surface characteristics of hot-rolled asphalt are produced by rolling in precoated chippings to clause 915.

Satisfactory skidding resistance is obtained by specifying the:

(1) Grading of chippings to BS 594[117] as shown in Table 23.14.
(2) Minimum rate of spread of chippings to BS 594 or the surface texture measured in accordance with either clause 921 of the *Specification for highway works*[72] (the sand patch test to BS 598, Part 3) or clause 922 (the TRRL mini texture meter to clause 929).
(3) Maximum aggregate abrasion values as required by TM H 16/76[116] and shown in Table 23.15.
(4) Minimum polished stone values again as required by TM H 16/76[116] and shown in Table 23.16.

The first three items determine the surface macrotexture. The aggregate abrasion value is a measure of the stones' resistance to wear and thus a maximum value is designed to maintain the projection of the chippings above the asphalt surface in order to retain the required texture in the long term.

The polished stone value is a measure of the long-term frictional property of the microtexture which is important to low-speed skid resistance.

Aggregate abrasion value and polished stone value limits are also specified for the exposed aggregate in surfacings other than hot-rolled asphalt.

23.9.3.5 Particular wearing course requirements

Dense tar surfacing is recommended for laybys, lorry parks, bus stations or any locations where diesel oil droppings are likely to occur.

The use of pervious macadam surfacing over an impermeable base course substantially reduces vehicle spray and reduces aquaplaning in heavy rain.

Pervious macadam also has the benefit of reducing dry surface L_{10} noise levels by between 4 and 5 dB(A) at 90 km/h and between 3 and 4 dB(A) at 70 km/h.[119]

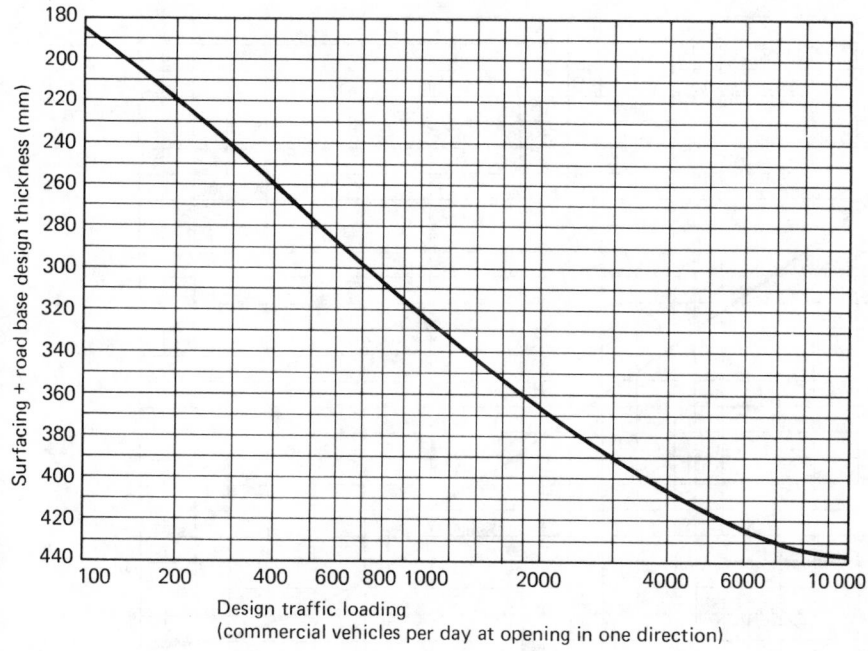

Design traffic loading
(commercial vehicles per day at opening in one direction)

Design traffic loading	Permitted materials and thickness	Special requirements		
		Binder grade	Coarse aggregate	Fine aggregate
100–3000 commercial vehicles per day	Wearing course (40 mm) rolled asphalt to C1.911	None	None	None
	Base course (60 mm) rolled asphalt to C1.905 dense bitumen macadam to C1.906 dense tarmacadam to C1.9907	50 pen bitumen 100 pen bitumen C54 grade tar	CRS CRS CRS	None None CRS
	Road base (remainder of thickness) rolled asphalt to C1.904 dense bitumen macadam to C1.903 dense tarmacadam to C1.902	50 pen bitumen 100 pen bitumen C58 or C54 grade tar	None None None	None None None
3000–10 000 commercial vehicles per day	Wearing course (40 mm) rolled asphalt to C1.911	None	None	None
	Upper road base (remainder of thickness) dense bitumen macadam to C1.903	100 pen bitumen	CRS	None
	Lower road base (125 mm) rolled asphalt to C1.904	50 pen bitumen	CRS	None

Notes:
(1) *Wearing course:* refer to BS594:1985 Part 1 Table 11 to select appropriate Marshall stability and flow values related to traffic loading.
(2) *Aggregate size:* refer to BS 594 and BS 4987 for guidance on the nominal size of coarse aggregate appropriate to the layer thickness and the level tolerance permitted on the top of the course.
(3) *Aggregate type:* CRS in the above table denotes that the aggregate shall be crushed rock or slag. For base course where there is local experience of the successful use of gravel it may be a permitted alternative coarse aggregate.

Figure 23.20 Design thickness of materials for flexible pavements with dense bitumen macadam, rolled asphalt or dense tarmacadam. (After Department of Transport (1987) *Structural design of new road pavements.* HMSO, London)

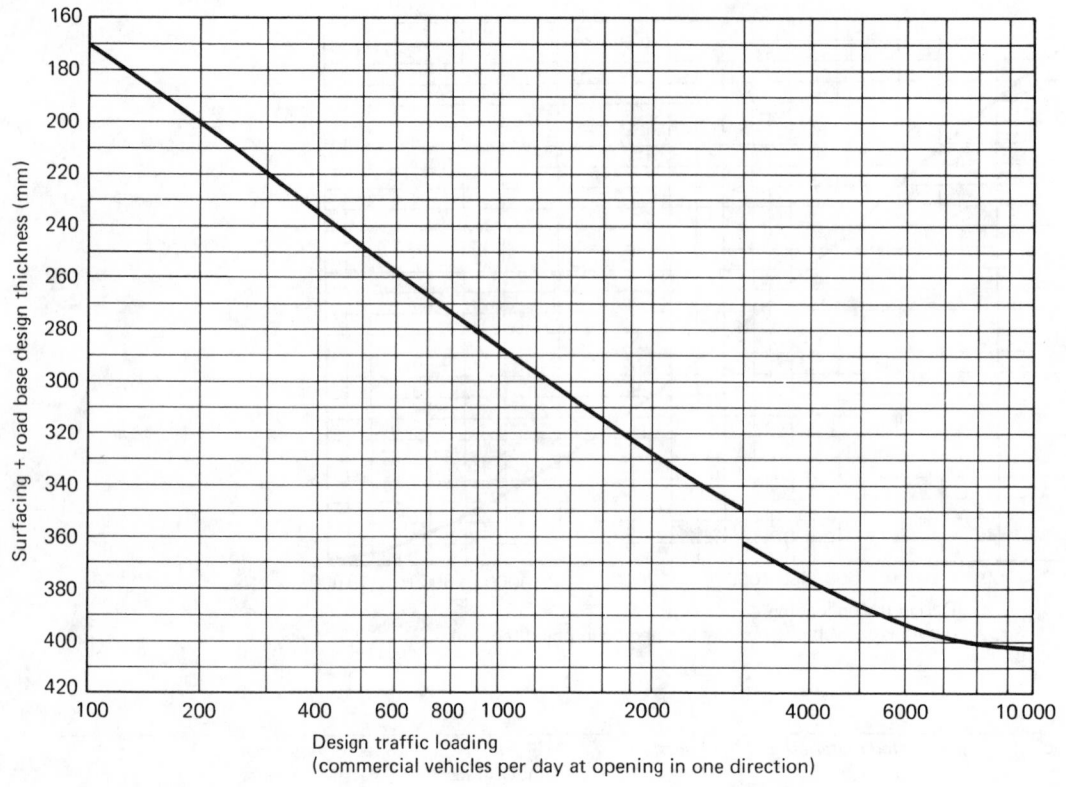

Design traffic loading
(commercial vehicles per day at opening in one direction)

Design traffic loading	Permitted materials and thickness	Special requirements for coarse aggregate
100–3000 commercial vehicles per day	Wearing course (40 mm) rolled asphalt to C1.911	None
	Base course (60 mm) dense bitumen macadam with 50 penetration grade binder to C1.934	CRS
	Road base (remainder of thickness) dense bitumen macadam with 50 penetration grade binder to C1.932	None
3000–10 000 commercial vehicles per day	Wearing course (40 mm) rolled asphalt to C1.911	None
	Upper road base (remainder of thickness) dense bitumen macadam with 50 penetration grade binder to C1.932	CRS
	Lower road base (125 mm) rolled asphalt to C1.904 (see note below)	CRS

Notes:
(1) *Wearing course:* refer to BS594:1985 Part 1 Table 11 to select appropriate Marshall stability and flow values related to traffic loading.
(2) *Aggregate size:* refer to BS 594 and BS 4987 for guidance on the nominal size of coarse aggregate appropriate to the layer thickness and the level tolerance permitted on the top of the course.
(3) *Aggregate type:* CRS in the above table denotes that the aggregate shall be crushed rock or slag. For base course where there is local experience of the successful use of gravel it may be a permitted alternative coarse aggregate.
(4) *Binder type:* for rolled asphalt lower road base the binder shall be 50 penetration grade bitumen.

Figure 23.21 Design thickness of materials for flexible pavements with dense bitumen macadam with 50-penetration grade binder. (After Department of Transport (1987) *Structural design of new road pavements*. HMSO, London)

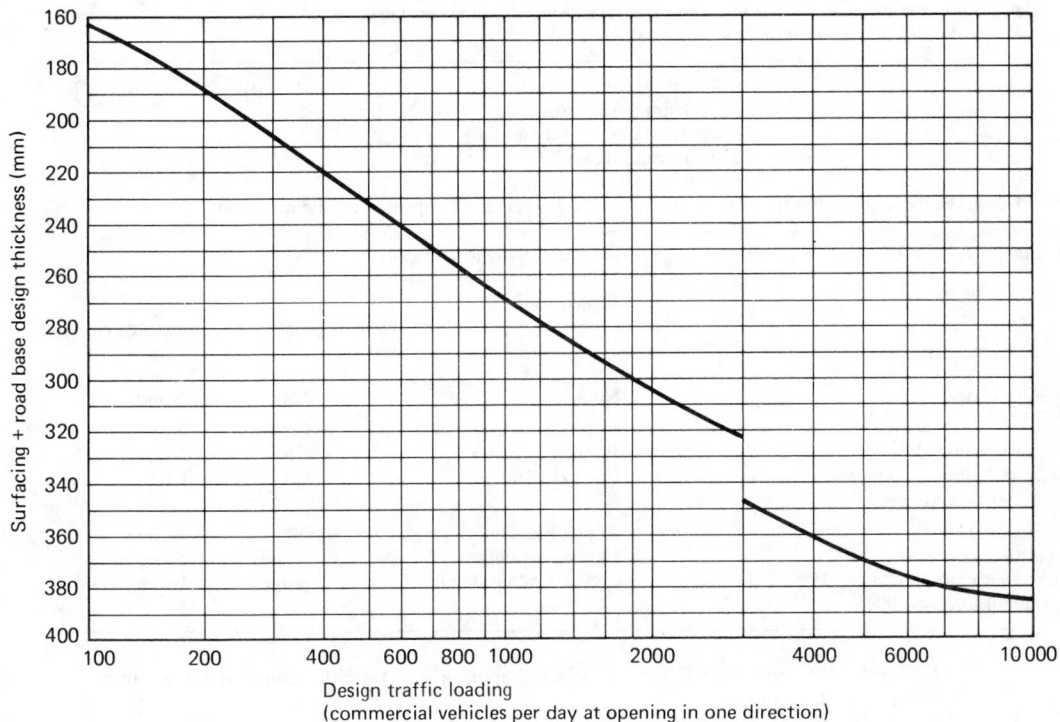

Design traffic loading
(commercial vehicles per day at opening in one direction)

Design traffic loading	Permitted materials and thickness	Special requirements for coarse aggregate
100–3000 commercial vehicles per day	Wearing course (40 mm) rolled asphalt to C1.911	None
	Base course (60 mm) heavy-duty macadam to C1.933	CRS
	Road base (remainder of thickness) heavy-duty macadam to C1.930	None
3000–10 000 commercial vehicles per day	Wearing course (40 mm) rolled asphalt to C1.911	None
	Upper road base (remainder of thickness) heavy-duty macadam to C1.930	CRS
	Lower road base (125 mm) rolled asphalt to C1.904 (see note below)	CRS

Notes:
(1) *Wearing course:* refer to BS 594:1985 Part 1 Table 11 to select appropriate Marshall stability and flow values related to traffic loading.
(2) *Aggregate size:* refer to BS 594 and BS 4987 for guidance on the nominal size of coarse aggregate appropriate to the layer thickness and the level tolerance permitted on the top of the course.
(3) *Aggregate type:* CRS in the above table denotes that the aggregate shall be crushed rock or slag. For base course where there is local experience of the successful use of gravel it may be a permitted alternative coarse aggregate.
(4) *Binder type:* for rolled asphalt lower road base the binder shall be 50 penetration grade bitumen.

Figure 23.22   Design thickness of materials for flexible pavements with heavy-duty macadam. (After Department of Transport (1987) *Structural design of new road pavements.* HMSO, London)

Table 23.13 Design thickness of materials for flexible composite pavements with indeterminable life. (After Department of Transport (1987) *Structural design of new road pavements*. HMSO, London)

40 mm wearing course
60 mm base course
100 mm road base (bituminous)
250 mm road base (CBM3)
or 225 mm road base (CBM4)
Design traffic loading 100 to 3000 commercial vehicles per day at opening in one direction

Permitted materials	Special requirements		
	Binder grade	*coarse aggregate*	*fine aggregate*
Wearing course			
Rolled asphalt to clause 911	None	None	None
Base course			
Rolled asphalt to clause 905	50 pen bitumen	CRS	None
Dense bitumen macadam to clause 906	100 pen bitumen	CRS	None
Dense tarmacadam to clause 907	C54 grade tar	CRS	CRS
Road base (bituminous)			
Rolled asphalt to clause 904	50 pen bitumen	None	None
Dense bitumen macadam to clause 903	100 pen bitumen	None	None
Dense tarmacadam to clause 902	C58 or C54 grade tar	None	None

Notes:
(1) *Wearing course:* Refer to BS 594: 1985, Part 1: Table 11 to select appropriate Marshall stability and flow values related to traffic loading.
(2) *Aggregate size:* Refer to BS 594 and BS 4987 for guidance on the nominal size of coarse aggregate appropriate to the layer thickness and the level tolerance permitted on the top of the course.
(3) *Aggregate type:* CRS in the above table denotes that the aggregate shall be crushed rock or slag. For base course where there is local experience of the successful use of gravel it may be a permitted alternative coarse aggregate.
(4) *CBM3 and CBM4:* CBM4 road base 225 mm thickness shall be included as an alternative where the CBM3 road base design thickness exceeds 225 mm.

Table 23.14 Grading of chippings. (After British Standards Institution (1985) *Hot-rolled asphalt for roads and other paved areas*, BS 594. BSI, Milton Keynes)

BS test sieve	Percentage by mass passing BS test sieve	
	20 mm nominal size	*14 mm nominal size*
28 mm	100	
20 mm	90–100	100
14 mm	0–25	90–100
10 mm	0–4	0–25
6.3 mm	—	0–4
75 μm	0–2	0–2

Table 23.15 Traffic loadings and maximum aggregate abrasion values for flexible surfaces. (After Department of Transport (1976) *Specification requirements for aggregate properties and texture depth for bituminous surfacings to new roads*. HMSO, London)

Traffic in commercial vehicles per lane per day	Under 250*	Up to 1000	Up to 1750	Up to 2500	Up to 3250	Over 3250
Maximum aggregate abrasion value for chippings	14	12	12	10	10	10
Maximum aggregate abrasion value for aggregate in coated macadam wearing courses	16	16	14	14	12	12

*For lightly trafficked roads carrying less than 250 commercial vehicles per lane per day aggregate of higher aggregate abrasion value may be used where experience has shown that satisfactory performance is achieved by aggregate from a particular source.

Table 23.16 Categories of sites and minimum polished stone values for flexible roads. (After Department of Transport (1976) *Specification requirements for aggregate properties and texture depth for bituminous surfacings to new roads*. HMSO, London)

Site		Definition	Minimum polished stone value	
A1 (difficult)	(i)	Approaches to traffic signals on roads with 85%ile speed of traffic greater than 40 mile/h (64 km/h)	Less than 250 commercial vehicles per lane per day:	60
			250 to 1000:	65
			1000 to 1750:	70
			More than 1750:	75
	(ii)	Approaches to traffic signals, pedestrian crossings and similar hazards on main urban roads		
A2 (difficult)	(i)	Approaches to and across major priority junctions on roads carrying more than 250 commercial vehicles per lane per day*	Less than 1750 commercial vehicles per lane per day:	60
			1750 to 2500:	65
			2500 to 3250:	70
			More than 3250:	75
	(ii)	Roundabouts and their approaches		
	(iii)	Bends with radius less than 150 m on roads with an 85%ile speed of traffic greater than 40 mile/h (64 km/h)		
	(iv)	Gradients of 5% or steeper, longer than 100 m		
B (average)		Generally straight sections of and large radius curves on:	Less than 1750 commercial vehicles per lane per day:	55
			1750 to 4000:	60
			More than 4000:	65
	(i)	Motorways		
	(ii)	Trunk and principal roads		
	(iii)	Other roads carrying more than 250 commercial vehicles per lane per day		
C (easy)	(i)	Generally straight sections of lightly trafficked roads, ie less than 250 commercial vehicles per day	45	
	(ii)	Other roads where wet skidding accidents are unlikely to be a problem		

*The 250 commercial vehicles per lane per day applies to each approach.

Where traffic flows at critical locations, such as the approaches to traffic signals on urban dual carriageways require a polished stone value of 75, consideration should be given to the use of high-skid-resistant surface treatment. The *Specification for highway works*[78] (clause 924) specifies such treatments consisting of calcined bauxite aggregate of high polished stone value with a resin-based binder. It is difficult to obtain chipping stones with a consistent polished stone value of 75 from UK sources.

23.9.4 Rigid (concrete) pavements

Concrete pavements are primarily constructed on large highway schemes where the layout permits economic working from the high production rates possible with concrete paving equipment associated with the availability of local materials. Older concrete pavements may exist which have now been overlaid with bituminous surfacing as part of normal maintenance.

Newly formed concrete pavements are of either rigid construction (i.e. concrete surface slabs or rigid composite construction) being continuously reinforced concrete road base with bituminous surfacing. Concrete surface slabs are either jointed unreinforced concrete pavement, jointed reinforced concrete pavement or continuously reinforced concrete pavement.[114]

23.9.4.1 Paving plant

Paving plant for constructing a concrete slab by machine is of two basic types:

(1) A fixed form paver supported on flat bottom rails laid to conform precisely with the required road profile.
(2) A slipform paver mounted on caterpillar tracks which travel on the sub-base outside the edges of the slab. The paving level of a slipform paver is controlled by tensioned guide wires.[120, 121]

23.9.4.2 Separation membrane

A separation membrane is required between the concrete slab and the sub-base to prevent loss of water from the fresh concrete. Polythene is used beneath jointed unreinforced and jointed reinforced concrete slabs to reduce the friction between the slab and the sub-base and inhibit the formation of mid-bay cracks. A bituminous spray is used for continuously reinforced concrete because a degree of restraint is required.

23.9.4.3 Concrete

The concrete for surface slabs and road bases is defined in clause 1001 of the *Specification for highway works*[72] as grade C40 with a minimum cement content of 320 kg/m³.

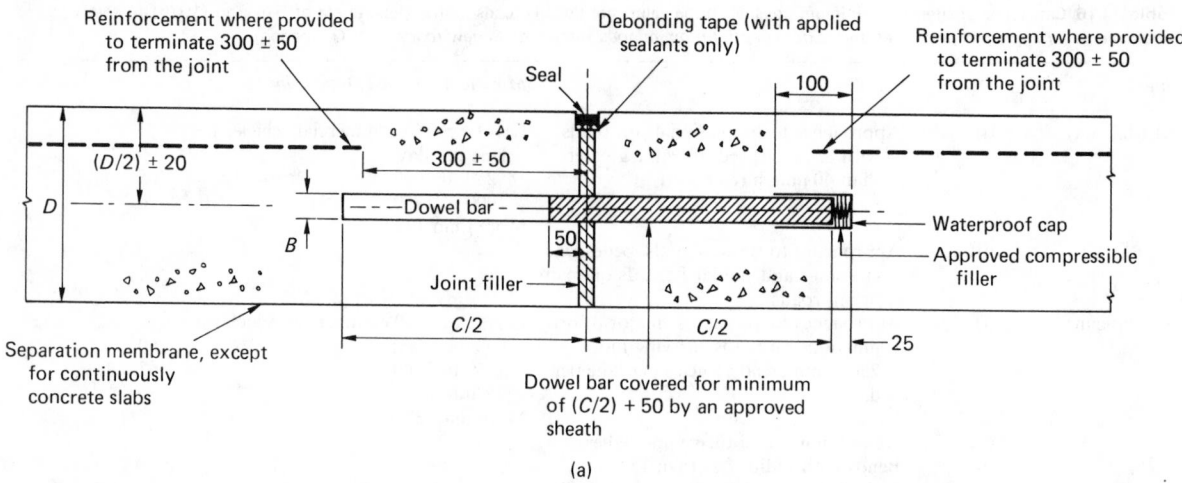

(a)

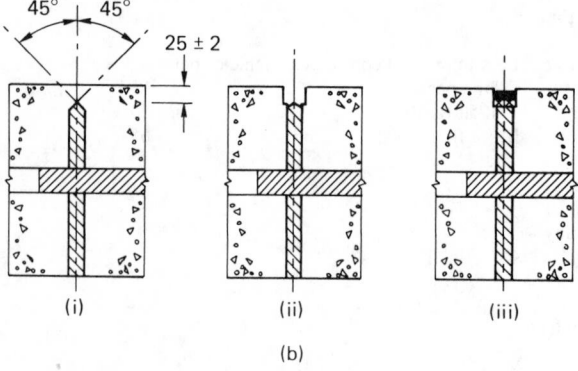

(i)   (ii)   (iii)

(b)

All dimensions in millimetres

Figure 23.23 Expansion joints: reinforced and unreinforced concrete slabs. (After Department of Transport (1987) *Highway construction details.* HMSO, London)

Dowel bar		
Slab thickness D	B	C
150–239	25	550
240 and over	32	650

Notes:

(1) The dowel bars shall be placed at 300 centres. This spacing shall be varied where necessary so that no dowel bar is within 150 of a slab edge or a joint parallel to the bars.

(2) Cover to all reinforcement to be 50 ± 5 for slabs less than 200 thick, 60 ± 10 for slabs 200 up to 270 thick and 70 ± 20 for slabs 270 thick or more.

23.9.4.4 Slab and joint details

Typical sections of slab and joint details are given in Series C of *Highway construction details.*[42] Figures 23.23 to 23.28 show a small selection of these details. Joints are formed either by presetting the joint assemblies on support cages or by insertion techniques. Discontinuities should be avoided with gullies and manholes positioned outside the main slab. Construction details are discussed in *Joints in concrete roads.*[122]

Jointed unreinforced concrete pavement. The design slab thickness and joint spacings are given in Chart 6 of HD 14/87[114] (Figure 23.29). The transverse joints are contraction, expansion or warping joints constructed such that the length:width ratio is not greater than 2. In summer periods, expansion joints may be replaced by contraction joints (clause 1009).[72] Warping joints are used for extra joints at manhole positions, alongside reinforced slabs or where required in long narrow or tapered slabs to reduce the length:width ratio to 2 or less. Longitudinal joints are positioned to limit the bay width to 4.2 m, or 5.0 m with limestone aggregates.

Jointed reinforced concrete pavement. The design slab thickness, longitudinal reinforcement and joint spacings are given in Chart 7 of HD 14/87[114] (Figure 23.30). The transverse joints are

contraction or expansion joints constructed as for an unreinforced pavement to limit the slab length:width ratio to 2. Expansion joints may be replaced by contraction joints in the summer period (clause 1009).[72] Longitudinal joints are positioned to limit the bay width to 4.2 m or 5.0 m with limestone aggregate.

Continuously reinforced concrete pavement. The design slab thickness and reinforcement are given in Chart 8 of HD 14/87[114] (Figure 23.31). Anchorages are provided at ends and at any discontinuities in the pavement.[42]

Continuously reinforced concrete road base with bituminous surfacing. The design slab thickness and reinforcement for the road base are given in Chart 9 of HD 14/87[114] (Figure 23.32). Anchorages are not required since the slab is protected from large thermal stresses by the 100 mm of flexible surfacing above it. This form of pavement is advocated where it is of paramount importance to minimize traffic disruption caused by future maintenance.

Surface texture. After final regulation of the surface of the slab and before the application of the curing membrane concrete surface slabs are brush textured across the carriageway (clause 1026).[72]

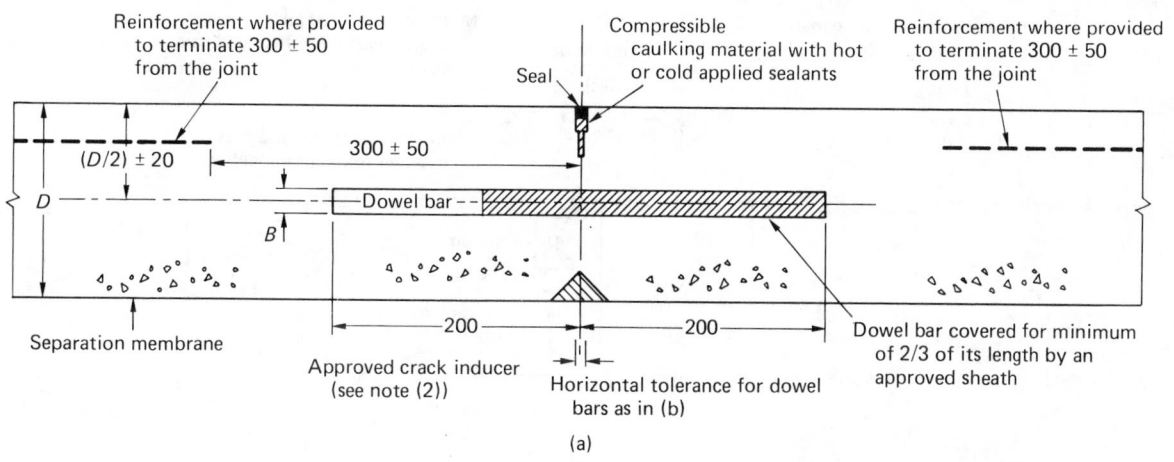

(a)

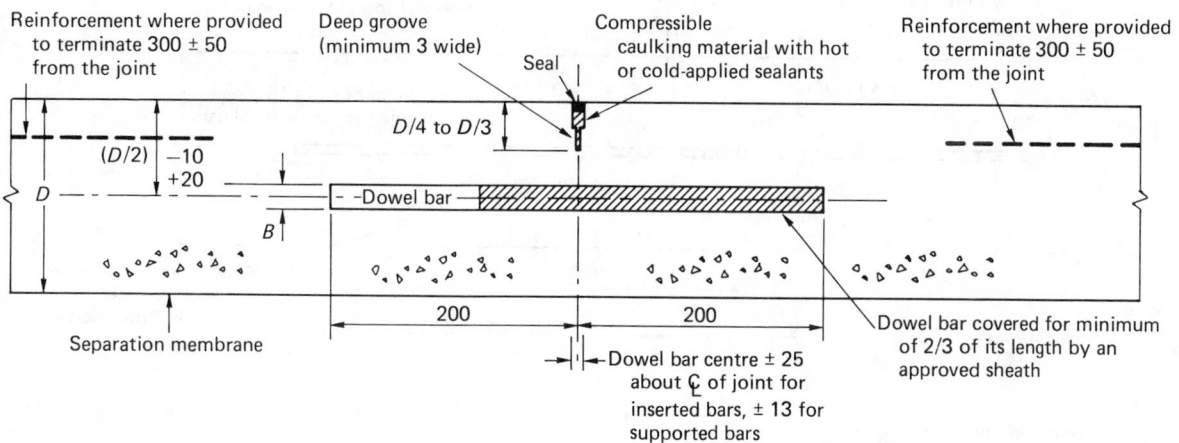

All dimensions in millimetres

(b)

Dowel bar (minimum dimensions)	
Slab thickness D	B
150 to 239	20
240 and over	25

Notes:
(1) The dowel bars shall be placed at 300 centres. This spacing shall be varied where necessary so that no dowel bar is within 150 of a joint parallel to the bars.

(2) Dimensions of crack inducer base shall not be greater than twice the height nor less than the height. The combined height of the crack inducer and the depth of the groove shall be between $D/4$ and $D/3$.

(3) Cover to all reinforcement to be 50 ± 5 for slabs less than 200 thick, 60 ± 10 for slabs 200 up to 270 thick and 70 ± 20 for slabs 270 thick or more.

(4) The bottom crack inducer may be omitted when construction takes place in summer from 21 April to 21 October and shall be omitted when the grooves are sawn. In these cases the groove shall be between $D/4$ and $D/3$ in depth.

Figure 23.24 Contraction joints. (a) With crack inducer; (b) with deep wet-formed or sawn groove. (After Department of Transport (1987) *Highway construction details.* HMSO, London)

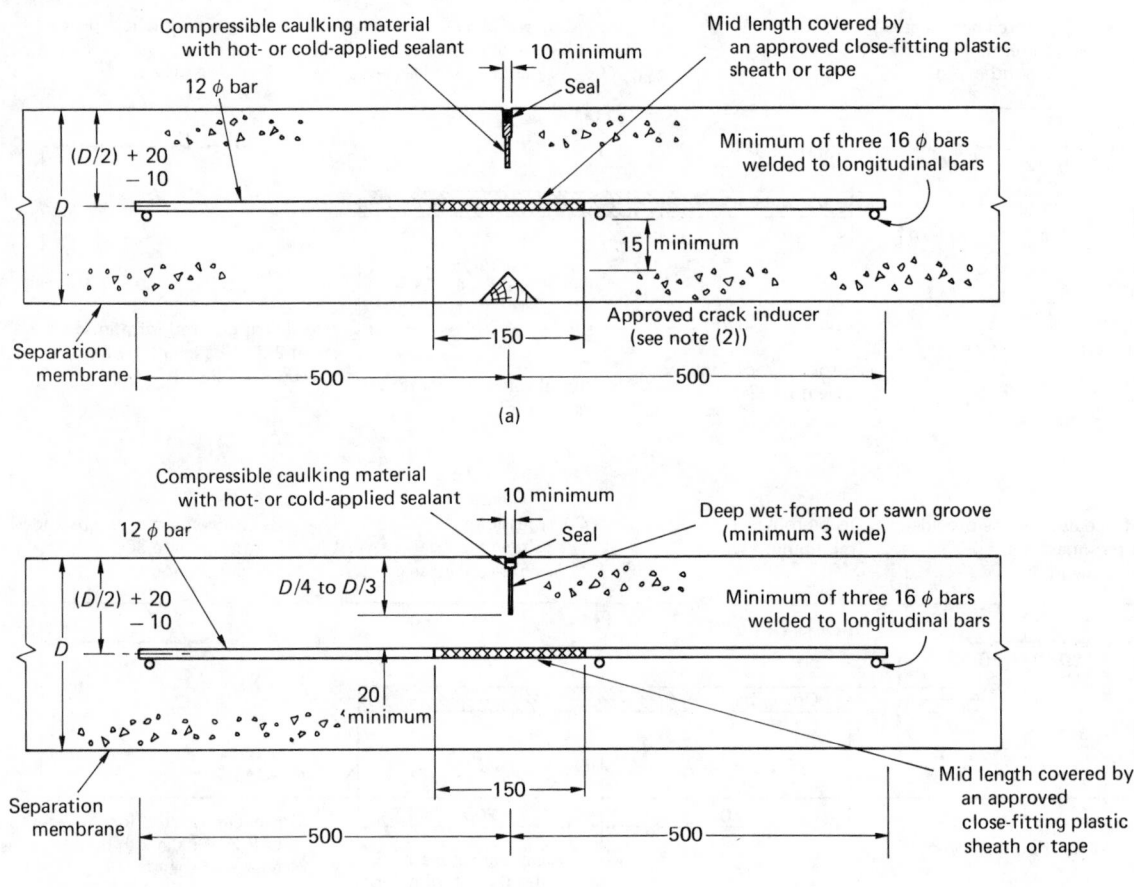

All dimensions in millimetres

(b)

Notes:
(1) Warping joints shall be
    constructed and sealed in
    accordance with the
    specification. The tie bar
    spacing shall be varied
    where necessary so that
    no tie bar is within 150
    of a slab edge or a joint
    parallel to the bars.
(2) *Dimensions of crack inducer:*
    base shall not be greater
    than twice the height nor
    less than the height.
    The combined height of the
    crack inducer and the depth
    of the groove shall be
    between $D/4$ and $D/3$.

(3) The bottom crack inducer
    may be omitted when
    construction takes place
    in summer from 21 April
    to 21 October and shall
    be omitted when the
    grooves are sawn. In
    these cases the groove
    shall be between $D/4$ and
    $D/3$ in depth.

Figure 23.25 Warping joints: unreinforced slabs only. (a) With
crack inducer; (b) with deep wet-formed or sawn groove. (After
Department of Transport (1987) *Highway construction details.*
HMSO, London)

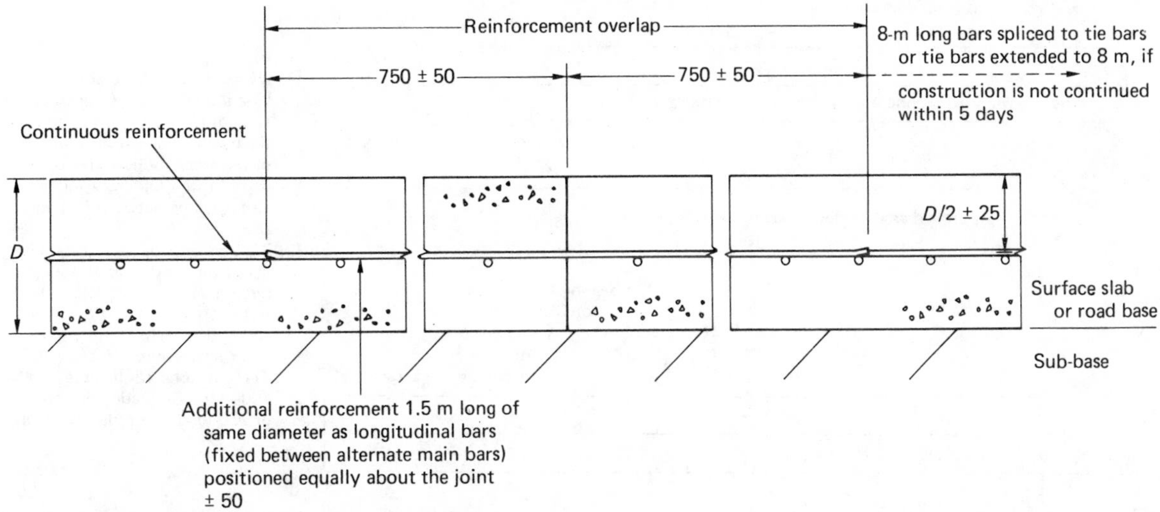

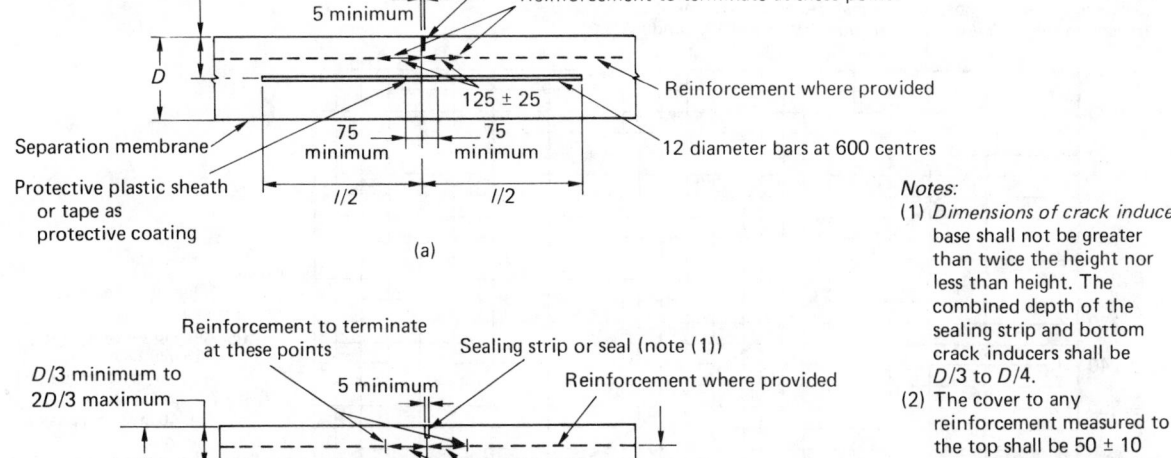

Direction of paving

Reinforcement overlap

750 ± 50

750 ± 50

8-m long bars spliced to tie bars or tie bars extended to 8 m, if construction is not continued within 5 days

Continuous reinforcement

D

D/2 ± 25

Surface slab or road base

Sub-base

Additional reinforcement 1.5 m long of same diameter as longitudinal bars (fixed between alternate main bars) positioned equally about the joint ± 50

All dimension in millimetres unless otherwise stated

Figure 23.26 Transverse construction joint for continuously reinforced concrete pavement or road base (longitudinal section). (After Department of Transport (1987) *Highway construction details*. HMSO, London)

Sealing strip or seal

$D/3$ minimum to $2D/3$ maximum

Reinforcement to terminate at these points

5 minimum

D

Reinforcement where provided

125 ± 25

Separation membrane

75 minimum

75 minimum

12 diameter bars at 600 centres

Protective plastic sheath or tape as protective coating

$l/2$

$l/2$

(a)

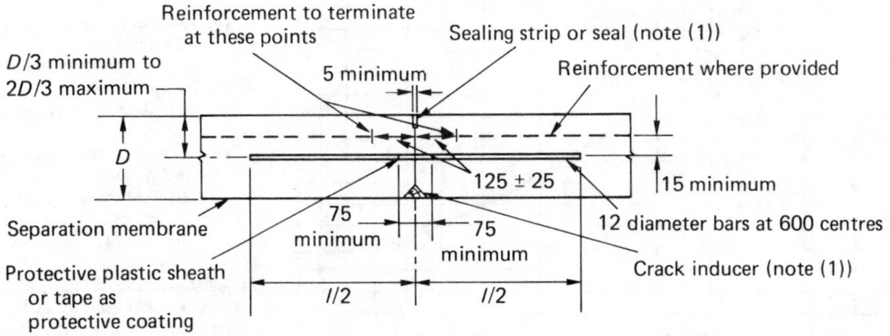

Reinforcement to terminate at these points

Sealing strip or seal (note (1))

$D/3$ minimum to $2D/3$ maximum

5 minimum

Reinforcement where provided

D

125 ± 25

15 minimum

Separation membrane

75 minimum

75 minimum

12 diameter bars at 600 centres

Protective plastic sheath or tape as protective coating

Crack inducer (note (1))

$l/2$

$l/2$

(b)

All dimensions in millimetres

Figure 23.27 Longitudinal joints for unreinforced concrete or jointed reinforced concrete slabs. (a) Type 1: longitudinal construction joint between two separately constructed unreinforced or jointed reinforced slabs; (b) Type 2: formed longitudinal joint for slabs of more than one lane width constructed in one operation. (After Department of Transport (1987) *Highway construction details*. HMSO, London)

Notes:
(1) *Dimensions of crack inducer:* base shall not be greater than twice the height nor less than height. The combined depth of the sealing strip and bottom crack inducers shall be $D/3$ to $D/4$.
(2) The cover to any reinforcement measured to the top shall be 50 ± 10 when D is less than 200, 60 ± 10 when D is 200 up to 270 and 70 ± 20 when D is 270 or greater.
(3) Tie bars length l to be 1000 for steel grade 250 750 for steel grade 460.

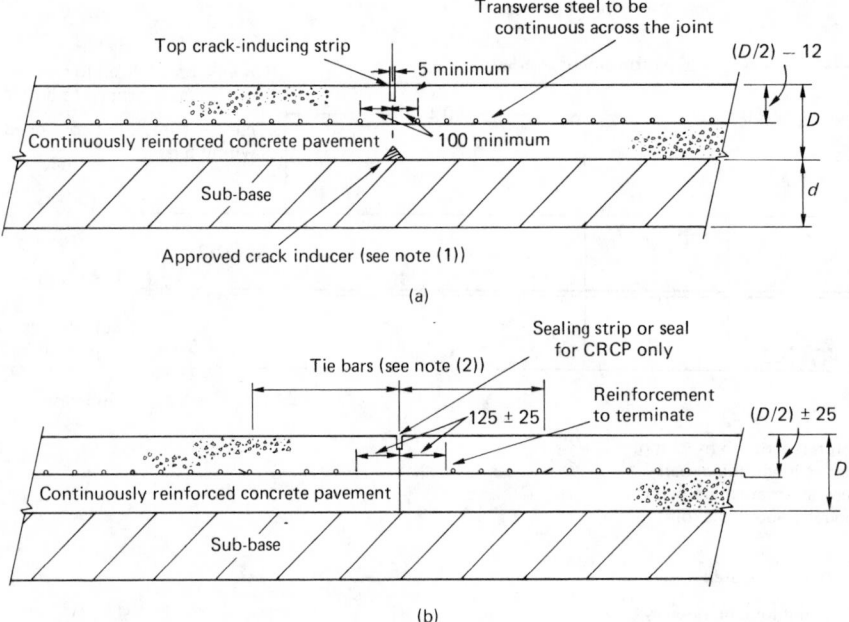

Notes:
(1) *Dimensions of the crack inducer:* base shall not be greater than twice the height nor less than the height. The combined height of the crack inducer and the depth of the top crack-inducing strip shall be between $D/4$ and $D/3$.

(2) Tie bars shall be placed equally about the joint ± 50 at the same spacing as and adjacent to the transverse reinforcement and tied to both longitudinal and transverse reinforcement. Tie bars to be 12 diameter 1000 long for grade 250 steel or 750 long for grade 460 steel.

All dimensions in millimetres unless otherwise stated

Figure 23.28 Longitudinal joint for continuously reinforced concrete pavement or road base. (a) A formed longitudinal joint for continuously reinforced concrete only (constructed in two or more lane widths in one operation): (b) butt-type construction joint (between separately constructed slabs). (After Department of Transport (1987) *Highway construction details*. HMSO, London)

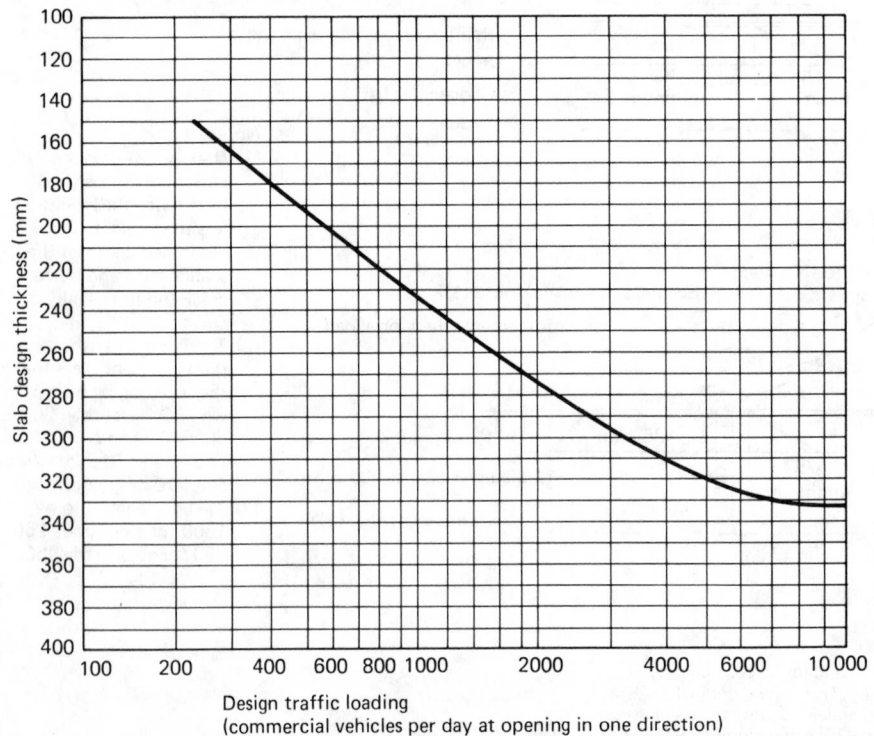

Notes:
(1) Maximum transverse joint spacings:
   (a) for slab thickness up to 225 mm:
      (i) 4 m for contraction joints,
      (ii) 40 m for expansion joints;
   (b) for slab thickness 225 mm and greater:
      (i) 5 m for contraction joints,
      (ii) 60 m for expansion joints.
(2) All transverse joint spacings may be increased by 20% if limestone coarse aggregate is used throughout the depth of the slab.
(3) Refer to the specification for highway works for periods when contraction joints may be substituted for expansion joints.

Figure 23.29 Design thickness for rigid pavements with jointed unreinforced concrete surface slabs. (After Department of Transport (1987) *Structural design of new road pavements*. HMSO, London)

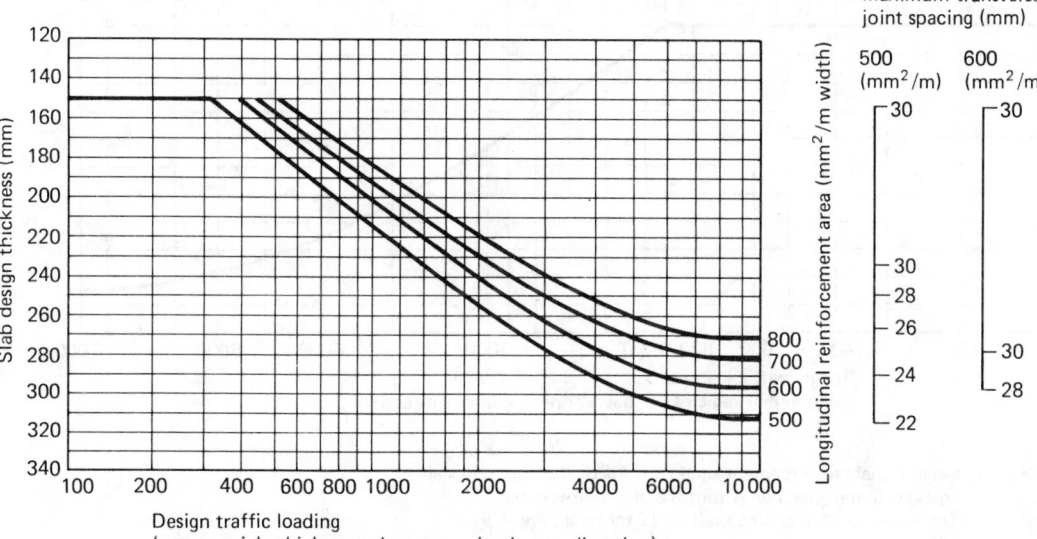

Notes:
(1) Intermediate values of slab thickness, longitudinal reinforcement area, and maximum transverse joint spacing may be interpolated.
(2) The maximum transverse joint spacing shall be 30 m, except for slabs with longitudinal reinforcement areas of 600 mm²/m and less where the maximum joint spacing corresponding to the slab thickness shall be as shown.
(3) Every third transverse joint shall be an expansion joint, with the remainder being contraction joints. Refer to the specification for highway works for periods when contraction joints may be substituted for expansion joints.
(4) All transverse joint spacings may be increased by 20% if limestone coarse aggregate is used throughout the depth of the slab.

Figure 23.30 Design thickness for rigid pavements with jointed reinforced concrete surface slabs. (After Department of Transport (1987) *Structural design of new road pavements.* HMSO, London)

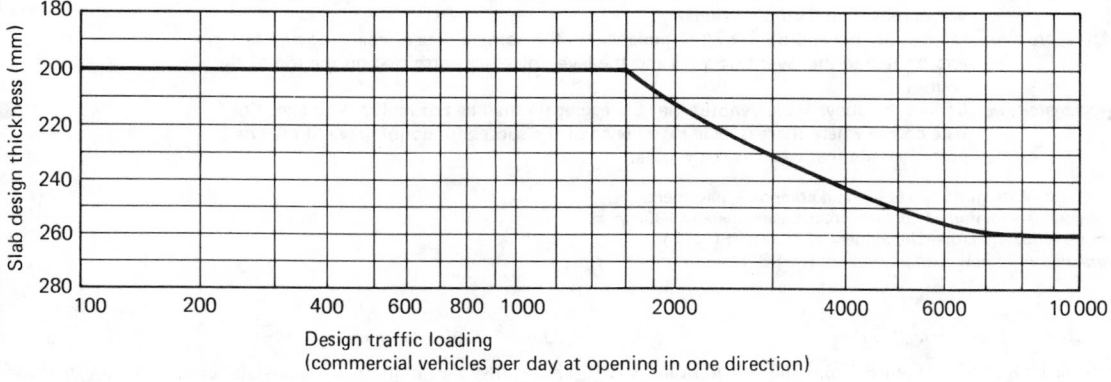

Notes:
(1) *Reinforcement:* longitudinal reinforcement shall be 0.6% of the concrete slab cross-sectional area, comprising 16 mm diameter bars. Transverse reinforcement shall be 12 mm diameter bars at 600 mm spacings.
(2) *Anchorages:* anchorages shall be provided at ends and any discontinuities in the pavement (ground beam or steel beam).

Figure 23.31 Design thickness for rigid pavements with continuously reinforced concrete surface slabs. (After Department of Transport (1987) *Structural design of new road pavements.* HMSO, London)

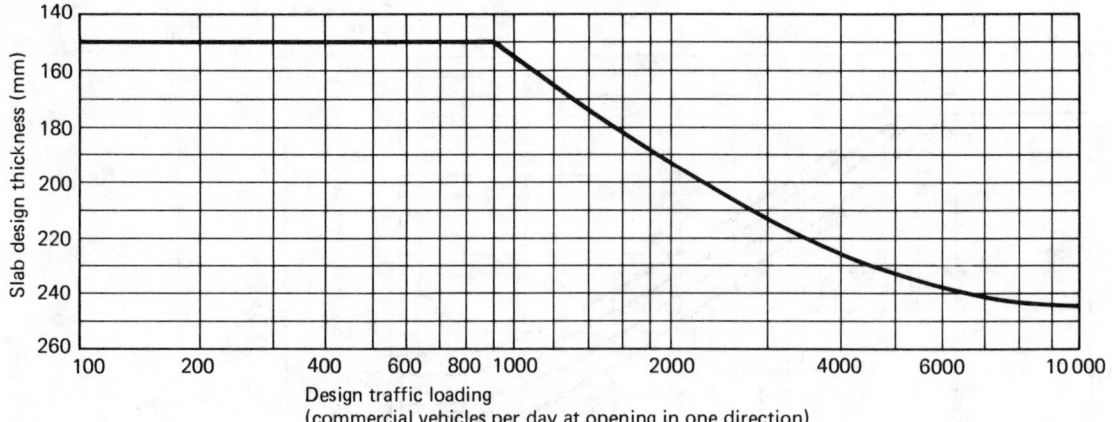

Notes:
(1) *Reinforcement:* longitudinal reinforcement shall be 0.4% of the concrete slab cross-sectional area, comprising 12 mm diameter bars. Transverse reinforcement shall be 12 mm diameter bars at 600 mm spacings.
(2) *Anchorages:* anchorages are not required.

Permitted bituminous materials and thickness Aggregate	Binder grade	Special requirements	
		Coarse aggregate	Fine
Wearing course (40 mm)			
rolled asphalt to C1.911	None	None	None
Base course (60 mm)			
rolled asphalt to C1.905	50 pen bitumen	CRS	None
dense bitumen macadam to C1.906	100 pen bitumen	CRS	None
dense tarmacadam to C1.907	C54 grade tar	CRS	CRS

Notes:
(1) *Wearing course:* refer to BS 594:1985 Part 1, Table 11 to select appropriate Marshall stability and flow values related to traffic loading.
(2) *Aggregate size:* refer to BS 594 and BS 4987 for guidance on the nominal size of coarse aggregate appropriate to the layer thickness and the level tolerance permitted on the top of the course.
(3) *Aggregate type:* CRS in the above table denotes that the aggregate shall be crushed rock or slag. For base course where there is local experience of the successful use of gravel it may be a permitted alternative coarse aggregate.

Figure 23.32 Design thickness for rigid composite pavements with continuously reinforced concrete road-base slabs and 100 mm bituminous surfacing. (After Department of Transport (1987) *Structural design of new road pavements.* HMSO, London)

23.9.4.5 Additional concrete slab thickness

Chart 10 of HD 14/87[114] (Figure 23.33) shows additional slab thicknesses for rigid and rigid composite pavements where the slab does not extend for 1 m or more beyond the edge of lanes carrying commercial vehicles.

23.9.4.6 Corrected design traffic loading for single carriageways

Chart 11 of HD 14/87[114] (Figure 23.34) shows corrected design traffic loading for single carriageways which applies to all types of construction (Figures 23.19 to 23.22 and Table 23.13; Figures 23.29 to 23.32).

23.9.4.7 Sealing of joints

All transverse joints in surface slabs except for construction joints in continuously reinforced concrete should be sealed (clauses 1016 and 1017).[72]

23.9.5 Designs for less than 100 commercial vehicles per day

Pavement designs for these roads should be taken from TRRL Research Report 87.[123]

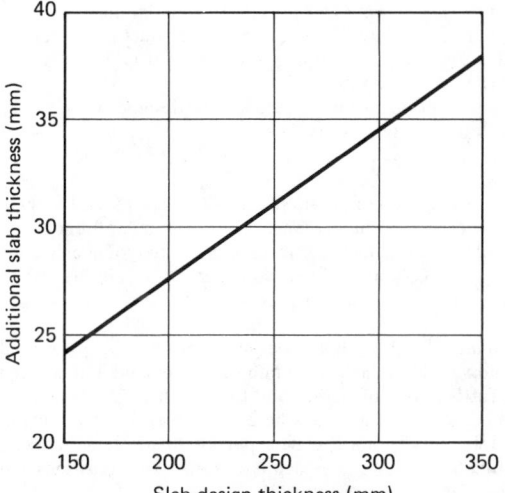

Figure 23.33 Additional concrete slab thickness for rigid and rigid composite pavements where the slab does not extend 1 m or more beyond the edge of lanes carrying commercial vehicles. (After Department of Transport (1987) *Structural design of new road pavements.* HMSO, London)

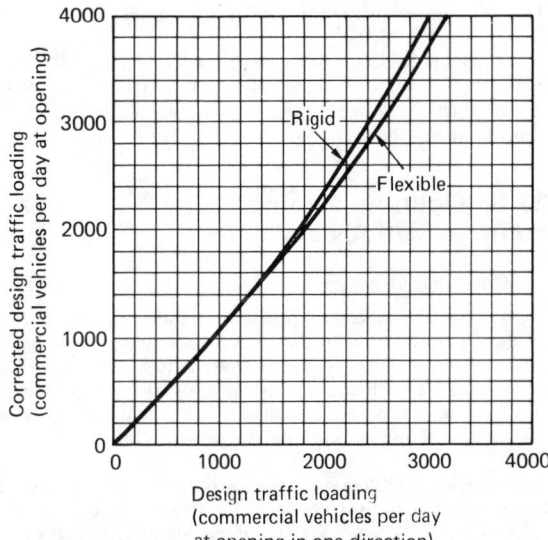

Figure 23.34 Corrected traffic loading for single carriageways. (After Department of Transport (1987) *Structural design of new road pavements.* HMSO, London)

23.9.6 Pavements for industrial usage

Road pavement design methods are inappropriate where axle loads are likely to greatly exceed those permitted on public highways. For example, a loaded front-lift truck capable of carrying containers may have a front axle loading of 90 t compared with the 8.2 t (80 kN) standard axle.

Concentrated loads, such as trailer jockey wheels, will cause deformation of bituminous surfacings. Stacked containers may result in loads of up to 100 t on an area of 0.5 m² at adjacent corners.

23.9.6.1 Construction types

A Concrete Society working party[124] considered the advantages and disadvantages of the following pavement types and gave an indication of comparative costs.

(1) Lean concrete sub base and base with a thin wearing course such as surface dressing.
(2) *In situ* concrete pavement (unreinforced, reinforced or pre-stressed) on a sub base.
(3) Precast concrete slabs on a cement-bound granular base.
(4) Precast concrete blocks on a cement-bound or bituminous-bound granular base.
(5) Clay paving bricks on a cement-bound or bituminous-bound granular base.
(6) Bituminous pavement comprising wearing course, base course and sub base.
(7) Bituminous pavement with specialist overlay wearing course.

The report, which also considered design methods, concluded that no particular form of construction could be universally applicable to all situations. However, on balance, a solution incorporating concrete in some form was judged to produce the most economic pavement.

23.9.6.2 Design methods

The Concrete Society working party considered that the British Ports Association *Design manual*[125] was, in relation to other heavy-duty pavement design methods, a step forward, particularly from the traffic loading viewpoint. This manual was based on the work by Barber and Knapton.[126] It caters for four types of surfacing including 80-mm thick block paving.

The Association have published block paving design guides for lightly and heavily trafficked roads and for specialized traffic.[127-130] However, these should be treated with caution since their later research established that the load-spreading abilities of block paving over a granular sub base had been seriously overestimated. See Gray's design guidance review.[131]

A code of practice for design of lightly trafficked pavements of clay pavers is provided in BS 6677.[132]

23.9.7 Block paving

Block paving, in addition to its industrial usage, is used for residential roads, pedestrianized areas and car parks. Specifications for clay and concrete blocks and codes of practice for laying such blocks are available from various sources.[131-135]

23.9.7.1 Edge details

The strength of the carriageway edge prevents lateral movement and local deterioration of the pavement under traffic loading. Edge restraint is particularly important to block paving both during construction and in its operational condition, in the latter case to preserve the interlock between blocks. Further control of surface water by properly designed edge detail is necessary to prevent ingress of water with consequent local softening of the subgrade.

The *Highway construction details*[42] provide edge-of-pavement details for motorways and trunk roads. Examples are shown in Figures 23.13 to 23.16.

A Concrete Society working party reported in 1974[136] on a wide range of edge details for all types of roads. The report covers function, design considerations, design practice, construction, maintenance and relative costs.

23.10 Lighting, signing, communications and safety

23.10.1 Lighting

Lighting can be expected to produce significant savings as a result of a reduction in night-time accidents, and upgrading old installations can also be expected to reduce accidents. The provision of street lighting in urban areas can deter criminals and there may also be an amenity value. There may be some environmental intrusion in rural areas, depending on the location.

The need for lighting on trunk roads and motorways should be appraised in accordance with TA 49/86.[137] A cost–benefit calculation and assessments of environmental and road-safety factors will generally be required and a framework approach can be adopted. Street lighting may be required for safety reasons alone at such locations as follows.

(1) In urban areas subject to a speed restriction of 30 mph.
(2) In semi-urban areas subject to a speed restriction of 40 mph.
(3) Where the road layout is substandard.
(4) Where junctions occur at frequent intervals.
(5) At roundabouts.
(6) At grade-separated junctions.
(7) On fog-prone sections of road.

23.10.1.1 Design principles

The design of all aspects of lighting is covered in BS 5489.[138] The aim of any road-lighting system is to illuminate all areas and aspects of the road and traffic which are of importance to all users, including pedestrians, in a manner which is aesthetically pleasing, especially in areas adjacent to areas of historical or visual interest. The main steps of the design process may be summarized as follows.

(1) The road category and preferred lighting arrangement (i.e. one side, staggered, etc.) should be decided.
(2) The data required for the calculation using BS 5489 and manufacturers' data should be compiled.
(3) The design spacing should be calculated.
(4) The lantern positions on the road plan with respect to fixed features such as junctions, pedestrian crossings, etc. should be plotted.
(5) The remaining lantern positions on uninterrupted stretches of road based on the design spacing but 'grading' into the fixed-feature locations should be plotted.
(6) The overall layout should be checked for misleading night-time arrays, general daylight appearance and 'line-of-sight' obstructions.
(7) The column locations should be checked on site for obstructions and visual acceptability. On the basis of the site survey, minor adjustments should be made as necessary.

23.10.1.2 Practical considerations

In addition to the design procedures described above, the following practical requirements relating to the installation of the lighting columns and associated equipment should be borne in mind.

Column foundations. The column foundations, whether planted or bolted-base, must always include duct routes for the incoming and outgoing cabling which should take account of the allowable bending radii of the cables.

The column foundations should also be designed to allow the column door to face away from oncoming traffic whenever practical.

Feeder pillars. Feeder pillars should always be located so as to allow safe access for maintenance. The feeder pillar foundations should also include ducts for incoming and outgoing cables, as described for columns, above. Spare ducts for future cables should be included wherever practical.

Cable routes. Cables should be buried 600 mm below ground level and should be laid on 75 mm of sieved sand and covered with a further layer of sieved sand to a depth of 75 mm. Cables should be run in soft ground wherever possible and the route of each cable must be marked with purpose-made cable markers indicating the cable voltage, depth of burial, joint positions and changes of direction.

Ducts. Where it is not possible to provide cable routes in soft ground, ducts should be provided, of minimum size 75 mm diameter. Typical locations where ducts are required are road crossings, cable routes within concrete structures, cable routes across vehicular accesses, etc.

Duct routes should take account of the permissible bending radii of cables and should include cable drawpits at all changes of direction and at intervals of 35 to 50 m, depending on the size of the cables involved.

Cast-in conduits. It is necessary in some instances (e.g. where lighting fittings are fixed on the underside of overbridges) to cast-in conduit routes from a location adjacent to the underground cabling to the light fitting(s) via a suitable chamber for location of the fused cutout unit normally located in the column base.

23.10.1.3 Maintenance and operation

The maintenance and operation of lighting systems is discussed in the British Standards.[138] Developmental Standard TD 23/86[139] deals with these aspects on trunk roads.

23.10.2 Traffic signs and road markings

A congress held in Vienna on 8 November 1968 resulted in the Convention on Road Signs and Signals. This included recommendations for signs and signals and set out standards for them. Further meetings were held at Geneva on 1 May 1971 and on 1 March 1973. These resulted in the European Agreement supplementing the Convention on Road Signs and Signals and the Protocol on Road Markings respectively.

These provisions have been accepted and agreed by the European Conference of Ministers of Transport and the *European rules concerning road traffic, signs and signals*[140] co-ordinates the provisions of the above.

The UK provisions are laid down in *Traffic signs regulations and general directions 1981*[53] and subsequent amendments and guidance on use, size, siting and illumination of signs and details of road markings are given in the *Traffic signs manual*[40] and *Circular roads* No. 7/75.[141] Further information is contained in Department of Transport circulars, standards, advice notes and a comprehensive bibliography,[142] which includes publications relating to working drawings for traffic sign design and post sizes, is produced also by the Department.

Permanent traffic signs, excluding signals, can be divided into:

(1) Warning signs, to advise of hazards
(2) Regulatory signs, to advise drivers of legal restrictions.
(3) Informatory signs giving information about directions, routes, places and facilities.

Road markings give information to help the drivers to select lanes and not as a guide base day and night. They may give a warning, a requirement, or a notice of restriction of prohibition.

For night-time visibility, signs may be illuminated externally or may contain reflective material, although in urban areas the use of one or other requires special consideration. The use of reflective material in road markings is a valuable aid to visibility as is the use of reflective road studs.

Variable message signs are also increasingly used as part of traffic control at locations where a message is not required to be displayed permanently or where alternative messages are required at different times. Examples of their use are as part of lane control, speed control, detection and warning to overheight vehicles and warning of adverse weather conditions. They can either be of the mechanical type (e.g. roller blind, prism or flap) or wholly electric where either an internal light source reveals the message or of the matrix type where an array of light sources create a range of messages, through the use of fibre optics.

Details of road signs and markings and a comparison of their use in many countries of the world were contained in the report of the Permanent International Association of Road Congresses[143] to the Geneva congress held in 1973.

23.10.3 Communications

23.10.3.1 Motorway standard equipment

All motorways in the UK are equipped with a standard Department of Transport specified communications system which includes electric cables, roadside equipment cabinets, signals, mains power supplies and electronic equipment.

Practice and presentation are consistent throughout the country. The system provides emergency telephones from the motorist's viewpoint, giving direct connection to the police in the local motorway control room, together with signals which indicate either a general need for caution or for specific actions, such as speed restrictions or changes of lane.

All motorways are each equipped with a multipair communications cable along one side only, taking a route as far from the carriageway as possible and which forms the longitudinal cable, which when connected to others at motorway-to-motorway junctions, provides a national cable network used by the motorway computer centres and control offices to communicate with each other and with the equipment to be controlled, such as signals, emergency telephones and certain lighting systems. These cables, in their various sizes, cater for all normal requirements in a motorway communications system.

23.10.3.2 Motorway emergency telephones

These are provided at approximately 1.5 km intervals along continuous motorway, normally in pairs, with one for each carriageway. This ensures that a motorist can reach a telephone by walking along the carriageway without attempting to cross it.

Extra pairs of telephones at motorway junctions are sited on the route but within the junction to accommodate motorists isolated by the sliproads. In the case of motorway-to-motorway interchanges, the link roads and main routes are provided with as many telephones – but not necessarily in pairs – as are required to meet the criteria stated above, which preclude the crossing of a carriageway or a spacing greater than approximately 1.5 km.

The telephone instrument is mounted in a standard housing, which has various reflective labels both on the outside facing the oncoming traffic and on the inside of the housing door. These labels show the reference number and letter of the telephone, which the motorist must give to the local motorway control office when reporting his situation. Various standard logos are also part of the telephone housing labels.

Telephone systems are normally, initially, a stand-alone system installed ready for motorway opening and comprising separate groups each of up to sixteen telephones and controlled by a telephone bridging unit near the roadside which sends the group's calls over a rented line to the motorway control office. The phase II systems which follow later operate the telephone under central computer control, switching them locally on to loaded circuits in the longitudinal communications cable and hence to the motorway control office.

23.10.3.3 Motorway signals

The most basic signal system, called Motorwarn, is that installed temporarily to cover from opening until the permanent signals are operational. Each signal is separate, consisting of a post set in concrete carrying a pair of amber lanterns, a radio receiver and a flasher mechanism which causes the lanterns to light alternately with a period of about 1 s. Power is supplied from a car battery on the ground below and the signal is switched on or off by radio transmission from an approaching police patrol vehicle.

Permanent motorway signals take two forms, depending on whether it is necessary to apply speed restrictions equally to all lanes of a two-lane or three-lane carriageway, or to apply different restrictions to each of any number of lanes. The former system is known as 'carriageway signalling' and employs post-mounted matrix-type indicators sited in the central reservation, while the latter is known as 'lane signalling' and mounts similar signal indicators on a gantry, with one over each lane. Access to a motorway at a junction with an all-purpose road, may be controlled by post-mounted matrix signals having additional red flashing lanterns.

All permanent signals discussed above have locally mounted units, either within the mounting post or on the gantry, known as 'distributors' and 'controllers' which are integral parts of signals. Mains supply is from Electricity Board interface cabinets usually set in the boundary fence and feeding signal installations over standard pattern mains-supply cable.

Control of signals and transmission over the motorway longitudinal cable network until the relevant responder is activated is from motorway control offices via the central computer. The responder is mounted in a cabinet near the roadside, local to the signals under its control, and acts on instructions bearing its electronic address. It also signals to the computer when any of its emergency telephones are off-hook and switches them to the motorway control office when instructed.

23.10.3.4 Current developments

All the above information refers to the National Motorway Communication System 1 (NMCS 1), but a new system, NMCS 2, is being brought into service using distributed computers associated with motorway control offices. The cabling to signals and other controlled devices is slightly different but uses similar cable with more cores in the longitudinal cable.

Mono-mode fibre-optic cable is coming into service as a standard cable for closed circuit television, but could easily be used to carry long-distance communications in the future. New services such as the automatic traffic surveillance system, fog detectors and ice detectors are coming into service with NMCS 2 and can be easily connected into the existing longitudinal cable at the nearest cabinet.

23.10.3.5 Traffic signals

The use of traffic signals is dealt with under sections 23.6 and 23.12.

23.10.4 Fencing

23.10.4.1 Boundary fencing

Motorway boundary fences are owned and maintained by the Department of Transport. It is normal, to simplify supply and maintenance, to use one type of wooden fence or one type of strained wire fence with droppers (Figures 23.35 and 23.36). The latter is less obtrusive than the former and would be used, for example, at the top of cutting slopes to reduce the effect on the skyline.

Other permanent fences generally become the responsibility of, and are chosen by, the landowner after erection by the Department of Transport. They are selected, if possible, from the accommodation fences included in the *Highway construction details*[42] or BS 1722.[144] Where appropriate, deer fencing is provided as shown in Figure 23.37.

Timber for use in the works and preservative treatment for timber fences should comply with BS 5589[145] subject to amendments detailed in the *Specification for highway works*.[72]

23.10.4.2 Noise barriers

Noise barriers can be made of glass reinforced polyester, glass-fibre cement, plastic-coated steel, aluminium, etc. as well as of conventional materials such as timber, brick and concrete. The material should be chosen to suit the surroundings. Grassed or planted earth mounds are aesthetically the most acceptable, particularly in rural areas. Criteria for design should be based on technical memorandum H 14/76 and Amendment No. 1.[146]

The main principles to be borne in mind are:

(1) The barrier should be sited close to either the noise source or the position to be protected for optimum effect. It should be long enough to obscure completely the noise source from view at the observation point.
(2) Normal range of barrier height is between 1 and 3 m. Less than 1 m is ineffective and more than 3 m usually unacceptably intrusive. Greater heights may be achieved by erecting a barrier on top of an earth mound.
(3) Comparatively light material is usually sufficient because

the limiting consideration is that the noise passing through the barrier should be less than that diffracted over or around it. The minimum mass required is given by the expression:

$$M = 3 \times \text{antilog} \frac{(A-10)}{14} \text{ kg/m}^2$$

where A (taken as positive) is the potential attenuation in dB(A) calculated by the path difference; and M is the mass in kg/m^2.

23.10.4.3 Safety fences and barriers

A safety fence absorbs some of the energy caused by a vehicle striking it and redirects the vehicle. A safety barrier provides containment and vehicle redirection without itself being deflected or deformed.

The main types of safety fences and barriers used in the UK are: (1) tensioned beam safety fence (corrugated or rectangular hollow section);[42] (2) untensioned beam safety fence (corrugated or open box);[42, 147] and (3) British concrete barrier (concrete profile barrier).[147]

Use of untensioned corrugated beam fences and concrete barriers is restricted to low speed roads (85 km/h or less) and in the case of British concrete barriers is limited to urban areas.

Safety fences are provided on the central reservations of all new motorways. They may also be used on dual-carriageway trunk roads and other all-purpose roads where there are high traffic flows or obstructions such as bridge piers and sign gantries. They are also used at the back of verges on embankments 6 m or more high, and on other embankments where there is a road, railway, water or other hazard below, the outsides of curves of less than 850 m radius on embankments over 3 m high, and at obstructions such as bridges, large signs and retaining walls. The requirements for safety fences on new and existing trunk roads are detailed in TD 19/85.[148]

23.10.5 User facilities

23.10.5.1 Motorway service areas

Department of Transport policy was to provide service areas at intervals of about 40 km, with potential infill sites midway between. The land, connections with the motorway, access roads, landscaping, parking areas, lighting and basic services are provided by the Department. The operator, who leases the site from the Department, pays for the buildings and equipment

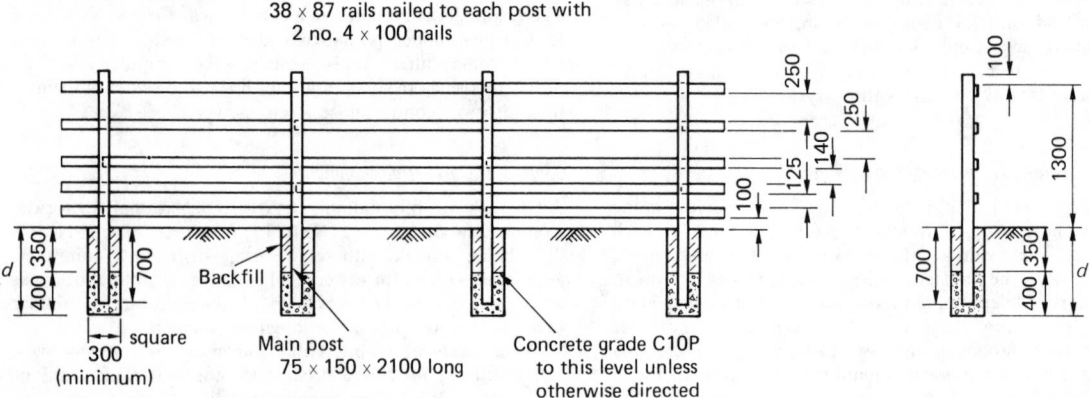

38 × 87 rails nailed to each post with 2 no. 4 × 100 nails

Backfill

Main post
75 × 150 × 2100 long

Concrete grade C10P
to this level unless
otherwise directed

300 square
(minimum)

All dimensions in millimetres

Figure 23.35 Wooden-post and five-rail fence. (After Department of Transport (1987) *Highway construction details*. HMSO, London)

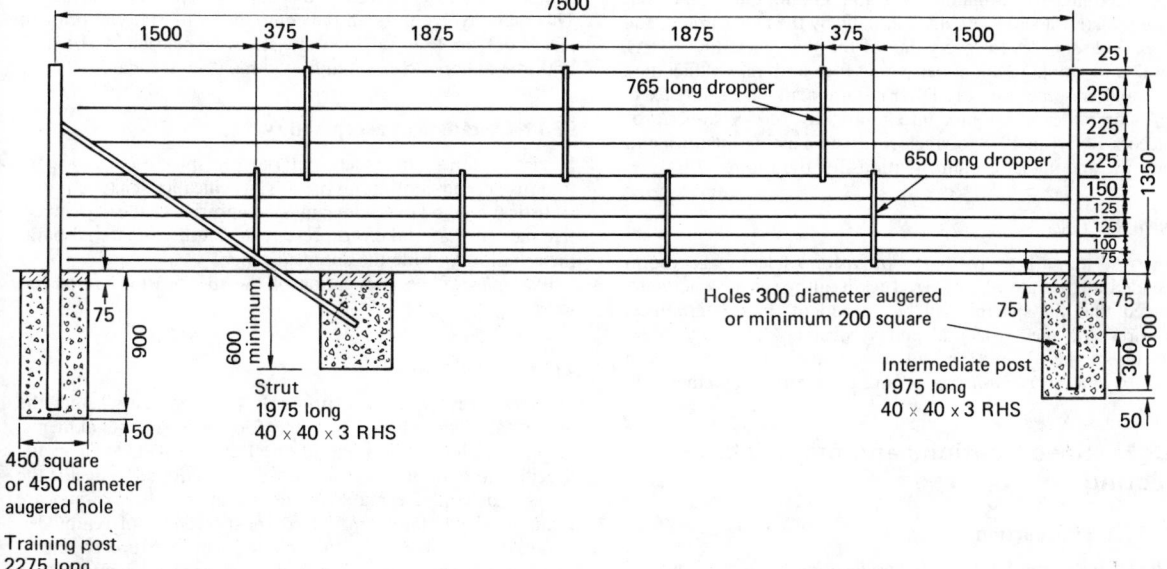

Figure 23.36 High-tensile strained-wire fence with droppers.
(After Department of Transport (1987) *Highway construction details*. HMSO, London)

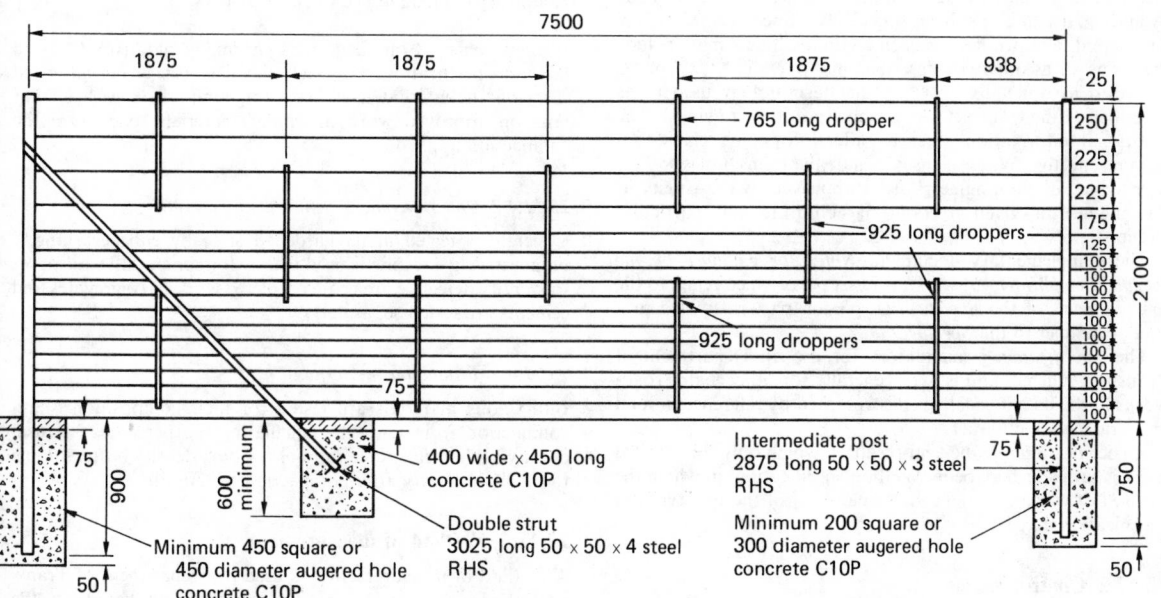

Figure 23.37 High-tensile strained-wire deer fence. (After
Department of Transport (1987) *Highway construction details*.
HMSO, London)

and is required to maintain the site and buildings in good repair, including the facilities provided initially by the Department, and to provide services for every day of the year, including toilets, food and drink retail outlets, and fuel and repair facilities, although in some cases this latter requirement has been relaxed.

A committee of inquiry into motorway service areas produced a report in 1978[149] which reviewed existing facilities and their use and made recommendations for the future.

23.10.5.2 Laybys

On rural all-purpose roads the provision of laybys is recommended at the rate of two per 1.6 km, although it is acknowledged that the spacing will be dependent on topographical features and the horizontal and vertical layout of the road. Details of their layout and advice on siting are contained in *Layout of roads in rural areas*[150] and the metric supplement.[151]

23.11 Specifications and materials testing

23.11.1 Introduction

The Department of Transport *Specification for highway works*[72] covers all aspects of works on highways (except signals), bridges and other associated structures. There are seven parts, and the first six are divided into twenty-six series, each covering a separate aspect of the work and an introduction. The final part covers accepted quality assurance schemes, certification-marked quality assurance schemes, approved lists of proprietary systems and materials, publications referred to in the *Specification*[72] and the manner in which variations to the specification can be made by adding, deleting or substituting clauses. Specific criteria may be inserted in appendices to each section. Clauses may include the terms 'or as otherwise described in Appendix *****' or 'as described in Appendix *****'. Other terminology used is 'as described in the Contract' or 'as shown on the drawings'. The additional information must be added for such clauses to become effective. Where clauses include the terms 'unless otherwise agreed by the Engineer' or 'as approved by the Engineer' the engineer is given freedom of action to suit particular circumstances.

The complementary document, *Notes for guidance on the specification for highway work*[73] is in six parts which relate to the first six parts of the *Specification*.[72] Each part is divided in a similar manner to the *Specification*.[72]

The *Specification* is mandatory for use on Department of Transport schemes and is used generally for other major roadworks in the UK; it usually is incorporated by reference in road and bridgework contracts.

A specification may be 'end-result' in which only the finished work is tested or may be by 'method specification' in which the method of working is defined. Some blend of the two types of specification may be adopted.

23.11.2 Control testing

Control testing may be defined as testing with the aim of monitoring compliance with the specification requirements for the material in question. It is principally carried out by a contractor or supplier and may be predictive in nature. The test may be the specified test itself or some other test which correlates sufficiently well with the specified test.

23.11.3 Acceptance testing

Acceptance testing is the testing carried out to ascertain whether compliance has been achieved within the specified test limits. This may involve tests on the components of material prior to placement, or tests on the material after placement checking both the material content and workmanship standard.

23.11.4 Statistical acceptability

In circumstances in which materials are produced as a continuous or semi-continuous process, specifications may require a statistical method of testing aimed at ensuring that early corrective measures are taken to prevent the required material parameter falling outside the specified range.

Examples of the application of the specification are as follows.

23.11.4.1 Earthworks

Cuttings Tests are taken primarily to determine whether the excavated materials are suitable for forming embankments. Acceptable limits for some of the tests are laid down in the specification but in others the designer is required to decide the limits. For cohesive materials, tests that may be required are grading, plastic limit, undrained shear strength of remoulded material and either moisture content or moisture condition value. A normal permitted moisture content for acceptability is not less than optimum moisture content (Test 12 of BS 1377)[74] and not greater than 1.2 times the plastic limit. The test for granular materials may be grading, uniformity coefficient, moisture content or moisture condition value, and for chalk, the saturation moisture content. A normal permitted moisture content for granular materials for acceptability is 1% from optimum to 2% below the optimum as determined by the laboratory compaction test (Test 12 of BS 1377).[74]

Embankments. Compaction is normally controlled by a method specification, although the engineer is permitted to carry-out field dry-density tests for comparison with similar tests on approved work in similar materials to confirm the compaction applied.

23.11.4.2 Concrete and cement-bound materials

Concrete is tested in its hardened state by cube crushing.[152] Nondestructive testing is used mainly for comparative purposes to identify defective areas of completed work.[152] Aggregates and cements are tested separately.[153-155]

23.11.4.3 Bituminous materials

Bituminous materials are tested for their composition[156] and compaction from cores by calculating the percentage refusal density as the percentage ratio of the bulk density of the sample to the final density after compaction to refusal.

23.11.5 Method of measurement

A method of measurement[157] is used on Department of Transport, and other, major roadworks contracts based on the Institution of Civil Engineers *Conditions of contract*.[158] It is also based on the *Specification for highway works*[72] and the *Highway construction details*.[42] The method allows bills of quantities to be prepared in a uniform manner for the benefit of engineers and others closely associated with highway works. The *Library of standard item descriptions*[159] is also provided.

Variations in the conditions of contract, the specification or highway construction details will require amendments to the method of measurement to suit the variation. Details of the amendments so made are required to be stated specifically.

23.12 Roads and traffic in urban areas

23.12.1 Introduction

Roads and traffic in urban areas are parts of the complex arrangement and use of the urban fabric. The growth of traffic has caused congestion, particularly in the peak hours and has had an impact on the environment and the people who live and work in urban areas with the noise and air pollution associated with traffic and increased danger.

Much has been done to combat these problems. Urban motorways and other major roads and improvements have been provided, but the general tendency is now to make better use of the existing fabric by providing traffic-management measures by a variety of means to reduce congestion and delays: (1) traffic-control measures have been introduced; (2) priority has been provided to public transport; (3) through traffic has been removed from residential areas and shopping centres; (4) pedestrianized areas and special facilities have been provided for cyclists and pedestrians as well as the disabled; and (5) safety has been improved. New roads can still contribute to the general good of the area and enable these improvements to take place more easily and, whilst new roads can have local adverse effects, they can still provide substantial overall benefits if properly sited and designed.

Roads and traffic in urban areas[12] provides a comprehensive review, and is intended as a guide to good practice in dealing with urban area problems.

23.12.2 Hierarchy of roads

It is most useful and desirable to establish a hierarchy of roads within an urban area, whereby a policy can be developed for the use of each road. There is an interaction between the highway network and land use, and the latter can be controlled in relation to its position in the road network.

Major through routes, for example, are primarily for traffic use and the number of accesses to it should be strictly controlled. Residential roads, at the other end of the spectrum, should have all through traffic excluded. The various levels of distributor roads can be decided and it may be possible to make the physical characteristics of the road more appropriate to its place in the hierarchy. It may be possible to channel more traffic on to main traffic routes thus allowing heavily used shopping areas to be pedestrianized.

23.12.3 Traffic management

Traffic management is aimed at improving an existing road network to meet set objectives, without having to resort to substantial new construction. Such objectives may include some, or all, of the following: (1) a reduction in road accidents; (2) environmental improvement; (3) improved access for people and goods; and (4) improved traffic flows on major routes.

These aims can be achieved by employing various traffic management measures involving combinations of:

(1) Improved road capacity.
(2) Giving priority to certain classes of vehicles, e.g. buses, emergency vehicles and cyclists.
(3) Restraining the demand for road space by introducing some form of restrictive measure.
(4) Improved facilities for pedestrians, cyclists and the disabled.

Traffic management measures may be limited to a small, localized area or may encompass an area-wide network. In the latter case, it should be recognized that measures regarded as providing solutions for one area can create problems elsewhere. It is therefore vital that all proposals be properly evaluated before implementation, and that adequate data collection and forecasting techniques be employed to examine effects over an appropriately wide area.

The Association of London Borough Engineers in association with the Department of Transport and the Metropolitan Police have produced a code of practice for traffic management in London.[160] The guidelines and advice contained within the code may have useful applications elsewhere.

23.12.4 Methods of assessment

The potential traffic throughput and effectiveness of alternative traffic management measures can be assessed by traffic simulation modelling techniques, such as CONTRAM,[18] TRAFFICQ,[19] SATURN,[17] and TRANSYT[58] referred to earlier in section 23.4. Such models typically require input data which includes link lengths and widths, junction type, traffic flows and estimates of the origins and destinations of the traffic pattern surrounding the study area.

Other computer models such as ARCADY,[49] PICADY,[49] MIDAS[50] and OSCADY[54] can be employed for the detailed assessment of individual junction designs within a scheme as outlined in section 23.6.

New roads may be proposed or new roads in conjunction with traffic management measures might be suggested as a package, and these proposals can be compared with other possible traffic-management measures.

The assessments should be made with the aid of a framework described earlier in this chapter so that all aspects may be considered. As with inter-urban roads, public consultation and participation is becoming more widespread and the results can be included; the impact on the environment and people is a major issue and both benefits and disbenefits will almost invariably arise in an urban area. All other appropriate factors should also be included. Any specific improvements in traffic and financial benefits in cost–benefit terms arising therefrom are, as in inter-urban roads, only two aspects of the comprehensive summary which must be produced to allow decision-making. The SACTRA report[9] and the government response[10] has been mentioned earlier.

23.12.5 Improvements in road capacity

The urban road network consists of a complex interaction of road links and the junctions which join them. Measures aimed at improving road capacity can therefore be assessed in terms of link or junction improvements, either separately or in tandem.

23.12.5.1 Road links

The treatment appropriate to road links between junctions should reflect the extent to which they serve the functions of through movement, local distribution and access to frontage premises.

Link improvements often may be achieved by straightforward measures such as road widening, or by the imposition of waiting and loading restraints[161] or turning restrictions. Other forms of link improvement commonly adopted include tidal flow arrangements and the creation of one-way streets.

23.12.5.2 Junctions

The traffic capacity of a road system is commonly constrained by its junctions. An important consideration when junction improvements are being investigated may be whether the existing junction type should be maintained, perhaps with modifications, or whether a different type of junction control would be more appropriate. Factors influencing the decision will include: (1) the existing and predicted traffic flow and composition; (2) the classification and function of the roads forming the junction; (3) available highway space; (4) proximity of adjacent junctions; (5) road safety record; and (6) other needs, such as public transport, pedestrians and cyclists. The four main types of junction in common use are:

(1) Major/minor.
(2) Signalled.
(3) Roundabouts and gyratory systems.
(4) Grade separation.

Major/minor junctions. Major/minor junctions are the most common type of junction in urban areas, their great advantage being that major road traffic is generally not delayed (except traffic waiting to turn right into the minor road). Improvements to heavily trafficked junctions may include the channelization of traffic to separate and clarify conflicting movements. This may be achieved by improved road markings and signing or by physical means, such as traffic islands. Departmental Advice Notes TA 23/81[42] and TA 20/84[48] give the main principles for layout and recommended design standards as stated previously.

Major/minor junctions tend to become unsuitable when minor road traffic cannot find suitable gaps in the main road traffic during times of peak traffic flow. These conditions can result in excessive queuing and delays in the minor road, often leading to accidents or diversionary routing. In these cases, an alternative form of junction control will be necessary and traffic signals often prove beneficial.

Signalled junctions. Traffic signal control is an important feature of junction control in urban areas, providing relatively efficient control within the confines of limited road space. Traffic signals initially were provided to reduce the police manpower required to control traffic. The use of traffic signals to control traffic movement can now bring about major reductions in congestion, improve road safety and enable specific strategies to be introduced which regulate the use of the road network. Such strategies might be: (1) to reinforce a designated route hierarchy; (2) to give priority to public service vehicles; (3) to provide crossing facilities for pedestrians and cyclists; and (4) to maximize traffic flow.

Section 23.6 includes details of computer programs used in connection with traffic signals and Department of Transport advice notes on this subject.

New signal controllers using microprocessors are now available for use at isolated junctions. Information on the traffic approaching from each leg of the junction can be processed at the signal controller and the signal timings adjusted to provide a more efficient use of the junction.

The numbers of traffic signal systems in urban areas may be extremely high, leading to adverse interaction between neighbouring installations. Recent advances in computer technology have made it possible to co-ordinate the operation of adjacent traffic signal sites by the use of area-wide traffic control.[12] This in turn has enabled vehicle movements to be controlled over a section of road network, which may result in a reduction in vehicle journey times, the number of stops, fuel consumption and environmental pollution. This is the basis of the majority of present-day urban traffic control schemes. Many different programs are in use: fixed-time programs such as TRANSYT[16, 58] adjust signal timings based on historical data, but SCOOT[162] both receives data on traffic and adjusts signal timings on real time. Other potential benefits which an urban traffic-control scheme may provide include: (1) the implementation of diversion schemes and variable message systems; (2) creating priorities for buses and bus routes; (3) priority for emergency vehicles

responding to incidents; and (4) provision of special signal timing plans to favour key routes from fire or ambulance stations.

Roundabouts and gyratory systems. Roundabouts provide a useful form of junction control which is generally conducive to 'free-flow' conditions when they are operating within their capacity limitations.[49] The different types of roundabouts, their capacities and designs have been discussed earlier in this chapter. They do not require control equipment and are therefore not prone to equipment failure. When approach speeds are very high or some of the approach links are dual carriageways, roundabouts are often the safest way of regulating traffic. In addition, they provide good opportunities for vehicles to turn right and allow U-turn manoeuvres, which can be very beneficial if restrictions exist elsewhere. However, roundabout solutions are unsuitable where linking of traffic flows between adjacent junctions is advantageous or where it is beneficial to change traffic priorities at different times of the day. The provision of pedestrian and cyclist facilities may also prove difficult.

In some circumstances (e.g. such as a measure to reduce accidents or during peak periods) there may be advantages in signalling one or more approach paths to a roundabout.

Gyratory systems – whereby a series of one-way streets are linked to form a circulatory system – can overcome the limitations of several small junctions. However, they can create access problems to properties within the central island.

Grade separation. As grade-separated junction facilities are quite expensive and can be visually intrusive they will generally only be used at junctions of major importance, where land is available and there is little or no adverse environmental impact.

23.12.5.3 Priority management measures

Traffic management measures may be undertaken to provide priority for particular classes of traffic, e.g. public transport vehicles, emergency vehicles, cyclists and pedestrians. Priority measures are normally achieved by allocating special facilities such as pedestrian[62] and cycle crossings, bus/emergency vehicle actuated signals[163] and bus and cycle lanes. Alternatively, certain classes of vehicles may be exempted from general traffic restrictions, e.g. exemption from turning bans or contraflow operations.

Priority measures are often applied where there is a general deficiency in the road network in terms of demand related to capacity. Any such measures should seek to ensure that the overall effects provide a net gain to the community as a whole and, where possible, should attempt to minimize the adverse effects on nonpriority road users.

23.12.6 The provision of information

The efficiency of traffic management schemes can be enhanced greatly by the careful provision of traffic signing and roadmarking systems.

Well-designed, clear and comprehensive signing should provide drivers with information on route choice well in advance of the approach to any junction and also warn of any turning prohibitions. Such systems are likely to be enhanced greatly by the emergence, in the foreseeable future, of electronic route guidance systems[163] such as AUTOGUIDE, a trial of which has been proposed in London, and ALI-SCOUT which has been installed in an area of West Berlin on a trial basis. A trial on route guidance, the CAC system,[143] was undertaken in Tokyo some years ago. In addition, the use of signing can provide information on items such as car-parking availability, diversion routes, guidance to tourist attractions, and direction signing to cyclists and pedestrians. The use of signs and markings has been dealt with earlier in this chapter.

23.12.7 Restrictive measures

Traffic restraint (or demand management) measures may be employed to control the level of traffic in an area or on a particular route. For example, in one area restraint may be aimed at through-traffic and in another area at commuter movements.

Restraint measures are designed to encourage those making trips to respond to imposed conditions in specific ways. Various responses may include: (1) a change in the mode of travel; (2) a change in the time of travel; (3) using a different route; (4) travelling to an alternative destination; or (5) not making the trip at all.

Very few restraint schemes have been implemented in the UK other than those involving some form of parking control[164] mainly effected through on-street waiting restrictions[165] or lorry bans.[166] However, more radical measures have been attempted in other countries such as Singapore, where there are restrictions on entering the central business district in the morning peak hours, and in Hong Kong,[163] where a trial on the feasibility of road pricing has been undertaken.

23.12.7.1 Facilities for pedestrians and cyclists

Pedestrians and cyclists can be very vulnerable to severe or fatal injuries on the roads and it is essential that traffic management measures take their safety into account. Typical facilities may include: (1) separate provision of footways or cycle routes; (2) guardrails; (3) pedestrian crossings; (4) subways or footbridges; and (5) pedestrianization schemes. In addition, consideration needs to be given to the handicapped.[167] Where space is limited, facilities for pedestrians can sometimes be shared with other users.

23.12.7.2 Residential areas

Previous reference has been made to residential areas. The residential precincts in The Netherlands (*Woonerf*) have been the subject of much interest.[163] The demonstration projects in The Netherlands[163] adopted three strategies for residential areas: (1) the exclusion of through-traffic by simple means; (2) the exclusion of through-traffic by simple means compiled with small-scale measures for reducing speed; and (3) the conversion to a residential precinct. It was felt that, for most areas, preference should be given to a redesign to the second strategy which gives relatively the best results for the costs involved.

23.12.7.3 Parking

Parking should be considered within the overall policy relating to the infrastructure and use of the urban area. The demand for parking, particularly in central areas, often exceeds the available space. The generous provision of parking space is likely to encourage the use of the private car and the limitation on the amount of parking to be provided may act as some deterrent to users of private vehicles.

Parking may be provided on-street, or off-street in ground-level or multistorey car parks. On-street parking may be controlled by indicated time limits free of charge or some payment may be required such as at parking meters. Parking charges, both on- and off-street, usually vary and depend on the location in relation to the central area.

Residents' parking schemes are often provided where parking spaces in an area would be otherwise occupied by the cars of visitors or commuters to the area. Charges may be levied or permits to park may be provided without charge.

Enforcement is a major part of parking policy. Fines are usually levied when the permitted parking time is exceeded. Vehicles may be removed in certain circumstances, and wheel clamps are now being introduced in certain areas which, in addition to incurring financial penalties, cause drivers delay and inconvenience before the cars are freed and can be driven away.

23.13 Highway maintenance

23.13.1 Introduction

As outlined in section 23.2, local highway authorities normally act as agents for the secretaries of state in maintaining trunk roads and motorways, in addition to maintaining their own highways.

The object of maintaining a highway is to preserve the fabric in such a condition that it provides safe passage for all traffic throughout its life. The *Report of the committee on highway maintenance*[168] proposed that maintenance operations be divided into: (1) structure; (2) aids to movement and safety; and (3) amenity.

It is not easy to ascribe priorities within those groups without assessing the type of traffic, its contribution to the community and the advantages to be gained by incurring the expenditure, or even the disadvantages, of not doing so.

The report[168] proposed national maintenance standards for the UK; good maintenance extends the life of a road pavement and adds to the convenience and safety of the public using the road. Good maintenance means:

(1) Day-to-day maintenance to maintain a road in proper condition for the traffic using it. This includes patching, surface dressing, gully emptying, repairs to drainage, kerb and footway maintenance, maintenance of bridges and other structures, embankments and verges, repair and maintenance of traffic lights, carriageway markings, street lighting and street furniture, snow and ice clearance.
(2) Structural work required to extend the life of the road or to enable a road to carry an increased volume or weight of traffic.

Tables 23.17 to 23.19 set out the needs in relation to structural works, aids to movement and amenity. Regular inspections and a system to determine priorities for short- and long-term attention (i.e. a maintenance rating system) will be required.

Most highway authorities carry out regular detailed inspections of highways at specified frequencies and have recording systems for defects which require maintenance action within a specified period of time.

The Department of Transport has instituted a code of practice[169] setting maintenance standards with a routine maintenance management system applying to motorways and trunk roads which sets out requirements for a computerized inspection system which is linked to an inventory of the highways and produces routine maintenance work schedules.

The Department also specifies the statement of service[170] and detailed standards to apply to the winter maintenance activities for motorways and trunk roads.

The local authorities associations in the UK have produced a code of good practice for highway maintenance[171] with model specifications of maintenance activities as a guide to the preparation of highway maintenance policies and standards.

Each highway authority produces particular maintenance policies to apply within the area, including those policies and standards relating to winter maintenance.

Table 23.17 Structural elements of highway maintenance

Element ( ) Item	Correction	Remark
Carriageway		

(1) Condition
    Shape
    Irregularity
	Reconstruction, resurfacing surface
	dressing
		Depending on degree of fault

(2) Strength	Reconstruction, resurfacing	Depending on degree of fault
Surface		
(1) Lack of skid resistance only	Surface dressing	This is corrected within other corrections operations as above
(2) Patching	Surface dressing	To maintain a watertight surface
(3) Defect through utility services		These should be masked during surface operations
	(Road markings are entered in Table 23.17)	
Drainage		
(1) Gully emptying		Dependent on local conditions – referred to also under sweeping and cleaning
(2) General performance	Width of running water in channels	Standards relate to the flow in channels reducing the effective width of carriageways
(3) Concentrated flow across carriageway	Install extra channels or gulleys	Not to be confused with normal flow across cambered roads
	(See also 'Structures' below)	
Footways (Urban)		
Irregularities related to safe use	Relay paving slabs	(1) Suggestions for height of projections are given and also rates of inspection in different areas
	Resurface flexible material	(2) Legal implications must be recognized
Footways (Rural)		
Irregularities and surface water related to safe use	Minimum maintenance in rural areas	Rate of inspection given
Kerbing		
(1) As drainage feature	Install to delineate drainage channel and support edge of carriageway	
(2) Edge of footway	Defective kerbing	A normal height of kerb in urban and rural conditions is given as safety feature for pedestrians
	Kerbs sunk to carriageway level or lower	Serves to support edge of carriageway
Structures		
(1) Bridges, culverts, walls	Necessity for programming of painting of steelwork	Rate of inspection to be assessed on local conditions

Table 23.17 – *continued*

Element ( ) Item	Correction	Remark
Safety of road user is paramount	Deterioration of concrete or other fabric Underwater inspection where necessary	
(2) Embankments and cuttings; to include ditches where appropriate	Regular inspection for incipient slips or failures Advice to adjoining landowners about structural or drainage defects	Legal implications are important

Table 23.18 Aids to movement and safety

Element ( ) Item	Correction	Remark
Road Markings (1) Advisory markings	Remake	General recognition should be possible
(2) Mandatory and statutory markings	Remake	These markings must lie within legal limits
(3) Reflecting studs	Loose or ineffective studs should be replaced. A wholesale change might be made at the end of effective life	An inspection before onset of winter conditions is advisable and equally an inspection should be made after winter maintenance operations are complete
Traffic signs and bollards (1) Illuminated signs and bollards	Regular inspection for (i) light failure (ii) drainage (iii) cleaning (iv) supports and frames	Particular attention to be paid to mandatory signs, e.g. 'Stop' signs, etc.
(2) Nonilluminated signs	Regular inspection for (i) drainage (ii) cleaning (iii) supports	See above
(3) Traffic signals	Regular inspection for (i) light failure (ii) general maintenance and cleaning (iii) phasing (iv) alignment (v) mechanism (vi) painting	(1) All highway personnel should report faults wherever discovered (2) Contract maintenance and guarantees to be operated. On-call arrangements to be made
Pedestrian Crossings (1) Beacons (2) Road markings	As for traffic lights (i) Slippery surfaces should be corrected (ii) Obscure markings should be made good	Legal implications are important
Road Lighting (1) Lanterns	Regular inspection for (i) illumination (ii) cleaning Conditions to be reported during inspection	The safety of the road user – driver and pedestrian – is paramount in this connection

Table 23.18 – *continued*

Element ( ) Item	Correction	Remark
(2) Columns	Conditions to be reported during inspection	
Guard Rails and safety fences Pedestrians and vehicular	To be included in regular inspections. Where risk to public is involved speedy action is necessary. Inspection to cover: (i) condition (ii) painting (iii) cleaning	Note legal implications
Winter maintenance (1) Precautionary salting	On receipt of frost warning roads should be treated The rate of salt-spread should be 14 g/m² Salt in accordance with BS 3247: 1970, Part 1	Treatment should be applied within a limited period after a warning: (a) Rural main roads and motorways – 2 h (b) Other important roads and accesses to emergency services – 2 h (c) Urban main roads – 1 h Crews of salting equipment should be on stand-by duty. Neat salt should be used. This should be mixed with grit in special circumstances only
(2) Snow and ice clearance	Use of specialist equipment	Major routes as in (a) above should never become impassable to traffic. This is related to traffic flows which have the effect along with salt of keeping snow from accumulating. Snow ploughs with blades effectively remove slush from main routes; crews should be on standby duty. Roads as in (c) above should not be impassable for longer than 1 h. Public transport is a major factor in the clearance of snow. Roads in other priorities should generally be cleared in 4 to 6 h unless conditions are exceptional. Pedestrian ways may require special treatment in town centres. Footpath clearance should be confined to busy areas, steep hills, etc.
(3) Snow fencing		Local knowledge is necessary to establish the siting and timing of snow-fence erection. Care must be taken to clear this with landowners
(4) Salt storage		Care must be taken to site salt heaps to avoid damage to local crops, watercourses

Table 23.19 Amenity items

Element ( ) Item	Correction	Remark
Grass cutting		
(1) Prevention of obstruction of sight lines	Grass cutting	Standards vary for urban or rural situations
(2) Maintenance on certain roads of reasonable pedestrian access		Rural (a) Major roads – 1.8 m of the verge should be kept below 15 cm. Elsewhere one or two cuts should be employed to keep grass to 30 cm long
(3) Control of noxious weeds		(b) Minor roads – one cut per year will normally suffice
		(c) Spraying of grassed areas can be used to control noxious weeds. Consideration must be given to width of road, nature of traffic and not least, the culture of the grass growth
		Urban
		(a) Major roads – grass should be kept down to 8 cm
		(b) Minor roads – minimum maintenance consistent with the environment.
	Use of chemical sprays	These should be used with caution where access for cutting is not available and for growth control generally. Examples are: around sign supports; central reserves with safety barriers; urban walkways
Hedge Trimming		
Prevention of obstruction of visibility at bends and at traffic signs	Tractor-mounted equipment might be employed	This is not normally a function of the highway authority. They have power to require land owners to reduce hedge heights and to control trees. Legal implications should be noted carefully
Trees		
This repeats very much the information under hedge trimming. Trees may call for specialist advice which is often available in the Parks Department of a Local Authority		Legal implications and ownership are important factors
Sweeping and Cleaning		
(1) Objects and material shed by vehicles	Heavy and dangerous items must be removed by hand This is material which can cause broken windscreens and mechanical damage	The rates of inspection and activity on this account vary with the weight of traffic, between daily inspection on motorways and weekly visits on less busy roads.
		All highway staff should be aware of the necessity to remove dangerous items whenever they see them

Table 23.19 – *continued*

Element ( ) Item	Correction	Remark
(2) Vegetation and detritus	This is material which can block drainage systems which can obscure road markings and cause dirty windscreens	Legal implications are important Again rates of inspection and activity are dictated by traffic but where traffic is heaviest, cleaning and scavenging is difficult. In rural areas carriageway sweeping is not necessary more often than at 2-monthly intervals. In town centres daily attendance is required, reducing to weekly attention in residential areas. This should include footpaths. The rate of gully emptying is dependent on the build-up of such material. In dry weather it may be necessary to top up gulleys with water to permit drainage traps to operate. This can fit in with road-washing operations

The structural needs of the road including resistance to skidding, surface irregularity, changes in traffic and usage of the road and a 'maintenance' rating[168] for, say, each 500 m length of road, requires to be derived. Computer programs have been developed to handle maintenance ratings for an extensive road network.

23.13.2 Assessment of structural needs

Two methods of assessing structural needs are the CHART[172, 173] and MARCH[174] systems. Maintenance assessment surveys are required by the Department of Transport in the determination of structural maintenance needs of motorways and trunk roads. The CHART and deflection surveys assist in the identification of priorities for structural carriageway maintenance.

The CHART is a visual and computerized inspection method of recording specific defects which, with the aid of a computer model, evaluates the condition of the road and recommends types of treatment with priorities.

Deflection measurements of flexible pavements are primarily taken by deflectograph, which is now used extensively on motorway and trunk road surveys. It is supplemented as necessary by deflection beam measurements.[175-177]

The deflection beam is usually used on short lengths of road of 1 km or less. The deflectograph provides a more rapid method of measurement than the beam and is better-suited for routine surveys of long lengths of road.

Skid resistance is checked in the UK using the sideways force coefficient routine investigation machine (SCRIM) developed by the TRRL. This provides a printout of sideways force coefficient (SFC) at 10 m intervals along a road and can test about 1500 km per year.[178-180]

23.13.3 Aids to movement and safety

Items falling under this heading are the routine requirements for signing, lighting and similar items and the winter maintenance requirements which have to be met as they arise.

23.13.3.1 Amenity items

Items included under the general heading of the amenity functions of a road also have a safety element in that grass-cutting, tree-lopping, etc. ensure that sight-lines at bends and the visibility of traffic signs are maintained.

23.13.3.2 Maintenance organization

The organization on the ground ranges from the outdated individual or lengthman method to the more cost-effective mobile gang/team. The latter method utilizes mechanical equipment with specialist teams for surface dressing, patching of carriageways, road markings, lighting, signing, bridge maintenance, motorway maintenance, etc.

23.13.3.3 Winter maintenance

Setting-up a winter maintenance system involves determining the standards to be adopted and then examining the meteorological data for the district so as to establish the period during which the winter maintenance organization will be expected to function. The organization with its resources and plant will undertake de-icing, gritting, snow clearance and similar work.

23.13.3.4 Signing for highway maintenance

Signing for maintenance work on UK highways is covered in the *Traffic signs manual*.[40] Information in the *Traffic signs manual*[40] for major roads is amended and supplemented by Departmental Standard TD 14/83.[181] This standard includes the TRW series of drawings which detail roadwork-signing requirements on motorways and dual carriageways including contraflow traffic control.

For minor roads (unclassified roads which form about half of the total road mileage in the UK), Department of Transport Advice Note TA 47/85[182] deals with the control of traffic at roadworks on single-carriageway roads and is complemented by

Department of Transport Standard TD 21/85[183] concerning portable traffic signals at roadworks.

All signs must comply with the *Traffic signs regulations and general directions*.[53]

Staffordshire County Council, with the assistance of eleven other county councils, have produced a useful 54-page booklet entitled *Safety at roadworks*.[184]

23.13.3.5 Staff for maintenance

The type of staff required for maintenance includes those: (1) for inspection, estimating and programming work; (2) for superintending day-to-day operations; and (3) for carrying out the physical work, either by direct labour or contract.

The officer controlling highways in a local authority will encompass the whole range of operations: (1) provision of new highways and bridges and their physical effects on the country through drainage arrangements, earthworks, etc.; (2) the maintenance of the roads, and its management in terms of control of traffic; and (3) surveillance of operations of public utilities and of security generally. He will probably require a second-in-command over the whole range of interests if the total length of highways is in excess of about 3000 km. Figure 23.38 illustrates a typical organization.

The Report of the Committee on Highway Maintenance[168] indicated that about 800 km of road would form a useful area unit; this could be increased significantly if comprehensive support services were readily available.

The area surveyor will be responsible for all maintenance in the area and will have to deal with some of the design problems met from time to time. It will be necessary to interpret survey data provided as a basis for work programming and to deal with the financial implications.

The area superintendent will need to organize labour and undertake ground surveys for work-programming purposes. The number of foremen must depend on the nature of the area and of the men under their control, e.g. rural conditions vary from urban conditions. Many of the inspections will be carried out by the foremen or inspectors and sometimes by members of their teams. The 'general services' available for maintenance work could include, for example: (1) a signs store, traffic signals, etc.; (2) a plant depot; (3) a winter maintenance depot and salt stores; and (4) workshops. An 'emergency' team will comprise a pool of skilled labour which can be used for special operations.

23.13.3.6 Alternative staffing structure

The Local Government Planning and Land Act 1980 requires local authority direct labour organizations to compete with contractors in the private sector in tendering for highway works contracts of maintenance and construction. Some local authorities have in consequence set up separate contracting organizations similar to the public works contractor within the local authority. The structure of such a highways department is divided into client and contracting organizations. On the client side, area surveyors/engineers with technical and administrative support staff work 'on the ground' and are responsible to the maintenance officer at headquarters. Contract works managers with direct-labour resources and staff, responsible to the works officer at headquarters, operate in the field to the client's requirements.

The contractors in these cases also provide the highways emergency services, including winter maintenance for the authority. The Department of Transport provides facilities for their agents.

23.14 Low-cost roads in developing countries

23.14.1 Introduction

This section comprises only an outline guide to the basic principles of road design and construction in developing countries. The nature of roads in these countries varies widely

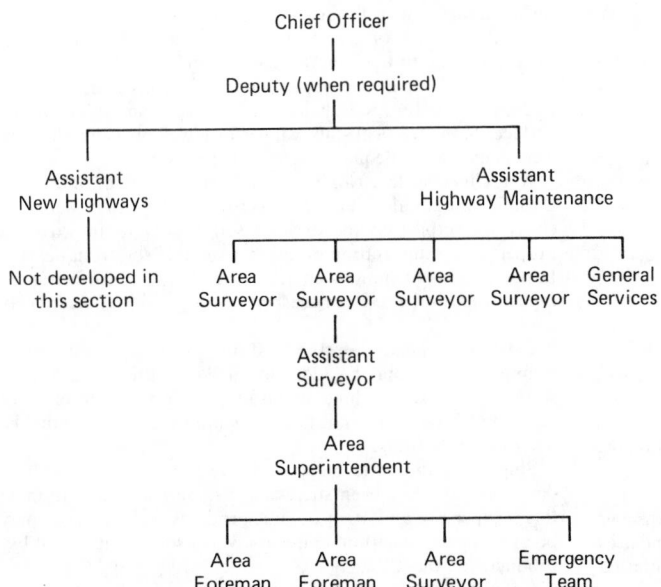

Figure 23.38 Typical operational structure for highway maintenance staff

depending upon the particular geographic, economic and political situation of the country concerned. Some countries follow British design and construction practices, some follow practices in the US and others those relating to yet other countries. Most countries will have developed road standards of their own, often based on a blend of one or more of the above but taking into account local conditions and requirements, availability of local materials and other local aspects, and it is essential that the designer should refer to the particular standards relating to the country in which he is operating. A publication produced for UNESCO[185] covers all aspects of design construction and maintenance of these roads.

Roads in developing countries have a number of common characteristics which influence the approach to design and generally require a philosophy different from that in the developed world. These characteristics include:

(1) Long lengths of road under consideration at one time.
(2) Low traffic volumes.
(3) Limited finance.
(4) Traffic composition (ox carts, bicycles, overloaded lorries, etc.).
(5) Tropical climate and geology (intense seasonal rainfall, lateritic soils, etc.).
(6) Poorly trained workforce and limited availability of construction materials.

The design and method of construction of a new road will vary significantly depending on the above characteristics and also on considerations of cost–benefit analysis, funding, and strategic and political factors. A solution should be selected (whatever the circumstances of a particular scheme) which is appropriate to the type and volume of traffic, geology, climate and availability of labour and machinery.

The construction of new roads in a developing country is a vital element in opening up the countryside and boosting industrial and agricultural development. Finance frequently is provided, at least in part, by an external funding agency. The economic return initially may be small in relation to the large capital expenditure and therefore it is generally of prime importance to keep the construction cost to the minimum. This can be achieved in a number of ways:

(1) *Phased development*: it may be appropriate to construct a gravel road in the first instance with the aim of adding a surfacing as traffic volumes and finance become available.
(2) *Appropriate construction techniques*: this may involve labour-intensive techniques in countries with a large labour force, but mechanization can often be a more appropriate solution.
(3) *Appropriate design standards*: these in respect of highway geometry, drainage and construction.

23.14.2 Route development

Routes for roads will often follow existing tracks but where a new alignment is to be provided the following principles will generally apply:

(1) A route should be chosen to avoid structural and drainage problems.
(2) A route should be chosen to avoid excessive earthworks.
(3) The alignment should be kept simple, as simple curves and straights are easier to set out in the field. Land boundaries and property acquisition will not generally be a limiting factor in developing countries.
(4) Where possible the vertical alignment should be slightly above ground level to aid drainage. Where this is not

possible drainage considerations are of paramount importance.
(5) The route should, where possible, be aligned near sources of locally available construction materials.

Suitable mapping often does not exist or is unavailable and it may be necessary to rely on aerial photography and walking the route to fix a line.

The road geometry will vary depending on the volume and type of traffic and the terrain. The cross-section, in general, should incorporate a verge at each side. This enables vehicles to pull off the road and also protects the edge of the pavement construction from erosion and damage. Figure 23.39 shows typical cross section standards adopted in tropical countries and also typical standards for low-volume roads.

23.14.3 Drainage

Water is the main enemy of road construction particularly in tropical countries where rainfall is frequently very intense. As already stated, drainage of the road formation is of paramount importance and is best achieved by setting the road slightly higher than the surrounding land and providing adequate side-drains as shown in Figure 23.40. Only in urban areas is positive drainage appropriate.

Drainage should be simple. Side-drains will generally discharge directly into streams or on to surrounding fields. Side-drains of appropriate cross-section are easily maintained by the use of a grader. Only on steep gradients need the ditch be lined with stone pitching to prevent erosion.

Cross-drainage should be kept to a minimum but where required is best achieved by the use of simple concrete pipes or corrugated steel structures. Complex reinforced concrete structures are not appropriate in developing countries except for long-span bridges.

A work by Watkins and Fiddes[186] deals with rainfall runoff and highway drainage.

23.14.4 Pavement construction

Many different types of pavement construction are used throughout the world and the type chosen will depend on the forecast volumes of traffic, the finance available and other factors specific to the country in question.

The pavement in its simplest form will consist of an earth track, the existing soil being graded to a smooth running surface. Many tropical soils will perform well provided they are well shaped and adequate drainage is provided.

A modified surface can be used if local soils are unsuitable or if traffic volumes demand a stronger road. This can be achieved either by stabilizing the existing soil *in situ* to improve its properties or importation of another suitable soil from nearby. Imported soils of suitable characteristics for road surfacing are typically natural gravels or crushed stones with a significant clay content.

Existing soil can be stabilized by many techniques. For example, cement may be added in small quantities (4 to 5%) to strengthen the soil, or lime can be added, which has the effect of reducing the plasticity index. The quantities to be used must be determined by trials.

Single-size and wind-blown soils are notoriously difficult to deal with but have been successfully stabilized using bitumen; this is expensive and other organic products such molasses may be appropriate in certain countries where such agricultural by-products are plentiful.

Gravel roads under heavy traffic flow require sealing to provide improved riding quality and protect the surface from the ingress of rainwater and axle load damage. The simplest

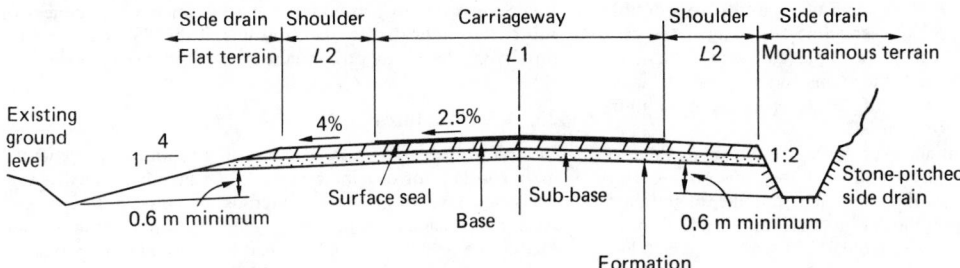

		Paved class			Gravel class	
	Average daily traffic	>1500	500–1500	50–500	50–500	<50
	Paved width L_1 (m)	7.3	6.7	6.1	6.1	5.5
Flat terrain	Shoulder width L_2 (m)	3	2–3	2	2	1
	Design speed (km/h)	100	100	100	80–100	60–80
	Maximum grade %	4	5	6	6	8
Rolling or hilly terrain	Shoulder width L_2 (m)	3	2	2	1	1
	Design speed (km/h)	100	80–100	80	60–80	50–60
	Maximum grade %	6	6	7	8	10
Mountainous terrain	Shoulder width L_2 (m)	1	1	1	—	—
	Design speed (km/h)	80	60–80	60	50–60	30–50
	Maximum grade %	7	8	8	10	12

Figure 23.39 Typical cross-section of a low-cost road

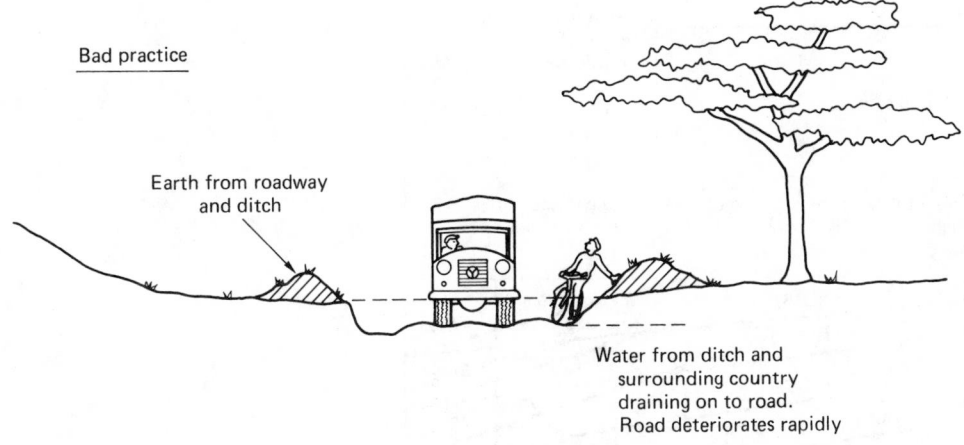

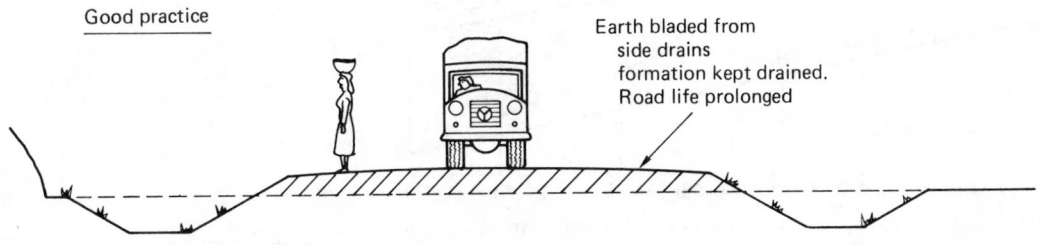

Figure 23.40 The importance of good drainage. (a) Good practice; (b) bad practice. (After Watkins and Fiddes (1984) *Highway and urban hydrology in the tropics*. Pentech Press, London).

form is a bituminous seal.[188] This can be upgraded to a double surface seal or a thin asphaltic concrete surfacing if the economic benefit can be shown. In some parts of the world, other materials such as waterbound macadams and penetration macadams, are employed. These can be appropriate if labour-intensive techniques are to be used.

Pavement design is undertaken by an empirical method as with other roads using the Californian bearing ratio of the subgrade in combination with the predicted number of standard axles using the road within the design life. Figure 23.41 shows a typical design chart from Road Note 31.[187] Care should be taken in design, since not only is the percentage of heavy vehicles likely to be very high in a developing country but overloading is common practice. Careful traffic forecasting, including field measurement of actual axle loads, is essential to a realistic design.

If it is desirable to provide, at the time of construction, a pavement capable of carrying more than 500 000 standard axles, the designer may choose either a 150 mm base with a 50 mm bituminous surfacing or a 200 mm base with a double surface dressing. For both of these alternatives, the recommended sub-base thickness is indicated by the broken line. Alternatively, a base 150 mm thick with a double surface dressing may be laid initially and the thickness increased when 500 000 standard axles have been carried. The extra thickness may consist of 50 mm of bituminous surfacing or at least 75 mm of crushed stone with a double surface dressing. The largest aggregate size in the crushed stone must not exceed 19 mm and the old surface must be prepared by scarifying to a depth of 50 mm. For this

stage construction procedure, the recommended thickness of sub-base is indicated by the solid line. Table 23.20 shows how rapidly the damaging power increases with increasing axle load.

23.14.5 Structures

Complex steel or reinforced concrete structures are generally inappropriate for developing countries since their construction will often involve imported materials and resultant high cost. Also the necessary expertise in construction techniques and quality control may not be readily available. Expertise in maintenance may also not be readily available and all structures should be designed with the ease of maintenance in mind.

The use of pontoon bridges, drifts or Irish bridges on low-trafficked routes may be appropriate. If a steel or reinforced concrete structure is unavoidable the designer should bear in mind the limitations of the contractor undertaking the work and the design should be kept simple.

Again, many countries will have their own locally developed bridging techniques which suit local conditions; hand-dug caisson foundations, for instance, are used widely in Asia.

23.14.6 Maintenance[185, 189, 190]

Where low-cost construction is used the road pavement is particularly susceptible to rapid deterioration and good maintenance is vital to prolong the life of the roads. Many maintenance activities must be carried out at frequent intervals.

Many loan agencies now put great emphasis on maintenance,

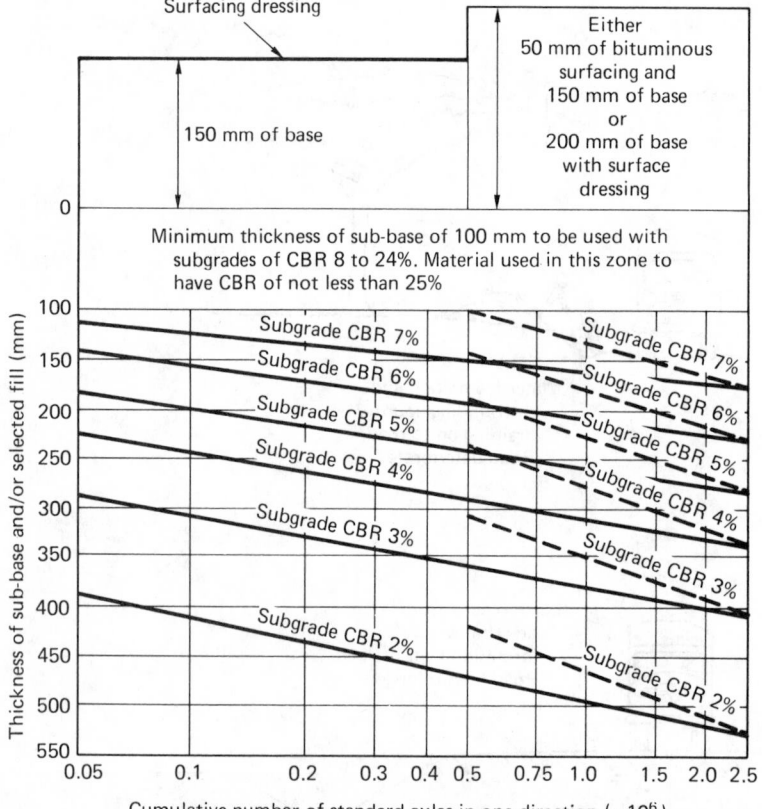

Figure 23.41 Pavement design chart for flexible pavements. (After Transport and Road Research Laboratory (1977) *A guide to the structural design of bitumen-surfaced roads in tropical and sub-tropical countries.* HMSO, London)

Table 23.20 Factor for converting numbers of axles to the eqivalent number of standard 8200 kg (18 000 lb) axles

Axle load (kg)	(lb)	Equivalence factor
910	2 000	0.0002
1 810	4 000	0.00025
2 720	6 000	0.01
3 630	8 000	0.04
4 540	10 000	0.08
5 440	12 000	0.2
6 350	14 000	0.3
7 260	16 000	0.6
8 160	18 000	1.0
9 970	20 000	1.6
9 980	22 000	2.4
10 890	24 000	3.6
11 790	26 000	5.2
12 700	28 000	7.2
13 610	30 000	9.9
14 520	32 000	13.3
15 430	34 000	17.6
16 320	36 000	22.9
17 230	38 000	29.4
18 140	40 000	37.3
19 070	42 000	47
19 980	44 000	58
20 880	46 000	72
21 790	48 000	87

since many low-cost roads built with financial aid have deteriorated prematurely because of the lack of maintenance. Many projects now include the setting-up and running of an ongoing maintenance management system as an integral part of the project.

Since the cost to the road user rises significantly as the standard of the pavement surface deteriorates, the economic advantage of maintenance is very marked. This is particularly true of low-cost roads with light construction where lack of maintenance can lead to the rapid development of corrugations, ruts and potholes.

Acknowledgements

The author acknowledges the patient help and assistance of his colleagues in the Croydon and Winchester offices of Mott, Hay & Anderson in the preparation of this chapter, and in particular H. Williams, J. Prince and Mrs P. Corston.

References

1 Cross, Charles A. (ed.) (var. dates) *Encyclopedia of highway law and practice*. Sweet and Maxwell, London.
2 Her Majesty's Stationery Office (1987) *Policy for roads in England*. HMSO, London.
3 Department of Transport (1982) *Traffic appraisal manual* and amendments. HMSO, London.
4 Ministry of Agriculture, Fisheries and Food (1966). *Agricultural land classification*. MAAF Technical Report No. 11. HMSO, London.
5 Transport and Road Research Laboratory (1976). *Preliminary sources of information for site investigation in Britain*. TRRL Laboratory Report No. 403. HMSO, London.
6 Department of Transport (1976) *Advisory committee on trunk road assessment*. HMSO, London.
7 Department of Transport (1979) *Trunk road proposals – a comprehensive framework of appraisal*. Standing Advisory Committee on Trunk Road Assessment. HMSO, London.
8 Department of Transport (1983) *Manual of environmental appraisal*. HMSO, London.
9 Standing Advisory Committee on Trunk Road Assessment (1986). *Urban road appraisal*. HMSO, London.
10 Department of Transport (1986) *The Government response to the SACTRA report on urban road appraisal*. HMSO, London.
11 The Standing Advisory Committee on Trunk Road Assessment (1979) *Forecasting traffic on trunk roads. A report on the regional highway traffic model project*. HMSO, London.
12 The Institution of Highways and Transportation and the Department of Transport (1987) *Roads and traffic in urban areas*. HMSO, London.
13 Department of Transport (1981) *Traffic surveys by roadside interview*. Departmental Advice Note TA 11/81. HMSO, London.
14 Wootton Jeffreys and Partners (1985) *'VISTA' Video System of Traffic Analysis*. Wooton Jeffreys, London.
15 Department of Transport (1981) *COBA user manual* and subsequent additions. HMSO, London.
16 Chard, B. M. and Lines, C. J. (1987) 'TRANSYT – the latest developments,' *Traffic Engng & Control*, July.
17 Van Vliet Institute for Transport Studies (1981) *SATURN, a user's guide*. University of Leeds.
18 Transport and Road Research Laboratory (1978) *CONTRAM: a traffic assignment model for predicting flows and queues during peak periods*, TRRL Laboratory Report LR 841. HMSO, London.
19 Transpotech (1983) *TRAFFICQ manual*, MVA Systematica.
20 Department of Transport (1984) *National road traffic forecasts (Great Britain)*, HMSO, London.
21 Department of Transport (1982) *QUADRO 2, Manual for assessing traffic-related roadworks costs*. HMSO, London.
22 Kelleway, R. C., White, C. D. and Matthews, D. H. (1984) *Delays caused by vehicle incidents and the development of CIDEL*. PTRC Summer Conference, Sussex University.
23 Committee on the Problems of Noise (1963) *Noise – final report*. Cmnd 2056. HMSO, London.
24 Department of the Environment (1972) *New housing and road traffic noise*. Design Bulletin No. 26. HMSO, London.
25 Department of the Environment (1972) *New roads in towns*. Urban Motorways Committee. HMSO, London.
26 Urban Motorways Project Team (1973) *Report on urban motorways*. HMSO, London.
27 Department of the Environment (1973) *Planning and noise*. Circular No. 10/73. HMSO, London.
28 Department of Transport (1976) *Route location with regard to environmental issues*. HMSO, London.
29 Council of the European Communities (1985) Directive of 27 June 1985 on the assessment of the effects of certain public and private projects on the environment. *J. of the European Communities*, **L175**, 40–48.
30 Transport and Road Research Laboratory (1971) *Survey of traffic-induced vibrations*. TRRL Laboratory Report No. LR 418. HMSO, London.
31 Transport and Road Research Laboratory (1987) *Traffic-induced ground-borne vibrations in dwellings*. TRRL Laboratory Report No. 102. HMSO, London.
32 Department of the Environment (1975) *Noise insulation regulations*. Statutory Instrument No. 1763. HMSO, London.
33 Department of the Environment (1975) *Calculation of road traffic noise*. HMSO, London.
34 British Standards Institution (1984/1986) *Noise control on construction and open sites*. BS 5228 Parts 1, 2 and 4. BSI, Milton Keynes.
35 Department of Transport (1981/1985) *Road layout and geometry – highway link design*. Departmental Standard No. TD 9/81 and Amendment No. 1. HMSO, London.
36 Department of Transport (1985) *Traffic flows and carriageway width assessment*. Department Standard No. TD 20/85. HMSO, London.
37 Department of Transport (1988) *Traffic flows and carriageway width assessment for rural roads*. Departmental Advice Note TA 46/85. HMSO, London.

38 Department of Transport (1982) *Choice between options for trunk road schemes*. Departmental Advice Note TA 30/82. HMSO, London.

39 The County Surveyors' Society (1969). *Highway transition curve tables*. CSS, London.

40 Department of Transport (var. dates) *Traffic signs manual*. HMSO, London.

41 Department of Transport (1988) *Cross-sections and headroom*. Departmental Standard No. TD 27/86. HMSO, London.

42 Department of Transport (1987) *Highway construction details*. Department of Transport Advice Note No. TA 23/81. HMSO, London.

43 Department of Transport (1977) *Residential roads and footpaths*. Design Bulletin No. 32. HMSO, London.

44 Hampshire County Council (1986) *Roads in residential areas* (2nd edn). HCC.

45 Savoy Computing (1981/1983) *TRACK library of vehicle-swept paths*. Crown and Travers Morgan Planning.

46 Freight Transport Association (1983) *Designing for deliveries*. FTA.

47 Department of Transport (1981) *Determination of size of roundabouts and major/minor junctions*. Departmental Advice Note No. TA 23/81. HMSO, London.

48 Department of Transport (1984) *The layout of major/minor junctions*. Departmental Advice Note No. TA 20/84. HMSO, London.

49 Department of Transport (1985) *ARCADY 2/PICADY 2, Capacities, queues and delays at major/minor junctions HCSL/R/30–31*. HMSO, London.

50 Department of Transport (1981) *MIDAS, assessment of delays at road junctions HECB/R/32*. HMSO, London.

51 Department of Transport (1984) *Geometric design of roundabouts*. Departmental Standard No. TD 16/84. HMSO, London.

52 Department of Transport (1984) *Geometric design of roundabouts*. Departmenal Advice Note No. TA 42/84. HMSO, London.

53 Department of Transport (1981) *Traffic signs regulations and general directions*. Statutory Instrument No. 859 and amendments. HMSO, London.

54 Department of Transport (1987) *OSCADY, Capacities, queues and delays at signal-controlled junctions HCSL/R/42*. HMSO, London.

55 Allsop, R. E. (1976) 'SIGCAP: a computer program for assesssing the traffic capacity of signal controlled junctions'. *Traffic Engng & Control*, **17**, 8 & 9.

56 Allsop, R. E. (1971) 'SIGSET: a computer program for calculating traffic signal settings'. *Traffic Engng & Control*, **13**, 2.

57 Simmonite, B. F. (1985) 'LINSIG: a program to assist traffic signal design and assessment'. *Traffic Engng & Control*, **26**, 6.

58 Transport and Road Research Laboratory (1980) *User guide to TRANSYT version 8*. TRRL Laboratory Report No. LR 888. HMSO, London.

59 Department of Transport (1981) *Traffic signals on high-speed roads*. Departmental Advice Note No. TA 12/81. HMSO, London.

60 Department of Transport (1981) *General principles of control by traffic signals*. Departmental Advice Note No. TA 16/81. HMSO, London.

61 Department of Transport (1981) *Junction layout for control by traffic signals*. Departmental Advice Note No. TA 18/81. HMSO, London.

62 Department of Transport (1981) *Pedestrian facilities at traffic signal installations*. Departmental Advice Note No. TA 15/81. HMSO, London.

63 Department of Transport (1986) *Layout of grade-separated junctions*. Departmental Advice Note No. TA 48/86. HMSO, London.

64 Department of Transport (1986) *Layout of grade-separated junctions*. Departmental Advice Note No. TA 22/86. HMSO, London.

65 British Standards Institution (1981) *Code of practice for earthworks*. BS 6031. HMSO, London.

66 Transport and Road Research Laboratory (1976) *Sources of information for site investigation*. TRRL Laboratory Report No. LR 403. HMSO, London.

67 Chandler, R. J. and Skempton, A. W. (1974) 'The design of permanent cutting slopes in stiff fissured clays', *Géotechnique*, **24**, 4, 457–466.

68 Bishop, A. W. (1955) 'The use of the slip circle in the stability analysis of slope', *Géotechnique*, **5**, 1, 7–17.

69 Bishop, A. W. and Morgenstern, N. (1960) 'Stability coefficients for earth slopes', *Géotechnique*, **10**, 4, 129–150.

70 Sarma S. K. (1979) 'Stability analysis of embankment and slope', *J. Am. Soc. Civ. Engrs, Geotech, Engng Div*. **105 GT** 12, 1522–1524.

71 Parsons, A. W. and Perry, J. (1985) 'Slope stability problems in ageing highway earthworks'. *Symposium on failures in earthworks*, Institution of Civil Engineers. Thomas Telford, London.

72 Department of Transport (1986) *Specification for highway works* (6th edn). HMSO, London.

73 Department of Transport (1986) *Notes for guidance on the specification for highway works*. HMSO, London.

74 British Standards Institution (1975) *Methods of test for soils for civil engineering purposes*, BS 1377, BSI, Milton Keynes.

75 Transport and Road Research Laboratory (1976) *The rapid measurement of the moisture condition of earthworks material*. TRRL Laboratory Report No. LR 750. HMSO, London.

76 Transport and Road Research Laboratory (1979) *The moisture condition test and its potential application in earthworks*, TRRL Supplementary Report No. SR 522. HMSO, London.

77 Transport and Road Research Laboratory (1987) *The precision of the moisture condition test*. TRRL Research Report No. 90. HMSO, London.

78 Transport and Road Research Laboratory (1986) *Assessing the quality of rockfill: a review of current practice for highways*. TRRL Research Report No. 60. HMSO, London.

79 Transport and Road Research Laboratory (1977) *The classification of chalk for use as a fill material*. TRRL Laboratory Report No. LR 806. HMSO, London.

80 Transport and Road Research Laboratory (1976) *Earthworks in soft chalk: a study of some factors affecting construction*. TRRL Laboratory Report No. LR 112. HMSO, London.

81 Institution of Civil Engineers (1978) 'Clay fills', *Proceedings, conference on clay fills*. Thomas Telford, London.

82 Imperial Chemical Industries (1986) *Lime stabilization manual*. ICI, London.

83 Transport and Road Research Laboratory (1982) *The effect of soil condition on earthmoving plant*. TRRL Laboratory Report No. LR 1034. HMSO, London.

84 Transport and Road Research Laboratory (1979) *The strength of clay fill sub-grades: its prediction in relation to road performance*. TRRL Laboratory Report No. LR 889. HMSO, London.

85 Transport and Road Research Laboratory (1984) *The structural design of bituminous roads*. TRRL Laboratory Report No. LR 1132. HMSO, London.

86 Transport and Road Research Laboratory (1968) *The properties of cement-stabilized materials*. TRRL Laboratory Report No. LR 205. HMSO, London.

87 British Standards Institution (1975) *Methods of test for stabilized soils*, BS 1924. BSI, Milton Keynes.

88 Rankilor, P. R. (ed.) (1986) *International directory of geotextiles and related products*. Manstock Geotechnical Consultancy Services, Manchester.

89 Transport and Road Research Laboratory (1963) *A guide for engineers to the design of storm sewer systems*. Road Note No. 35. HMSO, London.

90 Transport and Road Research Laboratory (1973) *Estimated rainfall for drainage calculations in the UK*. TRRL Laboratory Report No. LR 595. HMSO, London.

91 National Water Council Standing Technical Committee (1981) *Design and analysis of urban storm drainage – the Wallingford Procedure*. Reports STC 28–31.

92 American Iron and Steel Institute (1971) *Handbook of steel and highway construction products*. AISI, New York.

93 Bartlett, R. E. (1981) *Surface water sewerage* (2nd edn). Elsevier Applied Science, London.

94 Transport and Road Research Laboratory (1973) *The estimation of flood flows from natural catchments*. TRRL Laboratory Report No. LR 565. HMSO, London.

95 Natural Environment Research Council (1975) *The flood studies report*. NERC, London.

96 Institute of Hydrology (1978) *Methods of flood estimation – a guide to the* Flood Studies Report.

97 US Bureau of Public Roads (1965) *Hydraulic charts for the*

selection of highway culverts. Hydraulic Engineering Circular No. 5.

98 Construction Industry Research and Information Association (1980) *Guide to the design of storage ponds for flood control in partly urbanised catchments.* Technical Note No. 100. CIRIA, London.

99 Hydraulics Research Station (1983) *Charts for the hydraulic design of channels and pipes.* Hydraulics Research Paper No. 2. HMSO, London.

100 Hydraulics Research Station (1983) *Tables for the hydraulic design of storm drains, sewers and pipelines.* Hydraulics Research Paper No. 4. HMSO, London.

101 Colyer, D. J. (1977) 'The effect of surcharging on discharging through a pipe'. *Chartered Municipal Engnr*, **104**, 4, 60–62.

102 Transport and Road Research Laboratory (1967) *Subsoil drainage and the structural design of roads.* TRRL Laboratory Report No. LR 110. HMSO, London.

103 Transport and Road Research Laboratory (1968) *Depth of rainwater on road surfaces.* TRRL Laboratory Report No. LR 236. HMSO, London.

104 Transport and Road Research Laboratory (1969) *Hydraulic efficiency and spacing of BS road gullies.* TRRL Laboratory Report No. LR 277. HMSO, London.

105 Transport and Road Research Laboratory (1973) *Drainage of level or nearly level roads.* TRRL Laboratory Report No. LR 602. HMSO, London.

106 Building Research Establishment (1973) *Soakaways.* Digest No. 151. HMSO, London.

107 Transport and Road Research Laboratory (1986) *Simplified tables of external loads on buried pipelines.* HMSO, London.

108 Transport and Road Research Laboratory (1983) *A guide to design loadings for buried rigid pipes.* HMSO, London.

109 Department of Transport (1979) *Technical approval of highway structures on trunk roads (including motorways).* Departmental Standard No. BD 2/79. HMSO, London.

110 American Association of State Highway and Transportation Officials (1986). *Guide for design of pavement structures.* AASHTO, New York.

111 Shell International (1978) *Shell pavement design manual.* Shell International, London.

112 Shell International (1985) *Addendum, Shell pavement design manual.* Shell International, London.

113 Department of Transport (1987) *Structural design of new road pavements.* Departmental Advice Note No. HA 35/87. HMSO, London.

114 Department of Transport (1987) *Structural design of new road pavements.* Departmental Standard No. HD 14/87. HMSO, London.

115 Department of Transport (1983) *Deflection measurement of flexible pavements, analysis, interpretation and application of deflection measurements.* Departmental Advice Note No. HA 25/83. HMSO, London.

116 Department of Transport (1976) *Specification requirements for aggregate properties and texture depth for bituminous surfacings to new roads.* Technical Memorandum No. H16/76. HMSO, London.

117 British Standards Institution (1985) *Hot-rolled asphalt for roads and other paved areas.* BS 594 and amendments. BSI, Milton Keynes.

118 British Standards Institution (1974) *Sampling and examination of bituminous mixtures for roads and other paved areas.* BS 598, Parts 1 to 3. BSI, Milton Keynes.

119 Nelson, P. M. and Abbott, P. G. (1987) 'Low noise road surfaces'. *Applied Acoustics*, **21**.

120 Walker, B. J. and Beadle, D. (1975) *Mechanised construction of concrete roads.* Cement and Concrete Association, London.

121 Department of Transport (1978) *A guide to concrete road construction.* HMSO, London.

122 Concrete Society (1985) *Joints in concrete roads, aspects of construction and maintenance.* Technical Report No. 28.

123 Transport and Road Research Laboratory (1987) *Thickness design of concrete roads.* TRRL Research Report No. 87.

124 Concrete Society (1983) *Design of paved areas for industrial usage.* Technical Report No. 24.

125 British Ports Association (1982) *The structural design of heavy-duty pavements for ports and other industries.*

126 Barber, S. D. and Knapton, K. (1980) *Structural design of block*

pavements for ports. Proceedings, 1st international conference on concrete block paving.

127 Lilley, A. A. and Walker, B. J. (1978) *Concrete block paving for heavily trafficked roads and paved areas.* Cement and Concrete Association, London.

128 Cement and Concrete Association (1980) *A design method.* Advisory Data Sheet No. 36. C&CA, London.

129 Lilley, A. A. and Clark, A. J. (1980) *Concrete block paving for lightly trafficked roads and paved areas.* Cement and Concrete Association, London.

130 Clark, A. J. (1981) *Further investigations into the load-spreading of concrete block paving.* Technical Report No. 545. Cement and Concrete Association, London.

131 Gray, D. C. (1986) Block-paving – design guidance review. *Civ. Engng*, August.

132 British Standards Institution (1986) *Clay and calcium silicate pavers for flexible pavements.* BS 6677, Parts 1, 2, 3. BSI, Milton Keynes.

133 British Standards Institution (1986) *Precast concrete paving blocks.* BS 6717, Part 1. BSI, Milton Keynes.

134 County Surveyor's Society and Interpave (1980) *Specification for concrete paving blocks.* Cement and Concrete Association, London.

135 County Surveyor's Society and Interpave (1983) *Code of practice for laying precast concrete block pavements.* Cement and Concrete Association, London.

136 Concrete Society (1974) *A guide to good practice for road edge details.* Technical Report No. 10. CS, London.

137 Department of Transport (1986) *Appraisal of new and replacement lighting on trunk roads and trunk road motorways.* Department Advice Note No. TA 49/86. HMSO, London.

138 British Standards Institution (1980) *Code of practice for road lighting.* BS 5489 Part 1 to 9. BSI, Milton Keynes.

139 Department of Transport (1986) *Trunk roads and trunk road motorways – maintenance of road lighting.* Departmental Standard No. TD 23/86. HMSO, London.

140 Organization of Economic Co-operation and Development (1974) *European rules concerning road traffic signs and signals.* European Conference of Ministers of Transport, OECD, Brussels. Vienna, 1968. Geneva 1971–73.

141 Department of the Environment (1975) *Size design and mounting of traffic signs.* Circular 'Roads' No. 7/75. HMSO, London.

142 Department of Transport (1987) *Bibliography of publications relating to traffic signs, signals and road markings.* Circular 'Roads' No. 4/87. HMSO, London.

143 Permanent International Association of Road Congresses (1983) 'Technical committee report on roads in urban areas'. *Proceedings, 17th world road congress*, Sydney.

144 British Standards Institution (1972–1986) *Fences*, BS 1722, Parts 1 to 13. BSI, Milton Keynes.

145 British Standards Institution (1978) *Code of practice for preservation of timber*, BS 5589. BSI, Milton Keynes.

146 Department of Transport (1978) *Noise barriers – standards and materials.* Technical Memorandum H 14/76 and Amendment No. 1. HMSO, London.

147 British Standards Institution (1985) *Safety fences and barriers for highways.* BS 6579, Parts 1 to 8. BSI, Milton Keynes.

148 Department of Transport (1985) *Safety fences and barriers.* Department Standard No. TD 19/85. HMSO, London.

149 Department of Transport (1978) *Report of the committee of inquiry into motorway service areas.* HMSO, London.

150 Department of Transport (1968) *Layout of roads in rural areas.* HMSO, London.

151 Department of Transport (1974) *Layout of roads in rural areas, metric supplement.* HMSO, London.

152 British Standards Institution (1983) *Methods of testing concrete.* BS 188. BSI, Milton Keynes.

153 British Standards Institution (1985) *Testing aggregates.* BS 812. BSI, Milton Keynes.

154 British Standards Institution (1983) *Specification for natural resources for concrete.* BS 882. BSI, Milton Keynes.

155 British Standards Institution (1978) *Specification for ordinary and rapid-hardening Portland cement.* BS 12. BSI, Milton Keynes.

156 British Standards Institution (1974) *Sampling and examination of bituminous mixtures for roads and other paved areas.* BS 598. BSI, Milton Keynes.

157 Department of Transport (1987) *Method of measurement of highway works.* HMSO, London.

158 Institution of Civil Engineers (1986) *Conditions of contract and forms of tender, agreement and bond for use in connection with works of civil engineering construction* (5th edn). Thomas Telford, London.

159 Department of Transport (1987) *Library of standard item descriptions for highway works.* HMSO, London.

160 Association of London Borough Engineers (1985) *Highways and traffic management in London – a code of practice.* HMSO, London.

161 Department of Transport (1986) *Road Traffic Regulation (Parking) Act.* HMSO, London.

162 Transport and Road Research Laboratory (1981) *SCOOT – a traffic-responsive method of co-ordinating signals.* TRRL Laboratory Report No. LR 1014. HMSO, London.

163 Permanent International Association of Road Congresses (1987) 'Roads in urban areas'. *17th world congress.* Technical Committee Report No. 10. Brussels.

164 Transport and Road Research Laboratory (1977) *Nottingham zones and collar study – overall assessment.* TRRL Laboratory Report No. LR 805. HMSO, London.

165 Transport and Road Research Laboratory (1984) *The effects of wheel-clamping in central London.* TRRL Laboratory Report No. LR 1136. HMSO, London.

166 The Institution of Highway Engineers (1981) *Guidelines for lorry management schemes.*

167 The Institution of Highways and Transportation (1986) *Providing for people with a mobility handicap – guidelines.*

168 Department of Transport (1970) *Report of committee on highway maintenance.* HMSO, London.

169 Department of Transport (1985) *Code of practice for routine maintenance.* HMSO, London.

170 Department of Transport (1984) *Winter maintenance of motorways and trunk roads – statement of service and code of practice.* HMSO, London.

171 Association of County Councils, Association of District Councils and Association of Metropolitan Authorities (1983) *Highway maintenance – a code of good practice.*

172 Transport and Road Research Laboratory (1975) *The CHART system of assessing structural maintenance needs of highways.* TRRL Supplementary Report No. SR 153 UC. HMSO, London.

173 Transport and Road Research Laboratory (1983) *Improved data collection methods for CHART highway maintenance system.* TRRL Laboratory Report No. LR 1084, HMSO, London.

174 City Engineers' Group (1975) *The MARCH highway maintenance system.*

175 Department of Transport (1983) *Strength testing of flexible pavements by deflection measurement.* Departmental Standard No. HD 10/83. HMSO, London.

176 Department of Transport (1983) *Deflection measurement and flexible pavements – operational practice for the deflection beam and the deflectograph.* Departmental Advice Note No. HA 24/83. HMSO, London.

177 Department of Transport (1983) *Deflection measurement and flexible pavements – analysis, interpretation and application of deflection measurements.* Departmental Advice Note No. HA 25/83. HMSO, London.

178 Transport and Road Research Laboratory (1976) *Measurement of skidding resistance – guide to use of scrim.* TRRL Laboratory Report No. LR 737. HMSO, London.

179 Transport and Road Research Laboratory (1976) *Measurement of skidding resistance – three factors affecting scrim measurements.* TRRL Laboratory Report No. LR 739. HMSO, London.

180 Transport and Road Research Laboratory (1981) *Measurement of skidding resistance, Part V:* Precision of scrim measurements. TRRL Supplementary Report No. SR 642. HMSO, London.

181 Department of Transport (1983–1987) *Signing for traffic management at certain major road sites and amendments.* Departmental Standard No. TD 14/83. HMSO, London.

182 Department of Transport (1985) *Control of traffic at roadworks on single-carriageway roads.* Departmental Advice Note No. TA 47/85. HMSO, London.

183 Department of Transport (1985) *Portable traffic signals at roadworks on single-carriageway roads.* Departmental Standard No. TD 21/85. HMSO, London.

184 Staffordshire County Council (1985) *Safety at roadworks.* SCC, Stafford.

185 Odier, L., Millard, R., Dimertel Dossantos and Mehra, S. (1967) *Low-costs roads.* Butterworth.

186 Watkins, L., and Fiddes, D. (1984) *Highway and urban hydrology in the tropics.* Pentech Press, London.

187 Transport and Road Research Laboratory (1977) *A guide to the structural design of bitumen-surfaced roads in tropical and sub-tropical countries* (3rd edn). Road Note No. 31. HMSO, London.

188 Transport and Road Research Laboratory (1982) *A guide to surface dressing in tropical and sub-tropical countries.* Overseas Road Note No. 3. HMSO, London.

189 United Nations Economic Commission for Africa (1982) *Road maintenance handbook* (3 vols.) 'Practical guidelines for road maintenance in Africa'.

190 Transport and Road Research Laboratory (1981) *Road maintenance planning and management for developing countries.* ODA/TRRL One-day-seminar, 30 January. HMSO, London.

24

Airports

E V Finn CEng, FICE, FIStructE, FRSH,
MIWEM, MConsE
R H R Douglas BSc(Eng), CEng, FICE,
FIHT, MConsE and
D J Osborne BSc(Eng), CEng, FICE, FIHT,
MIWEM, MBIM
Sir Frederick Snow and Partners

Contents

24.1 Introduction

The planning and design of an airport is complex and involves specialists in airport planning, traffic forecasting, aeronautical ground lighting, telecommunications and navigational aids, air traffic control, baggage handling, and many other activities. The development of an airport will involve architects, structural, electrical, mechanical and telecommunications engineers, planners, economists, interior designers, quantity surveyors and other specialists, as well as civil engineers.

Traditionally, civil engineers have played a major role in the development of airports and the co-ordination and management of all the disciplines involved. This is, perhaps, because so many aspects of civil engineering have always been involved, such as the design of loadbearing pavements, access roads and car parks, surface water drainage, water supply, fire-fighting mains, foul drainage (including sewage treatment), as well as major building structures.

Airports have been required to cope with the increase in passenger traffic, the number of aircraft movements, and the size and weight of aircraft. The character of the airport has also changed, with greater emphasis on security, safety, comfort and convenience of passengers, efficiency and economical operation, and with the need for the involvement of more specialists in their planning and design.

In order to consider the civil engineering aspects of an airport in perspective, reference is made in this chapter to the location, standards and general concepts of airports, as well as to the other facilities which together make an airport. Only those aspects of civil engineering which are particular to airports are dealt with in detail.

24.2 Airport location

24.2.1 Basic considerations

The site selected for a new airport development must be capable of providing the longest possible useful life in order to secure the maximum return on the large investments which are required for its development. Many factors require examination in order to determine the most suitable site, but before consideration is given to the criteria involved, it is necessary to define the purpose for which the airport is required, and the size of the facilities to suit this requirement.

The need for an airport might be because: (1) none exists and it is believed air services will meet a specific physical or economic demand; (2) an existing airport cannot be expanded to meet growing traffic; or (3) an existing site has become environmentally unacceptable.

The facilities to be accommodated and considered will include the length and direction of the runway, the number of runways, the terminal building and apron, and ancillary requirements such as cargo handling, airport maintenance, catering and car parking. The scale of these facilities and, hence, the overall area of land needed for the airport site, will be assessed in relation to national or regional planning of airspace use (if such exists), traffic forecasts, and an assessment of aircraft types appropriate to predicted use.

24.2.2 Criteria for comparative analysis of sites

The essential factors to be considered in selection of an airport site include:

(1) Passenger catchment area.
(2) Environment.
(3) Economic appraisal.
(4) Financial appraisal.
(5) Airspace.
(6) Topography.
(7) Obstructions to aircraft operations.
(8) Meteorology.
(9) Construction problems.
(10) Utility services.

There is no particular order in which these should always be considered, and there are few fundamental criteria to provide a clear basis for rejection of a site from further consideration, other than perhaps the intrusion of unacceptable obstructions into the approach surfaces. There are clearly wide variations between what might be an acceptable site high in the Andes, in the desert of Jordan, on the southern tip of Shetland, or on the shores of Loch Neagh. An initial selection of sites for subsequent comparative analysis has to be made in the knowledge of these factors, but the final selection is made from an objective comparison of each.

24.2.2.1 Passenger catchment area

Where regional airports are concerned, a journey time of about 45 min from a centre of population is normally considered acceptable. In developed countries it will be necessary to assess the effect on journey time of any planned improvement or new highways. In less developed countries, it may be necessary to consider the effect the airport may have on the existing highway system.

A major international airport will attract passengers from a much wider catchment area, including those using feeder air routes from regional airports, and the proximity to a centre of population may be less critical.

24.2.2.2 Environment

An airport affects the environment in three major ways, through: (1) land use; (2) noise and (3) ecology.

In the UK most existing airports have been developed from wartime airfields. Where new sites have been sought, as for the third London Airport, there have been objections and lengthy inquiries, essentially on these environmental issues. In developing countries the emphasis is likely to be different.

The area required by an airport is large. A modest regional airport may occupy 450 ha; a major international airport might require 5000 ha. Unfortunately, one of the requirements for an airport site, namely relatively flat and well-drained land, is often also the best agricultural land in an area, or alternatively is an area suitably distant from a population centre to be designated for industrial use.

To avoid these conflicts, areas unsuitable for other use need to be looked at. Such sites may involve major earthwork problems as, for example, the site being considered for the new Bangkok Airport, which is largely waterlogged, or incur the possibility of disturbing the natural ecological balance, as was a major objection to proposals for the proposed development of the third London Airport at Maplin.

Noise became a major environmental issue in the 1960s and 1970s and is an important aspect of airport planning. Certification procedures introduced by the International Civil Aviation Organization (ICAO) in 1972 have resulted in a new generation of quieter aircraft, such as the Boeing 757, introduced into service by British Airways on domestic routes in the UK early in 1983. It is no longer permissible for earlier and noisier aircraft, such as the Trident and the BA 1-11, to be used in the UK. Such improvements and restrictions are unlikely to apply to developing countries for many years.

24.2.2.3 Economic appraisal

An economic appraisal compares the total cost of each site to the whole community.The comparison will take into account: (1) the capital cost of site acquisition and construction; (2) access to the airport by airport employees; (3) access for passengers and cargo; (4) noise and other environmental factors; and (5) operation of the airport. These costs will be offset by the revenue earned directly by the airport operator, the airlines, and airport-associated and airport-attached businesses. Many of these will be the same regardless of the site, but others may be affected considerably.

24.2.2.4 Financial appraisal

A financial appraisal compares alternative sites on the basis of the capital costs of development only, although it can be considered as including direct costs and revenues related to operating the airport, loan receipts, repayments and interest charges.

24.2.2.5 Airspace

All countries who are members of ICAO have a government authority responsible for Air Traffic Control. In the UK, National Air Traffic Services (NATS) is responsible and provides a combined service to both the Civil Aviation Authority (CAA) and the Ministry of Defence. The siting of an airport may be critical if there is the possibility of aircraft operations conflicting with operations from an adjacent airport, particularly if this is sited across a national border in another country. Otherwise, air traffic control services, and particularly landing and take-off procedures, can usually be adapted to meet the particular site requirements.

24.2.2.6 Topography

For the purpose of comparison of several sites it is not necessary, initially, to quantify the amount of work required to construct the airport on that site. It is necessary to compare the advantages and disadvantages and to identify any difficulties.

Ideally, an airport should be located on relatively flat ground, having effective natural drainage. The site should not be hemmed-in by hills, rivers, roads or development which may hinder future expansion, or form potential obstructions to aircraft approaching or departing.

The assessment can be made largely from examination of existing maps and aerial photographs, but an inspection of the site should be considered essential.

24.2.2.7 Obstructions to aircraft operations

Objects which project above the imaginary obstruction surfaces (Figure 24.1) are classified as obstructions and will need to be removed if possible, or marked, if a particular site is chosen and

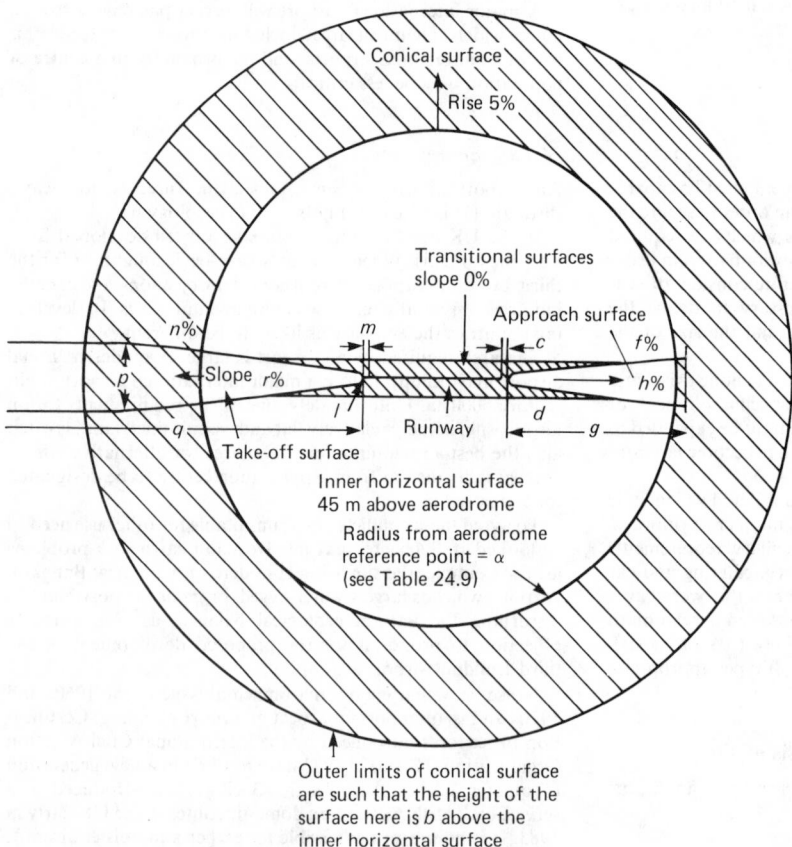

Figure 24.1 Plan view of obstruction surface (second and horizontal sections of approach surface for non-precision and precision approach are not shown for clarity)

Table 24.1 Aerodrome reference codes

Code Element 1			Code Element 2	
Code number	Aeroplane reference field length	Code letter	Wing span	Outer main gear* wheel span
1	2	3	4	5
1	Less than 800 m	A	Up to but not including 15 m	Up to but not including 4.5 m
2	800 m up to but not including 1200 m	B	15 m up to but not including 24 m	4.5 m up to but not including 6 m
3	1200 m up to but not including 1800 m	C	24 m up to but not including 36 m	6 m up to but not including 9 m
4	1800 m and over	D	36 m up to but not including 52 m	9 m up to but not including 14 m
		E	52 m up to but not including 60 m	9 m up to but not including 14 m

*Distance between the outside edges of the main gear wheels.

developed. At a stage of initial site appraisal, possibly before even the alignment of a runway has been determined, it is the potential of objects becoming obstructions which needs to be assessed, together with the degree of problems they could create in terms of removal or by inhibiting the location or alignment of a runway.

24.2.2.8 Meteorology

For any site to be appraised properly, meteorological records of wind direction, strength and frequency, together with visibility range and cloudbase height are necessary. This information provides the data for determining the runway alignment, and the need for and type of approach aids needed to provide the required level of usability.

There is usually sufficient data available in the general vicinity of an airport site in the UK for a valid interpolation to be made. This is frequently not the case in developing countries.

24.2.2.9 Construction problems

Any particularly difficult construction can usually be recognized in the initial stages of appraising a site. Such a problem in the UK is usually limited to the particular site characteristics, which may be poor soil conditions or bad drainage. In other countries these difficulties may extend to difficulties of access and lack of suitable materials for construction.

24.3 Standards

Details of international requirements for the layout of airfields are covered in the ICAO *Standards and recommended practices for aerodromes*,[1] Annex 14, and this publication is revised periodically. Any aerodrome (airfield or airport) requires a licence to accept a commercial service. The technical and other requirements for the licensing of a site on an aerodrome in the UK are incorporated in Civil Aviation Publication CAP 168, *Licensing of aerodromes*, published by the CAA.[2] In general, this conforms with and amplifies the information given in ICAO Annex 14, except for certain modifications which have been found appropriate to aerodromes in the UK.

The detailed standards and recommendations regarding airport layout, including recommendations for length, clearance and for the vertical alignment of runways and taxiways are given in Annex 14 with respect to the various airport reference codes. The following excerpts from Annex 14 are given for guidance only and reference should be made to Annex 14[1] or CAP 168[2] for full details.

24.3.1 Airport reference codes

From 24 November 1983, ICAO Annex 14[1] was subject to amendment. Two-element reference codes, incorporating numbers 1 to 4 together with letters A to E are now assigned to airports depending on the main runway length, aircraft wing span and outer main gear wheel span in accordance with Table 24.1.

24.3.2 Runway length

The actual runway length should be adequate to meet the operations requirements of the aeroplanes for which the runway is intended and should not be less than the longest length determined by applying the corrections for local conditions to the operations and performance characteristics of the relevant aeroplanes.

It may be noted that the actual runway length can be reduced within certain limits if a stopway or clearway is provided. Further comment on the design of runway length is made in sections 24.4.3 and 24.4.4.

24.3.3 Runway width

The width of a runway should not be less than the appropriate dimensions in Table 24.2.

Table 24.2 Runway widths (m)

Code number	Code letter				
	A	B	C	D	E
1	18	18	23	—	—
2	23	23	30	—	—
3	30	30	30	45	—
4	—	—	45	45	45

Note: The width of precision approach runway code number 1 or 2 should be not less than 30 m.

24.3.4 Runway vertical alignment

Recommendations in relation to the various components of vertical alignment are given in Table 24.3.

Table 24.3 Runway vertical alignment

	4	3	2	1
	Code letter			
Maximum effective slope	1%	1%	2%	2%
Maximum slope	1.25%	1.5%	2%	2%
Maximum change between consecutive slopes	1.5%	1.5%	2%	2%
Maximum rate of change of slope per 30 m	0.1%	0.2%	0.4%	0.4%
Minimum radius of curvature (m)	30 000	15 000	7500	7500
Minimum distance between successive points of intersection of vertical curves is the sum of the absolute numerical values of the corresponding slope changes multiplied by the factor given in metres	30 000	15 000	5000	5000

Notes: (1) The maximum slope for a runway code number 4 should not exceed 0.8% for the first and last quarters.
(2) The maximum slope for a runway code number 3 precision approach category II or III should not exceed 0.8% for the first and last quarters.

24.3.5 Runway transverse slopes

Recommendations for the transverse slopes are given in Table 24.4.

Table 24.4 Runway transverse slopes

Code letter				
E	D	C	B	A
1.5%	1.5%	1.5%	2%	2%

Note: The transverse slopes should not exceed 1.5 or 2% as applicable nor be less than 1% except at runway or taxiway intersections where flatter slopes may be necessary.

24.3.6 Taxiway widths

The width of a straight portion of a taxiway should be not less than that given in Table 24.5.

Table 24.5 Taxiway widths

Taxiway width (m)	Code letter
23	E or D and the taxiway is intended to be used by aeroplanes with an outer main gear wheel span equal to or greater than 9 m.
18	D and the taxiway is intended to be used by aeroplanes with an outer main gear wheel span of less than 9 m; C and the taxiway is intended to be used by aeroplanes with a wheel base equal to or greater than 18 m.
15	C and the taxiway is intended to be used by aeroplanes with a wheel base less than 18 m.
10.5	B
7.5	A

Note: The second subdivision of the 18 and the 15 m widths are defined by the wheel base, not the wheel span.

24.3.7 Taxiway vertical alignment

Recommendations in relation to the various components are given in Table 24.6.

Table 24.6 Taxiway vertical alignment

	E	D	C	B	A
	Code letter				
Maximum slope	1.5%	1.5%	1.5%	3%	3%
Maximum change of slope per 30 m	1%	1%	1%	—	—
Minimum radius of curvature (m)	3000	3000	3000	—	—
Minimum change of slope per 25 m	—	—	—	1%	1%
Minimum radius of curvature (m)	—	—	—	2500	2500
Maximum transverse slope	1.5%	1.5%	1.5%	2%	2%

24.3.8 Taxiway minimum separation distances

Recommendations for taxiway minimum separation distances are given in Table 24.7.

Table 24.7 Taxiway minimum separation distances

Code letter	Code number	Distance between taxiway centreline and runway centreline								Taxiway centreline to taxiway centreline	Taxiway & apron taxiway centreline to object	Aircraft stand taxilane centreline to object
		*Instrument runways**				*Other runways**						
		1	2	3	4	1	2	3	4			
A		82.5	82.5	—	—	37.5	47.5	—	—	21	13.5	12
B		87	87	—	—	42	52	—	—	31.5	19.5	16.5
C		—	—	168	—	—	—	93	—	46.5	28.5	24.5
D		—	—	176	176	—	—	101	101	68.5	42.5	36
E		—	—	—	180	—	—	—	105	76.5	46.5	40

*The separation distances shown represent ordinary combinations of runways and taxiways. The basis for development of these distances is given in the 'Aerodrome Design Manual, Part 2'.

24.3.9 Aprons – clearance distances

An aircraft stand should provide the clearances between an aircraft using the stand and any adjacent building, aircraft on another stand and other objects as shown in Table 24.8.

Table 24.8 Apron clearance distances

Code letter	Clearance (m)
A	3
B	3
C	4.5
D	7.5
E	7.5

Note: These clearances can be reduced in special circumstances where the code letter is D or E – for details reference should be made to ICAO Annex 14. Consideration must also be given to the provision of service roads and to manoeuvring and storage area for ground equipment.

24.3.10 Aprons – slopes

Slopes on an apron including those on an apron taxilane should be sufficient to prevent accumulation of water on the surface of the apron but should be kept as level as drainage requirements permit. On an aircraft stand the maximum slope should not exceed 1%.

24.3.11 Obstruction surfaces

Imaginary surfaces which extend over the area occupied by the airport and beyond its limits are defined. It is necessary to restrict the creation of new objects and to remove or mark existing objects (whether man-made or naturally occurring) which project above these imaginary surfaces. A plan view of them is shown in Figure 24.1 and dimensions are given in Tables 24.9 and 24.10. The main components are:

(1) An inner horizontal surface located 45 m above the airport elevation extending to a horizontal distance *a* measured from the aerodrome reference point.
(2) A conical surface with a slope of 5% above the horizontal, a lower edge coincident with the periphery of the inner horizontal surface and an upper edge located at a height *b* above the inner horizontal surface.
(3) Transitional surfaces along the side of the strip and part of the side of the approach surface *q* that slopes upwards and outwards at *c*% to the inner horizontal surface.
(4) Take-off surfaces established for each runway direction. The limits of the take-off surfaces are determined by an inner edge, two sides of which initially are diverging and then parallel and an outer edge, the inner and outer edges being perpendicular to the flight path. The inner edge has a length *l* and is at the end of the clearway if provided (and if it exceeds the specified distance) or at a distance *m* from the end of the runway. Each side diverges at a rate of *n*% relative to the extended centreline of the runway until a specified maximum width *p* is reached, continuing thereafter at that width to the outer edge. The distance between the inner and outer edges, or length of take-off surface, is *q* and the surface slopes up at *r*% to the horizontal.
(5) Approach surfaces established for each runway direction

Table 24.9 Approach runways: dimensions for obstacle limitation surfaces

Runway classification		Non-instrument code number				Non-precision approach code number			Precision approach		
									Category I code number		Category II or III code number
Surface and dimensions		4	3	2	1	4	3	2, 1	4, 3	2, 1	4, 3
Inner horizontal											
Height		45	45	45	45	45	45	45	45	45	45
Radius	*a*	4000	4000	2500	2000	4000	4000	3500	4000	3500	4000
Conical											
Slope		5%	5%	5%	5%	5%	5%	5%	5%	5%	5%
Height	*b*	100	75	55	35	100	75	60	100	60	100
Transitional											
Slope	*c*	14.3%	14.3%	20%	20%	14.3%	14.3%	20%	14.3%	14.3%	14.3%
Approach*											
Length of inner edge	*d*	150	150	80	60	300	300	150	300	150	300
Distance from threshold	*e*	60	60	60	30	60	60	60	60	60	60
Divergence (each side)	*f*	10%	10%	10%	10%	15%	15%	15%	15%	15%	15%
First section											
Length	*g*	3000	3000	2500	1600	3000	3000	2500	3000	3000	3000
Slope	*h*	2.5%	3.33%	4%	5%	2%	2%	3.33%	2%	2.5%	2%
Second section											
Length	*i*					3600†	3600†		3600†	12 000†	3000†
Slope	*j*					2.5%	2.5%		2.5%	3%	2.5%
Horizontal section											
Length						8400†	8400†		8400†		8400†
Total length						15 000	15 000	2500	15 000	15 000	15 000

*All dimensions are measured horizontally.
†Variable length. Under certain circumstances the length of the second section may be increased but the length of the horizontal section will be reduced by the same amount.

Table 24.10 Take-off runway: dimensions for obstacle limitation surfaces

Runway classification		Non-instrument code number				Non-precision approach code number			Precision approach		
										Category I code number	Category II or III code number
Surface and dimensions		4	3	2	1	4	3	2, 1	4, 3	2, 1	4, 3
Take-off climb											
Length of inner edge	l	180	180	80	60	180	180	60	180	80/60	180
Distance from runway end*	m	60	60	60	30	60	60	60/30	80	60/30	60
Divergence (each side)	n	12.5%	12.5%	10%	10%	12.5%	12.5%	10%	12.5%	10%	12.5%
Final width	p	1200	1200	580	380	1200	1200	580/380	1200	580/380	1200
		1800†	1800†			1800†	1800†		1800†		1800†
Length	q	15 000	15 000	2500	1600	15 000	15 000	2500	15 000	2500	15 000
								1600		1600	
Slope (%)	r	2	2	4	5	2	2	4/5	2	4/5	25

*The take-off climb surface starts at the end of the clearway if the clearway length exceeds the specified distance.
†1800 m when the intended track includes changes of heading greater than 15% for operations conducted in IMC, VMC by night.

used for the landing of aeroplanes. The limits of the approach surfaces are determined by an inner edge, two diverging sides (when viewed from the runway end) and an outer edge, the inner and outer edges being perpendicular to the flight path. The inner edge of length d is located at a distance e from the runway threshold. Each side diverges at a rate $f\%$ from the extended centreline of the runway to the outer edge and the length to the outer edge is g. The slope of the surface above the horizontal is $h\%$. Non-precision approach and precision approach runways have an approach surface in which the outer section length j is at a flatter slope $k\%$ and with a horizontal section beyond.

24.4 Airport concept and layout

24.4.1 General

Growth of aviation over recent years has been accompanied by a continuous process of change, and airport planners have increasingly become aware of the need to provide flexibility for future extensions and modifications of the facilities, bearing in mind that 10 to 15 years may elapse between master planning and commissioning of a major airport.

Conceptual planning has been influenced by the trend towards larger aircraft for handling the increasing numbers of passengers. The number of air passengers carried throughout the world on scheduled services by airlines of ICAO member states has risen from 111 million in 1961 to 639 million in 1981.

The size of aircraft has increased greatly; the Boeing 747, for example, is 70.5 m long and has a wingspan of 59.7 m and a maximum height of 19.4 m, whereas the earlier Boeing 707 had corresponding dimensions of 44.2, 39.8 and 12.7 m.

The ability of modern aircraft to land in crosswinds has resulted in a much reduced need for subsidiary runways in different directions. Also, the potential capacity of two independent runways means that few airports now need to be planned with more than two runways which can be parallel. Thus, a simple pattern of widely separated parallel runways has emerged which assists the planner to achieve a rational layout with the ground handling facilities located between the runways and served by a common access 'spine' as illustrated in Figure 24.2.

Examples of such layouts can be seen at Amman, Changi (Singapore), Munich and Athens. It will be noted that this type

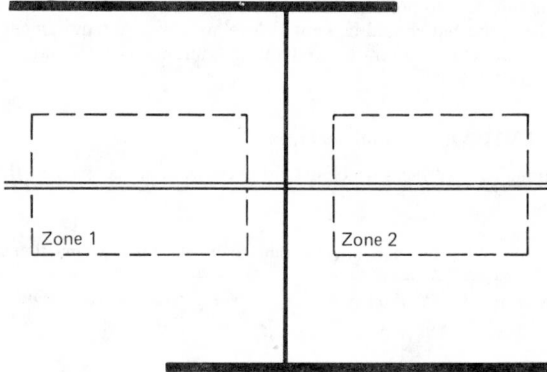

Figure 24.2 (1) Maintenance and cargo zones; (2) Terminal zones

of layout allows considerable scope for future extension of the airport facilities.

An airport is designed to meet many needs but compromise is inevitable since some of the most important requirements present varying degrees of incompatibility. The main factors are:

(1) Rapid and efficient handling of passengers.
(2) Minimum walking distances.
(3) Simple directional guidance for passengers.
(4) Maximum runway movement rates.
(5) Minimum taxiing times.
(6) Rapid aircraft turnround on the apron.

Whilst layouts of the airside facilities of runways, taxiways and aprons are governed by international standards described previously, no such standards exist for the design of passenger terminal buildings and other ground facilities. It is therefore in this part of the airport plan that the designer can exercise his individual skill.

No two terminal buildings are the same but modern airports for large- and medium-sized aircraft generally follow the pattern of parallel runways with the terminal facilities based on one of two principles, either: (1) centralized handling; or (2) decentralized handling. In the former, all the facilities such as check-in, baggage-handling, customs and immigration, restaurants, bars, concessions, banks, etc. are concentrated in one location, with associated car and aircraft parking facilities. There is, however, limited airside and landside frontage. Aircraft sometimes have to be parked away from the building with access by piers or apron buses and landside car parks tend to involve long walking distances.

Decentralization involves the distribution of these facilities over several centres in the terminal complex. The concept includes the range of variations from independent, or unit, terminals, each with the full complement of facilities, to the provision of facilities at the aircraft whereby passengers undergo a complete check-in (the gate check-in concept).

The small airport for light aircraft will almost certainly have centralized handling facilities and it may well require one or more cross-runways owing to inability of light aircraft to operate in strong crosswinds.

Various aspects of planning the airport layout follow in greater detail.

24.4.2 Runways

The number of runways at any airport, other than one for light aircraft, will be determined from the number of aircraft expected in a given period, usually an hour, but it is difficult to give general guidance since the capacity of any runway or runway system depends on a variety of factors such as:

(1) Aircraft types.
(2) Landing aids.
(3) Air traffic control techniques.
(4) Ground movement capability (e.g. taxiway and apron facilities).

There will be significant differences between the capacities under instrument flight rules (IFR) and visual flight rules (VFR) and the IFR capacities will be lower. The major airports handling high rates of commercial air transport movements operate under IFR even in good weather conditions.

As an indication, the capacity of a single runway handling a mixture of air transport and general aviation aircraft will be in the order of 37 movements per hour, assuming roughly equal numbers of landings and take-offs. The maximum figure may rise to about 50 movements per hour under VFR but, of course, VFR operations are entirely dependent on favourable weather conditions.

For parallel runways, maximum capacity is achieved when separation is sufficient to enable each runway to be operated independently with mixed landings and take-offs. The total capacity will then be in the order of 74 movements per hour. A minimum runway centreline spacing of 1800 m is required for this mode of operation.

The staggering of the parallel runways depicted in Figure 24.2 reduces taxiing distance at the expense of increased total land requirements.

24.4.3 Runway length

Runway length is dependent on the following main variables:

(1) Aircraft performance.
(2) Aircraft take-off or landing weight.
(3) Aircraft reference temperature.
(4) Airport elevation.
(5) Runway gradient.

Performance curves are published by aircraft manufacturers

and enable runway length to be computed for given sets of conditions.

At a specific airport runway take-off length will be determined by range considerations. Landing length is controlled by the maximum landing weight of an aircraft with allowance being made for the condition of the runway pavement in terms of braking ability.

It should be noted that runway lengths quoted in documents such as the 'UK Air Pilot' do not necessarily equate to actual physical lengths of pavement as account may be taken of the existence of a stopway, a clearway or a displaced threshold.

24.4.4 Temperature and elevation effect on runway length

The average daily temperature (over 24 h) for the hottest month of the year is of interest to the designer and it will be necessary to increase the length of the runways where high temperatures are recorded (see ICAO Annex 14).[1] The elevation of the airport has a like effect, and the basic length of a runway should be increased as also described in Annex 14.

24.4.5 Wind effect on alignment

The use of an airfield is controlled to a certain extent by the wind. Crosswind components may prevent safe usage of the runway and the direction of the runway should be aligned to keep instances of unacceptably high crosswinds to a minimum. To do this, a full summary of wind duration, speed and direction is required, taken over a period of years. From this a convenient graphical method of determining runway orientation as devised by Marwick is as follows.

The recorded hours (as percentage total) for each range of velocities are plotted in the sectors intercepted between concentric circles representing these velocities (Figure 24.3). The runway is then drawn in a trial direction through the centre of the circles and two parallel lines representing 13 knots (or any permissible crosswind component) to the same scale as the circles.

All winds falling outside these lines are in excess of the critical for that particular runway direction. Further trial and error establishes the desired pattern.

Alternatively, a computer may be employed to follow a similar process in order to establish the percentage usability of an airfield having one or several runways in various orientations.

In a multi-runway layout, the main runway may be set in the direction of the prevailing winds and the subsidiary runways are laid in the direction which yields the minimum crosswind component effect and the maximum percentage usability for the whole system. The present tendency is to aim for a single runway system with high permissible crosswind components.

The prevalence and nature of gusts and air turbulence in the area must be considered separately.

24.4.6 Taxiways

At busy airports there will certainly have to be a parallel taxiway for the full length of the runway and, at some of the more sophisticated airports, there may be double or even treble parallel taxiways. Exit taxiways linking the runway and parallel taxiway must be conveniently located so that landing aircraft can vacate the runway as soon as possible. The exit taxiways may either be perpendicular to the runway and parallel taxiway or, where particularly rapid turn-off from the runway is desirable, they may be angled up to 45° to the runway centreline for small aircraft although, for the larger aircraft, the maximum angle should be about 30° which will permit runway exit speeds

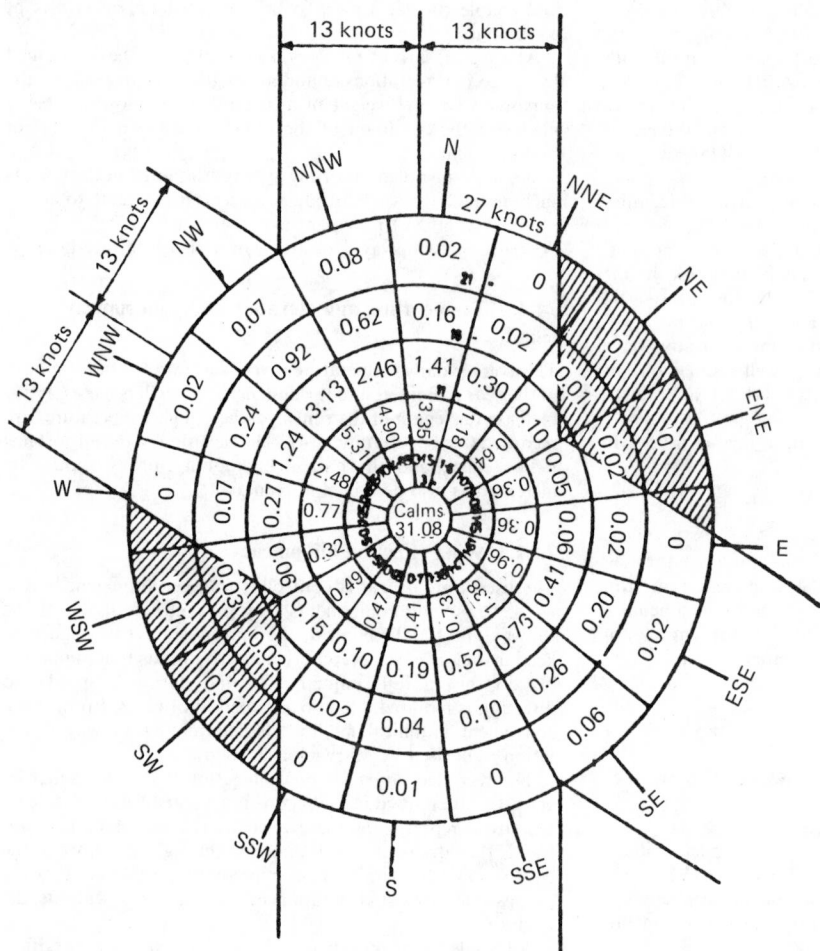

Figure 24.3 Graphical method to determine runway usability

up to 60 m.p.h. (96 km/h). At the other end of the scale, an airport with only low movement rates may not require a parallel taxiway, and back-tracking on the runway would be acceptable.

Taxiways should lead directly on to the end of the runway to enable aircraft to move rapidly into the take-off alignment with maximum occupancy of the runway, although again, at airports with low movement rates, taxiing along the runway may be acceptable to achieve economy in taxiway construction costs.

24.4.7 Terminal area

The terminal area has three main constituents: the aircraft apron, the terminal building and car parking with the associated road system. Their relation to each other will be determined in principle by whether the centralized or decentralized concept is adopted and, at major airports, by the method of internal surface transport. There are other factors which influence the relationship such as the pattern of airline operations, the ratio of domestic to international passengers, number of transfer passengers, etc.

The various centralized and decentralized concepts are illustrated in Figure 24.4.

24.4.8 Centralized concepts

The centralized concept may be considered to include the following variations: (1) simple terminals; (2) linear terminals; (3) finger terminals; (4) satellite terminals; and (5) mobile lounge terminals, although, depending on the extent of facilities provided in the satellite and mobile lounge terminals, these latter variations may tend towards the decentralized concept.

The simple terminal consists of a common area for all passenger handling facilities with several exits on to a small aircraft parking apron. It is only suitable for airports with low passenger and aircraft movements or is adaptable to general aviation operations whether located as a separate complex in a large airport or as an airport used exclusively by small general aviation aircraft.

The linear terminal concept is merely an extension of the simple terminal concept in which the latter is repeated to provide additional apron frontage and increased space for passenger processing which may feature a two-level arrangement for separating arriving and departing passengers. Passenger walking distance from set-down kerb to aircraft is relatively short. Linear terminals can easily be extended although this may destroy the advantage of short walking

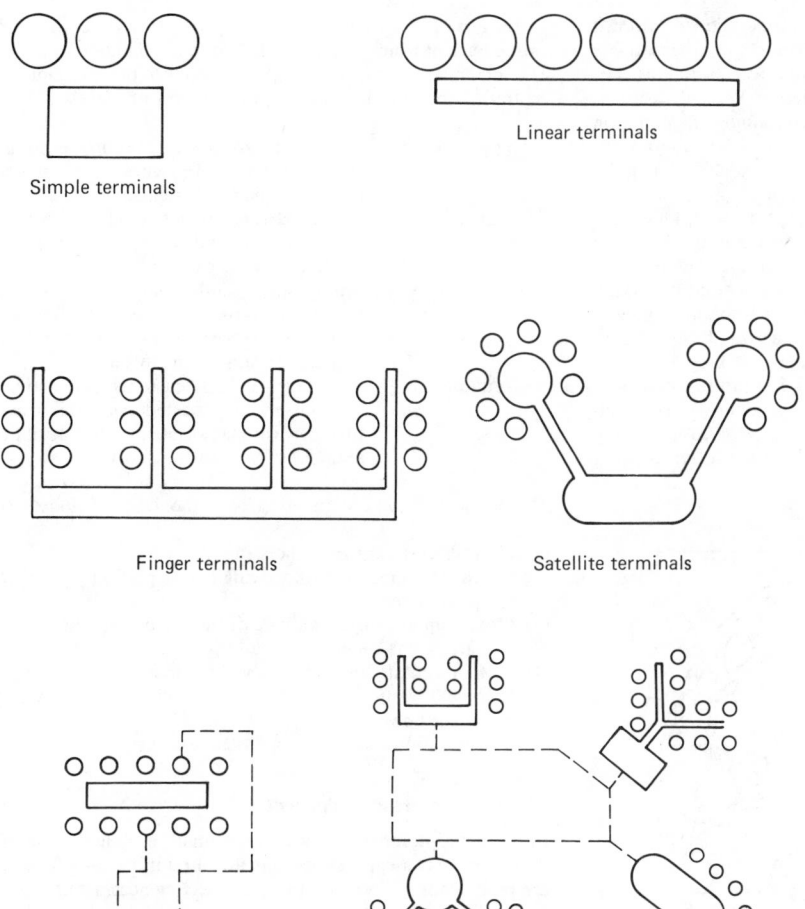

Simple terminals

Linear terminals

Finger terminals

Satellite terminals

Mobile lounge terminals

Unit terminals

Figure 24.4 Terminal area concepts

distance if directional signing is inadequate, and passengers cannot leave their cars opposite the appropriate aircraft departure gate with its adjacent passenger processing facilities.

The finger or pier terminal has evolved from the early provision of a covered walkway between the simple terminal and the aircraft such that later arrangements now incorporate holding lounges at the gate and vertical separation of departing and arriving passengers. A disadvantage of the concept is the long walking distance involved from the central processing facilities to the aircraft gate. There are many examples of this arrangement, that for Belfast Airport being illustrated in Figure 24.5. The necessity for provision of adequate space between fingers for manoeuvring aircraft is to be noted.

The features of the satellite concept are similar to those of the finger concept except that aircraft gates are located at the end of a long concourse rather than being spaced at intervals along it. Walking distances are relatively long and later developments have incorporated a people-mover system between the central terminal and satellites, as at London Gatwick. An advantage is that satellite gates can be served from a common holding lounge. The aircraft parking arrangement more readily allows the introduction of self-manoeuvring stands although the

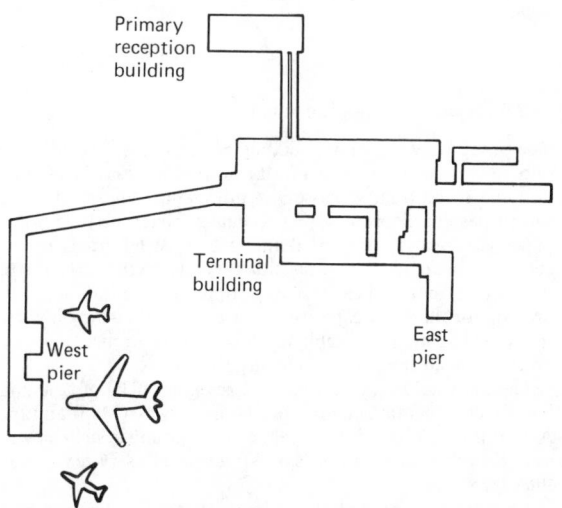

Primary reception building

Terminal building

West pier

East pier

Figure 24.5 Belfast airport

wedge-shaped stands tend to impair the operation of aircraft servicing equipment. Expansion is difficult with the satellite concept other than by introduction of additional satellites. The ultimate satellite arrangement is depicted at Paris Roissy (see Figure 24.6) where the main building containing the common facilities is completely surrounded by satellites containing waiting-lounges, access between terminal and satellite being by tunnel.

The mobile lounge or passenger transporter concept has been used at Dulles International Airport, Washington, D.C. The mobile lounges transport passengers between the common processing facilities in the central terminal and the aircraft parking apron or aprons, where they can be used as holding lounges. This arrangement reduces walking distances and allows considerable operational flexibility for aircraft parking-apron arrangements with excellent opportunities for future expansion. The cost of providing and operating independent service buildings and mobile lounges together with time involved in moving passengers by the mobile lounges will, however, often prove a disadvantage.

Figure 24.6 Roissy, Paris. Note 'drive-through' parking (A) car park

24.4.8.1 Heathrow Terminal Four

One of the largest terminal building projects in the world came into operation in 1986 at Heathrow Airport and is a good example of centralized passenger processing. The need for a fourth passenger terminal was recognized back in 1975 when passenger forecasts suggested that the three terminals in the central area would reach saturation capacity by the early 1980s. As there is not sufficient space within the central areas to provide for the extra capacity it was decided that the only site that could be made available for development to meet demand on time was to the south of the airport.

The Terminal Four complex occupies some 40 ha of land and has direct access to London's orbital motorway (M25) and the A30. It is also linked directly to the underground system as well as to the other three terminals by a frequent bus service through the cargo tunnel.

The designed annual throughput of Terminal Four is 8 million international passengers with one-way flow of 2000

passengers. The terminal is served by twenty wide-bodied aircraft stands of which sixteen are linked to the building by a pier. The planning of the building is based on the principle of centralized processing which is provided in three levels.

Upper level:	Departing passengers are processed on this level where after check-in, immigration and security passengers enter a common departures lounge which in effect is a pier 25 m wide and some 640 m long with satellite areas at each end.
Mezzanine level:	Immigration and health control are located on this level where arriving passengers are processed and then proceed to baggage reclaim at the lower level.
Ground level:	The baggage hall, customs and arrivals concourse are located on this level together with associated public facilities and access to road transport.

The major function criteria adopted in the design include:

(1) Centralized passenger processing.
(2) Complete segregation of arriving and departing passengers for security reasons.
(3) Maximum unassisted walking distance from the check-in to aircraft gate is 200 m.
(4) 75% of aircraft stands are served via loading bridges.
(5) Complete vertical separation of arriving and departing passenger flows.
(6) Maximization of non-aeronautical revenues.

24.4.9 Decentralized concept

In this concept, independent unit terminals, each incorporating the complete passenger processing and aircraft parking facilities are built around a system of interconnecting access and service roads. The separate terminals may take the form of any of the centralized concepts previously described and be built to the requirements of specific airlines or groups of airlines (as at Kennedy, New York) or may be split for operation by route type (as at Heathrow, London) into arrival and departure; alternatively, they may be split into domestic and international functions.

This concept is usually justifiable at high-volume airports where walking distances become excessive with finger terminals. It can, however, cause problems for transfer passengers unless a high level of inter-terminal connecting services is provided, as at Dallas, Forth Worth, US, illustrated in Figure 24.7. Future extension of the decentralized concept can be difficult because of the land requirements for each terminal. Development costs are high because similar facilities must be provided at each unit terminal.

Sophisticated developments of the centralized satellite terminal can result in this concept merging towards a decentralized system if each satellite contains complete passenger processing facilities. Complete decentralization is not achieved, however, if a central terminal is retained with common car-parking provision.

24.4.10 Apron layout

The overall size and layout of the aircraft apron will depend on the number and type of aircraft likely to be parked at any one time. The number of stands is derived from the aircraft standard busy rate (SBR, see section 24.5 for general definition) and is a complicated process often necessitating computer simulation studies. Small airports are treated empirically and a rough rule

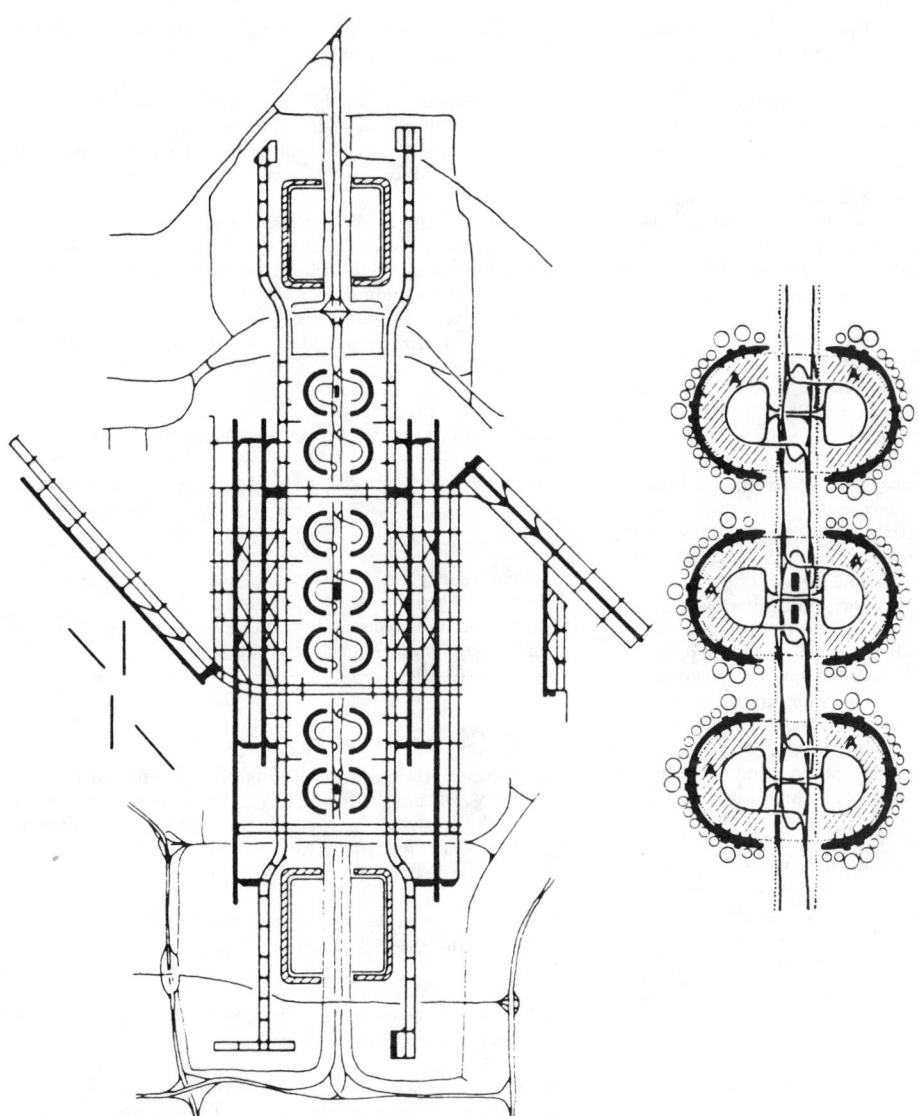

Figure 24.7 Dallas, Forth Worth (A) car park

is to increase the SBR by 10% and round up to the next whole number.

Minimum wing-tip clearances between adjacent aircraft and from aircraft to buildings must be maintained according to the standards previously referred to. The area of the stand will also be governed by the mode of parking. Nose-in parking, in which the aircraft must be mechanically pushed backwards on leaving the stand, requires special vehicles for this purpose but is more economical in overall area requirements than stands where the aircraft is self-manoeuvring. Most large and busy airports tend to adopt nose-in parking. Typical stand areas for various groups of aircraft are given in Table 24.11.

Access of aircraft to and from the parking stand is obtained by defined taxilanes on the apron surface. The width and other design parameters required for these taxilanes should be similar to those for independent taxiways as previously described such that the necessary wing-tip and obstacle clearance are maintained.

Table 24.11 Aircraft stand areas

	Nose-in parking (m)	Self-manoeuvring (m dia.)
Airbus	85×85	100
Long haul	65×65	90
Medium haul	50×50	60
Short haul	40×40	50
General aviation	—	30

Each stand position must be of sufficient area to accommodate the wide variety of mobile ground service equipment which is required for the modern aircraft. Generally, a minimum 3 m should be added to the apron depth to permit service access and 10 m additional depth may be required for operation of the push-out vehicle used in the case of nose-in parking.

A service road, typically 7 to 10 m wide, should be provided adjacent to the terminal building. Vertical clearance of 5 m should be available over the road.

A graphical design method for determining the separation of aircraft parking stands has been devised by the ICAO in the *Aerodrome design manual*,[3] Part 2 and, in *Airport aprons*,[4] the FAA has published graphs and equations for the determination of clearances for aircraft turning and taxiing out of a parking position. The *Apron and terminal building planning report*,[5] prepared for the FAA, provides scaled outlines for six groups of aircraft and gives general guidance for planning airport apron–terminal complexes.

24.4.11 Terminal building layout

The functions, flow pattern, accommodation, configuration and size of the terminal building or buildings need individual assessment for the factors of influence are many and differ in each case. Simulation and computer models have been developed to aid design of this most complex of buildings and are likely to be used for the larger terminals.

The usual approach to determining the required floor area is to estimate the requirement for each facility derived from the peak hour or SBR passenger demand (see section 24.5). After categorizing the peak hour passengers into international and domestic types and also into terminal and transit passengers, it is possible to estimate the number of passengers to be processed in each facility, such as check-in desks, lounges, customs and immigration, etc. and, hence, to determine the space requirement for each facility to ensure reasonable provision. Various guides are obtainable for estimating the space requirements of the different facilities. The Federal Aviation Administration and IATA have published the guidelines summarized in Tables 24.12 and 24.13. Perrett[6] gives the following approximate guide for the total capacity of the terminal:

(1) 1500 passengers per hour in each direction for every 15 000 m² of area available to the public.
(2) 1500 passengers per hour each way for every 25 000 m² of total terminal (excluding office accommodation).

Reductions of 30 to 40% could be made in the areas for terminals handling predominantly domestic traffic. Conversely, the space could be increased drastically if, for example, there were a high proportion of visitors.

Perrett gives additional useful data on terminal building design and the *Airport terminals reference manual*[7] (IATA) is also helpful.

Table 24.13 International Air Transport Association standards

Passenger requirements in any specific area	Space required per peak hour passenger (m²)
Standing passengers	1.0
Seated passengers	1.5
Plus 10% additional circulation and airline requirements space at lounges	

24.4.12 Car parking layout

The problems arising out of making provision for car parking are among the most difficult facing the airport designer. In general, the majority of passengers travel to and from airports by car. Visitors and airport workers must also be catered for. There are five main categories of car parking:

(1) Kerbside – for setting down and picking up.
(2) Short term – say up to 15 h.
(3) Long term.
(4) Staff – both airport and airline.
(5) Visitors – accompanying departing passengers, meeting arriving passengers and casual spectators.

The parking areas required to accommodate these various demands can be considerable and air travellers may comprise a small proportion of the total car users. No standard guidelines are available for determining the various parking requirements which are likely to differ from airport to airport. Estimates of traffic flow must be made by conventional methods such as census sampling. A decision must be made on the comparative proportion of short-term and long-term parking if, indeed, the alternatives are considered desirable. It is normal to price these facilities differentially to encourage rapid turnover in the short-term car park, which is usually located closest to the terminal building. On large airports, long-term parking may be extensive and the distance from the terminal building may necessitate a shuttle bus service.

24.4.13 Airport access

In addition to the terminal building, apron and car parking arrangement, the airport planner must consider and make provision for the alternative modes of surface access by which air passengers, airport workers and visitors may move to and from the airport.

Although road access for cars must invariably be provided, consideration must also be given to provision for taxis and

Table 24.12 Federal Aviation Administration standards

Domestic terminal space facility	Space required per peak hour passenger (m²)
Ticket lobby	1.0
Airline operational	4.8
Baggage claim	1.0
Waiting rooms	1.8
Restaurants	1.6
Kitchen and storage	1.6
Other concessions	0.5
Toilets	0.3
Circulation, mechanical and maintenance, walls	11.6
Total:	24.2

International terminal space facility (additional to domestic requirements)	Space required per peak hour passenger (m²)
Public health	1.5
Immigration	1.0
Customs	3.3
Agriculture	0.2
Visitors' waiting rooms	1.5
Circulation, baggage assembly, utilities, walls, partitions	7.5
Total:	15.0

public buses. Other modes of access such as railways may also be favoured – London Gatwick and Heathrow have surface and underground railway links respectively from the city centre which carry in excess of 42% of all persons passing through the airports.

The design of access road systems and other modes of airport access are outside the scope of this chapter.

24.4.14 Ancillary buildings

It has been customary to collect the remaining airport buildings under this heading but some, such as large hangars, cargo terminals, etc. may be major projects in their own right.

24.4.15 Control tower

This should give controllers a view of all the runways and is designed round the equipment required for air traffic control. Large areas of false floor to accommodate cabling may be needed. The tower is generally a separate building and not a part of the terminal building.

24.4.16 Apron control

Some airports include an apron control cabin located so that an apron controller can direct aircraft to the apron stands from the taxiways.

24.4.17 Aircraft catering building

This should preferably be located close to the terminal area and is a specialist catering building run by the airlines.

24.4.18 Cargo terminal building

This facility may be a simple framed building or a sophisticated terminal such as that of British Airways at Heathrow Airport which comprises transit sheds housing computer-controlled mechanical handling equipment, office blocks, vehicle parking, loading bays and circulation. There are no particular civil engineering requirements.

24.4.19 Maintenance hangars

These may range from a simple framed building to a major structure such as the British Airways hangar for the Boeing 747 at Heathrow Airport, London. The structure basically is a cladding for the maintenance requirements but consideration of large clear spans and door openings will determine the structural forms.

24.4.20 Buildings for electrical and electronic equipment

These, generally, are simple buildings designed to house particular items of equipment, some of which may require a controlled environment. The manufacturers advise on this point. The buildings for certain navigational aids cannot have ferrous metal above a specified level and the manufacturer's advice should be sought. Others may require special shielding. Generally speaking, there are no particular construction problems.

24.4.21 Airfield lighting

The extent of the approach and runway lighting provided not only depends on the airport classification but should be compatible with the radio and radar landing aids provided. It generally consists of high-intensity centreline and crossbar lighting for the approach areas, together with runway centreline and edge lighting. Taxiway lighting usually consists of green centreline lights supplemented with blue-edged lights at junctions and around the apron area.

For visual guidance in the angle of descent, visual approach slope indicators (VASIs), or precision approach path indicators (PAPIs) are provided.

24.4.22 Telecommunications

Telecommunications is a general term covering radio navigational aids and radar in addition to data and voice communications.

All modern airports require telecommunication services to some degree and in the case of a larger international airport those can be quite extensive. These services will consist of some or all of the following:

(1) Air–ground radio communication.
(2) Land mobile radio.
(3) Navigational aids.
(4) Final approach and landing aids.
(5) Radar.
(6) Direct speech communication.
(7) Direct data communication.
(8) Public communication services.

Recommendations and requirements concerning telecommunication requirements are given in the ICAO Annex 108.[8]

The positioning of the various telecommunication facilities is extremely important and must conform to the accepted ICAO recommendations in respect of siting and the grading of surrounding areas.

The following electronic services, not covered by the term 'telecommunications', are also required in most cases:

(1) Meteorological systems.
(2) Flight information display systems.
(3) Public address systems.
(4) Security systems.

24.4.23 Airport security

Attacks on civil aircraft for the furtherance of extreme political aims, both on the ground and in the air, have become a major feature of air travel since around 1970. Security on the ground at airports has therefore had to be developed to counter this trend.

New technology is playing a significant role in upgrading the standard of security at airports but there are some basic problems that remain unsolved. Airport security has made major advances since the early 1970s but it is only new terminals or airports that incorporate security as part of initial planning. In most cases, the attempt was to make secure an existing building, which in most cases proved very expensive and not 100% successful.

Computers play a prominent role in sophisticated security systems together with more advanced X-ray and electronic 'sniffer' equipment but the process is in a continuous state of evolution in order to cope with new types of explosives and plastic guns that are not easily detected on X-ray machines.

Apart from the severe high cost of security, the human factor is always at the centre of most security systems used at airports for screening passengers and their baggage.

24.5 Traffic forecasts

The capacity of the apron, terminal building and car parks is determined from the traffic forecasts. Such forecasts are normally made on an annual basis and are split into scheduled and charter flights for both domestic and international services.

These annual forecasts are determined by one of two main methods: firstly, extrapolation of historical data and, secondly, by analysis of such factors as future income levels, regional and national development plans, future population forecasts, tourist potential, etc. This analysis is usually carried out on a computer.

The annual figures are then converted to hourly flows, each called standard busy rate (SBR).

The SBR is defined as that rate which is exceeded 29 times in the year, and has been found to give a reasonable basis for design.

It is essential to obtain an estimate of the short-period flow rates for passengers and aircraft. This can be done by an analysis of monthly, weekly, daily and hourly aircraft movement patterns but, unless the relevant data is available, an assessment using ratios of standard busy hour passenger rates to annual movements is more likely to be the only suitable method. In general the ratios decrease with increasing annual movements and they tend to be higher at airports with high proportions of international leisure traffic. They also tend to be high where one route dominates the schedules, as occurs at many small airports, and therefore such airports need to be independently considered. Special consideration should clearly also be given at airports such as Aberdeen and Sumburgh where there is a high proportion of helicopter operations. Table 24.14 gives an indication of the ranges of ratios.

The passenger SBR is then used to determine the SBR of the aircraft movements estimating the likely mix of aircraft, capacity and load factors, taking future trends into account.

Table 24.14 Standard busy rate (SBR) values

Annual passenger movements	SBR/annual movement ratio
100 000	0.002–0.003
250 000	0.001–0.002
500 000	0.0007–0.0012
1 million	0.0006–0.0010
2–5 million	0.0004–0.0009

24.6 Aircraft pavements

24.6.1 General

Pavements suitable for the aircraft that will use them are required for runways, taxiways, aprons, maintenance areas, etc. The determination of pavement type and thickness is complex with many interacting variables involved which are often difficult to quantify. The first mathematical approach to airfield pavement design was introduced in 1945. Since then, there has been progressive refinement of the approach to suit increasing loads and complex landing gear configurations.

This section is intended to provide guidance to the principal considerations and methods used in aircraft pavement design.

24.6.2 Function of aircraft pavements

The general functions of aircraft pavements are as follows:

(1) Adequate strength for all aircraft types likely to use the airport.

(2) Adequate strength to resist the effects of repetitive loading.
(3) Absence of loose particles which could be sucked into aircraft engines.
(4) Imperviousness to water – resistance to jet blast.
(5) Resistance to fuel spillage (particularly on aprons and maintenance areas).
(6) Good surface drainage.
(7) Ability to accept temperature movements.
(8) Good skid resistance.
(9) Good riding surface for comfort in the aircraft.
(10) Economy in construction and maintenance.

24.6.3 General requirements of an aircraft pavement

From an operational point of view it is difficult at busy airports to close down a runway, taxiway or apron for the purposes of pavement maintenance or strengthening; indeed, routine maintenance may have to be carried out at night. Pavements designed to fulfil the needs of only the immediate future may prove to be expensive in the long term. A runway may be required to have a life of 20 years or more and the designer must anticipate requirements as far into the future as possible.

24.6.4 Construction

The three types of pavement construction may be grouped as follows: (1) rigid; (2) composite; and (3) flexible.

An example of each of the three types is shown in Figure 24.8.

In the UK it is usual practice to provide for all pavement types a 100 mm thick layer of dry lean concrete directly on the compacted subgrade. This gives immediate weather protection and acts as a working platform for placing subsequent layers. It is interesting to compare this with American practice where, under certain conditions, full-depth asphalt flexible pavements may be laid directly on the subgrade.

24.6.5 Choice of construction

The choice of type of construction and of materials depends on the location and function of the pavement, or overlay, the ground conditions or existing pavement and, very importantly, the cost. It is normal to carry out several designs, using all possible materials combinations, and to cost each design for comparison. A compromise between technical excellence and economy is often necessary.

24.6.6 Rigid pavements

Concrete surfacing is resistant to fuel spillage and to engine exhaust blast, has good friction characteristics and good resistance to scuffing. It is thus often preferred for aircraft parking and fuelling areas and for turning areas at runway ends. However, because of the need for construction in bays, with joints at regular intervals, it is often considered less suitable for runways and taxiways, where the uniform surface afforded by bituminous surfacing is of advantage.

A concrete pavement is usually considered as being rigid because the load is spread over a wide area of subgrade by virtue of its inherent flexural strength. The concrete can be reinforced or unreinforced and is divided into rectangular bays to restrict the tensile stresses which are induced by a combination of three factors:

(1) Contraction of the slab due to falling temperature and concrete shrinkage. This movement is restricted by the friction between the slab and the subgrade and as a result tensile stresses are induced in the slab.
(2) Warping of the slab due to a temperature gradient through

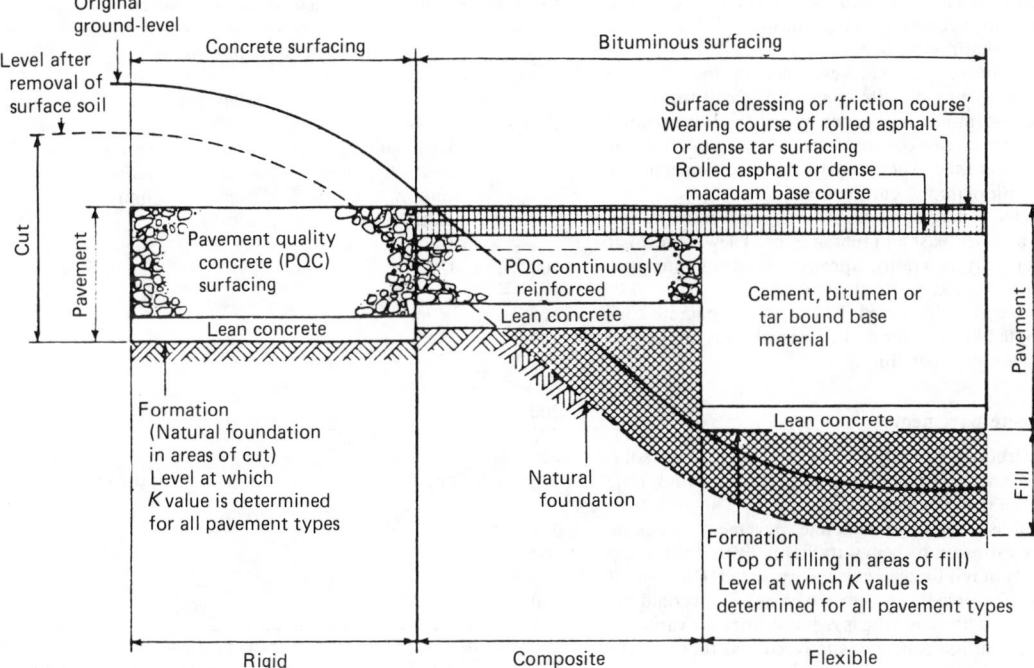

Figure 24.8 Alternative recommended types of aircraft pavements

the thickness of the slab. High surface temperatures cause the slab to dome until it is supported mainly at the edges, whilst low surface temperatures cause the corners to curl upwards.

(3) Loading. Slabs are usually most susceptible to loading near their corners which may cause cracks to form across the corner. Acute angles in slabs should therefore be avoided.

The bays are separated by contraction joints and the bay size depends on the slab thickness. The maximum bay sizes should be as indicated in Table 24.15.

Table 24.15 Maximum bay sizes of concrete runways

Slab thickness (mm)	Bay size (m)
150 or less	3
151–224	3.75
225–274	5.25
275 and over	6

The contraction joints may be formed by using crack inducers as shown in Figure 24.9. Load is usually transferred between adjacent slabs by aggregate interlock in which case no dowel bars are needed. If the aggregate particles in the concrete are not too hard, however, a more satisfactory solution is achieved by continuous casting of the slab, perhaps employing slipforming techniques, followed by sawing the joints after the concrete has set. Slots in the surface of concrete pavements, whether preformed or sawn, should be as narrow as possible; they should be filled with a semi-compressible material such as hardboard or fibreboard, depending upon the subgrade, and need not be sealed.

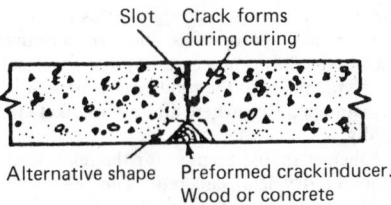

Figure 24.9 Contraction joint

Expansion joints may be provided in thin slabs but may be entirely omitted in slabs more than 250 mm in thickness.

Single butt construction joints as shown in Figure 24.10 are recommended since those incorporating a joggle are susceptible to cracking. Dowels may be omitted for slabs 275 mm thick and over.

The pavement quality concrete (PQC) used in rigid pavements should be designed on the basis of its flexural strength measured by loading 152×152 mm test beams, rather than on cube strength. It is the strength of the concrete when it is first loaded which is of importance so that age factors may be taken into account. The aggregate/cement ratio should not exceed 6.3:1 and the water/cement ratio should be less than 0.50.

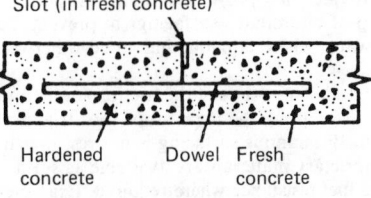

Figure 24.10 Construction joint

The PQC slabs should be placed over a layer of dry lean concrete having an aggregate/cement ratio of 15:1 and a minimum cube strength of 5.2 MN/m².

In order to improve the skid resistance of the concrete surface, the concrete may be wire combed or small transverse grooves may be cast into the wet concrete surface. It is essential that experiments are carried out to ensure that such treatment is applied at the correct time. Alternatively, the hardened concrete may be scored with diamond cutting drums.

Well-constructed concrete pavements show little cracking and are resistant to both jet blast and fuel spillage. They are ideal at runway ends, taxiway junctions, aprons and on maintenance areas where aircraft stand or are slow-moving.

Joints can be largely eliminated if prestressed concrete construction is adopted but this form of construction is unlikely to be economic under most conditions.

24.6.7 Composite pavements

Composite construction can often provide an economical solution, with the advantages of a bituminous surfacing without the disadvantages of a concrete pavement.

In a continuous reinforced concrete pavement, cracking (accentuated by exposure to heavy traffic) is likely to develop whatever quantity of reinforcement is incorporated. However, if the continuous reinforced concrete pavement is overlaid by bituminous surfacing, the cracking is reduced since the variation in the temperature in the concrete is lowered and those cracks which do form in the concrete are not subject to wear and are unlikely to be severe. While there is some tendency for cracks to form in the bituminous surfacing above those in the concrete they are usually minor and can be resealed easily.

The flexural strength of the concrete slab gives this form of construction good load-spreading properties, and a good riding quality surface can be obtained. It is not as resistant to jet blast, heat and fuel spillage compared with the rigid pavement so it is often used on runways and taxiways where aircraft are likely to be moving fairly rapidly.

Pavement-quality concrete should be used for the reinforced slab overlying 100 mm of dry lean concrete. The minimum cross-sectional areas specified for reinforcement for the appropriate concrete slab thickness, as recommended by Martin and Macrae,[9] are given in Table 24.16.

The surfacing, normally 100 mm in thickness, should be rolled Marshall asphalt or dense tar surfacing laid in two courses. This two-course work reduces the tendency to sympathetic cracking in the wearing course over cracks in the underlying slab.

24.6.8 Flexible pavements

The top structural layers of flexible aircraft pavements are usually of a hot rolled asphalt, with the mix designed and controlled by the Marshall method (bituminous concrete in American terminology). This achieves high density and stability and affords an excellent riding surface which has good friction characteristics in dry conditions. In wet weather, however, flat gradients and surface tension lead to retention of surface water. It is common to provide an open-textured, non-structural, friction course on top of bituminous surfacings to prevent the build up of surface water where aquaplaning could otherwise occur.

Water drains through the interstices of the friction course and passes to the runway edge along the impervious top structural layer of the pavement. Bituminous surfacing is not resistant to aviation fuel and proprietary materials are available as surface treatments to provide fuel resistance where required. Examples are 'Salviacim', an epoxy-based surfacing, and 'Jetseal', a tar-based surface sealant. Dense tar surfacing (DTS) by the Marshall method, where tar replaces bitumen as the binder, has also been successfully used in areas subject to fuel spillage.

A flexible pavement is one which depends on its thickness and elasticity to disperse the load to such an extent that the subgrade is not overstressed. It is made up of a number of layers of granular materials increasing in rigidity and decreasing in flexibility towards the surface. The lower materials may be unbound, or bound with bitumen or cement. The middle layers should be asphalt, bitumen or tarmacadam. The surface layers should be impervious and Marshall asphalt or dense-tar surfacing specifications are usual. The following factors have to be considered in relation to the design:

(1) The overall depth of pavement must be such that the strength of the subgrade is not exceeded.
(2) The strength of each individual layer of the pavement must be such as to resist the pressure at that level.
(3) The shearing strength of the surfacing and layers beneath must exceed the shear stresses produced by the tyre load.

For very light-duty pavements several layers may be omitted. When dry lean concrete is used on the subgrade to provide a good working surface it must be weak, otherwise cracks which form in this layer are likely to spread upwards towards the surface. An aggregate/cement ratio of 18:1 for gravel or 22:1 for crushed rock is usually suitable.

Well-designed flexible pavements have good riding qualities but some surfaces are susceptible to jet heat and fuel spillage may cause softening of the surface.

Relatively high landing and take-off speeds of modern aircraft, combined with the flat transverse slopes on runways, have led to the problem of aquaplaning.

24.6.9 Overlays of existing pavements

It is often necessary to overlay existing pavements to provide

Table 24.16 Steel reinforcement in rigid pavements

Slab thickness (mm)	Schedule of reinforcement			
	Main steel		Transverse steel	
	Minimum area (mm²/m width)	Spacing limits (mm)	Minimum area (mm²/m width)	Spacing limits (mm)
100	425		295	125–175
125				
150	530			
175				
200				
225	635	125–175		
250	740		170	150–225
275				
300				
325	825			
350				

Source: Martin, F. R. and Macrae, A. R. (1971) 'Current British pavement design', Paper 6, Proceedings, Conference on Airfield Pavement Design, Institution of Civil Engineers.

greater strength or to repair a damaged surface, to improve ride or friction characteristics, or to provide resistance to fuel spillage. Overlays are usually of bituminous materials, for ease of construction and potential for minimizing disruption of existing operations. However, some work has been carried out using concrete overlays bonded to original concrete slabs. Economic and practical considerations would generally mitigate against such treatment, except in cases where concrete surfacing might be considered essential.

24.6.10 Pavement design, UK method

24.6.10.1 Development

The construction of aircraft pavements did not commence until shortly before the outbreak of war in 1939 and design principles at this time were based on experience of highway construction.

The use by heavy bomber aircraft rapidly overstressed some of these early pavements, and led to investigations into the behaviour of paved surfaces and subgrades and the development of pavement design methods.

The first mathematical approach to airfield pavement design was made in 1945 when a design manual for concrete pavements was issued by the Air Ministry which contained design charts for single-wheel loads based on Westergaard's equations.

Investigations into the behaviour of pavements under increasing loads continued as heavier jet-powered military aircraft with larger and more complex landing gears came into service.

24.6.10.2 The load classification number (LCN) system

The principle of relating aircraft loads and pavement strength by means of a numerical scale, the load classification number (LCN), first established in 1945, remained the UK design and evaluation system until 1971.

The LCN was recognized by ICAO and incorporated into its 'Aerodrome Manual, Part 2' as a recommended method of aircraft and pavement classification in 1956.

In the late 1960s the LCN system was becoming increasingly difficult to apply to the heavy gear loads and a reappraisal of pavement design methods was undertaken utilizing both the latest analytical methods available at that time and the experience gained with the many heavy aircraft pavements constructed between 1950 and 1965. A revised system which introduced the concept of load classification groups (LCGs) for pavement evaluation replaced the LCN system in 1971 and is currently in use in the UK. This is set out in *Design and evaluation of aircraft pavements*[10] published in 1971 by the Department of the Environment, London.

24.6.10.3 The load classification group (LCG) system

The load classification group (LCG) system was published in 1971, and was recognized as a rigid pavement system by ICAO in 1974.

A coarse scale of seven groups was superimposed upon the old LCN scale, reflecting broadly the seven ICAO aircraft classification groups, as can be seen on the design and evaluation chart in Figure 24.11.

The seven groups are referenced by roman numerals in descending order as gear loads and pavement strengths increase, thus group VII is the group of lowest strength and group I is the highest.

Since it is UK practice to construct rigid pavements without load transfer devices at joints, provision is made at the design stage for the increased stresses due to edge and corner load cases by increasing the theoretical slab thickness and by providing a

100 mm subbase of rolled dry lean concrete. As the system was designed around the parameters of rigid pavements, the inclusion of flexible pavements into a common reporting system could only be accomplished by inserting in the group scale flexible pavement thicknesses derived empirically and from experience.

24.6.10.4 The LCG method for rigid pavements

The LCG method requires the following data: (1) aircraft LCG; (2) subgrade modulus; and (3) concrete flexural strength.

The highest LCG corresponding to the aircraft expected to use the airport, excepting the occasional visitor, is selected for the design. The soil subgrade is classified by its subgrade modulus or k value. The minimum flexural strength of the concrete is estimated for the time the pavement is to be loaded; this may be 6 months after construction. If no information is available it is reasonable to use the common value of $3.5 \, \text{MN/m}^2$.

The design chart is entered at the upper value of the LCG band and the pavement quality concrete thickness is then read off the corresponding band of the 'Rigid' column. An example is given on the published chart.

Some points should be noted:

(1) The LCG grouping for the aircraft or the LCN value must be taken from the corresponding column as the LCN values differ from that calculated by the original LCN method or which are given in the ICAO *Aerodrome design manual*,[11] Part 3.
(2) The LCG system and design method is basically related to UK practice and to the soils commonly found in the UK. Not only are these often clay soils with low strength but with all-the-year-round rainfall it is normally advantageous to prepare a working surface on which to lay the pavement-quality concrete. For these reasons the LCG method always incorporates a 100 mm layer of dry lean concrete. This could be omitted with certain suitable soils.
(3) The LCG system recognizes that over 95% of aircraft operate on the central 30 m of runway and almost all taxi along the centreline of taxiways. Thus, the central strips of runways and taxiways, to which must be added all the aprons or holding areas, can be considered as channelized areas. For 'non-channelized areas' one group lower can be selected to reduce the required design thickness.

24.6.10.5 The LCG method for composite pavements

The LCG method is also appropriate to design a pavement which is a composite of a reinforced concrete slab to spread the aircraft load with a bituminous surface. The design process is exactly similar to that of unreinforced rigid pavements except that the composite column of the chart is used.

24.6.10.6 The LCG method for flexible pavements

The simplest way of designing a flexible pavement is to use a similar process as for unreinforced rigid pavements except that the flexible column of the chart is used.

A flexible pavement constructed to such a design would be satisfactory, for the whole construction is in bound material. There is only one system of construction accepted which comprises a 100 mm layer of bitumen bound surfacing, a thick layer of cement, bitumen or tar-bound base material on the standard 100 mm of dry lean concrete.

Some confusion has existed between the LCN values and the LCG system since both use the common term LCN. The actual

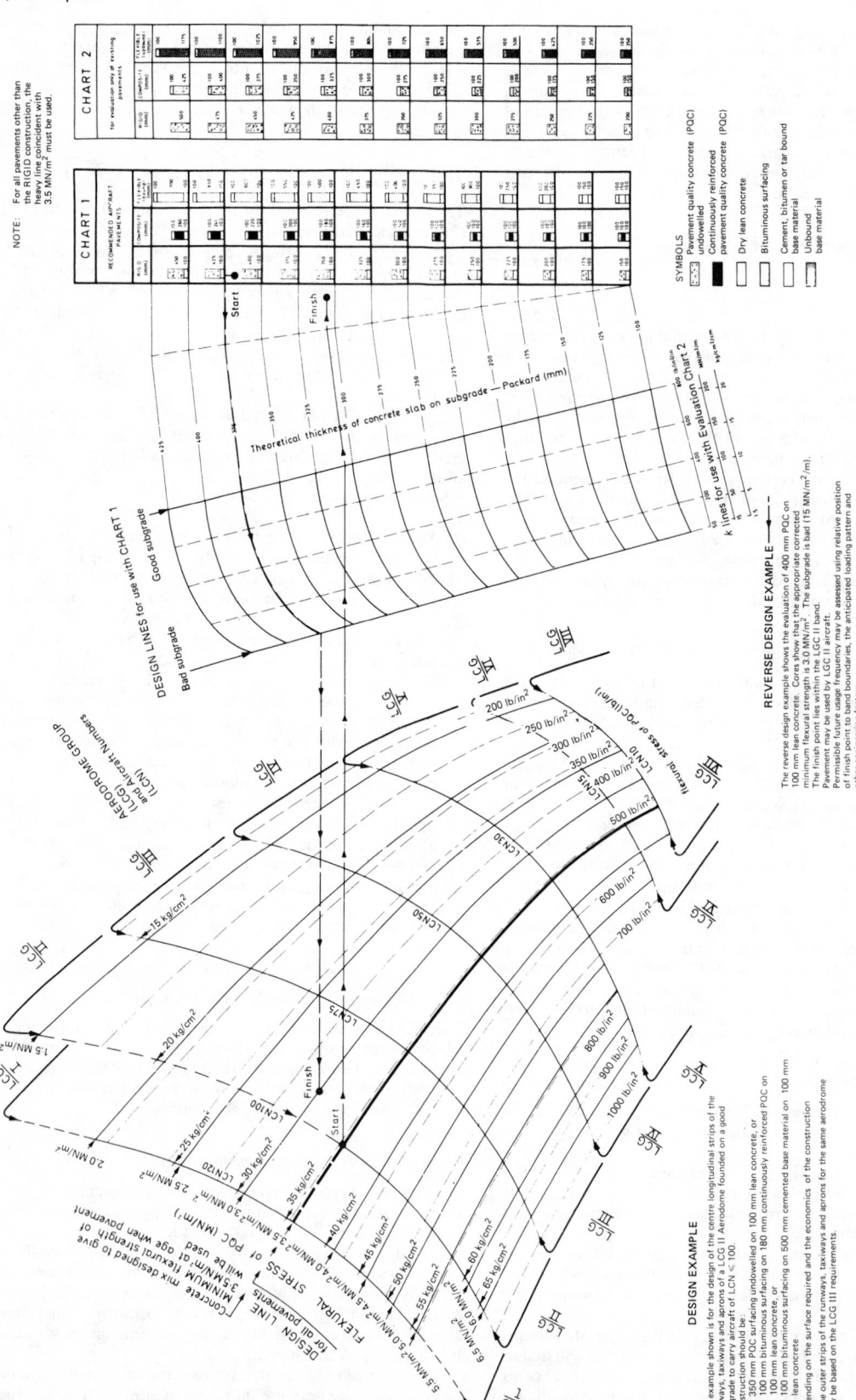

Figure 24.11 Design and evaluation chart for rigid, composite and flexible aircraft pavements

LCN values derived under the two systems are, however, different and unrelated and must not be confused or interchanged.

24.6.10.7 Evaluation of existing pavements

The majority of aircraft pavement works have consisted of strengthening and extending existing pavements. The evaluation of these pavements is rarely an easy matter as there is no mathematical basis on which a calculated evaluation can be made.

Pavement evaluations are normally made either by assessment or by physical testing. Assessments are made either by a 'reverse design' procedure or by the professional judgement of an experienced pavement engineer. Where a reasonably accurate assessment cannot be made, physical testing by means of plate bearing rigs or, more recently, by means of deflection measurements made with a heavy falling weight deflectometer, can be carried out.

24.6.10.8 The pavement classification number (PCN), and design system

The ICAO now requires airports to classify airfield pavements by means of the pavement classification number (PCN) and publishes in *Design manual*, Part 3, aircraft classification numbers (ACN) relating to types and thicknesses of pavement. The relationship between the ACN and PCN measures the ability of the aircraft to use the relevant pavement. It is a classification system but it is not a pavement design system.

The Airfield Pavements Branch of the Property Services Agency (PSA) has recently been developing a new design system for use in the UK and to be published in late 1987. It is expected to provide the design relationship to pavement classification numbers which is currently absent from the ACN/PCN system.

24.6.11 Pavement design – FAA method

24.6.11.1 General

The FAA design method is based on the gross weight of the critical aircraft operating at the maximum take-off weight. The areas of traffic concentration are considered as 'critical areas', which comprise the central portion of the runway, aprons, taxiways and runway ends, where departing traffic will load the pavement. The design charts produce a pavement thickness which is appropriate to critical areas; non-critical areas can have a reduced thickness.

24.6.11.2 Equivalent design aircraft departure

The design method requires the following initial steps:

(1) An estimate of the annual departures (half the total movements) of all aircraft forecast to use the pavement.
(2) Determination of the design aircraft.
(3) Calculation of the equivalent number of departures of the design aircraft.

It should be noted that arrivals are neglected since the landing weight of an aircraft is less than the take-off weight. The design method provides a pavement life of 20 years with the forecast annual departures.

24.6.11.3 Rigid pavement thickness design

The FAA advisory circular *Airport pavement design and evaluation*[12] gives design charts for groups and individual aircraft and use of the charts requires the following:

(1) Concrete flexural strength.
(2) Subgrade modulus.
(3) Gross weight.
(4) Number of the equivalent annual departures of the design aircraft.

The charts give the total concrete slab thickness for critical areas, which can then be reduced by the appropriate factors for non-critical areas.

24.6.11.4 Flexible pavement thickness design

The FAA advisory circular 'Airport Pavement and Design' also gives flexible pavement design charts for the same aircraft groups and aircraft as for the rigid pavements. Use of the figures requires the following:

(1) The CBR value of the subgrade.
(2) The gross weight.
(3) The annual departures of the design aircraft.

The figures give the total pavement thickness of a three-layer construction and the thickness for the critical and non-critical areas of the bituminous surface or wearing course, the thickness of the granular base course and, by deduction, the thickness of the granular subbase.

In addition to the design charts, the FAA present a further chart which shows the minimum base course thickness, and this has to be calculated as a check against the thickness determined from the main charts.

24.6.11.5 Pavement evaluation

The advisory circular contains separate charts for evaluating the strength of an existing pavement.

24.7 Surface water drainage design

24.7.1 General

As for all traffic-bearing pavements, a carefully designed water drainage system is a necessary requirement of an airport. Inadequate drainage may reduce the loadbearing capacity of the subgrade, decrease skid resistance on the surface and cause breakdown of surface vegetation.

In general, the same basic design methods for calculating runoff are used for airports as for highways or urban areas. On the 'landside' of an airport, the methods of dealing with the collection and disposal of surface water by way of gullies and piped systems are conventional. On the 'airside' there are problems which are particular to airports, largely related to the areas involved and the relatively flat grades, which are dealt with in this section.

24.7.2 Drainage for runways

A runway has longitudinal gradients limited to being not steeper than 1.25% on major runways and not steeper than 2% on minor runways. It is wide (up to 45 m), with transverse slopes, limited to between 1 and 1.5% on major, and between 1 and 2% on minor, runways. Gullies and gratings are not acceptable on the runway itself, nor are open ditches within the strip.

It is normal to have a shoulder about 3 m wide adjacent to the runway edge, sloping at 5% away from the runway, often with a subsoil drain under. Ideally, the strip will then slope away from the runway at a slope of about 1.5% to carry surface water runoff either to a storm drain system with grated inlets and manholes or, if the ground slopes away from the runway to the edge of the strip, to an open channel. Where the strip slopes towards the runway, as is permitted within the design standards, then a piped system with grated inlets has to be provided, preferably at the edge of the shoulder.

Because of the flat gradients and the potential for a film of water developing, friction courses have been used on many runways in the UK. These are a thin open texture of bituminous overlay, whereby the water flows through the interstices on the impervious surface of the runway, to the edge, where the collection and disposal is normal.

24.7.3 Taxiways

Permitted gradients on taxiways are such that these, also, have flat longitudinal and transverse gradients. They are dealt with in a similar way to runways.

24.7.4 Aprons

The maximum recommended gradient on an apron is 1%, sloping away from any buildings to minimize any risk arising from fuel spillage. Drainage is usually by continuous grated slot drains dividing the apron into drained areas such that drainage paths are not excessively long. It is undesirable to have frequent changes of gradient on an apron.

24.7.5 Subsoil drainage

Subsoil drainage may be necessary to drain low-lying water-logged areas, or to keep a fluctuating water table well below subgrade level. Open-jointed porous pipes laid in a 'herring-bone', 'parallel' or 'gridiron' system should be used. Depths should be as generous as possible, and should not be less than 0.6 m or greater than 1.2 m below the surface.

24.7.6 Stilling ponds

Airports are most frequently sited on low-lying relatively flat land, and therefore it is common for problems to arise in the discharge of surface water from the airport into the natural main drainage system, particularly when the flow is high in the latter. The use of stilling ponds is common. These provide storage until the level in the main drainage system has fallen, or relieve the peak flow in the main channel.

24.7.7 Main drainage channels

Another feature of airport sites is that there is often a natural major watercourse flowing across them. Where possible this should be diverted, but if this is not possible and a culvert has to be constructed, this should be sized generously, designed for aircraft loading, and be of sufficient length to pass under runway and strip.

24.8 Ancillary services

There are several ancillary services associated with the airport and the main ones are described in the following sections.

24.8.1 Aircraft sanitation

The aircraft toilets are emptied into vehicles and the contents are disposed of at airport sanitation buildings. These house macerators or comminutors which discharge into the foul drainage system. The buildings require an electrical power supply.

24.8.2 Fuel installation

The supply of fuel to aircraft is normally carried out by the fuel companies who contract for a specified period. There are two means of distributing fuel to the aircraft aprons: (1) by aircraft refuellers; and (2) by a hydrant system.

Aircraft refuellers are usually employed and they range from 2250 to 82 000 l. The larger is an articulated vehicle with an overall length of 21.5 m, a height of 3.65 m (including radio aerial), a width of 3.2 m, a turning circle of 21.4 m absolute minimum and a laden weight of 91 t.

Hydrant systems consisting essentially of a distribution network terminating in pits in the apron which contain hose couplings, have been installed at some airports but have, up to now, not been very popular for two main reasons. The first is the inherent inflexibility. The mixes of aircraft at any airport change rapidly and the parking stands on aprons have rarely remained constant for more than 2 or 3 years with the result that the hydrant point has frequently been in the wrong position almost as soon as it has been installed. The second is that the fuel companies' tenure is normally shorter than the hydrant system life and individual companies have not been prepared to finance the high capital cost. However, with the advent of very large aircraft of enormous fuel consumption, even larger fuel dispensers become less attractive and the hydrant system is likely to become of increasing interest in the future. Nose-in aircraft parking is becoming the standard with jet aircraft; this is tending to prolong the life of fixed apron stand positions and is a factor encouraging the greater use of hydrants.

Both the aircraft refuellers and hydrant systems incorporate safety features which prevent the pumping of fuel if the hosepipe should become disconnected.

24.8.4 Ground movement signs

These are placed adjacent to taxiways and aprons to direct the pilots. Details are given in ICAO Annex 14 and CAP 168. In addition, aircraft stand number signs are provided either free-standing or fixed to the terminal buildings or pier.

All these signs will require an electrical power supply.

24.8.5 Crash and rescue services

Fire engines and crash tenders are housed in buildings with quick and easy access to the aprons, taxiways and runway. The scale of provision for the UK is given in CAP 168 and the requirements are related to the heaviest aircraft in regular operation at the airport.

At some airports where a crash in water is possible, rescue boats should be provided.

24.8.6 Boundary and security fences, including crash access

Airports should be fenced properly and the choice of fence depends on availability and cost. Whilst a 1.2 m fence is adequate over most of the perimeter, security and customs may require a higher fence topped with barbed wire strands in the terminal area separating the landside from the airside. The airside/landside fence will require manned gates at all accesses, which should be kept to a minimum.

The perimeter fence should be provided with a number of frangible gates so that crash and rescue services can get quickly to the scene of any crashes which may occur outside the boundary.

24.9 Definitions

The following definitions are taken from these ICAO 'Standard and Recommended Practices for Aerodromes', Annex 14 and CAP 168, 'Licensing of Aerodromes'.

24.9.1 Aerodrome (airfield or airport)

Any area of land or water designed, equipped, set apart or commonly used for affording facilities for the landing and departure of aircraft and including any area or space, whether on the ground, on the roof of a building or elsewhere, which is designed, equipped or set apart for affording facilities for the landing and departure of aircraft capable of descending or climbing vertically, but shall not include any area the use of which for affording facilities for the landing and departure of aircraft has been abandoned and has not been resumed.

24.9.2 Aerodrome beacon

Aeronautical beacon used to indicate the location of an aerodrome.

24.9.3 Aerodrome elevation

The elevation of the highest point of the landing area.

24.9.4 Aerodrome reference point

The designated geographical location of an aerodrome.

24.9.5 Aerodrome reference field length

The minimum length required for take-off at maximum certificated take-off weight, sea-level, standard atmospheric conditions, still air and zero runway slope as described by the certificating authority or equivalent data from the aeroplane manufacturer.

24.9.6 Apron

A defined area on a land aerodrome, intended to accommodate aircraft for the purpose of loading or unloading of passengers or cargo, refuelling, parking or maintenance.

24.9.7 Barette

Three or more aeronautical ground lights closely spaced in a transverse line so that from a distance they appear as a short bar of light.

24.9.8 Clearway

A rectangular area at the end of the take-off run available and under the control of the aerodrome licensee, selected or prepared as a suitable area over which an aircraft may take a portion of its initial climb to a specified height.

24.9.9 Crosswind component

The velocity component of the wind measured at or corrected to a height of 10 m above ground-level at right angles to the direction of take-off or landing.

24.9.10 Instrument approach runway

A runway intended for the operation of aircraft using non-visual aids providing at least directional guidance in azimuth adequate for a straight-in approach. Those runways served by instrument landing systems (ILS) are designated precision approach runways and are further identified as either category I, II or III dependent on the sophistication of the ILS system and the ability to permit operations in various levels of reduced horizontal and vertical visibility.

24.9.11 Non-instrument runway

A runway intended for the operation of aircraft using visual approach procedures.

24.9.12 Obstacle

All fixed (whether temporary or permanent) and mobile objects, or parts thereof, that are located on an area intended for the surface movement of aircraft or that extend above a defined surface intended to protect aircraft in flight.

24.9.13 Runway effective slope

The slope computed by dividing the difference between the maximum and minimum elevations along the runway centreline by the runway length.

24.9.14 Shoulder

An area adjacent to the edge of a paved surface so prepared as to provide a transition between the pavement and the adjacent surface for aircraft running off the pavement.

24.9.15 Stopway

A defined rectangular area at the end of the take-off run available, prepared and designated as a suitable area in which an aircraft can be stopped in the case of an abandoned take-off.

24.9.16 Strip

An area of specified dimensions enclosing a runway to provide for the safety of aircraft operations.

24.9.17 Taxiway

A defined path, on a land aerodrome, selected or prepared for use of taxiing aircraft.

24.9.18 Threshold

The beginning of that portion of the runway usable for landing.

References

1 International Civil Aviation Organisation (1983) *International standards and recommended practices for aerodromes*, Annex 14 (8th edn). ICAO.
2 Civil Aviation Authority (1984) *Licensing of aerodromes*, CAP 168. CAA.
3 International Civil Aviation Organization (1977) *Aerodrome design manual: taxiways, aprons and holding bays*. (1st edn, Part 2) ICAO, Document 9157-AN/901, Montreal.
4 Federal Aviation Administration (1965) *Airport aprons*, Advisory Circular AC 150/5355-2, FAA.
5 Ralph M. Parsons Co. (1975) *The apron and terminal building planning report*, Report FAA-RD-75-191, FAA. (Rev. March 1976)

6   Perret, J. D. (1971) 'The capacity of airports – planning considerations', *Proc. Instn Civ. Engrs,* paper no. 7372, **50,** 435–450.

7   International Air Transport Association (1976) *Airport Terminals Reference Manual* (6th edn), IATA.

8   International Civil Aviation Organization (1972) *International standards and recommended practices: aeronautical telecommunications,* Annex 108, Vols I & II (3rd edn). ICAO.

9   Martin, F. R. and Macrae, A. R. (1971) 'Current British pavement design', paper 6, Proceedings, Conference on Aircraft Pavement Design, Institution Civil Engineers.

10 Department of the Environment (1971) *Design and evaluation of aircraft pavements,* DoE.

11 International Civil Aviation Organization (1977) *Aerodrome design manual* (1st edn) Part 4: Document 9157-AN/901, ICAO.

12 Federal Aviation Administration (1978) *Airport pavement design and evaluation,* FAA advisory circular AC150/5320-6C.

Bibliography

International Civil Aviation Organization (ICAO) publications

(a) 'International standards and recommended practices – environmental protection', Annex 16 (1st edn) Vol I.

(b) 'Aerodrome design manual' (1st edn) Part 4: 'Visual aids', Document 9157.

(c) 'Airport planning manual' (1st edn) Part 1: 'Master planning', Document 9134.

(d) 'Airport services manual' (1st edn) Part 1: 'Rescue & fire fighting', Document 9137.

(e) 'Airport services manual' (1st edn) Part 2: 'Pavement surface conditions', Document 9137.

(f) 'Manual on air traffic forecasting' (1st edn) Document 8991.

(g) 'Heliport manual' (1st edn) Document 9261.

(h) 'Stolport manual' (1st edn) Document 9150.

US Federal Aviation Administration advisory circulars

(a) 150/5300-6A 'Airport design standards, general aviation airports, basic & general transport' (2.24.81).

(b) 150/5200-8 'Planning and design criteria for metropolitan STOL ports' (11.5.70).

(c) 150/5325-2c 'Airport design standards – airport served by air carriers – surface gradient and line of sight (2.6.75).

(d) 150/5325-4 'Runway length requirements for airport design' (4.5.65).

(e) 150/5325-5B 'Aircraft data' (7.30.75).

(f) 150/5335-1A 'Airport design standards – airports served by air carriers – taxiways (5.15.70).

(g) 150/5335-4 'Airport design standards – airports served by air carriers – runway geometrics (7.21.75).

(h) 150/5340-4C 'Installation details for runway centreline & touchdown zone lighting systems (5.6.75).

(j) 150/5340-19 'Taxiway centreline lighting system' (11.4.68).

(k) 150/5340-24 'Runway & taxiway edge lighting system' (9.3.75).

(l) 150/5370-10 'Standards for specifying construction of airports (10.24.74).

(m) 150/5390-1B 'Heliport design guide'.

25

Railways

D S Currie FEng, FICE, MIMechE
Director of Civil Engineering
British Railways

Contents

25.1 Earthworks and drainage

The contours of the territory to be crossed by a railway are obviously decisive as to its average gradient but they are also the background to fixing the maximum permissible gradient within the limits of tractive and braking adhesion. The lower the difference between average gradient and maximum gradient the greater is the practicable train load and the lesser are the deviations from constant-speed running. The minimum curvature to be used also determines the differences between the line speed limit and local speed restrictions. Long curves of small radius on heavy gradients may involve derailment hazards for very long and heavy trains, arising from braking or tractive effort surges along the train. Both maximum gradient and minimum curvature have a large effect on the earthworks cost of constructing a railway and, because of this, the ideal of constant-speed running is often subject to heavy qualification. This is particularly the case in mountainous country.

Railway alignment needs to be planned to give a volume balance between excavation in cuttings and tipping in embankments, subject to the material excavated being suitable for tipping to form embankments and subject to minimizing the haul of the excavated soil. Recourse to borrow pits for embankments and spoil hauls for cuttings should be minimized.

The route may also be affected by other considerations which may be economic, environmental or technical. These would become apparent in a full site investigation, where such matters as previous mining activity, underground services, effect on neighbouring structures, nature of the groundwater table and watercourses would be taken into account. Guidance is available in relevant Codes of Practice, such as BS 6031 and BS 5930. (See also Chapters 9, 11 and 17.)

25.1.1 Site investigation

This will vary according to the extent of the problem. At the outset, a preliminary study may give adequate information to specify the route corridor from geological maps and memoirs, topographical maps and aerial photographs. Earth satellite imagery with interpretation of selective wavebands by specialists in remote sensing can indicate important features. Water table conditions may vary throughout the year from those obtaining at the time of exploration.

For more localized investigation, the type of equipment (augers, percussion and rotary tools, penetration heads, loading plates, pumps), instruments (piezometers, inclinometer tubes, seismometers, resistivity meters, gravimeters, etc.) must be chosen according to conditions. Relevant disturbed or undisturbed samples should be procured for testing. According to the type of ground, the construction and the design philosophy applied, it may be necessary to carry out full-scale site testing with long-term monitoring of instruments.

25.1.2 General

In the past, railways have been maintained over poor ground, with inadequate trackbed materials with a high input of labour time and at slow or moderate train speeds. Although the following concepts may be used in modifying old railways, the basic approach is to obtain a minimum maintenance high-speed railway accepting normal freight traffic on conventional sleepers.

25.1.2.1 Cutting slopes

Slopes in natural ground may be constructed at safe angles according to the properties of the soil.

25.1.2.2 Rock cuttings

British Railways has indicated safe slopes for rock cuttings and angles of repose for rock embankments in a chart reproduced as Table 3 in BS 6031. If steep cuttings in rock are essential, then it is necessary to apply engineering geology concepts to assess joint sets in relation to the direction of slope. Rock anchors, rock bolts and sprayed concrete may be used to assure stability before considering the use of mass-retaining walls. Chalk and certain other soft rocks are susceptible to weathering and frost action and may be protected by vegetation cover.

25.1.2.3 Soil cuttings

Where the slopes are in non-cohesive sands and gravels, the angle of repose is the limiting gradient (with a maximum value of 1:1). Usually, the gradient is shallower than this due to the presence of finer layers in the soil system or a silt or clay matrix around the gravel. The non-cohesive soils tend to be self-draining but erosion can occur when water springs part way up the slope. At such locations a non-woven geotextile filter can be placed and covered with uniform coarse stone (away from the sun's rays) which will hold the fabric in place.

Drainage measures should take the form of preventing water reaching the slope and of removing it from the slope. Unlined ditches behind the crest of the slope increase the hazard. Drainage trenches, whether behind or below the crest, should be designed to intercept water and have impermeable membranes below them and on the downfill face to prevent water once collected from re-entering the soil. Modern counterfort or buttress drains differ greatly from the original open forms. Like all drains they should be lined with a geotextile layer to prevent erosion behind and fouling within them. The top 1 or 2 m should be composed of impermeable material, either compacted clay fill or a system of stone and plastics membrane to prevent surface water reaching deep into the ground, where it could cause internal pore water pressure at likely slip surfaces.

In some cases, there is a mantle of more permeable silt or sand overlying the older clay and, if possible, a lateral interceptor drain should be placed about 20 m behind the crest to prevent the fast seepage of water.

In addition to the above rotational slides, translational slides can occur, usually as a shallow mass moving on a planar surface. They can take the form of slab or block slides, wedge failures, debris slides and flow slides. These possibilities are assessed in the site investigation. Mires are formed as peat is laid down in a specific sequence with variations in soil content, dimensions of fibre (roots, trees, etc.) and extent of humification. Peats reduce greatly in volume under the effects of loading and drainage. In making cuttings in a peat system, quite substantial waterflow can occur and a filter is advised to deal with fine particles otherwise carried from the slope.

In all cases of cuttings, other engineering works at ground level should be considered. Urban or industrial construction involves roads which act as catchments to deliver water to local drainage systems and also water services: if these are defective and near the slope, water can flow to increase pore pressure and precipitate a slip. Similarly, surcharging, especially if accompanied by dynamic loads from construction plant or the placing of storage containers, can seriously reduce the factor of safety.

Steep rock faces, chalk cliffs, and boulder-strewn hill slopes may present rockfall or chalkfall problems which may need special watchmen, signalling provisions, special fence, apron or tunnel protection.

25.1.2.4 Embankments

The nature of the natural soil to receive the embankment must be investigated. If it is too weak to receive the embankment loading at the rate of placing likely to be used by the contractor, failure could occur. A total stress analysis is applicable for this. The conventional technique is to place berms as counterweights at the position of heave. A modern alternative is the use of reinforcing geotextiles or meshes (usually of plastics) to resist tensile forces. The fabric is laid directly on the ground and covered with a granular layer at least 200 mm thick; this acts as a filter and permits construction plant to move easily over the site without sinking into the underlying soft soil. Further fabric or mesh is laid at higher levels, the number and spacing of layers depending upon the engineering properties of the specialist type of material chosen.

Slope angles. Embankments may be formed at angles varying from ratios of 1:1 horizontal to vertical for crushed rock and gravels to 2.5:1 or even shallower for clays and silts. The slope angles depend both on the material and rainfall. For modern railways, peat should not be used as fill material.

The surfaces of slopes should be protected from erosion. This may occur naturally as vegetation is established, followed by a protective topsoil. If surface erosion is a problem, various systems are available, such as spraying with a seed mulch, turving, placing filter fabrics held down by gravel layers or using a honeycomb mesh to hold seeded compost in place.

Rockfalls. Vertical or sloping rock faces may erode or topple to cause rocks of various sizes to fall towards the running line. Although vegetation may help by bonding superficially, protection must involve coping with the energy of rockfall and removing the debris regularly. If space is available, one or two berms are constructed between the base of the slope and the railway to collect scree. For some faces, a plastic geomesh is adequate.

Reinforced earth. The dimensions of embankments and of gravity-retaining structures can be reduced by the inclusion of various strengthening meshes, fabrics, strips and rods of metal, glass fibre or plastics. The tensile resistance of these elements is applied to the adjacent granular soil to produce a composite system permitting the construction of vertical faces in the fill. These external faces are protected by facing elements usually of concrete, resulting in a structure which is economic and which can accommodate settlement. The various qualities of creep, longevity of reinforcement, corrosion, etc. are still the subject of study but many such structures exist throughout the world, including railway environments. There are none so far beneath a high-speed running line and, in this particular case, such systems can be installed up to a horizontal distance of 5 m from the running line.

Trackbed designs. The thickness of subsleeper construction (trackbed) is a function of number and size of axle loadings and of the subgrade soil. Modern railways assess the subgrade soil in one of two approaches: (1) the classification of the soil according to its physical properties and taking account of the water table; or (2) correlating some measured strength or modulus of the subgrade with an empirical design chart. The thickness of the trackbed may also depend upon its component layers to protect against frost, water and particle movement. In very frost-prone areas, such as Central Europe, the thickness of ballast for frost protection exceeds that which might be required for prevention of bearing capacity failure.

The thickness of ballast (all dimensions are below bottom of sleeper) is a minimum of 200 mm to permit tamping machines to operate. Although new ballast injection machines would permit this to be reduced, a high-speed or high-axle-load railway would require 300 to 500 mm thickness for minimum maintenance.

Drainage. The control of water in the trackbed is a major factor in designing the construction layers in relation to the type of subgrade soil as discussed above. Soils such as sands and gravels which may be drained fairly easily do not present a great problem unless there is an artesian head: in this case, there can be a slow upward migration of fine or medium sand under track vibration combined with water flow, and a geotextile is necessary to hold down this sand. Cohesive soils cannot be drained easily and the installation of a channel or pipe will only reduce the pore water pressure to invert level in its immediate vicinity. It is not practical to attempt to remove water from clay in this way as such a large number of drains would be required. If water arrives through precipitation or by flow from adjacent areas, then a relatively small amount is sufficient to produce deleterious changes in pore pressure and so it is practical to provide drains to intercept and remove this free water. Most of the water in cohesive soil is held in capillary suction.

The relationship between the moisture content of cohesive soil and the water table is complex, depending upon the over consolidation ratio (OCR) of the soil. Weathering reduces the OCR effect at the surface.

One object in designing the system of drains and track is to produce a maintenance-free system or one needing minimal attention. Channel drains are readily accessible for cleaning and deal with rainwater, they can be laid at very slight gradients and can deliver at catchpits to deeper pipe carriers if necessary. Many pipe drains act both as collectors and carriers, allowing water to enter at open joints or through perforations. The various forms of pipe are glazed earthenware, galvanized corrugated steel, pitch fibre and, now coming more into use because of ease of handling, plain or perforated PVC pipe. Geotextiles are of great use in static drainage conditions and most railways report satisfactory results using commercially available filtering non-woven fabric. For normal purposes, a fully heat-bonded geotextile with a surface density in the range 100 to 200 g/m² is acceptable; needle-punched geotextiles should be of slightly heavier grade. Fabrics placed in quasi-static conditions near the track should be of heavier grade up to 350 g/m², or more if needle-punched. Such a geotextile would be placed in the sixfoot of a double track line after one track had been cleaned or blanketed. This would prevent slurry from the dirty adjacent track flowing across.

A modern standard design for a side drain is a trench lined with geotextile with a perforated pipe drain running along its base; above and around the pipe is placed uniform stone such as ballast, with the top of the geotextile lapped over the stone about 200 mm below the surface: more stone on the geotextile protects it from disturbance and from the effect of the ultraviolet rays of the sun. As perforated drains can release as well as collect water, possibly at susceptible locations, it is often the practice to place a polyethylene film to line the trench.

25.2 Ballast

Ballast, the material around and below the sleeper, is placed to provide support and lateral resistance to the sleeper. It permits adjustment of level and alignment as required and, if this is done manually, the maximum size of particles should be about 50 mm. Ballast should be free-draining, mainly of single size, of cubical shape but, above all, durable so that there is negligible volume change under track loading. The wet attrition value (WAV) gives the best correlation, with minimum maintenance requirements over the long term and this is determined by the test described in BS 812:1951, clause 27, which specifies the exact size and type of sample to be used. If particles of different size from the 50 to 37.5 mm required in the test are used, then a different WAV is obtained. The WAV should not exceed 4% for good ballast and it is possible to obtain stone with a WAV down to 1%. It is rare for synthetic stone, such as slag, to have adequate wet attrition properties and they may generally be grouped with most limestones as being unsuitable to be in contact with the sleeper. Hardness is difficult to define in relation to other tests but, if the ultrasonic pulse velocity of the homogeneous mineral is 6000 m/s or greater, the stone will be suitable. In severe climates, a freezing and thawing test may be applicable.

The dimensions should conform by weight to the values shown in Table 25.1; and the 1.18 mm limit effectively minimizes the amount of dust present.

Table 25.1

Square mesh sieve	% to pass
63	100
50	100–97
28	20–0
14	2–0
1.18	0.3–0

The stone should have a maximum flakiness index of 50%. For elongation qualities, not more than 2% by weight of particles should have a dimension exceeding 75 mm.

The effect of many tamping cycles is to break up ballast particles; stone as hard as possible is required to accommodate this. When a sleeper is tamped, a horizontal load – the major principal stress – is applied to the ballast, causing it to deform vertically and lift the sleeper. The minor principal stress is vertical with the subsequent arrangement of stone particles in the least favourable position to support track loading, even though the rail level is now correct. The maximum rate of rail settlement occurs after tamping, which reduces as the stone packs down under traffic, with the major principal stress becoming vertical. The trackbed becomes more stable under vertical loading and the best relevelling procedure is the manual or mechanical placing of measured quantities of small stones of nominal 20 mm size between the ballast bed and the sleeper, which is lifted to insert the stone. There is now a track-levelling machine, electronically controlled, which evaluates cant and level from an advancing inclinometer trolley and places the exact quantity of stone by pneumatic injection to obtain proper level. As stone more than 50 mm below the sleeper is not moved and brought with its dirty matrix up to bottom sleeper level, as with a tamper, the likelihood of pumping track is somewhat lessened; the beneficial effect of high-quality small stone is not reduced by mixing with existing worn stone and there is less attrition of the base of the sleeper.

25.2.1 Track profile

Under traffic, rail level, as measured by accurate optical or inclinometer-based machines, shows a profile which is repeated even after many successive tamping and loading cycles. This is related to the care with which the original trackbed layers were installed. When the subgrade is prepared initially, and as subsequent layers of blanket and ballast are placed and compacted, it is necessary to provide extra sighting instruments on compaction and grading plant so that there are no short variations in level. This can be successfully done by using a laser system aimed at the surface a fixed distance ahead and which causes the equipment to compensate for deviation from the required level by moving its instrumented blade up or down.

The ends of the sleepers need to be boxed-in with shoulder ballast to a minimum width of 150 mm for any tracks, 200 to 250 mm for running lines carrying moderate-speed, moderate-axle-load traffic, 300 mm minimum for welded track on the straight and 350 mm for welded track on curves. Generally, there is little advantage in extending shoulder ballasting beyond 300 to 350 mm. In the last decade or so, the practice of raising the shoulder ballast in a slope from about top of sleeper level at the rail to about top of rail level at the shoulder edge has become common on European railways. This practice not only increases the lateral stability of the track but provides a useful reserve of boxing ballast which can be temporarily utilized to make good the boxing ballast when the track is tamped. The angle of the shoulder should be about 55°.

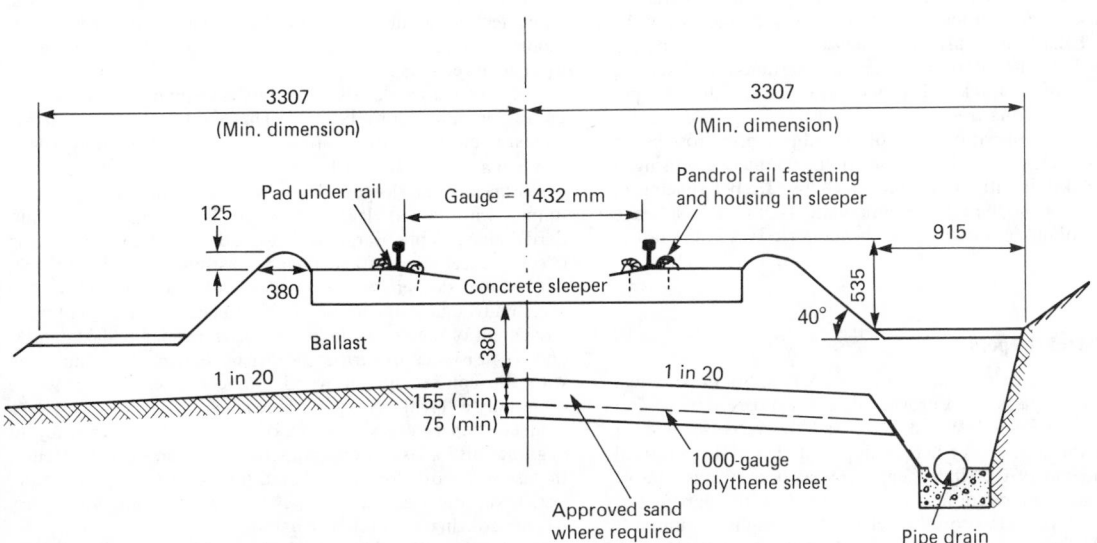

Figure 25.1 Components and formation for continuous welded rail (CWR) track (dimensions in millimetres)

25.3 Sleepers

25.3.1 Timber sleepers and sleeper spacing

The traditional track support since the early days of railways has been the timber sleeper, but now prestressed concrete sleepers are in use in many countries. In some countries the baseplated timber sleeper is already more expensive to install than a baseplateless prestressed concrete sleeper.

In Britain, the sleepers and point and crossing timbers have traditionally been of softwood, chiefly douglas fir from Canada, Maritime pine from southwest France and Corsica, and Baltic redwood from Poland and Russia. Homegrown fir sleepers have been used when they have been available. In continental Europe, beech and oak sleepers have been widely used. All softwood sleepers, and most hardwood sleepers that can be impregnated, are pressure creosoted by the Bethell or Ruping processes before being baseplated. The harder softwood sleepers such as douglas fir or scots fir and hardwood sleepers need to be incised prior to creosoting. Four to 14 litres of creosote per sleeper is a normal absorption, according to species. In Canada, the US and South America a mixture of mineral oil and creosote is used.

A timber sleeper may have a first or running line life of between 6 and 50 years with an average of about 20 years according to traffic loading, weather exposure, the nature and incidence of the indigenous vegetative and insect enemies of timber, the presence or absence of baseplates, the nature of the fastenings, the quality of the ballast and its maintenance, the species of timber and, significantly, the rate at which it has grown.

The British softwood timber sleeper is $250 \times 125 \times 2600$ mm with switch and crossing timbers of 300×150 mm section. In Europe, hardwood sleepers mainly of oak or beech are used with a depth of 160 mm. Sleeper spacing in Britain is 700 mm with 650 mm on curves or on timber sleepered continuous welded rail (CWR). In countries other than Britain, timber sleeper spacings of 600 mm are common with some as close as 550 mm in heavy-axle situations.

The sleeper spacing applied represents a compromise between overall sleeper cost and the lessened frequency of maintenance attention and reduced rail bending stress which results therefrom. In addition, a decrease in sleeper spacing may be appropriate where the formation is weak or where it is not practicable to increase the total track construction depth by increasing the depth of ballast. Since also increasing the density of sleepering increases both the lateral and vertical resistances to buckling movements of the track, it is general practice to reduce sleeper spacing when laying track intended to carry long welded rails. Where the rail formation is regularly subjected to frost heave distortions, there may be a case for deliberately choosing a rather weak rail with complementary close sleeper spacing in order to allow the track to accommodate itself to a frost heave contour without imposing undue bending stresses in the rails.

25.3.2 Steel sleepers

Steel sleepers are widely used in India, Africa, South America, Asia and in parts of Europe. Steel shortages during and following the Second World War, and the increased use of concrete, impeded the further development of steel sleepers until about the mid 1970s. Since then, there has been a considerable amount of research done in Australia, Europe, Japan and Britain to evolve improved steel sleeper designs. Computer-based design techniques, e.g. finite element analysis, and modern stress measuring and analysis techniques have been applied to both full-scale laboratory loading and fatigue testing

and to in-track service trials in order to prove the efficacy of new steel sleeper designs.

Many of the older steel sleeper designs gave average service lives of over 50 years before failing through cracking in, or adjacent to, the rail seat area. Generally, these cracks propagated from rail fastener holes or slots or from discontinuities in the rail seat area. Abrasive wear and/or corrosion in this area also contributed ultimately to sleeper failure. The steel sleeper design features, and especially those associated with the rail fastener mountings, together with the accumulated gross traffic tonnage carried, are the most important factors determining steel sleeper life. Recently improved steel sleeper designs have given due cognizance to these influences.

It is commonplace to recondition old steel sleepers which have already given a 30- to 50-year 'first life', by welding-on suitably designed reconditioning plates. If required, the reconditioning plates used may permit a change in rail fastener type or rail section to be made for the further 'second life' of the steel sleeper. When damaged in service, e.g. by derailments, steel sleepers can readily be repaired by pressing and/or welding.

Steel sleepers do not burn or suffer from exposure to dry heat. In tropical climates their immunity to insect or fungoidal damage is extremely beneficial.

Loss of metal section through corrosion is usually surprisingly low, but there may be a few very special sites subject to severe corrosion, where the use of steel sleepers would be inadvisable. Under-design of the rail seat, or inadequate support thereof, can greatly increase the localized corrosion and fatigue damage in this generally highly stressed area, and due allowances have been made in modern designs.

Steel sleepers pack neatly into bundles and thus simplify all handling and transport operations and greatly reduce these costs. The relatively low mass and convenient and uniform shape of steel sleepers is of benefit in both mechanical and manual handling.

The use of steel sleepers provides good resistance to lateral and longitudinal movement of the track and gives high consistency of gauge, both at installation and during subsequent service. The achievement of good alignment is necessary when steel sleepers are initially installed and when retamping after a brief 'running in' period. The additional care and attention at this early stage generally results in a reduced need for subsequent maintenance, realignment and retamping. Some of the older steel sleeper designs were difficult to install and adjust, but modern designs are now relatively easy to install and to use for 'spot replacements'.

The inverted trough shape and its entrapment of tamped-up ballast provides many benefits. The natural resistance of the ballast is more effectively utilized by steel sleepers, and ballast depth may be decreased by 75 to 100 mm as compared with wood or concrete sleepers. This can be of distinct advantage in tunnels with limited clearances and ballast depth. Where standard ballast depth is retained, the additional load-spreading effect of steel sleepers and their ballast-filled inverted trough, may enable sleeper spacing to be slightly increased. Also, the inverted trough shape has the in-built benefit that any enforced movement of the sleeper increases the entrapped ballast density and its resistance to further sleeper movement, in comparison with the solid shape of wood and concrete sleepers where eventually a 'break-free' point is reached.

Steel sleepers are similar to concrete in their tendency to degrade soft ballast more rapidly than wooden sleepers. Hence, the use of a hard stone is preferred for steel sleepers. Nevertheless, many old steel sleepers have given long and satisfactory service utilizing a gravel-type ballast.

Insulating pads, fastener shoes and/or washers must be incorporated in the steel sleepers designed for lines using track circuit signalling. In most cases the insulating components used are identical to those used with concrete sleepers. Modern plastic materials have improved the electrical security of these insulating components and the track circuits, but the high electrical conductivity of steel sleepers requires special attention to their design and maintenance.

The mass of individual steel sleepers is usually mid-way between that of wood and concrete, and may be considered a disadvantage with respect to resistance to the vertical buckling of continuously welded rail track. However, the entrapped ballast in the inverted trough of steel sleepers and their general shape (particularly the newer designs) greatly increases the 'effective in-track mass' of steel-sleepered track.

Most rail fastener types can be accommodated in steel sleeper designs, and there are only isolated fastener systems which cannot be economically incorporated in steel sleeper designs.

25.3.3 Concrete sleepers

The first experiments with reinforced concrete sleepers were made over 100 years ago. Concrete is attractively resistant to decay, insect attack and fire, but it is brittle, and reinforced concrete sleepers would not withstand the impact loads in main line track. A solution developed in France in the 1920s was to use pairs of reinforced concrete blocks connected by steel tie bars. This twin-block sleeper continues in use notably in France, Spain and North Africa but is of declining importance worldwide.

Under the pressures of timber shortages during the Second World War, a satisfactory monoblock concrete sleeper was developed in Britain in 1943 using the then new technique of pretensioned, prestressed concrete. This has led to a major industry in Britain (over 30 million pretensioned concrete sleepers have been supplied to British Rail) and many other countries (notably Norway, Sweden, USSR, Hungary, Czechoslovakia, Iraq, Japan, South Africa, Australia, Canada and the US).

In Britain and most of the other countries pretensioned concrete sleepers are made by the long-line method in which the tendons are fully bonded. This gives good distribution and control of prestress. Post-tensioned sleepers were developed in West Germany in the 1950s and have found limited favour (in West Germany, Italy, Spain and Mexico).

The current British Rail standard sleeper designated F27BS, is prestressed with six 9.3 mm, seven-wire strands (Figure 25.2). It is 2.515 m long, 203 mm deep under the rail, 264 mm wide at the base and weighs 280 kg. For more arduous conditions, a stronger version F27AS has been introduced, prestressed with eight strands in the same concrete envelope as the F27BS. Recently the F40, a 2.480 m long sleeper, has been developed to facilitate single-line mechanized track renewal within the tight British Rail structure gauge. The base width is increased to 285 mm to maintain the bearing area and the depth is 200 mm under the rail which gives the same weight as the F27BS. Prestressing is by six strands.

Spacing in British track is usually 700 or 650 mm, although 600 mm spacing is used in severe curves.

In the past 15 years, concrete sleepers have gained wider international acceptance for several reasons, including:

(1) Higher passenger train speeds and heavier freight axle loads necessitate higher quality and stronger track.
(2) High labour costs and intensive track use necessitate reduced frequency of maintenance cycles.
(3) Increased use of mechanized track-laying and renewal equipment has overcome the difficulty of handling heavy concrete sleepers.
(4) Softwood sleepers are often unable to withstand the stresses in modern main-line track and good-quality hardwood is increasingly expensive and difficult to obtain.
(5) With modern fastenings, only a resilient pad is required between the rail and sleeper thereby eliminating expensive base plates.

It is generally predicted that prestressed concrete sleepers will have an average life of 40 to 50 years. Some of the earlier designs of concrete sleepers have been removed from the track due to premature failure of the fastenings although the concrete remains structurally sound. Since 1964, British Rail have used the Pandrol fastening in which the less durable components can be renewed (Figure 25.3). At high speeds (over 110 km/h) wheel and rail surface defects can cause high stresses in concrete sleepers which, if allowed to persist, can cause cracking in the rail seat area of sleepers. Prestressed sleepers with such cracks can remain serviceable, providing the cause of high stress is promptly detected and rectified as necessary.

The brittle nature of concrete sometimes gives cause for concern in derailment damage. In practice, the enhanced stability of concrete-sleepered track reduces the incidence of derailments and their severity is reduced by modern fastenings which hold gauge well during derailments. Well-filled ballast cribs provide protection to the concrete and derailment damage to the track is not a significant problem with monoblock concrete sleepers. Twinblock sleepers, however, perform badly as the tie bar is relatively easily bent causing severe gauge narrowing.

Whilst concrete is a semiconductor, adequate insulation for track circuits can be obtained simply and cheaply by incorporating insulating components into the fastening.

Concrete sleepers can be cast into a concrete slab to provide ballastless track but it is advisable to use sleepers with projecting reinforcement to ensure permanent connection to the *in situ* concrete. In dry locations, especially tunnels under sensitive buildings where structure-borne vibration is objectionable, attenuation can be achieved by enclosing the sleepers partially (without projecting reinforcement) in rubber boots before casting into the slab.

Pretensioned concrete beams have been in use in Britain as switch and crossing bearers since the early 1970s. Use was inhibited by the practice of drilling the concrete to locate rail fastenings, but manufacturers have now developed methods of casting fastenings into the bearers, and increasing use is anticipated internationally.

Figure 25.2 British Rail F27 prestressed concrete sleeper (*Courtesy*: Costain Concrete Co. Ltd)

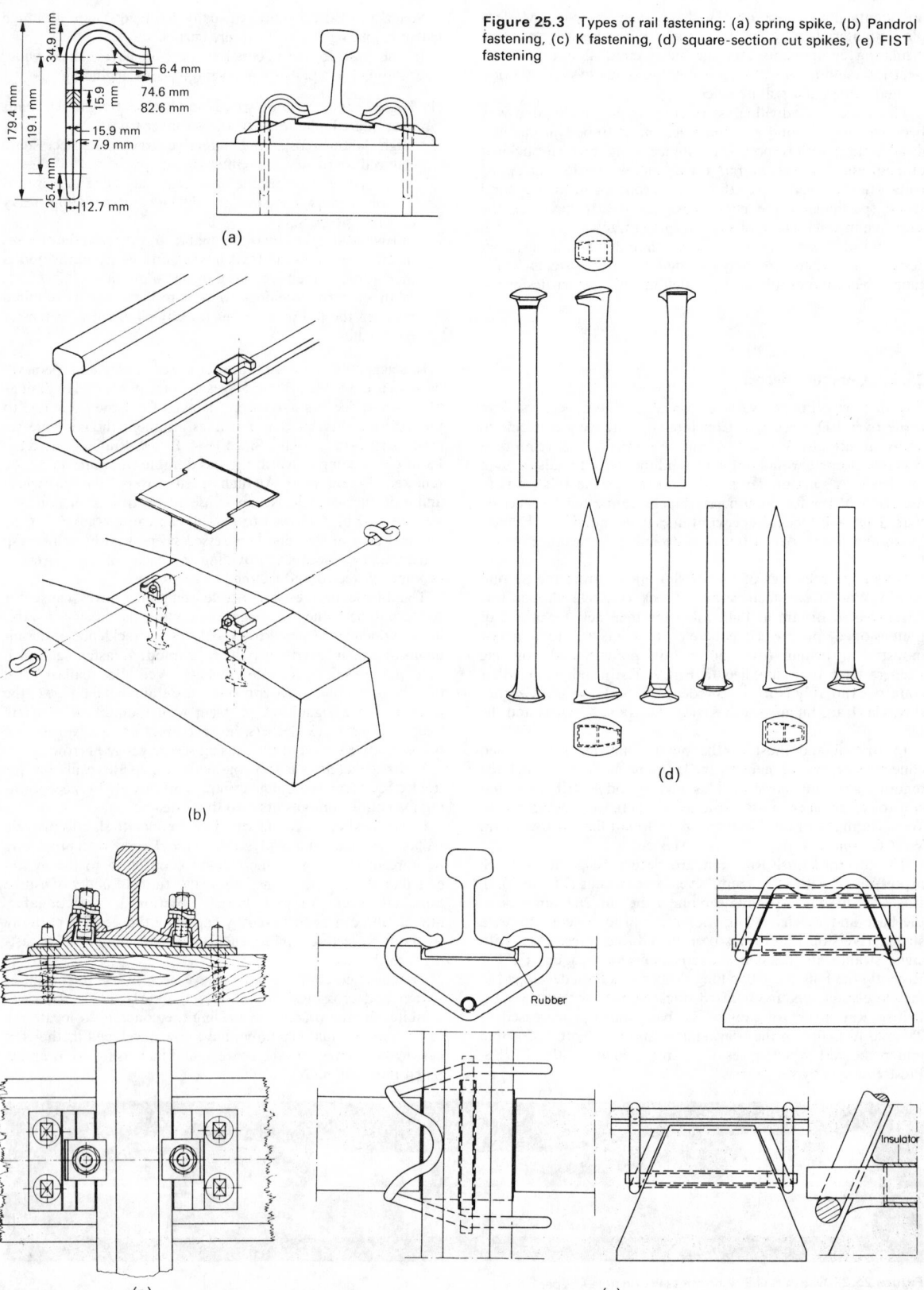

Figure 25.3 Types of rail fastening: (a) spring spike, (b) Pandrol fastening, (c) K fastening, (d) square-section cut spikes, (e) FIST fastening

25.4 Fastenings

Rail fastenings are divided into two categories: direct and indirect. These are further divided into elastic and rigid. Direct fastenings are those which fasten the rail directly to the sleeper, and indirect are those which fasten the rail to a chair or baseplate, which is fastened to the sleeper separately.

Both types of fastenings may be either elastic or rigid, depending on whether a spring element is incorporated, and also depending on whether they are adjustable or self-tensioning.

The square-section cut spike was the original fastening used for flat bottom rail (Figure 25.3), but in Europe it is no longer used except in sidings and light, narrow-gauge railways. In North America, however, and on American-built railways in various parts of the world, it is still in general use on lines carrying axle loads of up to 35 t. Resilient forms of the cut spike have, however, been used extensively outside Canada and the US, particularly in Europe.

The cut spike has a number of disadvantages which have led to the decline in its use by European railways. The major disadvantages are: (1) short sleeper life due to spike killing and baseplate cutting; (2) the need for high sleepering density (in America timber sleeper spacing of 500 mm is still regarded as normal); (3) the need for rail anchors either side of the sleeper with heavy-section rail; and (4) variable track gauge. The cut spike is a simple and robust system which lent itself well to rapid pioneer developments, but it is poorly suited to high speeds and welded rail, especially in countries where temperatures vary between extremes of heat and cold. It is also of little use in countries where timber is not readily available.

In Europe, the coachscrew direct fastening has been preferred and is still used extensively. Both cut spike and coachscrew have poor gauge-holding when used as direct fastenings, but when used with baseplates, their performance is improved and this can be improved further when additional screws or spikes are used to secure the baseplate to the sleeper. It is now general practice to fit baseplates on all curved track fastened with spikes or screws.

The main advantage of the screw over the cut spike is that it exerts a positive pressure on the rail foot whereas the spike allows the rail to float 'free'. It therefore goes some way toward resisting rail creep and rail expansion due to temperature change. In France and on French-built railways, the generally accepted standard fastening is a combination of the screw spike used with a spring clip (called the RN fastening).

In Germany and much of Central Europe, the K fastening of the Deutsche Bundesbahn has been the standard fastening since 1926. It consists of a rolled steel baseplate giving a 1:40 rail inclination with two heavy ribs forming a channel seating for the rail resting on a 5 mm compressed poplar or plastic pad and slotted to take a T-headed bolt each side, inverted rigid U-shaped clips being held by the T bolts to which heavy spring washers are fitted. Four screws are used to hold the baseplate down to timber sleepers in Germany but, in some other countries, e.g. Belgium, only two are used. In Germany, spring steel washers are commonly used on baseplate screws. A number of spring clips have been developed which can be driven into the rib of the baseplate to replace the rigid clips.

In Britain, the Pandrol resilient rail fastening has been the standard British Rail fastening since 1964 on both concrete and timber sleepers, and also in S and C work. It is also used extensively in Australia, Africa, Canada, the Middle East and many other countries throughout the world. It has been adopted by the railways of sixty countries since its introduction, and is generally regarded as one of the simplest and most economical fastenings yet developed. Other resilient fastening systems which have been used quite widely are the DE, Heyback, SHC and FIST – the SHC in the UK and the FIST in South Africa

Figure 25.4 Pandrol clips on a concrete sleeper (top) and a timber sleeper (bottom)

and Sweden. In many parts of the world, the increasing shortage of track maintenance labour and its growing cost have led to greater interest in 'fit and forget'-type fastenings such as the Pandrol fastening, particularly for continuously welded rail. The importance of the fastening in welded-rail track being able to maintain a constant clamping force on the rail foot is fundamental to the stability of the track. It is of even more significance if the ballast condition is below standard.

The clamping force generated by the rail fastening must be able to resist any tendency for the rail to creep through the fastening, but not so much as to allow the rail to push the sleepers through the ballast section. A nominal clip pressure of between 1.4 and 2 t per rail is generally sufficient to achieve this.

It is important that any self-tensioning clip is designed to have a large deflection when fitted in position in order to minimize the effects of manufacturing tolerances on clips, rail pad, insulator, rail clip housing and rail foot. For example, a clip with 6 mm of nominal deflection will lose 50% of its clamping force with 3 mm of wear or tolerances, whereas with 12 mm of deflection, the toe load loss will be only 25%. The ability to ensure positive and accurate track gauge retention is also an important feature of a rail fastening.

Resilient rail fastenings are now used in the harshest of operating conditions, such as heavy haul and high-speed passenger services, and particular attention is being devoted to improving the ability of fastening assemblies to attenuate the transmission of dynamic forces from rail to sleeper. This is proving to be of particular importance in the case of fastenings used with concrete sleepers in high-speed passenger operations. Considerable effort is being focused on improving rail pad characteristics.

A further area of development in rail fastenings is the need for mechanical installation in those areas of the world where track labour costs are relatively high, and machines are now available to drive some of the more widely-used fastening types.

25.5 Rails

The flat bottom, or Vignole, rail is now an almost universal standard so that the obsolescent bull head rail of past British practice need not be developed here.

Generally, rail sections or weights are derived from progressive experience, not because of analytical difficulties but because of inadequate quantified knowledge of what conditions of loading and support actually occur.

A simple rough guide to appropriate rail weight in kilograms per metre is to multiply the axle load in tonnes by 2. However, it is necessary to make allowance for a speed factor and the following formula given by Schramm does this:

$$\text{Rail weight in kilograms per metre} = 156 - \frac{10\,600}{(A\alpha + 67)} \quad (25.1)$$

where A = static axle load in tonnes and α = the speed factor. The speed factor is still under critical review but it can be evaluated by three equations which are widely accepted for practical use:

$$\alpha = 1 + (V^2/30\,000) \qquad \text{up to } 100\,\text{km/h} \quad (25.2)$$

$$\alpha = 1 + (4.5V^2/10^5) - (1.5V^3/10^7) \quad \text{up to } 140\,\text{km/h} \quad (25.3)$$

$$\alpha = 1.18 + (0.706V^3/10^7) \qquad \text{over } 140\,\text{km/h} \quad (25.4)$$

From the above expressions, Table 25.2 has been produced. The figures give an indication of appropriate weights within ± 15%.

British Standard specifications include rails from 25 to 56.5 kg/m; in Europe rails up to 60 kg/m are used and up to 70 kg/m in the US.

Table 25.2 Approximate rail weights, in kilograms per metre

Static axle weight (t)	Speed (km/h)				
	50	100	140	160	200
15	28	34	36	37	39
20	36	42	44	46	49
25	44	50	52	54	58
30	50	56	59	61	65
35	57	61	65	67	70

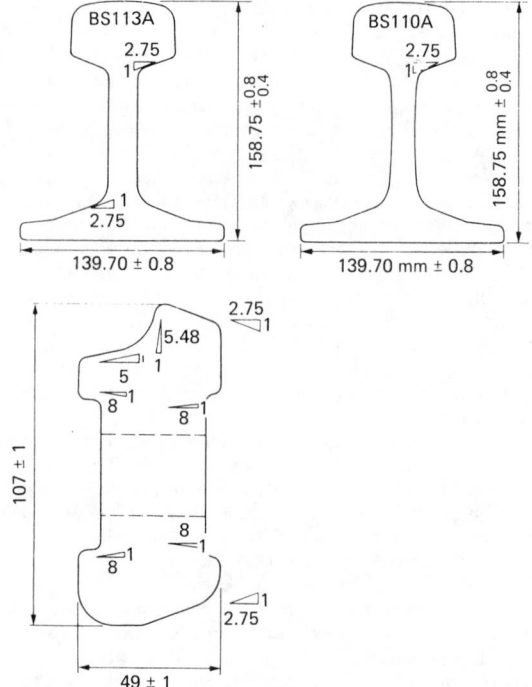

Figure 25.5 Sections of British Rail types 110A and 113A and fishplate (all dimensions in millimetres)

The present standard in Britain is the 56.5 kg/m (BS 113A) rail, adapted from the earlier British standard 109 lb/yd rail and its successor BS 110A section. Until relatively recently in Europe the standard section was the UIC 54 kg/m rail, but currently an increasing proportion of main line track is being constructed using UIC 60 kg/m section. In the USSR main line, heavy duty tracks incorporate 75 kg/m rail sections.

Table 25.2 relates to a representative sleeper spacing of 630 mm. If the spacing is wider than this, and until recently British practice utilized a sleeper spacing of 750 mm, the stress due to bending moment may be up to 9% greater, but in any case the occurrence of three loose sleepers can put up rail bending stress by 100% as can a heavy wheel flat at 30 km/h. At rail ends a 12 mm dipped joint can, with an unsprung mass of the order of 20% of the static load at a speed of 160 km/h, create a dynamic increment of load equal to the static load. Since the great majority of rail breaks occur at rail ends it can be stated that the rail weight selected must take into account the unsprung masses on the heavier axles and also the state of maintenance of joints which can be realized.

The weight of rail chosen has an influence on the shear loading of the subsoil. According to Eisenmann increasing the rail weight from 48 to 68 kg/m diminishes shear stress by 20%.

As rail weights increase, the breadth of the rail foot increases so that changing to a heavier-weight rail will generally involve changing the baseplates or, where cast-in inserts are used, respacing the inserts, unless the new rail is limited to new sleepers. Similarly, the administration of permanent way stocks is simplified, and the tied-up capital reduced, if the range of rail sections in use is limited.

The subsurface Hertzian stresses arising from wheel to rail contact must be very carefully considered, in view of their

Table 25.3

Country	Max. axle load (t)	Rail section (kg/m)	Max. speed (km/h)	Rail steel grade used Straight track	Curves
Britain	26	56.4	200	Normal	Wear-resist.
Europe	22.5	60	200	Wear-resist.	Premium
US	28	68	120–130	Wear-resist.	Premium

relationship with wheel load, wheel size and tensile strength of the rail steel. The Hertzian contact stresses are directly related to the square root of the wheel load and inversely related to the square root of the wheel radius. The nominal shape of the rail head and wheel tread profiles and the progressive wear of these two profiles in service, also influence the actual wheel–rail contact area and, hence, the Hertzian contact stresses. The maximum level of the Hertzian contact stress normally occurs a few millimetres below the contacting surfaces, and shelling and similar subsurface fatigue failures can develop if the shear stress in this contact zone exceeds about 50% of the ultimate tensile strength of the rail steel. Under conditions of excessive stressing, complete collapse of the rail head can occur.

The principal rail specifications in international use are BS11 of the British Standards Institution, UIC 860-0 of the Union Internationale des Chemins de Fer (UIC) and the *Manual* of the American Railway Engineering Association (AREA). A norm for railway rails (ISO 5003) has also been published by the International Standards Organization (ISO).

The most commonly used rail steels are defined in the above standard specifications and can be classified as follows:

(1) Normal grades, typically of 680 N/mm² minimum UTS as per BS 11 normal grade and UIC 860-0 grade 70.
(2) Wear resisting grades, typically of 880 N/mm² minimum UTS as per BS 11 wear resisting grades A and B, UIC 860 grades 90A and 90B, AREA standard carbon grade.
(3) Premium high strength grade, typically of 1080 N/mm² minimum UTS.

The premium high-strength rail grades are still the subject of much research and development, and are briefly referred to in only one specification – AREA.

These premium grade rails are produced either by the use of alloy steels, generally about 1% chromium and in some cases with micro-alloying additions of vanadium, molybdenum or niobium, or by the heat treatment of wear-resisting grade rails, e.g. BS wear resisting grade A, UIC 860 grade 90A, or AREA standard carbon grade. The heat treatment may consist of either full section hardening or localized head hardening only.

The rail steel grade selected for a particular track is dependent on various factors including track design, traffic and operating conditions, standards of maintenance and economic factors, etc.

Table 25.3 illustrates the different optimum strategies selected by different railroads. It is common practice to utilize a higher-strength, more wear-resistant, rail steel grade in tight curves, and a less expensive grade in straight track.

Broad recommendations for the selection of rail steel grade based on track curvature and traffic intensity are outlined in UIC Code 721-R.

A special grade of rail steel utilized throughout the world, albeit in relatively small quantities, is austenitic manganese steel (12–16% Mn) which is usually used in heavily trafficked turnouts, switches and crossings. Many crossing vees (frogs) are produced as austenitic manganese castings. Rolled austenitic manganese rails are produced in Britain in relatively small quantities being used in the fabrication of turnouts and S and C work, and occasionally in curved track. Austenitic manganese

steel surface hardens rapidly in service under stress and impact, and develops high wear resistance. However, it is expensive, needs great care in machining to avoid the risk of fatigue crack initiation and can only be welded to similar steel by normal rail welding techniques. Its higher thermal coefficient of linear expansion also produces manufacturing problems and service use limitations, but rolled austenitic manganese rails have been successfully welded and used in the relatively constant temperature conditions of underground railways.

Experimental work continues in the evaluation of more exotic and expensive rail steel chemistries for special applications, e.g. bainitic steels.

Rail joints are either suspended or supported. Joints are also square, i.e. opposite each other, or staggered by up to half the rail length. In some administrations, rail joints are square on the straight and staggered on the curves. Rail and fishplate sections need to be considered together in selecting a rail section, since if the ratio of the I_{xx} of the fishplate is much below 25% of the I_{xx} of the rail, broken fishplates and battered rail ends are likely to have a high incidence.

25.5.1 Rail failures

It is necessary in any railway system that a record should be kept of all cracked and broken rails removed from the track. In addition, the cause of failure must be noted in order to monitor any specific problems which may be developing. This is carried out with the use of a standard reporting form which gives details of track and conditions as well as the type of defect. This information is fed into a computer and frequent monitoring of the data gives any developing trends.

Whilst this latter information gives details of rails removed, it does not give any indication of those defects remaining in the track under observation. This information comes from the normal regular examination of the track which is carried out by means of an ultrasonic rail flaw detection train and also by hand-held ultrasonic rail flaw detectors used by pedestrian operators.

The ultrasonic test train is normally a self-propelled unit consisting of two vehicles. It operates at speeds up to 30 km/h and uses a series of probes which are applied to the rail head either in the form of sliding probes or wheel probes. The data from these probes are partially reduced by an on-board computer and the resulting information is stored on magnetic tape for subsequent off-line analysis. The defects detected by this system are sent to the appropriate maintenance engineers for action depending on the type of defect.

In addition to the train, hand-held ultrasonic units are used to carry out work in areas not covered by the test train and also to examine, in greater detail, defects which have been picked up by the test train but which require more detailed study. In addition, the hand units are also used for monitoring defects which are allowed to remain in track until such time as they can be repaired or removed.

Ultrasonic testing of rails is carried out at a frequency determined by the types and speeds of traffic carried. It ranges from 6 months to alternate years for some of the lightly used branch lines.

In addition to this routine testing, other specialized tests are carried out by both the test train and the hand operators which include gauge measurement, crack size estimation in special cases and weld testing.

25.6 Curved track

The main curves of the railway are nominally of constant radius, i.e. circular curves; curves made up of two or more circular curves of different radii curving in the same direction are called compound curves. Straights are generally and desirably connected to circular curves by transition curves of progressively varying radii; and adjoining circular curves of different radii are commonly joined together in a similar way if the difference of radii exceeds about 10%. Two adjoining circular curves curving in opposite directions comprise a reverse curve, and here, whatever the radius, the presence of connecting transitions is relatively more important than with circular or compound curves.

For practical purposes the cubic parabola $y = kx^3$ gives a uniform change of curvature between tangent point on the straight and tangent point on the curve, or between the tangent points of two curves, and is the most used form of transition curve. The versines, measured on half-chord overlapping chords along the transition, change with linear uniformity from zero to R although it is usual to smooth out the rate of increase and decrease at the start and end of the transition by putting on about one-sixth of the rate of the first versine at the zero station and reduce the increment at the final transition station to about five-sixths of the half-chord rate of increase. Versines are conveniently measured in millimetres.

The geometrical relation between a circular curve and a transition curve tangenting on to both the straight and the circular curve involves the moving inwards along a diameter normal to the straight of the circular curve by an amount termed the 'shift'. The transition curve bisects the shift at its midpoint measured along the straight tangent line and the offset from that line to the tangent point of the transition and circular curve is 4 times the shift or 8 times the offset at the mid-transition point. It follows that where no transition exists or where it is insufficient in length the institution or extension of a transition can be done only by sharpening the radius of the circular curve concerned or moving the tangent straight away from the curve, though this may be worthwhile since any transition is better than none.

The length of a transition curve is determined primarily by what is judged to be an acceptable rate of change of cant or cant deficiency. For standard gauge plain track a desirable rate may be 35 mm/s, with a maximum of, say, 55 mm/s to secure passenger comfort. In switches and crossings a rate as high as 80 mm/s may be applied but a good standard of switch and lead design is desirable for this rate of loss or gain of cant or cant deficiency.

Some limiting cant and cant deficiency values observed in British practice on a 1.432 m gauge are listed at top right.

Wherever space permits, curve design should be based on the desirable rate of change of cant or deficiency of 35 mm/s.

The amount of cant applied to a track depends upon consideration of the following factors:

(1) The line speed limit, i.e. the maximum speed at which traffic is allowed to run on a line or branch or section of a line or branch. This limit is usually fixed with reference to the value and distribution of permanent speed restrictions on the line or branch, or section thereof, involved.

Maximum cant:	
on curved track	150 mm
at station platforms	110 mm
maximum cant gradient	1 in 400
deficiency on plain line, CWR	110 mm
deficiency on plain line, jointed track	90 mm
deficiency on switches and crossings welded into CWR	
on through line	110 mm
on turnout	90 mm
with negative cant on turnout	90 mm
at switch toes	125 mm
deficiency on switches and crossings in jointed track	
on through line	90 mm
on turnout	90 mm
with negative cant on turnout	90 mm
at switch toes	125 mm
Maximum rate of change of cant	
on plain line	55 mm/s
on switches and crossings	55 mm/s
Maximum rate of change of cant deficiency	
on plain line	55 mm/s
on switches and crossings	55 mm/s
Maximum rate of change of cant deficiency	
on plain line	55 mm/s
on switches and crossings (inclined design)	55 mm/s
on switches and crossings (vertical design)	80 mm/s

(2) The proximity of permanent speed restrictions, junctions, stopping places, etc.
(3) Track gradients which may cause a reduction in the speed of freight or slow-moving passenger trains without having an appreciable effect on the speed of fast trains.
(4) The relative importance of the various types of traffic.

Generally, where fast and slow trains run on the same lines, an intermediate speed is selected to fix the cant. In this situation it may exceptionally be necessary to limit the cant and therefore the maximum speed to prevent surface damage to the low rail by heavy axles on slow-moving freight trains.

Each line of a double line should be separately assessed. In exposed locations subject to high winds it may be desirable to limit cant to below the 150 mm maximum.

Normally, cant or cant deficiency is uniformly gained or lost within the length of a transition curve or, where there is no transition curve, as may occur in switches and crossings, over the length of the virtual transition, which for practical purposes is taken as the shortest distance between the centres of bogies of coaching stock using the line. If the desired cant cannot be put on within this length observing a maximum cant gradient of 1 in 400, the cant loss or gain is continued on to the circular curve. The 1 in 400 cant gradient mentioned relates to axle twist derailment possibilities, especially where four-wheeled vehicles are concerned.

The maximum permissible speed on circular curves appropriate to the determination of permanent speed restrictions may, for standard gauge railways, be obtained from the following expressions:

Equilibrium (or theoretical) cant = $E = 11.82(Ve^2/R)$ mm

Equilibrium speed = $Ve = 0.29\sqrt{(RE)}$ km/h

Maximum speed = $Vm = 0.29\sqrt{R(E+D)}$ km/h

where R = radius in metres, E = cant, which may be either actual cant or equilibrium cant but in practice the difference is not likely to be significant, though the distinction has to be kept in mind in certain circumstances, and $D=$ maximum allowable cant deficiency in millimetres.

Desirable lengths of transition curves can be derived from the greater of the two values obtained from:

Length $= L = 0.0075EVm$ m
or $\qquad L = 0.0075DVm$ m

where $E=$ cant in millimetres, $D=$ deficiency of cant in millimetres and $Vm=$ maximum permissible speed in kilometres per hour.

If space is limited, the length of the transition may be reduced to two-thirds L subject to a minimum cant or cant deficiency gradient of 1 in 400.

On compound curves to be traversed at a uniform speed, the desirable length is obtained from the greater of the two following expressions:

$L=0.0075(E_1-E_2)Vm$ m or
$L=0.0075(D_1-D_2)Vm$ m

where E_1 and D_1 are the cant and cant deficiency conditions for one curve and E_2 and D_2 are the similar values for the other curve.

Similarly, on reverse curves the transition lengths are given by:

$L=0.0075(E_1+E_2)Vm$ m or
$L=0.0075(D_1+D_2)Vm$ m

It should be noted that in using the above formulae the constants used have reference to a standard gauge, to an assumed height of the centre of gravity of vehicles of about 1.5 m and a subjective passenger comfort assessment of the tolerable rate of change of cant or cant deficiency. To this extent the values are arbitrary rather than absolute and refer to standard gauge. Further qualifications are that the springing of vehicles may be such as to produce excessive lean under cant deficiency running to encroach significantly on side clearances or to diminish passenger comfort, whilst if the centre of gravity of a vehicle is higher or lower than that assumed the maximum permissible speed is affected in inverse ratio, so that certain vehicles may be permitted to run at a higher speed than other vehicles.

The versine or middle ordinate of a chord on to a curve is proportional to the curvature and is the basis of all railway curve alignment, checking and adjustment. Its value as determined from triangular analysis is given by:

$$2R=\{(C/2)^2/V\}+V \qquad (25.5)$$

where $R=$ radius of a curve, $C=$ length of chord on which the versine is measured, and $V=$ the versine, but since the value of V is very small in relation to R in railway situations, the final V of the expression may be disregarded so that for both field measurements and calculation purposes:

$$V=C^2/8R \qquad (25.6)$$

In Britain and on the Continent, railway curvature is usually described by the length of the radius measured in metres. The American practice is to describe a curve by the angle subtended at the centre of the curve by a chord of 30.48 m. For railway work it is sufficient in transposing degree units into radius units to divide 1746 by the degree of the curve to give the radius in

metres. Thus, what is described in American practice as a 10° curve would be described in Europe as a curve of 175 m radius.

Main railway curves are initially set out by theodolite generally on the basis that equal chords are subtended by equal angles but informal setting out by an offset from a tangent followed by the use of overlapping chords of convenient length using the versine appropriate to the radius can give a degree of accuracy sufficient for less important curves. A selection of curve formulae covering most calculations for the design or setting out of railway circular curves is set out in Table 25.4.

It is desirable to keep rail joints more or less square, or at a constant stagger where this is preferred, on curves by inserting short rails in the inner rail of curved track. The difference in rail length, D, for standard gauge and for 18.3 m rails is given by $D=(27.45/\text{radius in metres})$ m. Alternatively, the approximate difference in length can be obtained from the formula $D=\frac{2}{3}$ versine. It is normal to take a standard range of short rails from the makers and for 18.3 m rails, these are 18.25, 18.20 and 18.15 m.

The precise alignment of curves is vital to a smooth ride, stability of the track geometry and minimizing track wear. There are several alternative methods of adjusting railway curves as a maintenance operation but all are based on versine measurements and relate, either to a smoothing or averaging approach, in which the differences between a limited number of adjacent versines are adjusted to give a fairly uniform rate of change in transitions or to a fairly uniform value on circular curves; or to a design lining approach in which the whole of the versines of a curve are adjusted as one revision operation to give a precise rate of increase or decrease in transitions and a strictly uniform value on the circular curve portion, as far as clearances from structures, etc. will permit. It is generally convenient to use smoothing or local adjustment techniques after a curve has been aligned on a design basis. It is also normal practice to set markers or monuments or pegs in or beside the tracks when design lining is carried out.

Curve realignment can be based on manual measurement of versines or with less accuracy on the versine measurements of a track geometry recording car or of a track lining machine.

The side and interbody clearances shown on the diagram contained in the Department's Requirements (see Figure 25.15) relate to straight track and may need to be augmented on curved track for end throw or centre throw of vehicles, which are generally of the same magnitude.

Centre throw $= C = B^2/8R$

End throw $= E = (L^2-B^2)/8R$

where $B=$ wheelbase or bogie centres, $L=$ length of vehicle, and $R=$ radius.

To this increase of effective body width on curved track must be added a further allowance due to the tilting of the vehicle to the low side on canted track. In Britain, the loss of side clearance at the vehicle cornice is about 2.25 times the actual cant.

It is general practice slightly to widen the track on very sharp curves to allow all vehicles, especially locomotives with long rigid wheelbases, to pass round such curves without straining the track. The normal British practice is:

Curves over 200 m radius widened to 1435 gauge
200 to 110 m radius widened to 1439 gauge
110 to 70 m radius widened to 1451 gauge

Appropriate values of gauge widening for a given track radius depend on free play of wheelsets, length of half wheel flange below rail level and rigid wheelbase of the critical vehicle.

In the UK it is the Department of Transport's enjoined practice to install check rails on curves of 200 m or less radius.

Table 25.4 Curve formulae

Given	Sought	Formula	Given	Sought	Formula
E, M	C	$C = 2M\sqrt{\{(E+M)/(E-M)\}}$	T, α	M	$M = T\cot(\alpha/2)\operatorname{vers}(\alpha/2)$
E, R	C	$C = 2R\sqrt{\{E(2R+E)\}}/(R+E)$	D	R	$R = 50/\sin(D/2)$
E, T	C	$C = 2T(T^2-E^2)/(T^2+E^2)$	C, E	R	$0 = R^3 + R^2\{(4E^2-C^2)/8E\} - (RC^2/4) - (C^2E/8)$
E, α	C	$C = 2E\{\sin(\alpha/2)\}/\{\operatorname{ex sec}(\alpha/2)\}$	C, M	R	$R = \{(M^2+(C/2)^2)/2M\}$
M, T	C	$0 = C^3 - 2TC^2 + 4M^2C + 8M^2T$	C, α	R	$R = C/2\sin(\alpha/2)$
M, R	C	$C = 2\sqrt{\{M(2R-M)\}}$	E, α	R	$R = E/\operatorname{ex sec}(\alpha/2)$
M, α	C	$C = 2M\cot(\alpha/4)$	M, E	R	$R = EM/(E-M)$
R, T	C	$C = 2TR/\sqrt{(T^2+R^2)}$	M, α	R	$R = M/\operatorname{vers}(\alpha/2)$
R, α	C	$C = 2R\sin(\alpha/2)$	T, C	R	$R = CT/\sqrt{\{(2T+C)(2T-C)\}}$
T, α	C	$C = 2T\cos(\alpha/2)$	T, E	R	$R = \{(T+E)(T-E)\}/2E$
R	D	$\sin(D/2) = 50/R$	T, M	R	$0 = R^3 - R^2(M^2+T^2)/2M + RT^2 - MT^2/2$
α, L	D	$D = 100\alpha/L$ approx.	T, α	R	$R = T\cot(\alpha/2)$
C, M	E	$E = M(C^2+4M^2)/(C^2-4M^2)$	C, E	T	$0 = 2T^3 - T^2C - 2TE^2 - CE^2$
C, α	E	$E = C\{\operatorname{ex sec}(\alpha/2)\}/2\sin(\alpha/2)$	C, M	T	$T = C(C^2+4M^2)/2(C^2-4M^2)$
M, α	E	$E = M/\cos(\alpha/2)$	C, α	T	$T = C/\{2\cos(\alpha/2)\}$
R, C	E	$E = R^2/\sqrt{\{(R+C/2)(R-C/2)\}}$	E, α	T	$T = E\cot(\alpha/4)$
R, M	E	$E = RM/(R-M)$	M, E	T	$T = E\sqrt{\{(E+M)/(E-M)\}}$
R, T	E	$E = \sqrt{(T^2+R^2)} - R$	M, α	T	$T = M\{\tan(\alpha/2)\}/\{\operatorname{vers}(\alpha/2)\}$
R, α	E	$E = R\operatorname{ex sec}(\alpha/2)$	R, C	T	$T = CR/2\sqrt{\{(R+C/2)(R-C/2)\}}$
T, C	E	$E = T\sqrt{\{(2T-C)/(2T+C)\}}$	R, E	T	$T = \sqrt{\{E(2R+E)\}}$
T, M	E	$0 = E^3 + E^2M - ET^2 + MT^2$	R, M	T	$T = R\sqrt{\{M(2R-M)\}}/(R-M)$
T, α	E	$E = T\tan(\alpha/4)$	R, α	T	$T = R\tan(\alpha/2)$
R, α	E	$E = R\{1-\cos(\alpha/2)/\cos(\alpha/2)\}$	D, L	α	$\alpha = DL/100$ approx.
α, D	L	$L = 100\alpha/D$ approx.	M, C	α	$\tan(\alpha/4) = 2M/C$
C, E	M	$0 = M^3 + M^2E + (MC^2/4) - (C^2E/4)$	M, E	α	$\cos(\alpha/2) = M/E$
C, α	M	$M = (C/2)\tan(\alpha/4)$	R, C	α	$\sin(\alpha/2) = C/2R$
E, α	M	$M = E\cos(\alpha/2)$	R, E	α	$\operatorname{ex sec}(\alpha/2) = E/R$
R, C	M	$M = R - \sqrt{\{(R+C/2)(R-C/2)\}}$	R, M	α	$\operatorname{vers}(\alpha/2) = M/R$
R, E	M	$M = RE/(R+E)$	R, T	α	$\tan(\alpha/2) = T/R$
R, T	M	$M = R - \{R^2/\sqrt{(T^2+R^2)}\}$	T, C	α	$\cos(\alpha/2) = C/2T$
R, α	M	$M = R\operatorname{vers}(\alpha/2)$	T, E	α	$\tan(\alpha/4) = E/T$
T, C	M	$M = (C/2)\sqrt{\{(2T-C)/(2T+C)\}}$	R, α	L	$L = 0.017\,453\,292\,5\,R\alpha$
T, E	M	$M = E(T^2-E^2)/(T^2+E^2)$			

D = Degree of curve
R = Radius
α = Ext. angle = central angle
L = Length of curve

M = Mid-ordinate
T = Tangent
C = Long chord
E = External distance

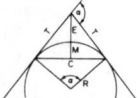

Note: $\operatorname{ex sec} A = \sec A - 1 = (1 - \cos A)/\cos A$ *and* $\operatorname{vers} A = 1 - \cos A$

Check rails are also sometimes installed in Britain on flatter curves to diminish side cutting of the high rail or as a protection against inadvertance or mishap at the bottom of a gradient. Wide flangeway check rails or guard rails may also be installed to protect bridge supports or girders against being struck by derailed vehicles. In the UK this provision is a Department of Transport Requirement.

25.7 Welded track

About two-thirds of the on-track maintenance work put into jointed permanent way is at or adjoining the rail joints and the great majority of rail failures occur at or near fishplated joints. For these and other reasons the tendency today is to use long welded or continuously welded rails.

This procedure entails the elastic containment of the longitudinal expansion and contraction stresses arising in the rails as a consequence of variation in the rail temperature with reference to the temperature at which the rails were fastened to the sleepers. If, as is usual, the rails are fastened down on installation within a prescribed narrow temperature range situated at or a little above the mean of the annual rail temperature range in a normal or stress-free condition, then the amount of tension in winter and compression in summer is limited to what is judged to give an optimum compromise between a cold season hazard of broken rails and a hot season hazard of track buckles.

The installation and maintenance of welded rail track need more technical insight and attention than ordinary track. The lateral strength of the track depends upon three main elements: (1) the I_{yy} of the two rails; (2) the framework stiffness of the assembled rails and sleepers as developed by the fastenings; and (3) the frictional loading of the ballast on the sides, ends and bottoms of the sleepers. The resistance of the track to vertical buckling, which generally occurs in combination with, or as a trigger to, lateral buckling, is determined mainly by the total weight of the track, and in this context concrete through-type sleepers show a considerable advantage. In a fully-ballasted track the ballast representatively accounts for about two-thirds of the total moment of lateral resistance. Loose sleepers and kinks in the alignment are of especial significance to the stability of welded track so that a high standard of maintenance is essential for this sort of permanent way.

25.7.1 Fastening welded rails

The need to fasten the rails in order to get a stress-free rail at about the mean of the annual temperature range would mean, in many parts of the world, a very short season for laying welded rails without artificial assistance. This assistance is normally supplied either by heating the rails by propane gas travelling heaters or by using hydraulic tensors, so that when fastened down after extension by heat or tensile pull, the rail has a length equivalent to a stress-free condition at a temperature between prescribed limits (21 to 27°C for the UK). In this way it is possible to lay welded track all the year round. The pull required to extend a 56 kg/m rail is of the order of 1.6 t/°C. The tensors have a capacity of about 70 t.

Figure 25.6 Hydraulic rail stresser. This has a 70-t pull and 380-mm stroke. For pushing action, the rail clamps are turned through 180°. (*Courtesy*: Permaquip Co. Ltd)

Welded rails are not normally laid in Europe in curves with a lesser radius than 600 m. Long welded rails are made up by flash butt welding of standard rail lengths into welded rails of 200 to 400 m length in depot and welded into continuous welded rails by Thermit welds, except in the USSR where a transportable flash butt welder is applied to join rails in the track.

It is normal practice in Britain and France to install sliding switches at the ends of long welded rails but in Germany this is not done, reliance being placed on very firm fastening of the welded rails where they connect with jointed track.

The advantages of welded track include an extension of rail lives by about a third, a reduction of on-track maintenance by about half, a dramatic reduction of rail breakages, an increase in running speeds, less damage to the formation, increased sleeper lives, improved ride comfort, a reduction of traction energy of up to 5%, and a reduction in train noise. In effect, welded rails, prestressed concrete sleepers and self-tensioning fastenings are complementary to each other in extending the life and reducing the long-term overall cost of permanent way.

25.8 Switches and crossings

All railway switch and crossing work is built up of three basic units: (1) switches; (2) common or acute (angle) crossings; and (3) obtuse (angle) or diamond crossings. Layouts of switches and crossings are illustrated in Figures 25.7 to 25.10.

25.8.1 Switches

The heel of a switch is the point from which it is free to move. Early switch designs had a loose heel formed by a semi-tight fishplated joint but this type is now rarely used outside of sidings. It is now the general practice for the heel of the switch to be formed by a bolted connection through a block between switchrail and stockrail so that the switch joint is well behind the heel. The length of the switch planing on the bead of the rail varies between about 1.6 and 11 m and most railway administrations have a range of two to eight standard switches to meet the requirements of short leads in depots and various speeds of traffic on the running lines.

Originally, ordinary or straight planing of switch blades was universally applied but during the last few decades curved planing, in which the switch rail is bent elastically whilst being planed on one side of the head, has become general. In this way the curve running through the lead is continued to the switch tip. A recent practice is to apply a larger radius to the planing of the switch than is applied to the switch beyond the end of the planing to give a more robust switch tip with a finite entry angle.

25.8.2 Crossings

Crossings are described by the angle at the crossing nose. In continental Europe it is common to state this in degrees but over the English-speaking world the angle is usually described as 1 in N. However, the angle of a crossing expressed as 1 in N varies slightly according to the measuring convention adopted. Thus if 1 in 8 is obtained by measuring 8 units along the centreline for 1 unit of symmetrical spread normal to the centreline (centreline measure), then measuring 8 units along one leg of the vee for a spread of 1 unit measured at right angles to that leg (right angle measure) will give 1 in 7.969, whilst if 8 units is measured along each leg to give a spread of 1 unit the crossing size will be 1 in 8.016 (isosceles measure). It is therefore necessary for any railway to have a specific mode of measuring crossings which is known to all supplying manufacturers of crossings, or to state the angle in degrees.

In the UK fixed diamond crossings are limited by Ministry requirement to be not flatter than 1 in 7.5. In continental Europe 1 in 10 is the general limit. Where a line crossing flatter than the allowed limit is necessary, the points of the diamond crossings are constructed as switches and have to be set to suit the train movement desired.

There is no official limit on the angle of common crossings but 1 in 28 is about the flattest angle at which the crossing nose has sufficient robustness. Crossings are either made up from rails or are cast in 12 to 14% manganese steel. Built-up crossings have the inherent limitations of bolted assemblies, and of recent years it has become general practice in the UK to weld the vee section and Huck-fasten the wings. In general, it is preferred practice to have cast crossings for high-speed or heavily worked lines using built-up crossings for less exacting duty. It is also preferred to use cast crossings for angles flatter than about 1 in 20. In the US it is common practice for the nose section of the crossing to be a manganese casting to which ordinary rail section wings are attached to form what is known as a rail bound cast crossing.

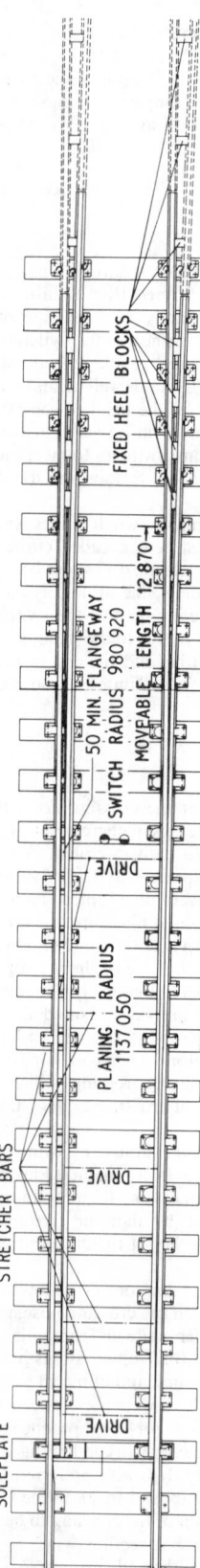

Figure 25.7 Timbered layout of long switches

Figure 25.8 Obtuse crossing

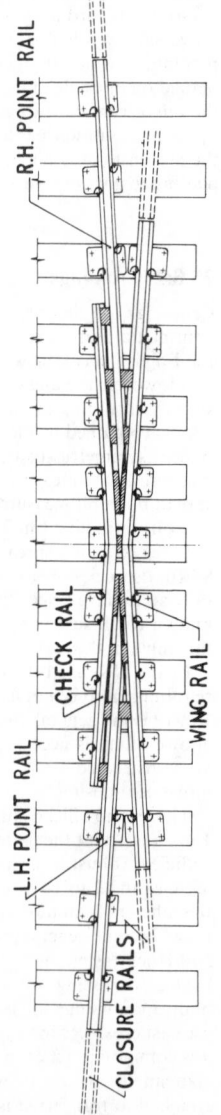

Figure 25.9 Cast common crossing. (*Courtesy*: Edgar Allen & Co. Ltd)

Figure 25.10 Part-welded crossing in lead on prestressed concrete bearers. (*Courtesy*: British Rail, London Midland Region)

25.8.3 Leads

The design of railway connections is based on standard leads, i.e. on specific combinations of switches and crossings forming turnouts of specific length giving specific departure angles. Exceptionally, as at important junctions where space is limited, leads are specially designed on the basis of standard switches with special crossings, e.g. instead of utilizing either a 1 in 8 or 1 in 9 crossing, a crossing is made to a fine decimal angle, such as 1 in 8.319, but this is avoided as far as possible in the interests of standardization. However, reference to a standard textbook on permanent way, such as *British Railway Track* by C. L. Heeler is recommended to civil engineers not familiar with fieldwork.

25.9 Slab track

Slab track installations have generally been constructed either as a continuous ribbon of reinforced concrete or as a series of precast units (usually specially designed sleepers) connected into a reinforced slab by *in situ* concrete. Both systems have been put into service in various parts of the world, but in general the continuously constructed slab (see Figure 25.11) has been favoured both because of its higher accuracy and its lower cost. The use of precast units has been restricted to sites with limited possession periods or to very short lengths.

The benefits of this type of track can be summarized as follows:

(1) It provides an accurate and stable track geometry. In tight locations it is not necessary to allow for the effects of track movement so that speed restrictions can be less onerous. Elsewhere, the high standards of accuracy that are attainable allow for high-speed running without the problems of maintaining alignment and top that are associated with conventional track, thus providing a considerable economic saving.

(2) The peak loading on the formation is reduced to something less than a quarter of that produced under conventional track. Consequently, it is possible to use the system to reduce formation problems, in many cases providing a good track geometry where most other track systems would require considerable and frequent effort to do so.

(3) The day-to-day maintenance cost of slab track is very low, requiring little more than standard patrolling. Additionally, however, experience has shown that periodic replacement costs are also greatly reduced since rates of wear and deterioration are much slower than for most other forms of track.

(4) The requirement of conventional track for considerable and recurring engineering possessions to permit the working of on-line maintenance equipment is eliminated.

(5) The hazard of track buckling, either vertically or laterally, is eliminated. Conventional sleepered track has a safety factor against lateral buckling which, at the high temperature limit, is typically of the order of 1.3. Some maintenance procedures can severely reduce this further. However, with slab track the full weight and stiffness of the slab is utilized and the safety factor is many times higher so that this form

of construction must be most attractive in areas having a large temperature range of, say, 70°C or more.

(6) Because of the high resistance to buckling of the slab, it is no longer imperative to restrict the maximum compressive force generated in the rails. The need for periodic restressing of the rail is eliminated and advantage can be taken to accept a stress-free condition at a lower temperature thereby reducing the likelihood of rail fractures during cold weather.

(7) Slab track, more especially continuous slab track, offers the possibility of continuous support to the rail. Even if the rail is supported on a series of discrete pads, the support is more stable and the elimination of the possibility of voids at support positions means that the rails are subjected to much lower bending stresses than those which they are designed to resist. There are grounds, therefore, for redesigning rails for use on slab track either to use a lighter section or to transfer more metal into the head to provide for extra wear before replacement.

Although techniques are available to restore the track geometry of slab track after subsidence of the slab, it is both difficult and costly in time and money to do so. It is far more preferable, and always possible, to carry out the construction of slab track, including its foundations, in such a way that the likelihood of permanent slab movement is very low.

The likely cost of installing slab track depends on the type of design adopted, but generally the cost of site preparation up to the sub-ballast level would be much the same as for conventional track whereas the provision of the remainder is likely to be 25 to 75% more than for conventional high-quality CWR track. However, this may be offset, and even reversed in many cases, by the reduced construction depth of the slab track which will reduce the ancillary operations, e.g. new tunnels can be driven with smaller cross-sections; overhead electrification clearances can be achieved without expensive bridge raising or formation lowering.

The laying of slab track is much slower than that of conventional track. The system which uses precast units surrounded by *in situ* concrete would typically be laid at about one-third of the rate of conventional track. The provision of a continuous slab may well be slower still, with the complication that the work cannot be subdivided and carried out in a series of short track possessions.

The main difference in logistics of supplying materials for slab track compared to conventional track is the need to place concrete soon after it has been mixed. However, this is only a problem of site organization and even in remote situations, such as in long tunnels, has not caused major problems.

25.9.1 Prefabricated slab track

Prefabricated slab tracks have been installed in a number of parts of the world. The major differences between these slabs and those described above are:

(1) Prefabricated systems are designed for ease of replacement of units during short possessions so that, except for the rail, there is little connection between adjacent units. Consequently, there is a much greater reliance on a strong, accurate formation.

(2) In practice, the installations using this type of system have suffered from the effects of dynamic forces generated as a wheel has passed over the junction between adjacent units. The need for maintenance of track geometry on a regular and frequent basis is usually encountered. This is often provided by the introduction or removal of height-adjusting pads located under baseplates mounted on the units.

Figure 25.11 Section of slab track laid in tunnel (to increase headroom for electrification, and to reduce maintenance costs)

Although these systems place a greater need on period maintenance than do the more continuous forms of slab track, the ease and lower cost of adjusting the track following major movement of the slab more than compensate for this extra effort in certain locations. In such locations, where the stability of slab track is required to allow high-speed traffic or the use of reduced clearances but where there is a strong possibility of significant movement of the track, e.g. as a result of an earthquake, this form of slab track construction has many advantages over others.

Estimates of the limit of speed at which trains can run on conventional tracks vary between about 250 and 400 km/h. The limits that are placed on speeds of trains by slab track are largely those of accuracy of measurement in order to detect and correct misalignments at the construction stage. If the call for higher speeds continues to be made, it may only be possible to meet it by using slab track.

Similarly, if the need to provide sinuous railway routes through highly congested areas continues, the use of slab track should allow this to be done using the least amount of space and with the least constraint on train speeds.

25.10 Track maintenance and renewal

Generally, the greater the strength and inherent stability of the design of the track construction and its foundation, the less demanding is the maintenance task. The stiffness of the rail, the sleeper spacing and the depth of ballast are important factors affecting the cost of maintenance.

The evolution of the automatic levelling lining and tamping machine in the 1960s (Figure 25.12) changed the general pattern of track maintenance. The original manual method of maintenance, using beater picks, was to drive individual pieces of ballast under the sleeper to obtain the required rail level and firmness of support. A later development, pioneered in France, was to jack the sleeper and spread small stone chippings over the bearing area. This system was known as measured shovel packing (MSP) and was extremely successful in producing good track although expensive in labour.

A recent development has been the mechanization of this process using a machine known as a 'Stoneblower'. This machine automatically measures the voids under the sleepers and, using sophisticated electronic controls, pneumatically injects the correct quantity of stone to produce the correct longitudinal, vertical track profile. A prototype machine using this process has recently been constructed (Figure 25.13). Development work on this principle has shown that track maintained by this system only deteriorates at approximately one-quarter of the rate of track maintained by conventional tamping. Very considerable savings are therefore possible in maintenance costs by using this new technique.

Other machines used in association with tamping and stoneblowing are ballast-regulating machines for positioning ballast

Figure 25.12 Tamping and lining machine suitable for use on switches and crossings. (*Courtesy*: Plasser & Theurer)

and providing the correct ballast profile and, for cases where ballast requires cleaning and renewal, ballast-cleaning machines which screen dirty ballast, returning usable stone and rejecting dirt which has accumulated over the years. Ballast cleaning is essential for maintaining track stability as it is vital to ensure good natural drainage in the ballast bed; this is particularly important in heavily used lines with high annual tonnages and axle loads.

With the reduction in manpower and greater reliance being placed on machines, methods additional to visual inspection by track walkers are necessary to ensure that track standards are maintained. This is achieved by the use of track geometry recording cars. These produce graphical records of the right- and left-hand top, track gauge variations, twist (the mutual angular deflection between two axles on a rigid frame), right- and left-hand versines and a location line showing reference points such as mileposts, bridges, stations, switches and crossings, etc. The data accumulated are recorded on an on-board computer and the information is presented in a printout and also, if required, in graphical form. This information is used to allocate priority to maintenance programmes, including tamping and lining of the track and also other remedial work.

A problem that has increased in recent years is rail corrugation. The cause of corrugation is not yet known but considerable research is being done by various academic and railway organizations on this subject. Meantime, civil engineers are faced with the problem of removing such corrugations. They occur both as shortwave corrugations of 40 to 180 m frequency with a depth of 0.1 to 0.4 mm and also as longwave corrugations of 200 to 3000 mm in frequency, with depths up to 3 mm.

If corrugation is not removed from the running surface, defects occur which can lead to metal fatigue. As a result, rails require replacing prematurely to avoid the risk of fractures. The extra vibration induced by corrugation causes the fastening elements between rails and sleepers to become loose, leading to excessive wear and early failure. The additional forces imposed by corrugation also increase the strain on sleepers and ballast causing deterioration and premature replacement. Corrugation generates noise, and in densely populated areas the result is an unacceptably high level of aggressive noise for local residents. It is also contrary to the aim of achieving increased standards of comfort for rail passengers.

Corrugations can be removed by means of rail grinding, and special trains operated by Messrs Speno provide this service worldwide. Grinding is carried out with the flat ends of rotating annular abrasive wheels and the trains operate groups of grinding wheels set at different angles over the running surface of the rails. These wheels are applied to the rail head and, after a number of passes at 5/7 km/h, a new rail head profile is produced eliminating the corrugation in the process. Various sizes of trains are in service ranging from small machines with sixteen grinding wheels to large trains operating more than a hundred. Recent research has highlighted the importance of maintaining the correct wheel–rail interaction involving both

Figure 25.13 Prototype 'Stoneblower' for packing and aligning track. (*Courtesy*: Plasser (GB) Ltd)

Figure 25.14 Track-laying gantries for use on single lines. (*Courtesy*: Geismar)

wheel and rail profiles. Rails often wear unevenly, particularly on curves, and past practice has been to transpose such rails once sidewear becomes excessive.

With speeds of modern passenger trains increasing, transposing of rails is not now an acceptable practice as the transposed rail usually presents a sharp edge to the wheel; this results in poor-quality running. It has therefore been necessary to develop an on-track rail planing machine which has the capacity to reprofile the rail head and thus provide an acceptable wheel–rail interaction suitable for high-speed running.

Mechanization of track-relaying has been developed over many years and present practice consists of laying sleepers by means of gantries and beams which carry up to sixty sleepers. These place the sleepers on the ballast and thereafter the rails (jointed or long-welded) are installed using further specialist machines. On new construction, it is possible to lay 3 to 4 km of track per day but this is dependent upon good organization for the supply of materials on a regular basis. On existing railways, different constraints apply, particularly relating to the passage traffic on adjoining lines. Speed of relaying on such tracks is much slower but can easily exceed 2 km per day if large relaying machines are used.

25.11 Railway structures

In the UK, all railway construction is subject to the approval of the Railway Inspectorate of the Department of Transport on behalf of the Secretary of State for Transport. Works of new construction, reconstruction, alteration or addition are accordingly governed by Requirements and Recommendations laid down by that Inspectorate; these are currently undergoing complete revision and metrication but the only section so far issued in the revised form is that relating to structural and electrical clearances. The following paragraphs detail the main requirements, being based as appropriate on the undernoted publications of Her Majesty's Stationery Office and such directives as have been given by the Railway Inspectorate since their issue: *Railway construction and operation: requirements for passenger lines and recommendations for goods lines* (1963 reprint), and *Railway construction and operation and structural and*

electrical clearances (1977). Where dimensions are given in the first of the above documents in imperial units, these have been converted to (rounded) S.I. units for the purposes of this chapter.

The Requirements are subject to some variations and relaxations in the case of light railways and lines of local interest and there are some modifications in the case of underground tube railways.

None of the Requirements is necessarily applicable in retrospect to existing railway structures.

25.11.1 General

All design, materials and standards of workmanship should accord with the appropriate current British Standard or Code of Practice, of which the most generally used are listed at the end of this chapter.

Where any of the Requirements cannot reasonably be met, the special dispensation of the Railway Inspectorate is necessary and may well be conditional upon additional stipulations.

25.11.2 Stations

The lines leading to passenger platforms should be arranged so that platform roads may be entered in the normal direction of movement without reversing.

Curvature of platform lines, and of station yards generally, should be minimized. Tracks through a station or serving a siding should not normally be on a gradient steeper than 1:260.

Platforms should be long enough to accommodate the longest passenger trains serving them and have a non-slip surface. The clear width of any platform throughout its length should not be less than 2 m; at important stations and for island platforms, it should be at least 4 m over the greater part of the length, but may be narrowed towards the ends to not less than 2 m..The descent at the ends of platforms is to be by ramps with a gradient not steeper than 1:8.

Columns for the support of roofs, together with other fixed works such as station buildings and passenger facilities, must not extend to closer than 2 m from the platform edge. A general clear headway of at least 2.5 m is required over platforms.

Platform levels depend upon the floor heights of passenger rolling stock using them, so that the gap between them is as small as practicable. In any case, the stepping distances between the platform edge and the footboards of the passenger rolling stock should not exceed 275 mm laterally and 250 mm vertically. At platform edges, coping should overhang the face walls by 300 mm and the recess so formed should be kept clear, so far as possible, of permanent obstruction.

Where a hazard exists at the rear of a platform or around the sides of stairwells or subway ramps, suitable protective fencing must be provided.

All station premises used by passengers or staff during the hours of darkness must be adequately illuminated, the minimum level of lighting on an open platform being 4 lx. Station names should be conspicuously displayed at intervals along platforms and illuminated at night.

Where passengers must cross the lines to reach a platform, a footbridge or subway should normally be provided.

Stairways and ramps must be of adequate width to avoid overcrowding (but not less than 1 m wide) and at no point narrower than at the top or contracted by any fixed obstruction. The steps of stairs should not be less than 280 mm in the tread nor more than 180 mm in the rise (with optimum dimensions of 300 and 150 mm respectively). Inclined slopes and ramps are not to be steeper than 1:8 (7.12°) while the angle of inclination of escalators should not exceed 30° to the horizontal. Intermediate landings should be provided between fixed flights of steps, with no single flight exceeding 3 m overall rise.

Stations should be constructed to be as fire resistant as practicable and extinguishing equipment provided for dealing with fires in station premises.

25.11.3 Bridges and viaducts

Bridges and viaducts may be constructed in steel, concrete (reinforced or prestressed), brickwork or masonry. The use of other materials such as light alloys will be considered on their merits, but cast iron must not be used in any portion of the structure of a bridge carrying a railway except when subject to direct compression only. Structural timber is usually restricted to light overline bridges.

New and reconstructed bridges must be built to provide at least the mandatory clearances to the kinematic envelope (described below) as illustrated in Figure 25.15. For British Rail, this requirement gives a standard structure gauge 4.640 mm high (above rail level) and 2.340 mm clear laterally from the centreline of the nearest track (when the latter is straight; additional lateral clearance is necessary alongside curved tracks).

Underline bridges must be provided with robust kerbs as part of the bridge structure and extending to a height of at least 305 mm above rail level to contain the wheels of any derailed vehicle. Such bridges must provide for a safe lineside walkway, with a substantial parapet or railing not less than 1.250 m in height above the walkway (unless the main girders themselves provide a sufficient parapet). On bridges and viaducts which are longer than 40 m or where visibility of approaching trains is limited, it may be necessary to incorporate recesses in the parapet design to afford the refuges described below.

Bridges and other structures built over or immediately adjacent to railway lines must be protected against the consequences of their supports being struck by railway vehicles in the event of a derailment. Such supports should be located as far from the tracks as practicable. If they can be kept 4.5 m or more clear of the nearest rail, there will normally be no requirement to design for impact forces. Where this clearance cannot be achieved, the supports must be designed with a degree of robustness and continuity to minimize the effects of contact by a derailed vehicle and to withstand nominal impact forces.

Overline bridges and elevated roads alongside railways must be provided with parapets complying with Department of Transport Technical Memorandum (Bridges) No. BE 5. These parapets are intended to protect pedestrians and/or to contain errant vehicles and prevent them from falling on to the railway; they may also be required to be solid, to prevent splash, reduce noise or screen railway electrification equipment.

For bridges carrying roads, such as motorways, from which pedestrians, animals and pedal cycles are excluded by order, the minimum height of parapet above the adjoining paved surface must be 1.250 m. The lower 600 mm of the parapet should have a mesh or solid infill panel, which may be mounted outside any longitudinal members.

For bridges carrying all-purpose roads and footbridges, the parapet is to have a traffic face which is smooth and without hand- or footholds. The minimum height of parapet above the adjoining paved surface must be 1.500 m. For bridges used frequently by equestrians, the height must be at least 1.800 m.

If a bridge over a railway which is electrified on the overhead system has a parapet with a plinth on its outer face, access must be denied to that plinth. Any parapet more than 100 mm thick is to be surmounted by a 'steeple-shaped' coping (i.e. with steep sides meeting at the top at an acute angle).

At locations where the likelihood of vehicle impact with the parapet and consequential damage resulting from its penetration outweigh the hazards resulting from the containment and redirection of errant vehicles within the traffic scheme, 'high containment' parapets must be provided.

Pipelines carrying liquids or gases over the railway must be incorporated in a bridge constructed for another purpose or be supported by a purpose-designed beam or service bridge. Such a bridge should preferably span the railway without intermediate supports. Consideration may be given to a free-standing design of pipe crossing only in the case of low-pressure water mains or similar pipes conveying non-hazardous materials.

The design loadings and other criteria for both underline and overline bridges are generally as set out in BS 5400. The major railway undertakings in Britain require underline bridges to be designed and constructed by their own staff or under their direct control.

Where bridges are to be constructed under or over existing railways, special techniques are usually necessary to minimize

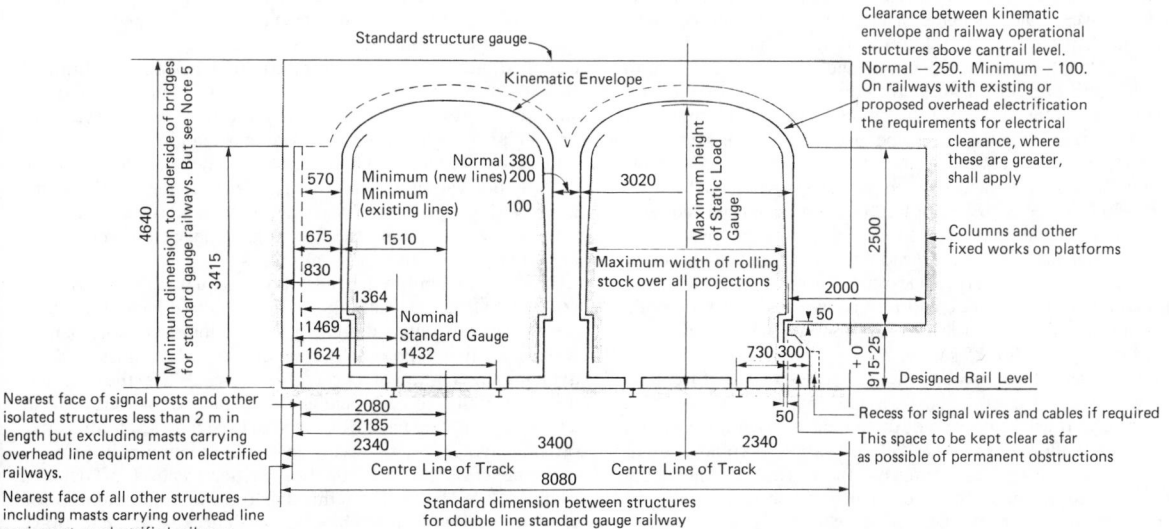

Figure 25.15 Department of Transport – standard structure gauge and kinematic load gauge

the interruption to the normal operation of such railways; the requirements of these techniques may significantly influence not only the temporary works but also the design of the permanent works. It is usually more economic to ensure that work adjacent to running lines, particularly excavation, is located sufficiently far away from them to avoid the necessity for the imposition of speed restrictions on railway traffic. The design and construction of all bridges and viaducts should also be such as to minimize future maintenance on grounds of both economy and avoidance of interference with railway traffic. Provision should be made to ensure that all parts of bridges are accessible for regular inspection and maintenance and that all exposed steel can be painted. Underline bridges should be designed to carry normal ballasted track.

25.11.4 Clearances

The structural and electrical clearances (to be provided on new lines and on existing railways where new structures are built or existing structures are modified or where clearances are otherwise altered) are specified in the revised part of the above-mentioned Department of Transport Requirements published in 1977.

Structural clearances are related to the 'kinematic envelope' of the stock which will use the line and are illustrated in Figure 25.14 in respect of lines carrying trains at speeds up to 200 km/h; that figure is based for the purpose of illustration on the kinematic envelope applicable to British Rail vehicles.

The kinematic envelope is derived from the static load gauge as follows: (1) the maximum permitted cross-sectional dimensions of vehicles and their loads when at rest and located centrally on straight and level track; (2) the static load gauge must be enlarged to allow for the maximum possible displacement of the vehicles when at rest or in motion, with respect to the rails, taking account of their suspension characteristics and making allowance for maximum permitted tolerances in the manufacture and maintenance of the vehicles, including wear; and (3) the resultant kinematic load gauge must then be further enlarged to take account of the maximum permitted tolerances in gauge, alignment, top- and cross-level of track including the effects of wear. The kinematic envelope so derived thus contains the full cross-section of vehicles and their loads under any permissible condition of operation and maintenance of both vehicles and of track. It does not, however, allow for the end-throw and centre-throw of vehicles on curved track, and additional allowance for these must be made as appropriate.

Electrical clearances are dependent upon the nominal voltage of the overhead conductor system and upon whether the category of clearance is to be normal, reduced or special reduced. They must include provision for tolerances in the installation and maintenance of the overhead equipment and of the track limiting values of the vertical clearance for normal, reduced and special reduced clearance conditions between the underside of a structure and the kinematic load gauge shown diagrammatically in Figure 25.15 (for 25 kV a.c.). Also shown are the corresponding clearances between the undersides of structures and the designed rail level, based on the normal British Rail static load gauge. It will be noted that the standard structure gauge on British Rail allows for overhead electrification at voltages up to 25 kV, albeit with 'close tolerances'; the vertical distance from rail level to the undersides of structures should ideally be increased to 4.780 m if this can be achieved with reasonable economy, and further increased where a bridge is adjacent to a level crossing, over which the overhead equipment has to be raised to give 5.600 m clearance over road levels.

25.11.5 Refuges

Refuges are to be provided in all tunnels and also on long bridges and viaducts and where the railway is enclosed by lineside structures unless lateral clearances over a continuous length of 40 m or more are such that staff can stand in safety during the passage of trains. Similarly, on embankments and in cuttings, wherever lineside clearances over a length of 40 m or more do not give safe standing room, refuge spaces are to be provided. The criteria as to what clearance allows men to stand at the lineside in safety are defined as 830 mm from the kinematic envelope at its widest point for train speeds of up to 160 km/h and 1.700 m for train speeds between 161 and 200 km/h.

Refuges and refuge spaces should be located on both sides of the line at a spacing not exceeding 40 m, staggered to give an effective spacing of 20 m or less. Refuges must be at least 2 m high, 1.400 m wide and 700 mm deep, with their floors substantially at the level of the adjacent lineside walkway. Handholds, to assist men in keeping their balance during the passage of a train, are to be provided in refuges and also elsewhere if lateral clearances only marginally exceed the criteria specified in the previous paragraph.

25.12 Inspection and maintenance of structures

In the interests of safety and economy, it is essential to have up-to-date and reliable information as to the current condition of all railway structures and to carry out any necessary repairs or renewals in due time.

The key to proper maintenance, therefore, lies in a system of regular routine and methodical examination by suitably trained and experienced staff in order to provide the responsible maintenance engineer with well-founded records, for each structure, as to: (1) its condition; (2) its current fitness for its intended purpose; (3) its rate of deterioration; and (4) the urgency for remedial action.

It is common practice to legislate for superficial examinations, undertaken at intervals of about 1 year, and less frequent detailed examinations, in which every element of the structure is carefully and thoroughly inspected at close range and the extent of any deterioration quantified. The frequency of detailed examinations will depend upon the nature, age and condition of the structure, but the interval between such examinations should rarely exceed 6 years unless there are special problems or unjustifiably high costs in providing adequate access; in the latter event, decisions as to the action to be taken following an examination should have particular regard to the length of time likely to elapse before the next one. Additional special examinations may be necessary after flooding, damage from impact, fire or vandalism and similar eventualities.

Proper examination depends upon adequate safe access to all parts of a structure. Suitable provision should have been made for this in the initial design by providing removable panels, walkways, fixing points for ladders or safety lines and by avoiding the creation of inaccessible spaces. Where the latter has not been done, some 'opening-out' should be undertaken, at least on a sampling basis. Recourse to expensive scaffolding may often be avoided by the use of mobile inspection platforms.

Examiners must compile full written reports of detailed examinations. To facilitate this, most railway undertakings use specially devised forms, which act as a checklist and encourage clear, concise and comprehensive coverage. Sketches or photographs are valuable for illustrating defects. Reports must record defects in quantitative form, e.g. the loss of section (or the amount of section remaining), the extent of any displacement, the magnitude of any movement, the width, length and orientation of cracks, the extent of hollow-sounding brickwork, the depth of perished mortar or the degree of water percolation. Where appropriate, 'tell tales' and plumbing/levelling points should be established and readings from them recorded. In some cases, more sophisticated techniques of investigation may be called for. The objective should be to ensure that the maintenance engineer has readily available data as to the nature, rate and extent of the development of any defect, i.e. the way in which deterioration is occurring over a period of time by comparing a succession of reports. Thus, he is assisted in diagnosing the cause of trouble and guided in his judgement as to the urgency for remedial action.

Financial constraints may well limit the number of structures which can receive maintenance attention in any year, and it is therefore necessary to establish some kind of priority rating system for defects, to ensure that available resources are directed where the need is most urgent. This may well involve some assessment of the residual strength of members. Calculations and possibly instrumental tests may be necessary to determine critical stresses; these should be based on the nature and magnitude of the loads actually having to be sustained and should have regard to the current condition of members, allowing for loss of original section, distortion and similar defects.

The nature, extent and timing of any repair will vary widely depending on the circumstances applicable in individual cases. Decisions hereon must rely heavily on the informed judgement of the experienced maintenance engineer. For repairs to be effective, it is essential that the cause of the defect is correctly diagnosed and that that cause, rather than its results, is properly dealt with. Desirably, this should be done before serious defects develop, by preventive maintenance of a stitch-in-time nature, e.g. painting of metalwork and timber, pointing of brickwork and masonry and sealing points of water percolation.

It is also a matter of professional judgement to determine when repairs are no longer economic and reconstruction is justified.

25.13 Research, development and international collaboration

The principal railway co-ordinating body in Europe is the UIC (Union Internationale des Chemins de Fer) whose headquarters is in Paris. Its primary purpose is the encouragement and development of standard railway policies and practices throughout Europe. It has a permanent secretariat but it works through committees covering the activities of most railway departments. Twenty-six nations are constituent members.

Its technical research co-ordinating arm is ORE (Office for Research and Experiments) which has its office at Utrecht. The UIC issues specifications, standards, and recommendations for railway practice to which railway administrations are expected or encouraged to adhere. The ORE is represented on all technical committees of UIC and issues reports made by working parties or by specially commissioned persons of academic or professional distinction to member administrations of UIC.

The field and laboratory work of ORE is assigned to member administrations or to universities or technical schools; such work is usually progressed by an appropriate working party composed of delegates from member administrations serviced by a technical secretary from ORE. Distinguished men from industry and universities are sometimes co-opted to serve on specific working parties; UIC and ORE reports are normally issued simultaneously in English, French and German, and these languages are used at UIC and ORE meetings, French being the primary procedural tongue.

Another European body which makes a significant contribution to railway technical literature in its periodical bulletins, is the International Railway Congress Association (IRCA) whose administration is in Brussels.

The principal technical railway co-ordinating organization in the US is the American Railway Engineering Association (AREA) of Chicago which issues a periodical journal covering minutes of meetings, reports of special committees and recommendations for specifications and technical procedures. The AREA also sponsors research and development projects which are mainly progressed by the Association of American Railroads (AAR) at its Research Centre at Chicago. The Pan American Railway Congress is a similar but less developed organization covering the countries of South America. Much valuable railway research and development work is done in Japan and the English version of the monthly *Journal* of the Permanent Way Society of Japan, founded in 1958 in Tokyo, provides a comprehensive contemporary reflection of this work.

Important permanent way research centres also exist in Prague, Moscow, Belgrade and Johannesburg but the results of their activities are normally limited to domestic distribution. The chief technical and development railway work carried out in the UK is associated with the Research and Development Centre at Derby which has a considerable international reputation. Its reports are generally limited to internal circulation within the British Railways Board but most of its railway research projects are reported through papers and articles for the British engineering institutions, chiefly the Institution of Civil Engineers, the Institution of Electrical Engineers, the Institution of Mechanical Engineers, and the technical press. It also accepts commissions for research from outside the British Railways Board.

The national railways of the UK, Canada, France, Germany and Japan are also associated with sponsored railway consultancy organizations which, whilst primarily concerned with the techniques of railway organization, also offer technical advice to clients.

Contemporary railway civil engineering practice and developments in the UK and elsewhere are more to be found in papers presented to the railway division of the Institution of Civil Engineers and to the Permanent Way Institution, in the bulletins and journals of ORE, IRC, and the AREA and in the railway technical press of the UK, France, Germany and America, rather than in the few standard textbooks on railway engineering which are necessarily less up to date.

Bibliography

Berridge, P. S. A. (1969) *The girder bridge*. Maxwell.
Eisenmann, J. (1969) 'Stress distribution in the permanent way due to heavy axleloads and high speeds,' American Railway Engineering Association *Proceedings*.
Fastenrath, F. (1981) *Railroad track – theory and practice*. New York.
Heeler, C. L. (1979) *British Railway Track*. Permanent Way Institution, London.
Prud'homme and Bentot (1969) 'The stability of tracks laid with long welded rails', International Railway Congress Association *Bulletin* (July 1969).
Schramm, G. (1961) *Permanent way technique and economy*. Darmstadt.
Turton, F. (1972) *Railway bridge maintenance*. Hutchinson, London.
Department of Transport (1983) *Bridge inspection guide*. HMSO, London.
Department of Transport (1982) 'Design of highway bridge parapets', Technical memorandum (bridges) No. BE 5. HMSO, London.
Institution of Civil Engineers (1984) *Proceedings of conference on track technology for the next decade* at University of Nottingham.

Permanent Way Institution, Journals and reports of proceedings, list of papers and authors 1884–1976.

Railway Construction and Operation: Requirements for Passenger Lines and Recommendations for Goods Lines (1963 reprint). HMSO, London.

Railway construction and operation: structural and electrical clearances (1977). HMSO, London.

Railway Construction and Operation: Level Crossings (1981). HMSO, London.

Railway engineering and maintenance encyclopaedia. Simmons Boardman, US.

Research and Development Division, British Rail and Association of American Railroads (1981) *Proceedings of conference on rail technology* at University of Nottingham.

British standards

BS 11:1978 (Amended 1984)
  'Specification for railway rails.'
BS 47:1959 (Amended 1979)
  'Steel fishplates for bullhead and flat bottom railway rails.'
BS 64:1946 (Amended 1960)
  'Steel fishbolts and nuts for railway rails.'
BS 144:1973 (Amended 1974)
  'Coal tar creosote for the preservation of timber.'
BS 500:1956 (Amended 1976)
  'Steel railway sleepers for flat bottom rails.'
BS 751:1959 (Amended 1976)
  'Steel bearing plates for flat bottom railway rails.'

BS 882:1983
  'Specification for aggregates from natural sources for concrete.'
BS 913:1973
  'Wood preservation by means of pressure creosoting.'
BS 1377:1975
  'Methods of test for soil for civil engineering purposes.'
BS 2589:1955 (1984)
  'Railway track spanners.'
BS 4521: Part I 1971: Part II 1975: Part III 1974: Part IV 1975
  'Turnouts using flat bottom rails.'
BS 4978:1975 (Amended 1984)
  'Timber grades for structural use.'
BS 5268:1984
  'Code of practice for the structural use of timber.'
BS 5400:1978 Parts 1 to 10
  'Steel, concrete and composite bridges.'
BS 5493:1977 (Amended 1984)
  'Code of practice for protective coating of iron and steel structures against corrosion.'
BS 5930:1981
  'Code of practice for site investigations.'
BS 6031:1981 (Amended 1983)
  'Code of practice for earthworks.'
BS 8110:1985: Parts I to III
  'The structural use of concrete.'
CP 2004:1972 (Amended 1975)
  'Foundations.'

26

Ports and Maritime Works

C J Evans MA(Cantab), FEng, FICE,
FIStructE
Wallace Evans and Partners

Contents

The function of a port is to provide an interface between two modes of transport – land and sea – for cargo and passengers. The requirements for sea transport are: (1) an adequate area of water of sufficient depth for navigation and berthing; and (2) adequate shelter so that berthing, loading and unloading can be carried out safely and efficiently. The requirements for the landside are: (1) adequate land area for working space, loading and unloading vessels and for handling and storage of cargoes; and (2) suitable access to areas served by the port.

26.1 Siting of ports and harbours

The siting of a port is generally dictated by commercial and economic requirements, particularly in relation to land transportation. A natural harbour is to be preferred in order to avoid the necessity of expensive breakwaters, even though some dredging may be required to provide the necessary area of deep water. If the material to be dredged is suitable, land reclamation may be possible using the dredged material to provide land for the shore facilities of a port.

If a natural harbour is not available, breakwaters will be required to provide adequate shelter. Breakwaters are normally very expensive however, and this must be weighed against any additional transport costs and compared with the expenditure incurred at a port where breakwaters are not required.

In planning a new harbour involving breakwaters, consideration must be given to the following factors, in addition to the design of the breakwater itself (see Chapter 31 for design of breakwaters): (1) waves; (2) littoral drift and sedimentation; (3) tides and currents; and (4) navigation.

26.1.1 Design of harbours

The main purpose of breakwaters is to provide protection from waves, and the biggest wave reduction is effected with the smallest entrance sited remote from the direction of approach of the waves. However, this can cause difficulty when approaching the entrance with heavy seas abeam the vessel. As harbours are normally designed to serve as a harbour of refuge, i.e. a protection to be sought by vessels during the height of a storm, it is common to site an entrance at a small angle to the heaviest sea, thereby improving accessibility at the expense of smoothness within the harbour.

Wave-height reduction within a harbour is improved as the distance from the entrance, and the width parallel to the shore, increase. It is desirable to have wave-spending beaches – or armoured slopes which absorb wave energy – facing the waves within the harbour, rather than vertical walls which reflect waves and could cause resonance resulting in significant increases of wave heights. Wave heights within a harbour are normally predicted using numerical models or a physical model; in both cases, various breakwater alignments can be tested to give the optimum alignment. An empirical method for assessing wave heights within a harbour is given in the Stevenson formula:

$$h_p = H\left[(b/B)^{\frac{1}{2}} - 0.027D^{\frac{1}{2}}\left(1 + \frac{b^{\frac{1}{2}}}{B}\right)\right] \qquad (26.1)$$

where h_p is the height of reduced wave at any point in the harbour, H is the height of wave at entrance, b is the breadth of entrance, B is the breadth of harbour at P, being length of arc with centre at midway of entrance and radius D and D is the distance from entrance to point P.

This formula does not take into account the result of any reflection of waves. For assessment of H, see Chapter 31.

26.1.2 Sedimentation

Sedimentation in a harbour can arise from three sources: (1) littoral drift; (2) tidal movements; and (3) where a harbour is located at a river mouth, from the river. The minimizing of sedimentation in navigation channels, at the entrance and within the harbour, is of prime importance in reducing the cost of maintenance dredging.

Littoral drift occurs to some extent along most coastlines. If the path of the drift is obstructed by a solid structure, the heavier particles will accumulate on the drift side and this accumulation may well extend round to the inside. The finer particles of the drift, which outside the harbour are kept in suspension by current velocities will, on entering the harbour, no longer be maintained in suspension and will settle out. Littoral drift normally occurs in one direction, but at certain times of the year or under some storm conditions, the direction of drift can be reversed. Littoral drift is discussed in more detail in Chapter 31.

Where a harbour is subjected to large tidal ranges, material in suspension will be brought into the harbour as the tide rises and, during periods of slack tide, material will settle on the sea-bed. Where a harbour is at a river mouth, the material carried down by the river is a further source of sedimentation. The interaction of river flows and movements of the sea makes for further complications, with the added difficulty of the difference in density between fresh and salt water.

Predictions of sedimentation are best carried out by numerical modelling. Physical models can also be used, but these can be less accurate – particularly with fine material in suspension – because of the difficulty of scaling-down the fine particle sizes to the scale of the model, and results should be treated with caution.

26.2 Port planning

The planning of a new port or expansion or improvement of an existing one requires many factors to be taken into consideration. Apart from passenger ferry terminals and cruise ship terminals, ports are primarily provided for the handling of cargo. Amongst the factors to be considered are:

(1) Nature of cargoes to be handled.
(2) Sizes and types of vessels to be catered for.
(3) Method of cargo handling.
(4) Land area and operations.
(5) Land access.

26.2.1 Types of cargoes

Between 1960 and 1980 a major revolution in the handling and carrying of maritime cargoes took place and this has led to new concepts in the design of ships, ports and land transportation systems. Generally speaking, during this period emphasis was given to handling and carrying cargoes in larger units, e.g. containers in the case of general cargo, and larger single shipments of bulk commodities such as wheat and oil, etc. Ship sizes also increased to obtain the benefits of the increased scale of operation.

26.2.1.1 General cargoes

Nonunitized (or break bulk) cargoes. These consist of small consignments requiring to be handled individually. The volumes now being conveyed by this method are rapidly diminishing and nonunitized working is practised only in areas where labour is plentiful.

Unitized cargoes. Unitization of cargoes permitting larger units of general cargo to be handled by mechanical equipment, so replacing labour, has become attractive. Unitized cargoes can be subdivided as follows:

(1) *Prepackaged.* Certain dry-bulk cargoes, of which sawn timber is one, are packaged into larger standard-sized units for handling in unit sizes ranging up to 5 t. Packaging is usually done using metal strapping.
(2) *Palletized cargoes.* These range from 1 t to 5 t and are suitable for handling by fork-lift trucks. Typical examples are bagged commodities such as cement and flour, and boxed products. Standard pallet sizes, in metres, are as follows:

    0.8 × 1.0
    0.8 × 1.2
    1.0 × 1.2
    1.2 × 1.6
    1.2 × 1.8

(3) *Flats.* These are usually 3.05 × 2.44 m and 6.10 × 2.44 m capable of carrying up to 10 t. Consignments can be of both regular or irregular shape but require lashing down to the flat. They can be handled by fork-lift trucks or a combination of fork-lifts and low-wheel trailers.
(4) *International Organization for Standardization (ISO) containers.* Standard sizes are usually quoted in tonnes equivalent units (TEUs), and those most commonly in use are:

    (a) 3.05 × 2.44 × 2.44 m (maximum load 10 t or 0.5 TEU);
    (b) 6.10 × 2.44 × 2.44 m (maximum load 20 t or 1 TEU);
    (c) 12.19 × 2.44 × 2.44 m (maximum load 40 t or 2 TEU).

These are sealed units, capable of being lifted from the bottom by fork-lift trucks or from the top at the ISO four-corner lock attachments by cranes and mobile equipment. They are also stackable. Specialized ISO containers have been developed as refrigerated and liquid tank units, but all are to the standardized overall dimensions and equipped with the ISO universal handling devices. Some of these, e.g. refrigerated units, require support services in the way of electrical power whilst in transit through the port.
(5) *Specialized forms.* The introduction of roll-on, roll-off (RoRo) ships allows cargoes in road trailers to be shipped either with or without the traction unit.

26.2.1.2 Bulk cargoes

Bulk cargoes fall into two categories: (1) dry; and (2) liquid. Commodities of these types, more often than not, are shipped in purpose-built vessels or carriers and are loaded and unloaded using specialized berths or terminals equipped with mechanical handling systems suitable for the commodity being handled. Typical commodities are grain, mineral ores, timber, sugar, vegetable oils, mineral oil and petroleum products, liquid chemicals, liquefied petroleum gases (LPG) and liquefied natural gas.

Some of these commodities are hazardous and have to be handled and stored under statutory regulations.

26.2.1.3 Miscellaneous trades

There are a number of cargo trades which do not fall readily into the above categories. An example of this is the advent of the car carrier solely handling cars for international distribution.

26.2.2 Sizes and types of vessels to be catered for

26.2.2.1 Classification

Ships are classified under a number of tonnages as follows.

(1) Gross registered tonnage (GRT): The value derived from dividing the total interior capacity of the vessel by 2.83 m^3, subject to the provisions of applicable laws and regulations.
(2) Net registered tonnage (NRT): The gross tonnage of the vessel minus the tonnage equivalent of crew cabins, engine-rooms, etc.
(3) Displacement tonnage: Indicates the total mass of the vessel, and is obtained by multiplying the volume of the displaced sea water by the density of sea water (1.03 t/m^3).
(4) Dead weight tonnage (DWT): Dead weight of a vessel is the weight equivalent of the displacement tonnage minus the ballasted weight of the vessel. Consequently, it indicates the weight of the cargo, fuel, water and all other items which can be loaded aboard the vessel.
(5) Tonne measurement: The value derived from dividing the cargo spaces of a vessel by 1.13 m^3.

The approximate relationships shown in Table 26.1 apply between the various tonnages. For port engineering purposes, DWT is the most significant although, for calculating berthing energies, the displacement of the vessel is required. The shipping industry uses the long ton. This is almost the same as the metric tonne and for planning purposes can be treated as being interchangeable.

26.2.3 Types of vessels

Vessels are generally categorized by the types of cargo they handle as follows.

(1) *General cargo.* These generally carry nonunitized (break-bulk) cargoes and/or unitized cargoes, but can also carry some containers. These range in size from small coasters (2000–3000 DWT) to long-distance vessels up to 30 000 DWT.
(2) *Container vessels.* These are specially designed ships for the purpose of carrying containers and can vary from small feeder vessels carrying perhaps 150 TEU up to the very large container vessels (used on long sea routes) carrying up to 4000 TEU and being of about 70 000 DWT.

Table 26.1

Vessel type	Approximate loaded displacement
Bulk carrier	GRT × 1.2–1.3
Container vessels	DWT × 1.4
Passenger liners	GRT × 1.0–1.1
General cargo	GRT × 2.0 or DWT × 1.4–1.6

(3) *Roll-on roll-off vessels.* These are specially designed to allow the movement of cargo through stern or bow ramps by vehicular movements without the need for cranes or other lifting devices, and are generally used on the shorter sea routes.

(4) *Bulk-cargo vessels.* These are normally designed specifically for a particular trade, such as iron ore, coal, grain sugar, etc. and can range from small vessels of 20 000 DWT up to large bulk carriers of up to 60 000 DWT.

(5) *Tankers.* These are designed for liquid bulk cargoes and can range from small vessels of 20 000 DWT up to the very large oil tanker of up to 1 million DWT.

Typical relationships of dimensions for various types of vessels are shown in Figures 26.1 to 26.4.

Certain characteristics of vessels may also need to be taken into account. Some vessels are equipped with bow thrusters for ease of manoeuvring, and these have been known to cause damage to quay walls. Problems can also occur with vessels that have bulbous bows, where the projecting bow located below water can cause damage to piled structures.

26.2.3.1 Vessel characteristics

In planning a port development, knowledge of the following characteristics of vessels likely to use the port is required in addition to the dimensions of vessels (length, beam and draft).

(1) Ship layouts, including the locations and dimensions of ramps and hatches, loaded and unloaded deck heights, superstructure positions and clearances for dockside cranes.
(2) Handling characteristics of ships for manoeuvring and turning operations.
(3) Windage areas of ships to assess forces on berths.
(4) Ship mooring line sizes and capacities for bollard pulls.
(5) Deck crane capacities and reaches.

26.2.4 Methods of cargo handling

These will depend largely on the nature of the cargoes and the types of vessels likely to use the port. The most important consideration is whether dockside cranes are required or whether ships' own lifting gear will be used for loading and unloading. Apart from cranes, cargo-handling equipment can range from fork-lift trucks, which can have a capacity from 30 to 400 kN for general cargo, to special container-handling equipment. The latter can be large fork-lift trucks (capacity 200 to 420 kN) straddle carriers and gantry cranes (rubber tyred or on rails).

26.2.5 Land area

This depends on: (1) throughput of cargo; (2) type of cargo; (3) methods of cargo handling; and (4) length of time cargo remains in the port. A modern general cargo berth is normally 200 m long and 200 m or more deep. Thus, an area of 200 × 200 m, or 4 ha, is required. With efficient cargo handling, this will handle approximately 250 000 t of cargo per year. A container berth requires more land behind the berth to maximize the throughput. Container berths are generally 300 m long or greater and with up to 200 to 800 m depth, although this can be reduced if containers are stacked. The area can therefore range up to about 20 ha which would handle up to about 1 million t of cargo per year. However, the land requirements must be investigated for individual cases according to the factors mentioned above. With a general cargo area, part of the land will be utilized by transit sheds and warehousing. In a container berth, the land area will largely be open for storage of

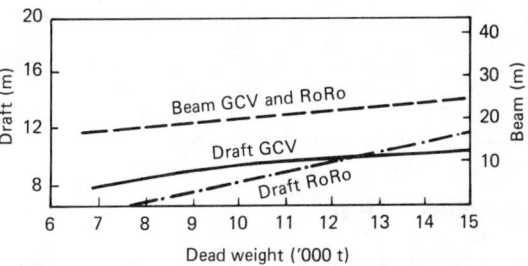

Figure 26.1 Typical general cargo and RoRo vessel dimensions

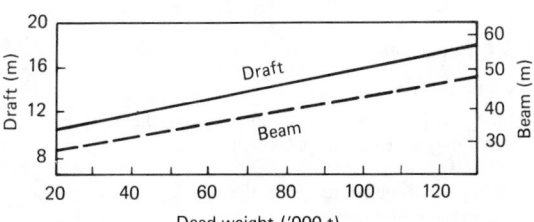

Figure 26.2 Typical oil tanker dimensions

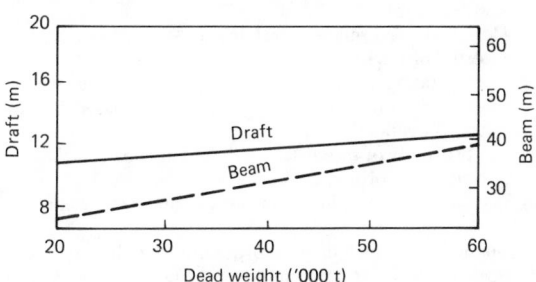

Figure 26.3 Typical ore carrier dimensions

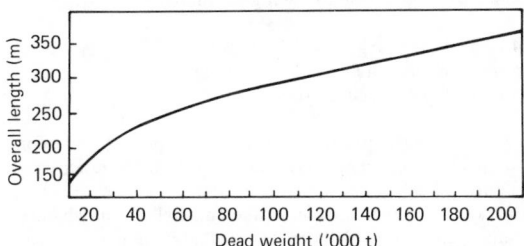

Figure 26.4 Typical lengths for general cargo RoRo vessels, tankers and bulk carriers

containers with sheds for filling and emptying containers, unless these operations are carried out at an inland depot away from the port.

26.2.6 Access

Access can be either by road, rail or both; or, in the case of liquid cargoes, by pipeline.

26.2.7 Other considerations

Other factors requiring consideration in the planning of port developments include:

(1) Tugs and pilotage.
(2) Security and policing services.

(3) Fuel bunkering facilities.
(4) Equipment maintenance facilities.
(5) Services to ships – water, electricity, sewerage, telephone.
(6) Rest rooms, canteens and offices, etc.
(7) Post offices.
(8) Customs and immigration arrangements.

26.3 Navigation

The navigation requirements of a harbour involve three aspects:
(1) the approach channel; (2) the entrance; and (3) the
manoeuvring area within the harbour.

26.3.1 Requirements

26.3.1.1 Channel width

Channel width is governed by many factors, the most important
of which may be summarized as follows:

(1) The vessel dimensions; in particular, the beam of the largest
    vessel using the port.
(2) The orientation and strength of the currents and the expo-
    sure to wind and wave action (which can cause vessels to
    yaw and crab).
(3) The speed and manoeuvrability of the vessels and the
    expertise of the pilots.
(4) The operating pattern of vessel movements, i.e. whether
    vessels are allowed to pass or whether a phased one-way
    system is operated.
(5) The proximity of the vessels to the channel banks (the effect
    of which is to promote additional yaw).
(6) The channel depth; in particular, the underkeel clearance.

Various methods, including ship deviation studies and scale-
model methods, have been employed to assess channel width
and various recommendations have been published. British
Standard 6349[1] gives the following recommendations.

(1) 4 to 6 × beam: large vessels, one-way traffic only.
(2) 6 to 8 × beam: smaller vessels passing.
(3) 5 to 7 × beam: large tankers.

Other studies have produced recommendations in the form
shown in Table 26.2, which gives an example for a design vessel
of length L of 260 m and breadth B of 40 m. The manoeuvring
lane is defined as that portion of the channel within which the
ship may manoeuvre without encroaching on the safe bank
clearance and without approaching another ship so closely that
dangerous interference between ships would occur. As vessels
pass each other, interactive hydrodynamic effects occur as
illustrated in Figure 26.5.

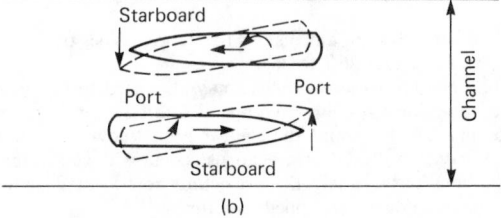

(a)

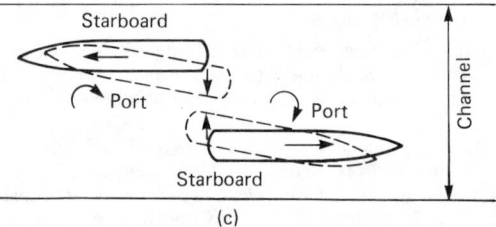

(b)

(c)

Figure 26.5 Hydrodynamic effects of ships passing in channels.
(a) Bows abreast: bows yaw away, but bank suction opposes this
tendency (sheer to starboard); (b) bows approach sterns: bows
yaw toward low water and the bank suction tends to reinforce this
movement (sheer to port); (c) sterns opposite each other: sterns
yaw toward low water at sterns but bank suction opposes this
tendency

The influence of depth of water on the channel width should
not be overlooked as a small underkeel clearance can have a
marked effect on the vessel's manoeuvrability and can increase
significantly the lane width required.

Where bends are unavoidable in the approach channel, the
channel width must be increased at the bend to take into
account the extra area swept by the ship during the turning
movement. It has not been possible to formulate precise rules
for this increased width, but it has been suggested that where the
change of heading is of the order of 30 to 45°, the channel width
should be increased by at least twice the largest vessel's beam.

Table 26.2

	Manoeuvring lane A	Bank clearance B	Shift clearance C	One-way traffic width	Two-way traffic width
Sheltered	A = 2.0 × beam	B = 1.5 × beam	C = 1.0 × beam	A + 2B	2A + 2B + C
Example (m)					
L = 260	80	60	40	200	320
B = 40					
Exposed location	A = 2 × beam + L sin 10°	B = 1.5 × beam	C = 1.0 × beam	A + 2B	2A + 2B + C
Example (m)	124	60	40	244	408

26.3.1.2 Channel depth

The depth of water available for shipping, whether natural or provided by dredging, is dependent on the variations in water level, the draught of the largest vessel, the change in salinity, the wave- and speed-induced vertical motion of the vessel and the required underkeel clearance. Account may also have to be taken of the accuracy of soundings, the sediment deposited between dredging operations and the dredging tolerances. These are shown diagrammatically in Figure 26.6. Much research has been carried out and recommendations published for minimum underwater keel clearances,[2] but it is advisable for general purposes to provide a depth below low water level of 1.15 times the maximum draught of the vessel, with a minimum gross underwater keel clearance of 1 m. Slightly greater clearances should be provided where the sea-bed is rock in order to increase the clearance for safety of the ship against grounding on a hard surface.

The depth alongside a berth can be slightly less than the channel depth, and in some ports (generally small ones) with high tidal ranges, provision is sometimes made for vessels to sit on the bottom during periods of low tide with access to the berth only during certain periods of the tidal range.

26.3.1.3 Turning circles

It is normally desirable for a ship to be able to manoeuvre within a harbour and to leave the harbour bow first; a sufficient turning area with the necessary depth of water must therefore be provided. For a vessel to turn unassisted in one circular movement the diameter required is ideally 4 times the length of the vessel. With the assistance of tugs a turning circle with a diameter of twice the vessel's length is acceptable. Where turning dolphins or other mooring arrangements, which enable the vessel to swing while partially moored, are provided, this requirement can be reduced further.

26.4 Design of maritime structures

The commonest types of maritime structures are:

(1) *Marginal berth (also termed quay or wharf).* A berth parallel to the shore and contiguous with it. Figure 26.7 shows a typical layout with three continuous marginal berths.
(2) *Pier.* A finger projection from the shore on which berths are provided (Figure 26.8).
(3) *Jetty.* A structure providing a berth or berths at some distance from the shore. It may be connected to the shore by an approach trestle or causeway, or the jetty may be of an island type (Figure 26.9).
(4) *Dolphin.* An isolated structure or strong point used for manoeuvring a vessel or to facilitate holding it in position at its berth (Figure 26.9).
(5) *Roll-on roll-off ramp.* A structure containing a fixed or adjustable ramp on to which a vessel's ramp is lowered to permit the passage of vehicles between vessel and shore (Figure 26.10).

26.5 Marginal berths

These require a vertical face against which the ship berths and a contiguous working area alongside for cargo-handling equipment and cargo storage. The vertical wall can be achieved by two main methods: (1) a solid wall – which can be a gravity wall or a sheet-piled wall; (2) an open type – piled structure. Both types are commonly used for marginal berths, the choice depending primarily on depth of water, the foundation conditions, and the availability of suitable material for filling behind the solid wall. Typical designs of quay walls for marginal berths are shown in Figure 26.11.

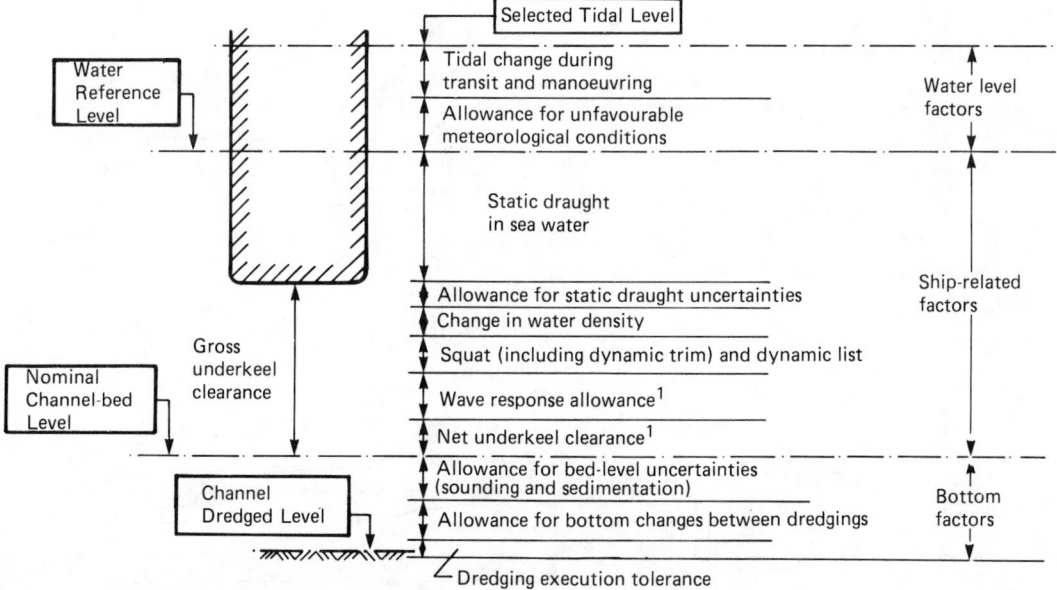

Note 1 Net underkeel clearance and wave response allowance contribute to the manoeuvrability margin

Figure 26.6 Factors determining the required underkeel clearance

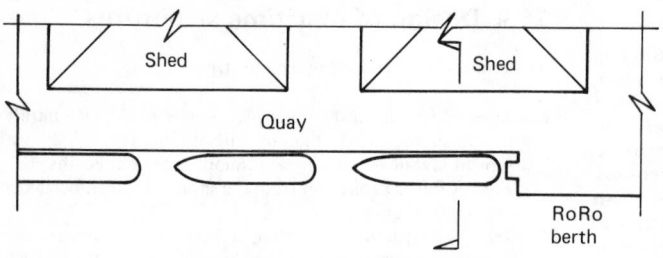
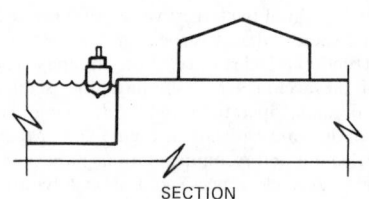

Figure 26.7 Marginal berths

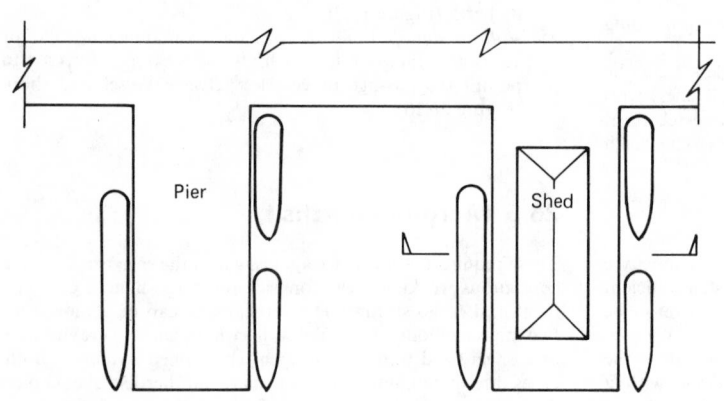
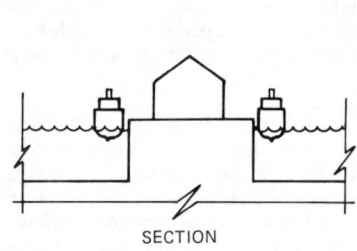

Figure 26.8 Pier berths

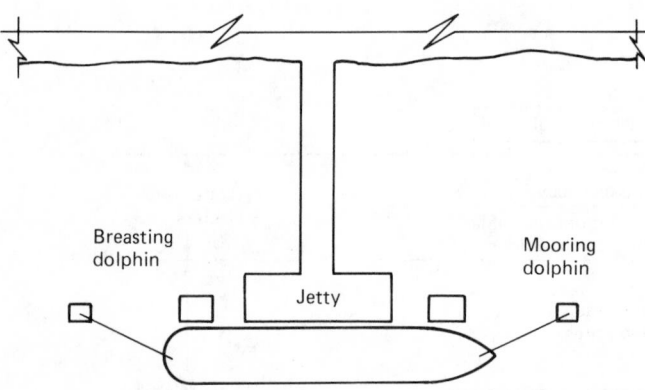
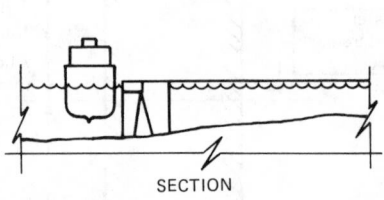

Figure 26.9 Jetty berth

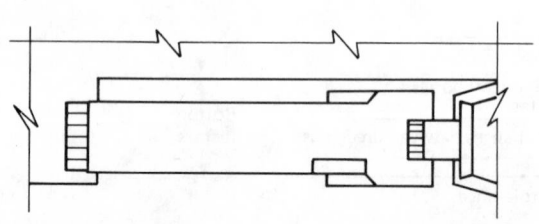
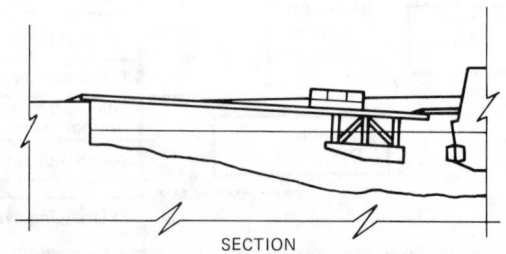

Figure 26.10 RoRo berth

26.6 Piers and jetties

26.6.1 Piers

A pier normally requires a vertical face on both sides against which ships are berthed, with the deck of the pier providing the working area for cargo handling and sometimes cargo storage. The methods of cargo handling and storage determine the width of the pier. If the pier is sufficiently wide, the seaward end of the pier can also be used for berthing ships.

As with marginal berths, the pier can be of a solid type or a suspended structure on piles. Because the pier extends into the seaway, particular consideration needs to be given to its effect on the hydraulic regime and littoral drift. The choice of whether the pier is solid or open will frequently depend on these considerations, although foundation conditions and availability of fill material may also affect the choice. Typical layout showing clearances required between adjacent piers is shown in Figures 26.12 and 26.13.

26.6.2 Jetties

A jetty is a structure providing a berth or berths at some distance from the shore where the required depth of water is available. It consists normally of a jetty head which provides the actual berth, which is connected to the shore by an approach trestle or causeway.

The jetty head should normally be aligned so that the vessel is berthed in the direction of the strongest currents, and is normally an open-piled structure although a solid 'island'-type structure is used occasionally. The approach section is generally built as an open-piled trestle type of structure mainly to avoid affecting the hydraulic regime, and often also on grounds of cost, although in shallow water a solid causeway may be cheaper. In some cases a causeway is used for the first section of the jetty approach from the shore, until the depth of water increases to the point where a piled structure becomes more economical. In determining this point, the life and maintenance costs of the open structures need to be taken into account, as a causeway generally requires very little maintenance.

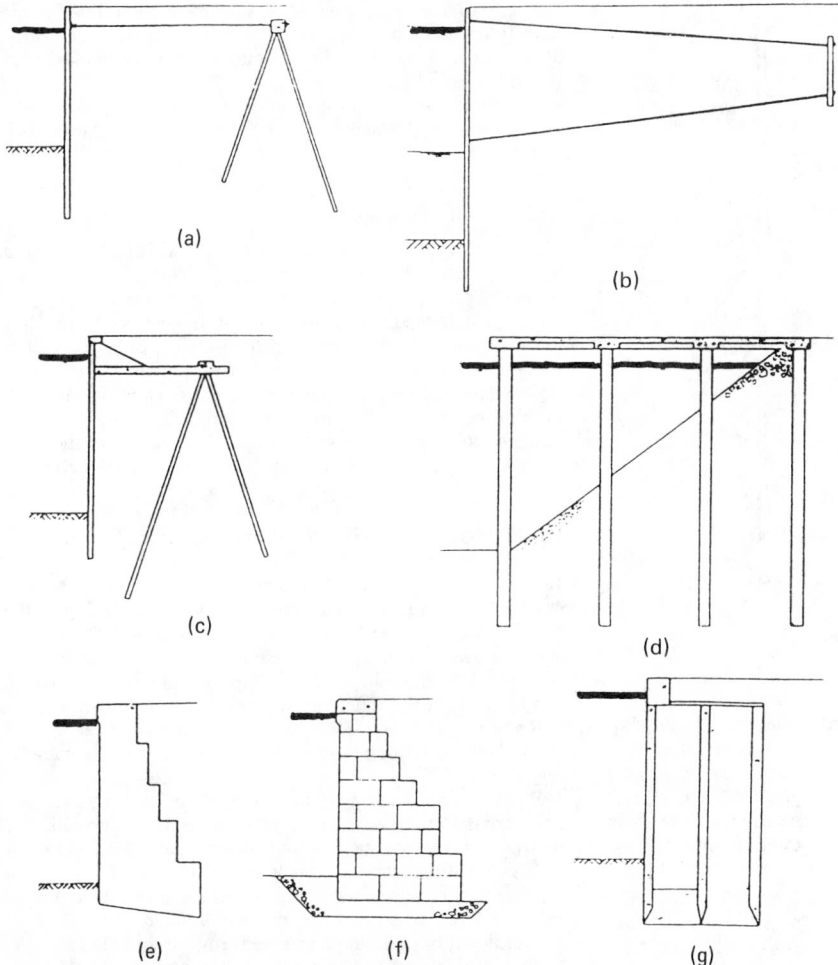

Figure 26.11 Types of quay walls. (a) Anchored sheet pile wall – single tie; (b) anchored sheet pile wall – two ties; (c) sheet pile wall with relieving platform; (d) open-piled construction with suspended deck; (e) concrete wall built in the dry; (f) concrete wall built in the wet; (g) monolith

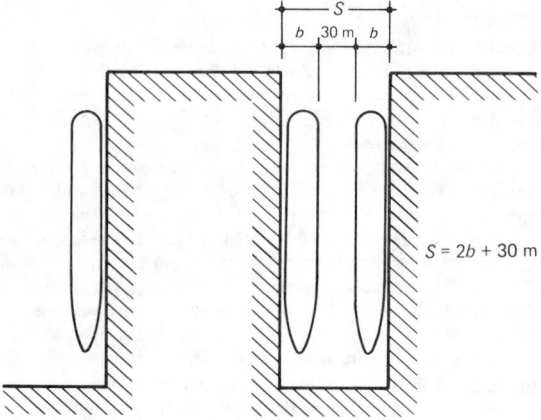

Figure 26.12 Clearances for single-berth piers

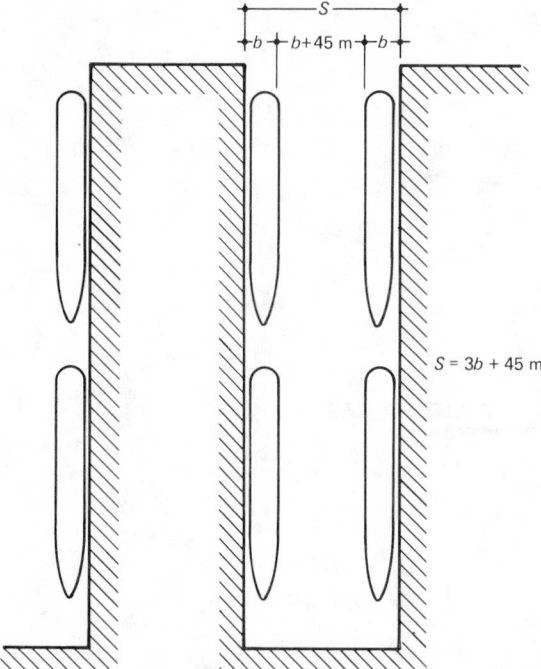

Figure 26.13 Clearances for multi-berth piers

The jetty head is normally smaller than the length of the ship it is designed to handle, and generally requires breasting dolphins and mooring dolphins (see sections 26.7.1 and 26.7.2) for berthing and for maintaining the vessel in position. In such cases the jetty head need not be designed to resist berthing impacts and can be of a lighter construction, designed primarily for vertical loads. A typical oil jetty terminal is shown in Figure 26.14.

26.7 Dolphins

Dolphins are of two kinds: (1) breasting dolphins; and (2) mooring dolphins. The use of both types is shown in Figure 26.14. Turning dolphins are also used occasionally to assist in berthing vessels.

26.7.1 Breasting dolphins

A breasting dolphin is an isolated structure designed to fulfil two distinct functions: (1) it must absorb the kinetic energy of the berthing vessel; and (2) it must assist in restraining the vessel at the berth. The optimum disposition of the breasting dolphins about the main service structure is critical to the design. Two berthing dolphins, one each side of the service platform, are generally sufficient, but if the berth is to be used by ships of widely varying size two additional dolphins closer to the platform may be required. The face line of the dolphins in relation to the front of the service platform will depend on the maximum deflection of the dolphins. To prevent impact between the unprotected platform and the vessel, a gap must be maintained between them when the dolphins are at maximum deflection. The gap must not be so large as adversely to affect the efficient operation of the cargo-handling equipment.

There are two basic types of breasting dolphin: (1) rigid; and (2) flexible. The former will be either a massive structure (such as a blockwork or caisson construction) or an open multiple-pile structure rigidly held together at the top by a massive deck or a steel jacket. Rigid structures simply withstand the berthing load relying on fenders to absorb the energy.

The flexible dolphin usually takes the form of parallel flexible steel tubes (or a single tube), with a high elastic limit, which absorb most of the energy by deflection up to a maximum of 1.5 to 2.0 m.

The choice between a rigid or flexible dolphin will usually be determined by the depth of water and the foundation conditions.

26.7.2 Mooring dolphins

Mooring dolphins are isolated structures to which mooring lines are attached to restrain the ship at the berth. They are not normally subject to impact from a berthing vessel and do not therefore need fendering or to be flexible to absorb energy. They must, however, be designed to resist the horizontal load from the mooring lines over a wide angular range, which arises from both wind and current load on the moored vessel and the ranging of the vessel from wave action. They must also be designed for uplift, to resist the vertical component of the force in the mooring line. They are normally rigid piled structures.

26.8 Roll-on roll-off berths

In parts of the world where tidal ranges are small, RoRo vessels can berth, offload, and load at any state of the tide, bridging the ship-to-shore gap with their own short ramps. Where tidal ranges are large, RoRo vessels must either use impounded docks or more elaborate ship-to-shore ramps must be provided which can tolerate greater differences in level between ship and quay.

Roll-on roll-off ramps can be of three types: (1) fixed at both ends; (2) fixed at one end – floating at ship end; and (3) completely afloat.

Buoyant RoRo ramps differ mainly from their nonbuoyant counterparts in the way their seaward ends are supported. The uplift of a submerged tank is used to carry the dead load instead of a conventional bridge foundation. From this fundamental difference have arisen many characteristics in the structures of buoyant ferry ramps that are uniquely different from those of conventional bridges; these may be seen in Figure 26.15.

The basic design parameters, which include range of water levels, freeboard, quay level, and limiting gradients, define the dimensions of the ramp with very little scope for variation.

The gradient of the ramp must allow vehicles and cargo-handling plant to drive over it at all states of the tide. Limiting gradients are usually dictated by the operators or ferry owners.

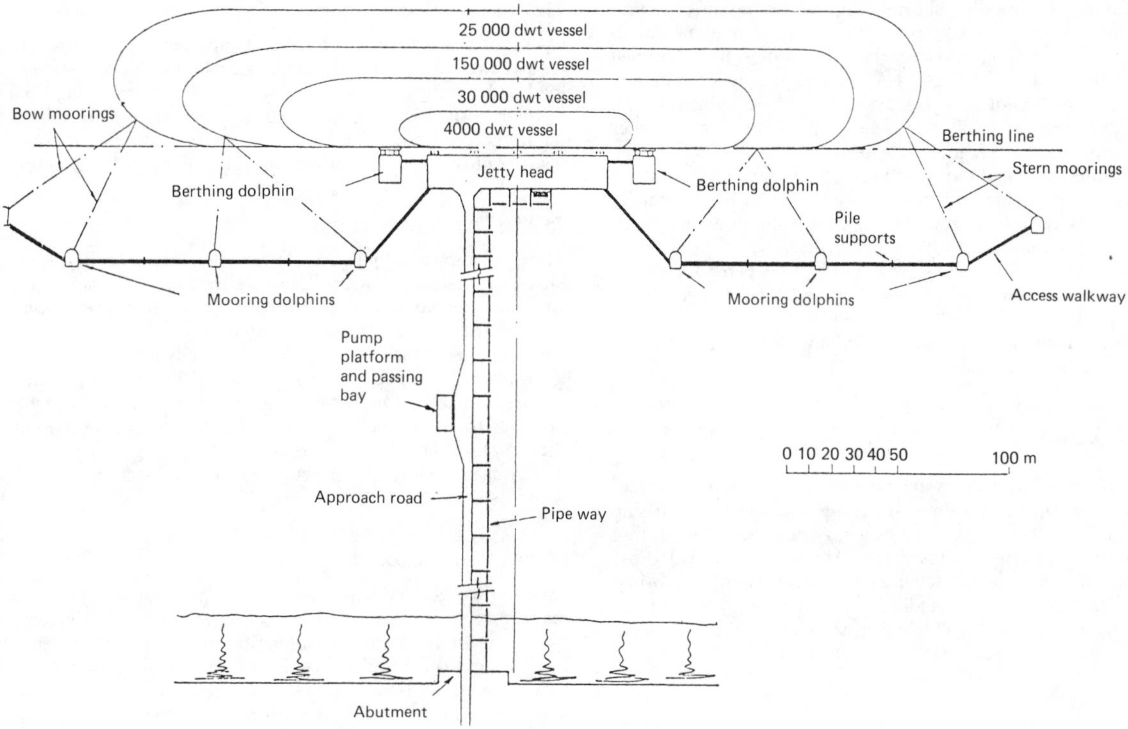

Figure 26.14 Typical plan of oil-jetty berth

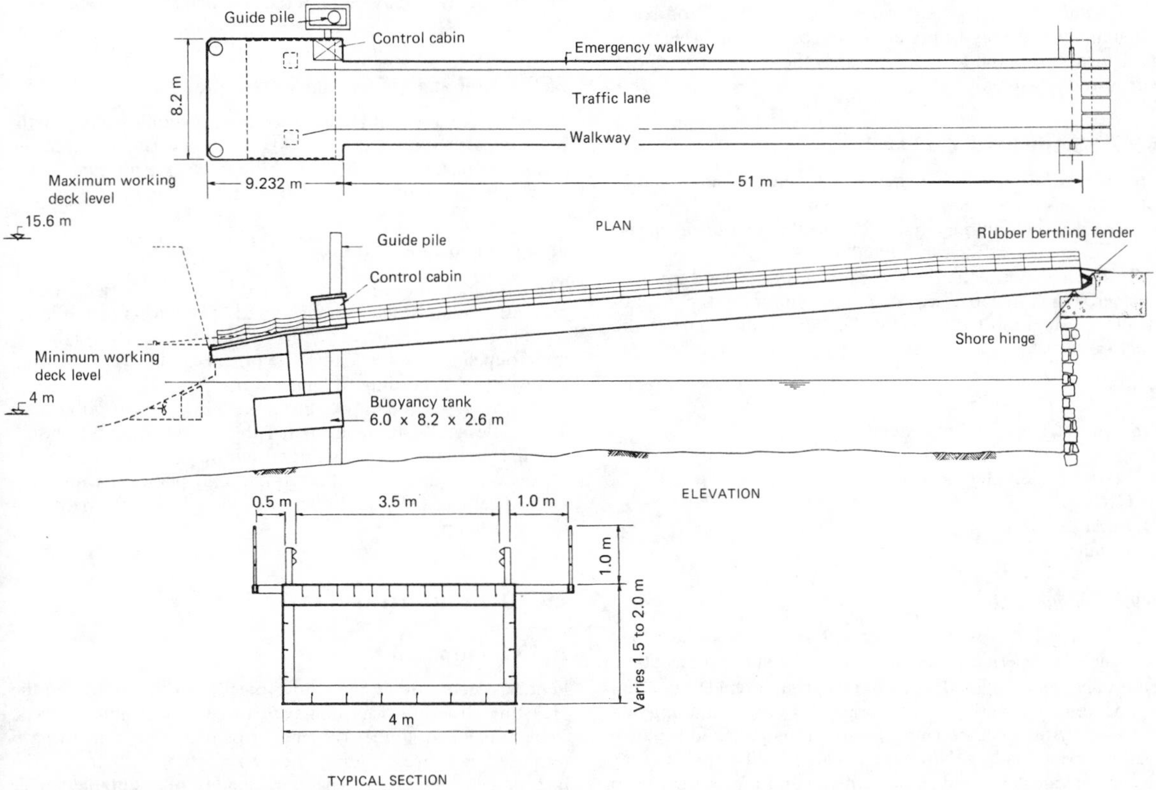

Figure 26.15 Buoyant RoRo ramp

Increased gradients allow shorter ramps with consequent economies of cost and a compromise must be struck. A maximum deck gradient of ± 10% is desirable, but circumstances have led to gradients of as high as 14% because the most economic solution may be not only to keep within the desirable limit on the most frequent tidal conditions, but also to allow steeper gradients on relatively rare occasions.

Considerable ingenuity is required to design ramps so that they are compatible with the very wide range of geometries adopted by ship designers. Harmonizing standards have been suggested, but unification has not yet arrived.

Roll-on roll-off terminals are dependent on being located in reasonably calm waters. In more exposed locations it can be very difficult to ensure that the dynamic wave-generated motion between a floating ramp and ship will not give rise to unacceptable working conditions.

26.9 Loads

In addition to dead loads and soil pressures, the other forces which may act on maritime structures are: (1) those arising from natural phenomena such as wind, snow, ice, temperature variations, currents, waves and earthquake; and (2) those imposed by operational activities such as berthing, mooring, cargo storage and cargo handling. These loads may be grouped conveniently into the following general categories: (1) dead; (2) superimposed dead; (3) imposed; (4) soil and differential water; and (5) environmental.

26.9.1 Dead load

Dead load is defined as the effective weight of the materials and parts of the structure that are structural elements, excluding soils, surfacing, fixed equipment and tracks, etc. For some design analyses it may be preferable to consider the weights of the elements in air and to treat separately the uplift forces due to hydrostatic pressures.

26.9.2 Superimposed dead load

Superimposed dead load is defined as the weight of all materials forming loads on the structure that are not structural elements and would include fill material on a relieving platform, surfacing, fixed equipment for cargo handling, etc.

The effect of removing the superimposed dead load must be considered in any analysis, as it may diminish the overall stability or diminish the relieving effect on another part of the structure.

26.9.3 Imposed load

Imposed loads may be subdivided into:

(1) Static and long-term cyclic.
(2) Cyclic.
(3) Impulsive.
(4) Random.

26.9.3.1 Static loads

Long-term cyclic loads are grouped with static loads where they have such long periods that they act on the structure as static loads. The main imposed loads in this group are: (1) superimposed live load covering cargo storage; (2) cargo handling and transport systems equipment; (3) current loading; and (4) time averaged wind loading. Normal methods of static analysis may be used to calculate the resulting stresses and movements from the imposed loads.

26.9.3.2 Cyclic loads

Cyclic loads are those which repeat in all essential features after regular intervals of time. The main cyclic loads are: (1) wave loading from regular trains of waves; (2) vortex shedding from circular sections in steady currents; (3) vibrations from vehicular traffic and tracked cranage; and (4) vibrating loads from heavy, out-of-balance, rotating machinery fixed to the structure.

26.9.3.3 Impulsive loads

The main impulsive loads are: (1) berthing forces; (2) release or failure of tensioned hawsers; (3) wave-slam forces on horizontal structural members due to the passage of the wave profile through the member; (4) crane snatch-loads when lifting cargo from moving vessels; and (5) vehicular impact and breaking loads from cranes and road and rail traffic. The most significant of these is likely to be berthing impact. Fendering is normally provided to absorb the energy of impact and reduce the load on the structure. The design of fendering is an integral part of the design of all structures subject to berthing impact and is dealt with in section 26.10.

26.9.3.4 Random loads

Random loads vary with time in a nonregular manner. The main random loads are: (1) normal wave loading; (2) loading from wave-induced motion of a moored vessel; (3) seismic loading; and (4) turbulent wind loading. Loading from wave-induced vessel motion is likely to be the most significant of these loads in open-sea conditions. In some cases these loads can be greater than those due to berthing, although they will always be less than berthing loads within a sheltered harbour.

Cyclic, impulsive and random loads are dynamic in value and dynamic analysis may be required to calculate the response of the structures.[3]

26.9.4 Soil and differential water load

These are the dominant loads affecting the stability of an earth-retaining structure. The soil loads should be derived from the properties of the soil. The disturbing forces are affected by the surcharge and imposed loads on the retained soil.

26.9.5 Environmental loads

These include the effects of temperature, snow, ice, waves, current, tide and time-averaged wind. The effects of the latter three are generally considered to be long-term cyclic loads and are grouped with static loads. The others can be cyclic, random or impulsive, according to their nature.

In most cases, the impact live loads are the most important. Typical design loads are given in Table 26.3. It should be borne in mind that many cargoes cause point loads, e.g. corner loads of containers, and also that cargo-handling plant can cause very high wheel loads. These should be obtained from the manufacturers of the plant.

26.10 Fendering

26.10.1 Introduction

Fender systems are designed to protect both the vessel and the breasting structure from damage caused by berthing impacts. They range from timber rubbing-strips fixed to a quay face, to purpose-built, free-standing energy-absorbing structures. The factors determining the type and capacity of a suitable fender system include the nature and size of the berthing vessels, the

Table 26.3

Load	(kN/m²)
Light traffic	5
General traffic	10
General cargo	20
Containers:	
empty, 4-high	15
2-high stacked	20
4-high stacked	30
RoRo cargo	30–50
Multipurpose	50
Offshore supply base	50–150
Paper	55
Forest products	70
Steel products	80
Coal	100–200
Ore	100–300

form of the structure to be protected, the environmental conditions (i.e. wind, waves, currents etc.,) the operational requirements and the consequences of damage to the vessel or structure.

The berthing force is often the predominant lateral load imposed on a quay or jetty structure and its effect is largely controlled by the fender system adopted. The design of the fendering system must therefore form an integral part of the structural design. The selection of the fendering system will often be the first step in the design process and can influence the shape, size and form of structure.

A fendering system may be defined as a structural element or a combination of elements which ensures the safe disposition of a vessel's kinetic energy whilst it berths in a controlled manner. Most systems incorporate an elastic energy-absorbing unit but occasionally a plastic or friable unit is also included, the deformation of which provides additional protection against marginal overloading.

26.10.2 Fendering systems

Fendering systems may be divided into two main categories:

(1) *Detached systems*, in which the berthing forces do not act on the main quay or jetty structure.
(2) *Attached systems*, in which the fender elements are attached to the main quay or jetty structure. The structure then provides the reactive force.

It will generally be found that when the operational, environmental and structural design parameters have been examined it is clear which category should be adopted.

26.10.2.1 Detached fender systems

The category of detached systems may be subdivided into; (1) detached quay fender systems; and (2) breasting dolphins. The main advantage of a detached system is that it permits a lighter main structure to be designed. Also, where the capacity of an existing berth is to be upgraded, a detached fender system may be used to advantage if the existing structures have insufficient strength to withstand the increased berthing loads.

A common example of a detached fender system is a row of free-standing piles (usually steel or timber) driven into the sea- or river-bed in front of the face of the main structure. Berthing energy is absorbed mainly by deflection, the capacity for energy absorption being determined by the size, length, penetration and material properties.

Breasting dolphins are berthing structures independent of the service structure provided for the vessel. Their disposition about the service quay or jetty head is such as to effect the most suitable compromise for the range of vessels envisaged. With flexible dolphins the energy absorption is provided in part by deflection of the entire structure and in part by energy-absorbing units attached to the dolphin face. With a rigid dolphin, all the energy dissipation is achieved by units similar to those used in attached fender systems.

26.10.2.2 Attached fender systems

An attached fender system normally consists of energy-absorbing units bolted to, or suspended from, a quay face or strong points on a jetty. Some types of energy-absorbing units are illustrated in Figure 26.16. Most of the units are made from synthetic rubber and the required energy absorption capacity is achieved by deformation in compression or shear. Some types, such as the hollow cylindrical rubber fenders, the arch type, and pneumatic fenders, can be allowed to make direct contact with the vessel's hull whereas others, such as Hidac, Raykin and cell types require face panels to reduce the contact pressure. The size of the face panel is determined by the permissible hull pressure. Typical attached fender systems are shown in Figure 26.16.

Gravity fenders are those in which kinetic energy is converted to potential energy by means of raising a large mass. They relieve the main structure of the berthing load but impose considerable dead load and a horizontal force depending on the movement of the fender block. The berthing beam principle combines the absorption capacities of cantilevered piles and gravity systems whilst avoiding the dead-load penalty of the latter.

26.10.3 Design of an attached fendering system

The design of an attached system must be integrated with the design of the structure to which it is attached and comprises:

(1) Calculation of energy to be absorbed.
(2) Investigation of alternative systems capable of absorption.
(3) The energy, and calculations of force (see sections 26.10.4 and 26.10.4.2 below), to be resisted by the structure for each alternative.
(4) Investigation of design of structure to resist force of berthing.
(5) Selection of fender system and structure.

In both cases, the calculation of energy to be absorbed is critical to the design.

26.10.4 The basic energy equation

The most generally accepted form of expressing the kinetic energy of a berthing vessel available for absorption by a fender system is:

$$E = 0.5 M v^2 \, C_F C_M C_S C_c \tag{26.2}$$

where E is the kinetic energy available for dissipation by the fender system, M is the mass of vessel (displacement tonnage), v is the velocity of vessel normal to fender at point of impact, C_E is the eccentricity factor, C_M is the mass factor (coefficient of hydrodynamic mass), C_S is the softness coefficient (stiffness factor) and C_c is the berth configuration coefficient.

26.10.4.1 Design velocity

As the energy varies with the square of the velocity of the

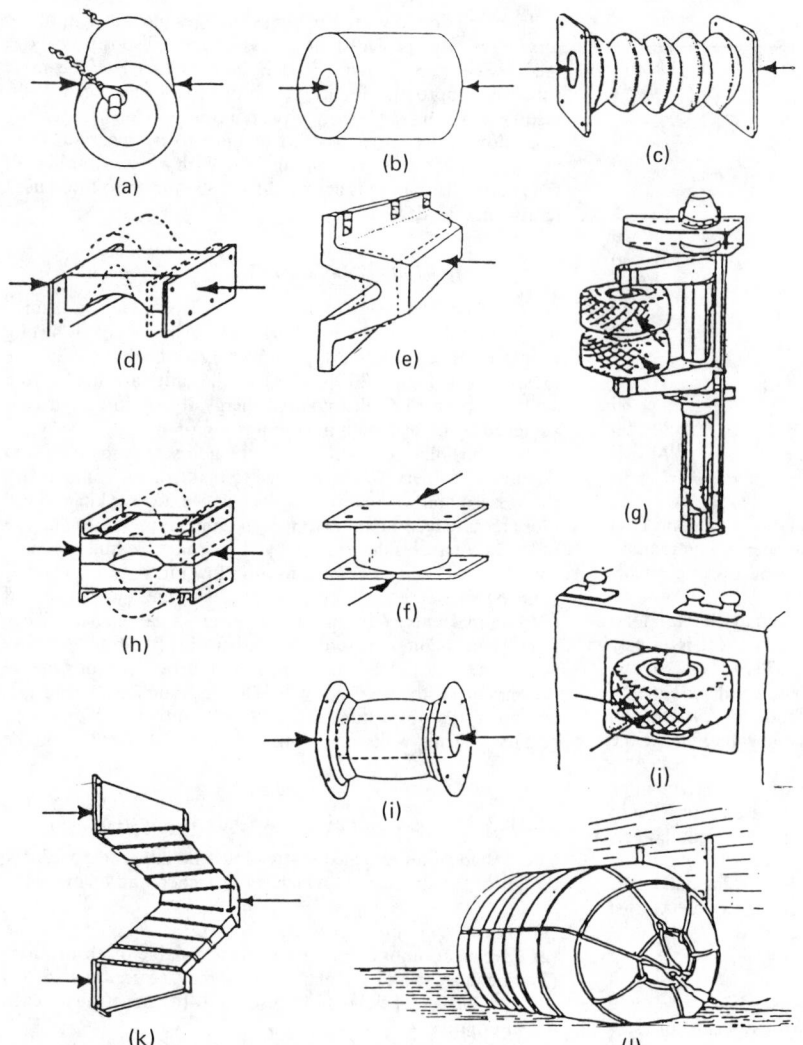

Figure 26.16 Typical replaceable energy-absorbing units

approaching vessel, the choice of the design velocity is most critical. It is, however, unfortunately, one of the most subjective choices in the design of maritime structures, depending as it does on: (1) ship size, type and frequency of arrival; (2) possible constraints on the ship's movement approaching berth; (3) wave conditions likely to be encountered at berthing; (4) current conditions likely to be encountered at berthing; (5) wind conditions likely to be encountered at berthing; (6) the use of tugs; and (7) whether or not speed-of-approach measuring equipment is fitted and used.

Figures suggested in BS 6349[1] are given in Table 26.4. These

Table 26.4

Displacement (t)	Transverse velocity (m/s)
Up to 2000	0.30
2000–10 000	0.18
10 000–125 000	0.16
Over 125 000	0.14

figures may however need to be modified as necessary to accommodate the local factors.

26.10.4.2 Eccentricity factor C_E

When a berthing vessel makes initial contact at a point remote from its centre of gravity, part of the energy is dissipated by the ensuing rotation. The consequent reduction in required energy absorption capacity in the fender is obtained by applying the eccentricity factor C_E. The eccentricity factor is normally calculated from Equation (26.3) below:

$$C_E = \frac{K^2}{K^2 + R^2} \tag{26.3}$$

where K is the radius of gyration of the ship (generally between $0.2L$ and $0.25L$ where L is the length of the vessel) and R is the distance of the point of impact from the centre of gravity of the vessel.

26.10.4.3 Mass factor C_M

The mass factor C_M takes account of the mass of water

entrained with the moving vessel. This is commonly referred to as the 'hydrodynamic mass'. The sum of the 'hydrodynamic mass' and the displacement gives the 'virtual mass' of the vessel. The mass factor C_M is the ratio of the virtual mass to the displacement. Many attempts have been made to define the hydrodynamic mass mathematically but none has been particularly successful. A value of 1.3 is generally used for C_M. Alternatively, the following simple relationship based on experimental results may be used:

$$C_M = 1 + \frac{2D}{B} \qquad (26.4)$$

where D is the vessel's draught and B is the vessel's beam.

26.10.4.4 Softness coefficient C_S

The softness coefficient allows for the portion of the impact energy that is absorbed by the ship's hull. Little research into energy absorption by ships' hulls has taken place, but it has been generally accepted that the value of C_S lies between 0.9 and 1.0. In the absence of more reliable information a figure of 1.0 for C_S is recommended when a soft fendering system is used, and between 0.9 and 1.0 for a hard fendering system.

26.10.4.5 Berth configuration coefficient C_c

The berth configuration coefficient allows for that portion of the ship's energy which is absorbed by the cushioning effect of water trapped between the ship's hull and quay wall. The value of C_c is influenced by the type of quay construction and its distance from the side of the vessel, the berthing angle, the shape of the hull and its underkeel clearance. A value of 1.0 for C_c should be used for piled jetty structures, and a value for C_c of between 0.8 and 1.0 is recommended for use with a solid quay wall.

26.10.5 The factor of safety

Two levels of energy of impact – 'normal' and 'abnormal' – should be established for the design of the fender system and the supporting berth structure.

The berthing energy as computed in accordance with the above formula is based on normal operations and may be exceeded for accidental occurrences such as: (1) engine failure of ship or tug; (2) breaking of mooring or towing lines; (3) sudden changes of wind or current conditions; and (4) human error. To provide a margin of safety against such unquantifiable risks it is recommended that the ultimate capacity of the fenders should be double that calculated for normal impacts.

Because of the nonlinear energy/reaction/deflection characteristics of most fender systems, the effects of both normal and abnormal impacts on the fender system and berth structures should be examined.

26.10.6 Structural considerations

The type of structure to which the fender system is attached is usually dictated by foundation conditions.

In situations where gravity structures such as blockwork walls or caissons are the economical solution it is unlikely that the overall design will be very sensitive to the berthing load, though more detailed factors such as the location of a service duct behind the berthing face or stability of the capping block, may be affected.

Where a piled structure is used, the berthing force may dictate the pile layout. Where a piled structure supports cargo-handling equipment or pipework, unacceptable deflection of the structure rather than overstressing is often found to be the limiting criterion.

In the design of fender support systems it is important that a robust means of restraint is provided against forces acting along the berth face. These forces are produced by friction between the hull and the fender and can reach 50% of the maximum fender reaction.

Consideration must also be given to the structural constraints imposed on the fender designs by the form of construction of the berthing vessel. Berthing loads are generally not considered as a basic design criterion by naval architects.

26.11 Locks

Maritime locks are used to allow vessels to pass between tidal waters and an impounded water area which can be a dock, harbour or ship canal, and enables the impounded water area to be maintained at a constant level, eliminating tidal effects.

In a port that has a large tidal range, locks may be essential to prevent vessels grounding while alongside berths. A constant water level behind a lock also has advantages in that the height of quays depends only on the ship's draught and no account need be taken of tides upstream of the lock. Loading and discharging is also simplified with a constant water level, and normally waves and currents can be disregarded.

26.11.1 Lock dimensions

The usable length of the lock chamber and the width and depth of the sill should be sufficient to ensure that all vessels entering the dock may be safely locked in and out. The level of the outer sill is normally dependent upon the dimension of the approach channel; the level of the inner sill is dependent upon the impounded water level and the maximum vessel draught. Normally a safety margin of 1 m is provided between the ship keel and the sill. A clearance of 10% of the maximum ship beam should be allowed on each side of the vessel. In determining the length of a lock it should be borne in mind that two to six tugs from 25 to 35 m in length may be required depending on the size of the ship and the ship's own manoeuvrability. Approximately 10 m additional length is required for towlines. The length and width of the lock also depends on the number of ships being locked simultaneously.

26.11.2 Lock gates

Lock gates can be caissons, mitre gates or sector gates. For smaller locks, single leaf gates (vertical or horizontal axis) are sometimes used, but these are less common.

Mitre gates are probably the most common type of gates for locks under 30 m wide, as they are generally more economical than other types, mainly because of the less extensive structural work required to house them in the lock heads, but also because in general terms they can be more easily handled for maintenance. Mitre gates obtain their strength largely from the head of water on one side holding them together. However, they have the disadvantage that they can only resist a head of water in one direction. They are not suitable when the tide level outside the lock rises above the impounded level, although it is sometimes found that in such a case it is cheaper to provide an additional pair of mitre gates rather than to use other types. Mitre gates also have the disadvantage that they are likely to vibrate where there is only a small hydraulic head holding them together. In smaller locks, such as those in marinas, it is generally found that radial sector gates or delta gates are the most suitable.

26.12 Pavements

Pavements in port areas which are used by cargo-handling plant are generally subjected to much higher loads and much greater repetition of loading than normal road pavements. They are also susceptible to damage from the plant itself, e.g. the lifting prongs attached to fork-lift trucks. The bearing capacity of the subsoil determines the pavement design to a large extent. Surfacing used for pavements can be either:

(1) *Flexible*: asphalt – normally expensive concrete blocks – easily maintained.

(2) *Rigid*: concrete – subject to cracking under high concentrated loads. Precast concrete rafts – if settlement occurs they can be lifted, the ground made up to level, and the rafts realigned.

The design of heavy-duty pavements is still semi-empirical. The British Ports Association has produced a manual of design charts[4] suitable for the design of pavements for cargo-handling plant currently in use with a wide range of soil conditions.

For low-cost storage areas for containers, gravel surfacings have been used in some ports.[5]

Grades and consequent drainage patterns should avoid excessive and frequent peaks and valleys. Large open expanses without grade breaks are needed for ground stacking. Pavement slopes should be held at 1 to 1.5% maximum.

26.13 Durability and maintenance

It is well known that the marine environment is one of the most severe as far as deterioration of materials is concerned. In addition to this, durability can be drastically affected by the location of the structure. Some of the factors affecting durability are:

(1) Temperature of the sea.
(2) Air pollution. It has been found, for example, that the rate of corrosion of galvanized steel can vary by a factor of 10:1 in different parts of the UK due to air pollutants.
(3) Pollution of the sea. This is sometimes dependent on the use of the berth, e.g. at a fertilizer terminal, where chemical compounds can find their way into the sea, and the rate of corrosion is much higher than in unpolluted sea water.

The factors affecting durability and repairs of reinforced concrete in the marine environment are well documented elsewhere[5, 6] but marine structures should be robust, avoiding thin sections and with a minimum cover to reinforcement of 50 mm. For greater protection against corrosion, galvanized or coated steel may be used or cathodic protection applied to the reinforcement.

The part of a structure most susceptible to corrosion or deterioration is the area in the tidal zone or splash zone, and special consideration needs to be given to this area. With steel piles, concrete muffs are frequently provided from the underside of the concrete deck to just below low-water level; steel sheet piling is frequently encased in concrete from cope level to below low-water level.

Durability is normally considered at the design stage, but maintenance has an equally important effect on the life of structures, and should also be considered at the design stage.

There are two kinds of maintenance: (1) preventive; and (2) remedial. Methods of both types of maintenance are well known. The principal preventative methods are protective coatings and cathodic production – both need to be considered and incorporated where appropriate at the design stage. From the designer's point of view it is important also to recognize how maintenance will be carried out and to design for simplicity in the provision of maintenance procedures. If the maintenance methods are straightforward they are far more likely to be carried out than if access is difficult, or if complicated plant or equipment is required.

References

1 British Standards Institution (1984–1985) *Code of practice for maritime structures*. BS 6349 Parts 1, 2 and 4. BSI, Milton Keynes.
2 Permanent International Association of Navigational Congresses (1985) *Underkeel clearance for large ships in maritime fairways with hard bottoms*. PIANC Supplement to Bulletin No. 51.
3 Construction Industry Research and Information Association (1977) *Dynamics of marine structures. Methods of calculating the dynamic response of fixed structures subject to wave and current action*. CIRIA Report No. UR8 Underwater Engineering Group, London.
4 British Ports Association (1982) *The structural design of heavy-duty pavements for ports and other industries*. BPA, London.
5 Institution of Civil Engineers (1986) *Maritime and offshore structure maintenance*. Thomas Telford, London.
6 Construction Industry Research and Information Association (1986) *Influence of methods and materials on the durability of repairs to concrete coastal and offshore structures*. UEG Publication No. UR36, Underwater Engineering Group, CIRIA, London.

Bibliography

American Association of Port Authorities (n.d.), *Port designs and construction*, AAPA, Washington, D.C.
American Iron and Steel Institution (1981) *Handbook of corrosion protection for steel pile structures in marine environments*. AISI, Washington, D.C.
American Society of Civil Engineers (1974) *Port structure costs: a report by the task committee on port structure costs*, Committee on Ports and Harbours, ASCE Waterways, Harbours and Coastal Engineering Division, New York.
Cornick, H. F. *Dock and harbour engineering*. Charles Griffin, London.
Department of Transport (1982) *Ship behaviour in ports and their approaches*. HMSO, London.
Fuhrer, M. and Romisch, K. (1983) Contribution to the design of fenders and dolphins. *8th International Harbour conference, Antwerp*.
Gulf Publishing Company (n.d.) *Port engineering*. GPC, Houston.
Ministry of Transport, Japan (1980) Technical standards for ports and harbour facilities in Japan. Bureau of Ports and Harbours.
National Ports Council (1920) *Port structures – an analysis of costs and designs of quay walls, locks and transit sheds*. NPC, London.
National Ports Council (1978) *Containers – their handling and transport. A survey of current practice*. NPC, London.
Permanent International Association of Navigational Congresses (1978) *Standardization of ro-ro ship and berth*. International Study Commission Report, PIANC, Geneva.
Permanent International Association of Navigational Congresses (1985) *Port maintenance handbook*. Supplement to Bulletin No. 50, PIANC, Geneva.
Permanent International Association of Navigational Congresses (1987) *Final report of the International Commission for the Study of Locks*. Supplement to Bulletin No. 55, PIANC, Geneva.
Permanent International Association of Navigational Congresses (1987) *Development of modern marine terminals*. Supplement to Bulletin No. 56, PIANC, Geneva.
Recommendations of the Committee for Waterfront Structures (1982) 4th English edition (translated from the 6th German Edition). Sohn.
United Nations (1985) *Port development – a handbook for planners in developing countries*. UN, New York.
US Army Corp of Engineers (1974) *Small craft harbours; design construction and operation*, Coastal Engineering Research Center, Special Report No. 2.
Vrijer, A. (1983) 'Fender forces caused by ship impacts'. *8th international harbour conference*, Antwerp.

27

Electrical Power Supply

Staff of Central Electricity Generating Board (Generation Development and Construction Division)

Contents

27.1 Fuels

by C W Maynard, BSc(Tech), CEng, FICE

In order to generate electricity in sufficient quantity for public supply, generators are normally driven by turbines which are themselves powered by steam raised from the combustion of coal or oil, or from nuclear fission (thermal energy), or by gas produced from the combustion of oil (also thermal energy) or by water pressure (hydraulic energy).

In England and Wales and the South of Scotland, practically all electricity generation is steam powered, although gas turbines are used to generate a small amount of electricity at times of peak demand. The generation of hydro-electric power is relatively small and is concentrated in Scotland.

On 31 March 1987 the Central Electricity Generating Board (CEGB) had 78 power stations with a declared net capability (DNC) of 52.5 GW nominally available from these stations for transmission via the CEGB's high-voltage system to its consumers. This output capacity is therefore described as 'sent out' and is measured in GWso (gigawatts sent out). The plant of each type and proportion of each type of fuel burnt during the year were as shown below in Table 27.1 demonstrating the heavy dependence on coal for electricity production.

27.1.1 Steam-powered generation

Electricity is produced from steam power by the association of two major plant items of matched output which, together with ancillary plant and services, form a unit:

(1) A boiler for raising steam, heated either by a coal- or oil-burning furnace or by the heat produced by the nuclear fission of radioactive material in a reactor.
(2) A steam turbine coupled to an electricity generator (a turbo-generator) in which the turbine converts the heat and pressure energy of steam into mechanical power to drive the generator.

Table 27.1 Existing plant and fuel burn

Type of plant	Declared net capability (DNC) at 31.3.87 GWso	%	Fuel burn 1981/82 mtce*	%
Coal	35.5**	67	77.00	77.0
Oil	6.8	13	6.77	7.0
Gas turbine	3.0	6		
Nuclear	5.0	10	16.20	16.0
Hydro/pumped storage	2.2	4		
	52.5	100	99.97	100

*mtce = million tonnes of coal equivalent.
**Includes plant capable of burning both oil or coal, and coal and gas.

Figure 27.1 shows diagrammatically the plant used to generate electricity from steam power, and the flow and losses of energy through the unit. The energy transmitted from the turbine to the generator is the electrical rating of the unit – typically 500 or 660 MW. The efficiency of a steam turbine increases with the drop in steam temperature across the turbine and, to achieve the lowest temperature at exhaust, the steam from the low-pressure stage of the turbine is discharged into condensers carrying water-cooled tubes. The condensate is returned to the boilers for recirculation, while the cooling water which has gained the heat lost by the steam in the condensers is either returned to source for the heat to be dissipated or is returned to cooling towers before recirculation through the condensers.

In the UK the electricity supply industry was nationalized in 1948, and in England and Wales the construction and operation of power stations and the transmission of electricity at high voltage is the responsibility of the CEGB. Privatization of the industry is planned for implementation in the early 1990s.

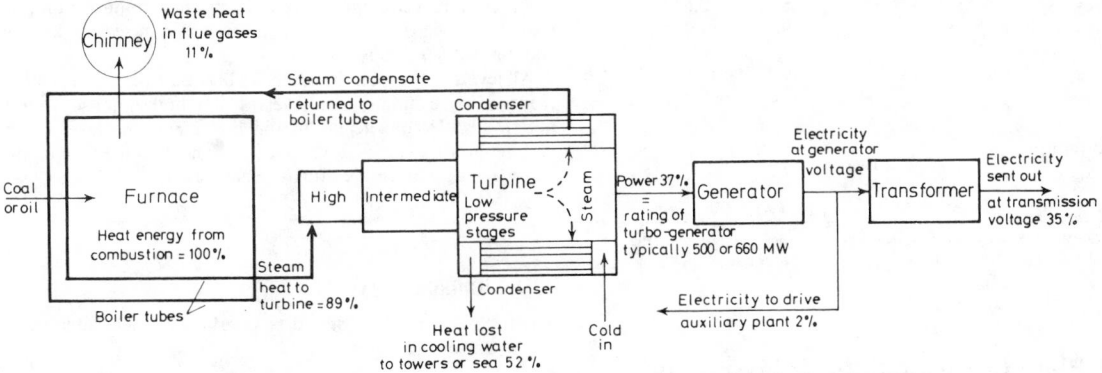

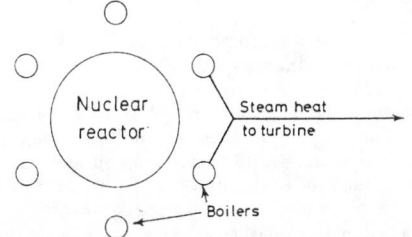

Figure 27.1 Flow of energy from plant used to generate electricity from steam power. Energy shown as percentage of heat in fuel

Fuel cost and the capital cost of the plant and buildings of a power station are the dominant costs of electrical power.

The operating and transmission costs are approximately the same in each case. The capital costs are approximately the same for coal and oil stations but are much higher for nuclear stations. The cost of coal is lower than oil but the cost of nuclear fuel is substantially the lowest. In total, the cheapest electricity is produced from nuclear fuel, followed by coal and with oil producing the most expensive electricity. There is greater potential for future technical improvement in the utilization of nuclear fuel than in the combustion of oil or coal, so that from technical considerations, nuclear fuel used in future power stations should produce still cheaper electricity under stable cost conditions.

The total fuel consumption for the generation of public supply electricity in England and Wales in 1986/7 was 99.97 million t (coal equivalent) in the following proportions:

Coal	77%
Oil	7%
Nuclear fuel	16%

The policy of the nationalized industry is to develop a flexible system capable of responding to the relative costs and availabilities of different fuels.

The size of generators in service at 31 March 1987 is shown in Table 27.2.

Table 27.2 Size of generators in service (31 March 1987)

Type	Nominal size (MW)	No.	Declared net capability (DNC) (GW_{50})
Large coal*	500/660	45	21.8
Large oil	500/660	17	6.2
Small coal	< 500	102	13.6
Small oil	< 500	5	0.6
Main gas turbine	50/90	29	1.4
Auxiliary gas turbine		70	1.5
Magnox	40/250	32	3.5
AGR	600/660	2	1.6
Hydro/pumped storage	0.2/90	32	2.2
Total			52.4

*Including 4 × 500 MW units at Kingsnorth Power Station capable of burning both coal and oil.

27.2 Plant layout, buildings layout and station siting

by D L McKie, CEng, MICE

27.2.1 Plant layout

For over a decade it has been the CEGB's normal practice to install 660 MW turbo-generators in fossil fuel and nuclear power stations. The fossil fuel stations have had between two and six machines and the nuclear stations two machines.

In the former type of station the boiler house and turbine house are built parallel to each other and are separated by the mechanical annexe which contains the deaerators and feed water tanks. Figure 27.2 shows in cross-section the main building of Drax coal-fired station which has 6 × 660 MW machines. On the side of the turbine house remote from the boilers the generator transformers are located and also the control room.

Beyond the boiler and bunker bays the flue gases are taken through precipitators before passing into the chimney which serves the six boilers. The layout of the Drax plant is shown in Figure 27.3.

The cross-section through an oil-fired station main building is virtually identical with the exception that the bunker bay is obviously eliminated and the boiler is slightly smaller.

Modern power stations are built on the unit principle in that each boiler serves one turbine. Thus, the boilers and turbo-generators are located at the same centres. The turbo-generators may be arranged longitudinally, transversely or diagonally in the turbine house. The first arrangement gives the minimum span for the turbine house and overhead cranes, whereas a transverse arrangement with the turbine end nearest to the boiler house, gives a minimum length for the high-pressure steam piping and the connections from the alternator to the generator transformer. A compromise between the two was adopted at Drax and the machines were set at 45° to the longitudinal axis of the turbine house.[1]

A turbine house is served by at least two overhead travelling cranes, one of which is capable of lifting the heaviest plant item. This is normally the stator which weighs about 300 t for a 660 MW machine. Due to the large span of the crane and in order to reduce the concentration of loading on the crane beams and columns it is common practice to provide two cranes each of approximately 150 t capacity and operate these in tandem to handle the heaviest load.

The power station levels should be determined so that either the turbine house operating floor or the basement floor, i.e. the floor below the operating floor, is at the same level as the permanent roads around the station. Various services, ducts and tunnels (other than the circulating water culverts) can with advantage be accommodated in a sub-basement up to about 3 m in depth. The choice of basement level is not influenced by the requirements of a recirculating water system (i.e. tower-cooled) but it may be economic to keep it low to minimize pumping costs in a once-through system, as discussed in section 27.6 on cooling-water systems.

All levels in the main building are derived from plant requirements. For example, the level of the boiler house roof is determined by the height of the boiler and the level of the turbine house crane is governed by the need to lift a generator stator over an assembled turbo-generator.

27.2.2 Buildings layout

The following matters should be considered when determining the layout of the station.

(1) The length of circulating water culverts (and tunnels) should be as short as practicable.
(2) The switch house (or compound if outdoor switchgear is used) should be on the same side of the main building as the outgoing transmission lines.
(3) The location and size of the fuel store for fossil-fired stations. Fuel may be delivered by rail, ship or oil pipeline. For delivery of coal by rail a track loop off the main line is required around the coal store which should be close to the bunkers to minimize the length of the conveyors.
(4) The main building and most heavily loaded structures should, where practicable and not in conflict with overriding

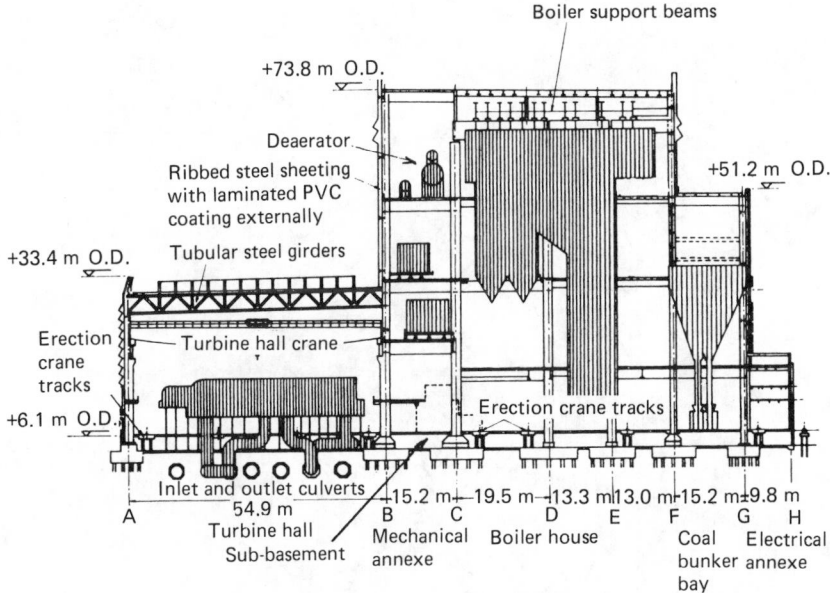

Figure 27.2 Section through main building of Drax power station

requirements, be located where ground conditions are most satisfactory.

(5) Environmental aspects.

27.2.3 Station siting

An important requirement of any power station site is proximity to a supply of river or sea water adequate for operation of the cooling water system. Within a given area of search, the best site is that which offers the best combination of desirable features: (1) little or no harm to amenity; (2) low capital cost; (3) low fuel cost; and (4) low transmission cost. Difficult foundation conditions may be accepted when these are outweighed by other factors.

The ideal size of site to provide ample construction, fabrication and storage areas for the various contractors, for site offices, canteens, hostel, bus and car parks, in addition to the areas to be occupied by the permanent station works, is frequently unavailable. Thus, a shortage of space has to be overcome by initially allocating for storage or fabrication those areas which will not be constructed until late in the programme, and also by the civil contractor handing areas over to the mechanical and electrical contractors as his workload decreases. Certain activities and storage may be off site as well as any hostel required due to a shortage of local labour.

In the Drax power stations six units were built in two stages of three units. Figure 27.4 shows the allocation of contractors' areas on the site during the first stage and gives an indication of the second stage permanent works which were occupied by the initial contractors.

In the case of a nuclear power station the basic differences lie in the source of steam supply and in the fuel and the methods by which it is handled. These matters are covered in section 27.9 on nuclear reactors and reactor buildings.

27.3 Power house steelwork

by N M Grieve, MIStructE

In a 2000 MW station with four 500 MW boiler and turbo-generator units each boiler measures about 24×46 m in plan and is suspended from girders about 60 m above ground level. Under operating conditions (i.e. filled with water and steam) it weighs about 12 000 t; other heavy items to be supported by the structure are the deaerator (300 t), the boiler drum (300 t) and two water tanks (660 t each). The boiler is surrounded by access floors at about 10 m intervals of height, and steelwork has proved to give the best structure for housing the boilers and ancillaries and carrying the loads. Generally, the support structure for one boiler only need be designed as the remainder will be a repetition of the first.

The turbine house framework may be of concrete or steel but, because the adjoining boiler house is invariably steel, there is no advantage to be gained from having a concrete-framed turbine house. The weight of the turbo-generators is carried directly to ground on concrete or steel structures (turbo-blocks) and is not carried by the turbine house framework.

The power house building therefore consists mainly of a boiler house and turbine house whose basic dimensions depend on the number and arrangement of boiler/turbo-generator units. In deciding the plant layout, it is important to have the boiler support columns arranged symmetrically around the boiler.

27.3.1 Design

After the plant layout has been finalized, it is necessary to ascertain if the boiler contractor has any special design requirements or limitations to be imposed on the boiler support structure. This is an important preliminary to design since the columns supporting the boiler will, with their connecting girders and bracing, form the core of the structure, and will provide

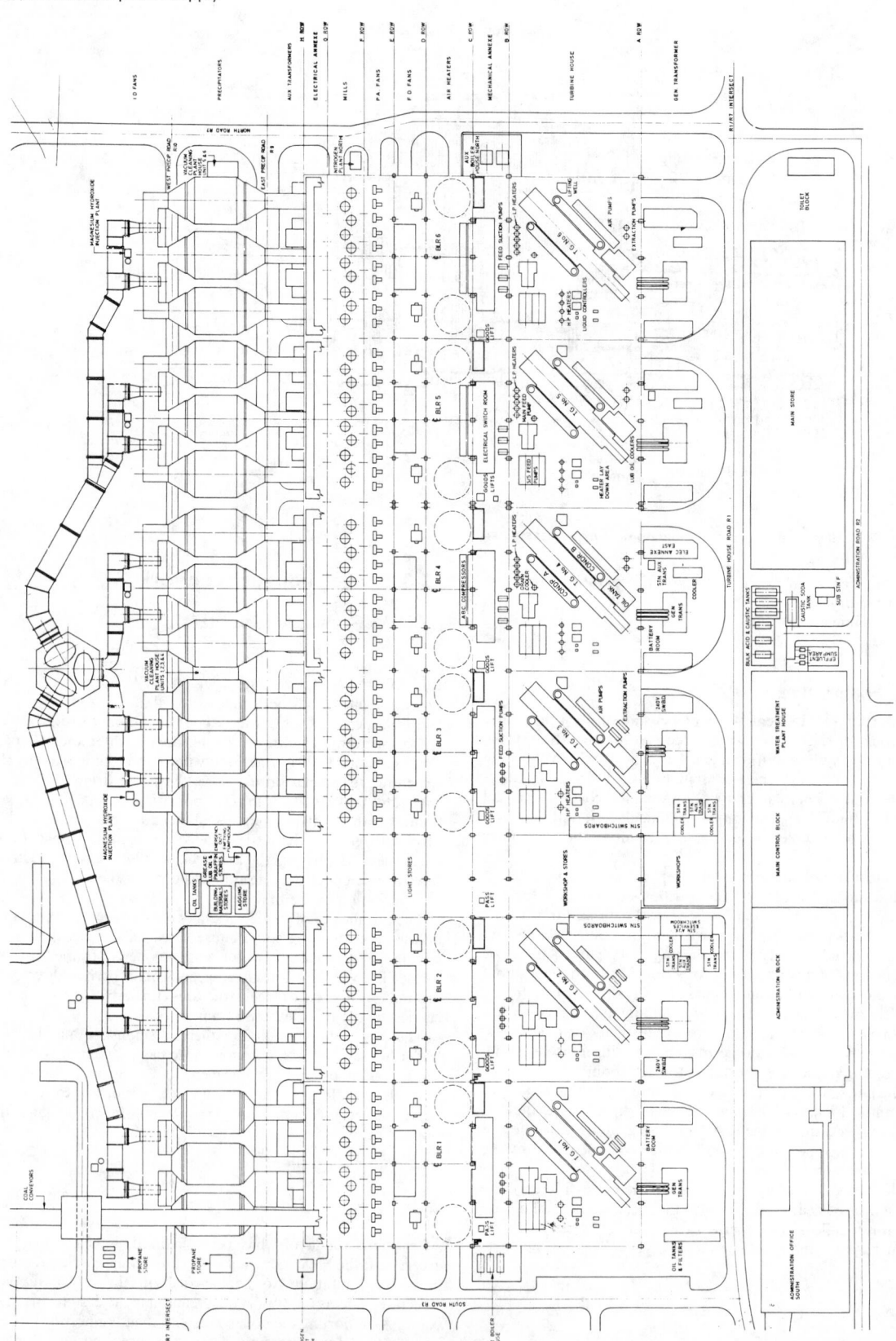

Figure 27.3 Layout of the main building of Drax power station

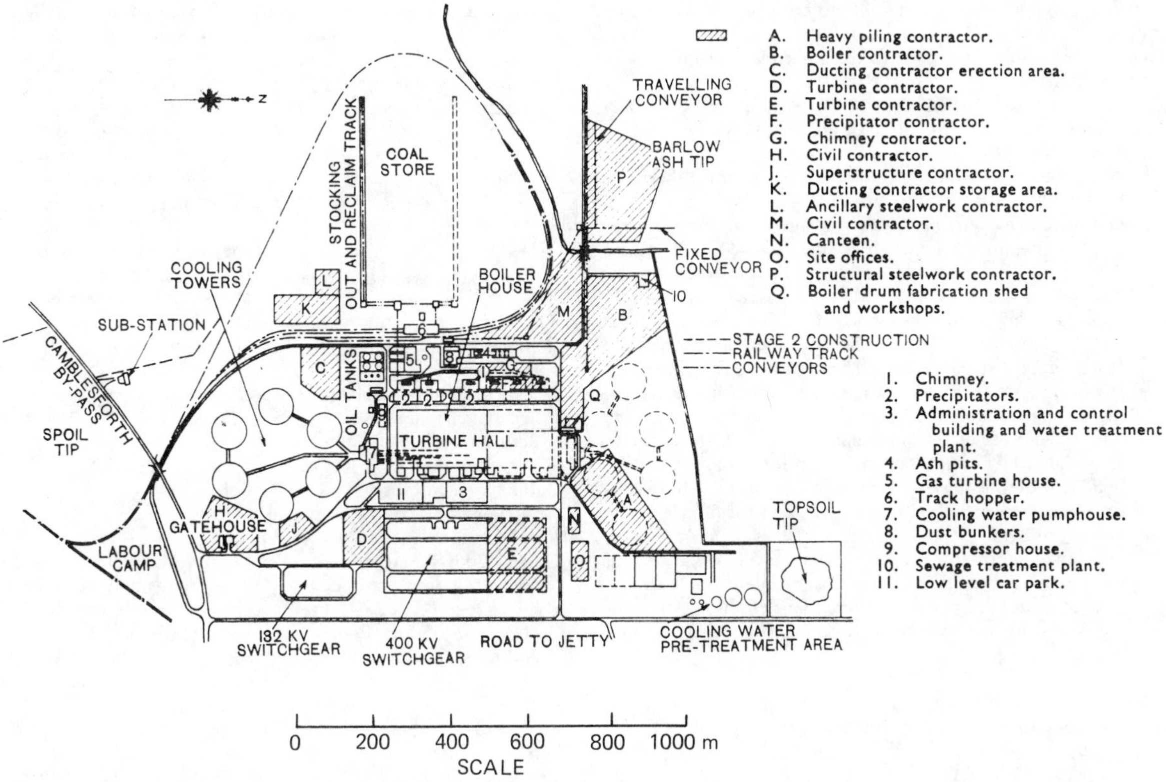
A. Heavy piling contractor.
B. Boiler contractor.
C. Ducting contractor erection area.
D. Turbine contractor.
E. Turbine contractor.
F. Precipitator contractor.
G. Chimney contractor.
H. Civil contractor.
J. Superstructure contractor.
K. Ducting contractor storage area.
L. Ancillary steelwork contractor.
M. Civil contractor.
N. Canteen.
O. Site offices.
P. Structural steelwork contractor.
Q. Boiler drum fabrication shed and workshops.
STAGE 2 CONSTRUCTION
RAILWAY TRACK
CONVEYORS
1. Chimney.
2. Precipitators.
3. Administration and control building and water treatment plant.
4. Ash pits.
5. Gas turbine house.
6. Track hopper.
7. Cooling water pumphouse.
8. Dust bunkers.
9. Compressor house.
10. Sewage treatment plant.
11. Low level car park.
TRAVELLING CONVEYOR
BARLOW ASH TIP
COAL STORE
STOCKING OUT AND RECLAIM TRACK
BOILER HOUSE
FIXED CONVEYOR
COOLING TOWERS
SUB-STATION
CAMBLESFORTH BY-PASS
SPOIL TIP
OIL TANKS
TURBINE HALL
GATEHOUSE
LABOUR CAMP
TOPSOIL TIP
132 KV SWITCHGEAR
400 KV SWITCHGEAR
ROAD TO JETTY
COOLING WATER PRE-TREATMENT AREA
0
200
400
600
800
1000 m
SCALE

Figure 27.4 Site layout of Drax power station

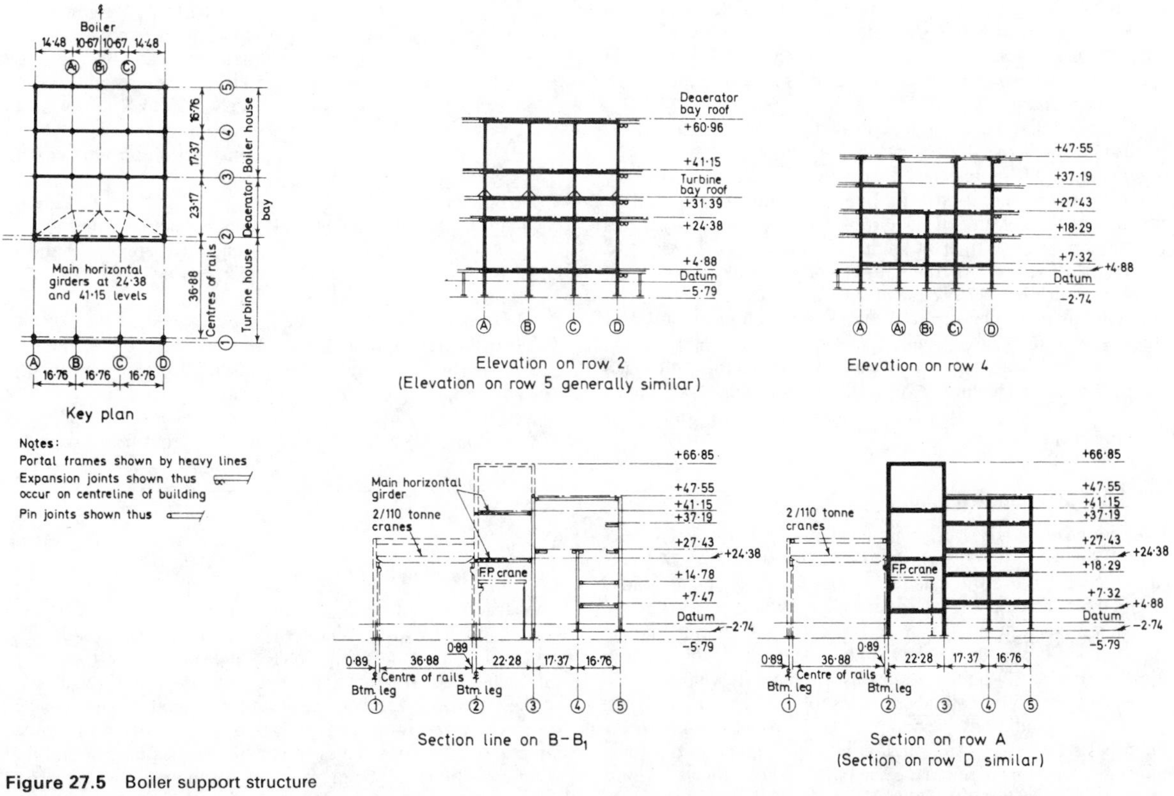
Boiler
Boiler house
Deaerator bay
Turbine house
Centres of rails
Main horizontal girders at 24·38 and 41·15 levels
Key plan
Notes:
Portal frames shown by heavy lines
Expansion joints shown thus occur on centreline of building
Pin joints shown thus
Deaerator bay roof
Turbine bay roof
Datum
Elevation on row 2
(Elevation on row 5 generally similar)
Elevation on row 4
Main horizontal girder
2/110 tonne cranes
F.P. crane
Centre of rails
Btm. leg
Section line on B–B1
Section on row A
(Section on row D similar)

Figure 27.5 Boiler support structure

transverse and longitudinal stability to both turbine and boiler house (see Figure 27.5). This is the reason for having a plant layout which allows a symmetrical arrangement of boiler support columns.

The main factors which depend on requirements generally obtained from the boiler contractor are as follows:

(1) The structure has to carry the wind load and must therefore be of sufficient stiffness to ensure that the combined maximum lateral movements under wind and operating temperatures are within the limits permissible for the high-pressure steam pipes and other connections to the boiler.
(2) The boiler drum of a 500 MW unit is a steel cylinder approximately 30 m long and 2 m in diameter; it weighs 260 t without water. It is fabricated off site and, when the boiler support structure is sufficiently stable, is hoisted into position above the boiler. Large open areas must be left in the floors to enable the boiler drum to pass through. The position and size of these openings must be established at an early date as it is important to design for continuity of the buildings during the drum-hoisting operation. Afterwards the open areas are generally filled in and form part of the flooring system.
(3) The boiler contractor may allow only limited deflection of the beams under important items of plant such as the deaerator vessel. Excessive differential deflections between the vessel supports can be detrimental to its efficient functioning.

Should the limitations on the horizontal movements of the structure due to wind load and temperature under item (1) be very stringent (say 50 mm in 60 m) then a braced frame system may be necessary to obtain the degree of stiffness required. This type of construction, whilst structurally effective, does not allow full flexibility of choice in the routing through the building of large-diameter pipework and ducting. Should the horizontal deflection requirements be less stringent, then a type of portal frame construction can be considered. This has the advantage of providing clear spaces between the girders and columns which form the frames.

The remaining part of the main building such as the floors, walls and roofs, are of conventional design. The turbine-house roof may require more detailed attention since the span can vary from approximately 30 to 55 m. Limitations may be required on its deflection to ensure effective roof drainage. The roof girders are sometimes used to erect the heavy-duty cranes.

Historically, the structural framework was designed in accordance with BS 449 *The use of structural steel in building*, but BS 449 has now been replaced by BS 5950. The members and their connections are required to take full account of the moments and stresses due to all possible combinations of loads and other effects including vertical loadings, wind loads, temperature effects and dynamic effects from cranes and other moving plant.

27.3.2 Superimposed loading

Superimposed loads are generally allowed for in accordance with BS 6399: Part 1 *Design loading for buildings*: Part 1, 'Code of Practice for dead and imposed loads', but all plant loads, loads from tanks, crane wheel loads, etc. should be the known weights of such items. Where accurate weights are not known because structural design has to be ahead of detailed plant design, then safely high assessments are made to enable steelwork design to proceed and estimates to be given for construction of the foundations. Temporary loads such as people, stored materials, movable plant and equipment, plant laid down during erection or maintenance, may be allowed for in various areas in accordance with Table 27.3.

Table 27.3 Temporary loading on floors and roofs

Operating floors in turbine and boiler houses	12.5 kN/m² (but 50 kN/m² in areas used to lay down turbo-generator parts). In the case of a large operating floor where there is little likelihood of the whole area being fully loaded at one time, a reduction of up to 20% of the temporary load can be taken on main beams and columns.
Loading bay in turbine house	Design for transporter with heaviest load.
Basement floors in turbine and boiler houses (i.e. floor immediately below operating floor)	25 kN/m²
Cable basement floor	7.5 kN/m²
Gallery floors (walkways only)	5 kN/m²
Gallery floors in large circulation areas in boiler house (e.g. between boilers), deaerator floors and tank floors	7.5 kN/m². The main beams and columns may be designed for a temporary load of 5 kN/m² but no reduction should be taken in the design of secondary beams.
Stairways in turbine and boiler houses	5 kN/m²
Flat metal-deck roof on turbine house	1.5 kN/m² to allow access for maintenance only. Load on main steelwork can be reduced to 0.75 kN/m².
Concrete shell roof on turbine house	0.75 kN/m² to allow access for maintenance only.
Flat metal-deck roof on boiler house	1.5 kN/m² for deck and main steelwork. 4 kN/m² for areas which have to take construction or maintenance loads.
Control-room floor	4 kN/m²
Switchgear floors	Design for actual loads (generally between 10 and 25 kN/m²).
Instrument workshop and laboratory floors	4 kN/m²
Electrical workshop floors and areas used for light engineering	7.5 kN/m² checked for point loading from heaviest plant load.
Stores and heavy-machine shop floors	15 kN/m² checked for point loading from heaviest plant load.

27.3.3 Wind loading

The structure should be designed for wind loading in accordance with CP3, Part 2, Chapter V: 'Loading – wind loads'. The effects of wind on the partly erected and clad structure as well as on the finished structure should be considered, and the absence of the stabilizing effect of heavy items of plant should be taken into account. From tests carried out on models of framed structures in various stages of erection, it has been found that wind conditions can be more severe during the erection period than on the completed structure. It may therefore be necessary during erection to stiffen temporarily certain main connections, and also provide temporary bracing to stabilize the structure against this more severe wind condition.

27.3.4 Turbine house crane loading

In the design of the crane support columns the transverse and longitudinal forces caused by the movements of the cranes should be taken into consideration as well as the station vertical loading. It is important that the crane support columns should be made stiff enough to reduce horizontal deflection due to roof loading and eccentric loading from the cranes to a minimum. Over-flexibility from the columns has in the past resulted in serious crane crabbing.

27.3.5 Thermal loading

The maximum temperature rise in the boiler house under certain conditions can be in excess of 35°C and the structure should be designed to accommodate thermal forces resulting from temperature changes of this magnitude. Although the high temperatures usually occur at the top of the boiler house, high differential temperatures have been noted at various levels in localized areas depending on the distance of the particular part of the structure from the boiler casing. Expansion joints are usually necessary to avoid excessive movements and stresses due to thermal forces.

27.3.6 Load combinations

The structure should be designed to cater for the worst combination of loading without exceeding the specified permissible stresses or the specified limits of vertical and horizontal deflection. It may also be considered necessary in the case of portal

frame construction to check the beam to column connections for possible 'pattern' loading due to alternate beam spans being loaded temporarily.

27.3.7 Construction

The total tonnage of structural steelwork in the turbine and boiler house of a 2000 MW power station can vary from 25 000 to 30 000 t depending on the plant layout and the height of the building. About 85% of this total tonnage is required for the support of the plant and the remaining 15% is required to provide the building framework.

The main columns in the turbine house are generally of welded box section about 2 m deep by 1.5 m wide with wall plates approximately 32 mm thick. A central division plate is generally required to reduce the unsupported width of the wall plates. Horizontal diaphragm plates are required with manholes positioned centrally for internal access during erection, and ladders are provided between each diaphragm plate. The span of the turbine house roof truss can be up to 55 m with a depth of 4.5 m.

The main boiler support columns are also generally of welded box section and vary from 0.9 to 1.4 m square with wall plates up to 50 mm thick. The columns are connected together by plate girders from 1.2 to 2.7 m deep with flange plates up to 75 mm thick. Where these members combine to form portal frames the girders are usually haunched to accommodate the large bending moments. Should diagonally braced frame construction be preferred, the bracing usually consists of heavy universal column sections.

The heavy loading and large spans at the ancillary plant levels in the boiler house demand the use of welded plate, box and lattice girders in the construction. Probably the most interesting of these girders are those which form the main boiler suspension

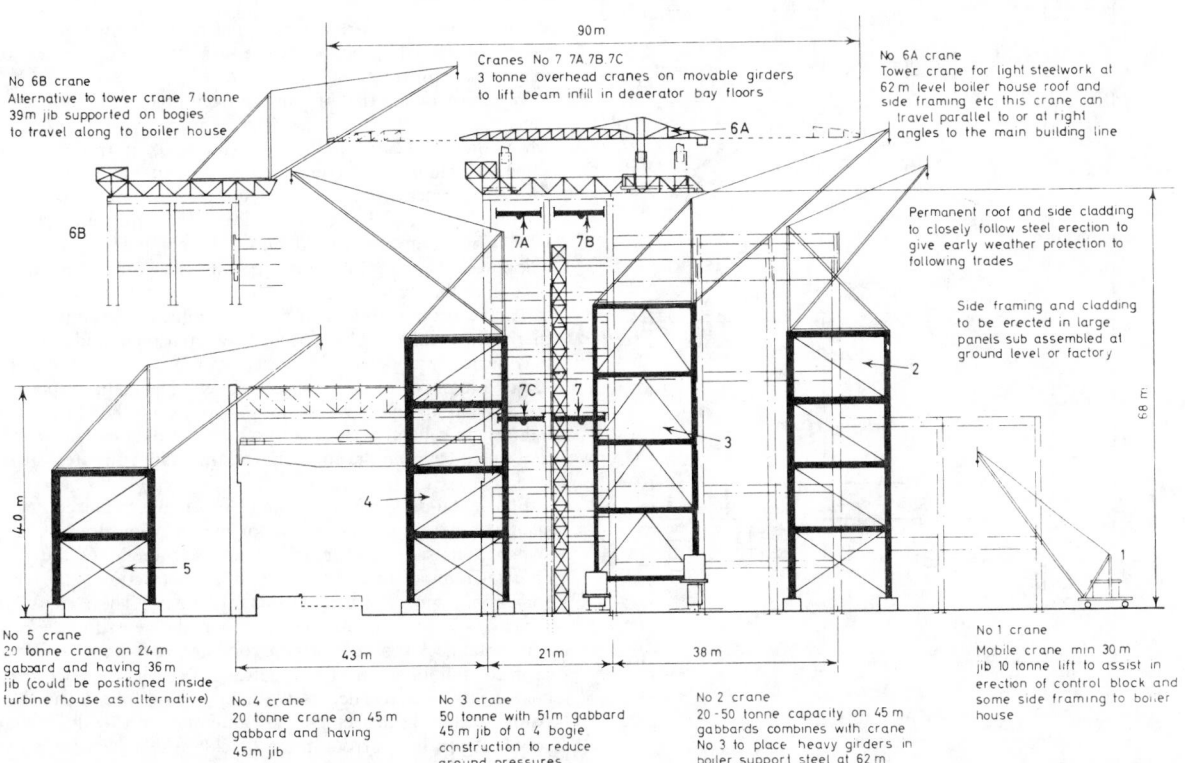

Figure 27.6 Steelwork erection scheme

steelwork. Some typical forms of suspension steelwork are as follows:

(1) Two principal lattice girders per boiler of up to 30 m span and weighing up to 100 t with smaller secondary cross-girders.
(2) Welded box girders varying from 15 to 21 m long and from 3 to 4.5 m deep, the flanges being from 0.9 to 1.2 m wide and up to 75 mm thick. They are stiffened internally by vertical and horizontal stiffeners. Where vertical diaphragm stiffening plates are used they are usually provided with access openings for the steel erectors. The box girders are fabricated in two lengths to suit the erection cranage and are connected together at site by high-strength friction grip bolts of diameters varying between 19 and 32 mm. The weight of a complete main box girder can be up to 160 t.
(3) A series of closely spaced plate girders about 27 m span and approximately 4.5 m deep with flanges up to 75 mm thick.

27.3.8 Erection

A typical steelwork erection scheme is shown in Figure 27.6. The average steelwork erection rate is in excess of 860 t/month. This average rate has to be increased considerably during the first 12 months in order to provide the necessary stability to the structure to enable the boiler contractor to install the first boiler drum. A sustained erection rate of 1400 t/month for a period of 12 consecutive months has been achieved. On another station 1040 t/month was achieved for a period of 31 consecutive months.

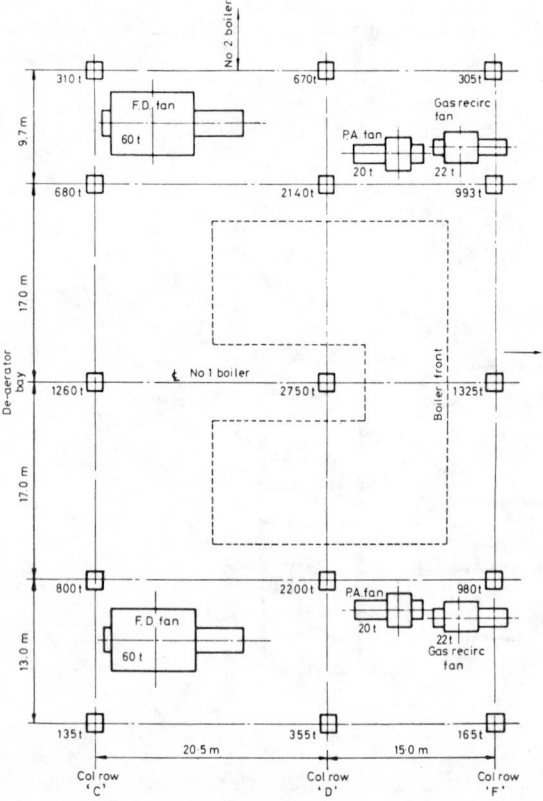

Figure 27.7 Boiler foundation loading plan (t=tonne). (*Courtesy:* Central Electricity Generating Board)

27.3.9 Foundations

The loads resulting from the boiler and supporting steelwork of a 500 MW unit are shown in Figure 27.7. The space between the boiler columns below ground (or basement) level is mainly occupied by cableways and pipe and drainage trenches, in addition to the bases shown for fans. It is seldom possible to carry these heavy column loads by direct bearing at a relatively shallow depth on the subsoil, and piled or cylinder foundations are fairly common. However, the use of large numbers of precast driven piles may result in excessive ground heave unless some preboring is undertaken.

The loads on the turbine house columns are considerably less than those on the boiler house columns although they may be of the order of 1000 t. Each row of the turbine house columns carries a crane beam and rail of the overhead electric travelling crane(s).

27.4 Roofs, walls, floors and ventilation

by V F Harman, BSc, CEng, MICE, MIStructE

Roofs, external walls and floors must be durable and capable of being erected quickly on the supporting steelwork to give support and protected conditions to allow the installation of plant to proceed with a minimum of delay.

27.4.1 Roofs

Power station roofs are generally flat roofs having a drainage slope not exceeding 5°, built of either reinforced concrete or metal-deck construction.

Where there is a risk of damage from objects falling from a higher level, such as on electrical or mechanical annexes, or where there is heavy traffic for maintenance purposes, as on some boiler houses, flat concrete roofs are suitable. They may be formed of precast units or of *in situ* concrete, or of a combination of these, and screeded to falls. They are waterproofed with two 10 mm layers of 10 mm white chippings. The use of chippings requires emphasis on regular inspection and maintenance since they tend to migrate and block rainwater outlets and can make location and repair of defects difficult. The differential thermal movements which occur between asphalt and concrete may be accommodated within a layer of insulating board or by use of a high-performance sheathing felt. The main purpose of the insulation layer, which may also be in the form of a lightweight cellular screed, is to reduce the differential thermal movement between the concrete and the supporting steel structure.

Flat metal-deck roofs are suitable for low-trafficked roofs, such as for turbine houses, reactor buildings and transmission switch houses; in each case the low self-weight of the roof covering is important for the economic construction of the long-span roof support structures of the building. These roofs are typically of the composite construction shown in Figure 27.8. The metal deck is secured to the roof steelwork and consists of profiled aluminium alloy or galvanized steel sheet which is coated with a suitable decorative and protective plastics finish or paint. Steel decks deflect less than aluminium decks of the same profile depth but corrode more quickly when their protective coating is damaged.

Aluminium alloy decks are favoured for turbine halls because

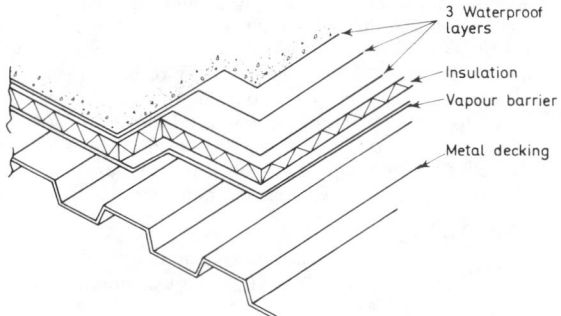

Figure 27.8 Metal-deck roof. (*Courtesy:* Central Electricity Generating Board)

their low collapse strength at elevated temperatures permits early venting of major fires which may occur with the high fire load of hydrogen gas and lubricating oil associated with the turbines. Venting of heat, smoke and fumes through the vent formed in the roof allows adjoining boiler houses to remain relatively free of smoke, the steelwork supporting the roof experiences less overheating and firefighting teams are better able to tackle the blaze. If a fire occurs under a steel roof deck, the main supporting steelwork may be extensively damaged since it is more highly stressed than the roof decking and therefore reaches collapse point earlier.

A vapour barrier of bituminous felt or other impervious sheeting is bonded to the metal deck and this is covered by a layer or layers of insulating board of thickness depending on the insulation value required. The insulation is bonded to the vapour barrier and mechanically fastened to the metal deck. On roofs where there is no particular requirement for thermal insulation, the main purpose of the insulation boards is to provide even support for the waterproofing membrane. The membrane usually comprises three layers of bituminous felt sheeting bonded with hot bitumen, first on the insulating layer and then on each other. The top layer of felt may be surfaced with mineral chippings or metal foil or painted to provide protection from solar radiation. The practice of adding a layer of 10 mm chippings in hot bitumen is going out of favour because of the problems of chipping migration and puncturing of the waterproof membrane on trafficked roofs. The insulating and waterproofing layers of metal deck roofs are susceptible to accidental damage, so insulation with a closed cellular structure is preferred to types which would absorb water and possibly deteriorate rapidly if water ingress occurred. Trafficked areas may be given additional protection by provision of a layer of cork above the insulation: (1) by use of felts based on high-tensile membranes which resist puncturing; (2) by use of additional layers of felt; or (3) by adding a protective layer of tiles bonded to the deck to prevent dislocation by wind forces.

In designing a built-up roof, it is most important to consider the system as a whole and ensure that all materials used are compatible not only with the duties to be imposed on them but also with the method of construction. Properly constructed metal-deck roofs will achieve an FAA fire-resistance rating (flat roof – grade AA) in accordance with BS 476: Part 3.

27.4.2 Walls

The lower part of external walls (the 'plinth' extending from ground level to about 2 m above operating floor level) is usually made of precast concrete panels in both boiler and turbine house to give reasonable resistance to impact damage. The external walls above the plinth are usually of profiled metal fixed to sheeting rails, the profiled cladding being formed from steel sheet finished with a plastics coating, or from aluminium sheet left with a mill finish or coated with plastics. Glazing can be incorporated to give natural lighting and to provide an architectural feature. In the turbine house, a glazing band is usually incorporated between the top of the plinth and the crane rail level to give natural lighting at operating level.

27.4.2.1 Insulating linings

In a boiler house the air is hot under normal working conditions owing to heat emission from the boilers and it is therefore unnecessary to insulate the metal clad walls to reduce loss of heat from the boiler house – indeed, loss of heat by conduction through the cladding is desirable. In a flat metal-deck roof of composite construction, the insulating layer is an essential component to support the waterproofing layers, so it must be retained although its insulating function is not required in a boiler house roof.

In a turbine house, incoming ventilation air acquires heat from the turbo-generator and also moisture from steam leakages. It is most undesirable to have moisture in this warm air condensing on cold surfaces of the metal-clad walls above crane level and of the roof, causing deterioration of paint films and damage by condensation dripping on to plant or cables beneath. Therefore metal-clad walls in turbine houses are normally insulated to have a thermal transmission coefficient ('U' value) not exceeding $2 \, W/m^2 \, °C$. Where there is a risk of impact damage from inside, sheet metal panels backed with mineral wool or glassfibre slab or quilt are used, while for areas less subject to impact damage, insulation board or plasterboard with a plastics or similar finish is used. Insulation lining, especially the type with metal trays, improves the sound-reduction factor of the walls.

Reactor buildings above charge face level are normally provided with insulation linings of a similar type to that used for turbine houses.

In transmission switch houses, the electrical equipment produces very little heat. Insulation is normally provided to reduce the risk of condensation inside the building.

27.4.2.2 Glazing

Fixed glazing may be incorporated in wall cladding as an architectural feature and to admit natural lighting, although in practice artificial lighting is continuous in power houses. Openable glazing is not recommended because the operating gear is susceptible to damage through misuse or becomes inoperable after long periods without operation or maintenance.

Glazing in the turbine house roof is not recommended because the cold surfaces induce condensation in cold weather. Glazing in the boiler house roof is unnecessary.

27.4.3 Floors

Open-grid flooring is used for stairways and galleries and for a large proportion of the floor areas inside the buildings. It has the advantage of low self-weight and also minimizes obstruction to the upward flow of air for the removal of heat in boiler houses. Sometimes imposed loadings require that solid suspended floors must be provided in certain locations; it is also necessary to use solid suspended floors in areas where liquid spillage requires to be collected, e.g. at deaerator level.

The floors at or about operating level and others adjacent to plant requiring regular maintenance, such as pulverized fuel pipework, burners and valves, have to support operating plant or heavy loads during maintenance. Where floors contain many perforations, such as in electrical annexes and control rooms, trimming is easier with *in situ* concrete construction.

27.4.4 Ventilation

Figure 27.9 shows a cross-section of a power house where the turbine house adjoins the boiler house and there is no wall between them. Heat is emitted by the boilers (and to a lesser extent by turbo-generators and other plant) and this raises the air temperature inside the buildings. The air is hottest around the boiler casing, so that the column of warm internal air for the height of the boiler house (about 50 m), being of lighter relative density than the cooler external air, causes the boiler house to act as a chimney, with sufficient air movement to ventilate adequately both the boiler house and turbine house if there is no wall between them. Cool air is drawn in through low-level louvres in the external walls of the boiler house and turbine house, and heated air escapes through outlets at roof level: some of the heated air is abstracted from the top of the boiler house by forced draught (FD) fans which feed it as combustion air to the boilers.

The turbine house inlet louvres contribute about one-third of the total air flow but the turbine house plant contributes only one-fifth of the total emitted heat. The temperature rise above outside ambient will therefore be less in the turbine house than in the boiler house. A turbine house which is not open to an adjoining boiler house (e.g. on a nuclear power station) must be mechanically ventilated, since natural ventilation of a turbine house without the chimney effect of the boiler house would demand excessive areas of inlet and outlet louvres to achieve the same limitation of temperature rise.

For a given rate of emission of heat, the air temperature in the power house is related to the rate at which air flows through it. The object in designing the ventilating system, therefore, is to provide inlet and outlet louvres in appropriate locations and of appropriate size to permit an adequate flow of air through the building.

The design sequence is as follows:

(1) Decide on the maximum permissible rise of inside temperature above outside temperature. This is derived from medical considerations of heat stress in personnel and of local climatic conditions; in the UK a maximum rise of 15°C is suitable.

(2) Assess the heat emission from plant and the solar gain or conduction loss through the fabric of the building to give the total quantity of heat to be removed by the ventilating air. For a 500 MW unit, about 12 MW of heat is emitted from the boiler plant and about 3 MW from the turbo-generator. The solar gain or conduction loss is relatively small and has little effect on the subsequent calculations.

(3) Calculate the total rate of air flow necessary to remove the heat; this gives the required air flow through the inlet ventilators.

$$\text{Quantity of air} = \frac{\text{heat to be removed by air}}{\text{specific heat of air} \times \text{temperature rise}}$$

For example, for a 2000 MW station of four 500 MW units allowing 15°C temperature rise:

$$Q = \frac{4 \times (12 + 3) \times 10^6 \, \text{J/s}}{1200 \, \text{J/m}^3 \, °\text{C} \times 15°\text{C}}$$

$$= 3300 \, \text{m}^3/\text{s}$$

(4) Determine the net area of the inlet ventilators to admit the total air flow using the method given in Clause 14 of BS 5925: 1980. The discharge coefficient should be taken from manufacturers' data but may be assumed to be 0.6 for preliminary design purposes. Louvred ventilators on power stations typically have a net opening area to gross ventilation area ratio of 0.5.

(5) The outlet ventilators must have the same total $A \times C_v$ value as the inlet ventilators to allow for the temporary condition at the end of shift when combustion ceases in the boilers. Then the FD fans stop running, but the boilers continue to emit heat which can only be removed by discharging the total air flow through the outlet ventilators. Under normal conditions, with the boilers and FD fans operating the fans may take 2000 m³/s of combustion air, equal to 60% of the total air flow. It should be possible to close an equal percentage of the area of the outlet ventilators, preferably around the FD fan intakes, so that a balance is maintained in

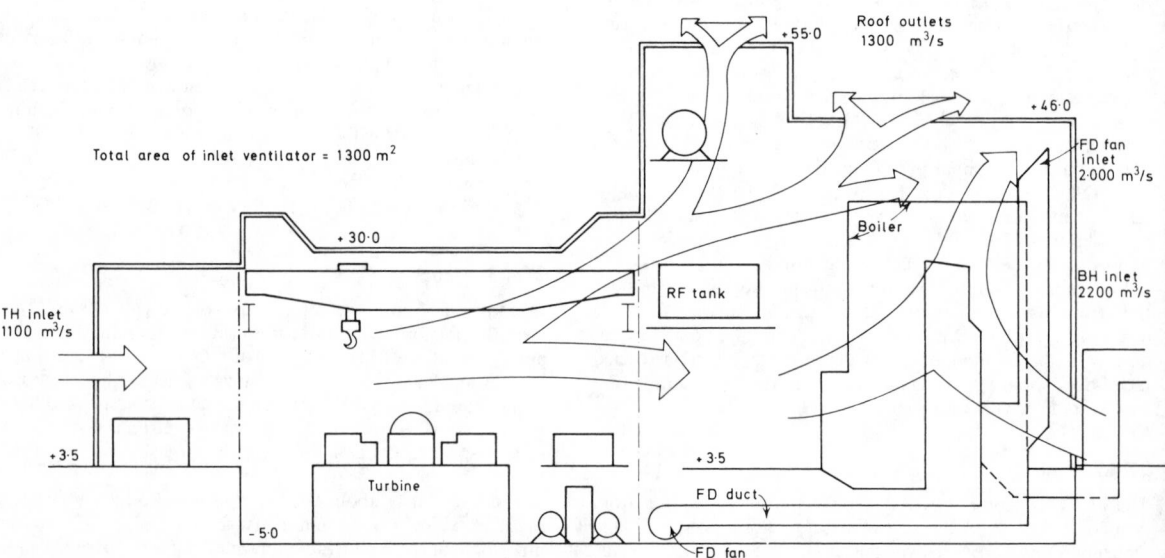

Figure 27.9 Cross-section through a 4×500 MW oil-fired power station showing ventilating air-flow pattern

the system. If the necessary proportion of the outlet venti-lators is not closed, the FD fans may draw air into the boiler house through the outlet ventilators.

In a nuclear power station, the size of the reactor building and the enclosure (by walls and floors) of the space around the reactor containment does not permit adequate natural ventila-tion of working places. The ventilation required in various parts of the reactor building also depends upon the operations carried out there, and in consequence most of the reactor building is mechanically ventilated. The natural ventilation of the turbine house of a nuclear power station cannot be augmented by the chimney effect of adjoining boiler house: consequently the turbine houses of nuclear stations are mechanically ventilated. In order to avoid large temperature variations, it is usual to provide ducting to introduce the incoming cold air at positions near to heat-emitting surfaces.

27.5 Turbo-generator support structures

by D J Maher, BSc, MSc, MICE

The turbo-generator support structure (the turbo-block) sup-ports the turbo-generator, steam chests and condensers and is constructed from a basement structure which is supported on a piled or raft foundation.

Modern turbine generators built for power stations in the UK are rated at 660 MW and are equipped with either pannier or underslung condensers. Pannier condensers are located either side of the low-pressure steam cylinders and were introduced to the UK at about the same time as turbo blocks were first generally constructed from structural steel (see Figures 27.10 and 27.11).

Underslung condensers have been preferred on the majority of recent units erected in the UK. An important criterion in their selection is fabrication, which is an activity primarily under-taken in the manufacturer's works with only a relatively small amount of fabrication required on the construction site. Thus, a saving on erection time may provide an overall benefit to the project programme and, for overseas projects, reduce the amount of skilled labour required on site.

Turbo blocks may be constructed in reinforced concrete or structural steel but recent practice in the UK has seen the use of structural steel as the preferred medium. The advantages of structural steel are:

(1) The slender dimensions of steel columns permit auxiliary equipment (particularly steam pipes and electrical connec-

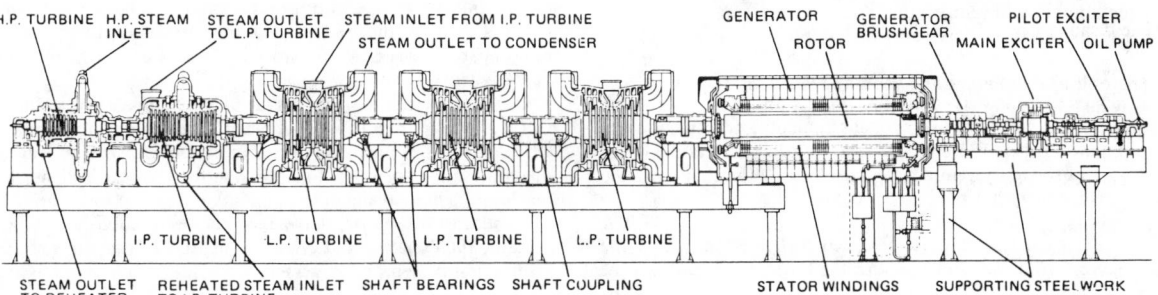

Figure 27.10 Arrangement of 660 MW turbo-generator on steel block with pannier condensers. (*Courtesy:* Central Electricity Generating Board)

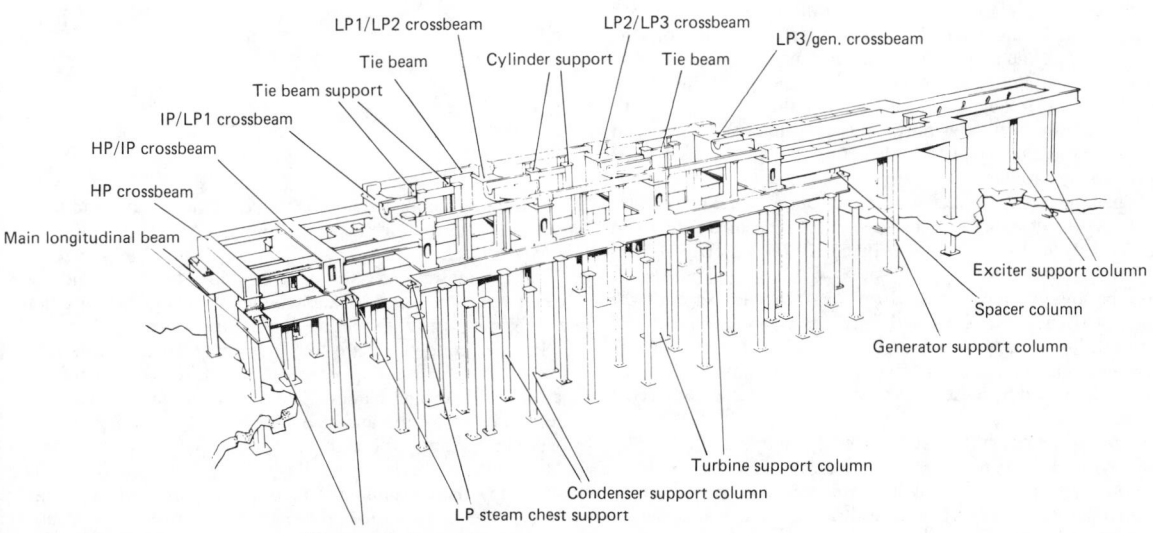

Figure 27.11 Steel block for turbo-generator with pannier condensers. (*Courtesy:* Central Electricity Generating Board)

tions to the generator) to be accommodated more conveniently.

(2) The improved ventilation maintains the steel columns at a more uniform temperature than concrete columns. There have been instances of large turbines being put out of alignment owing to temperature variations in the concrete columns at the turbine end.

(3) The behaviour under load of the steel block is consistent and maintained, whereas concrete blocks are subject to movement due to curing and creep. This has led to some machines requiring realignment after initial operation.

(4) Steel blocks have shown no deterioration by attrition under the turbine sole plates, whereas spalling of concrete surfaces has been occasionally observed.

(5) It is possible to make economies in construction by integrating the condenser and generator casings into the steel support system.

(6) Steel blocks can be readily adjusted to eliminate any local resonance.

(7) The time for erection of a steel block is much less than for a concrete block, although turbo-block erection time is generally not critical in the overall project programme.

(8) The steel block can be designed and supplied by the turbo-generator contractor thereby avoiding any division of responsibility for alignment between turbo-generator contractor and concrete block contractor.

The disadvantages of steel blocks are:

(1) Their capital cost is greater than that of concrete blocks.
(2) The steel columns require fire protection.
(3) The design of mountings for instruments and other attachments to the structure requires careful consideration since the absolute amplitude of steel block movement can be greater than that of concrete blocks.

The turbo block should be designed in accordance with the recognized Codes of Practice and the laws of mechanics. The final dimensions of the turbo block and its foundation will be determined following consideration of the vibratory forces acting on the supporting structures. The vibratory forces are mainly centrifugal forces developed from the elements of the rotating machine, and their vertical and horizontal components give rise to vertical and horizontal vibrations in the supporting structure. It is important therefore that the natural vibration frequencies of the supporting structure and its components in each direction do not approach the fundamental frequency, or the harmonic frequencies generated by the machine and so avoid resonance. It is generally a prerequisite of the design that the natural frequency of the supporting structure should differ from the operating speed by ± 20 to 30%.

When the vertical resonant frequency of the structure is above or below the machine running speed the turbo block is termed respectively 'high-tuned' or 'low-tuned'. In the majority of cases low-tuned blocks are made of steel, whereas concrete blocks may be low- or high-tuned. On a low-frequency support the machine must always pass through resonant levels of the support during start up and shut down, and the amplitude will depend on damping and machine speed. The transient dynamic forces developed at these levels should therefore be considered in the design analysis.

Concrete blocks have not been completely superseded, especially for smaller machines. They should, however, be built several months ahead of machine erection in order to reduce distortion caused by creep, drying shrinkage, and elastic deformation. Although a well-placed and good-quality concrete is required, a very high strength is not necessary; the mix design should aim at reducing shrinkage, a minimum of heat genera-

tion and a low temperature gradient within the pour. Other important points in the design of concrete blocks are the choice of positions for construction joints between lifts and the inclusion of adequate reinforcement (not less than 50 kg/m³) which should be placed in all three planes.

A site investigation should be completed before the design of the foundation is undertaken. This should include all tests necessary to assess the allowable bearing pressure of the soil and the immediate and long-term settlement of the foundation under static loads. The dynamic soil parameters should also be determined so that the soil response to the vibratory forces can be assessed and excessive settlements avoided. In order to reduce the transmission of vibratory forces to the operating floors of the main building the foundation should be isolated from the surrounding basement floor and, as far as practicable, from other structures and services, by the provision of movement joints.

One method of monitoring and measuring movement on the turbo support block is the manometric vertical movement monitoring system. This system detects small vertical movements along the length of the turbo-generator and utilizes the principle that a static head of fluid over an area is identical at all points, subjected to the same environment. The surface of the fluid may therefore be used as the datum from which vertical measurements can be made.

A programme of measuring and determining the causes of movement in turbo-generator blocks in use has been carried out by the CEGB so that the effect of movements on future turbo blocks can be assessed and satisfactory support structures can be designed.[2]

Useful guidance for the design of turbo blocks is given in German Standard DIN 4024:1955.[3] A Code of Practice, CP 2012: Part 1 is available for foundations for reciprocating machines and a complementary Code for rotating and impact machines is now available.

A number of papers have recently been published which describe the dynamic and economic aspects of turbo-generator foundation design and alignment.[4,5]

27.6 Cooling-water systems

by W G Jones, CGIA, MICE

27.6.1 Systems

27.6.1.1 Condensers

The function of a condenser is to reduce the steam condition at exhaust from the turbine to the lowest practicable temperature and pressure.

All condensers have the same function, types of condensers being named transverse, axial pannier or radial according to the arrangement of tubes relative to the turbine axis.

Pannier condensers are 'boxes' located on each side of the low-pressure turbine to receive the steam exhausted from the last row of turbine blades. Each box is about 3 m wide × 8 m deep in cross-section and extends for the full length of the low-pressure turbine (about 20 m). About 10 000 water tubes of 20 mm internal diameter run through the full length of each box and carry the continuous flow of cooling water (CW) (Figure 27.12). The condensate is returned to the boilers for reheating, while the CW, which has gained the heat lost by the steam, is either returned to source or is passed through cooling towers to dissipate the heat before re-use in the condensers.

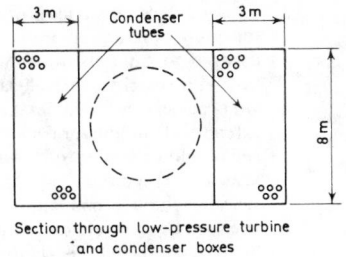

Section through low-pressure turbine
and condenser boxes

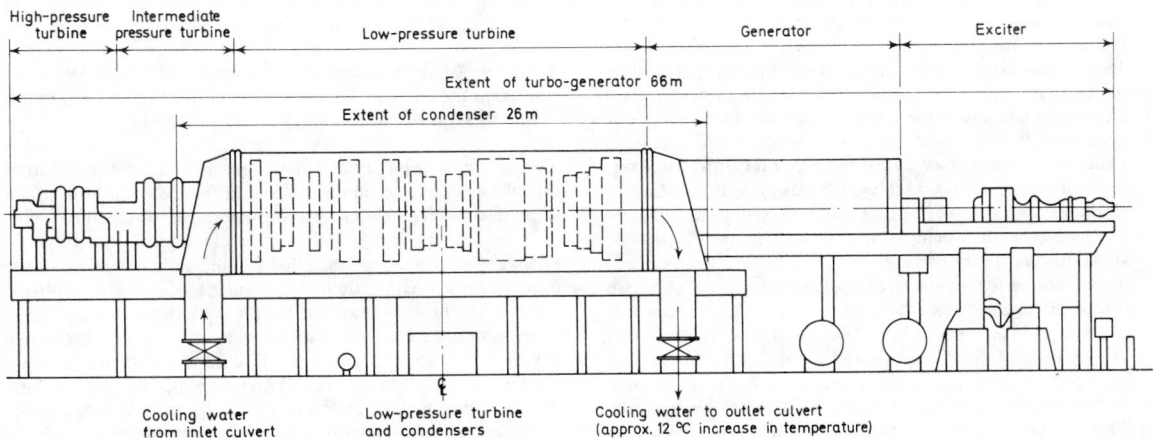

Figure 27.12 Turbo-generator with pannier condensers.
Condenser tubes total about 20 000 (tube diameter about 22 mm).
Sizes shown are typical for a 500 MW turbo-generator

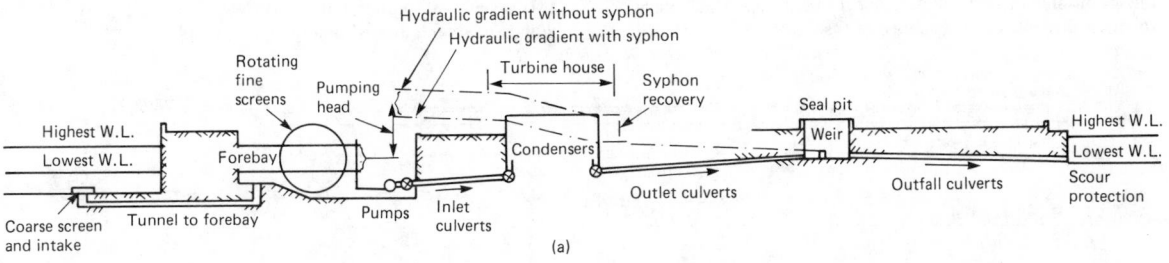

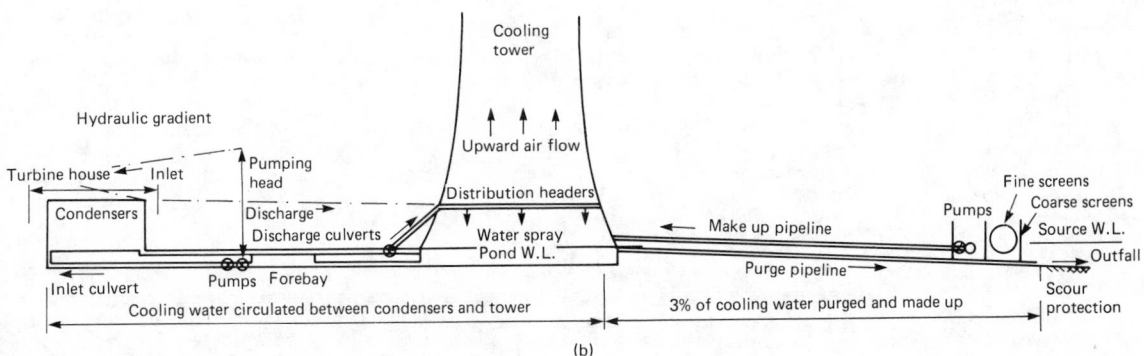

Figure 27.13 (a) Direct (once-through) cooling system; (b)
recirculating (tower-cooled) cooling system

27.6.1.2 Auxiliary plant

In the CW system, most of the water goes through the turbine condensers but about 5% is looped off to cool the turbine auxiliary plant. Water is taken from the inlet culvert and returned to the outlet culvert. The head loss across these auxiliaries matches the head loss across the tapping points. Auxiliary plant associated with boilers or reactors are not linked to the main CW system but cooled by separate pumps.

27.6.1.2 Once-through and recirculating systems

There are two alternative systems for maintaining the necessary flow of cold water through the condensers:

(1) Direct or once-through cooling (see Figure 27.13(a)) in which cold water is pumped from the sea or a river through the condensers and is then discharged back to the source in such a position that heated water is not taken into the intake again. In the case of CW taken from and returned to a lake, recirculation is unavoidable eventually, but the lake is channelled so that the heated water must flow a sufficient distance between discharge and intake for the heat to be dissipated before re-use.
(2) Recirculating or tower cooling (see Figure 27.13(b)) in which the heated water from the condensers is pumped to distribution channels at high level inside a cooling tower, from which it falls as a spray over a pack of timber slats or over parallel asbestos sheets. The cooling tower is designed to produce an upward flow of air so that heat in the descending water droplets is lost by evaporation and by contact with the rising air. The cooled water is collected in a shallow pond which forms the base area of the tower and is then recirculated by pumping through the condensers.

Direct (once-through) installation (Figure 27.13(a)). In passing from intake to outfall in a direct CW system, the water goes through the following sequence of installations.

Intake. This facilitates flow of water from source to the coarse screens. On rivers or protected coasts it is usually a channel, dredged to sufficient width and depth to carry the maximum flow at a velocity at which sand is not carried into the system. On exposed coasts, the intake is usually a concrete-lined tunnel extending from the forebay to a position offshore where the sea bed is below the effect of wave action, so that silt and sand is not drawn in (Figure 27.14). The tunnel terminates in a vertical shaft to the sea bed, the shaft being concrete lined, with a streamlined concrete cill around the shaft entrance at about a metre above bed level. There is usually a permanent headworks structure above the intake shaft to facilitate cleaning of coarse screens around the intake shaft and to facilitate closure of the intake shaft when necessary for maintenance.[6]

Coarse screens to exclude tree trunks and other large debris, are located at the entrance to tunnel intakes or at the landward end of channel intakes where water enters the forebay.

The forebay is a large open chamber sited on land to dissipate turbulence of the incoming water and induce good flow conditions to the screen chamber and pump house following.

Fine screens (where not installed in the headwork of tunnelled intakes) may be either travelling-band screens or rotating-drum screens. Drum screens are made of a layer of metal mesh 2.5 m wide, mounted circumferentially between a pair of rotating wheels about 15 m in diameter. Two concrete walls shaped as saddles lie under the wheels and divide the screen chamber into three compartments. Unscreened water from the forebay passes to the outer compartments and then into the space between the wheels. It then flows through the circumferential screen into the central compartment beneath the drum. From here the screened water flows through a draught tube to the pump. The rotating-drum carries trash retained inside the circumferential screen to trash removal equipment operating on the underside of the drum at the top. The drum dimensions have to be such that the trash removal equipment is above high tide, and there is sufficient screen area submerged at low tide.

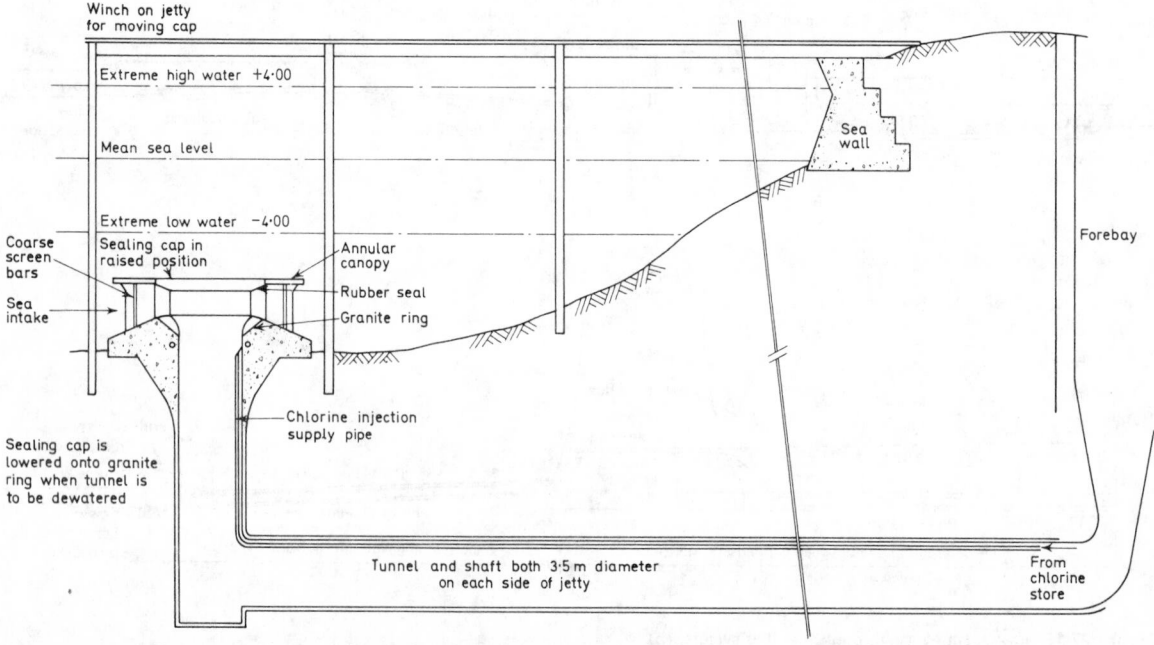

Figure 27.14 Sea-bed water intake

Pumps are located below lowest water level so that they can be started without priming. On a station generating 2000 MW, the cw pumps deal with 240 000 m³ of water per hour.

A non-return valve is located on the discharge side of the pump to prevent damage to the pump and motor resulting from reversal of flow from the condenser. The valve closes automatically within seconds if the pump is stopped by causes other than normal operating control.

Inlet culverts connect the pumps to the condenser and are subject to the maximum pumping head.

Outlet culverts take the cw to the seal pit after discharge from the condenser. They are subjected to little more pressure than the tidal head.

Seal pit: to limit the condenser syphon height. The flow of water through a condenser in a once-through system can be made to operate as a syphon, but the syphonic recovery of hydraulic head has to be limited because of the need to ensure stability of flow conditions. This is done by discharging the down leg of the syphon (i.e. the culvert that carries flow from the condenser) into a pit of water in which the surface is kept – by an overflow weir – at a level calculated to maintain optimum syphonic recovery in the condenser. The culvert discharge is kept wholly submerged to prevent backflow of air into the culvert. Normally, the weir level is not more than 9 m below the top of the condenser.

Outfall culverts take the cooling water from the seal pit to the sea.

For a 2000 MW station with four 500 MW turbo-generators, the cw system typically includes:

Intake for 240 000 m³ water per hour.
Coarse screens.
Forebay.
4 Rotating fine screens, 16 m diameter × 2.5 m wide.
4 Pumps each capable of pumping 60 000 m³ water per hour.
4 Inlet (pressure) culverts between pumps and condensers, interconnected for interchange facility in case of pump failure, each capable of carrying a flow of 120 000 m³/h at a velocity of 3 m/s.
4 Outlet (low-pressure) culverts between condensers and seal pit.
Outfall culvert(s) from seal pit to sea.

Recirculating (tower cooled) installation (Figure 27.13(b))
From the collecting ponds of the cooling towers, the cooled water gravitates along open channels to the pump house. It is pumped through the condensers and up to the distributors in the cooling towers about 14 m above ground level. From the distributors it trickles down over the surfaces of the slats stacked in the lower part of the tower, there being cooled by the natural updraught of air; it is finally collected in the pond for recirculation.

About 1% of the water passing through a tower evaporates and this causes an increase in the dissolved salts content of the recirculated water. The increase in dissolved salts is kept to an acceptable limit by purging – i.e. discharging to waste – about 2% of the circulating water, and both the purged and evaporated quantities are replaced with fresh supplies of water (make-up). Make-up water is drawn through coarse and fine screens from a nearby river and is pumped to one or more of the freely interconnected cooling tower ponds. From another point in the

recirculating system, a gravity pipeline is laid to take purged water back to the river at a point downstream of the intake screens.

For a 2000 MW station with four 500 MW turbo-generators, the cw system typically includes:

8 Natural draught cooling towers about 90 m in diameter at base and 115 m high.
4 Circulating pumps, each capable of pumping 60 000 m³ of water per hour against the head of the distributors.
Pressure culverts between pumps and condensers and between condensers and distributors, interconnected for interchange facility in case of pump failure and tower maintenance.
Gravity culverts between tower ponds and pumps forebay.

The make-up system includes:

River intake, with coarse and fine screens.
Three make-up pumps (giving 50% standby) each capable of pumping 5000 m³ of water per hour against the head of the cooling tower ponds.
Purge culvert to return 3300 m³ of water per hour to river or waste.

27.6.1.3 Comparison of systems

An important difference between tower and once-through systems is in their total pumping head. In a tower system the pumping head is the frictional head plus the static head (the height of the distributors above the pond surface, about 10 m); in a once-through system the pumping head is the frictional head plus the lesser static height above tide level of the highest point on the hydraulic gradient (see Figure 27.13(a)).

A typical pumping head for a once-through system is 15 m and for a tower-cooled system is 24 m. In both systems the quantity of water pumped through the condensers is approximately the same. The energy for pumping in the tower system takes 0.7% of the turbo-generator output, or reduces the station overall efficiency by 0.3%.

Generally, the water temperatures in a once-through system are from 5°C to 10°C lower than the temperatures in a recirculated system and this gives greater turbine efficiency.

Therefore, once-through systems have the financial advantage, but they are restricted in the UK to coastal or estuarine locations, since the heat rejection in cooling water from large stations (equivalent to about 1.3 × the electrical output) is not acceptable in rivers.

Effect of system on power house floor levels. Tower systems are effectively closed loops and hence are little affected economically by the height of the condensers above the river which is the source of make-up water. This cw system also has no effect on the basement and operating floor levels in the power station.

Pumping costs in a once-through system, however, are increased as the height of the condensers above sea-level increases. In cases where it is practicable to lower the floor levels of the turbine house by excavation, without risk of flooding, the resulting saving in pumping costs has to be considered relative to the cost of extra excavation.

27.6.2 Culverts

27.6.2.1 Layout of pumps and culverts

The inlet culverts are interconnected so that flow can be maintained to all condensers in the event of failure of any one pump. Figure 27.15 shows a diagrammatic layout of the pumps, valves, and inlet and outlet culverts for four turbo-generators with pannier condensers.

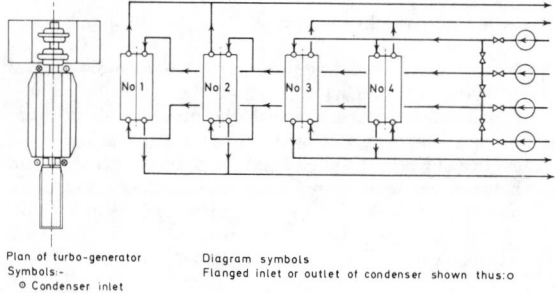

Plan of turbo-generator
Symbols:-
  ○ Condenser inlet
  ● Condenser outlet

Diagram symbols
Flanged inlet or outlet of condenser shown thus:○

Figure 27.15 Layout of pumps and culverts for transverse arrangement of turbo-generators

27.6.2.2 Sectional area of culverts

A culvert of small cross-section and high velocity tends towards high head loss and high pumping energy cost, but towards low capital installation cost. A larger culvert would have lower velocity and lower pumping energy cost, but higher installation cost.

The economic size of culvert is selected by calculating the cross-section of a number of culverts which give velocities of from 3 to 4 m/s in the inlet culvert. For each size the total cost of energy required to pump water through the culvert for the life of the station and the capital cost of installing the culvert are calculated. Energy and installation costs are added together for each size of culvert. The economic size is that which has the lowest combined cost. Culverts are normally about 3 m in diameter for inlet and outlet culverts and from 3 to 4.5 m in diameter for outfall culverts.

27.6.2.3 Hydraulic pressures

Cooling-water culverts are subjected to the following hydraulic pressures:

(1) Normal running pressures. These are as calculated from the hydraulic gradient.
(2) Starting pressure (closed-valve pressure). Pumping is started with the valves in the system closed. Each valve is opened in succession when the section before the valve has been filled. This procedure subjects the pressure culverts to the closed-valve pressure of the pumps.
(3) Sudden closure pressures (surge pressures and vacuum conditions). If the pump motor stops through other than normal operational control, the butterfly valve immediately downstream of the pump is designed to close automatically within seconds. This is to prevent damage to the pump and motor through reversal of flow from the condenser. When the valve closes, the water between it and the condenser briefly oscillates and causes alternatively high surge pressures and vacuum conditions in a length of culvert downstream of the closed valve.
(4) Culvert test pressure, applied during construction trials.

Culverts may be designed for the following hydraulic conditions:

Once-through system
  Inlet culvert   – closed-valve pressure of pump. Also length of culvert downstream of pump (say equal to 4 diameters of culvert) to withstand vacuum.

Outlet culvert   – normal working pressure.
Outfall culvert – pressure at entry to culvert.

Tower-cooled system
  Inlet culvert   – as for once-through system.
  Outlet culvert   – closed-valve pressure of pump (because of valve at entry to cooling tower).

Make-up pipelines can be quite long and surge and vacuum conditions can follow the normal shut-down of make-up pumps. The pipelines may therefore be designed for the closed valve head of the pumps, allowing an additional 10% for surge over the first 25% of length from the pumps; the same length should be designed to withstand vacuum conditions. Steel pipe is normally used for make-up pipelines so that the variations of pressure can be absorbed in the elasticity of the steel.

27.6.2.4 Loading on culverts

Culverts may be subjected to the following types of loading:

(1) Internal water pressure (normal running, closed valve, and surge).
(2) External air pressure (vacuum consequent on self-closure of valve).
(3) Handling stresses in steel pipes during installation.
(4) Ground load and superload.
(5) Self-weight of culvert and water.
(6) Buoyancy – the weight of the culvert when empty must withstand any upthrust from groundwater.

27.6.2.5 Concrete culverts

Concrete culverts in section may be circular or rectangular with chamfered corners, the latter being preferred for multiple culverts. They are extremely durable, but construction is slow and the open excavations may hinder traffic movement about the station site for long periods.

Present practice is to design for normal working pressure to CP 2007 (concrete not subject to surface cracking) and to check for surge conditions and closed-valve pressures to CP 114 which permits higher concrete stresses than CP 2007. In the design the internal pressures may be reduced by a safely low estimate of permanent external (ground) pressure where this is present. The culvert must also be designed to withstand a safely high external loading when it is empty or subject to vacuum. The test pressure for concrete culverts should be little above the normal working pressure, to avoid cracking inducing leakage unnecessarily.

27.6.2.6 Steel culverts

Steel culverts may be installed in the form of pipes made from steel plates with welded longitudinal and circumferential joints. These steel pipes, with a suitable external coating against corrosion when laid underground, are estimated to have a life of 40 to 60 yr which is adequate for the working life of the power station. Steel culvert offers speedier installation than concrete culvert and minimizes the duration of open-culvert excavations and, hence, restriction of movement on site.

Steel pipe has been installed under the following conditions:

(1) Laid overground with plate thickness not suitable for vacuum conditions.
(2) Laid overground with plate thickness suitable for vacuum conditions.
(3) Laid underground without concrete surround if the external loading (including vacuum conditions when applicable) can be carried by the pipe with lateral earth support; local

changes of diameter not exceeding 5% can be accepted without harm to the pipe.

(4) Laid underground with complete or partial concrete surround if the external loading exceeds that which can be carried by the pipe with earth support only.

If steel culverts were designed only to withstand internal pressures, the plate would probably be too thin to withstand the handling stresses during installation. There are therefore minimum practical thicknesses of plate from which pipes are made, these being 12 mm plate for pipes from 1.6 to 2.5 m in diameter, then thickening by 2 mm for each 0.5 m increase of diameter to 22 mm plate for 5 m diameter pipe. Such pipes are more than strong enough for internal pressures in CW culverts, but for diameters exceeding 1.6 m they are not suitable to resist vacuum conditions.

For steel pipe culverts subject to vacuum conditions, laid above ground without concrete surround, the plate thickness can be calculated from the formula for buckling:

$$p = \frac{2E}{1-\mu^2} \left(\frac{t}{d}\right)^3$$

where p is the external collapse pressure (taken as 2 atmospheres to include a factor of safety of 2 to allow for manufacturing tolerances = 0.2 N/mm²), E is the modulus of elasticity (taken as 200 000 N/mm²), μ is Poisson's ratio (taken as 0.3) and t/d is the ratio of plate thickness to pipe diameter.

From this, t should be not less than $d/130$, giving plate thicknesses varying from 16 mm for 2 m diameter pipe to 40 mm for 5 m diameter pipe. This takes into account only the external load due to vacuum conditions and applies only to steel pipes laid above ground without concrete surround. External loading and ground support must be considered where steel pipes subject to vacuum conditions are laid below ground.

27.6.2.7 Water velocities and mussel growth

Velocities are kept above the self-cleansing velocity of 0.8 m/s. The minimum velocity for sea water in once-through systems is considered to be 3.25 m/s, since below this velocity mussels can settle on the culvert walls and floor. When established, mussel growth can form a thick rough lining on the walls and floors of the inlet culverts, so restricting water flow. Mussels may also become detached and can cause damage if the shells are carried to the condenser to become lodged in the water tubes. However, it is impossible to maintain a velocity above 3.25 m/s at all times and provision is made to inject chlorine into the incoming cooling water during the spatting season to kill mussels or spat. A similar method is used to reduce algae growth on tower-cooled stations.

27.7 Natural-draught cooling towers

by K P Grubb, BSc(Eng), MICE

A natural-draught cooling tower consists essentially of a large reinforced concrete chimney or shell into which air is admitted around the base in such a manner that the induced flow of air intimately mixes with and cools a falling stream of water which has been heated in passing through the turbine condensers. This water is normally distributed uniformly across the area of the shell by a sprinkler system, supported at about 14 m above ground level.

Immediately below the sprinkler system a packing is situated, the main function of which is to increase the specific surface of the water stream so that maximum heat transfer takes place.

Currently, it takes the form of either: (1) splash packing, consisting generally of timber or plastic laths to transform the water stream into droplet form; or (2) film packing, consisting either of flat or corrugated asbestos cement sheets set on edge or of prefabricated plastic modules, both types creating a film of water on the sheeting surfaces such that maximum contact with the cooling air stream is created (Figure 27.16). The cooled water is finally collected in a concrete pond, which covers the base area of the tower, for recirculation to the condensers. Two cooling towers about 90 m in diameter and 115 m high are required to cool the water flowing through the condensers of a 500 MW unit.

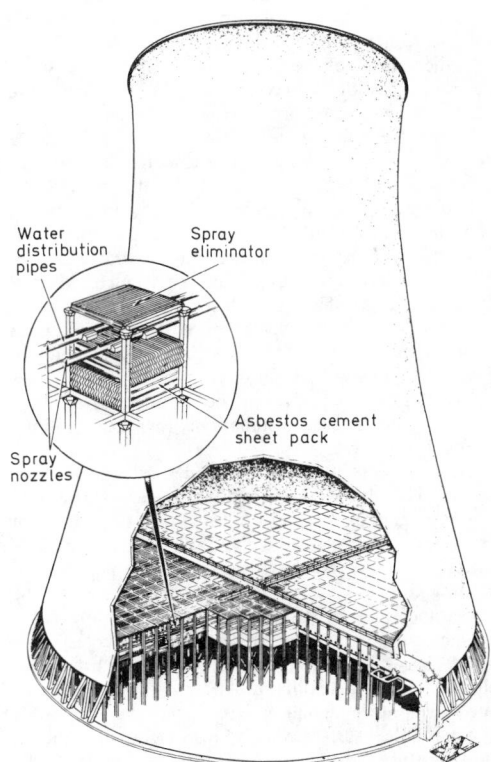

Figure 27.16 Natural-draught concrete cooling tower. (*Courtesy:* Central Electricity Generating Board)

The engineering design of a cooling tower thus presents a combination of hydraulic, aerodynamic, thermodynamic and structural problems. Only the last will be considered here.

The structure supporting the packing, distribution system and eliminator will generally be formed of precast concrete units designed in accordance with BS CP 110.

Two variations from this Code lie in the use of reduced cover to reinforcement in order to minimize the sectional size of the precast members and their impedance to cooling air, and in specifying tighter dimensional tolerances to avoid unacceptable cumulative errors during the erection of the large number of precast elements used in the pack support structure.

27.7.1 The tower shell

Hyperboloidal or near hyperboloidal shells have been used for cooling towers since 1918, when the shape was adopted to provide stability against the gross foundation settlement which was anticipated by van Itersen in his design for the Dutch State Mines. For equivalent strength and stability this shape has a substantial advantage in economy of both material and surface area over any equivalent cylindrical structure. When the further aesthetic advantage of the conic section is added, it is not surprising that the shape has remained virtually unchallenged since that date. It is now a shape uniquely associated with cooling towers.

The design of the shell presents the only unconventional structural design features in a cooling tower. In the UK, the only significant loading on a tower shell is that due to wind and in consequence the structure is very light when compared with other structural shells. As the wall thickness is much smaller than the other principal dimensions, the structure tends to resist wind loading predominantly by membrane and tangential shear stress resultants rather than by bending resistance.

However, the validity of the membrane theory as a basis of design has been questioned since the mid 1970s when finite element methods became generally available resulting in full bending analyses becoming more common. It is likely that a comparison of the two methods will not show significant design changes but the finite element methods can be of great value in more accurately modelling the geometry of the shell, ring beam, columns and foundations. Foundation settlement is one particular aspect that can be incorporated into a finite element analysis. Such an analysis will reveal the nonsymmetrical horizontal bending effects which the shell, if considered as a membrane, is not intrinsically well prepared to resist. Outside the UK there exists the possibility of seismic loadings being in excess of wind loadings. Where national codes require a nonlinear analysis of tower structures to be carried out in these circumstances, e.g. for nuclear installations, then again a finite element method is essential.

Several other aspects of shell design deserve mentioning. Under high-wind conditions the possibility of elastic instability exists. Although research work has indicated that this form of failure is significantly less probable than the overstressing mode in conventionally sized and designed shells, the presence of meridional cracking can cause a dramatic decrease in the buckling resistance of a hyperboloid. However, as towers increase in height and horizontal dimensions the need to estimate the buckling capacity of such towers becomes more important and requires that a geometrically nonlinear analysis should be carried out as part of the design process.[7]

Cracking in the shell surface can also result from operational thermal and moisture movement transients. These may well be intensified by ambient temperature variations. The effect of such cracking alone is only structurally significant in its consequential effect of rendering the shell more susceptible to buckling or vibrational failure although long-term corrosion of the shell reinforcement must also be taken into consideration. Hence, an estimation of temperature stresses in the tower shell should be made.[7]

Large constructional inaccuracies in the shape of hyperbolic shells can seriously reduce their resistance to wind loading. Hence precise control of shape is of great importance during construction.

27.7.2 Wind loading

The current British Standard[8] for the structural design of cooling tower shells requires that wind loading shall be assessed in accordance with BS CP 3: Chapter V, Part 2. The tower would therefore be regarded as a class C structure and would be designed to withstand the effects of a 15 s gust calculated at the cornice level and applied as a nonvariable quasi-static loading over the tower height.

A current draft amendment to BS CP 3, Chapter V, Part 2 suggests that this Code does not apply to buildings, or structures, with properties such as low natural frequency or low damping which make them susceptible to dynamic response. It goes on to say that the assessment of such response should be based upon published sources or experimental results such as properly conducted wind tunnel tests.

Wind tunnel testing of towers in groups carried out in the UK and abroad in the last decade have revealed that loadings in a tower caused by that portion of a wind spectrum at or near the natural frequency of the tower, i.e. due to the dynamic response, are of more significance than hitherto suspected. Hence, some allowance for resonance loadings should be made in assessing stress resultants in the shell although currently the range of tower sizes and of tower grouping to be considered only allows empirical expressions to be adopted.[7,9]

The distribution of pressure over the shell surface in any horizontal plane was measured at the time of the Ferrybridge investigation from wind tunnel tests and the results of those tests form the basis for the pressure coefficient distribution specified in the British Standard.[8]

27.7.3 Support columns and foundations

The design of foundations and diagonal support columns for tower shells is largely conventional, but any possibility of inducing uplift conditions on the upstream side of larger towers requires most careful consideration, in particular where the lifting of foundations under seismic loading might be used as a design aid for alleviating the uplift stresses in the shell and support columns.

Circumferential continuity of the foundations and/or pond wall is recommended in preference to discrete pad foundations below each support column node.

The transfer of load and horizontal shear from the shell into the support columns is effected through a thickening of the lower shell section, or ring beam. Very high maximum column loads, which may result from elastic analysis of the support complex, are likely to be modified or smoothed out by redistribution in practice, since it is clear that the loads are transient by nature and that the least degree of nonlinearity induced in the

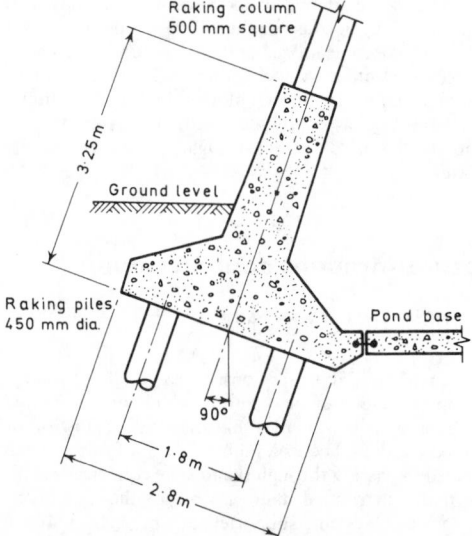

Figure 27.17 Cooling-tower foundation

columns by momentary overstress will permit such a redistribution. Some allowance for this effect is normally permitted in design.

Figure 27.17 shows a cross-section through a piled foundation for a cooling tower. The concrete ring beam (supported on raking piles) carries a triangulated system of raking support columns, which in turn support the ring beam created by the thickening of the lower section of the tower shell. The water in the pond and the tower packing are carried by the base slab.

In this design the pond wall forms an integral part of the pile cap. It can equally form part of a composite contact foundation. Some economies may be achieved by supporting the raking columns directly on the pile cap or contact foundation. In such cases the pond wall may be constructed outside the columns as the last item, hence providing easier access to the pond during the construction period.

27.8 Chimneys

by K P Grubb, BSc(Eng), MICE

The function of a chimney is to discharge flue gases to the atmosphere at such a height and velocity that the concentration of pollutants such as sulphur dioxide is kept within acceptable limits at ground level. Brickwork makes a suitable structure for free-standing chimneys up to about 60 m high but for taller chimneys the overturning moment due to increased wind load can be more economically resisted by a reinforced concrete shaft.

Fossil-fired stations will generally be equipped with either coal- or oil-fired boilers. Some coals have very high sulphur content, but generally the sulphur content of most British coal is reasonably low. With liquid fuels the sulphur content is more varied and depends in the main on the origin and subsequent processing. Hence, the flue gases from a large oil-fired boiler might have a sulphur content of up to 3% and a back-end temperature of about 150°C. At or near to dewpoint a dilute condensate of sulphuric acid might be produced on the flue surface if temperature conditions are favourable.

It is therefore necessary to provide a lining to protect the concrete shaft internally from heat and acid attack. The lining is normally in the form of free-standing acid-resisting brickwork one-half brick thick (say 100 mm) which is self-supporting for a maximum height of 10 m. Consequently, the lining is built as a series of truncated cones carried on corbels inside the concrete shaft at 10 m intervals (see the inner chimney detail at 'A' in Figure 27.18). There is a cavity 50 mm wide between the concrete shaft and the brickwork lining which may be filled with an insulating material or left as an air gap; the junction between successive sections of lining is sealed with glass fibre and lead to exclude flue gases from the cavity.

The lining is usually specified as dense acid-resisting brickwork to BS 3679, laid in potassium silicate mortar. Where alkaline or wet conditions may be experienced (as at the top of the chimney) a synthetic resin should be used instead of mortar to avoid softening of the joints which are normally kept as thin as possible, say 3 to 5 mm. Linings are usually one-half brick thick for flues up to 6 m in diameter, but in the lower levels of the chimney and around gas entry points, a lining one brick thick is provided.

27.8.1 Multi-flue chimneys

The column of hot gases rising inside a chimney continues to rise as a 'plume' without appreciable dispersion for some height after leaving the top of the chimney and so increases the effective height of emission; e.g. with a chimney 200 m high the effective height of emission could be 500 m.

From the expression:

$$C \propto \frac{Q}{H^2}$$

where C is the concentration of pollutant at ground level, Q is the rate of emission of flue gas and H is the effective height of emission

it can be seen that an increase in the effective height of emission has considerable effect in reducing the concentration of pollution at ground level. Research has shown that the plume rise (which governs the effective height of emission) is largely dependent upon the heat content of the plume; therefore for a power station with several boilers (each with its own flue to maintain effective emission) the plume rise can be maximized by closely grouping the flues to concentrate the heat into one plume. This has led to the development of the multi-flue chimney in which several flues are enclosed within a circular reinforced concrete windshield, and since the early 1960s all power stations with a capacity of 1500 MW or greater (i.e. with three or more units of 500 MW or larger) have been provided with a multi-flue chimney having one flue for each boiler.

Multi-flue chimneys are of two main types. Figure 27.18 shows a reinforced concrete windshield enclosing four free-standing reinforced concrete shafts with linings as already described. Floors are provided in the interspace between the chimneys and the windshield at approximately 40 m intervals for access and servicing of aircraft warning lights.

Figure 27.19 shows a reinforced concrete windshield which encloses flues formed only of lining brickwork. In this type the reinforced concrete chimney shafts are omitted, and the sections of flue brickwork are carried on a series of floors at about 10 m intervals. One feature of this type is the necessity to have deep beams supporting each load-bearing floor. In order to reduce the spans and deflections of these beams, various methods have been employed including the provision of a central column which can also serve as an access shaft, and the propping or tying of the floor beams at or near their centre points, to the windshield.

27.8.2 Flue design

The basic parameters for flue design are the height of the flue, the temperature, the efflux velocity and the rate of emission of the flue gases.

In the UK, the height of the flue and efflux velocity must be acceptable to H.M. Inspectorate of Pollutions which is concerned with the concentration of pollutants in the environment. The diameter of the top of the flue will be determined from the rate of emission and the efflux velocity, the latter being kept as high as practicable to minimize downwash of the emission. The flue gases are brought from the boiler furnace by the complementary action of the forced-draught and induced-draught fans through ducts as far as the base of the flue (see Figure 27.4): the pressure head which causes the flow of gases up the flue is the result of the difference in density between the flue gas and external atmosphere. It is good practice to maintain a slight negative pressure inside the flue to reduce gas leakage, and therefore a balance must be maintained between, on the one hand, the head available through density difference and, on the other hand, the losses caused at entry and exit and by friction in the flue. If the chimney is undersized or gas flow fluctuates excessively, a positive pressure can be created in the flue and cause gas leakage.

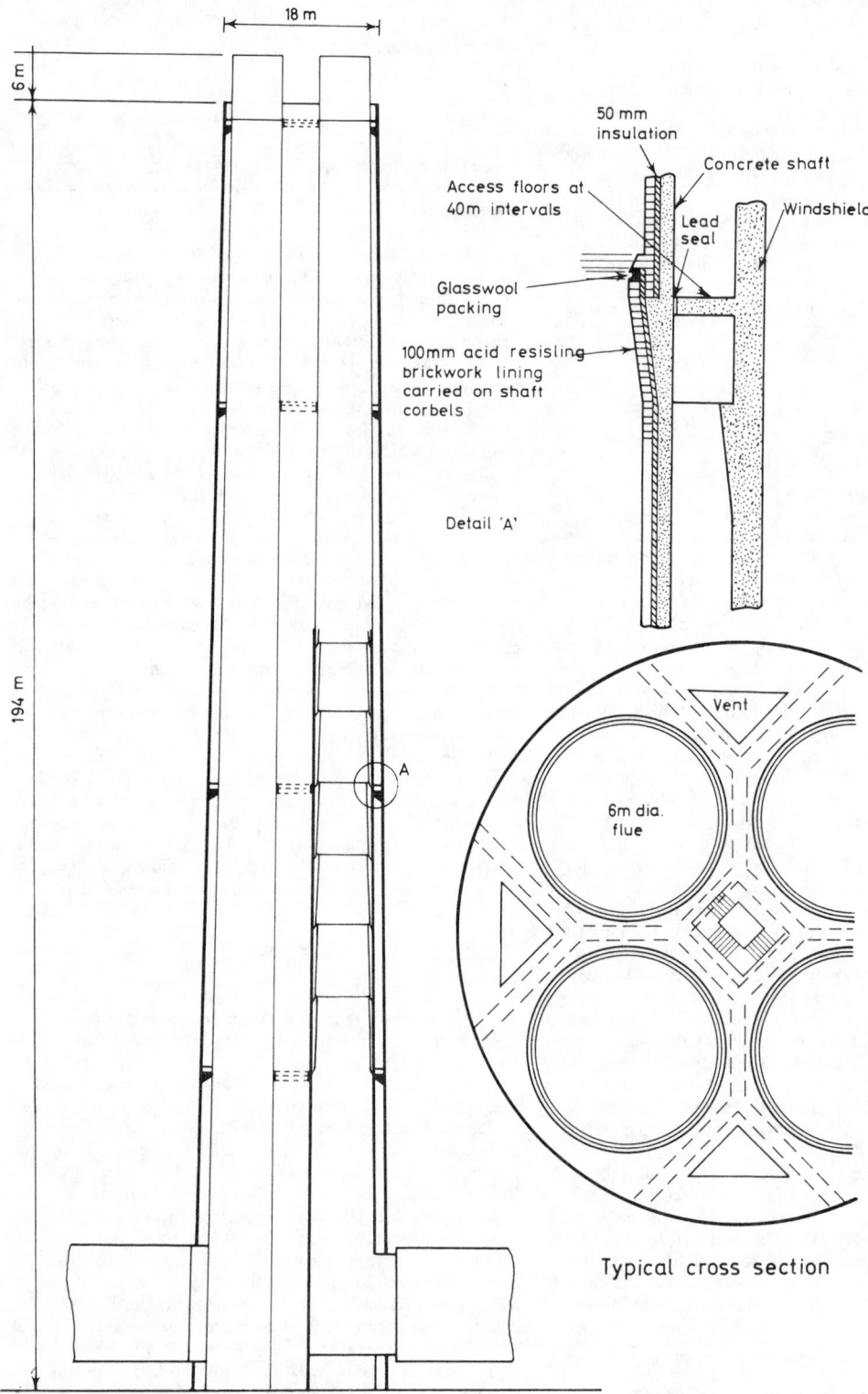

Figure 27.18 18 200 m multi-flue chimney with free-standing flues

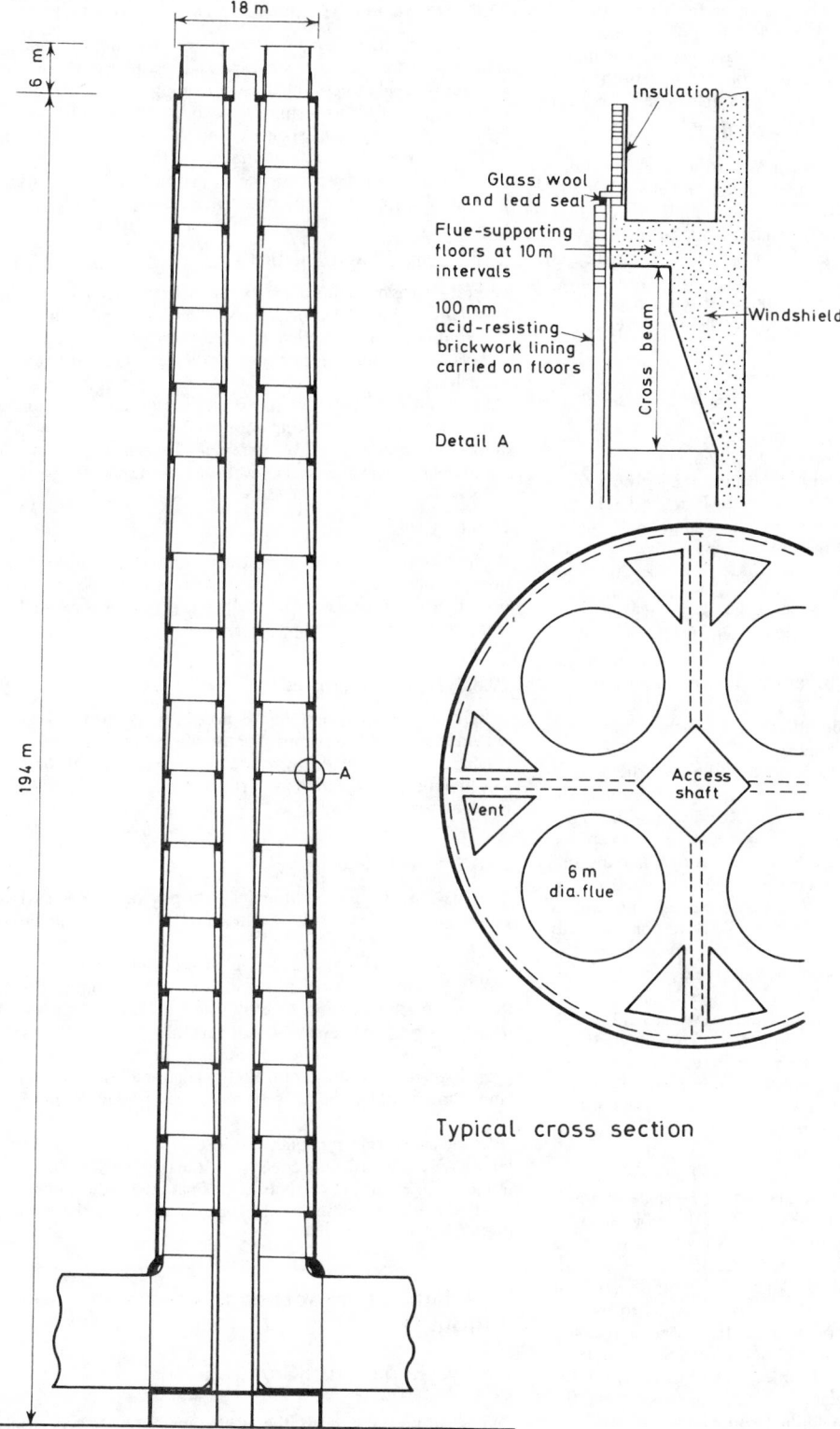

18 m

6 m

194 m

A

Insulation

Glass wool
and lead seal

Flue-supporting
floors at 10 m
intervals

100 mm
acid-resisting
brickwork lining
carried on floors

Detail A

Cross beam

Windshield

Access
shaft

Vent

6 m
dia. flue

Typical cross section

Figure 27.19 200 m multi-flue chimney with flues supported on
windshield

Details at flue entry vary; where entry is from beneath it is usual to provide a 'lobster-back' bend of steel plate in order to reduce friction and turbulence, in which case the flue lining commences at about 25 m above the chimney base. In the case of side entry it is usual to provide 'splitters' (i.e. deflector plates) over the area of entry to reduce turbulence in the gases due to the right angle change in direction.

The multi-flue chimney of a 2000 MW station would be about 200 m high with flues about 6 m in diameter and efflux velocity about 23 m/s; for a 4000 MW station the chimney would be about 260 m high and the efflux velocity about 26 m/s.

In the near future main boiler plant will be fitted with flue gas desulphurization equipment to reduce the toxicity of chimney emissions. This will affect chimney lining design and performance significantly, and it will be some time before these revised designs are proven.

27.8.3 Windshield design

The shafts of single-flue chimneys and the windshields of multi-flue chimneys must be designed to withstand the loadings from wind, self weight and temperature. Code of Practice 3, Chapter V, Part 2 requires that any structure whose greatest lateral or vertical dimension exceeds 50 m shall be designed to withstand a 15 s gust wind speed. However, the Code recognizes the limitations of applying quasi-static loadings to buildings or structures which are susceptible to dynamic response.

The basic design of the shaft or windshield as a cantilever resisting overturning moment under normal wind forces should take due account of buffeting in the natural wind and it is desirable to apply a factor to these forces which will adequately allow for dynamic downwind effects. Two such methods are proposed in current publications.[10,11]

Generally, this basic design will be carried out on the basis of an elastic analysis but model codes[11] normally require in addition consideration of an extreme wind condition with higher allowable stresses. Where a building or structure is susceptible to excitation by vortex shedding or other aerodynamic instability the maximum dynamic response may occur at wind velocities lower than normal wind force. It is therefore important to determine the conditions under which such responses could occur.[10-12]

It is necessary (particularly for a windshield enclosing free-standing shafts) to investigate the ovalling stresses caused by the varying pressure distribution around the windshield which result in positive and negative bending moments in the horizontal plane. Generally, the estimation of vertical and horizontal stress resultants are considered separately but with windshields having a ratio of mean diameter:shell thickness in excess of 50 it may be desirable to carry out a full bending and membrane analysis of the windshield as a thin shell.

Temperature stresses are calculated on the basis of the temperature differential which exists across the shaft or windshield walls and which causes tensile strain on the cooler face.[10,11]

In designing the floors inside the windshield their effect as stiffening diaphragms should be considered, otherwise the windshield could be of uneconomic thickness. The floor design must also include open areas, generally covered with open-mesh flooring, to allow sufficient upflow of air to cool the interspace in which the air temperature should not normally exceed 38°C.

Although it is usual to provide an expansion gap between the floors and any free-standing concrete shafts in a windshield, the floors might be brought into contact with the shafts and load transferred laterally owing to horizontal deflection of the windshield. Hence, the shafts should be designed to withstand a proportion of the total wind load based on the relative stiffness of shafts and windshield.

27.8.4 Protection of chimney top from acid attack

The top of a single- or multi-flue chimney for 10% of its height should be externally protected by acid-resisting paint or tiles. On multi-flue chimneys the flues projecting above the top floor are normally of acid-resisting brickwork only. Steel flues could be used, but they are expensive and difficult to erect. Glass-reinforced plastics are light and easy to erect but there is insufficient operating experience to justify their general use at present. The top floor may be covered with quarry tiles.

27.8.5 Aircraft warning lights

Aircraft warning lights must be provided in accordance with Regulations, usually at 50 m intervals vertically and at the top of the chimney. Three lights at 120° intervals (or four at 90° intervals, for a multi-flue chimney with four flues) are provided at each level.

On multi-flue chimneys, the lights are usually fixed on the outer face of inward-opening doors in the windshield, accessible at the appropriate floor level. On single-flue chimneys, lamps are maintained by steeplejacks and fittings are duplicated.

27.8.6 Access

Bronze sockets in which steeplejacks may insert ladder fixing hooks should be built in the external face of single-flue chimneys; they are unnecessary on multi-flue chimneys because there is internal access to the top.

27.8.7 Lightning protection

A lightning protection system is necessary. A coronal band is provided and CP 326 permits the use of steel reinforcement in a concrete structure as down conductors, provided that the reinforcement cage is adequately earthed and tested on completion for continuity.

27.8.8 Foundations

The varying subsoil conditions which may be experienced on power station sites results in an equal variety of foundation types.

The total weight of a 200-m high multi-flue chimney for a 2000 MW power station approaches 20 000 t. Combined with the resulting down-wind or cross-wind loadings the resulting maximum bearing pressure would generally preclude the use of contact foundations, solid or annular, unless sound rock is present at an economic depth. Both piled foundations and those employing cylinders or caissons have been provided on power stations.

In all cases, the dynamic response of a chimney will vary inversely as the natural frequency of the combined windshield/foundation. Hence, in calculating this parameter due allowance must be made for modelling the soil stiffness under the foundation.

27.9 Nuclear reactors and reactor buildings

by N G Eggleton, BSc, FICE

In nuclear power stations, the turbo-generators are very similar to those in coal- or oil-burning stations, but instead of heat being produced from the combustion of coal or oil in the furnace of a boiler, heat is produced by the nuclear fission of a radioactive material in a reactor. In the UK, except for the recent commitment to the pressurized water reactor at Sizewell 'B', all the reactors which have been built for the public supply industry use a circulating gas coolant for the nuclear fuel and are classified as gas-cooled reactors.

The reactor core consists of nuclear fuel elements housed in vertical channels formed in a graphite moderator which is assembled from accurately machined graphite blocks. The moderator slows down the neutrons radiated from the nuclear fuel in order to enhance the frequency of neutron collision and fission in the mass of nuclear fuel. The reactor is designed to contain sufficient mass of nuclear fuel to give a self-sustaining reaction. The reactor core is enclosed within a pressure vessel made of either steel or prestressed concrete, and heat exchangers ('boilers') are arranged around its circumference; the boilers are located outside steel pressure vessels, but are contained inside concrete vessels or are contained in cavities within the wall of concrete vessels. The reactor shown in Figure 27.20 is contained in a prestressed concrete pressure vessel. The gas coolant is circulated inside the pressure vessel; it flows up the channels in the moderator which houses the fuel elements and there it gains heat, then flows outwards from the core and down through the boilers, where it gives up heat to water and steam in the boiler–turbine steam circuit. It is then recirculated upwards through the core. In order to increase its capacity to carry heat, the gas is kept at high pressure.

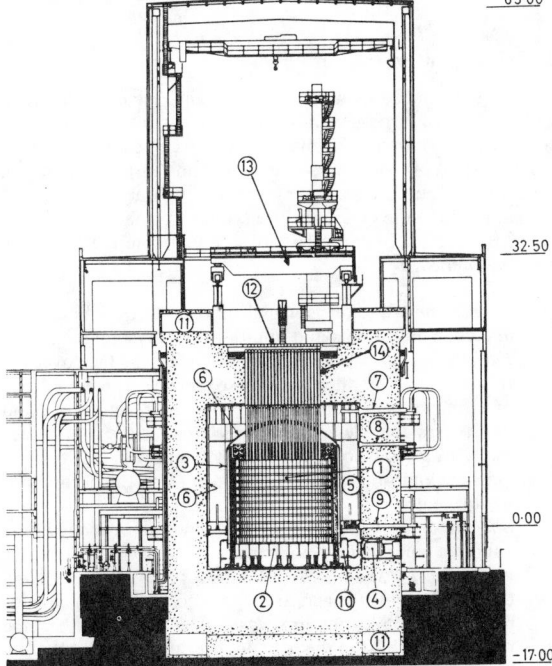

Figure 27.20 Advanced Gas-cooled Reactor. (1) Reactor core; (2) supporting grid; (3) gas baffle (steel cylinder 14 m diameter, without bottom and with torispherical dome); (4) gas circulators; (5) boilers; (6) thermal insulation; (7) reheat steam penetrations; (8) main steam penetrations; (9) boiler feed penetrations; (10) prestressed concrete pressure vessel (19 m diameter and 19.35 m high internally); (11) cable-stressing galleries; (12) charge-face level; (13) fuelling machine; (14) standpipes (one standpipe above every channel in core for refuelling or control)

The first nuclear power station programme in the UK consisted of nine Magnox stations commissioned between 1962 and 1968, and the second consists of seven Advanced Gas-cooled Reactor (AGR) stations commissioned between 1972 and 1978.

The Magnox reactors use natural uranium fuel elements in metal form, encased in magnesium alloy (magnox) finned containers about 25 mm in diameter and 0.5 to 1 m long to support the fuel and contain the radioactive products of fission; they are graphite-moderated and are cooled by carbon dioxide gas. The metal fuel and magnox cans limit the coolant temperature so that the steam temperatures and pressures reached in the boilers are suitable only for turbo-generators up to 300 MW capacity. The fuel irradiation is 3000 to 4000 MW days per tonne.

The AGRs use enriched uranium (uranium dioxide) fuel. The natural uranium is enriched in a separation process by increasing the proportion of the radioactive isotope present. The fuel elements consist of 36 stainless-steel ribbed tubes containing 14.5-mm diameter uranium dioxide pellets, the tube cluster being contained within a graphite sleeve 190 mm in diameter. Eight such elements each 1 m long, are linked together by a tie bar to form a fuel stringer, and each channel in the reactor core accommodates one stringer. Oxide fuel and stainless-steel cans permit higher operating temperatures and the steam conditions reached in the boilers (170 kgf/cm^2 and 540°C) are suitable for modern designs of 660 MW turbo-generators. The fuel irradiation is about 18 000 MW days per tonne. The moderator in the core remains graphite and the coolant remains carbon dioxide gas, but it is circulated at much higher pressure to increase its heat-transfer capacity.

The gas operating pressure in the series of Magnox reactors progressed from about 1000 kN/m^2 to 2000 kN/m^2 and the earlier reactors were contained in steel pressure vessels up to 90 mm thick. As reactors of increasing size were designed and coolant pressures increased, the fabrication of still larger and thicker steel vessels became impracticable, and prestressed concrete pressure vessels were used in the later Magnox reactors and in all the AGRs. The AGR pressure vessels operate at a coolant pressure of 4000–6000 kN/m^2.

27.9.1 Reactor pressure vessel

The prestressed concrete pressure vessel shown in Figure 27.20 is a vertical cylinder with helical multi-layer post-tensioned cables in the walls, so arranged that no cables are required across the top and bottom slabs. The pattern of cables is shown in Figure 27.21. The bottom slab is designed for a working pressure of 4200 kN/m^2 a.b.s. which is the outlet pressure from the gas circulators. The top slab and walls, however, are designed for the lower pressure of 3900 kN/m^2 following the coolant pressure drop in the core. The vessel inner surfaces are insulated and cooled to maintain concrete temperatures generally below 70°C.

The optimum angle of inclination of the helix is that for which the radial and vertical components of prestress are in the same ratio as the respective gas forces at ultimate load conditions. The prestressing cables are taken up into an annular extension of the cylinder wall beyond the flat slabbed end to such a height that the radial component of load from the extended cables provides sufficient force to restrain the end slabs. This arrangement permits any number of penetrations in the top slab without reducing the load-carrying capacity of the slab.

The prestressing design has a high degree of redundancy and the cable tensions are checked periodically; the pressure vessel is therefore very safe against rupture. The cables are not grouted-in and they can be removed individually for inspection, and, if necessary, replacement.

27.9.2 Reactor foundations

The main load to be carried by the reactor building foundations is the pressure vessel containing the reactor core. Because of the

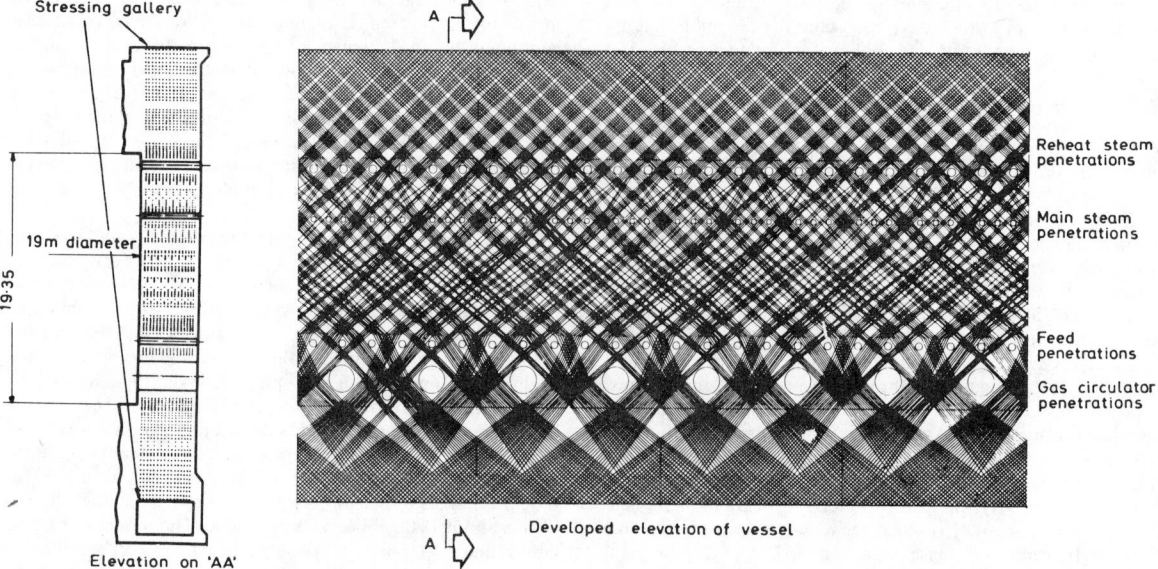

Figure 27.21 Prestressing system of a pressure vessel

load distribution, reactors in the UK have usually been supported by direct bearing on the subsoil at a depth of about 10 m, though in one instance where it was impracticable to spread the load sufficiently at a reasonable depth the reactor was carried on concrete shafts 2.3 m in diameter taken into rock about 40 m below ground level.

A contact foundation for an AGR with a prestressed concrete pressure vessel is shown in Figure 27.22. Here the total weight of the loaded vessel is transmitted to a rock stratum at a loading of about 1200 kN/m^2 through a mass concrete base 26 m in diameter. Surrounding the circular base is a reinforced concrete retaining wall, and the beams which support the considerably lighter external portions of the reactor building span the area between the pressure vessel and the retaining wall. The retaining wall is an independent structure and is only connected to the mass concrete base by a continuous rubber stop.

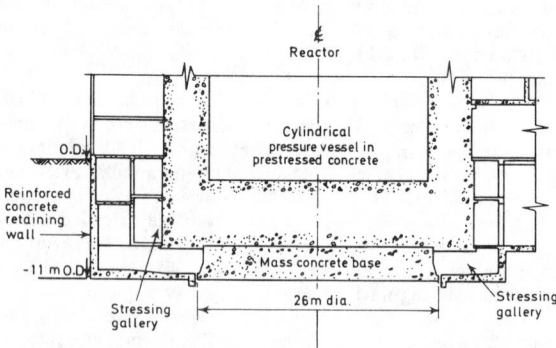

Figure 27.22 Reactor foundation. (*Courtesy:* Central Electricity Generating Board)

Stringent precautions are taken in the design and construction of reactor foundations to ensure that long-term overall settlement and differential settlement (both between buildings and that which results in tilt of the reactor) will be within acceptable limits.

27.9.3 Layout of reactor building

The Magnox and AGR nuclear stations each have two reactors, with one or two turbo-generators to each reactor. As with oil- or coal-burning stations, the layout has developed to an in-line arrangement of the units (reactor and turbo-generators), generally with transverse arrangement of the turbo-generators.

In addition to the pressure vessels and access facilities around them, the reactor building has to accommodate the following principal items.

(1) *Services annexe*, consisting of offices, stores, laboratories and changing rooms and ablutions for personnel.
(2) *Fuelling machine*, which traverses the charge face over the reactor core. Operated by remote control, it removes cans or stringers of irradiated fuel from the reactor core and inserts new fuel elements. Also located at the charge-face level are workshops for maintenance of the fuelling machine and other reactor equipment and the decontamination centre.
(3) *Shielded block*, which contains the new and irradiated fuel-handling equipment.
(4) *Active waste disposal*. After removal from an AGR core, the fuel element stringers are separated and certain components of the assembly are waste. This scrap has become radioactive whilst in the reactor core, and is disposed of by burial in deep vaults in the reactor building.
(5) *Cooling pond* in which irradiated fuel elements are stored for about 3 months after removal from the reactor core. This allows the radioactivity to decay to a level at which they can be transported away from the power station for processing. There is some 6 m depth of water in the cooling pond to shield operatives from radiation.
(6) *Cooling-pond water and active-effluent treatment plant*
(7) *Carbon dioxide treatment plant*
(8) *Ventilation plant*. Most of the reactor building is artificially ventilated, because the size of the reactor building leaves many working areas with insufficient natural ventilation and also because many areas must be kept under either forced or plenum ventilation conditions to prevent the possibility of spread of radioactive contamination.

Extracted air is also filtered before discharge to outside atmosphere. The ventilation occupies a substantial part of the reactor building.

(9) *Station control room*, which contains the instrumentation and control panels for the operation of the reactors, boilers, turbo-generators, and ancillary plant.
(10) *Electrical switchgear*
(11) *Emergency generators*. These are standby generators to safeguard against failure of the station supply and of the incoming supply from the public system.

The main difference between the layouts of various reactor buildings is in the location of the foregoing items, particularly of the services annexe. If the reactor spacing is kept to the minimum, the services annexe has to be located between the reactors and the turbo-generators, which results in long high-pressure pipework between boilers and turbines. By increasing the spacing of the reactors and locating the services annexe between them, the turbo-generators can be brought close to the reactors and the length of pipework reduced; this results in an increase in building volume and costs, but is more than offset by the saving on pipework.

27.10 Hydro-electric power and pumped storage

by D L McKie, CEng, MICE

27.10.1 Hydro-electric power

In hydro-electric power plants, turbines are powered by passing the greater part of the water flow of a river through the turbines. The plants fall into two categories according to the absence or use of storage:

(1) Run-of-river plants, which have insignificant reservoir capacity. At any particular time the power available for generation is limited according to the river flow at that time, and there is no reserve to meet high demand for electricity. These plants usually provide continuous generation (base load supply) and their output:capacity ratio (load factor) is high.
(2) Storage plants, which have a significant reservoir capacity sufficient to enable the water flow through the turbines to be regulated according to the demand for electricity. The load factor of these plants is usually low.

Thus, a run-of-river scheme will have a low dam (or weir) to raise the river level up to the intake works which give controlled conditions for flow of water to the pressure pipes or tunnels leading to the turbines. A storage scheme will have a high dam located in the best position, taking into account the volume impounded, elevation and construction cost. The intake works of a storage scheme are normally incorporated in the dam, or built as a tower either on the face of the dam, or free-standing in the reservoir.

For either run-of-river or storage schemes, the power station housing the turbines and generators is built at a lower level downstream of the dam, so that water flowing through the turbines is returned to the course of the river; the pressure head which operates the turbines is that due to the difference in level between the water surface at the intake and the tail-race level after leaving the turbines. The power station may be located at (or incorporated in) the base of the dam (Figure 27.23(a)) or it may be located at some distance from the dam and connected to

the intake by shafts and pressure tunnels (Figure 27.23(b)) or it may be located underground intermediately between the bottom of the shafts and the emergence of the tunnels (Figure 27.23(c)). In the latter case, the tunnels upstream of the turbines are pressure tunnels and the tunnels downstream of the turbines form the tail-race to the river.

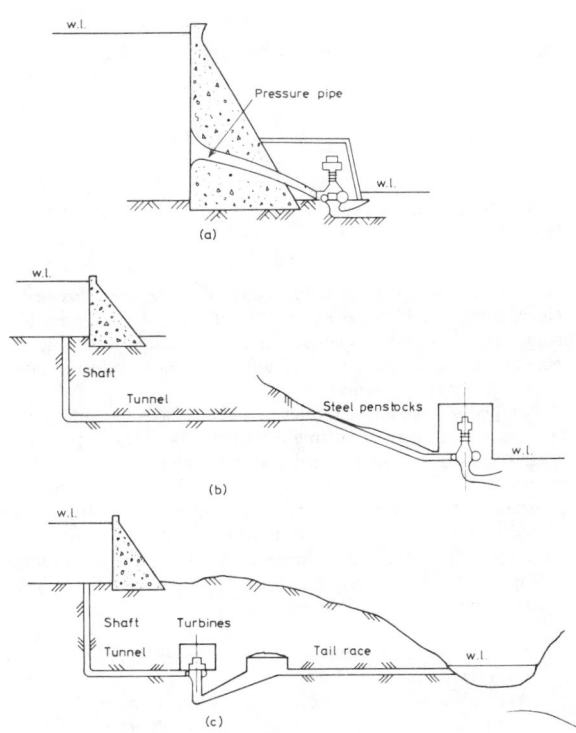

Figure 27.23 Locations for power station in hydro-electric scheme. (a) Low-to-medium head; (b) medium-to-high head; (c) high head

In the case of a scheme with a long pressure tunnel leading to the turbines, fluctuations of demand for electricity cause fluctuations in the speed of the turbines and of the flow of water through them, with consequent acceleration or deceleration of the mass of water in the tunnel and with pressure variations. A surge tank is constructed over and near the end of the tunnel to protect it from surges of pressure, and to facilitate the response of the plant to variations of demand.

According to the hydraulic operating head, different types of turbine are used, approximately as follows:

High head (1500–300 m)	Pelton wheels (horizontal or vertical shafts)
Medium/high head (550–30 m)	Francis turbines (horizontal or vertical shafts)
Low head (50–5 m)	Kaplan turbines (vertical shafts only)

27.10.2 Pumped storage

Demand for electricity varies throughout each 24 h, being greatest at mid morning and mid afternoon and least at night.

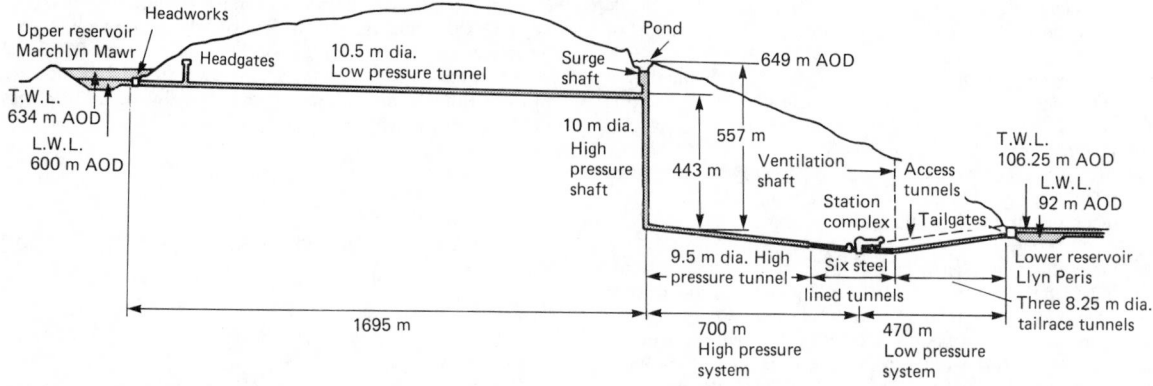

Figure 27.24 Pumped storage scheme, Dinorwic, Gwynedd, North Wales

Normally, generating capacity in excess of demand has to be closed down but, if electricity could be stored in large quantities, some of the more efficient steam plant could continue to generate at night and the stored output could be released later during the peaks of demand.

Pumped storage is an adjunct to highly efficient steam plants to achieve the effect of storing electricity. It operates by using their surplus generating capacity at the time of low demand to pump water from a low-level reservoir up to a high-level reservoir, and then at times of high demand using the water stored in the high-level reservoir to drive hydraulic turbines and generate electricity. The scheme is built as a conventional storage-type of hydro-electric installation, except for the following differences:

(1) Water after discharge from the turbines is stored in a low-level reservoir.
(2) The turbines must be reversible to act as pumps, or the pumps and turbines may be separate machines.
(3) The electrical alternators operate as motors to drive the pumps, or as generators when being driven by the turbines.

The Dinorwig pumped storage scheme[21] in North Wales is shown in Figure 27.24. The upper reservoir, Marchlyn Mawr, and the lower reservoir, Llyn Peris, each have a 'live' storage capacity of 7×10^6 m³. Both reservoirs were formed by enlarging existing lakes. Marchlyn dam was constructed with 1.85×10^6 m³ of zone graded and compacted slate rock fill which was surfaced with asphalt on the upstream face. The lower lake was restored to approximately its original size by excavating 4×10^6 m³ of waste which had been tipped into it during more than 150 yr of slate quarrying.

A maximum head of 542 m is available to drive the six 300 MW machines. Generation via all six vertical shaft Francis pump-turbines is possible for 5 h, whilst pumping the full 'live' storage capacity from the lower to the upper reservoir takes just over 6 h. A feature of the Dinorwig scheme is that an output from no load to 1320 MV can be achieved in 10 s.

The generating plant is housed in a series of caverns excavated in slate and is located at a level which provides a minimum submergence of 60 m to the pump-turbines, the depth which is necessary to prevent cavitation. The machine hall is the largest cavern and is 180 m long, 24 m wide and has a maximum depth of 60 m. Two tunnels give access from a road near the tailworks structure to the cavern complex.

Virtually all underground excavations were carried out by drill and blast methods. After excavation the walls and roofs of tunnels and caverns were given immediate support by applying

sprayed concrete with a minimum thickness of 25 mm. Rock bolts 25 mm in diameter and up to 4 m long were installed in tunnels and caverns and, for caverns only, further support for the walls and roofs was provided by tensioned double-corrosion-protected monobar anchors 36 mm in diameter and 12 m long.

From a manifold at the lowest point of the high-pressure tunnel a separate steel lined conduit 145 m long leads into each pump-turbine. These steel-lined lengths then continue for 110 m before being paired to form the three concrete tailrace tunnels. Apart from these six 255 m lengths all hydraulic tunnels have a smooth concrete lining reinforced only at bends and junctions. Pulverized fuel ash formed 25% of the cementitious content of the concrete to give added durability. The steel linings were pressure-grouted throughout but only limited grouting was carried out on the concrete-lined tunnels.

27.11 Overhead transmission lines and supports

by N G Eggleton, BSc, FICE

Electricity is distributed at high voltage in transmission lines from power stations through a sequence of substations, in which the voltage is transformed down to the consumer's voltage. Heat is generated in a conductor by the flow of current and the conductor size must be such that the temperature does not rise above the annealing point of the conductor material (approximately 75°C for hard-drawn copper and aluminium). For a given conductor, the heat generated (which is also the power lost in transmission) is proportional to the voltage, but the power transmitted is proportional to the square of the voltage, which makes transmission at high voltages desirable. In the UK, alternating current (three-phase, 50 Hz) is transmitted at 11, 33, 66, 275, and 400 kV, and higher voltages are under consideration. Conductors may be insulated cables laid underground or may be bare conductors carried at a safe height above ground between towers. The high degree of insulation and cooling, at high voltages, necessary on underground cables makes them expensive, and overhead conductors have a considerable economic advantage.

The range of overhead conductor sizes normally used is as shown in Table 27.4.

Table 27.4

Voltage limit	Conductor size (equivalent copper area mm²)
Up to 33	16–161
66	23–197
132	81–258
275	Twin 113–Twin 258
400	Twin 258

27.11.1 Design standards

The safety regulations covering the design of high-voltage lines in the UK may be summarized (using approximate SI equivalents) as:

(1) Conductors – minimum factor of safety = 2 (on breaking load) when at $-5°C$ they have a 10 mm radial thickness of ice and are subjected to an 80 km/h wind on the full projected area of the ice-coated conductor (equivalent to 384 N/m^2).
(2) Supports are to withstand the longitudinal, transverse and vertical forces imposed by the conductors under the above conditions of loading without damage and without movement in the ground. Wind pressure on supports = 384 N/m^2 on projected area, and with compound structures such as steel towers the pressure on the lee-side members may be taken as one half that on the windward side. Minimum factors of safety under these maximum working loads, calculated on the crippling load of struts and the elastic limit of tension members, are:

Iron or steel	2.5
Wood	3.5
Concrete	3.5

(3) Minimum height of conductors:

Maximum a.c. voltage (kV)	66	110	165	Exceeding 165
Ground clearance (m) at 50°C	6.1	6.4	6.7	7.0

27.11.2 Conductors

The three main conductor types are hard-drawn copper (BS 125), stranded aluminium (BS 215: Part 1) and steel-reinforced aluminium (BS 215: Part 2). Steel-reinforced aluminium is used for the majority of extra high-voltage transmission lines, because the high-strength conductors permit long spans between supports.

Conductors consist of three or more individual wires stranded together and are categorized according to the details of stranding (see Figure 27.25).

The mechanical characteristics of conductor materials are as shown in Table 27.5.

27.11.3 Conductor sags and tensions

Variations in conductor sags and tensions result from changes in temperature and loading, and conductors must be strung so that sags do not exceed that which ground clearance permits, nor tensions exceed the required factors of safety.

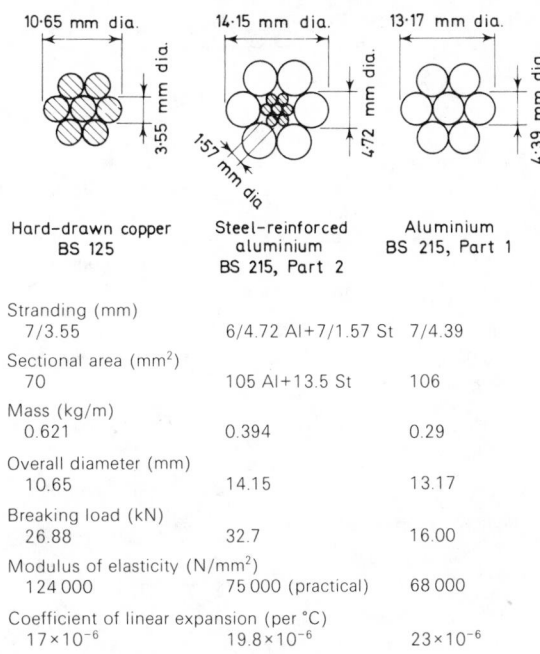

	Hard-drawn copper BS 125	Steel-reinforced aluminium BS 215, Part 2	Aluminium BS 215, Part 1
Stranding (mm)	7/3.55	6/4.72 Al + 7/1.57 St	7/4.39
Sectional area (mm²)	70	105 Al + 13.5 St	106
Mass (kg/m)	0.621	0.394	0.29
Overall diameter (mm)	10.65	14.15	13.17
Breaking load (kN)	26.88	32.7	16.00
Modulus of elasticity (N/mm²)	124 000	75 000 (practical)	68 000
Coefficient of linear expansion (per °C)	17×10^{-6}	19.8×10^{-6}	23×10^{-6}

Figure 27.25 Comparison of three conductors of 70 mm² equivalent copper area

Table 27.5

	Weight (kg/mm² m)	Tensile strength (N/mm²)	Coefficient of linear expansion (per °C)	Modulus of elasticity (N/mm²)
Copper (hard drawn)	0.008 9	415	17×10^{-6}	124 000
Aluminium (hard drawn)	0.002 7	160	23×10^{-6}	68 000
Steel	0.007 9	1340	11.5×10^{-6}	200 000

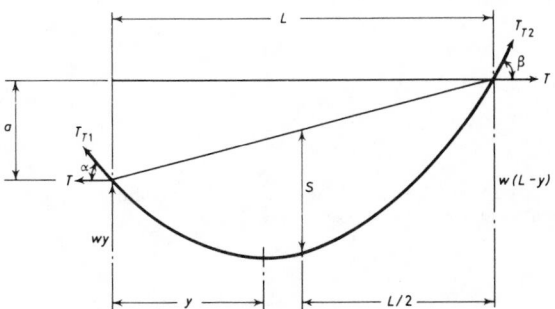

Figure 27.26 General case of the suspended conductor

The general case of a conductor suspended between supports at different levels is shown in Figure 27.26. Calculations within the practical requirements of accuracy, can, for simplicity, be made on the basis of a parabolic curve, although a catenary curve would be more accurate.

Symbols and definitions for Figure 27.26

General	Conductor data

$L=$ span (horizontal distance between supports)

$A=$ actual cross-sectional area of conductors

$S=$ sag (distance measured in the direction of the resultant load between the conductor and the midpoint of a straight line joining the two supports)

$d=$ overall diameter of conductor

$E=$ modulus of elasticity of complete conductor. Virtual modulus for composite conductor of materials a and b

$T=$ tension (horizontal component of the tension under load w, being uniform throughout any one span)

$=\dfrac{E_a m + E_b}{m+1}$ where

$m=\dfrac{\text{area of a}}{\text{area of b}}$

$T_T=$ tangential tension (actual tension at any given point in a conductor in the direction of the tangent to the curve)

$c=$ coefficient of linear expansion of complete conductor. Virtual coefficient for a composite conductor of materials a and b where E and m are as above

$w=$ resultant load per metre of conductor from vertical load of self-weight + ice (if present) and horizontal wind load

$=\dfrac{c_a E_a m + c_b E_b}{m E_a + E_b}$

$C_1=$ a constant $=\sqrt{EA/24}$

$C_2=$ a constant $=cEA$

$a=$ difference in level of adjacent supports

$y=$ horizontal distance of lowest point of curve from lower support

$t=$ temperature rise from initial to final conditions

Formulae for Figure 27.26

$$S=\frac{wL^2}{8T} \tag{27.1}$$

$$\text{Length of conductor} = L + \frac{8S^2}{3L} = L + \frac{w^2 L^3}{24 T^2} \tag{27.2}$$

$$\left(\frac{C_1 w_2 L}{T_2}\right)^2 - T_2 = \left(\frac{C_1 w_1 L}{T_1}\right)^2 - T_1 + C_2 t \tag{27.3}$$

where the suffixes to $w_1 T_1$ and $w_2 T_2$ denote conditions at initial and final temperature respectively

$$y = \frac{L}{2} - \frac{aT}{wL} \tag{27.4}$$

Vertical reaction at higher support $= w(L-y)$

Vertical reaction at lower support $= wy$

Example 27.1 To determine the maximum sag at 50°C of a steel-reinforced aluminium conductor of 100 mm² nominal aluminium area (6/4.72 mm aluminium + 7/1.57 mm steel) on a span of 200 m. Factor of safety = 2.0 on breaking load with 10 mm radial thickness of ice at −5°C and 80 km/h wind. The sequence of calculations is as follows:

(1) Evaluate the conductor data (see Figure 27.25)

$$A = 105 \text{ mm}^2 \text{ Al} + 13.5 \text{ mm}^2 \text{ Steel} = 118.5 \text{ mm}^2$$

$$d = 14.15 \text{ mm}; \quad \text{breaking load} = 32.7 \text{ kN}$$

$$C_1 = \left(\frac{EA}{24}\right)^{1/2} = \left(\frac{75\,000 \times 118.5}{24}\right)^{1/2} = 609$$

$$C_2 = cEA = 19.8 \times 10^{-6} \times 75\,000 \times 118.5 = 176$$

$$t = 50°C - (-5°C) = 55°C$$

(2) Calculate the initial tension T_1 in the cable at −5°C, the resultant load in the conductor w_1 at −5°C and w_2 at 50°C

$T_1 = $ initial tension at −5°C

$$= \frac{\text{breaking load}}{\text{factor of safety}} = \frac{32.7 \text{ kN}}{2} = 16\,350 \text{ N}$$

$$\text{Horizontal wind load} = 384 \left(\frac{20 + 14.15}{1000}\right)$$

$$= 13.2 \text{ N/m}$$

Weight of ice (at 0.91 g/cm³) on conductor at −5°C

$$= \frac{0.91}{1000} \times \frac{\bar{\Lambda}}{4} \left[\left(\frac{34.15}{10}\right)^2 - \left(\frac{34.15}{10}\right)^2\right] \times 100$$

$$= 0.686 \text{ kg/m}$$

Total weight = ice + self weight

$$= 0.686 + 0.394$$

$$= 1.08 \text{ kg/m}$$

Hence:

$$w_1 = [13.2^2 + (9.9 \times 1.08)^2]^{1/2}$$

$$= 17 \text{ N/m}$$

$w_2 = $ load due to self weight

$$= 9.9 \times 0.394 = 3.9 \text{ N/m}$$

(3) From Equation (27.3) find horizontal tension T_2 at 50°C

$$\left(\frac{C_1 w_2 L}{T_2}\right)^2 - T_2 = \left(\frac{C_1 w_1 L}{T_1}\right)^2 - T_1 + C_2 t$$

$$\left(\frac{609 \times 3.9 \times 200}{T_2}\right)^2 - T_2 = \left(\frac{609 \times 17 \times 200}{16\,350}\right)^2 - 16\,350 + 176 \times$$

$$\left(\frac{475\,000}{T_2}\right)^2 - T_2 = 9430$$

T_2 is solved by trial and error using a slide rule. In this case $T_2 = 4100$.

(4) From Equation (27.1):

$$\text{Sag } S \text{ at } 50°C = \frac{w_2 L^2}{8T_2}$$

$$= \frac{3.9 \times 200^2}{8 \times 4100}$$

$$= 4.75 \text{ m}$$

27.11.4 Supports

The configuration of supports (but not necessarily the form and material used) depends initially on the electrical requirements of number of circuits, conductor size and type, insulation and clearances, and on the arrangement of conductors and earth wires. The structural design of supports is related to the imposed loads: (1) horizontal transverse; (2) horizontal longitudinal; (3) vertical; and (4) wind and ice loads.

27.11.4.1 Horizontal transverse loads (P)

(1) *Wind on bare or ice-coated conductors.* Calculated on the support 'wind span' = half the sum of the adjacent span lengths (see Figure 27.27). (P_w)
(2) *Wind on supports.* For square-lattice structures, wind on the leeward face is taken as half that on the windward face; this shielding factor decreases with rectangular shapes until the full wind is taken on both faces. On cylindrical members, wind pressure is taken on 0.6 of projected area. (P_s)
(3) *Conductor tension at line deviations.*

$$\text{Transverse load} = 2T \sin \frac{\theta}{2}$$

where θ is the angle of deviation and T is the maximum conductor tension (see Figure 27.27) (P_a)

27.11.4.2 Horizontal longitudinal loads (T)

(1) Full conductor tension at line terminals. (T)
(2) Out-of-balance conductor tensions due to broken conductors or earthwires. At supports with suspension insulators, a reduced conductor tension, usually 70%, is allowed for the swing of insulators into the unbroken span. $(0.7 T \text{ or } T)$
(3) Out-of-balance conductor tension at angle or section positions. Only encountered in special cases, e.g. change from single to double earthwires. (T_x)

27.11.4.3 Vertical loads (V)

(1) Weight of bare or ice-coated conductors calculated on basis of support 'weight span' (from Equation (27.4)). (V_w)
(2) Weight of insulators, etc. (V_i)
(3) Support weight. (V_s)

27.11.4.4 Wind and ice loads

The relationship between wind velocity (V km/h) and pressure (P N/m²) may be taken as:

Flat surfaces $P = 0.1V^2$
Round surfaces (e.g. conductors) $P = 0.06V^2$

Figure 27.27 indicates the locations relative to the conductor in which the main types of support are used, i.e. intermediate, angle and terminal.

The structural form and materials of supports may be classified as:

Single or composite wooden poles
Single or composite reinforced concrete and prestressed concrete poles
Steel tubular poles
Narrow-base towers – lattice structures of rolled steel and tubular sections, with single block foundations, increasingly with the use of guy wires
Broad-base towers – steel lattice structures, with a separate foundation for each leg (Figures 27.28 and 27.33).

27.11.5 Design of broad-base towers

The following description applies chiefly to the design sequence for broad-base towers, but the principle applies equally to other types.

With the electrical requirements resolved, the first step in design of the supports is to decide upon the 'standard span', i.e. the most economic span assuming level ground. Exploratory design is concentrated on the intermediate supports (being the majority) and the following interdependent factors are taken

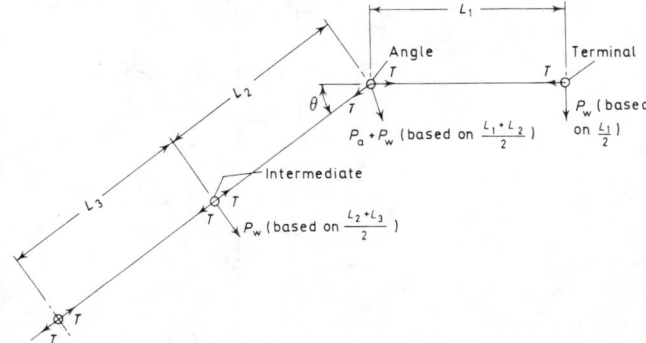

Figure 27.27 Horizontal loading relative to support positions

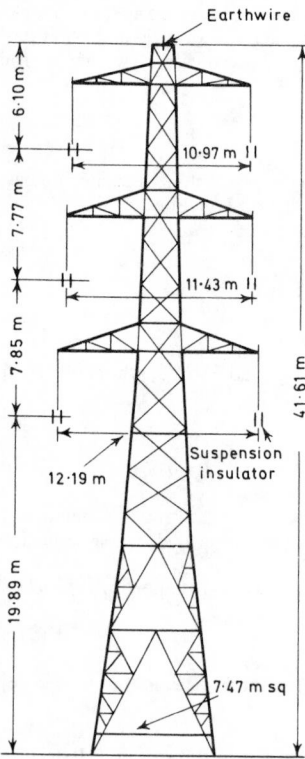

Figure 27.28 Broad-base tower

adequate midspan clearance dependent upon span length, sag and voltage, as well as factors such as ice shedding overcome by offsetting the conductors (Figure 27.28).

(b) Minimum live-metal-to-earth clearance, taking into account the maximum swing of suspension insulators related to horizontal (P_w and T) and vertical (V_w and V_i) conductor loading. Figure 27.29 shows a wire clearance diagram; the live-conductor-to-earthed-support clearances are decided according to the transmission voltage.

(c) Earthwire spacing. Protection against lightning is obtained by shielding the conductors with an overhead earthwire suitably earthed at the structures to intercept and earth-direct lightning strokes. The shielding angle ψ is preferred to be not greater than 30°. The earthwire sag should not exceed that of the conductor and the relative spacing is determined by the shielding angle (Figure 27.29).

When the standard span has been decided, the most economic tower to meet the prescribed conditions can be designed. For the intermediate tower, the basis of loading may be:

Wind span = greatest wind load = wind load on standard span + 10%
Weight span = greatest wind load = weight load from standard span + 100%
Maximum length of span = length of standard span + 40%

Final design is undertaken graphically by means of stress diagrams, usually on the basis of working loads. The factors of safety (e.g. in the UK 2.5 under normal, and 1.5 under broken-wire conditions) are applied when the individual member loads

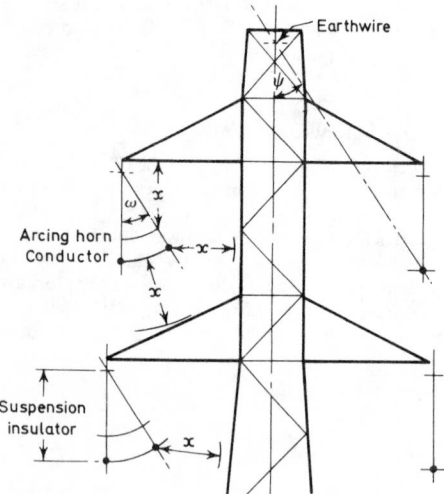

Figure 27.29 Wire clearance diagram

into consideration to determine the general outline and the arrangement and height of the cross arms:

(1) Height to bottom conductor, which is the minimum specified ground clearance, plus the maximum sag of the conductor.

(2) Conductor spacing:

   (a) Minimum horizontal and vertical spacing to provide

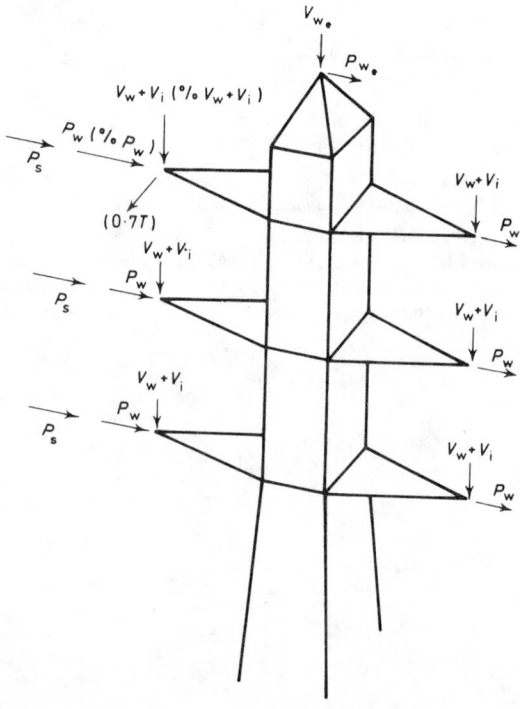

Figure 27.30 Loading diagram for a double-circuit intermediate tower

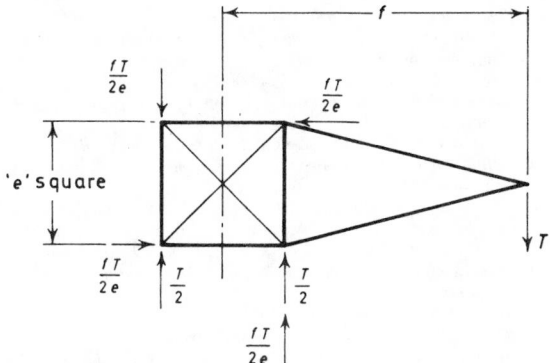

Figure 27.31 Torsion loading

are tabulated. Vertical loads are omitted from the stress diagrams, but at the tabulations are shared equally over the four main legs.

A typical loading diagram shown in Figure 27.30 includes a condition of any one conductor or the earthwire broken (shown in brackets for a top conductor but it must be considered individually at each conductor and earthwire position). Note the proportionate reductions of P_w and V_w for the broken-wire condition and the equal distribution at cross-arm level of the wind load P_s on the tower.

Stability under torsion loads due to broken wires depends on the tower being adequately braced in plan, essentially at cross-arm level but possibly at other levels dependent on particular design features. Figure 27.31 shows the reactions for a tower of square cross-section of the type shown in Figure 27.30.

27.11.6 Foundations

The forces to be resisted by overhead-line support foundations result largely from overturning moments, with a consequent emphasis on horizontal and uprooting forces. The types of foundation can be broadly classified as:

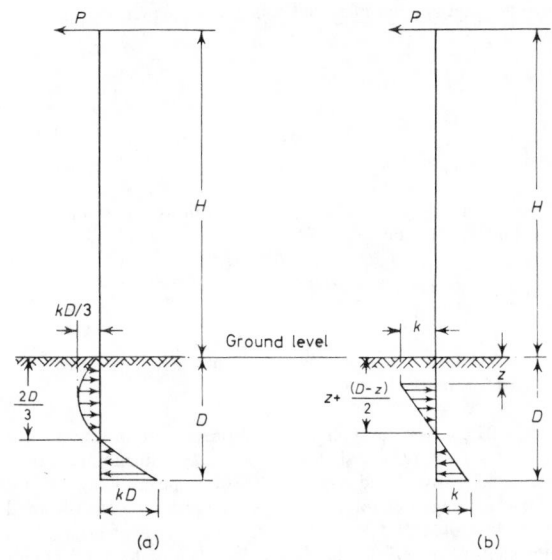

Figure 27.32 Side-bearing foundations

(1) Side bearing – resistance depends on horizontal soil reactions, i.e. single foundations used for unstayed poles and narrow-base towers.
(2) Uplift and compression – resistance depends on vertical soil reactions, i.e. in the case of broad-base towers where each of the four legs has a separate foundation, and in the case of stayed poles.

Figure 27.32 shows the pressure distribution assumed (neglecting the small values of direct horizontal shear) in two formulae used for pole and shallow concrete block side-bearing foundation. In Figure 27.32(a) (parabolic distribution) the pressure developed is based on the horizontal movement relative to the

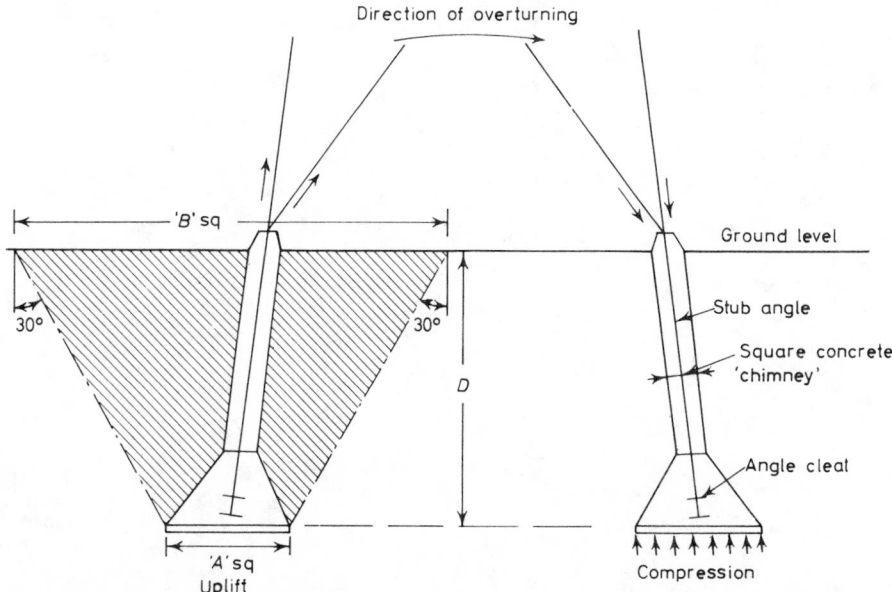

Figure 27.33 Uplift and compression foundations

pivotal point and assumes that soil resistance increases proportionately with depth:

$$P\left(H+\frac{2D}{3}\right)=\frac{kbD^3}{12}$$

where k is a constant and b is the breadth of the foundation.

Figure 27.32(b) involves similar assumptions but is based on constant soil resistance:

$$P\left(H+z+\frac{D-z}{2}\right)=\frac{kb(D-z)^2}{6}$$

where z is the amount of topsoil to be neglected, assumed to be at least 300 mm.

Figure 27.33 shows an uplift and compression foundation, each footing consisting of a shallow concrete pad surmounted by a truncated pyramid and chimney enclosing the stub angle. With an overturning moment, one pair of foundation blocks will tend to be uprooted and the other pair forced downwards. With intermediate towers the loading is largely due to wind and is reversible so that all four footings are identical. Ultimate uplift resistance is calculated on an assumed frustrum of earth above the foundation block.

References

1 Judson, J. C. and Morris, C. J. E. (1974) 'Drax power station.' *Proc. Instn Civ. Engrs*, **56**, 559–576.
2 Fitzherbert, W. A. and Barnett, J. H. (1967) 'Causes of movement in reinforced concrete turbo-blocks and developments in turbo-block design and construction.' *Proc. Instn Civ. Engrs*, **36**, 351–393.
3 *Supporting structures for rotary machines (especially pier foundations for steam turbines)*. German Standard DIN 4024. (English translation available.)
4 Lees, A. W. and Simpson, I. C. (1983) *The dynamics of turbo-alternator foundations*. Institution of Mechanical Engineers Conference.
5 Davies, W. G. R. and Pandley, P. C. (1983) *Economical optimisation of the alignment of turbine generators*. Institution of Mechanical Engineers Conference.
6 Chapman, E. K. J., Gibb, F. R. and Pugh, C. E. (1969) 'Cooling-water intakes at Wylfa power station.' *Proc. Instn Civ. Engrs*, **42**, 193–216.
7 Richtlinien, V. G. B. (1979) *Bautechnik bei Kühltürmen*. Kraftwerkstechnik, GmbH.
8 British Standards Institution (1975) *Structural design of cooling towers*. BS 4485, Part 4, BSI, Milton Keynes.
9 Armitt, J. (1980) 'Wind loading on cooling towers.' *J. Struct. Div. Am. Soc. Civ. Engrs*, **106**, ST3.
10 Pinfold, G. M. (1985) *Reinforced concrete chimneys and towers*. 2nd edn. Viewpoint Publications, (Scholium International), New York.
11 Comité International des Cheminées Industrielles (1984) *Model code for the design of chimneys*.
12 National Building Code of Canada (1980) *Supplement*. Commentary B.

Bibliography

Baker, L. H. (1970) 'Cockenzie and Longannet power stations: novel features in the design and construction.' *Proc. Instn Civ. Engrs*, **48**, 427–458.
Central Electricity Generating Board (1987) *Annual report and accounts*. HMSO, London.
Central Electricity Generating Board (1971) *Modern power station practice*. 8 vols. Pergamon, Oxford.
Rae, F. A. (1962) 'Design and construction of a reinforced concrete foundation block for a 200 MW turbo-generator.' *Proc. Instn Civ. Engrs*, **22**, 1962–2041.

28

Water Supplies

B H Rofe MA(Cantab), CEng, FICE, FIWEM, FGS
Rofe, Kennard and Lapworth,
Consulting Engineers

Contents

28.1 Organization and management

The organization of water supplies in Britain, as in other parts of the world, has evolved from initiatives by private companies and public corporations. This situation still exists in many parts of the world, but in Britain a framework of organization and management[1] has been created by successive Acts of Parliament starting with the Waterworks Clauses Act, 1874 through to the Water Act 1973 (England and Wales) and the Local Government (Scotland) Act 1973. Further modifications were introduced in the Water Bill 1983.

In England and Wales, the 1973 Act created a framework of river basin management, with ten regional authorities responsible for the complete hydrological cycle including conservation, water resources, treatment and distribution, land drainage, sewerage and sewage treatment. However, in Scotland water supply and sewage are the responsibility of the district council, except in the case of the Central Scotland Water Development Board who act as a bulk supply authority. River purification is the responsibility of separate boards except in the areas of the island district councils.

The Scottish pattern of management and organization of water supply, together with agent water companies (as in England and Wales) has been adopted in many English-speaking countries in local forms to comply with the form of government, but there is a movement towards the more logical format of total river basin management.

28.2 Present consumption and estimated demand

Demand for water varies according to the type of supply area but may generally be considered under the three headings of domestic, industrial and agricultural. Within the year, the monthly and weekly totals will vary considerably due to seasonal effect (wet or dry periods) and socio-economic effects (e.g. holiday periods, festivals, growth seasons). There will also be daily variations within each week and peak hours during each day. A peak hourly rate of around 3 times the average rate needs to be considered in the design of distribution systems, and in respect of local storage requirements.

Domestic consumption assessments in Britain during the period 1976–78 are summarized in Table 28.1, together with typical figures arising from studies in other countries. In Britain domestic supplies are not generally metered[2] but in most other countries domestic meters are a common feature. It is doubtful if metering offers any saving in cost, but with strict supervision and control it can assist in controlling demand using an increasing scale of charges at higher rates of consumption.

Table 28.1 Domestic consumption per head (1976–1978)

Area or country	Consumption (l/head/day) average	range
England & Wales	200	140–330
Scotland	275	240–350
London (urban)	260	—
Thames (rural)	210	—
Middle East (arid)	—	(standpipes) 50–450
Far East (tropical)		120–400

28.2.1 Domestic consumption (details)

A typical breakdown of the present consumption and an estimate of future demand was made for a group of six selected

Water Undertakings in south-east England by Sharp in 1967.[3,2] This is shown in Table 28.2.

Table 28.2 Breakdown of domestic consumption (1967)

Component	Estimated 1967 average consumption		Forecast of possible average consumption in 2000	
	gallons/ head/ day	litres/ head/ day	gallons/ head/ day	litres/ head/ day
Drinking and cooking	1	4.5	1	4.5
Dishwashing and cleaning	3	13.5	4	18
Laundry	3	13.5	5	22.5
Personal washing and bathing	10	45.5	13	59
Closet flushing and garbage disposal	11	50	14	63.5
Car washing	—	—	1	4.5
Garden use and recreation	1	4.5	6	27.5
Waste in distribution	5	22.5	8	36.5
Total	34	154	52	236

28.2.2 Industrial consumption

No generalization can be made as the industrial consumption in each town varies considerably according to the nature of the industry both in quantity and quality requirements. The water is generally required during the working day and this factor must be taken into account in the design of pumps, pipes and reservoirs as it affects the peak rates of flow. As general guidance the following examples are typical.

(1) For brewing, the quantity of water is substantially the amount brewed, but for beer the water is preferably hard; for stout, soft; cider must be made from pure soft water without iron.

(2) Canning is best done with hard water (except for peas), and iron must be less than 0.5 mg/l: anything between 20 and 40 l/kg canned.

(3) The dyeing industry requires soft, iron-free water, and about 100 l/kg, mercerizing textiles takes 250 l/kg.

(4) Industries such as distilling, ice-making and mineral-water-making require large amounts of water, plus that for power purposes in steam-raising.

(5) Leather requires 80 l/kg of raw hide tanned, water rich in sulphates being preferred. Rubber requires 70 l/kg processed.

(6) Paper or cardboard manufacture requires anything between 60 and 360 l/kg.

(7) A ton of soap requires about 2200 l of water in its manufacture.

(8) In the UK, sugar beet takes about 5 l/kg used in washing the beet, dissolving the sugar, transporting the material in the factory and in steam-raising.

(9) In the heavier industries the following quantities may be taken as approximate, e.g. railways take about 0.22 l/1000 kg of goods carried per kilometre. Cement takes 3 l/kg. Coke might consume 13 to 18 l of crude water per kilogram for cooling. Electricity works take 67 l of crude

water per kilowatt generated, for make up or loss in cooling towers, and 1.5 l of fresh water per kilowatt for boilers. Steelworks would consume some 9 l of mostly crude water per kilogram of steel manufactured.

(10) The cost of industrial water or metered supplies varies considerably but is generally in the range 2p to 10p per 1000.

28.2.3 Agricultural requirements

In addition to the human population, allowance must be made in a dairy farming district for the cow population. A cow requires as much as 135 to 180 l/day and there may be special requirements such as for bottling – 100 l milk means 200 l of water; manufacturing 455 kg dried milk needs 550 l, 455 kg alum needs 4500 l, 455 kg cheese needs 750 l. Where intensive fruit farming is practised a complete network of pipes is required throughout an orchard for treatment and irrigation. Similarly, considerable quantities of water are required to maintain bowling greens, golfcourses and racecourses, and overhead irrigation of crops by rotary sprinklers is on the increase. In the Thames Valley a total of 90 Ml/day has been estimated as the requirement for irrigation by the year 2000. Very high consumptions of the order of 50 000 to 100 000 l/ha per day would be required by a market gardener, and for tomatoes under glass approximately 200 000 l/ha per day. Again, the watercress industry[4] at certain times of the year consumes very large quantities of water and may require between 2.5 and 5.7 million l/ha per day. Paradoxical as it may seem, more water may be required in winter to keep watercress from freezing than in summer to keep it from scorching.[5]

28.2.4 Fire protection

Generally, hydrants are spaced not more than 130 m apart. Important buildings may require additional protection, i.e. more than two hydrants within 90 m. For less important buildings one hydrant within 140 m may suffice. Hydrants should be 6 m or more away from buildings, are best placed at crossings or corners, and are usually fixed on short 80 mm branches from the main which should not be less than 100 mm. Fire mains should deliver 550 l/min at each hydrant expected to be in use at the same time (generally two). As pressure in a main to command the highest buildings is generally impracticable, fire engines are used to deliver 1200 to 2000 l/min to a height of 50 m through a 24-mm dia. nozzle; the larger fire engines deliver up to 4500 l/min. A residual pressure of 3 m at the ground is desirable to avoid the engine creating a vacuum in the main on the suction side.

In towns the calculation for the distribution of water is based on very general assumptions of the amount of water required at any given moment, and it may not be practicable to design adequately for fire protection if the mains are assumed to be taking the maximum hour's domestic and industrial requirements as well. A good practical arrangement of valves and hydrants based on experience and checked occasionally by simple network analysis is of more value than any very exact calculations. Nowadays the fire authorities work closely with the water authorities to determine the positions of fire hydrants.

28.2.5 Waste

Some consumption of water by waste is inevitable and few statutory undertakers can seriously claim a figure of less than 10%, whilst in some areas where pressures are higher or the mains and services are old or in poor condition, or where efficient waste prevention methods are not applied, the wastage may amount to as much as 50% or more. Waste may be due to a number of factors including: (1) leakage from reservoirs, mains and other works of an undertaking, and from consumers' pipes and fittings through apertures, fractures, defective joints; (2) faulty washers and valve seatings; (3) bad design, failure to turn off taps; and (4) in all cases leakage and waste are intensified by unduly high pressures. Waste can be detected by detailed examination of the distribution system or house-to-house inspection, apart from a detailed check on the main reservoirs and aqueducts, etc.

28.2.5.1 Examination of the system

It is best to examine the water system section by section between midnight and 5.00 a.m. and check the night flow by a meter capable of reading small flows and recording them on a chart. A specific test on a 12 mm lead pipe under 3.2 kgf/cm^2 pressure gave a loss of 46 000 l/day for a 0.6 cm hole, 17 000 l/day for a 0.3 cm hole and 1600 l/day for a 0.15 cm hole. Tests on newly laid mains often call for a loss not exceeding 1 l/day per centimetre of diameter per kilometre of length. House-to-house inspections are probably in most cases the most effective way of checking waste. A dripping tap wastes up to 500 l/day and one running full as much as 10 000 l/day. The provision by the water authority of facilities for the rewashering, renewal and adjustment of taps, and repairs to service pipes at the lowest possible cost, undoubtedly encourages consumers to report leakages promptly and is an overall economy. In recent years several more sophisticated systems[6] have been developed to detect leaks in mains, and to identify the precise points of leakage, thus saving a lot of abortive exploratory excavation.

28.3 Transmission and distribution of water

Water may be transmitted under gravity along open or covered channels, through tunnels or through pipes.[7] Open channels are often used for catchwaters, waste-water channels or for river intakes to pumping stations in pumped storage schemes. Nowadays they are not generally used for the transmission of treated water due to the danger of pollution. In some cases canals are adapted as aqueducts for the transmission of water. Some large aqueducts have been constructed with sections of covered channel constructed by 'cut and cover' methods and modern practice is to construct these of plain or reinforced concrete. Where an aqueduct is required to pass through ground appreciably higher than the hydraulic gradient, tunnelling is necessary. The general principles of tunnelling are described elsewhere but for waterworks purposes the tunnels are usually lined, even in rock, partly to ensure that a fall does not block the waterway but also to reduce the friction. The use of pressure tunnels through the centre of a congested city with modern tunnelling methods is now becoming an economically satisfactory alternative to large trunk mains laid near the surface, provided the strata below the city are satisfactory. London is fortunate in this respect and several trunk aqueducts have been constructed in the London Clay.

28.3.1 Pipes

The major part of water transmission is through pipes and there has been a considerable increase in the numbers of new types of pipes and joints of all sizes in the last few years, including spun and cast iron, ductile iron, steel, concrete, asbestos cement, and their range of joints. The ducts also include unplasticized PVC and polythene pipes with their corresponding joints. Several technical factors affect the final choice of pipe material, including internal pressures, hydraulic and operating conditions, maximum permissible diameters, external and internal corrosion, and any special conditions of laying.

Joints may be classified into three categories, depending upon their capacity for movement, namely rigid, semirigid and flexible. Rigid joints are those which admit no movement at all and comprise flanged, welded and the now obsolete turned-and-bored joints. The semi-rigid joint is represented by the spigot-and-socket caulked lead joint which has given service for well over a century but is now largely obsolescent. Flexible joints are used where rigidity is undesirable and comprise mainly mechanical and rubber ring joints which permit some degree of deflection at each joint. Amongst the joints included in this category are the Tyton joint for cast iron and ductile iron pipes, the Johnson coupling and Fastite joint for steel pipes and the lock joint for prestressed concrete pipes and the detachable and Widnes joints for asbestos cement pipes. Victaulic joints are frequently used where longitudinal tension is required.

28.3.1.1 Cast-iron pipes (grey iron and ductile iron)

The use of vertical cast-iron pipes is now limited to the flanged pipes employed in connection to reservoirs, pumps and treatment plant, the bulk of the iron pipes in waterworks service being spun iron, centrifugally cast in metal or sand moulds. Such pipes may be of grey iron, the latter having the advantage of higher tensile strength and reduced tendency to fracture, but they are also thinner in section.

Grey iron pipes and fittings of sizes 80 to 700 mm are covered by British Standard (BS) 4622 and ductile iron pipes and fittings of sizes 800 to 1200 mm BS 4772. The former classes B, C and D have been replaced, in BS 4622, by classes 1, 2 and 3 which represent (for spun-iron pipes with socket and spigot joints) recommended maximum working pressures, inclusive of surge, of 10, 12.5 and 16 bar; maximum working pressures for flanged pipes and fittings in BS 4662 are lower than those for spun-iron pipes and where necessary the Standard advises the use of ductile iron or strengthened grey-iron fittings. Pressure ratings for ductile iron pipes (class K9) and fittings (class K12) vary with size: 40 bar up to 300 mm, 25 bar for 350 to 600 mm and 16 bar for 700 to 1200 mm (BS Code of Practice 2010: Part 3).

The standard length of spun-iron socket and spigot pipes to BS 4622 is 5.5 m, and available joints include Tyton (the most widely used) and mechanical flexible joints of bolted-gland type. The standard length of flanged pipes is 4 m.

The range of pipes likely to be available in future differs in some respects from that of BS 4622; as British, Japanese and American manufacturers have extended their range to meet demands for larger sizes on big supply schemes overseas in developing countries.

Protective coatings include bitumen sheathing and the application of centrifugally applied concrete or bitumen lining can be provided where conditions warrant these additional safeguards. Where aggressive soil conditions exist the pipe may be protected by a tubular polythene sleeve.[8] It is rare to install pipes without any protection, and where this has been done it has often proved a costly mistake.

28.3.1.2 Steel pipes

British Standard 534 covers the manufacturer of steel spigot and socket pipes and specials. Manufacturers of steel pipes are generally able to manufacture special pipes of any reasonable size, thickness or shape to suit customers' requirements. Pipes vary in size from 50 up to 1800 mm with wall thicknesses varying from 2.5 to approximately 20 mm. They may be jointed by welding with internal sleeve welds only, or internal sleeve welds and external sleeve welds to facilitate testing of butt welds. Alternatively, if greater flexibility is required, plain-ended pipes

are used in conjunction with Johnson couplings. The pipes may be protected with bitumen, concrete, or a sheathing of bitumen wrapped in hessian plus bitumen or coated in bitumen plus asbestos sometimes reinforced with woven glass. Where the surrounding groundwater is aggressive and the soil has a resistivity of less than 5000 Ω/cm^3 then cathodic protection is required which may be provided either by sacrificial anodes, or by the imposition of a protection current from a direct current source such as an accumulator or transformer rectifier unit.

28.3.1.3 Asbestos pipes

Asbestos cement pipe[5] is made of a mixture of asbestos and Portland cement to form a laminated material of great strength and density. The material is less subject to encrustation in soft-water districts and is not affected by electrolytic action. Flexible joints are used exclusively throughout the range of sizes up to 900 mm in diameter for working pressures up to 90 to 122 m head according to size. Special bends, tees and adaptors are not made in asbestos and those of cast iron are generally used for connections to asbestos pipes; BS 486 applies to A to C pressure pipes.

28.3.1.4 Concrete pipes

Standard concrete pipes of plain or reinforced concrete are made up to a diameter of about 2.3 m (or occasionally greater) and are chiefly used to convey liquids not under pressure. Sizes from 150 mm to 1.8 m are covered by BS 556. The joints are generally of the flexible type such as the Stanton–Cornelius. Prestressed concrete pipes can now be manufactured over a wide range of sizes, varying in diameter from 635 mm to 1.8 m (BS 4625 covers sizes from 400 mm to 1.8 m), and usually having a thin steel shell with a spun concrete interior lining stressed externally by prestressing wire on the outside of the steel shell, the whole then being protected by an outer covering of cement mortar. Working pressures of up to 120 m head of water can easily be obtained. In sizes over 1.2 m, longitudinal prestressing wires are normally employed and the steel cylinder is not used. The lock joints of the simple push-in selfcentring type are completely reliable provided that the manufacturers' jointing instructions are followed precisely.

28.3.1.5 Aluminium pipes

Aluminium pipes are available up to 700 mm diameter, manufactured by the helical method, but these have not been used to any great extent in water supply. The evidence would so far seem to indicate that this is a material which might be more widely used provided that suitable precautions are taken to protect the material similar to those adopted for steel. The main advantage is in the reduction in weight particularly for overground purposes.

28.3.1.6 Polyvinylchloride and fibreglass wrapped pipes

Polyvinylchloride pipes are light and easy to handle, corrosion resistant, and are generally available in sizes up to 600 mm in lengths of approximately 9 m. Larger pipes for waterworks purposes may require to be strengthened and this can be achieved by the use of glass fibre reinforcement. The joints are usually made by a push-on type of rubber ring joint or by a solvent welded joint, the latter only being practical where site conditions permit. It has also proved possible to mole plough long lengths of up to 200 m of this pipe up to a diameter of 300 mm underground without surface trenching. It should be noted that the coefficient of expansion of PVC is 8 times greater than that of steel and considerable movement can take place in long lengths of rigidly jointed pipelines.

28.3.1.7 Structural design

In considering the design of the pipeline the external loads generally arise from the weight of the pipe and its contents, the trench filling, superimposed loads including impact from traffic, and from subsidence. The design of pipelines and the strength of the pipes required has been considered empirically and the design method commonly used is that proposed by Marston and Spangler and described by Young and Smith.[9] When a pipeline has to be laid above ground over some obstruction it may either be carried on a pipe bridge or be designed as a selfsupporting arch. Special design and fabrication are necessary in these cases.

28.3.2 Flow in pipes

28.3.2.1 Streamline flow

Reynolds found by experiment that the average velocity below which streamline flow could be maintained for various diameters of pipes is approximately given by the equation:

$$\text{diameter (cm)} \times \text{mean velocity (m/s)} = \tfrac{2}{3} \text{ or less} \quad (28.1)$$

Hence, in a pipe of 300 mm diameter, for example, there should theoretically be a mean velocity of under 1/45 m/s, for maintaining streamline flow; in practice, however, it would be expected to carry water at a mean velocity of between 0.7 to 1.0 m/s, so that it is not economic to use this as a criterion for design.

28.3.2.2 Turbulent flow

Froude found empirically for turbulent flow that the friction consumed by water passing through a pipe varied: (1) almost as the square of the mean velocity of the water; (2) almost as the area of the wetted surface in contact with water, i.e. the circumference and length of the pipe; and (3) the nature of the surface inside the pipe.

28.3.2.3 Friction in pipes

Basic formulae have been derived by Bazin, d'Arcy, Chezy, Kutter, Ganguillet, Wisbert, Hazen, Manning, Flamant, Unwin, Barnes and others. The general principles are dealt with in Chapter 5.

The universal pipe friction diagram based on the Hazen Williams formula

$$v = 1.318C(D/4)^{0.63}(H/L)^{0.54}$$

was included in the *Manual of British water engineering practice* as chart D and it is reproduced in a simplified form in Figure 28.1. C is a constant depending upon the type, condition and diameter of the pipe, V is the mean velocity, D the internal diameter, H/L is the head loss per unit pipe length. This diagram is useful for preliminary design purposes, but for more accurate design the charts prepared by the Hydraulics Research Station,[10] based on the Colebrook–White equation, should be used.

28.3.2.4 Economic diameter of pumping main

Where water is to be pumped under pressure in a rising main, there is an economic diameter of pumping main to pass a given quantity of water. If the main is reduced in diameter the cost of the main will be less but the friction will be increased and the

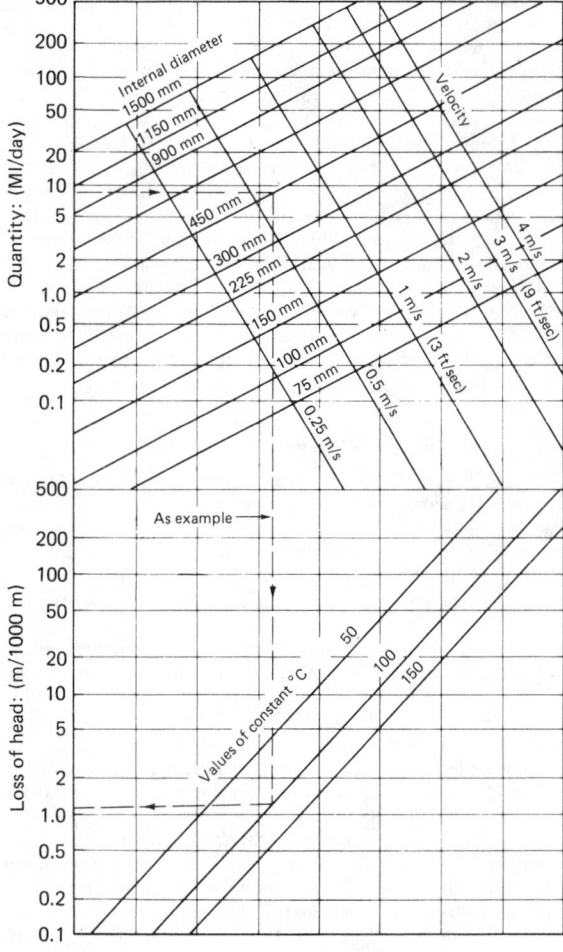

Figure 28.1 Pipe-friction diagram
Example of use
450-mm pipe is required to carry 9 Ml/day. What would be the velocity and head loss? On the top half of the diagram, read across from rate of flow to size–velocity; read off diagonal line giving 0.6 m/s. Strike down vertically to appropriate e value (assume 100). Read across horizontally to left scale.

Therefore head loss=1.2 m/100 m

Multiply by lengths of pipe to get total friction loss

cost of pumping, allowing also for larger machinery required, will be more. The converse also applies. Figure 28.2 shows the combined annual cost of the sinking fund taken at 35 yr on the main together with 15 yr on the machinery and pumping at a given unit rate of electricity. The curve is particularly instructive in showing that it is generally more economical to err by choosing too large a diameter than too small a diameter as the left-hand side of the curve rises more steeply than the right-hand side. Lea[11] has proved mathematically that the economic diameter lies between 0.535 and 0.675 times $Q^{1/2}$ (Q being the quantity pumped in m³/s). Hence the velocity for economic pumping to daily supply can be deduced as lying between 0.8 and 1.4 m/s. However, if the supply is only intermittent (as, for example, in a standby supply), then use of a higher velocity up to 2 m/s would be justified.

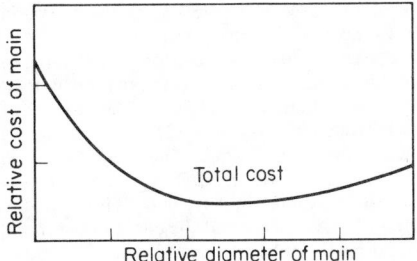

Figure 28.2 Variation of cost of main with diameter

28.3.3 Valves

28.3.3.1 Standard gate or sluice valve to BS 1218

These valves, of 37 to 300 mm diameter corresponding to cast-iron pipes, are to be had for class 1, working pressure 90 m and class 2 for a working pressure of 120 m. Valves up to 1.2 m diameter are available. Valves are usually flanged to enable them to be removed, repaired and re-inserted without disturbing the rest of the pipelines. The nonrising spindle type has the screw totally enclosed within a casing, is operated by a key, and usually opens by turning anticlockwise although this should be checked by looking at the arrow on the casing. For special purposes there are valves with spigots and sockets, double spigots, Victaulic joints, hand wheels, exposed screw rising spindle types, and anticlockwise opening or a combination of any of these.

Some sluice valves of the larger sizes, such as those situated in valve towers of impounding reservoirs, are often geared to facilitate operation by one man. Sluice valves may be provided with indicators to measure the amount of opening for both rising and nonrising spindles. There are locking devices. The larger sizes are also often provided with small bypasses to relieve pressures on opposite sides of the gate and the sizes of these bypasses are a matter of calculation, but the values in Table 28.3 may be taken as typical.

Table 28.3 Diameter of valves and bypasses

Main valve diameter (mm)	Bypass diameter (mm)
up to 200	10–20
225–300	25–30
350–525	50–75
550–900	75–100
900–1200	100–150

28.3.3.2 Butterfly valves

Butterfly valves in accordance with BS 3952 are now extensively used as they are easier to operate than gate valves, smaller in size and generally cheaper. They should not be operated at water velocities of over 5 m/s where rubber seatings are included.

28.3.3.3 Reflux valves

Reflux valves are also known as nonreturn, recoil, retaining, foot and flap valves. These are made up to 1.5 m diameter or more. There are many types, single door, multiple door, horizontal, vertical and tilting discs.

28.3.3.4 Air valves

Air valves are put at the highest points of mains and also on flat gradients of under 1 in 500 where the distances are more than 750 m. Those having large orifices let out air in large mains when being filled and those with small orifices are used for letting out air as it accumulates in coming out of the water. The double air valve has one large and one small orifice, a vulcanite ball being used for the large orifice and a rubber ball for the small 2.5 mm orifice. This is the type most generally in use but there are other modifications, e.g. that with and without an isolating valve to enable the balls to be inspected without emptying the main, or a refined type – kinetic – which prevents the balls slamming shut through a sudden rush of air and water. In this type the air is bypassed around the ball. For large mains a double air valve (isolating, kinetic type) is the most likely to be adopted, as it is capable of inspection and cleaning, which should be done at yearly intervals, and is not liable to slam shut when the mains are filled or refilled.

28.3.3.5 Hydrants

A hydrant consists usually of: (1) an 80 mm branch from the main with a duck foot on which rests the screw-down hydrant and stand pipe; (2) a screw-down hydrant and standpipe placed on the main itself; and (3) an 80 mm pipe and hose attachment.

28.3.3.6 Washouts

Washouts are usually branches of (say) 80 to 150 mm diameter, but may be larger, leading from the main to a ditch or river with an ordinary sluice valve control. Special branch tees having the invert of the branch coincident with that of the main are made to enable any sediment to be washed out of the main. Flap valves should be put on the ends of the branches. They are used to clear the main of contaminated water or sediment, and should be sized to achieve a flow of at least 0.5 m/s in the main if possible.

28.3.3.7 Valves for special purposes

The pressure-reducing valve. The ordinary sluice valve, half closed or throttled, is often used for reducing the pressure in a pipeline, but should not be, as special valves of various types are made for the purpose. The common form of pressure-reducing valve provides a constant pressure downstream at less than that upstream; the downstream water presses against a piston loaded with weights, and as the downstream pressure rises it forces the loaded piston upwards, and, through a system of levers, closes the valve. Modifications in this type of valve enable: (1) the downstream pressure to vary with the rate of flow through the valve; or (2) a reduction in the head through the valve by a constant amount.

The pressure-retaining valve. This type is sometimes called a sustaining valve and is an adaptation of the pressure-reducing valve. It enables a variable inlet pressure to be converted to a constant pressure or to prevent the upstream pressure from falling.

The pressure relief valve. This type is also sometimes called a sustaining valve and is an adaptation of the pressure-reducing valve. It may be a loaded spring affair or an adaptation of the pressure-retaining valve by which, for important installations, the times of opening can be regulated.

Flow control valves. Such valves are intended for controlling constant flows in pipelines irrespective of pressure, and modifications of them provide for dividing flows into two, or introducing other flows to make up the quantity. Some flow valves have balanced discs electrically or hydraulically operated; other forms, often of the needle type, are designed to close hydraulically in the event of a burst main, or to open and close hydraulically or electrically at stated hours, as in pumping stations.

28.3.4 Cost

The cost of water mains laid complete is of the order of £2.40 per centimetre diameter per metre with 10% to be added for valves and fittings (1985), including excavation (cover 1.2 m) and backfilling, but excluding road restoration.

28.3.5 Surge

A pressure transient or 'surge' is caused by a sudden alteration in the velocity of flow in a pipeline or aqueduct. This surge can be transmitted through the system causing an increase in pressure. Very high pressure can be generated leading to overstressing of the pipes, joints or pumps; likewise, low pressure can lead to the creation of a temporary vacuum causing infiltration, collapse or cavitation. These conditions may be caused by: (1) opening or closing a valve; (2) stopping or starting pumps or tankers; (3) the 'slamming' of a reflux valve or tidal flap; (4) on change of load demand on a hydro-electrical generator; (5) on vibration of guide vanes or impellers; or (6) generally by any situation causing a change in velocity of flow.

The phenomenon is often called 'water hammer' and can occur in small-diameter domestic plumbing. The general principles were reviewed by Lapworth[12] and a graphical analysis method was described by Lupton.[13] Alternatively, the mathematical 'method of characteristics', which utilizes the same equations as the graphical analysis, is more easily used on a computer and enables the pressure at several points of the system to be calculated simultaneously, whilst alternative solutions can be easily compared and refined.[14]

Control of surge is necessary if the rise or fall in pressure is found to be more than the components of the system are designed to withstand. The methods to be considered could include the following: (1) restricting rate of valve closure; (2) increasing pump inertia (e.g. by putting a layer flywheel); (3) providing a surge shaft; (4) fitting an air vessel; (5) providing air admission or release valves; (6) fitting a bypass round a pump; or (7) fitting a weighted surge suppressor valve or similar mechanical device.

Evaluation. The velocity of a pressure wave within a system may be calculated from the formula:

$$a = 1 / \sqrt{\frac{p}{g}\left[\frac{1}{K} + \frac{d(1-v^2)}{tE}\right]^{0.5}} \quad (28.2)$$

where p = density of water, g = acceleration due to gravity, K = bulk modulus of water, d = internal diameter of pipe, t = thickness of pipe, E = Young's modulus for the material of the pipe, and v = Poisson's ratio for the material of the pipe.

Typical values of a, resulting from the use of this formula, would be as follows: (1) small cast-iron pipe 1250 m/s; (2) large thin-wall steel pipe, 900 m/s; or (3) small µPVC pipe 300 to 400 m/s.

The time taken for the pressure wave to travel to the end of the system, length L distant, and back is $2L/a$ and is called the reflection time. If a valve is closed in less than the reflection time, pressure on the upstream side of the valve will rise above the stated pressure by a head H max. equal to a V_0/g where V_0 is the steady velocity before closure. After the end of the reflection time the pressure will drop by the same amount.

As an example, consider a simple system comprising a pump lifting water through a small cast-iron pumping main to a service reservoir against a static head of 80 m. The velocity of the pressure wave through the main is 1250 m/s, the initial velocity V_0 is 0.3 m/s and g is 9.8 m/s. Therefore the surge pressure on shut-off is $H_{max} = aV_0/g = \pm 38.4$ m. This gives a minimum pressure at the pump of $80 - 38.4 = 41.6$ m and a maximum pressure of 118.4 m. To avoid the cost of reinforcing the pipeline, it would probably be best to instal an air vessel.

The principles of this example can be applied to a more complex system and are fully described by Thorley and Enever.[14]

28.4 Measurement of flow in streams

For small streams or channels with flows up to about 50 000 l/hr, it is best to use a plate with a 90° Vee notch cut in, provided that there is not an excessive amount of debris and boulders (such as may be carried along in a steep mountain stream). The formula for a sharp-edged notch is $Q = 0.1573 \tan(\theta/2)H^{2.5}$, where Q = flow in litres per hour with coefficient of discharge 0.585; H is the head of water in millimetres measured 1 m upstream of the notch.

For permanent use the plate and notch should be made in stainless steel or bronze, but for temporary measurements a timber board is adequate.

For larger flows a wide rectangular notch is more suitable using the formula $Q = 0.2084LH^{1.5}$, using same definitions as for the Vee-notch formula except that the coefficient of discharge is 0.62 and L is the length of weir in millimetres ignoring end contractions. Although the reading of H should be taken 1 m upstream of the weir, it can be estimated by measuring it at the notch and then adding 10 mm to H if the velocity is about 0.5 m/s; or 20 mm if the velocity is 1 m/s.

For still larger flows, as over long overfall weirs in rivers and overflows from reservoirs, calibration by models is desirable and in any case the flow will be dependent on the particular profile adopted. However, approximate discharges can be assessed using the formula $Q = 0.0579H^{1.5}$; the coefficient can be increased to the order of 0.07 for an efficient crest profile. The approximate values of discharge for 90° Vee-notch, rectangular notch and the general long overflow are summarized in Table 28.4, evaluated in the usual zones of operation.

28.4.1 Current meter

The flow in a wide river is calculated by measuring the cross-sectional area of the river and the velocity by a current meter at several points across the section. A current meter is an instrument provided with a propeller screw which, when immersed, is turned by the velocity of the water, and the number of turns is a measure of that velocity; the mean of all the velocities multiplied by the cross-sectional area of the water is a measure of the quantity flowing. This method is often supplemented by dilution gauging for small rivers with high turbulence. Methods of dilution gauging are defined in BS 3680:Parts 2A and 2C. Further information on flow measurement is given in Chapters 22 and 31.

28.4.2 Crump weir

The Crump weir[15] has a sharp horizontal crest with a 1:2 slope on the upstream side and a 1:5 slope on the downstream side. Considerable research has been undertaken into the characteris-

Table 28.4 Discharges from weirs

H (mm)	90° Vee-notch (l/h)	Rectangular (long, per metre length) (l/h)	Overflow per metre length
10	50		
25	490	26 050	
50	2780	73 680	—
75	7650	135 350	—
100	15 750	208 400	—
130	30 300	308 900	—
160	50 950	421 750	in m³/s
200	89 000	589 450	0.16
250	—	823 750	—
400	—	—	0.46
600	—	—	0.85
800	—	—	1.31
1000	—	—	1.83
1500	—	—	3.36

tics of this type of weir, and this is summarized in Water Resources Board Publication TN8.[16] It is capable of measuring high discharges and can function in partially drowned conditions which means that the size of the structure can be kept down, thus avoiding objections by amenity and fishery interests.

Compound types with side section separated by piers can be used to improve discharge of low flows, and a 'flat Vee' type with the crest sloped in at 1:10 also provides for this. Each type should be individually calibrated for reliable results. The results can be read directly on a staff gauge or recorded on a chart or punch tape for later data processing.

28.5 Measurement of water in pipes

For measuring the flow of water within a pipe the most useful apparatus is the Venturi meter, named after the American inventor Herschel after an Italian of the eighteenth century who experimented on the flow of water in tapered pipes. As the heads in the pipe and at the throat vary with the velocity through the pipe the quantity passing through the meter is proportional to the square root of the difference of the pressure at the inlet and at the throat. By pressure pipes, connected to the upstream side of the meter and the throat, the rate of flow is recorded either visibly as in a manometer or transferred to a pen and paper chart. This rate of flow can also be integrated instrumentally to show the total quantity. Friction losses are 300 to 600 mm for good design. A truncated version called a Dall Tube is now generally used. On longer pipes fast electromagnetic flow meters are available with external fixings that provide no head loss and flows can easily be transmitted for remote reading.

28.6 Service reservoirs

An important item in the distribution of water is the covered service reservoir, not to be confused with the open impounding or pumped storage reservoir.

Whereas the function of an impounding reservoir is to store crude water for use in dry months or years, the service reservoir stores the drinking water for immediate use.

The covered service reservoir is an integral part of the distribution system and its object is to balance the daily fluctuations of demand on the one hand and the method of delivery

from the source, such as pumping, on the other. All service reservoirs must be at the highest possible level necessary to serve the houses which are to be supplied. If there is no natural ground sufficiently high for the purpose the water must be raised by being perpetually pumped (i.e. boosted), or the reservoir must be elevated and so becomes a water tower. At 1982 prices, the cost of pumping is about 2.5 p per 1000 l per 100 m lift.

28.6.1 Covered service reservoirs

All service reservoirs must be covered to keep the clean water which is put into them from being fouled by exposure to the atmosphere, which encourages algal growths (especially with hard water), by dirt from the air, or by vermin from the ground. Such reservoirs are usually built half in and half out of the ground, the material of excavation being used for banking material for the walls and covering the roof. About 300 mm of topsoil is usually placed over the whole expanse of roof to improve the appearance and also to keep the concrete of the roof at an even temperature to prevent expansion and cracking. Walls may be of mass, or reinforced, concrete, brick, masonry or puddled clay; columns may be of plain, or reinforced, concrete, brick, steel or masonry. Roofs may be of reinforced concrete, brick arched, or asbestos sheeting on mild steel trusses. Floors may be of plain, or reinforced, concrete, or bricks on puddled clay. Internally, reservoirs may be lined with asphalt. Different circumstances and different persons dictate the choice; mass or reinforced concrete walls, reinforced concrete roof and floors, concrete or coated steel columns being the author's usual choice.

The principles of design follow orthodox practice for masonry, concrete, or steel structures. For preliminary design purposes the dimensions, capacities and costs in Table 28.5 may be used.

Table 28.5 Service reservoirs on level ground

Capacity (Ml)	2.25	4.5	6.75	9.0	11.25	13.5
Dimensions depth (m)	3.7	4.6	5.5	6.4	7.3	8.2
Side of square (m)	25	32	35	38	39	41
or circular dia. (m)	28	36	40	42	44	46
Approx. cost (1987) (£000)	195	350	490	600	715	780

28.6.2 Water towers

The cost of water towers is balanced against the cost of boosting, i.e. the capital cost of the tower and its maintenance against the capital cost of boosting plant and its maintenance. If a reduction of capital expenditure is of paramount importance, boosting may be chosen, but in many cases greater security is felt with an elevated tank.

Towers are of several designs. For a given capacity the globular form, of steel, seen in the US, is the most economical in the weight of steel. The cylindrical form with 'dished' bottom is next-best. The rectangular form conveniently built up of steel plates is often used in the UK for industrial applications and in developing countries due to ease of design and fabrication. The majority of towers for waterworks purposes are of reinforced concrete, consisting of cylindrical structures, either on legs or totally enclosed; they are often lined with asphalt, and firms specializing in reinforced concrete are usually employed for their construction.

A form much favoured is the cylindrical steel tank surrounded with a thin external shell of reinforced concrete having a space of 1 or 1.3 m between the steel tank and the shell. The shell lends itself to architectural treatment to harmonize with the surroundings, can always be maintained to present a pleasing appearance, and can often be constructed by a local contractor. The steel tank inside the concrete shell can be inspected externally for leakage, and preserved by painting; this is not so easy with the reinforced concrete tank, particularly if it shows signs of leakage.

It is not economically justifiable to provide standby elevated storage and the largest towers seldom exceed 5 Ml in capacity or 30 m in height. They are subject to significant wind loading, and usually require substantial foundations. As a general guide a 1 Ml storage at 30 m height would cost approximately £600 000 (at 1985 prices).

28.6.3 Valves and fittings for service reservoirs

The appurtenances of a reservoir consist of suitable inlet, supply and washout valves, overflow arrangements, roof air inlets, a depth indicator and recorder, and an outlet meter and recorder. Flow through the reservoir is desirable, and to facilitate this the inlet is placed at one end and the outlet at the other. Air ventilators are provided to prevent accumulation of gases and accommodate the rise and fall of the water level. The overflow pipe is usually a standpipe with a bellmouth on top. In nine cases out of ten, and particularly where underground water is the source of supply, the service reservoir should rarely (e.g. 5 or 10 yr) need cleaning and, on these occasions, as the operation need not take more than a few hours, the supply can usually be maintained by bypassing perhaps through a small tank or at night; partition walls with all the necessary duplication of pipework are only required for the larger sizes where there is no alternative storage available during the period out of service. Some form of telltale or automatic level recorder is necessary, and there are many types, to inform the operator, at the source, of the water level. Although there may be a meter at the source to regulate the quantity of water flowing into the reservoir it is often desirable to have a meter on the outlet side to ascertain the rate of draw-off hour by hour in the day.

28.7 The underground scheme

28.7.1 Development of a new source

The origin of the water for an underground water scheme is rain which has fallen on the surface of the ground and has sunk in; to retrieve this water a well is dug in the fissured formation into which the rain has penetrated. If the strata are not sufficiently open to yield enough water an adit or heading may be driven in the bottom of the well. If the strata are well fissured and yield water readily, one or more boreholes are adopted. Whatever the 'hole' in the ground may be, its function is to accommodate a pump of the right dimensions for the economical pumping of water; this pump being low enough to reach the water when the water level in the well is at its lowest. Where the water level never falls or is never likely to fall more than 8 m below the surface, as in areas of alluvial estuarine plain with monsoon recharge occurring every year, it is usual to use surface pumps for ease of maintenance. However in most situations the fully submersible pump is now used. To drive the pump there are the alternatives of oil and electricity; oil is seldom used except as a source of power. Whatever the pumping machinery may be, however, for waterworks practice it must be absolutely reliable, and it is desirable to have provision for standby equipment and boreholes.

The capacity of the borehole pumps must be determined by the water requirements of the district and the capacity of the underground works. Some pumps may be left to themselves, others may require men in charge. Pumps may have to pump the whole daily supply of 24 h in 8 or 16 h to fit in with shifts, or even more during weekdays so as to close down at weekends. Hence the capacity of the pumping machinery, and any treatment works and pumping mains, if designed for 8 h pumping, must be 3 times as large as those designed for 24 h pumping, although the total daily or annual quantity pumped to supply is the same. The water, after being lifted by the well pumps to the surface, is pumped on either by the same pump or a modification thereof, or by a surface pump, with or without a balancing reservoir. Automatic pumping machinery for waterworks pumping is now generally adopted owing to its improved reliability and saving in cost.

The water, where necessary, may be softened, filtered (rare for underground supplies), or treated for removal of iron and/or manganese. The pumping main, which must be designed in accordance with the pumping rate, may be of asbestos, iron, steel or PVC with various coverings, and is best left free from tappings for house services and taken direct to the covered service reservoir. The service reservoir from which the water is distributed should hold at least 24 h and preferably 3 days' supply: the distribution mains from the reservoir should be designed to carry a rate of about 3 times the average consumption to allow for peak hourly demands.

28.7.2 Geology of source[17]

The source of water for a pumping scheme which depends upon obtaining water from an underground source, in England particularly, rather than in Wales or in Scotland, is based on the following geological formations, in order of merit.

28.7.2.1 Chalk[18]

This includes all three divisions, Upper, Middle, and Lower, with overlying gravels and crags, and occurs in Yorkshire, Lincolnshire and East Anglia. Chalk, which is associated with overlying porous gravels, Thanet Sand (and other Lower London tertiaries) is present in the Home Counties (Buckinghamshire, Berkshire, Surrey, Middlesex, Hertfordshire and Essex), London, Kent, Sussex, Hampshire, Isle of Wight, and with underlying Upper Greensand in Wiltshire and Dorset.

28.7.2.2 Bunter and Keuper Sandstones

These are present in South Lancashire, Cheshire, Yorkshire, Nottinghamshire, Staffordshire, Warwickshire and parts of Somerset and Dorset.

28.7.2.3 Oolites[19]

Lower oolites include somewhat arenaceous deposits in East Yorkshire; the Lincolnshire Limestone of Lincolnshire; thin beds in Northamptonshire and Oxfordshire; the Inferior oolites and Cotswold Sands of the Cotswold hills of Gloucestershire and Worcestershire and the somewhat arenaceous limestones of Somerset and Dorset. Upper oolites are found in Yorkshire, Oxfordshire and Wiltshire.

28.7.2.4 Lower Greensand

The chief development of the Lower Greensand is all around the foot of the Chalk beneath the Gault of the North and South Downs, which enclose the Weald of Kent and Sussex, and parts of Surrey and Hampshire. Some of the Lower Greensand provides useful soft water at a depth of over 300 m in the Slough area. Other divisions include spreads of Greensand at the northern foot of the Chalk in Bedfordshire and Hertfordshire, and in Lincolnshire.

28.7.2.5 Carboniferous

Under this general term may be included the Carboniferous Limestone which collects water in large fissures as in the Mendips of Somerset, Derbyshire and Yorkshire; the Millstone Grit and other grits with small fissures of the Yoredales and Coal Measures of Lancashire, Yorkshire and Wales; grits in the Upper Coal Measures for Coventry, and small supplies in the culm of Devonshire.

28.7.2.6 Permian and Magnesian Limestone

This occurs in the north-eastern counties.

28.7.2.7 Ashdown Sand and Tunbridge Wells Sand

These are present in the Kent and Sussex Wealds.

28.7.2.8 Old Red Sandstone

Old Red Sandstone is present in South Wales, the Forest of Dean and Herefordshire.

In addition there are water-bearing gravels in proximity to rivers or other so-called water-bearing formations in juxtaposition with many of the above water-bearing strata.

The following may be taken as a rough indication of the extent and quantities of water that are pumped from the main four formations (see Table 28.6).

Table 28.6 Underground supplies in the UK

Strata	Exposed surface area (km²)	Underground extent (km²)	Quantity pumped (Ml/day)
Chalk/Upper Greensand	13 000	18 000	1800
New Red Sandstone	4500	3000	400
Oolites	6500	2600	150
Lower Greensand	2600	13 000	120

28.7.3 Selection of source[20,21]

The natural factors governing the quantity of water obtainable at any one spot are: (1) the direction of flow of the water underground; (2) the general geological arrangement of the strata; (3) the area of strata exposed to the skies; (4) the thickness of formation; (5) the extent of the formation underground; (6) the nature of the fissures and formation and porosity of strata; (7) the amount of rainfall; and (8) the evaporated and dissolved salts from the strata.

The factors governing the final choice of site are those enumerated above, with conditions of proximity of buildings, proximity to places to be supplied, cost of development and other engineering considerations, elevation and acquisition of site. The direction of flow of water underground can be ascertained by plotting contours of underground water levels from the records of water levels, referred to Ordnance datum, in existing wells. The arrangement of the strata such as faults and/or rolls, anticlines, synclines, thinning out, and change in the lithology may modify profoundly the potential yield at any one spot and local depressions given an indication of possible over-extraction.

The proportion of rainfall which percolates into the underground strata is affected by surface land use and topography which dictate the amount of evaporation and surface runoff. As a general guide over porous strata, the average percolation is from 200 to 300 mm per year with an average rainfall of 600 to 900 mm per year. Lapworth has suggested that percolation in the Chalk is 0.9 of the average rainfall less 340 mm per year. The round figure of 250 mm per year is often assumed for a working basis and this is equivalent to 0.7 million l/day per square kilometre of gathering ground, the area being determined from the extent shown by the underground contours which may be assumed to flow (or be drawn) to the selected source. However, in a dry year with low rainfall and high evaporation, this could be reduced to almost nil. For an accurate assessment of the reliable yield of a particular aquifer, monthly or even daily figures should be assessed using the Penman[22] or similar formula which takes into account such factors as solar radiation, the drying power of the air, wind speed, vapour pressure and temperature. Application of these formulae have been programmed to incorporate a 'root' factor for different types of land use and soil moisture deficit, which can be obtained from the Meteorological Office in the UK.

With regard to natural factors affecting adversely the cost of underground water these are mainly: (1) hardness where rain acidified by vegetation dissolves calcium and magnesium from limestone; (2) chlorides, which come from contacts with rock salt as in Cheshire, or from rocks containing highly mineralized water from ancient seas or other sources and which have been protected throughout the ages by a clay layer; and (3) iron from acidified rain dissolved from the rocks, ground or peat through which it passes as in the ironstone of the Weald. The waters for new sources should be analysed chemically much more fully than those already in use, and it may be necessary, for example, to look for such a substance as fluorine which affects teeth.

Man-made factors affecting the quality of water are those of proximity of buildings, sewage, and noxious effluents, e.g. gas liquors on the gathering ground, refuse tips and manure, all objectionable in varying degrees according to their distance from, and the potential fissures underground leading to, the well. The site of the well should, of course, be chosen as near as possible to the place where the water is required; every kilometre of main adds greatly to the cost, e.g. a 300 mm main at a pumping rate of 5 million l/day costs £50 000 per kilometre. Among other engineering considerations, the highest underground water level when pumping is significant, because for every 30 m in elevation saved about 31p/1500 l is saved.

28.7.4 Construction of boreholes and adits

The diameter of the boreholes for water are much larger, and the depths are much less, than for oil wells. Boreholes under 300 mm diameter are seldom made for a permanent source of public water supply in the UK. Steel lining for boreholes is standardized to BS 879 with diameters up to 1.2 m. Bored wells, however, may be made up to 3.6 m diameter under water without pumping. Wells and adits when made by hand are dug in the dry, to rest water level, and by pumping below rest water level, but are very expensive and seldom justified in economic terms.

Boreholes are made by rotary drilling or by percussion; drilling enables exact cores to be obtained, whereas the percussion method pounds up the material. Both methods are used according to the conditions of hardness of the strata and any special requirements as to necessity of cores for the identification of strata. In sandy formations, boreholes are sunk by the mud flush system, the mud keeping back the sand as drilling proceeds through the soft sand; this process enables gravel to be inserted outside the perforated tubes, which is a very satisfactory way of keeping back the sand when the site is brought into commission. A modified method of drilling is known as the 'reversed' flow system where water is pumped under pressure into the borehole while being drilled. Sites are often developed by two boreholes so that duplicate machinery may be inserted in each. Often a sandstone site may be developed by several boreholes spread over the site with a small pumping unit in each in order not to pump too large a quantity at any one spot and so draw in sand.

Boreholes in the Chalk are risky, for although the site may be geologically a good one the borehole may quite easily miss water-bearing fissures. A geophysical survey of the site to ascertain where there is least electrical resistance to indicate the place most likely to be the most fissured may be useful and the use of aerial photographs and satellite images (LANDSAT) is useful in establishing fissure patterns and the presence of swallow holes.

The cost of drilling and testing a lined 600 mm diameter borehole over 50 m deep with a neighbouring observation hole would be approximately £400 per metre depth (1982).

28.7.5 Water-well casing

The lining of a borehole may be plain for lining out clay or other non-water-bearing material. Usually the top 15 m is lined out to prevent surface contamination. The water-bearing portion of the borehole, depending on the capacity of the rocks to stand up and not collapse, is often unlined. At some boreholes, even those in the Chalk which were thought to be safe, have collapsed it is often considered prudent to line a borehole throughout its depth with perforated or slotted tubes in the water-bearing horizons and plain tubes in the Clay or unstable sections.

The tubes in the water-bearing portion of the borehole may be perforated with holes at centres or by slots 3 to 12 mm wide, 150 mm or more long spaced at 100 to 150 mm centres. The British Standard casing (BS 879) for water wells includes lap welded and welded steel tubes. The main points of general difference are in the joints which may be screwed and socketed with: (1) V thread screwing; (2) square thread screwing; or (3) screw flush butt joints with square form thread parallel screwing. Nowadays µPVC tube and slotted lining is extensively used and is replacing the use of steel in many applications where structural strength is not required.

28.7.6 Testing boreholes and wells

The testing and development of newly developed sources is often complained of as being costly, but it is extremely necessary, as it is the basis upon which any pumping scheme rests. Some newly constructed holes show stationary pumping levels almost within a few minutes of commencement of the test; whereas in others the water levels are not stabilized for some time, even after 3 days or more. The frequent stopping and starting of pumping often improves the yield of a newly drilled borehole and the rate of rise at the end of the test gives the measure of the inflow into a well.

Yields may be increased in Chalk and other limestones by treating the boreholes with hydrochloric acid. Fissures, clogged by boring, are cleared and the well losses reduced. Development can also be achieved by hydraulic fracturing of the strata.

28.7.7 Yield of boreholes

For comparing the yield of one borehole with another it is convenient to compare the specific yield, the quantity pumped divided by the difference of level between rest and stabilized pumping level. Typical specific yields range from 75 million l/h per metre (poor) to 750 million l/h per metre of lowering (good).

Theoretically, the yield of wells varies as the logarithm of their diameters, but the mathematical theory of wells cannot be applied, except broadly, since conditions in the fissuring of strata are too variable and uncertain in practice. Geology does not lend itself to any mathematical treatment which presupposes that every cubic metre of strata over many square kilometres extent and many metres depth is of absolute uniform composition throughout. Measuring devices for the testing of wells include, of course, a flow recorder, or method for measuring the quantity, and a depth recorder for measuring the depth of water; both instruments are provided with charts or digital recorders connected directly into the logger system. Pumping tests must be continuous day and night, and the commonly accepted period for a public water scheme is 14 days. Frequently, longer periods are necessary where the water level is not stable or additional information about the aquifer is required, and this would normally include a series of 2 to 4 h step tests at different rates with equal rest periods between each test. The yield of a borehole can be assessed by plotting on a logarithmic scale and comparing with a number of theoretically produced curves prepared by Theis, Jacob and others which indicate if it is behaving as a confined, semi-confined or homogenous aquifer, and from which the sustained yield over annual or several months' pumping can be extrapolated from the base of the 7- to 14-day test.

28.8 Surface water schemes

28.8.1 Development of a new source and requirements[23]

A surface water scheme usually includes the construction of a reservoir to store river water at times of high flows for use in times of low flows in order to give a uniform daily rate during a design drought period consisting of consecutive drought years, with a given probability of occurrence – often taken as once in 50 or 100 years. Such a scheme basically depends on: (1) the quantity of rain falling on the gathering ground, and (2) the evaporation or loss in quantity after it falls on the gathering ground to the dam; (3) the storage of the reservoir consistent with cost and its reliable yield, and (4) the suitability of the geological conditions for safety of the submerged area and the dam site. Thus, meteorology, engineering and geology are interdependent for ensuring the safety and cost of a surface water scheme.

28.8.1.2 The main uses of reservoir conservation

The main uses of reservoir conservation are:

(1) Domestic and industrial water requirements.
(2) Hydro-electric generation.
(3) Irrigation.
(4) Regulation of the flow of a river by increasing dry weather flows and by reducing floods.
(5) Recreation (fishing, sailing).

28.8.1.3 The chief types of reservoirs

The chief types of reservoirs are:

(1) Impounding reservoirs, so called because the sides of a valley, with the dam, impound the natural flow of the river.
(2) Pumped-storage reservoirs, formed by a dam or bund

remote from the river from which they are filled only by pumping.

(3) Some impounding reservoirs may also be used partly for pumped storage.

(4) Both types could be used for river regulation, i.e. regulating the flow of a river to maintain abstractions lower down.

28.8.1.4 The chief types of modern dams

The chief types of modern dams are:

(1) Gravity concrete dams. Triangular in section where the line of pressure passes through the 'middle third' of the dam; so-called 'gravity' because any cross-section could stand without overturning.

(2) Modifications of concrete dams, i.e. curved gravity, pre-stressed, reinforced, thick arch, thin arch, double curvature.

(3) Buttress, multiple arch concrete dams.

(4) Earth dams with various forms of clay cores which are supported by sands, gravels, soft sedimentary and all other soft rocks. (Various types of construction, e.g. hydraulic fill dams, can be included in this category.)

(5) Rockfill dams with impervious cores, or faced with asphaltic concrete. Types (1), (2) and (3) concrete dams are usually adopted on rock foundations, while (4) and (5) earth rock dams are generally the best solution on soft alluvial or sedimentary deposits.

28.8.2 Geology of source

The chief valleys in Britain where surface waters are impounded are on the following geological formations, which are relatively impermeable:[17]

(1) The Ordovician, Silurian and Old Red Sandstone. The grits and shares of the Ordovician and Silurian in Wales present many developed sites in Wales (e.g. Claerwen, Llandegfedd) and in the Lake District, the artificially enlarged lakes, Thirlmere, Haweswater and Crummock.

(2) The Carboniferous Series, Yoredales, and other grits and shales below the Coal Measures underlie most of the sites which have been developed in Yorkshire, Lancashire and Cheshire (Scammonden, Erwood and Lamaload).

(3) Other Clay formations, e.g. the Keuper Marl (Chew) and Forest Marble of Somerset (Sutton Bingham), the Lias Clay of the Midlands (Eyebrook and Draycote), the Ashdown Sand and Weald Clay of Sussex (Weir Wood and Bough Beech).

(4) The granites of Dartmoor and Cornwall (e.g. Meldon, Siblyback) and the igneous and metamorphic rocks of Scotland, support the pre-stressed Allt na Lairige, the thin double curvature arch Monar and very many buttress dams for hydro-electric power.

The geographical requirements are that the valley should be wide and flat for the size of the reservoir but narrow at the site of the dam, sufficiently elevated to command the town and large enough to provide an adequate yield.

28.8.3 Economics of storage and yield of reservoirs

The storage of a reservoir is related to the annual runoff of the gathering ground.

If the required daily quantity of water is less than the driest weather runoff and can be acquired, there should be no necessity for a reservoir. If the required quantity is more than the daily runoff, the storage is based on a proportion of the annual runoff. Thus if the proportion is under 75 or 80% of the average annual runoff (generally known as the flow of the three driest consecutive years, which has been determined by records of gauging for 35 yr), economic reservoirs of reasonable size are assured.

For proportions greater than 80%, the size and cost of the reservoir may be doubled for only a very small increase in yield.

28.8.3.1 Storage and yield from river flow records

Where the flow of the river is known over a series of years the storage necessary for the different yields can best be calculated from plotting the flows cumulatively as a mass diagram. Figure 28.3 is a typical diagram for a small scheme in which the wavy line OX represents the cumulative runoff from a gathering ground of 425 ha for 2 yr. The straight line OX represents the uniform yield of 5.5 million l/day during 22 months. The vertical distance between the two lines represents the quantity by which the total actual runoff is below the average at any time.

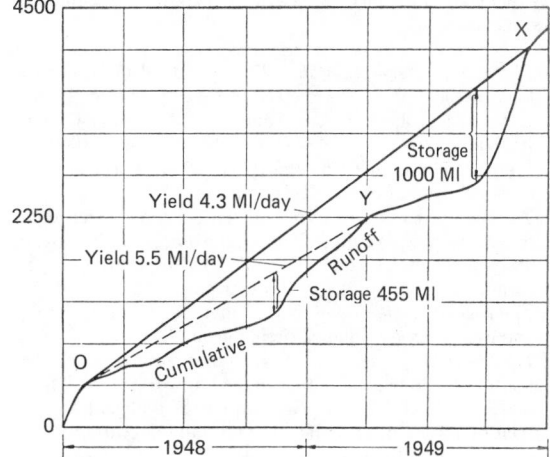

Figure 28.3 Cumulative runoff and yield

Thus the maximum deficiency during the 22 months is 100 million l, which occurs in September 1949, hence the storage required to maintain the guaranteed yield of 5.5 million l/day is 1000 million l.

Similarly, for a smaller yield, i.e. for 4.25 million l/day, represented by the line OY the storage required would be 450 million l, based on a period of 14 months.

28.8.3.2 Storage and yield from rainfall and evaporation records

Where the runoff of a stream or river is unknown, the relationship between storage and yield may be assessed for the gathering ground of the dam in the following steps:

(1) The average annual rainfall.

(2) The average annual evaporation and other losses.

(3) The average annual runoff ((1) − (2)).

(4) Various proportions of (3), known as yields.

(5) Storages corresponding with yields (4), from Table 28.7.

(6) Finally, the storage consistent with economy and site conditions nearest to giving the required yield is chosen.

Table 28.7 Lapworth yield–storage relationships

Average runoff from gathering ground (cm)	Storage (cm)					
	12.5	25	37.5	50	62.5	75
25	12.5	18.8	22.5	25.0	—	—
50	26.3	35.0	40.5	45.0	50.0	—
75	37.5	49.3	56.3	61.3	66.3	70.0
100	46.3	61.3	71.3	77.5	82.5	87.5
125	52.5	72.5	84.8	92.5	100.0	105.5
150	57.5	80.8	96.8	108.3	117.5	125.0
175	62.5	88.0	107.3	120.5	132.5	142.5
200	67.5	95.5	115.0	132.5	145.0	157.5

Conversions: Storage: 1 cm of water on 1 ha = 100 000 l
Runoff: 1 cm per annum on 1 ha = 274 l/day

The following notes may be useful in making a preliminary assessment.

(1) *The average annual rainfall.* The records of the Meteorological Office should be referred to for any area in Britain. Gathering grounds in the South and Midlands, 750 to 1000 mm, Wales, Lake District to Scotland 1000 to 1500 mm per annum. In other parts of the world best use must be made of often sparse data.
(2) *The evaporation or loss of the rainfall.* Penman[22] has compiled a map of Britain showing the average annual losses, and similar maps can be compiled by application of the formula using local factors in other parts of the world.
(3) *The average annual runoff* is the difference between rainfall and evaporation on the gathering ground.
(4) *For the yield–storage relation,* reference for a first approximation should be made either to the Lapworth chart,[24] from which Table 28.7 below gives extracts of the yield–storage relation for various runoffs. (The rainfall less evaporation is also known as the available yield.)

The determination of the size of a reservoir required to balance 50% or less of the available yield is difficult as the problem becomes sensitive to the dry weather flow (DWF) as in the first two examples above and special droughts may 'wreck' the calculations. The percentage of runoff taken for the yield and storage depends on the physical conditions of each site and on the daily requirements of each town.

Example of use of Lapworth Chart and Table 28.7.
Assume a gathering ground of 1000 ha with an average annual rainfall of 175 cm. Assume average annual evaporation is 50 cm so that average annual runoff = 175 − 50 = 125 cm.

From Table 28.7 for average annual runoff of 125 cm, consider storage of 25, 50 and 75 cm and tabulate as below.

Storage related to 125 cm	25	50	75
Corresponding yield (cm)	72.5	92.5	105.5
From conversion			
Storage (l per h)	2.5×10^6	5×10^6	7.5×10^6
Therefore Ml/1000 ha	2000	5000	7500
Yield (Ml/day)	20.0	25.3	28.9

26.8.4 Catchwaters

In order to augment the yield of a reservoir, catchwaters are often resorted to so that additional water is led into a reservoir from another gathering ground. As it is not economical to design catchwaters to take maximum floods, only up to 90% of the water available is taken. The proportion taken is regarded as the 'efficiency' of the catchwater and can be assessed by drawing up a flow frequency curve[25] relating to the stream at the intake point if such records are available. A catchwater may be a tunnel or, more often, an open channel graded to suit the contour of the land. This is the cheapest form of structure and, provided that the length is not too great, it is cheaper than a pipe.

The design of an open channel can be based on the Chézy formula.

28.9 Formation of reservoirs

Reservoirs may be formed with earthen (or rockfill) dams, concrete dams (gravity, arch, multiple arch, cupola, buttress,

Table 28.8 Dam categories and design flood factors

Category	Result of failure	Initial reservoir condition	Flood inflow (general standard)	Wind speed	Minimum wave surcharge (m)
A	Lives in a community endangered	Spilling	PMF	1 in 10 yr Max. h	0.6
B	Extensive damage, or lives not endangered in a community	Just full	0.5 PMF or 1/10 000 yr	1 in 10 yr Max. h	0.6
C	Negligible risk to life. Some damage	Just full	0.3 PMF	Av. annual	0.4
D	No loss of life likely. Limited damage	Spilling	0.2 PMF or 150-yr flood	Av. annual Max. h	0.3

reinforced, and prestressed), by raising existing lakes or by enclosing with artificial embankments, generally filled by pumping. In addition to the dam forming the reservoir, certain ancillary works such as drawoff tower, overflow and diversion works are normally required.

28.9.1 Valve towers

A water supply reservoir usually has a valve tower to contain pairs of valves at different levels, to draw off the water as it rises and falls and to ensure the water at that level is consistent with good quality. Recording instruments are often situated within the valve tower: (1) to record the level of the water below top water level; (2) to record the level of the water when the reservoir is overflowing; (3) to record the quantity of water taken to supply; (4) to record the quantity of water discharged for compensation. A rain gauge is usually placed in the vicinity of the dam.

28.9.2 Floods in reservoir practice

Earlier assessments of the size of floods were made based on valuable data recorded in the Institution of Civil Engineers (ICE) interim reports of 1933 and 1960, supplemented by local data when available and by the subjective judgement of experienced engineers. However, since 1975 these methods have been superseded by the *Flood studies report*[26] produced by the Institute of Hydrology with the ICE and the Meteorological Office. This introduced the concepts of probable maximum precipitation (PMP) and probable maximum flood (PMF). Probable maximum precipitation is defined as the theoretical greatest depth of precipitation (i.e. rain, sleet, snow or hail) for a given duration meteorologically possible for a given basin at a particular period. The flood hydrograph resulting from PMP is called the PMF and is assessed from local topographic and land use parameters. Detailed maps have been prepared for the report and these are continuously updated for Britain, and similar work is in hand for other parts of the world.

As the likelihood of a PMF occurring may be only once in 50 000 years, it would only be utilized in design if a failure arising from it would endanger lives in a community. For lesser consequences proportions of PMF are recommended in the ICE guide to *Floods and reservoir safety*[27] and these are summarized in Table 28.8 combined with associated design factors.

In Britain and Hong Kong, the decision on the category of dam and design flood is made by an engineer appointed under the Reservoirs (Safety Provisions) Act, 1930, or the Reservoirs Act 1975 when implemented in full, taking into account the effect of flood routing through the reservoir, the capacity of the spilling, available freeboard and the risk involved in the event of overtopping.

To obtain an approximate estimate of PMF for preliminary assessment of dam safety, typical curves have been prepared based on an undulating impermeable catchment. For other types of terrain add 15% for mountainous areas, 5% for hilly areas, deduct 5% for flatter areas and adjust total area for permeable zones. The reservoir soil master deficit (RSMD) is an index defined as the 1-day rainfall of 5-yr return period less effective mean soil master deficit; typical values given in the *Flood studies report* for Britain are 25 to 35 in the Midlands and Southeast, up to 70 in the Lake District, up to 90 in the Scottish Highlands. Similar figures can be computed from first principles for other parts of the world. Some typical figures are shown in Table 28.9 for general reference.

28.9.3 Overflow weirs

The function of the overflow is to carry the design peak flood safely over the dam. Overflow weirs may be formed along the crest of the dam, or as side weirs on one or both sides of the valley upstream of the dam, or as a series of siphons over the crest of the dam. Whatever form is adopted to fit in with the design of the dam, the weir should be capable of passing the design flood without overtopping of the main structure, though some limited overtopping can be accepted for lower categories of dam in the case of concrete dams or those with protected crest and downstream slope, often used in flood detention situations.

28.9.4 Drawoff and diversion culverts

Tunnels or culverts preferably constructed around the ends of the dam in solid strata are used for diverting streams during the construction of reservoirs, and their dimensions during construction depend upon the best approximation of the magnitude of the flood. In the case of bell-mouth overflows the tunnel is required to carry water which flows over the bellmouth permanently.

Frequently the same tunnel is used to carry water drawn off from the reservoir to supply or for regulation.

28.9.5 Earthen embankments and earth dams

28.9.5.1 General

The adoption of an *earthen* embankment for an impounding

Table 28.9 Approximate probable maximum flood assessment

Catchment area (km²)	RSMD (mm)	PMF peak flow (cumecs/km²)	Catchment area (km²)	RSMD (mm)	PMF peak flow (cumecs/km²)
1	25	9.2	10	25	5.7
1	40	14.5	10	50	11.5
1	70	26.0	10	75	17.7
5	25	6.5	10	100	23.6
5	50	13.5	50	25	4.1
5	75	20.5	50	50	8.4
5	100	27.5	50	75	12.9
	—		50	100	17.2

reservoir is largely a matter of choice, depending on the geological factors governing the site. If clay predominates at the site of the dam, an earthen embankment in all probability will be adopted, whereas in rocky country a masonry or concrete dam would be more suitable. This generalization is not rigid, for on the granite of Cornwall in the same valley there is a concrete dam upstream with an earthen embankment downstream. The earthen embankment usually depends upon a puddled or rolled clay core for watertightness, both in the trench below ground and in the body of the embankment above ground, although a concrete-filled cut-off trench was frequently used before grouting techniques were available. It is now generally ruled out on account of cost and permeable strata are sealed by grouting with cement or chemical mixtures from ground level or occasionally from a gallery or crest level after completion of construction.

Both upstream and downstream, earthen embankments depend for stability on adequate drainage by rubble and 'selected' material (known as 'filters', specially graded to certain rules) which is placed against the core or laid in layers in the ordinary filling. Broken rock, or material containing a large percentage of broken rock, permits steeper slopes to be adopted; thus for the upstream slope 1 (vertical) in 3 or 4 (horizontal) might suffice, and for that downstream 1 in 2 or 3; whereas, for the ordinary clayey materials so common in this country, 1 in 5 or 6 or more upstream, and 1 in 3 to 5 downstream are common. The guiding principle in such embankments is adequate drainage, as well as the application of the principles of soil mechanics[28] such as tests for shear strength and other properties of clays, and deduced slip planes and stability diagrams.[29] Bearing pressures of the subsoil below the embankment must be known if the weight of the embankment will be sustained without subsidence.

28.9.5.2 Cost of earth dams

It is useful to estimate the approximate cost of a dam when carrying out a comparative study of different sites and Mitchell[30] has suggested a formula, subsequently modified by Whincap as follows:

Cost (£ sterling) =
$3.6aLH^2 + 0.6bLH^2 + 2.0cLD + 66\,000H + 202\,000$
(1)　　　(2)　　　(3)　　(4)　　　(5)

where L is the crest length, H is the mean height from ground to crest (i.e. area of cross-section of valley divided by length of crest of chard across the valley), D is the mean depth of cutoff trench (all in metres).

The numbers in brackets below the items in the formula define the different sections as follows:

(1) Cost of forming the embankment, where a is mean rate per cubic metre.
(2) Represents the extra cost of rolled clay core where b is the rate per cubic metre. (Typical examples indicate that b varies from zero to one-third of a.)
(3) Cost of a concrete-filled cutoff trench 2 m wide, c is the rate per cubic metre; not often required. If rolled clay cutoff used, included in (2).
(4) Cost of stream diversion, overflow and valve shaft.
(5) Miscellaneous items such as pipework and valves, access to valve shaft, reinforcement and steelwork, recording instruments.

Items (4) and (5) are based on typical 1985 values, and appropriate 1985 values for factors a, b and c would be £16.50, zero to £5.50 and £124 respectively.

It should be noted that the formula does not include a number of common items in comparative studies such as contingencies, engineering design and supervision, land, diversion of utilities, site clearance, access roads and bridges, excavation of unsuitable material, grouting or construction camp accommodation. The first three of these items generally amount to approximately 25% of the formula cost, but the remainder needs to be priced separately as they are particular to a site.

28.9.5.3 Freeboard

For earthen dams and embankments it is important that the 'freeboard', the distance between top-water level and crest, should be adequate; it depends on the 'fetch' (the maximum distance the water is impounded at right angles to the dam), the density of trees and vegetation, the elevation of the site, intensity and direction of wind and, of course, a correct assessment of the floods.[27]

28.9.5.4 Design of earth dams

Some of the basic factors considered in design are:[31]

(1) To ascertain by site investigation and laboratory testing of the materials, available for constructing the dam, preferably those nearest to the site for economy.
(2) To ascertain the conditions and properties of the strata under the embankment to resist sliding or slipping.
(3) To analyse the factor of safety over *slipping* for the particular dam section by choosing a slip plane or circle through the probable weakest line of failure in and under the dam. This can now be done by one of several computer programs.

Common slopes of embankments vary between 1:1 and 1:6.

28.9.5.5 Slurry trench

Certain Bentonite clays or slurries have been used for sinking dam trenches in soft strata without timbering. The clays have the effect of remaining liquid when the trench is being dug but form a gel or colloid (slightly heavier than water) when not disturbed which has the effect of keeping the walls of the trench from falling in. Recently, Bentonite cement mixtures have been used to effect a permanent cutoff seal in porous strata below embankments as at Bewl Bridge,[32] in Kent.

28.9.5.6 Pore pressure under earth dams

Where embankments are constructed of, or rest on, cohesive materials containing water within the pores, a pore pressure is set up within the material when loaded either during construction or on the subsequent filling of the reservoir. The increase in pore pressure leads to a reduction in the shear strength of the material and a corresponding reduction in the factor of safety which can lead to failure of the embankment by slipping. To obviate this effect, sloping layers or 'blankets' of drainage material are included in the embankments to enable the pore pressure to be dissipated before it reaches a dangerous level.

A similar effect can be observed by infiltration of groundwater and if this is anticipated then relief wells or vertical sand drains should be incorporated under the embankment.

Pore pressure can be measured by installing piezometers, which are ceramic pots sealed in the strata and connected to a gauge by fine tubes.

28.9.5.7 Deformation of earth dams

Under this term are included the causes of the shapeless early nineteenth-century embankments often seen before the Safety Provisions Act, 1930.

These include: (1) subsidence of the crest – some have sunk nearly below the overflow top water level; (2) irregular shape of the embankment due to local slips; and (3) irregular toe lines due to slides.

Although sagging of the crest can be remedied by levelling and adding additional material, and irregular shapes and toes can be regraded and re-aligned, little or nothing is known of what is going on in the strata inside a dam and, since it is dangerous to take such measures without analysing the original cause of the deformation, instruments have been developed to indicate both horizontal and vertical movements.

Horizontal movements can now be measured by vibrations in an electrically stimulated wire (embedded between two concrete blocks) whose tension varies with the horizontal stress. For vertical movements, the relative displacement can be ascertained by lowering an induction coil down through a tube connecting two plates, or by various types of instruments using the U-tube principle. A fuller exposition on instrumentation in earth dams is given by Rofe and Tye.[33] For schemes including recent earthen embankments, see also works by Hallas and Titford,[34] Picken,[35] Walters and Walton,[36] Kennard and Kennard,[37] Kennard and Crann.[38]

28.9.5.8 Compaction of earth and rockfill dams

In earth dams it is important to have the impervious clay core well compacted either by heavy rollers or light vibrating rollers operating on layers of 250 to 500 m. The 'voids' percentage should be kept below 5%. The compaction of the rest of the 'fill' need not be to such a high specification. In a case such as Scammonden Dam (height about 80 m) an earth and rockfill dam near Huddersfield, a special investigation had to be made to ensure minimum settlement because a six-lane motorway goes over the crest.[40]

For this particular site (consisting of alternating grits, sandstones and shales of the Millstone Grit formation) extensive large-scale experiments revealed that the best compaction depended upon: (1) how the rock was quarried; (2) how the material was deposited on the embankment; and (3) the best type of compacting machines.

28.9.6 Concrete dams

The adoption of concrete for constructing a dam depends on the topography and geology of the site. The trench, as in the case of an earthen embankment, may be filled either with puddled clay or concrete, depending on the hardness or softness of the strata penetrated and the cost of filling it either with clay or concrete. Below the trench there may be a necessity for extensive grouting, to reduce leakage under the dam. The surface for the broad foundation for the superstructure must also be prepared. These foundation works may cost as much as the superstructure seen above ground.

For a straight gravity concrete dam 30 m high, the broad foundation would be about 18 m wide, and if rock or other stable formation is not found reasonably near (say 7 m below) the surface, considerable expense in foundation work may also be entailed. For a buttress or multiple-arch dam, it would be less, but the strata sustaining the buttresses would have to be stronger. Most dam failures may be attributed to faulty foundations.

Above ground, the *gravity* dam is so-called because any section can stand by itself because of its weight. Concrete is generally vibrated, in dam construction particularly, to eliminate air pockets, prevent leakage and increase speed of setting. Shuttering must be especially well constructed to withstand vibration during construction.

Curved-on-plan gravity dams are sometimes substituted for straight dams for aesthetic reasons, but the gravity section of the dam cannot be reduced.[41]

28.9.6.1 Cost of concrete gravity dams[31]

For comparative studies the cost can be illustrated by the following formula originally devised by Mitchell.

$$\text{Cost} = £(0.375xLH^2 + 0.675xLW^2 + 0.75yLD_1 + 2.0zLD_2 + 152\,000$$

$$\quad\quad (1) \quad\quad (2) \quad\quad (3) \quad\quad (4) \quad\quad (5)$$

where H is the mean height from broad foundation to crest in metres, i.e. area of cross-section of valley divided by length of crest or chord across valley, L is the length of crest, in metres, W the width of dam at crest in metres, D_1 the mean depth of broad foundation below ground level in metres and D_2 the mean depth of cutoff trench below broad foundation in metres.

Item (1) Cost of concrete in dam where x is rate per cubic metre (£82.50 at 1985 prices).

Item (2) Cost of crest road or footpath.

Item (3) Cost of excavating broad foundation where y is rate per cubic metre (£8 at 1985 prices).

Item (4) Cost of concrete-filled cutoff trench 2 m wide where z is rate per cubic metre (£99 at 1985 prices).

Item (5) Ancillary works such as pipework and valves, reinforcement and steelwork and recording instruments for water level, overflow and discharge below dam.

Note: Like the corresponding formula for earth dams, this formula is intended for comparing a number of sites when making a preliminary survey of alternative sources, and does not include the other factors described in that context.

28.9.6.2 Buttress and multiple-arch dams

The buttresses of a buttress dam form part of two adjacent halves of two arches (thick) which act as cantilevers and hence if the bearing pressure of one buttress differs relatively from the other, movement may occur at the centre of the arch, where an expansion joint is (or should be) inserted. The internal buttresses of the multiple-arch dam are merely blocks of concrete acting as abutments for supporting the two halves of two rigid (thin) arches. Hence, the foundations for the buttresses of the multiple-arch dam must have equal bearing capacity to avoid fracture of the true thin arches; whereas in those for the buttress dam, some inequality is taken care of by the expansion joint between the two cantilever arms of the buttresses.

Thin-arch dam. The true thin-arch concrete dam is suitable for the valley which has a good foundation and where the width at the level of the dam crest is not more than 3 times the proposed maximum height of the dam.

The volume of concrete in an arch dam is about half that in a comparable gravity dam.

Preliminary calculations are directed to finding the thickness t in metres of the dam at any depth in terms of the water pressure P (i.e. on a metre strip of dam), and the radius R of the upstream face in metres and the compressive strength S. If the abutment pressure is greater than the compressive strength of the strata on which the abutment rests, the concrete should be increased in width.

Thick-arch dam. The thick-arch dam lies between the thicknesses of the gravity and arch dams. It has been adopted in valleys with chord:height ratios up to 5 or 6. The theory of design involves doubtful assumptions, but nevertheless tests on models seem to confirm that these assumptions are reasonable. The saving in concrete and cost for all arch dams is well worthwhile but the supporting foundation strata must be above suspicion. Some notable failures have been attributed to inadequate treatment of the foundations.

Cupola, dome, or double curvature arch dam. This type of dam is generally suitable for valleys with a chord:height ratio of under 3. It is economical in concrete and its strength, for this thickness, is like an eggshell. Calculations are complex but models for testing to destruction are used with success. Foundations must be above reproach.

28.9.6.3 Prestressed concrete dam

Prestressed concrete dams have been developed and adopted in recent years for economy of concrete where good foundations are available. The thin concrete structure is anchored by steel cables embedded vertically in the concrete of the structure and with grout inserted in boreholes in hard strata below.

Existing dams have been raised successfully and for this purpose the use of prestressed steel enables the existing structure to be little interfered with beyond drilling vertical boreholes for the prestressed cables to be inserted.[42]

28.9.6.4 Special problems concerning concrete dams[32]

Floods over dams. The precise estimate of floods is not so important as those for earth dams but nevertheless overtopping the *crest* should not be permitted not only because of the extra weight of water pressure on the dam but particularly the risk of scouring the strata under the toe, unless these factors are taken into account in the design of the dam. It is true that an overflow of 100 m over the Vaiont dam caused no damage to the dam but the abutments were against hard dolomite limestone.

Rock testing by seismic methods. The velocity of sound through rock may give a valuable indication of its state below the surface, i.e. whether it is faulted, dry, wet, disturbed, open or revealing unsuspected faults or whether the density of concrete foundations is sufficient and the efficiency of grout curtains and contact grouting particularly for concrete dams which are on rock.

Pore pressure and uplift. Pore pressure is dangerous under a concrete dam as the pressure is upward and 'lightens' the dam tending to turn it over and make it slide.

In some cases pore pressure leads to leakage at the toe of the dam. In other cases it appears after a few years possibly from some kind of clogging and the pressure has to be reduced either by grouting or alternatively by drilling drains under the dam which, although it increases the leakage, reduces dangerous uplift.

Deformation of dam and strata. Strain gauges embedded in the dam give a measure of any untoward trouble going on in a solid concrete dam. They consist of electrically stimulated vibrating wires in which any change in vibration speed indicates change of stress in the dam, indicating degree of movement.

Other indicators of deformation are surface effects due to weather and temperature for which thermometers are inserted in the dam. Although movement of a dam can be ascertained by elaborate surveying equipment, pendulums and inverted pendulums inserted in boreholes inside the dam measure deformation more exactly.

Pendulums for high dams are used especially for showing the movement when the reservoir is filled and empty and if these values are the same or whether they change over the years. The relative movement of the strata with the dam can also be found.

28.9.7 Examples of raised lakes

There are two or three instances of the utilization of natural lakes (other than Thirlmere and Haweswater which have been developed by high dams), the chief of which is Loch Katrine, for Glasgow, where the natural surface of the lake has been raised 4.5 m to draw off 1 m below the original natural lake level, the total supply available being about 320 million l/day. Similarly, the Crummock Lake for Workington has been raised 600 mm; the drawoff pipe is 2.4 m below this level. This is estimated to give a gross supply of 59 million l/day. The utilization of existing lakes raises special methods of tunnelling to draw water from lower existing levels as well as raising the level of water and at the same time coping with storm water.

28.9.8 Pumped storage reservoirs

The largest examples of these reservoirs, constructed in Britain on clay and with clay cores and supported by any suitable material nearest the site, are those of the London metropolitan area. These reservoirs are of the order of 20 m in height and store water pumped from the Thames during periods of high flow.

28.10 Desalination[43]

Desalination should be regarded as a method of treatment to remove impurities, particularly salts, from a saline water.

It has come into use at an increasing rate during the last 30 yr; and in certain circumstances can compete with orthodox sources which depend on conventional treatment of water from boreholes, impounding reservoirs and river waters. However, generally the cost of desalination is at least double the cost of fresh water sources, and frequently a factor of 10 greater. The variation in cost arises from:

(1) The degree of salinity to be treated, e.g. sea water (chlorides 35 000 mg/l), brackish water (5000 down to 500 mg/l, which is tasteless to most palates).
(2) The location and availability and cost of power, heat, transport.
(3) The selection of plant, i.e. (a) multi-stage flash (MSF) and other variations of this distillation plant; (b) electrodialysis and the somewhat similar reverse osmosis plant; and (c) several other types used on ships and in factories and other special purposes.

28.10.1 Multi-stage flash distillation (MSF) (vacuum separation)

If sea water is evaporated, steam is condensed as pure water leaving solid salt behind, as in the Dead Sea region. In the mechanical MSF process the sea water is pumped through a pipe (sufficiently long to ensure that sand is not drawn in during rough weather), heated, and passed into a tank under partial or reduced vacuum. As water boils at a lower temperature than normal when at a lower pressure (as on a mountain), fresh water 'flashes off' as steam which is cooled by incoming pipes conveying the sea water and condenses to fresh water. This process is repeated in several stages to increase the efficiency. Many problems arise, apart from the multiplicity of pipes, particularly the elimination of alkaline and calcium sulphate scale which can be controlled by the addition of polysulphide, acid, and lime to increase the pH from 5 to 7. If temperatures could be used above 120°C the cost could be reduced.

28.10.2 Electrodialysis (membrane–electrode separation)

If brackish water is pumped through a tank between two membranes on each side of which is a positive electrode and a negative electrode, the electropositive sodium will go through one membrane to the negative electrode and the electronegative chlorine will go through the other membrane to the positive electrode. The water between the membranes, thus denuded of sodium chloride and other salts, is fresh. The method is only economic for water containing up to about 10 000 mg/l, and for reductions down to 500 mg/l, a cost of 20 to 50p per 1000 l is incurred (1983).

Operational plants with outputs of up to 22 million l/day have been installed in the Middle East, and the method is now well established.

28.10.3 Reverse osmosis (membrane pressure separation)

'Osmosis' may be envisaged as a natural flow of fresh water into sea water when in contact with each other; whereas 'reverse osmosis' acts when pressure is applied to the brine which, when pushed through a special membrane such as cellulose acetate, causes the fresh water to flow out of the brine for separate use.

Several plants are now in operation and act in the same range as electrodialysis plant. The main disadvantages are the high operating pressure and the fine limits involved in production of the membrane, but most of these problems have been overcome. A further disadvantage is the need for prefiltration and treatment to remove excess solids and biological impurities before the influent can be accepted through the membranes without early clogging.

28.10.4 Freezing

Two freezing processes are being evaluated – vacuum and secondary refrigerant – but neither are yet proven in a full-scale operation.

28.11 Treatment of water for potable supply [44,45]

The type of treatment required varies considerably according to the source of supply of the raw water, whether it be a surface water or from underground sources.

Surface waters may be divided into upland and lowland sources. Except in the case of very small supplies, upland waters are usually impounded in the catchment area and are good-quality waters, low in dissolved solids and with little organic contamination from the biological point of view, although they are frequently high in organic colour due to deposits of peat and can also contain, particularly in the Pennine area, iron, manganese, and aluminium in solution.

Although impounded water is generally of good quality it can be subject to disturbance due to stratification, thermal turnover if the water is deep, or by surface winds and flash runoff if the water is shallow. In these circumstances there is a marked and often very sudden deterioration in the quality of the water and, although this may be for only a short duration, it must be given full consideration when a treatment plant is being considered.

The increasing demands being made on upland sources, the difficulty of finding suitable reservoir sites and the extremely high cost of trunk main laying has led to a reappraisal and has indicated that in many cases a greater reliable yield can be established by using a reservoir for regulation of the river flow and abstracting direct from the river in its lower reaches, e.g. Clywedog.[46] Similarly, lowland pumped storage reservoirs (e.g. Grafham Water, Draycote and Empingham) are filled from low-quality river waters.

Therefore, increasing use is made of lowland waters taken from the lower reaches of comparatively slow-moving and turbid rivers. They present far greater problems from the point of view of organic and industrial pollution and the treatment aspect becomes far more complex.

Increasing sophistication in the equipment available for automatic control is leading to consideration of continuous monitoring of raw-water qualities for automatic control of the treatment process, and this now becomes established practice.

Underground supplies from aquifers such as Limestone, Chalk, Sandstone and Greensand are normally biologically pure, but are often very hard and can contain objectionable levels of iron, manganese, sulphates and chlorides, as well as excess carbon dioxide and hydrogen sulphide. In some circumstances, river gravel can also be used as an underground source, although this is not necessarily of the same degree of organic or biological purity. Although there are large quantities of water in old mine workings, it tends to be very high in dissolved solids, particularly sulphates and chlorides. Boreholes near the coast can also suffer from an infiltration of salinity with resultant brackish water.

28.11.1 Water characteristics

The main characteristics of a raw water which affect treatment processes are summarized in Table 28.10 follows:

Table 28.10 Chemical characteristics of raw water

Group	Main constituents affecting treatment
Gases	Oxygen, carbon dioxide, hydrogen sulphide, ammonia
Suspended solids	Clays, minerals, siliceous matter, vegetable debris
Organic matter	Organic acids, humus, peat, algae, faecal matter

Dissolved solids	(a) *Hardness salts:*
	    Permanent – calcium and magnesium sulphates, nitrates and chlorides
	    Temporary – calcium and magnesium bicarbonates
	(b) *Nonhardness salts:*
	    sodium sulphates, chlorides, nitrates or bicarbonates

A range of possible treatments for different characteristics of the raw water is summarized in Table 28.11.

The processes given in Table 28.11 are associated with the appropriate sedimentation and filtration plant and techniques. The full range of treatment for a poor-quality lowland river water may include storage, algal control, aeration, pH control, coagulation, precipitation softening, flocculation, sedimentation, filtration, chlorination, dechlorination, pH adjustment, and taste control.

The range of chemicals commonly used in treatment processes is summarized in Table 28.12.

The basic principles adopted in each stage of these treatment processes are briefly described in the following, with typical examples.

Table 28.11 Range of treatments for different characteristics

Characteristics	Possible treatment
Gases	Aeration
Dissolved impurities	Precipitation – (oxidation)
Suspended matter	Coagulation, settlement
Colloidal matter	Coagulation
Colour	Coagulation, activated carbon, ozone, chlorine
Odour	Aeration, activated carbon
Taste	Activated carbon, chlorine, chlorine dioxide, ozone
Acidity/free carbon dioxide	Aeration, control by alkali
Hardness	Lime and/or soda precipitation, or ion exchange methods
Iron and manganese	Aeration, precipitation and filtration with iron-removal media
Other metals	Coagulation and precipitation
Salinity (brackish)	Distillation, demineralization, reverse osmosis, freezing
Oil	Flotation, coagulation
Algae	Straining, copper sulphate, chlorine, cuprichloramine
Biological impurities	Storage, chlorine, chloramine, ozone, ultraviolet light
Industrial pollution	Combination of above as required

28.11.2 Storage

For a scheme using direct river abstraction a raw-water storage of 7 days is recommended to allow for settlement of heavy silt load to even out any rapid changes in water quality and to allow for rejection of water containing accidental and heavy pollution (in, for example, Oxford[47] and Nottingham[48]).

28.11.3 Algae

The growth of severe blooms of algae which would interfere with the treatment process can be inhibited by the use of an algicide (e.g. copper sulphate) and the exclusion of direct sunlight.

28.11.4 Aeration

The level of dissolved gases can be reduced substantially by an aeration system which in the order of ascending efficiency takes the form of cascades, sprays, and induced draft towers. If the water is particularly 'flat' in appearance the level of oxygen can be increased and the appearance of the water enhanced by similar means (in, for example, Ardleigh[49] and Oxford).

28.11.5 Coagulation

Any raw water containing colour or finely divided suspended solids needs the addition of a coagulant to neutralize the electrical charges causing dispersion and induce the impurities to coalesce and flocculate. This process is often assisted by slow-speed agitation to increase the collision between the particles. Normally, the reagents are delivered by road vehicle and taken into bulk storage at the treatment works. The method of adding the coagulant most usually adopted is the use of positive displacement ram-type metering pumps which can be controlled easily to vary the dose according to the treatment flow and, if required, to the water quality (as, for example, in Colchester and Swansea).

Table 28.12 Chemicals commonly used in water treatment

Substance	Formula	Purpose
Activated carbon	C	Taste and odour control
Aluminium sulphate (alum)	$Al_2(SO_4)_3$	Coagulant
Ammonia	NH_3	With chlorine for sterilization
Ammonium sulphate	$(NH_4)_2SO_4$	A source of ammonia for chloramine
Calcium carbonate (chalk)	$CaCO_3$	A source of bicarbonate alkalinity
Calcium hydroxide (slaked lime)	$Ca(OH)_2$	Softening and pH control
Calcium hypochlorite (bleach powder)	$Ca(OCl)Cl$	Disinfection
Calcium oxide (quick or burnt lime)	CaO	Softening and pH control
Chlorine	Cl_2	Disinfection
Chlorine dioxide	ClO_2	Disinfection, taste and odour removal
Copper sulphate (bluestone)	$CuSO_4 4H_2O$	Algal control
Ferrous sulphate (copperas)	$FeSO_4 7H_2O$	Coagulant
Ferric chloride	$FeCl_3$	Coagulant
Ozone	O_3	Disinfection, taste, odour and colour removal
Potassium permanganate	$KMnO_4$	Removal of iron, manganese, algal control
Sodium aluminate	$Na_2Al_2O_4$	Coagulant
Sodium carbonate (soda ash)	Na_2CO_3	Removal of permanent hardness and pH control
Sodium chloride (common salt)	$NaCl$	Regeneration of zeolites
Sodium hypochlorite (Chloros or Voxsan)	$NaOCl$	Disinfection
Sulphur dioxide	SO_2	Dechlorination

28.11.6 pH control

For efficient coagulation the pH value is critical and as the pH of the water with the added coagulant is unlikely to be at the required level it is necessary to correct this by the addition of acid or alkali, preferably under automatic control. Variation of pH from the level necessary for optimum coagulation can produce light fluffy and fragile flocs and high residual dissolved coagulant (as, for example, in Bradford and Londonderry[50]).

28.11.7 Precipitation

In any treatment process whch includes coagulation and sedimentation it is possible to precipitate the hardness salts by the addition of lime and/or soda and the precipitated salts are effectively removed in the general system and can in some circumstances increase the efficiency of the treatment although inevitably producing an increase in the volume of sludge to be discharged from the works (in, for example, northeast Lincolnshire[51] and Sheffield[52]).

28.11.8 Mixing

As the volume of reagent is small compared with the volume of water being treated it is critical to ensure that the chemical is fully dispersed into the body of the water and also that the reagents are added in the correct sequence according to the chemical requirements. Reagents can be diluted to ease the mixing problem, this is really two-stage mixing, and are then added in an area of turbulence induced either hydraulically or mechanically: hydraulically in the nappe of a weir or the standing wave of a Venturi flume or mechanically by high-speed mixing and sometimes by a pumped recirculating system (as, for example, in Bristol[53]).

28.11.9 Flocculation

After the reagents have been added and fully mixed it is normal to induce flocculation by passing the water through an area of slow agitation which, again, can be induced either hydraulically or mechanically (as, for example, in north Derbyshire).

28.11.10 Sedimentation

In its simplest form, sedimentation is the use of tanks giving a retention time that is long enough for the floc particles to settle and compact into sludge on the bottom of the tank, from which point the solids are discharged for disposal and the settled water is decanted to the following filters. However carefully such tanks are designed, the physical retention seldom exceeds 40% of the nominal retention and this has led to the development of the upflow type of treatment unit. After a flocculation zone the water is induced to flow vertically upwards through an area of suspended sludge where the floc particles have a large area of contact which greatly assists in the formation of denser agglomerates. Such tanks are designed so that the sludge can be withdrawn at the rate at which it is forming and provide a stable process that can be controlled over a varying range of duties. Construction can be in either steel or concrete, with the units either square or circular in plan, and as the capacities of treatment works increase it is generally more economical to consider a smaller number of circular-type tanks (as, for example, in Bradford in the Derwent Valley).

28.11.11 Filtration

After coagulation and sedimentation, the water still retains a quantity of suspended matter which is removed by filtration. It should be noted that the filtration process associated with the treatments being described is that known as 'rapid' as distinct from slow-sand filtration which is a biological process completely in itself.

The settled water passes through a layer of comparatively fine and specially graded sand supported on underbeds of graded pebbles with a piped header and lateral under-drain system, or supported on a flat floor with a system of closely spaced nozzles. In either design the clean water is collected from the base of the filter and as the resistance to flow increases, in proportion to the quantity of intercepted matter building up, the filter bed is cleaned, first by expanding the compacted bed, usually by the application of an air scour, which effectively loosens the intercepted impurities which are then flushed out to waste by a reverse flow of water.

The filter bed can be contained equally well in a steel pressure vessel or an open-topped gravity tank and the siting of the plant relative to the hydraulic gradient can determine which method is preferable. Where large flows are being considered the gravity-type filter does not have the same restriction on the size of individual units and it is unusual to follow sedimentation, requiring open-topped type tankwork, by pressure filters (as, for example, in the Lune Valley, West Glamorgan[53]).

Filtration technique is currently going through a period of change with advocates for downward, upward and sideways flow, for deep beds and shallow beds, for single media, multimedia, and multi-layer media.

All these variations have some application, however limited, and provide filtration in the depth of the bed rather than on the surface, with a greatly increased efficiency. Although there is very little, if any, long-term operating experience on some of the designs, the use of a two-layer downflow filter with a top stratum of graded anthracite resting on a shallower layer of conventional sand is a system that has been in use for a number of years in the UK and in gaining support as it has been proved that filter ratings can be increased and the length of filter runs extended.

28.11.12 Backwashing

Air for expanding and scouring the filter bed is normally provided by electrically driven blowers delivering large volumes of air at relatively low pressure direct to the induction system which is usually the underdrain system used for collecting the filtrate. Wash water is most often provided by direct pumping to the same common induction system (as, for example, in Wakefield[54]).

28.11.13 Chlorination

After the water has been clarified satisfactorily it is still necessary, if it is to become potable, for it to be fully disinfected, and chlorine is the most usual reagent for this duty. It is normally delivered as a liquid under pressure in either cylinders or drums, depending on the quantity required, and in some of the largest installations it is being delivered by bulk tanker and transferred into storage vessels at the treatment plant.

As it is considered prudent to carry a chlorine residual into the reticulation system as a measure of safety this has to be controlled at a level low enough to be unobjectionable to the consumer. Current practice is often to dose above the chlorine demand of the water and to control the residual passing into supply by adding sulphur dioxide to neutralize any excess. Sulphur dioxide is handled as a liquid under pressure in exactly the same way as chlorine and the dosing is normally under automatic control to ensure that the final chlorine residual is maintained at the correct level (as, for example, in West Surrey).

28.11.14 pH adjustment

Depending on the treatment adopted, the final water is unlikely to be at the pH required for distribution purposes and will need correction by the addition of acid or alkali. At the same time it is important to correct any undue corrosive tendencies which may be inherent in the treated water (see, for example, descriptions of works at Bedford[55] and in north Devon[56]).

28.11.15 Taste control

Taste which is objectionable as far as the consumer is concerned can be present in the raw water or can develop during the treatment process and activated carbon is often used to absorb the elements that are responsible. It can be added as a powder before the sedimentation process and removed with the sludge, directly as a powder or granule on to the filter beds and removed with the washwater, or as a granular filter medium in a separate filtration stage added to the end of the clarification treatment. In the first two applications the carbon is not recovered but if it is used as an additional filtration unit it can either be regenerated on site or returned to the manufacturer for this purpose (as, for example, in east Surrey[57] and Oxford[47]).

28.11.16 Waste products

Whatever the process used for clarification, or precipitation softening, the impurities removed are concentrated in the form of a sludge which under the best operating conditions is unlikely to be less than 95% water. In this form it can be fed to a filter press, or possibly a centrifuge for a softening sludge, to produce a dry solid suitable for mechanical handling and disposal. The filtrate, or centrate, has to be disposed of as a liquor.

In a few cases, attention is being given to the possibility of treating the sludge with acid to recover the coagulant but this process is not yet proven as commercially viable (as, for example, in the Fylde, mid Northamptonshire).

Sludge disposal in waterworks is not such a problem as sewage sludge. Local conditions can normally cope with the smaller quantities of waterworks sludge by distributing it on land, quarries, pits, river or sea which may be, and generally are, available.

Transport of sludge should be in closed containers in hilly districts, otherwise there is loss of sludge from well-filled open-top vehicles!

References

1 Institution of Water Engineers and Scientists (1979) 'The structure and management of the British water industry.' *Water practice manuals*, Book 1, p. 19.
2 Phillips, J. H. (1983) 'Water usage and the quantification of unaccounted water in a universally metered supply area.' *J. Instn W. Engrg*, 37, 4.
3 Sharp, R. G. (1967) 'Estimation of future demand on water resources in Britain.' *J. Instn W. Engrg* 21, 232.
4 Ministry of Agriculture, Fisheries and Food (1967) *Watercress growing*. Ministry of Agriculture Bulletin No. 136. HMSO.
5 Klein, R. L. (1959) 'The use of asbestos-cement pressure pipes in water supply practice.' *W. & W. Engrg*, 63, 356.
6 National Water Council (1980) *Leakage control policy and practice*. NWC.
7 Hydraulics Research Station, Hydraulic Design Charts Nos 1, 2 and 4.
8 Hayton, J. G. (1964) 'The use of polythene sleeving as a form of protection to spun iron water mains against external corrosion.' *J. Instn W. Engrg*, 18, 465.
9 Young, O. C. and Smith, J. H. (1970) *Simplified tables of external loads on buried pipelines*. Building Research Establishment, Garston.
10 Hydraulics Research Station (1983) *Charts for the design channels and pipes*. HRS.
11 Lea, F. C. (1938) *Hydraulics*. Edward Arnold.
12 Lapworth. C. F. (1944) *Surge control in pipelines*. *J. Instn W. Engrg*, 49, 29.
13 Lupton, H. R. (1953) 'Graphical analysis of pressure surge in pumping systems.' *J. Instn W. Engrg*, 7, 87.
14 Thorley, A. R. D. and Enever, K. J. (1979) *Control and suppression of pressure surges in pipes and tunnels*. Construction Industry Research and Information Association Report No. 84, CIRIA, London.
15 Crump, E. S. (1952) 'A new method of gauging flow.' *Proc. Instn. Civ. Engrs*, 1, 749.
16 Water Resources Board (1970) *Crump weir design*. WRB Technical Note No. 8,
17 British Geological Survey (n.d.) *Geological survey maps and memoirs*. British Geological Survey, London.
18 Walters, R. C. S. (1929) 'Hydro-geology of Chalk.' *J. Instn. W. Engrg*, 34, 79.
19 Walters, R. C. S. (1936) 'Oolites.' *J. Instn. W. Engrg*, 41, 134.
20 Ineson, J. (1970) 'Development of groundwater resources in England and Wales.' *J. Instn. W. Engrg*, 24, 155.
21 Rofe, B. H., Durrant, P. S. and Egerton, R. H. L. (1977) 'Some aspects of the use and management of groundwater resources.' *J. Instn. W. Engrs and Scientists*, 31.
22 Penman, H. L. (1954) *Evaporation over the British Isles*. Institution of Water Engineering, London.
23 Armstrong, R. B. and Clarke, K. F. (1972) 'Water resource planning in SE England.' *J. Instn W. Engrg*, 26, 11.
24 Lapworth, C. F. (1949) 'Reservoir storage and yield.' *J. Instn W. Engrg*, 3, 269.
25 Mansell-Moulin, M. (1966) *Flow frequency curves for design of catchwaters*. Institution of Water Engineering, London.
26 National Environmental Research Council (1975) *Flood studies report*.
27 Institution of Civil Engineers (1978) *Floods and reservoir safety – an engineering guide*. Thomas Telford, London.
28 Terzaghi, K. and Peck, R. B. (1967) *Soil mechanics in engineering practice*. New York.
29 Bishop, W. A. (1955) 'The use of the slipcircle in stability analysis of slopes.' *Géotechnique*, 5, 7.
30 Mitchell, P. B. (1951) 'Reservoir site investigation and economics.' *J. Instn W. Engrg*, 5, 445.
31 Walters, R. C. S. (1973) *Dam geology*. (Appendices by J. L. Knill). Butterworths, London.
32 Kennard, M. F. and Eden, W. H. (1978) *Bewl Bridge Reservoir: design and construction*. Thomas Telford, London.
33 Rofe, B. H. and Tye, P. F. (1971) 'Application of instrumentation to earth dams.' *J. Instn W. Engrg*, 25, 157.
34 Hallas, P. S. and Titford, A. R. (1971) 'Design and construction of Bough Beech Reservoir.' *J. Instn W. Engrg*, 25, 293.
35 Picken, J. A. (1957) 'The Chew Stoke Reservoir scheme.' *J. Instn W. Engrg*, 11, 33.
36 Walters, R. C. S. and Walton, R. J. C. (1957) 'Water supply for the Yeovil District (Sutton Bingham Scheme).' *Proc. Instn Civ. Engrs*, 8, 71.
37 Kennard, J. and Kennard, M. F. (1962) 'Selset Reservoir.' *Proc. Instn Civ. Engrs*, 21, 277.
38 Kennard, M. F. (1963) 'Balderhead Reservoir.' *Civ. Engng. & Publ. Wks Rev.*, 58, 633.
39 Crann, H. H. (1968) 'Design and construction of Llyn Celyn.' *J. Instn W. Engrg*, 22, 13.
40 Williams, H. and Stothard, J. N. (1967) 'Rock excavation and specification trials for the Lancashire–Yorkshire motorway – Yorkshire (West Riding) section.' *Proc. Instn Civ. Engrs*, 37, 607 and Discussion, 38, 135.
41 Farrar, R. E. S. (1972) 'Meldon Reservoir.' *Civ. Engng & Publ. Wks Rev.*, 67, 895.
42 Water and Water Engineering (1961) 'The heightening of Argal Dam for the Falmouth Corporation Water Undertaking.' *W. & W. Engrg*, 65, 537.
43 Silver, R. S. (1967) *Desalination*. HMSO, London.
44 Skeat, W. O. (ed.) (1969) *Manual of British water engineering practice*, 4th edn, vol. 3, Institution of Water Engineers.
45 Holden, W. S. (ed.) (1970) *Water treatment and examination*. Churchill.
46 Fordham, A. E., Cochrane, N. J., Kretschmer, J. M. and Baxter, R. S. (1970) 'The Clywedog Reservoir project.' *J. Instn W. Engrg*, 24, 17.
47 Cartwright, F. (1964) 'Design of Farmoor Treatment Works, Oxford Corporation Water Department.' *J. Instn W. Engrg*, 18, 381.
48 Adams, R. W., Robinson, R. D. and Kennett, C. A. (1973) 'The River Derwent Scheme of the Nottingham Corporation.' *J. Instn W. Engrg*, 27, 15.
49 Water and Water Engineering (1972) 'The Ardleigh Reservoir Scheme in North-East Essex.' *W. & W. Engrg*, 76, 3.
50 Wilcock, E. J. and Sard, B. A. (1964) 'Design and operation of the Carmoney Water Treatment Works: Faughan River scheme – Londonderry RDC.' *J. Instn. W. Engrg*, 18, 477.
51 Ashe, R. V. (1966) 'The Great Eau scheme: North-east Lincolnshire Water Board.' *J. Instn. W. Engrg*, 20, 435.
52 Earnshaw, F. (1962) 'Design of the Yorkshire Derwent headworks.' *J. Instn. W. Engrg*, 16, 139.

53 *Water and Water Engineering* (1955) 'The Usk Reservoir Scheme of the Swansea Corporation.' *W. & W. Engrg*, **59**, 377.
54 Collins, P. G. M. and Gibb, O. (1964) 'Design and construction of the Fixby Treatment Works of the Wakefield and District Water Board.' *J. Instn. W. Engrg*, **18**, 491.
55 *Water and Water Engineering* (1959) 'New water treatment works of the Borough of Bedford Water Undertaking.' *W. & W. Engrg*, **63**, 61.
56 *Water and Water Engineering* (1972) 'The Meldon Reservoir scheme of the North Devon Water Board.' *W. & W. Engrg*, **76**, 353.
57 Shinner, J. S. and Davison, A. S. (1971) 'The development of Bough Beech as a source of supply (The East Surrey Water Company).' *J. Instn. W. Engrg*, **25**, 243.

Bibliography

General Water Supply
The journals of the Institution of Water Engineers and Scientists (IWES).
The *Proceedings* of the Institution of Civil Engineers (ICE).
Institution of Water Engineering (1969) *Manual of British water engineering practice*. IWE, London.
Twort, A. C., Hoather, R. C. and Law, F. M. (1974) *Water supply*. Arnold, London.
Institution of Water Engineers and Scientists (1983) *Water practice manual*, Book 3: 'Water supply and sanitation in developing countries.' IWES, London.

29

Sewerage and Sewage Disposal

Staff of Watson Hawksley, Consulting Engineers

Contents

SEWERAGE

29.1 Introduction

29.1.1 Sewerage

The function of a sewerage system is to convey domestic and industrial wastewaters, and runoff from precipitation, safely and economically to a point of disposal.

Urban areas may be sewered by a combined system, a separate system, or a partially separate system. In a combined system, which is the most common in Britain, one network of sewers collects foul sewage and stormwater. In a separate system two sewer networks are used, one for foul sewage and the other for stormwater. A partially separate system is a compromise allowing some of the precipitation, e.g. from the backs of houses, to flow into the foul sewer; the second sewer carries the rest of the storm water.

29.1.2 Sewage

The term 'sewage' is applied to the contents of sewers carrying the waterborne wastes of a community. The network of sewers in which the wastes are conveyed is known as the sewerage system.

Domestic sewage is the discharge from water closets, sinks, baths, and washing machines in offices, schools, homes, factories, etc. Industrial effluent is the waterborne waste of industry. Infiltration is the unintended ingress of groundwater into the sewerage system. Foul sewage is a term commonly used for domestic sewage, but strictly includes any polluting wastewater, as distinct from stormwater. Storm sewage is foul sewage diluted by stormwater. It will readily be appreciated that combined and partially separate sewerage systems, carrying stormwater, must be designed for considerable variations in flow; in consequence it may be necessary to provide storm–sewage overflows as discussed below.

29.1.3 Disposal of stormwater and sewage

Runoff from precipitation, and certain other clean waters, is usually permitted by the pollution-control authorities to be discharged directly to the nearest watercourse.

Wastewaters collected by sewerage systems are usually delivered to a works for treatment before disposal to an appropriate receiving water. In combined and partially separate systems it is usually possible to limit the amount of wastewater passed forward for full treatment; the excess flow of storm sewage may, before disposal, require a lesser degree of treatment or even none at all if it has been sufficiently diluted with rainwater. In the latter case, separation of storm sewage may be effected at the most appropriate location within the sewerage system, the overflowed portion of the storm sewage passing directly, or via the stormwater sewerage network, to an adjacent watercourse.

Storm sewerage is needed to limit physical damage and financial loss caused by flooding.

29.1.4 Statutory control

The discharge of wastewaters to surface and underground waters in Britain is governed by Part 2 of the Control of Pollution Act 1974. This calls for the consent of the controlling authority before any wastewater may be discharged to a receiving water or to a public sewer, or before any change may be made in an existing discharge.

In England and Wales the control is exercised by the ten water authorities. In Scotland, discharge to receiving waters is subject to the consent of the ten river pollution prevention boards, while discharge to public sewers comes under the regional and island councils.

Normally, the consent will specify the quantity permitted and its quality. Industrial effluents, only, are subject to consent for discharge to sewer, and a reception and treatment charge will be made.

29.2 Design of storm sewers

29.2.1 'The Wallingford procedure'

A manual of practice[1] for the design and analysis of urban storm-drainage systems was published in 1981. This is known as the 'Wallingford procedure', and the five volumes not only describe the general procedure and choice of method of analysis and design, but also include maps of Britain with meteorological and soil data, and computer programs. In addition, one volume is devoted to the modified rational method, which is particularly suitable for small systems (not exceeding 100 to 150 ha in area or where pipe sizes are not larger than 600 to 1000 mm diameter).

29.2.2 Modified Rational method

This method is a development of the widely used Rational (or Lloyd-Davies) method; it gives the peak discharge from the equation:

$$Q = 2.78CiA \qquad (29.1)$$

where Q is the peak discharge in litres per second; C is a dimensionless coefficient; i is the average rainfall intensity during the time of concentration in millimetres per hour; and A is the contributing catchment area in hectares.

The coefficient C may be regarded as a combination of two separate coefficients – for volumetric runoff (C_v) and a dimensionless routing coefficient (C_r).

The duration of a storm to give peak rate of flow in the sewer is assumed to be equal to the time of concentration of the system. This is the sum of the time of entry and time of flow through the longest route of the system to the point under consideration.

In the detailed calculation it is necessary to consider the time of entry, which may vary from 3 to 10 min, according to size and slope of the catchment, and the severity of the storm. The Manual[1] gives values for time of entry which are shown in Table 29.1. The smaller values are applicable to subcatchments of less than 200 m² and with slope greater than 1 in 30, whilst the larger values are for subcatchments greater than 400 m² with slope less than 1 in 50.

Table 29.1 Time of entry

Return period	Time of entry (min)	
	Small subcatchments	*Large subcatchments*
5 years	3	6
2 years	4	7
1 year	4	8
1 month	5	10

The time of flow may be determined from pipe-full velocities obtained from design tables.[2] For the design of new systems, trial determinations are necessary to find the approximate size and gradient of pipe or channel, generally at the natural slope of the catchment.

The selection of the design return period is an economic, rather than a meteorological decision. Longer return periods will lead to systems with greater capacities, providing a higher standard of drainage at greater cost. At one time, design was frequently based upon storm-return periods of 1 year; this is still satisfactory where surface flooding during storms of greater severity is acceptable. Where inhabited basements in buildings are at risk, a design return period of once in 50 years or even once in 100 years should be considered.[3]

As a first approximation for a 1- and 5-year return period, the rainfall intensities (mm/h) in Table 29.2 could be used. Average rainfall intensities for a specific location in Britain and for different return periods may be obtained from the Meteorological Office, Bracknell, or may be derived from a simple manual calculation, which is set out in the appendix to Volume IV of the Manual.[1]

Table 29.2

Time of concentration (min)	Rainfall intensity (mm/h)	
	1–year	5–year
10	35–40	60–65
20	23–25	40–42
30	18–20	33–35

The volumetric runoff coefficient C_v may be defined as the proportion of the rain falling on the catchment which runs off into the storm-sewer system. The value is affected by whether the whole catchment (impervious and pervious areas) is considered, or the impervious areas alone. As a first approximation, if impervious areas alone are considered, the value of C_v could be taken as unity, although actual values may be within the range 0.6 to 0.9.

The routing coefficient (C_r) might be expected to vary with the shape of the catchment but examination of data led to the recommendation[1] of a constant value for C_r of 1.3 for both design and simulation.

29.2.3 Pollution from storm runoff

Urban storm runoff will be polluted to a greater or lesser extent. Several comprehensive studies of this pollution have been made, and are referred to in the Manual,[1] which includes a summary table to show the scale of the problem.

Accidental spillage of contaminants, e.g. in a road accident, can cause danger to watercourses, especially since it is common practice to remove such spillage by hosing into the surface-water drains. Where the result of such an accident can be particularly serious, e.g. in contaminating a potable water supply, special protective measures may be necessary in the drainage design.

29.3 Sewage

29.3.1 Introduction

The various types of sewage have been defined in section 29.1.2 above, and their polluting characteristics will be discussed in section 29.8 below. The current section is concerned with the volumes of flow for which the sewers must be designed, and with means for dealing with peak flows. It also looks at design flows for the treatment works.

29.3.2 Dry weather flow

The dry weather flow (DWF) is the rate of flow of sewage (together with infiltration if any) in a sewer in dry weather, usually defined as a period of 5 successive days and nights without measurable rain.

Different values of DWF will be obtained in summer and winter, as a result of changes in infiltration caused by variation in the level of the water table, or domestic holidays, or changes in the industrial pattern of operation.

The DWF of sewage in a sewer, on arriving at a sewage treatment works, is the sum of domestic flow, infiltration and industrial flow. Values for the average daily domestic water consumption should be ascertained from local records. A typical UK figure is 185 l per head·day but as little as 75 to 100 l per head·day may be appropriate in developing countries, and 400 to 500 l per head·day is often consumed in areas such as North America where air-conditioning, lawn watering, and automated car washes are in wide use.

Values of infiltration are best determined from sewer gauging at night, when domestic flow is almost zero, and industrial discharges are also least in number, and thus more readily calculable or measurable. Typical values might be 15 000 l per day per kilometre of sewer and house connections, for average conditions (sewer partly above water table and partly below).

Values of industrial discharge should be determined from metered records, or by reference to agreements with the local authority.

29.3.3 Storm sewage

Combined and partially separate sewers carry surface water in addition to the normal foul sewage. These sewers are designed to carry peak flows far in excess of the peak flow of foul sewage, when storms or long periods of heavy rainfall occur.

It is not necessary or economical to treat the full peak flow conveyed by such sewers. Provided the sewage-treatment works, downstream of storm separation (see below), has adequate capacity to treat fully the maximum contributory rate of flow of foul sewage, without bypassing in dry weather, it has been found that the remainder of the peak combined or partially separate flow can be separated using a storm-sewage overflow.

In the past it had been commonly assumed that dilution during storm periods would allow the excess storm flow to be bypassed to the nearest watercourse without serious detriment to its quality. This is not, however, the case and substantial pollution has been produced, not only from floating objects commonly seen caught by riverside bushes, but also because polluting sediments in the sewers are resuspended by the storm flush.

No complete solution has yet been achieved, but a combination of a suitably designed storm overflow structure with a storage basin, from which the first flush can be returned to the foul sewer for later treatment, has produced an improvement.

For small overflows (0.15 to 0.85 m³/s) storage-type overflows are suitable; control by throttle pipe and overflow weirs is preferred. At least manual screening of the overflow should be provided. For larger overflows only limited storage capacity is practicable, and design should be concentrated on avoiding overflow of the first storm flush. Control may be by throttle pipe, orifice, or flow regulator, and mechanically raked screens should be provided.

The overflow must be set to operate at a predetermined rate of flow, designed to mitigate, so far as possible, the pollution discharged with the excess storm sewage. In Britain the overflow setting (Q l per day) is given by the former Ministry of Housing:[4]

$$Q = \text{DWF} + 1360P + 2E,$$

where P is the tributary population and E is the industrial effluent flow.

29.3.4 Design flow for sewage treatment works

The volume of foul sewage flowing in the sewer, downstream of the last stormwater overflow, will be approximately 6 DWF. Not all of this can be fully treated, if the rainfall continues for long. It is recommended[4] that 3 DWF is fully treated (no allowance being made for an increase of infiltration), the remainder being bypassed to storm tanks to receive gravity settlement. It is usual, after the storm has ceased, to pump the contents of the storm tanks back to the works inlet for full treatment.

The rate of flow to the treatment works will vary over the day (and also weekly and seasonally, if the proportion of industrial effluent is substantial).

29.3.5 Pollution load

The inlet works, tanks, pumps, etc. on a sewage treatment works must be designed to deal with the design flow discussed in the previous subsection. In addition, the treatment processes, especially the secondary biological stage, must be designed for the pollution load, which is not necessarily affected by the actual fluid flow in the sewerage system.

It will be seen, from the discussion of sewage characteristics in section 29.8, that the principal parameter of pollution in domestic sewage is biochemical oxygen demand (BOD). The BOD load may be readily calculated by multiplying the average DWF by the average BOD concentration. In the absence of suitable measurements, the values given in Table 29.3 may be used as a first approximation.

Table 29.3 Strength of sewage

	Crude sewage (mg/l)			Settled sewage (% removals on crude)
	weak	medium	strong	
BOD$_5$	200	350	550	30–40
COD	350	600	950	30–40
SS	200	350	500	50–70
NH$_3$-N	25	35	60	—
Org. N.	10	15	20	15–20
Chloride (Cl)	70	100	130	—
Org. C	140	210	300	30–40

Hydraulic load and pollution load are important concepts in both the design of process units and treatment works, and in the determination of equitable charges to be applied to industrial users of sewers and sewage-treatment processes.

The balance between industrial and domestic waste will be important for any given sewage. Whilst many industrial wastes can readily be treated in admixture with domestic sewage, some industrial effluents prove difficult in terms of proportion, temperature, or BOD:N ratio. The concept of treatability is examined in more detail in section 29.8.

29.4 Design of sewerage systems

29.4.1 Introduction

The design of sewerage systems calls for the optimization of the hydraulic, structural and constructional aspects to suit the drainage area.

Very many factors affect this design and this section will concentrate mainly on hydraulic calculations, selection of materials, and structural design of the sewer line.

29.4.2 The Colebrook–White formula

Over past years many formulae have been developed for hydraulic design of pipes and channels. The equation derived by Colebrook in conjunction with White in 1939 is now regarded as the most satisfactory basis for hydraulic design. The Hydraulics Research Station at Wallingford has expressed the formula in tabular and graphical form more suited to the designer's needs.[5]

The tables and charts present flow rates (l/s), flow velocities (m/s) and hydraulic gradients for pipe sizes from 0.025 to 2.5 m diameter, and for roughness factors (k_s) from 0.003 to 600 mm. Recommended roughness factors are listed. The tabulated factors 'good', 'normal' and 'poor' relate only to the standard of uniformity of the surface of the pipeline or conduit when clean and new (unless otherwise stated). In the case of short pipelines, extra allowances must be made for discontinuities such as changes in direction, sizes, junctions and valves.

Pipelines and conduits may become fouled if not correctly designed and constructed. Physical fouling is caused by settlement of particulate matter in the invert; transport of biological matter present in wastewater results in sliming of pipeline surfaces below water, but both can be significantly reduced by maintaining high velocities. Grit need not be taken into consideration for new designs provided the pipeline has good self-cleansing characteristics.

Storm sewers may normally be considered as being in a clean state, whereas foul sewers become slimed, and necessitate the use of a roughness factor higher than that for storm sewers. Generally, the factors diminish as the velocity increases; this feature also applies to sewage rising mains.

Within a treatment works it is usual to assume that the main flow lines are 'sewers' until after the secondary stage of treatment. Gravity and pressure pipelines for sludge are special cases; friction factors will be dependent upon the characteristics of the sludge and may be up to 7 times that appropriate for sewage.

29.4.3 Design parameters

Experience has shown that a flow velocity of at least 0.75 m/s once a day for an hour or so is usually sufficient to keep gravity wastewater sewers clean. Designing for a higher daily peak velocity will also allow the use of a lower k_s factor and hence make possible decreased pipe size with improved conditions at minimum flow. Typical values for a concrete gravity slimed sewer are given in Table 29.4.

Table 29.4

Velocity (m/s)	Roughness k_s (mm)
0.5–1.0	6.0
1.0–1.5	1.5
>1.5	0.6

A pumping main always runs full, and flow may be discontinuous. Thus the cleansing velocity must be regularly achieved and sustained for a period sufficient to scour any settled solids.

Suggested minimum velocities are as shown in Table 29.5.

Typical roughness factors for coated steel and iron sewage-pumping mains are as shown in Table 29.6.

Table 29.5

	Settling velocity (m/s)	Pick-up velocity (m/s)		
		150 mm pipe	300 mm pipe	600 mm pipe
Probable particle size (any pipe diameter)				
Grit up to 5.0 mm dia.	1.50	1.2	1.5	1.8
Sand up to 2.5 mm dia.	0.45	0.6	0.6	0.6

Table 29.6

Velocity (m/s)	Suggested k_s factor (mm)
0.8–1.1	3.0
1.2–1.5	1.5
>1.5	0.3

29.4.4 Sewer materials

As usual, the selection of the most appropriate material is a compromise between first cost and service life. The costs of relaying, and of the upheaval caused during this process, are, however, so great that the first cost of the material cannot be the principal criterion for choice.

The material chosen must resist aggression by the liquid being carried (or outside the sewer) and by matter in suspension, and also by-products of biological degradation (e.g. sulphide). It must also be strong enough to withstand the internal and external loads. The following materials are in common use.

29.4.4.1 Clayware

Clay pipes are suitable for nonpressure applications, and are not generally available in diameters greater than 1 m. Their chemical inertness fits them for aggressive chemical wastes and sewage at high temperatures. Their main drawback is brittleness.

29.4.4.2 Cementitious

Pipes made from cementitious material are generally robust, reliable and relatively cheap. Unless expensive systems of protection are applied, however, such pipes are vulnerable to attack by sulphate in groundwaters, and to acid attack from industrial effluents or as a result of bacterial action in septic sewage.

Unreinforced concrete is available up to 1.4 m diameter, and is suitable only for gravity flow. Reinforced concrete pipes are widely used for gravity sewers in temperate climates, in diameters up to 3 m; they can also be used for pressure pipelines up to about 4 bar. Polyvinylchloride liners have been developed to protect the inner wall from septic sewage. Prestressed concrete pipes have been used up to 7 m diameter, and are particularly suitable for pressure sewers.

Asbestos cement pipes are widely available in diameters up to at least 2.5 m, and for pressures up to 32 bar. More recently glass and steel fibres have been used to reinforce concrete pipes.

29.4.4.3 Ferrous

Ductile iron is widely used for sewage-pumping mains up to 1.6 m diameter. Steel pipes are less widely used, but are available in larger diameters. Their suitability for welding means that joints capable of taking tensile loads can be made, making steel pipes suitable for long sea outfalls, river crossings, etc. Corrugated steel pipes have been widely developed and used in the US, particularly for storm and surface-water culverts. As complete pipes they are available up to 3 m diameter and, in sections for assembly on site, they can be made in spans of 10 m or more.

29.4.4.4 Plastics

There is a major distinction between thermoplastics, whose strength generally reduces markedly with temperature, and thermosetting resins (normally glass-fibre reinforced), whose strength falls much less with temperature. Both groups have very good chemical resistance, although this may be reduced when the pipe is stressed or strained.

There are two main groups of thermoplastics: the polyolefins, which include polyethylene (PE), polypropylene (PP) and polybutylene (PB), and the vinyls, which include polyvinylchloride (PVC) and acrylonitrile butadiene styrene (ABS). Polyethylene pipe is the most widely used of these for sewerage. It is available in medium and two main high-density forms (MDPE, HDPE1 and HDPE2). As extruded pipe it is made in diameters up to 1.6 m, suitable for pressures up to at least 12 bar at 20°C. In helically welded form it is available up to 3 m diameter for gravity sewers.

Polypropylene is available up to 1.2 m diameter, and for pressures to 15 bar (20°C). Polybutylene is made up to 600 mm diameter and 17 bar pressure (20°C). Polypropylene and PB have better high-temperature properties than PE, and PB is probably the best of all thermoplastic pipe materials, having particularly good high-temperature strength, environmental-stress cracking resistance, abrasion resistance and low creep. All the polyolefin plastics can be welded by thermal fusion, making them suitable for the pulling of outfalls and river crossings and for the slip-lining of old pipelines.

Of the vinyl-type thermoplastics PVC, in its unplasticized form, is the more common. It has been quite widely used since the late 1950s and its reputation has sometimes suffered as a result of its being the prototype for all plastic pipes. As a gravity sewer material, design can be carried out with confidence. For pressure applications it is important that the pipe should be derated not only for temperature, if appropriate, but also for fatigue effects where the pressure varies cyclically.[6] Acrylonitrile butadiene styrene pipes are available only up to 300 mm diameter.

Pitch-fibre pipes, which may be regarded as plastics, are limited to the even smaller diameter of 200 mm. Reinforced thermosetting resin pipes, variously known as GRP, FRP, RTRP, RPMP, etc. are now available in all sizes up to at least 4 m and for pressures up to at least 25 bar. For gravity and low-pressure applications the pipes often contain one or more layers of unreinforced sand and resin (RPM pipes). Pipes containing essentially only resin and glass fibre are known as GRP in Britain and FRP in the US.

29.4.5 Jointing materials

Flexible joints for rigid pipelines normally employ a socket (bell) and spigot arrangement, or a double collar or sleeve assembly. Both these jointing systems rely on an elastomeric sealing ring or gasket to ensure watertightness. Natural rubber has been used successfully for such sealing rings, but, in certain circumstances, may deteriorate as a result of microbial attack.

Work at the Water Research Centre in Britain[7] has led to ethylene-propylene rubber and styrene-butadiene rubber being the preferred materials for ordinary sewer use. Where industrial effluents are present consideration of other synthetic rubbers may be necessary, in order to obtain the appropriate chemical resistance.

29.4.6 Structural design of nonpressure pipes

Flexible joints and uniform beddings are used to minimize longitudinal bending moments, allowing structural design to consider only the two-dimensional case of the pipe cross-section. Two sources of loading are considered, that due to the backfill and that due to any surcharge loads on the surface. Backfill loads on all flexible pipes and on rigid pipes in trenches not wider than, say, 1.75 pipe diameters can be taken as the weight of the prism of soil vertically above the pipe. Rigid pipes in wider trenches may experience up to 1.5 times the 'prism load'. Surcharge loads are calculated according to Boussinesq, assuming a pattern of point loads due to vehicle wheels. Figure 29.1 shows the pressures at various depths, according to the usual British, American and German assumptions. Considerable care is required in the selection of factors of safety.[8]

Rigid pipes, e.g. clay, concrete and asbestos, are specified according to their crushing strength in a two- or three-edge line-load test. The load distribution and support provided by the bedding of a buried pipe enable it to carry a load greater than the test load by a factor (F_m) known as the bedding factor. (See Figure 29.2 for design values of F_m.) A factor of safety of 1.25 is normally taken and the enhanced pipe strength (i.e. $\times F_m$) must provide this.

Flexible pipes (metal and plastic) should be specified according to their stiffness ($Et^3/12D^3$, where E is Young's modulus, t is thickness of pipe wall, and D is pipe diameter). Table 29.7 lists values of the elastic constants for flexible pipes. (It should be

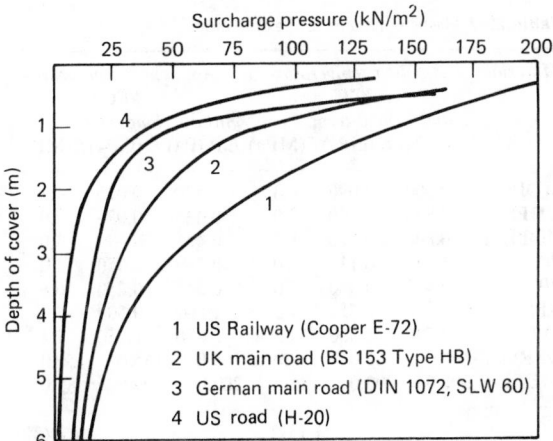

Figure 29.1 Pressures at various depths

1 US Railway (Cooper E-72)
2 UK main road (BS 153 Type HB)
3 German main road (DIN 1072, SLW 60)
4 US road (H-20)

noted that the Young's modulus of thermoplastic pipes reduces with time because of creep.)

The structural design of flexible pipes involves ensuring that the pipe neither collapses (by buckling) under the external load nor deflects to such an extent that it loses too much cross-sectional area, causes its joints to open or is overstressed or overstrained. Deflection is measured as percentage reduction of vertical diameter, and 5% deflection is normally regarded as the limit for loss of area or joint watertightness, with lower limits sometimes being imposed by stress or strain considerations. As for rigid pipes, the load-carrying capacity of flexible pipes is also influenced by the bedding. In this case it is the deformation modulus E' of the bedding, in resisting the outward deflection of

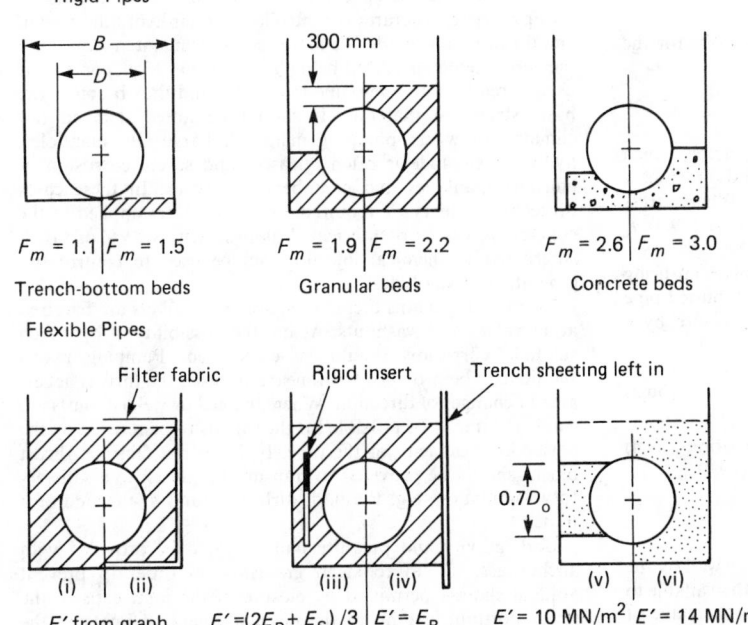

Figure 29.2 Bedding factors

Table 29.7 Elastic constants for flexible pipes

Material	Ambient temperature 20°C			Ambient temperature 40°C		
	initial E (GPa)	long-term E (GPa)	f (MPa)	initial E (GPa)	long-term E (GPa)	f (MPa)
MDPE	0.600	0.090	6.3	0.320	0.025	4.1
HDPE 1	0.875	0.130	5.0	0.435	0.030	2.0
HDPE 2	0.800	0.120	6.3	0.465	0.040	3.0
PP	1.150	0.115	5.0	0.760	0.050	3.0
PB	0.425	0.380	7.6	0.345	0.250	6.6
ABS	1.650	0.550	7.5	1.500	0.500	5.5
PVC	2.790	1.350	12.3	2.650	1.250	7.4
D.IRON	165.0	165.0	150.0	165.0	165.0	150.0
STEEL	200.0	200.0	85.0	200.0	200.0	85.0

			e (%)			e (%)
*			0.25			0.20
GRP †17.5–	10.0–		0.35	15.0–	8.0–	0.30
‡40.0	25.0		0.40	35.0	13.5	0.35
* 4.5–12.0	2.7–7.2	0.20		4.0–11.0	2.1–6.0	0.18
RPM 6.0–15.0	3.6–9.0	0.18		5.0–13.5	2.7–7.3	0.15
† 4.5–15.0	2.7–9.0	0.35		4.0–13.5	2.1–7.3	0.30
‡						

*ring tension †ring bending ††combined tension and bending

the sides of the pipe, which is used. Deflection is calculated as follows:

$$\text{Relative deflection } (\%) = F_{DL}\left(\frac{0.083 P_E}{8 S_p + 0.061 E'}\right) \quad (29.2)$$

where F_{DL} is deflection lag factor (increase in deflection with time – see Figure 29.3), P_E is the external pressure (backfill-+surcharge) and S_p the pipe stiffness, as from Table 29.7.

The deflections corresponding to stress or strain limits for the pipe material are calculated as follows:

$$\text{strain-limited deflection} = e_L D / F_G t \quad (29.3)$$

where e_L is the limiting strain and F_G the strain factor, which takes account of the geometry of the distortion, and for which a value of 6.0 can be taken for design purposes. Where stress limits apply, f_L / E may be substituted for e_L in Equation (29.3), f_L being the limiting stress.

Design of flexible pipes to resist buckling involves ensuring that the critical pressure (P_{CR}) which will cause the buried pipe to buckle, exceeds the actual external loading pressure by a suitable factor of safety.

$$P_{CR} = (\sqrt{32 E' S_p})(1 - 3 \times \text{relative deflection}) \quad (29.4)$$

The value of relative deflection inserted in Equation (29.4) should be that calculated according to Equation (29.2).

29.4.7 Structural design of pressure pipes

The circumferential tensile stresses set up in the pipe wall by the pressure within the pipe reduce the effective strength available to resist the external loads. With rigid pressure pipes the value of the crushing strength must be increased by a factor F_p given by Equation (29.5):

$$F_p = 1/(1 - P_i/P_u)^n \quad (29.5)$$

where P_i is the design internal pressure, P_u is the ultimate pressure capacity of the pipe, and n is 1 for reinforced concrete pipes, 1/2 for asbestos-cement pressure pipes, 1/3 for prestressed concrete pipes.

When designing flexible pressure pipes the stress or strain in the pipe wall, produced by the internal pressure, is added to the stress or strain induced by deflection, as indicated above. Such total stress or strain must not exceed the limit given in Table 29.7.

Where flexible pressure pipes may be subjected to sub-atmospheric pressures, e.g. during surging following pump shutdown, the vacuum pressure should be added to the external pressure and the factor of safety against buckling rechecked using Equation (29.4) above.

29.4.8 River crossings and submerged outfalls

For these types of installations the stresses and strains in the completed pipeline are seldom great. Because, however, they are often constructed by assembling long strings of pipes on land and then towing or pulling these into position, high stresses and strains may be set up during construction. These stresses and strains are likely to be due either to direct tension or to curvature of the pipeline as it passes over supports. The tensile loading depends on the pulling force involved in the particular method of construction. Thus, a maximum pulling load may be calculated, for given pipe properties, and this must be specified as not to be exceeded in construction. Curvature of the pipeline may occur both during construction and in its final position. The radius of curvature should not be less than a critical value, which will be governed by stress, strain or buckling.

29.4.9 Ancillary structures

Other than pumping stations, which are dealt with elsewhere, the structures associated with underground drainage systems include manholes, drop chambers and stormwater overflows. Sizing of these structures is controlled by their hydraulic design, and the need to provide adequate access for maintenance. Since they are often constructed below groundwater level, concrete is usually required to overcome buoyancy, and thus becomes the basic structural material. It should be noted that, in hot climates, or where pumping mains discharge into manholes, hydrogen sulphide is often released, and severe corrosion of concrete manholes and chambers can occur. In these cases protective systems are required – either coatings applied to the concrete *in situ* or prefabricated linings such as PVC or GRP. Alternatively, chemical injection can be used to control the generation of sulphides.[9]

On pumping mains themselves, access chambers are required at air valves and washouts. Again, the possibility of hydrogen sulphide corrosion should be considered. Pumping mains should also be provided with means to resist the thrusts generated at changes of direction. Where flanged or welded joints are employed it may be possible for the thrusts to be resisted by the tensile-load capacity of the pipe. If this is not feasible, thrust blocks should be provided to transmit the thrust to a satisfactory foundation, e.g. the undisturbed ground at the side of a trench.[10]

Both gravity and pressure mains should be provided with anchorages, if laid to steep gradients, in order to prevent gradual sliding, permitted by closure of the joint gaps in the lower portion, leading to the disengagement of joints in the upper portion of the pipeline.

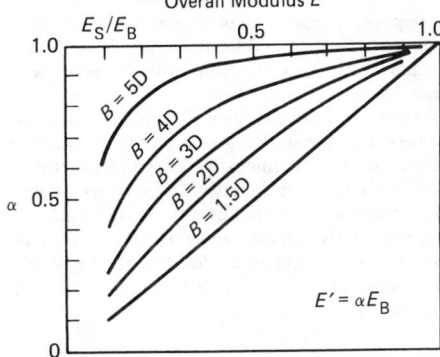

Overall Modulus E'

$E' = \alpha E_B$

Bedding Modulus E_B
In terms of depth in metres (H)

Material	Compaction	E_B (MN/m^2)
Gravel	95% MPD	$8 + 1.7H$
	90	$4.5 + 1.4H$
	85	$3 + H$
Coarse sand	95	$7 + 0.85H$
	90	$5 + 0.85H$
	85	$3 + 0.85H$
	80	$2 + 0.85H$
Fine sand	90	$3 + 0.285H$
	85	$2 + 0.285H$

Native Soil Modulus E_S

Material	Peat	Clay	Silt	Sand	Gravel	Rock
E_o (MN/m^2)	<0.3	0.5-5.0	2.5-4.5	4.0-10.0	10.0-20.0	>30
Creep F_{DL}	4.0-2.0	2.5-1.5	2.25-1.5	1.75-1.25	1.5-1.25	1.0

Figure 29.3 Modulus E'

29.5 Pumping sewage

29.5.1 Introduction

Pumping sewage presents a particular problem in the need to handle the solids contained therein. It is common practice to assume that the smallest sewer within a sewerage system is 100 mm diameter and therefore may pass solids of almost this size. Therefore pumps are specified as being capable of passing a 90-mm diameter sphere, and the inlet and discharge connections must not be less than 100 mm diameter, and this is true for pumping mains. This limitation precludes satisfactory pumping at rates below about 15 l/s.

The need to handle solids also dictates that end-suction single-stage pumps are used, thus limiting the possible head that can be generated to about 75 m.

Sewage pumps are normally centrifugal or mixed-flow machines. In smaller sizes, submersible sewage pumps are manufactured. These pumps have a close-coupled, fully submersible electric motor fitted to the pump and are designed to be lowered into the sewage.

If the flow rate required falls below 15 l/s, special devices are required. Various manufacturers can supply these and they depend either on comminuting the solids or on some method of handling solids without passing them through a pump (e.g. solids diverter (R) or compressed-air ejector).

For lifting duties at sewage works large Archimedean screws are being increasingly used. These devices are suitable for lifting sewage, but not for feeding into pumping mains under pressure.

Sewage sludges are of various consistencies. The ability of centrifugal sewage pumps to be used satisfactorily or the need to use positive displacement pumps are covered in a publication issued by the Water Research Centre.[11]

29.5.2 Sewage pumping stations

Sewage and drainage installations differ from almost all others in one important point. This is that once the installation has been commissioned it is virtually impossible to close it down; sewage continues to flow in the sewers. All sewage installations must be designed with this in mind, particularly pumping stations. For whilst it may be possible to bypass a part of the treatment process, it may be essential to continue pumping in all circumstances.

There are two facets of the need to maintain a pumping station so that it is continuously available for service or running. The first is that it should be possible for all routine maintenance, including major overhauls, to be carried out with the station operating, and the second is that machine breakdowns or other similar circumstances should be 'fail-safe'. The various ways of meeting these requirements underlie the remainder of this section.

Sewage-pumping stations are almost always equipped with electrically driven automatic pumps, operated from level-measuring devices or switches in the reception sump, enabling them to operate without full-time pump attendants.

The sump should be designed to allow easy flow to the pump suctions, and with sufficient benching to avoid undue settlement of solids.[12] In major installations, two interconnecting sumps are often provided to enable either to be drained for cleaning or maintenance, without closing down the installation. Some intakes to sumps are fitted with screens. However, there are conflicting views on the fitting of screens. If screens are fitted there is the need to dispose of screenings; if they are not disposed of there is a risk of items reaching the pumps and blocking or damaging them. Current practice may follow either view.

Major pumping stations are normally designed to be similar to that shown in Figure 29.4. Of particular note as being current good practice are the following features:

(1) The pump casings are below the invert of the incoming sewer, thus ensuring that the pumps require no special priming equipment.
(2) The nonreturn valves and the entries to the rising main are both horizontal, thus avoiding some of the problems caused by deposition of solids when the pumps are not running.
(3) The electrical equipment is at a high level, thus obviating damage in the event of the pump well being flooded. It is also normal to fit automatic cellar-drainage pumps to this well.
(4) Sufficient access and cranage is provided to ease maintenance as far as possible. Most designs provide access stairs to the pump well rather than ladders to encourage maintenance staff to inspect the machinery regularly.

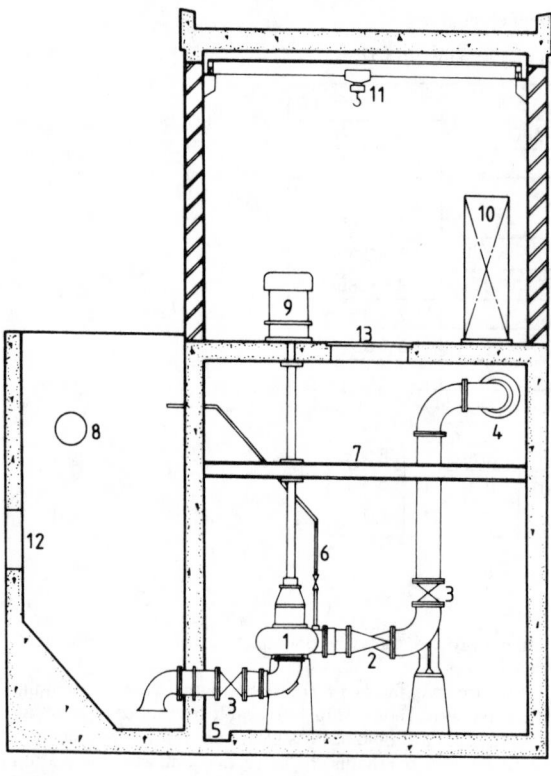

KEY

(1) Pump
(2) Nonreturn valve
(3) Isolating valves
(4) Rising main
(5) Drainage channel
(6) Air release pipework
(7) Intermediate shaft support
(8) Overflow
(9) Electric motor
(10) Switchboard
(11) Overhead crane
(12) Sewer inlet
(13) Machinery access cover

Figure 29.4 Pumping station

The switchboard should be of such a design that individual pump starters, controls, etc. can be isolated for maintenance, whilst the board is live, and other pumps are running or available for service.

The ability to continue to pump or bypass sewage under all circumstances is normally provided by several features. At least one standby pump will be provided, and will have suitable automatic controls for it to take over the duties of any pump which fails from whatever cause. The number of pumps provided will depend on the expected flow variation, length of pumping main, lift and other similar design parameters.

The electricity supply may need to be secured, either by duplicating the connections to the public supply, or by providing standby generating plant within the pumping station, or both.

Despite these precautions it is wise to provide a high-level overflow to avoid flooding if there is a total breakdown.

The design of stations using close-coupled submersible sewage pumps is similar but normally rather simpler. It is not necessary to provide a building, so long as there is good access to the well containing the pumps, and the electrical switchgear is housed in a suitable weatherproof kiosk.

29.5.3 Rising mains

Sewage-pumping mains differ from those containing most other fluids in that whenever the sewage therein becomes stationary, or falls to a low velocity, deposition of the solids will occur. To ensure that this deposition is not cumulative, it is good practice to design the pipeline so that a velocity at which solids are picked up is achieved on some occasion daily. This velocity should be at least 1 m/s.

As in other systems, hydraulic surge will occur whenever the velocity in a sewage pipeline is changed. Suitable precautions should be taken to ensure that this surge does not generate a pressure which is likely to cause damage. The velocity of such hydraulic surges is materially lowered by any dissolved gases in the fluid;[13] sewage normally contains gases. However, since on some occasions the system might be filled with water, the precautions taken should be effective with surge velocities both for water and for sewage.

29.6 Construction

29.6.1 Introduction

Construction should aim to achieve the design objectives with the greatest economy. The choice of materials may make differing demands on the installation costs of buried pipelines. The achievement of the necessary pipe-bedding standard (see Figure 29.2) is crucial and, because of the greater dependence of flexible pipes on their bedding, may invalidate cost comparisons based on material prices only. The influence of trench width and native ground conditions on the design of flexible pipelines is ignored in many published 'design methods'. The information provided in Figure 29.2 is intended to remedy this situation, and shows how trench width, bedding material and its degree of compaction must be considered together. The data relating bedding moduli to the degree of compaction applied to various materials are based on empirical relationships obtained from a consensus of various published sources.

Gravel beds are frequently preferred because they can achieve 90 to 95% MPD with minimal compaction. Where gravels are used they must be prevented from acting as groundwater drains, by the use of regular impermeable barriers, e.g. polyethylene sheeting. In some cases, e.g. where the native soil is fine sand or silt, it may be necessary to enclose the whole of the bedding in an impermeable membrane or filter fabric to prevent groundwater flows leaching out fine material and forming voids at the trench side. A high-modulus bedding material may, however, still not achieve a high overall modulus if the native soil is soft. The use of wide trenches will improve this, but may be impracticable for large-diameter pipes, in which case resort should be made to one of the special beddings.

For all types of pipeline the provision of uniform support by the bedding is essential to prevent the development of unacceptably high shear forces or longitudinal bending moments. Where this cannot be achieved, e.g. where pipes are built into the walls of underground structures, in areas of mining subsidence or abrupt transitions from rock to soil, closely spaced mechanical joints should be specified. Flexible pipes are often supplied in long lengths, very long in the cases of welded steel or polyolefins, and the basic flexibility of the pipe may not be sufficient to accommodate differential settlements without the use of such flexible joints.

Careful attention must be given to the manner of joining lateral sewers and house connections to main sewers.[14] The Construction Industry Research and Information Association has recently issued an authoritative report on trenching.[15]

In the construction of above-ground pipelines, proper consideration must be given to possible thermal movements, and to even load distribution at supports.

The construction of river and estuarine crossings, and of submerged outfalls, favours the use of materials which can be joined into long strings on land and then pulled into position. Steel and polyolefin plastics with welded joints are therefore often used. Glass reinforced plastic pipes, with hand lay-up overwrap joints, have also been successfully used in this manner, as also has prestressed concrete.

29.6.2 Renovation of pipelines

The construction of new underground pipelines by trenching is expensive in established urban areas, not only in direct cost but also in the indirect costs of disruption. This has encouraged development of so-called 'nondisruptive' construction methods. The techniques which have received most attention are miniaturized tunnelling, and sewer renovation.

Quite apart from the fact that many old urban sewers require renovation because they are structurally unsafe, construction based on renovation of the old sewer has the additional advantages that a pipeline route clear of other underground services is automatically provided, and also that all lateral connections are automatically located.

Renovation techniques have been reviewed extensively by the Water Research Centre,[16] which has adopted the following system of categorization:

Type 1: Included in this category are lining systems which are bonded to the fabric of the old sewer so as to form a composite rigid structure. Examples are glass reinforced concrete segmental linings, and GRP segmented, or complete pipe, linings roughened to provide the required bond.

Type 2: Lining systems in this category do not rely on the formation of a bond to the old sewer structure. The liner usually consists of a polyethylene pipe inserted by sliplining, a reinforced thermosetting resin liner, installed by inversion and cured *in situ*, or a plastic pipe liner formed by individual insertion of GRP or polyolefin pipes. As with Type 1 linings, the annulus is grouted, but no reliance is placed upon the formation of a bond, so that the liner is regarded as acting as a flexible pipe.

Type 3: Linings of this type, thin-walled GRP or *in situ* resin, are not regarded as fulfilling any permanent structural role. Rather, they are considered as formwork left-in, with the annulus grout providing the structural element.

Most of the renovation systems result in lateral connections being temporarily blocked, but several ingenious methods of reopening laterals have been developed, e.g. by remote cutting from the sewer, by remote cutting from the lateral or by 'minimum excavation' techniques from the surface.

Pipe 'renovation' has developed to the stage that one technique, polyethylene pipe sliplining, can provide an increase in the diameter of the sewer. In a 1985 example, a 225 mm diameter clay pipe was 'relined' with a 350 mm diameter polyethylene pipe, inserted behind an impact mole which split the old pipe.

29.7 Maintenance

Given good design and construction, the correct choice of materials should reduce sewer maintenance to the clearance of occasional blockages. Although good hydraulic design should minimize blockages, the fact that they may never be completely avoided requires proper provision for maintenance to be included in the design. Easy and safe access for men and equipment is essential, but further consideration should be given to the possible need to extricate injured workers. Thus space for at least two men should be provided in all manholes, together with openings permitting unobstructed lifts to the surface.

Since access arrangements in deep manholes should preclude the possibility of long falls, intermediate platforms are required. Since such platforms might interrupt full height vertical lifting, manholes on deep sewers should preferably have two surface access openings.

It is essential to ensure proper ventilation of sewer systems to minimize the generation of hydrogen sulphide and to dispose of this and other toxic gases.

SEWAGE TREATMENT

29.8 Introduction

29.8.1 Characteristics of sewage

Municipal sewage is mainly the wastewaters from homes, offices and shops, and, therefore, consists of human wastes and of the discharges of man's domestic activities. Many industries use large quantities of water, which must also be disposed of after use. In industrialized countries, a very large proportion of their industrial effluents is discharged to the municipal sewers, and treated with the domestic sewage; this may demand some pretreatment to ensure that it does not interfere with the normal treatment process (especially the biological stage) or with the disposal of sludge.

A partial analysis of a typical British domestic sewage is given in Table 29.3. (The strength of sewage depends somewhat on the diet and other living habits of the contributory population, and markedly on the quantity of water used.) If this were to be discharged to an inland stream in quantity it would cause substantial pollution. The principal polluting matters in sewage are suspended solids (SS) and organic matter.

The suspended solids would be unsightly, and, being at least partly organic, would reduce the dissolved oxygen in the receiving water.

The organic matter is partly carbonaceous and partly nitrogenous. Both are oxidized by naturally occurring microorganisms in the receiving water, and so reduce the dissolved oxygen, which is essential for fish and other animal life in the water. Since we are primarily concerned with oxygen demand, carbonaceous organic matter is normally measured as biochemical oxygen demand (BOD), the oxygen consumed in 5 days at 20°C by microorganisms consuming the organic matter, or as chemical oxygen demand (COD), a purely chemical parameter which approximates to the ultimate oxygen demand.

Nitrogen is commonly analysed in its various forms, and we are mainly concerned with ammoniacal nitrogen. This will be oxidised in the receiving water, and so will increase the oxygen demand. At high pH values, ammonia can also be poisonous to fish.

It has been found, in Britain, that each person contributes the following pollution loads (grams per day): BOD 60, suspended solids 60, ammoniacal nitrogen 8. In the absence of more specific data, these values may be used to assess the pollution load to be removed in sewage treatment.

29.8.2 Sampling and analysis

Before selecting the method of disposal, and the appropriate treatment of the sewage to permit disposal without pollution, it is necessary to sample and analyse the sewage.

Sampling should usually be carried out over the full 24 h, since flow varies greatly over the day, and the individual samples must be bulked in such a way as to give a properly weighted representative sample. It is desirable that sampling should be carried out over the various seasons and in a range of weather conditions, but this is often not possible.

The analyses should be carried out by the standardized methods,[17,18] which have been laid down in the UK and US, and which are generally used throughout the world.

Although suspended solids, oxygen demand and ammonia are the most important design parameters, it is essential, in these preliminary analyses, to seek also a wide range of substances that might cause danger or damage to sewer workers, to the sewerage system or to the treatment processes, and to ensure that these are at acceptable levels.

For operational control the sewage works operator will analyse routinely (probably daily on the larger works) for SS, COD, and ammoniacal and oxidized nitrogen in the raw sewage and in the effluent after various stages of treatment. (Although it tells little about the biological effect of organic matter, COD is routinely used in place of BOD because the analysis for COD can be carried out in about 2 h as compared to 5 days for BOD.)

29.8.3 Ease of treatment

The final method of disposal, and the proper degree of treatment to allow this without environmental nuisance, is discussed in the next section. It is important, at an early stage, to be able to estimate the ease, or otherwise, with which the pollution can be removed.

Generally, much of the suspended solids in domestic sewage can be readily removed by gravity (sedimentation); in this process some 60% of the SS may be removed, with a consequent reduction in the BOD of about 30%.

In normal sewage treatment the majority of the organic matter is removed from the aqueous stream by biological means during the secondary stage (see section 29.12). It is essential, therefore, to be able to assess the ease with which the organic matter can be oxidized, and to estimate the possible interference of toxic and other substances with the biological oxidation.

Actual tests of the 'treatability' of an effluent may be made by bench- or pilot-scale tests, the latter being the more reliable. Often, however, such tests are not possible, and judgement must be based on experience.

A crude estimate of 'treatability' in relation to a strictly domestic sewage may be obtained from the COD:BOD ratio. For raw domestic sewage this ratio is usually about 2; the organic matter in a wastewater will be *more* easily degraded biologically if this ratio is *less than* 2, and less easily broken down the more the ratio exceeds 2. The most easily degraded substances are broken down first and, in consequence, the COD:BOD ratio of the aqueous stream increases as it passes through biological treatment.

29.8.4 Possible effects of industrial effluents

Many industries concerned with the manufacture and processing of food and drink use great quantities of water, and produce correspondingly large quantities of effluent. The pollution carried by such an effluent is of much the same nature as domestic sewage (mainly organic matter), and can also be treated by biological processes, some being more suitable than others. It is frequently very much 'stronger' than sewage, and due allowance must be made in design and operation.

Other industries discharge a very wide range of substances into their effluents, and many of these can interfere with treatment processes. In such cases it is essential to pretreat the industrial effluent to remove or neutralize the interfering substance. The treatment of industrial effluents is far too extensive a topic to treat in an introductory chapter, and reference should be made to recent books.[19]

Discharge of dangerous or inhibitory wastes is controlled by local ordinance, and local regulatory authorities have wide powers in respect of consent to discharge, inspection, with-

drawal of consent, or the penalizing of offenders. Discharge to sewers of petrol or cyanides, for example, is forbidden for safety reasons.

Discharge of inhibitory matter, such as certain metallic ions or phenol, has to be very closely controlled if treatment processes are not to be upset, and watercourses put at risk by contamination. Trade-waste control is thus an extremely important factor in the day-to-day operation of sewers and sewage treatment works.

29.9 Effluent disposal

29.9.1 Introduction

A wastewater can be finally 'disposed of' only into water, on to land and into the ground. The last of these is available in practice only for small quantities of hazardous materials that cannot be safely dealt with in any other way, e.g. into worked-out salt mines.

Until the early years of this century, discharge on to land (sewage farming) was the only method in Britain acceptable to the Local Government Board. Generally, now, however, sewage farming is of no more than historic interest, although the irrigation of growing crops with fully treated sewage is of great interest in parts of the world where water is short.

This chapter is therefore largely concerned with the disposal of sewage into surface waters, and this section briefly considers the degree of treatment necessary to avoid danger, damage or nuisance resulting from such disposal.

29.9.2 Effects of water pollution

Wastewaters may contain substances which are poisonous to man and to plant and animal life in the water; sewage ought not to hold such substances, and it is essential to ensure that toxic matters are not allowed to enter municipal sewers from industry.

More important for municipal sewage are its power to use up the small amount of oxygen dissolved in the water, as the organic matter and nitrogen are oxidized by aqueous microorganisms, and the possible aesthetic effect of floating and suspended substances.

Nitrogen is of significance in a number of ways. Organic and ammoniacal nitrogen are oxidized in the receiving water, and so also use up dissolved oxygen. The fully oxidized form (nitrate) in sufficiently high concentrations may cause methaemoglobinaemia in very young infants. Above all, nitrogen is a plant fertilizer, and its presence in water, especially in standing bodies of water, may promote undesirable weed growth.

Sewage will, of course, also contain faecal microorganisms, and some of these may be pathogenic. In urban, industrialized communities the main protection against waterborne diseases is water treatment, with its accompanying disinfection. Nevertheless, as will be seen from Table 29.8, full normal sewage treatment reduces sewage bacteria very substantially.

It is usual to treat sewage to remove, partly or fully, suspended solids, oxygen demand (measured as BOD or COD – see section 29.8) and nitrogen, to prevent the kinds of polluting effects mentioned in previous paragraphs.

29.9.3 Degree of treatment necessary

The effects of pollution outlined in the previous section are closely related to the dilution available for the effluent discharged. The lower the dilution the greater will be the damage caused. For this reason it is usual to prescribe the quality of

Table 29.8 Removal efficiencies of sewage treatment

| | Percentage removal of | | |
	SS	BOD	Bacteria
Primary sedimentation	40–70	25–40	25–75
Chemical precipitation	70–90	50–85	40–80
Sedimentation + trickling filters + final sed.	70–92	80–95	90–95
Sedimentation + activated sludge + final sed.	85–95	80–95	90–98
Chlorination following full biological treatment			98–99

Table 29.10 Treatment of organic wastewaters

BOD (mg/l)	Method		BOD loading
< 500	Single filtration or activated sludge		0.1 kg/m³·day 0.2 kg/kg MLSS·day
500 +	Filtration with recirculation or alternating double filtration		0.15 kg/m³·day 0.2 kg/m³·day
1000	Extended aeration		0.05–0.15 kg/kg MLSS·day
1000–1500	High-rate filtration (with recirculation) followed by percolating filters or ADF or activated sludge		Up to 5 kg/m³·day 0.1 kg/m³·day 0.2 kg/m³·day 0.2 kg/kg MLSS·day
1500 +	Anaerobic treatment followed by one or two stages of aerobic treatment		1–5 kg/m³·day (depending on degree of removal)
Any	Oxidation ponds (multi-stage)	anaerobic first stage aerobic final stage	7000 kg/ha·day 250 kg/ha·day

effluent required for discharge to various types of receiving water; some suggested values are shown in Table 29.9.

The various methods of treatment that may be given to municipal sewage are indicated diagrammatically in Figure 29.5. The degree of removal that may be achieved by various combinations of treatment process is shown in Table 29.10.

It may be seen, by considering Tables 29.9 and 29.10 in conjunction, that full primary, biological and final treatment will be necessary for discharge to inland rivers; nitrification followed by denitrification may be required, in addition, for low dilutions. Equally effective treatment is likely also to be necessary for discharge to lakes, together with, in some cases, removal of the other important plant nutrient, phosphorus. Preliminary treatment alone is likely to be sufficient for ocean discharge, although it may sometimes be desirable also to provide primary settlement.

During full treatment not more than about one-third of the incoming pollution is converted to relatively harmless substances; the rest remains on the treatment works as solids for disposal (sludge). This is a major problem and expense in sewage treatment, and is discussed in sections 29.15 and 29.16.

29.10 Preliminary treatment

29.10.1 Introduction

The principal objective of preliminary treatment is to protect subsequent treatment processes, by preventing blockage and

Table 29.9 Typical concentrations of pollution for discharge to various receiving waters (all values in mg/l, except pH)

Parameter	Inland dilution less than 8[1]	River dilution more than 8[1]	Estuary	Open sea[2]
BOD	10	20	150	—[3]
SS	15	30	200	250[4]
Ammoniacal nitrogen	10	—	—	
pH	5–9	5–9	5–9	

Notes:
(1) With clean water.
(2) The ocean outfall must be carried sufficiently far out to sea to ensure that pollution is not brought back to the bathing beaches, etc. The end of the outfall must be provided with a properly designed diffuser and be located in a sufficient depth of water to ensure thorough mixing and dilution before the effluent reaches the surface.
(3) Not usual to specify a BOD limit.
(4) A SS standard is not always specified, but it is always desirable that floating matter should be reduced to a practicable minimum.

damage to the plant, and to increase the reliability and efficiency of the treatment process. Those objectives are achieved by removal of large solids by screening, removal of grit, removal of oil and grease, balancing of flow and/or load, pH control, and nutrient addition.

29.10.2 Screening

The quantity and nature of screenings vary, often substantially, between one plant and another. Relevant factors are social conditions and habits, industrial contributions, the type of sewerage system and the design of the screening plant. The following guidelines may be used to estimate quantities where there is no previous experience of local conditions.

The volume of screenings depends more on the character of the sewage than on the bar spacing; average volumes for domestic sewage are in the range 1 to 3 m³ per day per 100 000 population. Where there is an industrial effluent, the nature of the industry may suggest the additional screening load. The peak hourly rate of screenings removed is likely to be 4 to 6 times the average, and with combined sewers the peak rate during a storm following a dry period may be 10 to 20 times the average.

There are two basic approaches to handling large solids:

(1) To comminute in the flow or to remove, disintegrate and return to flow.
(2) To remove from the flow and dispose of elsewhere.

Comminutors are clean, innoffensive and relatively trouble-free machines, which are normally left unattended. However, rags tend to be shredded rather than cut up and may 'ball up' in later treatment stages, and scum volumes are increased by comminution. Similar problems are encountered with disintegrators.

The alternative of permanent removal and separate disposal of screenings is preferable in relation to the operation of the remainder of the works, but the problems of handling, transporting and disposing of the screenings are not easily resolved.

For most applications, mechanically raked bar screens are preferred for removing large solids from the flow. Curved bar screens are restricted to use in shallower channels, but the vertical or inclined bar types may be used in deeper channels and may use front or back raking.

Bar spacing (clear opening between bars) depends on the required removal efficiency. To protect pumps or fine screen units, a spacing of up to 150 mm is acceptable, although 90 mm is usual for all but the largest pumps. Pump clearway dimensions must obviously be considered. For normal sewage works a spacing of 20 mm is common although the need for individual design to suit local sewage characteristics must be considered.

Screenings are normally transported from the screen by mechanical conveyor or by launders. Whilst belt conveyors cannot be duplicated for standby they are attractive for smaller works. Launders are favoured for their simplicity, wherever the addition of transport water can be accepted, and are preferred on large works where a number of screens is to be installed. The use of a launder implies the need for a subsequent dewatering stage unless the screenings are disintegrated and returned to the sewage flow.

In the past, screenings were allowed to drain and were then carted away for burial, but the trend now is towards mechanical dewatering followed by transport to a screenings burial area, a refuse disposal site or an incinerator. The dewatering of screenings by ram or roller press not only reduces the volume, but often also improves the appearance and odour. Generally, the drier the product the more unrecognizable and acceptable it becomes.

On-site burial of screenings is normally practicable only on small to medium works. Specialized small-capacity incinerators can be considered for large works, but they are an expensive option and transport to a controlled refuse tip may be preferable. Incineration in combination with municipal refuse can be considered, but difficulties may arise from the different combustion characteristics of screenings and refuse.

29.10.3 Grit removal

Grit is a mixture of heavy mineral particles in sewage, such as sand, gravel, cinders, glass, etc. and when removed from sewage it contains some organic matter. The quantity of grit in normal municipal sewage varies according to area, degree of separation of surface water, time of year, etc. Typical yields range from 0.02 to 0.2 m³/1000 m³ of sewage, with most data tending towards the lower values.

Grit removal is normally preceded by screening, but comminutors, if used, are best installed after grit removal, with coarse bar screening upstream. Four methods of grit removal are commonly used: (1) detritors; (2) constant velocity channels; (3) spiral flow aerated channels; and (4) vortex-type chambers.

The detritor, which is a short-period settlement tank with a mechanical scraper for grit removal, is a satisfactory method of grit removal, but in general requires more complex civil engineering structures and larger land areas than the spiral flow channel.

Constant velocity channels are a simple design suitable for small and medium-sized works. They consist of long channels with a cross-section approximately parabolic. This maintains a relatively constant velocity over the full range of flow, when controlled by a flume or similar device. Settled grit is removed either manually or by suction dredgers.

The spiral flow aerated channel combines the constant velocity principle of differential settlement with the washing action of air turbulence. Air injected into a rectangular channel at one side induces a rotational motion, which sweeps grit into a floor

hopper and maintains lighter particles in suspension. Spiral flow aerated channels are the most suitable method for large works.

In vortex-type chambers a mechanically induced vortex in a conical tank produces secondary currents, which maintain the organic matter in suspension. A deep hopper at the bottom of the tank is used for grit collection and as a sump for grit pumping. The vortex-type trap has not given consistently good results, but has applications in some situations.

Grit washing after extraction is essential with a detritor, and desirable with constant velocity channels. Grit from spiral flow channels should normally be sufficiently clean. Grit from vortex traps varies in quality. Various washing mechanisms are available, often combined with dewatering classifiers to ease handling of the final product.

29.10.4 Skimming, flocculation and preaeration

Other pretreatment operations have been used to remove grease, oil and scum from sewage prior to primary sedimentation (see section 29.11) and to improve the treatability of wastewater. Skimming, flocculation and preaeration have been used for this purpose, although these techniques, other than perhaps preaeration, are not commonly practised in the UK.

Skimming tanks to remove floating matter may be designed to provide retention periods of 1 to 15 min. The outlet, which is submerged, is opposite the inlet and at a lower elevation to assist in flotation and to allow any solids that may settle to pass on to subsequent treatment stages.

Flocculation of sewage by mechanical or air agitation, although not commonly used, is sometimes given consideration when it is desired to increase the removal of suspended solids and BOD in primary sedimentation facilities, to condition wastewater containing certain industrial wastes, and to improve the performance of secondary sedimentation tanks following biological treatment processes, particularly the activated sludge process.

Preaeration, i.e. aeration of sewage prior to primary sedimentation, may be practised to provide grease separation. It may also be used for a variety of other purposes including grit removal and flocculation, to prevent septicity and hence to control odours, to promote uniform distribution of suspended and floating solids to treatment units, and to increase BOD removals. Preaeration may be practised in purpose-designed tanks, sometimes as an extension to aerated grit channels. Depending on the particular objective, it may also be carried out using aerated channels, which also serve to distribute flows to subsequent treatment units.

29.10.5 Flow/load balancing

Balancing of variations of flow and pollution load involves the damping of flowrate fluctuations so that a constant or reduced peak flow and pollution-load rate is achieved. It reduces the size of subsequent treatment units and reduces the risk of shock loadings (including elevated temperatures) affecting the performance of the main treatment units. If practised at all in sewage treatment it is normally coupled with primary sedimentation. Separate balancing with mixing facilities to maintain solids in suspension is normally considered only for industrial waste treatment applications or where the sewage has a high industrial waste content. Combined sedimentation/balancing tanks are discussed further in section 29.11.1.

29.10.6 pH control

The pH of a wastewater is a key factor in the growth of organisms within biological treatment processes. Most organisms cannot tolerate pH levels above 9.5 or below 4.0. Generally, the optimum pH for growth lies between 6.5 and 7.5.

Where the sewage contains a high proportion of industrial wastes and pH control is not practised at the factory premises, pH correction facilities may have to be installed at the sewage treatment works to ensure the satisfactory operation of biological processes.

Control of pH is also frequently necessary where chemical flocculation is practised (see section 29.11.2). pH correction, in the form of alkali addition, may also be necessary where a high degree of biological nitrification (ammonia oxidation) is required for sewages having low alkalinities and hence insufficient buffering capacity to absorb the reduction in alkalinity associated with this process.

Chemicals used for pH correction are common inorganic acids and alkalis, including sulphuric or hydrochloric acid, lime, caustic soda or soda ash. Dosing can be adjusted automatically by pH probes linked to pH controllers.

29.10.7 Nutrient addition

If a biological treatment system is to function correctly, nutrients must be available in adequate amounts. The principal nutrients are nitrogen and phosphorus. Assuming an average composition of cell tissue of $C_5H_7NO_2$, about 12.4% by weight of nitrogen will be required. The phosphorus requirement is normally assumed to be about one-fifth of this value. These are typical values rather than fixed quantities, since the distribution of nitrogen and phosphorus in cell tissue varies with the age of the cell and with environmental conditions. An alternative approach is to relate nutrient requirements to the waste BOD requiring removal; BOD:N:P ratios of 100:5:1 are often quoted for aerobic biological treatment, with lower N and P requirements being appropriate for anaerobic processes.

Domestic sewage generally contains more than adequate nitrogen and phosphorus concentrations to satisfy all sewage treatment requirements. However, where a high proportion of industrial wastes is also present, a nutrient deficiency may result and consideration then has to be given to the addition of nutrient chemicals. Soluble ammoniacal and phosphate salts, urea or combinations of these chemicals may be used.

29.11 Primary treatment

29.11.1 Sedimentation

Primary sewage treatment, following preliminary treatment, commonly takes the form of sedimentation for the removal of readily settleable suspended solids (SS) and associated BOD. Sedimentation tanks may be circular (radial flow), rectangular (horizontal flow) or pyramidal (upward flow) and are normally designed to remove 60 to 70% SS together with 30 to 40% associated BOD. They normally operate on a continuous flow basis, and include hoppers or troughs for collection of sludge and, in the case of circular and rectangular tanks, power-driven scrapers to move the sludge across the floor to the outlet. Facilities are also usually provided for collecting and removing surface scum and other floating material for subsequent treatment and/or disposal with the settled sludge.

The design criteria for primary sedimentation are based primarily on the maximum flow to receive treatment which should be established, taking full account of the effects of the return of secondary sludge and works liquors. The two principal design parameters are surface hydraulic loading ($m^3/m^2 \cdot h$ or $m^3/m^2 \cdot day$) and retention period (h). Maximum surface loadings in the range 1.5 to $2 \, m^3/m^2 \cdot h$ and minimum retentions of 1.5 to 2.0 h are commonly selected, due account being taken of the

need to provide a sufficient number of tanks, so that any one tank may be taken out of service without greatly affecting the sedimentation process. Consequently, even with a small works there should ideally be a minimum of two tanks.

Somewhat higher primary-tank loadings than specified above (e.g. up to $3 \, m^3/m^2 \cdot h$ and 0.75 to 1 h minimum retention) are sometimes adopted, at the expense of slightly lower pollution removals, depending on the particular application; this is particularly applicable in warmer climates, where the retention of sewage and sludge can usefully be minimized to reduce risk of septicity, rising sludge and associated problems. In the extreme, high-rate sedimentation tanks (8 to $12 \, m^3/m^2 \cdot h$) are sometimes considered as an alternative to fine-mesh screens for removal of gross solids (including floating material) plus associated BOD (15 to 30% SS, 5 to 10% BOD) prior to discharge to sea (see Table 29.11).

Table 29.11 Suggested design upflow velocities (surface loading) for sedimentation tanks

For settlement of	Design upflow velocity (surface loading)
Sewage – primary	2 m/h at peak flow
– primary before plastic (structural medium)	3 m/h
Sewage – final	1.5 m/h at peak
– final following extended aeration	1 m/h
With chemical treatment	Up to 2.5 m/h
With ferric salts or alum	1.5–2 m/h
Difficult solids following lime treatment	1 m/h

Note: The design upflow velocity can vary quite widely for various types of industrial wastes. It will not usually be suitable to use sedimentation for light solids that have upflow velocities less than 1 m/h; in such cases alternative clarification processes, such as flotation, should be considered.

Settled sewage usually leaves the sedimentation tank via a weir. Maximum weir loadings in the range 300 to $450 \, m^3/m^2 \cdot day$ are commonly specified by designers, although the significance of this parameter on the performance of primary sedimentation tanks is a subject of debate; weir loadings are of greater importance in the case of secondary sedimentation tanks following biological treatment.

Where extreme diurnal variations in flow and strength of sewage reach a treatment works, it is sometimes necessary to consider combined sedimentation/balancing tanks to equalize the flow and pollution load passing to subsequent treatment units. Such tanks are not as efficient for removal of suspended solids as tanks designed for sedimentation only, but are to be preferred to the installation of separate balancing tanks.

Sedimentation tanks are commonly sized to accept up to about 3 × DWF, with the higher flows arising during wet weather being diverted to storm tanks, which are normally allowed to overflow directly to the receiving watercourse when full. When flows reduce, storm-tank contents, including sludge, are then pumped back to the head of the works for further treatment. With small works, however, storm tanks are sometimes dispensed with and the capacity of the sedimentation tanks is increased in order to receive flows up to 6 × DWF or even higher.

29.11.2 Chemical treatment

During the earlier part of this century, it was a common practice in the UK, especially for sewages containing a high industrial waste content, to add one or more chemical flocculants (often at controlled pH) in order to improve the removal of suspended solids and associated BOD. While such chemical treatment is sometimes practised today throughout the world, chemical treatment, because of high cost, is now normally used in the UK only as a temporary measure to relieve an overloaded works by improving the efficiency of primary sedimentation.

However, chemical treatment may also be necessary where there is a need to reduce phosphate levels to meet strict standards for nutrients or as part of a tertiary treatment facility to produce a high-quality effluent for discharge to watercourse or for reuse. In the former case, flocculation with lime, iron or aluminium salts, often with polymer addition as a flocculant aid, is appropriate prior to primary sedimentation and subsequent treatment. Alternatively, variations on this practice may be adopted including chemical treatment following primary and secondary (biological) treatment or dosing directly into the biological treatment stage.

29.11.3 Flotation

Flotation is a process used to separate solids or liquid particles from a liquid phase by introducing fine bubbles of gas (usually air) into the liquid phase. The bubbles attach to the particulate matter and the buoyant force of the combined particle and gas bubble causes the particles to rise to the surface. In sewage treatment, flotation may be used to remove suspended solids either at the primary solids-removal stage or following biological treatment; in practice, however, flotation has greater interest as a method of thickening biological sludge. In all cases, chemical flocculant addition is required prior to the flotation process. Flotation thickening for waste sludges is discussed more fully in section 29.15.4.2.

29.11.4 Septic tanks

Cesspools and septic tanks are frequently used for receiving sewage from houses and other premises which are too isolated

for connection to a public sewer. A cesspool is simply a storage tank which should be emptied at regular intervals by a suitable tank-emptying vehicle.

A septic tank is a continuous, horizontal-flow tank in which settled sludge is retained sufficiently long for the organic content to undergo anaerobic digestion. When sludge is eventually removed (maybe once or twice yearly), a portion is left in the tank as a seed to initiate further digestion. A septic tank therefore combines the operations of sedimentation and sludge digestion.

29.12 Biological treatment

29.12.1 Introduction

Preliminary and primary treatment may remove two-thirds of the suspended matter in sewage, but no more than perhaps one-third of the organic pollution (BOD). If further reduction of BOD is necessary – and it usually is – it is essential to provide secondary treatment, in which the organic matter is oxidized by microorganisms, part being removed as such end-products as carbon dioxide and water, and part being converted to new microorganisms, which must be removed from the aqueous stream by final settlement.

Normal biological treatment of domestic sewage, together with final sedimentation, can consistently achieve an effluent with less than 20 mg/l BOD and with most of the ammoniacal nitrogen oxidized to nitrate. Very much better effluents can be produced only with additional treatment (see section 29.13). Biological processes may also be used for partial treatment where appropriate.

Biological oxidation may be achieved with the microorganisms held in a fixed film, or suspended in the sewage, or within a fluidized bed (see Figure 29.5). Biological breakdown may also be carried out by anaerobic processes (in the absence of air).

There is a very wide range of aerobic processes which may be

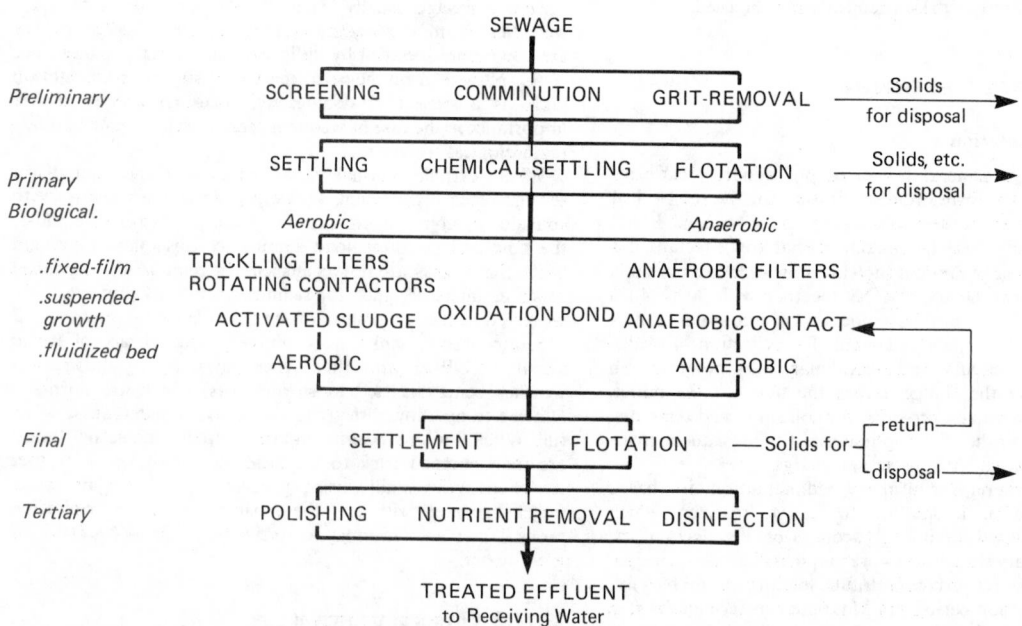

Figure 29.5 Treatment of municipal sewage

used for treating organic wastewaters of a correspondingly wide range of strength. An indication of suitable processes is given in Table 29.10 (p. 29/13).

29.12.2 Percolating filters

The oldest, and the most common, fixed-bed process is the percolating filter, which consists of a solid medium over which the sewage is sprinkled, and through which air passes freely, as the source of the essential oxygen. The medium may be stone, slag, gravel or specially made random-pack plastic; structural plastic media of various kinds have also been used.

For treatment to take place both BOD and oxygen must diffuse into the biological film, whose active depth must, therefore, be relatively limited. Surface area is, accordingly, an important design requirement, and may be increased by using a small medium. On the other hand, the film increases in volume, as the microorganisms multiply, and so may block the interstices and, in turn, prevent the access of oxygen. Although the film continually drops off, and there may also be a seasonal sloughing, it is essential to select the size and nature of medium with care. Guidance is given in Table 29.12.

It will be seen that single filtration removes the lowest amount of BOD, although it will produce an effluent of high quality. Performance may be improved by recirculating treated and settled effluent. In addition, filters may be used in-series, with the order of filtration being alternated. Both these modifications are useful for preventing the excessive increase of film.

The more recently introduced plastic media, both structural and random pack, have at least two advantages over stone, slag, etc. They provide a very much greater specific surface, and their lower weight can allow a lighter containing structure, which may be less costly.

Filters using the heavier media are commonly about 2 m deep. They must be contained in brickwork or concrete walls, and are underdrained in such a way that the ingress of air is encouraged. Plastic-packed filters may be much deeper.

Settled sewage may be applied to the filter beds by fixed sprays or moving sprinklers. Many filters are circular, and their rotating sprinklers are often driven by the reaction of the sewage itself; this calls for intermittent application controlled by a siphon dosing chamber.

The biological film is normally colonized by a number of animals and insects, which may assist in controlling the build-up of biomass. Filter flies, however, can be an unpleasant local nuisance. Odour, too, can be troublesome, especially with high-rate filters, and special steps, e.g. by exhausting the air and passing it through a scrubber, may be called for.

29.12.3 Rotating biological contactors

The rotating biological contactor (RBC) is a secondary biological treatment system. It consists of a large-diameter corrugated plastic medium, or set of discs, mounted on a horizontal shaft and placed in a concrete or steel tank. The medium is slowly rotated while approximately 40% of the surface area is submerged in the wastewater. During operation, biological growths adhere to the surface of the medium and form a slime layer over the entire wetted surface area. The rotation of the medium brings about oxygen transfer, and maintains the biomass in an aerobic condition as it absorbs and degrades the soluble organic constituents of the wastewater. The rotation is also a mechanism by which excess slime growth (or humus) is sloughed off into the tank, while at the same time maintaining the solids in suspension before they are carried from the unit to a subsequant settlement tank.

The attached biomass typically contains of the order of 50 000 mg/l suspended solids. If these solids were removed and placed in the mixed liquor, the resulting mixed liquor-suspended solids concentration would be 10 000 to 20 000 mg/l, i.e. very significantly higher than is practicable for an activated sludge plant. Hence, a high degree of treatment is possible for a relatively short retention period; however, in view of this short retention period, suspended solids in the influent to the plant are bioflocculated in the mixed liquor tank but otherwise pass through to the final settlement tank largely unchanged.

For industrial waste treatment applications, rotation of the plastic medium is achieved by direct motor drive. For domestic sewage applications, the alternative of diffused air, released into the mixed liquor near the periphery of the immersed medium, is sometimes used to create rotation, while at the same time helping to maintain aerobic conditions, but air drive units are not normally recommended for treatment of industrial waste. The use of periodic supplementary aeration is, however, beneficial, particularly for industrial waste treatment, for controlling the thickness of the slime layer on the medium. This control

Table 29.12 Biological filtration

Mode of operation	Overall filter loading (kg BOD/m³·d)		Medium size (mm)	Specific surface area (m²/m³)	Strength of sewage feed BOD (mg/l)
	without nitrification	with nitrification			
Straight filtration	0.09–0.12	0.06–0.09	40	120	150–350
Straight filtration with recirculation	0.15–0.18	0.09–0.15	40–60	80–120	150–350
Double filtration	0.18–0.24	0.15–0.21	40–60	80–120	>350
Alternating double filtration	0.18–0.24	—	40–60	80–120	>350
High-rate (slag)	Up to 2.7	—	100–150	33–60	200–400
High-rate (structural plastic) with recirculation if necessary	Up to 8	—	—	85	Up to 1500
High-rate (random pack plastic) with recirculation if necessary	Up to 10	—	—	180	Up to 1500

helps to reduce overall power costs by minimizing the weight of biomass adhering to the rotating medium.

Rotating biological contactor units are sized primarily on soluble BOD loadings per square metre of effective surface per day. Choice of loading will vary according to the type, strength and temperature of the wastewater, as well as on the degree of BOD removal required. When comparing differences among the various units marketed by RBC manufacturers, due account must be taken of differences in effective surface area of medium among the various designs. Rotating biological contactor plants are often installed within a building to protect the plant against adverse weather conditions.

29.12.4 Activated sludge

In the activated sludge process the microorganisms responsible for breaking down the organic matter are suspended in the sewage. There are at least two basic requirements: (1) oxygen for the metabolism of the organisms; and (2) good mixing to ensure that the organic matter, the organisms and the oxygen are in intimate contact, and that the solids are kept in suspension.

The agitation and oxygenation are usually combined. The two common processes are:

(1) *Diffused air*, in which air is blown into the sewage through porous diffusers or orifices.
(2) *Mechanical aeration*, in which the mixture of sewage and microorganisms is vigorously agitated mechanically, thus dissolving air.

It is essential to return active microorganisms to the aeration stage (returned activated sludge).

An important design parameter is the sludge loading rate (often called the food:microorganism ratio). The higher this rate, the lower will be the quality of the effluent. For a given BOD load, therefore, it is necessary to have a specified weight of active organisms, known as mixed liquor suspended solids (MLSS). There is a limit to which the concentration of MLSS can be increased in the aeration tank, which is governed by the process requirements. Suitable design parameters are shown in Table 29.13.

During the aeration stage, part of the incoming BOD is converted to new biological cells, up to 1 kg or more of new cells being produced for each kg of BOD removed. As indicated

above (and as shown in column 3 of Table 29.13) some of this must be returned to the reaction vessel (aeration tank), but most of it must be discarded after final settlement. The quantities involved are shown in the last column of Table 29.13.

It will also be seen from the table that the MLSS concentration is normally maintained at no more than 3000 to 4000 mg/l, with a consequent relatively low sludge loading rate. This is largely because transfer of oxygen to the microorganisms is limited with normal aeration systems. Therefore, the sludge loading rate must be relatively low to maintain a residual dissolved oxygen concentration in the mixed liquor of about 1 to 2 mg/l. Much higher oxygen-transfer rates are possible if pure oxygen is used in place of air, or if a deep shaft (50 to 150 m deep) is used as the aeration vessel. In both cases, higher oxygen transfer rates allow higher MLSS and sludge loading rates. The parameters for these two important modifications of the activated sludge process are shown in the bottom two lines of Table 29.13.

The figures given in the table are especially applicable to settled domestic sewage (say, BOD 200 to 300 mg/l and SS 100 to 150 mg/l), but the processes are suitable for a wide range of organic wastewaters, the sludge loading rates given in column 4 governing the size of aeration tank.

Pumps for returning the activated sludge to the aeration tank must be carefully selected to avoid excess shearing forces on the sludge flocs. It is usual, therefore, to use axial flow, torque flow or screw pumps.

The extended aeration modification, shown in the table, can be applied simply in an oxidation ditch, a process first developed in the Netherlands. An aeration rotor both circulates the mixed liquor and aerates it. The ditch is sometimes also used as the final settling tank by intermittently discontinuing the operation of the rotor. Surplus sludge must be periodically withdrawn from a collecting sump.

Retention periods will vary from 1 to 2 h for the high-rate process, through 6 to 10 h for the conventional process to 1 or more days for extended aeration.

The removal of organic and ammoniacal nitrogen is becoming increasingly important. It will seem from the table that nitrification (i.e. oxidation of the ammoniacal nitrogen) calls for a low sludge loading rate, and hence a large aeration tank.

Even oxidized nitrogen (NO_3) is coming under fire for health reasons, as well as because it is an important plant nutrient which may produce eutrophic conditions in lakes. Recent

Table 29.13 Activated sludge

Mode of operation	MLSS (mg/l)	Sludge return (%)	Sludge-loading rate (kg BOD/ kg MLSS·day)	Average oxygen requirement (kg O_2/ kg BOD)	Excess sludge (kg/kg BOD)
High-rate	1500–2500	25 +	1.0–2.0	0.5	1.2
Conventional (30/20 effluent)	1500–2500	33 +	0.35–0.45	0.8	0.8
Conventional (10/10 effluent)	2500–3000	50 +	0.20–0.30	1.2	0.7
Conventional with nitrification	3000–4000	50 +	0.15	1.2 + 4.5 × NH_3 N-oxidized	0.6
Extended aeration	2500–6000	100 +	0.05–0.15	2.0	0.8–1.0
Oxygen activated sludge	4500–8000	25 +	0.4–1.0	0.8–1.2	0.4–0.75
Deep shaft	4000–6000	100	0.4–1.0	1.3	0.5–0.85

modifications of the activated sludge process have introduced anoxic zones (oxygen free) in which the oxidized nitrogen is denitrified to gaseous nitrogen, and is so removed.

29.12.5 Oxidation ponds

Where land is cheap, and especially where sunshine is plentiful and temperatures high, sewage may be inexpensively treated in oxidation ponds. These are large lagoons, normally up to 4 ha in area, and usually about 1.0 to 2.5 m deep.

The biological oxidation may be carried out by the same sort of organisms as act in the aerobic processes described in sections 29.12.2 to 29.12.4 above. Anaerobic organisms may also be used, as may a combination of the two commonly called 'facultative'. Some typical design parameters are given in Table 29.14. There may be odour problems from the anaerobic and facultative ponds.

Table 29.14 Some typical design parameters for oxidation ponds

Operation	Anaerobic Recirculation of aerated effluent series	'Facultative' Mixed surface layer series or parallel	Aerobic Intermittently mixed series or parallel
Pond size (ha)	0.2–1	1–4	4
Depth (m)	2.5–5	1–2.5	1–1.5
Retention time (days)	20–50	7–20	10–40
Temperature range (°C)	6–50	0–50	0–30
optimum (°C)	30	20	20
BOD reduction (%)	50–85	80–95	80–95
Effluent SS (mg/l)	80–160	40–60	80–140

(After Metcalf and Eddy, Inc. (1979) *Wastewater engineering: treatment, disposal, reuse*, (2nd edn) McGraw-Hill)

Note: Examples quoted relate to pond systems in North America. In tropical climates, anaerobic pond loadings may be considerably higher in the range 100 to 400 g BOD/m³·day giving 1 to 2 days' retention. Similarly, facultative pond loadings up to 300 to 400 kg BOD/ha·day may be appropriate.

29.12.6 Anaerobic treatment

Organic matter may also be broken down, usually less completely, in the absence of oxygen by anaerobic microorganisms. The end-products are mainly carbon dioxide, methane and water, and the gas produced has a high calorific value and can be a useful source of energy.

Anaerobic processes have been used for very many years for treatment of sewage sludge before final disposal (see section 29.15.5). Recently such processes have been successfully applied to the treatment of strong organic effluents, e.g. from production of starch from wheat.

29.12.7 Fluidized beds

Fluidized beds have long been used in chemical engineering to promote rapid and complete reactions. They are now beginning to be applied to the much more dilute conditions of sewage treatment, using both aerobic and anaerobic organisms.

29.12.8 Final sedimentation

It cannot be too strongly emphasized that about one-half of the BOD applied to the biological treatment stage is converted to

new microorganisms, and must effectively be removed if an effluent of high quality is required. Final settlement is, therefore, essential. It is also required in order to separate the activated sludge solids for return to the aeration tank.

Activated sludge does not always settle easily. For example, high concentrations of carbohydrates, e.g. in brewery effluents, may cause special problems. It is essential, therefore, to consider the requirements for final sedimentation in designing the biological treatment stage.

The solids to be settled in final tanks are somewhat lighter than those removed in primary settlement. Accordingly, the design upflow velocity is commonly limited to 1.5 m/h at peak flow. It may be desirable to reduce this even further to 1 m/h following extended aeration.

29.13 Tertiary treatment

29.13.1 Introduction

It is sometimes necessary to produce a final effluent quality significantly better than the 30 mg/l SS, 20 mg/l BOD, which may be achieved with biological treatment methods. Such further treatment, commonly referred to as tertiary treatment or 'polishing', usually involves reduction of residual suspended solids, and hence associated BOD. Sand filtration, upflow clarification, microstraining or lagooning are commonly employed for this purpose; irrigation over grassland is also feasible where sufficient land is available.

29.13.2 Sand filters

Filtration may be carried out using slow sand filters, rapid downflow filters or upward flow filters. Slow sand filters are not often installed on modern works since they occupy large land areas and maintenance costs are high. Rapid downflow filters involve passing the sewage effluent downwards through a bed of graded sand (up to 1.5 m deep), under the influence of gravity or pressure, and these can cope with loadings up to 50 times that of slow sand filters. The gravity type, operated with 3 to 4 m head and at loadings up to 10 m³/m²·h, is normally used in the sewage treatment field, rather than fully enclosed pressure filters. Periodic backwashing with treated effluent (1 to 3 times daily), typically at 10 l/m²·s, is required to prevent clogging of the filter medium, backwash flows being returned to the head of the treatment works for further treatment. An air scour is also commonly used to aid the backwashing process. During backwashing, flows to the filter are usually diverted to standby or other duty filters. However, package rapid downflow filters are available on the market, where filtration through each filter can be maintained without interruption, sand from the base of the filter being air scoured during continuous airlifting to the top where washing takes place prior to the clean sand particles re-entering the main filtration zone. Removals of 80% SS and 70% associated BOD can be obtained with rapid downflow filters.

Rather higher loading rates can be achieved with upward flow sand filters (up to 17 m³/m²·h) for similar pollution removals. Backwashing involves increasing the effluent feed rate to achieve expansion of the bed which has first been loosened by an air scour. Whichever type of filter is employed, performance improves the more stable the suspended solids in the effluent feed are.

29.13.3 Upward flow clarifiers

Upward flow clarifiers are sometimes used at small sewage works for effluent polishing. The secondary effluent passes through a shallow bed of pea gravel, a mesh screen or other suitable medium which removes suspended solids by flocculation and settlement. The filter medium is normally incorporated within the final clarifier(s) and is kept clean by simple backwashing procedures. Maximum hydraulic loadings are normally in the range 1.3 to 1.8 $m^3/m^2 \cdot h$.

29.13.4 Microstrainers

Microstrainers (or microscreens) involve the passage of effluent through a stainless steel fabric which is retained around the periphery of a rotating drum. Flow enters the drum through a single open end and flows radially out through the steel mesh. Solids collected on the inside of the drum are then removed by a continuous filtrate spray applied along the drum at the highest point, the washings being collected in a trough and returned to the works' inlet. Best performance is achieved when the secondary effluent is well oxidized. Suspended solids removals of 30 to 80% and 25 to 70% associated BOD can be achieved depending on the mesh size employed (65 to 15 μm openings) and the hydraulic loading applied (typically 12 to 30 m^3/m^2 fabric·day).

29.13.5 Lagoons

Provision of lagoons as a final stage of treatment allows further flocculation and settlement of suspended solids and removal of associated BOD. Some further biological oxidation, reoxygenation and removal of bacteria also take place; a reduction in nitrate content due to denitrification, and of phosphate due to uptake by algae may also occur. In order to minimize the growth of algae, retention times should not normally exceed about 2 days, divided among several lagoons in-series. For typical depths of the order of 1 m, algal growths tend to increase the SS content of the final effluent, if retention periods exceed about 2.5 to 3 days.

Lagoons often support a fish population and attract wildlife and, hence, have appreciable amenity value. However, the availability and cost of land normally determines whether they are to be preferred to alternative tertiary-treatment methods.

Improved performance may be expected with a number of lagoons rather than a single lagoon of the same capacity. More than one lagoon is in any case desirable to provide flexibility of operation, particularly if solids deposition exceeds sludge solubilization and degradation, and desludging is eventually required.

29.13.6 Irrigation over grassland

This method of tertiary treatment is practised at many small and some medium-sized UK works. The grass plot retains suspended solids, thus improving effluent quality. Loadings are typically in the range 2000 to 5000 $m^3/ha \cdot day$ depending on climate and soil structure for secondary effluents of consistent quality. Effluent is applied via feed channels or by spraying. Periodical rest periods for the plots in rotation are necessary to reaerate the soil, and regular grass cutting is desirable to encourage fresh growth. Removals of 60 to 80% SS and 50 to 75% BOD can be achieved together with some removal of bacteria, nitrogen and phosphorus and some increase in dissolved oxygen.

29.13.7 Disinfection

The disinfection of sewage effluents (e.g. by chlorination or ozonation) is not extensively practised in the UK, although it is widely practised in many countries around the world, particularly where waterborne diseases or parasites are prevalent, and where effluent reuse for irrigation or other purposes is required.

Chlorination and ozonation both provide a very effective means of destroying bacteria and viruses in wastewaters and, in addition, have a remarkable effect on the colour and clarity of the effluent.

Chlorination tends to be significantly cheaper than ozonation. However, its suitability for treating sewage effluents should take full account of the final destination of the effluent, since certain residual compounds can combine with chlorine to form more objectionable compounds.

29.14 Advanced treatment

29.14.1 Introduction

The object of advanced treatment is to remove those pollutants which persist after conventional secondary treatment, but need to be removed to render the water suitable for reuse. Such pollutants include suspended, colloidal or dissolved, organic and inorganic compounds, as well as bacteria and viruses. Many processes exist for advanced treatment, and a few of these are outlined in the following paragraphs.

29.14.2 Chemical coagulation and flocculation

The removal of fine suspended solids may be improved by the addition of chemical coagulants. These act by destabilizing the colloidal particles, causing them to settle out more rapidly. In addition, the coagulant may also precipitate out soluble salts by chemically combining or physically adsorbing on to the floc.

Chemical coagulants include alum (aluminium sulphate), lime (calcium hydroxide), and ferric chloride, but only the former two can be recovered from the coagulation step. Organic polymers (polyelectrolytes) may be used in conjunction with, or in the place of, the inorganic coagulants in order to increase flocculation.

Addition of coagulants to wastewater takes place in rapid mix basins in which very high velocities are obtained and retention times are of the order 0.5 to 2 min. Flocculation is achieved by inducing velocity gradients (using mechanical stirrers or diffused air systems), into the wastewater to increase floc collisions. Typical retention times for flocculation are 15 to 40 min.

29.14.3 Ammonia stripping

This unit operation is one of mass transfer in a packed tower, using a countercurrent flow of air and water. Ammonia is transferred from the water to the airstream and may be exhausted to the atmosphere or recovered in an absorption column with acid to precipitate the corresponding salt. However, there are two limitations to the system: (1) the efficiency of the process is significantly reduced as the ambient air temperature falls below 10°; and (2) calcium carbonate scale may deposit on the tower packing and reduce the mass transfer efficiency. Scale formation has been found to be readily removed by frequent light hosing with water.

Ammonia-stripping towers have been successfully employed with ammoniacal removals of over 90%, achieved in 7 m towers with an air throughput of 1560 m^3/m^3 water. For higher removals greater airflows are necessary.

29.14.4 Recarbonation

The addition of lime to wastewaters (e.g. in phosphorus removal) increases the pH to a value of 10 or higher, which is not only outside usual discharge limits, but is also detrimental to the operation of downstream units such as carbon adsorption. Recarbonation is employed to lower the pH of wastewater and render it suitable for discharge. The pH is lowered by introducing carbon dioxide into solution. Generally the source of CO_2 is stack, or oven, gas, although underwater burners are also used.

Two forms of recarbonation are in use: single- and two-stage. Since the solubility of calcium carbonate is a minimum at pH 9.3, much of the $CaCO_3$ present precipitates out in the first tank, and enables the lime to be recalcined and reused. Single-stage recalcination does not afford this saving, since little $CaCO_3$ is precipitated. However, substantial savings can be made in capital and operating costs. In the former type the pH is lowered to about 7 in one tank, whereas in the latter case, two tanks are employed with possible intermediate settling. In this case, the pH is reduced to around 9.3 and 7 in the first and second tanks respectively. Single-stage recarbonation systems have typical residence times of 5 min and two-stage systems 15 and 45 min for the first and second stages respectively.

29.14.5 Granular activated carbon

Removal of organic materials from wastewaters may be successfully achieved by adsorption on to a granular activated carbon (GAC) bed. Powdered activated carbon (PAC) is seldom employed, since it can usually be used only once and is therefore costly, and presents dust and disposal problems. The advantage of GAC systems is that they can accept shock hydraulic and organic loadings without a reduction in removal efficiency, and may be regenerated in an on-site furnace. Operational experience has shown that this is a simple and effective process, giving reductions of total organic carbon of more than 70%. A secondary, but desirable application of activated carbon, is as a pretreatment for membrane processes.

29.14.6 Membrane processes

The introduction of membrane processes into wastewater treatment is relatively recent. The technique is one of selective removal of molecules or ions by a semi-permeable membrane. Three variations of this technique are discussed below.

29.14.6.1 Reverse osmosis

Reverse osmosis (or hyperfiltration) is the most popular of the membrane separation processes. The technique employs a pressure difference greater than the wastewater's osmotic pressure to force the solvent (water) through the membrane, producing a very clean water on one side of the membrane and a concentrated solution of impurities on the other. Although a small amount of dissolved solids will remain, bacteria, viruses and other pathogens are effectively removed. The water quality obtained from a RO pilot shows a reduction in excess of 90% in total dissolved solids and chemical oxygen demand.

The major problem with membranes is their potential to clog with suspended solids, chemical deposition or biological growths. Cleaning the membranes is accomplished by air/water scouring, although some chemical treatment may be necessary too. With domestic sewage a large proportion of the fouling matter is present as fats and oils, which have been successfully removed using detergents in conjunction with scouring.

29.14.6.2 Ultrafiltration

Ultrafiltration uses a membrane to discriminate between the size and shape of molecules, whereas the membrane in reverse osmosis operates in a selective manner for the transport of water. This is the fundamental difference between the techniques. Operationally the difference lies in the applied pressure – here it is significantly lower (0.5 to 7 bar) with typical water fluxes of 1 to 2.25 $m^3/m^2 \cdot day$. This method is applicable for the separation of solutes of high molecular weight (e.g. 500) such as bacteria, viruses, starch and proteins.

29.14.6.2 Electrodialysis

Electrodialysis is a technique for separating the ions of dissolved salts in solution. The ions are separated according to charge by the application of a current between two electrodes immersed in the solution. As the ions migrate to their respective electrodes they pass through a series of semi-permeable membranes which are ion-selective. The membranes are arranged with alternative selectivity, forming alternating zones of clean water and ion concentrate. The wastewater is pumped through the cell to achieve a retention time of 10 to 20 s. Scaling on the membrane may occur, but can be reduced by maintaining a lower pH with sulphuric acid. As in all the membrane processes, pretreatment is often necessary and prudent to avoid membrane fouling.

29.14.7 Ion exchange

The soluble ions present in wastewaters may be effectively removed by substitution by insoluble ions of a different chemical species. This is the basis of ion exchange which is frequently used to demineralize water.

The resin, an insoluble natural or synthetic material, needs to be regenerated once it becomes saturated with ions from the aqueous phase. Chemical treatment is employed to restore the resin.

Two methods of operation are currently in use: batch and continuous exchange. In the batch mode of operation, the resin is added to the water and stirred until the reaction is complete. Spent resin is allowed to settle out and is removed. In the continuous mode, the resin is packed in columns and the wastewater is passed through the bed. It is often prudent to have two columns so that one may be regenerated without halting the flow of water.

Ion exchange has applications in the treatment of municipal wastes in the removal of nitrates and phosphates. Using a strong-base anion exchanger, phosphate removals greater than 97% can be achieved, producing an effluent level of 0.2 p.p.m. Similarly, nitrate levels may be reduced by 90%, and COD by 45% with typical water throughputs of 200 × bed volume.

29.15 Sludge treatment

29.15.1 Introduction

The volume of liquid sludge produced at a sewage treatment works typically amounts to some 1 to 2% of the total sewage flow. However, its treatment and disposal are usually major operations, accounting for as much as 50% of the operating costs of the works. The purpose of sludge treatment is to render the sludge more amenable to disposal, and to minimize the cost of disposal.

29.15.2 Character and amount of sludge

Where primary sedimentation is practised, as at most sewage treatment works, about 60 to 70% of the suspended solids is normally removed, together with 30 to 40% of the associated biochemical oxygen demand (BOD). This sludge normally contains an average of about 5 to 6% dry solids, although the concentration at any one works will vary according to the frequency of sludging practised. This is commonly between once and three times daily, but where primary sludge is subsequently thickened in continuous-flow tanks, sludging will be more frequent (e.g. every hour) giving thinner sludges of the order of 2% dry solids.

Biological processes will contribute to the overall sludge volume produced in the works, but quantities generated depend on the type of biological treatment process operated. Yields of biological sludge are expressed as kilograms dry solids per kilogram BOD removed per day; examples, including typical dry solids contents are given in Table 29.15. In practice, actual yields of sludge requiring treatment and/or disposal will be within the ranges given in Table 29.15, less the dry weight of suspended solids discharged in the secondary sedimentation tank effluent; however, this loss will be relatively small in the case of secondary effluents conforming to the common 30 mg/l SS, 20 mg/l BOD standard. Biological sludges are frequently returned to the primary sedimentation tanks (where these are provided) and co-settled with primary sludge, although recent trends have been towards separate thickening of surplus activated sludge.

Table 29.15 Yields of biological sludge

Process	Biological sludge yield (kg dry solids/kg BOD removed·day)	Dry solids (%)
Biological filters		
low-rate	0.25–0.5	0.5–2.0
high-rate	up to 1.0	0.5–2.0
Conventional activated sludge	0.6–0.8	0.5–0.8
Extended-aeration activated sludge	0.8–1.0	0.5–1.0
Rotating biological contactors		
following removal of gross primary solids only	0.7–0.8	1.0–3.0
following conventional primary sedimentation	0.5–0.6	1.0–3.0

Where tertiary treatment is practised, the weight of dry solids removed at this stage, which is returned for treatment and disposal, should be added to the quantity of biological solids produced. Additional quantities of sludge will also be produced where chemical treatment is practised, e.g. for the removal of phosphorus.

29.15.3 Screening

Sewage is normally screened or comminuted on arrival at the sewage treatment works. However, since such screening is never completely effective, and because shredded rags can tend to 'ball up' and cause blockages in sludge pipelines and in treatment equipment, sludge is also sometimes screened before treatment. Manually cleaned screens (19 mm bar spacings) or, for biological sludge alone, mesh screens (3 to 5 mm) may be used, e.g. rotary brush type, are suitable; screenings should be disposed of together with other works sludges. The alternative of sludge comminution is also sometimes practised.

29.15.4 Sludge thickening

Thickening of sludge is frequently carried out to reduce the volume of liquid sludge requiring subsequent treatment and/or disposal. The resultant liquors produced are normally returned for further treatment with the main sewage flow. The commonest form of thickening is by gravity, but centrifuging or, particularly in the case of surplus activated sludges, gas flotation or the use of a belt thickener, are sometimes employed.

29.15.4.1 Gravity thickening

Gravity thickening, often aided by a rotating picket fence, is normally carried out in the UK as a batch process, three 1-day retention tanks commonly being provided, one each for filling, quiescent settlement and emptying respectively; in this case, supernatant liquors are removed by floating or swivel arm, telescopic weir or a series of controlled outlets set at different levels. However, the tanks may also be designed for continuous flow operation and may be used as storage tanks. Continuous flow tanks are primarily designed on the basis of solid loadings per unit area per day; examples of some typical figures are given in Table 29.16.

The degree of thickening achieved by fill-and-draw and continuous flow tanks is dependent on the type and age of the sludge, and on its initial dry solids content, as well as on the design of the thickening tank provided. Increases in dry solids contents as a result of gravity thickening are commonly in the range 1 to 3% dry solids.

Table 29.16 Tank loadings – continuous-flow sludge thickeners

Sludge type	Solids loading (kg dry solids per m²/day)
Raw primary	100–150
Mixed primary/humus	60–100
Mixed primary/surplus activated	40–80

29.15.4.2 Flotation thickening

Sludge thickening by gas flotation is sometimes considered, where separate thickening of surplus activated sludge is required. Activated sludges are difficult to concentrate by gravity alone, but flotation thickening to 4 to 5% dry solids can be achieved by dissolved air or electrolytic flotation, the former being the more common of the two processes. Polyelectrolyte flocculant (1 to 4 kg/t dry solids) is usually added to achieve efficient thickening in the dissolved air process, but owing to charge neutralization effects, flocculant aids are not normally necessary for the electrolytic process.

Dissolved air flotation includes dissolving air into a water stream under pressure, usually the subnatant liquors or works final effluent, and introducing a mixture of this stream and the raw surplus activated sludge into the bottom of a flotation tank. The associated release of pressure generates very small air bubbles and the air/sludge mixture rises to the surface of the tank, whence it is removed by surface scraper. The clarified liquor is returned to the main treatment works or to the recycle stream for saturation with air and reuse.

Electrolytic flotation is achieved by passing the sludge between a grid of electrodes operating with an applied d.c. potential. The resulting electrolysis produces very small gas bubbles which enable thickening effects similar to those achieved with dissolved air flotation to be achieved. The sizing of both systems is based on solids loading per unit area of tank surface per hour, a figure of 10 kg dry solids/m²·h commonly being used for design purposes.

29.15.4.3 Centrifuges

Two types of centrifuge are available for thickening of sewage sludges; the nozzle bowl or disc stack type for thickening surplus activated sludge, and the solid-bowl type for either activated sludge or mixed co-settled sludges.

The nozzle- or disc-type centrifuge is a vertical spindle machine rotating at 900 to 3300 rev./min. The internal bowl contains a stack of conical discs which provide a large surface area to aid settling of solid particles. To minimize wear and risk of blockage, screening of the sludge before pumping through the centrifuge is required. Centrifuges of this type can achieve thickened sludges of up to 6% dry solids without polyelectrolyte flocculant addition giving a 90% solids recovery.

The solid bowl centrifuge comprises a horizontally mounted cylinder with a tapered section forming a 'beach' for the solids, up which they are conveyed by the scroll rotating at a speed slightly higher than that of the bowl. The performance of this type of machine depends on: (1) the type of sludge; (2) the type of polyelectrolyte being used; (3) the bowl/screw conveyor speed difference; and (4) the liquid level maintained in the centrifuge. It is more commonly used, in conjunction with polyelectrolytes, to dewater sludge to a stackable cake, but can be used to produce a thickened slurry if the liquid is maintained above the discharge port. Polyelectrolyte dosing is normally essential to achieve efficient solids recovery.

29.15.4.4 Belt thickening

Gravity thickening of surplus activated sludge, with addition of polyelectrolyte, has recently been developed using belt machines such as the Aquabelt manufactured by Simon Hartley Ltd. Thickened sludges of 4 to 5% dry solids can be produced. This type of machine incorporates a travelling belt on to which flocculated sludge is discharged. As the sludge travels along the length of the machine, thickening occurs with sludge liquors passing through the porous belt cloth. This development is a modification of the free-draining zone incorporated at the front end of belt presses which are commonly used for dewatering sludges to give a sludge cake.

29.15.5 Anaerobic sludge digestion

Anaerobic digestion of raw primary, or mixed primary/secondary sludges, is carried out in the absence of oxygen, and results in conversion of organic matter into soluble and gaseous products. The process changes a malodorous sludge into one which is relatively inoffensive, destroying grease and reducing numbers of certain pathogenic organisms. The process has developed from cold digestion in open tanks to the modern mesophilic system with covered tanks operated at 30 to 35°C, utilizing the gas for heating or power generation. The sludge gas normally consists of about 65 to 70% methane, the remainder being mainly carbon dioxide with a small amount of other gases; typically the gross calorific value is 24 000 to 26 000 kJ/m³.

Process efficiency depends on the type of sludge and the extent to which inhibitory substances may be present (e.g. heavy metals), the method of addition, the degree of internal mixing within the digester, the retention period and the operating pH and temperature. For a 20-day retention at 30 to 35°C the organic content of the sludge is reduced by 40 to 50% (equivalent to 30 to 40% total solids reduction), and of the order of 1 m³ sludge gas per kilogram volatile matter·day is produced (equivalent to about 0.03 m³/hd·day). In the past, many heated digesters have been designed for longer retention periods (25 to 30 days) to provide a margin of safety, although in theory a digestion period of about 10 days is adequate. In practice, some safety margin is desirable to deal with inadequate mixing or sludge-loading variations, but with improvements in available mixing and heating systems, retentions of 15 to 20 days can be quite satisfactory.

Cold digestion is practised at many small treatment works. Mesophilic digestion at 30 to 35°C is, however, preferred for improved efficiency. Thermophilic digestion (40 to 50°C) is not practised in the UK. Digester mixing may be carried out by mechanical mixing, by internally or externally mounted gas-lift pumps (confined mixers), or by passing digester gas through floor-mounted diffusers (unconfined mixers). Mixing systems, including use of the gas, combine the advantages of good mixing efficiency with the benefit of having no moving parts within the digester, thus avoiding maintenance problems.

Heat-exchange units can be incorporated within the sludge-mixing system or as a separate circuit. Heat is supplied from a boiler which burns the methane gas or an alternative fuel. Gasholding capacity may be provided by means of floating covers on the digesters or by means of separate, normally floating-roof, gasholders.

Sludge gas, surplus to that required for heating the digester contents, is commonly used for space heating or for power production. Sludge gas has also been used for incineration of screenings, sludge drying and, in an emergency, for vehicle propulsion. At large sewage treatment works, the total yield of sludge gas produced during anaerobic digestion is often used for power generation, the waste heat from the engines being recovered in the form of hot water for heating the sludge. The power produced may be used to generate electricity or for driving air blowers or pumps. Gas consumption typically lies in the range 0.45 to 0.55 m³/kWh.

Following heated digestion, the digested sludge is generally passed to storage tanks or lagoons (commonly referred to as secondary digesters) prior to dewatering and/or off-site disposal. They enable the sludge to cool and allow some thickening to take place, supernatant liquors being returned to the sewage stream for further treatment. Retention periods vary from a few days to often more than 30 days. Rotating picket-fence mechanisms are often incorporated.

29.15.6 Aerobic sludge digestion

The alternative of aerobic sludge digestion involves partial oxidation of sludges utilizing aerobic microorganisms supported by aeration. It serves the function of stabilizing the sludge to minimize odour nuisance and of reducing the solids content. The application of the process is limited within the UK, tending to be confined to use with small package sewage treatment plants.

Under UK climatic conditions, surplus activated sludges and mixed primary/secondary sludges require a minimum aeration period of 10 to 15 days and around 20 days respectively to achieve a solids reduction of 30% and a relatively odour-free sludge. However, the resultant digested sludges often do not thicken readily and the overall dewatering properties of the sludge may also deteriorate.

Aeration is normally carried out using mechanical or diffused-air systems. Electrical power consumption is high and solids concentrations should not exceed 1.5 to 2% to avoid problems of poor mixing and inadequate aeration. Oxidation rates increase with increasing temperature. Thermophilic aerobic digestion has been considered for certain applications, particularly where efficient removal of pathogenic organisms is required, but data currently available on this process are limited.

29.15.7 Sludge dewatering

Sludge dewatering produces a readily handleable cake, normally containing at least 15% dry solids and possibly up to 40% dry solids or more, depending on the type of sludge and method of dewatering concerned.

In the early part of this century, drying beds, consisting of a sand and gravel bed overlying tile underdrainage, were in widespread use for dewatering sludge. In the UK, changing disposal strategies and the development of a range of mechanical dewatering equipment have resulted in land-intensive drying beds becoming less attractive in favour of direct disposal of sludge in liquid or mechanically dewatered form.

29.15.7.1 Filter presses

Mechanical dewatering processes include filter plate and filter belt pressing, centrifuging and vacuum filtration. A filter plate press comprises a series of chambers formed between recessed plates. An appropriate filter cloth is fitted over the surfaces of each plate. The plates are closed together and held hydraulically, or by screws, to withstand the applied filtration pressure supplied by positive-displacement pumps. Filtrate passes out of the press via ports and is returned to the sewage stream for further treatment, leaving the sludge solids retained within the chambers as a cake. Lime with an iron salt, aluminium chlorohydrate or polyelectrolyte are commonly used to condition the sludge prior to feeding the press on a batch basis. Sludge cakes of 30 to 40% dry solids are normally achieved. With the press mounted in a building at first-floor level, cakes can readily be dropped into a lorry or skip.

29.15.7.2 Filter belt presses

Filter belt presses (or band filters) involve the continuous discharge of liquid sludge, normally conditioned with polyelectrolyte (1 to 4 kg/t dry solids) on to a moving open mesh, endless roller-mounted belt (0.5 to 2.5 mm wide) which acts as filtering medium. This belt then converges with a second belt to produce pressure on the sludge layer. At the discharge end, a doctor blade lifts the sludge cake from the belt which is then washed with high-pressure water, or effluent sprays, before travelling back to the sludge inlet end of the machine. Since the early machines, more rollers have been introduced and, hence, better performance can now be achieved. Some designs include vertical belts, spring loaded rollers and caterpillar tracks. Sludge cakes typically lie in the range 15 to 35% dry solids, depending on the type of sludge, polyelectrolyte dose and machine design.

29.15.7.3 Centrifuges

The use of centrifuges has already been described as a method of thickening sludges prior to further processing and/or disposal. The solid bowl (or decanter-type) centrifuge is most commonly used for dewatering sewage sludges. With polyelectrolyte dosing (2 to 5 kg/t dry solids) of mixed primary/secondary and digested sludges, sludge cakes averaging 20 to 25% dry solids can usually be achieved associated with solids recovery efficiencies of 90 to 99%. As is frequently the case with sludge-dewatering processes, performance improves the thicker and fresher the sludge feed.

29.15.7.4 Vacuum filters

Rotary drum vacuum filters consist of a cylindrical drum covered with a filtration medium, normally cloths or stainless steel coils, and rotating partially submerged in a tank of sludge. With chemically conditioned sludge in contact with the external face of the filtration medium and the application of a vacuum to the internal face, liquor is drawn through leaving a dewatered sludge retained on the revolving outer surface, which can then be removed by a tine bar or doctor blade. Lime with an iron salt,

aluminium chlorohydrate or a polyelectrolyte can be used for conditioning the sludge, either of the latter two being more common today, since problems of scale formation giving increased wear and reduced efficiency of the vacuum are then minimized. Sludge cakes from 20 to 25% dry solids, and occasionally higher, can be produced depending on the particular application. A large number of rotary vacuum filters were installed in the UK during the 1960s; however, they are less popular today, important limitations being the relatively low dry solids content of the cake compared with plate presses and modern belt presses, high maintenance costs and the need for experienced operators to optimize performance.

Vacuum disc filters differ in design from rotary vacuum filters, but operate on similar principles, and are simpler to operate. They are suitable for use at small sewage treatment works. In the UK, aluminium chlorohydrate has been the more usual conditioning aid, and using mixed raw sludge as a feed, sludge cakes typically in the range 15 to 18% dry solids can be achieved.

29.15.8 Other sludge treatment processes

Elutriation, as a sludge conditioning process, is sometimes employed, although it is not common today for new works. The process involves washing the sludge to remove fine suspended solids which reduces the chemical conditioner requirements prior to mechanical dewatering. It also reduces the alkalinity of sludge (particularly by removing ammoniacal compounds) prior to anaerobic digestion.

Heat treatment of raw sludge at temperatures up to 250°C, with or without the introduction of air, is sometimes practised as a conditioning process. Although very effective, major disadvantages include the emission of strong odours, production of strong sludge liquors, which are difficult to treat, and high operational and maintenance costs.

Following dewatering of sewage sludge to produce a cake (using drying beds or mechanical processes), thermal drying can be practised, where a granular product containing 85 to 90% dry solids is required. Oil, digester gas or other fuel is used to produce hot air for the drying process. The process is expensive and is normally considered only where the dried product can be sold, e.g. as a fertilizer or soil conditioner.

29.16 Sludge disposal

Following treatment, or in a small number of situations without any treatment, sludge residues must be disposed of off-site. The method of disposal selected will usually depend upon the outlets available and on the characteristics of the sludge, in particular its toxic metal content. It is normal for the ultimate disposal route to influence the type of treatment practised prior to disposal, rather than vice versa. Sludge treatment and disposal are costly, and the objective is to obtain a balance between treatment and disposal, to discharge the sludge to the environment safely and at the minimum possible cost.

The principal methods of sludge disposal are: (1) as liquid to agricultural land, usually and preferably after digestion; (2) as a liquid at sea, often after digestion; (3) as a cake to agricultural land or to tip; and (4) as cake to incineration, the ashes produced being dumped to tip. In the UK, disposal as a liquid to agricultural land is favoured where enough land is within reasonable reach of the treatment works and providing the sludge is of an acceptable quality. Disposal of liquid sludge to sea is also favoured, particularly for large conurbations (i.e. large amounts of sludge and more distant agricultural land) with access to a sea terminal. In both cases it is common practice to thicken the sludge prior to disposal to reduce its volume, and, hence, the disposal costs. The disposal of sludge cake to agricultural land is less favoured because of the cost of dewatering to a cake, and the difficulty of spreading it. Incineration is costly and is generally practised only when the sludge is unsuitable for land disposal because of toxicity or due to the absence of a suitable dumping site. Dumping on sacrificial land is usually preferred to incineration for toxic sludges because it is cheaper. However, it is not always possible to establish environmentally safe dumping sites of adequate capacity within economical transport distance.

In the past, sludge was disposed of to the environment with little concern for any adverse effects other than, perhaps, odour nuisance. An increasing level of environmental awareness, coupled with a fuller understanding of potential hazards, led to the introduction of controls upon sludge disposal in many countries. In Britain, sludge disposal to agricultural land is not subject to national legally enforced control at present. However, Regional Water Authorities and other responsible bodies do apply their own control measures working to guidelines laid down by the Department of the Environment/National Water Council[20] and by the Ministry of Agriculture, Fisheries and Food.[21,22] The guidelines specify the maximum load of toxic elements that can be added to agricultural land over a 30-year period. The load is calculated as individual elements and as 'zinc equivalent'; the latter concept permits the relative toxicity of the three most common and harmful toxic metals to be allowed for and quantified in a single figure. Zinc equivalent is the sum of the zinc content, twice the copper content and 8 times the nickel content of a sludge.

The dumping of sludge at sea is practised according to two international Conventions as embodied in the Disposal at Sea Act 1974. Sludge discharged to sea via pipelines is effectively not controlled at present, although mechanisms exist. All sludge dumping activities are controlled by licence. The licence imposes pragmatically based limitations depending on the nature of the dumping grounds and the quantity of the sludge in mind. It is not possible to give typical standards, but the most stringent control is on persistent and recalcitrant materials, in particular mercury and cadmium.

The dumping of sludge by tipping on sacrificial land is subject to normal solid waste disposal legislation in Britain, under which tipping sites are licensed and controlled by Waste Disposal Authorities. Controls on incineration relate to the dumping of the ash as a solid waste and the discharge of gaseous emissions to the atmosphere.

Other countries have their own control mechanisms which may or may not impose stricter constraints upon sludge disposal activities. It is not possible to generalize. It is however, important to note that measures are being taken by the European Community to standardize the control of sludge disposal to agricultural land and to the sea. It may be expected that EEC Directives will be issued and that they will impose limitations and controls at least as stringent as those now applied in Britain.

29.17 Intermediate technology

In many of the poorer developing countries there are large areas where not even a piped water supply exists, let alone a sewerage system. Sanitation practices are often based on open air defecation sites, which may be fields, waste land, or alleyways; it appears that no understanding exists of the link between faecal contamination and health. It is to these areas that the UN International Drinking Water Supply and Sanitation Decade is principally addressed. There is no possibility that conventional sanitation facilities could be introduced in all of these areas: the cost is unaffordable. In the more densely populated conurbations, some form of waterborne sewerage system could be

shown to be essential; elsewhere, consideration must be given to low-cost solutions to allow the benefits of sanitation to be enjoyed by the maximum number of people.

The most common low-cost sanitation system in the past was the earth closet, usually associated with a nightsoil collection system in urban areas. Whilst this system is still widely used it cannot be commended. It offers a health hazard to those handling the containers and, as is often the case, when the crude excreta are used as fertilizer. Other methods, such as squatting huts built over fish ponds or excreta chutes to pig pens may be less unhygienic but do carry hazards and, in any case, cannot have a wide application.

Biogas installations, which were originally developed in India in 1938, are being used in some countries with variable success. Here, biodegradable human, animal and vegetable wastes are anaerobically decomposed to produce a relatively harmless fertilizer and digester gas. The gas can be used for domestic cooking, heating and lighting. The concept is attractive, but has not gained widespread international favour, because of the relative complexity of design, the cost of the installation, the risk of process instability and the possibility of explosion or fire.

In recent years much attention has been paid to privies of simple design and affordable construction. Privies vary from simple pits or shafts with a squatting plate above, to composting chambers into which both excreta and vegetable waste may be disposed. Refinement may be introduced by substituting the squatting plate by a water-sealed unit, hand-flushed with small volumes of water. Where pour-flushing cannot be afforded, or availability of water is inadequate, fly and odour nuisance should be minimized. This is best done by ensuring that light is excluded from the privy and that the collection pit is ventilated by a pipe. The pipe should be at least 75, and preferably about 200 mm, in diameter, should be painted black and sited in the sunniest position to encourage an upflow of air, and should contain a mesh filter to trap flies, since any entering the pit will tend to escape towards the light at the top of the pipe. Properly designed and maintained, ventilated pit latrines can be aesthetically and hygienically acceptable.

There are usually two pits associated with these simple latrines, one in use and one standing full. Each pit should have at least 1 year's capacity and up to 15 years is used in practice. The full pit should not be emptied until it has been standing for a year or more, by which time its contents are a relatively harmless compost. Whether or not a pour-flush facility is installed it is desirable to ensure that liquid can soak away from the pit; otherwise it will fill too quickly and the contents will be less easy to dispose of safely. Where liquid can drain from the pit the risk of groundwater pollution must be avoided.

Latrines are designed to accept only human excreta; sullage ('grey water') must be disposed of separately. Often it may be discharged to a soakaway, or used for irrigation. However, in other than the most rural situation it is common to discharge sullage haphazardly to surface water drains. Since surface water systems are often crude and poorly designed the practice frequently leads to situations where health is at risk and which are aesthetically objectionable. When considering low-cost sanitation, safe sullage disposal must be given equal emphasis.

Pit latrines are suitable only where water usage is relatively low; they should not be associated with normal toilet flushing practices. Where water is available for normal flushing, and the facility can be afforded, the lowest cost acceptable system is the septic tank.

Septic tanks are designed to discharge an effluent usually to a soakaway. Again, groundwater pollution must be avoided. Where septic tanks exist, or are installed in relatively densely populated areas, they can be upgraded by collecting their effluents into a small-bore sewerage system for central treatment. However, whether or not this is done, a septic tank must be sludged periodically, ideally once or twice a year, and the facility must exist to meet this relatively costly need.

The overall concept of upgrading is important and must be borne in mind at design stage. As water availability improves, and a society becomes more affluent, the need will be perceived to move, perhaps, from simple pit latrines, to pour-flush latrines, to septic tanks and even eventually to main drainage. A long-term goal should be set appropriate to the area and each step towards it should be taken with the sequential development of the sanitation plan in mind.

The need to generate reliable comparative costs is basic to the final evaluation of the options that are judged to be viable in technical terms. The World Bank has published a summary for the technologies of total annual economic cost per household;[22,23] their data are reproduced in Table 29.17, which shows not only total costs but also a breakdown into on-site, collection and treatment costs. As the World Bank points out, the figures do not divide clearly into community and individual systems as might have been expected. However, it is evident that the choice among the three groups of technologies is clearcut, with large buffer areas available for up-grading any system within any one group.

Table 29.17 Average annual on-site, collection and treatment costs per household[23]

	Average annual costs per household ($US 1978)			
	Mean total costs	On-site costs	Collection costs	Treatment costs
Low cost				
Pour-flush toilet	18.7	18.7	—	—
Pit privy	28.5	28.5	—	—
Communal toilet	34.0	34.0	—	—
Vacuum truck cartage	37.5	16.8	14.0	6.6
Low-cost septic tanks	51.6	51.6	—	—
Composting toilets	55.0	47.0	—	8.0
Bucket cartage	64.9	32.9	26.0	6.0
Medium cost				
Sewered aquaprivy	159.2	89.8	39.2	30.2
Aquaprivy	168.0	168.0	—	—
High cost				
Septic tanks	269.2	332.3	25.6	11.3
Sewerage	400.3	201.6	82.8	115.9

Intermediate technology is an important aspect of environmental engineering. Considerable practical experience now exists, and it is possible to select and design the system that will most closely meet local economic, geographic and ethnic needs. However, it must be emphasized that whilst low-cost sanitation systems may also be low-technology systems, this does not mean that optimum solutions can be established inexpertly. Low-cost systems must be selected with the same care and expertise that are required when working with more conventional technology.

References

1 Department of the Environment and National Water Council (1981) *Design and analysis of urban storm drainage*, vols 1–5. National Water Council, London.

2 Hydraulics Research Station (1977) *Tables for the hydraulic design of pipes* (3rd edn). HMSO, London.

3 Transport and Road Research Laboratory (1976) *A guide for engineers to the design of storm sewer systems*. Road Note No. 35 (2nd edn). HMSO, London.

4 Ministry of Housing and Local Government (1970) *Technical committee on storm overflows and the disposal of storm sewage. Final report*. HMSO, London.

5 Hydraulics Research Station (1978) *Charts for the hydraulic design of channels and pipes* (4th edn). HMSO, London.

6 Moore, D. R. and Gotham, K. V. (1978) 'The mechanical properties of uPVC in relation to pressure pipes.' *The Pub. Health Engr*, **6**, 239.

7 Kirby, P. C. and Ridgway, J. W. (1982) 'Recent developments in rubber joint rings for water mains.' *Proceedings of conference on the use of plastics and rubber in water and effluents*, 6.1.–6.15. Plastic and Rubber Institute, London.

8 Olliff, J. L. (1982) *Factors of safety in the structural design of large sewers*. Proceedings of the 1st International Seminar on Urban Drainage Systems, Southampton University.

9 Griffiths, I.W. 'Sulphide control in rising mains.' *Water Pollution Control*, **80**, 5, 654–647.

10 Manganaro, C. A. (1968) 'Design for unbalanced thrust for buried water conduits.' *J. Amer. Wat. Wks Assoc*, **60**, 6, 705–716.

11 Johnson, M. (1981) *First report on the WRC sewage sludge pumping project*. Technical Report No. TR162. Water Research Centre/British Hydromechanics Research Association.

12 Prosser, M. J. (1977) *The hydraulic design of pumps, sumps and intakes*. British Hydromechanics Research Association/Construction Industry Research and Information Association, London.

13 Frost, R. C. (1983) *How to design sewage sludge pumping systems*. Technical Report No. TR 185, Water Research Centre, Swindon.

14 Young, O. C. (1978) *A review of practice and recommendations in making connections to pipe sewers*. Occasional Technical Paper No. 1, National Water Council, London.

15 Irving, D. J. and Smith, R. J. H. (1983) *Trenching practice*. Report No. 97. Construction Industry Research and Information Association, London.

16 Parkinson, R. W. and Giles, R. G. (1983) Water Research Centre and Water Authorities Association. Sewerage rehabilitation manual. WRC Engineering. Swindon.

17 Department of the Environment and National Water Council (1977) *Methods for the examination of waters and associated materials*. London (1977 and continuing).

18 American Public Health Association, American Water Works Association and Water Pollution Control Federation (1981) *Standard methods for the examination of water and wastewaters* (15th edn) ALPHA, New York.

19 Callely, A. G., Forster, C. F. and Stafford, D. A. (eds) (1977) *Treatment of industrial effluents*. Hodder and Stoughton, London; Koziorowski, B. and Kucharski, J. (1972) *Industrial waste disposal*. Pergamon Press, Oxford.

20 Department of the Environment and National Water Council (1981) *Report of the sub-committee on the disposal of sewage sludge to land*. Standing Committee on the Disposal of Sludge, Report No. 20.

21 Ministry of Agriculture, Fisheries and Food (1971) *Permissible levels of toxic metals in sewage used on agricultural land*. ADAS Advisory Paper No. 10.

22 Ministry of Agriculture, Fisheries and Food (1978) *The use of sewage sludge as a fertiliser*. AF 51.

23 International Bank for Reconstruction and Development (1979) *Appropriate sanitation alternatives: a technical and economic appraisal*. World Bank, Washington DC.

30

Irrigation, Drainage and River Engineering

W Pemberton BSc, FICE
Head of Irrigation and Drainage Department
Sir Murdoch MacDonald and Partners

C E Rickard BSc, CEng, MICE, MIWEM
Head of River Engineering Department
Sir Murdoch MacDonald and Partners

Contents

PART A: IRRIGATION AND DRAINAGE

30.1 Irrigation – fundamental concepts

30.1.1 Introduction

Irrigation is desirable where natural rainfall does not meet the plant water requirements for all or part of the year. Irrigation is essential for agriculture in the desert but even in areas such as northern Europe it can improve the yield of crops normally grown under rainfall conditions only.

30.1.2 Soil moisture

The soil can be considered a moisture reservoir. Soils can be classified under the International Soil Science Association (ISSA) system as follows:

Fraction	Particle size (mm)
Coarse sand	2–0.2
Fine sand	0.2–0.02
Silt	0.02–0.002
Clay	<0.002

Water is held by the soil in the soil pores. The amount of water held can be defined as follows:

(1) Saturation: the state of complete soil wetness when no further water may be added to the soil.
(2) Field capacity (FC): the condition reached after water has drained from the soil by gravity.
(3) Permanent wilting point (PWP): the condition reached after plants have extracted all the moisture they can from the soil.
(4) Available water: defined as (FC − PWP), the amount of water held by the soil that plants can use.

Plants respond to how tightly the water is held by the soil which is defined as soil moisture tension. Generally, it is assumed that the soil moisture tension at field capacity is 0.3 bar pressure. Soil moisture tension at PWP is assumed to be 15 bar.

Typical moisture contents for various soils are shown in Table 30.1.

Table 30.1

Soil type	Moisture content (percentage by weight)		
	Field capacity	Permanent wilting point	Available moisture
Coarse sand	8	4	4
Fine sand	15	8	7
Silt	28	18	10
Clay	45	30	15

With knowledge of the crop rooting depth, the available soil moisture and the crop water requirements, it is possible to select a suitable irrigation interval (time between irrigations). Not all water in the root zone is readily available to the crop. It is normal to allow the crop to deplete only 50% of the available moisture before irrigating again. More detailed guidelines are given by the Food and Agricultural Organization.[1]

30.1.3 Crop water requirements

Crop water requirements are defined as the depth of water required to meet the water loss through evapotranspiration (ET_{crop}) of a crop. The effect of climate on crop water requirements is given by the reference crop evapotranspiration (ET_0) which is defined as the rate of evapotranspiration from an extensive surface of green grass of uniform height (8 to 15 cm):

$$ET_{crop} = k_c \times ET_0 \tag{30.1}$$

where k_c is the crop coefficient which varies with crop, growth stage, growing period and prevailing weather conditions

The most reliable method of estimating ET_0 is generally considered to be the PENMAN method. This method is best described by Doorenbos and Pruitt[1] which also gives details of crop coefficients for a wide range of crops. Values of ET_{crop} are normally calculated for 10-day periods. A typical crop coefficient curve is shown in Figure 30.1.

A simpler method was proposed by Blaney and Criddle[2] in

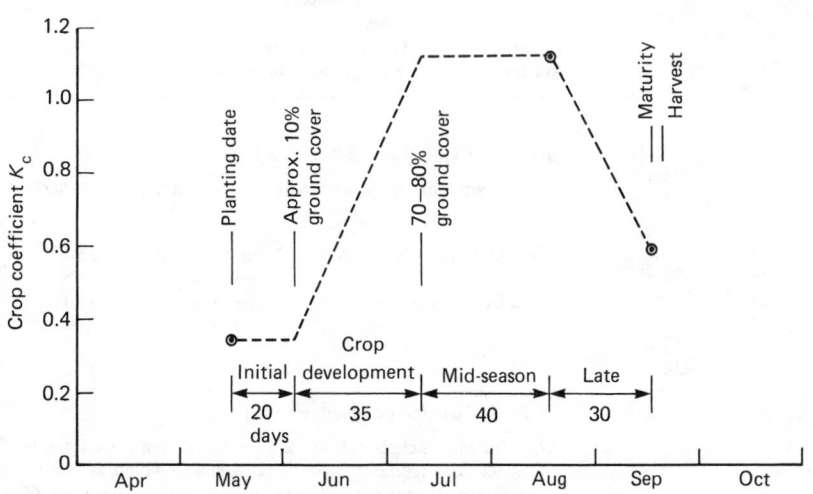

Figure 30.1 Example of crop coefficient curve.
(After J. Doorenbos and W. O. Pruitt (1977) *Crop water requirements.* Food and Agriculture Organization Irrigation and Drainage Paper No. 24.)

Table 30.2 Monthly percentage of annual daytime hours (*p*) for different latitudes

Latitude North South	Jan. Jul.	Feb. Aug.	Mar. Sep.	Apr. Oct.	May Nov.	Jun. Dec.	Jul. Jan.	Aug. Feb.	Sep. Mar.	Oct. Apr.	Nov. May	Dec. Jun.
40°	6.76	6.72	8.33	8.95	10.02	10.08	10.22	9.54	8.29	7.75	6.72	7.52
42°	6.63	6.65	8.31	9.00	10.14	10.22	10.35	9.62	8.40	7.69	6.62	6.37
44°	6.49	6.58	8.30	9.06	10.26	10.38	10.49	9.70	8.41	7.63	6.49	6.21
46°	6.34	6.50	8.29	9.12	10.39	10.54	10.64	9.79	8.42	7.57	6.36	6.04
48°	6.17	6.41	8.27	9.18	10.53	10.71	10.80	9.89	8.44	7.51	6.23	5.86
50°	5.98	6.30	8.24	9.24	10.68	10.91	10.99	10.00	8.46	7.45	6.10	5.65
52°	5.77	6.19	8.21	9.29	10.85	11.13	11.20	10.12	8.49	7.39	5.93	5.43
54°	5.55	6.08	8.18	9.36	11.03	11.38	11.43	10.26	8.51	7.30	5.74	5.18
56°	5.30	5.95	8.15	9.45	11.22	11.67	11.69	10.40	8.53	7.21	5.54	4.89
58°	5.01	5.81	8.12	9.55	11.46	12.00	11.98	10.55	8.55	7.10	5.04	4.56
60°	4.67	5.65	8.08	9.65	11.74	12.39	12.31	10.70	8.57	6.98	4.31	4.22

Note: Southern latitudes apply 6-month difference as shown.

which the monthly crop water requirements ET_{crop} (in millimetres) are found by multiplying the mean monthly temperature T_m (°C) by the monthly percentage of annual daytime hours p and a monthly crop coefficient k:

$$ET_{crop} = (0.46T_m + 8)kp$$

Table 30.2 shows the monthly percentage of p for different latitudes.

A sample calculation of water requirements for maize planted mid May near Saskatoon (latitude 52°N) is shown in Table 30.3.

A simple calculation of gross irrigation requirements (I_{gross}) can be made as follows:

$$I_{gross} = (ET_{crop} - R_e)/E_a \qquad (30.2)$$

where ET_{crop} is the crop water requirements, R_e is the effective rainfall and E_a is the field application efficiency

Table 30.3

Period		Mean monthly temp. (°C)	% annual daytime (p)	Crop coefficient (k)	Water require- ments (mm)
	Days				
15/5–31/5	17	11.6	5.95	0.35	27.8
1/6– 3/6	3	19.7	1.11	0.35	6.6
4/6–30/6	27		10.02	0.96	164.1
1/7– 8/7	8	19.3	2.89	1.05	51.2
9/7–31/7	23		8.31	1.14	159.9
1/8–17/8	17	21.3	5.55	1.14	112.6
18/8–31/8	14		4.57	1.02	83.0
1/9–16/9	16	8.7	4.53	0.75	40.8
					646.0

30.1.4 Irrigation efficiency

It is necessary to account for losses of water incurred during conveyance and application to the field. Efficiencies can be divided into three parts: (1) field application; (2) field canal; and (3) distribution efficiency.

30.1.4.1 Field application efficiency (E_a)

E_a is dependent on soil type and type of irrigation system used. Typical values are given in Table 30.4.

Table 30.4

Irrigation method	Application practices	E_a, % water application efficiency	
		Soil texture heavy	light
Sprinkler	— Daytime application, moderately strong wind	60	60
	— Night application	70	70
Trickle		80	80
Basin	— Poorly levelled and shaped	60	45
	— Well levelled and shaped	75	60
Furrow	— Poorly graded and sized	55	40
Border	— Well graded and sized	65	50

30.1.4.2 Field canal efficiency (E_f)

E_f is dependent on type of field channel used and area served.

Blocks larger than 20 ha	Unlined canals	0.90
	Lined or piped	0.95
Blocks up to 20 ha	Unlined canals	0.80
	Lined or piped	0.90

30.1.4.3 Distribution efficiency (E_d)

Distribution efficiency (E_d) is dependent on area served, the level of water management, and canal seepage which is the main component. Canal seepage can be calculated separately from seepage rates given in Table 30.11 (page 30/9).

Typical overall values for E_d are given below:

Table 30.5 Average monthly effective rainfall as related to average monthly ET_{crop} and mean monthly rainfall

	12.5	25	37.5	50	62.5	75	87.5	100	112.5	125	137.5	150	162.5	175	187.5	200
Monthly mean rainfall (mm)																
Average monthly ET_{crop} (mm)																
25	8	16	24													
50	8	17	25	32	39	46										
75	9	18	27	34	41	48	56	62	69							
100	9	19	28	35	43	52	59	66	73	80	87	94	100			
125	10	20	30	37	46	54	62	70	76	85	92	98	107	116	120	
150	10	21	31	39	49	57	66	74	81	89	97	104	112	119	127	133
175	11	23	32	42	52	61	69	78	86	95	103	111	118	126	134	141
200	11	24	33	44	54	64	73	82	91	100	109	117	125	134	142	150
225	12	25	35	47	57	68	78	87	96	106	115	124	132	141	150	159
250	13	25	38	50	61	72	84	92	102	112	121	132	140	150	158	167

Where net depth of water that can be stored in the soil at time of irrigation is greater or smaller than 75 mm, the correction factor to be used is:

Effective storage	20	25	37.5	50	62.5	75	100	125	150	175	200
Correction factor	0.73	0.77	0.86	0.93	0.97	1.00	1.02	1.04	1.06	1.07	1.08

Continuous supply with no substantial change in flow	0.90
Rotational supply in projects of 3000 to 7000 ha and rotation areas of 70 to 300 ha, with effective management	0.80
Rotational supply in large schemes (> 10 000 ha) and small schemes (< 1000 ha) with respective problematic communication and less effective management:	
based on predetermined schedule	0.70
based on advance request	0.65

The total irrigation requirements at the head of the system (I_{sys}) can be calculated from:

$$I_{sys} = \frac{I_{gross}}{E_d E_f} \tag{30.3}$$

30.1.5 Effective rainfall (R_e)

All rainfall is not effective in providing water for crop use. The calculation of effective rainfall is discussed in detail by Dastane.[3] A simple method has been developed by the US Soil Conservation Service which relates effective rainfall with ET_{crop} and mean monthly rainfall (see Table 30.5).

For example, with a monthly rainfall of 50 mm, an ET_{crop} of 100 mm and an effective soil storage of 100 mm, the correction factor is 1.02 and the effective rainfall is $1.02 \times 35 = 36$ mm.

30.1.6 Salinity and leaching requirement

All irrigation water contains some dissolved salts. If no effort is made to move salts through and beyond the root zone, the soil salinity level will increase to make it unfit for plant growth. The process of dissolving and transporting soluble salts downwards to below the root zone is known as leaching.

The maximum leaching requirements can be calculated from:

Leaching requirement (LR) for surface or sprinkler irrigation

$$LR = \frac{EC_w}{5EC_e - EC_w}$$

For drip and high frequency sprinklers (almost daily)

$$LR = \frac{EC_w}{2MaxEC_e}$$

EC_w = electrical conductivity of irrigation water, mmho/cm

EC_e = electrical conductivity of the soil saturation extract for a given crop to the tolerable degree of yield reduction (see Table 30.6)

Max EC_e = maximum tolerable electrical conductivity of the soil saturation extract for a given crop (see Table 30.7)

Alkalinity and toxicity may also affect soil permeability and crop growth. For further details, Ayes and Westcott[4] can be consulted.

Salinity hazard has been classified by the US Department of Agriculture (USDA) as shown in Table 30.6.

Table 30.6

Salinity of water (mmho/cm)	Salinity hazard	
<0.25	Low	Suitable for most crops and soils
0.25 to 0.75	Medium	Suitable for moderately salt tolerant crops
0.75 to 2.25	High	Not suitable for low permeability soils
>2.25	Very high	Generally not suitable for irrigation

In many instances, the usual inefficiencies of water application satisfy the leading requirements, but it is sometimes necessary to allow additional irrigation water for leaching. The leaching efficiency varies with soil type and may vary from 100% for sandy soils to perhaps as low as 30% for swelling heavy clay soils.

30.2 Irrigation methods

30.2.1 Introduction

The choice of method of irrigation is dependent on technical feasibility and economics. Normal methods fall into four main categories: (1) surface; (2) sprinkler; (3) trickle; and (4) sub-irrigation.

30.2.2 Surface irrigation

Surface irrigation is still the most common method of irrigation employed and is suitable for the irrigation of most soils with an infiltration rate of less than 150 mm/h and for lands with a flat topography with an overall slope of less than 3%, although these limitations are exceeded in some situations.

There are four main types of surface irrigation: (1) basin; (2) border strip; (3) furrow; and (4) corrugation irrigation.

30.2.2.1 Basin irrigation

Water is applied from a small canal by gravity to fill a level basin surrounded by earth bunds. In practice, these basins are often small but the most efficient irrigation is obtained by using large basins, at least a hectare in area and accurately levelled. Water should be applied to those basins at a rate of at least 2 to 4 times the infiltration rate of the soil.

Basin irrigation is most suitable for very flat and level land and soils with low infiltration rates. When adopted for uneven topography the basin size must be kept small in order to limit the quantity of land levelling required. Land levelling rates in excess of 1000 m³/ha should be avoided. The cultivation of paddy rice is normally done using basin irrigation. Basin irrigation is illustrated in Figure 30.2.

30.2.2.2 Border strip irrigation

The land is divided into strips separated by earth bunds which run generally down the slope, and water is applied at the head of the strip and allowed to flow down the slope infiltrating the soil as it flows across it (see Figure 30.3). The strip is graded at an even slope along its length in the direction of flow and level across the strip. Efficient irrigation is obtained by choosing the strip width, length and discharge to meet the soil infiltration rate

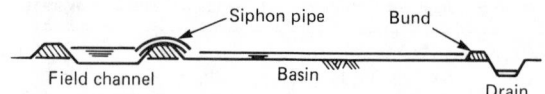

Figure 30.2 Basin irrigation

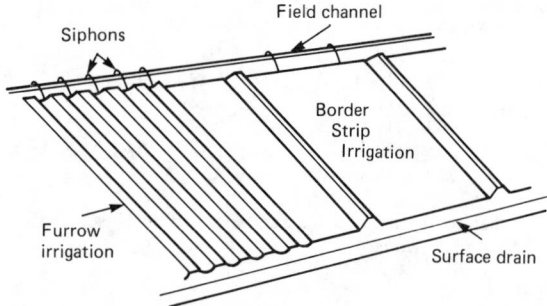

Figure 30.3 Surface irrigation methods

and land slope conditions to give as constant a depth of water as possible infiltrated over the length of the strip. Typical border strip designs from the USDA *Yearbook* are given in Table 30.8. Border strips are suitable for land with a more pronounced existing slope, thus reducing the amount of land levelling necessary. (For more information on design of sprinklers, see work by Barrs.[6])

30.2.2.3 Furrow irrigation

Furrow irrigation is used for the irrigation of row crops or crops grown on beds between furrows. Furrow irrigation usually implies sloping land although horizontal furrows can be used for row crops within level basins. Water is applied to the upper end of each furrow and flows down the furrow with water infiltrating into the beds between the furrows on which the crop is grown.

Furrow spacings are a function of crop and type of tillage machinery used. Typically furrows are spaced 0.75 to 1.05 m apart. Table 30.9 gives recommended maximum furrow lengths in metres for various soil types, furrow slopes and average depth of water applied over the whole field.

Furrow slopes should be checked for erodability. The maximum non-erosive flow in furrows (Q_m) can be estimated from:

$$Q_m = 0.60/S \qquad (30.3)$$

Table 30.7 Crop salt tolerance for selected crops

Crop	Yield potential 100% (EC_e)	(EC_w)	90% (EC_e)	(EC_w)	75% (EC_e)	(EC_w)	50% (EC_e)	(EC_w)	Max EC_e
Barley	8.0	5.3	10.0	6.7	13.0	8.7	18.0	12.0	28
Wheat	6.0	4.0	7.4	4.9	9.5	6.4	13.0	8.7	20
Typical vegetables (beans, carrots, lettuce, onions)	1.0	0.7	1.7	1.1	2.8	1.9	4.6	3.1	8
Forage, grasses	4.6	3.1	5.9	3.9	7.9	5.3	11.1	7.4	18
Fruit trees	1.7	1.1	2.3	1.6	3.3	2.2	4.8	3.2	8
Date palm	4.0	2.7	6.8	4.5	10.9	7.3	17.9	12.0	32

Table 30.8 Typical border strip designs

Soil type	Slope ((%)	Depth applied (mm)	Strip width (m)	Strip length (m)	Flow (l/s)
	0.25	50	15	150	240
		100	15	250	210
		150	15	400	180
	1.00	50	12	100	80
Coarse		100	12	150	70
		150	12	250	70
	2.00	50	10	60	35
		100	10	100	30
		150	10	200	30
	0.25	50	15	250	210
		100	15	400	180
		150	15	400	100
	1.00	50	12	150	70
Medium		100	12	300	70
		150	12	400	70
	2.00	50	10	100	30
		100	10	200	30
		150	10	300	30
	0.25	50	15	400	120
		100	15	400	70
		150	15	400	40
	1.00	50	12	400	70
Fine		100	12	400	35
		150	12	400	20
	2.00	50	10	320	30
		100	10	400	30
		150	10	400	20

where Q_m is in litres per second and S, the furrow slope, is in per cent. Generally, cross-slopes in furrow irrigation should be less than the major ground slope down the furrows to limit the furrow flows breaking out.

30.2.2.4 Corrugation irrigation

Corrugation irrigation is a variant of furrow irrigation in which the furrows are very small. It is suitable for medium soils only and is used for close-growing crops such as wheat. The corrugations are some 10 cm deep and spaced 40 to 75 cm apart. Because the corrugation flows are small, slopes up to 8% have been used. This method of irrigation is not widely used outside the US.

For more details of surface irrigation methods, Booher[5] can be consulted.

30.2.3 Sprinkler irrigation

30.2.3.1 Types of sprinkler

The application of water by overhead sprinklers takes many forms which include the following.

(1) Permanent and solid set – a network of pipes and sprinklers which covers the whole area to be irrigated. No movement of equipment within a season is necessary. This is the most expensive form of sprinkler irrigation.
(2) Lateral move sprinklers – sprinklers on a lateral line that is moved by hand after each irrigation application to the next area to be irrigated. This is the most widely used system.
(3) Traveller systems – these are motorized methods of moving sprinklers and include:

    (a) sideroll – lateral pipe and sprinklers on wheels, pushed by hand or small motor from one position to next irrigation position;
    (b) mobile rain gun – single gun winched across field whilst irrigating and fed from a hose reel;
    (c) centre pivot – overhead lateral with sprinklers which rotates about centre whilst irrigating;
    (d) linear move – similar to centre pivot but moves laterally across the field.

The most common system used in developing countries, where labour is inexpensive, is the lateral move sprinkler system. In developed countries where labour is expensive, various forms of travellers are more common. Sideroll is suitable for low crops as the lateral is not normally more than 1.8 m above the ground. Rain guns with their high water-pressure requirements and, hence, high energy costs are best used for supplementary irrigation. Centre pivot and linear move are becoming the most popular traveller systems in arid areas. Permanent and solid sets are very expensive and hence used only on high-value crops.

30.2.3.2 Sprinkler design

Individual sprinklers provide a cone of precipitation so that

Table 30.9 Maximum recommended furrow lengths (m). (After Booher (1974) *Surface irrigation*. Food and Agriculture Organization Land and Water Development Series No. 3.)

Soil type	Fine				Medium				Coarse			
Furrow slope (%)	average depth of water applied (cm)											
	7.5	15	22.5	30	5	10	15	20	5	7.5	10	12.5
0.05	300	400	400	400	120	270	400	400	60	90	150	190
0.1	340	440	470	500	180	340	440	470	90	120	190	220
0.2	370	470	530	620	220	370	470	530	120	190	250	300
0.3	400	500	620	800	280	400	500	600	150	220	280	400
0.5	400	500	560	750	280	370	470	530	120	190	250	300
1.0	280	400	500	600	250	300	370	470	90	150	220	250
1.5	250	340	430	500	220	280	340	400	80	120	190	220
2.0	220	270	340	400	180	250	300	340	60	90	150	190

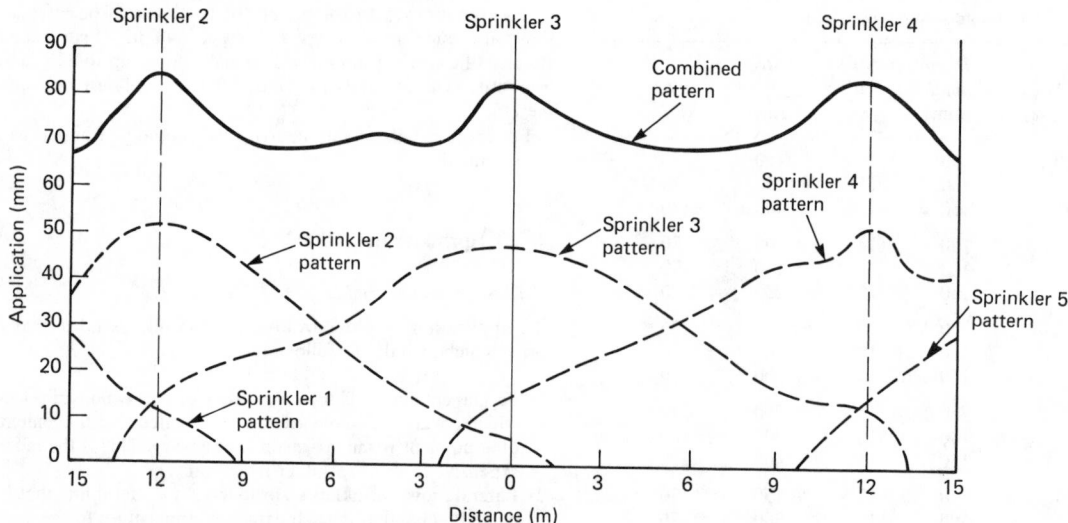

Figure 30.4 Sprinkler application patterns

overlap of sprinkler patterns is necessary to give a reasonable uniform application as shown in Figure 30.4.

The discharge of a sprinkler Q in cubic metres per second:

$$Q = CF\sqrt{2gH} \qquad (30.4)$$

where: C = the contraction coefficient varying between 0.79 and 0.98, F = the cross-sectional area of the nozzle in square metres, g = 9.81 m/s^2 and H = the height of hydraulic head behind the nozzle in metres

To give a particular precipitation rate over a field there is a range of solutions of sprinkler spacing, nozzle diameter and pressure as shown in Table 30.10.

Table 30.10 Spacing and precipitation rates of single-nozzle sprinklers. (After Baars (1973) *Design of sprinkler installations*. Department of Irrigation, Civil Engineering, Agricultural University, Wageningen)

Details of sprinkler				Square and rectangular spacing of sprinklers (m)							
Nozzle size	Pressure	Discharge	Diameter coverage	12 × 12	12 × 18	18 × 18	18 × 24	24 × 24	24 × 30	30 × 30	30 × 36
(mm)	(atm)	(m³/h)	(m)	Precipitation rate (mm/h)							
4.0	3.0	1.02	30	7.1	4.7						
	3.5	1.11	31	7.7	5.1						
	4.0	1.19	32	8.3	5.5						
5.0	3.0	1.63	33		7.5	5.0					
	3.5	1.76	34		8.1	5.4					
	4.0	1.88	35		8.7	5.8					
6.0	3.5	2.56	36			7.9	5.9				
	4.0	2.74	36			8.5	6.3				
	4.5	2.90	37			9.0	6.7				
7.0	3.5	3.48	40				8.1	6.0			
	4.0	3.73	41				8.6	6.5			
	4.5	3.96	42				9.2	6.9			
8.0	3.5	4.44	43				10.3	7.7			
	4.0	4.74	43				11.0	8.2			
	4.5	5.04	44				11.7	8.8			
9.0	3.5	5.67	44					9.8	7.9		
	4.0	6.06	45					10.5	8.4		
	4.5	6.42	46					11.1	8.9		
10.0	3.5	7.12	46					12.4	9.9		
	4.0	7.60	47					13.2	10.6		
	4.5	8.06	48					14.0	11.2		
11.0	3.5	8.63	48						12.0	9.6	
	4.0	9.23	49						12.8	10.3	
	4.5	9.79	50						13.6	10.9	
12.0	3.5	10.18	49						14.1	11.3	
	4.0	10.88	50						15.1	12.1	
	4.5	11.55	52						16.0	12.8	

Note: Exceeding the line is not recommended for ideal irrigation.

The variation of head between sprinklers is normally limited to $\pm 0.2H$ where H is the design head at the sprinkler, taking into account any difference in ground level at sprinklers. It is this criterion which limits the lateral pipe diameter and length.

Movable laterals are normally made of aluminium or galvanized steel, the aluminium being lighter and hence easier to handle, and the galvanized steel being cheaper and more easily repaired.

The supply pipeline can be designed using the normal pipe friction formulae described in Chapter 5.

The calculation of head loss in sprinkler lines having sprinklers at constant spacing can be calculated using the Christiansen formula:

$$h_z = \frac{hfan}{100} \tag{30.5}$$

where h_z = head loss in the sprinkler line in metres, h = head loss in 100 m line in metres, through which a quantity of water flows which corresponds to the total discharge of all sprinklers on the line, n = number of sprinklers on the sprinkler line, a = spacing of the sprinklers and f = factor which varies with the number of sprinklers, n, as follows:

n	f	n	f	n	f	n	f
2	0.625	8	0.398	14	0.370	20	0.359
3	0.518	9	0.391	15	0.367	25	0.354
4	0.469	10	0.385	16	0.365	30	0.350
5	0.440	11	0.380	17	0.363	40	0.345
6	0.421	12	0.376	18	0.361		
7	0.408	13	0.373	19	0.360		

30.2.4 Trickle irrigation

The basis of trickle irrigation is to provide irrigation water to individual plants. A plastic pipe is run along the ground at the base of a row of plants and water is carried to each plant through orifices in the pipe or using an emitter. Trickle irrigation is more accurately described as localized irrigation as it includes a wide range of emitters such as micro-sprinklers and bubblers.

Trickle irrigation is most suitable for row crops and trees and is generally able to use more saline water supplies than surface irrigation or sprinkler irrigation. The design of localized irrigation systems is described by Vermeirei and Jobling.[7]

30.2.5 Sub-irrigation

Sub-irrigation is only suitable for specialized soil conditions. High horizontal permeability and low vertical permeability are

Table 30.11 Seepage rates from canals. (After Etcheverry (1915) *Irrigation practice and engineering*. McGraw-Hill)

Type of soil	Seepage losses (m^3/s per million m^2)
Impervious clay loam	0.8–1.2
Medium clay loam	1.2–1.7
Clay loam or silty soil	1.7–2.7
Gravelly clay loam or sandy clay or gravel cemented with clay	2.7–3.5
Sandy loam	3.5–5.2
Sandy soil	5.2–6.4
Sandy soil with gravel	6.4–8.6
Pervious gravelly soil	8.6–10.4
Gravel with some earth	10.4–20.8

required, or a barrier layer beneath the root zone. Water is passed to the crop from open feeder ditches via buried perforated pipes. Control of the water level in the ditches determines the quantity of water available to the crops. A combined system of irrigation and drainage is common with the ditches and pipes doubling for both irrigation and drainage.

30.2.6 Irrigation canal design

The basic and most common method of designing a rigid boundary channel is the Manning equation.

The design method and values of Manning's n are described in section 30.8. Earth canals which transport significant quantities of sediment can be designed, using a regime method or one of the sediment transport formulae described in sections 30.6 and 30.7.

30.2.6.1 Freeboard

Freeboard is defined as the distance between the design water level and the canal bank top level.

Minimum freeboard above design water level for earth canals can be defined by:

$$Fb = 0.2 + 0.235Q^{1/3} \tag{30.6}$$

with a minimum value of 0.3 m

where Fb is the freeboard in metres and Q is the design discharge in cubic metres per second

30.2.6.2 Canal seepage

The quantity of water that will seep from the canal is normally measured in cubic metres per second per million square metres of wetted perimeter. Seepage rates for various materials in which the canals are constructed are given in Table 30.11.[8]

30.3 Drainage of agricultural land

30.3.1 Introduction

Agricultural drainage is necessary to remove excess water from the soil to improve the agricultural potential.

The benefits of drainage may include:

(1) Seed germination – excess moisture associated with low temperatures impairs germination. Waterlogging may cause seeds to rot and not germinate.
(2) Crop growth – most crops require air in the root zone to grow.
(3) Control of water table – high water tables will limit depth of root zone.
(4) Disease – waterlogged crops are more susceptible to disease.
(5) Yield gain – generally higher crop yields are experienced from drained land.
(6) Poaching – wet soil that carries stock experiences surface damage by grazing animals.
(7) Cultivation – improved drainage will allow easier access for cultivation machinery.
(8) Salinity – control of salinity in crop root zone.

Drainage systems can be defined as subsurface and surface. Surface drains are designed to remove excess runoff from the land which would otherwise cause localized flooding. Subsurface drainage is designed to remove excess water from the soil mass. It is discussed in the following sections.

30.3.2 Sub-surface drainage of irrigated land

Sub-surface drainage for irrigated lands in arid areas is normally associated with the control of the water table depth. Most crops grow best with the water table below their root depth although crops may not be affected by a higher water table for a short period. Rice is an exception since it grows well in totally waterlogged conditions.

Recommended minimum water table depths are shown in Table 30.12. The necessary drainage is frequently achieved by providing perforated drainage pipes below ground at regular intervals. It is necessary to install the drains below the desired design water table depth.

Table 30.12 Minimum water table depths

Crop	Water table depth below ground level (m) Fine textured (permeable soil)	Light textured soil
Field crops	1.2	1.0
Vegetables	1.1	1.0
Tree crops	1.6	1.2

The shallowest drain depth for water table control is:

$H + 0.5h + 0.1$ m

where H = design water table depth given above and h = rise in water table resulting from the maximum individual recharge from a water application.

30.3.3 Drainable surplus

The quantity of water to be removed by a subsurface drainage system can be estimated from a water balance:

$$Q_s = R_f + S_c + S_i - D_n \qquad (30.7)$$

where Q_s = water to be removed by drainage, R_f = recharge to the water table from rainfall or irrigation, S_c = seepage from canals or rivers, S_i = groundwater flow into the area and D_n = groundwater flow out of the area.

Recharge (R_f) to the water table will vary with soil type, irrigation method and efficiency of water management.

Food and Agriculture Organization Paper No. 38[9] Drainage design factors gives the estimated recharge for various conditions as shown in Table 30.13. Seepage from canals can be estimated using Table 30.11.

Groundwater inflow and outflow can be calculated from data on groundwater slope, flow cross-section and soil permeability using Darcy's law, which states that:

$$V = \frac{Kh}{L} \qquad (30.8)$$

where V = flow velocity in metres per day, K = hydraulic conductivity of the soil in metres per day, and h/L = hydraulic gradient.

And $Q = VA$

where Q = flow in cubic metres per day and A = area of flow in square metres

Table 30.13 Estimated recharge to watertable as related to irrigation method and soil type

Irrigation method	Application practices	Average recharge as percentage of irrigation water delivered to the field Soil texture heavy	light
Sprinkler	Daytime application, moderately strong wind	30	30
	Night application	25	25
Trickle		15	15
Basin	Poorly levelled and shaped	30	40
	Well levelled and shaped	20	30
Furrow, border	Poorly graded and sized	30	40
	Well graded and sized	25	35

Approximate design drainage rates are likely to be in the following ranges:

Less than 1.5 mm/day	For soils having a low infiltration rate.
1.5 to 3.9 mm/day	For most soils, with the higher rate for more permeable soils and where cropping intensity is high.
3.0 to 4.5 mm/day	For extreme conditions of climate, crop and salinity management, and under poor irrigation practices.
More than 4.5 mm/day	For special conditions, e.g. rice irrigation on lighter textured soils.

30.3.4 Drainage of lands subject to excess rainfall

The drainage of irrigated land in arid areas is described above. However, many areas require drainage due to an excess of rainfall. The drain discharge due to rainfall rises to a peak following a rainstorm and then recedes.

For the design of a buried pipe-drainage system, the discharge is often based directly on rainfall data. For instance, in the UK field drainage design is based on 5-day rainfall divided by 5 to give the daily drainage rate with return periods as shown in Table 30.14. Typical drainage rates in northwest Europe would be of the order of 7 to 10 mm/day. Drainage systems incorporating mole drainage are normally based on a 1-day rainfall value, because of the shorter response.

The design depth to water table is often taken at 0.5 m for shallow rooted crops and 0.75 to 1 m for deep-rooted and high-value crops. Drains in the UK, in practice, are usually laid at

Table 30.14

Crop	Design rainfall exceedance
Specialist high value crops	1 yr in 25
Horticultural	1 yr in 10
Roots	1 yr in 5
Intensive grass, cereals	1 yr in 2
Grassland	1 yr in 1

depths ranging from 0.75 m in low permeability soils to 1.25 to 1.5 m in permeable soils. A more detailed discussion of drainage discharge design is given by Smedema and Rycroft.[10]

30.3.5 Drain spacing

The required spacing of drains can be calculated using the Hooghoudt equation:

$$L^2 = \frac{8K_b dh}{q} + \frac{4K_a h^2}{q} \tag{30.9}$$

where K_a = hydraulic conductivity above the drain in metres per day, K_b = hydraulic conductivity below the drain in metres per day, h = height of water table above the drain level midway between the drains in metres, q = drain discharge in metres per day and d = equivalent depth – function of depth to impermeable barrier (D) and drain spacing (L) (see Table 30.15)

Table 30.15 Equivalent depths (d) for 80 mm corrugated PVC pipe drains

L (m) D (m)	5	10	15	20	25	30	35	40	45
0.25	0.25	0.25	0.25	0.25	0.25	0.25	0.25	0.25	0.25
0.5	0.43	0.46	0.47	0.48	0.48	0.49	0.49	0.49	0.49
0.75	0.53	0.62	0.66	0.68	0.69	0.70	0.71	0.71	0.72
1.00	0.59	0.74	0.81	0.85	0.88	0.90	0.91	0.92	0.93
1.25	0.62	0.83	0.93	1.00	1.04	1.07	1.09	1.11	1.12
1.50	0.63	0.89	1.03	1.12	1.18	1.22	1.26	1.28	1.30
1.75	0.64	0.94	1.11	1.22	1.30	1.36	1.40	1.44	1.47
2.0	0.64	0.97	1.17	1.31	1.41	1.48	1.54	1.58	1.62

The Hooghoudt equation allows two layers of soil with differing hydraulic conductivity (K_a, K_b) (see Figure 30.5). Values of hydraulic conductivity can be measured in the field using the auger hole method. Alternatively, the designer can use values measured on similar soils elsewhere. The single auger hole method requires a hole some 80 mm in diameter to be bored below the water table. The water in the hole is then pumped or baled out and the rate at which it refills is measured. From these measurements the value of hydraulic conductivity can be calcu-

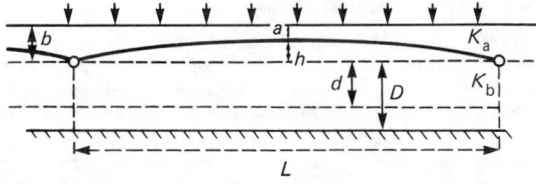

Figure 30.5 The Hooghoudt equation (definitions)

Table 30.16 Hydraulic conductivity (m/day)

	K (m/day)
Coarse gravelly sand	10–50
Medium sand	1–5
Sandy loam/fine sand	1–3
Loam/clay loam/clay, well structured	0.5–2.0
Very fine sandy loam	0.2–0.5
Clay loam/clay, poorly structured	0.02–0.2
Dense clay, not cracked and no bio-pores	< 0.002

lated. For more details, see van Beers's work.[11] Typical values of hydraulic conductivity are given in Table 30.15. A more detailed explanation of the calculation of drain spacings is given in ILRI *Bulletin* no. 8.[12]

30.3.6 Drain flow

Drain pipe sizes can be calculated using the Darcy–Weisbach equation for smooth pipes and Chezy–Manning for corrugated pipes. For which have a constant discharge along their length:

$$Q = 89\varphi^{2.71} i^{0.57} \quad \text{smooth pipes}$$
$$Q = 38\varphi^{2.67} i^{0.50} \quad \text{corrugated pipes}$$

where Q = discharge in pipe, in cubic metres per second, φ = pipe internal diameter, in metres and i = hydraulic gradient, in metres per metre

It is common to 'over design' the pipe to allow for some siltation with the drain capacity normally increased by some 30%. It is normal to assume that the hydraulic gradient line coincides with the pipe soffit, i.e. the pipe flows full.

If the drains are installed in hydraulically unstable soils they will require to be surrounded by a gravel envelope. Generally, soils with a high clay content will be stable and will not require an envelope. Granular envelopes are normally 50 to 100 mm thick. The gradation of the filter should be designed using the US Bureau of Reclamation method.[13]

30.3.7 Drainage layouts

Typical layouts of a buried drainage system, regular and irregular are shown in Figure 30.6.

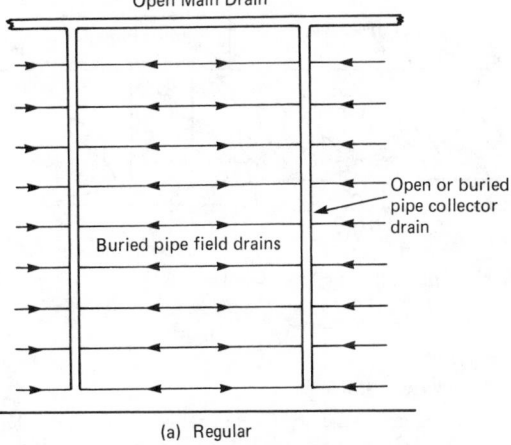

(a) Regular

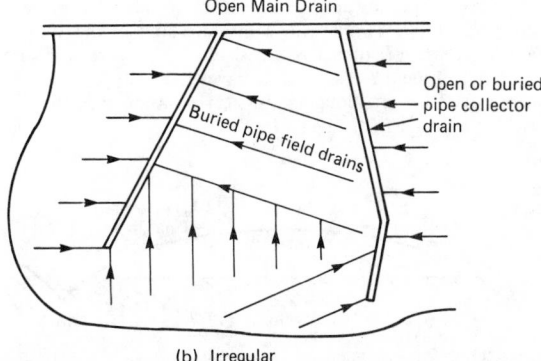

(b) Irregular

Figure 30.6 Typical layouts of buried drainage systems

Collector drains can be open ditches or buried pipes. Buried pipe collectors are to be preferred where sufficient ground slope is available. Pipe drain slopes should not be less than 0.0005 whilst open collector slopes can be as low as 0.0001. To allow drains to be cleaned, they should not exceed 300 m in length without a manhole or outfall into an open channel.

30.3.8 Drainage of heavy soils

For soils with very low permeability it becomes uneconomic to install drainage systems with buried field drains at spacings of between 1 and 5 m as are indicated by the use of the Hooghoudt equation. A common solution is the use of mole drainage. Moles are installed by using a mole plough that draws a 75-mm diameter bullet through the soil at a depth between 400 and 600 mm. The mole forms a tunnel in the soil and some fissuring in the upper soil area. Mole drains are normally spaced at 1 to 3 m and drawn across the line of the collector drains which have permeable fill in the pipe drain trench above the drain (see Figure 30.7).

Collector drains are normally spaced at 20 to 60 m.

Moling is best suited to clay soils with a minimum clay content of 30% and the moles have a relatively long life in stable calcareous clays. However, remoling will be necessary on average every 5 yr or so.

Efforts have been made to increase the life of mole drains by filling the tunnels with gravel. However, this is very expensive and for normal field cropping is not economic.

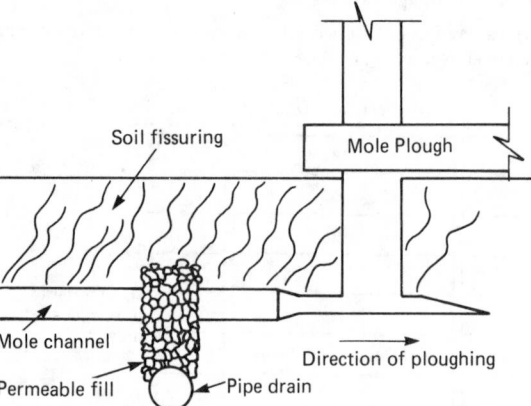

Figure 30.7 Mole drainage

30.3.9 Bedding systems

Bedding is a common method for the drainage of flat heavy land subject to excess rainfall. Wide beds are most suitable for mechanized agriculture and are up to 30 m in width. Drainage is mainly by surface runoff with some interflow in the topsoil region as shown in Figure 30.8. The shallow drains are normally some 0.5 m in depth. The raised crowned beds are normally built up over time by ploughing in such a way to turn the soil towards the centre of the bed.

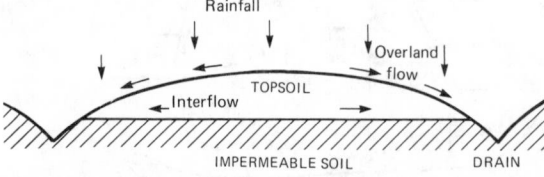

Figure 30.8 Wide bedding

30.3.10 Surface drainage for irrigated land

Surface drainage is often provided to irrigated land to collect excess irrigation supplies and runoff from rainfall. For surface irrigation typical surface drain capacities can be based on 24 h, 1 in 5 yr rainfall with 24 to 48 h storage on the field. For rice drainage, the drain capacity should be sufficient to allow the drawdown of water in the paddies where this is part of the cultivation pattern.

Typical values of surface drainage capacity are in the range of 2 to 4 l/s per hectare.

PART B: LAND DRAINAGE AND RIVER ENGINEERING

30.4 Land drainage and flood alleviation

30.4.1 Objectives of land drainage

The drainage of agricultural lands has already been discussed in the first part of this chapter. To the river engineer the term 'land drainage' has a broader interpretation, encompassing both the removal of excess water and the prevention of flooding of the urban as well as the rural environment.

In general terms, the problem of ineffective land drainage occurs when inflow into the system exceeds outflow, so that there is a build-up of water over a period of time. This may occur rapidly over a few hours in response to heavy rainfall, or it may be a gradual rise in water table during wet periods. Flooding occurs when a channel has inadequate capacity to convey the amounts of water flowing into it, or when flood defence works fail. Thus, the solutions to land drainage problems invariably involve either control of inflow into the system or works to improve the capability of the drainage channels to carry flows through the system. The basic objective is to reduce the frequency and/or the intensity of inundation to acceptable levels, appropriate for the situation.

30.4.2 Rivers as natural drains

Rivers are the Earth's natural drainage channels, conveying surface flow from the land to the sea or to inland lakes and marshes. Some rivers are essentially ephemeral (wadis) and flow for only very brief periods, often with very high discharges and consequently devastating erosive power. Others are seasonal, being dry for part of the year, but flowing steadily during the wetter months. Others still are perennial, generally flowing throughout the year but with varying intensity. Most rivers in Europe fall into this latter category.

No two rivers are the same, but rivers exhibit similarities which, to a certain extent, can be defined mathematically, thus enabling engineers to assess the problems with which they are faced. Perhaps the most fundamental property of a river is its flow or discharge. However, as has been indicated above, this is not a fixed property – the flow varies both spatially and with time. There are ways in which the flow in a river can be controlled or reduced, but often the engineer is faced with the problem of designing a structure or a scheme which is capable of withstanding the flow which passes through a specific point or reach of the river. It is therefore necessary to estimate the river flow for which the scheme or structure must be designed and this involves an exercise in statistics which is described later in this chapter.

30.4.3 Economic issues

Since funds are limited and there is always competition from other potential schemes, it is necessary to undertake some form of economic evaluation of proposed drainage improvement works. Such an evaluation requires the estimation of the *benefits* which might accrue from the scheme and the *costs* of its implementation.

For an urban flood relief scheme, some of the benefits are obvious and can be evaluated in a straightforward manner. Elimination of the physical damage caused by flooding is one such benefit, which can be assessed by counting the cost of replacement or repair of goods and property so damaged. In addition, there are less tangible costs of flooding which must be evaluated, such as loss of production due to flooding of industrial properties and disruption to traffic resulting from flooded roads. These too must be estimated. Finally it is now common practice to evaluate the intangible factors such as the distress caused to the public by flooding, particularly to those people in a high-risk area. From a knowledge of the frequency of flooding the present value of all the 'damage' likely to occur during the lifetime of the proposed works can be estimated. The benefits so derived should then be compared with the estimated costs of the works so that competing schemes can be compared on a similar basis or to determine the most economic level of protection which could be provided.

For agricultural lands it is possible to estimate the increased value of production generated by improved drainage, although this can involve some fairly subjective assessments. In general, the agricultural benefit will accrue as a result of either a lowered water table and/or a reduced risk of periodic flooding, both enabling a wider range of crops to be grown and/or better yields to be achieved as well as extending the period for which agricultural operations are possible and improving 'traffickability' of the land. Thus, an estimate of the increased value of annual production is made possible by the drainage works and this figure is capitalized over the life of the scheme to determine the benefits. As with the urban scheme the benefits are then compared with costs as a means of evaluating schemes.

30.5 Hydrology

30.5.1 Introduction

The design of river engineering and land drainage works is based on hydrological criteria, predominantly estimates of channel flow and its variation with time.

The ideal basis for the calculation of design parameters is a long period of recorded data which can then be analysed using statistical methods. Such data are often not available, but a record from a neighbouring catchment may be, and this can be corrected for use in the area concerned. Even short periods of data are useful, but if no records exist or their reliability is doubtful, empirical techniques of parameter estimation can be employed.

30.5.2 Measurement

The measurement of channel flow (discharge) is most commonly undertaken by velocity–area methods or at flow-measuring structures. Flows are measured over a range of stages (water levels) so that a stage–discharge relationship can be developed. Velocity–area methods depend upon the use of a current meter to record velocities and a knowledge of the cross-sectional area to which the velocity measurements can be applied, the product of these two variables being discharge. Flow-measuring structures are operated on the principle that there is a unique

relationship between level upstream of a structure and discharge. Flumes and weirs are commonly used on small rivers whereas velocity–area methods are usually applied to medium- and large-sized channels.

In recent years, permanent flow measurement installations using electromagnetic or ultrasonic gauging techniques have been developed as an alternative to weirs and flumes. Essentially, these methods measure velocity at a defined section.

Where records of channel flow are not available, rainfall records may be used to estimate likely flows. Within the UK rainfall records are maintained by the Meteorological Office, which operates over 5500 gauges, and flow data are archived by the Institute of Hydrology.

30.5.3 Statistics

Methods of statistics[14] are frequently used in hydrology to estimate the return periods of natural events. A flood flow is said to have a return period of, for example, 50 yr if, on average over a long period of time, that flow is equalled or exceeded once in a 50-yr period. Frequency analyses are required so that standards of protection can be met, risk assessments made and economic analyses undertaken.

The Fisher–Tippett type 1 extremal distribution (commonly known as the Gumbel distribution) is often used to analyse annual maxima series of discharge and rainfall. For a series of data values, Q_M, the magnitude of the event of return period T yr, Q_T, is given by:

$$Q_T = Q_A + \sigma(0.78y - 0.45) \qquad (30.10)$$

where $Q_A = \dfrac{\sum Q_M}{n}$, $\qquad \sigma = \dfrac{n}{n-1}\left(\dfrac{\sum Q_M^2}{n} - Q_A^2\right)$,

$y = -\log_e\left(-\log_e\left(1 - \dfrac{1}{T}\right)\right)$ and n = number of years of data

The probability, P_N, of an event of return period T yr being equalled or exceeded during a period of N yr, is given by:

$$P_N = 1 - [1 - (1/T)]^N \qquad (30.11)$$

Typical acceptable frequencies of flooding within the UK are once a year for grassland, once in 10 yr for arable land and once in 100 yr for urban areas. Where the risk to life is high, as in coastal situations, standards of protection even higher than those for urban areas may be considered.

30.5.4 Flood flow calculation methods

If statistical methods cannot be used, the design flow may be estimated from empirical equations relating rainfall and/or catchment characteristics to runoff. Early formulae to evaluate the 'maximum flood', Q_{max}, were of the form:

$$Q_{max} = CA^u \qquad (30.12)$$

where C = coefficient depending on the type of climate and catchment, A = catchment area and n = an index, usually between 0.5 and 1.2

The rational formula is representative of the many formulae developed to relate rainfall to peak runoff and is of the form:

$$Q = CiA \qquad (30.13)$$

where Q = peak discharge, C = runoff coefficient, depending on the characteristics of the catchment, i = rainfall intensity and A = catchment area

This approach is based on the assumption that maximum flow occurs as the result of the maximum rainfall intensity to be expected within the 'time of concentration' of the catchment. 'Time of concentration' is the time taken for rainfall falling on the most remote part of the catchment to reach the part of the drainage network under consideration. Nowadays the rational formula is generally only used for urban drainage design, for which the assumptions are valid.

Rainfall formulae have also been developed relating the intensity, duration and frequency of events. The Bilham formula[15] related these for the UK by the following equation:

$$r = 25.4[(1.25T/N)^{0.282} - 0.1] \qquad (30.14)$$

where r = total rainfall in millimetres, T = duration of storm in hours and N = probable number of occurrences in 10 yr

30.5.5 Hydrographs

A more accurate approach to flood peak and volume prediction, which has been widely accepted and developed since its inception by Sherman, is the unit hydrograph concept. A hydrograph is a plot of discharge versus time (Figure 30.13(b)) and, as such, gives the engineer much more information than just a peak flow estimate. The unit hydrograph of a catchment is defined as the hydrograph of direct runoff resulting from a unit of effective rainfall generated uniformly over the catchment at a uniform rate during a specified period of time.

Where runoff and rainfall records are available the catchment's unit hydrograph can be derived from these data and used to produce hydrographs of runoff for any given rainfall profile. Since such data are often not available, synthetic unit hydrograph techniques have been developed which relate catchment characteristics to the parameters of the unit hydrograph. The US Soil Conservation Service (USSCS)[16] and the Environmental Research Council[17] have presented such techniques.

30.5.6 Curve number method

The curve number method developed by the USSCS uses the following methodology:

(1) Rainfall is converted to discharge using a curve number graph based on catchment characteristics.
(2) Discharge is developed into a basin hydrograph using the USSCS unit hydrograph.
(3) The design drainage rate is taken from the peak of the hydrograph.

This method is described in detail in the USDA *National engineering handbook*, section 4.

30.5.7 The *Flood studies report*

In the UK, the *Flood studies report* provides a comprehensive guide to the estimation of maximum floods, return periods and flood volumes for any site. The methods described apply to both gauged and ungauged catchments.

Of the two methods developed for ungauged catchments – the mean annual flood plus growth curve and the synthetic unit

hydrograph plus design storm – the latter is given more weighting in the report. However, for preliminary estimates of peak discharge the simplicity of the former technique is very appealing. Use of the *Flood studies report* is described in detail by Sutcliffe.[18]

30.6 Channel regime

30.6.1 Regime flow

A channel is said to be in regime when, over a hydrological cycle, the channel shows no appreciable change in its width, depth or gradient. Regime theory postulates that for a stable channel there is a relationship between the channel parameters of width, depth, gradient and flow. Thus if any one of these four parameters is artificially (or naturally) changed, the channel will adjust itself so that regime conditions are re-established.

Also fundamental to river regime is sediment transport, which is discussed in more detail later. Sediment is material which is picked up, transported and deposited by the river. It can vary from very fine clay particles, often referred to as wash load, to large cobbles and boulders which can be moved by a river in flood. A regime channel is generally transporting sediment which is similar in size to the material which forms the bed and banks of the channel.

Many regime theories have been developed from a study of irrigation canals and the application of regime theory to natural rivers raises the problem of what flow should be taken as significant in determining the dimensions of the stable channel. The low flows which go on for most of the time cause little or no change in the channel section, while the maximum flood flows which occur for a few hours at intervals of several years cause rapid but temporary changes which the river will subsequently tend to restore. The 'bank-full' flow is commonly used in the UK, where it has been quantified as the flow which is exceeded for 0.6% of the time (i.e. about 2 days per year on average).

30.6.2 Regime formulae

Of the many regime formulae postulated, that of Lacey is probably one of the most useful. The Lacey formulae can be expressed as:

$$S = \frac{Ke^{\frac{1}{3}}f^x E}{Q^{1/6}}$$
$$W_s = 4.83eQ^{1/2}$$
$$d_m = \frac{2.46V^2}{f}$$

where Q = design flow in cubic metres per second, S = channel slope, W_s = water surface width in metres, d_m = mean depth = A/W_s in metres, A = cross sectional area of flow in square metres, V = flow velocity = Q/A in metres per second, E = shape factor = P/W_s, P = channel wetted perimeter in metres, and e = width factor and B = bed width (generally taken as W_s).

The values of K and x vary with the size of sediment present in the channel bed, as shown in Table 30.17. The shape factor E takes account of the differences between wide shallow channels and narrow channels. For the former, the value of E approaches unity.

The width factor e varies with soil type, in response to erodibility. Its value may range between 0.7 for stiff soils (clays) to 1.0 for erodible soils (sands and silts).

The silt factor f depends on the size of sediment which will form the channel bed in the long term. Indicative values of f are

Table 30.17

Sediment median grain size, D_{50} (mm)	K	x
$D_{50} < 0.2$	0.000206	$\frac{2}{3}$
$0.2 < D_{50} < 0.6$	0.000274	$\frac{7}{6}$
$0.6 < D_{50} < 2.0$	0.000303	$\frac{5}{3}$
$D_{50} > 2.0$	0.000188	$\frac{13}{6}$

0.4 to 0.6 for clays and 1.0 for sands. Actual values adopted should be based on local experience of stable channels.

In the context of rivers in the UK Nixon[19] carried out an interesting study of channel characteristics in 1959. He examined twenty-nine British rivers and attempted to establish regime formulae which were independent of bed material properties. His formulae are thus a simplification of reality, since they imply that the channel cross-section is independent of the material from which it is formed. Nevertheless, they can be used as a guide to channel dimensions for British rivers, although it should be noted that they are not applicable to channels with gravel beds. The Nixon formulae (converted to metric units) are:

$$W_s = 3.0Q_b^{0.5}$$
$$d_m = 0.55Q_b^{0.33}$$
$$V = 0.61Q_b^{0.17} \text{ and}$$
$$A = 1.65Q_b^{0.83}$$

where W_s, A and V are defined as above, d_m = mean depth = A/W_s in metres and Q_b = bank-full flow in cubic metres per second

30.6.3 Practical applications

Regime equations can be used as a guide in the design of new channels or of remodelling works to existing channels. In the absence of any other information the characteristics of a representative reach of existing channel will give a good indication of what is appropriate for new works in terms of channel width and depth, bed slope and bend radius. It is worth remembering that, if it is necessary to improve the conveyance of a river channel (as part of a flood improvement scheme) it is better to increase the depth of flow and/or the slope rather than the channel width, because natural adjustments of the former tend to occur at a much slower rate.

A channel may be capable of carrying the required flow with smaller dimensions and a steeper gradient than the regime values but the higher velocities generated will increase the width and depth by scour and will reduce the gradient by deposition of the eroded material at the lower end of the reach. Similarly, many channels have been made with excess width to carry the maximum flood flows, but sediment deposition has subsequently resulted in a narrower meandering deep channel within the main channel. The most common error arising from lack of knowledge of the regime theory is when a river is straightened by cutting across meanders without considering whether the resultant shortening of the course and increase of slope will produce scouring velocities.

30.7 Sediment transport

30.7.1 Basic concepts

Sediment load is generally divided into two categories depend-

ing on the mechanism of its transport. The 'wash load' comprises relatively fine material and the rate of wash load transport is mainly determined by its rate of supply from the drainage basin rather than the transport capacity of the stream. This material settles out rather slowly and can be maintained in suspension in large quantities by relatively slow-flowing water. Particles of 0.06 mm or finer are often considered as the wash load fraction. More accurately, wash load can be defined as that fraction of sediment which is finer than the D_{10} size of material (D_{10} = size for which 10% is finer) found on the channel bed.

In contrast, the 'bed material load' transported is almost entirely a function of the transporting capacity of the flow. The use of the term 'bed material' indicates that this is what the sediment load mainly comprises. It should not be confused with 'bed load' which has been used in the past to describe the larger particles transported near the bed of the channel.

Sediment loads are normally expressed as parts per million (p.p.m.) by weight (i.e. 1 g of sediment in 1 g of water = 1 p.p.m.). Transport rates are generally expressed in tonnes per day.

30.7.2 Sediment transport estimates

Sediment transport estimates are most commonly required so that the rate of scour or deposition in a channel can be predicted. In northern Europe problems of sediment transport are generally modest and concentrations of less than 1000 p.p.m. are common. However, in tropical or arid zones, sediment loads in rivers in flood can exceed 20 000 p.p.m., and much higher loads have been recorded in extreme cases. Clearly, if such highly charged water is diverted from a river into, for example, a slow-flowing irrigation canal, there will inevitably be extensive deposition in the canal.

Sediment transport estimates may also be required in the design of river regrading works where it is necessary to check whether the changes imposed (such as steepening the river gradient) are likely to lead to excessive erosion.

Of course, wherever possible, attempts should be made to measure sediment transport rates *in situ*. This is a notoriously difficult operation and even the most carefully controlled sampling can yield widely differing results. This is not only because the sampling technique is prone to error, but also because extreme sediment loads occur in relatively short-lived floods which are unpredictable.

30.7.3 Sediment transport equations

There are many sediment transport equations, none of which can claim a high degree of predictive accuracy. An estimate which is in the range 0.5 to 2 times the actual value is all that can be reliably expected. The recent Ackers–White[20] equations are amongst the most accurate although these are rather cumbersome. The more simple Engelund and Hansen[21] equation yields similar levels of accuracy.

The Engelund and Hansen equation for bed material load can be expressed as:

$$X = \frac{16\,000sVd^{\frac{1}{2}}S^{1.5}}{(s-1)^2 D_{50}}$$

where X = sediment concentration in parts per million, s = sediment specific gravity (normally 2.65), V = average flow velocity in metres per second, d = average channel depth in metres, S = channel slope and D_{50} = median sediment size of bed material in metres

Thus, a channel flowing at 1.0 m/s at a depth of 1 m with a slope of 0.5 m/km and a median sediment size of 1 mm would transport:

$$\frac{16\,000 \times 2.65 \times 1.0 \times 1.0^{\frac{1}{2}}(0.0005)^{1.5}}{1.65^2 \times 0.001} = 174 \text{ p.p.m.}$$

30.7.4 Stable channel design

Recent research by White, Paris and Bettess[20] has resulted in the publication of a set of *Tables for the design of stable alluvial channels*. These tables have been derived from the results of extensive flume experiments, and list values of sediment size, sediment concentration, channel flow, flow velocity, channel slope, depth of flow, channel width and friction factor. Given a sediment size and any two of the other parameters, it is possible to estimate values for all the other variables.

30.8 Channel design

30.8.1 Channel flow formulae

The most commonly used and universally accepted channel flow equation is that of Manning, which can be expressed in terms of flow as:

$$Q = \frac{AR^{\frac{2}{3}}S^{1/2}}{n}$$

where Q = channel flow in cubic metres per second, A = channel cross-sectional area below water level in square metres, R = hydraulic radius, A/P in metres, P = wetted perimeter in metres, S = channel slope and n = Manning's roughness coefficient

The selection of an appropriate value for Manning's roughness coefficient, which in normal engineering practice lies in the range 0.010 to 0.150, is a matter of judgement and experience. Ven Te Chow[22] gives comprehensive guidelines, with values for all common situations supported by photographs of typical channels. Table 30.18 gives some illustrative values.

In selecting an appropriate value of Manning's n the effect of future changes in the nature of the channel must be considered. Perhaps the most significant factor is vegetation which, if left

Table 30.18 Values of Manning's n*

Surface/channel	Normal range of n (design value)
Concrete lined channel (smooth finish)	0.012–0.017 (0.015)
Brick-lined channel	0.012–0.018 (0.015)
Mortared rubble masonry	0.017–0.030
Earth channel: clean, uniform	0.020–0.030 (0.025)
Earth channel: very overgrown with weeds, etc.	0.050–0.120
Minor stream: clean, straight	0.025–0.033 (0.030)
Minor stream: sluggish, weedy with deep pools	0.050–0.080
Major stream: regular section	0.025–0.060
Floodplain	0.025–0.150†

Notes: *Assumes channel flowing at or near full stage, lower flows may result in higher *n* values because of the relative significance of obstructions.
†On floodplains the value of *n* depends very much on the type of vegetation, its height and the season. A typical value for short grass might be 0.030, whereas dense brush in summer might be 0.100.

unchecked, can reduce the capacity of a channel to a fraction of its design capacity. Thus, whereas it is feasible to construct an earth channel which will have an *n* value of 0.025, experience has shown that even modest vegetative growth will increase this to 0.035. The problem is worse in tropical zones where plant growth is prolific, and irrigation canals can require clearance every few months. It is therefore recommended that, for such channels, a minimum value of $n = 0.030$ is adopted, with higher values if it is known that maintenance will be infrequent.

30.8.2 Channel stability

Manning's equation gives us a simple tool for determining the channel size, but gives us no information as to the long-term stability of the channel. It is most important, therefore, that a check is made on likelihood of sedimentation or scour occurring. Sedimentation has already been discussed. To assess the scouring potential of a stream it is common to determine the *tractive force*, defined as:

$$\tau_0 = C\gamma RS \tag{30.15}$$

where τ_0 = unit tractive force in newtons per square metre, γ = water specific weight (9810 newtons per cubic metre), S = water surface slope, R = hydraulic radius in metres and C = coefficient depending on the shape of the channel and the part of the channel considered

Unless the channel is particularly narrow, values of C of 1.0 for the bed and 0.76 for the banks are usually assumed.

For non-cohesive soils there are recommended values of tractive force on the channel bed for a range of soil types, recommended limiting velocities of flow are also given. Guidelines are given in Table 30.19.

Table 30.19 Suggested limiting tractive force (τ_0) and flow velocity (V) values

Material	Clear water		Water transporting colloidal silts	
	V (m/s)	τ_0 (N/m²)	V (m/s)	τ_0 (N/m²)
Fine sands and non-colloidal silts	0.55	1.9	0.85	4.0
Firm loam	0.75	3.6	1.00	7.2
Stiff clay and colloidal silts	1.15	12.5	1.50	22.0
Fine gravel	0.75	3.6	1.50	15.3
Graded colloidal silts to gravel	1.20	20.6	1.65	38.3
Coarse gravel	1.20	14.4	1.50	32.0
Cobbles and shingles	1.50	43.6	1.65	52.7

Source: Etcheverry.(1915) *Irrigation practice and engineering*. McGraw-Hill.

The figures in Table 30.18 can be used as a guide to determining limiting slopes for channels in terms of the movement of bed material. For the channel sides, even though the tractive force is lower, the banks may be less stable because of the effect of gravity. In practice it is found that, for fine non-cohesive material, small amounts of sediment in the water tend to cement the particles together and the use of tractive force theory is conservative. However, side slope stability should be considered for coarse non-cohesive materials (medium-sized gravels and above).

30.8.3 Other considerations

The channel shape, as well as its size, is also important. For irrigation canals and drainage channels bed width to depth ratios ($B:d$) of between 3 and 4 are often used. Channels designed using the Lacey regime equations tend to have larger $B:d$ ratios. Such channels have lower tractive force values but this principle cannot be extended too far since drains with beds which are too wide will tend to form sub-channels at lower flows leading to a higher local tractive force and, hence, erosion.

Channel side slopes depend mainly on the nature of the ground in which they are cut. Slopes of 1:1.5 (vertical:horizontal) and 1:2 are quite commonly adopted, with flatter slopes of 1:3 or even 1:5 if the bank material is highly erodible.

Other considerations may dictate the choice of side slope such as use of the slopes for grazing (where flow is intermittent) and ecological factors (e.g. desire to re-establish reed growth, hence shallow slopes).

For bends in channels a rough guide to the appropriate bend radius is $10 \times W_s$ (W_s = water surface width), but this takes no account of the erodibility of the bank material. Lacey proposed design radii of $128\sqrt{Q}$ (Q = design flow in cubic metres per second) which is recommended where space is available for channels in fine alluvial soils. If possible, the designer should measure the radii of other stable channels in the area to give a guide to acceptable values. For lined channels (concrete, brick, masonry, etc.) a minimum radius of $3W_s$ is recommended, with $5W_s$ being used where possible.

30.9 Channel improvements

30.9.1 Channel clearance

In northern Europe the clearance of trees, brushwood and weeds offers greater improvement in reduced flood levels in relation to cost than any other form of channel improvement. Such works should be executed with a sympathetic approach since our rivers are frequently areas of great natural beauty and are well used by anglers and for other recreational interests (see Figure 30.9). Some very valuable advice in this respect can be found in a handbook published by the Royal Society for the Protection of Birds and the Royal Society for Nature Conservation.[27] The aim should be to retain an appearance which is as natural as possible without prejudicing the aim of improved channel conveyance.

As an alternative to channel clearance, consideration should

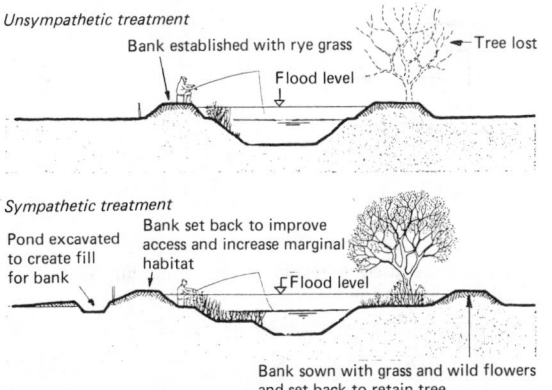

Unsympathetic treatment

Sympathetic treatment

Figure 30.9 Conservation in river engineering
Courtesy: Nature Conservancy Council (1983) *Nature Conservation and River Engineering*

be given to forming a bypass channel which would leave a sensitive reach of channel untouched. The bypass would be set at a level where it only operated in flood conditions, thus preserving the main stream for all normal flows and allowing use of the land taken up by the bypass for grazing.

30.9.2 Realignment

Natural rivers and streams often follow a meandering course, and lower flood levels in a particular area can be obtained by straightening the course downstream by diversions across meanders or by a completely new channel. Realignment will eliminate sharp bends where erosion takes place and will produce a more stable course, but in rivers of high amenity value long straight reaches will lead to complaints that the river has been converted to a 'canal' and in such cases a course with sweeping curves is preferable.

With any realignment exercise it is essential to check that the increased slope which will result will not lead to instability. In extreme cases the shortening of a river reach can result in upstream progressing degradation which could undermine bridge foundations or cause the collapse of river frontages. Where realignment results in a gradient steeper than the stable regime gradient, the provision of weirs may be necessary to limit erosion. In small channels, or in short lengths of larger channels, bed scour may be prevented by the use of some form of flexible bed protection such as dumped stone or gabion mattress.

30.9.3 Revetments and lining

30.9.3.1 Introduction

A revetment is any means of protecting a channel bank from erosion or undermining. Revetments are frequently required on river bends, in the vicinity of structures (where flow may be more turbulent), where the natural bank material is unstable, and where the wash from boats causes progressive collapse of banks. Revetments may be rigid (e.g. sheet piling) or flexible (e.g. dumped stone).

Channel lining is used where both bed and bank scour are to be prevented or where it is necessary to streamline the channel. Linings, too, can be rigid (e.g. reinforced concrete) or flexible (e.g. gabion mattress). Flexible linings are often preferable because they will accept some settlement or damage whilst retaining their appearance and integrity.

Typical details of rigid and flexible revetment systems are illustrated in Figures 30.10 and 30.11. The most important considerations in revetment design are:

(1) Adequate scour protection at the toe to prevent undermining.
(2) Flexibility if settlement is likely.
(3) Adequate weephole provision if the revetment is impermeable and rapid drawdown of river level is possible.
(4) Cost (taking into account locally available materials, especially in developing countries, and availability of inexpensive labour).
(5) Environmental acceptability.

30.9.3.2 Traditional methods

One of the earliest methods of bed and bank protection was the use of brushwood faggots or fascines and these are still used in certain conditions. For revetment purposes the brushwood is made up into tight bundles laid side by side end-on to the channel, with successive layers stepped back to conform with the bank slope. The bundles are held down by stakes but, after a

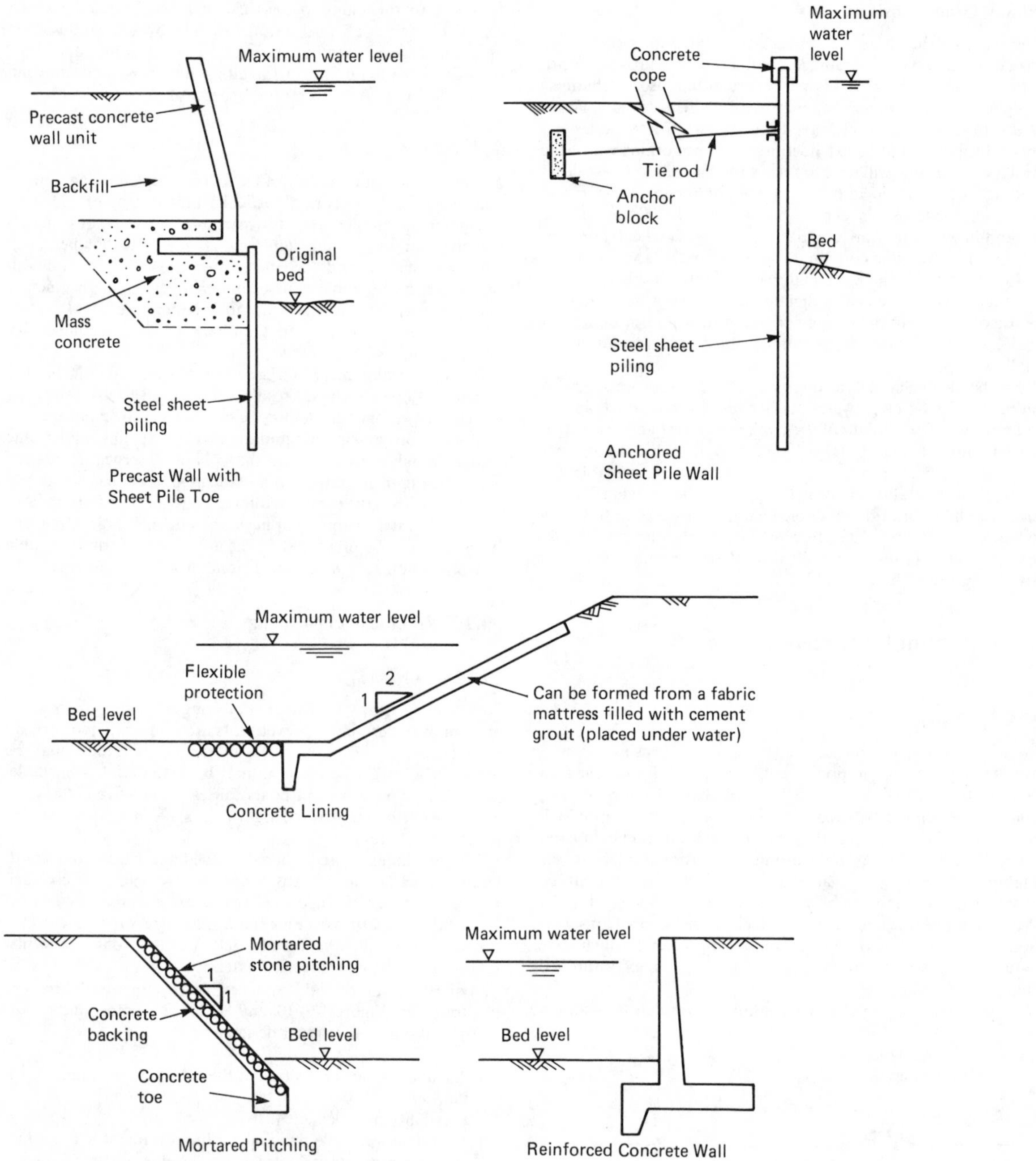

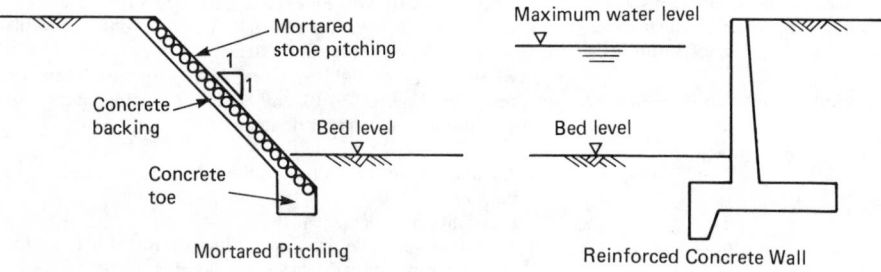

Figure 30.10 Typical rigid revetments

few floods or tides the whole mass is impregnated with silt which holds and preserves the brushwood. Brushwood is also used in longer lengths to build up large mattresses which are floated into position and sunk by loading with stone to protect the river bed or the lower bank slopes.

Stone is also much used for revetment work. This may be in the form of dumped stone (usually machine placed), dry stone pitching (hand placed), pitching grouted with bitumen, or mortared stone pitching, the latter being rigid. The following formula may be used to estimate the minimum stone size required for a given flow velocity:

$$W = \frac{0.011 V^6 s}{(s-1)^3 \sin^3(p-\alpha)} \tag{30.16}$$

where W = critical weight of stone in kilograms (two-thirds of stones to be heavier), V = flow velocity in metres per second, s = stone specific gravity (assume 2.5 if no other information),

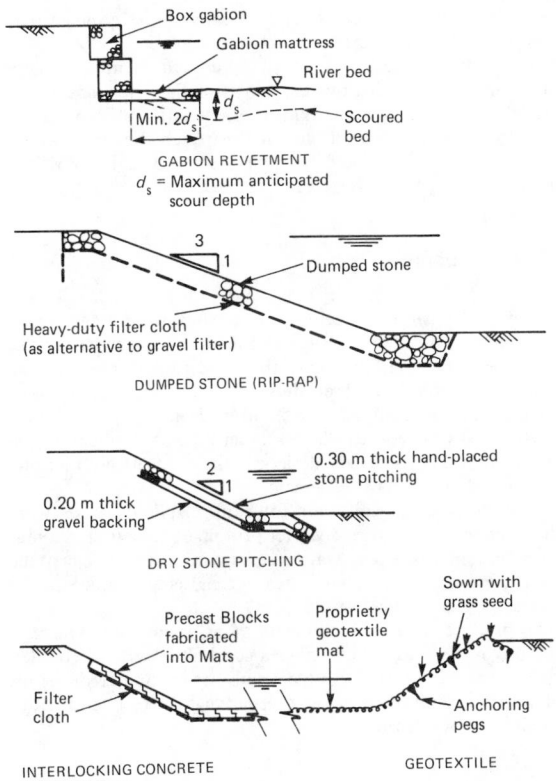

Figure 30.11 Typical flexible revetments

Table 30.20 Gabion mattress thickness

Clays, heavy cohesive soils:			
maximum water velocity (m/s)	2	3	4.5
minimum mattress thickness (mm)	170	230	300
Silts, fine sands:			
maximum water velocity (m/s)	2	3	N/A
minimum mattress thickness (mm)	230	300	N/A
Shingle with gravel:			
maximum water velocity (m/s)	3.5	5	6
minimum mattress thickness (mm)	170	230	300

p = a stability factor, 70° for random stone (riprap) and α = slope of the bank (i.e. angle with horizontal, in degrees)

Note: Impinging velocity may be assumed to be 1.25 × average velocity on the outside of bends. Thickness of stone should be at least 1.5*D* where *D* is the effective diameter of the normal size of rock specified. Bank slopes should not exceed 1:1.5 (vertical to horizontal). A filter of graded gravel or geotextile should be provided under the stone to prevent the leaching of fines from the bank.

30.9.3.3 Gabions

Gabions and gabion mattresses have been used for many years for revetment and lining work. The gabions are crates formed from wire or plastic mesh and filled with stone. Common sizes are $2 \times 1 \times 1$ and $2 \times 1 \times 0.5$ m. The gabion mattress is similar but comes in units of 6×2 m with a range of thickness (up to 500 mm). The gabion crates/mattress are subdivided by diaphragms and are packed with stones of a size generally just larger than the mesh size. Good-quality control during filling is essential to achieve the desired effect. The end-product is a flexible permeable structure of considerable erosion resistance. It is normal to use a filter beneath the gabions to prevent the washing out of fines. Table 30.20 gives indicative mattress thicknesses for a range of conditions. The mattress can be used alone as a flexible lining, or in conjunction with gabion boxes as illustrated in Figure 30.11.

30.9.3.4 Concrete, geotextile and other methods

In recent years the use of concrete block systems and geotextile revetments has become popular and there are many proprietary systems on the market. Most of these systems are flexible, permeable and allow the growth of grass and waterside plants through them. Most of the concrete block systems have mechanical interlocks which prevent the lifting of individual units, some are connected together by polypropylene strands or glued to geotextile mats so that they can be placed in large units. The incorporation of a geotextile fabric filter under the armouring is common practice. This prevents fines being washed through the armouring in the same way that a graded gravel filter does under stone protection.

Various forms of geotextile reinforcement are also available. These can be placed on the soil surface and sown with an appropriate grass seed to give a natural looking erosion-proof surface (in limited erosion conditions). They can even be provided with grass already established. The use of jute fibres for this purpose is also coming into vogue because of its environmental acceptability (it biodegrades within 1 or 2 yr leaving an established grass cover).

Steel sheet piling is commonly used in the neighbourhood of weirs, locks or sluices, where there is wave action or heavy turbulence. In fine bed material such as silts, corrugated asbestos cement sheets have been used as an inexpensive form of sheet piling. This material is only suitable to support up to 1 m faces where damage by boats, etc. is not expected.

There are many other proprietary revetment systems, any one of which may be appropriate in certain circumstances. These include:

(1) Grout-filled mattress (rigid, permeable, can be placed under water).
(2) Fabric with pockets which can be filled with soil (allows rapid planting of appropriate waterside plants).
(3) Plastic webbing spanning between vertical supports.

30.10 Embankments

30.10.1 Introduction

Embankments are provided along river channels to prevent flooding. They are normally set back from the river so that during floods they provide the necessary increase in the waterway section by providing both extra width and extra depth without overflow. In urban areas the land required to set back the embankments may not be available. In fact, there may not be space for the wide-based embankment at all and a flood wall, or a 'half-bank' supported by a wall may have to be used (see Figure 30.12).

Most flood embankments in the UK and northern Europe are of moderate dimensions, not exceeding 4 m in height and many not exceeding 2 m. They are constructed of the best material available, preferably containing some clay to make the banks watertight. Pure clays are not ideal because they tend to crack on drying. Low flood embankments are often constructed from material dredged from the river channel, frequently sands and gravels with varying silt content. This achieves two objectives simultaneously by enlarging the channel and forming flood banks. Such banks may be designed to overtop every 5 or 10 yr or so when providing protection to agricultural land. Provided that the land-side slopes are relatively flat and a good soil and turf cover is provided, overtopping of long reaches of such embankment will cause little damage.

Clay cores are not normally provided in embankments although these can be specified where residential properties are protected and even small amounts of seepage would therefore be unacceptable. Steel sheet piling driven down the middle of an embankment has also been used to cure a local seepage problem.

30.10.2 Design

Bank top levels should provide a freeboard above the design flood level to allow for settlement and damage by cattle or pedestrians. The freeboard also allows for any inaccuracies in the estimation of flood level. A minimum freeboard of 0.5 m for embankments is normal, with the lower figure of 0.3 m adopted for walls.

The bank top width is often about 2 m but may be increased to permit the passage of a tractor along the top with various maintenance equipment.

Side slopes on the river face should not be steeper than 1:2 (vertical:horizontal) and on the landward face may be somewhat flatter. Where space permits and fill is available the bank may be constructed with very flat slopes on the landward side so that the area may be mown as part of the adjoining field. Arable cultivation may also be permitted on such slopes provided that the crest of the embankment is clearly demarcated to prevent gradual lowering by farming operations. The planting of trees or shrubs on the embankment should be discouraged. The ideal surface treatment is good turf grazed by sheep or cut mechanically several times per year.

Embankment slopes of 1:2.5 can be mown by tractor, but flatter slopes are preferred. Slopes of 1:3 or flatter may be grazed by cattle, but cows tend to cause more damage than sheep so careful management will be necessary.

The hydraulic gradient through the embankment is the slope of a line from the high-water mark on the river side to the landward toe of the embankment and this should not exceed 1:4. In important cases a 'flow-net' through the embankment and its foundation should be calculated to reveal any risk of 'piping' due to excessive rates of seepage.

30.10.3 Stability

The subsoil adjoining many rivers in their lower reaches consists of recent deposits of alluvium and is often waterlogged. Such material may be unable to support the increase of superimposed load due to the construction of the flood bank. Unless the soft material is a thin layer, the solution is to construct an embankment of reduced height with a flood wall on top of it. Alternatively, the use of geotextile layers under the embankment may be considered. Appropriately designed, these effectively reinforce the subsoil and reduce settlement.

The most serious condition for bank instability arises when the flood level drops rapidly after prolonged retention at a high level. The increased pore water pressure in the embankment has insufficient time to dissipate, shear strengths are reduced and a classic slip failure may result.

In practice, with embankments of no more than 4 m high, stability is rarely a problem given a good fill material adequately compacted. However, if problem soils are encountered, the use of geogrid reinforcement or alternative forms of constuction should be considered.

30.10.4 Construction

The area of the base of the embankment should be stripped of turf and topsoil to a depth of 0.25 m. This may be stockpiled for later use on the surfacing of the bank. Where the subsoil is weak, or the embankment high, compaction of the subgrade before placing fill should be considered. This will reduce settlements experienced subsequently. The fill should then be deposited and rolled down in layers, generally by bulldozer or tractor shovel.

Finally, the embankment will be cased with topsoil and sown with grass seed. The height of the newly completed bank should make allowance for settlement which is likely during the first year at the rate of about one-tenth of the height of the bank, depending on the type of material and the degree of compaction.

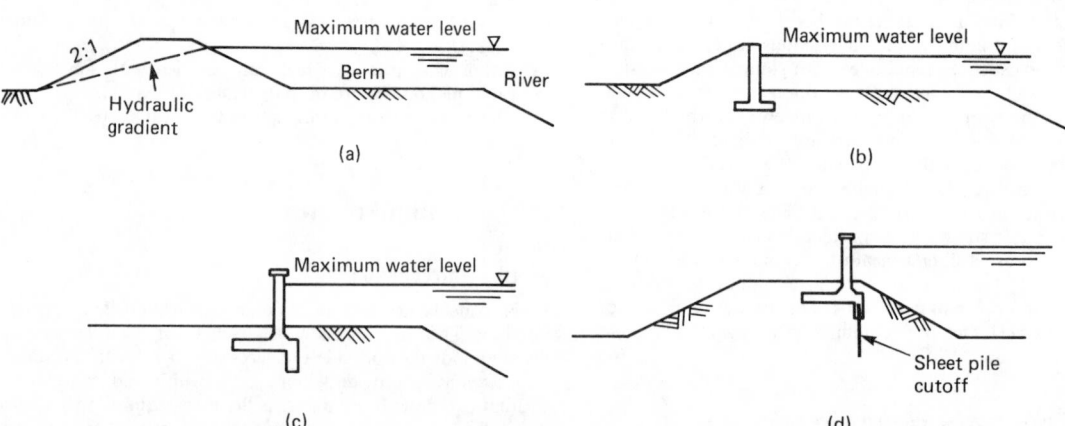

Figure 30.12 (a) Flood embankment (b) Half-bank with wall (c) Flood wall (d) Flood wall on embankment

30.10.5 Flood walls

Where space or foundation conditions do not permit the construction of embankments, flood walls may be used. To meet amenity requirements, the wall may have to be cased in brick, or natural or artificial stone, or a half bank may be formed behind it which will provide support, seal any leakage and overcome objections to an exposed concrete face.

Where it is necessary to allow access through a flood wall in residential or industrial areas and steps over the wall are impracticable, flood gates may be incorporated. These are side-hung hinged steel gates with a rubber seal and stout locking device. Normally left open, these gates are closed by the residents when flood levels get dangerously high. Similar flood gates may be provided on vehicular access ways in which case they may be bottom hinged or housed in a slot below road level and raised when necessary.

30.11 Detention basins, washlands and catchwater drains

30.11.1 Detention basins

The peak discharge in a river or stream can be reduced by storing some of the flow in a detention basin temporarily. The flow into the basin should be controlled by banks and weirs so that flows up to the bank-full capacity can pass down the stream leaving the basin empty. In flood the excess flow can then be spilled into the basin over a weir or through a sluice so that flow down the channel is still restricted to a safe value (Figure 30.13). When the peak of the flood has passed the basin can be emptied through an escape sluice back into the river. The escape sluice may be manual or automatic and should be designed to allow rapid emptying of the basin so that the storage is available in the event of a second flood in quick succession.

Provided that an adequate area of suitable land exists in the river valley, the maximum flow passing downstream may be reduced to any desired extent above the normal flow, but the volume of storage required increases in greater proportion than the reduction of residual flow, and economy may require a compromise between the provision of storage and channel improvements downstream (Figure 30.14).

In order to spill a substantial part of the flow into the basin when the safe residual flow is reached without a further increase in the flow passing downstream, a long side weir may be used separating the basin from the normal channel, or an automatic sluice of adequate capacity designed to maintain a constant water level on the main channel side may be used. The long side weir may take the form of a low embankment suitably protected against erosion by stone, gabion, concrete block or geotextile reinforcement.

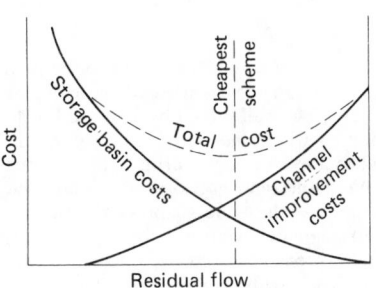

Figure 30.14 Combined channel improvement and storage – economic assessment

30.11.2 Washlands

The use of washlands is common in the Fens of the southeast UK and elsewhere. Flood embankments are constructed well back from the river enclosing an area which may be more than 1 km wide and 30 km long. This area acts both as a detention basin and as a flood channel. Large quantities of water can be temporarily stored and because of the large cross-sectional area of the waterway at flood level, velocities and surface gradients are very small. The area of land involved is considerable and its use is usually restricted to summer grazing or the production of hay or lucerne. If the washland is only required for major floods, lesser floods may be excluded by a lower bank alongside the river and the washland may be cultivated. For small schemes uncontrolled flow on to the washlands may be acceptable but large schemes rely on sluices to regulate the flow into and out of the washlands, thus optimizing their use.

30.11.3 Catchwater drains

In many cases the water from areas of high ground runs down into low-lying areas to create or accentuate drainage problems. The upland area may be more extensive than the lowland and also produces a greater runoff per unit area. The upland flow may be diverted away from the lowland area by a catchwater drain. In the case of lowland pumped drainage schemes the diversion of the upland water will reduce both the capital cost of the pumping plant and its subsequent running cost.

The catchwater drain is located on the edge of the upland and is designed to intercept the streams running down into the lowland area. Control structures are necessary to allow some flow to follow the original course in dry periods to avoid a shortage of water in the lowland area. In flood periods the whole

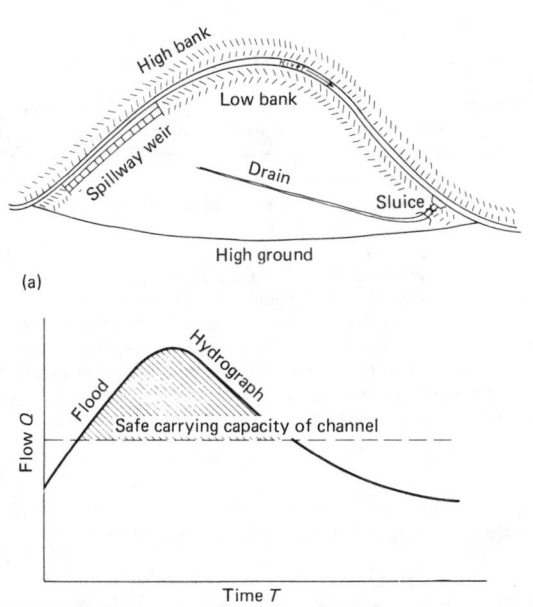

(a)

(b)

Figure 30.13 Flood detention basin. (a) layout, (b) estimation of storage capacity

flow may be diverted down the catchwater to a suitable outlet away from the lowland area.

30.12 Structures

30.12.1 Introduction

The design of river structures must involve an understanding of the fundamentals of channel hydraulics, sediment transport and the associated problems of scour and deposition. It must be appreciated that any structure in a river is likely to interfere with the natural regime of the channel, the consequences of which can be far-reaching unless the process is fully understood and is catered for in the design.

Some of the structures commonly encountered in river engineering are described below.

30.12.2 Retaining walls

The riverside retaining wall is used to support the river bank or a road or building adjacent to it and is used where a sloping revetment is inappropriate. The wall should be designed for the rapid drawdown condition where the water table behind the wall remains high when the river level drops quickly after prolonged flooding. Weepholes and gravel drains behind the wall may be provided to relieve the hydraulic pressure, but these must not be assumed to eliminate the pressure differential, only to reduce it. It is normal to provide a cutoff (see Figure 30.12(d)) on the base of the wall. This member serves three functions:

(1) It provides a key to improve the factor of safety against sliding.
(2) It reduces seepage under the wall when river levels are high.
(3) It provides stability in the event of bed erosion in front of the wall.

Various alternative walls have been described in section 30.9.3. Steel sheet piling is very common in river work, particularly for wall heights in excess of 3 m. Sheet piling has the considerable advantages of speed of erection and no need for dewatering.

30.12.3 Bridges

30.12.3.1 General

Bridges spanning rivers affect channel regime inasmuch as the presence of abutments and piers changes the flow pattern causing local acceleration of flow. The simplest way to avoid problems is for the bridge to span the entire river channel, although this is often uneconomic or impractical for large rivers. Even if the bridge abutments are outside the main river channel it will be necessary to check that the approaches do not unduly restrict floodplain flow.

As a guideline, a bridge should be designed to have an open area of not less than 80% of the channel cross-sectional flow area for bank-full conditions, although smaller values may be appropriate where the natural river is very slow-flowing.

30.12.3.2 Scour at bridges

Bridge piers and abutments must be designed such that they are not undermined by any scouring of the river bed or banks. Scour may be caused locally by the presence of the piers, or it may be a feature of the natural channel, occurring during floods.

The following equations, put forward by Holmes[24] for rivers

in New Zealand, may be used to give indicative scour depths. The local scour (d_s) due to the piers must be *added to* the general bed scour (D_s).

$$d_s = 0.8\sqrt{Vb}$$

where d_s = local scour depth in metres, V = flow velocity in metres per second and b = projected pier width in metres

$$D_s = (yVK)/\sqrt{A/W}$$

where D_s = depth of scour measured from *flood water level* in metres, y = rise in water level to flood level upstream of the bridge measured from normal water level in metres, W = waterway width at bridge, no deduction for piers, in metres, A = waterway area at bridge, no deduction for piers, in square metres, V = mean flow velocity upstream, no allowance for scour, in metres and $K = \sqrt{(W/4.83Q^{1/2})}$, maximum value $K = 1$, Q = discharge in cubic metres per second

30.12.3.3 Afflux at bridges

Unless a bridge is designed with very streamlined approach and exit conditions, head losses are likely to be significant during flood flows. The head loss at a bridge may be estimated from the following equation:

$$h_L = \frac{C}{2g} (V_2^2 - V_1^2) \tag{30.17}$$

where h_L = head loss (drop in water level through bridge) in metres, V_1 = flow velocity upstream in metres per second, V_2 = flow velocity through bridge in metres per second, g = acceleration due to gravity (9.81 metres per squared second) and C = coefficient depending on the degree of streamlining (1.2 for normal bridges)

30.12.4 Weirs

Weirs are designed for one or more of the following roles:

(1) Measurement of flow (as part of a hydrological network).
(2) Control of channel depth for navigation.
(3) Reducing the slope of steep, erosive channels.
(4) Creation of ponds to improve river fisheries.
(5) Control of rivel level to allow perennial irrigation abstractions.

The design of weirs is discussed in Chapters 5 and 22.

For flow measurement the Crump weir is often adopted. This is a robust structure offering minimal obstruction to the passage of sediment and debris. It is modular up to a drowning ratio of 80% and can be used for measurement in any state of drowning if a crest tapping is provided (i.e. if both upstream and crest water levels are measured). For modular flow the equation is:

$$Q = 1.96Wh^{1.5} \tag{30.18}$$

where Q = discharge in cubic metres per second, W = weir width in metres and h = upstream head over the weir in metres

Where it is necessary to record both high and low flows accurately it is normal to provide a compound weir. A typical structure might comprise a Crump weir for high flows with a parallel Vee-notch weir for low flows, the crest of the Crump weir being above that of the Vee-notch.

For controlling navigation depths a wide range of weir types have been used. The weir will, of course, incorporate a lock to permit the passage of vessels and often sluice gates are provided to pass flood flows (these effectively reduce the range of water levels between normal and flood flow conditions and they may be essential to avoid flooding upstream). The typical weir comprises a crest and a stilling basin, the design of which is discussed in Chapter 22. Provision must be made to reduce seepage under the weir and to resist uplift on the downstream apron under all flow conditions.

Wherever the head loss across the weir is more than about 1 m, or where the foundations comprise permeable soils, a check on underseepage, uplift and exit gradient should be made. All of these can be determined by plotting a flow net using an electrical analogue device or an appropriate computer program.

For larger weirs the use of a physical model is recommended to test the design before the prototype is built. This is particularly important where entry and exit flow conditions are not straightforward or where heavy sediment loads are expected.

For rivers which are used by migratory fish, weirs should be provided with a fish pass. The most common type is a series of pools connected by small weirs or submerged orifices. The pools should not be less than 3 m long and 2.5 m wide with not less than 1 m depth below the connecting openings. The difference in level between successive pools should not exceed 500 mm.

30.12.5 Gated control structures

Gated control structures are required to maintain a design water level or range of water levels throughout the range of flow conditions experienced. For an unregulated weir the water level upstream rises with increasing flow, but the introduction of a water control gate can enable a relatively constant upstream water level to be maintained. In effect, the gates (which have a sill level lower than the weir crest) allow the waterway area to be increased, thus permitting the flow to increase without raising upstream water level.

The most common form of water control gate is the vertical lift sluice gate, but radial gates and hinged tilting gates are also used extensively. Figure 30.15 illustrates the three types diagrammatically.

30.12.5.1 Vertical lift gates

For small sizes the vertical lift gate may be a standard penstock gate fabricated from cast iron or steel sliding in cast iron or steel frames and normally operated by screwed spindles and simple reduction gearing.

Larger vertical lift gates are fabricated from welded steel plates and sections having wheels and guide rollers running on wheel tracks built into recesses in the piers and abutments. Such gates are usually suspended from the gearing by plate link chains or steel wire ropes passing over sprockets or grooved winding drums mounted on the overhead steel superstructure.

To decrease the loading on the gearing and so increase the manual speed of operation, the gates are often counterweighted, the counterweights being connected directly to the chains or ropes. Modern vertical gates have wheels, bushed with self-lubricating bearing material, bearing on stainless steel axles fixed to the gate structure and having a fairly low coefficient of friction.

Opening and closing of vertical lift gates may be by manual means or electric power. When electric power is used the opening and closing of the gates may be made automatic, controlled by the upstream water level usually by incremental movements of the gate as a result of variations in water level of about 100 mm. Operation of the gate under automatic control is purposely made slow with time delays between each incremental

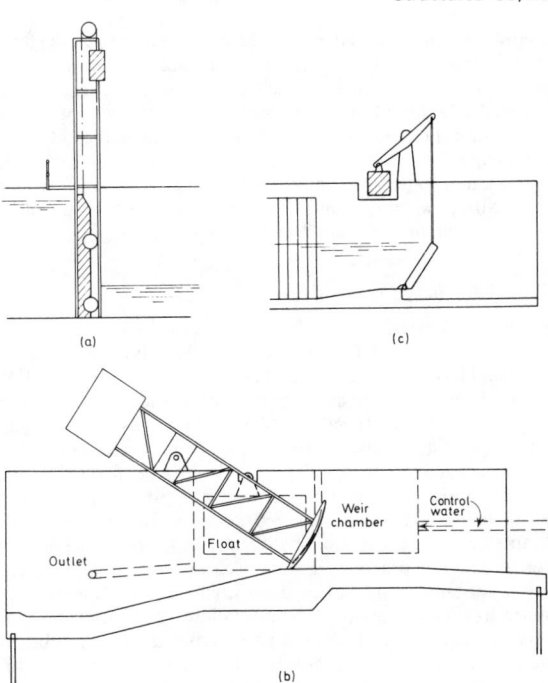

Figure 30.15 (a) Vertical sluice gate
(b) Float operated radial gate
(c) Tilting gate

movement in order to prevent 'hunting' which can result if the incremental movement causes large changes in upstream water level.

30.12.5.2 Radial gates

Radial gates are constructed so that the resultant reaction of the water loading passes through the centre of rotation of the gate and thus there is no component of water load to be handled by the gate-lifting mechanism. The water loading is transmitted to the pivot bearings by gate arms and thence to the concrete work. Unlike vertical lift gates, there is no need for groove recesses in the piers and abutments and the side members are required simply to provide seal bearing surfaces. With automatic-type radial gates the weight of the gate structure is balanced by counterweights on extensions to the arms which support the gate water load.

The lower edge of the gate closes on to a sill and is sealed by a strip of rubber or similar material. The ends of the gate have rubber or leather seals sliding on the metal plates set in the abutments. Radial gates are normally operated by electric motors but there is a form of radial gate operated by floats. The float operated radial gate (see Figure 30.15(b)) is a special type of radial gate designed to maintain a constant upstream water level automatically. It can be very useful in remote locations where the power supply is unreliable or non-existent.

The floats are located in chambers in the abutments or piers. Water from the river upstream passes over an adjustable weir and into the float chamber, from which it escapes at a constant rate through a valve. If the upstream water level rises, water flows into the float chamber at a greater rate than it escapes, causing the float to rise in response to the increased level. The float acts on the gate arm causing the gate to open, which in turn increases the flow through the structure and lowers the upstream water level. Conversely, if the level falls, flow over the control weir does not balance the escape of water from the float chamber, and the gate closes until normal level is restored. Again, the design must be such as to avoid hunting but in practice it has been found possible to control the level of a large river within 12 mm of a set level. In major floods the gate rises completely out of the water and there is a possibility that the downstream water level will take control and prevent the gate from closing when it should. This situation can be avoided by proper design of the control arrangements.

30.12.5.3 Tilting gates

Radial and vertical gates have the disadvantage that flow takes place under them and floating debris of all kinds collects against them and has to be removed manually. This is avoided by tilting gates which are hinged at the bottom edge and allow the water to pass over them. The gates are lowered by links or chains and may be operated manually or automatically by electricity. The gate when fully open lies flat in a shallow pit formed in the foundation to give maximum discharge, or in some cases still forms a low weir in the lowered position. In rivers carrying coarse sediment there may be difficulty in lowering the gate to bed level, but in practice the water weiring over the gate usually keeps the area immediately downstream clear. This type of sluice has the advantage that any failure of power supply or other operating failure will not allow water levels to fall below the gate level. As an adjustable weir it is particularly suitable for small installations with manual operation.

30.12.6 Tidal outfalls

Tidal outfalls are required where drainage channels discharge through a sea wall or tidal embankment. Their function is to allow discharge at low tide but to prevent the tidal water from flowing back into the drainage system during the high tide. Essentially they consist of a culvert through the tidal embankment with a tidal flap or door, which may be at the outer end of the culvert or in a chamber in the embankment (Figure 30.16). Where beach levels near the sea wall are relatively high the culvert may be extended for some distance out on the foreshore and a door on the outer end could be subject to severe wave action. At the wall itself the door can be sheltered by wing walls and breakwaters, and still greater protection is obtained by the use of a chamber in the embankment.

The door is usually circular or square, of cast iron, steel, or plastic, hung from the top by double hinges which allow the door to seat freely and to accommodate small obstructions such as weed or sticks on the seat. There should be sufficient space between the bottom of the door and the apron on to which the water falls to avoid debris from being trapped behind the door and preventing closure. There should also be adequate clearance between the sides of the door and the wing walls or sides of the chamber. When the door is in a chamber it must be mounted on the upstream wall and the chamber itself must be built up above high tide level. It is not advisable for the chamber to be sited inland of the tidal embankment as the intervening culvert will be under pressure at high tides. If the door is some distance out on

the foreshore there is the possibility of tidal water breaking into the culvert behind the door, and it is therefore advisable to provide a sluice or penstock capable of shutting off the culvert at the tidal embankment or at the inland end.

For small drainage channels discharging into a river a similar arrangement is used. The culvert through the river bank is fitted with a flap gate, which is a small version of the tidal door, generally circular and of diameter 300 mm to 1.2 m. Larger flap gates may be rectangular and can be counterweighted to minimize the head loss required to open the gate. Counterweighted gates would not be used in tidal situations where waves could cause cyclical movements leading to damage.

30.13 Pumping

30.13.1 Single or multiple pumps

Pump capacities for land drainage installations are commonly in the range 500 to 2000 l/s. It is usual to provide some standby capacity in pumping stations but, in the case of pumps which operate infrequently, a single pump may be appropriate for small stations. For larger stations where two or more pumps are required it may be appropriate to omit a standby unit, but where the design capacity is regularly achieved, one standby pump should be provided. Pumps of variable capacity allow flexibility in running but add to capital costs and complicate automatic running.

30.13.2 Motive power

Most modern pumping installations are powered by electricity. This provides the considerable advantages of automatic operation and reduced maintenance. In some rural areas failures of supply are not uncommon but these are not often long enough to cause difficulties.

Where electrical power is not readily available, particularly in developing countries, direct diesel-driven pumps are normally used. For long-life applications the diesel is of the straight vertical cylinder, slow speed, nonautomotive type and generally requires continuous attendance.

30.13.3 Pumps

Land drainage pumps are required to produce large outputs at low heads, generally between 3 and 7 m but occasionally up to 10 m. This flow–head characteristic falls into the axial and mixed flow bowl range. For the lower heads the axial-flow pump is used but typically the mixed-flow bowl pump of the vertical spindle configuration is used. A typical pump station layout is shown in Figure 30.17.

With axial- and mixed-flow bowl pumps, small head variations have a relatively large effect on the quantity of water pumped. Consequently, careful selection of pump and prime mover is required to cope with all demands. Also, because of the head–flow characteristics, the discharge pipe usually has a submerged termination to provide syphonic recovery. To reduce

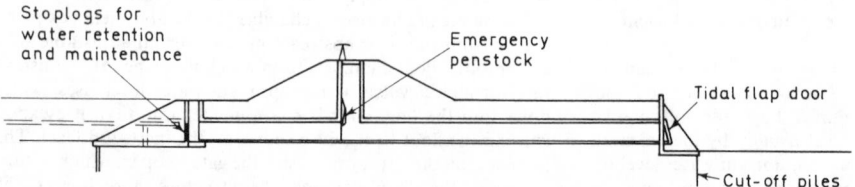

Figure 30.16 Tidal outfall

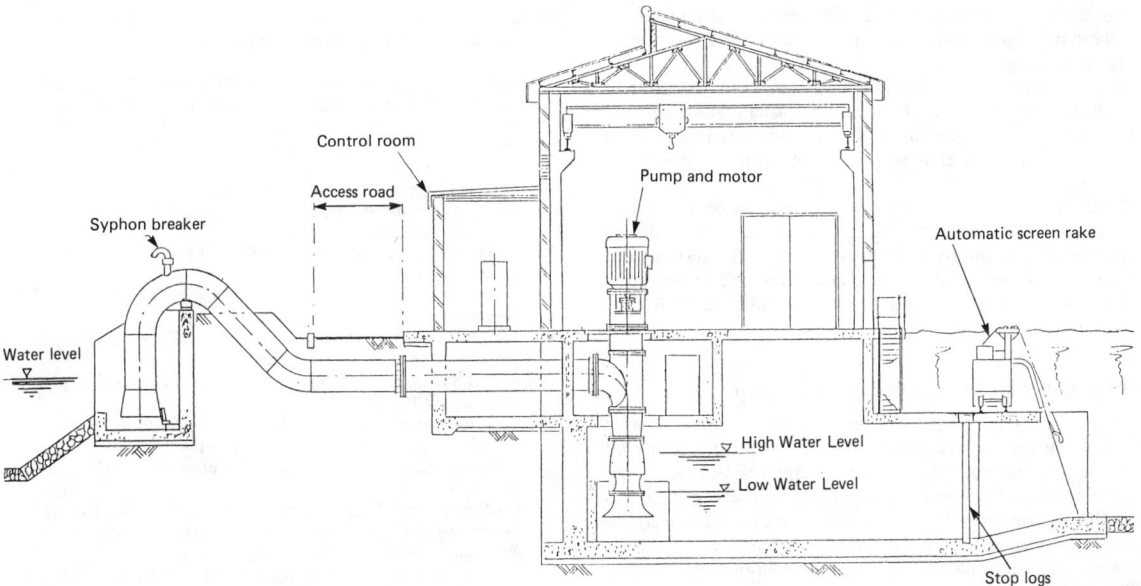

Figure 30.17 Typical mixed-flow bowl pump installation

the risk of reverse flow a syphon breaker is installed. This comprises a paddle-operated butterfly valve which is held shut by forward flow and opened to atmosphere during reverse flow, thus breaking the syphon.

30.13.4 Control

Unmanned electrically powered pumps are generally stopped and started automatically by preset level sensors located in the suction sump. The level-sensory equipment normally comprises mercury float switches but there is a move to use more sophisticated sensory equipment such as ultrasonics because of their low maintenance, ease with which preset levels can be changed, and

because they are more compatible with remote instrumentation and automation.

Fully automatic stations are now common in the developed countries. Details of water levels and pump operation can be conveyed automatically by a telemetry system to a central monitoring station. The same system can carry warning alarms to signal pump failure or other maloperation.

30.13.5 Pump station building

Vertical spindle axial- and mixed-flow bowl pumps are suspended in sumps close to the back wall which is curved to reduce the risk of swirl and vortices occurring. To maximize on sump

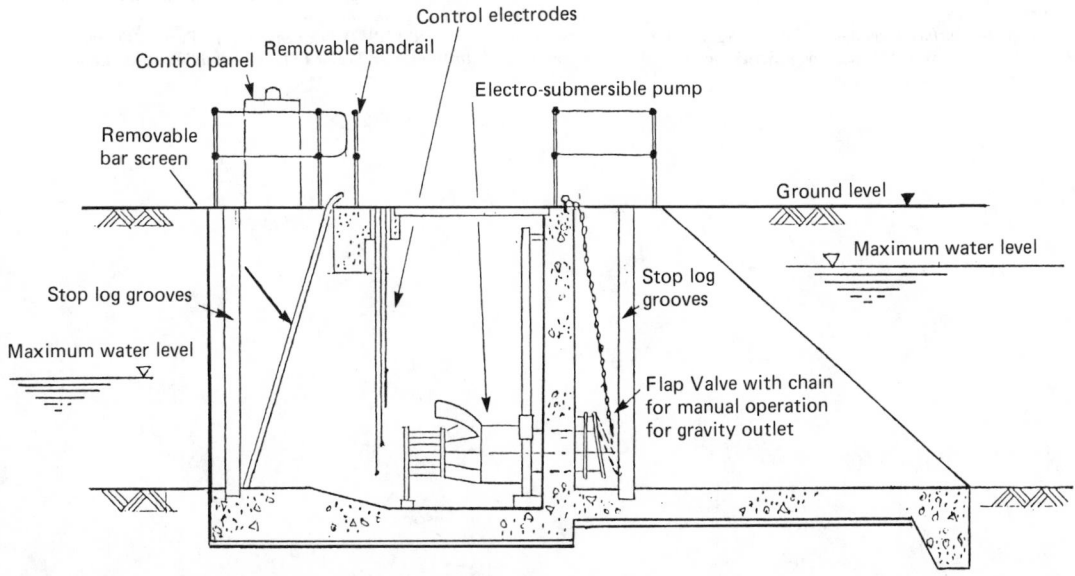

Figure 30.18 Typical electro-submersible pump installation

configuration with minimum civil substructure, particularly for non-standard applications, sump model tests are often carried out prior to design.

The main walls of the sump may be of reinforced concrete or steel sheet piling. Incoming channels are normally screened with sloping steel bars to prevent weeds and large items of debris entering the sump. Screens require raking either manually or automatically.

Buildings are generally required to give protection against the elements for plant and operating staff, to avoid damage due to vandalism, and to improve the appearance of the station. For remote locations and in developing countries little or no super-structure is provided and equipment is suitably rated for the outside locations.

30.13.6 Other types of pumping installation

For rivers carrying a high sediment load conventional channel off-takes can become quickly blocked, so requiring continual dredging. One method which has been employed to overcome this is the use of floating pontoons to carry the pumps, or to carry the suction pipes from land-based pumps. Such installations are also useful where the water level varies to the extent that the water's edge recedes leaving a conventional station dry.

The most recent development towards changing the pump type is the electro-submersible pumpset. This is being favoured because it can be located below ground level requiring little or no ground equipment or superstructure. A typical arrangement is shown in Figure 30.18.

References

1 Doorenbos, J. and Pruitt, W. O. (1977) *Crop water requirements.* Food and Agriculture Organization Irrigation and Drainage Paper No. 24.

2 Blaney, H. F. and Criddle, W. D. (1950) *Determining water requirements in irrigated areas from climatological and irrigation data.* US Department of Agriculture – SCS TP 96.

3 Dastane, N. G. (1974) *Effective rainfall in irrigated agriculture.* Food and Agriculture Organization Irrigation and Drainage Paper No. 25.

4 Ayes, R. S. and Wescot, D. W. (1976) *Water quality for agriculture.* Food and Agriculture Organization Irrigation and Drainage Paper No. 29.

5 Booher, L. J. (1974) *Surface irrigation.* Food and Agriculture Organization Land and Water Development Series No. 3.

6 Baars, C. (1973) *Design of sprinkler installations.* Department of Irrigation, Civil Engineering, Agricultural University, Wageningen.

7 Vermeirei, L. and Jobling, G. A. (1980) *Localized irrigation,* Food and Agriculture Organization Irrigation and Drainage Paper No. 36.

8 Etcheverry, B. A. (1915) *Irrigation practice and engineering.* McGraw-Hill.

9 Food and Agriculture Organization (1980) *Drainage design factors.* Food and Agriculture Organization Irrigation and Drainage Paper No. 38.

10 Smedema, L. K. and Rycroft, D. W. (1983) *Land drainage.* Batsford Academic.

11 Van Beers, W. J. F. (1963) *The Auger hole method,* ILRI, Wageningen.

12 Van Beers, W. J. F. (1979) *Some nomographs for the calculation of drain spacings,* ILRI, Bulletin No. 8, Wageningen.

13 United States Bureau of Reclamation (1978) *Drainage manual.* US Government Printing Office.

14 Ven Te Chow and Yevjevich, V. M. (1964) *Statistical and probability applied hydrology.* McGraw-Hill.

15 Bilham, E. G. (1962) *The classification of heavy falls of rain in short periods.* HMSO.

16 US Department of Agriculture (1968) *A method of estimating volume and rate of runoff in small watersheds.* US Department of Agriculture, Soil Conservation Service.

17 National Environment Research Council (1975) *Flood studies report.* NERC, London.

18 Sutcliffe, J. V. (1978) *Methods of flood estimation – a guide to the* Flood studies report. IOH.

19 Nixon, M. (1959) 'A study of the bank-full discharges of rivers in England and Wales'. *Proc. Instn. Civ. Engrs,* **12**, p. 157.

20 White, W. R., Paris, E. and Bettess, R. (1981) *Tables for the design of stable alluvial channels.* Hydraulics Research Station.

21 Vanoni, V. A. (ed) (1975) *Sedimentation engineering.* American Society of Civil Engineers, New York.

22 Ven Te Chow (1959) *Open channel hydraulics.* McGraw-Hill.

23 Royal Society for the Protection of Birds and the Royal Society for Nature Conservation (1984) *Rivers and wildlife handbook.*

24 Holmes, P. S. (1974) 'Analysis and prediction of scour at railway bridges in New Zealand.' *New Zealand Engineering,* (Nov) p. 313.

Suggested further reading

Withers, B. and Vipond, S. (1983) *Irrigation: design and practice.*
United Nations (Economic Commission for Asia and the Far East) (1953) *River training and bank protection.*
Brandon, T. W. (ed.) (1987) *River engineering,* Part 1: 'Design principles'. Institution of Water Engineers and Scientists, London.

31

Coastal and Maritime Engineering

F L Terrett MEng, CEng, FICE, MConsE
Posford Duvivier

Contents

31.1 Tides

31.1.1 Tide-raising forces

The alternate rising and falling of sea-level is caused by the attractive forces of the Moon and the Sun on the rotating Earth. The predominant effect, that of the Moon, can be explained in a simplified form by omitting in the first place the rotation of the Earth and Moon about their own axes and considering the relative motion of the two bodies about their common centre of rotation G (Figure 31.1). They revolve about G independently, not as a single rigid body, and points P_1 and P_2 on the Earth's surface rotate about G_1 and G_2 in which GG_1 and GG_2 are respectively parallel to CP_1 and CP_2, and P_1G_1 and P_2G_2 are parallel to CG. The attractive force of the Moon on a particle of mass m at the centre of the Earth is gmM_1/L^2 in which M_1 is the mass of the Moon, L the distance between the centres of the Moon and the Earth and g is the gravitational constant. This attractive force is the centripetal force F which restrains the particle in its circular orbit round G.

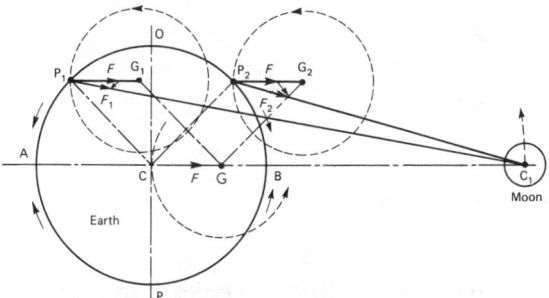

Figure 31.1 Tide-raising forces

If particles of water, also of mass m, are to remain in position at points P_1 and P_2 they must also be acted upon by forces F towards their centres of rotation G_1 and G_2. The attraction of the Moon on these particles is respectively F_1 which is less than F, and F_2 which is greater than F since P_1C_1 and P_2C_1 are respectively greater and less than L. The vector differences shown in the figure F minus F_1, and F_2 minus F, are the tide-raising forces. The vertical components of these forces are small in relation to the Earth's gravity and are of little importance; the horizontal components which are towards A and B, respectively, generate the tidal wave. They are zero at points A and B in line with the Moon and near points O and P at right angles to AB, and are a maximum midway between these points. Their directions, indicated by the circumferential arrows in the figure, cause two high waters, one directly under the Moon and the other on the opposite side of the Earth.

The Sun produces similar tide-raising forces but of barely half the magnitude of those due to the Moon. As the Earth rotates, the tides are phased with the apparent motion of the Moon so that the interval between successive high waters is approximately half the lunar day of about 24 h 50 min.

31.1.2 Tidal variations – effects of declination

Variations in tide level result from the varying positions of the Sun and the Moon relative to the Earth; at times of new and full moon the tide-raising forces of the Sun reinforce those of the Moon giving spring tides and when the Moon is at the first and third quarter the Sun's tide-raising forces counteract those of the Moon giving neap tides.

In many places there is a marked inequality in the height and

range of succeeding tides which is largely due to the angles between the plane of rotation of the Earth about its axis and the planes of the orbit of the Moon round the Earth and of the Earth round the Sun. These varying angles, which are the declination of the Moon and the Sun, introduce a diurnal component which combines with the semidiurnal tides. It is possible for one high water to be suppressed altogether and for an inequality in time also to be caused by declination so that the interval from high to low water may not be the same as from low to high water.

In the waters of northwestern Europe and the eastern seaboard of America the tides are essentially semi-diurnal, the tidal pattern, which is readily explained, being one large set of spring tides and one smaller set each lunar month. The tidal range varies from month to month with the varying distance of the Earth from the Sun, the largest range being at the equinoxes (March and September).

In the Pacific and many other places away from the Atlantic Ocean the tides have a strong diurnal inequality; generally tides of large and small range alternate, the largest tides occurring in December and June at the solstice when the diurnal component of the tide-raising force most nearly coincides with the semi-diurnal component.

In these areas the tides cannot be classified simply as springs or neaps; the highest and lowest water levels do not precede or succeed one another and there are significant changes in mean sea-level from week to week. At times the diurnal component predominates and at others the semidiurnal component, so that the tidal pattern is extremely complex varying from one to two high- or low-water levels per day and from small to large tidal range in either mode.

It is important to note that in these areas the greatest rate of change in level from high to low water or vice versa does not necessarily coincide with tides of greatest range nor is there an obvious relationship between tidal current and tidal range.

31.1.3 Tidal currents – coastal effects – reflection and resonance

In the open oceans the tide generated by the attractive forces of the Moon and Sun takes the form of a progressive wave in which the associated currents are in the direction of wave propagation below the crest and in the opposite direction in the trough. The maximum current velocities are at the crest and trough, i.e. at high and low water, and zero at half tide on both rising and falling tides.

This simple description of the tidal motion is, however, much altered by many factors, in particular by the shape and disposition of the land masses and the depth of the seas around them. Reflection of the tidal wave from the shores and resonance effects in enclosed or partially enclosed gulfs and straits result in standing oscillations in which the tidal current is zero at high and low water and a maximum at half tide or thereabouts. An example of such a standing oscillation is found in the eastern half of the English Channel where high water between the Isle of Wight and Dover occurs within about 10 min at all places along the English and French coasts.

As the tidal wave enters shallow water, in an estuary, for example, it is distorted: the speed of propagation is reduced and the wave crest tends to overtake the preceding trough. Thus the time interval from low to high water is reduced and from high to low water increased, and the flood current becomes stronger than the ebb. Also, the height of the tide may increase as the estuary narrows inland.

31.1.4 The Coriolis force

The ocean currents, whether tidal or wind-generated, or the

result of density gradients due to salinity and temperature differences, are affected by the rotation of the Earth, being deflected to the right in the northern hemisphere and to the left in the southern. This is known as the Coriolis effect after the French scientist of that name (1792–1843). In a narrow sea, such as the English Channel, deflection of the flood current to the right is inhibited by the proximity of the shore lines and the Coriolis force leads to higher tides along the French than the English coast.

In a more open sea the tidal wave and its associated currents may become rotary about an amphidromic point at which the currents are zero and there is no tidal variation in level. There are three such amphidromic points in the North Sea (Figure 31.2).

In small tidal inlets the Coriolis force is not significant, but where the width exceeds about 20 km it has an important effect on the currents which flush pollutants from these waters and erode and transport fine sediments.

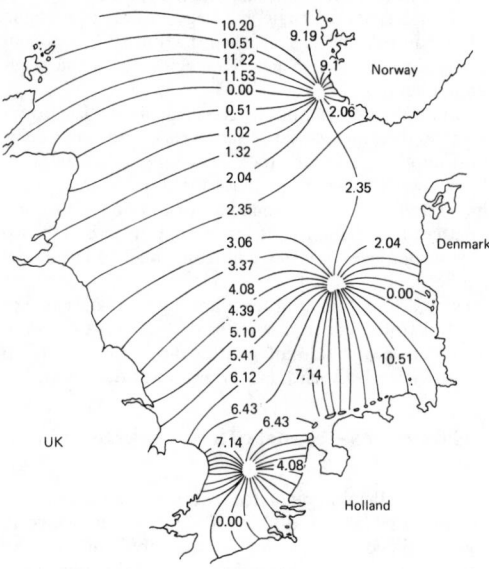

Figure 31.2 Amphidromic points in the North Sea

31.1.5 Prediction of tides

The astronomical tide-raising forces create semidiurnal and longer frequencies in the tidal cycle; shallow-water effects introduce higher frequencies. The recorded tidal curve at any place can be broken down into these various frequencies and the individual constituents recombined to give tidal predictions.

For most places in the world, such predictions are made by national government agencies for their own territorial waters, and where available, should be used in preference to the worldwide tables prepared by Her Majesty's Stationery Office, as they are often more detailed and likely to be related to a local land datum.

However, the quality of tidal predictions varies widely and should be checked by examination of the basic data before starting work at any unfamiliar coastal site.

In the absence of published tables, predictions can now be readily prepared for any site for which a month or more of good-quality record is available, using a small digital computer following the method of working set out in the *Admiralty Manual of Tides*.[1]

31.2 Waves

31.2.1 General

Coastal and estuarine processes are complex and it is seldom possible to find solely analytical solutions to practical problems. A great deal of theoretical work has, however, been carried out and the more important results are given below with notes on their significance and application. For their derivation see Ippen.[2]

In Figure 31.3, T is the wave period (= time interval for motion to recur at a fixed point), c is the velocity of wave propagation or wave celerity, $\eta(x, t)$ is surface elevation at position x and time t, u is the horizontal component of instantaneous velocity of fluid element, v is the vertical component of instantaneous velocity of fluid element, p the instantaneous 'static' pressure, H the wave height (= $2a$), ρ the density (mass per unit volume) and v the kinematic viscosity.

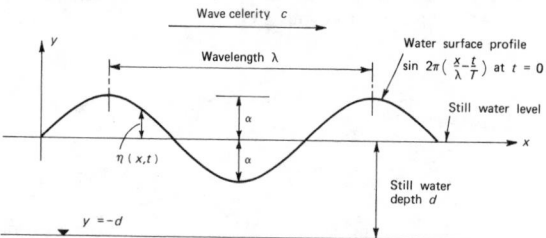

Figure 31.3 Coordinate system

31.2.2 Wave length, celerity and period as functions of depth

$$c^2 = \frac{g\lambda}{2\pi} \tanh\left(\frac{2\pi d}{\lambda}\right) \tag{31.1}$$

$$\lambda = cT \tag{31.2}$$

Equation (31.1), derived from theoretical work by Stokes,[3] is strictly accurate only for waves of small amplitude but the error resulting from its application to practical problems is small and the theory may be used with confidence. When a wave moves from deep to shallow water (or vice versa), c and λ both change, while T necessarily remains constant. Published tables and graphs[2,3] relating the variables are available so that manipulation of the equations is not necessary.

Waves are conveniently classified into types according to the relative depth d/λ as follows:

Shallow-water (long) waves $d/\lambda = 0$ to $1/20$
         $\tanh(2\pi d/\lambda) \simeq 2\pi d/\lambda$

Deep-water (short) waves $d/\lambda = 1/2$ to ∞
         $\tanh(2\pi d/\lambda) \simeq 1$

Intermediate waves $d/\lambda = 1/20$ to $1/2$

Using these approximations, Equations (31.1) and (31.2) become:

(1) *For shallow water:*

$$\left[c = (gd)^{1/2} \right]^{1} \tag{31.3}$$

$$\left[\lambda = T(gd)^{1/2} \right]^{2} \tag{31.4}$$

(2) *For deep water:*

$$\left[c = gT/2\pi \right]^3 \tag{31.5}$$

$$\left[\lambda = gT^2/2\pi \right]^4 \tag{31.6}$$

Within the limits indicated above, Equations (31.3) to (31.6) resulted in errors of less than 1%. For many purposes such precision is unnecessary and the limits may be widened.

31.2.3 Fluid velocity and pressure

The elements of fluid in a wave move in nearly closed orbits. If u is the instantaneous horizontal velocity and v the corresponding vertical component then:

$$u = \frac{agT}{\lambda} \frac{\cosh\left[\frac{2\pi}{\lambda}(d+y)\right]}{\cosh(2\pi d/\lambda)} \sin\left[2\pi\left(\frac{x}{\lambda} - \frac{t}{T}\right)\right] \tag{31.7}$$

$$v = -\frac{agT}{\lambda} \frac{\sinh\left[\frac{2\pi}{\lambda}(d+y)\right]}{\cosh(2\pi d/\lambda)} \cos\left[2\pi\left(\frac{x}{\lambda} - \frac{t}{T}\right)\right] \tag{31.8}$$

For any given phase angle (the sine and cosine terms) the velocities diminish with increasing depth, and where $|y| \geqslant -\lambda/2$ there is no appreciable motion.

If the water depth is less than $\lambda/2$, v becomes zero at the bed and the fluid above the bed is constrained to move in elliptical orbits in which A, the major axis, and B, the minor axis, are given by:

$$A = 2a \frac{\cosh\left[\frac{2\pi}{\lambda}(d+y)\right]}{\sinh(2\pi d/\lambda)} \tag{31.9}$$

$$B = 2a \frac{\sinh\left[\frac{2\pi}{\lambda}(d+y)\right]}{\sinh(2\pi d/\lambda)} \tag{31.10}$$

For shallow water, as defined on page 31/4, A and B become:

$$A = a\lambda/\pi d \tag{31.11}$$

$$B = 2a(d+y)/d \tag{31.12}$$

For deep water:

$$A = B = 2a \exp(2\pi y/\lambda) \tag{31.13}$$

These results which are shown schematically in Figure 31.4 are of value in estimating the depths at which sediment may be disturbed by wave action. In practice, however, particularly in the case of shallow-water waves of large amplitude, the trajectories of the water particles are not closed orbits, the velocities in the direction of wave propagation being greater and those in the reverse direction less than predicted by the theory. There is a resultant net movement of fluid in the direction of wave propagation at the surface and, in the case of shallow-water waves, close to the bed also, which in a confined system must be balanced by a return flow, normally at mid depth.

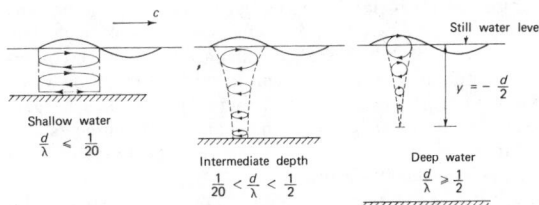

Figure 31.4 Water particle trajectories

Pressure fluctuations under a wave also diminish with depth and the instantaneous hydrostatic pressure p is given by:

$$p = \rho g \left[a \sin\left[2\pi\left(\frac{x}{\lambda} - \frac{t}{T}\right)\right] \frac{\cosh\left[\frac{2\pi}{\lambda}(d+y)\right]}{\cosh(2\pi d/\lambda)} - y \right] \tag{3.14}$$

The sine term has limiting values of ± 1 under the crest and trough of the wave, and at the bed where $d = -y$ the maximum and minimum pressures are:

$$\rho g \left[d \pm \frac{a}{\cosh(2\pi d/\lambda)} \right] \tag{3.15}$$

A pressure-sensing instrument located at a moderate depth may thus be used for the measurement of wave heights. It should also be noted that the pressure gradient will impose forces on any solid object within the fluid in addition to the drag forces arising from the fluid velocities u and v.

31.2.4 Superposition

Most of the properties arising from separate wave trains, i.e. surface elevation, particle velocity and instantaneous pressure, may be superposed, provided that both the amplitude of the component trains and the amplitude of the combined trains remain small. For example:

Combined surface elevation at position x and time t
$$\eta_\tau = \eta_1 + \eta_2 + \eta_3 + \ldots \eta_n$$

Combined instantaneous horizontal velocity
$$u_\tau = u_1 + u_2 + u_3 + \ldots u_n$$

Combined instantaneous vertical velocity
$$v_\tau = v_1 + v_2 + v_3 + \ldots v_n$$

The resulting surface elevations can be most readily determined graphically; when considering the interaction of wave trains travelling in different directions the surface profiles at, say, quarter-period intervals should be sketched to give a visual appreciation of the motion. Wave energy is proportional to the square of the wave amplitude a and cannot therefore be superposed.

The mathematically derived results for special cases are given by Ippen,[2] Stokes[3] and Wiegel,[4] and the following are of particular interest:

(1) Two wave trains moving in the same direction with periods T_1 and T_2, wave lengths λ_1 and λ_2 and phase angles δ_1 and δ_2:

$$\eta_\tau = a_1 \sin\left[2\pi\left(\frac{x}{\lambda_1} - \frac{t}{T_1} + \delta_1\right)\right] + a_2 \sin\left[2\pi\left(\frac{x}{\lambda_2} - \frac{t}{T_2} + \delta_2\right)\right] \tag{31.16}$$

Equation (31.16) is not harmonic but it may be periodic. The constants δ_1 and δ_2 are the phase angles at the arbitrary origin where $t=0$, $x=0$, but the phase difference will change continuously, and an alternative origin may be found at which $\delta_1=\delta_2=0$ and η_τ is zero. If after an interval of time T, η_τ is again zero and $T=mT_1=nT_2$ in which m and n are integers, the resultant wave train is periodic, the period T being the smallest value that will satisfy $T=mT_1=nT_2$.

If $a_1=a_2$ and η_τ is periodic, nodes will occur at periods T, $2T$, $3T\ldots$, the envelope of the crest being another wave of period T, amplitude $2a_1$ and celerity

$$\frac{\lambda_1\lambda_2}{T_1T_2}\left(\frac{T_2-T_1}{\lambda_2-\lambda_1}\right)$$

This special case is known as 'pure beat'.

(2) For two progressive waves moving in opposite directions:

$$\eta_\tau=a_1\sin\left[2\pi\left(\frac{x}{\lambda_1}-\frac{t}{T_1}+\delta_1\right)\right]$$
$$+a_2\sin\left[2\pi\left(\frac{x}{\lambda_2}+\frac{t}{T_2}+\delta_2\right)\right]\tag{31.17}$$

If the origin of coordinates is chosen so that $\delta_1=0$ and if in addition $T_1=T_2$ $(\lambda_2=\lambda_1)$, then:

$$\eta_\tau=a_1\sin\left[2\pi\left(\frac{x}{\lambda}-\frac{t}{T}\right)\right]$$

$$+a_2\cos\delta_2\sin\left[2\pi\left(\frac{x}{\lambda}+\frac{t}{T}\right)\right]$$

$$+a_2\sin\delta_2\cos\left[2\pi\left(\frac{x}{\lambda}+\frac{t}{T}\right)\right]\tag{31.18}$$

Equation (31.18) applies to an incoming wave which is totally or partially reflected by a structure such as a breakwater when $a_2=K_ra_1$ in which K_r is the reflection coefficient. In practice K_r is close to 1 for small waves impinging on a vertical wall.

(a) *Perfect reflection* $(K_r=1)$

If a wave train is perfectly reflected $(K_r=1$ and $a_2=a_1)$ by a vertical barrier at $x=x_1$, then:

$$\eta_\tau=2a\sin\left[2\pi\left(\frac{x_1}{\lambda}-\frac{t}{T}\right)\right]\cos\left[\frac{2\pi}{\lambda}(x-x_1)\right]\tag{31.19}$$

It will be noted that η_τ is the product of two harmonic terms, one a function of x only and the other a function of t only. Thus, there are certain times when $\eta_\tau=0$ for *all* values of x, i.e. the water surface is flat, and certain positions where $\eta_\tau=0$ for all values of t, i.e. where there is no vertical displacement of the surface at any time; the latter points are called nodes and will be located where:

$$\cos\left[\frac{2\pi}{\lambda}(x-x_1)\right]=0$$

i.e.:

$$x_{\text{node}}=x_1-\frac{(2n+1)\lambda}{4}\tag{31.20}$$

in which n can have any of the values 0, 1, 2, 3, ... Thus, the nodes will occur at $\lambda/4$, $3\lambda/4$, $5\lambda/4$ ... from the barrier. This

condition of stationary nodes is known as a 'standing wave' or 'clapotis'. The instantaneous horizontal and vertical water particle velocities in a standing wave are:

$$u=\frac{2agT}{\lambda}\frac{\cosh\left[\frac{2\pi}{\lambda}(d+y)\right]}{\cosh(2\pi d/\lambda)}$$

$$\times\cos\left[2\pi\left(\frac{x_1}{\lambda}-\frac{t}{T}\right)\right]\sin\left[\frac{2\pi}{\lambda}(x-x_1)\right]\tag{31.21}$$

$$v=-\frac{2agT}{\lambda}\frac{\sinh\left[\frac{2\pi}{\lambda}(d+y)\right]}{\cosh(2\pi d/\lambda)}$$

$$\times\cos\left[2\pi\left(\frac{x_1}{\lambda}-\frac{t}{T}\right)\right]\cos\left[\frac{2\pi}{\lambda}(x-x_1)\right]\tag{31.22}$$

Since the nodes occur where $\cos[(2\pi/\lambda)(x-x_1)]=0$ there are horizontal motions only under the nodes and vertical motion only under the antinodes. This result is important to the understanding of the relationship between tidal level and tidal current in coastal waters where the reflection or interaction of the tidal wave may result in a partial or complete standing wave.

The pressure p within a standing wave is given by:

$$p=\rho g\left(\eta_\tau\frac{\cosh\left[\frac{2\pi}{\lambda}(d+y)\right]}{\cosh(2\pi d/\lambda)}-y\right)\tag{31.23}$$

and is hydrostatic under the nodes where $\eta_\tau=0$ and there is no vertical movement.

(b) *Imperfect reflection* $(K_r<1)$

If partial reflection takes place at a vertical barrier at $x=x_1$, $a_2=K_ra_1$

$$\eta_\tau=a\sin\left[2\pi\left(\frac{x-x_1}{\lambda}-\frac{t}{T}\right)\right]$$

$$-K_ra\sin\left[2\pi\left(\frac{x-x_1}{\lambda}+\frac{t}{T}\right)\right]\tag{31.24}$$

The maximum and minimum values for this expression occur in the same positions as the antinodes and nodes for the case of perfect reflection, the maximum and minimum amplitudes being:

$$a_{\min}=a_1-a_2$$

$$a_{\max}=a_1+a_2$$

and

$$K_r=\frac{a_{\max}-a_{\min}}{a_{\max}}$$

31.2.5 Wave trains and wave energy

31.2.5.1 Group celerity C_G

For a pure beat with waves travelling in the same direction the node and, hence, the wave group between each pair of nodes progress at a celerity of:

$$\frac{\lambda_1 \lambda_2}{T_1 T_2} \left(\frac{T_2 - T_1}{\lambda_2 - \lambda_1} \right)$$

(see page **31/6**). The group of waves between any pair of nodes may be considered separately from the preceding and succeeding groups and it can be shown that as T_1 approaches T_2 the 'group celerity' C_G becomes:

$$\frac{c}{2} \left[1 + \frac{4\pi d/\lambda}{\sinh (4\pi d/\lambda)} \right] \tag{31.25}$$

Thus, in deep water the group celerity is half the celerity of the individual waves in the group while in shallow water it approaches the celerity of the individual waves.

In a finite group of waves travelling in otherwise undisturbed water, wave crests will form at the back of the group, travel through it, in deep water at twice the speed of the group but at decreasing relative velocity as the water becomes shallow, and disappear at the front. It is evident that the energy within the wave train travels at the group celerity, not the wave celerity, and that the time taken for waves to reach a location distant from the area in which they have been generated is a function of C_G.

31.2.5.2 Energy

The average potential energy density (average potential energy per unit surface area) which is attributable to the presence of a progressive wave on the free surface is $\rho g a^2/4$; the average kinetic energy density is also $\rho g a^2/4$ and the total average energy density E is given by:

$$E = \rho g a^2 / 2 \tag{31.26}$$

For a two-component composite wave train with both waves travelling in the same direction the average potential and kinetic energy densities are both $(\rho g/4)(a_1^2 + a_2^2)$, and

$$E = \frac{\rho g}{2} (a_1^2 + a_2^2) \tag{31.27}$$

For a standing wave, $E = \rho g a^2$ where a is the amplitude of the incident and reflected waves.

The proportion of the total energy which is carried along with a progressive wave train is given by:

$$E \times \tfrac{1}{2} \left(1 + \frac{(4\pi d/\lambda)}{\sinh (4\pi d/\lambda)} \right) = E \times (C_G/c) \tag{31.28}$$

In deep water this is half of the total energy, while it approaches the total energy in shallow water.

31.2.6 Transformation of waves

When waves travel from deep into shallow water there will be no reflection of energy if the bed slope does not exceed 1 in 20, and the energy flux across any two planes parallel to the wave crests will remain constant provided no energy is dissipated or generated between the two planes. Using this principle of energy conservation, and allowing for changes in channel width b, or crest length due to wave refraction, changes in wave length, height and celerity can be calculated.

It is usual to refer these parameters to the corresponding values in deep water to which is ascribed the suffix '0', the basic transformation expressions being:

$$\frac{c}{c_0} = \frac{\lambda}{\lambda_0} = \tanh (2\pi d/\lambda) \tag{31.29}$$

and

$$\frac{H}{H_0} = \left(\frac{b_0^{\frac{1}{2}}}{b} \times \left[\frac{2 \cosh^2 (2\pi d/\lambda)}{(4\pi d/\lambda) + \sinh (4\pi d/\lambda)} \right]^{1/2} \right) \tag{31.30}$$

Change in wave steepness H/λ is obtained by combining Equations (31.29) and (31.30).

Tables[5] of the various wave functions are available, from which typical values are given in Table 31.1.

It should be noted that at a certain depth, depending upon wave height and length, the wave will start to break and the above relationships then become invalid.

Table 31.1 Wave transformation functions

d/λ_0	1.0	0.5	0.1	0.05	0.01	0.005
$c/c_0 = \lambda/\lambda_0$	1.0	0.99	0.71	0.53	0.25	0.15
C_G/c	0.5	0.52	0.81	0.91	1.0	1.0
H/H_0	0.98	0.92	0.90	1.0	1.1	1.15

31.2.7 Reflection coefficients

Reflection coefficients for abrupt changes in geometry must normally be determined by experiment but some guidance may be obtained from published results.[6] It should be noted that the reflection coefficient is a function of both the incident wave steepness and the geometry of the solid boundaries.

31.2.8 Dissipation of wave energy

The rate at which energy is dissipated as a wave train travels through deep water is exceedingly small. The resulting reduction of wave amplitude with distance and time can be derived from:

$$a = a_0 e^{-\alpha x} = a_0 e^{-\alpha C_G t} \tag{31.31}$$

in which the damping modulus α is given by:

$$\alpha = \frac{4v}{c} \left(\frac{2\pi}{\lambda} \right)^2$$

This expression gives the typical times and distances for the wave height to be reduced to half of its original value as shown in Table 31.2.

Table 31.2 Distance and time for 50% reduction in wave height

λ	30 m	3 m	0.3 m
t	1700 h	17 h	10 min
$x = C_G t$	22 000 km	64 km	210 m

For shallow water the theoretical solution is inaccurate owing to turbulence near the bed. Experimental work has shown α to be considerably larger than given above, the best fit to the available data being

$$\alpha = \frac{13.5\pi^{3/2} (Tv)^{1/2}}{(4\pi d/\lambda) + \sinh (4\pi d/\lambda)} \tag{31.32}$$

These results cannot be applied to very long waves or intermediate waves where the extent of the turbulence is uncertain. If the bed is permeable there will be an energy loss due to wave-induced flow and the wave heights will be less than predicted.

31.2.9 Finite amplitude theory – breaking of waves

The foregoing results are strictly applicable only to waves of small amplitude. They are, however, sufficiently precise for many purposes and it is usually only necessary to have recourse to the more difficult finite-amplitude theory to obtain an appreciation of the processes which limit the maximum possible height of wave.

It should be noted that waves of finite amplitude have longer, shallower troughs and shorter, steeper crests than the sine wave assumed in small-amplitude theory, and this departure should be taken into account when determining the height of structures above mean sea-level and the forces on them.

The wave celerity is insensitive to second- and higher-order effects but waves of finite height travel faster than small waves.

In the finite theory developed by Stokes it is assumed that if the water particle velocity at the crest of the wave exceeds the celerity of the wave it will 'topple over' or 'spill'. The crest angle determined for this condition in deep water is 120° when the wave steepness $H/\lambda = 1/7$. From this the height of breaking waves H_b in deep water is given by $H_b/gT^2 = 0.0272$ which fits experimental data when $d/T^2 > 1$ m/s². The crest height above mean sea-level a_c reaches a maximum value of $0.68H$.

In shallow water, long waves can be looked upon as 'solitary waves' (see below). This applies when $d/T^2 < 0.1$ m/s² when the ratio a_c/H approaches 1 and $H_b/d = 0.78$.

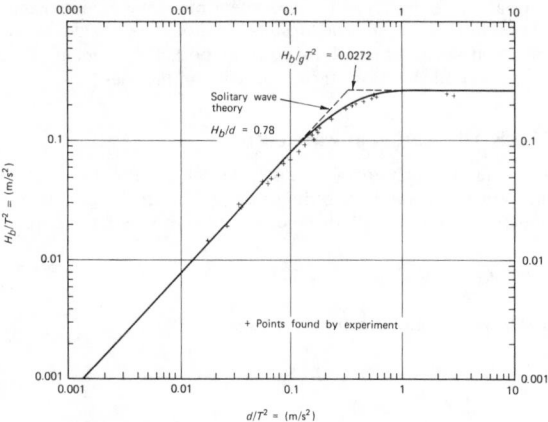

Figure 31.5 Breaking index curve. (After Reid and Bretschneider (1952) 'Revised wave forecasting relationships.' *Proceedings, 2nd Conference on Coastal Engineering*

These results are shown in Figure 31.5 in which the curve in the range $0.1 < d/T^2 < 1.0$ has been fitted empirically to the available data.

31.2.10 The solitary wave

It is possible for a single wave, lying entirely above the still-water level, to be generated; such a wave propagates at constant velocity and is unaltered in form. In nature, waves generated by landslides or earthquakes may approximate to this type and, as already mentioned, long oscillatory waves moving into shallow water.

The surface profile of such a wave is given by the relation:

$$\eta = H \left(\operatorname{sech} \left[\left(\frac{3H}{4d^3} \right)^{\frac{1}{2}} (x - ct) \right] \right)^2 \tag{31.33}$$

and

$$c = [g(H + d)]^{1/2} \tag{31.34}$$

the origin of x being at the wave crest.

31.2.11 Wave generation

When wind blows across a free water surface at a very low velocity the interface remains perfectly stable and the mirror calm is undisturbed. If the velocity increases slightly, ripples appear; these are capillary waves and have a length of about 17 mm and a period of 0.07 s. With further increase in wind speed the ripples start to grow and become gravity (rather than capillary) waves. Although great interest has been shown in these critical wind speeds, they are of little importance in engineering practice.

Once gravity waves have formed, energy is transferred from the air to the water in three ways:

(1) By shear at the interface.
(2) By pressure differences due to the form resistance of the waves.
(3) By random pressure fluctuations associated with the turbulent air stream.

Initially the second of these will be dominant but as the wave length increases, shear at the interface becomes more important. If the wind blows for sufficient time over a long enough 'fetch' the energy input from the wind will become equal to the losses within the wave motion and further growth ceases. This is known as a fully developed sea for which Bretschneider gives:

$$c/U = 1.95$$

Table 31.3 Relationship between wave height, wind speed, time and fetch

Wind speed		Wave spectrum method (fully developed sea)			Significant wave method (90% developed sea)			JONSWAP		
(m/s)	$H_{\frac{1}{3}}$ (m)	Time (h)	Fetch (km)	$H_{\frac{1}{3}}$ (m)	Time (h)	Fetch (km)	$H_{\frac{1}{3}}$ (m)	Time (h)	Fetch (km)
5	0.4	2	18	0.7	17	150	0.6	10	60
10	2.1	9	125	2.6	33	600	2.5	20	230
15	6.7	22	500	5.9	50	1400	5.6	30	520
20	13.0	40	1250	10.5	66	2550	9.9	40	930
25	22.2	65	2400	16.2	83	3850	15.5	50	1450
28	29.2	82	3500	20.1	92	4700	19.4	56	1820

and

$$\frac{gH}{U^2} = 0.283$$

where U is the wind speed. It will be noted that the waves are travelling at nearly twice the wind speed. However, other workers give significantly different results, for the physical processes involved are not sufficiently understood and so-called wave forecasting methods which relate wave height, period and length to wind speed and fetch, cannot therefore be other than semi-empirical. The approximate time and distance, or 'fetch', for which a steady wind has to blow in one direction over deep water, and the corresponding significant wave height as assessed by the 'wave spectrum' method of Pierson, Neuman and James[7] for a fully developed sea, and by the 'significant wave' method of Sverdrup, Munk and Bretschneider[8] for a 90% developed sea, are compared in Table 31.3 with results derived from the more recent JONSWAP experiment.[9]

The 'significant wave' method gives such large values for time and fetch for a fully developed sea as to be unattainable in reality. For short fetches and high wind speeds the following expressions are derived from the work of Bretschneider:

$$H_{\frac{1}{3}} = 0.024 \, (U^2 F)^{1/2} \tag{31.35}$$

$$T_{\frac{1}{3}} = 0.6 \, (U^2 F)^{1/4} \tag{31.35}$$

$$\frac{F_{min}}{t_{min}} = 1.26 \, (U^2 F)^{1/4} \tag{31.36}$$

$H_{\frac{1}{3}}$ is the significant wave height in metres (mean of the highest one-third of the waves), $T_{\frac{1}{3}}$ is the significant wave period in seconds (mean period of the highest one-third of the waves), U is the windspeed in metres per second, F the fetch length in kilometres, F_{min} the minimum fetch for the wave condition to develop, and, t_{min} the minimum duration for the wave condition to develop.

Since in real storms the wind is not constant in speed or direction nor unlimited in extent, the practical application of these results is complicated.

Wave heights are randomly distributed about some mean value and it is not possible to define a 'highest' wave from a recording of a wave train with a given mean height, or the energy level; it is, however, possible to determine the probability of a certain height being equalled or exceeded in a given sample. It appears to be generally agreed that in deep water the spectrum of wave heights fits approximately to a Rayleigh distribution and the following relationships have been derived:

$$H_{\frac{1}{3}} = 1.6 \times H_{mean} \tag{31.37}$$

$$H_{\frac{1}{10}} = 2.03 \times H_{mean} \tag{31.38}$$

($H_{\frac{1}{10}}$ = mean of the highest 10% of the waves)

About 1% of the waves will equal or exceed about $2.8 H_{mean}$, about one wave in 10 000 will equal $3.4 H_{mean}$ and about 16% of the waves will exceed $H_{\frac{1}{3}}$; it is usual to measure $H_{\frac{1}{3}}$ when examining wave records. These results for deep water, based on the Rayleigh distribution, cannot be applied to waves in shallow water.

31.2.12 Wave generation in shallow water

The physical processes involved in the generation of waves in shallow water are the same as those in deep water except that when $d/T^2 < 0.75$ m/s the waves 'feel the bottom' and the growth of the longer-period waves is restricted. If the fetch and duration are unlimited the relationship between wind speed, depth and significant wave height, based on the work of Bretschneider, is as shown in Table 31.4.

Figure 31.6 gives forecasting curves after Thijsse and Schijf [10]

Table 31.4 Wave heights in shallow water

Wind speed (m/s)	Water depth (m)					
	1	2	3	5	7	9
	Wave heights (m)					
10	0.30	0.48	0.63	0.90	1.13	1.35
15	0.38	0.60	0.78	1.14	1.44	1.73
20	0.45	0.71	0.92	1.35	1.72	2.04
25	0.52	0.81	1.07	1.53	1.95	2.32
28	0.56	0.88	1.14	1.64	2.06	2.45

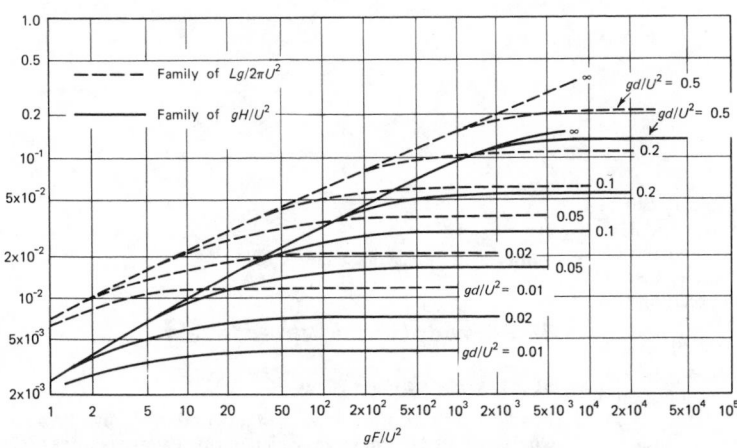

Figure 31.6 Growth of waves in limited depth. (After Thijsse and Schijf (1949) 'Report on waves.' *Proceedings, 17th International Congress*. Section II: Communication, page 4)

for wave height and length when both fetch and depth are restricted.

31.2.13 Wave decay

When waves generated in deep water travel out of the generating area, or when the generating wind abates, they lose their extreme irregularities and diminish in height. The processes which result in this decay of wave height are principally the following:

(1) Lateral diffraction of energy.
(2) Selective attenuation; the long-period part of the wave spectrum travels faster than the short-period part and the energy is spread out in the direction of wave propagation.
(3) Air resistance or directly opposing winds.
(4) Viscous damping in the water.

A conclusion derived empirically by the Admiralty in 1942 is that waves lose one-third of their height each time they travel a distance in nautical miles equal to their length in feet. Charts based on the work of Sverdrup, Munk and Bretschneider are available.[8]

31.2.14 Propagation of waves into shallow water – refraction

When waves travel into shallow water $(d < \lambda/2)$ their speed diminishes, their form alters and if the wavefronts are long and not parallel to the contours they are refracted and become curved. Since any wave train consists of a number of components of varying length the bottom will thus have a sorting effect, the longer components being affected sooner and therefore to a greater overall extent than the shorter components.

If the direction of wave propagation is represented by rays (orthogonals – see Figure 31.7), refraction of the waves is precisely analogous to the bending of light rays passing from one medium into another of greater density, and Snell's law, $\sin i / \sin r = c_1/c_2$, applies.

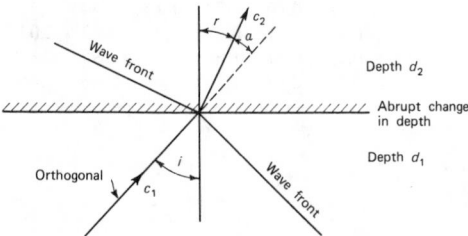

Figure 31.7 Wave refraction

If the wave period and depth are known, c_1 and c_2 can be calculated from the relationships given on pages **31**/4 and **31**/7, or more expeditiously from published tables. Starting from the assumed direction of the orthogonals in deep water, it is possible to plot them as they are refracted, using a templet relating c_1/c_2 to a to facilitate this tedious operation.

This forward tracking procedure may result in orthogonals crossing, to avoid which, computer programs now available project the orthogonals in reverse, i.e. from the area of interest seaward into deep water.

It will be noted that a sea-bottom ridge will cause the wavefronts to become concave and the orthogonals to converge so that the wave energy and wave height increase locally. A valley will have the reverse effect.

Wavefronts approaching a shoreline at an angle will be convex with diverging orthogonals.

It is usual, and nearly correct, to assume that no energy crosses the orthogonals so that if the wave height in deep water is known the wave height at any other point in the diagram can be calculated from Equation (31.34).

31.2.15 Wave forecasting

Using the principles set out on pages **31**/8 to **31**/10 it should in theory be possible to compute from the meteorological synoptic charts a wave spectrum for a given return period for any coastal site. The practical application of the various methods available is, however, complex and beyond the scope of this chapter. For further information the reader is referred to publications dealing specifically with forecasting methods.[7-9] Moreover, the theoretical and empirical basis for the methods is far from perfect and errors will be introduced at each stage; if at all possible, design decisions should therefore be based on statistical analysis of wave records for the site. It is then doubtful whether consideration of the physical processes generating the waves is necessary except to establish the typicality of the sampling period.

31.2.16 Diffraction

If waves pass the end of a breakwater or through a hole in a barrier they spread into the water behind the obstruction, and like refraction, the phenomenon is analogous to the diffraction of light. Mathematical solutions of problems involving diffraction are difficult but have been carried out for a few simple cases, e.g. for a train of uniform low waves in uniform depth passing the end of a breakwater for which the diffraction pattern is shown in Figure 31.8.

The application of diffraction theory to harbour design is limited since in a real harbour basin the diffraction pattern will normally be curtailed by the boundaries of the basin and refraction and reflection will occur simultaneously; therefore the study of problems of diffraction usually requires a physical model investigation. In principle a mathematical model is possible but has not yet been generally accepted.[11]

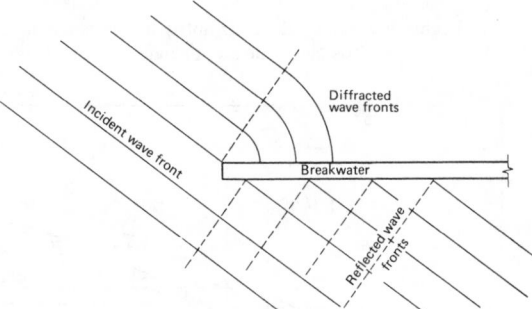

Figure 31.8 Wave difraction

31.3 Exceptional water levels

31.3.1 Long waves – surge

In addition to the tidal wave with a period of about 12.5 h and wind-generated storm waves which normally have periods in the range 4 to 15 s, sea-level is frequently disturbed by long waves of intermediate or longer periods, by solitary waves and by long-term secular changes. These disturbances of the sea surface are of importance in the design of harbour basins, from which long waves cannot be excluded and may be amplified by resonance effects, and in the determination of the crest levels of sea walls and flood defences. They can also cause sea-level to fall below predicted tide levels with effect on shipping movements and the operation of sea-water intakes. The main causes of these long period oscillations of the sea surface are as follows:

(1) Seismic activity (earthquakes).
(2) Wind shear on the water surface in shallow seas or lakes which causes not only short period waves but also an inclination of the water surface.
(3) Swell waves of slightly different periods reaching the locality at the same time from separate storms and producing 'surf beats'.
(4) Rapid changes in barometric pressure which cause the sea-level to rise or fall, 34 mb change in pressure causing 300 mm change in water level. A rapidly moving depression or cyclonic front may generate long waves as well as a general rise in mean sea-level.
(5) Ice falling from the end of a glacier.
(6) Long-term and seasonal weather changes including advance or recession of the polar ice-caps.
(7) Geological effects such as movement of the continental land masses and local settlement of coastal lands following, for instance, the pumping of an aquifer.
(8) Seasonal changes in salinity in coastal and estuarine waters.

In different parts of the world, one or other of these phenomena will generally be dominant and the others need not be considered; for instance in the North Pacific, particularly around the Japanese Islands, the main concern is the seismically generated wave known there as a tsunami which has been the subject of several papers published in the *Proceedings* of the Coastal Engineering Conferences.

There are numerous examples of areas where the effects of wind shear are important, notably the Gulf of Mexico where hurricane winds tend to pile up the shallow water of the Gulf towards the coast and blow the water out of inshore lakes and lagoons. Along most sub-tropical coasts, surges caused by tropical storms (hurricanes) are amplified when they enter coastal inlets. These areas have been the subject of intensive study and, since hurricane wind speeds and rates of travel do not vary greatly, attempts have been made to understand the physical processes involved and to predict the effects of storms of this kind.

In the North Sea, which is very shallow, wind shear again plays a dominant role particularly when combined with a fall in barometric pressure. Here prolonged northerly winds may generate a surge or surges as in 1953 when it reached a height of 2.7 m along the East Anglian coast and exceeded 3.0 m in the Delta area of The Netherlands. According to the Admiralty tide tables, depression of the sea-level or negative surges of 0.6 to 0.9 m occur several times a year in the southern North Sea; levels 2.1 m below tidal prediction were recorded at Southend in 1967.

The longer-period disturbances of significant amplitude will show up on tide gauge recordings which can be used for the prediction of such sea-level variations.

The simplest method is to abstract from the records all abnormally high and low water levels and plot them as log probability against level; the rarer values will normally plot as a straight line which can be extrapolated with confidence. This procedure is strictly correct if there is no secular trend in sea-level.

Other statistical methods which have been used are extreme value analysis of annual maximum and minimum levels,[12,13] surge residuals at each high water[14] and the joint probabilities of tide level and surge.[15] Annual maxima have also been used to deduce secular trends.[16]

The annual maximum method can be criticized for discarding much of the available data, while the use of surge residual, although probably conservative in assuming independence of surge and tide (whereas there is evidence of surge–tide interaction in some localities) is sensitive to the accuracy of the tidal predictions.

The duration of a surge at any particular locality is generally no more than a few hours and the probability that the peak of an exceptionally high surge will coincide with a very high astronomical tide is small; return periods for the highest conceivable water levels are therefore very long, and the choice of crest level for flood protection requires careful judgement in balancing the cost of the works against the risk of damage and perhaps loss of life if they are overtopped.

Well known examples of enclosed waters in which long-period wind-induced oscillations are set up are Loch Ness with a period of 33 min, Lough Neagh (period 45 min) and the Baltic (period 15 h).

Waves of appreciable height and periods in the range of 20 s to a few minutes, which cause surging in tidal dock basins, have been observed in many places and may be expected in any coastal waters open to the ocean. They are particularly prevalent in the southern hemisphere.

Analytical results which are of help in considering long-wave problems are given below.

31.3.2 Wind set-up

For a wind of constant direction and speed U the wind set-up S above still-water level can be determined from the water slope:

$$\frac{dS}{dx} = \frac{KU^2}{g(d+S)} \qquad (31.39)$$

in which the constant K depends on surface stress, which is a function of the wave state and the current structure associated with shear and roughness of the bottom. From a study of Lake Okeechobee in Florida, K has been evaluated for enclosed waters as 3.3×10^{-6}.

For an open coastline with the wind perpendicular to the shore (Figure 31.9) an approximation is obtained by assuming constant depth; then Equation (31.39) can be solved giving:

$$S = d\left[\left(\frac{2KU^2x}{gd^2}+1\right)^{\frac{1}{2}} - 1\right] \qquad (31.40)$$

In order to allow for bottom slope it is suggested that $K = 3.0 \times 10^{-6}$ be used instead of 3.3×10^{-6}.

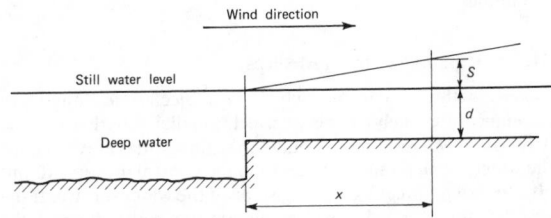

Figure 31.9 Wind set-up

31.3.3 Wave set-up

In addition to wind set-up the breaking of waves on a beach also raises the mean sea-level locally and this may be as much as 10 to 20% of the incident wave height. Consequently, tide gauges on open beaches will give misleading results while a partly sheltered beach will be subject to littoral currents flowing from the exposed to the sheltered region.

31.3.4 Resonance in harbour basins

Resonant standing wave systems can be demonstrated in a simple manner if an open-topped tank part filled with water is moved to and fro with the correct frequency; if the tank is mounted on rollers and driven by a variable-speed drive it can be excited in several different modes, the first or slowest of which occurs when the length of the tank is equal to half the wavelength of the progressive gravity wave which would occur in that depth of water. The tank then contains one 'cell' of a standing wave system with one 'node' at constant level, but with maximum horizontal velocities at its centre, and maximum vertical movement and vertical velocity at its ends.

Resonance in a rectangular basin of length a, width b and depth d will occur when the period T of the varying exciting force coincides with one of the modes of oscillation of the basin, i.e.:

$$T = \frac{2}{(gd)^{\frac{1}{2}}} \left[\left(\frac{n}{a}\right)^2 + \left(\frac{m}{b}\right)^2 \right]^{-1/2} \tag{31.41}$$

where n and m are integers representing the various modes of oscillation in directions a and b respectively. For oscillation in one direction only:

$$T_a = \frac{2a}{n(gd)^{1/2}} \tag{31.42}$$

$$T_b = \frac{2b}{m(gd)^{1/2}} \tag{31.43}$$

Periods of oscillation for circular and elliptical basins can be evolved analytically while numerical solutions may be used for irregular basins.[2]

In a harbour basin this kind of oscillation, often referred to as *ranging* or *scend*, can be initiated by long-wave activity in the approaches to the harbour. If the opening to the basin through which the forcing wave enters is at one end, oscillation in the first mode will be encouraged; if it is at the centre this mode of oscillation will be suppressed, but a second mode oscillation may occur with two nodes at one-quarter of the basin length from either end. The determination of the mode of oscillation for basins of irregular shape, with openings of appreciable width interconnected with one another and the open sea, is complex and requires the use of either a hydraulic model or a mathematical model solved on a computer or a combination of both techniques.

31.3.5 Ranging of moored ships

Vessels within a harbour which is subjected to long-wave resonance are likely to be of much smaller length than the waves, and will respond not only to the horizontal movement of the water which is much greater than the vertical movement, but also to the continually changing slope of the water surface. If the vessel is unrestrained it will accelerate down the slope of the wave in one direction, and then decelerate to rest before accelerating again in the opposite direction as the water slope changes. For a vessel which is restrained by elastic moorings, resonance will occur if $T^2 = 4\pi^2 M/k$ in which M is the 'virtual mass', i.e. the mass of the vessel plus the mass of water associated with the motion, and k the stiffness of the mooring. In practice, resonance may occur if the moorings are moderately

stiff. If the moorings are very stiff but with some slack, which is the most usual case, the motion is irregular and may become extremely violent. In either case large forces may be imposed on the moorings and lines may be broken. For further information the reader is referred to the Oil Companies International Marine Forum.[17]

31.4 Sea-bed and littoral sediments

31.4.1 Sources of material

The primary sources of sea-bed and littoral sediments are the adjacent land masses, from which the material is derived either by the normal processes of subaerial denudation and transported to the coast by streams and rivers, or from erosion of the coastline under wave attack. There is little evidence of any significant transport of material to the shore from deep water, apart from silt which finds its way into some of the estuaries in the UK,[18,19] and in some cases sand, e.g. from the Irish Sea into Liverpool Bay and Morecambe Bay. There is some evidence that shingle may be moved from off-lying shoals in shallow water on to the shore but as a source of beach-building material the quantities so moved are of little importance.

31.4.2 Modes of transport – currents and waves

Movement of sediment may be caused by currents alone, depending upon grain size and current velocity (Figure 31.10).[20] Where the currents are strong, as is frequently found close inshore at headlands and in the entrances to rivers and tidal inlets, they may have a significant effect on the sea-bed profile and configuration of the shore. Along the greater part of the coast, movement of material is initiated by wave action and the resulting direction of littoral transport is dependent upon the relative strengths and directions of the wave induced and tidal currents and the grain size of the material. Along the foreshore and in the breaker zone, wave action predominates and the material is moved inshore and offshore with changes in wave height and period. The angle between the wave crests and the shore line determines the direction of the along-shore component of the wave-induced current and the along-shore transport direction. Within and to seaward of the breaker zone, tidal and other along-shore currents have an increasing effect and as the depth increases may become dominant. It is thus possible for transport outside the breaker zone to be in a contrary direction to transport along the foreshore. The mechanics of these modes of transport are not precisely known and it is not possible from theoretical considerations to determine the quantity of material moved along the coast. The direction of movement may, however, be deduced if the dominant wave direction and littoral currents are known.

31.4.3 Wave direction

Determination of the tidal currents is dealt with on page **31**/21. The dominant wave direction cannot be determined directly other than by observation over a prolonged period for which time and the substantial funds required may not be available except for large and important projects. Forecasting or 'hindcasting' procedures from synoptic weather charts as developed by Bretschneider and others or more recently based on the JONSWAP results may be adopted to assess wave height and period or wave spectrum, and refraction analysis to indicate direction of approach to the shore. In many locations the direction of wave approach to the shore is closely correlated with the local wind directions so that simple observation com-

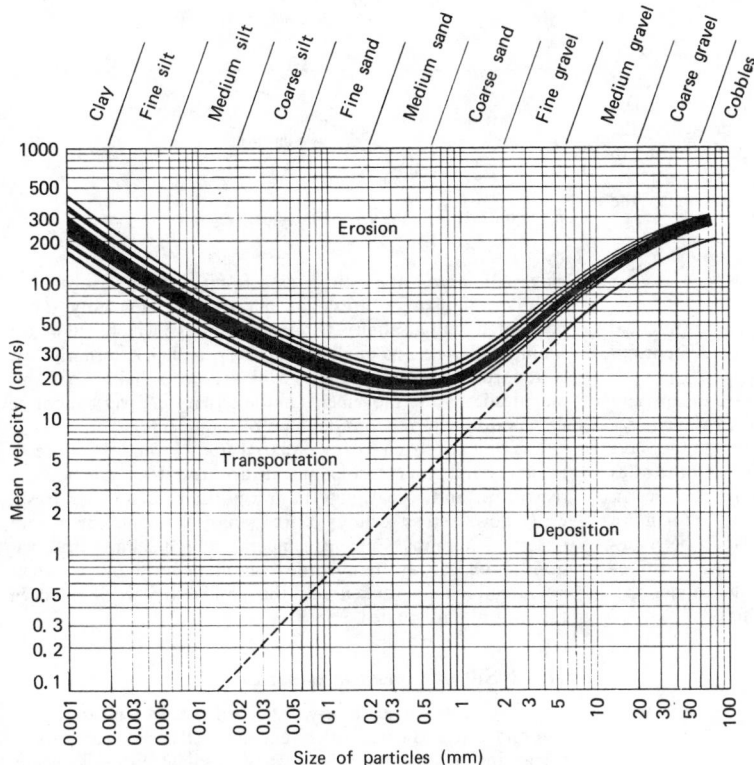

Figure 31.10 Erosion, transportation and deposition curves.
(After Kuenen (1950) *Marine geology*. Wiley, Chichester)

bined with a study of wind records, which are widely available, and a visual assessment of wave height will meet many engineering needs.

31.4.4 Effect of size of beach material

It is commonly found that the coarsest material comprising the foreshore and adjacent sea-bed is at the crest of the beach, i.e. at the limit of wave uprush, and gradually becomes finer towards and below low water mark. This sorting of beach material is generally ascribed to the varying velocities in the oscillating wave currents. Whereas in deep water the velocity of the water particles, in the direction of wave propagation below the wave crest and in the opposite direction below the trough, are equal, this no longer holds when the waves begin to feel the bottom and the wave form becomes distorted; the crests steepen and shorten in relation to the trough length and the forward velocities become higher and for a shorter duration below the crest than the reverse velocities in the trough. Thus the coarsest material which rolls along the bed or quickly falls out of suspension as the current slackens, moves only shoreward with the wave crests while the finer material returns seaward below the troughs. A balance is achieved where the beach slope is sufficiently steep for gravity to counteract the current effect, but seldom holds for long owing to changing tide levels and wave conditions. Short, steep seas tend to draw the beach down forming a steep upper slope and flat lower slope, while long-swell waves restore it to a more uniform gradient.

31.4.5 Erosion and accretion

On an open coastline the waves will approach with varying severity and from varying directions and the movement of material on the shore will consequently vary in quantity and direction. The dominant direction is that which prevails in the long term. The most reliable estimation of the amount of the littoral drift is provided by a study of past records, if available, or prolonged observation or experiment (see pages **31/23**). It can be most readily determined if there is a complete artificial barrier to the drift where it can be measured as accretion on the updrift side of the barrier or erosion on the downdrift side. In the absence of such a barrier, erosion or accretion within any specified lengths of the coastline will depend upon the balance of transport into and out of the area, and it is to be noted that a stable coastline does not indicate that there is no littoral transport or that the reversals of direction of transport with the varying direction of wave approach, and the direction and strength of wave-induced and tidal currents are equal. Apart from the movement of material into the area from the updrift of the length under study and out of the area to the downdrift beaches, account must be taken of other sources of material supply and losses from other causes. Supply to the area will be augmented by material eroded from the back-land and of this the finer material will quickly be washed out to sea and finally deposited in deep water. Attrition will reduce the size of The coarser material until it also may be lost in the same way and there may be direct loss of coarse material in the event of deep water, or deep gullies existing close to the shore. Sand may be lost by wind action resulting in the formation of dunes behind the foreshore. Exceptionally, material may find its way on to the beaches from offlying shoals and banks. Estimation of these sources of supply and loss can only be made by comparison of surveys of the shore and sea-bed.

31.4.6 Computation of littoral drift

It is not possible in the present state of knowledge to quantify all of the various coastal processes which determine the rate of littoral transport but it is commonly accepted as more or less logical that it should relate to the wave energy approaching the coast. Empirical relationships have been developed resulting in the generally accepted formula for longshore transport developed by the Coastal Engineering Research Center in the US.[5]

$$S = KH_b^2 c \sin \phi_b \cos \phi_b$$

There are, however, a number of fundamental and practical objections to this formula and its use can only provide approximate results.

Nevertheless, it is now possible to measure wave energy and direction of approach or to determine the required wave parameters by hind-casting methods and hence to compute the wave energy parallel and perpendicular to the coastline and the rates of sediment transport at any point. For the more simple cases an analytical solution of the equations is possible while for more complicated situations numerical procedures have been developed. For large and important investigations the construction and operation of the necessary large computer model may be justified in the absence of more reliable methods.

31.5 Stratification and densimetric flow

31.5.1 Saline wedge in estuaries

The difference in density of fresh and sea water has an important effect on the movement of sediments in estuaries and estuarial dock systems and on the behaviour of effluent discharges and industrial cooling water intakes and outfalls.

In the absence of turbulence, liquids of different densities do not readily mix and if brought together tend to form layers, the less dense rising to the surface and the more dense remaining below. In turbulent flow the density difference required to maintain a stable interface between the two liquids depends upon the degree of turbulence but is commonly quite small, e.g. the 2.5% difference in density between fresh water and sea water is more than sufficient to inhibit mixing, and the interface in the case of a river or stream discharging into the open sea, unless it does so via a long tidal estuary, is frequently visible as a sharp line on the surface.

If the two liquids of density ρ_1 and ρ_2, which in nature may be salt and fresh water, or silty and clear water, are initially separated by a vertical gate which is then removed, the more dense or silty water will flow in one direction as a wedge under the less dense fresh or clear water while the fresh or clear water flows in the opposite direction forming an upper wedge. The velocity of the interface at the surface and the bed (see Figure 31.11) is given by:

$$V = 2/3 \left(\frac{\rho_1 - \rho_2}{\rho_1 + \rho_2} gd \right)^{1/2} \tag{31.44}$$

For salt and fresh water and a depth $d = 12$ m, which is not unusual in estuarial dock systems, $V = 0.8$ m/s which is sufficient to erode and transport fine bed material. Examples of this phenomenon leading to a siltation problem in impounded docks are found on the Thames and the Mersey.

In estuaries and tidal rivers the fresh water flows over the sea water, the interface moving up- and down-river with the rising and falling tide. The interface will generally be found to lie

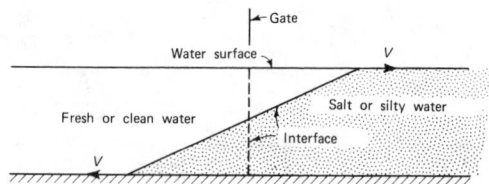

Figure 31.11 Saline wedge in estuaries

diagonally across the estuary as a result of the Coriolis force with partial mixing of fresh and salt water in its vicinity; it can be located with a salinity–temperature bridge which will detect salinity differences as small as 0.2 parts per 1000. Alternatively it is reported that dye patches will follow isohalines of this magnitude. Water movements in the estuary or river cannot be understood without consideration of these factors.

The first approach to any estuarial study should therefore be to survey and plot the isohalines at various states of the tide and river flow, and to calculate the proportional sea-water–fresh-water flows across sections of the estuary for comparison with observed dilutions. Current, temperature and salinity observations should always be recorded simultaneously and no other investigations undertaken until this preliminary work has been carried out and studied.

31.5.2 Silt movement in estuaries

Fine sediments carried by rivers toward the sea tend to flocculate and settle out when they encounter the salt water to be carried inshore again with the advancing saline wedge. Typically the inflow of sea water in a tidal estuary may be 10 to 100 times the fresh-water flow and the fine sediments are trapped; their distribution in the estuary is then a guide to the haline circulations which need to be appreciated in selecting spoil grounds for disposal of dredgings, sewage sludge and other waste products if the return of the unwanted polluting material is to be avoided. They will also indicate where dredging is likely to be effective and where siltation rates are so high as to make dredging uneconomic.

31.5.3 Effluent outfalls

Studies and experimental work by Abraham[21] have shown that, for sewage discharged into the North Sea off The Netherlands coast a dilution of 50 times, which reduces the density difference to 0.05%, is sufficient to avoid a stable sewage 'slick', but this result does not apply to discharge into calm, shallow and stratified water. Dilutions of this order can often be achieved, however, within the buoyant plume which rises from an effluent outfall discharging close to the sea-bed in moderate depth, and can be calculated from data published by various workers.[21-23] The dilution depends almost entirely on the ratio of the diameter z of the discharge port to the water depth d, from approximately 10 for $z/d = 10$ to approximately 150 for $z/d = 100$. The initial jet angle and velocity of efflux v are not significant except for very small values of z/d. It should be noted that the densimetric Froude number given by $v [gz(\rho_0 - \rho_s)/(\rho_s)]^{-1/2}$ in which ρ_0 and ρ_s are respectively the density of the receiving body of water and effluent, cannot in practice be less than 1; if the selected area produces a value less than 1, then the port will discharge only 'part full' as an inverted weir, or in the case of a multiport system the flow will be concentrated in only a few of the ports.

A current flowing across the point of discharge will lengthen the plume with a corresponding increase in dilution. Field studies by Agg and Wakeford[24] have shown that a cross-current of velocity u increases the dilution by $10 \times u/v$ but this result should not be applied outside the range of the experiments which was small.

31.5.4 Density and turbidity currents

If silt-laden water, brine or some other fluid having a density greater than sea-water ($\rho_0 = 1.025$ g/cm³) is discharged on to a sloping sea-bed it will flow down the slope as a discrete and coherent stream until the density difference is eliminated either by dilution due to turbulent mixing with the ambient fluid or the deposition of the silt load. Streams of this kind can also occur naturally when, for example, a disturbance on a sloping sea-bed puts fine material into suspension and initiates a flow down the slope which scours further material and gains increasing momentum like an avalanche. These density or turbidity currents are capable of carrying bed material, or solid waste matter dumped on a sloping sea-bed, long distances into deep water.

Calculation of the velocity of a turbidity current requires an assumption as to the way the current will spread laterally unless it is confined by the sea-bed topography and it is necessary also to establish that the current will be stable, which will only be the case on moderate slopes.

The results of the experimental work by Tesaker[25] relate velocity v to a number θ developed by Keulegan for density currents, i.e.:

$$\theta = \frac{1}{v} \left(\frac{vg \ (\rho_s - \rho_0)}{\rho_s} \right)^{-1/3}$$

where ρ_0 is the density of ambient fluid and ρ_s the density and v are kinematic viscosity of the fluid forming the current.

θ is a function of slope only and is found by Tesaker[25] to be 0.02 for a slope of 1 in 10 and 0.027 for a slope of 1 in 20. For a theoretical treatment of the subject the reader is referred to Elliston and Turner's paper.[26]

31.6 Wave and current forces

31.6.1 Forces on a circular cylinder or pile

The simplest approach to the assessment of wave and current forces on piled and braced structures or pipelines in the sea is by the summation of 'drag' and 'inertia' components due respectively to the velocities and accelerations of the water particles in the motion. The horizontal force on an elementary length ds of a cylinder or pipe of diameter D is then:

$$dF = \left(\tfrac{1}{2} C_d \rho D u \bar{u} + C_m \rho \frac{\pi D^2}{4} \frac{du}{dt} \right) ds \qquad (31.45)$$

in which C_d is the drag coefficient, C_m the inertia or mass coefficient, and $\bar{u}$ is the absolute value of u.

Using the small-amplitude theory (pages **31**/4 and **31**/5) the required values of u and du/dt are obtained from Equation (31.7). (Similarly, vertical forces can be obtained using v and dv/dt from Equation (31.8).

It is convenient to transfer the origin of the coordinates x and t to the wave crest and to assume that the elementary length ds of the cylinder is located at $x = 0$ and elevation s above the sea-bed ($d + y = s$). By substituting from Equations (31.1) and (31.2) the equations for velocity and acceleration can be reduced to:

$$u = \frac{\pi H}{T} \frac{\cosh (2\pi s/\lambda)}{\sinh (2\pi d/\lambda)} \cos \left(\frac{2\pi t}{T} \right) \qquad (31.46)$$

and

$$\frac{du}{dt} = \frac{2\pi^2 H}{T^2} \frac{\cosh (2\pi s/\lambda)}{\sinh (2\pi d/\lambda)} \sin \left(\frac{2\pi t}{T} \right) \qquad (31.47)$$

In the case of breaking waves, it is suggested that the solitary wave[27-29] theory should be used with graphical and tabulated values of u and du/dt as given by Munk.[30]

For steady flow the theoretical value of $C_m = 2.0$ while C_d is related to Reynolds number (uD/kinematic viscosity). For orbital flow in waves the coefficients have been determined experimentally on model and full-scale piles.[31,32] A mean value of 2.5 is reported for C_m but it is to be noted that 20% of the results were above 3.5 and 10% above 4.0. Attempts to relate C_d to Reynolds number were unsatisfactory (see Figure 31.12). It has, however, been demonstrated that more consistent results can be obtained if the data are analysed using the higher-order wave theory; in that case values of $C_m = 2.0$ and $C_d = 0.7$ can be justified. For an approximate assessment of forces using the linear wave theory the values $C_m = 2.5$ and C_d from an envelope of the experimental data in Figure 31.12 are recommended.

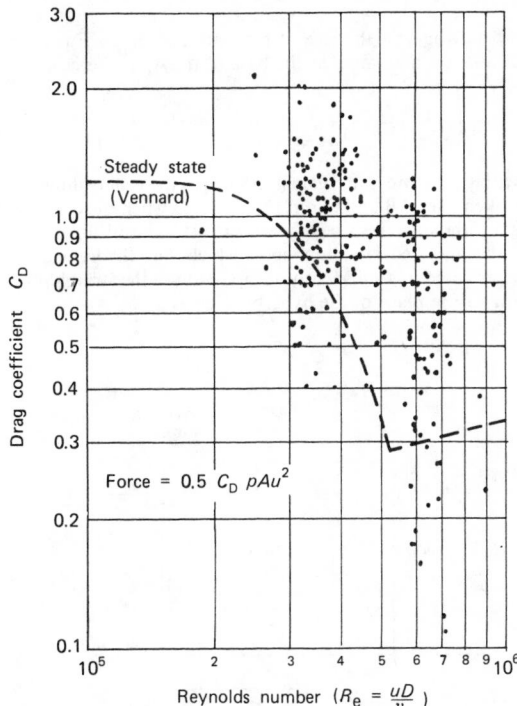

Figure 31.12 Coefficient of drag C_d as a function of Reynolds number R_e for waves higher than 3m. A is the projected area, u the particle velocity, D the pile diameter, uD the kinematic viscosity and ρ the density of water. (After Wiegel, Beebe and Moon (1957) 'Ocean wave forces on circular cylindrical piles.' *J. Hydraulics Div. Am. Soc. Civ. Engrs*, **83**, No. HY2)

This quasi-static analysis is not adequate if the natural period of oscillation of the structure or individual elements is of the same order as the frequency of vortex shedding or wave action. If the ratio $V:ND$ is greater than 1, V being incident flow velocity, N the natural frequency of structure or element and D typical cross-sectional dimension (e.g. diameter of circular member) the reader should consult Hallam.[33]

31.6.2 Forces on sea walls and breakwaters

In an analysis of the forces exerted by waves on a structure it is first necessary to decide if the waves will break against, or immediately in front of, the structure (see page **31**/8). Forces due to nonbreaking waves will be essentially hydrostatic, whereas breaking waves exert additional pressures due to the dynamic effects of turbulent water motion and entrapped air pockets; breakwaters should not be designed for unbroken waves unless the circumstances are exceptional.

31.6.2.1 Nonbreaking waves – Sainflou method

With complete reflection from a vertical face a 'standing wave' or 'clapotis' will be set up (see page **31**/6, $K_r = 1$). The wave height at the wall becomes twice the incident wave height H, the mean level or orbit centre being a height h_0 above still water level where:

$$h_0 = \frac{\pi H^2}{\lambda} \frac{1}{\tanh (2\pi d/\lambda)} \tag{31.48}$$

The pressure against the face of the wall is given by Equation (31.23) and is a maximum at the base of the wall where:

$$p_{max} = \rho g \left[\frac{\pm H}{\cosh (2\pi d/\lambda)} + d \right] \tag{31.49}$$

The variation in pressure against the face of the wall is shown by the broken lines AB and GF in Figure 31.13 when the crest and trough of the wave respectively are against the wall. Bearing in mind the idealized situation upon which this theoretical approach is based it is sufficiently accurate to use the straight lines AB and GF in place of the hyperbolic curves.

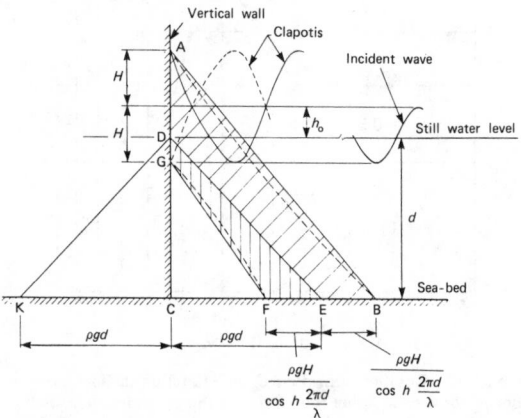

Figure 31.13 Sanflou diagram

If there is still water behind the wall there will be an outward pressure represented by the triangle DKC and the resultant force on the wall is then given by the area ABED in the landward direction when the wave crest is at the wall and by the area DEFG in the seaward direction when the wave trough is at the wall.

If the crest of the wall is less than $(H + h_0)$ above still-water level the usual procedure is to assume the full clapotis but to omit the part of area ABED which lies above the crest level. Clearly, this will give a conservative estimate of the total force.

31.6.2.2 Breaking waves

Evidence from various sources, including analysis of existing structures, suggests that the forces due to breaking waves on a plane vertical surface will be in the range 100 to 1000 kN/m² but no satisfactory analytical methods have been devised whereby they can be predicted with any great confidence.

The following expression for the maximum dynamic pressure P_{dm}, based on experimental work by Bagnold,[34] has been proposed for the case in which the sea-bed shelves steeply (in excess of 1:20) from deep water up to the face of a vertical wall:

$$P_{dm} = \pi \rho g H_b \frac{d}{\lambda_0} \left(1 + \frac{d}{d_0} \right) \tag{31.50}$$

in which H_b is the height of the breaking wave, d the depth at the wall, d_0 the depth away from the wall, and λ_0 the wavelength in depth d_0.

It is suggested that the maximum pressure P_{dm} at still-water level falls off parabolically to zero at the crest and trough of the wave and is added to the hydrostatic pressure as indicated in Figure 31.14.

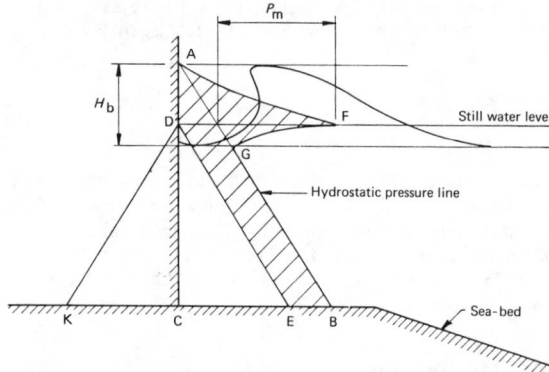

Figure 31.14 Breaking wave diagram. (After Bagnold (1939) 'Interim report on wave pressure research.' *J. Instn Civ. Engrs,* **12**)

With still-water pressure behind the structure the resultant pressure is given by the shaded area AFGBED.

An alternative approach is to assume that, immediately before breaking, the water mass in the wave moves forward with the velocity of wave propagation c. Using the approximate relationship for shallow water $c = (gd)^{1/2}$ (Equation (31.3)), the dynamic pressure $P_{dm} = \rho c^2 = \rho g d$ which is added to the hydrostatic pressure as shown in Figure 31.15.

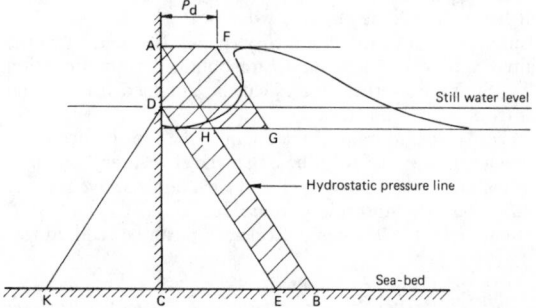

Figure 31.15 Breaking wave diagram – Momentum method

With still water behind the structure the resultant pressure is given by the shaded area AFGHBED.

Neither of these methods on its own can be relied upon and in any particular case the results should be checked by comparison with the performance of existing structures and with the results of relevant experimental work available from hydraulic institutes specializing in this work. Large and important projects should be model-tested in random waves with a statistical analysis of the results to determine the forces and moments resulting from rare events as described by Terrett, Ganly and Stubbs.[35]

31.6.2.3 Sloping walls and breakwaters

The analysis given in the preceding paragraphs can be applied to walls with sloping faces (Figure 31.16).

According to Miche[36] the maximum steepness of waves which are completely reflected is given by:

$$\frac{H}{\lambda} = \frac{\cos^2 \beta}{\pi} \left(1 - \frac{2\beta}{\pi}\right)^{1/2}$$

For breaking waves the dynamic pressure is reduced by $\cos^2 \beta$.

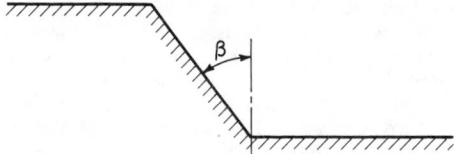

Figure 31.16 Sloping wall

Wave forces on sloping rubble-mound breakwaters or sea walls are the most difficult to treat analytically, for the hydrodynamics are almost hopelessly complicated by the highly variable characteristics of such structures. Therefore the calculation of the required weight of armour unit has been invariably based on model testing. The most-used formula is that of Irribarren as later modified by Hudson,[37] namely:

$$W_r = \frac{\gamma_r H^3}{K_D (\gamma_r/\gamma_w - 1)^3 \cot \alpha}$$

in which W_r is the weight of armour units, γ_r the specific weight of armour units, γ_w the specific weight of water, H the wave height, α the angle of breakwater slope to the horizontal, and K_D an empirical coefficient (see section 31.12.2).

It should be noted that this is only a statement that the rock size (linear dimension of cubical rock) is proportional to wave height for a given slope and density. A similar design approach is used for 'riprap' slopes[38] and is satisfactory for rock armoured slopes not steeper than about 1:2. In recent years it has, however, been adopted for steeper slopes armoured with very large shaped precast concrete blocks, and has led to some spectacular failures.

31.7 Scaling laws and models

31.7.1 General

Although an understanding of the physical processes involved is essential to the solution of most problems in coastal engineering, few are amenable to mathematical analysis alone, and it is therefore necessary to rely on past experience aided where

appropriate by physical or mathematical model studies. In some cases, e.g. studies of wave diffraction and refraction, model study results will give an accurate interpretation of the prototype behaviour, but in others, such as studies of siltation in large estuaries, both the operation of the model and the interpretation of the results will require a measure of judgement based on experience and observation.

31.7.2 Scaling and similarity

According to Buckingham's Π theorem, which provides the most general and elegant approach to the problem of similarity and scaling, if a_0 is a function of n independent dimensional parameters $a_1, a_2, a_3 \ldots a_n$, then the variables can be combined so that the dependent dimensionless group A_0 is a function of $n - 3$ independent dimensionless groups $A_1, A_2, A_3 \ldots A_{n-3}$

where

$$A_1 = (a_1^{x_1} \quad a_2^{y_1} \quad a_3^{z_1}) a_4$$

$$A_2 = (a_1^{x_2} \quad a_2^{y_2} \quad a_3^{z_2}) a_5$$

$$A_0 = (a_1^{x_0} \quad a_2^{y_0} \quad a_3^{z_0}) a_0$$

the indices being shown by trial and error to make the relationships dimensionally correct.

This approach yields directly, and in only a few lines, to many well-known results, e.g. the drag coefficient of a smooth object in a fluid stream which is a function of Reynolds number alone.

In general, apart from geometrical similarity, it is necessary for the following parameters to be effectively the same:

Froude number	$= V/(Lg)^{1/2}$
Reynolds number	$= VL/v$
Weber number	$= V \Big/ \left(\dfrac{\phi}{\rho L}\right)^{1/2}$
Mach number	$= V \Big/ \left(\dfrac{E}{\rho}\right)^{1/2}$
Euler number	$= V \Big/ \left(\dfrac{2\Delta p}{\rho}\right)^{1/2}$

in which V is the velocity in metres per second, L the length in metres, v the kinematic viscosity in square metres per second, ϕ the surface tension in newtons per metre, E the elastic modulus in newtons per square metre, ρ the density in kilograms per cubic metre, and Δp the pressure difference in newtons per square metre.

In many cases, similarity in these parameters is only possible if the model is the same size as the prototype or, using a special fluid, nearly the same size. The greatest difficulty occurs with the modelling of the sediments which form the mobile beds of seas and estuaries; the particle size cannot be reduced to scale and in other situations also, similarity cannot always be maintained; models of tidal estuaries and rivers, for example, are almost invariably distorted. Nevertheless, it has been shown repeatedly that satisfactory reproduction of tidal flows, currents and salinity differences can be obtained in such models.

It can be shown by mathematical analysis that a distorted model may be able to reproduce prototype behaviour, or it can be justified in more general terms as follows. If a vertical sided basin is connected to a tidal sea through a long channel of appreciable hydraulic resistance, the tidal response of the basin (provided it is short compared with the length of the tidal wave) depends only on the area of the basin and the hydraulic resistance of the channel, and a deep narrow channel can be made with the same hydraulic properties as a wide shallow one.

31.7.3 Tidal models

Models of estuaries or other large tidal inlets have frequently been built, and they commonly have vertical scales of the order of 1:100 and horizontal scales of the order of 1:500. When the roughness of the bed has been adjusted with reference to prototype data, they reproduce accurately the tidal levels and currents in the main channels, but not currents across mudflats or other areas which are uncovered at low water. Such models may therefore be used with confidence to predict the effect on currents and tide or surge levels of, for instance, dredging, training works or a barrage in the upper estuary. Salt-water intrusion and the related phenomena of dispersion and flushing of wastes (either effluents or warm water from power station discharges) will only be correctly reproduced if the scale is undistorted and the Coriolis effect is negligible but, with careful calibration, correspondence in a particular reach is sufficiently good to justify the use of the model in the case of important projects. The movement and deposition of silt is the least satisfactory aspect of the tidal model, and requires the most elaborate and careful calibration and very extensive prototype data. Since, however, the cost of dredging the approaches to many ports is very large there have been frequent attempts to build and operate models of the type, and from time to time exceptional claims have been made as to the ability of particular mobile bed models to predict changes in the prototype; all such claims should be regarded with some reserve and the report on a model study of this nature should not be expected to be more than a qualified subjective opinion as to whether proposed changes by construction works or in dredging procedures would lead to accretion or erosion.

An estuary model investigation together with the associated field studies will usually require the employment of teams of four or five engineers and assistants in the laboratory and the field for at least a year and the cost will be substantial.

31.7.4 Harbour models

Provided that the model is large enough for the waves to be unaffected by surface tension forces and bed friction, wave disturbance models of harbours with unbroken waves obey Froude's law precisely. Although the procedure adopted for the operation of a particular model (e.g. whether the input should be an irregular wave train or a series of single period waves, or whether it is sufficient to assume that the response of the harbour will be linear so that only one wave height at sea for each wave period need be considered), is a matter on which opinions differ but the construction and testing of a model in some form is usually justified before embarking on major harbour works. The linear scale adopted is commonly of the order of from 1:50 to 1:100, the time-scale being the square root of the linear scale. Judgement and experience are required in assessing the acceptable wave height at any particular berth within the harbour because of the lack of basic data and difficulty in determining the response of moored vessels to waves of specified periods and height.

The response of the harbour to long waves, i.e. those which may cause the harbour basins to resonate in the first or second modes, is difficult to model because the waves are reflected and rereflected from the boundaries of the basin which contains the model. These extraneous reflections can be suppressed by the provision of wide, flat beaches (their width being large compared with the wavelength) along the boundaries of the basin but this results in either an impractically large basin or an unacceptably small model. For this type of study it is therefore necessary to have recourse to a mathematical model to solve the whole problem or to specify the wave characteristics at the harbour entrance, and then to reproduce them in the model.

31.7.5 Forces on structures

Forces on structures, i.e. breakwaters, piles, pipelines, etc. and on moored ships also obey Froude's law provided that the scale is large enough to ensure fully developed turbulence and that air entrainment is not important. The first proviso is easily satisfied as model scales of about 1:20 are usually not inconvenient, while the second applies only to structures in breaking waves. With these qualifications it is generally accepted that forces measured on model structures may be scaled up and used for design purposes. For rigid structures the forces may be measured by pressure cells, dynamometers or strain gauges recording electronically on to paper or magnetic tape. For rock-mound breakwaters and similar structures for which the stability of the individual armour units is the main consideration, observation of the percentage of the units which rock or are displaced during each test run provides the basis for design.

In the past it was customary to use only regular trains of waves in the laboratory, testing with a variety of heights and periods, and from these to produce the necessary design figures. While the procedure is probably sound for unbroken waves, it is necessary to use irregular waves if the waves break in front of the structure, which may be due partly to the reflection of the previous wave from the structure itself. The irregular wave train may be either a synthetic combination of a number of sine waves or a 'facsimile copy' of real waves recorded at sea. Since the magnitude of the force on a structure caused by a breaking wave depends very greatly on the form of the wave and exactly how it breaks, and since even regular waves produce a very irregular pattern of forces, the largest forces from a train of irregular waves will almost certainly not be caused by the highest wave and a statistical approach to the analysis of the results becomes essential.

31.7.6 Overtopping

The crest level required for a breakwater to provide adequate protection, or for a sea wall to be safe from scour on its landward side, requires a knowledge of tide and surge levels and wave heights from which it can be assessed by experience or can be estimated from a model study. The amount of water thrown over the top of a sea wall or the size of the regenerated wave on the harbour side of a breakwater can be measured in a wind–wave flume but, unless the crest level is abnormally low relative to the sea-level, air entrainment and water droplet size will be of great importance. These are not susceptible to scaling; the size of the drop of water blown from the surface will be the same in both model and prototype but thereafter the prototype drop will be exposed to a stronger wind and be broken up into fine spray whereas the model drop will fall relatively quickly on to the rear slope of the wall or into the harbour. Overtopping tests should therefore only be used to establish the relative merits of various types of structure and their crest levels and one of known performance should be included as a basis of comparison.

31.7.7 Digital numerical models

With the exception of breaking waves and fluid–solid interaction at an erodible boundary, equations of motion can be written for all free surface engineering hydraulic phenomena. For example, for two dimensions in plan the equations are:

(1) Momentum equations:

$$\frac{\delta u}{\delta t} + \underbrace{\left(u\frac{\delta u}{\delta x} + v\frac{\delta u}{\delta y} \right)}_{\text{convective terms}} + g\frac{\delta \eta}{\delta x} - fv + \tau u = \psi_x \qquad (31.51)$$

$$\frac{\delta v}{\delta t} + \left(u\frac{\delta v}{\delta x} + v\frac{\delta v}{\delta y} \right) + g\frac{\delta \eta}{\delta y} + fu + \tau v = \psi_y \qquad (31.52)$$

(2) Continuity equation:

$$\frac{\delta \eta}{\delta t} + \frac{\delta(hu)}{\delta x} + \frac{\delta(hv)}{\delta y} = 0 \qquad (31.53)$$

where u, v is velocity in x and y directions; η is water surface elevation above MSL; $h = d + \eta$ is total depth of water; f is Coriolis parameter $2\Omega \sin \theta$; Ω is angular velocity of Earth; θ is latitude; $\tau = g\frac{(u^2 + v^2)^{1/2}}{C^2 h}$; g is gravitational acceleration; C is Chezy coefficient; and ψ_x, ψ_y are wind stress and atmospheric pressure terms in x, y directions.

These equations can be solved using a numerical implicit finite difference scheme on a large computer provided the fluid and solid boundaries can be defined. Turbulent diffusion in the x and y directions and friction at the bed should be adjusted separately until the velocities are correctly reproduced. The computer should be programmed to plot velocity vectors for each state of the tide so that the flow can be visualized. The model can then be used with considerable confidence to predict the effect on the flow of changes to the boundaries.

If the flow through the fluid boundaries is known, the model can be elaborated to reproduce the dispersion and movement of suspended solids, dissolved solids, heat, pollutants and dissolved oxygen.

A two-dimensional model will only be correct if the velocity through each boundary of each element follows the logarithmic velocity distribution law and can be represented by a single vector: if there are significant vertical salinity or temperature gradients this will not be the case and a multilayer model is required. In a strongly stratified estuary a two- or three-layer model will be an adequate representation of reality and is practicable. An estuary with significant vertical density gradients can only be properly represented by a three-dimensional model and the computational difficulties are formidable.

The successful application of a numerical flow model to an engineering problem requires:

(1) Knowledge of the nature of the water body so that a one-, two- or three-dimensional model can be selected as appropriate.
(2) An understanding of the computational scheme in order to select grid sizes and time steps and avoid spurious numerical diffusion and instability.
(3) Data with which to define the boundaries and calibrate the model.
(4) A precise definition of the objectives.

31.7.8 Littoral processes

The part played by wave action in the movement of sediments can only be modelled if it is secondary to the tidal currents, i.e. only in the circumstances in which waves initiate movement by putting the sediment into suspension but do not control the direction in which it is moved.

Small-scale experiments with waves and beaches have been of value in providing an understanding of littoral processes but these are not 'models' of real beaches and cannot be used for the solution of particular problems.

31.8 Surveys and data collection

31.8.1 Sources of information

Hydrographic surveys are costly and every effort should therefore be made to obtain the results of previous surveys and measurements which may be relevant. In the UK, information may be obtained from:

(1) Published Admiralty charts which also contain information on sea-bed materials and tidal currents, and the unpublished originals which contain more information than the published charts.
(2) Geological and Ordnance Survey maps.
(3) Port authorities' charts and records.
(4) Authorities or engineers responsible for the construction of works in the area.
(5) Research laboratories concerned with coastal processes, e.g. the National Institute of Oceanographic Sciences, the Hydraulics Research Station, the Fisheries Laboratory and the Water Pollution Research Laboratory.

31.8.2 General

The problems associated with the collection of information for the design of maritime works increases with distance from high-water mark and the degree of exposure of the coastline. Between tides, conventional land surveying techniques are frequently adequate and can be extended a short distance beyond low-water mark with the assistance of a diver or chainman in a wading suit, using a marked line of 'corlene', or other fibre which floats and does not shrink when wet, for measuring distances, and a sounding chain or marked pole.

Conventional subsoil investigation procedures are also appropriate for work on the foreshore or in shallow water. A lightweight boring rig can be set up quickly and a useful amount of work done in the few hours the foreshore is uncovered, and the working time can be extended, if the coastline is not too exposed, by using a small scaffold working platform.

Beyond low-water mark, floating craft or other specialized equipment are necessary to form a working platform. In all of this work position fixing and level are most important; accurate work requires practice and trained surveyors should be employed where possible.

31.8.3 Position fixing

Methods currently in use for fixing the positions of a survey vessel, drill barge or diving boat are:

(1) Resection of two angles measured by sextant to co-ordinated stations on shore, the position being plotted by circle diagram, station pointer or computer.
(2) Intersection by two theodolite observers on shore.
(3) By angle and range from a station on shore.
(4) As (3), but operating in a dynamic mode, and continuously transmitting the information to a computer, logger, display and plotter on the vessel.

(5) A laser beam set up on shore and a laser range finder on the vessel.
(6) High-frequency radio ranging with a master unit on board and two co-ordinated slave stations ashore, the data being continuously logged by computer and the boat's position displayed on VDUs to the surveyor and the helmsman.
(7) Sonar beacons on the sea-bed.
(8) Satellite.

The most usual method now employed is radio ranging by one or other of a number of proprietary systems. These have a range accuracy of about ± 1 m up to about 60 km and accuracy in position dependent on the angle to the shore stations. The data may be logged by hand, but it is preferable to use a computer which will reject erratic readings, average the remainder and compute the vessel's position in rectangular coordinates. This is then logged on magnetic disc with paper back-up, and the vessel's track plotted by the machine; for accurate work account must be taken of the distance travelled by the vessel in the time taken to reduce and log the data.

31.8.4 Bathymetry

31.8.4.1 Sea surface level

Sea-bed and sub-bed levels are measured from a survey boat with reference to sea-level which must be recorded and related to a fixed datum. This may be done by any of the following means:

(1) Observation of a graduated staff.
(2) A pressure gauge attached to a bellows or bubbler unit fixed below the surface.
(3) A float in a stilling well.

Float-operated autographic gauges are best and are available with the telemetry required for the control of dredgers which is useful for other survey work. Pressure gauges should not be used if the water density can change from salt to fresh as in an estuary.

31.8.4.2 The level of the sea-bed

Isolated soundings can readily be taken with a lead and chain but for a survey of any extent an echo sounder should be used. This instrument consists of a 'transducer' which generates acoustic pulses, focuses them as a beam downwards to the sea-bed and picks up the reflected waves. The time taken for the sound wave to travel from the surface to the sea-bed is recorded and gives a measure of the water depth. The frequency of the pulse and the spread of the beam can range from 20 kHz with a 7 to 9° beam to 700 kHz with a 1° beam. The latter gives the highest resolution and may be used to resolve small features on the sea-bed, but the instrument is large and expensive and has to be mounted on gimbals or used only in calm weather. Instruments called 'trench profilers' employing a very high-frequency narrow beam are used to plot cross-sections directly from a survey boat; they can pick up features as small as 200 mm and are useful for 'inspecting' such items as armour blocks at the toe of a breakwater, a pipe in a trench, or damage to the face of a quay wall. It has a range of only 40 m and must be mounted on a very stable support.
Standard survey instruments, which record in analogue form on dry paper and digitally on magnetic disc for automatic plotting by computer can operate at 20 kHz, 200 kHz or both frequencies together; to obtain a realistic profile of an irregular rocky bottom requires the higher frequency and a 3° beam. As the speed of sound propagation through water varies with the water density, frequent calibration by reading the depth to a reflector plate suspended below the transducer is essential.

31.8.4.3 Horizons recorded

Sound waves are reflected at any abrupt change in density and echo sounders may therefore record any of the following:

(1) Density gradients in a stratified estuary.
(2) Sub-surface floating rubbish and fish.
(3) The surface of fluid mud layers which may have densities as low as 1.05 and no detectable shear strength, but need to be recognized as they do not interfere with the movement of vessels.
(4) The surface of sediments deposited since the last dredging operation.
(5) The surface of marine deposits disturbed by dredging.
(6) The surface of undisturbed marine deposits.

An echo sounder may record any one or more of the sea-bed horizons if they are present, and they will all be visible on a good dual-frequency record from which much can be learned by careful scrutiny.
The depth to a soft bottom measured by sounding pole, lead line or diver will extend a little below the undisturbed surface and is unlikely to coincide with any of the horizons recorded by an echo sounder.

31.8.4.4 Accuracy of soundings

It is normally a reasonable assumption that the accuracy of a single sounding observation will be ± 150 mm of the true value, but may be reduced to perhaps ± 100 mm if the best equipment has been used with great care and sea conditions are favourable. By averaging a large number of readings, as may be required for the calculation of dredging quantities or siltation studies, the probable error is much reduced. If successive surveys are to be compared for any purpose the same instruments and procedures should be used for each one.
Accurate determination of the level of hard submarine features are best made by direct observation from the shore on to a staff or by a diver with a sensitive pressure cell.

31.8.5 Nature of the sea-bed

A general survey of sea-bed features can be made with 'transit' or 'side-scan' sonar equipment which transmits a fan-shaped acoustic beam very narrow in the horizontal plane. The equipment is towed behind or alongside the survey vessel and scans an area of the sea-bed normally up to 500 m wide to one side of the vessel's track so that large areas can be covered quickly. The record is made to scale on a wide strip of wet paper. A uniform flat surface will produce little return except directly under the boat, but any upstanding feature, (e.g. rock outcrop, wreck, pile stump, etc.) will be recorded as a dark mark on the paper with a white shadow behind. The equipment gives good resolution and a clear record of any upstanding features larger than about 300 mm across.
It is possible to discriminate between mud, flat sand, rippled sand, sand waves, gravel, exposed clays and rock by the 'texture' of the record. When these regions have been mapped they may be further investigated by sampling, which can be done from the surface using a grab, gravity corer or vibro corer, depending upon the sea-bed materials. The latter is of particular value, as an economical alternative to boring, in extracting samples to a depth of 3 to 6 m below the sea-bed for dredging investigations and in prospecting for minerals.

Divers may then be employed to collect samples of exposed rock or hard clay and to investigate anomalies. The diving boat should carry survey equipment so that the observations can be co-ordinated with the same accuracy as the rest of the survey.

31.8.6 Nature of material below the sea-bed

Offshore borings are expensive and may take up a great deal of time. It is therefore important to keep their number to a minimum and to ensure that they are located so as to give the maximum amount of data for the expenditure involved. All possible alternative sources of information should be explored before the borings are undertaken, e.g. by examination of exposures of the strata on shore, from onshore borings and by superficial inspection and sampling of the sea-bed (see page **31**/20). If further information is required, a seismic survey can give a great deal of information in a short time. The equipment used is similar in principle to an echo sounder but the pulse energy is very much greater and high-gain amplifiers are required to process the reflected signals. The sound pulses from the transducer are reflected from the sea-bed as with an echo sounder, but also penetrate the bed below and are reflected from any interfaces between materials through which the velocity of sound differs. With high-resolution 'boomer'-type equipment it is possible to distinguish base of marine mud, base of alluvium, surface of weathered rock and surface of sound rock. With 'sparker'-type equipment which provides more energy but less resolution, the interface between such strata as sedimentary and igneous rocks, and sedimentary rocks and coal seams will be detected with faults and other important geological features to a depth of 100 to 150 m.

It should be noted that the equipment measures (and plots) the travel time of the pulse; the determination of depth to the reflecting horizons requires a knowledge of the velocity of sound through the materials. For a precise calculation of depth, borehole cores are required for direct measurement of the velocity but, for a preliminary interpretation, figures for comparable strata may be employed as, for example, in Table 31.5.

Table 31.5 Wave velocity for comparable strata

	Wave velocity (m/s)
Sand	300–800
Glacial moraine	1500–2100
Chalk	2000–2600
Lower Greensand	1600–2000
Gault (Clay)	1800–2200
Jurassic Limestone	1800–3600
Carboniferous	2400–3600
Ordovician (N. Wales)	2400–3600
Cambrian (N. Wales)	4000–6500
Limestones	3500–6500
Granites	4600–7000

Unless the strata are dipping steeply the interpretation of the record at more than twice the water depth below the sea surface is difficult owing to multiple sound reflections from the strata interfaces and the surface.

When the information obtained above has been considered it is usual to prepare tentative designs of the proposed works from which the most suitable positions and depths of the offshore borings, having regard to construction requirements and the seismic interpretation, can be decided.

Three methods for setting up the boring are available:

(1) Above the sea surface on a piled platform or jack-up barge.
(2) On a barge or selfpropelled craft.
(3) On the sea-bed.

If methods (1) or (2) are used the drilling procedure and equipment will be the same as would be employed on land. In reasonably sheltered water, shell and auger boring to moderate depths can be carried out rapidly at reasonable cost from a platform erected over the side of a small coaster or large fishing vessel. As the exposure and/or the depth of the hole increases, and for core boring, a jack-up barge or a boring vessel with a central well is required, or in depths not exceeding about 50 m diver-operated rigs set up on the sea-bed can be used to drill up to about 300 m.

If cores are not recovered from a borehole, identification of the strata and information on the permeability and density of the rock can be obtained from *in situ* measurement of the electrical resistivity and the scatter and absorption of gamma radiation by instruments lowered down the hole or acoustic tomography may be used in which an airgun is fired in one borehole and the time for the sound waves to reach an array of geophones in another is recorded.

31.8.7 Fluid mud layer

Where the bottom of a navigation channel or dock consists of fluid mud, port authorities now define the 'navigation depth' as the depth to the horizon at which a particular density, in the range 1.12 and 1.20, is reached. The chosen horizon will not, in general, be detectable on an echo sounder record but instruments are being developed to measure the fluid density and other fluctuating properties in these layers such as acoustic velocity and gamma-ray back scatter. No commercial instrument has emerged as yet but at least two port authorities are using prototype equipment which is towed by a survey vessel.

31.8.8 Current measurement

The measurement of currents at a fixed point in the sea is usually carried out with a current meter or by repeatedly timing floats over short distances while the movement and mixing of masses of water over large areas is carried out by tracking floats or mapping in plan and depth the dispersion of indicators such as radioactive, bacterial or fluorescent tracers.

31.8.8.1 Floats and tracers

Floats less than 30 cm deep are susceptible to wind and wave action and their paths will not be representative of the general current movement; floats made from poles ballasted at their lower end, so as to float vertically, are better but surface effects will still be significant and unpredictable. Deep pole floats or 'logslips' (see the Admiralty *Manual*[39]) are a useful guide, however, to the effect of the currents on the movement of a ship. Floats consisting of a substantial ballasted cruciform drogue attached by a line to a small surface float with a marked flag, radar reflector, or a lamp for night observation, are preferable for most purposes and under favourable conditions one boat can track several floats at a time. Drogue floats tracked for a whole tidal cycle give a clear measure of the residual current not obtainable in any other way.

If information on the turbulent dispersion of a neutrally buoyant effluent is required, measurements can be made of the dilution of indicators such as Rhodamine B dye which can be detected in a fluorimeter at concentrations up to 1 part in 10^9. Radioactive tracers (e.g. Bromine 82) have also been used, and bacteriological tracers, e.g. Serratia and Coliform bacteria from sewer outfalls. Radioactive tracers are also used to study the dispersal of dredgings, sludge and other waste materials dumped on the sea-bed. Experiments with tracers in the open sea are, however, expensive and difficult and should not be attempted until as much information as possible has been obtained from float tracking and other sources.

31.8.8.2 Current meters

Modern current meters consist of a rotor or electromagnetic detector to measure speed, a vane to align the instrument to the current and a compass to record direction. They also incorporate some form of data-recording system and usually a salinity/temperature bridge.

The electromagnetic meter should be the most reliable as it has no moving parts to clog with weed or debris but is expensive and requires a powerful electricity supply. The vertical spindle meter with Savonius-type rotor is liable to over-register the mean current in fluctuating flow but is robust and popular with practical surveyors. The horizontal spindle meter with propeller-type rotor is the most liable to clog with weed and debris and likely to under-register the mean current in varying flow.

The instruments can be arranged to be read directly or the information, in the form of revolutions per minute in the case of rotor meters or converted to velocity, together with compass heading, temperature and salinity can be relayed by cable to a display screen on a survey vessel or to shore if not too distant. Instruments of this latter kind can be coupled to a radio buoy which transmits the data to a receiver on shore where they can be recorded either automatically or manually. Completely self-contained instruments which record the data on magnetic tape are now more popular.

Current meters must be secured at a fixed level and not attached directly to a floating buoy or survey boat. The laying and recovery of the meters has to be carried out with great care to ensure that they are not damaged during these operations, and their location must be chosen with due regard to the risk of damage by shipping and fishing nets or fouling by weed and rubbish. As a consequence of the difficulties (e.g. possible malfunction or loss of an instrument) current meter data is seldom complete; on some occasions as little as 50% may have to be accepted and as much as 90% will only be achieved by experienced surveyors with the best equipment and guard boats or lighted buoys over each instrument. It has also to be remembered that the recording will be affected by wave action, depending upon the wave height and length and water depth, and on the response of the instrument to variations in strength and direction of the current.

31.8.9 Water properties

Instruments which give direct readings of salinity, temperature and dissolved oxygen, and are quick and easy to operate, are available commercially so that information on these characteristics can be obtained readily while a survey team is in the field.

Domestic sewage can be detected in sea water up to concentrations of about 1 p.p.m. by culturing samples and counting the number of colonies of Coliform bacteria which develop. These measurements, which require the services of a bacteriological laboratory, are the best means of judging the degree of pollution from an existing outfall.

Tests for the presence of other pollutants and 'heavy' metals such as cadmium, zinc and mercury may also be required when considering the effects of industrial outfalls; very small concentrations are difficult to identify but filter-feeding shellfish effectively concentrate many of the more objectionable pollutants and it may be worthwhile collecting them for analysis with the water samples.

Other properties that may need to be measured for the study of the effect of effluent discharges are biological and chemical oxygen demand (BOD and COD) and the concentration of nutrients such as nitrates, nitrites and phosphates, and chlorine demand for the specification of chlorinators for sterilizing process and cooling water. Conductivity, required for the design of cathodic protection systems, can be extracted from the salinity/temperature bridge.

Suspended solids measurements are most difficult to interpret for the following reasons:

(1) Standard laboratory tests measure total filterable solids including organic and clay particles which may not contribute to siltation.
(2) Silt and clay particles with low settling velocities in fresh water tend to flocculate and settle rapidly on encountering salt water.
(3) Solid concentrations can increase many times in thin layers near the bed.
(4) Silt meters measure 'turbidity' or transmission of light which may not correlate with filterable or settleable solids.

Water samples may be taken by Nansen bottle, pump or by diver and as many tests as practicable should be carried out in the survey boat to reduce the large number of samples which otherwise have to be transported and stored, including dissolved oxygen, chlorine demand, BOD and COD which alter in storage.

31.8.10 Waves and tides

The practice of estimating wave heights by eye either from the shore or from a ship is unreliable and data obtained in this way or by observation against a graduated pole, although of value for some purposes, are not suitable for numerical analysis; wave data have to be analysed statistically and a very large number of systematic observations are required.

Instruments for measuring wave height are of four kinds:

(1) Double-integrating accelerometers housed in a moored ship or a buoy designed to rise and fall with the water surface.
(2) Pressure gauges which are held at a fixed level below the surface.
(3) Wave staffs which pass through the sea surface and record its level directly.
(4) Inverted, narrow-beam, high-frequency transducers (echo sounders) supported at fixed level below the sea surface.

Instruments of the first kind have been installed in lightships but it is clear that in this situation they will not record the true waveform and the amplitude may also be distorted by the response of the ship. If installed in a small, shallow-draft buoy with a light mooring, instruments of this kind, of which the Datawell 'Wave Rider' is the most used, produce satisfactory records. Long waves and tides are not recorded.

Instruments of the second kind can be either completely self-contained or the pressure cell can be laid on the sea-bed and connected by a cable or pressure hose to a recording instrument on shore. The drawback of this type of instrument is that the pressure variation caused by a wave of given height decreases with increasing value of the ratio depth:wave length; since in tidal waters both quantities vary continuously and at any one time waves of various periods will be present, the interpretation of the records is laborious and not always easy. Shorter waves tend to be lost owing to the high attenuation. If the instrument is arranged to record on paper rolls the tidal variation should be compensated within the instrument or the record will be too cramped for satisfactory interpretation. If information on waves with periods of over, say, 15 s is required, hydraulic filters can be inserted between the cell and the measuring instrument to suppress the pressure variations due to the shorter-period storm waves and the record can then be scanned visually for long-wave activity.

Instruments of the third and fourth types require a fixed support and a power supply, and therefore can only be used close to the shore. They provide a complete and undistorted record of all changes in the water level; the acoustic type is probably more reliable than the wave staff which obtains its signal from variation of the capacitance of a taut insulated wire passing through the water surface. Acoustic gauges can use standard echo sounder electronics and digitizers and are now the commonest type for inshore measurements.

Wave staffs and acoustic gauges can be mounted on tall pillar buoys which, if they are correctly designed, barely respond to storm waves; if the buoy does so respond its movement can be measured with a double-integrating accelerometer incorporated in it, and combined with the record from the wave staff.

Owing to the vast amount of data that would otherwise be collected (about 6×10^6 waves per annum) wave recorders do not usually run continuously and an arbitrary sampling period, e.g. 15 min/2 h, is usually adopted. It would be better to have a trigger circuit arranged to sample waves of increasing height with increasing intensity, e.g. 0 to 1 m, no measurements; 1 to 3 m, 5 min/h; over 3 m, continuous recording.

The measurement of very long-period waves (e.g. the tide) presents little difficulty provided short-period waves can be excluded from the record; either the second, third or fourth types of wave gauge described above can be adapted for the purpose. The traditional tide gauge consists of a float in a stilling well mechanically connected to a clockwork-driven drum recorder. Wave direction can be determined by observation or by radar.

31.8.11 Meteorological data

Tide levels are affected by wind and barometric pressure and waves are generated by the wind which also affects the currents; therefore the recording of any wave, tide, or current data should be accompanied by concurrent and immediate antecedent wind and barometric pressure records. This information can frequently be obtained from an existing meteorological station, but failing this a recording anemometer and barograph should be set up. Portable direct-reading cup anemometers should be carried in the survey boat for checking local variations in wind speed during the survey.

31.8.12 Coastal stability – movement of beach and sea-bed sediments

If the shore or sea-bed consists of soft or loose material evidence will be required as to its stability, both in the long and short term. The questions to be considered are the quantity of material in transport across any section of the shore and sea-bed, the extent of short-term fluctuations in level, whether there is on balance, erosion or accretion, to what depth may scour or 'fluidization' of the material occur and what may be the response to any construction works contemplated.

These questions can only be answered on the basis of experience and judgement following examination of relevant records supplemented by survey and observation, although mathematical modelling (see page 31/14) can assist in the absence of more positive physical evidence.

Typical sources of information are:

(1) Comparison of old and current topographical and bathymetric plans.
(2) Evidence of accretion (e.g. sand dunes, salt marsh or grove) and erosion (e.g. exposed clay and rock in cliffs and foreshore) and the accumulation or scour of littoral material against natural or artificial obstructions to the drift.
(3) The depth below the foreshore or sea-bed to rock, clay or sediments deposited under previous geological conditions.
(4) The presence or absence of surface forms associated with high transport rates (e.g. dunes and plunge bars) or moderate transport, e.g. ripples.
(5) Dredging records.
(6) The presence of fluid or very soft mud layers which can be detected by an echo sounder, or organic mud with bubbles which absorb acoustic waves.

It should be noted that almost complete stability does not preclude the possibility that very large quantities of sediment are passing through an area and the investigation should be extended along the coast in both directions to points where the passage of sediment is barred by topographical features.

Estimates of the quantity of material moving along a coastline can be made by labelling the sediment with radioactive or fluorescent tracers but, since transport rates may vary widely from year to year and one exceptional storm can have as much effect as many months of typical weather, such experiments are of doubtful value unless continued over a long period; experimental groynes or dredged channels may well yield more information for an equivalent cost and effort.

Along coastlines exposed to the open sea, where sediments are moved primarily by wave action as bed load, material finer than sand will be dispersed rapidly and settle in deep water so that measurements of suspended solids are not normally necessary. In sheltered estuaries the situation is quite different and measurements of suspended solids may be required.

Evidence of scour obtained by the procedures outlined above must not be assumed to show the limit to which disturbance can occur; fluctuating pressures and flow in conditions of extreme flood or storm may fluidize sediments to much greater depth. Reported examples are the recovery of pottery and glass shards at a depth of 30 m below the normal bed of an estuary channel, and the settlement of a pipeline to over 9 m in the surf zone on an exposed beach.

31.9 Design parameters and data analysis

31.9.1 Ground conditions

The inspection, testing and analysis of superficial and subsoil samples of sea-bed and beach materials and of the underlying strata following the site investigations (outlined on pages 31/20 and 31/21) will follow normal soils and rock-testing procedures, including grading of sediments and laboratory tests for compressive strength, internal angle of friction, cohesion, etc.

The work may need to be extended, however, to include a study of sea-bed and foreshore stability. For this purpose all available charts and surveys should be plotted or reproduced to a common scale and datum, and carefully scrutinized for changes between one survey and the next. The significance of any observed variation can be determined by superimposing cross-sections from successive surveys or by plotting accretion and erosion contours from which changes in successive periods of time in the volume of material within a given area can be calculated.

31.9.2 Waves, tides and currents

The other measurements which may have been made, for instance of wave height and period, tides, currents, salinity, silt content and similar phenomena, will almost always be scattered over a wide range and it will not normally be possible to make direct use of them. In some cases, particularly if the data are of limited extent or quality, their use may be confined to corroborating, or modifying in some particular, decisions based on previous experience. Alternatively, the data may be analysed statistically with a view to choosing values for the required design parameters to give a selected low probability of their being exceeded within the estimated lifespan of the works to be constructed. In this case the method of analysis adopted will be limited by the amount of data available but within this limit it has to be remembered that a very elaborate processing procedure will only be justified if techniques are available whereby the results can be applied rigorously in the subsequent design and subjective judgements avoided at a later stage.

As far as possible, therefore, certain decisions should be made before data collection is started, e.g.:

(1) The type of structure to be built, or range of possible types, or the kind of solution most likely to be adopted for some particular problem.
(2) The methods by which the data will be processed and the designs carried out, which may be by numerical analysis by hand or by computer, by a model study or by a combination of these and, hence, the form in which the data are to be recorded. These decisions will be influenced by questions of cost and time and by the availability of computer and model study facilities.

Suppose, for example, a breakwater is being considered; if the design is to be based upon empirical formula, e.g. the Hudson equation for a rubble mound or the Sainflou method for a mass concrete or blockwork breakwater, then a straightforward record of wave height for a relatively short period, say one winter, to enable a value for $H_{\frac{1}{3}}$ or $H_{\frac{1}{10}}$ (page **31**/9) to be estimated may be all that is required, with visual observations of the direction of wave approach. On the other hand, if some less traditional type of structure is envisaged, e.g. the cellular structure adopted for the Brighton Marina Breakwaters (see Figure 31.26, page **31**/32) for which model testing in random waves is necessary (see page **31**/18) then the wave record should include all wave characteristics, namely, height period and shape, preferably in a form that can be used directly to control the model wave generator, or be simulated by a number of sinusoidal components following Fourier analysis of the prototype wave train; the measurements made in the model should, in turn, be in a form which can be analysed statistically to produce the worst combination of wave energy, water level, wind speed and direction of wave approach.

31.10 Materials

The maritime environment is among the most aggressive in which civil engineering works are constructed and the question of durability, therefore, is one of the main factors in the choice of materials. In this context four zones can be identified as follows:

(1) The splash zone above high-water level where surfaces may become coated with salts due to alternate wetting by spray and evaporation, and abrasion may be caused by blown sand and shingle thrown up by the sea.
(2) The beach zone between high- and low-water marks where severe abrasion by wave-driven shingle or gravel may override other considerations.
(3) The tidal zone away from a beach where corrosion of metals is likely to be most severe.

(4) The submerged zone below low water which is the least aggressive of the four.

Most of the common engineering materials can be employed in maritime works, e.g. rock, mass concrete, reinforced concrete, timber, iron, steel, special steels, bronzes, plastics and bitumen. Some of these can be used in most situations while others have more limited application as discussed below.

31.10.1 Rock

Most of the harder and more durable types of rock are suitable for any of the zones defined above and may be in the form of masonry or random placed blocks in revetments or rubble mounds. Usually the quality of the stone used is dependent upon what is economically available; there is no recognized testing procedure to determine the suitability of a rock specifically for maritime works but tests developed for other purposes are useful.[40] The rocks most commonly used are of the igneous and metamorphic types but also some of the harder sedimentary rocks. It is to be noted that some of the softer sedimentary rocks may be attacked by piddock or pholas near or below low water mark.

31.10.2 Brickwork

Engineering brickwork is suitable for use in Zone (1) and has from time to time been used with successful results in Zones (2) and (3). In all instances it is essential to ensure that the joints are completely filled with strong durable mortar.

31.10.3 Concrete[41]

Concrete is suitable for use in all four zones but special attention must be given to questions of durability and protection of embedded steelwork. These considerations require the mix to be designed for maximum density and minimum permeability rather than strength.

The required properties can usually be achieved using uncontaminated sound aggregates and ordinary Portland cement by specifying a mix which can be fully compacted in the prevailing circumstances and having a minimum cement content in the range 270 to 300 kg/m³ for plain concrete and 320 to 360 kg/m³ for reinforced concrete, with a free water:cement ratio not exceeding 0.5.

For *in situ* work large pours are desirable and special attention should be paid to construction joints which are vulnerable to attack by the sea. Replacement of cement by up to 35% of pulverized fuel ash (PFA) or 70% of ground granulated blast furnace slag (GGBFS) is a worthwhile economy in large structures and reduces the heat of hydration and associated shrinkage cracking, while the introduction of the slag improves durability but reduces the rate at which strength is developed and may impair early resistance to abrasion. If abrasion is a potential problem the coarse aggregate should be the largest size and hardest rock available.

Special precautions are necessary with reinforced concrete. The cover to reinforcing steel should be not less than 50 mm with additional cover of 100 mm or more in Zones (1) and (2) if abrasion is likely to occur. Underwater *in situ* construction should be avoided if possible, and where necessary, cover should be increased by an appropriate tolerance for the assembly and fixing of the reinforcement. It should be noted, however, that excessive cover may not be beneficial if it results in reduced control over surface cracking.

For placing underwater, concrete needs to be cohesive in texture with good workability and freedom from segregation; it must be of a consistency which will allow it to flow into position and be consolidated by its own mass, compaction by vibration being seldom practicable. These properties may be achieved by designing the mix as for work to be constructed in the dry and adding an extra 25% of cement with appropriate increase in added water or by the addition of plasticizers or retarders.

If conditions are warm or the water polluted, concrete may be attacked by the sulphates in sea water and this can be serious if the concrete is porous; sulphate-resisting cement in which tricalcium alluminate is limited may be used as a precaution but not as a substitute for sound impervious concrete. There is, however, evidence[42] that in the tidal and splash zones the use of sulphate-resisting cement may increase the risk of corrosion of embedded steel as a result of chloride ion diffusion which can be equally serious if the concrete is porous. It has been suggested that PFA or GGBFS replacement increases liability to sulphate attack and that a cement with 5% C_3Al is probably the optimum for concrete in the marine environment.[43]

31.10.4 Timber

Timber is widely used in coastal engineering works notably for jetties, slipways, groynes and outfall pipe support trestles. Although subject to various forms of deterioration, an acceptable lifespan can generally be assured by suitable choice of species and/or preservative treatment. Above water level where the timber will be continuously damp, or alternately wet and dry, decay by normal wet or dry rot will be most prevalent; in the beach zone abrasion may be a more important consideration while near and below mean tide level attack by marine borers, of which the Teredo and Limnoria are the most common, will be of greatest concern.

A number of naturally resistant hardwoods are available of which greenheart is the best-known and outstanding as regards strength and resistance to attack by marine borers; others in common use are jarrah, ekki, and opepe. Softwoods need to be treated by impregnation with creosote or one of the proprietary preservatives which are mostly based on copper, chromium or arsenic salts. The usual specification for this is impregnation by the full cell process at a pressure of $1.2 \, MN/m^2$ for a period of 4 h or to refusal; alternatively a retention of 8 kg of preservative salt per cubic metre of timber. Softwoods most commonly used in marine work are pitchpine and douglas fir. These timbers must be incized in order to achieve adequate penetration of the preservative but are outstanding in respect of resistance to abrasion in the beach zone.

The more important properties of the above timbers are given in Table 31.6.

Table 31.6 Timber properties

Species	Specific gravity at 50% moisture content	Ultimate stresses parallel to grain (N/mm²)			Modulus of elasticity (N/mm²)
		Bending	Compression	Shear	
Greenheart	1.32	135	70	9	18 000
Jarrah	1.01	68	36	9	10 000
Ekki	1.32	120	68	16	14 000
Opepe	0.95	90	50	12	12 000
Basralocus	1.07	82	41	11	12 000
Karri	1.04	77	37	10	14 000
Turpentine	1.03	77	41	10	11 000
Pitchpine	0.74	54	25	6	10 000
Douglas fir	0.64	53	26	6	11 000

Working stresses will depend upon the quality of the timber and the location and size of blemishes such as knots and shakes but will usually be from one-fifth to one-third of the ultimate stresses given in Table 31.6.

31.10.5 Iron

In the past cast iron, in pipes and jetty or pier columns, was frequently used and has a long life in the marine environment except where subjected to abrasion. Apart from abrasion, deterioration occurs by graphitization in which the iron corrodes leaving a residue of soft and porous iron oxide and graphite. In modern practice the use of cast iron is limited to pipe specials and valves, spun iron or ductile iron now being more usual for pipes.

Wrought iron is most durable and outstanding for many purposes. It is not now readily obtainable and its use is confined to special small items such as pipe straps and fittings in inaccessible situations.

31.10.6 Steel

On grounds of economy and ease of construction, steel in the form of piles and structural sections is frequently employed in maritime structures in spite of its relatively high rate of corrosion in many situations. The normal rate of atmospheric corrosion tends to be accelerated particularly in the tidal and splash zones and precautions are needed to give an acceptable life span. Tests over a long period by the Sea Action Committee of the Institution of Civil Engineers and measurements on existing structures indicate an average loss of thickness of 0.08 mm per annum in sea water. Protective coatings of tar, bitumastic, or other paints, wrapping with coal tar or synthetic tapes or galvanizing will retard commencement of corrosion and with modern methods of preparation and application, these and coatings developed especially for tidal and splash zones, can have an effective life of 10 years or so. Below low-water level, coating or wrapping systems combined with cathodic protection by impressed current or sacrificial anodes, as used frequently on steel jetties and submerged pipelines, can prolong the life of the structure indefinitely. If an impressed current system is used care should be taken to prevent stray currents from causing corrosion of steel in adjacent concrete. These protective measures are seldom economical for steel sheet piling for which it is more usual to make allowance for corrosion in the design; it is usual to assume that a wall of steel sheet piling will have reached the end of its useful life by the time 50% of the metal has been lost by corrosion, when the initial stresses in the material will have doubled. For the sections of piling most commonly used, this criterion gives a life of from 60 to 100 years.

Steel has very poor resistance to abrasion and on shingle beaches may have a life of not more than one-tenth of normal; steel sheet piling in groynes, for example, is known to have worn through in 5 years or less in some situations. For these conditions concrete encasement of the vulnerable parts of a steel structure is probably the best form of protection, or cladding with timber which can be renewed from time to time.

31.10.7 Corrosion-resistant metals

Metals in this category such as stainless steel, bronzes and monel metal are expensive and their use is generally confined to small special items.

Of the various types of stainless steel available, austenitic steel containing nickel and chromium has the most suitable strength and corrosion resistance for use in structural and civil engineering work. It is durable in most situations encountered in maritime engineering with the exception of anaerobic conditions which may occur due to marine growth or below the sea-bed, particularly in estuarine muds, and in stagnant conditions where the oxygen supply is low. Under these circumstances, stainless steel is subject, owing to the breakdown of the protective oxide film, to pitting and crevice corrosion, a tendency which is increased in the presence of the chlorides in the sea water. Care is therefore needed to avoid laps and crevices at joints.

Duplex and molybdenum stabilized austenitic stainless steels are less prone to pitting and crevice attack, and may also be chosen, but at extra cost, to avoid superficial rust staining where appearance is important.

Monel metal, an alloy of 30% copper and 66% nickel with 4% iron and manganese, combines high strength with maximum resistance to corrosion in all maritime situations.

For use in sea water, bronzes must be zinc-free. They are normally used only for such items as valve seatings and trims.

31.10.8 Synthetic materials

This category of materials includes sealing compounds such as polysulphides and epoxies which are used in marine work although some are difficult to apply in damp conditions. A recent development is an epoxy coating which can be applied under water. Polyvinylchloride pipes are suitable as liners in outfall and intake structures or, strengthened with a glass reinforced polyester or epoxy coating, as pipelines for various purposes; polythene is outstanding for pipes as it can be supplied in long lengths, is easily floated or towed into position and is a good material for abrasive conditions. All these materials are corrosion-resistant, but expensive, and apart from polythene not very tough. It has, however, been reported that PVC has been attacked by marine borers and sealing compounds in sewage works by fungal and bacterial organisms; such materials may be similarly susceptible in the marine environment. Their use is generally restricted to special applications where there are problems with the more traditional materials.

31.10.9 Bitumen

Bitumen[44-46] is widely used in the form of sand mastic (sand, filler and bitumen) as the binder in asphaltic concrete, as a jointing material for stone or concrete blockwork revetments, as asphaltic carpets above or below water and for grouting rubble revetments, groynes or breakwaters. These uses have largely been developed for coast defence work in The Netherlands but are now finding an increasing use in other countries. This form of construction has the advantage of being sufficiently flexible to accommodate long-term settlement without fracture while being rigid enough to withstand the large but short-term dynamic forces of breaking waves.

In order to achieve good durability, asphaltic concrete requires a higher bitumen content than normal paving mixes but limited so as to resist flow down the slopes on which it is to be laid and to avoid segregation during mixing and handling. Usually 6 to 9% of bitumen of penetration grade 40/50, 60/70 or 80/100 will be used depending upon the situation of the work and climatic conditions. A graded aggregate of maximum size 25 mm is commonly used, although it can be gap-graded, with a filler content (usually ground limestone or cement) between 8 and 13%. Mixes of this type require to be compacted and their use is therefore confined to work above water level.

For the jointing of blockwork revetments the filler may be replaced with a fibrous material such as asbestos and a typical composition for the mastic would be: bitumen 40/50 45%, sand 50%, asbestos fibre 5%. The composition for any specific work depends upon the size of the joints, the angle of slope and the range of ambient temperature and should be determined by trial

on the basis of the preceding figures which are for a pouring mastic. If the joints are to be trowelled, a typical mix would be: bitumen 40/50 25%, sand 70%, asbestos fibre 5%.

The inert materials in sand mastic (sand and filler) are overfilled with bitumen so that the mixture is pourable when hot and does not require compaction. It can therefore be used both above and below water level. The composition can vary widely depending upon the requirements which may be an asphalt carpet for scour protection as used in the Delta Works in The Netherlands, either placed *in situ* or as prefabricated mattresses reinforced with wire mesh, or as a material for grouting rubble revetments or breakwaters either above or below water level. The materials and their proportions should be chosen to give a viscosity which is low enough at the placing temperature to give adequate spread or penetration but not so high as to allow excessive flow on slopes; it will normally be in the range 10 to 10^2 Ns/m^2. At ambient temperature the viscosity must be high enough to prevent excessive long-term flow and will normally be in the range 10^6 to 10^9 Ns/m^2

For use above water, a bitumen of penetration grade 20/30 to 80/100 is likely to be chosen, and for underwater work one of the softer grades, e.g. 180/200 or 280/320. A typical mix is: sand 75%, filler 10%, bitumen 15%. For underwater work the temperature of the mastic should be relatively low (160 to 180°C) so as to reduce heat losses while placing and the production of excessive steam which otherwise could cause the mastic to harden as a spongy mass with a tendency to float.

31.11 Sea-defence and coast protection works

31.11.1 Sea walls

Sea walls may be required for the protection of land which is being eroded by the sea, of low-lying land against flooding by the sea, or for the purpose of reclaiming land from the sea. The type of wall to be built in any particular case is very much dependent upon circumstances such as the exposure of the site, the material forming the foundation upon which the wall will be built, whether or not there is a beach in front of the wall and type of beach, and the cost of the work in relation to the value of the property or land to be protected. They thus range from heavy mass concrete or masonry structures founded upon rock or other firm foundation, and frequently used for the protection of urban areas, to light timber breastworks or pitched revetments for the protection of open land.

Within the range of types indicated above, the following are typical examples.

Figure 31.17 shows a cross-section of a mass concrete vertical sea wall founded upon shale bedrock underlying a flat sandy beach. The exposure is not severe, being in a sheltered bay within which there is little littoral drift. The foundations of the wall are protected against erosion of the sand by a concrete apron and toe wall in a trench in the rock, but scour is not a serious risk at this particular site and the fact that waves are reflected from this type of wall has not had any significant effect upon beach conditions.

Walls of this type have been used with varying degrees of success in places where the exposure is severe. They are eminently suitable in two extreme conditions: (1) where there is no beach, the foreshore on which the wall is founded being exposed bedrock; and (2) at the other extreme, where large quantities of beach material in transit with the littoral drift provide an ample supply of material to make good scour which is likely to take place in front of such a wall during storm conditions.

This type of wall may be the first choice in situations where abrasion by wave-driven shingle or gravel is severe, since,

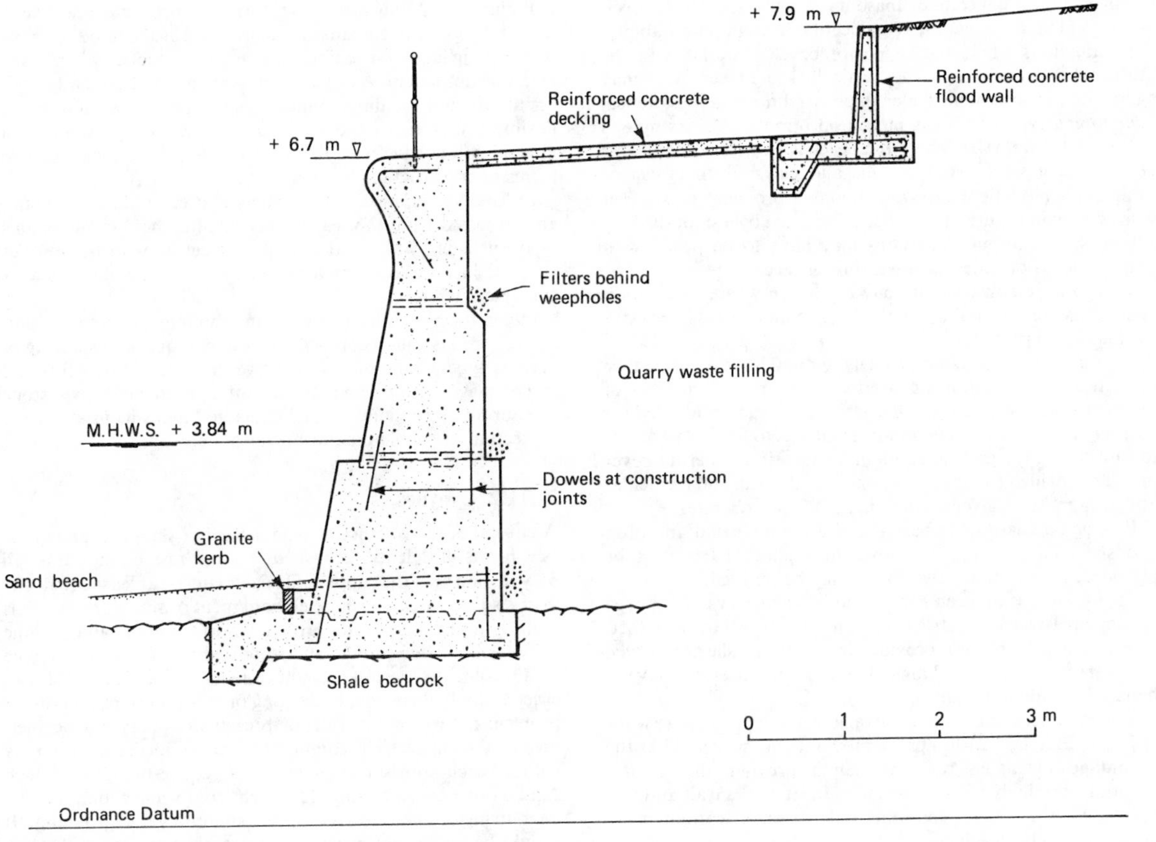

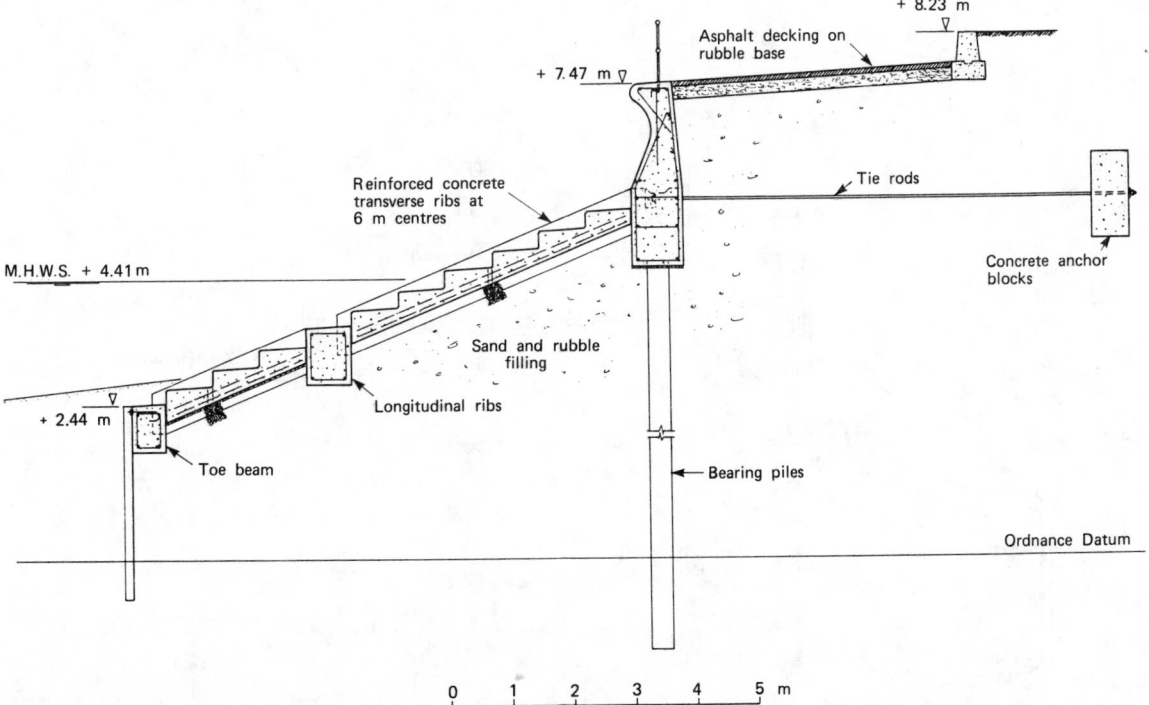

+ 7.9 m ▽

Reinforced concrete flood wall

Reinforced concrete decking

+ 6.7 m ▽

Filters behind weepholes

Quarry waste filling

M.H.W.S. + 3.84 m

Dowels at construction joints

Granite kerb

Sand beach

Shale bedrock

0   1   2   3 m

Ordnance Datum

Figure 31.17 Mass concrete sea wall at Tenby, Pembrokeshire

+ 8.23 m ▽

Asphalt decking on rubble base

+ 7.47 m ▽

Reinforced concrete transverse ribs at 6 m centres

Tie rods

Concrete anchor blocks

M.H.W.S. + 4.41 m

Sand and rubble filling

+ 2.44 m ▽

Longitudinal ribs

Toe beam

Bearing piles

Ordnance Datum

0   1   2   3   4   5 m

Figure 31.18 Sloping sea wall at Crosby, Lancashire

whether built in concrete or masonry, it can suffer an appreciable loss of material without significant effect upon its stability.

In situations where the supply of beach material with the littoral drift is not high the vertical type of sea wall may aggravate the situation. Under these conditions the less reflective sloping type of wall is more appropriate. An example is shown in Figure 31.18 where the beaches consist of a variable depth of fine sand overlying silts and clays. With gravel or shingle beaches where abrasion may be a problem the stepped work shown in Figure 31.18 is better replaced by a smooth slope which under extreme conditions may have to be paved with granite blocks or other hard-wearing surface.

This form of construction may be chosen where foundation conditions are unsuitable for the high loading associated with the massive vertical type of wall.

Figure 31.19 is an example of a lighter and less expensive type of vertical wall which has proved satisfactory in a number of different situations. Being a piled structure it can be adapted for a variety of foundation conditions from silts to firm clays. If it is to withstand any significant amount of abrasion, generous cover must be provided to the reinforcement in the wall and particular attention must be given to the quality of the concrete.

If large fluctuations in beach level are anticipated an apron and sheet piled cutoff, as shown in Figure 31.19, must be provided, but in other situations it may be omitted.

Figure 31.20 is an even lighter and less expensive type of sea wall or breastwork constructed of greenheart or jarrah timber. Provision against scour is provided by the timber sheeters driven below beach level. In shingle beaches the sheeters may be replaced by underplanking.

In areas where heavy stone is available the type of breastwork shown in Figure 31.21 has proved to be economical and has the advantage of being less reflective than the previous illustrations. It is thus less likely to cause scour in front of the wall and if it does occur the rubble blocks are able to settle without being drawn down the beach.

Figure 31.22 illustrates the form of protection which is frequently given to embankments provided as the flood defences of low-lying land in situations such as river estuaries where wave action is not severe. As such embankments are frequently built on a soft and yielding foundation the protection has to be flexible and stone rubble or concrete blocks with bitumen run into the joints have been found to be one of the most suitable forms of construction for this work.

In The Netherlands, where extensive areas of low-lying land are protected from the sea by very large embankments of this type, surfacing with a sand mastic carpet or with bituminous concrete has proved to be a satisfactory alternative to stone or concrete pitching.

Other forms of defence that are suitable in some circumstances are fascines, rubble-filled mesh gabions, and sheeting of various types, none of which, however, are able to withstand severe wave action, and at the other extreme heavy stone armouring or specially shaped concrete blocks as used for the protection of rubble-mound breakwaters.

31.11.2 Groynes

A natural beach whether of sand, gravel or shingle is one of the best forms of defence against the sea but in the long term will only remain so if the supply of beach material at least equals the losses, so that an alongshore embankment of shingle or gravel is built up above high-water level, or in the case of sand, a dune belt which, if wide enough and stabilized by vegetation, may be an adequate defence on account of its bulk and width. If losses exceed supply there will, in the long or short term, be an erosion problem for which sea walls or breastworks may not on their own be the answer. The solution is either to increase the supply with a beach nourishment scheme, or reduce the losses, which can be done by protecting the beach from wave action, e.g. by an offlying breakwater, or more commonly by groyning. It should be noted that groynes will not prevent the beach from

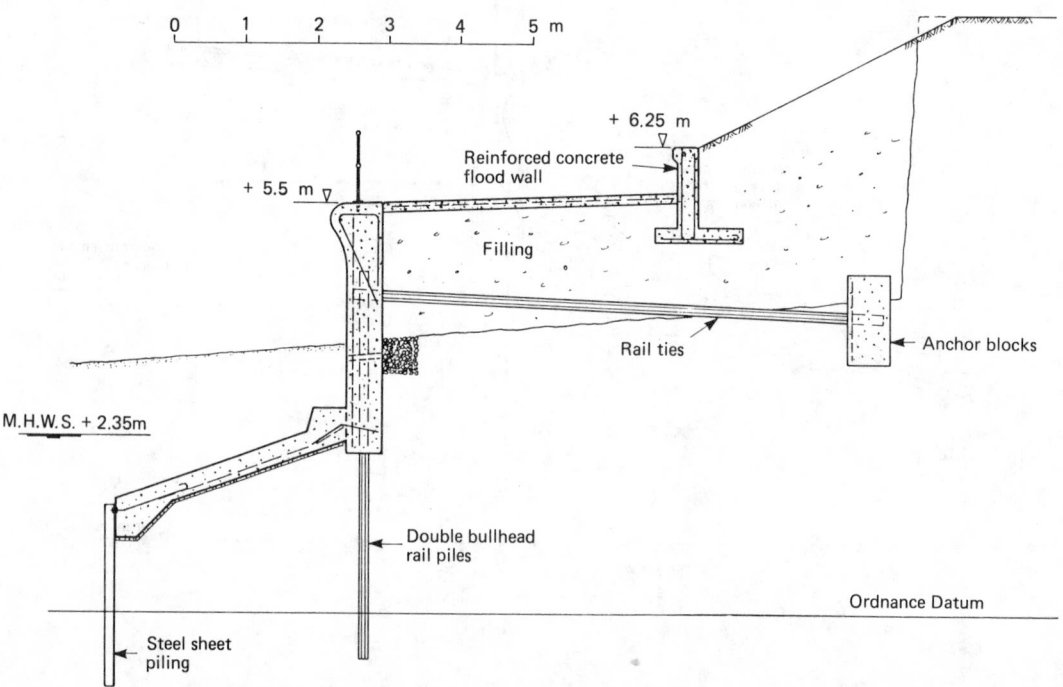

Figure 31.19 Light reinforced concrete sea wall

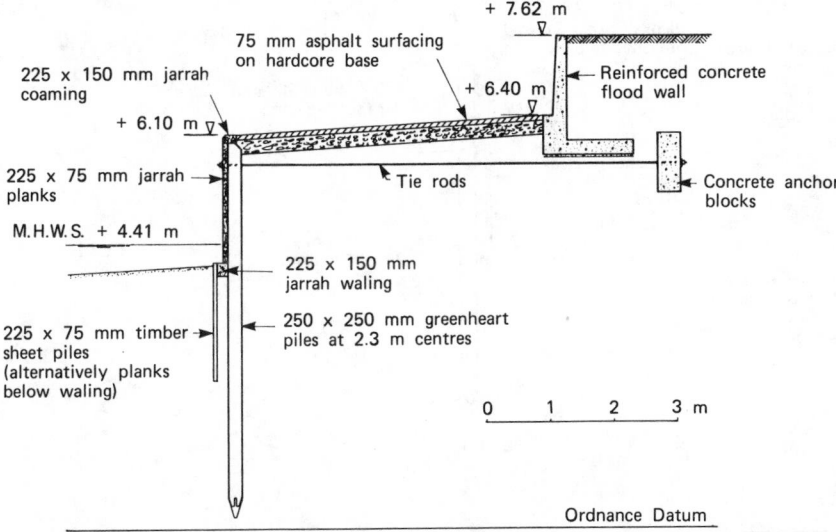

225 x 150 mm jarrah
coaming

75 mm asphalt surfacing
on hardcore base

+ 7.62 m

+ 6.40 m

Reinforced concrete
flood wall

+ 6.10 m

225 x 75 mm jarrah
planks

M.H.W.S. + 4.41 m

Tie rods

Concrete anchor
blocks

225 x 150 mm
jarrah waling

250 x 250 mm greenheart
piles at 2.3 m centres

225 x 75 mm timber
sheet piles
(alternatively planks
below waling)

0   1   2   3 m

Ordnance Datum

Figure 31.20   Timber breastwork

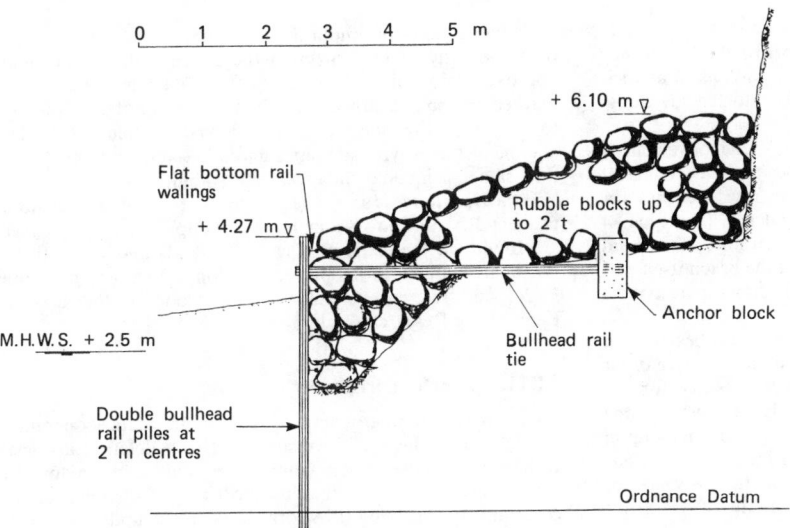

0   1   2   3   4   5 m

+ 6.10 m

Flat bottom rail
walings

+ 4.27 m

Rubble blocks up
to 2 t

Anchor block

M.H.W.S. + 2.5 m

Bullhead rail
tie

Double bullhead
rail piles at
2 m centres

Ordnance Datum

Figure 31.21   Rail pile and rubble breastwork

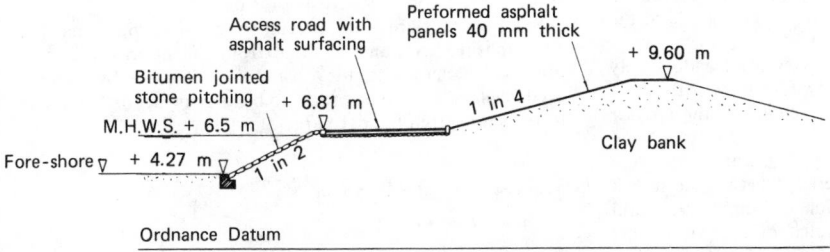

Access road with
asphalt surfacing

Preformed asphalt
panels 40 mm thick

+ 9.60 m

Bitumen jointed
stone pitching

+ 6.81 m

M.H.W.S. + 6.5 m

1 in 4

Fore-shore   + 4.27 m

1 in 2

Clay bank

Ordnance Datum

0   2   4   6   8   10 m

Figure 31.22   Revetment to estuary embankments: Hurditches sea
wall. (*Courtesy*: Somerset River Authority)

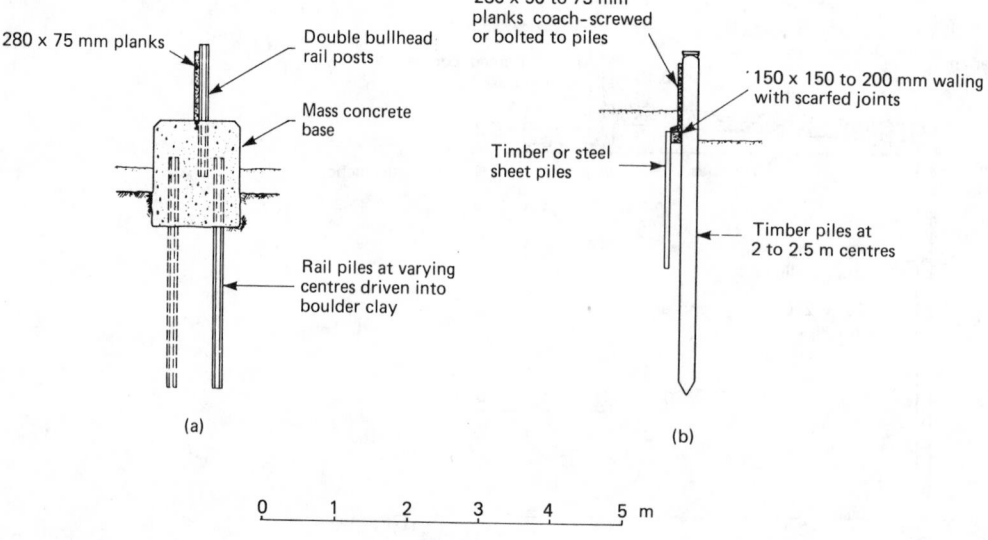

Figure 31.23 (a) Concrete groyne; (b) timber piled groyne

being drawn down under storm conditions, so that intermittent erosion of the shoreline or the stratum below the beach may continue and in time the groynes may be outflanked at their inner ends. It is normally necessary, therefore, to provide along-shore defences as well as groynes.

Although the physical processes involved in the movement of beach material are not fully understood there is sufficient experience in the control of shingle and gravel beaches for systems of groynes to be built with confidence. The type of construction adopted in any particular situation will depend upon the exposure of the site, the nature of the beach itself and the foundation conditions. Figure 31.23 shows two typical examples.

As the slopes associated with shingle and gravel beaches are steep, fluctuations in level can be very considerable and, in order to control these fluctuations, it is usually necessary for the groynes for such beaches to be built at fairly close centres, the actual spacing in any particular case being dependent upon direction of wave approach and volume of littoral drift. It has been found from experience in the UK that the provision of approximately 1 m of groyne for every metre of sea front protected is required.

For the flat gradients of sand beaches the groynes can be much more widely spaced but need also to be much longer, and again the provision of approximately 1 m length of groyne for every metre of frontage protected is a reasonable guide to the probable requirement.

For gravel beaches it is usually necessary to provide widely spaced groynes across the lower sandy foreshore with short intermediate groynes to control the movement of the coarser material at the top of the beach.

Since groynes can only serve their purpose by interrupting the along-shore movement of beach material, whether the movement is predominantly in one direction or not, a successful system of groynes along any one stretch of the coast must inevitably lead to a reduction in the amount of material available to the downdrift beaches. In general, therefore, a system of groynes should not be made so efficient as to form a complete barrier to the littoral drift. As far as is practicable, the groynes should be designed so that initially they can be kept low relative to the beach level and only built up gradually as material accumulates. The alignment of the groynes relative to the

coastline is also of importance; if the littoral drift is not predominantly in one direction the groynes should be built approximately normal to the coastline but where there is a marked drift in one direction it is better to lay off the groynes up to 10° from the normal on the downdrift side. With this alignment the groynes are less subject to scour on their down-drift side which could endanger their stability.

If the tidal range is small the groynes, to be effective, will need to extend beyond low-water mark. Examples of such groynes are found on the coast of The Netherlands and in southern Portugal. Suitable forms of construction are shown in Figure 31.24, and reference should also be made to the Coastal Engineering Research Center.[5]

31.11.3 Beach nourishment

As an alternative to groyning, or as a complementary operation in the maintenance of a satisfactory beach defence, artificial recharging is undertaken in many places. This has been done by excavating shingle from the downdrift end of a stretch of the coast and transporting it by lorry back to the updrift end and repeating the operation as often as necessary. Elsewhere, beaches have been recharged with quarried stone, or gravel excavated from inland gravel pits, and with colliery waste, the shore being a convenient disposal site.

Sand beaches are more difficult to renourish but it has been accomplished by pumping sand from the updrift side of an obstruction across to the downdrift side, and by pumping dredged sand from offshore on to beaches or depositing it close to the shore from hopper barges.

31.11.4 Cliff stabilization[47]

31.11.4.1 General

The need for cliff stabilization frequently arises as a result of coastal erosion and it is then necessary to look far ahead in planning coast protection works.

A cliff which has been steepened gradually as the toe has been eroded may not appear to present an urgent problem but the long-term angle of stability may be considerably flatter than that which pertains.

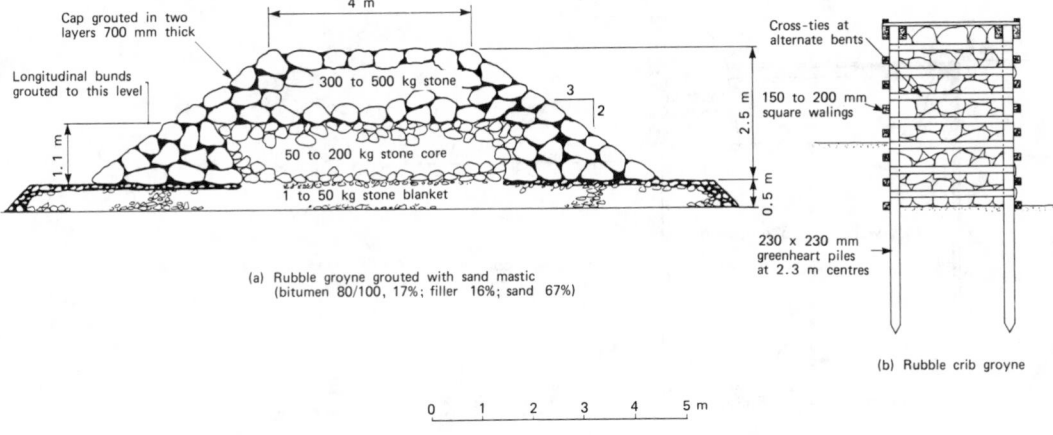

Cap grouted in two
layers 700 mm thick

4 m

300 to 500 kg stone

Longitudinal bunds
grouted to this level

3
2

50 to 200 kg stone core

1.1 m

1 to 50 kg stone blanket

(a) Rubble groyne grouted with sand mastic
(bitumen 80/100, 17%; filler 16%; sand 67%)

Cross-ties at
alternate bents

150 to 200 mm
square walings

2.5 m

0.5 m

230 x 230 mm
greenheart piles
at 2.3 m centres

(b) Rubble crib groyne

0   1   2   3   4   5 m

Figure 31.24 Rubble groynes

Cliffs in hard clay present one of the more difficult problems as they remain relatively stable for many years at an average slope between 1:1 and 1:3 after toe erosion has been halted but will ultimately degrade by gradual surface slipping or by movement on a deep-seated slip plane to a much flatter angle depending upon the type of clay, the height of the cliff and the presence of waterbearing strata. In London Clay, for example, the ultimate slope is unlikely to be steeper than 10° to the horizontal (1:5.5) and, where deep slips and mud runs have formed, an even flatter angle may eventually prevail. Boulder clay and Lias Clay may generally have long-term stability at 1:2 to 1:2.5 but in places will need to be drained to maintain this angle against surface creep.

Mixed strata of sand, gravel, silts and clays can present a difficult problem as water may be flowing at levels which are too deep to intercept by conventional drainage systems.

The angle at which cliffs will stand in rock depends upon their hardness and durability and very largely upon the orientation of joint and bedding planes. If they have to be trimmed a batter not more than 20° to the vertical will prevent water from collecting on the surface and reduce the effects of sun and frost which are the principal agents of denudation. If flatter slopes are required it is best to trim back to not more than 50° to the horizontal so that vegetation can become established; it is difficult to soil and seed a slope steeper than 40° to the horizontal. In some rocks, such as soft sandstone, marl and chalk, ultimate stability may not be achieved at slopes greater than 1:1.5.

In such situations, sea walls should not normally be constructed to stop the erosion without stabilization works to prevent the wall from being overloaded by slipping or falling material.

31.11.4.2 Methods of stabilization

The main methods employed in cliff stabilization are: (1) modifying the slope by filling or excavation; (2) drainage; (3) control of seepage erosion; and (4) provision of retaining structures or toe protection. Regrading of the slope frequently involves adding weight at the toe by filling with material excavated from the upper parts of the cliff. Drainage has an important effect by reducing pore water pressures in the ground and may consist of french drains, herringbone drains, or sub-horizontal bored drains installed from ground level or from shafts or tunnelled galleries. In all cases the drains must be protected by suitable filters to prevent clogging; geofabrics are now widely used for this purpose. Surface drains should be taken down to undisturbed ground where practicable and filled

to the surface with durable free-draining material. Herringbone drains of this type have the additional effect of buttressing the surface layers of the cliff and improving its overall stability.

Seepage erosion, which occurs in fine sands and silts when soil particles are washed out by groundwater seepage leading to undercutting of the slope, can be controlled by the provision of suitable filters.

For rock cliffs, alternatives to trimming back are rock bolting and grouting. Clay slopes have also been stabilized by grouting.

31.11.4.3 Toe protection

The toe protection provided for coastal cliffs need only vary from that provided in other situations to take account of potential instability.

With clay cliffs the slope of the wall should not be significantly greater than the clay slope behind and may require a wide decking or berm, possibly at two levels in order to attain the required height. It may also incorporate sheet piling driven to intersect deep-seated slip planes.

Drainage behind the walls will usually be necessary and should be at a low level in order to reduce water pressures in the cliff. As it is difficult to keep the sea out of drains below tide level and the outfalls clear of beach material and seaweed it may be necessary to pump the flow.

If complete cliff stability is not certain, a flexible form of construction as shown in Figures 31.20 and 31.21 is economical to construct and to repair if damaged by cliff falls or movement.

31.12 Breakwaters

Breakwaters can be designed either to reflect or absorb wave energy but for reasons of construction or economy many are composite structures which reflect waves of certain characteristics or at particular water levels while under other circumstances they cause the waves to break and dissipate their energy.

31.12.1 Vertical-faced structures

The traditional mass concrete, masonry or concrete blockwork breakwater, of which there are numerous examples in the UK and elsewhere, is of the reflecting type, a typical example being the breakwaters at Dover.[48] These structures differ mainly in the method adopted for the construction of the foundations; at Dover this was done by excavating a trench in the chalk sea-bed working from a diving bell, and in it the blockwork was placed

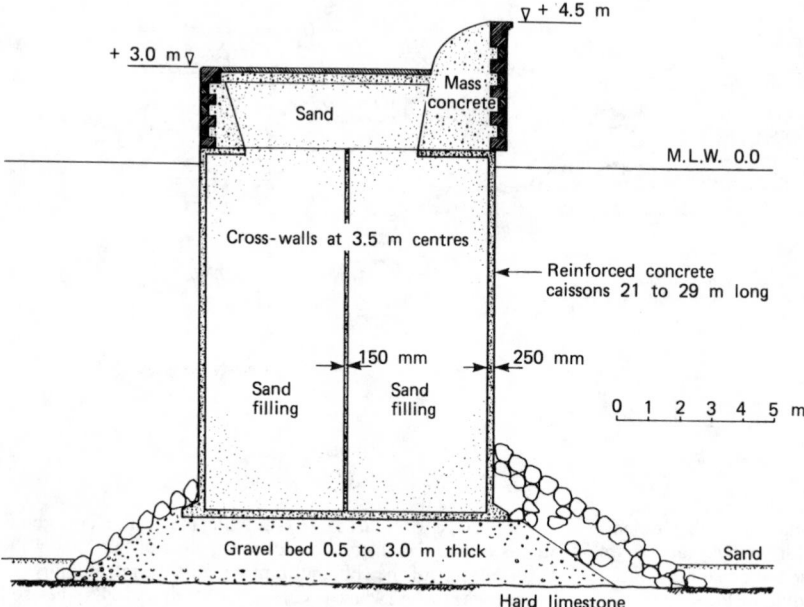

Figure 31.25 Caisson breakwater at Helsingborg, Sweden

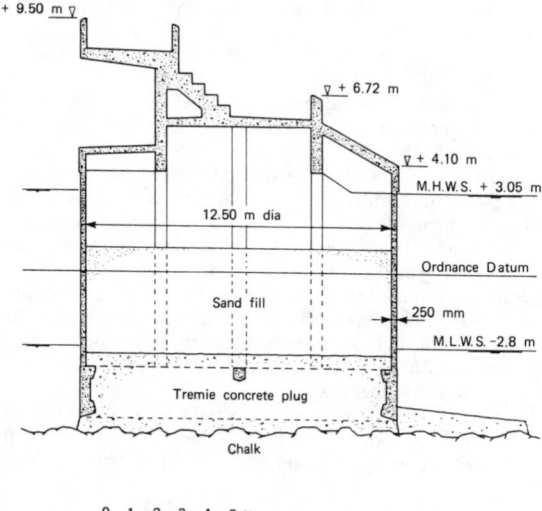

Figure 31.26 Cellular breakwater at Brighton

by goliath crane. Variations of this form of construction are the slicework breakwater, e.g. at Colombo,[49] in which the blocks are able to settle without disrupting the structure, and breakwaters with outer walls of blockwork or masonry and rubble or concrete heartings. At Newhaven[50] the breakwater is of mass concrete cast *in situ* on a foundation of 100-t concrete sacks deposited from a special bottom-opening vessel. A similar form of construction at Whitby[51] was carried out within shutters erected on the sea-bed from a gantry which was designed to 'walk' ahead of the work.

An alternative form of construction for this type of break-water is the caisson, constructed partly on shore and partly afloat in a building dock, floated into position and sunk on to a

prepared foundation. The breakwater at Helsingborg (Figure 31.25) is a typical example.

The main advantage of caissons is that very large units can be handled and with adequate space for building them progress can be rapid. However, their use is limited by the depth in which they can be floated.

A caisson system which does not have this limitation was used for two breakwaters at Hanstholm[52] (Denmark) and adapted for the harbour arms for the Brighton Marina. The caissons are 12-m diameter concrete cylindrical units weighing up to 600 t each, and open at their lower ends. They are constructed on shore and transported to the end of the completed part of the breakwater on a selfpropelled trolley where they were suspended in position clear of the sea-bed by a special crane while tremied concrete was placed in the bottom to form a massive foundation plug. A cross-section of the Brighton breakwater is shown in Figure 31.26.

31.12.2 Rubble-mound breakwaters

The rubble mound is the traditional form of energy-absorbing breakwater but there are numerous variations. In most cases quarried rock forms the core of the mound and, when economically available in suitable sizes, has frequently been used for the armour layers and capping. Where large rock is difficult to obtain, concrete armour units of various shapes have been used, among the best-known being the Tetrapod, Accropode, Tribar, Akmon, Stabit and Dolos. For rock armour, K_D values in Hudson's modification of Irribarren's equation given on page **31**/17 between 2.0 and 2.8 are suggested, depending upon the angularity of the rock and the position in the breakwater. For concrete units, values of K_D from about 8.0 to 18.0 have been used, but these are based on model tests which take no account of the strength of the individual units. In the event, failures have been initiated by breaking of the units, and the whole approach to the design of sloping structures armoured with concrete units is being reviewed.

It is usual to provide at least two layers of 'primary' armour

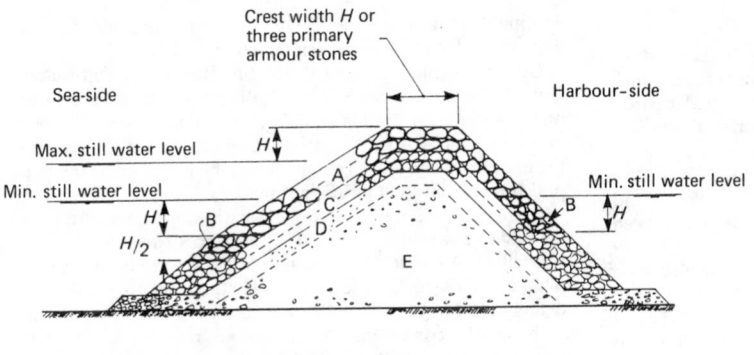

A – Primary armour 1 to 1.5 W_r
B – Primary armour 0.5 to 1 W_r
C - Secondary armour 0.6 to 0.1 W_r
D – Filter layer if required
E – Core-run of quarry max. size
0.05 W_r with not more than 20%
smaller than 10 kg

Figure 31.27 Type section for rubble-mounted breakwater.
A, primary armour 1 to 1.5 W_r; B, primary armour 0.5 to 1 W_r;
C, secondary armour 0.066 to 0.100 W_r; D, filter layer if required;

E, core – run of quarry. Maximum size 0.050 W_r with not more than
20% smaller than 10 kg

and below this one or more 'filter' or 'secondary armour' layers of a range of sizes of stone depending upon the voids in the primary armour and the size of the core rock. Figure 31.27 is a type section for this form of construction in which W_r is the weight of primary armour as determined above.

The choice of side slopes and armour size requires a considerable amount of judgement and will depend upon the availability of the required rock sizes, and the plant and equipment to be used for placing it. Generally a batter of 1 vertical to 1.25 horizontal is the steepest that can be used, this being the natural slope of the tipped rock core. However, the exposed side of the breakwater is liable to be flattened by wave action during construction and batters of 1 to 1.5 near the base and up to 1 to 3 for the upper slope are typical.

Variations of the rubble-mound breakwater include the replacement of the cap stones by a concrete deck and parapet, an example being the breakwaters at Port Talbot,[53] and the use of a sand core placed between successive pairs of rubble mounds built up as the work proceeds.

At a number of places, where overtopping of the breakwater, and the resulting wave action behind it, can be tolerated, the crest level may be close to or even below water level. The new North Mole at the Hook of Holland is in this category (Figure 31.28).

31.12.3 Rock-filled crib breakwaters

A type of breakwater intermediate between the fully reflecting vertical wall and the mainly absorbent rubble mound is the rock-filled crib. It can be constructed with piles and walings, as the crib-groyne in Figure 31.24, or the cribs can be fabricated on shore and floated into position.

Composite construction. Apart from the above distinct types of construction numerous combinations of them have been built and, for very deep water, mass concrete, blockwork and caissons have been founded on rubble mounds as at Casablanca, Bari (Italy), Marseilles and Funchal.

31.12.4 Piled breakwaters

Timber, concrete and steel sheet piling have been used in breakwater construction in a number of ways. Where wave action is not severe a single line of piles supported by a jetty structure may be adequate, but in more exposed situations a double line will be necessary, tied together with or without cross-walls at intervals and filled with sand, gravel, rubble or concrete. Straight web cellular sheet piled construction is also suitable if conditions are not too severe.

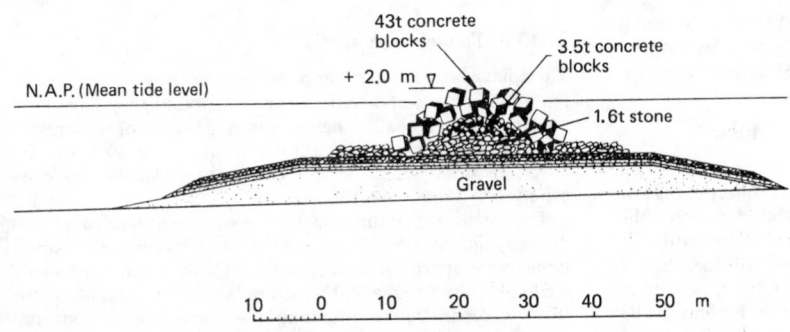

Figure 31.28 North mole at the Hook of Holland

31.12.5 Experimental breakwaters

A number of attempts have been made to design vertical breakwaters which absorb rather than reflect wave energy thus reducing the forces on the structure. The perforated caisson breakwater at Baie Comeau on the St Lawrence River is a notable example but it is not subjected to very severe wave action. Others have been built in Sicily and as a protective screen to an oil storage tank for the Ekofisk oilfield, and experimental work[54] aimed at extending their range has been carried out.

Experimental work and some prototype trials have also been made on pneumatic, hydraulic and floating breakwaters, but none of these has been fully developed and, for a number of reasons, it is doubtful if they will have much practical application.

31.13 Sea-water intakes and outfalls

Apart from pipes for gas and oil which are outside the scope of this chapter, pipelines are constructed in coastal waters for the disposal and dispersion of effluents, and the abstraction of sea water for cooling purposes or industrial processes and its return to the sea.

In order that effluents can be dispersed without nuisance and that relatively sediment-free water can be obtained at all times from an intake, which requires the entry ports to be below the wave troughs at lowest low water and the intake pipes or culverts to be below water level up to the pump suctions, works of considerable magnitude may be required. They fall into three main categories: (1) jointed pipelines; (2) pipelines towed or floated into position; and (3) tunnels and shafts.

31.13.1 Jointed pipelines

These include pipes laid and jointed by diver in a trench excavated in the sea-bed and backfilled with imported or dredged material, or partially or wholly with concrete in the case of a rocky sea-bed, or supported above the sea-bed on piled trestles or concrete saddles.

The most usual type of pipe for this form of construction is cast or spun iron with flexible joints, but steel, plastics, concrete and aluminium have all been used. The choice of pipe material should be made for each project, having regard to durability and method of installation; the cost of the pipe itself is usually a less important consideration.

The design of the work should take account of methods of construction and the type of equipment to be used, which may be conventional plant working on the shore between tides, from a gantry, from floating craft, or from a jack-up spud platform.

Allowance must be made for possible changes in sea-bed level to ensure that the pipe or its supports are not undermined, and if the pipes are supported above the sea-bed they must be designed to resist forces due to wave action and currents (see page **31**/15).

31.13.2 Pipelines towed or floated into position

Pipes installed by these methods, in which they are fabricated into long lengths on shore and towed into position by tugs or winches, have usually been of steel manufactured by either the longitudinal or spiral weld process and complying with the American Petroleum Institute Specification, APISpec5L, for line pipe and Standard, API Std1104, for welding. Protection to the pipe may be a mortar, epoxy or polyethylene lining, and externally a bitumen or coal tar impregnated glass fibre wrapping; the pipe is then encased in gunite or concrete to protect the wrapping against mechanical damage and provide negative buoyancy at the sea-bed when sealed and full of air.

Usually the pipes are joined into long 'strings' by butt welding and, if floated and sunk into position, the strings are joined under water with flexible couplings. For lines towed into position along the sea-bed the strings are butt welded together and the lining and wrappings completed at the joints as the pipes are pulled out from the fabricating site.

In this bottom pull method the negative buoyancy of the pipe is critical and small construction tolerances can have a significant effect. A check by on-site weighing of a length of prepared pipe is advisable to determine whether or not additional, detachable buoyancy needs to be added. Pulling forces vary with sea-bed conditions and can range from 0.5 to 1.0 of the negative buoyancy of the pipeline.

Maximum stresses in the pipe invariably occur during construction and arise from pipe curvature and spanning, pulling forces and external hydrostatic pressure while the pipe is full of air. Timoshenko[55] gives the maximum pressure, p_y, before yield of an imperfect cylinder as:

$$p_y^2 - \left[\frac{\sigma_y t}{R} + \left(1 + 6\frac{u_o}{t} \right) p_c \right] p_y + \frac{\sigma_y p_c t}{R} = 0 \tag{31.54}$$

p_c = critical collapse pressure of a perfect cylinder

$$= \frac{Et^3}{4(1-\mu^2)R^3} \tag{31.55}$$

where E is Young's modulus; σ_y is yield stress; μ is Poisson's ratio; R is pipe radius; t is pipe wall thickness; and u_o is maximum initial radial deviation.

For pipelines not exceeding 1 m in diameter high-density polyethylene has been used and in the smaller sizes has advantages in flexibility and ease of handling. In the larger sizes the flexibility becomes a problem and difficulties may also occur due to temperature changes, site welding in wet conditions and shape deformation. Because they are light in weight, polyethylene pipe must be anchored if it is to be left exposed on the sea-bed.

Concrete pipe has also been used for long outfalls using the techniques developed for submerged tube tunnels. For this operation the foundation or bedding conditions are of great importance.

There is some evidence that a pipe left exposed on a sandy sea-bed will, at least partially, embed itself naturally, but in exposed situations where the bed level is liable to fluctuate it is necessary to bury the pipe. This is usually done by dredging a trench prior to launching the pipe but ploughs and jetting equipment have been used for this purpose.

For further information on this subject the reader should consult Det Norske Veritas *Rules*[56] and Reynolds.[57]

31.13.3 Tunnels and shafts

There have been many instances of tunnels being constructed as sea-water intakes and outfalls. In soft ground they have been confined to fairly shallow depths allowing the use of compressed air. In rock there is more latitude in the choice of depth and cover, fissured strata being treated ahead of the work by grouting.

The connection of the tunnel to the sea has in some cases been made by flooding the tunnel and breaking through with a single large round of explosive, but more usually by one or more shafts leading to some permanent structure (a tower or caisson) on the sea-bed, or drilled from floating plant or a jack-up platform. For more detail of this method of construction see Moore and Osorio.[58]

References

1 Doodson, A. T. and Warburgh, H. D. (1973), *Admiralty manual of tides*. HMSO, London.

2 Ippen, A. T. (1966) *Estuary and coastline hydrodynamics*. McGraw-Hill, Maidenhead.

3 Stokes, G. C. 'On the theory of oscillating waves'. *Trans. Cambridge Phil. Soc.*, **8** and Supplement, *Sci. Papers*, **1**.

4 Wiegel, R. L. *Gravity waves, tables and functions*. Council on Wave Research, The Engineering Foundation, London.

5 Coastal Engineering Research Center (1984) 'Shore protection manual' (4th edn). US Government Printing Office, Washington.

6 Bourodimos, E. L. and Ippen, A. T. (1968) 'Wave reflection and transmission in channels of variable section'. *Proceedings, 11th Conference on Coastal Engineering*, vol. 1, London.

7 Pierson, W. J. Jr, Gerhard Neuman and James, R. W. (1955) *Practical methods for observing and forecasting ocean waves by means of wave spectra and statistics*. Navy Hydrographic Office, Publication No. 603. NHO.

8 Bretschneider, C. L. (1952) 'Revised wave forecasting relationships', *Proceedings, 2nd Conference on Coastal Engineering*, Houston.

9 Carter, D. J. J. (1982) 'Predictions of wave height and period from a constant wind velocity using the JONSWAP results'. *Ocean Engng*, **9**, 1.

10 Thijsse, T. Th. and Schijf, J. B. (1949) 'Report on waves', *Proceedings, 17th International Navigation Congress*, Section II: 'Communication'. 4. Lisbon.

11 Rottman, W. and Zielke, W. (1983) 'FEM analysis of combined diffraction and refraction in a harbour'. *Proceedings, International Conference on Coastal and Port Engineering in Developing Countries*, Colombo.

12 Suthons, C. T. (1963) 'Frequency of occurrence of abnormally high sea-levels on the east and south coasts of England'. *Proc. Instn Civ. Engrs*, **25**.

13 Lennon, G. W. (1963) 'A frequency investigation of abnormally high tidal levels at certain west coast ports', *Proc. Instn Civ. Engrs*, **25**.

14 Ackers, P. and Ruxton, T. D. (1974) 'Extreme levels arising from meteorological surges'. *Proceedings, 14th Coastal Engineering Conference*, Copenhagen.

15 Pugh, D. T. and Vassie, J. M. (1980) 'Applictions of the joint probability method for extreme sea-level'. *Proc. Instn Civ. Engrs*, Part 2, **69**.

16 Blackman, D. L. and Graff, J. (1978) 'The analysis of extreme sea-levels at certain ports in southern England'. *Proc. Instn Civ. Engrs*, Part 2, **65**.

17 Oil Companies International Marine Forum (1978) *Guidelines and recommendations for the safe mooring of large ships at piers and sea islands*. Witherby.

18 Inglis, Sir Claude and Kestner, F. J. T. (1958) 'Changes in the washes as affected by training walls and reclamation works', *Proc. Instn Civ. Engrs*, **11**, 435–466.

19 Price, W. A. and Kendrick, M. P. (1963) 'Field and model investigation into reasons for siltation in the Mersey estuary'. *Proc. Instn Civ. Engrs*, **24**.

20 Kuenen, Ph.H. (1950) *Marine geology*, Wiley.

21 Abraham, G. (1963) *Jet diffusion in stagnant ambient fluid*. Delft, Hydrological Laboratory Publication No. 29. DHL, Delft.

22 Frankel, R. F. and Cummings, J. D. (1965) 'Turbulent mixing phenomena of ocean outfalls'. *Proc. Am. Soc. Civ. Engrs*, **91SA**.

23 Anwar, H. O. (1969) 'Behaviour of buoyant jet in a calm fluid'. *Proc. Am. Soc. Civ. Engrs*, **95HY4**.

24 Agg, A. R. and Wakeford, A. C. (1972) 'Field studies of jet dilution of sewage at sea outfalls'. *J. Inst. Pub. Health Engrg*.

25 Tesaker, E. (1969) 'Uniform turbidity current experiments', *Proceedings, 13th Congress of International Association Hydraulic Research*.

26 Elliston, T. H. and Turner, J. S. (1959) 'Turbulent entrainment in stratified flow'. *J. Fluid Mech.*, **6**.

27 Wiegel R. L. and Beebe, K. E. (1956) 'The design wave in shallow water'. *J. Waterways Div., Proc. Am. Soc. Civ. Engrs*, **82**, WWI.

28 Wiegel, R. L. and Skjei, R. E. (1958) 'Breaking wave force prediction'. *J. Waterways and Harbors Div. Proc. Am. Soc. Civ. Engrs*, **84**, WW2.

29 Bretschneider, C. L. (1958) 'Selection of design wave for offshore structures. *J. Waterways and Harbors Div., Proc. Am. Soc. Civ. Engrs*, **84**, WW2.

30 Munk, W. H. (1949) 'The solitary wave theory and its application to surf problems, ocean surface waves' (*Ann. N.Y. Acad. Sci.*), **51**.

31 Wiegel R. L., Beebe, K. E. and Moon, J. (1957) 'Ocean wave forces on circular cylindrical piles'. *J. Hydraulics Div. Proc. Am. Soc. Civ. Engrs*, **83**, HY2.

32 Morison, J. R., Johnson, J. W. and O'Brien, M. P. (1953) 'Experimental studies of forces on piles'. *Proceedings, 4th Conference on Coastal Engineering*.

33 Hallam, M. G., Heaf, N. J. and Wooton, L. R. (1977) *Dynamics of marine structures*. Construction Industry Research and Information Association, Report No. UR8, CIRIA, London.

34 Bagnold, R. A. (1939) 'Interim report on wave pressure research'. *J. Inst. Civ. Engrs*, **12**.

35 Terrett, F. L., Ganly, P. and Stubbs, S. B. (1979) 'Harbour works at Brighton Marina: investigations and design'. *Proc. Instn Civ. Engrs*, Part 1, **66**.

36 Miche, R. (1951) 'La pouvoir réfléchissant des ouvrage maritimes exposés à l'action de la houle'. *Ann. Ponts Chauss.*, 121.

37 Hudson, R. Y. (1959) 'Laboratory investigation of rubble mound breakwaters'. *Waterways Harbors Div., Proc. Am. Soc. Civ. Engrs*, Paper No. 2171.

38 Kitt, J. D. and Ackers, P. (1982) *Review of field and laboratory tests on rip rap*. Construction Industry Research and Information Association Report No. 94. CIRIA, London.

39 The Admiralty (1969) *Admiralty manual of hydrographic surveying*, vol. 2, Ch. 2. NP 134b(2), HMSO, London.

40 Allsop, N. W. H., Bradbury, A. B., Poole, A. B., Dibb, T. E. and Hughes, D. W. (1985) *Rock durability in the marine environment*. Hydraulics Research Limited Report No. SR11.

41 Allen, R. T. L. and Terrett, F. L. (1968) 'Durability of concrete in coast protection works'. *Proceedings, 11th Conference on Coastal Engineering*, London.

42 Buenfeld, N. A. and Newman, J. B. (1985) 'Permeability of marine concrete'. *Proceedings, International Conference on Concrete in the Maritime Environment*. The Concrete Society, London.

43 Burdall, A. C. and Sharp, J. V. (1985) 'Some aspects of revisions to the UK guidance notes for offshore structures'. *Proceedings, International Conference on Concrete in the Maritime Environment*. The Concrete Society, London.

44 Van Asbeck, W. F. (1959) *Bitumen in hydraulic engineering*, vol. 1, Shell International Petroleum Company, London; vol. II (1964), Elsevier, Amsterdam.

45 Visser, W. (n.d.) *Coast protection with bitumen*, Shell Bitumen Reprint No. 20, Shell International Petroleum Company, London.

46 Kerkhoven, R. E. (n.d.) *Hydraulic applications in The Netherlands*. Company Report No. 110 F, Shell International Petroleum Company, London.

47 Hutchinson, J. N. (1982) 'The geotechnics of cliff stabilization'. *Proceedings, Institution Civil Engineers Conference on Shore Protection*, Southampton.

48 Wilson, M. F. G. (1921) 'Admiralty harbour, Dover'. *Minutes, Proc. Instn Civ. Engns*, **209**.

49 Kyle, J. (1886/87) 'Colombo harbour works, Ceylon'. *Minutes, Proc. Instn Civ. Engrs*, **87**, Part 1.

50 Carey, A. E. 'Harbour improvements at Newhaven, Sussex'. *Minutes, Proc. Instn Civ. Engineers*, **87**, Part I (1886/87).

51 Mitchell, J. (1921) 'Whitby harbour improvement'. *Minutes, Proc. Instn Civ. Engrs*, **209**.

52 Lundgren, H. (1962) 'A new type of breakwater for exposed positions'. *Dock and Harbour* (Nov.).

53 McGarey, D. G. and Fraenkel, P. M. (1970) 'Port Talbot harbour: planning and design'. *Proc. Instn Civ. Engrs*, **45**.

54 Terrett, F. L., Osorio, J. D. C. and Lean, G. H. (1968) 'Model studies of a perforated breakwater'. *Proceedings, 11th Conference on Coastal Engineering*, London.

55 Timoshenko, S. P. (1941) *Strength of materials* (2nd edn), Part 2. Macmillan, London.

56 Det Norske Veritas (1976) *Rules for the design, construction and inspection of submarine pipelines and pipeline risers*.

57 Reynolds, J. M. (1980) 'Design and construction of seabed outfall'. *Proceedings, Institution Civil Engineers Conference on Coastal Discharges*, London.

58 Moore, K. H. and Osorio, J. D. C. (1980) 'Tunnel outfall design and construction'. *Proceedings, Institution Civil Engineers Conference on Coastal Discharges*, London.

32

Tunnelling

A M Muir Wood FRS, FEng, FICE
Consultant, Sir William Halcrow & Partners

Contents

Many concerned with tunnelling continue, at their own peril and that of others, to underestimate the need for practical understanding of the behaviour of the ground, the essence of good tunnelling. No two tunnels are the same; experience, and real insight of the value of that experience, are necessary to transmute particular experience to more general understanding and thus to transmit the experience of one tunnel appropriately to another.

Advances in tunnelling usually arise not from research so much as from innovations in methods of design and construction. Monitoring of the results in the field may then follow, supported by research where existing knowledge fails to explain the findings.

Two essential elements to economic tunnelling are:

(1) The tunnel (unless permanently unlined) must be considered as a composite ground-lining structure. Not only does the lining support the ground but the ground in its turn supports the lining.
(2) The design of the permanent tunnel must be considered in association with the methods of construction. The overall cost of the process requires to be minimized and the finished geometry is only one of many factors.

There are many barriers to a full understanding of the behaviour of the ground around a tunnel: (1) the three-dimensional, time-dependent nature of the problem; (2) the complexity of the stress–strain relations in soft ground; (3) the effects of the initial state of stress, discontinuities and joints upon the behaviour of a rock; (4) the dependence upon the method of excavation; (5) the standard of workmanship and (6) inhomogeneity of the ground.

Full-scale tunnels provide the one reliable laboratory for testing theory against practice.

32.1 The options for a tunnel route

The ground is the principal determinant of the cost of a tunnel of a given size. For this reason great economic benefits may derive from the capability of selecting a favourable and relatively consistent type of ground for tunnelling. Until the geological structure is known, the object should be to keep the options for a tunnel route as open as possible.

For each type of tunnel there are certain geometrical constraints and other specific factors affecting cost. For a road tunnel, for example, acceptable gradients and curves will be related to the design speed and, hence, to traffic costs.[1] For a pressure tunnel, on the other hand, there is little direct geometrical constraint and the differential cost of construction in relation to the ground would need to be considered against the capitalized head losses.

A general knowledge of the geological structure will indicate whether or not the most direct route conforms to a favourable geological horizon or whether, on the contrary, it may encounter unstable ground such as squeezing rock, running sand, major fault zones, decomposed rock, karstic limestone or similar hazards which may only be penetrated at great expense.

Where there is a possibility of adopting an economic method of tunnelling, related specifically to a type of ground with limited variation, there may be the greatest benefit from diverging from the most direct route, in order to situate the tunnel throughout in such ground.

At the earliest stage in planning, such factors should be considered so that the options may be described, systematically tested and reduced as information arises from the first stage of site investigation.

32.2 Costs of tunnelling

32.2.1 Principal factors

Attempts are made periodically to set out tunnelling costs in a systematized form, with costs per unit length of a certain size of tunnel related to a few generalized ground types and to a few other simplified categories of accessibility and tunnel length. Except for specific areas in which the ground can be reliably depended upon, there is no valid way of expressing tunnelling costs on a simple unit cost basis.

From a knowledge of the ground a system of tunnelling may be selected and the costs evaluated on an assumed average rate of progress. The rate of progress may be assessed from experience in similar ground elsewhere, taking account of any innovation in the tunnelling method, and not forgetting the costs of ground treatments or similar ancillary operations. In general, the extent of variability in the cost of tunnelling is increasing for these principal reasons:

(1) Tailor-made tunnel systems to suit a particular type of ground permit increasing economies in construction.
(2) The cost of labour-intensive tunnelling systems adopted for difficult ground or in congested circumstances will naturally reflect the trend of labour costs including incentive payments.
(3) The demand for tunnels in urban development tends to reduce the options available for a tunnel route.

As the result of these factors, at the present time there is at least an order of magnitude between the unit cost of constructing the cheapest and the most expensive tunnel of the same size. Hence, there is an increasing benefit to be derived from undertaking studies appropriate to choosing the most economic expedient in each situation.

A feature that may be overlooked in comparing the costs of tunnels concerns the means of access during construction. While a shallow urban tunnel or a short tunnel through a hill may be approached directly from the ground surface, long and subaqueous tunnels usually require working shafts and access headings, adding not only to the direct cost of the project but also to the cost of all the consequent tunnelling operations.

32.2.2 Effect of tunnel size

The cost of a given tunnel is specific to its situation and its timing, on account of the varying differences in prices, varying local skills and technical capabilities. There is therefore no simple factor to be applied to the cost of a tunnel in order to determine its hypothetical cost at a different place or time.

Neither is there a simple formula to determine the cost of a tunnel by consideration of another tunnel in the same ground and conditions but of different size. As a simplification, where variation in size does not entail a change in basic techniques, we may consider each factor in construction as entailing a unit cost U expressed as:

$$U = A + Bd + Cd^2 \tag{32.1}$$

where A, B and C are constants and d is the finished diameter

For a highly mechanized system, A will be high, while for a labour-intensive system C will be high. For excavation there will be an appreciable element in spoil disposal costs for which C will predominate while for temporary tunnel supports A and B will be the principal factors.

As the size of tunnel is reduced, the increasing congestion leads to reduced efficiency in working. In consequence, there is a size of tunnel for which the costs will be a minimum; the greater the degree of mechanization, the greater will be the size d_{min} for minimum cost (i.e. B and $C \to 0$ as $d \to d_{min}$). For a long length of tunnel in the London Clay the minimum cost is obtained for a tunnel diameter of about 2.5 m while for certain machine-driven tunnels in soft rock the optimum diameter has been found to be about 3 m, and about 2 m for a hand-driven tunnel in hard rock.

32.3 Systematic site investigation

32.3.1 Geological data

The scheme for determining the geological conditions should work from the general towards the particular. This will entail a study of geological maps and papers, first on a regional and then on a local basis. In the UK there are normally available sheets at scales of 1:50 000 and 1:10 000 with explanatory memoirs, produced by the Institute of Geological Sciences. Where geological maps do not exist, aerial photographs often provide useful information on the geological structure.

32.3.2 Objects

According to the apparent options for the tunnel the scheme of site investigation may then be designed with these main objects:

(1) To test geological data at doubtful points.
(2) To explore particular areas of tunnelling difficulty.
(3) To obtain information necessary to complement available data on important aspects of geology and geohydrology.
(4) To obtain samples for testing and to undertake *in situ* tests in order to establish the suitability of ground for alternative methods of tunnelling.
(5) To determine design and construction parameters.

Far too often a site investigation is undertaken without adequate thought to its purpose; in consequence, information vital for good tunnelling is overlooked at the expense of acquiring much irrelevant material. The site investigation should be supervised by those with a direct practical understanding of the associated techniques of tunnel design and construction.

32.3.3 Means

A few large boreholes or adits may be justified for direct examination, *in situ* testing and for subsequent inspection by tendering contractors and others.

There is no general rule on the spacing between boreholes. At one extreme, for sedimentary rocks of a uniform character it may only be necessary to be able to establish general continuity of the geological sequence by identification of marker beds or horizons. At the other extreme, igneous intrusions and metamorphosed rocks may present so complex a pattern as to necessitate a tunnelling method highly tolerant to change, however well the ground may be investigated. A good general rule is to establish during site investigation a set of hypotheses, concerning the geological structure and the properties of the ground to be tested so that when a conflicting anomaly is indicated by a new borehole its significance is appreciated, i.e. is the benefit of an additional borehole likely to exceed its cost? Where there is doubt concerning the practicability of adopting a mechanical system of tunnelling, special care is required to ensure exploration of the ground in sufficient detail to determine the feasibility of the scheme.[2]

Geophysical methods of exploration may serve not only to extend the data from individual boreholes in the second and third dimension but also to reveal specific features such as faults and igneous intrusions. Without adequate 'fixes' geophysical results may permit widely different possible interpretations.

Benefits are usually to be found in undertaking a site investigation in two or more stages, depending on the initial knowledge of the terrain, the magnitude of the project and the diversity of possible options. The investigation should be designed initially to investigate those features most likely to determine the tunnel location; otherwise money and time are wasted on investigation too far from the selected line to be of great value. However, in ground variable to a common pattern, information obtained away from the tunnel route may yet be relevant; the validity of such transference needs careful assessment.

Water constitutes a hazard encountered in many forms. The site investigation should, as appropriate to the circumstances, be designed to provide information about water-bearing fault zones, fault zones with a weak filling, open joints and the effect of tunnelling upon aquicludes whose rupture may expose the tunnel to water from aquifers. The geological structure and the possible head of water will control the zone of ground around the tunnel which calls for investigation.

32.4 Tunnelling methods related to the ground

32.4.1 Historical background

The history of tunnelling is one of increasing diversification of methods with an increasing capability to explore and to understand the ground.

While Brunel used the first tunnelling shield for the Thames Tunnel in 1825–28, tunnels throughout the nineteenth century continued generally to be constructed by means of one of the traditional methods of excavation and timbered support.[3] Although these are now largely of historical interest only, the English method, widely and successfully used, sometimes in soft ground and in broken jointed rocks where other methods had failed, merits mention.

An essential feature of the English method concerned the use of longitudinal crown bars, supported at the forward end on props and sill and at the rearward end on the last section of completed permanent lining, which might be brickwork or masonry. In this way continuous support was provided to the ground over the tunnel from the time of first excavation and, in principle, the method may be considered as the forerunner of the tunnelling shield.

32.4.2 Shield tunnelling

Shield tunnelling is strongly associated with the name of Greathead. He worked with the first circular shield designed by Barlow for the Tower Subway beneath the River Thames in 1869. Greathead designed a shield (Figure 32.1) for the South London Railway in 1886–90 incorporating most of the essential features which have survived to the present day.[4] Greathead not only recognized that a shield reduced the risks in tunnelling in water-bearing ground but he was also one of the few of his time to appreciate that it permitted faster and cheaper tunnelling in good ground.

The first shield with a mechanical cutting head was the Price excavator used in 1897 for the Central London Railway.

Since this time there have been rapid developments, predominantly in Japan, followed by Western Germany, the US and Canada, of mechanical shields provided with means for (partially) balancing soil and groundwater pressures.[5] For the most open-textured grounds, sands and gravels, the choice may be between a bentonite (Figure 32.2) or a type of earth-balance shield. This latter may use a fully plated head with controllable

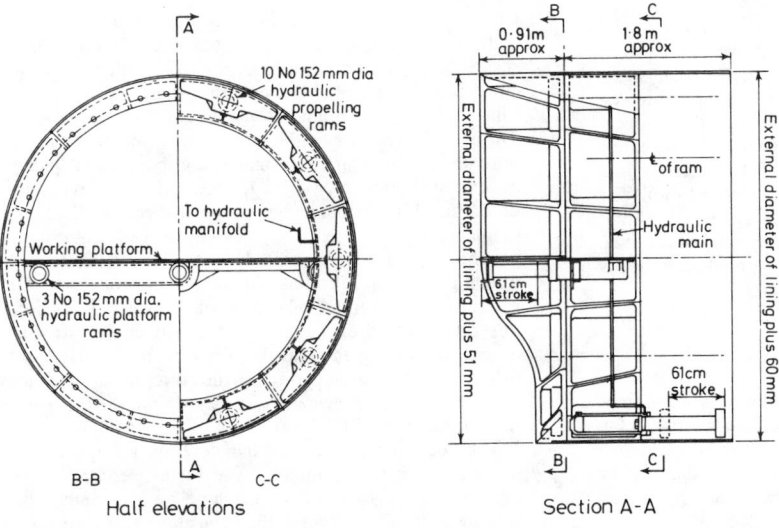

Figure 32.1 Hooded Greathead shield with platform rams suitable for 3.5-m diameter tunnel

slots or (Figure 32.3), alternatively, the shield may be open-faced and the soil extracted at a controlled rate by an archimedean screw conveyor as the shield advances. For finer-textured ground, a slurry shield may be used with natural clay mixed with water and used for pumping the spoil to the surface. The hydro shield for yet finer ground maintains the face of the shield under hydraulic pressure with a supply of water mixed with the spoil which, again, is transported by pumping. Developments in these directions have incorporated a number of novel features. One particular contribution has been the perfection of seals between the shield tail and the enclosed lining; another, for the pressure balancing shield, has been the use of an air-vessel in the pressurized face of the shield to dampen pressure fluctuations. All such special shields are designed for limited variability of the ground and contingency measures may need to be incorporated to deal with departures, such as the presence of large boulders. In Japan, developments are proceeding towards full automation so that all operation of advancing and steering the shield, excavating the ground and transporting spoil are controlled from the surface.

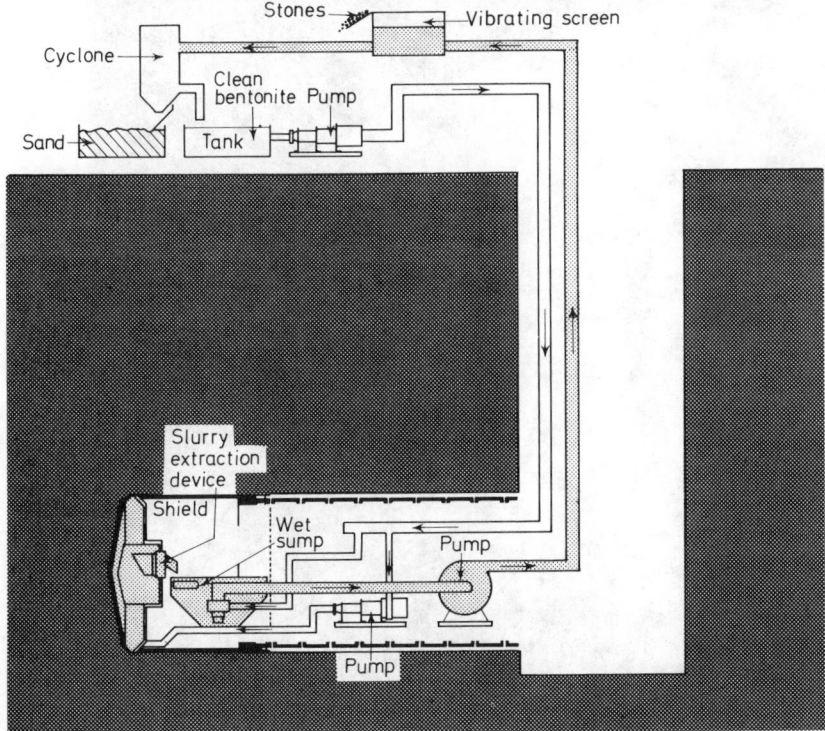

Figure 32.2 Bentonite tunnelling machine. (After National Research Development Council)

32.4.3 The bentonite shield

It is possible, where frequent access is not required to the face of a shield, to provide for ground support by compressed air, water or mud confined to the face of the shield. Air is not, however, recommended for this purpose. The use of mud in this application offers considerable benefits in permitting an approximately balanced pressure over the full height of the face and in providing a suitable medium for the pumping away of spoil. The first such shield was used in Mexico City,[5] utilizing the mud formed from the natural montmorillonitic clay spoil of the area. A true bentonite tunnelling machine was first used successfully in London (Figure 32.2) in sands and gravels.[6]

32.4.4 Rock-tunnelling machines

Many tunnelling machines for rock have been evolved since 1956, although here again the prototype machine belongs to the last century, usually attributed to Beaumont,[7] used for Channel Tunnel heading in 1881–82 and subsequently for the Mersey Railway Tunnel.[8] There are several features of such machines[9] which merit differentiation as shown below.

(1) *Cutters* (see Figure 32.4). For the softest rock the cutters are fixed picks which chisel the rock out as a succession of grooves. For harder rocks, generally in ascending order of hardness, machines make use of single or multiple-disc cutters, toothed cutters, roller cutters or cutters with tungsten carbide insert buttons.

(2) *Cutter heads*. For the smaller machines a single full-diameter rotary head is adopted (Figure 32.5). As the machine size increases so there is a tendency to introduce planetary cutters to share the work between cutters and reduce the range between minimum and maximum cutter speeds.

(3) *Thrust of machine*. For the softest rock, machines receive purchase by rams thrusting against a gripper ring expanded against the periphery. With a few exceptions the remainder of the machines obtain their forward thrust by means of diametrically opposed thrust pads jacked against the ground. All such machines advance by periodically withdrawing and repositioning the thrust ring or pads. One machine uses a central pilot drill which is firmly anchored into a hole ahead of the face. Not only does this provide a means for pulling the machine forward but it also establishes a firm forward bearing for the cutter head which may be a valuable feature where rock variation in the face causes uneven loading on the head.

The road header, a machine developed originally for mining, has a rotary milling head on a telescopic boom, attached to the body of the machine by a universal joint (Figure 32.6). Thus a typical machine may excavate a gallery up to 4.5 m high and 5.8 m wide. The cutter head usually mounts picks in the pattern of a conical scroll. Operation is by means of a 'sump' formed in the face, extended by lateral pressure on the rotating head. Generally the loading on each pick will be less, and less controllable, than for a full-face machine and hence for the limitations on rock strength for effective application.

Figure 32.3 Shield with fully-plated head with controllable slots

A

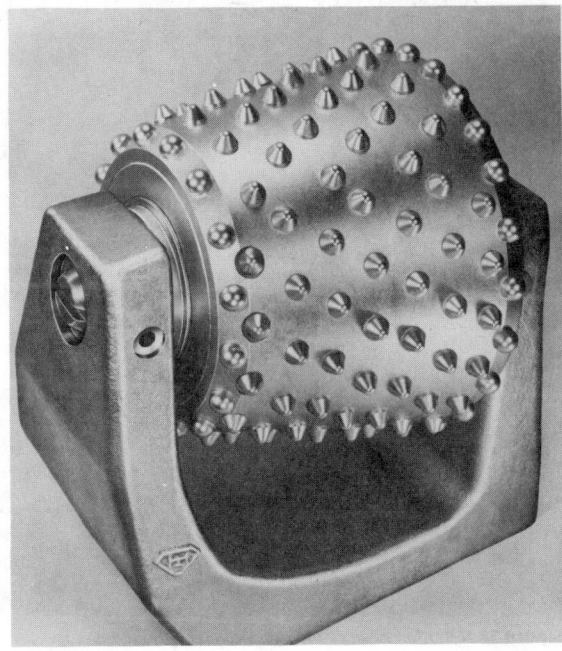

B

C

Figure 32.4 Types of cutters for tunnelling machines
32.4a 'Series 12' tooth cutter–MNX
32.4b 'Series 12' HHIX cutter
32.4c Bolt-on-disc cutter, type DGX (*Courtesty*: Hughes Tool Co.)

The selection of a tunnelling machine must take account of the rock types to be encountered along the length of drive. The efficiency of the machine is related to the inherent properties of the rock, principally to strength, the extent of jointing, strain modulus and abrasion. The aim is to minimize the specific energy needed to fragment a rock by keeping the size of particle high and by provoking brittle fracture. The cutter action aims to set up a high local difference in principal stresses and high-enough tensile stresses to induce cracking. Each type of cutter and pattern of cutters operates most efficiently in a rock whose properties lie within a limited range. Difficulties may arise from several causes, e.g.: (1) from excessive wear of the cutters in hard rock which grooves instead of fragmenting; (2) from inefficient fracture of rock too soft or plastic for the type of cutter; (3) from excessive wear due to overheating in the presence of high content of silica; (4) from excessive bearing loads where rock varies appreciably in the face; and (5) from jamming of the head where hard rock is heavily jointed and tends to collapse on to the machine or at the face.[10]

A simple criterion for the economics of machine excavation concerns the cost of repairing and replacing cutters. In sound soft rock this will be found to be a trivial sum in relation to other costs of excavation. For the hardest rock, the cost will be found to climb to a figure of several pounds per cubic metre, with stoppages every few metres for replacement, and this stage represents the present economic limit.

The tolerance of a machine to the full range of rock types to be encountered should be considered in weighing the overall merits and costs of its introduction. Another essential question concerns prediction of the need for temporary support close to the face, a process that presents greater difficulty for the full face machines and for which object certain machines make special provision in their design.[11]

Figure 32.5 Cutter head of hard-rock machine (*Courtesy*: The Robbins Co.)

Figure 32.6 Road header in iron mine (*Courtesy*: Anderson Mavor Ltd)

32.5 Tunnel construction

32.5.1 Drilling and blasting

The traditional scheme of advancing rock tunnels has been by drilling and blasting and this method continues to be generally adopted for short tunnels, hard rock tunnels and for tunnels in variable ground. At the present date, for example, machine tunnelling is unlikely to be economic in shattered rock or in rock of strengths greater than 200 MN/m².

The principle behind blasting in a tunnel is to obtain the greatest 'pull' for the minimum explosive charge and for the minimum damage of the rock around the tunnel. Secondary objectives are: (1) to fragment the rock adequately; and (2) to form a compact stock pile against the face.

The pattern of drill holes is designed to suit the rock and the explosive. Cut holes are arranged towards the centre of the face, usually inclined towards each other in order to remove a cone or wedge. One or more central unloaded holes of larger diameter may be used to assist the cut. The remainder of the holes are drilled parallel to the tunnel axis. Delays of a few milliseconds are used between groups of drill holes, from the cut outwards, so that the excavation is enlarged with the travel of the shock wave.

Considerable effort has been applied to establishing the neatest periphery to the excavation by the trimming holes, which may be charged or uncharged. In the technique of pre-splitting, the trimming holes are fired before the remainder with distributed charges to cause cracking around the periphery between adjacent holes. Another technique which has been used in tunnelling is termed smooth-blasting, whereby the line of trimmer holes is required to coincide with the periphery of the excavation, each being loaded with a reduced distributed charge and fired with a short delay after the remainder. It may be well worth considering means for reducing overbreak by careful control of the spacing, line and charging of the trimmer holes. The geometry of the drills or the drill carriage should be designed to permit the trimming holes to be drilled as parallel as possible to the tunnel axis. Care in these respects may show considerable benefit not only in reduction of direct overbreak but also in the reduced extent of the surrounding zone of cracking and displacement of the rock, with consequent savings in the extent of temporary support.

Sectional drawings of tunnels have often indicated the periphery of the 'minimum section' and the 'payment line' which allows payment for overbreak to be assessed in relation to the volume of excavation and the volume of concrete lining. Occasionally a 'limit line' is also shown, beyond which a leaner concrete mix may be used for filling. Overbreak in a tunnel is frequently expressed as a percentage of sectional area but, without knowledge of the size of tunnel, this designation has little significance.

The present tendency[12] is to indicate surface areas of different sizes of tunnel, possibly subdivided into different types of ground, in order to facilitate translation of overbreak into the corresponding additional volumes of ground to be excavated.

For small tunnels, hand-operated drills are used on telescopic air-legs. For larger tunnels there is usually a wider choice, including ladder drills, light mobile boom-mounted drills or heavier drills mounted on a jumbo. The latter may provide advantage in controlling the drill pattern and with the speed of drilling, also in protection close to the face for other operations; the main disadvantage arises from inflexibility in the event of departure from full-face driving. For the Mont Blanc Tunnel, for example, it was fortunate that a jumbo was used only from the French end, since difficulties encountered along the Italian drive compelled the enlargement from headings over a considerable length of tunnel.[13]

A more recent development has been the introduction of the hydraulic drill, offering a rate of drilling some 50 to 100% greater than the corresponding pneumatic rotary percussion drill, at a considerable reduction in noise level of 10 to 15 dB.

32.5.2 Spoil handling

The handling of spoil from the face cannot be considered separately from the method of excavation. Mechanical shields and tunnelling machines have built-in chain or belt conveyors loading to a hopper or to another conveyor. The same operation is achieved in a drill-and-blast tunnel by means of a mechanical loader, often with composite face shovel and conveyor. The general trend is to use rail wagons for transport for tunnels up to about 7 m diameter and for tunnels worked from vertical shafts and to use dump trucks for large tunnels directly accessible from the surface or for tunnels at a gradient of more than about 2.5%.

Many solutions to the problem of loading rail cars at the face have been adopted. One currently used in relatively small tunnels (say 3 m diameter) is to use a long transit car with an armoured conveyor floor so that spoil loaded at one end may be evenly distributed. Another system for rather larger tunnels (say 4 to 5 m diameter) uses an overhead conveyor capable of loading in turn each of a train of six or more (or fewer) rail cars, preferably to contain the spoil from a complete round. An alternative uses a long sliding platform with rail track and turnouts in consequence maintained close to the working face.

Conveyors are also used for dry materials and where access is by inclined shaft. The pulverizing of spoil and its discharge by pipe as a slurry has been adopted for suitable soft rock. Frequently the bottle-neck in materials handling is found to occur at the foot of a working shaft and here mining practice has introduced the use of automatic tipping of tunnel wagons into large hoppers from which shaft skips are rapidly loaded. The entire process of excavation and removal of spoil merits considerable study at an early stage as to its adequacy, with contingency plans to overcome foreseeable causes of break-down.

32.5.3 Tunnel lining

The method of tunnel lining is essentially related to the nature of the ground and to the scheme of excavation. General-purpose tunnel lining has economic application to small tunnels in variable ground. Recent progress and attendant economy have been demonstrated to result from the capability for designing the lining specifically to the condition of the ground and the overall tunnelling system.

The first subdivision in type of lining results from whether or not the need exists for an immediate support at the face. In North America the common practice in tunnelling in soft ground has been to tunnel by hand, to erect continuous support in timber sets or steel liner plates and subsequently to place an *in situ* concrete lining. In the UK, and generally throughout Europe, shields have been more widely used together with permanent primary segmental linings.

The traditional lining over more than 100 yr has been the ring of bolted cast-iron segments built within the protection of the tail of the shield, with the external annulus often grouted with lime or cement. Improvements in site investigation procedure have allowed the development of alternative types of lining which can be adopted in certain restricted types of soft ground.

Reinforced concrete segments[14] have been preferred to cast-iron segments for reasons of cost since 1938 except where loading is heavy or where watertightness is an essential object.

Another general type of tunnel lining is built in rings of segments immediately behind the shield. Each ring is then expanded directly against the ground with elimination of the procedures of bolting and grouting. Evidently the system can only be used where the ground around the tunnel is self-supporting over the width of a ring for a short period and thus a certain minimum apparent cohesion of the ground is necessary. Such techniques were developed predominantly in London Clay, sufficiently stiff (i.e. with a low enough stability ratio – see page **32**/18) and homogeneous for a specifically designed system.[15] Two types of lining based on this principle merit mention.

The Donseg lining is created from rings of tapered segments, expansion against the ground being achieved by the process of inserting alternate segments, as longitudinally tapered keys, into the ring by the shield rams (Figure 32.7). This is a highly economic method, limited to tunnels of diameter not exceeding about 3 m, because of the geometry of the lining.

For larger tunnels, the Halcrow lining provides for articulating joints between segments. In this way, a part ring of segments may be assembled clear of the extrados. The insertion and expansion of jacks between special segments cause the ring to expand against the ground, accompanied by relative rotation between adjacent segments (Figure 32.8). A special feature of a lining of this type is that secondary stresses are limited to a low level with consequent savings in the structural thickness of the lining. For the Cargo Tunnel at Heathrow[11] a lining 300 mm thick has been used for a 10.3-m diameter tunnel.

One of the most highly developed linings of this type has been developed by Holzmann and used *inter alia* by Wayss and Freytag for metro tunnels in Antwerp in open water-bearing ground, using a slurry shield and necessitating high standards of water tightness.[16] Each ring (Figure 32.9) comprises eight longitudinally tapered segments, the width of the ring itself being tapered so that all segments are built without packings, corrections to line and level being achieved by relative rolling of the ring. This is one example of recent concrete and (ductile) iron

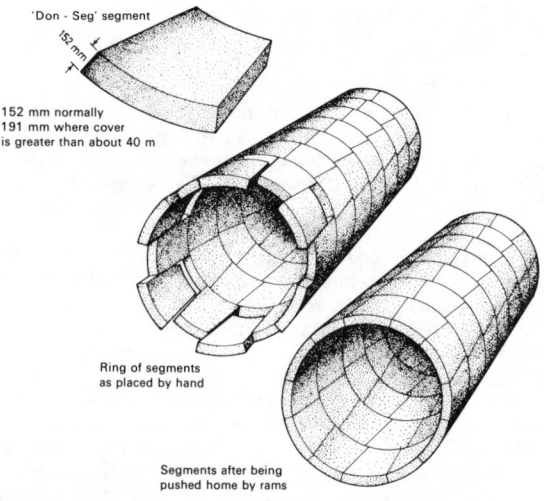

Figure 32.7 Donseg tunnel lining

rings which depends on extruded plastic seals compressed into recesses to achieve watertightness.[17]

Steel linings of two basic types are used for soft-ground tunnels. Pressed liner plates with a maximum sheet thickness of about 8 mm serve as a primary lining for hand-driven tunnels.[18] Such a lining is inadequate for accepting the thrust from a shield but, for particularly arduous conditions, fabricated steel linings may be used here. These conditions may arise from excessive variation in loading around the lining, on account of the nature

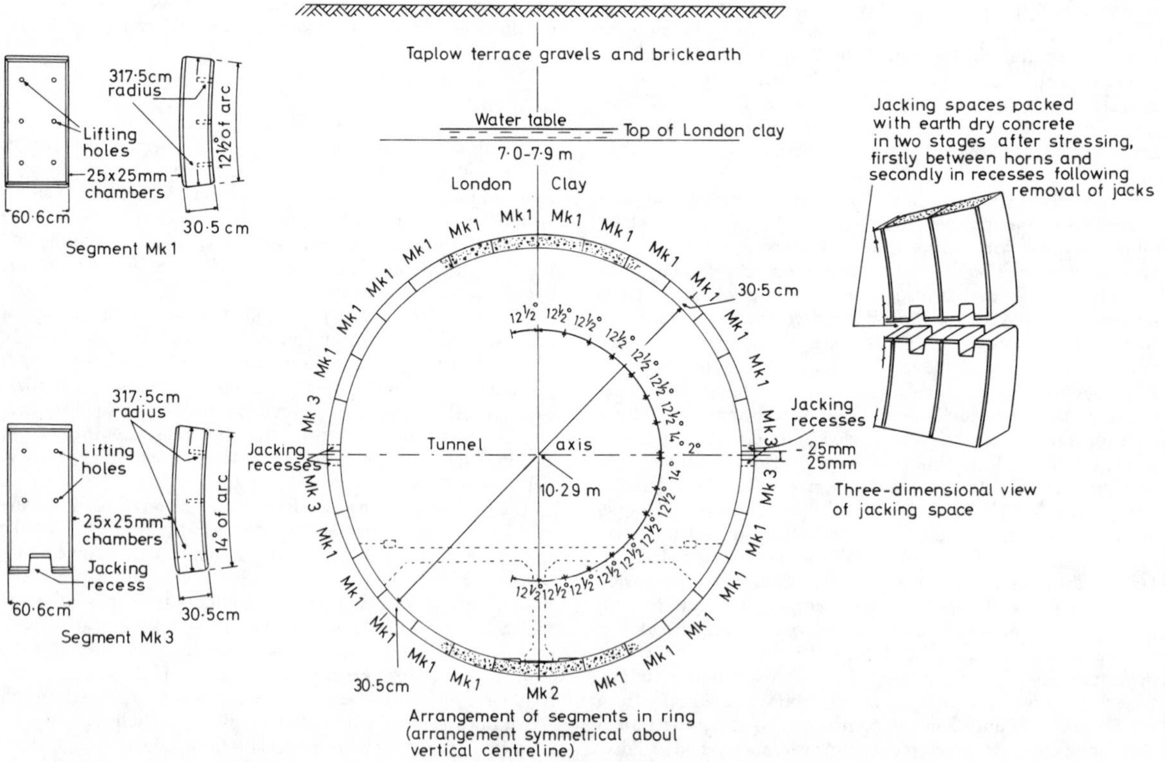

Figure 32.8 Lining for cargo tunnel at Heathrow Airport, London. (After Muir Wood and Gibb (1971) 'Design and construction of the cargo tunnel at Heathrow Airport, London.' *Proc. Instn. Civ. Engrs*, **48**, 11–34)

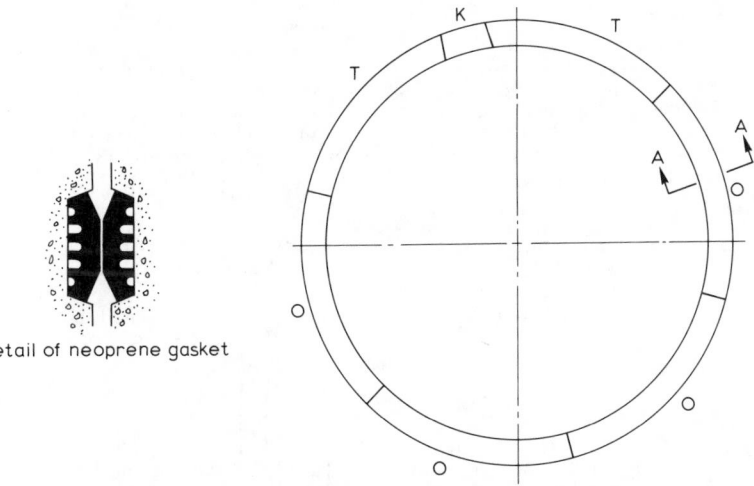

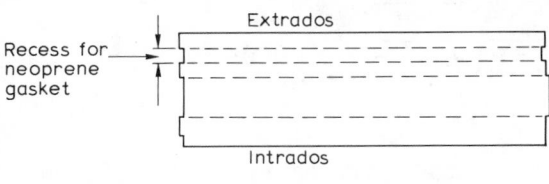

Detail of neoprene gasket

Fixing and lifting details omitted

Recess for neoprene gasket

Extrados

Intrados

Longitudinal section of segment
on A - A

Figure 32.9 Precast lining with neoprene gasket seal

of the ground, low top cover, confined side clearance or proximity to foundations.

Several types of flush lining have been designed for initial erection around a central spider but these are suitable only for small-diameter tunnels. For the Mersey Tunnels 3A and 3B, lining segments (Figure 32.10) were made in mass concrete with an internal steel face.[10] Each ring is attached to the previous ring by means of long bolts inserted into threaded sleeves. Waterproofing of the lining is achieved by welding cover plates across the joints. The concrete expanded linings result in a flush interior surface, which may be beneficial for tunnels serving as conduits.

Where a lining is built in any but very weak ground, bolting between segments has no permanent structural significance. Fastenings are therefore required primarily to control shape during erection prior to filling the extrados, the space between ground and lining. One effective cheap system uses tapered elm dowels to achieve alignment of adjacent circumferential joints while each radial joint is located by a longitudinal tube in semicircular channels; the tube collapses under load to avoid excessive local pressures.

Ductile (spheroidal graphite) iron has been used for tunnel linings. While this material allows a considerable saving in weight by comparison with grey iron, the reduced depth of segment is a disadvantage for obtaining purchase for the thrust rams but generally such linings are found to offer economic benefits as alternatives to steel linings where high loading and appreciable tensile stresses are expected.

Tunnel segments are erected in rings and the width of the ring determines the stroke of the propelling ram and hence the length of the shield. In the UK the tendency has been, in soft ground,

to maintain tunnel linings to a width of no more than 70 cm while on the Continent segment width is generally greater, with 1 m as a common standard and this trend is generally extended as new shields are built.

Rock tunnels are usually lined *in situ* with concrete placed behind shutters. The lining may be cast in discrete lengths or continuously behind shutters travelled forward in a retracted mode. Concrete is usually pumped, with placers used with decreasing frequency for filling the crown. Subsequent contact grouting is usually necessary to fill shrinkage cracks and voids between lining and rock.

For many years, attempts have been made to form a continuous *in situ* lining immediately behind a shield in soft ground. Success requires synchronization of advancing the shield and filling the concrete annulus. This has been achieved by Hochtief, Holzmann and Wayss and Freytag in Hamburg for the metro. The shield thrust is transmitted through internal shutters to avoid pressure on newly placed concrete.[19]

32.5.4 Thrust boring

Thrust boring of tunnels[20] has developed from pipe jacking, whereby lengths of steel pipe are pushed through the ground, from a jacking pit, with the addition of a new length of pipe at the rearward end after each extension of the jack. Thrust-bored tunnels are frequently in the form of reinforced concrete elements or layered materials incorporating fibre reinforced plastic. The limiting distance of thrust boring depends upon the ground, the geometry of the tunnel and the capacity of the jacks. This may be extended by the use of an external lubricant such as

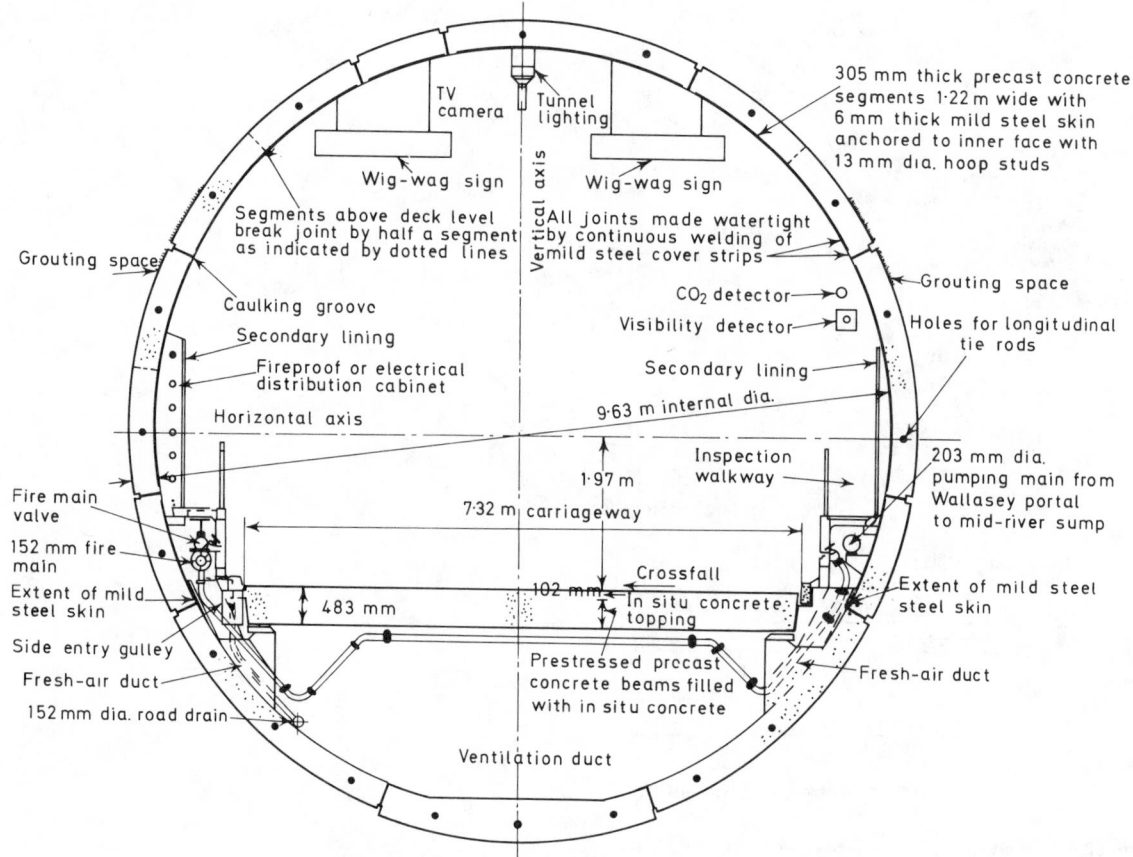

Figure 32.10 Cross-section of Mersey Kingsway Tunnel. (After McKenzie and Dodds (1972) 'Mersey Kingsway Tunnel: construction' *Proc. Instn. Civ. Engrs,* **51**, 503–533)

bentonite or by using intermediate jacking points to control the maximum length to be advanced at a time. For small pipes excavation is often by continuous-flight auger; for larger tunnels, excavation may be by hand or by small mechanical excavator.

Various types of cutting head or shield are used with jacked tunnels which, for longer drives, may be designed to help correct errors in alignment. The extent of support to the face will compare with that necessary for shield tunnelling in similar ground. Thus, very weak silt may be extruded into the tunnel through a ported head.

The most variable features of jacked tunnels concerns the joint between elements. Traditionally, for a concrete lining, a spigot was used with clearance between the internal parts of the joint which would subsequently be sealed. Such an arrangement reduces the surface area available to transmit thrust. Several types of flush external sleeve are now used in association with a butt joint, usually associated with an external annular seal to exclude the ground.

For relatively short lengths of tunnel through soft ground thrust-boring offers the benefit of erecting all lining at the thrust pit directly accessible from the ground surface, in lengths of 2 m or more, thus reducing manufacturing costs and the aggregate lengths of joints to be sealed. Furthermore, in weak ground the pipe form provides improved circumferential strength. Tunnels may be jacked in continuous easy curves using tapered pipes (or tapered packings) at the expense of increased thrust. Experience shows that a well designed and engineered jacked tunnel, built

to fine tolerances, considerably reduces thrust loads and in consequence extends the total length of tunnel, or spacing between intermediate jacks. Measurement of the build up of thrusts in the initial period of jacking will help to establish appropriate spacing between jacking stations.

32.5.5 Waterproofing

The availability of new sealing materials provides a wide choice of waterproofing systems for the joints between preformed tunnel elements.[23] Selection will normally be on the basis of cost and durability, to meet particular criteria concerning:

(1) Capacity to tolerate relative movement between elements.
(2) Hydraulic pressure.
(3) Application to wet surfaces and under pressure.

The first barrier is normally provided by annular or contact grouting. Thereafter there are fundamentally three choices: (1) a sealant provided in a liquid or plastic state; (2) a material caulked into the joint space; and (3) a preformed gasket compressed between elements. The latter has found wide acceptance as a high-performance seal used with segments cast to fine dimensional tolerances. Where practicable, seals should be formed at a radius beyond that occupied by bolts or other fastenings so that these do not require separate treatment for the exclusion of water.

32.5.6 Temporary support

In rock tunnelling, the permanent lining cannot be considered separately from the scheme of excavation and temporary support. The initial stability of the excavated ground depends not only upon the inherent quality of the rock but also on the method and quality of the excavation process. Generally, mechanical excavation will not only provide a better shaped arch around the tunnel but, more important, also much less disturbance of the surrounding rock. Recent studies have indicated that blasting may cause cracking of the rock up to a diameter outside the tunnel.

The essence of good tunnelling in jointed rock is to provide adequate support to incipiently collapsing rock as soon as possible. The means for achieving this end are directly related to the nature of rock and its jointing. The situation may be summarized thus:

(1) Where the rock is highly shattered or with frequent open joints, effective support may require the use of heavy arches. These must be provided with adequate foot supports to avoid punching into the invert and must be blocked off the rock sufficiently frequently to avoid excessive bending stresses.[22] One means of achieving an even or virtually continuous blocking is by the use of porous bolsters placed behind the arch into which a weak element/flyash grout is pumped.[23] Arches of the yielding type,[23] designed originally for colliery support, are now widely used in tunnels in recognition of their ease in erection and the virtual equivalence of their major and minor second moments of area, and hence greatly reduced tendency to distort, in conditions in which their higher cost may be justified.

(2) Where the rock is subject to progressive deterioration or to surface weathering, an immediate application of concrete or mortar may provide great benefit. A thin application of pneumatically applied mortar (gunite) or fine concrete (shotcrete)[24] will often serve in this respect, applied preferably to enter open crevices between blocks so that an adequate arch is provided around the tunnel. Shotcrete is frequently reinforced with steel mesh attached to the rock face by rockbolts or pins. Alternatively, the shotcrete may be applied with a wire staple or fibre reinforced content. A somewhat heavier and more expensive version with the same general object may be provided by an initial concrete lining placed against the newly exposed rock, possibly behind perforated steel sheeting supported by arches.[26] There are often great advantages in the reduction of overbreak if support of this nature can be applied so close to the face as to receive benefit of the three-dimensional dome that occurs here. There is also a certain time dependence of the tendency for collapse from a tunnel roof; thus a great deal of the barring down of an unstable tunnel roof can frequently be avoided by immediate support.

(3) The action of rock bolts in supporting the ground around a tunnel depends upon the nature of the jointing.[27-29]

For a regular pattern of sets of joints, the areas around a tunnel arch may be identified from which unsupported blocks may tend to fall or slide. Rock bolting will be designed in a regular pattern to create a reinforced rock arch, taking account of the strength characteristics of the joints or the stress–strain behaviour of the rock mass, where unacceptable rock convergence may otherwise develop. Special circumstances for rock support may arise from:

(1) High horizontal ground stresses, recognizing the need for appropriate disposition of support.
(2) Strongly laminated rock, for which rock bolts may serve to tie together the laminations.

(3) Presence of dominant pattern of open joints, necessitating great care in design of support.
(4) Weak filling of joints which may further weaken or be eroded as a result of tunnelling.

Large blocks of rock adjacent to rock caverns, bounded by joints of low strength, have called for special measures of anchorage in order to ensure stability, by means of anchored tendons and cables.

There are many types of rock bolt but these may conveniently be considered in two groups: (1) those which rely upon end anchorage, usually by some method of mechanical expansion of the end of the bolt, and (2) those which are keyed along their length. The latter type may be deformed bolts, set in cement or in epoxy resin or similar adhesive. The cement is introduced either as a mortar introduced in a split expanded metal cage, or as a grout through a perforated bolt or a separate tube. The epoxy resin is usually in the form of cartridges inserted ahead of the bolt, with twisting of the bolt used to burst the cartridges and mix the two-part resin. The form of keying depends, *inter alia*, on the ability to drill true regular holes in the rock. Where this is possible a bolt in the form of a hollow split sleeve may be driven into the rock.

The end anchorage bolt is generally the cheaper expedient and is more readily stressed but, in soft or weathered rock the anchoring should be achieved by a resin bonding. The head of the bolt should be fitted with plate washers or a short length of channel to spread the load adequately over the surface of a soft rock. Progressive failure of a jointed rock may be controlled by a wire mesh between bolts acting as a containing cage which is also a useful safety measure. Evidently the effective depth of a bolted rock arch or slab depends upon the bolt size, length and spacing, the length usually requiring, for overall economy, to be twice the spacing or more.[28]

Rock bolting and shotcreting are often used in association, the former providing the major support, the latter controlling surface deterioration without which aid the bolts would be effective for a short time only. Surface cracking of shotcrete provides early warning of continuing movement of the ground.

Dowels generally represent reinforcement placed in the rock without tensioning. They may be preferred to bolts in the following circumstances: (1) where subsequent movement of the rock will suffice to stress the dowel but would be otherwise liable to overstress or dislodge a prestressed bolt; (2) where light (possibly temporary) support only is required; and (3) where subsequent tunnelling or mining will excavate through the supported area, favouring the use of wooden (bamboo) or reinforced plastic dowels. Spiles are driven into the ground, ahead of the face or around the tunnel periphery, providing support by shear stress mobilized along the spile. In consequence, spiles are frequently of rolled steel sections providing a high superficial area per unit weight. They may be driven obliquely ahead of the tunnel face, first to support the face itself and, subsequently, by successive redriving as the face advances, to support the periphery of the tunnel.

There has been much development in recent years in tunnelling with support designed to reinforce the rock (or stiff soil) such that minimum applied support is required to achieve stability. Such support is often in the form of rock bolts and projected concrete (shotcrete) but may also include the use of grouting, compressed air and similar expedients. The principle of such a method is that observations should be undertaken to confirm that the support is adequately stabilizing the surrounding ground, so that the rate of convergence towards the tunnel is perceived to be approaching an asymptotic value. Where, after initial support, this is not assured, further support may be applied incrementally. Thus, the principle of such a system may be described as that of 'incremental support'. The most widely known application of this approach is the New Austrian Tunnelling Method (NATM). Confusion has arisen by the incorrect application of the term NATM to all forms of support by rock bolts and shotcrete or, more widely yet, to forms of tunnelling which do not adopt formal linings.

The essence of design of a system of incremental support is to understand the stress–strain properties of the ground and of the tunnel support since, essentially, controlled strain of the ground has to occur in order to develop a changed stress field in the ground around the tunnel compatible with the degree of support.

32.5.7 Advance by full face or by heading

A first consideration in excavating a large tunnel concerns the practicability of full-face excavation. This will depend upon the stability of the rock in relation to the tunnel size and upon the need for any advance heading to explore the ground and to provide an opportunity for undertaking ground treatment ahead of the main excavation. In very large tunnels it may be economic to excavate a top heading first in order to insert supports for the crown and then subsequently to work the invert section as a vertical bench; a variation to such a method may utilize a bottom heading in addition, serving for drainage and for removal of spoil.[30]

In swelling ground (i.e. in rock containing a montmorillonitic clay) the difficulty in support may be roughly expressed as proportional to the area of the tunnel. In consequence there may be considerable benefits in utilizing a series of small headings or drifts around the periphery of the tunnel in which the permanent lining for the full arch and invert is cast section by section.

32.6 Aids to tunnelling

32.6.1 Compressed air

As a tunnel advances, relaxation of the ground in the vicinity of the face will induce dilation. In fine-grained soils this can occur only at the rate at which water can be drawn into the soil. As the soil dilates, effective stress between the grains is reduced and the soil may flow or ravel. The period during which the face remains stable is known as the stand-up time and, in any particular circumstances, the dominant controlling features are the soil permeability and swelling modulus. A first aim of any one of the several aids to tunnelling will be to extend the stand-up time.

The application of the use of compressed air to soft ground tunnelling is another development associated with Greathead and the South London Railway (1886–90).[4]

In soft clays, compressed air will provide direct support to the ground. In silts and sands the compressed air displaces the greater part of the pore water and causes cohesion between grains of the soil by surface tension. The effect allows running sands to be treated in excavation as a soft rock. Another side-effect of compressed air in the ground is to reduce its permeability to the flow of water (by as much as an order of magnitude for silts).

The use of compressed air necessitates a considerable outlay in low-pressure compressors, air coolers, air locks (including a medical lock) and the associated control and monitoring system. Even a momentary loss in air pressure might have fatal consequences and, hence, the need for a high degree of duplication and standby equipment. The working conditions in compressed air owe a great deal to pioneering studies by Professor J. S. Haldane, leading to a set of recommendations by the Institution of Civil Engineers later revised and issued as a set of regulations under the Factory Inspectorate.[31] Comparable standards have been evolved in other countries which undertake tunnelling or caisson work in compressed air.

Compressed air introduces increased direct and indirect costs, the latter arising from the reduction in effective working time and the period spent in 'locking out' which may for instance increase from 25 to 45 min for a 6-h shift as the working pressure (measured above atmospheric) rises from 1 to 2 atmospheres. The upper limit, without special air mixtures, is about 3 atmospheres. It is recognized that it is necessary for strict medical supervision to be provided for workmen in compressed air.[32] For many years it has been known that the amount of nitrogen dissolved in the blood is related to the period of exposure to a given pressure so that if the pressure is lowered too rapidly bubbles are formed, particularly at the joints, leading to the condition known as 'the bends'. More recently, compressed air has become associated with a more serious complaint, that of bone necrosis, which may leave the victim crippled. One reaction to this discovery has been to resort to other forms of aid in order to dispense totally with the use of compressed air. A more reasonable attitude appears to be to discover the causative process and to eliminate the offending factor, since alternatives for compressed air may not only entail high cost but also introduce new hazards. In the past, before bone necrosis was associated with compressed air, medical inspection of workmen passed them fit to work in compressed air without attention being given to any latent defect of the bones or joints, an oversight that should not recur for future compressed-air working.

Compressed air has been used for many subaqueous tunnels in soft ground. The problem of balancing the external water pressure increases with the depth, as well as the size, of the tunnel. The depth below the water surface and the texture of the ground will determine the quantity of air required. In coarse sand a rule of thumb for determining the maximum demand in relation to losses through the face has been stated as $7.5D^2\,\mathrm{m}^3/$ min where D is the tunnel[32] diameter in metres. To face losses need to be added losses through the lining and through airlocks and bulkheads. Lining losses, in the absence of special care in sealing and caulking, can for a long length of tunnel represent a high fraction of total losses.

Where the ground comprises clay interbedded with thin layers of sand or silt it has frequently been found that a relatively low ratio between the pressure of air and the external head of water is adequate to provide greatly improved stability to a tunnel face.

In open ground, air losses may be reduced by locally sealing the exposed face, for which purpose bentonite dust has been used. A further problem area arises, in the construction of a segmental tunnel lining behind a tunnelling shield, in the avoidance of collapse of the ground on to the lining immediately behind the tail of the shield. This has been countered by grouting with bentonite through the skin of the shield in order to increase the capability of supporting the ground by compressed air. An alternative method has been to fill the annular space with pea gravel as the shield advances, by no means easy to perform satisfactorily.

32.6.2 Ground treatments

A wide choice of grouting media[33] is now available for consolidating weak or water-bearing ground:

(1) Setting grouts containing cement, bentonite, fly ash and other materials may be selected, at the lowest cost compatible with adequate travelling capability for the dimensions of pores and joints to be filled. Bentonite may also be used on its own as a lubricant for the extrados of the shield skin,

for thrust-bored tunnelling and for shaft sinking. Bentonite mixtures are thixotropic, i.e. they form a gel in the absence of shearing motion.

(2) Chemical grouts are used in medium to fine sands, single chemical systems having a time-dependent control of setting and two-chemical systems, of which the Joosten is the most familiar process, depending on contact between the two components.

(3) For silty sands, resin grouts may be used, low viscosity grouts being available for permeabilities down to about 10^{-5} m/s.

Generally, the finer the ground and the lower its permeability the more expensive the grouting process. The principle in grouting variable ground is therefore one of working through the available grouts from cements, clays, chemicals and resins as appropriate, so that the cheaper grouts are used to confine the travel and hence the 'take' of the more expensive grouts (Figure 32.11). In fine material, electrochemical grouting may be used in tunnels in the future. It has already been used for foundations.[34] In this process, electro-osmosis accelerates the rate of penetration of the chemical agent through the ground.

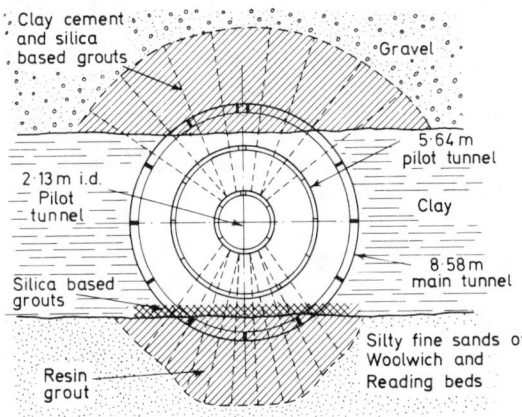

Figure 32.11 Pattern of ground treatment for north end of second Blackwall Tunnel, London (*J. Instn Civ. Engrs*, **35**, 19, October 1966 – with acknowledgement to the Council of the Institution of Civil Engineers)

32.6.3 Freezing

Freezing has had a longer history as an aid to sinking mine shafts than for civil engineering applications.[35] It has been used in civil engineering works predominantly for situations of unusual difficulty and for installations using vertical freezing-holes sunk from the surface. In each hole is inserted a U-tube or, alternatively, a composite freezing tube with concentric inner and outer tube through which cold brine is circulated, usually down the inner and up the outer tube. The brine is usually used at a temperature of about $-20°C$ but this may be reduced to $-35°C$.

Freezing was widely used for construction of the Moscow Underground in the early 1930s, for vertical shafts and for inclined escalator tunnels. For several lengths of recent sewer tunnel in Germany freezing has been adopted with horizontal freezing holes. A freezing operation with the use of brine usually occupies several weeks after the installation of the tubes and equipment.

Freezing has been used for six shafts for the Ely–Ouse water

scheme,[36] each requiring control of groundwater to a depth of 25–65 m below its surface. The cost of freezing (1969) was about £560/m for a 4.5-m internal diameter shaft and £740/m for a 7.5-m internal diameter shaft.

A new development has entailed the use of liquid nitrogen as the freezing agent. Since the operating temperature may then be lowered to $-150°C$ the freezing operation occurs rapidly and the process has frequently been used for penetrating relatively thin bands of water-bearing ground during the sinking of shafts. Freezing by liquid nitrogen has been used in tunnels in Switzerland and South Africa in conjunction with shotcrete, to provide a secure, if expensive, temporary support with low subsidence in weak ground.

32.6.4 Dewatering

Control of water for tunnelling may be achieved by lowering the water table by pumping or by diverting the water as a tunnel is lined. The first requires no further explanation here beyond the observation that pumping continues to be adopted in association with urban tunnelling with inadequate appreciation of the risks of settlement to adjacent buildings, particularly where organic soils are concerned. In certain circumstances recharging wells may be used to control the extent of the depression of the water table.

If the water is permitted to flow freely into a tunnel there may be a risk of ground settlement but this is not generally an important consideration in rock tunnels. Exceptions to this general rule occur in crushed or altered fault zones, where weak joint filling may be softened or washed out or where the rock is incompetent in relation to the pressure of groundwater. Particular care in controlling water is demanded where weak, jointed, rock is associated with stronger rock serving as aquifers. Gypsiferous rocks in the presence of water may continue to swell over a long period. Provisions may be made to permit continued swelling without excessive pressure on the tunnel; another expedient may be to exclude water from the area by sealing or drainage. In the Seelisberg Tunnel, Switzerland,[36] such expedients were employed conjointly.

While major flow of water will require to be controlled in consideration of pumping capacity and deterioration of the ground, minor flow will only present problems for the lining operation. Over a period of many years several expedients have been devised for the diversion and control of water to allow placing of the lining. One of the successful methods has been to provide a continuous protection of plastic sheeting around the tunnel supported on panels of steel mesh with longitudinal french drains along each side of the invert which are grouted up as a final operation. An alternative arrangement, where water flow is general but not great, will use a quick-set mortar pneumatically applied on to steel mesh with pipes inserted at intervals to concentrate the water flow. The pipes are stopped off on completion of lining or, occasionally, allowed to flow where permanent drainage and pressure relief are intended.

Where water is confined locally to joints it may be adequate to form a stopping of flash-set mortar around a flexible tubular former to provide a drainage path.

32.7 Ground movements

Excavation for a tunnel may give rise to associated ground movements for two principal reasons. These may either be caused by over-excavation, leaving cavities beyond the space occupied by the lined tunnel, or by release of original stresses in the ground, giving rise to elastic or plastic deformation towards the tunnel.

In rock, over-excavation may occur from roof collapses or from failure to line solidly against the ground. Rock falls may develop domes, arches or chimneys depending upon the nature of the ground and the pattern of initial stresses. High horizontal stresses, for example, will normally tend to limit the extent of the cavity provided the rock is sufficiently competent in relation to the maximum resultant stress around the periphery of the tunnel. Crush zones in a homogeneous rock around a tunnel may indicate high stress; the same phenomenon occurs in the release of strain energy by rock bursts when thin slabs of rock become violently detached from the periphery, in strong rocks at depth and in weaker rocks nearer the surface. As rock fractures it increases in bulk; in consequence, once a plug is provided at the base of a cavity its upward extent will be limited and may be approximately calculated.

Even small cavities immediately behind the lining are serious in that they may lead to uneven loading on the lining and consequential failure; hence the need for systematic contact grouting. In soft incompetent ground, over-excavation will usually be transmitted in full to the surface approximately to equate to the volume of surface settlement. However, in dense sand the total settlement at the surface will be reduced; in loose sand, settlement may occur as a result of disturbance by tunnelling even in the absence of any over-excavation. It is often impossible in soft ground to subdivide the effects of over-excavation and of changes in the stress pattern, the latter tending to give rise to loss of ground towards the exposed face.

The shape of the 'trough' of settlement at the surface will usually be influenced by loss of ground along a length of tunnel somewhat greater than its depth below surface, the influence factors being highly dependent upon the geological structure. In homogeneous soft ground and for a tunnel advanced with consistent standards of design, workmanship and progress, a characteristic depression will develop over the tunnel which may be described approximately in terms of the shape of statistical normal distribution curves.[37] Approximately half the total settlement will have occurred immediately above the advancing tunnel face, for tunnels at no great depth.

Tests undertaken during the construction of tunnels in London Clay indicate that with increasing depth, there is a greater tendency for loss of ground arising from deformation towards the advancing face. In general, the contribution to loss of ground may be as set out in Table 32.1.

A special cause for settlement over a tunnel may occur where a shield in soft ground can only be kept to correct line by means of maintaining an appreciable 'look up' on account of a tendency to settle at the cutting edge. This loss may be countered to some extent by grouting above the shield as it advances, with fly ash or similar material.

32.8 Tunnel design

32.8.1 Stresses around a tunnel

The state of stress in real ground around a full-size tunnel during the course of construction is too complex to analyse fully. A more rewarding process is to idealize the problem to a certain degree and then, by inference and judgement, determine the significance of inadequacies of the conceptual model.

We start by considering the two-dimensional problem of a long unlined circular tunnel pierced instantaneously at great depth in perfectly elastic ground. We can in this instance build up the overall stress pattern around the tunnel by superposition of its constituents.[38] The initial vertical loading will be redistributed and will set up the tangential and radial principal stresses σ_θ and σ_r shown in Figure 32.12 for the vertical and horizontal axes. At the periphery,

$$\sigma_r = 0 \tag{32.2}$$

and at axis and crown level,

$$\sigma_\theta = 3\sigma^* \quad \text{and} \quad \sigma_r = -\sigma^* \tag{32.3}$$

respectively where σ^* was the final vertical loading in the ground.

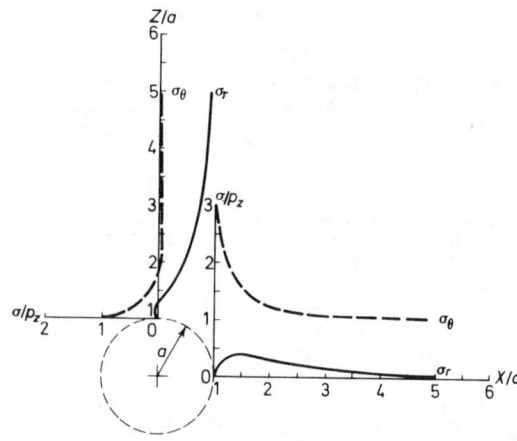

Figure 32.12 Stresses around circular tunnel in elastic ground initially stressed in vertical direction only

Table 32.1 Contribution to loss of ground around a shield-driven tunnel

Nature of ground loss	Computation	Normal limits (%)
Ground loss at face	$\pi d^2 h/4$	0.1–?
Ground loss behind cutting edge	πdt	0.1–0.5
Ground loss along the shield	$\pi l v/8$	0–1
Ground loss behind the tail	$\pi d(d-d_0)/2$	0–4
	$(\pi d(d-d_0)/4$ above water table)	0–2

Where the loss per unit length is expressed as a percentage of area of tunnel face and d is the diameter of the shield, d_0 is the external diameter of the lining, t is the relief behind the cutting edge, v is the 'look up' of the shield measured as the extent of out of plumb on vertical diameter, l is the length of shield and h is the horizontal movement of ground at the face per unit length of advance of shield.

A similar set of relationships may be obtained for the horizontal loads $N\sigma^*$. For ground loaded from above and laterally constrained, it can readily be shown that $N = v/(1-v)$ where v is Poisson's ratio. For ground loaded and then subjected to reduction of vertical loading, N may be greater than unity and, indeed, in over-consolidated ground where appreciable surface erosion has occurred N, according to the circumstances, may be 2, 3 or more. Evidently if $N = 1$, Equation (32.3) indicates by superposition that $\sigma_\theta = 2\sigma^*$ around the periphery. The factor N may vary in azimuth and be influenced, *inter alia*, by tectonic forces.

The most important departures from this simple model may be caused by:

(1) Nonelastic behaviour of the ground.
(2) Limiting ultimate strength of the ground.
(3) Inability of the ground to accept tension.
(4) Discontinuities in the ground.

The simplest nonelastic model is that for ground assumed to behave elastically up to certain limiting differences between maximum and minimum principal stresses (generally the stress parallel to the tunnel axis may be considered as intermediate between the other two) and thereafter to deform perfectly plastically. For example, a jointed rock might be considered as elastic for stresses lying within the Mohr's envelope with plastic deformation occurring at the limiting shear stress of:

$$\tau = c' + \sigma_N \tan \varphi' \qquad (32.4)$$

The stress pattern around a circular tunnel[39] might then be represented as Figure 32.13. It will be noted that full development of the plastic zone will entail appreciable movement of ground into the tunnel and theoretical considerations suggest delay in supporting ground to reduce to a minimum the load on tunnel supports (see page 32/13). In most tunnels, the object is to provide support as rapidly as possible and then to consider merits of systems that will yield noncatastrophically at excess loads. A diagram such as Figure 32.13 permits examination of the reduction in plastic movement as a result of increased σ_r at the periphery of the tunnel by means of ground support.

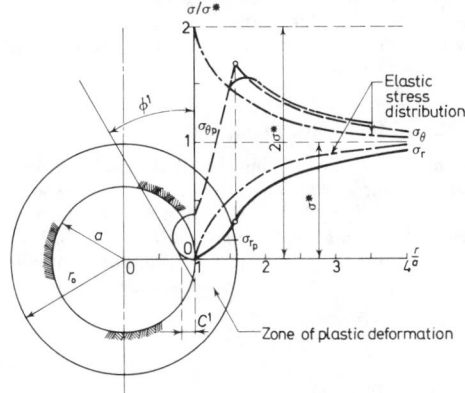

Figure 32.13 State of stresses around a circular tunnel in ground yielding to plastic and elastic strains ($N=1$). (After Kastner (1971) *Statik des Tunnel- und Stollenbaues*, (2nd edn) Springer-Verlag)

Many of the recent advances in economic rock tunnelling have required the behaviour of the rock to be better understood in two particular respects: (1) the determination of the approximate shape of the Mohr or other form of yield envelope, determining the permissible changes of stress within the rock, i.e. the stress history, which may be tolerated without inevitable damage of the rock structure;[29] (2) an understanding of the stress–strain behaviour of the rock when it is stressed beyond yield. Leaving aside questions of nonhomogeneity and anisotropy, knowledge of these two characteristics would permit prediction of the (two-dimensional) behaviour of the rock with specific schemes of support. In fact, the problem is not so simple on account of the rotation of principal stresses which occurs in the ground in the vicinity of a tunnel. While, therefore, such an

approach provides considerable insight into the behaviour of the rock, this cannot be in a fully quantitative sense, on account of the sheer complexity of the problem even as simplified to two dimensions, and neglecting time-dependent effects. There are nevertheless great economic benefits in devising schemes of tunnelling which utilize the rock in the vicinity of the tunnel to support a high fraction of the ground load. The technique is then one of making predictions, on the basis of experience and on tests on the particular rocks or types of rock, on expectations of behaviour, and then systematically to monitor specific predicted features. The simplest and most effective set of measurements concerns the convergence of the rock face as the tunnel is advanced, possibly supplemented by measurements of internal movements up to a diameter or so from the tunnel. Monitoring implies that if movements are observed to be beyond tolerable limits in magnitude, velocity or distribution, designed countermeasures may be introduced as reinforcement, to re-establish tolerable conditions. This is the application of the observational method, at the heart of all techniques of tunnelling which use incremental support. The best-publicized but by no means unique exponent of the technique is the NATM; while it has attracted mythological accretions, most of the applications have been successful and many of these economic. Contributions to the tunnelling techniques based on the observational method have evolved separately in numerous countries and continents.[27] One essential feature lies in the recognition that design and construction become inseparable processes; for successful application the contractual relationships must reflect this interaction.

The strength of the rock in true plastic yielding may ultimately be represented as a purely frictional material, giving a limiting strength line, as in Figure 32.14, to be reproduced in an analysis of the failure of the ground around a tunnel.

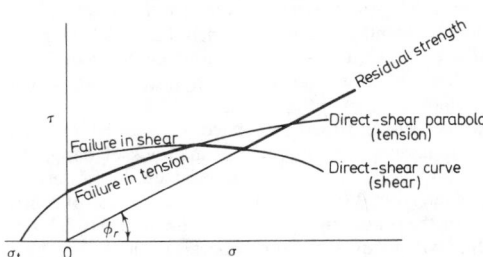

Figure 32.14 Failure limits for rock. (After Lajtai (1969) 'Shear strength of weakness planes in rock.' *Inst. J. Rock Mech. Min. Sci.*, **6**, 5, 499–515

From Equation (32.3) it is readily seen that $3 > N > 1/3$ will cause no tension in elastic ground. Outside these limits, tension zones will occur for a circular tunnel (Figure 32.15).

The behavioural implications of an elementary feature of this nature needs to be covered satisfactorily by any effective method of stress analysis applied to a rock tunnel. A single discontinuity may have a considerable influence upon the stress distribution in an otherwise sound rock.

The degree of knowledge of the ground, its homogeneity and the extent of the tunnel will determine the length to which simple analysis, models and numerical methods may appropriately be applied to defining the economic basis for tunnel design. The method of finite elements[40] has almost unlimited potential for solving this class of problem but the cost of the solution increases rapidly with increasing complexity. Great care is required for the nonlinear stress–strain conditions, since the result depends upon the loading sequence.

The objectives of any such analysis should be stated at the

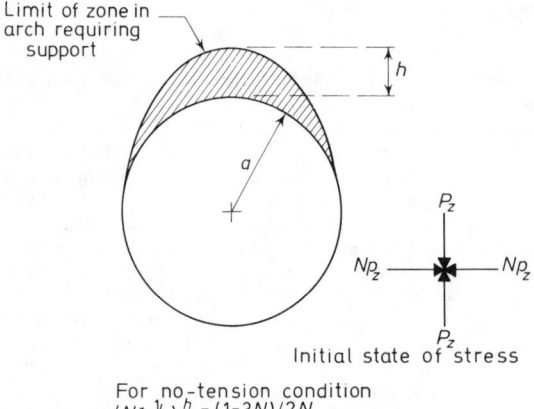

Limit of zone in
arch requiring
support

h

a

P_z

Np_z —— Np_z

P_z

Initial state of stress

For no-tension condition
$(N \geqslant \frac{1}{3}), \frac{h}{a} = (1-3N)/2N$

Figure 32.15 Circular tunnel in no-tension material

outset. Often a small fraction of the cost of elaborate analysis would have been better spent on a better understanding of the rock structure without which the results of analysis are unreliable. Increasing use is made of boundary integral methods which, for linear problems, entails superposition of familiar patterns of stress distribution and will, hence, provide solutions of complex geometry, including discontinuities and nonhomogeneities, at modest cost. Often, the variability of the rock and the imprecision of construction methods renders it appropriate only to make qualitative assessments of stress distribution. The notion of 'stress flow' around a cavity[41] helps to identify areas of concentration; important 'stress raisers' around a cavity may also be readily identified. Simple models have yielded much insight on the behaviour of nonuniform or jointed rock[27] and, for qualitative solutions, inexpensive physical models should not be overlooked.

Laboratory tests and their associated analyses have contributed to the development of rational methods of design for tunnels in soil,[42] especially in determining factors of safety for stressing of the ground beyond elastic limits. The use of centrifuge testing is then effective in scaling-up from model to full size, for which gravity assumes a more important role.[40]

At the present day, many techniques are available to the engineer for the analytical aspects of design; the matter is much more a question of selection of the appropriate technique in relation to reliable knowledge of the ground. For rock, the most ubiquitous problem is that of presenting the rock properties in a form assailable by analysis and in identifying the relevant factors.

32.8.2 Stability ratio, rock competence and classification

A simple failure mechanism at the face of a tunnel in soil derives a stability ratio N as the ratio (in consistent units) of nett overburden stress $(\gamma D - P_i)$ to undrained cohesive strength c_u. Generally stability, in the absence of special measures, requires N to be no greater than 4 or 5. It will be noted that N may be reduced by increasing P_i, the internal pressure in the face of the tunnel, by some such expedient as air or liquid under pressure.

The competence of a rock is a measure of its capacity to resist deformation under a given loading. Since the loading is usually directly related to the overburden, it appears helpful, in classifying the ground from the view point of tunnelling, to define a competence factor[43] as:

$$F_c = q_u/\gamma^D \qquad (32.5)$$

where q_u is unconfined compressive strength of the ground and γ is density of overburden of depth D above tunnel

(Note that, if $q_u = 2c_u$ and $P_i = 0$, $F_c = 2/N$

Where $F_c < 2$, immediate support is required. Where $10 > F_c > 2$, stability of unsupported ground will depend on the initial state of stress and on stress–strain–time characteristics. Where $F_c > 10$, the ground will be competent, the strength of the rock structure may become largely irrelevant, and the real problem concerns discontinuities and joints, pre-existing or caused by tunnelling.

The problem then beomes that of describing a jointed rock in a form which can lead to rational designation of requirements for support. Several systems of rock classification have been devised for this purpose which embrace selected factors affecting stability. There remain two major limitations. First, that the dominant factors vary from situation to situation; e.g. for weak jointed rock, rock strength may be important. Second, that each factor has such a wide degree of variability on account of geological history and structure that we can scarcely expect to find unique combinations of two or more factors reliably represented by a single classification index. The system of Barton[43] is the most comprehensive and is helpful in making a first assessment; thereafter, for any particular situation of reasonably consistent rock type, a local classification system may be devised, as a basis of different degrees of rock support. Often, the most difficult feature to be represented in any classification system concerns joint quality, including such characteristics as openness, continuity, nature (shear, tension, etc.), roughness, planarity, filling (and how liable to be affected by movement, water, exposure) which cannot hope to be covered by one or two numerical factors.

32.8.3 Stiffness of the tunnel lining

The tunnel lining and the surrounding ground should be treated as a composite structure when considering states of stress and deformations. A question of first importance concerns the relative stiffness of the lining and the ground it displaces. For elastic conditions for the simplest, 'elliptical', mode of deformation of a circular tunnel[44] we may express this stiffness ratio as:

$$R_s = 3EI/a^3\lambda \qquad (32.6)$$

coefficient of ground reaction

$$\lambda = \frac{3E_c}{(1+v)(5-6v)a} \qquad (32.7)$$

Likewise, the compressibility factor

$$R_c = aE_c(1-v_c^2)/AE(1+v) \qquad (32.8)$$

where A represents lining area per unit length

The average hoop stress and the maximum bending moment in the lining are then, respectively:

$$P_0 = p/(1+R_c) \text{ where } p = (p_v + p_h)/2 \qquad (32.9)$$

$$M = \tfrac{1}{6}(p_v - p_h)a^2[R_s/(1+R_s)] \qquad (32.10)$$

where p_v and p_h are initial vertical and horizontal pressures in the ground

The maximum and minimum hoop stresses are then:

$$p/(1 + R_c) \pm M/a$$

and the radial departures from the original circle of radius a are found to be, on the vertical and horizontal diameters, respectively:

$$P_0 a/E \pm Ma^2/3E\,I$$

For most applications, R_s is considerably less than unity and lining stiffness then makes little contribution to the reduction of deflections. An objective in economic design in such circumstances will be to minimize stiffness. Likewise, compressible linings need only to support a reduced fraction of hoop loading.

32.8.4 Towards a better understanding

The advance in techniques of tunnelling is a continuing process and the results of monitoring the behaviour of a tunnel by comparison with prediction provide valuable correction to design techniques and assumptions. Great skill is required in determining the data to be acquired and in interpreting the quantities of data that may be obtained in a comprehensible manner without introducing the risks of oversimplification. Such monitoring will normally be concerned with strains in the ground and with deformations of, and stresses in, the lining.[56] Generally, the simplest and most robust instruments should be used to avoid disappointment from incomplete results arising from damage.

Improvements in understanding the behaviour of the ground and in tunnelling techniques bring economies in cost and time. These also lead to better ground control and reduced subsidence over the tunnel, a feature which has received much attention.[45]

One example of the specialized development of tunnelling relates to storage of oil and liquefied petroleum gas (LPG). This application requires selection of sites of reliable rock quality, the confidence to excavate caverns of great size and span with little or no ground support and the ability to ensure containment of the product stored, by natural tightness of the rock, possibly supplemented by a water curtain to ensure that any flow across the boundary of the cavern is inwards not outwards.[46]

Where a tunnelling option exists, its optimization depends upon its assessment by those knowledgeable in the potential of tunnelling at an early stage in planning. Too often the tunnel option is considered too late to be optimized, where it might otherwise have provided the most economic solution with associated social and environmental benefits.

References

1 Kell, J. (1963) 'The Dartford Tunnel.' *Proc. Instn Civ. Engrs,* **24,** 359–372.

2 Grange, A. and Muir Wood, A. M. (1970) 'The site investigations for a Channel Tunnel, 1964–1965.' *Proc. Instn Civ. Engrs,* **45,** 103–123.

3 Szechy, K. (1966) *The art of tunnelling,* 1st edn (in English). Akadémiai Kiadó, Budapest, p.891.

4 Greathead, J. H. (1895) 'The City and South London Railway: with some remarks upon subaqueous tunnelling by shield and compressed air.' *Instn Civ. Engrs Papers on London Underground Railways, 1885–1929,* Paper 2872, 39–73.

5 Harries, D. A. (1971) 'Constructing the deep-level drainage system of Mexico City.' *Tunnels and Tunnelling,* **3,** 35–42.

6 Stack, Barbara (1982) *Handbook of mining and tunnelling machinery.* Wiley, Chicester.

7 *The Engineer* (1883) 'English boring machine.' **55,** 455.

8 Fox, F. (1886) 'The Mersey Railway.' *J. Instn Civ. Engrs,* **86,** 40–49.

9 Nicholson, W. E. (1967) 'Big bits drill big.' *World mining.* Part I, 41–48 (September); Part II, 44–51 (October).

10 McKenzie, J. C. and Dodds, G. S. (1972) 'Mersey Kingsway Tunnel: construction.' *Proc. Instn Civ. Engrs,* **51,** 503–533.

11 Muir Wood, A. M. and Gibb, F. R. (1971) 'Design and construction of the cargo tunnel at Heathrow Airport, London.' *Proc. Instn Civ. Engrs,* **48,** 11–34.

12 Construction Industry Research and Information Association (1978) *Tunnelling: improved contract practices.* CIRIA Report No. 79.

13 Sandström, Gosta, E. (1963) *The history of tunnelling.* Barrie and Rockcliff, p.427.

14 Groves, G. L. (1943) 'Tunnel linings with special reference to a new form of reinforced concrete lining.' *J. Instn Civ. Engrs,* **20,** 29–42.

15 Craig, R. N. and Muir Wood, A. M. (1978) 'A review of tunnel lining practice in the United Kingdom.' Transport and Road Research Laboratory, Supplementary Report No. 335.

16 Engelmann, E. (1981) 'Tunnelling beneath the old port of Spandau: tunnelling with the Hydroshield' (section H110: Underground railway construction) in: *Construction underground: present and future.* (In German.) Stuva, Cologne, pp.212–221.

17 Glang, S. (1981) 'Elastomer joint tapes and profiles.' *Tunnel (Tiefban).* (In German and English.) **3,** 195–206.

18 Apel, F. (1968) *Tunnel mit Schildvortrieb.* Werner-Verlag, p.293.

19 Martin, D. and Braun, W. M. (1982) 'Blade shield tunnelling machine extrudes its own lining, *Tunnels and Tunnelling,* **3,** 54–56; **6,** 55.

20 Craig, R. N. (1983) *Pipejacking: state of the art review.* Construction Industry Research and Information Association, Technical Note No. 12.

21 Construction Industry Research and Information Association (1979) Tunnelling waterproofing. CIRIA Report No. 81, (rev. 1981).

22 Proctor, R. V. and White, T. L. (1946) *Rock tunnelling with steel supports.* Commercial Shearing and Stamping Co., Ohio.

23 Craig, R. N. (1979) 'The Lewes road tunnel, Sussex, England.' *Tunnelling.* (1979 Symposium.)

24 Cunliffe, J. and Johnston, A. G. (1958) 'Roadway support with special reference to yielding arches.' *Trans. Instn Min. Engrs,* **117,** 805–818.

26 Alberts, C. and Bäckström, S. (1971) 'Instant shotcrete support in rock tunnels.' *Tunnels and Tunnelling,* **3,** 1, 29–32.

27 Wöhlbier, H. (1969) 'Der Ausbau unterirdischer Hohlräume mit S und A Bleche System.' *Bergbauwissenschaften,* **16,** 117–126.

28 Douglas, T. H. and Arthur, L. J. (1983) *A guide to the use of rock reinforcement in underground excavations.* Construction Industry Research and Information Association, Report No. 101.

29 Hoek, E. and Brown, E. T. (1980) *Underground excavation in rock.* Institute Minerals and Metals.

30 Anderson, D. (1936) 'The construction of the Mersey Tunnel.' *J. Instn Civ. Engrs,* **2,** 473–516.

31 Construction Industry Research and Information Association (1982) *Medical code of practice in compressed air.* CIRIA Report No. 44.

32 McCallum, R. I. (ed.) (1967) *Decompression of compressed air workers in civil engineering.* Oriel Press, p.329.

33 Camberfort, H. (1977) 'The principles and application of grouting.' *Q. J. Eng. Geol.,* **10,** 57–95.

34 Caron, C. (1971) 'Consolidation des terrains argileux par électro-osmose.' *Ann. de l'Inst. Tech. du Bâtiment et des travaux publics,* **285,** 75–91.

35 Mussche, H. E. and Waddington, J. C. (1946) 'Applications of the freezing process to civil engineering works.' *Inst. Civ. Engrg Works Contr.* Paper 5.

36 Buri, F. (1977) 'Seelisburg middle-section construction.' *Tunnels and Tunnelling,* **9,** 5, 40–44.

37 Muir Wood, A. M. (1970) 'Soft ground tunnelling.' *Technology and Potential of Tunnelling,* **1,** 167–174; **11,** 72–75, Johannesburg.

38 Terzaghi, K. and Richart, F. E. (1953) 'Stresses in rock about cavities.' *Géotechnique,* **3,** 2, 57–90.

39 Kastner, H. (1971) *Statik des Tunnel- und Stollenbaues.* (2nd edn). Springer-Verlag, Berlin, p.269.

40 Davis, E. H., Gunn, M. J., Mair, R. J. and Seneviratne, H. N. (1980) 'The stability of shallow tunnels and underground openings in cohesive material.' *Géotechnique*, **30**, 4, 397–416.

41 Wilson, A. H. (1983) 'The stability of underground workings in the soft rocks of the Coal Measures.' *Inst. J. Min. Engrg*, **1**, 2, 91–187.

42 Schofield, A. N. (1980) 'Cambridge geotechnical centrifuge operations.' *Géotechnique*, **30**, 3, 227–268.

43 Barton, N. N., Lien, R. and Lunde, J. (1974) 'Engineering classification of rock masses for the design of tunnel support.' *Rock Mechanics*, **6**, 4, 189–236.

44 Muir Wood, A. M. (1975) 'The circular tunnel in elastic ground.' *Géotechnique*, **25**, 1, 115–127.

45 Attewell, P. B., Yeates, J. and Selby, A. R. (1986) *Soil movements induced by tunnelling and their effects on pipelines and structures.* Blackie, London.

46 Bergman, M. (ed.) (1980) 'Subsurface space.' *Proceedings*, Pergamon Press, Oxford (3 vols).

33

Project and Contract Management

P A Thompson BSc(Eng), MSc, CEng, FICE, MIWES
University of Manchester Institute of Science and Technology

Contents

33.1 Introduction

Management is concerned with the setting and achievement of realistic objectives for the project or contract. This will demand effort – it will not happen as a matter of course – and it will require the dedication and motivation of people. The provision and training of an adequate management team is therefore an essential prerequisite for a successful job for it is their drive and judgement, their ability to persuade and lead, which will ensure that the project objectives are achieved.

Managers of projects and contracts involving engineering construction will frequently encounter a mixture of technical, environmental, logistical and physical problems. The style of management required for such work will therefore differ in many ways from that required in the relatively static surroundings of line management in a factory. The temporary nature of the organization and the considerably greater element of uncertainty associated with construction projects will be particularly significant.

Uncertainty is the source of many of the problems encountered in construction work and will influence project appraisal, estimating, planning, the form of contract and the procedures for contractual measurement and valuation. Excessive uncertainty which leads to continuous or multiple changes to design will reduce productivity and will almost certainly affect adversely the morale of the workforce. It will also result in extra cost to both client and contractor. All parties involved in construction projects and contracts would therefore benefit greatly from reduction in uncertainty prior to financial commitment. Effort should be devoted to risk management and the person(s) ultimately responsible for the financial outcome should be appraised fully of the full risk spectrum before committing his or her organization to the project or contract.

When considering planning techniques, construction contracts and their valuation and other aspects of management in this chapter there may be a tendency for the reader to think of each system as a separate entity. He or she is advised to remember that the purpose of all these techniques and procedures is to help people make the judgements and decisions and to perform the administrative functions which are necessary for the successful accomplishment of the project or contract. The project or contract manager is primarily concerned with the direction and motivation of other human beings.

33.2 Project and contract organization

33.2.1 Introduction

There are many steps between the inception of a new capital project and its successful operation and maintenance. For efficient project management it is helpful to group these steps together into the following project stages, which are also identified in Figure 33.1:

APPRAISAL	Assess alternative strategies for meeting needs
	Establish technical and economic feasibility
	Derive the master plan
DEFINITION	Statement of project objectives
	Conceptual design and associated cost estimates
	Design review
	Arrange project funding
	Sanction
DESIGN	Detailed design
	Design review
	Contract strategy report and definition of contract packages
	Detailed cost estimates
	Procurement/tendering
	Contract award
CONSTRUCTION	Site construction
	Offsite fabrication and manufacture
	Installation
	Quality control

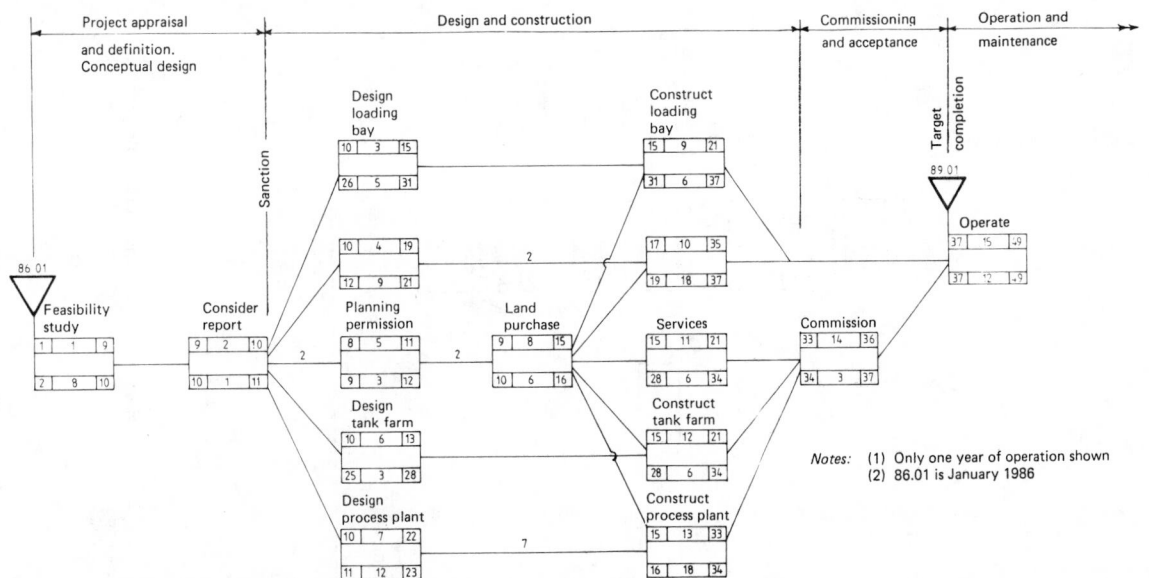

Figure 33.1 Precedence diagram for new manufacturing plant. This simple network will later be developed into a time-and-money model for appraisal of the project

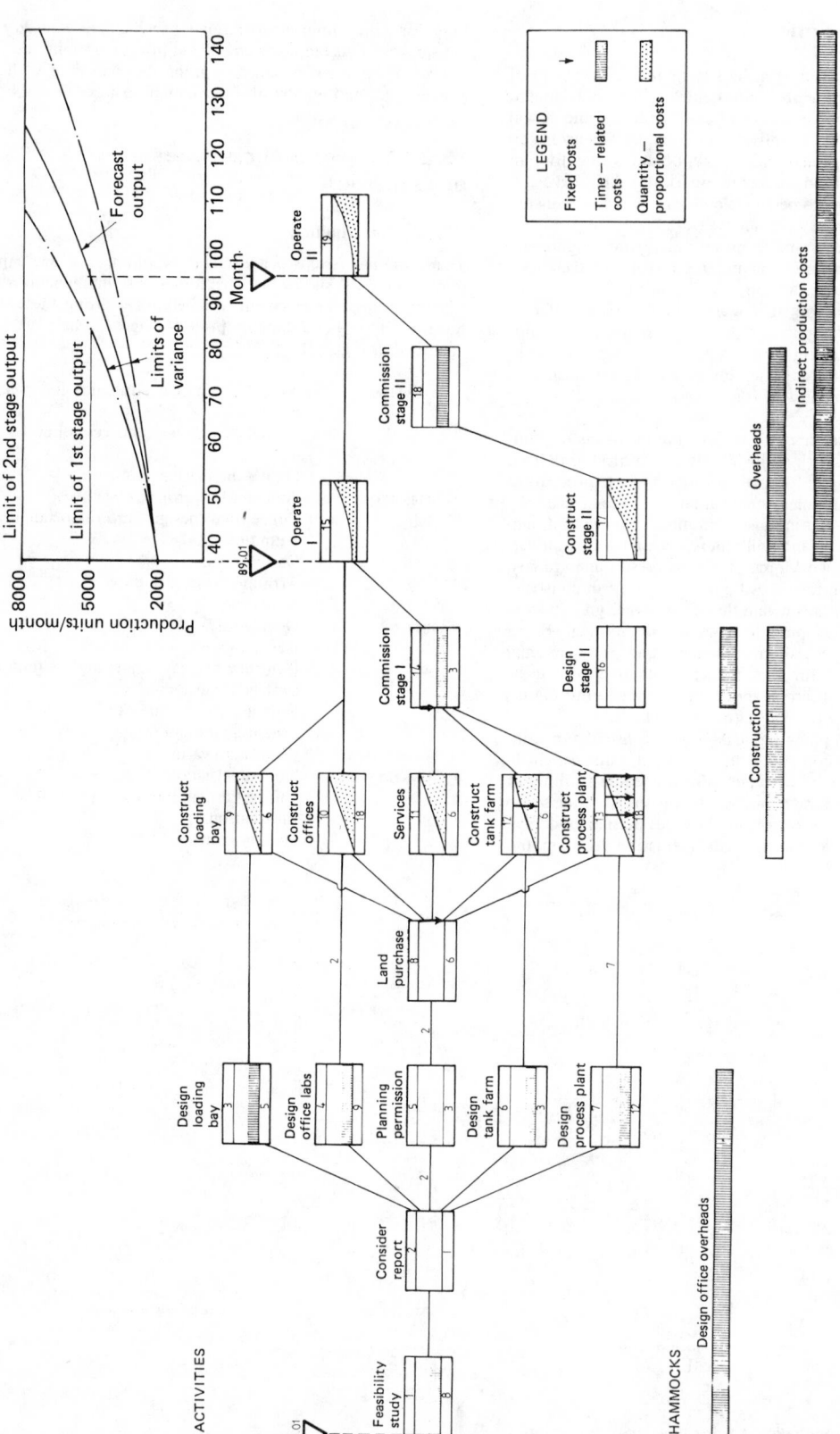

Figure 33.2 Flow chart for new manufacturing plant. The precedence diagram is developed and extended as the framework for cost/benefit analysis

COMMISSIONING   Expediting
Construction management
Contract administration
Engineering and performance tests
Acceptance

OPERATION   Organization for operation and maintenance
Project review

These stages may frequently overlap and their relative durations can vary greatly. For public works projects the early stages of appraisal and definition are likely to extend over many years and the facility, say a road or a water-treatment plant, will also be utilized over a period frequently exceeding 25 years. Commercial projects, such as the new manufacturing plant illustrated in Figure 33.2, are more likely to be appraised, sanctioned and implemented rapidly before a competitor enters the market.

33.2.2 Organization of design and construction

In the traditional system for the procurement of civil engineering works[1] it is normal for the client to employ an experienced consulting engineer to assist him with project development and implementation. The consultant is likely to play an important role in project appraisal and definition, to undertake the design and contract documentation, to oversee tendering by contractors, to supervise construction work on site and administer the construction contracts. A contractor or group of contractors complete the fabrication and construction of the works. Management of design and construction is therefore the responsibility of different organizations.

This well-established system is still widely used but many variations are adopted to meet the particular requirements of the client, particularly in the private sector. In general, these reflect alternatives used in the related process plant, offshore and building industries[2] and frequently respond to pressure for quicker and timely completion to reduce the payback period. There is also increasing emphasis on effective project management and on the overall management of design and construction.

33.2.3 Decision making and control

Control is concerned with regulation of the future. This implies the ability to predict the consequences of specific courses of action and necessitates decision making under conditions of uncertainty. That is to say that the decision maker must choose a specific course of action from those available to him even though the consequences of the possible courses of action will depend on events that cannot be predicted with certainty. Decision making on construction projects is inextricably linked with uncertainty and also frequently encounters conditions of urgency and constraint. Many decisions are required during each of the stages listed above and may be influenced by a need to keep options open or a need to reduce uncertainty.

The scope for control diminishes as the project proceeds. The key events for the client are *sanction* – when he commits himself to a project of particular characteristics – and *contract award* – when he commits himself to specific contractors and to major cost expenditure. The contractor's commitment is, of course, made in his tender: thereafter he will exert control mainly through the allocation and use of resources.

33.2.4 Project management

The role of project management is to exercise overall control of the project from its inception through to the completion of commissioning. Thus, the ultimate responsibility for project management lies squarely with the client. His primary function is to define the parameters of the project and thereafter to provide decisions, approvals and guidance. Several reports on the performance of the construction industry over the last 10 to 20 years have concluded that good project management by the client is an essential ingredient for a successful project.[3] Regrettably, it was found that many projects lacked an appropriate project management structure, resources and expertise.

For each project, a single individual from within the client's organization should be named as project manager, and given the necessary authority and responsibility.[4] His role, therefore, is to manage the client's investment, and he should have sufficient seniority to exercise effective control both within and outside the client organization. For most projects he will need the support of a small project management team.

Brief guidelines for effective project management are given in section 33.9.

33.2.5 Project objectives

The client will have a number of overall objectives for undertaking the project. These may be commercial, reflect the perceived needs of society and/or have political overtones. Specific project management objectives must be compatible with the overall objectives and should be clearly formulated early in the definition stage of project development.

The dominant considerations must be fitness for purpose of the completed project and safety during both the construction and operational phases. Thereafter cost, time and functional performance form a minimum set of values from which the primary objectives will be drawn. The potential for conflict between these objectives, as problems arise during project implementation, is obvious.[5] The disasters which beset the Montreal Olympics stadium – a prestige project of novel design with an unrealistic budget and a fixed time constraint – offer salutary reading to all project managers.[6]

Consequently, it is necessary for the project objectives to be ranked in terms of their relative importance. Tolerances must also be specified – as range of acceptable variation in performance, float in the programme and tolerance and contingency allowances in the estimate. The greater the perceived uncertainty, the more flexible these criteria must be. Unclear or ill-defined objectives will have a detrimental effect on decision making and progress.

Thereafter, the monitoring of progress and performance against these objectives will determine the need for replanning, revision of estimates and changes in project scope and specifications. The latter are normally reductions in work content or quality which are accepted in order to meet stringent financial constraints. This approach is prevalent in public works projects,[7] is disruptive, has an adverse effect on morale, and is likely to lead to client or public dissatisfaction with the project however well the remainder of the work may be completed. Far better to expose the uncertainties, allow for them in the estimate and adopt realistic objectives in the first place.

Quality control demands effective liaison between design office and the project management team both in terms of specification and verification of client requirements. Effective inspection and testing procedures should be established and agreed by all parties. There is a tendency for some clients to economize in the allocation of engineering staff to inspect fabrication and supervise construction. When this follows the award of a contract to a low bidder without prequalification it does not surprise the Author that the client may be dissatisfied with the completed works.

Quality Assurance systems can assist project management in the setting and achievement of project objectives. According to a recent Construction Industry Research and Information Association (CIRIA) report:[8]

> Quality Assurance is a systematic way of ensuring that organised activities happen the way they are planned. It is a management discipline concerned with anticipating problems and with creating the attitudes and controls which prevent problems arising.

The report continues:

> Quality Assurance is concerned with systematically providing evidence to the client that all reasonable actions have been taken to achieve the required quality. But it is also concerned with spelling out the risks involved in any civil engineering project, and advising the client on operation and maintenance.

Properly practised, the system requires precision of communication and this, in the Author's opinion, is its greatest value to project management.

Quality Assurance is usually associated with manufactured products and with complex multi-disciplinary projects where safety and quality of plant operation are the primary objectives. The application of such systems to any project should depend on whether there is benefit. In the civil engineering context, care must be taken to ensure that the adoption of a Quality Assurance system does not result in rigid adherence to unnecessarily demanding specifications. Neither must the system inhibit the flexibility and judgement required for the management of the uncertainties associated with the one-off job.

33.2.6 Safety

The nature of construction work makes it vulnerable to accidents and project management must at all times enforce safety procedures. This is an area where great benefit to the physical well-being of staff, to morale and to the progress of the works could be achieved by the adoption of effective safety Quality Assurance programmes.[9,10]

33.3 Commercial considerations and cashflow

Projects and contracts involving engineering construction are commercial ventures. Both the client promoting a project and a contractor employed by him are investing money and taking financial risks in order to achieve some desired return. Project and contract management is concerned with the control of both investment and risk with the aim of achieving this return.

The client invests money in the realization of the project to provide either a service or the production of goods. The project is conceived and developed to meet a predicted demand for the services or goods: a motorway, a hospital, a power station and a shoe factory are all examples of projects. In the first two cases, the return obtained from the investment is represented by benefit to the users, whereas the unit price charged for the product of a commercial factory will be calculated to ensure an attractive profit to the investors. The power station must both provide a statutory service and make a small profit.

The client may also be called the promoter or the employer. He is concerned with the flow of money to and from the project – the project cashflow – throughout the life of the project from its conception to the end of some defined period of operation, a period that may extend over many years. Projects are consequently capital intensive, i.e. a large amount of investment is required over a long period of time before a substantial benefit is achieved. This is well illustrated in Figure 33.3. Even in that manufacturing project with a relatively short life cycle of 12 years, the client will have capital committed for almost 7 years before a positive balance is shown in the project account.

He may also enter into a legal agreement or contract with a contractor, or a number of contractors, for the construction of the project. Contractors are commonly specialists in a particular field and may also undertake the detailed design of their work if so required by the client.

A contractor's pattern of investment is very different and the time-scale of his involvement is much shorter. His cashflow may well comprise a substantial investment in construction plant,

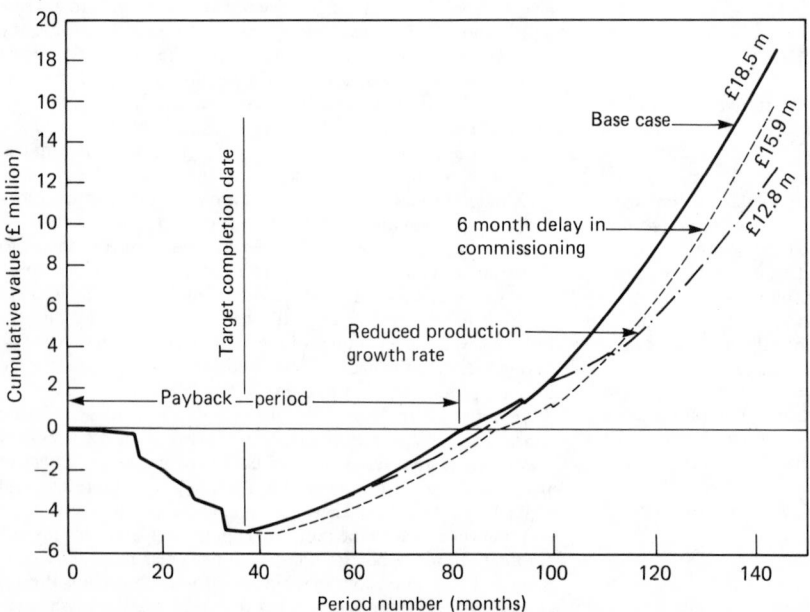

Figure 33.3 Cashflow diagram for new manufacturing plant. Delay in commissioning the plant generates a significant increase in capital investment

labour, and materials, particularly in the early stages of a contract, offset by regular monthly payments received from the client for work completed. The contractor must apply his expertise to produce a method statement for construction, assess the risks, and submit a tender that is both realistic and competitive. His targets are defined in his estimate and tender; thereafter he controls against them. The plan on which the tender is based will represent the most efficient use of the contractor's resources to achieve the work defined in the contract documents. It follows that any delay or change in the work content may also affect the timing of the flow of money to and from the contractor, and may involve additional funding from his capital as illustrated in the contract cashflow example given in section 33.3.2 below. If the disruption is caused by the client, the contractor will expect to be recompensed.

At any time, a contractor will probably be employed on a variety of contracts in different locations for different clients and will also be tendering for future work. His commercial skill is to utilize a relatively small amount of capital and to 'turn it over' as many times as possible by employing it to finance several contracts. The corporate cashflow and total financial risk will be very sensitive to changes in the timing of flows of money. It is important to appreciate the very different investment patterns experienced by clients and contractors as these greatly influence both attitudes and procedures adopted within the contractual relationship.

33.3.1 Cashflow

In order to quantify both the demand for money to meet the project or contract costs and the pattern of income it will generate it is necessary to predict the cashflow. A cashflow is a financial model of the project or contract which quantifies the actual flow of money, i.e. it takes account of delays between incurring a commitment and making a money transaction. The model is compiled simply by adding the costs and revenues to every activity on a bar chart programme which extends over the entire life-cycle of the project or contract.

When developing cashflow it is vital to distinguish between different categories of charge:

(1) Fixed charges which occur when certain stages are reached, e.g. a mobilization payment at the start of a construction contract.
(2) Time-related charges which are paid or received in regular increments over a period of time, e.g. the cost of site overheads, the weekly hire charge for a crane and payments to employees.
(3) Quantity proportional charges which are related to the quantity of material used or to the number of units of production of a factory or power station.

Only in this way will it be possible to use the model to predict realistically the effect on investment or return of such diverse factors as variations in output, delay, disruption, cost or wastage of materials.

33.3.2 Contract cashflow

The significance to the contractor's cashflow of different patterns of payment from the client and of delays is illustrated in Figure 33.4 by consideration of a small hypothetical contract of 2 years' duration.

The estimated cost curve (1) is seen in Figure 33.4(a) to be a flat 'S' due to the dominance of time-related costs for resources and overheads. The anticipated predicted revenue (2) assumes monthly payments for work completed and a 4-week delay between certification and payment by client. The shaded area

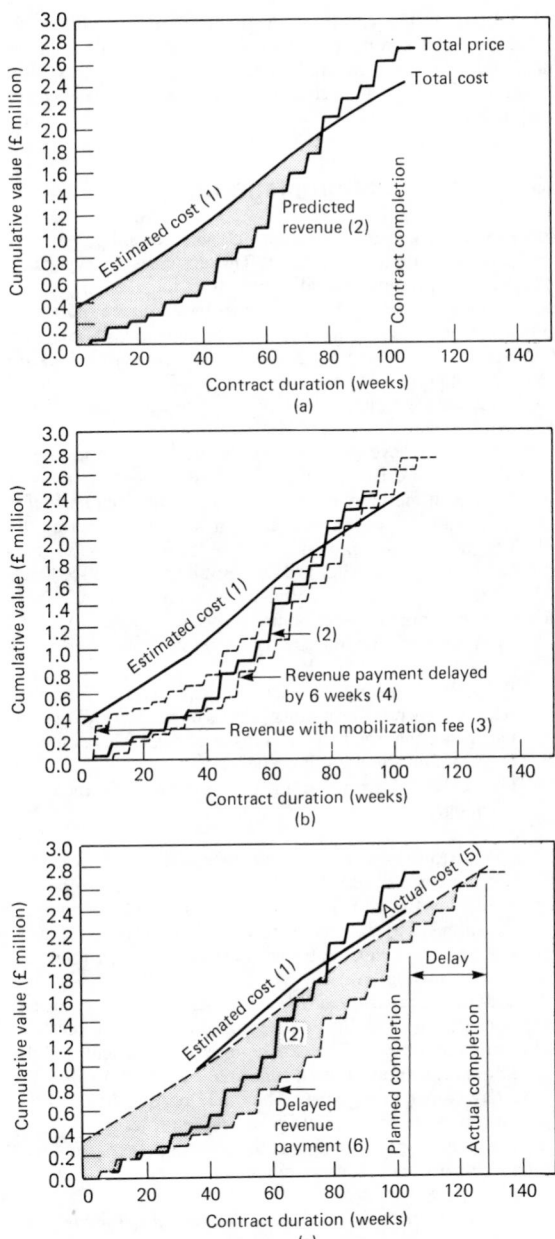

Figure 33.4 Contract cashflow. The contractors' investment is sensitive to change in payment and to delay

between the curves represents the funding to be provided by the contractor. This would be considerably reduced if the client agreed to a 10% advance mobilization payment (3), or increased if payments were received late (4), as illustrated in Figure 33.4(b).

The effect of delay in completion is also serious as cost increases and payment for completed work is delayed (see Figure 33.4(c)). If the delay was caused by factors outside the contractor's control, an extension of time may be granted and payment increased but this will depend on the circumstances and the contract conditions.

In all cases, the effect on the contractor's investment is significant. The cumulative effects of reduction of credit by suppliers, delay in payment by the client, perhaps on several contracts, can be that a contractor will be led quickly into bankruptcy!

33.4 Construction planning

The success of a project or contract will depend greatly on careful and continuous planning. The activities of designers, manufacturers, suppliers and contractors must be organized and integrated to meet the objectives set by the client and/or the contractor. Sequences of activities will be defined and linked on a time-scale to form the programme to ensure that priorities are identified and that efficient use is made of expensive and/or scarce resources within the perceived physical constraints affecting the job.

Remember, however, that because of the uncertain nature of construction work it should be expected that the plan will change. It must therefore be updated quickly and regularly if it is to remain a guide to the most efficient way of completing the job. The programme should therefore be simple – so that updating is straightforward and does not demand the feedback of large amounts of data from busy men – and flexible, so that all alternative courses of action are obvious.

The purposes of planning are therefore:

(1) To persuade people to perform their tasks before they delay the operations of other groups of people, and in such a sequence that the best use is made of available resources.
(2) To communicate with those people.
(3) To provide a framework for decision making in the event of change.

It is difficult to enforce a plan which is conceived in isolation, and it is therefore essential continually to involve the people responsible for the constituent operations and to encourage their commitment as the plan is developed and revised. Ideally, it should provide a flexible framework within which they can exercise their own initiative.

Programmes are required at various stages in the contract; when considering feasibility or sanction, at the precontract stage and during the contract. They are required by the client and the contractors. They may be used for initial budget control or for day-to-day construction work. They may pertain to one contract, or a number of contracts in one large project.

Before compiling a programme, the planner is therefore faced with a number of decisions and must decide on:

(1) *The appropriate level of detail for the programme.* The golden rule is to keep it simple. A programme of 100 activities is easy to comprehend; one of 1000 activities is not. Initially, divide the job into the minimum number of large work packages and develop detail later only in specific areas where it is required due to the complexity of the work or when there is need to determine precise resource requirements.
(2) *The choice of programming technique.* Particularly important in this choice is the level of management at which the programme is to be used, and the level of detail required. The main programming techniques are given in section 33.4.3 below.

33.4.1 Compiling a programme

This is a process of repeated refinement. Before sketching out the first draft, the planner must be familiar with the objectives and priorities defined for the job and be aware of constraints. These will include: (1) restrictions on access to parts of the works; (2) likely periods of bad weather which may inhibit particular operations; and (3) availability of resources. Although the efficient use of key resources will be considered in detail at a later stage of the refining process, awareness of levels of skills, machines and materials likely to be available will, of course, aid the preparation of a realistic first draft.

Assumptions are invariably made as the plan is developed and it is essential that these should be stated clearly so that everyone using it is aware of its validity.

When compiling the programme the planner is concerned with:

(1) *Resources:* with the allocation and utilization of people with expertise and skills, fabrication facilities, construction plant and materials.
(2) *Activities:* packages of work which *consume resources* and are defined by considerations of:
    (a) the type of work (and therefore the type of resource required);
    (b) the location of the work;
    (c) any restraints on the continuity of the activity.
    Each activity will be identified and appear as a bar or other symbol on the programme.
(3) *Logic:* with the relationships and links between activities which will be represented by lines or arrows on the programme. Most programmes will contain obvious sequences of activities which will provide the basic shape of the diagram. It is also advantageous to identify any opportunity to *overlap* activities, i.e. an activity may start before the preceding one is completed. In this way the overall duration of the job may be minimized, frequently with consequential savings in cost. Initially each activity should be shown at its *earliest possible start.*
(4) *Duration* of each activity which is a function of the quantity of work to be done, the number of units of resource allocated to the activity, and their predicted output.
    It is important to think of duration in this way as all these factors may be variable. The quantity of work, e.g. volume of excavation or number of drawings, may be more or less than originally estimated, one or more teams may be allocated to the activity, and their predicted production may or may not be achieved. Any change will affect the activity duration and in turn alter the overall demand for resources, the total duration of the job, the cost and the cashflow.

The initial allocation of resources to an activity will be a matter of judgement and is quite likely to be changed subsequently.

(5) *Potential problems and uncertainties* must be identified and the implications and possible responses considered. The greater the uncertainty, the more flexible the programme must be in order to provide alternative courses of action. This may be achieved either by allocating additional resources or by extending the contract duration. In either case, the estimated cost will increase and it is therefore essential to link the programme with the cost forecast.
(6) *Overall duration of the job* calculated when all the activities at their assumed durations have been assembled within the overriding constraints which form the framework of the programme.

33.4.2 Resource scheduling

Whichever programming technique is used, the next important

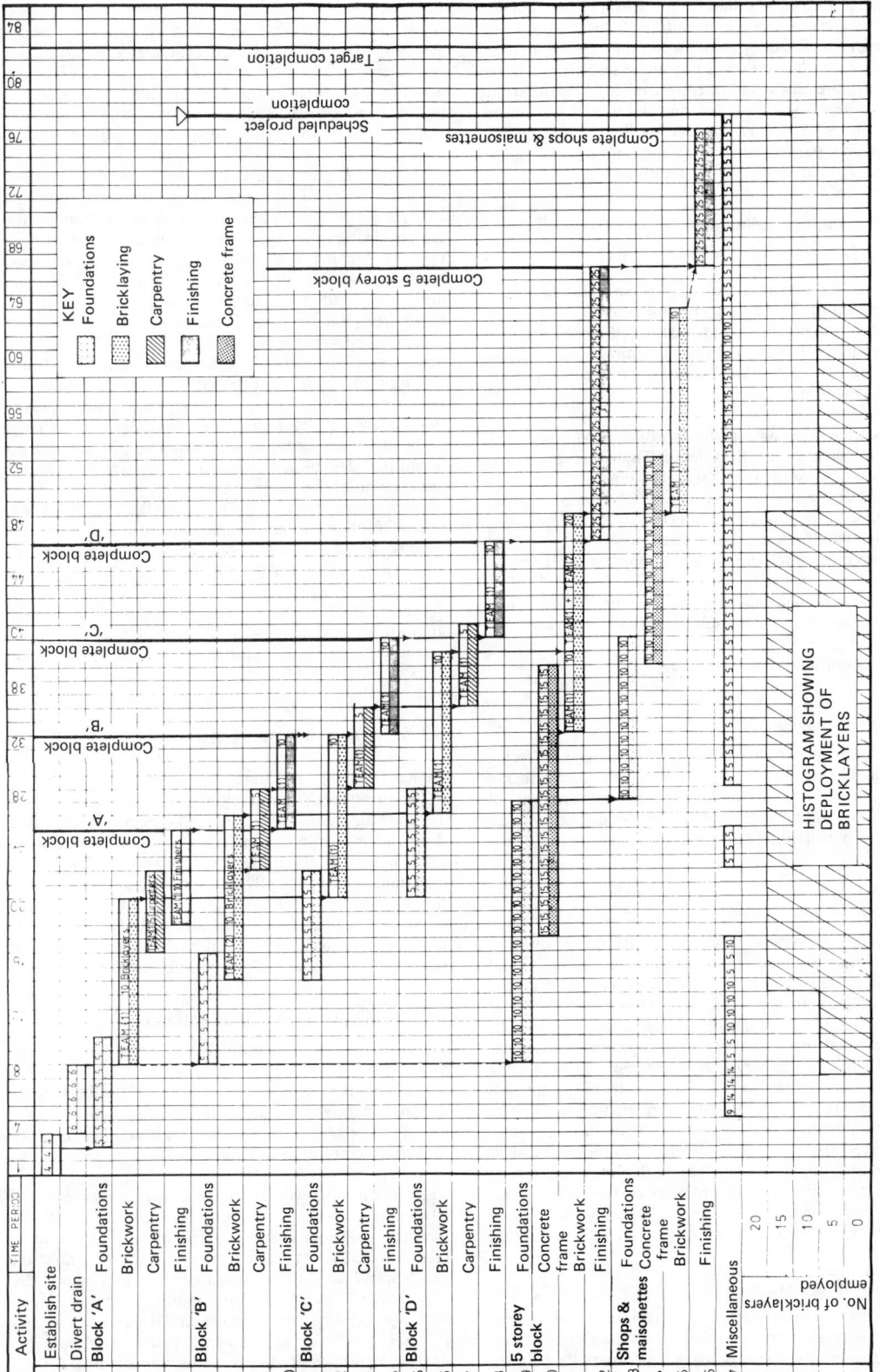

Figure 33.5 Bar chart programme. The work is organized to
achieve the project objectives and efficient use of resources

step in refining the plan would be to consider the overall demand for key resources. The definition of key resources is likely to differ for different types of project and particularly for their location. Consideration will be given especially to those resources which are *scarce* and/or *expensive*. It is clear that the adjustment, or levelling, of one resource will have an effect on the usage of others. Generally, resource levelling is only applied to a few resources, particularly if it is being done manually. The use of a computer can allow greater sophistication.

33.4.3 Programming techniques

33.4.3.1 Bar charts

The best form of plan for site use is the bar chart (a simple example is shown in Figure 33.5). In this example the planner was required to give priority to construction of blocks A and B and to complete the housing development in 82 weeks. Bricklayers were seen to be the key resource. Note the following aspects of the diagram:

(1) Each activity is shown in its scheduled position which results in efficient use of resources. The histogram shows continuous use of the same teams of bricklayers.
(2) The logic of the inter-relationship of activities and movement of resource teams is shown clearly.
(3) The space within the bars can be used for figures of output or resource demand, and there is room beneath to mark actual progress.
(4) Important constraints and key dates are marked to ensure that these are clearly communicated to all concerned with the project.
(5) Five weeks of float have been allowed.

33.4.3.2 Line of balance

This simple technique was developed for house building and is also useful for other forms of repetitive work, such as producing and distributing prefabricated units. The axes are the number of completed units and time: the work of each gang appears as an inclined line, the inclination being related to the output of the gang.[11]

33.4.3.3 Location–time diagram

In cross-country jobs such as major roadworks, the erection of transmission lines, or pipelaying, the performance of individual activities will be greatly affected by their location and the various physical conditions encountered. Restricted access to the works, the relative positions of cuttings and embankments, sources of materials from quarries and temporary borrowpits, the need to provide temporary or permanent crossings for watercourses, roads and railways, and the nature of the ground, will all influence the continuity of the construction work. The output achieved by similar resources of men and machines may vary when they are working in different locations. In such circumstances the programme may best be developed on axes of location and time on which all obstacles or features may be marked. The work of each resource team will again appear as an inclined bar with inclination related to the output of the gang.[12]

33.4.3.4 Network analysis

These techniques are used to evaluate programmes where complex or multiple relationships exist between a number of activities. The greatest benefit is the discipline imposed on the planner when compiling the logic of the network, i.e. when specifying the individual activities and the links between them. A further major advantage is that the computer may be utilized to explore many possible combinations of the timing of individual activities in order to achieve really efficient use of resources. Networks are rarely the best method of communication with the people actually responsible for performing the constituent activities, and the results of the analyses will normally be translated into some other form of working programme such as a bar chart.

Precedence diagrams and activity networks. The Author advocates the use of a simple precedence system which is illustrated in Figure 33.1. This system utilizes preprinted node-sheets and thereby enables the engineer to devote less time to drawing and more to planning. It also offers a simple method of overlapping activities and is easy to update and revise. Individual activities are represented by rectangular nodes which are linked by dependency lines to form the network.

Several overlaps are shown in the example. The figure 7 on the dependency line between activities 7 and 13 indicates that construction of the plant should start a maximum of 7 months before completion of design. It has been the Author's experience that it is the activity durations and overlaps which change most frequently when updating a programme. Consequently, he has found that definition of overlaps in this simple manner simplifies revision of the programme.

Float is also identified quickly from the diagram. The imposition of fixed start and completion dates in this example has resulted in all activities displaying at least 1 month of total float – as calculated from the difference between earliest and latest start of any activity.

Precedence diagrams provide an excellent basis for the development of cost models as illustrated when Figure 33.1 is extended to cover the operational phase of the project in Figure 33.2.

33.5 Cost estimating

Estimates of cost and time are prepared and revised at many stages throughout the development of a project or contract. They are all *predictions* of the final outcome of the job and the degree of realism and confidence achieved will depend on the level of definition of the work and the extent of risk and uncertainty. Consequently, the accuracy of successive estimates should improve as the project or contract develops. The most important estimates prepared are probably for a project, at sanction, and for a contract, at tender, for it is at these points that the client and contractor become committed.

33.5.1 Requirements of an estimate

The requirements of an estimate are:

(1) To predict the most probable cost of the works and also to define the range within which the final cost is likely to lie.
(2) To produce a forecast of expenditure: the cashflow based on the project programme.

These predictions will be influenced by factors peculiar to the particular project under consideration. Location, logistics, weather, availability and capacity of resources and market factors will all affect the final price. The estimate must therefore be compiled with the circumstances of the project clearly in mind and all assumptions, uncertainties and exclusions should be stated. Ideally, any estimate should be presented as a most probable value and a tolerance together with a range of less likely values to emphasize that it is an estimate.

It is important to realize that the precise value of a specific single-figure estimate made at an early stage of the project or contract is most unlikely to be achieved due to the uncertainty associated with civil engineering work.

33.5.2 Cost-estimating techniques

All these techniques rely on historical data of some kind and it is prudent to note the following points:

(1) Ideally, the data should be from a sufficiently large sample of similar work in a similar location and constructed in similar circumstances. Unfortunately, this is rarely the case and corrections have to be applied.
(2) Cost data needs to be related to a specific historical date, chosen with care. Historical costs must be corrected for inflation, changes in exchange rates and market factors.

The five basic estimating techniques available to the estimator are summarized below.

(1) *Global*. This term describes the 'broadest brush' category of technique which is derived from libraries of achieved costs of similar projects related to the overall size or capacity of the asset provided. This technique may also be known as 'rule-of-thumb' or 'ballpark' estimating. Examples are:
   (a) cost per square metre of building floor area or per cubic metre of building volume;
   (b) cost per megawatt capacity of power stations;
   (c) cost per kilometre of roads;
   (d) cost per tonne of output for process plants.
   The technique relies entirely on historical data and therefore must be used in conjunction with inflation indices and a judgement of the influence of the construction market appropriate to the envisaged timing of the project. Global estimates can only be used to give a rough indication of the order of cost in the appraisal and definition stages of project development.
(2) *Factorial*. These techniques are used widely for early estimates of the cost of process plants, power stations, etc. where the core of the project consists of major items of plant which can be identified and for which current budget prices may be obtained from suppliers.
   The techniques provide factors for a comprehensive list of peripheral costs such as pipework, electrics, instruments, structures and foundations. The estimate for each peripheral will be the product of its factor and the estimate for the main plant item.[13]
   A detailed programme is not a necessity but it is recommended that one is prepared. This will be valuable particularly in identifying problems of construction which will go undetected if the technique is applied in a purely arithmetical way, and is required for cashflow prediction. The technique has the considerable advantage of being predominantly based on current costs, thereby taking account of market conditions and needing little, if any, reliance on inflation indices. Factorial techniques are not normally reliable for site works.
(3) *Manhours*. This technique is only suitable for labour-intensive construction, design-office activities, and operations such as mechanical erection work where reliable records of hourly productivity of different trades are available. The total manhours estimated for a given operation are then costed at the current labour rates and added to the costs of materials and equipment. The advantages of working in current costs is obtained.
(4) *Unit rates*. The estimator selects historical rates or prices for each item in the bill of quantities using either information from recent similar contracts, or published informa-

tion, or 'built-up' rates from his own analysis. As the technique relies on historical data it is subject to the general dangers outlined above.

The technique is most appropriate to building and repetitive work where the allocation of costs to specific operations is reasonably well-defined. It is essential that the rates are selected from an adequate sample of similar work with reasonably constant levels of productivity and limited distortions arising from construction risks and uncertainties, e.g. access problems. The technique is less appropriate for civil engineering where the method of construction is more variable and where the uncertainties of ground conditions are more significant.

The unit rate technique does not demand an examination of the programme or method of construction and the estimate is frequently compiled by the direct application of historical 'prices'. It therefore does not require an analysis of the real costs of the work, neither does it encourage consideration of the peculiarities, constraints and risks affecting the particular project. Nevertheless, unit rate estimating is probably the most frequently used technique. It can result in reliable estimates when practised by experienced estimators with good, intuitive judgement and the ability to assess the realistic programme and circumstances of the work.

(5) *Operational (resource-cost)*. This is the fundamental estimating technique wherein the total cost of the work is compiled from consideration of the constituent operations or activities defined in the construction method statement and programme and from the accumulated demand for resources. Labour, plant and materials are costed at current rates. The advantage of working in current costs is obtained.

The most difficult data to obtain are the productivities of labour and construction plant in the geographical location and special circumstances of the project under consideration. Claimed outputs of plant are obtainable from suppliers' handbooks but these need to be reviewed in the light of actual experience. Labour productivities will vary from site to site depending on management, organization, industrial relations, site conditions, etc. and also from country to country.

The operational technique is, by far, the best method of evaluating uncertainties and risks, particularly those likely to cause delay.[12] Because the technique exposes the basic sources of costs, the sensitivities of the estimate to alternative assumptions/methods can be investigated easily and the reasons for variations in cost appreciated. It also provides a detailed current cost/time basis for the application of inflation forecasts and, hence, the compilation of a project cashflow.

This is the most reliable estimating technique for civil engineering work. Compilation is relatively painstaking and time consuming compared with other techniques, but when preparing an operational estimate the estimator will gain a realistic appreciation of the risk and special circumstances of the project.

33.6 Project appraisal

Project appraisal is a process of investigation, review, and evaluation undertaken as the project or alternative concepts of the project are defined. This study is designed to assist the client to reach informed and rational choices concerning the nature and scale of investment in the project. The core of the process is an economic evaluation – based on a cashflow analysis of all costs and benefits which can be valued in money terms – which is also therefore called cost/benefit analysis.

Appraisal is likely to be a cyclical process repeated as new ideas are developed, additional information received and uncertainty reduced until the client is able to make the critical decision to sanction implementation of the project and commit the investment in anticipation of the predicted return.

33.6.1 Risk and uncertainty

The greatest degree of uncertainty about the future is encountered early in the life of a new project. Decisions taken during the appraisal stage can have a very large impact on final cost, duration and benefits. The extent and effects of change are frequently underestimated during this phase although these are often considerable, particularly in developing countries and remote locations.

At this stage, the engineering and project management input will normally concentrate on providing:

(1) Realistic estimates of capital and running costs.
(2) Realistic time-scales and programmes for project implementation.
(3) Appropriate specifications for performance standards.

At appraisal, the level of project definition is likely to be low and therefore risk response should be characterized by a broad-brush approach.[14] It is recommended that effort should be concentrated on:

(1) Seeking solutions which avoid/reduce risk.
(2) Considering whether the extent or nature of the major risks are such that the normal transfer routes may be unavailable or particularly expensive.
(3) Outlining any special treatments which may need to be considered for risk transfer, e.g. for insurance or unconventional contractual arrangements.
(4) Setting realistic contingencies and estimating tolerances consistent with the objective of preparing the best estimate of anticipated total project cost.
(5) Identifying comparative differences in the riskiness of alternative project schemes.

Engineers/project managers will usually have less responsibility for identifying the revenues and benefits from the project: this is usually the function of marketing or development planning departments. The involvement of engineers/project managers in the planning team is recommended as the appraisal is essentially a multi-disciplinary brainstorming exercise through which the client seeks to evaluate all alternative ways of achieving his objectives.

For many projects this assessment is complex, as not all the benefits/disbenefits may be quantifiable in monetary terms. For others it may be necessary to consider the development in the context of several different scenarios (or views of the future). In all cases, the predictions are concerned with the future needs of the customer or community. They must span the overall period of development and operations of the project which is likely to range from a minimum of 8 or 10 years for a plant manufacturing consumer products, to 30 years for a power station and much longer for public works projects. Phasing of the development should always be considered.

33.6.2 Project evaluation

The process of economic evaluation and the extent of uncertainty associated with project development is illustrated by the appraisal of a hypothetical new manufacturing plant.

The simple precedence diagram (Figure 33.1) has been developed into a flow chart (Figure 33.2) and extended to include 9 years of plant operation. The diagram also gives some indication of the patterns of costs and revenue.

The plant is designed for development in two stages, the first with a manufacturing capacity of 5000 units and the second raising this to the maximum of 8000 units. During an appraisal study, uncertainty frequently exists with regard to the demand for the product and is indicated by the definition of a range of possible output. The forecast curve assumes a 20% per annum growth in the market but a range of 15 or 25% is considered possible. If the growth rate fell below the favoured 20% it is likely that stage 2 would not be implemented.

If all the predictions of costs, revenues, markets and programme over the 12-year project life were correct, the project would require maximum investment of £4.96 million and would ultimately generate a surplus of £18.5 million as shown in the basecase cumulative cashflow curve (the full line in Figure 33.3).

Other parameters used to quantify this investment could be:

Payback period	80 months
Net present value @ 10% discount rate	£6.13 million
Internal rate of return (i.r.r.)	27.6%

and it is strongly advised that a similar range of criteria are employed when determining any investment.

It is, of course, most unlikely that those precise values will be achieved due to all the risks and uncertainties which exist at the early stage of project development. The chain line in Figure 33.3 indicates that, should the market growth rate be only 15%, the surplus would be reduced to £12.8 million and i.r.r. to 22.9%. The obvious effect of a 6-month delay in completion of the plant, shown by the broken line in Figure 33.3 would be to reduce the surplus to £15.9 million and i.r.r. to 23.7%. A far more serious consequence could be loss of the market to a competitor.

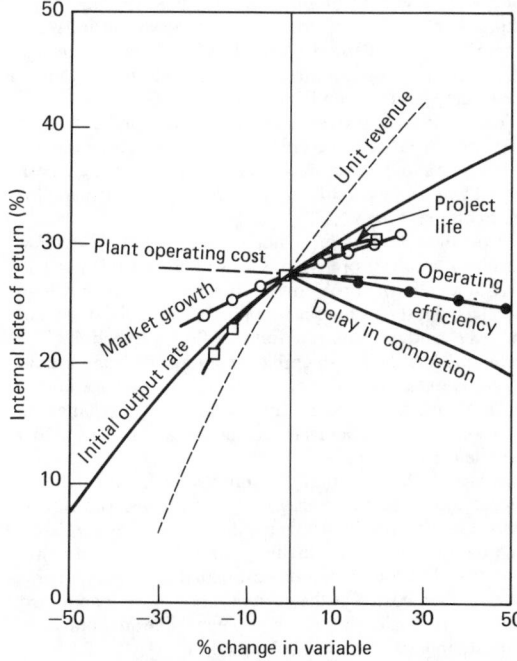

Figure 33.6 New manufacturing plant: sensitivity diagram.
This 'spider' diagram communicates the relative effect of major variables on the viability of the project

The effect on the investment of variation in a whole range of factors is well shown in the sensitivity diagram (Figure 33.6). Each variable is considered independently and it is obvious that market factors are dominant. Delay in completion is also shown to have a serious effect on the return obtained from the investment. The great value of this diagram is that it indicates where further effort is needed to reduce uncertainty, perhaps by additional market surveys in this case. It also suggests that management policy must give priority to timely completion of the plant.

In practice, a combination of all these uncertainties and risks is likely to be experienced and a better prediction of the probable range of outcome of this project can be obtained from the cumulative probability diagram (Figure 33.7). This diagram is generated by substituting 1000 combinations of these factors in the basic model on a random basis in a Monte Carlo simulation.[14,16] The base case prediction is seen to be optimistic when uncertainties are taken into account as there is a 77% probability that i.r.r. will be less than 27.6%. It is predicted that there is 50/50 chance of achieving an i.r.r. of 21% but that extreme values of zero and 40% are just possible. Although analyses of this type require judgement to be made on the likely range and probability distribution of each variable, the Author strongly recommends that this discipline of a rigorous risk analysis is adopted for all major projects.

The output of power and market factors is also seen to dominate the sensitivity diagram for a real project, the proposed Severn tidal power scheme,[15] (see Figure 33.8). Again, the most sensitive engineering factors are delays in completion of the works and installations of the turbines.

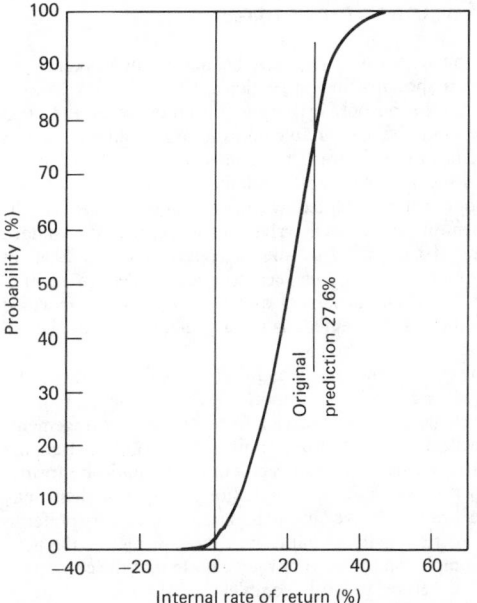

Figure 33.7 New manufacturing plant: cumulative probability diagram. There is a 77% probability that the internal rate of return will be less than the base case prediction of 27.6%

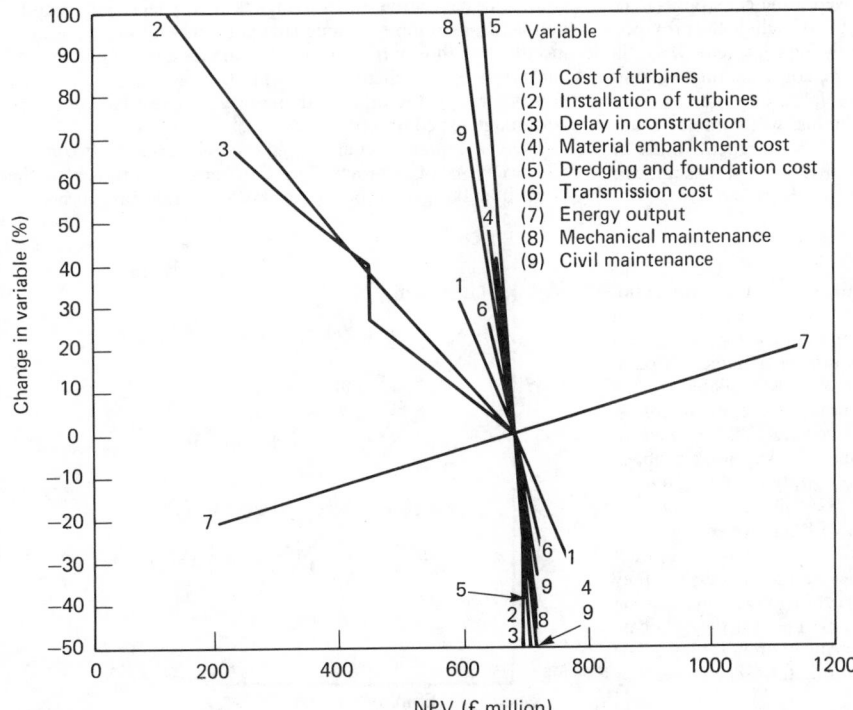

Figure 33.8 Severn tidal power scheme: sensitivity analysis. The amount of power produced, the selling price, and the delay in completion of the project again have the most serious effect on project viability

33.7 Engineering contracts

Construction work of all types is normally undertaken by a contractor, a specialist in the particular field of work, who is employed for this purpose by the client. In most cases, the client will invite a number of suitable contractors to submit tenders and subsequently will award the contract on the basis of the lowest realistic and acceptable tendered price.

This approach is adopted worldwide and has led to the development of well-defined systems of working and Standard Conditions of Contract for different types of work. Each of these traditional procedures has been developed to meet a particular set of circumstances and will work well, provided their limitations are accepted, as the associated case law is well established.

It is, however, important for engineers to realize that, because of the diversity of both construction work and clients' requirements, no single uniform approach to contractual arrangements can be specified or advocated. A number of alternative strategies are available to the client and each contract should be formulated with the specific job in mind. For example, a client may wish to be directly involved in site management or may prefer to delegate this responsibility entirely to the contractor, the need for early completion of the work may dictate that the contractor is appointed before design is completed, risks may be apportioned in various ways between the parties, the contractor may be required to undertake the detailed design or to provide varying amounts of finance – these and many other considerations will all influence the client's contract strategy.

Obviously, this strategy will also be greatly affected by the nature of the work to be completed under the contract. The fabrication and positioning of an offshore oil production platform is a high-risk venture which may involve advanced or new technology and will be subject to severe time constraints. The contract for building a chemical plant may include the provision of unique process know-how offered by the contractor who will consequently undertake detailed design, construction and commissioning of the plant. Tunnelling implies uncertainty – and therefore risk – about ground conditions, whilst minor roadworks or house building are likely to involve repetitive use of traditional techniques with relatively little financial risk and the overriding requirement of minimum cost.[16]

33.7.1 Contract strategy

The term 'contract strategy' is used to describe the organizational and contractual policies chosen for the execution of a specific project.[17] The development of a contract strategy is an important task for the client or his project manager. It comprises a thorough assessment of the choices available for the implementation and management of design and construction. A pattern of inter-related decisions is required which seeks to maximize the likelihood of achievement of key project objectives. The selected strategy is likely to be optimal in that it must satisfy a variety of constraints and be sufficiently robust to withstand the uncertainty associated with the project.

The decisions taken during the development of a contract strategy affect: (1) the responsibilities of the parties; (2) they influence the control of design, construction and commissioning and, hence, the co-ordination of the parties; (3) they allocate risk and define policies for risk management; and (4) they define the extent of control transferred to contractors. Therefore, they affect cost, time and quality.

The first step in the development of a contract strategy is to identify the areas which constitute the strategic choices. These are:

(1) The project management objectives as defined by the client.

(2) The organizational system for design and construction.
(3) The type of contract.
(4) The Conditions of Contract and other contract documents.
(5) The tendering procedure.

The project manager must then choose from the options available within each of these five strategic areas.

33.7.2 Choice of contract type

There are three essential requirements of any contract:

(1) *Incentive.* The aim is to provide an adequate incentive for efficient performance from the contractor. This must be reflected by an incentive for the client to provide appropriate information and support in a timely manner.
(2) *Flexibility.* The aim is to provide the client with sufficient flexibility to introduce change which can be anticipated but not defined at the tender stage. An important and related requirement is that the contract should provide for systematic and equitable evaluation of such changes.
(3) *Risk sharing.* The aim should be to allocate all risk between client and contractor. This must take account of the management and control of the effects of risks which materialize. The contractor will include a risk contingency sum in his tender as protection against the risks he has been asked to carry.

The inter-relationship of these requirements with the type of contract is demonstrated in Figure 33.9 in which the requirements are expressed in terms of contractor's incentive, client's flexibility and exposure to risk. It is apparent that, generally, a contractor's incentive and a client's flexibility tend to be incompatible. For example, a lump-sum contract imposes maximum incentive on the contractor but also implies a very high level of constraint on the client against introducing change. The converse is true at the other extreme of a cost-reimbursable plus percentage fee contract.

There are many detailed points of difference between the various types of contract. Those of most importance to the client in making an appropriate choice are summarized below.

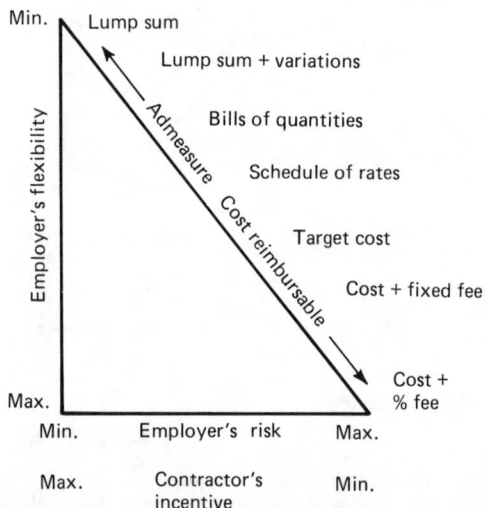

Figure 33.8 Characteristics of different types of construction contract

33.7.3 The main types of contract

Types of contract are virtually classified by their payment system: (1) *price-based* – lump sum and admeasurement (prices or rates are submitted by the contractor in his tender); and (2) *cost-based* – cost-reimbursable and target cost (the actual costs incurred by the contractor are reimbursed, together with a fee for overheads and profit).

33.7.2.1 Price-based contracts

Lump-sum contracts. Lump-sum contracts are based on a single price tendered for the whole works. Payment may be staged at time intervals or related to achieved milestones. Lump sum does not necessarily imply a fixed price; in particular, price may be adjusted for cost escalation.

The implications are that complete, final design is available at tender and that minimal changes or variations are expected. No contractual mechanism is specified for price adjustment and such contracts are therefore rarely used in engineering construction.

A high degree of tender competition may be achieved and the price may include a high level of financing by contractor.

Admeasurement contracts. These are based on bills of quantities or schedules of rates in which items of work are specified with quantities. Contractors then tender unit rates or prices against each item. Payment is usually monthly and is derived from measuring quantities of completed work and valuing at rates in the tender, or new rates negotiated from tender rates.

Mechanisms are provided for adjusting both price and time, as discussed in section 33.8 in the likely event of change. This facility to introduce limited variation is frequently abused and design may be only partially complete at tender. Extensive change and delay will generate claims and the final price is invariably different from the tender total. The price will include an allowance for any financing required by the contractor and a risk contingency.

Cost-reimbursable contracts. These are based on payment of actual cost incurred by the contractor plus a specified fee for overheads and profit. The contractor's cost accounts are open to audit by the client (Openbook Accounting). Payments may be monthly in advance, in arrears, or from an imprest account.

Cost-reimbursible contracts are normally used when the client wishes to employ a contractor at an early stage of project definition and design, or when there are major risks associated with the contract. There is little contractual incentive for the contractor to perform and the final price will depend both on the extent to which risks materialize and on the efficiency of the contractor. The client carries the risk and will therefore require to participate in contract management.

Target-cost contracts. Target cost is based on the setting of a probable (or target) cost for the work. The target cost will subsequently be adjusted for major changes in the work and cost inflation. The contractor's actual costs are monitored and reimbursed as in a cost-reimbursable contract. Any difference between actual cost and target cost is shared in a specified way between the client and the contractor. There is a separate fee for overheads and profit.

By using target-cost contracts, it has been possible to achieve a high degree of collaboration between the parties. They are most suitable for high-risk contracts where the work content is well defined, such as tunnels.[18]

33.7.4 Management contracting

Various forms of management contract are widely used in building construction but are rarely appropriate to civil engineering works. In these systems, an external organization – the management contractor or construction manager – is employed specifically to manage and co-ordinate design and construction on behalf of the client. The systems are frequently adopted to achieve early completion of the project by 'fast tracking' or by overlapping and integrating design and construction. They are most appropriate for construction which can be split into a series of well-defined contract packages, each of which is awarded immediately the relevant design is completed.

The management contractor is normally employed on a cost-reimbursable basis and, although the construction contracts are of the familiar admeasurement form, it is important to realize that the allocation of risk between the parties may be considerably changed.[2]

33.8 Contractual measurement and valuation

The type of contract, the level of detail required in contractual measurement and the method of valuation of the work are interrelated.

Several procedures are provided in admeasurement contracts for civil engineering works for valuation and payment in respect of changes to the work defined in the contract documents.

(1) Changes to quantities of measured work listed in the bill of quantities are priced on a pro rata basis, the actual quantity of work completed being substituted for the original estimated figure in the final account. Under the Institution of Civil Engineers[19] and Federation Internationale des Ingénieuers Confeils (FIDIC). Conditions of Contract there is, in theory, no limit to the adjustment permitted to any quantity but it should be noted that under Clause 56(2) of the former (fifth edition) the tendered rate may be varied following such adjustment. In some other Conditions of Contract, a definite range of adjustment to quantity, over which the tendered rate is deemed to be valid, is specified.

(2) Variations to the work defined in the contract may be ordered by the Engineer who will issue a written variation order. The Engineer is empowered to fix the value of the work covered by the variation order after consultation with the contractor and, wherever possible, utilizing tendered rates in the bill of quantities. Should the contractor dispute the engineer's valuation, the contractor may claim additional payment.

(3) The contractor may also claim additional payment and/or extension of time should he incur additional cost on account of 'unforeseen conditions' or delay. The Engineer will assess the value of each claim from evidence submitted by the contractor.

The Engineer is empowered to settle all disputes between the client and the contractor. If either party is dissatisfied with his decision they may then resort to arbitration.

33.8.1 Bills of quantities

The conventional British and international civil engineering contracts are of the admeasurement type wherein the contract price is accumulated in the bills of quantities. These list and quantify the constituent items of work, each of which is priced individually by a tendering contractor. The quantities of work are stated to be the best estimate which can be made by the Engineer prior to tender: all work items are subsequently remeasured during the course of the contract and valued at the rates tendered by the successful contractor.

The bill of quantities is one of the contract documents and has several main functions:

(1) To itemize and quantify the elements of work to be completed within the contract.
(2) To facilitate comparison of tender prices.
(3) Interim and final valuation of completed work.
(4) Evaluation of change and variation.

33.8.2 The concepts incorporated in the traditional bills of quantities

The admeasurement contract developed from the lump-sum contract in order to provide essential flexibility as jobs become larger and more complex and the traditional admeasurement procedures and bills of quantities have been developed from the following simple concepts:

(1) All prices are deemed to be proportional to the quantity of work completed: all quantities are remeasured on completion of the contract.
(2) The client will pay only for completed permanent works.
(3) The payment lines are specified.
(4) The contractor can price the component items in any way he wishes.
(5) The tender price is to be the total price for completing the works specified in the contract documents.

33.8.3 Development of the bills of quantities

The simple and expedient procedures outlined above proved acceptable and equitable for repetitive and labour-intensive work. They are rarely adequate nowadays for the evaluation of plant-intensive work or for contracts including significant items of temporary work when there is a significant incidence of variation or delay. There are several reasons for this. The prices entered in the traditional bill of quantities rarely represent the true cost of completing the work defined in the individual items as the contractor's costs are not all directly related to the quantity of work completed. It follows that any adjustment of price resulting from a change in quantity of a particular item is unlikely to represent the true variation in cost.

A significant part of a contractor's costs are time-related and it is these costs which are affected by disruption or delay. Time-related charges are not, however, separated in the traditional bill of quantities and it is therefore not possible to evaluate systematically the very issues which are a major source of contractual claims.

As the incidences of change and variation have increased, various attempts to improve measurement and valuation procedures have included:

(1) Additional 'preliminary' items for overheads and specified facilities to be provided by the contractor.
(2) Separation of major temporary works items.
(3) Limits on the range of permissible variation for individual billed items and for the total extent of variation within the contract.
(4) Separation of 'method-related' charges.

33.8.4 Method-related charges

Systematic evaluation of a range of changes and variations, including delays, may be achieved by the separation of method-related charges in the bill of quantities. This approach was introduced in the British Civil Engineering Standard Method of Measurement (CESMM) in 1976[20] and moves away from the concept that all charges are proportional to quantities of completed work. Method-related charges are introduced to permit tenderers to enter their own items in the first section of the bill of quantities for any operations whose costs are not directly linked to the quantities of permanent works. The rates entered against the bill of quantities are consequently more realistic and are dominated by material costs.

The principal improvements in measurement procedures are seen to be:

(1) Items and prices take account of method of construction.
(2) Systematic evaluation of variations and claims arising from disruption and delay.
(3) Greater similarity between cost and price.

The production of estimates in an operational form which is directly related to his programme has greatly facilitated cost forecasting by the contractor. Similarly the separation of method-related charges in the bill of quantities permits the meaningful correlation of price and time which is essential for cashflow forecasts by the contractor, budget forecasts by the client and the mutual evaluation of contractual changes.

Important points to note are:

(1) Method-related charges are specified as either fixed or time-related and are entered in the *Preliminary and General section* of the bill of quantities. They are all priced as *Lump Sums*.
(2) Where the contractor enters such method-related items arising from his method of construction, they must be defined in sufficient detail for the Engineer to be able to identify the particular resource.
(3) It is most important to note what method-related charges are not subject to remeasurement. They may, however, be adjusted where relevant under a variation order.

33.8.5 Pricing and tendering policy

The contractor may translate cost into the priced items in the bill of quantities in any way he wishes, i.e. he may separate all or some of his method-related charges and 'weight' items to improve cashflow. In doing so, he must consider both the effect on his investment and the possible consequences of variations to the work defined in the bill of quantities. The incentive for a contractor to separate method-related charges normally stems from an improvement in cashflow; in consequence, he provides less investment and may reduce his tender price.

33.9 Project management

The responsibility of the project manager normally spans design, construction and commissioning. His function is to control the sequence of events and decisions leading to the completion of the project. Indecision is costly, as resources – design teams, manufacturers and contractors – are employed and will require compensation if their work is disrupted. Nevertheless, change is a characteristic of the engineering phase of projects involving construction and the project manager must be prepared to take the necessary corrective action.

If he is to fulfil his task of control of the realization of the project on behalf of the client, the project manager cannot divorce decisions taken on engineering matters from all other factors affecting the investment. Control may only be achieved by regular reappraisal of the project as a whole so that the current situation in the design office, on fabrication, on the supply of materials, and on site may be related to the latest market predictions. If this is done the advantage to be gained, say, from early access of land may be equated with any additional costs in full knowledge of the value of early or timely completion. The continual updating of a simple 'time and money' model of the project originally compiled for appraisal as illustrated in section 33.6 will greatly facilitate effective control during the engineering phase.[21]

33.9.1 Guidelines for project management

The project management process is briefly summarized below:

(1) The success of a project or contract depends greatly on the management effort expended by the client prior to sanction and by both parties prior to award of a contract.

(2) The client commits himself to investment in the project on the basis of the appraisal completed prior to sanction. The appraisal must be realistic and identify all risk, uncertainties and potential problem areas. Single-figure estimates are misleading and should be supported by figures showing the range of likely outcome of the investment.

(3) The client has a crucial role to play during implementation of the project, and the early appointment of an experienced project manager to pursue his interests is essential. He must be supported by an adequate project team set up in good time. The function of this team is co-ordination of all aspects of the project and particularly the contribution of the client organization.

The prime roles of the project manager are to drive the project forward and to think ahead; he must therefore delegate routine functions and concern himself with any problem areas.

(4) It is essential that project management ensures that the client clearly defines the project objectives together with the ranking of their relative importance. The likelihood of a successful project is greatly improved when all key managers of design, construction and supporting groups are fully informed and committed to these objectives. The project objectives should also be communicated to the other parties (contractors, consultants, etc.) involved in project implementation.

The dominant considerations must be fitness for purpose of the completed project and safety during both the construction and operation phases. Thereafter, the normal primary project objectives are concerned with cost, time and quality. These are inter-related and may conflict.

The fact that the client does not see any return on his investment until the project is commissioned suggests that timely completion should be a prime objective.

(5) Engineering projects are normally of short duration and are completed against a demanding time-scale. Adequate staff of the right quality must therefore be appointed and given training in the appropriate techniques and procedures. All staff concerned with contract management must be familiar with the contractual procedures employed.

(6) Although the scope of the project will be agreed at sanction it is probable that conceptual design, which will determine the final layout and size of the functional units, will follow early in the engineering phase. It is recommended by the Author that the conceptual design is rigorously reviewed as this is the main opportunity both for cost saving and for ensuring that the proposals meet the client's objectives. Particular attention should be given at this stage to subsequent operation and maintenance of the project.

(7) Effective control of the project or contract will only be achieved through continual planning and replanning. Management effort should be concentrated on the present and the future: time devoted to the reporting and collection of historical data should be kept to a minimum.

In his planning, the client must take a broad view of the project and aim to co-ordinate design, construction, commissioning and subsequent operation and maintenance. Interaction of contractors, access, statutory requirements and public relations must all be considered.

A contractor will plan in more detail and aim to achieve continuous and efficient deployment of his resources. Because of the greater likelihood of change, the contractor's programme should be flexible and subject to constant review.

(8) The plan used by senior staff should clearly show the financial consequences of alternative courses of action and of indecision. It is therefore convenient to develop the plan as a time-and-money model of the project or contract which will react realistically to changes in timing, method, content and cost of work. Realism is largely dependent on the correct definition and allocation of costs and revenues as either fixed, time-related or quantity-proportional charges.

Time-related costs are significant in all types of construction work and predominate in many civil engineering projects. Adherence to the programmed time schedule for the work will therefore also control both cost and investment.

(9) Time lost at the beginning of a project can rarely be recovered: particular attention must therefore be given to the start-up of the project. Similarly, sufficient time must be allowed for mobilization by each contractor.

(10) Consideration of alternative contract strategies will frequently focus attention on deficiency of information and on the problems which will hinder the project objectives. Selection of an appropriate contract strategy at an early stage of project implementation is perhaps the most important single activity of the project management team.

(11) Appointment of a contractor on the sole criterion of lowest bid price will not necessarily lead to a harmonious contractual relationship. The lowest tender may not produce the lowest contract price.

Both parties are making their commitment at this point and should be fully aware of both the client's objectives and the contractual responsibilities.

Selective tendering followed by rigorous bid appraisal, including study of the contractor's programme and resource allocation, will do much to ensure that the contractor has not misjudged the job and that his price is realistic. The production of his own operational type of cost estimate will greatly aid the project manager in this appraisal.

The client must check that all his obligations can be honoured before award of the contract.

(12) The items in the bill of quantities or other contractual financial document should reflect the method of construction. Similarity between the tendered prices and the contractor's costs will greatly aid evaluation of change and equitable adjustment of price for inflation.

(13) Throughout the implementation period of the project the client or his representative will inspect and approve the quality of workmanship of contractors and manufacturers. Again, an adequate number of staff with relevant experience must be employed. Prior definition and agreement of acceptable standards is essential and all parties should be aware of tolerances. There is a tendency for design engineers to specify unnecessarily high standards, the achievement of which may prove difficult and/or expensive. The desired quality of workmanship must always be considered in relation to the client's other prime

objective, usually timely completion and economical cost.

(14) Clients frequently underestimate both the extent and consequence of change. The project manager should assess rigorously the cost and benefit of all design changes before they are implemented. Priority should be given to timely completion of the project.

The better organized the contractor, the more likely it is that he is working to a tight, well-resourced programme. The disruptive effect of variation may therefore be serious.

Modifications to manufacturing plant are sometimes best implemented during some future shutdown of the plant for maintenance.

(15) Involvement in prolonged bargaining over claims is a sign of failure. Evaluate and agree payment for variations and claims as the job progresses. The valuation should be based on prices, resource output and efficiencies similar to those incorporated in the contractor's tender.

(16) Projects and contracts are managed by people who are directing and communicating continuously with other human beings. Great attention must be paid to the selection and motivation of staff. Personality and ability to think ahead are as important as technical know-how.

Project and contract management staff must be given adequate authority to manage in their dynamic working environment without continual reference to head office.

For both the client and contractor, one man in each organization – the project/contract manager – must ultimately be responsible and be known to be responsible, for the realization of the project or contract. Each must be identified with, and committed to, the project.

References

1 Institution of Civil Engineers (1979) *Civil engineering procedure.* Thomas Telford, London.

2 Hayes, R. W., Perry, J. G. and Thompson, P. A. (1983) *Management contracting.* Construction Industry Research and Information Association, Report No. 100, CIRIA, London.

3 National Economic Development Office (1983) *Faster building for industry.* HMSO, London.

4 Ninos, F. E. and Wearne, S. H. (1986) 'Control of projects during construction, *Proc. Instn Civ. Engrs,* **80**, 931.

5 Barnes, N. M. L. (1978) 'Human factors in cost control.' *Proc. 5th International Cost Engineering Congress.* Danish Association of Cost Engineers, Utrecht, pp. 244–248.

6 Neil, J. M. (1982) *Construction cost estimating for project control.* Prentice-Hall, Englewood Cliffs, New Jersey.

7 National Economic Development Office (1975) *The public client and the construction industries.* HMSO, London.

8 Power, R. D. (1985) *Quality Assurance in civil engineering.*

Construction Industry Research and Information Association, Report No. 109, CIRIA, London.

9 Davies, V. J., Dodd, L. A., Mills, T. R., Tietz, S. B., Woods, D. R., Barker, J. E., King, F. W., Harlow, W. A., Summersgill, I., Rust, C. E., Eye, D. and Beagley, C. (1986) 'Conference on Engineering for safety.' *Proc. Instn Civ. Engrs,* **80**, 13–119; (discussion, February 1987).

10 *New Civil Engineer* (1986) 'ICE to press for greater safety enforcement', *New Civ. Engr,* no. 718.

11 Lumsden, P. (1968) *The line of balance method.* Pergamon Press, Oxford.

12 Thompson, P. A. (1981) *Organization and economics of construction.* McGraw-Hill, Maidenhead.

13 Institution of Chemical Engineers (1979) *A new guide to capital cost estimating.* Institution of Chemical Engineers, London.

14 Perry, J. G. and Hayes, R. W. (1985) 'Risk and its management in construction projects'. *Proc. Instn Civ. Engrs,* **78**, 499–520.

15 University of Manchester Institute of Science and Technology (1980) *Severn tidal power: sensitivity and risk analysis.* Project Management Group. UMIST, Manchester.

16 Science and Engineering Research Council (1987) *Risk management in engineering construction.* SERC, Thomas Telford, London.

17 Perry, J. G. (1988) *Contract strategies for construction.* Collins, London.

18 Perry, J. G., Thompson, P. A. and Wright, M. (1982) *Target and cost-reimbursable construction contracts.* Construction Industry Research and Information Association, Report No. 85, CIRIA, London.

19 Institution of Civil Engineers (1979) *Conditions of contract,* 5th edn. ICE, London.

20 Institution of Civil Engineers (1985) *Civil engineering standard method of measurement.* Thomas Telford, London.

21 Thompson, P. A. and Willmer, G. (1985) 'Caspar – a program for engineering project appraisal and management. *Proc. Civ. Comp.,* **1**, 83.

Bibliography

Abrahamson, M. W. (1984) 'Risk management.' *Intnl. Const. Law Rev.* **1**, 241–264.

Darnell, H. et al. (1986) *Total project management.* Booklets 1 to 3, The Asset Management Group, British Institute of Management, London.

Derrington, J. A. and Barnett, M. J. N. (1986) 'Civil engineering projects – what is value for money?' *Proc. Instn Civ. Engrs,* **80**, 1589–1596.

Franks, J. (1984) *Building procurement systems.* Chartered Institute of Building, London.

Gaisford, R. W. (1986) 'Project management in the North Sea. *Intnl J. Project Mangmnt,* **4**, 1.

Harris, F. and McCaffer, R. (1977) *Modern construction management.* Crosby Lockwood Staples, London.

Harrison, F. L. (1983) *Advanced project management.* Gower, London.

Kennawry, A. (1984) 'Errors and failures in building: why they happen and what can be done to reduce them.' *Intnl Const. Law Rev.* **2**, 5779.

Pilcher, R. (1985) *Project cost control in construction.* Collins, London.

34

Setting Out on Site

D W Quinion BSc(Eng), FICE, FIStructE
Tarmac Construction Ltd

Contents

34.1 Principles

'Setting out', as practised on civil engineering and building sites, is the locating of the works to be constructed, ensuring that they are dimensionally within permissible tolerances and correctly constructed. This service is essentially an aid to the labour force and must necessarily be provided in a form that is easy for them to use and understand; the information must be reliable and must be available as and when required. Errors in setting out will in most cases result in remedial works which will be expensive. Whatever lines or levels are provided should be checked to be sure of their accuracy, and they should be provided to the foreman efficiently so that he can have the necessary confidence in them.

Clause 17 of the Institution of Civil Engineers' Conditions of Contract states:

The Contractor shall be responsible for the true and proper setting out of the Works and for the correctness of the position, levels, dimensions and alignment of all parts of the Works and for the provision of all necessary instruments, appliances and labour in connection therewith. If at any time during the progress of the works any error shall appear or arise in the position, levels, dimensions or alignment of any part of the Works the Contractor, on being required so to do by the Engineer, shall, at his own expense, rectify such error to the satisfaction of the Engineer, unless such error is based on incorrect data supplied in writing by the Engineer or Engineer's representative, in which case the expense of rectifying same shall be borne by the Employer. The checking of any setting out or of any line or level by the Engineer or the Engineer's representative shall not in any way relieve the Contractor of his responsibility for the correctness thereof and the Contractor shall carefully protect and preserve all bench marks, sight rails, pegs and other things used in setting out works.

In this chapter, the initials SOE (setting-out engineer) are used to identify whoever undertakes the setting out. This function is performed by engineers, surveyors, technicians and foremen. 'The Engineer' is used to define the Client's technical representative.

Since the previous edition there has been continuing development of surveying instruments and aids. In particular there is ready availability of instruments which reduce the risk of user error and of instruments with electronic direct reading and computational facilities. The work required to be 'set out' has changed little and thus the basic principles are unaltered although complex setting out can be greatly simplified by the use of computer calculations and instruments.

34.2 Surveying instruments and their use in setting out

The usual instruments employed in setting out are a 20-s theodolite and a quick-set level. The theodolites and levels will be complete in a box containing the recommended tools for adjustments and an operating booklet, and will have an accompanying tripod. Theodolites will generally permit optical plumbing over setting-out points. They should have a loose-fitting hood for protection from dust or rain between measurements. For work over long distances, a 1-s theodolite or electronic distance measuring (EDM) instruments are frequently used. These are available with microprocessor units permitting the use of programmed setting out data or the transcribing of measurements into a prescribed printout. It is important that staff are trained in the use of such instruments beforehand. Careful use

of instruments gives greater accuracy than the mere use of a more sophisticated one.

The instruments should be checked upon receipt and their accuracy tested. The level staff should be examined to ensure all the graduation marks are clearly visible and, if the staff has been repaired, that no errors have been introduced.

A range of laser beam equipment is used to provide alignment beams, and reference planes can be provided by rotating beams. The uses commonly made of these are described later.

The reader is referred to Chapter 6 for more detailed information on the use of surveying instruments and methods, but a number of practical points will be emphasized here.

(1) Theodolites and levels are delicate and easily damaged or strained. They should, therefore, be treated with great care. They should not be erected on potentially slippery surfaces. They should not be left unattended and when not in use should be carefully and correctly replaced in their boxes and the fasteners secured. They should be checked for accuracy and alignment at least once a week and whenever there is any reason for doubt.

(2) When instruments have to be moved on their tripods they should be carried with the tripod legs straddling the shoulder such that the instrument is sitting alongside the head of the bearer in the normal vertical position. When instruments become wet they should be carefully dried by the SOE and should always be kept clean. It is preferable that they are replaced in their boxes for moving between locations.

(3) A 20-s error with a theodolite at 33 m gives an error of 3.2 mm.

34.3 Working procedures

The SOE will have a kit of setting-out equipment usually carried in a shoulder bag which is large enough to also carry and keep dry a survey book, reference book or papers, and folded drawings. The kit typically comprises a 30-m steel tape, 1-m folding rule, graduated scale, 30-m fine string line, 500-g plumb bob, triplicate book, club hammer, claw hammer, nails, centre punch, hardened steel point for scribing lines on steel or concrete, knife, spirit level, crayon, pencils and cloths. Spirit levels are now available which can give alignments other than horizontal and vertical. The setting out equipment is usually cared for by the chainman who, only after he has received careful instructions, may also take care of the instruments and will generally transport them in their boxes about the site. The chainman should be allocated to one or more SOEs and should be instructed in a signalling system which the SOE will use to indicate his requirements when out of hearing. The chainman should be instructed as to the correct method of holding a level staff and the correct use of the measuring tape. A well-instructed chainman will greatly ease the work of the SOE, whereas a poorly instructed or indifferent one will cause errors and delays. The chainman should be provided with the necessary tools and be capable of making and erecting profiles, sight rails, batter rules, and boning rods. He should have access to suitable timber, which may well be scrap from general site use, for such purposes. After use, the materials should be recovered. The profiles and pegs should be painted to be clearly visible and identifiable.

When taping distances it is usually more accurate to measure from the 1-m mark on the tape, with the end of the tape held clear of the starting marker. The SOE should always make it clear to his chainman what starting position he requires. Allowances should be made for measuring errors which occur due to slackness in the tape. The catenary error between the ends of the tape can be greatly reduced by providing intermediate support. Without such support a 33-m steel tape weighing 0.0219 kg/m with a 5 kg tension will give an inaccuracy of 28.8 mm and with a 10 kg tension an inaccuracy of 7.2 mm. Likewise, correction should be made for measuring on slopes. The measurement of 33 m on a 1 in 50 slope means a 7.5 mm error on the horizontal measurement, and on a 1 in 10 slope would mean an error of 164 mm. The best method of slope correction is by taking levels at each end and calculating the correct length which is the taped

length along the slope less the sum of $\frac{H^2}{2l} + \frac{H^4}{83^3}$ where H is the difference in height.

Appropriate setting out should be completed as a closed traverse and any cumulative errors traced and eliminated by recognized surveying techniques. When closed traverses cannot be used, try to check your setting out using different reference points so that any mistakes are unlikely to be repeated. Unless it is impractical, fore- and back-sight distances for levelling should be roughly equal as this will reduce inaccuracies if the instrument is in need of adjustment.

All tapes and bands should be kept clean and lightly oiled to avoid rusting, but not so oily as to pick up dirt.

The marking out of a right angle without an instrument is quickly achieved by using a 3:4:5 triangle of measurements. Rapid but more approximate results can be obtained by standing over the offset point on the baseline with arms out sideways at shoulder height in the line of the baseline. As the hands are brought together in front of one they will indicate the line at right angles. Use can also be made of optical squares.

When establishing a route across difficult country with bushes or other features obscuring required lines, it is often quicker and simpler to locate the positions of markers in clearings where they can be seen and to transfer lines locally.

When transferring a mark to concrete or steel make a mark each side of the required position and scribe a line between them with a hard steel point. On concrete this can be stencilled-in with indelible pencil and on steel the required point can be emphasized with a centre punch.

Try to identify setting-out pegs by writing on them or colour coding but first remove any previous references.

When possible, check your initial setting out using a different set of references to avoid inadvertently repeating an error.

Temporary bench marks (TBMs) are usually established on hopefully immovable features of the site. They can be scribed on to the sides of walls, on the tops of foundations or kerbs, or on to piles or bases constructed for the purpose. They must all be levelled-in from the main site bench mark and regularly checked to ensure reliability. Finished work should not be permanently damaged by marking on setting-out points. Make sure that a level staff can be held truly vertical above the level mark. Sometimes a piece of steel angle iron, perhaps 1 m long, driven into the ground will meet the requirement.

Setting out usually involves knocking in timber pegs or steel pins to mark the extremities or centrelines of the excavation or area concerned. Offcuts of steel reinforcement painted white serve well and can be re-used many times. When the setting-out lines are required more accurately and are required for several operations, then timber profiles are usually employed. Commonly, these consist of low timber rails fixed to two square timber pegs. Nails are lined in on the top to denote the required centre, building or other setting-out lines. The foremen usually extend string lines between profiles. Profiles need not necessarily be accurately at right angles to the setting-out lines but reasonable accuracy makes offsetting of the line, by the foreman, very much easier. The rails are often painted and the positions of the nails referenced on them in pencil.

Profiles for levelling excavations are usually set much closer to make sighting between them that much easier.

Setting-out stations are usually square timber pegs or reinforcement steels knocked well into the ground and protected with a surround of concrete. Nails or marks locate the true line or intersection point. These points can also be scribed directly on to suitable existing concrete or other surfaces. If likely to be damaged they should be clearly marked and where necessary protected by a simple fence or guard.

In special cases it may be justifiable to erect a small rigid platform above a setting-out point on which to set up the theodolite and gain a clear view not only across the site but in some cases also down into excavations.

Setting-out work has to serve the foremen and they should be given diagrams clearly indicating how the points and levels relate to the work they have to do and they should be shown the pegs, profiles, etc. from which they will work. Interference with these must not be tolerated.

Colour coding may be necessary where a profile is used with a different length of boning rod on each side or different lengths of rod for various purposes.

A number of mistakes frequently give rise to common errors. It is easy to transfer offset dimensions from drawings to notebook to site and set out bases, etc. on the wrong side of the main setting-out lines. It is easy to give some pegs and markers in offset positions and others on-line. It is easy to set up profiles accurately but set the wrong length for the boning rods. Errors of a unit can easily occur in reading tapes and staffs. The chainman can make simple errors when erecting profiles or holding markers. Simple errors usually arise, not from calculation mistakes but from lack of attention; straightforward setting out should always be checked, as well as the apparently more complex. The SOE must always be alert to pegs and profiles which have been disturbed and may have been replaced without his knowledge.

34.4 Site survey and preparations

Before the commencement of a contract it is necessary to establish a survey of the site as it currently exists, picking up all natural features and locating the site in relation to Ordnance Survey datums, local authority building lines, kerb lines of main roads, or other features that can be regarded as permanent. A principal bench mark should be established on site and agreed as a datum with the engineer. Likewise, basic lines must be agreed for the location and orientation of the works as a whole and about which they will be set out. In cases where there is the possibility of the construction of the works having an effect on adjacent properties due to construction up to the site boundary or as a result of possible ground movements or vibration, it may be necessary to survey and record features of those properties. This may comprise the recording of levels, inclinations to vertical, positions of cracks. Supporting photographs are valuable in recording the state of such properties.

The SOE now has a basis for proceeding with the setting out. He is frequently faced with the need to set out the first stages of site construction for an immediate start on the 'access to site date' and the simultaneous need to establish main setting-out lines which may have to last the length of the contract and be installed with considerable accuracy. Initial construction operations usually consist of site clearance and levelling and approximate setting-out methods can usually enable these operations to commence without delay. In some cases, the SOE may not be able to establish the principal datum lines he requires until features of the site, such as old buildings, trees, mounds, etc. have been removed. The SOE will establish his principal datum lines in positions where they can be of use for as long as possible. Positions for the principal points should be where they can preferably remain undisturbed and free from construction operations. The ground conditions and importance of these datum lines may justify the casting of a concrete block or even a pile on which to secure a stable setting-out point. The SOE will usually find it desirable to prepare a master plan indicating his principal setting-out lines, points and his key bench marks for the site. This will relate his principal setting-out lines to the building lines, centrelines of buildings, roads and principal services required on the site. This information should be used to check the dimensions given on the engineer's drawings and copies should be supplied to the engineer with a request that he confirms that the dimensions are as required. This will also enable the engineer to satisfy himself or herself on the accuracy of the SOE's setting out. In some cases, to preserve the principal setting-out points, it may be desirable for these to be located right on the site boundaries, or even outside it if permission can be obtained from the adjoining landowner. It should be borne in mind that for a large site, the Ordnance Survey bench marks around the site will not necessarily correspond within the accuracy with which the site levelling will be done, hence the need to establish a single bench mark on the site for the purpose of the works. From this single key bench mark a number of TBMs will be established and used locally. It is a sensible precaution to check at intervals of not more than 1 month that these have not moved or been damaged.

Before commencing their setting out, SOEs should discuss with the foremen the methods to be used in the construction of the works. Foremen will require pegs, profiles, batter rules and other information in locations which will not interfere with the movement of machines, men and materials. They may require offset pegs to be provided by the SOE or may decide to make their own offset measurements. The SOE must determine not only when the setting-out pegs and lines are required for use, so that he can anticipate these times, but also the accuracy with which the information is required in relation to the purpose for which it will be used. Where considerable accuracy is required it is customary to provide timber pegs or rails and use nails for the precise position of the line. Lesser accuracy, but greater speed, can frequently be obtained by knocking in steel pins, particularly if the ground is difficult to penetrate.

Checking is all-important and, having established setting-out points and checked that they are in the right position, it is still essential to check as the work is carried out, to ensure that the original setting-out pegs and profiles have not been disturbed during the progress of the work. So, the SOE must stay in constant contact with the construction operations and provide constant services to those operations. It is wise to check that boning rods are being used properly and that the work conforms to the dimensions on the drawings and not assume that the foremen and operatives will necessarily be working accurately.

For some large sites, such as motorways, the original site survey will have been performed with reference to early established survey stations and the information incorporated in a computer program. It is possible to use such data to calculate setting-out instructions which can be fed into programmed EDM instruments which, set up over the survey stations, can enable the principal setting-out points to be established quickly without tedious prior calculations.

34.5 Setting out for excavation and grading works

For these operations a lesser degree of accuracy is needed than for the setting out of foundations and building works. The SOE should bear in mind the likelihood that positions will need to be established and re-established with speed. The initial marking out of the areas to be excavated and those to be filled will be disturbed when soil stripping takes place. Either long pegs clearly visible from earthmoving machines or smaller pegs with ranging rods to identify them should be used. Attendance will be required by the SOE to provide what is needed. As soon as it is practical to do so, lines and profiles should be established around the areas in question and batter rules set up to give guidance for the forming of slopes to cuttings and embankments. It will frequently be necessary as the work proceeds to provide additional profiles and points within the excavation or on the embankments. A typical situation is shown in Figure 34.1.

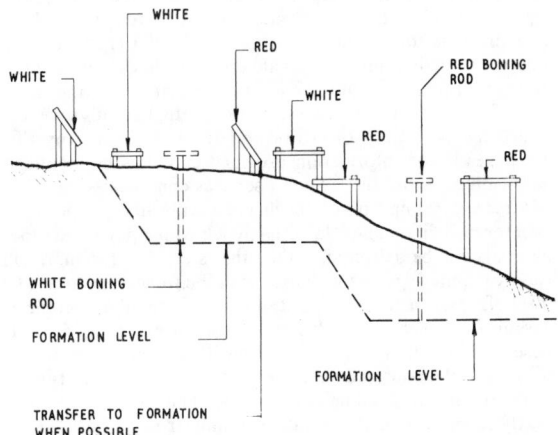

Figure 34.1

Where several levels have to be established, a colour-coding system on the pegs and profiles should be adopted and this should be carefully explained to the foreman and the machine operators and the foreman should be provided with diagrams and explanations from the triplicate book. It is important to discuss the method of setting out with the earthmoving manager or foreman so that the information is provided to suit his intended plant operations when he needs it. The cost per hour of large earthmoving and excavation plant is high and its utilization is an important factor on the cost and programme of this work which is very susceptible to adverse weather conditions. As the various levels are established, new setting-out points should be provided so that deeper individual foundations and local requirements can be quickly marked out for work to proceed without delay using bulk earthmoving equipment to the best economical advantage.

It is important at an early stage to locate the toes of batters and tops of slopes to ensure that the process of shaping and trimming is carried out quickly and easily the first time. It is better to provide a few too many pegs or batter rules than to provide too little information. It may be necessary for the SOE to attend on the excavating machines as they approach formation levels literally to level them in as they proceed using a level staff attached to the side of a scraper. Lasers giving a constant plane of reference can also be very useful by providing a visible indicator to the operator as to his working level. On motorway and aerodrome contracts in areas of intersections this attendance by the SOE often saves a lot of secondary grading.

The SOE should take into account whether the bulk excavations are to be taken straight down to formation level or left high to protect the formation until on exposure it can be blinded immediately or sealed. Likewise with embankments or fill areas, allowance is usually required for consolidation and settlement. The allowances made in each case, and how the levels given correspond to finished or initial levels, should be made clear in writing to the foreman. Surfaces should be formed with self-draining falls to avoid ponding of water upon them. Where batter rules are set up on varying ground levels to give a continuous finished sloping cut or fill line these can be quickly checked by eye for alignment.

34.6 Trenching and pipelaying

Pipelines, culverts, serviceways and the like are usually tied to specific positions and levels where they enter and leave buildings, pass under roads and intersect with each other. It is therefore wise to set out the entire length of trench between consecutive tie-in points and locate the centrelines and essential levels at all junctions, horizontal and vertical bends and manholes. The treating of several sections together in this way will reduce the possibility of errors or late alterations. Any discrepancies in the information provided can thus be identified and resolved before the laying of services commences.

When excavating a trench the machine will usually deposit the spoil for backfill on one side of the trench whilst pipes and other materials will be delivered to the other side. The foreman will usually require pegs on the centreline of the trench and a specific offset of, say, 3 m at all key positions. He will require a profile as close to the trench as possible at centres not exceeding 45 m. If these positions are likely to interfere with trench excavation or movement of labour and materials then he may require a further profile offset from the line of the trench. The profiles should be clearly labelled with the length of boning rod to be used for excavation. The length of the boning rod should be marked on it. It is not usual to mark out the width of the trench as this will be determined by the bucket of the excavating machine, which would have been selected as the most appropriate, bearing in mind the construction width required and available bucket sizes.

Within a length of pipework between manholes there may be junctions for lead-in pipes from gulleys or other items not requiring a manhole connection. The positions of these will need to be marked by pegs installed at the side of the trench as the excavation proceeds. It will be necessary to indicate on which side the connection will be made and at what relative angle to the horizontal the junction pipe should be set.

Where it is known that existing services have to be crossed these should be marked ahead of excavations and, if necessary, exploratory work should be carried out to locate them and confirm that there will be no clash between them and the new services. Many SOEs are able to trace existing services in the ground by 'dousing' methods. In addition, there is equipment available for the location of underground services, and it may well be worth while getting such an instrument on site to avoid the complications or charges which occur when existing services are damaged. At manhole positions there may be changes in level or line and the SOE will be required to provide further information to the foreman in order that the manholes can be constructed quickly and economically. It is more economic for the main excavator to take out the required enlargement at these positions as it reaches them rather than for them to be trimmed out afterwards with more expensive removal of the spoil.

Where pipes have to be laid within trenches the SOE should clearly determine with the foreman the level the latter requires, bearing in mind that he may require to dig out locally for collars if the barrel of the pipe is laid directly on virgin ground, or he may require a different relative level for other circumstances.

It is now quite common to use laser beams in trenches. These can be set up at manhole positions or in the ends of previous pipe runs to provide a beam to the right alignment at a defined relationship to the work such as pipe centreline. It is possible for the excavator driver to trim to formation if he can identify the beam to a reference point on his bucket or a boning rod used by the banksman. Any pipe bedding can be laid by reference to the beam and followed by the pipes. The use of the pipe centreline for the beam enables all operations to be controlled. The laser, being expensive, should be protected from accidental impact in use and for security should be returned to the office between uses.

As excavation proceeds, the trench should be checked periodically to ensure that it is being excavated to sufficient but not excessive width. With large-diameter pipes the wrong diameter can easily be used and diameters should be checked. Care should be taken that cracked or damaged pipes are not used and connections should be temporarily sealed off to maintain cleanliness. The provision of draw wires, where specified, should not be overlooked.

The SOE should be aware of the dangers associated with trench work and should be familiar with the Construction Industry Research and Information Association guide *Trenching practice.*

34.7 Foundations

Foundations are commonly set out by establishing a series of profiles around the excavations with the location on these profiles of specific setting-out lines notified to the foreman. Typically these may be as shown in Figure 34.2.

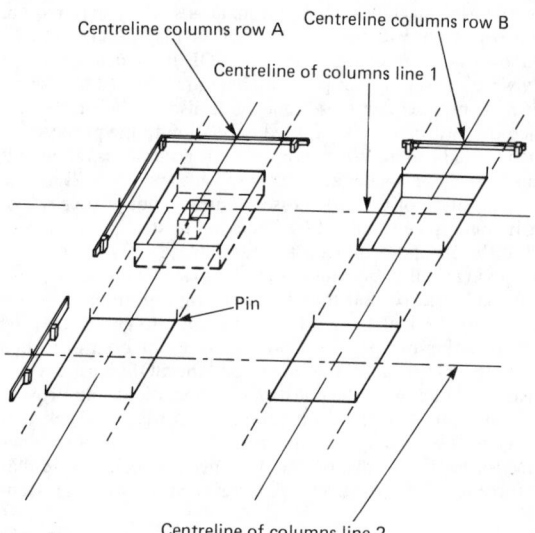

Centreline columns row A

Centreline columns row B

Centreline of columns line 1

Pin

Centreline of columns line 2

Figure 34.2

These profiles are usually set just above ground level with the rails horizontal but not necessarily at any particular level. Nails inserted into the rails locate the required setting-out lines and the foreman can offset these to move from, say, column centreline to the outside column face or exterior face of the brickwork to suit his requirements. He will normally stretch cord lines or piano wire between the nails and from these he can plumb-down using a spirit level or plumb bob. For level purposes he may require level profiles but more usually a series of specific level points on the works can be transferred by the foreman using straight edges and spirit levels or, more commonly on building contracts, with a water level. Once the foreman has been provided with setting-out profiles he can usually get by with little further assistance from the SOE other than in checking the various stages of construction. In many cases it is helpful for the SOE to set up a theodolite over a setting-out point and transfer a line directly into an excavation or on to a foundation at a number of points for the operatives' easier use.

Once the foundations have been correctly installed it is comparatively simple to transfer lines and levels up and through the building or structure. On the other hand, inaccuracy in the foundations will be difficult to overcome with later construction, and it is necessary that key items such as holding-down bolts for steel frames, reinforcement starter bars for *in situ* concrete construction and pockets for precast concrete columns are correctly positioned and levelled. The SOE should consider at an early stage the tolerances which are appropriate in comparison with the cost of remedying any inaccuracies later. The SOE must be aware of the extent by which the accuracy of the works may deviate from the positions given on the drawings. These tolerable deviations govern what is acceptable. General guidance on tolerances is contained in British Standards and in particular in BS 5606 but the contract specification may set particular requirements.

Holding-down bolts are usually assembled in a frame slung from a template. The bolts usually have the bottom head under a washer plate with a sleeve tube above to provide an annular space around the bolts after concreting. The threaded end protrudes through the template and a top nut can be adjusted to set the bolts to the required level. It is important that the bolts should hang vertically and the sleeves should provide the bolts with some play after concreting. On many occasions the washer plates are replaced with steel members joining two or more bolts and this helps correct installation. These bolt assemblies have to be supported within the proposed concrete foundation. Supports, spanning across the base excavation or shutters, are frequently used. Since they will be subject to deflections and dislodgements they must be set firmly into position. Where the assembly is difficult to locate and suspend accurately, a frame should be made and supported on the concrete blinding. The supporting legs may or may not be lost in the concreting of the base but any parts above the concrete level can be re-used. This method is particularly appropriate if the bolts have to be built solidly without sleeving into the foundations.

When the bolts are sleeved they should be tapped with a hammer to make sure they are free as the concrete is setting. After concreting the threads should be cleaned, regreased and wrapped with sacking for protection and the sleeves covered to prevent stones from entering.

Starter bars protruding from bases into columns or walls usually extend one lap length above the height of a small concrete kicker. Starter bars and kickers need to be set accurately and restrained there to provide the correct concrete cover within the column or wall shutter and to lap correctly with the lower end of the column or wall main reinforcement. It is essential to check starter bars before, during and on completion of concreting. Where the kicker is cast integrally with the base it will be checked with the level and positioning of the reinforcement.

Where pockets are to be formed in the foundations for precast concrete or steel sections, they should be sized for a reasonable clearance all round so that this can finally be effectively filled with concrete using a slim poker vibrator. Such a clearance can also be useful for removing any debris from the pocket. An excessive clearance, on the other hand, can affect the size of column base and will increase the temporary wedging and guying used to position the column accurately. Again, it is important to locate the pocket accurately. The pocket formwork will be subject to an uplift from the fluid concrete which

must be resisted. Rather, the box should be set low as it is easier to pack up than remove concrete to deepen the pocket. The material forming the pocket should be so constructed that it can easily be removed.

Once column bases have been concreted, it is usual to scribe the centrelines each side of the column position on the concrete. This serves to check that they are correct and is very useful to the erectors of the steel or concrete framework for rapid erection.

Structural steel framework erection should almost set itself out if the columns are properly aligned and levelled on their baseplates and holding-down bolts. It is desirable to set the steelwork from the centre of any building and work outwards. This will halve any creep which might occur by fixing from one end. It should not be checked finally and the baseplates concreted-in until a securely braced section of the structure has been completed to ensure correct fitting of other essential members at higher levels. The SOE should check the fitting of members, that the correct members and bolts are used and that they are installed in the manner indicated upon the drawings. The SOE should give attention to the needs for structural safety during erection and the safety of the workforce engaged upon and below it. The SOE may have the duty of ensuring that mating or other surfaces which will afterwards be inaccessible, are painted first. Before any cladding is attached to the roof or sides of the building, the trusses, purlins or other members to which such cladding is to be attached should be correctly aligned and, if necessary, temporarily braced until the cladding has been attached.

With transmission towers and similar multi-legged structures it is usually necessary to use large templates to set the starter lengths of the tower legs. The four legs of such towers are usually inclined to the vertical in two directions. The excavation for the four legs are marked out with steel pins in the usual way. The starter lengths of the tower legs are attached to the corners of a square assembly template of four trusses with cross-ties between them, and the assembly is levelled across the excavations with the legs extending down into them. After checking for level and line the excavations are concreted.

34.8 Work within buildings

Within a building envelope there are the two categories of activity associated with building finishes and with plant. The building finishes comprise internal walls, doors, floor surfaces, ceilings and fitting-out works. The plant is associated with building services and the operations of the building user.

For building finishes, the foreman and tradesmen require reference lines and levels to which they can refer easily.

Lines and levels can be provided by marking on to floor and wall surfaces in ways which can be removed later. Rotating laser beams can provide reference planes. It is essential that the light references can be re-established quickly and accurately if they need to be removed between uses. The rotating laser reference plane has advantages by servicing many tradesmen in a large area but is of less value with a number of isolated areas each requiring levels. In such cases water levels are very easy to use for transferring level datums between rooms.

It is customary to specify the level of a floor in terms of tolerable deviations from the specified level and over any length of 3 m. When goods are to be stocked to nearly ceiling height, great accuracy is required. The SOE should consult with the foreman as to the most appropriate method of construction to achieve the specified standard and should control closely the levels of the work and consistency of the materials used.

For plant there is the need to locate the positions of plinths and fixings as well as the openings required in walls and floors for ducts and pipes. It is important to minimize the interference of one trade with another and to ensure that holes are incorporated in walls and floors as they are built. The SOE should therefore study drawings of plant and services and try to ensure that such openings are detailed on the structural drawings for incorporation during the main floor and wall construction. It is easy with some plant drawings to mistake the orientation of plant requiring a number of plinths and fixings and some care and checking is necessary.

Where items have to be built-in or are in modular units, a check is necessary to determine whether the items will fit readily with some tolerance or whether cutting will be necessary. In the latter case, the cutting procedure should be agreed with the engineer or architect as early as possible.

34.9 Piles and diaphragm walls

It is customary for the main contractor to provide centrelines for each base or pile group to a specialist piling subcontractor who then locates each pile position. Before piling commences it is often necessary to provide a firm working surface (pile carpet) for the specialist plant, and agreement must be reached as to the disposal of any spoil from the piling operations. These considerations make it difficult to establish and maintain the setting-out points and it is important for the SOE to monitor and check the piling operations. The SOE must so establish setting-out lines and references that he can readily check on the pile positions. The piling subcontractor will establish each pile position and then set up casing, auger, precast pile or pile shell over that position. It is easy for this casing or preformed pile to be displaced as it penetrates the pile carpet and overburden layers which can frequently contain hard pieces. It is necessary that the SOE checks the position of the casing or pile when it has penetrated a metre or so and becomes set on its course. It may be necessary to extract, fill the void and start again if the pile is displaced outside the permitted deviation. The permissible deviation from the specified pile position is usually up to 76 mm. There is also usually a tolerance on verticality of 1 in 85 which applies to the rake of inclined piles as well. The setting out of the pile must take account of these rakes if the finished tops of the piles are located at a different level to that of the setting out at the piling carpet level.

A record log should be kept of the installation of each pile, whatever type of pile it might be. When the required depth for bored piles has been reached, a light should be lowered to the bottom to assess the alignment and the condition at the bottom of the pile, and the depth should be measured. When full depth casing is not used with tripod rigs, the boring tool can be displaced by hard inclusions in the ground so producing a 'banana-like' shape. This, if it occurs, should be reported to the engineer. With precast piles and continuous casings displacement can also occur and this can throw the alignment and top position out of tolerance and so this fact must also be reported. Displacement piles have usually to be installed to a sequence and timing in order that damage or uplift of completed piles does not occur as subsequent ones are installed. It is the duty of the SOE to see that the sequence and timing are observed and to record and report the occurrence of any disruptions of adjacent piles, buildings or services. When driven piles are installed to a specified set, the SOE should check that this is achieved and at the level expected by the engineer. Any variation should be reported as the pile may be held on a boulder or other intrusion. To be assured as to the quality of a pile it is essential to have a reliable record as to the installation process and have the pile installed correctly and the SOE must see that this happens.

Sheet piling must be started by the accurate driving of the first panel of piles. Should there be uncertainty over obstructions in the upper layers of the ground it may be wise to excavate a shallow trench and place the sheet piles in it with some backfill around them. This will steady the piles and ensure a vertical start. When driving piles through gravels, in particular, there is the risk of declutching and piles going off-line and the SOE should watch for signs of this. When driving cofferdams, the lengths of the sides should equate to the width of piles to be driven and provide working tolerances around the specified dimensions of the permanent works. The cofferdam perimeter should be completed with a number of piles at one corner undriven to ensure correct interlocking.

Diaphragm walls are formed by digging under bentonite through guide trenches. A pair of 'inverted L'-shaped reinforced concrete walls form the sides of the trench. The width of the trench is 50 to 75 mm greater than the width of the digging bucket. The trench depth of up to 18 m depends on the stability of the upper ground strata in relation to the disruption expected from the digging operation. The verticality of the wall excavation must be checked regularly. When steel reinforcement cages, steel or precast concrete members are set in the trench prior to concreting it is necessary that they are suspended freely and vertically as well as to line. The suspension points for the cages and members must be designed to enable any necessary adjustments to be made under the direction of the SOE.

34.10 Tall buildings and structures

The ease with which tall buildings and structures can be erected depends on the accuracy of the foundations from which they rise. The formwork needs to be set accurately horizontal so that corners and other such features will be cast truly vertical, reinforcement is aligned vertically in relation to the formwork, and fixings are provided in their correct positions. It becomes increasingly difficult to correct misaligned lifts of construction work as the structure rises. For these reasons it is also important for the starter reinforcement to be set out correctly for position, cover and verticality. The larger diameters of steel reinforcing bars may not be truly straight or easy to 'push over' if a misalignment of, say, 5 mm in the cover over a 1 m lift of concrete is to be corrected over the remaining vertical lifts. Likewise in the case of falsework systems erected as two-, three-, or four-legged framed towers, it is worth the care of starting correctly so that the load is correctly distributed between the members of the tower and eccentricity from the vertical does not result in avoidable lateral reactions: 1.5 degrees out of plumb represents 25 mm in a height of 1 m and a restraining horizontal force of 2.5% of the vertical load in the member is needed. In the case of structural steel frames, provision is usually made to adjust the levels of the baseplates with packers to ensure the baseplates are correctly level before they are grouted or concreted solid after the holding-down bolts have been finally tightened. Having ensured that the initial lifts of construction rise correctly, they should receive any permanent or temporary horizontal bracing and lacing as the specified positions are attained. This lacing and bracing is needed for the lateral stability of the structure and makes erection easier if it is attached at the right time. It is frequently difficult to insert such members later than the designer intended, and can result in distortion of connections and the 'building'-in of locating forces which the structure was not designed to resist.

Tall buildings are commonly clad with storey-height panels and designed with movement joints at storey heights. It is therefore necessary to check that those storey heights are controlled and accumulations of errors do not occur such that the higher panels will not fit with the fixings or the movement joints will not perform as intended. Before construction commences a check should be made on the tolerances expected on the cladding and other components and, hence, the deviations allowed in the construction of the work on site to which the components will be attached.

It is possible to check and control verticality in several ways. Theodolites set up at ground level can project lines up the faces of a building or structure. Autoplumbs can be used through openings formed for lift wells and staircases. Laser beams can be projected upwards from ground stations or piano wires with heavy weights suspended in or around the building. The weights are usually hung in a barrel of water or oil, and the location of the suspending wires easily checked against two reference lines. It is important to ensure that the edges of the floor slabs are cast accurately to the required details so that nibs and edge fixings will perform as intended in relation to the cladding. It is important to maintain cavity widths in cavity brickwork to ensure that the ties are correctly installed.

Before slipformed concrete construction commences to silos, towers and cores of tall buildings, the platform arrangements should be checked and all the critical components aligned. Misalignment can result in the platform being urged into a distorted slide which is often difficult to correct, and if corrected, often leaves poor concrete in the affected areas. The jacking rods should be vertical and the platform horizontal. The shutters should be inclined in accordance with the design and the projecting vertical bars restrained temporarily in their correct alignment. The cover to the reinforcement must be correct. Fixings and openings in the height of the first lengths of vertical reinforcement should be correctly inserted. Level controls should be established at each jacking position for water level or electronic recording. It is always useful to have some water levels even if electronics are used. A uniform rate of jacking without unnecessary stops will avoid set and snatch conditions.

The SOE should see that the supplies of steel reinforcement, inserts and fixings are provided to the sliding platform as needed and should keep level marks available on the vertical reinforcement to aid the operatives. A periodic check should be made on the true height of the work and on alignment to ensure that the platform is not twisting in plan.

34.11 Marine structures

The initial work for the SOE may concern the initial survey of the area of the works and the contouring of sea-bed, river-bed or marshy areas. When the area is extensive it is most practical to use aerial photography and echo-sounding methods to obtain the information quickly. When the area is more limited, conventional surveying practices can be adopted.

For dredging works, a relationship must be followed between dredge levels and location. Markers or buoys can be established for position alignment in shallow locations and the dredger can usually operate to a chosen depth. In offshore locations the problems are more complex but electronic systems exist to deal with them.

To locate a pile or a structure just offshore one usually employs the intersecting-line method from two known points on a baseline. Depending on the accuracy required, the SOE may line-in the object with two theodolites or establish two pairs of markers for guidance. Depending on the distance apart of the base stations, these methods will be more-or-less accurate. When the required location is a considerable distance offshore, more sophisticated equipment, such as Deccafix, will be used; here, an instrument offshore can be adjusted into a position that is a required distance away from two known base stations where electronic signallers have been sited.

Usually, once an offshore location has been established and centrelines marked, the remaining setting out is simple. It is not always so easy to transfer a level datum from shore to the offshore structure. In most cases there will not be the need for the accuracy used with onshore setting out, and transfer with care by normal surveying methods over long sights may be sufficiently reliable provided that fore and back distances are roughly equal. Astro methods may be necessary over considerable distances offshore.

For the transfer of levels about an offshore structure, a water-level system is very convenient.

When lines or levels need to be transferred to divers working on the sea- or river-bed, vertical measuring rods are used. These will be difficult to control in flowing water conditions, but the conditions can be improved by working within a sleeve pipe 1.2 or 1.8 m in diameter into which clear water is introduced under a small pressure head to improve visibility.

When tunnels are being driven to meet with a control structure or diffuser offshore, either a small-diameter bore to which the tunnel can locate is sunk or a bore is raised with an emitter marker from the tunnel and its position located from the surface.

34.12 Tunnelling

Working underground is restricting to the SOE since access to the works for surveying is generally limited. Usually, surveying is undertaken outside normal working hours which may mean only on Saturday night and Sunday along with maintenance operations. Because the survey works can be checked only infrequently, they must be reliable and firmly established. The most common underground works are tunnels and shafts and often the only access to the former is by the latter and this might involve projecting a long tunnel from a 6-m diameter shaft and consequently something less than a 6-m baseline.

Under these conditions, theodolites of 1-s accuracy are necessary and much patience is required. It may be necessary for a number of engineers to undertake the setting-out procedure several times and the mean of their lines used. It may mean that the shafts will have to be watched in case they are moving – in soft ground conditions they can lean towards large adjacent excavations as the surrounding ground readjusts to these operations.

Where considerable accuracy must be maintained, any interference arising from traffic or other vibrations, heat and pollution hazes, and general surrounding activities needs to be minimal. So even the preliminary ground level surveying in busy areas is usually carried out under more peaceful conditions at weekends or at night. It is usual to establish, across access shafts, the centrelines of the tunnels below. These centrelines are established by means of piano wires suspended down the shafts. The piano wires are wound around screw adjusters at ground level and at their lower ends have heavy 9 to 15 kg weights in buckets of water or oil. They need to be close to the shaft linings to secure as long a baseline as possible.

At the bottom of the shafts the centreline given by these wires has to be picked up by instruments and established on markers rigidly attached to plates in the crown of the tunnel. A number of surveying methods are employed for this purpose but probably the use of Weisbach's triangle is as reliable as any. The established centreline is projected forward whenever the opportunity occurs. An instrument reference point can never be too near to the face, particularly if curves are being negotiated. The use of laser reference beams giving the tunnel centreline is now common practice and is a great boon to the tunnel boss.

For negotiating tunnel curves, information is usually provided in the form of offsets at chord lengths with instrumental checks as necessary and as possible. It is important to offset on the correct side of the tunnel! Before reaching tangent points it is necessary to work out with the tunnel boss how the tunnel shield will be adjusted to negotiate the curve and whether it can negotiate at the radius required for the finished work. It may be necessary to enlarge the tunnel at these positions using hand methods and such possibilities should be considered from the outset so that the permanent works can, if possible, be adjusted to accommodate the construction practice. Great care is required with horizontal and vertical changes in alignment or level to ensure that the work is accurately set for the new course.

Another important feature to check with tunnels in soft ground is the amount of squat or wander. Using tunnel segments, the true diameter may easily be reduced vertically and extended horizontally. It may not be a uniform distortion and considerable difficulty may be experienced with the final tunnel lining to achieve the required finished accuracy. Measurements of the tunnel-segment diameters are kept and checked for every ring both horizontally and vertically and plotted in relation to the required tunnel centreline – the result is often termed a 'wriggle diagram'.

When the tunnel is being driven under compressed air the centreline has to be transferred through an airlock and this is done by setting up the instrument within the lock, aligning it first with the outside door open, and then after closing it and compressing the lock, opening the inner door and transferring the line ahead.

When there is access above the tunnel it may be desirable to sink a borehole ahead of the tunnel and through it pick up a check on the centreline. Special precautions will be required when tunnelling using compressed air.

Increasing use is being made of full face and pipe jacking tunnel machines which can be controlled remotely. With such machines, laser beams are used in conjunction with 'targets' on the backs of the machines. By television viewing of the target and console control of the machine hydraulics, the operator in the shaft or above it can steer the machine forward and often negotiate bends.

When reliance is placed on laser beams for alignment, it is necessary to use lasers which indicate by beam oscillation when they have been disturbed.

Larger works underground employ an extension of these methods. When tunnels have to cross or deliberately connect with other services it is wise to locate these intersection points at an early stage. Positions and levels of older services are rarely accurate and adjustments to the new works may be more readily accommodated some distance from the intersection point. The SOE may need to undertake a complicated survey to locate accurately existing services by working within them. The Health and Safety Regulations should be observed when working in or adjacent to existing underground services.

Bibliography

British Standard Institution (1978) *Accuracy in building.* BS 5606. BSI, Milton Keynes.

Clark, D. (1969) *Plane and geodetic surveying,* vols I and II. Constable, London.

Construction Industry Research and Information Association (1983) *Trenching practise.* Report No. 97. CIRIA, London.

Construction Industry Research and Information Association (in press) *Setting-out procedures.* CIRIA, London.

Richardson, H. W. and Mayo, R. F. (1941) *Practical tunnel driving* (rev. 1975) McGraw-Hill, New York.

35

Temporary Works

C J Wilshere OBE, BA, BAI, FICE
Laing Engineering and Temporary Works Office

Contents

35.1 The legal position

In the normal contractual arrangement, the engineer provides all the information about the permanent structure. However, on many occasions a temporary structure of some type is needed in order to reach the final position. The design and construction of this is a matter solely for the contractor. In other types of contract, things may be somewhat different. For example, in a direct labour situation the design of all temporary and permanent works is likely to be in the same hands; or the engineer may choose to design the temporary works because of their close interaction with the structure. However, in contracts undertaken under the Institution of Civil Engineers conditions, it is firmly a matter for the contractor, though the submission of details to the engineer is normally required. This submission in no way relieves the contractor of responsibility, but does provide a further check on practicability and safety.

The engineer has no legal responsibility to the contractor for approving or 'not objecting to' these drawings. He must be careful not to attract responsibility to himself unintentionally. But he has a responsibility towards his client, to ensure that the temporary works will be satisfactory; this means that they must serve this purpose without delaying the work, or cause the client to be in difficulties because his structure has in some way interfered with others. If the structure takes longer to build through inadequate temporary works design or a failure, the client suffers.

That very briefly outlines the basic position under English contract law. But the position in common law is slightly different. Everyone who has a direct supervisory position on the site and who has the ability to make appropriate judgements has a responsibility. Thus, if the engineer is aware that the temporary works are not all they should be and some harm befalls, he may well share responsibility with the contractor should a court of law award costs arising out of such an incident.

Other legal requirements arise from problems such as preserving amenities presently enjoyed by neighbours or the public. Requirements are not specifically laid down but follow from this.

The Health and Safety Executive lays down general requirements for health and welfare which guide the designer, especially in connection with access scaffolding.

The main legislation in the UK is the Health and Safety at Work Act 1974. Under its authority, regulations[1-4] are laid down giving general, and in the case of access scaffolding, detailed requirements. These are enforced by the Factory Inspectorate, under the Health and Safety Commission. They also publish a series of *Guidance notes*,[5] some of which are relevant to construction. There is a variety of approach in different countries, based on different traditions and attitudes to life. In the EEC, there are pressures for harmonization, but this will take many years.

35.2 The temporary works condition

Design and construction of temporary works comprise a particular case of construction in general but there are certain items which are different. These are:

(1) The time for which the structure is in use will be measured not in decades but in months or possibly only hours.
(2) Because of this short duration it is easier to predict what loadings will actually have to be carried, which may enable a slightly lower safety factor to be used. Conversely, unless the site is well organized and controlled, unpredicted loads of considerable magnitude can arise.
(3) In some cases a collapse of temporary works would be merely a costly nuisance; in others it will be catastrophic, both to life and property. It may be appropriate to make adjustments to the design parameters depending on circumstances.
(4) The available design facilities may be different from the conventional. For example, some temporary works structures may be required at very short notice and the design therefore must be carried out by whoever is available at the time, and checking of a normal standard may not be possible.
(5) Similarly, the materials may be somewhat different. They may be unusual, and frequently they are not new. This is discussed below under particular materials.
(6) Because of the short-term nature of the works, and possible financial advantage, there is a strong bias to take risks with life and property which would not be contemplated elsewhere.

35.2.1 Limit state design

At the time of writing, some structural codes have become available using partial or gamma factors, commonly called limit state philosophy. This calculation method is based on characteristic values, which are not available for temporary works loadings, and thus cannot be used for such calculations at present. But the philosophy, which involves many safety factors more directly related to the various aspects of uncertainty and risk, lends itself admirably to the design of temporary works, and enables an engineer to adjust factors to the conditions of his particular case. As they are revised, structural codes are being transformed into limit state philosophy, and Eurocodes are similarly based.

35.3 Materials

Materials common in construction are used and some particular notes are given below. Those which have already been fabricated to form equipment are described in section 35.4 following.

35.3.1 Stresses

For virtually all materials in use for temporary works, there is a Code of Practice which lays down the stresses which are applicable in permanent construction. Higher stresses may be appropriate in some cases, where conditions of loading are more accurately known. If materials have deteriorated, it may be possible to use lower stresses, rather than rejecting them out of hand.

35.3.2 Steel

There are several grades of steel available, with different properties. As there is no system of permanent marking, great care should be exercised to ensure that a basic steel is not used instead of a higher grade steel. As the stiffness of all steel grades is the same, no advantage is obtained by using a higher grade, unless the controlling factor in the design is strength. Steel which has been used is often damaged; it may be straightened for reuse. Rectification should only carried out by experienced personnel, as there are pitfalls. Over-enthusiastic straightening may leave cracks which are not noticed. Heat treatment can cause changes in the properties of steel. The final accuracy may be rather wide of what is desirable, and if it is critical as, for example, in a strut, lower working stresses may be appropriate. Rust-pitted steel should also be used with lower stresses and deeply pitted steel should be discarded. Particular modes of failure with steel joists too frequently overlooked, are instability and web buckling.

35.3.2.1 Steel scaffolding

Scaffold tube is 48 mm outside diameter. The steel tube used is of various thicknesses, from 2.9 to 4.6 mm, and several grades of steel. A small amount is aluminium. Attempts to standardize throughout the EEC are proceeding, but a variety of types will persist for at least a decade or two. Other sizes, thicknesses and grades are often used for prefabricated scaffold.

In the UK, steel tube is 4 mm thick with a ductile steel of yield stress 210 N/m^2, and at present there is no need to check thickness and grade as there is only this type in use. In other countries a check must be made when design is based on any but the lowest grade tube in circulation there.

Table 35.1 gives the safe working loads for steel tube complying with BS 1139:1982, Part 1[6] when loaded concentrically. The effective length l is a function of the end conditions. In Figure 35.1 and Table 35.1, the effective length l of the top cantilever and the adjacent section are given by:

$$l = L_1 + 2mL_1 \qquad (35.1)$$

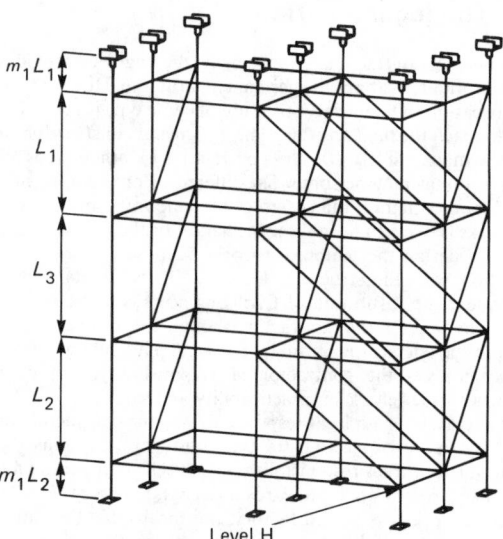

Figure 35.1 Effective lengths of scaffolding

Table 35.1 Maximum permissible axial loads in steel scaffold tubes BS 1139:1982, Part 1 *Specification for tubes for use in scaffolding*

Effective length, l (mm)	Slenderness ratio l/r	New tubes Permissible axial load (kN)	Used tubes Permissible axial load (kN)
0		70.7	60.1
250	15.9	68.5	58.2
500	31.8	66.2	56.3
750	47.8	63.0	53.6
1000	63.7	57.7	49.1
1250	79.6	50.3	42.8
1500	95.5	42.0	35.7
1750	111.5	34.2	29.1
2000	127.4	27.9	23.7
2250	143.3	22.8	19.4
2500	159.2	18.9	16.1
2750	175.2	16.0	13.6
3000	191.1	13.5	11.5
3250	207.0	11.6	9.9
3500	222.9	10.1	8.6
3750	238.8	8.8	7.5
4000	254.8	7.9	6.7

Note: Effective lengths above 3000 mm are undesirable.

The bottom is similar unless lateral stability is provided at level H. The value for intermediate sections may be taken as $1.0 \times L$.

35.3.3 Timber

Timber has been available for many years, classified by 'commercial' grading, which is basically aesthetic. Some stress-graded timber is now available, which means that minimum strengths can be reliably anticipated. The timber Code[7] gives information about designs using such timber. This provides data on the strengths of various species, as well as on strength classes, a new concept classifying all timber into nine classes of

Table 35.2 Properties of steel scaffold tube to BS 1139:1982, Part 1, *Specification for tubes for use in scaffolding*

External diameter	48.3 mm
Thickness of wall	4 mm
Cross-sectional area	557 mm^2
Weight	4.37 kg/m
Radius of gyration r	15.7 mm
Modulus of cross-section z	5700 mm^3
Minimum yield stress	210 N/mm^2
Maximum allowable compressive stress	127 N/mm^2

ascending strength. Classes SC3, 4 and 5 are the three most likely to be used in temporary works. A variety of modification factors are given, of which the most important is for moisture content. Outdoors in the UK, timber should always be considered to be 'wet', and the appropriate factor used. For the short-term applications of temporary works, the duration of loading factor may be applied. Provided the timber can recover between uses, time should not be considered as cumulative. A new factor relating to depth of section is now included. The falsework Code[8] gives two other factors which may be used in that context, and has tables of stresses. Table 35.3 gives stresses for three strength classes, appropriate for most falsework applications. Where it is impractical to obtain stress-graded timber, this Code also gives some guidance on what action to take. With a rejection of the worst pieces, a parcel of a particular commercial grade may be used with level of stress equal to SC3.

To ensure reused timber is in adequate condition, inspection must be carried out. Guidance may be found in a leaflet by the Timber Research and Development Association (TRADA).[9]

35.3.3.1 Plywood and man-made timbers

Waterproof plywood of structural thickness is now available in three main types: (1) Douglas fir from North America; (2) birch from Finland; and (3) tropical hardwood from a variety of sources. From structural considerations, there is little to choose between them. But where used as shuttering, care must be taken to choose appropriately and to apply suitable treatment (see

Table 35.3 Permissible stresses and moduli of elasticity: timber

Strength class	Bending stress parallel to grain (N/mm²)	Tension stress parallel to grain (N/mm²)	Compression stress perpendicular to grain (N/mm²)	Shear stress perpendicular to grain (N/mm²)	Modulus of elasticity E	
					mean (N/mm²)	minimum (N/mm²)
SC3	6.29	3.73	2.63	1.32	7600	5000
SC4	8.90	5.24	2.87	1.40	8600	5700
SC5	11.87	6.90	3.35	1.97	9200	6100

section 35.5 on formwork). Large sheets can be obtained to special order with scarfed joints, producing concrete virtually free from joint lines. Data for calculations are available from the manufacturers or suppliers.

Chipboard. For many years, chipboard has been a cheap but moisture-sensitive material. Recently, a few chipboards which are adequately moisture-resistant for formwork have become available, and have been used extensively for reactor construction in Germany. Its stiffness is not as good as virgin timber, and so it is often necessary to use a thicker section than the equivalent plywood.

Other man-made timber materials. Due to low stiffness and poor moisture resistance, they have relatively little application in temporary works.

35.3.4 Aluminium

The lightness of aluminium is countered by its flexibility, and so aluminium has only limited structural applications. As scaffold tube, it is about 3 times as flexible as ordinary tube and in consequence its use in conjunction with steel tubes must be exercised with care to prevent unexpected distortions. Thus, it is desirable to make any one structure of either steel or aluminium. Because of its high scrap value, it is very susceptible to theft. It can be extruded into complex sections (see section 35.4.2 on soldiers and walings).

35.3.5 The ground

In many cases, naturally occurring soil or rock is fundamental to the success of the work. Its properties must be established to enable the design to proceed. However, there are cases where a rigorous, classical analysis will indicate capacities so small as to be impracticable; but for small loads, and short periods, it may still be possible to achieve an acceptable result. Reference may be made to the falsework Code.[8] In all cases, an examination should be made of the ground conditions, at least using a spade. For the situation where the ground must be supported laterally, see section 35.9 on temporary excavations.

35.3.6 Bricks and the like

Where such materials can remain permanently they can form an economical material for formwork, especially if they bring in a trade at that time under-utilized. Brickwork can also be an economic way of forming support towers. Where existing structures contribute to the strength of temporary works, careful investigation, supplemented by tests if necessary, is needed before relying on them.

Concrete, precast or *in situ*, has similar applications.

35.3.7 Plastics

The uses of plastics in temporary works are still few and far between. The difficulties which must be overcome to make it a successful proposition are cost and the low modulus of elasticity. To use plastic effectively, fabricated sections built up in some way to give a large effective depth must be used and this process is expensive in labour. The alternative of thick, solid sections of plastic is considerably more expensive than traditional means, and so it will be some time before this material is in use in large quantities.

There are three main groups: (1) a thermoplastic group in which heat will render the plastic flexible so that it can be shaped again; (2) a thermosetting group in which the plastic once set by the manufacturing process, cannot be altered; and (3) plastic reinforced with fibres, usually glass fibre. As a structural material, only the third class need be considered (see page 35/10). Thermoplastic materials can be used to make textured formwork surfaces and these are discussed by Blake.[10] Small items such as tie cones are very effectively made in plastic, easy stripping from the concrete being an important reason. Bulk with low weight and relatively low strength can be obtained fairly cheaply with expanded plastic such as polystyrene used, for example, as a permanent form for a void, or for forming holes.

35.4 Equipment

Materials are often fabricated to make equipment designed especially for the job in hand, to be scrapped thereafter, but much equipment is conceived for continuing use. There are a number of firms supplying equipment of various kinds useful in temporary works, available for either sale or hire. When these are proven items, a saving of time and design cost are immediately available and it is more likely to be economic to write-off only a part on the job in question. For example, a set of equipment may be purchased, and the job completed in a given time. Alternatively, by hiring twice as much equipment for a total cost of perhaps half the outlay, the job can be completed in half the time.

Data for design should be obtained from the manufacturer. It should be noted that there are no standard tests except for telescopic centres,[11] props,[12] and heavy-duty support towers.[13] Factors of safety are discussed in the falsework Code.[8] The items below are the more common.

35.4.1 Formwork panels

The face of the concrete will often be formed by panels which may be of steel or of a steel frame with a ply face. This latter is normally a component in a system designed primarily either for walls or soffits. In general, these will serve the alternative purpose but not quite so efficiently. These are available commercially, or panels can be purpose-made.

35.4.2 Soldiers and walings

The majority of large shutters today rely on soldiers or strongbacks. A number are available to suit most problems: they can be used in other ways, e.g. as walings, and some are designed for this purpose.

The big advantage of aluminium is the possibility of extruding complex shapes, and sections are now available for formwork applications. One face is designed to accommodate a timber for fixing the sheeting while the other provides simplified fixing for the main outer strength member. Due to the lightness, spans larger than the timber it displaces are practical.

35.4.3 Centres

Telescopic beams are available in a variety of sizes to support slab formwork. These may present problems of end support owing to the small area in contact with any timber they sit on. Deflections may also need care. In the largest types, spans of 10 m are possible (see also section 35.4.9 on heavy-duty girders).

35.4.4 Form ties

Today, form ties in many varieties are available. The simplest approach is a threaded bolt with a cardboard or plastic sleeve. Many of these are based on the Dividag reinforcing bar, with special nut units; alternatively, a plain rod with various proprietary types of friction grip at each end may be used. The form tie will normally be withdrawn and the hole filled. The oldest type of proprietary form tie is the coil tie in which two coils are connected together by rods. The shutter is fixed with special bolts which engage with these coils. On withdrawal of the bolts the holes are made good.

A comparable group of form ties perform the same function using a she bolt and a threaded rod as the expendable piece. With this system it is more difficult to obtain accurate spacing or thickness of the wall, and correct location of the portion left in the wall. Note that threads are not standard and mixing ties and bolts can result in failure.

There are also snap ties in which a complete assembly is cast into the wall and the ends broken off.

35.4.5 Clamps

Clamps have been available for many years for constructing columns. They are also available in a variety of styles for clamping beam shutters.

35.4.6 Props and struts

Telescopic props consist of telescopic tubes between which the load is transferred by a pin, which sits on an adjustable screwed collar. Its capacity is related to the accuracy of its use, as was demonstrated by research carried out on props to BS 4074:1982.[14] It is practical on site to limit out of plumb to 1.5°, and the location of the load at the top to within 25 mm of the ideal position. In these cases, the lower pair of lines in Figure 35.2 are appropriate. Where concentricity can be guaranteed – by means of pins on the beams above locating in the holes in the prop – the upper pair of lines may be used.

The standard prop is not designed for tension, but a variety of push–pull props of varying capacity are available. The ends must be of a type to permit connection to take the tension. They are useful for aligning formwork and for locating precast units in position until they are secured permanently.

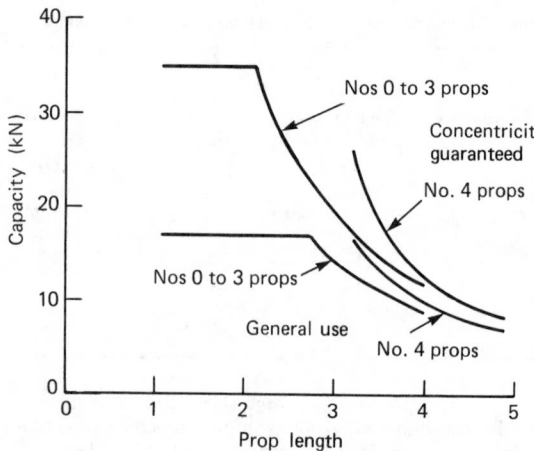

Figure 35.2 Safe working loads for telescopic props to BS 4074

Struts for trenching are of the same basic design, but are much shorter and have clawed ends.

35.4.7 Prefabricated scaffolding

As an alternative to traditional tubular scaffolding, prefabricated units are available. While most were designed principally for access, they can also be used for soffit support purposes. There are two main types: (1) in which frames are used to construct a series of independent towers or a line; and (2) in which components of linear nature connect together to provide a grid of vertical supports, or when used for access, in a line. When used as falsework, leg loads range from 3 to 6 t.

35.4.8 Heavy-duty support equipment

There are a few specialist towers, and Bailey bridging[15] is available. For even larger loads, military trestling is appropriate with a capacity of over 250 t per tower.

35.4.9 Heavy-duty girders

Bailey bridging and its modern derivatives can be used, as well as other girder systems built up from modular components, with the additional possibility of under-trussing.

35.4.10 Piles

The two normally available sheet piles, Larssen and Appleby Frodingham, each come in a variety of types, accompanied by appropriate specials. For data, contact British Steel Corporation. Lighter trench sheeting, some of which is interlocking, is also available usually in lengths up to 5 m. Some identical sections are sold under different names by different suppliers.

For kingposts, deadmen and the like, box piles are available. In addition, some foundation piles are suitable.

35.5 Formwork

(*Note*: BS 4340:1968 gives a glossary of formwork terms. This is being incorporated into BS 6100, section 6.5.[16])

35.5.1 Purpose

There are three main aspects of formwork: (1) it is important to have formwork which is of the appropriate quality, to produce both satisfactory dimensions and surface appearance; (2) it is necessary that the formwork should be safe and that the risk of damage to people and property should be a minimum; and (3) it is important that the cost should be as low as possible. The most exhaustive textbook currently available is *Formwork: a guide to good practice*.[17]

35.5.2 Design data

There are two principal loading cases in formwork: the horizontal and the vertical.

35.5.2.1 Horizontal loading

For anything greater than the smallest of lifts, a proper assessment of lateral pressure is needed. This can most reliably be carried out in accordance with the tables of the Construction Industry Research and Information Association (CIRIA)

method given in Clear and Harrison's[18] report *Concrete pressure on formwork*. This takes into account the following factors:

(1) *Plan dimensions.* Where the dimensions of the piece to be concreted are both less than 2 m, it is classified as a column. If either dimension is greater, the table for walls should be used. Experience shows this gives too high a figure for very small columns.
(2) *Type of cement and admixture.* There are two main classes.
(3) *Form height.* This is the height of the form, even though concrete may not reach this height.
(4) *Rate of rise.* The known volume of concrete to be supplied per hour is divided by the plan area.
(5) *Concrete temperature.* This is the temperature as the concrete goes into the formwork.

The tables in the CIRIA report *Concrete pressure on formwork* are based on a concrete weight density of 25 kN/m³.

Table 35.4 Design pressures on formwork, in kilonewtons per square metre

			Walls and Bases — A wall or base is a section where at least one of the plan dimensions is greater than 2 m								Columns — A column is a section where both plan dimensions are less than 2 m					
Concrete group	Conc. temp. (°C)	Form height (m)	Rate of rise (m/h)							Form height (m)	Rate of rise (m/h)				
			0.5	1.0	1.5	2.0	3.0	5.0	10		2	4	6	10	15
1) OPC, RHPC or SRPC without admixtures	5	2	40	45	50	50	50	50	50	3	75	75	75	75	75
		3	50	55	60	65	70	75	75	4	85	100	100	100	100
		4	60	65	65	70	75	85	100	6	95	115	125	145	150
		6	70	75	80	80	90	100	115	10	115	135	145	170	190
		10	85	90	95	100	105	115	135	15	130	150	165	190	210
2) OPC, RHPC or SRPC with any admixture except a retarder	10	2	35	40	45	45	50	50	50	3	65	75	75	75	75
		3	40	45	50	55	60	70	75	4	75	90	100	100	100
		4	45	50	55	60	65	75	90	6	80	100	115	130	150
		6	50	55	60	65	75	85	105	10	95	115	130	150	175
		10	60	70	75	80	85	95	115	15	105	125	140	165	190
	15	2	30	35	40	45	50	50	50	3	60	75	75	75	75
		3	35	40	45	50	55	65	75	4	65	85	95	100	100
		4	35	45	50	50	60	70	90	6	75	90	105	130	150
		6	40	50	55	60	65	75	95	10	80	100	115	140	165
		10	50	55	60	65	75	85	105	15	90	110	125	150	175
3) OPC, RHPC or SRPC with a retarder	5	2	50	50	50	50	50	50	50	3	75	75	75	75	75
		3	65	70	75	75	75	75	75	4	100	100	100	100	100
		4	75	80	85	90	95	100	100	6	120	130	140	150	150
		6	95	100	105	105	110	110	135	10	145	160	175	195	215
		10	120	125	130	130	140	150	165	15	170	190	205	225	245
4) LHPBFC, PBFC, PPFAC or a blend containing less than 70% g.g.b.f.s. or 40% p.f.a. without admixtures	10	2	40	45	50	50	50	50	50	3	75	75	75	75	75
		3	50	55	60	65	70	75	75	4	85	95	100	100	100
		4	60	60	65	70	75	85	100	6	95	110	125	145	150
		6	70	75	80	80	90	100	115	10	115	130	145	170	190
		10	85	90	95	100	105	115	135	15	130	150	165	190	210
5) LHPBFC, PBFC, PPFAC or a blend containing less than 70% g.g.b.f.s. or 40% p.f.a. with any admixture except a retarder (see Note)	15	2	35	40	45	45	50	50	50	3	65	75	75	75	75
		3	40	45	50	55	60	70	75	4	75	90	100	100	100
		4	45	50	55	60	65	75	90	6	80	100	115	135	150
		6	50	60	65	65	75	85	105	10	95	115	130	155	175
		10	65	70	75	80	85	100	120	15	105	125	140	165	190

Notes:
(1) The maximum pressures are in units of kN/m² to the nearest 5 kN/m². They were calculated assuming a concrete weight density of 25 kN/m². Pressures for lightweight or heavyweight concretes should be calculated in a proportion to their densities.
(2) The pressures in italic are outside recorded experience. The highest recorded pressures on site were 90 kN/m² for walls and 166 kN/m² for columns.
(3) The tables do not include the use of concretes which contain a retarder in combination with LHPBFC, PBFC, PPFAC or any cement blend. Guidance on these combinations is given in CIRIA Report 108.
(4) New types of superplasticizers have been introduced recently, referred to as 'extended life superplasticizers'. They also act as retarders and therefore relate to Concrete Group 3.

Table 35.4 sets out design pressures for OPC, RHPC and SRPC with any admixture other than a retarder. For further information, see Clear and Harrison's *Concrete pressure on formwork*.

35.5.3 Vertical loading

The dead load is straightforward to calculate, but care must be taken to establish the type of concrete in use. The loading figures to be adopted for live loads are somewhat less certain. For concrete placed by wheelbarrow, a design load of 1.5 kN/m^2 should be taken. This will also be appropriate when concreting is by crane skip, and heaping of concrete is carefully limited. With newer methods of placing, such as pumps and crane skips, care must be taken to establish that the figure is appropriate to the case. As far as deflection is concerned, this is unlikely to be a serious problem. For example, the impact loadings from a crane skip are of very short duration and it is most probable that the formwork will recover, so that any deflections are caused by the dead load only. However, the strength aspect is important.

There may be other environmental loads such as wind, or the possible risk of damage from passing vehicles, which should be considered as possible loads.

35.5.4 Other cases

Wind frequently produces significant load on wall formwork, and should be given due consideration. Other forces may arise from snow and ice (see also section 35.6.1).

35.5.5 Design considerations

Formwork may be divided broadly into two cases: (1) vertical surfaces; and (2) horizontal surfaces. In the former, failure will result at least in local deformation requiring remedial work to the concrete. It is very infrequent amongst the failures of wall shutters that there is movement sufficient to cause damage to people working with them. However, a slab, when it fails, is often the cause of more serious damage, partly because materials and any person concerned will almost inevitably have a significant distance to fall. Thus a slightly more conservative approach is appropriate to soffit shuttering, while wall shuttering stresses may be taken a little higher.

The safety of formwork may be equated more-or-less directly with its strength. One aspect of quality is a function of the deflection. Specifications call for standards of flatness over an area, normally for visual or aesthetic reasons, on occasion because something has to fit against them. It is also desirable to have the correct cross-section of concrete. Too little will result in loss of strength and reinforcement cover or displacement; too much will add to the weight.

With typical formwork constructed in the traditional manner of a sheeting material spanning a comparatively short distance between framing members which in their turn span a greater distance, the question of deflections involves an analysis both of the individual deflections of individual structural components and consideration of the structural system as a whole. If a given figure is arrived at for a total deflection from the theoretical plane, part of this will be attributable to each stage of the structural system. The point which sums up the various deflections to the greatest degree (note that this is not a straight arithmetical addition) should be at the maximum deviation figure permitted. However, the shape of the whole area is important.

If a flexible sheeting is used in conjunction with closely spaced framing members, concrete can result which is markedly rippled but still within the overall tolerance. In addition to the straight dimensional tolerance provided, it is usual to limit the deflection to a fraction of the span between the supporting members. This figure should be 1/270 or, if the quality of the work is particularly important, somewhat smaller. It should also be noted that, particularly with timber, there is a reserve of stiffness in almost all cases and so the theoretical deflection is seldom likely to be reached.

35.5.6 Concrete finish

The other main aspect of quality is the actual surface finish. Concrete is an amorphous material which will mould itself relatively easily to the shape of the containing mould, although certain characteristics (sometimes regarded as blemishes) will always be present in some degree. The most difficult to eliminate is the blowhole. This is because the air mixed in to the concrete cannot get away but is worked to the edge of the concrete by the compaction and frequently remains at the face of the shutter.

One method which helps to avoid this is to have a shutter surface of absorbent material, although this will produce another characteristic variation in colour. While concrete has a given colour in given circumstances, these circumstances do not remain sensibly constant. The absorbency of the formwork, differing pressures of the concrete, variations in the actual constituents of the mix from batch to batch and varying compaction will all lead to differently coloured patches on the face of the concrete.

It is difficult to eliminate these completely, though the use of impervious shutters tends to even-up the colour. However, a high gloss finish on the forms such as plastic may create a further problem of dark markings on the concrete. It will thus be clear that it is difficult to have formwork which will give a very smooth concrete, an evenly coloured concrete and a concrete free from blowholes. The problems of surface finish are discussed at more length by Monks.[19-23]

While the above discussion relates to a typical piece of concrete against a typical shutter, it is necessary to have joints between one day's concrete and another and between two pieces of formwork which will be dismantled and re-erected on a subsequent occasion. Both of these give rise to variations in the appearance of the concrete. There may be a leakage at the shutter joint, albeit very small, which will change the mix locally and so create a difference of colour. There may be a physical step on such a joint because the formwork was not aligned with enough accuracy. In general, these are natural characteristics of the concreting process.

In accepting this basic difficulty, many designers take advantage of it by designing the appearance of the concrete so that these minor disabilities are turned to advantage. For example, along joint lines, a fillet is fixed to the face of the formwork, thus creating a groove. If there is a step, or a discoloration, it is inconspicuous by comparison with the groove and the arrangement or pattern of these can be designed so that the appearance of the structure is considerably enhanced.

Apart from a straightforward patterning as suggested above, modern plastic materials in particular can provide shaped, textured or patterned concrete, which can provide an alternative way of making the concrete appearance an adjunct to the structure. This is discussed, along with a number of other approaches to surface treatment, by Gage[24] in the *Guide to exposed concrete finishes*. The present state of the art does not admit of precise statements on a number of these points. For example, the effect of different mixes and different materials for the mixes will vary the incidence of dark markings and the tendency to leak through a shutter, but there is as yet no known and accepted index to either of these.

35.5.7 The form face

To ensure the satisfactory detachment of the form from the

concrete, a release agent is applied to the face before concreting. This is usually an oil, which serves to prevent the concrete physically sticking to the form face. Various types of oil gave different effects and reference should be made to Monks[19-23] for information. A chemical release agent acts by actual chemical combination with a very thin layer of the outer skin of the concrete to ease stripping, thus reducing damage to shutter face and concrete alike.

Claims have been made that the use of various materials eliminates the need for a release agent; but in all cases a release agent increases the life and makes it easier to strip. Plastic is the most likely material to be acceptable without oil but there is usually a build-up of laitance.

For steel, a suitable oil is essential to reduce rust. With timber or plywood, it is often appropriate to paint on a coating to make it more durable before applying mould oil. Provided that the dry and clean application conditions needed to ensure its effective adhesion exist, polyurethane paints give good service, prolonging the life of the face. Many paints react with the alkali in concrete, and these should not be used.

35.5.8 Basic philosophy

The basic design and planning of formwork should take account of the following points.

(1) *Strength and stiffness.* The whole structure must be such that there are no weak links, ideally no over-strong sections, and the material content is at a minimum.
(2) *Repetition.* In general, formwork will be used a number of times, in straightforward designs in a precisely repetitive way. Elsewhere, components of the formwork may be re-erected in a different way. As the capital cost is often a considerable part of the total cost, repetition always requires careful analysis, to reduce the total amount of equipment.
(3) *Durability.* As formwork is expected to last, owing to its cost, careful consideration should be given to using materials that are reasonably durable so that the complete assembly can be used and handled without undue wear.

(4) *Stripability.* Many items of formwork have to be set up in re-entrant situations and it is essential to have a design which can be stripped without damaging the formwork and which is not excessively time-consuming.
(5) *Cost.* The sum of the costs of forming an area of concrete are made up of, firstly, the cost of providing the formwork. This is usually the cost of buying or making the equipment, but in some cases it may be hired. The second part of the cost is the labour in erecting, stripping, cleaning and carrying forward to a later use. The cost of any expendable components such as form ties is the final part. It is essential that the total of these three is kept to a minimum. In today's rising-price situation, studying the labour content carefully will repay effort best.

35.5.8.1 Details

The success of formwork schemes depends on the basic scheme. But failure to cope satisfactorily with any of the details may cause much trouble.

35.5.8.2 Construction joints

No formwork is required for a horizontal joint but vertical joints in slabs and walls present considerable difficulty. The traditional approach is to use timber or plywood cut in small pieces to surround the projecting reinforcement. This is tedious, though it is not normally necessary to make a very concrete-proof shutter. In many cases a slot at the level of the reinforcement is acceptable. Expanded polystyrene can also be used. An alternative method is to use a suitable type of expanded metal set on a timber frame. This can be used in slabs or walls of considerable depth, and the leakage through the material is very small. If it is carried right out to the edge of the concrete member, rusting on the surface is almost inevitable.

35.5.8.3 Kickers

To build a wall or the like on a slab, it is possible merely to stand two shutters on top of the concrete. Two main difficulties arise:

Figure 35.3 Vertical and horizontal formwork being erected

(1) the slab is most unlikely to be sufficiently flat to provide a seal at the bottom of the shutter and grout will leak out; (2) the location and thickness of this wall will not be reliable, as the least knock will move the shutters from the position in which they were set up. To get over this problem, it is usual to cast a small height of structure of the same cross-section, between 50 and 150 mm high, called a kicker. This can be cast with the slab below, or alternatively may be cast-on afterwards, generally enabling greater accuracy to be achieved. Where it is felt more desirable to ensure that the wall above the slab is directly over the wall below, rather than being in the correct theoretical position, a kicker device based on the wall below has a considerable advantage. This may be done by making precast blocks which can be set in the top of the shutter below, acting as a spacer to the shutters, and projecting sufficiently high so that when the wall above is to be cast it will act as a spacer at the bottom of that wall. It may be desirable to use this line of blocks to suspend a pair of timbers to form a separate *in situ* kicker. Where slabs are fairly accurate, it is practical to place timbers outside the wall shutter, and nail these down to the slabs.

35.5.8.4 *Voids*

There are a number of occasions where a completely enclosed empty space is required, and this problem gives scope for considerable ingenuity. Where it is cylindrical, proprietary products are available in expanded polystyrene or sheet metal. It is imperative that adequate steps are taken to tie them down, as the buoyancy is considerable. Where the space is rectangular it may be possible to obtain a suitable cardboard void former; alternatively, light timber may be used. When the void is deep, it will often be sensible to cast the concrete in two lifts. Thus, all formwork but the soffit will be recovered. Asbestos cement sheet and its substitutes are useful soffit material; the makers should be consulted for data on strength. For voids, the design criteria may be relaxed somewhat, as clearly the aesthetic consideration does not apply.

35.5.8.5 *Top shutters*

It is required in clause 15(2) of the *Standard method of measurement of building works*[25] that surfaces at an angle of more than 15° to the horizontal shall be formed with a shutter. However, this is a somewhat arbitrary requirement, because the various considerations may produce other sensible answers. The factors concerned are: (1) the angle of the slope; (2) the concrete mix; (3) the thickness of the slab; (4) the amount of reinforcement; and (5) the degree of accuracy of the surface required. In some cases, it is possible to construct a slope at an angle as steep as 60°, and in the case of gunite this can go to beyond 90°. Where a top shutter is felt to be necessary, it is essential that it is both supported off and tied to the lower shutter and designed for the full theoretical hydrostatic pressure. An alternative approach is briefly described in section 35.5.10 on sliding formwork or slipform, below.

35.5.8.6 *Curved shapes*

From a structural point of view, the circle is an ideal shape. For example, a circular column form can be designed in pure tension, provided it has enough stiffness for handling. Any kind of curved surface provides a stiffness greater than the equivalent plane one. It is fairly straightforward to use traditional materials for parts of a cylinder, but forming three-dimensional curves is more difficult. Boatbuilding techniques can be applied to them but, in many cases, the advantage of plastic which can be moulded is paramount. While plastic is a very flexible material, this only matters when it is being used in a situation where bending is important. When used in tension the movement is not of significance. Thus reinforced plastics can be used to good advantage in forming cylindrical shutters and also in forming more complex shapes. Circular column shutters are also made in steel.

35.5.8.7 *Stripping*

The removal of the forms should only take place when the concrete is strong enough. For sections which carry their own weight, the strength required will be considerable. But for other cases, it is merely necessary to consider accidental damage, wind and frost. Harrison[26] gives figures which give security, based on information about the actual circumstances of the case. Forms which are left in position will aid curing.

35.5.9 Particular types of formwork

35.5.9.1 *Walls*

In the typical wall, it is necessary to have two form surfaces which are held rigidly at the required distance apart. The reaction from the concrete pressure on one side is normally taken by means of form ties through the concrete to the opposite face which is thus balanced. This tie will usually act as a spacer so that the shutter can be set up accurately in the first place. There are a variety of methods within this broad idea. The initial subdivison of wall types is based on handling.

It may be that traditional methods are in use in which case it will be man-handled; alternatively, a crane may be available to lift the shutter in relatively large pieces without fully dismantling it. As an alternative to this, some form of wheels or skids may be provided if the form is moving in a horizontal plane. Where the form is to be crane-handled, it is practical to use relatively large structural sections and thereby reduce the number of form ties considerably. This has the effect of reducing labour. On the other hand, where man-handling is essential, the form must be dismantlable into fairly small and light pieces, and it is then more practical to have lighter form ties connecting the two forms at many more points, and thus eliminating the need for heavy structural members. Inevitably, this leads to a somewhat higher labour content but the capital cost will be less. In addition, it is much more versatile, as the components can be built up differently for each successive use. The conditions applicable to the problem in hand will dictate which is the more sensible solution.

The actual forms may be of basic materials, or proprietary equipment or a combination of both. Proprietary equipment is particularly suitable for the piece small approach, but such components can also be assembled into a large form, to be employed as a single unit for a number of uses.

The piece small approach almost inevitably leads to a patterning on the concrete from the individual components that is fairly small. This may be acceptable, or considered desirable, but there are other occasions when large unmarked areas are needed. In these cases the crane approach permits fabricating a form which will remain as a unit with satisfactory joints over a period. Where a wall has to be waterproof, it is very desirable that any form ties have a part cast into the wall and lost to further use. This reduces the risk of leakage which inevitably arises when a number of open holes have to be sealed after the removal of the forms. Occasionally it is required to fix a disc on the tie to lengthen any possible leakage path, but the value of this is not undisputed. If possible, avoid tie-holes and their problems – aesthetic and waterproofing – by tying above or outside the concrete.

35.5.9.2 Columns

Where forms to a vertical structural member can be tied round the ends, from the construction point of view the member is a column (see Figure 35.4). The problems of tying through the concrete disappear, and all the equipment is reusable. The typical solution is the use of column clamps which are available from numerous manufacturers. For smaller columns the use of steel strapping is also a very satisfactory solution. Larger columns are frequently constructed using pairs of soldiers tied at their ends, but often special equipment is designed and fabricated for this purpose. For repetitive columns, two-piece crane-handled shutters with permanently attached access and plumbing devices are appropriate.

35.5.9.3 Soffits

The form surface needs a support. This comes either from the ends of the span which will subsequently support the slab itself, or through a number of propping points from the floor or ground directly below. The former solution has obvious structural advantages, the disadvantages being that the formwork itself is often rather flexible and so the quality of the soffit may be poor. If the span is large, the cost of providing and handling the centring beams may be considerable.

While the actual surface of a soffit shutter in the past was timber boarding and in Europe is still often specially formed timber panels, in the UK plywood or ply-faced panels are the most favoured methods. In the case of plywood, the support

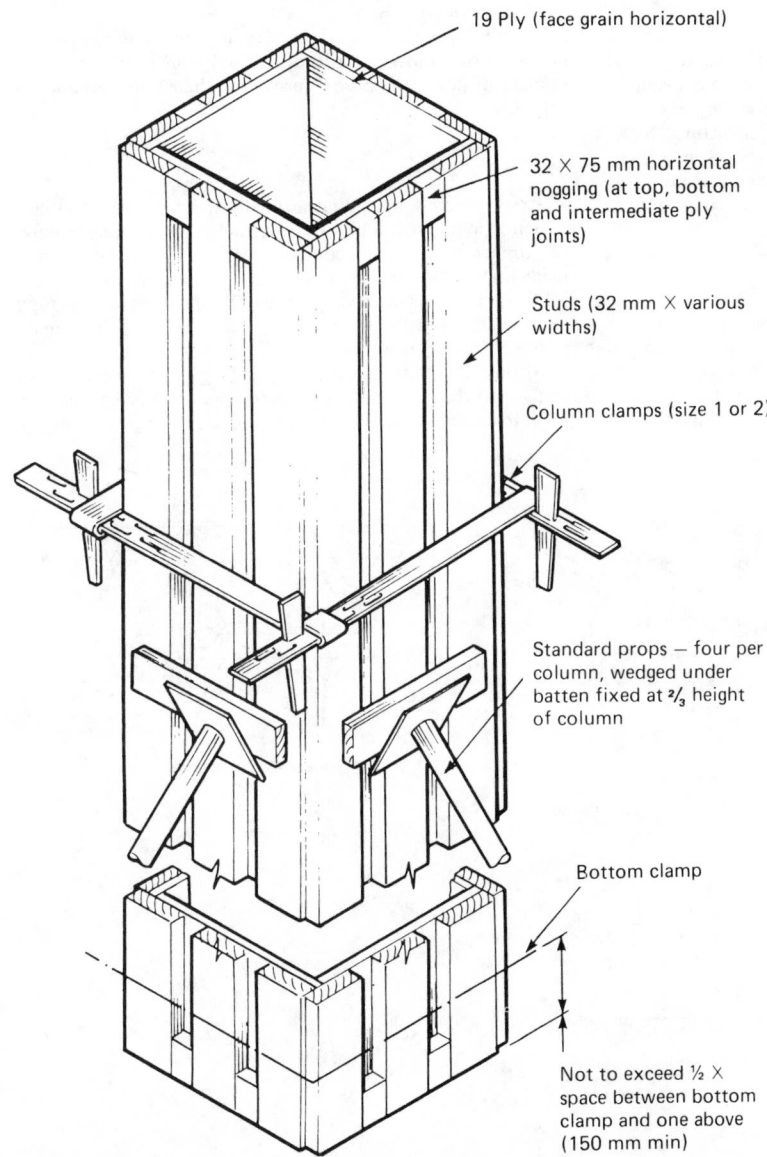

19 Ply (face grain horizontal)

32 × 75 mm horizontal nogging (at top, bottom and intermediate ply joints)

Studs (32 mm × various widths)

Column clamps (size 1 or 2)

Standard props — four per column, wedged under batten fixed at ²/₃ height of column

Bottom clamp

Not to exceed ½ × space between bottom clamp and one above (150 mm min)

Figure 35.4 Rectangular-column formwork using plywood face, suitable for fair-faced work. (After Powell (ed.) (1979) *House builder's reference book*. Butterworth, London)

may be timber beams of either one or two layers, supported in turn on props or scaffold from below, or perhaps some form of centring beam supported by the walls (see Figure 35.5). For panels, similar support may be provided but in most cases the suppliers offer a system of support designed to work with the panels, providing a low-labour method of assembling and dismantling.

Another variant is the use of table forms or flying forms. This is the term given to a unit of formwork, comprising an area of soffit, complete with its supports. In a tall building these are invariably handled by crane. They are pushed to the edge of the building after lowering on to wheels, or dropped and skidded out. The crane is then attached to them by slings or preferably by a lifting hook device, and they are lifted to the new position.

In buildings with a large floor area they may be pushed to the new position on the same level.

35.5.10 Sliding formwork or slipform

The use of continuously moving formwork is applicable to any structure of constant cross-section and appreciable height – 30 m or so. While the traditional application of sliding has been to the strictly constant cross-section, tapered structures have been successfully formed, and those with changes in cross-section at one or more points in their height. The system is most suitable for structures with a simple outline and so any such complication must be carefully assessed.

35.5.10.1 Brief description

Instead of having two shutter panels held together with ties, they are located with a framework over the top of the shutter called a yoke. Fixed to the centre of each of these yokes is a jack and this jack operates by climbing a rod or tube through the middle of the wall. Traditionally, this jack was a screw-jack

manually operated but this has now almost completely been superseded by a mechanically operated jack, normally from a central hydraulic supply. The form itself should have a smooth face to the concrete and be set up so that there is a very small taper from bottom to top, preventing the concrete from being caught in the shutter and dragged up with it. The access to the job is obtained from platform or platforms at the level of the top of the formwork, and carried on it. Forms are connected across spaces in the structure, e.g. the bin of a silo, by a framework which doubles as the support for such a platform.

In addition, platforms are suspended below, so that as the concrete emerges from the forms, it may be inspected and, if necessary, smoothed over to present a more uniform appearance. It is normal to use a concrete of fairly high strength, with a high cement content. This helps the workability, so that the compacting of the concrete presents little difficulty, and lubricates the form so that it slides more easily. It is usual to operate 24 h/day, climbing from 75 to 500 mm/h.

More information is given by Hunter[27] in *Construction with moving forms* and in standard textbooks on formwork.[28,29] A number of firms specialize in providing the equipment and the expertise.

35.5.10.2 Pros and cons

A structure built by sliding normally has no horizontal construction joints. It will be built in a fraction of the time required for conventional construction. Unless some subsequent treatment makes it necessary, there need be no conventional scaffolding. There is a saving of overheads due to the rapid construction and the whole exercise catches people's imagination, with consequent benefits.

Against this is the need to organize gangs to work on two shifts, and the availability of expertise and competent management to ensure success in the operation. If the height is not large,

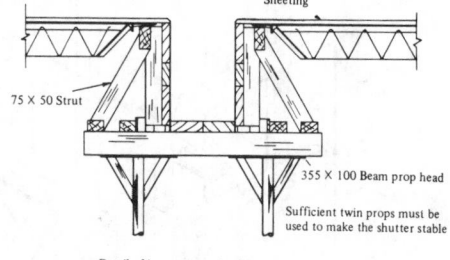

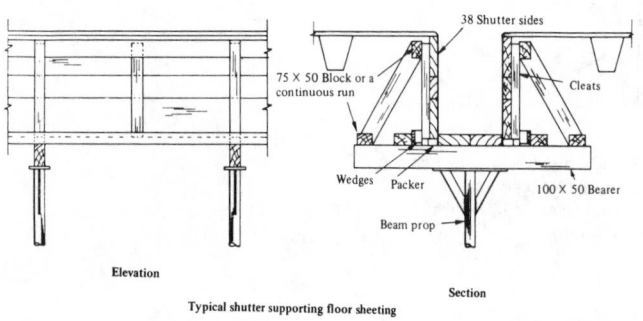

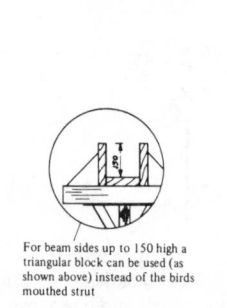

Figure 35.5 Small beam formwork. Where the stripping of the side shutter will not precede the stripping of the soffit, the packers below the beam sides should be omitted. All dimensions are in millimetres

the cost of making and setting up will not be offset by the lower cost of operation, though a succession of low structures may be viable. Because it is essential to preplan much of the work, especially services – instead of dealing with it in the traditional hand-to-mouth manner – time may not permit its use.

35.5.10.3 *Variants*

A comparable technique is used for concrete roads, but it can also be used to advantage on slopes. In this case, winches are used to pull up forms, which are weighted to resist the concrete pressure.

35.6 Falsework

This is the subject of BS 5975.[8] It is temporary support work in construction, needed for some part of the structure until it is capable of supporting itself. This might be wet concrete, precast units or steelwork. The basic requirement for the design is to carry the loads and forces down to a firm foundation. Where practical, this will be the foundation of the ultimate structure, but there will be many cases where such an arrangement is not possible. An introduction, primarily for site engineers, is given by Wilshere.[30]

35.6.1 Loads

The dead weight which falsework must support is straightforward to calculate. But care must be taken to consider the sequence of loading and to include in the calculations all the other loads which may occur. These are shock loads due to placing the item, the use of machinery on top of the falsework, e.g. a dumper truck delivering concrete, or a crane placing the next unit, and other environmental loads, such as wind. While the accuracy of loads is known more precisely than is typical in a permanent structure, as the time concerned is so much shorter, the care with which people treat such a structure is very much less, and the unexpected is more likely to happen. Normally, lateral concrete pressure is not supported by falsework, but sometimes it is necessary to arrange the falsework to act as a tie between opposing form faces above.

35.6.2 Stresses

There are many different circumstances in which falsework may

Figure 35.6 A straightforward falsework arrangement

be used. These differ not only in the order of magnitude of the loadings, but in the approach taken to the design and construction. In the ideal case, where the complete design is carried out in full detail, and the site supervision ensures that it is constructed exactly in this manner, it is possible to use stresses greater than those used for permanent construction. Conversely, where design work is sketchy and supervision not over-efficient, it is desirable to use stresses lower than those adopted in permanent work. Basic design stresses are to be found in the various codes of practice but except for timber – BS 5268[7] – guidance for any possible variations is not given.

35.6.3 Typical construction

Falsework may be constructed out of any of the usual materials as used in a permanent structure. For vertical loads, where any horizontal load is a small part, tubular scaffolding may be employed to advantage. Where leg loads in excess of 3 t are required, there are other specialized towers available as well. Bailey bridging on end or military trestling may be used. Box piles or other structural-steel sections are appropriate, but their secondhand value is not good.

Other constructions may be made of brickwork or concrete, and on occasion the ground itself can be used. In the past and today in timber-growing countries, wood forms an ideal shoring material. Where the falsework spans horizontally, steel joists are very suitable, and proprietary adjustable beam units are available in various sizes.

Falsework often comprises a structure set upon a foundation built by a different group of people and supporting in its turn the components of the permanent structure. Where more than one group of people are involved, it is particularly important that the responsibility of each is clearly understood by the other. For example, the foundation for falsework can only be designed satisfactorily when exact information on the loadings is available. This is important as the foundation is often the least satisfactory part of a falsework structure and to improve its bearing capacity involves considerable expense. Similarly, at the top of the scaffold or falsework, the arrangement in detail of how the load is to be supported is of very considerable importance and is a point where failures have occurred in the past (see Figure 35.6).

35.7 Handling and erection of precast units

The advantage of casting concrete other than in its ultimate position is well known, but this leads to the problem of handling it. Where a unit weighs more than 20 kg, and today this may reach 3000 t, there is a need to consider techniques for moving it into position. The mechanical equipment may include cranes, sticks, gantries, rollers and sliding ways. The range of units will include blocks, beams, columns, cladding panels and walls. The problem, apart from appropriate design, is that the industry has been handling this type of unit for a relatively short time and there is no traditional approach or 'feel' for it. All too often the labour which is asked to deal with it is comparatively inexperienced.

35.7.1 Moving units

Where units can be slid or rolled into position they do not need to be crane-handled; jacks underneath can be arranged to do all the necessary vertical movement. The horizontal ways may be bullhead rails with steel balls or well-greased timber. But cranes or gantries may well be needed to handle some items. Because the capacity of cranes falls off rapidly as the radius increases, it

is sometimes convenient to use two cranes. Effective co-ordination between two cranes is very difficult, so schemes which involve moving the cranes under load should not be used.

To connect the concrete item to the lifting device requires some form of equipment. Ideally, this could consist of slings or a forklift device as both of these would be easy to attach to the unit and require no special provision made in the unit itself. But often it is difficult to set the unit down. A traditional approach is to cast loops of reinforcing steel projecting from the top of the concrete and connect to them the hooks of a sling with two or more legs. Steel should be notch–ductile, which is difficult to obtain in small quantities.

An alternative method is to put pins of appropriate diameter horizontally through the unit, usually in conjunction with suitable yokes. If a single pin is used, a column can be up-ended ready for placing.

It may be necessary to have some form of anchorage cast into the unit, e.g. a screwed socket. While this provides a neat appearance in the finished structure, threads are relatively unsatisfactory in this type of work. It is very difficult to keep them clean and thus the thread device which will connect to the crane may not be screwed home properly. The bolt threads wear. A strictly limited number of sizes – ideally one – should be used on any one site. Various special devices can be made for lifting items and there are a number of specialized proprietary items available. The width and shape of the unit may well limit the types of fixing which can be placed in it.

Any cast-in metal work on a unit to be exposed to the weather is a potential source of rusting or staining. The initial decision should always take this into account and any metal work should either be adequately protected by galvanizing or be of a non-staining nature. Alternatively, they could be protected with mortar to prevent the weather causing trouble. Holes in the top of the unit may well fill with water and freeze in cold weather, with the risk of splitting it.

35.7.2 Lifting gear

The gear which goes between the unit and the crane is a piece of lifting equipment. Because of this it must be tested to comply with the Regulations[2] and the certificate must be available. The test will normally be an overload depending on the total weight, varying from 10% at a large load to a double load at 1 t. Guidance is given in the Docks Regulations.[3] In addition to the initial test, there is a need for continuing inspection, and the Regulations lay down the minimum. However, badly treated equipment can very quickly become unserviceable and dangerous.

If a unit can be picked up at a central point, it is a convenient arrangement requiring virtually no lifting gear. But usually units have to be picked up from points near their ends. Similar points of support must be used when they are put down temporarily. If slings are used, the unit will act as a strut. This is clearly advantageous as the cost and weight of a separate spreader beam is eliminated, but the unit may not be strong enough. Where it is not very obvious, the top of a unit should be so marked. In many cases inverted lifting will cause overstressing or immediate failure.

In the design of all lifting gear, careful consideration must be given to ensuring that all the parts will mate with one another. For example, a large crane will have a large hook which will not go through the ring of size appropriate to the chain sling in use. This will normally be solved if the crane driver has acquired some large shackles but it can prove very embarrassing. Always mark units to be lifted with their weight.

35.7.3 Erection

Many units are completely stable when placed in position.

Where tall units are being used, it may be possible to balance them on edge but they could be dangerous. This is a situation which is not obvious, as a concrete unit looks very solid and stable. Thus, units should not be released from the lifting gear unless it is certain that they cannot fall. It is normal to use a push–pull prop to give stability. An anchorage will be required in the floor or ground at one end and a fixing in the unit itself. The actual placing of the units will normally be on shims – thin steel or plastic plates of various thicknesses to build up the required total thickness. After positioning, the gap will then be filled with a nearly dry mortar packed hard to take the weight of the unit down to its support. In some cases stressing may be used to make vertical units secure but until they have adequate strength in themselves, the prop should not be removed.

35.7.4 Damage

This may occur because crane drivers are not careful. But it is also caused by poor lifting schemes and badly detailed arrangements. Pins in holes must be designed not to impose a load at their extreme ends. Fixings only suitable for a direct load must not be used for an angled pull.

The handling of units may involve turning them from a horizontal to a vertical position, as it is often much more convenient to transport them from the casting place to the point of erection flat. It will normally be necessary to arrange that the bottom corner about which the unit turns should be placed in a bed of sand or a turning frame. It is very difficult for the crane driver to lift so that the heel does not slide as it is raised and this could cause damage to the unit and to anything on which it slides.

35.8 Access scaffolding

In the UK the Regulations[1] made under the various Acts make two principal requirements for scaffolding: (1) a safe working place; and (2) a safe means of access. The main contractor has an overall responsibility for scaffolding but it is also the duty of each employer to satisfy himself that it is in an appropriate condition before he sends his employees on to it. Although the erection may be subcontracted, the responsibility cannot be so simply delegated. A weekly inspection by a competent person is required by the Regulations partly because scaffolding is equipment which all construction operatives feel capable of adapting. Thus, the careful examination of it at frequent intervals is highly necessary.

The design and layout of scaffold falls into two headings. It is necessary to have sufficient space on them and sufficient convenient access, so that the operations to be done from them may be carried out quickly. Certain minimum platform widths are laid down in the Regulations referred to above.

Secondly, it is necessary to design the scaffold to cope with its own deadload and the liveload of the construction traffic, windloads, and any other environmental loads. While this is frequently done from tables for simple scaffolds, it is necessary on large scaffolds to do calculations. The stability of such a scaffold will almost invariably depend on support from the structure being constructed or maintained. Ties are fixed to it, for tube and fitting scaffold, at not more than 8.5 m spacings in both horizontal and vertical directions. It is essential that where ties have to be taken out temporarily, other ties are put in instead before this happens. Information is now available in *Guidance Notes*[5] produced by the Health and Safety Executive on some aspects and there is a works construction guide on access scaffolding.[31]

35.8.1 Types

Many scaffolds in the UK are constructed of tube and fittings, although more are now being built with proprietary modular scaffold, often called unit scaffold. Timber scaffolding was in use for many centuries, and the terminology has largely been derived from it. In the design of scaffolding it is usual to assume that the scaffold has pin joints, and that any stiffness they have does not materially strengthen the scaffold. When this is so, diagonal bracing is used on all scaffolds irrespective of size.

An advantage of the unit scaffold is the reduction of labour. It is possible to erect such scaffolds quickly, but the material costs are greater.

Where a building is being constructed of brick, or stone, a putlog scaffold can be used. A single set of standards supports the outer ends of putlogs, whose flattened inner ends rest in the brick joints. But most scaffolds are independent, using two sets of standards to support the decks. These are not truly independent as they rely on the ties described above. Information on tubular scaffolding is given in BS 5973:1981[32] (Tables 1 and 2 and Figure 1 of that document).

For maintenance or where the frame of the building is built rapidly, e.g. in steel, a suspended scaffold is used. While such scaffolds have traditionally been manually operated, there are now power-operated models available. Comparable suspended scaffolds are provided for window cleaning operations. Information may be found in BS 5974.[33]

There are now many tower scaffolds, particularly for use indoors. These are typically made of aluminium and are mobile. Almost all comply with BS 1139:1983, Part 3.[34]

The term 'scaffold' is usually taken to mean a static arrangement, as described above. But there is an ever-increasing range of alternatives of a mechanical nature. These include hydraulic platforms, scissor platforms, and mast-supported platforms. They may be an economic alternative because they are in use in one location for only a short time. Because of the ability, particularly of the hydraulic platforms, to reach awkward places, their high cost can be quite acceptable.

35.9 Temporary excavations

A great deal of construction work takes place below ground level. This may be done by digging a hole with sloping sides; by making these sides vertical less excavation is needed; however, in some cases adjacent buildings prevent the open-cut approach. Where the ground is not good rock, some form of sheeting and a supporting arrangement may be installed to achieve this. Traditionally, such sheeting was supported by a maze of struts either raking to the floor of the excavation or spanning across to the other side. Invariably, such an arrangement obstructs construction, making it more tedious and expensive. Modern trends in excavation of such holes tend toward providing a completely free working space by the use of anchorage systems in the ground immediately outside the excavation.

While digging such holes is the contractor's responsibility, and normally it is up to him to propose and carry out the method, there are many cases where an integration of the ultimate structure with the temporary problem of support to the sides can produce a much better solution. There are many modern examples where either the design *ab initio* has assumed that the structure will form part of the temporary support works as well as becoming the permanent structure in the long term, or where the contractor has decided to adopt this approach and the necessary modifications to the structure have been possible.

35.9.1 Materials

The structural framework for groundwork supports divides into three parts:

(1) The sheeting.
(2) The framing directly supporting individual sheets.
(3) The members at right angles to the sheeting which carry the reaction from the earth to some point where it is contained satisfactorily.

Consider first the sheeting. The traditional material is timber planks 37 to 75 mm thick, and this is often the most satisfactory material to use. Additionally, today steel trench sheeting is used as well as interlocking steel sheet piles. For small jobs, timber is preferred, for big ones, sheet piling. Timber is often in short lengths of 900 to 1200 mm fixed vertically.

Trench sheets and sheet piles are also fixed vertically, trench sheets being normally in the range 2 to 6 m, whereas piles may be longer, up to 13 m. Where necessary another section will be welded on to create an even longer pile. The limitation of pile size is often a function of the difficulty of driving. Depending on the type of ground encountered and the type of hammer a lighter or heavier section will be appropriate.

The framing, either horizontal or vertical members supporting the sheeting, may likewise be of timber, traditionally 225×225 mm, 300×300 mm or larger. It may be rolled-steel sections, box piles and, on occasion, precast and *in situ* concrete beams. The choice of these depends upon availability, weight, the size of the job and the strength required.

In addition to this approach, concrete diaphragm walls are often used today, becoming part of the permanent structure and being both sheeting and framing in one single structural unit. While the diaphragm wall has considerable strength and possible height, it is normally constructed in very short lengths of not more than 6 m, whose joints are typically articulated and thus can carry no moment. The support to the diaphragm wall must take this into account. There are more elaborate methods, which permit moment to be carried.

Another approach is to drive a series of soldier piles. If these are placed in prebored holes, the noise of piledriving can be eliminated. Horizontal members are then placed between them as excavation proceeds. While such soldier piles are normally steel, they can also be concrete, e.g. bored piles. If these are bored to touch each other or intersect (secant piling), a complete wall can be built though it will require walings and supports unless it can be designed as a cantilever.

The loads created by the ground are taken through the sheeting and framing to some reaction member. This may be any of the materials which have already been described. In some cases an arch or ring effect is created with a part or whole circle of piles.

A very useful modern development is the ground anchor drilled into either soft ground or rock and providing significant holding power, with no interference whatsoever with subsequent operations (see Figure 35.7). This is very similar to the older method of using a 'dead man', in which an anchorage such as a large block of concrete or a pile or two is put some distance back from the face and a rod ties the sheeting back to it. Useful information for designing with sheet piling is given in the British Steel Corporation's *Piling handbook*.[35]

35.9.2 Typical problems – a trench

The large number of trench accidents, in which men are killed and hurt through the sides collapsing, is all too well known. The treacherousness of any ground must not be underrated. In a trench, the traditional approach with hand digging is to use vertical timbers which are let down rather than driven as a trench is deepened. Individual boards are wedged from the walings to ensure they are tight against the ground. Where the trench is greater than the length of a single plank, a stepping arrangement is required whereby the lower part of the trench is narrower than the upper; alternatively, the planks slope outward as they go down. A waling is normally provided inside the sheeting, and pairs are strutted apart across the trench. When complete this forms an ideal arrangement, but the process of reaching this situation during modern mechanical excavation often leaves room for considerable improvement in safety. Various minor differences on this theme involve using trench sheeting as the facing material.

In some cases it is not necessary to have the boards touching each other and individual pairs of boards may be propped apart. It is normally necessary to prop at more than one level, though with short boards of perhaps 1 m only a single line of props may be adequate.

Trenching has been an art rather than a science. But attempts to codify existing experience and to relate it to a more engineering approach have been made and resulted in a report on trenching practice by Irving and Smith[36] which provides guidance for much of the range of trench work.

As an alternative to the traditional techniques, there are now available a number of trench-support devices. One class comprises a frame which is used to support the area being dug, and is big enough to enable pipelaying to be carried out safely. It is dragged forward as work proceeds. There are also several designed to remain unmoved until no longer needed, thus permitting a length of trench to be kept open safely.

35.9.3 Wider excavations

A trench is normally dug to construct a pipeline or duct but it may also be used to construct the retaining wall of the building. Once this is constructed and stable, the remaining earth on the inside of the structure can then be simply dug away. Constructing such a retaining wall in a trench which has many struts is time-consuming. It may also present problems of watertightness and quality on the face of the wall.

As an alternative, wider excavations will often be chosen instead of a trench. In this case, sheet piling may be driven before excavation commences, or timbering or sheeting placed as it is in progress. The support necessary to retain these will be provided by a raking shore system to the foot of the excavated hole, or by the ground anchor method mentioned above. The ground anchors will be installed as the excavation proceeds, as soon as it is possible to have access to the points on the piling where these are required.

In some cases the traditional kingpost solution will be appropriate, where horizontal struts are carried from one side of the excavation to the other but at least at one intermediate point they are supported by a vertical post fixed in the base of the excavation. It is essential to anchor this post down, as otherwise if it rises owing to any lack of horizontality in the struts, the entire cofferdam will collapse. With a diaphragm wall, the same general considerations apply. As this will invariably become part of the structure, it is often appropriate to use the floor beams of the ultimate structure as the strutting holding it in position while the building is constructed. Where access to the surrounding ground is possible, ground anchors are advantageous.

35.9.4 Loads

The design of works such as these requires consideration of the earth pressures, active and passive, because in many cases the

Figure 35.7 Ground anchors permit the clear working area shown in this picture

toe at the bottom of the sheeting is a significant part of the support. Any superimposed load which exists adjacent to the excavation must be assessed. For example, a building may already stand beside it, or it may be possible for heavy lorries to drive along close to it delivering goods to the site. The presence of water must be taken into account. While it is often difficult to make sheet piling waterproof it is equally unrealistic to expect wet conditions on the assumption that the piling will leak. Where there is water such as a river, the situation is clearly simpler, though consideration should be given to any shipping knocking the sheet piles. A large ship cannot be expected to be resisted, but where this is a risk, additional fenders may well be used to give the cofferdam some protection.

Another factor which must be taken into account is temperature change. This question has been raised on many occasions but there are no recorded cases where this has produced any significant problems. Undoubtedly, the temperature range

between extremes can be considerable, but this is taken up in elasticity of the struts and of the ground at the ends. Where conditions are extremely cold, the earth behind the cofferdam may freeze and create problems.

35.9.5 Newer approaches

Even with the best-devised schemes for digging out a hole and then building the structure within it, the time involved is great. If any part of the temporary work can be part of the permanent, time is potentially saved. This can be both the outer face against the earth and the strutting. It is also possible first to construct the piles for a building, designing the upper parts as columns, build the ground floor on these, and then construction proceeds both upwards and downwards simultaneously. In this case, the piles become the basement columns of the permanent building.

35.9.6 Water

Sheet piling is fairly watertight and is quite effective in keeping water out in normal ground conditions, though the possibility of blowouts should be checked. With a small amount of waterproofing it can be satisfactory for temporary works in a river or the sea. As an alternative to sheet piling, a pumped dewatering system may be put in, chemical treatment can be used or, where expense is little object, freezing can be adopted.

In all cases, the pressure from the water must be taken into account in the design.

References

1 Health and Safety Commission (1966) *The construction (working places) regulations*. SI 94. HSC, London.
2 Health and Safety Commission (1961) *The construction (lifting operations) regulations*. SI 1581. HSC, London.
3 Health and Safety Commission (1934) *The docks regulations*. SRO 279. HSC, London.
4 Health and Safety Commission (1984) *The construction (metrication) regulations*. SI 1593. HSC, London.
5 Health and Safety Commission (various dates) *Guidance notes*. HMSO, London.
   General series:

   GS10,   *Roofwork: prevention of falls.*
   GS15,   *General access scaffolds.*
   GS28/1, *Safe erection of structures*, Part 1: 'Initial planning and design.'
   GS28/2, *Safe erection of structures*, Part 2: 'Site management and procedures.'
   GS29/1, *Health and safety in demolition work*, Part 1: 'Preparation and planning.'
   GS29/2, *Health and safety in demolition work*, Part 2: 'Legislation.'
   GS29/3, *Health and safety in demolition work*, Part 3: 'Techniques.'
   GS29/4, *Health and safety in demolition work*, Part 4: 'Health hazards.'
   GS31,   *Safe use of ladders, step ladders and trestles.*

   Plant and machinery series:

   PM30, *Suspended access equipment.*
   PM54, *Lifting gear standards.*
6 British Standards Institution (1982a) *Metal scaffolding*, BS 1139, Part 1: 'Specification for tubes for use in scaffolding.' BSI, Milton Keynes.
7 British Standards Institution (1984) *Code of Practice for the structural use of timber*, BS 5268, Part 2: 'Permissible stress design, materials and workmanship.' BSI, Milton Keynes.
8 British Standards Institution (1982b) *Code of Practice for falsework*. BS 5975. BSI, Milton Keynes.
9 Timber Research and Development Association (1981) *Simplified rules for the inspection of secondhand timber for load-bearing use.* TRADA, High Wycombe.
10 Blake, L. S. (1967) *Recommendations for the production of high-quality concrete surfaces.* Cement and Concrete Association, London.
11 British Standards Institution (1977) *Methods of test for falsework equipment*, BS 5507, 'Floor centres.' BSI, Milton Keynes.
12 British Standards Institution (1982c) *Methods of test for falsework equipment*, BS 5507, Part 3: 'Props.' BSI, Milton Keynes.
13 British Standards Institution (1983) *Methods of testing and assessing the performance of prefabricated heavy-duty support towers*. DD89. BSI, Milton Keynes.
14 British Standards Institution (1982d) *Specification for metal props and struts*. BS 4074. BSI, Milton Keynes.
15 Hathrell, Major J. A. (1966) *The Bailey and uniflote handbook*, 2nd edn. Acrow Press, London.
16 British Standards Institution (1987) *Glossary of building and civil engineering terms*. BS 6110, Part 6, Section 6.5: 'Formwork.' BSI, Milton Keynes.
17 The Concrete Society (1986) *Formwork: a guide to good practice.* The Concrete Society, London.
18 Clear, C. A. and Harrison, T. A. (1985) *Concrete pressure on formwork*. Cement and Concrete Association, London.
19 Monks, W. (1980) *Visual concrete: design and production. Appearance Matters*, no. 1. Cement and Concrete Association, London.
20 Monks, W. (1981) *The control of blemishes in concrete. Appearance Matters*, no. 3. Cement and Concrete Association, London.
21 Monks, W. (1986) *Textured and profiled concrete finishes. Appearance Matters*, no. 7. Cement and Concrete Association, London.
22 Monks, W. (1985) *Exposed aggregate concrete finishes. Appearance Matters*, no. 8. Cement and Concrete Association, London.
23 Monks, W. (1985) *Tooled concrete finishes. Appearance Matters*, no. 9. Cement and Concrete Association, London.
24 Gage, M. (1970) *Guide to exposed concrete finishes.* The Architectural Press and The Cement and Concrete Association, London.
25 Royal Institute of Chartered Surveyors and the National Federation of Building Trades Employers (1979) *Standard method of measurement of building works*, 6th edn. RICS and NFBTE, London.
26 Harrison, T. A. (1980) *Tables of minimum striking times for soffit and vertical formwork*. Construction Industry Research and Information Association, Report No. 67. CIRIA, London.
27 Hunter, L. E. (1951) *Construction with moving forms*. Concrete Publications, London.
28 Hurd, M. K. (ed.) (1981) *Formwork for concrete*. The American Concrete Institute, Michigan.
29 Wynn, A. E. and Manning, C. P. (1974) *Design and construction of formwork for concrete structures*, 6th edn. Concrete Publications, London.
30 Wilshere, C. J. (1983) *Falsework*. Institution of Civil Engineers Works construction guide. Thomas Telford, London.
31 Wilshere, C. J. (1981) *Access scaffolding*. Institution of Civil Engineers Works construction guide. Thomas Telford, London.
32 British Standards Institution (1981) *Code of Practice for access and working scaffolds and special scaffold structures in steel*. BS 5973. BSI, Milton Keynes.
33 British Standards Institution (1982) *Code of Practice for temporarily installed suspended scaffolds and excess equipment*. BS 5974. BSI, Milton Keynes.
34 British Standards Institution (1983) BS 1139: *Metal scaffolding*, BS 1139, Part 3: 'Specification for prefabricated access and working towers.' BSI, Milton Keynes.
35 British Steel Corporation (1984) *Piling handbook*, 4th edn. BSC, Scunthorpe.
36 Irving, D. J. and Smith, R. J. H. (1983) *Trenching practice*. Construction Industry Research and Information Association, Report No. 97, CIRIA, London.

36

The Selection and Operation of Construction Plant and Equipment

Hugh C Wylde
Independent management consultant
specializing in contractors' plant

Contents

36.1 Introduction

The effective selection of construction equipment for use on any construction project relies on the proper analysis of three principal considerations:

(1) Technical efficiency, i.e. the requirement that the particular construction task be completed to the correct specification within the project timetable, by using the correct machines. In short, this identifies the plant with the ability to perform the job.
(2) Commercial and financial viability, i.e. that the cost of the equipment falls within the estimates for the specific project. In addition, where purchase of equipment is involved, the selection must meet the overall financial criteria required by the construction company as a whole.
(3) Availability – equipment can be supplied from a number of sources, i.e. existing internal holdings, the hire market or by additional purchase.

Within the UK, the last 15 years have seen the continued development of a sophisticated and efficient plant hire industry, giving contractors the option of hiring-in their plant requirements, as an alternative to buying and providing their own machines. This development has now reached the stage where it is estimated that 50% of all contractors' requirements are met by the plant hire industry. Therefore, the commercial analysis element of selection defined in (2) above should consider the relative cost of owning or hiring in any particular situation and most contractors will use a mixture of internal and external resources.

Outside the UK, there are very few countries with a comparable hire industry, with the result that, on overseas contracts, most requirements will be met by purchase. However, staff responsible for selecting and procuring plant for overseas work, should be aware of the plant hire alternative, albeit that local resources may be limited. There is no doubt that a hire market will develop gradually in other countries.

Within any construction group, the main financial assets consist of plant and equipment. Therefore, site staff should be aware that the correct selection of these items not only affects the profitability of their own contract but is also fundamental to the longer-term success of the company. As a consequence, the above selection considerations should be analysed by the appropriate level of management, as there may well be a need to resolve a conflict between what is technically desirable and what is financially viable. Although the young engineer on a particular project is concerned mainly that the equipment fulfils the technical requirements, the plant manager and/or the construction manager need to reconcile the technical/commercial factors so that the best commercial decision is taken in the overall interests of the company. On many occasions this will involve compromise.

This chapter, therefore, is designed to indicate how the selection and procurement of the correct piece of plant should be approached. There is no substitute for dealing with this selection process in a formal and organized manner. Although there is a tendency to make decisions on an informal basis, often because of time constraints, this can be dangerous. Before any organization commits itself to expenditure, especially of a capital nature, there should be an established procedure for arriving at a decision to ensure that no options are overlooked. This procedure should be applied whatever the size of company. Clearly, in the smaller firm the decision may be the responsibility of at most one or two people but in the larger construction organizations there are usually specialist departments which may be involved in different aspects of the decision. Most major contractors have a separate plant department whose primary function is to serve the plant requirements of the construction sites whilst protecting the longer-term position of the company. It is normal in the UK for the ownership of plant within a construction group to be vested in the plant department, whether it be a division or a separate operating company. It is therefore the responsibility of this department to ensure that the correct selection procedures are followed particularly where purchase is involved.

Having selected and procured the right piece of equipment, there is then a responsibility to ensure that it is used in an efficient, economic and safe manner in order to support the original decision. This chapter, therefore, also looks at the operation and control of plant in the field situation.

Figure 36.1 A Caterpillar dozer stripping a site

36.2 Plant selection

36.2.1 The principles of selection

The need for selection of construction equipment can arise from a number of situations, which vary according to the nature and size of the organization.

36.2.1.1 Requirements for new contracts or groups of contracts in the UK

During any construction work, there will be requirements for items of plant and equipment in order to carry out the work in a more cost-effective manner. This involves site staff initially in making a technical selection using the following criteria:

(1) Comparing mechanization with other more labour-intensive methods of working. For example, the degree of mechanization may be higher in the UK where labour costs are greater, than in other countries where there is an ample supply of more economic labour. However, the nature of modern construction methods, materials and components means that even overseas contracts are nowadays substantially mechanized.

(2) Comparing alternative plant methods for a particular operation. Bulk earthmoving may be carried out either with tractors and scrapers or with lorries and loading shovels/excavators depending on the outputs required.

Although site management will ultimately be responsible for selecting the appropriate type of equipment in line with the above criteria, it must be stressed that the analysis should start at the time of tender. Estimators and planners should take steps to define the scope of the construction work and to make a preliminary technical selection in order that the work can be priced. As far as cost and availability are concerned, for smaller items of general plant, especially of the nonoperated variety, the estimator will probably be aware of internal and external resources and the relevant charging structures. Therefore, at this stage there is unlikely to be a great deal of further commercial

analysis of these items. The situation will be different for larger specific pieces of equipment, e.g. tower cranes, earthmoving plant and similar operated machines. Because of limited availability internally and/or externally, a more detailed commercial analysis will be necessary to identify the source of probable supply which complies with the programme. In addition, consideration must be given to the financial aspect, which may have to include the cost of a number of factors such as erection and dismantling, transport, etc. This is particularly important if the contract is likely to entail a substantial investment in the purchase of plant, in order to establish at the outset that the financial resources are available and approved.

The estimators' information should ideally be made available to site management, who, of course, must then be in a position to review and modify the initial projections to suit conditions on the ground. At that stage, however, greater attention must be paid to the commercial aspects of selection over the whole range of plant to ensure that the best deal is achieved.

36.2.1.2 Requirements for major new overseas contracts

This is a somewhat specialized problem. For practical and financial reasons it is often impractical to move plant and equipment from one country to another. Apart from the cost of shipping, import restrictions and duty, etc. may inhibit the movement of plant. Therefore, requirements are usually met by purchasing new or secondhand, specially for the particular project. This equipment will then be maintained and operated totally at the expense of that project and will be sold on completion.

In the period when overseas work was plentiful and profitable, price tended to be a secondary consideration behind technical specification. However, the present decline in overseas work, particularly in areas such as the Middle East, together with increased competition, means that just as much emphasis must be placed on commercial considerations. The purchase price, the shipping cost, the cost of spares and the ultimate resale value are of critical importance if work is to be competitive.

Having said this, the importance of correct technical selection

Figure 36.2 A Caterpillar wheeled loader working in a quarry

should not be underestimated. Overseas contracts are often carried out in remote areas where the contractor is required to be self-sufficient. Site investigations must be intensive to determine the conditions under which the plant will operate. Problems such as extreme heat, dust, high humidity, and inexperienced operators must be allowed for both in the initial selection and in the provision of spares and maintenance facilities. A serious breakdown cannot be overcome by phoning the nearest dealer or plant hire organization for a replacement. Therefore, new models or developments should not be considered for an overseas environment until they have been thoroughly proven.

36.2.1.3 Existing machine replacement

Taking the UK, this is most likely to occur in: (1) a major construction company; (2) a plant hire organization; (3) a public authority; and (4) a process contractor, i.e. open-cast mining. Each of these organizations will own a permanent fleet of equipment and will be able to generate sufficient utilization to warrant ownership. As a result of age or obsolescence, this fleet will have to be replaced periodically. However, since the replacement will almost certainly be of the same type as the original and may even be the identical model, there need be only a limited technical assessment. An examination of modifications and improvements should be sufficient, although admittedly there is the opportunity to look at similar machines of a different manufacture. Nevertheless, one of the most important commercial factors affecting equipment selection is standardization, as substantial ownership of the same make and model of plant can achieve savings in ownership costs. For example, maintenance and operating familiarity are important considerations, as is the level of spares holdings.

It is clear, therefore, that in this particular situation, technical considerations may well be secondary to commercial ones, assuming that the original machine has been successful in operation. In fact, the construction company or public authority may well decide to divest itself, either partly or wholly, of a particular type of plant if it feels that its requirements can be more economically met from the plant hire market. The process contractor is unlikely to have this opportunity because of the specialized nature of the plant involved.

36.2.1.4 New developments

From time to time, manufacturers produce new types of machines or new models within existing ranges. Investment in new developments should be treated with great caution until it has been possible to carry out a full technical and commercial appraisal on the same basis as any other selection situation. It is important to establish that the new equipment is either technically or commercially superior to previous alternatives. The benefits should also outweigh the costs of change including items such as spare parts and retraining, etc.

Having done a theoretical appraisal, it is imperative that practical equipment tests are carried out, preferably on more than one machine and over a lengthy period. A manufacturer's demonstration is not suitable and neither is a free loan or trial as this often has strings attached. The most reliable method is to hire-in a similar unit so that it can be tested under normal field conditions over a proper trial period.

From the foregoing paragraphs it will be seen that the selection process can be summarized under the following stages:

(1) Selection of the best type of machine to do a particular site task.
(2) Selecting the source of supply.
(3) Selecting the right make and model, should purchase be necessary either as new or replacement.

Each of these stages involves a technical and commercial investigation to varying degrees and in the next subsection we look at the procedures and methods to be adopted for completing each stage.

36.2.2 Selection methods and procedures

36.2.2.1 Selection to meet contract requirements

This selection process can be broken down into six separate stages: (1) task identification; (2) preliminary selection; (3) machine output estimation; (4) machine matching; (5) output costing; and (6) final selection. One of the dangers of any analytical process is that the work involved may not be justified by the end-result. Therefore, an objective view must be taken at the outset to ensure that the project warrants detailed consideration of the alternative machines and methods available. It may well be possible, in the case of straightforward requirements, to shorten this selection process, particularly if site management have previous similar experience on which to base a decision.

Task identification. When selecting the most appropriate equipment to meet a contract requirement, consideration must firstly be given to the nature of the particular site task before the actual, alternative plant methods can be examined. To enable us to demonstrate a practical construction situation, in all the following examples we have used earthmoving as an illustration, as it is one of the main components in civil engineering work. Therefore an earthmoving job should be looked at from the following aspects: (1) duration and programme; (2) location; (3) material specification; (4) distances and site conditions, i.e. gradients, etc.; (5) weather; (6) special conditions, i.e. safety; and (7) legal and contractual. The site engineer will generally be responsible for analysing these factors, often taking into account any surveys done at tender stage.

Preliminary selection. The next step is for the site engineer to identify the various types of plant which would be suitable for carrying out the above task. At this stage, all options would be considered in relation to factors such as the nature of digging, i.e. hard or soft ground, and the travelling conditions, i.e. whether tracked or wheeled machines are most appropriate. Preliminary enquiries would also be made concerning the availability either internally or externally of various groups of machines.

Machine output estimation. Calculations must then be made of the outputs which can be achieved using two or more alternative methods. Where operations lose money because of inefficiency, it is very often because over-optimistic assessments have been made of machine outputs, usually based on unrealistic figures provided by equipment manufacturers. In this area there is no substitute for practical production experience and in a later section we shall look at outputs in more detail.

Machine matching. Logical selection implies that the machine/s finally selected will be matched to the task, i.e. the machine capacity will match the job requirement. For example, concrete-mixer output should match the placing team production, and crane capacity and speed should match the other plant with which it will work. To take a simple example, if the company owns/operates three 5-m³ tippers matched with a loading shovel with a potential output of 30 m³/h, then the shovel can complete six loads per hour. A decision must then be made whether the distance to the tipping point is such that three tippers can cope with this output. If not, more vehicles or larger ones may be needed.

Output costing. Once the outputs of various alternatives have been established, these outputs must be converted into costs. Taking our example of the loader with an output of 30 m³/h, if we assume that the hire rate or provision cost for that machine is £12/h, each metre cube will cost 40 p to load plus vehicle cost.

Final selection. The final selection is best achieved by tabulating the cost of the various alternatives, so that the most economical can be identified (see Figure 36.3). This cost comparison may also reflect other commercial factors, i.e. the charges for transporting the various alternatives together with the differing fuel costs, if it is felt that these will influence the final decision. This also assumes that all the alternatives will complete the task within the programme period. However, this is not always the case and there may have to be some compromise between cost and time, particularly if there are penalty clauses in operation. Finally, the availability of the chosen method must be checked to ensure that the equipment can be supplied to meet the programme starting date.

In the above analysis of the technical, commercial and availability factors, clearly the site management initiate the investigation and must take the final decision on machine selection. However, it is essential that they draw on the accumulated experience of the company plant organization. This department should be able to supply detailed information in the following areas:

(1) Provision of machine data from their technical library.
(2) Provision of weekly or hourly hire rates for machines based either on their charging system for internal plant or on the external market.

(3) Costs of ancillary items such as transport, erection and dismantling, and fuel consumption.
(4) The availability of equipment from internal resources or from the hire market. This includes information on the delivery of new purchases.

There has been a tendency on the part of contract management over the years to underestimate the role of the experienced plant manager and of the plant department. On a civil engineering contract, the plant content of the job value can be as high as 40%, particularly on major overseas projects, and the efficient management and operation of this plant are therefore critical to success. The field plant manager and the central plant department should therefore have clearly defined functions which are accorded a proper position in the overall management team.

36.2.2.2 *Selection for purchase purposes*

Within a construction organization, equipment will be purchased either as a replacement or as an addition to existing holdings. This investment must be preceded by a selection process to determine the most suitable make and model from a technical point of view for the applications to which the equipment will be put. In addition, the commercial implications of ownership must be calculated:

(1) In order to establish a hire rate for the machine so that it can be charged to individual sites on the basis of usage. Most construction organizations now treat their plant and equipment holdings as a separate profit centre with the result that plant is charged out internally on an equitable

Start date	Preliminary selection	Volume to be moved/ excavated	Estimated output	Time allowed for task completion (h)	Time required for task completion (h)	Total cost of machine per h	Cost per yd³/m³ per h	Length of time on site	Final selection
	(1) 'A' machine								
	(2) 'B' machine								
	(3) 'C' and 'D' machines								

Figure 36.3 Plant selection analysis chart

basis to each site. In addition, this method ensures that the full costs of ownership are recovered over the life of the machine.

(2) To enable the company to make a comparison between the cost of ownership and the cost of hiring-in the equivalent item. If an internal hire rate is calculated based on ownership costs, then this can be directly compared with the market-place to ascertain whether it would be more economical to hire.

In the UK, on a large project, in particular a joint venture, it may be necessary to purchase special equipment specifically for that job, with a view to either disposing of it at the end or alternatively writing it off over that contract. A tunnelling machine could be an example of such equipment. Even then, a technical appraisal will be carried out, although the commercial appraisal will be concerned principally with the cost of financing the purchase. It will be necessary to define the method and timing of the recovery of the ownership costs from the client.

Therefore, although the selection process which accompanies a purchase still takes into consideration the technical and commercial factors, the actual procedures and methods of selection vary from those adopted when making a site selection. In fact, they are usually an extension to that situation, i.e. the site selects the machine it requires given certain performance and charging rates; the plant department then either supplies the machine from within existing resources, or purchases new or arranges external hire.

Technical evaluation. In any purchase situation, the buyer is likely to find that there is a number of alternative makes and models of machine on the market which may meet the required specification. It is therefore essential to establish a formal procedure for evaluating these alternatives to ensure that the most satisfactory purchase is made. In the book *Construction plant – management and investment decisions*, Frank Harris discusses a systematic approach to evaluation of equipment alternatives developed by the American consultants, Kepner and Tregoe which is suitable for carrying out a detailed analysis on a major investment. However, there is again a danger that the analytical work involved may not be justified by the result. The buyer should look at each situation and decide what degree of investigation is warranted. For example, greater effort would be devoted to the selection of a large crawler crane than to a two-tool compressor. For practical purposes, there are certain steps that should be undertaken as a minimum requirement, whatever the scale of the equipment. These are as follows:

(1) Define the minimum machine characteristics required to fulfil the type of work required of it. This definition should certainly include:
   - (a) performance capabilities i.e. digging depth, speed, etc.;
   - (b) physical dimensions and weight;
   - (c) engine specification;
   - (d) chassis, i.e. wheeled or tracked;
   - (e) statutory requirements; and
   - (f) safety features.

   Also, try to define any characteristics which may not be essential but which would be desirable. For example, you may prefer automatic lubrication to be fitted, but this need not be a critical factor.

(2) Make an initial selection of the makes and models which appear to come nearest to meeting the above characteristics and obtain manufacturers' detailed specification sheets and data.

(3) The basic information must then be analysed in a chart form under the headings mentioned in (1) above so that there is an easy visual comparison, with the object of finding those models that correspond most nearly to the overall requirements. On this chart you will also indicate any special features possessed by each machine together with details of any optional extras offered by the manufacturer. All serious contenders should match all of the essential requirements and as many of the preferred requirements as possible.

(4) Contact the various manufacturers and ask to see their representatives so that you can discuss the technical specification in more detail. If you are not familiar with the firm and their products, ask for information on other existing users whom you can contact to discuss whether they are satisfied with that model. Manufacturers should be quite happy to give you this sort of information and other users are usually very willing to talk about their experiences. In addition, seek information from the manufacturer on any units owned by hire firms, in order that you can arrange for a machine to be hired-in for a trial period.

Commercial evaluation. At the time you obtained the technical information from the manufacturers, you should also have obtained an initial quotation showing the cost of the machine to the basic specification together with the price of any optional extras. However, the commercial evaluation does not end with the price. Therefore, in discussion with the representative, satisfy yourself regarding the following additional factors:

(1) The delivery period.
(2) Delivery costs.
(3) Spare parts availability and cost.
(4) Service support from dealers/manufacturers.
(5) Operator and fitter training if necessary.
(6) Fuel consumption and running costs.
(7) Driver comfort.
(8) The longer-term reliability of the company.
(9) Payment terms, i.e. discounts and deferred payments.
(10) Special financing arrangements, i.e. leasing, hire purchase.
(11) Buy-back terms against future replacement.

This may seem a mammoth exercise, but any of these items are neglected at your peril. It is no use having the machine with the best technical specification if the supplier goes out of business and spare parts become unobtainable.

One other aspect which is particularly relevant to overseas plant is that of secondhand equipment. A contractor working overseas may well find that there is plant available from other previous projects in the same area. Clearly, if it is possible to obtain good secondhand machines locally, then there must be a substantial saving on the new price, particularly if this avoids shipping costs. However, it is essential that, firstly, a full inspection and technical evaluation is carried out. In an overseas environment, it is far more difficult and expensive to rectify selection mistakes so the purchaser must be absolutely sure that: (1) the equipment is right for the job; and (2) it is mechanically sound. In most cases, a full physical examination is a necessity, possibly carried out by the manufacturers or their representatives.

Final selection. At this point the management must make a decision based on the importance they attach to the technical and commercial factors and in many cases the result will be a compromise. In the situation where a machine is being bought to carry out a specific duty on certain well-defined contracts then, clearly, the technical specification will be of overriding importance, but if standard items of contractor's plant are being purchased as part of the general fleet, then commercial considerations become more important.

Communications. In a construction company of any size, the selection of plant for a job either at the tender or operational stage requires the co-operation and communication of a number of different people from different departments working towards a common objective. Good communications are therefore imperative. If basic disciplines are to be observed, then a good deal of these communications are best conducted in a formal manner. It is worth highlighting at this stage certain essential examples:

(1) The plant department representative should be involved in tender meetings. Where the estimators lack information and costs on plant, the advice of the plant personnel should be sought in obtaining this information. Finally, the estimators/planners should produce a list of plant requirements in bar-chart form.

(2) Once a tender looks promising, the plant department should be sent a copy of the plant list, so that they can start making certain preliminary enquiries on availability, etc. Clearly, when the contract manager arrives on site this may well alter, but certain elements will remain unchanged.

(3) Once the contract has been awarded, there is often a pre-contract meeting of everyone likely to be involved. It is imperative that plant is seen to be an important part of the agenda and that the plant department is represented. Certain decisions may have to be made at this early stage on, for example, site accommodation.

(4) Once the site is under way, it is vital that plant is ordered from the plant department in writing in a formal manner using a requisition or order form, so that the essential information is clearly set out and so that the site can progress requirements in an organized manner. This form should include information on the delivery date and the estimated period of the requirement, as this is a vital factor in the commercial assessment. In addition, it should clearly specify any attachments or modifications that are required. It is amazing the attention which is paid to the specification, ordering and scheduling of materials, when substantial plant requirements can be ordered over the telephone in a haphazard fashion.

36.3 Plant hire rates

36.3.1 Hire rate philosophy

The decision to own plant and equipment involves the contractor in accepting the costs of that ownership. In addition, these costs must be recovered over the life of the machine in order that a fund is created to buy the replacement. Assuming that the equipment is only used on internal work, then the ownership costs must be paid for out of the construction work. Most plant managers are all too familiar with the site management which expects internal machines at rock-bottom prices, in order to secure work. However, there are serious dangers for the financial well-being of the company as a whole, in the philosophy that plant rates should be cut to secure work. Any failure to recover the true ownership costs from the user contracts means that the shortfall must be made good as a company from the overall construction profits.

It is also fair that each contract should bear the proportion of the ownership costs which relate to the period of their machine use. As mentioned earlier, most construction groups nowadays operate their plant activity as a separate profit centre which charges the internal sites for the machines they use at an agreed internal rate, on a similar basis to the charge received from external hire firms. This method not only overcomes the problems mentioned above but it also simplifies tendering and

internal charging. At the same time it is also possible to make a straightforward comparison between the alternative of supplying requirements from internal resources or from the external hire market.

Therefore, all levels of construction management should be aware of the basis of calculation of hire rates and the importance of ensuring that they reflect the true ownership costs. The cheapest is often not the best, even as far as outside hire firms are concerned, as it is likely that the firm will be offering unreliable plant and poor service. In the context of overall contract costs, a small saving in hire rate can easily be more than offset by the cost of lost production when plant fails to perform satisfactorily.

36.3.2 The calculation of a hire rate

The accompanying chart (Figure 36.4) defines:

(1) The factors which should be considered when calculating a rate.
(2) The method of calculation to arrive at an hourly, daily or weekly rate depending on how the machine is charged.

Machine details: Smiths Model 36 Thumper c/w spare skip
Machine cost: £5000 incl. spare skip and delivery charge

Rate factors
Economic life: 5 years
Residual value: £500
Utilization: 70% based on a 49-week working year
Interest charge: 15% per annum on reducing capital cost
Repair cost: 12% per annum on capital cost
Overheads: 10% per cent on total machine costs

Rate calculation	*Annual cost*
Depreciation – cost less residual over 5 years	900.00
Interest charge	375.00
Repair cost	600.00
Total machine cost	1875.00
Overhead cost	187.50
Total cost	£2062.50
Cost recovery period 49 weeks × 70%	34 weeks
Hire rate per week necessary to recover cost £2062.50	£60.66 per week
Hire rate based on a 60% utilization, i.e. 29 weeks	£71.12 per week

---

Driver and fuel costs are not included in the above rates. However, where appropriate an annual cost can be included in the calculation. Licensing and insurance can be calculated separately but for the purposes of this exercise have been included in the overheads.

The above rates do not include any profit element, only costs.

Assuming normal inflation, the basic purchase price of the replacement will rise over the 5-year period. Rates should therefore be reviewed annually in line with inflation.

Figure 36.4 Hire rate calculation

The costs of fuel and operator have not been included in the calculation as these are usually paid for by the site when plant is supplied internally. External hire firms will, however, provide an inclusive rate which covers these items. One factor which is often overlooked by construction managers is the element of finance. Before plant is purchased, money must be available to pay for it. The requisite cash can be provided from a variety of sources, i.e. loans, shareholders' investment, leasing or spare funds generated in the course of trading. Whatever the source, the provider of the money will expect a return or interest. Alternatively, if it is internal money, then it could be invested in a deposit account to yield a regular interest or could be used to finance further construction work. Therefore, it is important that the hire rate reflects the cost of that money. A leasing company will expect the customer to pay an interest charge in addition to paying back the capital.

The other factors in the hire calculation are mainly self-explanatory. In the case of repair and overhead costs, it is clearly difficult to quantify these accurately for each individual machine. However, by looking at historical ratios, it is usually possible to establish an acceptable method of assessing future costs in these areas.

The principal factor determining the hire rate on any machine is the utilization. Once the costs of ownership have been quantified, then these costs must be recovered over the actual working or chargeable hours as the equipment will not be earning whilst it is standing off-hire. Some of the cost elements in the hire rate, i.e. depreciation, finance, insurance and to a certain extent overheads, etc. are of a fixed nature irrespective of how much work the machine does, and can be predicted reasonably accurately over the life of the machine at the time the rate is calculated. However, the other costs, such as maintenance, are variable depending on the level of utilization. Any drop in utilization therefore increases the proportion of fixed cost in the rate so that costs overall do not decline in proportion to income. Conversely, if the targeted utilisation is achieved, then at that point most of the ownership costs have been covered and any additional utilization over and above target is extremely profitable. A hire rate is therefore very sensitive to changes in utilization, a factor which is to a certain degree outwith the control of the owner. Prediction of future utilization levels is therefore difficult and should be treated with caution unless there is a very clear historical pattern. The impact of a change in the predicted utilization is illustrated in the hire rate calculation shown in Figure 36.4.

If contract management recommend the purchase of a particular machine on the basis that they can find work for it for on average 36 weeks a year out of a possible 48 weeks, then that equates to a 75% utilization. On this basis the company may decide that purchase is warranted because the ownership costs can be recovered by the plant department over that sort of period using a hire rate which is more economical than the external rate. Suddenly the contract management finds that certain work does not materialize and the utilization level drops below the anticipated level. In this situation not only is there the danger that the company will not recover its costs at the right rate but it has also been led into a decision in favour of ownership when external hire could have been cheaper. Once a commitment to purchase has been made, it is usually expensive to try to reverse that situation.

Over recent years the construction market in the UK has been declining and future workloads have been difficult to predict. This partly explains the growth of the hire market in an environment where the contractor is reluctant to invest in plant on the basis of an uncertain utilization. Hence, the importance of understanding the hire rate calculation.

Earlier, it was stated that the estimator/contract manager requires hire rates for potential machines in order to be able to calculate the costs of output. It cannot be emphasized too strongly that equipment selection should not be made on the criteria of hire rate alone. What matters are production costs and the cheapest machine may not be the most efficient or reliable.

36.3.3 The hire v. buy decision

Several times in preceding paragraphs, mention has been made of the necessity to make the decision between buying and hiring as one of the last stages in the selection process. Normally, the decision to hire will be taken for the basic commercial reason that the external hire rate is cheaper, assuming that both internal and external sources have equipment of similar specification. However, there are a number of other practical considerations that could influence that decision in favour of external hire as follows:

(1) Short-term requirements where the utilization does not justify purchase.
(2) The distance of the site from the company operating base making transport costs on plant uneconomical, particularly in relation to short-term requirements.
(3) Bad ground conditions. The company may be reluctant to use its own machines on ground which will cause undue wear and fatigue.
(4) The supply of company machine operators could present problems, with the result that it is easier to hire a unit complete with operator.

Despite these factors in favour of external hire, every reasonable effort should be made to use internal company resources providing the cost differentials will not have a material effect on the success of a contract. Apart from the basic commercial reason that the company is recovering some of the costs incurred on that investment, there are also a number of practical reasons why internal hire may be favourable, i.e.:

(1) Better company image and presentation.
(2) Certain clients insist at time of tender that the contractor should show evidence of its plant resources.
(3) It may be possible to exert greater control over the quality and performance of internal plant.

To summarize, contract management and plant management should make a joint decision on the sourcing of equipment requirements which takes into account the overall best interests of the company.

36.3.4 The plant hire market

Apart from the decision to hire for clear commercial reasons, there may also be situations, even on overseas contracts, where hiring is the only reasonable solution. For example, very few contractors own the largest heavy lift cranes because of the massive investment involved to meet what may be intermittent requirements. In addition, this type of equipment requires a high degree of specialist back-up and operating expertise which is unlikely to be present within the construction organization. In this situation, it will still be necessary to carry out a technical appraisal to identify the most suitable plant method but the commercial appraisal will be limited to a comparison of alternative hire companies to identify the most economical package. Availability may also be critical in this selection, as there is only a limited number of machines of this size in the market.

The decision to hire any item of plant involves the contractor in entering into a formal legal contract with the hire company. In the UK most hires are conducted under the conditions laid down in the document known as the *Model conditions for the hiring of plant* published by the Contractors' Plant Hire Association. These conditions place certain clearly defined responsibilities on the hirer regarding, for example, the control of the driver. Therefore, within most construction organizations, the ordering of hired equipment is conducted by the plant department on behalf of the individual contract. Apart from giving the plant department the opportunity to look at the alternative of supplying internally, this also ensures that the company does not enter into a contract where the terms and conditions could prove onerous or unfair. Although site management are inclined to hire-in plant on their own initiative, there are many examples of situations where this has proved disastrous, particularly when something has gone wrong, i.e. in the case of an accident involving the plant. Contract management should avail themselves of the experience of the plant organization in dealing with the hire of plant and, in fact, in many organizations it is mandatory that all requirements are channelled through the plant organization.

36.4 Plant operation

36.4.1 Equipment output and production

We have established that the first stage in the selection process requires the site engineer to choose the right type of machine for the job. In other words, the method will be selected which achieves the task within the allotted contract programme. At this stage the engineer has three sources of information, on which to make the appropriate calculations and decision. These are:

(1) The parameters of the particular task as set out in section 36.2.2.1.
(2) The machine data supplied by the manufacturer.
(3) The practical experience of performing that sort of work.

In efforts to secure work it is not unknown for estimators and planners to take an over-optimistic view of the outputs which can be achieved on site because they have either relied on manufacturers' information which is unrealistic or they have neglected to take account of previous experience and the effect of adverse site conditions. Certain manufacturers' literature will even quote outputs which can be expected from machines, but these are invariably based on certain assumptions on site conditions, which may not necessarily be relevant to the task under review. Therefore, there is no substitute for carrying out your own calculations of output. Obviously, the time spent on this sort of exercise must relate to the importance and value of that particular task. However, in any situation where the progress of the job depends on materials production or handling, the output should be looked at in two distinct phases:

(1) The calculation of optimum outputs based on the operating cycle and on the manufacturers' machine data.
(2) The assessment of realistic outputs taking into account the effect of site conditions.

As a simple example, on a house-building job, the success of the brickwork operation can depend on the size and output of the mixer used. A 5/3.5 mixer will take around half a minute to mix a 0.1m³ batch of concrete but the actual production time will take longer than this, depending on the efficiency of the operative and the availability of materials, etc.

36.4.1.1 *Calculating optimum output*

Taking an earthmoving operation, as this represents a major component in civil engineering work, let us consider a task involving loading lorries using a loading shovel. The first step is to calculate the operating cycle for the loader which is as follows:

(1) Dig into spoil heap.
(2) Reverse and raise bucket.
(3) Slew and travel to vehicle.
(4) Dump material.
(5) Slew and return to dig whilst lowering bucket.

Timings for all these elements can be obtained from a combination of the manufacturers' specification details and the physical site circumstances. The next step is to calculate the size of a bucket load of material, i.e. the heaped capacity of the bucket. In this context it is most important to remember that the loose material may represent a different volume from the same material in its solid banked form, i.e. soil increases in volume when dug. In its undisturbed state, the soil is measured in 'bank' units. The moment it is disturbed, the soil swells and, when further break-up occurs, e.g. the transfer from the bucket to the vehicle, the volume increases still further. As might be expected, materials consisting of small grains such as sand or sand/gravel

Figure 36.5 An Akerman excavator loading a dumptruck

Figure 36.6 An Akerman excavator on a pipelaying contract

mixes do not increase in volume as much as, say, clay. To change bank measures into loose measures, multiply by the swell or banking factor. Examples of typical swell factors are given in Table 36.1. From the calculated cycle time and the assessed loading of the bucket vehicle it is possible to calculate the output of the loader assuming that there is a constant supply of vehicles.

Table 36.1 Typical material characteristics

Material	Swell factor
Clay	1.40
Earth loam	1.25
Gravel	1.12
Gypsum	1.74
Iron ore	1.33
Limestone	1.67
Sand	1.22
Sandstone	1.54
Trap rock	1.65

The weight and load size will vary with factors such as moisture content, degree of compaction, etc. A test must be carried out to determine an exact material characteristic.

When calculating the cycle time and output of plant or vehicles, where the essential function is transport of materials, information must be available on the anticipated routes, analysed into the following elements: (1) distance; (2) gradient; (3) rolling resistance, i.e. ground conditions; (4) average speed; and (5) the nature of the route, i.e. curved or straight. These elements may change as the job progresses, i.e. different tipping areas can be employed. There may also be variations in the elements due to seasonal factors, i.e. a hard road in the summer may become a muddy one in the winter, thereby altering the rolling resistance and the speed.

36.4.1.2 Reducing factors

To convert optimum outputs into realistic outputs, there are a number of reducing factors which are as follows:

(1) The mechanical reliability of the machine. Breakdowns can never be completely eliminated and some time must be lost for this factor, based on previous experience.
(2) It should be possible to carry out routine servicing and refuelling outside normal working hours, so that machine availability is not affected. However, this may not always be possible and it may be necessary to carry out certain major servicing jobs such as oil changes during the working day.
(3) Operator efficiency. An operator who fails to fill his bucket, who moves his bucket further than the job requires or whose co-ordination of the controls is poor, will not produce maximum output. It is difficult to quantify operator efficiency without some specific on-the-job measurement, but some assessment must be made if realistic outputs are to be achieved.
    It must also be appreciated that operators cannot work 60 min in every hour. Stops must be made for personal reasons, for refreshment, to check work, etc. To accommodate this factor, some managers work on a 50 min/h machine operating time based on European yardsticks. Clearly, in some countries there is a shortage of skilled operators and training is needed on the job. This efficiency factor may then decline.
(4) Climate and weather conditions will also affect output. If a machine is scheduled to work in a region where there are extreme climatic conditions, then allowance will be made for this when compiling the technical specification, e.g. special filters may be necessary to cope with dusty conditions or low ground pressure tracks fitted for work in bad ground conditions. However, there will still be situations where work will be stopped by the weather, as in Britain where rain can bring everything to a halt. An estimated reduction must be made in output to reflect these conditions. If necessary, particularly when working in an unfamiliar area, detailed meteorological reports may be necessary.

Having made provision for all the above factors, it should be possible to arrive at a realistic output which may be in some cases as low as 50 to 60% of the optimum. Where a number of machines are working as a team on a specific task, there is the risk that the failure of one of the components can have a serious cumulative affect on the whole operation. In this case it might be considered prudent to carry spare units or relief drivers. The extra cost may be more than offset by the improved output.

36.4.1.3 Human errors

Whatever steps are taken to determine machine outputs for costing purposes, it is worth reflecting that the tender seldom makes any provision for the human errors which can arise in the course of the actual site operations. Failure to achieve the theoretical outputs is generally the result of poor discipline in one or more of the following areas:

(1) The wrong choice of plant due to lack of prior investigation.
(2) Overloading and overspeeding without regard to long-term consequences, possibly because of ill-conceived bonus schemes.
(3) Bad layout planning.
(4) Poor site control resulting in machine queueing and shortage.
(5) Inadequate maintenance and repair programmes and facilities.
(6) Poor fuel supplies.
(7) Bad maintenance of haul roads and working areas.
(8) Poor drainage.
(9) Ineffective communication and supervision.

Clearly, similar patterns may not occur in other types of operation such as concreting and cranage which tend to be less complex. However, anyone who has been involved in the installation of a tower crane for materials handling will appreciate that this requires a high degree of planning and control, if the right machine is to be selected and operated successfully.

36.4.2 Plant maintenance

All the work put into selecting and procuring the right piece of equipment is so much wasted effort if the machine fails to achieve the estimated outputs and exceeds the target operating costs which have been incorporated in the hire rate. In order to ensure that this does not happen, site staff must, first of all, aim towards maximum availability. All equipment requires servicing and refuelling and even breaks down from time to time. At the beginning of the first shift of the day, it has to be started and checked over and at the end of the working day it has to be parked up in a secure place.

Although a number of these duties rest with the operator of the plant, most large civil engineering sites will have their own facilities for maintaining and repairing the plant, so that a high level of utilization is achieved. A consideration of site maintenance and repair falls into three categories: (1) the systems; (2) the maintenance equipment including buildings; and (3) the maintenance staff. Each of these categories must be clearly defined initially at the time of tender, particularly if the site is large enough to warrant its own facilities. Obviously, there are situations on smaller sites, when it makes sense for all but the routine servicing to be carried out off site or by visiting fitters. This may be the case where there is a plant depot nearby. However, we will consider the situation that will arise on a major contract which justifies its own maintenance facilities, which is usually the situation on overseas jobs.

36.4.2.1 Systems

This category can be divided into the following specific activities:

(1) Daily servicing which is carried out in the field.
(2) Periodic maintenance which is done in the workshops.
(3) Specialist maintenance, e.g. tyre repairs.
(4) Refuelling.

Although the daily servicing is often the operator's responsibility, this does mean that part of his time has to be set aside for doing this work, usually at the start or finish of the day. There are then the practical problems, i.e. that the operator has to start and finish half an hour before and after the normal working hours and also that the work may entail the use of specialized equipment, i.e. an air compressor for tyres or heavy-duty greasing equipment. For this reason on most large projects, particularly where earthmoving is involved, it is usually found to be more efficient to establish a special servicing team equipped with specialist gear, which can travel from machine to machine in rotation making use of working breaks during the day. This team will include one or more 'greasers' who will work from a special greasing truck. Refuelling will be dealt with in a similar manner using a fuel bowser. It is imperative that routine servicing and refuelling are carried out to a predetermined programme under the overall control of the site plant manager or foreman. It will then be his responsibility to ensure that programmes are adhered to and to liaise with the contract management.

The major periodic maintenance, i.e. oil changes, etc. is best done under workshop conditions, especially as far as major items of plant are concerned. Site staff are often reluctant to release machines for maintenance when they are under pressure to finish the job, e.g. during a period of good weather. However, this is an extremely shortsighted view. It is essential that maintenance is carried out in accordance with the timetables laid down by the manufacturer and failure to do this can result in breakdowns which cause far more disruption to the work than any time spent on maintenance. Therefore, the plant manager for the job needs to establish a programme of preventative maintenance in conjunction with site management. As the word implies, a 'preventative' programme is designed to prevent breakdowns and lost production by ensuring that the machine is kept in good running order and that any problems are identified and rectified before they reach the point of failure. This enables the workshop to plan its repair programme and to order spare parts so that they are available at the time the work is ready to be carried out. Apart from considerations of production, a planned approach also ensures that proper safety standards are maintained. The workshop manager should adopt a simple system for recording when periodic maintenance is due and when it has been completed on each machine, so that the overall position can be seen at a glance at any time. This can be done quite easily on a maintenance chart similar to that illustrated in Figure 36.7.

From time to time unplanned work will be necessary to deal with sudden breakdowns and, clearly, this will place a strain on site resources. In order to minimize the problems, steps should be taken at the start of the contract to establish the identity of the major plant manufacturers and the location of their nearest agents. Not only may they be responsible for supplying spare parts but they may be able to assist with machine repairs. A visit by the plant manager to the manufacturer at the start of the job can be invaluable in ascertaining what back-up facilities are available in the event of unscheduled difficulties. The supply and storage of spare parts is particularly important and will be covered separately.

On any civil engineering contract, there is bound to be a variety of machines and vehicles fitted with tyres. On certain larger items of earthmoving equipment such as scrapers, they are an extremely expensive item with a high wear factor. Therefore constant attention to tyre condition and replacement

Week service due – /
Service completed – X

Machine description	Plant no.	Month & week no.								
		January					February			
		1	2	3	4	5	1	2	3	4	5
23cwt dumper	A1234		X								
23cwt dumper	A1235							/			
JCB excavator	B1367			X							
JCB excavator	B1369								/		

Figure 36.7 Service record chart

is a fundamental requisite of any major civil engineering site, even if it affects only smaller machines like dumpers. Breakdown through punctures or tyre failure can be extremely costly in terms of lost production. In the UK, on all but the largest sites, it is normal to rely on a specialist tyre firm who are experienced in the repair, supply and fitting of earthmoving tyres. An agreement should be reached with one of these firms for them to carry out regular tyre inspections and to change round tyres to prevent uneven wear. This agreement should include breakdown repairs. If necessary, there should be negotiations for the firm to base a tyre fitter and breakdown truck permanently on the site. The fitting of large tyres is a skilled job which requires strict safety precautions and should under no circumstances be undertaken by inexperienced personnel. Detailed records should be kept on site of the tyre numbers and

the machines they are fitted to, as lax control of earthmoving tyres can cost a contractor thousands of pounds. In an overseas situation, where there is unlikely to be a local network of experienced tyre dealers, the contractor may well have to undertake this task, and in this case, care should be taken to employ experienced tyre fitters, if necessary expatriates, and to provide the appropriate tyre-fitting equipment.

The provision of proper refuelling facilities entails the installation of tanks with meters and locking devices for security purposes. Alternatively, a fuel bowser may be used if it is more economical to take the fuel to the equipment rather than bring the equipment to the fuel. Issues and deliveries of fuel should then be tightly controlled. Within the UK, very few problems are experienced with the quality of fuel. However, overseas local fuel supplies can vary widely in quality; for example, they can have a higher sulphur content than normal. This factor should be taken into account when planning the job and even when selecting the plant, as it can adversely affect the normal servicing and maintenance intervals.

36.4.2.2 The equipment

We have already mentioned the use of special trucks for servicing, tyre fitting and refuelling. If their use is warranted, then they should be specialist units designed for the purpose and fitted with the correct ancillary equipment. The adaptation of older secondhand vehicles may seem the cheapest method of supply but it must be remembered that reliability is critical. The right vehicles are therefore likely to be more cost-effective in the long run.

In addition, the site will need a fitting shop, a plant stores and a compound where large tyres, lubricants, etc. can be held in secure conditions. As far as buildings are concerned, the main requirement is that they should be of the sectional re-locatable variety, so that the capital cost can be spread over a number of contracts. There are now folding workshop buildings available, which can be brought on to the site on a flat-bed lorry, lifted off by crane and placed in position on a prepared concrete base with the minimum of assembly work. This workshop should then be fitted-out properly with fitters' benches, lifting equip-

Figure 36.8 A Caterpillar scraper on an earthmoving operation

ment, compressed air, light and heat so that the staff can deal with the range of equipment to be used on the job. The stores building must be fitted out with suitable racking, counter and office facilities. Many civils jobs nowadays carry a large fleet of personnel vehicles which are used by the resident engineer and by the contractor, e.g. cars, vans and landrovers. Depending on local facilities, this may necessitate a special transport fitting shop which is equipped to prepare and maintain vehicles to comply with statutory requirements. Provision for all of this back-up equipment should have been made in the tender based as far as possible on standard layouts which are applicable in modular form to a range of civil engineering contracts.

36.4.2.3 The staff

A typical major civil engineering contract will have a requirement for the following grades of maintenance staff: (1) fitters; (2) electricians; (3) welders; (4) greasers; and (5) storemen. The numbers of each will depend on the type of contract and the nature and volume of the equipment. Tunnelling work will, for example, require a high level of electrical personnel. Within any civil engineering company there should be staff ratios relevant to particular types of work using the experience of previous contracts. Personnel must be of the highest calibre as they will be expected to be adaptable and flexible with the skill to improvise. If necessary, assistance should be sought from the leading manufacturers in training fitters, especially in the types of plant they are likely to meet.

The subject of maintenance on civil engineering contracts is one that in the past has tended to be given a low priority. However, this is a recipe for disaster. The plant presence on site, in terms of a site plant manager/engineer with the proper support is vital to successful contracting. This is particularly true on overseas contracts where the equipment may be operating under arduous conditions.

36.4.3 Plant spares and stock control

Maintenance costs are a major component of the total ownership costs for any group of plant and spare parts will account for around 50% of maintenance. Labour is fairly predictable in the short term, but the use of spare parts can vary widely depending on the level of unforeseen breakdowns. It is therefore difficult to control and requires constant management attention. This is particularly true on overseas contracts where supplies are difficult to procure locally or take time to import. In this situation, it becomes necessary to carry a much higher level of spares stocks than would normally be expected on a large UK contract. In fact, there have been several cases where an overseas job has been nearing the end of the contract period, when somebody has suddenly realized that they are still carrying a large stock of spares which are unlikely to be used. Not only has cash been tied up unnecessarily in surplus stocks but there is also a situation where a forced disposal of those stocks will make it impossible to recover anything like their real value. Therefore, on any major civils contract, the ordering and control of spares is an important part of the plant operations.

36.4.3.1 Spares categories

Spare parts can be classified as follows:

(1) Cheap fast-moving items which are used regularly for routine maintenance and repairs. Filters, spark plugs, bulbs, etc. are typical examples. Because they are fast-moving, a stock must be kept so that fitters can obtain them on demand. Generally, because of their value, they cannot be identified with any particular machine repair.

However, a tight control must be kept on ordering and stock levels.

(2) Specific spares which can be readily identified to a particular machine and repair job. Generally, they are ordered only as required, but it may be necessary to carry a small stock if there is a pattern of certain repairs recurring.

(3) Consumable items such as lubricants, welding rods, gas, etc. which are fast-moving and which are part of the workshop overhead costs. It is impractical to allocate the cost of these items to any particular repair.

Wherever possible, spares should be ordered for a specific repair and the items should be costed against the individual machine so that there is a financial control not only of the overall level of purchases but also of the cost of spares consumed by any particular machine.

At the start of any major contract, plant management should investigate the local availability of spares in each of the above categories for each of the types of plant likely to be used on that job, with a view to deciding on whether it is necessary to carry site stocks and at what level. In fact, on a major project, the manufacturer can be asked to suggest a level of spares appropriate to support the plant. As an extension to this, the manufacturer may be prepared to supply a stock of basic spares on a consignment basis, i.e. they are paid for only as used. In any event, it may be wise to have a buy-back agreement on spares so that you can recover most of the cost on any items which are returned in good condition at the end of the job. However, consignment stocks and buy-back agreements do not absolve the contractor from the responsibility to store and document stocks properly.

36.4.3.2 Spares stocks

In theory, spares stocks should be kept to an absolute minimum, but in practice manufacturers have become less efficient in this area and it may be necessary to hold buffer stocks. This will depend to a large extent on the lead time when ordering. Money tied up in stocks is dead money and requires tight control. To start with, spares stocks must relate to current machines, as obsolescence can present a problem. When plant becomes obsolete or there is a change of manufacturer, the stocks of spare parts should be run down in good time so that sites and depots are not left holding dead stock.

Inevitably, spares will be ordered for a specific repair job but will not be used. In this case, they must be returned to the supplier for immediate credit or they must be put into stock. Unless this is done, there is a real danger that each fitter will build up under his bench a heap of unidentified spares over which there is no control.

36.4.3.3 Stock control

The most basic form of stock control is the bin card. A card is issued for each item held in stock and it is designed to show the part identification number, the stock code number and details of receipts and issues of that item. From the simple card it is possible to move to a computerized system, where similar information is recorded on the computer files for easy reference on screen. The type of system adopted depends on the scale of operations, but if in doubt start with a manual system until a need for something more sophisticated can be proven.

The requirements for any stock control system are ideally that it should provide information on:

(1) The numbers of each item theoretically in stock at any one time. This can be compared with physical stocks by a

count. If there is a major discrepancy, there should be an investigation.

(2) The movements into and out of stock.
(3) Minimum and maximum stock levels to control re-ordering.
(4) Suppliers and purchase prices.

The fixing of minimum and maximum stock levels is an important part of stores control. Based on the following factors: (1) purchase lead time; (2) historical consumption; (3) future consumption; and (4) economic order quantity, it should be possible to define these levels. Any person in the stores then has sufficient information to know when and how much to re-order. It is even possible to arrange automatic re-ordering when minimum levels are reached. However, this has dangers in a construction environment where circumstances can change quite quickly.

36.4.3.4 Stocktaking procedures

Regular stocktaking is vital not only as part of the annual financial accounting procedures, but also as a management control on the security and level of stocks. It can be done annually as a one-off exercise when the stores are closed and a total physical check is carried out. However, this can be difficult to organize and it may be preferable to opt for a system of progressive stocktake in which different sections of the stores are checked progressively throughout the year, so that the whole stores are completed within the 12-month period. If it is decided to alter the basis of stocktaking, this should not be done without consulting the accountants.

36.4.4 Licensing and insurance

36.4.4.1 Licensing

At one time or another, it is necessary to move heavy civil engineering plant and vehicles on the public roads, particularly dumptrucks, lorries and similar items which are designed for carrying materials. Within the UK, there are laws covering road traffic in general including various licensing arrangements. Although the UK statutory requirements are being brought into line with EEC Regulations, in other overseas countries there will clearly be other sets of legal requirements, which may or may not follow European lines. However, it is sensible to define briefly vehicle licensing as it applies in the UK. This will then provide a reminder or checklist against which overseas requirements relative to any particular country can be checked.

Licensing of drivers. Basically, in the UK no person may drive on the public roads without a licence as a result of passing a driving test. There are different classes of licence depending on the nature of the vehicle to be driven. Also, the minimum age at which a person may drive on the public roads varies with the class of vehicle. Further details of these regulations and requirements can be obtained from local licensing offices.

However, in order to drive a heavy goods vehicle a person must hold a separate HGV licence. This licence covers vehicles with a gross weight of over 7.5 t as shown in the plate fitted to the chassis. There are various classes of licence depending on the classification of the vehicle and in order to obtain the appropriate class of HGV licence a driver must pass a separate test. However, the drivers of certain types of construction equipment are exempt from the HGV requirements. This includes specialist vehicles such as road rollers, engineering plant, works trucks and digging machines.

Operator's licensing. In 1968, under the Transport Act, licensing was introduced for all firms operating goods vehicles exceeding 3.5 t gross vehicle weight. The licence includes all such goods vehicles owned by the holder of the licence together with all vehicles temporarily in his possession on loan or hire. The UK is divided into a number of traffic areas, each of which will issue a licence for the user's operating centres contained within that area. Before granting a licence, the traffic authority will take into account the user's ability to maintain and operate the vehicles on the licence in a safe condition. They must be satisfied that the user has formal procedures for inspecting and maintaining the vehicles on a regular basis although this work can be subcontracted out to a competent garage. Nevertheless, the user still carries the overall responsibility. However, under this and other legislation there are certain exemptions where vehicles perform less than 9.66 km per week on the public roads.

As far as major civil engineering sites are concerned, this whole aspect of operator's licensing has particular significance. If qualifying goods vehicles from the site are to be used on the public roads, it then becomes an operating centre and must comply with the legislation. If this entails a new licence or additions to an existing licence in that area, account must be taken of the time delay in obtaining the necessary authority. Under an operator's licence the firm is required to keep certain minimum records for a period of at least 15 months available for scrutiny by Department of Transport officials.

All heavy goods vehicles are now fitted with a plate following an inspection at a goods vehicle testing station within certain time limits. This plate sets out the vehicle's prescribed axle and gross weights. At the same time there will also be a test of roadworthiness. Every 12 months thereafter, the vehicle must be submitted for retesting for roadworthiness.

Drivers' records and hours. Within the European community including the UK there are now clearly laid down limits to the hours a person can drive a vehicle over 3.5 t gross vehicle weight both on a daily and a weekly basis. These limits should be checked for any particular country of operation. Also, the use of tachographs, i.e. automatic recording devices, is a statutory requirement.

Construction and Use Regulations. The Motor Vehicles (Construction and Use) Regulations control the specification of vehicles which can be used on the public roads. However, certain items of heavy equipment, i.e. mobile cranes and similar construction equipment which are outwith the general Regulations can be authorized under the Motor Vehicles (Authorization of Special Types) General Order 1979.

The foregoing paragraphs are intended only as a general guide. Legislation can and does change. Therefore, where a site activity involves the use of public roads, the local statutory requirements should be checked beforehand.

36.4.4.2 Insurance

Within any construction organization, insurance will be taken out to cover a wide variety of risks, some of which will relate to the operation of the plant and equipment. The need for insurance can arise for the following reasons:

(1) A statutory or legal requirement. For example, in the UK vehicles must have minimum third party cover.
(2) Contractual requirements. Certain conditions of contract, i.e. Institution of Civil Engineers conditions, will place insurance requirements on the contractor, particularly as far as plant is concerned. However, it is worth noting that, where plant is hired-in under a CPA Hire Agreement, there is a footnote to the terms in that Agreement which states that, unless otherwise agreed, the hirer is responsible for

insuring against the liabilities set out in clauses 8 and 13. This is not the same thing as a contractual obligation to insure, as insurance is not specifically mentioned in the actual conditions of hire.

(3) Commercial risk. Some companies will extend their third party vehicle insurance to cover comprehensive risks including damage to the vehicle itself. Others will consider it more economical to carry this risk themselves.

The various types of insurance found within a construction company and their effect on plant risks are set out in the following paragraphs.

Employers' Liability and Public Liability policies. These insurances have only a very limited impact on plant operations, insofar as they provide cover in respect of injury to employees and to the public. However, where plant is hired-in under the CPA Agreement with an operator, the hirer is usually asked to indemnify the owner against claims from the operator or from the public.

Contractors' All Risk policy. This policy provides cover for the contractor against loss or damage affecting the contract works. This includes any plant, equipment, tools and temporary buildings whilst they are on the site enabling the contractor to recover the full cost of anything which is damaged. The policy will also protect the contractor against any indemnity claims from plant hire firms where their plant is on hire under a CPA-type agreement. Certain specialist types of plant, e.g. tower cranes, may have to be declared separately in view of the risk involved. Also marine plant may be the subject of a separate marine policy.

Plant insurance. Major contractors are unlikely to take out blanket damage cover for their general plant as it can normally be dealt with under another policy. Similarly, mechanical breakdown is not normally insured, as the contractor is prepared to stand any risks of that nature. However, steps should be taken to ensure that equipment, including that which is hired-in, is covered for damage, whilst it is not on site under the umbrella of the Contractors' All Risk policy. For example, there is the risk when it is being stored whilst in the plant depot or is in transit between locations other than sites.

Plant and transport licensed for use on the public highway will also be covered by the Motor Vehicle policy.

Engineering insurance. Certain high-risk types of plant, i.e. cranes, hoists and pressure vessels, are subject to regular statutory inspection and testing and are covered under a separate Engineering policy. This inspection, etc. can be carried out by qualified insurance company engineers under the policy, which will then also include cover against the risks of operating that type of plant.

Motor vehicle insurance. As mentioned earlier, plant and transport used on the public road requires motor vehicle insurance. By law, this must consist of minimum third party cover for personal injury. Additional comprehensive insurance is available but many commercial firms prefer to accept their own risks where damage to their own vehicles is concerned.

This has been a brief general guide to insurance as it affects plant and equipment used in the UK. In other countries, the statutory requirements and the commercial risks may warrant an altogether different approach.

36.5 Plant control

In section 36.4.2 I highlighted the contribution made by good maintenance systems to the efficient operation of plant. In addition, there are other controls which must be exercised by the prudent contractor in order to achieve effective plant selection and operation. These can be divided into: (1) operational controls; and (2) financial controls. There is no use selecting the right plant, if it is not then used under controlled conditions. Furthermore, there must be some effective measure of the operational costs, so that management can review the financial performance of different items or types of equipment, to ensure that it compares with the projections made at the time of original selection.

36.5.1 Operational controls

On any civil engineering contract, the plant and equipment requirements should be channelled through one member of the site management team, whether this be the site plant engineer, site agent or even office manager. In this way, one person is responsible for:

(1) Issuing requisitions and orders to the company plant department or to outside hire companies.
(2) Progressing orders to ensure that requirements are delivered on time.
(3) Making arrangements for plant to be checked over on receipt. This is essential to ensure that items are in good working order and safe to operate and is of particular importance where external plant is concerned.
(4) Maintaining a schedule of items located on site. This schedule should be reviewed weekly with the site management team so that equipment is released off-hire as soon as it is finished with, thereby minimizing hire charges. It should also be possible to identify losses of, and damage to, plant and equipment at the earliest possible moment. It is quite commonplace for management to discover that quantities of equipment cannot be accounted for when the site has been finally cleared, with the result that unnecessary hire charges have been incurred.
(5) Ensuring that details are kept of working time, breakdowns and standing time. In the case of operated plant, the drivers, both internal and external, must complete weekly time-sheets which will show this information in detail and the reasons for plant not working. From these sheets, site management can weed out under-utilized plant together with items which are unreliable.
(6) Suppliers' invoices must also be checked in detail to ensure that the site is not paying for plant when it is unavailable due to breakdown, etc.
(7) Maintaining documentary records to include copies of test certificates and inspection reports relating to statutory requirements on cranes, hoists, pressure vessels and general lifting tackle. Also arranging where appropriate for the necessary tests and inspections to be carried out.
(8) Controlling the issue of fuel and maintaining fuel consumption records.

Unless this system of control is adopted, there is real danger that one section of the contract will be obtaining fresh plant whilst similar items lie idle on other parts of the site.

36.5.2 Financial control

It is essential that all items of capital plant and equipment owned by the contractor can be identified by means of a separate plant identity number. This applies to all mechanically driven items even down to electric tools. However, it is clearly impossible to identify separately items of nonmechanical plant such as scaffolding. From a financial point of view, they will therefore be treated on a group number basis. Once plant items and groups have been identified, these numbers can be used as cost centres against which it is then possible to allocate not only the internal hire income but also the direct operating costs. This entails:

(1) The completion of fitters' and drivers' time-sheets in order that they can allocate their time to the particular item of equipment on which they are working. This time can then be converted into a labour cost using standard labour rate.
(2) The allocation of purchase orders for all spares and materials to specific cost codes. Wherever possible, purchases will be costed direct to individual machines or equipment groups except where they are for stock or cannot be identified specifically.
(3) The recording of the issue from stock of fuel, etc. to individual machines.

Within a construction company, the income and direct costs against each plant number can be gathered and presented as a profit/loss situation over a defined period, i.e. annually. In this way, management can see if the hire rate used has effectively covered the operating costs and provided sufficient surplus to pay for overheads and the indirect costs.

The plant department will therefore be responsible for accumulating all the costs incurred against the company-owned equipment and for monitoring its financial performance. Some of these costs will arise at site level and there must, therefore, be an agreed system for ensuring that all costs are fed back to the central plant cost-control point. It has been found that modern computer systems are ideal for gathering and reporting plant financial information. Once the plant items and their identifica-tion code numbers have been stored on the computer together with the site location codes, it is comparatively simple for the computer to be programmed to deal with plant administration. Once the cost data has also been added to the computer files, the following information can be regularly produced: (1) the plant and equipment asset register and hire rates; (2) internal hire invoices; (3) accumulated plant costs and profitability; and (4) plant utilization. Where major sites have their own terminals linked to a central computer, they can have direct access to plant information subject to certain operating restrictions.

On large overseas contracts where firms own and operate their own plant, it is possible, with the aid of a micro-computer based on the site, to identify and control all the equipment which is supplied to that site.

Bibliography

Harris, F. and McCaffer, R. (1982) *Construction plant – management and investment decisions*. Granada.

Harris, F. (1981) 'Construction plant', *Excavating and materials handling, equipment and methods*. Granada.

Harris, F. and McCaffer, R. (1982) *Modern construction management* (2nd edn). Granada.

Powell-Smith, V. (1981) *Contractors' guide to the model conditions for the hiring of plant* (1979 edn) IPC Building and Contract Journals.

Horner, P. C. (1981) *ICE works construction guides – earthworks*. Thomas Telford.

Higgins, L. R. (1979) *Handbook of equipment maintenance*. McGraw-Hill, US.

Eaglestone, F. N. (1979) *Insurance for the construction industry*. George Godwin.

Health & Safety Executive. *Guidance notes*. This series of guidance notes includes a section on plant and machinery.

See also various government publications relating to: The Health and Safety at Work Act 1974; Road Traffic Act 1972; and Construction Regulations 1961 and 1966.

37

Concrete Construction

Keith M Brook BSc, FICE, FIHT
Wimpey Laboratories Ltd

Contents

37.1 Introduction

Concrete in its simplest basic form consists of a mixture of cement, sand, stone (aggregate) and water. Its rapid development in this century has made it one of the principal construction materials in use throughout the world.

The Romans have left evidence of their skill in making an earlier form of concrete in the remains of buildings such as the Pantheon in Rome, although evidence of the use of concrete goes back to well before Roman times.

The production of the world's first Portland cement dates back to 1824 when Joseph Aspdin of Leeds took out a patent on the cement he had produced. This cement was formed by heating a mixture of finely divided clay and limestone or chalk in a furnace to a temperature sufficiently high to drive off all the carbon dioxide. He called it Portland cement because he thought it resembled Portland stone in colour.

Since then, there have been many developments in the manufacture of cement and modern Portland cement is now made to high standards in most parts of the world.

Although concrete usually has a reasonably high compressive strength, it always has a relatively low tensile strength. Steel reinforcing bars are therefore generally provided in structural concrete to take the tensile stresses which occur. Hence the term 'reinforced concrete' which is in general use.

37.2 Concrete production

37.2.1 Storage of aggregates

Except in those cases where the components of concrete are readily available at short notice from stockpiles at pits and quarries, ground or bin storage of at least several hours' supply of all materials is called for. In some cases, where concreting operations are concentrated into a relatively small proportion of the total contract time as, for instance, in concrete road construction, the building of large stockpiles adjacent to the batching plant will normally be required.

The equipment required for building and drawing from these large ground stockpiles is expensive and the planning of site operations must be directed towards achieving a sound economic balance between the rate of consumption and the rate of supply. It is essential to avoid ground storage of excessive amounts of materials at batching plants and to rely, as far as is possible, on current production from a number of pits and quarries.

For concretes where the standards of control required are high, provision must be made in the vicinity of batching plants for storing at least two sizes of coarse aggregates and, generally, two days' supply of fine aggregate. This latter requirement stems from the fact that fine aggregates, which are normally washed, can contain up to 15% water which will drain away at a fairly rapid rate, allowing the moisture content to stabilize at a much lower figure over a period of perhaps 24 h.

Large stockpiles should be so constructed that segregation of the larger fractions from graded materials, which could cause embarrassing variations in the overall grading of the concrete aggregate, is avoided to the greatest extent possible. Where large stockpiles of coarse aggregates are to be built, consideration should be given to the use of controlled tipping of lorry loads.

Segregation of the coarser fractions of sands from the finer is not normally a major problem since the moisture content of the whole will be sufficient to prevent particle separation. It is always desirable and will often prove economic to provide adequate paved areas, laid to falls, round the aggregate stockpiles so that access to and drawing from them does not lead to contamination by dirt from the site. It could happen that the loss of aggregate into the ground beneath the stockpile exceeds the cost of providing sufficient hardstanding for aggregate storage.

This paved area will preferably extend well outside the range of any stockpile reclaiming devices and give clean access to, and room for, manoeuvre round the stockpile. However, it will not normally be necessary to pave right up to the batching plant since with most reclaiming units there will be a certain amount of dead storage which is not removed until concreting operations are virtually complete.

A requirement of any binning arrangements made round batching plants is that their walls should be built high enough to avoid overspilling of one grade of material into another; it is also necessary to ensure that overfilling, which leads to spillage and mixing of different grades of aggregate, does not take place.

It is quite often required that concretes be made from aggregates with special properties, e.g. lightweight (for low-density concretes used for better thermal insulation and lower structural weight) or heavyweight (used for such purposes as radiation shielding to nuclear reactors). Rather than complicate binning arrangements round batching plants used for the major part of the concrete it may be found advisable to use a separate batching/mixing plant of sufficient capacity to meet the requirements of the special concrete.

The sizes of stockpiles and the rates at which materials will need to be taken from them are matters which will need to be given proper consideration by job planners and site management. Among the types of equipment most frequently seen operating in Britain are:

(1) Ground hoppers fed with materials drawn from stockpiles by, for example, forward-loading shovels and transferred thence by conveyor belts to short-term storage bins above the batching plants.
(2) Drag shovels, draglines or grabs mounted atop stockpiles located at the batching plant, feeding into short-term storage bins or pulling materials into live storage areas whence they feed by gravity into weighing equipment.

With high-capacity plants these storage bins will, of necessity, hold materials for only a small number of batches, hence the need for adequate ground storage.

37.2.2 Storage of cement

It is generally required that cement be stored after grinding in high-capacity silos at the works. This is done so that the quality of the cement can be assessed for compliance with the appropriate British Standard before delivery and also to ensure, so far as is possible, that the high temperatures attained during the grinding process can be to some extent dissipated before delivery to the site.

The consequences of using physically hot cement as opposed to cooler cement are not thought to be serious, but nevertheless provision must sometimes be made for ensuring that cement temperatures are limited. In some instances where large masses of concrete are to be placed and where overheating due to the exothermic reaction of the cement must be reduced so far as possible, it may be required that the cement be circulated through coolers or kept in circulation through a number of storage units.

Storage of large quantities of cement might be called for to smooth out irregularities in delivery when construction sites are remote from cement works. The movement of bulk supplies of cement from mills to works by special cement trains rather than by road delivery offers the opportunity for short-term site bulk storage in special high-capacity wagons. However, it will generally be necessary for at least part of the journey to be made by road and facilities for pneumatic transfer from rail direct to road vehicles and then into short-term storage at the batching plant, will have to be provided.

On site, overhead cement storage silos of capacities up to about 150 t, either singly or in interconnected groups, are quite widely used. These are generally charged pneumatically by dry-air blowers capable of lifting cement to a height of 30 m or so above the ground, mounted on transport vehicles.

Some cement companies are willing to provide storage facilities on sites and their availability should be investigated.

Although bulk storage of cement in silos is generally preferable to the use of cement in bags, nevertheless there are many small construction jobs, particularly work by small builders, where bagged cement is used. Although the bags are made of strong 3-ply paper, they are not waterproof and it is therefore important that they are protected from the weather. This should preferably be done by storage in dry, well-ventilated sheds or in the case of very small amounts, the bags may be stored outside on a dry platform raised above the ground and covered with plastic sheeting, tarpaulins or similar covers.

Whether cement is stored in bulk in silos or in bags, it should be used in the order in which it has been delivered to the site. Failure to follow this elementary precaution can result in some cement being stored for too long and consequently becoming lumpy, i.e. 'air-set' and unsuitable for use.

When a cement replacement material such as pulverized fuel ash (PFA) or ground blast-furnace slag is included in the concrete being produced, it will usually be necessary to provide another silo to store it. To avoid confusion and the possible misdirection of PFA or ground blast-furnace slag into the cement silo, it is important where bulk deliveries are made, either by road or rail, that the connections at the silos are clearly marked. Where 'split' silos are used, inspection of the diaphragm to check that it is not perforated or damaged in any way is a necessary precaution to ensure that the cement and the cement replacement material stay in their respective compartments.

37.2.3 Water storage

On average, each cubic metre of concrete produced will require between 100 and 140 l of water. In addition, large quantities will be used round the plant at the end of a concreting session for the thorough cleaning down of the whole of the mixing plant and the lorries, skips, pumps or other devices used for the transport of the wet concrete.

These large quantities of water can sometimes be drawn from a water authority's mains, but quite often some local source, such as a stream, will have to be used. Permission to extract water from a stream will generally have to be sought from the local river or water authority.

Overhead storage of water in steel tanks to give a sufficient head to supply a concrete plant can be quite expensive, particularly when drawing from mains is restricted to night-time only and adequate storage has to be provided for all day-time operations. In view of this high cost of water storage a practice sometimes adopted is to excavate a hole of sufficient capacity close to the batching plant and to provide a waterproof lining, generally 1000-gauge polythene sheeting, held round the top edge by embedment in concrete or some other means. From this, stored water can be drawn by pump to supply a small header tank above the batching plant. When this mode of storage is used every precaution must be taken to avoid damage to the lining since repair will be difficult.

The large amounts of water used for cleaning down cannot be discharged into any local water course or sewer without first allowing the cement and aggregates to settle out to such an extent that the effluent is acceptable to the authority. This calls

for the building and keeping clean of a comprehensive system of settling tanks from which clear water can be decanted at the end of the line. Provision can be made to recirculate this water into the supply system, but it is doubtful, except where water is very expensive, if recovery is a practical proposition.

37.2.4 Admixtures

The use of admixtures to modify properties of unhardened and hardened concrete in one way or another is becoming increasingly common practice in the construction industry. Generally, these materials are added in very small amounts in relation to the size of a batch and it is usual to measure them by volume and feed them into the mixing water supply line from gauge tanks to the mixer.

When using liquid admixtures it is essential to maintain an adequate supply and to ensure that each batch of concrete has its proper dosage added at the correct time. This calls for the full interlocking of water and admixture supplies so that underdosing or overdosing cannot take place.

37.2.5 Batching concrete

The gauging of concrete to give mixes either of specified proportions or to meet strengths or other requirements is carried out in batching plants using weight as the unit of measurement. In almost all cases, batching plants incorporate a facility for mixing the concrete. However, many ready-mixed concrete plants have a batching facility only, mixing being carried out in truck mixers.

Although there is evidence that given proper control over operations, particularly with regard to the measurement of cement, volume batching of concrete can give high standards of quality control, this system has been almost completely superseded by weight batching. In these plants it is usual to weigh the various aggregates cumulatively in one hopper whilst the cement, any bulky additive such as PFA and water will be measured separately.

Water can be batched into a concrete mix either by weight or volume, but with the increasing tendency to use fully automatic batching/mixing plants in which the moisture content of the fine aggregate is monitored continuously and its batch weight adjusted accordingly, there is rather more emphasis on weight batching.

A small header tank is generally provided and this in turn is kept continuously charged direct from the supply mains, or by pump when the mains pressure is inadequate or supplies have to be drawn from ground storage.

Computer controls are becoming available on modern batching and mixing plants and they are designed to facilitate accurate batching of the materials and the production of concrete mixers of uniform workability. The use of computer control systems also provides a ready means of keeping accurate records of the quantities of materials used and of the weights of the constituents in each batch of concrete.

37.2.6 Mixing concrete

To achieve the full potential strength of a concrete mix it is most important that there should be a proper dispersal of the various constituents within each element of concrete. The speed at which this dispersal takes place will depend upon a number of factors amongst which are:

(1) Type of mixer and its speed of rotation.
(2) Size of charge put into the mixer in relation to the volume of the mixer drum.
(3) Degree of wear on paddles and blades.
(4) Order of charging materials.

Figure 37.1 Production of concrete for motorway base. *Note*: large aggregate stockpiles; ground storage of cement in wheeled bulk silos; small overhead cement storage; groundwater storage (in background); continuous proportioning and mixing of 4 No. aggregates and cement

Various types and sizes of mixer are available and the following are commonly used in British practice:

(1) Rotating-drum mixers:
    (a) tilting drum;
    (b) nontilting drum, including reversing drum.
(2) Split-drum mixers.
(3) Pan and annular-ring mixers.
(4) Trough mixers.
(5) Continuous mixers.

Figure 37.2 shows these mixers in general outline.

Types (1), (2) and (3) are commonly described as 'free fall' mixers since their action is derived from the falling within the drum of elements of concrete materials lifted from the bottom towards the top by a series of blades. In types (4) and (5) the mixing action is more vigorous and it is claimed that this both improves the efficiency of mixing and increases the speed at which a sufficiently high degree of uniformity can be attained.

The largest-capacity batch mixer of any type used to date in British practice has been a 6 m³ tilting-drum type with an hourly throughput of about 200 m³. Sizes of the various types available range from a few litres to about 3 m³ but there are, as noted, some exceptionally large mixers.

37.2.6.1 Rotating-drum mixers

In type (1) mixers which are normally rotated at speeds up to about 20 rpm, mixing is achieved by carrying the ingredients from the bottom of the mixer to the top by a series of paddles of differing form mounted inside the drum. As the paddles approach the top of the mixer, materials are spilled from them and fall to the bottom of the mixer whence they are again lifted

towards the top. The speed of the mixer is important in that a slow rotation extends the mixing time while too fast a rate will reduce the efficiency by tending to carry materials over.

A type (1a) tilting-drum mixer is charged at the open end with the axis of rotation of the mixer inclined upwards at an angle of about 45°. Mixing takes place whilst the drum is in this attitude. Discharge is accomplished by moving the axis of rotation through an angle of about 180°. Depression of the axis below the horizontal is carefully controlled to avoid too high a rate of discharge.

With nontilting drum mixers of type (1b), charging is via a retractable or nonretractable chute at one side of the mixer, depending on the loading arrangements, whilst discharge is brought about by inserting an inclined retractable chute at the opposite side. This chute intercepts the 'free-falling' materials within the drum and causes them to be discharged into a receiving hopper or other device. An alternative design is the reversing-drum mixer. In this the concrete is discharged after mixing by reversing the drum; thus, no chutes are needed.

37.2.6.2 Split-drum mixers

This type of mixer had its origin in Belgium but has found a good deal of favour in Britain, largely because of its simplicity, its ability to mix efficiently all types of concrete and its rapid clean discharge. The mixer drum, which rotates on a horizontal axis, is split vertically into two approximately equal volume sections. These sections are closed together during charging and mixing and retracted one from the other for discharging. Mixes are carried from the bottom to the top of the drum by cohesion and a small number of cleats secured to the inside of the drum. It has no blades or paddles of the form usually seen in mixers. Because of the large area of the gap between the two sections,

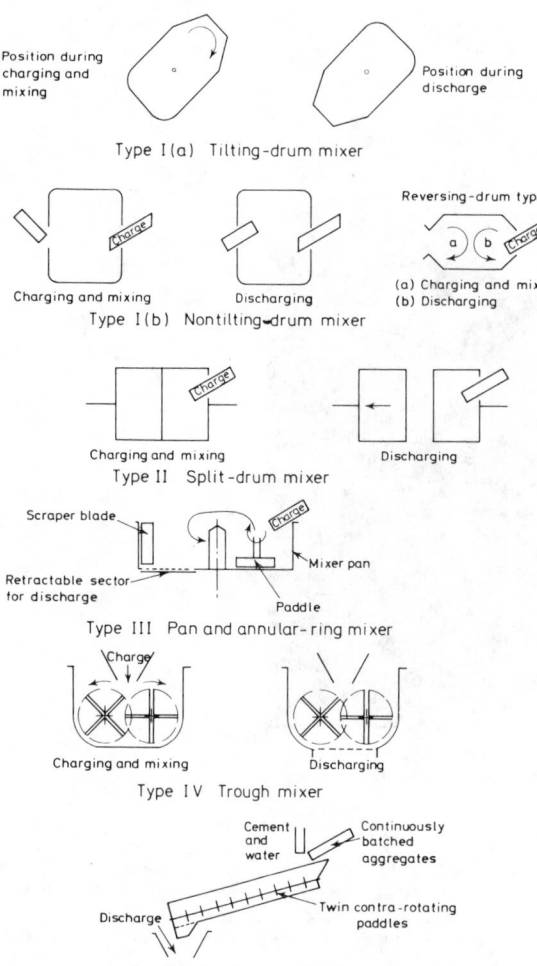

Position during charging and mixing

Position during discharge

Type I(a) Tilting-drum mixer

Reversing-drum type

Charge

Charging and mixing

Discharging

(a) Charging and mixing
(b) Discharging

Type I(b) Nontilting-drum mixer

Charge

Charging and mixing

Discharging

Type II Split-drum mixer

Scraper blade

Charge

Retractable sector for discharge

Mixer pan

Paddle

Type III Pan and annular-ring mixer

Charge

Charging and mixing

Discharging

Type IV Trough mixer

Cement and water

Continuously batched aggregates

Discharge

Twin contra-rotating paddles

Type V Continuous mixer

Figure 37.2 Concrete mixers. (*Courtesy*: John Laing and Sons Ltd)

Figure 37.3 Small reversing-drum mixer (Winget 400R). *Note*: drag shovel loading: charge and discharge at opposite ends of the mixer drum; means of controlling point of discharge

when retracted, discharge is very rapid. Mixing cycles are relatively short, in particular because of this rapid discharge, and their efficiency is said to be quite high.

37.2.6.3 Pan and annular-ring mixers

To speed up the mixing cycles and at the same time achieve a higher degree of uniformity of the mixed concrete a series of different types of pan mixer have come to be much more widely used in recent years. Mixers of this type, generally in the smaller sizes, have been used in precast concrete works for many years and have proved very efficient. Because a satisfactory degree of uniformity of the mixed concrete can be achieved with this type of mixer in a good deal shorter time than with type (1) their use has now extended to heavy civil engineering work and there has been a steady increase in the sizes of batches which can be mixed in them.

Since the concrete contained in the pans of these mixers is moved round the pan by a series of paddles whose action and speed varies with the design adopted, they have come to be known as 'forced-action' mixers as opposed to the 'free-falling' type described at (1), (2) and (3) above. The rotating paddles which mix the materials in the pan have varying forms and action and are driven from either above or below the pan. In some types the paddles rotate on their own axis as well as round the mixer pan and, with these particularly, a high degree of uniformity of the mixed concrete is claimed after only a very short period of mixing.

Because of the large diameter of these mixers, which for efficient mixing will have an average of about 150 to 250 mm depth of concrete over the area of the pan, they are seldom made with capacities greater than 2 m³. A mixer of this capacity will have a pan diameter of about 3 m.

All mixers of this type are quickly discharged by retracting a section of the bottom of the pan to allow concrete to be swept from the pan into a receiver. This feature results in their being mounted quite high in a batching/mixing plant. Each different make of this general type of mixer is claimed to have advantages over its rivals, but it is probable that the final choice will depend more upon such matters as the service afforded by the maker and delivery times. Wear and tear in pan mixers will generally prove to be higher than in a 'free fall' type but wear-resistant metals are coming to be more widely used.

Power consumption is somewhat higher than for 'free-fall' mixers, but this is compensated for by the higher hourly output in relation to the size of the mixer and its accompanying batching arrangements.

37.2.6.4 Trough mixers

This very heavy and robust type of forced-action mixer is more widely used in the production of asphalt and coated materials than for concrete, though a number of them have been built into batching/mixing plants in recent years. They can be used for mixing in much larger batches than can pan mixers since the effective depth of the concrete in them can be greater.

Mixers of this type may have a single or two contra-rotating shafts carrying blades which are so shaped and disposed as to move the constituents of the mix longitudinally along the axis of the mixer as well as round it. Mixing efficiency for short mixing cycles and hourly outputs in relation to the batch sizes are high for all types of concrete.

37.2.6.5 Continuous mixers

Recent developments in machines of this type include proportioning of all the individual materials over a series of weight feeders together with the interlocks which close down propor-

Figure 37.4 Medium-size (30m³/h) plant (Benford PB 40). *Note*: aggregate stockpiles; overhead cement storage; ground batching; annular-ring mixer.

tioning and mixing operations on the malfunctioning of any one of the feeders. These developments have resulted in there now being available batching/mixing plants which are capable of outputs approaching 300 m³/h. The continuous flow of accurately proportioned materials through an inclined-trough type of mixer with high-speed contra-rotating paddles ensures very effective mixing with a relatively low power consumption in both proportioning and mixing units of the plant. The rate of feed of the materials can be varied over wide ranges and the retention time in the mixer can be varied by changing the angle of tilt. All types of concrete are quite effectively mixed since the feeders batch constituent materials in their correct proportions on to a continuous moving collecting belt. This ensures that each element of concrete passed through the mixer is correctly proportioned.

37.2.7 Sizes of batching/mixing plants

A decision which will have to be made by a contractor at a very early stage in a contract is that concerning the size of plant to be installed. No doubt early planning of a job whilst preparing a tender will have given a clear indication of what the concreting programme is likely to be and to have set guidelines as to the number, capacity and siting of the batching/mixing plant needed to meet it.

In general it is considered preferable to concentrate concrete production at one central location in the area of maximum demand rather than at a number of dispersed points. This simplifies the problem of supply of materials and of storage. However, over sites to which access can be gained at a number of points it might be thought preferable to locate a number of smaller plants in strategic areas. The early provision of a network of substantially built temporary and permanent site roads will thus be of the greatest value. A further alternative is

to provide more than one concrete production unit of smaller size at a central point. Although it will inevitably cost a good deal more to provide a number of small plants with a combined capacity sufficient to meet peak demands than it will one large one, the ability to continue operations when one section is out of commission is particularly attractive. A further point to be considered is that high-capacity single plants producing concrete in large batches need means for carrying these heavy loads around a site and the provision of adequate means for handling the concrete into position.

In many applications of batching and mixing plants – concrete road construction is one – a very obvious requirement of such a plant is that the time occupied in erecting, dismantling, transporting and re-erecting in a new location, and any heavy cranage required for this, should be reduced to a minimum. For this reason a number of manufacturers are building into their plants either self-erecting facilities or features which will result in low costs and short downtimes for these operations. The use of harnessed electrical wiring assembled by no more involved a process than inserting heavy-duty plugs into sockets is a further aid to cost reduction.

For ease of transport it is sought to proportion units of the plant so that they are within acceptable loading gauges on public roads and can be moved without being broken down into smaller units.

37.2.8 Mixing efficiency

In order that concrete should meet the requirements of a specification and be in accordance with mix design developed in the laboratory it is necessary that mixing should continue for such length of time that all the materials are uniformly dispersed throughout a batch. This time will vary with the intensity of the mixing action. Thus it is considered that the 'forced-action' types of mixer are able to achieve a higher degree of uniformity in a shorter time than can 'free-fall' mixers. However, for the reasons given, the capacity of 'forced-action' mixers is limited; hence the largest mixers are usually of the 'free-fall' type.

The time of a mixing cycle is governed by the speed at which the operations of charging, mixing and discharging can be carried out.

Of these three components of the mixing cycle the lengthiest is likely to be the second, though mixing is, to some extent, being carried out throughout the three.

Some specifications require that there should be a minimum lapsed time between completion of charging and discharging, this time being dependent upon the measured time required to achieve the degree of uniformity within the mix suggested in BS 3963:1974 *Method for testing the mixing performance of concrete mixers.*

37.3 Ready-mixed concrete

Over 80% of site-placed concrete in the UK is produced by the ready-mixed concrete industry and this accounts for about 45% of the total cement sales for all purposes. Similar figures apply to other industrialized countries and they demonstrate the increasing tendency of contractors to rely on the supply of a concrete from a ready-mixed concrete plant as opposed to the use of site mixers.

'Building' sites are a major use of ready-mixed concrete since their demand for concrete is usually intermittent and therefore the provision of site mixing facilities is probably not economic. Also there may be a problem of finding space for site mixers and stockpiles of aggregates on many building projects, particularly in congested urban areas.

On 'civil engineering' sites, the choice between the use of ready-mixed concrete or site-mixed concrete is dependent on a number of factors. They include costs (which means a detailed study of the whole concreting operation, not just material costs), continuity of supply, rate of concrete production, site access, availability of concrete with special cements or admixtures and control of quality.

A big advantage arising from the use of ready-mixed concrete on civil engineering works is that deliveries can be brought to locations at, or very close to, the point of use. The unit price quoted will normally be to site and, provided that a reasonable standard of access road to various points is available, this will include anywhere on the site. Thus, the means of transporting large quantities of mixed concrete from a central plant to any point on the site are provided by the ready-mixed concrete operator.

There is often justification for using both ready-mixed concrete and site-mixed concrete on a site. For example, in the early stage of a contract, concrete may be required before the site mixer has been installed. It will therefore be appropriate to use ready-mixed concrete for the early work before the site-mixed concrete can be produced. There is also the situation when the quantities of concrete required in a given time are beyond the capacity of the site mixing plant. In such cases, ready-mixed concrete can be used with advantage to supplement the site production.

There are certain situations where the use of ready-mixed concrete becomes an obvious choice. The placing of large pours of concrete for large foundations in one continuous operation is an example of a construction requirement which demands large-scale production facilities for a short period. The full resources of one or possibly two large ready-mixed concrete plants may be required for a period of up to about 24 h in such a situation. There are other examples of high placing rates which call for the

use of ready-mixed concrete such as the concreting of large-bored piles.

On certain road construction sites both in the UK and abroad, ready-mixed concrete operators have entered into sub-contracts with main contractors to supply the aggregates and the cement, mix the concrete and distribute it over the site in agitator trucks or other vehicles. The merits of this arrangement depend on a number of factors, not least being the question of whether or not the main contractor can himself find the necessary amounts of aggregates in reasonable proximity to the site.

37.3.1 Plant

Ready-mixed concrete is supplied to jobs in a number of different ways depending largely on the preference of the operators, but perhaps on the terms of the specification in use.

The most widely used method in the UK, since it dispenses with the need for an expensive central mixing plant, is to dry-batch all the materials in two or three operations into the truck mixer whilst the latter is turning at 10 to 15 rpm below the batcher. On completion of the charging of the materials, water is added as required and mixing continues at the plant for at least 100 revolutions of the drum (10 to 15 min). Mixing then continues at about 1 to 2 rpm while the truck mixer travels to the site.

A second 'dry' method is to batch in the same way, but to charge the water required into a tank carried on the mixer. Only on arrival at the site is the calculated amount of water added to the materials in the mixer drum and mixing carried out for a minimum of 100 revolutions (10 to 15 min) before discharge takes place.

Most batching plants now in current use are the two-stage type of plants (i.e. a first stage where the aggregates are weighed and then taken by belt conveyor to the second stage where the cement is weighed and added to the aggregates). Compared with the earlier plants, they have the advantage of reduced height, flexibility of layout, smaller individual structures and ribbon feeding of the batched materials on conveyor belts to the truck mixers. A typical plant will provide for two cement compartments with a total of 60 to 80 t capacity and five aggregates compartments with a total capacity of 250 to 500 t.

Plant mixers are used for a small proportion of central plants supplying ready-mixed concrete; in the UK, probably 10 to 20% of suppliers' depots have central plant mixers. However, in several European countries, especially Germany and France,

Figure 37.5 Truck mixers awaiting loading at ready-mixed concrete depot. *Note*: conveyor-belt attachment on the truck mixer to the left.(*Courtesy*: British Ready Mixed Concrete Association)

the approach is different and the majority of plants are provided with central mixers.

The main advantage of a central mixing plant is that it facilitates better control over the quality of the concrete produced. However, this method of concrete production involves higher capital costs for the plant since a high-capacity mixer capable of completely charging a truck mixer or agitator truck in one, two or three batches is necessary (see Figure 37.5). The concrete is fully mixed at the plant and charged into the truck mixer whilst the latter is rotating at high speed. After complete charging, the concrete is 'kept alive' and prevented from settling and compacting in the drum by rotating at 1 to 2 rpm en route.

Most types of mixer have been used for producing ready-mixed concrete, but the type which is now used most widely in the UK is the tilting drum mixer of 3 m³ and 6 m³ capacity. This mixer can be charged and discharged quickly, it occupies low headroom, and because it revolves at relatively low speeds, it uses less power and involves less maintenance than most other types.

37.3.2 Semi-mobile plant

Construction projects of relatively short-term duration are sometimes supplied with concrete by semi-mobile plant. Whereas a mobile plant is one that may easily be moved from place to place, a semi-mobile plant may be defined as one which can only be moved from place to place with a little difficulty.

This type of plant is a compromise between the requirements of total mobility on the one hand and the efficient and economic production of concrete on the other. It is desirable to aim for a 6 m³ batch size and reasonable quantities of cement and aggregate storage. As a general guide, 60 t of cement and 150 t of overhead aggregate storage should be provided as a minimum requirement.

37.3.3 Truck mixers

Truck mixers are basically free-fall mixers mounted on a truck chassis. They can be used either for mixing and transporting the dry-batched concrete, with water added at the supplier's depot or site, or they can be used for transporting centrally mixed concrete, i.e. as agitators.

The size of drum is largely governed by the official weight restrictions on public roads which results in a nominal capacity of 6 m³ for the majority of truck mixers used in the UK.

Truck mixers fall broadly into two main types namely: (1) the drum is driven by a separate donkey engine; and (2) the drum is hydraulically driven by power take-off from the truck engine. The use of the separate donkey engine is, however, decreasing because it adds weight to the vehicle and is consequently a handicap to the need to optimize the payload.

On arrival at site, truck mixers generally discharge directly into the required location (for ground and foundation works) or into hoppers (for elevation to a structure) (see Figure 37.6).

Figure 37.6 Foundation slab being constructed with concrete fed from truck mixers. The concrete is directed down chutes at the side of the excavation. (*Courtesy*: Cement and Concrete Association)

There is an increasing tendency for truck mixers to work in conjunction with mobile pumps, and in this case the concrete is discharged into the receiving hopper of the pump. The use of a pump means that specific attention must be given to the mix design of the concrete to ensure that it is pumpable, and resulting changes to the cement or sand content can result in the imposition of a cost surcharge by the supplier.

Articulated belt conveyors for mounting beneath the discharge chute of a truck mixer are now available and these provide added height and distance when placing concrete. However, since these conveyors are mounted on the truck mixer, there is a loss of payload and consequently their use is very limited.

Whichever method of placing is adopted, safe and adequate site access must be provided for the truck mixers and the method of placing carefully organized to ensure the best use of labour. The supplier normally aims to discharge the concrete at a rate of not more than 5 min/m³ and therefore expects a truck to spend about 30 min on site. Excessive delay on site can incur an additional charge by the supplier.

37.3.4 Quality control

The principles of quality control for ready-mixed concrete are the same as those for site-mixed concrete and one of the main requirements is that the concrete should meet the specific compressive strength requirement. The strength achieved depends largely on the cement content which is the most expensive ingredient of the concrete. In the face of strong competition, each supplier aims to keep cement content to the minimum possible and there is consequently a risk that the more unscrupulous suppliers will provide a cement content which is too low. Hence the need for tight control by the customer and the regular casting of concrete cubes for testing to check that the required strength has been achieved. In cases where the customer is concerned about the durability of the concrete, he should specify a minimum cement content.

Another primary factor is the water content of the concrete which affects both the strength and workability. The required workability is normally quoted at the ordering stage and the customer is entitled to expect this degree of workability, within certain tolerances, at the time of discharge on site. If the workability at this stage is too low, the onus is on the supplier to add more water to the concrete to bring it to the specified workability. On the other hand, if the workability is correct, but the customer's site staff request the addition of more water to increase the workability, then the supplier will expect a signature to this effect since he cannot be held responsible if the required strength is not then achieved.

In the UK, quality assurance procedures for ready-mixed concrete are controlled by a national certification organization called the Quality Scheme for Ready Mixed Concrete (QSRMC). The scheme establishes technical requirements, including standards of production and quality control by manufacturers of ready-mixed concrete and provides a system for the accreditation of ready-mixed concrete plants.

The main objectives of the QSRMC can be defined as:

(1) Defining technical requirements for the quality systems to be operated by member companies in accordance with BS 5750 'Quality systems' and BS 5328 *Methods for specifying concrete including ready-mixed concrete.*
(2) Providing independent assessments of plants and their associated quality records by qualified specialists in concrete production and technology.
(3) Issuing certificates of accreditation of plants and their associated quality records.

(4) Assuring, by continuing surveillance, that the standards of the scheme are being maintained.
(5) Providing information to the construction industry on the status of accredited plants.

The QSRMC is very similar to other schemes which are now well established in Europe, particularly in Belgium, France, Germany, Holland and Sweden.

37.4 Distribution of concrete

37.4.1 General observations

Whatever the scale of the work, the problem of distributing concrete around a site will arise and a contractor has a wide range of equipment available from which to make his selection.

Where concrete is delivered from a mixing plant to the work by one mode only – e.g. truck mixer – and chuted direct into the work, this can be regarded as primary distribution only. Where it is conveyed for part of the distance by one means and for the rest by another as, for instance, an agitator truck, transferring concrete into a pump hopper, thence into the work, this can be regarded as primary and secondary distribution.

Generally, primary distribution will be by means of wheeled transport of one kind or another, but other methods for moving concrete in bulk over fairly long distances – as, for instance, pumps and conveyor belts – are coming to be used more widely as their design, versatility and standards of reliability are improved.

One problem which sometimes arises, particularly in warm weather or in hot climates, is that changes in workability can take place during transfer and occasionally cause difficulties in placing. However, the use is becoming widespread of plasticizers and set retarders in mixes as means by which a consistency suited to proper placing can be maintained for quite long periods.

In the following sections all the commonly used methods for distribution and handling are briefly described. However, as is so often the case, the personal choice of job planners based on their own experience and evidence from previous works, and, of course, availability of suitable equipment, will play a major part in determining the precise form that primary and secondary distribution are to take.

Amongst the methods of distribution which will be considered here are:

(1) Wheeled transport – for mainly horizontal movement (H).
(2) Hoists – for vertical movement (V) only.
(3) Cranes of different types for H and V.
(4) Concrete pumps for H and V.
(5) Pneumatic methods for H and V.
(6) Conveyors mainly H, but also V.
(7) Cableways for H and V.

37.4.2 Wheeled transport

37.4.2.1 Lorries

The cheapest mode of transport for concrete is undoubtedly the tipping lorry but in general the volume carried and the way in which discharge takes place makes them unsuitable for most purposes. However, they are widely used in paving operations and as flat-bottom transporters for concrete hoppers.

Widely used forms of wheeled transport used for both primary and secondary distribution are dumpers of a range of different types and sizes, flat-bottom tipping lorries (mainly for

Figure 37.7 1-t turntable dumper capable of depositing concrete over arc of 180°. (*Courtesy*: Benford Ltd)

paving operations) and lorry- or trailer-mounted concrete hoppers.

37.4.2.2 Dumpers

Small, hand-propelled dumpers – wheelbarrows and prams – still have their occasional use on even major works. Mechanization of a simple form improves the rate at which concrete can be distributed by this means, but it should be noted that the increase in volume carried, which mechanization allows, does call for rather more sophisticated means of access than just a series of planks serving for a barrow-run.

Mechanical dumpers are supplied in a range of different sizes, generally geared to multiples of the batch size of the mixer they are serving. They are often plagued by the fact that the more workable concretes which are readily discharged from them are very easily spilled unless the haul roads along which they are used are of a reasonably good standard and straight. Clean discharge of the less workable concretes, which are not so readily spilled, is more difficult to achieve and may entail a good deal of stripping out. They also have the disadvantage that concrete is literally dumped from them in one mass; this can pose a number of problems such as the effect of sudden heavy impact loading on formwork and the displacement of reinforcing steel.

To overcome these problems a range of dumpers whose rate of discharge can be controlled hydraulically by the dumper driver adjusting the angle of tilt of the body, has been introduced. A further variant of the simple end-tipping dumper is that which allows rotation of the body to feed concrete over an arc of about 180° (see Figure 37.7).

When used for loading a device for secondary distribution, such as crane skips of various forms, a high discharge level for the concrete is a marked advantage. Equipment of this type is available with a discharge height of up to 2 m and can be used for filling a large range of skips and hoist buckets.

37.4.2.3 Lorry- and trailer-mounted transporters

For the distribution of concrete from the larger sizes of batching plant – say upwards of 25 m³/h – it is necessary to consider units which are capable of carrying several cubic metres at one time. This implies the use of a chassis of not less than 5 t carrying capacity. High discharge trucks from which discharge is assisted and controlled by the angle of tilt of the body and by a

hydraulically driven paddle which propels the concrete towards the outlet, are widely used on sites.

A good deal of concrete is now moved around sites from central mixing plants by truck mixers. By this means, mixing of the concrete is continued whilst in transit, in just the same way as ready-mixed concrete would be.

In those cases where concrete is moved into its final position via crane skips of varying capacity such as, for instance, in dam construction, it is common practice to load the skips themselves at the mixer and to transport these on towed trailer bodies to the pick-up point closest to the concrete's final position. These trailer bodies will often be fitted with cradles into which the crane skips can be easily located, since fully loaded skips will have a high centre of gravity and might well prove unstable on the trailers if difficult ground conditions have to be negotiated.

37.4.3 Hoists

Hoists of various types are used solely for vertical movement of concrete and, despite the competition of a vast range of different crane types, they still play an important role in building and civil engineering construction. They range from the rudimentary platform hoist capable of lifting two or three barrows or a pram full of concrete over a short distance to an automatic hoist of up to about 2 t capacity with skip discharge into receiving hoppers at pre-set levels up to about 200 m. Where concrete in substantial volumes has to be lifted to a considerable height, over which travelling times are likely to be extended, the use of twin hoists is sometimes called for. Apart from giving better continuity of flow of concrete such an arrangement can ensure that work will not cease in the event of breakdown of a single unit.

Whilst most hoists currently in use call for the erection of a substantial tower which, for adequate stability may need to be tied-in to the structure being erected at frequent intervals, there is for some applications, such as concrete chimney construction, a trend towards the use of rope-guided hoisting systems. Here the bottom works including hoist winches, etc. and the headgear are connected only by tensioned guide ropes. Since these guide ropes pass through eyelets on the outside of the hoist cage they serve to constrain the hoist cage, within close limits, to a vertical path.

Building operations generally call for the vertical transportation of personnel as well as the materials of construction. Major movement of personnel will generally be confined to fixed periods but materials of all types will be required throughout the day. For this reason some hoist systems have the facility for on and off loading of special wheeled concrete skips as required.

The design and operation of all loading devices are subject to rigorous regulations to ensure safety in operation and it is incumbent on site management to ensure that these are enforced and that detailed inspection of the whole of the mechanism is carried out at frequent intervals.

37.4.4 Cranes

Handling materials by means of crane and skips is probably one of the oldest construction techniques known to the industry. It would be unwise for any contractor to price work on the basis of using recently developed handling methods without giving full consideration to the use of well-tried equipment such as the crane and skip.

The range of types of crane available to the construction industry is a wide one, consisting as it does of: (1) derricks; (2) tower cranes; (3) cranes mounted on crawler base machines; (4) lorry-mounted cranes and wheeled cranes; and (5) hydraulic cranes.

The lifting capacities of each type also cover a wide range.

Categories (4) and (5) have probably seen the greatest number of very recent advances but since their introduction some years ago both (2) and (3) have seen considerable changes in design and stepping-up of capacity. Advances have been made, too, in derrick design but basically, apart from changes mainly concerned with prime movers, they have remained unchanged for decades. The steam crane is now all but a museum piece; however, its ruggedness, simplicity and general freedom from breakdown still assures its place where electric power cannot be provided at reasonable cost; nor is the standard of maintenance available likely to be adequate for diesel power.

37.4.4.1 Derricks

These fall into three basic types: (1) stiff leg derricks; (2) guyed derricks; and (3) mono-tower derricks. Type (2) are mainly used for special work such as the handling of very heavy, indivisible steelwork loads. They would rarely, if ever, be used for handling concrete.

Stiff leg derricks with carrying capacities up to about 15 t and jib lengths to 45 m are, like all other lifting devices, restricted to loads much below their maxima when their radius of action approaches the maximum. For example, a derrick as above will only handle the maximum load at a radius of less than 30 m; above this figure it will be much reduced. The stays restrict the effective working arc to rather less than 270°, but this restriction rarely rules out their use, particularly when their range can be increased by mounting them on three bogie trucks set on two parallel pairs of rails.

The range of working height can be increased by mounting the lying legs, with their sole plates, on gabbards, which can again be either fixed or on rail-mounted bogies. Ground conditions are always an important factor in considering the stability of derrick cranes and they become of even greater significance when travelling derricks, mounted on tall gabbards, are called for. The need for soundly constructed rail tracks, laid to close tolerances, cannot be overstressed.

Where full-circle operation of a derrick is needed, then the mono-tower crane can be the answer, but such a machine will be restricted in working area swept by the jib from a fixed point. Mono-towers can be made to travel, but problems of the stability of a single moving tower makes this version much less attractive.

37.4.4.2 Tower cranes

Tower cranes were not widely used as an aid to building and civil engineering work in Britain until 1953 but they were coming to be widely used on the continent, particularly in France, for many years before this. Tower cranes, which can be mounted on fixed or movable towers are of two basic types: (1) luffing jib; and (2) saddle jib. Both of them can be mounted on tracked base machines, which can be self-stabilizing, or on lorries fitted with hydraulically actuated outriggers whereby the verticality of the tower can be ensured.

With both the luffing-jib and saddle-jib types of machines, load-carrying capacity at increasing radius is greatly restricted and manufacturers' specifications should be studied before making a selection. Where very tall towers are to be used, these must be tied-in to the erected structure at intervals of about 20 m. When the radius of action of a tower crane is increased by mounting the tower on rail-borne bogies, the same attention to accuracy in track laying – and of course its maintenance – (as for derricks on gabbards) is essential.

Lorry- and crawler-base-mounted tower cranes have generally lower lifting capacities than have the static and rail-mounted types, but they fulfil a useful purpose. However, their high capital cost may restrict their use to those sites where a high

degree of manoeuvrability from one working place to another, coupled with a reasonable speed of movement, is of greater significance than is high load-carrying capacity.

Restrictions need to be placed on the operation of any lifting device during high winds and this is particularly the case with tower cranes; this might be considered a serious handicap to their use. However, examination of meteorological and down-time records for the majority of sites shows availability to be generally upwards of 90%. The location of the work will of course have some bearing on this availability factor.

37.4.4.3 Crawler-base-mounted cranes

The attractiveness of this type of machine is that cranage is only one of a range of functions which the base machine can perform. Within a short space of time and with the appropriate equipment, it can be re-rigged as a face shovel or a dragline. However, to ensure stability when ground conditions are difficult, it may be necessary to use a machine with wider and longer tracks than would be needed for excavating functions alone.

The range of base machine available ensures their versatility as means of handling a range of capacities of concrete buckets, and their fast rate of slewing gives good output figures in this activity. The versatility of cranes of this type is much increased when the main jib is supplemented by a fly jib since their working range when operating close into a structure is thereby greatly increased.

The higher-capacity mobile cranes are normally those available on crawler bases but there seems to be a certain amount of rivalry between makers of both crawler-mounted and lorry-mounted cranes to achieve the accolade of highest capacity.

37.4.4.4 Lorry-mounted and wheel-mounted cranes

The former is an independent self-contained crane unit mounted on a multi-wheeled, multi-axle chassis of appropriate size which can be moved under its own power along public roads. The motive power and controls of the crane are completely separated from those of the lorry chassis on which it is mounted.

A wheel-mounted crane is one in which both the travelling of the crane over the ground and the various motions of the crane unit are controlled from one central point. Because of their rather low ground speed these units are normally to be found working in such sections of construction sites as pre-casting

Figure 37.8 Hydraulic crane handling roll-over skip. (*Courtesy*): Coles Cranes Ltd)

yards, plant yards, etc.; they are moved from site to site on low-loading lorries.

The larger sizes of lorry-mounted crane with extended or heavy-duty jibs will normally be moved in at least two units, one carrying the crane mechanism and the other, sections of the jib and fly jib to make up the required mast length.

With all but the smallest loads for which these cranes are used, it is necessary to use the in-built hydraulically actuated outriggers to achieve the necessary degree of stability and to relieve the lorry axles of the crane burden.

37.4.4.5 *Hydraulic cranes*

The rapid advances which have taken place in the design of hydraulically actuated cranes with their multi-section telescopic jibs, have possibly outstripped those of almost any other type of lifting device during recent years. Their adaptability seems to be endless and new versions, and the uses to which they can be put, are encountered with astonishing frequency.

In this type of crane the only nonhydraulically actuated function is that of hoisting, but even here hydraulic motors can be used for the cable-drum drive. Slewing, luffing, jib extension and outrigger control are all performed by hydraulic cylinders and rams alone or in combination with a form of cable drive. The speed at which the requirements for a particular type of operation can be met make hydraulic cranes one of the most favoured items of equipment at the commencement of work on many sites.

37.4.5 Concrete skips and buckets

Whilst cranes of the various types already described can be and are used for a multitude of purposes on all construction sites, their role in the handling of concrete is that of moving containers charged in one way or another from point to point on a site. These containers are known variously as skips or buckets and there are several different designs to meet differing loading and placing conditions. In size, they range from capacities of about 400 l to 9 m³ which latter size has been used abroad in concrete dam construction, mainly in conjunction with cableways. Skips are of two basic types: (1) roll-over; and (2) constant attitude. The first (see figure 37.8) is normally charged whilst lying on the ground in a horizontal position, close to the mixer or concrete transporter; it assumes a vertical position when hoisted by a lifting device. Concrete is released through the skip discharge in a controlled flow by means of a simple flap, actuated by a lever which is locked in position during transit on the crane hook to the point of deposit (see Figure 37.9). These skips can be fitted with variously shaped outlets and deflectors which permit their being used to good advantage in filling columns and walls.

The design of a skip which will give clean discharge of concrete of a range of consistencies without recourse to hammering and rodding has been the objective of all manufacturers of this type of equipment but it is probable that the answer to the problem of clean discharge lies just as much with the user as with the designer of the equipment. Cleanliness and freedom from build up of concrete in either its wet or hardened state are essential if discharge problems are to be avoided.

Constant-attitude skips are charged with concrete whilst in the same attitude as they will be during transport and discharge. Generally they are of larger capacity than the roll-over skips, but there is no well-defined range of sizes used for one type or the other.

In the larger sizes, where the weight of concrete above the outlet might be quite substantial, some form of mechanically operated device will be needed to open the gate for discharge. Simple geared clamshell gates are sometimes used at the lower

Figure 37.9 Concrete being placed by skip and compacted by poker vibrator. (*Courtesy*: Cement and Concrete Association)

end of the size range, but they tend to be rather slow in operation. For a faster and more positive gate action, pneumatically or hydraulically actuated rams are built into the structure of the bucket. Such mechanisms add substantially to skip weights.

For air operation, it is essential to have a pressure airline available at the point of discharge and, of course, time is spent in coupling up and building up sufficient pressure to actuate the mechanism. Recently developed built-in hydraulically actuated gate-opening devices whose sources of power are hydraulic accumulators charged during hoisting, are said to give very effective control over the discharge of concrete, whether or not the whole batch is to be deposited in one place.

It is clear that the self-weight of a concrete skip or bucket is a factor of much importance to the user since the effective rate of operation of a lifting device will be controlled, in part at least, by this. For this reason, manufacturers have from time to time experimented with different materials for skip construction. Very light weight has been achieved by fabricating in glass-reinforced plastic but because of inability to withstand the inevitable rough usage on sites, these skips have been far from successful. Significantly better concrete weight to total weight ratios are achieved by using magnesium in the manufacture of buckets, but only at very considerable additional capital cost. It should be noted that because of the chemical reaction between concrete and aluminium neither this metal nor its alloys are suitable for this purpose.

37.4.6 Concrete pumping

Production of concrete pumps in the UK by the Concrete Pump Company started in 1932 and by 1939 upwards of seventy machines had been built, many for export.

In these first pumps, the concrete was moved down the pipeline by a piston, actuated mechanically by a diesel engine or electric motor. The energy-consuming process of accelerating a column of concrete in the pipeline, allowing it to come to rest whilst a new charge was fed into the pump cylinder and then re-accelerating the whole, soon came to be recognized as a marked disadvantage of the process. Better continuity of movement along the line was achieved by the German device, developed independently by Torkret and Schwing, of using twin cylinders, one charging whilst the other was discharging. Torkret machines used water as the operating medium, bleeding a small proportion away at each stroke to lubricate the pump cylinder.

Schwing adopted oil as the operating medium, but of necessity used water as cylinder lubrication.

There is now a wide choice of equipment available from both the longer established manufacturers and from newcomers.

The basic differences between various makes of pump are in the actuating medium – oil or water – and in the type of valve used – gate or flapper. So far as can be seen, all types of pump, when operated in the correct manner with suitable concretes, are capable of satisfactory performance but some of the plant servicing departments associated with contractors appear to have their own preferences.

Most of the early single-cylinder ram pumps were of 150 mm bore and suited to pumping concretes with aggregate up to about 38 mm maximum size, at rates approaching 12 to 15 m³/h, dependent upon the extent to which the pump cylinder was fully charged at each stroke. However, smaller, 75 and 100 mm bore pumps were also available for pumping concrete with aggregate up to about 19 mm at rates of around 6 to 8 m³/h.

Concurrently with the development of hydraulically actuated pumps in Germany and elsewhere, Challenge–Cook Brothers in America introduced an entirely new concept of a concrete pump which they designated the Squeez-Crete pump. In this machine a short length of flexible but highly abrasion-resistant 100 mm diameter circular tube in the form of a U is charged with concrete from a hopper, this charging being assisted by maintaining a high vacuum in the surrounding chamber. Concrete is expressed from the flexible tube by rollers which, as they rotate on an axis and round the U from inlet to outlet, depress the tube, pushing the contained concrete forward towards the outlet. Since the rotating of the rollers along the U length of flexible tube is continuous, the very desirable objective of continuous flow along the flexible pipe and hence the delivery line, is achieved.

Although 150, 200 mm and even larger diameter pumps are in use, by far the majority of concrete is pumped through pipelines of 75 and 100 mm diameter. One of the main reasons for the use of the smaller sizes is that lengths of large-diameter pipe charged with concrete are very heavy and difficult to move on site. Thus, where large-diameter lines are used they will generally be associated with a semi-permanent pump and pipeline installation so that the advantage of the favourable area:wetted perimeter ratio of the large diameter line can be exploited fully. For flexibility of movement around construction sites, the smaller lorry-mounted and thus fully mobile equipment with pumping mains mounted on hydraulically actuated articulated booms is much favoured (Figure 37.10).

Generally, small-bore pumping equipment is used to handle concrete supplied either from ready-mix plants via truck mixers or concrete mixed on site and distributed in agitator trucks. By this means, the flow of concrete from the carrying unit into the re-mixer and feed hopper of the pump can be accurately controlled so as to keep the head of concrete approximately constant.

Where bigger pumping units are used they often form an

Figure 37.10 Third-floor slab being concreted using mobile pump fitted with folding boom carrying the pipeline. The concrete is supplied to the pump by truck mixer. (*Courtesy*: Cement and Concrete Association)

integral part of a combined batching/mixing/distribution unit where facilities may be provided for alternative methods of distribution as desired.

As would be expected, the effort required to pump concrete vertically is a good deal greater than that for horizontal movement – in ratios varying between 10:1 and 6:1 according to the make. In addition, extra effort is required to negotiate bends, which, when they are unavoidably incorporated into a pipeline, should be of large radius.

When pumping at their maximum range, which is usually claimed as being from 250 to 500 m horizontally and up to 80 m vertically, the output from all pumps tends to fall away, sometimes quite seriously.

It is generally preferred to use small-bore lines for vertical and larger ones for horizontal transport. However, it is not considered advisable to change diameter along a length of pipe. Where a long horizontal movement of concrete is to be followed by considerable vertical movement it might be thought advisable to transfer into a smaller-bore pipeline, through a supplementary pump.

In view of these limitations in the scope of pumping operations, the scale of the work involved should be carefully studied in advance to determine what is likely to be the most appropriate equipment. The layout of pumping points in relation to reception points should also be arranged to keep distances as short as possible.

When concreting operations include the use of expensive pumping equipment, adequate planning to ensure that utilization is as high as possible is essential. This implies that an aim should always be to have a sufficiency of work available to fully use semi-permanent installations on construction sites or to exploit the capabilities of hired-in pumps to the maximum.

37.4.7 Pneumatic placing of concrete

The use of compressed air to convey concrete from a container vessel – generally referred to as a pneumatic placer – along a pipeline to the desired location is little used now. It was, however, quite a popular technique some years ago, being used mainly for such work as tunnel linings for which it may still have some limited application. In the case of the latter, the end of the pipeline is inserted into the space between the tunnel form and the excavation rock face, and the concrete is blown in. The force with which the concrete is blown out of the pipe ensures that it is adequately compacted and this is an advantage where the restricted space prevents compaction by other means.

A particular problem with using pneumatic placers is the tendency of the concrete to segregate at the discharge point. The costs and dangers of using compressed air also tend to distract from their use. In view of these limitations and the availability of more effective modern types of concrete pumps, the use of pneumatic placers has declined.

37.4.8 Conveyor belts

Conveyor belts are very widely used for moving a vast range of materials cheaply and effectively and so it is natural that there should be interest in using them for moving concrete. The

Figure 37.11 Cableway with constant-attitude concrete bucket used in dam construction. (*Courtesy*: J. M. Henderson & Co. Ltd)

throughput of a conveyor system in relation to the power used in moving it is probably more favourable than with any other method of distribution. This is probably as true with concrete as with any other materials, but there are problems associated with conveying concrete which do not apply with other materials, e.g. the tendency for coarse aggregates in the wetter mixes to separate out from the matrix and the need for elaborate and effectively maintained belt-cleaning equipment. There is, in practical terms, no limit to the speed at which a belt carrying concrete can be run so that quite narrow ones achieve high rates of delivery.

Concrete transported on a conveyor belt can have a wide range of slump values, though it is likely that the wetter mixes could cause more problems than the drier ones. There are examples of concrete having been conveyed over long distances, but it should be borne in mind that they are generally open to the weather and that rain and strong sunshine even over a short time can affect the properties of the wet concrete.

By far the widest application of conveyor belts has been in America. There, the fundamental requirements of a conveyor system – a high degree of flexibility and lightness of individual sections – seem to have been met and a high standard of acceptability achieved.

It is an essential feature of effective conveyor-belt operation that arrangements for belt scraping should ensure complete emptying of the belt, down to the rubber, at each discharge point, whether from one belt to another or into the work. This implies the use of vulcanized rubber joints only and the immediate replacement or repair of any section of belting which is damaged for any reason. As mentioned previously, high-speed movement of wet concrete can bring about a tendency to segregation of the coarse aggregate from the matrix; to avoid this, each discharge point should be fitted with a hooded funnel within which the separated elements of the mix can be recombined before discharge on to the next section of belt or into the work.

Evenness of flow on to the conveyor is an essential to effective operation and a form of belt feeder to give this uniformity is a worthwhile investment, even though it is another piece of equipment to be maintained.

In some classes of work the final length of belting is a short one to give discharge over a fairly limited area without of necessity moving the whole system. In others, scraper blades are used to sweep a belt clean of concrete at any point along its length and then to direct it into the work. Yet another method is to use a pivoted conveyor on to which concrete can be deposited at any point along its length. This equipment gives a very wide range of placing facilities by the movement of one belt only of a system.

37.4.9 Cableways

Cableways, working singly or in pairs, have been one of the principal methods used for placing concrete in many of the world's major dams across valleys of various profiles. To give good coverage of the plan area of a dam, cableways are often set out with a head mast fixed in one position but capable of being pivoted to an angle of about 10° from the vertical. The tail mast is normally mounted on a rail-borne carriage which moves over an arc. The stability of both head and tail masts need special consideration in the light of the load to be carried – concrete and containing skip.

To ensure a steady discharge of concrete from the skip and thus gradual return of the cableway to its unloaded bucket condition, special devices for controlling the rate of discharge are essential. These have been discussed.

Cableways will normally be controlled by an operator from the head-mast position, acting under radio guidance from loading and unloading points; discharge of concrete will be controlled locally at deposit points.

37.5 Placing and compacting concrete

37.5.1 Placing

Wet concrete is set into the position in which it is to harden with the aid of crane skips, pump lines, conveyor belts, etc., and the primary objective of placing techniques should be to avoid the need for extensive subsequent movement from the point of deposit to its final position. This will normally be within shutters of one form or another.

In the majority of work, the first pours of concrete will need to be set on or against excavated or filled ground and, in order to avoid contamination of the structural concrete, a thin veneer – up to 100 mm thick – of rather lower-quality blinding concrete is laid over the formation. Besides preventing contamination, the blinding concrete, which can be set to reasonably accurate levels, can be used as a working platform for the erection of reinforcement where required. When the risk of contamination is low, for instance where concrete is to be built up on a rock formation, reasonably effective cleaning of the rock surface should be carried out but over-insistence on a high degree of cleanliness should be avoided. It is most unlikely that any structural design will include the need for a substantial degree of bond between the formation and the structure.

Wherever it is allowable, the cheapest way of filling concrete into deep excavations is via inclined chutes so positioned as to take concrete direct from a transporter into the work. Purpose-made chuting of lightweight but adequate stiffness, achieved by having a narrow but deep section, and supported as necessary, should always be used and generally it should be at an angle greater than about 30° to the horizontal. The range over which concrete can be chuted is sometimes increased by raising the transporter on a specially built movable platform, if necessary, approached by a ramp.

Frequent resiting of the chutes to avoid too great a build-up of concrete at the discharge point should be aimed at – hence the need for light weight. Alternatively, a number of chutes and loading points can be arranged around the work and used in suitable sequence.

Where heavily laden concrete transporters are brought close-in to the sides of an excavation to give maximum range for chuting operations the ability of the excavation support to sustain the heavy surcharge should be checked.

When concrete is to be filled into excavations in which water is rising, it will often be necessary to conduct this water through channels around the periphery of the work to low points or sumps from which it can be raised clear of the work.

Concrete carried out in series of lifts should have the top surface of each prepared by removing laitance whilst the concrete is still relatively unhardened yet not at risk from the action of water jets or other devices used in surface preparation.

With reinforced work, it is particularly important to remove all scraps of tying wire and other debris from joint planes by means of compressed air/water lances or other devices such as suction pipes prior to concreting lest unsightly joints additionally marred by rust stains should develop.

Concrete should be fed into shutters by a method which will give uniform distribution along a section in layers some 350 to 500 mm thick, each layer being placed and compacted before a succeeding layer is placed. By this means, the forming of unsightly segregation planes in the concrete as coarse aggregate separates from the matrix will be avoided. There will also be a reduced risk of displaced reinforcement resulting from unbalanced local concrete pressures. As successive layers are placed, each should be properly merged with the preceding layer by shallow penetration of the compacting device, generally an immersion vibrator.

37.5.2 Placing in deep lifts

At one time, the placing of concrete in lifts of more than 2 m was regarded as unsatisfactory mainly because of segregation troubles with the concrete and the difficulty of obtaining good compaction. However, the continuous placing of concrete in relatively thin sections of concrete walling and columns in heights of up to 10 to 12 m has become an accepted construction technique in recent years.

When pours of this height are adopted, in thin sections it is not usually practicable to use often recommended devices such as full-depth trunking through which to place the concrete, but it may be found advantageous to use special deflector plates, either on the discharge from the concrete skip or at the top of the shutter to direct the bulk of the concrete vertically downwards. There is always a risk that chuting concrete directly into walls between reinforcement will result in a certain amount of aggregate separation. This should be of little consequence provided the concrete is designed to be as cohesive as possible and to contain a slight excess of sand.

When placing concrete in deep lifts, it may be necessary in the case of thin sections to provide openings in the formwork at one side to permit better control over placing and compacting, especially at the bottom of the lift. These openings, which are sometimes referred to as access doors, usually range in size from about 0.3 m square to about 1 m long by 0.5 m deep and they are usually provided at about one-third to one-half the way up and at 2 to 3 m centres laterally. By feeding concrete through these access doors, any problems which arise by dropping the concrete through the full height of the lift are avoided or, at least, substantially reduced. More important, however, the access doors enable the concrete to be seen more clearly and the poker vibrators to be controlled more easily.

Access doors are especially useful for thin walls and columns where the concentration of steel reinforcement is high and where it would not be possible to see the bottom of the lift from the top. They are indispensable for sloping columns and walls because it is very difficult to pass the concrete down from the top satisfactorily with such members.

An alternative to the provision of access doors is to leave out a number of panels of formwork on one side and secure these in place as the concrete rises.

The surface appearance of the concrete may be marred by access doors since there is invariably a clear indication of their positions in the concrete face when the formwork is stripped. With care, a reasonably neat job may be obtained but, whenever possible, it is preferable to position access doors in a side of the member which will not be readily seen.

As far as appearance is concerned, an advantage of deep lift construction is the absence of horizontal construction joints. At best, it is not easy to disguise these joints and sometimes they can be very unsightly. It is for this reason that many columns and walls are now concreted in one lift.

The absence of horizontal construction joints is not the only factor influencing the improved appearance of deep lifts. For example, blowholes on the surface of concrete are generally much more numerous towards the top of a lift regardless of whether the lift is only 2 or 3 m high or whether it is very deep. Consequently, a deep concrete member which is cast in a number of successive lifts may exhibit a band of blowholes at the top of each of these lifts; if only one deep lift is used, the band of blowholes will occur only once.

When walls are poured in deep lifts there is always a risk of excessive water gain at the top surface due to the rising of water from concrete in the lower levels under the high pressures existing there. This water gain, which can often result in a lower-quality concrete marred by less completely closed surfaces and perhaps colour change, is best countered by using somewhat drier batches of concrete as work approaches the top of the lift.

37.5.3 Joints in concrete structures

Joggles set into the top of a concrete lift ostensibly to give a key for succeeding layers are a potent cause of trouble at horizontal construction joints since they are difficult to clean properly and hold water in excess; they should therefore not be used. Instead, joint planes should be slightly crowned to shed any water used in washing down prior to concreting.

Whether or not to use a cement grout or a cement/sand mortar of creamy consistency at joint planes has long been a subject of controversy. Some engineers prefer to use a thin layer of mortar over the area of a joint prior to concreting; others prefer to have the surface dampened but to use no mortar. If mortar is used it is difficult to ensure that it is only thinly spread and worked well into the top of the previous lift; the excessive amounts which may accumulate can give rise to an undesirable amount of shrinkage at the joint plane and perhaps slight colour differences. It is probably wiser to omit mortar – except where a thin (6 to 12 mm) layer can be properly brushed into the hardened concrete of the previous lift – and to rely on complete compaction of the lowest layer of concrete to give the desired quality of construction joint with as high a degree of bond between the two as it is possible to obtain.

The extravagant use of water bars of different types in concrete construction below ground is perhaps indicative of engineers' and architects' lack of confidence that joints can be made to resist the passage of water without their use. They may be desirable and perhaps necessary where water pressures are high but in many instances it has been found that joints incorporating water bars have tended to show traces of water seepage, sometimes severe, whereas those without such a device are much less prone to trouble.

When such faults are investigated it is generally found that the mode of installation has been faulty rather than that the water bar has been inadequate. They must be installed in properly designed movement joints built precisely in accordance with the drawings; construction methods must be such as to ensure that this is done. Horizontal joints in particular, call for a method which will ensure that correct positioning of the bar is maintained; in vertical joints they must be so secured in position that they remain sensibly normal to the joint plane.

Where joints are built in accordance with known good practice it will rarely be necessary to install water bars in horizontal joints when the effective head is less than 5 to 6 m; they may, however, be needed in vertical joints as indicated. Joints should be detailed as shown in Figure 37.12. To avoid the need for water bars in the vertical joints of, say, a basement perimeter wall it will be preferable either to cast *in situ* or precast sections some 5 to 6 m long leaving 1 m gaps between sections. These gaps will be concrete with as dry a mix as can be effectively placed after an interval of more than about 3 weeks. An effective flexible seal built on to the pressure side of the wall should then ensure water tightness.

37.5.4 Underwater concreting

In many civil engineering works it will be necessary to place concrete in situations where it will be under water either all or most of the time. Offshore work between tidal limits is an instance in which concrete may have to be placed during a short period of slack water and protected from the effects of scour very soon after placing and whilst still unhardened. Other work

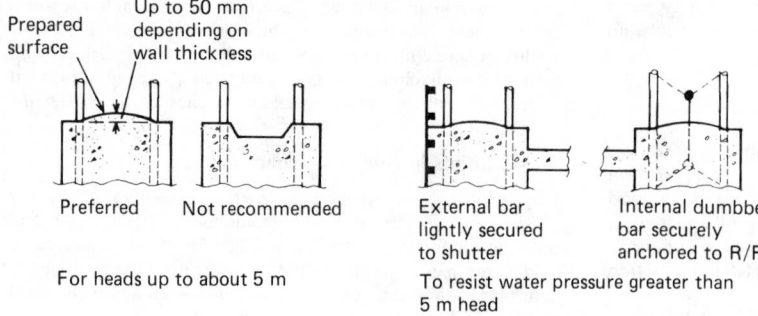

(i) Horizontal Construction Joints

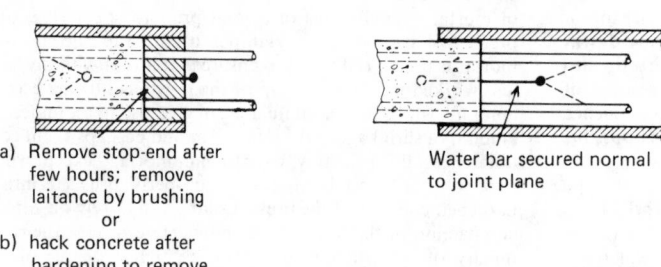

(a) Remove stop-end after few hours; remove laitance by brushing
or
(b) hack concrete after hardening to remove laitance

Water bar secured normal to joint plane

(ii) Vertical Joint to Resist Water Pressure in Continuous Construction

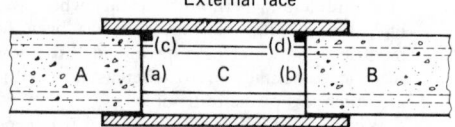

(1) Prepare (a) and (b) as in continuous construction
(2) Interval between A-B and C not less than 21 days
(3) Seal vertical joints at (c) and (d) under favourable conditions

(iii) Vertical Joint for Non-continuous Construction
Pressures up to 5 m Water Bars Installed above this Pressure

Figure 37.12 Details of construction joints

carried out in tidal and nontidal waters will call for the placing of concrete in parts of structures which are permanently under water. This section concerns itself with this type of work.

Since the hardening of concrete is a purely chemical reaction which can take place only in the presence of water, its setting under water is in no way inhibited. What can happen, however, is that improper methods for placing the concrete into position can cause a serious reduction in quality by virtue of the leaching-out of some of the cement. The objective of underwater concreting techniques is therefore to protect the concrete both during placing and whilst still unhardened, against undue cement loss.

Concrete cannot be placed in fast-running water without recourse to devices such as cofferdamming, but where current velocities are low then it can be placed satisfactorily provided certain precautions are taken both in the proportioning of the mix and in its placing. Some protection against scour or too wide spreading of the concrete mass can be provided by setting the concrete into steel shutters or perhaps walls built up with bagged concrete. Shutters will normally be used if reasonably

accurate concrete levels and shapes are to be achieved by rough screeding carried out under water by divers. Where the shape of the mass of concrete is of no particular concern, its only function being to provide a firm base from which the structure proper can be erected, then shuttering is often dispensed with and the concrete mass is allowed to adopt its own angle of repose.

As an alternative to using concrete, it might sometimes be preferred to set single-size coarse stone into a heap of the approximate shape required and then to bind the mass together with a sand–cement mortar fed into the interstices between stones.

This particular technique is generally referred to as the grouted aggregate process and various patented methods, for which the proprietors put forward their various claims, are available. Whichever method is selected it is important that the work be carried out by competent staff, well trained in the art of producing a solid mass of 'concrete' by this apparently simple process.

Whether the choice of an underwater concreting technique

falls on the use of conventional concrete or on a grouting method, the clearing of unsuitable material from beneath the structure will be essential. It can be carried out by means of a diver-directed grab, suctionpipe or airlift pump. In those situations where resilting will rapidly take place, concreting should be started as soon as possible after the base has been cleared.

37.5.4.1 Methods for placing concrete

Concrete may be placed under water through a steel tremie tube which will have a diameter some 5 or 6 times the maximum size of the aggregate in it. It will preferably be very workable so that it will flow easily down the tremie pipe and over a large area – up to a radius of about 2.5 m – once it leaves the bottom of the pipe. Where bigger areas than those are required it is normally considered better practice to use more than one pipe or to repeat the whole tremie concreting process at a number of points rather than attempt to cover a large area by moving a single pipe from place to place during one operation.

In operation, the tremie tube, which will often be made of a number of sections to allow easy adjustment of its outlet height above the base of the work, will have a receiving hopper at the top. This hopper, together with the tremie tube, will preferably be slung from a crane or overhead structure and fed with concrete from either a crane skip or a pump.

At the start of operations, the tremie pipe is set hard down on the base of the work, water rising up the pipe. A travelling plug of one of a variety of types is set into the outlet from the concrete hopper. As concrete is poured into the hopper it forces the plug down, displacing water from the tremie and preventing the concrete from falling directly through the water. When the plug has been driven down to the bottom of the tube, the tremie is lifted slightly whereupon the weight of concrete finally forces the plug from the pipe, allowing concrete to well out in all directions, including upwards. Further concrete passing down the tremie is encouraged to flow outwards by slightly raising the pipe, but without allowing its end to lift clear of the mass of concrete.

When an area has been brought to the required height, the tremie should be cleared and moved to a new location, where the sequence of operations will be repeated.

For concreting in deep underwater lifts, which will require reducing tremie lengths, provision must be made for supporting the lower lengths of tremie tube whilst adjustments are made.

An alternative to the use of a tremie tube is a special type of bottom-opening skip. This will be fitted with a canvas cover to protect the concrete during lowering through water and with an enveloping metal shroud which will restrain a sudden surge of wet concrete as the flap-type doors are unlatched, only allowing it to flow as the skip is raised slowly off the bottom or away from previously laid concrete.

Generally speaking, skip placing, because of its intermittent nature, will be technically less satisfactory than tremie placing. Also because of the high cost of such skips and the relatively slower rate of placing, the economics of the method may be less favourable than for tremie work. However, where tolerably accurate surface levels without recourse to heavy underwater screeding are required, work may well be easier with skips. Whatever method is adopted, a consistent supply of concrete which has a sufficiently high cement content to cater for the inevitable loss through leaching is essential to the successful completion of this type of work.

Grouted aggregate method. A useful material for this class of work will be graded 75 to 40 mm and have not more than 10% fines; this is often available as rejects from concrete aggregate processing screens. It should be free from any clay and dusty coatings.

Setting stone into position on a prepared base and grouting should be carried out as quickly as possible, lest silt be deposited over the mass of aggregate or other forms of contamination, such as algae growth which would prevent adequate bond between matrix and stone, should develop. An advantage of using this technique in flowing water is that in passing through the stone mass its velocity will be lowered; this will reduce the likelihood of serious cement teaching.

When using this method of underwater construction the base of the structure should be cleared of any deposits of silt, with an airlift, immediately prior to laying a 75 mm thick bed of sand or pea gravel. Grout pipes are set vertically into the area to be concreted at intervals before stone is tipped round them. These pipes are subsequently coupled through flexible hosing to the grout pumps.

As soon as stone setting is completed – no compaction of any kind will be required – a cement grout of creamy consistency is pumped in turn into the grout pipes and followed with a 1 : 1.5 cement–sand mortar, again of an easy-flowing creamy consistency. Grouting will normally start at the lowest point in the mass and as this area is seen to be filled, then pipes can be moved further into the mass of stone until such time as the stone and matrix become a solid mass. When grout tubes are lifted in the work to ease pump pressure, care should be taken to ensure that they are always kept at a level slightly below the grout plane. Underwater work carried out by this method alone is not capable of producing a fair surface, but in combination with normal tremie- or skip-placed concrete it will make a first-class base.

37.5.5 Compacting concrete

The relationship between the degree of compaction of concrete and its compressive strength has been indicated in a previous section and the need for achieving a densely compacted mix will be apparent. What is not sufficiently appreciated is that even highly workable concrete mixes need to have some work done on them in order that they should have an adequate standard of compaction.

The effort required to achieve a high degree of compaction with highly workable concrete is minimal, and can be obtained satisfactorily by hand-punning. However, for mixes of medium and low workability, vibration is needed to attain good compaction quickly and the use of vibrators has therefore replaced the earlier hand methods.

Internal vibrators, or poker vibrators as they are more usually called, are the most common type of vibrator in use. They consist of a vibrating tube at the end of a flexible drive and the usual sizes vary from 25 to 75 mm in diameter. The power for the flexible drive can be provided by small petrol engines or electric motors. Although vibrators should not be used for moving concrete (other than the downward movement during compaction), some movement horizontally will inevitably result from their use.

Vibrators fixed to the outside of the formwork are sometimes used, especially where there is a heavy congestion of reinforcement in a wall or deep beam web. In such cases it is not possible to insert a poker vibrator. However, the application of external vibrators is limited by the need to provide heavy formwork which can resist the stresses and shaking produced.

Concrete having been set in the shutters as described in continuous layers about 350 to 500 mm thick along the length of a section, it should be compacted by feeding the vibrator vertically down through the depth of the layer and not more than about 70 mm into the next lower layer. When it is clear that all air has been sensibly expelled from the area being compacted, the vibrator should be slowly withdrawn and plunged into the next section of the work, this process being repeated along the length of the section.

The proper compaction of concrete into highly reinforced zones such as the ends of prestressed concrete beams has always been a problem. The concentration of reinforcing bars and hoop steel round cable anchorages is sometimes such that it is virtually impossible to insert even the smallest-diameter poker vibrator. In such cases external vibrators securely fixed to stiff shutters should be used and moved up the work as concreting proceeds. Honeycombed concrete in the areas of highest steel amounts will be avoided if concrete of a suitable workability is filled into the shutter at a very slow rate only, so that visual inspection will be able to reveal areas of inadequate compaction.

In some situations where suitable lifting devices can be made available, it might be considered to precast the anchorage block on end so that immersion vibrators can be used in the direction of the main reinforcing steel and ducting and not across them. Proper treatment of the top surface of these end blocks, as described, should ensure their proper bonding into the beams with no risk of loss of structural strength of the whole.

37.5.6 Curing concrete

In order to achieve full potential strength and to reduce the amount of drying shrinkage and moisture movement to the lowest possible levels it is necessary to carry out the 'curing' process. This connotes a method whereby the amount of water in the newly placed concrete, which is always higher than that required to fully hydrate the cement present, will be retained over a period of at least several days.

Curing large flat areas such as road slabs and monoliths in plain and reinforced concrete work presents no great problem since, as is indicated in later section, the former can be effectively cured with an impermeable membrane sprayed on to the surface after finishing and the latter with hessian or similar material maintained in a damp condition by means of water sprinklers or occasional drenching with water.

Vertical construction, however, presents a problem which is more difficult to solve. With shuttered vertical faces, the evaporation of moisture from the concrete is effectively barred by the shutters themselves but the economics of construction will generally require that these be used as frequently as possible; this involves removal for re-erection as soon as the concrete has achieved a strength adequate for self-support. At this stage, the concrete is liable to lose moisture to an extent which can result in a good deal of surface crazing and shrinkage cracking. Particularly is this the case if the concrete attains a temperature within the mass which is a good deal higher than the ambient.

Thus for vertical construction, curing of some form should be undertaken as soon as possible after the shutters have been removed. This will preferably consist in draping sections with hessian which can be kept damp by frequent spraying with water. Alternatively a spray bar of plastic tubing perforated at intervals with small-diameter holes can be laid along the top of the concrete and connected to a supply point. It is unnecessary to use an excessive amount of water in this process and, in fact, it is not advisable, since conditions underfoot can thereby be made more unsatisfactory. Polythene sheeting is often used as a barrier to the movement of moisture from concrete but unless it is properly secured to the concrete members with adequate ties it may well be less effective than making no attempt at curing whatsoever, since it encourages the more rapid circulation of air around the member.

In some instances, vertical surfaces are cured by spraying a membrane-forming curing compound on to the concrete surface. The curing compound should be applied, using a hand spray, as soon as the forms have been removed. If the surface

has dried out, it should first be sprayed with water and allowed to assume a uniformly damp appearance before the curing membrane is applied.

The application of curing membranes to vertical surfaces suffers from the disadvantage that they stain the concrete surface and do not always disintegrate and fall away after a short time, as claimed by the manufacturers. It must also be remembered that curing membranes should not be used on surfaces that are to receive applied finishes such as plaster or cement renderings, paint, tiles, etc. that require a positive bond with the concrete.

When it is considered desirable for the sake of expedience to cure concrete surfaces which are to receive finishings with a membranous compound, then all traces should be removed by heavy wire brushing or, as has sometimes proved necessary, a more drastic treatment of sand blasting or bush hammering.

37.6 Construction of concrete roads and airfields

37.6.1 General observations

In the last 15 years or so, about 20% of all major roads constructed in the UK have been built in concrete. This figure indicates an increase over the proportion previously constructed (about 6% before 1969) and is attributed to changes in pavement design, the effects of government policy and the rising price of bituminous materials for alternative asphalt-surfaced roads. However, the proportion of concrete roads is still considerably less than in the US where about 50% of major roads are built of concrete.

Important design changes were introduced in the UK in 1969 which permitted the use of unreinforced concrete paving for major roads. Although the omission of reinforcement fabric from the concrete means that many more contraction joints are required, the overall result is a considerable reduction in cost relative to reinforced concrete paving. This cost reduction is one of the main reasons for the increase in the proportion of roads built in concrete.

Machine-laid concrete roads, which have all but replaced hand-laid work throughout the world, are constructed by two different methods. The first uses forms both to contain the unhardened concrete slab and to support rails on which the road-building machines – spreaders, compactors and finishers – are mounted. In Britain, this is sometimes referred to as conventional construction; it is often favoured where reinforced slabs are required by the specification. The second method is slipform paving and, as the term implies, the formwork in which the unhardened concrete assumes the required shape of the road slab moves forward as an integral part of the paving machines, leaving the concrete unsupported after only a very short period of time.

In Britain and Europe, conventional construction has held sway for many years, but in recent years about 30% of major concrete roads in Britain have been constructed by slipform pavers. In America, slipform paving is relentlessly taking over from conventional work and it seems likely that before long there will be little other than this type of concrete road paving work there.

A high proportion of all the runways, taxiways and loading and parking areas for heavy transport aircraft throughout the world are built of concrete of varying thicknesses, reinforced and unreinforced as the design requires. The same methods as are used for building roads can be used for runways, etc. but as thicknesses are often a good deal greater, construction problems may be somewhat different. For example, slipform paving which is successfully used in road construction may not be so attractive a method for building thicker runway slabs because of the greater risk of serious edge slumping. The greater thicknesses involved may also require that compaction from the surface be supplemented by immersion vibrators.

Records of surface profiles of roads laid in recent years indicate that there is now little to choose between the best of the bituminous work and the best of the concrete.

37.6.2 Conventional construction

This class of work is described under the following headings:

(1) Forms and form setting.
(2) Trimming of base and laying sliding membrane.
(3) Setting of expansion and dummy joint assemblies.
(4) Concrete production and transport.
(5) Spreading concrete.
(6) Laying reinforcement – where required.
(7) Compacting and finishing concrete.
(8) Joint forming and sealing.
(9) Texturing of road surfaces.
(10) Curing of road surfaces.

37.6.2.1 Forms and form setting

All formwork for machine-laid concrete roads should be made from steel plate of at least 5 mm gauge and have an adequate number of stiffeners. They should be fitted with heavy-gauge rails on which construction machines, weighing up to about 12 t when loaded, can be run and firmly fixed to a concrete base about 100 mm thick. This base is preferably laid with a wire-guided datum laying machine or similar, capable of laying a strip of good-quality concrete as wide as the form base, to which the formwork can be secured. Normally, laying these base strips to high standards of accuracy should ensure that the forms and rails on which the machines are to run are accurately positioned. Since finishing machines are required to be of the articulated floating beam type (see section 37.6.2.7) the accuracy of the top edge of the formwork is not of great importance, but it is often an advantage to have a square, rather than a rounded, edge to the formwork. When the forms available are shallower than the slab thickness they can still be used since the concrete base on which they are laid can be any thickness greater than about 100 mm. To achieve a balance between adequate rigidity and lowest cost, form depths will generally be about 175 to 200 mm.

In those instances where the design of the road calls for the use of a lean concrete base it might be considered preferable to dispense with the concrete form base and use full-depth forms.

It should be borne in mind that very high loads are imposed on formwork when spreaders are loaded by side-discharge trucks. If these are used care should be taken to ensure adequate strength of the form–base combination so that there is no deflection under these impact loads.

37.6.2.2 Trimming of base and laying sliding membrane

Bases to concrete slabs should be laid accurately to ensure that slab thicknesses are correct. Where the bases are of a granular nature they can be set slightly low and trimmed upwards with fine material to be compacted by roller. If the stronger lean concrete bases are required then they should be laid as accurately as possible with no positive tolerance. An accurately laid base will result in lower shrinkage and temperature stresses developing in the slab than would otherwise be the case. These stresses are sometimes required to be reduced by laying a membrane of polythene or heavy gauge kraft paper. A membrane is perhaps not quite so important when a granular base, trimmed with fines, is used.

37.6.2.3 Setting of expansion and dummy joint assemblies

Under British conditions, expansion joints are not required in concrete roads laid between 21 April and 21 October. It is a natural assumption that contractors will hope to avoid laying concrete in the period October to April, and so expansion joints will not normally be required. However, a full complement of contraction and warping joints is required in the designs. It is essential to have these extensively prefabricated so that setting up at the correct spacing can be carried out expeditiously.

In recent years a dowel bar setting device has become available. This machine provides the means for accurately positioning bars at contraction joints and forcing them down to their correct position within the road slab by vibratory means. The performance of this machine is said to be highly satisfactory as regards both the accuracy with which bars are placed in position and in cost comparisons with methods used hitherto.

37.6.2.4 Concrete production and transport

Concrete for road construction can be produced by any of the methods previously described. The method of construction will dictate the means of transport of concrete but it should be noted that high-capacity end-tipping lorries are likely to result in the lowest overall transport costs. Where it is elected to use box spreaders it has generally been considered necessary to use high-discharge side-tipping lorries whose capacity will be no higher than that of the spreader they are loading. However, various devices have been used by contractors to enable them to use the more economical end-tipper lorries for this purpose.

A matter of some importance when operating high-discharge side-tipping lorries is that, in order to achieve rapid and clean discharge, the load must be quite high above the ground. Concrete laden lorries might be rather unstable at speed unless haul roads are well maintained.

The exposed surface of a lorry load of concrete can be noticeably dried out or wetted by exposure to weather during long hauls and a cover should be provided with each lorry so that it is available for use as required.

The overall rate of paving will depend upon a number of factors, including the programmed time for paving with due allowance for unfavourable weather, and the rate at which supplies of aggregate and cement can be made available, supplemented as necessary by materials drawn from stockpiles accumulated during nonpaving work.

The size of batching and mixing plant required will be governed by the expected rate of forward progress, slab thickness and width of road to be built.

Various alternative combinations of lane widths are allowable and often the overall design, on motorways at least, is simplified to the extent that the surfacing of hard shoulders can be concrete rather than a contrasting bituminous material. Where the design and specification of the road structure allows, it will often be found that construction costs can be reduced by paving in equal widths so that batching, mixing and transporting concrete is carried out at the same tempo throughout the work, with constant width placing and finishing equipment.

The prime requisite for high-quality paving work has been found to be a steady rather than a high rate of progress, but there may be some advantages, so far as quality is concerned, in aiming at a higher rather than lower overall speed of construction. What has been found from past experience is that the irregularity index (q) for the road surface profile is closely linked with the average rate of construction, low q values being associated with high average rates of uninterrupted progress and high values with sporadic working. This implies that the quality of work will be improved if preventive maintenance is carried out diligently to reduce breakdown time of plant to the minimum; the effect of this on the economics of concrete laying will be clear.

37.6.2.5 Spreading concrete

The importance of proper spreading of concrete as a factor contributing to the production of a good riding quality of a concrete road with the minimum effort in finishing cannot be overstated. Ideally the equipment used should be capable of so spreading concrete that its loose density as spread over the formation does not vary by more than about $35 \, \text{kg/m}^3$ from point to point.

Concrete is almost universally spread by means of hopper spreaders fed from side-tipping lorries of various designs. The outlet from the hopper may be controlled by a helmet gate or by contraction to a narrow opening through which concrete is passed by vibration. With some models, concrete is tipped into the hopper whilst this is in a roughly horizontal position with the outlet remote from the point of loading. For spreading, the hopper is brought into the erect position and only when this is done does concrete flow from the outlet.

Certain hopper spreaders can be made to function with the hoppers mounted square to the line of the road as opposed to along the line of the road. They can then be loaded by means of end-tipping lorries backed up to the formation; concrete is spread in lanes along the line of the road rather than in bands transverse to it.

This latter system involves the use of a joint assembly which can be very readily and securely set into position on the formation without interfering with the movement of concrete lorries to and from the spreader. It should also be noted that these lorries are liable to cause damage to the formation and any sliding membrane laid on it.

37.6.2.6 Laying reinforcement

When reinforcement is used in road slabs it usually takes the form of welded mesh and is set at about 60 mm below the running surface. Three methods for laying are commonly adopted:

(1) Concrete to the level of the reinforcement is laid and compacted. This operation is normally carried out by the first of two spreaders and compacting beams. After reinforcement has been laid as required, surface concrete is laid and compacted by the second pair of machines. Note that the second spreader will have only about one-third to one-quarter of the throughput of the first.

   As an alternative to this method the reinforcing mesh is laid on the uncompacted lower concrete and depressed into its surface by the beam of the compacting machine before laying the surfacing.

(2) The reinforcement is laid in advance of concreting operations, supported at the correct level on closely spaced stirrups set on the formation, the rate of stirrupping being about $1/\text{m}^2$. When this method is adopted, the spacing of the main longitudinal bars should be adequate to allow the free passage of concrete through them – a minimum of 100 mm is suggested. Various proprietary systems are available for this purpose and they can be used for either mesh or bar reinforcement.

(3) Loose concrete is spread to the depth required for the full slab thickness. Reinforcement is then laid over the area. Prior to compaction in depth the mesh is depressed into the loose concrete to the required depth by vibrating tines. Besides forcing the mesh into the concrete, these vibrating tines make a contribution to the compaction of the concrete but compaction in depth is carried out from the

Figure 37.13 Conventional reinforced concrete road construction. Photograph shows from foreground: concrete edge strip carrying rails; sliding membrane and joint assembly; spreading base concrete; laying reinforcement; compacting base concrete through reinforcement; spreading surface concrete; finishing surface with diagonal finishing; texturing and curing (tenting in background)

surface only by compacting beams. This method is fairly widely used in America.

It should be noted that with method (1) it is allowable to use any quality of concrete in the base (provided of course that it complies with the specification) irrespective of the nature of the aggregate; it need not be air-entrained. The surfacing concrete, which in Britain and many other countries must be air-entrained, can then be laid about 60 mm thick and be made of such materials as will cause the concrete to comply with skid-resistance requirements. Often this smaller volume of concrete for top layer, with air entrainment and a selected aggregate, will be produced in a smaller mixing plant than the rest of the concrete.

With methods (2) and (3) all concrete must be of the same type complying with requirements regarding strength, air entrainment and skid resistance of abraded samples.

In methods (1) and (2) it is usual to lay out the reinforcing mats along the line of the road against the position they will occupy in the slab. As an alternative they can be off-loaded from the supplier's vehicle in bulk on to a wheeled mesh cart which straddles the slab and is either self-propelled or can be towed by one of the leading road-building machines. Mesh is dragged from the pile on the cart as it moves forward with the road slab.

It is claimed that method (1), which requires the final spreading and separate compaction of only 60 mm of surfacing concrete above the base concrete and reinforcement, can lead to easier final finishing to the required surface tolerances. Some of the best work has been carried out by this method. On the other hand good work has been carried out when the whole depth of the slab has been compacted and finished as one layer.

37.6.2.7 Compacting and finishing concrete

Compacting and surface finishing of concrete were at one time carried out by the same machine but, currently, compaction and partial finishing is undertaken by one or two machines of the same type, depending upon whether compaction is carried out to the full depth of the slab or in two distinct layers below and above reinforcement as described. Final finishing is carried out by a further machine which imparts very little compactive effort to the concrete but strikes the surface to a true profile by a to-and-fro screeding action across the slab.

Where two-layer work is adopted, the leading compacting machine will be used to impart a beam finish to the base layer, prior to laying the reinforcement. The surface layer will be compacted and partially finished by the second compacting machine which may have a floating oscillating finishing beam carried on an articulated chassis. Final truing up of the surface will be done with an articulated finishing machine with an angled oscillating single- or double-acting beam. The beam is mounted at an angle of about 50 to 60° to the line of the road and is of particular value in that it can encourage the quick removal of any excess of 'fat' gathered in front of the beam during final screeding.

In some reinforced road construction work, transverse joints are formed by vibratory or other means in the wet concrete immediately behind the second finisher and kept open prior to sealing by a removable insert of one form or another. Here, an angled beam has a marked advantage over a square beam in that it advances steadily from one end of the joint to the other and does not exert a disturbing pressure on any but a short length of the insert at any one time; this ensures that the joint insert remains in its correct position. Furthermore, the angling of the beam extends the wheelbase of the machine considerably and it has been found that this tends to much reduce the 'yaw' of the machine when the resistance to forward movement is greater on one side than the other.

For greatest effectiveness a smoothing beam should be so heavy that there is no tendency to ride over rather than plane off high spots in the concrete. It is considered that for maximum efficiency a beam should weigh about 150 kg/m.

37.6.2.8 Joint forming and sealing

Transverse joints. As noted previously, expansion joints are unlikely to be used in concrete roads except where, inadvertently, construction has been delayed or advanced so that work is in progress during the period mid October to mid April. However, contraction and/or warping joints are necessary and these will be provided with the specified number and sizes of load-transfer bars. Weakening of the slab at these contracting joints to ensure cracking there rather than elsewhere within the slab is provided by a wood or plastic fillet secured to the base and by a groove sawn or formed from the surface directly above the fillet. The reduction in slab thickness by fillet and groove should amount to at least 20% and preferably 33% to ensure adequate weakening.

Joint grooves can be formed in the unhardened concrete by means of vibrating blades or wobbly wheels, either of which will displace sufficient concrete to allow a temporary strip of adequate depth and width to be inserted and held secure in the concrete whilst the disturbed surface is being retrued. This refinishing of the surface is most effectively carried out by the angled finishing machine as described.

As an alternative to forming grooves in the unhardened

concrete, they may be sawn when the concrete is sufficiently hard to allow this being done without disruption of the concrete along the joint line.

Concrete saws are power driven and the cutting blades are tipped with various grades of silicon carbide or diamonds. The ease or otherwise of cutting concrete depends much more on the kind of aggregate used than on its strength. Limestone concretes are by far the easiest to cut, next in order are other types of crushed rock whilst hardest are quartzites and flints. The last-named are particularly difficult and expensive to cut.

A difficulty that often arises in cutting partially hardened concrete by saw is that cracking may be induced at the surface before the concrete is hard enough to be sawn. Limestone concretes crack less readily than other types and can be sawn fairly soon after hardening; they are therefore much to be preferred, provided the skid resistance of the concrete can be made to meet specified requirements.

Longitudinal joints. Longitudinal joints may be either full depth at a slab edge or part depth formed by a bottom fillet and a surface groove in the centre of a slab. These surface grooves can be built into the unhardened concrete behind a surface-finishing machine by means of an attachment thereto which displaces concrete and, at the same time, inserts a preformed sealing strip. As with the forming of transverse joints, there is some disturbance of the surface when this insertion is carried out; this is best corrected by using the angled finisher.

Since concrete inevitably shrinks away from the preformed sealer its efficiency is somewhat in doubt. But also in doubt is the real need for a completely efficient longitudinal joint seal.

Full-depth longitudinal joints are made against the formwork and will need tie bars across them, as specified. The ties are most readily positioned by cranking to 90° and laying one arm against the form for later recovery and bending into the adjacent lane. These joints are best sealed by sticking a preformed sealer on to the top of the hardened slab just prior to concreting an adjoining slab.

The adequate sealing of joints in concrete paving is a problem which does not appear to have been solved. It is open to question whether or not sealing against water penetration is necessary when the base, as is generally the case, has been built in such a way as to make it much less susceptible to damage from seeping water. However, there is a need to avoid spalling at joints caused by intrusion of stones and perhaps grit in sealing grooves. If joint edges are left completely unsupported, there is also a risk that traffic will, in time, break away some section of joint edge.

37.6.2.9 Texturing of road surfaces

Recent research into the high-speed skid resistance of various types and textures of concrete surfaces has indicated quite clearly that a deeply textured one, which causes discontinuities in the surface water film in wet weather, markedly improves this characteristic.

It is now the practice on all concrete roads which will be used by high-speed traffic to score the surface, after final finishing, with mechanically operated brushes which impart this texture in the thin layer of surface mortar found on all of them. The result of this scoring is often to produce a drumming similar to that caused by running over heavily surface-dressed black-top roads.

Also available is a grooving machine which consists of a series of vibrating blades at irregular spacings. These make incisions into the unhardened concrete. It is claimed that the wholly random spacing of the grooves formed by this machine does much to reduce the noise nuisance which is sometimes induced where too regular and heavy texturing is a feature of the surface.

37.6.2.10 *Curing of road surfaces*

Adequate curing of concrete road slabs is essential, particularly in periods of low relative humidity, if damaging cracking and other surface defects such as widespread crazing are to be avoided.

Currently, practice throughout the world is to spray the concrete immediately after finishing with a solution of resin in a volatile carrier, or some other liquid, which will retain a high proportion of the water in the concrete for the vital first few days yet will not remain to reduce the skid resistance of the surface, once it is used by traffic. Where hot sunshine prevails for much of the day, a metallic or white reflecting pigment added to the membrane will reduce heat absorption into the slab.

British requirements in respect of curing are currently that the first 2 or 3 h production of concrete slabs should be protected by tenting which supplements the curing effect of the membrane. This does not pose any problems where rail-mounted equipment is used; but it becomes impracticable when slabs are laid at high speed by slipform paving techniques – in fact, it is most uncommon to see any form of tenting used when slipform paving methods are adopted, except a small amount carried on or close to the machine, for emergency use only.

37.6.3 Slipform paving

A method of building concrete roads at high speed without the use of prefixed formwork – slipform paving – has made a major impact on construction techniques, not only in America but over the rest of the world where concrete roads find favour. Slipform paving is essentially concerned with laying, on a prepared base, a section of road slab in plastic concrete, within a moving form. The sides of the section are defined by formwork which is rigidly mounted on the machine itself at the limit of width of the slab, whilst its upper surface is formed either by a truly flat 'conforming' plate extending to the full width of the slab over a length of several feet or by oscillating screeds which strike off the surface of the concrete at a controlled level. Some machines use both.

The side forms may vary in length between about 4 and 9 m with different machines. Since they generally operate at forward speeds in excess of 1.5 m/min, the concrete slab is supported at its vertical edges for a matter of a few minutes only, in contrast to the normal practice of maintaining support by formwork for several hours before striking.

Slipform pavers are normally positioned as to line and level by means of electrohydraulic equipment. This in turn is controlled by sensors which detect the line and level of wires or cords supported well outside but parallel to and above one or both sides of the slab. Where one wire only is used it is accurately set up to control the position of the nearest slab edge, that of the remote edge being governed by a cross-levelling device on the machine. The locating of this is dictated by the position of the machine in the road, which governs the sense and degree of cross-fall.

Figure 37.14 Slipform paving with G and Z machine. Unreinforced concrete slab on stabilized base. Transverse joints sawn in hardened concrete of left-hand carriageway after 10 to 24 h

The concrete for slipform pavers may be deposited on the formation from end-tipping trucks or discharged into a spreading device in front of the machine. It is then struck off approximately level across the slab by various means, e.g. shuttle spreaders, auger screws and transverse paddles. Compaction to its maximum density by a battery of immersion vibrators, supplemented as necessary by surface vibrators, follows. As the machine moves forward, a conforming plate or other levelling device is passed over the highly fluidized concrete and causes it to take up the shape of the slab. When a conforming plate is used, the amount of concrete passed below the plate will depend to some extent on the head of concrete in front; thus the level at its rear may be subject to slight variations. The action of oscillating screeds, on the other hand, is to strike off excess concrete or to make good deficiencies with concrete carried in front of them. For this reason there may be some technical advantage in using a machine with oscillating screeds rather than a long conforming plate.

It is often preferred to lay concrete road slabs on a very accurately prepared base of cement- or lime-stabilized material, asphalt, sand – asphalt, lean concrete, etc., which will be adequately strong to carry the weight of the paving equipment. When this is done, the paver works at a fixed height above the base tending, by virtue of its long track base, to smooth out irregularities thereon. Direction has to be controlled by means of a line sensor on the machine but apart from this all levels of the finished road slab are dictated by those of the base on which it is being built.

Slipform pavers are best suited to laying unreinforced concrete slabs built without joints but subdivided – when the concrete has been laid a few hours and hardened sufficiently to resist damage – into short lengths by gang saws. However, by deploying other machines such as spreaders and mesh depressors, it is possible to build slabs which are, to all intents and purposes, the equivalent of our reinforced concrete slabs. When this type of road is built, the basic simplicity of slipform paving is lost and the train of equipment becomes not unlike that used for rail-mounted work as previously described. The capital cost of the machines employed is much higher than for rail-mounted equipment but its potential throughput is also much higher.

In America the final running surface behind a slipform paver is often trued-up to a very regular profile by a tube finisher. This simple device consists of a 200 mm diameter tube of length about 1.5 times the slab width, mounted on a wheeled chassis which straddles the newly laid concrete slab.

The profiles of concrete slabs finished by either a slipform paver or a combination of slipform paver and tube finisher are generally very uniform and hence should comply adequately with specification requirements.

Although there do not appear to be any insuperable technical problems associated with the successful operation of slipform pavers, there are those of logistics. These are mainly that the machines operate most economically with high throughputs of concrete. Under normal conditions in Britain this necessarily involves stockpiling large amounts of concrete aggregates and perhaps special arrangements for the supply of corresponding amounts of cement. Building and drawing from stockpiles are expensive operations and significant reductions in construction costs have to be achieved to effect the desired savings in overall costs.

37.7 Concrete floors

37.7.1 Construction procedures

There are many types of toppings and surface finishes that are applied to concrete ground-floors and the choice will depend on the specific requirements in each case. However, in the majority of cases, good finishing techniques on the basic concrete itself will be perfectly adequate.

Floors can be laid in the same way as concrete paving for roads and airfields, although it is customary to see manual or semi-manual methods being used as opposed to machine methods. The latter are limited for economic reasons to the construction of large floors where uninterrupted lengths of at least 80 m are required.

A traditional way of laying concrete floors has been by the 'chequer-board' method of construction in which individual bays are cast alternatively within stop-ends forming the joints. Infill bays are usually specified to be placed no earlier than 7 days afterwards, the basis of this requirement being to allow a considerable proportion of the shrinkage movement of the earlier bays to occur. Since, however, shrinkage of concrete takes place over a period of several months, it is now recognized that this latter requirement is of dubious value. Moreover, this method of working is not efficient and access for constructing the infill bays is poor.

A more modern method of floor construction which has found favour is the so-called 'long-strip' procedure which is basically the same approach as that adapted for concrete road construction. Alternate strips – usually not more than 4.5 m wide – are first laid continuously the full length of the floor area and divided into bays by means of induced joints. These joints may be formed either in the plastic concrete or by sawing shallow slots in the surface 2 or 3 days after the concrete has been laid. The infill strips of concrete are placed a few days later when the first strips of concrete have hardened sufficiently to withstand the effects of the compacting beam without damage to the edges.

Again, as in the case of roads, reinforcement is often provided in the slab near the top surface. This entails laying the concrete in two courses.

37.7.2 Finishing techniques

For many industrial applications, the concrete used for the floor slabs can be directly finished to provide a suitable wearing surface. A 28-day strength of 30 N/mm^2 or more should be specified for the quality of the concrete and it is also desirable that the concrete should have a cement content of at least 330 kg/m^3 in order to provide good durability and resistance to abrasion.

Traditionally, concrete floors are finished by hand trowelling using steel trowels. Each trowelling operation follows the previous one after an interval of about 1 h (during which time further moisture will have evaporated from the concrete surface).

This trowelling process has become mechanized in recent years and it is now usual to see the concrete finished by mechanical means using power floats and power trowels. The former uses a rotating solid circular disc whereas the latter is provided with three or four rotating blades. However, the terminology for these processes varies and the term 'power float' is often taken to cover both power floats and power trowels.

Power floating and power trowelling can, with care, produce very good finishes. In comparison with hand trowelling they enable the work to be done up to 6 times more quickly and their use permits a slightly drier concrete to be used (with consequent advantages in enhanced strength and resistance to wear). However, it is still necessary to resort to hand trowelling in small confined areas, such as floors of domestic dwellings, where it is not practicable to use a power float efficiently.

The power float does not, as is sometimes thought, provide compaction of the concrete. The concrete must have been compacted and screeded to level before the power float is used.

The time at which the power float is brought into use on the floor surface depends on various factors, notably the ambient temperature and the workability of the concrete. It is recommended that under average conditions it should be used about 1 h after the concrete has been laid. A good guide which can be used to determine the time at which to use the power float is the depth of the impression left on the surface when a man stands on it; if the footprints are about 2 mm deep, the concrete is ready for treatment.

In cool conditions, there may be a considerable delay before the power float can be brought on to the concrete and this may entail overtime working by the operators. To avoid this, a vacuum dewatering process is sometimes applied to the concrete surface soon after it has been compacted and has received its initial surface finish. The process involves the use of a flexible vacuum mat, provided with a fine filter sheet, which is connected to a vacuum generator. When the mat is laid on the concrete surface, a vacuum is created underneath the mat and this causes water to be drawn out of the concrete. The vacuum is usually applied for about 20 min after which the concrete will be sufficiently stiff to receive the first surface treatment with the power float.

37.7.3 Surface hardeners

The use of surface hardeners on concrete floors is rather a controversial subject. On the one hand, a properly laid concrete floor should give a satisfactory performance without any further treatment. On the other hand, the quality of many concrete floors, when laid, is far from perfect and a surface hardening treatment can be beneficial in these cases. However, in the case of weak concrete floors, a surface hardening treatment will not be effective at all.

It follows that the use of a surface hardener should not be regarded as a substitute for producing a good-quality concrete in the first place.

The most commonly used surface hardener is sodium silicate which combines with the lime in the concrete to form a hard glassy substance. Magnesium silicofluoride and certain other salts are also used as surface hardeners.

37.8 Other forms of concrete construction

37.8.1 Dam construction

Circumstances arise from time to time in the construction of reservoirs for either water supply or hydro-electric schemes when it is advantageous to build a concrete dam rather than an earthen or rock-fill embankment.

The design and specifications for the dam will give details of the lengths of each section of the dam, the depth of each lift and the interval which must lapse between the concreting of successive lifts; it will also place a limitation on the depth of concrete which can be poured in the lifts and time interval between adjacent monoliths.

Section lengths are generally of the order of 15 to 18 m and joint planes will normally be at 1.5 to 2 m.

Where it is intended that several monoliths in the same area be brought up together, this can be done provided an adequate gap is left between each for subsequent infilling. Gaps of about 2 m have been used and filled some time after the main lengths of the wall, when major drying and thermal shrinkage of the main blocks has taken place. Proper provision for water bars must be left in each side of this gap so that the method involves somewhat heavier expenditure on joint preparation and sealing.

A number of factors have to be taken into account in deciding the method of construction to be adopted.

First will be the duration of the contract and the likely time to be spent on preparation of the foundations before dam construction can begin. Another major factor will be the profile of the valley across which it will be built. In Britain, the sites on which long and high dams can be built are very few so that a contractor is usually faced with building either a small number of high monoliths or a larger number of low ones.

The required number and dimensions of the monoliths will determine the method by which concrete will best be handled into place. Where a long, low dam is required to be built across a shallow valley, it might be considered to carry the concrete from a central mixing plant in crane skips mounted on flat-bottomed lorries or trailer units, thence into the work via a crawler or other type of crane. Flat-bottomed bogies carried on a narrow-gauge rail track along a low-level gantry might also be considered. To build a shorter but higher dam across a narrow steep-sided valley it might be thought preferable to handle concrete in crane skips via a series of derrick cranes mounted on temporary concrete pillars. These could well be sited on the upstream side of the dam, since the greater volume of concrete will be there.

All possible ways of handling concrete are discussed in a previous chapter and the contractor, in preparing his scheme for the work, will make decisions on the rate of concreting and the means by which it is to be placed in position.

It is common practice to build the main mass of a dam wall with a low cement content and hence fairly low-strength hearting concrete, but to use a higher quality having greater durability at the upstream and downstream faces. This facing must be placed within a short period of placing the hearting concrete so that there shall be a complete bond between the two.

Probably the easiest method is to build up the hearting in a 400 to 500 mm lift over the whole area to within about 400 to 500 mm of the dam faces – or whatever thickness surfacing concrete is called for in the design. The facing is then filled in to the same depth between hearting and the shuttered upstream and downstream faces. The richer concrete is not normally required to be poured against the shuttering to transverse joints.

Rock suitable for good-quality concrete aggregates is often available at the site of work and is generally quarried as required to make concrete on site. Since sections are normally large and there is a need to keep cement contents low, aggregate up to 150 mm maximum is often used. The production of lean and sufficiently workable mixes for low-strength hearting concrete is facilitated by using large-size aggregate. However, the use of plums or displacers is not economic – nor is it good practice.

In some instances where only poorer qualities of rock – as regards their suitability as aggregate – are to be found at site it might be necessary to import the better-quality material for exposed concrete but to use the inferior aggregate for the mass of low-strength concrete in the hearting. Proper mix design and placing techniques are quite capable of producing concretes satisfactory for this work from the most unlikely materials.

A disfiguring feature of many dams in the past has been the tendency to seep water through transverse joints and, more rarely, along horizontal joint planes. The former faults can be avoided by proper detailing of joints to facilitate their building according to the intended design. With horizontal joints it is essential to avoid the presence of laitence, downward joggles and deep indentations which will hold water. Laitence can be removed by air–water blast when the concrete is hardened sufficiently to avoid damage. Treatment of concrete with water sprayed on to hessian strips will ensure its proper curing and avoid any shrinkage cracks which might contribute to failure; it will also keep the whole area clean for subsequent operations.

A recent development is the use of 'rolled concrete' for dam construction (referred to as 'Rollcrete' in the US). The concrete in this case has a low cement content and has a relatively dry consistency which permits compaction by rolling. It may be compared with the dry lean concrete which has been used successfully for the construction of road bases, but the concrete mix used for dams benefits from the inclusion of a fairly high proportion of PFA, the proportion of which has generally ranged from 30 to 75% by volume of cementitious material. The use of rolled concrete for dams has the advantage that conventional earthmoving plant may be used for handling and compacting the concrete and that therefore large outputs can be quickly and economically achieved. Construction can proceed in layers, about 200 to 300 mm thick, laid continuously from one side of the dam to the other.

37.8.2 Tunnel linings

In situ concrete linings may be called-for in hard-ground tunnels to give support to rock which will deteriorate in the course of time or to improve hydraulic characteristics.

The dimensions of the tunnel – length, diameter, number of access points and other factors – will need to be taken into consideration in deciding the method to be used and the order in which work is to be carried out. Mention has already been made of the use of pneumatic placers but concrete pumping is coming to be more widely favoured. A variety of methods for getting the concrete to the working face have been used in the past and new ideas coming from time to time are adequately described in technical literature.

37.8.3 Mass plain and reinforced concrete sections

Concrete will be placed into sections of this type, which will most frequently be found in power stations, foundations to large buildings and other heavy work, by various combinations of methods already described. The sizes of bay to be concreted will be dictated by the output of the batching plant available for the particular operation in hand. Owing to the complexity of shuttering work in reinforced concrete it might be found advantageous to restrict the depth of pour so as to increase the area to be concreted at any one time – shuttering costs are then likely to be rather lower.

In heavily reinforced concrete foundations there is a strong financial incentive to dispense as far as possible with costly construction joints and to place the concrete continuously in large pours. Placing of volumes of concrete of the order of 200 to 300 m³ is quite common for such foundations and very much larger pours of the order of 3000 m³ have been placed where the plant facilities and access have permitted such a large-scale approach.

In unreinforced concrete foundations, however, where there is no steel reinforcement to restrain the subsequent shrinkage of the concrete, it is necessary to restrict the areas of concrete cast in one pour in order to avoid cracking resulting from the concrete shrinkage. Due to the absence of reinforcement, the provision of shuttering for the construction joints will be relatively straightforward in these circumstances.

The tendency to crack when the concrete contracts is related to the value of the maximum temperature reached by the concrete soon after it has been placed. The cracking is, of course, caused by the subsequent cooling down to ambient temperature. In large foundation slabs, where cooling is generally in a vertical direction, the maximum temperature rise attained is related to the lift height. However, due to a combination of self-insulating effects and the reduction in the adiabatic rate of heat generation after 24 h, increase in lift height above 2 m causes very little further increase in the maximum temperature reached.

37.8.4 Vertical construction with sliding formwork

For many years, certain types of vertical construction in concrete, e.g. materials storage silos and the service cores of tall buildings, both of which have either a constant cross-section throughout their height or only a small number of variations in wall thickness, have been carried out by using formwork which was moved continuously upwards as the concrete was placed within the forms. The infrequent changes in plan have been accomplished by altering the dimensions of the moving shutter as required; this has involved completion of a section and re-erection of one or both faces of the shutter before work recommenced.

More recent developments have seen modifications to the original conception of moving formwork to allow slight and gradual changes in cross-section as the work proceeded upwards. The construction of reinforced concrete chimneys and the erection of tall towers serving as supports for TV aerials and amenity buildings are examples of work in which both external dimensions and wall thickness have decreased as the height above ground increased.

Sliding formwork, though now widely used, particularly by companies who have become specialists through mastering the techniques involved, is not a new art, there being reference to such work as long as 60 years ago. However, as the use of the method has become more widespread so have improvements been made in the mode of operation.

Basically the method consists in continuously raising formwork of the correct plan dimensions into which concrete with the designed amount of reinforcement is placed in a series of narrow bands up to 150 to 200 mm depth, proceeding over the whole plan area. The formwork to the inner and outer faces is connected at close intervals by a series of straddling yokes, each having a device by means of which the yoke and the attached shutter can be moved upwards in relation to jacking rods at each point. These jacking rods are carried from bottom to top of the structure and are located in circular cavities formed within the walls by tubes which surround the rods below the jacking points.

Apart from the external and internal shutters and the jacking systems and their control, it is necessary to provide a working platform which will generally cover the whole plan area of the structure. From this men can operate and on it can be stored such materials as reinforcement and blocking-out pieces for door openings, etc.

Concrete will normally be raised by a hoist or tower crane and distributed around the outside of the structure by hand barrows, light skips, monorail or other devices considered appropriate.

An essential feature of a sliding form is a platform connected with and below the inner and outer faces of the main shutter, from which operatives can carry out work to impart a sufficiently high standard of surface to the concrete emerging from the shutter. Occasionally, when faults have developed in the work, these will be corrected from the same platforms.

In the earliest examples of sliding formwork, raising of the shutter with reference to the jacking rods was carried out by hand-actuated screw jacks. However, the specially designed jacks used now are almost universally hydraulically operated from a central control panel. The mode of operation is basically that the jaws on the jacks grip the jacking rod firmly whilst the body of the jacks, attached to the yoke, are moved upwards in about 12 mm intervals, the rate of travel varying between 150 and 500 mm/h according to circumstances.

The forming of the narrow cavity round the jacking rods ensures that whilst adequate support is given against buckling under vertical loading as they are lengthened, the rods can be recovered for re-use after completion of the slide.

To ensure success of sliding operations it is essential to

Figure 37.15 Continuous vertical construction of circular silos with sliding formwork. *Note*: sliding form; jacking points; working platform; monorail transporter distributing hoisted concrete. (*Courtesy*: John Laing and Sons Ltd)

provide a high standard of control over the quality of the concrete used and the rate of sliding so that when it emerges from the shutter the concrete is capable of selfsupport without slumping. The concrete must also be amenable to surface floating, etc., to remove minor blemishes.

Sliding operations involve 24-h working, often under very variable weather conditions, and a high standard of job organization is necessary to ensure that there is continuity of all operations involved. Breakdown of equipment which could result in long delays and perhaps in the extreme, the abandonment of a slide, is best guarded against by either duplication of vital items or constant survey to reduce the risk of untimely failure.

37.8.5 Gunite (shotcrete)

Gunite consists essentially of a mixture of cement, sand and water which is sprayed from a nozzle into the required position. In some cases, coarse aggregate of about 10 mm size may also be added. The first step in the process is the mixing of cement and sand in the required proportions. The mixture is then fed into a piece of plant called the 'gun' which consists of one or more chambers connected to a compressed-air supply. This gun feeds the material in a continuous flow into a pipeline, along which it is conveyed pneumatically, until it reaches the nozzle at the placing end. At the nozzle, a spray of water is introduced, under pressure, into the passing material and the resulting mix emerges from the nozzle at high velocity on to the required surface. The optimum distance between the nozzle and the surface is 1 to 1.5 m (Figure 37.16).

The condition of the sand, before mixing with cement, should be damp (a moisture content of 3 to 5% is generally considered desirable) in order that the particles can retain a coating of cement. Sand which is too wet, however, may cause a blockage to develop in the system.

The amount of water added at the nozzle is controlled by a valve operated by the 'nozzleman'. The amount is critical since too wet a mix will result in slumping off the surface, whilst a mix which is too dry will lack cohesion and will result in a considerable loss of material due to excessive rebound off the surface.

This basic guniting procedure is sometimes referred to as the 'dry process' to distinguish it from the so-called 'wet process'.

In the 'wet process', the materials are mixed initially, with the required amount of water (as in the case of normal concrete) before they are fed into the pipeline. The mix is forced along the pipeline by the positive displacement action of a concrete pump or, alternatively, by pneumatic means. At the nozzle end, compressed air is introduced to provide momentum to force the material out in the form of a spray.

Figure 37.16 Gunite being sprayed from a nozzle to increase the thickness of existing concrete on a cooling tower

The amount of water added in the 'wet process' is predetermined to a controlled amount and is not dependent on the judgement of the nozzleman, as in the case of gunite. However, the nature of the wet-mix process requires a mix with a higher water content than that produced in the gunite process, and accordingly, the strength and allied properties of the material will be inferior. There is also more difficulty in clearing any blockages which may occur in the pipeline.

In the US, both the 'dry process' and the 'wet process' are generally referred to as 'shotcrete'.

Gunite has been in use for about 50 years and its main application has been in the repair of deteriorated concrete structures where a layer of good-quality mortar is needed to reinstate the surface. More extensive use of the process has been limited by economics since normal concreting procedures are cheaper. It has, however, been used successfully in constructing swimming pools, culverts, retaining walls, tunnel linings, and intricate curved structures. It has also been used to strengthen existing structures by increasing the thickness of the concrete.

37.8.6 No-fines concrete

As the name implies, no-fines concrete contains no sand or fine aggregate. It is therefore characterized by uniformly distributed voids throughout the mass, which give it a relatively low density. No-fines concrete may therefore be considered to be a particular form of lightweight concrete.

The main applications of no-fines concrete are in the construction of loadbearing walls for low- and medium-rise housing. It has also been used extensively for the infilling panels on high-rise framed structures. Other uses include the provision of drainage layers in civil engineering works and the paving of free-draining parking areas.

When used for building purposes, the optimum size of the aggregate is 10 to 20 mm. It is usual to specify an aggregate of which not more than 5% is retained on a 20 mm mesh sieve and not more than 10% passes the 10 mm sieve.

In order to achieve a satisfactory cellular structure with adequate strength, it is found that mix proportions with a cement:aggregate ratio of 1:8 by volume give the optimum result. In cases where strengths higher than the normal requirements are called for as, for example, on loadbearing no-fines concrete used for four- and five-storey buildings, cement:aggregate ratios of 1:7 by volume, or even slightly richer, may be required.

Cube strengths obtained with 1:8 mixes vary from about 4 to 9 N/mm^2 at 28 days, the corresponding densities being about 1600 to 1850 kg/m^3.

Where strength is less important, as in the case of drainage layers, cement:aggregate ratios can be reduced to 1:10 by volume or leaner.

The water content of no-fines concrete should be the minimum necessary to ensure that each particle of aggregate is coated with a shining film of cement paste. If insufficient water is used, there is a lack of cohesion between the particles giving a friable appearance and loss of strength. Too much water causes the cement paste to run and separate from the aggregate. A water:cement ratio of about 0.40 is usually satisfactory for 1:8 mixes when using dense aggregate.

The main advantages of no-fines concrete, in comparison with normal dense concrete, when used for building construction are:

(1) Lightness in weight.
(2) Low thermal conductivity.
(3) Capillary absorption of water is virtually eliminated.
(4) Light formwork can be used.
(5) The open texture provides an excellent surface for the application of a rendered finish.

As a building process, the no-fines concrete construction technique has the particular benefits of being simple, economical and fast.

A further benefit of using no-fines concrete is that it will not segregate and it can therefore be readily placed in deep lifts of up to three storeys high in one operation, if required. It is important to maintain a level head of no-fines concrete in the formwork along the wall under construction, since localized full-height pouring may cause inclined planes of weakness (pour planes).

Unlike normal concrete it is not necessary to compact no-fines concrete, but some rodding should be given to it to ensure that the formwork is evenly filled. Also careful rodding should be carried out whenever obstacles such as window openings and lintel bearings occur.

No-fines concrete presents some difficulty in the fixing of various fittings and it is necessary to embed nailing blocks of timber which are attached to the formwork prior to pouring. Provision should also be made for suitable openings and chases before pouring since it is difficult to cleanly cut away the no-fines concrete for services.

37.8.7 Concrete diaphragm walls

During the last 15 years or so, there has been a spectacular growth in the construction of concrete diaphragm walls in the UK.

The technique, which was initially developed in Italy to prevent the seepage of water below dams, has subsequently been extended to the construction of retaining walls and loadbearing elements at the sides of deep basements, underpasses, etc. Diaphragm wall construction is invariably carried out by specialist contractors.

The basic process consists of excavating a trench in the ground which is filled with a slurry of bentonite mud to stabilize the sides of the trench as the excavation proceeds. A reinforcement cage is lowered through the bentonite into the trench and concrete is then placed by tremie pipe, gradually displacing the bentonite as it fills the trench.

The key to the process is the use of bentonite which is a thixotropic clay. When it is mixed with water to form a slurry, it has the useful property of forming a membrane of low permeability at the sides of the excavation. The face stabilization is improved as the density of the bentonite slurry is increased but in general the aim should be the achievement of a density of between 1.02 and 1.04 g/cm³. This may be achieved with a slurry containing 4 to 6% bentonite and 1% fine sand. The gel membrane which is formed along the sides of the trench allows the bentonite slurry to exert a hydrostatic head in excess of the *in situ* head of groundwater and the lateral earth pressures. Additional stabilization is obtained by limited penetration of the bentonite slurry into the adjacent soil.

In the initial stages of diaphragm wall construction, it is necessary to construct concrete guide walls which are usually about 1 m deep and about 300 mm wide. These walls serve the purpose of fixing the line of the wall, controlling the direction of the trenching tool and retaining the soil near the surface.

Percussive, rotary or excavating tools may be used for forming the trench, the first two types being necessary where excavation in rock is required. Excavating tools may be of the auger, bucket, shovel or clamshell grab type which cut the soil in bulk and bring it up above ground level for discharge.

As the excavation proceeds, the bentonite slurry is simultaneously pumped into the trench. Since it is not possible to avoid the mixing of the bentonite with detritus arising from the excavation process, there is inevitably a settling of this sludge to the bottom of the trench and this must be removed before the reinforcement cage is positioned in the trench. If this is not done, it is likely that the concrete which is poured into the bottom of the trench will flow over the sludge and not displace it. Clearly, the presence of such a soft layer would seriously reduce the loadbearing properties of the wall.

After the bottom of the trench has been cleaned out the reinforcement cages are lifted into position and supported at the right level. Concrete is then tremied into the trench and the bentonite slurry is gradually displaced as the level of concrete rises. The displaced slurry is pumped into settling tanks for re-use or else removed from site.

The difference in density between the bentonite slurry and the concrete is generally enough to prevent intermixing except for a layer of about 300 to 600 mm in the interface zone. The concrete should be very workable in order that it can flow readily and the aim should therefore be a concrete having a slump of 150 to 250 mm so that it behaves like a heavy viscous fluid. The coarse aggregate should preferably be rounded gravel of 20 mm maximum size to enhance the flow properties and plasticizing admixtures are normally recommended. A cement content of not less than 400 kg/m³ is necessary to provide adequate strength in the concrete.

The degree of compaction achieved by gravity in a very workable concrete placed by tremie pipe is generally adequate. Vibration is not required and in any case would be undesirable since it would cause segregation in a very workable concrete.

Diaphragm walls are normally constructed in panels ranging from 700 mm to 1 m wide, 3 to 6 m long and 6 to 30 m deep. The length of a panel is mainly determined by the soil stability and this leads to a practicable maximum length of 6 m in practice. The distance of lateral flow of concrete from the bottom of a tremie pipe should not exceed 3 m in order to ensure a uniform flow. For long panels, therefore, two or more tremies should be used.

Continuity of concrete placing is essential and a continuous rate of at least 20 m³ concrete per hour is desirable. If serious delays occur in the supply of concrete, difficulties will arise with achieving a correct tremie technique and this can lead to undesirable trapping of bentonite slurry in the body of the concrete wall.

Since diaphragm walls are cast in a series of panels, it is necessary to ensure that the resulting joints between panels are watertight and that they provide an effective key between the panels. This is generally achieved by means of a steel tube installed vertically as a stop end at the end of a panel. This steel tube is removed after the concrete has set to leave a semi-circular-shaped joint at the end of the panel.

One of the problems is the complete removal of the bentonite slurry from the surfaces of the reinforcement bars. Although the tremie technique enables the concrete to displace the bentonite slurry as a mass movement, it cannot be expected to remove completely the coating of bentonite around the reinforcement bars. Since such a coating will adversely affect the bond between the concrete and the steel, it is necessary to use deformed bar reinforcement.

In the case of diaphragm walls constructed round the perimeter of deep basements, the excavation of the ground will follow the completion of the diaphragm wall. Some form of support system will then be necessary to resist the lateral pressure of the earth behind and this may be achieved by props or by ground anchors. The technique whereby the wall is tied back with ground anchors has the particular merit of allowing a clear space free of obstructing props and struts in the excavated area.

37.9 Precast concrete

Precasting of concrete is widely practised in all branches of civil engineering but perhaps the most spectacular is in maritime work. For example, units weighing thousands of tons to be linked together to form submerged vehicle tunnels across narrow waters are frequently built in docks and made temporarily buoyant by adding bulkheads. They are then floated out to their permanent location where a number are strung together on a prepared base below the sea-bed, to make a complete tunnel.

37.9.1 Bridges

Many concrete structures can be built either *in situ* or by using a number of precast units which when assembled together, often by *in situ* work, will form an equivalent structure; here, the emphasis will be on bridge work.

Design studies carried out by the engineer and influenced in large measure by his past experience will indicate which method is the more likely to result in lower cost, simplicity and speed of building. The findings of these studies will be incorporated in the contract designs.

When prestressed concrete beams of various types are a feature of the design then the contractor may have to consider either buying-in or making within his own organization. He will rarely consider setting up equipment to produce long-span box-section beams designed for production by the fully bonded (long-line) system because of the high capital cost involved; but where the beams incorporate their own inbuilt anchorages for prestressing tendons it is often open to consideration whether the beams should be factory made and hauled to site by road and rail or built on the job.

In some instances, for instance where very heavy long-span beams, which could not be brought to the site because of road or rail restrictions, are required, there will be no alternative to site casting. For the smaller, readily transportable, handleable units, such factors as the cost of preparation of suitable casting beds, the cost of concrete production and of providing the high degree of supervision over production need to be considered. Quite often, bridges are built in locations where access is difficult even for small units. Here the alternatives of building on site or waiting until better access can be provided for brought-in beams and the cranes for handling them need to be considered.

Precast concrete units weighing up to about 130 t have formed a substantial part of several overhead urban road works.

37.9.2 Tunnel works

Not perhaps so much in the public eye have been schemes in which the roadways in bored two-lane vehicular tunnels have been precast and set in position somewhat below the axis of the tunnel. In this arrangement the large area below the road is used primarily for ventilation but also for services of all kinds.

Tunnel road deck units will normally be formed in lengths of up to 6 to 8 m in precasting yards at one or both ends of the tunnel, according to the number required and the time available between the completion of driving and lining and opening to traffic.

A high standard of dimensional accuracy and surface finishes is called for, particularly if the upper surface is to form the running surface of the road without recourse to an applied bituminous or further concrete finish; additionally, accurate and sufficient bedding of the units is called for to ensure good performance.

Where ground conditions are appropriate, precast concrete units may be used for lining tunnels. These units may be either solid sections which are stressed into contact with the ground (often manhandled units in the smaller-diameter tunnels), or ribbed sections, both of which are put into position with mechanical erectors. Precast tunnel linings are used as widely as possible because of their low cost as compared with that of cast-iron tubbing.

37.9.3 Cladding panels

Precast concrete cladding panels have been used extensively for the façades of buildings for many years and a great variety of shapes, sizes and surface finishes may be seen throughout the world.

The size of a cladding panel is often determined by the site cranage and the aim should be to limit the weight of a cladding unit to a value which is no more than the other site loads that the crane will have to lift. If this aim is disregarded, it will be necessary to hire-in a heavy mobile crane for lifting the cladding into place.

Although large cladding panels are more expensive to transport and handle, the merits of using them should be considered in each case. The advantages of large panels include faster construction, a reduction in the number of fixings and fewer joints. The latter is a pertinent point as far as leakage problems are concerned.

The weight of the panels will also be influenced by the density of the concrete and there is an obvious benefit in using lightweight concrete. The latter may take the form of concrete made with lightweight aggregate or it may be made from concrete having a cellular structure. An alternative approach to minimizing the weight is to use normal concrete mixes in thin panels, but this introduces problems of providing adequate cover to the steel reinforcement. The consequences of inadequate cover can be seen in the unsightly cracking and corrosion staining that has occurred in those cases where this basic requirement has been given scant attention. For very thin panels where it is not possible to obtain the necessary cover, stainless steel reinforcement should be used.

The development of glass reinforced cement (GRC) in recent years had led to this material being successfuly used as a cladding material. Glass reinforced cement is basically formed in thin sheets which not only gives it the advantage of light weight but also enables it to be provided in panels to a wide variety of shapes. One particular application is the production of panels having a light insulating material (such as polystyrene) sandwiched between two sheets of GRC.

Whatever the material used for cladding panels, dimensional accuracy is an essential requirement. Normal standards of accuracy are covered by the tolerances given in CP 297:1972, 'Precast concrete cladding'. If tighter tolerances are sought, it is necessary to bear in mind that costs increase dramatically as the degree of accuracy is increased.

There are various means of fixing precast concrete cladding panels to the supporting structure and the choice of a suitable fixing arrangement must, in particular, take account of adequate supporting strength, the need for adjustment to accommodate any inaccuracies and the long-term corrosion resistance of the metal used.

37.10 Concrete construction in hot arid countries

37.10.1 Introduction

Concreting in hot arid countries presents difficulties not usually encountered in areas with temperate climates. The prime difficulty is dealing with the adverse effect of the high temperatures and solar radiation not only on the concrete in the handling stages but also on the concrete in the early hardening state.

However, another factor which can be equally difficult in hot arid areas, if not more so, is the presence of aggressive salts in the ground and in sources of fine and coarse aggregates. Sources of water can also contain undesirable levels of salts and even the atmosphere contains wind-borne salts which can cause trouble.

When no allowance is made for these factors in areas of rapid development, such as in parts of the Middle East, the result has been poor-quality concrete work lacking in durability.

37.10.2 Mixing and handling concrete in hot weather

The effects of high ambient temperatures, intense solar radiation and variable humidity pose several problems when planning and carrying out concreting operations. In the Middle East, the annual temperature range in the shade generally varies from about 10 to about 50°C, the latter being experienced during the period June to September. Moreover, the problems caused by high temperatures during this period are increased by the rapid changes in relative humidity which can vary from 25 to 100% within a daily cycle.

This harsh environment influences the behaviour of concrete in both the plastic and the hardened state. In the former, the rapid loss of water by evaporation results in a corresponding loss of workability, and concrete which has been carefully designed to give the right workability at the mixer may well be too stiff for proper compaction by the time it reaches the point of placing. Allowance must therefore be made for this effect by increasing the water content at the mixer above that required. This approach calls for careful judgement since there are several factors which influence the loss of workability, not least being the method of transporting, and the distance to the placing point. It is generally considered preferable to convey the concrete in truck mixers which provide protection from the direct heat of the sun; they may be kept relatively cool by spraying with water periodically. If open transport is used, the concrete should be covered with damp canvas or similar material.

It is beneficial to paint all concreting plant white since the exposed surfaces will more readily reflect the solar radiation. Temperature reductions on former dark surfaces can amount to 10 to 17°C by this straightforward expedient.

Care is required to ensure that the water adjustment at the mixer is not excessive since segregation of the over-wet concrete may otherwise occur during transporting. To avoid this situation, the use of admixtures is recommended. Water-reducing and retarding admixtures have become very popular for concrete work in hot countries since they permit the amount of water in the concrete to be reduced and, moreover, they extend the restricted period of time during which the concrete remains sufficiently workable for handling purposes.

The rate of loss of workability increases as the temperature of the concrete increases and therefore it is advantageous to start with the concrete at as low a temperature as possible. Measures to keep the stockpiles of aggregates cool are therefore of benefit. These measures include shading the aggregates from the sun and spraying with cool water. The latter should be used with discretion because there is a danger that the water content of the aggregates may become very variable and consequently cause difficulties in controlling the water content of the concrete. A fine spray of water uniformly applied two or three times during the day is probably the best approach.

The temperature of the concrete when it is first produced at the mixer will depend on the temperatures, the specific heats, and the proportions of the constituent materials. The temperature of the fresh concrete can be estimated from the following formula:

$$T_c = \frac{(t_c + AT_a + 5Wt_w)}{(1 + A + 5W)}$$

where T_c, t_c, t_a and t_w are the temperatures of the concrete, cement, aggregate and water respectively, A is the aggregate:cement ratio, and W is the water:cement ratio.

This formula is based on a specific heat for water of unity and on a specific heat for cement and aggregates of 0.2 (this figure being used for simplicity instead of the more correct figure of 0.22).

This formula indicates that the initial water temperature may be more important than the aggregate temperature since water has a high specific heat. It is therefore important to keep the water cool and consequently it should preferably be stored in tanks below ground; if this is not done, the tanks should be shaded and painted white. All supply pipes should be buried in the ground and insulated.

In severe conditions, the use of ice will cool the water very significantly. However, ice is frequently not available in the quantities required and also it can be expensive. Ice-making plant or refrigeration equipment is sometimes installed at concrete-mixing plants which provide large quantities of concrete. Generally, it is not considered that ice is really necessary, although it may be desirable in some cases to add blocks of ice to the tank(s) of water early in the morning well before concreting starts.

37.10.3 Maximum temperature of the concrete

To ensure that good-quality concrete is produced, many specifications impose a maximum temperature on the concrete produced. Based on experience in the hot arid zones of the US this maximum temperature is often stated at 32°C.

The practical limitations of this value in the Middle East have led to a more realistic approach there as more experience has been obtained. Depending on circumstances, the maximum temperature allowed in the concrete is now usually between 35 and 38°C. Certainly, a very considerable amount of good-quality concrete has been placed with a maximum temperature requirement of 38°C. However, it must be added that it would not be prudent to relax the temperature requirement to more than 40°C since loss of workability, handling difficulties and the tendency of the concrete to crack after placing increase very significantly above this temperature level.

Clearly, there is no one unique value for a maximum concrete temperature since the behaviour of the concrete will be influenced by the relative humidity and amount of wind as well as the temperature. In some of the coastal regions of the Middle East, the high relative humidity will tend to alleviate the drying effect of the high temperatures. Conversely, where relative humidity is low, it is preferable to be cautious and limit the concrete temperature to a relatively low value, especially where drying winds are prevalent.

The concrete temperature can, of course, be influenced by the time of day during which the construction work takes place. Ideally, it is best to start work either in the early morning or early evening and avoid concreting during the high midday temperatures. Apart from the effect on the concrete, there is the important benefit that the operatives will be able to work more efficiently during the cooler parts of the day. Working at night is sometimes a way of avoiding high daytime temperatures, although it may not be so popular and, moreover, supervision is generally not as good as that in daylight.

37.10.4 Protection and curing of the concrete

Once the concrete has been placed, compacted and finished, adequate protection from the sun and adequate curing is

essential. This aspect of the work certainly requires far more attention in hot climates than in temperate ones.

If exposed concrete surfaces are not protected from the sun directly following the finishing operations, continuing evaporation of water is likely to lead to plastic shrinkage cracking. This type of cracking on paved areas is characterized by a series of cracks parallel to each other, often running at about 45° to the line of greatest slope. Although they are usually relatively shallow, they may well penetrate to the steel reinforcement and provide a path for corrosive salts to reach the reinforcement with a risk of consequent steel corrosion.

More severe cracking can occur subsequently in the first 2 or 3 days if the concrete is not protected and is allowed to reach a high temperature before it eventually cools and tries to contract.

The risk of cracking is related to the initial concrete temperature and this adverse effect is one of the main reasons for specifying a maximum concrete temperature at the time of placing, as mentioned above.

The Portland Cement Association[1] has produced a chart (Figure 37.17) which enables the rate of evaporation of water from the concrete to be determined from known values of air

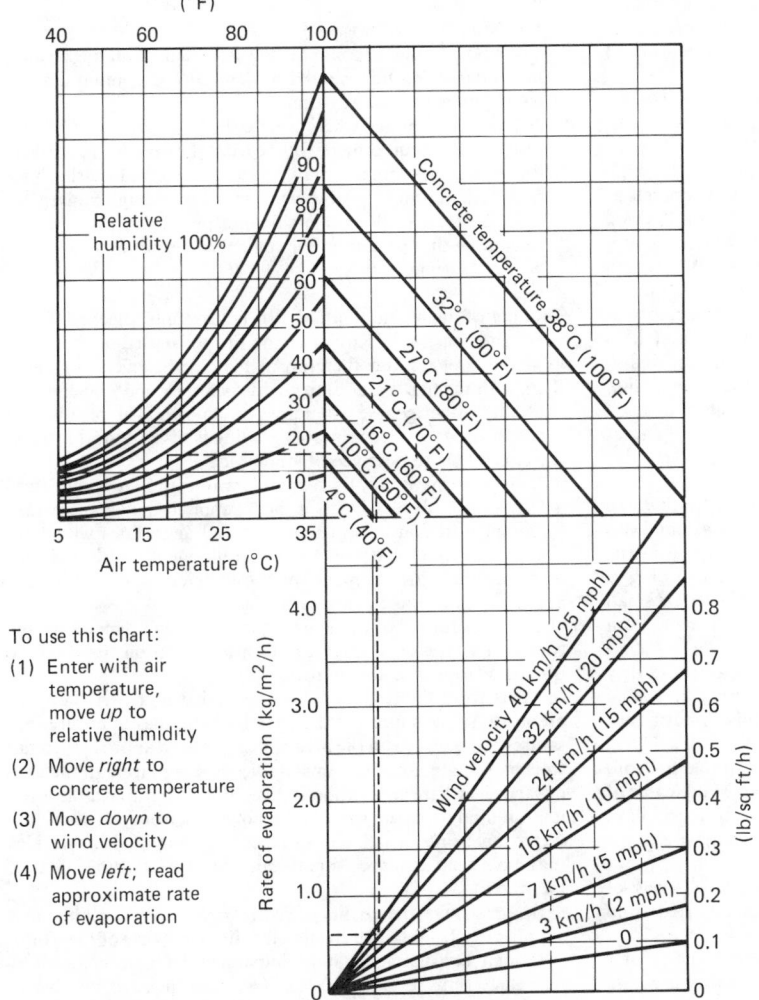

To use this chart:

(1) Enter with air temperature, move *up* to relative humidity

(2) Move *right* to concrete temperature

(3) Move *down* to wind velocity

(4) Move *left*; read approximate rate of evaporation

Figure 37.17 Effect of concrete and air temperature, relative humidity, and wind velocity on the rate of evaporation of surface moisture from concrete

temperature, relative humidity, concrete temperature and wind speed. If it is found that the rate of evaporation approaches 1 kg/m²/h, precautions against plastic shrinkage working are deemed necessary for paving work. These precautions invariably will be required when working in areas with hot arid climates.

Curing procedures in hot conditions are the same as those already described in earlier sections for concrete work in general. However, for hot climates, curing should be started earlier and be applied more diligently. Exposed surfaces, such as paving, should not be left exposed for more than 20 min and preferably less. Curing membranes which are sprayed on to the fresh concrete surfaces should contain a pigment to reflect solar radiation, and two coats should be applied in very hot conditions. It may also be necessary to provide covers or suitable shading over the concrete for several hours to protect it from the excessive heat of the sun. It is important that provision is made for adequate ventilation under these covers.

Where damp hessian is used to cover concrete for several days after placing, the water which is sprayed periodically on to the hessian should be relatively free from salts. If it is not, the repeated application will build up chlorides in the concrete which may reach a level where corrosion of the reinforcement will be initiated. This is an important point to watch in hot arid regions where suitable water supplies are limited. In coastal areas, the temptation to use seawater for curing reinforced concrete must be firmly resisted.

37.10.5 Strength development in hot weather

It is well known that concrete which is placed and cured at high temperatures will achieve higher early strengths than concrete at standard laboratory conditions of 20°C. At an age of 28 days, however, the situation is reversed, and concrete which is maintained at a temperature of 40°C will have a 28-day strength which is nearly 20% less than concrete maintained at 20°C.

Research has shown that concrete made in the laboratory at 38°C produced cube results at 28 days which were about 15% lower than concrete made at 18°C. This reduction occurred in spite of the fact that after 1 day, all the cubes in the investigation were stored in water at 14 to 19°C before testing at 28 days.

37.10.6 Concreting materials

Rigorous quality control is the key to producing trouble-free concrete in hot arid areas, particularly the Middle East, and this applies to the choice and handling of materials as well as the construction process. On major projects, extensive testing is required both before and during construction and even on small projects it is advisable to ensure that the materials have been checked by some basic tests.

Aggregates are probably the main cause for concern and they should be carefully assessed by a full programme of testing before they are approved for use. Natural sands in hot desert countries, especially beach sands, can have very high salt (i.e. sodium chloride) contents and are clearly undesirable for concrete unless the chloride is removed by washing. Many cases of corroded reinforcement and cracked concrete in structures have been due primarily to the use of a sand containing too much salt.

Coarse aggregates may also contain undesirable levels of chloride, particularly some crushed limestones quarried from near the ground surface.

Sulphates must also be checked in both fine and coarse aggregates since undesirable levels can cause expansion within the concrete.

The following recommendations apply to the permissible chloride and sulphate contents of fine and coarse aggregates.

Table 37.1 Limits of chloride content for aggregate used in reinforced concrete

Aggregate	Maximum chloride content (as Cl) (by weight of aggregate)	
	Using ordinary Portland cement (%)	Using sulphate-resisting Portland cement (%)
Fine aggregate (sand)	0.06	0.04
Coarse aggregate	0.03	0.02

Chlorides (for reinforced concrete)
Small adjustments may be made to these limits, if necessary, subject to the overriding requirement that the acid-soluble chloride present in the concrete does not exceed 0.30% by weight of ordinary Portland cement or 0.20% weight of sulphate-resisting Portland cement.

Sulphates (for all classes of concrete)
The acid-soluble sulphate (as SO_3) content of all aggregates, both coarse and fine, should not exceed a maximum limit of 0.40% by weight of aggregate.

This limit applies when using either ordinary or sulphate-resisting Portland cement. In the case of mixes with relatively low cement contents, it may be necessary to reduce this limit of 0.40% in order to comply with the overriding requirement that the total acid-soluble sulphate present in the concrete (including that present in the cement) does not exceed 4% by weight of cement.

Grading of aggregates should also be carefully checked. Dune sands and beach sands are often too fine to be used on their own as fine aggregate and therefore it may be beneficial to blend them with the fine material from crushed rock and gravel which is usually rather coarse when used on its own. The blending of sands to produce an acceptable grading is often desirable when the concrete is to be placed by pumping.

Dust content is another consideration and in some areas where water is expensive or in short supply it may be necessary to accept dust contents in the processed aggregates which are slightly in excess of BS or ASTM requirements. Although this increases the water demand of the concrete made with such aggregates, the effect is not serious.

The importance of other aggregate tests will depend on the source and type of aggregate and information on these can be obtained from specialist papers.

Apart from the tests, it is recommended that periodic visits are made to the source of the aggregate supplies to ensure that sand is being dug from the area which has been approved or that work is being quarried from suitable working faces. In the case of sandpits where excavation is taken down to water-table level, it is essential to ensure that sand is not taken from just above the water table since it is likely to be contaminated with chlorides that have been absorbed due to upward capillary movement of the water.

In some areas, the quality of cement can be very variable since supplies can be obtained from many different parts of the world. Although consignments may sometimes be accompanied by manufacturers' test certificates, it is nevertheless advisable to check the quality of the cement by tests on samples, especially if the cement has been in transit or store for some months and is no longer fresh. Some form of testing is essential whenever a

Table 37.2 Local criteria for reinforced concrete in the Gulf region

Exposure Conditions	Range of Specification Limits		
	Minimum cement content for 20 mm aggregates (kg/m³)	Maximum water:cement ratio	Minimum cover for reinforcement (mm)
(1) Superstructures, inland with no risk of wind-borne salts and ground level well above capillary rise zone	300–320	0.52–0.50	30
(2) Superstructures, in areas of saltflats exposed to wind-borne salts. Ground level within capillary rise zone	320	0.50	40
(3) Parts of structures in contact with the soil, well above capillary rise zone and with no risk of water introduced at the surface	320–350	0.50–0.45	40–50
(4) Parts of structures in contact with soil within the capillary rise zone, below groundwater level, or where water may be introduced at the surface:			
(a) soil and groundwater free from significant contamination	300–320	0.50	40–50
(b) soil and groundwater contaminated with sulphates and/or chlorides	320–400*	0.50–0.42*	40–50
(5) Marine structures	370–400	0.45–0.42	75–100
(6) Water-retaining structures	400	0.50	40

Note:
* The wide range of these requirements reflects the range of sulphate concentrations in the soil or groundwater, but takes no account of chloride concentrations. (The five levels of significant sulphate concentration adopted in BRE Digest 250 and BS 8110 are used in local specifications.) A tanking membrane is required for the more severe conditions.

change of source of cement occurs, otherwise there is a risk of low concrete strengths being obtained if no allowance is made for a lower-quality cement.

The importance of good concreting materials and the quality of the resulting concrete has been well covered in the CIRIA *Guide to concrete construction in the Gulf region*.[2]

Table 37.2, taken from information in the CIRIA guide, gives the basic requirements of the concrete as based on specifications in use in the Gulf region. These requirements are related to material exposure conditions which can be very severe in some cases. Particular attention needs to be given to ensuring that the minimum cover to the reinforcement is achieved since lack of adequate cover has been responsible for early corrosion and concrete cracking in many cases.

The recommended type of cement in these various exposure conditions depends on the amount of sulphates and chlorides present. Where resistance is needed against sulphate attack and there is no risk of chloride-induced corrosion, sulphate-resisting cement to BS 4027 or ASTM type V should be used. Where there is no significant exposure to sulphates but there is a risk of chloride-induced corrosion, cement with a medium to high C_3A content is preferred (as found with ordinary Portland cement or ASTM type I). Where resistance is needed against both sulphates and chlorides, a compromise has to be made on the type of cement used. Generally a cement containing at least 3.5, but not more than 9%, C_3A is preferred.

Steel reinforcing bars can also be supplied from a wide variety of sources and therefore checks on quality are advisable. In a saline atmosphere, the reinforcement must be stored under covers to prevent the deposition of salts which will cause corrosion. On important projects it is sometimes considered advisable to clean the steel by grit-blasting or rotary wire-brushing to ensure that it is free from salts and corrosion before being fixed. When reinforcement bars are left projecting from the initial lifts of concrete, they should be covered with polythene sheeting as protection from atmospheric salts if any delay in placing subsequent concrete lifts is expected.

References

1 American Concrete Institute (1977) 'Hot weather concreting'. ACI Committee 305. *Proc. Conc. Inst.* 74, 8, 317–332.
2 Construction Industry Research and Information Association (1984) *The CIRIA guide to concrete construction in the Gulf region*. CIRIA Special Publication Number 31. London.

Bibliography

American Concrete Institute (1981) *Manual of concrete practice* Parts 1 to 5. ACI, Detroit, Michigan.
Blake, L. S. (1974) *Recommendations for the production of high-quality concrete surfaces*. Cement and Concrete Association, London, 40 pp.
United States Department of the Interior (1975) *Concrete manual* (8th edn). A Water Resources Technical Publication US Department of the Interior, Bureau of Reclamation. Denver, Colorado.
Concrete Society (1985) *Pumping concrete*. Concrete Society Digest Number 1. Laing Design and Development Centre.
Deacon, R. C. (1975) *Concrete ground floors – their design, construction and finish* (2nd edn). Cement and Concrete Association.
Illingworth, J. R. (1972) *Movement and distribution of concrete*. McGraw-Hill.

Institution of Civil Engineers (1975) 'Diaphragm walls and anchorages'. *Proceedings,* Conference 18–20 September 1974.

Muir Wood, A. M. and Gibb, F. R. S. (1971) 'Cargo tunnel at Heathrow Airport'. *Proc. Instn. Civ. Engrs.*

Murdock, L. J. and Brook, K. M. (1979) *Concrete materials and practice* (5th edn). Edward Arnold, London.

Orchard, D. F. (1979) *Concrete technology* (4th edn). Vol. II, 'Practice'. Applied Science Publishers, London.

Quality Scheme for Ready Mixed Concrete (1984) Regulations.

Sharp, D. R. (1970) *Concrete in highway engineering.* Pergamon, Oxford.

Short, A. and Kinninburgh, W. (1978) *Lightweight concrete* (3rd edn). Applied Science Publishers, London.

Stein, J. and Donaldson, P. K. (1966) *Techniques and formwork for continuous vertical construction.* Concrete Society, London.

Concrete Society (1971) *Underwater concreting.* Technical Report No. 3. Concrete Society, London.

38

Heavy-welded Structural Fabrication

J L Pratt BSc(Eng), MIEE, FWeldl
formerly Research Manager,
Braithwaite & Co. Engineers Ltd

Contents

The vast majority of fabrications in steel are now welded and it is rare to see a new fabrication that is joined by any other method. The steel used is generally to BS 4360 *Specification for weldable structural steels*; the revised 1986 edition includes the weathering steels and the whole range covers steels with yield stresses ranging from 230 to 450 N/mm².

38.1 Welding processes

Figure 38.1 shows the welding processes most commonly used in steel fabrications; in all cases an arc is struck between the electrode or electrode wire and the workpiece resulting in a high arc temperature which melts off the electrode and deposits it in the joint which has to be made. The manual metal arc (MMA)[1] is the most common process and the electrode is deposited manually with the operator controlling the direction of the weld and the build-up of the weld metal. The flux extruded around the core wire when melted by the heat of the arc provides a gaseous shroud which protects the molten pool and arc from atmospheric contamination and controls the weld metal reactions; it can also be the vehicle for supplying certain alloying constituents to the weld metal. The fused slag around the deposited weld metal also helps to form the weld bead shape. There are several types of electrode coverings which function in different capacities and are classified in BS 639 'Covered electrodes for the manual metal-arc welding of carbon and carbon manganese steels'.

Gas-shielded arcs with bare wire or cored wire, can be of the semi-automatic or automatic type.[1-3] The semi-automatic process utilizes a power source, a wire drive unit, incorporating the necessary control units, and a 'gun' which is held by the operator and manipulated manually; the wire is driven through a flexible tube to the gun and a suitable designed nozzle concentric to the gun orifice supplies the gas to the arc. The automatic process usually has a heavier 'gun' or head with the wire (also known as electrode wire or feed wire) fed directly through the gun without the intervening flexible tube; the whole apparatus travels automatically for longitudinal welds or may be stationary for circular fabrications. Higher welding currents and deposition rates are generally used with subsequent water cooling of the head being necessary. The weld metal and arc is protected from the atmosphere by the shroud but a bare wire must contain deoxidizers such as silicon, manganese and sometimes aluminium; these are necessary to prevent some oxidizing processes which occur within the arc atmosphere. A cored wire has the flux enfolded within the electrode wire as typified in the cross-section shown in Figure 38.1; it may be used in semi- or fully-automatic processes. The necessary deoxidants are carried in the flux which may also be the vehicle for additional alloys to be added to the weld; the flux allows for higher welding currents than that in solid or bare wire shielded welds, with the slag allowing better bead shapes to be obtained, and is generally more tolerant to rusty plate conditions which could otherwise lead to porosity.

Shielding gases used for structural steels are carbon dioxide or argon with addition of oxygen or with carbon dioxide and oxygen, the cheapest being carbon dioxide.

Another type of semi-automatic welding popular in the US and now being used in the UK is the self-shielded arc where the continuous electrode contains in its core flux ingredients which vaporize in the arc, shielding the arc and forming a thin slag around the metal droplets as they transfer across the arc gap; deoxiding materials also form part of the flux. The process requires no gas shield and is therefore better-suited for outdoor operations when windy conditions prevail.

The submerged arc (SA) is a process which feeds a bare wire into the arc and the arc is covered by a granulated flux which is

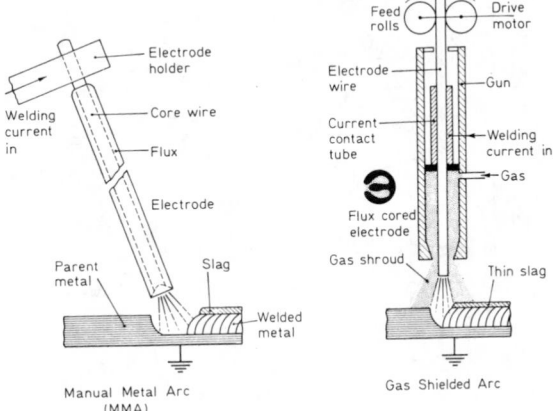

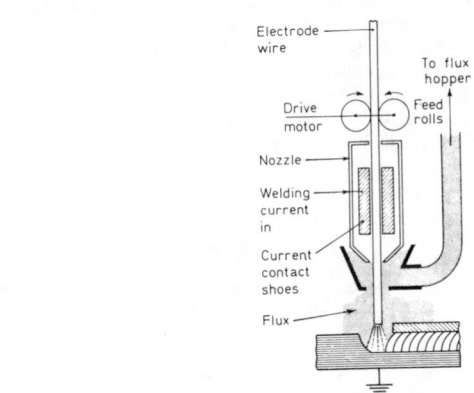

Figure 38.1 Common welding processes

also automatically fed; some of the flux is melted to cover the weld pool as slag and to provide the arc with a gaseous shield. Again alloys can be added to the weld either via the arc or the flux; very high currents can be used in this process[1] and very smooth bead-contour shapes can be obtained. The arc is completely shrouded by the flux and thus it cannot be seen; this gives a total absence of arc glare but, correspondingly, guiding is that much more difficult. This process is more susceptible to rusty or dirty plate conditions than MMA but less susceptible than metal arc inert gas (MIG) and for very heavy weld metal depositions on thick plate, multiple electrodes may be used in the same weld. A semi-automatic form using a small diameter wire can also be obtained.

Figure 38.2 shows schematically two other processes, electro-slag and electro-gas welding; both are completely automatic. In electro-slag welding, the plates to be welded are mounted vertically with the edges of the plate square or unprepared; watercooled copper shoes are mounted either side of the weld seam to contain the molten metal. An arc is struck on the starting block with a little granulated flux added to the weld pool; as the wire or electrode burns off, the temperature of the slag bath increases and the slag becomes electrically conducting; from then on the electrode protrudes into the bath, the arc extinguishes, and the wire metals off due to the I^2R heating of the current. It is thus not an arc process but a continuous cast process used on plate thicknesses over 25 mm and certainly up to 450 mm; for the greater thicknesses three electrodes are fed

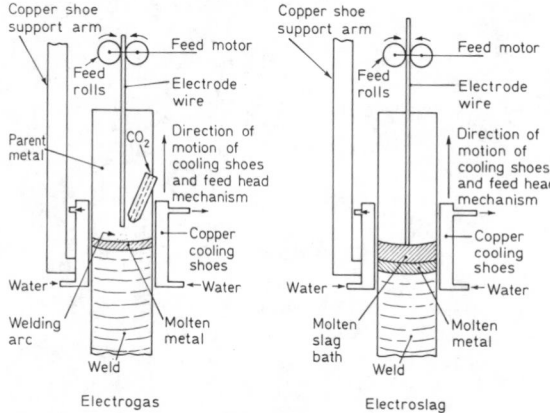

Figure 38.2 Electro-gas and electro-slag processes

simultaneously into the slag bath with, in one application, the whole assembly oscillating across the width of the joint.

There are two methods of applying this process known as electrode or consumable guide (see Figure 38.3); in the electrode method the feed head is at the side of the plate being welded and moves up with the copper cooling shoes as the weld is made. In the consumable guide method the feed head is stationary at the

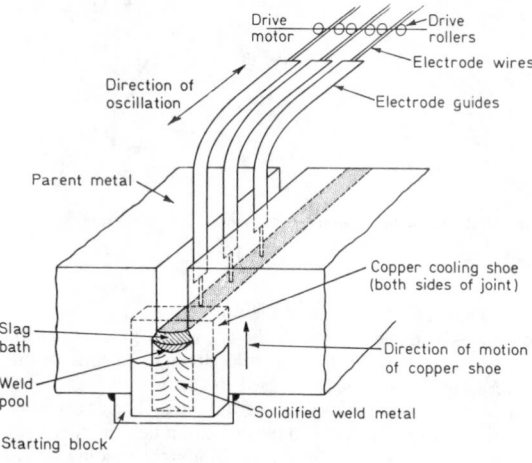

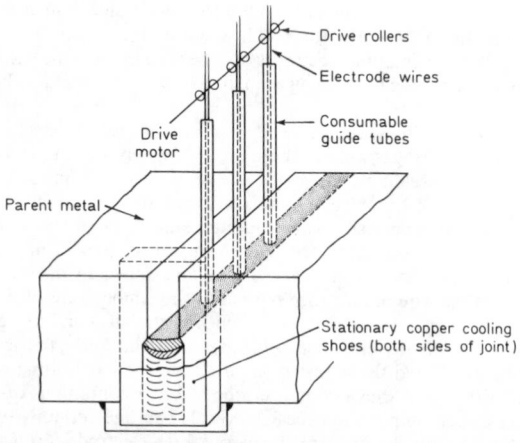

Figure 38.3 Electro-slag welding process

top of the joint to be welded and the wire(s) are fed down to the slag bath by a consumable guide which is insulated from the workpiece by fusible spacers.

As the wire(s) melt so does the bottom of the consumable guide and the copper shoes can be stationary of a length equal to the length of the welded joint; this method requires less sophisticated machinery than the former and is therefore, where it can be applied, cheaper. It cannot obviously be oscillated across the width of the joint.

The electro-gas process is similar to that of electro-slag welding in that the weld metal is contained by watercooled shoes but different in that the weld metal is deposited by a true arc with a thin slag from the flux in the cored electrode; the weld metal and arc is protected by a stream of CO_2.

Since both processes can evolve large heat inputs to the heat-affected zone of the weld and to the weld itself with resultant large grain microstructure, poor fracture toughness may result. Some improvement may be obtained by postweld normalizing treatment, but where fracture toughness may be a problem, expert opinion is advisable.

38.2 Weld details

The two main types of welds used in fabrication are the fillet and butt welds. Fillet welds are shown in Figure 38.4; BS 5400[4] and 449[5] lay down that the allowable stress in a fillet weld is based on the throat thickness, t, or '$0.70 \times L$' where L is the leg length, because t for a stated leg length L will vary according to the included angle, $\gamma°$, between the fusion faces. Table 38.1 gives the values of t for varying angle γ.

Table 38.1

Angle between fusion faces	60 to 90°	91 to 100°	101 to 106°	107 to 113°	114 to 120°
Factor by which fillet size is multiplied to give throat thickness	0.70	0.65	0.60	0.55	0.50

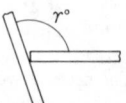

Some typical butt welds are shown in Figure 38.5; these are generally for manual welding.[4] The root-run is usually back-grooved (except where a backing strip is used) so that clean weld metal from the previous root is obtained (Figure 38.6); this ensures homogeneity of weld metal at the root area. These same preparations may be used for semi-automatic welding with no root gaps where root gaps are shown, or with a root run of manual weld to seal the root before applying any semi-automatic process for the rest of the weld. The root run on the second side does not generally require back-grooving since the penetration is enough to ensure weld metal homogeneity.

Submerged arc welding is a high-deposition welding process with deep-penetration characteristics although with a direct current electrode negative power source the burn-off or deposition rate increases with a large diminution in penetration; multiple wires or electrodes may be used in the same weld with

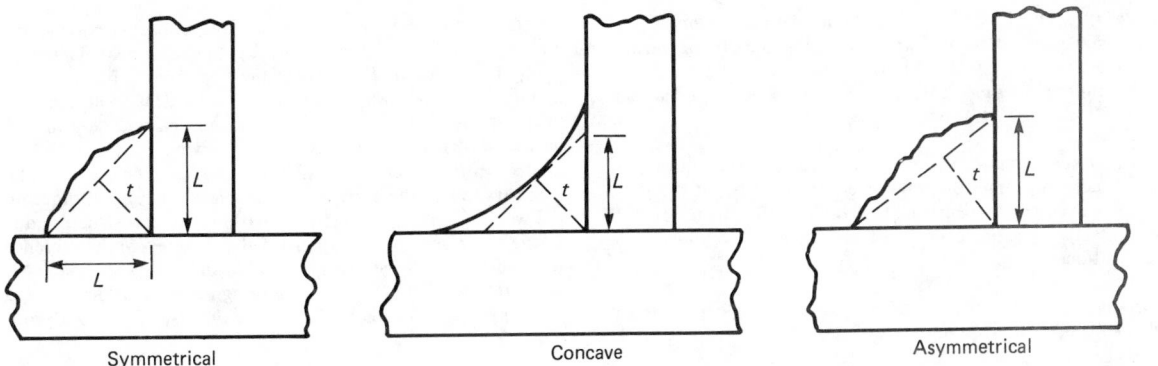

Figure 38.4 Fillet welds. *Note*: minimum length of both legs to be measured for *L*; for concave weld $t \neq 0.7\,L$.

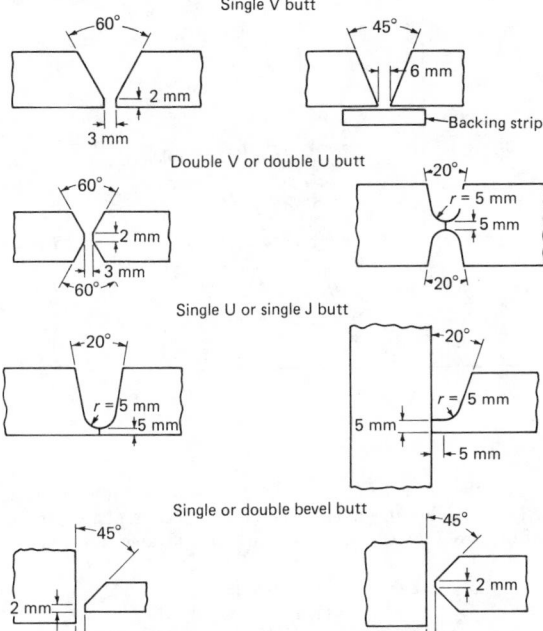

Figure 38.5 Typical butt welds. *Note*: angles and dimensions of root gaps and root faces may be altered to suit welding technique and position of weld, the above being suitable for flat-position welding. Welding is carried out from both sides of all preparations except where a backing strip is employed. To achieve complete penetration, back-gouging (back-grooving) may be employed

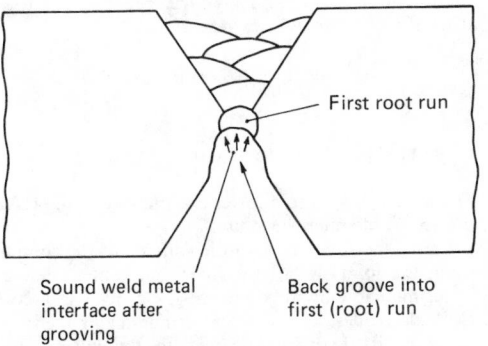

Figure 38.6 Butt weld back groove

the electrodes sharing, in parallel, the same power source or each electrode connected to a separate power source. Weld preparations for such a process are infinite and reference should be made to the suppliers of electrodes and fluxes for their advice; for notch ductile materials, basic fluxes and an increase in the number of runs may be necessary.

38.3 Weld defects

Some typical weld defects are shown in Figure 38.7.

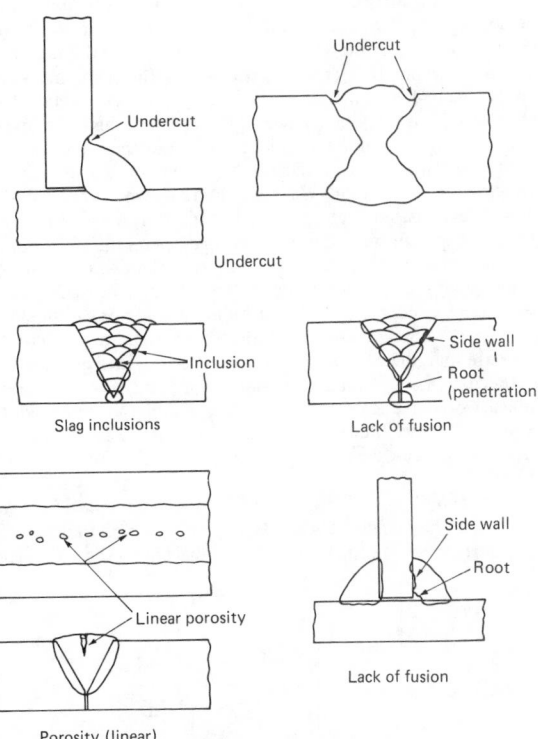

Figure 38.7 Some weld defects

'*Undercut.*' A groove melted into the parent metal at the toe of a fillet weld or root of a butt-weld – produced by the arc but left unfilled by the filler metal. Undercut may be due to incorrect angle between the electrode and the workpiece, too high an arc voltage or travel speed or scaled parent material.

'Porosity.' Due to dirty or rusty parent material surface, damp consumables, arc instability (as evidenced by the stop or starting of the arc when using MMA), gas entrapment from air due to inefficient shielding gas, grease on filler wires. When linear can denote 'lack of penetration'.

'Lack of penetration' for butt welds. Inclusive angle of prepared faces too small to allow the electrode to get at the root, insufficient current to penetrate the landing or landing too small for the set arc parameters or root gap too small to allow more penetration.

'Lack of fusion.' Incorrect manipulation and angle of electrode to ensure side-wall fusion, or root fusion.

'Slag inclusion.' Nonmetallic solid material entrapped between runs of weld metal or between weld metal and parent metal. Due to inefficient clearing of the slag between each run which in turn may be due to wrong weld parameters giving the wrong-shaped interpass bead and positional requirements.

'Spatter.' Small metallic particles ejected from the weld area and forming on the parent material adjacent to the weld. Spatter varies with the arc process and within that particular process may increase or decrease with differing arc parameters.

'Hot-cracking' (solidification cracking). Cracks appearing in the central region of the weld (Figure 38.8) where segregation of sulphur and phosphorus, the lowest melting-point constituents of the weld metal, occurs; thus at temperatures in the region of the solidus thin films of the liquid segregates occur along the grain boundaries (intergranular). The weld metal may thus become susceptible to cracking because of the high shrinkage stresses generated during the cooling of the weld metal. The effect of sulphur may be reduced by obtaining a higher manganese:sulphur ratio in the weld metal whilst at the same sulphur and phosphorus levels an increased carbon content may cause cracking; likewise silicon. Weld metal on its own is low in all the above elements but high 'pick-up' or dilution may be derived from the parent metal; thus any deep-penetration welding process could lead to hot-cracking. A weld bead or nugget whose depth is greater than its width can, in such deep-penetration processes such as the submerged arc or gas-shielded arcs, promote such cracks when the above metallurgical conditions are marginally operative. In this case a wider preparation or the use of more than one run of weld with lower current values would reduce the dilution factor and the depth:width ratio, to decrease the risk of cracking. Guidance is given in BS 1535 on such cracking.

'Cold-cracking' (underbead or HAZ or hydrogen-induced cracking). The heat-affected zone (HAZ) of a weld is that (generally) narrow zone in the parent metal adjacent to the weld bead

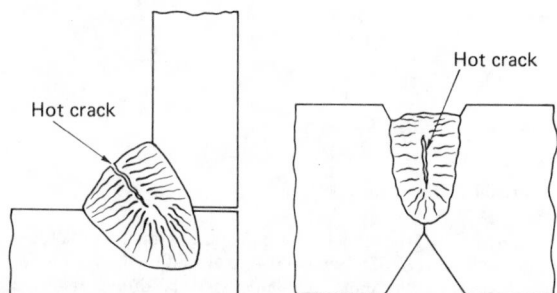

Figure 38.8 Hot-cracking in SA welds

affected by the heat input of the weld and whose microstructure and physical properties might be affected by that heat. This zone is rapidly cooled by the mass of the surrounding parent metal and if this cooling rate is high enough a hardened (martensitic) microstructure may be formed. Cracking may develop in this hardened structure (see Figure 38.9) owing to: (1) the alloy content of the parent material increasing; (2) high cooling rate; (3) restraint and therefore higher residual stresses resulting from the weld contraction; (4) stresses within the microstructure due to the transformation to a hardened structure; (5) the presence of absorbed hydrogen from the weld diffusing into the HAZ when that weld cools and contributing to the creation of micro fissures; and (6) for fillet welds, where the fit-up is bad with root gaps.

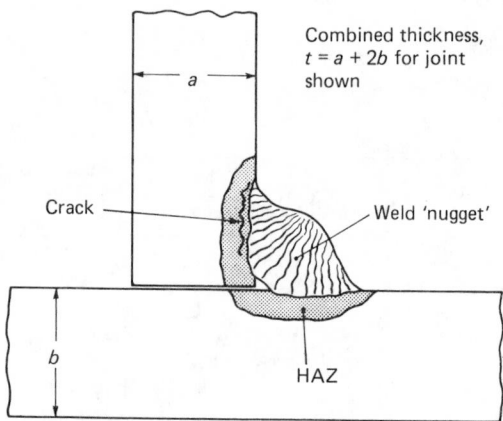

Figure 38.9 HAZ crack

In (1) the presence of alloys in increasing amounts increases the hardenability in the HAZ and their effect can be related to that of carbon by the following carbon equivalent formula:

$$\text{Carbon equivalent}\% = \text{C}\% + \frac{\text{Mn}}{6}\% + \frac{\text{Cr}+\text{Mo}+\text{V}}{5}\% + \frac{\text{Ni}+\text{Cu}}{15}\%$$

Thus any increase of carbon equivalent due to the increase in any of the above alloys will increase the hardenability of the steel. This formula only applies to those steels in BS 4360.

In (2) the cooling rate is assessed partly by the combined thickness t of the joint being welded (Figure 38.9) and partly by the heat input from the weld and any given preheat. The total heat input from any arc may be expressed as:

$$H \text{ (joules/mm)} = \frac{\text{arc voltage} \times \text{current (amps) } XT}{L}$$

where T is the time in seconds to deposit L mm of weld.

Lower t and higher H lead to a lower cooling rate in the HAZ with a less hardenable microstructure.

In (3) the restraint increases with the stiffness of the components making the joint. In (5) the hydrogen content can be reduced by using a low-hydrogen process, e.g. hydrogen-controlled MMA electrodes. The MMA electrode may have to be baked to reduce its hydrogen content to the lowest level possible; in SA welding the flux would have to be dry and preferably of the agglomerate rather than fused flux. With all automatic wires or electrodes no wire drawing compound contaminates should be present; gas-shielded arc processes with solid wire could prove to give weld metals with the lowest hydrogen content.

Preheat curves necessary for combined thicknesses and size of weld deposit (and thus heat input) are given in BS 5135.

Preheat, when applied, reduces the rate of the cooling of the weld and allows more hydrogen to be evolved by the weld metal to the surrounding atmosphere; therefore to be effective it must be applied to the correct temperature and over a sufficient width of the plate. British Standard 5135 indicates that the width of the preheat zone on each side of the weld should be at least 75 mm in any direction from the joint preparation. In practice the temperature is measured by using thermo indicating crayons or paints, the former melting and the latter changing colour when the correct temperature is achieved, and to make certain that the heat has penetrated the full thickness it is customary to heat the far side of the plate with the temperature indicator on the near side, or by heating the near side until the required temperature is indicated on the same side for 2 min for each 25 mm of steel thickness after the heat source has been removed. Although the heats applied are generally low (on the average 100°C), the wide area over which they are used can lead to more distortion than that of the weld itself; it is therefore preferable to use a higher heat input weld source to reduce the preheat required. It is also more economical.

38.4 Distortion

Distortion due to welding is dependent on the heat input from the weld; such heat is concentrated in a narrow zone around the weld area. The subsequent contraction of the heated weld metal and parent metal produce undue stresses in the fabricated part; if unrestrained the fabrication will distort and, if restrained against distortion, residual stresses up to the yield point of the material may occur. The parts being welded may in themselves have residual stresses due to their shape and size and thus their manufacture; these stresses or some of these stresses may be relieved or increased by the local welding heat and thus their distortion due to welding may be difficult to predict. Metals with differing expansion coefficients, thermal conductivities and physical properties will produce different distortion levels with the same weld heat input.

Figure 38.10 shows distortions for typical welds. Figure 38.11 shows joint preparations, welding procedures and some typical plate presetting to compensate for weld distortions. For the heavier type of fabrication it is generally better to fabricate all subsections prior to incorporating them into the main structure but to control the increased distortions for thin-walled constructions it may be preferable to assemble and tack the whole assembly to give a much stiffer structure more able to withstand distortion.

38.4.1 Correction of distortion

For a dished plate (e.g. a dish resulting from an area of plate welded all round the periphery of that area on one side of the plate only) the amount of dishing resulting from such a weld depends on the heat input of the weld and the thickness of the plate. To flatten such an area, spot heat from a heating torch can be applied in several places within the dished area on the outside (convex side) of the bulge; this will increase the amount of dishing on heating but on contracting that side will shrink and reduce the dish. Heat can be up to red heat (600 to 650°C) but does depend on the thickness of the plate; for very thin plate the applied heat may heat both sides to an equal temperature

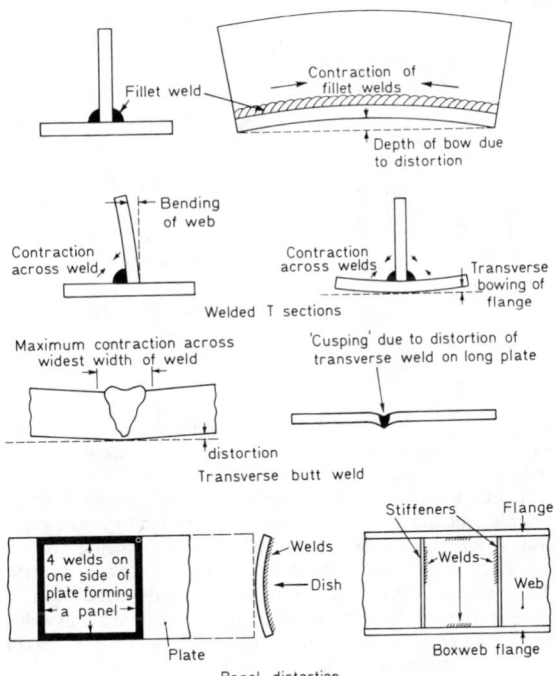

Figure 38.10

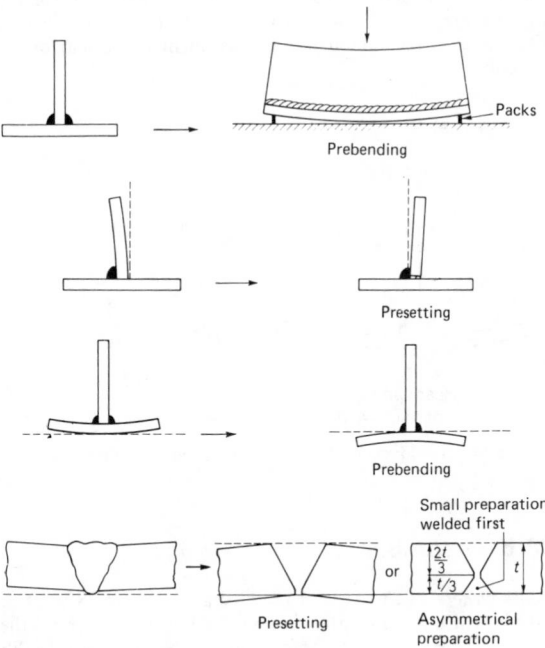

Figure 38.11   Methods to reduce distortion

resulting in equal contractions on both sides of the plate with no decrease in the dish.

Triangular heating on the web and bar heating on the flange of a plate girder or section (Figure 38.12) will increase the camber and can also be applied, within certain limits, to box sections. It is important to note that heating the flange and not

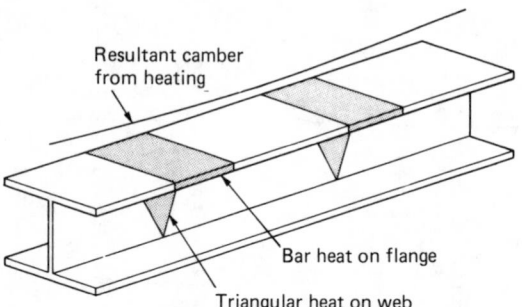

Figure 38.12 Correction of camber

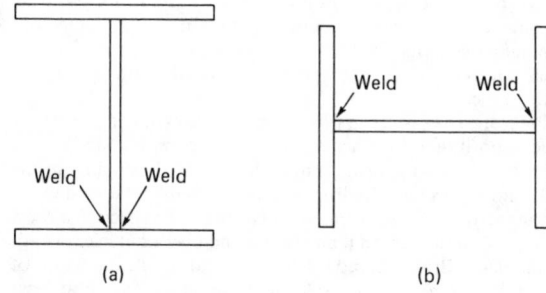

Figure 38.14 Plate girder welds

the web may shrink the flange and increase the camber of the girder but the web, not being heated, cannot shrink to accommodate the increased camber and may therefore buckle.

Angular bending of a flange plate due to the two fillet welds attaching the web to the flange may be corrected by heating in a straight line (Figure 38.13). The effect of introducing heat to shrink areas and to introduce distortions to reduce others, must of a necessity induce stresses into the fabrication; the effect of these stresses and the subsequent increased load on some welds must be watched carefully and if necessary those welds increased in size to accommodate the increased load. All welds or any other form of localized high heats give high residual tensile stresses local to that heat; these stresses in turn generate compressive stresses outside those tensile stress areas. Stress relieving (at about 600 to 650°C) may relieve the structure of any induced stress but in turn must lead to increased or different distortions to accommodate the subsequent movement of the structure.

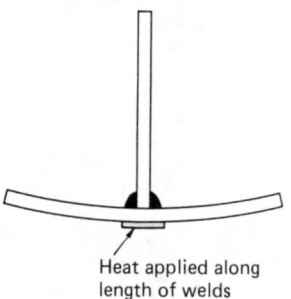

Heat applied along
length of welds

Figure 38.13 Correction for transverse flange distortion

38.5 Assembly

'Plate girders.' These may be welded as in Figure 38.14(a) or (b) by MMA or automatic welding. Tack welding to hold the assembly together must conform to the requirements of BS 5135, with minimum root gaps; large root gaps may lead to HAZ cracks as described previously or 'burn through' when using high current density automatic welds. For girders with top and bottom flanges of differing thicknesses or with top-flange-to-web and bottom-flange-to-web welds of different sizes, different shrinkages may occur in each flange and hence alter the camber of the girder; in such cases it may be necessary to induce triangular heating as described above or to increase deliberately the camber if the web plate is cut to camber in the preparatory

stage. For thin flanges it may be found necessary to prebend the plates as shown in Figure 38.11.

For crane girders it may be necessary to make full penetration welds (Figure 38.15) for the web-to-top-flange welds. When using automatic welds such as SA, care must be taken that hot-cracking does not occur; this can happen when trying to achieve penetration and the dilution of the weld metal by the parent metal is high. To reduce dilution back-grooving may be used (but is difficult in this situation) or the web preparation made wider; the increase in the ratio $p{:}w$ in Figure 38.15 indicates a bigger dilution of the weld metal by the patent metal and the possibility of hot-cracking increases. Hot-cracking invariably occurs on the second side of the joint to be welded since the first weld has made the web–flange assembly rigid or constrained, and it should be noted that hot-cracking may be contained below the surface of the weld and thus not be visible (Figure 38.8).

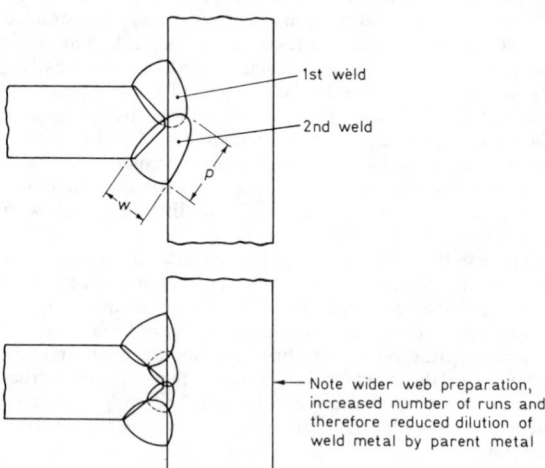

Figure 38.15 Full penetration butt weld, submerged arc

'Box girders.' These are invariably assembled on one flange as the base fabrication plate and must lie perfectly flat on the assembly stallage or a twist in the box may result; all diaphragms are then placed in position after being subassembled and the two webs then tack-welded to diaphragms and flange. As much internal welding as possible is then made before the fourth closing flange plate is placed in position and tack-welded; the four longitudinal web-to-flange welds are then made.

The same comments about differing flange thicknesses or web-to-flange welds in plate girders can apply to box girders.

The choice of flange-to-web longitudinal weld detail may be dictated by the camber and or curvature required in the box. For a large box where it may be difficult to rotate during fabrication (a) in Figure 38.16 may be preferable; where there is camber, (b) is easier to assemble with the flanges outside the webs than (a) where the box-closing flange coming inside the webs can only sit on the diaphragms unless backing strips on the webs are installed between the diaphragms to maintain the closing flange profile.

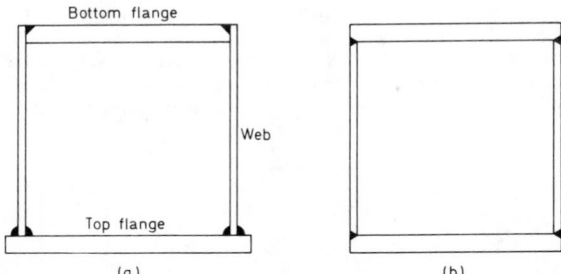

Figure 38.16 Box-girder assembly

Where boxes are to be jointed on site the open ends of adjacent boxes should be stiffened if there are no diaphragms close to the open ends to hold those ends to the required square or trapezoidal profile; for all such ends they would tend, without such stiffening, to have inward bows on all flanges and webs although the four corners are dimensionally correct (Figure 38.17).

In boxes all stiffeners are subassembled on the webs and flanges before the main assembly is completed; to keep all resultant distortion, shrinkage, etc to a minimum it is better that intermittent welding be used on such items if the design requirements can be met with such welds. Stiffeners may be of the bulb, angle or T-type; the latter two may present difficulties for blast or other type of cleaning after welding and also for the subsequent painting.

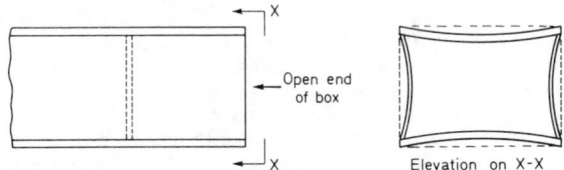

Figure 38.17 Box open-end distortion

38.6 Stud welding

Stud welding shear connectors on the top flange for bridge girders for composite concrete decks is now a widespread practice; the diameters of the studs usually range from 12 to 25 mm and generally vary from 100 to 150 mm in length although 250 mm studs have been welded. The form of the stud is shown in Figure 38.18 and the head of the stud fixes into a gripping chuck in the operator's gun which in turn is placed vertical over the spot to be welded and which rests firmly on a three-point support on the steel surface. When the trigger is pulled an electronic timing device controls the following sequence: The chuck is lifted about 3 mm by an electromagnet and

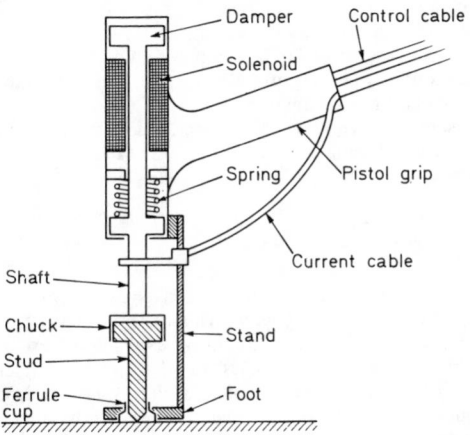

Figure 38.18 Stud-welding gun

a pilot arc is formed about the tapered point (Figure 38.19) which then develops into the main arc, the main arc current being drawn usually from a drooping characteristic transformer–rectifier power source. This arc melts the end of the stud with a resultant melted area on the workpiece and after a preset time the solenoid is de-energized and the stud is plunged by a return spring on to the workpiece while the arc current is still flowing.

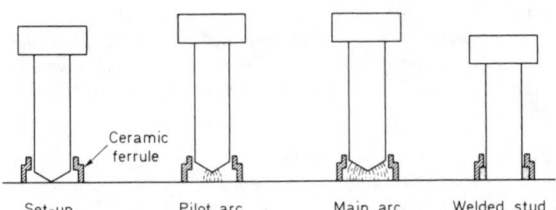

Figure 38.19 Sheer-connector stud

The stud when correctly welded should be of a correct length after welding with a formed upset fillet around it with no undercut; such undercut may be due to incorrect welding parameters or arc blow and when present can lead to easy fracture of the stud from the plate surface. Arc blow because of the high, though transient, currents used may be prevalent when the studs are placed near to the edge of the plate; in such cases an edge plate to extend the magnetic field of the current in the main plate may be utilized (see Figure 38.20).

Studs greater than 22 mm φ are difficult to weld, leading to erratic arcs and sometimes unsound welds; the plate surface on which the studs are being welded should be free of all oily

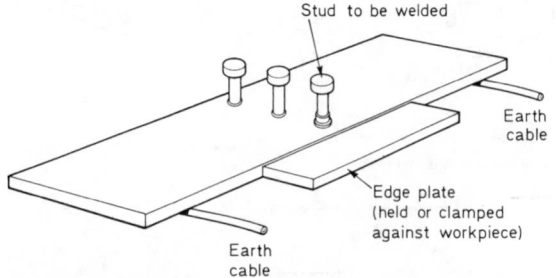

Figure 38.20 Magnetic field edge plate

contaminants, millscale and deep rust. A light surface grinding in the stud area is recommended. The tip of the stud is either sprayed with aluminium or holds a 'slug' of aluminium which acts as a deoxidant when vaporized in the arc; it is important that this deoxidant is not damaged.

One test sometimes employed to ensure the soundness of the steel weld is to bend some to an angle of 30° and to hit or 'ring' the others with a hammer.

38.7 Testing

'Methods.' These may include nondestructive testing methods such as X or gamma radiography, ultrasonics, dye penetrant or magnetic particle testing. Radiography is used almost exclusively on butt welds and ultrasonics on butt and some fillet welds; site welds are invariably tested by ultrasonics and/or gamma radiography employing in general iridium as the source of gamma rays. The standard and scope of testing required is usually determined by the customer and should be ascertained at the enquiry stage.

Before most contracts are commenced some welding procedures may have to be approved by the customer, i.e. a weld joint simulating the thickness, preparation, etc. of an actual weld configuration used in the fabrication must be welded to prove that the proposed welding consumable and method of welding is satisfactory. Such welds may be subsequently tested by nondestructive methods and then physically tested by means of transverse tensile and bend tests, Charpy impact tests, nick-break tests (for fillet welds) and cross-section macrostructures (see BS 709).

The skill of the welders may be approved by the customer's own specific test or by BS 4872 *'Approval testing of welders when welding procedure approval is not required'* or any other subsequent standard with any appropriate nondestructive test requirements.

'Laminations.' Where a plate is laminated or where a plate must be tested for laminations before being incorporated in a fabrication, ultrasonics is the only method by which any such lamination may be detected and the extent measured. Material may be supplied to a standard of lamination testing by the steelmakers; the details of such standards and the appropriate costs may be obtained from the steel supplier. The effect of any lamination on the stability of the structure must be referred to the designer, e.g. the effect of a lamination in a compression member is generally more severe than one in a tension member. It is probably true to say that most structures can tolerate a fairly large degree of lamination in a member before repair is required; the repair of such a lamination is shown in Figure 38.21.

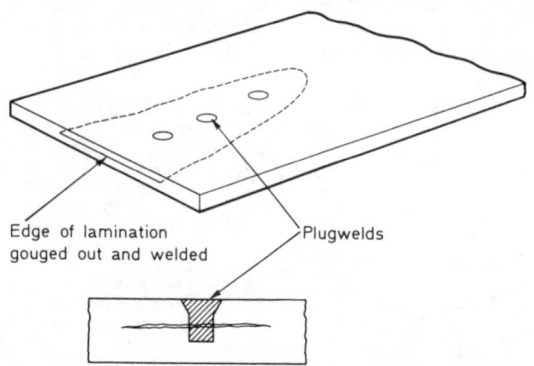

Edge of lamination
gouged out and welded

Plugwelds

Figure 38.21 Repair to lamination

'Lamellar tearing.' This is a result of nonmetallic inclusions in the steel in the plane of rolling merging into a tear due to the stress imparted by the weld (Figure 38.22) or other derived stresses normal to the plane of inclusions. Because these inclusions are very small and scattered throughout the thickness of the material they are not detected radiographically, and, up to now, although detectable by ultrasonic inspection, they cannot be quantified to assess potential cracking. The tear, when it occurs, is of a ductile nature, fibrous and stepped or ragged as shown schematically in Figure 38.22, the steps resulting from the inclusions in different planes being joined to form the tear; the presence of such inclusions decreases the ability of the material to withstand extension under a load applied across the thickness of the plate. One destructive method of assessing the material for tearing is to machine a small transverse tensile test piece and to measure its reduction in area at failure of applied tensile load.

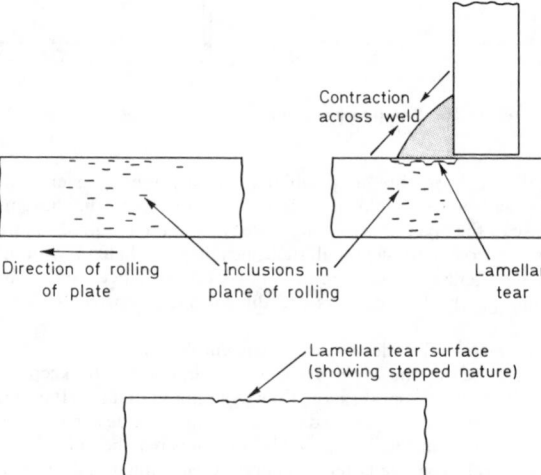

Contraction across weld

Direction of rolling of plate

Inclusions in plane of rolling

Lamellar tear

Lamellar tear surface (showing stepped nature)

Figure 38.22 Lamellar tearing

Joint details that can promote tearing are shown in Figure 38.23; generally the welds must be large to cause contraction across the welds on cooling to create sufficient tensile stresses across the plate to produce a tear in a susceptible material. For fillet welds to generate such a tear the weld in almost every occasion would have to be greater than 12 mm, although many materials can tolerate much bigger welds, and any details which have both large weld and imposed load stresses across their thickness should be avoided; some preferential details are shown in Figure 38.23. On some suspect material it may be helpful to reduce the risk of tearing by buttering the weld fusion face with MMA or SA welds as also shown.

38.8 Significance of defects

Sudden and catastrophic failures in some steel fabrications has led to the further development of Griffiths'[6] classical work on linear elastic fracture mechanics (LEFM) by Wells[7] and Cottrell,[8] who independently proposed the crack-opening displacement test for determining the fracture toughness of engineering materials where crack propagation ahead of a crack was accompanied by a large plastic deformation at the tip of the crack.

Fracture toughness gives a measure of the material's resistance to failure by cracking due to the stress intensity around the

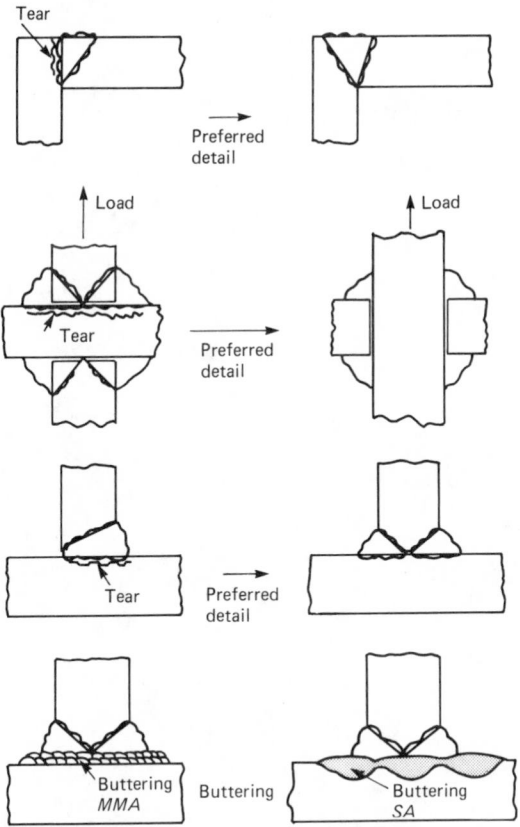

Figure 38.23 Lamellar tearing. Preferred details and buttering

tip of the crack in the presence of an applied and/or residual stress in the material. The measurement of fracture toughness has led to the publication of the BS document PD 6493[9] which gives guidance on the methods for determining acceptance levels of weld defects.

Fracture mechanics can also be applied to the growth of a fatigue crack under load-cycle conditions and the prediction of fatigue failure is largely due to Maddox[10] and Gurney.[11]

Factors which affect fracture toughness of steels are material thickness, service temperature, residual stress due to welding, applied stress, material type and strain rate. Lower service temperatures, thicker material, higher-yield material will adversely affect the fracture toughness; strain rate increase will not permit the normal mechanism of fracturing by slip along the atomic planes in time, and the material behaves elastically. The total stress across a weld defect may be the sum of both applied and residual, the latter due to the stresses set up by the contraction of the weld when cold; it is not unknown for brittle fracture to occur under residual stress only. Once the material has been selected for welding, the correct electrode must be used for yield, ultimate and required impact values; ultimately the level of defects in the weld will govern the service performance of the structure. Crack-like defects normal to principal stresses are most at risk, such as heat-affected zone cracks and lack of penetration; cracks in the outer quarters of a weld are more significant than those in the middle half of the weld. Rounded defects such as porosity and slag inclusions can be ignored in many cases but they themselves may, if profuse or long, indicate possible cracking of some form; such profuse defects may shadow more serious defects when ultrasonic or radiographic testing is applied. However, they indicate bad welding practice and a limit should be applied to the extent of their occurrence.

The fitness for purpose concept, i.e. defects are acceptable in structures if their size, orientation and type do not affect the integrity of the structure, indicate that where stress and design criteria require it an acceptable level of defects must be tabled. A typical statement that no weld should contain defects is inadmissible since all welds contain some defects, however small; with the advent of ultrasonic techniques with improved discriminating powers many small, structurally insignificant defects can be found. Repairs of such defects are not necessary and are expensive, with the possibility of reintroducing more significant defects upon repairing.

Burdekin[12] has proposed acceptance criteria for welded joints in crane girders and proposals are in hand for bridge structures to BS 5400.

Some limited guidance is given in BS5135 related to welders' tests but to relate these tests to a structure's service requirements would be unrealistic. Design detail concepts to minimize the risk of brittle fracture include the obvious step of avoiding welded joints in areas of high stress concentration or high tensile stress and, where this is impossible, to lay down acceptable weld-defect levels. Keep all weld sizes to a minimum commensurate with design criteria to reduce the incidence of weld defects and to reduce weld residual stress. Recognize that a higher-yield material will reduce the acceptable weld defect size and will not be superior to lower strength steel in fatigue life.

To select the correct notch-ductile material related to material thickness and service temperature. A welded joint that has to be tested should be fully accessible to the testing method chosen and to the arc process used to make that joint; poor welding access will inevitably lead to more significant defects. To achieve defect size, location, orientation and type, the use of ultrasonics is mandatory; it lacks a permanent record (compared to radiography) and hence the skill and integrity of the operator is paramount.

Generally, building frames in steel are at little risk to brittle fracture but bridges, pressure vessels and oil platforms all have their areas where care in design, correct steel selection, proven fabrication skill and erection procedures will reduce the possibility of such failures.

References

1 Gourd, L. M. (1980) *Principles of welding technology*. Edward Arnold, London.
2 Houldcroft, P. T. (1967) *Welding processes*. Cambridge University Press, Cambridge.
3 *A.W.S. Welding handbook*, Vol 2. American Welding Society.
4 British Standards Institution (1978–1980) *Steel, concrete and composite bridges*. BS 5400. BSI, London.
5 British Standards Institution (1969–1970) *The use of structural steel in building*. BS 449. BSI, London.
6 Griffiths, A. A. (1920) *Phil. trans. Roy. Soc.* Series A, **221**, 163.
7 Wells, A. A. (1963) 'Application of fracture mechanics at and beyond general yielding.' *British Welding Journal* 573/570 (Nov.)
8 Cottrell, A. H. (1961) International Standards Institute, Report No. 69, pp. 281–291.
9 British Standards Institution (1980) *Guidance on some methods for the derivation of acceptance levels for defects in fusion-welded joints.* PD 6493. BSI, London.
10 Maddox, S. J. (1972) Welding Institute Report No. E49.
11 Gurney, T. R. (1979) Welding Institute Report No. 91.
12 Burdekin, F. M. (1981) 'Practical aspects of fracture mechanics in engineering design.' *Proc. Inst. Mech. Engrs.*, **195**, 12.

39

Steelwork Erection

W H Arch BSc(Eng), MICE

Contents

39.1 Introduction

It is not entirely by chance that the French refer to a bridge as a work of art. For it is indeed a very close liaison between science, and the art of its application, that makes for success in bridge building. This is particularly true for bridges, but is also applicable – though perhaps to a lesser degree – in other types of structures.

A structure, of any type, is essentially designed for its completed condition, and the act of achieving its successful completion is that of coping with logistics of its components, with the partially erected frame, and with the people who will be involved in its erection.

Many variable factors affect the choice of erection method which will be used in any particular case, but on the other hand there are a number of common elements connected with any erection plan which do not vary from job to job.

For any project to be successful it is essential that the designer of the structure has clearly in mind how he visualizes the job will be built. It may well be that the final scheme may vary from that which was originally envisaged due to modifications and improvements, but at least the original conception would have been based on a completely thought out, fully integrated design.

The problems posed by the logistics of a job, the control of materials flow, and also of the organization of the manpower who will build the structure, are common.

Two examples of design points which the original conception should be able to accommodate, immediately spring to mind. The effects of temporary points of support and possible reversal of stresses during bridge building, and secondly the orientation of column webs and flanges to enable the closing beams to be erected in the final stages of completion of a multi-storey frame.

39.2 Effects of the site on a project

The shape and location of the site on which a project is to be built will have had its effect on the basic design of the structure to be erected.

For a tall building, for example, the size of the site, as well as the eventual use of the building will have determined the column centres; the height or number of storeys will have determined the section of the column and, therefore, the weight. In a bridge the configuration of the area to be spanned together with the ground conditions will have determined the spans. Only in uncomplicated cases will the economics of the actual design of the spans themselves be the sole determining factor in the choice of span – or indeed of the method of construction.

Thus, the site through its bearing capacity, or through existing structures or services, has a profound effect on the design of the structure to be erected upon it. That the design of the structure is affected by site conditions is thus clear, but the effect those same conditions have on the erection method to be adopted is even more profound.

A restriction on access for the materials, or the presence of underground services can each affect the placing of cranes and thus the whole approach to the erection plan.

Time on a site costs money and a good scheme should, therefore, aim to reduce as far as is practicable the number of manhours required on that site by the maximum use of prefabrication and subassembly off site. If this is taken to extremes then the components requiring to be lifted could become too large to transport or too heavy to lift. There is thus an optimum balance for any one job, and in major construction projects it is not often that what is right for one job is also right for another.

The ground conditions will have been fully evaluated before the foundations of the structure are designed. It is of equal importance that the ground conditions under temporary foun-

Figure 39.1 This view of the deck of the Forth Road Bridge during erection shows the changes in profile which can occur during the erection process. Arrangements must be made to accommodate and follow such changes in the crane support arrangements. The connections in the structure itself must also accommodate such changes and be designed accordingly

dations are properly evaluated. In a bridge erection scheme, for example, the whole stability of the project can depend upon the adequacy of a temporary pier and, therefore, on the foundations under that pier.

All temporary loads – even of such a transient nature as those under the outriggers of a mobile crane – are important. Only by full attention to such matters can the safety of the men working on the job be safeguarded, and the safety of the structure itself be assured against collapse.

39.3 The effect of plant on design

Everyone is familiar with the contractor's plant yard. Items go there from one job, to be refurbished, tested as required, often given a coat of paint, and there they lie ready to be sent out on to the next job.

The choice of erection plant which is to be used on the next job will, therefore, mean that, for economic reasons, existing tackle will be used where possible. There is, therefore, a need for plant which is as universally useful as possible. Mobile cranes, either mounted on wheels or tracks are one example. At the other end of the scale are the special 'one-off' erection devices built to suit the peculiarities of a particular job. The chances of all the peculiarities repeating exactly on another job are remote, and, therefore, the 'one-off' device is only used on the really big jobs where the very high cost can be written off against that job.

Whatever the type of plant which is being used to erect the steel the economics of that job will be vastly affected by the

Figure 39.2 The mobile crane is a standard piece of equipment for bridge erection. Points to note are the proper provision and blocking of the outriggers and the use of the minimum radius possible for handling the lifts

variance between the actual average weight of each piece to be lifted and the capacity of the crane at each relevant radius. Thus, in the ideal situation the weight of every piece should equal the capacities given by the safe-load indicator on the crane. Clearly, this is an ideal unlikely to be experienced, but a job on which a conscious effort has been made in that direction will be a more economic job to build.

Where a contractor's plant yard and operational area is remote from access to plant hire companies, greater emphasis must be placed on a careful choice of universally useful pieces of equipment. Little emphasis can be placed on being able to hire a major piece of equipment which may be required for a particular project. An approach often used by international contractors, building major projects in remote areas, is to buy all the equipment needed for a job and then to sell it on the local market at the end of the contract. This option is, generally speaking, not available to a local contractor working his local market by carrying out a large number of smaller contracts. He is in the same position as a small local contractor anywhere, but without the option of hiring the major plant he may need.

When making a choice between one mobile crane and another, consideration must be given to the road conditions and the distances to the likely locations on which the plant will be used. A truck-mounted crane can be driven on the roads, but a crane mounted on caterpillar tracks requires a low-loader and special equipment to move it from one location to another. The existence of suitable maintenance and testing facilities may affect the choice between a hydraulically operated crane or one operated by winches.

If the operational area is large and the likely workload is predominantly at the light end of the market, conditions may be such that the planning emphasis should move towards a more labour-intensive method using easily transportable equipment.

A light lattice pole, guyed to the ground at suitable anchorage points can be used to erect most single-storey frames and girders.

Special techniques are needed where the height of the frame being erected requires that the pole – or the crane – be supported on the frame itself as it grows upwards. The winch providing the power to work the crane can either be arranged at ground level, or to climb with the frame.

The choice of plant and the location of that plant during the various stages of building a complex frame will have a significant effect on the stress generated in the structure during erection. Very careful thought must, therefore, be given to the need for temporary material in additional supports and to the need for additional permanent material to transmit the transient additional erection loads, stresses and deflections.

The use of models can be an invaluable help in the positioning of plant for the best use to be made of the capacity available. The increase in the availability of computer facilities is enabling the optimization of erection methods, and the resultant requirements in terms of additional material, manpower and plant, to be achieved more easily. In addition, a large number of alternative possibilities can be considered. A scale model – made of balsa or of cardboard – can also be invaluable as a method of explaining to the erection squad exactly what they have to do, and why.

The temporary supports and the erection cranes are arranged to safeguard the safety of the structure during erection. It is of equal importance that full consideration is given to the many working places where men will have to work during that erection. The provision of safe means of access to these working places will have the dual benefit of improving the safety of the individual, and at the same time of speeding the flow of the work. Careful thought must be given to the use and location of

Figure 39.3 An erection pole is much lighter and is more easily broken down into small sections for transport than is a mobile crane with an equivalent lifting capacity. This pole is being used to erect the heavy girders of an overhead electric travelling crane. An erection pole is guyed at the top and bottom and should always be used in as vertical position as possible. By the use of suitable tackles in the overhead and ground guys it is possible to 'walk' the pole from one erection to the next

safety nets since a badly used net can be a hindrance to the job and can indeed be a safety hazard rather than providing the protection intended.

39.4 Tolerances

There are four main areas in which the question of tolerances can affect the engineer. These are: (1) the accuracy of the geometry of the structure in the horizontal and (2) the vertical planes; (3) the accuracy with which the components match at joints and splices; and (4) lastly – but by no means least – the control of deformations in the structure, particularly where welding is involved.

There is only one way of achieving a specified tolerance, and that is to control the operation right from the start. Begin with accurate setting out, follow with accurate fabrication and finish with careful erection. Work which is allowed to slip below standard can seldom be improved at a later stage.

Whilst coarser tolerances can be accepted at the time the ground is being excavated, careful centrelines should be marked on the foundations of columns or piers to ensure that when the steelwork is positioned on them it will be accurate. Levels of both concrete and steelwork can be controlled by the careful use of a dumpy level. It is far easier to place a column on to levelled pads than it is to pull the whole frame into position after it has been erected.

The specification for each job will determine the tolerances which can be accepted on that job, and these will reflect the purposes for which it is to be used. There are a number of standards which give quantitative guidance on acceptable tolerances and these may be quoted in a specification for any particular job.

39.4.1 Camber and vertical curvature of a bridge

Where a bridge is being built using large prefabricated components there is always a possibility that local dimensional variations and twists can occur across the section. Measures must be adopted to accommodate these effects so that force to bring the bridge into contact with its bearings is not necessary. In some cases it may be necessary to provide a temporary support on the centreline of the bridge to allow the structure to take up its natural shape before setting the bearings to suit. If this is necessary, care has to be taken to ensure that the required vertical curvature and camber are maintained as closely as possible. It is important to ensure that controllable errors are not allowed to remain uncorrected and that the adjustment to the bearings is not used merely as a means of compensation for careless work.

The use of machine levels and adjusting screws for levelling the bearings will enable extreme accuracy to be attained in setting the upper bearing surfaces of the bearings. There should be a small air gap between the upper and lower halves of the bearing during final setting to ensure an even distribution of load across the bearing before the load is finally transferred.

Great care must be taken to ensure that the bearing is correctly placed below any diaphragm or stiffener in the structure above.

There should be no need to emphasize the effects which can be caused by loads being applied at points not designed to receive them, or of loads eccentric to their designed position.

The specification of the bridge will lay down the precise tolerances permitted for position and level of the bearings in relation to the structure they are supporting. In multi-span continuous girders the tolerances are tighter than those for simply supported members so far as level is concerned. The position of bearings should be such that they are aligned accurately with the centreline of diaphragms and stiffeners shown on the drawing.

The procedures in the fabricating shop, and the shop assembly of adjacent sections which the specification may have called for, will ensure that these components will recreate the camber called for when they are erected on the bearings we have been discussing.

It is essential to ensure that the methods used for handling, stacking and subsequent erection do not cause permanent distortion of the components. Means of protection should be provided not only to avoid damage to the slings used, but also to prevent damage to the component being lifted or stacked. Incorrect handling can cause deformations and it can also damage any protective treatment which has been previously applied. Many of the points made in this section apply equally to the erection of a building frame.

39.4.2 Tolerances across site butts or joints

The jigs and shop erection procedures used to ensure accuracy of shop fabrication will have minimized the errors in positioning of stiffeners, etc. and will have kept distortions within practicable limits. The responsibility of those on site is then to reassemble the components and erect them into place.

The tolerances allowable in the butt joints between the components are dictated by the need to limit the eccentricity of adjoining members and the secondary stresses those eccentricities would produce. The nature of the stresses, compression or tension, to which the members will be subjected in service, affects the acceptable limits. The compression condition requires a more severe limitation than does a connection loaded in tension. A temporary or transient loading condition, say, during erection, can of course reverse the permanent condition and this must be taken into account when the connections are being designed and fabricated.

If a butt in the flange plate of a girder is to be made by welding then care must be taken to limit the amount of 'cusping' that can occur when the butt is completed. It may be necessary to arrange to have the material preset in order to be able to compensate for the effects of shrinkage and distortion as the welded area cools.

The tolerances both of alignment at butts, and of straightness, will be based on measurement taken over a gauge length dependent on the spacing of stiffeners, and the amounts of allowable deviations will in general be based on the thickness of the plate or flange being measured. Reference has already been made to the relaxation that is possible in connections subject only to tension loading.

There is clearly a balance between the maintenance of practical limits on the accuracy of workmanship and the achievement of very high standards approaching perfection. The fabrication industry has always been keenly aware of the need to maintain a high standard of workmanship in its products whilst at the same time ensuring that it remains competitive in the market places of the world. Careful site planning and control can ensure that these high standards are maintained in the completed structure.

39.5 The interaction of design and erection factors

In section 39.3 consideration was given to the choice of plant which would be used in a particular project. The weight of the pieces to be lifted is clearly of paramount importance in this consideration.

The weight of a component is a function of its size. Piece weights, therefore, affect the location of the splices connecting the components. In addition there are local limitations, and associated regulations, which limit the size of component that can be transported. Thus, the site and any limitations imposed by its access limitations can affect the design just as much as the capacity of the available crane on the job. All these points must be borne in mind in addition to the purely design considerations.

The 'nesting' of components must be considered during the design phase in order to minimize the volume required to be occupied by a given weight of material, particularly where transportation by sea is involved. The cost penalties for shipping a bulky lightweight component are very heavy.

The restrictions referred to, where size must be limited, can result in components which are uneconomical to erect. In these cases, consideration should be given to ways of enabling subassembly or prefabrication to be carried out at site before finally lifting the component into place. An example of this in the tall-building field is shown in Figure 39.4. In this example the floor beam stubs have been welded to the column shaft after arrival on site, but before erection. There were a number of advantages in this procedure. The plain components would 'nest' well for transport, thus reducing the cost of transportation. The depth of the floor girder sections could be increased at the columns in order to be able better to transmit the wind loads, and lastly the splices remaining to be made up in the air are in a much more accessible location. Splices should be located just above a floor level to give safe access for bolting up. The location of the splices between sections of columns is limited both by the capacity of the crane available for the erection, and also by the climbing cycle which has been decided upon. Three floor levels in one piece is normal, with a reduction towards the lower sections where heavier scantlings will be used.

39.6 Foundations and temporary supports

The design of the permanent foundations for a structure will have been considered in very great detail when the structure was in the design office. From the point of view of the erector, however, the temporary condition must be examined. This applies more particularly to bridge building, but the principle applies in any erection plan. Not only can transient loadings and moments which will be applied to foundations during erection exceed very considerably those which can occur after the structure is complete, but temporary foundations may also be required.

Where temporary foundations are required, as much consideration should be given to their design and construction as their importance to the safety of the erection scheme dictates. The main foundations will not have been designed without reference being made to the results of test borings. Where loads are significant or the effects of settlement on the safety of the

Figure 39.4 A site-subassembled column section for a high-rise building being erected. Site-subassembly enables more easily transportable components to be built up into economic units for erection. In this case also moment connections could be more easily welded up in optimum conditions and the stub-beam connection reduced to an easily jigged and readily accessible connection

structure could be dangerous, then a trial boring should be put down local to the temporary foundation to ensure that adequate precautions are taken.

The additional loads that must be considered when designing temporary foundations should include those that arise from the incompleteness of the structure that they are supporting. The effects of wind loading and aerodynamic instability must be considered. The weights of items of plant, plus the loads they are carrying and any resultant uplift and dynamic loads due to movement, can affect the design.

Particularly in bridgework there will be temporary piers to place on the foundations discussed earlier, and also temporary extensions to the permanent piers. These are commonly constructed of standard components which can be stored and re-used on future work. In the event that specially designed and fabricated temporary supports are not being used, a check calculation should be made to ensure the adequacy of the standard components planned for use.

The points at which it is required to make connections between the permanent structure and the temporary erection structures should be given careful consideration. They must be incorporated in the original design conception. They must be adequate to carry the various loads and stresses to which they will be subjected during the progress of the work. It must be remembered that temperature changes will result in the movement of the structure, and that these movements have to be accommodated by the design of the temporary works. Remember too, that the temporary structure will deflect when the load comes on to it. There must be provision for relieving this load before the structure can be removed.

In buildings, the temporary structures most commonly used are, again, props or columns used to support girder structures at intermediate points during the construction, for example, large lattice girders spanning distances too great to permit erection of the component in one subassembly. It is important here to check the adequacy of the beam carrying the prop, or of the lower column length being temporarily extended. It is not uncommon where the temporary erection loads have to be accommodated for permanent additional material to be incorporated to provide the additional carrying capacity required.

39.7 The partially completed structure

We have considered the need for temporary supports during the erection of a structure, and these considerations have dealt principally with the need to shorten cantilevers during bridge erection or to prop long girders during assembly in a building.

However, all structures are designed to have a minimum amount of redundancy and it follows, therefore, that until an erection procedure is complete the structure is at risk. It is to eliminate this risk and to take due account of the changing stress patterns that will arise during an erection procedure that the engineer must be concerned. It has already been stressed that an erection scheme should be borne in mind when the original design is being made, since only by that means can a feasible, safe and well regulated design be produced.

If a lattice girder structure is to be built out by cantilevering from a pier it is clear that the stresses in the members will be reversed until the next pier is reached and the girder spans from pier to pier. Similarly, only when all the spans are completed will any continuity assumed by the designer be achieved.

Where the lattice girder is supported by piers or, in the case of an arch rib, may be supported from above by temporary cables (see Figure 39.5), very special stress patterns can develop. It may be necessary to carry out a number of case studies at progressive stages of erection in order to ensure that the most critical condition for each member and for each connection has been considered and adequately dealt with.

A building designed to derive its rigidity from shear walls or from the composite action of the floor slab will be unstable until these features are completed. Temporary means of providing this rigidity must be arranged.

The location of these temporary members, and their stiffness relative to adjacent members of the partially completed frame must be carefully considered. It must be possible to construct the permanent members without removing the temporary load-carrying members, and to be able to remove the temporary members afterwards.

The permanent structure must be capable of absorbing the loads induced by the temporary members. These can in many cases frame into locations not intended to carry those loads when the frame is completed. Cases of eccentricity need special attention in view of the secondary stresses which can be induced.

39.8 Stockyards and transport

We have seen that prefabrication enables manhours to be expended on the ground rather than at a height, and thus reduces to the absolute minimum the time that a site is occupied. It is this factor that enables a steel frame to be put up so quickly, and to provide an instant support for the follow-up trades.

Inevitably, components must be transported from the factory to the site. Nothing can ever be perfectly controlled, and a stockyard is therefore necessary to absorb the differences between materials arriving on site, and those actually required from day to day for erection. The effects of delays in the transport system must be cushioned, otherwise expensive erection equipment and manpower would be kept idle.

It has been very truly said that a good stockyard control can make a successful job. The reverse is certainly true: if components and fasteners are not sent out to the erection front as required there will be delays.

Incoming material must be recorded correctly, and located in order that it can be relocated and transported out when required without resorting to double handling. It is often convenient to colour-code the material to conform either to types or area of

Figure 39.5 In the erection of an arch bridge, temporary supports are required to assist in cantilevering the two halves of the arch out from the abutments. Transient loads, deformations and stress reversals which occur during the erection process must all be calculated carefully for all erection stages and constant checks made of the actual behaviour of the structure as it grows out of the crown

the project. Cranage will be provided to handle the weights of components involved, with the same rules applying as were discussed in section 39.2. A light crane with large coverage will be required to handle the light components, and all the heavy components will be placed under the heavy crane. A Goliath-type crane is often used for stockyard duties since it occupies little ground area for its tracks and can easily cover a large area without reduction of capacity.

Adequate roads – and rail tracks where required – must be provided to give all-weather service. Axle loads of the vehicles using a stockyard are heavy and maintenance time on broken roads is time lost for erection.

Vehicles bringing material into a stockyard will be the normal form of fixed axle or semi-trailer type of transport common on the public roads. Only in exceptional cases will it be necessary to bring components on to site on special transport vehicles requiring police escort. There are limitations imposed by law on both the weight and dimensions of loads that can be moved on the public highway and a knowledge of those restrictions is essential. Additionally, many sites are located in places where the road layout, or low bridges, impose restrictions and a study of these must be undertaken before the component gets stuck and the site is disrupted.

On-site transport is not restricted in the same way, and the size of load is similarly subject to different controls. Site subassembly can lead to awkward loads, and the safe loading and fixing of these on the site transport is important.

The proper control and distribution of fasteners is important. Bolts should be rebagged into 'sets' required for particular connections or areas. This saves time at the splice, and also cuts down the waste of bolts when an excessive number are issued by the store. The proper storage of welding electrodes, their issue and conditioning while being held at the point of use are also functions often delegated to the stockyard stores control function.

39.9 Manpower and safety

None of the operations discussed in this chapter can be carried out without an adequate and sufficiently skilled number of men. The proper use of subassembly techniques can reduce the number of manhours which have to be worked at a height, but work has still to be done up on the open steel, and this must be properly planned.

Steel erection is a task that can only be learned by experience. Formal training can teach the basic skills of rigging, scaffolding, slinging, crane driving, burning and welding, but only experience can enable these skills to be applied on the job. It is the individual erector who brings his expertise and skill; it is the employer who must provide the planned erection method and adequate equipment to enable the erector to safely apply his skill. He needs not only the crane and the spanners, but also a safe place in which to work. Figure 39.6 shows a man working on an unsafe platform. Not only is it dangerous for a person working in such conditions, but the safety of the whole job could be put at risk. The provision of boxes to contain loose tools, and good housekeeping in the handling of small components in general can each help to ensure a smooth running, and a safe, job.

In the UK, a large number of regulations control the manner of the ordering of work on a construction site, and these, together with summaries of them are available from HMSO and from RoSPA. These cover such major items as the testing and retesting of lifting equipment and the dimensional requirements for working platforms and access ways. Many accidents, though, are caused not by major problems, but by insufficient attention being given to the small items.

Thought given to the design and detailing of the steelwork itself can result in a real reduction in the risk of an accident occurring during erection.

Accidents will always happen, but careful planning from the

Figure 39.6 An example of an unsafe working platform. To ask a man to work under such conditions endangers not only the man on the platform but also the safety of the other men with whom he is working, and the safety of the whole job could be endangered

start will prevent a simple accident from escalating into a major calamity.

39.10 Construction management

The construction manager is a man whose function is becoming increasingly important. The basic object of management on a construction site is to have the right component at the right place at the right time with adequate equipment and men to handle it, and to repeat this over and again to complete the structure on time and within a cost target.

To this basic requirement must be added the co-ordination of a number of contractors, each with his own problems and targets, together with an ability to overcome problems created by other influences, often outside his direct control.

There are a number of contractual forms within which construction work can be carried out, but one important feature of all these is that the roles of the parties to the contract are clearly defined. It is essential that responsibilities are clearly defined in this way since we have seen how closely the original design affects the erection process and vice versa. These interconnecting factors must be appreciated by all the parties or disputes, and perhaps tragedies, can result.

There is a variety of ways in which a prospective owner can have the management of the construction of his project organized. One factor common to all methods, however, is that the earlier in the time-scale all the parties who are to be involved can add their particular expertise to the planning of that project, then the sooner the project can be started, and the sooner it can be completed. The attainment of this ideal situation is made easier with some forms of contract and management organization than with others.

In what is being presented here as the ideal situation, a project team is brought together at the inception of the planning of the project. This team will comprise those who will be directly responsible for each of the phases of the project, from the design, not only of the frame, but of all the associated services, right through to the site erection functions, at the other end of the planning and construction time-scale.

The other end of the spectrum is the job where all the functions are carried out in watertight compartments, the designs being prepared in a number of engineers' offices, and the contractors tendering in isolation. With this system no meaningful discussions can take place between those who either are, or will be, most intimately involved with the project until so late a time in the construction process that their contribution can only be limited.

39.11 Summary

The more complex a project becomes and the more services are involved – a tall building is an excellent example – the greater the need for early consultation and planning. In a tall building where the lower floors can be approaching occupancy while those above are having services installed and those at the top are still being framed, the whole complex construction pattern is seen in microcosm.

Accesses for foundation construction must integrate with the demolition of what was there before. Material arriving on site requires close control and schedules if traffic chaos and a hopelessly congested site are to be avoided.

The positioning of cranes and the stability of the partially completed frame and of any temporary works must be very carefully thought through.

The safety of those working on the structure and of those working, and perhaps also living, in the vicinity of the new structure must be constantly considered.

Only by the co-operation of everybody who will be involved in the project, be it bridge or building, can the best design be made, the best construction method adopted and the earliest completion date obtained.

All these factors – design, method of construction and construction time – must be optimized if the owner is to be able to make use of his new facility at the earliest possible time and so be able to generate revenue from his investment.

Bibliography

Arch, W. H. (1985) *Structural steelwork erection*. British Council for Steel Erection, London

Barron, T. (1963) *Erection of constructional steelwork*. Iliffe, London.

Constrado (1972) *Steel designers' manual* (4th edn). Crosby Lockwood, London.

Leech, L. V. (1972) *Structural steelwork for students*. Butterworths, London.

O'Connor, C. (1971) *Design of bridge superstructures*. Wiley-Interscience, Chichester.

Ward, F. C., Bryant, E. G. and Pound, R. P. (1970) 'Simply supported bridges in composite construction'. British Council for Steel Erection, London

Journal

Building with steel: controlled circulation, British Steel Corporation, address: 9 Albert Embankment, London SE1 7SN

40

Buried Pipelines and Sewer Construction

D J Irvine BSc, CEng, FICE
Technical Services Manager,
Tarmac Construction Ltd

Contents

Buried pipelines are used to transport fluids which may be gases, liquids or slurries. The pipes may operate at high or low pressure. The pipelines may form part of extensive structures such as large national grids or be limited in extent, such as culverts under roads. Consequently, a wide range of skill and engineering expertise is needed to construct and maintain the structures which represent a considerable capital investment and form a significant proportion of any developed country's infrastructure.

40.1 Routing

The choice of a pipeline route is usually governed by economic considerations but practical and legal factors will have a considerable influence which varies upon the substance being conveyed. A useful checklist of matters to be considered is provided in Part 1 of *Pipelines in land*.[1]

The development of pipelines in the UK is regulated by a number of Acts of Parliament which must be strictly followed by the pipeline promoter. The principal legislation comprises:

Water Acts 1945 and 1948
Gas Acts 1948 and 1965
Public Health Acts 1936 and 1961
Requisitioned Land and Waterworks Acts 1945 and 1948
Land Powers Act 1958

Other legislation which is relevant is:

Public Utilities Street Works Act 1950
(for pipes laid along or across streets)

Coast Protection Act 1949
(for parts of pipelines laid below high water)

Land Drainage Act 1961
(for pipelines crossing rivers and streams)

The legal situation is frequently complex and it is essential to obtain expert advice. The Pipelines Inspectorate, the Health and Safety Executive and the relevant local and statutory authorities should be approached at an early stage in the development of any large project. It should be noted that a planning consent or pipeline authorization does not confer any rights to enter or carry out works in land and it is up to the pipeline promoter to obtain the necessary rights from the owners or occupiers. If these cannot be obtained voluntarily a Compulsory Rights Order may be required.

The greatest problems, both practical and legal, will be encountered in congested urban areas. Pressure pipes for oil, gas and water can be laid at relatively shallow depths and within limits may follow the ground surface profile. Sewers, however, are generally laid to falls so that the contents flow under gravity and this may result in pipelines being constructed at greater depths with potentially greater problems for adjacent owners.

40.2 Materials

The material for a pipeline must be selected to suit its type and purpose (see Table 40.1). The manufacture and use of all materials listed are covered by British Standards which should be consulted to ensure that both the pipe material and the jointing system will comply with the desired service conditions. For work outside the UK, other relevant national standards may apply and should be consulted.

Most pipe manufacturers have devised their own pattern of joint and modern practice uses flexible joints rather than rigid.

These are easier to fix and allow a limited degree of relative settlement between pipes without the risk of leakage or fracture. 'Push in' joints using rubber sealing rings are used for low-pressure work. At higher pressures the seal may be compressed by loose flanges and bolts or screwed fittings. Steel and some plastic pipes can also be jointed by welding and this process is usual where 100% watertightness is required for safety reasons. Heat-shrink plastic and other proprietary joints are gaining favour particularly for repair work in the smaller diameters of pipe.

It should be noted that most flexible joints do not resist longitudinal forces. Therefore, pressure mains utilizing such joints should be provided with anchors at bends, tees and end caps to resist thrusts arising from the internal pressure.

40.3 Pipes in trench

40.3.1 Structural design

A buried pipe and the soil surrounding it are interactive structures. The extent of the interaction and hence the magnitude of the pipe loads arising depends on the relative stiffnesses. As a result two separate traditions of design have been evolved; one for 'rigid', and one for 'flexible', pipes.

40.3.2 Rigid and flexible pipes

A rigid pipe may be defined as likely to fracture under very small deformation. The pipe wall resists the external load by circumferential bending and the tensile strength of the pipe material is the usual limiting factor. The tensile bending stress due to the applied loads will be affected by the circumferential tensile stress resulting from any internal pressure caused by the pipe contents.

A flexible pipe will not necessarily crack under slow deformation even when this becomes large. It will fail under vertical loading by buckling or flattening if it is not adequately supported laterally. A circular flexible pipe reacts to external loading above it by deflecting downwards at the crown and outwards at the sides. The latter movement induces a passive resistance in the adjacent soil. The pipe should be capable of sustaining a crown deflection of at least 10% of the original diameter without damage, although it is usual to limit the working deflection to 5% or even less if protective linings are used.[2]

It should be noted that in reality the behaviour of any soil-pipe system varies with the pipe diameter : wall thickness ratio, its stiffness and the modulus of the soil. Consequently, there is an intermediate category of 'semi-flexible' pipes which for convenience are normally designed as rigid pipes.[2]

40.3.3 Longitudinal bending

This is also known as axial bending or beam effect and arises due to differential settlement along the pipeline (which is likely in most soils) or from local concentration of support. The use of flexible joints and good workmanship will assist in reducing the effects of longitudinal bending but special attention should be paid to small rigid pipes laid at shallow depths under heavy wheel loads, and where pipes intersect with structures such as buildings and manholes. A flexible joint as close to the face as possible and a short 'rocker' pipe should be used to accommodate differential movement at structures. If concrete beddings are used for the pipe it is essential to retain flexibility at pipe joints by forming flexible joints through the bedding.

In poor ground it may be necessary to use a piled foundation with a continuous reinforced concrete capping beam to support the pipeline at the required line and grade.

Table 40.1 General pipeline applications (based on BS CP 2010)

Pipe material	Design method	Crude oil and petroleum products	Liquefied petroleum gases	Natural gas	Town gas	Industrial gases	Chemicals	Brine	Sludges and slurries	Water	Sewage and trade effluent	Remarks
						Application						
Asbestos	R							A	A	A	A	Larger diameters are reinforced
Clay	I							A	A	A	A	
Concrete	G						A	A	A	A	A	
Grey cast iron	I / D			A^	A^		A	A	A	A*	A	
Ductile cast iron	F			A^	A^		A	A	A	A*	A	
Pitch fibre	L										A	
Steel with butt-welded joints	E / X / I	A	A	A	A	A	A*	A*	A	A*	A*	
Steel with other than butt-welded joints	B / L			A^	A^		A*	A*	A	A*	A*	
Plastics	E			A	A		A	A	A	A	A	Lightweight; some types can be chemically or heat welded

Notes:
(1) *Indicates pipeline may need special lining to prevent corrosion.
(2) ^Used at lower operating pressures.
(3) It is necessary to ensure that any jointing compounds are resistant to the fluid being carried.

40.4 Imposed loads on pipes in trench

The most widely used method of estimating external loads on a buried pipeline was pioneered by Marston, Spangler and Schlick in the US. It was further developed and extended for UK practice by Clark and Young and is generally termed the 'Marston' or 'computed load' method (Table 40.2).

The method is to some extent empirical and uses a soil model based on Rankine's theories of soil behaviour. This has been the cause of some criticism but considerable testing and successful practical experience has demonstrated its merits, and it remains the standard method in the UK, Europe and the US.

40.4.1 Installation conditions considered for design

40.4.1.1 Narrow trench condition

This is the case when the trench is narrow and deep compared to the width of the pipe. The upper limit of narrow trench width for a given combination of trench depth and pipe diameter is known as the 'transition width' (see below). The narrow trench condition results in smaller soil loads on pipes than other conditions but there are practical construction difficulties which can prevent the design assumptions being realized.

The theory is based on an analogy with the Jansen theory of

Table 40.2 Notes on imposed loads on pipes in trench

Source of load	Comment
(1) Soil overburden	The magnitude of the vertical load on the pipe is estimated using the Marston model and is influenced by: (a) the depth of the fill and its nature (b) the width of the trench (c) whether negative or positive projection (d) when the trench sheeting is removed (e) level of water table
(2) Superimposed loads on surface (a) uniformly distributed load of large extent, e.g. temporary filling	(a) the load is expressed as an equivalent additional depth of fill in the Marston model but ignores any shear forces induced in the surcharge due to differential settlement (b) for narrow trench conditions an appreciable error may result when the notional increase in depth approaches the same magnitude as the original depth
(b) uniformly distributed loads of limited extent (permanent), e.g. foundations of structures, stacking of construction materials, ground loads from caterpillar tracks	(a) the load on the pipe is estimated using Newmark's integration of the Boussinesq equation (b) the soil stress at the pipe crown is calculated and assumed to be constant over 1 m run of pipe and across the diameter (c) the calculated load is added to the soil loads
(c) concentrated loads e.g. vehicle loads	(a) the load on the pipe is estimated by the Boussinesq method for the distribution of stress in a semi-infinite homogeneous elastic medium due to a point load at the surface (b) the method results in a peaked load along the pipe. This is converted to an average load over a length of 0.9 m or less if appropriate (c) the load thus calculated is added to the soil loads (d) the loading conditions are shown in Table 40.3 together with appropriate impact factors (e) all pipes should be checked for the worst case arising during construction as well as the permanent works service condition
(3) Fluid load, i.e. load of pipe contents	(a) weight of pipe contents causes circumferential bending in the pipe wall. (Note self-weight of pipe usually neglected) (b) the magnitude of the bending moment depends on the manner in which the pipe is bedded and whether the pipe is running full (c) the effect is allowed by adding an 'equivalent water load' to the other loads on the pipe. The value can be obtained from charts or can be estimated as 0.75 times water load

(4) Internal pressure	(a) pipes should be designed for the worst internal surge or test pressure which is likely to arise
	 (b) in gravity pipelines the maximum static head (ignoring surge) occurs when the velocity of flow is zero
	 (c) in pumping mains the maximum head will be either:
	   (i) the sum of the maximum static head plus friction head at maximum flow plus any other loss of head, or
	   (ii) maximum surge pressure due to sudden stoppage of the pumps and closing of non-return valves particularly if no surge suppression is provided in the system
	 (d) partial vacuum conditions arising from inefficient air valves, etc. can be treated as an additional temporary external water pressure

Table 40.3 Concentrated vehicle loads on pipelines

Condition	Reference		Loading
Main roads	BS 5400:1975 type HB		Eight wheel loads of 112.5 kN each (including impact factor of 1.25) distributed over circular or square contact area at effective pressure of 1100 kN/m²
Light roads	Ministry of Housing and Local Government (1967) *Working party on the design and construction of underground pipe sewers*, 2nd Report, HMSO, London, as quoted in: Young and O'Reilly (1983) *A guide to design loadings for buried rigid pipes*. Transport and Road Research Laboratory, HMSO, London		Two wheel loads of 105 kN each (i.e. 70 kN static weight with impact factor of 1.5) and contact pressure of 700 kN/m²
Field loading	Ministry of Housing and Local Government (1967) *Working party on the design and construction of underground pipe sewers*, 2nd Report, HMSO, London, as quoted in: Young and O'Reilly (1983) *A guide to design loadings for buried rigid pipes*. Transport and Road Research Laboratory, HMSO, London		Two wheel loads of 60 kN each (i.e. 30 kN static weight with impact factor of 2.0) and contact pressure of 400 kN/m²
Railway loading RU (mainline railways of 1.4 m gauge and above)	BS 5400:1978	Loads are static and must be increased by impact factor of 2.0 for pipes up to 3 m diameter	
Railway loading RL (reduced loading for passenger rapid transit systems where main line locomotives and rolling stock do not operate)	BS 5400:1978		

Construction vehicles	Trott and Gaunt (1976) *Experimental pipelines under a major road: performance during and after road construction*. Report LR 692, Transport and Road Research Laboratory, Crowthorne	Manufacturers' loading data for actual plant should be used wherever possible. Alternatively, loads can be estimated from following:

Plant	Total mass (t)	Static wheel load (t)	Tyre inflation pressure (KN/m²)
Small scraper	23.2	6	200–400
Large scraper	110.3	28	500–600
Small dump truck	24.3	4	350–700
Large dump truck	80.4	20	up to 650
Ready mix truck (6 m³ capacity)	24.0	5.5	up to 750

pressures within silos (see also Table 40.3). In a vertically sided trench (Figure 40.1) the load on the pipe is the weight of the prism of fill at level X–X minus the friction of the fill on the adjacent soil. The theory ignores the effect of cohesion on the shear surface and, when rigid pipes are used, any support provided by the fill below level X–X.

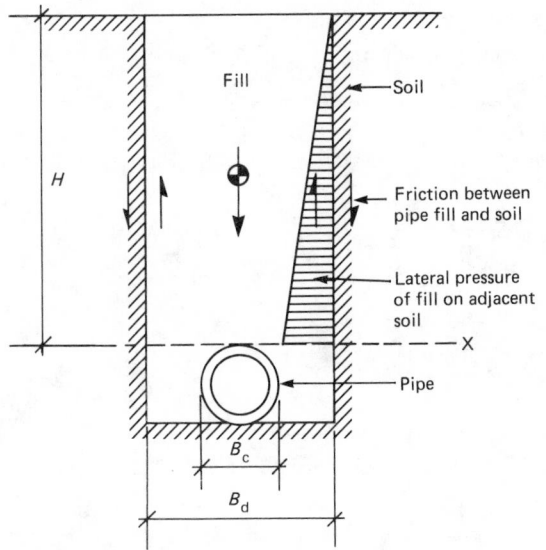

Figure 40.1   Narrow trench conditions: diagram of forces

The load on the pipe is given by the expression:

$$W_c = \gamma B_d^2 \left[\frac{1 - e^{-2K\mu'}}{2K\mu'} \times (H/B_d) \right]$$

$$= \gamma B_d^2 C_d$$

where W_c = fill load on pipe in narrow conditions in kilonewtons per metre; γ = unit weight of soil in kilonewtons per cubic metre; B_d = effective width of trench in metres; e = base of Napierian logarithms; H = height of ground surface above top of pipe in metres; K = Rankine active earth pressure coefficient; and μ' = coefficient of sliding friction of backfill against trench sides

$K\mu'$ is a semi-empirical constant for particular soil types and it is common to use design charts to determine values of C_d for different soil types; μ' is used to distinguish the coefficient of sliding friction from the coefficient of internal friction for backfill used elsewhere in Young's and O'Reilly's calculations.[3]

It has been shown that the narrow trench condition can exist in battered and stepped trenches (Figure 40.2) and the value of $K\mu'$ appropriate to the material in which shear occurs should be used for design.

40.4.1.2 Late removal of trench support sheeting in the narrow trench condition

The narrow trench loading condition assumes that the pipe load will be reduced by friction at interface between the fill and the soil. However, in close-sheeted trenches it is frequently not practicable for reasons of safety to remove the sheets before placing and compacting the backfill. In such cases it should be

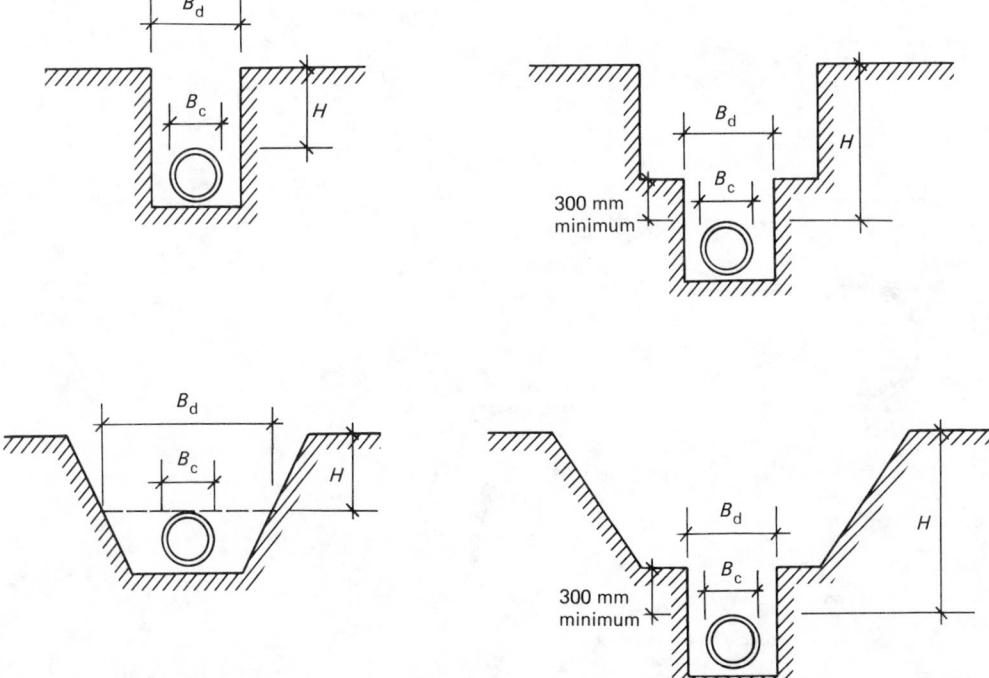

Figure 40.2   Effective trench widths in battered and stepped trenches

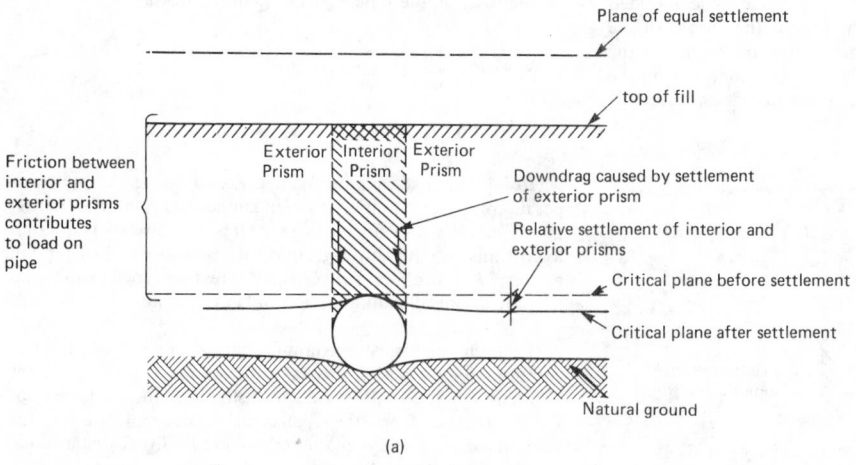

(a)

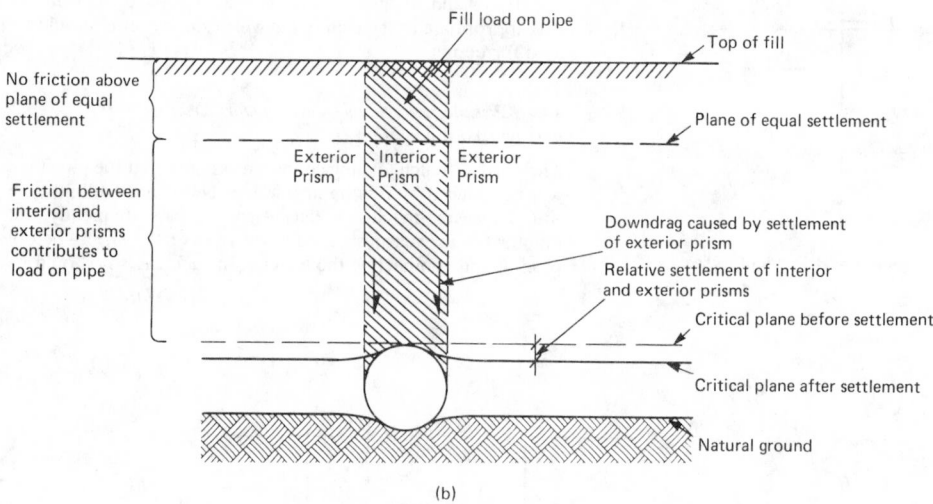

(b)

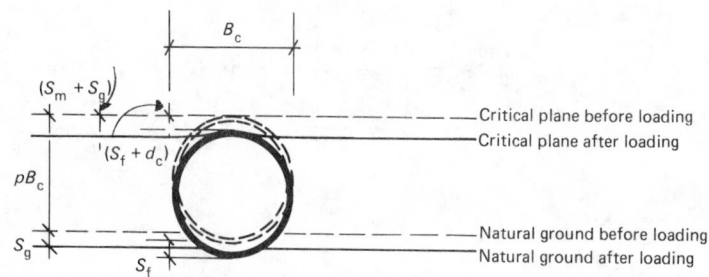

Settlement ratio r_{sd}
$$= \frac{(S_m + S_g) - (S_f + d_c)}{S_m}$$

p = projection ratio
pB_c = projection height
S_m = compression of soil in projection height due to overlying fill
S_f = total settlement of pipe invert
S_g = settlement of natural ground surface
d_c = vertical deflection (shortening) of the pipe

(Note: It is not practical to pre-determine the settlement ratio and it is always assessed from experimental load measurements and practical experience: see Young and O' Reilly (1983) *A guide to design loadings for buried rigid pipes.* Transport and Road Research Laboratory, HMSO, London.)

(c)

Figure 40.3 Positive projection condition under valley fill or embankment. (a) Complete projection, i.e. shear stresses between interior and exterior prisms extend to top of fill; (b) incomplete projection, i.e. shear stresses between interior and exterior prisms do not extend to top of fill; (c) settlement ratio

assumed that the total weight of backfill will be exerted on the pipe.

40.4.1.3 Embankment or valley fill with positive projection condition

As the trench width increases, the prism of soil over the pipe will be bounded on either side by deeper prisms of fill down to formation level (Figure 40.3). The latter are likely to settle relative to the pipe and, hence, will exert downward forces on the sides of the centre prism and increase the load on the pipe.

In a low embankment the shear stresses on the centre prism will extend up to the surface of the fill. This is known as the 'complete projection condition'.

In a sufficiently high embankment, the shear stresses do not extend to the surface but cease at an intermediate level known as the 'plane of equal settlement'. This is the 'incomplete projection condition' and no frictional forces are exerted on the centre prism above the plane of equal settlement.

It should be noted that the amount of deformation of the critical plane in Figures 40.3(a) and (b) depends on, among other things, the 'projection ratio', i.e. the extent to which the pipe projects above the natural ground.

The total deformation of the critical plane is made up of:

(1) The settlement of the fill in the projection height pB_c.
(2) The total settlement of the pipe invert.
(3) The settlement of the natural ground surface.
(4) The vertical shortening of the pipe.
(5) The 'settlement ratio' defined in Figure 40.3(c). (Actual values of settlement ratio have been derived semi-empirically.[3])

The total load on the pipe is given by the expression:

$$W_c' = C_c \gamma B_c^2$$

where $W_c' =$ fill load on pipe in kilonewtons per metre; $C_c =$ fill load coefficient for positive projection case; and $B_c =$ external diameter of pipe

C_c is derived from complex expressions depending on soil and

fill characteristics and geometry, and it is usual to obtain design values from standard charts.[3]

40.4.1.4 Wide-trench condition

It has been shown that if a trench is progressively widened and other conditions do not change then the narrow trench loading does not continue to increase but reaches a limiting value given by the appropriate positive projection equation. The trench width at which this limit is reached is known as the 'transition width'.

The same logic applies to a narrow trench being reduced in depth. The depth at which the narrow trench load equals the positive projection load is known as the 'transition depth'. For depths deeper than the transition depth the narrow trench condition applies.

40.4.1.5 Embankment or valley fill with negative projection condition

This case differs from positive projection in that the pipe is laid in a subtrench below the natural ground level (Figure 40.4). The middle prism of soil will tend to settle relative to the adjacent prisms on either side thus reducing the potential fill load on the pipe.

The concepts of the plane of equal settlement and settlement ratios are also used in this case.

The total load on the pipe is given by:

$$W_c' = C_n \gamma B_d^2$$

where $W_c' =$ fill load on pipe in kilonewtons per metre; $C_n =$ fill load coefficient for negative projection case; and $B_d =$ width of trench

As in the case of positive projection, the fill load coefficient C_n is dependent on soil and fill characteristics and geometry, and values may be obtained from standard charts.[3]

40.4.1.6 Loads on flexible pipes

The calculation of fill loads on flexible pipes can be simplified by

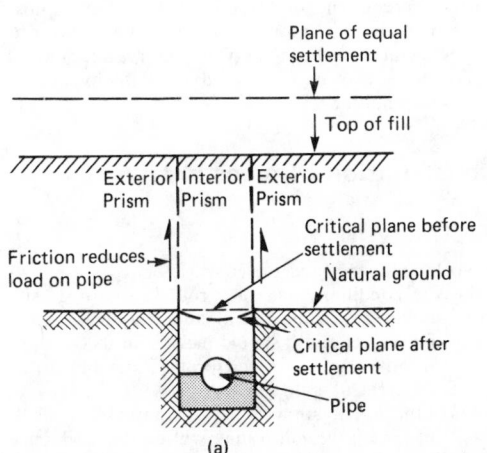

(a)

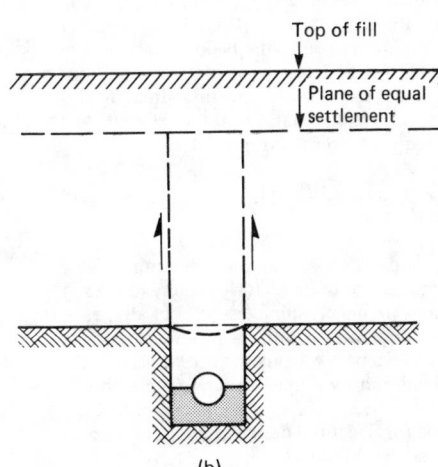

(b)

Figure 40.4 Negative projection condition under embankment or valley fill. (a) Complete projection, i.e. shear stresses between interior and exterior prisms extend to top of fill; (b) incomplete projection, i.e. shear stresses do not extend to top of fill

taking the total effective weight of the prism above the pipe of width equal to the pipe horizontal diameter[4] and extending up to the ground surface. To this may be added the self-weight of the pipe above the midplane and the buoyancy caused by any groundwater above the level of the midplane although these loads will be minimal in the case of small pipes.

The effect of concentrated vehicle loads may be calculated using a Boussinesq distribution or by reference to design curves in Compston *et al.*[4] and Nath.[5] The pressures due to vehicle loads will tend to be greatest at the crown and those from external water pressure greatest at the invert. The maximum radial pressure is selected for design and is assumed to be distributed as ring compression around the pipe wall[4] in addition to the fill load.

For the design of corrugated buried steel structures under roadways in the UK, the Department of Transport[6] should be consulted.

40.5 Pipe strength

40.5.1 Rigid pipes

The design of a rigid pipe wall is based on its resistance to the circumferential bending moments induced by external loads plus any circumferential tensile stress caused by internal pressure.

The effect of external loads on a pipe can be calculated by mathematical theory. However, due to the difficulties in accurately modelling the conditions which apply in practice, it is usual to design rigid pipes using an empirical method linked to standard crushing values in the British Standards Specification.

The empirical design approach takes account of the benefits gained by using specific bedding methods immediately around the pipe (Figure 40.5). These distribute the foundation reaction around the lower pipe periphery thus reducing the circumferential bending moment in the pipe wall.

The 'bedding factor' is the ratio by which the ultimate strength of the pipe in the ground is increased compared to its strength in the standard 'three-edge bearing test'. Standard tables of bedding factors are published[2,3] and they normally ignore the effects of lateral restraint. When pipes are laid under embankments with positive projection, the soil will exert an active soil pressure acting horizontally to the pipe. Since this pressure tends to produce a bending moment in the pipe ring which opposes that produced by the vertical load, the net effect is to produce an apparent increase in the bedding factor. The enhanced value of bedding factor can be calculated by methods presented by Young and Trott.[7] The supporting strength of a pipe must be equal to or greater than the total external load (W_c). Therefore, the required strength of pipe W_T is:

$$W_T = \frac{\text{External load } (W_c) \times \text{factor of safety}}{\text{Bedding factor}}$$

The factor of safety can be chosen at the discretion of the designer. It is usual practice to use a factor of safety of 1.25 in conjunction with the maximum crushing test load of a pipe, or factor of safety = 1.0 in conjunction with the proof test load for an unreinforced concrete pipe. (All pipes must withstand the specified maximum load and reinforced concrete pipes should not crack by more than a specified amount under proof load.)

Tables are published for structural design of different types of pipe under various load conditions.[8–10]

40.5.1.1 Internal pressure in rigid pipes

As previously noted, internal pressure produces circumferential

tension which reduces the available strength for resisting circumferential bending tension due to the external load. In these cases, Young and O'Reilly[3] and Young and Trott[7] should be used for design.

40.5.2 Flexible pipes

The design of flexible pipes is based on the yield strength of the pipe material in compression and on its resistance to circumferential buckling and distortion while being restrained by the surrounding ground. In pressure pipes the yield strength in tension must also be considered.

A significant element of the strength of a flexible pipe is derived from the adjacent soil and consequently the adequacy of the surrounding material and the workmanship used will have a significant effect on the performance of the pipe.

The pipe wall thickness for external loading is derived from ring compression theory.[4,11] It should be noted that a minimum ratio of pipe diameter to wall thickness must be provided to avoid damage to the pipe during handling.

40.5.3 Practical pipe design

The design and construction of pipes for most uses are covered by appropriate national standards. These usually give preferred methods of analysis and allowable stresses and deal with practical matters. Minimum cover depths (usually not less than 0.9 m) and the need to provide anchorages, air valves and other venting arrangements may have significant effects on design loads.

Nevertheless, the designer will be required to make a number of engineering judgements and it is essential that he should understand the basic theories involved and their limitations. Variability in soil, pipe strength and site workmanship means that extreme precision in design is not appropriate. Design tables are available[8–10] which reduce tedious calculation in a large number of situations. In situations where site conditions vary and the tables do not apply, the works by Young and O'Reilly[3] and Young and Trott[7] should be used.

The worst conditions for a length of pipeline should be used for design. Sources of extreme stress should be avoided rather than strengthening the pipe to resist them. The designer's assumptions should not demand excessively high qualities of workmanship or supervision on site. In particular, the designer must recognize that trenches need to be wide enough for access and support during construction. Experience shows that in most circumstances it is cheaper and more satisfactory in service to use strong pipes with cheap beddings (Class B for sewer pipes,[12]) and to employ the least expensive method of installation appropriate to the site conditions.

40.6 Construction of pipes in trench

40.6.1 Site investigation

Both the pipeline designer and the constructor require knowledge of the site conditions and the ground. A proper site investigation should be carried out at an early stage in the development of a project and it should include all the relevant topographical information and a soil profile extending to at least 1.5 times the trench depth.

The results of the investigation should be reported formally.[13] It is unlikely that extensive soil testing will be required since visual descriptions will be sufficient for most methods of design for pipe strength and trench support. Particular attention should be given to groundwater tables and to potentially corrosive conditions.

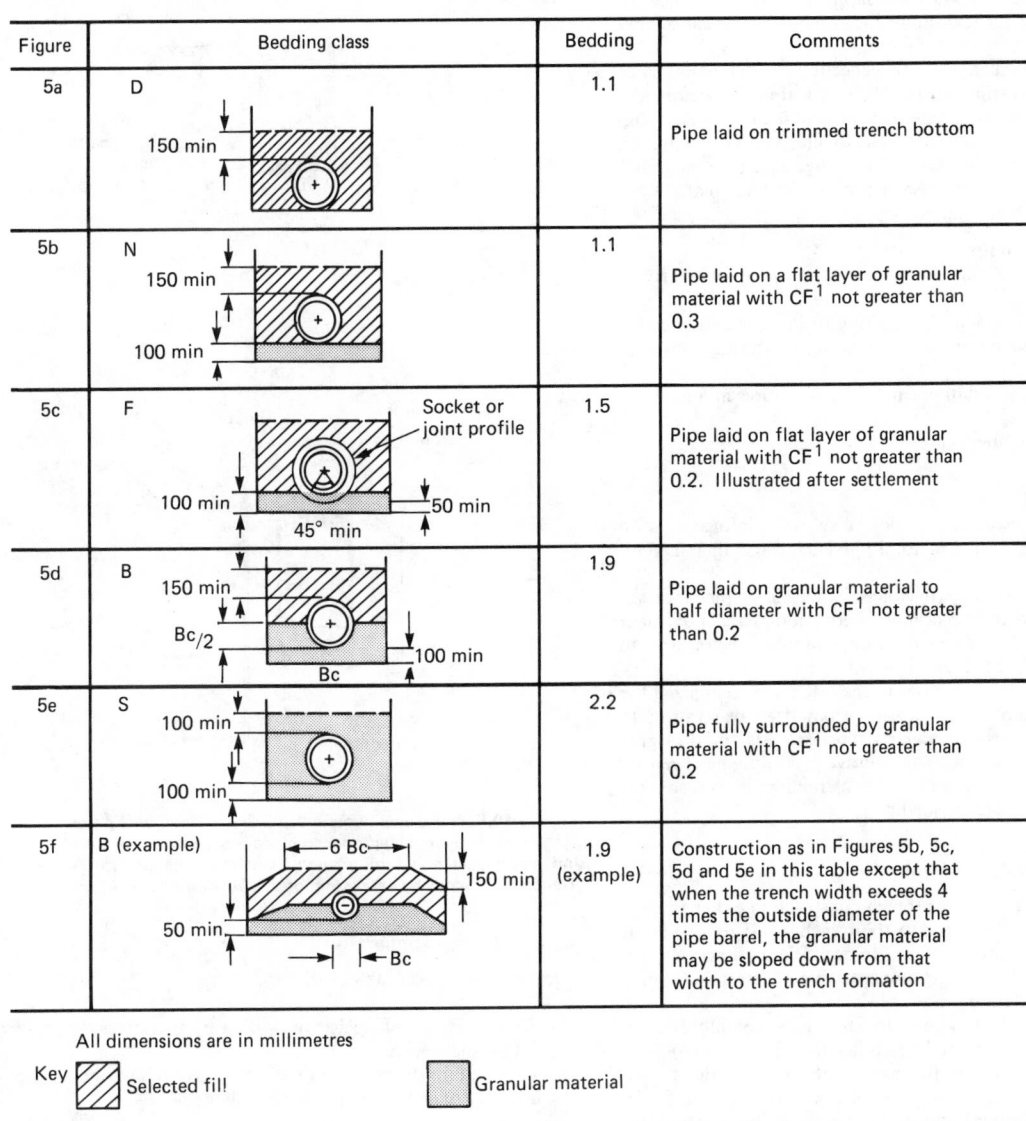

Figure	Bedding class		Bedding	Comments
5a	D — 150 min		1.1	Pipe laid on trimmed trench bottom
5b	N — 150 min, 100 min		1.1	Pipe laid on a flat layer of granular material with CF[1] not greater than 0.3
5c	F — 100 min, 50 min, 45° min, Socket or joint profile		1.5	Pipe laid on flat layer of granular material with CF[1] not greater than 0.2. Illustrated after settlement
5d	B — 150 min, Bc/2, Bc, 100 min		1.9	Pipe laid on granular material to half diameter with CF[1] not greater than 0.2
5e	S — 100 min, 100 min		2.2	Pipe fully surrounded by granular material with CF[1] not greater than 0.2
5f	B (example) — 6 Bc, 150 min, 50 min, Bc		1.9 (example)	Construction as in Figures 5b, 5c, 5d and 5e in this table except that when the trench width exceeds 4 times the outside diameter of the pipe barrel, the granular material may be sloped down from that width to the trench formation

All dimensions are in millimetres

Key ▨ Selected fill ▦ Granular material

Notes: (1) CF = Compaction fraction (see Appendix D, BS 8005: Sewerage, Part 1)

(2) Recommendations for granular pill and selected fill are given in clauses 5.5.1 and 5.5.2 of BS 8005.

(3) Where pipes with sockets are used these sockets should not be less than 50 mm above the floor of the trench.

Figure 40.5 Beddings for rigid pipes

40.6.2 Ground movement

All ground has existing *in situ* stresses and the balance of these stresses can be disturbed by the construction of the pipeline. In a trench excavation the sides and bottom lose the restraint of the excavated spoil and deform until either sufficient shear strength is mobilized in the ground to restore equilibrium or failure occurs. The insertion of the trench support assists in reducing this potential movement and preventing collapse. However, it should be recognized that ground movement cannot be eliminated.

The stress–strain relationships of soils are very complex. The degree of strain is dependent not only on the magnitude of the original stress and the properties of the soil but also on time and other factors. In weak soils, e.g. soft clay, light loading can cause large deformations. Wide trenches in soft clay can heave and fail at the base even when side support has been installed. In strong soils, e.g. cemented sands, the same loads will cause only small deformations with no failure.

The deformations are shear deflections which commence immediately excavation starts. There is a time lag before the maximum potential movement is achieved at each level of dig. For this reason soil can continue to creep after the trench support has been installed since it may have gaps between it and the soil and will in any case deflect under the load. Other movements may occur due to:

(1) Inadequately compacted backfill.
(2) Volume changes due to moisture variations particularly in fine grained soils.
(3) Short-term consolidation resulting from increased effective stresses caused by groundwater lowering during construction.
(4) Long-term consolidation caused by structures and geological processes.
(5) Ground heave due to groundwater pressures beneath clay.
(6) Frost heave.
(7) Traffic loading.

Symons[14] and Attewell and Taylor[15] give useful information on the possible components of total movement and their magnitude.

Soil movement is a cause of concern since buried pipelines will move with the soil and hence must accommodate the movement without distress. Pipelines with flexible joints are better able to resist such movements than pipes with rigid joints.

The water and gas industries in the UK have recognized the risk caused by movements arising from deep excavation to adjacent buried pipes, particularly those manufactured of grey cast-iron. For this reason a consultative procedure has been set up between them to investigate construction works so that potential problems are avoided.[16]

40.6.3 Groundwater control

This can be required during trench construction either to cope with groundwater flowing into the trench or to reduce uplift pressures on the bottom of the trench. Water flows may be controlled either by driving impermeable sheets to obtain a cutoff or by dewatering.

Dewatering of trenches down to 6 m is most commonly done by sump pumping and well pointing (see Figure 40.6). The effectiveness depends on the nature of the soil, the trench geometry and the degree and rate of lowering required. Sump pumping is the cheapest and simplest method but draws water inwards towards the base of the trench which may cause instability due to seepage forces and erosion of fine particles.

Well points have the advantage of drawing water away from the trench bottom which tends to stabilize the ground by increasing its shear resistance. Well points are most effective in sands; clays and fine silts are usually too impermeable.

In these cases exclusion methods are generally employed, although ground freezing or electro-osmosis has been used successfully at great cost.[17]

40.6.4 Battered trenches

Where sufficient space is available, battered trenches are often the quickest and cheapest method of construction. Large excavators can be employed and the need for side support is eliminated by cutting the sides of the excavation at a safe slope.[13] Slips due to subsequent deterioration of the exposed faces must be prevented by suitable measures where they are likely to cause problems.

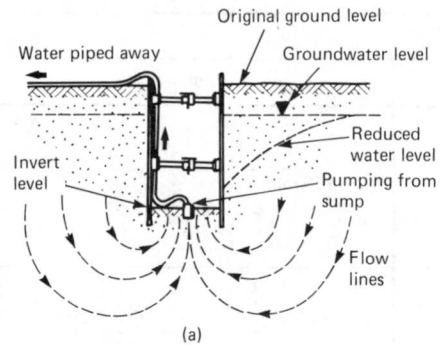

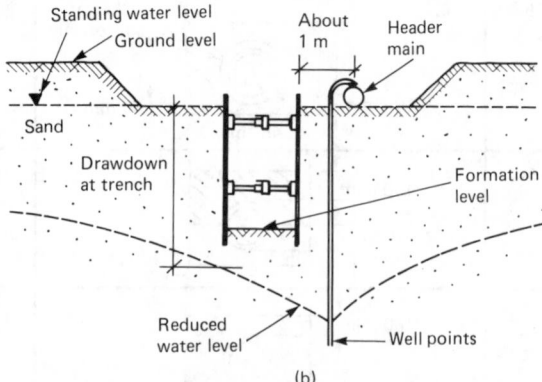

Figure 40.6 (a) Sump pumping; (b) single-sided well point system. (This figure is reproduced by courtesy of the Construction Industry Research and Information Association, Report No. 97 *Trenching practice* by permission of the Director of CIRIA)

40.6.5 Trench support

Trench support has two functions:

(1) To provide a safe place of work for operatives working within the trench.
(2) To maintain the stability of adjacent ground and hence the integrity of structures supported by it.

Trench support systems can be considered in two groups: (1) those based on traditional timbering methods but now frequently employing steel sheets and adjustable struts; and (2) those methods which use proprietary components or complete proprietary systems.[13,18,19]

In the traditional system, the sheets may be either predriven or inserted as excavation proceeds, and then supported by walings and adjustable struts (Figure 40.7).

The sizes of the components are estimated by simple calculation.[13] The use of highly mathematical theories is not justified since soil characteristics are generally not accurately known and the reuse of simple components keeps material costs low. The system can be adopted readily to suit changes in conditions or geometry but it should be installed only by men experienced in its use.

40.6.6 Proprietary systems

These may be described[13,18] as:

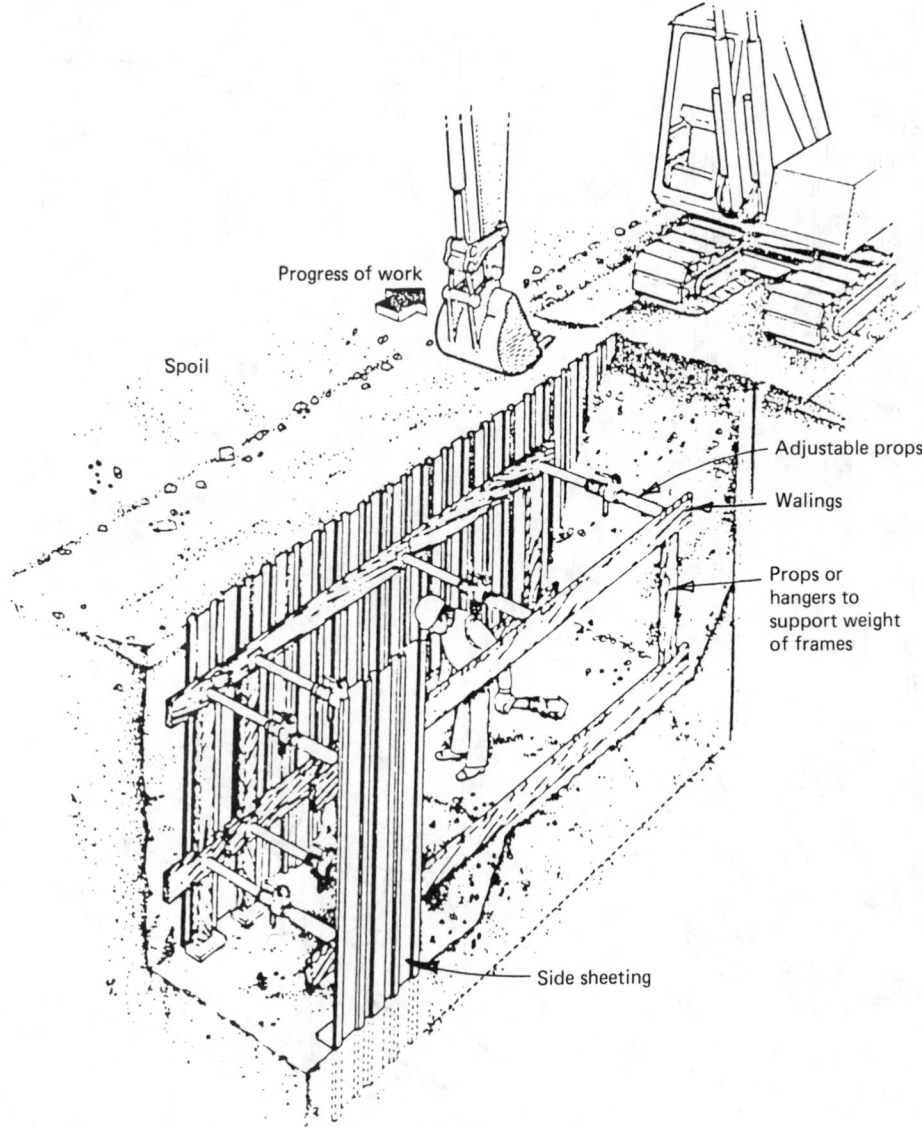

Progress of work

Spoil

Adjustable props

Walings

Props or
hangers to
support weight
of frames

Side sheeting

Figure 40.7 Typical trench support system with struts, walings
and side sheeting. (This figure is reproduced by courtesy of the
Construction Industry Research and Information Association,
Report No. 97 *Trenching practice* by permission of the Director of
CIRIA)

(1) Hydraulic frames and shores.
(2) Boxes.
(3) Slide rail systems.
(4) Shields.
(5) Piling frames.

Hydraulic frames (Figure 40.8) have adjustable struts permanently fixed to walings and may be used in place of conventional
screw struts and separate walings. The advantage of the hydraulic system is that in some conditions it can be placed and
tightened without the need for operatives to enter the trench.
The vertical shores are of similar construction and can be used
as pinchers where full face sheeting is not required.

Boxes are used to form modular strutted support walls
(Figure 40.8(b)). They are either lowered into a predug trench to
provide a protective box for operatives or are buried using the
dig-and-push technique. In this case, the box is progressively
pushed down and the spoil dug out between the walls. Where
efforts are made to limit overbreak, the latter method will result
in the box being a tighter fit in the ground thus giving more
positive support to the trench sides. It is normal practice when
laying pipes to use three or four boxes in line.

Slide rail systems have vertical slide rails or soldiers strutted
apart with horizontal walls spanning between them (Figure
40.8(c)). The dig-and-push installation method is used and the
wall panels slide vertically in the rails.

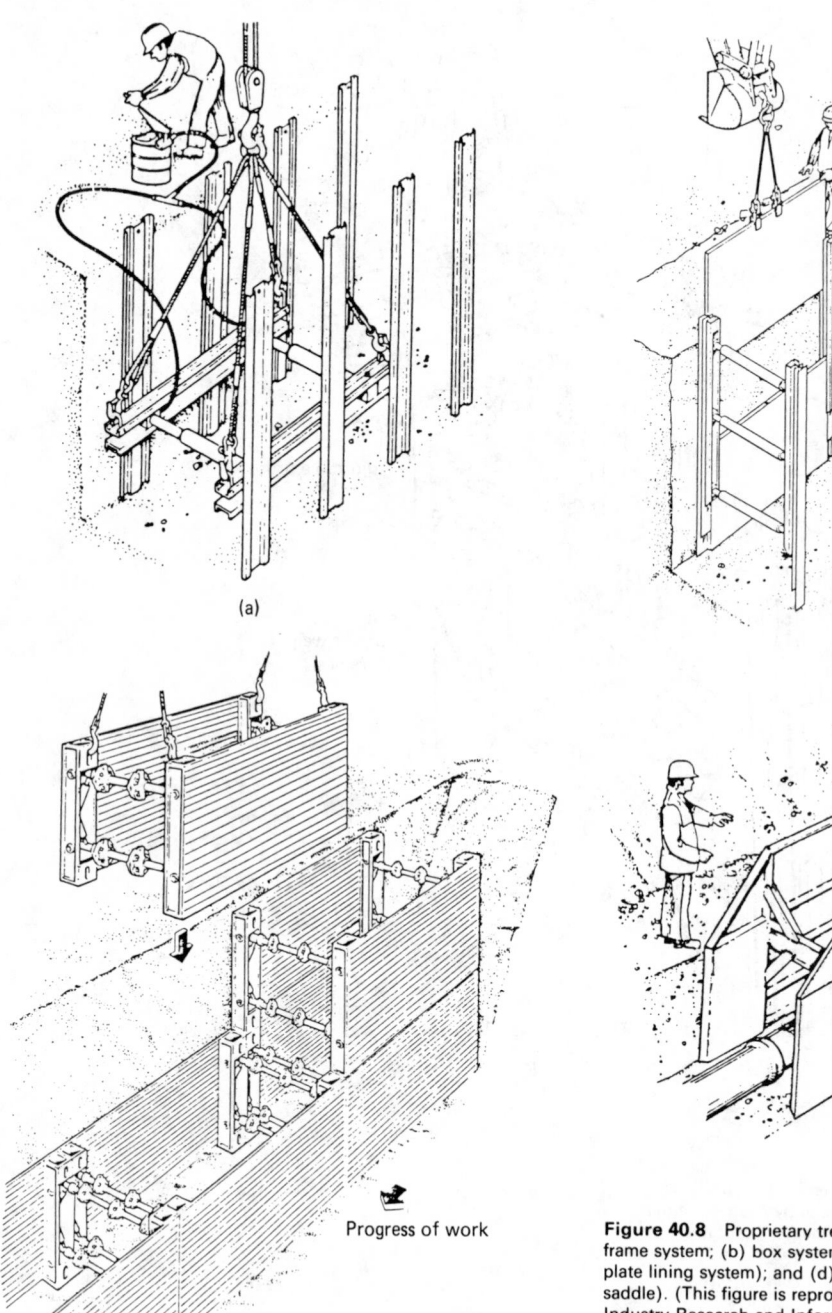

(a)

(b)

(c)

(d)

Progress of work

Progress of work

Figure 40.8 Proprietary trench support systems. (a) Hydraulic frame system; (b) box system; (c) slide rail system (also known as a plate lining system); and (d) drag box (also known as a shield or saddle). (This figure is reproduced by courtesy of the Construction Industry Research and Information Association, Report No. 97 *Trenching practice* by permission of the Director of CIRIA)

Shields, also called drag boxes or saddles (Figure 40.8(d)), are vertical support walls permanently strutted apart. The shield is lowered into the excavation and dragged forward by the excavator as the trench is extended. It is a loose fit in the ground and provides a protected place of work.

The presence of existing crossing services causes practical difficulties for boxes, shields and slide rail systems. Consequently, short lengths of traditional trench support may be needed in these areas.

Proprietary piling frames can be used to install vertical sheeting in the ground (Figure 40.9). The non-powered type consists of two sets of parallel walings forming a gate through which the sheets are inserted. The walings are permanently strutted apart and dig-and-push methods are used to force the sheets into the ground. In the powered version, the waling gates are equipped with hydraulic rams which enable the machine to push individual sheets up or down. The sheets are specially made for the machine and can be used to depths greater than 6 m if intermediate waling frames are used.

A further development for installing small-diameter pipes is

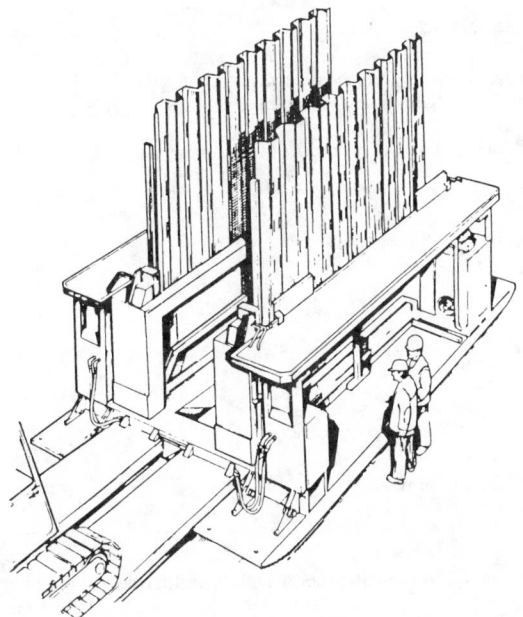

Figure 40.9 Proprietary rail-mounted piling frame. (This figure is reproduced by courtesy of the Construction Industry Research and Information Association, Technical Note 95 *Proprietary trench support systems*)

to use a self-propelled machine which cuts a continuous slot in the ground. It carries horizontal shields to keep the slot open so that the pipes can be jointed continuously at the surface and installed down a ladder guide into the bottom of the trench.

40.6.7 Submarine crossings

Pipes to be laid under water can be installed by several methods depending on circumstances:

(1) In cofferdam.
(2) Floating and sinking.
(3) Bottom pulling.
(4) Lay barge.

Cofferdams are costly and not used for pipelines of great length. Floating and sinking requires the pipeline to be fabricated on shore and fitted with buoyancy tanks before it is towed into position and sunk. Care must be taken not to bend the pipe and its protective coating beyond their allowable stresses.

Bottom pulling is more frequently used since it is easier to control stresses. The pipe is fabricated on shore in line with the trench and then pulled into the trench using a winch. In the lay barge method the pipe lengths are jointed on the barge as it is winched forward. The pipe spans between the barge and the trench bottom and frequently a ladder is used to give intermediate support.

In all cases, sufficient weight must be provided to anchor the pipe in place when it is empty.

40.6.8 Trench backfilling and reinstatement

The backfill forms part of the load-bearing system of a pipeline structure and must be constructed with care. The fill around the pipe needs particular attention to achieve an adequate bedding factor while ensuring no damage is caused by over-compaction.

The materials used for backfill should be capable of densification to the required standard without undue effort. The reuse of poor material from the excavation can be a false economy particularly under trafficked areas. The permanent reinstatement of road surfacing should be done as soon as practicable. The use of temporary surfacing may cause more problems than it is intended to cure, particularly where it does not give adequate support to the adjacent pavement or protect the subgrade.

40.6.9 Protection

Pipelines may require protection both internally and externally against a wide range of aggressive conditions. Steel and ductile iron pipes will tend to corrode in the presence of moisture and air. Consequently, they may require protection both internally and externally using a wide range of materials such as reinforced bitumen, coal tars, plastics, epoxy resin-based coatings and Portland cement mortars. Factory-applied protection is to be preferred but, since accidental damage is likely to occur during installation, ease of repair must be an important consideration.

The corrosion of ferrous materials is an electro-chemical process and cathodic protection can be used to provide external protection in addition to coatings. This may be particularly necessary in aggressive ground conditions and where clay soil could promote bacteriological corrosion. The pipeline is made cathodic by using either an impressed direct current or by connecting the pipeline to sacrificial anodes which corrode in preference.[20] It is important that cathodic protection systems are properly engineered since corrosion may be accelerated by an inadequate design.

Concrete pipes can be damaged due to sulphate in the groundwater and septicity in sewage. The latter generates hydrogen sulphide which can be reduced to sulphuric acid by certain types of bacteria. Septicity can to some degree be prevented by careful design of the sewer system to ensure an adequate oxygen supply.

40.6.10 Testing

Pipelines should be inspected for line level and material defects as they are constructed. Welded joints in steel pipelines can be checked using radiographic or other non-destructive tests. Internal proof pressure tests should be applied, preferably before joints have been backfilled. Air pressure is usually quicker and more convenient than water pressure; smoke tests can be used to discover leaks. However, air-pressure tests can be dangerous if failure should occur on large-diameter pipes. Water testing is to be preferred in these cases. Special arrangements may be necessary to anchor unbalanced forces particularly when testing large pipes. Also, the practical problems of obtaining sufficient water and its subsequent disposal can be considerable.

40.6.11 Safety

The principal legislation dealing with construction is the Health and Safety at Work Act 1974[21] which defines duties and responsibilities of employers, controllers of premises, manufacturers and suppliers of equipment and employees. This Act is supported by the Construction Regulations[22,23] together with other legislation. Practical advice on safety can be found elsewhere.[13,19,24,25]

The object of the legislation is to provide a safe place of work for operatives with safe means of entry and egress and a safe system of work. At the same time the safety of persons outside the site must also be ensured.

By law, trenches must be supported properly and inspected regularly.[22] Excavators used in the work can be used as cranes

Table 40.4 Safety checks for confined spaces

Before work starts:
(1) Check ground conditions for hazards and sources of gas such as organic strata, refuse, sewers, gas mains, industrial pipelines and the interaction between carbonate and acid (particularly in chalk, limestone and greensands)
(2) Ensure personnel are fit and properly trained
(3) Ensure breathing apparatus, lifelines and safety equipment is available. Define the procedures for contact with the emergency services
(4) Check gas monitoring equipment is available and working

Before entering:
(5) Check atmosphere for oxygen deficiency and explosive or toxic gases
(6) Check arrangements for ventilation
(7) Use breathing apparatus if the atmosphere is dangerous
(8) Check arrangements for entry and egress and lighting are adequate

While working:
(9) Continuously monitor the atmosphere
(10) Ensure space is properly ventilated
(11) Do not smoke
(12) Ensure direct communication is maintained between everyone involved in the work
(13) Ensure all equipment is maintained in good order and properly used

In emergencies:
(14) Do not enter the space without proper equipment or without an attendant at the entrance. A lifeline and harness should be used when entering and pulling out victims.
(15) Do not attempt to purge dangerous atmospheres with pure oxygen which will cause an explosive hazard

---

only if special exemption is obtained[26] and should be inspected weekly.

It should be noted that trenches may in some circumstances constitute a confined space and appropriate safety measures should be applied (see Table 40.4).

The legal responsibilities for safety cannot easily be subcontracted and the Swan Hunter case[27] clarified further duties for providing information. Contractors working alongside other contractors (whether subcontractors, tertiary contractors or independent contractors) have a duty to provide sufficient information so that each employer is aware how his operations affect others and how the operations of others affect his own employees. This duty applies not only downwards from main contractor to subcontractor but upwards also.

40.7 Trenchless pipelaying

In urban areas the use of trenchless techniques can reduce to a large extent the disruption caused by trench construction and also allow obstructions such as rivers, railways, major roads, etc. to be negotiated conveniently. The techniques currently used may be roughly divided into microtunnelling (formation of a new bore smaller than man-entry size) and conventional tunnelling (larger than man-entry size). The lower limit of man-entry size is generally accepted to be about 900 mm diameter. This is the smallest diameter in which a man can work effectively (albeit with difficulty). Where pipes are to be inspected only, men can enter down to 600 mm diameter if proper precautions are taken.

Since 95% of the UK sewer system is 900 mm diameter or smaller, microtunnelling is frequently used for sewer refurbishment as well as new construction.

40.7.1 Microtunnelling design

In recent years there has been a considerable emphasis on the development of microtunnelling techniques, and methods are available for producing bores from 50 to 900 mm diameter.

Figure 40.10 indicates the current techniques, which are based mainly on pipe-jacking methods.

Little research has been done on the vertical loads exerted by the ground on pipes installed by jacking. It is usually assumed that a slight overbreak of soil around the pipe may occur which will create a soil stress system similar to a Marston narrow trench condition. Therefore, soil load can be estimated by the equation by Young and O'Reilly (page 40/7) where B_d = effective width of trench. This assumption is likely to be conservative in stable soil.[7] Other soil theories may also be used, in particular Terzaghi's theory for buried pipes[28] which is based on assumptions similar to those used by Marston.

The loads on the pipe due to ground surcharge may also be calculated using the methods derived for pipes in trench.

The pipes must be strong enough to cope with the axial loads produced by the jacking operation as well as the external and service loads. Experience shows that pipes which are designed to cope with the jacking forces are generally adequate for the ground loads. Once the ground forces have been established, the pipe strength can be checked either using the bedding factor method (a value of 1.9 is generally accepted for pipes installed by jacking) or by using elastic theories of ring compression.[29]

The jacking loads arise mainly from friction. This varies with different ground conditions from 5 to 25 kN/m² although more extreme figures have been known. Sands and gravels cause higher values than cohesive soils and friction can be reduced by lubricating the outside of the pipe using bentonite slurry or other methods.

Jacking pipes of different materials and joint design vary in the load accepted in end bearing, the different materials available including steel, concrete, clay and GRP. Angular variations arising between pipes during installation tend to reduce the failure load and many types of joint have been tried to reduce this danger.[28] The jacking forces must be resisted at the thrust pit by a thrust wall or a heavy foundation. These are designed by conventional soil mechanics methods.

40.7.2 Site investigation

A proper site investigation is an indispensable requirement. It is

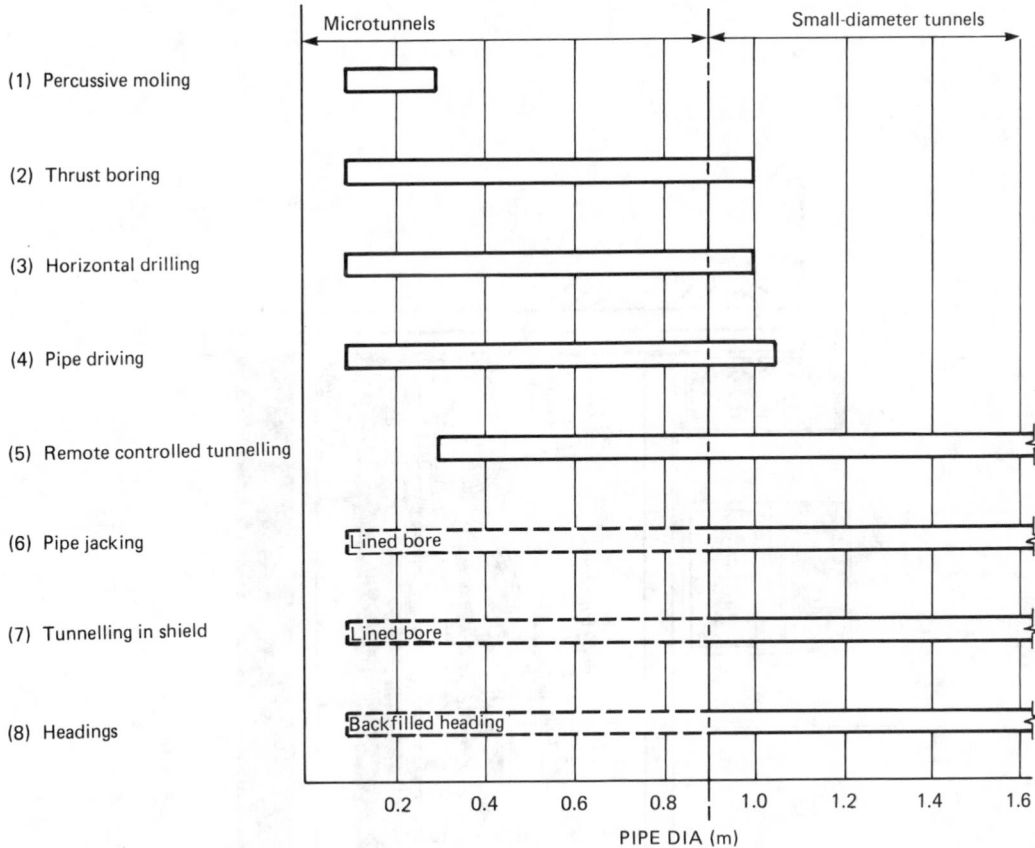

Figure 40.10 Trenchless pipelaying techniques for new pipelines

essential to choose the right construction method for the conditions, since there is no access to the end of the bore and the tunnelling machine may be lost if unacceptable ground conditions are encountered. If this happens it will probably require a shaft or second tunnel to be dug to recover the machine. In extreme cases the machine and the original tunnel must be abandoned.

A full topographical survey is required including the accurate location of all existing buried structures. The soil investigation should extend to a generous depth below the invert of the pipe. Boreholes and trial pits should not be located exactly on the centreline of the pipe to avoid weakening the ground and they should be properly grouted up and backfilled for the same reason. Soil testing is required to ascertain the description, density and strength of the ground. Cobbles and boulders are a particular hazard and small-diameter boreholes may be inadequate for identifying such conditions.

Information on groundwater levels and permeabilities must be provided. Poor conditions may need treatment before the tunnel excavation starts. Suitable techniques might include groundwater lowering with well points or deep wells, grouting or ground freezing.

The detection of existing services may be done by a variety of methods. Magnetic field location is commonly used. Passive locators are used to locate buried conductors by detecting the natural magnetic field surrounding it. The method is quick and simple but only locates. It is not appropriate to trace or identify a buried service. If this is required, an active locator should be

used. This comprises a transmitter, which applies a signal of known frequency to the pipe or cable and a receiver which is tuned to the signal and used to trace the service. The method can be time-consuming and requires a skilled operator but gives greater accuracy than the passive technique.

Non-metallic pipes can be traced by the use of small self-contained transmitters known as sondes. These are moved through the pipes using rods or high-pressure water or towing cables, and the signal is traced from ground level.

Other location methods are available including ground-probing radar, seismic reflection techniques and the detection of variations in static magnetic fields. Also dowsing methods have been used with success.

A common fault of existing service maps is lack of accuracy. The digitization of records and production of digital maps based on Ordnance Survey maps is more accurate and is gaining favour. This should improve the situation in the future and ease the exchange of information between utility companies and others.

40.7.3 Ground movement

The installation techniques for microtunnels may be classified as 'convergent', where the volume of excavation exceeds the volume of the pipe installed, or 'expansive' where the opposite is the case.

In the convergent case, ground may be lost due to face and peripheral encroachment (similar to overdig in larger tunnels),

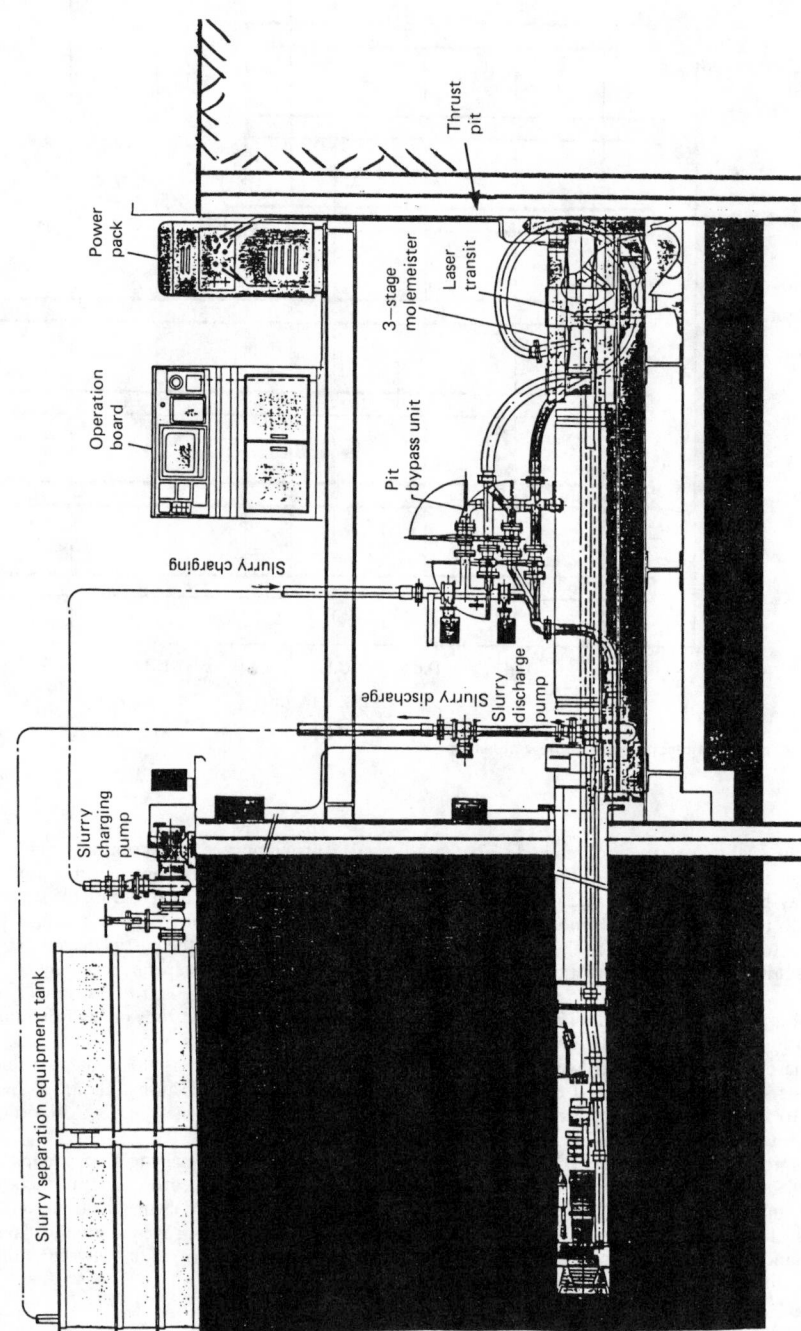

Figure 40.11 Microtunnelling with a bentonite slurry shield. (Courtesy: Iseti Poly-Tech Inc.)

consolidation due to changes in drainage patterns, and local instabilities during driving. Because of the small diameter of the hole the absolute volume lost tends to be small and, hence, the potential movement is usually limited compared to man-accessible tunnels.

Expansive techniques include percussive moling, pipe driving and on-line pipe replacement and may disturb adjacent buried pipes or cause ground heave at shallow depths. This problem is discussed by Howe and Hunter[30] and O'Rouke.[31]

40.7.4 Pipe jacking with slurry shields

These machines are essentially miniaturized versions of shields used for conventional tunnelling and employ a full-face cutting head within a shield (Figure 40.11). The cutter can be moved relative to the shield, allowing it to exert a constant pressure on the ground irrespective of the rate of advance of the pipe. The groundwater pressure is balanced by flooding the cutting face with water or a bentonite slurry. In many machines the slurry is used to transport the spoil back to the surface.

The ability to balance soil and water pressure gives good control of ground movement at the face. The alignment of the machine is monitored by using a laser beam focused on a target on the shield. The target is observed by an operator at ground level via closed-circuit TV and he can adjust the alignment by remote control using steering jacks within the shield.

The shield is jacked forward from the jacking pit leading a pipe string through which its power and slurry lines run. As new pipes are added to the pipe string the power and other lines must

be broken and reconnected. One manufacturer avoids this by using a temporary lining behind the shield which has a recess to accommodate the lines. The temporary lining is replaced by the permanent pipe in a second-stage operation.

Slurry shield machines can operate in most soft ground conditions although the cutter type and the slurry separation systems may need adjustment for different materials which can cause practical difficulties in variable ground conditions. The size of cobble which can be excavated depends on the size and type of machine used. One manufacturer has machines which can crush stones measuring up to about one-third of the diameter of the shield. In ground containing large stones it will be necessary to install a larger machine which can cope with the expected conditions.

40.7.5 Pipe jacking with steerable borers

This category covers a wide range of machines. The steering may be by means of jacks on the shield or by rotating a special cam behind the articulated cutting collar. Alignment is monitored by visual surveying from the launch pit or by using a laser system observed by closed-circuit TV. Spoil is usually removed back to the launch pit by screw conveyor which also forms the drive shaft. Water or slurry may be used to cool the cutting head and assist in spoil transport. Other systems form a pilot hole which is then reamed out by passing spoil forward to the receiving pit.

Steerable borers can deal with a wide range of ground conditions by selecting an appropriate cutting head. Auger bits

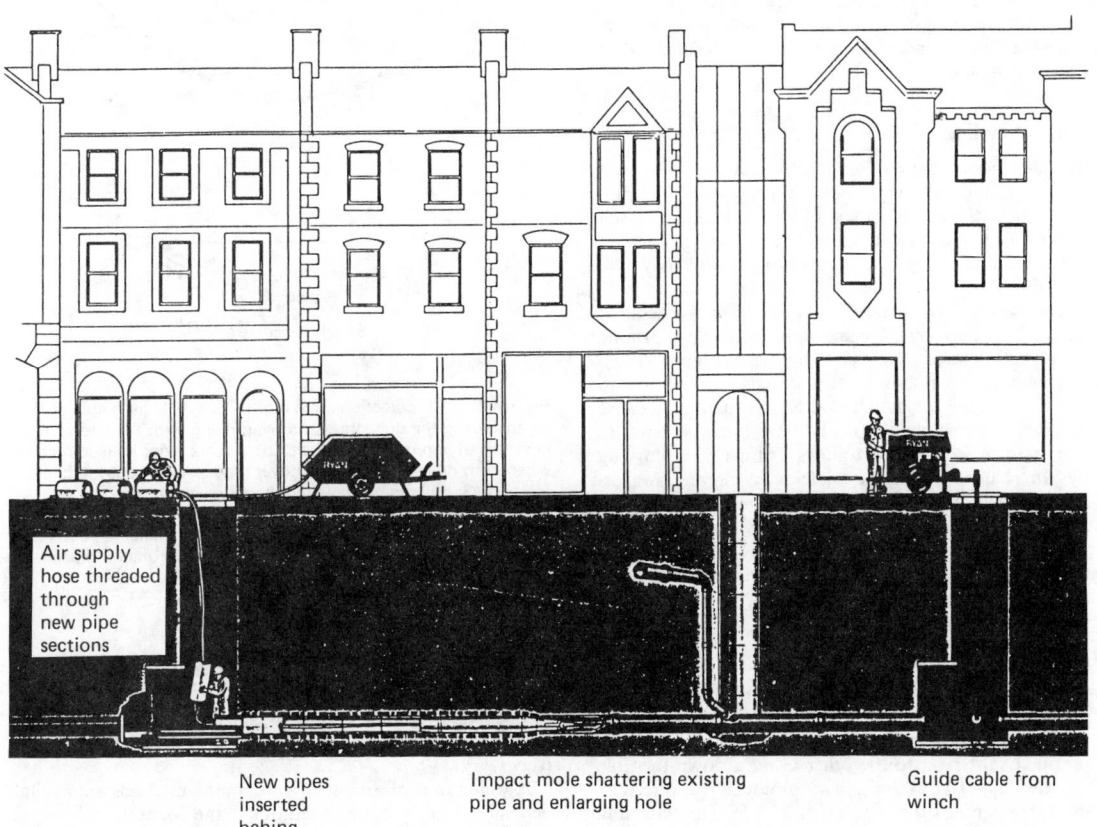

New pipes inserted behing impact mole

Impact mole shattering existing pipe and enlarging hole

Guide cable from winch

Air supply hose threaded through new pipe sections

Figure 40.12 Impact mole used for upsizing an existing sewer

can be used in soft strata while rock roller bits can be used in harder materials. Various methods are employed for dealing with water pressure including short pitch augers or pressurizing the face with water or compressed air.

40.7.6 Pipe jacking with non-steerable augers

There are less complicated tools than those described above although they do incorporate similar features. The non-steering auger uses a simple cutting head rotated by an auger screw. The rotational power and thrust for the auger is provided by a drive unit at the launch pit. The cutting head is difficult to control accurately once boring has progressed a few metres and often alignment is not monitored between the launch and receiving pits during excavation. Large deviations may occur if the bore obliquely crosses hard strata. In long drives steady bearings have been installed in intermediate pits and reasonable accuracy achieved. A wide range of ground conditions can be excavated but problems may arise in water-bearing sands and gravels.

40.7.7 Pipe ramming

This technique, also known as pipe driving or pipe pushing, is analagous to pile driving. It normally utilizes a steel pipe which is driven into the ground from a launch pit using percussive loading at the trailing end. Small-diameter pipes are driven closed-ended but larger diameters are left open and cleaned out using augers or water jets.

Pipe ramming can be carried out in most types of ground and the percussive action tends to break up cobbles and other obstructions. There is little control of accuracy once the drive is under way and obstructions can cause deviations. The final accuracy depends on ground conditions and lengths of drive do not usually exceed 30 m.

40.7.8 Impact moling

This technique relies on displacing and compacting soil to form a void in the ground. The mole is usually powered by compressed air which causes a hammer piston to strike a chisel-headed anvil. This pierces the ground and the tool moves forward. The machines are generally reversible to aid recovery if refusal conditions are encountered.

Moles can be used to make pilot holes for subsequent enlargement or alternatively they can tow in pipes while forming the bore if polyethelene or plastic pipes are used (Figure 40.12).

Impact moles can operate in most materials and are able to break up isolated stones or boulders although these may cause some deviation in line. In very soft conditions the mole tends to sink under its own weight and follows a downward curving trajectory. In shallow bores the mole may curve upwards. Therefore to prevent ground heave, and to control accuracy, a minimum cover of 9 times the diameter should be used. Accuracy depends on ground conditions and is typically about 1%.

Impact moles are also used widely in replacing and up-sizing gas and sewer pipes (see below).

40.7.9 Directional drilling

This method employs techniques used in drilling oil- and gaswells, and is particularly suited to the construction of river crossings, etc. A surface-mounted rig is used to drill a pilot hole from one side of the river to the other using a down-the-hole motor at the end of drill rods. The motor is powered by bentonite slurry pumped through the drill rods. The returning slurry cools the drill bit and pushes the cuttings back to the surface. The curved profile is achieved by steering the drill rods using a special fitting mounted behind the motor.

The pilot hole is cased with a steel pipe called a washover pipe which is then used as a draw string for a reaming tool when the pilot hole is complete. The reaming tool is a circular cutter which is attached to the leading end of the washover pipe. The rig then rotates the pipe and pulls it back through the pilot hole (Figure 40.13). A further washover pipe is towed behind the reaming tool. Bentonite slurry is used together with the second washover pipe to stabilize the enlarged hole. The permanent pipe is then pulled into place using the second washover pipe. Further reaming and smoothing tools may be used in this operation to ensure a proper fit for the permanent pipe.

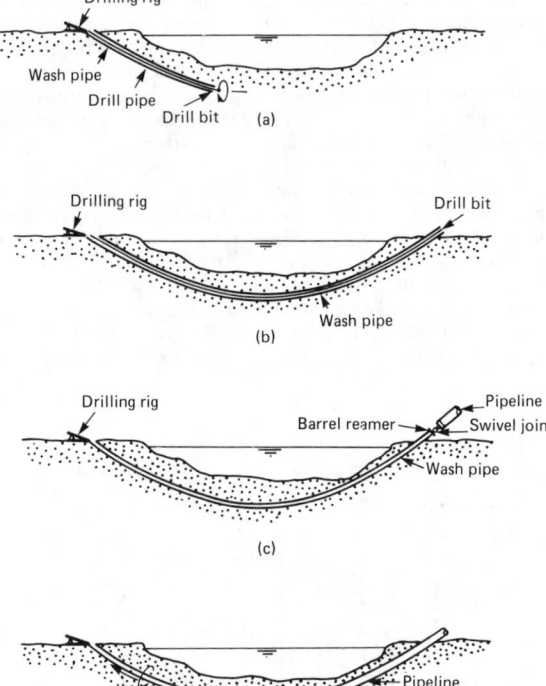

Figure 40.13 Directional drilling. (a) Stage 1: pilot hole drilled by advancing the drill string and overdrilling with the washover pipe in stages of approximately 80 m; (b) stage 2: pilot hole completed when both drill string and washover pipe exit on opposite bank; (c) stage 3: drill pipe is removed, barrel reamer connected to washover pipe which is in turn connected to the pipeline pulling head by a swivel joint; (d) stage 4: the barrel reamer is pulled back and rotated by the drill rig positioning the non-rotating pipeline into the formed hole

40.8 Man-accessible tunnels

Man-accessible tunnels are defined as having diameters greater than 900 mm which is the smallest in which a man can effectively work.

A wide range of excavation and lining methods are available depending on ground conditions and operational requirements.[32] In soft ground timbered headings, segmental linings and pipe jacking are the principal construction methods used.

Excavation is usually done by hand in timber heading. In

segmental tunnels and pipe jacking hand excavation may be used but more usually a variety of mechanized methods are employed, including backhoe excavators within a shield, and full face tunnelling machines. The latter include bentonite slurry shields and mechanical earth pressure machines intended to minimize ground loss and movement at the excavation face.

Soft ground tunnels are invariably lined and the choice of lining depends on a number of factors including ground conditions, operational requirements and the construction method to be used. The relationship of these factors is complex and the final choice should be based on mutual agreement between the employer and the contractor.

In hard rock the traditional method of excavation is drill and blast which has the advantage of coping with a wide range of ground conditions. Machine excavation can be much quicker and hence cheaper, but it is more sensitive to changing ground conditions which may be outside its range.

Linings for bores in hard competent rock may be unnecessary for structural reasons but could be desirable to provide a smooth hydraulically efficient surface. In this case an *in situ* sprayed or poured concrete lining may be used. In less competent ground the rock may be stabilized with rock bolts prior to lining with *in situ* concrete or a bolted precast concrete lining may be used.

In many cases the primary lining supporting the excavation may be too large for efficient hydraulic design and the tunnel must be lined with a small carrier pipe or a channel invert. Timbered headings are invariably backfilled around the carrier pipe which must be designed for the loads arising.[3]

40.8.1 Site investigation

The main factor affecting construction cost of a tunnel is the nature of the ground. The site investigation will be similar in nature to that for microtunnelling but usually will be more extensive due to the larger works being constructed. When assessing suitable construction methods and the loads on the lining, more extensive *in situ* and laboratory testing will be required. Full descriptions must be given for the strata together with a geotechnical interpretive report since the ground structure has increasing significance as the tunnel size increases. When large tunnels are contemplated, consideration should be given to the possibility of constructing a pilot tunnel as a separate contract. This will yield valuable information for the design and construction of the larger tunnel.

A tunnel lining is interactive with the surrounding soils in a manner which depends on their relative stiffnesses. Measurements in existing tunnels show that frequently the lining does not carry the whole overburden load.

The fundamental design problem is predicting the behaviour of the ground and the way it is varied by the construction process. The current trend is towards designing flexible linings which will deflect to produce virtual equality of radial stress in the ground, usually by the horizontal axis extending as the vertical axis becomes shorter. The process of design must consider the construction method and the operational requirements of the lining, and then check that it is adequate for the ground loads expected. Handling stresses in the lining elements often give the worst design conditions particularly in tunnels for shallow depths.

A variety of methods of designing lining thickness are available.[32,33] Empirical techniques based on field observations allow the designer to estimate the likely ground loads and deformation of the lining when similar construction methods are used in similar ground conditions. Other design procedures are based on mathematical analysis and use closed form elastic solutions or numerical methods. These require the soil behaviour to be modelled in terms of its elastic, plastic and time-dependent

characteristics to predict the changes in the *in situ* ground stresses. It is difficult to determine representative values for these factors and to assess the *in situ* state of stress of the ground. Hence, the methods are frequently used to model a range of possibilities to assess the sensitivity of each parameter and allow an engineering judgement to be made on likely stable or unstable configurations.

Pipes to be installed by jacking methods are designed as rigid structures as previously described for microtunnelling. Particular attention must be paid to the leading pipe and the trailing pipes at intermediate jacking stations since these are repeatedly subject to the full ram load during the jacking operation.

40.8.2 Headings

Headings are rectangular or square in section and usually dug by hand. The ground is supported by timber frames which are erected as the excavation progresses (Figure 40.14). The timber sizes are either estimated by rule of thumb or designed as temporary works with appropriate factors of safety.[19] The dimensions of the heading must allow sufficient working room for the construction of the carrier pipe and the compaction of the backfilling around it. The smallest practical size is about 1.15 × 0.7 m clear between supports.

The technique requires experienced and skilled operatives and is relatively expensive. As a result its use is restricted to special circumstances.

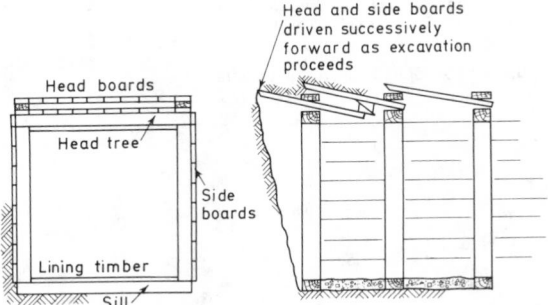

Figure 40.14 Timbered heading

40.8.3 Steel liner plates

Steel liner plates may be used for lining small-diameter tunnels in firm ground. Typically, four plates are used to construct the ring and the plates are bolted on the longitudinal and circumferential joints. The plates are back-grouted to ensure even bearing of the ground. Joints can be welded if necessary to prevent water ingress.

The finished lining behaves as a flexible pipe and is generally given a secondary lining.

40.8.4 Precast concrete segmental linings

These are available for either bolted or boltless construction. Bolted segments have flanges on all four sides and tend to be used in poor ground and where there are problems with water ingress (Figure 40.15). The segments are erected within the protection of a shield and are back-grouted every ring or every shift. The method is particularly useful in conditions requiring compressed air and for hydraulic tunnels with a moderate pressure.

Where a smooth internal finish is required it is necessary to use a secondary lining. To avoid this, smooth bore, bolted

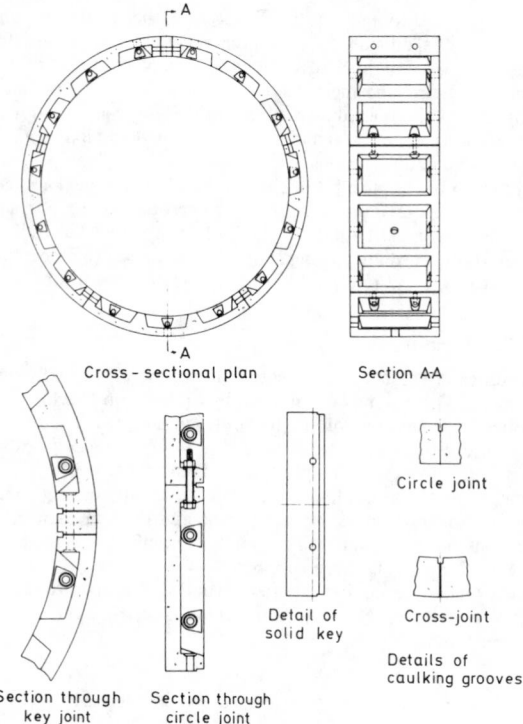

Cross - sectional plan Section A-A

Circle joint

Section through key joint   Section through circle joint

Detail of solid key

Cross-joint

Details of caulking grooves

Figure 40.15 Bolted concrete segmental ring

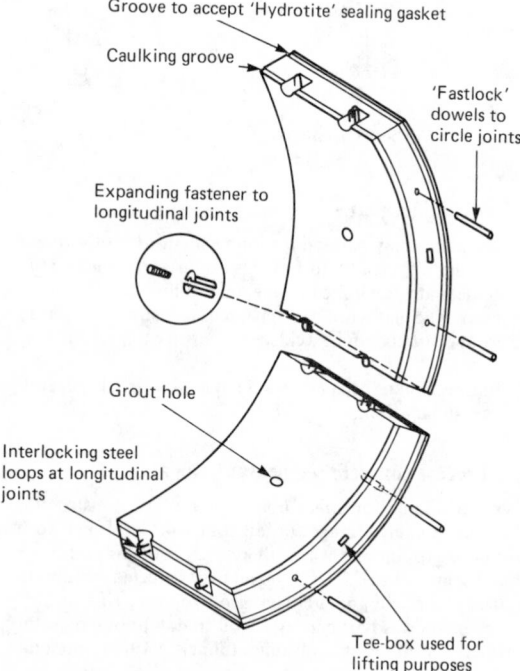

Groove to accept 'Hydrotite' sealing gasket

Caulking groove

'Fastlock' dowels to circle joints

Expanding fastener to longitudinal joints

Grout hole

Interlocking steel loops at longitudinal joints

Tee-box used for lifting purposes

Figure 40.16 Bolted smooth bore concrete segmental ring. (*Courtesy*: Charcon Tunnels Ltd)

segments have been developed and their use is generally known as one pass construction (Figure 40.16).

Boltless segments also produce a smooth bore tunnel and are used in ground which has a reasonable stand-up time. The segments are erected on a temporary steel former ring using bolted connections. After the annulus behind the concrete ring is grouted the steel former is demounted and the process repeated (Figure 40.17).

Other smooth bore linings are of the expanded type intended for use in clay. The linings are erected on an erector arm or bars and expanded by using a wedge-shaped block or by cirumferential jacks which form spaces to be filled with dry pack or special blocks.

Grey cast iron has great strength and traditionally was used where heavy ground loads were expected. More recently it has been replaced by spheroidal cast iron which has a better tensile strength. Cast iron has been used for both bolted and expanded systems. The segments can be manufactured to a better tolerance than concrete and, hence, produce a better seal against water ingress in bad conditions.

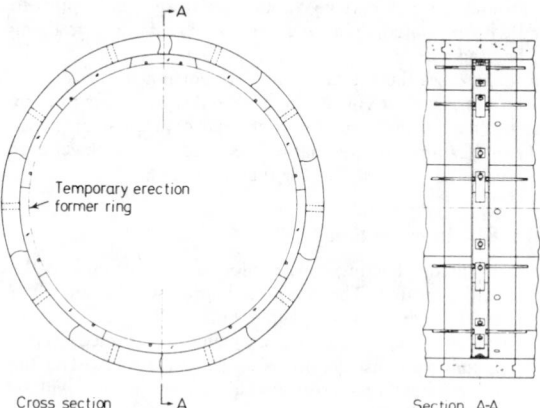

Temporary erection former ring

Cross section

Section A-A

Figure 40.17 Boltless smooth bore concrete segmental ring. (*Courtesy*: Charcon Tunnels Ltd)

40.8.5 Triple segmental block (minitunnel system)

This proprietary system has been developed to provide smooth-bore tunnels in the range 1.0 to 1.3 m diameter. The lining consists of three segments which can be erected without the use of a former ring (Figure 40.18). Each segment has longitudinal V-shaped grooves inside and outside which function as stress raisers to ensure that the lining deforms and acts with forces in compression.

The segments are erected within the protection of a shield which is used to dig an oversize hole. The overbreak is filled with gravel injected through an orifice in the rear of the shield tail. The gravel packs around the completed segment ring and supports the ground. The main longitudinal and circumferential joints are sealed with a rubber bitumen strip compound during erection and the gravel annulus may be grouted up to provide additional waterproofing.

40.8.6 Pipe jacking

As with the microtunnelling technique, the pipes are forced into the ground using hydraulic jacks at a working shaft (Figure 40.19). A steel or concrete cutting edge is used at the leading end of the pipe string and miners working within the pipe excavate

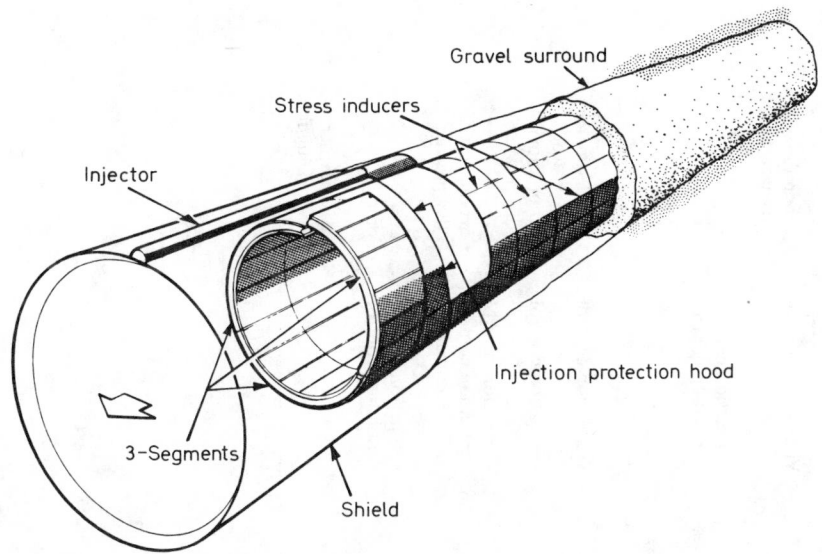

Figure 40.18 Mini tunnel system. (*Courtesy*: William F. Rees Ltd)

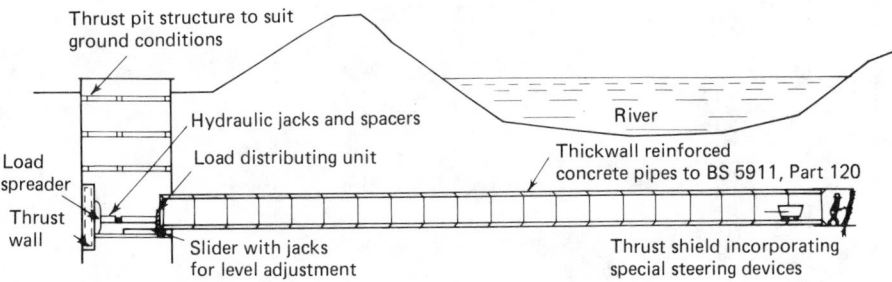

Figure 40.19 Section through a typical pipe jacking operation

material as in a normal tunnel shield operation. The accuracy of the drive can be improved by using trimming jacks or steering fins to adjust the attitude of the cutting shield. These also allow pipes to be jacked round bends.

The maximum length of drive can be increased by injecting bentonite or other lubricants to reduce the friction on the outside of the pipe. Also intermediate jacking stations can be built into the pipe string to increase overall jacking capacity and to restrict the length of pipe to be pushed by each set of jacks.

40.8.7 Ground movement

This is caused by ground loss during tunnel excavation and by additional consolidation as the porewater pressure is reduced due to drainage induced by the presence of the tunnel. Observations of various tunnels in the past have provided a basis for estimating the magnitude of potential ground movements caused by tunnelling.[15,34] Efficient management and good workmanship are critical factors in reducing movement to a practicable minimum.

In soft, loose and very permeable ground some pretreatment may be necessary to limit the ground loss and to improve the safety of the construction phase. Typical methods include dewatering, grouting and freezing.

The use of compressed air may be considered to overcome technical difficulties during excavation in soft clays or water-bearing ground. It acts in three ways, by: (1) balancing the water head; (2) providing a direct reaction to field forces; and (3) drying the skin of the soil, thus increasing its effective stress. However, it also has significant physiological risks for operatives and must be regarded as a last resort.

40.9 Working in confined spaces

A confined space may be defined as a workplace which does not have the benefit of natural ventilation. Consequently, the danger exists that the atmosphere in such places may become either deficient in oxygen due to the build up of gases or vapours, or hazardous due to concentrations of toxic or flammable gases or vapours.

The construction and maintenance of buried pipelines involves working in a variety of confined spaces. Examples are manholes, shafts, tunnels, pits, boreholes and, in some situations, trenches. Precautions must be taken in these cases to ensure that safe working conditions exist and the work being carried out does not give rise to hazards. Much specialized advice is available[24-27,35-37] and tunnels and boreholes have Bri-

Table 40.5 Summary of the most commonly encountered dangerous gases in confined spaces

Gas		Specific gravity	Hazard	TLV* (p.p.m.)	Explosive limits Lower (%)	Upper (%)	Principal sources	Prevention	In case of fire
Carbon monoxide	CO	0.97	Toxic flammable explosive	50	—	—	Explosives/engines	Ventilation control	Carbon dioxide dry powder
Carbon dioxide	CO_2	1.53	Asphyxiant	5000	—	—	Natural/engines	Ventilation control	—
Nitrogen oxides	NO	1.04	Toxic	25	—	—	Explosives/engines	Ventilation control	—
	NO_2	1.60	Extremely toxic	5	—	—			
Nitrogen	N_2	0.80	Asphyxiant liquid causes burns	5	—	—	Freezing processes	Ventilation control	—
Methane	CH_4	0.60	Explosive and asphyxiant	—	5.3	14	Natural	Ventilation control	Carbon dioxide dry powder
Hydrogen sulphide	H_2S	1.70	Toxic and explosive	10	4.3	46	Natural	Ventilation control	Carbon dioxide dry powder
Sulphur dioxide	SO_2	2.30	Toxic	5	—	—	Natural	Ventilation control	—
Propane		1.55	Explosive and asphyxiant	1000	2.2	9.5	Leakages	Ventilation control	Carbon dioxide dry powder
Butane		2.10		600	1.5	8.5	Leakages	} Cylinder care	
Acetylene		0.91		—	2.5	81.0	Leakages		
Deoxygenated air	N_2		Asphyxiant	—	—	—	Natural/induced	Ventilation control	—
Petrol/diesel vapour			Explosive	—	1.3	7.5.	Spillage	Ventilation control	Foam carbon dioxide dry powder vaporizing liquid

*Threshold limit value.

tish Standard requirements.[38,39] The Construction Regulations and Factory Acts also give guidance.

Work in confined spaces should be undertaken only by persons properly trained for the job. Under the Health and Safety at Work Act 1974 the employer has a duty to provide such information, instruction, training and supervision as is necessary to ensure, so far as is reasonably practicable, the health and safety at work of his employees. The employer should also check that employees who are expected to work in confined spaces are physically and mentally suitable.

Safe systems of work must be laid down and strictly observed. The minimum requirements will be:

(1) To test the atmosphere before entering the confined space.
(2) To monitor the atmosphere while people are working.
(3) To maintain contact between the operatives and an attendant in free air. The attendant must be trained to carry out emergency procedures.

Adequate means of access and egress must be provided. Rectangular or oval holes of 458×407 mm and circular holes of 407 mm diameter are the minimum sizes laid down by the Factories Act.

Larger access should be provided where feasible so that rescue equipment can be used more easily. Permit to Enter and Permit to Work systems should be established.[24] These must be administered by a responsible person who will keep records of the safety measures taken and will sign a certificate accordingly.

Gas testing of the atmosphere may be done using simple, rugged equipment. Potential hazards can often be anticipated by an intelligent assessment of local conditions. Works near waste tips are likely to encounter methane or hydrogen sulphide generated by the decomposition of organic material as well as other industrial compounds. Carbon dioxide, which is heavier than air, may be produced by the action of acidic water on limestone or other calcareous rock. It may also arise from the exhausts of nearby internal combustion engines and will be mixed with carbon monoxide. Table 40.5 summarizes the commonly encountered dangerous gases. Oxygen deficiency can arise due to rusting processes or a fall in atmospheric pressure causing deoxygenated gas to seep out of the surrounding ground.

In sewers additional hazards arise due to the danger of being swept away or drowned by fast-flowing streams. Dangerous conditions can arise quickly and may be noticed by increasing movement of air or water through the sewer or the noise of approaching water. Attention should be paid to weather forecasts as a preliminary precaution.

Sewers can also give rise to infection from a number of causes and all persons entering should be inoculated. Personal hygiene must be meticulous and any injury or abrasion should receive proper medical attention.

Confined spaces can rapidly become contaminated from the work processes being used. Often these processes are themselves intended to avoid other hazards. For instance, ground freezing and chemical grouting for excluding water, and pipe and cable freezing to isolate sections of pipe for maintenance or repair. Burning and thermal welding processes use up oxygen and give rise to toxic fumes as do many solvents and paints. Pipes which have contained flammable materials must be purged thoroughly before any work is carried out.

Dust and noise must also be regarded as dangerous contaminants in confined spaces. Lighting must be efficient and provide illumination to at least 20 lx using flameproof equipment and no more than 24 V. An emergency lighting system must be available for immediate use in the event of failure of the main system.

Rescue from confined spaces should not be attempted without proper equipment. The rescue team must be a minimum of two so that one person can attend at the entrance while the other enters. Non-observance of these requirements is likely to make the situation worse.

40.10 Rehabilitation of pipes and sewers

Trenchless methods for the rehabilitation of buried pipes offer the advantages of reduced disruption and cost. The gas, water supply and water disposal industries in particular have developed new methods of construction, particularly for non-man-entry pipes, and further innovation can be expected.

The methods are divided broadly into those incorporating the original pipe structure, e.g. relining, and those removing or destroying the existing structure, e.g. pipe bursting with percussive moles. The choice of method is heavily influenced by the size of new pipe required and the access available. In non-man-entry sizes work must be carried out either by remote methods or by open trench.

40.10.1 Inspection and basic strategy

Before repair and refurbishing operations can be undertaken the pipe must be inspected. Manual inspection is generally preferred where possible but in dangerous atmospheres and small-diameter pipes closed-circuit television is used routinely. It should be recognized that in such cases breathing apparatus or forced ventilation may be required when inserting and recovering the camera.

The results of the inspection should be recorded in a standard manner to reduce the subjective nature of assessing defects. In sewers the recommendations of the National Water Council[40] should be used.

After inspection, a repair strategy must be decided. In sewers,

Table 40.6 Definitions of remedial works in sewers

Rehabilitation:	All aspects of upgrading performance of existing sewers. Structural rehabilitation includes repair, renovation and renewal. Hydraulic rehabilitation includes replacement, reinforcement, flow reduction or attenuation and occasionally renovation
Maintenance:	Minor repairs not involving reconstruction of main sewer or alteration of dimensions
Repair:	Rectification of damage to the structural fabric of the sewer and reconstruction of short lengths but not reconstruction of the whole pipeline.
Renovation:	Improvement of performance of sewer by incorporating the original sewer fabric
Reinforcement:	Provision of an additional pipeline which, in conjunction with the existing sewer, increases overall flow capacity
Renewal:	Construction of a new sewer on or off the line of an existing sewer. The function and capacity of the new sewer is similar to the old
Replacement:	Construction of a new sewer on or off the line of an existing sewer. The function of the new sewer will incorporate that of the old but may also include improvement or development work.

Table 40.7 Pipe stabilization and relining methods

Technique	Brief description	Original effective pipe diameter	Advantage	Disadvantage
Cleaning				
(1) Scaling	High-pressure water jetting to remove scale and encrustation	All	Can be used in all diameters of pipe. In small diameter pipes the work can be done remotely and monitored by closed-circuit TV	High-pressure equipment is expensive and requires trained operators
(2) Removal of protrusions	In small-diameter pipes cutting is done by remotely controlled mechanical or high-pressure cutters operated under observation by closed-circuit TV	All		
Stabilization				
(3) Repointing	Hand- or pressure-pointing used to renew joints in masonry or brick sewers where these have not closed up	0.9 m upwards	Minimal disruption to existing service. Materials and equipment inexpensive	Man-access necessary and existing sewer must be structurally sound. Infiltration must be controlled and flow must be diverted when working in the invert. Quality control difficult due to poor working conditions
(4) Cement grouting	Cement grout injected to seal leaks, fill voids and strengthen ground	All	Materials and equipment inexpensive. PFA and other additives can be added to grout to improve properties. Can be injected from within pipe or from ground surface	May not be suitable where infiltration is actually flowing
(5) Chemical grouting	Low-viscosity chemicals injected into ground generally to seal leaks. Some increase in soil strength may be achieved	All	Can be used for man-entry and non-man-entry pipes. Fast gel time allows method to be used where infiltration is occurring. Low flows tolerated within pipe. Remote method for small pipes allows joints to be air tested as work proceeds	Expensive particularly where large voids exist. High flows must be diverted. Restricted number of specialists for non-man-entry system. Difficult to seal packers where pipe surface is irregular
Pipe lining				
(6) Pipelining generally	New pipes are inserted in existing pipes		Can improve structural stability, hydraulic capacity and chemical resistance. Structural design will require annulus between old and new pipes to be grouted up. Existing pipe may then form part of the structure	Reduces cross-sectional area but may improve hydraulic performance to compensate. Problems may arise where laterals occur
(7) Sliplining	After cleaning and proving the diameter of the existing pipe the new pipe (usually polyolefin) is jointed at ground level using butt fusion welding and then towed or pushed into the existing pipe via a lead in trench	0.1–0.9 m	Quick and large-diameter bends may be accommodated	Lateral connections must be re-made quickly. This could take a lot of time or considerable excavation in non-man-access pipes and services could be disrupted. Pipe may tend to float during grouting. Lead in trench disruptive. Circular cross-section only available

Method	Description	Diameter	Features	Limitations/notes
(8) Sectional lining pipes	Lining pipes are jointed within a working pit and then jacked or towed into the existing pipe	All	Various materials may be used (e.g. µPVC, reinforced plastic mortar, GRP, GRC). Quick large-diameter bends may be accommodated. Joints can be screwed, socket and spigot or welded depending on material. Cost effective for short deep lengths. May be possible to vary shape of cross-section with some materials	Generally needs working pit and large number of joints. Relatively high loss of cross-sectional area. Pipe may tend to float during grouting. Lateral connections may cause a problem in non-man-accessible diameters
(9) Inversion: polyester lining cured in place	After cleaning the existing pipe a special polyester felt liner impregnated with polyester thermosetting resin is inserted into the pipe through an existing manhole. The liner is initially inside out and unrolled into the pipe by flooding with water. The liner is specially tailored to suit the diameter and length of the pipe. The resin is then cured by circulating hot water within the lined pipe	0.1–1.2 m	Rapid installation using existing manholes. Bends and minor deformations in the existing pipe can be accommodated. Thickness is approximately 3mm minimum but can be increased to 19 mm. Smooth finish improves flow characteristics. Resins can be specially formulated to give chemical-resistant finish	Full over-pumping required during installation. Site set-up cost is relatively high on small jobs. Monopoly supplier
(10) Segmental relining	These methods are frequently used for non-circular man-accessible pipes			
(a) GRC segments	Segments are usually produced in two sections with lap joints. Connections are made using bolts, pop rivets, self-tapping screws, etc. The invert is laid first and bedded on mortar or wooden blocks. The crown is then installed and centralized with wedges. The annulus is grouted up and the laterals reconnected as work proceeds	0.6 m upwards	Variety of cross-sections available. Material easily cut to form connections	Jointing is labour-intensive. Strutting often required during grouting operation
(b) GRP segments	Segments can be either one piece with a single longitudinal joint or two pieces with lap joints	0.6 m upwards	Deforms easily to suit cross-section. High strength : weight ratio	Jointing is labour-intensive. Support usually required during grouting. Care must be taken not to overstrain or otherwise damage the barrier layer of resin
(c) Preshot gunite segments	Segments are installed by laying invert and then jacking crown into position. Reinforcement is tied and in situ gunite used to make joints. Annulus then pressure grouted	See note	Variety of cross-sections available. No strutting required. Segments can support earth load	Segments are heavy. Jointing is labour-intensive. Not suitable for pipe clearances less than 1070 × 760 mm
(d) Precast resin concrete segments	Similar to (c) but lap jointed using epoxy-modified mortar seals	See note	Similar to (c)	Similar to (c). Not suitable for pipe clearances less than 900 × 600 mm

Table 40.7 Pipe stabilization and relining methods (*continued*)

In situ *coatings*		All methods need thorough cleaning of the pipe and dewatering of the pipe to prevent infiltration		
(11) *In situ* gunite	0.9 m upwards	Standard gunite process used in man-accessible pipes to rebuild inside of pipe. Small-diameter rod or mesh required for structural applications	Cheap material and simple process produces a jointless lining. Connections and changing cross-section easily accommodated. Access through existing manholes	Dusty and difficult to supervise. Very dependent on operator skill. Must remove rebound material from invert. Overpumping required
(12) Centrifugal cement mortar lining	All	The wet mortar mix is pumped through a revolving mortar dispenser on to the walls of the pipe. The lining machine is towed through the pipe at a specified rate followed by rotary or drag trowels which give a smooth finish	Existing manholes can be used for access. Suitable for a wide range of pipe diameters. Cheap material	Non-structural. Laterals difficult to deal with. Flow must be diverted
(13) Polyurethane or cold-cured resin		Similar to (12) and currently undergoing development		

the pipeline is likely to form only one part of a larger reticulation and changes in its flow characteristics may have adverse effects elsewhere. The *Sewerage rehabilitation manual*[41] sets out the requirements of a proper management strategy for sewer rehabilitation and gives definitions of the various terms to be used when describing the works (Table 40.6).

40.10.2 Methods of rehabilitation

Some methods of rehabilitation are based on microtunnelling techniques while others are based on more traditional procedures (Table 40.7).

The practical requirements common to gas, water and sewer pipes are: (1) the need to cope with existing flows; (2) the need to work through existing manholes wherever possible; and (3) the ability to cope with lateral connections.

When relining gas and sewer pipes it is sometimes possible to use the annulus between the old and new pipes to maintain existing service connections. The annulus is grouted up once the service connections have been remade and the new pipe is on stream. In other cases bypasses and overpumping must be used with the possibility of interrupted services for the consumer.

Lateral connections present a considerable problem in sewers. They must be located accurately prior to any work being carried out and all live connections identified. The position and angle of entry must be recorded with great accuracy. If the connection protrudes into the sewer then it may be necessary to cut the projection back even before the inspection phase can be completed.

Various remote methods of reconnection have been invented,[42,43] but local excavation is often the only practical solution. On occasions, relined sewers can be reconnected using a 'down the drain' remote-controlled cutter observed by closed-circuit TV within the new sewer or using a cutter working within the new sewer. Accurate location of the cutting tool is required to ensure coincidence within the existing branch and several specialized devices are available to achieve this. Once the new pipe has been holed through, special sleeves or grouting techniques are used to make a watertight joint usually working from within the pipe.

In cases where excavation is unavoidable for reconnection special 'keyhole' techniques have been evolved which require a minimum of spoil to be removed so that special pipe-fitting tools can be used from ground level.

40.11 Costs

The total cost of constructing any buried pipeline is made up of the costs incurred by its promoter and the social costs borne by the community at large. The social costs can arise in a variety of ways. Examples are traffic delays, additional wear on unsuitable diversion roads, business losses, etc. which are not direct charges to the promoter. There is increasing concern in some critical urban situations that the social cost may be greater than the initial construction cost of the pipeline, particularly where trenched construction is used.[44] Thus, concern has given added impetus to the development of trenchless techniques which may offer significant benefits in congested areas by reducing disruption.

40.11.1 Project cost appraisal

The prudent promoter of a new pipeline will consider its total lifetime costs which simply stated are:

$$\begin{array}{cccc} \text{Initial construction} + & \text{Running} + & \text{Maintenance} + & \text{Replacement} \\ \text{cost} & \text{cost} & \text{cost} & \text{cost} \end{array}$$

It is likely that several different routes and types of construction appear feasible and the promoter will undertake cost/benefit studies to ascertain which combination offers the best advantage within given financial limits. The studies will require a number of estimates to be made of future events. Consequently, the results are unlikely to be accurate in absolute terms but will offer a relative assessment of the choices available. A further factor to be considered in cost/benefit analysis is the possibility of maximizing benefit for a small increase in construction cost. An example of this is the enlargement of ducts and tunnels to permit them to be used by two or more separate services.

The degree to which social costs are considered in such studies will depend on legal requirements and the nature of the promoter. The law gives some third parties entitlement to recompense for financial damage suffered as a result of the works. In other cases, for instance, where diverted traffic uses more fuel, the promoter is unlikely to have to foot the bill and will not consider such costs in the feasibility study unless political requirements or public policy dictate otherwise. Table 40.8

Table 40.8 Typical direct and indirect costs associated with sewer construction in trench

	Direct costs	Indirect costs
Engineering costs		
Planning and design	Yes	No
Site investigations	Yes	No
Construction	Yes	No
Supervision	Yes	No
Service diversions	Yes	No
Reinstatement and maintenance	Initial reinstatement	Increased long-term maintenance (damage due to trenching)
Social costs		
Other service replacements	For example, gas mains, water mains, etc.	No
Traffic diversions and delays	Traffic lights, signs	Cost to road users of delay and extra fuel, additional policing, wear and tear on alternative routes
Loss of business	Compensation	Any costs not met by compensation
Environmental	Include in permanent works	Yes
Dislocation of supply	Temporary works	Consumer inconvenience

indicates typical direct and indirect costs which may arise in the construction of a sewer in trench.

40.11.2 Construction costs

The estimation of costs for civil engineering construction is a skilled procedure. If prices are required to be within a reasonable order of accuracy then it is necessary to build them up from first principles. The variation of conditions for design and construction will tend to make the use of historical overall unit costs unreliable except on the simplest jobs.

Ground conditions have a major influence on both the design and the risks arising during construction and in any type of construction the cost of a proper site investigation should be regarded as money well spent. The report should consider the problems likely to arise in both the permanent and temporary conditions and the report should be made available to the tenderers.

The promoter must decide at an early stage what the contractual basis will be for the design and the construction of the works. The contracts with the designer and the constructor must place the obligations and responsibilities for the risks in an equitable manner. The promoter must recognize that by bearing some part of the potential risks himself, e.g. unexpected ground conditions, then the prices tendered are likely to be more realistic.

Decisions made during the design phase may result in some methods of construction being precluded and the effect of this must be assessed. Also different construction methods impress their own individual characteristics on the serviceability and life of the completed works which must be recognized by the designer.

The total construction cost of a pipeline is made up of:

(1) Labour cost.
(2) Plant cost.
(3) Materials cost.
(4) Constructor's on-costs including temporary works.
(5) Constructor's profit.
(6) Design cost.
(7) Promoter's supervision cost.

In most cases, the cost of the pipe and other permanent materials is not affected significantly by the depth of construc-

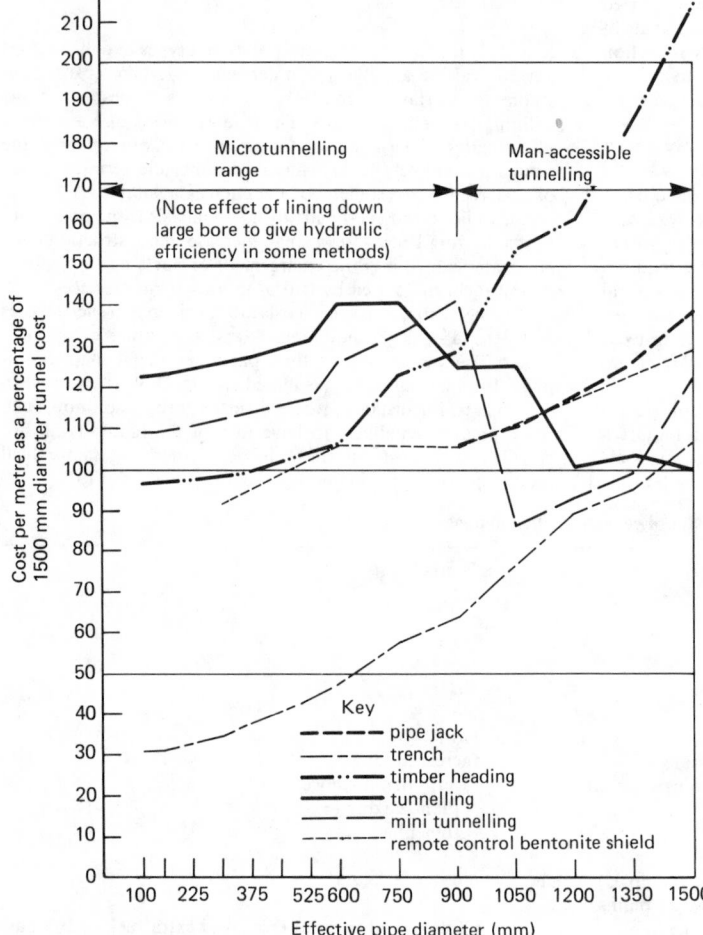

Note: Based on 600-m long sewer including manholes

Figure 40.20 Comparison of direct engineering costs of various tunnelling methods in good ground

tion. For a given diameter of pipe constructed in trench the cost of labour and plant increases as a power function of the depth. The cost of pipelines laid by non-disruptive methods is largely unaffected by depth.

The comparison of costs for different types of construction must be made on the basis of the same ground and site conditions. Where conditions change, then the relativities and ranking of the various methods are also likely to change. Some construction methods can cope readily with unexpected ground conditions, whereas others may need to be supplemented with expensive secondary operations such as extensive grouting or ground freezing. The potential cost of such risks must be assessed at the feasibility stage of a project.

The total job size will also have an influence on the unit costs of construction especially when using methods with a high initial set-up cost. Figure 40.20 illustrates the relative costs for installing 600-m long sewers by different methods in good ground.

References

1 British Standards Institution (var. dates) *Pipelines in land*: Parts 1 to 5, CP 2010. BSI, Milton Keynes.
2 British Standards Institution (n.d.) BS 8005 (In press) BSI, Milton Keynes.
3 Young, O. C. and O'Reilly M. P. (1983) *A guide to design loadings for buried rigid pipes.* Transport and Road Research Laboratory, Crowthorne.
4 Compston, D. G., Cray, P., Schofield, A. N. and Shann, C. D. (1978) *Design and construction of buried thin wall pipes.* Construction Industry Research and Information Association Report No. 78. CIRIA London.
5 Nath, P. (1981) *Pressures on buried pipelines due to revised H.B. loading.* Transport and Road Research Laboratory Report No. LR 977, Crowthorne. Report No. LR 977.
6 Department of Transport (1982) *Corrugated steel buried structures.* Departmental Standard BD12/82. HMSO, London.
7 Young, O. C. and Trott, J. J. (1984) *Buried rigid pipes.* Elsevier Applied Science, London.
8 Young, O. C., Brennan, G., O'Reilly M. P. (1986) *Simplified tables of external loads on buried pipelines.* Transport and Road Research Laboratory, Crowthorne.
9 Concrete Pipe Association (1983) *Loads on buried pipelines*: Part 1 'Tables of total design loads in trench.' Concrete Pipe Association Technical Bulletin No. 2 (1st rev.). CPA, Leicester.
10 Bland C. E. G. (1983) *Design tables for determining the bedding construction of vitrified clay pipelines.* Clay Pipe Development Association, London.
11 American Iron and Steel Institute (1971) *Handbook of steel drainage and highway construction products.* AISI.
12 Water Research Council (1982) *Economics of minimum pipe beddings in sewer construction.* WRC External Report No. SWM2.82. WRC, Medmenham.
13 Irvine, D. J. and Smith R. J. H. (1983) *Trenching practice.* Construction Industry Research and Information Association Report No. 97. CIRIA, London.
14 Symons, I. F. (1978) 'Ground movements and their influence on shallow buried pipes.' *Pub. Health Engr* **8**,4, 149–153.
15 Attewell, P. B. and Taylor, R. K. (eds) (1984) *Ground movements and their effects on structures.* Surrey University Press, Glasgow.
16 Water Authorities Association/British Gas Corporation (1984). *Model consultative procedure for pipeline constructing involving deep excavation* WAA/BGC, London.
17 Somerville, S. H. (1986) *Control of groundwater for temporary works.* Construction Industry Research and Information Association Report No. 113. CIRIA, London.
18 Mackay, E. B. (1986) *Proprietary trench support systems.* Construction Industry Research and Information Association Technical Note No. 95 (3rd edn). CIRIA, London.
19 Timber Research and Development Association (1981) *Timber in excavations.* TRADA, High Wycombe.
20 British Standards Institution (1973) *Code of practice for cathodic protection,* CP 1021 BSI, Milton Keynes.
21 Health and Safety at Work Act 1974. HMSO, London.
22 *Construction (general provisions) regulations,* 1961. HMSO, London.
23 *Construction (lifting operations) regulations,* 1966. HMSO, London. Research and Development Association (1981) op. cit.
24 Building Employers Confederation (n.d.) *Construction safety.* BAS Management Services, London (regularly revised).
25 The Royal Society for the Prevention of Accidents (1976) *Construction regulations handbook.* RoSPA, Birmingham.
26 Health and Safety Executive (1987) *Safety in construction work: excavations* HMSO, London (new edition in preparation).
27 R. vs Swan Hunter Shipbuilding Ltd (1981) *The Times,* 6 July, (ICR 831).
28 Craig, R. N. (1983) *Pipe jacking: a state of the art review.* Construction Industry Research and Information Association Technical Note No. 112. CIRIA, London.
29 Roark, R. J. and Young, W. C. (1975) *Formulae for stress and strain.* McGraw-Hill, Maidenhead.
30 Howe, M. and Hunter, P. (1985) 'Trenchless mainlaying within British Gas.' *Proceedings, 1st international conference on trenchless construction for utilities,* No Dig '85, London.
31 O'Rouke, T. D. (1985) 'Ground movements caused by trenchless construction.' *Proceedings, 1st international conference on trenchless construction for utilities.* No Dig '85, London.
32 Craig, R. N. and Muir Wood, A. M. (1978) *A review of tunnel lining practice in the United Kingdom.* Transport and Road Research Laboratory Supplementary Report No. 335, TRRL, Crowthorne.
33 O'Rouke, T. D. (ed.) (1984) *Guidelines for tunnel lining design.* American Society of Civil Engineers, New York.
34 Attewell, P. B., Yeats J. and Selby, A. R. (1986) *Soil movements induced by tunnelling.* Blackie, London.
35 Institution of Civil Engineers (1972) *Safety in wells and boreholes.* Thomas Telford, London.
36 Water Authorities Association (1979) *Safe working in sewers and at sewage works.* WAA Publication No. 2. WAA, London.
37 Health and Safety Executive (1900) *Entry into confined spaces.* Guidance Note No. G.5. HMSO, London.
38 British Standards Institution (1982) BS 6164: *Safety in tunnelling in the construction industry,* BSI, Milton Keynes.
39 British Standards Institution (1978) BS 5573: *Code of practice for safety precautions in the construction of large diameter boreholes for piling and other purposes.* BSI, Milton Keynes.
40 Water Authorities Association/Department of the Environment Manual of sewer conditions classification. Standing Technical Committee Report No. 24. WAA, London.
41 Water Research Centre (1983) *Sewerage rehabilitation manual.* WRC, Medmenham.
42 Gale, J. (1982) *Drain connections in small diameter sewer renovation.* Water Research Centre External Report No. 54E. WRC, Medmenham.
43 Cox, G. C. and Knott, G. E. (1984) 'Innovative achievements underground in the UK water industry.' Pipetech Conference, London.
44 Glennie, E. B. and Reed, K. (1985) 'Social costs: trenchless vs trenching.' *Proceedings, 1st international conference on trenchless construction for utilities.* No Dig '85, London.

Bibliography

Binnie and Partners (1985) *Trenchless construction for new pipelines: a review of current methods and developments.* Water Research Centre External Report No. 168E, WRC, Medmenham.
British Standards Institution (1966) *Protection of iron and steel structures from corrosion.* CP 2008. BSI, Milton Keynes.
British Standards Institution (1981). *Code of practice for site investigations,* BS 5930, BSI, Milton Keynes.
Clarke, J. R. G. (1984) 'Pipeline renovation.' Pipetech, '84 Conference, London.
Clarke, N. W. B. (1968) *Buried pipelines.* Maclaren, London.
Irvine, D. J. and Wishart, J. Tunnelling for Sewers, The Institute of Water Pollution Control, London, May 1986.
Lo, K. Y. (ed.) (1984) *Tunnelling in soil and rock.* American Society of Civil Engineers, New York.
Ministry of Housing and Local Government (1967) *Working party on the design and construction of underground pipe sewers, 2nd Report.* HMSO, London.

Peabody, A. W. (1978) *Control of pipeline corrosion.* National Association of Corrosion Engineers, Houston.

Stanton and Staveley (1979) *Ductile iron pipelines – embedment design.* S and S, Nottingham.

Tarmac Construction (1985) *Report on the comparative costs of various* *methods of laying and renovating sewer pipes.* Water Research Centre External Report No. 168E. WRC, Medmenham.

Watson, T. J. (1987 *Trenchless construction for underground services.* Construction Industry Research and Information Association Technical Note No. 127, CIRIA London.

41

Dredging

John H Sargent CEng, FICE, FGS
Costain Group plc, London

Contents

41.1 History

It is believed that dredging works originated in Egypt at about 4000 BC when canals were excavated using massive labour forces and the simplest of tools. During the sixteenth and seventeenth centuries dredging was often carried out both to improve navigation and obtain material to be utilized as ballast for outgoing ships, often undertaken using the 'bag-and-spoon' method.

An important historical point in the development of dredging plant was reached about 1590, with the invention in Holland of the 'mud mill'. This machine superficially resembled the modern bucket ladder dredger in which the mud was scooped up a chute suspended below a moving chain and discharged overboard or into barges.

The introduction of the steam engine and the development of the centrifugal dredge pump in the nineteenth century marked a dramatic step forward and led directly to the development of the modern dredger. Improvement in the design of 'traditional' dredging plant is still taking place, in particular in relation to improved production and in the ability of the dredger to work in more adverse sea conditions. In the 1970s in the Middle East a self-elevating platform with a cutter suction installation was in use, enabling work to take place in considerable sea swell. The introduction of modern electronics, especially in the field of survey control and on-board construction, is now in full application and dredging plant is becoming increasingly sophisticated.

41.2 Dredging plant and techniques

41.2.1 General comments

For convenience, dredging plant can be subdivided into: (1) mechanically operated dredgers; and (2) hydraulically operated dredgers.

The following description of plant and techniques is related primarily to the use and application of plant from a civil engineering viewpoint, and does not deal exhaustively with the mechanical design or performance of equipment.

Likewise, the aspects of cost and dredged material production have only been considered in a general manner. These aspects will vary over a very wide range since they are dependent on a number of factors which cannot be considered without a much fuller discussion and will in any case vary from project to project. Such factors will include: (1) the physical conditions imposed by the site and its geology; (2) the working hours and tides available; (3) the water depths existing prior to and during dredging; (4) the location in relation to authorized disposal areas at sea or on land; (5) the hydraulic conditions (allied to the exposure of the site); (6) the availability and skill of local staff, tradesmen and labour; (7) the proximity of the project to navigation; and (8) probable interference from other water-source traffic.

In addition to these factors will be the local attitude to private or public employment and the state of the market in the dredging industry at the time of tender and contract discussion and agreement. As for civil engineering projects, each job must be viewed separately and costed for its own particular set of circumstances.

Dredging works can also be considered conveniently in the categories of: (1) capital projects; and (2) maintenance works. Capital projects include the construction of new ports, harbours, waterways and canals and normally involve private contractors due often to the size of the project and the need to obtain competitive bids.

Many of the world's ports need some maintenance dredging,

either continuously or at frequent intervals, sometimes called 'campaign dredging'. The demand which maintenance dredging places on the port budget can be onerous and may directly affect the economic viability of the port operation, especially when dredging work is a constant need.

In this respect there is a frequent discussion regarding the use of private contractors versus publicly or locally owned dredging organizations. When a port has a continuous demand, can reliably predict the geology, working conditions, and motivate staff to attain acceptable production and be certain of the mechanical efficiency of its plant, there is a case for the purchase of specific dredging equipment. If these factors are not assured and difficult or uncertain geological conditions may be encountered, then the private contractor is often preferred and is likely to be more efficient and cost-effective. Criticism is sometimes levelled at port authorities who operate their own plant that the true operational cost is hidden by adopting 'noncommercial' accounting procedures. Port authorities should therefore ensure that calculations are based on realistic economic parameters and a comparison with private industry should be sought whenever possible prior to decision on purchase.

A useful report which outlines the types of dredging plant commonly available was issued by the Permanent International Association of Navigational Congresses (PIANC)[1]

41.3 Mechanically operated dredgers

41.3.1 Bucket ladder dredging

The design of the basic bucket ladder dredger comprises a chain of continuously revolving buckets moving along a steel ladder inclined into the sea or river bed, excavated material being carried to the top of the ladder where it is gravity-discharged into a chute (see Figure 41.1). Historically, the majority of bucket ladder dredgers were dumb, i.e. requiring towage to and from site, and discharged through the chute into a self-propelled

Figure 41.1 Bucket dredger with self-propelled hopper alongside. (*Courtesy:* Westminster Dredging Co. Ltd)

hopper or dumb barge for disposal. Later designs include dredgers which are self-loading hoppers and, as in this case, may be self-propelled.

Since the bucket ladder dredger needs positioning in its working area and in particular requires a reaction when excavating into the face of material to be dredged, the vessel is usually held by several moorings, the headline (working through a strong-head winch) being of utmost importance to provide the reaction into the digging face. In practice, the headline may extend for some hundreds of metres for fully effective operation. Short headlines can be used, but result in loss of efficiency. Side lines (winch operated) are used to traverse the dredger across the working face.

The bucket ladder dredger is commonly of poor stability when moving or under towage from site to site and, consequently, towage can be slow and expensive. Insurance costs are also likely to be high. Depending on its size, the dredger will normally require light tug or workboat assistance when on site, both to run lines and to help move the dredger from area to area.

As with most dredging craft based on a pontoon design, the average bucket ladder dredger is susceptible to heavy swell and may suffer badly in swell conditions in excess of about 1 m. An important feature in such work will be the presence of hoppers or barges alongside the dredger. Serious problems can occur under swell conditions owing to the differential movement between the craft, with consequent danger to both craft and men.

The advantage of the heavy bucket ladder dredger as a dredging tool is its ability to dredge a wide variety of materials ranging from soft alluvium to soft rock. The lighter units can tackle a range of soil types but will be very restricted in respect of rock and the heavier clay soils.

Foreign material (such as scrap iron, wires, etc.) can also be removed by the bucket ladder dredger, although production (compared with dredging 'normal' soils) will suffer seriously.

Bucket types can be changed to suit the ground conditions: it is common to use a heavy bucket in soil, and a bucket fitted with strengthened lip and fitted with digging teeth when working in soft rock. Bucket capacities vary from 0.5 to 1.2 m³, using bucket chain speeds at 18 to 30 buckets per minute depending on the material and the conditions. Dredging depths vary with the size of plant but commonly extend to 20 m or so with extensions to allow greater depths.

Environmental problems may occur when using bucket ladder dredgers owing to the noise of the operation. Disturbance is particularly common where work is in progress at night and the site is close to residential property, but 'silent' bucket chains are now available which can help to solve this problem.

Outside the civil engineering industry the bucket ladder dredger is much used as a basic tool in mining for rare earths and precious metals. Perhaps best-known is the extensive use of large dredgers of this type for tin-mining in the Far East.

41.3.2 Grab dredging

This type of plant (often called a clamshell dredger in North America) is nearest to normal civil engineering equipment employing a prime mover mounted on a pontoon or vessel loading either into attendant hoppers or barges, or into its own hopper. In the latter case the vessel may also be self-propelled.

Grab dredger sizes can vary over a wide range from very small

Figure 41.2 Grab dredger mounted on pontoon, loading hopper barge

units with small cranes mounted on pontoons or barges (such as the small hydraulic cranes used in narrow barges for canal dredging) to large machines capable of handling grabs in excess of 30m³.

Simple grab dredgers can be built easily and quickly by running a crawler-mounted excavator on to a pontoon, although in this case production may suffer unless work is being carried out in very quiet waters owing to the instability of the design (see Figure 41.2).

Small and medium-sized grab dredgers have found much favour in harbours and ports because of the limited capital expenditure required, their relative ease of maintenance and their particular ability to dredge close to quay walls, jetties and in basins without damage. Further, the quantity of 'foreign' material (scrap iron, etc.) within port basins is often very considerable and a grab can handle much of this material with reasonable ease. Commonly occurring harbour materials such as silts or soft silty clays, can also be removed easily with standard grabs, making the unit particularly useful to the port authority.

Dredging depths will be limited by the size of the available winch drum with a practical limitation imposed by the hoisting speed of the grab and the effect of sea or river currents on the grab when working at or near maximum depths. Working depths up to 25 m or so below water level are common although greater depths have been undertaken. An example where dredging was used for purposes other than producing navigational depth occurred in the Hong Kong Plover Cove earth dam, where work was undertaken by a specially designed cantilever grab dredger to considerable depths to excavate the cutoff trench for the dam.

Grab dredgers are able to operate in a swell condition, but production is affected badly over about 1.5 m swell, and under all conditions the profiling of the bottom must be carefully supervised. Single grab dredging requires a fairly simple pattern of attack, working forward after dredging through an arc. With several cranes mounted on the unit the dredger must be moved diagonally to obtain optimum coverage.

41.3.3 Dipper and backhoe dredgers

As with the grab dredger, the dipper and backhoe dredger have evolved through the adaptation of land-based excavators. Advances in design, in particular with the backhoe (which excavates with the bucket movement towards the hull) have been rapid in recent years and a very high output and mobility is possible when this plant is used on 'dry' land jobs. Outputs when used for dredging cannot normally match 'dry' land work, but the tool is still of considerable value since it can excavate effectively not only the softer sediments but also cemented sands, clays and soft rock (see Figure 41.3).

Limitation is generally in depth capability and hopper barges will be needed to remove the dredged soil.

The dipper dredger, which is usually of very heavy construction, is still in use but has never gained general favour in Europe owing to its relatively low production and limited depth of digging. The dipper dredger can, however, be valuable when dredging clays in a firm or stiff condition, or soft and friable rock. The tool has also been usefully employed for the demolishing of old stone defence works, jetties, locks and similar 'heavy' jobs, which other dredging plant could not handle. Dipper dredge plant is normally controlled by a spudded arrangement, often using two forward spuds to pin up the hull and stern spuds for walking and advancing the dredger.

Figure 41.3 Backhoe on spudded pontoon. (*Courtesy*: Westminster Dredging Co. Ltd)

41.4 Hydraulically operated dredgers

41.4.1 General comments

This class of dredger comprises the most widely used unit in the world's dredging fleets. Basically, the design comprises a centrifugal dredge pump installed in a pontoon or hull with a suction pipe lowered on a ladder to the bed. Where the bed material is loose and granular in nature, such as a running sand, it will be drawn under suction directly through the pipe and dredge pump into a discharge line. If the bed material is compact or dense, such as a firm clay or cemented sand, it is necessary to fit a rotating cutter on to the end of the suction line. The cutter will be turned by a separate cutter motor which is commonly mounted at the head of the suction line ladder. For deep dredging this may now be placed close to the suction head.

The hydraulic technique of dredging is often claimed to be highly efficient because it combines excavating and transporting waterborne dredged material in one continuous operation. Improvements over the last few years have occurred primarily through better pump design, and dredging pumps now represent a compromise with high efficiency against low power consumption matched against reliability and accessibility. Additionally, the development of the submerged dredge pump has produced a greater efficiency.

A problem with the technique is inherent in the movement of abrasive materials such as sands and gravels through a pipeline system, since this may cause extremely high wear and tear. Pump liners and pipes must be designed to stand this mixture, and in some cases double-walled pumps are now installed.

Attention has also been given to the use of water jets at or close to the suction mouth to assist in the break up of *in situ* material and its movement into the suction pipe. Suction heads comprise a wide variety of shapes and designs often linked to the dredge manufacturer's or contractor's own individual beliefs and theories.

Suction dredgers have evolved into several classes of plant which for simplicity may be listed as: (1) plain suction dredgers; (2) cutter suction dredgers; (3) hopper suction dredgers; and (4) pump-ashore units.

41.4.2 Plain suction dredgers

Plain suction dredgers (sometimes called profile dredgers) may be defined as having a plain 'mouth' to the suction tube, the suction mouth or head being buried into the material to be excavated. This unit is much used where there is clean free-running granular material that is used as fill on reclamation sites.

After dredging, the material may be pumped through side arms and discharged into barges moored alongside the dredger or pumped directly from the dredger through a pipeline system, parts of which will usually be floating, to shore.

Some use has been made with this type of dredger for the removal of soft, cohesive materials such as silts or silty clays in river and estuary conditions. Although the dredger is useful in these conditions, problems will arise if a suitable deposit area is not available nearby for the disposal of the dredged material, which will normally be unsuitable for use on a reclamation site.

The plain suction dredger design is usually based on a dumb pontoon which requires either an anchoring system or stern spuds, to allow walking when moving forward, plus anchors.

A variation of this type of dredger, developed in North America, is the 'dustpan' dredger based on the use of a suction head which often has a width equal to the hull and is fitted with high-velocity water jets and capable of dredging alluvial materials at very high volumes. In this case the dredger is often self-propelled and has been employed to keep open river navigation, pumping unconsolidated sands or alluvial soft silts and silty clays away from the navigable channel into a nearby area (outside the channel) where the discharged material is distributed or spread thinly by erosion currents.

41.4.3 Cutter suction dredgers

This type of plant is the most commonly used dredging unit. Dredger sizes range widely from small designs with 150 mm suction pipe diameters and limited power, to very large units with suction pipe diameters of 1000 mm and installed power in excess of 15 000 kW (see Figure 41.4).

Basically, the cutter suction dredger is pontoon- or hull-built. It is normally dumb, requiring towage and assistance on site (as for the majority of plain suction dredgers and bucket ladder dredgers) although an increasing number of large units are now self-propelled and capable of seagoing journeys.

A qualifying feature is the incorporation of a rotating cutter around the suction pipe mouth, powered by a separate electric or hydraulic cutter motor which turns the cutter into the excavation face. The cutter in this case lies with its axis of rotation parallel and in line with the cutter ladder.

When working, the unit usually has a spud driven to the river or sea bed mounted in the stern and the vessel pivots about the

Figure 41.4 Large cutter suction dredger. (*Courtesy*: Boskalis Westminster Baggeren b.v.)

spud, pulled by side anchors in order that the cutter head can be hauled across the working face. A second stern spud known as the 'walking spud' is employed to move the dredger forward. In the majority of cutter suction dredgers, the side lines are led through sheaves fixed to the suction ladder and run to anchors along or just above the sea or river bed. Side booms are often fitted which allow the facility of lifting and moving anchors forward without the necessity to use a separate workboat.

Depending on the structural strength of the design (in particular the weight of the ladder and the available power on the cutter coupled to the power available from the centrifugal dredge pump) the cutter suction dredger can excavate and remove most 'normal' soils and soft rock formations. With regard to large units, it may be considered as a rough guide that the softer sedimentary rocks such as chalk, marl, coral and weathered sandstones are within the scope of direct excavation.

After excavation, the material can either be pumped through a pipeline system (in which case suitable material will be used as reclamation fill and unsuitable material placed into a deposit area) or pumped through the dredge pump and loaded into hoppers or barges which lie alongside the dredger. The discharge distance through the pipeline system from the dredger will be dictated by the installed pump power, unless separate booster units are used.

Working depths vary considerably with size of plant but are commonly up to 30 m or so with the larger units. Much research

and development is in hand in this respect and greater depths can be achieved with specific units.

The design of the cutter is a subject with innumerable individual ideas and opinions. When working in hard and abrasive materials it is essential that cutter replacement can be effected at high speed, and in order to assist with this problem many rock cutters have replaceable teeth. A very wide variety of cutters have been designed for different soil and rock conditions but contractors usually employ their own preferences based on experience.

Modern cutter suction dredgers are both expensive to build and operate and are becoming increasingly more sophisticated. In order to work at optimum efficiency, it is necessary to provide adequate control and instrumentation, and to do this it is essential to provide a system to optimize output by co-ordinating the functions of vacuum (on the dredger pump), cutter torque and the swing speed of the cutter ladder via side winches.

Of particular interest for small-project work is the dismantleable cutter suction dredger, typified by a standard range from a major manufacturer with units with installed powers ranging from 100 to 1120 kW. Although these units can be broken down into component parts for road, rail or water transport, the size and weight of the individual pontoons with the larger units should not be underestimated.

All conventional dredging plant are vulnerable to swell conditions which restrict working. In answer to the problems of

Figure 41.5 Self-elevating walking heavy-duty cutter suction dredger. (*Courtesy*: Gulf-Cobla (Private) Ltd)

operating under severe sea conditions on a project in the Middle East, a self-elevating walking heavy duty cutter suction dredger was designed (see Figure 41.5). The vessel could be raised above the sea and was able to work continuously in swell over 4 m and in high winds.

41.4.3.1 Bucket wheel dredger

A recent development of the cutter suction dredger has been adaption of the bucket wheel in the place of the cutter head. Very large bucket wheel excavators have been used in open-cast mining on land for many years, but recently small bucket wheel dredgers have been developed for underwater work. It is claimed that the bucket wheel gives a greater efficiency than the normal cutter head in certain soil types and different designs are being developed to extend its use. Designs so far have been fairly limited in size, dredgers typically being small to medium in relation to cutter suction dredging plant.

41.4.4 Hopper suction dredgers

In this category of 'suction dredging' plant, two types are best known; the trailer dredger and the stationary dredger, of which the self-propelled trailer hopper suction dredger has become a dominant member of the dredging family in recent years, particularly in its role of deepening or maintaining navigable waterways under exposed conditions.

41.4.4.1 Trailer dredger

The trailer dredger is designed as a self-contained vessel equipped with a suction pipe or pipes trailed along the bottom whilst the dredger is moving forward under its own propulsion. The dredged material is excavated by a suction head and taken through the pipe and dredge pump and passed into the hopper. After loading, the dredger will either (in the case of poor-quality material) sail to sea to deposit the cargo through bottom doors or (if the material is suitable) sail to a pump-out installation for reclamation fill. In the latter instance an alternative method is to deposit the material in a prepared location for double-handling, probably by stationary suction or cutter suction dredger and pipeline system.

A considerable improvement in the bottom discharge of certain types of soil (especially in relation to material such as boulder clay) has been effected through the successful development of split-hull vessels. Originally confined to dumb hopper barges, the design has been extended to self-propelled hoppers and to trailer dredgers (see Figure 41.6).

A trailer dredger operates in exactly the same manner as a ship without wires, spuds or other restrictions, and is consequently highly manoeuvrable. It does, however, require sufficient room to work both in terms of waterway and water depth. Manoeuvrability is often improved by the provision of twin-propulsion units and sometimes by the addition of bow thrusters.

Sizes of trailer dredgers vary greatly as with all dredging

Figure 41.6 Seagoing split trailer hopper dredger. (*Courtesy*: Costain-Blankevoort (UK) Dredging Co. Ltd)

plant, but typically the medium-sized trailer dredger in present-day use will have about 2500 m³ hopper capacity and an overall length of about 110 m. Dredging depths vary but are frequently up to 25 m with an increasing number of vessels able to dredge to 35 m.

Because the trailer dredger is frequently used to produce and maintain waterways for modern shipping, particularly large tanker transport, the dredging depth and capacity of the dredger are geared to the optimum waterway depth required by ports on a global scale.

Close control of a sophisticated and costly ship such as the modern trailer dredger is vital and the control systems for both the suction pipe and operation of the sand pump are usually centred on the bridge together with the usual controls for propulsion and other navigational equipment. Further automation and dredging aids now provided include the use of concentration, production and loading meters together with multi-channel recorders and on-board computer systems, both to provide automatic dredging and to give instant read-out facilities of survey position and production to aid the dredgemaster.

A limited number of trailer dredgers have pump-ashore facilities. To effect this operation, a suction line is installed in the hopper well together with an upper set of doors plus the necessary discharge lines. Designs provide either for the sand pump to be connected directly to a shore line or for an additional pressure pump to be included which may be driven by one of the trailer's propulsion engines.

The trailer dredger has a high production characteristic in soft or loose alluvial soils but due to the need to drag the suction head, it is more difficult for the dredger to excavate stiff or hard clays, cemented sands or similar material. The work can be undertaken with adapted dragheads, but low productions have normally to be accepted. Very fine material such as silt can be loaded easily, but because of its poor settlement characteristics in the hopper it is usual to take only part loads during each cycle.

A dredging technique where hydraulic conditions are known to be suitable and in which the soil is a fine freely moving sediment, such as fluid silt, is agitation dredging. One method is to use the side-cast or boom trailer dredger. This involves a special design of trailer dredger with a revolving deck-mounted boom that will allow discharge of the dredged material outside the limits of the navigable channel. It follows that the hydraulic conditions must be suitable to remove the discharged material away from the navigable channel with the minimum of resiltation.

Adaptation of plant is typified by the development of a trailer dredger as an alternative oil-spill recovery vessel. The economics of such vessels must, however, be closely checked since, as is usual with engineering plant, multi-purpose tools invariably suffer in cost effectiveness.

Recent research and development has taken place regarding the development of a silt draghead for use in trailer dredging. It has a useful application in specific cases and should improve the results of maintenance dredging in harbours and waterways.

An important feature of the trailer dredger when considering the dredging requirements of a new project is its ability to operate in exposed locations, often in swell in excess of several metres and in wind, weather and sea conditions totally beyond the capacity of other types of dredging plant. There is little doubt that it will continue to be of major importance for work in sea or estuary approach channels.

41.4.4.2 Stationary dredger

The stationary hopper dredger is of the same basic design as the trailer dredger but does not move from the working location while dredging and often has the suction pipe placed in a forward direction. Vessels of this type have been much employed for the dredging of offshore sand and gravel deposits to obtain aggregates.

Offloading of the material will be in the same manner as the trailer dredger with the exception that, for sand and gravel extraction operation, the cargo may often be rehandled directly from the hopper by a grab to stockpiles or to screening and washing plant ashore. In some instances screening may be carried out on the vessel with reject material pumped overboard, where this is permitted by local legislation.

41.4.4.3 Scraper dredger

A recent development by a Finnish company has involved the construction of a scraping hopper dredger in which the complete hopper is hinged and lowered to the sea-bed, to be filled mechanically by moving forward at low level prior to the hopper being lifted in place for transport. Such plant, although interesting in conception, has a very small hopper capacity and its application will be restricted.

41.4.5 Pump-ashore plant

Pump-ashore units are normally used in conjunction with other dredging plant, the purpose of the installation being to empty a filled barge or hopper by suction means and to move the dredged material ashore through a pipeline system to a reclamation area or stockpile.

In essence, the technique provides an alternative in the cycle of the total dredging operation and avoids the use of lengthy floating pipeline systems (in the case of suction or cutter suction or bucket wheel dredgers) and is applicable in particular where the distance from dredging location to reclamation site is too great or is uneconomic for direct pumping by the dredger.

41.5 Ancillary plant and equipment

In conjunction with the major items of dredging plant discussed so far, the provision of waterborne ancillary plant and craft to support the dredging operation is essential and includes barges, tugs, workboats, survey launches and small marine craft.

Of prime importance are dumb barges and hoppers, used extensively with bucket ladder dredgers and other plant for carrying dredged material to disposal grounds or to pump-ashore plant for reclamation fill.

The size of project and quantities of dredged material will dictate the capacity and type of such craft to be used, e.g. self-propelled hoppers are used where large volumes and long sailing distances are involved.

41.6 Reclamation works

A major contribution of the dredging industry to development is the use of dredging plant to enable land reclamation (in the sense of building new land) to take place. Particular examples of this are the immense reclamation schemes in Holland, the US and Japan and schemes in many other countries, including the UK (see Figure 41.7).

Reclamation methods using dredging plant as the prime mover may be subdivided into reclamation by polder method or reclamation using pumped fill material.

In the case of the former (polder) method, normal practice is to construct dykes around a water area and artificially drain the area. The resulting dry land (frequently below sea- or groundwater level) is in Holland termed a 'polder' which gives its name to the method.

Figure 41.7 Reclamation at port using cutter suction dredger

In this instance it will be necessary to maintain the water levels inside the polder by pumping through a series of interconnecting drainage ditches and systems.

A major problem where land is open to the sea in tidal areas will be the final closure of the dyke where the final gap will pass large quantities of tidal flow at increasing velocities.

In the case of the second method, the general level will be raised by pumping-in soil dredged from elsewhere. It is normal practice to elevate the surface above high water mark for safety.

When considering the economics of the two methods, a major cost of polder reclamation will be the dyke construction and its subsequent protection. A project covering a large area will therefore often be more attractive since the cost of the boundary dyke will be lower in cost in relation to the reclaimed area.

In considering reclamation using fill material, a significant cost will be the availability of suitable material. Ideally, the fill material should be easily dredgeable, close to the site and have good pumping characteristics with low wear and tear during pumping and high compaction and load-bearing characteristics when in place. In Holland, where reclamation works are extensive, the predominant soil is a fine sand which meets several (though not all) of these needs. In the UK an exceedingly wide range of naturally occurring soils is encountered and a balance will have to be obtained in deciding the choice of fill material.

Fill by dredging methods normally means that the material is pumped hydraulically through a pipeline system as a mixture with water as the transport medium. Reclamation works using dredging methods therefore involve very large quantities of water which must either be drained away or recirculated within the system (see Figure 41.8).

Figure 41.8 Reclamation in progress using hydraulic fill

From a geotechnical viewpoint, materials such as sand and gravel will make excellent earthworks fill, but granular materials with increasing particle size (from coarse sand grading upwards) become increasingly expensive to pump.

Soft clays and silts are easily dredged and pumped but are unsatisfactory fill materials and are usually avoided when possible. However, use has been made of such material, both in Holland and the UK, where deposit grounds can be husbanded and redeveloped for agriculture.

In reclamation works clays with a firm or stiff cohesion characteristic usually 'ball' if passed through a dredge pump, and although they may form satisfactory fills, care is needed to analyse and predict the behaviour of such material.

Prior to the commencement of reclamation projects, site investigation into the virgin land conditions (which will become the subsoil below the dredged fill) is important, in particular to provide the basic information for the engineering design of substructures.

41.6.1 Construction of 'islands'

The construction of offshore 'islands' by dredging methods has developed rapidly in recent years and designs ranging from very small islands for pleasure purposes to jumbo projects for airports or waste disposal have been researched.

Such islands, except where they are of a temporary nature, will require effective (and usually expensive) protection against wave and wind attack.

A new dredging technology has developed in order to construct from silt or fine-grained materials temporary islands beyond the Arctic Circle in seas which become icebound for much of the year. The purpose of such islands is to enable oil and gas exploration to be effected on a year-long basis rather than be confined to a narrow weather window. Hybrid dredgers with very high production capacities have been developed and used for such work.

41.7 Other important considerations

In the final analysis technical factors relating to the type of dredging plant or which system to employ are one aspect of implementing effective dredging work. Equally important factors can be identified as:

(1) Exposure of the port and harbour to weather and sea and the hydraulic conditions in the vicinity.
(2) Volume and frequency of dredging requirement.
(3) Environmental aspects such as availability of dredged material disposal grounds or land-fill areas.
(4) Availability of dredging plant from private contractors or publicly-owned organizations.

Where a port is to be built from new, or where developments take place at an existing port or harbour, the designer should take into account a wide range of engineering criteria to determine the design and location of waterways, structures and land reclamation in order to minimize future dredging and limit the chance of siltation.

In relation to the dredging aspects, it cannot be over-emphasized that the most comprehensive site investigation (involving a study of the geological, climatic and hydraulic conditions) in conjunction with an assessment of the dredging quantities and available time-scales is needed in order to decide on the most suitable plant and *modus operandi* for the works.

Where a port is already in operation and it is necessary to plan the dredging maintenance, a more specific site investigation matched to a review of all earlier data collected over the years from the previous construction and dredging works is needed.

As implied previously, the nature of the materials to be dredged and the quantities involved are vital statistics when deciding on the correct dredging plant. For efficiency it is important to use a common scientific language when describing soil and rocks and it is recommended that a standard classification method should be adopted. In this respect, the best approach is to use the PIANC report[2] which provides a suitable classification system for dredging purposes and gives recommendations regarding the best type of soil and rock tests and investigation procedures to be adopted.

When major dredging works are being considered, the possibility of trial dredging should not be overlooked. Clearly, because of the cost of mobilization it is likely to be economic only if suitable equipment is available locally, or is passing the location. If trial dredging can be arranged, the measurement of the trial work must be to a high standard and recorded in a suitable manner for use in the future by the port engineer, his civil consultant, and by dredging contractors required to bid for the work.

Increasingly over the last few years, instrumentation has been installed in dredging plant. The majority of dredging plant now has instrumentation which indicates the dredged material density, pipeline velocity and the hopper loading. The larger and more sophisticated plant may have advanced data-collection systems based on computer technology to guide the operator in the search for high production. Where it is designed and installed efficiently the introduction of such a system should lead to labour, energy and time saving, lowering of training time, improved recording and an increased dredging efficiency.

41.8 Environmental aspects

In common with civil engineering projects, present-day dredging works need great care to evaluate environmental factors and to meet legislation which is local, national and international.

Each country will have its own regulations concerning aspects of dredging work. For example, in the UK the disposal of dredged material at sea must conform to the Food and Environmental Protection Act (Part II) 1985. Furthermore, in the UK the Oslo and London (Dumping) Conventions[3,4] regulate disposal and often impose strict requirements. Developments in 1986 in the London Dumping Convention[5] have produced guidelines for dredged materials which is a first rational step to consider dredged materials separately from industrial wastes and sewage sludge. Although the greater proportion of material deposited at sea comprises dredged materials, the percentage of polluted material is very small and has been estimated as probably only 5% of the total volume.

While every care must be taken to comply with environmental restrictions when using dredging plant, the greatest attention is required at the disposal situation. Disposal at sea should be considered as an equal option with disposal on land. When the dredged material is coarse-grained and clean, its use causes no problem and it may be ideal for landfill. Fine-grained materials (such as silts and clays) are much more difficult to use and disposal to lagoons on land or disposal areas at sea would be the usual method. In these cases, environmental studies will often be needed if a new site is under consideration.

Greater attention is now being given to the beneficial use of dredged material, including widespread application for beach creation or nourishment. A further recent example is processing of silt material into a useful topsoil, although such special treatment is rarely economic except in specific circumstances.

At the present time, views on the impact of dredged material disposal both at sea and on land sites vary widely. Several authorities maintain that (provided the material for deposit is not itself severely contaminated) disposal at sea causes no harm if the site is properly selected. Even where some contamination is present in the material, a well-argued case has been presented that the natural 'binding' properties of the material will contain the pollutant.

Deposition of polluted material on land, apart from being very expensive, may well be detrimental if leaching of pollutants into the groundwater system occurs.

The sensible approach to the selection of a disposal site appears to be that of evaluating the 'choice of least detriment' (accepting that in most instances there will be no detrimental and even beneficial effects to be gained) and allowing disposal at sea to be on an equal basis with land disposal.

For an overview on this developing subject reference should be made to a PIANC report[6] issued in 1986 which is concerned with disposal at sea. The Association is currently preparing a further report which will deal with disposal on land.

41.9 Hydrographic surveys and geotechnical investigations

A vital feature of the assessment of dredging conditions to enable plant selection to be made involves the early hydrographic survey and investigation work.

In order to calculate quantities of material for excavation, accurate hydrographic surveys are required. Additionally, during the dredging operations, surveys may need to be carried out to assess interim payments, particularly if these are made on an *in situ* quantity basis. After the completion of dredging work, a particularly accurate survey will be needed to check that the specified depth has been reached, side slopes produced and to agree the final quantities for payment.

With regard to the setting out of dredging work, stationary plant such as bucket ladder dredgers, suction and cutter suction dredgers and grabs can normally be positioned by relatively simple hydrographic survey methods, using beacons, shore markers, theodolite systems and sextant observations. Further offshore and, where visibility problems occur (which may often be the case for trailer dredging) radio position fixing systems have been used for many years and range-range or range-bearing may be applicable. Laser-based systems have been developed, some involving simultaneous measurement of range and bearing. Due to the continuing development of the electronics and telecommunications industries, improved survey and control systems for dredging works are constantly under review. Satellite-based systems, for example, may be applicable in the future.

Investigations to check the nature and occurrence of the materials to be dredged are within the field of marine geotechnical investigations and will include boring in soils and drilling in rock together with the recovery of samples or cores for inspection, logging and testing. Laboratory work on the recovered samples and cores is invaluable in determining the properties of the materials and provides essential information for the selection of plant and the costing of the project (see Figure 41.9).

The application of geophysics during the site investigation is proving increasingly useful, provided sufficient correlation information, possibly in the form of boreholes, is available and is used.

Survey specification can vary widely but in the UK it is recommended that reference is made to a joint publication by the Institution of Civil Engineers and Royal Institution of Chartered Surveyors[7] published in 1985.

Of particular importance is the delineation of rockhead for a dredging contract. The difference in production by a specific dredger in soil and the same dredger in rock is often so considerable that the unexpected occurrence of rock on a

Figure 41.9 Sea-bed sampler designed for dredging investigations. (*Courtesy*: Osiris-Seaway Ltd)

project will be costly to all concerned! In many cases the dredger mobilized for a soil job may be totally unsuitable for rock work and the mobilization of a new item of plant can be very expensive and time-wasting.

At the present time, it is usual to apply soil and rock mechanics practice to the evaluation of materials but an increasing value is the recovering of as much *in situ* data as possible, especially with respect to investigations into rock conditions.

The need to estimate whether any form of pretreatment (such as drilling and blasting) is needed is particularly relevant to rock. At present the softer sedimentary rocks (soft sandstones, limestones, corals, etc.) can frequently be dredged direct with large and heavy plant whilst the unweathered igneous and metamorphic rocks will require pretreatment before removal.

In many dredging projects it will be necessary to determine the hydraulic conditions at the dredging site, the disposal area and the project site (especially if reclamation works are to be undertaken). In some cases the construction of a physical hydraulic model will be needed in order to study the influence of the various hydraulic conditions. A study of current patterns will be very important, in particular to check that navigation problems are not caused, siltation is kept to a minimum and wave action is reduced. Information on the latter is also vital to determine the method of protection to be afforded to dykes and embankments.

If conditions are not too complex, it may be possible to evaluate the hydraulic conditions using a mathematical model without having to construct a physical model which will usually be much more expensive and time-consuming.

41.10 Organizations

It may be useful to those seeking further information about dredging and its technology to note that the following organizations have considerable involvement or access to such knowledge.

41.10.1 World Association of Dredging Organizations (WODA)

The World Association of Dredging Organizations consists of three linked organizations:

(1) Western Dredging Association (WEDA) responsible for North America.
(2) Central Dredging Association (CEDA) responsible for Europe, Middle East and Africa.
(3) Eastern Dredging Association (EADA) responsible for Far East, China, Japan and Australasia.

It embraces the total field of dredging including operators, designers, manufacturers and research bodies and is responsible for frequent world conferences (see bibliography).

41.10.2 Permanent International Association of Navigational Congresses (PIANC)

This long-established (1885) organization is based in Brussels and is an important professional international body concerned with the design, operation and maintenance of ports, harbours, waterways and canals.

An important feature is its four-yearly congress (see bibliography). The organization is supported by governments and has national sections in many countries (including a British section).

Dredging has frequently been a subject for consideration at the congresses and valuable reports on specific subject areas related to dredging have been produced both at the congresses and by international working groups.

41.10.3 International Association of Dredging Companies (IADC)

This Hague-based organization (which is in effect a trade association for operators) has a worldwide membership. It often assists in study work with its special expertise and also produces an informative and valuable publication entitled *Terra et Aqua* related to dredging (see bibliography).

References

1 Permanent International Association of Navigational Congresses (1977) *Study of environmental effects of dredging and disposal of dredged materials*. Report of the International Commission, annexe to Bulletin No. 27. PIANC, Brussels.
2 Permanent International Association of Navigational Congresses (1984) *Classification of soil and rocks to be dredged*. Report of the International Commission. Supplement to Bulletin No. 47. PIANC, Brussels.
3 Oslo Commission (1974) *Convention for the prevention of marine pollution by dumping from ships and aircraft*, (Oslo Convention). Oslo Commission, London.
4 International Maritime Organisation (1972) *Convention on the prevention of maritime pollution by dumping of wastes and other matters*, (London Dumping convention). IMO, London.
5 International Maritime Organisation (1986) *Guidelines for the application of the annexes to the disposal of dredged material*. Report of the 10th consultative meeting of constructing parties to the London Dumping convention, annexe 2, IMO, London.
6 Permanent International Association of Navigational Congresses (1986) *Disposal of dredged material at sea*. Supplement to Bulletin No. 52. PIANC, Brussels.
7 Institution of Civil Engineers and Royal Institute of Chartered Surveyors (1985) *Guidelines for the preparation of hydrographic surveys for dredging*. ICE and RICS, Thomas Telford, London.

Bibliography

Central Dredging Association *Proceedings* world dredging conferences (various dates from June 1968). CEDA, Delft.
Permanent International Association of Navigational Congresses *Proceedings* (various dates from 1885). PIANC, Brussels.
International Association of Dredging Companies *Terra et Aqua – Journal of International Association of Dredging Companies* (from 1972). IADC, The Hague.
Dredging and Port Construction (monthly journal) Published by Industrial and Marine Publications, Redhill.
World Dredging (monthly journal) Published by Symcon Co., California.

42

Underwater Work

Cdr H Wardle, Subsea Consultant,
Sovereign Oil and Gas PLC

R W Barrett MSc (Eng)
Research Manager, UEG
(the underwater engineering group of CIRIA)

Contents

42.1 Introduction

Underwater work involves divers, submersibles, remotely operated vehicles (ROVs) or tools from above the surface. The operation of submersibles and ROVs is a highly specialized field which is at present mainly confined to work on offshore structures, pipelines and cables laid in open water. Surface-operated 'tools' include dredging and drilling operations (see Chapter 41), pile driving, etc. which are essentially surface-controlled activities extending underwater. The aim in this chapter is to discuss some of the problems associated with the utilization of divers by the construction industry and to describe some of the diver alternative systems available.

42.1.1 Diver employment

Work for divers has increased considerably in the last three decades in line with man's increasing exploitation of the oceans. Some supertankers have a laden draught of over 30 m, with a consequent increase in the depth of water for berths and moorings. Increasing world population and improving environmental standards demand the disposal of waste by the construction of long sea effluent outfalls; land reclamation schemes are being carried out in many parts of the world. By far the most important development has been the production of oil and gas from wells many miles offshore in water of increasing depth; oilfield divers may be required to work in excess of 200 m of water.

42.2 Diving operations

42.2.1 Decompression

Common to all diving operations is the concept of decompression which is the controlled procedure by which a diver returns to surface pressure from the ambient pressure of his working depth. As the working depth increases so does the time taken to decompress; in addition, the longer the diver stays under pressure the more the decompression time is needed.

Decompression is necessary because the diver breathing under pressure absorbs inert gas into his tissues and blood. The decompression procedure allows this gas to be released gradually and thereby avoid decompression sickness. Decompression sickness (the bends) is directly attributable to a too-rapid reduction of pressure which can cause gas bubbles to form in the body tissues from the dissolved gases. Decompression schedules or tables have therefore been developed which allow the diver to be brought back to the surface in predetermined steps that avoid decompression sickness. Typical air-decompression tables allow for 80 min work at 15 m before 'stops' are incurred. Stops increase rapidly as the depth and working time increase. Divers and others who work under pressure may suffer from bone necrosis, i.e. damage to small areas of the bone. Some divers may be more susceptible than others. Strict compliance with decompression procedures will minimize this risk.

42.2.2 Surface-orientated diving

In this type of diving, the diver enters the water from the surface, proceeds to the working depth, carries out his work and then returns to the surface. Decompression, if necessary, can be carried out either by stopping at intervals in the water or by returning directly to the surface to be decompressed in a compression chamber. The choice depends on circumstances at the work site and other factors such as the decompression time required.

Commercial diving of this type generally uses compressed air as the breathing gas, which is pumped down to the diver from the surface via an umbilical hose. Because the nitrogen in air has a narcotic effect when breathed under pressure, compressed air should not be used in any diving operations at depths greater than 50 m. As most underwater construction takes place in less than this depth – about nine in ten dives are carried out in depths of 10 m or less – this constraint very rarely affects diving associated with conventional civil engineering works.

Surface-orientated diving can also be undertaken from simple diving stages. These act as lifts to carry the divers to and from the work site. Decompression can again be carried out by stopping the stage at intervals or in a surface compression chamber. A variation of this technique is the 'wet bell' which is an open-sided bell carrying the divers' air supply and having an upper canopy in which air is trapped thus providing a refuge for the divers' heads.

42.2.3 Bell diving

A diving bell is normally cylindrical or spherical in shape, has a bottom hatch, and is designed to withstand internal and external pressure. It can accommodate two or more divers and is provided with breathing gas from the surface.

Bell diving is a technique in which divers are transported to the work site in a diving bell from where they carry out their operations on completion of their work; they return to the bell which is then sealed (to maintain the pressure as it was at the work site) and winched to the surface.

At the surface, it is 'locked on' to a deck compression chamber via a transfer-under-pressure (TUP) system, and the divers are decompressed. If the decompression period is short, decompression can be completed in the bell.

Bell diving is most commonly used in support of offshore oil and gas fieldwork, and is obligatory in UK and Norwegian waters when diving to depths greater than 50 m. It is sometimes used for work in shallower depths especially for long duration tasks when the additional support it affords the divers may be necessary. Bell diving is now well developed and practical operations in excess of 300 m have been carried out. Exceptionally, operational dives have taken place to 450 m and experimental dives down to 600 m have been achieved.

There are two basic diving techniques associated with bell diving – 'bounce' diving and 'saturation' diving. Bounce diving means that the diver is not exposed to pressure long enough for the dissolved gas in his body tissues to reach saturation. On completion of his work, he is returned to the deck compression chamber for immediate decompression. In practical terms, this type of bell diving is generally used for simple tasks requiring a short bottom time.

Saturation diving, as its name implies, involves maintaining the diver at pressure for a long enough time for the dissolved gas to reach saturation. Saturation means that his body tissues and blood cannot absorb any more inert gas from his lungs and once this condition has been reached the time need for decompression is the same, no matter how long he remains saturated. Saturation diving avoids the need for decompression to atmospheric pressure at the end of each working period; the divers live in the deck compression chamber (maintained at the same pressure as the working depth) for periods of up to 30 days. For this type of diving the divers usually breathe a mixture of oxygen and another inert gas, usually helium. They are transferred to and from the work site by the diving bell.

The choice between bounce or saturation diving is decided primarily by the depth of water and the expected duration of the dive, which in turn is governed by the nature of the work to be carried out. For continuous work or work that will take a relatively long time, saturation diving has clear advantages. As the depth of water, and therefore the decompression time, increase, the technique becomes more attractive. In relatively deep water even quite short dives require such a long decompression that a saturation facility is likely to be the most cost-effective way of doing the work.

42.3 Diving equipment

As over 90% of all diving is carried out on work in less than 50 m using air as the breathing gas, this chapter deals mainly with surface-orientated diving. Some factors apply equally to the deep diver but, should specific information be required on deep work, reference should be made to one of the specialist companies providing deep-diving services.

The following is a summary of diving apparatus in general commercial use for air diving.

42.3.1 Standard diving apparatus

The traditional standard diving apparatus consists of a heavy tinned copper helmet and corselet (attachment between helmet and the heavy twill diving dress), lead-soled boots and weights. Air is supplied via an armoured air hose from the surface to the helmet and escapes from the helmet through a relief valve which maintains the pressure at slightly above the ambient (water) pressure.

The apparatus is used for relatively static and heavy work. Although very comfortable for the diver when on the bottom, it is cumbersome and difficult to work in when operating in a swell and when mobility and vision are restricted.

Its use is now limited to 'old hands' in the diving profession, but it still has a place for localized heavy work and in protecting the diver from cold or pollution.

It is sometimes referred to as 'helmet' or 'hard hat' diving gear (Figure 42.1).

42.3.2 The aqualung or SCUBA

The aqualung or self-contained underwater breathing apparatus (SCUBA) normally consists of twin high-pressure cylinders supplying air via a manifold and demand valve (reducer) which automatically adjusts the air breathed by the diver to ambient pressure. This automatic adjustment is achieved by the water pressure acting on a diaphragm which, in turn, operates a tilt valve supplying air to the diver. Thus air is provided to the diver at the correct pressure and 'on demand', the action of breathing operating the valve. The diver wears a rubber suit, fins, mask and a releasable weight belt.

This equipment gives a very high degree of mobility but low endurance. It has inherent limitations for sustained hard work at any depth and as depth increases it becomes increasingly difficult for the diver to make a reliable assessment of gas consumption under varying work conditions. It is not therefore generally recommended except for short-duration observation work. It is widely used for survey work.

42.3.3 Surface-demand diving equipment

The principle of this apparatus is the same as the aqualung except that the air is piped from the surface to the diver (Figure 42.2). It is used where a combination of endurance and mobility is required. Some aqualungs have a dual capability, with the diver normally supplied from the surface but carrying 'emergency' air, usually in a cylinder strapped to his back.

42.3.4 The fibreglass helmet

This equipment is an attempt to provide the diver with the

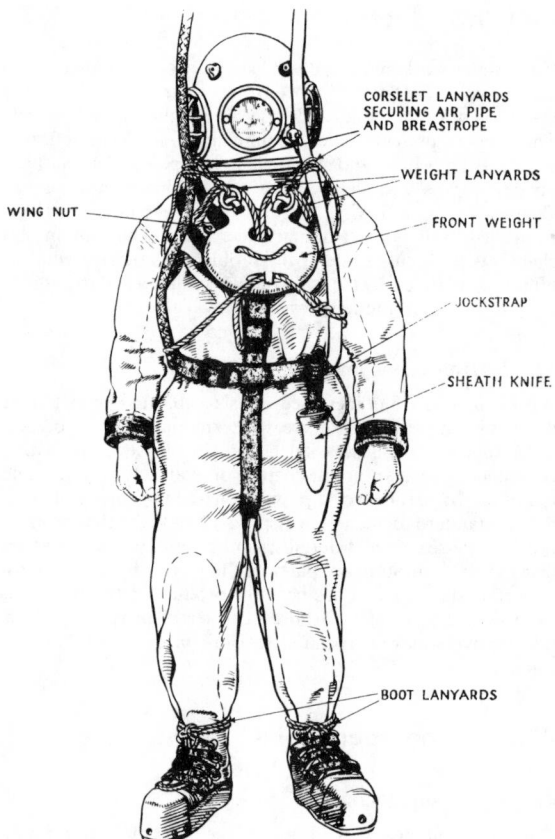

CORSELET LANYARDS SECURING AIR PIPE AND BREASTROPE

WEIGHT LANYARDS

WING NUT

FRONT WEIGHT

JOCKSTRAP

SHEATH KNIFE

BOOT LANYARDS

Figure 42.1 Standard diver (front view)

comfort and good communications of the helmet of the standard diving apparatus while retaining the mobility of the frogman. The principle of operation is the same as the standard helmet but with the fibreglass helmet attached directly to the frogman-type suit. The volume of water displaced by this apparatus is less than with the standard diving apparatus and therefore the weight required to 'sink' the diver is less.

42.3.5 Bell diving system

A bell diving system can be designed for either bounce or saturation diving. A basic system consists of one or more deck compression chambers, a diving bell complete with its handling system and equipment necessary for life-support, environmental control and communications. A saturation system normally has more or larger compression chambers and a more developed life-support and environmental-control system. It may be capable of accommodating only one saturation diving team or a number of teams or relays of divers in order to carry out work around the clock (Figures 42.3 and 42.4).

Life-support and environmental-control systems ensure that the correct breathing mixtures are provided to the deck compression chamber and the diving bell, and that the divers are maintained in a safe environment and in thermal balance.

42.3.6 Diver heating

It is important that each diver's thermal balance is kept within safe limits, neither too hot nor too cold. When diving operations are carried out at depths greater than 50 m, equipment must be

Figure 42.2. Diver wearing typical surface-demand diving equipment

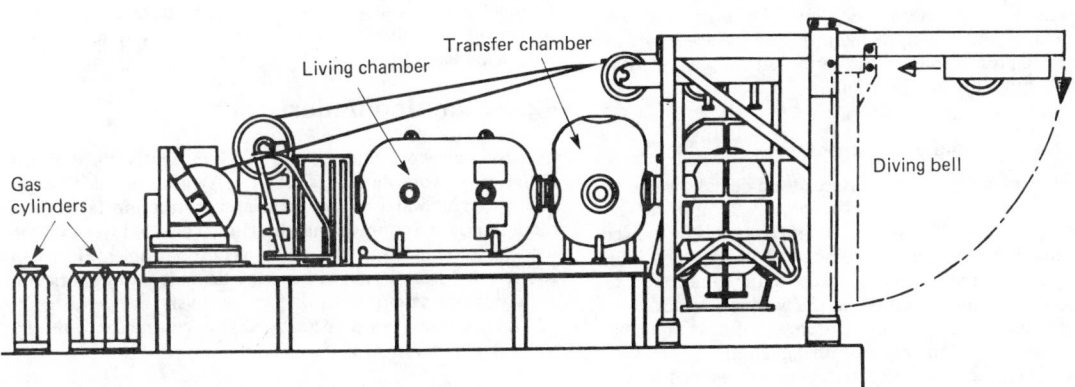

Figure 42.3 Basic saturation diving system

provided for heating the diver's body. At depths greater than 150 m the diver's breathing mixture must also be heated. This is due to the fact that a diver breathing oxy-helium mixtures, loses body heat very much faster than a diver breathing air. Currently, the most popular diver heating systems use hot water fed from the surface to the diver via the main bell umbilical and then distributed throughout the diver's suit.

Heating is not usually necessary for air diving but it should be considered in special circumstances, e.g. when diving in very low-temperature waters.

42.4 Air-supply systems

42.4.1 Rigid free-flow helmet diving system

A helmet diver requires approximately 42 l of free air per atmosphere, e.g. 84 l at 10 m, 126 at 20 m, etc. The airflow is calculated on the basis of ensuring sufficient ventilation to prevent the diver suffering from carbon-dioxide poisoning.

The output of the compressor should be sufficient to provide air for a standby diver to go to the assistance of the working

Figure 42.4 Diver in lightweight equipment working from a diving bell

diver. Because the output of compressors drops with wear, it is suggested that a safety factor of 50% be applied when calculating the required output. A receiver should be in the line and have sufficient capacity to allow the divers to surface safely in the event of compressor failure. The correct air supply is of vital importance for safe diving operations.

42.4.2 Surface-demand diving apparatus

This apparatus requires gas at pressure at the demand valve of not less than 4.5 bar above ambient before it will supply at the correct rate to the diver. The volume of air consumed by a diver varies in a number of factors including his work rate and lung capacity; however, an allowance should be made for an average consumption of 55 l/min measured at the working depth.

Supply pressure is critical and compressor output must be carefully considered for the planned diving depth.

A reserve emergency high-pressure cylinder is carried by the diver.

42.4.3 The aqualung or SCUBA

The air supply required for SCUBA divers is not less than 65 l/min measured at the working depth. High-pressure compressors operating between 140 and 280 bar are used to charge the cylinders. For major surveys, or other works where aqualungs are used extensively, high-pressure storage banks are used and the aqualung cylinders are charged by decanting from the banks. If the diver's SCUBA bottles were charged for more than 6 h before a dive they should be checked for correct pressure immediately prior to diving.

42.4.4 Compressors

All compressors used in connection with diving must supply air suitable for breathing purposes. Tool or similar compressors should on no account be used. To avoid contamination of the compressed air, compressors and cylinders must be maintained properly. It is also essential that the air intake of the compressors should be positioned to prevent foul or contaminated air reaching the cylinders.

42.4.5 Mixed-gas supply

Supply of mixed gas is normally from banks of cylinders containing the appropriate mixture dependent on the water depth in which work is being carried out. Normally, the gas mix is of helium and oxygen with the oxygen content reducing as the depth increases.

42.5 Size of the diving team

The diving contractor should set up a diving team with at least the minimum number of divers and support personnel to operate the plant and equipment necessary to undertake the diving operation safely.

A diving supervisor is responsible for the diving operation and for all the members of the diving team.

A minimum number of personnel in the diving team is usually legally specified but, typically (and as would be applicable to most civil engineering work) the team for air diving in less than 30 m of water where no decompression is planned would be: the diving supervisor, the diver in the water and a standby diver. The standby diver should be dressed with his equipment immediately to hand and ready to enter the water. In the UK, an additional diver would be required on the surface if the air-diving operation was connected with offshore installations or pipelines, if decompression stops were required, or if there was a special hazard, e.g. where a diver might be endangered by a current, become trapped or his equipment become entangled. Also in the UK, if the diving operation takes place in less than 1.5 m of water, and there are no special hazards, the diving supervisor may also act as the attendant thereby reducing the size of the team to two.

42.6 Planning underwater works

Construction work under water is self-evidently more difficult than similar work on land. Clearly, it is unreasonable to expect anyone under water to carry out all construction tasks to the same standard as a tradesman on land. The employer of diving services for simple recovery work has little problem. For more complicated work he can either redesign underwater fabrication so it is basic assembly work for the divers or, if this is not possible, can employ a specialist diving contractor with engineering backup.

42.7 Diver qualification

Standards of diver training in the UK are laid down by the Health and Safety Executive and minimum qualifications have been established in four grades:

(1) *Part I*: air divers who are trained to perform a wide range of air-diving operations and diving techniques including surface decompression and who have received basic training in the performance of work tasks using tools underwater at depths of up to 50 m.
(2) *Part II*: divers who are trained in deep diving using diving bells and mixed gas and saturation techniques, involving open-water experience of these techniques to 100 m.

(3) *Part III*: divers who need to perform only a limited range of air diving using both types of surface-orientated equipment. The training covers operations to 30 m but after appropriate work-up dives, operations to 50 m can be undertaken by this category of diver provided they do not exceed 20 min decompression time. This standard is often adequate for those working in inland/inshore locations involving underwater inspection of visual survey work but not for heavy manual tasks.

(4) *Part IV*: as for Part III but restricted to the use of self-contained SCUBA apparatus. This category includes those employed as scientists, archaeologists, photographers, scallop fishermen, etc.

(*Only divers qualified in Parts I and II may be employed in the offshore oil and gas industry.*)

42.8 The importance of an accurate underwater survey

Underwater surveying may be a highly specialized, expensive and painstaking task, but it is a vital preliminary to underwater construction or remedial works.

A practical example can be quoted of how a comprehensive but expensive survey resulted in a cheap and completely successful repair being carried out to the clapping face and sill of a large dock gate which was losing water. On the other hand, in the past, inspections of vital bridge foundations have been known to have been carried out by a two-man diving team after a tender – at a price which could only be described as ridiculous. All that can be achieved in a case like this is the classic 'yard arm' clearer: 'The foundations were inspected by . . . on. . . .'

Probably the worst results of all come when a well-meaning but misguided employer tries to get, as he sees it, maximum value for money from the survey diving contractors with requests like: Whilst you are there you can fix this, and: I want all the diving team to be working divers, and so on. The end result is that nothing gets done properly. The diving contractor must, of course, carry out the wishes of his employer, but it cannot be emphasized too strongly that surveying is a job requiring a special approach. Accurate records are vital and, basically, the surveying supervisor must be on the surface, diving only to clarify specific points. Work and surveying are not mixed on the surface nor should they be under water.

42.9 Factors affecting the diver's work

42.9.1 Effect of tidal flow

The most important limiting factor which adversely affects a diver's work is the velocity of water flow. A flow of 1 knot has roughly the same effect on a diver as that of an 80 km/h wind acting on a man on land. The maximum tidal flow in which a diver can work effectively is, not surprisingly, about 1 knot. The ability to work in a strong tidal flow is dependent on the work task, the work location and adjacent physical support available to the diver.

42.9.2 Visual inspection

A diver looking, as he does, through layers of air, glass and water observes objects apparently larger and closer than they actually are. He is therefore liable to report incorrect dimensions if he relies solely on observation, and he should always use a sea-bed ruler. A spirit-level may be used to establish levels in shallow water.

42.9.3 Poor visibility

In rivers which run through highly populated or industrialized areas, visibility is often very poor. The converse usually applies in sparsely populated areas. Similarly, in coastal areas near river estuaries, the outflow of polluted rivers may adversely affect visibility many miles offshore although the prospect of good visibility improves further offshore. Apart from pollution caused by man, sand from the sea-bed brought into suspension by storms can reduce visibility to a few centimetres even well offshore. After a few days of good weather this can change to give a visibility in excess of 30 m. In poor visibility, high-candlepower lights illuminate only the particles in suspension so the diver sees myriads of bright reflections from the particles. Special low-candlepower lights are available for close inspection in poor visibility. In areas of permanent low visibility, tactile measuring devices are invaluable and the diver's fingers become his eyes.

Underwater floodlights are likely to be useless in shallow water where daylight has not penetrated. The effective use of floodlights is primarily limited to nightwork or at intakes and other areas where natural light cannot penetrate.

42.9.4 The underwater season

In the summer months, weeds and other marine growth are at their most prolific, particularly in shallow coastal waters. Inspection of outfalls and other structures is therefore best carried out in the early spring.

42.9.5 Fatigue

Breathing underwater involves appreciable effort. The muscles of the rib cage which draw air into the lungs have to work harder to ventilate the lungs with the denser air. With the demand-valve systems, some effort is also required to activate the tilt valve. The effort required to swim underwater is heavy and trials have shown that a diver in standard diving apparatus uses approximately the same effort when walking as the underwater swimmer. Two hours is considered the maximum time a diver can work efficiently using a demand-valve breathing system. Provided that he is working in one area, 4 h duration is possible in helmet gear.

42.9.6 Sickness

Of necessity, divers are required to have a high degree of physical fitness and, generally, they are very rarely ill. However, working in cold water and experiencing temperature changes exposes them to the common cold. This can be serious since the presence of mucus in the eustachian tubes can prevent 'clearing his ears'. He is then unable to balance the pressure across his ear drums and the drums are forced inwards. The forcing can cause damage or infection of the ear. Consequently, divers should not dive with a head cold.

42.10 Preparation for underwater work

42.10.1 General

Recognizing the difficulties under which divers work, great care is necessary in briefing the diving team before underwater work begins. Basic information should also be established. For example, before diving commences on a bridge survey: (1) record a fixed datum point above water level (if required, it can be tied to Ordnance Datum at a later date); (2) on a tidal river, fix a temporary tide board from which direct readings can be taken during the survey; (3) establish chainages along the face of piers and abutments; and (4) establish the river-bed profile in relation to the water level and therefore the datum.

Dimensioned elevation drawings of the structure to be surveyed can be produced from this basic information. As the divers proceed with their inspection, findings can be related to the chainage position and the water level or bed level.

When working in tidal rivers and estuaries, reference should be made to the Admiralty tide tables to establish the time of high and low water predicted for the nearest primary or secondary port. Strong onshore winds can cause higher water levels than those predicted in the tables. The converse applies with strong offshore winds.

42.10.2 Offshore

When working offshore, divers can normally work from a vessel in conditions in which vessels can moor and work. The sea states generated by a force 5 wind normally prevent operations unless land protection is provided as with an offshore wind.

In most coastal areas, diving can take place only during slack water between high and low water. Large-scale tidal charts are available with times, velocities and directions tabulated for fixed locations. Times are related to the time of high water at a primary port in the tide tables. It is unlikely that work will be at a tabulated location so interpolation is necessary between the nearest locations for which information is available. A good knowledge of chart work and tide tables is essential for the offshore diver.

The following two important factors should be remembered when assessing tidal speed and direction. The tabulated information refers only to surface tides (which affect shipping) – on the sea-bed the tidal flow could be in the opposite direction to that of the surface. Tabulated information can only give an indication of what may be expected and some adjustment will need to be established during the first day's work. The second factor is that the direction of tidal flow moves around through 180° during the change of the tide. Thus a ship at single anchor may start off over the site of work and swing steadily away as the tide slackens, just when the diver is experiencing the best conditions for working on the bottom.

42.11 Construction work under water – preplanning

The work the diver carries out underwater usually involves a steel or concrete structure. As a general rule, underwater structures are designed in the same way as similar structures on land and little consideration is given to the problems and slow progress associated with underwater construction. It is common to find vital bolts missing simply because the tolerances were too tight or access so difficult that the diver could not insert or set up the bolt.

When planning construction work to be carried out by divers the following factors should be taken into account:

(1) Visibility may be poor and can be virtually nil, therefore units should be designed in such a way that the diver can work by touch.
(2) The diver's vision is distorted, making fine assembly difficult. Wider tolerances should be adopted.
(3) The diver and the structure he is working on are affected by tidal flow. Simple connections should be used between new and existing works.
(4) Water flow accelerates around massive units placed underwater. The final velocity depends on the size of the unit and,

of course, on the basic flow. Consideration should be given to means of protecting the diver from strong currents.

(5) It is difficult to place heavy units accurately into position from surface floating craft. A diver can normally handle a unit that weighs approximately 50 kg in air. The use of controlled buoyancy is practical for units weighing up to approximately 250 kg in air but it must be remembered that tidal forces act on the buoyancy unit as well as the unit being placed. A decision on the method of placement is an essential part of design work.
(6) A diver is virtually weightless underwater; he can apply little downward force. Providing he can establish a footing he can lift his full strength, although some of his energy may be diverted to overcoming tidal currents. The one way in which a diver can exert his full strength is with his arms and shoulders – using one hand as a reaction point, he can pull with the other. When preparing the detailed design of an underwater structure, the use of this force should be exploited. Modifications to the design of concrete blocks to facilitate diver assembly is likely to increase design and casting costs, but these should be more than offset by increased output from the divers.
(7) A diver's 'feel' when setting up bolts underwater cannot be relied upon. By breathing compressed air he is, in effect, breathing enriched air due to the higher partial pressure of oxygen. As a result of this stimulant he may apply too much force to the bolts. The use of a torque wrench is therefore recommended.
(8) In many cases the diver needs one hand to hold himself in position against current flow and so, effectively, he has only one hand free for working. Where a large number of bolts have to be set up, provision should be made to lock the bolt heads.
(9) Whenever possible, divers should be allowed to examine units to be installed under water while at the surface. This familiarization will reduce the time needed to complete the underwater task.

42.12 Use of tools

It is difficult for the diver, with the many restraints on his performance, to use tools effectively. He has a number of tools at his disposal but careful planning is necessary to ensure their proper use. For example, if concrete saddles are to be fixed to a rock bed by rock bolts it is sensible to cast the lower section of the saddle block so that it forms a template for the drilling operation, i.e. with its hole of sufficient diameter to accept the drill bit and its thickness related to drill steel lengths to give the correct penetration.

As a general principle, the use of tools should be kept to a minimum. By careful design, it may be possible to construct underwater by a combination of interlocking bolted concrete blocks and/or units. The diver then has a repetitive 'Meccano-like' assembly to carry out.

42.12.1 Pneumatic tools

Pneumatic tools are frequently used by divers. Generally, all surface air-diving tools can be used effectively to about 30 m depth. It will be obvious that extra care is necessary to maintain the tools. Unsatisfactory operation is often caused by lack of consideration of the following basic facts:

(1) Pneumatic tools usually require an effective pressure 'at the tool' of approximately 4.5 bar. At 30 m the back-pressure of the sea-water is approximately 3 bar and the minimum pressure required would therefore be 7.5 bar.

(2) The diver is normally a long way from the compressor and the use of normal 19 mm tool hose causes a large pressure drop in the line. This must be corrected by the use of a larger-bore hose, the introduction of a 'pig' (receiver) in the air line near the diver, or a combination of both (which is recommended).

The use of exhaust hose to the surface to remove back-pressure appears attractive, but the extra resistance to airflow and the drag on the exhaust hose removes this apparent advantage in carrying the exhaust air back to surface-pressure.

42.12.2 Airlifting

The simplest form of airlift consists of a 150 to 300 mm diameter length of PVC pipe with an air connection and control valve inserted approximately 500 mm from the lower end. The diver locates the lower pipe end close to the material to be excavated and opens up the air supply. The air travelling up the tube expands as it ascends, drawing water and sea-bed material to the surface.

With an adequate air supply a 10-m long by 200-mm diameter airlift will easily lift 150-mm stones approximately 2 m above the surface. A diver can operate and move such an airlift manually, like a vacuum cleaner, to clear an area of sea-bed.

A PVC airlift can be operated entirely underwater where the materials removed from a cofferdam on the sea-bed can be distributed on the surrounding sea-floor. In a water depth of 25 m a short airlift will work quite effectively if the discharge point is less than 15 m underwater. At this depth the air does not expand as much as it would if the lift extended to the surface. It does not therefore have its full lifting capacity and a greater flow of air is thus required to deal with a given quantity of spoil.

Purpose-built steel airlifts deployed from surface barges may also be used. Generally, where the 'lifted' materials are to be removed from the site, normal dredger techniques should be used.

42.12.3 Underwater television

Underwater television is frequently used for instruction and control in offshore work. Because most inshore diving work is done in shallow water and the greater proportion of this work is carried out where visibility is poor, television is suitable only for limited localized inspection. However, manufacturers are developing new and more sensitive systems with complex lighting which improves the performance. This development, together with the gradual cleaning up of our rivers, should increase the suitability of television for routine inspection in shallow water.

The use of television is justifiable where major repairs are considered necessary to an important structure. A videotape recording made in the best conditions, which the engineer can study at any time, has obvious attractions.

42.12.4 Underwater photography

Again, owing to poor visibility in, for example, rivers, docks and harbours, photography is seldom used. As with underwater television, improvements in photographic techniques under water increase the possibility of obtaining useful photographic records.

42.12.5 Production of drawings

For structural surveys there are obvious advantages in using engineers or draughtsmen who can dive to produce adequate drawings illustrating findings.

42.12.6 Hydraulic tools

Hydraulic tools are available but the logistic problems of special pumps limit their use to specialized tasks. In addition to hydraulic rotary tools, a number of hand-pumped hydraulic tools are available for underwater use, especially those used for cutting cables, hawsers, etc.

42.12.7 Underwater thermal cutting

There are three basic types:

(1) *The oxy-hydrogen torch* which is little used commercially. Divers need regular practice.
(2) *Oxy-arc equipment*, consisting of a 'gun' which holds a hollow carbon or steel rod through which oxygen is supplied. Power to the rod is provided by a welding generator. An arc is struck with the rod, as for welding, with the oxygen jet removing molten metal in addition to providing fuel.
(3) *Thermal arc*, which is similar in principle to the thermic lance but with a flexible plastic hose which incorporates a metal 'fuel'. The hose is expensive and the oxygen consumption high. The thermal arc is a quick if crude cutting system which is attractive in that no generator is needed.

42.12.8 Cutting using explosives

The cutting of steel plate or tubular section using explosives is limited in underwater construction work owing to the possibility of damage to adjacent structures. For offshore work in deep or exposed waters, specially designed and shaped charges have been developed. These use the minimum weight of explosive to cut through a plate or tube of a given thickness. They are expensive but worth considering where, for example, divers' working time on the bottom is limited by depth or tide.

42.12.9 Underwater welding

Simple underwater welding has been carried out for many years but quality has been poor in the past and applications limited. Highly sophisticated and therefore expensive systems have been developed for use in the oil industry, such as welding within specially designed underwater environments and automatic underwater welding machines.

42.12.10 Concreting underwater

Placing concrete underwater by tremie tube has been practised for many years. The difficulties in handling the tremie tube and pouring concrete to avoid passing the mix through water (with the consequent loss of cement) are well known. As an alternative, the present-day efficiency of the concrete pump allows direct placement.

42.12.11 Grouting underwater

Unlike concrete with its high weight per unit volume and viscosity, grout, consisting of sand, cement, water and retarding additives, can be pumped over long distances. It is used particularly to 'tie together' precast concrete units. It is very effective where both the paths and connections for supplying and controlling the grout flow can be built into the structure. The problems of grouting the foundations of structures in water are much the same as on the surface, i.e. the possible excessive loss of grout through coarse gravel or other open material must be guarded against.

42.12.12 Underwater bolt-firing gun

These guns are capable of fixing bolts into bricks, steel or concrete. It is not normally practical to fix bolts into rock owing to its variable consistency.

42.12.13 Nondestructive testing

Underwater nondestructive testing techniques are similar to those used on the surface. Time must be allocated for preparing and cleaning surfaces where, for example, metal thickness readings are required. Where potential differences are to be measured, the diver may carry only the instrument probe connected by cable to the instrument on the surface. In this way the divers are controlled by telephone from the surface where readings are taken and recorded. Encapsulated complete instruments are also available so that a trained diver/inspector may take his own reading underwater but, where practical, surface control is recommended.

42.13. Diver-alternative systems

The most widely used diver-alternative system prior to 1960 was the observation chamber. This consisted of a watertight cylindrical chamber with a removable upper access door which could accommodate one man. Observation portholes allowed the occupant to observe the underwater scene in relative comfort at atmospheric pressure. Life support was simple – a manually controlled oxygen cylinder, whilst carbon dioxide was removed by breathing out through an absorbent canister. This system was widely used on salvage operations in deep water with the observer controlling a grab deployed from the surface. Probably the most well-known operation was the recovery of gold from the SS *Egypt* which sank in about 130 m of water in 1932, a depth which then was well beyond divers' capacity.

Attempts to protect man from water pressure resulted in many versions of 'iron men'. These were articulated armoured suits (Figure 42.5) which had little success until fairly recent times when development of materials and joint design gave the 'diver' greater mobility. The most significant early development in diver alternatives prior to the start of major offshore oil exploration was the introduction of underwater television by the Royal Navy in 1951 during the search for the submarine HMS *Affray*, lost in the English Channel.

A major boost in demand for sophisticated underwater services took place in the UK in 1956. Progress since that date has been enormous; diving techniques have improved and depth capabilities extended. Although the safety record of diving companies has been good, the danger to human life has remained. This, plus the fact that exploration was taking place in water depths greater than divers had ever previously reached led to a surge in investment in systems which could replace divers.

The introduction of manned submersibles to carry out surveys and inspection work using television with video recording was a fairly straightforward development. Some were fitted with a diver lock-out capability. Manned sea-bed crawler-type vehicles for working on pipelines were introduced, with both submarines and crawlers being fitted with hydraulically powered arms and specialized tools of increasingly sophisticated design (Figures 42.6–42.8).

The systems still posed an element of risk to personnel. Associated with the production of oil offshore has been the enormous technical development in every branch of technology relating to the remote control of instruments, tools, machines and installations. These, plus a similar advance in navigational systems for guidance of submersibles, led the development of many types of unmanned underwater vehicles controlled entirely from the surface (Figure 42.9).

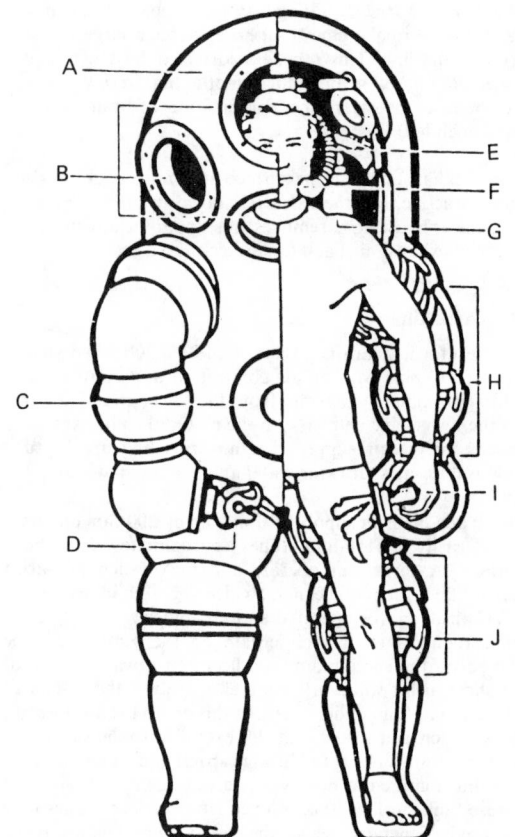

Figure 42.5 An atmospheric diving suit. A. Magnesium alloy pressure hull. B. View ports. C. Ballast weight. D. Special-purpose manipulator. E. Breathing hose. F. Oral-nasal breathing mask. G. Carbon dioxide scrubber. H. Articulated fluid-supported joints. I. Manipulator hand-levers. J. Articulated fluid-supported joints

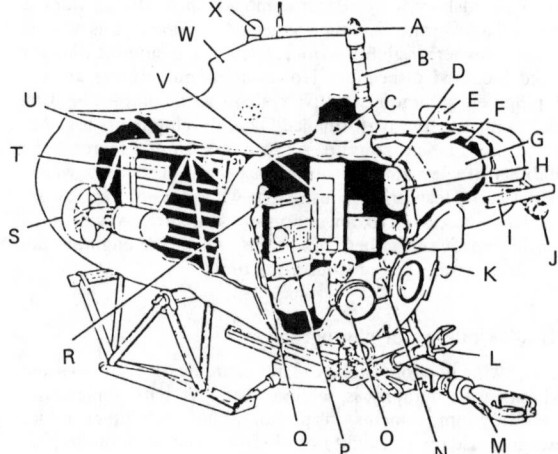

Figure 42.6 A two-man minisub. A. Surface radio antennae. B. Transponder. C. Hatch. D. Crew sphere. E. Speed log. F. Trim sphere. G. Oilbag. H. Oxygen bottle. I. Badge bar. J. Light. K. Sonar. L. Telechiric arm. M. Torpedo recovery arm. N. Control console. O. Viewing port. P. Receiver. Q Air-purification unit. R. Videotape recorder. S. Propulsion motor. T. Batteries. U. Machinery sphere. V. Main fuse panel. W. Sail. X. Emergency release buoy

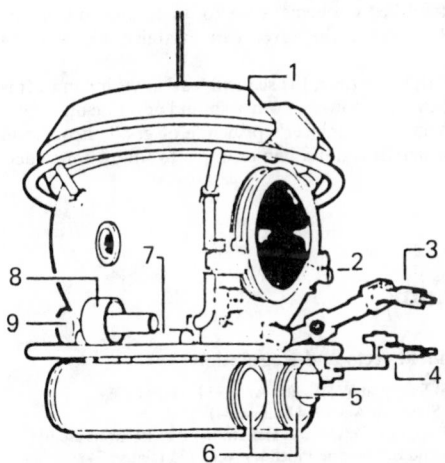

Figure 42.7 A manned mobile observation chamber fitted with manipulators. 1. Flotation material. 2. Television camera and light. 3. Force feedback manipulator. 5. Sonar dome. 6. Battery pods. 7. Oxygen supply. 8. Thruster. 9. Hydraulic power supply

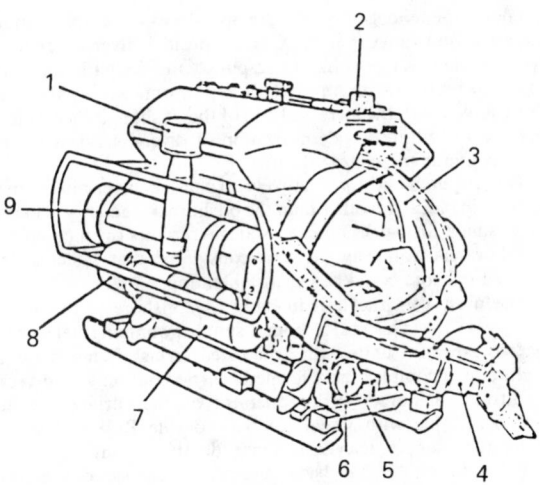

Figure 42.8 A one-man minisub. 1. Vertical thruster. 2. Acoustic communications transducers. 3. Forward acrylic viewport. 4. Manipulator. 5 Directional transducer for locating acoustic 'pinger'. 6. Lamp. 7. Variable buoyancy compressed-air tank. 8. Horizontal thruster. 9. Electronics container for thruster control

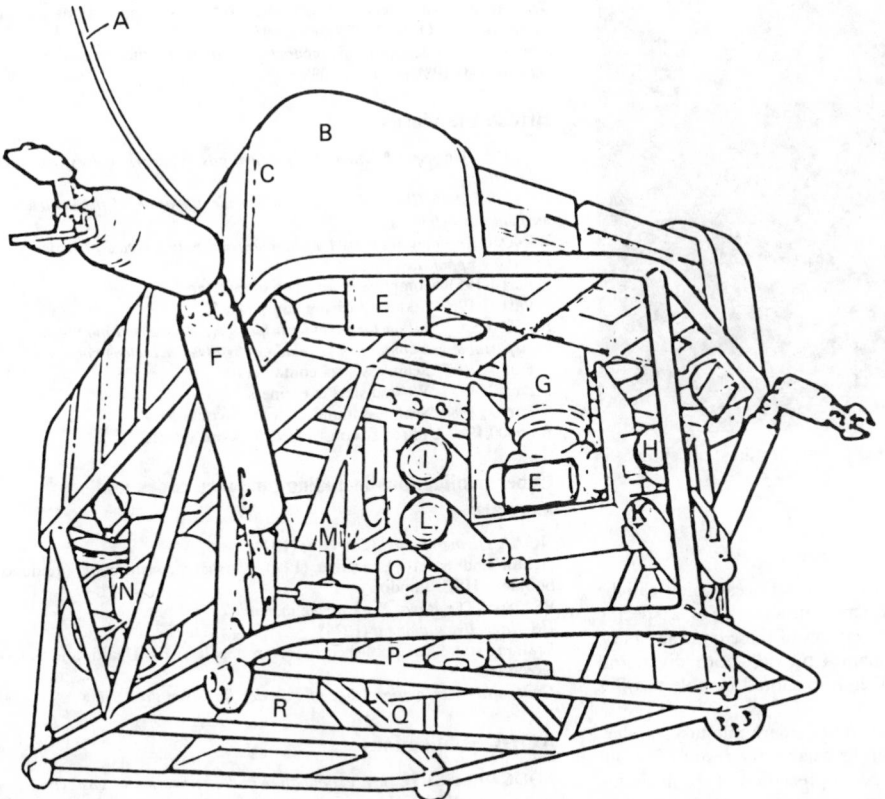

Figure 42.9 A typical advanced remotely operated vehicle. A. Umbilical cable. B. Syntactic foam buoyancy. C. Foam attachment straps. D. Trim foam. E. Lamp. F. Manipulator. G. Pan and tilt unit. H. Strobe lamp. I. Cine camera. J. Vertical thruster. K. Still camera. L. Colour television. M. Lateral thruster. N. Longitudinal thruster. O. Pan unit. P. Side-scan sonars. Q. Sector-scan sonar. R. Sub-bottom profiler

Today, the tendency in offshore operations in depths beyond the air diving range, i.e. 50 m, is to consider diver-alternative systems whenever possible. In response, the diving companies have not been slow to improve their performance capabilities: this coupled with the increased size of the offshore industry, has ensured a continued wide utilization of divers' services where these continue to be cost-effective.

The requirement for relatively expensive diver-alternative systems in industries other than the offshore oil industry is likely to be small; for works on deep-water reservoirs or in contaminated waters, for example, it is considered that systems developed offshore could be worth evaluating.

The most versatile tool in daily use offshore is the ROV (Figure 42.10). This varies from a simple swimming 'television eye' to powerful units with articulated 'arms' designed for a multitude of complex tasks formerly carried out only by divers. Control is achieved by an operator at the surface via an umbilical cable with visual feedback of the ROV's position using, for example, television, sonar, depth, heading, etc.

Work by ROVs has been successfully carried out in the northern sectors of the North Sea in water depths greater than 600 m with a 2-knot effective current in addition. Navigational systems are available by which the vehicle can return repeatedly to a given position.

Figure 42.10 The 'Pioneer' ROV with features as shown in Figure 42.9

42.14 Conclusion

This chapter has outlined some of the services which can be provided by specialist underwater contractors. For a specific task it is recommended that the services of these contractors be employed. A number of companies provide both diver and diver-alternative systems and therefore should be able to offer an objective project analysis.

Divers can, and do, carry out some remarkable tasks underwater. Such works are, however, invariably the result of sound practical engineering and a proper appreciation of the maritime problems associated with underwater working. Divers need to be used intelligently as part of the construction team.

It is not sensible to ask the diver to carry out heavy manual work where this can be avoided. Fatigue reduces the mental faculties of the diver in the same way as it would any other person. Diving is hard work which, to some extent, accounts for the loss of memory associated with the job in hand.

Having established a scheme, stick to it and aim to standardize assembly so that the diver can establish his working techniques.

It is said that, given time, the sea will 'eat' anything in it. It is certainly true that the power of water should never be underestimated. The Victorian engineer's principle of good, strong and durable construction is not a bad example to follow for underwater work.

Bibliography

UK Government Acts and Regulations

The Health and Safety at Work, etc. Act 1974 (clause 37)
The Merchant Shipping Act 1974 (clause 34)
The Mineral Workings (Offshore Installations) Act 1971 (clause 61)
The Petroleum and Submarine Pipelines Act 1975 (clause 74)

Statutory Instruments

The Construction (General Provisions) Regulations 1961 (number 1580)
The Diving Operations at Work Regulations 1981 (number 399)
The Merchant Shipping (Diving Operations) Regulations 1975 (number 116)
The Submarine Pipelines (Diving Operations) Regulations 1976 (number 923)
The Petroleum (Production) Regulations 1976 (number 1129)
The Health and Safety at Work, etc. Act 1974 (Applications outside Great Britain) Order 1977 (number 1232)
The Merchant Shipping (Submersible Craft Construction and Survey) Regulations 1981 (number 1098)

British Standards

BS 1319:1976, *Specifications for medical gas cylinders, valves and yoke connections*
BS 1319C:1976 *Chart of colours for the identification of the contents of medical gas cylinders*
BS 4001 *Recommendations for the care and maintenance of underwater breathing apparatus:*
    Part I 1961 'Compressed air open-circuit type'
    Part II 1967 'Standard diving equipment'
BS 5430 *Specification for periodic inspection, testing and maintenance of transportable gas containers (excluding dissolved-acetylene containers):*
    Part 1: 1977 'Seamless steel containers'
    Part 2: 1977 'Welded steel containers'
    Part 3: 1980 'Seamless aluminium alloy containers'
BS 5500:1982 Unfired fusion-welded pressure vessels

Other publications including guidance codes and manuals:

NOAA *Diving manual* (2nd edn) (1979)
Health and Safety Executive (1980) *Offshore construction* (guidance booklet), HSE, London
BR 2806: *The Royal Navy diving manual*
US Navy diving manual (1974)
Training standards published by or on behalf of the Health and Safety Executive
Underwater Association Code of Practice for Scientific Diving

AODC publications

AODC 010:1983 *Testing, examination and certification of gas cylinders*
AODC 014:1983 *Guidance note on minimum quantities of gas required offshore*
AODC 015:1983 *Guidance note on surface orientation (air) diving from DP vessels*
AODC 016:1983 *Guidance on colour coding and marking of diving gas cylinders and banks*
AODC 027:1984 *Oil-lubricated compressors*
AODC 035:1985 *Code of practice for the safe use of electricity underwater*

UEG publications

UR34: *Control and monitoring of carbon dioxide in diving bells*
UR31: *Tables for saturation and excursion diving on nitrogen–oxygen mixtures*
UR23: *The principles of safe diving practice*
UR28: *Thermal stress on divers in oxy-helium environments*

UR18: *Handbook of underwater tools*
UTN26: *Procedures and language for underwater communication*
UTN25: *Aseptic bone necrosis in commercial divers*
UR14: *Underwater electrical safety – some guidance on protection against shock*
UR11: *Oxy-helium saturation diving tables prepared by RNPL*
UR7: *RNPL metric air diving tables*

43

Demolition

T R Mills CEng, MICE, FIDE
Chartered Engineer

Contents

43.1 Introduction

The demolition industry has its origin among those small contractors who specialized in the removal and resale of architectural items, building materials and structural components. In the past, their operations were labour-intensive and time-consuming and often the salvaged items remained on their hands for considerable periods.

Development over the years has determined that demolition contractors organize their work and equip themselves to meet critical programmes. The size of firms undertaking demolition and associated fringe activities, such as excavation and temporary works, varies from organizations offering a complete package, through to small firms undertaking specialist works such as concrete-cutting or demolition blasting.

The capability of contractors varies from those thoroughly organized with managerial and technical facilities backing up a skilled labour force, down to those who work in excess of their capacity to the detriment of their clients in particular and the industry in general.

It is important that demolition contracts or the demolition phase of major projects, should receive the same careful consideration as those for rebuilding and civil engineering.

Much can be done to eliminate delays to works by ensuring that obligations to third parties and statutory undertakings are negotiated before commencing work.

Details of the buildings or structures to be demolished or altered, together with a careful assessment of the restraints imposed by location, should be discussed by clients and prospective contractors at an early stage to permit realistic and fair tendering.

'Demolition' embraces a wide spectrum of activities – from simple cottage demolition to the dismantling and removal of industrial complexes. No broad boundaries are found between demolition styles in different localities. Certain types of structure lend themselves towards a preferred method, but of the many techniques available few can be applied individually and most successful jobs employ a combination of one or more methods.

43.2 Organization and planning

The person inviting tenders or negotiating for demolition work should ensure that it will be undertaken by experienced contractors who have an awareness of the problems likely to be encountered on and off the site.

Forms of contract normally used for building or civil engineering may consist of documentation made unnecessarily complex by alterations, omissions and additions. A simple form of contract such as that published by the National Federation of Demolition Contractors in London may be acceptable.

The specification should be as simple as possible while adequately defining the work required. Items such as shoring, weatherproofing and accommodation works may be impossible to determine until parts of the redundant building or structure have been removed and assessments made with the professional adviser(s) appointed by owners of the adjoining premises. Where this is the case, 'provisional sums' should be included.

In partial demolition or refurbishment, the structural detail may have a critical effect on the proposals. Even though original drawings may exist, it is prudent to open up relevant parts of the superstructure to confirm what has, in fact, been built. This preliminary work will often be needed by the consulting engineers charged with modifying the structure, but in any event the structural survey should be done prior to inviting tenders for demolition in order that prospective contractors may have the benefit of seeing the structural detail.

Where original drawings exist an allowance must be made for alterations and additions which have not been formally recorded.

Parts of the premises which are inaccessible should be carefully opened up. It should not be a prospective contractor's responsibility to survey below the surface; indeed, the practicality of a number of such contractors bringing equipment on to a site which is not formally in their responsibility normally precludes this.

The likelihood of toxic or hazardous material either being built in to the structure or arising from previous use should be assessed and the information forwarded to tenderers. It may be prudent to decontaminate premises prior to the start of demolition works or carefully phase a programme so that demolition does not produce a hazardous environment by scattering residues. The discovery of toxic and hazardous material at a late stage can have the serious result of closing down work until residues are removed.

The specification should state, and drawings should clearly indicate, the levels to which buildings are to be demolished. If ground slabs and foundations are to be removed, approximate quantities with provisional sums for excesses, or a schedule of rates, should be included for an item which is basically excavation. Any architectural features or sections of the building which are to be retained should be defined clearly and removed early in the job if possible.

Once the decision to demolish has been taken, there are many matters that can be put in hand alongside preparation of contract documents.

The so-called party wall awards, with their very specific details of what can and cannot be done to adjacent properties, need considerable time for practical assessment and preparation of documents. Whilst it may be impossible to agree fully the extent of the party wall award prior to opening up and exposing an adjacent property, it is better to have made some progress, thus allowing a demolition contractor to work normally and logically rather than be prohibited from approaching a site boundary.

The removal of statutory undertakings' equipment should be organized as soon as the premises become vacant, in order to avoid delays to the works and accidents caused by live services.

Demolition work may involve the diversion or protection of services where there is a wayleave across the site and such matters should be considered early, as sometimes extremely complex planning has to be done to avoid disconnecting adjoining premises.

In industrial premises which have existed harmoniously with their neighbours, demolition, even over a short time, can cause concern to residents, or hazards to adjoining premises. There needs to be liaison between the client and concerned neighbours in order that prospective contractors are aware of restraints upon their methods of work.

43.2.1 Insurance

Clients and their professional advisers should make certain that specified insurance requirements are adequate, but at the same time realistic, for the risks arising from their projects.

Policies on behalf of the client, the contractor and sometimes jointly, should be in force prior to commencement of the works.

The amount of cover required can vary considerably, dependent upon location. Demolition work on an active mainline rail terminal, for example, would require considerably more cover than work on a remote and disused factory premises.

Employer's liability insurance is a statutory requirement, the premium reflecting the hazardous nature of demolition work and the record of the contractor.

43.2.2 Surveys and method statements

The object of a demolition survey is to establish sufficient detail about the premises to allow a decision on the method of demolition and to identify any restraints affecting proposed activities on the site.

Problems may arise due to characteristics of the building or structure, or to the previous use of the premises or the location.

Drawings of the building are not always available, and even where they are, the survey should be comprehensive enough to determine the extent of alterations from, and additions to, the original plan.

The sequence of building or erection, together with an assessment of any temporary works or equipment used, can be usefully compared with the proposals for demolition.

The need to provide shoring either on adjacent premises or within the structure being demolished should emerge from the survey; also, the need for weatherproofing either temporary or permanent (see sections 43.2.3 and 43.2.4).

On any type of building or structure, the condition can, to a greater or lesser extent, determine the method of demolition to be used.

Traditional brick, masonry and timber structures left unattended and open to the atmosphere, may have deteriorated to the extent that they cannot be relied upon to provide safe access.

Industrial uses producing corrosive atmospheres and long-term lack of maintenance may have affected structural steelwork, particularly connections, to the extent that preweakening of a structure would be hazardous.

Persons undertaking demolition surveys should be aware of, and experienced in, finding such conditions.

In addition to surveying the building, the adjoining buildings should be assessed in terms of structure and equipment contained to determine any restraints. For example, in urban areas the increase in use of computers, which are very sensitive to vibration, has added to the problems traditionally arising from dust and vibration.

Following the demolition survey, a detailed method statement should be prepared prior to the start of work.

The presentation of this statement has been widely debated: BS 6187:1982 *Code of practice for demolition*[1] recommends that a programme should be drawn up in which the proposed sequence and the method of operation is indicated clearly. The Health and Safety Executive Guidance Note GS29/1[2] goes further by advising that a detailed method statement should identify problems and their solutions and form a reference for site supervision.

The degree of difficulty in the work must determine the detail on the method statement, but in any event, it must be in terms that can be understood by the supervisor and labour force. It may be written with annotated drawings, or instructions may be painted on to the structure.

43.2.3 Shoring

The need for shoring-up adjacent buildings or, on occasions, the temporary shoring of the structure being demolished, should be investigated prior to the start of demolition. In cases where this is not possible a procedure for checking the requirements as the job progresses should be formalized.

The simplest approach is to leave parts of adjacent walls or part of the structural frame as buttresses if they are in an acceptable condition – provided their retention would not hinder subsequent building.

If it is decided that raking shores are necessary, it is sometimes acceptable to erect them after demolition, but if the durability of the exposed wall is suspect in the short term, the shores may have to be inserted prior to demolition, with all the

attendant difficulties of preparing the foundation block and handling the sections in confined spaces.

When a free area of land is required, and where there is a convenient support, a flying shore can be connected horizontally to an adjacent and sufficiently robust premises, provided the owner of the supporting building agrees.

On flying shores of long spans a central prop may be necessary and, although this to some extent defeats the object of obtaining a clear area, it can be less inconvenient in the centre of a site, where it may interfere only with floor construction, than raking shores at the edge, interfering with walls.

Raking and flying shores are inserted to prevent a sideways movement of a wall. It is sometimes necessary to effect vertical support to a wall, either because its foundations have failed, or it is to be underpinned, or is to have additional openings made in it.

In these circumstances, needle shores, in which short horizontal members are placed through a wall and given their own support, are used. Vertical support is referred to as 'dead shoring'.

The loading on the shoring must be assessed and the system designed and detailed in a manner which takes into account the circumstances of its erection and later dismantling.

Shoring was traditionally of timber, but examples from scaffold tubes and fittings, and from fabricated steelwork are more common today, with the use of traditional carpentry skills becoming uneconomical.

For dealing with dangerous structures, say a building which has been hit by a vehicle or suffered a severe fire, shores made from scaffolding are normally used, often erected not by the demolition contractor but by scaffolding companies who have arrangements with the authorities for rapid turnout to deal with such emergencies.

For retaining façades, scaffolds may be used, but fabricated steelwork providing quick erection and a compact design is popular, particularly in congested city areas.

43.2.4 Weatherproofing

Walls previously protected by buildings which are being demolished, will quickly absorb moisture and deteriorate if weatherproofing is not provided.

Walls exposed on a long-term basis can be rendered or skimmed with brickwork if this is desirable from a visual viewpoint. Walls in good condition can be treated with water-repellent compounds.

It is more usual for bitumen or polythene sheeting to be used on a temporary basis to stand until the new adjacent walls provide protection. Such sheeting must be fixed to the wall with vertical and horizontal battens in a regular pattern and in a workmanlike manner if it is not to be torn by the wind and become ineffective. It is both annoying and costly to have to return to repair small areas of weatherproofing as the access from the original building may not be available and scaffolds or elevating platforms will have to be obtained. The sheeting must be overlapped and horizontal faces must be weatherproofed to ensure that the arrangement is completely watertight.

43.2.5 Hoardings

It is usual for the contractor to provide a hoarding, sometimes to his own design, sometimes to that of others.

As a precaution against trespassers, it is in the contractor's best interests to secure the site at all times, but this may not be practicable, particularly during the early stages of work.

The provision of a hoarding can be a substantial item. Erection should be done by experienced personnel in order to ensure that it will resist overturning by the wind, and be robust enough to resist the ravages of subsequent building activity where it will be cut about to complete sight lines or have panels removed to give access for materials.

Hoardings should not go across manholes and valve covers. Any openings or doors which are at ground level should not open on to open basements or trial holes.

43.3 Demolition by hand

Hand demolition involves workmen using pneumatic or hand tools to remove part of a building or structure prior to completion by mechanical means (Figure 43.1), or all of the structure where mechanical means cannot be used.

This work may involve opening up floors to allow debris to be dropped to the lowest hard level. Opening up must not be allowed to make the building unstable and material must be cleared to avoid lateral pressure on lower walls.

If material is being dropped inside, it will be deflected a considerable distance if it hits other parts of the structure. Windows and apertures should be boarded up, or sheeting erected, to prevent debris being ejected outside.

Suitable working platforms must be provided where work cannot be done safely from the structure. These can be formed from scaffolding or the use of man-riding skips or hydraulic elevating platforms.

Scaffolds should be sheeted and fans provided over adjacent public areas. As the height of the building is reduced the scaffold should follow; ties should not be removed because this would render the scaffold unsafe.

It is normally preferable to demolish floor by floor. To preserve adjacent properties, shoring may have to be built either before or during demolition. The removal of roofs will normally be done by hand prior to other activities.

43.3.1 Preweakening

Preweakening is the removal, cutting or partial cutting of structural members as an activity prior to final collapse being achieved by some other means.

The preweakening of structures has for many years been a preliminary to wire-rope-pulling, but more recently the use of explosives has involved preweakening to reduce the amount of explosive needed or to form 'hinges' to encourage structures to collapse in a predictable manner, thus allowing economic removal of debris.

There is argument about the relationship between the need for expert engineering advice and reliance on experience and intuition to decide the style of preweakening.

It is accepted that most structures, when relieved of sheeting, industrial plant and bulk material, can be considerably preweakened. Most engineered solutions use simplifications of the structure to arrive at the selected procedures.

There is a school of thought which encourages the retention of loads at height in order to provoke movement once the structure is, by whatever means, made unstable. This method requires very careful analysis and supervision.

Notwithstanding an empirical or an engineered approach after any surplus load has been removed, a number of cuts will be made to finally preweaken the structure.

Cuts that allow members to 'sit', or 'hinge', have characteristic styles, but great care is required to monitor movement as preweakening progresses to ensure that the structure is behaving as intended and not about to collapse prematurely from inbuilt stresses, wind or impact.

Failure of structures to respond to the final drama of removing the key parts can produce extremely complex problems involving time and expense.

43.4 Mechanical demolition

There are comparatively few machines which are specifically designed for demolition, equipment generally being obtained through the normal plant markets. With few exceptions, demolition attachments to base machines designed primarily for excavation and material handling are used.

Demolition duties impose considerable wear and tear on machinery. Good maintenance procedures therefore need to be provided if the machinery is not to deteriorate until it becomes unsafe and beyond economic repair.

Some contractors elect to improve the durability of their machinery by fitting heavy-duty buckets and protection to cabs and engines to guard against damage by debris.

Figure 43.1 Manual demolition work using pneumatic breakers on a city centre roof

43.4.1 Wire-rope pulling

This apparently simple and effective method of demolition requires careful planning and organization if it is to be successful, especially when used on tall structures.

Static winches or vehicle recovery trucks may be used to effect the pull, as may the drag rope of an excavator or a winch 'dozer.

The distances available must be such that there is no risk of the resultant debris falling or being projected on to the winch or pulling vehicle.

The pulling ropes must be of steel wire, be in good condition and have proper fittings at their ends.

Workmen and spectators must be well clear of the area in which a breaking rope may whip.

More than one rope should be attached in advance, in order to avoid the risk of approaching the structure which may be unstable or partially collapsed after the first pull.

The load should be applied gradually, and it will become apparent at an early stage if the equipment has sufficient pulling capacity.

In the event that it has been decided to preweaken the structure in advance of ordering final collapse by pulling, expert engineering advice may be necessary to ensure that structural stability is retained. An unsuccessful pull may weaken a structure.

43.4.2 Hydraulic excavators and tractor shovels

Most demolition sites will eventually reach the stage where a hydraulic excavator or a tractor shovel will be used for digging out basements or loading debris.

These machines may be used successfully for pulling or pushing over low-level buildings by using their buckets. Brick and masonry structures will collapse readily, but steelwork may need some preweakening if the section resists the capacity of the machine.

A Ripper tooth attachment can be used to drag down low-level steelwork and industrial plant.

When adopting this approach, there is a tendency to start on the structure before the building has been properly stripped out, the result being mixed debris (containing brickwork, masonry, steelwork, trunking and fittings) which is unsaleable and which may need sorting prior to dumping.

43.4.3 Balling machines

This well-publicized method is popular with demolition contractors because it is economic, quick and safe, provided certain simple precautions are taken.

Balling machines need space in which to operate and should be outside rather than within the confines of the building being demolished (Figure 43.2).

It may be necessary to remove roofs by hand prior to balling, and it is sometimes necessary to remove storeys to bring a building within the capacity of a balling machine. Areas of floors may need to be removed to facilitate the dropping of debris.

If vibration is likely to be a problem, a split should be made between the building being demolished and adjacent properties.

The demolition ball should generally not be allowed to swing free on the hoist rope, but should at all times be tethered by a drag-rope. In order to bring the ball into contact with the building, it may be dropped, pulled and released in line with the jib, or slewed. Slewing the jib is particularly hazardous as an excessive slew angle, rotational acceleration, and braking can impose excessive stresses encouraging jib failures.

Balling at heights in excess of 30 m, known as 'high balling' is now undertaken regularly. Great skill and experience is required for this work, and it does not follow that a good crane driver is a good balling-machine operator.

It is important to ensure that the ball does not become trapped or that the machine becomes overloaded by sudden collapse of large parts of the structure. A technique of 'little-and-often' is preferred.

Many recommendations concerning the use of balling machines and other mechanical methods are given in BS 6187:1982.[1]

43.4.4 Pusher arm

These devices were formally rigid arms fitted to hydraulic excavators which exerted a horizontal thrust to demolish walls

Figure 43.2 A balling machine working in a confined area

or structures. More recently, a telescopic demolition arm allowing a pushing or pulling motion has become available, considerably extending the adaptability of this method.

Some preparation may be necessary, either in removing the roof, reducing the height or isolating the building from adjacent premises. The building should be attacked in a logical manner without leaving slender sections free-standing. Walls should not be pushed more than 600 mm below their top level and, once again, a 'little-and-often' technique is required both to bring the building down safely and avoid the necessity of breaking-up large slabs of brickwork or masonry as a secondary operation.

Pusher arms are not effective against steel-framed or heavily reinforced concrete structures although they may be used to remove infill panels or sheeting.

43.4.5 Impact breakers

This tool is essentially a larger development of the traditional 'jack hammer'. It may be pneumatic, but is more regularly hydraulically operated from the power pack of the excavator on which it is mounted. These devices will achieve considerably better and more economical output than hand-held tools.

Large impact breakers are normally used for breaking mass concrete or old foundations. When used on reinforced concrete, it is sometimes necessary to cut manually the reinforcement enabling the pick of the impact breaker to remove the fractured concrete prior to restarting work on a fresh area.

Alongside the development of small and mini hydraulic excavators has been the development of the so-called mini breaker. These machines can be lifted or manoeuvred into buildings for breaking floors or be placed in confined areas for breaking basements and foundations, which would previously have required handwork.

43.4.6 Hydraulic shears

Large hydraulic shears mounted on the boom of an excavator and powered from the machine can be used instead of manual cutting methods. These machines have the advantage of reducing the risks associated with hot cutting, be it of fire or risks to health by lead fumes from paint or heated chemicals.

Some shears have the ability to both elevate and swivel allowing them to be used for cutting down structures rather than simply reducing to pieces the structures that have been felled by other methods.

When working on areas of steelwork, it is important to establish a sequence which maintains the stability of the structure by leaving braced bays in position until a collapse can be engineered and the shear used to complete cutting at ground level.

The shears should always be worked in an attitude which allows cut sections to fall free without overloading the base machine. The shears can, by careful manipulation, pick and load cut sections directly into skips or vehicles.

43.4.7 Heavy-duty grabs

Heavy-duty multi-reeved rope-operated grabs are normally used for rehandling debris at ground level. They can be used directly to remove brick and masonry structures, but this method is not popular in the UK. The rate of production is often determined by the availability of transport which, particularly in urban locations, can be unreliable.

43.4.8 Diamond drilling and sawing

Diamond drilling and sawing or a combination of both may have to be used to remove parts of structures where other

Figure 43.3 A small 'skidsteer' rubber-tyred loader working on an intermediate floor

methods, which would produce a nuisance through dust, noise or vibration, are not acceptable. It should always be borne in mind, however, that if mains power is not available, diesel prime movers in themselves produce noise and exhaust fumes.

Tipped-core drills can be used to stitch drill, i.e. form overlapping holes around a section to be moved, be it horizontal or vertical, but it is necessary to ensure that there is no movement before the work is complete, as this may stop the core.

Cores may have to be taken out at the start and finish of sawcut lines to accommodate the crescent left by the circular blade.

Great care and experience is required to ensure that premature movement of parts being cut away does not stop the rotating sawblades.

Accurate work with some judicious angle cutting is necessary to ensure that a piece, especially the first one of a pattern, will move clear without jamming once the cutting is complete. For large blocks or slabs inside buildings, complex systems for moving pieces both vertically and horizontally may be needed.

43.4.9 Thermic lancing

Thermic lancing is a process in which a reaction between oxygen and iron is used to generate an intense heat at the point of a lance which is used to cut through, for example, heavy sections of concrete, and is useful where noise and vibration are prohibited. The volume of fumes given off in confined spaces can be a problem, as can the heat and the need for the slag to be drained away without causing a fire hazard. In the open, however, it is used regularly where time is 'of the essence', e.g. in an alteration to busy public areas during nights or at weekends.

43.4.10 Expanding cement

The drilling of holes, filling them with a cement which expands on setting – thus breaking off pieces of concrete – is sometimes advocated as a method of demolition. It is used regularly in quarrying and rock excavation, but has not found favour with demolition contractors in the UK.

43.4.11 Cutting with water jets

The cutting of concrete using high-pressure water jets is offered as a service by a small number of specialist firms, but it has not been seen in general use alongside more traditional methods.

43.4.12 Hydraulic bursting

Mass or lightly reinforced concrete can be broken by drilling a series of holes near an open face and inserting a burster from which plungers are forced out by hydraulic pressure, thus cracking the concrete.

A more recent application of this technique is the removal of a core into which a small hydraulic jack is placed. When the jack is operated, the concrete will crack from the core to the open face.

Bursting techniques are relatively quiet, but tend to be slow. In an effort to speed the process, large areas of concrete may be broken out, but their subsequent removal, particularly if in awkward areas inside the building, may be a problem.

43.5 Demolition blasting

The use of explosives for demolition attracts a great deal of publicity, but it should be considered as another demolition method requiring careful selection based on suitability, location, competent practitioners and careful planning.

For all demolition blasting, good protection and carefully selected clear zones must be achieved. Brick, masonry or mass concrete, when drilled and charged with explosive, will rupture or disintegrate when detonated. The effect will vary and be dependent upon the amount and type of explosive, the weight of material charged in relation to the amount of explosive, the pattern and position of the charges in relation to open faces and the integrity of the material.

Taking these many variables into account, successful blasts can be achieved and used to remove key parts to fell structures or simply fracture banks of material to allow removal by mechanical excavation.

With reinforced concrete, the problem becomes more complex. It is easy enough to disintegrate the concrete but bursting open the steel and thus allowing the debris to disperse is not quite so simple. The firing of test shots may be necessary, and the manual cutting of reinforcement will (subject to advice from engineers) assist towards a successful demolition.

Recent years have seen more use of explosives on the demolition of steel structures. The cutting of steelwork was done traditionally by using substantial charges to blast a section apart — a procedure which tended to have a limited application due to noise and the risk of shrapnel. However, the development of linear or cutting charges, previously well established for military or underwater operations, with the proviso that proprietary products are available, has meant that such charges, coupled with the careful use of delay detonators and propellant charges to shift cut sections, are now regularly used to fell substantial steel structures.

With all types of demolition blasting, no matter what type of structure, it is essential that those personnel involved understand or have access to advice concerning the 'mechanism of collapse'. A structure which has failed to respond to the use of explosives, or has partly collapsed, can be extremely dangerous with all the attendant problems of safe access for completion of the task.

43.6 Preferred methods of demolition

Some types of structures lend themselves to particular methods of demolition, subject, of course, to local restraints. The basis of these methods is that the sequence of work retains the stability of the remaining part of the structure.

Simple examples of preferred demolition methods concern arches which, if demolished by hand, should be in strips parallel to the rings forming the span of the arch. Likewise, reinforced concrete floors should be demolished in strips parallel to the main reinforcement.

When working on a filler joist floor, it should be tackled progressively by releasing individual filler joists, whilst ensuring that the workmen have the protection of a well-secured temporary deck spanning the main floor beams.

The principle of removing as much deadload as possible also should be applied to bridges, bearing in mind that temporary support may be necessary to support the skeleton before removal by crane or dropping on to a prepared area.

The procedures for demolishing chimneys by explosives revolve around leaving the chimney standing on pillars created by taking 'windows' out of the base in a pattern arranged to encourage a particular direction of fall when the pillars are blasted clear. If chimneys are to be demolished by hand, access must be provided to the top. This is normally by a laddered scaffold on the outside. It is normal for material to be dropped inside the chimney and removed through a doorway cut in the side of the base. The debris must be cleared to avoid pressure on the walls of the chimney, and regular checks must be made to ensure that material has not arched inside the flue.

Brick or reinforced concrete chimneys can be treated as described, but in prestressed and post-tensioned structures, expert advice from engineers should be sought to deal with the particular problems.

On sites in confined spaces it will be necessary to dismantle pylons and masts by hand, but in open areas, these can be felled accurately by preweakening the legs against the direction of fall, cutting the rear legs and pulling over the structure with a previously secured steel-wire rope.

For steel flues, the guylines at 45° on plan against the line of fall should be maintained or, if absent, provided and made taut prior to cutting the remote guys and severing the base of the flue in a pattern which allows it to hinge over.

Above ground, storage tanks must be free of fire and explosion risks before structural dismantling takes place. If the tank is too large for the roof to be lifted clear as one piece, then the plates should be removed exposing the structural steel 'spider' which should be removed in a regular pattern until a simple cruciform exists which finally can be cut down.

Some tanks have roofs which float on the contents. Such roofs should be lowered to their supports and removed once the tank shell has been dealt with. Tank shells should be removed in horizontal bands rather than vertical strips as the behaviour of plates cut down vertically is unpredictable.

British Standard 6187:1982[1] includes a guide to typical methods of demolition from which it can be noted that on the more widely found buildings and structures, more than one method may be applied successfully.

There is little experience to date with the demolition of prestressed concrete. There is, however, a recognition that uncontrolled release of stressed wires, cables or bars may provoke the ejection of anchorages, particularly if the stressed member is unbonded or the bonding is faulty.

Pretensioned members are rarely a problem during demolition. This is not the case with post-tensioned members, which are sometimes incrementally tensioned as successive loading is applied. Most post-tensioned members are bonded by grouting, but some unbonded examples must remain and it is difficult to identify the difference other than by physically exposing the stressing wires. Expert engineering advice should be sought when dealing with such members, as uncontrollable release of the tension can lead to catastrophic collapse and/or ejection of debris from anchorages which will fly a considerable distance at great speed.

Incorrect unloading of incrementally tensioned beams with-

Figure 43.4 Example of temporary steelwork supporting a five-storey façade

out partial release of the stressing can lead to hogging and/or bursting with unpredictable results.

The technology for dealing with prestressed concrete exists, but it is important to recognize the problem at the outset. It is also important to realize that, complex as the problems may be, the actual method for destressing and removing prestressed concrete beams, say, may pall into insignificance in comparison with the problems posed by their location (which is likely to be permanently encapsulated in working structures).

43.7 Salvage and recycling

On some premises, particularly those of historic interest, architectural items, such as weathercocks, clocks, door porticos and boundary marks may have to be recovered ahead of the demolition work for restoration and later repositioning.

Internal fittings, staircases and banisters may also be of interest, particularly in listed buildings. The demolition contractor has traditionally been interested in metal, be it nonferrous or ferrous scrap. During demolition, these will need to be graded into type and size for marketing to processors for resmelting or export. The removal and resale of builders' items, doors, window frames, flooring and china is no longer attractive as it is time-consuming and prohibited by tight schedules.

The practice of recycling debris to produce a single-size crushed concrete or an all-in material has developed in recent years. Although it has proved difficult to maintain a specific grading because of the variable raw material, these materials are often acceptable as backfill, subgrade for estate roads or car parks.

Research aimed at producing a specification for crushed concrete and subsequently a wider acceptance of its use is currently being undertaken. In the UK this is at an early stage, but in Europe it is more advanced, boosted in part by a lack of acceptable landfill sites and in some areas by a lack of roadmaking aggregate.

The recycling of timber as a fuel has been developed in The Netherlands where charcoal is obtained with some gas production. Such plants are extremely expensive and do not, as yet, exist in the UK.

43.8 Health and safety

For the demolition industry, the risks to health and safety range from those generally associated with construction through to chemical hazards normally found on manufacturing plants. Logical work planning, an analysis of the structural condition of buildings or structures as they are being dismantled, and the provision of good access and working platforms can greatly reduce the risks from premature collapse, falling materials and falls from heights. Such measures are specified for particular circumstances in UK legislation.

Items needed in ensuring personal protection against physical injury such as safety helmets, goggles, ear muffs and dust and fume respirators are, in many circumstances, required by legislation. However, no matter how rigidly the use of such equipment is enforced, much needs to be done to improve its design. Although it may prove satisfactory in manufacturing conditions, it very quickly becomes uncomfortable when used for the manual tasks and long duration of use associated with the work.

The standard of respiratory protection during burning of lead-painted steelwork or cleaning of lead dust will depend on the natural ventilation around the workman. The use of an auri-nasal mask may be sufficient in the open air, whereas a full facepiece with a high-efficiency filter or even airline breathing apparatus may be needed indoors. The difference in comfort and visibility is apparent, and workmen have a tendency to discard uncomfortable equipment.

The shrouding of work areas, the full respiratory protection and the decontamination procedures required for asbestos removal have to be maintained stringently; most workmen take seriously the publicity given to the risks from this material.

Apart from these two well-documented hazards, the demolition contractor often encounters others associated with the structure or its previous use. Timber used for building may be treated with preservative requiring its removal to a licensed tip, as burning will produce a toxic smoke. Heavy metals such as cadmium or mercury may be found engrained in slabs or even the subsoil of factory premises. Sometimes these contaminants surface from processes used many years prior to the final use of the premises.

43.8.1 Fire and explosion risks

Flame-cutting through structures and within buildings can create a serious fire hazard if precautions are not taken to isolate areas and clear burnable debris from underneath working areas.

When bonfires are lit they should be carefully sited, of modest size and be extinguished before the end of the shift.

At the other end of the scale is the devastation caused by explosions due to the ignition of vapours from small quantities of residues in closed vessels.

Residues in petroleum, oil, gas or chemical tanks, pipes and vessels remain almost indefinitely after the tanks have been drained and abandoned. Even when containers have been degassed, pockets of residues may remain and vaporize.

If it is proposed to dismantle such containers using flame-cutters or tools which produce sparks, such as abrasive disc cutters, the contents must be made inert.

Water can be used to fill a tank to exclude vapour, but it does not render the contents inert; indeed, flammable liquid may float on top of the water. Nitrogen, an inert gas, is recommended in BS 6187:1982[1] for purging tanks. Nitrogen is, however, an asphyxiant and should be handled carefully by specialists. 'Dry ice' – frozen carbon dioxide – can also be used as an inhibitor.

As well as the problems of neutralizing fuel tanks, contractors are increasingly involved with the dismantling of factories which have used chemicals in their working processes. On some occasions, the plants which produce or convert such chemicals are themselves demolished.

Notwithstanding the operator may have purged the process equipment, pipework, tanks and vessels, they may nevertheless have accumulated volatile residues over a considerable period which present health risks from toxicity or irritants.

These problems will be anticipated by experienced contractors who will know where to seek the necessary expert advice.

43.9 Environmental matters

The demolition contractor's activities will attract the attention of the authorities required to look after the interests of the public both physically and environmentally. Local authorities will be concerned particularly about: (1) noise; (2) vibration; and (3) dust.

Restrictions on working hours, e.g. placing limits on early starts or imposing early finishes, or the enforcement of silent periods through the day, can seriously affect site production.

The level of noise acceptable at a site boundary may be settled in advance by agreement with the local authority, but few developers opt for this procedure, preferring to leave the negotiations to the contractor once the job has started.

The careful siting of diesel-driven plant, shielding of static items and use of electrically driven equipment (if mains power is available) can reduce noise considerably. Good maintenance of engines and bodywork, and the avoidance of unnecessary dropping when loading, can also contribute to this.

Vibration may be impossible to avoid; indeed, efforts to eradicate it by changing methods may produce other problems. However, the intermittent use of a demolition ball may be more acceptable to neighbours than the constant noise of pneumatic breakers.

The demolition method quite obviously must take into account the susceptibility of adjacent or neighbouring structures to damage from vibration, and the increasing use of sophisticated office equipment – which is highly susceptible to vibration – means that work must be organized to provide as early as possible a gap between occupied buildings and the building being demolished. Even then, there may be a problem if office equipment is in basements, picking up vibration from falling debris or moving plant and vehicles.

The major problem is dust. Sheeting and dampening down can reduce the inconvenience but not eliminate it completely. Contractors face an increasing number of complaints about blocked filters which purify the air circulation to computer rooms; air-conditioning units are also affected. Much can be done to relieve this problem if the contractor allows for the time and cost of additional sheeting or occasionally the use of a prefilter at the installation.

In general, the contractor can do much to prevent environmental difficulties by good housekeeping around his sites, responding to complaints from the authorities and individuals, and keeping local people informed of the reasons for temporary inconvenience.

43.10 Future developments

The demolition industry, as conservative as any section of the construction industry, watches new techniques and machinery with interest as it continues to move toward greater mechanization, encouraged both by economics and legislation, effectively bringing about methods which allow a reduced workforce.

On a broader front, contractors recognize that their managerial and supervisory skills will be assessed thoroughly prior to obtaining a share of the markets becoming available, e.g. decommissioning offshore oil installations and nuclear power stations. Many governmental policy decisions will affect the procedures.

References

1 British Standards Institution (1982) *Code of practice for demolition*, BS 6187. BSI, Milton Keynes.
2 Health and Safety Executive, *Health and safety in demolition work*, Part 1: 'Preparation and planning'. Guidance note, GS29/1, HSE, London.

44

Offshore Construction

Goodfellow Associates Ltd

Contents

44.1 Introduction

Understanding offshore construction operations requires some familiarity with the type and form of the structures involved. For readers who are not familiar with such structures, section 44.2 briefly describes their general form and function.

Offshore structures are dominated by oil and gas production facilities as exploration for hydrocarbons extended from land to shallow waters and moved to deeper and more hostile environments such as the North Sea.

Other types of offshore structures include cargo and offloading terminals, offshore wind turbines, ocean thermal energy facilities, military and defence-related structures and some novel floating structures for leisure or other purposes.

The construction and installation techniques vary depending on the types of structures involved, but in this chapter some typical examples, mostly related to the oil industry, are introduced and give a good representation of the methods and activities involved.

Construction methods for both steel and concrete structures are described. Reference is made to the general factors affecting the techniques with particular reference to cost, safety and practicality of operations.

Offshore operations involve a well-planned programme of work and project organization with effective control and management. This subject is briefly discussed to demonstrate its importance in a multidisciplinary operation of great complexity.

Finally, reference is made to codes and regulations and operations involving inspection, maintenance and repair of offshore structures.

It is hoped that the reader will gain a general understanding of offshore construction techniques and their impact on various fields of engineering by the examples given.

The subject has been addressed purely as an introduction to this topic and readers who are interested in extending their knowledge further have access to numerous publications including those mentioned in the bibliography at the end of this chapter.

44.2 Offshore structures

The oil industry only began to move offshore in the late 1940s. Offshore operations were first carried out in the US, where a gradual move could be made from the swamps of Louisiana. Exploration results there indicated that the oil area extended offshore into the shallow waters of the Gulf of Mexico. The mobile jack-up drilling unit was originally developed for this region.

44.2.1 Jack-up rigs

The jack-up unit is a barge fitted with movable legs (Figure 44.1). The unit can be towed or self-propelled from site to site with the legs in an elevated position. Once at a drilling location, the legs can be lowered to the sea-bed and the barge can 'jack' itself up the legs so that it comes out of the water, clear of any anticipated wave action, ready for drilling. When the well is finished, the operation is reversed to make the barge ready for moving to its next location. The length of the legs determines the water depth in which the jack-up can be used, but they are commonly designed for use in up to 75 m of water and occasionally as much as 105 m. Reasonably calm weather is required when the units are being jacked up and down.

In order to enable offshore drilling to be carried out in the deeper waters (e.g. in the Gulf of Mexico), semisubmersible and drill-ship drilling units were developed.

44.2.2 Fixed platforms

Once exploration drilling has confirmed the existence of an oil- or gasfield, appraisal drilling is usually required to show if it is large enough to be developed commercially. Field development calls for the drilling of a series of production wells and the installation of equipment to control the production. The usual method is to install a fixed platform and to drill deviated production wells from it. Deviated wells are drilled inclined from the vertical and in a direction away from the platform to reach parts of the reservoir as far away from the platform as possible. Sometimes satellite wells are drilled up to 10 km away and tied back to the platform by pipeline. Both steel and concrete platforms have been used in the North Sea in a variety of designs. The first fixed platforms installed in UK waters were relatively small uncomplicated steel structures for the southern North Sea gasfields in water depths up to 45 m. These have become dwarfed by those subsequently installed in the northern North Sea oil- and gasfields, in water depths of up to 180 m. These have overall heights of around 275 m from the sea-bed and are able to withstand storm waves 30 m high and winds of 240 km/h.

A steel platform consists of a framework called a 'jacket' on which a deck is mounted (Figure 44.2). The jacket is fabricated onshore and towed out to sea on its side, either afloat or on a large barge. On reaching its location, it is carefully up-ended and secured by piles driven into the sea-bed. Once this has been completed, the deck is installed and modules containing the drilling, production and accommodation facilities are added.

Concrete platforms vary considerably in design and consequently in method of construction (Figure 44.3). Normally, a buoyant base is built in a dry dock and floated into progressively deeper water as the structure is built up from it. This requires sheltered, deep water close to shore. The weight of a concrete platform is several hundred thousand tonnes greater than a steel platform. A concrete platform is frequently designed with chambers for oil storage. When completed, with a superstructure containing drilling, production and accommodation facilities, it is towed out by a number of large tugs to its location. It is then ballasted down until it rests on the sea-bed where it remains secure under its own weight. Concrete platforms are consequently called gravity platforms. All the fixed platforms are, therefore, bottom-supported structures.

Another approach developed for deeper waters is the guyed tower (Figure 44.4). The platform deck is supported by a lightweight steel compliant tower, held upright by guy lines radiating outwards. This type of platform has been used for a field in the Gulf of Mexico.

44.2.3 Floating platforms

Because of the very large cost of fixed platforms and the possibility of finding oil in waters which are so deep that fixed platforms would neither be technically feasible nor economical, considerable attention has been given to developing oilfields by other methods.

One approach is to use a floating production platform. However, it is necessary to restrict lateral and vertical movements to a minimum, so as to avoid unacceptable loads on the high-pressure vertical pipes known as 'risers' which provide the link between the platform and the wells on the sea-bed.

The semisubmersible rig is a floating platform with the deck supported by vertical columns on submerged pontoons which provide its buoyancy (Figures 44.5 and 44.6). By varying the quantity of ballast water in the pontoons, the rig can be raised or lowered in the water. The lower the pontoons lie beneath the water the less they are influenced by wave action. This reduces vertical movement and allows drilling or production to continue

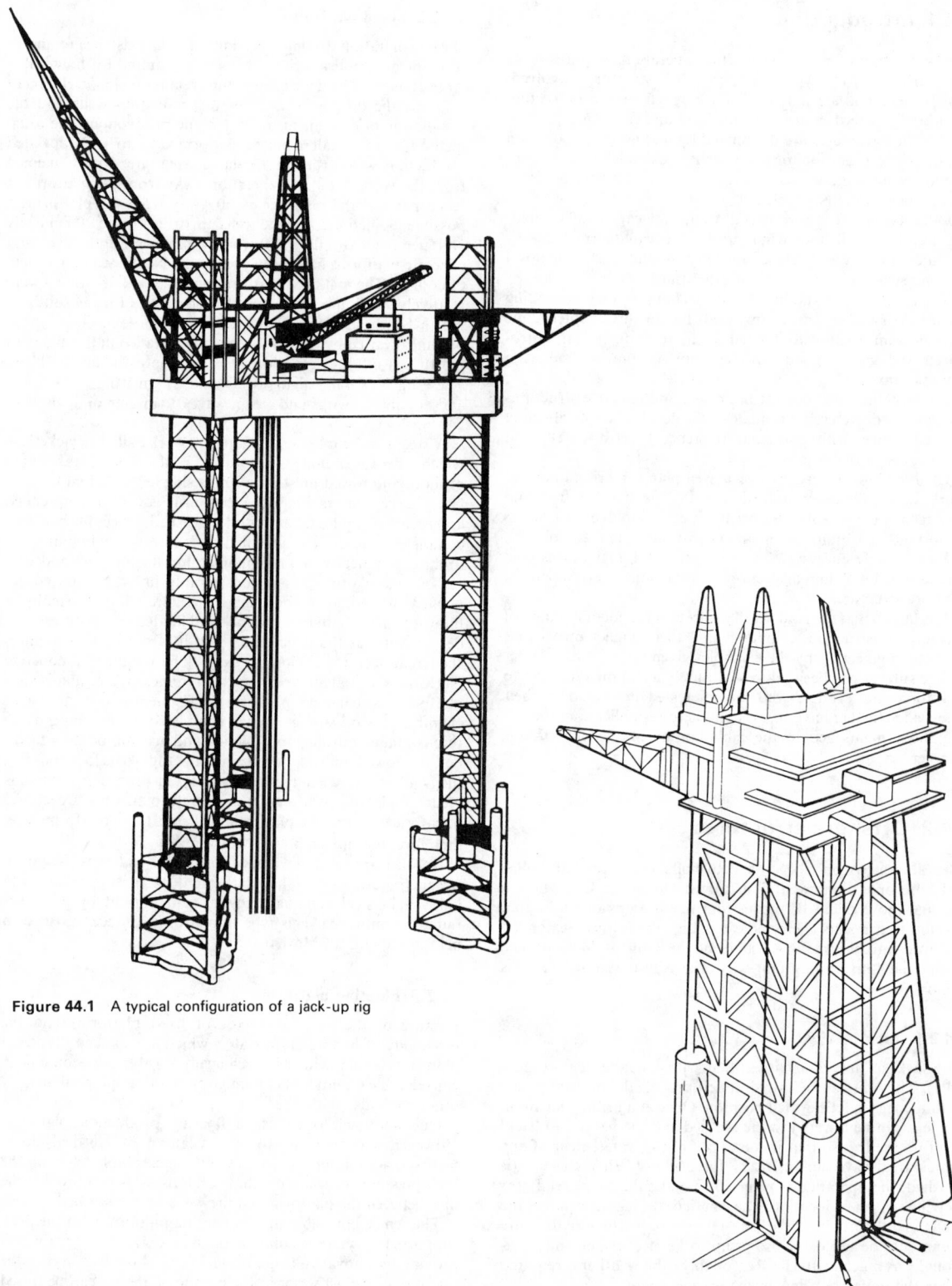

Figure 44.1 A typical configuration of a jack-up rig

Figure 44.2 A typical configuration of a conventional fixed steel jacket (platform)

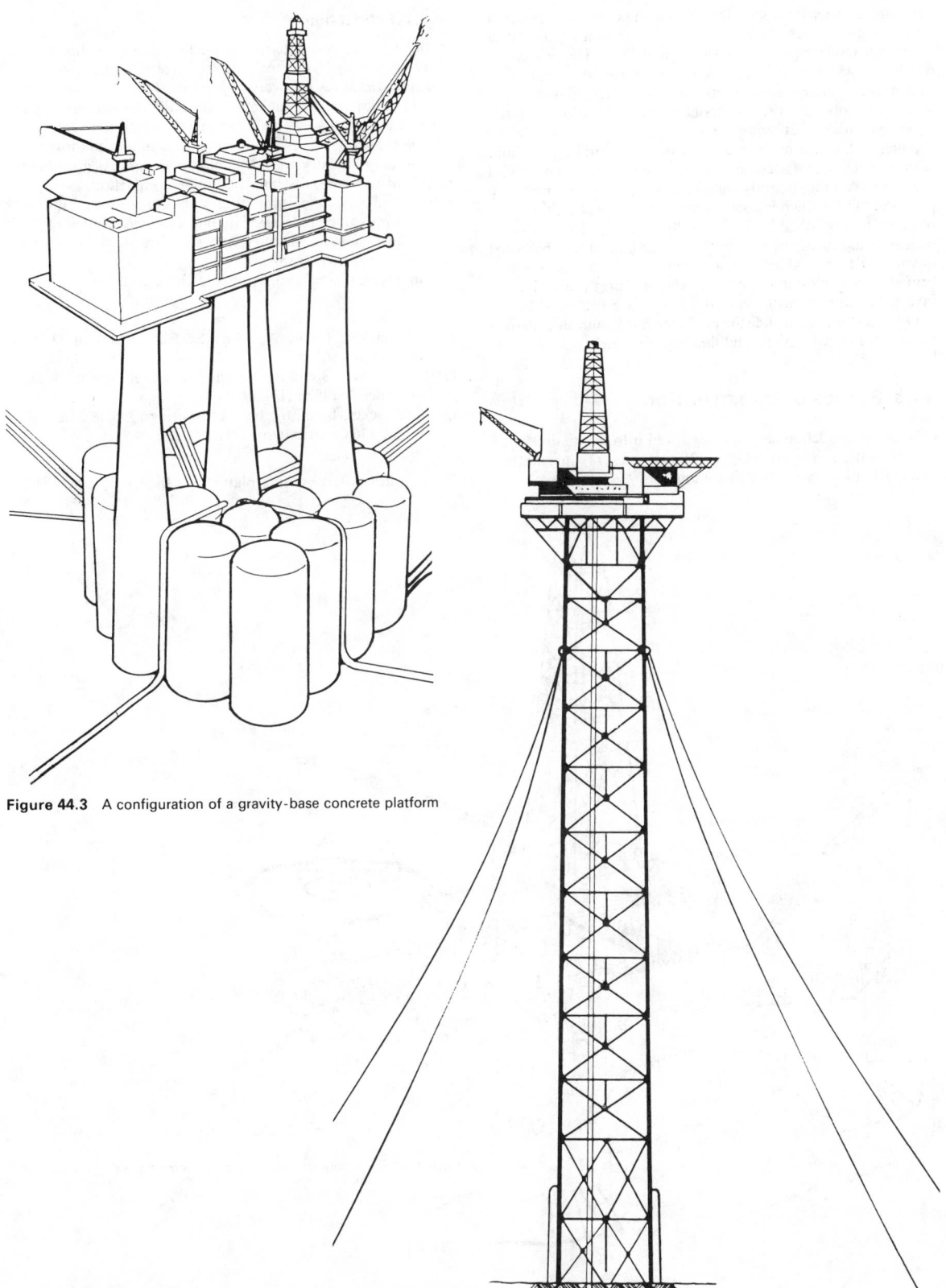

Figure 44.3 A configuration of a gravity-base concrete platform

Figure 44.4 A typical configuration of a guyed tower

in rough seas. A semisubmersible rig is normally held in position by up to twelve very large anchors. The design of the latest semisubmersible rigs enables them to drill in UK waters at depths of 450 m and over, all the year round, despite the exceptionally high waves experienced in winter. Semisubmersible platforms can also be designed as production facilities equipped with process equipment.

Anchored semisubmersible units used for drilling or built with production and accommodation facilities are in use around the world. Another floating technique is the use of a tension leg platform (TLP) which is a semisubmersible type of unit, held in place by tensioned cables anchored to the sea-bed immediately beneath each corner of the platform. The platform is ballasted down while the cables are attached and then deballasted, bringing the cables under tension. The platform moves like an inverted pendulum, with very little heave. See Figure 44.8.

Other techniques include the use of specially built ship-shaped vessels, converted tankers and floating concrete platforms.

44.3 Stages of construction

Offshore construction can be categorized into five main stages: (1) fabrication; (2) launching; (3) tow-out; (4) installation; and (5) hook-up and commissioning.

44.3.1 Fabrication

In this section, construction of steel structures is discussed in order to highlight the main tasks involved. Construction of concrete structures is covered in section 44.5.

Fabrication of steel jackets is generally carried out in land-based fabrication yards which have access to waterways, or the open sea. Such facilities are in some ways similar to those in the shipbuilding industry with dry docks and slipways allowing the vessels to be eventually launched upon completion.

Size and weight of structures vary considerably and as a result, some can be fabricated in only a limited number of yards which have suitable facilities with sufficient draught along the waterways for their transport.

Some typical sizes and weights of the jackets are:

(1) Steel jacket, Thistle A, North Sea: jacket weight 31 396 t, water depth 161 m.
(2) Steel jacket, Brent A, North Sea: jacket weight 14 225 t, water depth 140 m (Figure 44.7).
(3) Steel jacket, Indefatigable CD, southern North Sea: jacket weight 536 t, water depth 29 m.

The world's tallest existing platform is the Cognac steel jacket

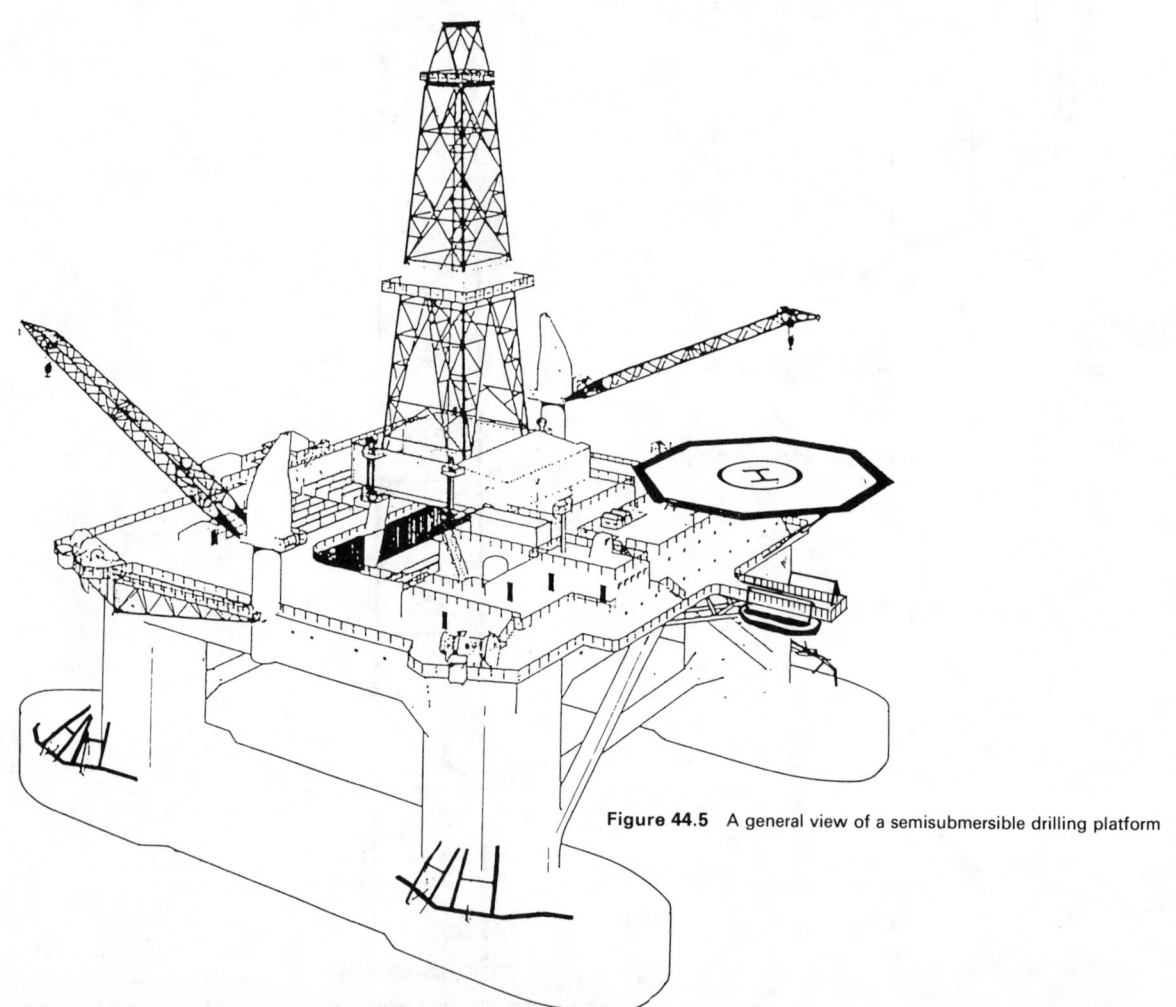

Figure 44.5 A general view of a semisubmersible drilling platform

platform with a height of 385 m. However, the Bullwinkle platform, which is of a similar design to the Cognac, will be 492 m tall when installed in 1988. This platform will then be 49 m taller than the world's tallest building. This record will no doubt be broken again in future years.

Limited dimensions and handling capacities of fabrication yards and dry docks may result in the need to fabricate the structures in more than one piece. In addition, parts of the platform may be fabricated separately in other yards. The parts will then be brought together and mated under separate operations. Deck structures of jackets and modules are often fabricated and assembled separately. These modules, which could weigh from under 50 t to a few thousand tonnes are transported and are lifted and installed on the deck of the platforms, using crane barges, when the deck is installed. As an example, the total topside weight of the *BP Magnus* platform, which consists of a multistorey deck 75 m square and 32 m high, is in the order of 31 000 t.

To minimize cost, the maximum possible work on fabrication, assembly, testing, inspection and installation of various components is carried out inland. Costs of offshore construction operations are significantly higher than the land-based work and are therefore limited to essential tasks which cannot be carried out in any other way.

For fabrication of steel structures, welding tubulars ranging from 300 mm to 2 m diameter or more, and with varying thickness of up to 80 mm is involved; an example is the *BP Magnus* platform in which two of its four legs each has a diameter of 10.5 m. Welding such large structures requires efficient automatic welding techniques with quality control and stress-relieving in many cases.

Fabrication of nodes consisting of several tubular members of different sizes is one of the most complex parts of the welding operation. Techniques of casting nodes have been developed which enhance their load-carrying capacities by eliminating high welding stresses and streamlining and strengthening the joint structure. The design of tubular joints is discussed in a publication by the Underwater Engineering Group of the Construction Industry Research and Information Association (see Bibliography).

Covered fabrication facilities are available to allow work to be independent of weather conditions.

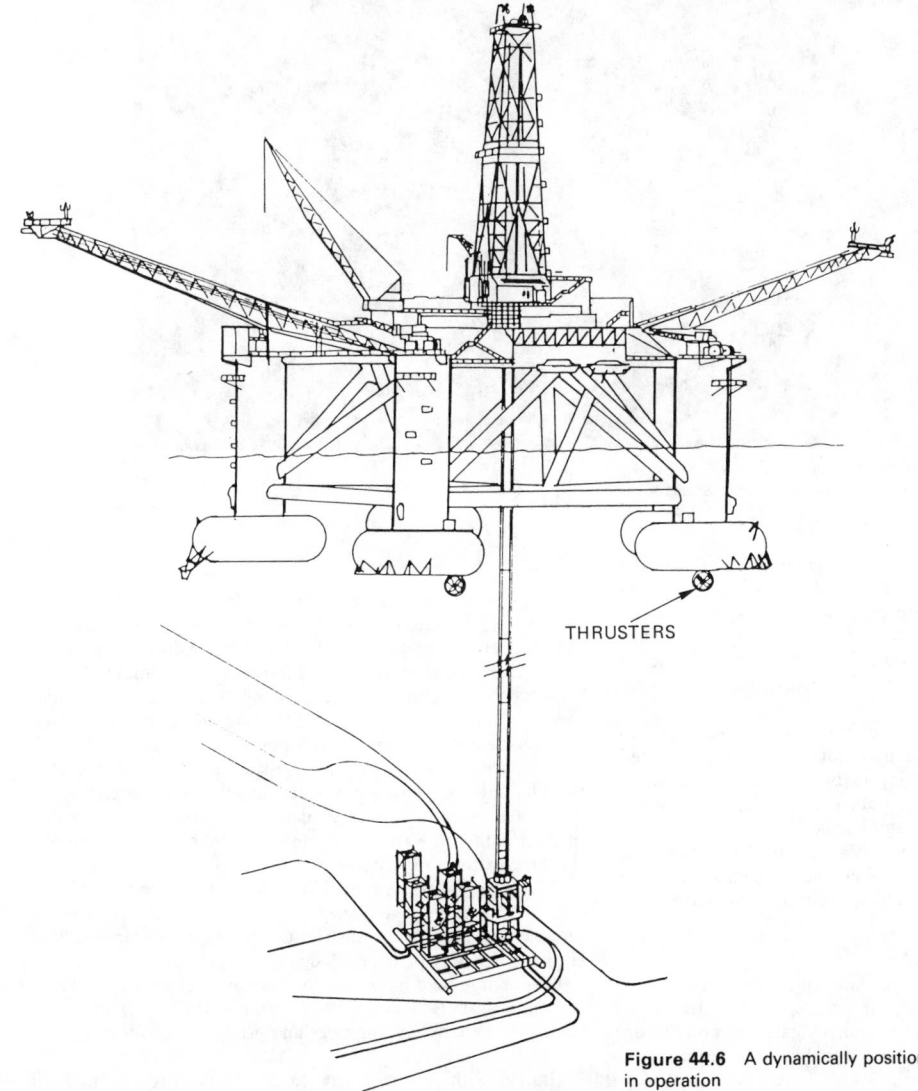

THRUSTERS

Figure 44.6 A dynamically positioned semisubmersible drilling rig in operation

Figure 44.7 Steel production platform Brent A operated by Shell and Esso in 140 m of water in the North Sea. (*Courtesy*: Shell)

Steel structures are fabricated in sections which can be accommodated and handled in the yard. Close tolerances are required to enable mating with other sections. Inspection and quality control become integral parts of the fabrication operations, as these structures are required to withstand high loading conditions with theoretical fatigue life equivalent to 10 times their service life.

Failures of welds resulting from bad workmanship, unpredicted loading conditions and poor tolerances have provided lessons to the industry, resulting in bringing about improvements in welding techniques, more extensive nondestructive testing and attention to detailing of structures.

Large-size structures are generally fabricated and assembled on pre-installed trestles and rails to enable the next stage of the operation, which is launching and tow-out, to take place.

44.3.2 Launching

When fabrication is complete on land, the structure is transferred to waterways for towing and transportation to its offshore destination. The method of launching depends on the size and weight of the structure and the facilities used for its construction.

44.3.2.1 Load-out from quays

Lighter structures, or those which, because of the draught limitations of the waterways, are fabricated on quays, are moved on to flat-top barges (moored against the load-out quays) for transporting to sea. Limitations of cranes in fabrication yards to handle weights ranging from a few hundred to several thousand tonnes require the completed structures to be transported slowly on rails or bogies and loaded on to the barges. Alternatively, they can be supported on pads, each of which floats on a cushion of water or oil, using the principle of hydraulic or air flotation. Reduction in friction, as the result of pads floating on water or oil cushions, enables the structure to be winched on to the barge with relatively small pulling loads.

Modular trailers with over 700 wheels and capacities of up to 12 000 t or more have been used for this purpose. Bogies also enable the load to be distributed to levels within the load-bearing capacity of the quayside which is often below 5.5 t/m².

The barge-loading operation requires powerful ballasting facilities on barges so that they maintain their level against the quayside under changing tides and gradual transfer of the load on to their flat decks.

Barges with sufficient deck capacities need to be fitted with sea-fastenings to secure the structure during transportation. Extensive deck stiffening is sometimes required to enable the barge to hold its load safely.

44.3.2.2 Load-out from dry docks

Load-out from dry docks requires flooding of the docks allowing the structure to float. Limitation in the draught in the dry docks often requires the operation to take place within the limited period of high tide. The floating structure is then towed out of the dock by tugs for transport to sea.

44.3.2.3 Launching from slipways

The completed structure rests on a number of rails which extend along the slipway into water, similar to the method used in the shipbuilding industry. The structure is freed from its trestles for launching, and is gradually winched and allowed to move into the water until it floats. This technique is particularly suitable for structures which are too heavy to be transported on barges, or have excessive draught and need to be fabricated and loaded-out in yards closer to the open sea. The *BP Magnus*, a self-floating jacket and piles weighing 42 000 t, was launched in this way from the Highland Fabricators' yard in Scotland in 1982.

44.3.3 Transportation at sea: marine operations

Transportation of structures, whether floating or transported on flat-top barges, is carried out by a number of tugs. The tugs position themselves in a 'star' formation, providing the power and controlling the movement of the structure along its pre-determined path.

Suitable weather windows are required to ensure the safety of the structure during transportation. The speed of the tow is limited to a few knots and, depending on the distance, may take anything from several hours to a few days. At the destination and prior to its installation, intensive survey and inspection of the sea-bed, subsea template and other structures is made.

Back-up facilities are mobilized, and trial runs are undertaken to ensure that the final crucial stages of the operation pass without difficulties.

At this stage, support vessels carrying power, personnel, equipment, divers and inspectors are at the site to carry out the highly controlled and co-ordinated operation of setting the structure in its final position. Sonar systems and satellites are used to monitor the position of the structure and help to hold it in a position within the small allowed tolerances, which vary from a few metres for the first stage of station-keeping to a few millimetres for the final setting stage into the support structures or templates.

44.3.4 Installation

44.3.4.1 Installation of the main structure

The method of installation varies and depends on the type of structure. For floating, bottom-supported, steel jackets a controlled up-ending operation is carried out followed by further ballasting; the structure is then lowered on to the sea-bed.

Maintaining structural safety and stability during the up-ending operation is crucial. This stage is therefore a well-investigated and tested operation during which the movement of the structure is also helped by lines from tugs and crane vessels.

Structures transported on barges may be lifted either by heavy-lift crane barges and lowered on to the sea-bed, or submersible barges may be used if they remain afloat. Submersible barges can, by a process of ballasting, have their draught increased until the structures they carry float freely and are towed clear.

An alternative method is to launch the structure directly from a barge equipped with a launching frame at its stern.

Crane barges are used to drive piles around the legs of the jackets; piles are guided by pre-installed sleeves around the legs. Pile-driving techniques now allow the use of underwater pile hammers with high driving capacities of 200 tonf and beyond.

Piles driven for the *BP Magnus* jacket are a typical example of the support system, being 100 m long with 2.1 m diameter and 63 mm plate thickness. The 36 piles have been driven by two of Menck's (MHU 1700) underwater hammers, delivering a striking energy of 170 tonf. Each pile has been designed to take loads of up to 6000 t.

The techniques explained in this section are examples of installing fixed jackets. The installation of other types of structures such as templates, articulated columns, floating structures such as semisubmersibles, and TLPs are all different, with differing levels of complexity.

In the case of TLPs (Figure 44.8), for example, installing and tying the tethers to their templates on the sea-bed is a complex and lengthy operation. It can be carried out from the platform itself or, alternatively, by pre-installing the tethers using crane barges and finally mating and tying them to the main structure as a second operation, has been shown to be more economical.

44.3.4.2 Installation of secondary components (topside)

For installing other components or parts of a production platform such as the deck structure and some subsea components, heavy-lift cranes are used. Fixed jackets could have deck structures weighing several thousand tonnes which are transported separately on flat-top barges. The deck is lifted by one or two cranes and is installed on top of the support structure. Various modules, part of the hydrocarbon production facilities on the deck, and each weighing from a few hundred to a few thousand tonnes, are also transported separately and, using a crane barge, are installed on the deck.

Early installation of all modules on the deck is not often feasible because of weight limitations of barges and stability problems during transportation at sea.

In the *BP Magnus* platform, the total topside payload of the structure was 31 000 t divided into 19 modules each weighing up to 2200 t, some 40 to 50 m long.

44.3.4.3 Installation of secondary components (subsea)

For installation of subsea modules such as templates or manifolds, often accurate positioning and mating with existing subsea structures are required. In such conditions, a guidance system is required in addition to cranes to control their lowering and positioning. Tensioned guide wires combined with guide posts are examples of the methods used for the controlled lowering and positioning of the modules subsea. The guide wires are tensioned by winches from the installation vessel and the lines are tied subsea to the guide posts. The component which is being installed is equipped with funnel-shaped guidance sleeves which are engaged on to the guide wires and which enable the unit to be lowered to its position guided by the tensioned lines.

All operations are closely monitored by divers or remotely operated vehicles (ROVs) carrying underwater television cameras.

44.3.5 Hook-up and commissioning

The term 'hook-up' refers to the operations which link the various components and parts of the offshore facilities when they are all installed.

Tying the subsea pipelines to the platform risers, installing and tying umbilical lines, cabling and pipework to complete the

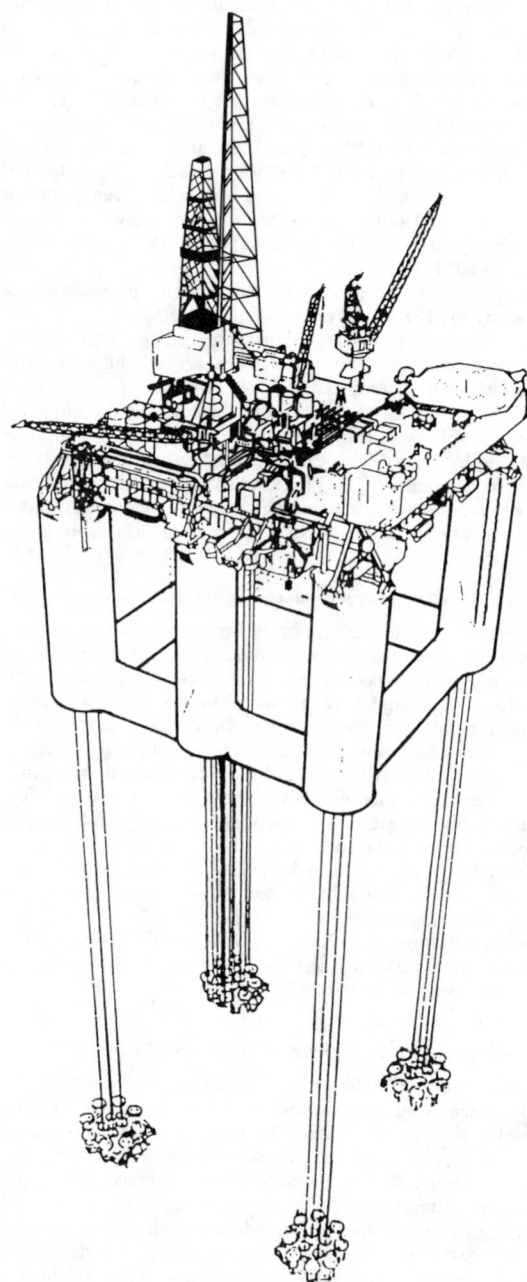

Figure 44.8 A general configuration of a tension leg platform

linking of the topside modules which have been transported and installed on the deck separately, are examples of the hook-up operation. This work requires a well-co-ordinated multidisciplinary taskforce with back-up facilities such as cranes, service vessels, divers and ROVs.

Several thousand man-hours are required to complete this stage of the work and commission the facilities. With offshore man hour rates ranging from 5 to 10 times that of land-based operations, hook-up and commissioning are costly operations which need to be minimized as much as possible.

The manpower required for such an operation offshore could run to over 1000 men for major projects. Temporary accommodation and transportation offshore are required for such a large number of people who may stay in accommodation vessels moored close to the platform. Part of the operation is sensitive to sea state and may result in significant delay (downtime) in completion. Selection of suitable vessels which can operate safely close to the platform at more severe sea states, although more costly, is more economical in the long run for severe environments such as those in the North Sea.

44.4 General factors affecting construction techniques

Selection of suitable techniques for fabrication and installation of offshore structures are influenced by many factors which include:

(1) Material (steel, concrete or hybrid structure, and other new materials).
(2) Economic factors such as the need to bring the field to partial production early and improve overall cashflow.
(3) Cost.
(4) Environmental conditions: sea states, wind, current.
(5) Water depth.
(6) Safety.
(7) Constraints imposed by regulatory authorities, such as vessel operation constraints, pollution control, navigation restrictions, etc.
(8) Existence of suitable fabrication yards/dry docks with sufficient space and load capacity and available draught in the waterways for transport of the structures.
(9) Socio-political factors which may influence selection of yards and even the type and form of structures.

In addition, with the development of novel techniques and new equipment and tools, traditional methods have been replaced by new methods and are likely to continue changing.

The introduction of high-pressure flexible lines, subsea trenching crawlers for trenching and laying pipelines, dynamically positioned vessels capable of maintaining position at more severe sea states, ROVs, crane barges with heavy lifting capacities of 8000 t or more all influence not only fabrication and installation techniques but have played major roles in changes to the form and design of the structures.

The handling capacity of crane barges enable bigger modules to be built onshore and provide a reduction in the cost of offshore hook-up operations. The use of underwater high-capacity hammers has allowed the sizes of piles to be increased, resulting in reductions in numbers and, therefore, savings in material costs and offshore operation costs.

Some of the developments and trends have been described in publications listed in the bibliography.

44.5 Concrete structures

44.5.1 Types

Concrete structures, by their nature are, in general, bulkier and heavier than those constructed in steel and involve different construction techniques.

In order to understand and appreciate the differences, it is helpful to refer to a number of major concrete structures and their functions, as listed below.

(1) Concrete gravity platforms (resting on the sea-bed with no piling involved).

(2) Floating concrete structures (semisubmersible, TLPs or ship-shape structures).
(3) Arctic caissons.
(4) Concrete pontoons, supporting various types of structures.
(5) Articulated buoyant columns.

In this section, construction of concrete gravity platforms is discussed to demonstrate the tasks involved.

44.5.2 Major requirements

For all such structures the prime considerations are:

(1) Suitable facilities and locations for their construction.
(2) Offshore construction (if applicable), mooring and support facilities.
(3) Marine operations involving transport, mooring, mating of components and installation.
(4) Foundations and scour prevention.

The weight and size of concrete gravity structures increased substantially as their application to deeper waters of 100 to 300 m was introduced, and resulted in the construction of structures weighing in excess of 800 000 t with topside loads of over 30 000 t.

The Brent platform in the North Sea consists of a cellular base of 90 m square and 54 m high. The four towers rise some 107 m above the base to support a deck with a total area of around 39 000 m² and weighing 31 000 t (Figure 44.9).

Figure 44.9 One of the concrete platforms under construction by McAlpine Sea Tank at Ardoyne Point, for use in the Brent field

The platform displaces 436 300 t of water. Construction to deck level required over 257 000 t of concrete and 15 000 t of reinforcing steel.

Construction of such large platforms in existing dry docks has been impractical because of the limitation in the size and weight capacity of the docks and the draught available for tow-out. For these reasons the practice has been to construct dry basins with access to deeper waters and to construct part of the base to a height at which the available water depth allows flotation, tow-out and transportation.

Construction of such a basin at Ardoyne required the removal of some 900 000 m³ of material.

Limitations in water depth of 10 to 15 m in many coastal areas and waterways leaves only a few locations suitable in the UK for such operations. Norway with its sheltered deep fjords, however, offers good surroundings for construction of concrete structures. Stability requirements during transport dictate the depth to which the floating structure should be submerged. Such requirements acknowledge the need for a deep sheltered site where the partly completed base can be moved, and be moored, and where the remainder of the construction work offshore can be completed. It should be remembered that draughts of 100 to 150 m are often required for major platforms.

44.5.3 Concrete construction

High-grade sulphate-resisting cement concrete (grade 50 or more) is used for offshore construction work. Durability in hostile sea environments requires high grades of cement, aggregate and good workmanship. The large quantities involved pose supply and storage problems. Concrete production plants with high output capacity in excess of 100 m³/h are often required. This can be achieved by using more than one plant to ensure continuity of supply during breakdowns.

Concrete is pumped, or moved by trucks, within the site. For offshore construction, several pumps are used, each with capacities in excess of 300 m³/h. The concrete production plants can be located on pontoons, moored against the platform. Long pumping distance often requires the addition of plasticizers and retarders to the concrete.

Slipforming is the common method of placing concrete, with rates in excess of 50–100 mm/h for the caissons and higher rates of 100–200 mm/h for the main towers. Slipforming of inclined surfaces has also been developed and has proved to be practical.

Thicknesses of concrete slabs and walls vary from 500 mm to a few metres. The ducts are introduced within the thick members to help in the dissipation of heat to cope with the high heat of hydration.

Both reinforcing bars and prestressing tendons similar to those used in land-based structures are used.

For the Brent offshore platform, 1000 jacks of 3 t capacity were used and required 1100 m³ of concrete to achieve a 1 m lift. The base slab required 20 000 m³ of grade 50 cement concrete. Several tower cranes with the capacity of 10 to 15 t were required for concreting and handling reinforcement and formwork.

The effects of creep and temperature changes require thorough investigation for both construction and service life when, during oil production, parts of the cellular base space are used for storage of crude oil at temperatures of 30 to 40°C above the surrounding sea-water temperature.

44.5.3.1 Deck installation

Following completion of the concrete platform it is ballasted-down to enable the deck structure to be lifted and positioned on the towers by heavy-lift crane barges. Other modules for the deck are brought into their positions and installed. It is also sometimes possible to ballast-down the platform until only a few metres of the towers are above water. The deck may then be transported on pontoons, each with a clearance to enable the deck to be moved over the towers. By gradually deballasting the platform, the deck can be aligned with the towers.

Winches installed on the towers are used to perform the final pulling stages of the deck over the towers, and the final few millimetres of the positioning is completed with the help of jacks.

This operation requires delicate control of the platform and the deck, continuous monitoring of the movements and a powerful ballasting system to cope with the ballasting rates required.

44.5.3.2 Towing to the final position

The significant draught of the structure is often in excess of 100 metres; it is therefore necessary to select and survey a towing route in order to ensure that sufficient water depth exists along the total distance. The effect of current, waves and wind are studied to ensure tugs have sufficient reserve power to cope with towing under specified adverse weather conditions. Towing speed can be as low as 0.5 kn, increasing to 2 to 2.5 kn in safer passages.

Navigation, towing and monitoring of the operation may require a crew of from 30 to 50 men.

When the structure reaches its destination, tugs in star formation hold it in position while, by gradual ballasting, the structure is lowered on to the sea-bed.

44.5.3.3 Foundation considerations

In addition to the common requirements for load-bearing, long- and short-term settlement, stability and keying against shear forces, it is important to note that, owing to the action of waves, loads on the foundation are cyclic and affect the drainage of the soil underneath both in the short and long term. The direct effects of waves on soil, particularly in shallow waters of up to 50 m, could also be significant. Variations in pore pressure depend, among other things, on the densities of the oil to be stored.

Problems of scour around the perimeter of the base require careful consideration. Various methods, varying from dumping stone to the installation of manmade mattresses filled with grout, sand or stone, have been used with varying degrees of success.

44.6 Construction in the arctic

Oil in arctic zones was first discovered in the MacKenzie delta and Arctic islands in North America. Further studies in the US in the late 1970s showed that there were substantial potential resources offshore in the Arctic zones, particularly in the Bering, Beaufort and Chukchi Seas.

The first field was developed in water depths of 1 to 20 m. Future discoveries in the lease sale areas involved operating in depths of 20 to 50 m.

Structures suitable for such relatively shallow depths but extremely hostile environments are therefore different from the conventional offshore structures. The environmental conditions, particularly the presence of ice packs, play dominant roles and are worth mentioning.

44.6.1 Environmental conditions

The expected maximum wind and wave conditions in arctic areas of immediate interest are less severe than those of the North Sea. The 100-year expected maximum wave height is in the range of 12 to 15 m for water depths of 15 to 30 m. Storm surges in excess of 6 m are, however, significant for the design of arctic structures.

Ice criteria dominate the design of the structures. The main features of the arctic ice are:

(1) *First-year ice.* The thickness of ice formed within 1 year could be up to 2 m, depending on the area.
(2) *Multi-year ice.* This is the ice which has lasted more than one melt season and has resulted in the build-up of an ice sheet into a thickness of 6 m or over, with a diameter of 3 to 5 km being typical.

Collision of two large sheets of ice may result in the formation of pressure ridges several metres above the water level as ice mountains and their coves could extend several metres into the soft sea-bed.

Multi-year ice-floes could travel at velocities of up to 2 m/s and their impact with any structure would result in an effective total load of several thousand tonnes, depending on the form of ice and details of the structure.

Ambient temperature reaches a low of $-50°C$.

So far as ground conditions are concerned, the new features particular to arctic zones are permafrost and gas hydrates. The permafrost table could vary from a few feet below the mud line to several metres. Gas hydrates are ice-like pockets of natural gas which fit into the structural voids in the lattice of water molecules.

Freezing and thawing of soil columns are other features which affect ground conditions to support gravity base structures.

44.6.2 Types of arctic structures

The most common types of structures considered as arctic platforms for drilling or production of hydrocarbons are: (1) artificial islands; (2) hybrid islands; (3) cone structures; (4) tower structures; and (5) floating structures.

44.6.2.1 Artificial islands

Since early 1970s, a number of artificial islands have been constructed in water depths of 1 to 20 m. Most of these islands are in the MacKenzie delta, in northern Canada. The construction method has varied from over-the-ice construction to dredging the loose soil and filling with dredged sand and armour stone. Armouring is particularly the cause of high cost because of lack of quarry stone in the nearby areas.

Artificial islands are attractive economically for shallow waters below 10 to 20 m depth. For depth ranges of above 10 m, other types could become more economical.

Ice pads are another type of structure which consist of layers of ice formed on top of one another by pumping water from the lower depth of water to the surface of the ice-pack. The thickness of each layer is in the order of 6 m. The ice-pack covers the entire water depth forming a platform for the operation.

44.6.2.2 Hybrid islands

They include caisson-retained islands in which sand-filled barges or ship hulls form the central core of the island and rest on beams which extend 4 to 5 m below the water level. The benefit of this type of island is the reduction in volume of fill and short construction time.

44.6.2.3 Cone structures

The most common types of island developed are cone-shaped

structures. Cone-shaped gravity platforms vary in form and shape and are constructed of steel, concrete or a hybrid of steel and concrete structures. The outer walls are inclined to break ice on impact in the most effective way. The main structure of the cone consists generally of cellular form.

44.6.2.4 Tower structures

Other types of platforms are braced-steel structures and concrete gravity platforms with cylindrical towers. These structures are suitable for areas with light ice conditions.

44.6.2.5 Floating structures

Floating structures vary in form and include ship-shaped structures, floating concrete caissons and conical floating platforms. Most structures in this category are suitable for deeper waters and arctic areas where ice surveillance and management is practical and economical. These structures are basically moored to the sea-bed with several mooring lines.

44.6.3 Construction

With temperatures down to $-50°C$, the presence of ice floes and limited open water restrict the working season to 1–3 months. Construction operations are costly and, for both economic and practical reasons, most structures are designed to be constructed in easier conditions and are towed to location for installation.

Construction of artificial islands using arctic dredgers has proved possible. Use of support vessels, ice-breakers for towing and management of ice-packs, and tugs enable the platforms to be fabricated in several sections and be brought together for final mating and setting on location.

Concrete cone structures with total displacement in excess of 500 000 t are fabricated in segments, using conventional techniques of concreting. Similar to concrete platforms used for other offshore locations, limitations in draught for towing the structure dictate the location for fabrication and the construction techniques.

Ice loadings on arctic cone structures are not known precisely but could vary in intensity from 1000 to 2000 kN/m^2 global loading and to 12 000 kN/m^2 local pressure. These require concrete structures to withstand high punching shears as well as high bending and shear forces. The structures are therefore heavily reinforced with high-strength temperature-compatible steel as well as prestressing tendons.

Concrete has been shown to gain strength with time in low-temperature conditions. This includes compressive strength, tensile strength, bond strength, impact resistance and modulus of elasticity. Application of concrete for arctic structures is therefore a viable solution.

Low temperature and presence of ice have been used as an aid for construction purposes, e.g. ice roads several kilometres long and 10 to 20 m deep. These roads stretch into the sea and form access routes to artificial islands. Artificial ice-platforms for drilling in high arctic areas are another example.

Offshore construction in hostile arctic areas has therefore led to the development of novel ideas and use of special equipment suitable for such conditions. Arctic engineering has become a specialized field involving the development of material, equipment and better understanding of environmental loads such as ice loads and soil conditions.

44.7 Fabrication/construction facilities

Major facilities suitable for fabrication and construction of offshore structures are: (1) land-based fabrication yards; (2) dry docks; (3) slipways; and (4) offshore floating facilities.

44.7.1 Fabrication yards

Land-based yards are close to waterways with loadout quays for transporting the structures to sea. Main features of such yards are:

(1) Covered areas for weather-independent work such as steel-rolling, fabrication, assembly and painting.
(2) Cranes with sufficient reach and capacity.
(3) Quays with surface load capacity of 50 to 150 kN/m^2 to cope with heavy loads of several thousand tonnes.
(4) Access to deep water and open sea for towing out structures.

Such facilities are often required to be approved by certifying authorities to ensure that they provide conditions needed to meet the necessary standards of workmanship and quality control. Fabrication yards are used primarily for fabrication of steel jackets, deck structure of the platforms and a variety of modules for installation on the decks of offshore structures.

44.7.2 Dry docks

Dry docks for fabrication of offshore structures are, in general, larger structures than those used for shipbuilding. These facilities are equipped with cranes and other support facilities required for fabrication or construction of large and heavy structures, which are outside the capacities of the fabrication yards, and can be floated out for transport to offshore locations. The dimensions of Kishorn dry dock, Scotland, are $180 \times 170 \times 11.5$ m deep. This facility, with its deep-water mooring site and various fabrication and paint shops, is a typical example of the dry dock suitable for fabrication of large steel jackets.

44.7.3 Slipways

These facilities are similar to shipbuilding slipways and allow the fabrication in land-based environments. When the construction is completed, the structure is loaded-out on rails on to the water in a similar manner to launching a ship. Purpose-built slipways, with direct access to open seas, suit large-size structures which are outside the handling range of available fabrication yards and dry docks.

44.7.4 Offshore fabrication platforms

Large floating pontoons made of steel or concrete have been developed and moored offshore as fabrication yards. The use of such platforms is justified when other conventional facilities are not available, or there are specific restrictions such as depth of water for transport to the sea.

Those countries involved in the oil industry, such as the UK, France, Norway, Holland, the US, have developed and, at times, maintained such facilities with government assistance.

44.7.5 Back-up facilities

A vital key to success is the use of suitable equipment for efficient and cost-effective execution of work. Speed of operation, completion of work on time and achievement of high standards of workmanship demand that the most up-to-date equipment is available for these purposes. Well-equipped covered areas, automatic welding equipment, nondestructive testing facilities, all backed-up by computer services, are examples of what are needed.

A host of equipment and services is needed offshore to carry out the various stages of fabrication, mating, transport and installation of structures. The following is a list of some of the major facilities required.

(1) Heavy-lift crane barges with capacities ranging from a few hundred to over 12 000 t. Some semisubmersible crane vessels available at present have two cranes with total lift capacities of up to 10 to 12 000 t.
(2) Support vessels for specialized work, such as diving support vessels, inspection vessels, and vessels for carrying power and control equipment.
(3) Accommodation vessels or semisubmersibles as offshore hotels for engineers, inspectors and fitters.
(4) Remotely operated vehicles for subsea operations.
(5) Tugs for towing or station-keeping floating structures.
(6) Anchor-handling vessels.

Involvement of such vessels and associated equipment is a costly stage of the installation operation because of the high daily rates involved in their deployment.

44.8 Analysis

Analysis is by no means restricted to the behaviour of the completed structures in their installed condition. The structures either as part or complete units are subjected to loads different from their normal service condition during fabrication, launching, tow-out and installation.

Static and dynamic loading conditions are involved which require analysis for various purposes, including:

(1) Checking stresses (local and global).
(2) Static stability of the floating structure at various stages of installation.
(3) Dynamic behaviour and stability of the structure subjected to wind, wave and current loads.
(4) Load cycles experienced during transport and installation and their effect on the fatigue life of the structure.
(5) Deflections and deformations of structures, particularly pipelines and risers during installation.
(6) Behaviour of the guidance systems, such as tensioned guide-wires, if used for lowering and locating components subsea.
(7) Behaviour and response of floating structures and vessels which are used during the installation operations.

For analysis of the conditions listed here, computer programs have been developed and are used for both static and dynamic analysis. There are, however, many cases where computer programs and techniques are insufficient and model tests are needed to verify predicted behaviours of structures.

Model-testing in water tanks is an example of the type of tests carried out for the oil industry.

44.9 Schedule of work: cost factor

Fabrication, assembly and installation of the various components and modules, as well as the main platform structure, are all complex multidisciplinary tasks. The work often involves acquisition of some long lead items which need to be ordered and manufactured well ahead of time.

So far as the offshore operations are concerned, many facilities such as heavy-lift crane barges, support vessels and tugs, are required. These require mobilization, modification and installation of equipment.

A well-detailed programme of work is required in order to carry out all such tasks. Complex civil engineering projects are no exception, and readers familiar with the programmes of work involved in conventional civil engineering will appreciate the additional complexity of offshore construction.

In offshore work, sensitivity to weather and seasonal sea conditions, involvement of high-cost facilities, such as heavy lift cranes, support vessels and the like, create a demand for thorough planning of the operation.

Bar charts and critical path analysis techniques are used to develop the following key areas of the operation:

(1) Duration of each operation.
(2) Order of work to be carried out and identification of critical activities.
(3) Equipment and facilities required, together with specifications for performance.
(4) Materials needed.
(5) Site/plant requirements.
(6) Requirements and restrictions imposed by regulatory authorities.
(7) Manpower requirements.
(8) Tendering and selection of contractors and subcontractors.
(9) Quality assurance and quality-control requirements.
(10) Route survey and selection for transport.
(11) Co-ordination of work.
(12) Planning for completion and transport of various modules.
(13) Approval and certification for all stages of the operation.
(14) Management system and cost control.

Complex offshore structures often take more than a year to complete and cost several million pounds in capital expenditure. The high rates of cost involved in deployment of these facilities and the use of skilled personnel mean that delays or miscalculations are likely to incur high cost penalties.

The total capital cost of developing offshore hydrocarbon fields varies significantly depending on the depth of water, complexity of the structure and the production system involved, typical costs being, for example, £500 million for the Fulmar field and £1250 million for the Magnus field. These compare with multimillion pound civil engineering projects such as the Thames Barrier at £430 million (1976 price level).

The cost of the development of the oilfields includes drilling, pipelines, production and export facilities. The capital cost of the platform and the construction and installation operations are therefore only one part of a large capital investment in the development of a hydrocarbon field.

44.10 Codes and regulations

There are many codes which apply to the design and fabrication of offshore systems. Specific codes related to the design of offshore structures have been issued by various authorities in the UK, the US and other countries, such as Norway. There are also regulations relating to marine and other offshore operations, some of which are specific to particular countries or areas.

The facilities require to be certified as fit for the purposes specified for offshore structures, whether for production of hydrocarbons or other purposes. The certificates confirm the safety of the operation, safety of the crew, structural and environmental requirements.

There are organizations which assess and issue such certificates. These bodies have set out guidelines and rules with reference to codes and acts which are to be followed. Adherence to such codes and regulations is essential and is one of the requirements for all stages of the project development, from conceptual design to commissioning.

The major certifying authorities are:

(1) American Bureau of Shipping	(US)
(2) Bureau Veritas	(France)
(3) Det Norske Veritas	(Norway)
(4) Germanischer Lloyd	(W. Germany)
(5) Lloyd's Register of Shipping	(UK)

The codes and guidelines cover a broad area ranging from environmental conditions, loads to be considered, allowable stresses, stability, fatigue requirements, methods of analysis, lifting operations, corrosion protection, material specification, fabrication and associated quality control and testing and installation operations.

There are codes and guidelines issued by a number of organizations in the UK including the Department of Energy, Lloyd's Register of Shipping and the British Standards Institution.

In the US, the codes issued by the American Petroleum Institute, the American Bureau of Shipping, the American Concrete Institute, the American Society of Mechanical Engineers and the American National Standard Institute are the main guidelines.

The following is a shortlist of some of the codes and regulations currently in use.

(1) American Petroleum Institute API RP2A: *Recommended practice for planning, design and constructing fixed offshore platforms.*
(2) British Standard 6235: *Code of Practice for fixed offshore structures.*
(3) Det Norske Veritas: *Rules for the design, construction and inspection of offshore structures.*
(4) Department of Energy: *Offshore installations – guidance on design and construction.*
(5) Lloyd's Register of Shipping: *Code for lifting appliances in a marine environment.*

44.11 Organization and management of offshore projects

Interdependency of design, construction and installation techniques plays an important part in the development of offshore structures. Integration of multidisciplinary tasks at all stages demands well-organized management, co-ordination and control of the work.

44.11.1 Project requirements

Like many other complex projects, the main groups or organizations involved are: (1) the client(s); (2) the designers and consultants; (3) the contractors and subcontractors; (4) suppliers of materials and components; (5) inspectors and approving authorities; (6) finance organizations; and (7) insurance companies.

Management requires a project execution plan and an organized team to carry out the tasks of planning, organization and manpower control, contract administration, quality control, expediting, cost control and liaison and co-ordination.

44.11.2 Project organization

Management can be carried out with varying emphases on decision-making, delegation and construction. The matrix would therefore be different for each approach. The most common approaches for the form of project organization are:

(1) Owner project management.
(2) Owner partial involvement plus project services contractor.
(3) Management contractor.
(4) Prime contractor.

44.11.2.1 Owner project management

In owner project management, the owner parcels out various parts of the work to contractors and subcontractors and manages the entire work directly using his own project team. This approach requires a vast team of engineers and planners from the owners who do not often have such a pool of experts.

44.11.2.2 Owner project services contractor

In the project services contractor approach, the owner still has an active role in the management and decision-making processes but selects a contractor to carry out all or most of the project management services.

44.11.2.3 Management contractor

In the management contractor approach, the management contractor acts on behalf of the owner and carries out all management tasks with the main work being contracted out to selected engineering, procurement and construction subcontractors. The owner's role in this case is top-level management and surveillance of the management contractor using his selected project team.

44.11.2.4 Prime contractor

In the prime contractor approach, work is carried out on a 'turnkey' basis by a contractor on a design/construct basis. The contractor is, in this case, responsible for the management and execution of the work, which he may undertake partly himself while subcontracting many other parts of the work to other subcontractors.

In addition to the above approaches, there are cases where combinations of these methods are used. Each approach has its benefits and weaknesses. Selection of the right approach depends on the capabilities of the owner to manage the work, the type of project, country and location.

The number of project managers, planners, engineers, construction inspectors and contract, purchasing, estimating, safety and administration staff varies significantly, and runs from a few hundred to a few thousand depending on the project and method of management. The team operates in various locations, i.e. the central office, land-based sites and offshore sites.

The project management of the Fulmar field in the North Sea involved an in-house team of 95 and a site team of 170 people.

44.12 Inspection, maintenance and repair

The emergence of certification for offshore structures has meant that requirements have been established for inspection, maintenance and repair at regular intervals during the life of the platform.

The inspection of offshore structures presents difficulties because they are being placed in ever deeper and more hostile and turbulent waters. A steel platform can weigh 25 000 t or more and have a total weld length of over 1 km, distributed over some 900 weld points. Templates, manifolds, wellheads, christmas trees, risers, pipelines, flowlines and loading facilities all require regular inspection.

A typical inspection routine for an offshore structure would include the following:

(1) General inspection.
(2) Marine growth inspection.
(3) Debris survey and mapping.
(4) Sea-bed, scour and structure stability inspection.
(5) Corrosion damage inspection.
(6) Cathodic protection potential surveys.
(7) Anode inspection.
(8) Still photography and photo formatting.
(9) Videography.
(10) Nondestructive testing inspection which may include: (a) magnetic particle inspection (crack detection); (b) eddy current inspection; (c) ultrasonics; (d) AC–PD methods; (e) Harwell ultrasonic torch technique; (f) radiography; (g) vibrodetection; and (h) photogrammetry.

Once a defect has been located, there are numerous repair possibilities to consider depending on the type of structure. Steel structures are generally repaired by cutting out the defective area and re-welding or by strengthening, using grouted clamps. Reference should also be made to Chapter 42 for further information on inspection and repair underwater. Typical operations involved in the repair of offshore structures are as follows.

44.12.1 Underwater cutting

There are four underwater cutting processes generally in use: (1) oxy-arc; (2) thermic cutting; (3) gas cutting; and (4) shielded metal arc. Of these, oxy-arc is probably the most widely used. Shielded metal arc can cut steels resistant to oxidization and corrosion and nonferrous materials, and is useful where no oxygen is available. Oxy-hydrogen cutting is performed with a torch rather than with a cutting electrode and an experienced operator can achieve a very neat cut in thick metal. Thermic cutting will burn through almost any material, including reinforced concrete.

44.12.2 Underwater welding

There are three underwater welding techniques:

(1) *Dry hyperbaric welding.* Using either the semiautomatic or manual metal arc-welding processes, the weld area can be enclosed in three ways: (a) full-sized habitat; (b) mini habitat; and (c) portable dry box.
(2) *One-atmosphere welding.* This technique uses an underwater chamber in which the environment is maintained at one atmosphere. The dry hyperbaric welding and one-atmosphere welding are the same except that dry hyperbaric welding is conducted under pressure.
(3) *Shielded metal arc wet welding.* Basically the same equipment is used as for surface welding, but with insulated cable joints and a torch with waterproof electrodes.

44.12.3 Grouted clamps

Grouted clamps are used to strengthen nodes and braces on existing steel platforms. The clamps are bolted together and then filled with grout. They have been extensively used to repair defective nodes on older North Sea platforms.

44.12.4 Concrete repair

A discussion of methods of repair for concrete structures is given by the Underwater Engineering Group of CIRIA (see Bibliography). Techniques for inspection, maintenance and repair operations vary depending on the water depth and many other features of the platforms. Divers are used to perform some of these operations in shallow waters within the range of their safe operation which, in most cases, is up to 15 m. In deeper waters, remotely operated vehicles or remotely operated equipment is used.

Many techniques for remote inspection and maintenance operations have been developed recently. These have resulted in the need to modify details of the structures so that such operations can be carried out successfully. Design and construction methods are therefore influenced by inspection, maintenance and repair requirements during the platform's service life. These operations, apart from having to be practical, need to be safe and economical as, in most cases, the costs of divers, service support vessels and other equipment for offshore use are very high, compared with inspection, maintenance and repair operations for land-based structures.

44.13 Cathodic protection

Cathodic protection is the most commonly used corrosion protection method for steel offshore structures. It is normally used in combination with an insulating or protective coating, where the coating forms the first line of defence. Where the coating is damaged, however, corrosion can occur and cathodic protection is used to provide protection at such locations.

Cathodic protection uses either sacrificial anodes or impressed current to make the structure cathodic. This causes an electrical current to flow from the anodes, through the electrolyte, to the cathode (the structure) thereby opposing the natural electrical current arising from the flow of electrically charged ions away from the surface that is corroding.

In a sacrificial anode system, the current can be generated by the use of sacrificial anodes, such as zinc or aluminium. These will corrode instead of the structure, by virtue of their stronger anodic reaction with respect to the environment. They corrode at known rates, which means that their life expectancies can be estimated and maintenance replacement programmes specified so that new anodes can be installed before the older ones are entirely used up. A large platform anode can produce around 4 A d.c. at about 25 V. The current is transmitted over only relatively short distances.

The current alternatively can be generated by an impressed current system where an outside electrical power supply (e.g. a transformer rectifier) supplies a current to an auxiliary anode of some highly resistant material, such as platinum-coated titanium. An electric field is established which inhibits current flows out of the protected metal. Typical operating power for a single impressed current anode may be around 50 A, 20 V d.c. so that high power levels can be achieved with only a few anodes and long distances can be covered.

44.14 Removal of platforms

Most North Sea fixed platforms have planned lives of 20 to 30 years. A few platforms will therefore be decommissioned in the 1990s with a bunching of decommissioning dates between 2000 and 2010. The latest estimates put the decommissioning cost of all the 250 existing platforms at about $20 bn.

There are no set laws regarding the removal of offshore platforms at present. A consultative document recently issued by the UK Department of Energy envisages complete removal of platforms to a depth of 50 m in the southern North Sea, partial removal providing a minimum clearance of 55 m below the surface in the central North Sea and partial removal to a water depth of 75 m in the northern North Sea.

This is expected to be challenged by the US and the Soviet Union, whose strategic concerns are to minimize the hazards for submarine navigation and require total removal.

Only a few small offshore platforms in the shallow southern North Sea have been removed to date at relatively low cost. The major removal problems will be associated with the 40 or so large platforms located in the central and northern North Sea which are located in water of 100 m or more. While most of these platforms are steel jackets weighing up to 40 000 t, there are also 18 large concrete gravity-based structures with base weights of up to 800 000 t and topside weights of up to 50 000 t.

Suitable dumping sites for the platforms have been investigated by government and offshore contractors. In the US and Japan, the creation of artificial reefs in shallow waters, which could enhance the fish population, have been considered.

However, oil companies are presently trying to increase the life of existing platforms by bringing new fields on stream using subsea completions or unmanned platforms and linking them back to the existing platforms.

Several detailed studies have been carried out to develop cost-effective and safe techniques for removal of such large structures. Many of these techniques involve using some of the methods established for handling such structures during their installation. The methods involve removal of the topside equipment and the deck structure by floating crane barges and the use of underwater explosives to cut the steel structure from its piled foundation.

Acknowledgements

Goodfellow Associates wish to acknowledge contributions to this chapter by M. M. Sarshar, H. D. Parker, L. E. Clarke and R. E. Lawrence.

Bibliography

Det Norske Veritas (1983) *Guidelines for the Preparation of Underwater Inspection Procedures.* DNV, Oslo.

Donovan, J. F., Bearn, W. T. and Nash, N. W. (1987) 'The Balmoral subsea production template.' Paper OTC 5431, *Proceedings offshore technology conference*, 27–30 April, Houston.

Goodfellow Associates Ltd (1986) *Offshore Engineering Development of Small Oilfields.* Graham and Trotman, London.

Graff, W. J. (1981) *Introduction to Offshore Structures: Design/Fabrication/Installation.* Gulf Publishing Company, Houston.

Institution of Civil Engineers (1977) *Proceedings, conference on design and construction of offshore structures,* 27–28 October. ICE, London.

Jones, M. E. (1981) *Deepwater Oil Production and Manned Underwater Structures.* Graham and Trotman, London.

Leniham, J. E., Austin, R. T. C. and Flanagan, P. J. (1984) 'The rapid installation of a large North Sea jacket over a subsea template.' Paper OTC 4759, *Proceedings, offshore technology conference*, 7–9 May, Houston.

Mahoney, T. R. (1987) 'Balmoral, conception to production.' Paper OTC 5430, *Proceedings offshore technology conference*, 27–30 April, Houston.

Myers, J. J., Holm, C. H. and McAllister, R. F. (1969) *Handbook of Ocean and Underwater Engineering.* McGraw Hill, New York.

North Sea Platform Guide (1985) Oilfield Publications, Ledbury, Herefordshire, England.

North Sea Subsea Construction Guide (1986). Oilfield Publications, Ledbury, Herefordshire, England.

Ranney, M. W. (1979) *Offshore Oil Technology: Recent Developments.* Noyes Data Corporation, New Jersey.

Society of Underwater Technology (1985) *The Design and Installation of Subsea Systems.* Volume 2 of Proceedings, subsea international conference on advances in underwater technology and offshore engineering, London, 15–16 January. Graham and Trotman, London.

Thomas, D. B. J. (1981), 'Offshore steel structure repair and maintenance.' In: D. Faulkner, M. J. Cowling and P. A. Fieze (eds), *Integrity of Offshore Structures.* Applied Science Publishers, London.

Underwater Engineering Group (1985) *Design of Tubular Joints for Offshore Structures,* UEG UR30 (3 vols). CIRIA, London.

Underwater Engineering Group (1986) *The Influence of Methods and Materials on the Durability of Repairs to Concrete Coastal and Offshore Structures,* UEG UR36. CIRIA, London.

Underwater Technology (1984) *Proceedings, international conference on deepwater technology.* Bergen, Norway.

Walker, D. B. L. (1980) 'The design and installation of the Buchan Field subsea equipment.' Paper EUR 174, *Proceedings, European Offshore petroleum conference*, 21–24 October, Society of Petroleum Engineers, London.

45

Units, Conversions and Symbols

Contents

45.1 Introduction

45.1.1 Units

The Système International (SI) system of units used throughout this book is the standard system used throughout Europe and many other countries in the world. It was first accepted at an international conference in 1960 and, in 1971, a directive by the European Economic Community required the existing imperial and metric CGS systems to be replaced by SI.

The definitions of, and the symbols for, SI units are given in sections 2.1.1 to 2.1.5.

Although, in time, it can be expected that there will be strict adherence to the SI units given in sections 2.1.1 to 2.1.5, there are at present some cases in which, for convenience or because of previously established practice, the units are varied or auxiliary units are introduced. The most common variations are given in section 2.1.6 and it should be noted that they do not represent a serious departure from the SI system.

45.1.2 Conversion factors

Section 2.2.1 and, in particular, Table 2.4 gives conversion factors between most common imperial and SI units, together with the reciprocals. Section 2.2.2 draws attention to some differences between imperial and US units.

45.1.3 Properties of materials

For information on the properties of materials, the reader is referred to the appropriate earlier chapter, as follows:

Aluminium	Chapter 14
Bituminous materials	Chapters 23 and 24
Concrete	Chapters 4, 12 and 37
Masonry	Chapter 15
Paint	Chapter 4
Plastics	Chapter 4
Reinforcement	Chapter 12
Rock	Chapter 10
Rubber	Chapter 4
Soil	Chapter 9
Timber	Chapter 16

45.1.4 Mathematical relations and trigonometrical functions

Mathematical relations commonly employed in civil engineering work, including those for statistical applications, are given in Chapter 1.

The reader is referred to any of the standard works which evaluate mathematical relations and provide tables of trigonometrical functions for more detailed information.

45.2 International unit system (SI)

The SI is a metric system giving a fully coherent set of units for science, technology and engineering, involving no conversion factors. The starting point is the selection and definition of a minimum set of independent 'base' units. From these, 'derived' units are obtained by forming products or quotients in various combinations, again without numerical factors. For convenience, certain combinations are given shortened names. A single SI unit of energy (joule = kilogram metre-squared per second-squared) is, for example, applied to energy of any kind, whether it be kinetic, potential, electrical, thermal, chemical . . . thus unifying usage throughout science and technology.

The SI system has seven *base* units, and two *supplementary* units of angle. Combinations of these are *derived* for all other units.

45.2.1 Base units

Definitions of the seven base units have been laid down in the following terms. The quantity symbol is given in italic, the unit symbol (with its standard abbreviation) in roman type. As measurements become more precise, changes are occasionally made in the definitions.

(1) *Length*: l, metre (m)   The metre was defined in 1983 as the length of the path travelled by light in a vacuum during a time interval of $1/299\,792\,458$ of a second.
(2) *Mass*: m, kilogram (kg).   The mass of the international prototype (a block of platinum preserved at the International Bureau of Weights and Measures, Sèvres).
(3) *Time*: t, second (s)   The duration of $9\,192\,631\,770$ periods of the radiation corresponding to the transition between the two hyperfine levels of the ground state of the caesium-133 atom.
(4) *Electric current*: i, ampere (A)   The current which, maintained in two straight parallel conductors of infinite length, of negligible circular cross-section and 1 m apart in vacuum, produces a force equal to 2×10^{-7} N/m length.
(5) *Thermodynamic temperature*: T, kelvin (K)   The fraction $1/273.16$ of the thermodynamic (absolute) temperature of the triple point of water.
(6) *Luminous intensity*: l, candela (cd)   The luminous intensity in the perpendicular direction of a surface of $1/600\,000$ m^2 of a black body at the temperature of freezing platinum under a pressure of $101\,325$ N/m^2.
(7) *Amount of substance*: Q, mole (mol)   The amount of substance of a system which contains as many elementary entities as there are atoms in 0.012 kg of carbon-12. The elementary entity must be specified and may be an atom, a molecule, an ion, an electron . . . or a specified group of such entities.

45.2.2 Supplementary units

Plane angle: : a, β, . . . radian (rad)   The plane angle between two radii of a circle which cuts on the circumference of the circle an arc of length equal to the radius.

Solid angle: Ω, steradian (sr)   The solid angle which, having its vertex at the centre of a sphere, cuts off an area of the surface of the sphere equal to a square having sides equal to the radius.

45.2.3 Notes

Temperature   At 0 K, bodies possess no thermal energy. Specified points (273.16 and 373.16 K) define the Celsius (centigrade) scale (0 and 100°C). In terms of *intervals*, 1°C = 1 K. In terms of *levels*, a scale Celsius temperature θ corresponds to $(\theta + 273.16)$ K.

Force   The SI unit is the newton (N). A force of 1 N endows a mass of 1 kg with an acceleration of 1 m/s^2.

Weight   The weight of a mass depends on gravitational effect. The standard weight of a mass of 1 kg at the surface of the Earth is 9.807 N.

45.2.4 Derived units

All physical quantities have units derived from the base and supplementary SI units, and some of them have been given names for convenience in use. Base, supplementary and some of the derived units are listed in Table 45.1.

Table 45.1 Système International base, supplementary and derived units

Quantity	Unit name	Derivation	Unit symbol
Base			
length	Metre		m
mass	Kilogram		kg
time	Second		s
electric current	Ampere		A
thermodynamic temperature	Kelvin		K
luminous intensity	Candela		cd
amount of substance	Mole		mol
Supplementary			
plane angle	Radian		rad
solid angle	Steradian		sr
Derived			
force	Newton	$kg\,m/s^2$	N
pressure, stress	Pascal	N/m^2	Pa
energy	Joule	N m, W s	J
power	Watt	J/s	W
electric charge, flux	Coulomb	A s	C
magnetic flux	Weber	V s	Wb
electric potential	Volt	J/C	V
magnetic flux density	Tesla	Wb/m^2	T
resistance	Ohm	V/A	Ω
inductance	Henry	Wb/A, V s/A	H
capacitance	Farad	C/V, A s/V	F
conductance	Siemens	A/V	S
frequency	Hertz	s^{-1}	Hz
luminous flux	Lumen	cd sr	lm
illuminance	Lux	lm/m^2	lx
radiation activity	Becquerel	s^{-1}	Bq
absorbed dose	Gray	J/kg	Gy
mass density	Kilogram per cubic metre		kg/m^3
dynamic viscosity	Pascal-second		Pa s
concentration	Mole per cubic metre		mol/m^3
linear velocity	Metre per second		m/s
linear acceleration	Metre per second-squared		m/s^2
angular velocity	Radian per second		rad/s
angular acceleration	Radian per second-squared		rad/s^2
torque	Newton metre		N m
current density	Ampere per square metre		A/m^2
resistivity	Ohm metre		Ω m
conductivity	Siemens per metre		S/m
thermal capacity	Joule per kelvin		J/K
specific heat capacity	Joule per kilogram kelvin		J/(kg K)
thermal conductivity	Watt per metre kelvin		W/(m K)
luminance	Candela per square metre		cd/m^2

42.2.5 Decimal multiples and submultiples

Decimal multiples and submultiples of SI units are indicated by the prefix letters given in Table 45.2. Thus, MN is meganewton and μs is microsecond. Prefixes for the kilogram are expressed in terms of the gram, i.e. 1000 kg = 1 Mg, not 1 kkg. There is a preference to express stress as $1\,N/mm^2$ instead of $1\,MN/mm^2$.

Table 45.2 Decimal prefixes

Factor by which unit is multiplied	Prefix Name	Prefix Symbol
10^{18}	exa	E
10^{15}	peta	P
10^{12}	tera	T
10^9	giga	G
10^6	mega	M
10^3	kilo	k
10^2	hecto	h
10^1	deca	da
10^{-1}	deci	d
10^{-2}	centi	c
10^{-3}	milli	m
10^{-6}	micro	μ
10^{-9}	nano	n
10^{-12}	pico	p
10^{-15}	femto	f
10^{-18}	atto	a

45.2.6 Common variations and auxiliary units

The main variations that are commonly applied to civil engineering are:

Stress expressed as N/mm instead of pascals (Pa) (1 N/mm = 1 MN/m = 1 MPa)

Pressure, e.g. underwater, expressed as bar instead of pascals (1 bar = 100 kPa, 1 mbar = 0.1 kPa)

Temperature expressed as °C (Celsius or centigrade) instead of K (kelvin) (0°C = 273.16 K, 100°C = 373.16 K)

Mass expressed as tonne instead of kilograms (kg) (1 t = 1000 kg)

Some of the common variations from the strict SI system are listed in Table 45.3 and are here termed 'auxiliary' units.

Table 45.3 Auxiliary units

Quantity	Symbol	SI	
Angle			
degree	(°)	$\pi/180$	rad
minute	(′)	—	—
second	(″)	—	—
Area			
acre	a	100	m^2
hectare	ha	0.01	km^2
barn	barn	10^{-28}	m^2
Energy			
erg	erg	0.1	μJ
calorie	cal	4.186	J
electron-volt	eV	0.160	aJ
gauss-oersted	Ga Oe	7.96	$μJ/m^3$
Force			
dyne	dyn	10	μN
Length			
Ångstrom	Å	0.1	μm

Table 45.3 (*continued*)

Quantity	Symbol	SI	
Mass			
tonne	t	1000	kg
Nucleonics, Radiation			
becquerel	Bq	1.0	s^{-1}
gray	Gy	1.0	J/kg
curie	Ci	3.7×10^{10}	Bq
rad	rd	0.01	Gy
roentgen	R	2.6×10^{-4}	C/kg
Pressure			
bar	b	100	kPa
torr	Torr	133.3	Pa
Time			
minute	min	60	s
hour	h	3600	s
day	d	86 400	s
Volume			
litre	l or L	1.0	dm^3

45.3 Conversion factors

45.3.1 Système International and imperial units

Although SI is now the standard system in use throughout Europe and much of the rest of the world, imperial units are used occasionally in some specialized areas and many publications prior to about 1980 were in imperial units.

Conversion factors between SI and imperial units are given in Table 45.4. Column 1 gives the imperial units, column 2 the SI equivalent and column 3 the reciprocal.

Table 45.4 Conversion factors: imperial to SI

Imperial	SI	Reciprocal
Length (m)		
1 in	25.40 mm	0.0394
1 ft	0.3048 m	3.2800
1 yd	0.9144 m	1.0940
1 fathom	1.829 m	0.5470
1 mile	1.6093 km	0.6210
1 nautical mile	1.852 km	0.5400
Area (m^2)		
1 in^2	645.2 mm^2	1.5500×10^{-3}
1 ft^2	0.0929 m^2	10.7600
1 yd^2	0.8361 m^2	1.2000
1 acre	4047 m^2	0.2470×10^{-3}
1 $mile^2$	2.590 km^2	0.3860
Volume (m^3)		
1 in^3	16.39×10^3 mm^3	0.0610×10^{-3}
1 ft^3	0.0283 m^3	35.300
1 yd^3	0.7646 m^3	1.310
1 UK gal	4.546 dm^3	0.220
Second moment of area (m^4)		
1 in^4	416×10^3 mm^4	2.40×10^{-6}
Velocity (m/s, rad/s)		
Acceleration (m/s^2, rad/s^2)		
1 ft/s	0.3048 m/s	3.2800

Table 45.4 (*continued*)

Imperial	SI	Reciprocal
1 mile/h	0.4470 m/s	2.2370
1 knot	0.5144 m/s	1.9440
1 deg/s	17.45 mrad/s	0.0573
1 rev/s	6.283 rad/s	0.1590
1 ft/s^2	0.3048 m/s^2	3.2810
Mass (kg)		
1 oz	28.35 g	0.0353
1 lb	0.454 kg	2.2000
1 cwt	50.80 kg	0.0197
1 UK ton	1016 kg	0.9840×10^{-3}
Energy (J), Power (W)		
1 ft lbf	1.356 J	0.737
1 Btu	1055 J	0.948×10^{-3}
1 therm	105.5 kJ	9.478×10^{-3}
1 kW h	3.60 MJ	0.278
1 Btu/h	0.293 W	3.413
1 ft lbf/s	1.356 W	0.737
1 hp	745.9 W	1.341×10^{-3}
Thermal quantities (W, J, kg, K)		
1 $Btu/(ft^2 h)$	3.155 W/m^2	0.3170
1 $Btu/(ft^3 h)$	10.35 W/m^3	0.9660
1 Btu/(ft h °F)	1.731 W/(m K)	0.5780
1 ft lbf/lb	2.989 J/kg	0.3340
1 Btu/lb	2326 J/kg	0.4300×10^{-3}
1 Btu/ft^3	37.26 KJ/m^3	0.0268
1 ft lbf/(lb °F)	5.380 J/(kg K)	0.1860
1 Btu/(lb °F)	4.187 kJ/(kg K)	0.2390
1 $Btu/(ft^3 °F)$	67.07 kJ/m^3 K	0.0149
Density (kg/m^3)		
1 lb/in^3	27.68 Mg/m^3	0.0361
1 lb/ft^3	16.02 kg/m^3	0.0624
1 ton/yd^3	1329 kg/m^3	0.7520×10^{-3}
Flow rate (kg/s, m^3/s)		
1 lb/h	0.1260 g/s	7.9360
1 ton/h	0.2822 kg/s	3.5440
1 lb/s	0.4536 kg/s	2.2046
1 ft^3/h	7.866 cm^3/s	0.1270
1 ft^3/s	0.0283 m^3/s	35.3360
1 gal/h	1.263 cm^3/s	0.7920
1 gal/min	75.77 cm^3/s	0.0312
1 gal/s	4.546 dm^3/s	0.2200
Force (N), pressure (Pa)		
1 dyn	10.0 μN	0.1000
1 lbf	4.445 N	0.2250
1 tonf	9.964 kN	0.1004
1 lbf/ft^2	47.88 Pa	0.0209
1 lbf/in^2	6.895 kPa	0.1450
1 $tonf/ft^2$	107.2 kPa	9.3280×10^{-3}
1 $tonf/in^2$	15.44 MPa	0.0648
1 inHg	3.386 kPa	0.2950
1 inH_2O	149.1 Pa	6.7070×10^{-3}
Torque (N m)		
1 lbf in	0.113 N m	8.8490
1 lbf ft	1.356 N m	0.7370
1 tonf ft	3.307 kN m	0.3020
Inertia ($kg\ m^2$)		
Momentum (kg m/s, kg m^2/s)		
1 $lb\ in^2$	0.293 $g\ m^2$	3.4130

Table 45.4 (*continued*)

Imperial	SI	Reciprocal
1 lb ft^2	0.0421 kg m^2	23.7530
1 ton ft^2	94.30 kg m^2	0.0106
1 lb ft/s	0.138 kg m/s	7.2460
1 lb ft^2/s	0.042 kg m^2/s	23.8100
Viscosity (Pa s, m^2/s)		
1 poise	9.807 Pa s	0.1020
1 lbf s/ft^2	47.88 Pa s	0.0209
1 lbf h/ft^2	172.4 kPa s	5.8000×10^{-3}
1 stokes	1.0 cm^2/s	1.0000
1 in^2/s	6.452 cm^2/s	0.1550
1 ft^2/s	929.0 cm^2/s	1.0760×10^{-3}
Illumination (cd, lm)		
1 lm/ft^2	10.76 lm/m^2	0.0929
1 cd/ft^2	10.76 cd/m^2	0.0929
1 cd/in^2	1550 cd/m^2	0.645×10^{-3}

45.3.2 Système International and US units

United States units differ from imperial units in respect of liquid measurement and mass.

1 US gal.	= 0.8332 imperial gal. (reciprocal 1.200)
	= 3.788 dm^3 (reciprocal 0.264)
	= 3.788 l
1 US long ton	= 1.020 t (SI)
1 US short ton	= 0.909 t (SI)
(1 imperial ton	= 1.016 t (SI)

45.4 Symbols

45.4.1 The Greek alphabet

Although very little use is made of Greek letters for symbols in SI, the Greek alphabet is, of course, widely used in mathemati-cal and other applications in civil engineering. Table 45.5 gives the Greek alphabet in the form used throughout the text of this book.

Table 45.5 The Greek alphabet

Capital	Lower case	Name	English transliteration
A	α	alpha	a
B	β	beta	b
Γ	γ	gamma	g
Δ	δ	delta	d
E	ε	epsilon	e
Z	ζ	zeta	z
H	η	eta	ē
Θ	θ	theta	th
I	ι	iota	i
K	κ	kappa	k
Λ	λ	lambda	l
M	μ	mu	m
N	ν	nu	n
Ξ	ξ	xi	x
O	o	omicron	o
Π	π	pi	p
P	ρ	rho	r
Σ	σ	sigma	s
T	τ	tau	t
Y	υ	upsilon	u
Φ	ϕ	phi	ph
X	χ	chi	kh
Ψ	ψ	psi	ps
Ω	ω	omega	ō

Index